Providing the Steadfast Support Yo

This text is written to engage and support you in your le
both the conceptual understanding and the analytical skills necessary for success.

SUPPORT FOR LEARNING CONCEPTS

Examples and step-by-step solutions include side comments and section references to previously covered material. Pointers in the examples provide on-the-spot reminders.

Example/Solution videos in MyMathLab offer a detailed solution process for every example in this textbook.

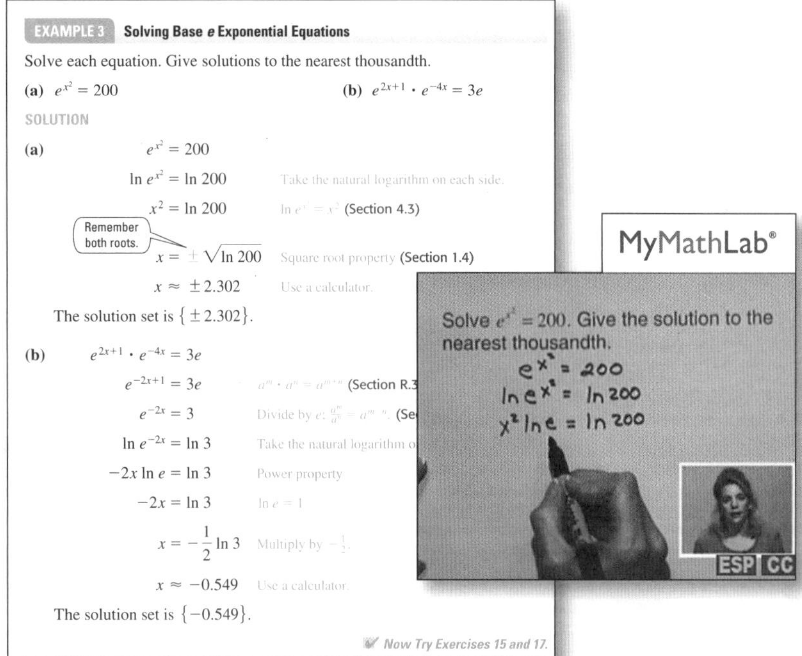

EXAMPLE 3 **Solving Base *e* Exponential Equations**

Solve each equation. Give solutions to the nearest thousandth.

(a) $e^{x^2} = 200$ (b) $e^{2x+1} \cdot e^{-4x} = 3e$

SOLUTION

(a)

$e^{x^2} = 200$

$\ln e^{x^2} = \ln 200$ Take the natural logarithm on each side.

$x^2 = \ln 200$ $\ln e^{x^2} = x^2$ (Section 4.3)

Remember both roots.

$x = \pm\sqrt{\ln 200}$ Square root property (Section 1.4)

$x \approx \pm 2.302$ Use a calculator.

The solution set is $\{\pm 2.302\}$.

(b)

$e^{2x+1} \cdot e^{-4x} = 3e$

$e^{-2x+1} = 3e$ $a^m \cdot a^n = a^{m+n}$ (Section R.3

$e^{-2x} = 3$ Divide by e; $\frac{a^m}{a^n} = a^{m-n}$. (Se

$\ln e^{-2x} = \ln 3$ Take the natural logarithm o

$-2x \ln e = \ln 3$ Power property

$-2x = \ln 3$ $\ln e = 1$

$x = -\frac{1}{2}\ln 3$ Multiply by $-\frac{1}{2}$.

$x \approx -0.549$ Use a calculator.

The solution set is $\{-0.549\}$.

Now Try Exercises 15 and 17.

Page 441

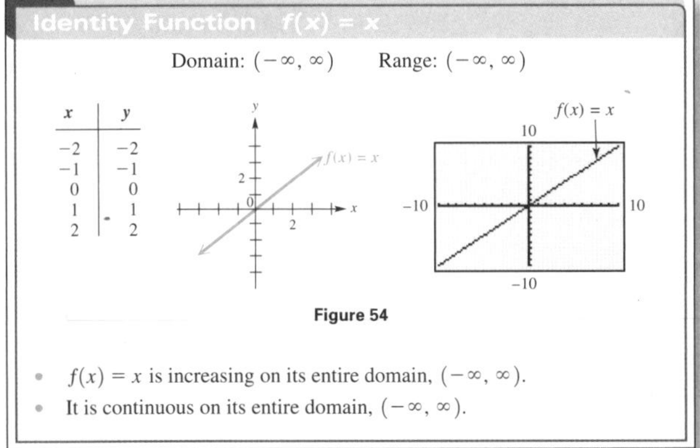

AVAILABLE IN MyMathLab®

Identity Function $f(x) = x$

Domain: $(-\infty, \infty)$ Range: $(-\infty, \infty)$

x	y
-2	-2
-1	-1
0	0
1	1
2	2

Figure 54

- $f(x) = x$ is increasing on its entire domain, $(-\infty, \infty)$.
- It is continuous on its entire domain, $(-\infty, \infty)$.

Page 232

Function boxes offer a comprehensive, visual introduction to each class of function and also serve as an excellent resource for your reference and review throughout the course. Each function box includes a table of values alongside traditional and calculator graphs, as well as the domain, range, and other specific information about the function.

Interactive animations in MyMathLab explore the connection between plotted points and their graphs, bringing the text's function boxes to life.

Real-life applications in the examples and exercises draw from fields such as business, entertainment, sports, life sciences, and environmental studies to show the relevance of algebra to daily life.

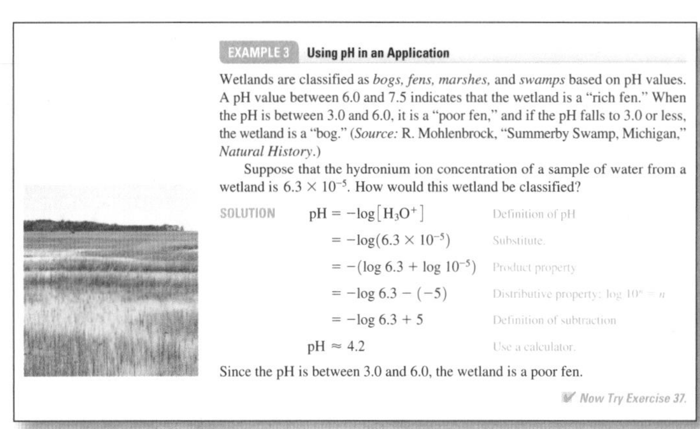

EXAMPLE 3 **Using pH in an Application**

Wetlands are classified as *bogs, fens, marshes,* and *swamps* based on pH values. A pH value between 6.0 and 7.5 indicates that the wetland is a "rich fen." When the pH is between 3.0 and 6.0, it is a "poor fen," and if the pH falls to 3.0 or less, the wetland is a "bog." (*Source:* R. Mohlenbrock, "Summerby Swamp, Michigan," *Natural History*.)

Suppose that the hydronium ion concentration of a sample of water from a wetland is 6.3×10^{-5}. How would this wetland be classified?

SOLUTION

$\text{pH} = -\log[\text{H}_3\text{O}^+]$ Definition of pH

$= -\log(6.3 \times 10^{-5})$ Substitute.

$= -(\log 6.3 + \log 10^{-5})$ Product property

$= -\log 6.3 - (-5)$ Distributive property; $\log 10^a = a$

$= -\log 6.3 + 5$ Definition of subtraction

$\text{pH} \approx 4.2$ Use a calculator.

Since the pH is between 3.0 and 6.0, the wetland is a poor fen.

Now Try Exercise 37.

Page 429

Continued next page

SUPPORT FOR PRACTICING AND REVIEWING CONCEPTS

AVAILABLE IN MyMathLab®

Chapter 4 Quiz (Sections 4.1–4.4)

1. For the one-to-one function $f(x) = \sqrt[3]{3x-6}$, find $f^{-1}(x)$.
2. Solve $4^{2x+1} = 8^{3x-6}$.
3. Graph $f(x) = -3^x$. Give the domain and range.
4. Graph $f(x) = \log_4(x+2)$. Give the domain and range.
5. *Future Value* Suppose that $15,000 is deposited in a bank certificate of deposit at an annual rate of 3.5% for 8 yr. Find the future value if interest is compounded as follows.
   (a) annually **(b)** quarterly **(c)** monthly **(d)** daily (365 days)
6. Use a calculator to evaluate each logarithm to four decimal places.
   (a) log 34.56 **(b)** ln 34.56
7. What is the meaning of the expression $\log_6 25$?
8. Solve.
   (a) $x = 3^{\log_3 4}$ **(b)** $\log_x 25 = 2$ **(c)** $\log_4 x = -2$
9. Assuming all variables represent positive real numbers, use properties of logarithms to rewrite

Page 439

Mid-chapter quizzes allow you to periodically check your understanding of the material. Quizzes and cumulative tests in MyMathLab provide unlimited opportunity for practice and mastery.

AVAILABLE IN MyMathLab®

Relating Concepts

For individual or collaborative investigation *(Exercises 117–122)*

Assume $f(x) = a^x$, where $a > 1$. ***Work these exercises in order.***

117. Is f a one-to-one function? If so, based on **Section 4.1,** what kind of related function exists for f?

118. If f has an inverse function f^{-1}, sketch f and f^{-1} on the same set of axes.

119. If f^{-1} exists, find an equation for $y = f^{-1}(x)$ using the method described in **Section 4.1.** You need not solve for y.

120. If $a = 10$, what is the equation for $y = f^{-1}(x)$? (You need not solve for y.)

121. If $a = e$, what is the equation for $y = f^{-1}(x)$? (You need not solve for y.)

122. If the point (p, q) is on the graph of f, then the point ________ is on the graph of f^{-1}.

Page 413

Relating Concepts Exercises help you tie together topics and develop problem-solving skills as you compare and contrast ideas, identify and describe patterns, and extend concepts to new situations. In-chapter **Summary Exercises** provide mixed topic review problems.

AVAILABLE IN MyMathLab®

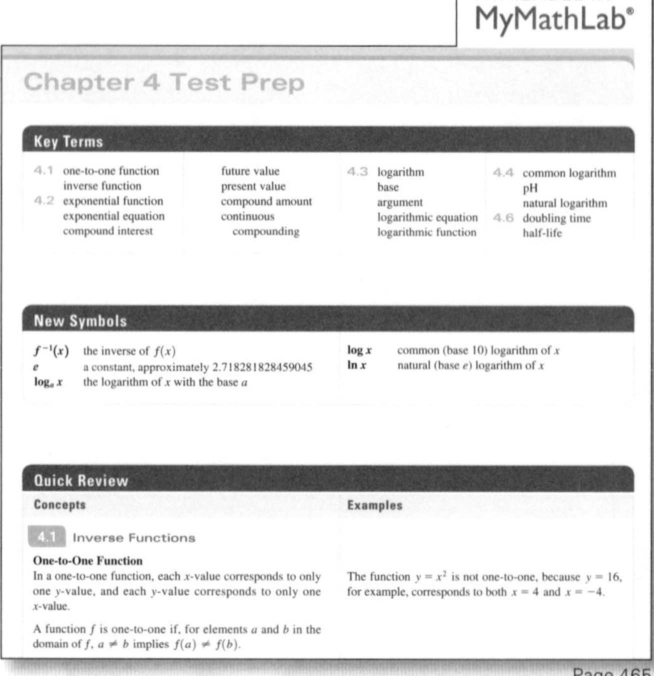

Chapter 4 Test Prep

Key Terms

4.1 one-to-one function, inverse function
4.2 exponential function, exponential equation, compound interest, future value, present value, compound amount, continuous compounding
4.3 logarithm, base, argument, logarithmic equation, logarithmic function
4.4 common logarithm, pH, natural logarithm
4.6 doubling time, half-life

New Symbols

$f^{-1}(x)$	the inverse of $f(x)$
e	a constant, approximately 2.718281828459045
$\log_a x$	the logarithm of x with the base a
$\log x$	common (base 10) logarithm of x
$\ln x$	natural (base e) logarithm of x

Quick Review

Concepts	Examples
4.1 Inverse Functions	
One-to-One Function In a one-to-one function, each x-value corresponds to only one y-value, and each y-value corresponds to only one x-value.	The function $y = x^2$ is not one-to-one, because $y = 16$, for example, corresponds to both $x = 4$ and $x = -4$.
A function f is one-to-one if, for elements a and b in the domain of f, $a \neq b$ implies $f(a) \neq f(b)$.	

Page 465

Chapter Test Prep provides Key Terms, New Symbols, and a Quick Review of important concepts, with corresponding examples. Review Exercises and Chapter Tests are also provided to make test preparation easy.

Quick Review videos in MyMathLab cover key definitions and procedures from each section.

Interactive Chapter Summaries in MyMathLab allow you to quiz yourself via interactive examples, key vocabulary, symbols, and concepts.

Precalculus

FIFTH EDITION

Annotated Instructor's Edition

Margaret L. Lial
American River College

John Hornsby
University of New Orleans

David I. Schneider
University of Maryland

Callie J. Daniels
St. Charles Community College

PEARSON

Boston Columbus Indianapolis New York San Francisco Upper Saddle River
Amsterdam Cape Town Dubai London Madrid Milan Munich Paris Montréal Toronto
Delhi Mexico City São Paulo Sydney Hong Kong Seoul Singapore Taipei Tokyo

Editor-in-Chief: Anne Kelly
Executive Content Editor: Christine O'Brien
Editorial Assistant: Judith Garber
Senior Managing Editor: Karen Wernholm
Senior Production Project Manager: Kathleen A. Manley
Digital Assets Manager: Marianne Groth
Associate Media Producer: Stephanie Green
Software Development: Kristina Evans, MathXL; Mary Durnwald, TestGen
Marketing Manager: Peggy Sue Lucas
Marketing Assistant: Justine Goulart
Senior Author Support/Technology Specialist: Joe Vetere
Rights and Permissions Advisor: Michael Joyce
Image Manager: Rachel Youdelman
Procurement Manager: Evelyn Beaton
Procurement Specialist: Debbie Rossi
Media Procurement Specialist: Ginny Michaud
Associate Director of Design: Andrea Nix
Senior Designer: Beth Paquin
Text Design, Production Coordination, Composition: Cenveo Publisher Services/Nesbitt Graphics, Inc.
Illustrations: Cenveo Publisher Services/Nesbitt Graphics, Inc., Laserwords, Network Graphics
Cover Design: Jenny Willingham
Cover Image: Agatha Brown/Fotolia

Library of Congress Cataloging-in-Publication Data

Precalculus/Margaret L. Lial … [et al.].—5th ed.
p. cm.
Includes index.
ISBN-13: 978-0-321-78380-6
ISBN-10: 0-321-78380-8
1. Algebra. 2. Trigonometry. I. Lial, Margaret L.
QA154.3.L56 2013
512—dc22 2011013067

1 2 3 4 5 6 7 8 9 10—CRK—16 15 14 13 12

Annotated Instructor's Edition
ISBN-10: 0-321-78606-8
ISBN-13: 978-0-321-78606-7

ISBN-10: 0-321-78380-8
ISBN-13: 978-0-321-78380-6

www.pearsonhighered.com

To my friend, Inkie Landry

E.J.H.

To my U of O professors, Buddy Smith and Noel Rowbotham

C.J.D.

Contents

Preface

WELCOME TO THE 5TH EDITION

As authors, we have called upon our classroom experiences, use of MyMathLab, suggestions from users and reviewers, and many years of writing to provide tools that will support learning and teaching. This new edition of *Precalculus* continues our effort to provide a sound pedagogical approach through logical development of the subject matter. This approach forms the basis for all of the Lial team's instructional materials available from Pearson Education, in both print and technology forms.

Our goal is to produce a textbook that will be an integral component of the student's experience in learning precalculus mathematics. With this in mind, we have provided a textbook that students can read more easily, which is often a difficult task, given the nature of mathematical language. We have also improved page layouts for better flow, provided additional side comments, and updated many figures.

We realize that today's classroom experience is evolving and that technology-based teaching and learning aids have become essential to address the ever-changing needs of instructors and students. As a result, we've worked to provide support for all classroom types—traditional, hybrid, and online. In the 5th edition, text and online materials are more tightly integrated than ever before. This enhances flexibility and ease of use for instructors and increases success for students. See pages xix–xxi for descriptions of these materials.

NEW TO THE 5TH EDITION

- In **Chapter 1** we have rewritten the introduction to the set of complex numbers and have included a new diagram illustrating the relationships among its subsets. We have also expanded the discussion of solving absolute value equations and inequalities.

- In **Chapter 2** we have prepared new art for shrinking and stretching graphs, added new exercises on translations, and expanded the exercises involving the difference quotient to include additional greater-degree functions.

- **Chapter 3** has undergone a particularly extensive revision. We have updated the opening discussion by providing a table of examples of polynomial functions (constant, linear, quadratic, cubic, quartic) and have identified the degree and leading coefficient. There is an increased emphasis on displaying all possibilities for positive, negative, and nonreal complex zeros in a table format, with graphical representations to illustrate them. We now have a new summary exercise set on solving both equations and inequalities (linear, quadratic, polynomial, rational, and miscellaneous).

- In **Chapter 4** we have added more work with translations of exponential and logarithmic functions, and we have included additional examples and exercises on solving equations that involve these functions.

- In **Chapter 5** we have provided a revised visual in the figure accompanying the explanation of the ranges of the sine and cosine functions (ratios) in conjunction with right triangles. There are new exercises for evaluating trigonometric expressions with function values of special angles. We have expanded the discussion on using inverse trigonometric functions to find angle measures using a calculator, and we have updated examples of solving right triangles to prepare students for the more challenging exercises in the exercise sets.

- In **Chapter 6** we have added a figure clarifying the concept of radian measure, and there are new exercises involving application of the formula for the area of a sector of a circle. We now provide an expanded explanation of what values a calculator returns for inverse trigonometric functions. We have added a new figure in explaining how to use the concept of inverse functions as well as a revised figure and discussion relating the sine graph to the unit circle. We have expanded the discussion of sketching graphs of translated trigonometric functions and have updated the guidelines for these sketches. There are new examples of connecting graphs with equations, and we have included new exercises of this type in the exercise sets and chapter review exercises.

- In **Chapter 7** we have updated the discussion on finding inverse function values using a calculator. We have included updated examples and many more exercises in which trigonometric equations are solved for *all solutions* in both degrees and radians. Several new figures are included relating solutions of trigonometric equations with angle measures and arc lengths on the unit circle. For equations involving inverse trigonometric functions that are solved for a specified variable, restrictions are now given so that each equation provides a one-to-one correspondence, and a new figure is given to provide conceptual understanding.

- In **Chapter 8** we have updated and improved many of the illustrations in the examples and exercises. New figures illustrate operations with vectors geometrically, and we have also included the justification for the geometric interpretation of the dot product. We include a new example of vectors applied to a navigation problem. There is a new introduction to the set of complex numbers, and we have included a revised diagram illustrating the relationships among its subsets. Polar graphs now include underlying grids for easier placement of polar coordinates.

- In **Chapter 9** we now present the determinant theorems within the exposition, and we have written new examples and exercises showing how they are used.

- **Chapter 10** now offers more discussion on graphing ellipses and hyperbolas centered away from the origin.

- The discussion of binomial probability in **Chapter 11** has been rewritten to provide more detail and greater clarity.

- For visual learners, numbered **Figure** and **Example** references within the text are set using the same typeface as the figure and bold print for the example. This makes it easier for the students to identify and connect them. We also have increased our use of a "drop down" style, when appropriate, to distinguish between simplifying expressions and solving equations, and we have added many more explanatory side comments. Interactive figures with accompanying exercises and explorations are now available and assignable in MyMathLab.

- Enhancing the already well-respected exercises, nearly 1300 are new or modified, and hundreds present updated real-life data. In addition, the MyMathLab course has expanded coverage of all exercise types appearing in the exercise sets, as well as the mid-chapter Quizzes and Summary Exercises.

FEATURES OF THIS TEXT

SUPPORT FOR LEARNING CONCEPTS

We provide a variety of features to support students' learning of the essential topics of precalculus mathematics. Explanations that are written in understandable terms, figures and graphs that illustrate examples and concepts, graphing technology that supports and enhances algebraic manipulations, and real-life applications that enrich the topics with meaning all provide opportunities for students to deepen their understanding of mathematics. These features help students make mathematical connections and expand their own knowledge base.

- **Examples** Numbered examples that illustrate the techniques for working exercises are found in every section. We use traditional explanations, side comments, and pointers to describe the steps taken—and to warn students about common pitfalls. Some examples provide additional graphing calculator solutions, although these can be omitted if desired.
- **Now Try Exercises** Following each numbered example, the student is directed to try a corresponding odd-numbered exercise (or exercises). This feature allows for quick feedback to determine whether the student has understood the principles illustrated in the example.
- **Real-Life Applications** We have included hundreds of real-life applications, many with data updated from the previous edition. They come from fields such as business, entertainment, sports, biology, astronomy, geology, and environmental studies.
- **Function Boxes** Beginning in Chapter 2, functions provide a unifying theme throughout the text. Special function boxes (for example, see page 232) offer a comprehensive, visual introduction to each type of function and also serve as an excellent resource for reference and review. Each function box includes a table of values, traditional and calculator-generated graphs, the domain, the range, and other special information about the function. These boxes are now assignable in MyMathLab.
- **Figures and Photos** Today's students are more visually oriented than ever before, and we have updated the figures in this edition to a greater extent than in our previous few editions. Interactive figures with accompanying exercises and explorations are now available and assignable in MyMathLab.
- **Use of Graphing Technology** We have integrated the use of graphing calculators where appropriate, although ***this technology is completely optional and can be omitted without loss of continuity.*** We continue to stress that graphing calculators support understanding but that students must first master the underlying mathematical concepts. Exercises that require their use are marked with an icon.
- **Cautions and Notes** Text that is marked **CAUTION** warns students of common errors, and NOTE comments point out explanations that should receive particular attention.
- **Looking Ahead to Calculus** These margin notes offer glimpses of how the topics currently being studied are used in calculus.

SUPPORT FOR PRACTICING CONCEPTS

This text offers a wide variety of exercises to help students master precalculus mathematics. The extensive exercise sets provide ample opportunity for practice, and the exercise problems increase in difficulty so that students at every level of understanding are challenged. The variety of exercise types promotes understanding of the concepts and reduces the need for rote memorization.

- **Exercise Sets** We have revised many drill and application exercises for better pairing of corresponding even and odd exercises, and answers to the odd exercises are provided in the Student Edition. In addition to these, we include writing exercises, optional graphing calculator problems, and multiple-choice, matching, true/false, and completion exercises. Those marked *Concept Check* focus on conceptual thinking. *Connecting Graphs with Equations* exercises challenge students to write equations that correspond to given graphs. Finally, MyMathLab offers Pencast solutions for selected Connecting Graphs with Equations problems.

- **Relating Concepts Exercises** Appearing in selected exercise sets, these groups of exercises are designed so that students who work them in numerical order will follow a line of reasoning that leads to an understanding of how various topics and concepts are related. All answers to these exercises appear in the student answer section, and these exercises are now assignable in MyMathLab.

- **Complete Solutions to Selected Exercises** Exercise numbers marked indicate that a full worked-out solution appears at the back of the text. These are often exercises that extend the skills and concepts presented in the numbered examples.

SUPPORT FOR REVIEW AND TEST PREP

Ample opportunities for review are found within the chapters and at the ends of chapters. Quizzes that are interspersed within chapters provide a quick assessment of students' understanding of the material presented up to that point in the chapter. Chapter "Test Preps" provide comprehensive study aids to help students prepare for tests.

- **Quizzes** Students can periodically check their progress with in-chapter quizzes that appear in all chapters, beginning with Chapter 1. All answers, with corresponding section references, appear in the student answer section. These quizzes are now assignable in MyMathLab.

- **Summary Exercises** These sets of in-chapter exercises give students the all-important opportunity to work *mixed* review exercises, requiring them to synthesize concepts and select appropriate solution methods. The summary exercises are now assignable in MyMathLab.

- **End-of-Chapter Test Prep** Following the final numbered section in each chapter, the Test Prep provides a list of **Key Terms,** a list of **New Symbols** (if applicable), and a two-column **Quick Review** that includes a section-by-section summary of concepts and examples. This feature concludes with a comprehensive set of **Review Exercises** and a **Chapter Test.** The Test Prep, Review Exercises, and Chapter Test are assignable in MyMathLab.

- **Glossary** A comprehensive glossary of important terms drawn from the entire book follows Appendix C.

Student Supplements

Student's Solutions Manual
By Beverly Fusfield
- Provides detailed solutions to all odd-numbered text exercises

ISBN: 0-321-79136-3 & 978-0-321-79136-8

Video Lectures with Optional Captioning
- Feature Quick Reviews and Example Solutions:
  Quick Reviews cover key definitions and procedures from each section.
  Example Solutions walk students through the detailed solution process for every example in the textbook.
- Ideal for distance learning or supplemental instruction at home or on campus
- Include optional text captioning
- Available in MyMathLab®

Additional Skill and Drill Manual
By Cathy Ferrer, Valencia Community College
- Provides additional practice and test preparation for students

ISBN: 0-321-52928-6 & 978-0-321-52928-2

MyNotes
- Available in MyMathLab and offer structure for student reading and understanding of the textbook
- Include textbook examples along with ample space for students to write solutions and notes
- Include key concepts along with prompts for students to read, write, and reflect on what they have just learned
- Customizable so that instructors can add their own examples or remove examples that are not covered in their courses

MyClassroomExamples
- Available in MyMathLab and offer structure for classroom lecture
- Include Classroom Examples along with ample space for students to write solutions and notes
- Include key concepts along with fill in the blank opportunities to keep students engaged
- Customizable so that instructors can add their own examples or remove Classroom Examples that are not covered in their courses

Instructor Supplements

Annotated Instructor's Edition
- Provides answers in the margins to almost all text exercises, as well as helpful Teaching Tips and Classroom Examples
- Includes sample homework assignments indicated by problem numbers underlined in blue within each end-of-section exercise set
- Sample homework problems assignable in MyMathLab

ISBN: 0-321-78606-8 & 978-0-321-78606-7

Online Instructor's Solutions Manual
By Beverly Fusfield
- Provides complete solutions to all text exercises
- Available in MyMathLab or downloadable from Pearson Education's online catalog

Online Instructor's Testing Manual
By Christopher Mason, Community College of Vermont
- Includes diagnostic pretests, chapter tests, final exams, and additional test items, grouped by section, with answers provided
- Available in MyMathLab or downloadable from Pearson Education's online catalog

TestGen®
- Enables instructors to build, edit, print, and administer tests
- Features a computerized bank of questions developed to cover all text objectives
- Available in MyMathLab or downloadable from Pearson Education's online catalog

Online PowerPo int Presentation, Active Learning Questions, and Classroom Example PowerPoints
- Written and designed specifically for this text
- Include figures and examples from the text
- Provide active learning questions for use with classroom response systems, including multiple-choice questions to review lecture material (available in MyMathLab only)
- Provide Classroom Example PowerPoints that include full worked-out solutions to all Classroom Examples
- Available in MyMathLab or downloadable from Pearson Education's online catalog

MEDIA RESOURCES

MyMathLab® Online Course (access code required)

MyMathLab delivers **proven results** in helping individual students succeed.

- MyMathLab has a consistently positive impact on the quality of learning in higher education math instruction. MyMathLab can be successfully implemented in any environment—lab-based, hybrid, fully online, or traditional—and demonstrates the quantifiable effect that integrated usage has on student retention, subsequent success, and overall achievement.
- MyMathLab's comprehensive online gradebook automatically tracks students' results on tests, quizzes, and homework and in the study plan. The gradebook can be used to quickly intervene if students have trouble or to provide positive feedback on a job well done. The data within MyMathLab are easily exported to a variety of spreadsheet programs, such as Microsoft Excel. Instructors can determine which points of data they want to export, and then analyze the results to determine student success.

MyMathLab provides **engaging experiences** that personalize, stimulate, and measure learning for each student.

- **Tutorial Exercises** Homework and practice exercises in MyMathLab are correlated with the exercises in the textbook, and they regenerate algorithmically to give students unlimited opportunities for practice and mastery. The software offers immediate, helpful feedback when students enter incorrect answers.
- **Multimedia Learning Aids** Exercises include guided solutions, sample problems, animations, videos, and eText clips for extra help at point-of-use.
- **Expert Tutoring** Although many students describe MyMathLab itself as "like having your own personal tutor," students using MyMathLab have access to live tutoring from Pearson, in the form of qualified math instructors who provide tutoring sessions for students via MyMathLab.

And MyMathLab comes from a **trusted partner** with educational expertise and an eye on the future.

Using a Pearson product means using quality content. This means that our eTexts are accurate, our assessment tools work, and that our exercises and answers are carefully checked. And whether instructors are just getting started with MyMathLab or have a question along the way, we're here to help them learn about our technologies and how to incorporate them into their courses.

To learn more about how MyMathLab combines proven learning applications with powerful assessment, visit **www.mymathlab.com** or contact your Pearson representative.

MyMathLab® Ready to Go Course (access code required)

These new Ready to Go courses provide students with all the same great MyMathLab features that they are used to, but the courses make it easier for instructors to get started. Each course includes preassigned homework and quizzes to make creating a course even simpler. Ask your Pearson representative about the details for this particular course, or request a copy of the course.

MyMathLab® Plus

MyLabsPlus combines proven results and engaging experiences from MyMathLab® with convenient management tools and a dedicated service team. Designed to support growing math and statistics programs, it includes additional features such as

- **Batch Enrollment** Schools can create the login name and password for every student and instructor so that everyone can be ready to start class on the first day. Automation of this process is also possible through integration with the school's Student Information System.
- **Login From Your Campus Portal** Instructors and their students can link directly from their campus portal into their MyLabsPlus courses. A Pearson service team works with the institution to create a single sign-on experience for instructors and students.
- **Advanced Reporting** MyLabsPlus's advanced reporting enables instructors to review and analyze students' strengths and weaknesses by tracking their performance on tests, assignments, and tutorials. Administrators can review grades and assignments across all courses on a MyLabsPlus campus for a broad overview of program performance.
- **24/7 Support** Students and instructors receive 24/7 support, 365 days a year, by phone, email, or online chat.

MyLabsPlus is available to qualified adopters. For more information, visit our website at www.mylabsplus.com or contact your Pearson representative.

MathXL® Online Course (access code required)

MathXL® is the homework and assessment engine that runs MyMathLab. (MyMathLab is MathXL plus a learning management system.) With MathXL, instructors can

- Create, edit, and assign online homework and tests using algorithmically generated exercises correlated at the objective level with the textbook.
- Create and assign their own online exercises and import TestGen tests for added flexibility.
- Maintain records of all student work tracked in MathXL's online gradebook.

With MathXL, students can

- Take chapter tests in MathXL and receive personalized study plans and/or personalized homework assignments based on their test results.
- Use the study plan and/or the homework to link directly to tutorial exercises for the objectives they need to study.
- Access supplemental animations and video clips directly from selected exercises.

MathXL is available to qualified adopters. For more information, visit our website at www.mathxl.com or contact your Pearson representative.

ACKNOWLEDGMENTS

We wish to thank the following individuals who provided valuable input into this edition of the text.

Mark Burtch – Austin Community College – Rio Grande
Tilak De Alwis – Southeastern Louisiana University
Elaine Fitt – Bucks County Community College
Audrey Gillant – SUNY Maritime
Richard G. Goldthwait – Youngstown State University
Laura Hoye – Trident Tech
Rene Lumampao – Austin Community College – Rio Grande
Eric Matsuoka – Leeward Community College
Pat Miceli – Wright City College
Preeti Parikh – SUNY Maritime
Carole Phipps – Sampson Community College
Lucia Riderer – Citrus College
Kristina Sampson – Lonestar College – CyFair
David Schweitzer – University of Central Florida
Mayada Shahrokhi – Lonestar College – CyFair
Linda Sturges – SUNY Maritime
Jinhua Tao – Central Missouri University
Magdalena Toda – Texas Tech University

Our sincere thanks to those individuals at Pearson Education who have supported us throughout this revision: Greg Tobin, Anne Kelly, Kathy Manley, and Christine O'Brien. Terry McGinnis continues to provide behind-the-scenes guidance for both content and production. We have come to rely on her expertise during all phases of the revision process. Marilyn Dwyer of Nesbitt Graphics, Inc., with the assistance of Carol Merrigan, provided excellent production work. Special thanks go out to Abby Tanenbaum for updating data in applications, and to Chris Heeren and Paul Lorczak for their excellent accuracy-checking. We thank Lucie Haskins, who once again provided an accurate index. We appreciate the valuable suggestions for Chapter 9 that Mary Hill of *College of Dupage* made during our meeting with her in March 2010. Further thanks go out to Dr. Mohammed Alaimia, KFUPM, Saudi Arabia, who assisted us in revising the exercises in Section 7.7.

As an author team, we are committed to providing the best possible precalculus mathematics course to help instructors teach and students succeed. As we continue to work toward this goal, we welcome any comments or suggestions you might send, via e-mail, to math@pearson.com.

Margaret L. Lial
John Hornsby
David I. Schneider
Callie J. Daniels

R Review of Basic Concepts

Positive and negative numbers, used to represent gains and losses on a board such as this one, are examples of *real numbers* encountered in applications of mathematics.

R.1 Sets

- Basic Definitions
- Operations on Sets

Basic Definitions A **set** is a collection of objects. The objects that belong to a set are called the **elements,** or **members,** of the set. In algebra, the elements of a set are usually numbers. Sets are commonly written using **set braces,** $\{\ \}$. For example, the set containing the elements 1, 2, 3, and 4 is written as follows.

$$\{1, 2, 3, 4\}$$

Since the order in which the elements are listed is not important, this same set can also be written as $\{4, 3, 2, 1\}$ or with any other arrangement of the four numbers.

To show that 4 is an element of the set $\{1, 2, 3, 4\}$, we use the symbol $\in$.

$$4 \in \{1, 2, 3, 4\}$$

Since 5 is *not* an element of this set, we place a slash through the symbol $\in$.

$$5 \notin \{1, 2, 3, 4\}$$

It is customary to name sets with capital letters. If S is used to name the set above, then we write it as follows.

$$S = \{1, 2, 3, 4\}$$

Set S was written by listing its elements. Set S might also be described as

"the set containing the first four counting numbers."

In this example, the notation $\{1, 2, 3, 4\}$, with the elements listed between set braces, is briefer than the verbal description.

The set F, consisting of all fractions between 0 and 1, is an example of an **infinite set,** one that has an unending list of distinct elements. A **finite set** is one that has a limited number of elements. The process of counting its elements comes to an end. Some infinite sets can be described by listing. For example, the set of numbers N used for counting, called the **natural numbers,** or the **counting numbers,** can be written as follows.

$$N = \{1, 2, 3, 4, \ldots\} \quad \text{Natural (counting) numbers}$$

The three dots (*ellipsis points*) show that the list of elements of the set continues according to the established pattern.

Sets are often written using a variable to represent an arbitrary element of the set. For example,

$$\{x \mid x \text{ is a natural number between 2 and 7}\} \quad \text{Set-builder notation}$$

(which is read "the set of all elements x such that x is a natural number between 2 and 7") uses **set-builder notation** to represent the set $\{3, 4, 5, 6\}$. The numbers 2 and 7 are *not* between 2 and 7.

Classroom Example 1
Identify each set as *finite* or *infinite*. Then determine whether 8 is an element of the set.

(a) $\{5, 6, 7, \ldots, 10\}$

(b) $\{1, \frac{1}{2}, \frac{1}{3}, \frac{1}{4}, \ldots\}$

(c) $\{x \mid x \text{ is a fraction between 9 and 10}\}$

(d) $\{x \mid x \text{ is a natural number between 7 and 9}\}$

Answers:
(a) finite; 8 is an element.
(b) infinite; 8 is not an element.
(c) infinite; 8 is not an element.
(d) finite; 8 is an element.

EXAMPLE 1 Using Set Notation and Terminology

Identify each set as *finite* or *infinite*. Then determine whether 10 is an element of the set.

(a) $\{7, 8, 9, \ldots, 14\}$

(b) $\{1, \frac{1}{4}, \frac{1}{16}, \frac{1}{64}, \ldots\}$

(c) $\{x \mid x \text{ is a fraction between 1 and 2}\}$

(d) $\{x \mid x \text{ is a natural number between 9 and 11}\}$

SOLUTION

(a) The set is finite, because the process of counting its elements 7, 8, 9, 10, 11, 12, 13, and 14 comes to an end. The number 10 does belong to the set, and this is written as follows.

$$10 \in \{7, 8, 9, \ldots, 14\}$$

(b) The set is infinite, because the ellipsis points indicate that the pattern continues forever. In this case,

$$10 \notin \left\{1, \tfrac{1}{4}, \tfrac{1}{16}, \tfrac{1}{64}, \ldots\right\}.$$

(c) Between any two distinct natural numbers there are infinitely many fractions, so this set is infinite. The number 10 is not an element.

(d) There is only one natural number between 9 and 11, namely 10. So the set is finite, and 10 is an element.

Now Try Exercises 1, 3, 5, and 7.

Classroom Example 2
Use set notation, and write the elements belonging to each set.
(a) $\{x \mid x$ is a natural number between 8 and 12$\}$
(b) $\{x \mid x$ is a natural number greater than 6 and less than 8$\}$

Answers:
(a) $\{9, 10, 11\}$ (b) $\{7\}$

EXAMPLE 2 Listing the Elements of a Set

Use set notation, and write the elements belonging to each set.

(a) $\{x \mid x$ is a natural number less than $5\}$

(b) $\{x \mid x$ is a natural number greater than 7 and less than $14\}$

SOLUTION

(a) The natural numbers less than 5 form the set $\{1, 2, 3, 4\}$.

(b) This is the set $\{8, 9, 10, 11, 12, 13\}$.

Now Try Exercise 15.

When we are discussing a particular situation or problem, the **universal set** (whether expressed or implied) contains all the elements included in the discussion. The letter U is used to represent the universal set. The **null set,** or **empty set,** is the set containing no elements. We write the null set by either using the special symbol $\emptyset$, or else writing set braces enclosing no elements, $\{\ \}$.

CAUTION Do not combine these symbols. **$\{\emptyset\}$ *is not the null set.***

Every element of the set $S = \{1, 2, 3, 4\}$ is a natural number. S is an example of a *subset* of the set N of natural numbers, and this is written

$$S \subseteq N.$$

By definition, set A is a **subset** of set B if every element of set A is also an element of set B. For example, if $A = \{2, 5, 9\}$ and $B = \{2, 3, 5, 6, 9, 10\}$, then $A \subseteq B$. However, there are some elements of B that are not in A, so B is not a subset of A, which is written

$$B \not\subseteq A.$$

By the definition, every set is a subset of itself. Also, by definition, $\emptyset$ is a subset of every set.

If A is any set, then $\emptyset \subseteq A$.

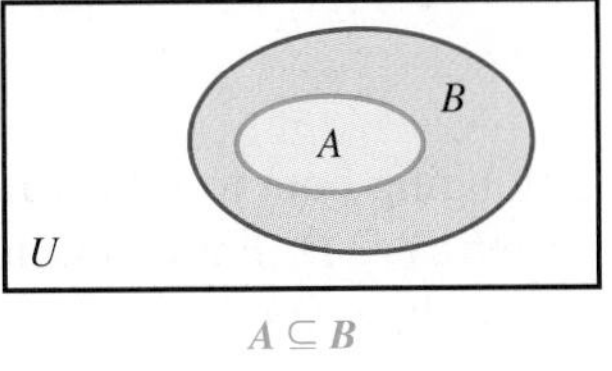

$A \subseteq B$

Figure 1

Figure 1 shows a set A that is a subset of set B. The rectangle in the drawing represents the universal set U. Such diagrams are called **Venn diagrams.**

Two sets A and B are equal whenever $A \subseteq B$ and $B \subseteq A$. Equivalently, $A = B$ if the two sets contain exactly the same elements. For example,

$$\{1, 2, 3\} = \{3, 1, 2\}$$

is true, since both sets contain exactly the same elements. However,

$$\{1, 2, 3\} \neq \{0, 1, 2, 3\},$$

since the set $\{0, 1, 2, 3\}$ contains the element 0, which is not an element of $\{1, 2, 3\}$.

Classroom Example 3
Let $U = \{3, 9, 15, 21, 27, 33, 39\}$,
$A = \{3, 9, 15, 21, 27, 33\}$,
$B = \{3, 9, 21, 27\}$,
$C = \{9, 27, 33\}$,
and $D = \{3, 27\}$.
Determine whether each statement is *true* or *false*.
(a) $B \subseteq D$ (b) $D \subset B$
(c) $D \nsubseteq A$ (d) $U \neq A$

Answers:
(a) false (b) true
(c) false (d) true

EXAMPLE 3 Examining Subset Relationships

Let $U = \{1, 3, 5, 7, 9, 11, 13\}$, $A = \{1, 3, 5, 7, 9, 11\}$, $B = \{1, 3, 7, 9\}$, $C = \{3, 9, 11\}$, and $D = \{1, 9\}$. Determine whether each statement is *true* or *false*.

(a) $D \subseteq B$ **(b)** $B \subseteq D$ **(c)** $C \nsubseteq A$ **(d)** $U = A$

SOLUTION

(a) All elements of D, namely 1 and 9, are also elements of B, so D is a subset of B, and $D \subseteq B$ is true.

(b) There is at least one element of B (for example, 3) that is not an element of D, so B is *not* a subset of D. Thus, $B \subseteq D$ is false.

(c) C is a subset of A, because every element of C is also an element of A. Thus, $C \subseteq A$ is true, and as a result, $C \nsubseteq A$ is false.

(d) U contains the element 13, but A does not. Therefore, $U = A$ is false.

Now Try Exercises 43, 45, 53, and 55.

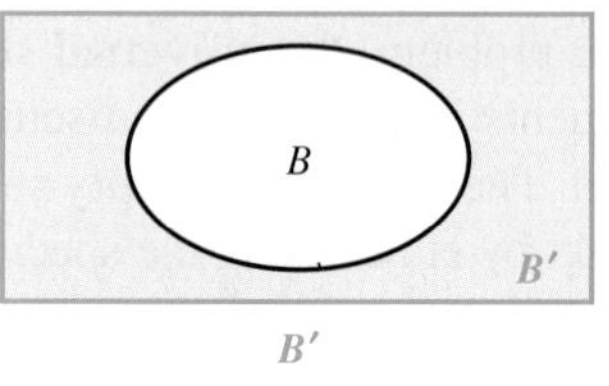

B′

Figure 2

Operations on Sets Given a set A and a universal set U, the set of all elements of U that do not belong to set A is called the **complement** of set A. For example, if set A is the set of all students in your class 30 years old or older, and set U is the set of all students in the class, then the complement of A would be the set of all the students in the class younger than age 30. The complement of set A is written A' (read **"A-prime"**). The Venn diagram in **Figure 2** shows a set B. Its complement, B', is in color.

Classroom Example 4
Let $U = \{1, 2, 3, 4, 5, 6, 7, 8, 9\}$,
$A = \{2, 4, 6, 8\}$,
and $B = \{3, 6, 9\}$.
Find each set.
(a) A' (b) B'
(c) U' (d) $\emptyset'$

Answers:
(a) $\{1, 3, 5, 7, 9\}$
(b) $\{1, 2, 4, 5, 7, 8\}$
(c) $\emptyset$
(d) U

EXAMPLE 4 Finding the Complement of a Set

Let $U = \{1, 2, 3, 4, 5, 6, 7\}$, $A = \{1, 3, 5, 7\}$, and $B = \{3, 4, 6\}$. Find each set.

(a) A' **(b)** B' **(c)** $\emptyset'$ **(d)** U'

SOLUTION

(a) Set A' contains the elements of U that are not in A. Thus, $A' = \{2, 4, 6\}$.

(b) $B' = \{1, 2, 5, 7\}$ **(c)** $\emptyset' = U$ **(d)** $U' = \emptyset$

Now Try Exercise 79.

Given two sets A and B, the set of all elements belonging both to set A ***and*** to set B is called the **intersection** of the two sets, written $A \cap B$. For example, if $A = \{1, 2, 4, 5, 7\}$ and $B = \{2, 4, 5, 7, 9, 11\}$, then we have the following.

$$A \cap B = \{1, 2, 4, 5, 7\} \cap \{2, 4, 5, 7, 9, 11\} = \{2, 4, 5, 7\}$$

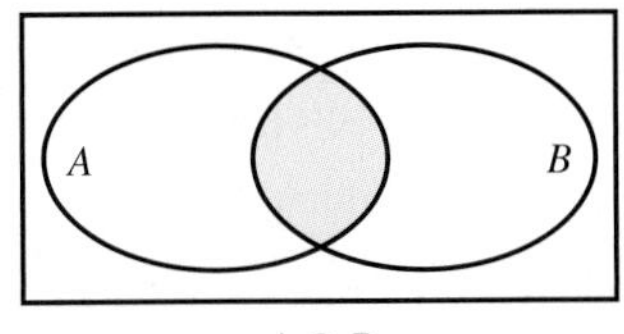

$A \cap B$

Figure 3

The Venn diagram in **Figure 3** shows two sets A and B. Their intersection, $A \cap B$, is in color.

Two sets that have no elements in common are called **disjoint sets.** If A and B are any two disjoint sets, then $A \cap B = \emptyset$. For example, there are no elements common to both $\{50, 51, 54\}$ and $\{52, 53, 55, 56\}$, so these two sets are disjoint.

$$\{50, 51, 54\} \cap \{52, 53, 55, 56\} = \emptyset$$

Classroom Example 5

Find each of the following.

(a) $\{15, 20, 25, 30\} \cap \{12, 18, 24, 30\}$

(b) $\{3, 6, 9, 12, 15, 18\} \cap \{6, 12, 18, 24\}$

(c) $\{6, 7, 8\} \cap \{678\}$

Answers:

(a) $\{30\}$ (b) $\{6, 12, 18\}$

(c) $\emptyset$

EXAMPLE 5 Finding the Intersection of Two Sets

Find each of the following.

(a) $\{9, 15, 25, 36\} \cap \{15, 20, 25, 30, 35\}$

(b) $\{2, 3, 4, 5, 6\} \cap \{1, 2, 3, 4\}$

(c) $\{1, 3, 5\} \cap \{2, 4, 6\}$

SOLUTION

(a) $\{9, 15, 25, 36\} \cap \{15, 20, 25, 30, 35\} = \{15, 25\}$
The elements 15 and 25 are the only ones belonging to both sets.

(b) $\{2, 3, 4, 5, 6\} \cap \{1, 2, 3, 4\} = \{2, 3, 4\}$

(c) $\{1, 3, 5\} \cap \{2, 4, 6\} = \emptyset$ Disjoint sets

Now Try Exercises 59, 65, and 75.

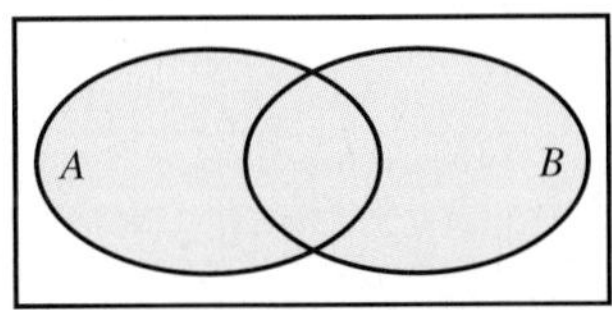

$A \cup B$

Figure 4

The set of all elements belonging to set A ***or*** to set B (or to both) is called the **union** of the two sets, written $A \cup B$. For example, if $A = \{1, 3, 5\}$ and $B = \{3, 5, 7, 9\}$ then we have the following.

$$A \cup B = \{1, 3, 5\} \cup \{3, 5, 7, 9\} = \{1, 3, 5, 7, 9\}$$

The Venn diagram in **Figure 4** shows two sets A and B. Their union, $A \cup B$, is in color.

Classroom Example 6

Find each of the following.

(a) $\{1, 3, 5, 7, 9\} \cup \{3, 6, 9, 12\}$

(b) $\{9, 10, 11, 12\} \cup \{10, 12, 14, 16\}$

(c) $\{2, 6, 10, 14, \ldots\} \cup \{4, 8, 12, 16, \ldots\}$

Answers:

(a) $\{1, 3, 5, 6, 7, 9, 12\}$

(b) $\{9, 10, 11, 12, 14, 16\}$

(c) $\{2, 4, 6, 8, 10, \ldots\}$

EXAMPLE 6 Finding the Union of Two Sets

Find each of the following.

(a) $\{1, 2, 5, 9, 14\} \cup \{1, 3, 4, 8\}$

(b) $\{1, 3, 5, 7\} \cup \{2, 4, 6\}$

(c) $\{1, 3, 5, 7, \ldots\} \cup \{2, 4, 6, \ldots\}$

SOLUTION

(a) Begin by listing the elements of the first set, $\{1, 2, 5, 9, 14\}$. Then include any elements from the second set that are not already listed.

$$\{1, 2, 5, 9, 14\} \cup \{1, 3, 4, 8\} = \{1, 2, 3, 4, 5, 8, 9, 14\}$$

(b) $\{1, 3, 5, 7\} \cup \{2, 4, 6\} = \{1, 2, 3, 4, 5, 6, 7\}$

(c) $\{1, 3, 5, 7, \ldots\} \cup \{2, 4, 6, \ldots\} = N$ Natural numbers

Now Try Exercises 73 and 77.

The **set operations** are summarized below.

Set Operations

Let A and B be sets, with universal set U.

The **complement** of set A is the set A' of all elements in the universal set that do *not* belong to set A.

$$A' = \{x \mid x \in U,\quad x \notin A\}$$

The **intersection** of sets A and B, written $A \cap B$, is made up of all the elements belonging to both set A *and* set B.

$$A \cap B = \{x \mid x \in A \text{ and } x \in B\}$$

The **union** of sets A and B, written $A \cup B$, is made up of all the elements belonging to set A *or* to set B.

$$A \cup B = \{x \mid x \in A \text{ or } x \in B\}$$

R.1 Exercises

1. finite; yes 2. finite; yes
3. infinite; no 4. infinite; yes
5. infinite; no 6. infinite; yes
7. infinite; no 8. infinite; yes

9. $\{12, 13, 14, 15, 16, 17, 18, 19, 20\}$
10. $\{8, 9, 10, 11, 12, 13, 14, 15, 16, 17\}$
11. $\{1, \frac{1}{2}, \frac{1}{4}, \frac{1}{8}, \frac{1}{16}, \frac{1}{32}\}$
12. $\{3, 9, 27, 81, 243, 729\}$
13. $\{17, 22, 27, 32, 37, 42, 47\}$
14. $\{74, 68, 62, 56, 50, 44, 38\}$
15. $\{8, 9, 10, 11, 12, 13, 14\}$
16. $\{1, 2, 3, 4\}$

17. $\in$ 18. $\in$ 19. $\notin$
20. $\notin$ 21. $\in$ 22. $\in$
23. $\notin$ 24. $\notin$ 25. $\notin$
26. $\notin$ 27. $\notin$ 28. $\notin$

Identify each set as finite *or* infinite. *Then determine whether* 10 *is an element of the set.* ***See Example 1.***

1. $\{4, 5, 6, \ldots, 15\}$
2. $\{1, 2, 3, 4, 5, \ldots, 75\}$
3. $\{1, \frac{1}{2}, \frac{1}{4}, \frac{1}{8}, \ldots\}$
4. $\{4, 5, 6, \ldots\}$
5. $\{x \mid x$ is a natural number greater than $11\}$
6. $\{x \mid x$ is a natural number greater than or equal to $10\}$
7. $\{x \mid x$ is a fraction between 1 and $2\}$
8. $\{x \mid x$ is an even natural number$\}$

Use set notation, and list all the elements of each set. ***See Example 2.***

9. $\{12, 13, 14, \ldots, 20\}$
10. $\{8, 9, 10, \ldots, 17\}$
11. $\{1, \frac{1}{2}, \frac{1}{4}, \ldots, \frac{1}{32}\}$
12. $\{3, 9, 27, \ldots, 729\}$
13. $\{17, 22, 27, \ldots, 47\}$
14. $\{74, 68, 62, \ldots, 38\}$
15. $\{x \mid x$ is a natural number greater than 7 and less than $15\}$
16. $\{x \mid x$ is a natural number not greater than $4\}$

Insert $\in$ or $\notin$ in each blank to make the resulting statement true. ***See Examples 1 and 2.***

17. 6 ____ $\{3, 4, 5, 6\}$
18. 9 ____ $\{3, 2, 5, 9, 8\}$
19. -4 ____ $\{4, 6, 8, 10\}$
20. -12 ____ $\{3, 5, 12, 14\}$
21. 0 ____ $\{2, 0, 3, 4\}$
22. 0 ____ $\{0, 5, 6, 7, 8, 10\}$
23. $\{3\}$ ____ $\{2, 3, 4, 5\}$
24. $\{5\}$ ____ $\{3, 4, 5, 6, 7\}$
25. $\{0\}$ ____ $\{0, 1, 2, 5\}$
26. $\{2\}$ ____ $\{2, 4, 6, 8\}$
27. 0 ____ $\emptyset$
28. $\emptyset$ ____ $\emptyset$

29. false 30. false 31. true
32. true 33. true 34. true
35. true 36. true 37. false
38. false 39. true 40. true

41. true 42. true 43. true
44. true 45. false 46. false
47. true 48. true 49. true
50. false 51. false 52. false

53. $\subseteq$ 54. $\subseteq$ 55. $\not\subseteq$
56. $\not\subseteq$ 57. $\subseteq$ 58. $\subseteq$

59. true 60. true 61. false
62. false 63. true 64. false
65. false 66. true 67. true
68. false 69. true 70. true

71. $\{0, 2, 4\}$
72. M, or $\{0, 2, 4, 6, 8\}$
73. $\{0, 1, 2, 3, 4, 5, 6, 7, 8, 9, 11, 13\}$
74. $\{0, 1, 2, 3, 4, 6, 8\}$
75. $\emptyset$; M and N are disjoint sets.
76. N, or $\{1, 3, 5, 7, 9, 11, 13\}$
77. $\{0, 1, 2, 3, 4, 5, 7, 9, 11, 13\}$
78. Q, or $\{0, 2, 4, 6, 8, 10, 12\}$
79. Q, or $\{0, 2, 4, 6, 8, 10, 12\}$
80. N, or $\{1, 3, 5, 7, 9, 11, 13\}$
81. $\{10, 12\}$
82. $\{6, 8, 10, 12\}$
83. $\emptyset$; $\emptyset$ and R are disjoint.
84. $\emptyset$; $\emptyset$ and Q are disjoint.
85. N, or $\{1, 3, 5, 7, 9, 11, 13\}$; N and $\emptyset$ are disjoint.
86. R, or $\{0, 1, 2, 3, 4\}$; R and $\emptyset$ are disjoint.
87. R, or $\{0, 1, 2, 3, 4\}$
88. $\{0, 2, 4\}$
89. $\{0, 1, 2, 3, 4, 6, 8\}$
90. N, or $\{1, 3, 5, 7, 9, 11, 13\}$
91. R, or $\{0, 1, 2, 3, 4\}$
92. M, or $\{0, 2, 4, 6, 8\}$
93. $\emptyset$; Q' and $(N' \cap U)$ are disjoint.
94. U, or $\{0, 1, 2, 3, 4, 5, 6, 7, 8, 9, 10, 11, 12, 13\}$

Determine whether each statement is true *or* false. **See Examples 1–3.**

29. $3 \in \{2, 5, 6, 8\}$
30. $6 \in \{-2, 5, 8, 9\}$
31. $1 \in \{3, 4, 5, 11, 1\}$
32. $12 \in \{18, 17, 15, 13, 12\}$
33. $9 \notin \{2, 1, 5, 8\}$
34. $3 \notin \{7, 6, 5, 4\}$
35. $\{2, 5, 8, 9\} = \{2, 5, 9, 8\}$
36. $\{3, 0, 9, 6, 2\} = \{2, 9, 0, 3, 6\}$
37. $\{5, 8, 9\} = \{5, 8, 9, 0\}$
38. $\{3, 7, 12, 14\} = \{3, 7, 12, 14, 0\}$
39. $\{x \mid x \text{ is a natural number less than } 3\} = \{1, 2\}$
40. $\{x \mid x \text{ is a natural number greater than } 10\} = \{11, 12, 13, \ldots\}$

Let $A = \{2, 4, 6, 8, 10, 12\}$, $B = \{2, 4, 8, 10\}$, $C = \{4, 10, 12\}$, $D = \{2, 10\}$, *and* $U = \{2, 4, 6, 8, 10, 12, 14\}$.

Determine whether each statement is true *or* false. **See Example 3.**

41. $A \subseteq U$
42. $C \subseteq U$
43. $D \subseteq B$
44. $D \subseteq A$
45. $A \subseteq B$
46. $B \subseteq C$
47. $\emptyset \subseteq A$
48. $\emptyset \subseteq \emptyset$
49. $\{4, 8, 10\} \subseteq B$
50. $\{0, 2\} \subseteq D$
51. $B \subseteq D$
52. $A \subseteq C$

Insert $\subseteq$ *or* $\not\subseteq$ *in each blank to make the resulting statement true.* **See Example 3.**

53. $\{2, 4, 6\}$ ____ $\{3, 2, 5, 4, 6\}$
54. $\{1, 5\}$ ____ $\{0, -1, 2, 3, 1, 5\}$
55. $\{0, 1, 2\}$ ____ $\{1, 2, 3, 4, 5\}$
56. $\{5, 6, 7, 8\}$ ____ $\{1, 2, 3, 4, 5, 6, 7\}$
57. $\emptyset$ ____ $\{1, 4, 6, 8\}$
58. $\emptyset$ ____ $\emptyset$

Determine whether each statement is true *or* false. **See Examples 4–6.**

59. $\{5, 7, 9, 19\} \cap \{7, 9, 11, 15\} = \{7, 9\}$
60. $\{8, 11, 15\} \cap \{8, 11, 19, 20\} = \{8, 11\}$
61. $\{2, 1, 7\} \cup \{1, 5, 9\} = \{1\}$
62. $\{6, 12, 14, 16\} \cup \{6, 14, 19\} = \{6, 14\}$
63. $\{3, 2, 5, 9\} \cap \{2, 7, 8, 10\} = \{2\}$
64. $\{8, 9, 6\} \cup \{9, 8, 6\} = \{8, 9\}$
65. $\{3, 5, 9, 10\} \cap \emptyset = \{3, 5, 9, 10\}$
66. $\{3, 5, 9, 10\} \cup \emptyset = \{3, 5, 9, 10\}$
67. $\{1, 2, 4\} \cup \{1, 2, 4\} = \{1, 2, 4\}$
68. $\{1, 2, 4\} \cap \{1, 2, 4\} = \emptyset$
69. $\emptyset \cup \emptyset = \emptyset$
70. $\emptyset \cap \emptyset = \emptyset$

Let $U = \{0, 1, 2, 3, 4, 5, 6, 7, 8, 9, 10, 11, 12, 13\}$, $M = \{0, 2, 4, 6, 8\}$, $N = \{1, 3, 5, 7, 9, 11, 13\}$, $Q = \{0, 2, 4, 6, 8, 10, 12\}$, *and* $R = \{0, 1, 2, 3, 4\}$.

Use these sets to find each of the following. Identify any disjoint sets. **See Examples 4–6.**

71. $M \cap R$
72. $M \cap U$
73. $M \cup N$
74. $M \cup R$
75. $M \cap N$
76. $U \cap N$
77. $N \cup R$
78. $M \cup Q$
79. N'
80. Q'
81. $M' \cap Q$
82. $Q \cap R'$
83. $\emptyset \cap R$
84. $\emptyset \cap Q$
85. $N \cup \emptyset$
86. $R \cup \emptyset$
87. $(M \cap N) \cup R$
88. $(N \cup R) \cap M$
89. $(Q \cap M) \cup R$
90. $(R \cup N) \cap M'$
91. $(M' \cup Q) \cap R$
92. $Q \cap (M \cup N)$
93. $Q' \cap (N' \cap U)$
94. $(U \cap \emptyset') \cup R$

R.2 Real Numbers and Their Properties

- Sets of Numbers and the Number Line
- Exponents
- Order of Operations
- Properties of Real Numbers
- Order on the Number Line
- Absolute Value

Sets of Numbers and the Number Line As mentioned in the previous section, the set of **natural numbers** is written in set notation as follows.

$$\{1, 2, 3, 4, \ldots\} \quad \text{Natural numbers (Section R.1)}$$

Including 0 with the set of natural numbers gives the set of **whole numbers.**

$$\{0, 1, 2, 3, 4, \ldots\} \quad \text{Whole numbers}$$

Including the negatives of the natural numbers with the set of whole numbers gives the set of **integers.**

$$\{\ldots, -3, -2, -1, 0, 1, 2, 3, \ldots\} \quad \text{Integers}$$

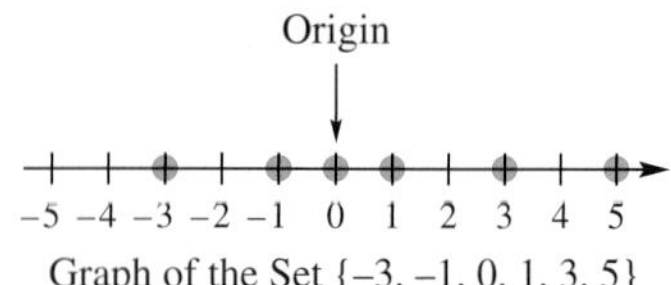

Graph of the Set {–3, –1, 0, 1, 3, 5}

Figure 5

Integers can be **graphed** on a **number line.** See **Figure 5**. Every number corresponds to one and only one point on the number line, and each point corresponds to one and only one number. The number associated with a given point is called the **coordinate** of the point. This correspondence forms a **coordinate system.**

The result of dividing two integers (with a nonzero divisor) is called a *rational number,* or *fraction.* A **rational number** is an element of the set defined as follows.

$$\left\{\frac{p}{q} \,\middle|\, p \text{ and } q \text{ are integers and } q \neq 0\right\} \quad \text{Rational numbers}$$

The set of rational numbers includes the natural numbers, the whole numbers, and the integers. For example, the integer -3 is a rational number because it can be written as $\frac{-3}{1}$. Numbers that can be written as repeating or terminating decimals are also rational numbers. For example, $0.\overline{6} = 0.66666\ldots$ represents a rational number that can be expressed as the fraction $\frac{2}{3}$.

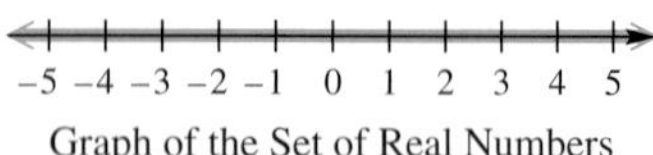

Graph of the Set of Real Numbers

Figure 6

The set of all numbers that correspond to points on a number line is the **real numbers,** shown in **Figure 6**. Real numbers can be represented by decimals. Since every fraction has a decimal form—for example, $\frac{1}{4} = 0.25$—real numbers include rational numbers.

Some real numbers cannot be represented by quotients of integers. These numbers are **irrational numbers.** The set of irrational numbers includes $\sqrt{3}$ and $\sqrt{5}$. Another irrational number is π, which is *approximately* equal to 3.14159. The numbers in the set $\left\{-\frac{2}{3}, 0, \sqrt{2}, \sqrt{5}, \pi, 4\right\}$ can be located on a number line, as shown in **Figure 7**.

$\sqrt{2}$, $\sqrt{5}$, and π are irrational. Since $\sqrt{2}$ is approximately equal to 1.41, it is located between 1 and 2, slightly closer to 1.

Figure 7

The sets of numbers discussed so far are summarized as follows.

Sets of Numbers

Set	Description
Natural numbers	$\{1, 2, 3, 4, \ldots\}$
Whole numbers	$\{0, 1, 2, 3, 4, \ldots\}$
Integers	$\{\ldots, -3, -2, -1, 0, 1, 2, 3, \ldots\}$
Rational numbers	$\{\frac{p}{q} \mid p \text{ and } q \text{ are integers and } q \neq 0\}$
Irrational numbers	$\{x \mid x \text{ is real but not rational}\}$
Real numbers	$\{x \mid x \text{ corresponds to a point on a number line}\}$

Classroom Example 1
Repeat **Example 1** for
$S = \left\{-12, -7.25, -\sqrt{3}, -\frac{5}{6}, 0, 0.\overline{45}, \sqrt{9}, \frac{21}{3}, 999\right\}$.

Answers:
(a) $\sqrt{9}$ (or 3), $\frac{21}{3}$ (or 7), 999
(b) 0, $\sqrt{9}$ (or 3), $\frac{21}{3}$ (or 7), 999
(c) -12, 0, $\sqrt{9}$ (or 3), $\frac{21}{3}$ (or 7), 999
(d) All elements of S are rational except $-\sqrt{3}$.
(e) $-\sqrt{3}$
(f) All elements of S are real numbers.

EXAMPLE 1 Identifying Sets of Numbers

Let $A = \left\{-8, -6, -\frac{12}{4}, -\frac{3}{4}, 0, \frac{3}{8}, \frac{1}{2}, 1, \sqrt{2}, \sqrt{5}, 6\right\}$. List the elements from A that belong to each set.

(a) Natural numbers **(b)** Whole numbers **(c)** Integers
(d) Rational numbers **(e)** Irrational numbers **(f)** Real numbers

SOLUTION

(a) Natural numbers: 1 and 6 **(b)** Whole numbers: 0, 1, and 6

(c) Integers: -8, -6, $-\frac{12}{4}$ (or -3), 0, 1, and 6

(d) Rational numbers: -8, -6, $-\frac{12}{4}$ (or -3), $-\frac{3}{4}$, 0, $\frac{3}{8}$, $\frac{1}{2}$, 1, and 6

(e) Irrational numbers: $\sqrt{2}$ and $\sqrt{5}$

(f) All elements of A are real numbers.

✔ *Now Try Exercises 1, 11, and 13.*

Figure 8 shows the relationships among the subsets of the real numbers.

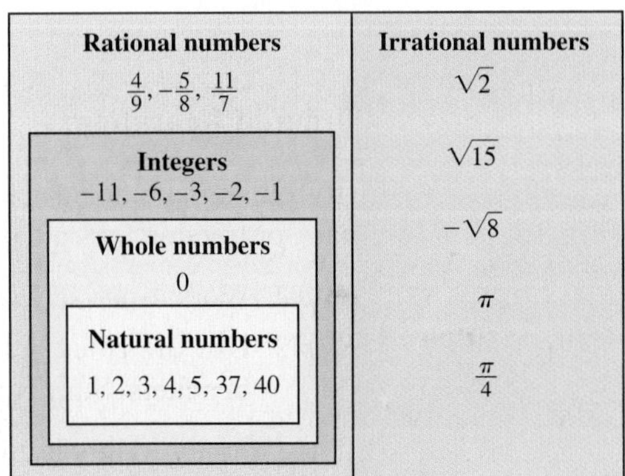

Figure 8

Exponents The product $2 \cdot 2 \cdot 2$ can be written as 2^3, where the 3 shows that three factors of 2 appear in the product.

Teaching Tip Point out that when an exponent is not indicated, it is understood to be 1. This fact will be used in **Section R.3** when we discuss polynomials.

Exponential Notation

If n is any positive integer and a is any real number, then the nth power of a is written using exponential notation as follows.

$$a^n = \underbrace{a \cdot a \cdot a \cdot \ldots \cdot a}_{n \text{ factors of } a}$$

That is, a^n means the product of n factors of a. The integer n is the **exponent,** a is the **base,** and a^n is a **power** or an **exponential expression** (or simply an **exponential**). Read a^n as **"a to the nth power,"** or just **"a to the nth."**

Classroom Example 2
Evaluate each exponential expression, and identify the base and the exponent.

(a) 10^3 (b) $(-3)^4$ (c) -3^4
(d) $2 \cdot 5^2$ (e) $(2 \cdot 5)^2$

Answers:
(a) 1000; base: 10; exponent: 3
(b) 81; base: -3; exponent: 4
(c) -81; base: 3; exponent: 4
(d) 50; base: 5; exponent: 2
(e) 100; base: 10; exponent: 2

Teaching Tip Focus on the difference between **Examples 2(b) and (c).** One of the most common mistakes students make is to treat them as the same.

EXAMPLE 2 Evaluating Exponential Expressions

Evaluate each exponential expression, and identify the base and the exponent.

(a) 4^3 **(b)** $(-6)^2$ **(c)** -6^2 **(d)** $4 \cdot 3^2$ **(e)** $(4 \cdot 3)^2$

SOLUTION

(a) $4^3 = \underbrace{4 \cdot 4 \cdot 4}_{\text{3 factors of 4}} = 64$ The base is 4 and the exponent is 3.

(b) $(-6)^2 = (-6)(-6) = 36$ The base is -6 and the exponent is 2.

(c) $-6^2 = -(6 \cdot 6) = -36$
The base is 6 and the exponent is 2.

Notice that parts (b) and (c) are different.

(d) $4 \cdot 3^2 = 4 \cdot 3 \cdot 3 = 36$ The base is 3 and the exponent is 2.

$3^2 = 3 \cdot 3$, NOT $3 \cdot 2$

(e) $(4 \cdot 3)^2 = 12^2 = 144$
The base is $4 \cdot 3$, or 12, and the exponent is 2.

$(4 \cdot 3)^2 \neq 4 \cdot 3^2$

Now Try Exercises 17, 19, 21, and 23.

Order of Operations When a problem involves more than one operation symbol, we use the following order of operations.

Teaching Tip Have students identify the common **errors** contained in these examples.

(a) $32 - 8 \div 2 = 24 \div 2 = 12$

(b) $3 \cdot 4^2 - 7 \cdot 5 = 12^2 - 35$
$= 144 - 35$
$= 109$

(c) $\dfrac{3^3 - 5 \cdot 3}{36 \div 12} = \dfrac{27 - 5 \cdot 3}{3}$
$= 27 - 5$
$= 22$

Order of Operations

If grouping symbols such as parentheses, square brackets, absolute value bars, or fraction bars are present, begin as follows.

Step 1 Work separately above and below each **fraction bar.**

Step 2 Use the rules below within each set of **parentheses** or **square brackets.** Start with the innermost set and work outward.

If no grouping symbols are present, follow these steps.

Step 1 Simplify all **powers** and **roots.** ***Work from left to right.***

Step 2 Do any **multiplications** or **divisions** in order. ***Work from left to right.***

Step 3 Do any **negations, additions,** or **subtractions** in order. ***Work from left to right.***

Classroom Example 3
Evaluate each expression.

(a) $3 \cdot 9 - 2^5 \div 4$
(b) $(30 - 5) \cdot 3 \div 15 + 7$
(c) $\dfrac{2^4 - 11}{9 + 3 \cdot 2}$
(d) $\dfrac{-7^2 - (-9)}{6(-3) - 1(-2)}$

Answers:
(a) 19 (b) 12
(c) $\frac{1}{3}$ (d) $\frac{5}{2}$

EXAMPLE 3 Using Order of Operations

Evaluate each expression.

(a) $6 \div 3 + 2^3 \cdot 5$ **(b)** $(8 + 6) \div 7 \cdot 3 - 6$

(c) $\dfrac{4 + 3^2}{6 - 5 \cdot 3}$ **(d)** $\dfrac{-(-3)^3 + (-5)}{2(-8) - 5(3)}$

SOLUTION

(a) $6 \div 3 + 2^3 \cdot 5 = 6 \div 3 + 8 \cdot 5$ Evaluate the exponential.
$= 2 + 8 \cdot 5$ Divide.
$= 2 + 40$ Multiply.
$= 42$ Add.

Multiply or divide in order from left to right.

(b) $(8+6) \div 7 \cdot 3 - 6 = 14 \div 7 \cdot 3 - 6$ Work inside parentheses.

Be careful to divide *before* multiplying here.

$= 2 \cdot 3 - 6$ Divide.

$= 6 - 6$ Multiply.

$= 0$ Subtract.

(c) $\dfrac{4+3^2}{6-5\cdot 3} = \dfrac{4+9}{6-15}$ Evaluate the exponential and multiply.

$= \dfrac{13}{-9}$, or $-\dfrac{13}{9}$ Add and subtract; $\frac{a}{-b} = -\frac{a}{b}$.

(d) $\dfrac{-(-3)^3+(-5)}{2(-8)-5(3)} = \dfrac{-(-27)+(-5)}{2(-8)-5(3)}$ Evaluate the exponential.

$= \dfrac{27+(-5)}{-16-15}$ Multiply.

$= \dfrac{22}{-31}$, or $-\dfrac{22}{31}$ Add and subtract; $\frac{a}{-b} = -\frac{a}{b}$.

Now Try Exercises 25, 27, and 33.

Classroom Example 4
Evaluate each expression for $x=-4$, $y=3$, and $z=-6$.
(a) $6x^2+5y-3z$
(b) $\dfrac{4y-3(x-1)^2}{z+9}$
(c) $\dfrac{\frac{x}{4}+\frac{y}{3}}{\frac{z}{2}-\frac{x}{2}}$

Answers:
(a) 129 (b) -21 (c) 0

Teaching Tip Include an explanation of why

$$5-2(x+y)$$

is *not* equivalent to

$$3(x+y).$$

EXAMPLE 4 Using Order of Operations

Evaluate each expression for $x=-2$, $y=5$, and $z=-3$.

(a) $-4x^2-7y+4z$ **(b)** $\dfrac{2(x-5)^2+4y}{z+4}$ **(c)** $\dfrac{\frac{x}{2}-\frac{y}{5}}{\frac{3z}{9}+\frac{8y}{5}}$

SOLUTION

Use parentheses around substituted values to avoid errors.

(a) $-4x^2-7y+4z = -4(-2)^2-7(5)+4(-3)$ Substitute: $x=-2$, $y=5$, and $z=-3$.

$= -4(4)-7(5)+4(-3)$ Evaluate the exponential.

$= -16-35-12$ Multiply.

$= -63$ Subtract.

(b) $\dfrac{2(x-5)^2+4y}{z+4} = \dfrac{2(-2-5)^2+4(5)}{-3+4}$ Substitute: $x=-2$, $y=5$, and $z=-3$.

$= \dfrac{2(-7)^2+20}{1}$ Work inside parentheses. Then multiply and add.

$= 2(49)+20$ Evaluate the exponential.

$= 98+20$ Multiply.

$= 118$ Add.

(c) $\dfrac{\frac{x}{2}-\frac{y}{5}}{\frac{3z}{9}+\frac{8y}{5}} = \dfrac{\frac{-2}{2}-\frac{5}{5}}{\frac{3(-3)}{9}+\frac{8(5)}{5}}$ Substitute: $x=-2$, $y=5$, and $z=-3$.

$= \dfrac{-1-1}{-1+8}$, or $-\dfrac{2}{7}$ Simplify the fractions.

Now Try Exercises 35, 43, and 45.

Properties of Real Numbers The following basic properties can be generalized to apply to expressions with variables.

Properties of Real Numbers

Let a, b, and c represent real numbers.

Property	Description
Closure Properties $a + b$ is a real number. ab is a real number.	The sum or product of two real numbers is a real number.
Commutative Properties $a + b = b + a$ $ab = ba$	The sum or product of two real numbers is the same regardless of their order.
Associative Properties $(a + b) + c = a + (b + c)$ $(ab)c = a(bc)$	The sum or product of three real numbers is the same no matter which two are added or multiplied first.
Identity Properties There exists a unique real number 0 such that $a + 0 = a$ and $0 + a = a$. There exists a unique real number 1 such that $a \cdot 1 = a$ and $1 \cdot a = a$.	The sum of a real number and 0 is that real number, and the product of a real number and 1 is that real number.
Inverse Properties There exists a unique real number $-a$ such that $a + (-a) = 0$ and $-a + a = 0$. If $a \neq 0$, there exists a unique real number $\frac{1}{a}$ such that $a \cdot \frac{1}{a} = 1$ and $\frac{1}{a} \cdot a = 1$.	The sum of any real number and its negative is 0, and the product of any nonzero real number and its reciprocal is 1.
Distributive Properties $a(b + c) = ab + ac$ $a(b - c) = ab - ac$	The product of a real number and the sum (or difference) of two real numbers equals the sum (or difference) of the products of the first number and each of the other numbers.

Teaching Tip A common **error** students make is to "distribute" multiplication over multiplication. For instance, they write

$$2(xy) = 2x \cdot 2y.$$

CAUTION Notice that with the commutative properties, the ***order*** changes from one side of the equality symbol to the other. With the associative properties the order does not change, but the ***grouping*** does.

Commutative Properties	Associative Properties
$(x + 4) + 9 = (4 + x) + 9$	$(x + 4) + 9 = x + (4 + 9)$
$7 \cdot (5 \cdot 2) = (5 \cdot 2) \cdot 7$	$7 \cdot (5 \cdot 2) = (7 \cdot 5) \cdot 2$

Classroom Example 5
Simplify each expression.
(a) $(12 + 2x) + 18$
(b) $\left(\frac{4}{7}\right)(-35t)$
(c) $(-54s)\left(-\frac{4}{9}\right)$
Answers:
(a) $30 + 2x$ (b) $-20t$
(c) $24s$

EXAMPLE 5 Simplifying Expressions

Use the commutative and associative properties to simplify each expression.

(a) $6 + (9 + x)$ **(b)** $\frac{5}{8}(16y)$ **(c)** $-10p\left(\frac{6}{5}\right)$

SOLUTION

(a) $6 + (9 + x) = (6 + 9) + x$ Associative property
$= 15 + x$ Add.

(b) $\frac{5}{8}(16y) = \left(\frac{5}{8} \cdot 16\right)y$ Associative property
$= 10y$ Multiply.

(c) $-10p\left(\frac{6}{5}\right) = \frac{6}{5}(-10p)$ Commutative property
$= \left[\frac{6}{5}(-10)\right]p$ Associative property
$= -12p$ Multiply.

✔ Now Try Exercises 63 and 65.

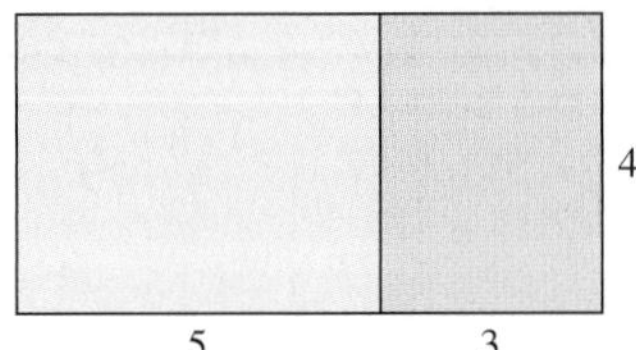

Geometric Model of the Distributive Property

Figure 9

Figure 9 helps to explain the distributive property. The area of the entire region shown can be found in two ways, as follows.

$$4(5 + 3) = 4(8) = 32$$

or

$$4(5) + 4(3) = 20 + 12 = 32$$

The result is the same. This means that

$$4(5 + 3) = 4(5) + 4(3).$$

Classroom Example 6
Rewrite each expression using the distributive property and simplify, if possible.
(a) $8(m - 2n)$
(b) $-(-3r + 5s)$
(c) $\frac{3}{4}\left(\frac{5}{6}p + \frac{1}{2}q - 28\right)$
(d) $22t - 55$
Answers:
(a) $8m - 16n$
(b) $3r - 5s$
(c) $\frac{5}{8}p + \frac{3}{8}q - 21$
(d) $11(2t - 5)$

EXAMPLE 6 Using the Distributive Property

Rewrite each expression using the distributive property and simplify, if possible.

(a) $3(x + y)$ **(b)** $-(m - 4n)$ **(c)** $\frac{1}{3}\left(\frac{4}{5}m - \frac{3}{2}n - 27\right)$ **(d)** $7p + 21$

SOLUTION

(a) $3(x + y) = 3x + 3y$

(b) $-(m - 4n) = -1(m - 4n)$
$= -1(m) + (-1)(-4n)$
$= -m + 4n$

Be careful with the negative signs.

(c) $\frac{1}{3}\left(\frac{4}{5}m - \frac{3}{2}n - 27\right) = \frac{1}{3}\left(\frac{4}{5}m\right) + \frac{1}{3}\left(-\frac{3}{2}n\right) + \frac{1}{3}(-27)$
$= \frac{4}{15}m - \frac{1}{2}n - 9$

(d) $7p + 21 = 7p + 7 \cdot 3$
$= 7(p + 3)$ Distributive property in reverse

✔ Now Try Exercises 67, 69, and 71.

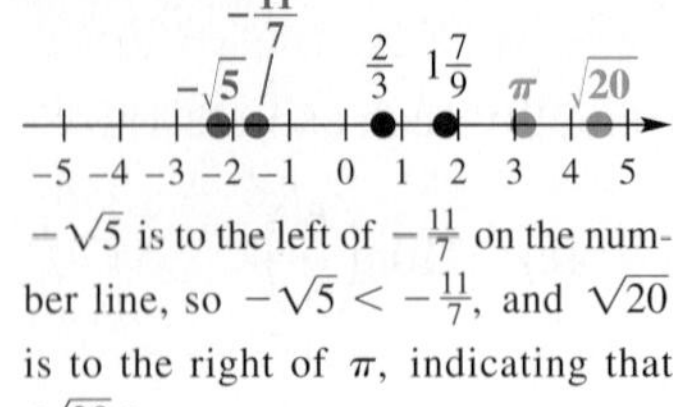

$-\sqrt{5}$ is to the left of $-\frac{11}{7}$ on the number line, so $-\sqrt{5} < -\frac{11}{7}$, and $\sqrt{20}$ is to the right of π, indicating that $\sqrt{20} > \pi$.

Figure 10

Order on the Number Line If the real number a is to the left of the real number b on a number line, then

$$a \text{ is less than } b, \quad \text{written} \quad a < b.$$

If a is to the right of b, then

$$a \text{ is greater than } b, \quad \text{written} \quad a > b.$$

The inequality symbol must point toward the lesser number.

Figure 10 illustrates this with several pairs of numbers. Statements involving these symbols, as well as the symbols less than or equal to, $\leq$, and greater than or equal to, $\geq$, are called **inequalities.** The inequality $a < b < c$ says that b is ***between*** a and c since $a < b$ and $b < c$.

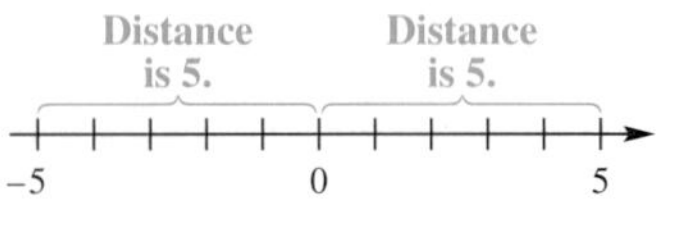

Figure 11

Absolute Value The distance on the number line from a number to 0 is called the **absolute value** of that number. The absolute value of the number a is written $|a|$. For example, the distance on the number line from 5 to 0 is 5, as is the distance from -5 to 0. See **Figure 11.** Therefore, both of the following are true.

$$|5| = 5 \quad \text{and} \quad |-5| = 5$$

NOTE ***Since distance cannot be negative, the absolute value of a number is always positive or 0.***

The algebraic definition of absolute value follows.

Absolute Value

Let a represent a real number.

$$|a| = \begin{cases} a & \text{if } a \geq 0 \\ -a & \text{if } a < 0 \end{cases}$$

That is, the absolute value of a positive number or 0 equals that number, while the absolute value of a negative number equals its negative (or opposite).

Classroom Example 7
Evaluate each expression.
(a) $|-6.85|$ (b) $-|50|$
(c) $-\left|-\frac{2}{3}\right|$
(d) $|y|$, for $y = \sqrt{2}$

Answers:
(a) 6.85 (b) -50
(c) $-\frac{2}{3}$ (d) $\sqrt{2}$

EXAMPLE 7 Evaluating Absolute Values

Evaluate each expression.

(a) $\left|-\frac{5}{8}\right|$ **(b)** $-|8|$ **(c)** $-|-2|$ **(d)** $|2x|$, for $x = \pi$

SOLUTION

(a) $\left|-\frac{5}{8}\right| = \frac{5}{8}$ **(b)** $-|8| = -(8) = -8$

(c) $-|-2| = -(2) = -2$ **(d)** $|2\pi| = 2\pi$

Now Try Exercises 83 and 87.

Teaching Tip In Example 7(c), compare $-|-2|$ with $-(-2)$, which is positive.

Absolute value is useful in applications where only the *size* (or magnitude), not the *sign,* of the difference between two numbers is important.

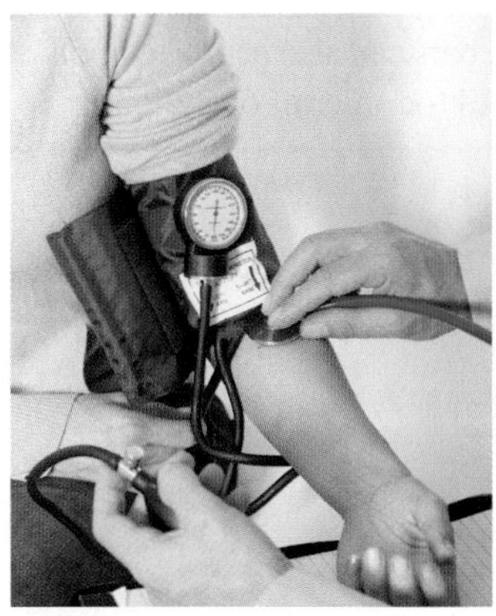

Classroom Example 8
Refer to **Example 8.** Find P_d for a patient with a systolic pressure, P, of 146.

Answer: 26

EXAMPLE 8 Measuring Blood Pressure Difference

Systolic blood pressure is the maximum pressure produced by each heartbeat. Both low blood pressure and high blood pressure may be cause for medical concern. Therefore, health care professionals are interested in a patient's "pressure difference from normal," or P_d.

If 120 is considered a normal systolic pressure, then

$$P_d = |P - 120|, \quad \text{where } P \text{ is the patient's recorded systolic pressure.}$$

Find P_d for a patient with a systolic pressure, P, of 113.

SOLUTION

$$P_d = |P - 120|$$
$$= |113 - 120| \quad \text{Let } P = 113.$$
$$= |-7| \quad \text{Subtract.}$$
$$= 7 \quad \text{Definition of absolute value}$$

✔ Now Try Exercise 89.

Teaching Tip Replace a and b with numerical values (both positive and negative) to illustrate the properties of absolute value. When explaining the triangle inequality, use $a = 4$ and $b = 10$, $a = -4$ and $b = -10$, and $a = -4$ and $b = 10$.

Properties of Absolute Value

Let a and b represent real numbers.

Property	Description
1. $\|a\| \geq 0$	The absolute value of a real number is positive or 0.
2. $\|-a\| = \|a\|$	The absolute values of a real number and its opposite are equal.
3. $\|a\| \cdot \|b\| = \|ab\|$	The product of the absolute values of two real numbers equals the absolute value of their product.
4. $\frac{\|a\|}{\|b\|} = \left\|\frac{a}{b}\right\| \quad (b \neq 0)$	The quotient of the absolute values of two real numbers equals the absolute value of their quotient.
5. $\|a + b\| \leq \|a\| + \|b\|$ (the triangle inequality)	The absolute value of the sum of two real numbers is less than or equal to the sum of their absolute values.

LOOKING AHEAD TO CALCULUS
One of the most important definitions in calculus, that of the **limit,** uses absolute value. (The symbols ϵ (epsilon) and δ (delta) are often used to represent small quantities in mathematics.)

Suppose that a function f is defined at every number in an open interval I containing a, except perhaps at a itself. Then the limit of $f(x)$ as x approaches a is L, written

$$\lim_{x \to a} f(x) = L,$$

if for every $\epsilon > 0$ there exists a $\delta > 0$ such that $|f(x) - L| < \epsilon$ whenever $0 < |x - a| < \delta$.

To illustrate these properties, see the following.

$$|-15| = 15 \text{ and } 15 \geq 0 \quad \text{Property 1}$$

$$|-10| = 10 \text{ and } |10| = 10, \text{ so } |-10| = |10|. \quad \text{Property 2}$$

$$|5x| = |5| \cdot |x| = 5|x| \text{ since 5 is positive.} \quad \text{Property 3}$$

$$\left|\frac{2}{y}\right| = \frac{|2|}{|y|} = \frac{2}{|y|}, \quad y \neq 0 \quad \text{Property 4}$$

To illustrate the triangle inequality, we let $a = 3$ and $b = -7$.

$$|a + b| = |3 + (-7)| = |-4| = 4$$
$$|a| + |b| = |3| + |-7| = 3 + 7 = 10$$

Thus, $\quad |a + b| \leq |a| + |b|. \quad$ Property 5

NOTE As seen in **Example 9(b),** absolute value bars can also act as symbols of inclusion. Remember this when applying the rules for order of operations.

Classroom Example 9
Let $m = 13$ and $n = -9$. Evaluate each expression.
(a) $|3m + 5n|$
(b) $\dfrac{|2m| - 3|n|}{|m + n|}$

Answers:
(a) 6 (b) $-\frac{1}{4}$

EXAMPLE 9 Evaluating Absolute Value Expressions

Let $x = -6$ and $y = 10$. Evaluate each expression.

(a) $|2x - 3y|$ **(b)** $\dfrac{2|x| - |3y|}{|xy|}$

SOLUTION

(a) $|2x - 3y| = |2(-6) - 3(10)|$ Substitute.

$= |-12 - 30|$ Work inside absolute value bars. Multiply.

$= |-42|$ Subtract.

$= 42$ Definition of absolute value

(b) $\dfrac{2|x| - |3y|}{|xy|} = \dfrac{2|-6| - |3(10)|}{|-6(10)|}$ Substitute.

$= \dfrac{2 \cdot 6 - |30|}{|-60|}$ $|-6| = 6$; multiply.

$= \dfrac{12 - 30}{60}$ Multiply. $|30| = 30$, $|-60| = 60$

$= \dfrac{-18}{60}$ Subtract.

$= -\dfrac{3}{10}$ Write in lowest terms; $\frac{-a}{b} = -\frac{a}{b}$.

✔ *Now Try Exercises 93 and 95.*

Distance between Points on a Number Line

If P and Q are points on a number line with coordinates a and b, respectively, then the distance $d(P, Q)$ between them is given by the following.

$$d(P, Q) = |b - a| \quad \text{or} \quad d(P, Q) = |a - b|$$

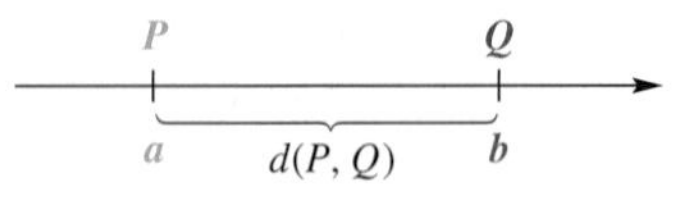

Figure 12

***That is, the distance between two points on a number line is the absolute value of the difference between their coordinates in either order. See* Figure 12.**

Classroom Example 10
Find the distance between -8 and 14.

Answer: 22

EXAMPLE 10 Finding the Distance between Two Points

Find the distance between -5 and 8.

SOLUTION Use the first formula above, with $a = -5$ and $b = 8$.

$$|b - a| = |8 - (-5)| = |8 + 5| = |13| = 13$$

Alternatively, for $a = 8$ and $b = -5$, we obtain the same result.

$$|b - a| = |(-5) - 8| = |-13| = 13$$

✔ *Now Try Exercise 105.*

R.2 Exercises

Complete solutions to all exercises denoted with a green are given at the back of the book for the student.

1. (a) B, C, D, F
(b) A, B, C, D, F
(c) D, F
(d) A, B, C, D, F
(e) E, F
(f) D, F

3. false; Some are whole numbers, but negative integers are not.
4. true
5. false; No irrational number is an integer.
6. true
7. true
8. false; No rational number is irrational.
9. true
10. true

11. 1, 3
12. 0, 1, 3
13. $-6, -\frac{12}{4}$ (or -3), 0, 1, 3
14. $-6, -\frac{12}{4}$ (or -3), $-\frac{5}{8}, 0, \frac{1}{4}$, 1, 3
15. $-\sqrt{3}, 2\pi, \sqrt{12}$
16. All are real numbers.

17. -16 18. -243
19. 16 20. 64
21. -243 22. -32
23. -162 24. -500

25. -6 26. 23
27. -60 28. 18
29. -12 30. 16
31. $-\frac{25}{36}$ 32. $-\frac{5}{8}$
33. $-\frac{6}{7}$ 34. $\frac{6}{5}$

35. 28 36. -42
37. $-\frac{1}{2}$ 38. $-\frac{7}{2}$
39. $-\frac{23}{20}$ 40. $-\frac{71}{20}$
41. $-\frac{25}{11}$ 42. 3
43. $-\frac{11}{3}$ 44. 2
45. $-\frac{13}{3}$ 46. $\frac{1}{2}$
47. 6 48. 2

1. *Concept Check* Match each number from Column I with the letter or letters of the sets of numbers from Column II to which the number belongs. There may be more than one choice, so give all choices.

I		II	
(a) 0	**(b)** 34	**A.** Natural numbers	**B.** Whole numbers
(c) $-\frac{9}{4}$	**(d)** $\sqrt{36}$	**C.** Integers	**D.** Rational numbers
(e) $\sqrt{13}$	**(f)** 2.16	**E.** Irrational numbers	**F.** Real numbers

2. Explain why no answer in **Exercise 1** can contain both D and E as choices.

Concept Check Decide whether each statement is true *or* false. *If it is false, tell why.*

3. Every integer is a whole number.
4. Every natural number is an integer.
5. Every irrational number is an integer.
6. Every integer is a rational number.
7. Every natural number is a whole number.
8. Some rational numbers are irrational.
9. Some rational numbers are whole numbers.
10. Some real numbers are integers.

Let set $A = \left\{-6, -\frac{12}{4}, -\frac{5}{8}, -\sqrt{3}, 0, \frac{1}{4}, 1, 2\pi, 3, \sqrt{12}\right\}$. *List all the elements of A that belong to each set.* ***See Example 1.***

11. Natural numbers
12. Whole numbers
13. Integers
14. Rational numbers
15. Irrational numbers
16. Real numbers

Evaluate each expression. ***See Example 2.***

17. -2^4
18. -3^5
19. $(-2)^4$
20. $(-2)^6$
21. $(-3)^5$
22. $(-2)^5$
23. $-2 \cdot 3^4$
24. $-4 \cdot 5^3$

Evaluate each expression. ***See Example 3.***

25. $-2 \cdot 5 + 12 \div 3$
26. $9 \cdot 3 - 16 \div 4$
27. $-4(9 - 8) + (-7)(2)^3$
28. $6(-5) - (-3)(2)^4$
29. $(4 - 2^3)(-2 + \sqrt{25})$
30. $(5 - 3^2)(\sqrt{16} - 2^3)$
31. $\left(-\frac{2}{9} - \frac{1}{4}\right) - \left[-\frac{5}{18} - \left(-\frac{1}{2}\right)\right]$
32. $\left[-\frac{5}{8} - \left(-\frac{2}{5}\right)\right] - \left(\frac{3}{2} - \frac{11}{10}\right)$
33. $\dfrac{-8 + (-4)(-6) \div 12}{4 - (-3)}$
34. $\dfrac{15 \div 5 \cdot 4 \div 6 - 8}{-6 - (-5) - 8 \div 2}$

Evaluate each expression for $p = -4$, $q = 8$, *and* $r = -10$. ***See Example 4.***

35. $-p^2 - 7q + r^2$
36. $-p^2 - 2q + r$
37. $\dfrac{q + r}{q + p}$
38. $\dfrac{p + r}{p + q}$
39. $\dfrac{3q}{r} - \dfrac{5}{p}$
40. $\dfrac{3r}{q} - \dfrac{2}{r}$
41. $\dfrac{5r}{2p - 3r}$
42. $\dfrac{3q}{3p - 2r}$
43. $\dfrac{\frac{q}{2} - \frac{r}{3}}{\frac{3p}{4} + \frac{q}{8}}$
44. $\dfrac{\frac{q}{4} - \frac{r}{5}}{\frac{p}{2} + \frac{q}{2}}$
45. $\dfrac{-(p + 2)^2 - 3r}{2 - q}$
46. $\dfrac{-(q - 6)^2 - 2p}{4 - p}$
47. $\dfrac{3p + 3(4 + p)^3}{r + 8}$
48. $\dfrac{5q + 2(1 + p)^3}{r + 3}$

49. distributive
50. distributive
51. inverse 52. inverse
53. identity 54. identity
55. commutative
56. commutative
57. associative
58. associative
59. closure 60. closure

63. $20z$ 64. $-9r$
65. $m + 11$
66. $15 + a$, or $a + 15$
67. $\frac{2}{3}y + \frac{4}{9}z - \frac{5}{3}$
68. $-5m - 2y + 8z$

69. $(8 - 14)p = -6p$
70. $(15 - 10)x = 5x$
71. $-4z + 4y$
72. $-3m - 3n$

73. 1700 74. 3200
75. 150 76. 174

77. false; $|6 - 8| = |8| - |6|$
78. false; $|(-3)^3| = 3^3$
79. true
80. true
81. false; $|a - b| = |b| - |a|$
82. true

83. 10 84. 15
85. $-\frac{4}{7}$ 86. $-\frac{7}{2}$
87. -8 88. -12

89. 9 90. 113 or 147

Identify the property illustrated in each statement. Assume all variables represent real numbers. ***See Examples 5 and 6.***

49. $6 \cdot 12 + 6 \cdot 15 = 6(12 + 15)$

50. $8(m + 4) = 8m + 32$

51. $(t - 6) \cdot \left(\frac{1}{t - 6}\right) = 1$, if $t - 6 \neq 0$

52. $\frac{2 + m}{2 - m} \cdot \frac{2 - m}{2 + m} = 1$, if $m \neq 2$ or -2

53. $(7.5 - y) + 0 = 7.5 - y$

54. $1 \cdot (3x - 7) = 3x - 7$

55. $5(t + 3) = (t + 3) \cdot 5$

56. $-7 + (x + 3) = (x + 3) + (-7)$

57. $(5x)\left(\frac{1}{x}\right) = 5\left(x \cdot \frac{1}{x}\right)$

58. $(38 + 99) + 1 = 38 + (99 + 1)$

59. $5 + \sqrt{3}$ is a real number.

60. 5π is a real number.

61. Is there a commutative property for subtraction? That is, in general, is $a - b$ equal to $b - a$? Support your answer with examples.

62. Is there an associative property for subtraction? That is, does $(a - b) - c$ equal $a - (b - c)$ in general? Support your answer with examples.

Simplify each expression. ***See Examples 5 and 6.***

63. $\frac{10}{11}(22z)$

64. $\left(\frac{3}{4}r\right)(-12)$

65. $(m + 5) + 6$

66. $8 + (a + 7)$

67. $\frac{3}{8}\left(\frac{16}{9}y + \frac{32}{27}z - \frac{40}{9}\right)$

68. $-\frac{1}{4}(20m + 8y - 32z)$

Use the distributive property to rewrite sums as products and products as sums. ***See Example 6.***

69. $8p - 14p$

70. $15x - 10x$

71. $-4(z - y)$

72. $-3(m + n)$

Concept Check *Use the distributive property to calculate each value mentally.*

73. $72 \cdot 17 + 28 \cdot 17$

74. $32 \cdot 80 + 32 \cdot 20$

75. $123\frac{5}{8} \cdot 1\frac{1}{2} - 23\frac{5}{8} \cdot 1\frac{1}{2}$

76. $17\frac{2}{5} \cdot 14\frac{3}{4} - 17\frac{2}{5} \cdot 4\frac{3}{4}$

Concept Check *Decide whether each statement is* true *or* false. *If false, correct the statement so it is true.*

77. $|6 - 8| = |6| - |8|$

78. $|(-3)^3| = -|3^3|$

79. $|-5| \cdot |6| = |-5 \cdot 6|$

80. $\frac{|-14|}{|2|} = \left|\frac{-14}{2}\right|$

81. $|a - b| = |a| - |b|$, if $b > a > 0$

82. If a is negative, then $|a| = -a$.

Evaluate each expression. ***See Example 7.***

83. $|-10|$

84. $|-15|$

85. $-\left|\frac{4}{7}\right|$

86. $-\left|\frac{7}{2}\right|$

87. $-|-8|$

88. $-|-12|$

Solve each problem. ***See Example 8.***

89. *Blood Pressure Difference* Calculate the P_d value for a woman whose actual systolic pressure is 116 and whose normal value should be 125.

90. *Systolic Blood Pressure* If a patient's P_d value is 17 and the normal pressure for his gender and age should be 130, what are the two possible values for his systolic blood pressure?

91. 16 92. 18
93. 20 94. 14
95. −1 96. 6
97. −5 98. −2

99. Property 2 100. Property 1
101. Property 3 102. Property 5
103. Property 1 104. Property 4

105. 3 106. 12
107. 9 108. 13

109. x and y have the same sign.
110. y must be positive.
111. x and y have different signs.
112. x must be negative.
113. x and y have the same sign.
114. x and y have different signs.

115. 17; This represents the number of strokes between their scores.
116. 17,648 yd; No, it is not the same, because the sum of the absolute values is 17,660.

117. 0.031 118. 0.035
119. 0.024; 0.023; Increased weight results in lower BACs.

Let $x = -4$ and $y = 2$. Evaluate each expression. ***See Example 9.***

91. $|3x - 2y|$ **92.** $|2x - 5y|$ **93.** $|-3x + 4y|$ **94.** $|-5y + x|$

95. $\dfrac{2|y| - 3|x|}{|xy|}$ **96.** $\dfrac{4|x| + 4|y|}{|x|}$ **97.** $\dfrac{|-8y + x|}{-|x|}$ **98.** $\dfrac{|x| + 2|y|}{-|x|}$

Justify each statement by giving the correct property of absolute value from this section. Assume all variables represent real numbers.

99. $|m| = |-m|$ **100.** $|-k| \geq 0$

101. $|9| \cdot |-6| = |-54|$ **102.** $|k - m| \leq |k| + |-m|$

103. $|12 + 11r| \geq 0$ **104.** $\left|\dfrac{-12}{5}\right| = \dfrac{|-12|}{|5|}$

Find the given distances between points P, Q, R, and S on a number line, with coordinates −4, −1, 8, and 12, respectively. ***See Example 10.***

105. $d(P, Q)$ **106.** $d(P, R)$ **107.** $d(Q, R)$ **108.** $d(Q, S)$

Concept Check *Determine what signs on values of x and y would make each statement true. Assume that x and y are not 0. (You should be able to work these mentally.)*

109. $xy > 0$ **110.** $x^2y > 0$ **111.** $\dfrac{x}{y} < 0$

112. $\dfrac{y^2}{x} < 0$ **113.** $\dfrac{x^3}{y} > 0$ **114.** $-\dfrac{x}{y} > 0$

Solve each problem.

115. *Golf Scores* Phil Mickelson won the 2010 Masters Golf Tournament with a total score that was 16 under par, and Zach Johnson won the 2007 tournament with a total score that was 1 above par. Using −16 to represent 16 below par and +1 to represent 1 over par, find the difference between these scores (in either order) and take the absolute value of this difference. What does this final number represent? (*Source:* www.masters.org)

116. *Total Football Yardage* During his 16 years in the NFL, Marcus Allen gained 12,243 yd rushing, 5411 yd receiving, and −6 yd returning fumbles. Find his total yardage (called *all-purpose yards*). Is this the same as the sum of the absolute values of the three categories? Explain. (*Source: The Sports Illustrated Sports Almanac.*)

Blood Alcohol Concentration *The blood alcohol concentration (BAC) of a person who has been drinking is approximated by the following expression.*

number of oz × % alcohol × 0.075 ÷ body weight in lb − hr of drinking × 0.015

(*Source:* Lawlor, J., *Auto Math Handbook: Mathematical Calculations, Theory, and Formulas for Automotive Enthusiasts,* HP Books.)

117. Suppose a policeman stops a 190-lb man who, in 2 hr, has ingested four 12-oz beers (48 oz), each having a 3.2% alcohol content. Calculate the man's BAC to the nearest thousandth. Follow the order of operations.

118. Find the BAC to the nearest thousandth for a 135-lb woman who, in 3 hr, has drunk three 12-oz beers (36 oz), each having a 4.0% alcohol content.

119. Calculate the BACs in **Exercises 117 and 118** if each person weighs 25 lb more and the rest of the variables stay the same. How does increased weight affect a person's BAC?

120. Decreased weight will result in higher BACs; 0.040; 0.053

120. Predict how decreased weight would affect the BAC of each person in **Exercises 117 and 118.** Calculate the BACs if each person weighs 25 lb less and the rest of the variables stay the same.

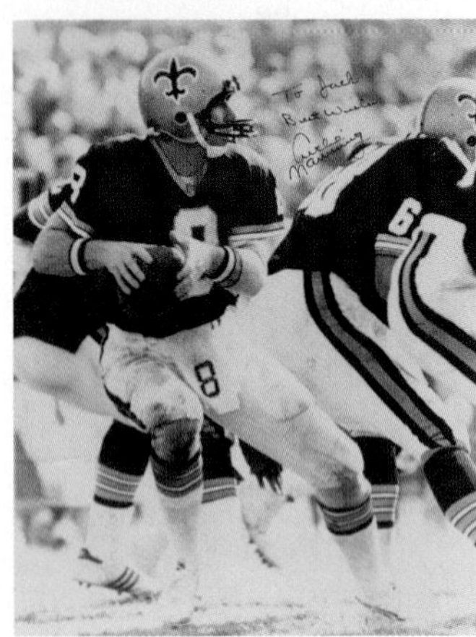

Archie Manning, father of NFL quarterbacks Peyton and Eli, signed this photo for author Hornsby's son, Jack.

Passer Rating for NFL Quarterbacks The current system of rating passers in the National Football League, adopted in 1973, is based on four performance components: completions, touchdowns, yards gained, and interceptions, as percentages of the number of passes attempted. It uses the following formula.

$$\text{Rating} = \frac{\left(250 \cdot \frac{C}{A}\right) + \left(1000 \cdot \frac{T}{A}\right) + \left(12.5 \cdot \frac{Y}{A}\right) + 6.25 - \left(1250 \cdot \frac{I}{A}\right)}{3},$$

where $A =$ attempted passes, $C =$ completed passes, $T =$ touchdown passes, $Y =$ yards gained passing, and $I =$ interceptions.

In addition to the weighting factors appearing in the formula, the four category ratios are limited to nonnegative values with the following maximums.

$$0.775 \text{ for } \frac{C}{A}, \quad 0.11875 \text{ for } \frac{T}{A},$$

$$12.5 \text{ for } \frac{Y}{A}, \quad 0.095 \text{ for } \frac{I}{A}$$

121. 111.0 122. 101.8
123. 101.2 124. 100.2
125. 97.0 126. 95.9
127. 93.6 128. 93.0
129. 92.0 130. 91.9
131. 91.0 132. 90.9

133. 121.1 134. 158.3

Exercises 121–132 give the 2010 regular season statistics for the top twelve quarterbacks. Use the formula to determine the rating for each. Round each answer to the nearest tenth.

Quarterback, Team	*A* Att	*C* Comp	*T* TD	*Y* Yds	*I* Int
121. Tom Brady, NE	492	324	36	3900	4
122. Philip Rivers, SD	541	357	30	4710	13
123. Aaron Rodgers, GB	475	312	28	3922	11
124. Michael Vick, PHI	372	233	21	3018	6
125. Ben Roethlisberger, PIT	389	240	17	3200	5
126. Josh Freeman, TB	474	291	25	3451	6
127. Joe Flacco, BAL	489	306	25	3622	10
128. Matt Cassel, KC	450	262	27	3116	7
129. Matt Schaub, HOU	574	365	24	4370	12
130. Peyton Manning, IND	679	450	33	4700	17
131. Matt Ryan, ATL	571	357	28	3705	9
132. Drew Brees, NO	658	448	33	4620	22

Source: www.nfl.com

133. Steve Young, of the San Francisco 49ers, set a full season rating record of 112.8 in 1994 and held that record until Peyton Manning surpassed it in 2004. (As of 2011, Manning's all-time record holds.) If Manning had 336 completions, 49 touchdowns, 10 interceptions, and 4557 yards, for 497 attempts, what was his rating in 2004?

134. Refer to the passer rating formula and determine the highest rating possible (considered a "perfect" passer rating).

R.3 Polynomials

- Rules for Exponents
- Polynomials
- Addition and Subtraction
- Multiplication
- Division

Teaching Tip Help students understand the product rule by showing that

$$2^3 \cdot 2^4 = 2^{3+4}.$$

Write 2^3 as $2 \cdot 2 \cdot 2$ and 2^4 as $2 \cdot 2 \cdot 2 \cdot 2$ and then combine them.

$$2^3 \cdot 2^4 = 2 \cdot 2 \cdot 2 \cdot 2 \cdot 2 \cdot 2 \cdot 2 = 2^7$$

Similar illustrations will help students with the three power rules.

Teaching Tip Stress that all rules for exponents can also be used from right to left. For instance,

$$a^{m+n} = a^m \cdot a^n$$

is used extensively in **Section R.4.**

Rules for Exponents

From **Section R.2,** the notation a^m (where m is a positive integer and a is a real number) means that a appears as a factor m times. In the same way, a^n (where n is a positive integer) means that a appears as a factor n times. In the product $a^m \cdot a^n$, the base a would appear $m + n$ times, so the **product rule** states the following.

$$a^m \cdot a^n = a^{m+n} \quad \text{Product rule}$$

Also consider the expression $(2^5)^3$, which can be written as follows.

$$(2^5)^3 = 2^5 \cdot 2^5 \cdot 2^5 \quad \text{Definition of exponent}$$

$$= 2^{5+5+5}, \text{ or } 2^{15} \quad \text{Generalization of the product rule}$$

The exponent 15 could have been obtained by multiplying 5 and 3. This example suggests the first of the **power rules** below. The others are found in a similar way.

1. $(a^m)^n = a^{mn}$ **2.** $(ab)^m = a^m b^m$ **3.** $\left(\frac{a}{b}\right)^m = \frac{a^m}{b^m}$ $(b \neq 0)$

Power rules for positive integers m and n and real numbers a and b

Rules for Exponents

For all positive integers m and n and all real numbers a and b, the following rules hold.

Rule	Description
Product Rule $a^m \cdot a^n = a^{m+n}$	When multiplying powers of like bases, keep the base and add the exponents.
Power Rule 1 $(a^m)^n = a^{mn}$	To raise a power to a power, multiply the exponents.
Power Rule 2 $(ab)^m = a^m b^m$	To raise a product to a power, raise each factor to that power.
Power Rule 3 $\left(\frac{a}{b}\right)^m = \frac{a^m}{b^m}$ $(b \neq 0)$	To raise a quotient to a power, raise the numerator and the denominator to that power.

Classroom Example 1
Find each product.
(a) $m^6 \cdot m^8$
(b) $(-5r^3)(6r^4)(-3r)$

Answers:
(a) m^{14} (b) $90r^8$

EXAMPLE 1 Using the Product Rule

Find each product.

(a) $y^4 \cdot y^7$ **(b)** $(6z^5)(9z^3)(2z^2)$

SOLUTION

(a) $y^4 \cdot y^7 = y^{4+7} = y^{11}$ Product rule: Keep the base and add the exponents.

(b) $(6z^5)(9z^3)(2z^2) = (6 \cdot 9 \cdot 2) \cdot (z^5 z^3 z^2)$ Commutative and associative properties (Section R.2)

$= 108z^{5+3+2}$ Multiply. Apply the product rule.

$= 108z^{10}$ Add.

✔ *Now Try Exercises 3, 5, and 7.*

Classroom Example 2
Simplify. Assume all variables represent nonzero real numbers.

(a) $(7^3)^5$ (b) $(2^5y^3)^4$

(c) $\left(\frac{4^3}{z^2}\right)^5$ (d) $\left(\frac{-3a^3}{bc^4}\right)^2$

Answers:
(a) 7^{15} (b) $2^{20}y^{12}$
(c) $\frac{4^{15}}{z^{10}}$ (d) $\frac{9a^6}{b^2c^8}$

EXAMPLE 2 Using the Power Rules

Simplify. Assume all variables represent nonzero real numbers.

(a) $(5^3)^2$ **(b)** $(3^4x^2)^3$ **(c)** $\left(\frac{2^5}{b^4}\right)^3$ **(d)** $\left(\frac{-2m^6}{t^2z}\right)^5$

SOLUTION

(a) $(5^3)^2 = 5^{3(2)} = 5^6$ Power rule 1

(b) $(3^4x^2)^3 = (3^4)^3(x^2)^3$ Power rule 2

$= 3^{4(3)}x^{2(3)}$ Power rule 1

$= 3^{12}x^6$

(c) $\left(\frac{2^5}{b^4}\right)^3 = \frac{(2^5)^3}{(b^4)^3}$ Power rule 3

$= \frac{2^{15}}{b^{12}}$ Power rule 1

(d) $\left(\frac{-2m^6}{t^2z}\right)^5 = \frac{(-2m^6)^5}{(t^2z)^5}$ Power rule 3

$= \frac{(-2)^5(m^6)^5}{(t^2)^5z^5}$ Power rule 2

$= \frac{-32m^{30}}{t^{10}z^5}$, or $-\frac{32m^{30}}{t^{10}z^5}$ Evaluate $(-2)^5$. Then use Power rule 1.

Now Try Exercises 11, 17, and 19.

CAUTION The expressions mn^2 and $(mn)^2$ are ***not*** equivalent. The second power rule can be used only with the second expression:

$$(mn)^2 = m^2n^2.$$

A zero exponent is defined as follows.

Zero Exponent

For any nonzero real number a, $\mathbf{a^0 = 1}$.

***That is, any nonzero number with a zero exponent equals* 1.**

To illustrate why a^0 is defined to equal 1, consider the product

$$a^n \cdot a^0, \quad \text{for} \quad a \neq 0.$$

We want the definition of a^0 to be consistent so that the product rule applies. Now apply this rule.

$$a^n \cdot a^0 = a^{n+0} = a^n$$

The product of a^n and a^0 must be a^n, and thus a^0 is acting like the identity element 1. So for consistency, we *define* a^0 to equal 1. (**0^0 is undefined.**)

Classroom Example 3
Evaluate each power.
(a) 8^0 (b) -8^0
(c) $(-8)^0$ (d) $-(-8)^0$
(e) $(-3b^8)^0$

Answers:
(a) 1 (b) -1
(c) 1 (d) -1
(e) $1, b \neq 0$

EXAMPLE 3 Using the Definition of a^0

Evaluate each power.

(a) 4^0 **(b)** $(-4)^0$ **(c)** -4^0 **(d)** $-(-4)^0$ **(e)** $(7r)^0$

SOLUTION

(a) $4^0 = 1$ Base is 4. **(b)** $(-4)^0 = 1$ Base is -4.

(c) $-4^0 = -(4^0) = -1$ Base is 4. **(d)** $-(-4)^0 = -(1) = -1$ Base is -4.

(e) $(7r)^0 = 1,\ r \neq 0$ Base is $7r$.

Now Try Exercise 23.

Polynomials Any collection of numbers or variables joined by the basic operations of addition, subtraction, multiplication, or division (except by 0), or the operations of raising to powers or taking roots, formed according to the rules of algebra, is an **algebraic expression.**

$$-2x^2 + 3x, \quad \frac{15y}{2y - 3}, \quad \sqrt{m^3 - 64}, \quad (3a + b)^4 \quad \text{Algebraic expressions}$$

The product of a real number and one or more variables raised to powers is a **term.** The real number is the **numerical coefficient,** or just the **coefficient,** of the variables. The coefficient of the variable in $-3m^4$ is -3, while the coefficient in $-p^2$ is -1. **Like terms** are terms with the same variables each raised to the same powers.

$$-13x^3, \quad 4x^3, \quad -x^3 \quad \text{Like terms} \qquad 6y, \quad 6y^2, \quad 4y^3 \quad \text{Unlike terms}$$

A **polynomial** is defined as a term or a finite sum of terms, with only positive or zero integer exponents permitted on the variables. If the terms of a polynomial contain only the variable x, then the polynomial is a **polynomial in x.**

$$5x^3 - 8x^2 + 7x - 4, \quad 9p^5 - 3, \quad 8r^2, \quad 6 \quad \text{Polynomials}$$

The terms of a polynomial cannot have variables in a denominator.

$$9x^2 - 4x + \frac{6}{x} \quad \text{Not a polynomial}$$

The **degree of a term** with one variable is the exponent on the variable. For example, the degree of $2x^3$ is 3, and the degree of $17x$ (that is, $17x^1$) is 1. The greatest degree of any term in a polynomial is the **degree of the polynomial.** For example,

$$4x^3 - 2x^2 - 3x + 7$$

has degree 3, because the greatest degree of any term is 3. A nonzero constant such as -6, equivalent to $-6x^0$, has degree 0. (The polynomial 0 has no degree.)

A polynomial can have more than one variable. A term containing more than one variable has degree equal to the sum of all the exponents appearing on the variables in the term. For example, $-3x^4y^3z^5$ has degree $4 + 3 + 5 = 12$. The degree of a polynomial in more than one variable is equal to the greatest degree of any term appearing in the polynomial. By this definition, the polynomial

$$2x^4y^3 - 3x^5y + x^6y^2$$

has degree 8, because the x^6y^2 term has the greatest degree, 8.

A polynomial containing exactly three terms is a **trinomial.** A two-term polynomial is a **binomial.** A single-term polynomial is a **monomial.**

Classroom Example 4
Fill in the blank entries with the appropriate response.

	Polynomial	Degree	Type
(a)	$x^4 + 3x^3 + 5$		
(b)	$16x^2y^5$		
(c)	$m^3 - n^3$		

Answers:
(a) 4; trinomial (b) 7; monomial
(c) 3; binomial

EXAMPLE 4 Classifying Polynomials

The table classifies several polynomials.

Polynomial	Degree	Type
$9p^7 - 4p^3 + 8p^2$	7	Trinomial
$29x^{11} + 8x^{15}$	15	Binomial
$-10r^6s^8$	14	Monomial
$5a^3b^7 - 3a^5b^5 + 4a^2b^9 - a^{10}$	11	None of these

Now Try Exercises 27, 29, 33, and 35.

Addition and Subtraction Since the variables used in polynomials represent real numbers, a polynomial represents a real number. This means that all the properties of the real numbers mentioned in **Section R.2** hold for polynomials. In particular, the distributive property holds.

$$3m^5 - 7m^5 = (3 - 7)m^5 = -4m^5 \quad \text{Distributive property}$$

Thus, polynomials are added by adding coefficients of like terms, and they are subtracted by subtracting coefficients of like terms.

Classroom Example 5
Add or subtract, as indicated.
(a) $(17x^3 - 10x^2 + x) + (-9x^3 + 10x^2 - 5x)$
(b) $(-6m^4 - 11m^2 + 21) - (m^4 - 6m^2 + 35)$
(c) $(10r^3s^6 + 5r^6s^3) + (25r^3s^6 - 15r^6s^3)$
(d) $6(z^2 - 5z + 3) - 4(3z^2 - 2z + 9)$

Answers:
(a) $8x^3 - 4x$
(b) $-7m^4 - 5m^2 - 14$
(c) $35r^3s^6 - 10r^6s^3$
(d) $-6z^2 - 22z - 18$

EXAMPLE 5 Adding and Subtracting Polynomials

Add or subtract, as indicated.

(a) $(2y^4 - 3y^2 + y) + (4y^4 + 7y^2 + 6y)$
(b) $(-3m^3 - 8m^2 + 4) - (m^3 + 7m^2 - 3)$
(c) $(8m^4p^5 - 9m^3p^5) + (11m^4p^5 + 15m^3p^5)$
(d) $4(x^2 - 3x + 7) - 5(2x^2 - 8x - 4)$

SOLUTION

(a) $(2y^4 - 3y^2 + y) + (4y^4 + 7y^2 + 6y)$ $y = 1y$

$= (2 + 4)y^4 + (-3 + 7)y^2 + (1 + 6)y$ Add coefficients of like terms.

$= 6y^4 + 4y^2 + 7y$ Work inside parentheses.

(b) $(-3m^3 - 8m^2 + 4) - (m^3 + 7m^2 - 3)$

$= (-3 - 1)m^3 + (-8 - 7)m^2 + [4 - (-3)]$ Subtract coefficients of like terms.

$= -4m^3 - 15m^2 + 7$ Simplify.

(c) $(8m^4p^5 - 9m^3p^5) + (11m^4p^5 + 15m^3p^5) = 19m^4p^5 + 6m^3p^5$

(d) $4(x^2 - 3x + 7) - 5(2x^2 - 8x - 4)$

$= 4x^2 - 4(3x) + 4(7) - 5(2x^2) - 5(-8x) - 5(-4)$ Distributive property (Section R.2)

$= 4x^2 - 12x + 28 - 10x^2 + 40x + 20$ Multiply.

$= -6x^2 + 28x + 48$ Add like terms.

Now Try Exercises 39 and 41.

Teaching Tip In **Example 5(b)**, rewrite
$(-3m^3 - 8m^2 + 4) - (m^3 + 7m^2 - 3)$
as $(-3m^3 - 8m^2 + 4) + (-1)(m^3 + 7m^2 - 3)$,
and then apply the distributive property to show how subtraction affects each term of the second polynomial.

As shown in **Examples 5(a), (b), and (d),** polynomials in one variable are often written with their terms in **descending order** (or descending degree). Thus, the term of greatest degree is first, the one with the next greatest degree is next, and so on.

Multiplication One way to find the product of two polynomials, such as $3x - 4$ and $2x^2 - 3x + 5$, is to treat $3x - 4$ as a single expression and use the distributive property.

$$\begin{aligned}
&(3x - 4)(2x^2 - 3x + 5)\\
&= (3x - 4)(2x^2) - (3x - 4)(3x) + (3x - 4)(5)\\
&= 3x(2x^2) - 4(2x^2) - 3x(3x) - (-4)(3x) + 3x(5) - 4(5)\\
&= 6x^3 - 8x^2 - 9x^2 + 12x + 15x - 20\\
&= 6x^3 - 17x^2 + 27x - 20
\end{aligned}$$

Another method is to write such a product vertically, similar to the method used in arithmetic for multiplying whole numbers.

Place like terms in the same column.

$$\begin{array}{rl}
2x^2 - 3x + 5 & \\
3x - 4 & \\
\hline
-8x^2 + 12x - 20 & \leftarrow -4(2x^2 - 3x + 5)\\
6x^3 - 9x^2 + 15x \quad\quad & \leftarrow 3x(2x^2 - 3x + 5)\\
\hline
6x^3 - 17x^2 + 27x - 20 & \text{Add in columns.}
\end{array}$$

Classroom Example 6
Multiply $(4t - 5)(3t^2 - 2t + 7)$.

Answer: $12t^3 - 23t^2 + 38t - 35$

EXAMPLE 6 Multiplying Polynomials

Multiply $(3p^2 - 4p + 1)(p^3 + 2p - 8)$.

SOLUTION

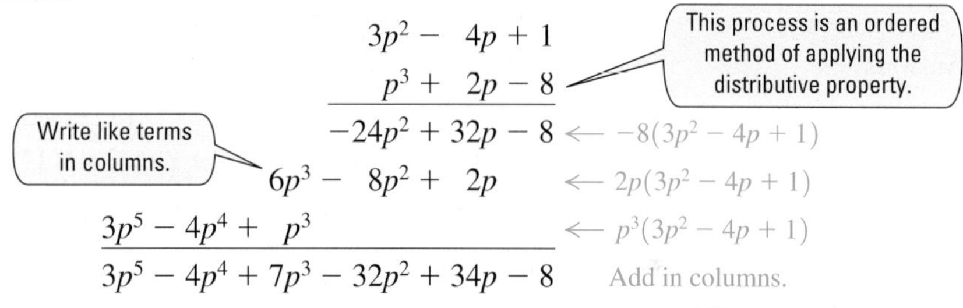

This process is an ordered method of applying the distributive property.
Write like terms in columns.

$$\begin{array}{rl}
3p^2 - 4p + 1 & \\
p^3 + 2p - 8 & \\
\hline
-24p^2 + 32p - 8 & \leftarrow -8(3p^2 - 4p + 1)\\
6p^3 - 8p^2 + 2p \quad\quad & \leftarrow 2p(3p^2 - 4p + 1)\\
3p^5 - 4p^4 + p^3 \quad\quad\quad\quad & \leftarrow p^3(3p^2 - 4p + 1)\\
\hline
3p^5 - 4p^4 + 7p^3 - 32p^2 + 34p - 8 & \text{Add in columns.}
\end{array}$$

Teaching Tip Remind students to put each polynomial in descending order before multiplying vertically.

✔ *Now Try Exercise 53.*

The **FOIL method** is a convenient way to find the product of two binomials. The memory aid **FOIL** (for **F**irst, **O**utside, **I**nside, **L**ast) gives the pairs of terms to be multiplied to find the product, as shown in the next example.

Classroom Example 7
Find each product.
(a) $(7y + 3)(4y - 5)$
(b) $(6p + 11)(6p - 11)$
(c) $x^3(2x - 5)(2x + 5)$

Answers:
(a) $28y^2 - 23y - 15$
(b) $36p^2 - 121$
(c) $4x^5 - 25x^3$

EXAMPLE 7 Using the FOIL Method to Multiply Two Binomials

Find each product.

(a) $(6m + 1)(4m - 3)$ **(b)** $(2x + 7)(2x - 7)$ **(c)** $r^2(3r + 2)(3r - 2)$

SOLUTION

(a) $(6m + 1)(4m - 3) = \overset{F}{6m(4m)} + \overset{O}{6m(-3)} + \overset{I}{1(4m)} + \overset{L}{1(-3)}$

$= 24m^2 - 14m - 3$ $\qquad -18m + 4m = -14m$

(b) $(2x + 7)(2x - 7) = 4x^2 - 14x + 14x - 49$ FOIL

$= 4x^2 - 49$ Combine like terms.

(c) $r^2(3r + 2)(3r - 2) = r^2(9r^2 - 6r + 6r - 4)$ FOIL

$= r^2(9r^2 - 4)$ Combine like terms.

$= 9r^4 - 4r^2$ Distributive property

✔ *Now Try Exercises 45 and 47.*

In **Example 7(a),** the product of two binomials is a trinomial, while in **Examples 7(b) and (c),** the product of two binomials is a binomial. ***The product of two binomials of the forms $x + y$ and $x - y$ is always a binomial.*** The squares of binomials, $(x + y)^2$ and $(x - y)^2$, are also special products.

Special Products

Product of the Sum and Difference of Two Terms	$(x + y)(x - y) = x^2 - y^2$
Square of a Binomial	$(x + y)^2 = x^2 + 2xy + y^2$
	$(x - y)^2 = x^2 - 2xy + y^2$

Classroom Example 8
Find each product.
(a) $(7m - 10)(7m + 10)$
(b) $(4r^2 + 9)(4r^2 - 9)$
(c) $(5x^2 - 8y^4)(5x^2 + 8y^4)$
(d) $(8z + 3)^2$
(e) $(5z - 12q^3)^2$

Answers:
(a) $49m^2 - 100$
(b) $16r^4 - 81$
(c) $25x^4 - 64y^8$
(d) $64z^2 + 48z + 9$
(e) $25z^2 - 120zq^3 + 144q^6$

EXAMPLE 8 Using the Special Products

Find each product.

(a) $(3p + 11)(3p - 11)$ **(b)** $(5m^3 - 3)(5m^3 + 3)$

(c) $(9k - 11r^3)(9k + 11r^3)$ **(d)** $(2m + 5)^2$

(e) $(3x - 7y^4)^2$

SOLUTION

$(3p)^2 = 3^2p^2$, *not* $3p^2$

(a) $(3p + 11)(3p - 11) = (3p)^2 - 11^2$ $\quad (x + y)(x - y) = x^2 - y^2$

$= 9p^2 - 121$ $\quad$ Power rule 2

(b) $(5m^3 - 3)(5m^3 + 3) = (5m^3)^2 - 3^2$ $\quad (x - y)(x + y) = x^2 - y^2$

$= 25m^6 - 9$ $\quad$ Power rules 2 and 1

(c) $(9k - 11r^3)(9k + 11r^3) = (9k)^2 - (11r^3)^2$

$= 81k^2 - 121r^6$

(d) $(2m + 5)^2 = (2m)^2 + 2(2m)(5) + 5^2$ $\quad (x + y)^2 = x^2 + 2xy + y^2$

$= 4m^2 + 20m + 25$ $\quad$ Power rules 2 and 1; Multiply.

Remember the middle term, $2(2m)(5)$.

(e) $(3x - 7y^4)^2 = (3x)^2 - 2(3x)(7y^4) + (7y^4)^2$ $\quad (x - y)^2 = x^2 - 2xy + y^2$

$= 9x^2 - 42xy^4 + 49y^8$ $\quad$ Power rule 2; Multiply.

✔ ***Now Try Exercises 59, 61, 63, and 65.***

CAUTION See **Examples 8(d) and (e).** ***The square of a binomial has three terms.*** Do **not** give $x^2 + y^2$ as the result of expanding $(x + y)^2$, or $x^2 - y^2$ as the result of expanding $(x - y)^2$.

$$(x + y)^2 = x^2 + 2xy + y^2$$
$$(x - y)^2 = x^2 - 2xy + y^2$$

Remember to include the middle term.

Classroom Example 9
Find each product.
(a) $[(4x-3)+7y]\cdot[(4x-3)-7y]$
(b) $(a-b)^4$
(c) $(s+4t)^3$

Answers:
(a) $16x^2-24x+9-49y^2$
(b) $a^4-4a^3b+6a^2b^2-4ab^3+b^4$
(c) $s^3+12s^2t+48st^2+64t^3$

EXAMPLE 9 Multiplying More Complicated Binomials

Find each product.

(a) $[(3p-2)+5q][(3p-2)-5q]$ **(b)** $(x+y)^3$ **(c)** $(2a+b)^4$

SOLUTION

(a) $[(3p-2)+5q][(3p-2)-5q]$

$= (3p-2)^2-(5q)^2$ Product of the sum and difference of two terms

$= 9p^2-12p+4-25q^2$ Square both quantities.

(b) $(x+y)^3 = (x+y)^2(x+y)$

This does *not* equal x^3+y^3.

$= (x^2+2xy+y^2)(x+y)$ Square $x+y$.

$= x^3+2x^2y+xy^2+x^2y+2xy^2+y^3$ Multiply.

$= x^3+3x^2y+3xy^2+y^3$ Combine like terms.

(c) $(2a+b)^4 = (2a+b)^2(2a+b)^2$

$= (4a^2+4ab+b^2)(4a^2+4ab+b^2)$ Square each $2a+b$.

$= 16a^4+16a^3b+4a^2b^2+16a^3b+16a^2b^2 + 4ab^3+4a^2b^2+4ab^3+b^4$ Distributive property

$= 16a^4+32a^3b+24a^2b^2+8ab^3+b^4$ Combine like terms.

Now Try Exercises 69, 73, and 75.

Division The quotient of two polynomials can be found with an algorithm (that is, a step-by-step procedure) for long division similar to that used for dividing whole numbers. ***Both polynomials must be written in descending order to use this algorithm.***

Classroom Example 10
Divide $12n^3+11n^2+5n-8$ by $3n+2$.

Answer: $4n^2+n+1+\frac{-10}{3n+2}$

EXAMPLE 10 Dividing Polynomials

Divide $4m^3-8m^2+5m+6$ by $2m-1$.

SOLUTION

$4m^3$ divided by $2m$ is $2m^2$.
$-6m^2$ divided by $2m$ is $-3m$.
$2m$ divided by $2m$ is 1.

$$\begin{array}{r} 2m^2-3m+1 \\ 2m-1\overline{)4m^3-8m^2+5m+6} \\ \underline{4m^3-2m^2} \\ -6m^2+5m \\ \underline{-6m^2+3m} \\ 2m+6 \\ \underline{2m-1} \\ 7 \end{array}$$

$\leftarrow 2m^2(2m-1)=4m^3-2m^2$
$\leftarrow$ Subtract. Bring down the next term.
$\leftarrow -3m(2m-1)=-6m^2+3m$
$\leftarrow$ Subtract. Bring down the next term.
$\leftarrow 1(2m-1)=2m-1$
$\leftarrow$ Subtract. The remainder is 7.

To subtract, add the opposite.

Teaching Tip Point out that the process stops when the remainder is zero or when the degree of the remainder is less than the degree of the divisor. This is analogous to long division where the process stops when the remainder is less than the divisor.

Thus, $\dfrac{4m^3-8m^2+5m+6}{2m-1} = 2m^2-3m+1+\dfrac{7}{2m-1}$.

Remember to add $\frac{\text{remainder}}{\text{divisor}}$.

Now Try Exercise 91.

When a polynomial has a missing term, we allow for that term by inserting a term with a 0 coefficient for it.

Classroom Example 11
Divide $8x^4 + 12x^2 + 7x - 18$ by $x^2 + 2$.

Answer: $8x^2 - 4 + \frac{7x - 10}{x^2 + 2}$

EXAMPLE 11 Dividing Polynomials with Missing Terms

Divide $3x^3 - 2x^2 - 150$ by $x^2 - 4$.

SOLUTION Both polynomials have missing first-degree terms. Insert each missing term with a 0 coefficient.

$$\begin{array}{r} 3x - 2 \\ x^2 + 0x - 4 \overline{)3x^3 - 2x^2 + 0x - 150} \\ \underline{3x^3 + 0x^2 - 12x} \\ -2x^2 + 12x - 150 \\ \underline{-2x^2 + 0x + 8} \\ 12x - 158 \end{array}$$

Missing term (0x in dividend); Missing term (0x in divisor); Insert placeholders for missing terms.; $12x - 158$ ← Remainder

The division process ends when the remainder is 0 or the degree of the remainder is less than that of the divisor. Since $12x - 158$ has lesser degree than the divisor $x^2 - 4$, it is the remainder. Thus, the entire quotient is written as follows.

$$\frac{3x^3 - 2x^2 - 150}{x^2 - 4} = 3x - 2 + \frac{12x - 158}{x^2 - 4}$$

Now Try Exercise 93.

R.3 Exercises

Simplify each expression. ***See Example 1.***

1. $(-4x^5)(4x^2)$ 2. $(3y^4)(-6y^3)$ 3. $n^6 \cdot n^4 \cdot n$

4. $a^8 \cdot a^5 \cdot a$ 5. $9^3 \cdot 9^5$ 6. $4^2 \cdot 4^8$

7. $(-3m^4)(6m^2)(-4m^5)$ 8. $(-8t^3)(2t^6)(-5t^4)$ 9. $(5x^2y)(-3x^3y^4)$

10. *Concept Check* Decide whether each expression has been simplified correctly. If not, correct it. Assume all variables represent nonzero real numbers.

(a) $(mn)^2 = mn^2$ (b) $y^2 \cdot y^5 = y^7$ (c) $\left(\frac{k}{5}\right)^3 = \frac{k^3}{5}$ (d) $3^0y = 0$

(e) $4^5 \cdot 4^2 = 16^7$ (f) $(a^2)^3 = a^5$ (g) $cd^0 = 1$ (h) $(2b)^4 = 8b^4$

Simplify each expression. Assume variables represent nonzero real numbers. ***See Examples 1–3.***

11. $(2^2)^5$ 12. $(6^4)^3$ 13. $(-6x^2)^3$

14. $(-2x^5)^5$ 15. $-(4m^3n^0)^2$ 16. $-(2x^0y^4)^3$

17. $\left(\frac{r^8}{s^2}\right)^3$ 18. $\left(\frac{p^4}{q}\right)^2$ 19. $\left(\frac{-4m^2}{tp^2}\right)^4$

20. $\left(\frac{-5n^4}{r^2}\right)^3$ 21. $-\left(\frac{x^3y^5}{z}\right)^0$ 22. $-\left(\frac{p^2q^3}{r^3}\right)^0$

Answers:

1. $-16x^7$ 2. $-18y^7$
3. n^{11} 4. a^{14}
5. 9^8 6. 4^{10}
7. $72m^{11}$ 8. $80t^{13}$
9. $-15x^5y^5$
10. (a) incorrect; $(mn)^2 = m^2n^2$
(b) correct
(c) incorrect; $\left(\frac{k}{5}\right)^3 = \frac{k^3}{5^3}$
(d) incorrect; $3^0y = 1 \cdot y = y$
(e) incorrect; $4^5 \cdot 4^2 = 4^{5+2} = 4^7$
(f) incorrect; $(a^2)^3 = a^{2(3)} = a^6$
(g) incorrect; $cd^0 = c \cdot 1 = c$
(h) incorrect; $(2b)^4 = 2^4b^4 = 16b^4$
11. 2^{10} 12. 6^{12}
13. $(-6)^3x^6$, or $-216x^6$
14. $(-2)^5x^{25}$, or $-32x^{25}$
15. -4^2m^6, or $-16m^6$
16. -2^3y^{12}, or $-8y^{12}$
17. $\frac{r^{24}}{s^6}$ 18. $\frac{p^8}{q^2}$
19. $\frac{(-4)^4m^8}{t^4p^8}$, or $\frac{256m^8}{t^4p^8}$
20. $\frac{(-5)^3n^{12}}{r^6}$, or $-\frac{125n^{12}}{r^6}$
21. -1 22. -1

23. (a) B (b) C (c) B (d) C
24. (a) D (b) E (c) B (d) B

27. polynomial; degree 11; monomial
28. polynomial; degree 5; monomial
29. polynomial; degree 4; binomial
30. polynomial; degree 3; binomial
31. polynomial; degree 5; trinomial
32. polynomial; degree 4; trinomial
33. polynomial; degree 11; none of these
34. polynomial; degree 12; none of these
35. not a polynomial
36. not a polynomial
37. polynomial; degree 0; monomial
38. polynomial; degree 0; monomial

39. $x^2 - x + 2$
40. $m^3 - 4m^2 + 10$
41. $12y^2 + 4$
42. $9p^2 - 5p - 20$
43. $6m^4 - 2m^3 - 7m^2 - 4m$
44. $-6x^3 - 3x^2 - 4x + 4$

45. $28r^2 + r - 2$
46. $15m^2 + 2m - 24$
47. $15x^4 - \frac{7}{3}x^3 - \frac{2}{9}x^2$
48. $6m^5 + \frac{1}{4}m^4 - \frac{1}{8}m^3$
49. $12x^5 + 8x^4 - 20x^3 + 4x^2$
50. $2b^5 - 8b^4 + 6b^3$
51. $-2z^3 + 7z^2 - 11z + 4$
52. $-3w^3 + 10w^2 - w - 6$
53. $m^2 + mn - 2n^2 - 2km + 5kn - 3k^2$
54. $2r^2 - 7rs + 3s^2 + 3rt - 4st + t^2$
55. $16x^4 - 72x^2 + 81$
56. $81y^4 - 450y^2 + 625$
57. $x^4 - 2x^2 + 1$
58. $t^4 - 32t^2 + 256$

59. $4m^2 - 9$ 60. $64s^2 - 9t^2$
61. $16x^4 - 25y^2$ 62. $4m^6 - n^2$
63. $16m^2 + 16mn + 4n^2$
64. $a^2 - 12ab + 36b^2$
65. $25r^2 - 30rt^2 + 9t^4$
66. $4z^8 - 12z^4y + 9y^2$
67. $4p^2 - 12p + 9 + 4pq - 6q + q^2$
68. $16y^2 - 8y + 1 + 8yz - 2z + z^2$
69. $9q^2 + 30q + 25 - p^2$
70. $81r^2 - 18rs + s^2 - 4$

Match each expression in Column I with its equivalent in Column II. ***See Example 3.***

I	II	I	II
23. (a) 6^0	**A.** 0	**24. (a)** $3p^0$	**A.** 0
(b) -6^0	**B.** 1	**(b)** $-3p^0$	**B.** 1
(c) $(-6)^0$	**C.** -1	**(c)** $(3p)^0$	**C.** -1
(d) $-(-6)^0$	**D.** 6	**(d)** $(-3p)^0$	**D.** 3
	E. -6		**E.** -3

25. Explain why $x^2 + x^2$ is not equivalent to x^4.

26. Explain why $(x + y)^2$ is not equivalent to $x^2 + y^2$.

Identify each expression as a polynomial *or* not a polynomial. *For each polynomial, give the degree and identify it as a* monomial, binomial, trinomial, *or* none of these. ***See Example 4.***

27. $-5x^{11}$ **28.** $-4y^5$ **29.** $6x + 3x^4$

30. $-9y + 5y^3$ **31.** $-7z^5 - 2z^3 + 1$ **32.** $-9t^4 + 8t^3 - 7$

33. $15a^2b^3 + 12a^3b^8 - 13b^5 + 12b^6$ **34.** $-16x^5y^7 + 12x^3y^8 - 4xy^9 + 18x^{10}$

35. $\frac{3}{8}x^5 - \frac{1}{x^2} + 9$ **36.** $\frac{2}{3}t^6 + \frac{3}{t^5} + 1$

37. 5 **38.** 9

Find each sum or difference. ***See Example 5.***

39. $(5x^2 - 4x + 7) + (-4x^2 + 3x - 5)$

40. $(3m^3 - 3m^2 + 4) + (-2m^3 - m^2 + 6)$

41. $2(12y^2 - 8y + 6) - 4(3y^2 - 4y + 2)$

42. $3(8p^2 - 5p) - 5(3p^2 - 2p + 4)$

43. $(6m^4 - 3m^2 + m) - (2m^3 + 5m^2 + 4m) + (m^2 - m)$

44. $-(8x^3 + x - 3) + (2x^3 + x^2) - (4x^2 + 3x - 1)$

Find each product. ***See Examples 6–8.***

45. $(4r - 1)(7r + 2)$ **46.** $(5m - 6)(3m + 4)$

47. $x^2\left(3x - \frac{2}{3}\right)\left(5x + \frac{1}{3}\right)$ **48.** $m^3\left(2m - \frac{1}{4}\right)\left(3m + \frac{1}{2}\right)$

49. $4x^2(3x^3 + 2x^2 - 5x + 1)$ **50.** $2b^3(b^2 - 4b + 3)$

51. $(2z - 1)(-z^2 + 3z - 4)$ **52.** $(3w + 2)(-w^2 + 4w - 3)$

53. $(m - n + k)(m + 2n - 3k)$ **54.** $(r - 3s + t)(2r - s + t)$

55. $(2x + 3)(2x - 3)(4x^2 - 9)$ **56.** $(3y - 5)(3y + 5)(9y^2 - 25)$

57. $(x + 1)(x + 1)(x - 1)(x - 1)$ **58.** $(t + 4)(t + 4)(t - 4)(t - 4)$

Find each product. ***See Examples 8 and 9.***

59. $(2m + 3)(2m - 3)$ **60.** $(8s - 3t)(8s + 3t)$ **61.** $(4x^2 - 5y)(4x^2 + 5y)$

62. $(2m^3 + n)(2m^3 - n)$ **63.** $(4m + 2n)^2$ **64.** $(a - 6b)^2$

65. $(5r - 3t^2)^2$ **66.** $(2z^4 - 3y)^2$

67. $[(2p - 3) + q]^2$ **68.** $[(4y - 1) + z]^2$

69. $[(3q + 5) - p][(3q + 5) + p]$ **70.** $[(9r - s) + 2][(9r - s) - 2]$

71. $9a^2 + 6ab + b^2 - 6a - 2b + 1$
72. $4m^2 + 28m + 49 - 4mn - 14n + n^2$
73. $y^3 + 6y^2 + 12y + 8$
74. $z^3 - 9z^2 + 27z - 27$
75. $q^4 - 8q^3 + 24q^2 - 32q + 16$
76. $r^4 + 12r^3 + 54r^2 + 108r + 81$

71. $[(3a + b) - 1]^2$ **72.** $[(2m + 7) - n]^2$ **73.** $(y + 2)^3$

74. $(z - 3)^3$ **75.** $(q - 2)^4$ **76.** $(r + 3)^4$

Perform the indicated operations. ***See Examples 5–9.***

77. $(p^3 - 4p^2 + p) - (3p^2 + 2p + 7)$ **78.** $(x^4 - 3x^2 + 2) - (-2x^4 + x^2 - 3)$

79. $(7m + 2n)(7m - 2n)$ **80.** $(3p + 5)^2$

81. $-3(4q^2 - 3q + 2) + 2(-q^2 + q - 4)$ **82.** $2(3r^2 + 4r + 2) - 3(-r^2 + 4r - 5)$

83. $p(4p - 6) + 2(3p - 8)$ **84.** $m(5m - 2) + 9(5 - m)$

85. $-y(y^2 - 4) + 6y^2(2y - 3)$ **86.** $-z^3(9 - z) + 4z(2 + 3z)$

77. $p^3 - 7p^2 - p - 7$
78. $3x^4 - 4x^2 + 5$
79. $49m^2 - 4n^2$
80. $9p^2 + 30p + 25$
81. $-14q^2 + 11q - 14$
82. $9r^2 - 4r + 19$
83. $4p^2 - 16$
84. $5m^2 - 11m + 45$
85. $11y^3 - 18y^2 + 4y$
86. $z^4 - 9z^3 + 12z^2 + 8z$

Perform each division. ***See Examples 10 and 11.***

87. $\dfrac{-4x^7 - 14x^6 + 10x^4 - 14x^2}{-2x^2}$ **88.** $\dfrac{-8r^3s - 12r^2s^2 + 20rs^3}{-4rs}$

89. $\dfrac{4x^3 - 3x^2 + 1}{x - 2}$ **90.** $\dfrac{3x^3 - 2x + 5}{x - 3}$

91. $\dfrac{6m^3 + 7m^2 - 4m + 2}{3m + 2}$ **92.** $\dfrac{10x^3 + 11x^2 - 2x + 3}{5x + 3}$

93. $\dfrac{x^4 + 5x^2 + 5x + 27}{x^2 + 3}$ **94.** $\dfrac{k^4 - 4k^2 + 2k + 5}{k^2 + 1}$

87. $2x^5 + 7x^4 - 5x^2 + 7$
88. $2r^2 + 3rs - 5s^2$
89. $4x^2 + 5x + 10 + \frac{21}{x - 2}$
90. $3x^2 + 9x + 25 + \frac{80}{x - 3}$
91. $2m^2 + m - 2 + \frac{6}{3m + 2}$
92. $2x^2 + x - 1 + \frac{6}{5x + 3}$
93. $x^2 + 2 + \frac{5x + 21}{x^2 + 3}$
94. $k^2 - 5 + \frac{2k + 10}{k^2 + 1}$

Many exercise sets will contain groups of exercises under the heading Relating Concepts. These exercises are provided to illustrate how the concepts currently being studied relate to previously learned concepts. In most cases, they should be worked sequentially. We provide the answers to all such exercises, both even- and odd-numbered, in the Answer Section at the back of the student book.

Relating Concepts

For individual or collaborative investigation *(Exercises 95–98)*

The special products can be used to perform selected multiplications. On the left, we use $(x + y)(x - y) = x^2 - y^2$. *On the right,* $(x - y)^2 = x^2 - 2xy + y^2$.

$51 \times 49 = (50 + 1)(50 - 1)$	$47^2 = (50 - 3)^2$
$= 50^2 - 1^2$	$= 50^2 - 2(50)(3) + 3^2$
$= 2500 - 1$	$= 2500 - 300 + 9$
$= 2499$	$= 2209$

Use special products to evaluate each expression.

95. 99×101 **96.** 63×57 **97.** 102^2 **98.** 71^2

95. 9999 96. 3591
97. 10,404 98. 5041

Solve each problem.

99. *Geometric Modeling* Consider the figure, which is a square divided into two squares and two rectangles.

(a) The length of each side of the largest square is $x + y$. Use the formula for the area of a square to write the area of the largest square as a power.

(b) Use the formulas for the area of a square and the area of a rectangle to write the area of the largest square as a trinomial that represents the sum of the areas of the four figures that comprise it.

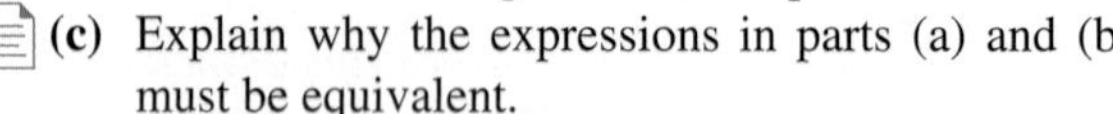

(c) Explain why the expressions in parts (a) and (b) must be equivalent.

(d) What special product formula from this section does this exercise reinforce geometrically?

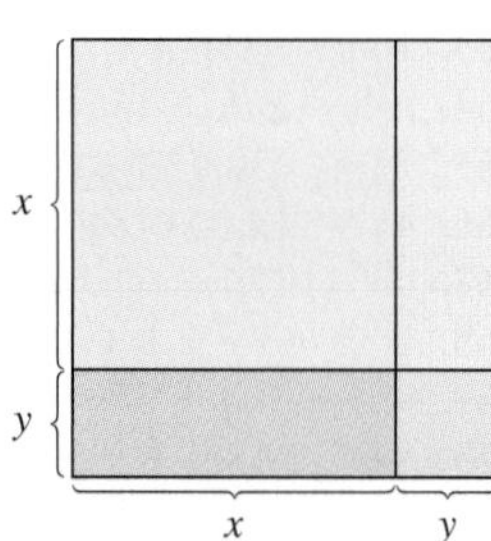

99. (a) $(x + y)^2$
(b) $x^2 + 2xy + y^2$
(d) the special product for squaring a binomial

101. (a) approximately 60,501,000 ft^3
(b) The shape becomes a rectangular box with a square base, with volume $V = b^2h$.
(c) If we let $a = b$, then $V = \frac{1}{3}h(a^2 + ab + b^2)$ becomes $V = \frac{1}{3}h(b^2 + bb + b^2)$, which simplifies to $V = hb^2$. Yes, the Egyptian formula gives the same result.

102. (a) $V = \frac{1}{3}hb^2$ (This is the correct formula.)
(b) approximately 91.6 million ft^3; The pyramid is slightly smaller.
(c) approximately 13.1 acres

103. 6.3; exact
104. 3.1; 0.2 high
105. 2.1; 0.3 low
106. 2.1, 0.1 low

107. 1,000,000
108. 144
109. 32
110. 81

100. *Geometric Modeling* Use the reasoning process of **Exercise 99** and the accompanying figure to geometrically support the distributive property. Write a short paragraph explaining this process.

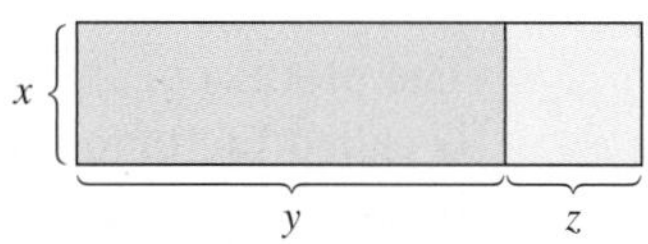

101. *Volume of the Great Pyramid* An amazing formula from ancient mathematics was used by the Egyptians to find the volume of the frustum of a square pyramid, as shown in the figure. Its volume is given by

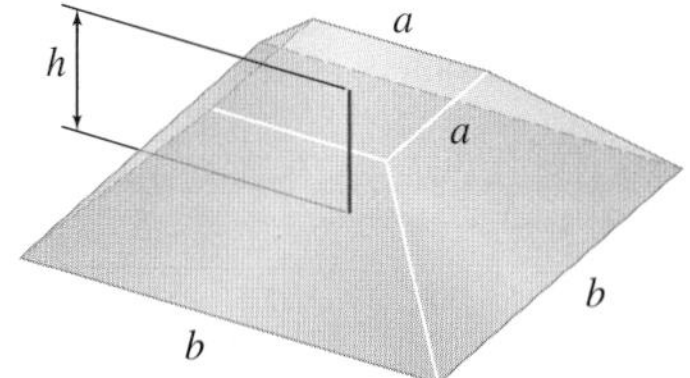

$$V = \frac{1}{3}h(a^2 + ab + b^2),$$

where b is the length of the base, a is the length of the top, and h is the height. (*Source:* Freebury, H. A., *A History of Mathematics,* Macmillan Company, New York.)

(a) When the Great Pyramid in Egypt was partially completed to a height h of 200 ft, b was 756 ft, and a was 314 ft. Calculate its volume at this stage of construction.
(b) Try to visualize the figure if $a = b$. What is the resulting shape? Find its volume.
(c) Let $a = b$ in the Egyptian formula and simplify. Are the results the same?

102. *Volume of the Great Pyramid* Refer to the formula and the discussion in **Exercise 101.**

(a) Use $V = \frac{1}{3}h(a^2 + ab + b^2)$ to determine a formula for the volume of a pyramid with square base of length b and height h by letting $a = 0$.
(b) The Great Pyramid in Egypt had a square base of length 756 ft and a height of 481 ft. Find the volume of the Great Pyramid. Compare it with the 273-ft-tall Superdome in New Orleans, which has an approximate volume of 100 million ft^3. (*Source: Guinness Book of World Records.*)
(c) The Superdome covers an area of 13 acres. How many acres does the Great Pyramid cover? (*Hint:* 1 acre = 43,560 ft^2)

(Modeling) Number of Farms in the United States The graph shows the number of farms in the United States since 1940 for selected years. The polynomial

$$0.001147x^2 - 4.5905x + 4595$$

provides a good approximation of the number of farms for these years by substituting the year for x and evaluating the polynomial. For example, if $x = 1960$, *the value of the polynomial is approximately* 3.9, *which differs from the data in the bar graph by only* 0.1.

Evaluate the polynomial for each year and then give the difference from the value in the graph.

103. 1940

104. 1970

105. 1990

106. 2009

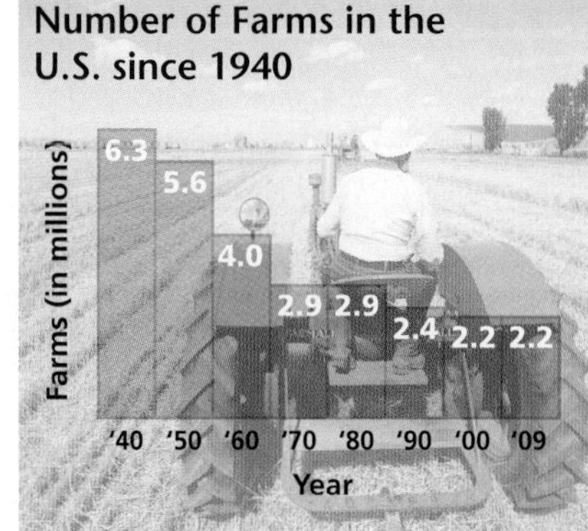

Source: U.S. Department of Agriculture.

Concept Check Perform each operation mentally.

107. $(0.25^3)(400^3)$ **108.** $(24^2)(0.5^2)$ **109.** $\frac{4.2^5}{2.1^5}$ **110.** $\frac{15^4}{5^4}$

R.4 Factoring Polynomials

- Factoring Out the Greatest Common Factor
- Factoring by Grouping
- Factoring Trinomials
- Factoring Binomials
- Factoring by Substitution

The process of finding polynomials whose product equals a given polynomial is called **factoring.** Unless otherwise specified, we consider only integer coefficients when factoring polynomials. For example, since

$$4x + 12 = 4(x + 3)$$

both 4 and $x + 3$ are **factors** of $4x + 12$ and $4(x + 3)$ is a **factored form** of $4x + 12$.

A polynomial with variable terms that cannot be written as a product of two polynomials of lower degree is a **prime polynomial.** A polynomial is **factored completely** when it is written as a product of prime polynomials.

Factoring Out the Greatest Common Factor

To factor $6x^2y^3 + 9xy^4 + 18y^5$, we look for a monomial that is the greatest common factor (GCF) of the three terms.

$$6x^2y^3 + 9xy^4 + 18y^5 = 3y^3(2x^2) + 3y^3(3xy) + 3y^3(6y^2) \quad \text{GCF} = 3y^3$$

$$= 3y^3(2x^2 + 3xy + 6y^2) \quad \text{Distributive property (Section R.2)}$$

Classroom Example 1
Factor out the greatest common factor from each polynomial.
(a) $6a^2 - 18a^4$
(b) $14x^3y^2 - 28x^2y^3 + 21x^2y^2$
(c) $24(x-2)^3 - 16(x-2)^2 + 6(x-2)$

Answers:
(a) $6a^2(1 - 3a^2)$
(b) $7x^2y^2(2x - 4y + 3)$
(c) $2(x-2)(12x^2 - 56x + 67)$

EXAMPLE 1 Factoring Out the Greatest Common Factor

Factor out the greatest common factor from each polynomial.

(a) $9y^5 + y^2$ **(b)** $6x^2t + 8xt + 12t$

(c) $14(m + 1)^3 - 28(m + 1)^2 - 7(m + 1)$

SOLUTION

(a) $9y^5 + y^2 = y^2(9y^3) + y^2(1)$ GCF $= y^2$

$= y^2(9y^3 + 1)$ Distributive property

Remember to include the 1.

CHECK Multiply out the factored form: $y^2(9y^3 + 1) = \underbrace{9y^5 + y^2}_{\text{Original polynomial}}$. ✓

(b) $6x^2t + 8xt + 12t = 2t(3x^2 + 4x + 6)$ GCF $= 2t$

CHECK $2t(3x^2 + 4x + 6) = 6x^2t + 8xt + 12t$ ✓

(c) $14(m + 1)^3 - 28(m + 1)^2 - 7(m + 1)$

$= 7(m + 1)[2(m + 1)^2 - 4(m + 1) - 1]$ GCF $= 7(m + 1)$

$= 7(m + 1)[2(m^2 + 2m + 1) - 4m - 4 - 1]$ Square $m + 1$ (Section R.3); distributive property

Remember the middle term.

$= 7(m + 1)(2m^2 + 4m + 2 - 4m - 4 - 1)$ Distributive property

$= 7(m + 1)(2m^2 - 3)$ Combine like terms.

Now Try Exercises 3, 9, and 15.

Teaching Tip In **Example 1(c),** students may have trouble seeing $(m + 1)$ as a factor. Use substitution to replace $(m + 1)$ with u, factor using the GCF, then replace u with $(m + 1)$ to get the final result. This factoring technique is used later in this section.

CAUTION In **Example 1(a),** the 1 is essential in the answer, since

$$y^2(9y^3) \neq 9y^5 + y^2.$$

Factoring can always be checked by multiplying.

Factoring by Grouping When a polynomial has more than three terms, it can sometimes be factored using **factoring by grouping.** Consider this example.

$$ax + ay + 6x + 6y = \overbrace{(ax + ay)}^{\text{Terms with common factor } a} + \overbrace{(6x + 6y)}^{\text{Terms with common factor } 6}$$ Group the terms so that each group has a common factor.

$$= a(x + y) + 6(x + y)$$ Factor each group.

$$= (x + y)(a + 6)$$ Factor out $x + y$.

It is not always obvious which terms should be grouped. In cases like the one above, group in pairs. Experience and repeated trials are the most reliable tools.

Classroom Example 2
Factor each polynomial by grouping.
(a) $r^2s + 3r^2 - 5s - 15$
(b) $4m^2 - m^2n + 4n - n^2$
(c) $9y^3 - 15y^2 + 6y - 10$

Answers:
(a) $(r^2 - 5)(s + 3)$
(b) $(m^2 + n)(4 - n)$
(c) $(3y^2 + 2)(3y - 5)$

EXAMPLE 2 Factoring by Grouping

Factor each polynomial by grouping.

(a) $mp^2 + 7m + 3p^2 + 21$ **(b)** $2y^2 + az - 2z - ay^2$

(c) $4x^3 + 2x^2 - 2x - 1$

SOLUTION

(a) $mp^2 + 7m + 3p^2 + 21 = (mp^2 + 7m) + (3p^2 + 21)$ Group the terms.

$= m(p^2 + 7) + 3(p^2 + 7)$ Factor each group.

$= (p^2 + 7)(m + 3)$ $p^2 + 7$ is a common factor.

CHECK $(p^2 + 7)(m + 3) = mp^2 + 3p^2 + 7m + 21$ FOIL (Section R.3)

$= mp^2 + 7m + 3p^2 + 21$ ✓ Commutative property (Section R.2)

Teaching Tip Explain to students that grouping the terms in pairs will lead to success only after the binomials are grouped so that each one has a common factor.

(b) $2y^2 + az - 2z - ay^2 = 2y^2 - 2z - ay^2 + az$ Rearrange the terms.

$= (2y^2 - 2z) + (-ay^2 + az)$ Group the terms.

Be careful with signs here. $= 2(y^2 - z) - a(y^2 - z)$ Factor out 2 and $-a$ so that $y^2 - z$ is a common factor.

$= (y^2 - z)(2 - a)$ Factor out $y^2 - z$.

(c) $4x^3 + 2x^2 - 2x - 1 = (4x^3 + 2x^2) + (-2x - 1)$ Group the terms.

$= 2x^2(2x + 1) - 1(2x + 1)$ Factor each group.

$= (2x + 1)(2x^2 - 1)$ Factor out $2x + 1$.

✓ ***Now Try Exercises 19 and 21.***

Factoring Trinomials As shown here, factoring is the opposite of multiplication.

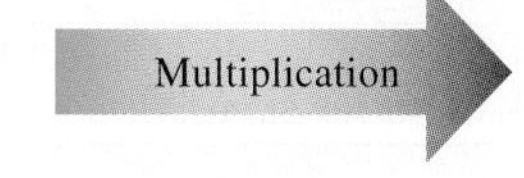

$$(2x + 1)(3x - 4) = 6x^2 - 5x - 4$$

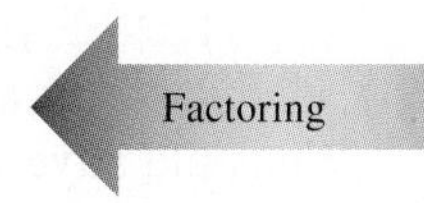

One strategy in factoring trinomials requires using the FOIL method in reverse.

Classroom Example 3
Factor each trinomial, if possible.
(a) $5z^2 + 4z - 12$
(b) $12t^2 - 5t - 3$
(c) $3x^2 - 15x + 16$
(d) $24x^2 + 42x + 15$

Answers:
(a) $(5z - 6)(z + 2)$
(b) $(4t - 3)(3t + 1)$
(c) prime
(d) $3(4x + 5)(2x + 1)$

Teaching Tip Tell students to always attempt to factor out the GCF first. Once this has been done, explain that the terms of a binomial factor cannot have a common factor (i.e., $(2y - 6)$ cannot be a factor once the GCF has been factored out). This will help lessen the amount of trial and error.

EXAMPLE 3 Factoring Trinomials

Factor each trinomial, if possible.

(a) $4y^2 - 11y + 6$　　**(b)** $6p^2 - 7p - 5$

(c) $2x^2 + 13x - 18$　　**(d)** $16y^3 + 24y^2 - 16y$

SOLUTION

(a) To factor this polynomial, we must find integers a, b, c, and d such that

$$4y^2 - 11y + 6 = (ay + b)(cy + d). \quad \text{FOIL}$$

Using FOIL, we see that $ac = 4$ and $bd = 6$. The positive factors of 4 are 4 and 1 or 2 and 2. Since the middle term has a negative coefficient, we consider only negative factors of 6. The possibilities are -2 and -3 or -1 and -6.

Now we try various arrangements of these factors until we find one that gives the correct coefficient of y.

$$(2y - 1)(2y - 6) = 4y^2 - 14y + 6 \quad \text{Incorrect}$$
$$(2y - 2)(2y - 3) = 4y^2 - 10y + 6 \quad \text{Incorrect}$$
$$(y - 2)(4y - 3) = 4y^2 - 11y + 6 \quad \text{Correct}$$

Therefore, $4y^2 - 11y + 6$ factors as $(y - 2)(4y - 3)$.

CHECK $(y - 2)(4y - 3) = 4y^2 - 3y - 8y + 6$ FOIL
$= 4y^2 - 11y + 6$ ✓ Original polynomial

(b) Again, we try various possibilities to factor $6p^2 - 7p - 5$. The positive factors of 6 could be 2 and 3 or 1 and 6. As factors of -5 we have only -1 and 5 or -5 and 1.

$$(2p - 5)(3p + 1) = 6p^2 - 13p - 5 \quad \text{Incorrect}$$
$$(3p - 5)(2p + 1) = 6p^2 - 7p - 5 \quad \text{Correct}$$

Thus, $6p^2 - 7p - 5$ factors as $(3p - 5)(2p + 1)$.

(c) If we try to factor $2x^2 + 13x - 18$ as above, we find that none of the pairs of factors gives the correct coefficient of x.

$$(2x + 9)(x - 2) = 2x^2 + 5x - 18 \quad \text{Incorrect}$$
$$(2x - 3)(x + 6) = 2x^2 + 9x - 18 \quad \text{Incorrect}$$
$$(2x - 1)(x + 18) = 2x^2 + 35x - 18 \quad \text{Incorrect}$$

Additional trials are also unsuccessful. Thus, this trinomial cannot be factored with integer coefficients and is prime.

(d) $16y^3 + 24y^2 - 16y = 8y(2y^2 + 3y - 2)$ Factor out the GCF, $8y$.
$= 8y(2y - 1)(y + 2)$ Factor the trinomial.

Remember to include the common factor in the final form.

✔ ***Now Try Exercises 25, 27, 29, and 31.***

NOTE In **Example 3,** we chose positive factors of the positive first term. We could have used two negative factors, but the work is easier if positive factors are used.

Each of the special patterns for multiplication given in **Section R.3** can be used in reverse to get a pattern for factoring. Perfect square trinomials can be factored as follows.

Factoring Perfect Square Trinomials

$$x^2 + 2xy + y^2 = (x + y)^2$$
$$x^2 - 2xy + y^2 = (x - y)^2$$

Classroom Example 4
Factor each trinomial.
(a) $49x^2 + 28xy + 4y^2$
(b) $81a^2b^2 - 90ab + 25$

Answers:
(a) $(7x + 2y)^2$ (b) $(9ab - 5)^2$

EXAMPLE 4 Factoring Perfect Square Trinomials

Factor each trinomial.

(a) $16p^2 - 40pq + 25q^2$ **(b)** $36x^2y^2 + 84xy + 49$

SOLUTION

(a) Since $16p^2 = (4p)^2$ and $25q^2 = (5q)^2$, we use the second pattern shown in the box with $4p$ replacing x and $5q$ replacing y.

$$16p^2 - 40pq + 25q^2 = (4p)^2 - 2(4p)(5q) + (5q)^2$$
$$= (4p - 5q)^2$$

Make sure that the middle term of the trinomial being factored, $-40pq$ here, is twice the product of the two terms in the binomial $4p - 5q$.

$$-40pq = 2(4p)(-5q)$$

Thus, $16p^2 - 40pq + 25q^2$ factors as $(4p - 5q)^2$.

CHECK $(4p - 5q)^2 = 16p^2 - 40pq + 25q^2$ ✓ Multiply.

(b) $36x^2y^2 + 84xy + 49$ factors as $(6xy + 7)^2$ ← $2(6xy)(7) = 84xy$

CHECK Square $6xy + 7$: $(6xy + 7)^2 = 36x^2y^2 + 84xy + 49$. ✓

Now Try Exercises 41 and 45.

Factoring Binomials Check first to see whether the terms of a binomial have a common factor. If so, factor it out. The binomial may also fit one of the following patterns.

Factoring Binomials

Difference of Squares	$x^2 - y^2 = (x + y)(x - y)$
Difference of Cubes	$x^3 - y^3 = (x - y)(x^2 + xy + y^2)$
Sum of Cubes	$x^3 + y^3 = (x + y)(x^2 - xy + y^2)$

CAUTION ***There is no factoring pattern for a sum of squares in the real number system.*** In particular, $x^2 + y^2$ **does not factor as** $(x + y)^2$, for real numbers x and y.

Classroom Example 5
Factor each polynomial.
(a) $64r^2 - 49$
(b) $169u^6 - 144v^4$
(c) $(2c - 3d)^2 - 16f^2$
(d) $x^2 + 18x + 81 - 25y^2$
(e) $4x^2 - y^2 - 10y - 25$

Answers:
(a) $(8r + 7)(8r - 7)$
(b) $(13u^3 + 12v^2)(13u^3 - 12v^2)$
(c) $(2c - 3d + 4f)(2c - 3d - 4f)$
(d) $(x + 9 + 5y)(x + 9 - 5y)$
(e) $(2x + y + 5)(2x - y - 5)$

EXAMPLE 5 Factoring Differences of Squares

Factor each polynomial.

(a) $4m^2 - 9$ **(b)** $256k^4 - 625m^4$ **(c)** $(a + 2b)^2 - 4c^2$

(d) $x^2 - 6x + 9 - y^4$ **(e)** $y^2 - x^2 + 6x - 9$

SOLUTION

(a) $4m^2 - 9 = (2m)^2 - 3^2$ Write as a difference of squares.

$= (2m + 3)(2m - 3)$ Factor.

Check by multiplying.

(b) $256k^4 - 625m^4 = (16k^2)^2 - (25m^2)^2$ Write as a difference of squares.

Don't stop here. $= (16k^2 + 25m^2)(16k^2 - 25m^2)$ Factor.

$= (16k^2 + 25m^2)(4k + 5m)(4k - 5m)$ Factor $16k^2 - 25m^2$.

CHECK $(16k^2 + 25m^2)(4k + 5m)(4k - 5m)$

$= (16k^2 + 25m^2)(16k^2 - 25m^2)$ Multiply the last two factors.

$= 256k^4 - 625m^4$ ✓ Original polynomial

(c) $(a + 2b)^2 - 4c^2 = (a + 2b)^2 - (2c)^2$ Write as a difference of squares.

$= [(a + 2b) + 2c][(a + 2b) - 2c]$ Factor.

$= (a + 2b + 2c)(a + 2b - 2c)$

Check by multiplying.

(d) $x^2 - 6x + 9 - y^4 = (x^2 - 6x + 9) - y^4$ Group terms.

$= (x - 3)^2 - y^4$ Factor the trinomial.

$= (x - 3)^2 - (y^2)^2$ Write as a difference of squares.

$= [(x - 3) + y^2][(x - 3) - y^2]$ Factor.

$= (x - 3 + y^2)(x - 3 - y^2)$

Check by multiplying.

(e) $y^2 - x^2 + 6x - 9 = y^2 - (x^2 - 6x + 9)$ Factor out the negative sign and group the last three terms.

Be careful with signs. This is a perfect square trinomial.

$= y^2 - (x - 3)^2$ Write as a difference of squares.

$= [y - (x - 3)][y + (x - 3)]$ Factor.

$= (y - x + 3)(y + x - 3)$ Distributive property

Check by multiplying.

✔ ***Now Try Exercises 51, 55, 57, and 59.***

CAUTION When factoring as in **Example 5(e),** be careful with signs. Inserting an open parenthesis following the minus sign requires changing the signs of all of the following terms.

Classroom Example 6
Factor each polynomial.
(a) $t^3 + 1000$
(b) $r^3 - 8s^3$
(c) $125u^9 - 216v^{12}$

Answers:
(a) $(t + 10)(t^2 - 10t + 100)$
(b) $(r - 2s)(r^2 + 2rs + 4s^2)$
(c) $(5u^3 - 6v^4) \cdot (25u^6 + 30u^3v^4 + 36v^8)$

EXAMPLE 6 Factoring Sums or Differences of Cubes

Factor each polynomial.

(a) $x^3 + 27$ **(b)** $m^3 - 64n^3$ **(c)** $8q^6 + 125p^9$

SOLUTION

(a)
$$\begin{aligned} x^3 + 27 &= x^3 + 3^3 && \text{Write as a sum of cubes.} \\ &= (x + 3)(x^2 - 3x + 3^2) && \text{Factor.} \\ &= (x + 3)(x^2 - 3x + 9) && \text{Apply the exponent.} \end{aligned}$$

(b)
$$\begin{aligned} m^3 - 64n^3 &= m^3 - (4n)^3 && \text{Write as a difference of cubes.} \\ &= (m - 4n)[m^2 + m(4n) + (4n)^2] && \text{Factor.} \\ &= (m - 4n)(m^2 + 4mn + 16n^2) && \text{Simplify.} \end{aligned}$$

(c)
$$\begin{aligned} 8q^6 + 125p^9 &= (2q^2)^3 + (5p^3)^3 && \text{Write as a sum of cubes.} \\ &= (2q^2 + 5p^3)[(2q^2)^2 - 2q^2(5p^3) + (5p^3)^2] && \text{Factor.} \\ &= (2q^2 + 5p^3)(4q^4 - 10q^2p^3 + 25p^6) && \text{Simplify.} \end{aligned}$$

✔ ***Now Try Exercises 61, 63, and 65.***

Factoring by Substitution We introduce a new technique for factoring.

Classroom Example 7
Factor each polynomial.
(a) $8(3x + 1)^2 + 10(3x + 1) - 25$
(b) $(3x + 1)^3 - 27$
(c) $15m^4 - m^2 - 6$

Answers:
(a) $(12x - 1)(6x + 7)$
(b) $(3x - 2)(9x^2 + 15x + 13)$
(c) $(3m^2 - 2)(5m^2 + 3)$

EXAMPLE 7 Factoring by Substitution

Factor each polynomial.

(a) $10(2a - 1)^2 - 19(2a - 1) - 15$ **(b)** $(2a - 1)^3 + 8$

(c) $6z^4 - 13z^2 - 5$

SOLUTION

(a)
$$\begin{aligned} &10(2a - 1)^2 - 19(2a - 1) - 15 && \text{Replace } 2a - 1 \text{ with } u \text{ so that } (2a-1)^2 \text{ becomes } u^2. \\ &= 10u^2 - 19u - 15 \\ &= (5u + 3)(2u - 5) && \text{Factor.} \\ &= [5(2a - 1) + 3][2(2a - 1) - 5] && \text{Replace } u \text{ with } 2a - 1. \\ &= (10a - 5 + 3)(4a - 2 - 5) && \text{Distributive property} \\ &= (10a - 2)(4a - 7) && \text{Simplify.} \\ &= 2(5a - 1)(4a - 7) && \text{Factor out the common factor.} \end{aligned}$$

Don't stop here. Replace u with $2a - 1$.

(b)
$$\begin{aligned} (2a - 1)^3 + 8 &= u^3 + 8 && \text{Replace } 2a - 1 \text{ with } u. \\ &= u^3 + 2^3 && \text{Write as a sum of cubes.} \\ &= (u + 2)(u^2 - 2u + 4) && \text{Factor.} \\ &= [(2a - 1) + 2][(2a - 1)^2 - 2(2a - 1) + 4] && \text{Replace } u \text{ with } 2a - 1. \\ &= (2a + 1)(4a^2 - 4a + 1 - 4a + 2 + 4) && \text{Add, and then multiply.} \\ &= (2a + 1)(4a^2 - 8a + 7) && \text{Combine like terms.} \end{aligned}$$

Teaching Tip Some students prefer to factor the type of trinomial in **Example 7(c)** directly using trial and error with FOIL.

(c)
$$\begin{aligned} 6z^4 - 13z^2 - 5 &= 6u^2 - 13u - 5 && \text{Replace } z^2 \text{ with } u. \\ &= (2u - 5)(3u + 1) && \text{Use FOIL to factor.} \\ &= (2z^2 - 5)(3z^2 + 1) && \text{Replace } u \text{ with } z^2. \end{aligned}$$

Remember to make the final substitution.

✔ ***Now Try Exercises 79 and 83.***

R.4 Exercises

Factor out the greatest common factor from each polynomial. **See Examples 1 and 2.**

1. $12m + 60$
2. $15r - 27$
3. $8k^3 + 24k$
4. $9z^4 + 81z$
5. $xy - 5xy^2$
6. $5h^2j + hj$
7. $-4p^3q^4 - 2p^2q^5$
8. $-3z^5w^2 - 18z^3w^4$
9. $4k^2m^3 + 8k^4m^3 - 12k^2m^4$
10. $28r^4s^2 + 7r^3s - 35r^4s^3$
11. $2(a + b) + 4m(a + b)$
12. $6x(a + b) - 4y(a + b)$
13. $(5r - 6)(r + 3) - (2r - 1)(r + 3)$
14. $(4z - 5)(3z - 2) - (3z - 9)(3z - 2)$
15. $2(m - 1) - 3(m - 1)^2 + 2(m - 1)^3$
16. $5(a + 3)^3 - 2(a + 3) + (a + 3)^2$

17. *Concept Check* When directed to completely factor the polynomial $4x^2y^5 - 8xy^3$, a student wrote $2xy^3(2xy^2 - 4)$. When the teacher did not give him full credit, he complained because when his answer is multiplied out, the result is the original polynomial. Give the correct answer.

18. *Concept Check* Kurt factored $16a^2 - 40a - 6a + 15$ by grouping and obtained $(8a - 3)(2a - 5)$. Callie factored the same polynomial and gave an answer of $(3 - 8a)(5 - 2a)$. Which answer is correct?

Factor each polynomial by grouping. **See Example 2.**

19. $6st + 9t - 10s - 15$
20. $10ab - 6b + 35a - 21$
21. $2m^4 + 6 - am^4 - 3a$
22. $15 - 5m^2 - 3r^2 + m^2r^2$
23. $p^2q^2 - 10 - 2q^2 + 5p^2$
24. $20z^2 - 8x + 5pz^2 - 2px$

Factor each trinomial, if possible. **See Examples 3 and 4.**

25. $6a^2 - 11a + 4$
26. $8h^2 - 2h - 21$
27. $3m^2 + 14m + 8$
28. $9y^2 - 18y + 8$
29. $15p^2 + 24p + 8$
30. $9x^2 + 4x - 2$
31. $12a^3 + 10a^2 - 42a$
32. $36x^3 + 18x^2 - 4x$
33. $6k^2 + 5kp - 6p^2$
34. $14m^2 + 11mr - 15r^2$
35. $5a^2 - 7ab - 6b^2$
36. $12s^2 + 11st - 5t^2$
37. $12x^2 - xy - y^2$
38. $30a^2 + am - m^2$
39. $24a^4 + 10a^3b - 4a^2b^2$
40. $18x^5 + 15x^4z - 75x^3z^2$
41. $9m^2 - 12m + 4$
42. $16p^2 - 40p + 25$
43. $32a^2 + 48ab + 18b^2$
44. $20p^2 - 100pq + 125q^2$
45. $4x^2y^2 + 28xy + 49$
46. $9m^2n^2 + 12mn + 4$
47. $(a - 3b)^2 - 6(a - 3b) + 9$
48. $(2p + q)^2 - 10(2p + q) + 25$

49. *Concept Check* Match each polynomial in Column I with its factored form in Column II.

I	II
(a) $x^2 + 10xy + 25y^2$	**A.** $(x + 5y)(x - 5y)$
(b) $x^2 - 10xy + 25y^2$	**B.** $(x + 5y)^2$
(c) $x^2 - 25y^2$	**C.** $(x - 5y)^2$
(d) $25y^2 - x^2$	**D.** $(5y + x)(5y - x)$

50. *Concept Check* Match each polynomial in Column I with its factored form in Column II.

I	II
(a) $8x^3 - 27$	**A.** $(3 - 2x)(9 + 6x + 4x^2)$
(b) $8x^3 + 27$	**B.** $(2x - 3)(4x^2 + 6x + 9)$
(c) $27 - 8x^3$	**C.** $(2x + 3)(4x^2 - 6x + 9)$

1. $12(m + 5)$ 2. $3(5r - 9)$
3. $8k(k^2 + 3)$ 4. $9z(z^3 + 9)$
5. $xy(1 - 5y)$ 6. $hj(5h + 1)$
7. $-2p^2q^4(2p + q)$
8. $-3z^3w^2(z^2 + 6w^2)$
9. $4k^2m^3(1 + 2k^2 - 3m)$
10. $7r^3s(4rs + 1 - 5rs^2)$
11. $2(a + b)(1 + 2m)$
12. $2(a + b)(3x - 2y)$
13. $(r + 3)(3r - 5)$
14. $(3z - 2)(z + 4)$
15. $(m - 1)(2m^2 - 7m + 7)$
16. $(a + 3)(5a^2 + 31a + 46)$
17. The *completely* factored form is $4xy^3(xy^2 - 2)$.
18. Both are correct.
19. $(2s + 3)(3t - 5)$
20. $(5a - 3)(2b + 7)$
21. $(m^4 + 3)(2 - a)$
22. $(3 - m^2)(5 - r^2)$
23. $(p^2 - 2)(q^2 + 5)$
24. $(5z^2 - 2x)(4 + p)$
25. $(2a - 1)(3a - 4)$
26. $(4h - 7)(2h + 3)$
27. $(3m + 2)(m + 4)$
28. $(3y - 4)(3y - 2)$
29. prime
30. prime
31. $2a(3a + 7)(2a - 3)$
32. $2x(6x - 1)(3x + 2)$
33. $(3k - 2p)(2k + 3p)$
34. $(7m - 5r)(2m + 3r)$
35. $(5a + 3b)(a - 2b)$
36. $(4s + 5t)(3s - t)$
37. $(4x + y)(3x - y)$
38. $(5a + m)(6a - m)$
39. $2a^2(4a - b)(3a + 2b)$
40. $3x^3(3x - 5z)(2x + 5z)$
41. $(3m - 2)^2$
42. $(4p - 5)^2$
43. $2(4a + 3b)^2$
44. $5(2p - 5q)^2$
45. $(2xy + 7)^2$
46. $(3mn + 2)^2$
47. $(a - 3b - 3)^2$
48. $(2p + q - 5)^2$
49. (a) B (b) C (c) A (d) D
50. (a) B (b) C (c) A

51. $(3a + 4)(3a - 4)$
52. $(4q + 5)(4q - 5)$
53. $(x^2 + 4)(x + 2)(x - 2)$
54. $(y^2 + 9)(y + 3)(y - 3)$
55. $(5s^2 + 3t)(5s^2 - 3t)$
56. $9(2z + 3y^2)(2z - 3y^2)$
57. $(a + b + 4)(a + b - 4)$
58. $(p - 2q + 10)(p - 2q - 10)$
59. $(p^2 + 25)(p + 5)(p - 5)$
60. $(m^2 + 36)(m + 6)(m - 6)$
61. $(2 - a)(4 + 2a + a^2)$
62. $(3 - r)(9 + 3r + r^2)$
63. $(5x - 3)(25x^2 + 15x + 9)$
64. $(2m - 3n)(4m^2 + 6mn + 9n^2)$
65. $(3y^3 + 5z^2)(9y^6 - 15y^3z^2 + 25z^4)$
66. $(3z^3 + 4y^4)(9z^6 - 12z^3y^4 + 16y^8)$
67. $r(r^2 + 18r + 108)$
68. $b(b^2 + 9b + 27)$
69. $(3 - m - 2n)(9 + 3m + 6n + m^2 + 4mn + 4n^2)$
70. $(5 - 4a + b)(25 + 20a - 5b + 16a^2 - 8ab + b^2)$

71. B   72. C

73. $(x - 1)(x^2 + x + 1) \cdot (x + 1)(x^2 - x + 1)$
74. $(x - 1)(x + 1)(x^4 + x^2 + 1)$
75. $(x^2 - x + 1)(x^2 + x + 1)$
76. additive inverse property (0 in the form $x^2 - x^2$ was added on the right.); associative property of addition; factoring a perfect square trinomial; factoring a difference of squares; commutative property of addition
77. They are the same.
78. $(x^4 - x^2 + 1)(x^2 + x + 1) \cdot (x^2 - x + 1)$

Factor each polynomial. ***See Examples 5 and 6.***

51. $9a^2 - 16$ **52.** $16q^2 - 25$ **53.** $x^4 - 16$

54. $y^4 - 81$ **55.** $25s^4 - 9t^2$ **56.** $36z^2 - 81y^4$

57. $(a + b)^2 - 16$ **58.** $(p - 2q)^2 - 100$

59. $p^4 - 625$ **60.** $m^4 - 1296$

61. $8 - a^3$ **62.** $27 - r^3$

63. $125x^3 - 27$ **64.** $8m^3 - 27n^3$

65. $27y^9 + 125z^6$ **66.** $27z^9 + 64y^{12}$

67. $(r + 6)^3 - 216$ **68.** $(b + 3)^3 - 27$

69. $27 - (m + 2n)^3$ **70.** $125 - (4a - b)^3$

71. *Concept Check* Which of the following is the correct complete factorization of $x^4 - 1$?

A. $(x^2 - 1)(x^2 + 1)$ **B.** $(x^2 + 1)(x + 1)(x - 1)$
C. $(x^2 - 1)^2$ **D.** $(x - 1)^2(x + 1)^2$

72. *Concept Check* Which of the following is the correct factorization of $x^3 + 8$?

A. $(x + 2)^3$ **B.** $(x + 2)(x^2 + 2x + 4)$
C. $(x + 2)(x^2 - 2x + 4)$ **D.** $(x + 2)(x^2 - 4x + 4)$

Relating Concepts

For individual or collaborative investigation *(Exercises 73–78)*

The polynomial $x^6 - 1$ can be considered either a difference of squares or a difference of cubes. ***Work Exercises 73–78 in order,*** *to connect the results obtained when two different methods of factoring are used.*

73. Factor $x^6 - 1$ by first factoring as a difference of squares, and then factor further by using the patterns for a sum of cubes and a difference of cubes.

74. Factor $x^6 - 1$ by first factoring as a difference of cubes, and then factor further by using the pattern for a difference of squares.

75. Compare your answers in **Exercises 73 and 74.** Based on these results, what is the factorization of $x^4 + x^2 + 1$?

76. The polynomial $x^4 + x^2 + 1$ cannot be factored using the methods described in this section. However, there is a technique that enables us to factor it, as shown here. Supply the reason why each step is valid.

$$x^4 + x^2 + 1 = x^4 + 2x^2 + 1 - x^2 \quad \_\_\_\_\_\_\_\_\_\_$$
$$= (x^4 + 2x^2 + 1) - x^2 \quad \_\_\_\_\_\_\_\_\_\_$$
$$= (x^2 + 1)^2 - x^2 \quad \_\_\_\_\_\_\_\_\_\_$$
$$= (x^2 + 1 - x)(x^2 + 1 + x) \quad \_\_\_\_\_\_\_\_\_\_$$
$$= (x^2 - x + 1)(x^2 + x + 1) \quad \_\_\_\_\_\_\_\_\_\_$$

77. Compare your answer in **Exercise 75** with the final line in **Exercise 76.** What do you notice?

78. Factor $x^8 + x^4 + 1$ using the technique outlined in **Exercise 76.**

79. $9(7k-3)(k+1)$
80. $2(8z-3)(6z-5)$
81. $(3a-7)^2$
82. $(10x+17)^2$
83. $(m^2-5)(m^2+2)$
84. $(a^2-8)(a^2+6)$

85. $(2b+c+4)(2b+c-4)$
86. $(2y-3)^2$
87. $(x+y)(x-5)$
88. prime
89. $(m-2n)(p^4+q)$
90. $(6a+5)^2$
91. $(2z+7)^2$
92. $(3p^2-1)(2p^2+3)$
93. $(10x+7y)(100x^2-70xy+49y^2)$
94. $(b+4+a)(b+4-a)$
95. $(5m^2-6)(25m^4+30m^2+36)$
96. $(q+3+p)(q+3-p)$
97. $9(x+2)(3x^2+4)$
98. $(6p+5q)(36p^2-30pq+25q^2)$
99. $2y(3x^2+y^2)$
100. $(10r+13s)(10r-13s)$
101. prime
102. $(3a-4)^2$
103. $4xy$
104. $(4z^2+5)(z^2-3)$

107. $\left(7x+\frac{1}{5}\right)\left(7x-\frac{1}{5}\right)$
108. $\left(9y+\frac{1}{7}\right)\left(9y-\frac{1}{7}\right)$
109. $\left(\frac{5}{3}x^2+3y\right)\left(\frac{5}{3}x^2-3y\right)$
110. $\left(\frac{11}{5}y^2+7x\right)\left(\frac{11}{5}y^2-7x\right)$

111. ± 36 112. ± 30
113. 9 114. 25

Factor each polynomial by substitution. ***See Example 7.***

79. $7(3k-1)^2+26(3k-1)-8$ **80.** $6(4z-3)^2+7(4z-3)-3$

81. $9(a-4)^2+30(a-4)+25$ **82.** $4(5x+7)^2+12(5x+7)+9$

83. m^4-3m^2-10 **84.** a^4-2a^2-48

Factor by any method. ***See Examples 1–7.***

85. $4b^2+4bc+c^2-16$ **86.** $(2y-1)^2-4(2y-1)+4$

87. $x^2+xy-5x-5y$ **88.** $8r^2-3rs+10s^2$

89. $p^4(m-2n)+q(m-2n)$ **90.** $36a^2+60a+25$

91. $4z^2+28z+49$ **92.** $6p^4+7p^2-3$

93. $1000x^3+343y^3$ **94.** $b^2+8b+16-a^2$

95. $125m^6-216$ **96.** $q^2+6q+9-p^2$

97. $64+(3x+2)^3$ **98.** $216p^3+125q^3$

99. $(x+y)^3-(x-y)^3$ **100.** $100r^2-169s^2$

101. $144z^2+121$ **102.** $(3a+5)^2-18(3a+5)+81$

103. $(x+y)^2-(x-y)^2$ **104.** $4z^4-7z^2-15$

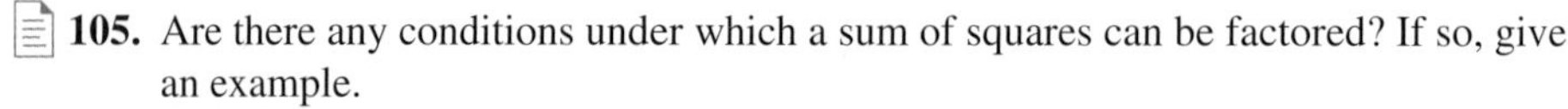

105. Are there any conditions under which a sum of squares can be factored? If so, give an example.

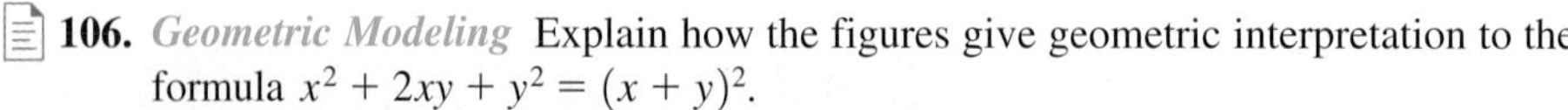

106. *Geometric Modeling* Explain how the figures give geometric interpretation to the formula $x^2+2xy+y^2=(x+y)^2$.

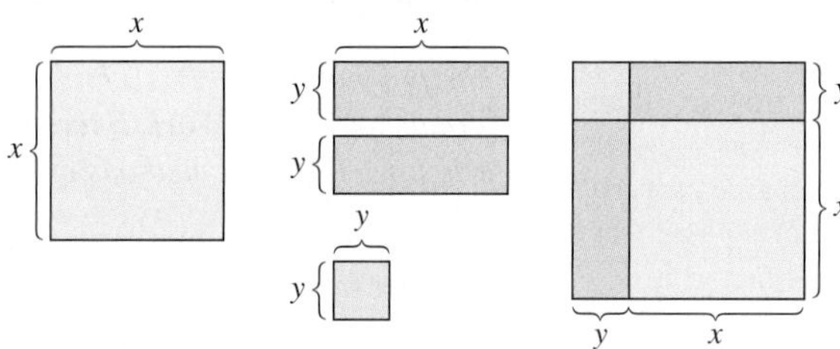

Factor each polynomial over the set of rational number coefficients.

107. $49x^2-\frac{1}{25}$ **108.** $81y^2-\frac{1}{49}$

109. $\frac{25}{9}x^4-9y^2$ **110.** $\frac{121}{25}y^4-49x^2$

Concept Check Find all values of b or c that will make the polynomial a perfect square trinomial.

111. $4z^2+bz+81$ **112.** $9p^2+bp+25$

113. $100r^2-60r+c$ **114.** $49x^2+70x+c$

R.5 Rational Expressions

- Rational Expressions
- Lowest Terms of a Rational Expression
- Multiplication and Division
- Addition and Subtraction
- Complex Fractions

Rational Expressions The quotient of two polynomials P and Q, with $Q \neq 0$, is a **rational expression.**

$$\frac{x+6}{x+2}, \quad \frac{(x+6)(x+4)}{(x+2)(x+4)}, \quad \frac{2p^2+7p-4}{5p^2+20p} \qquad \text{Rational expressions}$$

The **domain** of a rational expression is the set of real numbers for which the expression is defined. Because the denominator of a fraction cannot be 0, the domain consists of all real numbers except those that make the denominator 0. We find these numbers by setting the denominator equal to 0 and solving the resulting equation. For example, in the rational expression

$$\frac{x+6}{x+2},$$

the solution to the equation $x + 2 = 0$ is excluded from the domain. Since this solution is -2, the domain is the set of all real numbers x not equal to -2, or

$$\{x \mid x \neq -2\}. \qquad \text{Set-builder notation (Section R.1)}$$

Teaching Tip Give a quick review of solving simple linear equations such as

$$x + 2 = 0,$$

and of the zero-factor property. (See **Chapter 1.**)

If the denominator of a rational expression contains a product, we determine the domain with the **zero-factor property** (covered in more detail in **Section 1.4**), which states that $ab = 0$ if and only if $a = 0$ or $b = 0$.

Classroom Example 1
Find the domain of the rational expression.

$$\frac{(x-7)(x-1)}{(x+3)(x-1)}$$

Answer: $\{x \mid x \neq -3, 1\}$

EXAMPLE 1 Finding the Domain

Find the domain of the rational expression.

$$\frac{(x+6)(x+4)}{(x+2)(x+4)}$$

SOLUTION

$$(x+2)(x+4) = 0 \qquad \text{Set the denominator equal to zero.}$$

$$x+2=0 \quad \text{or} \quad x+4=0 \qquad \text{Zero-factor property}$$

$$x=-2 \quad \text{or} \quad x=-4 \qquad \text{Solve each equation.}$$

The domain is the set of real numbers *not equal to* -2 or -4, written

$$\{x \mid x \neq -2, -4\}.$$

✔ *Now Try Exercises 1, 3, and 7.*

Lowest Terms of a Rational Expression A rational expression is written in **lowest terms** when the greatest common factor of its numerator and its denominator is 1. We use the following **fundamental principle of fractions.**

Fundamental Principle of Fractions

$$\frac{ac}{bc} = \frac{a}{b} \quad (b \neq 0, c \neq 0)$$

Classroom Example 2
Write each rational expression in lowest terms.

(a) $\frac{12x^2 - 30x}{4x^2 - 25}$

(b) $\frac{x^2 - 8x + 16}{8x - 2x^2}$

Answers:
(a) $\frac{6x}{2x+5}$
(b) $\frac{-(x-4)}{2x}$, or $\frac{4-x}{2x}$

Teaching Tip Show that

$$\frac{6+3x}{x^2+x-6},$$

which is equivalent to

$$\frac{3(2+x)}{(x+3)(x-2)},$$

does not contain a common factor greater than 1 in the numerator and denominator.

EXAMPLE 2 Writing Rational Expressions in Lowest Terms

Write each rational expression in lowest terms.

(a) $\frac{2x^2 + 7x - 4}{5x^2 + 20x}$ **(b)** $\frac{6 - 3x}{x^2 - 4}$

SOLUTION

(a) $\frac{2x^2 + 7x - 4}{5x^2 + 20x} = \frac{(2x-1)(x+4)}{5x(x+4)}$ Factor. (Section R.4)

$= \frac{2x-1}{5x}$ Fundamental principle and lowest terms

To determine the domain, we find values of x that make the *original* denominator $5x^2 + 20x$ equal to 0, and exclude them.

$5x^2 + 20x = 0$ Set the denominator equal to 0.

$5x(x+4) = 0$ Factor.

$5x = 0$ or $x + 4 = 0$ Zero-factor property

$x = 0$ or $x = -4$ Solve each equation.

The domain is $\{x \mid x \neq 0, -4\}$. ***From now on, we will assume such restrictions when writing rational expressions in lowest terms.***

(b) $\frac{6 - 3x}{x^2 - 4} = \frac{3(2-x)}{(x+2)(x-2)}$ Factor.

$= \frac{3(2-x)(-1)}{(x+2)(x-2)(-1)}$ $2 - x$ and $x - 2$ are opposites. Multiply numerator and denominator by -1.

$= \frac{3(2-x)(-1)}{(x+2)(2-x)}$ $(x-2)(-1) = -x + 2 = 2 - x$

Be careful with signs.

$= \frac{-3}{x+2}$ Fundamental principle and lowest terms

Working in an alternative way would lead to the equivalent result $\frac{3}{-x-2}$.

Now Try Exercises 13 and 17.

LOOKING AHEAD TO CALCULUS
A standard problem in calculus is investigating what value an expression such as $\frac{x^2-1}{x-1}$ approaches as x approaches 1. We cannot do this by simply substituting 1 for x in the expression since the result is the indeterminate form $\frac{0}{0}$. When we factor the numerator and write the expression in lowest terms, it becomes $x + 1$. Then by substituting 1 for x, we get $1 + 1 = 2$, which is called the **limit** of $\frac{x^2-1}{x-1}$ as x approaches 1.

CAUTION The fundamental principle requires a pair of common *factors,* one in the numerator and one in the denominator. ***Only after a rational expression has been factored can any common factors be divided out.***

For example, $\frac{2x+4}{6} = \frac{2(x+2)}{2 \cdot 3} = \frac{x+2}{3}$. Factor first, and then divide.

Multiplication and Division We now multiply and divide fractions.

Multiplication and Division

For fractions $\frac{a}{b}$ and $\frac{c}{d}$ $(b \neq 0, d \neq 0)$, the following hold.

$$\frac{a}{b} \cdot \frac{c}{d} = \frac{ac}{bd} \quad \text{and} \quad \frac{a}{b} \div \frac{c}{d} = \frac{a}{b} \cdot \frac{d}{c} \quad (c \neq 0)$$

That is, to find the product of two fractions, multiply their numerators to find the numerator of the product. Then multiply their denominators to find the denominator of the product. To divide two fractions, multiply the* dividend *(the first fraction) by the reciprocal of the* divisor *(the second fraction).

Classroom Example 3
Multiply or divide, as indicated.

(a) $\frac{6z^6}{7} \cdot \frac{28}{9z^2}$

(b) $\frac{4n^2+3n-10}{2n^2+3n-2} \cdot \frac{2n-1}{n+4}$

(c) $\frac{5z^2-16z+3}{z^2+z-12} \div \frac{30z^2-6z}{2z^3+8z^2}$

(d) $\frac{x^2-1}{x^3-1} \cdot \frac{xy-2y+3x-6}{xy+3x+y+3}$

Answers:

(a) $\frac{8z^4}{3}$ (b) $\frac{4n-5}{n+4}$

(c) $\frac{z}{3}$ (d) $\frac{x-2}{x^2+x+1}$

EXAMPLE 3 Multiplying or Dividing Rational Expressions

Multiply or divide, as indicated.

(a) $\frac{2y^2}{9} \cdot \frac{27}{8y^5}$ **(b)** $\frac{3m^2-2m-8}{3m^2+14m+8} \cdot \frac{3m+2}{3m+4}$

(c) $\frac{3p^2+11p-4}{24p^3-8p^2} \div \frac{9p+36}{24p^4-36p^3}$

(d) $\frac{x^3-y^3}{x^2-y^2} \cdot \frac{2x+2y+xz+yz}{2x^2+2y^2+zx^2+zy^2}$

SOLUTION

(a) $\frac{2y^2}{9} \cdot \frac{27}{8y^5} = \frac{2y^2 \cdot 27}{9 \cdot 8y^5}$ Multiply fractions.

$= \frac{2 \cdot 9 \cdot 3 \cdot y^2}{9 \cdot 2 \cdot 4 \cdot y^2 \cdot y^3}$ Factor.

$= \frac{3}{4y^3}$ Fundamental principle

Although we usually factor first and then multiply the fractions (see parts (b)–(d)), we did the opposite here. Either order is acceptable.

(b) $\frac{3m^2-2m-8}{3m^2+14m+8} \cdot \frac{3m+2}{3m+4}$

$= \frac{(m-2)(3m+4)}{(m+4)(3m+2)} \cdot \frac{3m+2}{3m+4}$ Factor.

$= \frac{(m-2)(3m+4)(3m+2)}{(m+4)(3m+2)(3m+4)}$ Multiply fractions.

$= \frac{m-2}{m+4}$ Fundamental principle

(c) $\frac{3p^2+11p-4}{24p^3-8p^2} \div \frac{9p+36}{24p^4-36p^3}$

$= \frac{(p+4)(3p-1)}{8p^2(3p-1)} \div \frac{9(p+4)}{12p^3(2p-3)}$ Factor.

$= \frac{(p+4)(3p-1)}{8p^2(3p-1)} \cdot \frac{12p^3(2p-3)}{9(p+4)}$ Multiply by the reciprocal of the divisor.

$= \frac{12p^3(2p-3)}{9 \cdot 8p^2}$ Divide out common factors. Multiply fractions.

$= \frac{3 \cdot 4 \cdot p^2 \cdot p(2p-3)}{3 \cdot 3 \cdot 4 \cdot 2 \cdot p^2}$ Factor.

$= \frac{p(2p-3)}{6}$ Fundamental principle

(d) $\frac{x^3 - y^3}{x^2 - y^2} \cdot \frac{2x + 2y + xz + yz}{2x^2 + 2y^2 + zx^2 + zy^2}$

$= \frac{(x - y)(x^2 + xy + y^2)}{(x + y)(x - y)} \cdot \frac{2(x + y) + z(x + y)}{2(x^2 + y^2) + z(x^2 + y^2)}$ Factor. Group terms and factor.

$= \frac{(x - y)(x^2 + xy + y^2)}{(x + y)(x - y)} \cdot \frac{(2 + z)(x + y)}{(2 + z)(x^2 + y^2)}$ Factor by grouping. (Section R.4)

$= \frac{x^2 + xy + y^2}{x^2 + y^2}$ Multiply fractions; fundamental principle

✔ *Now Try Exercises 23, 33, and 37.*

Addition and Subtraction We now add and subtract fractions.

Addition and Subtraction

For fractions $\frac{a}{b}$ and $\frac{c}{d}$ $(b \neq 0, d \neq 0)$, the following hold.

$$\frac{a}{b} + \frac{c}{d} = \frac{ad + bc}{bd} \quad \text{and} \quad \frac{a}{b} - \frac{c}{d} = \frac{ad - bc}{bd}$$

That is, to add (or subtract) two fractions in practice, find their least common denominator (LCD) and change each fraction to one with the LCD as denominator. The sum (or difference) of their numerators is the numerator of their sum (or difference), and the LCD is the denominator of their sum (or difference).

Finding the Least Common Denominator (LCD)

Step 1 Write each denominator as a product of prime factors.

Step 2 Form a product of all the different prime factors. Each factor should have as exponent the ***greatest*** exponent that appears on that factor.

Classroom Example 4
Add or subtract, as indicated.

(a) $\frac{3}{10z^4} + \frac{2}{15z^2}$

(b) $\frac{7}{m - 5} + \frac{2m}{5 - m}$

(c) $\frac{4}{(x - 3)(x + 5)} - \frac{6}{(x + 5)(x - 5)}$

Answers:

(a) $\frac{9 + 4z^2}{30z^4}$

(b) $\frac{7 - 2m}{m - 5}$, or $\frac{2m - 7}{5 - m}$

(c) $\frac{-2x - 2}{(x - 3)(x + 5)(x - 5)}$

EXAMPLE 4 Adding or Subtracting Rational Expressions

Add or subtract, as indicated.

(a) $\frac{5}{9x^2} + \frac{1}{6x}$ **(b)** $\frac{y}{y - 2} + \frac{8}{2 - y}$

(c) $\frac{3}{(x - 1)(x + 2)} - \frac{1}{(x + 3)(x - 4)}$

SOLUTION

(a) $\frac{5}{9x^2} + \frac{1}{6x}$

Step 1 Write each denominator as a product of prime factors.

$$9x^2 = 3^2 \cdot x^2$$

$$6x = 2^1 \cdot 3^1 \cdot x^1$$

Step 2 For the LCD, form the product of all the prime factors, with each factor having the greatest exponent that appears on it.

Greatest exponent on 3 is 2. Greatest exponent on x is 2.

$$\text{LCD} = 2^1 \cdot 3^2 \cdot x^2$$
$$= 18x^2$$

Write the given expressions with this denominator, and then add.

$$\frac{5}{9x^2} + \frac{1}{6x} = \frac{5 \cdot 2}{9x^2 \cdot 2} + \frac{1 \cdot 3x}{6x \cdot 3x} \qquad \text{LCD} = 18x^2$$

$$= \frac{10}{18x^2} + \frac{3x}{18x^2} \qquad \text{Multiply.}$$

$$= \frac{10 + 3x}{18x^2} \qquad \text{Add the numerators.}$$

Always check to see that the answer is in lowest terms.

(b) $\dfrac{y}{y-2} + \dfrac{8}{2-y}$ We arbitrarily choose $y - 2$ as the LCD.

$$= \frac{y}{y-2} + \frac{8(-1)}{(2-y)(-1)} \qquad \text{Multiply the second expression by } -1 \text{ in both the numerator and the denominator.}$$

$$= \frac{y}{y-2} + \frac{-8}{y-2} \qquad \text{Simplify.}$$

$$= \frac{y-8}{y-2} \qquad \text{Add the numerators.}$$

We could use $2 - y$ as the common denominator instead of $y - 2$.

$$\frac{y(-1)}{(y-2)(-1)} + \frac{8}{2-y} \qquad \text{Multiply the first expression by } -1 \text{ in both the numerator and the denominator.}$$

$$= \frac{-y}{2-y} + \frac{8}{2-y}, \quad \text{or} \quad \frac{8-y}{2-y}$$

This equivalent expression results.

(c) $\dfrac{3}{(x-1)(x+2)} - \dfrac{1}{(x+3)(x-4)}$ The LCD is $(x-1)(x+2)(x+3)(x-4)$.

$$= \frac{3(x+3)(x-4)}{(x-1)(x+2)(x+3)(x-4)} - \frac{(x-1)(x+2)}{(x+3)(x-4)(x-1)(x+2)}$$

$$= \frac{3(x^2 - x - 12) - (x^2 + x - 2)}{(x-1)(x+2)(x+3)(x-4)} \qquad \text{Multiply in the numerators, and then subtract them.}$$

Be careful with signs.

$$= \frac{3x^2 - 3x - 36 - x^2 - x + 2}{(x-1)(x+2)(x+3)(x-4)} \qquad \text{Distributive property (Section R.2)}$$

$$= \frac{2x^2 - 4x - 34}{(x-1)(x+2)(x+3)(x-4)} \qquad \text{Combine like terms in the numerator.}$$

Now Try Exercises 47, 51, and 59.

CAUTION ***When subtracting fractions where the second fraction has more than one term in the numerator, as in Example 4(c), be sure to distribute the negative sign to each term.*** Use parentheses as in the second step to avoid an error.

Complex Fractions The quotient of two rational expressions is a **complex fraction.** There are two methods for simplifying a complex fraction.

Classroom Example 5
Simplify each complex fraction.

(a) $\dfrac{3+\frac{4}{x^2}}{6-\frac{1}{x^2}}$ (b) $\dfrac{\frac{1}{z+1}-\frac{1}{z-1}}{\frac{1}{z}+\frac{1}{z+1}}$

Answers:

(a) $\frac{3x^2+4}{6x^2-1}$ (b) $\frac{-2z}{2z^2-z-1}$

EXAMPLE 5 Simplifying Complex Fractions

Simplify each complex fraction. In part (b), use two methods.

(a) $\dfrac{6-\frac{5}{k}}{1+\frac{5}{k}}$ **(b)** $\dfrac{\frac{a}{a+1}+\frac{1}{a}}{\frac{1}{a}+\frac{1}{a+1}}$

SOLUTION

(a) Method 1 for simplifying uses the identity property for multiplication. We multiply both numerator and denominator by the LCD of all the fractions, k.

$$\frac{6-\frac{5}{k}}{1+\frac{5}{k}}=\frac{k\left(6-\frac{5}{k}\right)}{k\left(1+\frac{5}{k}\right)}=\frac{6k-k\left(\frac{5}{k}\right)}{k+k\left(\frac{5}{k}\right)}=\frac{6k-5}{k+5}$$

Distribute k to *all* terms within the parentheses.

(b) $$\frac{\frac{a}{a+1}+\frac{1}{a}}{\frac{1}{a}+\frac{1}{a+1}}=\frac{\left(\frac{a}{a+1}+\frac{1}{a}\right)a(a+1)}{\left(\frac{1}{a}+\frac{1}{a+1}\right)a(a+1)}$$

For Method 1, multiply both numerator and denominator by the LCD of all the fractions, $a(a+1)$.

(Method 1)

$$=\frac{\frac{a}{a+1}(a)(a+1)+\frac{1}{a}(a)(a+1)}{\frac{1}{a}(a)(a+1)+\frac{1}{a+1}(a)(a+1)}$$ Distributive property

$$=\frac{a^2+(a+1)}{(a+1)+a}$$ Multiply.

$$=\frac{a^2+a+1}{2a+1}$$ Combine like terms.

$$\frac{\frac{a}{a+1}+\frac{1}{a}}{\frac{1}{a}+\frac{1}{a+1}}=\frac{\frac{a^2+1(a+1)}{a(a+1)}}{\frac{1(a+1)+1(a)}{a(a+1)}}$$

For Method 2, find the LCD, and add terms in the numerator and denominator of the complex fraction.

(Method 2)

$$=\frac{\frac{a^2+a+1}{a(a+1)}}{\frac{2a+1}{a(a+1)}}$$ Combine terms in the numerator and denominator.

$$=\frac{a^2+a+1}{a(a+1)}\cdot\frac{a(a+1)}{2a+1}$$ Definition of division

$$=\frac{a^2+a+1}{2a+1}$$ Multiply fractions, and write in lowest terms.

The result is the same as in Method 1.

✔ Now Try Exercises 61 and 73.

R.5 Exercises

Find the domain of each rational expression. ***See Example 1.***

1. $\dfrac{x+3}{x-6}$ **2.** $\dfrac{2x-4}{x+7}$ **3.** $\dfrac{3x+7}{(4x+2)(x-1)}$

4. $\dfrac{9x+12}{(2x+3)(x-5)}$ **5.** $\dfrac{12}{x^2+5x+6}$ **6.** $\dfrac{3}{x^2-5x-6}$

7. $\dfrac{x^2-1}{x+1}$ **8.** $\dfrac{x^2-25}{x-5}$ **9.** $\dfrac{x^3-1}{x-1}$

10. *Concept Check* Use specific values for x and y to show that in general, $\frac{1}{x}+\frac{1}{y}$ is not equivalent to $\frac{1}{x+y}$.

Write each rational expression in lowest terms. ***See Example 2.***

11. $\dfrac{8x^2+16x}{4x^2}$ **12.** $\dfrac{36y^2+72y}{9y^2}$ **13.** $\dfrac{3(3-t)}{(t+5)(t-3)}$

14. $\dfrac{-8(4-y)}{(y+2)(y-4)}$ **15.** $\dfrac{8k+16}{9k+18}$ **16.** $\dfrac{20r+10}{30r+15}$

17. $\dfrac{m^2-4m+4}{m^2+m-6}$ **18.** $\dfrac{r^2-r-6}{r^2+r-12}$ **19.** $\dfrac{8m^2+6m-9}{16m^2-9}$

20. $\dfrac{6y^2+11y+4}{3y^2+7y+4}$ **21.** $\dfrac{x^3+64}{x+4}$ **22.** $\dfrac{y^3-27}{y-3}$

Find each product or quotient. ***See Example 3.***

23. $\dfrac{15p^3}{9p^2}\div\dfrac{6p}{10p^2}$ **24.** $\dfrac{8r^3}{6r}\div\dfrac{5r^2}{9r^3}$ **25.** $\dfrac{2k+8}{6}\div\dfrac{3k+12}{2}$

26. $\dfrac{5m+25}{10}\div\dfrac{6m+30}{12}$ **27.** $\dfrac{x^2+x}{5}\cdot\dfrac{25}{xy+y}$ **28.** $\dfrac{y^3+y^2}{7}\cdot\dfrac{49}{y^4+y^3}$

29. $\dfrac{4a+12}{2a-10}\div\dfrac{a^2-9}{a^2-a-20}$ **30.** $\dfrac{6r-18}{9r^2+6r-24}\div\dfrac{4r-12}{12r-16}$

31. $\dfrac{p^2-p-12}{p^2-2p-15}\cdot\dfrac{p^2-9p+20}{p^2-8p+16}$ **32.** $\dfrac{x^2+2x-15}{x^2+11x+30}\cdot\dfrac{x^2+2x-24}{x^2-8x+15}$

33. $\dfrac{m^2+3m+2}{m^2+5m+4}\div\dfrac{m^2+5m+6}{m^2+10m+24}$ **34.** $\dfrac{y^2+y-2}{y^2+3y-4}\div\dfrac{y^2+3y+2}{y^2+4y+3}$

35. $\dfrac{x^3+y^3}{x^3-y^3}\cdot\dfrac{x^2-y^2}{x^2+2xy+y^2}$ **36.** $\dfrac{x^2-y^2}{(x-y)^2}\cdot\dfrac{x^2-xy+y^2}{x^2-2xy+y^2}\div\dfrac{x^3+y^3}{(x-y)^4}$

37. $\dfrac{xz-xw+2yz-2yw}{z^2-w^2}\cdot\dfrac{4z+4w+xz+wx}{16-x^2}$

38. $\dfrac{ac+ad+bc+bd}{a^2-b^2}\cdot\dfrac{a^3-b^3}{2a^2+2ab+2b^2}$

39. *Concept Check* Which of the following rational expressions is equivalent to -1? In choices A, B, and D, $x\neq -4$, and in choice C, $x\neq 4$. (*Hint:* There may be more than one answer.)

A. $\dfrac{x-4}{x+4}$ **B.** $\dfrac{-x-4}{x+4}$ **C.** $\dfrac{x-4}{4-x}$ **D.** $\dfrac{x-4}{-x-4}$

40. Explain how to find the least common denominator of several fractions.

1. $\{x\,|\,x\neq 6\}$
2. $\{x\,|\,x\neq -7\}$
3. $\{x\,|\,x\neq -\frac{1}{2}, 1\}$
4. $\{x\,|\,x\neq -\frac{3}{2}, 5\}$
5. $\{x\,|\,x\neq -2, -3\}$
6. $\{x\,|\,x\neq -1, 6\}$
7. $\{x\,|\,x\neq -1\}$
8. $\{x\,|\,x\neq 5\}$
9. $\{x\,|\,x\neq 1\}$

10. For example, let $x=4$ and $y=2$. Then $\frac{1}{4}+\frac{1}{2}=\frac{3}{4}$, $\frac{1}{4+2}=\frac{1}{6}$, and $\frac{3}{4}\neq\frac{1}{6}$.

11. $\frac{2x+4}{x}$ **12.** $\frac{4y+8}{y}$
13. $\frac{-3}{t+5}$ **14.** $\frac{8}{y+2}$
15. $\frac{8}{9}$ **16.** $\frac{2}{3}$
17. $\frac{m-2}{m+3}$ **18.** $\frac{r+2}{r+4}$
19. $\frac{2m+3}{4m+3}$ **20.** $\frac{2y+1}{y+1}$
21. $x^2-4x+16$
22. y^2+3y+9

23. $\frac{25p^2}{9}$ **24.** $\frac{12r^3}{5}$
25. $\frac{2}{9}$ **26.** 1
27. $\frac{5x}{y}$ **28.** $\frac{7}{y}$
29. $\frac{2a+8}{a-3}$, or $\frac{2(a+4)}{a-3}$
30. $\frac{2}{r+2}$
31. 1 **32.** $\frac{x-4}{x-5}$
33. $\frac{m+6}{m+3}$ **34.** $\frac{y+3}{y+4}$
35. $\frac{x^2-xy+y^2}{x^2+xy+y^2}$ **36.** $x-y$
37. $\frac{x+2y}{4-x}$ **38.** $\frac{c+d}{2}$

39. B, C

Perform each addition or subtraction. ***See Example 4.***

41. $\dfrac{3}{2k}+\dfrac{5}{3k}$ 42. $\dfrac{8}{5p}+\dfrac{3}{4p}$ 43. $\dfrac{1}{6m}+\dfrac{2}{5m}+\dfrac{4}{m}$

44. $\dfrac{8}{3p}+\dfrac{5}{4p}+\dfrac{9}{2p}$ 45. $\dfrac{1}{a}-\dfrac{b}{a^2}$ 46. $\dfrac{3}{z}+\dfrac{x}{z^2}$

47. $\dfrac{5}{12x^2y}-\dfrac{11}{6xy}$ 48. $\dfrac{7}{18a^3b^2}-\dfrac{2}{9ab}$ 49. $\dfrac{17y+3}{9y+7}-\dfrac{-10y-18}{9y+7}$

50. $\dfrac{7x+8}{3x+2}-\dfrac{x+4}{3x+2}$ 51. $\dfrac{1}{x+z}+\dfrac{1}{x-z}$ 52. $\dfrac{m+1}{m-1}+\dfrac{m-1}{m+1}$

53. $\dfrac{3}{a-2}-\dfrac{1}{2-a}$ 54. $\dfrac{4}{p-q}-\dfrac{2}{q-p}$

55. $\dfrac{x+y}{2x-y}-\dfrac{2x}{y-2x}$ 56. $\dfrac{m-4}{3m-4}-\dfrac{5m}{4-3m}$

57. $\dfrac{4}{x+1}+\dfrac{1}{x^2-x+1}-\dfrac{12}{x^3+1}$ 58. $\dfrac{5}{x+2}+\dfrac{2}{x^2-2x+4}-\dfrac{60}{x^3+8}$

59. $\dfrac{3x}{x^2+x-12}-\dfrac{x}{x^2-16}$ 60. $\dfrac{p}{2p^2-9p-5}-\dfrac{2p}{6p^2-p-2}$

Simplify each expression. ***See Example 5.***

61. $\dfrac{1+\frac{1}{x}}{1-\frac{1}{x}}$ 62. $\dfrac{2-\frac{2}{y}}{2+\frac{2}{y}}$ 63. $\dfrac{\frac{1}{x+1}-\frac{1}{x}}{\frac{1}{x}}$

64. $\dfrac{\frac{1}{y+3}-\frac{1}{y}}{\frac{1}{y}}$ 65. $\dfrac{1+\frac{1}{1-b}}{1-\frac{1}{1+b}}$ 66. $\dfrac{2+\frac{2}{1+x}}{2-\frac{2}{1-x}}$

67. $\dfrac{\frac{1}{a^3+b^3}}{\frac{1}{a^2+2ab+b^2}}$ 68. $\dfrac{\frac{1}{x^3-y^3}}{\frac{1}{x^2-y^2}}$ 69. $\dfrac{m-\frac{1}{m^2-4}}{\frac{1}{m+2}}$

70. $\dfrac{y+\frac{1}{y^2-9}}{\frac{1}{y+3}}$ 71. $\dfrac{\frac{3}{p^2-16}+p}{\frac{1}{p-4}}$ 72. $\dfrac{\frac{6}{x^2-25}+x}{\frac{1}{x-5}}$

73. $\dfrac{\frac{y+3}{y}-\frac{4}{y-1}}{\frac{y}{y-1}+\frac{1}{y}}$ 74. $\dfrac{\frac{x+4}{x}-\frac{3}{x-2}}{\frac{x}{x-2}+\frac{1}{x}}$

75. $\dfrac{\frac{1}{x+h}-\frac{1}{x}}{h}$ 76. $\dfrac{\frac{-2}{x+h}-\frac{-2}{x}}{h}$

77. $\dfrac{\frac{1}{(x+h)^2+9}-\frac{1}{x^2+9}}{h}$ 78. $\dfrac{\frac{2}{(x+h)^2+16}-\frac{2}{x^2+16}}{h}$

41. $\frac{19}{6k}$ 42. $\frac{47}{20p}$
43. $\frac{137}{30m}$ 44. $\frac{101}{12p}$
45. $\frac{a-b}{a^2}$ 46. $\frac{3z+x}{z^2}$
47. $\frac{5-22x}{12x^2y}$ 48. $\frac{7-4a^2b}{18a^3b^2}$
49. 3 50. 2
51. $\frac{2x}{(x+z)(x-z)}$
52. $\frac{2m^2+2}{(m-1)(m+1)}$
53. $\frac{4}{a-2}$, or $\frac{-4}{2-a}$
54. $\frac{6}{p-q}$, or $\frac{-6}{q-p}$
55. $\frac{3x+y}{2x-y}$, or $\frac{-3x-y}{y-2x}$
56. $\frac{6m-4}{3m-4}$, or $\frac{4-6m}{4-3m}$
57. $\frac{4x-7}{x^2-x+1}$
58. $\frac{5x-18}{x^2-2x+4}$
59. $\frac{2x^2-9x}{(x-3)(x+4)(x-4)}$
60. $\frac{p^2+8p}{(2p+1)(p-5)(3p-2)}$
61. $\frac{x+1}{x-1}$ 62. $\frac{y-1}{y+1}$
63. $\frac{-1}{x+1}$ 64. $\frac{-3}{y+3}$
65. $\frac{(2-b)(1+b)}{b(1-b)}$
66. $\frac{x^2+x-2}{x^2+x}$
67. $\frac{a+b}{a^2-ab+b^2}$
68. $\frac{x+y}{x^2+xy+y^2}$
69. $\frac{m^3-4m-1}{m-2}$
70. $\frac{y^3-9y+1}{y-3}$
71. $\frac{p^3-16p+3}{p+4}$
72. $\frac{x^3-25x+6}{x+5}$
73. $\frac{y^2-2y-3}{y^2+y-1}$
74. $\frac{x^2-x-8}{x^2+x-2}$
75. $\frac{-1}{x(x+h)}$
76. $\frac{2}{x(x+h)}$
77. $\frac{-2x-h}{(x^2+9)[(x+h)^2+9]}$
78. $\frac{-4x-2h}{(x^2+16)[(x+h)^2+16]}$

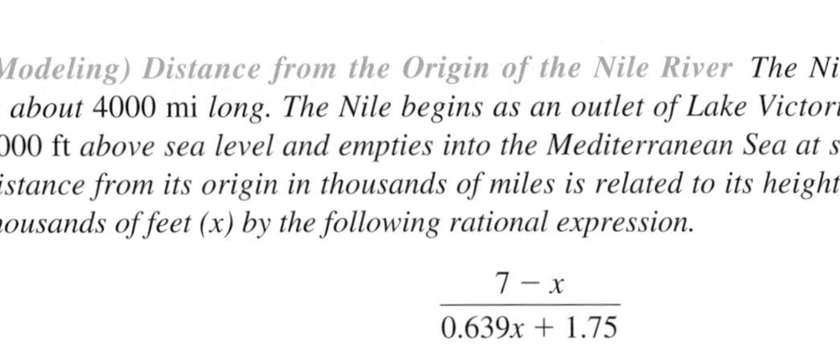

(Modeling) Distance from the Origin of the Nile River *The Nile River in Africa is about* 4000 mi *long. The Nile begins as an outlet of Lake Victoria at an altitude of* 7000 ft *above sea level and empties into the Mediterranean Sea at sea level* (0 ft). *The distance from its origin in thousands of miles is related to its height above sea level in thousands of feet (x) by the following rational expression.*

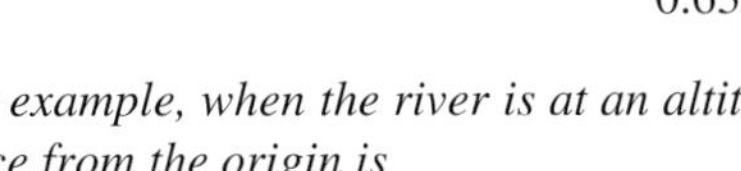

$$\frac{7-x}{0.639x+1.75}$$

For example, when the river is at an altitude of 600 ft, $x = 0.6$ *(thousand), and the distance from the origin is*

$$\frac{7-0.6}{0.639(0.6)+1.75} \approx 3, \quad \textit{which represents } 3000 \text{ mi.}$$

(Source: World Almanac and Book of Facts.)

79. 0 mi
80. about 2305 mi

79. What is the distance from the origin of the Nile when the river has an altitude of 7000 ft?

80. What is the distance from the origin of the Nile when the river has an altitude of 1200 ft?

81. 20.1 (thousand dollars)
82. 127.3 (thousand dollars)

(Modeling) Cost-Benefit Model for a Pollutant *In situations involving environmental pollution, a* ***cost-benefit model*** *expresses cost in terms of the percentage of pollutant removed from the environment. Suppose a cost-benefit model is expressed as*

$$y = \frac{6.7x}{100-x},$$

where y is the cost in thousands of dollars of removing x percent of a certain pollutant. Find the value of y for each given value of x.

81. $x = 75$ (75%)

82. $x = 95$ (95%)

R.6 Rational Exponents

- Negative Exponents and the Quotient Rule
- Rational Exponents
- Complex Fractions Revisited

Negative Exponents and the Quotient Rule In **Section R.2,** we justified the definition $a^0 = 1$ for $a \neq 0$ using the product rule for exponents. Suppose that n is a positive integer, and we wish to define a^{-n} to be consistent with the application of the product rule. Consider the product $a^n \cdot a^{-n}$, and apply the rule.

$$a^n \cdot a^{-n} = a^{n+(-n)} \quad \text{Product rule}$$
$$= a^0 \quad n \text{ and } -n \text{ are additive inverses.}$$
$$= 1 \quad \text{Definition of } a^0$$

The expression a^{-n} acts as the *reciprocal* of a^n, which is written $\frac{1}{a^n}$. Thus, these two expressions must be equivalent.

Negative Exponent

Suppose that a is a nonzero real number and n is any integer.

$$a^{-n} = \frac{1}{a^n}$$

Classroom Example 1
Evaluate each expression. In parts (d) and (e), write the expression without negative exponents. Assume all variables represent nonzero real numbers.

(a) 10^{-3} (b) -5^{-1} (c) $\left(\frac{4}{9}\right)^{-2}$

(d) mn^{-4} (e) $(mn)^{-4}$

Answers:
(a) $\frac{1}{1000}$ (b) $-\frac{1}{5}$
(c) $\frac{81}{16}$ (d) $\frac{m}{n^4}$
(e) $\frac{1}{(mn)^4}$, or $\frac{1}{m^4n^4}$

EXAMPLE 1 Using the Definition of a Negative Exponent

Evaluate each expression. In parts (d) and (e), write the expression without negative exponents. Assume all variables represent nonzero real numbers.

(a) 4^{-2} **(b)** -4^{-2} **(c)** $\left(\frac{2}{5}\right)^{-3}$ **(d)** $(xy)^{-3}$ **(e)** xy^{-3}

SOLUTION

(a) $4^{-2} = \frac{1}{4^2} = \frac{1}{16}$ **(b)** $-4^{-2} = -\frac{1}{4^2} = -\frac{1}{16}$

(c) $\left(\frac{2}{5}\right)^{-3} = \frac{1}{\left(\frac{2}{5}\right)^3} = \frac{1}{\frac{8}{125}} = 1 \div \frac{8}{125} = 1 \cdot \frac{125}{8} = \frac{125}{8}$

Multiply by the reciprocal of the divisor.

(d) $(xy)^{-3} = \frac{1}{(xy)^3}$, or $\frac{1}{x^3y^3}$ **(e)** $xy^{-3} = x \cdot \frac{1}{y^3} = \frac{x}{y^3}$

Base is xy. Base is y.

Now Try Exercises 3, 5, 7, 9, and 11.

CAUTION ***A negative exponent indicates a reciprocal, not a sign change of the expression.***

Example 1(c) showed the following.

$$\left(\frac{2}{5}\right)^{-3} = \frac{125}{8} = \left(\frac{5}{2}\right)^3$$

We can generalize this result. If $a \neq 0$ and $b \neq 0$, then for any integer n, the following is true.

$$\left(\frac{a}{b}\right)^{-n} = \left(\frac{b}{a}\right)^n$$

The **quotient rule** for exponents follows from the definition of exponents.

Quotient Rule

Suppose that m and n are integers and a is a nonzero real number.

$$\frac{a^m}{a^n} = a^{m-n}$$

That is, when dividing powers of like bases, keep the same base and subtract the exponent of the denominator from the exponent of the numerator.

CAUTION When applying the quotient rule, be sure to subtract the exponents in the correct order. Be careful especially when the exponent in the denominator is negative, and avoid sign errors.

Classroom Example 2
Simplify each expression. Assume all variables represent nonzero real numbers.

(a) $\frac{15^8}{15^3}$ (b) $\frac{y^4}{y^{-9}}$

(c) $\frac{35r^6}{25r^{-4}}$ (d) $\frac{34a^8b^{11}}{51a^{12}b^5}$

Answers:
(a) 15^5 (b) y^{13}
(c) $\frac{7r^{10}}{5}$ (d) $\frac{2b^6}{3a^4}$

EXAMPLE 2 Using the Quotient Rule

Simplify each expression. Assume all variables represent nonzero real numbers.

(a) $\frac{12^5}{12^2}$ (b) $\frac{a^5}{a^{-8}}$ (c) $\frac{16m^{-9}}{12m^{11}}$ (d) $\frac{25r^7z^5}{10r^9z}$

SOLUTION

(a) $\frac{12^5}{12^2} = 12^{5-2} = 12^3$

(b) $\frac{a^5}{a^{-8}} = a^{5-(-8)} = a^{13}$ Use parentheses to avoid errors.

(c) $\frac{16m^{-9}}{12m^{11}} = \frac{16}{12} \cdot m^{-9-11}$

$= \frac{4}{3}m^{-20}$

$= \frac{4}{3} \cdot \frac{1}{m^{20}}$

$= \frac{4}{3m^{20}}$

(d) $\frac{25r^7z^5}{10r^9z} = \frac{25}{10} \cdot \frac{r^7}{r^9} \cdot \frac{z^5}{z^1}$

$= \frac{5}{2}r^{-2}z^4$

$= \frac{5z^4}{2r^2}$

✔ *Now Try Exercises 15, 21, 23, and 25.*

The rules for exponents in **Section R.3** were stated for positive integer exponents and for zero as an exponent. Those rules continue to apply in expressions involving negative exponents, as seen in the next example.

Classroom Example 3
Simplify each expression. Write answers without negative exponents. Assume all variables represent nonzero real numbers.

(a) $5x^3(2^{-1}x^4)^{-3}$

(b) $\frac{30r^4s^{-9}}{45r^{-6}s^3}$

(c) $\frac{(4b^3)^{-2}(4b^{-1})^{-3}}{(4^{-1}b^3)^{-4}}$

Answers:
(a) $\frac{40}{x^9}$ (b) $\frac{2r^{10}}{3s^{12}}$ (c) $\frac{b^9}{4^9}$

EXAMPLE 3 Using the Rules for Exponents

Simplify each expression. Write answers without negative exponents. Assume all variables represent nonzero real numbers.

(a) $3x^{-2}(4^{-1}x^{-5})^2$ (b) $\frac{12p^3q^{-1}}{8p^{-2}q}$ (c) $\frac{(3x^2)^{-1}(3x^5)^{-2}}{(3^{-1}x^{-2})^2}$

SOLUTION

(a) $3x^{-2}(4^{-1}x^{-5})^2 = 3x^{-2}(4^{-2}x^{-10})$ Power rules **(Section R.3)**

$= 3 \cdot 4^{-2} \cdot x^{-2+(-10)}$ Rearrange factors; product rule **(Section R.3)**

$= 3 \cdot 4^{-2} \cdot x^{-12}$ Simplify the exponent on x.

$= \frac{3}{16x^{12}}$ Write with positive exponents.

(b) $\frac{12p^3q^{-1}}{8p^{-2}q} = \frac{12}{8} \cdot \frac{p^3}{p^{-2}} \cdot \frac{q^{-1}}{q^1}$

$= \frac{3}{2} \cdot p^{3-(-2)}q^{-1-1}$ Quotient rule

$= \frac{3}{2}p^5q^{-2}$ Simplify the exponents.

$= \frac{3p^5}{2q^2}$ Write with positive exponents.

(c) $\dfrac{(3x^2)^{-1}(3x^5)^{-2}}{(3^{-1}x^{-2})^2} = \dfrac{3^{-1}x^{-2}3^{-2}x^{-10}}{3^{-2}x^{-4}}$ Power rules

$= \dfrac{3^{-1+(-2)}x^{-2+(-10)}}{3^{-2}x^{-4}} = \dfrac{3^{-3}x^{-12}}{3^{-2}x^{-4}}$ Product rule

$= 3^{-3-(-2)}x^{-12-(-4)} = 3^{-1}x^{-8}$ Quotient rule

Be careful with signs.

$= \dfrac{1}{3x^8}$ Write with positive exponents.

✔ *Now Try Exercises 29, 33, 35, and 37.*

CAUTION Notice the use of the power rule $(ab)^n = a^nb^n$ in **Example 3(c):**

$$(3x^2)^{-1} = 3^{-1}(x^2)^{-1} = 3^{-1}x^{-2}.$$

***Remember to apply the exponent to the numerical coefficient* 3.**

Rational Exponents The definition of a^n can be extended to rational values of n by defining $a^{1/n}$ to be the nth root of a. By one of the power rules of exponents (extended to a rational exponent),

$$(a^{1/n})^n = a^{(1/n)n} = a^1 = a,$$

which suggests that $a^{1/n}$ is a number whose nth power is a.

The Expression $a^{1/n}$

$a^{1/n}$, n Even If n is an *even* positive integer, and if $a > 0$, then $a^{1/n}$ is the positive real number whose nth power is a. That is, $(a^{1/n})^n = a$. (In this case, $a^{1/n}$ is the principal nth root of a. See **Section R.7.**)

$a^{1/n}$, n Odd If n is an *odd* positive integer, and a is *any nonzero real number,* then $a^{1/n}$ is the positive or negative real number whose nth power is a. That is, $(a^{1/n})^n = a$.

For all positive integers n, $0^{1/n} = 0$.

Classroom Example 4
Evaluate each expression.

(a) $49^{1/2}$ (b) $-144^{1/2}$
(c) $-(144)^{1/2}$ (d) $64^{1/6}$
(e) $(-64)^{1/6}$ (f) $-64^{1/6}$
(g) $(-125)^{1/3}$ (h) $-64^{1/3}$

Answers:
(a) 7 (b) -12
(c) -12 (d) 2
(e) $(-64)^{1/6}$ is not a real number.
(f) -2 (g) -5
(h) -4

EXAMPLE 4 Using the Definition of $a^{1/n}$

Evaluate each expression.

(a) $36^{1/2}$ **(b)** $-100^{1/2}$ **(c)** $-(225)^{1/2}$ **(d)** $625^{1/4}$

(e) $(-1296)^{1/4}$ **(f)** $-1296^{1/4}$ **(g)** $(-27)^{1/3}$ **(h)** $-32^{1/5}$

SOLUTION

(a) $36^{1/2} = 6$ because $6^2 = 36$. **(b)** $-100^{1/2} = -10$

(c) $-(225)^{1/2} = -15$ **(d)** $625^{1/4} = 5$

(e) $(-1296)^{1/4}$ is not a real number. **(f)** $-1296^{1/4} = -6$

(g) $(-27)^{1/3} = -3$ **(h)** $-32^{1/5} = -2$

✔ *Now Try Exercises 39, 41, and 45.*

The notation $a^{m/n}$ must be defined in such a way that all the previous rules for exponents still hold. For the power rule to hold, $(a^{1/n})^m$ must equal $a^{m/n}$. Therefore, $a^{m/n}$ is defined as follows.

The Expression $a^{m/n}$

Let m be any integer, n be any positive integer, and a be any real number for which $a^{1/n}$ is a real number.

$$a^{m/n} = (a^{1/n})^m$$

Classroom Example 5
Evaluate each expression.
(a) $81^{3/4}$ (b) $25^{3/2}$
(c) $-4^{5/2}$ (d) $(-64)^{2/3}$
(e) $216^{-2/3}$ (f) $(-100)^{3/2}$

Answers:
(a) 27 (b) 125
(c) -32 (d) 16
(e) $\frac{1}{36}$
(f) $(-100)^{3/2}$ is not a real number.

EXAMPLE 5 Using the Definition of $a^{m/n}$

Evaluate each expression.

(a) $125^{2/3}$ **(b)** $32^{7/5}$ **(c)** $-81^{3/2}$ **(d)** $(-27)^{2/3}$ **(e)** $16^{-3/4}$ **(f)** $(-4)^{5/2}$

SOLUTION

(a) $125^{2/3} = (125^{1/3})^2 = 5^2$, or 25

(b) $32^{7/5} = (32^{1/5})^7 = 2^7$, or 128

(c) $-81^{3/2} = -(81^{1/2})^3 = -9^3$, or -729

(d) $(-27)^{2/3} = [(-27)^{1/3}]^2 = (-3)^2$, or 9

(e) $16^{-3/4} = \frac{1}{16^{3/4}} = \frac{1}{(16^{1/4})^3} = \frac{1}{2^3}$, or $\frac{1}{8}$

(f) $(-4)^{5/2}$ is not a real number. This is because $(-4)^{1/2}$ is not a real number.

Now Try Exercises 49, 53, and 55.

NOTE For all real numbers a, integers m, and positive integers n for which $a^{1/n}$ is a real number, $a^{m/n}$ can be interpreted as follows.

$$a^{m/n} = (a^{1/n})^m \quad \text{or} \quad a^{m/n} = (a^m)^{1/n}$$

So $a^{m/n}$ can be evaluated either as $(a^{1/n})^m$ or as $(a^m)^{1/n}$.

$$27^{4/3} = (27^{1/3})^4 = 3^4 = 81$$

or $$27^{4/3} = (27^4)^{1/3} = 531{,}441^{1/3} = 81$$

The result is the same.

Teaching Tip Remind students that a common base is necessary in order to apply the product and quotient rules. Encourage students to verify the definitions and rules for exponents by writing out exponential terms as repeated factors. For example,

$$(x^2)^3 = x^2 \cdot x^2 \cdot x^2 = (x \cdot x) \cdot (x \cdot x) \cdot (x \cdot x) = x^6$$

shows how the first power rule can be verified.

The earlier results for integer exponents also apply to rational exponents.

Definitions and Rules for Exponents

Suppose that r and s represent rational numbers. The results here are valid for all positive numbers a and b.

Product rule $a^r \cdot a^s = a^{r+s}$

Quotient rule $\frac{a^r}{a^s} = a^{r-s}$

Negative exponent $a^{-r} = \frac{1}{a^r}$

Power rules $(a^r)^s = a^{rs}$

$(ab)^r = a^r b^r$

$\left(\frac{a}{b}\right)^r = \frac{a^r}{b^r}$

Classroom Example 6
Simplify each expression. Assume all variables represent positive real numbers.

(a) $\frac{18^{1/2} \cdot 18^{7/2}}{18^3}$

(b) $100^{3/2} \cdot 16^{-3/4}$

(c) $4z^{3/4} \cdot 5z^{2/5}$

(d) $\left(\frac{5m^{4/3}}{n^{2/3}}\right)^2\left(\frac{m^4}{8n^5}\right)^{1/3}$

(e) $y^{3/7}(y^{4/7} - 5y^{11/7})$

Answers:
(a) 18 (b) 125
(c) $20z^{23/20}$ (d) $\frac{25m^4}{2n^3}$
(e) $y - 5y^2$

EXAMPLE 6 Using the Rules for Exponents

Simplify each expression. Assume all variables represent positive real numbers.

(a) $\frac{27^{1/3} \cdot 27^{5/3}}{27^3}$ **(b)** $81^{5/4} \cdot 4^{-3/2}$ **(c)** $6y^{2/3} \cdot 2y^{1/2}$

(d) $\left(\frac{3m^{5/6}}{y^{3/4}}\right)^2\left(\frac{8y^3}{m^6}\right)^{2/3}$ **(e)** $m^{2/3}(m^{7/3} + 2m^{1/3})$

SOLUTION

(a) $\frac{27^{1/3} \cdot 27^{5/3}}{27^3} = \frac{27^{1/3+5/3}}{27^3}$ Product rule

$= \frac{27^2}{27^3}$ Simplify.

$= 27^{2-3}$ Quotient rule

$= 27^{-1}$, or $\frac{1}{27}$ Negative exponent

(b) $81^{5/4} \cdot 4^{-3/2} = (81^{1/4})^5(4^{1/2})^{-3}$

$= 3^5 \cdot 2^{-3}$

$= \frac{3^5}{2^3}$, or $\frac{243}{8}$

(c) $6y^{2/3} \cdot 2y^{1/2} = 12y^{2/3+1/2}$

$= 12y^{7/6}$

(d) $\left(\frac{3m^{5/6}}{y^{3/4}}\right)^2\left(\frac{8y^3}{m^6}\right)^{2/3} = \frac{9m^{5/3}}{y^{3/2}} \cdot \frac{4y^2}{m^4}$ Power rules

$= 36m^{5/3-4}y^{2-3/2}$ Quotient rule

$= 36m^{-7/3}y^{1/2}$ Simplify the exponents.

$= \frac{36y^{1/2}}{m^{7/3}}$ Simplify.

(e) $m^{2/3}(m^{7/3} + 2m^{1/3}) = m^{2/3} \cdot m^{7/3} + m^{2/3} \cdot 2m^{1/3}$ Distributive property (Section R.2)

Do *not* multiply the exponents.

$= m^{2/3+7/3} + 2m^{2/3+1/3}$ Product rule

$= m^3 + 2m$ Simplify.

Now Try Exercises 61, 63, 67, 69, and 75.

Classroom Example 7
Factor out the least power of the variable or variable expression. Assume all variables represent positive real numbers.

(a) $28y^{-5} + 21y^{-2}$

(b) $18n^{4/3} - 12n^{1/3}$

(c) $(x + 3)^{-2/5} - (x + 3)^{3/5}$

Answers:
(a) $7y^{-5}(4 + 3y^3)$
(b) $6n^{1/3}(3n - 2)$
(c) $(x + 3)^{-2/5}(-2 - x)$

EXAMPLE 7 Factoring Expressions with Negative or Rational Exponents

Factor out the least power of the variable or variable expression. Assume all variables represent positive real numbers.

(a) $12x^{-2} - 8x^{-3}$ **(b)** $4m^{1/2} + 3m^{3/2}$ **(c)** $(y - 2)^{-1/3} + (y - 2)^{2/3}$

SOLUTION

(a) The least exponent on $12x^{-2} - 8x^{-3}$ is -3. Since 4 is a common numerical factor, factor out $4x^{-3}$.

$12x^{-2} - 8x^{-3} = 4x^{-3}(3x^{-2-(-3)} - 2x^{-3-(-3)})$ Factor.

$= 4x^{-3}(3x - 2)$ Simplify the exponents.

Check by multiplying on the right.

LOOKING AHEAD TO CALCULUS
The technique of **Example 7(c)** is used often in calculus.

(b) $4m^{1/2} + 3m^{3/2} = m^{1/2}(4 + 3m)$ Factor out $m^{1/2}$.

To *check*, multiply $m^{1/2}$ by $4 + 3m$.

(c) $(y-2)^{-1/3} + (y-2)^{2/3} = (y-2)^{-1/3}[1 + (y-2)]$
$= (y-2)^{-1/3}(y-1)$

Now Try Exercises 83, 89, and 93.

Complex Fractions Revisited Negative exponents are sometimes used to write complex fractions. Recall that complex fractions are simplified either by first multiplying the numerator and denominator by the LCD of all the denominators, or by performing any indicated operations in the numerator and the denominator and then using the definition of division for fractions.

Classroom Example 8
Simplify $\frac{x^{-1} + y^{-1}}{x^{-2} - y^{-2}}$. Write the result with only positive exponents.

Answer: $\frac{xy}{y - x}$

EXAMPLE 8 Simplifying a Fraction with Negative Exponents

Simplify $\frac{(x+y)^{-1}}{x^{-1} + y^{-1}}$. Write the result with only positive exponents.

SOLUTION
$$\frac{(x+y)^{-1}}{x^{-1} + y^{-1}} = \frac{\frac{1}{x+y}}{\frac{1}{x} + \frac{1}{y}}$$ Definition of negative exponent

$$= \frac{\frac{1}{x+y}}{\frac{y+x}{xy}}$$ Add fractions in the denominator. (Section R.5)

$$= \frac{1}{x+y} \cdot \frac{xy}{x+y}$$ Multiply by the reciprocal of the denominator of the complex fraction.

$$= \frac{xy}{(x+y)^2}$$ Multiply fractions.

Now Try Exercise 99.

CAUTION Remember that if $r \neq 1$, then $(x+y)^r \neq x^r + y^r$. In particular, this means that $(x+y)^{-1} \neq x^{-1} + y^{-1}$.

R.6 Exercises

1. (a) B (b) D (c) B (d) D
2. (a) C (b) D (c) D (d) C

Concept Check In Exercises 1 and 2, match each expression in Column I with its equivalent expression in Column II. Choices may be used once, more than once, or not at all.

I	II
1. (a) 4^{-2}	**A.** 16
(b) -4^{-2}	**B.** $\frac{1}{16}$
(c) $(-4)^{-2}$	**C.** -16
(d) $-(-4)^{-2}$	**D.** $-\frac{1}{16}$

I	II
2. (a) 5^{-3}	**A.** 125
(b) -5^{-3}	**B.** -125
(c) $(-5)^{-3}$	**C.** $\frac{1}{125}$
(d) $-(-5)^{-3}$	**D.** $-\frac{1}{125}$

3. $\frac{1}{(-4)^3}$, or $-\frac{1}{64}$
4. $\frac{1}{(-5)^2}$, or $\frac{1}{25}$
5. $-\frac{1}{5^4}$, or $-\frac{1}{625}$
6. $-\frac{1}{7^2}$, or $-\frac{1}{49}$
7. 3^2, or 9
8. $\left(\frac{3}{4}\right)^3$, or $\frac{27}{64}$
9. $\frac{1}{16x^2}$ 10. $\frac{1}{125t^3}$
11. $\frac{4}{x^2}$ 12. $\frac{5}{t^3}$
13. $-\frac{1}{a^3}$ 14. $-\frac{1}{b^4}$

15. 4^2, or 16 16. 5^2, or 25
17. x^4 18. y^4
19. $\frac{1}{r^3}$ 20. $\frac{1}{y^4}$
21. 6^6 22. 7^8
23. $\frac{2r^3}{3}$ 24. $3s^4$
25. $\frac{4n^7}{3m^7}$ 26. $\frac{3}{5a^3b^5}$
27. $-4r^6$ 28. $-2m^5$
29. $\frac{5^4}{a^{10}}$ 30. $\frac{3^2}{p^{11}}$
31. $\frac{p^4}{5}$ 32. $\frac{m^3}{9}$
33. $\frac{1}{2pq}$ 34. $-\frac{2}{x^4}$
35. $\frac{4}{a^2}$ 36. $2k^5$
37. $\frac{5}{x^2}$ 38. $\frac{1}{y^{10}}$

39. 13 40. 11
41. 2 42. 5
43. $-\frac{4}{3}$ 44. $-\frac{2}{3}$
45. This expression is not a real number.
46. This expression is not a real number.

47. (a) E (b) G (c) F (d) F
48. (a) D (b) A (c) B (d) C

49. 4 50. 81
51. 1000 52. 512
53. -27 54. $\frac{1}{16}$

Write each expression with only positive exponents and evaluate if possible. Assume all variables represent nonzero real numbers. ***See Example 1.***

3. $(-4)^{-3}$ 4. $(-5)^{-2}$ 5. -5^{-4}

6. -7^{-2} 7. $\left(\frac{1}{3}\right)^{-2}$ 8. $\left(\frac{4}{3}\right)^{-3}$

9. $(4x)^{-2}$ 10. $(5t)^{-3}$ 11. $4x^{-2}$

12. $5t^{-3}$ 13. $-a^{-3}$ 14. $-b^{-4}$

Perform the indicated operations. Write each answer using only positive exponents. Assume all variables represent nonzero real numbers. ***See Examples 2 and 3.***

15. $\frac{4^8}{4^6}$ 16. $\frac{5^9}{5^7}$ 17. $\frac{x^{12}}{x^8}$ 18. $\frac{y^{14}}{y^{10}}$

19. $\frac{r^7}{r^{10}}$ 20. $\frac{y^8}{y^{12}}$ 21. $\frac{6^4}{6^{-2}}$ 22. $\frac{7^5}{7^{-3}}$

23. $\frac{4r^{-3}}{6r^{-6}}$ 24. $\frac{15s^{-4}}{5s^{-8}}$ 25. $\frac{16m^{-5}n^4}{12m^2n^{-3}}$ 26. $\frac{15a^{-5}b^{-1}}{25a^{-2}b^4}$

27. $-4r^{-2}(r^4)^2$ 28. $-2m^{-1}(m^3)^2$ 29. $(5a^{-1})^4(a^2)^{-3}$ 30. $(3p^{-4})^2(p^3)^{-1}$

31. $\frac{(p^{-2})^0}{5p^{-4}}$ 32. $\frac{(m^4)^0}{9m^{-3}}$ 33. $\frac{(3pq)q^2}{6p^2q^4}$ 34. $\frac{(-8xy)y^3}{4x^5y^4}$

35. $\frac{4a^5(a^{-1})^3}{(a^{-2})^{-2}}$ 36. $\frac{12k^{-2}(k^{-3})^{-4}}{6k^5}$ 37. $\frac{(5x)^{-2}(5x^3)^{-3}}{(5^{-2}x^{-3})^3}$ 38. $\frac{(8y^2)^{-4}(8y^5)^{-2}}{(8^{-3}y^{-4})^2}$

Evaluate each expression. ***See Example 4.***

39. $169^{1/2}$ 40. $121^{1/2}$ 41. $16^{1/4}$ 42. $625^{1/4}$

43. $\left(-\frac{64}{27}\right)^{1/3}$ 44. $\left(-\frac{8}{27}\right)^{1/3}$ 45. $(-4)^{1/2}$ 46. $(-64)^{1/4}$

Concept Check *In Exercises 47 and 48, match each expression from Column I with its equivalent expression from Column II. Choices may be used once, more than once, or not at all.*

I		II	
47. (a) $\left(\frac{4}{9}\right)^{3/2}$	48. (a) $\left(\frac{8}{27}\right)^{2/3}$	A. $\frac{9}{4}$	B. $-\frac{9}{4}$
(b) $\left(\frac{4}{9}\right)^{-3/2}$	(b) $\left(\frac{8}{27}\right)^{-2/3}$	C. $-\frac{4}{9}$	D. $\frac{4}{9}$
(c) $-\left(\frac{9}{4}\right)^{3/2}$	(c) $-\left(\frac{27}{8}\right)^{2/3}$	E. $\frac{8}{27}$	F. $-\frac{27}{8}$
(d) $-\left(\frac{4}{9}\right)^{-3/2}$	(d) $-\left(\frac{27}{8}\right)^{-2/3}$	G. $\frac{27}{8}$	H. $-\frac{8}{27}$

Perform the indicated operations. Write each answer using only positive exponents. Assume all variables represent positive real numbers. ***See Examples 5 and 6.***

49. $8^{2/3}$ 50. $27^{4/3}$ 51. $100^{3/2}$

52. $64^{3/2}$ 53. $-81^{3/4}$ 54. $(-32)^{-4/5}$

55. $\frac{256}{81}$ 56. $\frac{1000}{1331}$
57. 9 58. 36
59. 4 60. 25
61. y 62. r
63. $k^{2/3}$ 64. $z^{3/2}$
65. x^3y^8 66. rs^{10}
67. $\frac{1}{x^{10/3}}$ 68. $\frac{1}{p^{7/4}}$
69. $\frac{6}{m^{1/4}n^{3/4}}$ 70. $\frac{10}{a^{1/8}b^{3/2}}$
71. p^2 72. $z^{1/3}$

73. (a) approximately 250 sec
(b) $2^{-1.5} \approx 0.3536$
74. (a) approximately 0.56 hr or almost 34 min
(b) no

75. $y - 10y^2$
76. $3p^3 + 9p^6$
77. $-4k^{10/3} + 24k^{4/3}$
78. $-15y^{19/10} - 20y^{13/10}$
79. $x^2 - x$
80. $2z - z^{3/2} - z^2$
81. $r - 2 + r^{-1}$, or $r - 2 + \frac{1}{r}$
82. $p - p^{-1}$, or $p - \frac{1}{p}$

55. $\left(\frac{27}{64}\right)^{-4/3}$
56. $\left(\frac{121}{100}\right)^{-3/2}$
57. $3^{1/2} \cdot 3^{3/2}$
58. $6^{4/3} \cdot 6^{2/3}$
59. $\frac{64^{5/3}}{64^{4/3}}$
60. $\frac{125^{7/3}}{125^{5/3}}$
61. $y^{7/3} \cdot y^{-4/3}$
62. $r^{-8/9} \cdot r^{17/9}$
63. $\frac{k^{1/3}}{k^{2/3} \cdot k^{-1}}$
64. $\frac{z^{3/4}}{z^{5/4} \cdot z^{-2}}$
65. $\frac{(x^{1/4}y^{2/5})^{20}}{x^2}$
66. $\frac{(r^{1/5}s^{2/3})^{15}}{r^2}$
67. $\frac{(x^{2/3})^2}{(x^2)^{7/3}}$
68. $\frac{(p^3)^{1/4}}{(p^{5/4})^2}$
69. $\left(\frac{16m^3}{n}\right)^{1/4}\left(\frac{9n^{-1}}{m^2}\right)^{1/2}$
70. $\left(\frac{25^4a^3}{b^2}\right)^{1/8}\left(\frac{4^2b^{-5}}{a^2}\right)^{1/4}$
71. $\frac{p^{1/5}p^{7/10}p^{1/2}}{(p^3)^{-1/5}}$
72. $\frac{z^{1/3}z^{-2/3}z^{1/6}}{(z^{-1/6})^3}$

Solve each applied problem.

73. *(Modeling) Holding Time of Athletes* A group of ten athletes were tested for isometric endurance by measuring the length of time they could resist a load pulling on their legs while seated. The approximate amount of time (called the **holding time**) that they could resist the load was given by the formula

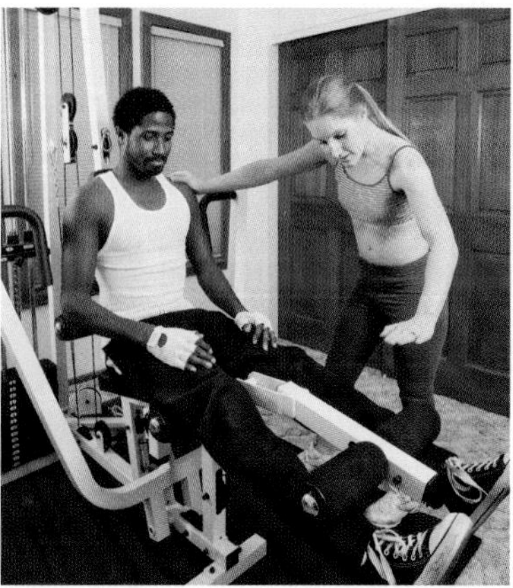

$$t = 31{,}293w^{-1.5},$$

where w is the weight of the load in pounds and the holding time t is measured in seconds. (*Source:* Townend, M. Stewart, *Mathematics in Sport,* Chichester, Ellis Horwood Limited.)

(a) Determine the holding time for a load of 25 lb.
(b) When the weight of the load is doubled, by what factor is the holding time changed?

74. *Duration of a Storm* Suppose that meteorologists approximate the duration of a particular storm by using the formula

$$T = 0.07D^{3/2},$$

where T is the time in hours that a storm of diameter D (in miles) lasts.

(a) The National Weather Service reports that a storm 4 mi in diameter is headed toward New Haven. How long can the residents expect the storm to last?
(b) After weeks of dry weather, a thunderstorm is predicted for the farming community of Apple Valley. The crops need at least 1.5 hr of rain. Local radar shows that the storm is 7 mi in diameter. Will it rain long enough to meet the farmers' need?

Find each product. Assume all variables represent positive real numbers. ***See Example 6(e) in this section and Example 7 in Section R.3.***

75. $y^{5/8}(y^{3/8} - 10y^{11/8})$
76. $p^{11/5}(3p^{4/5} + 9p^{19/5})$
77. $-4k(k^{7/3} - 6k^{1/3})$
78. $-5y(3y^{9/10} + 4y^{3/10})$
79. $(x + x^{1/2})(x - x^{1/2})$
80. $(2z^{1/2} + z)(z^{1/2} - z)$
81. $(r^{1/2} - r^{-1/2})^2$
82. $(p^{1/2} - p^{-1/2})(p^{1/2} + p^{-1/2})$

83. $k^{-2}(4k+1)$
84. $y^{-5}(1-3y^2)$
85. $4t^{-4}(t^2+2)$
86. $5r^{-8}(r^2-2)$
87. $z^{-1/2}(9+2z)$
88. $m^{-1/3}(3m-4)$
89. $p^{-7/4}(p-2)$
90. $r^{-5/3}(6r-5)$
91. $4a^{-7/5}(-a+4)$
92. $-3p^{-7/4}(p+10)$
93. $(p+4)^{-3/2}(p^2+9p+21)$
94. $3(3r+1)^{-2/3}(3r^2+3r+1)$
95. $6(3x+1)^{-3/2}(9x^2+8x+2)$
96. $-35(5t+3)^{-5/3}(15t^2+16t+4)$
97. $2x(2x+3)^{-5/9}(-16x^4-48x^3-30x^2+9x+2)$
98. $2y(4y-1)^{-3/7}(115y^2-60y+8)$
99. $b+a$
100. $q-p$
101. -1
102. $\frac{y^2+x^2}{(y-x)(x-y)}$
103. $\frac{y(xy-9)}{x^2y^2-9}$
104. $\frac{b(ab-16)}{a^2b^2-16}$
105. $\frac{2x(1-3x^2)}{(x^2+1)^5}$
106. $\frac{3y(y^2+2-6y^3)}{(y^2+2)^3}$
107. $\frac{1+2x^3-2x}{4}$
108. $\frac{2-20x^3+45x}{3(4x^2-9)^4}$
109. $\frac{3x-5}{(2x-3)^{4/3}}$
110. $\frac{4(5t+2)}{(3t+1)^{3/2}}$
111. 27
112. 4
113. 4
114. 27
115. $\frac{1}{100}$
116. 8

Factor, using the given common factor. Assume all variables represent positive real numbers. ***See Example 7.***

83. $4k^{-1}+k^{-2}$; k^{-2} **84.** $y^{-5}-3y^{-3}$; y^{-5} **85.** $4t^{-2}+8t^{-4}$; $4t^{-4}$

86. $5r^{-6}-10r^{-8}$; $5r^{-8}$ **87.** $9z^{1/2}+2z^{1/2}$; $z^{-1/2}$ **88.** $3m^{2/3}-4m^{-1/3}$; $m^{-1/3}$

89. $p^{-3/4}-2p^{-7/4}$; $p^{-7/4}$ **90.** $6r^{-2/3}-5r^{-5/3}$; $r^{-5/3}$

91. $-4a^{-2/5}+16a^{-7/5}$; $4a^{-7/5}$ **92.** $-3p^{-3/4}-30p^{-7/4}$; $-3p^{-7/4}$

93. $(p+4)^{-3/2}+(p+4)^{-1/2}+(p+4)^{1/2}$; $(p+4)^{-3/2}$

94. $(3r+1)^{-2/3}+(3r+1)^{1/3}+(3r+1)^{4/3}$; $(3r+1)^{-2/3}$

95. $2(3x+1)^{-3/2}+4(3x+1)^{-1/2}+6(3x+1)^{1/2}$; $2(3x+1)^{-3/2}$

96. $7(5t+3)^{-5/3}+14(5t+3)^{-2/3}-21(5t+3)^{1/3}$; $7(5t+3)^{-5/3}$

97. $4x(2x+3)^{-5/9}+6x^2(2x+3)^{4/9}-8x^3(2x+3)^{13/9}$; $2x(2x+3)^{-5/9}$

98. $6y^3(4y-1)^{-3/7}-8y^2(4y-1)^{4/7}+16y(4y-1)^{11/7}$; $2y(4y-1)^{-3/7}$

Perform all indicated operations and write each answer with positive integer exponents. ***See Example 8.***

99. $\frac{a^{-1}+b^{-1}}{(ab)^{-1}}$ **100.** $\frac{p^{-1}-q^{-1}}{(pq)^{-1}}$

101. $\frac{r^{-1}+q^{-1}}{r^{-1}-q^{-1}}\cdot\frac{r-q}{r+q}$ **102.** $\frac{x^{-2}+y^{-2}}{x^{-2}-y^{-2}}\cdot\frac{x+y}{x-y}$

103. $\frac{x-9y^{-1}}{(x-3y^{-1})(x+3y^{-1})}$ **104.** $\frac{a-16b^{-1}}{(a+4b^{-1})(a-4b^{-1})}$

Simplify each rational expression. Use factoring, and refer to ***Section R.5*** *as needed. Assume all variable expressions represent positive real numbers.*

105. $\frac{(x^2+1)^4(2x)-x^2(4)(x^2+1)^3(2x)}{(x^2+1)^8}$ **106.** $\frac{(y^2+2)^5(3y)-y^3(6)(y^2+2)^4(3y)}{(y^2+2)^7}$

107. $\frac{4(x^2-1)^3+8x(x^2-1)^4}{16(x^2-1)^3}$ **108.** $\frac{10(4x^2-9)^2-25x(4x^2-9)^3}{15(4x^2-9)^6}$

109. $\frac{2(2x-3)^{1/3}-(x-1)(2x-3)^{-2/3}}{(2x-3)^{2/3}}$ **110.** $\frac{7(3t+1)^{1/4}-(t-1)(3t+1)^{-3/4}}{(3t+1)^{3/4}}$

Concept Check Answer each question.

111. If the lengths of the sides of a cube are tripled, by what factor will the volume change?

112. If the radius of a circle is doubled, by what factor will the area change?

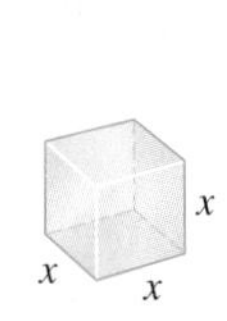

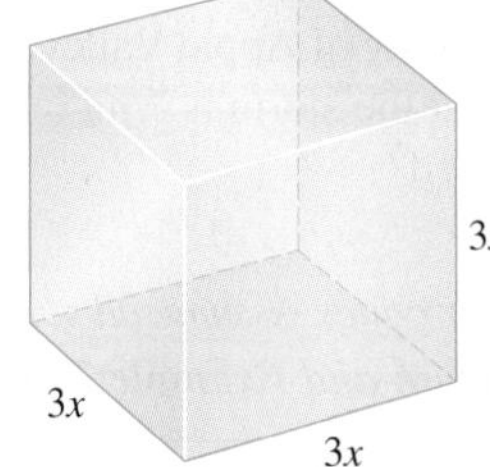

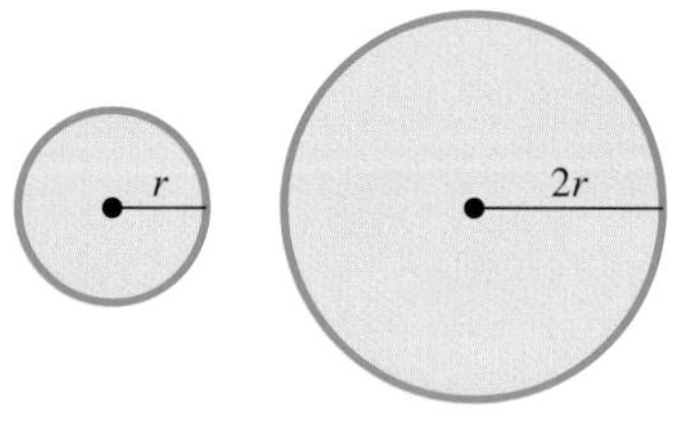

Concept Check Calculate each value mentally.

113. $0.2^{2/3}\cdot 40^{2/3}$ **114.** $0.1^{3/2}\cdot 90^{3/2}$ **115.** $\frac{2^{2/3}}{2000^{2/3}}$ **116.** $\frac{20^{3/2}}{5^{3/2}}$

R.7 Radical Expressions

- Radical Notation
- Simplified Radicals
- Operations with Radicals
- Rationalizing Denominators

Radical Notation In **Section R.6** we used rational exponents to express roots. An alternative notation for roots is **radical notation.**

Radical Notation for $a^{1/n}$

Suppose that a is a real number, n is a positive integer, and $a^{1/n}$ is a real number.

$$\sqrt[n]{a} = a^{1/n}$$

Teaching Tip Emphasize that the denominator of the exponent is always the root. Students sometimes mistake $x^{m/n}$ for the mth root of x^n.

Radical Notation for $a^{m/n}$

Suppose that a is a real number, m is an integer, n is a positive integer, and $\sqrt[n]{a}$ is a real number.

$$a^{m/n} = \left(\sqrt[n]{a}\right)^m = \sqrt[n]{a^m}$$

In the radical $\sqrt[n]{a}$, the symbol $\sqrt[n]{\ }$ is a **radical symbol,** the number a is the **radicand,** and n is the **index.** We use the familiar notation $\sqrt{a}$ instead of $\sqrt[2]{a}$ for the square root.

For even values of n (square roots, fourth roots, and so on), when a is positive, there are two nth roots, one positive and one negative. In such cases, the notation $\sqrt[n]{a}$ represents the positive root, the **principal nth root.** We write the **negative root** as $-\sqrt[n]{a}$.

Classroom Example 1
Write each root using exponents and evaluate.

(a) $\sqrt[3]{27}$ (b) $-\sqrt[4]{10{,}000}$
(c) $\sqrt[3]{-216}$ (d) $\sqrt[4]{-81}$
(e) $\sqrt[3]{\frac{125}{512}}$ (f) $-\sqrt[5]{-243}$

Answers:
(a) $27^{1/3} = 3$
(b) $-10{,}000^{1/4} = -10$
(c) $(-216)^{1/3} = -6$
(d) $\sqrt[4]{-81}$ is not a real number.
(e) $\left(\frac{125}{512}\right)^{1/3} = \frac{5}{8}$
(f) $-(-243)^{1/5} = 3$

EXAMPLE 1 Evaluating Roots

Write each root using exponents and evaluate.

(a) $\sqrt[4]{16}$ **(b)** $-\sqrt[4]{16}$ **(c)** $\sqrt[5]{-32}$

(d) $\sqrt[3]{1000}$ **(e)** $\sqrt[6]{\frac{64}{729}}$ **(f)** $\sqrt[4]{-16}$

SOLUTION

(a) $\sqrt[4]{16} = 16^{1/4}$
$= 2$

(b) $-\sqrt[4]{16} = -16^{1/4}$
$= -2$

(c) $\sqrt[5]{-32} = (-32)^{1/5}$
$= -2$

(d) $\sqrt[3]{1000} = 1000^{1/3}$
$= 10$

(e) $\sqrt[6]{\frac{64}{729}} = \left(\frac{64}{729}\right)^{1/6}$
$= \frac{2}{3}$

(f) $\sqrt[4]{-16}$ is not a real number.

✔ *Now Try Exercises 1, 3, 7, and 11.*

Classroom Example 2
Write in radical form and simplify. Assume all variable expressions represent positive real numbers.

(a) $16^{3/4}$ (b) $(-64)^{2/3}$
(c) $-121^{3/2}$ (d) $y^{7/8}$
(e) $7z^{4/5}$ (f) $12q^{1/4}$
(g) $(5x + 2y)^{1/6}$

Answers:
(a) $(\sqrt[4]{16})^3 = 8$
(b) $(\sqrt[3]{-64})^2 = 16$
(c) $-(\sqrt{121})^3 = -1331$
(d) $\sqrt[8]{y^7}$ (e) $7\sqrt[5]{z^4}$
(f) $12\sqrt[4]{q}$ (g) $\sqrt[6]{5x + 2y}$

EXAMPLE 2 Converting from Rational Exponents to Radicals

Write in radical form and simplify. Assume all variable expressions represent positive real numbers.

(a) $8^{2/3}$ **(b)** $(-32)^{4/5}$ **(c)** $-16^{3/4}$ **(d)** $x^{5/6}$

(e) $3x^{2/3}$ **(f)** $2p^{1/2}$ **(g)** $(3a + b)^{1/4}$

SOLUTION

(a) $8^{2/3} = (\sqrt[3]{8})^2 = 2^2 = 4$

(b) $(-32)^{4/5} = (\sqrt[5]{-32})^4 = (-2)^4 = 16$

(c) $-16^{3/4} = -(\sqrt[4]{16})^3 = -(2)^3 = -8$

(d) $x^{5/6} = \sqrt[6]{x^5}$

(e) $3x^{2/3} = 3\sqrt[3]{x^2}$

(f) $2p^{1/2} = 2\sqrt{p}$

(g) $(3a + b)^{1/4} = \sqrt[4]{3a + b}$

Now Try Exercises 13, 15, and 17.

LOOKING AHEAD TO CALCULUS
In calculus, the "power rule" for derivatives requires converting radicals to rational exponents.

CAUTION It is not possible to "distribute" exponents over a sum, so in **Example 2(g),** $(3a + b)^{1/4}$ *cannot be written as* $(3a)^{1/4} + b^{1/4}$.

$$\sqrt[n]{x^n + y^n} \quad \textit{is not equivalent to} \quad x + y.$$

(For example, let $n = 2$, $x = 3$, and $y = 4$ to see this.)

Classroom Example 3
Write in exponential form. Assume all variable expressions represent positive real numbers.

(a) $\sqrt[7]{n^3}$ (b) $\sqrt[4]{10x}$
(c) $15(\sqrt[3]{r})^4$ (d) $-2\sqrt[5]{(3x^2)^8}$
(e) $\sqrt[3]{r^2 + s^4}$

Answers:
(a) $n^{3/7}$ (b) $(10x)^{1/4}$
(c) $15r^{4/3}$ (d) $-2 \cdot 3^{8/5}x^{16/5}$
(e) $(r^2 + s^4)^{1/3}$

EXAMPLE 3 Converting from Radicals to Rational Exponents

Write in exponential form. Assume all variable expressions represent positive real numbers.

(a) $\sqrt[4]{x^5}$ **(b)** $\sqrt{3y}$ **(c)** $10(\sqrt[5]{z})^2$

(d) $5\sqrt[3]{(2x^4)^7}$ **(e)** $\sqrt{p^2 + q}$

SOLUTION

(a) $\sqrt[4]{x^5} = x^{5/4}$

(b) $\sqrt{3y} = (3y)^{1/2}$

(c) $10(\sqrt[5]{z})^2 = 10z^{2/5}$

(d) $5\sqrt[3]{(2x^4)^7} = 5(2x^4)^{7/3} = 5 \cdot 2^{7/3}x^{28/3}$

(e) $\sqrt{p^2 + q} = (p^2 + q)^{1/2}$

Now Try Exercises 19 and 21.

We *cannot* simply write $\sqrt{x^2} = x$ for all real numbers x. For example, if $x = -5$, then

$$\sqrt{x^2} = \sqrt{(-5)^2} = \sqrt{25} = 5 \neq x.$$

To take care of the fact that a negative value of x can produce a positive result, we use absolute value. For any real number a, the following holds.

$$\sqrt{a^2} = |a|$$

For example, $\sqrt{(-9)^2} = |-9| = 9$ and $\sqrt{13^2} = |13| = 13$.

We can generalize this result to any *even* nth root.

Evaluating $\sqrt[n]{a^n}$

If n is an *even* positive integer, then $\sqrt[n]{a^n} = |a|$.

If n is an *odd* positive integer, then $\sqrt[n]{a^n} = a$.

Classroom Example 4

Simplify each expression.

(a) $\sqrt{z^6}$ (b) $\sqrt[7]{t^7}$

(c) $\sqrt{81r^8s^{10}}$ (d) $\sqrt[4]{(-3)^4}$

(e) $\sqrt[5]{m^{10}}$ (f) $\sqrt{(3x-4)^2}$

(g) $\sqrt{x^2-10x+25}$

Answers:

(a) $|z^3|$ (b) t

(c) $9r^4|s^5|$ (d) 3

(e) m^2 (f) $|3x-4|$

(g) $|x-5|$

EXAMPLE 4 Using Absolute Value to Simplify Roots

Simplify each expression.

(a) $\sqrt{p^4}$ (b) $\sqrt[4]{p^4}$ (c) $\sqrt{16m^8r^6}$

(d) $\sqrt[6]{(-2)^6}$ (e) $\sqrt[5]{m^5}$ (f) $\sqrt{(2k+3)^2}$

(g) $\sqrt{x^2-4x+4}$

SOLUTION

(a) $\sqrt{p^4} = \sqrt{(p^2)^2} = |p^2| = p^2$

(b) $\sqrt[4]{p^4} = |p|$

(c) $\sqrt{16m^8r^6} = |4m^4r^3| = 4m^4|r^3|$

(d) $\sqrt[6]{(-2)^6} = |-2| = 2$

(e) $\sqrt[5]{m^5} = m$

(f) $\sqrt{(2k+3)^2} = |2k+3|$

(g) $\sqrt{x^2-4x+4} = \sqrt{(x-2)^2} = |x-2|$

Now Try Exercises 27, 29, and 31.

NOTE When working with variable radicands, we will *usually* assume that all variables in radicands represent only nonnegative real numbers.

The following rules for working with radicals are simply the power rules for exponents written in radical notation.

Rules for Radicals

Suppose that a and b represent real numbers, and m and n represent positive integers for which the indicated roots are real numbers.

Rule	Description
Product rule $\sqrt[n]{a} \cdot \sqrt[n]{b} = \sqrt[n]{ab}$	The product of two roots is the root of the product.
Quotient rule $\sqrt[n]{\frac{a}{b}} = \frac{\sqrt[n]{a}}{\sqrt[n]{b}}$ $(b \neq 0)$	The root of a quotient is the quotient of the roots.
Power rule $\sqrt[m]{\sqrt[n]{a}} = \sqrt[mn]{a}$	The index of the root of a root is the product of their indexes.

Classroom Example 5
Simplify. Assume all variable expressions represent positive real numbers.

(a) $\sqrt{5} \cdot \sqrt{45}$ (b) $\sqrt[5]{n^3} \cdot \sqrt[5]{n^2}$

(c) $\sqrt{\frac{11}{169}}$ (d) $\sqrt[6]{\frac{a}{b^{12}}}$

(e) $\sqrt[5]{\sqrt[4]{17}}$ (f) $\sqrt{\sqrt[6]{8}}$

Answers:

(a) 15 (b) n

(c) $\frac{\sqrt{11}}{13}$ (d) $\frac{\sqrt[6]{a}}{b^2}$

(e) $\sqrt[20]{17}$ (f) $\sqrt[12]{8}$

EXAMPLE 5 Simplifying Radical Expressions

Simplify. Assume all variable expressions represent positive real numbers.

(a) $\sqrt{6} \cdot \sqrt{54}$ **(b)** $\sqrt[3]{m} \cdot \sqrt[3]{m^2}$ **(c)** $\sqrt{\frac{7}{64}}$

(d) $\sqrt[4]{\frac{a}{b^4}}$ **(e)** $\sqrt[7]{\sqrt[3]{2}}$ **(f)** $\sqrt[4]{\sqrt{3}}$

SOLUTION

(a) $\sqrt{6} \cdot \sqrt{54} = \sqrt{6 \cdot 54}$ Product rule
$= \sqrt{324}$, or 18

(b) $\sqrt[3]{m} \cdot \sqrt[3]{m^2} = \sqrt[3]{m^3}$
$= m$

(c) $\sqrt{\frac{7}{64}} = \frac{\sqrt{7}}{\sqrt{64}}$ Quotient rule
$= \frac{\sqrt{7}}{8}$

(d) $\sqrt[4]{\frac{a}{b^4}} = \frac{\sqrt[4]{a}}{\sqrt[4]{b^4}}$
$= \frac{\sqrt[4]{a}}{b}$

(e) $\sqrt[7]{\sqrt[3]{2}} = \sqrt[21]{2}$ Power rule

(f) $\sqrt[4]{\sqrt{3}} = \sqrt[4 \cdot 2]{3} = \sqrt[8]{3}$

Now Try Exercises 37, 41, 45, and 65.

NOTE Converting to rational exponents shows why these rules work.

$$\sqrt[7]{\sqrt[3]{2}} = (2^{1/3})^{1/7} = 2^{(1/3)(1/7)} = 2^{1/21} = \sqrt[21]{2} \quad \text{Example 5(e)}$$

Simplified Radicals In working with numbers, we prefer to write a number in its simplest form. For example, $\frac{10}{2}$ is written as 5, and $-\frac{9}{6}$ is written as $-\frac{3}{2}$. Similarly, expressions with radicals can be written in their simplest forms.

Simplified Radicals

An expression with radicals is simplified when all of the following conditions are satisfied.

1. The radicand has no factor raised to a power greater than or equal to the index.
2. The radicand has no fractions.
3. No denominator contains a radical.
4. Exponents in the radicand and the index of the radical have greatest common factor 1.
5. All indicated operations have been performed (if possible).

Classroom Example 6
Simplify each radical.

(a) $\sqrt{288}$ (b) $-8\sqrt[3]{125}$

(c) $\sqrt[3]{128a^6b^8c^{10}}$

Answers:

(a) $12\sqrt{2}$ (b) -40

(c) $4a^2b^2c^3\sqrt[3]{2b^2c}$

EXAMPLE 6 Simplifying Radicals

Simplify each radical.

(a) $\sqrt{175}$ **(b)** $-3\sqrt[5]{32}$ **(c)** $\sqrt[3]{81x^5y^7z^6}$

SOLUTION

(a) $\sqrt{175} = \sqrt{25 \cdot 7}$
$= \sqrt{25} \cdot \sqrt{7}$
$= 5\sqrt{7}$

(b) $-3\sqrt[5]{32} = -3\sqrt[5]{2^5}$
$= -3 \cdot 2$
$= -6$

(c) $\sqrt[3]{81x^5y^7z^6} = \sqrt[3]{27 \cdot 3 \cdot x^3 \cdot x^2 \cdot y^6 \cdot y \cdot z^6}$ Factor. (Section R.4)

$= \sqrt[3]{(27x^3y^6z^6)(3x^2y)}$ Group all perfect cubes.

$= 3xy^2z^2\sqrt[3]{3x^2y}$ Remove all perfect cubes from the radical.

✔ ***Now Try Exercises 33 and 51.***

Operations with Radicals Radicals with the same radicand and the same index, such as $3\sqrt[4]{11pq}$ and $-7\sqrt[4]{11pq}$, are **like radicals.** On the other hand, examples of **unlike radicals** are as follows.

$2\sqrt{5}$ and $2\sqrt{3}$ Radicands are different.

$2\sqrt{3}$ and $2\sqrt[3]{3}$ Indexes are different.

We add or subtract like radicals by using the distributive property. ***Only like radicals can be combined.*** Sometimes we need to simplify radicals before adding or subtracting.

Classroom Example 7
Add or subtract, as indicated. Assume all variables represent positive real numbers.

(a) $14\sqrt{5pq} - 11\sqrt{5pq}$

(b) $\sqrt{75ab^3} - b\sqrt{12ab}$

(c) $\sqrt[3]{81x^5y^7} + \sqrt[3]{24x^8y^4}$

Answers:

(a) $3\sqrt{5pq}$ (b) $3b\sqrt{3ab}$

(c) $(3xy^2 + 2x^2y)\sqrt[3]{3x^2y}$

Teaching Tip Use the distributive property to justify the result in **Example 7(a).**

$3\sqrt[4]{11pq} + (-7\sqrt[4]{11pq})$
$= (3-7)\sqrt[4]{11pq}$
$= -4\sqrt[4]{11pq}$

Explain that this is similar to simplifying the expression

$3x + (-7x)$.

EXAMPLE 7 Adding and Subtracting Radicals

Add or subtract, as indicated. Assume all variables represent positive real numbers.

(a) $3\sqrt[4]{11pq} + \left(-7\sqrt[4]{11pq}\right)$ **(b)** $\sqrt{98x^3y} + 3x\sqrt{32xy}$

(c) $\sqrt[3]{64m^4n^5} - \sqrt[3]{-27m^{10}n^{14}}$

SOLUTION

(a) $3\sqrt[4]{11pq} + \left(-7\sqrt[4]{11pq}\right) = -4\sqrt[4]{11pq}$

(b) $\sqrt{98x^3y} + 3x\sqrt{32xy} = \sqrt{49 \cdot 2 \cdot x^2 \cdot x \cdot y} + 3x\sqrt{16 \cdot 2 \cdot x \cdot y}$ Factor.

$= 7x\sqrt{2xy} + 3x(4)\sqrt{2xy}$ Remove all perfect squares from the radicals.

$= 7x\sqrt{2xy} + 12x\sqrt{2xy}$ Multiply.

$= (7x + 12x)\sqrt{2xy}$ Distributive property (Section R.2)

$= 19x\sqrt{2xy}$ Add.

(c) $\sqrt[3]{64m^4n^5} - \sqrt[3]{-27m^{10}n^{14}} = \sqrt[3]{(64m^3n^3)(mn^2)} - \sqrt[3]{(-27m^9n^{12})(mn^2)}$

$= 4mn\sqrt[3]{mn^2} - (-3)m^3n^4\sqrt[3]{mn^2}$

$= 4mn\sqrt[3]{mn^2} + 3m^3n^4\sqrt[3]{mn^2}$

$= (4mn + 3m^3n^4)\sqrt[3]{mn^2}$

This *cannot* be simplified further.

✔ ***Now Try Exercises 69 and 73.***

If the index of the radical and an exponent in the radicand have a common factor, we can simplify the radical by first writing it in exponential form. We simplify the rational exponent, and then write the result as a radical again, as shown in **Example 8** on the next page.

Classroom Example 8
Simplify each radical.
(a) $\sqrt[10]{2^5}$ (b) $\sqrt[3]{a^9b^{18}}$
(c) $\sqrt[6]{\sqrt[3]{4^2}}$

Answers:
(a) $\sqrt{2}$ (b) a^3b^6 (c) $\sqrt[9]{4}$

EXAMPLE 8 Simplifying Radicals

Simplify each radical. Assume all variables represent positive real numbers.

(a) $\sqrt[6]{3^2}$ **(b)** $\sqrt[6]{x^{12}y^3}$ **(c)** $\sqrt[9]{\sqrt{6^3}}$

SOLUTION

(a) $\sqrt[6]{3^2} = 3^{2/6}$
$= 3^{1/3}$
$= \sqrt[3]{3}$

(b) $\sqrt[6]{x^{12}y^3} = (x^{12}y^3)^{1/6}$
$= x^2y^{3/6}$
$= x^2y^{1/2}$
$= x^2\sqrt{y}$

(c) $\sqrt[9]{\sqrt{6^3}} = \sqrt[9]{6^{3/2}}$
$= (6^{3/2})^{1/9}$
$= 6^{1/6}$
$= \sqrt[6]{6}$

Now Try Exercises 63 and 67.

In **Example 8(a),** we simplified $\sqrt[6]{3^2}$ as $\sqrt[3]{3}$. However, to simplify $(\sqrt[6]{x})^2$, the variable x must represent a nonnegative number. For example, consider the statement

$$(-8)^{2/6} = [(-8)^{1/6}]^2.$$

This result is not a real number, since $(-8)^{1/6}$ is not a real number. On the other hand,

$$(-8)^{1/3} = -2.$$

Here, even though $\frac{2}{6} = \frac{1}{3}$,

$$(\sqrt[6]{x})^2 \neq \sqrt[3]{x}.$$

If a is nonnegative, then it is always true that $a^{m/n} = a^{mp/(np)}$. Simplifying rational exponents on negative bases should be considered case by case.

Classroom Example 9
Find each product.
(a) $(\sqrt{11} + \sqrt{17})(\sqrt{11} - \sqrt{17})$
(b) $(5 + \sqrt{32})(3 - \sqrt{2})$

Answers:
(a) -6 (b) $7 + 7\sqrt{2}$

EXAMPLE 9 Multiplying Radical Expressions

Find each product.

(a) $(\sqrt{7} - \sqrt{10})(\sqrt{7} + \sqrt{10})$ **(b)** $(\sqrt{2} + 3)(\sqrt{8} - 5)$

SOLUTION

(a) $(\sqrt{7} - \sqrt{10})(\sqrt{7} + \sqrt{10}) = (\sqrt{7})^2 - (\sqrt{10})^2$ Product of the sum and difference of two terms (Section R.3)
$= 7 - 10$ $(\sqrt{a})^2 = a$
$= -3$ Subtract.

(b) $(\sqrt{2} + 3)(\sqrt{8} - 5) = \sqrt{2}(\sqrt{8}) - \sqrt{2}(5) + 3\sqrt{8} - 3(5)$ FOIL (Section R.3)
$= \sqrt{16} - 5\sqrt{2} + 3(2\sqrt{2}) - 15$ Multiply; $\sqrt{8} = 2\sqrt{2}$.
$= 4 - 5\sqrt{2} + 6\sqrt{2} - 15$ Simplify.
$= -11 + \sqrt{2}$ Combine like terms.

Now Try Exercises 77 and 83.

Rationalizing Denominators The third condition for a simplified radical requires that no denominator contain a radical. We achieve this by **rationalizing the denominator**—that is, multiplying by a form of 1.

Classroom Example 10
Rationalize each denominator.

(a) $\frac{2}{\sqrt{7}}$ (b) $\sqrt[3]{\frac{4}{9}}$

Answers:

(a) $\frac{2\sqrt{7}}{7}$ (b) $\frac{\sqrt[3]{12}}{3}$

Teaching Tip In **Example 10(b)**, caution students *not* to simplify $\frac{\sqrt[4]{375}}{5}$ as $\sqrt[4]{75}$.

EXAMPLE 10 Rationalizing Denominators

Rationalize each denominator.

(a) $\frac{4}{\sqrt{3}}$ **(b)** $\sqrt[4]{\frac{3}{5}}$

SOLUTION

(a) $\frac{4}{\sqrt{3}} = \frac{4}{\sqrt{3}} \cdot \frac{\sqrt{3}}{\sqrt{3}}$ Multiply by $\frac{\sqrt{3}}{\sqrt{3}}$ (which equals 1).

$= \frac{4\sqrt{3}}{3}$ $\sqrt{a} \cdot \sqrt{a} = a$

(b) $\sqrt[4]{\frac{3}{5}} = \frac{\sqrt[4]{3}}{\sqrt[4]{5}}$ Quotient rule

The denominator will be a rational number if it equals $\sqrt[4]{5^4}$. That is, four factors of 5 are needed under the radical. Since $\sqrt[4]{5}$ has just one factor of 5, three additional factors are needed, so multiply by $\frac{\sqrt[4]{5^3}}{\sqrt[4]{5^3}}$.

$\frac{\sqrt[4]{3}}{\sqrt[4]{5}} = \frac{\sqrt[4]{3} \cdot \sqrt[4]{5^3}}{\sqrt[4]{5} \cdot \sqrt[4]{5^3}}$ Multiply by $\frac{\sqrt[4]{5^3}}{\sqrt[4]{5^3}}$.

$= \frac{\sqrt[4]{3 \cdot 5^3}}{\sqrt[4]{5^4}}$ Product rule

$= \frac{\sqrt[4]{375}}{5}$ Simplify.

✔ *Now Try Exercises 55 and 59.*

Classroom Example 11
Simplify each expression. Assume all variables represent positive real numbers.

(a) $\frac{\sqrt[3]{a^5 b}}{\sqrt[3]{a^2 b^5}}$ (b) $\sqrt[4]{\frac{6}{x^8}} - \sqrt[4]{\frac{3}{x^{16}}}$

Answers:

(a) $\frac{a\sqrt[3]{b^2}}{b^2}$ (b) $\frac{x^2\sqrt[4]{6} - \sqrt[4]{3}}{x^4}$

EXAMPLE 11 Simplifying Radical Expressions with Fractions

Simplify each expression. Assume all variables represent positive real numbers.

(a) $\frac{\sqrt[4]{xy^3}}{\sqrt[4]{x^3y^2}}$ **(b)** $\sqrt[3]{\frac{5}{x^6}} - \sqrt[3]{\frac{4}{x^9}}$

SOLUTION

(a) $\frac{\sqrt[4]{xy^3}}{\sqrt[4]{x^3y^2}} = \sqrt[4]{\frac{xy^3}{x^3y^2}}$ Quotient rule

$= \sqrt[4]{\frac{y}{x^2}}$ Simplify the radicand.

$= \frac{\sqrt[4]{y}}{\sqrt[4]{x^2}}$ Quotient rule

$= \frac{\sqrt[4]{y}}{\sqrt[4]{x^2}} \cdot \frac{\sqrt[4]{x^2}}{\sqrt[4]{x^2}}$ Rationalize the denominator.

$= \frac{\sqrt[4]{x^2 y}}{x}$ $\sqrt[4]{x^2} \cdot \sqrt[4]{x^2} = \sqrt[4]{x^4} = x$

LOOKING AHEAD TO CALCULUS

Another standard problem in calculus is investigating the value that an expression such as $\frac{\sqrt{x^2+9}-3}{x^2}$ approaches as x approaches 0. This cannot be done by simply substituting 0 for x, since the result is $\frac{0}{0}$. However, by **rationalizing the numerator,** we can show that for $x \neq 0$ the expression is equivalent to $\frac{1}{\sqrt{x^2+9}+3}$. Then, by substituting 0 for x, we find that the original expression approaches $\frac{1}{6}$ as x approaches 0.

(b) $\sqrt[3]{\frac{5}{x^6}} - \sqrt[3]{\frac{4}{x^9}} = \frac{\sqrt[3]{5}}{\sqrt[3]{x^6}} - \frac{\sqrt[3]{4}}{\sqrt[3]{x^9}}$ Quotient rule

$= \frac{\sqrt[3]{5}}{x^2} - \frac{\sqrt[3]{4}}{x^3}$ Simplify the denominators.

$= \frac{x\sqrt[3]{5}}{x^3} - \frac{\sqrt[3]{4}}{x^3}$ Write with a common denominator. **(Section R.5)**

$= \frac{x\sqrt[3]{5} - \sqrt[3]{4}}{x^3}$ Subtract the numerators.

✔ ***Now Try Exercises 85 and 87.***

In **Example 9(a),** we saw that the product

$$(\sqrt{7}-\sqrt{10})(\sqrt{7}+\sqrt{10}) \quad \text{equals} \quad -3, \text{ a rational number.}$$

This suggests a way to rationalize a denominator that is a binomial in which one or both terms is a square root radical. The expressions $a-b$ and $a+b$ are **conjugates.**

Classroom Example 12
Rationalize the denominator.

$$\frac{2}{3+\sqrt{5}}$$

Answer: $\frac{3-\sqrt{5}}{2}$

EXAMPLE 12 Rationalizing a Binomial Denominator

Rationalize the denominator of $\frac{1}{1-\sqrt{2}}$.

SOLUTION

$\frac{1+\sqrt{2}}{1+\sqrt{2}} = 1$

$\frac{1}{1-\sqrt{2}} = \frac{1(1+\sqrt{2})}{(1-\sqrt{2})(1+\sqrt{2})}$ Multiply numerator and denominator by the conjugate of the denominator: $1+\sqrt{2}$.

$= \frac{1+\sqrt{2}}{1-2}$ $(x-y)(x+y) = x^2-y^2$

$= -1-\sqrt{2}$ $\frac{1+\sqrt{2}}{-1} = -1-\sqrt{2}$

✔ ***Now Try Exercise 93.***

R.7 Exercises

1. 5 2. 6
3. 3 4. 4
5. −5 6. −7
7. This expression is not a real number.
8. This expression is not a real number.
9. 2 10. 2
11. 2 12. 7

Write each root using exponents and evaluate. ***See Example 1.***

1. $\sqrt[3]{125}$ **2.** $\sqrt[3]{216}$ **3.** $\sqrt[4]{81}$

4. $\sqrt[4]{256}$ **5.** $\sqrt[3]{-125}$ **6.** $\sqrt[3]{-343}$

7. $\sqrt[4]{-81}$ **8.** $\sqrt[4]{-256}$ **9.** $\sqrt[5]{32}$

10. $\sqrt[7]{128}$ **11.** $-\sqrt[5]{-32}$ **12.** $-\sqrt[3]{-343}$

13. (a) F (b) H (c) G (d) C
14. (a) B (b) D (c) A (d) E
15. $\sqrt[3]{m^2}$, or $(\sqrt[3]{m})^2$
16. $\sqrt[4]{p^5}$, or $(\sqrt[4]{p})^5$
17. $\sqrt[3]{(2m+p)^2}$, or $(\sqrt[3]{2m+p})^2$
18. $\sqrt[7]{(5r+3t)^4}$, or $(\sqrt[7]{5r+3t})^4$
19. $k^{2/5}$ 20. $z^{5/4}$
21. $-3 \cdot 5^{1/2}p^{3/2}$ 22. $-2^{1/2}my^{5/2}$
23. A
24. 3, 5, 7, ... (odd positive integers greater than or equal to 3)
25. $x \geq 0$ 26. D; $\frac{\sqrt{3}}{2}$
27. $|x|$ 28. $|x|$
29. $5k^2|m|$ 30. $3|p^3q|$
31. $|4x-y|$ 32. $|5+2m|$
33. $3\sqrt[3]{3}$ 34. $5\sqrt[3]{2}$
35. $-2\sqrt[4]{2}$ 36. $-3\sqrt[4]{3}$
37. $\sqrt{42pqr}$ 38. $\sqrt{35xt}$
39. $\sqrt[3]{14xy}$ 40. $\sqrt[3]{36xy}$
41. $-\frac{3}{5}$ 42. $-\frac{4}{7}$
43. $-\frac{\sqrt[3]{5}}{2}$ 44. $-\frac{\sqrt[4]{3}}{2}$
45. $\frac{\sqrt[4]{m}}{n}$ 46. $\frac{\sqrt[6]{r}}{s}$
47. -15 48. -35
49. $32\sqrt[3]{2}$ 50. $15\sqrt[3]{75}$
51. $2x^2z^4\sqrt{2x}$ 52. $2m^3n^2\sqrt{6n}$
53. This expression cannot be simplified further.

Concept Check In Exercises 13 and 14, match the rational exponent expression in Column I with the equivalent radical expression in Column II. Assume that x is not 0. ***See Example 2.***

I		II	
13. (a) $(-3x)^{1/3}$	**14. (a)** $-3x^{1/3}$	**A.** $\frac{3}{\sqrt[3]{x}}$	**B.** $-3\sqrt[3]{x}$
(b) $(-3x)^{-1/3}$	**(b)** $-3x^{-1/3}$	**C.** $\frac{1}{\sqrt[3]{3x}}$	**D.** $\frac{-3}{\sqrt[3]{x}}$
(c) $(3x)^{1/3}$	**(c)** $3x^{-1/3}$	**E.** $3\sqrt[3]{x}$	**F.** $\sqrt[3]{-3x}$
(d) $(3x)^{-1/3}$	**(d)** $3x^{1/3}$	**G.** $\sqrt[3]{3x}$	**H.** $\frac{1}{\sqrt[3]{-3x}}$

If the expression is in exponential form, write it in radical form. If it is in radical form, write it in exponential form. Assume all variables represent positive real numbers. ***See Examples 2 and 3.***

15. $m^{2/3}$ **16.** $p^{5/4}$ **17.** $(2m+p)^{2/3}$ **18.** $(5r+3t)^{4/7}$

19. $\sqrt[5]{k^2}$ **20.** $\sqrt[4]{z^5}$ **21.** $-3\sqrt{5p^3}$ **22.** $-m\sqrt{2y^5}$

Concept Check Answer each question.

23. For which of the following cases is $\sqrt{ab} = \sqrt{a} \cdot \sqrt{b}$ a true statement?

A. a and b both positive **B.** a and b both negative

24. For what positive integers n greater than or equal to 2 is $\sqrt[n]{a^n} = a$ always a true statement?

25. For what values of x is $\sqrt{9ax^2} = 3x\sqrt{a}$ a true statement? Assume $a \geq 0$.

26. Which of the following expressions is *not* simplified? Give the simplified form.

A. $\sqrt[3]{2y}$ **B.** $\frac{\sqrt{5}}{2}$ **C.** $\sqrt[4]{m^3}$ **D.** $\sqrt{\frac{3}{4}}$

Simplify each expression. ***See Example 4.***

27. $\sqrt[4]{x^4}$ **28.** $\sqrt[6]{x^6}$ **29.** $\sqrt{25k^4m^2}$

30. $\sqrt[4]{81p^{12}q^4}$ **31.** $\sqrt{(4x-y)^2}$ **32.** $\sqrt[4]{(5+2m)^4}$

Simplify each expression. Assume all variables represent positive real numbers. ***See Examples 1, 4–6, and 8–11.***

33. $\sqrt[3]{81}$ **34.** $\sqrt[3]{250}$ **35.** $-\sqrt[4]{32}$

36. $-\sqrt[4]{243}$ **37.** $\sqrt{14} \cdot \sqrt{3pqr}$ **38.** $\sqrt{7} \cdot \sqrt{5xt}$

39. $\sqrt[3]{7x} \cdot \sqrt[3]{2y}$ **40.** $\sqrt[3]{9x} \cdot \sqrt[3]{4y}$ **41.** $-\sqrt{\frac{9}{25}}$

42. $-\sqrt{\frac{16}{49}}$ **43.** $-\sqrt[3]{\frac{5}{8}}$ **44.** $-\sqrt[4]{\frac{3}{16}}$

45. $\sqrt[4]{\frac{m}{n^4}}$ **46.** $\sqrt[6]{\frac{r}{s^6}}$ **47.** $3\sqrt[5]{-3125}$

48. $5\sqrt[3]{-343}$ **49.** $\sqrt[3]{16(-2)^4(2)^8}$ **50.** $\sqrt[3]{25(-3)^4(5)^3}$

51. $\sqrt{8x^5z^8}$ **52.** $\sqrt{24m^6n^5}$ **53.** $\sqrt[4]{x^4+y^4}$

54. $\sqrt[3]{27 + a^3}$ 55. $\sqrt{\frac{2}{3x}}$ 56. $\sqrt{\frac{5}{3p}}$

57. $\sqrt{\frac{x^5y^3}{z^2}}$ 58. $\sqrt{\frac{g^3h^5}{r^3}}$ 59. $\sqrt[3]{\frac{8}{x^4}}$

60. $\sqrt[3]{\frac{9}{16p^4}}$ 61. $\sqrt[4]{\frac{g^3h^5}{9r^6}}$ 62. $\sqrt[4]{\frac{32x^5}{y^5}}$

63. $\sqrt[8]{3^4}$ 64. $\sqrt[9]{5^3}$ 65. $\sqrt[3]{\sqrt{4}}$

66. $\sqrt[4]{\sqrt{25}}$ 67. $\sqrt[4]{\sqrt[3]{2}}$ 68. $\sqrt[5]{\sqrt[3]{9}}$

Simplify each expression. Assume all variables represent positive real numbers. ***See Examples 7, 9, and 11.***

69. $8\sqrt{2x} - \sqrt{8x} + \sqrt{72x}$

70. $4\sqrt{18k} - \sqrt{72k} + \sqrt{50k}$

71. $2\sqrt[3]{3} + 4\sqrt[3]{24} - \sqrt[3]{81}$

72. $\sqrt[3]{32} - 5\sqrt[3]{4} + 2\sqrt[3]{108}$

73. $\sqrt[4]{81x^6y^3} - \sqrt[4]{16x^{10}y^3}$

74. $\sqrt[4]{256x^5y^6} + \sqrt[4]{625x^9y^2}$

75. $5\sqrt{6} + 2\sqrt{10}$

76. $3\sqrt{11} - 5\sqrt{13}$

77. $(\sqrt{2} + 3)(\sqrt{2} - 3)$

78. $(\sqrt{5} + \sqrt{2})(\sqrt{5} - \sqrt{2})$

79. $(\sqrt[3]{11} - 1)(\sqrt[3]{11^2} + \sqrt[3]{11} + 1)$

80. $(\sqrt[3]{7} + 3)(\sqrt[3]{7^2} - 3\sqrt[3]{7} + 9)$

81. $(\sqrt{3} + \sqrt{8})^2$

82. $(\sqrt{5} + \sqrt{10})^2$

83. $(3\sqrt{2} + \sqrt{3})(2\sqrt{3} - \sqrt{2})$

84. $(4\sqrt{5} + \sqrt{2})(3\sqrt{2} - \sqrt{5})$

85. $\frac{\sqrt[3]{mn} \cdot \sqrt[3]{m^2}}{\sqrt[3]{n^2}}$

86. $\frac{\sqrt[3]{8m^2n^3} \cdot \sqrt[3]{2m^2}}{\sqrt[3]{32m^4n^3}}$

87. $\sqrt[3]{\frac{2}{x^6}} - \sqrt[3]{\frac{5}{x^9}}$

88. $\sqrt[4]{\frac{7}{t^{12}}} + \sqrt[4]{\frac{9}{t^4}}$

89. $\frac{1}{\sqrt{2}} + \frac{3}{\sqrt{8}} + \frac{1}{\sqrt{32}}$

90. $\frac{2}{\sqrt{12}} - \frac{1}{\sqrt{27}} - \frac{5}{\sqrt{48}}$

91. $\frac{-4}{\sqrt[3]{3}} + \frac{1}{\sqrt[3]{24}} - \frac{2}{\sqrt[3]{81}}$

92. $\frac{5}{\sqrt[3]{2}} - \frac{2}{\sqrt[3]{16}} + \frac{1}{\sqrt[3]{54}}$

Rationalize the denominator of each radical expression. Assume all variables represent nonnegative numbers and that no denominators are 0. ***See Example 12.***

93. $\frac{\sqrt{3}}{\sqrt{5} + \sqrt{3}}$ 94. $\frac{\sqrt{7}}{\sqrt{3} - \sqrt{7}}$ 95. $\frac{\sqrt{7} - 1}{2\sqrt{7} + 4\sqrt{2}}$

96. $\frac{1 + \sqrt{3}}{3\sqrt{5} + 2\sqrt{3}}$ 97. $\frac{p - 4}{\sqrt{p} + 2}$ 98. $\frac{9 - r}{3 - \sqrt{r}}$

99. $\frac{3m}{2 + \sqrt{m + n}}$ 100. $\frac{a}{\sqrt{a + b} - 1}$ 101. $\frac{5\sqrt{x}}{2\sqrt{x} + \sqrt{y}}$

102. *Concept Check* What should the numerator and denominator of

$$\frac{1}{\sqrt[3]{3} - \sqrt[3]{5}}$$

be multiplied by in order to rationalize the denominator? Write this fraction with a rationalized denominator.

54. This expression cannot be simplified further.
55. $\frac{\sqrt{6x}}{3x}$ 56. $\frac{\sqrt{15p}}{3p}$
57. $\frac{x^2y\sqrt{xy}}{z}$ 58. $\frac{gh^2\sqrt{ghr}}{r^2}$
59. $\frac{2\sqrt[3]{x^2}}{x^2}$ 60. $\frac{\sqrt[3]{36p^2}}{4p^2}$
61. $\frac{h\sqrt[4]{9g^3hr^2}}{3r^2}$ 62. $\frac{2x\sqrt[4]{2xy^3}}{y^2}$
63. $\sqrt{3}$ 64. $\sqrt[3]{5}$
65. $\sqrt[3]{2}$ 66. $\sqrt[4]{5}$
67. $\sqrt[12]{2}$ 68. $\sqrt[15]{9}$
69. $12\sqrt{2x}$ 70. $11\sqrt{2k}$
71. $7\sqrt[3]{3}$ 72. $3\sqrt[3]{4}$
73. $3x\sqrt[4]{x^2y^3} - 2x^2\sqrt[4]{x^2y^3}$
74. $4xy\sqrt[4]{xy^2} + 5x^2\sqrt[4]{xy^2}$
75. This expression cannot be simplified further.
76. This expression cannot be simplified further.
77. -7 78. 3
79. 10 80. 34
81. $11 + 4\sqrt{6}$
82. $15 + 10\sqrt{2}$
83. $5\sqrt{6}$
84. $11\sqrt{10} - 14$
85. $\frac{m\sqrt[3]{n^2}}{n}$ 86. $\frac{\sqrt[3]{4}}{2}$
87. $\frac{x\sqrt[3]{2} - \sqrt[3]{5}}{x^3}$ 88. $\frac{\sqrt[4]{7} + t^2\sqrt[4]{9}}{t^3}$
89. $\frac{11\sqrt{2}}{8}$ 90. $-\frac{7\sqrt{3}}{36}$
91. $-\frac{25\sqrt[3]{9}}{18}$ 92. $\frac{13\sqrt[3]{4}}{6}$
93. $\frac{\sqrt{15} - 3}{2}$ 94. $\frac{-\sqrt{21} - 7}{4}$
95. $\frac{-7 + 2\sqrt{14} + \sqrt{7} - 2\sqrt{2}}{2}$
96. $\frac{3\sqrt{5} - 2\sqrt{3} + 3\sqrt{15} - 6}{33}$
97. $\sqrt{p} - 2$ 98. $3 + \sqrt{r}$
99. $\frac{3m(2 - \sqrt{m + n})}{4 - m - n}$
100. $\frac{a(\sqrt{a + b} + 1)}{a + b - 1}$
101. $\frac{5\sqrt{x}(2\sqrt{x} - \sqrt{y})}{4x - y}$
102. $\sqrt[3]{9} + \sqrt[3]{15} + \sqrt[3]{25}$; $\frac{\sqrt[3]{9} + \sqrt[3]{15} + \sqrt[3]{25}}{-2}$

103. 17.7 ft per sec
104. 19.1 ft per sec

105. $-12°$F
106. 19° F

107. 2
108. 3
109. 2
110. 2
111. 3
112. 2

113. It gives six decimal places of accuracy.
114. It first differs in the fourth decimal place.
115. It first differs in the fourth decimal place.

(Modeling) Solve each problem.

103. *Rowing Speed* Olympic rowing events have one-, two-, four-, or eight-person crews, with each person pulling a single oar. Increasing the size of the crew increases the speed of the boat. An analysis of Olympic rowing events concluded that the approximate speed, s, of the boat (in feet per second) was given by the formula

$$s = 15.18\sqrt[9]{n},$$

where n is the number of oarsmen. Estimate the speed of a boat with a four-person crew. (*Source:* Townend, M. Stewart, *Mathematics in Sport,* Chichester, Ellis Horwood Limited.)

104. *Rowing Speed* **See Exercise 103.** Estimate the speed of a boat with an eight-person crew.

(Modeling) Windchill *The National Weather Service has used the formula*

$$\text{Windchill temperature} = 35.74 + 0.6215T - 35.75V^{0.16} + 0.4275TV^{0.16},$$

where T is the temperature in °F and V is the wind speed in miles per hour, to calculate windchill. (Source: National Oceanic and Atmospheric Administration, National Weather Service.) Use the formula to calculate the windchill to the nearest degree given the following conditions.

105. 10°F, 30 mph wind

106. 30°F, 15 mph wind

Concept Check *Simplify each expression mentally.*

107. $\sqrt[4]{8} \cdot \sqrt[4]{2}$

108. $\dfrac{\sqrt[3]{54}}{\sqrt[3]{2}}$

109. $\dfrac{\sqrt[5]{320}}{\sqrt[5]{10}}$

110. $\sqrt{0.1} \cdot \sqrt{40}$

111. $\dfrac{\sqrt[3]{15}}{\sqrt[3]{5}} \cdot \sqrt[3]{9}$

112. $\sqrt[6]{2} \cdot \sqrt[6]{4} \cdot \sqrt[6]{8}$

The screen in **Figure A** *seems to indicate that π and $\sqrt[4]{\frac{2143}{22}}$ are exactly equal, since the eight decimal values given by the calculator agree. However, as shown in* **Figure B** *using one more decimal place in the display, they differ in the ninth decimal place.* ***The radical expression is a very good approximation for π, but it is still only an approximation.***

```
π
          3.14159265
⁴√2143/22
          3.14159265
```

Figure A

```
π
          3.141592654
⁴√2143/22
          3.141592653
```

Figure B

Use your calculator to answer each question. Refer to the display for π in **Figure B**.

113. The Chinese of the fifth century used $\frac{355}{113}$ as an approximation for π. How many decimal places of accuracy does this fraction give?

114. A value for π that the Greeks used circa A.D. 150 is equivalent to $\frac{377}{120}$. In which decimal place does this value first differ from π?

115. The Hindu mathematician Bhaskara used $\frac{3927}{1250}$ as an approximation for π circa A.D. 1150. In which decimal place does this value first differ from π?

Chapter R Test Prep

Key Terms

R.1 set, elements (members), infinite set, finite set, Venn diagram, disjoint sets

R.2 number line, coordinate system, coordinate, power or exponential expression (exponential), exponent, base, absolute value

R.3 algebraic expression, term, coefficient, like terms, polynomial, polynomial in x, degree of a term, degree of a polynomial, trinomial, binomial, monomial, descending order, FOIL method

R.4 factoring, factored form, prime polynomial, factored completely, factoring by grouping

R.5 rational expression, domain of a rational expression, lowest terms, complex fraction

R.7 radicand, index of a radical, principal nth root, like radicals, unlike radicals, rationalizing the denominator, conjugates

New Symbols

$\{\ \}$	set braces
$\in$	is an element of
$\notin$	is not an element of
$\{x \mid x \text{ has property } p\}$	set-builder notation
U	universal set
$\emptyset$, or $\{\ \}$	null (empty) set
$\subseteq$	is a subset of
$\nsubseteq$	is not a subset of
A'	complement of a set A
$\cap$	set intersection
$\cup$	set union
a^n	n factors of a
$<$	is less than
$>$	is greater than
$\leq$	is less than or equal to
$\geq$	is greater than or equal to
$\lvert a \rvert$	absolute value of a
$\sqrt{\ }$	radical symbol

Quick Review

Concepts	Examples
R.1 Sets	
Set Operations	
For all sets A and B, with universal set U: The **complement** of set A is the set A' of all elements in U that do not belong to set A. $A' = \{x \mid x \in U, x \notin A\}$	Let $U = \{1, 2, 3, 4, 5, 6\}$, $A = \{1, 2, 3, 4\}$, and $B = \{3, 4, 6\}$. $A' = \{5, 6\}$
The **intersection** of sets A and B, written $A \cap B$, is made up of all the elements belonging to both set A and set B. $A \cap B = \{x \mid x \in A \text{ and } x \in B\}$	$A \cap B = \{3, 4\}$
The **union** of sets A and B, written $A \cup B$, is made up of all the elements belonging to set A or to set B. $A \cup B = \{x \mid x \in A \text{ or } x \in B\}$	and $A \cup B = \{1, 2, 3, 4, 6\}$

Concepts	Examples
R.2 Real Numbers and Their Properties	
Sets of Numbers	
Natural numbers $\{1, 2, 3, 4, \ldots\}$	$5, 17, 142$
Whole numbers $\{0, 1, 2, 3, 4, \ldots\}$	$0, 27, 96$
Integers $\{\ldots, -3, -2, -1, 0, 1, 2, 3, \ldots\}$	$-24, 0, 19$
Rational numbers $\{\frac{p}{q} \mid p \text{ and } q \text{ are integers and } q \neq 0\}$	$-\frac{3}{4}, -0.28, 0, 7, \frac{9}{16}, 0.66\overline{6}$
Irrational numbers $\{x \mid x \text{ is real but not rational}\}$	$-\sqrt{15}, 0.101101110\ldots, \sqrt{2}, \pi$
Real numbers $\{x \mid x \text{ corresponds to a point on a number line}\}$	$-46, 0.7, \pi, \sqrt{19}, \frac{8}{5}$
Properties of Real Numbers	
For all real numbers a, b, and c:	
Closure Properties	
$a + b$ **is a real number.**	$1 + \sqrt{2}$ is a real number.
ab **is a real number.**	$3\sqrt{7}$ is a real number.
Commutative Properties	
$a + b = b + a$	$5 + 18 = 18 + 5$
$ab = ba$	$-4 \cdot 8 = 8 \cdot (-4)$
Associative Properties	
$(a + b) + c = a + (b + c)$	$[6 + (-3)] + 5 = 6 + (-3 + 5)$
$(ab)c = a(bc)$	$(7 \cdot 6)20 = 7(6 \cdot 20)$
Identity Properties	
There exists a unique real number 0 such that $a + 0 = a$ and $0 + a = a$.	$145 + 0 = 145$ and $0 + 145 = 145$
There exists a unique real number 1 such that $a \cdot 1 = a$ and $1 \cdot a = a$.	$-60 \cdot 1 = -60$ and $1 \cdot (-60) = -60$
Inverse Properties	
There exists a unique real number $-a$ such that $a + (-a) = 0$ and $-a + a = 0$.	$17 + (-17) = 0$ and $-17 + 17 = 0$
If $a \neq 0$, there exists a unique real number $\frac{1}{a}$ such that $a \cdot \frac{1}{a} = 1$ and $\frac{1}{a} \cdot a = 1$.	$22 \cdot \frac{1}{22} = 1$ and $\frac{1}{22} \cdot 22 = 1$
Distributive Properties	
$a(b + c) = ab + ac$	$3(5 + 8) = 3 \cdot 5 + 3 \cdot 8$
$a(b - c) = ab - ac$	$6(4 - 2) = 6 \cdot 4 - 6 \cdot 2$

(continued)

Concepts	Examples
Order $a > b$ if a is to the right of b on a number line. $a < b$ if a is to the left of b on a number line.	$7 > -5$ $0 < 15$
Absolute Value $$\lvert a\rvert = \begin{cases} a & \text{if } a \geq 0 \\ -a & \text{if } a < 0 \end{cases}$$	$\lvert 3\rvert = 3$ and $\lvert -3\rvert = 3$

R.3 Polynomials

Concepts	Examples
Special Products	
Product of the Sum and Difference of Two Terms $(x + y)(x - y) = x^2 - y^2$	$(7 - x)(7 + x) = 7^2 - x^2 = 49 - x^2$
Square of a Binomial $(x + y)^2 = x^2 + 2xy + y^2$	$(3a + b)^2 = (3a)^2 + 2(3a)(b) + b^2$ $= 9a^2 + 6ab + b^2$
$(x - y)^2 = x^2 - 2xy + y^2$	$(2m - 5)^2 = (2m)^2 - 2(2m)(5) + 5^2$ $= 4m^2 - 20m + 25$

R.4 Factoring Polynomials

Concepts	Examples
Factoring Patterns	
Difference of Squares $x^2 - y^2 = (x + y)(x - y)$	$4t^2 - 9 = (2t + 3)(2t - 3)$
Perfect Square Trinomial $x^2 + 2xy + y^2 = (x + y)^2$ $x^2 - 2xy + y^2 = (x - y)^2$	$p^2 + 4pq + 4q^2 = (p + 2q)^2$ $9m^2 - 12mn + 4n^2 = (3m - 2n)^2$
Difference of Cubes $x^3 - y^3 = (x - y)(x^2 + xy + y^2)$	$r^3 - 8 = (r - 2)(r^2 + 2r + 4)$
Sum of Cubes $x^3 + y^3 = (x + y)(x^2 - xy + y^2)$	$27x^3 + 64 = (3x + 4)(9x^2 - 12x + 16)$

R.5 Rational Expressions

Concepts	Examples
Operations Let $\frac{a}{b}$ and $\frac{c}{d}$ $(b \neq 0, d \neq 0)$ represent fractions.	
$\frac{a}{b} \pm \frac{c}{d} = \frac{ad \pm bc}{bd}$	$\frac{2}{x} + \frac{5}{y} = \frac{2y + 5x}{xy}$ $\quad$ $\frac{x}{6} - \frac{2y}{5} = \frac{5x - 12y}{30}$
$\frac{a}{b} \cdot \frac{c}{d} = \frac{ac}{bd}$ and $\frac{a}{b} \div \frac{c}{d} = \frac{ad}{bc}$ $(c \neq 0)$	$\frac{3}{q} \cdot \frac{3}{2p} = \frac{9}{2pq}$ $\quad$ $\frac{z}{4} \div \frac{z}{2t} = \frac{z}{4} \cdot \frac{2t}{z} = \frac{2zt}{4z} = \frac{t}{2}$

Concepts	Examples
R.6 Rational Exponents	
Rules for Exponents Let r and s be rational numbers. The following results are valid for all positive numbers a and b.	
$a^r \cdot a^s = a^{r+s}$ $\quad (ab)^r = a^r b^r \quad (a^r)^s = a^{rs}$	$6^2 \cdot 6^3 = 6^5 \quad (3x)^4 = 3^4 x^4 \quad (m^2)^3 = m^6$
$\frac{a^r}{a^s} = a^{r-s} \quad \left(\frac{a}{b}\right)^r = \frac{a^r}{b^r} \quad a^{-r} = \frac{1}{a^r}$	$\frac{p^5}{p^2} = p^3 \quad \left(\frac{x}{3}\right)^2 = \frac{x^2}{3^2} \quad 4^{-3} = \frac{1}{4^3}$
R.7 Radical Expressions	
Radical Notation Suppose that a is a real number, n is a positive integer, and $a^{1/n}$ is defined.	
$\sqrt[n]{a} = a^{1/n}$	$\sqrt[4]{16} = 16^{1/4} = 2$
Suppose that m is an integer, n is a positive integer, and a is a real number for which $\sqrt[n]{a}$ is defined.	
$a^{m/n} = \left(\sqrt[n]{a}\right)^m = \sqrt[n]{a^m}$	$8^{2/3} = \left(\sqrt[3]{8}\right)^2 = \sqrt[3]{8^2} = 4$
Operations Operations with radical expressions are performed like operations with polynomials.	$\sqrt{8x} + \sqrt{32x} = 2\sqrt{2x} + 4\sqrt{2x} = 6\sqrt{2x}$ $(\sqrt{5} - \sqrt{3})(\sqrt{5} + \sqrt{3}) = 5 - 3 = 2$ $(\sqrt{2} + \sqrt{7})(\sqrt{3} - \sqrt{6}) = \sqrt{6} - 2\sqrt{3} + \sqrt{21} - \sqrt{42}$ FOIL; $\sqrt{12} = 2\sqrt{3}$
Rationalize the denominator by multiplying numerator and denominator by a form of 1.	$\frac{\sqrt{7y}}{\sqrt{5}} = \frac{\sqrt{7y}}{\sqrt{5}} \cdot \frac{\sqrt{5}}{\sqrt{5}} = \frac{\sqrt{35y}}{5}$

Chapter R Review Exercises

1. Use set notation to list all the elements of the set $\{6, 8, 10, \ldots, 20\}$.
2. Is the set $\{x \mid x \text{ is a decimal between 0 and 1}\}$ finite or infinite?
3. *Concept Check* *True* or *false:* The set of negative integers and the set of whole numbers are disjoint sets.
4. *Concept Check* *True* or *false:* 9 is an element of $\{999\}$.

Determine whether each statement is true *or* false.

5. $1 \in \{6, 2, 5, 1\}$
6. $7 \notin \{1, 3, 5, 7\}$
7. $\{8, 11, 4\} = \{8, 11, 4, 0\}$
8. $\{0\} = \emptyset$

Answers:
1. $\{6, 8, 10, 12, 14, 16, 18, 20\}$
2. infinite
3. true
4. false
5. true
6. false
7. false
8. false

9. true
10. true
11. true
12. false

13. $\{2, 6, 9, 10\}$
14. $\{4, 8\}$
15. $\emptyset$
16. $\{1, 3, 5, 7\}$, or C
17. $\emptyset$
18. $\{2, 4, 6, 8\}$, or B
19. $\{1, 2, 3, 4, 6, 8\}$
20. $\{3, 4, 5, 6, 7, 8, 9, 10\}$
21. $\{1, 2, 3, 4, 5, 6, 7, 8, 9, 10\}$, or U

22. true

23. $-12, -6, -\sqrt{4}$ (or -2), 0, 6
24. $-12, -6, -0.9, -\sqrt{4}$ (or -2), $0, \frac{1}{8}, 6$

25. irrational number, real number
26. none of these
27. whole number, integer, rational number, real number
28. integer, rational number, real number

35. commutative
36. distributive
37. associative
38. inverse
39. identity
40. identity

41. 3550

Let $A = \{1, 3, 4, 5, 7, 8\}$, $B = \{2, 4, 6, 8\}$, $C = \{1, 3, 5, 7\}$, $D = \{1, 2, 3\}$, $E = \{3, 7\}$, *and* $U = \{1, 2, 3, 4, 5, 6, 7, 8, 9, 10\}$.

Determine whether each statement is true *or* false.

9. $\emptyset \subseteq A$ **10.** $E \subseteq C$ **11.** $D \not\subseteq B$ **12.** $E \not\subseteq A$

Refer to the sets given for Exercises 9–12. Specify each set.

13. A' **14.** $B \cap A$ **15.** $B \cap E$

16. $C \cup E$ **17.** $D \cap \emptyset$ **18.** $B \cup \emptyset$

19. $(C \cap D) \cup B$ **20.** $(D' \cap U) \cup E$ **21.** $\emptyset'$

22. *Concept Check* *True* or *false:* For all sets A and B, $(A \cap B) \subseteq (A \cup B)$.

For Exercises 23 and 24, let set $K = \{-12, -6, -0.9, -\sqrt{7}, -\sqrt{4}, 0, \frac{1}{8}, \frac{\pi}{4}, 6, \sqrt{11}\}$. *List all elements of K that belong to each set.*

23. Integers **24.** Rational numbers

For Exercises 25–28, choose all words from the following list that apply.

natural number whole number integer
rational number irrational number real number

25. $\frac{4\pi}{5}$ **26.** $\frac{\pi}{0}$ **27.** 0 **28.** $-\sqrt{36}$

Write each algebraic identity (true statement) as a complete English sentence without using the names of the variables. For instance, $z(x + y) = zx + zy$ *can be stated as "The multiple of a sum is the sum of the multiples."*

29. $\frac{1}{xy} = \frac{1}{x} \cdot \frac{1}{y}$ **30.** $a(b - c) = ab - ac$

31. $(ab)^n = a^n b^n$ **32.** $a^2 - b^2 = (a + b)(a - b)$

33. $\left(\frac{a}{b}\right)^n = \frac{a^n}{b^n}$ **34.** $|st| = |s| \cdot |t|$

Identify by name each property illustrated.

35. $8(5 + 9) = (5 + 9)8$ **36.** $4 \cdot 6 + 4 \cdot 12 = 4(6 + 12)$

37. $3 \cdot (4 \cdot 2) = (3 \cdot 4) \cdot 2$ **38.** $-8 + 8 = 0$

39. $(9 + p) + 0 = 9 + p$ **40.** $\frac{1}{\sqrt{2}} \cdot \frac{\sqrt{2}}{\sqrt{2}} = \frac{\sqrt{2}}{2}$

41. *Ages of College Undergraduates* The following table shows the age distribution of college students in 2008. In a random sample of 5000 such students, how many would you expect to be over 19?

Age	Percent
14–19	29
20–24	43
25–34	16
35 and older	12

Source: U.S. Census Bureau.

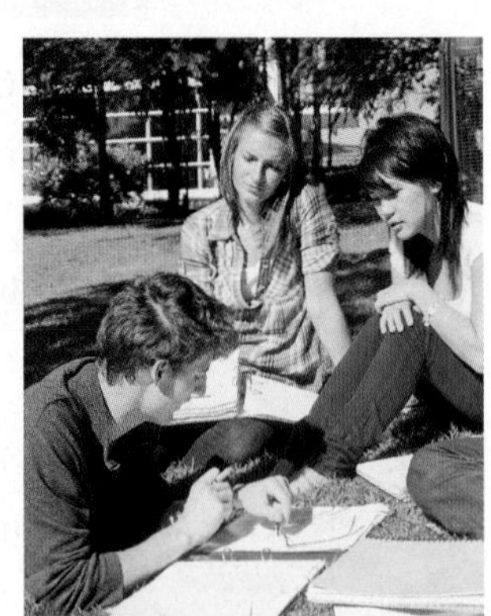

42. 29

43. 32 44. $-\frac{27}{4}$

45. $-\frac{37}{18}$ 46. $-\frac{37}{20}$

47. $-\frac{12}{5}$ 48. $-\frac{19}{42}$

49. -32 50. $\frac{26}{5}$

51. -13 52. $-\frac{5}{2}$

53. $7q^3 - 9q^2 - 8q + 9$
54. $y^6 - 18y^2 + 8y$
55. $16y^2 + 42y - 49$
56. $8r^2 + 26rs - 99s^2$
57. $9k^2 - 30km + 25m^2$
58. $16a^2 - 24ab + 9b^2$

59. $6m^2 - 3m + 5$
60. $9r + 4$
61. $3b - 8 + \frac{2}{b^2 + 4}$
62. $5m - 7 + \frac{10m}{m^2 - 2}$

63. $3(z - 4)^2(3z - 11)$
64. $z(7z - 9z^2 + 1)$
65. $(z - 8k)(z + 2k)$
66. $(r + 7p)(r - 6p)$
67. $6a^6(4a + 5b)(2a - 3b)$
68. $(3m + 1)(2m - 5)$
69. $(7m^4 + 3n)(7m^4 - 3n)$
70. $(13y^2 + 1)(13y^2 - 1)$
71. $3(9r - 10)(2r + 1)$
72. $8(y - 5z^2)(y^2 + 5yz^2 + 25z^4)$
73. $(x - 1)(y + 2)$
74. $(5p + 3q)(3m - 2n)$

75. $(3x - 4)(9x - 34)$
76. $(7x - 8)^2(121 - 56x)$

42. *Counting Marshmallows* In early 2011, there were media reports about students providing a correction to the following question posed on boxes of Swiss Miss Chocolate: *On average, how many mini-marshmallows are in one serving?*

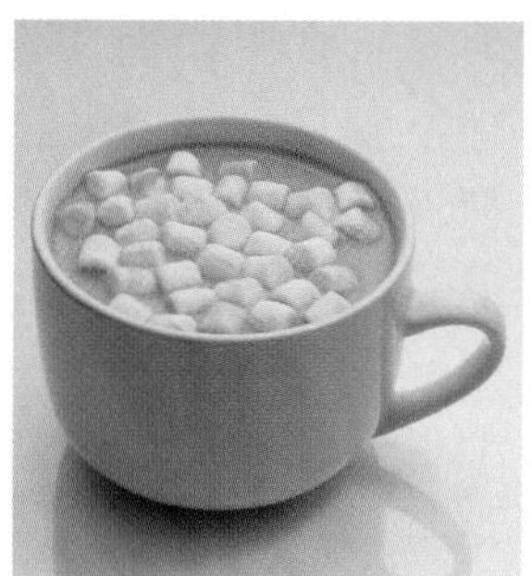

$$3 + 2 \times 4 \div 2 - 3 \times 7 - 4 + 47 = \_\_\_\_\_\_$$

The company provided 92 as the answer. What is the *correct* calculation provided by the students? (*Source:* Swiss Miss Chocolate box.)

Simplify each expression.

43. $(-4 - 1)(-3 - 5) - 2^3$

44. $(6 - 9)(-2 - 7) \div (-4)$

45. $\left(-\frac{5}{9} - \frac{2}{3}\right) - \frac{5}{6}$

46. $\left(-\frac{2^3}{5} - \frac{3}{4}\right) - \left(-\frac{1}{2}\right)$

47. $\dfrac{6(-4) - 3^2(-2)^3}{-5[-2 - (-6)]}$

48. $\dfrac{(-7)(-3) - (-2^3)(-5)}{(-2^2 - 2)(-1 - 6)}$

Evaluate each expression for $a = -1$, $b = -2$, and $c = 4$.

49. $-c(2a - 5b)$

50. $(a - 2) \div 5 \cdot b + c$

51. $\dfrac{9a + 2b}{a + b + c}$

52. $\dfrac{3|b| - 4|c|}{|ac|}$

Perform the indicated operations.

53. $(3q^3 - 9q^2 + 6) + (4q^3 - 8q + 3)$

54. $2(3y^6 - 9y^2 + 2y) - (5y^6 - 4y)$

55. $(8y - 7)(2y + 7)$

56. $(2r + 11s)(4r - 9s)$

57. $(3k - 5m)^2$

58. $(4a - 3b)^2$

Perform each division.

59. $\dfrac{30m^3 - 9m^2 + 22m + 5}{5m + 1}$

60. $\dfrac{72r^2 + 59r + 12}{8r + 3}$

61. $\dfrac{3b^3 - 8b^2 + 12b - 30}{b^2 + 4}$

62. $\dfrac{5m^3 - 7m^2 + 14}{m^2 - 2}$

Factor as completely as possible.

63. $3(z - 4)^2 + 9(z - 4)^3$

64. $7z^2 - 9z^3 + z$

65. $z^2 - 6zk - 16k^2$

66. $r^2 + rp - 42p^2$

67. $48a^8 - 12a^7b - 90a^6b^2$

68. $6m^2 - 13m - 5$

69. $49m^8 - 9n^2$

70. $169y^4 - 1$

71. $6(3r - 1)^2 + (3r - 1) - 35$

72. $8y^3 - 1000z^6$

73. $xy + 2x - y - 2$

74. $15mp + 9mq - 10np - 6nq$

Factor each expression. (These expressions arise in calculus from a technique called the product rule that is used to determine the shape of a curve.)

75. $(3x - 4)^2 + (x - 5)(2)(3x - 4)(3)$

76. $(5 - 2x)(3)(7x - 8)^2(7) + (7x - 8)^3(-2)$

77. $\frac{1}{2k^2(k-1)}$ 78. $\frac{3}{8r}$

79. $\frac{x+1}{x+4}$

80. $(3m+n)(3m-n)$

81. $\frac{p+6q}{p+q}$ 82. $\frac{37}{20y}$

83. $\frac{2m}{m-4}$, or $\frac{-2m}{4-m}$

84. $\frac{x+9}{(x-3)(x-1)(x+1)}$

85. $\frac{q+p}{pq-1}$

86. $\frac{3m^2+2m-12}{5(m+2)}$

87. $\frac{16}{25}$ 88. $\frac{1}{12}$

89. $-10z^8$ 90. $-\frac{16p^7}{q}$

91. 1 92. $\frac{y^6}{36x^4z^4}$

93. $-8y^{11}p$ 94. $\frac{1}{a^{23}}$

95. $\frac{1}{(p+q)^5}$ 96. $\frac{1}{p^2(m+n)}$

97. $-14r^{17/12}$ 98. $\frac{a^{11/8}}{b^{1/6}}$

99. $y^{1/2}$ 100. $\frac{n^{1/2}}{5m^{5/2}}$

101. $\frac{1}{p^2q^4}$

102. $-8m^{5/4}-4m^{-3/4}$, or $-8m^{5/4}-\frac{4}{m^{3/4}}$

103. $10\sqrt{2}$ 104. $2\sqrt[3]{2}$

105. $5\sqrt[4]{2}$ 106. $-\frac{4\sqrt{3}}{3}$

107. $-\frac{\sqrt[3]{50p}}{5p}$ 108. $\frac{8y^4\sqrt{2m}}{m^2}$

109. $\sqrt[12]{m}$ 110. $2q$

111. 66 112. $\frac{23\sqrt{5}}{30}$

113. $-9m\sqrt{2m}+5m\sqrt{m}$, or $m\left(-9\sqrt{2m}+5\sqrt{m}\right)$

114. $\frac{7+\sqrt{3}}{23}$

115. $\frac{6(3+\sqrt{2})}{7}$

116. $\frac{k(\sqrt{k}+3)}{k-9}$

In Exercises 117–125, we give only the corrected right-hand sides of the equations.

117. x^3+5x 118. -9

Perform the indicated operations.

77. $\frac{k^2+k}{8k^3}\cdot\frac{4}{k^2-1}$

78. $\frac{3r^3-9r^2}{r^2-9}\div\frac{8r^3}{r+3}$

79. $\frac{x^2+x-2}{x^2+5x+6}\div\frac{x^2+3x-4}{x^2+4x+3}$

80. $\frac{27m^3-n^3}{3m-n}\div\frac{9m^2+3mn+n^2}{9m^2-n^2}$

81. $\frac{p^2-36q^2}{p^2-12pq+36q^2}\cdot\frac{p^2-5pq-6q^2}{p^2+2pq+q^2}$

82. $\frac{1}{4y}+\frac{8}{5y}$

83. $\frac{m}{4-m}+\frac{3m}{m-4}$

84. $\frac{3}{x^2-4x+3}-\frac{2}{x^2-1}$

85. $\frac{p^{-1}+q^{-1}}{1-(pq)^{-1}}$

86. $\frac{3+\frac{2m}{m^2-4}}{\frac{5}{m-2}}$

Simplify each expression. Write the answer with only positive exponents. Assume all variables represent positive real numbers.

87. $\left(-\frac{5}{4}\right)^{-2}$

88. $3^{-1}-4^{-1}$

89. $(5z^3)(-2z^5)$

90. $(8p^2q^3)(-2p^5q^{-4})$

91. $(-6p^5w^4m^{12})^0$

92. $(-6x^2y^{-3}z^2)^{-2}$

93. $\frac{-8y^7p^{-2}}{y^{-4}p^{-3}}$

94. $\frac{a^{-6}(a^{-8})}{a^{-2}(a^{11})}$

95. $\frac{(p+q)^4(p+q)^{-3}}{(p+q)^6}$

96. $\frac{[p^2(m+n)^3]^{-2}}{p^{-2}(m+n)^{-5}}$

97. $(7r^{1/2})(2r^{3/4})(-r^{1/6})$

98. $(a^{3/4}b^{2/3})(a^{5/8}b^{-5/6})$

99. $\frac{y^{5/3}\cdot y^{-2}}{y^{-5/6}}$

100. $\left(\frac{25m^3n^5}{m^{-2}n^6}\right)^{-1/2}$

101. $\frac{(p^{15}q^{12})^{-4/3}}{(p^{24}q^{16})^{-3/4}}$

102. Simplify the product $-m^{3/4}(8m^{1/2}+4m^{-3/2})$. Assume the variable represents a positive real number.

Simplify. Assume all variables represent positive real numbers.

103. $\sqrt{200}$

104. $\sqrt[3]{16}$

105. $\sqrt[4]{1250}$

106. $-\sqrt{\frac{16}{3}}$

107. $-\sqrt[3]{\frac{2}{5p^2}}$

108. $\sqrt{\frac{2^7y^8}{m^3}}$

109. $\sqrt[4]{\sqrt[3]{m}}$

110. $\frac{\sqrt[4]{8p^2q^5}\cdot\sqrt[4]{2p^3q}}{\sqrt[4]{p^5q^2}}$

111. $(\sqrt[3]{2}+4)(\sqrt[3]{2^2}-4\sqrt[3]{2}+16)$

112. $\frac{3}{\sqrt{5}}-\frac{2}{\sqrt{45}}+\frac{6}{\sqrt{80}}$

113. $\sqrt{18m^3}-3m\sqrt{32m}+5\sqrt{m^3}$

114. $\frac{2}{7-\sqrt{3}}$

115. $\frac{6}{3-\sqrt{2}}$

116. $\frac{k}{\sqrt{k}-3}$

Concept Check *Correct each* ***INCORRECT*** *statement by changing the right-hand side of the equation.*

117. $x(x^2+5)=x^3+5$

118. $-3^2=9$

119. m^6
120. $9xy$
121. $\frac{a}{2b}$
122. $\frac{mn}{r^2}$
123. $(-2)^{-3}$
124. 5^2
125. $\frac{7b}{8b + 7a}$

119. $(m^2)^3 = m^5$

120. $(3x)(3y) = 3xy$

121. $\dfrac{\left(\frac{a}{b}\right)}{2} = \dfrac{2a}{b}$

122. $\dfrac{m}{r} \cdot \dfrac{n}{r} = \dfrac{mn}{r}$

123. $\dfrac{1}{(-2)^3} = 2^{-3}$

124. $(-5)^2 = -5^2$

125. $\left(\dfrac{8}{7} + \dfrac{a}{b}\right)^{-1} = \dfrac{7}{8} + \dfrac{b}{a}$

Chapter R Test

[R.1]
1. false
2. true
3. false
4. true

[R.2]
5. (a) $-13, -\frac{12}{4}$ (or -3), 0, $\sqrt{49}$ (or 7)
(b) $-13, -\frac{12}{4}$ (or -3), 0, $\frac{3}{5}$, 5.9, $\sqrt{49}$ (or 7)
(c) All are real numbers.
6. 4
7. (a) associative
(b) commutative
(c) distributive
(d) inverse
8. 109.6

[R.3]
9. $11x^2 - x + 2$
10. $36r^2 - 60r + 25$
11. $3t^3 + 5t^2 + 2t + 8$
12. $2x^2 - x - 5 + \frac{3}{x-5}$

Let $U = \{1, 2, 3, 4, 5, 6, 7, 8\}$, $A = \{1, 2, 3, 4, 5, 6\}$, $B = \{1, 3, 5\}$, $C = \{1, 6\}$, *and* $D = \{4\}$. *Tell whether each statement is* true *or* false.

1. $B' = \{2, 4, 6, 8\}$

2. $C \subseteq A$

3. $(B \cap C) \cup D = \{1, 3, 4, 5, 6\}$

4. $(A' \cup C) \cap B' = \{6, 7, 8\}$

5. Let $A = \left\{-13, -\frac{12}{4}, 0, \frac{3}{5}, \frac{\pi}{4}, 5.9, \sqrt{49}\right\}$. List the elements of A that belong to the given set.

(a) Integers **(b)** Rational numbers **(c)** Real numbers

6. Evaluate the expression $\left|\dfrac{x^2 + 2yz}{3(x + z)}\right|$ for $x = -2$, $y = -4$, and $z = 5$.

7. Identify each property illustrated. Let a, b, and c represent any real numbers.

(a) $a + (b + c) = (a + b) + c$

(b) $a + (c + b) = a + (b + c)$

(c) $a(b + c) = ab + ac$

(d) $a + [b + (-b)] = a + 0$

8. *Passer Rating for NFL Quarterbacks* Approximate the quarterback rating (to the nearest tenth) of Drew Brees of the New Orleans Saints during the 2008 regular season. He attempted 514 passes, completed 363, had 4388 total yards, threw for 34 touchdowns, and had 11 interceptions. (*Source: World Almanac and Book of Facts.*)

$$\text{Rating} = \frac{\left(250 \cdot \frac{C}{A}\right) + \left(1000 \cdot \frac{T}{A}\right) + \left(12.5 \cdot \frac{Y}{A}\right) + 6.25 - \left(1250 \cdot \frac{I}{A}\right)}{3},$$

where A = attempted passes, C = completed passes, T = touchdown passes, Y = yards gained passing, and I = interceptions.

In addition to the weighting factors appearing in the formula, the four category ratios are limited to nonnegative values with the following maximums.

$$0.775 \text{ for } \frac{C}{A}, \quad 0.11875 \text{ for } \frac{T}{A}, \quad 12.5 \text{ for } \frac{Y}{A}, \quad 0.095 \text{ for } \frac{I}{A}$$

Perform the indicated operations.

9. $(x^2 - 3x + 2) - (x - 4x^2) + 3x(2x + 1)$

10. $(6r - 5)^2$

11. $(t + 2)(3t^2 - t + 4)$

12. $\dfrac{2x^3 - 11x^2 + 28}{x - 5}$

(Modeling) Adjusted Poverty Threshold *The adjusted poverty threshold for a single person between the years 1995 and 2009 can be approximated by the polynomial*

$$4.872x^2 + 170.2x + 7787,$$

where $x = 0$ corresponds to 1995, $x = 1$ corresponds to 1996, and so on, and the amount is in dollars. According to this model, what was the adjusted poverty threshold, to the nearest dollar, in each given year? (Source: U.S. Census Bureau.)

13. 2000 **14.** 2008

Factor completely.

15. $6x^2 - 17x + 7$ **16.** $x^4 - 16$

17. $24m^3 - 14m^2 - 24m$ **18.** $x^3y^2 - 9x^3 - 8y^2 + 72$

19. $(a - b)^2 + 2(a - b)$ **20.** $1 - 27x^6$

Perform the indicated operations.

21. $\dfrac{5x^2 - 9x - 2}{30x^3 + 6x^2} \div \dfrac{x^4 - 3x^2 - 4}{2x^8 + 6x^7 + 4x^6}$ **22.** $\dfrac{x}{x^2 + 3x + 2} + \dfrac{2x}{2x^2 - x - 3}$

23. $\dfrac{a + b}{2a - 3} - \dfrac{a - b}{3 - 2a}$ **24.** $\dfrac{y - 2}{y - \dfrac{4}{y}}$

Simplify or evaluate as appropriate. Assume all variables represent positive real numbers.

25. $\sqrt{18x^5y^8}$ **26.** $\sqrt{32x} + \sqrt{2x} - \sqrt{18x}$

27. $(\sqrt{x} - \sqrt{y})(\sqrt{x} + \sqrt{y})$ **28.** $\dfrac{14}{\sqrt{11} - \sqrt{7}}$

29. $\left(\dfrac{x^{-2}y^{-1/3}}{x^{-5/3}y^{-2/3}}\right)^3$ **30.** $\left(-\dfrac{64}{27}\right)^{-2/3}$

31. *Concept Check* *True* or *false*: For all real numbers x, $\sqrt{x^2} = x$.

32. *(Modeling) Period of a Pendulum* The period t, in seconds, of the swing of a pendulum is given by the equation

$$t = 2\pi\sqrt{\frac{L}{32}},$$

where L is the length of the pendulum in feet. Find the period of a pendulum 3.5 ft long. Use a calculator.

13. \$8760
14. \$10,823

[R.4]
15. $(3x - 7)(2x - 1)$
16. $(x^2 + 4)(x + 2)(x - 2)$
17. $2m(4m + 3)(3m - 4)$
18. $(x - 2)(x^2 + 2x + 4) \cdot (y + 3)(y - 3)$
19. $(a - b)(a - b + 2)$
20. $(1 - 3x^2)(1 + 3x^2 + 9x^4)$

[R.5]
21. $\dfrac{x^4(x + 1)}{3(x^2 + 1)}$
22. $\dfrac{x(4x + 1)}{(x + 2)(x + 1)(2x - 3)}$
23. $\dfrac{2a}{2a - 3}$, or $\dfrac{-2a}{3 - 2a}$
24. $\dfrac{y}{y + 2}$

[R.6, R.7]
25. $3x^2y^4\sqrt{2x}$
26. $2\sqrt{2x}$
27. $x - y$
28. $\dfrac{7(\sqrt{11} + \sqrt{7})}{2}$
29. $\frac{y}{x}$
30. $\frac{9}{16}$
31. false
32. approximately 2.1 sec

1 Equations and Inequalities

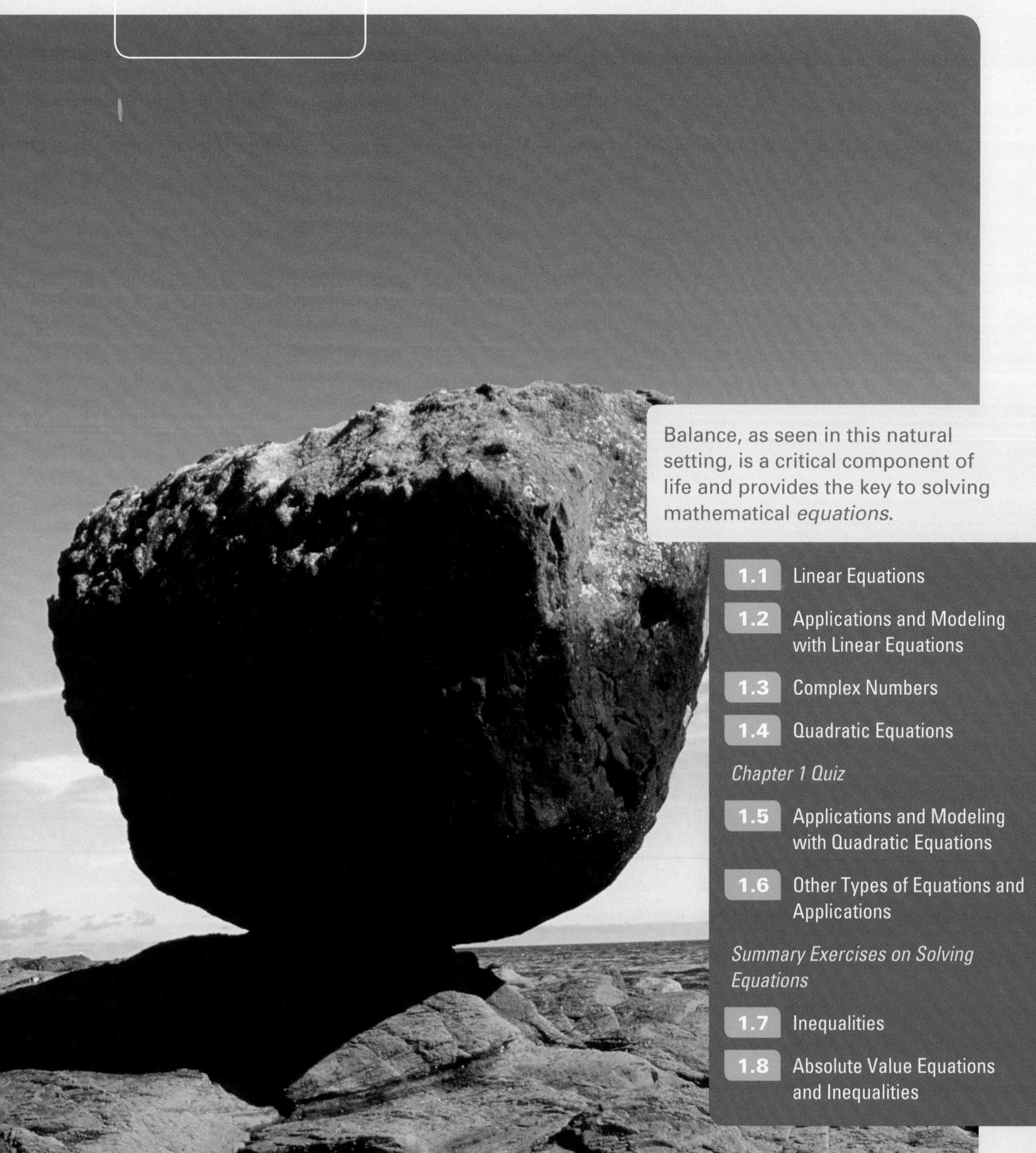

Balance, as seen in this natural setting, is a critical component of life and provides the key to solving mathematical *equations*.

1.1 Linear Equations

- Basic Terminology of Equations
- Solving Linear Equations
- Identities, Conditional Equations, and Contradictions
- Solving for a Specified Variable (Literal Equations)

Basic Terminology of Equations An **equation** is a statement that two expressions are equal.

$$x + 2 = 9, \quad 11x = 5x + 6x, \quad x^2 - 2x - 1 = 0 \quad \text{Equations}$$

To *solve* an equation means to find all numbers that make the equation a true statement. These numbers are the **solutions,** or **roots,** of the equation. A number that is a solution of an equation is said to *satisfy* the equation, and the solutions of an equation make up its **solution set.** Equations with the same solution set are **equivalent equations.** For example,

$$x = 4, \quad x + 1 = 5, \quad \text{and} \quad 6x + 3 = 27 \quad \text{are equivalent equations}$$

because they have the same solution set, $\{4\}$. However, the equations

$$x^2 = 9 \quad \text{and} \quad x = 3 \quad \text{are } \textit{not} \text{ equivalent,}$$

since the first has solution set $\{-3, 3\}$ while the solution set of the second is $\{3\}$.

One way to solve an equation is to rewrite it as a series of simpler equivalent equations using the **addition and multiplication properties of equality.**

Teaching Tip Emphasize this important "Golden Rule" of algebra: "Do unto one entire side of the equation (on one side of the equals symbol) as you do unto the other" to keep sides equal.

Addition and Multiplication Properties of Equality

Let a, b, and c represent real numbers.

If $a = b$, then $a + c = b + c$.

That is, the same number may be added to each side of an equation without changing the solution set.

If $a = b$ and $c \neq 0$, then $ac = bc$.

That is, each side of an equation may be multiplied by the same nonzero number without changing the solution set. (Multiplying each side by zero leads to $0 = 0$.)

These properties can be extended: The same number may be subtracted from each side of an equation, and each side may be divided by the same nonzero number, without changing the solution set.

Teaching Tip To help students recognize *linear* equations, list several equations and ask which ones are linear. Here are some possible examples.

(a) $3x + 5 = 0$ [yes]

(b) $(x - 2)(2x + 3) = 0$ [no]

(c) $2(4x + 1) = 3 - (2x + 4)$ [yes]

(d) $3\sqrt{x - 1} = 6$ [no]

Solving Linear Equations We use the properties of equality to solve *linear equations.*

Linear Equation in One Variable

A **linear equation in one variable** is an equation that can be written in the form

$$ax + b = 0,$$

where a and b are real numbers with $a \neq 0$.

A linear equation is a **first-degree equation** since the greatest degree of the variable is 1.

$$3x + \sqrt{2} = 0, \quad \frac{3}{4}x = 12, \quad 0.5(x + 3) = 2x - 6 \qquad \text{Linear equations}$$

$$\sqrt{x} + 2 = 5, \quad \frac{1}{x} = -8, \quad x^2 + 3x + 0.2 = 0 \qquad \text{Nonlinear equations}$$

Classroom Example 1
Solve $-4(3x - 5) = 3 - (8x + 7)$.

Answer: $\{6\}$

EXAMPLE 1 Solving a Linear Equation

Solve $3(2x - 4) = 7 - (x + 5)$.

SOLUTION

$$3(2x - 4) = 7 - (x + 5)$$

Be careful with signs.

$$6x - 12 = 7 - x - 5 \qquad \text{Distributive property (Section R.2)}$$

$$6x - 12 = 2 - x \qquad \text{Combine like terms. (Section R.3)}$$

$$6x - 12 + x = 2 - x + x \qquad \text{Add } x \text{ to each side.}$$

$$7x - 12 = 2 \qquad \text{Combine like terms.}$$

$$7x - 12 + 12 = 2 + 12 \qquad \text{Add 12 to each side.}$$

$$7x = 14 \qquad \text{Combine like terms.}$$

$$\frac{7x}{7} = \frac{14}{7} \qquad \text{Divide each side by 7.}$$

$$x = 2$$

Teaching Tip In the solution for **Example 1**, emphasize that the final line of the check does *not* give the answer, only a confirmation that the answer found is correct. Also, give an example of how a check can show that an answer is incorrect.

CHECK

A check of the solution is recommended.

$$3(2x - 4) = 7 - (x + 5) \qquad \text{Original equation}$$

$$3(2 \cdot 2 - 4) \stackrel{?}{=} 7 - (2 + 5) \qquad \text{Let } x = 2.$$

$$3(4 - 4) \stackrel{?}{=} 7 - (7) \qquad \text{Work inside the parentheses.}$$

$$0 = 0 \ ✓ \qquad \text{True}$$

Since replacing x with 2 results in a true statement, 2 is a solution of the given equation. The solution set is $\{2\}$.

✔ *Now Try Exercise 11.*

Classroom Example 2
Solve $\frac{3x + 6}{10} - \frac{1}{2}x = \frac{2}{5}x + \frac{33}{5}$.

Answer: $\{-10\}$

EXAMPLE 2 Solving a Linear Equation with Fractions

Solve $\frac{2x + 4}{3} + \frac{1}{2}x = \frac{1}{4}x - \frac{7}{3}$.

SOLUTION

$$\frac{2x + 4}{3} + \frac{1}{2}x = \frac{1}{4}x - \frac{7}{3}$$

Distribute to *all* terms within the parentheses.

$$12\left(\frac{2x + 4}{3} + \frac{1}{2}x\right) = 12\left(\frac{1}{4}x - \frac{7}{3}\right) \qquad \text{Multiply by 12, the LCD of the fractions. (Section R.5)}$$

$$12\left(\frac{2x + 4}{3}\right) + 12\left(\frac{1}{2}x\right) = 12\left(\frac{1}{4}x\right) - 12\left(\frac{7}{3}\right) \qquad \text{Distributive property}$$

$$4(2x + 4) + 6x = 3x - 28 \qquad \text{Multiply.}$$

$$8x + 16 + 6x = 3x - 28 \qquad \text{Distributive property}$$

$$14x + 16 = 3x - 28 \qquad \text{Combine like terms.}$$

$$11x = -44 \qquad \text{Subtract } 3x. \text{ Subtract 16.}$$

$$x = -4 \qquad \text{Divide each side by 11.}$$

CHECK

$$\frac{2x+4}{3}+\frac{1}{2}x=\frac{1}{4}x-\frac{7}{3} \quad \text{Original equation}$$

$$\frac{2(-4)+4}{3}+\frac{1}{2}(-4)\stackrel{?}{=}\frac{1}{4}(-4)-\frac{7}{3} \quad \text{Let } x=-4.$$

$$\frac{-4}{3}+(-2)\stackrel{?}{=}-1-\frac{7}{3} \quad \text{Simplify.}$$

$$-\frac{10}{3}=-\frac{10}{3} \checkmark \quad \text{True}$$

The solution set is $\{-4\}$.

✔ *Now Try Exercise 19.*

Identities, Conditional Equations, and Contradictions An equation satisfied by every number that is a meaningful replacement for the variable is an **identity.**

$$3(x+1)=3x+3 \quad \text{Identity}$$

An equation that is satisfied by some numbers but not others is a **conditional equation.**

$$2x=4 \quad \text{Conditional equation}$$

The equations in **Examples 1 and 2** are conditional equations. An equation that has no solution is a **contradiction.**

$$x=x+1 \quad \text{Contradiction}$$

Classroom Example 3
Determine whether each equation is an *identity,* a *conditional equation,* or a *contradiction.* Give the solution set.

(a) $6x-9=4x+13$
(b) $10+14x=7(2x-5)$
(c) $-3(2x-1)+5x=3-x$

Answers:
(a) conditional equation; $\{11\}$
(b) contradiction; $\emptyset$
(c) identity; {all real numbers}

EXAMPLE 3 Identifying Types of Equations

Determine whether each equation is an *identity,* a *conditional equation,* or a *contradiction.* Give the solution set.

(a) $-2(x+4)+3x=x-8$ **(b)** $5x-4=11$ **(c)** $3(3x-1)=9x+7$

SOLUTION

(a) $-2(x+4)+3x=x-8$

$$-2x-8+3x=x-8 \quad \text{Distributive property}$$

$$x-8=x-8 \quad \text{Combine like terms.}$$

$$0=0 \quad \text{Subtract } x. \text{ Add } 8.$$

When a *true* statement such as $0=0$ results, the equation is an identity, and the solution set is **{all real numbers}.**

(b) $5x-4=11$

$$5x=15 \quad \text{Add 4 to each side.}$$

$$x=3 \quad \text{Divide each side by 5.}$$

This is a conditional equation, and its solution set is $\{3\}$.

(c) $3(3x-1)=9x+7$

$$9x-3=9x+7 \quad \text{Distributive property}$$

$$-3=7 \quad \text{Subtract } 9x.$$

When a *false* statement such as $-3=7$ results, the equation is a contradiction, and the solution set is the **empty set,** or **null set,** symbolized $\boldsymbol{\emptyset}$.

✔ *Now Try Exercises 29, 31, and 35.*

Teaching Tip Remind students that an equation with no solution, such as the one in **Example 3(c)**, has the empty (or null) set as its solution set.

Identifying Types of Linear Equations

1. If solving a linear equation leads to a true statement such as $0 = 0$, the equation is an **identity.** Its solution set is **{all real numbers}. (See Example 3(a).)**
2. If solving a linear equation leads to a single solution such as $x = 3$, the equation is **conditional.** Its solution set consists of a single element. **(See Example 3(b).)**
3. If solving a linear equation leads to a false statement such as $-3 = 7$, the equation is a **contradiction.** Its solution set is $\emptyset$. **(See Example 3(c).)**

Solving for a Specified Variable (Literal Equations) A formula is an example of a **literal equation** (an equation involving letters).

Classroom Example 4
Solve for the specified variable.
(a) $d = rt$, for t
(b) $S = kr^2 + kr\ell$, for k
(c) $11y + 8 = 2(4y + 5w) - 6z$, for y

Answers:
(a) $t = \frac{d}{r}$
(b) $k = \frac{S}{r^2 + r\ell}$
(c) $y = \frac{10w - 6z - 8}{3}$

EXAMPLE 4 Solving for a Specified Variable

Solve for the specified variable.

(a) $I = Prt$, for t **(b)** $A - P = Prt$, for P

(c) $3(2x - 5a) + 4b = 4x - 2$, for x

SOLUTION

(a) This is the formula for **simple interest** I on a principal amount of P dollars at an annual interest rate r for t years. Here, treat t as if it were the only variable, and the other variables as if they were constants.

$$I = Prt \quad \text{Goal: Isolate } t \text{ on one side.}$$

$$\frac{I}{Pr} = \frac{Prt}{Pr} \quad \text{Divide each side by } Pr.$$

$$\frac{I}{Pr} = t, \quad \text{or} \quad t = \frac{I}{Pr}$$

Teaching Tip When solving a formula for a specified variable, encourage students to circle or underline the specified variable.

(b) The formula $A = P(1 + rt)$, which can also be written $A - P = Prt$, gives the **future value,** or **maturity value,** A of P dollars invested for t years at annual simple interest rate r.

$$A - P = Prt \quad \text{Goal: Isolate } P\text{, the specified variable.}$$

$$A = P + Prt \quad \text{Transform so that all terms involving } P \text{ are on one side.}$$

$$A = P(1 + rt) \quad \text{Factor out } P. \text{ (Section R.4)}$$

Pay close attention to this step.

$$\frac{A}{1 + rt} = P, \quad \text{or} \quad P = \frac{A}{1 + rt} \quad \text{Divide by } 1 + rt.$$

(c)
$$3(2x - 5a) + 4b = 4x - 2 \quad \text{Solve for } x.$$
$$6x - 15a + 4b = 4x - 2 \quad \text{Distributive property}$$
$$6x - 4x = 15a - 4b - 2 \quad \text{Isolate the } x\text{-terms on one side.}$$
$$2x = 15a - 4b - 2 \quad \text{Combine like terms.}$$
$$x = \frac{15a - 4b - 2}{2} \quad \text{Divide each side by 2.}$$

✔ *Now Try Exercises 39, 47, and 49.*

Classroom Example 5
Jared North borrowed \$2580 to buy new kitchen appliances for his home. He will pay off the loan in 9 months at an annual simple interest rate of 6.0%. How much interest will he pay?

Answer: \$116.10

EXAMPLE 5 Applying the Simple Interest Formula

Becky Brugman borrowed \$5240 for new furniture. She will pay it off in 11 months at an annual simple interest rate of 4.5%. How much interest will she pay?

SOLUTION Use the simple interest formula $I = Prt$.

$$I = Prt = 5240(0.045)\left(\frac{11}{12}\right) = \$216.15$$

$P = 5240$, $r = 0.045$, and $t = \frac{11}{12}$ (year)

She will pay \$216.15 interest on her purchase.

Now Try Exercise 59.

1.1 Exercises

1. true 2. true
3. false 4. false
7. B
9. $\{-4\}$ 10. $\{-5\}$
11. $\{1\}$ 12. $\{-1\}$
13. $\{-\frac{2}{7}\}$ 14. $\{\frac{5}{12}\}$
15. $\{-\frac{7}{8}\}$ 16. $\{-\frac{6}{11}\}$
17. $\{-1\}$ 18. $\{-2\}$
19. $\{10\}$ 20. $\{-5\}$
21. $\{75\}$ 22. $\{3\}$
23. $\{0\}$ 24. $\{0\}$
25. $\{12\}$ 26. $\{-24\}$
27. $\{50\}$ 28. $\{20\}$
29. identity; {all real numbers}
30. identity; {all real numbers}

Concept Check In Exercises 1–4, decide whether each statement is true *or* false.

1. The solution set of $2x + 5 = x - 3$ is $\{-8\}$.
2. The equation $5(x - 8) = 5x - 40$ is an example of an identity.
3. The equations $x^2 = 4$ and $x + 2 = 4$ are equivalent equations.
4. It is possible for a linear equation to have exactly two solutions.
5. Explain the difference between an identity and a conditional equation.
6. Make a complete list of the steps needed to solve a linear equation. (Some equations will not require every step.)
7. *Concept Check* Which one is not a linear equation?
   A. $5x + 7(x - 1) = -3x$ **B.** $9x^2 - 4x + 3 = 0$
   C. $7x + 8x = 13x$ **D.** $0.04x - 0.08x = 0.40$
8. In solving the equation $3(2x - 8) = 6x - 24$, a student obtains the result $0 = 0$ and gives the solution set $\{0\}$. Is this correct? Explain.

Solve each equation. **See Examples 1 and 2.**

9. $5x + 4 = 3x - 4$
10. $9x + 11 = 7x + 1$
11. $6(3x - 1) = 8 - (10x - 14)$
12. $4(-2x + 1) = 6 - (2x - 4)$
13. $\frac{5}{6}x - 2x + \frac{4}{3} = \frac{5}{3}$
14. $\frac{7}{4} + \frac{1}{5}x - \frac{3}{2} = \frac{4}{5}x$
15. $3x + 5 - 5(x + 1) = 6x + 7$
16. $5(x + 3) + 4x - 3 = -(2x - 4) + 2$
17. $2[x - (4 + 2x) + 3] = 2x + 2$
18. $4[2x - (3 - x) + 5] = -6x - 28$
19. $\frac{1}{14}(3x - 2) = \frac{x + 10}{10}$
20. $\frac{1}{15}(2x + 5) = \frac{x + 2}{9}$
21. $0.2x - 0.5 = 0.1x + 7$
22. $0.01x + 3.1 = 2.03x - 2.96$
23. $-4(2x - 6) + 8x = 5x + 24 + x$
24. $-8(3x + 4) + 6x = 4(x - 8) + 4x$
25. $0.5x + \frac{4}{3}x = x + 10$
26. $\frac{2}{3}x + 0.25x = x + 2$
27. $0.08x + 0.06(x + 12) = 7.72$
28. $0.04(x - 12) + 0.06x = 1.52$

Determine whether each equation is an identity, *a* conditional equation, *or a* contradiction. *Give the solution set.* **See Example 3.**

29. $4(2x + 7) = 2x + 22 + 3(2x + 2)$
30. $\frac{1}{2}(6x + 20) = x + 4 + 2(x + 3)$

31. conditional equation; $\{0\}$
32. conditional equation; $\{0\}$
33. identity; $\{\text{all real numbers}\}$
34. identity; $\{\text{all real numbers}\}$
35. contradiction; $\emptyset$
36. contradiction; $\emptyset$

39. $l = \frac{V}{wh}$
40. $P = \frac{I}{rt}$
41. $c = P - a - b$
42. $w = \frac{P - 2l}{2}$, or $w = \frac{P}{2} - l$
43. $B = \frac{2\mathcal{A} - hb}{h}$, or $B = \frac{2\mathcal{A}}{h} - b$
44. $h = \frac{2\mathcal{A}}{B + b}$
45. $h = \frac{S - 2\pi r^2}{2\pi r}$, or $h = \frac{S}{2\pi r} - r$
46. $g = \frac{2s}{t^2}$
47. $h = \frac{S - 2lw}{2w + 2l}$

Answers in Exercises 49–58 exist in equivalent forms.
49. $x = -3a + b$
50. $x = 2a + 5c$
51. $x = \frac{3a + b}{3 - a}$
52. $x = \frac{4a - 3b}{b + a}$
53. $x = \frac{3 - 3a}{a^2 - a - 1}$
54. $x = \frac{1 - 2a^2}{1 - 4a}$
55. $x = \frac{2a^2}{a^2 + 3}$
56. $x = \frac{a^2 + b^2}{b - a}$
57. $x = \frac{m + 4}{2m + 5}$
58. $x = -\frac{3k + 1}{5k + 2}$

59. (a) \$63 (b) \$3213
60. (a) \$33,449.25 (b) \$2549.25

31. $2(x - 8) = 3x - 16$

32. $-8(x + 5) = -8x - 5(x + 8)$

33. $0.3(x + 2) - 0.5(x + 2) = -0.2x - 0.4$

34. $-0.6(x - 5) + 0.8(x - 6) = 0.2x - 1.8$

35. $4(x + 7) = 2(x + 12) + 2(x + 1)$

36. $-6(2x + 1) - 3(x - 4) = -15x + 1$

37. A student claims that the equation $5x = 4x$ is a contradiction, since dividing both sides by x leads to $5 = 4$, a false statement. Explain why the student is incorrect.

38. If $k \neq 0$, is the equation $x + k = x$ a contradiction, a conditional equation, or an identity? Explain.

Solve each formula for the indicated variable. Assume that the denominator is not 0 if variables appear in the denominator. **See Examples 4(a) and (b).**

39. $V = lwh$, for l (volume of a rectangular box)

40. $I = Prt$, for P (simple interest)

41. $P = a + b + c$, for c (perimeter of a triangle)

42. $P = 2l + 2w$, for w (perimeter of a rectangle)

43. $\mathcal{A} = \frac{1}{2}h(B + b)$, for B (area of a trapezoid)

44. $\mathcal{A} = \frac{1}{2}h(B + b)$, for h (area of a trapezoid)

45. $S = 2\pi rh + 2\pi r^2$, for h (surface area of a right circular cylinder)

46. $s = \frac{1}{2}gt^2$, for g (distance traveled by a falling object)

47. $S = 2lw + 2wh + 2hl$, for h (surface area of a rectangular box)

48. Refer to **Exercise 45.** Why is it not possible to solve this formula for r using the methods of this section?

Solve each equation for x. **See Example 4(c).**

49. $2(x - a) + b = 3x + a$

50. $5x - (2a + c) = 4(x + c)$

51. $ax + b = 3(x - a)$

52. $4a - ax = 3b + bx$

53. $\frac{x}{a - 1} = ax + 3$

54. $\frac{x - 1}{2a} = 2x - a$

55. $a^2x + 3x = 2a^2$

56. $ax + b^2 = bx - a^2$

57. $3x = (2x - 1)(m + 4)$

58. $-x = (5x + 3)(3k + 1)$

Work each problem. **See Example 5.**

59. *Simple Interest* Elmer Velasquez borrowed \$3150 from his brother Julio to pay for books and tuition. He agreed to repay Julio in 6 months with simple annual interest at 4%.
   (a) How much will the interest amount to?
   (b) What amount must Elmer pay Julio at the end of the 6 months?

60. *Simple Interest* Levada Qualls borrows \$30,900 from her bank to open a florist shop. She agrees to repay the money in 18 months with simple annual interest of 5.5%.
   (a) How much must she pay the bank in 18 months?
   (b) How much of the amount in part (a) is interest?

61. 104°F
62. 392°F
63. 10°C
64. 25°C
65. 37.8°C
66. 176.7°C
67. 463.9°C
68. −128.9°F
69. −13.9°C
70. 82.6°F

Celsius and Fahrenheit Temperatures In the metric system of weights and measures, temperature is measured in degrees Celsius (°C) instead of degrees Fahrenheit (°F). To convert between the two systems, we use the equations

$$C = \frac{5}{9}(F - 32) \quad \text{and} \quad F = \frac{9}{5}C + 32.$$

In each exercise, convert to the other system. Round answers to the nearest tenth of a degree if necessary.

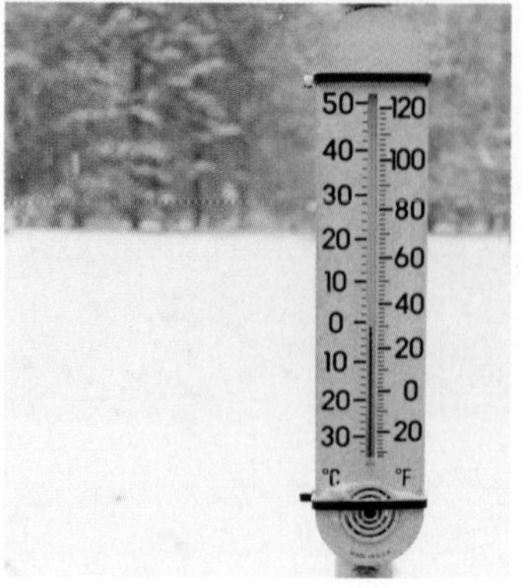

61. 40°C
62. 200°C
63. 50°F
64. 77°F
65. 100°F
66. 350°F

Work each problem.

67. *Temperature of Venus* Venus is the hottest planet with a surface temperature of 867°F. What is this temperature in Celsius? (*Source: World Almanac and Book of Facts.*)

68. *Temperature at Soviet Antarctica Station* A record low temperature of −89.4°C was recorded at the Soviet Antarctica Station of Vostok on July 21, 1983. Find the corresponding Fahrenheit temperature. (*Source: World Almanac and Book of Facts.*)

69. *Temperature in Montreal* The average low temperature in Montreal, Canada, for the date January 30 is 7°F. What is the corresponding Celsius temperature to the nearest tenth of a degree? (*Source:* www.wunderground.com)

70. *Temperature in Haiti* The average annual temperature in Port-au-Prince, Haiti, is approximately 28.1°C. What is the corresponding Fahrenheit temperature to the nearest tenth of a degree? (*Source:* www.climatetemp.info/haiti)

1.2 Applications and Modeling with Linear Equations

- Solving Applied Problems
- Geometry Problems
- Motion Problems
- Mixture Problems
- Modeling with Linear Equations

Solving Applied Problems One of the main reasons for learning mathematics is to be able use it to solve application problems. While there is no one method that enables us to solve all types of applied problems, the following six steps provide a useful guide.

Solving an Applied Problem

Step 1 **Read** the problem carefully until you understand what is given and what is to be found.

Step 2 **Assign a variable** to represent the unknown value, using diagrams or tables as needed. Write down what the variable represents. If necessary, express any other unknown values in terms of the variable.

Step 3 **Write an equation** using the variable expression(s).

Step 4 **Solve** the equation.

Step 5 **State the answer** to the problem. Does it seem reasonable?

Step 6 **Check** the answer in the words of the original problem.

Geometry Problems

Classroom Example 1
The length of a rectangle is 2 in. more than the width. If the length and width are each increased by 3 in., the perimeter of the new rectangle will be 4 in. less than 8 times the width of the original rectangle. Find the dimensions of the original rectangle.

Answer: width: 5 in.; length: 7 in.

EXAMPLE 1 Finding the Dimensions of a Square

If the length of each side of a square is increased by 3 cm, the perimeter of the new square is 40 cm more than twice the length of each side of the original square. Find the dimensions of the original square.

SOLUTION

Step 1 **Read** the problem. We must find the length of each side of the original square.

Step 2 **Assign a variable.** Since the length of a side of the original square is to be found, let the variable represent this length.

Let $x =$ the length of a side of the original square in centimeters.

The length of a side of the new square is 3 cm more than the length of a side of the old square.

Then $x + 3 =$ the length of a side of the new square.

See **Figure 1.** Now write a variable expression for the perimeter of the new square. The perimeter of a square is 4 times the length of a side.

Thus, $4(x + 3) =$ the perimeter of the new square.

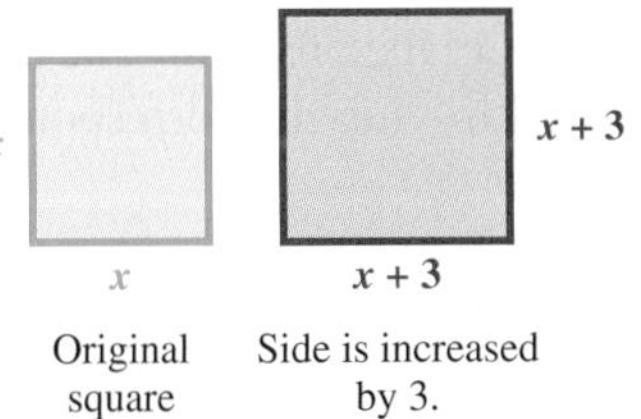

x and $x + 3$ are in centimeters.

Figure 1

Step 3 **Write an equation.** Translate the English sentence that follows into its equivalent algebraic equation.

The new perimeter	is	40	more than	twice the length of a side of the original square.
$4(x + 3)$	$=$	40	$+$	$2x$

Step 4 **Solve** the equation.

$$4x + 12 = 40 + 2x \quad \text{Distributive property (Section R.2)}$$

$$2x = 28 \quad \text{Subtract } 2x \text{ and } 12. \text{ (Section 1.1)}$$

$$x = 14 \quad \text{Divide by 2. (Section 1.1)}$$

Teaching Tip With geometric applications, students should not only draw diagrams but also label them with the appropriate variable expressions.

Step 5 **State the answer.** Each side of the original square measures 14 cm.

Step 6 **Check.** Go back to the words of the original problem to see that all necessary conditions are satisfied. The length of a side of the new square would be $14 + 3 = 17$ cm. The perimeter of the new square would be $4(17) = 68$ cm. Twice the length of a side of the original square would be $2(14) = 28$ cm. Since $40 + 28 = 68$, the answer checks.

Now Try Exercise 13.

Motion Problems

LOOKING AHEAD TO CALCULUS
In calculus the concept of the **definite integral** is used to find the distance traveled by an object traveling at a *non-constant* velocity.

PROBLEM-SOLVING HINT In a motion problem, the components *distance, rate,* and *time* are denoted by the letters d, r, and t, respectively. (The *rate* is also called the *speed* or *velocity*. Here, rate is understood to be constant.) These variables are related by the equation

$$d = rt, \quad \text{and its related forms} \quad r = \frac{d}{t} \quad \text{and} \quad t = \frac{d}{r}.$$

Classroom Example 2
Krissa drove to her grandmother's house for a weekend visit. She averaged 40 mph driving there on Friday. As she returned home on Sunday, traffic was lighter, so she was able to average 48 mph, and her driving time was 1 hr less. What is the distance between Krissa's home and that of her grandmother?

Answer: 240 mi

EXAMPLE 2 Solving a Motion Problem

Maria and Eduardo are traveling to a business conference. The trip takes 2 hr for Maria and 2.5 hr for Eduardo, since he lives 40 mi farther away. Eduardo travels 5 mph faster than Maria. Find their average rates.

SOLUTION

Step 1 **Read** the problem. We must find Maria's and Eduardo's average rates.

Step 2 **Assign a variable.** Since average rates are to be found, we let the variable represent one of these rates.

Let $\quad x =$ Maria's rate.

Because Eduardo travels 5 mph faster than Maria, we can express his average rate using the same variable.

Then $\quad x + 5 =$ Eduardo's rate.

The table below organizes the information given in the problem. The expressions in the last column were found by multiplying the corresponding rates and times.

	r	t	d
Maria	x	2	$2x$
Eduardo	$x + 5$	2.5	$2.5(x + 5)$

Summarize the given information in a table.

Use $d = rt$.

Step 3 **Write an equation.** Eduardo's distance traveled exceeds Maria's distance by 40 mi. Translate this relationship into its algebraic form.

$$\underbrace{2.5(x+5)}_{\text{Eduardo's distance}} \underbrace{=}_{\text{is}} \underbrace{2x + 40}_{\text{40 more than Maria's.}}$$

Step 4 **Solve.**

$$2.5x + 12.5 = 2x + 40 \quad \text{Distributive property}$$

$$0.5x = 27.5 \quad \text{Subtract } 2x \text{ and } 12.5.$$

$$x = 55 \quad \text{Divide by } 0.5.$$

Step 5 **State the answer.** Maria's rate of travel is 55 mph, and Eduardo's rate is

$$55 + 5 = 60 \text{ mph.}$$

Step 6 **Check.** The diagram below shows that the conditions of the problem are satisfied.

Distance traveled by Maria: $\quad 2(55) = 110$ mi

Distance traveled by Eduardo: $\quad 2.5(60) = 150$ mi

$150 - 110 = 40$

✔ *Now Try Exercise 19.*

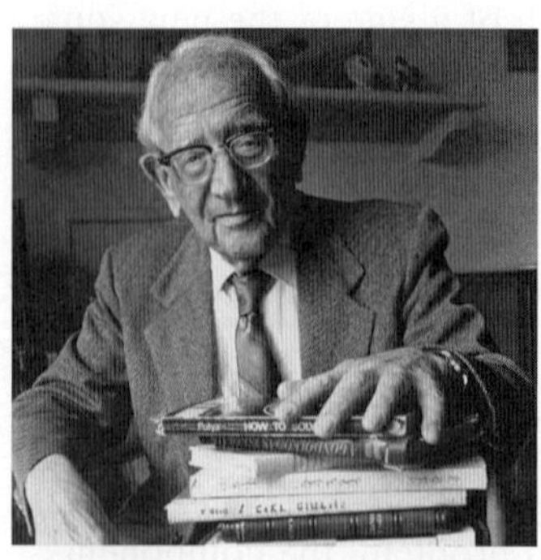

George Polya (1887–1985)

Polya, a native of Budapest, Hungary, wrote more than 250 papers and a number of books. He proposed a general outline for solving applied problems in his classic book, *How to Solve It.*

Mixture Problems Problems involving mixtures of two types of the same substance, salt solution, candy, and so on, often involve percent.

PROBLEM-SOLVING HINT In mixture problems involving solutions, the rate (percent) of concentration is multiplied by the quantity to get the amount of pure substance present. ***The concentration of the final mixture must be between the concentrations of the two solutions making up the mixture.***

Classroom Example 3
How many liters of a 25% antifreeze solution should be added to 5 L of a 10% solution to obtain a 15% solution?

Answer: 2.5 L

EXAMPLE 3 Solving a Mixture Problem

Lisa Harmon is a chemist. She needs a 20% solution of alcohol. She has a 15% solution on hand, as well as a 30% solution. How many liters of the 15% solution should she add to 3 L of the 30% solution to obtain her 20% solution?

SOLUTION

Step 1 **Read** the problem. We must find the required number of liters of 15% alcohol solution.

Step 2 **Assign a variable.**

Let $x =$ the number of liters of 15% solution to be added.

Figure 2 and the table show what is happening in the problem. The numbers in the last column were found by multiplying the strengths and the numbers of liters.

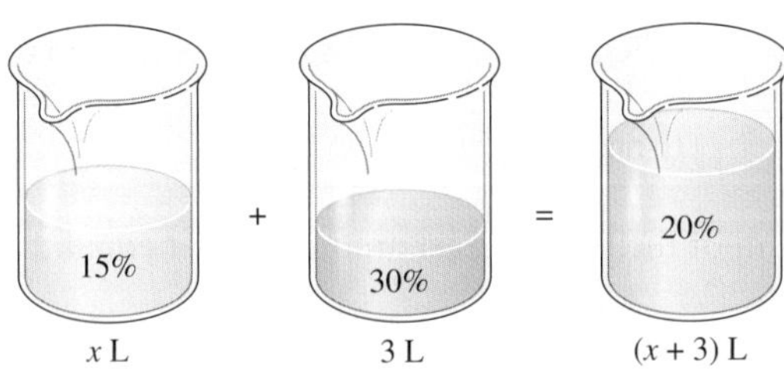

Figure 2

Strength	Liters of Solution	Liters of Pure Alcohol
15%	x	$0.15x$
30%	3	$0.30(3)$
20%	$x + 3$	$0.20(x + 3)$

Sum must equal

Step 3 **Write an equation.** The number of liters of pure alcohol in the 15% solution plus the number of liters in the 30% solution must equal the number of liters in the final 20% solution.

$$\underbrace{0.15x}_{\text{Liters of pure alcohol in 15\%}} + \underbrace{0.30(3)}_{\text{Liters of pure alcohol in 30\%}} = \underbrace{0.20(x+3)}_{\text{Liters of pure alcohol in 20\%}}$$

Step 4 **Solve.**

$$0.15x + 0.30(3) = 0.20(x + 3)$$

$$0.15x + 0.90 = 0.20x + 0.60 \quad \text{Distributive property}$$

$$0.30 = 0.05x \quad \text{Subtract 0.60 and } 0.15x.$$

$$6 = x \quad \text{Divide by 0.05.}$$

Step 5 **State the answer.** Thus, 6 L of 15% solution should be mixed with 3 L of 30% solution, giving $6 + 3 = 9$ L of 20% solution.

Step 6 **Check.** The answer checks, since the amount of alcohol in the two solutions is equal to the amount of alcohol in the mixture.

$$0.15(6) + 0.9 = 0.9 + 0.9 = 1.8 \quad \text{Solutions}$$

$$0.20(6 + 3) = 0.20(9) = 1.8 \quad \text{Mixture}$$

Now Try Exercise 29.

PROBLEM-SOLVING HINT In mixed investment problems, multiply each principal by the interest rate and the time in years to find the amount of interest earned.

Classroom Example 4
Last year, Owen Reid earned a total of \$784 in interest from two investments. He invested a total of \$28,000, part of it at 2.4% and the rest at 3.1%. How much did he invest at each rate?

Answer: \$12,000 at 2.4%; \$16,000 at 3.1%

EXAMPLE 4 Solving an Investment Problem

An artist has sold a painting for \$410,000. He needs some of the money in 6 months and the rest in 1 yr. He can get a Treasury bond for 6 months at 2.65% and one for a year at 2.91%. His broker tells him the two investments will earn a total of \$8761. How much should be invested at each rate to obtain that amount of interest?

SOLUTION

Step 1 **Read** the problem. We must find the amount to be invested at each rate.

Step 2 **Assign a variable.**

Let $x =$ the dollar amount to be invested for 6 months at 2.65%.

$410{,}000 - x =$ the dollar amount to be invested for 1 yr at 2.91%.

Invested Amount	Interest Rate (%)	Time (in years)	Interest Earned
x	2.65	0.5	$x(0.0265)(0.5)$
$410{,}000 - x$	2.91	1	$(410{,}000 - x)(0.0291)(1)$

Summarize this information in a table using the formula $I = Prt$.

Step 3 **Write an equation.** The sum of the two interest amounts must equal the total interest earned.

$$\underbrace{0.5x(0.0265)}_{\text{Interest from 2.65\% investment}} + \underbrace{0.0291(410{,}000 - x)}_{\text{Interest from 2.91\% investment}} = \underbrace{8761}_{\text{Total interest}}$$

Step 4 **Solve.**

$$0.5x(0.0265) + 0.0291(410{,}000 - x) = 8761$$

$$0.01325x + 11{,}931 - 0.0291x = 8761 \quad \text{Distributive property}$$

$$11{,}931 - 0.01585x = 8761 \quad \text{Combine like terms.}$$

$$-0.01585x = -3170 \quad \text{Subtract 11,931.}$$

$$x = 200{,}000 \quad \text{Divide by } -0.01585.$$

Step 5 **State the answer.** The artist should invest \$200,000 at 2.65% for 6 months and

$$\$410{,}000 - \$200{,}000 = \$210{,}000$$

at 2.91% for 1 yr to earn \$8761 in interest.

Step 6 **Check.** The 6-month investment earns

$$\$200{,}000(0.0265)(0.5) = \$2650,$$

while the 1-yr investment earns

$$\$210{,}000(0.0291)(1) = \$6111.$$

The total amount of interest earned is

$$\$2650 + \$6111 = \$8761, \quad \text{as required.}$$

✔ ***Now Try Exercise 35.***

Modeling with Linear Equations A **mathematical model** is an equation (or inequality) that describes the relationship between two quantities. A **linear model** is a linear equation. The next example shows how a linear model is applied.

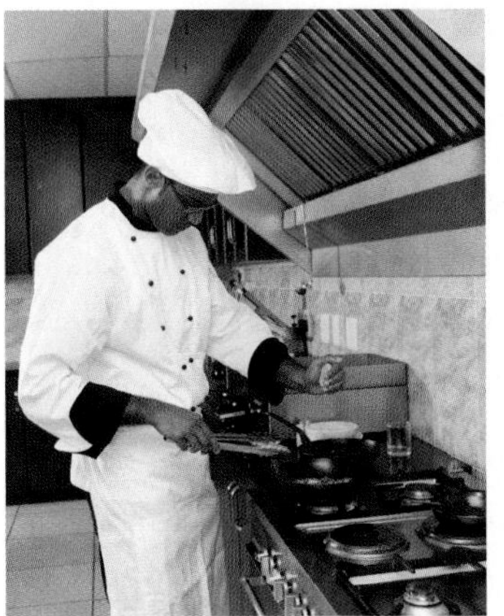

Classroom Example 5
Refer to the model in **Example 5.** What flow F must a range hood have to remove 70% of the contaminants from the air?

Answer: approximately 59.26 L of air per second

EXAMPLE 5 Modeling Prevention of Indoor Pollutants

If a vented range hood removes contaminants such as carbon monoxide and nitrogen dioxide from the air at a rate of F liters of air per second, then the percent P of contaminants that are also removed from the surrounding air can be modeled by the linear equation

$$P = 1.06F + 7.18, \quad \text{where } 10 \leq F \leq 75.$$

What flow F must a range hood have to remove 50% of the contaminants from the air? (*Source: Proceedings of the Third International Conference on Indoor Air Quality and Climate.*)

SOLUTION Replace P with 50 in the linear model, and solve for F.

$$P = 1.06F + 7.18 \quad \text{Given model}$$
$$50 = 1.06F + 7.18 \quad \text{Let } P = 50.$$
$$42.82 = 1.06F \quad \text{Subtract 7.18.}$$
$$F \approx 40.40 \quad \text{Divide by 1.06.}$$

Therefore, to remove 50% of the contaminants, the flow rate must be approximately 40.40 L of air per second.

Now Try Exercise 41.

Classroom Example 6
Refer to the linear model in **Example 6.**
(a) What were the per capita health care expenditures in the year 2007?
(b) When did the per capita expenditures reach \$8500?

Answers:
(a) \$6913 (b) 2011

EXAMPLE 6 Modeling Health Care Costs

The projected per capita health care expenditures in the United States, where y is in dollars, and x is years after 2000, are given by the following linear equation.

$$y = 343x + 4512 \quad \text{Linear model}$$

(*Source:* Centers for Medicare and Medicaid Services.)

(a) What were the per capita health care expenditures in the year 2010?

(b) If this model continues to describe health care expenditures, when will the per capita expenditures reach \$9200?

SOLUTION In part (a) we are given information to determine a value for x and asked to find the corresponding value of y, whereas in part (b) we are given a value for y and asked to find the corresponding value of x.

(a) The year 2010 is 10 yr after the year 2000. Let $x = 10$ and find the value of y.

$$y = 343x + 4512 \quad \text{Given model}$$
$$y = 343(10) + 4512 \quad \text{Let } x = 10.$$
$$y = 7942 \quad \text{Multiply and then add.}$$

In 2010, the estimated per capita health care expenditures were \$7942.

(b) Let $y = 9200$ in the given model, and find the value of x.

$$9200 = 343x + 4512 \quad \text{Let } y = 9200.$$
$$4688 = 343x \quad \text{Subtract 4512.}$$
$$x \approx 13.7 \quad \text{Divide by 343.}$$

13 corresponds to $2000 + 13 = 2013$.

The x-value of 13.7 indicates that per capita health care expenditures are projected to reach \$9200 during the 13th year after 2000—that is, 2013.

Now Try Exercise 45.

1.2 Exercises

1. 20 mi
2. 90 L
3. \$40
4. \$70
5. A
6. D
7. D
8. A

9. 90 cm
10. length: 14 ft; width: 8 ft
11. 6 cm
12. 7.6 cm

Concept Check Exercises 1–8 should be done mentally. They will prepare you for some of the applications found in this exercise set.

1. If a train travels at 80 mph for 15 min, what is the distance traveled?

2. If 120 L of an acid solution is 75% acid, how much pure acid is there in the mixture?

3. If a person invests \$500 at 2% simple interest for 4 yr, how much interest is earned?

4. If a jar of coins contains 40 half-dollars and 200 quarters, what is the monetary value of the coins?

5. *Acid Mixture* Suppose two acid solutions are mixed. One is 26% acid and the other is 34% acid. Which one of the following concentrations cannot possibly be the concentration of the mixture?

   A. 24% **B.** 30% **C.** 31% **D.** 33%

6. *Sale Price* Suppose that a computer that originally sold for x dollars has been discounted 60%. Which one of the following expressions does not represent its sale price?

   A. $x - 0.60x$ **B.** $0.40x$ **C.** $\frac{4}{10}x$ **D.** $x - 0.60$

7. *Unknown Numbers* Consider the following problem.

   One number is 3 less than 6 times a second number. Their sum is 46. Find the numbers.

   If x represents the second number, which equation is correct for solving this problem?

   A. $46 - (x + 3) = 6x$ **B.** $(3 - 6x) + x = 46$
   C. $46 - (3 - 6x) = x$ **D.** $(6x - 3) + x = 46$

8. *Unknown Numbers* Consider the following problem.

   The difference between seven times a number and 9 is equal to five times the sum of the number and 2. Find the number.

   If x represents the number, which equation is correct for solving this problem?

   A. $7x - 9 = 5(x + 2)$ **B.** $9 - 7x = 5(x + 2)$
   C. $7x - 9 = 5x + 2$ **D.** $9 - 7x = 5x + 2$

***Note:* Geometry formulas can be found on the back inside cover of this book.**

Solve each problem. ***See Example 1.***

9. *Perimeter of a Rectangle* The perimeter of a rectangle is 294 cm. The width is 57 cm. Find the length.

10. *Perimeter of a Storage Shed* Michael Gomski must build a rectangular storage shed. He wants the length to be 6 ft greater than the width, and the perimeter will be 44 ft. Find the length and the width of the shed.

11. *Dimensions of a Puzzle Piece* A puzzle piece in the shape of a triangle has perimeter 30 cm. Two sides of the triangle are each twice as long as the shortest side. Find the length of the shortest side. (Side lengths in the figure are in centimeters.)

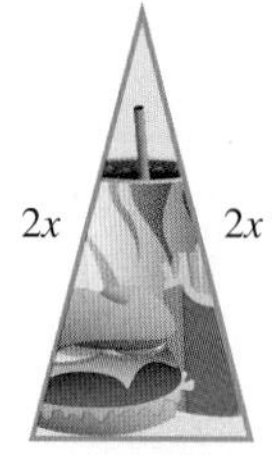

12. *Dimensions of a Label* The length of a rectangular label is 2.5 cm less than twice the width. The perimeter is 40.6 cm. Find the width. (Side lengths in the figure are in centimeters.)

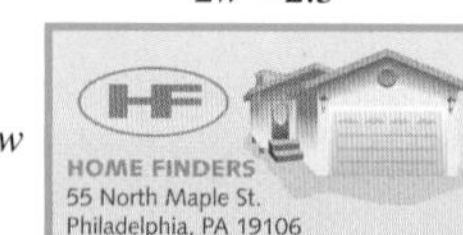

13. *Perimeter of a Plot of Land* The perimeter of a triangular plot of land is 2400 ft. The longest side is 200 ft less than twice the shortest. The middle side is 200 ft less than the longest side. Find the lengths of the three sides of the triangular plot.

14. *World Largest Ice Cream Cake* The world's largest ice cream cake, made at the Baxy ice cream factory in Beijing, China, on January 16, 2006, had length 5.9 ft greater than its width. Its perimeter was 51 ft. What were the length and width of this 8-ton cake? (*Sources:* www.chinadaily.com.cn, www.foodmall.org)

15. *World's Smallest Quran* The world's smallest Quran, published in Cairo, Egypt, in 1982 has width 0.42 cm shorter than its length. The book's perimeter is 5.96 cm. What are the width and length of this Quran, in centimeters? (*Source:* www.guinnessworldrecords.com)

16. *Cylinder Dimensions* A right circular cylinder has radius 6 in. and volume 144π in.^{3}. What is its height? (In the figure, h = height.)

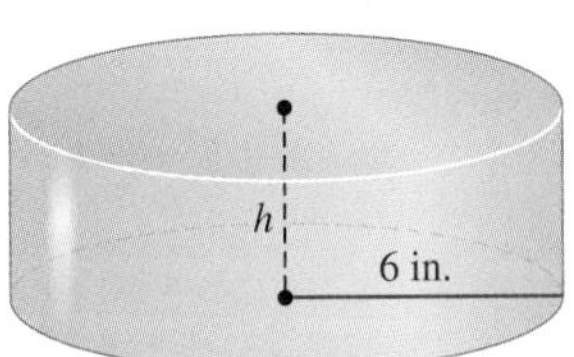

h is in inches.

17. *Recycling Bin Dimensions* A recycling bin is in the shape of a rectangular box. Find the height of the box if its length is 18 ft, its width is 8 ft, and its surface area is 496 ft^2. (In the figure, h = height. Assume that the given surface area includes that of the top lid of the box.)

18. *Concept Check* Which one or more of the following cannot be a correct equation to solve a geometry problem, if x represents the length of a rectangle? (*Hint:* Solve each equation and consider the solution.)

A. $2x + 2(x - 1) = 14$ **B.** $-2x + 7(5 - x) = 52$

C. $5(x + 2) + 5x = 10$ **D.** $2x + 2(x - 3) = 22$

13. 600 ft, 800 ft, 1000 ft
14. width: 9.8 ft; length: 15.7 ft
15. width: 1.28 cm; length: 1.7 cm
16. 4 in.
17. 4 ft
18. B, C
19. 50 mi
20. about 840 mi

Solve each problem. ***See Example 2.***

19. *Distance to an Appointment* Margaret drove to a business appointment at 50 mph. Her average speed on the return trip was 40 mph. The return trip took $\frac{1}{4}$ hr longer because of heavy traffic. How far did she travel to the appointment?

	r	t	d
Morning	50	x	
Afternoon	40	$x + \frac{1}{4}$	

20. *Distance between Cities* On a vacation, Elwyn averaged 50 mph traveling from Denver to Minneapolis. Returning by a different route that covered the same number of miles, he averaged 55 mph. What is the distance between the two cities if his total traveling time was 32 hr?

	r	t	d
Going	50	x	
Returning	55	$32 - x$	

21. 2.7 mi
22. 250 mph; 300 mph
23. 45 min
24. 25 min
25. 1 hr, 7 min, 34 sec; It is about $\frac{1}{2}$ the world record time.
26. 1 hr, 6 min, 48 sec; It is about $\frac{1}{2}$ the world record time.
27. 35 km per hr
28. about 6.2 mph
29. $7\frac{1}{2}$ gal
30. 120 mL

21. *Distance to Work* David gets to work in 20 min when he drives his car. Riding his bike (by the same route) takes him 45 min. His average driving speed is 4.5 mph greater than his average speed on his bike. How far does he travel to work?

22. *Speed of a Plane* Two planes leave Los Angeles at the same time. One heads south to San Diego, while the other heads north to San Francisco. The San Diego plane flies 50 mph slower than the San Francisco plane. In $\frac{1}{2}$ hr, the planes are 275 mi apart. What are their speeds?

23. *Running Times* Mary and Janet are running in the Apple Hill Fun Run. Mary runs at 7 mph, Janet at 5 mph. If they start at the same time, how long will it be before they are 1.5 mi apart?

24. *Running Times* If the run in **Exercise 23** has a staggered start, and Janet starts first, with Mary starting 10 min later, how long will it be before Mary catches up with Janet?

25. *Track Event Speeds* At the 2008 Summer Olympics in Beijing, China, Usain Bolt (Jamaica) set a new Olympic and world record in the 100-m dash with a time of 9.69 sec. If this pace could be maintained for an entire 26-mi marathon, what would his time be? How would this time compare to the fastest time for a marathon, which is 2 hr, 3 min, 59 sec, also set in 2008? (*Hint:* 1 m $\approx$ 3.281 ft.) (*Source: World Almanac and Book of Facts.*)

26. *Track Event Speeds* On August 19, 2009, at the World Track and Field Championship in Berlin, Usain Bolt broke his own record in the 100-m dash with a time of 9.58 sec. Refer to **Exercise 25** and answer the questions using Bolt's 2009 time. (*Source:* www.sports.espn.go.com)

27. *Boat Speed* Callie took 20 min to drive her boat upstream to water-ski at her favorite spot. Coming back later in the day, at the same boat speed, took her 15 min. If the current in that part of the river is 5 km per hr, what was her boat speed?

28. *Wind Speed* Joe traveled against the wind in a small plane for 3 hr. The return trip with the wind took 2.8 hr. Find the speed of the wind if the speed of the plane in still air is 180 mph.

Solve each problem. ***See Example 3.***

29. *Acid Mixture* How many gallons of a 5% acid solution must be mixed with 5 gal of a 10% solution to obtain a 7% solution?

Strength	Gallons of Solution	Gallons of Pure Acid
5%	x	
10%	5	
7%	$x + 5$	

30. *Acid Mixture* Marin Caswell needs 10% hydrochloric acid for a chemistry experiment. How much 5% acid should she mix with 60 mL of 20% acid to get a 10% solution?

Strength	mL of Solution	mL of Pure Acid
5%	x	
20%	60	
10%	$x + 60$	

31. 2 L
32. $2\frac{2}{3}$ gal
33. 4 mL
34. 7.2 L
35. short-term note: $100,000; long-term note: $140,000
36. $70,000 for land that made a profit; $50,000 for land that produced a loss
37. $10,000 at 2.5%; $20,000 at 3%
38. $20,000 at 3%; $80,000 at 2.75%
39. $50,000 at 1.5%; $90,000 at 4%
40. $20,000 at 3.25%; $14,560 at 1.75%
41. (a) $52 (b) $2500 (c) $5000

31. *Alcohol Mixture* Beau Glaser wishes to strengthen a mixture from 10% alcohol to 30% alcohol. How much pure alcohol should be added to 7 L of the 10% mixture?

32. *Alcohol Mixture* How many gallons of pure alcohol should be mixed with 20 gal of a 15% alcohol solution to obtain a mixture that is 25% alcohol?

33. *Saline Solution* How much water should be added to 8 mL of 6% saline solution to reduce the concentration to 4%?

34. *Acid Mixture* How much pure acid should be added to 18 L of 30% acid to increase the concentration to 50% acid?

Solve each problem. ***See Example 4.***

35. *Real Estate Financing* Cody Westmoreland wishes to sell a piece of property for $240,000. He wants the money to be paid off in two ways—a short-term note at 2% interest and a long-term note at 2.5%. Find the amount of each note if the total annual interest paid is $5500.

Note Amount	Interest Rate (%)	Time (in years)	Interest Paid
x	2	1	$x(0.02)(1)$
$240{,}000 - x$	2.5	1	$(240{,}000 - x)(0.025)(1)$

36. *Buying and Selling Land* Roger bought two plots of land for a total of $120,000. When he sold the first plot, he made a profit of 15%. When he sold the second, he lost 10%. His total profit was $5500. How much did he pay for each piece of land?

37. *Retirement Planning* In planning her retirement, Janet Karrenbrock deposits some money at 2.5% interest, with twice as much deposited at 3%. Find the amount deposited at each rate if the total annual interest income is $850.

38. *Investing a Building Fund* A church building fund has invested some money in two ways: part of the money at 3% interest and four times as much at 2.75%. Find the amount invested at each rate if the total annual income from interest is $2800.

39. *Lottery Winnings* Linda won $200,000 in a state lottery. She first paid income tax of 30% on the winnings. She invested some of the rest at 1.5% and some at 4%, earning $4350 interest per year. How much did she invest at each rate?

40. *Cookbook Royalties* Becky Schantz earned $48,000 from royalties on her cookbook. She paid a 28% income tax on these royalties. The balance was invested in two ways, some of it at 3.25% interest and some at 1.75%. The investments produced $904.80 interest per year. Find the amount invested at each rate.

(Modeling) Solve each problem. ***See Examples 5 and 6.***

41. *Warehouse Club Membership* Membership warehouse clubs offer shoppers low prices, along with rewards of cash back on club purchases. If the yearly fee for a warehouse club membership is $100 and the reward rate is 2% on club purchases for the year, then the linear equation

$$y = 100 - 0.02x$$

models the actual yearly cost of the membership y, in dollars. Here x represents the yearly amount of club purchases, also in dollars.

(a) Determine the actual yearly cost of the membership if club purchases for the year are $2400.

(b) What amount of club purchases would reduce the actual yearly cost of the membership to $50?

(c) How much would a member have to spend in yearly club purchases to reduce the yearly membership cost to $0?

42. (a) $26 (b) $3125
(c) For yearly club purchases over $3125, the model would yield a negative value for y, the actual yearly cost of the membership. Essentially, the cash-back reward has exceeded the initial fee of $50, creating a positive gain for the member.

43. (a) $F = 14{,}000x$
(b) about 1.9 hr

44. (a) $V = 900x$ (b) $A = 0.06x$
(c) $A = 2.4$ ach
(d) Ventilation should be increased by $3\frac{1}{3}$. (Smoking areas require more than triple the ventilation.)

42. *Warehouse Club Membership* Suppose that the yearly fee for a warehouse club membership is $50 and that the reward rate on club purchases for the year is 1.6%. Then the actual yearly cost of a membership y, in dollars, for an amount of yearly club purchases x, in dollars, can be modeled by the following linear equation.

$$y = 50 - 0.016x$$

(a) Determine the actual yearly cost of the membership if club purchases for the year are $1500.

(b) What amount of club purchases would reduce the actual yearly cost of the membership to $0?

(c) If club purchases for the year exceed $3125, how is the actual yearly membership cost affected?

43. *Indoor Air Pollution* Formaldehyde is an indoor air pollutant formerly found in plywood, foam insulation, and carpeting. When concentrations in the air reach 33 micrograms per cubic foot ($\mu g/ft^3$), eye irritation can occur. One square foot of new plywood could emit 140 μg per hr. (*Source:* A. Hines, *Indoor Air Quality & Control.*)

(a) A room has 100 ft^2 of new plywood flooring. Find a linear equation F that computes the amount of formaldehyde, in micrograms, emitted in x hours.

(b) The room contains 800 ft^3 of air and has no ventilation. Determine how long it would take for concentrations to reach 33 $\mu g/ft^3$.

44. *Classroom Ventilation* According to the American Society of Heating, Refrigerating and Air-Conditioning Engineers, Inc. (ASHRAE), a nonsmoking classroom should have a ventilation rate of 15 ft^3 per min for each person in the room.

(a) Write an equation that models the total ventilation V (in cubic feet per hour) necessary for a classroom with x students.

(b) A common unit of ventilation is air change per hour (ach). 1 ach is equivalent to exchanging all of the air in a room every hour. If x students are in a classroom having volume 15,000 ft^3, determine how many air exchanges per hour (A) are necessary to keep the room properly ventilated.

(c) Find the necessary number of ach (A) if the classroom has 40 students in it.

(d) In areas like bars and lounges that allow smoking, the ventilation rate should be increased to 50 ft^3 per min per person. Compared to classrooms, ventilation should be increased by what factor in heavy smoking areas?

45. *College Enrollments* The graph shows the projections in total enrollment at degree-granting institutions from fall 2009 to fall 2018.

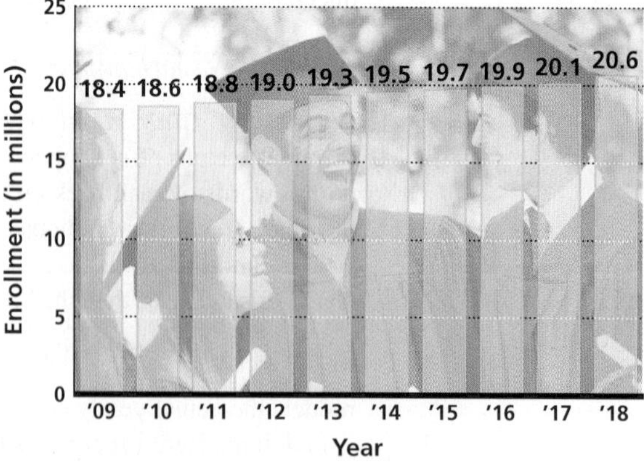

Source: U.S. Department of Education, National Center for Education Statistics.

The following linear model provides the approximate enrollment, in millions, between the years 2009 and 2018, where $x = 0$ corresponds to 2009, $x = 1$ to 2010, and so on, and y is in millions of students.

$$y = 0.2309x + 18.35$$

45. (a) 19.5 million
(b) 2016
(c) They are quite close.
(d) 15.6 million

(a) Use the model to determine projected enrollment for fall 2014.
(b) Use the model to determine the year in which enrollment is projected to reach 20 million.
(c) How do your answers to parts (a) and (b) compare to the corresponding values shown in the graph?
(d) The actual enrollment in fall 1997 was 14.5 million. The model here is based on data from 2009 to 2018. If you were to use the model for 1997, what would the projected enrollment be?
(e) Why do you think there is such a discrepancy between the actual value and the value based on the model in part (d)? Discuss the pitfalls of using the model to predict enrollment for years preceding 2009.

1.3 Complex Numbers

- Basic Concepts of Complex Numbers
- Operations on Complex Numbers

Basic Concepts of Complex Numbers The set of real numbers does not include all the numbers needed in algebra. For example, there is no real number solution of the equation

$$x^2 = -1,$$

since no real number, when squared, gives -1. To extend the real number system to include solutions of equations of this type, the number i is defined to have the following property.

The Imaginary Unit *i*

$$i = \sqrt{-1}, \quad \text{and therefore,} \quad i^2 = -1.$$

(Note that $-i$ is also a square root of -1.)

Square roots of negative numbers were not incorporated into an integrated number system until the 16th century. They were then used as solutions of equations and later (in the 18th century) in surveying. Today, such numbers are used extensively in science and engineering.

Complex numbers are formed by adding real numbers and multiples of i.

Complex Number

If a and b are real numbers, then any number of the form $a + bi$ is a **complex number.** In the complex number $a + bi$, a is the **real part** and b is the **imaginary part.***

Two complex numbers $a + bi$ and $c + di$ are equal provided that their real parts are equal and their imaginary parts are equal; that is, they are equal if and only if $a = c$ and $b = d$.

* In some texts, the term bi is defined to be the imaginary part.

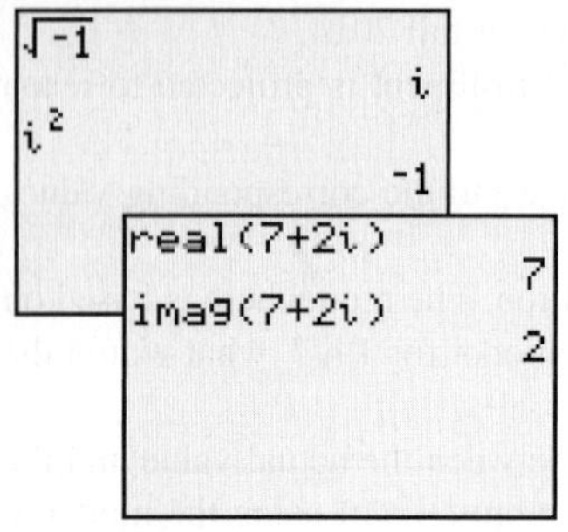

The calculator is in complex number mode. The top screen supports the definition of i. The bottom screen shows how the calculator returns the real and imaginary parts of the complex number $7 + 2i$.

Figure 3

Some graphing calculators, such as the TI 83/84 Plus, are capable of working with complex numbers, as seen in **Figure 3.** ■

For a complex number $a + bi$, if $b = 0$, then $a + bi = a$, which is a real number. Thus, the set of real numbers is a subset of the set of complex numbers. If $a = 0$ and $b \neq 0$, the complex number is said to be a **pure imaginary number.** For example, $3i$ is a pure imaginary number. A pure imaginary number, or a number such as $7 + 2i$ with $a \neq 0$ and $b \neq 0$, is a **nonreal complex number.** A complex number written in the form $a + bi$ (or $a + ib$) is in **standard form.** (The form $a + ib$ is used to write expressions such as $i\sqrt{5}$, since $\sqrt{5}i$ could be mistaken for $\sqrt{5i}$.)

The relationships among the subsets of the complex numbers are shown in **Figure 4.**

Complex Numbers $a + bi$, for a and b Real

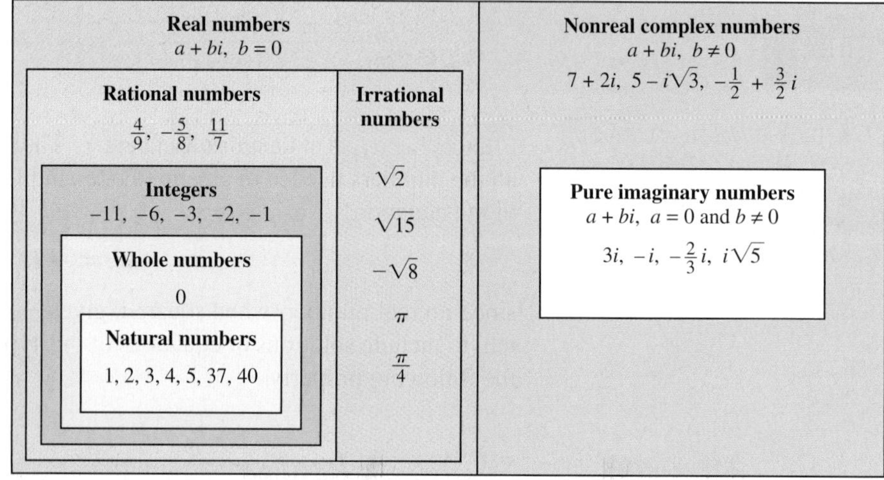

Figure 4

For a positive real number a, the expression $\sqrt{-a}$ is defined as follows.

The Expression $\sqrt{-a}$

If $a > 0$, then $\sqrt{-a} = i\sqrt{a}.$

Classroom Example 1
Write as the product of a real number and i, using the definition of $\sqrt{-a}$.
(a) $\sqrt{-81}$ (b) $\sqrt{-55}$
(c) $\sqrt{-98}$

Answers:
(a) $9i$ (b) $i\sqrt{55}$
(c) $7i\sqrt{2}$

EXAMPLE 1 Writing $\sqrt{-a}$ as $i\sqrt{a}$

Write as the product of a real number and i, using the definition of $\sqrt{-a}$.

(a) $\sqrt{-16}$ **(b)** $\sqrt{-70}$ **(c)** $\sqrt{-48}$

SOLUTION

(a) $\sqrt{-16} = i\sqrt{16} = 4i$ **(b)** $\sqrt{-70} = i\sqrt{70}$

(c) $\sqrt{-48} = i\sqrt{48} = i\sqrt{16 \cdot 3} = 4i\sqrt{3}$ Product rule for radicals (Section R.7)

Now Try Exercises 17, 19, and 21.

Operations on Complex Numbers Products or quotients with negative radicands are simplified by first rewriting $\sqrt{-a}$ as $i\sqrt{a}$ for a positive number a. Then the properties of real numbers and the fact that $i^2 = -1$ are applied.

CAUTION ***When working with negative radicands, use the definition $\sqrt{-a} = i\sqrt{a}$ before using any of the other rules for radicals.*** In particular, the rule $\sqrt{c} \cdot \sqrt{d} = \sqrt{cd}$ is valid only when c and d are *not* both negative. For example,

$$\sqrt{-4} \cdot \sqrt{-9} = 2i \cdot 3i = 6i^2 = -6 \quad \text{is correct,}$$

while $$\sqrt{-4} \cdot \sqrt{-9} = \sqrt{(-4)(-9)} = \sqrt{36} = 6 \quad \text{is incorrect.}$$

Classroom Example 2
Multiply or divide, as indicated. Simplify each answer.

(a) $\sqrt{-21} \cdot \sqrt{-21}$

(b) $\sqrt{-5} \cdot \sqrt{-30}$

(c) $\dfrac{\sqrt{-42}}{\sqrt{-3}}$ (d) $\dfrac{\sqrt{-63}}{\sqrt{21}}$

Answers:
(a) -21 (b) $-5\sqrt{6}$
(c) $\sqrt{14}$ (d) $i\sqrt{3}$

EXAMPLE 2 Finding Products and Quotients Involving $\sqrt{-a}$

Multiply or divide, as indicated. Simplify each answer.

(a) $\sqrt{-7} \cdot \sqrt{-7}$ **(b)** $\sqrt{-6} \cdot \sqrt{-10}$ **(c)** $\dfrac{\sqrt{-20}}{\sqrt{-2}}$ **(d)** $\dfrac{\sqrt{-48}}{\sqrt{24}}$

SOLUTION

(a) $\sqrt{-7} \cdot \sqrt{-7} = i\sqrt{7} \cdot i\sqrt{7}$ First write all square roots in terms of i.

$= i^2 \cdot (\sqrt{7})^2$

$= -1 \cdot 7$ $\quad i^2 = -1$

$= -7$

(b) $\sqrt{-6} \cdot \sqrt{-10} = i\sqrt{6} \cdot i\sqrt{10}$

$= i^2 \cdot \sqrt{60}$

$= -1\sqrt{4 \cdot 15}$

$= -1 \cdot 2\sqrt{15}$

$= -2\sqrt{15}$

(c) $\dfrac{\sqrt{-20}}{\sqrt{-2}} = \dfrac{i\sqrt{20}}{i\sqrt{2}} = \sqrt{\dfrac{20}{2}} = \sqrt{10}$ Quotient rule for radicals **(Section R.7)**

(d) $\dfrac{\sqrt{-48}}{\sqrt{24}} = \dfrac{i\sqrt{48}}{\sqrt{24}} = i\sqrt{\dfrac{48}{24}} = i\sqrt{2}$

✔ ***Now Try Exercises 25, 27, 29, and 31.***

Teaching Tip Point out that the result in **Example 2(a)** follows directly from the definition of the square root of -7—namely, it is the number whose square is -7.

Classroom Example 3
Write $\dfrac{15 - \sqrt{-75}}{5}$ in standard form $a + bi$.

Answer: $3 - i\sqrt{3}$

EXAMPLE 3 Simplifying a Quotient Involving $\sqrt{-a}$

Write $\dfrac{-8 + \sqrt{-128}}{4}$ in standard form $a + bi$.

SOLUTION

$$\frac{-8 + \sqrt{-128}}{4} = \frac{-8 + \sqrt{-64 \cdot 2}}{4}$$

$$= \frac{-8 + 8i\sqrt{2}}{4} \qquad \sqrt{-64} = 8i$$

Be sure to factor before simplifying.

$$= \frac{4(-2 + 2i\sqrt{2})}{4} \qquad \text{Factor. (Section R.4)}$$

$$= -2 + 2i\sqrt{2} \qquad \text{Lowest terms (Section R.5)}$$

✔ ***Now Try Exercise 37.***

With the definitions $i^2 = -1$ and $\sqrt{-a} = i\sqrt{a}$ for $a > 0$, all properties of real numbers are extended to complex numbers. As a result, complex numbers are added, subtracted, multiplied, and divided using real number properties and the definitions on the following pages.

Teaching Tip Point out the similarities between addition of complex numbers and addition of linear polynomials.

Addition and Subtraction of Complex Numbers

For complex numbers $a + bi$ and $c + di$,

$$(a + bi) + (c + di) = (a + c) + (b + d)i$$

and

$$(a + bi) - (c + di) = (a - c) + (b - d)i.$$

That is, to add or subtract complex numbers, add or subtract the real parts and add or subtract the imaginary parts.

Classroom Example 4
Find each sum or difference.
(a) $(4 - 5i) + (-5 + 8i)$
(b) $(-10 + 7i) - (5 - 3i)$

Answers:
(a) $-1 + 3i$
(b) $-15 + 10i$

EXAMPLE 4 Adding and Subtracting Complex Numbers

Find each sum or difference.

(a) $(3 - 4i) + (-2 + 6i)$ **(b)** $(-4 + 3i) - (6 - 7i)$

SOLUTION

(a) $(3 - 4i) + (-2 + 6i) = \underbrace{[3 + (-2)]}_{\text{Add real parts.}} + \underbrace{[-4 + 6]}_{\text{Add imaginary parts.}}i$ Commutative, associative, distributive properties (Section R.2)

$= 1 + 2i$

(b) $(-4 + 3i) - (6 - 7i) = (-4 - 6) + [3 - (-7)]i$

$= -10 + 10i$

Now Try Exercises 43 and 45.

The product of two complex numbers is found by multiplying as though the numbers were binomials and using the fact that $i^2 = -1$, as follows.

$(a + bi)(c + di) = ac + adi + bic + bidi$ FOIL (Section R.3)

$= ac + adi + bci + bdi^2$ Associative property

$= ac + (ad + bc)i + bd(-1)$ Distributive property; $i^2 = -1$

$= (ac - bd) + (ad + bc)i$ Group like terms.

Multiplication of Complex Numbers

For complex numbers $a + bi$ and $c + di$,

$$(a + bi)(c + di) = (ac - bd) + (ad + bc)i.$$

This definition is not practical in routine calculations. To find a given product, it is easier just to multiply as with binomials.

Classroom Example 5
Find each product.
(a) $(5 + 3i)(2 - 7i)$
(b) $(4 - 5i)^2$
(c) $(9 - 8i)(9 + 8i)$

Answers:
(a) $31 - 29i$
(b) $-9 - 40i$
(c) 145, or $145 + 0i$

EXAMPLE 5 Multiplying Complex Numbers

Find each product.

(a) $(2 - 3i)(3 + 4i)$ **(b)** $(4 + 3i)^2$ **(c)** $(6 + 5i)(6 - 5i)$

SOLUTION

(a) $(2 - 3i)(3 + 4i) = 2(3) + 2(4i) - 3i(3) - 3i(4i)$ FOIL

$= 6 + 8i - 9i - 12i^2$ Multiply.

$= 6 - i - 12(-1)$ Combine like terms; $i^2 = -1$

$= 18 - i$ Standard form

(b) $(4+3i)^2 = 4^2 + 2(4)(3i) + (3i)^2$ Square of a binomial (Section R.3)

Remember to add twice the product of the two terms.

$= 16 + 24i + 9i^2$ Multiply.

$= 16 + 24i + 9(-1)$ $i^2 = -1$

$= 7 + 24i$ Standard form

(c) $(6+5i)(6-5i) = 6^2 - (5i)^2$ Product of the sum and difference of two terms (Section R.3)

$= 36 - 25(-1)$ Square 6 and 5; $i^2 = -1$.

$= 36 + 25$ Multiply.

$= 61$, or $61 + 0i$ Standard form

✔ *Now Try Exercises 51, 55, and 59.*

```
(2-3i)(3+4i)
            18-i
(4+3i)²
           7+24i
(6+5i)(6-5i)
              61
```

This screen shows how the TI–83/84 Plus displays the results found in **Example 5.**

Example 5(c) showed that $(6+5i)(6-5i) = 61$. The numbers $6+5i$ and $6-5i$ differ only in the sign of their imaginary parts and are called **complex conjugates.** ***The product of a complex number and its conjugate is always a real number.*** This product is the sum of the squares of the real and imaginary parts.

Property of Complex Conjugates

For real numbers a and b,

$$(a+bi)(a-bi) = a^2 + b^2.$$

To find the quotient of two complex numbers in standard form, we multiply both the numerator and the denominator by the complex conjugate of the denominator.

Classroom Example 6
Write each quotient in standard form $a + bi$.

(a) $\frac{5-5i}{3+i}$

(b) $\frac{15}{-i}$

Answers:
(a) $1 - 2i$
(b) $15i$, or $0 + 15i$

EXAMPLE 6 Dividing Complex Numbers

Write each quotient in standard form $a + bi$.

(a) $\frac{3+2i}{5-i}$ (b) $\frac{3}{i}$

SOLUTION

(a) $\frac{3+2i}{5-i} = \frac{(3+2i)(5+i)}{(5-i)(5+i)}$ Multiply by the complex conjugate of the denominator in both the numerator and the denominator.

$= \frac{15 + 3i + 10i + 2i^2}{25 - i^2}$ Multiply.

$= \frac{13 + 13i}{26}$ Combine like terms; $i^2 = -1$

$= \frac{13}{26} + \frac{13i}{26}$ $\frac{a+bi}{c} = \frac{a}{c} + \frac{bi}{c}$ (Section R.5)

$= \frac{1}{2} + \frac{1}{2}i$ Write in lowest terms and standard form.

CHECK $\left(\frac{1}{2} + \frac{1}{2}i\right)(5 - i) = 3 + 2i$ ✓ Quotient × Divisor = Dividend

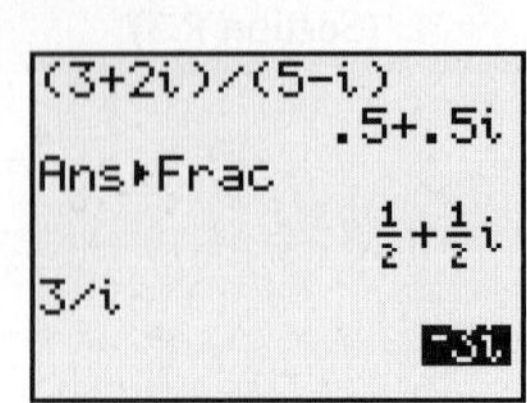

This screen supports the results in **Example 6.**

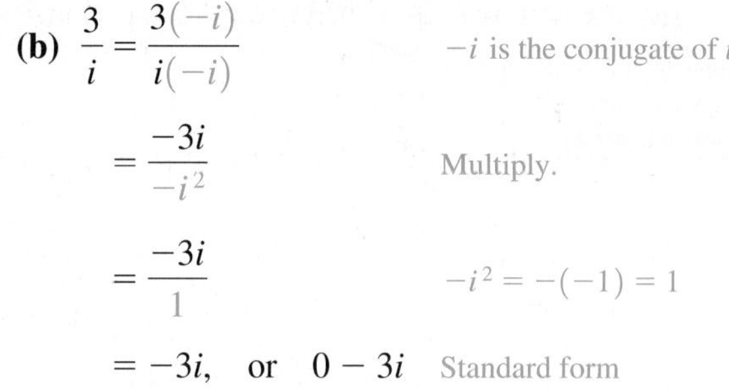

(b) $\frac{3}{i} = \frac{3(-i)}{i(-i)}$ $\quad -i$ is the conjugate of i.

$= \frac{-3i}{-i^2}$ $\quad$ Multiply.

$= \frac{-3i}{1}$ $\quad -i^2 = -(-1) = 1$

$= -3i, \quad \text{or} \quad 0 - 3i$ $\quad$ Standard form

✔ *Now Try Exercises 69 and 75.*

Powers of i can be found on the TI–83/84 Plus calculator.

Powers of i can be simplified using the facts

$$i^2 = -1 \quad \text{and} \quad i^4 = (i^2)^2 = (-1)^2 = 1.$$

Consider the following powers of i.

$i^1 = i$	$i^5 = i^4 \cdot i = 1 \cdot i = i$
$i^2 = -1$	$i^6 = i^4 \cdot i^2 = 1(-1) = -1$
$i^3 = i^2 \cdot i = (-1) \cdot i = -i$	$i^7 = i^4 \cdot i^3 = 1 \cdot (-i) = -i$
$i^4 = i^2 \cdot i^2 = (-1)(-1) = 1$	$i^8 = i^4 \cdot i^4 = 1 \cdot 1 = 1$ and so on.

Powers of i cycle through the same four outcomes (i, −1, −i, and 1) since i^4 has the same multiplicative property as 1. Also, any power of i with an exponent that is a multiple of 4 has value 1. As with real numbers, $i^0 = 1$.

Classroom Example 7
Simplify each power of i.
(a) i^{33} (b) i^{-14}

Answers:
(a) i (b) -1

EXAMPLE 7 Simplifying Powers of *i*

Simplify each power of i.

(a) i^{15} **(b)** i^{-3}

SOLUTION

(a) Since $i^4 = 1$, write the given power as a product involving i^4.

$$i^{15} = i^{12} \cdot i^3 = (i^4)^3 \cdot i^3 = 1^3(-i) = -i$$

(b) Multiply i^{-3} by 1 in the form of i^4 to create the least positive exponent for i.

$$i^{-3} = i^{-3} \cdot 1 = i^{-3} \cdot i^4 = i \qquad i^4 = 1$$

✔ *Now Try Exercises 85 and 93.*

1.3 Exercises

1. true 2. true
3. true 4. true
5. false; *Every* real number is a complex number.
6. true

Concept Check *Determine whether each statement is* true *or* false. *If it is false, tell why.*

1. Every real number is a complex number.

2. No real number is a pure imaginary number.

3. Every pure imaginary number is a complex number.

4. A number can be both real and complex.

5. There is no real number that is a complex number.

6. A complex number might not be a pure imaginary number.

7. real, complex
8. real, complex
9. pure imaginary, nonreal complex, complex
10. pure imaginary, nonreal complex, complex
11. nonreal complex, complex
12. nonreal complex, complex
13. real, complex
14. real, complex
15. pure imaginary, nonreal complex, complex
16. pure imaginary, nonreal complex, complex

17. $5i$ 18. $6i$
19. $i\sqrt{10}$ 20. $i\sqrt{15}$
21. $12i\sqrt{2}$ 22. $10i\sqrt{5}$
23. $-3i\sqrt{2}$ 24. $-4i\sqrt{5}$

25. -13 26. -17
27. $-2\sqrt{6}$ 28. $-5\sqrt{3}$
29. $\sqrt{3}$ 30. $\sqrt{10}$
31. $i\sqrt{3}$ 32. $i\sqrt{2}$
33. $\frac{1}{2}$ 34. $\frac{1}{3}$
35. -2 36. -3

37. $-3 - i\sqrt{6}$ 38. $-3 - i\sqrt{2}$
39. $2 + 2i\sqrt{2}$ 40. $10 + i\sqrt{2}$
41. $-\frac{1}{8} + \frac{\sqrt{2}}{8}i$ 42. $-\frac{1}{2} + \frac{\sqrt{2}}{2}i$

43. $12 - i$ 44. $12 + 4i$
45. 2 46. 1
47. $1 - 10i$ 48. $-10 + i$
49. $-13 + 4i\sqrt{2}$ 50. $-3\sqrt{7} + 2i$

51. $8 - i$ 52. $-2 + 16i$
53. $-14 + 2i$ 54. $17 + i$
55. $5 - 12i$ 56. $3 + 4i$
57. 10 58. 26
59. 13 60. 52
61. 7 62. 18
63. $25i$ 64. $53i$
65. $12 + 9i$ 66. $-120 - 35i$
67. $20 + 15i$ 68. $20 - 60i$

69. $2 - 2i$ 70. $4 - i$
71. $\frac{3}{5} - \frac{4}{5}i$ 72. $\frac{7}{25} - \frac{24}{25}i$
73. $-1 - 2i$ 74. $-2 + i$
75. $5i$ 76. $6i$
77. $8i$ 78. $12i$
79. $-\frac{2}{3}i$ 80. $-\frac{5}{9}i$

Identify each number as real, complex, pure imaginary, *or* nonreal complex. *(More than one of these descriptions will apply.)*

7. -4 8. 0 9. $13i$ 10. $-7i$ 11. $5 + i$
12. $-6 - 2i$ 13. π 14. $\sqrt{24}$ 15. $\sqrt{-25}$ 16. $\sqrt{-36}$

Write each number as the product of a real number and i. ***See Example 1.***

17. $\sqrt{-25}$ 18. $\sqrt{-36}$ 19. $\sqrt{-10}$ 20. $\sqrt{-15}$
21. $\sqrt{-288}$ 22. $\sqrt{-500}$ 23. $-\sqrt{-18}$ 24. $-\sqrt{-80}$

Multiply or divide, as indicated. Simplify each answer. ***See Example 2.***

25. $\sqrt{-13} \cdot \sqrt{-13}$ 26. $\sqrt{-17} \cdot \sqrt{-17}$ 27. $\sqrt{-3} \cdot \sqrt{-8}$
28. $\sqrt{-5} \cdot \sqrt{-15}$ 29. $\dfrac{\sqrt{-30}}{\sqrt{-10}}$ 30. $\dfrac{\sqrt{-70}}{\sqrt{-7}}$
31. $\dfrac{\sqrt{-24}}{\sqrt{8}}$ 32. $\dfrac{\sqrt{-54}}{\sqrt{27}}$ 33. $\dfrac{\sqrt{-10}}{\sqrt{-40}}$
34. $\dfrac{\sqrt{-8}}{\sqrt{-72}}$ 35. $\dfrac{\sqrt{-6} \cdot \sqrt{-2}}{\sqrt{3}}$ 36. $\dfrac{\sqrt{-12} \cdot \sqrt{-6}}{\sqrt{8}}$

Write each number in standard form a + bi. ***See Example 3.***

37. $\dfrac{-6 - \sqrt{-24}}{2}$ 38. $\dfrac{-9 - \sqrt{-18}}{3}$ 39. $\dfrac{10 + \sqrt{-200}}{5}$
40. $\dfrac{20 + \sqrt{-8}}{2}$ 41. $\dfrac{-3 + \sqrt{-18}}{24}$ 42. $\dfrac{-5 + \sqrt{-50}}{10}$

Find each sum or difference. Write the answer in standard form. ***See Example 4.***

43. $(3 + 2i) + (9 - 3i)$ 44. $(4 - i) + (8 + 5i)$
45. $(-2 + 4i) - (-4 + 4i)$ 46. $(-3 + 2i) - (-4 + 2i)$
47. $(2 - 5i) - (3 + 4i) - (-2 + i)$ 48. $(-4 - i) - (2 + 3i) + (-4 + 5i)$
49. $-i\sqrt{2} - 2 - (6 - 4i\sqrt{2}) - (5 - i\sqrt{2})$
50. $3\sqrt{7} - (4\sqrt{7} - i) - 4i + (-2\sqrt{7} + 5i)$

Find each product. Write the answer in standard form. ***See Example 5.***

51. $(2 + i)(3 - 2i)$ 52. $(-2 + 3i)(4 - 2i)$ 53. $(2 + 4i)(-1 + 3i)$
54. $(1 + 3i)(2 - 5i)$ 55. $(3 - 2i)^2$ 56. $(2 + i)^2$
57. $(3 + i)(3 - i)$ 58. $(5 + i)(5 - i)$ 59. $(-2 - 3i)(-2 + 3i)$
60. $(6 - 4i)(6 + 4i)$ 61. $(\sqrt{6} + i)(\sqrt{6} - i)$ 62. $(\sqrt{2} - 4i)(\sqrt{2} + 4i)$
63. $i(3 - 4i)(3 + 4i)$ 64. $i(2 + 7i)(2 - 7i)$ 65. $3i(2 - i)^2$
66. $-5i(4 - 3i)^2$ 67. $(2 + i)(2 - i)(4 + 3i)$ 68. $(3 - i)(3 + i)(2 - 6i)$

Find each quotient. Write the answer in standard form a + bi. ***See Example 6.***

69. $\dfrac{6 + 2i}{1 + 2i}$ 70. $\dfrac{14 + 5i}{3 + 2i}$ 71. $\dfrac{2 - i}{2 + i}$ 72. $\dfrac{4 - 3i}{4 + 3i}$
73. $\dfrac{1 - 3i}{1 + i}$ 74. $\dfrac{-3 + 4i}{2 - i}$ 75. $\dfrac{-5}{i}$ 76. $\dfrac{-6}{i}$
77. $\dfrac{8}{-i}$ 78. $\dfrac{12}{-i}$ 79. $\dfrac{2}{3i}$ 80. $\dfrac{5}{9i}$

81. $E = 2 + 62i$
82. $E = 260 + 20i$
83. $Z = 12 + 8i$
84. $I = 8 + 3i$

85. i 86. i
87. -1 88. -1
89. $-i$ 90. $-i$
91. 1 92. 1
93. $-i$ 94. -1
95. $-i$ 96. 1

(Modeling) Alternating Current Complex numbers are used to describe current I, voltage E, and impedance Z (the opposition to current). These three quantities are related by the equation

$$E = IZ, \quad \text{which is known as } \textbf{Ohm's Law.}$$

Thus, if any two of these quantities are known, the third can be found. In each exercise, solve the equation $E = IZ$ for the remaining value.

81. $I = 5 + 7i, Z = 6 + 4i$

82. $I = 20 + 12i, Z = 10 - 5i$

83. $I = 10 + 4i, E = 88 + 128i$

84. $E = 57 + 67i, Z = 9 + 5i$

Simplify each power of i. ***See Example 7.***

85. i^{25} **86.** i^{29} **87.** i^{22} **88.** i^{26}

89. i^{23} **90.** i^{27} **91.** i^{32} **92.** i^{40}

93. i^{-13} **94.** i^{-14} **95.** $\frac{1}{i^{-11}}$ **96.** $\frac{1}{i^{-12}}$

97. Suppose that your friend, Kathy Strautz, tells you that she has discovered a method of simplifying a positive power of i. "Just divide the exponent by 2. Your answer is then the simplified form of i^2 raised to the quotient times i raised to the remainder." Explain why her method works.

98. Explain why the following method of simplifying i^{-42} works.

$$i^{-42} = \frac{1}{i^{42}} = \frac{1}{(i^2)^{21}} = \frac{1}{(-1)^{21}} = \frac{1}{-1} = -1$$

99. Show that $\frac{\sqrt{2}}{2} + \frac{\sqrt{2}}{2}i$ is a square root of i.

100. Show that $\frac{\sqrt{3}}{2} + \frac{1}{2}i$ is a cube root of i.

101. Show that $-2 + i$ is a solution of the equation $x^2 + 4x + 5 = 0$.

102. Show that $-3 + 4i$ is a solution of the equation $x^2 + 6x + 25 = 0$.

1.4 Quadratic Equations

- Solving a Quadratic Equation
- Completing the Square
- The Quadratic Formula
- Solving for a Specified Variable
- The Discriminant

Teaching Tip To help students see the difference between linear and quadratic equations, list several equations and have students tell whether each is linear, quadratic, or neither. Remind them that a quadratic equation has degree two.

A *quadratic equation* is defined as follows.

Quadratic Equation in One Variable

An equation that can be written in the form

$$ax^2 + bx + c = 0,$$

where a, b, and c are real numbers with $a \neq 0$, is a **quadratic equation.** The given form is called **standard form.**

A quadratic equation is a **second-degree equation**—that is, an equation with a squared variable term and no terms of greater degree.

$$x^2 = 25, \quad 4x^2 + 4x - 5 = 0, \quad 3x^2 = 4x - 8 \quad \text{Quadratic equations}$$

Solving a Quadratic Equation The factoring method of solving a quadratic equation depends on the **zero-factor property.**

Zero-Factor Property

If a and b are complex numbers with $ab = 0$, then $a = 0$ or $b = 0$ or both equal zero.

Classroom Example 1
Solve $10x^2 + x - 2 = 0$ by factoring.

Answer: $\left\{-\frac{1}{2}, \frac{2}{5}\right\}$

EXAMPLE 1 Using the Zero-Factor Property

Solve $6x^2 + 7x = 3$.

SOLUTION

Don't factor out x here.

$$6x^2 + 7x = 3$$

$$6x^2 + 7x - 3 = 0 \quad \text{Standard form}$$

$$(3x - 1)(2x + 3) = 0 \quad \text{Factor. (Section R.4)}$$

$$3x - 1 = 0 \quad \text{or} \quad 2x + 3 = 0 \quad \text{Zero-factor property}$$

$$3x = 1 \quad \text{or} \quad 2x = -3 \quad \text{Solve each equation. (Section 1.1)}$$

$$x = \frac{1}{3} \quad \text{or} \quad x = -\frac{3}{2}$$

Teaching Tip Students often make the mistake of writing the equation in **Example 1** as

$$x(6x + 7) = 3,$$

and then setting $x = 3$ and $6x + 7 = 3$. Emphasize that the right side of the equation must be 0 before factoring on the left.

CHECK $6x^2 + 7x = 3$ Original equation

$$6\left(\frac{1}{3}\right)^2 + 7\left(\frac{1}{3}\right) \stackrel{?}{=} 3 \quad \text{Let } x = \tfrac{1}{3}.$$

$$\frac{6}{9} + \frac{7}{3} \stackrel{?}{=} 3$$

$$3 = 3 \checkmark \text{ True}$$

$$6\left(-\frac{3}{2}\right)^2 + 7\left(-\frac{3}{2}\right) \stackrel{?}{=} 3 \quad \text{Let } x = -\tfrac{3}{2}.$$

$$\frac{54}{4} - \frac{21}{2} \stackrel{?}{=} 3$$

$$3 = 3 \checkmark \text{ True}$$

Both values check, since true statements result. The solution set is $\left\{\frac{1}{3}, -\frac{3}{2}\right\}$.

Now Try Exercise 15.

Teaching Tip Remind students that a polynomial of degree two can have as many as two solutions. Solve the equation

$$x^2 - 16 = 0$$

by factoring as a difference of squares. This will help make the point that 4 and -4 are *both* solutions of the equation $x^2 = 16$, while at the same time making the square root property more understandable.

A quadratic equation of the form $x^2 = k$ can also be solved by factoring.

$$x^2 = k$$

$$x^2 - k = 0 \quad \text{Subtract } k.$$

$$\left(x - \sqrt{k}\right)\left(x + \sqrt{k}\right) = 0 \quad \text{Factor.}$$

$$x - \sqrt{k} = 0 \quad \text{or} \quad x + \sqrt{k} = 0 \quad \text{Zero-factor property}$$

$$x = \sqrt{k} \quad \text{or} \quad x = -\sqrt{k} \quad \text{Solve each equation.}$$

This proves the **square root property.**

Teaching Tip Mention that

$$x^2 - k = 0$$

is a special form of

$$ax^2 + bx + c = 0,$$

where the coefficient b of the linear term is zero.

Square Root Property

If $x^2 = k$, then $\boldsymbol{x = \sqrt{k}}$ **or** $\boldsymbol{x = -\sqrt{k}}$.

That is, the solution set of $x^2 = k$ is

$$\left\{\sqrt{k}, -\sqrt{k}\right\}, \quad \textit{which may be abbreviated} \quad \left\{\pm\sqrt{k}\right\}.$$

Both solutions $\sqrt{k}$ and $-\sqrt{k}$ are real if $k > 0$, and both are pure imaginary if $k < 0$. If $k < 0$, then we write the solution set as

$$\{\pm i\sqrt{|k|}\}.$$

If $k = 0$, then there is only one distinct solution, 0, sometimes called a **double solution.**

Classroom Example 2
Solve each quadratic equation using the square root property.
(a) $x^2 = 29$
(b) $x^2 = -144$
(c) $(x - 8)^2 = 24$

Answers:
(a) $\{\pm\sqrt{29}\}$
(b) $\{\pm 12i\}$
(c) $\{8 \pm 2\sqrt{6}\}$

EXAMPLE 2 Using the Square Root Property

Solve each quadratic equation.

(a) $x^2 = 17$ **(b)** $x^2 = -25$ **(c)** $(x - 4)^2 = 12$

SOLUTION

(a) By the square root property, the solution set of $x^2 = 17$ is $\{\pm\sqrt{17}\}$.

(b) Since $\sqrt{-1} = i$, the solution set of $x^2 = -25$ is $\{\pm 5i\}$.

(c)

$$(x - 4)^2 = 12$$

$$x - 4 = \pm\sqrt{12} \quad \text{Generalized square root property}$$

$$x = 4 \pm \sqrt{12} \quad \text{Add 4.}$$

$$x = 4 \pm 2\sqrt{3} \quad \sqrt{12} = \sqrt{4 \cdot 3} = 2\sqrt{3} \text{ (Section R.7)}$$

CHECK $(x - 4)^2 = 12$ Original equation

$$(4 + 2\sqrt{3} - 4)^2 \stackrel{?}{=} 12 \quad \text{Let } x = 4 + 2\sqrt{3}.$$

$$(2\sqrt{3})^2 \stackrel{?}{=} 12$$

$$2^2 \cdot (\sqrt{3})^2 \stackrel{?}{=} 12$$

$$12 = 12 \quad \checkmark \text{ True}$$

$$(4 - 2\sqrt{3} - 4)^2 \stackrel{?}{=} 12 \quad \text{Let } x = 4 - 2\sqrt{3}.$$

$$(-2\sqrt{3})^2 \stackrel{?}{=} 12$$

$$(-2)^2 \cdot (\sqrt{3})^2 \stackrel{?}{=} 12$$

$$12 = 12 \quad \checkmark \text{ True}$$

The solution set is $\{4 \pm 2\sqrt{3}\}$.

✔ Now Try Exercises 27, 29, and 31.

Completing the Square Any quadratic equation can be solved by the method of **completing the square,** summarized as follows.

Teaching Tip Students may find this method tedious, but it is useful later in being able to quickly graph (or analyze the graphs of) parabolas.

If the linear coefficient is even after dividing by a, using this method is often easier than using the quadratic formula.

Solving a Quadratic Equation by Completing the Square

To solve $ax^2 + bx + c = 0$, where $a \neq 0$, by completing the square, use these steps.

Step 1 If $a \neq 1$, divide both sides of the equation by a.

Step 2 Rewrite the equation so that the constant term is alone on one side of the equality symbol.

Step 3 Square half the coefficient of x, and add this square to each side of the equation.

Step 4 Factor the resulting trinomial as a perfect square and combine like terms on the other side.

Step 5 Use the square root property to complete the solution.

Classroom Example 3
Solve $x^2 + 10x - 20 = 0$ by completing the square.

Answer: $\{-5 \pm 3\sqrt{5}\}$

EXAMPLE 3 Using Completing the Square ($a = 1$)

Solve $x^2 - 4x - 14 = 0$.

SOLUTION $x^2 - 4x - 14 = 0$

Step 1 This step is not necessary since $a = 1$.

Step 2 $x^2 - 4x = 14$ Add 14 to each side.

Step 3 $x^2 - 4x + 4 = 14 + 4$ $\left[\frac{1}{2}(-4)\right]^2 = 4$; Add 4 to each side.

Step 4 $(x - 2)^2 = 18$ Factor. (Section R.4) Combine like terms.

Step 5 $x - 2 = \pm\sqrt{18}$ Square root property

Take *both* roots.

$x = 2 \pm \sqrt{18}$ Add 2 to each side.

$x = 2 \pm 3\sqrt{2}$ Simplify the radical.

The solution set is $\{2 \pm 3\sqrt{2}\}$.

Now Try Exercise 41.

Teaching Tip Point out that the -2 in Step 4 of **Example 3** is one-half the coefficient of x in Step 3.

Classroom Example 4
Solve $4x^2 + 6x + 5 = 0$ by completing the square.

Answer: $\left\{-\frac{3}{4} \pm \frac{\sqrt{11}}{4}i\right\}$

EXAMPLE 4 Using Completing the Square ($a \neq 1$)

Solve $9x^2 - 12x + 9 = 0$.

SOLUTION $9x^2 - 12x + 9 = 0$

$x^2 - \frac{4}{3}x + 1 = 0$ Divide by 9. (Step 1)

$x^2 - \frac{4}{3}x = -1$ Subtract 1 from each side. (Step 2)

$x^2 - \frac{4}{3}x + \frac{4}{9} = -1 + \frac{4}{9}$ $\left[\frac{1}{2}\left(-\frac{4}{3}\right)\right]^2 = \frac{4}{9}$; Add $\frac{4}{9}$ to each side. (Step 3)

$\left(x - \frac{2}{3}\right)^2 = -\frac{5}{9}$ Factor. Combine like terms. (Step 4)

$x - \frac{2}{3} = \pm\sqrt{-\frac{5}{9}}$ Square root property (Step 5)

$x - \frac{2}{3} = \pm\frac{\sqrt{5}}{3}i$ $\sqrt{-\frac{5}{9}} = \frac{\sqrt{-5}}{\sqrt{9}} = \frac{i\sqrt{5}}{3}$, or $\frac{\sqrt{5}}{3}i$ (Section 1.3)

$x = \frac{2}{3} \pm \frac{\sqrt{5}}{3}i$ Add $\frac{2}{3}$ to each side.

The solution set is $\left\{\frac{2}{3} \pm \frac{\sqrt{5}}{3}i\right\}$.

Now Try Exercise 47.

The Quadratic Formula If we start with the equation, $ax^2 + bx + c = 0$, for $a > 0$, and complete the square to solve for x in terms of the constants a, b, and c, the result is a general formula for solving any quadratic equation.

$ax^2 + bx + c = 0$

$x^2 + \frac{b}{a}x + \frac{c}{a} = 0$ Divide each side by a. (Step 1)

$x^2 + \frac{b}{a}x = -\frac{c}{a}$ Subtract $\frac{c}{a}$ from each side. (Step 2)

Square half the coefficient of x: $\left[\frac{1}{2}\left(\frac{b}{a}\right)\right]^2 = \left(\frac{b}{2a}\right)^2 = \frac{b^2}{4a^2}$.

$$x^2 + \frac{b}{a}x + \frac{b^2}{4a^2} = -\frac{c}{a} + \frac{b^2}{4a^2}$$ Add $\frac{b^2}{4a^2}$ to each side. (Step 3)

$$\left(x + \frac{b}{2a}\right)^2 = \frac{b^2}{4a^2} + \frac{-c}{a}$$ Factor. Use the commutative property. (Step 4)

$$\left(x + \frac{b}{2a}\right)^2 = \frac{b^2}{4a^2} + \frac{-4ac}{4a^2}$$ Write fractions with a common denominator. **(Section R.5)**

$$\left(x + \frac{b}{2a}\right)^2 = \frac{b^2 - 4ac}{4a^2}$$ Add fractions. **(Section R.5)**

$$x + \frac{b}{2a} = \pm\sqrt{\frac{b^2 - 4ac}{4a^2}}$$ Square root property (Step 5)

$$x + \frac{b}{2a} = \frac{\pm\sqrt{b^2 - 4ac}}{2a}$$ Since $a > 0$, $\sqrt{4a^2} = 2a$. **(Section R.7)**

$$x = \frac{-b}{2a} \pm \frac{\sqrt{b^2 - 4ac}}{2a}$$ Subtract $\frac{b}{2a}$ from each side.

Quadratic Formula This result is also true for $a < 0$. →

$$x = \frac{-b \pm \sqrt{b^2 - 4ac}}{2a}$$ Combine terms on the right.

Teaching Tip Point out that the equation in **Example 5** can be thought of as

$$1x^2 + (-4)x + 2 = 0.$$

Therefore, $a = 1$, $b = -4$, and $c = 2$.

Quadratic Formula

The solutions of the quadratic equation $ax^2 + bx + c = 0$, where $a \neq 0$, are given by the quadratic formula.

$$x = \frac{-b \pm \sqrt{b^2 - 4ac}}{2a}$$

Classroom Example 5
Solve $x^2 + 6x = 3$ using the quadratic formula.

Answer: $\{-3 \pm 2\sqrt{3}\}$

EXAMPLE 5 Using the Quadratic Formula (Real Solutions)

Solve $x^2 - 4x = -2$.

SOLUTION

$$x^2 - 4x + 2 = 0$$ Write in standard form. Here $a = 1$, $b = -4$, and $c = 2$.

$$x = \frac{-b \pm \sqrt{b^2 - 4ac}}{2a}$$ Quadratic formula

$$x = \frac{-(-4) \pm \sqrt{(-4)^2 - 4(1)(2)}}{2(1)}$$ Substitute $a = 1$, $b = -4$, and $c = 2$.

The fraction bar extends *under* $-b$.

$$x = \frac{4 \pm \sqrt{16 - 8}}{2}$$ Simplify.

$$x = \frac{4 \pm 2\sqrt{2}}{2}$$ $\sqrt{16 - 8} = \sqrt{8} = \sqrt{4 \cdot 2} = 2\sqrt{2}$ **(Section R.7)**

$$x = \frac{2(2 \pm \sqrt{2})}{2}$$ Factor out 2 in the numerator. **(Section R.5)**

Factor first, then divide.

$$x = 2 \pm \sqrt{2}$$ Lowest terms

The solution set is $\{2 \pm \sqrt{2}\}$.

Now Try Exercise 53.

CAUTION ***Remember to extend the fraction bar in the quadratic formula under the $-b$ term in the numerator.***

Throughout this text, unless otherwise specified, we use the set of complex numbers as the domain when solving equations of degree 2 or greater.

Classroom Example 6
Solve $4x^2 = 3x - 5$ using the quadratic formula.

Answer: $\left\{\frac{3}{8} \pm \frac{\sqrt{71}}{8}i\right\}$

EXAMPLE 6 Using the Quadratic Formula (Nonreal Complex Solutions)

Solve $2x^2 = x - 4$.

SOLUTION

$$2x^2 - x + 4 = 0 \quad \text{Write in standard form.}$$

$$x = \frac{-(-1) \pm \sqrt{(-1)^2 - 4(2)(4)}}{2(2)} \quad \text{Quadratic formula with } a = 2, b = -1, c = 4$$

Use parentheses and substitute carefully to avoid errors.

$$x = \frac{1 \pm \sqrt{1 - 32}}{4}$$

$$x = \frac{1 \pm \sqrt{-31}}{4} \quad \text{Simplify.}$$

$$x = \frac{1 \pm i\sqrt{31}}{4} \quad \sqrt{-1} = i \text{ (Section 1.3)}$$

The solution set is $\left\{\frac{1}{4} \pm \frac{\sqrt{31}}{4}i\right\}$.

Now Try Exercise 57.

Teaching Tip Emphasize that the values of a, b, and c are determined only after the quadratic equation is written in standard form.

The equation $x^3 + 8 = 0$ is a **cubic equation** because the greatest degree of the terms is 3.

Classroom Example 7
Solve $x^3 - 125 = 0$ using factoring and the quadratic formula.

Answer: $\left\{5, -\frac{5}{2} \pm \frac{5\sqrt{3}}{2}i\right\}$

EXAMPLE 7 Solving a Cubic Equation

Solve $x^3 + 8 = 0$ using factoring and the quadratic formula.

SOLUTION

$$x^3 + 8 = 0$$

$$(x + 2)(x^2 - 2x + 4) = 0 \quad \text{Factor as a sum of cubes. (Section R.4)}$$

$$x + 2 = 0 \quad \text{or} \quad x^2 - 2x + 4 = 0 \quad \text{Zero-factor property}$$

$$x = -2 \quad \text{or} \quad x = \frac{-(-2) \pm \sqrt{(-2)^2 - 4(1)(4)}}{2(1)} \quad \text{Quadratic formula with } a = 1, b = -2, c = 4$$

$$x = \frac{2 \pm \sqrt{-12}}{2} \quad \text{Simplify.}$$

$$x = \frac{2 \pm 2i\sqrt{3}}{2} \quad \text{Simplify the radical.}$$

$$x = \frac{2\left(1 \pm i\sqrt{3}\right)}{2} \quad \text{Factor out 2 in the numerator.}$$

$$x = 1 \pm i\sqrt{3} \quad \text{Lowest terms}$$

The solution set is $\left\{-2, 1 \pm i\sqrt{3}\right\}$.

Now Try Exercise 67.

Solving for a Specified Variable To solve a quadratic equation for a specified variable, we usually apply the square root property or the quadratic formula.

Classroom Example 8
Solve for the specified variable. Use $\pm$ when taking square roots.

(a) $V = \frac{1}{3}\pi r^2 h$, for r

(b) $2my^2 - ny = 3p \ (m \neq 0)$, for y

Answers:

(a) $r = \frac{\pm\sqrt{3V\pi h}}{\pi h}$

(b) $y = \frac{n \pm \sqrt{n^2 + 24mp}}{4m}$

EXAMPLE 8 Solving for a Quadratic Variable in a Formula

Solve for the specified variable. Use $\pm$ when taking square roots.

(a) $\mathcal{A} = \frac{\pi d^2}{4}$, for d **(b)** $rt^2 - st = k \ (r \neq 0)$, for t

SOLUTION

(a) $\mathcal{A} = \frac{\pi d^2}{4}$ Goal: Isolate d, the specified variable.

$4\mathcal{A} = \pi d^2$ Multiply each side by 4.

$\frac{4\mathcal{A}}{\pi} = d^2$ Divide each side by π.

$d = \pm\sqrt{\frac{4\mathcal{A}}{\pi}}$ Square root property

See the Note following this example.

$d = \frac{\pm\sqrt{4\mathcal{A}}}{\sqrt{\pi}} \cdot \frac{\sqrt{\pi}}{\sqrt{\pi}}$ Multiply by $\frac{\sqrt{\pi}}{\sqrt{\pi}}$. (Section R.7)

$d = \frac{\pm\sqrt{4\mathcal{A}\pi}}{\pi}$ Multiply numerators. Multiply denominators.

$d = \frac{\pm 2\sqrt{\mathcal{A}\pi}}{\pi}$ Simplify the radical.

(b) Because $rt^2 - st = k$ has terms with t^2 and t, use the quadratic formula.

$rt^2 - st - k = 0$ Write in standard form.

$t = \frac{-b \pm \sqrt{b^2 - 4ac}}{2a}$ Quadratic formula

$t = \frac{-(-s) \pm \sqrt{(-s)^2 - 4(r)(-k)}}{2(r)}$ Here, $a = r$, $b = -s$, and $c = -k$.

$t = \frac{s \pm \sqrt{s^2 + 4rk}}{2r}$ Simplify.

✔ *Now Try Exercises 71 and 77.*

NOTE In **Example 8,** we took both positive and negative square roots. However, if the variable represents time or length in an application, we consider only the *positive* square root.

The Discriminant The quantity under the radical in the quadratic formula, $b^2 - 4ac$, is called the **discriminant.**

$$x = \frac{-b \pm \sqrt{b^2 - 4ac}}{2a} \leftarrow \text{Discriminant}$$

When the numbers a, b, and c are *integers* (but not necessarily otherwise), the value of the discriminant can be used to determine whether the solutions of a quadratic equation are rational, irrational, or nonreal complex numbers. The number and type of solutions based on the value of the discriminant are shown in the following table.

Discriminant	Number of Solutions	Type of Solutions	
Positive, perfect square	Two	Rational	
Positive, but not a perfect square	Two	Irrational	← As seen in **Example 5**
Zero	One (a double solution)	Rational	
Negative	Two	Nonreal complex	← As seen in **Example 6**

CAUTION ***The restriction on a, b, and c is important.*** For example,

$$x^2 - \sqrt{5}x - 1 = 0 \quad \text{has discriminant} \quad b^2 - 4ac = 5 + 4 = 9,$$

which would indicate two rational solutions *if the coefficients were integers.* By the quadratic formula, the two solutions $\frac{\sqrt{5} \pm 3}{2}$ are *irrational* numbers.

Classroom Example 9
Determine the number of distinct solutions, and tell whether they are *rational, irrational*, or *nonreal complex* numbers.
(a) $4x^2 - 12x + 9 = 0$
(b) $3x^2 + x = -5$
(c) $2x^2 = 6x + 7$

Answers:
(a) one distinct rational solution
(b) two distinct nonreal complex solutions
(c) two distinct irrational solutions

EXAMPLE 9 Using the Discriminant

Determine the number of distinct solutions, and tell whether they are *rational, irrational,* or *nonreal complex* numbers.

(a) $5x^2 + 2x - 4 = 0$ **(b)** $x^2 - 10x = -25$ **(c)** $2x^2 - x + 1 = 0$

SOLUTION

(a) For $5x^2 + 2x - 4 = 0$, use $a = 5$, $b = 2$, and $c = -4$.

$$b^2 - 4ac = 2^2 - 4(5)(-4) = 84 \leftarrow \text{Discriminant}$$

The discriminant 84 is positive and not a perfect square, so there are two distinct irrational solutions.

(b) First, write the equation in standard form as $x^2 - 10x + 25 = 0$. Thus, $a = 1$, $b = -10$, and $c = 25$.

$$b^2 - 4ac = (-10)^2 - 4(1)(25) = 0 \leftarrow \text{Discriminant}$$

There is one distinct rational solution, a double solution.

(c) For $2x^2 - x + 1 = 0$, use $a = 2$, $b = -1$, and $c = 1$.

$$b^2 - 4ac = (-1)^2 - 4(2)(1) = -7 \leftarrow \text{Discriminant}$$

There are two distinct nonreal complex solutions. (They are complex conjugates.)

✔ *Now Try Exercises 83, 85, and 89.*

1.4 Exercises

1. G 2. A 3. C 4. E
5. H 6. B 7. D 8. F

Concept Check Match the equation in Column I with its solution(s) in Column II.

I		II	
1. $x^2 = 25$	**2.** $x^2 = -25$	**A.** $\pm 5i$	**B.** $\pm 2\sqrt{5}$
3. $x^2 + 5 = 0$	**4.** $x^2 - 5 = 0$	**C.** $\pm i\sqrt{5}$	**D.** 5
5. $x^2 = -20$	**6.** $x^2 = 20$	**E.** $\pm\sqrt{5}$	**F.** -5
7. $x - 5 = 0$	**8.** $x + 5 = 0$	**G.** ± 5	**H.** $\pm 2i\sqrt{5}$

9. D; $\left\{\frac{1}{3}, 7\right\}$

10. B; $\left\{\frac{-5 \pm \sqrt{7}}{2}\right\}$

11. C; $\{-4, 3\}$ 12. A; $\left\{-\frac{1}{3}, 6\right\}$

13. $\{2, 3\}$ 14. $\{-4, 2\}$

15. $\left\{-\frac{2}{5}, 1\right\}$ 16. $\left\{-\frac{5}{2}, 3\right\}$

17. $\left\{-\frac{3}{4}, 1\right\}$ 18. $\left\{-\frac{5}{6}, 2\right\}$

19. $\{\pm 10\}$ 20. $\{\pm 8\}$

21. $\left\{\frac{1}{2}\right\}$ 22. $\left\{\frac{2}{3}\right\}$

23. $\left\{-\frac{3}{5}\right\}$ 24. $\left\{-\frac{5}{6}\right\}$

25. $\{\pm 4\}$ 26. $\{\pm 11\}$

27. $\{\pm 3\sqrt{3}\}$ 28. $\{\pm 4\sqrt{3}\}$

29. $\{\pm 9i\}$ 30. $\{\pm 20i\}$

31. $\left\{\frac{1 \pm 2\sqrt{3}}{3}\right\}$ 32. $\left\{\frac{-1 \pm 2\sqrt{5}}{4}\right\}$

33. $\{-5 \pm i\sqrt{3}\}$

34. $\{4 \pm i\sqrt{5}\}$

35. $\left\{\frac{3}{5} \pm \frac{\sqrt{3}}{5}i\right\}$ 36. $\left\{\frac{5}{2} \pm i\sqrt{2}\right\}$

37. $\{1, 3\}$ 38. $\{3, 4\}$

39. $\left\{-\frac{7}{2}, 4\right\}$ 40. $\left\{-\frac{5}{4}, 2\right\}$

41. $\{1 \pm \sqrt{3}\}$ 42. $\{5 \pm \sqrt{7}\}$

43. $\left\{-\frac{5}{2}, 2\right\}$ 44. $\left\{-\frac{5}{3}, 1\right\}$

45. $\left\{\frac{2 \pm \sqrt{10}}{2}\right\}$ 46. $\left\{\frac{3 \pm 2\sqrt{6}}{3}\right\}$

47. $\left\{1 \pm \frac{\sqrt{3}}{2}i\right\}$ 48. $\left\{\frac{3}{2} \pm \frac{\sqrt{3}}{6}i\right\}$

49. He is incorrect because $c = 0$.
50. She is incorrect because $b = 0$.

51. $\left\{\frac{1 \pm \sqrt{5}}{2}\right\}$ 52. $\left\{\frac{3 \pm \sqrt{17}}{2}\right\}$

53. $\{3 \pm \sqrt{2}\}$ 54. $\{2 \pm \sqrt{3}\}$

55. $\{1 \pm 2i\}$ 56. $\{1 \pm 3i\}$

57. $\left\{\frac{3}{2} \pm \frac{\sqrt{2}}{2}i\right\}$ 58. $\left\{-\frac{1}{4} \pm \frac{\sqrt{39}}{12}i\right\}$

59. $\left\{\frac{-1 \pm \sqrt{97}}{4}\right\}$ 60. $\left\{\frac{-3 \pm 3\sqrt{129}}{16}\right\}$

61. $\left\{\frac{-2 \pm \sqrt{10}}{2}\right\}$ 62. $\left\{\frac{1 \pm \sqrt{13}}{2}\right\}$

63. $\left\{\frac{-3 \pm \sqrt{41}}{8}\right\}$ 64. $\left\{\frac{2 \pm \sqrt{10}}{3}\right\}$

65. $\{5\}$

67. $\{2, -1 \pm i\sqrt{3}\}$

68. $\left\{3, -\frac{3}{2} \pm \frac{3\sqrt{3}}{2}i\right\}$

69. $\left\{-3, \frac{3}{2} \pm \frac{3\sqrt{3}}{2}i\right\}$

Concept Check *Use Choices A–D to answer each question in Exercises 9–12.*

A. $3x^2 - 17x - 6 = 0$ **B.** $(2x + 5)^2 = 7$

C. $x^2 + x = 12$ **D.** $(3x - 1)(x - 7) = 0$

9. Which equation is set up for direct use of the zero-factor property? Solve it.

10. Which equation is set up for direct use of the square root property? Solve it.

11. Only one of the equations does not require Step 1 of the method for completing the square described in this section. Which one is it? Solve it.

12. Only one of the equations is set up so that the values of a, b, and c can be determined immediately. Which one is it? Solve it.

Solve each equation by the zero-factor property. ***See Example 1.***

13. $x^2 - 5x + 6 = 0$ **14.** $x^2 + 2x - 8 = 0$ **15.** $5x^2 - 3x - 2 = 0$

16. $2x^2 - x - 15 = 0$ **17.** $-4x^2 + x = -3$ **18.** $-6x^2 + 7x = -10$

19. $x^2 - 100 = 0$ **20.** $x^2 - 64 = 0$ **21.** $4x^2 - 4x + 1 = 0$

22. $9x^2 - 12x + 4 = 0$ **23.** $25x^2 + 30x + 9 = 0$ **24.** $36x^2 + 60x + 25 = 0$

Solve each equation by the square root property. ***See Example 2.***

25. $x^2 = 16$ **26.** $x^2 = 121$ **27.** $27 - x^2 = 0$

28. $48 - x^2 = 0$ **29.** $x^2 = -81$ **30.** $x^2 = -400$

31. $(3x - 1)^2 = 12$ **32.** $(4x + 1)^2 = 20$ **33.** $(x + 5)^2 = -3$

34. $(x - 4)^2 = -5$ **35.** $(5x - 3)^2 = -3$ **36.** $(-2x + 5)^2 = -8$

Solve each equation by completing the square. ***See Examples 3 and 4.***

37. $x^2 - 4x + 3 = 0$ **38.** $x^2 - 7x + 12 = 0$ **39.** $2x^2 - x - 28 = 0$

40. $4x^2 - 3x - 10 = 0$ **41.** $x^2 - 2x - 2 = 0$ **42.** $x^2 - 10x + 18 = 0$

43. $2x^2 + x = 10$ **44.** $3x^2 + 2x = 5$ **45.** $-2x^2 + 4x + 3 = 0$

46. $-3x^2 + 6x + 5 = 0$ **47.** $-4x^2 + 8x = 7$ **48.** $-3x^2 + 9x = 7$

49. *Concept Check* Francisco claimed that the equation $x^2 - 8x = 0$ cannot be solved by the quadratic formula since there is no value for c. Is he correct?

50. *Concept Check* Francesca, Francisco's twin sister, claimed that the equation $x^2 - 19 = 0$ cannot be solved by the quadratic formula since there is no value for b. Is she correct?

Solve each equation using the quadratic formula. ***See Examples 5 and 6.***

51. $x^2 - x - 1 = 0$ **52.** $x^2 - 3x - 2 = 0$ **53.** $x^2 - 6x = -7$

54. $x^2 - 4x = -1$ **55.** $x^2 = 2x - 5$ **56.** $x^2 = 2x - 10$

57. $-4x^2 = -12x + 11$ **58.** $-6x^2 = 3x + 2$ **59.** $\frac{1}{2}x^2 + \frac{1}{4}x - 3 = 0$

60. $\frac{2}{3}x^2 + \frac{1}{4}x = 3$ **61.** $0.2x^2 + 0.4x - 0.3 = 0$ **62.** $0.1x^2 - 0.1x = 0.3$

63. $(4x - 1)(x + 2) = 4x$ **64.** $(3x + 2)(x - 1) = 3x$ **65.** $(x - 9)(x - 1) = -16$

66. Explain why the following equations have the same solution set. (Do not actually solve.)

$$-2x^2 + 3x - 6 = 0 \quad \text{and} \quad 2x^2 - 3x + 6 = 0$$

Solve each cubic equation using factoring and the quadratic formula. ***See Example 7.***

67. $x^3 - 8 = 0$ **68.** $x^3 - 27 = 0$ **69.** $x^3 + 27 = 0$ **70.** $x^3 + 64 = 0$

Solve each equation for the indicated variable. Assume no denominators are 0. **See Example 8.**

71. $s = \frac{1}{2}gt^2$, for t
72. $\mathcal{A} = \pi r^2$, for r
73. $F = \frac{kMv^2}{r}$, for v
74. $E = \frac{e^2k}{2r}$, for e
75. $r = r_0 + \frac{1}{2}at^2$, for t
76. $s = s_0 + gt^2 + k$, for t
77. $h = -16t^2 + v_0t + s_0$, for t
78. $S = 2\pi rh + 2\pi r^2$, for r

For each equation, **(a)** solve for x in terms of y, and **(b)** solve for y in terms of x. **See Example 8.**

79. $4x^2 - 2xy + 3y^2 = 2$
80. $3y^2 + 4xy - 9x^2 = -1$
81. $2x^2 + 4xy - 3y^2 = 2$
82. $5x^2 - 6xy + 2y^2 = 1$

Evaluate the discriminant for each equation. Then use it to predict the number of distinct solutions, and whether they are rational, irrational, *or* nonreal complex. *Do not solve the equation.* **See Example 9.**

83. $x^2 - 8x + 16 = 0$
84. $x^2 + 4x + 4 = 0$
85. $3x^2 + 5x + 2 = 0$
86. $8x^2 = -14x - 3$
87. $4x^2 = -6x + 3$
88. $2x^2 + 4x + 1 = 0$
89. $9x^2 + 11x + 4 = 0$
90. $3x^2 = 4x - 5$
91. $8x^2 - 72 = 0$

92. Show that the discriminant for the equation

$$\sqrt{2}x^2 + 5x - 3\sqrt{2} = 0$$

is 49. If this equation is completely solved, it can be shown that the solution set is $\left\{-3\sqrt{2}, \frac{\sqrt{2}}{2}\right\}$. We have a discriminant that is positive and a perfect square, yet the two solutions are irrational. Does this contradict the discussion in this section? Explain.

93. Is it possible for the solution set of a quadratic equation with integer coefficients to consist of a single irrational number? Explain.

94. Is it possible for the solution set of a quadratic equation with real coefficients to consist of one real number and one nonreal complex number? Explain.

Find the values of a, b, and c for which the quadratic equation

$$ax^2 + bx + c = 0$$

has the given numbers as solutions. (Hint: Use the zero-factor property in reverse.)

95. 4, 5
96. $-3, 2$
97. $1 + \sqrt{2}, 1 - \sqrt{2}$
98. $i, -i$

Answers:

70. $\{-4, 2 \pm 2i\sqrt{3}\}$
71. $t = \frac{\pm\sqrt{2sg}}{g}$
72. $r = \frac{\pm\sqrt{\mathcal{A}\pi}}{\pi}$
73. $v = \frac{\pm\sqrt{FrkM}}{kM}$
74. $e = \frac{\pm\sqrt{2Erk}}{k}$
75. $t = \frac{\pm\sqrt{2a(r - r_0)}}{a}$
76. $t = \frac{\pm\sqrt{(s - s_0 - k)g}}{g}$
77. $t = \frac{v_0 \pm \sqrt{v_0^2 - 64h + 64s_0}}{32}$
78. $r = \frac{-\pi h \pm \sqrt{\pi^2h^2 + 2\pi S}}{2\pi}$
79. (a) $x = \frac{y \pm \sqrt{8 - 11y^2}}{4}$ (b) $y = \frac{x \pm \sqrt{6 - 11x^2}}{3}$
80. (a) $x = \frac{2y \pm \sqrt{31y^2 + 9}}{9}$ (b) $y = \frac{-2x \pm \sqrt{31x^2 - 3}}{3}$
81. (a) $x = \frac{-2y \pm \sqrt{10y^2 + 4}}{2}$ (b) $y = \frac{2x \pm \sqrt{10x^2 - 6}}{3}$
82. (a) $x = \frac{3y \pm \sqrt{5 - y^2}}{5}$ (b) $y = \frac{3x \pm \sqrt{2 - x^2}}{2}$
83. 0; one rational solution (a double solution)
84. 0; one rational solution (a double solution)
85. 1; two distinct rational solutions
86. 100; two distinct rational solutions
87. 84; two distinct irrational solutions
88. 8; two distinct irrational solutions
89. -23; two distinct nonreal complex solutions
90. -44; two distinct nonreal complex solutions
91. 2304; two distinct rational solutions

In Exercises 95–98, there are other possible answers.
95. $a = 1, b = -9, c = 20$ 96. $a = 1, b = 1, c = -6$ 97. $a = 1, b = -2, c = -1$
98. $a = 1, b = 0, c = 1$

Chapter 1 Quiz (Sections 1.1–1.4)

1. Solve the linear equation $3(x - 5) + 2 = 1 - (4 + 2x)$.

2. Determine whether each equation is an *identity*, a *conditional equation*, or a *contradiction.* Give the solution set.

(a) $4x - 5 = -2(3 - 2x) + 3$ **(b)** $5x - 9 = 5(-2 + x) + 1$
(c) $5x - 4 = 3(6 - x)$

3. Solve the equation $ay + 2x = y + 5x$ for y. Assume $a \neq 1$.

Answers:

[1.1]
1. $\{2\}$
2. (a) contradiction; $\emptyset$ (b) identity; {all real numbers} (c) conditional equation; $\left\{\frac{11}{4}\right\}$
3. $y = \frac{3x}{a - 1}$

[1.2]
4. \$10,000 at 2.5%; \$20,000 at 3%
5. \$5.62; The model predicts a wage that is \$0.47 greater than the actual wage.

[1.3]
6. $-\frac{1}{2} + \frac{\sqrt{6}}{4}i$
7. $\frac{3}{10} - \frac{8}{5}i$

[1.4]
8. $\left\{\frac{1}{6} \pm \frac{\sqrt{11}}{6}i\right\}$
9. $\left\{\pm\sqrt{29}\right\}$
10. $r = \frac{\pm\sqrt{2\mathcal{A}\theta}}{\theta}$

4. *Earning Interest* Johnny Ramistella deposits some money at 2.5% annual interest and twice as much at 3.0%. Find the amount deposited at each rate if his total annual interest income is \$850.

5. *(Modeling) Minimum Hourly Wage* One model for the minimum hourly wage in the United States for the period 1978–2009 is

$$y = 0.126x - 246.25,$$

where x represents the year and y represents the wage, in dollars. (*Source:* Bureau of Labor Statistics.) The actual 1999 minimum wage was \$5.15. What does this model predict as the wage? What is the difference between the actual wage and the predicted wage?

6. Write $\frac{-4 + \sqrt{-24}}{8}$ in standard form $a + bi$.

7. Write the quotient $\frac{7 - 2i}{2 + 4i}$ in standard form $a + bi$.

Solve each equation.

8. $3x^2 - x = -1$

9. $x^2 - 29 = 0$

10. $\mathcal{A} = \frac{1}{2}r^2\theta$, for r

1.5 Applications and Modeling with Quadratic Equations

- Geometry Problems
- Using the Pythagorean Theorem
- Height of a Projected Object
- Modeling with Quadratic Equations

Geometry Problems To solve these applications, we continue to use the problem-solving strategy discussed in **Section 1.2.**

Classroom Example 1
Repeat **Example 1** if the length of the rectangle is twice the width, the squares cut from the corners measure 10 cm on a side, and the volume of the box is to be 7500 cm^3.

Answer: 35 cm by 70 cm

EXAMPLE 1 Solving a Problem Involving Volume

A piece of machinery produces rectangular sheets of metal such that the length is three times the width. Equal-sized squares measuring 5 in. on a side can be cut from the corners so that the resulting piece of metal can be shaped into an open box by folding up the flaps. If specifications call for the volume of the box to be 1435 in.^{3}, find the dimensions of the original piece of metal.

SOLUTION

Step 1 **Read** the problem. We must find the dimensions of the original piece of metal.

Step 2 **Assign a variable.** We know that the length is three times the width.

Let x = the width (in inches) and thus, $3x$ = the length.

The box is formed by cutting $5 + 5 = 10$ in. from both the length and the width. See **Figure 5. Figure 6** indicates that the width of the bottom of the box is $x - 10$, the length of the bottom of the box is $3x - 10$, and the height is 5 in. (the length of the side of each cut-out square).

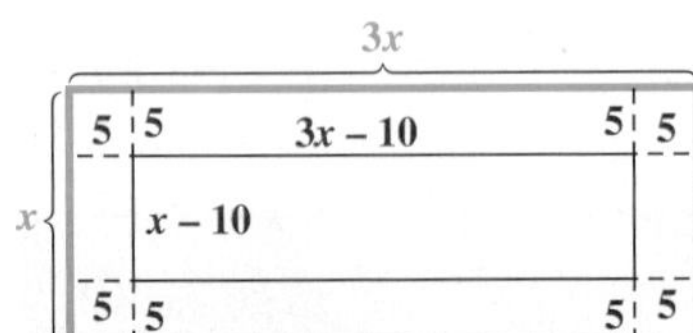

Figure 5

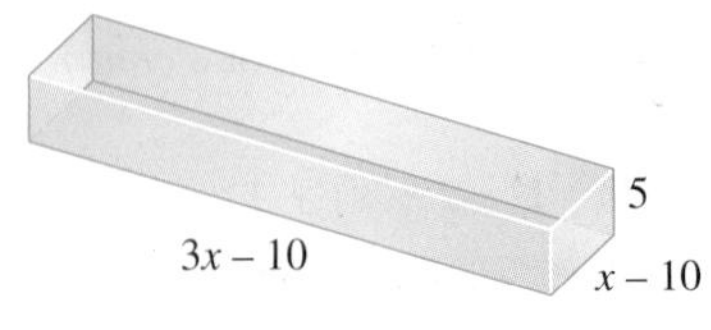

Figure 6

Step 3 **Write an equation.** The formula for volume of a box is $V = lwh$.

$$\begin{array}{ccc} \text{Volume} & = & \text{length} \times \text{width} \times \text{height} \\ \downarrow & & \downarrow \quad\quad \downarrow \quad\quad \downarrow \\ 1435 & = & (3x - 10)(x - 10)(5) \end{array}$$

(Note that the dimensions of the box must be positive numbers, so $3x - 10$ and $x - 10$ must be greater than 0, which implies $x > \frac{10}{3}$ and $x > 10$. These are both satisfied when $x > 10$.)

Step 4 **Solve** the equation.

$1435 = 15x^2 - 200x + 500$		Multiply. (Section R.3)
$0 = 15x^2 - 200x - 935$		Subtract 1435 from each side.
$0 = 3x^2 - 40x - 187$		Divide each side by 5.
$0 = (3x + 11)(x - 17)$		Factor. (Section R.4)
$3x + 11 = 0$	or $x - 17 = 0$	Zero-factor property (Section 1.4)
$x = -\frac{11}{3}$	or $x = 17$	Solve each equation. (Section 1.1)

The width cannot be negative.

Step 5 **State the answer.** Only 17 satisfies the restriction $x > 10$. Thus, the dimensions of the original piece should be 17 in. by $3(17) = 51$ in.

Step 6 **Check.** The length of the bottom of the box is $51 - 2(5) = 41$ in. The width is $17 - 2(5) = 7$ in. The height is 5 in. (the amount cut on each corner), so the volume of the box is

$$V = lwh = 41 \times 7 \times 5 = 1435 \text{ in.}^3, \quad \text{as required.}$$

✔ *Now Try Exercise 23.*

PROBLEM-SOLVING HINT We may get a solution that does not satisfy the physical constraints of a problem. As seen in **Example 1,** a dimension of a box cannot be a negative number such as $-\frac{11}{3}$. Therefore, we reject this outcome.

Using the Pythagorean Theorem **Example 2** requires the use of the **Pythagorean theorem** for right triangles. Recall that the **legs** of a right triangle form the right angle, and the **hypotenuse** is the side opposite the right angle.

Pythagorean Theorem

In a right triangle, the sum of the squares of the lengths of the legs is equal to the square of the length of the hypotenuse.

$$a^2 + b^2 = c^2$$

Hypotenuse c
Leg a
Leg b

Classroom Example 2
Repeat **Example 2** if the longer leg is 10 m shorter than twice the length of the shorter leg and the hypotenuse is 20 m longer than the longer leg.

Answer: 80 m, 150 m, 170 m

EXAMPLE 2 Applying the Pythagorean Theorem

A piece of property has the shape of a right triangle. The longer leg is 20 m longer than twice the length of the shorter leg. The hypotenuse is 10 m longer than the length of the longer leg. Find the lengths of the sides of the triangular lot.

SOLUTION

Step 1 **Read** the problem. We must find the lengths of the three sides.

Step 2 **Assign a variable.**

Let $x =$ the length of the shorter leg (in meters).

Then $2x + 20 =$ the length of the longer leg, and

$(2x + 20) + 10$, or $2x + 30 =$ the length of the hypotenuse.

See **Figure 7.**

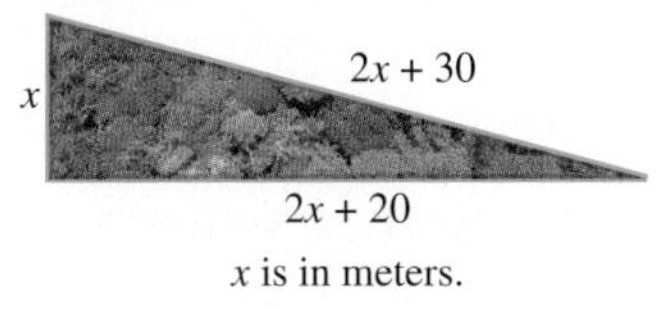

x is in meters.

Figure 7

Step 3 **Write an equation.**

$$\begin{array}{ccccc} a^2 & + & b^2 & = & c^2 \\ \downarrow & & \downarrow & & \downarrow \\ x^2 & + & (2x+20)^2 & = & (2x+30)^2 \end{array}$$

The hypotenuse is c.

Substitute into the Pythagorean theorem.

Step 4 **Solve** the equation.

Remember the middle terms.

$$x^2 + (4x^2 + 80x + 400) = 4x^2 + 120x + 900 \quad \text{Square the binomials. (Section R.3)}$$

$$x^2 - 40x - 500 = 0 \quad \text{Standard form}$$

$$(x - 50)(x + 10) = 0 \quad \text{Factor.}$$

$$x - 50 = 0 \quad \text{or} \quad x + 10 = 0 \quad \text{Zero-factor property}$$

$$x = 50 \quad \text{or} \quad x = -10 \quad \text{Solve each equation.}$$

Step 5 **State the answer.** Since x represents a length, -10 is not reasonable. The lengths of the sides of the triangular lot are 50 m, $2(50) + 20 = 120$ m, and $2(50) + 30 = 130$ m.

Step 6 **Check.** The lengths 50, 120, and 130 satisfy the words of the problem and also satisfy the Pythagorean theorem.

Now Try Exercise 31.

Galileo Galilei (1564–1642)

According to legend, Galileo dropped objects of different weights from the Leaning Tower of Pisa to disprove the Aristotelian view that heavier objects fall faster than lighter objects. He developed the formula $d = 16t^2$ for freely falling objects, where d is the distance in feet that an object falls (neglecting air resistance) in t seconds, regardless of weight.

Height of a Projected Object If air resistance is neglected, the height s (in feet) of an object projected directly upward from an initial height of s_0 feet, with initial velocity v_0 feet per second, is given by the following equation.

$$s = -16t^2 + v_0t + s_0$$

Here t represents the number of seconds after the object is projected. The coefficient of t^2, -16, is a constant based on the gravitational force of Earth. This constant varies on other surfaces, such as the moon and other planets.

Classroom Example 3

If a projectile is launched upward from the ground with an initial velocity of 73.5 m per sec, neglecting air resistance, its height (in meters) above the ground t seconds after projection is given by

$$s = -4.9t^2 + 73.5t.$$

(a) After how many seconds will the projectile be 100 m above the ground?

(b) How long will it take for the projectile to return to the ground?

Answers:

(a) after 1.51 sec and after 13.49 sec

(b) 15 sec

EXAMPLE 3 Solving a Problem Involving Projectile Height

If a projectile is launched vertically upward from the ground with an initial velocity of 100 ft per sec, neglecting air resistance, its height s (in feet) above the ground t seconds after projection is given by $s = -16t^2 + 100t$.

(a) After how many seconds will it be 50 ft above the ground?

(b) How long will it take for the projectile to return to the ground?

SOLUTION

(a) We must find value(s) of t so that height s is 50 ft.

$$50 = -16t^2 + 100t \quad \text{Let } s = 50 \text{ in the given equation.}$$

$$0 = -16t^2 + 100t - 50 \quad \text{Standard form}$$

$$0 = 8t^2 - 50t + 25 \quad \text{Divide by } -2.$$

Substitute carefully.

$$t = \frac{-(-50) \pm \sqrt{(-50)^2 - 4(8)(25)}}{2(8)} \quad \text{Quadratic formula (Section 1.4)}$$

$$t = \frac{50 \pm \sqrt{1700}}{16} \quad \text{Simplify.}$$

$$t \approx 0.55 \quad \text{or} \quad t \approx 5.70 \quad \text{Use a calculator.}$$

Both solutions are acceptable, since the projectile reaches 50 ft twice—once on its way up (after 0.55 sec) and once on its way down (after 5.70 sec).

LOOKING AHEAD TO CALCULUS

In calculus, you will need to be able to write an algebraic expression from the description in a problem like those in this section. Using calculus techniques, you will be asked to find the value of the variable that produces an optimum (a maximum or minimum) value of the expression.

(b) When the projectile returns to the ground, the height s will be 0 ft.

$$0 = -16t^2 + 100t \qquad \text{Let } s = 0.$$

$$0 = -4t(4t - 25) \qquad \text{Factor.}$$

$$-4t = 0 \quad \text{or} \quad 4t - 25 = 0 \qquad \text{Zero-factor property}$$

$$t = 0 \quad \text{or} \quad t = 6.25 \qquad \text{Solve each equation.}$$

The first solution, 0, represents the time at which the projectile was on the ground prior to being launched, so it does not answer the question. The projectile will return to the ground 6.25 sec after it is launched.

✔ *Now Try Exercise 43.*

Modeling with Quadratic Equations

Classroom Example 4
Based on figures from the period 1992–2003, the quadratic equation

$$S = 0.00579x^2 + 0.2579x + 0.9703$$

models sales of SUVs, where S represents sales in millions, and $x = 0$ represents 1992, $x = 1$ represents 1993, and so on.

(a) Use the model to determine sales in 2002. Compare the results to the actual figure of 4.2 million.

(b) According to the model, in what year after 1992 did sales reach 3.5 million? (Round down to the nearest year.)

Answers:

(a) The prediction of 4.1 million is 0.1 million less than the actual figure of 4.2 in 2002.

(b) 2000

EXAMPLE 4 Analyzing Trolley Ridership

The I-Ride Trolley service carries passengers along the International Drive Resort Area of Orlando, Florida. The bar graph in **Figure 8** shows I-Ride Trolley ridership data in millions. The quadratic equation

$$y = -0.0072x^2 + 0.1081x + 1.619$$

models ridership from 2000 to 2010, where y represents ridership in millions and $x = 0$ represents 2000, $x = 1$ represents 2001, and so on.

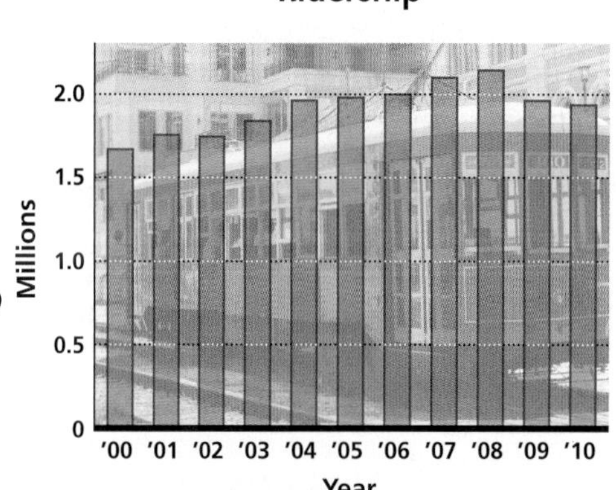

Source: I-Ride Trolley, International Drive Master Transit, www.itrolley.com

Figure 8

(a) Use the model to determine ridership in 2008. Compare the result to the actual ridership figure of 2.1 million.

(b) According to the model, in what year did ridership reach 1.9 million?

SOLUTION

Teaching Tip Point out that the quadratic model provided in **Example 4** is valid only for 2000–2010. Discuss with students the danger of using the quadratic model to extrapolate too far beyond the given years.

(a) Since $x = 0$ represents the year 2000, use $x = 8$ to represent 2008.

$$y = -0.0072x^2 + 0.1081x + 1.619 \qquad \text{Given model}$$

$$y = -0.0072(8)^2 + 0.1081(8) + 1.619 \qquad \text{Let } x = 8.$$

$$y \approx 2.0 \text{ million} \qquad \text{Use a calculator.}$$

The prediction is about 0.1 million (that is, 100,000) less than the actual figure of 2.1 million.

Solve this equation for x.

(b)

$$1.9 = -0.0072x^2 + 0.1081x + 1.619 \qquad \text{Let } y = 1.9 \text{ in the model.}$$

$$0 = -0.0072x^2 + 0.1081x - 0.281 \qquad \text{Standard form}$$

$$x = \frac{-0.1081 \pm \sqrt{(0.1081)^2 - 4(-0.0072)(-0.281)}}{2(-0.0072)} \qquad \text{Quadratic formula}$$

$$x \approx 3.3 \quad \text{or} \quad x \approx 11.7 \qquad \text{Use a calculator.}$$

The year 2003 corresponds to $x = 3.3$. Thus, according to the model, ridership reached 1.9 million in the year 2003. This outcome closely matches the bar graph and seems reasonable.

The year 2011 corresponds to $x = 11.7$. Round down to the year 2011 because 11.7 yr from 2000 occurs during 2011. There is no value on the bar graph to compare this to, because the last data value is for the year 2010. Always view results that are *beyond* the data in your model with skepticism, and realistically consider whether the model will continue as given. The model *predicts* that ridership will be 1.9 million again in the year 2011.

✔ Now Try Exercise 47.

1.5 Exercises

1. A 2. C
3. D 4. B

5. 7, 8 or −8, −7
6. 10, 11 or −11, −10

Concept Check *Answer each question.*

1. *Area of a Parking Lot* For the rectangular parking area of the shopping center shown, which one of the following equations says that the area is 40,000 yd^2?

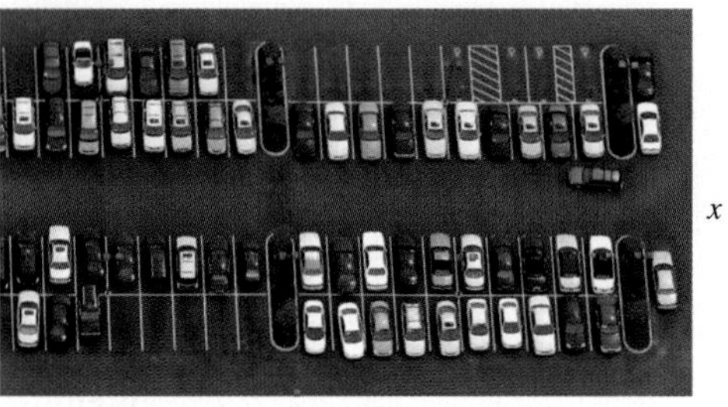

A. $x(2x + 200) = 40{,}000$
B. $2x + 2(2x + 200) = 40{,}000$
C. $x + (2x + 200) = 40{,}000$
D. $x^2 + (2x + 200)^2 = 40{,}000^2$

2. *Diagonal of a Rectangle* If a rectangle is r feet long and s feet wide, which one of the following expressions is the length of its diagonal in terms of r and s?

r
Diagonal
s

A. $\sqrt{rs}$ **B.** $r + s$
C. $\sqrt{r^2 + s^2}$ **D.** $r^2 + s^2$

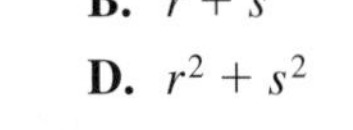

3. *Sides of a Right Triangle* To solve for the lengths of the right triangle sides, which equation is correct?

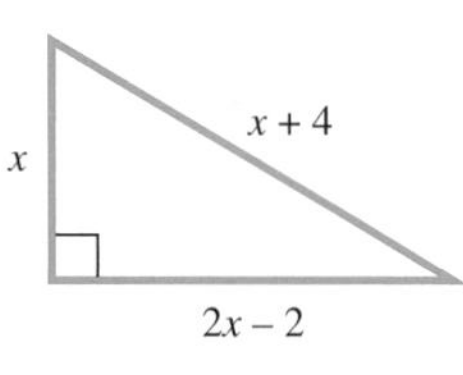

A. $x^2 = (2x - 2)^2 + (x + 4)^2$
B. $x^2 + (x + 4)^2 = (2x - 2)^2$
C. $x^2 = (2x - 2)^2 - (x + 4)^2$
D. $x^2 + (2x - 2)^2 = (x + 4)^2$

4. *Area of a Picture* The mat around the picture shown measures x inches across. Which one of the following equations says that the area of the picture itself is 600 $in.^2$?

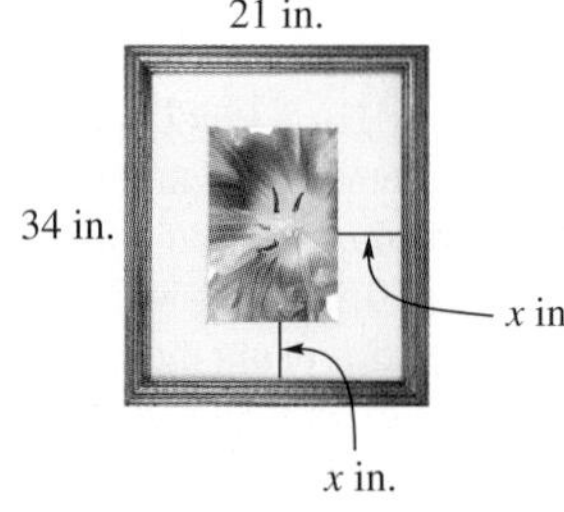

A. $2(34 - 2x) + 2(21 - 2x) = 600$
B. $(34 - 2x)(21 - 2x) = 600$
C. $(34 - x)(21 - x) = 600$
D. $x(34)(21) = 600$

To prepare for the applications that come later, work the following basic problems that lead to quadratic equations.

Unknown Numbers *In Exercises 5–14, use the following facts.*

If x represents an integer, then $x + 1$ represents the next consecutive integer.

If x represents an even integer, then $x + 2$ represents the next consecutive even integer.

If x represents an odd integer, then $x + 2$ represents the next consecutive odd integer.

5. Find two consecutive integers whose product is 56.

6. Find two consecutive integers whose product is 110.

7. 12, 14 or $-14, -12$
8. 14, 16 or $-16, -14$
9. 7, 9 or $-9, -7$
10. 11, 13 or $-13, -11$
11. 9, 11 or $-11, -9$
12. 4, 6 or $-6, -4$
13. 20, 22
14. 7, 9
15. 6, 8, 10
16. 3, 4, 5
17. 7 in., 10 in.
18. 7 in., 12 in.
19. 100 yd by 400 yd
20. 50 m by 100 m
21. 9 ft by 12 ft

7. Find two consecutive even integers whose product is 168.

8. Find two consecutive even integers whose product is 224.

9. Find two consecutive odd integers whose product is 63.

10. Find two consecutive odd integers whose product is 143.

11. The sum of the squares of two consecutive odd integers is 202. Find the integers.

12. The sum of the squares of two consecutive even integers is 52. Find the integers.

13. The difference of the squares of two positive consecutive even integers is 84. Find the integers.

14. The difference of the squares of two positive consecutive odd integers is 32. Find the integers.

15. *Dimensions of a Right Triangle* The lengths of the sides of a right triangle are consecutive even integers. Find these lengths. (*Hint:* Use the Pythagorean theorem.)

16. *Dimensions of a Right Triangle* The lengths of the sides of a right triangle are consecutive positive integers. Find these lengths. (*Hint:* Use the Pythagorean theorem.)

17. *Dimensions of a Square* The length of each side of a square is 3 in. more than the length of each side of a smaller square. The sum of the areas of the squares is 149 in.^{2}. Find the lengths of the sides of the two squares.

18. *Dimensions of a Square* The length of each side of a square is 5 in. more than the length of each side of a smaller square. The difference of the areas of the squares is 95 in.^{2}. Find the lengths of the sides of the two squares.

Solve each problem. **See Example 1.**

19. *Dimensions of a Parking Lot* A parking lot has a rectangular area of 40,000 yd^2. The length is 200 yd more than twice the width. Find the dimensions of the lot. (**See Exercise 1.**)

20. *Dimensions of a Garden* An ecology center wants to set up an experimental garden using 300 m of fencing to enclose a rectangular area of 5000 m^2. Find the dimensions of the garden.

21. *Dimensions of a Rug* Zachary Daniels wants to buy a rug for a room that is 12 ft wide and 15 ft long. He wants to leave a uniform strip of floor around the rug. He can afford to buy 108 ft^2 of carpeting. What dimensions should the rug have?

22. 1 ft
23. 20 in. by 30 in.
24. 20 in. by 40 in.
25. 1 ft
26. $\frac{1}{3}$ ft, or 4 in.
27. 4
28. 6, 12
29. 3.75 cm
30. length: 10.6 in.; width: 7.4 in.
31. 5 ft

22. *Width of a Flower Border* A landscape architect has included a rectangular flower bed measuring 9 ft by 5 ft in her plans for a new building. She wants to use two colors of flowers in the bed, one in the center and the other for a border of the same width on all four sides. If she has enough plants to cover 24 ft^2 for the border, how wide can the border be?

23. *Volume of a Box* A rectangular piece of metal is 10 in. longer than it is wide. Squares with sides 2 in. long are cut from the four corners, and the flaps are folded upward to form an open box. If the volume of the box is 832 $in.^3$, what were the original dimensions of the piece of metal?

24. *Volume of a Box* In **Exercise 23,** suppose that the piece of metal has length twice the width, and 4-in. squares are cut from the corners. If the volume of the box is 1536 $in.^3$, what were the original dimensions of the piece of metal?

25. *Manufacturing to Specifications* A manufacturing firm wants to package its product in a cylindrical container 3 ft high with surface area 8π ft^2. What should the radius of the circular top and bottom of the container be? (*Hint:* The surface area consists of the circular top and bottom and a rectangle that represents the side cut open vertically and unrolled.)

26. *Manufacturing to Specifications* In **Exercise 25,** what radius would produce a container with a volume of π times the radius? (*Hint:* The volume is the area of the circular base times the height.)

27. *Dimensions of a Square* What is the length of the side of a square if its area and perimeter are numerically equal?

28. *Dimensions of a Rectangle* A rectangle has an area that is numerically twice its perimeter. If the length is twice the width, what are its dimensions?

29. *Radius of a Can* A can of Blue Runner Red Kidney Beans has surface area 371 cm^2. Its height is 12 cm. What is the radius of the circular top? Round to the nearest hundredth.

30. *Dimensions of a Cereal Box* The volume of a 10-oz box of corn flakes is 180.4 $in.^3$. The width of the box is 3.2 in. less than the length, and its depth is 2.3 in. Find the length and width of the box to the nearest tenth.

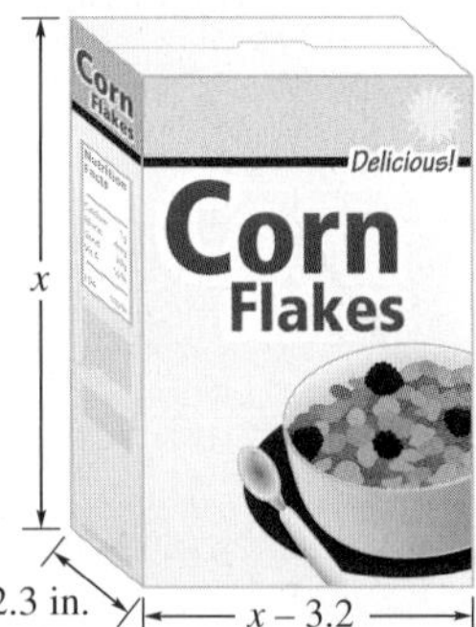

Solve each problem. ***See Example 2.***

31. *Height of a Dock* A boat is being pulled into a dock with a rope attached to the boat at water level. When the boat is 12 ft from the dock, the length of the rope from the boat to the dock is 3 ft longer than twice the height of the dock above the water. Find the height of the dock.

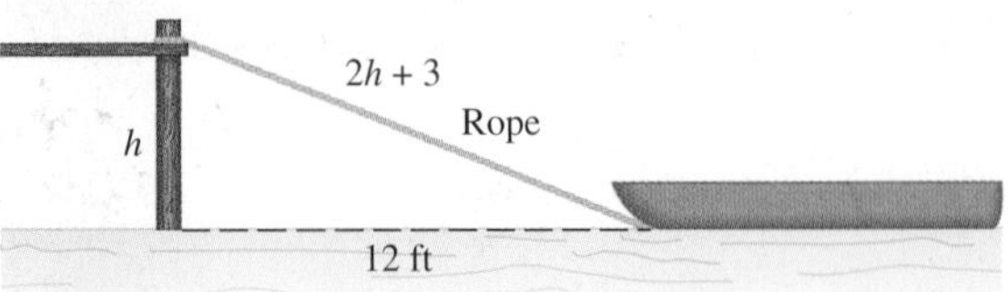

32. horizontal: 30 ft; vertical: 45 ft
33. $10\sqrt{2}$ ft
34. longer leg: 23.3 in.; shorter leg: 11.6 in.
35. 16.4 ft
36. 61 min
37. 3000 yd

32. *Height of a Kite* Grady is flying a kite on 50 ft of string. Its vertical distance from his hand is 10 ft more than the horizontal distance from his hand. Assuming that the string is being held 5 ft above ground level, find its horizontal distance from Grady and its vertical distance from the ground.

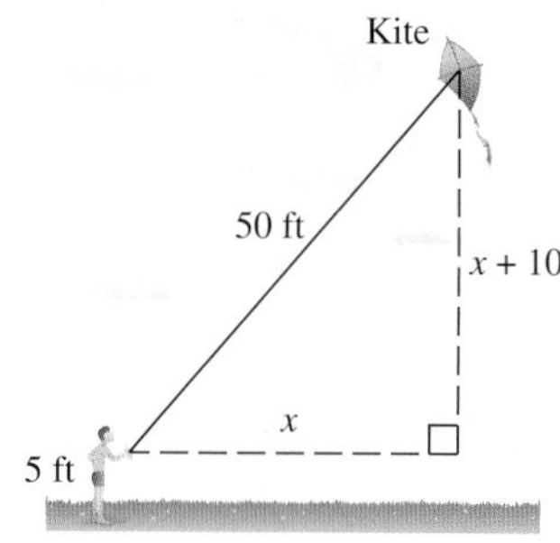

33. *Radius Covered by a Circular Lawn Sprinkler* A square lawn has area 800 ft^2. A sprinkler placed at the center of the lawn sprays water in a circular pattern as shown in the figure. What is the radius of the circle?

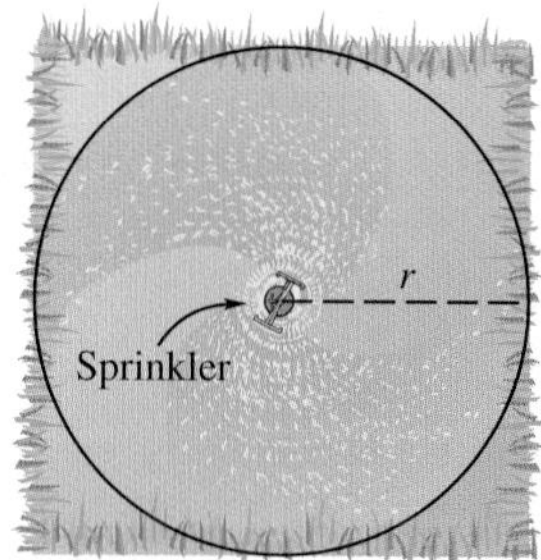

34. *Dimensions of a Solar Panel Frame* Molly has a solar panel with a width of 26 in. To get the proper inclination for her climate, she needs a right triangular support frame that has one leg twice as long as the other. To the nearest tenth of an inch, what dimensions should the frame have?

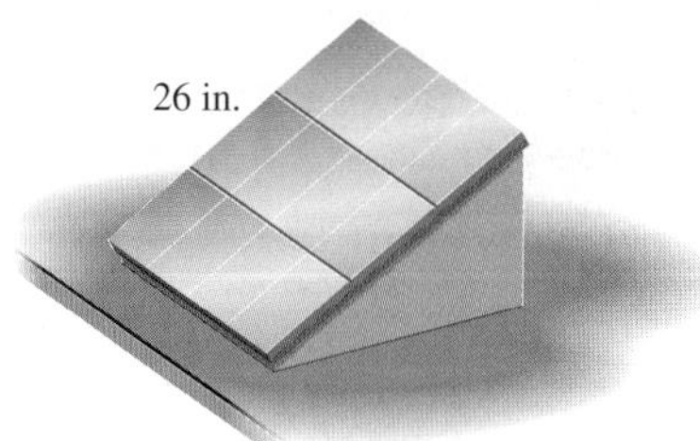

35. *Length of a Ladder* A building is 2 ft from a 9-ft fence that surrounds the property. A worker wants to wash a window in the building 13 ft from the ground. He plans to place a ladder over the fence so it rests against the building. (See the figure.) He decides he should place the ladder 8 ft from the fence for stability. To the nearest tenth of a foot, how long a ladder will he need?

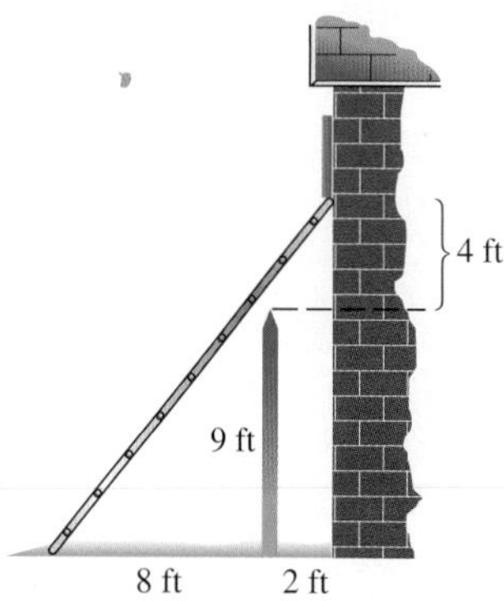

36. *Range of Receivers* Tanner Jones and Sheldon Furst have received communications receivers for Christmas. If they leave from the same point at the same time, Tanner walking north at 2.5 mph and Sheldon walking east at 3 mph, how long will they be able to talk to each other if the range of the communications receivers is 4 mi? Round your answer to the nearest minute.

37. *Length of a Walkway* A nature conservancy group decides to construct a raised wooden walkway through a wetland area. To enclose the most interesting part of the wetlands, the walkway will have the shape of a right triangle with one leg 700 yd longer than the other and the hypotenuse 100 yd longer than the longer leg. Find the total length of the walkway.

38. 4.55 ft

39. (a) 1 sec, 5 sec
(b) 6 sec

40. (a) 0.68 sec, 7.32 sec
(b) 8 sec

41. (a) It will not reach 80 ft.
(b) 2 sec

42. (a) It will not reach 80 ft.
(b) 1 sec

43. (a) 0.19 sec, 10.92 sec
(b) 11.32 sec

44. When the quadratic formula is applied to $-2.7t^2 + 30t + 6.5 = 100$, the discriminant is negative.

45. (a) approximately 19.2 hr
(b) 84.3 ppm (109.8 is not in the interval $[50, 100]$.)

46. (a) 10.3 hr
(b) 722.2 ppm (857.8 is not in the interval $[500, 800]$.)

38. *Broken Bamboo* Problems involving the Pythagorean theorem have appeared in mathematics for thousands of years. This one is taken from the ancient Chinese work, *Arithmetic in Nine Sections:*

> *There is a bamboo 10 ft high, the upper end of which, being broken, reaches the ground 3 ft from the stem. Find the height of the break.*

(Modeling) Solve each problem. **See Examples 3 and 4.**

Height of a Projectile A projectile is launched from ground level with an initial velocity of v_0 feet per second. Neglecting air resistance, its height in feet t seconds after launch is given by

$$s = -16t^2 + v_0 t.$$

*In Exercises 39–42, find the time(s) that the projectile will **(a)** reach a height of* 80 ft *and **(b)** return to the ground for the given value of v_0. Round answers to the nearest hundredth if necessary.*

39. $v_0 = 96$ **40.** $v_0 = 128$ **41.** $v_0 = 32$ **42.** $v_0 = 16$

43. *Height of a Projected Ball* An astronaut on the moon throws a baseball upward. The astronaut is 6 ft, 6 in. tall, and the initial velocity of the ball is 30 ft per sec. The height s of the ball in feet is given by the equation

$$s = -2.7t^2 + 30t + 6.5,$$

where t is the number of seconds after the ball was thrown.

(a) After how many seconds is the ball 12 ft above the moon's surface? Round to the nearest hundredth.

(b) How many seconds will it take for the ball to return to the surface? Round to the nearest hundredth.

44. The ball in **Exercise 43** will never reach a height of 100 ft. How can this be determined algebraically?

45. *Carbon Monoxide Exposure* Carbon monoxide (CO) combines with the hemoglobin of the blood to form carboxyhemoglobin (COHb), which reduces transport of oxygen to tissues. Smokers routinely have a 4% to 6% COHb level in their blood, which can cause symptoms such as blood flow alterations, visual impairment, and poorer vigilance ability. The quadratic model

$$T = 0.00787x^2 - 1.528x + 75.89$$

approximates the exposure time in hours necessary to reach this 4% to 6% level, where $50 \le x \le 100$ is the amount of carbon monoxide present in the air in parts per million (ppm). (*Source: Indoor Air Quality Environmental Information Handbook: Combustion Sources.*)

(a) A kerosene heater or a room full of smokers is capable of producing 50 ppm of carbon monoxide. How long would it take for a nonsmoking person to start feeling the above symptoms?

(b) Find the carbon monoxide concentration necessary for a person to reach the 4% to 6% COHb level in 3 hr. Round to the nearest tenth.

46. *Carbon Monoxide Exposure* Refer to **Exercise 45.** High concentrations of carbon monoxide (CO) can cause coma and death. The time required for a person to reach a COHb level capable of causing a coma can be approximated by the quadratic model

$$T = 0.0002x^2 - 0.316x + 127.9,$$

where T is the exposure time in hours necessary to reach this level and $500 \le x \le 800$ is the amount of carbon monoxide present in the air in parts per million (ppm). (*Source: Indoor Air Quality Environmental Information Handbook: Combustion Sources.*)

(a) What is the exposure time when $x = 600$ ppm?

(b) Estimate the concentration of CO necessary to produce a coma in 4 hr.

47. (a) 781.8 million metric tons
(b) 2005
48. 2010
49. $13,094 million
50. 99.9 million

47. *Methane Gas Emissions* The table gives methane gas emissions from all sources in the United States, in millions of metric tons. The quadratic model

$$y = 1.493x^2 + 7.279x + 684.4$$

approximates the emissions for these years. In the model, x represents the number of years since 2004, so $x = 0$ represents 2004, $x = 1$ represents 2005, and so on.

Year	Millions of Metric Tons of Methane
2004	686.6
2005	691.8
2006	706.3
2007	722.7
2008	737.4

Source: U.S. Energy Information Administration.

(a) According to the model, what would be the emissions in 2010? Round to the nearest tenth of a million metric tons.

(b) Find the year beyond 2004 for which this model predicts that the emissions reached 700 million metric tons.

48. *Cost of Public Colleges* The average cost, in dollars, for tuition and fees for in-state students at four-year public colleges over the period 2000–2010 can be modeled by the equation

$$y = 3.026x^2 + 377.7x + 3449,$$

where $x = 0$ corresponds to 2000, $x = 1$ corresponds to 2001, and so on. Based on this model, for what year after 2000 was the average cost $7605? (*Source:* The College Board, *Annual Survey of Colleges.*)

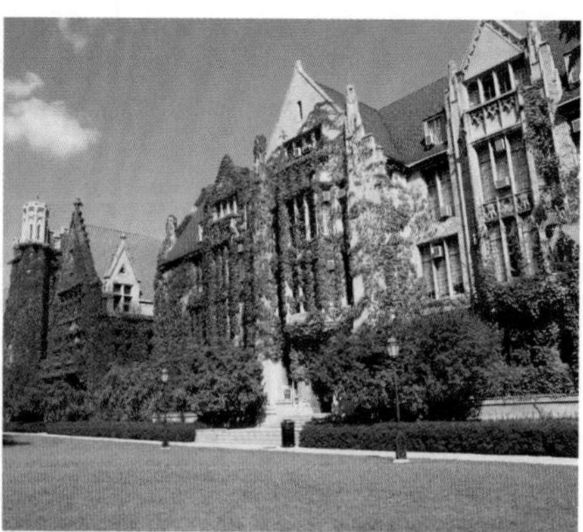

49. *Internet Publishing and Broadcasting* Estimated revenue from Internet publishing and broadcasting in the United States during the years 2004 through 2008 can be modeled by the equation

$$y = 318.4x^2 + 1612x + 8596,$$

where $x = 0$ corresponds to the year 2004, $x = 1$ corresponds to 2005, and so on, and y is in millions of dollars. Approximate the revenue from Internet publishing and broadcasting in 2006 to the nearest million. (*Source:* U.S. Census Bureau.)

50. *Cable TV Subscribers* The number of U.S. households subscribing to cable TV for the period 2000 through 2010 can be modeled by the equation

$$y = -0.0746x^2 + 3.146x + 79.52,$$

where $x = 0$ corresponds to 2000, $x = 1$ corresponds to 2001, and so on, and y is in millions. Based on this model, approximately how many U.S. households, to the nearest tenth of a million, subscribed to cable TV in 2008? (*Source:* Nielsen Media Research.)

Relating Concepts

For individual or collaborative investigation *(Exercises 51–54)*

If p units of an item are sold for x dollars per unit, the revenue is $R = px$. Use this idea to analyze the following problem, ***working Exercises 51–54 in order.***

Number of Apartments Rented *The manager of an 80-unit apartment complex knows from experience that at a rent of $300, all the units will be full. On the average, one additional unit will remain vacant for each $20 increase in rent over $300. Furthermore, the manager must keep at least 30 units rented due to other financial considerations. Currently, the revenue from the complex is $35,000. How many apartments are rented?*

(continued)

51. $80 - x$
52. $300 + 20x$
53. $R = (80 - x)(300 + 20x)$ $= 24{,}000 + 1300x - 20x^2$
54. 10, 55; Because of the restriction, only $x = 10$ is valid. The number of apartments rented is 70.
55. 80
56. 15, 65; Because of the restriction, only $x = 15$ is valid. The number of passengers is 85.
57. 4
58. 5

51. Suppose that x represents the number of \$20 increases over \$300. Represent the number of apartment units that will be rented in terms of x.

52. Represent the rent per unit in terms of x.

53. Use the answers in **Exercises 51 and 52** to write an equation that defines the revenue generated when there are x increases of \$20 over \$300.

54. According to the problem, the revenue currently generated is \$35,000. Substitute this value for revenue into the equation from **Exercise 53.** Solve for x to answer the question in the problem.

*Solve each problem. (**See Relating Concepts Exercises 51–54.**)*

55. *Number of Airline Passengers* The cost of a charter flight to Miami is \$225 each for 75 passengers, with a refund of \$5 per passenger for each passenger in excess of 75. How many passengers must take the flight to produce a revenue of \$16,000?

56. *Number of Bus Passengers* A charter bus company charges a fare of \$40 per person, plus \$2 per person for each unsold seat on the bus. If the bus holds 100 passengers and x represents the number of unsold seats, how many passengers must ride the bus to produce revenue of \$5950? (*Note:* Because of the company's commitment to efficient fuel use, the charter will not run unless filled to at least half-capacity.)

57. *Harvesting a Cherry Orchard* The manager of a cherry orchard wants to schedule the annual harvest. If the cherries are picked now, the average yield per tree will be 100 lb, and the cherries can be sold for 40 cents per pound. Past experience shows that the yield per tree will increase about 5 lb per week, while the price will decrease about 2 cents per pound per week. How many weeks should the manager wait to get an average revenue of \$38.40 per tree?

58. *Recycling Aluminum Cans* A local group of scouts has been collecting old aluminum cans for recycling. The group has already collected 12,000 lb of cans, for which they could currently receive \$4 per hundred pounds. The group can continue to collect cans at the rate of 400 lb per day. However, a glut in the old-can market has caused the recycling company to announce that it will lower its price, starting immediately, by \$0.10 per hundred pounds per day. The scouts can make only one trip to the recycling center. How many days should they wait in order to get \$490 for their cans?

1.6 Other Types of Equations and Applications

- Rational Equations
- Work Rate Problems
- Equations with Radicals
- Equations with Rational Exponents
- Equations Quadratic in Form

Rational Equations A **rational equation** is an equation that has a rational expression for one or more terms. To solve a rational equation, multiply each side by the least common denominator (LCD) of the terms of the equation to eliminate fractions, and then solve the resulting equation.

A value of the variable that appears to be a solution after each side of a rational equation is multiplied by a variable expression (the LCD) is called a **proposed solution.** Because a rational expression is not defined when its denominator is 0, **proposed solutions for which any denominator equals 0 are excluded from the solution set.**

Be sure to check all proposed solutions in the original equation.

Classroom Example 1
Solve each equation.

(a) $\frac{2x-3}{2}+\frac{5x}{x+1}=x$

(b) $\frac{x}{x-5}+5=\frac{5}{x-5}$

Answers:

(a) $\{\frac{3}{7}\}$ (b) $\emptyset$

EXAMPLE 1 Solving Rational Equations That Lead to Linear Equations

Solve each equation.

(a) $\frac{3x-1}{3}-\frac{2x}{x-1}=x$ **(b)** $\frac{x}{x-2}=\frac{2}{x-2}+2$

SOLUTION

(a) The least common denominator is $3(x-1)$, which is equal to 0 if $x=1$. Therefore, 1 cannot possibly be a solution of this equation.

$$\frac{3x-1}{3}-\frac{2x}{x-1}=x$$

$$3(x-1)\left(\frac{3x-1}{3}\right)-3(x-1)\left(\frac{2x}{x-1}\right)=3(x-1)x$$ Multiply by the LCD. $3(x-1)$, where $x \neq 1$. **(Section R.5)**

$$(x-1)(3x-1)-3(2x)=3x(x-1)$$ Divide out common factors.

$$3x^2-4x+1-6x=3x^2-3x$$ Multiply. **(Section R.3)**

$$1-10x=-3x$$ Subtract $3x^2$. Combine like terms.

$$1=7x$$ Solve the linear equation. **(Section 1.1)**

$$x=\frac{1}{7}$$ Proposed solution

The proposed solution $\frac{1}{7}$ meets the requirement that $x \neq 1$ and does not cause any denominator to equal 0. Substitute to check for correct algebra.

CHECK $$\frac{3x-1}{3}-\frac{2x}{x-1}=x$$ Original equation

$$\frac{3\left(\frac{1}{7}\right)-1}{3}-\frac{2\left(\frac{1}{7}\right)}{\frac{1}{7}-1}\stackrel{?}{=}\frac{1}{7}$$ Let $x=\frac{1}{7}$.

$$-\frac{4}{21}-\left(-\frac{1}{3}\right)\stackrel{?}{=}\frac{1}{7}$$ Simplify the complex fractions. **(Section R.5)**

$$\frac{1}{7}=\frac{1}{7}\ \checkmark$$ True

The solution set is $\{\frac{1}{7}\}$.

(b) $$\frac{x}{x-2}=\frac{2}{x-2}+2$$

$$(x-2)\left(\frac{x}{x-2}\right)=(x-2)\left(\frac{2}{x-2}\right)+(x-2)2$$ Multiply by the LCD, $x-2$, where $x \neq 2$.

$$x=2+2(x-2)$$ Divide out common factors.

$$x=2+2x-4$$ Distributive property **(Section R.2)**

$$-x=-2$$ Solve the linear equation.

$$x=2$$ Proposed solution

The proposed solution is 2. However, the variable is restricted to real numbers except 2. If $x=2$, then not only does it cause a zero denominator, but also multiplying by $x-2$ in the first step is multiplying both sides by 0, which is not valid. Thus, the solution set is $\emptyset$.

Teaching Tip When teaching students to solve rational equations, introduce the restriction(s) of the variable in the **first step** as you multiply both sides of the equation by the LCD. This will help students remember to check restrictions on proposed solutions at the end of the process.

Now Try Exercises 7 and 9.

Classroom Example 2
Solve each equation.

(a) $\frac{x-5}{x-3} + \frac{1}{x} = \frac{-7}{x^2-3x}$

(b) $\frac{x}{x+5} + \frac{5}{x-5} = \frac{50}{x^2-25}$

Answers:
(a) $\{2\}$ (b) $\emptyset$

EXAMPLE 2 Solving Rational Equations That Lead to Quadratic Equations

Solve each equation.

(a) $\frac{3x+2}{x-2} + \frac{1}{x} = \frac{-2}{x^2-2x}$ **(b)** $\frac{-4x}{x-1} + \frac{4}{x+1} = \frac{-8}{x^2-1}$

SOLUTION

(a)

$$\frac{3x+2}{x-2} + \frac{1}{x} = \frac{-2}{x^2-2x}$$

$$\frac{3x+2}{x-2} + \frac{1}{x} = \frac{-2}{x(x-2)} \quad \text{Factor the last denominator. (Section R.4)}$$

$$x(x-2)\left(\frac{3x+2}{x-2}\right) + x(x-2)\left(\frac{1}{x}\right) = x(x-2)\left(\frac{-2}{x(x-2)}\right) \quad \text{Multiply by } x(x-2),\ x \neq 0, 2.$$

$$x(3x+2) + (x-2) = -2 \quad \text{Divide out common factors.}$$

$$3x^2 + 2x + x - 2 = -2 \quad \text{Distributive property}$$

$$3x^2 + 3x = 0 \quad \text{Standard form (Section 1.4)}$$

$$3x(x+1) = 0 \quad \text{Factor.}$$

Set *each* factor equal to 0.

$$3x = 0 \quad \text{or} \quad x+1 = 0 \quad \text{Zero-factor property (Section 1.4)}$$

$$x = 0 \quad \text{or} \quad x = -1 \quad \text{Proposed solutions}$$

Because of the restriction $x \neq 0$, the only valid proposed solution is -1. Check -1 in the original equation. The solution set is $\{-1\}$.

(b)

$$\frac{-4x}{x-1} + \frac{4}{x+1} = \frac{-8}{x^2-1}$$

$$\frac{-4x}{x-1} + \frac{4}{x+1} = \frac{-8}{(x+1)(x-1)} \quad \text{Factor.}$$

The restrictions on x are $x \neq \pm 1$. Multiply by the LCD, $(x+1)(x-1)$.

$$(x+1)(x-1)\left(\frac{-4x}{x-1}\right) + (x+1)(x-1)\left(\frac{4}{x+1}\right) = (x+1)(x-1)\left(\frac{-8}{(x+1)(x-1)}\right)$$

$$-4x(x+1) + 4(x-1) = -8 \quad \text{Divide out common factors.}$$

$$-4x^2 - 4x + 4x - 4 = -8 \quad \text{Distributive property}$$

$$-4x^2 + 4 = 0 \quad \text{Standard form}$$

$$x^2 - 1 = 0 \quad \text{Divide by } -4.$$

$$(x+1)(x-1) = 0 \quad \text{Factor.}$$

$$x+1 = 0 \quad \text{or} \quad x-1 = 0 \quad \text{Zero-factor property}$$

$$x = -1 \quad \text{or} \quad x = 1 \quad \text{Proposed solutions}$$

Neither proposed solution is valid, so the solution set is $\emptyset$.

Now Try Exercises 15 and 17.

Work Rate Problems If a job can be completed in 3 hr, then the rate of work is $\frac{1}{3}$ of the job per hr. After 1 hr the job would be $\frac{1}{3}$ complete, and after 2 hr the job would be $\frac{2}{3}$ complete. In 3 hr the job would be $\frac{3}{3}$ complete, meaning that 1 complete job had been accomplished.

PROBLEM-SOLVING HINT If a job can be completed in t units of time, then the rate of work, r, is $\frac{1}{t}$ of the job per unit time.

$$r = \frac{1}{t}$$

The amount of work completed, A, is found by multiplying the rate of work, r, and the amount of time worked, t. This formula is similar to the distance formula $d = rt$.

Amount of work completed = rate of work × amount of time worked

or $$A = rt$$

Classroom Example 3
Lisa and Keith are raking the leaves in their backyard. Working alone, Lisa can rake the leaves in 5 hr, while Keith can rake them in 4 hr. How long would it take them to rake the leaves working together?

Answer: $2\frac{2}{9}$ hr

Teaching Tip Remind students that many applications ask for *two or more* answers, as in **Example 3.** Also, caution students to carefully define variables.

For example, if the expression for the faster printer's work rate in **Example 3** was given by $\frac{1}{2x}$ and the slower printer's work rate was given by $\frac{1}{x}$, then the solutions would have been reversed. A check of the answer would reveal that the solutions don't make sense.

EXAMPLE 3 Solving a Work Rate Problem

One printer can do a job twice as fast as another. Working together, both printers can do the job in 2 hr. How long would it take each printer, working alone, to do the job?

SOLUTION

Step 1 **Read** the problem. We must find the time it would take each printer, working alone, to do the job.

Step 2 **Assign a variable.** Let x represent the number of hours it would take the faster printer, working alone, to do the job. The time for the slower printer to do the job alone is then $2x$ hours.

Therefore, $\frac{1}{x}$ = the rate of the faster printer (job per hour)

and $\frac{1}{2x}$ = the rate of the slower printer (job per hour).

The time for the printers to do the job together is 2 hr. Multiplying each rate by the time will give the fractional part of the job accomplished by each.

	Rate	Time	Part of the Job Accomplished
Faster Printer	$\frac{1}{x}$	2	$2\left(\frac{1}{x}\right) = \frac{2}{x}$
Slower Printer	$\frac{1}{2x}$	2	$2\left(\frac{1}{2x}\right) = \frac{1}{x}$

← $A = rt$

Step 3 **Write an equation.** The sum of the two parts of the job accomplished is 1, since one whole job is done.

Part of the job done by the faster printer + Part of the job done by the slower printer = One whole job

$$\frac{2}{x} + \frac{1}{x} = 1$$

Step 4 **Solve.**

$$x\left(\frac{2}{x} + \frac{1}{x}\right) = x(1) \quad \text{Multiply each side by } x \text{, where } x \neq 0.$$

$$x\left(\frac{2}{x}\right) + x\left(\frac{1}{x}\right) = x(1) \quad \text{Distributive property}$$

$$2 + 1 = x \quad \text{Multiply.}$$

$$3 = x \quad \text{Add.}$$

Step 5 **State the answer.** The faster printer would take 3 hr to do the job alone, and the slower printer would take $2(3) = 6$ hr. Be sure to give *both* answers here.

Step 6 **Check.** The answer is reasonable, since the time working together (2 hr, as stated in the problem) is less than the time it would take the faster printer working alone (3 hr, as found in Step 4).

Now Try Exercise 29.

NOTE **Example 3** can also be solved by using the fact that the sum of the rates of the individual printers is equal to their rate working together. Since the printers can complete the job together in 2 hr, their combined rate is $\frac{1}{2}$ of the job per hr.

$$\frac{1}{x} + \frac{1}{2x} = \frac{1}{2}$$

$$2x\left(\frac{1}{x} + \frac{1}{2x}\right) = 2x\left(\frac{1}{2}\right) \quad \text{Multiply each side by } 2x.$$

$$2 + 1 = x \quad \text{Distributive property}$$

$$x = 3 \quad \text{Same solution found earlier}$$

Equations with Radicals

To solve an equation such as

$$x - \sqrt{15 - 2x} = 0,$$

in which the variable appears in a radicand, we use the following **power property** to eliminate the radical.

Power Property

If P and Q are algebraic expressions, then every solution of the equation $P = Q$ is also a solution of the equation $P^n = Q^n$, for any positive integer n.

When the power property is used to solve equations, the new equation may have *more* solutions than the original equation. For example, the equation

$$x = -2 \quad \text{has solution set} \quad \{-2\}.$$

If we square each side of the equation $x = -2$, we obtain the new equation

$$x^2 = 4, \quad \text{which has solution set} \quad \{-2, 2\}.$$

Since the solution sets are not equal, the equations are not equivalent. ***When we use the power property to solve an equation, it is essential to check all proposed solutions in the original equation.***

CAUTION ***Be very careful when using the power property.*** It does *not* say that the equations $P = Q$ and $P^n = Q^n$ are equivalent. It says only that each solution of the original equation $P = Q$ is also a solution of the new equation $P^n = Q^n$.

Solving an Equation Involving Radicals

To solve an equation containing radicals, follow these steps.

Step 1 Isolate the radical on one side of the equation.

Step 2 Raise each side of the equation to a power that is the same as the index of the radical so that the radical is eliminated.

If the equation still contains a radical, repeat Steps 1 and 2.

Step 3 Solve the resulting equation.

Step 4 Check each proposed solution in the *original* equation.

Classroom Example 4
Solve $x - \sqrt{4x + 12} = 0$.

Answer: $\{6\}$

EXAMPLE 4 Solving an Equation Containing a Radical (Square Root)

Solve $x - \sqrt{15 - 2x} = 0$.

SOLUTION

$$x - \sqrt{15 - 2x} = 0$$

$$x = \sqrt{15 - 2x} \quad \text{Isolate the radical. (Step 1)}$$

$$x^2 = \left(\sqrt{15 - 2x}\right)^2 \quad \text{Square each side. (Step 2)}$$

$$x^2 = 15 - 2x \quad \left(\sqrt{a}\right)^2 = a, \text{ for } a \ge 0. \text{ (Section R.7)}$$

$$x^2 + 2x - 15 = 0 \quad \text{Solve the quadratic equation. (Step 3)}$$

$$(x + 5)(x - 3) = 0 \quad \text{Factor.}$$

$$x + 5 = 0 \quad \text{or} \quad x - 3 = 0 \quad \text{Zero-factor property}$$

$$x = -5 \quad \text{or} \quad x = 3 \quad \text{Proposed solutions}$$

CHECK $x - \sqrt{15 - 2x} = 0$ Original equation (Step 4)

Let $x = -5$.	Let $x = 3$.
$-5 - \sqrt{15 - 2(-5)} \stackrel{?}{=} 0$	$3 - \sqrt{15 - 2(3)} \stackrel{?}{=} 0$
$-5 - \sqrt{25} \stackrel{?}{=} 0$	$3 - \sqrt{9} \stackrel{?}{=} 0$
$-5 - 5 \stackrel{?}{=} 0$	$3 - 3 \stackrel{?}{=} 0$
$-10 = 0$ False	$0 = 0$ ✓ True

As the check shows, only 3 is a solution, so the solution set is $\{3\}$.

Now Try Exercise 35.

Classroom Example 5
Solve $\sqrt{3x + 1} - \sqrt{x + 4} = 1$.

Answer: $\{5\}$

EXAMPLE 5 Solving an Equation Containing Two Radicals

Solve $\sqrt{2x + 3} - \sqrt{x + 1} = 1$.

SOLUTION

$$\sqrt{2x + 3} - \sqrt{x + 1} = 1$$

Isolate one of the radicals on one side of the equation.

$$\sqrt{2x + 3} = 1 + \sqrt{x + 1} \quad \text{Isolate } \sqrt{2x + 3}. \text{ (Step 1)}$$

$$\left(\sqrt{2x + 3}\right)^2 = \left(1 + \sqrt{x + 1}\right)^2 \quad \text{Square each side. (Step 2)}$$

$$2x + 3 = 1 + 2\sqrt{x + 1} + (x + 1)$$

Don't forget this term when squaring.

Be careful: $(a + b)^2 = a^2 + 2ab + b^2$ (Section R.3)

$$x + 1 = 2\sqrt{x + 1} \quad \text{Isolate the remaining radical. (Step 1)}$$

$$(x+1)^2 = \left(2\sqrt{x+1}\right)^2 \quad \text{Square again. (Step 2)}$$

$$x^2 + 2x + 1 = 4(x+1) \quad (ab)^2 = a^2b^2 \quad \text{Apply the exponents.}$$

$$x^2 + 2x + 1 = 4x + 4 \quad \text{Distributive property}$$

$$x^2 - 2x - 3 = 0 \quad \text{Solve the quadratic equation. (Step 3)}$$

$$(x-3)(x+1) = 0 \quad \text{Factor.}$$

$$x - 3 = 0 \quad \text{or} \quad x + 1 = 0 \quad \text{Zero-factor property}$$

$$x = 3 \quad \text{or} \quad x = -1 \quad \text{Proposed solutions}$$

CHECK $\sqrt{2x+3} - \sqrt{x+1} = 1$ Original equation (Step 4)

Let $x = 3$.	Let $x = -1$.
$\sqrt{2(3)+3} - \sqrt{3+1} \stackrel{?}{=} 1$	$\sqrt{2(-1)+3} - \sqrt{-1+1} \stackrel{?}{=} 1$
$\sqrt{9} - \sqrt{4} \stackrel{?}{=} 1$	$\sqrt{1} - \sqrt{0} \stackrel{?}{=} 1$
$3 - 2 \stackrel{?}{=} 1$	$1 - 0 \stackrel{?}{=} 1$
$1 = 1$ ✓ True	$1 = 1$ ✓ True

Both 3 and -1 are solutions of the original equation, so $\{-1, 3\}$ is the solution set.

Teaching Tip Students may be tempted to reject the negative solution $x = -1$ in **Example 5** because of the radical sign in the original equation. Explain that -1 is a valid solution because the expressions under the radical sign are nonnegative when -1 is substituted for x. Emphasize checking solutions. Point out that the equation

$$-x = \sqrt{15 - 2x}$$

has proposed solutions $x = -5$ and $x = 3$, and the *positive* solution $x = 3$ must be rejected.

Now Try Exercise 47.

CAUTION Remember to isolate a radical in Step 1. It would be incorrect to square each term individually as the first step in **Example 5.**

EXAMPLE 6 Solving an Equation Containing a Radical (Cube Root)

Solve $\sqrt[3]{4x^2 - 4x + 1} - \sqrt[3]{x} = 0$.

Classroom Example 6
Solve

$$\sqrt[3]{5x^2 - 12x + 6} - \sqrt[3]{x} = 0.$$

Answer: $\left\{\frac{3}{5}, 2\right\}$

SOLUTION

$$\sqrt[3]{4x^2 - 4x + 1} - \sqrt[3]{x} = 0$$

$$\sqrt[3]{4x^2 - 4x + 1} = \sqrt[3]{x} \quad \text{Isolate a radical. (Step 1)}$$

$$\left(\sqrt[3]{4x^2 - 4x + 1}\right)^3 = \left(\sqrt[3]{x}\right)^3 \quad \text{Cube each side. (Step 2)}$$

$$4x^2 - 4x + 1 = x \quad \text{Apply the exponents.}$$

$$4x^2 - 5x + 1 = 0 \quad \text{Solve the quadratic equation. (Step 3)}$$

$$(4x - 1)(x - 1) = 0 \quad \text{Factor.}$$

$$4x - 1 = 0 \quad \text{or} \quad x - 1 = 0 \quad \text{Zero-factor property}$$

$$x = \frac{1}{4} \quad \text{or} \quad x = 1 \quad \text{Proposed solutions}$$

CHECK $\sqrt[3]{4x^2 - 4x + 1} - \sqrt[3]{x} = 0$ Original equation (Step 4)

Let $x = \frac{1}{4}$.	Let $x = 1$.
$\sqrt[3]{4\left(\frac{1}{4}\right)^2 - 4\left(\frac{1}{4}\right) + 1} - \sqrt[3]{\frac{1}{4}} \stackrel{?}{=} 0$	$\sqrt[3]{4(1)^2 - 4(1) + 1} - \sqrt[3]{1} \stackrel{?}{=} 0$
$\sqrt[3]{\frac{1}{4}} - \sqrt[3]{\frac{1}{4}} \stackrel{?}{=} 0$	$\sqrt[3]{1} - \sqrt[3]{1} \stackrel{?}{=} 0$
$0 = 0$ ✓ True	$0 = 0$ ✓ True

Both are valid solutions, and the solution set is $\left\{\frac{1}{4}, 1\right\}$.

Now Try Exercise 59.

Equations with Rational Exponents An **equation with a rational exponent** contains a variable, or variable expression, raised to an exponent that is a rational number. For example, the radical equation

$$\left(\sqrt[5]{x}\right)^3 = 27 \quad \text{can be written with a rational exponent as} \quad x^{3/5} = 27$$

and solved by raising each side to the reciprocal of the exponent, with care taken regarding signs as seen in **Example 7(b).**

Classroom Example 7
Solve each equation.
(a) $x^{3/2} = 216$
(b) $(x + 7)^{2/5} = 4$

Answers:
(a) $\{36\}$ **(b)** $\{-39, 25\}$

EXAMPLE 7 Solving Equations with Rational Exponents

Solve each equation.

(a) $x^{3/5} = 27$ **(b)** $(x - 4)^{2/3} = 16$

SOLUTION

(a)

$$x^{3/5} = 27$$

$$\left(x^{3/5}\right)^{5/3} = 27^{5/3} \quad \text{Raise each side to the power } \tfrac{5}{3}\text{, the reciprocal of the exponent of } x.$$

$$x = 243 \quad 27^{5/3} = \left(\sqrt[3]{27}\right)^5 = 3^5 = 243$$

CHECK Let $x = 243$ in the original equation.

$$x^{3/5} = 243^{3/5} = \left(\sqrt[5]{243}\right)^3 = 27 \ ✓ \ \text{True}$$

The solution set is $\{243\}$.

(b)

$$(x - 4)^{2/3} = 16 \quad \text{Raise each side to the power } \tfrac{3}{2}\text{. Insert } \pm \text{ since this involves an even root, as indicated by the 2 in the denominator.}$$

$$\left[(x - 4)^{2/3}\right]^{3/2} = \pm 16^{3/2}$$

$$x - 4 = \pm 64 \quad \pm 16^{3/2} = \pm\left(\sqrt{16}\right)^3 = \pm 4^3 = \pm 64$$

$$x = 4 \pm 64 \quad \text{Add 4 to each side.}$$

$$x = -60 \quad \text{or} \quad x = 68 \quad \text{Proposed solutions}$$

CHECK $(x - 4)^{2/3} = 16$ Original equation

$(-60 - 4)^{2/3} \stackrel{?}{=} 16$ Let $x = -60$.	$(68 - 4)^{2/3} \stackrel{?}{=} 16$ Let $x = 68$.
$(-64)^{2/3} \stackrel{?}{=} 16$	$64^{2/3} \stackrel{?}{=} 16$
$16 = 16$ ✓ True	$16 = 16$ ✓ True

Both proposed solutions check, so the solution set is $\{-60, 68\}$.

✔ ***Now Try Exercises 65 and 69.***

Equations Quadratic in Form The equation

$$(x + 1)^{2/3} - (x + 1)^{1/3} - 2 = 0$$

is not a quadratic equation in x. However, with the substitutions

$$u = (x + 1)^{1/3} \quad \text{and} \quad u^2 = \left[(x + 1)^{1/3}\right]^2 = (x + 1)^{2/3},$$

the equation becomes

$$u^2 - u - 2 = 0,$$

which is a quadratic equation in u. This quadratic equation can be solved to find u, and then $u = (x + 1)^{1/3}$ can be used to find the values of x, the solutions to the original equation.

Equation Quadratic in Form

An equation is said to be **quadratic in form** if it can be written as

$$au^2 + bu + c = 0,$$

where $a \neq 0$ and u is some algebraic expression.

Classroom Example 8
Solve each equation.
(a) $(x-3)^{1/2} - 6(x-3)^{1/4} + 8 = 0$
(b) $15x^{-2} - 4x^{-1} = 3$

Answers:
(a) $\{19, 259\}$ (b) $\{-3, \frac{5}{3}\}$

EXAMPLE 8 Solving Equations Quadratic in Form

Solve each equation.

(a) $(x+1)^{2/3} - (x+1)^{1/3} - 2 = 0$ **(b)** $6x^{-2} + x^{-1} = 2$

SOLUTION

(a) $(x+1)^{2/3} - (x+1)^{1/3} - 2 = 0$ Since $(x+1)^{2/3} = [(x+1)^{1/3}]^2$, let $u = (x+1)^{1/3}$.

$u^2 - u - 2 = 0$

$(u-2)(u+1) = 0$ Factor.

$u - 2 = 0$ or $u + 1 = 0$ Zero-factor property

$u = 2$ or $u = -1$ Solve each equation.

Don't forget this step.

$(x+1)^{1/3} = 2$ or $(x+1)^{1/3} = -1$ Replace u with $(x+1)^{1/3}$.

$[(x+1)^{1/3}]^3 = 2^3$ or $[(x+1)^{1/3}]^3 = (-1)^3$ Cube each side.

$x + 1 = 8$ or $x + 1 = -1$ Apply the exponents.

$x = 7$ or $x = -2$ Proposed solutions

CHECK $(x+1)^{2/3} - (x+1)^{1/3} - 2 = 0$ Original equation

Let $x = 7$.	Let $x = -2$.
$(7+1)^{2/3} - (7+1)^{1/3} - 2 \stackrel{?}{=} 0$	$(-2+1)^{2/3} - (-2+1)^{1/3} - 2 \stackrel{?}{=} 0$
$8^{2/3} - 8^{1/3} - 2 \stackrel{?}{=} 0$	$(-1)^{2/3} - (-1)^{1/3} - 2 \stackrel{?}{=} 0$
$4 - 2 - 2 \stackrel{?}{=} 0$	$1 + 1 - 2 \stackrel{?}{=} 0$
$0 = 0$ ✓ True	$0 = 0$ ✓ True

Both proposed solutions check, so the solution set is $\{-2, 7\}$.

(b) $6x^{-2} + x^{-1} = 2$

$6x^{-2} + x^{-1} - 2 = 0$ Subtract 2 from each side.

$6u^2 + u - 2 = 0$ Let $u = x^{-1}$. Then $u^2 = x^{-2}$.

$(3u+2)(2u-1) = 0$ Factor.

$3u + 2 = 0$ or $2u - 1 = 0$ Zero-factor property

Don't stop here. Remember to substitute for u.

$u = -\frac{2}{3}$ or $u = \frac{1}{2}$ Solve each equation.

$x^{-1} = -\frac{2}{3}$ or $x^{-1} = \frac{1}{2}$ Replace u with x^{-1}.

$x = -\frac{3}{2}$ or $x = 2$ x^{-1} is the reciprocal of x. **(Section R.6)**

Both proposed solutions check, so the solution set is $\{-\frac{3}{2}, 2\}$.

Now Try Exercises 83 and 87.

CAUTION ***When using a substitution variable in solving an equation that is quadratic in form, do not forget the step that gives the solution in terms of the original variable.***

Classroom Example 9
Solve $18x^4 - 29x^2 + 3 = 0$.

Answer: $\left\{\pm\frac{1}{3}, \pm\frac{\sqrt{6}}{2}\right\}$

EXAMPLE 9 Solving an Equation Quadratic in Form

Solve $12x^4 - 11x^2 + 2 = 0$.

SOLUTION

$$12x^4 - 11x^2 + 2 = 0$$

$$12(x^2)^2 - 11x^2 + 2 = 0 \qquad x^4 = (x^2)^2$$

$$12u^2 - 11u + 2 = 0 \qquad \text{Let } u = x^2. \text{ Then } u^2 = x^4.$$

$$(3u - 2)(4u - 1) = 0 \qquad \text{Solve the quadratic equation.}$$

$$3u - 2 = 0 \quad \text{or} \quad 4u - 1 = 0 \qquad \text{Zero-factor property}$$

$$u = \frac{2}{3} \quad \text{or} \quad u = \frac{1}{4} \qquad \text{Solve each equation.}$$

$$x^2 = \frac{2}{3} \quad \text{or} \quad x^2 = \frac{1}{4} \qquad \text{Replace } u \text{ with } x^2.$$

$$x = \pm\sqrt{\frac{2}{3}} \quad \text{or} \quad x = \pm\sqrt{\frac{1}{4}} \qquad \text{Square root property (Section 1.4)}$$

$$x = \frac{\pm\sqrt{2}}{\sqrt{3}} \cdot \frac{\sqrt{3}}{\sqrt{3}} \quad \text{or} \quad x = \pm\frac{1}{2} \qquad \text{Simplify radicals. (Section R.7)}$$

$$x = \pm\frac{\sqrt{6}}{3}$$

Check that the solution set is $\left\{\pm\frac{\sqrt{6}}{3}, \pm\frac{1}{2}\right\}$.

Now Try Exercise 77.

NOTE Some equations that are quadratic in form are simple enough to avoid using the substitution variable technique. For example, to solve

$$12x^4 - 11x^2 + 2 = 0, \qquad \text{Equation from Example 9}$$

we could factor $12x^4 - 11x^2 + 2$ directly as $(3x^2 - 2)(4x^2 - 1)$, set each factor equal to zero, and then solve the resulting two quadratic equations. ***Which method to use is a matter of personal preference.***

1.6 Exercises

1. $-\frac{3}{2}, 6$ 2. $-1, \frac{2}{5}$
3. $2, -1$ 4. $-3, 1$
5. 0 6. 0

Decide what values of the variable cannot possibly be solutions for each equation. Do not solve. ***See Examples 1 and 2.***

1. $\frac{5}{2x+3} - \frac{1}{x-6} = 0$

2. $\frac{2}{x+1} + \frac{3}{5x-2} = 0$

3. $\frac{3}{x-2} + \frac{1}{x+1} = \frac{3}{x^2-x-2}$

4. $\frac{2}{x+3} - \frac{5}{x-1} = \frac{-5}{x^2+2x-3}$

5. $\frac{1}{4x} - \frac{2}{x} = 3$

6. $\frac{5}{2x} + \frac{2}{x} = 6$

7. $\{-10\}$ 8. $\left\{\frac{3}{5}\right\}$
9. $\emptyset$ 10. $\emptyset$
11. $\emptyset$ 12. $\emptyset$
13. $\{-9\}$ 14. $\{3\}$
15. $\{-2\}$ 16. $\left\{-\frac{1}{4}\right\}$
17. $\emptyset$ 18. $\emptyset$
19. $\left\{-\frac{5}{2}, \frac{1}{9}\right\}$ 20. $\left\{-\frac{1}{3}, \frac{7}{2}\right\}$
21. $\left\{\frac{3}{4}, 1\right\}$ 22. $\left\{\frac{9}{4}, \frac{4}{3}\right\}$
23. $\{3, 5\}$ 24. $\left\{-\frac{3}{2}, 4\right\}$
25. $\left\{-2, \frac{5}{4}\right\}$ 26. $\left\{-1, \frac{2}{3}\right\}$
27. $1\frac{7}{8}$ hr 28. $3\frac{3}{7}$ hr
29. 78 hr 30. 16 hr
31. $13\frac{1}{3}$ hr

Solve each equation. ***See Example 1.***

7. $\dfrac{2x+5}{2} - \dfrac{3x}{x-2} = x$

8. $\dfrac{4x+3}{4} - \dfrac{2x}{x+1} = x$

9. $\dfrac{x}{x-3} = \dfrac{3}{x-3} + 3$

10. $\dfrac{x}{x-4} = \dfrac{4}{x-4} + 4$

11. $\dfrac{-2}{x-3} + \dfrac{3}{x+3} = \dfrac{-12}{x^2-9}$

12. $\dfrac{3}{x-2} + \dfrac{1}{x+2} = \dfrac{12}{x^2-4}$

13. $\dfrac{4}{x^2+x-6} - \dfrac{1}{x^2-4} = \dfrac{2}{x^2+5x+6}$

14. $\dfrac{3}{x^2+x-2} - \dfrac{1}{x^2-1} = \dfrac{7}{2x^2+6x+4}$

Solve each equation. ***See Example 2.***

15. $\dfrac{2x+1}{x-2} + \dfrac{3}{x} = \dfrac{-6}{x^2-2x}$

16. $\dfrac{4x+3}{x+1} + \dfrac{2}{x} = \dfrac{1}{x^2+x}$

17. $\dfrac{x}{x-1} - \dfrac{1}{x+1} = \dfrac{2}{x^2-1}$

18. $\dfrac{-x}{x+1} - \dfrac{1}{x-1} = \dfrac{-2}{x^2-1}$

19. $\dfrac{5}{x^2} - \dfrac{43}{x} = 18$

20. $\dfrac{7}{x^2} + \dfrac{19}{x} = 6$

21. $2 = \dfrac{3}{2x-1} + \dfrac{-1}{(2x-1)^2}$

22. $6 = \dfrac{7}{2x-3} + \dfrac{3}{(2x-3)^2}$

23. $\dfrac{2x-5}{x} = \dfrac{x-2}{3}$

24. $\dfrac{x+4}{2x} = \dfrac{x-1}{3}$

25. $\dfrac{2x}{x-2} = 5 + \dfrac{4x^2}{x-2}$

26. $\dfrac{3x^2}{x-1} + 2 = \dfrac{x}{x-1}$

Solve each problem. ***See Example 3.***

27. *Painting a House* (This problem appears in the 1994 movie *Little Big League*.) If Joe can paint a house in 3 hr, and Sam can paint the same house in 5 hr, how long does it take them to do it together?

28. *Painting a House* Repeat **Exercise 27,** but assume that Joe takes 6 hr working alone, and Sam takes 8 hr working alone.

29. *Pollution in a River* Two chemical plants are polluting a river. If plant A produces a predetermined maximum amount of pollutant twice as fast as plant B, and together they produce the maximum pollutant in 26 hr, how long will it take plant B alone?

	Rate	Time	Part of Job Accomplished
Pollution from A	$\frac{1}{x}$	26	$\frac{1}{x}(26)$
Pollution from B		26	

30. *Filling a Settling Pond* A sewage treatment plant has two inlet pipes to its settling pond. One pipe can fill the pond 3 times as fast as the other pipe, and together they can fill the pond in 12 hr. How long will it take the faster pipe to fill the pond alone?

31. *Filling a Pool* An inlet pipe can fill Blake's pool in 5 hr, while an outlet pipe can empty it in 8 hr. In his haste to surf the Internet, Blake left both pipes open. How long did it take to fill the pool?

32. $4\frac{5}{8}$ hr more
33. 10 min
34. $4\frac{1}{2}$ min

35. $\{3\}$ 36. $\{6\}$
37. $\{-1\}$ 38. $\{3\}$
39. $\{5\}$ 40. $\{-1, 3\}$
41. $\{9\}$ 42. $\{8\}$
43. $\{9\}$ 44. $\{16\}$
45. $\emptyset$ 46. $\emptyset$
47. $\{\pm 2\}$ 48. $\{\frac{10}{9}, 2\}$
49. $\{0, 3\}$ 50. $\{8\}$
51. $\{-2\}$ 52. $\{27\}$
53. $\{-\frac{2}{9}, 2\}$ 54. $\{3\}$
55. $\{4\}$ 56. $\{0, 9\}$
57. $\{-2\}$ 58. $\{-\frac{2}{3}\}$
59. $\{\frac{2}{5}, 1\}$ 60. $\{\frac{4}{3}, 2\}$
61. $\{31\}$ 62. $\{0\}$
63. $\{-3, 1\}$ 64. $\{-8, 2\}$

65. $\{25\}$ 66. $\{16\}$
67. $\{-27, 3\}$ 68. $\{-\frac{64}{3}, 4\}$
69. $\{-29, 35\}$ 70. $\{-416, 16\}$
71. $\{\frac{3}{2}\}$ 72. $\{5\}$
73. $\{\frac{1}{4}, 1\}$ 74. $\{1, 9\}$
75. $\{0, 8\}$ 76. $\{0, \frac{1}{81}\}$

77. $\left\{\pm 1, \pm \frac{\sqrt{10}}{2}\right\}$
78. $\left\{\pm \frac{\sqrt{6}}{2}, \pm \frac{\sqrt{2}}{2}\right\}$
79. $\left\{\pm \sqrt{3}, \pm i\sqrt{5}\right\}$
80. $\left\{\pm \frac{\sqrt{15}}{3}, \pm i\sqrt{5}\right\}$
81. $\{-63, 28\}$ 82. $\{-13, 1\}$
83. $\{0, 31\}$ 84. $\{-130, 59\}$
85. $\left\{\frac{-6 \pm 2\sqrt{3}}{3}, \frac{-4 \pm \sqrt{2}}{2}\right\}$
86. $\left\{\frac{8 \pm \sqrt{2}}{2}, \frac{8 \pm \sqrt{3}}{2}\right\}$
87. $\{-\frac{2}{7}, 5\}$ 88. $\{-\frac{7}{4}, \frac{1}{2}\}$
89. $\{-\frac{1}{27}, \frac{1}{8}\}$ 90. $\{-32, 1\}$
91. $\{\pm\frac{1}{2}, \pm 4\}$ 92. $\{\pm\frac{5}{2}, \pm 5\}$

32. *Filling a Pool* Suppose Blake discovered his error (see **Exercise 31**) after an hour-long surf. If he then closed the outlet pipe, how much more time would be needed to fill the pool?

33. *Filling a Sink* With both taps open, Robert can fill his kitchen sink in 5 min. When full, the sink drains in 10 min. How long will it take to fill the sink if Robert forgets to put in the stopper?

34. *Filling a Sink* If Robert (see **Exercise 33**) remembers to put in the stopper after 1 min, how much longer will it take to fill the sink?

Solve each equation. ***See Examples 4–6.***

35. $x - \sqrt{2x + 3} = 0$

36. $x - \sqrt{3x + 18} = 0$

37. $\sqrt{3x + 7} = 3x + 5$

38. $\sqrt{4x + 13} = 2x - 1$

39. $\sqrt{4x + 5} - 6 = 2x - 11$

40. $\sqrt{6x + 7} - 9 = x - 7$

41. $\sqrt{4x} - x + 3 = 0$

42. $\sqrt{2x} - x + 4 = 0$

43. $\sqrt{x} - \sqrt{x - 5} = 1$

44. $\sqrt{x} - \sqrt{x - 12} = 2$

45. $\sqrt{x + 7} + 3 = \sqrt{x - 4}$

46. $\sqrt{x + 5} + 2 = \sqrt{x - 1}$

47. $\sqrt{2x + 5} - \sqrt{x + 2} = 1$

48. $\sqrt{4x + 1} - \sqrt{x - 1} = 2$

49. $\sqrt{3x} = \sqrt{5x + 1} - 1$

50. $\sqrt{2x} = \sqrt{3x + 12} - 2$

51. $\sqrt{x + 2} = 1 - \sqrt{3x + 7}$

52. $\sqrt{2x - 5} = 2 + \sqrt{x - 2}$

53. $\sqrt{2\sqrt{7x + 2}} = \sqrt{3x + 2}$

54. $\sqrt{3\sqrt{2x + 3}} = \sqrt{5x - 6}$

55. $3 - \sqrt{x} = \sqrt{2\sqrt{x} - 3}$

56. $\sqrt{x} + 2 = \sqrt{4 + 7\sqrt{x}}$

57. $\sqrt[3]{4x + 3} = \sqrt[3]{2x - 1}$

58. $\sqrt[3]{2x} = \sqrt[3]{5x + 2}$

59. $\sqrt[3]{5x^2 - 6x + 2} - \sqrt[3]{x} = 0$

60. $\sqrt[3]{3x^2 - 9x + 8} = \sqrt[3]{x}$

61. $\sqrt[4]{x - 15} = 2$

62. $\sqrt[4]{3x + 1} = 1$

63. $\sqrt[4]{x^2 + 2x} = \sqrt[4]{3}$

64. $\sqrt[4]{x^2 + 6x} = 2$

Solve each equation. ***See Example 7.***

65. $x^{3/2} = 125$

66. $x^{5/4} = 32$

67. $(x^2 + 24x)^{1/4} = 3$

68. $(3x^2 + 52x)^{1/4} = 4$

69. $(x - 3)^{2/5} = 4$

70. $(x + 200)^{2/3} = 36$

71. $(2x + 5)^{1/3} - (6x - 1)^{1/3} = 0$

72. $(3x + 7)^{1/3} - (4x + 2)^{1/3} = 0$

73. $(2x - 1)^{2/3} = x^{1/3}$

74. $(x - 3)^{2/5} = (4x)^{1/5}$

75. $x^{2/3} = 2x^{1/3}$

76. $3x^{3/4} = x^{1/2}$

Solve each equation. ***See Examples 8 and 9.***

77. $2x^4 - 7x^2 + 5 = 0$

78. $4x^4 - 8x^2 + 3 = 0$

79. $x^4 + 2x^2 - 15 = 0$

80. $3x^4 + 10x^2 - 25 = 0$

81. $(x - 1)^{2/3} + (x - 1)^{1/3} - 12 = 0$

82. $(2x - 1)^{2/3} + 2(2x - 1)^{1/3} - 3 = 0$

83. $(x + 1)^{2/5} - 3(x + 1)^{1/5} + 2 = 0$

84. $(x + 5)^{2/3} + (x + 5)^{1/3} - 20 = 0$

85. $6(x + 2)^4 - 11(x + 2)^2 = -4$

86. $8(x - 4)^4 - 10(x - 4)^2 = -3$

87. $10x^{-2} + 33x^{-1} - 7 = 0$

88. $7x^{-2} - 10x^{-1} - 8 = 0$

89. $x^{-2/3} + x^{-1/3} - 6 = 0$

90. $2x^{-2/5} - x^{-1/5} - 1 = 0$

91. $16x^{-4} - 65x^{-2} + 4 = 0$

92. $625x^{-4} - 125x^{-2} + 4 = 0$

93. $\{16\}$; $u = -3$ does not lead to a solution of the equation.
94. $\{16\}$; 9 does not satisfy the equation.
96. $\{4\}$
97. $h = \frac{d^2}{k^2}$
98. $y = \pm(a^{2/3} - x^{2/3})^{3/2}$
99. $m = (1 - n^{3/4})^{4/3}$
100. $R = \frac{r_1 r_2}{r_1 + r_2}$
101. $e = \frac{Er}{R + r}$
102. $b = \pm\sqrt{c^2 - a^2}$

Relating Concepts

For individual or collaborative investigation *(Exercises 93–96)*

In this section we introduced methods of solving equations quadratic in form by substitution and solving equations involving radicals by raising each side of the equation to a power. Suppose we wish to solve

$$x - \sqrt{x} - 12 = 0.$$

We can solve this equation using either of the two methods. **Work Exercises 93–96 in order** *to see how both methods apply.*

93. Let $u = \sqrt{x}$ and solve the equation by substitution. What is the value of u that does not lead to a solution of the equation?

94. Solve the equation by isolating $\sqrt{x}$ on one side and then squaring. What is the value of x that does not satisfy the equation?

95. Which one of the methods used in **Exercises 93 and 94** do you prefer? Why?

96. Solve $3x - 2\sqrt{x} - 8 = 0$ using one of the two methods described.

Solve each equation for the indicated variable. Assume all denominators are nonzero.

97. $d = k\sqrt{h}$, for h

98. $x^{2/3} + y^{2/3} = a^{2/3}$, for y

99. $m^{3/4} + n^{3/4} = 1$, for m

100. $\frac{1}{R} = \frac{1}{r_1} + \frac{1}{r_2}$, for R

101. $\frac{E}{e} = \frac{R + r}{r}$, for e

102. $a^2 + b^2 = c^2$, for b

Summary Exercises on Solving Equations

1. $\{3\}$
2. $\{-1\}$
3. $\{-3 \pm 3\sqrt{2}\}$
4. $\{2, 6\}$
5. $\emptyset$
6. $\{-31\}$
7. $\{-6\}$
8. $\{6\}$
9. $\{\frac{1}{5} \pm \frac{2}{5}i\}$
10. $\{-2, 1\}$
11. $\{-\frac{1}{243}, \frac{1}{3125}\}$
12. $\{-1\}$
13. $\{\pm i, \pm 2\}$
14. $\{-2.4\}$
15. $\{4\}$
16. $\{\frac{1}{3} \pm \frac{\sqrt{2}}{3}i\}$
17. $\{\frac{15}{7}\}$
18. $\{4\}$
19. $\{3, 11\}$
20. $\{1\}$
21. $\{x \mid x \neq 3\}$
22. $a = \pm\sqrt{c^2 - b^2}$

This section of miscellaneous equations provides practice in solving all the types introduced in this chapter so far. Solve each equation.

1. $4x - 3 = 2x + 3$

2. $5 - (6x + 3) = 2(2 - 2x)$

3. $x(x + 6) = 9$

4. $x^2 = 8x - 12$

5. $\sqrt{x + 2} + 5 = \sqrt{x + 15}$

6. $\frac{5}{x + 3} - \frac{6}{x - 2} = \frac{3}{x^2 + x - 6}$

7. $\frac{3x + 4}{3} - \frac{2x}{x - 3} = x$

8. $\frac{x}{2} + \frac{4}{3}x = x + 5$

9. $5 - \frac{2}{x} + \frac{1}{x^2} = 0$

10. $(2x + 1)^2 = 9$

11. $x^{-2/5} - 2x^{-1/5} - 15 = 0$

12. $\sqrt{x + 2} + 1 = \sqrt{2x + 6}$

13. $x^4 - 3x^2 - 4 = 0$

14. $1.2x + 0.3 = 0.7x - 0.9$

15. $\sqrt[6]{2x + 1} = \sqrt[6]{9}$

16. $3x^2 - 2x = -1$

17. $3[2x - (6 - 2x) + 1] = 5x$

18. $\sqrt{x + 1} = \sqrt{11 - \sqrt{x}}$

19. $(14 - 2x)^{2/3} = 4$

20. $2x^{-1} - x^{-2} = 1$

21. $\frac{3}{x - 3} = \frac{3}{x - 3}$

22. $a^2 + b^2 = c^2$, for a

1.7 Inequalities

- Linear Inequalities
- Three-Part Inequalities
- Quadratic Inequalities
- Rational Inequalities

An **inequality** says that one expression is greater than, greater than or equal to, less than, or less than or equal to another (**Section R.2**). As with equations, a value of the variable for which the inequality is true is a solution of the inequality, and the set of all solutions is the solution set of the inequality. Two inequalities with the same solution set are equivalent.

Inequalities are solved with the properties of inequality, which are similar to the properties of equality in **Section 1.1.**

Properties of Inequality

Let a, b, and c represent real numbers.

1. If $a < b$, then $a + c < b + c$.

2. If $a < b$ and if $c > 0$, then $ac < bc$.

3. If $a < b$ and if $c < 0$, then $ac > bc$.

Replacing $<$ with $>$, $\leq$, or $\geq$ results in similar properties. (Restrictions on c remain the same.)

NOTE Multiplication may be replaced by division in Properties 2 and 3. ***Always remember to reverse the direction of the inequality symbol when multiplying or dividing by a negative number.***

Linear Inequalities The definition of a *linear inequality* is similar to the definition of a linear equation.

Linear Inequality in One Variable

A **linear inequality in one variable** is an inequality that can be written in the form

$$ax + b > 0,$$

where a and b are real numbers, with $a \neq 0$. (Any of the symbols $\geq$, $<$, and $\leq$ may also be used.)

Classroom Example 1
Solve $-2x + 7 < -5$.

Answer: $\{x \mid x > 6\}$

EXAMPLE 1 **Solving a Linear Inequality**

Solve $-3x + 5 > -7$.

SOLUTION

$$-3x + 5 > -7$$

$$-3x + 5 - 5 > -7 - 5 \quad \text{Subtract 5.}$$

$$-3x > -12 \quad \text{Combine like terms.}$$

$$\frac{-3x}{-3} < \frac{-12}{-3} \quad \text{Divide by } -3. \text{ Reverse the direction of the inequality symbol when multiplying or dividing by a negative number.}$$

Don't forget to reverse the inequality symbol here.

$$x < 4$$

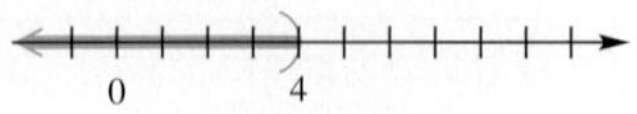

Figure 9

Thus, the original inequality $-3x + 5 > -7$ is satisfied by any real number less than 4. The solution set can be written $\{x \mid x < 4\}$. A graph of the solution set is shown in **Figure 9**, where the parenthesis is used to show that 4 itself does not belong to the solution set. Note that testing values from the solution set in the original inequality will produce true statements, while testing values outside the solution set produces false statements.

The solution set of the inequality,

$$\{x \mid x < 4\}, \quad \text{Set builder notation (Section R.1)}$$

is an example of an **interval.** We use a simplified notation, called **interval notation,** to write intervals. With this notation, we write the interval as

$$(-\infty, 4). \quad \text{Interval notation}$$

The symbol $-\infty$ does not represent an actual number. Rather it is used to show that the interval includes all real numbers less than 4. The interval $(-\infty, 4)$ is an example of an **open interval,** since the endpoint, 4, is not part of the interval. A **closed interval** includes both endpoints. A square bracket is used to show that a number *is* part of the graph, and a parenthesis is used to indicate that a number *is not* part of the graph.

✔ *Now Try Exercise 13.*

In the table that follows, we assume that $a < b$.

Teaching Tip Have students translate English phrases such as

"all values of x such that x is greater than -3 and less than or equal to 4"

into set notation and interval notation.

Type of Interval	Set	Interval Notation	Graph
Open interval	$\{x \mid x > a\}$	(a, ∞)	a
	$\{x \mid a < x < b\}$	(a, b)	a b
	$\{x \mid x < b\}$	$(-\infty, b)$	b
Other intervals	$\{x \mid x \geq a\}$	$[a, \infty)$	a
	$\{x \mid a < x \leq b\}$	$(a, b]$	a b
	$\{x \mid a \leq x < b\}$	$[a, b)$	a b
	$\{x \mid x \leq b\}$	$(-\infty, b]$	b
Closed interval	$\{x \mid a \leq x \leq b\}$	$[a, b]$	a b
Disjoint interval	$\{x \mid x < a \text{ or } x > b\}$	$(-\infty, a) \cup (b, \infty)$	a b
All real numbers	$\{x \mid x \text{ is a real number}\}$	$(-\infty, \infty)$	

Classroom Example 2
Solve $3 - 4x \geq 2x + 8$. Give the solution set in interval notation.

Answer: $\left(-\infty, -\frac{5}{6}\right]$

EXAMPLE 2 Solving a Linear Inequality

Solve $4 - 3x \leq 7 + 2x$. Give the solution set in interval notation.

SOLUTION

$$4 - 3x \leq 7 + 2x$$

$$4 - 3x - 4 \leq 7 + 2x - 4 \quad \text{Subtract 4.}$$

$$-3x \leq 3 + 2x \quad \text{Combine like terms.}$$

$$-3x - 2x \leq 3 + 2x - 2x \quad \text{Subtract } 2x.$$

$$-5x \leq 3 \quad \text{Combine like terms.}$$

$$\frac{-5x}{-5} \geq \frac{3}{-5} \quad \text{Divide by } -5. \text{ Reverse the direction of the inequality symbol.}$$

$$x \geq -\frac{3}{5}$$

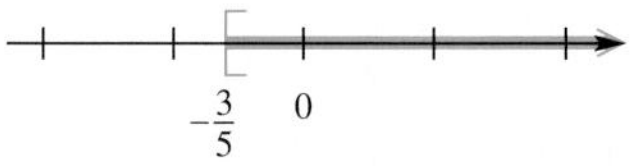

Figure 10

In interval notation the solution set is $\left[-\frac{3}{5}, \infty\right)$. See **Figure 10** for the graph.

Now Try Exercise 15.

A product will break even, or begin to produce a profit, only if the revenue from selling the product at least equals the cost of producing it. If R represents revenue and C is cost, then the **break-even point** is the point where $R = C$.

Classroom Example 3
If the revenue and cost of a certain product are given by

$R = 45x$ and $C = 30x + 5250$,

where x is the number of units produced and sold, at what production level does R at least equal C?

Answer: $[350, \infty)$

EXAMPLE 3 Finding the Break-Even Point

If the revenue and cost of a certain product are given by

$$R = 4x \quad \text{and} \quad C = 2x + 1000,$$

where x is the number of units produced and sold, at what production level does *R at least equal C*?

SOLUTION Set $R \geq C$ and solve for x.

At least equal to translates as $\geq$.

$$R \geq C$$

$$4x \geq 2x + 1000 \quad \text{Substitute.}$$

$$2x \geq 1000 \quad \text{Subtract } 2x.$$

$$x \geq 500 \quad \text{Divide by 2.}$$

The break-even point is at $x = 500$. This product will at least break even if the number of units produced and sold is in the interval $[500, \infty)$.

Now Try Exercise 25.

Three-Part Inequalities The inequality $-2 < 5 + 3x < 20$ says that $5 + 3x$ is *between* -2 and 20. This inequality is solved using an extension of the properties of inequality given earlier, working with all three expressions at the same time.

Classroom Example 4
Solve $1 \leq 6x - 8 \leq 4$.

Answer: $\left[\frac{3}{2}, 2\right]$

EXAMPLE 4 Solving a Three-Part Inequality

Solve $-2 < 5 + 3x < 20$.

SOLUTION

$$-2 < 5 + 3x < 20$$

$$-2 - 5 < 5 + 3x - 5 < 20 - 5 \quad \text{Subtract 5 from each part.}$$

$$-7 < 3x < 15 \quad \text{Combine like terms in each part.}$$

$$\frac{-7}{3} < \frac{3x}{3} < \frac{15}{3} \quad \text{Divide each part by 3.}$$

$$-\frac{7}{3} < x < 5$$

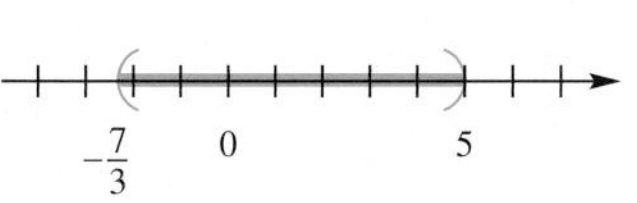

Figure 11

The solution set, graphed in **Figure 11,** is the interval $\left(-\frac{7}{3}, 5\right)$.

Now Try Exercise 29.

Quadratic Inequalities Solving *quadratic inequalities* is more complicated than solving linear inequalities and depends on finding solutions of quadratic equations.

Quadratic Inequality

A **quadratic inequality** is an inequality that can be written in the form

$$ax^2 + bx + c < 0,$$

for real numbers a, b, and c, with $a \neq 0$. (The symbol $<$ can be replaced with $>$, $\leq$, or $\geq$.)

One method of solving a quadratic inequality involves finding the solutions of the corresponding quadratic equation and then testing values in the intervals on a number line determined by those solutions.

Teaching Tip Some instructors use the term **critical numbers** to refer to the solutions of the corresponding equation in Step 1.

Solving a Quadratic Inequality

Step 1 Solve the corresponding quadratic equation.

Step 2 Identify the intervals determined by the solutions of the equation.

Step 3 Use a test value from each interval to determine which intervals form the solution set.

Classroom Example 5
Solve $x^2 - 2x - 15 \leq 0$.

Answer: $[-3, 5]$

EXAMPLE 5 Solving a Quadratic Inequality

Solve $x^2 - x - 12 < 0$.

SOLUTION

Step 1 Find the values of x that satisfy $x^2 - x - 12 = 0$.

$$x^2 - x - 12 = 0 \quad \text{Corresponding quadratic equation}$$

$$(x + 3)(x - 4) = 0 \quad \text{Factor. (Section R.4)}$$

$$x + 3 = 0 \quad \text{or} \quad x - 4 = 0 \quad \text{Zero-factor property (Section 1.4)}$$

$$x = -3 \quad \text{or} \quad x = 4 \quad \text{Solve each equation.}$$

Step 2 The two numbers -3 and 4 cause the expression $x^2 - x - 12$ to ***equal*** zero and can be used to divide the number line into three intervals, as shown in **Figure 12.** The expression $x^2 - x - 12$ will take on a value that is either ***less than*** zero or ***greater than*** zero on each of these intervals. Since we are looking for x-values that make the expression ***less than*** zero, use open circles at -3 and 4 to indicate that they are not included in the solution set.

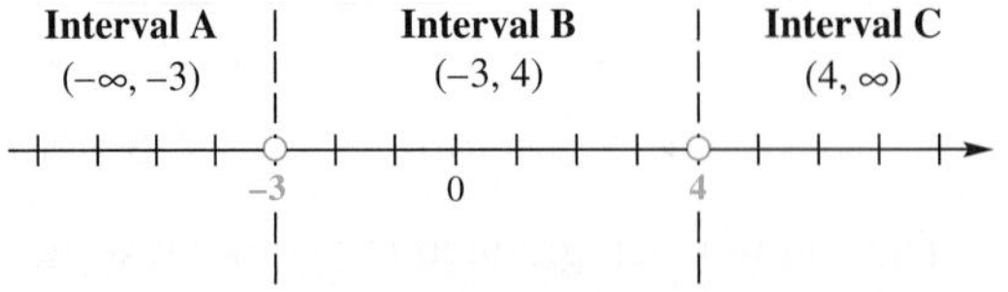

Figure 12

Use open circles since the inequality symbol does not include *equality*. -3 and 4 do not satisfy the *inequality*.

Step 3 Choose a test value in each interval to see whether it satisfies the original inequality, $x^2 - x - 12 < 0$. If the test value makes the statement true, then the entire interval belongs to the solution set.

Interval	Test Value	Is $x^2 - x - 12 < 0$ True or False?
A: $(-\infty, -3)$	-4	$(-4)^2 - (-4) - 12 \overset{?}{<} 0$ $8 < 0$ False
B: $(-3, 4)$	0	$0^2 - 0 - 12 \overset{?}{<} 0$ $-12 < 0$ True
C: $(4, \infty)$	5	$5^2 - 5 - 12 \overset{?}{<} 0$ $8 < 0$ False

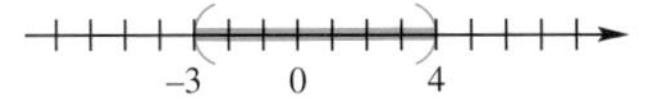

Figure 13

Since the values in Interval B make the inequality true, the solution set is $(-3, 4)$. See **Figure 13.**

Now Try Exercise 41.

Classroom Example 6
Solve $3x^2 - 11x - 4 > 0$.

Answer: $\left(-\infty, -\frac{1}{3}\right) \cup (4, \infty)$

EXAMPLE 6 Solving a Quadratic Inequality

Solve $2x^2 + 5x - 12 \geq 0$.

SOLUTION

Step 1 Find the values of x that satisfy $2x^2 + 5x - 12 = 0$.

$$2x^2 + 5x - 12 = 0 \quad \text{Corresponding quadratic equation}$$

$$(2x - 3)(x + 4) = 0 \quad \text{Factor.}$$

$$2x - 3 = 0 \quad \text{or} \quad x + 4 = 0 \quad \text{Zero-factor property}$$

$$x = \frac{3}{2} \quad \text{or} \quad x = -4 \quad \text{Solve each equation.}$$

Step 2 The values $\frac{3}{2}$ and -4 cause the inequality $2x^2 + 5x - 12$ to equal 0 and can be used to form the intervals $(-\infty, -4)$, $\left(-4, \frac{3}{2}\right)$, and $\left(\frac{3}{2}, \infty\right)$ on the number line, as seen in **Figure 14.**

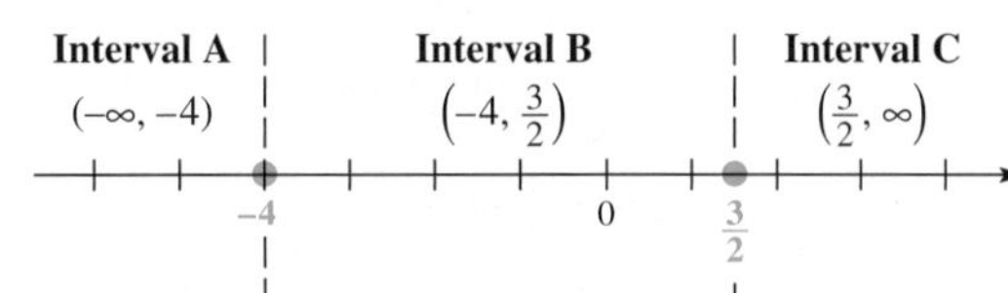

Use closed circles since the inequality symbol includes *equality*. -4 and $\frac{3}{2}$ satisfy the *inequality*.

Figure 14

Step 3 Choose a test value from each interval.

Interval	Test Value	Is $2x^2 + 5x - 12 \geq 0$ True or False?
A: $(-\infty, -4)$	-5	$2(-5)^2 + 5(-5) - 12 \overset{?}{\geq} 0$ $13 \geq 0$ True
B: $\left(-4, \frac{3}{2}\right)$	0	$2(0)^2 + 5(0) - 12 \overset{?}{\geq} 0$ $-12 \geq 0$ False
C: $\left(\frac{3}{2}, \infty\right)$	2	$2(2)^2 + 5(2) - 12 \overset{?}{\geq} 0$ $6 \geq 0$ True

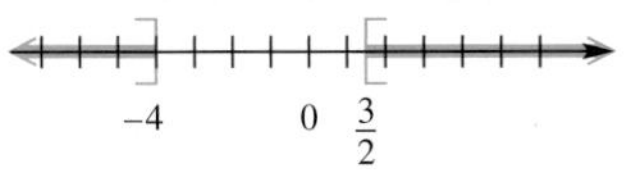

Figure 15

The values in Intervals A and C make the inequality true, so the solution set is the *union* (**Section R.1**) of the two intervals, written $(-\infty, -4] \cup \left[\frac{3}{2}, \infty\right)$. The graph of the solution set is shown in **Figure 15.**

Now Try Exercise 39.

NOTE Inequalities that use the symbols $<$ and $>$ are **strict inequalities,** while $\leq$ and $\geq$ are used in **nonstrict inequalities.** The solutions of the equation in **Example 5** were not included in the solution set since the inequality was a *strict* inequality. In **Example 6,** the solutions of the equation *were* included in the solution set because of the nonstrict inequality.

Classroom Example 7
If an object is launched from ground level with an initial velocity of 144 ft per sec, its height s in feet t seconds after launching is

$$s = -16t^2 + 144t.$$

When will the object be greater than 128 ft above ground level?

Answer: The object is greater than 128 ft above ground level between 1 and 8 sec after it is launched.

EXAMPLE 7 Finding Projectile Height

If a projectile is launched from ground level with an initial velocity of 96 ft per sec, its height s in feet t seconds after launching is given by the following equation.

$$s = -16t^2 + 96t$$

When will the projectile be greater than 80 ft above ground level?

SOLUTION

$$-16t^2 + 96t > 80 \quad \text{Set } s \text{ greater than 80.}$$

$$-16t^2 + 96t - 80 > 0 \quad \text{Subtract 80.}$$

$$t^2 - 6t + 5 < 0 \quad \text{Divide by } -16.$$

Reverse the direction of the inequality symbol.

Now solve the corresponding *equation.*

$$t^2 - 6t + 5 = 0$$

$$(t-1)(t-5) = 0 \quad \text{Factor.}$$

$$t - 1 = 0 \quad \text{or} \quad t - 5 = 0 \quad \text{Zero-factor property}$$

$$t = 1 \quad \text{or} \quad t = 5 \quad \text{Solve each equation.}$$

Use these values to determine the intervals $(-\infty, 1)$, $(1, 5)$, and $(5, \infty)$. Since we are solving a strict inequality, solutions of the equation $t^2 - 6t + 5 = 0$ are ***not*** included. See **Figure 16.**

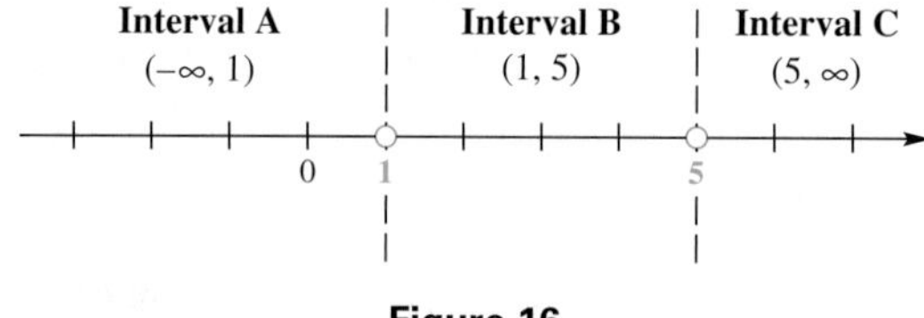

Figure 16

Choose a test value from each interval and use the procedure of **Examples 5 and 6** to determine that values in Interval B, $(1, 5)$, satisfy the inequality. The projectile is greater than 80 ft above ground level between 1 and 5 sec after it is launched.

Now Try Exercise 95.

Rational Inequalities Inequalities involving one or more rational expressions are **rational inequalities.**

$$\frac{5}{x+4} \geq 1 \quad \text{and} \quad \frac{2x-1}{3x+4} < 5 \quad \text{Rational inequalities}$$

Teaching Tip As with quadratic inequalities, some instructors use the term **critical numbers** to refer to the values found in Step 2.

Solving a Rational Inequality

Step 1 Rewrite the inequality, if necessary, so that 0 is on one side and there is a single fraction on the other side.

Step 2 Determine the values that will cause either the numerator or the denominator of the rational expression to equal 0. These values determine the intervals on the number line to consider.

Step 3 Use a test value from each interval to determine which intervals form the solution set.

A value causing a denominator to equal zero will never be included in the solution set. If the inequality is strict, any value causing the numerator to equal zero will be excluded. If the inequality is nonstrict, any such value will be included.

CAUTION Solving a rational inequality such as

$$\frac{5}{x+4} \geq 1$$

by multiplying each side by $x + 4$ to obtain $5 \geq x + 4$ requires considering *two cases,* since the sign of $x + 4$ depends on the value of x. If $x + 4$ is negative, then the inequality symbol must be reversed. The procedure described in the preceding box and used in the next two examples eliminates the need for considering separate cases.

Classroom Example 8

Solve $\frac{6}{x-3} \geq 4$.

Answer: $\left(3, \frac{9}{2}\right]$

EXAMPLE 8 Solving a Rational Inequality

Solve $\frac{5}{x+4} \geq 1$.

SOLUTION

Step 1

$$\frac{5}{x+4} - 1 \geq 0 \quad \text{Subtract 1 so that 0 is on one side.}$$

$$\frac{5}{x+4} - \frac{x+4}{x+4} \geq 0 \quad \text{Use } x+4 \text{ as the common denominator.}$$

Note the careful use of parentheses.

$$\frac{5-(x+4)}{x+4} \geq 0 \quad \text{Write as a single fraction. (Section R.5)}$$

$$\frac{1-x}{x+4} \geq 0 \quad \text{Combine like terms in the numerator, being careful with signs.}$$

Step 2 The quotient possibly changes sign only where x-values make the numerator or denominator 0. This occurs at

$$1 - x = 0 \quad \text{or} \quad x + 4 = 0$$

$$x = 1 \quad \text{or} \quad x = -4.$$

These values form the intervals $(-\infty, -4)$, $(-4, 1)$, and $(1, \infty)$ on the number line, as seen in **Figure 17** on the next page.

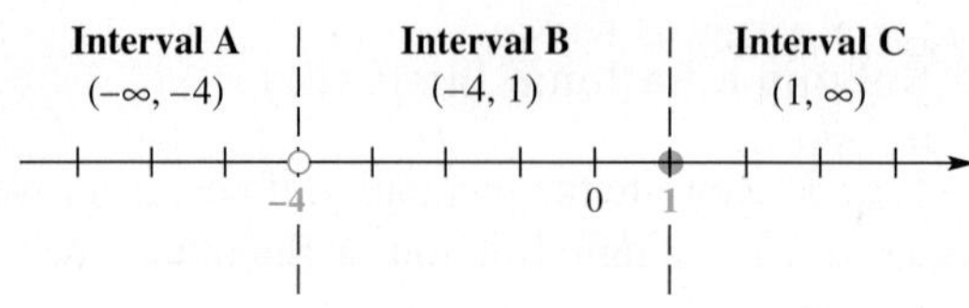

Figure 17

Use a solid circle on 1, since the symbol is ≥. The value −4 cannot be in the solution set since it causes the denominator to equal 0. Use an open circle on −4.

Step 3 Choose test values.

Teaching Tip Explain that it is also possible to substitute test values into the equivalent inequality

$$\frac{1-x}{x+4} \geq 0.$$

Interval	Test Value	Is $\frac{5}{x+4} \geq 1$ True or False?
A: $(-\infty, -4)$	-5	$\frac{5}{-5+4} \overset{?}{\geq} 1$ $-5 \geq 1$ False
B: $(-4, 1)$	0	$\frac{5}{0+4} \overset{?}{\geq} 1$ $\frac{5}{4} \geq 1$ True
C: $(1, \infty)$	2	$\frac{5}{2+4} \overset{?}{\geq} 1$ $\frac{5}{6} \geq 1$ False

The values in the interval $(-4, 1)$ satisfy the original inequality. The value 1 makes the nonstrict inequality true, so it must be included in the solution set. Since -4 makes the denominator 0, it must be excluded. The solution set is $(-4, 1]$.

Now Try Exercise 73.

CAUTION ***Be careful with the endpoints of the intervals when solving rational inequalities.***

Classroom Example 9
Solve $\frac{3x+1}{2x-3} < 4$.

Answer: $\left(-\infty, \frac{3}{2}\right) \cup \left(\frac{13}{5}, \infty\right)$

EXAMPLE 9 Solving a Rational Inequality

Solve $\frac{2x-1}{3x+4} < 5$.

SOLUTION

$$\frac{2x-1}{3x+4} - 5 < 0 \quad \text{Subtract 5.}$$

$$\frac{2x-1}{3x+4} - \frac{5(3x+4)}{3x+4} < 0 \quad \text{Common denominator is } 3x+4.$$

$$\frac{2x-1-5(3x+4)}{3x+4} < 0 \quad \text{Write as a single fraction.}$$

Be careful with signs.

$$\frac{2x-1-15x-20}{3x+4} < 0 \quad \text{Distributive property (Section R.2)}$$

$$\frac{-13x-21}{3x+4} < 0 \quad \text{Combine like terms in the numerator.}$$

Set the numerator and denominator equal to 0 and solve the resulting equations to find the values of x where sign changes may occur.

$$-13x - 21 = 0 \quad \text{or} \quad 3x + 4 = 0$$

$$x = -\frac{21}{13} \quad \text{or} \quad x = -\frac{4}{3}$$

Use these values to form intervals on the number line, as seen in **Figure 18.**

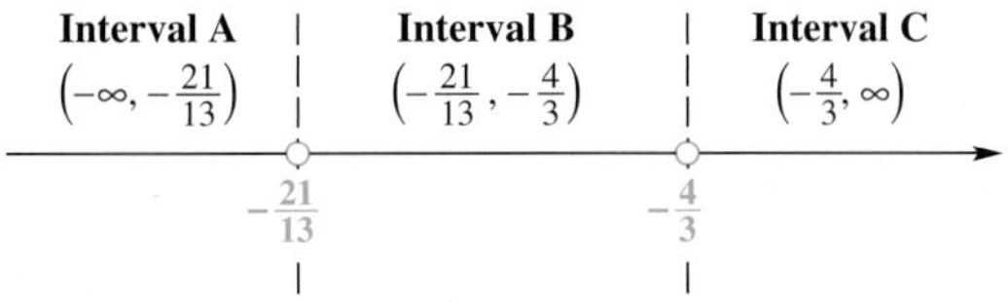

Figure 18

Use an open circle at $-\frac{21}{13}$ because of the strict inequality, and use an open circle at $-\frac{4}{3}$ since it causes the denominator to equal 0.

Now choose test values from the intervals in **Figure 18.** Verify that

-2 from Interval A makes the inequality true.

-1.5 from Interval B makes the inequality false.

0 from Interval C makes the inequality true.

Because of the $<$ symbol, neither endpoint satisfies the inequality, so the solution set is

$$\left(-\infty, -\frac{21}{13}\right) \cup \left(-\frac{4}{3}, \infty\right).$$

✔ *Now Try Exercise 85.*

1.7 Exercises

1. F 2. J
3. A 4. H
5. I 6. D
7. B 8. G
9. E 10. C

12. D

Concept Check *Match the inequality in each exercise in Column I with its equivalent interval notation in Column II.*

I	II
1. $x < -6$	**A.** $(-2, 6]$
2. $x \leq 6$	**B.** $[-2, 6)$
3. $-2 < x \leq 6$	**C.** $(-\infty, -6]$
4. $x^2 \geq 0$	**D.** $[6, \infty)$
5. $x \geq -6$	**E.** $(-\infty, -3) \cup (3, \infty)$
6. $6 \leq x$	**F.** $(-\infty, -6)$
7. 	**G.** $(0, 8)$
8. 	**H.** $(-\infty, \infty)$
9. 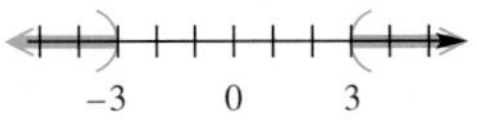	**I.** $[-6, \infty)$
10. 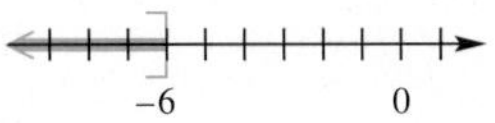	**J.** $(-\infty, 6]$

11. Explain how to determine whether to use a parenthesis or a square bracket when graphing the solution set of a linear inequality.

12. *Concept Check* The three-part inequality $a < x < b$ means "a is less than x and x is less than b." Which one of the following inequalities is not satisfied by some real number x?

A. $-3 < x < 10$ **B.** $0 < x < 6$
C. $-3 < x < -1$ **D.** $-8 < x < -10$

13. $[-4, \infty)$ 14. $[-5, \infty)$
15. $[-1, \infty)$ 16. $(-\infty, 1]$
17. $(-\infty, \infty)$ 18. $(-\infty, \infty)$
19. $(-\infty, 4)$ 20. $(-\infty, \frac{15}{7})$
21. $[-\frac{11}{5}, \infty)$ 22. $(-\infty, \frac{1}{2}]$
23. $(-\infty, \frac{48}{7}]$ 24. $[-4, \infty)$
25. $[500, \infty)$ 26. $[15, \infty)$
27. The product will never break even.
28. The product will never break even.
29. $(-5, 3)$ 30. $(-3, 1)$
31. $[3, 6]$ 32. $[-\frac{3}{2}, 3]$
33. $(4, 6)$ 34. $(\frac{1}{6}, 1)$
35. $[-9, 9]$ 36. $[-12, 6]$
37. $(-16, 19]$ 38. $(-\frac{13}{4}, \frac{3}{4}]$
39. $(-\infty, -2) \cup (3, \infty)$
40. $(-\infty, 2) \cup (5, \infty)$
41. $[-\frac{3}{2}, 6]$ 42. $[-\frac{4}{3}, 1]$
43. $(-\infty, -3] \cup [-1, \infty)$
44. $(-4, -2)$
45. $[-2, 3]$ 46. $(-4, 3)$
47. $[-3, 3]$
48. $(-\infty, -4) \cup (4, \infty)$
49. $\emptyset$ 50. $\emptyset$
51. $[1 - \sqrt{2}, 1 + \sqrt{2}]$
52. $(-\infty, -2 - \sqrt{3}) \cup (-2 + \sqrt{3}, \infty)$
53. A 54. D
55. $\{\frac{4}{3}, -2, -6\}$
56. –6 –2 0 $\frac{4}{3}$

Solve each inequality. Write each solution set in interval notation. **See Examples 1 and 2.**

13. $-2x + 8 \le 16$
14. $-3x - 8 \le 7$
15. $-2x - 2 \le 1 + x$
16. $-4x + 3 \ge -2 + x$
17. $3(x + 5) + 1 \ge 5 + 3x$
18. $6x - (2x + 3) \ge 4x - 5$
19. $8x - 3x + 2 < 2(x + 7)$
20. $2 - 4x + 5(x - 1) < -6(x - 2)$
21. $\dfrac{4x + 7}{-3} \le 2x + 5$
22. $\dfrac{2x - 5}{-8} \le 1 - x$
23. $\dfrac{1}{3}x + \dfrac{2}{5}x - \dfrac{1}{2}(x + 3) \le \dfrac{1}{10}$
24. $-\dfrac{2}{3}x - \dfrac{1}{6}x + \dfrac{2}{3}(x + 1) \le \dfrac{4}{3}$

Break-Even Interval Find all intervals where each product will at least break even. **See Example 3.**

25. The cost to produce x units of picture frames is $C = 50x + 5000$, while the revenue is $R = 60x$.
26. The cost to produce x units of baseball caps is $C = 100x + 6000$, while the revenue is $R = 500x$.
27. The cost to produce x units of coffee cups is $C = 105x + 900$, while the revenue is $R = 85x$.
28. The cost to produce x units of briefcases is $C = 70x + 500$, while the revenue is $R = 60x$.

Solve each inequality. Write each solution set in interval notation. **See Example 4.**

29. $-5 < 5 + 2x < 11$
30. $-7 < 2 + 3x < 5$
31. $10 \le 2x + 4 \le 16$
32. $-6 \le 6x + 3 \le 21$
33. $-11 > -3x + 1 > -17$
34. $2 > -6x + 3 > -3$
35. $-4 \le \dfrac{x + 1}{2} \le 5$
36. $-5 \le \dfrac{x - 3}{3} \le 1$
37. $-3 \le \dfrac{x - 4}{-5} < 4$
38. $1 \le \dfrac{4x - 5}{-2} < 9$

Solve each quadratic inequality. Write each solution set in interval notation. **See Examples 5 and 6.**

39. $x^2 - x - 6 > 0$
40. $x^2 - 7x + 10 > 0$
41. $2x^2 - 9x \le 18$
42. $3x^2 + x \le 4$
43. $-x^2 - 4x - 6 \le -3$
44. $-x^2 - 6x - 16 > -8$
45. $x(x - 1) \le 6$
46. $x(x + 1) < 12$
47. $x^2 \le 9$
48. $x^2 > 16$
49. $x^2 + 5x + 7 < 0$
50. $4x^2 + 3x + 1 \le 0$
51. $x^2 - 2x \le 1$
52. $x^2 + 4x > -1$
53. *Concept Check* Which one of the following inequalities has solution set $(-\infty, \infty)$?
 A. $(x - 3)^2 \ge 0$ **B.** $(5x - 6)^2 \le 0$
 C. $(6x + 4)^2 > 0$ **D.** $(8x + 7)^2 < 0$
54. *Concept Check* Which one of the inequalities in **Exercise 53** has solution set $\emptyset$?

57. In the interval $(-\infty, -6)$, choose $x = -10$, for example. It satisfies the original inequality. In the interval $(-6, -2)$, choose $x = -4$, for example. It does not satisfy the inequality. In the interval $\left(-2, \frac{4}{3}\right)$, choose $x = 0$, for example. It satisfies the original inequality. In the interval $\left(\frac{4}{3}, \infty\right)$, choose $x = 4$, for example. It does not satisfy the original inequality.

58. –6 –2 0 $\frac{4}{3}$

$(-\infty, -6] \cup \left[-2, \frac{4}{3}\right]$

59. $\left[-2, \frac{3}{2}\right] \cup [3, \infty)$

60. $[-5, -2] \cup \left[\frac{4}{3}, \infty\right)$

61. $(-\infty, -2] \cup [0, 2]$

62. $(-\infty, -4] \cup [0, 4]$

63. $(-\infty, -1) \cup (-1, 3)$

64. $(-\infty, -1)$

65. $[-4, -3] \cup [3, \infty)$

66. $(-\infty, -4] \cup [-3, 4]$

67. $(-\infty, \infty)$

68. $(-\infty, \infty)$

69. $(-5, 3]$

70. $(-\infty, -1) \cup (4, \infty)$

71. $(-\infty, -2)$

72. $(-2, 2)$

73. $(-\infty, 6) \cup \left[\frac{15}{2}, \infty\right)$

74. $(-\infty, 2) \cup (5, \infty)$

75. $(-\infty, 1) \cup \left(\frac{9}{5}, \infty\right)$

76. $\left(-\infty, \frac{2}{3}\right] \cup \left(\frac{5}{3}, \infty\right)$

77. $\left(-\infty, -\frac{3}{2}\right) \cup \left[-\frac{1}{2}, \infty\right)$

78. $\left(-2, -\frac{5}{3}\right]$

79. $(-2, \infty)$

80. $(-\infty, -1)$

81. $\left(0, \frac{4}{11}\right) \cup \left(\frac{1}{2}, \infty\right)$

82. $\left(-\infty, -\frac{2}{3}\right) \cup \left[-\frac{1}{2}, 0\right)$

83. $(-\infty, -2] \cup (1, 2)$

84. $(-\infty, -5) \cup (-3, -1)$

85. $(-\infty, 5)$

86. $\left(-\infty, -\frac{3}{2}\right) \cup \left[-\frac{13}{9}, \infty\right)$

87. $\left[\frac{3}{2}, \infty\right)$ 88. $\left(-\infty, \frac{8}{9}\right)$

89. $\left(\frac{5}{2}, \infty\right)$ 90. $\left(-\infty, \frac{3}{5}\right]$

91. $\left[-\frac{8}{3}, \frac{3}{2}\right] \cup (6, \infty)$

92. $\left(-\frac{7}{2}, \frac{11}{9}\right) \cup \left(\frac{8}{3}, \infty\right)$

Relating Concepts

For individual or collaborative investigation *(Exercises 55–58)*

Inequalities that involve more than two factors, such as

$$(3x - 4)(x + 2)(x + 6) \le 0,$$

can be solved using an extension of the method shown in ***Examples 5 and 6. Work Exercises 55–58 in order,*** *to see how the method is extended.*

55. Use the zero-factor property to solve $(3x - 4)(x + 2)(x + 6) = 0$.

56. Plot the three solutions in **Exercise 55** on a number line, using closed circles because of the nonstrict inequality, $\le$.

57. The number line from **Exercise 56** should show four intervals formed by the three points. For each interval, choose a number from the interval and decide whether it satisfies the original inequality.

58. On a single number line, graph the intervals that satisfy the inequality, including endpoints. This is the graph of the solution set of the inequality. Write the solution set in interval notation.

Use the technique described in ***Relating Concepts Exercises 55–58*** *to solve each inequality. Write each solution set in interval notation.*

59. $(2x - 3)(x + 2)(x - 3) \ge 0$

60. $(x + 5)(3x - 4)(x + 2) \ge 0$

61. $4x - x^3 \ge 0$

62. $16x - x^3 \ge 0$

63. $(x + 1)^2 (x - 3) < 0$

64. $(x - 5)^2 (x + 1) < 0$

65. $x^3 + 4x^2 - 9x \ge 36$

66. $x^3 + 3x^2 - 16x \le 48$

67. $x^2(x + 4)^2 \ge 0$

68. $-x^2(2x - 3)^2 \le 0$

Solve each rational inequality. Write each solution set in interval notation. ***See Examples 8 and 9.***

69. $\dfrac{x - 3}{x + 5} \le 0$

70. $\dfrac{x + 1}{x - 4} > 0$

71. $\dfrac{1 - x}{x + 2} < -1$

72. $\dfrac{6 - x}{x + 2} > 1$

73. $\dfrac{3}{x - 6} \le 2$

74. $\dfrac{3}{x - 2} < 1$

75. $\dfrac{-4}{1 - x} < 5$

76. $\dfrac{-6}{3x - 5} \le 2$

77. $\dfrac{10}{3 + 2x} \le 5$

78. $\dfrac{1}{x + 2} \ge 3$

79. $\dfrac{7}{x + 2} \ge \dfrac{1}{x + 2}$

80. $\dfrac{5}{x + 1} > \dfrac{12}{x + 1}$

81. $\dfrac{3}{2x - 1} > \dfrac{-4}{x}$

82. $\dfrac{-5}{3x + 2} \ge \dfrac{5}{x}$

83. $\dfrac{4}{2 - x} \ge \dfrac{3}{1 - x}$

84. $\dfrac{4}{x + 1} < \dfrac{2}{x + 3}$

85. $\dfrac{x + 3}{x - 5} \le 1$

86. $\dfrac{x + 2}{3 + 2x} \le 5$

Solve each rational inequality. Write each solution set in interval notation.

87. $\dfrac{2x - 3}{x^2 + 1} \ge 0$

88. $\dfrac{9x - 8}{4x^2 + 25} < 0$

89. $\dfrac{(5 - 3x)^2}{(2x - 5)^3} > 0$

90. $\dfrac{(5x - 3)^3}{(25 - 8x)^2} \le 0$

91. $\dfrac{(2x - 3)(3x + 8)}{(x - 6)^3} \ge 0$

92. $\dfrac{(9x - 11)(2x + 7)}{(3x - 8)^3} > 0$

93. (a) 1995 (b) 2007
94. (a) 2005 (b) 2006–2008
95. between (and inclusive of) 4 sec and 9.75 sec
96. between (and inclusive of) 1 sec and 1.75 sec
97. between -1.5 sec and 4 sec

*(Modeling) Solve each problem. For Exercises 95 and 96, **see Example 7**.*

93. *Box Office Receipts* U.S. movie box office receipts, in billions of dollars, are shown in 5-year increments from 1989 to 2009. (*Source:* www.boxofficemojo.com)

Year	Receipts
1989	5.033
1994	5.396
1999	7.413
2004	9.381
2009	10.595

These receipts R are reasonably approximated by the linear model

$$R = 0.3022x + 4.542,$$

where $x = 0$ corresponds to 1989, $x = 5$ corresponds to 1994, and so on. Using the model, calculate the year in which the receipts first exceed each amount.

(a) \$6.5 billion **(b)** \$10 billion

94. *Recovery of Solid Waste* The percent W of municipal solid waste recovered is shown in the bar graph. The linear model

$$W = 0.68x + 30.6,$$

where $x = 1$ represents 2004, $x = 2$ represents 2005, and so on, fits the data reasonably well.

(a) Based on this model, when did the percent of waste recovered first exceed 32%?
(b) In what years was it between 33% and 34%?

Municipal Solid Waste Recovered

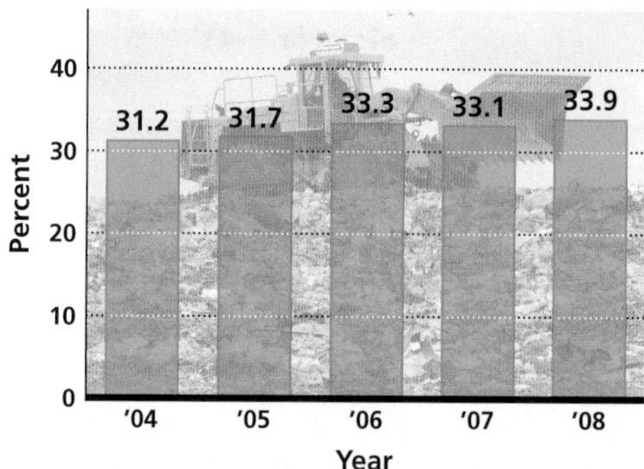

Source: U.S. Environmental Protection Agency.

95. *Height of a Projectile* A projectile is fired straight up from ground level. After t seconds, its height above the ground is s feet, where

$$s = -16t^2 + 220t.$$

For what time period is the projectile at least 624 ft above the ground?

96. *Height of a Baseball* A baseball is hit so that its height, s, in feet after t seconds is

$$s = -16t^2 + 44t + 4.$$

For what time period is the ball at least 32 ft above the ground?

97. *Velocity of an Object* Suppose the velocity, v, of an object is given by

$$v = 2t^2 - 5t - 12,$$

where t is time in seconds. (Here t can be positive or negative.) Find the intervals where the velocity is negative.

98. between 2 sec and 4 sec

98. *Velocity of an Object* The velocity of an object, v, after t seconds is given by

$$v = 3t^2 - 18t + 24.$$

Find the interval where the velocity is negative.

99. A student attempted to solve the inequality

$$\frac{2x - 3}{x + 2} \leq 0$$

by multiplying each side by $x + 2$ to get

$$2x - 3 \leq 0$$

$$x \leq \frac{3}{2}.$$

He wrote the solution set as $\left(-\infty, \frac{3}{2}\right]$. Is his solution correct? Explain.

100. A student solved the inequality $x^2 \leq 144$ by taking the square root of each side to get $x \leq 12$. She wrote the solution set as $(-\infty, 12]$. Is her solution correct? Explain.

1.8 Absolute Value Equations and Inequalities

- Basic Concepts
- Absolute Value Equations
- Absolute Value Inequalities
- Special Cases
- Absolute Value Models for Distance and Tolerance

Basic Concepts Recall from **Section R.2** that the **absolute value** of a number a, written $|a|$, gives the distance from a to 0 on a number line. By this definition, the equation $|x| = 3$ can be solved by finding all real numbers at a distance of 3 units from 0. As shown in **Figure 19,** two numbers satisfy this equation, 3 and -3, so the solution set is $\{-3, 3\}$.

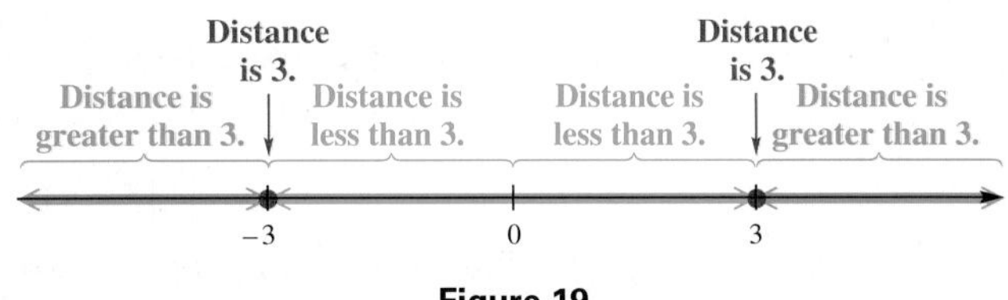

Figure 19

Similarly, $|x| < 3$ is satisfied by all real numbers whose distances from 0 are less than 3—that is, the interval

$$-3 < x < 3, \quad \text{or} \quad (-3, 3).$$

See **Figure 19.** Finally, $|x| > 3$ is satisfied by all real numbers whose distances from 0 are greater than 3. These numbers are less than -3 or greater than 3, so the solution set is

$$(-\infty, -3) \cup (3, \infty).$$

Notice in **Figure 19** that the union of the solution sets of $|x| = 3$, $|x| < 3$, and $|x| > 3$ is the set of real numbers.

These observations support the cases for solving absolute value equations and inequalities summarized in the table on the next page. If the equation or inequality fits the form of Case 1, 2, or 3, change it to its equivalent form and solve. The solution set and its graph will look similar to those shown.

LOOKING AHEAD TO CALCULUS

The precise definition of a **limit** in calculus requires writing absolute value inequalities.

A standard problem in calculus is to find the "interval of convergence" of something called a **power series,** by solving an inequality of the form

$$|x - a| < r.$$

This inequality says that x can be any number within r units of a on the number line, so its solution set is indeed an interval—namely the interval $(a - r, a + r)$.

For each equation or inequality in Cases 1–3 in the table, assume that $k > 0$.

Solving Absolute Value Equations and Inequalities

Absolute Value Equation or Inequality	Equivalent Form	Graph of the Solution Set	Solution Set
Case 1: $\lvert x \rvert = k$	$x = k$ or $x = -k$	$-k$ k	$\{-k, k\}$
Case 2: $\lvert x \rvert < k$	$-k < x < k$	$-k$ k	$(-k, k)$
Case 3: $\lvert x \rvert > k$	$x < -k$ or $x > k$	$-k$ k	$(-\infty, -k) \cup (k, \infty)$

In Cases 2 and 3, the strict inequality may be replaced by its nonstrict form. Additionally, if an absolute value equation takes the form $|a| = |b|$, then a and b must be equal in value or opposite in value.

Thus, the equivalent form of $|a| = |b|$ ***is*** $a = b$ **or** $a = -b$.

Teaching Tip Students will sometimes write $|x| > k$ in Case 3 as $-k > x > k$. Caution them against this.

Absolute Value Equations Because absolute value represents distance from 0 on a number line, solving an absolute value equation requires solving two possibilities, as shown in the examples that follow.

Classroom Example 1
Solve each equation.
(a) $|9 - 4x| = 7$
(b) $|3x + 2| = |x - 5|$

Answers:
(a) $\left\{\frac{1}{2}, 4\right\}$ (b) $\left\{-\frac{7}{2}, \frac{3}{4}\right\}$

EXAMPLE 1 Solving Absolute Value Equations (Case 1 and the Special Case $|a| = |b|$)

Solve each equation.

(a) $|5 - 3x| = 12$ **(b)** $|4x - 3| = |x + 6|$

SOLUTION

(a) For the given expression $5 - 3x$ to have absolute value 12, it must represent either 12 or -12. This equation fits the form of Case 1.

$$|5 - 3x| = 12$$

Don't forget this second possibility.

$$5 - 3x = 12 \quad \text{or} \quad 5 - 3x = -12 \qquad \text{Case 1}$$

$$-3x = 7 \quad \text{or} \quad -3x = -17 \qquad \text{Subtract 5.}$$

$$x = -\frac{7}{3} \quad \text{or} \quad x = \frac{17}{3} \qquad \text{Divide by } -3.$$

Check the solutions $-\frac{7}{3}$ and $\frac{17}{3}$ by substituting them in the original absolute value equation. The solution set is $\left\{-\frac{7}{3}, \frac{17}{3}\right\}$.

(b) If the absolute values of two expressions are equal, then those expressions are either equal in value or opposite in value.

$$|4x - 3| = |x + 6|$$

$$4x - 3 = x + 6 \quad \text{or} \quad 4x - 3 = -(x + 6) \qquad \text{Consider both possibilities.}$$

$$3x = 9 \quad \text{or} \quad 4x - 3 = -x - 6 \qquad \text{Solve each linear equation. (Section 1.1)}$$

$$x = 3 \quad \text{or} \quad 5x = -3$$

$$x = -\frac{3}{5}$$

CHECK $|4x - 3| = |x + 6|$ Original equation

$\left|4\left(-\frac{3}{5}\right) - 3\right| \stackrel{?}{=} \left|-\frac{3}{5} + 6\right|$ Let $x = -\frac{3}{5}$.	$|4(3) - 3| \stackrel{?}{=} |3 + 6|$ Let $x = 3$.
$\left|-\frac{12}{5} - 3\right| \stackrel{?}{=} \left|-\frac{3}{5} + 6\right|$	$|12 - 3| \stackrel{?}{=} |3 + 6|$
$\left|-\frac{27}{5}\right| = \left|\frac{27}{5}\right|$ ✓ True	$|9| = |9|$ ✓ True

Both solutions check. The solution set is $\left\{-\frac{3}{5}, 3\right\}$.

✔ *Now Try Exercises 9 and 19.*

Absolute Value Inequalities

Classroom Example 2
Solve each inequality.
(a) $|4x - 6| < 10$
(b) $|4x - 6| > 10$

Answers:
(a) $(-1, 4)$
(b) $(-\infty, -1) \cup (4, \infty)$

EXAMPLE 2 Solving Absolute Value Inequalities (Cases 2 and 3)

Solve each inequality.

(a) $|2x + 1| < 7$ **(b)** $|2x + 1| > 7$

SOLUTION

(a) This inequality fits Case 2. If the absolute value of an expression is less than 7, then the value of the expression is *between* -7 and 7.

$$|2x + 1| < 7$$

$$-7 < 2x + 1 < 7 \quad \text{Case 2}$$

$$-8 < 2x < 6 \quad \text{Subtract 1 from each part. (Section 1.7)}$$

$$-4 < x < 3 \quad \text{Divide each part by 2.}$$

The final inequality gives the solution set $(-4, 3)$.

(b) This inequality fits Case 3. If the absolute value of an expression is greater than 7, then the value of the expression is either less than -7 *or* greater than 7.

$$|2x + 1| > 7$$

$$2x + 1 < -7 \quad \text{or} \quad 2x + 1 > 7 \quad \text{Case 3}$$

$$2x < -8 \quad \text{or} \quad 2x > 6 \quad \text{Subtract 1 from each side.}$$

$$x < -4 \quad \text{or} \quad x > 3 \quad \text{Divide each side by 2.}$$

The solution set is $(-\infty, -4) \cup (3, \infty)$.

✔ *Now Try Exercises 27 and 29.*

Teaching Tip Now would be a good time to talk to students about the difference between a solution set, such as $\left\{-\frac{3}{5}, 3\right\}$ in **Example 1(b),** of an absolute value *equation*, and a solution set, such as $(-4, 3)$ in **Example 2(a),** of an absolute value *inequality*. The set braces { } in **Example 1(b)** are used to *list* the two solutions in the solution set, while the interval notation (,) in **Example 2(a)** is used to denote infinitely many solutions *between* -4 and 3.

Cases 1, 2, and 3 require that the absolute value expression be isolated on one side of the equation or inequality.

Classroom Example 3
Solve $|5 - 8x| + 6 \geq 14$.

Answer: $\left(-\infty, -\frac{3}{8}\right] \cup \left[\frac{13}{8}, \infty\right)$

EXAMPLE 3 Solving an Absolute Value Inequality (Case 3)

Solve $|2 - 7x| - 1 > 4$.

SOLUTION

$$|2 - 7x| - 1 > 4$$

$$|2 - 7x| > 5 \quad \text{Add 1 to each side.}$$

$$2 - 7x < -5 \quad \text{or} \quad 2 - 7x > 5 \quad \text{Case 3}$$

$$-7x < -7 \quad \text{or} \quad -7x > 3 \quad \text{Subtract 2.}$$

$$x > 1 \quad \text{or} \quad x < -\frac{3}{7} \quad \text{Divide by } -7\text{. Reverse the direction of each inequality. (Section 1.7)}$$

The solution set is $\left(-\infty, -\frac{3}{7}\right) \cup (1, \infty)$.

✔ *Now Try Exercise 51.*

Special Cases The three cases given in this section require the constant k to be positive. ***When $k \leq 0$, use the fact that the absolute value of any expression must be nonnegative, and consider the truth of the statement.***

Classroom Example 4
Solve each equation or inequality.
(a) $|7x + 28| = 0$
(b) $|6x - 9| > -2$
(c) $|2 - 5x| \leq -5$

Answers:
(a) $\{-4\}$ (b) $(-\infty, \infty)$
(c) $\emptyset$

EXAMPLE 4 Solving Special Cases

Solve each equation or inequality.

(a) $|2 - 5x| \geq -4$ **(b)** $|4x - 7| < -3$ **(c)** $|5x + 15| = 0$

SOLUTION

(a) Since the absolute value of a number is always nonnegative, the inequality

$$|2 - 5x| \geq -4 \text{ is always true.}$$

The solution set includes all real numbers, written $(-\infty, \infty)$.

(b) There is no number whose absolute value is less than -3 (or less than *any* negative number).

$$\text{The solution set of } |4x - 7| < -3 \text{ is } \emptyset.$$

(c) The absolute value of a number will be 0 only if that number is 0. Therefore, $|5x + 15| = 0$ is equivalent to

$$5x + 15 = 0, \quad \text{which has solution set} \quad \{-3\}.$$

CHECK Substitute -3 into the original equation.

$$|5x + 15| = 0 \quad \text{Original equation}$$
$$|5(-3) + 15| \stackrel{?}{=} 0 \quad \text{Let } x = -3.$$
$$0 = 0 \ \checkmark \quad \text{True}$$

✔ *Now Try Exercises 55, 57, and 59.*

Teaching Tip Substitute real numbers for a and b and model this concept on a number line.

Absolute Value Models for Distance and Tolerance If a and b represent two real numbers, then the absolute value of their difference,

$$\text{either} \quad |a - b| \quad \text{or} \quad |b - a|, \quad \text{(Section R.2)}$$

represents the distance between them.

Classroom Example 5
Write each statement using an absolute value inequality.
(a) m is no more than 9 units from 3.
(b) t is within 0.02 unit of 5.8.

Answers:
(a) $|m - 3| \leq 9$
(b) $|t - 5.8| < 0.02$

EXAMPLE 5 Using Absolute Value Inequalities to Describe Distances

Write each statement using an absolute value inequality.

(a) k is no less than 5 units from 8. **(b)** n is within 0.001 unit of 6.

SOLUTION

(a) Since the distance from k to 8, written $|k - 8|$ or $|8 - k|$, is no less than 5, the distance is greater than or equal to 5. This can be written as

$$|k - 8| \geq 5, \quad \text{or, equivalently,} \quad |8 - k| \geq 5. \quad \text{Either form is acceptable.}$$

(b) This statement indicates that the distance between n and 6 is less than 0.001.

$$|n - 6| < 0.001, \quad \text{or, equivalently,} \quad |6 - n| < 0.001$$

✔ *Now Try Exercises 83 and 85.*

Classroom Example 6
Suppose $y = 5x - 2$ and we want y to be within 0.001 unit of 6. For what values of x will this be true?

Answer: $(1.5998, 1.6002)$

EXAMPLE 6 Using Absolute Value to Model Tolerance

In quality control and other applications, we often wish to keep the difference between two quantities within some predetermined amount, called the **tolerance.** Suppose $y = 2x + 1$ and we want y to be within 0.01 unit of 4. For what values of x will this be true?

SOLUTION

$\lvert y - 4 \rvert < 0.01$	Write an absolute value inequality.
$\lvert 2x + 1 - 4 \rvert < 0.01$	Substitute $2x + 1$ for y.
$\lvert 2x - 3 \rvert < 0.01$	Combine like terms.
$-0.01 < 2x - 3 < 0.01$	Case 2
$2.99 < 2x < 3.01$	Add 3 to each part.
$1.495 < x < 1.505$	Divide each part by 2.

Reversing these steps shows that keeping x in the interval $(1.495, 1.505)$ ensures that the difference between y and 4 is within 0.01 unit.

Now Try Exercise 89.

1.8 Exercises

1. F 2. B
3. D 4. E
5. G 6. A
7. C 8. H

9. $\{-\frac{1}{3}, 1\}$ 10. $\{-\frac{7}{4}, \frac{3}{4}\}$
11. $\{\frac{2}{3}, \frac{8}{3}\}$ 12. $\{\frac{4}{3}, \frac{10}{3}\}$
13. $\{-6, 14\}$ 14. $\{-16, 12\}$
15. $\{\frac{5}{2}, \frac{7}{2}\}$ 16. $\{\frac{1}{8}, \frac{7}{8}\}$
17. $\{-\frac{4}{3}, \frac{2}{9}\}$ 18. $\{\frac{1}{5}, 7\}$
19. $\{-\frac{7}{3}, -\frac{1}{7}\}$ 20. $\{0, 1\}$
21. $\{1\}$ 22. $\{2\}$
23. $(-\infty, \infty)$

Concept Check Match each equation or inequality in Column I with the graph of its solution set in Column II.

I	II
1. $\lvert x \rvert = 7$	A. −7 0 7
2. $\lvert x \rvert = -7$	B. −7 0 7 ∅
3. $\lvert x \rvert > -7$	C. −7 0 7
4. $\lvert x \rvert > 7$	D. −7 0 7
5. $\lvert x \rvert < 7$	E. −7 0 7
6. $\lvert x \rvert \geq 7$	F. −7 0 7
7. $\lvert x \rvert \leq 7$	G. −7 0 7
8. $\lvert x \rvert \neq 7$	H. −7 0 7

*Solve each equation. See **Example 1.***

9. $\lvert 3x - 1 \rvert = 2$
10. $\lvert 4x + 2 \rvert = 5$
11. $\lvert 5 - 3x \rvert = 3$
12. $\lvert 7 - 3x \rvert = 3$
13. $\left\lvert \frac{x - 4}{2} \right\rvert = 5$
14. $\left\lvert \frac{x + 2}{2} \right\rvert = 7$
15. $\left\lvert \frac{5}{x - 3} \right\rvert = 10$
16. $\left\lvert \frac{3}{2x - 1} \right\rvert = 4$
17. $\left\lvert \frac{6x + 1}{x - 1} \right\rvert = 3$
18. $\left\lvert \frac{2x + 3}{3x - 4} \right\rvert = 1$
19. $\lvert 2x - 3 \rvert = \lvert 5x + 4 \rvert$
20. $\lvert x + 1 \rvert = \lvert 1 - 3x \rvert$
21. $\lvert 4 - 3x \rvert = \lvert 2 - 3x \rvert$
22. $\lvert 3 - 2x \rvert = \lvert 5 - 2x \rvert$
23. $\lvert 5x - 2 \rvert = \lvert 2 - 5x \rvert$

24. The equation $|5x - 6| = 3x$ cannot have a negative solution. Why?

25. The equation $|7x + 3| = -5x$ cannot have a positive solution. Why?

26. *Concept Check* Determine the solution set of each equation by inspection.

(a) $-|x| = |x|$ **(b)** $|-x| = |x|$ **(c)** $|x^2| = |x|$ **(d)** $-|x| = 9$

Solve each inequality. Give the solution set using interval notation. ***See Example 2.***

27. $|2x + 5| < 3$ **28.** $|3x - 4| < 2$ **29.** $|2x + 5| \geq 3$

30. $|3x - 4| \geq 2$ **31.** $\left|\frac{1}{2} - x\right| < 2$ **32.** $\left|\frac{3}{5} + x\right| < 1$

33. $4|x - 3| > 12$ **34.** $5|x + 1| > 10$ **35.** $|5 - 3x| > 7$

36. $|7 - 3x| > 4$ **37.** $|5 - 3x| \leq 7$ **38.** $|7 - 3x| \leq 4$

39. $\left|\frac{2}{3}x + \frac{1}{2}\right| \leq \frac{1}{6}$ **40.** $\left|\frac{5}{3} - \frac{1}{2}x\right| > \frac{2}{9}$ **41.** $|0.01x + 1| < 0.01$

42. Explain why the equation $|x| = \sqrt{x^2}$ has infinitely many solutions.

Solve each equation or inequality. ***See Examples 3 and 4.***

43. $|4x + 3| - 2 = -1$ **44.** $|8 - 3x| - 3 = -2$ **45.** $|6 - 2x| + 1 = 3$

46. $|4 - 4x| + 2 = 4$ **47.** $|3x + 1| - 1 < 2$ **48.** $|5x + 2| - 2 < 3$

49. $\left|5x + \frac{1}{2}\right| - 2 < 5$ **50.** $\left|2x + \frac{1}{3}\right| + 1 < 4$ **51.** $|10 - 4x| + 1 \geq 5$

52. $|12 - 6x| + 3 \geq 9$ **53.** $|3x - 7| + 1 < -2$ **54.** $|-5x + 7| - 4 < -6$

Solve each equation or inequality. ***See Example 4.***

55. $|10 - 4x| \geq -4$ **56.** $|12 - 9x| \geq -12$ **57.** $|6 - 3x| < -11$

58. $|18 - 3x| < -13$ **59.** $|8x + 5| = 0$ **60.** $|7 + 2x| = 0$

61. $|4.3x + 9.8| < 0$ **62.** $|1.5x - 14| < 0$ **63.** $|2x + 1| \leq 0$

64. $|3x + 2| \leq 0$ **65.** $|3x + 2| > 0$ **66.** $|4x + 3| > 0$

Relating Concepts

For individual or collaborative investigation *(Exercises 67–70)*

To see how to solve an equation that involves the absolute value of a quadratic polynomial, such as $|x^2 - x| = 6$, ***work Exercises 67–70 in order.***

67. For $x^2 - x$ to have an absolute value equal to 6, what are the two possible values that it may be? (*Hint:* One is positive and the other is negative.)

68. Write an equation stating that $x^2 - x$ is equal to the positive value you found in **Exercise 67,** and solve it using factoring.

69. Write an equation stating that $x^2 - x$ is equal to the negative value you found in **Exercise 67,** and solve it using the quadratic formula. (*Hint:* The solutions are not real numbers.)

70. Give the complete solution set of $|x^2 - x| = 6$, using the results from **Exercises 68 and 69.**

Use the method described in ***Relating Concepts Exercises 67–70,*** *if applicable, and properties of absolute value to solve each equation or inequality. (Hint: Exercises 77 and 78 can be solved by inspection.)*

71. $|3x^2 + x| = 14$ **72.** $|2x^2 - 3x| = 5$ **73.** $|4x^2 - 23x - 6| = 0$

---

26. (a) $\{0\}$ (b) $(-\infty, \infty)$ (c) $\{-1, 0, 1\}$ (d) $\emptyset$

27. $(-4, -1)$ **28.** $\left(\frac{2}{3}, 2\right)$

29. $(-\infty, -4] \cup [-1, \infty)$

30. $\left(-\infty, \frac{2}{3}\right] \cup [2, \infty)$

31. $\left(-\frac{3}{2}, \frac{5}{2}\right)$ **32.** $\left(-\frac{8}{5}, \frac{2}{5}\right)$

33. $(-\infty, 0) \cup (6, \infty)$

34. $(-\infty, -3) \cup (1, \infty)$

35. $\left(-\infty, -\frac{2}{3}\right) \cup (4, \infty)$

36. $(-\infty, 1) \cup \left(\frac{11}{3}, \infty\right)$

37. $\left[-\frac{2}{3}, 4\right]$ **38.** $\left[1, \frac{11}{3}\right]$

39. $\left[-1, -\frac{1}{2}\right]$

40. $\left(-\infty, \frac{26}{9}\right) \cup \left(\frac{34}{9}, \infty\right)$

41. $(-101, -99)$

43. $\left\{-1, -\frac{1}{2}\right\}$ **44.** $\left\{\frac{7}{3}, 3\right\}$

45. $\{2, 4\}$ **46.** $\left\{\frac{1}{2}, \frac{3}{2}\right\}$

47. $\left(-\frac{4}{3}, \frac{2}{3}\right)$ **48.** $\left(-\frac{7}{5}, \frac{3}{5}\right)$

49. $\left(-\frac{3}{2}, \frac{13}{10}\right)$ **50.** $\left(-\frac{5}{3}, \frac{4}{3}\right)$

51. $\left(-\infty, \frac{3}{2}\right] \cup \left[\frac{7}{2}, \infty\right)$

52. $(-\infty, 1] \cup [3, \infty)$

53. $\emptyset$ **54.** $\emptyset$

55. $(-\infty, \infty)$ **56.** $(-\infty, \infty)$

57. $\emptyset$ **58.** $\emptyset$

59. $\left\{-\frac{5}{8}\right\}$ **60.** $\left\{-\frac{7}{2}\right\}$

61. $\emptyset$ **62.** $\emptyset$

63. $\left\{-\frac{1}{2}\right\}$ **64.** $\left\{-\frac{2}{3}\right\}$

65. $\left(-\infty, -\frac{2}{3}\right) \cup \left(-\frac{2}{3}, \infty\right)$

66. $\left(-\infty, -\frac{3}{4}\right) \cup \left(-\frac{3}{4}, \infty\right)$

67. -6 or 6

68. $x^2 - x = 6$; $\{-2, 3\}$

69. $x^2 - x = -6$; $\left\{\frac{1}{2} \pm \frac{\sqrt{23}}{2}i\right\}$

70. $\left\{-2, 3, \frac{1}{2} \pm \frac{\sqrt{23}}{2}i\right\}$

71. $\left\{-\frac{7}{3}, 2, -\frac{1}{6} \pm \frac{\sqrt{167}}{6}i\right\}$

72. $\left\{-1, \frac{5}{2}, \frac{3}{4} \pm \frac{\sqrt{31}}{4}i\right\}$

73. $\left\{-\frac{1}{4}, 6\right\}$

74. $\left\{-\frac{7}{2}, -\frac{1}{3}, 0\right\}$
75. $\{-1, 1\}$
76. $\left\{-3, -\frac{2}{3}, \frac{2}{3}, 3\right\}$
77. $\emptyset$
78. $\emptyset$
79. $\left(-\infty, -\frac{1}{3}\right) \cup \left(-\frac{1}{3}, \infty\right)$
80. $\left(-\infty, -\frac{7}{8}\right) \cup \left(-\frac{7}{8}, \infty\right)$

In Exercises 81–88, the expression in absolute value bars may be replaced by its additive inverse. For example, in Exercise 81, $p - q$ may be written $q - p$.

81. $|p - q| = 2$
82. $|r - s| = 6$
83. $|m - 7| \leq 2$
84. $|z - 4| \geq 5$
85. $|p - 9| < 0.0001$
86. $|k - 10| < 0.0002$
87. $|r - 29| \geq 1$
88. $|q - 22| \leq 8$
89. $(0.9996, 1.0004)$
90. $(0.3998, 0.4002)$
91. $[6.7, 9.7]$
92. $[-140, -28]$
93. $|F - 730| \leq 50$
94. $|x - 123| \leq 25$; $|x - 21| \leq 5$
95. $25.33 \leq R_L \leq 28.17$; $36.58 \leq R_E \leq 40.92$
96. $5699.25 \leq T_L \leq 6338.25$; $8230.5 \leq T_E \leq 9207$

74. $|6x^3 + 23x^2 + 7x| = 0$ **75.** $|x^2 + 1| - |2x| = 0$ **76.** $\left|\frac{x^2 + 2}{x}\right| - \frac{11}{3} = 0$

77. $|x^4 + 2x^2 + 1| < 0$ **78.** $|x^2 + 10| < 0$

79. $\left|\frac{x - 4}{3x + 1}\right| \geq 0$ **80.** $\left|\frac{9 - x}{7 + 8x}\right| \geq 0$

81. *Concept Check* Write an equation involving absolute value that says the distance between p and q is 2 units.

82. *Concept Check* Write an equation involving absolute value that says the distance between r and s is 6 units.

Write each statement as an absolute value equation or inequality. ***See Example 5.***

83. m is no more than 2 units from 7. **84.** z is no less than 5 units from 4.

85. p is within 0.0001 unit of 9. **86.** k is within 0.0002 unit of 10.

87. r is no less than 1 unit from 29. **88.** q is no more than 8 units from 22.

89. *Tolerance* Suppose that $y = 5x + 1$ and we want y to be within 0.002 unit of 6. For what values of x will this be true?

90. *Tolerance* Repeat **Exercise 89,** but let $y = 10x + 2$.

(Modeling) Solve each problem. ***See Example 6.***

91. *Weights of Babies* Dr. Tydings has found that, over the years, 95% of the babies he has delivered weighed x pounds, where

$$|x - 8.2| \leq 1.5.$$

What range of weights corresponds to this inequality?

92. *Temperatures on Mars* The temperatures on the surface of Mars in degrees Celsius approximately satisfy the inequality $|C + 84| \leq 56$. What range of temperatures corresponds to this inequality?

93. *Conversion of Methanol to Gasoline* The industrial process that is used to convert methanol to gasoline is carried out at a temperature range of 680°F to 780°F. Using F as the variable, write an absolute value inequality that corresponds to this range.

94. *Wind Power Extraction Tests* When a model kite was flown in crosswinds in tests to determine its limits of power extraction, it attained speeds of 98 to 148 ft per sec in winds of 16 to 26 ft per sec. Using x as the variable in each case, write absolute value inequalities that correspond to these ranges.

(Modeling) Carbon Dioxide Emissions When humans breathe, carbon dioxide is emitted. In one study, the emission rates of carbon dioxide by college students were measured during both lectures and exams. The average individual rate R_L (in grams per hour) during a lecture class satisfied the inequality

$$|R_L - 26.75| \leq 1.42,$$

whereas during an exam the rate R_E satisfied the inequality

$$|R_E - 38.75| \leq 2.17.$$

(*Source:* Wang, T. C., *ASHRAE Trans.,* 81 (Part 1), 32.)

Use this information in Exercises 95–96.

95. Find the range of values for R_L and R_E.

96. The class had 225 students. If T_L and T_E represent the total amounts of carbon dioxide in grams emitted during a one-hour lecture and exam, respectively, write inequalities that model the ranges for T_L and T_E.

Chapter 1 Test Prep

Key Terms

1.1 equation
solution or root
solution set
equivalent equations
linear equation in one variable
first-degree equation
identity
conditional equation
contradiction
simple interest
literal equation
future or maturity value

1.2 mathematical model
linear model

1.3 imaginary unit
complex number
real part
imaginary part
pure imaginary number
nonreal complex number
standard form
complex conjugate

1.4 quadratic equation
standard form
second-degree equation
double solution
cubic equation
discriminant

1.5 leg
hypotenuse

1.6 rational equation
proposed solution
equation quadratic in form

1.7 inequality
linear inequality in one variable
interval
interval notation
open interval
closed interval
break-even point
quadratic inequality
strict inequality
nonstrict inequality
rational inequality

1.8 tolerance

New Symbols

Symbol	Meaning
$\emptyset$	empty or null set
i	imaginary unit
∞	infinity
(a, b), $(-\infty, a]$, $[a, b)$	interval notation
$\lvert a \rvert$	absolute value of a

Quick Review

Concepts / **Examples**

1.1 Linear Equations

Addition and Multiplication Properties of Equality
Let a, b, and c represent real numbers.

If $a = b$, then $a + c = b + c$.

If $a = b$ and $c \neq 0$, then $ac = bc$.

Solve. $5(x + 3) = 3x + 7$

$5x + 15 = 3x + 7$ Distributive property

$2x = -8$ Subtract $3x$. Subtract 15.

$x = -4$ Divide by 2.

Solution set: $\{-4\}$

1.2 Applications and Modeling with Linear Equations

Problem-Solving Steps
Step 1 Read the problem.

How many liters of 30% alcohol solution and 80% alcohol solution must be mixed to obtain 50 L of 50% alcohol solution?

Concepts	Examples
Step 2 Assign a variable.	Let $x =$ the number of liters of 30% solution. $50 - x =$ the number of liters of 80% solution. Summarize the information of the problem in a table.

Strength	Liters of Solution	Liters of Pure Alcohol
30%	x	$0.30x$
80%	$50 - x$	$0.80(50 - x)$
50%	50	$0.50(50)$

Step 3 Write an equation.

The equation is $0.30x + 0.80(50 - x) = 0.50(50)$.

Step 4 Solve the equation.

Solve the equation to obtain $x = 30$.

Step 5 State the answer.

Therefore, 30 L of the 30% solution and $50 - 30 = 20$ L of the 80% solution must be mixed.

Step 6 Check.

CHECK $\quad 0.30(30) + 0.80(50 - 30) \stackrel{?}{=} 0.50(50)$

$$25 = 25 \checkmark \quad \text{True}$$

1.3 Complex Numbers

Definition of *i*

$$i = \sqrt{-1} \quad \text{and} \quad i^2 = -1$$

Definition of Complex Number (*a* and *b* real)

$$\underset{\substack{\uparrow \\ \text{Real} \\ \text{part}}}{a} + \underset{\substack{\uparrow \\ \text{Imaginary} \\ \text{part}}}{bi}$$

In the complex number $-6 + 2i$, the real part is -6 and the imaginary part is 2.

Definition of $\sqrt{-a}$

For $a > 0$, $\quad \sqrt{-a} = i\sqrt{a}.$

Simplify.

$$\sqrt{-4} = 2i$$

$$\sqrt{-12} = i\sqrt{12} = 2i\sqrt{3}$$

Adding and Subtracting Complex Numbers

Add or subtract the real parts, and add or subtract the imaginary parts.

$$(2 + 3i) + (3 + i) - (2 - i)$$
$$= (2 + 3 - 2) + (3 + 1 + 1)i$$
$$= 3 + 5i$$

Multiplying and Dividing Complex Numbers

Multiply complex numbers as with binomials, and use the fact that $i^2 = -1$.

$$(6 + i)(3 - 2i) = 18 - 12i + 3i - 2i^2 \quad \text{FOIL}$$
$$= (18 + 2) + (-12 + 3)i \quad i^2 = -1$$
$$= 20 - 9i$$

Divide complex numbers by multiplying the numerator and denominator by the complex conjugate of the denominator.

$$\frac{3 + i}{1 + i} = \frac{(3 + i)(1 - i)}{(1 + i)(1 - i)} \quad \text{Multiply by } \frac{1 - i}{1 - i}.$$
$$= \frac{3 - 3i + i - i^2}{1 - i^2} \quad \text{Multiply.}$$
$$= \frac{4 - 2i}{2} \quad \text{Combine like terms; } i^2 = -1$$
$$= \frac{2(2 - i)}{2} \quad \text{Factor.}$$
$$= 2 - i \quad \text{Divide out the common factor.}$$

(continued)

Examples

1.4 Quadratic Equations

Zero-Factor Property
If a and b are complex numbers with $ab = 0$, then $a = 0$ or $b = 0$ or both equal zero.

Solve. $6x^2 + x - 1 = 0$

$(3x - 1)(2x + 1) = 0$ Factor.

$3x - 1 = 0$ or $2x + 1 = 0$ Zero-factor property

$x = \frac{1}{3}$ or $x = -\frac{1}{2}$

Solution set: $\left\{-\frac{1}{2}, \frac{1}{3}\right\}$

Square Root Property
The solution set of $x^2 = k$ is

$$\left\{\sqrt{k}, -\sqrt{k}\right\}, \text{ abbreviated } \left\{\pm\sqrt{k}\right\}.$$

Solve. $x^2 = 12$

$x = \pm\sqrt{12} = \pm 2\sqrt{3}$

Solution set: $\left\{\pm 2\sqrt{3}\right\}$

Quadratic Formula
The solutions of the quadratic equation $ax^2 + bx + c = 0$, where $a \neq 0$, are given by the quadratic formula.

$$x = \frac{-b \pm \sqrt{b^2 - 4ac}}{2a}$$

Solve. $x^2 + 2x + 3 = 0$

$x = \frac{-2 \pm \sqrt{2^2 - 4(1)(3)}}{2(1)}$ $a = 1, b = 2, c = 3$

$= \frac{-2 \pm \sqrt{-8}}{2}$ Simplify.

$= \frac{-2 \pm 2i\sqrt{2}}{2}$ Simplify the radical.

$= \frac{2(-1 \pm i\sqrt{2})}{2}$ Factor out 2 in the numerator.

$= -1 \pm i\sqrt{2}$ Lowest terms

Solution set: $\left\{-1 \pm i\sqrt{2}\right\}$

1.5 Applications and Modeling with Quadratic Equations

Pythagorean Theorem
In a right triangle, the sum of the squares of the lengths of legs a and b is equal to the square of the length of hypotenuse c.

$$a^2 + b^2 = c^2$$

In a right triangle, the shorter leg is 7 in. less than the longer leg, and the hypotenuse is 2 in. greater than the longer leg. What are the lengths of the sides?

Let $x =$ the length of the longer leg.

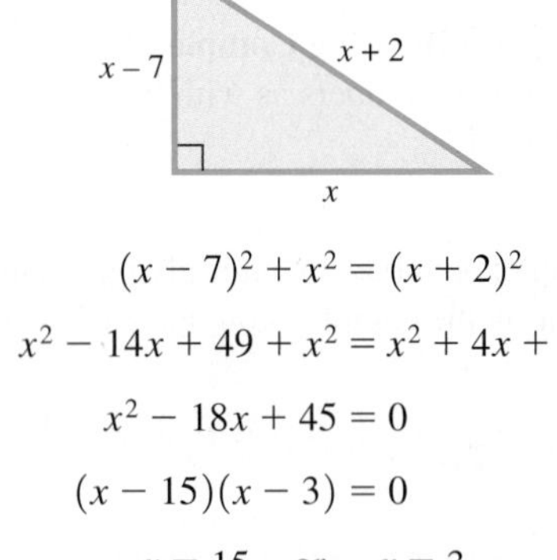

$$(x - 7)^2 + x^2 = (x + 2)^2$$
$$x^2 - 14x + 49 + x^2 = x^2 + 4x + 4$$
$$x^2 - 18x + 45 = 0$$
$$(x - 15)(x - 3) = 0$$
$$x = 15 \quad \text{or} \quad x = 3$$

The value 3 must be rejected. The lengths of the sides are 15 in., 8 in., and 17 in. Check to see that the conditions of the problem are satisfied.

Height of a Projected Object
The height s (in feet) of an object projected directly upward from an initial height of s_0 feet, with initial velocity v_0 feet per second, is

$$s = -16t^2 + v_0 t + s_0,$$

where t is the number of seconds after the object is projected.

Concepts	Examples

1.6 Other Types of Equations and Applications

Power Property
If P and Q are algebraic expressions, then every solution of the equation $P = Q$ is also a solution of the equation $P^n = Q^n$ for any positive integer n.

Quadratic in Form
An equation in the form $\boldsymbol{au^2 + bu + c = 0}$, where u is an algebraic expression, can often be solved by using a substitution variable.

If the power property is applied, or if both sides of an equation are multiplied by a variable expression, check all proposed solutions.

Solve. $(x+1)^{2/3} + (x+1)^{1/3} - 6 = 0$

$$u^2 + u - 6 = 0 \quad \text{Let } u = (x+1)^{1/3}.$$

$$(u+3)(u-2) = 0$$

$$u = -3 \quad \text{or} \quad u = 2$$

$$(x+1)^{1/3} = -3 \quad \text{or} \quad (x+1)^{1/3} = 2$$

$$x + 1 = -27 \quad \text{or} \quad x + 1 = 8 \quad \text{Cube.}$$

$$x = -28 \quad \text{or} \quad x = 7 \quad \text{Subtract 1.}$$

Both solutions check. The solution set is $\{-28, 7\}$.

1.7 Inequalities

Properties of Inequality
Let a, b, and c represent real numbers.

1. **If $a < b$, then $a + c < b + c$.**
2. **If $a < b$ and if $c > 0$, then $ac < bc$.**
3. **If $a < b$ and if $c < 0$, then $ac > bc$.**

Solve. $-3(x+4) + 2x < 6$

$$-3x - 12 + 2x < 6$$

$$-x < 18$$

$$x > -18 \quad \text{Multiply by } -1. \text{ Change } < \text{ to } >.$$

Solution set: $(-18, \infty)$

Solving a Quadratic Inequality

Step 1 Solve the corresponding quadratic equation.

Solve. $x^2 + 6x \le 7$

$$x^2 + 6x - 7 = 0 \quad \text{Corresponding equation}$$

$$(x+7)(x-1) = 0 \quad \text{Factor.}$$

$$x = -7 \quad \text{or} \quad x = 1 \quad \text{Zero-factor property}$$

Step 2 Identify the intervals determined by the solutions of the equation.

Step 3 Use a test value from each interval to determine which intervals form the solution set.

The intervals formed are $(-\infty, -7)$, $(-7, 1)$, and $(1, \infty)$. Test values show that values in the intervals $(-\infty, -7)$ and $(1, \infty)$ do not satisfy the original inequality, while those in $(-7, 1)$ do. Since the symbol $\le$ includes equality, the endpoints are included.
Solution set: $[-7, 1]$

Solving a Rational Inequality

Step 1 Rewrite the inequality so that 0 is on one side and a single fraction is on the other.

Solve. $\dfrac{x}{x+3} \ge \dfrac{5}{x+3}$

$$\frac{x}{x+3} - \frac{5}{x+3} \ge 0$$

$$\frac{x-5}{x+3} \ge 0$$

Step 2 Find the values that make either the numerator or the denominator 0.

The values -3 and 5 make either the numerator or the denominator 0. The intervals formed are

$$(-\infty, -3), (-3, 5), \text{ and } (5, \infty).$$

Step 3 Use a test value from each interval to determine which intervals form the solution set.

The value -3 must be excluded and 5 must be included. Test values show that values in the intervals $(-\infty, -3)$ and $(5, \infty)$ yield true statements.

Solution set: $(-\infty, -3) \cup [5, \infty)$

(continued)

Concepts	Examples
1.8 Absolute Value Equations and Inequalities	
Solving Absolute Value Equations and Inequalities For each equation or inequality in Cases 1-3, assume that $k > 0$.	
Case 1: To solve $\lvert x \rvert = k$, use the equivalent form $x = k \quad \text{or} \quad x = -k.$	Solve. $\lvert 5x - 2 \rvert = 3$ $5x - 2 = 3 \quad \text{or} \quad 5x - 2 = -3$ $5x = 5 \quad \text{or} \quad 5x = -1$ $x = 1 \quad \text{or} \quad x = -\frac{1}{5}$ Solution set: $\left\{-\frac{1}{5}, 1\right\}$
Case 2: To solve $\lvert x \rvert < k$, use the equivalent form $-k < x < k.$	Solve. $\lvert 5x - 2 \rvert < 3$ $-3 < 5x - 2 < 3$ $-1 < 5x < 5$ $-\frac{1}{5} < x < 1$ Solution set: $\left(-\frac{1}{5}, 1\right)$
Case 3: To solve $\lvert x \rvert > k$, use the equivalent form $x < -k \quad \text{or} \quad x > k.$	Solve. $\lvert 5x - 2 \rvert \geq 3$ $5x - 2 \leq -3 \quad \text{or} \quad 5x - 2 \geq 3$ $5x \leq -1 \quad \text{or} \quad 5x \geq 5$ $x \leq -\frac{1}{5} \quad \text{or} \quad x \geq 1$ Solution set: $\left(-\infty, -\frac{1}{5}\right] \cup [1, \infty)$

Chapter 1 Review Exercises

1. $\{6\}$ 2. $\{5\}$
3. $\left\{-\frac{11}{3}\right\}$ 4. $x = \frac{11(k+p)}{10-a}$
5. $f = \frac{AB(p+1)}{24}$
6. B, C

Solve each equation.

1. $2x + 8 = 3x + 2$

2. $\frac{1}{6}x - \frac{1}{12}(x - 1) = \frac{1}{2}$

3. $5x - 2(x + 4) = 3(2x + 1)$

4. $9x - 11(k + p) = x(a - 1)$, for x

5. $A = \frac{24f}{B(p + 1)}$, for f (approximate annual interest rate)

6. *Concept Check* Which of the following cannot be a correct equation to solve a geometry problem, if x represents the measure of a side of a rectangle? (*Hint:* Solve the equations and consider the solutions.)

A. $2x + 2(x + 2) = 20$ **B.** $2x + 2(5 + x) = -2$

C. $8(x + 2) + 4x = 16$ **D.** $2x + 2(x - 3) = 10$

7. A, B
8. (a) no
   (b) AirTran, Alaska/Horizon, Southwest
9. 13 in. on each side
10. 8 mi
11. $3\frac{3}{7}$ L
12. \$55,000 at 5.5%; \$35,000 at 6%
13. 15 mph
14. 9 hr
15. (a) $A = 36.525x$
    (b) 2629.8 mg

7. *Concept Check* If x represents the number of pennies in a jar in an applied problem, which of the following equations cannot be a correct equation for finding x? (*Hint:* Solve the equations and consider the solutions.)

A. $5x + 3 = 11$ **B.** $12x + 6 = -4$

C. $100x = 50(x + 3)$ **D.** $6(x + 4) = x + 24$

8. *Airline Carry-On Baggage Size* Carry-on rules for domestic economy-class travel differ from one airline to another, as shown in the table.

Airline	Size (linear inches)
AirTran	55
Alaska/Horizon	51
American	45
Delta	45
Southwest	50
United	45
USAirways	45

Source: Individual airline websites.

To determine the number of linear inches for a carry-on, add the length, width, and height of the bag.

(a) One Samsonite rolling bag measures 9 in. by 12 in. by 21 in. Are there any airlines that would not allow it as a carry-on?

(b) A Lark wheeled bag measures 10 in. by 14 in. by 22 in. On which airlines does it qualify as a carry-on?

Solve each problem.

9. *Dimensions of a Square* If the length of each side of a square is decreased by 4 in., the perimeter of the new square is 10 in. more than half the perimeter of the original square. What are the dimensions of the original square?

10. *Distance from a Library* Becky Anderson can ride her bike to the university library in 20 min. The trip home, which is all uphill, takes her 30 min. If her rate is 8 mph faster on her trip there than her trip home, how far does she live from the library?

11. *Alcohol Mixture* Alan wishes to strengthen a mixture that is 10% alcohol to one that is 30% alcohol. How much pure alcohol should he add to 12 L of the 10% mixture?

12. *Loan Interest Rates* A realtor borrowed \$90,000 to develop some property. He was able to borrow part of the money at 5.5% interest and the rest at 6%. The annual interest on the two loans amounts to \$5125. How much was borrowed at each rate?

13. *Speed of an Excursion Boat* An excursion boat travels upriver to a landing and then returns to its starting point. The trip upriver takes 1.2 hr, and the trip back takes 0.9 hr. If the average speed on the return trip is 5 mph faster than on the trip upriver, what is the boat's speed upriver?

14. *Toxic Waste* Two chemical plants are releasing toxic waste into a holding tank. Plant I releases waste twice as fast as Plant II. Together they fill the tank in 3 hr. How long would it take the slower plant to fill the tank working alone?

15. *(Modeling) Lead Intake* As directed by the "Safe Drinking Water Act" of December 1974, the EPA proposed a maximum lead level in public drinking water of 0.05 mg per liter. This standard assumed an individual consumption of two liters of water per day.

(a) If EPA guidelines are followed, write an equation that models the maximum amount of lead A ingested in x years. Assume that there are 365.25 days in a year.

(b) If the average life expectancy is 72 yr, find the EPA maximum lead intake from water over a lifetime.

16. $333.24 billion
17. (a) $4.31; The model gives a figure that is $0.51 more than the actual figure of $3.80.
(b) 47.6 yr after 1956, which is mid-2003. This is close to the minimum wage changing to $5.85 in 2007.
18. (a) 1965; 38.2 million; 1975: 39.3 million; 1985: 42.6 million; 1995: 43.9 million; 2005: 46.8 million

16. *(Modeling) Online Retail Sales* Projected e-commerce sales (in billions of dollars) for the years 2007–2012 can be modeled by the equation

$$y = 32.13x + 172.59,$$

where $x = 0$ corresponds to 2007, $x = 1$ corresponds to 2008, and so on. Based on this model, what would you expect the retail e-commerce sales to be in 2012? (*Source:* Forrester Research, Inc.)

17. *(Modeling) Minimum Wage* Some values of the U.S. minimum hourly wage, in dollars, for selected years from 1956 to 2009 are shown in the table. The linear model

$$y = 0.1132x + 0.4609$$

approximates the minimum wage during this time period, where x is the number of years after 1956 and y is the minimum wage in dollars.

Year	Minimum Wage	Year	Minimum Wage
1956	1.00	1996	4.75
1963	1.25	1997	5.15
1975	2.10	2007	5.85
1981	3.35	2008	6.55
1990	3.80	2009	7.25

Source: Bureau of Labor Statistics.

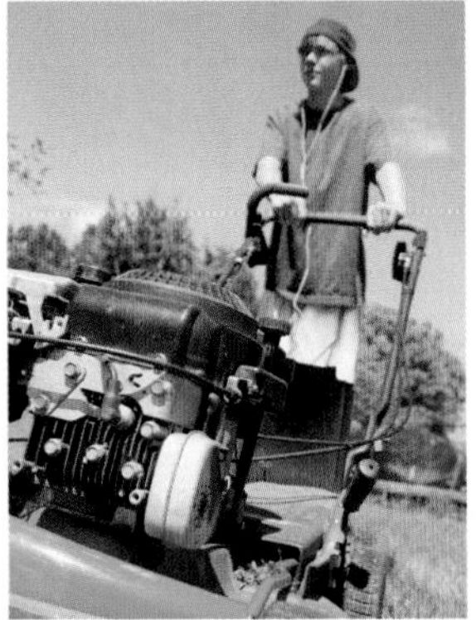

(a) Use the model to approximate the minimum wage in 1990. How does it compare to the data in the table?

(b) Use the model to approximate the year in which the minimum wage was $5.85. How does your answer compare to the data in the table?

18. *(Modeling) Mobility of the U.S. Population* The U.S. population, in millions, is given in the table. The figures are for the midyear of each decade from the 1960s through the 2000s. If we let the midyear shown in the table represent each decade, we can use the population figures of people moving with the data given in the graph.

Year	Population
1965	194
1975	216
1985	238
1995	263
2005	296

Source: U.S. Census Bureau.

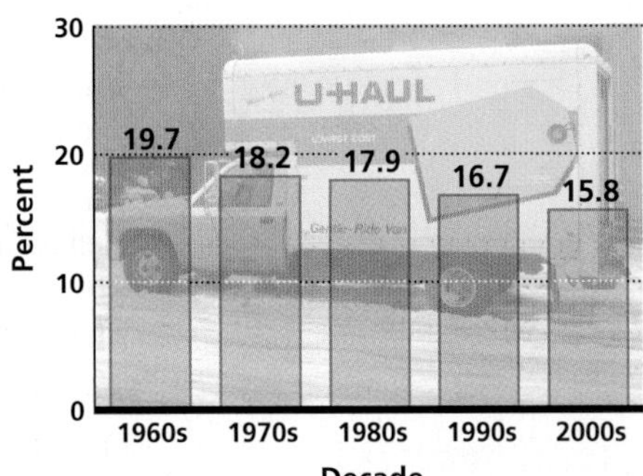

Source: U.S. Census Bureau.

(a) Find the number of Americans, to the nearest tenth of a million, moving in each year given in the table.

(b) The percents given in the graph decrease each decade, while the populations given in the table increase each decade. From your answers to part (a), is the number of Americans moving each decade increasing or decreasing? Explain.

19. $13 - 3i$
20. $-19 + 9i$
21. $-14 + 13i$
22. $-4 - 4i$
23. $19 + 17i$
24. $6 - 10i$
25. 146
26. $7 - 24i$
27. $-30 - 40i$
28. $-32 + 36i$
29. $1 - 2i$
30. $3 - 4i$
31. $-i$
32. 1
33. i
34. -1
35. i
36. $-i$
37. $\{-7 \pm \sqrt{5}\}$
38. $\left\{\frac{2 \pm 2\sqrt{2}}{3}\right\}$
39. $\left\{-3, \frac{5}{2}\right\}$
40. $\left\{\frac{1}{6}, \frac{1}{2}\right\}$
41. $\left\{-\frac{3}{2}, 7\right\}$
42. $\left\{-\frac{1}{3} \pm \frac{\sqrt{14}}{3}i\right\}$
43. $\{2 \pm \sqrt{6}\}$
44. $\{\sqrt{2} \pm 1\}$
45. $\left\{\frac{\sqrt{5} \pm 3}{2}\right\}$
46. $\{-2 \pm 2i\}$
47. D
48. B, C
49. A
50. -188; two distinct nonreal complex solutions
51. 76; two distinct irrational solutions
52. 484; two distinct rational solutions
53. -124; two distinct nonreal complex solutions
54. 0; one rational solution (a double solution)
55. 0; one rational solution (a double solution)
57. 6.25 sec and 7.5 sec
58. 10 in. by 13 in.
59. $\frac{1}{2}$ ft

Perform each operation and write answers in standard form.

19. $(6 - i) + (7 - 2i)$ **20.** $(-11 + 2i) - (8 - 7i)$

21. $15i - (3 + 2i) - 11$ **22.** $-6 + 4i - (8i - 2)$

23. $(5 - i)(3 + 4i)$ **24.** $(-8 + 2i)(-1 + i)$

25. $(5 - 11i)(5 + 11i)$ **26.** $(4 - 3i)^2$ **27.** $-5i(3 - i)^2$

28. $4i(2 + 5i)(2 - i)$ **29.** $\dfrac{-12 - i}{-2 - 5i}$ **30.** $\dfrac{-7 + i}{-1 - i}$

Find each power of i.

31. i^{11} **32.** i^{60} **33.** i^{1001} **34.** i^{110} **35.** i^{-27} **36.** $\dfrac{1}{i^{17}}$

Solve each equation.

37. $(x + 7)^2 = 5$ **38.** $(2 - 3x)^2 = 8$ **39.** $2x^2 + x - 15 = 0$

40. $12x^2 = 8x - 1$ **41.** $-2x^2 + 11x = -21$ **42.** $-x(3x + 2) = 5$

43. $(2x + 1)(x - 4) = x$ **44.** $\sqrt{2}x^2 - 4x + \sqrt{2} = 0$

45. $x^2 - \sqrt{5}\,x - 1 = 0$ **46.** $(x + 4)(x + 2) = 2x$

47. *Concept Check* Which one of the following equations has two real, distinct solutions? Do not actually solve.

A. $(3x - 4)^2 = -9$ **B.** $(4 - 7x)^2 = 0$

C. $(5x - 9)(5x - 9) = 0$ **D.** $(7x + 4)^2 = 11$

48. *Concept Check* Which equations in **Exercise 47** have only one distinct, real solution?

49. *Concept Check* Which one of the equations in **Exercise 47** has two nonreal complex solutions?

Evaluate the discriminant for each equation, and then use it to predict the number and type of solutions.

50. $8x^2 = -2x - 6$ **51.** $-6x^2 + 2x = -3$ **52.** $16x^2 + 3 = -26x$

53. $-8x^2 + 10x = 7$ **54.** $25x^2 + 110x + 121 = 0$ **55.** $x(9x + 6) = -1$

56. Explain how the discriminant of $ax^2 + bx + c = 0$ $(a \neq 0)$ is used to determine the number and type of solutions.

Solve each problem.

57. *(Modeling) Height of a Projectile* A projectile is fired straight up from ground level. After t seconds its height s, in feet above the ground, is given by

$$s = 220t - 16t^2.$$

At what times is the projectile exactly 750 ft above the ground?

58. *Dimensions of a Picture Frame* Zach Levy went into a frame-it-yourself shop. He wanted a frame 3 in. longer than it was wide. The frame he chose extended 1.5 in. beyond the picture on each side. Find the outside dimensions of the frame if the area of the unframed picture is 70 in.^{2}.

59. *Kitchen Flooring* Paula Story plans to replace the vinyl floor covering in her 10-ft by 12-ft kitchen. She wants to have a border of even width of a special material. She can afford only 21 ft^2 of this material. How wide a border can she have?

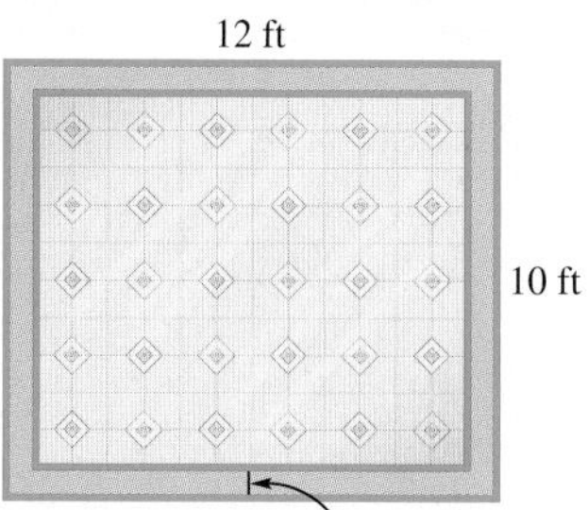

60. approximately 104.5 ft per sec
61. \$565.9 billion
62. 5 in., 12 in., 13 in.
63. $\{\pm i, \pm\frac{1}{2}\}$ 64. $\left\{0, \pm\frac{\sqrt{2}}{2}\right\}$
65. $\{-\frac{7}{24}\}$ 66. $\{-\frac{1}{2}, 3\}$
67. $\emptyset$ 68. $\{\frac{5}{2}, 4\}$
69. $\{-\frac{7}{4}\}$ 70. $\emptyset$
71. $\{-15, \frac{5}{2}\}$ 72. $\{-4, -\frac{80}{27}\}$
73. $\{3\}$ 74. $\{-1\}$
75. $\{-2, -1\}$ 76. $\{1\}$
77. $\emptyset$ 78. $\{8\}$
79. $\{-4, 1\}$ 80. $\{-2\}$
81. $\{-1\}$ 82. $\{1, 4\}$
83. $(-\frac{7}{13}, \infty)$ 84. $[-\frac{8}{9}, \infty)$
85. $(-\infty, 1]$ 86. $(-\infty, \frac{2}{5}]$
87. $[4, 5]$ 88. $(-\frac{7}{3}, -1)$
89. $[-4, 1]$
90. $(-\infty, -7) \cup (3, \infty)$
91. $(-\frac{2}{3}, \frac{5}{2})$
92. $\left(-\infty, \frac{3-\sqrt{29}}{2}\right] \cup \left[\frac{3+\sqrt{29}}{2}, \infty\right)$
93. $(-\infty, -4] \cup [0, 4]$
94. $(-\infty, -1) \cup (0, \frac{5}{2})$
95. $(-\infty, -2) \cup (5, \infty)$
96. $(-\infty, -\frac{1}{2}) \cup [6, \infty)$

60. *(Modeling) Airplane Landing Speed* To determine the appropriate landing speed of a small airplane, the formula

$$D = 0.1s^2 - 3s + 22$$

is used, where s is the initial landing speed in feet per second and D is the length of the runway in feet. If the landing speed is too fast, the pilot may run out of runway. If the speed is too slow, the plane may stall. If the runway is 800 ft long, what is the appropriate landing speed? Round to the nearest tenth.

61. *(Modeling) U.S. Government Spending on Medical Care* The amount spent in billions of dollars by the U.S. government on medical care during the period 1990–2010 can be approximated by the equation

$$y = 1.717x^2 + 0.8179x + 167.3,$$

where $x = 0$ corresponds to 1990, $x = 1$ corresponds to 1991, and so on. According to this model, about how much was spent by the U.S. government on medical care in 2005? Round to the nearest tenth of a billion. (*Source*: *U.S. Office of Management and Budget*.)

62. *Dimensions of a Right Triangle* The lengths of the sides of a right triangle are such that the shortest side is 7 in. shorter than the middle side, while the longest side (the hypotenuse) is 1 in. longer than the middle side. Find the lengths of the sides.

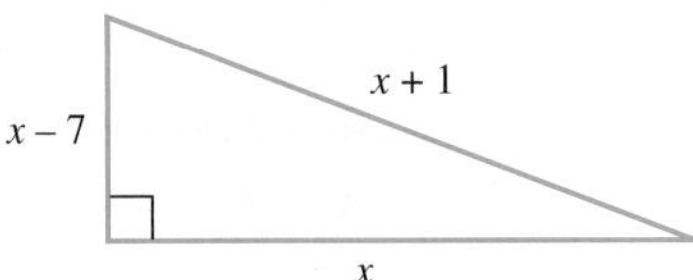

Solve each equation.

63. $4x^4 + 3x^2 - 1 = 0$ **64.** $x^2 - 2x^4 = 0$ **65.** $\frac{2}{x} - \frac{4}{3x} = 8 + \frac{3}{x}$

66. $2 - \frac{5}{x} = \frac{3}{x^2}$ **67.** $\frac{10}{4x - 4} = \frac{1}{1 - x}$ **68.** $\frac{13}{x^2 + 10} = \frac{2}{x}$

69. $\frac{x}{x + 2} + \frac{1}{x} + 3 = \frac{2}{x^2 + 2x}$ **70.** $\frac{2}{x + 2} + \frac{1}{x + 4} = \frac{4}{x^2 + 6x + 8}$

71. $(2x + 3)^{2/3} + (2x + 3)^{1/3} - 6 = 0$ **72.** $(x + 3)^{-2/3} - 2(x + 3)^{-1/3} = 3$

73. $\sqrt{4x - 2} = \sqrt{3x + 1}$ **74.** $\sqrt{2x + 3} = x + 2$

75. $\sqrt{x + 2} - x = 2$ **76.** $\sqrt{x} - \sqrt{x + 3} = -1$

77. $\sqrt{x + 3} - \sqrt{3x + 10} = 1$ **78.** $\sqrt{5x - 15} - \sqrt{x + 1} = 2$

79. $\sqrt{x^2 + 3x} - 2 = 0$ **80.** $\sqrt[5]{2x} = \sqrt[5]{3x + 2}$

81. $\sqrt[3]{6x + 2} - \sqrt[3]{4x} = 0$ **82.** $(x - 2)^{2/3} = x^{1/3}$

Solve each inequality. Write each solution set using interval notation.

83. $-9x + 3 < 4x + 10$ **84.** $11x \geq 2(x - 4)$

85. $-5x - 4 \geq 3(2x - 5)$ **86.** $7x - 2(x - 3) \leq 5(2 - x)$

87. $5 \leq 2x - 3 \leq 7$ **88.** $-8 > 3x - 5 > -12$

89. $x^2 + 3x - 4 \leq 0$ **90.** $x^2 + 4x - 21 > 0$

91. $6x^2 - 11x < 10$ **92.** $x^2 - 3x \geq 5$

93. $x^3 - 16x \leq 0$ **94.** $2x^3 - 3x^2 - 5x < 0$

95. $\frac{3x + 6}{x - 5} > 0$ **96.** $\frac{x + 7}{2x + 1} - 1 \leq 0$

97. $(-2, 0)$
98. $\left(-\frac{1}{3}, 0\right)$
99. $(-3, 1) \cup [7, \infty)$
100. $(-2, 4) \cup (16, \infty)$
101. (a) 79.8 ppb (b) 87.7 ppb
102. $[300, \infty)$
103. (a) 20 sec (b) between 2 sec and 18 sec
104. 7.1 yr after 2000, which is 2007. This is very close to the bar graph.

97. $\dfrac{3x-2}{x} - 4 > 0$

98. $\dfrac{5x+2}{x} < -1$

99. $\dfrac{3}{x-1} \le \dfrac{5}{x+3}$

100. $\dfrac{3}{x+2} > \dfrac{2}{x-4}$

(Modeling) Solve each problem.

101. *Ozone Concentration* Guideline levels for indoor ozone are less than 50 parts per billion (ppb). In a scientific study, a Purafil air filter was used to reduce an initial ozone concentration of 140 ppb. The filter removed 43% of the ozone. (*Source:* Parmar and Grosjean, *Removal of Air Pollutants from Museum Display Cases,* Getty Conservation Institute, Marina del Rey, CA.)

(a) What is the ozone concentration after the Purafil air filter is used?

(b) What is the maximum initial concentration of ozone that this filter will reduce to an acceptable level?

102. *Break-Even Interval* A company produces earbuds. The revenue from the sale of x units of these earbuds is $R = 8x$. The cost to produce x units of earbuds is $C = 3x + 1500$. In what interval will the company at least break even?

103. *Height of a Projectile* A projectile is launched upward from the ground. Its height s in feet above the ground after t seconds is given by

$$s = 320t - 16t^2.$$

(a) After how many seconds in the air will it hit the ground?

(b) During what time interval is the projectile more than 576 ft above the ground?

104. *Social Security* The total amount paid by the U.S. government to individuals for Social Security retirement and disability insurance benefits during the period 2000–2010 can be approximated by the linear model

$$y = 29.4x + 391.2,$$

where $x = 0$ corresponds to 2000, $x = 1$ corresponds to 2001, and so on. The variable y is in billions of dollars. Based on this model, during what year did the amount paid by the government first exceed $600 billion? Round your answer to the nearest year. Compare your answer to the bar graph.

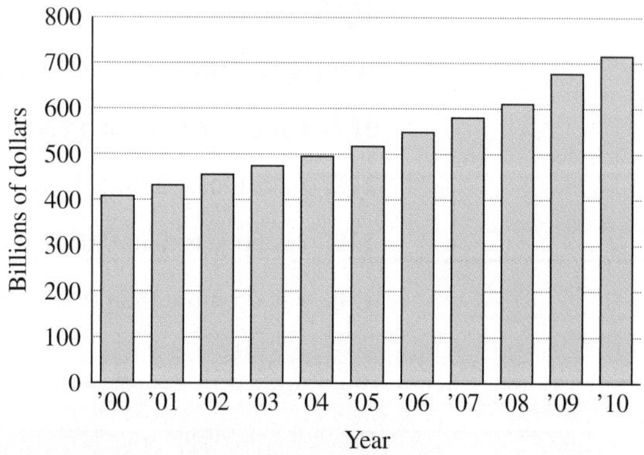

Source: U.S. Office of Management and Budget.

105. Without actually solving the inequality, explain why 3 cannot be in the solution set of $\frac{14x+9}{x-3} < 0$.

106. Without actually solving the inequality, explain why -4 must be in the solution set of $\frac{x+4}{2x+1} \ge 0$.

107. $c \geq 30$
108. $h \geq 12$; $g \geq 3.2$
109. $x > 14{,}000$
110. $d < 42$; $p < 5$
111. $\{-11, 3\}$
112. $\{-1, 5\}$
113. $\left\{\frac{11}{27}, \frac{25}{27}\right\}$
114. $\left\{-\frac{15}{13}, -\frac{13}{29}\right\}$
115. $\left\{-\frac{2}{7}, \frac{4}{3}\right\}$
116. $\left\{\frac{1}{2}\right\}$
117. $[-6, -3]$
118. $\left(-\infty, \frac{6}{5}\right] \cup [2, \infty)$
119. $\left(-\infty, -\frac{1}{7}\right) \cup (1, \infty)$
120. $\left(-\frac{22}{3}, \frac{14}{3}\right)$
121. $\left\{-4, -\frac{2}{3}\right\}$
122. $\left(-\infty, -\frac{11}{7}\right) \cup \left(-\frac{5}{7}, \infty\right)$
123. $(-\infty, \infty)$
124. $\emptyset$
125. $\{0, -4\}$
126. $(-\infty, -4) \cup (-4, 0) \cup (0, \infty)$
127. $|k - 6| = 12$ (or $|6 - k| = 12$)
128. $|p - 1| \geq 3$ (or $|1 - p| \geq 3$)
129. $|t - 5| \geq 0.01$ (or $|5 - t| \geq 0.01$)
130. $|s - 100| \leq 0.001$ (or $|100 - s| \leq 0.001$)

Rewrite the numerical part of each statement as an inequality using the indicated variable.

107. *Teacher Retirement* A teacher in the Public School Retirement System of Missouri can retire at any age with at least 30 yr of credit. (*Source:* Public School Retirement System of Missouri.) Let c represent the number of years of credit that a retiree has earned.

108. *Honor Roll* To make the honor list for the 2010 fall semester, students at Missouri University of Science and Technology had to carry a minimum of 12 credit hours and have a grade point average of 3.2 or above. (*Source: Warren County Record.*) Let h represent the number of credit hours, and let g represent the grade point average of a student on the honor list.

109. *Blood Donation* Severe winter weather throughout much of the eastern half of the United States caused the cancellation of more than 14,000 blood and platelet donations through the American Red Cross. (*Source: O'Fallon Community News.*) Let x represent the number of donation cancellations.

110. *Blood Donation* Blood is perishable and has no substitute. Red blood cells have a shelf life of 42 days, and platelets have a shelf life of just 5 days. (*Source: O'Fallon Community News.*) Let d represent the number of days that red blood cells may be stored prior to use. Let p represent the number of days that platelets may be stored prior to use.

Solve each equation or inequality.

111. $|x + 4| = 7$

112. $|2 - x| - 3 = 0$

113. $\left|\frac{7}{2 - 3x}\right| - 9 = 0$

114. $\left|\frac{8x - 1}{3x + 2}\right| = 7$

115. $|5x - 1| = |2x + 3|$

116. $|x + 10| = |x - 11|$

117. $|2x + 9| \leq 3$

118. $|8 - 5x| \geq 2$

119. $|7x - 3| > 4$

120. $\left|\frac{1}{2}x + \frac{2}{3}\right| < 3$

121. $|3x + 7| - 5 = 0$

122. $|7x + 8| - 6 > -3$

123. $|4x - 12| \geq -3$

124. $|7 - 2x| \leq -9$

125. $|x^2 + 4x| \leq 0$

126. $|x^2 + 4x| > 0$

Write as an absolute value equation or inequality.

127. k is 12 units from 6 on the number line.

128. p is at least 3 units from 1 on the number line.

129. t is no less than 0.01 unit from 5.

130. s is no more than 0.001 unit from 100.

Chapter 1 Test

[1.1]
1. $\{0\}$ 2. $\{-12\}$

[1.4]
3. $\left\{-\frac{1}{2}, \frac{7}{3}\right\}$ 4. $\left\{\frac{-1 \pm 2\sqrt{2}}{3}\right\}$

Solve each equation.

1. $3(x - 4) - 5(x + 2) = 2 - (x + 24)$

2. $\frac{2}{3}x + \frac{1}{2}(x - 4) = x - 4$

3. $6x^2 - 11x - 7 = 0$

4. $(3x + 1)^2 = 8$

5. $\left\{-\frac{1}{3} \pm \frac{\sqrt{5}}{3}i\right\}$

[1.6]
6. $\emptyset$
7. $\left\{-\frac{3}{4}\right\}$
8. $\{4\}$
9. $\{-3, 1\}$
10. $\{-2\}$
11. $\{\pm 1, \pm 4\}$
12. $\{-30, 5\}$

[1.8]
13. $\left\{-\frac{5}{2}, 1\right\}$ 14. $\left\{-6, \frac{4}{3}\right\}$

[1.1]
15. $W = \frac{S - 2LH}{2H + 2L}$

[1.3]
16. (a) $5 - 8i$ (b) $-29 - 3i$
(c) $55 + 48i$ (d) $6 + i$
17. (a) -1 (b) i
(c) i

[1.2]
18. (a) $A = 806{,}400x$
(b) 24,192,000 gal
(c) $P = 40.32x$; approximately 40 pools
(d) approximately 24.8 days
19. length: 200 m; width: 110 m
20. cashews: $23\frac{1}{3}$ lb; walnuts: $11\frac{2}{3}$ lb
21. 560 km per hr

5. $3x^2 + 2x = -2$

6. $\dfrac{12}{x^2 - 9} = \dfrac{2}{x - 3} - \dfrac{3}{x + 3}$

7. $\dfrac{4x}{x - 2} + \dfrac{3}{x} = \dfrac{-6}{x^2 - 2x}$

8. $\sqrt{3x + 4} + 5 = 2x + 1$

9. $\sqrt{-2x + 3} + \sqrt{x + 3} = 3$

10. $\sqrt[3]{3x - 8} = \sqrt[3]{9x + 4}$

11. $x^4 - 17x^2 + 16 = 0$

12. $(x + 3)^{2/3} + (x + 3)^{1/3} - 6 = 0$

13. $|4x + 3| = 7$

14. $|2x + 1| = |5 - x|$

15. *Surface Area of a Rectangular Solid* The formula for the surface area of a rectangular solid is

$$S = 2HW + 2LW + 2LH,$$

where S, H, W, and L represent surface area, height, width, and length, respectively. Solve this formula for W.

16. Perform each operation. Give the answer in standard form.

(a) $(9 - 3i) - (4 + 5i)$ **(b)** $(4 + 3i)(-5 + 3i)$

(c) $(8 + 3i)^2$ **(d)** $\dfrac{3 + 19i}{1 + 3i}$

17. Simplify each power of i.

(a) i^{42} **(b)** i^{-31} **(c)** $\dfrac{1}{i^{19}}$

Solve each problem.

18. *(Modeling) Water Consumption for Snowmaking* Ski resorts require large amounts of water in order to make snow. Snowmass Ski Area in Colorado plans to pump between 1120 and 1900 gal of water per minute at least 12 hr per day from Snowmass Creek between mid-October and late December. (*Source:* York Snow Incorporated.)

(a) Determine an equation that will calculate the *minimum* amount of water A (in gallons) pumped after x days during mid-October to late December.

(b) Find the minimum amount of water pumped in 30 days.

(c) Suppose the water being pumped from Snowmass Creek was used to fill swimming pools. The average backyard swimming pool holds 20,000 gal of water. Determine an equation that will give the minimum number of pools P that could be filled after x days. How many pools could be filled each day?

(d) In how many days could a minimum of 1000 pools be filled?

19. *Dimensions of a Rectangle* The perimeter of a rectangle is 620 m. The length is 20 m less than twice the width. What are the length and width?

20. *Nut Mixture* To make a special mix, the owner of a fruit and nut stand wants to combine cashews that sell for \$7.00 per lb with walnuts that sell for \$5.50 per lb to obtain 35 lb of a mixture that sells for \$6.50 per lb. How many pounds of each type of nut should be used in the mixture?

21. *Speed of a Plane* Mary Lynn left by plane to visit her mother in Louisiana, 420 km away. Fifteen minutes later, her mother left to meet her at the airport. She drove the 20 km to the airport at 40 km per hr, arriving just as the plane taxied in. What was the speed of the plane?

[1.5]
22. (a) 1 sec and 5 sec
(b) 6 sec

[1.2, 1.5]
23. B

[1.7]
24. $(-3, \infty)$
25. $[-10, 2]$
26. $(-\infty, -1] \cup [\frac{3}{2}, \infty)$
27. $(-\infty, 3) \cup (4, \infty)$

[1.8]
28. $(-2, 7)$
29. $(-\infty, -6] \cup [5, \infty)$
30. $\{-\frac{7}{3}\}$

22. *(Modeling) Height of a Projectile* A projectile is launched straight up from ground level with an initial velocity of 96 ft per sec. Its height in feet, s, after t seconds is given by the equation

$$s = -16t^2 + 96t.$$

(a) At what time(s) will it reach a height of 80 ft?
(b) After how many seconds will it return to the ground?

23. *(Modeling) U.S. Airline Passengers* The number of fliers on scheduled flights of U.S. airline companies was approximately 548 million in 1995, 739 million in 2005, and 704 million in 2009. (*Source:* Air Transport Association of America.) Here are three possible models for these data, where $x = 0$ corresponds to the year 1995.

Year	A	B	C
0	562	547.8	544.5
10	688.8	738.6	726.9
14	739.5	703.9	708.5

A. $y = 12.68x + 562.0$
B. $y = -1.983x^2 + 38.91x + 547.8$
C. $y = -1.632x^2 + 34.56x + 544.5$

The table shows each equation evaluated at the years 1995, 2004, and 2009. Decide which equation most closely models the data for these years.

Solve each inequality. Give the answer using interval notation.

24. $-2(x - 1) - 12 < 2(x + 1)$

25. $-3 \leq \frac{1}{2}x + 2 \leq 3$

26. $2x^2 - x \geq 3$

27. $\frac{x + 1}{x - 3} < 5$

28. $|2x - 5| < 9$

29. $|2x + 1| - 11 \geq 0$

30. $|3x + 7| \leq 0$

2 Graphs and Functions

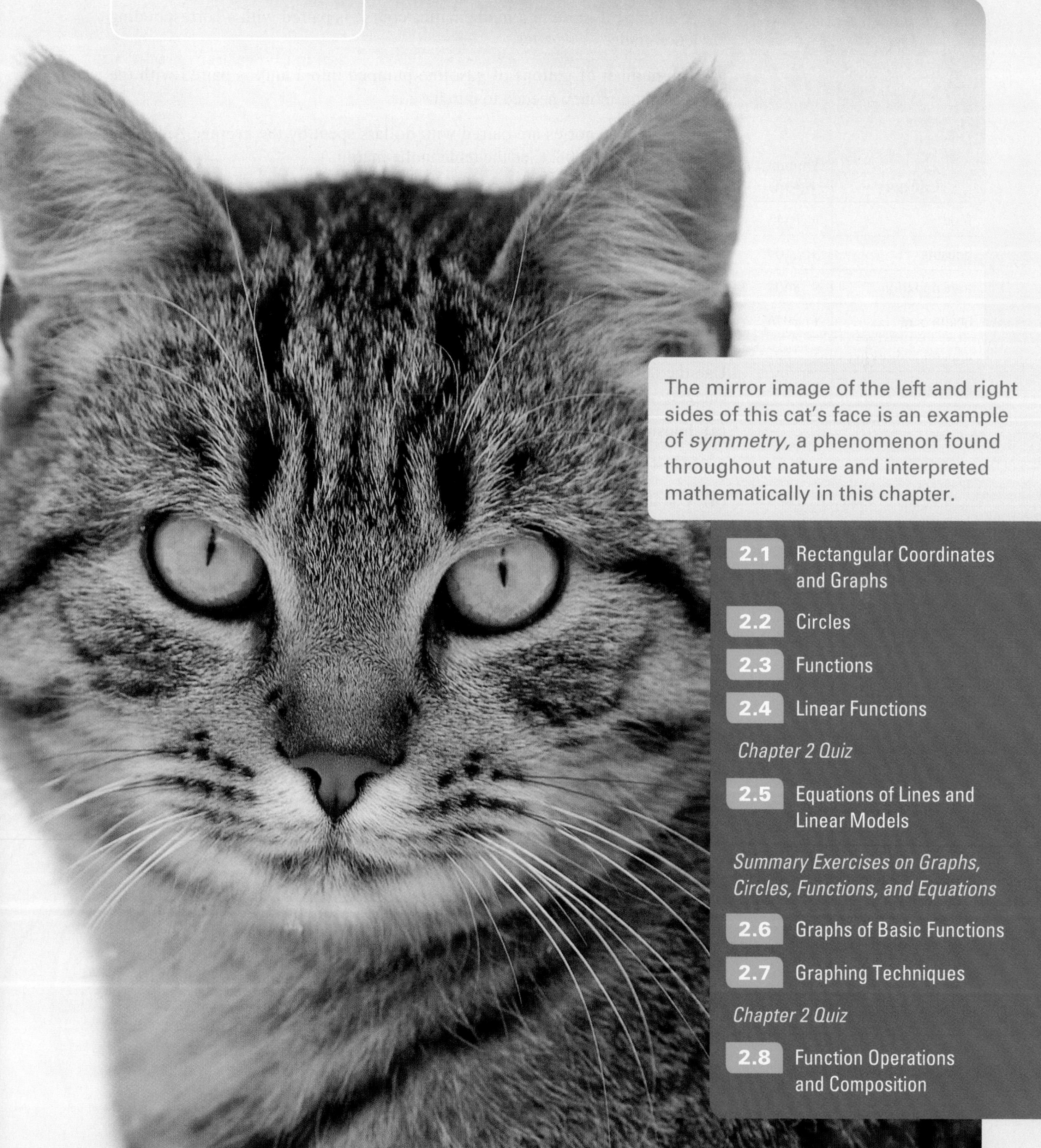

The mirror image of the left and right sides of this cat's face is an example of *symmetry,* a phenomenon found throughout nature and interpreted mathematically in this chapter.

2.1 Rectangular Coordinates and Graphs

- Ordered Pairs
- The Rectangular Coordinate System
- The Distance Formula
- The Midpoint Formula
- Graphing Equations in Two Variables

Category	Amount Spent
food	\$ 6443
housing	\$17,109
transportation	\$ 8604
health care	\$ 2976
apparel and services	\$ 1801
entertainment	\$ 2835

Source: U.S. Bureau of Labor Statistics.

Ordered Pairs The idea of pairing one quantity with another is often encountered in everyday life.

- A numerical score in a mathematics course is paired with a corresponding letter grade.
- The number of gallons of gasoline pumped into a tank is paired with the amount of money needed to purchase it.
- Expense categories are paired with dollars spent by the average American household in 2008. (See the table in the margin.)

Pairs of related quantities, such as a 96 determining a grade of A, 3 gallons of gasoline costing \$10.50, and 2008 spending on food of \$6443, can be expressed as *ordered pairs:* (96, A), (3, \$10.50), (food, \$6443). An **ordered pair** consists of two components, written inside parentheses.

EXAMPLE 1 Writing Ordered Pairs

Use the table to write ordered pairs to express the relationship between each category and the amount spent on it.

(a) housing **(b)** entertainment

SOLUTION

(a) Use the data in the second row: (housing, \$17,109).

(b) Use the data in the last row: (entertainment, \$2835).

Now Try Exercise 9.

Classroom Example 1
Use the table to write ordered pairs to express the relationship between each category and the amount spent on it.
(a) transportation
(b) health care

Answers:
(a) (transportation, \$8604)
(b) (health care, \$2976)

In mathematics, we are most often interested in ordered pairs whose components are numbers. The ordered pairs (a, b) and (c, d) are equal provided that $a = c$ *and* $b = d$.

NOTE Notation such as $(2, 4)$ was used in **Chapter 1** to show an interval on the number line, and the same notation is used to indicate an ordered pair of numbers. The intended use is usually clear from the context of the discussion.

The Rectangular Coordinate System As mentioned in **Section R.2,** each real number corresponds to a point on a number line. This idea is extended to ordered pairs of real numbers by using two perpendicular number lines, one horizontal and one vertical, that intersect at their zero-points. This point of intersection is called the **origin.** The horizontal line is called the ***x*-axis,** and the vertical line is called the ***y*-axis.**

The x-axis and y-axis together make up a **rectangular coordinate system,** or **Cartesian coordinate system** (named for one of its coinventors, René Descartes. The other coinventor was Pierre de Fermat). The plane into which the coordinate system is introduced is the **coordinate plane,** or ***xy*-plane.** See **Figure 1.** The x-axis and y-axis divide the plane into four regions, or **quadrants,** labeled as shown. The points on the x-axis and y-axis belong to no quadrant.

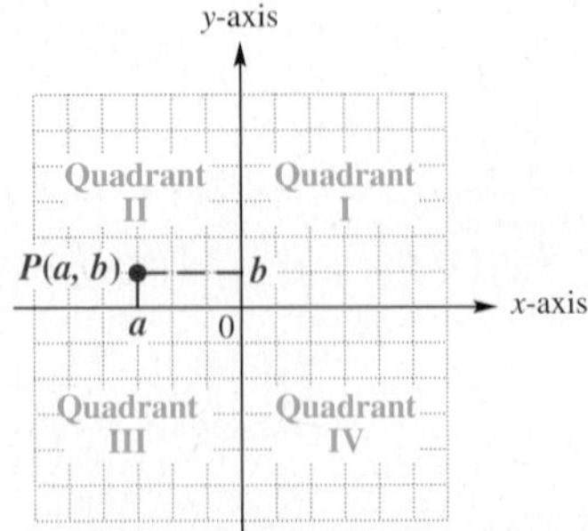

Figure 1

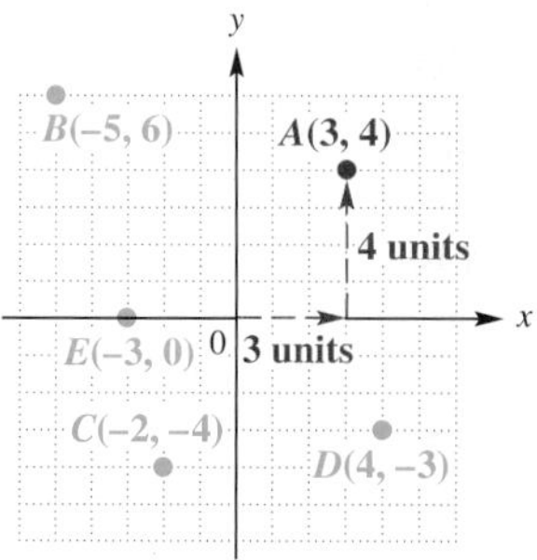

Figure 2

Each point P in the xy-plane corresponds to a unique ordered pair (a, b) of real numbers. The point P corresponding to the ordered pair (a, b) often is written $P(a, b)$ as in **Figure 1** and referred to as "the point (a, b)." The numbers a and b are the **coordinates** of point P. To locate on the xy-plane the point corresponding to the ordered pair $(3, 4)$, for example, start at the origin, move 3 units in the positive x-direction, and then move 4 units in the positive y-direction. See **Figure 2.** Point A corresponds to the ordered pair $(3, 4)$.

The Distance Formula Recall that the distance on a number line between points P and Q with coordinates x_1 and x_2 is

$$d(P, Q) = |x_1 - x_2| = |x_2 - x_1|.$$ Definition of distance **(Section R.2)**

By using the coordinates of their ordered pairs, we can extend this idea to find the distance between any two points in a plane.

Figure 3 shows the points $P(-4, 3)$ and $R(8, -2)$. We complete a right triangle as in the figure. This right triangle has its 90° angle at $Q(8, 3)$. The legs have lengths

$$d(P, Q) = |8 - (-4)| = 12$$

and

$$d(Q, R) = |3 - (-2)| = 5.$$

By the Pythagorean theorem, the hypotenuse has length

$$\sqrt{12^2 + 5^2} = \sqrt{144 + 25} = \sqrt{169} = 13.$$ **(Section 1.5)**

Thus, the distance between $(-4, 3)$ and $(8, -2)$ is 13.

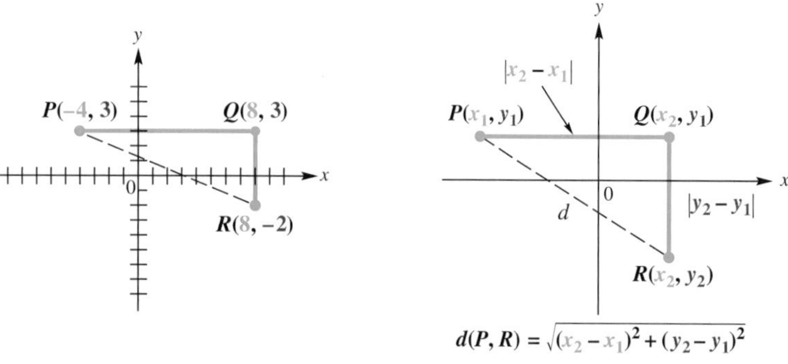

Figure 3

Figure 4

To obtain a general formula, let $P(x_1, y_1)$ and $R(x_2, y_2)$ be any two distinct points in a plane, as shown in **Figure 4.** Complete a triangle by locating point Q with coordinates (x_2, y_1). The Pythagorean theorem gives the distance between P and R.

$$d(P, R) = \sqrt{(x_2 - x_1)^2 + (y_2 - y_1)^2}$$

Absolute value bars are not necessary in this formula, since for all real numbers a and b,

$$|a - b|^2 = (a - b)^2.$$

The **distance formula** can be summarized as follows.

René Descartes (1596–1650)

The initial flash of *analytic geometry* may have come to Descartes as he was watching a fly crawling about on the ceiling near a corner of his room. It struck him that the path of the fly on the ceiling could be described if only one knew the relation connecting the fly's distances from two adjacent walls.

Source: An Introduction to the History of Mathematics by Howard Eves.

Distance Formula

Suppose that $P(x_1, y_1)$ and $R(x_2, y_2)$ are two points in a coordinate plane. The distance between P and R, written $d(P, R)$, is given by the following formula.

$$d(P, R) = \sqrt{(x_2 - x_1)^2 + (y_2 - y_1)^2}$$

LOOKING AHEAD TO CALCULUS
In analytic geometry and calculus, the distance formula is extended to two points in space. Points in space can be represented by **ordered triples.** The distance between the two points (x_1, y_1, z_1) and (x_2, y_2, z_2) is given by the following expression.

$$\sqrt{(x_2 - x_1)^2 + (y_2 - y_1)^2 + (z_2 - z_1)^2}$$

The distance formula can be stated in words.

The distance between two points in a coordinate plane is the square root of the sum of the square of the difference between their x-coordinates and the square of the difference between their y-coordinates.

Although our derivation of the distance formula assumed that P and R are not on a horizontal or vertical line, the result is true for any two points.

Classroom Example 2
Find the distance between

$P(3, -5)$ and $Q(-2, 8)$.

Answer: $\sqrt{194}$

EXAMPLE 2 Using the Distance Formula

Find the distance between $P(-8, 4)$ and $Q(3, -2)$.

SOLUTION Use the distance formula.

$$\begin{aligned} d(P, Q) &= \sqrt{(x_2 - x_1)^2 + (y_2 - y_1)^2} && \text{Distance formula} \\ &= \sqrt{[3 - (-8)]^2 + (-2 - 4)^2} && x_1 = -8,\ y_1 = 4,\ x_2 = 3,\ y_2 = -2 \\ &= \sqrt{11^2 + (-6)^2} && \text{Be careful when subtracting a negative number.} \\ &= \sqrt{121 + 36} \\ &= \sqrt{157} \end{aligned}$$

✔ Now Try Exercise 11(a).

A statement of the form "If p, then q" is called a **conditional statement.** The related statement "If q, then p" is called its **converse.** In **Section 1.5** we studied the Pythagorean theorem. Its *converse* is also a true statement.

If the sides a, b, and c of a triangle satisfy $a^2 + b^2 = c^2$, then the triangle is a right triangle with legs having lengths a and b and hypotenuse having length c.

We can use this fact to determine whether three points are the vertices of a right triangle.

Classroom Example 3
Are the points

$R(0, -2)$, $S(5, 1)$, and $T(-4, 3)$

the vertices of a right triangle?

Answer: no

EXAMPLE 3 Applying the Distance Formula

Are points $M(-2, 5)$, $N(12, 3)$, and $Q(10, -11)$ the vertices of a right triangle?

SOLUTION A triangle with the three given points as vertices, shown in **Figure 5,** is a right triangle if the square of the length of the longest side equals the sum of the squares of the lengths of the other two sides. Use the distance formula to find the length of each side of the triangle.

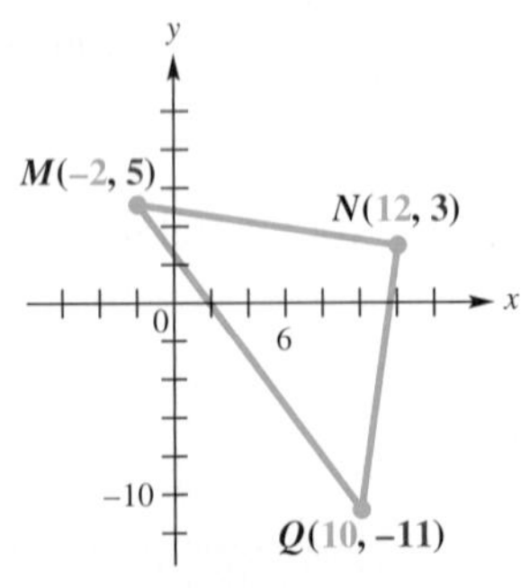

Figure 5

$$d(M, N) = \sqrt{[12 - (-2)]^2 + (3 - 5)^2} = \sqrt{196 + 4} = \sqrt{200}$$

$$d(M, Q) = \sqrt{[10 - (-2)]^2 + (-11 - 5)^2} = \sqrt{144 + 256} = \sqrt{400} = 20$$

$$d(N, Q) = \sqrt{(10 - 12)^2 + (-11 - 3)^2} = \sqrt{4 + 196} = \sqrt{200}$$

The longest side, of length 20 units, is chosen as the possible hypotenuse. Since

$$\left(\sqrt{200}\right)^2 + \left(\sqrt{200}\right)^2 = 400 = 20^2$$

is true, the triangle is a right triangle with hypotenuse joining M and Q.

✔ Now Try Exercise 19.

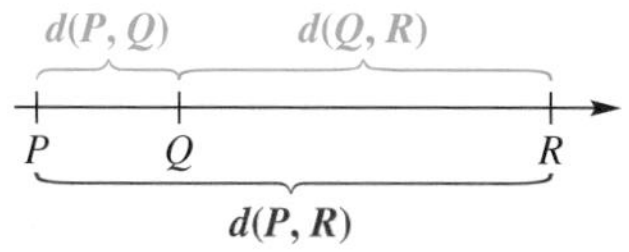

d(P, Q) d(Q, R)

P Q R

d(P, R)

$d(P, Q) + d(Q, R) = d(P, R)$

Figure 6

Using a similar procedure, we can tell whether three points are **collinear** (that is, lying on a straight line). ***Three points are collinear if the sum of the distances between two pairs of the points is equal to the distance between the remaining pair of points.*** See **Figure 6.**

EXAMPLE 4 Applying the Distance Formula

Are the points $P(-1, 5)$, $Q(2, -4)$, and $R(4, -10)$ collinear?

SOLUTION

$$d(P, Q) = \sqrt{(-1-2)^2 + [5-(-4)]^2} = \sqrt{9+81} = \sqrt{90} = 3\sqrt{10}$$

(Section R.7)

$$d(Q, R) = \sqrt{(2-4)^2 + [-4-(-10)]^2} = \sqrt{4+36} = \sqrt{40} = 2\sqrt{10}$$

$$d(P, R) = \sqrt{(-1-4)^2 + [5-(-10)]^2} = \sqrt{25+225} = \sqrt{250} = 5\sqrt{10}$$

Because $3\sqrt{10} + 2\sqrt{10} = 5\sqrt{10}$ is true, the three points are collinear.

✔ *Now Try Exercise 25.*

Classroom Example 4
Are the points $P(-2, 5)$, $Q(0, 3)$, and $R(8, -5)$ collinear?

Answer: yes

NOTE In **Exercises 76–80** of **Section 2.5,** we examine another method of determining whether three points are collinear.

The Midpoint Formula The midpoint of a line segment is equidistant from the endpoints of the segment. The **midpoint formula** is used to find the coordinates of the midpoint of a line segment. To develop the midpoint formula, let $P(x_1, y_1)$ and $Q(x_2, y_2)$ be any two distinct points in a plane. (Although **Figure 7** shows $x_1 < x_2$, no particular order is required.) Let (x, y) be the midpoint of the segment joining P and Q. Draw vertical lines from each of the three points to the x-axis, as shown in **Figure 7.**

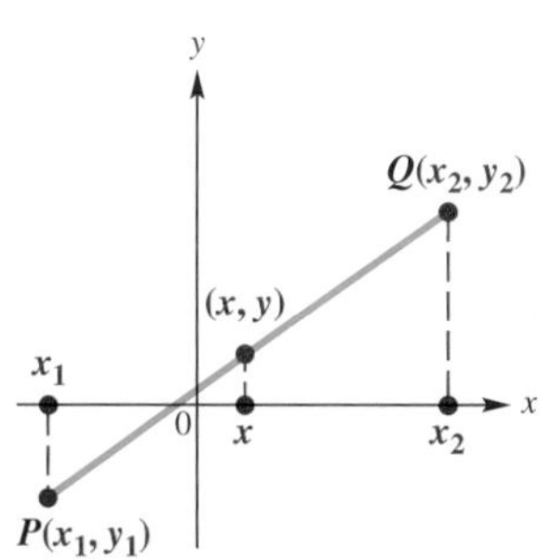

Figure 7

Since (x, y) is the midpoint of the line segment joining P and Q, the distance between x and x_1 equals the distance between x and x_2.

$$x_2 - x = x - x_1$$

$$x_2 + x_1 = 2x \qquad \text{Add } x \text{ and } x_1. \text{ (Section 1.1)}$$

$$x = \frac{x_1 + x_2}{2} \qquad \text{Divide by 2 and rewrite.}$$

Similarly, the y-coordinate is $\frac{y_1 + y_2}{2}$, yielding the following formula.

Midpoint Formula

The midpoint M of the line segment with endpoints $P(x_1, y_1)$ and $Q(x_2, y_2)$ has the following coordinates.

$$\left(\frac{x_1 + x_2}{2}, \frac{y_1 + y_2}{2}\right)$$

That is, the x-coordinate of the midpoint of a line segment is the* average *of the x-coordinates of the segment's endpoints, and the y-coordinate is the* average *of the y-coordinates of the segment's endpoints.

Classroom Example 5
Use the midpoint formula to do each of the following.
(a) Find the coordinates of the midpoint M of the segment with endpoints $(-7, -5)$ and $(-2, 13)$.
(b) Find the coordinates of the other endpoint Q of a segment with one endpoint $P(8, -20)$ and midpoint $M(4, -4)$.

Answers:
(a) $\left(-\frac{9}{2}, 4\right)$ (b) $(0, 12)$

EXAMPLE 5 Using the Midpoint Formula

Use the midpoint formula to do each of the following.

(a) Find the coordinates of the midpoint M of the segment with endpoints $(8, -4)$ and $(-6, 1)$.

(b) Find the coordinates of the other endpoint Q of a segment with one endpoint $P(-6, 12)$ and midpoint $M(8, -2)$.

SOLUTION

(a) The coordinates of M are found using the midpoint formula.

$$\left(\frac{8+(-6)}{2}, \frac{-4+1}{2}\right) = \left(1, -\frac{3}{2}\right) \quad \text{Substitute in } \left(\frac{x_1+x_2}{2}, \frac{y_1+y_2}{2}\right).$$

The coordinates of midpoint M are $\left(1, -\frac{3}{2}\right)$.

(b) Let (x, y) represent the coordinates of Q. Use the midpoint formula twice.

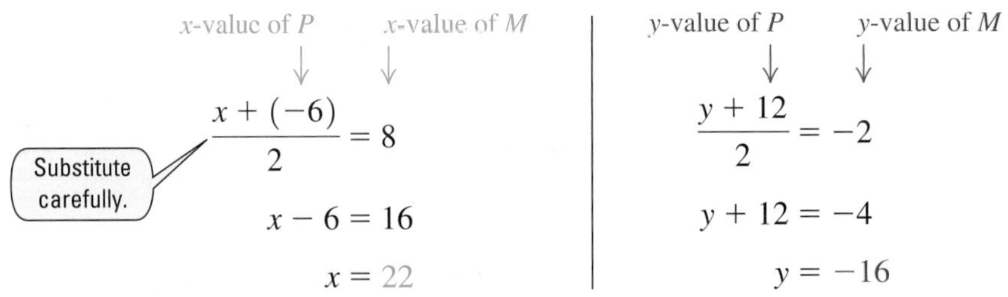

$$\frac{x+(-6)}{2} = 8 \qquad \frac{y+12}{2} = -2$$
$$x - 6 = 16 \qquad y + 12 = -4$$
$$x = 22 \qquad y = -16$$

The coordinates of endpoint Q are $(22, -16)$.

Now Try Exercises 11(b) and 31.

Classroom Example 6
Total revenue of full-service restaurants in the United States increased from \$117.8 billion in 1998 to \$184.2 billion in 2010 (projected). Use the midpoint formula to estimate the total revenue for 2004, and compare this to the actual figure of \$156.9 billion. (*Source:* National Restaurant Association.)

Answer: \$151.0 billion; This amount is \$5.9 billion less than the actual figure.

EXAMPLE 6 Applying the Midpoint Formula

Figure 8 depicts how a graph might indicate the increase in the revenue generated by fast-food restaurants in the United States from \$69.8 billion in 1990 to \$164.8 billion in 2010. Use the midpoint formula and the two given points to estimate the revenue from fast-food restaurants in 2000, and compare it to the actual figure of \$107.1 billion.

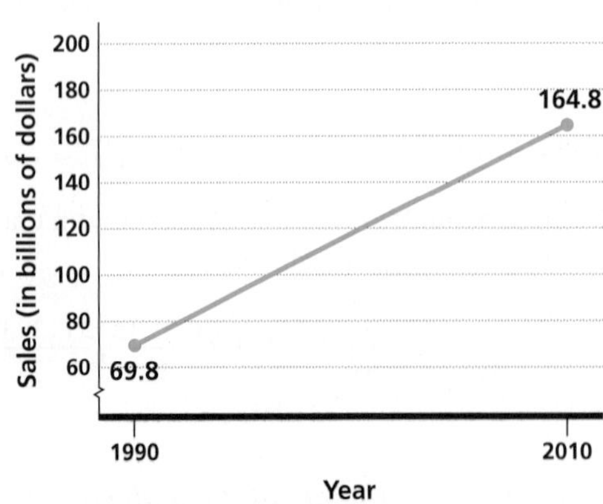

Source: National Restaurant Association.

Figure 8

SOLUTION The year 2000 lies halfway between 1990 and 2010, so we must find the coordinates of the midpoint of the segment that has endpoints

$$(1990, 69.8) \quad \text{and} \quad (2010, 164.8).$$

(Here, the second component is in billions of dollars.)

$$\left(\frac{1990 + 2010}{2}, \frac{69.8 + 164.8}{2}\right) = (2000, 117.3) \quad \text{Use the midpoint formula.}$$

Thus, our estimate is \$117.3 billion, which is greater than the actual figure of \$107.1 billion. (This discrepancy is due to actual revenue with an average increase of \$3.73 billion between 1990 and 2000, and then of \$5.77 billion between 2000 and 2010. Graphs such as this can sometimes be misleading.)

✔ *Now Try Exercise 37.*

Graphing Equations in Two Variables Ordered pairs are used to express the solutions of equations in two variables. When an ordered pair represents the solution of an equation with the variables x and y, the x-value is written first. For example, we say that

$$(1, 2) \quad \text{is a solution of} \quad 2x - y = 0,$$

since substituting 1 for x and 2 for y in the equation gives a true statement.

$$2x - y = 0$$
$$2(1) - 2 \stackrel{?}{=} 0 \quad \text{Let } x = 1 \text{ and } y = 2.$$
$$0 = 0 \ ✓ \quad \text{True}$$

Classroom Example 7
For each equation, find at least three ordered pairs that are solutions.
(a) $y = -2x + 5$
(b) $x = \sqrt[3]{y + 1}$
(c) $y = -x^2 + 1$

Answers:
(a) $(-1, 7), (0, 5), (3, -1)$
(b) $(1, 0), (-1, -2), (2, 7), (0, -1), (-2, -9)$
(c) $(0, 1), (-1, 0), (1, 0), (-2, -3), (2, -3)$
(Other answers are possible.)

EXAMPLE 7 Finding Ordered-Pair Solutions of Equations

For each equation, find at least three ordered pairs that are solutions.

(a) $y = 4x - 1$ **(b)** $x = \sqrt{y - 1}$ **(c)** $y = x^2 - 4$

SOLUTION

(a) Choose any real number for x or y and substitute in the equation to get the corresponding value of the other variable. For example, let $x = -2$ and then let $y = 3$.

$y = 4x - 1$		$y = 4x - 1$	
$y = 4(-2) - 1$	Let $x = -2$.	$3 = 4x - 1$	Let $y = 3$.
$y = -8 - 1$	Multiply.	$4 = 4x$	Add 1.
$y = -9$	Subtract.	$1 = x$	Divide by 4.

This gives the ordered pairs $(-2, -9)$ and $(1, 3)$. Verify that the ordered pair $(0, -1)$ is also a solution.

(b)

$$x = \sqrt{y - 1} \quad \text{Given equation}$$
$$1 = \sqrt{y - 1} \quad \text{Let } x = 1.$$
$$1 = y - 1 \quad \text{Square each side. (Section 1.6)}$$
$$2 = y \quad \text{Add 1.}$$

One ordered pair is $(1, 2)$. Verify that the ordered pairs $(0, 1)$ and $(2, 5)$ are also solutions of the equation.

(c) A table provides an organized method for determining ordered pairs. Here, we let x equal -2, -1, 0, 1, and 2 in $y = x^2 - 4$ and determine the corresponding y-values.

x	y	
-2	0	$(-2)^2 - 4 = 4 - 4 = 0$
-1	-3	$(-1)^2 - 4 = 1 - 4 = -3$
0	-4	$0^2 - 4 = -4$
1	-3	$1^2 - 4 = -3$
2	0	$2^2 - 4 = 0$

Five ordered pairs are $(-2, 0)$, $(-1, -3)$, $(0, -4)$, $(1, -3)$, and $(2, 0)$.

Now Try Exercises 43(a), 47(a), and 49(a).

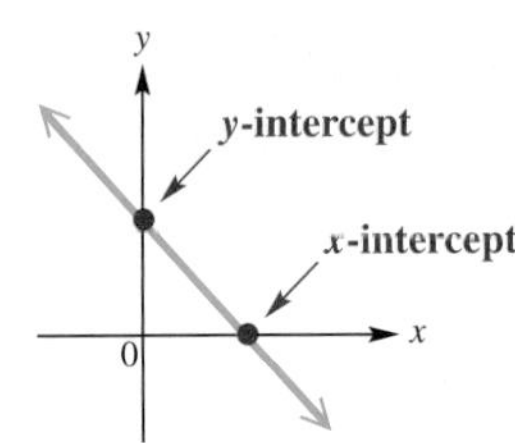

The **graph** of an equation is found by plotting ordered pairs that are solutions of the equation. The **intercepts** of the graph are good points to plot first. An ***x*-intercept** is an x-value where the graph intersects the x-axis. A ***y*-intercept** is a y-value where the graph intersects the y-axis. In other words, the x-intercept is the x-coordinate of an ordered pair where $y = 0$, and the y-intercept is the y-coordinate of an ordered pair where $x = 0$.

A general algebraic approach for graphing an equation using intercepts and point-plotting follows.

Graphing an Equation by Point Plotting

Step 1 Find the intercepts.

Step 2 Find as many additional ordered pairs as needed.

Step 3 Plot the ordered pairs from Steps 1 and 2.

Step 4 Join the points from Step 3 with a smooth line or curve.

Classroom Example 8
Graph each of the equations here, from **Classroom Example 7.**

(a) $y = -2x + 5$

(b) $x = \sqrt[3]{y + 1}$

(c) $y = -x^2 + 1$

Answers:

(a)

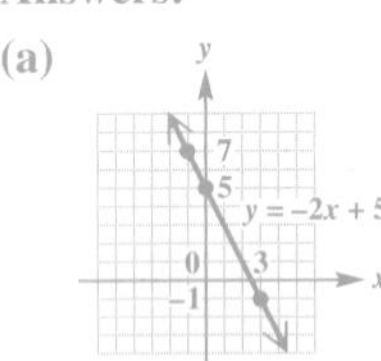

(b)

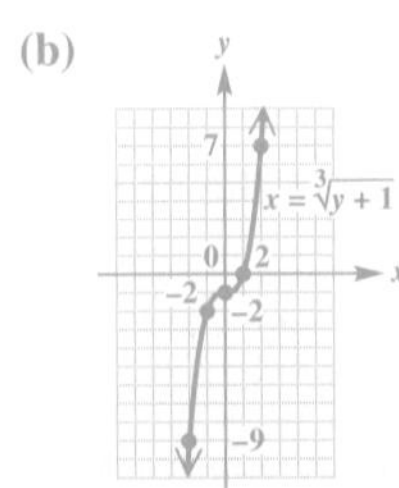

(c)

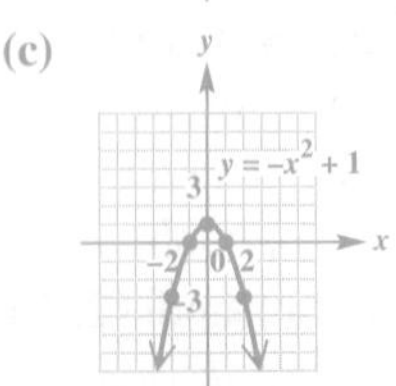

EXAMPLE 8 Graphing Equations

Graph each of the equations here, from **Example 7.**

(a) $y = 4x - 1$ **(b)** $x = \sqrt{y - 1}$ **(c)** $y = x^2 - 4$

SOLUTION

(a) ***Step 1*** Let $y = 0$ to find the x-intercept, and let $x = 0$ to find the y-intercept.

$y = 4x - 1$		$y = 4x - 1$	
$0 = 4x - 1$	Let $y = 0$.	$y = 4(0) - 1$	Let $x = 0$.
$1 = 4x$		$y = 0 - 1$	
$\frac{1}{4} = x$	x-intercept*	$y = -1$	y-intercept*

These intercepts lead to the ordered pairs $\left(\frac{1}{4}, 0\right)$ and $(0, -1)$. Note that the y-intercept yields one of the ordered pairs we found in **Example 7(a).**

*The intercepts are sometimes defined as ordered pairs, such as $\left(\frac{1}{4}, 0\right)$ and $(0, -1)$ instead of numbers, such as x-intercept $\frac{1}{4}$ and y-intercept -1. In this text, we define them as numbers.

Step 2 We use the other ordered pairs found in **Example 7(a):**

$$(-2, -9) \quad \text{and} \quad (1, 3).$$

Step 3 Plot the four ordered pairs from Steps 1 and 2 as shown in **Figure 9.**

Step 4 Join the points plotted in Step 3 with a straight line. This line, also shown in **Figure 9,** is the graph of the equation $y = 4x - 1$.

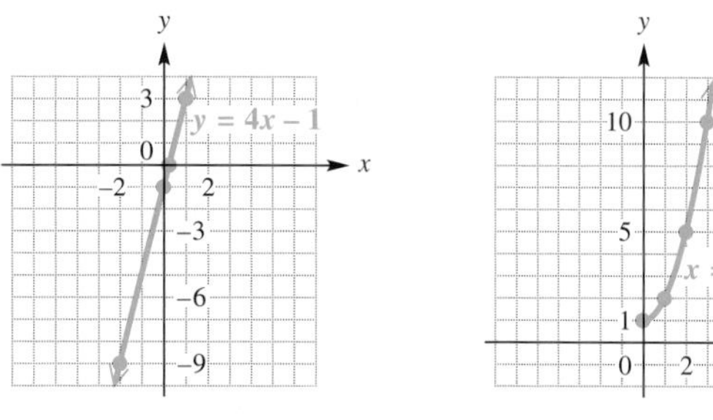

Figure 9 **Figure 10**

(b) For $x = \sqrt{y - 1}$, the y-intercept 1 was found in **Example 7(b).** Solve

$$x = \sqrt{0 - 1} \quad \text{Let } y = 0.$$

for the x-intercept. Since the quantity under the radical is negative, there is no x-intercept. In fact, $y - 1$ must be greater than or equal to 0, so y must be greater than or equal to 1.

We start by plotting the ordered pairs from **Example 7(b)** and then join the points with a smooth curve as in **Figure 10.** To confirm the direction the curve will take as x increases, we find another solution, $(3, 10)$. (Point plotting for graphs other than lines is often inefficient. We will examine other graphing methods later.)

(c) In **Example 7(c),** we made a table of five ordered pairs that satisfy the equation $y = x^2 - 4$.

$$(-2, 0), \quad (-1, -3), \quad (0, -4), \quad (1, -3), \quad (2, 0)$$

x-intercept ↑ (−2, 0); y-intercept ↑ (0, −4); x-intercept ↑ (2, 0)

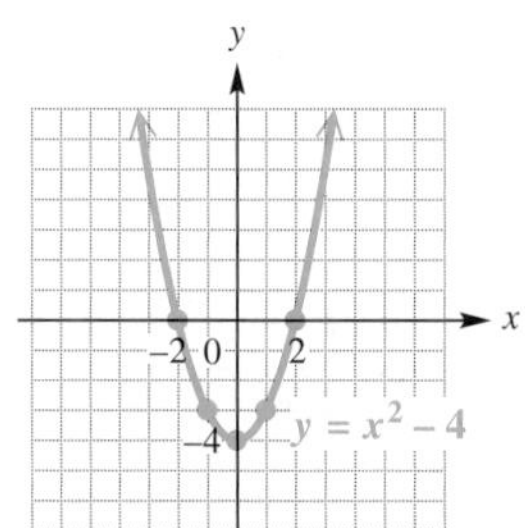

Figure 11

Plotting the points and joining them with a smooth curve gives the graph in **Figure 11.** This curve is called a **parabola.**

✔ *Now Try Exercises 43(b), 47(b), and 49(b).*

To graph an equation on a calculator, such as

$$y = 4x - 1, \quad \text{Equation from } \textbf{Example 8(a)}$$

we must first solve it for y (if necessary). Here the equation is already in the correct form, $y = 4x - 1$, so we enter $4x - 1$ for Y_1.

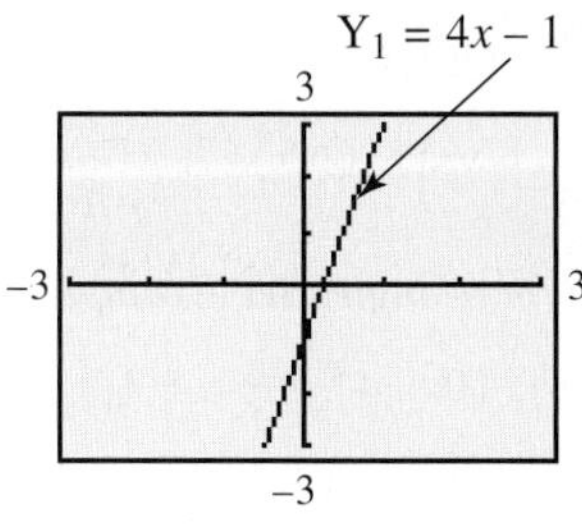

Figure 12

The intercepts can help determine an appropriate window, since we want them to appear in the graph. A good choice is often the **standard viewing window** for the TI-83/84 Plus, which has X minimum $= -10$, X maximum $= 10$, Y minimum $= -10$, Y maximum $= 10$, with X scale $= 1$ and Y scale $= 1$. (The X and Y scales determine the spacing of the tick marks.) Since the intercepts here are very close to the origin, we have chosen the X and Y minimum and maximum to be -3 and 3 instead. See **Figure 12.** ■

2.1 Exercises

1. false; $(-1, 3)$ lies in quadrant II.
2. false; The expression should be $\sqrt{(x_2 - x_1)^2 + (y_2 - y_1)^2}$.
3. true
4. true
5. true
7. any three of the following: $(2, -5), (-1, 7), (3, -9), (5, -17), (6, -21)$
8. any three of the following: $(3, 3), (-5, -21), (8, 18), (4, 6), (0, -6)$
9. any three of the following: $(1997, 36), (1999, 35), (2001, 29), (2003, 22), (2005, 23), (2007, 20)$
10. any three of the following: $(1998, 90.0), (2000, 88.5), (2002, 86.8), (2004, 89.8), (2006, 90.7), (2008, 97.4), (2010, 106.5)$
11. (a) $8\sqrt{2}$ (b) $(-9, -3)$
12. (a) 10 (b) $(-1, -1)$
13. (a) $\sqrt{34}$ (b) $\left(\frac{11}{2}, \frac{7}{2}\right)$
14. (a) $\sqrt{202}$ (b) $\left(-\frac{5}{2}, -\frac{1}{2}\right)$
15. (a) $3\sqrt{41}$ (b) $\left(0, \frac{5}{2}\right)$
16. (a) $2\sqrt{17}$ (b) $(5, 2)$
17. (a) $\sqrt{133}$ (b) $\left(2\sqrt{2}, \frac{3\sqrt{5}}{2}\right)$
18. (a) $3\sqrt{55}$ (b) $\left(2\sqrt{7}, \frac{7\sqrt{3}}{2}\right)$
19. yes 20. yes
21. no 22. no
23. yes 24. yes

Concept Check *Decide whether each statement in Exercises 1–5 is* true *or* false. *If the statement is false, tell why.*

1. The point $(-1, 3)$ lies in quadrant III of the rectangular coordinate system.

2. The distance from $P(x_1, y_1)$ to $Q(x_2, y_2)$ is given by

$$d(P, Q) = \sqrt{(x_1 - y_1)^2 + (x_2 - y_2)^2}.$$

3. The distance from the origin to the point (a, b) is $\sqrt{a^2 + b^2}$.

4. The midpoint of the segment joining (a, b) and $(3a, -3b)$ has coordinates $(2a, -b)$.

5. The graph of $y = 2x + 4$ has x-intercept -2 and y-intercept 4.

6. In your own words, list the steps for graphing an equation.

In Exercises 7–10, give three ordered pairs from each table. **See Example 1.**

7.

x	y
2	−5
−1	7
3	−9
5	−17
6	−21

8.

x	y
3	3
−5	−21
8	18
4	6
0	−6

9. *Percent of High School Students Who Smoke*

Year	Percent
1997	36
1999	35
2001	29
2003	22
2005	23
2007	20

Source: Centers for Disease Control and Prevention.

10. *Number of U.S. Viewers of the Super Bowl*

Year	Viewers (millions)
1998	90.0
2000	88.5
2002	86.8
2004	89.8
2006	90.7
2008	97.4
2010	106.5

Source: www.tvbythenumbers.com

For the points P and Q, find **(a)** *the distance d(P, Q) and* **(b)** *the coordinates of the midpoint of the segment PQ.* **See Examples 2 and 5(a).**

11. $P(-5, -7),\ Q(-13, 1)$
12. $P(-4, 3),\ Q(2, -5)$
13. $P(8, 2),\ Q(3, 5)$
14. $P(-8, 4),\ Q(3, -5)$
15. $P(-6, -5),\ Q(6, 10)$
16. $P(6, -2),\ Q(4, 6)$
17. $P(3\sqrt{2}, 4\sqrt{5}),\ Q(\sqrt{2}, -\sqrt{5})$
18. $P(-\sqrt{7}, 8\sqrt{3}),\ Q(5\sqrt{7}, -\sqrt{3})$

Determine whether the three points are the vertices of a right triangle. **See Example 3.**

19. $(-6, -4),\ (0, -2),\ (-10, 8)$
20. $(-2, -8),\ (0, -4),\ (-4, -7)$
21. $(-4, 1),\ (1, 4),\ (-6, -1)$
22. $(-2, -5),\ (1, 7),\ (3, 15)$
23. $(-4, 3),\ (2, 5),\ (-1, -6)$
24. $(-7, 4),\ (6, -2),\ (0, -15)$

25. yes 26. yes
27. no 28. no
29. no 30. yes

31. $(-3, 6)$ 32. $(-5, 3)$
33. $(5, -4)$ 34. $(-2, 7)$
35. $(2a - p, 2b - q)$
36. (a, c)

37. 25.35%; This estimate is very close to the actual figure of 25.2%.
38. \$487
39. \$20,666
40. (a) 12,366.5 thousand (b) 13,476 thousand

42. $d = [(x_2 - x_1)^2 + (y_2 - y_1)^2]^{1/2}$

Other ordered pairs are possible in Exercises 43–54.

43. (a)

x	y
0	−2
4	0
2	−1

(b)
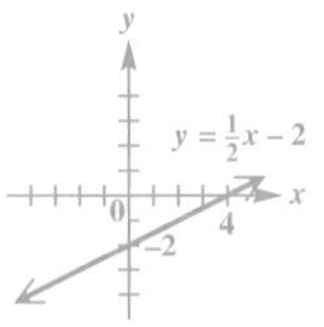

44. (a)

x	y
0	3
3	0
1	2

(b)
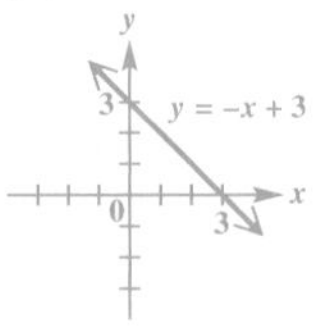

45. (a)

x	y
0	$\frac{5}{3}$
$\frac{5}{2}$	0
4	−1

(b)
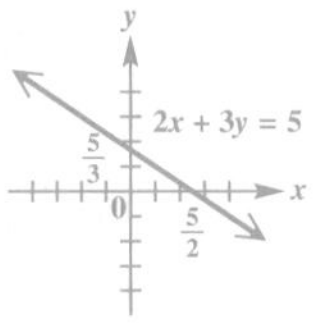

46. (a)

x	y
0	−3
2	0
4	3

(b)
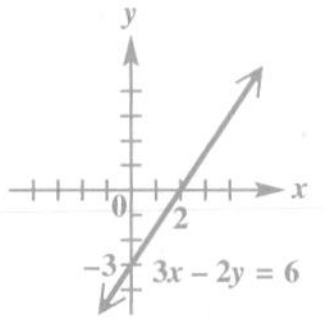

47. (a)

x	y
0	0
1	1
−2	4

(b)
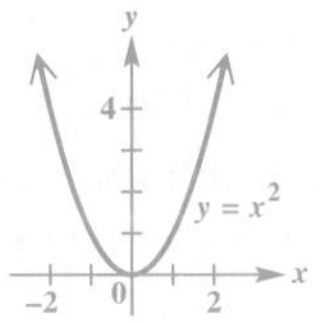

48. (a)

x	y
0	2
−1	3
2	6

(b)
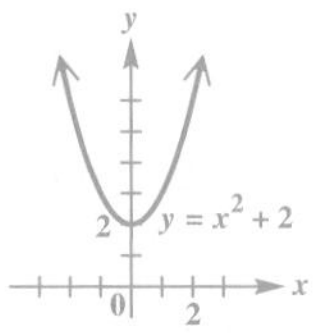

Determine whether the three points are collinear. **See Example 4.**

25. $(0, -7), (-3, 5), (2, -15)$
26. $(-1, 4), (-2, -1), (1, 14)$
27. $(0, 9), (-3, -7), (2, 19)$
28. $(-1, -3), (-5, 12), (1, -11)$
29. $(-7, 4), (6, -2), (-1, 1)$
30. $(-4, 3), (2, 5), (-1, 4)$

Find the coordinates of the other endpoint of each segment, given its midpoint and one endpoint. **See Example 5(b).**

31. midpoint $(5, 8)$, endpoint $(13, 10)$
32. midpoint $(-7, 6)$, endpoint $(-9, 9)$
33. midpoint $(12, 6)$, endpoint $(19, 16)$
34. midpoint $(-9, 8)$, endpoint $(-16, 9)$
35. midpoint (a, b), endpoint (p, q)
36. midpoint $\left(\frac{a+b}{2}, \frac{c+d}{2}\right)$, endpoint (b, d)

Solve each problem. **See Example 6.**

37. ***Bachelor's Degree Attainment*** The graph shows a straight line that approximates the percentage of Americans 25 years and older who had earned bachelor's degrees or higher for the years 1990–2008. Use the midpoint formula and the two given points to estimate the percent in 1999. Compare your answer with the actual percent of 25.2.

Source: U.S. Census Bureau.

38. ***Temporary Assistance for Needy Families (TANF)*** The graph shows an idealized linear relationship for the average monthly payment per recipient to needy families in the TANF program. Based on this information, what was the average payment to families in 2004?

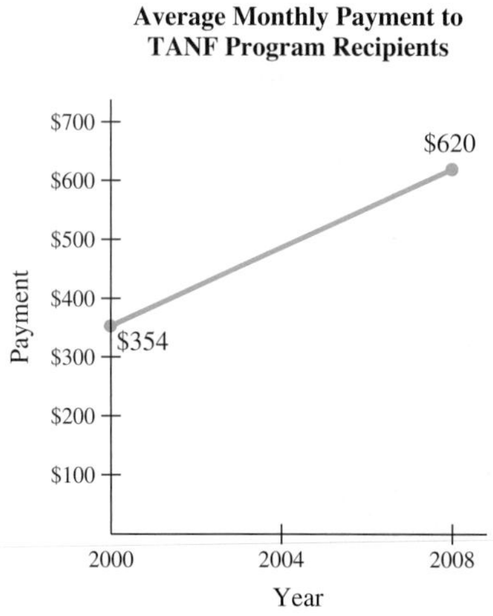

Source: U.S. Department of Health and Human Services.

39. ***Poverty Level Income Cutoffs*** The table lists how poverty level income cutoffs (in dollars) for a family of four have changed over time. Use the midpoint formula to approximate the poverty level cutoff in 2006 to the nearest dollar.

Year	Income (in dollars)
1980	8414
1990	13,359
2000	17,604
2004	19,307
2008	22,025

Source: U.S. Census Bureau.

49. (a) (b)

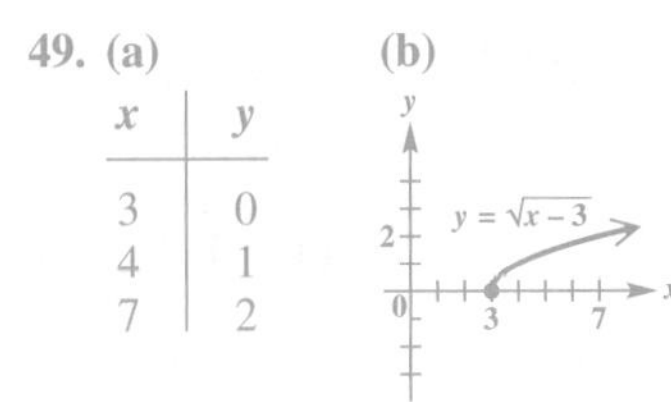

50. (a) (b)

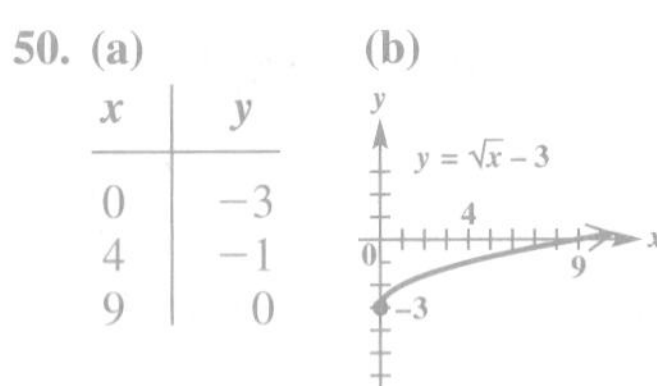

51. (a) (b)

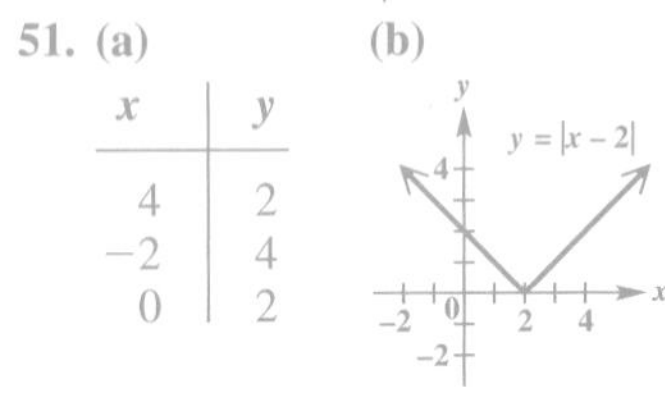

52. (a) (b)

x	y
−2	−2
−4	0
0	−4

$y = -|x+4|$

53. (a) (b)

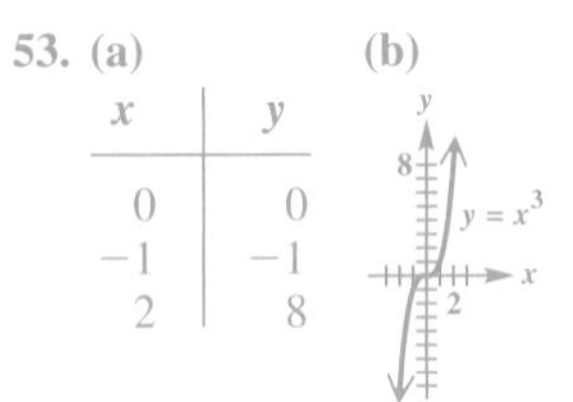

54. (a) (b)

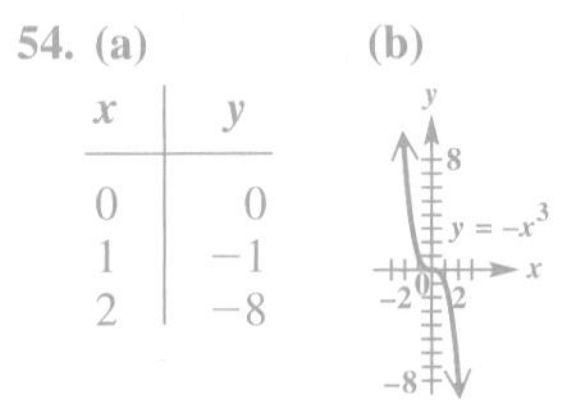

55. (4, 0) 56. (0, 3)

57. III; I; IV; IV 59. yes; no 60. (6, 8) and (8, 11)

40. *Public College Enrollment* Enrollments in public colleges for recent years are shown in the table. Assuming a linear relationship, estimate the enrollments for **(a)** 2002 and **(b)** 2006. Give answers to the nearest tenth of thousands if applicable.

Year	Enrollment (in thousands)
2000	11,753
2004	12,980
2008	13,972

Source: U.S. Census Bureau.

41. Show that if M is the midpoint of the line segment with endpoints $P(x_1, y_1)$ and $Q(x_2, y_2)$, then $d(P, M) + d(M, Q) = d(P, Q)$ and $d(P, M) = d(M, Q)$.

42. Write the distance formula $d = \sqrt{(x_2 - x_1)^2 + (y_2 - y_1)^2}$ using a rational exponent.

*For each equation, **(a)** give a table with at least three ordered pairs that are solutions, and **(b)** graph the equation. **See Examples 7 and 8.***

43. $y = \frac{1}{2}x - 2$ **44.** $y = -x + 3$ **45.** $2x + 3y = 5$

46. $3x - 2y = 6$ **47.** $y = x^2$ **48.** $y = x^2 + 2$

49. $y = \sqrt{x - 3}$ **50.** $y = \sqrt{x} - 3$ **51.** $y = |x - 2|$

52. $y = -|x + 4|$ **53.** $y = x^3$ **54.** $y = -x^3$

Concept Check Answer the following.

55. If a vertical line is drawn through the point $(4, 3)$, at what point will it intersect the x-axis?

56. If a horizontal line is drawn through the point $(4, 3)$, at what point will it intersect the y-axis?

57. If the point (a, b) is in the second quadrant, in what quadrant is $(a, -b)$? $(-a, b)$? $(-a, -b)$? (b, a)?

58. Show that the points $(-2, 2)$, $(13, 10)$, $(21, -5)$, and $(6, -13)$ are the vertices of a rhombus (all sides equal in length).

59. Are the points $A(1, 1)$, $B(5, 2)$, $C(3, 4)$, and $D(-1, 3)$ the vertices of a parallelogram (opposite sides equal in length)? of a rhombus (all sides equal in length)?

60. Find the coordinates of the points that divide the line segment joining $(4, 5)$ and $(10, 14)$ into three equal parts.

2.2 Circles

- Center-Radius Form
- General Form
- An Application

Center-Radius Form By definition, a **circle** is the set of all points in a plane that lie a given distance from a given point. The given distance is the **radius** of the circle, and the given point is the **center.**

We can find the equation of a circle from its definition by using the distance formula. Suppose that the point (h, k) is the center and the circle has radius r, where $r > 0$. Let (x, y) represent any point on the circle. See **Figure 13.**

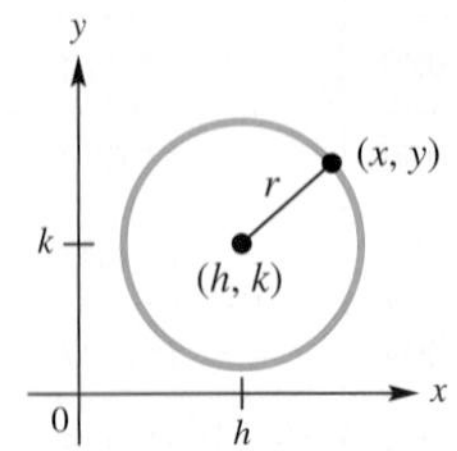

Figure 13

$$\sqrt{(x_2 - x_1)^2 + (y_2 - y_1)^2} = r \quad \text{Distance formula (Section 2.1)}$$

$$\sqrt{(x - h)^2 + (y - k)^2} = r \quad (h, k) = (x_1, y_1) \text{ and } (x, y) = (x_2, y_2)$$

$$\mathbf{(x - h)^2 + (y - k)^2 = r^2} \quad \text{Square each side. (Section 1.6)}$$

LOOKING AHEAD TO CALCULUS

The circle $x^2 + y^2 = 1$ is called the **unit circle.** It is important in interpreting the *trigonometric* or *circular* functions that appear in the study of calculus.

Center-Radius Form of the Equation of a Circle

A circle with center (h, k) and radius r has equation

$$(x - h)^2 + (y - k)^2 = r^2,$$

which is the **center-radius form** of the equation of the circle. As a special case, a circle with center $(0, 0)$ and radius r has the following equation.

$$x^2 + y^2 = r^2$$

Classroom Example 1
Find the center-radius form of the equation of each circle described.
(a) center at $(1, -2)$, radius 3
(b) center at $(0, 0)$, radius 2

Answers:
(a) $(x - 1)^2 + (y + 2)^2 = 9$
(b) $x^2 + y^2 = 4$

EXAMPLE 1 Finding the Center-Radius Form

Find the center-radius form of the equation of each circle described.

(a) center at $(-3, 4)$, radius 6 **(b)** center at $(0, 0)$, radius 3

SOLUTION

(a)
$$(x - h)^2 + (y - k)^2 = r^2 \quad \text{Center-radius form}$$
$$[x - (-3)]^2 + (y - 4)^2 = 6^2 \quad \text{Substitute. Let } (h, k) = (-3, 4) \text{ and } r = 6.$$
Watch signs here.
$$(x + 3)^2 + (y - 4)^2 = 36 \quad \text{Simplify.}$$

(b) The center is the origin and $r = 3$.

$$x^2 + y^2 = r^2 \quad \text{Special case of the center-radius form}$$
$$x^2 + y^2 = 3^2 \quad \text{Let } r = 3.$$
$$x^2 + y^2 = 9 \quad \text{Apply the exponent.}$$

Now Try Exercises 1(a) and 7(a).

Classroom Example 2
Graph each circle discussed in **Classroom Example 1.**
(a) $(x - 1)^2 + (y + 2)^2 = 9$
(b) $x^2 + y^2 = 4$

Answers:
(a) $(x - 1)^2 + (y + 2)^2 = 9$
(b) $x^2 + y^2 = 4$

EXAMPLE 2 Graphing Circles

Graph each circle discussed in **Example 1.**

(a) $(x + 3)^2 + (y - 4)^2 = 36$ **(b)** $x^2 + y^2 = 9$

SOLUTION

(a) Writing the given equation in center-radius form

$$[x - (-3)]^2 + (y - 4)^2 = 6^2$$

gives $(-3, 4)$ as the center and 6 as the radius. See **Figure 14.**

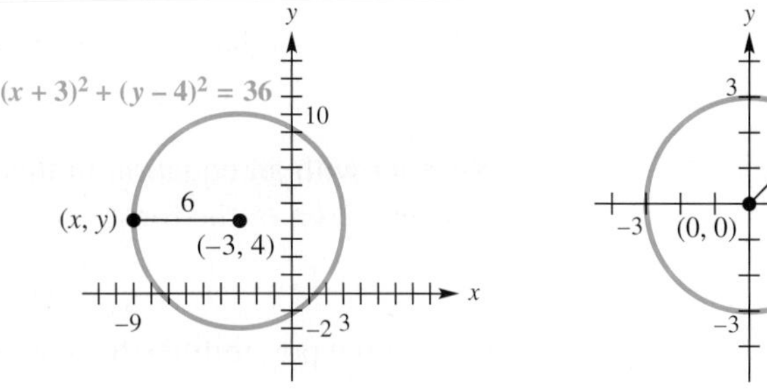

Figure 14 **Figure 15**

(b) The graph with center $(0, 0)$ and radius 3 is shown in **Figure 15.**

Now Try Exercises 1(b) and 7(b).

The circles graphed in **Figures 14 and 15** of **Example 2** can be generated on a graphing calculator by first solving for y and then entering two functions y_1 and y_2. See **Figures 16 and 17.** In both cases, the plot of y_1 yields the top half of the circle while that of y_2 yields the bottom half.

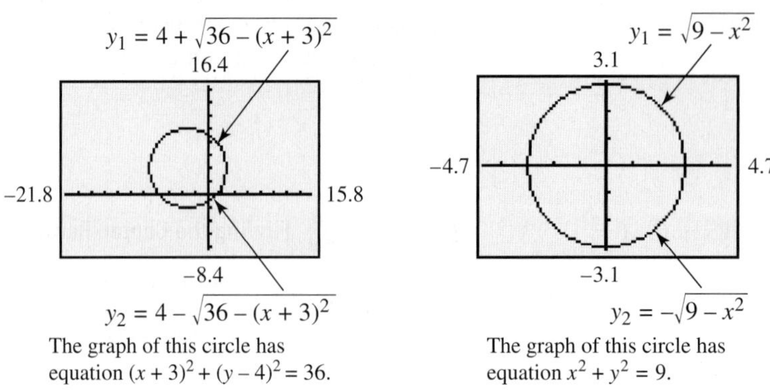

The graph of this circle has equation $(x + 3)^2 + (y - 4)^2 = 36$.

Figure 16

The graph of this circle has equation $x^2 + y^2 = 9$.

Figure 17

Because of the design of the viewing window, it is necessary to use a **square viewing window** to avoid distortion when graphing circles. Refer to your owner's manual to see how to do this. ■

General Form Consider the center-radius form of the equation of a circle, and rewrite it so that the binomials are expanded and the right side equals 0.

$$(x - h)^2 + (y - k)^2 = r^2 \quad \text{Center-radius form}$$

Don't forget this term when squaring.

Remember this term.

$$x^2 - 2xh + h^2 + y^2 - 2yk + k^2 - r^2 = 0 \quad \text{Square each binomial, and subtract } r^2. \text{ (Section R.3)}$$

$$x^2 + y^2 + \underset{c}{(-2h)}x + \underset{d}{(-2k)}y + \underset{e}{(h^2 + k^2 - r^2)} = 0 \quad \text{Properties of real numbers (Section R.2)}$$

If $r > 0$, then the graph of this equation is a circle with center (h, k) and radius r, as seen earlier. This is the **general form of the equation of a circle.**

General Form of the Equation of a Circle

For some real numbers c, d, and e, the equation

$$\mathbf{x^2 + y^2 + cx + dy + e = 0}$$

can have a graph that is a circle or a point, or is nonexistent.

Starting with an equation in this general form, we can complete the square to get an equation of the form

$$(x - h)^2 + (y - k)^2 = m, \quad \text{for some number } m.$$

There are three possibilities for the graph based on the value of m.

1. If $m > 0$, then $r^2 = m$, and the graph of the equation is a circle with radius $\sqrt{m}$.
2. If $m = 0$, then the graph of the equation is the single point (h, k).
3. If $m < 0$, then no points satisfy the equation, and the graph is nonexistent.

Classroom Example 3
Show that

$$x^2 + 4x + y^2 - 8y - 44 = 0$$

has a circle as its graph. Find the center and radius.

Answer: $m = 64$, and $64 > 0$, so the graph is a circle; center: $(-2, 4)$; radius: 8

EXAMPLE 3 Finding the Center and Radius by Completing the Square

Show that $x^2 - 6x + y^2 + 10y + 25 = 0$ has a circle as its graph. Find the center and radius.

SOLUTION We complete the square twice, once for x and once for y. Begin by subtracting 25 from each side.

$$x^2 - 6x + y^2 + 10y + 25 = 0$$

$$(x^2 - 6x \qquad) + (y^2 + 10y \qquad) = -25$$

Think: $\left[\frac{1}{2}(-6)\right]^2 = (-3)^2 = 9$ and $\left[\frac{1}{2}(10)\right]^2 = 5^2 = 25$

Add 9 and 25 on the left to complete the two squares, and to compensate, add 9 and 25 on the right.

$$(x^2 - 6x + 9) + (y^2 + 10y + 25) = -25 + 9 + 25 \quad \text{Complete the square. (Section 1.4)}$$

Add 9 and 25 on *both* sides.

$$(x - 3)^2 + (y + 5)^2 = 9 \quad \text{Factor. (Section R.4) Add on the right.}$$

$$(x - 3)^2 + (y - (-5))^2 = 3^2 \quad \text{Center-radius form}$$

Since $3^2 = 9$ and $9 > 0$, the equation represents a circle with center at $(3, -5)$ and radius 3.

✔ *Now Try Exercise 19.*

Classroom Example 4
Show that

$$2x^2 + 2y^2 + 2x - 6y = 45$$

has a circle as its graph. Find the center and radius.

Answer: $m = 25$, and $25 > 0$, so the graph is a circle; center: $\left(-\frac{1}{2}, \frac{3}{2}\right)$; radius: 5

EXAMPLE 4 Finding the Center and Radius by Completing the Square

Show that $2x^2 + 2y^2 - 6x + 10y = 1$ has a circle as its graph. Find the center and radius.

SOLUTION To complete the square, the coefficients of the x^2- and y^2-terms must be 1. In this case they are both 2, so begin by dividing each side by 2.

$$2x^2 + 2y^2 - 6x + 10y = 1$$

$$x^2 + y^2 - 3x + 5y = \frac{1}{2} \quad \text{Divide by 2.}$$

$$(x^2 - 3x \qquad) + (y^2 + 5y \qquad) = \frac{1}{2} \quad \text{Rearrange and regroup terms in anticipation of completing the square.}$$

$$\left(x^2 - 3x + \frac{9}{4}\right) + \left(y^2 + 5y + \frac{25}{4}\right) = \frac{1}{2} + \frac{9}{4} + \frac{25}{4}$$

Complete the square for *both* x and y; $\left[\frac{1}{2}(-3)\right]^2 = \frac{9}{4}$ and $\left[\frac{1}{2}(5)\right]^2 = \frac{25}{4}$

$$\left(x - \frac{3}{2}\right)^2 + \left(y + \frac{5}{2}\right)^2 = 9 \quad \text{Factor and add.}$$

$$\left(x - \frac{3}{2}\right)^2 + \left(y - \left(-\frac{5}{2}\right)\right)^2 = 3^2 \quad \text{Center-radius form}$$

The equation has a circle with center at $\left(\frac{3}{2}, -\frac{5}{2}\right)$ and radius 3 as its graph.

✔ *Now Try Exercise 23.*

Classroom Example 5
The graph of the equation
$$x^2 - 6x + y^2 + 2y + 12 = 0$$
either is a point or is nonexistent. Which is it?

Answer: $m = -2$ and $-2 < 0$, so the graph is nonexistent.

EXAMPLE 5 Determining Whether a Graph Is a Point or Nonexistent

The graph of the equation $x^2 + 10x + y^2 - 4y + 33 = 0$ either is a point or is nonexistent. Which is it?

SOLUTION

$$x^2 + 10x + y^2 - 4y + 33 = 0$$

$$x^2 + 10x + y^2 - 4y = -33 \quad \text{Subtract 33.}$$

Think: $\left[\frac{1}{2}(10)\right]^2 = 25$ and $\left[\frac{1}{2}(-4)\right]^2 = 4$ Prepare to complete the square for both x and y.

$$(x^2 + 10x + 25) + (y^2 - 4y + 4) = -33 + 25 + 4 \quad \text{Complete the square.}$$

$$(x + 5)^2 + (y - 2)^2 = -4 \quad \text{Factor on the left, and add.}$$

Since $-4 < 0$, there are *no* ordered pairs (x, y), with x and y both real numbers, satisfying the equation. The graph of the given equation is nonexistent—it contains no points. (If the constant on the right side were 0, the graph would consist of the single point $(-5, 2)$.)

✔ *Now Try Exercise 25.*

Classroom Example 6
If three receiving stations at
$$(1, 4), \quad (-6, 0), \quad \text{and} \quad (5, -2)$$
record distances to an earthquake epicenter of 4 units, 5 units, and 10 units, respectively, show algebraically that the epicenter lies at $(-3, 4)$.

Answer: The point $(-3, 4)$ lies on all three circles:
$$(x - 1)^2 + (y - 4)^2 = 16,$$
$$(x + 6)^2 + y^2 = 25,$$
and $(x - 5)^2 + (y + 2)^2 = 100.$

An Application Seismologists can locate the epicenter of an earthquake by determining the intersection of three circles. The radii of these circles represent the distances from the epicenter to each of three receiving stations. The centers of the circles represent the receiving stations.

EXAMPLE 6 Locating the Epicenter of an Earthquake

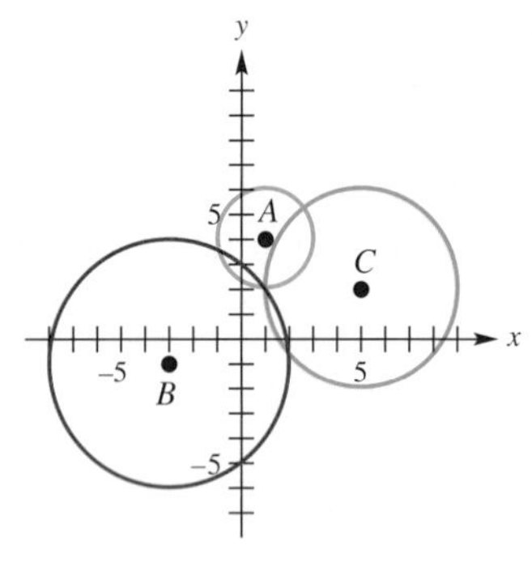

Figure 18

Suppose receiving stations A, B, and C are located on a coordinate plane at the points $(1, 4)$, $(-3, -1)$, and $(5, 2)$. Let the distances from the earthquake epicenter to these stations be 2 units, 5 units, and 4 units, respectively. Where on the coordinate plane is the epicenter located?

SOLUTION Graph the three circles as shown in **Figure 18.** From the graph it appears that the epicenter is located at $(1, 2)$. To check this algebraically, determine the equation for each circle and substitute $x = 1$ and $y = 2$.

Station A:	Station B:	Station C:
$(x - 1)^2 + (y - 4)^2 = 4$	$(x + 3)^2 + (y + 1)^2 = 25$	$(x - 5)^2 + (y - 2)^2 = 16$
$(1 - 1)^2 + (2 - 4)^2 \stackrel{?}{=} 4$	$(1 + 3)^2 + (2 + 1)^2 \stackrel{?}{=} 25$	$(1 - 5)^2 + (2 - 2)^2 \stackrel{?}{=} 16$
$0 + 4 \stackrel{?}{=} 4$	$16 + 9 \stackrel{?}{=} 25$	$16 + 0 \stackrel{?}{=} 16$
$4 = 4$	$25 = 25$	$16 = 16$

The point $(1, 2)$ lies on all three graphs. Thus, we can conclude that the epicenter of the earthquake is at $(1, 2)$.

✔ *Now Try Exercise 41.*

2.2 Exercises

1. (a) $x^2 + y^2 = 36$
   (b) $x^2 + y^2 = 36$
2. (a) $x^2 + y^2 = 81$
   (b) $x^2 + y^2 = 81$
3. (a) $(x - 2)^2 + y^2 = 36$
   (b) $(x - 2)^2 + y^2 = 36$
4. (a) $(x - 3)^2 + y^2 = 9$
   (b) $(x - 3)^2 + y^2 = 9$
5. (a) $x^2 + (y - 4)^2 = 16$
   (b) $x^2 + (y - 4)^2 = 16$
6. (a) $x^2 + (y + 3)^2 = 49$
   (b) $x^2 + (y + 3)^2 = 49$
7. (a) $(x + 2)^2 + (y - 5)^2 = 16$
   (b) $(x + 2)^2 + (y - 5)^2 = 16$

*In Exercises 1–12, **(a)** find the center-radius form of the equation of each circle, and **(b)** graph it. **See Examples 1 and 2.***

1. center $(0, 0)$, radius 6
2. center $(0, 0)$, radius 9
3. center $(2, 0)$, radius 6
4. center $(3, 0)$, radius 3
5. center $(0, 4)$, radius 4
6. center $(0, -3)$, radius 7
7. center $(-2, 5)$, radius 4
8. center $(4, 3)$, radius 5
9. center $(5, -4)$, radius 7
10. center $(-3, -2)$, radius 6
11. center $(\sqrt{2}, \sqrt{2})$, radius $\sqrt{2}$
12. center $(-\sqrt{3}, -\sqrt{3})$, radius $\sqrt{3}$

Connecting Graphs with Equations *In Exercises 13–16, use each graph to determine the equation of the circle in **(a)** center-radius form and **(b)** general form.*

13.

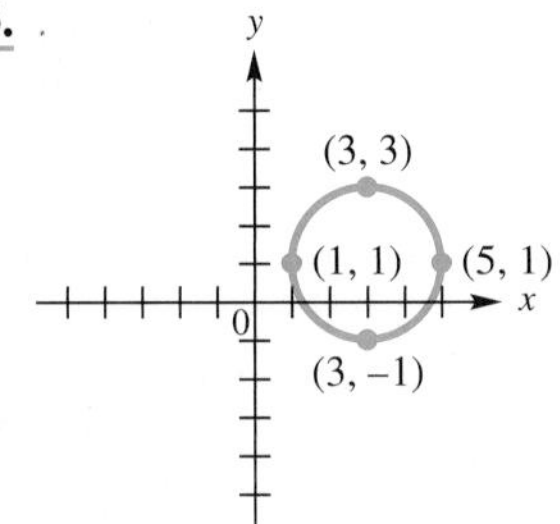

14.

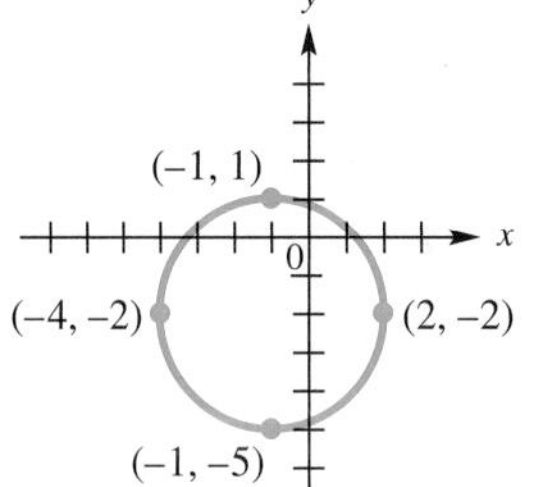

15.

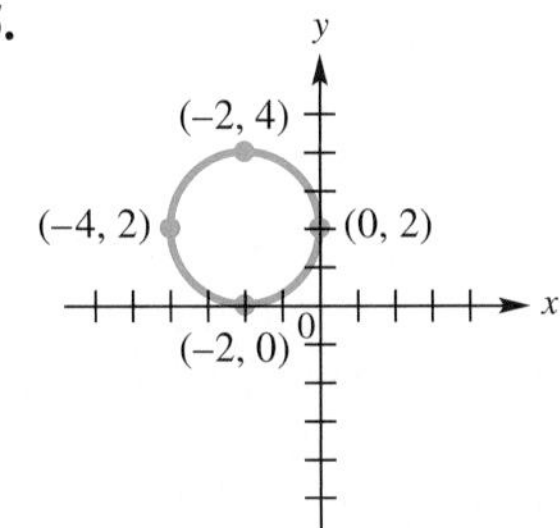

16.

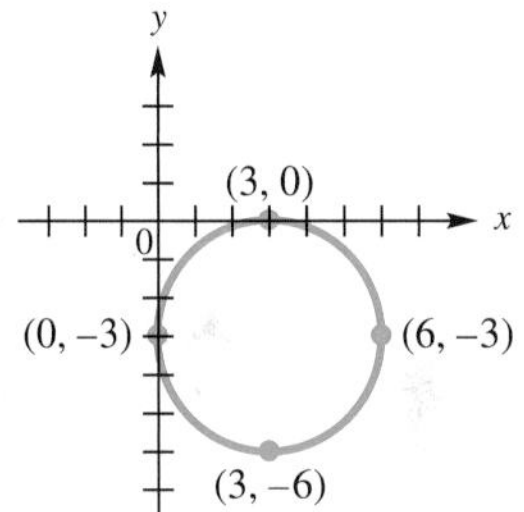

17. *Concept Check* Which one of the two screens is the correct graph of the circle with center $(-3, 5)$ and radius 4?

A.

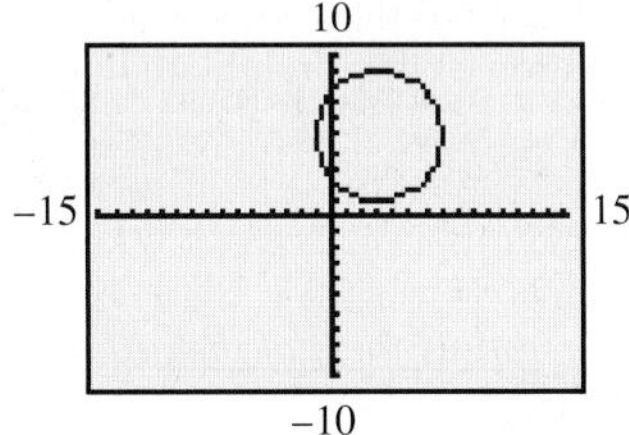

B.

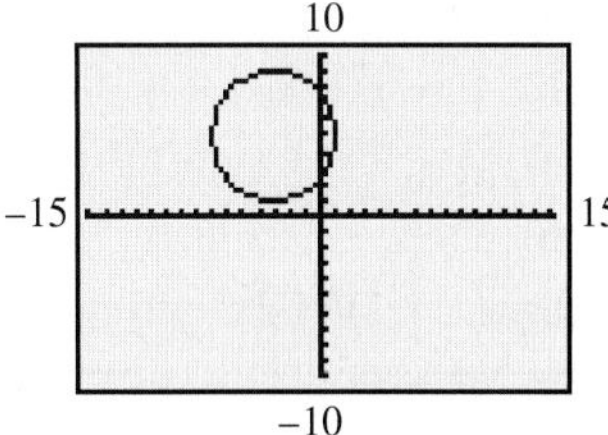

18. When the equation of a circle is written in the form

$$(x - h)^2 + (y - k)^2 = m,$$

how does the value of m indicate whether the graph is a circle, is a point, or is nonexistent?

*Decide whether or not each equation has a circle as its graph. If it does, give the center and the radius. If it does not, describe the graph. **See Examples 3–5.***

19. $x^2 + y^2 + 6x + 8y + 9 = 0$
20. $x^2 + y^2 + 8x - 6y + 16 = 0$
21. $x^2 + y^2 - 4x + 12y = -4$
22. $x^2 + y^2 - 12x + 10y = -25$
23. $4x^2 + 4y^2 + 4x - 16y - 19 = 0$
24. $9x^2 + 9y^2 + 12x - 18y - 23 = 0$

8. **(a)** $(x-4)^2+(y-3)^2=25$
(b)

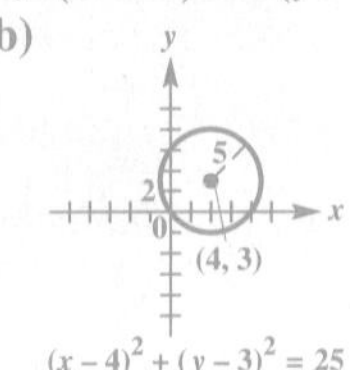

9. **(a)** $(x-5)^2+(y+4)^2=49$
(b)

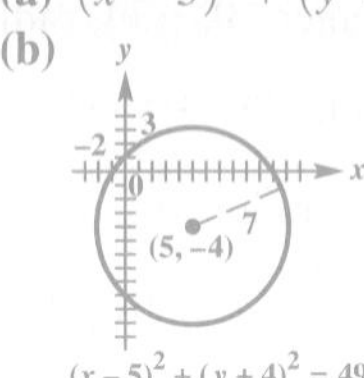

10. **(a)** $(x+3)^2+(y+2)^2=36$
(b)

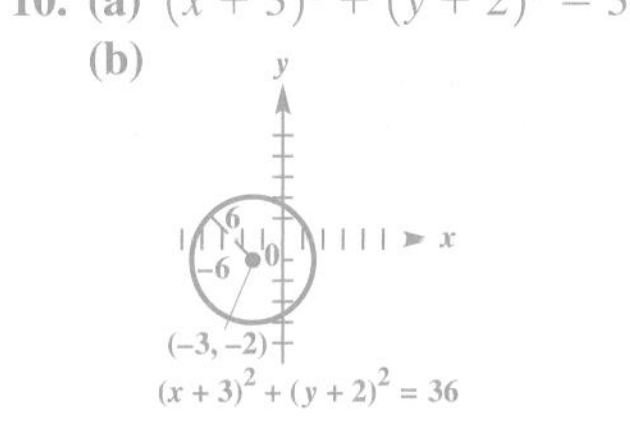

11. **(a)** $\left(x-\sqrt{2}\right)^2+\left(y-\sqrt{2}\right)^2=2$
(b)

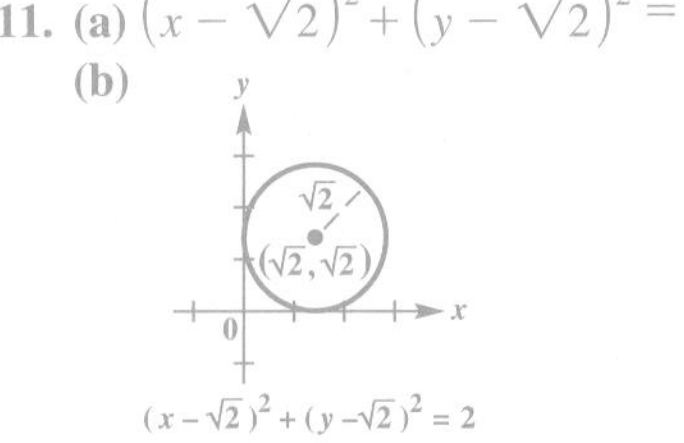

12. **(a)** $\left(x+\sqrt{3}\right)^2+\left(y+\sqrt{3}\right)^2=3$
(b)

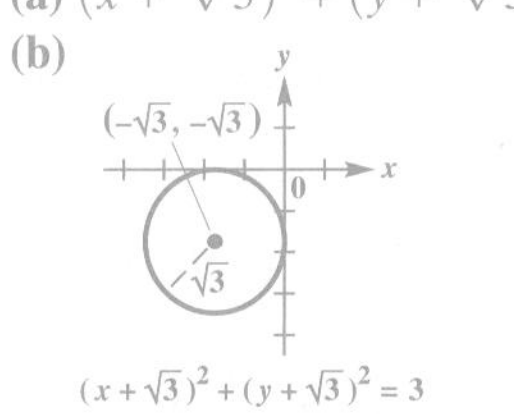

13. **(a)** $(x-3)^2+(y-1)^2=4$
(b) $x^2+y^2-6x-2y+6=0$

14. **(a)** $(x+1)^2+(y+2)^2=9$
(b) $x^2+y^2+2x+4y-4=0$

15. **(a)** $(x+2)^2+(y-2)^2=4$
(b) $x^2+y^2+4x-4y+4=0$

16. **(a)** $(x-3)^2+(y+3)^2=9$
(b) $x^2+y^2-6x+6y+9=0$

17. B

19. yes; center: $(-3, -4)$; radius: 4
20. yes; center: $(-4, 3)$; radius: 3
21. yes; center: $(2, -6)$; radius: 6
22. yes; center: $(6, -5)$; radius: 6
23. yes; center: $\left(-\frac{1}{2}, 2\right)$; radius: 3
24. yes; center: $\left(-\frac{2}{3}, 1\right)$; radius: 2
25. no; The graph is nonexistent.

25. $x^2+y^2+2x-6y+14=0$
26. $x^2+y^2+4x-8y+32=0$
27. $x^2+y^2-6x-6y+18=0$
28. $x^2+y^2+4x+4y+8=0$
29. $9x^2+9y^2-6x+6y-23=0$
30. $4x^2+4y^2+4x-4y-7=0$

Relating Concepts

For individual or collaborative investigation *(Exercises 31–36)*

The distance formula, the midpoint formula, and the center-radius form of the equation of a circle are closely related in the following problem.

A circle has a diameter with endpoints $(-1, 3)$ and $(5, -9)$. Find the center-radius form of the equation of this circle.

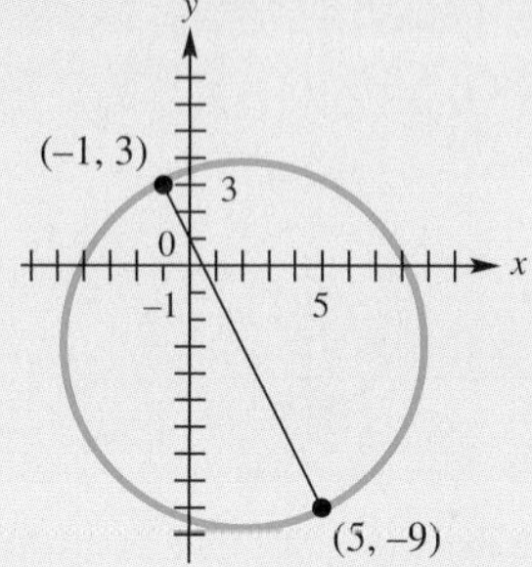

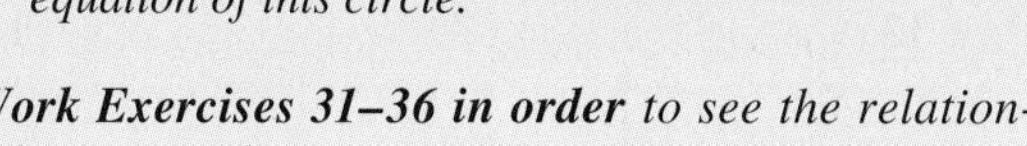

Work Exercises 31–36 in order *to see the relationships among these concepts.*

31. To find the center-radius form, we must find both the radius and the coordinates of the center. Find the coordinates of the center using the midpoint formula. (The center of the circle must be the midpoint of the diameter.)

32. There are several ways to find the radius of the circle. One way is to find the distance between the center and the point $(-1, 3)$. Use your result from **Exercise 31** and the distance formula to find the radius.

33. Another way to find the radius is to repeat **Exercise 32,** but use the point $(5, -9)$ rather than $(-1, 3)$. Do this to obtain the same answer you found in **Exercise 32.**

34. There is yet another way to find the radius. Because the radius is half the diameter, it can be found by finding half the length of the diameter. Using the endpoints of the diameter given in the problem, find the radius in this manner. You should once again obtain the same answer you found in **Exercise 32.**

35. Using the center found in **Exercise 31** and the radius found in **Exercises 32–34,** give the center-radius form of the equation of the circle.

36. Use the method described in **Exercises 31–35** to find the center-radius form of the equation of the circle with diameter having endpoints $(3, -5)$ and $(-7, 3)$.

Find the center-radius form of the circle described or graphed. ***(See Relating Concepts Exercises 31–36.)***

37. a circle having a diameter with endpoints $(-1, 2)$ and $(11, 7)$

38. a circle having a diameter with endpoints $(5, 4)$ and $(-3, -2)$

39.

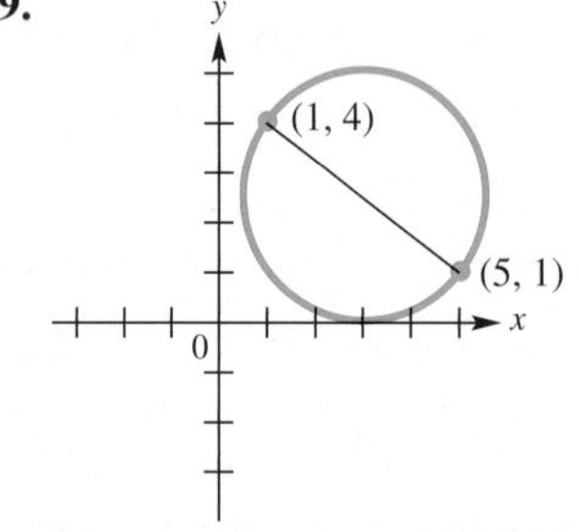

40.

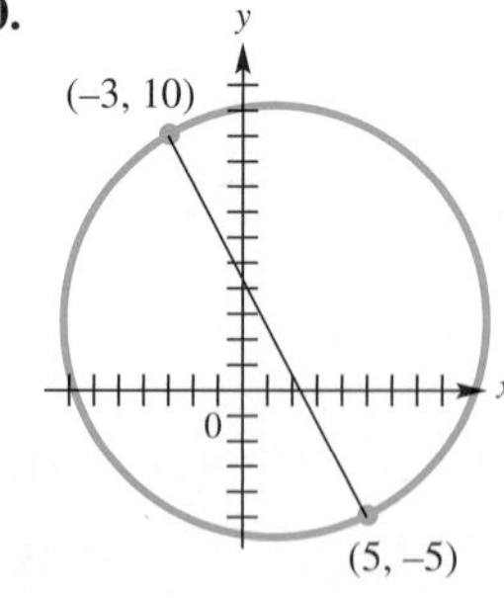

26. no; The graph is nonexistent.
27. no; The graph is the point $(3, 3)$.
28. no; The graph is the point $(-2, -2)$.
29. yes; center: $\left(\frac{1}{3}, -\frac{1}{3}\right)$; radius: $\frac{5}{3}$
30. yes; center: $\left(-\frac{1}{2}, \frac{1}{2}\right)$; radius: $\frac{3}{2}$
31. $(2, -3)$
32. $3\sqrt{5}$
33. $3\sqrt{5}$
34. $3\sqrt{5}$
35. $(x-2)^2 + (y+3)^2 = 45$
36. $(x+2)^2 + (y+1)^2 = 41$
37. $(x-5)^2 + \left(y - \frac{9}{2}\right)^2 = \frac{169}{4}$
38. $(x-1)^2 + (y-1)^2 = 25$
39. $(x-3)^2 + \left(y - \frac{5}{2}\right)^2 = \frac{25}{4}$
40. $(x-1)^2 + \left(y - \frac{5}{2}\right)^2 = \frac{289}{4}$
41. at $(3, 1)$
42. at $(5, 2)$
43. at $(-2, -2)$
44. at $(5, 0)$
45. $(x-3)^2 + (y-2)^2 = 4$
46. $(x+4)^2 + (y-3)^2 = 106$
47. $\left(2 + \sqrt{7}, 2 + \sqrt{7}\right)$, $\left(2 - \sqrt{7}, 2 - \sqrt{7}\right)$
48. $\left(\frac{-5 + \sqrt{127}}{2}, \frac{5 - \sqrt{127}}{2}\right)$, $\left(\frac{-5 - \sqrt{127}}{2}, \frac{5 + \sqrt{127}}{2}\right)$
49. $(2, 3)$ and $(4, 1)$
50. $(x+1)^2 + (y-3)^2 = 5$
51. $9 + \sqrt{119}, 9 - \sqrt{119}$
52. $\left(x + \sqrt{2}\right)^2 + \left(y + \sqrt{2}\right)^2 = 2$
53. $\sqrt{113} - 5$
54. $\left(3 - \sqrt{3}, 1\right), \left(3 + \sqrt{3}, 1\right)$

Epicenter of an Earthquake *Solve each problem. To visualize the situation, use graph paper and a pair of compasses to carefully draw the graphs of the circles.* ***See Example 6.***

41. Suppose that receiving stations X, Y, and Z are located on a coordinate plane at the points $(7, 4)$, $(-9, -4)$, and $(-3, 9)$, respectively. The epicenter of an earthquake is determined to be 5 units from X, 13 units from Y, and 10 units from Z. Where on the coordinate plane is the epicenter located?

42. Suppose that receiving stations P, Q, and R are located on a coordinate plane at the points $(3, 1)$, $(5, -4)$, and $(-1, 4)$, respectively. The epicenter of an earthquake is determined to be $\sqrt{5}$ units from P, 6 units from Q, and $2\sqrt{10}$ units from R. Where on the coordinate plane is the epicenter located?

43. The locations of three receiving stations and the distances to the epicenter of an earthquake are contained in the following three equations: $(x-2)^2 + (y-1)^2 = 25$, $(x+2)^2 + (y-2)^2 = 16$, and $(x-1)^2 + (y+2)^2 = 9$. Determine the location of the epicenter.

44. The locations of three receiving stations and the distances to the epicenter of an earthquake are contained in the following three equations: $(x-2)^2 + (y-4)^2 = 25$, $(x-1)^2 + (y+3)^2 = 25$, and $(x+3)^2 + (y+6)^2 = 100$. Determine the location of the epicenter.

Concept Check *Work each of the following.*

45. Find the center-radius form of the equation of a circle with center $(3, 2)$ and tangent to the x-axis. (*Hint:* A line **tangent** to a circle touches it at exactly one point.)

46. Find the equation of a circle with center at $(-4, 3)$, passing through the point $(5, 8)$. Write it in center-radius form.

47. Find all points (x, y) with $x = y$ that are 4 units from $(1, 3)$.

48. Find all points satisfying $x + y = 0$ that are 8 units from $(-2, 3)$.

49. Find the coordinates of all points whose distance from $(1, 0)$ is $\sqrt{10}$ and whose distance from $(5, 4)$ is $\sqrt{10}$.

50. Find the equation of the circle of least radius that contains the points $(1, 4)$ and $(-3, 2)$ within or on its boundary.

51. Find all values of y such that the distance between $(3, y)$ and $(-2, 9)$ is 12.

52. Suppose that a circle is tangent to both axes, is in the third quadrant, and has radius $\sqrt{2}$. Find the center-radius form of its equation.

53. Find the shortest distance from the origin to the graph of the circle with equation $x^2 - 16x + y^2 - 14y + 88 = 0$.

54. Find the coordinates of the points of intersection of the line $y = 1$ and the circle centered at $(3, 0)$ with radius 2.

55. Phlash Phelps is the morning radio personality on SiriusXM Satellite Radio's *Sixties on Six* Decades channel. Phlash is an expert on U.S. geography and loves traveling around the country to strange, out-of-the-way locations. The photo shows Curt Gilchrist (standing) and Phlash (seated) visiting a small Arizona settlement called *Nothing*. (Nothing is so small that it's not named on current maps.) The sign indicates that Nothing is 50 mi from Wickenburg, AZ, 75 mi from Kingman, AZ, 105 mi from Phoenix, AZ, and 180 mi from Las Vegas, NV. Discuss how the concepts of **Example 6** can be used to locate Nothing, AZ, on a map of Arizona and southern Nevada.

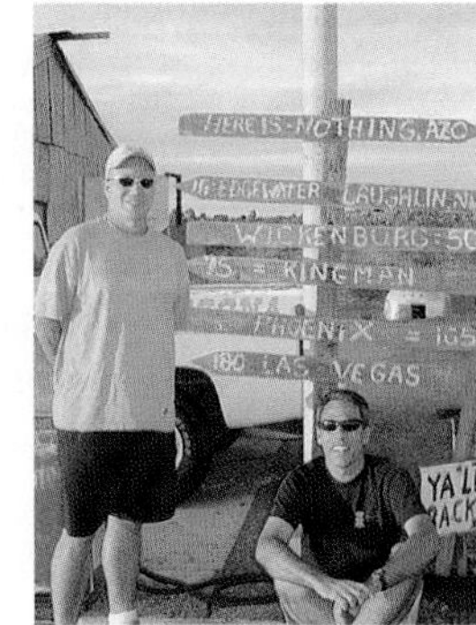

2.3 Functions

- Relations and Functions
- Domain and Range
- Determining Whether Relations Are Functions
- Function Notation
- Increasing, Decreasing, and Constant Functions

Relations and Functions Recall from **Section 2.1** how we described one quantity in terms of another.

- The letter grade you receive in a mathematics course depends on your numerical scores.
- The amount you pay (in dollars) for gas at the gas station depends on the number of gallons pumped.
- The dollars spent on household expenses depends on the category.

We used ordered pairs to represent these corresponding quantities. For example, (3, \$10.50) indicates that you pay \$10.50 for 3 gallons of gas. Since the amount you pay *depends* on the number of gallons pumped, the amount (in dollars) is called the *dependent variable,* and the number of gallons pumped is called the *independent variable.*

Generalizing, if the value of the second component y depends on the value of the first component x, then y is the **dependent variable** and x is the **independent variable.**

Independent variable ↘ ↙ Dependent variable

$$(x, y)$$

A set of ordered pairs such as $\{(3, 10.50), (8, 28.00), (10, 35.00)\}$ is called a *relation.* A special kind of relation called a *function* is very important in mathematics and its applications.

Relation and Function

A **relation** is a set of ordered pairs. A **function** is a relation in which, for each distinct value of the first component of the ordered pairs, there is *exactly one* value of the second component.

NOTE The relation from the beginning of this section representing the number of gallons of gasoline and the corresponding cost is a function since each x-value is paired with exactly one y-value.

Teaching Tip An excellent classroom illustration of the *function* concept is the activity of pumping gasoline. For every number of gallons that appears on the pump display, there is one and only one price displayed. The same illustration can be used to explain *domain, range, components, independent variable,* and *dependent variable.*

EXAMPLE 1 Deciding Whether Relations Define Functions

Decide whether each relation defines a function.

$$F = \{(1, 2), (-2, 4), (3, 4)\}$$
$$G = \{(1, 1), (1, 2), (1, 3), (2, 3)\}$$
$$H = \{(-4, 1), (-2, 1), (-2, 0)\}$$

SOLUTION Relation F is a function, because for each different x-value there is exactly one y-value. We can show this correspondence as follows.

$$\{1, -2, 3\} \quad x\text{-values of } F$$
$$\downarrow \quad \downarrow \quad \downarrow$$
$$\{2, \ 4, \ 4\} \quad y\text{-values of } F$$

Classroom Example 1
Decide whether each relation defines a function.

$M = \{(-4, 0), (-3, 1), (3, 1)\}$
$N = \{(2, 3), (3, 2), (4, 5), (5, 4)\}$
$P = \{(-4, 3), (0, 6), (2, 8), (-4, -3)\}$

Answer: M is a function; N is a function; P is not a function.

As the correspondence below shows, relation G is not a function because one first component corresponds to *more than one* second component.

$$\{1, 2\} \quad x\text{-values of } G$$

$$\{1, 2, 3\} \quad y\text{-values of } G$$

In relation H the last two ordered pairs have the same x-value paired with two different y-values (-2 is paired with both 1 and 0), so H is a relation but not a function. ***In a function, no two ordered pairs can have the same first component and different second components.***

Different y-values

$$H = \{(-4, 1), (-2, 1), (-2, 0)\} \quad \text{Not a function}$$

Same x-value

Now Try Exercises 1 and 3.

F

x-values y-values

1 → 2
−2 → 4
3 → 4

F is a function.

H

x-values y-values

−4 → 1
−2 → 1
−2 → 0

H is not a function.

Figure 19

Relations and functions can also be expressed as a correspondence or *mapping* from one set to another, as shown in **Figure 19** for function F and relation H from **Example 1.** The arrow from 1 to 2 indicates that the ordered pair $(1, 2)$ belongs to F—each first component is paired with exactly one second component. In the mapping for relation H, which is not a function, the first component -2 is paired with two different second components, 1 and 0.

Since relations and functions are sets of ordered pairs, we can represent them using tables and graphs. A table and graph for function F are shown in **Figure 20.**

x	y
1	2
−2	4
3	4

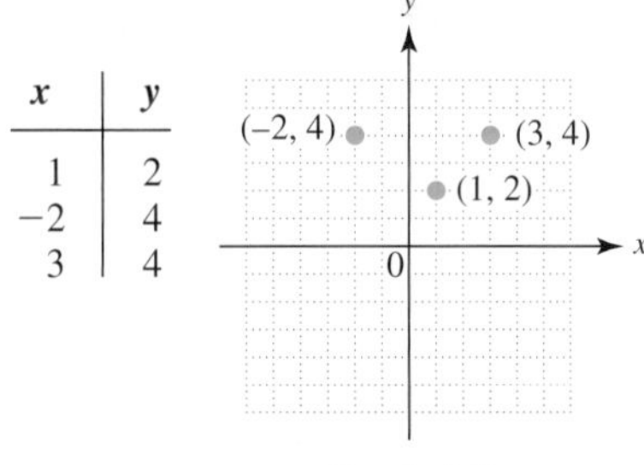

Graph of F

Figure 20

Finally, we can describe a relation or function using a rule that tells how to determine the dependent variable for a specific value of the independent variable. The rule may be given in words: for instance, "the dependent variable is twice the independent variable." Usually the rule is an equation, such as the one below.

$$\text{Dependent variable} \longrightarrow y = 2x \longleftarrow \text{Independent variable}$$

Teaching Tip In the gasoline-pumping illustration, the *price* of the amount pumped *depends* on the number of gallons pumped.

NOTE Another way to think of a function relationship is to think of the independent variable as an input and the dependent variable as an output. This is illustrated by the input-output (function) machine for the function defined by $y = 2x$.

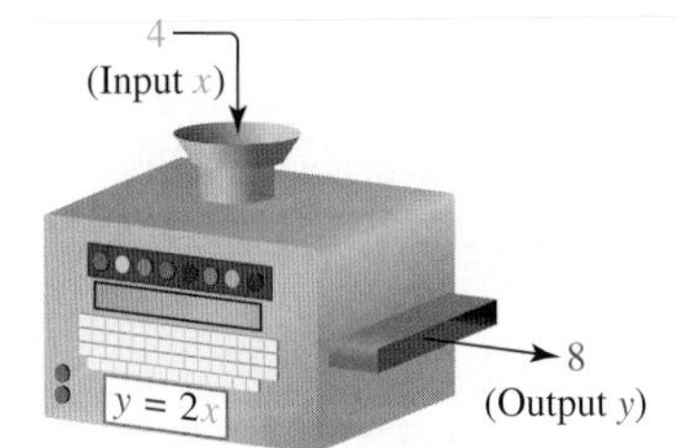

Function machine

In a function, there is exactly one value of the dependent variable, the second component, for each value of the independent variable, the first component.

On this particular day, an *input* of pumping 7.870 gallons of gasoline led to an *output* of \$29.58 from the purchaser's wallet. This is an example of a function whose domain consists of numbers of gallons pumped, and whose range consists of amounts from the purchaser's wallet. Dividing the dollar amount by the number of gallons pumped gives the exact price of gasoline that day. Use your calculator to check this. Was this pump fair? (Later we will see that this price is an example of the slope m of a linear function of the form $y = mx$.)

Domain and Range For every relation there are two important sets of elements called the *domain* and *range*.

Domain and Range

In a relation consisting of ordered pairs (x, y), the set of all values of the independent variable (x) is the **domain.** The set of all values of the dependent variable (y) is the **range.**

EXAMPLE 2 Finding Domains and Ranges of Relations

Give the domain and range of each relation. Tell whether the relation defines a function.

(a) $\{(3, -1), (4, 2), (4, 5), (6, 8)\}$

(b)

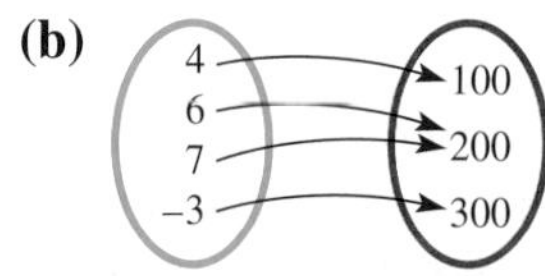

(c)

x	y
-5	2
0	2
5	2

SOLUTION

(a) The domain is the set of x-values, $\{3, 4, 6\}$. The range is the set of y-values, $\{-1, 2, 5, 8\}$. This relation is not a function because the same x-value, 4, is paired with two different y-values, 2 and 5.

(b) The domain is $\{4, 6, 7, -3\}$ and the range is $\{100, 200, 300\}$. This mapping defines a function. Each x-value corresponds to exactly one y-value.

(c) This relation is a set of ordered pairs, so the domain is the set of x-values $\{-5, 0, 5\}$ and the range is the set of y-values $\{2\}$. The table defines a function because each different x-value corresponds to exactly one y-value (even though it is the same y-value).

✔ *Now Try Exercises 9, 11, and 13.*

Classroom Example 2

Give the domain and range of each relation. Tell whether the relation defines a function.

(a) $\{(-4, -2), (-1, 0), (1, 2), (3, 5)\}$

(b)

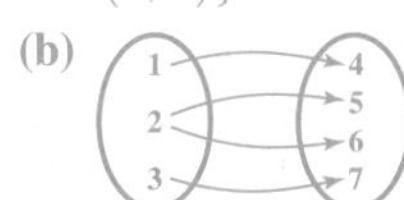

(c)

x	y
-3	5
0	5
3	5
5	5

Answers:

(a) domain: $\{-4, -1, 1, 3\}$; range: $\{-2, 0, 2, 5\}$; function

(b) domain: $\{1, 2, 3\}$; range: $\{4, 5, 6, 7\}$; not a function

(c) domain: $\{-3, 0, 3, 5\}$; range: $\{5\}$; function

EXAMPLE 3 Finding Domains and Ranges from Graphs

Give the domain and range of each relation.

(a)

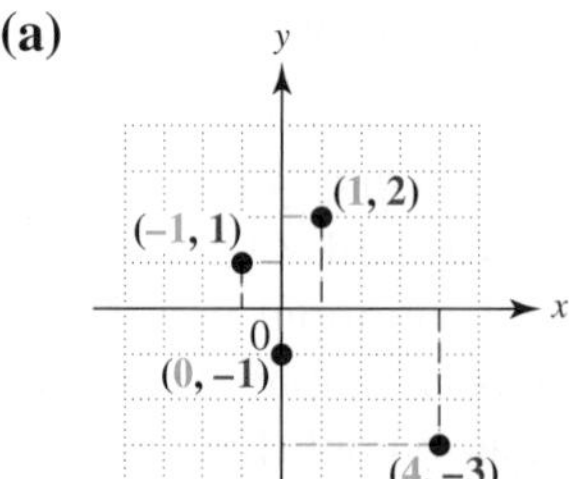

(b)

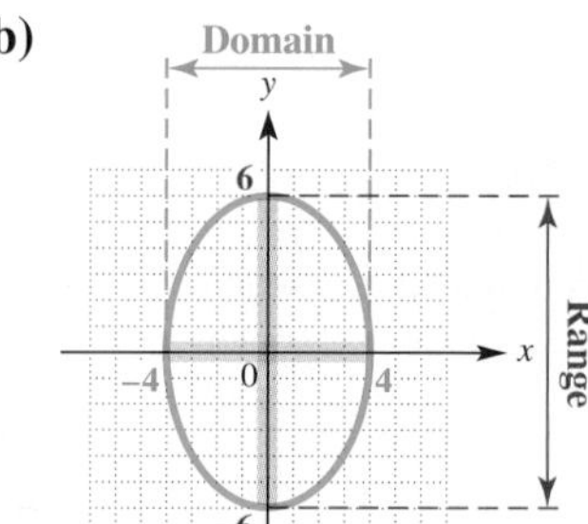

(c)

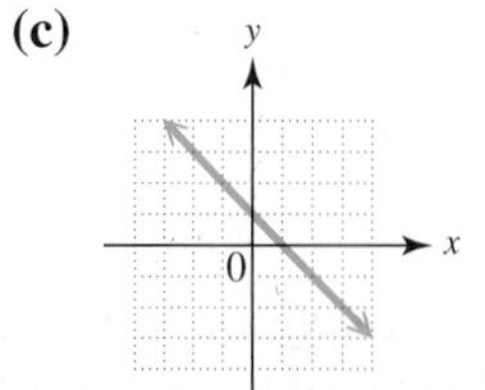

(d)

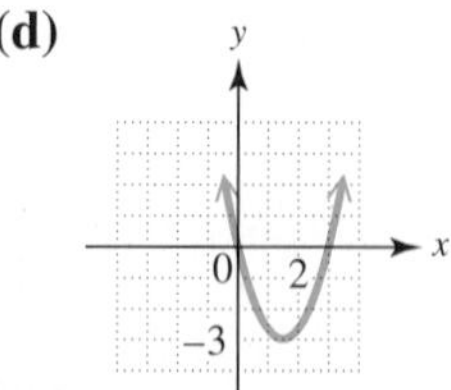

Classroom Example 3
Give the domain and range of each relation.

(a) (b)

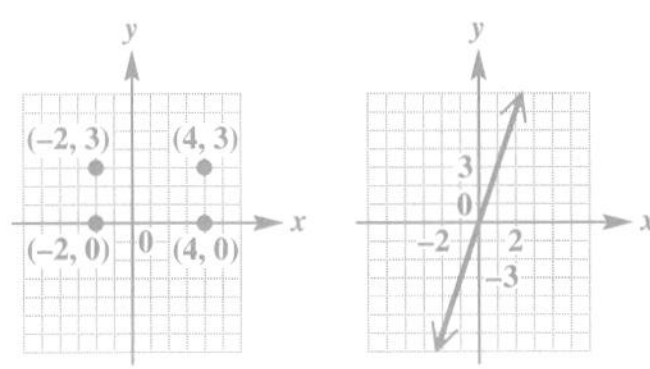

(c) (d)

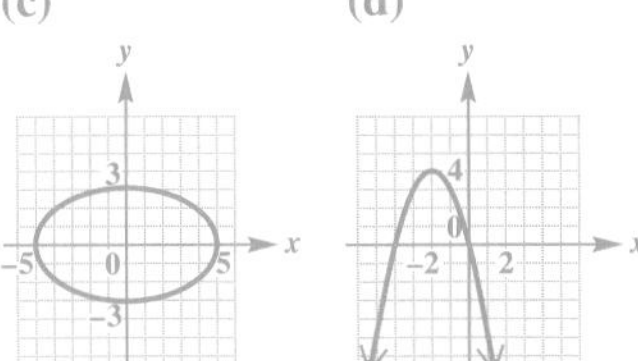

Answers:

(a) domain: $\{-2, 4\}$; range: $\{0, 3\}$

(b) domain: $(-\infty, \infty)$; range: $(-\infty, \infty)$

(c) domain: $[-5, 5]$; range: $[-3, 3]$

(d) domain: $(-\infty, \infty)$; range: $(-\infty, 4]$

Teaching Tip Use graphs of vertical and horizontal lines, parabolas, hyperbolas, and square root functions to show how to find domain and range. Give results in interval notation, paying special attention to the use of parentheses or brackets.

SOLUTION

(a) The domain is the set of x-values, $\{-1, 0, 1, 4\}$. The range is the set of y-values, $\{-3, -1, 1, 2\}$.

(b) The x-values of the points on the graph include all numbers between -4 and 4, inclusive. The y-values include all numbers between -6 and 6, inclusive.

The domain is $[-4, 4]$. Use interval notation.

The range is $[-6, 6]$. (Section 1.7)

(c) The arrowheads indicate that the line extends indefinitely left and right, as well as up and down. Therefore, both the domain and the range include all real numbers, which is written $(-\infty, \infty)$.

(d) The arrowheads indicate that the graph extends indefinitely left and right, as well as upward. The domain is $(-\infty, \infty)$. Because there is a least y-value, -3, the range includes all numbers greater than or equal to -3, written $[-3, \infty)$.

✔ *Now Try Exercises 17 and 19.*

Since relations are often defined by equations, such as $y = 2x + 3$ and $y^2 = x$, we must sometimes determine the domain of a relation from its equation. In this book, we assume the following agreement on the domain of a relation.

Agreement on Domain

Unless specified otherwise, the domain of a relation is assumed to be all real numbers that produce real numbers when substituted for the independent variable.

To illustrate this agreement, since any real number can be used as a replacement for x in $y = 2x + 3$, the domain of this function is the set of all real numbers. As another example, the function defined by $y = \frac{1}{x}$ has all real numbers *except* 0 as domain, since y is undefined if $x = 0$.

In general, the domain of a function defined by an algebraic expression is all real numbers, except those numbers that lead to division by 0 or to an even root of a negative number.

(There are also exceptions for logarithmic and trigonometric functions. They are covered in further treatment of precalculus mathematics.)

Determining Whether Relations Are Functions Since each value of x leads to only one value of y in a function, any vertical line must intersect the graph in at most one point. This is the **vertical line test** for a function.

Vertical Line Test

If every vertical line intersects the graph of a relation in no more than one point, then the relation is a function.

The graph in **Figure 21(a)** represents a function because each vertical line intersects the graph in no more than one point. The graph in **Figure 21(b)** is not the graph of a function since a vertical line intersects the graph in more than one point.

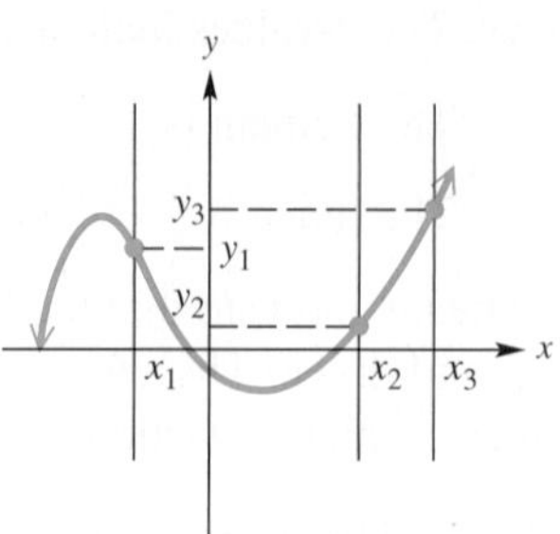

This is the graph of a function. Each x-value corresponds to only one y-value.

(a)

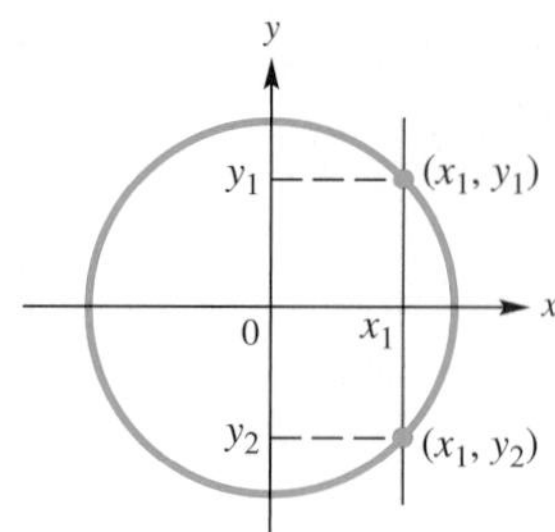

This is not the graph of a function. The same x-value corresponds to two different y-values.

(b)

Figure 21

Classroom Example 4
Use the vertical line test to determine whether each relation graphed in **Classroom Example 3** is a function.

Answers:
(a) not a function
(b) function
(c) not a function
(d) function

EXAMPLE 4 Using the Vertical Line Test

Use the vertical line test to determine whether each relation graphed in **Example 3** is a function.

SOLUTION We repeat each graph from **Example 3,** this time with vertical lines drawn through the graphs.

(a)

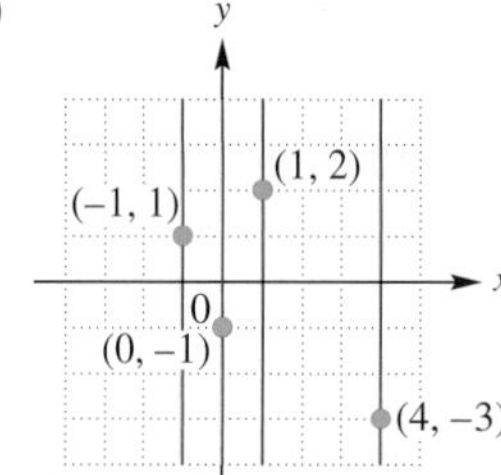

(b)

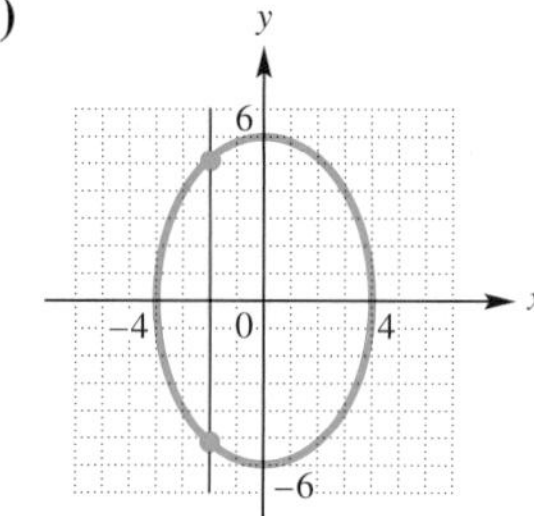

(c)

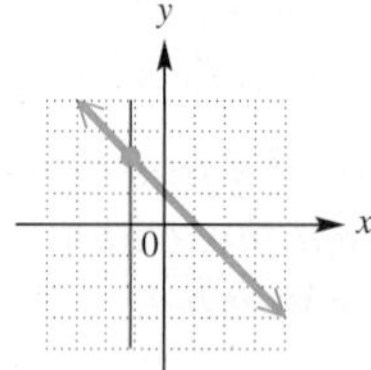

(d)

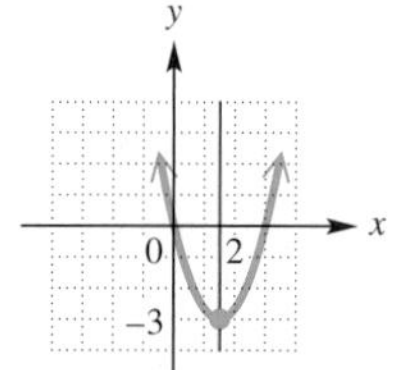

- The graphs of the relations in parts (a), (c), and (d) pass the vertical line test, since every vertical line intersects each graph no more than once. Thus, these graphs represent functions.
- The graph of the relation in part (b) fails the vertical line test, since the same x-value corresponds to two different y-values. Therefore, it is not the graph of a function.

Now Try Exercises 17 and 19.

The vertical line test is a simple method for identifying a function defined by a graph. Deciding whether a relation defined by an equation or an inequality is a function, as well as determining the domain and range, is more difficult. The next example gives some hints that may help.

Classroom Example 5
Decide whether each relation defines a function and give the domain and range.

(a) $y = 2x - 5$ (b) $y = x^2 + 3$
(c) $x = |y|$ (d) $y \geq -x$
(e) $y = \frac{3}{x+2}$

Answers:

(a) function; domain: $(-\infty, \infty)$; range: $(-\infty, \infty)$
(b) function; domain: $(-\infty, \infty)$; range: $[3, \infty)$
(c) not a function; domain: $[0, \infty)$; range: $(-\infty, \infty)$
(d) not a function; domain: $(-\infty, \infty)$; range: $(-\infty, \infty)$
(e) function; domain: $(-\infty, -2) \cup (-2, \infty)$; range: $(-\infty, 0) \cup (0, \infty)$

EXAMPLE 5 Identifying Functions, Domains, and Ranges

Decide whether each relation defines a function and give the domain and range.

(a) $y = x + 4$ **(b)** $y = \sqrt{2x - 1}$ **(c)** $y^2 = x$

(d) $y \leq x - 1$ **(e)** $y = \frac{5}{x - 1}$

SOLUTION

(a) In the defining equation (or rule), $y = x + 4$, y is always found by adding 4 to x. Thus, each value of x corresponds to just one value of y, and the relation defines a function. The variable x can represent any real number, so the domain is

$$\{x \mid x \text{ is a real number}\}, \quad \text{or} \quad (-\infty, \infty).$$

Since y is always 4 more than x, y also may be any real number, and so the range is $(-\infty, \infty)$.

(b) For any choice of x in the domain of $y = \sqrt{2x - 1}$, there is exactly one corresponding value for y (the radical is a nonnegative number), so this equation defines a function. Since the equation involves a square root, the quantity under the radical sign cannot be negative.

$$2x - 1 \geq 0 \quad \text{Solve the inequality. (Section 1.7)}$$
$$2x \geq 1 \quad \text{Add 1.}$$
$$x \geq \frac{1}{2} \quad \text{Divide by 2.}$$

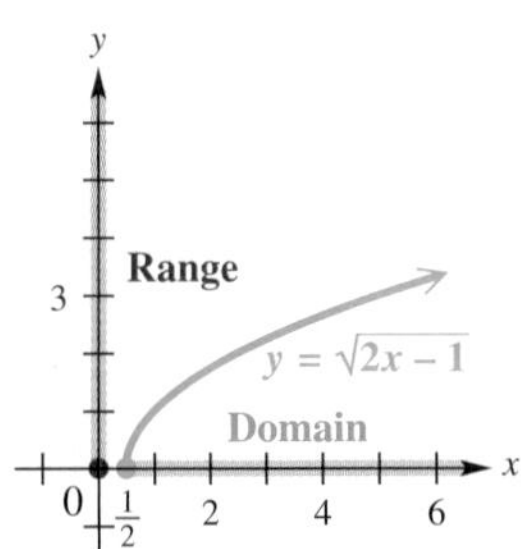

Figure 22

The domain of the function is $\left[\frac{1}{2}, \infty\right)$. Because the radical must represent a nonnegative number, as x takes values greater than or equal to $\frac{1}{2}$, the range is $\{y \mid y \geq 0\}$, or $[0, \infty)$. See **Figure 22.**

(c) The ordered pairs $(16, 4)$ and $(16, -4)$ both satisfy the equation $y^2 = x$. Since one value of x, 16, corresponds to two values of y, 4 and -4, this equation does not define a function.

Because x is equal to the square of y, the values of x must always be nonnegative. The domain of the relation is $[0, \infty)$. Any real number can be squared, so the range of the relation is $(-\infty, \infty)$. See **Figure 23.**

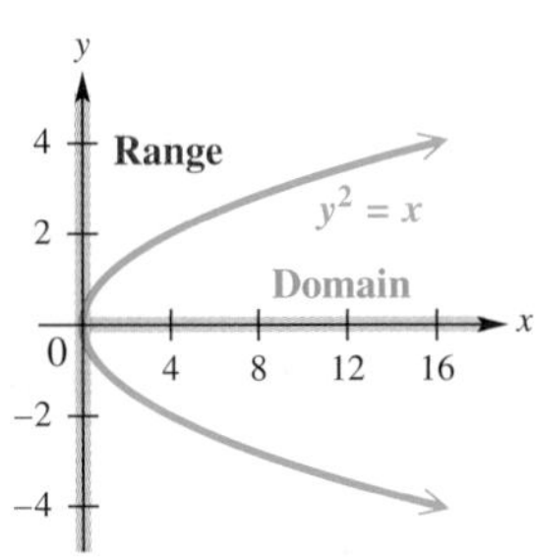

Figure 23

(d) By definition, y is a function of x if every value of x leads to exactly one value of y. Substituting a particular value of x, say 1, into $y \leq x - 1$ corresponds to many values of y. The ordered pairs

$$(1, 0), \quad (1, -1), \quad (1, -2), \quad (1, -3), \quad \text{and so on}$$

all satisfy the inequality, so y is not a function of x here. Any number can be used for x or for y, so the domain and the range of this relation are both the set of real numbers, $(-\infty, \infty)$.

(e) Given any value of x in the domain of

$$y = \frac{5}{x - 1},$$

we find y by subtracting 1 from x, and then dividing the result into 5. This process produces exactly one value of y for each value in the domain, so this equation defines a function.

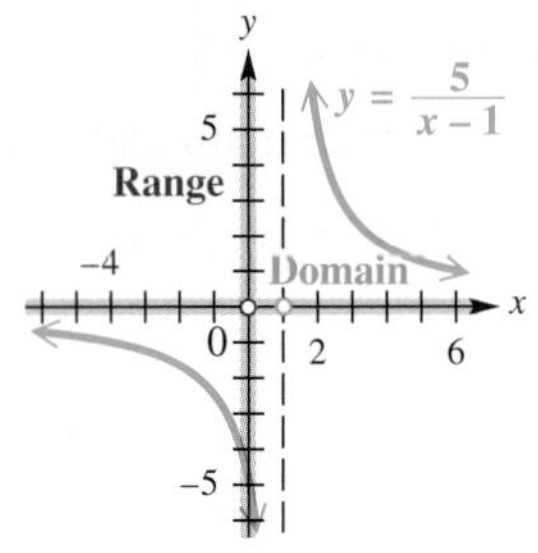

Figure 24

The domain of $y = \frac{5}{x-1}$ includes all real numbers except those that make the denominator 0. We find these numbers by setting the denominator equal to 0 and solving for x.

$$x - 1 = 0$$

$$x = 1 \quad \text{Add 1. (Section 1.1)}$$

Thus, the domain includes all real numbers except 1, written as the interval $(-\infty, 1) \cup (1, \infty)$. Values of y can be positive or negative, but never 0, because a fraction cannot equal 0 unless its numerator is 0. Therefore, the range is the interval $(-\infty, 0) \cup (0, \infty)$, as shown in **Figure 24.**

✔ *Now Try Exercises 27, 29, and 35.*

Variations of the Definition of Function

1. A **function** is a relation in which, for each distinct value of the first component of the ordered pairs, there is exactly one value of the second component.
2. A **function** is a set of ordered pairs in which no first component is repeated.
3. A **function** is a rule or correspondence that assigns exactly one range value to each distinct domain value.

LOOKING AHEAD TO CALCULUS
One of the most important concepts in calculus, that of the **limit of a function,** is defined using function notation:

$$\lim_{x \to a} f(x) = L$$

(read "the limit of $f(x)$ as x approaches a is equal to L") means that the values of $f(x)$ become as close as we wish to L when we choose values of x sufficiently close to a.

Function Notation When a function f is defined with a rule or an equation using x and y for the independent and dependent variables, we say, "y is a function of x" to emphasize that y *depends on* x. We use the notation

$$y = f(x),$$

called **function notation,** to express this and read $f(x)$ as **"f of x."** The letter f is the name given to this function.

For example, if $y = 3x - 5$, we can name the function f and write

$$f(x) = 3x - 5.$$

Note that $f(x)$ is just another name for the dependent variable y. For example, if $y = f(x) = 3x - 5$ and $x = 2$, then we find y, or $f(2)$, by replacing x with 2.

$$f(2) = 3 \cdot 2 - 5 \quad \text{Let } x = 2.$$

$$f(2) = 1 \quad \text{Multiply, and then subtract.}$$

The statement "In the function f, if $x = 2$, then $y = 1$" represents the ordered pair $(2, 1)$ and is abbreviated with function notation as follows.

$$f(2) = 1$$

The symbol $f(2)$ is read "f of 2" or "f at 2."

These ideas can be illustrated as follows.

Name of the function

Defining expression

$$y = f(x) = 3x - 5$$

Value of the function

Name of the independent variable

CAUTION ***The symbol $f(x)$ does not indicate "f times x,"*** but represents the y-value for the indicated x-value. As just shown, $f(2)$ is the y-value that corresponds to the x-value 2.

Classroom Example 6
Let $f(x) = -x^2 - 6x + 4$ and $g(x) = 3x + 1$. Find and simplify each of the following.

(a) $f(-3)$
(b) $f(r)$
(c) $g(r + 2)$

Answers:
(a) 13
(b) $-r^2 - 6r + 4$
(c) $3r + 7$

EXAMPLE 6 Using Function Notation

Let $f(x) = -x^2 + 5x - 3$ and $g(x) = 2x + 3$. Find and simplify each of the following.

(a) $f(2)$ **(b)** $f(q)$ **(c)** $g(a + 1)$

SOLUTION

(a) $f(x) = -x^2 + 5x - 3$

$f(2) = -2^2 + 5 \cdot 2 - 3$ Replace x with 2.

$= -4 + 10 - 3$ Apply the exponent and multiply. (Section R.2)

$= 3$ Add and subtract.

Thus, $f(2) = 3$, and the ordered pair $(2, 3)$ belongs to f.

(b) $f(x) = -x^2 + 5x - 3$

$f(q) = -q^2 + 5q - 3$ Replace x with q.

(c) $g(x) = 2x + 3$

$g(a + 1) = 2(a + 1) + 3$ Replace x with $a + 1$.

$= 2a + 2 + 3$ Distributive property (Section R.2)

$= 2a + 5$ Add.

The replacement of one variable with another variable or expression, as in parts (b) and (c), is important in later courses.

Now Try Exercises 41, 49, and 55.

Functions can be evaluated in a variety of ways, as shown in **Example 7.**

Classroom Example 7
For each function, find $f(-1)$.

(a) $f(x) = 2x^2 - 9$
(b) $f = \{(-4, 0), (-1, 6), (0, 8), (2, -2)\}$
(c)

(d)

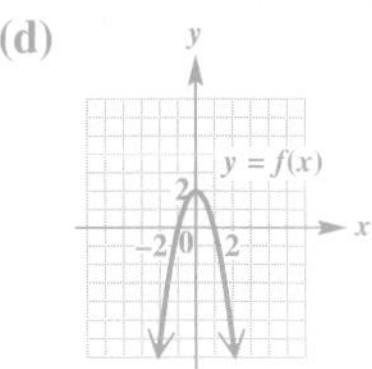

Answers:
(a) -7 (b) 6
(c) 5 (d) 0

EXAMPLE 7 Using Function Notation

For each function, find $f(3)$.

(a) $f(x) = 3x - 7$ **(b)** $f = \{(-3, 5), (0, 3), (3, 1), (6, -1)\}$

(c)

Domain	f	Range
-2	→	6
3	→	5
10	→	12

(d)

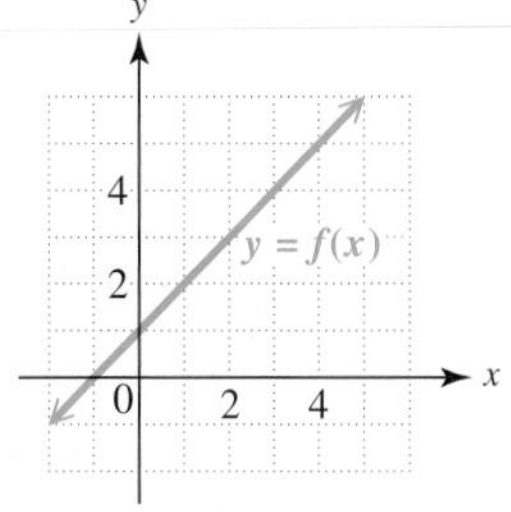

SOLUTION

(a) $f(x) = 3x - 7$

$f(3) = 3(3) - 7$ Replace x with 3.

$f(3) = 2$ Simplify.

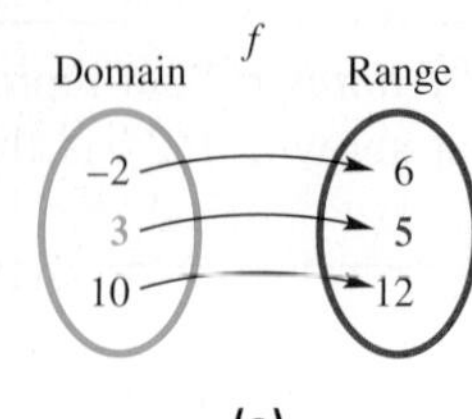

(a)

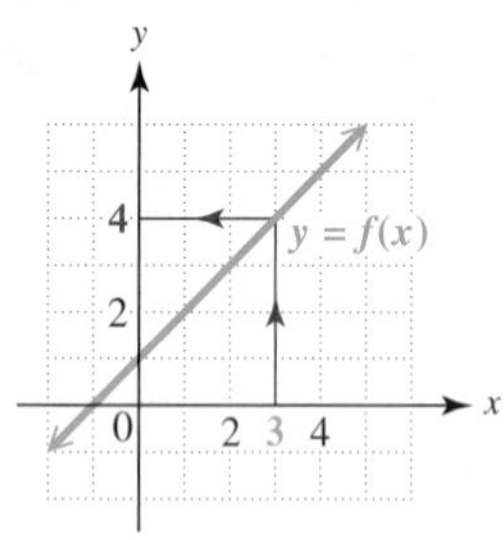

(b)

Figure 25

(b) For $f = \{(-3, 5), (0, 3), (3, 1), (6, -1)\}$, we want $f(3)$, the y-value of the ordered pair where $x = 3$. As indicated by the ordered pair $(3, 1)$, when $x = 3$, $y = 1$, so $f(3) = 1$.

(c) In the mapping, repeated in **Figure 25(a),** the domain element 3 is paired with 5 in the range, so $f(3) = 5$.

(d) To evaluate $f(3)$ using the graph, find 3 on the x-axis. See **Figure 25(b).** Then move up until the graph of f is reached. Moving horizontally to the y-axis gives 4 for the corresponding y-value. Thus, $f(3) = 4$.

Now Try Exercises 57, 59, and 61.

If a function f is defined by an equation with x and y (and not with function notation), use the following steps to find $f(x)$.

Finding an Expression for $f(x)$

Consider an equation involving x and y. Assume that y can be expressed as a function f of x. To find an expression for $f(x)$, use the following steps.

Step 1 Solve the equation for y.

Step 2 Replace y with $f(x)$.

Classroom Example 8
Assume that y is a function f of x. Rewrite each equation using function notation. Then find $f(-5)$ and $f(b)$.

(a) $y = x^2 + 2x - 3$

(b) $2x - 3y = 6$

Answers:

(a) $f(x) = x^2 + 2x - 3$; $f(-5) = 12$; $f(b) = b^2 + 2b - 3$

(b) $f(x) = \frac{2}{3}x - 2$; $f(-5) = -\frac{16}{3}$; $f(b) = \frac{2}{3}b - 2$

EXAMPLE 8 Writing Equations Using Function Notation

Assume that y is a function f of x. Rewrite each equation using function notation. Then find $f(-2)$ and $f(a)$.

(a) $y = x^2 + 1$ **(b)** $x - 4y = 5$

SOLUTION

(a)

$$y = x^2 + 1 \quad \text{Equation is already solved for } y. \text{ (Step 1)}$$

$$f(x) = x^2 + 1 \quad \text{Let } y = f(x). \text{ (Step 2)}$$

Now find $f(-2)$ and $f(a)$.

$$f(-2) = (-2)^2 + 1 \quad \text{Let } x = -2.$$
$$f(-2) = 4 + 1$$
$$f(-2) = 5$$

$$f(a) = a^2 + 1 \quad \text{Let } x = a.$$

(b)

$$x - 4y = 5 \quad \text{Solve for } y. \text{ (Section 1.1) (Step 1)}$$

$$-4y = -x + 5$$

$$y = \frac{x - 5}{4} \quad \text{Multiply by } -1. \text{ Divide by 4.}$$

$$f(x) = \frac{1}{4}x - \frac{5}{4} \quad \text{Let } y = f(x). \text{ (Step 2)} \quad \frac{a-b}{c} = \frac{a}{c} - \frac{b}{c} \text{ (Section R.5)}$$

Now find $f(-2)$ and $f(a)$.

$$f(-2) = \frac{1}{4}(-2) - \frac{5}{4} \quad \text{Let } x = -2.$$
$$f(-2) = -\frac{7}{4}$$

$$f(a) = \frac{1}{4}a - \frac{5}{4} \quad \text{Let } x = a.$$

Now Try Exercises 63 and 67.

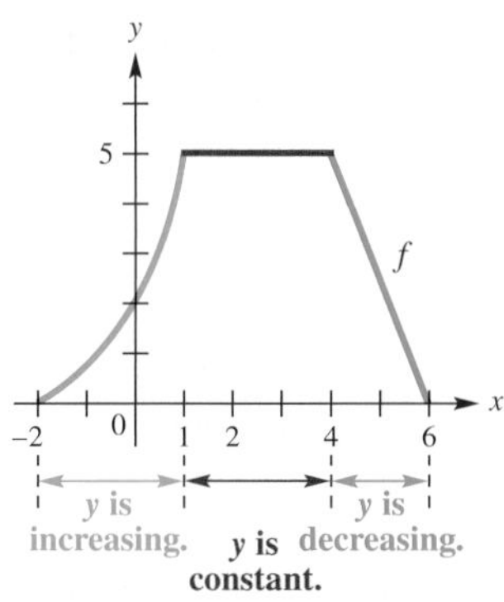

Figure 26

Increasing, Decreasing, and Constant Functions Informally speaking, a function *increases* on an interval of its domain if its graph rises from left to right on the interval. It *decreases* on an interval of its domain if its graph falls from left to right on the interval. It is *constant* on an interval of its domain if its graph is horizontal on the interval.

For example, consider **Figure 26.** The function increases on the interval $[-2, 1]$ because the y-values continue to get larger for x-values in that interval. Similarly, the function is constant on the interval $[1, 4]$ because the y-values are always 5 for all x-values there. Finally, the function decreases on the interval $[4, 6]$ because there the y-values continuously get smaller. ***The intervals refer to the x-values where the y-values either increase, decrease, or are constant.***

The formal definitions of these concepts follow.

Teaching Tip Point out that "whenever $x_1 < x_2$, $f(x_1) < f(x_2)$" means that for every two numbers x_1 and x_2 in the interval with x_1 to the left of x_2, the point on the graph at x_1 is lower than the point on the graph at x_2. Refer students to **Figure 27(a)** to see this.

Increasing, Decreasing, and Constant Functions

Suppose that a function f is defined over an interval I and x_1 and x_2 are in I.

(a) f **increases** on I if, whenever $x_1 < x_2$, $f(x_1) < f(x_2)$.

(b) f **decreases** on I if, whenever $x_1 < x_2$, $f(x_1) > f(x_2)$.

(c) f is **constant** on I if, for every x_1 and x_2, $f(x_1) = f(x_2)$.

Figure 27 illustrates these ideas.

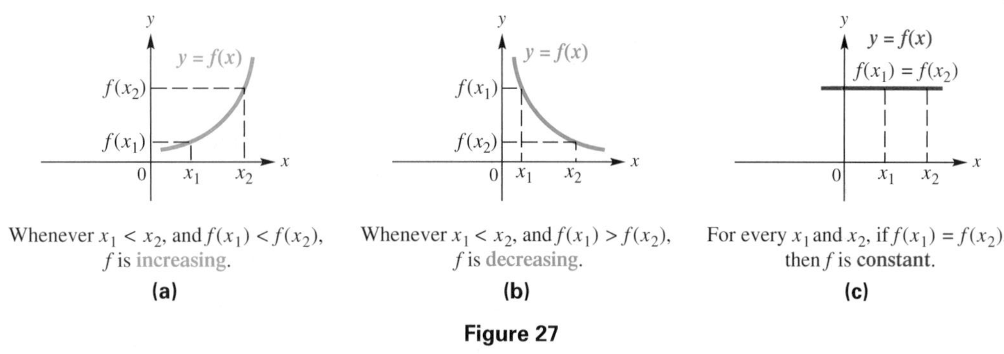

Whenever $x_1 < x_2$, and $f(x_1) < f(x_2)$, f is increasing.
(a)

Whenever $x_1 < x_2$, and $f(x_1) > f(x_2)$, f is decreasing.
(b)

For every x_1 and x_2, if $f(x_1) = f(x_2)$, then f is constant.
(c)

Figure 27

NOTE To decide whether a function is increasing, decreasing, or constant on an interval, ask yourself, ***"What does y do as x goes from left to right?"***

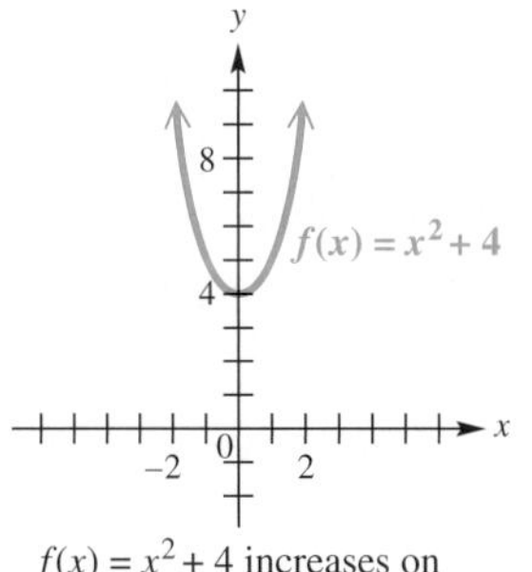

$f(x) = x^2 + 4$ increases on $[0, \infty)$ and decreases on $(-\infty, 0]$.

Figure 28

There can be confusion regarding whether endpoints of an interval should be included when determining intervals over which a function is increasing or decreasing. For example, consider the graph of $y = f(x) = x^2 + 4$, shown in **Figure 28.** Is it increasing on $[0, \infty)$ or just on $(0, \infty)$?

The definition of increasing and decreasing allows us to include 0 as a part of the interval I over which this function is increasing, because if we let $x_1 = 0$, then $f(0) < f(x_2)$ whenever $0 < x_2$. Thus, $f(x) = x^2 + 4$ is increasing on $[0, \infty)$. A similar discussion can be used to show that this function is decreasing on $(-\infty, 0]$. Do not confuse these concepts by saying that f both increases and decreases at the point $(0, 0)$.

The concepts of increasing and decreasing functions apply to intervals of the domain, not to individual points.

It is not incorrect to say that $f(x) = x^2 + 4$ is increasing on $(0, \infty)$—there are infinitely many intervals over which it increases. However, we generally give the largest possible interval when determining where a function increases or decreases. (*Source:* Stewart J., *Calculus,* Fourth Edition, Brooks/Cole Publishing Company, p. 21.)

EXAMPLE 9 Determining Intervals over Which a Function Is Increasing, Decreasing, or Constant

Figure 29 shows the graph of a function. Determine the intervals over which the function is increasing, decreasing, or constant.

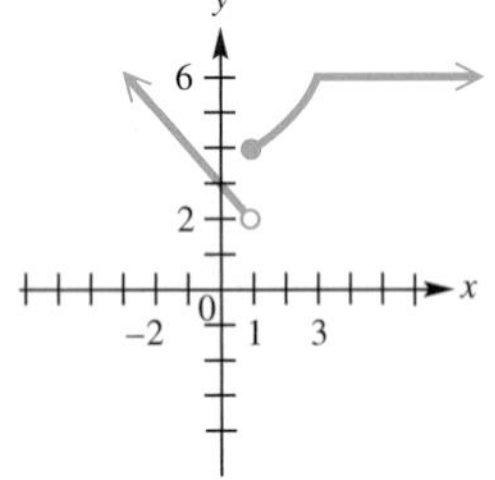

Figure 29

SOLUTION We should ask, "What is happening to the y-values as the x-values are getting larger?" Moving from left to right on the graph, we see the following:

- On the interval $(-\infty, 1)$, the y-values are *decreasing*.
- On the interval $[1, 3]$, the y-values are *increasing*.
- On the interval $[3, \infty)$, the y-values are *constant* (and equal to 6).

Therefore, the function is decreasing on $(-\infty, 1)$, increasing on $[1, 3]$, and constant on $[3, \infty)$.

✔ *Now Try Exercise 77.*

Classroom Example 9
The figure shows the graph of a function. Determine the intervals over which the function is increasing, decreasing, or constant.

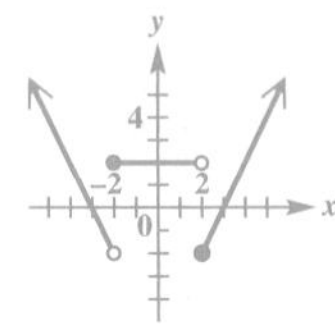

Answer:
decreasing on $(-\infty, -2)$;
constant on $[-2, 2)$;
increasing on $[2, \infty)$

EXAMPLE 10 Interpreting a Graph

Figure 30 shows the relationship between the number of gallons, $g(t)$, of water in a small swimming pool and time in hours, t. By looking at this graph of the function, we can answer questions about the water level in the pool at various times. For example, at time 0 the pool is empty. The water level then increases, stays constant for a while, decreases, and then becomes constant again. Use the graph to respond to the following.

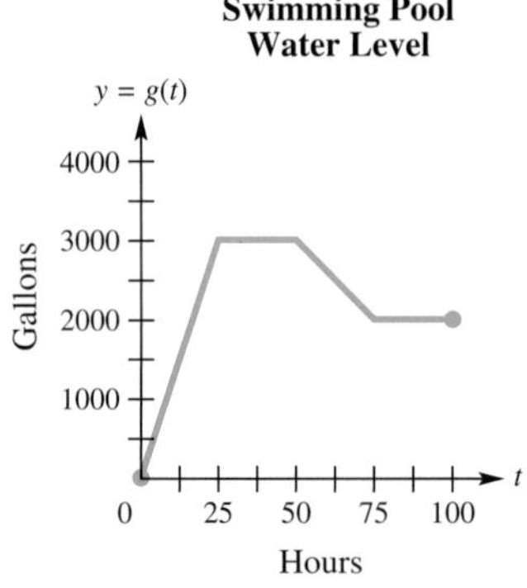

Figure 30

(a) What is the maximum number of gallons of water in the pool? When is the maximum water level first reached?

(b) For how long is the water level increasing? decreasing? constant?

(c) How many gallons of water are in the pool after 90 hr?

(d) Describe a series of events that could account for the water level changes shown in the graph.

SOLUTION

(a) The maximum range value is 3000, as indicated by the horizontal line segment for the hours 25 to 50. This maximum number of gallons, 3000, is first reached at $t = 25$ hr.

(b) The water level is increasing for $25 - 0 = 25$ hr and is decreasing for $75 - 50 = 25$ hr. It is constant for

$$(50 - 25) + (100 - 75) = 25 + 25, \quad \text{or} \quad 50 \text{ hr}.$$

(c) When $t = 90$, $y = g(90) = 2000$. There are 2000 gal after 90 hr.

Classroom Example 10
Use the graph in **Figure 30** to respond to the following.

(a) Over what period of time is the water level changing most rapidly?

(b) After how many hours does the water level start to decrease?

(c) How many gallons of water are in the pool after 75 hr?

Answers:
(a) from 0 to 25 hr
(b) 50 hr
(c) 2000 gal

(d) Looking at the graph in **Figure 30,** we might write the following description:

The pool is empty at the beginning and then is filled to a level of 3000 gal during the first 25 hr. For the next 25 hr, the water level remains the same. At 50 hr, the pool starts to be drained, and this draining lasts for 25 hr, until only 2000 gal remain. For the next 25 hr, the water level is unchanged.

Now Try Exercise 83.

2.3 Exercises

1. function 2. function
3. not a function
4. not a function
5. function 6. function
7. function 8. function
9. not a function; domain: $\{0, 1, 2\}$; range: $\{-4, -1, 0, 1, 4\}$
10. not a function; domain: $\{2, 3, 5\}$; range: $\{5, 7, 9, 11\}$
11. function; domain: $\{2, 3, 5, 11, 17\}$; range: $\{1, 7, 20\}$
12. function; domain: $\{1, 2, 3, 5\}$; range: $\{10, 15, 19, 27\}$
13. function; domain: $\{0, -1, -2\}$; range: $\{0, 1, 2\}$
14. function; domain: $\{0, 1, 2\}$; range: $\{0, -1, -2\}$
15. function; domain: $\{2005, 2006, 2007, 2008\}$; range: $\{63.5, 60.4, 62.3, 61.2\}$
16. function; domain: $\{2006, 2007, 2008, 2009\}$; range: $\{10{,}878{,}322, 11{,}120{,}822, 11{,}160{,}293, 11{,}134{,}738\}$

Decide whether each relation defines a function. ***See Example 1.***

1. $\{(5, 1), (3, 2), (4, 9), (7, 8)\}$
2. $\{(8, 0), (5, 7), (9, 3), (3, 8)\}$
3. $\{(2, 4), (0, 2), (2, 6)\}$
4. $\{(9, -2), (-3, 5), (9, 1)\}$
5. $\{(-3, 1), (4, 1), (-2, 7)\}$
6. $\{(-12, 5), (-10, 3), (8, 3)\}$
7.

x	y
3	-4
7	-4
10	-4

8.

x	y
-4	$\sqrt{2}$
0	$\sqrt{2}$
4	$\sqrt{2}$

Decide whether each relation defines a function and give the domain and range. ***See Examples 1–4.***

9. $\{(1, 1), (1, -1), (0, 0), (2, 4), (2, -4)\}$
10. $\{(2, 5), (3, 7), (3, 9), (5, 11)\}$

11. 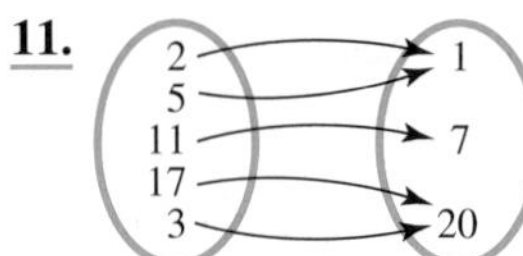

12. 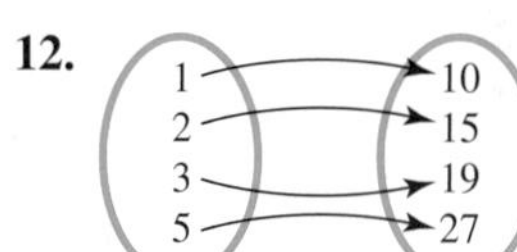

13. 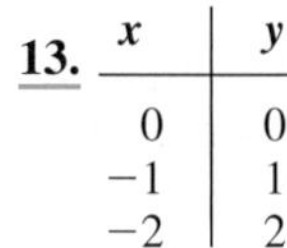

x	y
0	0
-1	1
-2	2

14.

x	y
0	0
1	-1
2	-2

15. *Number of Visits to U.S. National Parks*

Year (x)	Number of Visits (y) (millions)
2005	63.5
2006	60.4
2007	62.3
2008	61.2

Source: National Park Service.

16. *Attendance at NCAA Women's College Basketball Games*

Season* (x)	Attendance (y)
2006	10,878,322
2007	11,120,822
2008	11,160,293
2009	11,134,738

Source: NCAA.

*Each season overlaps the starting year (given) with the following year.

17. function;
domain: $(-\infty, \infty)$;
range: $(-\infty, \infty)$
18. function;
domain: $(-\infty, \infty)$;
range: $(-\infty, 4]$
19. not a function;
domain: $[3, \infty)$;
range: $(-\infty, \infty)$
20. not a function;
domain: $[-4, 4]$;
range: $[-3, 3]$
21. function;
domain: $(-\infty, \infty)$;
range: $(-\infty, \infty)$
22. function;
domain: $[-2, 2]$;
range: $[0, 4]$

23. function;
domain: $(-\infty, \infty)$;
range: $[0, \infty)$
24. function;
domain: $(-\infty, \infty)$;
range: $(-\infty, \infty)$
25. not a function;
domain: $[0, \infty)$;
range: $(-\infty, \infty)$
26. not a function;
domain: $[0, \infty)$;
range: $(-\infty, \infty)$
27. function;
domain: $(-\infty, \infty)$;
range: $(-\infty, \infty)$
28. function;
domain: $(-\infty, \infty)$;
range: $(-\infty, \infty)$
29. not a function;
domain: $(-\infty, \infty)$;
range: $(-\infty, \infty)$
30. not a function;
domain: $(-\infty, \infty)$;
range: $(-\infty, \infty)$
31. function;
domain: $[0, \infty)$;
range: $[0, \infty)$
32. function;
domain: $[0, \infty)$;
range: $(-\infty, 0]$
33. function;
domain: $(-\infty, 0) \cup (0, \infty)$;
range: $(-\infty, 0) \cup (0, \infty)$

17.

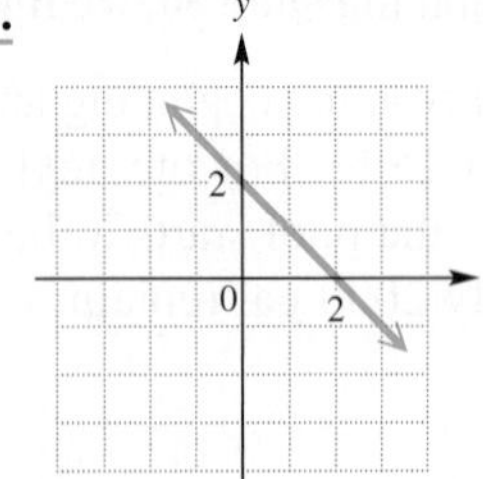

18.

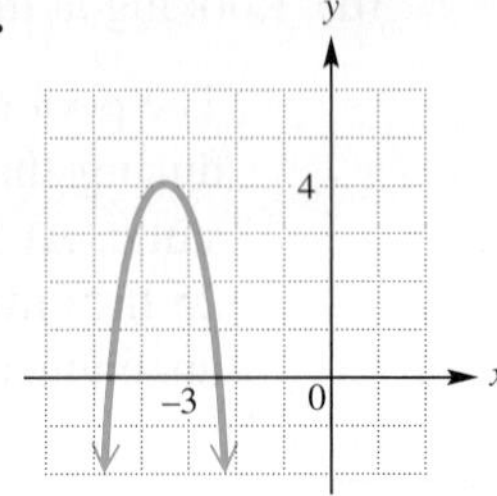

19.

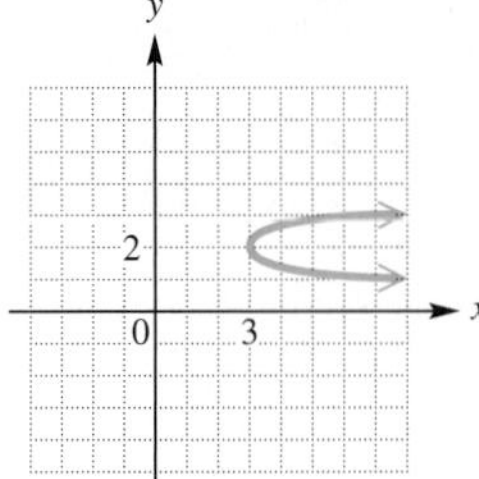

20.

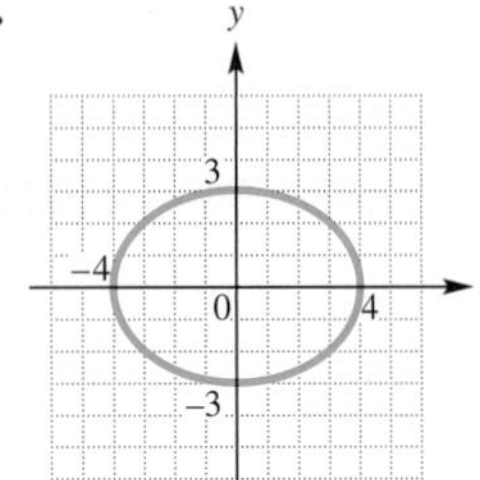

21.

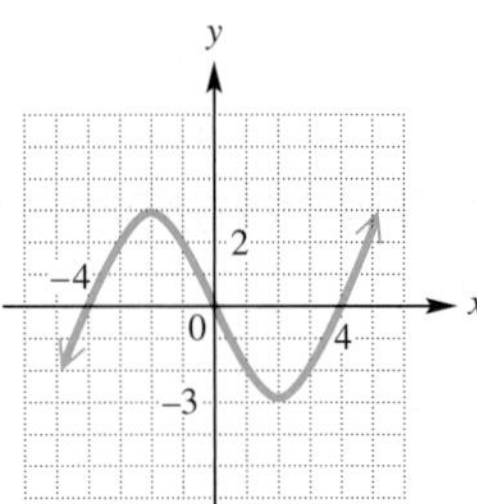

22.

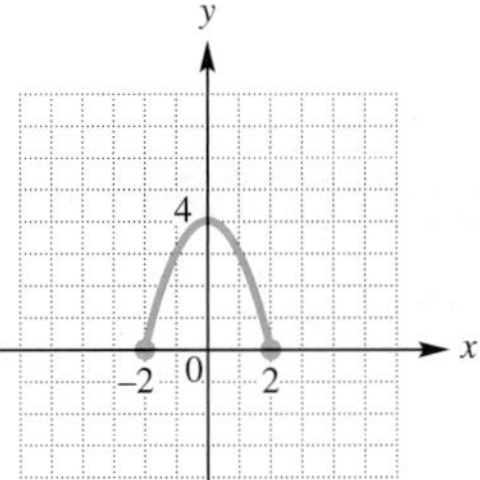

Decide whether each relation defines y as a function of x. Give the domain and range. ***See Example 5.***

23. $y = x^2$ **24.** $y = x^3$ **25.** $x = y^6$ **26.** $x = y^4$

27. $y = 2x - 5$ **28.** $y = -6x + 4$ **29.** $x + y < 3$ **30.** $x - y < 4$

31. $y = \sqrt{x}$ **32.** $y = -\sqrt{x}$ **33.** $xy = 2$ **34.** $xy = -6$

35. $y = \sqrt{4x + 1}$ **36.** $y = \sqrt{7 - 2x}$ **37.** $y = \dfrac{2}{x - 3}$ **38.** $y = \dfrac{-7}{x - 5}$

39. *Concept Check* Choose the correct answer: For function f, the notation $f(3)$ means

A. the variable f times 3, or $3f$.

B. the value of the dependent variable when the independent variable is 3.

C. the value of the independent variable when the dependent variable is 3.

D. f equals 3.

40. *Concept Check* Give an example of a function from everyday life. (*Hint:* Fill in the blanks: _____ depends on _____, so _____ is a function of _____.)

Let $f(x) = -3x + 4$ *and* $g(x) = -x^2 + 4x + 1$. *Find and simplify each of the following.* ***See Example 6.***

41. $f(0)$ **42.** $f(-3)$ **43.** $g(-2)$ **44.** $g(10)$

45. $f\left(\frac{1}{3}\right)$ **46.** $f\left(-\frac{7}{3}\right)$ **47.** $g\left(\frac{1}{2}\right)$ **48.** $g\left(-\frac{1}{4}\right)$

49. $f(p)$ **50.** $g(k)$ **51.** $f(-x)$ **52.** $g(-x)$

53. $f(x + 2)$ **54.** $f(a + 4)$ **55.** $f(2m - 3)$ **56.** $f(3t - 2)$

For each function, find ***(a)*** $f(2)$ *and* ***(b)*** $f(-1)$. ***See Example 7.***

57. $f = \{(-1, 3), (4, 7), (0, 6), (2, 2)\}$ **58.** $f = \{(2, 5), (3, 9), (-1, 11), (5, 3)\}$

59.

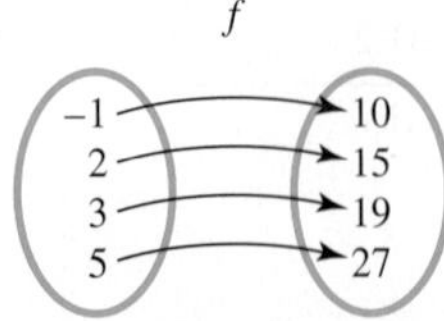

60.

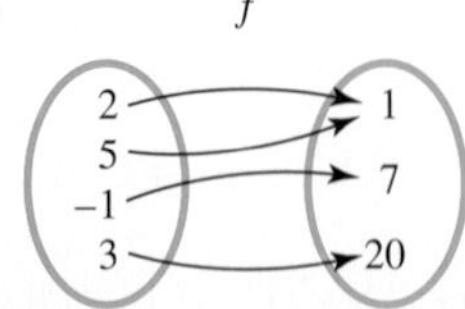

34. function;
domain: $(-\infty, 0) \cup (0, \infty)$;
range: $(-\infty, 0) \cup (0, \infty)$

35. function;
domain: $\left[-\frac{1}{4}, \infty\right)$;
range: $[0, \infty)$

36. function;
domain: $\left(-\infty, \frac{7}{2}\right]$;
range: $[0, \infty)$

37. function;
domain: $(-\infty, 3) \cup (3, \infty)$;
range: $(-\infty, 0) \cup (0, \infty)$

38. function;
domain: $(-\infty, 5) \cup (5, \infty)$;
range: $(-\infty, 0) \cup (0, \infty)$

39. B

40. Here is one example: The cost of gasoline; number of gallons used; cost; number of gallons

41. 4 **42.** 13
43. -11 **44.** -59
45. 3 **46.** 11
47. $\frac{11}{4}$ **48.** $-\frac{1}{16}$
49. $-3p + 4$
50. $-k^2 + 4k + 1$
51. $3x + 4$
52. $-x^2 - 4x + 1$
53. $-3x - 2$ **54.** $-3a - 8$
55. $-6m + 13$ **56.** $-9t + 10$

57. **(a)** 2 **(b)** 3
58. **(a)** 5 **(b)** 11
59. **(a)** 15 **(b)** 10
60. **(a)** 1 **(b)** 7
61. **(a)** 3 **(b)** -3
62. **(a)** -3 **(b)** 2

63. **(a)** $f(x) = -\frac{1}{3}x + 4$
(b) 3
64. **(a)** $f(x) = \frac{1}{4}x - 2$
(b) $-\frac{5}{4}$
65. **(a)** $f(x) = -2x^2 - x + 3$
(b) -18
66. **(a)** $f(x) = 3x^2 + x + 2$
(b) 32
67. **(a)** $f(x) = \frac{4}{3}x - \frac{8}{3}$
(b) $\frac{4}{3}$
68. **(a)** $f(x) = \frac{2}{5}x + \frac{9}{5}$
(b) 3

69. $f(3) = 4$
70. 0.164 square unit
71. -4
72. -3

61.

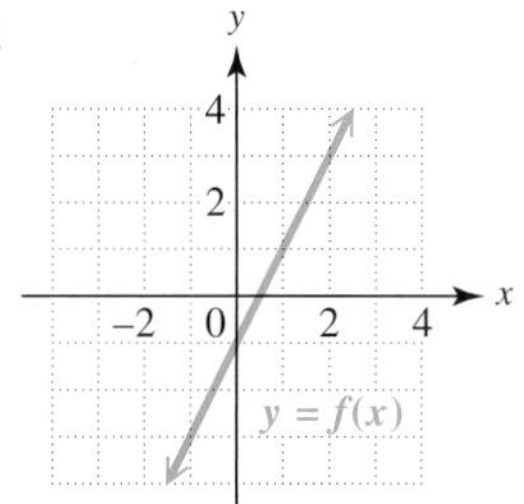

62.

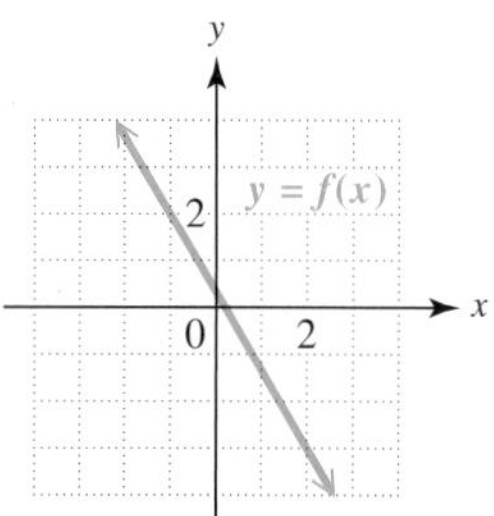

An equation that defines y as a function of x is given. ***(a)*** *Solve for y in terms of x and replace y with the function notation $f(x)$.* ***(b)*** *Find $f(3)$.* ***See Example 8.***

63. $x + 3y = 12$ **64.** $x - 4y = 8$ **65.** $y + 2x^2 = 3 - x$

66. $y - 3x^2 = 2 + x$ **67.** $4x - 3y = 8$ **68.** $-2x + 5y = 9$

Concept Check *Answer each question.*

69. If $(3, 4)$ is on the graph of $y = f(x)$, which one of the following must be true: $f(3) = 4$ or $f(4) = 3$?

70. The figure shows a portion of the graph of $f(x) = x^2 + 3x + 1$ and a rectangle with its base on the x-axis and a vertex on the graph. What is the area of the rectangle? (*Hint:* $f(0.2)$ is the height.)

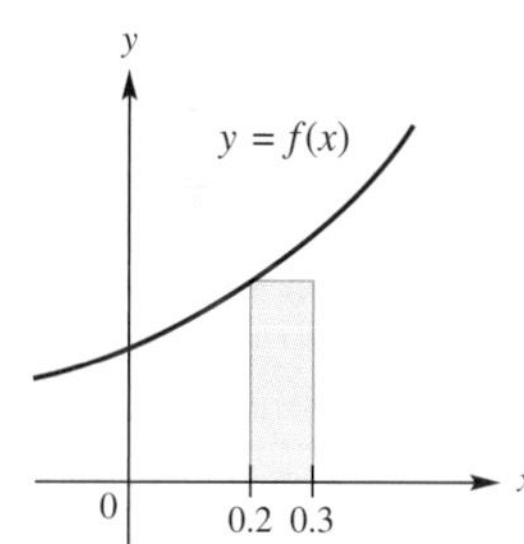

71. The graph of $Y_1 = f(X)$ is shown with a display at the bottom. What is $f(3)$?

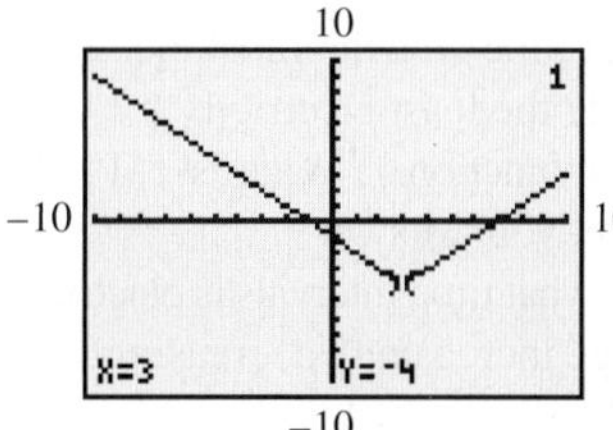

72. The graph of $Y_1 = f(X)$ is shown with a display at the bottom. What is $f(-2)$?

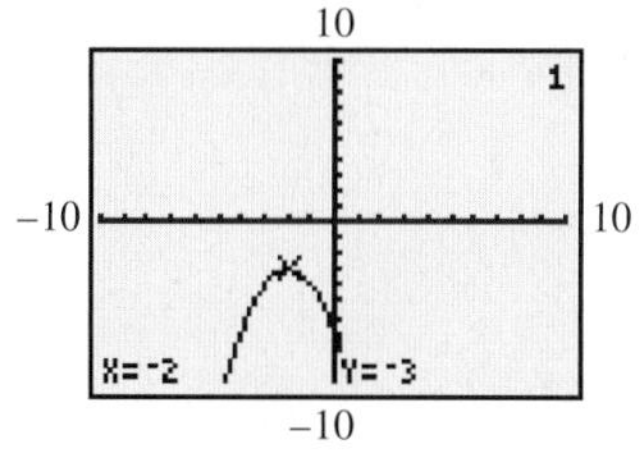

In Exercises 73–76, use the graph of $y = f(x)$ to find each function value: ***(a)*** *$f(-2)$,* ***(b)*** *$f(0)$,* ***(c)*** *$f(1)$, and* ***(d)*** *$f(4)$.* ***See Example 7(d).***

73.

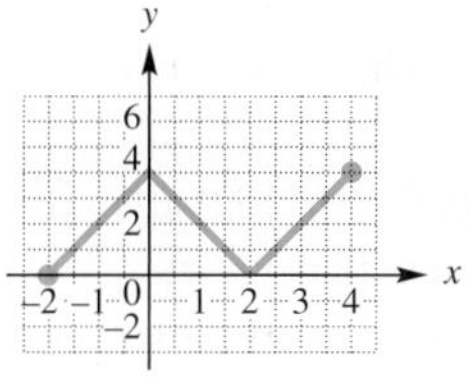

74.

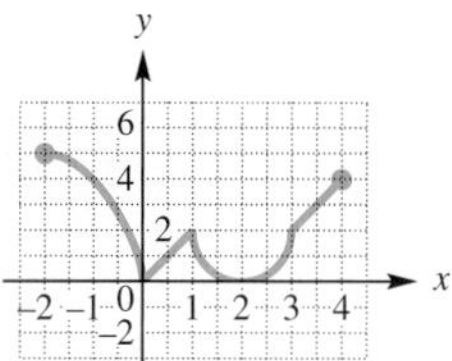

75.

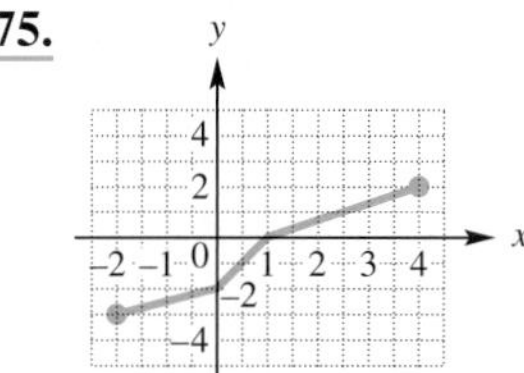

76.

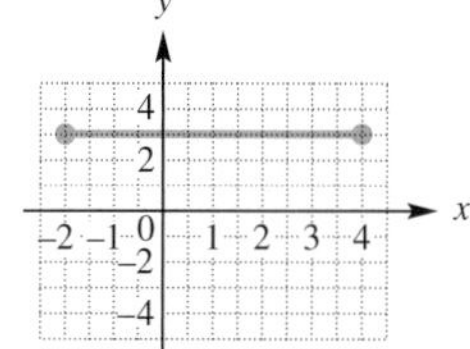

73. (a) 0 (b) 4 (c) 2 (d) 4
74. (a) 5 (b) 0 (c) 2 (d) 4
75. (a) −3 (b) −2 (c) 0 (d) 2
76. (a) 3 (b) 3 (c) 3 (d) 3

77. (a) $[4, \infty)$ (b) $(-\infty, -1]$
(c) $[-1, 4]$
78. (a) $(-\infty, 1]$ (b) $[4, \infty)$
(c) $[1, 4]$
79. (a) $(-\infty, 4]$ (b) $[4, \infty)$
(c) none
80. (a) none (b) $(-\infty, \infty)$
(c) none
81. (a) none
(b) $(-\infty, -2]$; $[3, \infty)$
(c) $(-2, 3)$
82. (a) $(3, \infty)$ (b) $(-\infty, -3)$
(c) $(-3, 3]$

83. (a) yes
(b) $[0, 24]$
(c) 1200 megawatts
(d) at 17 hr or 5 P.M.; at 4 A.M.
(e) $f(12) = 1900$; At 12 noon, electricity use is 1900 megawatts.
(f) increasing from 4 A.M. to 5 P.M.; decreasing from midnight to 4 A.M. and from 5 P.M. to midnight
84. (a) 240 ft
(b) at 1 sec and at 5 sec
(c) from 0 to 3 sec; from 3 to 7 sec
(d) 256 ft; at 3 sec
(e) 7 sec

*Determine the intervals of the domain for which each function is **(a)** increasing, **(b)** decreasing, and **(c)** constant.* ***See Example 9.***

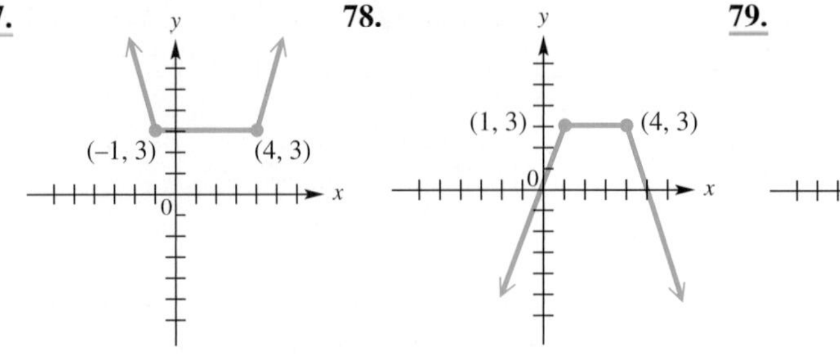

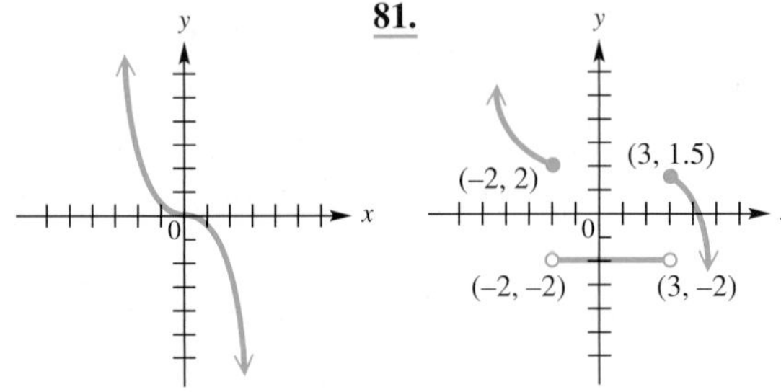

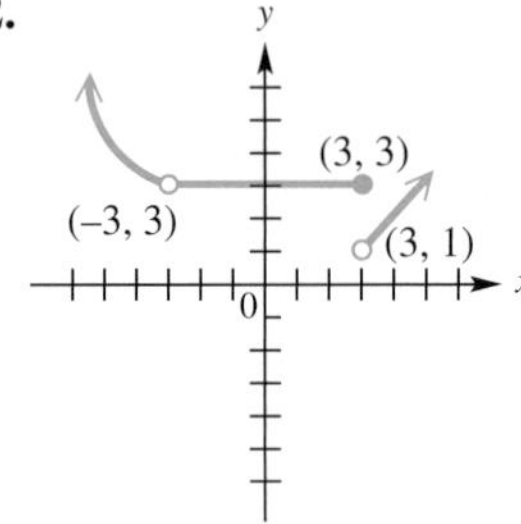

Solve each problem. ***See Example 10.***

83. *Electricity Usage* The graph shows the daily megawatts of electricity used on a record-breaking summer day in Sacramento, California.

(a) Is this the graph of a function?
(b) What is the domain?
(c) Estimate the number of megawatts used at 8 A.M.
(d) At what time was the most electricity used? the least electricity?
(e) Call this function f. What is $f(12)$? Interpret your answer.
(f) During what time intervals is electricity usage increasing? decreasing?

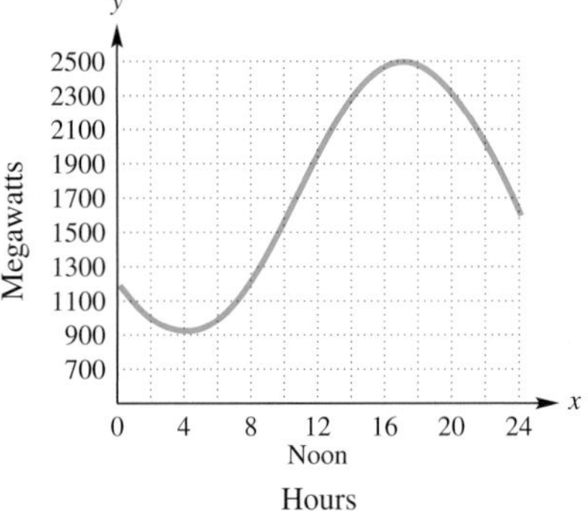

Source: Sacramento Municipal Utility District.

84. *Height of a Ball* A ball is thrown straight up into the air. The function defined by $y = h(t)$ in the graph gives the height of the ball (in feet) at t seconds. (*Note:* The graph does *not* show the path of the ball. The ball is rising straight up and then falling straight down.)

(a) What is the height of the ball at 2 sec?
(b) When will the height be 192 ft?
(c) During what time intervals is the ball going up? down?
(d) How high does the ball go, and when does the ball reach its maximum height?
(e) After how many seconds does the ball hit the ground?

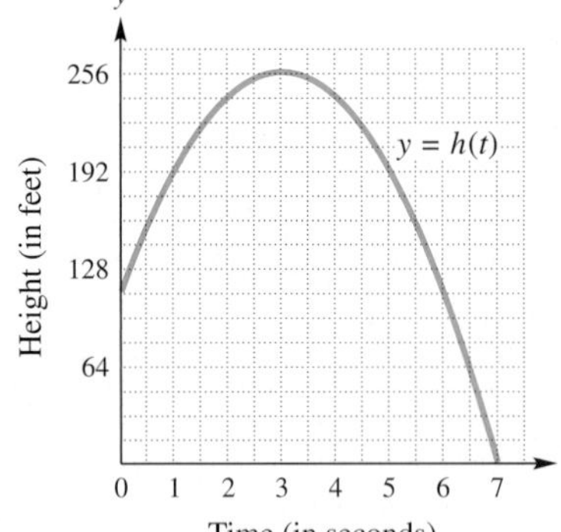

85. (a) about 12 noon to about 8 P.M.
(b) from midnight until about 6 A.M. and after 10 P.M.
(c) about 10 A.M. and 8:30 P.M.
(d) The temperature is just below 40° from midnight to 6 A.M., when it begins to rise until it reaches a maximum of just below 65° at 4 P.M. It then begins to fall until it reaches just under 40° again at midnight.

86. (a) 24 units
(b) from 0 to 2 hr after it is taken; from 2 to 12 hr after it is taken
(c) 2 hr; 64 units
(d) 8 hr
(e) When the drug is administered, the level is 0 units. The level begins to rise quickly for 2 hr until it reaches a maximum of 64 units. The level then begins to decrease gradually until it reaches a level of 12 units 12 hr after it was administered.

85. *Temperature* The graph shows temperatures on a given day in Bratenahl, Ohio.

(a) At what times during the day was the temperature over 55°?

(b) When was the temperature below 40°?

(c) Greenville, South Carolina, is 500 mi south of Bratenahl, Ohio, and its temperature is 7° higher all day long. At what time was the temperature in Greenville the same as the temperature at noon in Bratenahl?

(d) Use the graph to give a word description of the 24-hr period in Bratenahl.

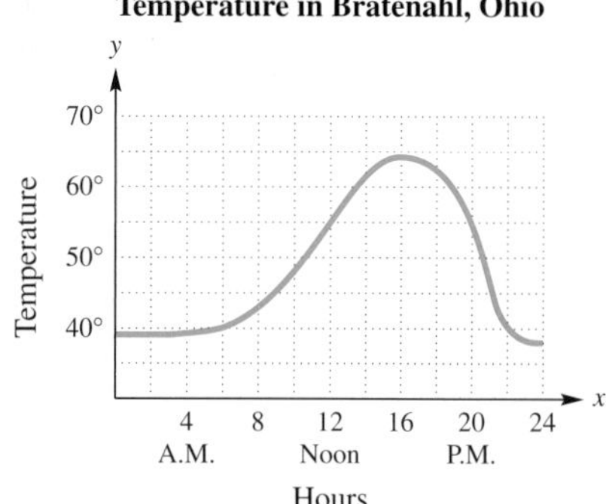

86. *Drug Levels in the Bloodstream* When a drug is taken orally, the amount of the drug in the bloodstream after t hours is given by the function defined by $y = f(t)$, as shown in the graph.

(a) How many units of the drug are in the bloodstream at 8 hr?

(b) During what time interval is the drug level in the bloodstream increasing? decreasing?

(c) When does the level of the drug in the bloodstream reach its maximum value, and how many units are in the bloodstream at that time?

(d) When the drug reaches its maximum level in the bloodstream, how many additional hours are required for the level to drop to 16 units?

(e) Use the graph to give a word description of the 12-hr period.

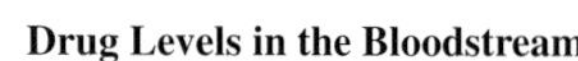

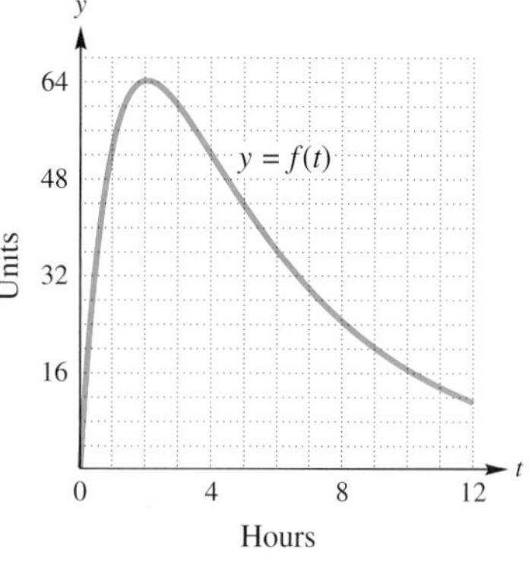

2.4 Linear Functions

- Graphing Linear Functions
- Standard Form $Ax + By = C$
- Slope
- Average Rate of Change
- Linear Models

Graphing Linear Functions We begin our study of specific functions by looking at *linear functions.*

Linear Function

A function f is a **linear function** if

$$f(x) = ax + b,$$

for real numbers a and b. If $a \neq 0$, the domain and the range of a linear function are both $(-\infty, \infty)$.

In **Section 2.1,** we graphed lines by finding ordered pairs and plotting them. Although only two points are necessary to graph a linear function, we usually plot a third point as a check. The intercepts are often good points to choose for graphing lines.

Classroom Example 1
Graph $f(x) = \frac{3}{2}x + 6$. Give the domain and range.

Answer:

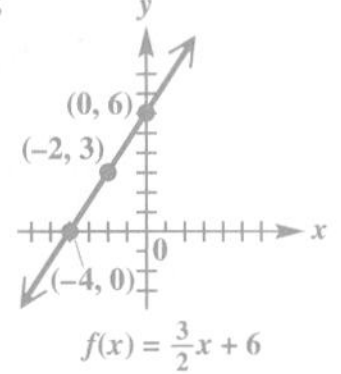

domain: $(-\infty, \infty)$;
range: $(-\infty, \infty)$

EXAMPLE 1 **Graphing a Linear Function Using Intercepts**

Graph $f(x) = -2x + 6$. Give the domain and range.

SOLUTION The x-intercept is found by letting $f(x) = 0$ and solving for x.

$$f(x) = -2x + 6$$

$$0 = -2x + 6 \quad \text{Let } f(x) = 0.$$

$$x = 3 \quad \text{Add } 2x \text{ and divide by 2. (Section 1.1)}$$

The x-intercept is 3, so we plot $(3, 0)$. The y-intercept is found by evaluating $f(0)$.

$$f(0) = -2(0) + 6 \quad \text{Let } x = 0.$$

$$f(0) = 6 \quad \text{Simplify.}$$

Therefore, another point on the graph is $(0, 6)$. We plot this point and join the two points with a straight-line graph. We use the point $(2, 2)$ as a check. See **Figure 31.** The domain and the range are both $(-\infty, \infty)$.

The corresponding calculator graph is shown in **Figure 32.**

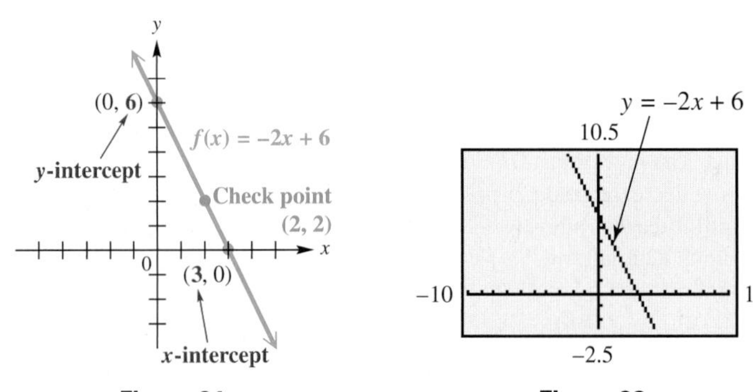

Figure 31 **Figure 32**

Now Try Exercise 9.

If $a = 0$ in the definition of linear function, then the equation becomes $f(x) = b$. In this case, the domain is $(-\infty, \infty)$ and the range is $\{b\}$. A function of the form $\boldsymbol{f(x) = b}$ is a **constant function,** and its graph is a horizontal line.

Classroom Example 2
Graph $f(x) = 2$. Give the domain and range.

Answer:

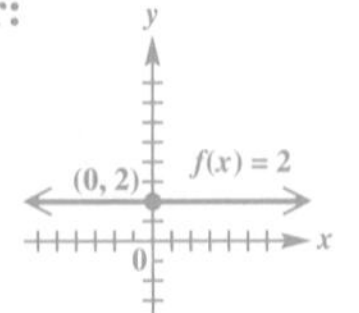

domain: $(-\infty, \infty)$; range: $\{2\}$

EXAMPLE 2 **Graphing a Horizontal Line**

Graph $f(x) = -3$. Give the domain and range.

SOLUTION Since $f(x)$, or y, always equals -3, the value of y can never be 0. This means that the graph has no x-intercept. The only way a straight line can have no x-intercept is for it to be parallel to the x-axis, as shown in **Figure 33.** The domain of this linear function is $(-\infty, \infty)$ and the range is $\{-3\}$. **Figure 34** shows the calculator graph.

Teaching Tip Write $f(x) = -3$ in **Example 2** as

$$f(x) = 0 \cdot x - 3$$

to illustrate that x can take on any value and y remains constant.

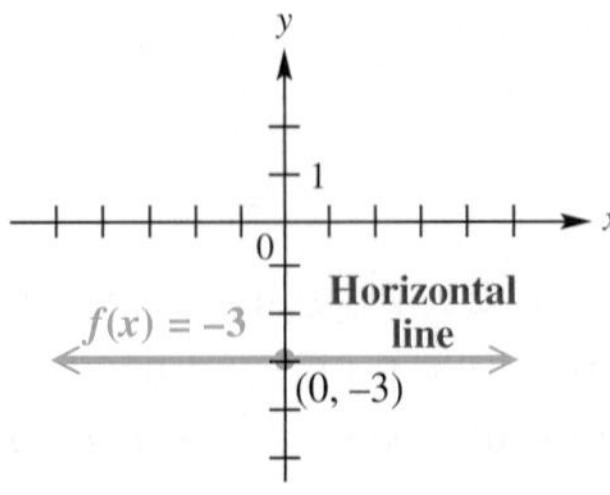

Figure 33

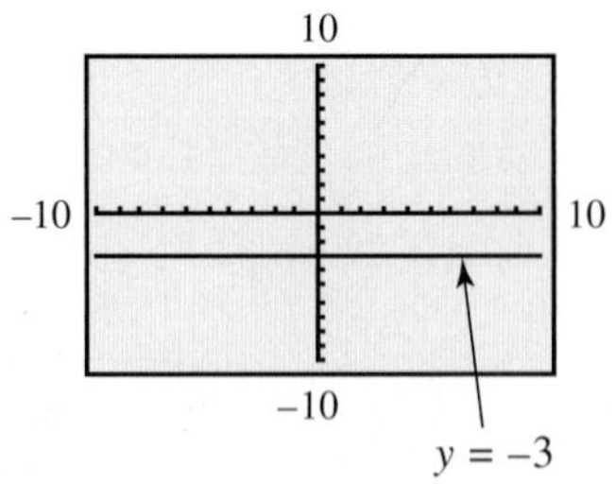

Figure 34

Now Try Exercise 13.

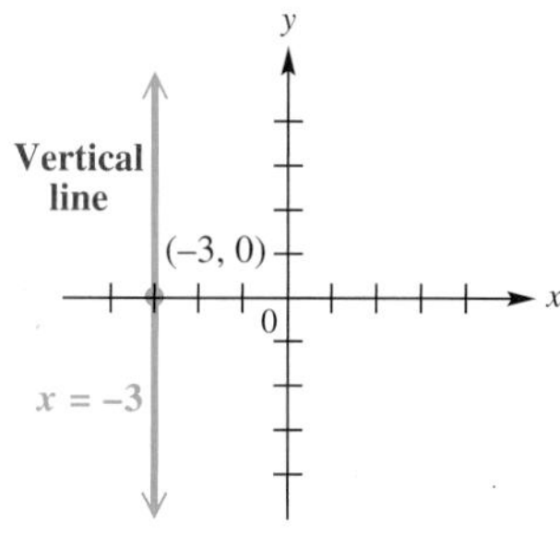

Figure 35

EXAMPLE 3 Graphing a Vertical Line

Graph $x = -3$. Give the domain and range of this relation.

SOLUTION Since x always equals -3, the value of x can never be 0, and the graph has no y-intercept. Using reasoning similar to that of **Example 2,** we find that this graph is parallel to the y-axis, as shown in **Figure 35.** The domain of this relation, which is *not* a function, is $\{-3\}$, while the range is $(-\infty, \infty)$.

Now Try Exercise 19.

Classroom Example 3
Graph $x = 5$. Give the domain and range of this relation.

Answer:

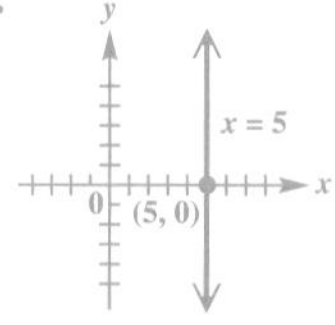

domain: $\{5\}$; range: $(-\infty, \infty)$

Standard Form $Ax + By = C$ Equations of lines are often written in the form $Ax + By = C$, which is called **standard form.**

NOTE The definition of "standard form" is, ironically, not standard from one text to another. Any linear equation can be written in infinitely many different, but equivalent, forms. For example, the equation $2x + 3y = 8$ can be written equivalently as

$$2x + 3y - 8 = 0, \quad 3y = 8 - 2x, \quad x + \frac{3}{2}y = 4, \quad 4x + 6y = 16,$$

and so on. In this text we will agree that if the coefficients and constant in a linear equation are rational numbers, then we will consider the standard form to be $Ax + By = C$, where $A \geq 0$, A, B, and C are integers, and the greatest common factor of A, B, and C is 1. If $A = 0$, then we choose $B > 0$. (If two or more integers have a greatest common factor of 1, they are said to be **relatively prime.**)

In **Example 8(a)** of **Section 2.1,** we graphed $y = 4x - 1$ (or, equivalently, $4x - y = 1$) by finding the x- and y-intercepts. The following example is similar.

Classroom Example 4
Graph $3x + 4y = 0$. Give the domain and range.

Answer:

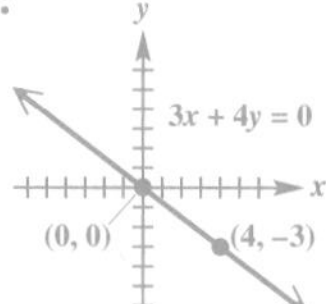

domain: $(-\infty, \infty)$;
range: $(-\infty, \infty)$

EXAMPLE 4 Graphing $Ax + By = C$ ($C = 0$)

Graph $4x - 5y = 0$. Give the domain and range.

SOLUTION Find the intercepts.

$$4(0) - 5y = 0 \quad \text{Let } x = 0.$$
$$y = 0 \quad y\text{-intercept}$$

$$4x - 5(0) = 0 \quad \text{Let } y = 0.$$
$$x = 0 \quad x\text{-intercept}$$

The graph of this function has just one intercept: the origin $(0, 0)$. We need to find an additional point to graph the function by choosing a different value for x (or y).

$$4(5) - 5y = 0 \quad \text{We choose } x = 5.$$
$$20 - 5y = 0 \quad \text{Multiply.}$$
$$4 = y \quad \text{Add } 5y. \text{ Divide by 5.}$$

This leads to the ordered pair $(5, 4)$. Complete the graph using the two points $(0, 0)$ and $(5, 4)$, with a third point as a check. The domain and range are both $(-\infty, \infty)$. See **Figure 36.**

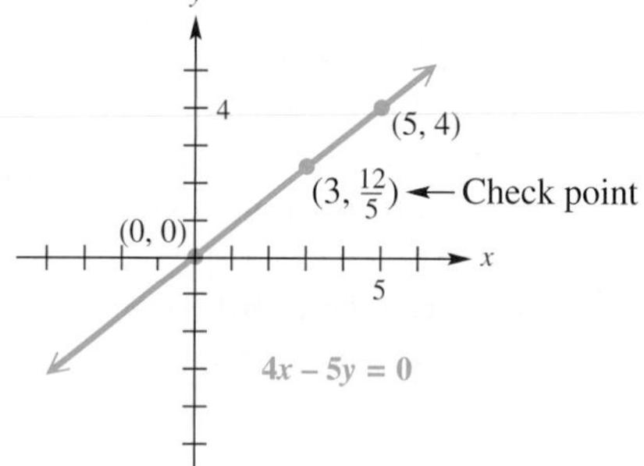

Figure 36

Now Try Exercise 17.

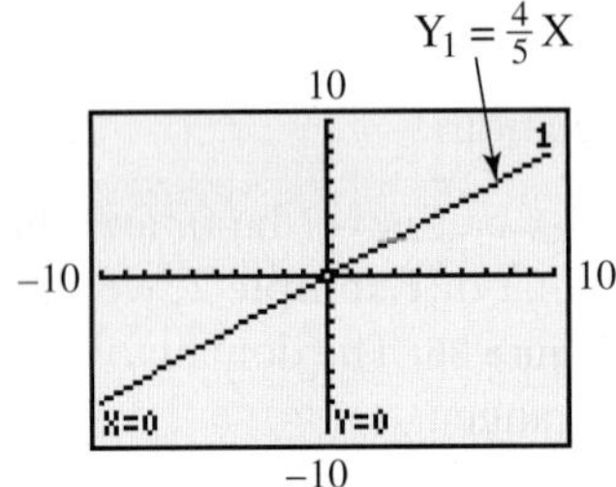

Figure 37

To use a graphing calculator to graph a linear function given in standard form, we must first solve the defining equation for y.

$$4x - 5y = 0 \quad \text{Equation from Example 4}$$

$$y = \frac{4}{5}x \quad \text{Subtract } 4x. \text{ Divide by } -5.$$

We graph $Y_1 = \frac{4}{5}X$. See **Figure 37.** ■

Slope An important characteristic of a straight line is its *slope,* a numerical measure of the steepness and orientation of the line. (Geometrically, this may be interpreted as the ratio of **rise** to **run.**) The slope of a highway (sometimes called the *grade*) is often given as a percent. For example, a 10% $\left(\text{or } \frac{10}{100} = \frac{1}{10}\right)$ slope means the highway rises 1 unit for every 10 horizontal units.

To find the slope of a line, start with two distinct points (x_1, y_1) and (x_2, y_2) on the line, as shown in **Figure 38,** where $x_1 \neq x_2$. As we move along the line from (x_1, y_1) to (x_2, y_2), the horizontal difference

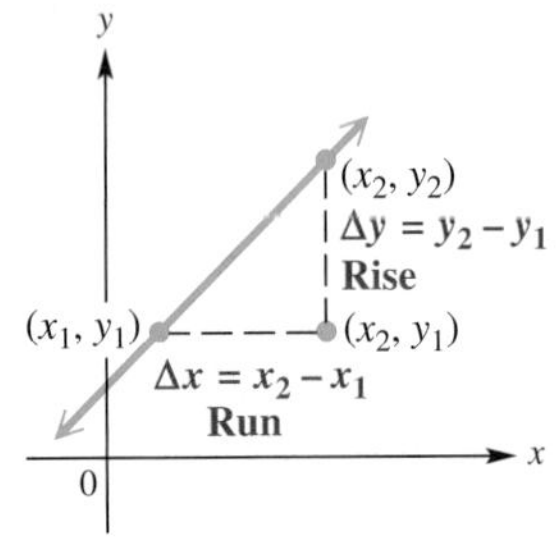

Figure 38

$$\Delta x = x_2 - x_1$$

is called the **change in x,** denoted by Δx (read **"delta x"**), where Δ is the Greek letter **delta.** The vertical difference, called the **change in y,** can be written

$$\Delta y = y_2 - y_1.$$

The *slope* of a nonvertical line is defined as the quotient (ratio) of the change in y and the change in x, as follows.

Slope

The **slope** m of the line through the points (x_1, y_1) and (x_2, y_2) is given by the following.

$$m = \frac{\text{rise}}{\text{run}} = \frac{\Delta y}{\Delta x} = \frac{y_2 - y_1}{x_2 - x_1}, \quad \text{where } \Delta x \neq 0$$

That is, the slope of a line is the change in y divided by the corresponding change in x, where the change in x is not 0.

LOOKING AHEAD TO CALCULUS

The concept of slope of a line is extended in calculus to general curves. The **slope of a curve at a point** is understood to mean the slope of the line tangent to the curve at that point.

The line in the figure is tangent to the curve at point P.

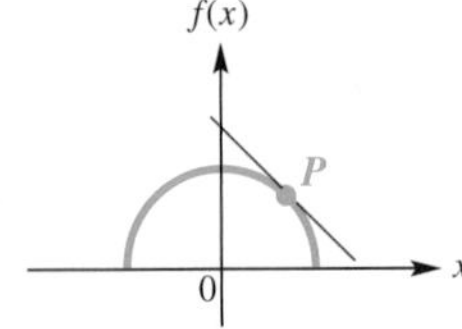

CAUTION ***When using the slope formula, it makes no difference which point is (x_1, y_1) or (x_2, y_2). However, be consistent.*** Start with the x- and y-values of *one* point (either one) and subtract the corresponding values of the *other* point.

$$\textit{Use} \quad \frac{y_2 - y_1}{x_2 - x_1} \quad \text{or} \quad \frac{y_1 - y_2}{x_1 - x_2}, \quad \textit{not} \quad \frac{y_2 - y_1}{x_1 - x_2} \quad \text{or} \quad \frac{y_1 - y_2}{x_2 - x_1}.$$

Be sure to write the difference of the y-values in the numerator and the difference of the x-values in the denominator.

The slope of a line can be found only if the line is nonvertical. This guarantees that $x_2 \neq x_1$ so that the denominator $x_2 - x_1 \neq 0$.

Undefined Slope

The slope of a vertical line is undefined.

Classroom Example 5
Find the slope of the line through the given points.
(a) $(-2, 4), (2, -6)$
(b) $(-3, 8), (5, 8)$
(c) $(-4, -10), (-4, 10)$

Answers:
(a) $-\frac{5}{2}$ (b) 0
(c) undefined

EXAMPLE 5 Finding Slopes with the Slope Formula

Find the slope of the line through the given points.

(a) $(-4, 8), (2, -3)$ **(b)** $(2, 7), (2, -4)$ **(c)** $(5, -3), (-2, -3)$

SOLUTION

(a) Let $x_1 = -4$, $y_1 = 8$, and $x_2 = 2$, $y_2 = -3$. Then the slope is

$$m = \frac{\text{rise}}{\text{run}} = \frac{\Delta y}{\Delta x} \quad \text{Definition of slope}$$

$$= \frac{-3 - 8}{2 - (-4)} \quad \text{Substitute carefully.}$$

$$= \frac{-11}{6}, \text{ or } -\frac{11}{6}.$$

We can also subtract in the opposite order, letting $x_1 = 2$, $y_1 = -3$ and $x_2 = -4$, $y_2 = 8$. The same slope results.

$$m = \frac{8 - (-3)}{-4 - 2} = \frac{11}{-6}, \text{ or } -\frac{11}{6}$$

Teaching Tip Mention the case where the run is 1. That is, if you start at any point of the line and move 1 unit to the right, you must move $|m|$ units in the y-direction to return to the line. The distance you move in the y-direction is up if m is positive and down if m is negative.

(b) If we attempt to use the formula with the points $(2, 7)$ and $(2, -4)$, we get a zero denominator.

$$m = \frac{-4 - 7}{2 - 2} = \frac{-11}{0} \quad \text{Undefined}$$

The formula is not valid here because $\Delta x = x_2 - x_1 = 2 - 2 = 0$. A sketch would show that the line through $(2, 7)$ and $(2, -4)$ is vertical. As mentioned above, the slope of a vertical line is undefined.

(c) For $(5, -3)$ and $(-2, -3)$, the slope is

$$m = \frac{-3 - (-3)}{-2 - 5} = \frac{0}{-7} = 0.$$

A sketch would show that the line through $(5, -3)$ and $(-2, -3)$ is horizontal.

✔ *Now Try Exercises 35, 41, and 43.*

The results in **Example 5(c)** suggest the following generalization.

Zero Slope

The slope of a horizontal line is 0.

LOOKING AHEAD TO CALCULUS

The **derivative** of a function provides a formula for determining the slope of a line tangent to a curve. If the slope is positive on a given interval, then the function is increasing there. If it is negative, then the function is decreasing. If it is 0, then the function is constant.

Theorems for similar triangles can be used to show that the slope of a line is independent of the choice of points on the line. ***That is, slope is the same no matter which pair of distinct points on the line are used to find it.***

If the equation of a line is in the form $y = ax + b$, we can show that the slope of the line is a. To do this, we use function notation and the definition of slope. Let $f(x) = ax + b$, $x_1 = x$, and $x_2 = x + 1$.

$$m = \frac{f(x_2) - f(x_1)}{x_2 - x_1}$$

$$= \frac{[a(x+1) + b] - (ax + b)}{(x+1) - x} \quad \text{Substitute in the slope formula.}$$

$$= \frac{ax + a + b - ax - b}{x + 1 - x} \quad \text{Distributive property (Section R.2)}$$

$$= \frac{a}{1} \quad \text{Combine like terms.}$$

$$= a \quad \text{The slope is } a.$$

This discussion enables us to find the slope of the graph of any linear equation by solving for y and identifying the coefficient of x, which is that slope.

Classroom Example 6
Find the slope of the line $2x - 5y = 10$.

Answer: $\frac{2}{5}$

EXAMPLE 6 Finding the Slope from an Equation

Find the slope of the line $4x + 3y = 12$.

SOLUTION Solve the equation for y.

$$4x + 3y = 12$$

$$3y = -4x + 12 \quad \text{Subtract } 4x.$$

$$y = -\frac{4}{3}x + 4 \quad \text{Divide by 3.}$$

The slope is $-\frac{4}{3}$, the coefficient of x.

Now Try Exercise 49(a).

Since the slope of a line is the ratio of vertical change (rise) to horizontal change (run), if we know the slope of a line and the coordinates of a point on the line, we can draw the graph of the line.

Classroom Example 7
Graph the line passing through $(-2, -3)$ and having slope $\frac{4}{3}$.

Answer:

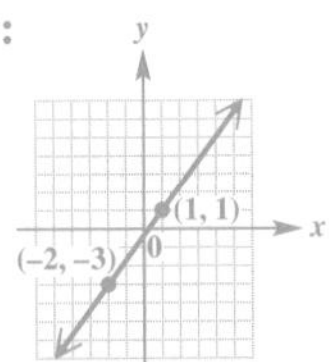

EXAMPLE 7 Graphing a Line Using a Point and the Slope

Graph the line passing through $(-1, 5)$ and having slope $-\frac{5}{3}$.

SOLUTION First locate the point $(-1, 5)$ as shown in **Figure 39.** Since the slope of this line is $\frac{-5}{3}$, a change of -5 units vertically (that is, 5 units *down*) corresponds to a change of 3 units horizontally (that is, 3 units to the *right*). This gives a second point, $(2, 0)$, which can then be used to complete the graph.

Because $\frac{-5}{3} = \frac{5}{-3}$, another point could be obtained by starting at $(-1, 5)$ and moving 5 units *up* and 3 units to the *left*. We would reach a different second point, $(-4, 10)$, but the graph would be the same. Confirm this in **Figure 39.**

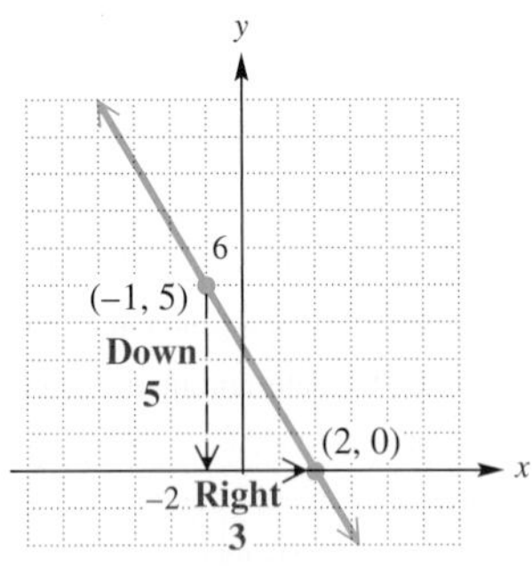

Figure 39

Now Try Exercise 53.

Classroom Example 8
In 2001, the number of basic cable TV subscribers in the United States was 66.7 million. In 2009, the number was 62.9 million. Assume a linear relationship, and find the average rate of change in the number of subscribers per year. Graph as a line segment and interpret the result. (*Source:* SNL Kagan.)

Answer: -0.475; The graph confirms that the line through the ordered pairs falls from left to right and therefore has negative slope. Thus, the number of basic cable subscribers *decreased* by an average of 0.475 million (or 475,000) each year from 2001 to 2009.

Figure 40 shows lines with various slopes. *Notice the following important concepts:*

- A line with a positive slope rises from left to right. The corresponding linear function is increasing on its entire domain.
- A line with a negative slope falls from left to right. The corresponding linear function is decreasing on its entire domain.
- A line with 0 slope neither rises nor falls. The corresponding linear function is constant on its entire domain.

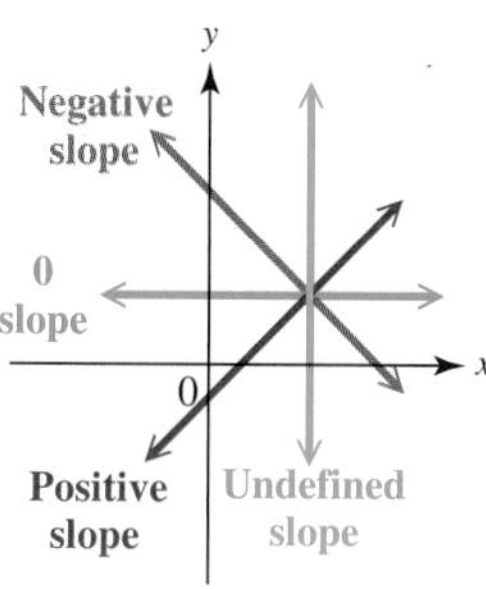

Figure 40

Average Rate of Change We know that the slope of a line is the ratio of the vertical change in y to the horizontal change in x. ***Thus, slope gives the average rate of change in y per unit of change in x, where the value of y depends on the value of x.*** If f is a linear function defined on $[a, b]$, then we have the following.

$$\textbf{Average rate of change on } [a, b] = \frac{f(b) - f(a)}{b - a}$$

This is simply another way to write the slope formula, using function notation.

EXAMPLE 8 Interpreting Slope as Average Rate of Change

In 1980, the average monthly rate for basic cable TV in the United States was \$7.69. In 2008, the average monthly rate was \$46.13. Assume a linear relationship, and find the average rate of change in the monthly rate per year. Graph as a line segment and interpret the result. (*Source:* SNL Kagan.)

SOLUTION To use the slope formula, we need two ordered pairs. Here, if $x = 1980$, then $y = 7.69$, and if $x = 2008$, then $y = 46.13$. This gives the ordered pairs (1980, 7.69) and (2008, 46.13). (Note that y is in dollars.)

$$\text{Average rate of change} = \frac{46.13 - 7.69}{2008 - 1980} = \frac{38.44}{28} \approx 1.37$$

The graph in **Figure 41** confirms that the line through the ordered pairs rises from left to right and therefore has positive slope. Thus, the monthly rate for basic cable *increased* by an average of about \$1.37 each year from 1980 to 2008.

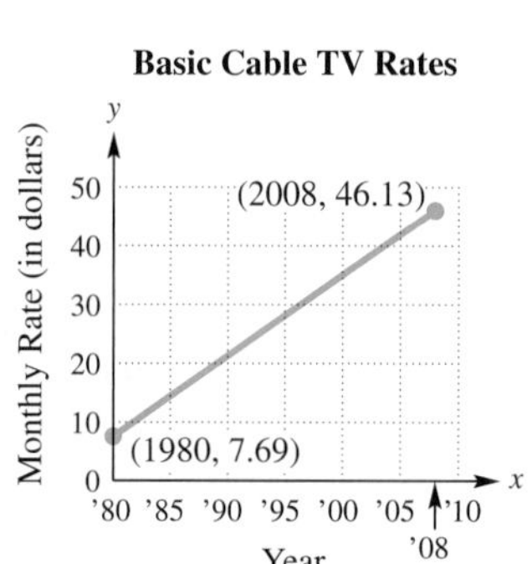

Figure 41

✔ *Now Try Exercise 75.*

Linear Models In **Example 8,** we used the graph of a line to approximate real data. Recall from **Section 1.2** that this is called **mathematical modeling.** Points on the straight-line graph model, or approximate, the actual points that correspond to the data.

A **linear cost function** has the form

$$C(x) = mx + b, \quad \text{Linear cost function}$$

where x represents the number of items produced, m represents the **cost** per item, and b represents the **fixed cost.** The fixed cost is constant for a particular product and does not change as more items are made. The value of mx, which increases as more items are produced, covers labor, materials, packaging, shipping, and so on.

The **revenue function** for selling a product depends on the price per item p and the number of items sold x. It is given by the following function.

$$R(x) = px \quad \text{Revenue function}$$

Profit is found by subtracting cost from revenue and is described by the **profit function.**

$$P(x) = R(x) - C(x) \quad \text{Profit function}$$

Classroom Example 9
Repeat **Example 9** if the fixed cost is \$2400, the variable cost per item is \$120, and the item sells for \$150.

Answers:
(a) $C(x) = 120x + 2400$
(b) $R(x) = 150x$
(c) $P(x) = 30x - 2400$
(d) at least 81 items

EXAMPLE 9 Writing Linear Cost, Revenue, and Profit Functions

Assume that the cost to produce an item is a linear function and all items produced are sold. The fixed cost is \$1500, the variable cost per item is \$100, and the item sells for \$125. Write linear functions to model

(a) cost, **(b)** revenue, and **(c)** profit.

(d) How many items must be sold for the company to make a profit?

SOLUTION

(a) Since the cost function is linear, it will have the following form.

$$C(x) = mx + b \quad \text{Cost function}$$
$$C(x) = 100x + 1500 \quad \text{Let } m = 100 \text{ and } b = 1500.$$

(b) The revenue function is defined by the product of 125 and x.

$$R(x) = px \quad \text{Revenue function}$$
$$R(x) = 125x \quad \text{Let } p = 125.$$

(c) The profit function is found by subtracting the cost function from the revenue function.

$$\begin{aligned} P(x) &= R(x) - C(x) \\ &= 125x - (100x + 1500) && \text{Use parentheses here.} \\ &= 125x - 100x - 1500 && \text{Distributive property} \\ P(x) &= 25x - 1500 && \text{Combine like terms.} \end{aligned}$$

ALGEBRAIC SOLUTION

(d) To make a profit, $P(x)$ must be positive.

$$P(x) = 25x - 1500 \quad \text{Profit function from part (c)}$$

Set $P(x) > 0$ and solve.

$$\begin{aligned} P(x) &> 0 \\ 25x - 1500 &> 0 && P(x) = 25x - 1500 \\ 25x &> 1500 && \text{Add 1500 to each side. (Section 1.7)} \\ x &> 60 && \text{Divide by 25.} \end{aligned}$$

Since the number of items must be a whole number, at least 61 items must be sold for the company to make a profit.

GRAPHING CALCULATOR SOLUTION

(d) Define Y_1 as $25X - 1500$ and graph the line. Use the capability of your calculator to locate the x-intercept. See **Figure 42.** As the graph shows, y-values for x less than 60 are negative, and y-values for x greater than 60 are positive, so at least 61 items must be sold for the company to make a profit.

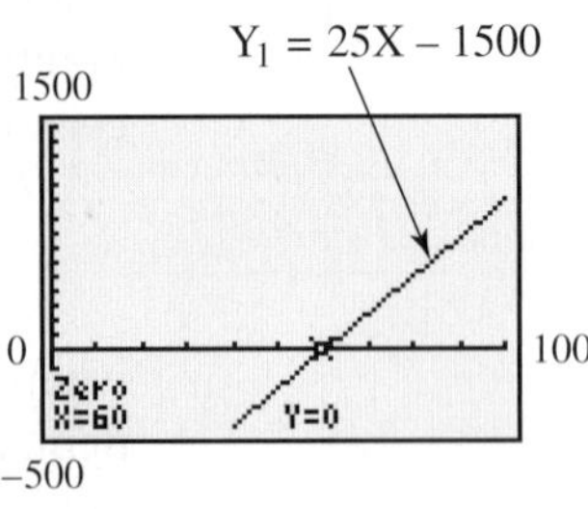

Figure 42

Now Try Exercise 87.

2.4 Exercises

1. B 2. H 3. C
4. G 5. A 6. D

In Exercises 7–24, we give the domain first and then the range.

7. $(-\infty, \infty)$; $(-\infty, \infty)$

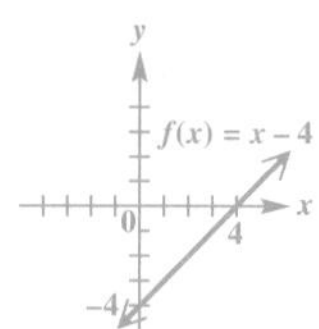

8. $(-\infty, \infty)$; $(-\infty, \infty)$

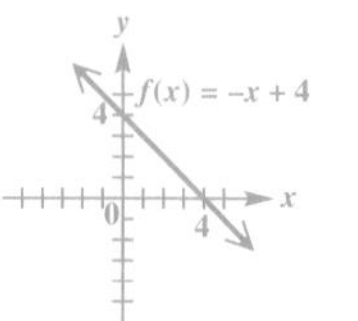

9. $(-\infty, \infty)$; $(-\infty, \infty)$

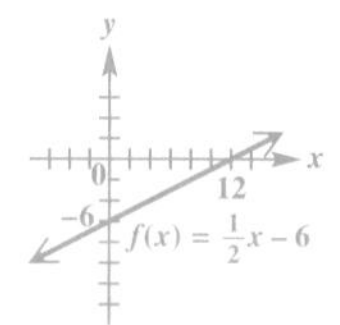

10. $(-\infty, \infty)$; $(-\infty, \infty)$

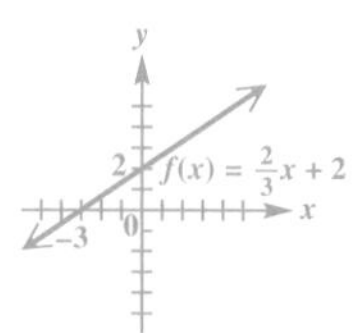

11. $(-\infty, \infty)$; $(-\infty, \infty)$

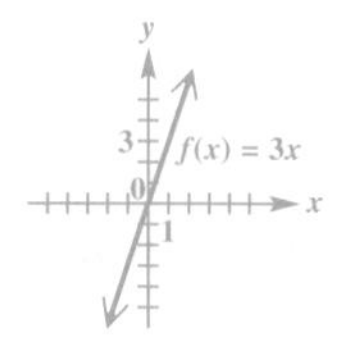

12. $(-\infty, \infty)$; $(-\infty, \infty)$

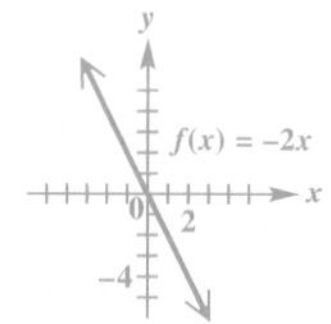

13. $(-\infty, \infty)$; $\{-4\}$; constant function

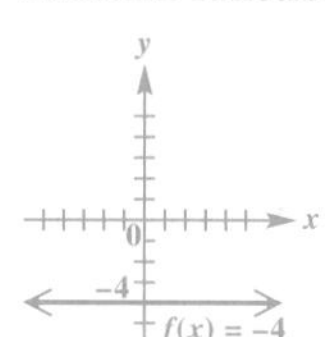

Concept Check *Match the description in Column I with the correct response in Column II. Some choices may not be used.*

I	II
1. a linear function whose graph has y-intercept 6	**A.** $f(x) = 5x$
2. a vertical line	**B.** $f(x) = 3x + 6$
3. a constant function	**C.** $f(x) = -8$
4. a linear function whose graph has x-intercept -2 and y-intercept 4	**D.** $f(x) = x^2$
	E. $x + y = -6$
5. a linear function whose graph passes through the origin	**F.** $f(x) = 3x + 4$
	G. $2x - y = -4$
6. a function that is not linear	**H.** $x = 9$

Graph each linear function. Identify any constant functions. Give the domain and range. **See Examples 1 and 2.**

7. $f(x) = x - 4$ **8.** $f(x) = -x + 4$ **9.** $f(x) = \frac{1}{2}x - 6$

10. $f(x) = \frac{2}{3}x + 2$ **11.** $f(x) = 3x$ **12.** $f(x) = -2x$

13. $f(x) = -4$ **14.** $f(x) = 3$

Graph each line. Give the domain and range. **See Examples 3 and 4.**

15. $-4x + 3y = 12$ **16.** $2x + 5y = 10$

17. $3y - 4x = 0$ **18.** $3x + 2y = 0$

19. $x = 3$ **20.** $x = -4$

21. $2x + 4 = 0$ **22.** $-3x + 6 = 0$

23. $-x + 5 = 0$ **24.** $3 + x = 0$

Match each equation with the sketch that most closely resembles its graph. **See Examples 2 and 3.**

25. $y = 5$ **26.** $y = -5$ **27.** $x = 5$ **28.** $x = -5$

A. **B.** **C.** **D.**

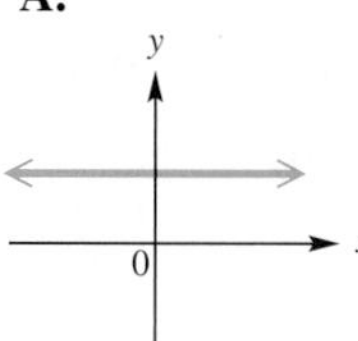

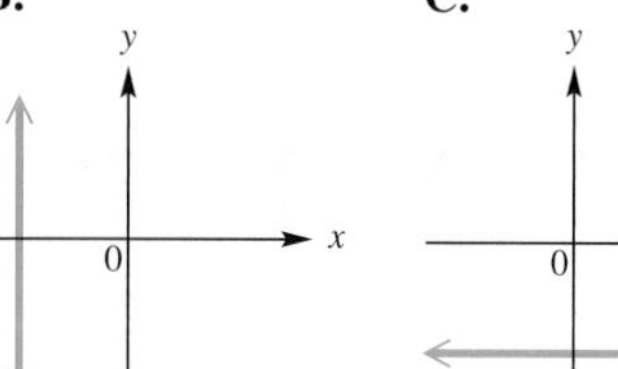

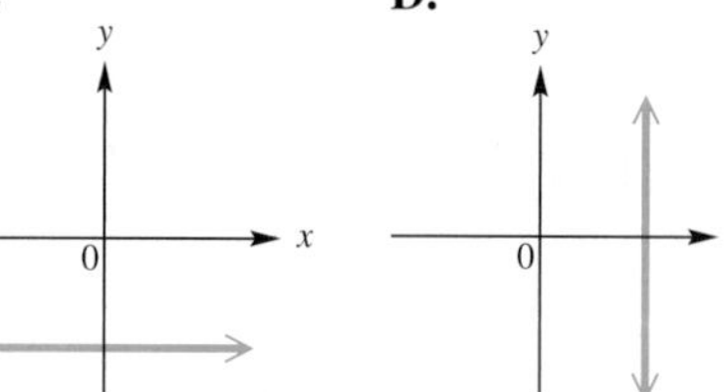

Use a graphing calculator to graph each equation in the standard viewing window. **See Examples 1, 2, and 4.**

29. $y = 3x + 4$ **30.** $y = -2x + 3$

31. $3x + 4y = 6$ **32.** $-2x + 5y = 10$

14. $(-\infty, \infty)$; $\{3\}$; constant function

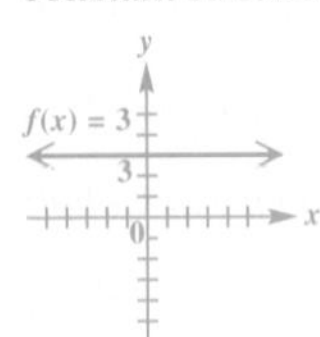

15. $(-\infty, \infty)$; $(-\infty, \infty)$

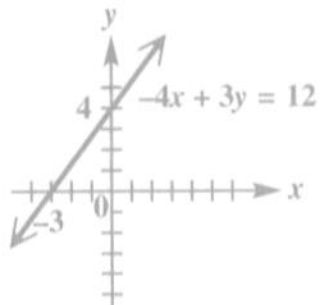

16. $(-\infty, \infty)$; $(-\infty, \infty)$

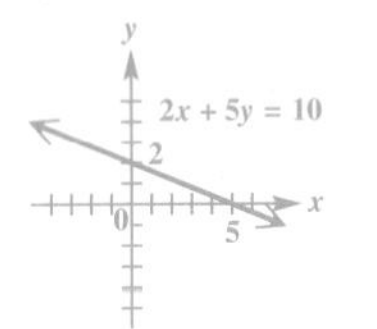

17. $(-\infty, \infty)$; $(-\infty, \infty)$

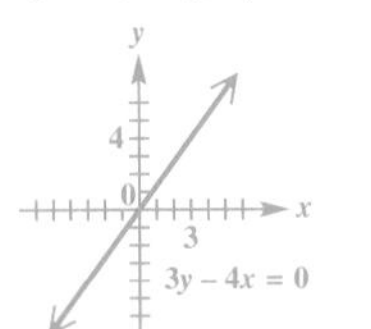

18. $(-\infty, \infty)$; $(-\infty, \infty)$

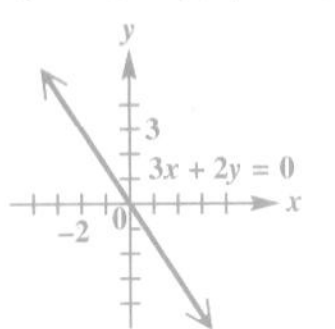

19. $\{3\}$; $(-\infty, \infty)$

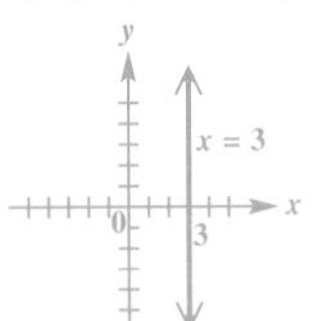

20. $\{-4\}$; $(-\infty, \infty)$

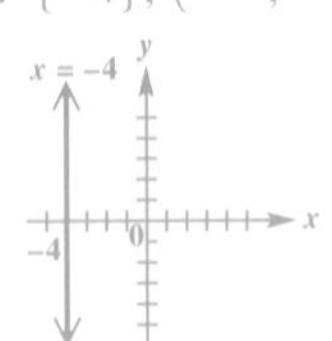

21. $\{-2\}$; $(-\infty, \infty)$

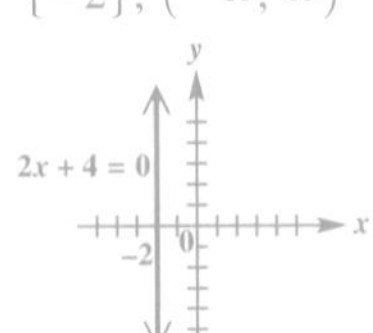

22. $\{2\}$; $(-\infty, \infty)$

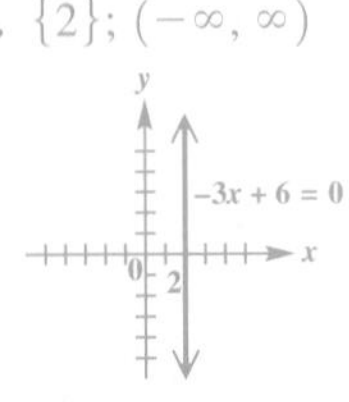

33. *Concept Check* If a walkway rises 2.5 ft for every 10 ft on the horizontal, which of the following express its slope (or grade)? (There are several correct choices.)

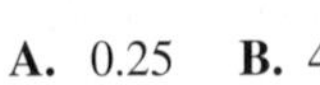

A. 0.25 B. 4 C. $\frac{2.5}{10}$ D. 25%

E. $\frac{1}{4}$ F. $\frac{10}{2.5}$ G. 400% H. 2.5%

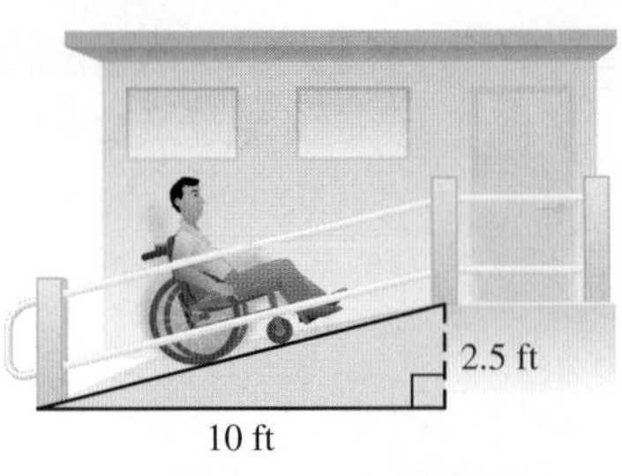

34. *Concept Check* If the pitch of a roof is $\frac{1}{4}$, how many feet in the horizontal direction correspond to a rise of 4 ft?

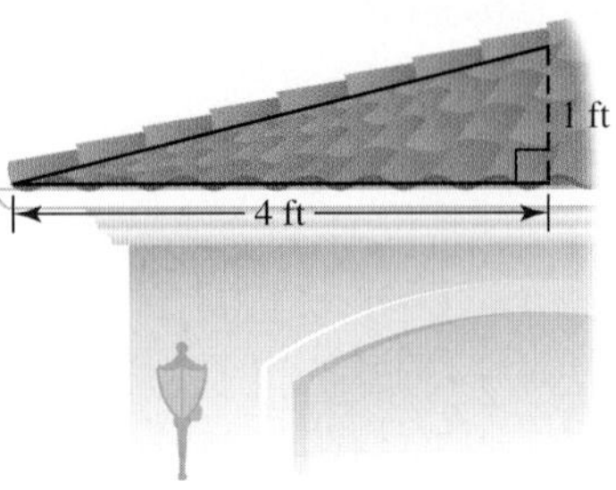

Find the slope of the line satisfying the given conditions. ***See Example 5.***

35. through $(2, -1)$ and $(-3, -3)$

36. through $(-3, 4)$ and $(2, -8)$

37. through $(5, 8)$ and $(3, 12)$

38. through $(5, -3)$ and $(1, -7)$

39. through $(5, 9)$ and $(-2, 9)$

40. through $(-2, 4)$ and $(6, 4)$

41. horizontal, through $(5, 1)$

42. horizontal, through $(3, 5)$

43. vertical, through $(4, -7)$

44. vertical, through $(-8, 5)$

For each line, ***(a)*** *find the slope and* ***(b)*** *sketch the graph.* ***See Examples 6 and 7.***

45. $y = 3x + 5$

46. $y = 2x - 4$

47. $2y = -3x$

48. $-4y = 5x$

49. $5x - 2y = 10$

50. $4x + 3y = 12$

Graph the line passing through the given point and having the indicated slope. Plot two points on the line. ***See Example 7.***

51. through $(-1, 3)$, $m = \frac{3}{2}$

52. through $(-2, 8)$, $m = \frac{2}{5}$

53. through $(3, -4)$, $m = -\frac{1}{3}$

54. through $(-2, -3)$, $m = -\frac{3}{4}$

55. through $\left(-\frac{1}{2}, 4\right)$, $m = 0$

56. through $\left(\frac{3}{2}, 2\right)$, $m = 0$

57. through $\left(-\frac{5}{2}, 3\right)$, undefined slope

58. through $\left(\frac{9}{4}, 2\right)$, undefined slope

Concept Check For each given slope in Exercises 59–64, identify the line in A–F that could have this slope.

59. $\frac{1}{3}$

60. -3

61. 0

62. $-\frac{1}{3}$

63. 3

64. undefined

A.

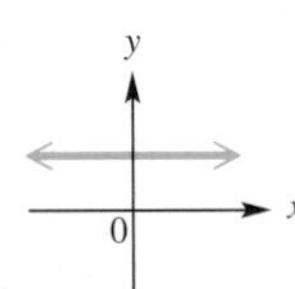

B.

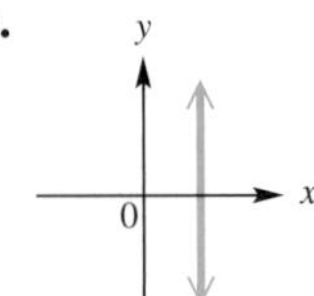

C.

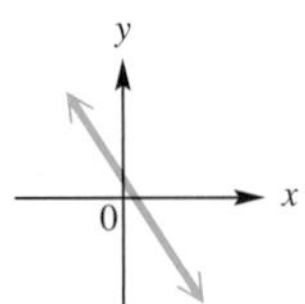

D.

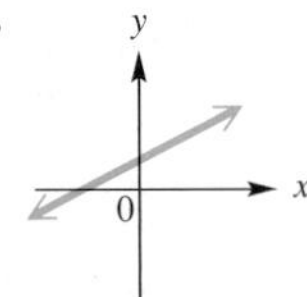

E.

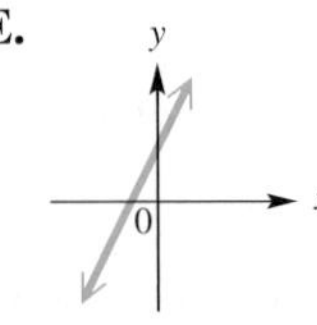

F.

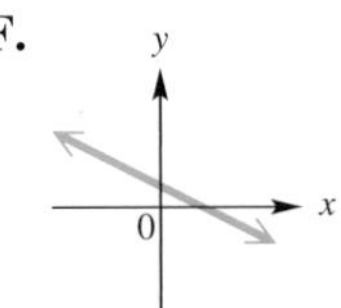

23. $\{5\}$; $(-\infty, \infty)$

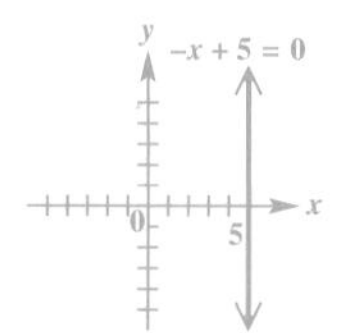

24. $\{-3\}$; $(-\infty, \infty)$

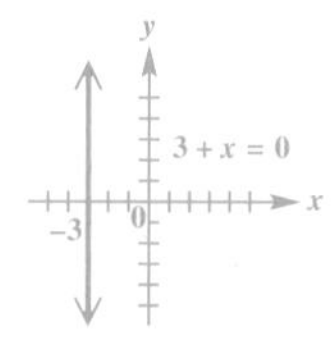

25. A 26. C 27. D 28. B

29. $y = 3x + 4$

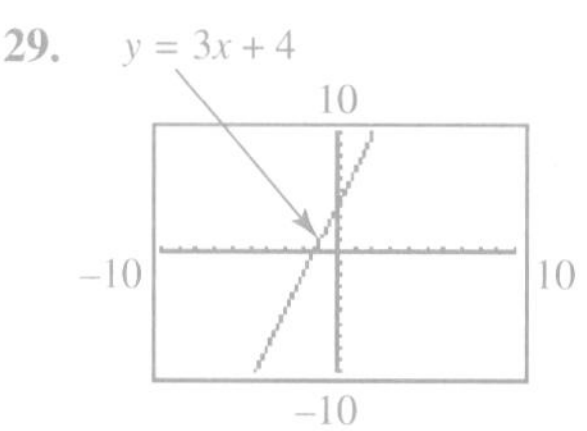

30. $y = -2x + 3$

10

-10 10

-10

31. $3x + 4y = 6$

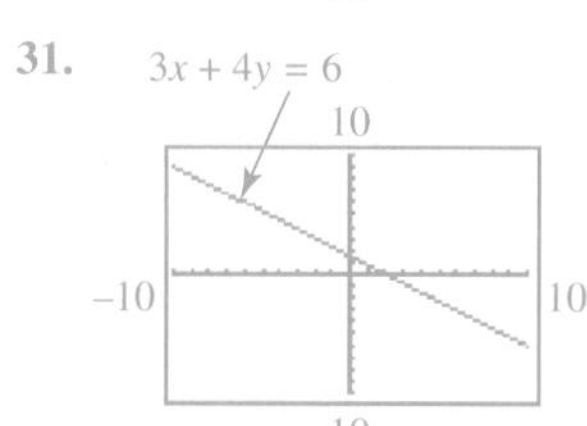

32. $-2x + 5y = 10$

10

-10 10

-10

33. A, C, D, E 34. 16 ft

35. $\frac{2}{5}$ 36. $-\frac{12}{5}$

37. -2 38. 1

39. 0 40. 0

41. 0 42. 0

43. undefined 44. undefined

45. (a) $m = 3$

(b)

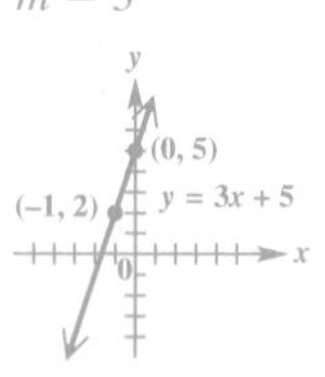

Concept Check *Find and interpret the average rate of change illustrated in each graph.*

65.

y

Value of Machine (in thousands of dollars)

20 16 12 8 4

0 1 2 3 4 x

Year

66.

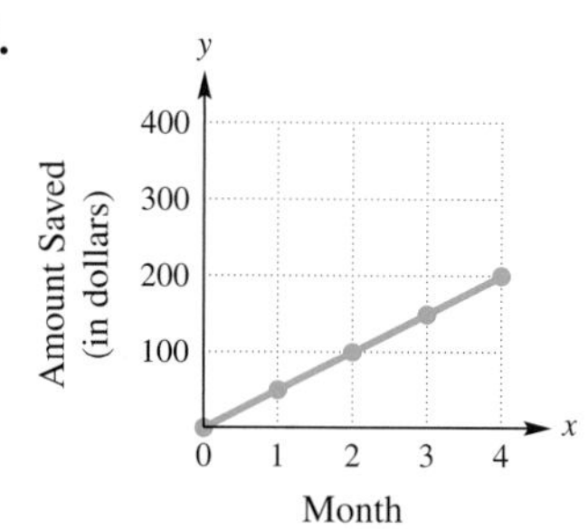

67.

y

Percent of Pay Raise

8 6 4 2

0 1 2 3 4 x

Year

68.

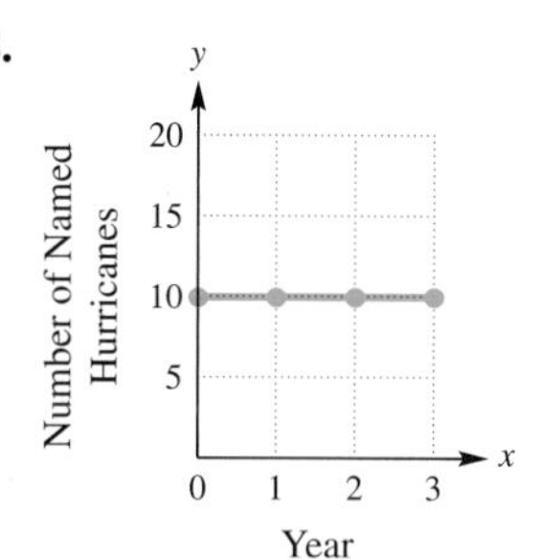

69. *Concept Check* For a constant function, is the average rate of change positive, negative, or zero? (Choose one.)

Solve each problem. ***See Example 8.***

70. *(Modeling) Olympic Times for 5000-Meter Run* The graph shows the winning times (in minutes) at the Olympic Games for the men's 5000-m run, together with a linear approximation of the data.

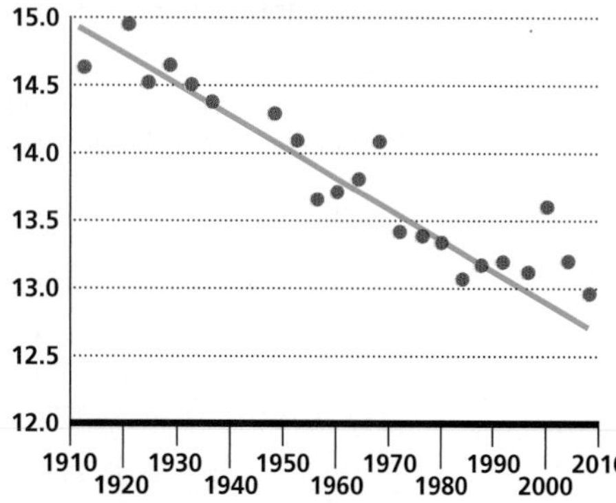

Source: World Almanac and Book of Facts.

(a) An equation for the linear model, based on data from 1912–2008 (where x represents the year), is

$$y = -0.0193x + 51.73.$$

Determine the slope. (**See Example 6.**) What does the slope of this line represent? Why is the slope negative?

(b) Can you think of any reason why there are no data points for the years 1916, 1940, and 1944?

(c) The winning time for the 1996 Olympic Games was 13.13 min. What does the model predict? How far is the prediction from the actual value?

46. (a) $m = 2$
(b)
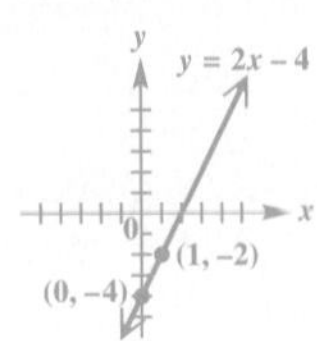

47. (a) $m = -\frac{3}{2}$
(b)
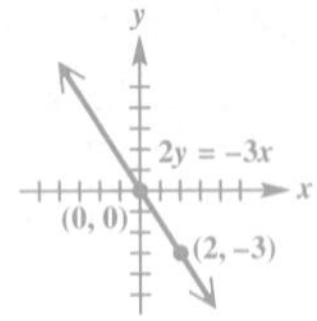

48. (a) $m = -\frac{5}{4}$
(b)
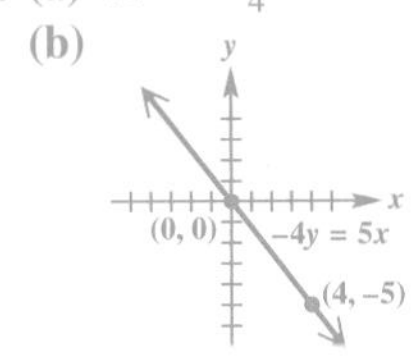

49. (a) $m = \frac{5}{2}$
(b)
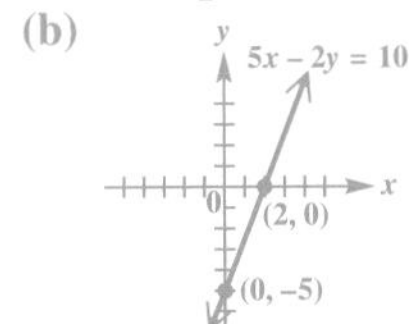

50. (a) $m = -\frac{4}{3}$
(b)
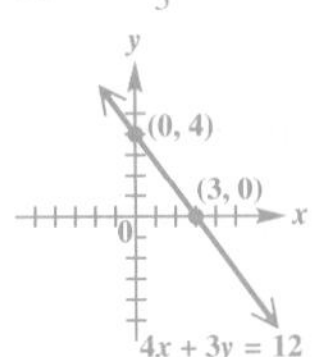

51.
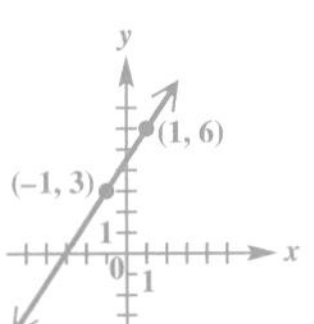

52.
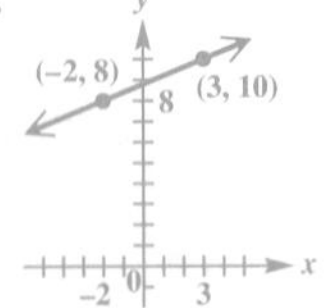

53.
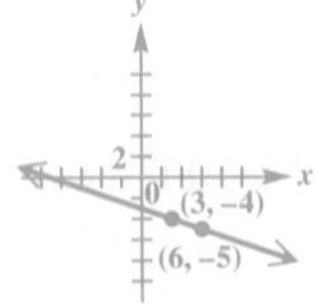

54.
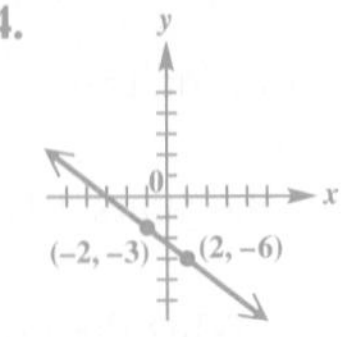

71. *(Modeling) U.S. Radio Stations* The graph shows the number of U.S. radio stations on the air, along with the graph of a linear function that models the data.

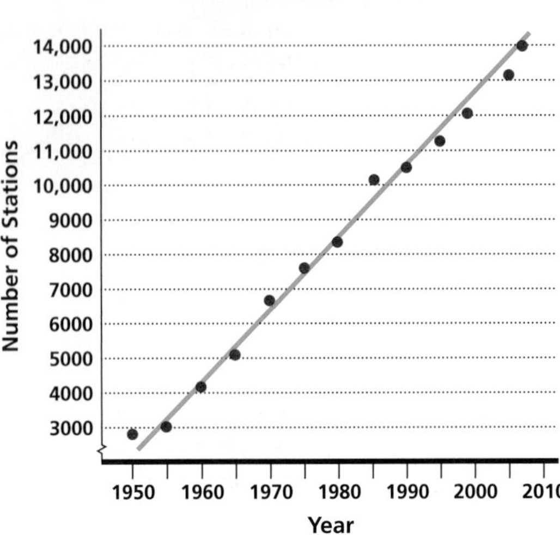

Source: National Association of Broadcasters.

(a) Discuss the predictive accuracy of the linear function.

(b) Use the two data points $(1950, 2773)$ and $(2007, 13{,}977)$ to find the approximate slope of the line shown. Interpret this number.

72. *Cellular Telephone Subscribers* The table gives the number of cellular telephone subscribers in the U.S. (in thousands) from 2004 through 2009.

(a) Find the change in subscribers for 2004–2005, 2005–2006, and so on.

(b) Are the changes in successive years approximately the same? If the ordered pairs in the table were plotted, could an approximately straight line be drawn through them?

Year	Subscribers (in thousands)
2004	182,140
2005	207,896
2006	233,041
2007	255,396
2008	270,334
2009	285,646

Source: CTIA-The Wireless Association.

73. *Mobile Homes* The graph provides a good approximation of the number of mobile homes (in thousands) placed in use in the United States from 1999 through 2009.

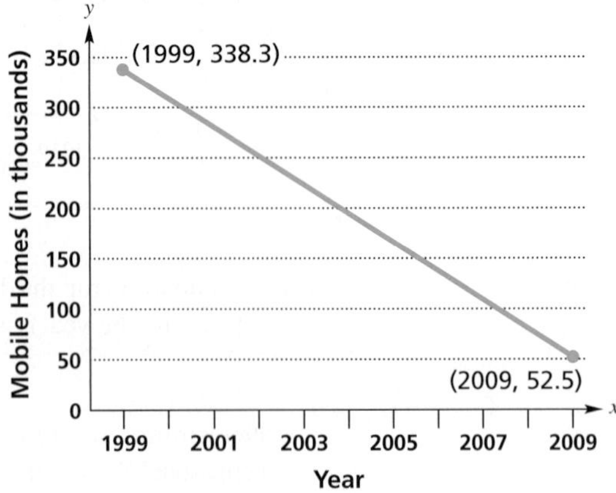

Source: U.S. Census Bureau.

(a) Use the given ordered pairs to find the average rate of change in the number of mobile homes per year during this period.

(b) Interpret what a negative slope means in this situation.

74. *Olympic Times for 5000-Meter Run* **Exercise 70** showed the winning times for the Olympic men's 5000-m run. Find and interpret the average rate of change for the following periods.

(a) The winning time in 1912 was 14.6 min. In 1996 it was 13.1 min.

(b) The winning time in 1912 was 14.6 min. In 2008 it was 13.0 min.

75. *High School Dropouts* In 1980, the number of high school dropouts in the United States was 5212 thousand. By 2008, this number had decreased to 3118 thousand. Find and interpret the average rate of change per year in the number of high school dropouts. (*Source:* U.S. Census Bureau.)

76. *Plasma Flat-Panel TV Sales* The total amount spent on plasma flat-panel TVs in the United States changed from \$5302 million in 2006 to \$2907 million in 2009. Find and interpret the average rate of change in sales, in millions of dollars per year. Round your answer to the nearest hundredth. (*Source:* Consumer Electronics Association.)

Relating Concepts

For individual or collaborative investigation *(Exercises 77–86)*

The table shows several points on the graph of a linear function. ***Work Exercises 77–86 in order,*** *to see connections between the slope formula, the distance formula, the midpoint formula, and linear functions.*

x	y
0	-6
1	-3
2	0
3	3
4	6
5	9
6	12

77. Use the first two points in the table to find the slope of the line.

78. Use the second and third points in the table to find the slope of the line.

79. Make a conjecture by filling in the blank: If we use any two points on a line to find its slope, we find that the slope is ______ in all cases.

80. Find the distance between the first two points in the table. (*Hint:* Use the distance formula.)

81. Find the distance between the second and fourth points in the table.

82. Find the distance between the first and fourth points in the table.

83. Add the results in **Exercises 80 and 81,** and compare the sum to the answer you found in **Exercise 82.** What do you notice?

84. Fill in the following blanks, basing your answers on your observations in **Exercises 80–83:** If points A, B, and C lie on a line in that order, then the distance between A and B added to the distance between ______ and ______ is equal to the distance between ______ and ______.

85. Find the midpoint of the segment joining $(0, -6)$ and $(6, 12)$. Compare your answer to the middle entry in the table. What do you notice?

86. If the table were set up to show an x-value of 4.5, what would be the corresponding y-value?

55.

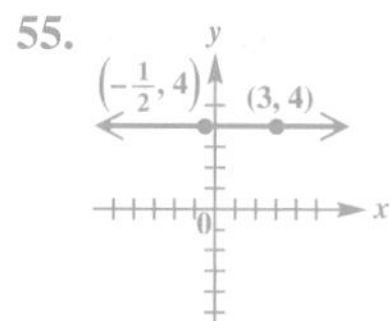

56.

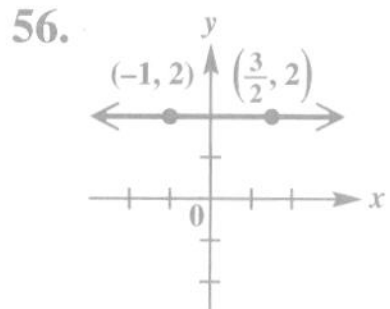

57.

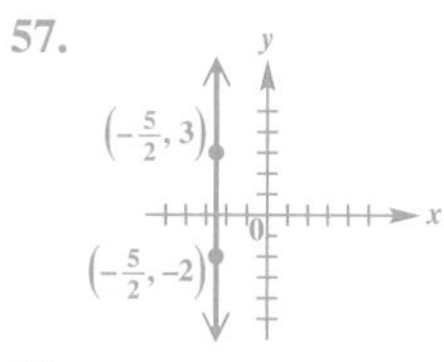

58.

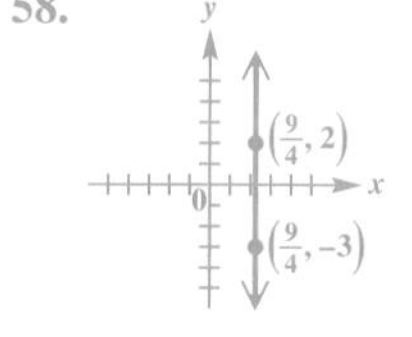

59. D **60.** C **61.** A
62. F **63.** E **64.** B

65. $-\$4000$ per yr; The value of the machine is decreasing \$4000 each year during these years.
66. \$50 per month; The amount saved is increasing \$50 each month during these months.
67. 0% per yr (or no change); The percent of pay raise is not changing—it is 3% each year during these years.
68. 10 named hurricanes each year (or no change); For four consecutive years, there were 10 named hurricanes.
69. zero

70. **(a)** The slope of -0.0193 indicates that the average rate of change per year of the winning time for the 5000-m run is 0.0193 min less (faster). It is negative because the times are generally decreasing as time progresses.
(b) The Olympics were not held during World Wars I and II.
(c) about 13.21 min; The times differ by 0.08 min.

71. **(b)** 196.6; This means that the average rate of change in the number of radio stations per year is an increase of 196.6.

72. **(a)** 25,756 thousand; 25,145 thousand; 22,355 thousand; 14,938 thousand; 15,312 thousand **(b)** no; no **73.** **(a)** -28.58 thousand mobile homes per yr
(b) The negative slope means that the number of mobile homes *decreased* by an average of 28.58 thousand each year from 1999 to 2009. **74.** **(a)** -0.0179 min per yr; The winning time decreased an average of 0.0179 min each year from 1912 to 1996.
(b) -0.0167 min per yr; The winning time decreased an average of 0.0167 min each year from 1912 to 2008.

75. -74.8 thousand per yr; The number of high school dropouts decreased by an average of 74.8 thousand per yr from 1980 to 2008.
76. $-\$798.33$ million per yr; Sales of plasma flat-panel TVs decreased by an average of \$798.33 million per yr from 2006 to 2009.

77. 3
78. 3
79. the same
80. $\sqrt{10}$
81. $2\sqrt{10}$
82. $3\sqrt{10}$
83. The sum is $3\sqrt{10}$, which is equal to the answer in Exercise 82.
84. *B*; *C*; *A*; *C*
85. The midpoint is $(3, 3)$, which is the same as the middle entry in the table.
86. 7.5

87. (a) $C(x) = 10x + 500$ (b) $R(x) = 35x$ (c) $P(x) = 25x - 500$ (d) 20 units; do not produce
88. (a) $C(x) = 150x + 2700$ (b) $R(x) = 280x$ (c) $P(x) = 130x - 2700$ (d) 20.77 or 21 units; produce
89. (a) $C(x) = 400x + 1650$ (b) $R(x) = 305x$ (c) $P(x) = -95x - 1650$

(Modeling) Cost, Revenue, and Profit Analysis A firm will break even (no profit and no loss) as long as revenue just equals cost. The value of x (the number of items produced and sold) where $C(x) = R(x)$ is called the ***break-even point.*** *Assume that each of the following can be expressed as a linear function. Find*

(a) *the cost function,* ***(b)*** *the revenue function, and* ***(c)*** *the profit function.*

(d) *Find the break-even point and decide whether the product should be produced, given the restrictions on sales.*

See Example 9.

	Fixed Cost	Variable Cost	Price of Item	
87.	\$ 500	\$ 10	\$ 35	No more than 18 units can be sold.
88.	\$2700	\$150	\$280	No more than 25 units can be sold.
89.	\$1650	\$400	\$305	All units produced can be sold.
90.	\$ 180	\$ 11	\$ 20	No more than 30 units can be sold.

(Modeling) Break-Even Point The manager of a small company that produces roof tile has determined that the total cost in dollars, $C(x)$, of producing x units of tile is given by

$$C(x) = 200x + 1000,$$

while the revenue in dollars, $R(x)$, from the sale of x units of tile is given by

$$R(x) = 240x.$$

91. Find the break-even point and the cost and revenue at the break-even point.

92. Suppose the variable cost is actually \$220 per unit, instead of \$200. How does this affect the break-even point? Is the manager better off or not?

89. (d) $R(x) < C(x)$ for all positive x; don't produce, impossible to make a profit
90. (a) $C(x) = 11x + 180$ (b) $R(x) = 20x$ (c) $P(x) = 9x - 180$ (d) 20 units; produce 91. 25 units; \$6000 92. The break-even point is 50 units instead of 25 units. The manager is not better off, because twice as many units must be sold before beginning to show a profit.

Chapter 2 Quiz (Sections 2.1–2.4)

[2.1]
1. $\sqrt{41}$
2. 2002: 6.25 million; 2004: 6.76 million
3.

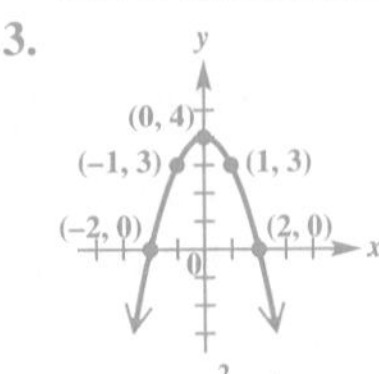

1. For $A(-4, 2)$ and $B(-8, -3)$, find $d(A, B)$, the distance between A and B.

2. *Two-Year College Enrollment* Enrollments in two-year colleges for selected years are shown in the table. Use the midpoint formula to estimate the enrollments for 2002 and 2006.

Year	Enrollment (in millions)
2000	5.95
2004	6.55
2008	6.97

Source: U.S. Center for Education Statistics.

3. Sketch the graph of $y = -x^2 + 4$ by plotting points.

4. Sketch the graph of $x^2 + y^2 = 16$.

5. Determine the radius and the coordinates of the center of the circle with equation

$$x^2 + y^2 - 4x + 8y + 3 = 0.$$

[2.2]

4.

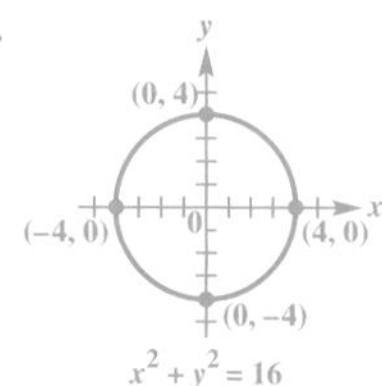

5. radius: $\sqrt{17}$; center: $(2, -4)$

[2.3]

6. 2

7. domain: $(-\infty, \infty)$; range: $[0, \infty)$

8. (a) $(-\infty, -3]$ (b) $[-3, \infty)$ (c) none

[2.4]

9. (a) $\frac{3}{2}$ (b) 0 (c) undefined

10. -1711 thousand per yr; The number of new motor vehicles sold in the United States decreased by an average of 1711 thousand per yr from 2005 to 2009.

For Exercises 6–8, refer to the graph of $f(x) = |x + 3|$.

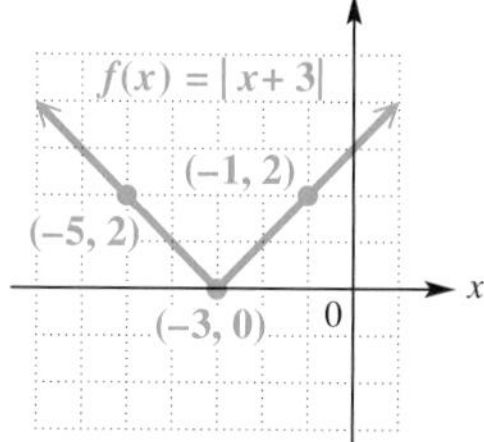

6. Find $f(-1)$.

7. Give the domain and the range of f.

8. Give the largest interval over which the function f is **(a)** decreasing, **(b)** increasing, **(c)** constant.

9. Find the slope of the line through the given points.

(a) $(1, 5)$ and $(5, 11)$ **(b)** $(-7, 4)$ and $(-1, 4)$ **(c)** $(6, 12)$ and $(6, -4)$

10. *Motor Vehicle Sales* The graph shows a straight line segment that approximates new motor vehicle sales in the United States from 2005 to 2009. Determine the average rate of change from 2005 to 2009, and interpret the results.

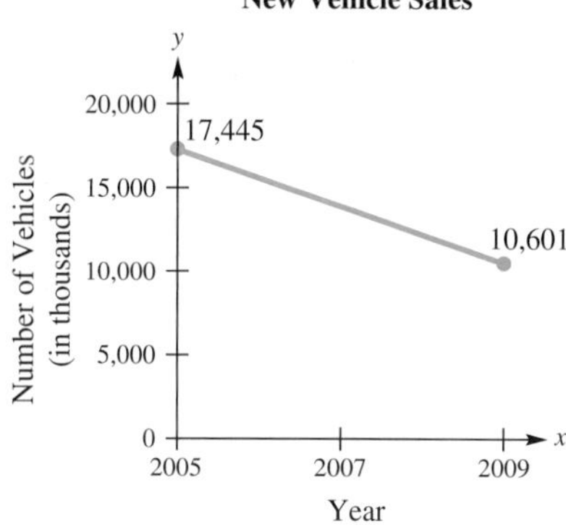

Source: U.S. Bureau of Economic Analysis.

2.5 Equations of Lines and Linear Models

- Point-Slope Form
- Slope-Intercept Form
- Vertical and Horizontal Lines
- Parallel and Perpendicular Lines
- Modeling Data
- Solving Linear Equations in One Variable by Graphing

Point-Slope Form The graph of a linear function is a straight line. We now develop various forms for the equation of a line.

Figure 43 shows the line passing through the fixed point (x_1, y_1) having slope m. (Assuming that the line has a slope guarantees that it is not vertical.) Let (x, y) be any other point on the line. Since the line is not vertical, $x - x_1 \neq 0$. Now use the definition of slope.

$$m = \frac{y - y_1}{x - x_1} \quad \text{Slope formula (Section 2.4)}$$

$$m(x - x_1) = y - y_1 \quad \text{Multiply each side by } x - x_1.$$

or

$$y - y_1 = m(x - x_1) \quad \text{Interchange sides.}$$

This result is the *point-slope form* of the equation of a line.

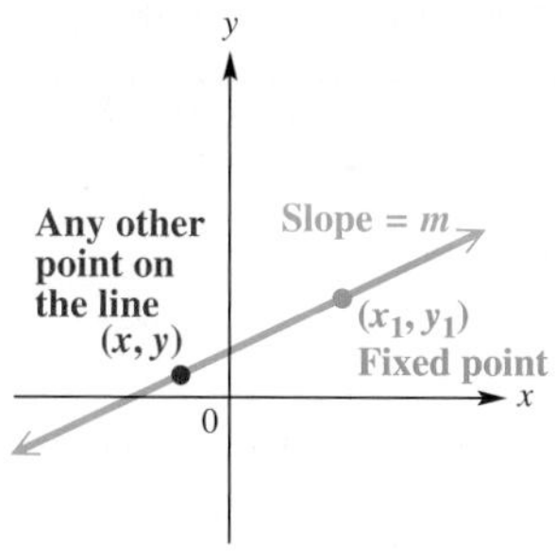

Figure 43

Point-Slope Form

The **point-slope form** of the equation of the line with slope m passing through the point (x_1, y_1) is

$$y - y_1 = m(x - x_1).$$

LOOKING AHEAD TO CALCULUS

A standard problem in calculus is to find the equation of the line tangent to a curve at a given point. The derivative (see *Looking Ahead to Calculus* in **Section 2.4**) is used to find the slope of the desired line, and then the slope and the given point are used in the point-slope form to solve the problem.

EXAMPLE 1 Using the Point-Slope Form (Given a Point and the Slope)

Write an equation of the line through $(-4, 1)$ having slope -3.

SOLUTION Here $x_1 = -4$, $y_1 = 1$, and $m = -3$.

$$y - y_1 = m(x - x_1) \quad \text{Point-slope form}$$

$$y - 1 = -3[x - (-4)] \quad x_1 = -4,\ y_1 = 1,\ m = -3$$

$$y - 1 = -3(x + 4) \quad \text{Be careful with signs.}$$

$$y - 1 = -3x - 12 \quad \text{Distributive property (Section R.2)}$$

$$y = -3x - 11 \quad \text{Add 1. (Section 1.1)}$$

Now Try Exercise 23.

Classroom Example 1
Write an equation of the line through $(3, -5)$ having slope -2.

Answer: $y = -2x + 1$

EXAMPLE 2 Using the Point-Slope Form (Given Two Points)

Write an equation of the line through $(-3, 2)$ and $(2, -4)$. Write the result in standard form $Ax + By = C$.

SOLUTION Find the slope first.

$$m = \frac{-4 - 2}{2 - (-3)} = -\frac{6}{5} \quad \text{Definition of slope}$$

The slope m is $-\frac{6}{5}$. Either $(-3, 2)$ or $(2, -4)$ can be used for (x_1, y_1). We choose $(-3, 2)$.

$$y - y_1 = m(x - x_1) \quad \text{Point-slope form}$$

$$y - 2 = -\frac{6}{5}[x - (-3)] \quad x_1 = -3,\ y_1 = 2,\ m = -\tfrac{6}{5}$$

$$5(y - 2) = -6(x + 3) \quad \text{Multiply by 5.}$$

$$5y - 10 = -6x - 18 \quad \text{Distributive property}$$

$$6x + 5y = -8 \quad \text{Standard form (Section 2.4)}$$

Verify that we obtain the same equation if we use $(2, -4)$ instead of $(-3, 2)$ in the point-slope form.

Now Try Exercise 13.

Classroom Example 2
Write an equation of the line through $(-4, 3)$ and $(5, -1)$. Write the result in standard form $Ax + By = C$.

Answer: $4x + 9y = 11$

NOTE The lines in **Examples 1 and 2** both have negative slopes. Keep in mind that a slope of the form $-\frac{A}{B}$ may be interpreted as either $\frac{-A}{B}$ or $\frac{A}{-B}$.

Slope-Intercept Form As a special case of the point-slope form of the equation of a line, suppose that a line passes through the point $(0, b)$, so the line has y-intercept b. If the line has slope m, then using the point-slope form with $x_1 = 0$ and $y_1 = b$ gives the following.

$$y - y_1 = m(x - x_1) \quad \text{Point-slope form}$$

$$y - b = m(x - 0) \quad x_1 = 0,\ y_1 = b$$

$$y = mx + b \quad \text{Solve for } y.$$

Slope (m) — y-intercept (b)

Since this result shows the slope of the line and the y-intercept, it is called the *slope-intercept form* of the equation of the line.

Teaching Tip Mention that for many applications, equations of lines must first be put into slope-intercept form.

Slope-Intercept Form

The **slope-intercept form** of the equation of the line with slope m and y-intercept b is

$$y = mx + b.$$

Classroom Example 3 Find the slope and y-intercept of the line with equation $3x - 4y = 12$.

Answer: slope: $\frac{3}{4}$; y-intercept: -3

EXAMPLE 3 Finding the Slope and y-Intercept from an Equation of a Line

Find the slope and y-intercept of the line with equation $4x + 5y = -10$.

SOLUTION Write the equation in slope-intercept form.

$$4x + 5y = -10$$

$$5y = -4x - 10 \quad \text{Subtract } 4x.$$

$$y = -\frac{4}{5}x - 2 \quad \text{Divide by 5.}$$

$$\uparrow m \qquad \uparrow b$$

The slope is $-\frac{4}{5}$ and the y-intercept is -2.

Now Try Exercise 35.

NOTE Generalizing from **Example 3,** the slope m of the graph of

$$Ax + By = C$$

is $-\frac{A}{B}$, and the y-intercept b is $\frac{C}{B}$.

Classroom Example 4 Write an equation of the line through $(-2, 4)$ and $(2, 2)$. Then graph the line using the slope-intercept form.

Answer: $y = -\frac{1}{2}x + 3$

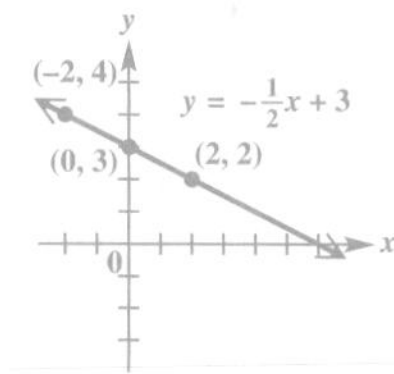

EXAMPLE 4 Using the Slope-Intercept Form (Given Two Points)

Write an equation of the line through $(1, 1)$ and $(2, 4)$. Then graph the line using the slope-intercept form.

SOLUTION In **Example 2,** we used the *point-slope form* in a similar problem. Here we show an alternative method using the *slope-intercept form.* First, find the slope.

$$m = \frac{4 - 1}{2 - 1} = \frac{3}{1} = 3 \quad \text{Definition of slope}$$

Now substitute 3 for m in $y = mx + b$ and choose one of the given points, say $(1, 1)$, to find the value of b.

$$y = mx + b \quad \text{Slope-intercept form}$$

$$1 = 3(1) + b \quad m = 3, x = 1, y = 1$$

$$y\text{-intercept} \rightarrow b = -2 \quad \text{Solve for } b.$$

The slope-intercept form is

$$y = 3x - 2.$$

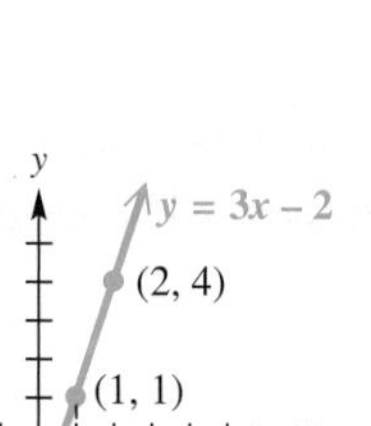

Figure 44

The graph is shown in **Figure 44.** We can plot $(0, -2)$ and then use the definition of slope to arrive at $(1, 1)$. Verify that $(2, 4)$ also lies on the line.

Now Try Exercise 13.

Classroom Example 5
Use the graph of the linear function f shown here to complete the following.

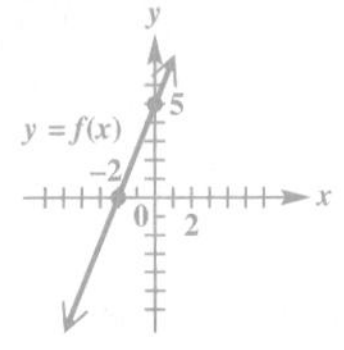

(a) Find the slope, y-intercept, and x-intercept.
(b) Write the equation defining f.

Answers:
(a) slope: $\frac{5}{2}$; y-intercept: 5; x-intercept: -2
(b) $f(x) = \frac{5}{2}x + 5$

EXAMPLE 5 Finding an Equation from a Graph

Use the graph of the linear function f shown in **Figure 45** to complete the following.

(a) Find the slope, y-intercept, and x-intercept.

(b) Write the equation that defines f.

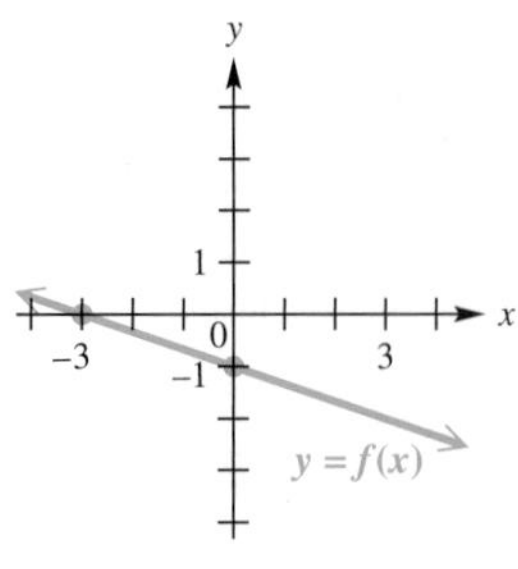

Figure 45

SOLUTION

(a) The line falls 1 unit each time the x-value increases by 3 units. Therefore, the slope is $\frac{-1}{3} = -\frac{1}{3}$. The graph intersects the y-axis at the point $(0, -1)$ and intersects the x-axis at the point $(-3, 0)$. Therefore, the y-intercept is -1 and the x-intercept is -3.

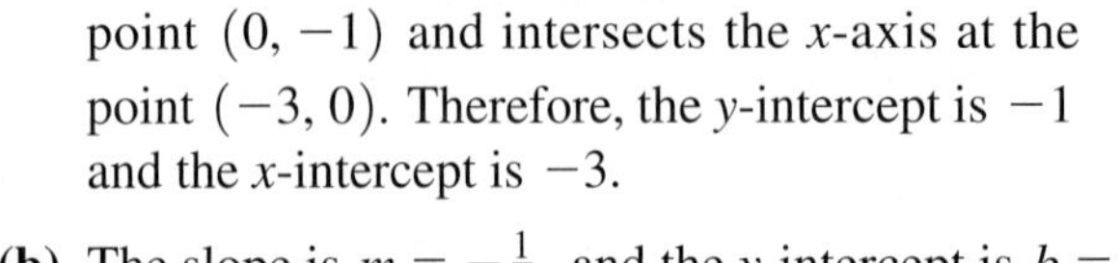

(b) The slope is $m = -\frac{1}{3}$, and the y-intercept is $b = -1$.

$$y = f(x) = mx + b \qquad \text{Slope-intercept form}$$

$$f(x) = -\frac{1}{3}x - 1 \qquad m = -\tfrac{1}{3},\ b = -1$$

Now Try Exercise 39.

Vertical and Horizontal Lines We first saw graphs of vertical and horizontal lines in **Section 2.4.** The vertical line through the point (a, b) passes through all points of the form (a, y), for any value of y. Consequently, the equation of a vertical line through (a, b) is $x = a$. For example, the vertical line through $(-3, 1)$ has equation $x = -3$. See **Figure 46(a).** Since each point on the y-axis has x-coordinate 0, ***the equation of the y-axis is x = 0.***

The horizontal line through the point (a, b) passes through all points of the form (x, b), for any value of x. Therefore, the equation of a horizontal line through (a, b) is $y = b$. For example, the horizontal line through $(1, -3)$ has equation $y = -3$. See **Figure 46(b).** Since each point on the x-axis has y-coordinate 0, ***the equation of the x-axis is y = 0.***

Vertical line
(−3, 1)
$x = -3$
(a)

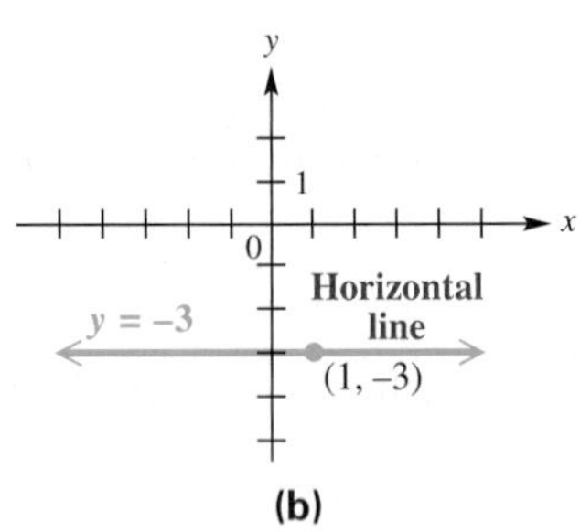

(b)

Figure 46

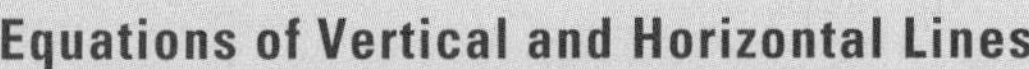

Equations of Vertical and Horizontal Lines

An equation of the **vertical line** through the point (a, b) is $x = a$.

An equation of the **horizontal line** through the point (a, b) is $y = b$.

Parallel and Perpendicular Lines Since two parallel lines are equally "steep," they should have the same slope. Also, two distinct lines with the same "steepness" are parallel. The following result summarizes this discussion. (The statement "p if and only if q" means "if p then q *and* if q then p.")

Parallel Lines

Two distinct nonvertical lines are parallel if and only if they have the same slope.

When two lines have slopes with a product of -1, the lines are perpendicular.

Perpendicular Lines

Two lines, neither of which is vertical, are perpendicular if and only if their slopes have a product of -1. Thus, the slopes of perpendicular lines, neither of which is vertical, are *negative reciprocals.*

For example, if the slope of a line is $-\frac{3}{4}$, the slope of any line perpendicular to it is $\frac{4}{3}$, since $-\frac{3}{4}\left(\frac{4}{3}\right) = -1$. (Numbers like $-\frac{3}{4}$ and $\frac{4}{3}$ are **negative reciprocals** of each other.) A proof of this result is outlined in **Exercises 69–75.**

NOTE Because a vertical line has *undefined* slope, it does not follow the *mathematical* rules for parallel and perpendicular lines. We intuitively know that all vertical lines are parallel and that a vertical line and a horizontal line are perpendicular.

Classroom Example 6
Write the equation in both slope-intercept and standard form of the line that passes through the point $(2, -4)$ and satisfies the given condition.

(a) parallel to the line $3x - 2y = 5$

(b) perpendicular to the line $3x - 2y = 5$

Answers:

(a) $y = \frac{3}{2}x - 7$; $3x - 2y = 14$

(b) $y = -\frac{2}{3}x - \frac{8}{3}$; $2x + 3y = -8$

EXAMPLE 6 Finding Equations of Parallel and Perpendicular Lines

Write the equation in both slope-intercept and standard form of the line that passes through the point $(3, 5)$ and satisfies the given condition.

(a) parallel to the line $2x + 5y = 4$

(b) perpendicular to the line $2x + 5y = 4$

SOLUTION

(a) Since we know that the point $(3, 5)$ is on the line, we need only find the slope to use the point-slope form. We find the slope by writing the equation of the given line in slope-intercept form. (That is, we solve for y.)

$$2x + 5y = 4$$

$$5y = -2x + 4 \quad \text{Subtract } 2x.$$

$$y = -\frac{2}{5}x + \frac{4}{5} \quad \text{Divide by 5.}$$

The slope is $-\frac{2}{5}$. Since the lines are parallel, $-\frac{2}{5}$ is also the slope of the line whose equation is to be found. Now substitute this slope and the given point $(3, 5)$ in the point-slope form.

$$y - y_1 = m(x - x_1) \quad \text{Point-slope form}$$

$$y - 5 = -\frac{2}{5}(x - 3) \quad m = -\frac{2}{5}, x_1 = 3, y_1 = 5$$

$$y - 5 = -\frac{2}{5}x + \frac{6}{5} \quad \text{Distributive property}$$

$$\text{Slope-intercept form} \rightarrow y = -\frac{2}{5}x + \frac{31}{5} \quad \text{Add } 5 = \frac{25}{5}.$$

$$5y = -2x + 31 \quad \text{Multiply by 5.}$$

$$\text{Standard form} \rightarrow 2x + 5y = 31 \quad \text{Add } 2x.$$

(b) There is no need to find the slope again, because in part (a) we found that the slope of the line $2x + 5y = 4$ is $-\frac{2}{5}$. The slope of any line perpendicular to it is $\frac{5}{2}$.

$$y - y_1 = m(x - x_1) \quad \text{Point-slope form}$$

$$y - 5 = \frac{5}{2}(x - 3) \quad m = \tfrac{5}{2},\ x_1 = 3,\ y_1 = 5$$

$$y - 5 = \frac{5}{2}x - \frac{15}{2} \quad \text{Distributive property}$$

$$\text{Slope-intercept form} \rightarrow y = \frac{5}{2}x - \frac{5}{2} \quad \text{Add } 5 = \tfrac{10}{2}.$$

$$2y = 5x - 5 \quad \text{Multiply by 2.}$$

$$\text{Standard form} \rightarrow 5x - 2y = 5 \quad \text{Subtract } 2y \text{, add 5, and rewrite.}$$

✔ ***Now Try Exercises 45 and 47.***

We can use a graphing calculator to support the results of **Example 6.** In **Figure 47(a),** we graph the equations of the parallel lines

$$y = -\frac{2}{5}x + \frac{4}{5} \quad \text{and} \quad y = -\frac{2}{5}x + \frac{31}{5}$$

from **Example 6(a).** The lines appear to be parallel, giving visual support for our result. We must use caution, however, when viewing such graphs, as the limited resolution of a graphing calculator screen may cause two lines to *appear* to be parallel even when they are not. For example, **Figure 47(b)** shows the graphs of the equations

$$y = 2x + 6 \quad \text{and} \quad y = 2.01x - 3$$

in the standard viewing window, and they appear to be parallel. This is not the case, however, because their slopes, 2 and 2.01, are different.

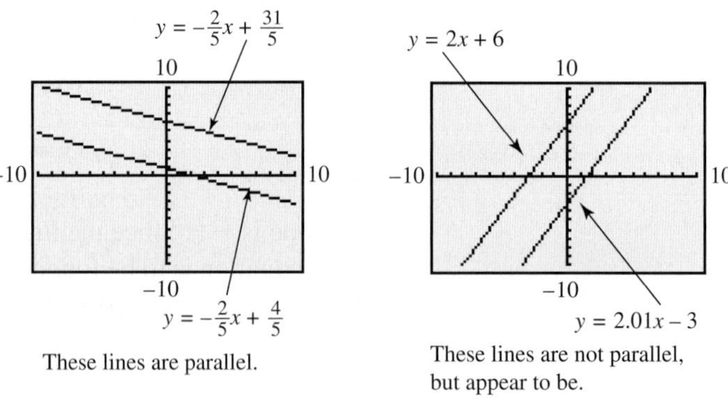

Figure 47

To support the result of **Example 6(b),** we graph the equations of the perpendicular lines

$$y = -\frac{2}{5}x + \frac{4}{5} \quad \text{and} \quad y = \frac{5}{2}x - \frac{5}{2}.$$

If we use the standard viewing window, the lines do not appear to be perpendicular. See **Figure 48(a).** To obtain the correct perspective, we must use a square viewing window, as in **Figure 48(b).**

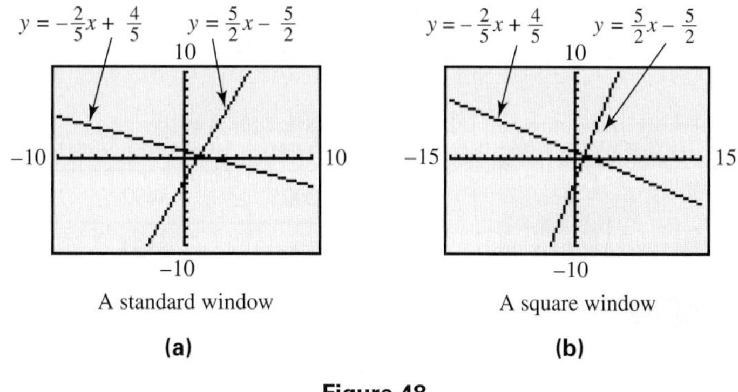

Figure 48

A summary of the various forms of linear equations follows.

Equation	Description	When to Use
$y = mx + b$	**Slope-Intercept Form** Slope is m. y-intercept is b.	The slope and y-intercept can be easily identified and used to quickly graph the equation. This form can also be used to find the equation of a line given a point and the slope.
$y - y_1 = m(x - x_1)$	**Point-Slope Form** Slope is m. Line passes through (x_1, y_1).	This form is ideal for finding the equation of a line if the slope and a point on the line or two points on the line are known.
$Ax + By = C$	**Standard Form** (If the coefficients and constant are rational, then A, B, and C are expressed as relatively prime integers, with $A \geq 0$.) Slope is $-\frac{A}{B}$ $(B \neq 0)$. x-intercept is $\frac{C}{A}$ $(A \neq 0)$. y-intercept is $\frac{C}{B}$ $(B \neq 0)$.	The x- and y-intercepts can be found quickly and used to graph the equation. The slope must be calculated.
$y = b$	**Horizontal Line** Slope is 0. y-intercept is b.	If the graph intersects only the y-axis, then y is the only variable in the equation.
$x = a$	**Vertical Line** Slope is undefined. x-intercept is a.	If the graph intersects only the x-axis, then x is the only variable in the equation.

Modeling Data We can write equations of lines that mathematically describe, or model, real data if the data change at a fairly constant rate. In this case, the data fit a linear pattern, and the rate of change is the slope of the line.

Classroom Example 7
Repeat **Example 7** using the data points $(1, 5804)$ and $(4, 7050)$.

Answers:
(a) $y = 415.3x + 5389$
(b) \$8296

EXAMPLE 7 Finding an Equation of a Line That Models Data

Average annual tuition and fees for in-state students at public four-year colleges are shown in the table for selected years and graphed as ordered pairs of points in **Figure 49,** where $x = 0$ represents 2005, $x = 1$ represents 2006, and so on, and y represents the cost in dollars. This graph of ordered pairs of data is called a **scatter diagram.**

Year	Cost (in dollars)
2005	5492
2006	5804
2007	6191
2008	6591
2009	7050
2010	7605

Source: Trends in College Pricing 2010, The College Board.

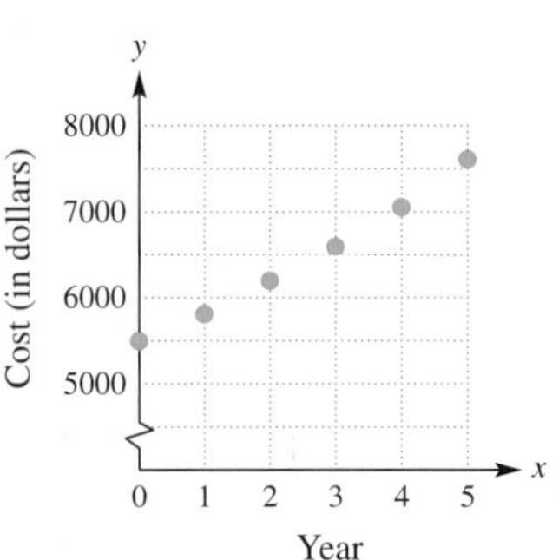

Figure 49

(a) Find an equation that models the data.

(b) Use the equation from part (a) to predict the cost of tuition and fees at public four-year colleges in 2012.

SOLUTION

(a) Since the points in **Figure 49** lie approximately on a straight line, we can write a linear equation that models the relationship between year x and cost y. We choose two data points, $(0, 5492)$ and $(5, 7605)$, to find the slope of the line.

$$m = \frac{7605 - 5492}{5 - 0} = \frac{2113}{5} = 422.6$$

The slope 422.6 indicates that the cost of tuition and fees for in-state students at public four-year colleges increased by about \$423 per year from 2005 to 2010. We use this slope, the y-intercept 5492, and the slope-intercept form to write an equation of the line.

$$y = mx + b \qquad \text{Slope-intercept form}$$
$$y = 422.6x + 5492 \qquad \text{Substitute.}$$

(b) The value $x = 7$ corresponds to the year 2012, so we substitute 7 for x.

$$y = 422.6x + 5492 \qquad \text{Model from part (a)}$$
$$y = 422.6(7) + 5492 \qquad \text{Let } x = 7.$$
$$y = 8450.2 \qquad \text{Multiply, and then add.}$$

The model predicts that average tuition and fees for in-state students at public four-year colleges in 2012 would be about \$8450.

Now Try Exercise 57(a) and (b).

NOTE In **Example 7,** if we had chosen different data points, we would have gotten a slightly different equation.

Guidelines for Modeling

Step 1 Make a scatter diagram of the data.

Step 2 Find an equation that models the data. For a line, this involves selecting two data points and finding the equation of the line through them.

A technique from statistics called **linear regression** provides the line of "best fit." **Figure 50** shows how a TI-83/84 Plus calculator can accept the data points, calculate the equation of this line of best fit (in this case, $Y_1 = 420.1X + 5405$), and plot both the data points and the line on the same screen.

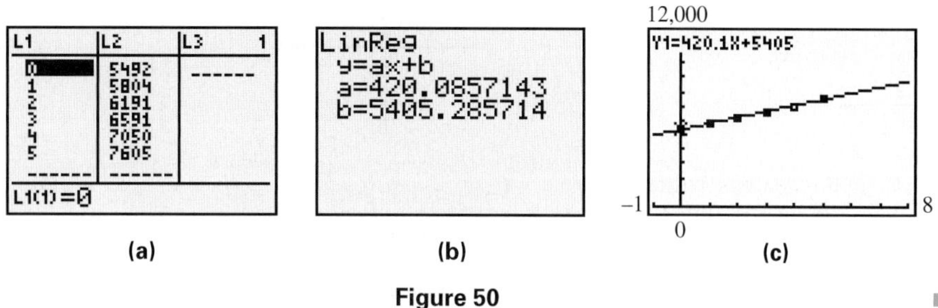

(a) (b) (c)

Figure 50

Solving Linear Equations in One Variable by Graphing Suppose that Y_1 and Y_2 are linear polynomials in x, and we want to solve the equation $Y_1 = Y_2$. Assuming that the equation has a unique solution, the equation can be found graphically as follows and is illustrated in **Example 8.**

1. Rewrite the equation as $Y_1 - Y_2 = 0$.
2. Graph the linear function $Y_3 = Y_1 - Y_2$.
3. Find the x-intercept of the graph of the function Y_3. This is the solution of $Y_1 = Y_2$.

Classroom Example 8
Use a graphing calculator to solve

$$-3x + 2(5 - x) = 2x + 38.$$

Answer: $\{-4\}$

EXAMPLE 8 Solving an Equation with a Graphing Calculator

Use a graphing calculator to solve $-2x - 4(2 - x) = 3x + 4$.

SOLUTION We write an equivalent equation with 0 on one side.

$$-2x - 4(2 - x) - 3x - 4 = 0 \quad \text{Subtract } 3x \text{ and } 4.$$

Then we graph $Y = -2X - 4(2 - X) - 3X - 4$ to find the x-intercept. The standard viewing window cannot be used because the x-intercept does not lie in the interval $[-10, 10]$.

As seen in **Figure 51,** the x-intercept of the graph is -12, and thus the solution of the equation is -12. (-12 is the zero of the function Y.) The solution set is $\{-12\}$.

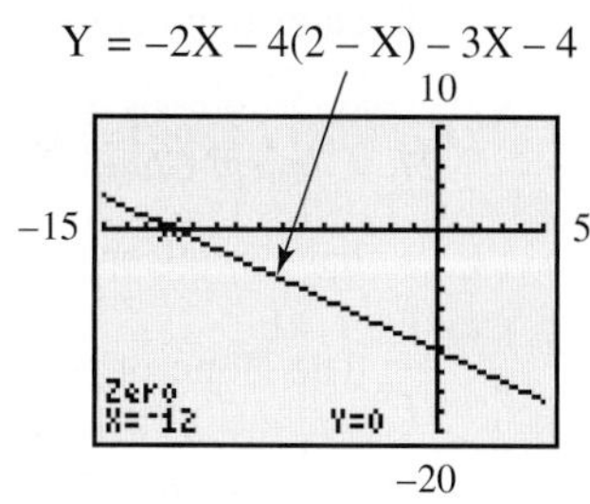

Figure 51

Now Try Exercise 63.

2.5 Exercises

1. D 2. B 3. C 4. A
5. $2x + y = 5$
6. $x + y = 6$
7. $3x + 2y = -7$
8. $3x - 4y = -24$
9. $x = -8$
10. $x = 5$
11. $y = -8$
12. $y = 12$
13. $x - 4y = -13$
14. $x - 3y = -7$
15. $y = \frac{2}{3}x - 2$
16. $y = \frac{3}{4}x + 3$
17. $x = -6$ (cannot be written in slope-intercept form)
18. $x = 2$ (cannot be written in slope-intercept form)
19. $y = 4$
20. $y = -2$
21. $y = 5x + 15$
22. $y = -2x + 12$
23. $y = -4x - 3$
24. $y = -2x + 1$
25. $y = \frac{3}{2}$
26. $y = -\frac{5}{4}$
27. -2; does not; undefined; $\frac{1}{2}$; does not; 0
28. (a) D (b) B (c) A (d) C
29. slope: 3; y-intercept: -1

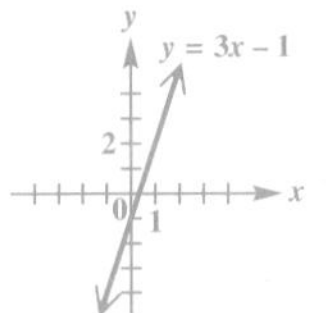

30. slope: -2; y-intercept: 7

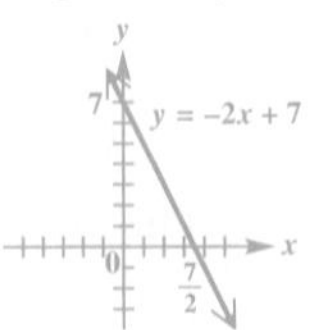

31. slope: 4; y-intercept: -7

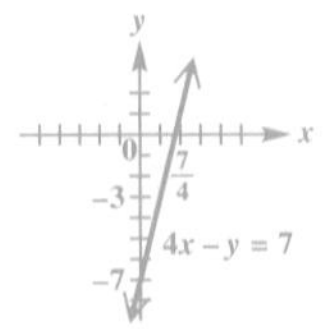

Concept Check *Match each equation in Exercises 1–4 to the correct graph in A–D.*

1. $y = \frac{1}{4}x + 2$
2. $4x + 3y = 12$
3. $y - (-1) = \frac{3}{2}(x - 1)$
4. $y = 4$

A.

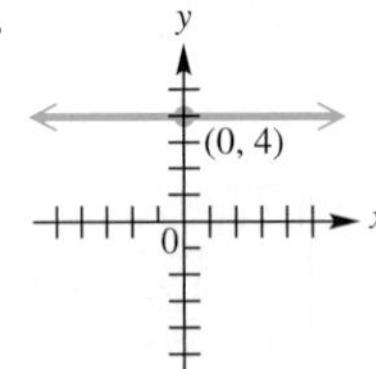

B.

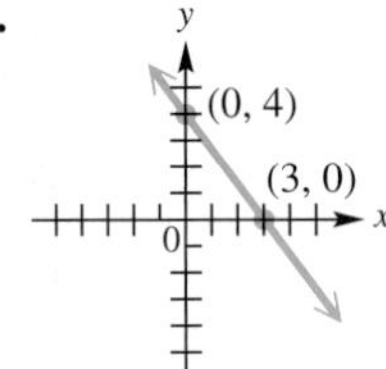

C.

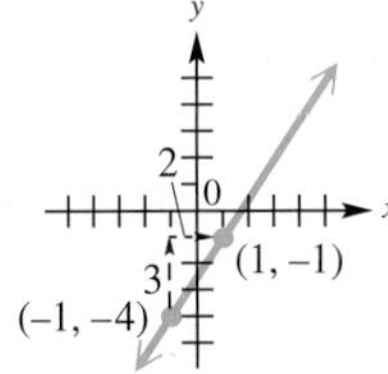

D.

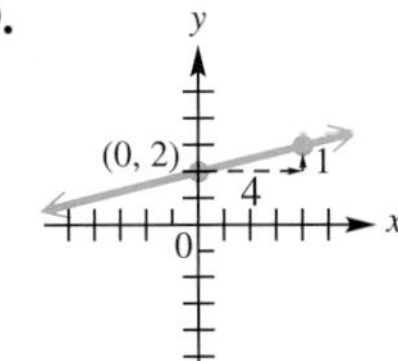

In Exercises 5–26, write an equation for the line described. Give answers in standard form for Exercises 5–14 and in slope-intercept form (if possible) for Exercises 15–26. ***See Examples 1–4.***

5. through $(1, 3)$, $m = -2$
6. through $(2, 4)$, $m = -1$
7. through $(-5, 4)$, $m = -\frac{3}{2}$
8. through $(-4, 3)$, $m = \frac{3}{4}$
9. through $(-8, 4)$, undefined slope
10. through $(5, 1)$, undefined slope
11. through $(5, -8)$, $m = 0$
12. through $(-3, 12)$, $m = 0$
13. through $(-1, 3)$ and $(3, 4)$
14. through $(2, 3)$ and $(-1, 2)$
15. x-intercept 3, y-intercept -2
16. x-intercept -4, y-intercept 3
17. vertical, through $(-6, 4)$
18. vertical, through $(2, 7)$
19. horizontal, through $(-7, 4)$
20. horizontal, through $(-8, -2)$
21. $m = 5$, $b = 15$
22. $m = -2$, $b = 12$
23. through $(-2, 5)$ having slope -4
24. through $(4, -7)$ having slope -2
25. slope 0, y-intercept $\frac{3}{2}$
26. slope 0, y-intercept $-\frac{5}{4}$

27. *Concept Check* Fill in each blank with the appropriate response: The line $x + 2 = 0$ has x-intercept ________. It ________ (does/does not) have a y-intercept. The slope of this line is ________ (0/undefined). The line $4y = 2$ has y-intercept ________. It ________ (does/does not) have an x-intercept. The slope of this line is ________ (0/undefined).

28. *Concept Check* Match each equation with the line that would most closely resemble its graph. (*Hint:* Consider the signs of m and b in the slope-intercept form.)

(a) $y = 3x + 2$ **(b)** $y = -3x + 2$ **(c)** $y = 3x - 2$ **(d)** $y = -3x - 2$

A.

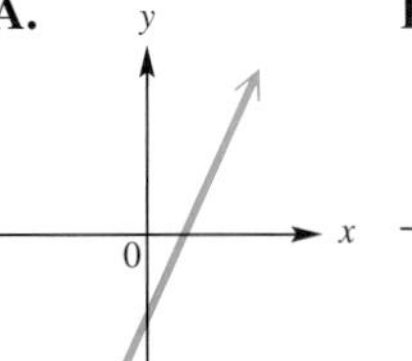

B.

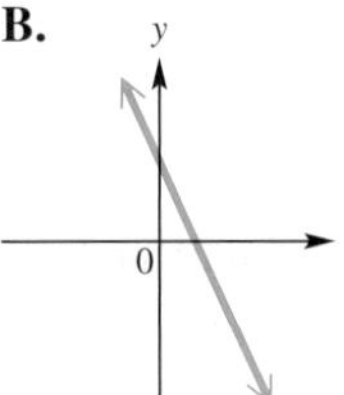

C. **D.**

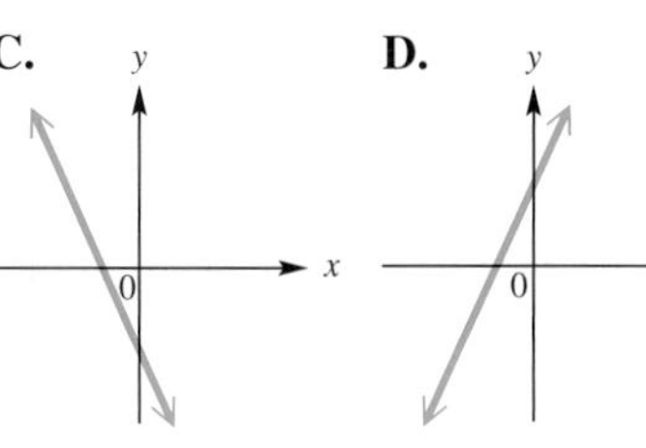

Give the slope and y-intercept of each line, and graph it. ***See Example 3.***

29. $y = 3x - 1$ **30.** $y = -2x + 7$ **31.** $4x - y = 7$

32. $2x + 3y = 16$ **33.** $4y = -3x$ **34.** $2y = x$

35. $x + 2y = -4$ **36.** $x + 3y = -9$ **37.** $y - \frac{3}{2}x - 1 = 0$

38. *Concept Check* The table represents a linear function f.

(a) Find the slope of the line defined by $y = f(x)$.
(b) Find the y-intercept of the line.
(c) Find the equation for this line in slope-intercept form.

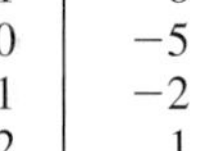

x	y
−2	−11
−1	−8
0	−5
1	−2
2	1
3	4

Connecting Graphs with Equations The graph of a linear function f is shown. ***(a)*** *Identify the slope, y-intercept, and x-intercept.* ***(b)*** *Write the equation that defines f.* ***See Example 5.***

39.

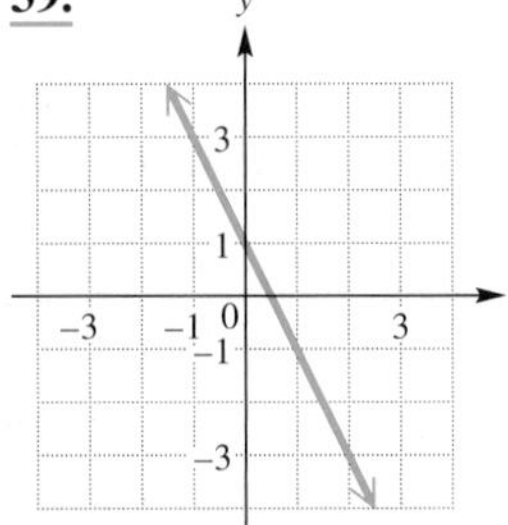

40.

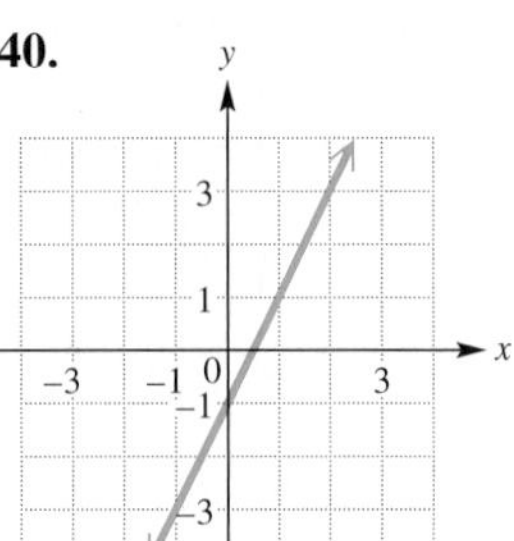

41.

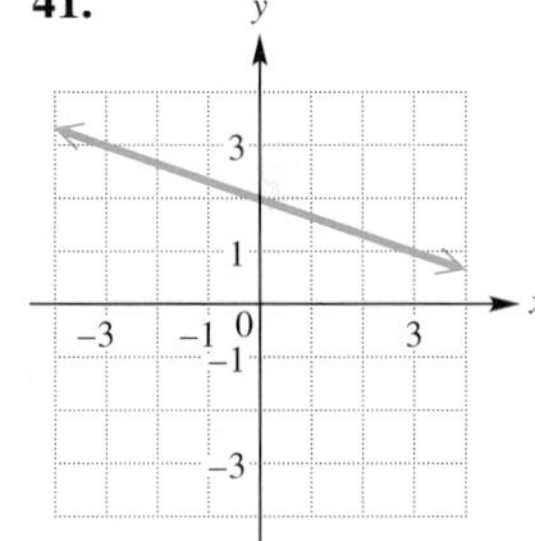

42.

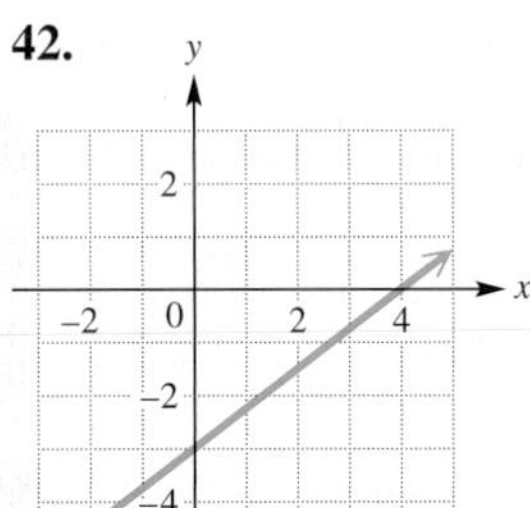

43.

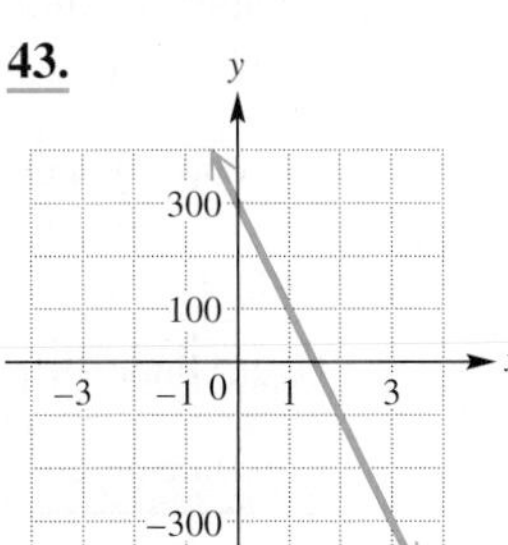

44.

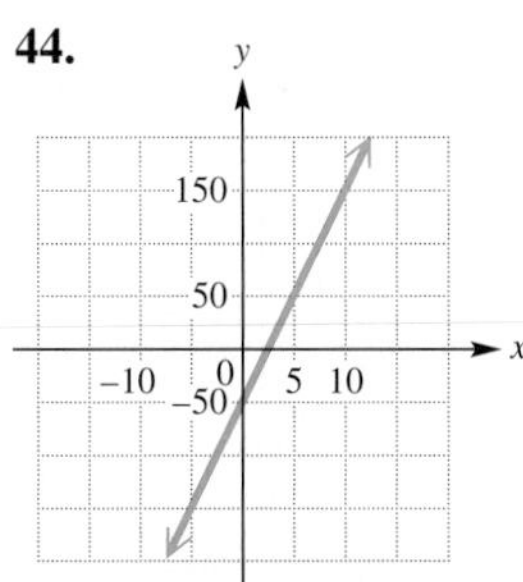

In Exercises 45–52, write an equation ***(a)*** *in standard form and* ***(b)*** *in slope-intercept form for the line described.* ***See Example 6.***

45. through $(-1, 4)$, parallel to $x + 3y = 5$

46. through $(3, -2)$, parallel to $2x - y = 5$

47. through $(1, 6)$, perpendicular to $3x + 5y = 1$

48. through $(-2, 0)$, perpendicular to $8x - 3y = 7$

---

32. slope: $-\frac{2}{3}$; y-intercept: $\frac{16}{3}$

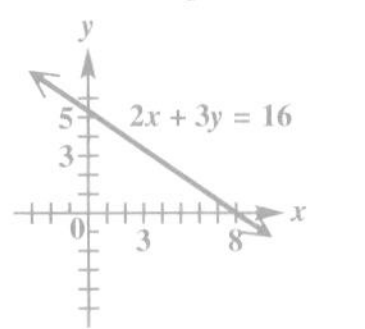

33. slope: $-\frac{3}{4}$; y-intercept: 0

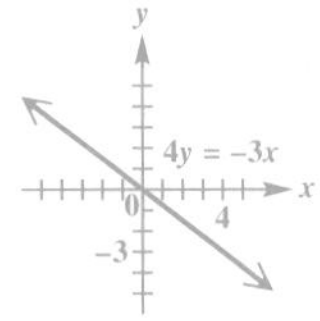

34. slope: $\frac{1}{2}$; y-intercept: 0

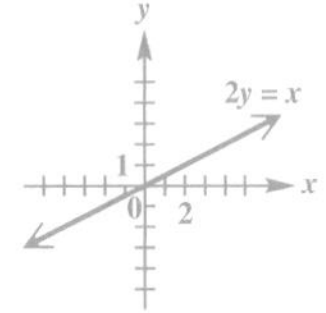

35. slope: $-\frac{1}{2}$; y-intercept: -2

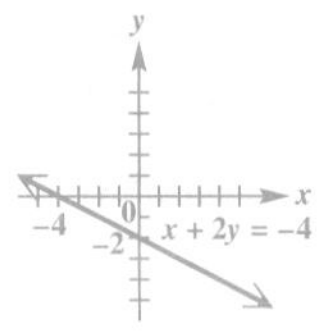

36. slope: $-\frac{1}{3}$; y-intercept: -3

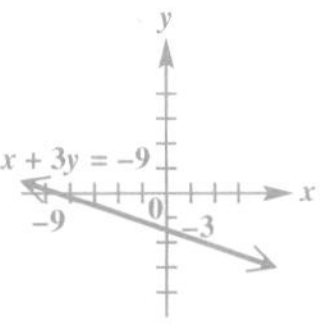

37. slope: $\frac{3}{2}$; y-intercept: 1

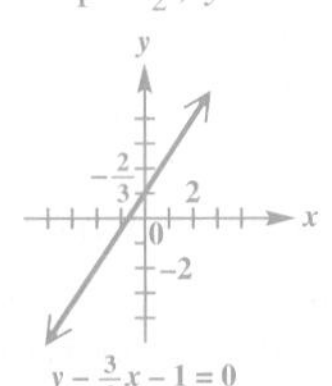

$y - \frac{3}{2}x - 1 = 0$

38. (a) 3 **(b)** -5 **(c)** $y = 3x - 5$

39. (a) -2; 1; $\frac{1}{2}$
(b) $f(x) = -2x + 1$

40. (a) 2; -1; $\frac{1}{2}$
(b) $f(x) = 2x - 1$

41. (a) $-\frac{1}{3}$; 2; 6
(b) $f(x) = -\frac{1}{3}x + 2$

42. (a) $\frac{3}{4}$; -3; 4
(b) $f(x) = \frac{3}{4}x - 3$

43. (a) -200; 300; $\frac{3}{2}$
(b) $f(x) = -200x + 300$

44. (a) 20; -50; $\frac{5}{2}$
(b) $f(x) = 20x - 50$

49. through $(4, 1)$, parallel to $y = -5$

50. through $(-2, -2)$, parallel to $y = 3$

51. through $(-5, 6)$, perpendicular to $x = -2$

52. through $(4, -4)$, perpendicular to $x = 4$

53. Find k so that the line through $(4, -1)$ and $(k, 2)$ is

(a) parallel to $3y + 2x = 6$; **(b)** perpendicular to $2y - 5x = 1$.

54. Find r so that the line through $(2, 6)$ and $(-4, r)$ is

(a) parallel to $2x - 3y = 4$; **(b)** perpendicular to $x + 2y = 1$.

(Modeling) Solve each problem. ***See Example 7.***

55. *Annual Tuition and Fees* Use the data points $(0, 5492)$ and $(4, 7050)$ to find a linear equation that models the data shown in the table accompanying **Figure 49** in **Example 7.** Then use it to predict the average annual tuition and fees for in-state students at public four-year colleges in 2010. How does the result compare to the actual figure given in the table, \$7605?

56. *Annual Tuition and Fees* Repeat **Exercise 55** using the data points for the years 2006 and 2008 to predict the average annual tuition and fees for 2010. How does the result compare to the actual figure given in the table, \$7605?

57. *Cost of Private College Education* The table lists the average annual cost (in dollars) of tuition and fees at private four-year colleges for selected years.

Year	Tuition and Fees (in dollars)
1996	12,881
1998	13,973
2000	15,470
2002	16,826
2004	18,604
2006	20,517
2008	22,449

Source: The College Board, *Annual Survey of Colleges.*

(a) Determine a linear function $f(x) = mx + b$ that models the data, where $x = 0$ represents 1996, $x = 1$ represents 1997, and so on. Use the points $(0, 12{,}881)$ and $(12, 22{,}449)$ to graph f and a scatter diagram of the data on the same coordinate axes. (You may wish to use a graphing calculator.) What does the slope of the graph of f indicate?

(b) Use this function to approximate tuition and fees in 2007. Compare your approximation to the actual value of \$21,979.

(c) Use the linear regression feature of a graphing calculator to find the equation of the line of best fit.

58. *Distances and Velocities of Galaxies* The table lists the distances (in megaparsecs; 1 megaparsec $= 3.085 \times 10^{24}$ cm, and 1 megaparsec $= 3.26$ million light-years) and velocities (in kilometers per second) of four galaxies moving rapidly away from Earth.

Galaxy	Distance	Velocity
Virgo	15	1600
Ursa Minor	200	15,000
Corona Borealis	290	24,000
Bootes	520	40,000

Source: Acker, A., and C. Jaschek, *Astronomical Methods and Calculations,* John Wiley and Sons. Karttunen, H. (editor), *Fundamental Astronomy,* Springer-Verlag.

(a) Plot the data using distances for the x-values and velocities for the y-values. What type of relationship seems to hold between the data?

(b) Find a linear equation in the form $y = mx$ that models these data using the points $(520, 40{,}000)$ and $(0, 0)$. Graph your equation with the data on the same coordinate axes.

---

45. (a) $x + 3y = 11$ **(b)** $y = -\frac{1}{3}x + \frac{11}{3}$

46. (a) $2x - y = 8$ **(b)** $y = 2x - 8$

47. (a) $5x - 3y = -13$ **(b)** $y = \frac{5}{3}x + \frac{13}{3}$

48. (a) $3x + 8y = -6$ **(b)** $y = -\frac{3}{8}x - \frac{3}{4}$

49. (a) $y = 1$ **(b)** $y = 1$

50. (a) $y = -2$ **(b)** $y = -2$

51. (a) $y = 6$ **(b)** $y = 6$

52. (a) $y = -4$ **(b)** $y = -4$

53. (a) $-\frac{1}{2}$ **(b)** $-\frac{7}{2}$

54. (a) 2 **(b)** -6

55. $y = 389.5x + 5492$; \$7440; The result is \$165 less than the actual figure.

56. $y = 393.5x + 5410.5$; \$7378; The result is \$227 less than the actual figure.

57. (a) $f(x) = 797.3x + 12{,}881$

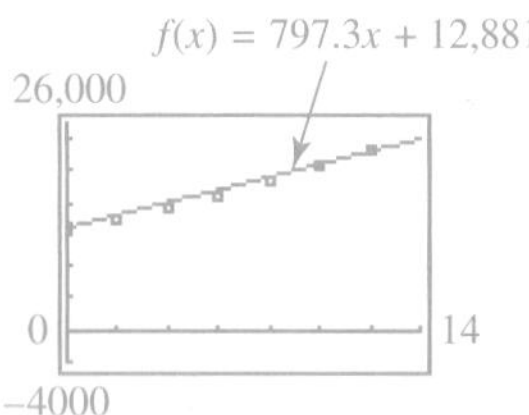

The average tuition increase is about \$797 per year for the period, because this is the slope of the line.

(b) $f(11) = \$21{,}651$; This is a fairly good approximation.

(c) $f(x) = 802.3x + 12{,}432$

58. (a) See the graph in the answer to part (b). It seems to be linear.

(b) $y = 76.9x$

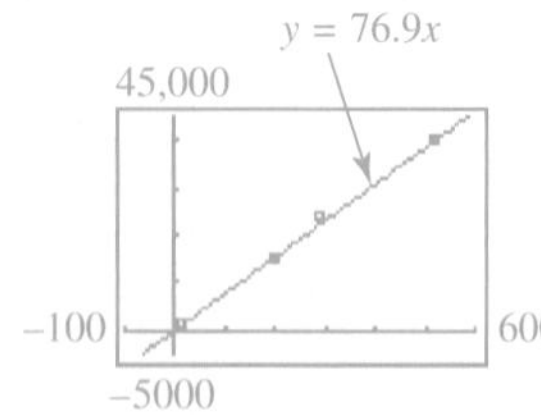

58. **(c)** approximately 780 megaparsecs
(d) approximately 12.35 billion yr
(e) $A(50) \approx 1.9 \times 10^{10}$, $A(100) \approx 9.5 \times 10^{9}$; between 9.5 billion and 19 billion yr

59. **(a)** $F = \frac{9}{5}C + 32$
(b) $C = \frac{5}{9}(F - 32)$
(c) $-40°$

60. **(a)** $p(x) = 0.0292x + 1$
(b) approximately 2.75 atmospheres

61. **(a)** $C = 0.8636I - 296.3$
(b) 0.8636

62. $\{3\}$

63. $\{1\}$ **64.** $\{-0.5\}$

65. $\{4\}$ **66.** D

67. **(a)** $\{12\}$

68. The graph is a horizontal line that does not intersect the x-axis. The solution set is $\emptyset$.

69. $\sqrt{x_1^2 + m_1^2x_1^2}$

70. $\sqrt{x_2^2 + m_2^2x_2^2}$

71. $\sqrt{(x_2 - x_1)^2 + (m_2x_2 - m_1x_1)^2}$

73. $-2x_1x_2(m_1m_2 + 1) = 0$

74. Since $x_1 \neq 0$, $x_2 \neq 0$, we have $m_1m_2 + 1 = 0$, implying that $m_1m_2 = -1$.

75. If two nonvertical lines are perpendicular, then the product of the slopes of these lines is -1.

77. yes **78.** yes

79. no **80.** no

Answers to the Summary Exercises following Section 2.5

1. **(a)** $\sqrt{65}$ **(b)** $\left(\frac{5}{2}, 1\right)$
(c) $y = 8x - 19$

2. **(a)** $\sqrt{29}$ **(b)** $\left(\frac{3}{2}, -1\right)$
(c) $y = -\frac{2}{5}x - \frac{2}{5}$

3. **(a)** 5 **(b)** $\left(\frac{1}{2}, 2\right)$
(c) $y = 2$

4. **(a)** $\sqrt{10}$ **(b)** $\left(\frac{3\sqrt{2}}{2}, 2\sqrt{2}\right)$
(c) $y = -2x + 5\sqrt{2}$

5. **(a)** 2 **(b)** $(5, 0)$
(c) $x = 5$

6. **(a)** $4\sqrt{2}$ **(b)** $(-1, -1)$
(c) $y = x$

(c) The galaxy Hydra has a velocity of 60,000 km per sec. How far away is it according to the model in part (b)?

(d) The value of m is called the **Hubble constant.** The Hubble constant can be used to estimate the age of the universe A (in years) using the formula

$$A = \frac{9.5 \times 10^{11}}{m}.$$

Approximate A using your value of m.

(e) Astronomers currently place the value of the Hubble constant between 50 and 100. What is the range for the age of the universe A?

59. *Celsius and Fahrenheit Temperatures* When the Celsius temperature is 0°, the corresponding Fahrenheit temperature is 32°. When the Celsius temperature is 100°, the corresponding Fahrenheit temperature is 212°. Let C represent the Celsius temperature and F the Fahrenheit temperature.

(a) Express F as an exact linear function of C.
(b) Solve the equation in part (a) for C, thus expressing C as a function of F.
(c) For what temperature is $F = C$ a true statement?

60. *Water Pressure on a Diver* The pressure p of water on a diver's body is a linear function of the diver's depth, x. At the water's surface, the pressure is 1 atmosphere. At a depth of 100 ft, the pressure is about 3.92 atmospheres.

(a) Find the linear function that relates p to x.
(b) Compute the pressure at a depth of 10 fathoms (60 ft).

61. *Consumption Expenditures* In Keynesian macroeconomic theory, total consumption expenditure on goods and services, C, is assumed to be a linear function of national personal income, I. The table gives the values of C and I for 2004 and 2009 in the United States (in billions of dollars).

Year	2004	2009
Total consumption (C)	\$8285	\$10,089
National income (I)	\$9937	\$12,026

Source: U.S. Bureau of Economic Analysis.

(a) Find the formula for C as a function of I.
(b) The slope of the linear function is called the **marginal propensity to consume.** What is the marginal propensity to consume for the United States from 2004–2009?

Use a graphing calculator to solve each linear equation. ***See Example 8.***

62. $2x + 7 - x = 4x - 2$

63. $7x - 2x + 4 - 5 = 3x + 1$

64. $3(2x + 1) - 2(x - 2) = 5$

65. $4x - 3(4 - 2x) = 2(x - 3) + 6x + 2$

66. The graph of $y = f(x)$ is shown in the standard viewing window. Which is the only value of x that could possibly be the solution of the equation $f(x) = 0$?

A. -15 **B.** 0 **C.** 5 **D.** 15

10
−10 10
−10

67. **(a)** Solve $-2(x - 5) = -x - 2$ using the methods of **Chapter 1.**
(b) Explain why the standard viewing window of a graphing calculator cannot graphically support the solution found in part (a). What minimum and maximum x-values would make it possible for the solution to be seen?

68. Using a graphing calculator, try to solve $-3(2x + 6) = -4x + 8 - 2x$. Explain what happens. What is the solution set?

7. (a) $4\sqrt{3}$ (b) $(4\sqrt{3}, 3\sqrt{5})$
(c) $y = 3\sqrt{5}$

8. (a) $\sqrt{34}$ (b) $\left(\frac{3}{2}, -\frac{3}{2}\right)$
(c) $y = \frac{5}{3}x - 4$

9. $y = -\frac{1}{3}x + \frac{1}{3}$

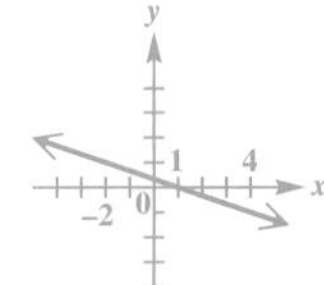

10. $y = 3$

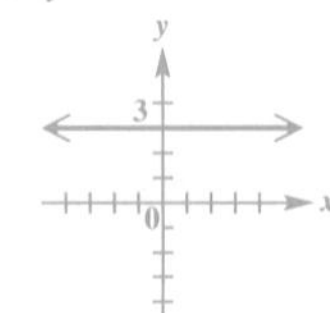

11. $(x-2)^2 + (y+1)^2 = 9$

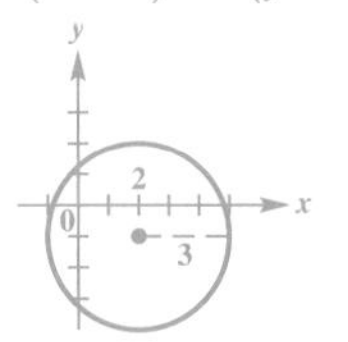

12. $x^2 + (y-2)^2 = 4$

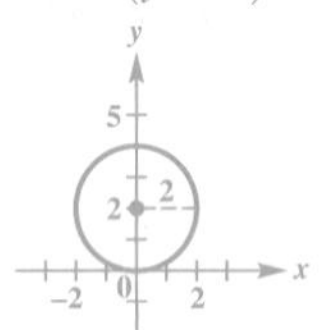

13. $y = -\frac{5}{6}x - \frac{5}{2}$

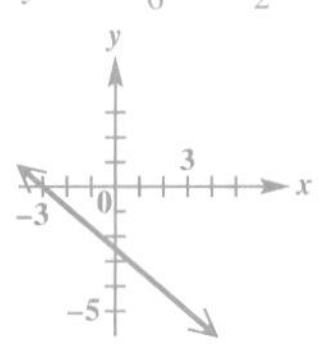

14. $y = -\frac{4}{3}x$

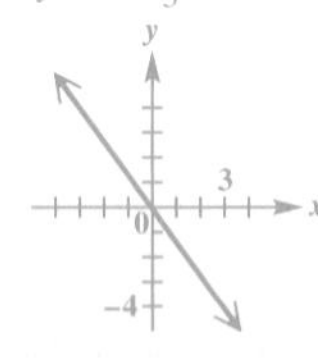

15. $y = -\frac{2}{3}x$

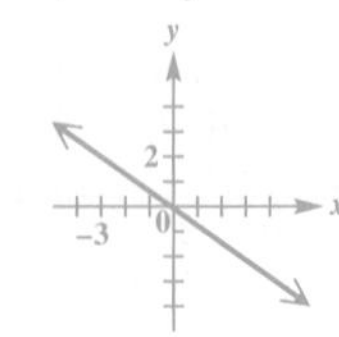

Relating Concepts

For individual or collaborative investigation *(Exercises 69–75)*

In this section we state that two lines, neither of which is vertical, are perpendicular if and only if their slopes have a product of -1*. In Exercises 69–75, we outline a partial proof of this for the case where the two lines intersect at the origin.* ***Work these exercises in order,*** *and refer to the figure as needed.*

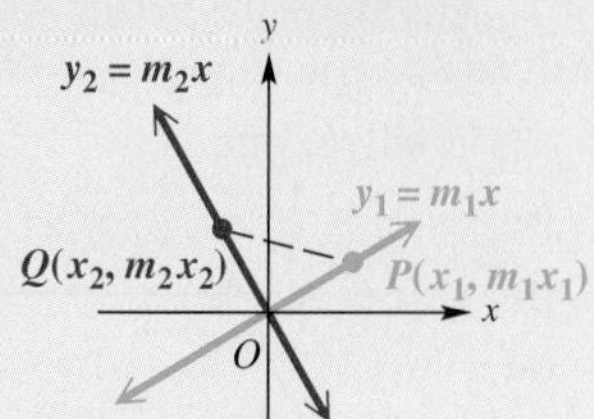

By the converse of the Pythagorean theorem, if

$$[d(O,P)]^2 + [d(O,Q)]^2 = [d(P,Q)]^2,$$

then triangle POQ is a right triangle with right angle at O.

69. Find an expression for the distance $d(O, P)$.

70. Find an expression for the distance $d(O, Q)$.

71. Find an expression for the distance $d(P, Q)$.

72. Use your results from **Exercises 69–71,** and substitute into the equation from the Pythagorean theorem. Simplify to show that this leads to the equation

$$-2m_1m_2x_1x_2 - 2x_1x_2 = 0.$$

73. Factor $-2x_1x_2$ from the final form of the equation in **Exercise 72.**

74. Use the property that if $ab = 0$ then $a = 0$ or $b = 0$ to solve the equation in **Exercise 73,** showing that $m_1m_2 = -1$.

75. State your conclusion based on **Exercises 69–74.**

76. Refer to **Example 4** in **Section 2.1,** and prove that the three points are collinear by taking them two at a time and showing that in all three cases, the slope is the same.

Determine whether the three points are collinear by using slopes as in ***Exercise 76.*** *(Note: These problems were first seen in* ***Exercises 25–28*** *in* ***Section 2.1.****)*

77. $(-1, 4)$, $(-2, -1)$, $(1, 14)$

78. $(0, -7)$, $(-3, 5)$, $(2, -15)$

79. $(-1, -3)$, $(-5, 12)$, $(1, -11)$

80. $(0, 9)$, $(-3, -7)$, $(2, 19)$

Summary Exercises on Graphs, Circles, Functions, and Equations

These summary exercises provide practice with some of the concepts from ***Sections 2.1–2.5.***

For the points P and Q, find ***(a)*** *the distance* $d(P, Q)$, ***(b)*** *the coordinates of the midpoint of the segment PQ, and* ***(c)*** *an equation for the line through the two points. Write the equation in slope-intercept form if possible.*

1. $P(3, 5)$, $Q(2, -3)$

2. $P(-1, 0)$, $Q(4, -2)$

3. $P(-2, 2)$, $Q(3, 2)$

4. $P(2\sqrt{2}, \sqrt{2})$, $Q(\sqrt{2}, 3\sqrt{2})$

16. $x = -4$

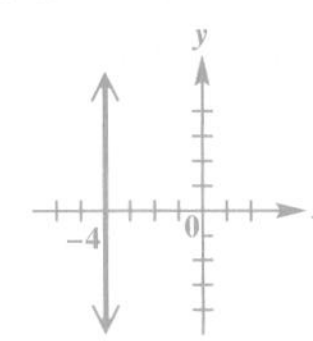

17. yes; center: $(2, -1)$; radius: 3
18. no
19. yes; center: $(6, 0)$; radius: 4
20. yes; center: $(-1, -8)$; radius: 2
21. no
22. yes; center: $(0, 4)$; radius: 5
23. $(4 - \sqrt{7}, 2), (4 + \sqrt{7}, 2)$
24. 8
25. (a) domain: $(-\infty, \infty)$; range: $(-\infty, \infty)$
(b) $f(x) = \frac{1}{4}x + \frac{3}{2}$; 1
26. (a) domain: $[-5, \infty)$; range: $(-\infty, \infty)$
(b) y is not a function of x.
27. (a) domain: $[-7, 3]$; range: $[-5, 5]$
(b) y is not a function of x.
28. (a) domain: $(-\infty, \infty)$; range: $[-\frac{3}{2}, \infty)$
(b) $f(x) = \frac{1}{2}x^2 - \frac{3}{2}$; $\frac{1}{2}$

5. $P(5, -1)$, $Q(5, 1)$
6. $P(1, 1)$, $Q(-3, -3)$
7. $P(2\sqrt{3}, 3\sqrt{5})$, $Q(6\sqrt{3}, 3\sqrt{5})$
8. $P(0, -4)$, $Q(3, 1)$

Write an equation for each of the following, and sketch the graph.

9. the line through $(-2, 1)$ and $(4, -1)$
10. the horizontal line through $(2, 3)$
11. the circle with center $(2, -1)$ and radius 3
12. the circle with center $(0, 2)$ and tangent to the x-axis
13. the line through $(3, -5)$ with slope $-\frac{5}{6}$
14. the line through the origin and perpendicular to the line $3x - 4y = 2$
15. the line through $(-3, 2)$ and parallel to the line $2x + 3y = 6$
16. the vertical line through $(-4, 3)$

Decide whether or not each equation has a circle as its graph. If it does, give the center and the radius.

17. $x^2 + y^2 - 4x + 2y = 4$
18. $x^2 + y^2 + 6x + 10y + 36 = 0$
19. $x^2 + y^2 - 12x + 20 = 0$
20. $x^2 + y^2 + 2x + 16y = -61$
21. $x^2 + y^2 - 2x + 10 = 0$
22. $x^2 + y^2 - 8y - 9 = 0$
23. Find the coordinates of the points of intersection of the line $y = 2$ and the circle with center at $(4, 5)$ and radius 4.
24. Find the shortest distance from the origin to the graph of the circle with equation

$$x^2 + y^2 - 10x - 24y + 144 = 0.$$

*For each relation, **(a)** find the domain and range, and **(b)** if the relation defines y as a function f of x, rewrite the relation using function notation and find $f(-2)$.*

25. $x - 4y = -6$
26. $y^2 - x = 5$
27. $(x + 2)^2 + y^2 = 25$
28. $x^2 - 2y = 3$

2.6 Graphs of Basic Functions

- Continuity
- The Identity, Squaring, and Cubing Functions
- The Square Root and Cube Root Functions
- The Absolute Value Function
- Piecewise-Defined Functions
- The Relation $x = y^2$

Continuity Earlier in this chapter we graphed linear functions. The graph of a linear function, a straight line, may be drawn by hand over any interval of its domain without picking the pencil up from the paper. In mathematics we say that a function with this property is *continuous* over any interval. The formal definition of continuity requires concepts from calculus, but we can give an informal definition at the college algebra level.

Continuity (Informal Definition)

A function is **continuous** over an interval of its domain if its hand-drawn graph over that interval can be sketched without lifting the pencil from the paper.

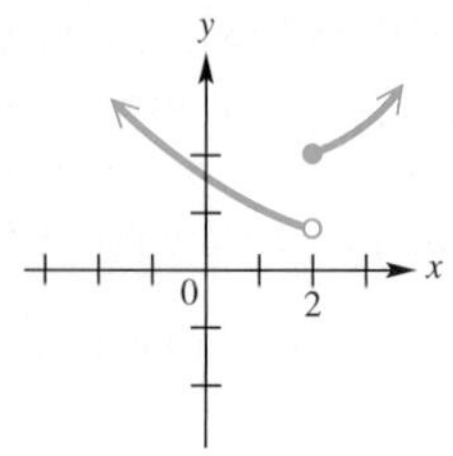

The function is discontinuous at $x = 2$.

Figure 52

If a function is not continuous at a *point,* then it has a *discontinuity* there. **Figure 52** shows the graph of a function with a discontinuity at the point where $x = 2$.

Classroom Example 1
Describe the intervals of continuity for each function.

(a)

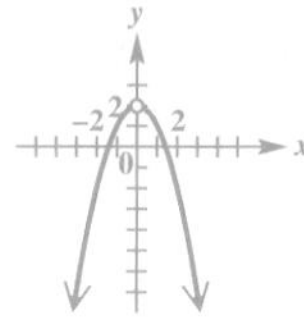

(b)

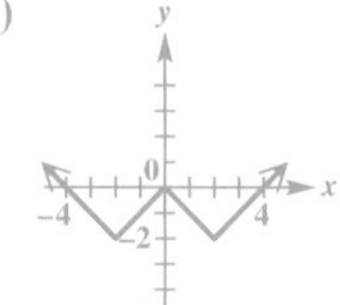

Answers:
(a) $(-\infty, 0)$ and $(0, \infty)$
(b) $(-\infty, \infty)$

EXAMPLE 1 Determining Intervals of Continuity

Describe the intervals of continuity for each function in **Figure 53.**

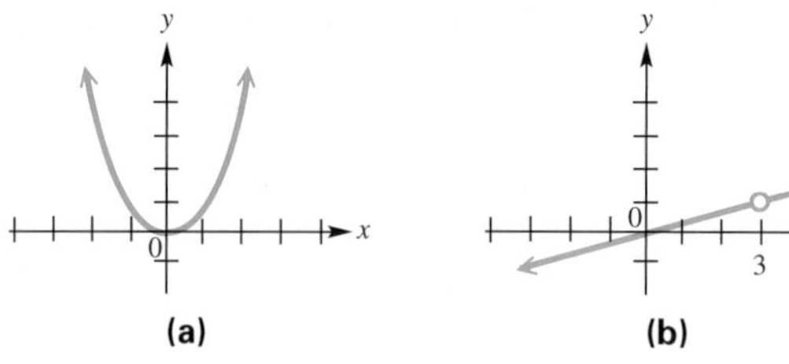

Figure 53

SOLUTION The function in **Figure 53(a)** is continuous over its entire domain, $(-\infty, \infty)$. The function in **Figure 53(b)** has a point of discontinuity at $x = 3$. Thus, it is continuous over the intervals

$$(-\infty, 3) \quad \text{and} \quad (3, \infty).$$

Now Try Exercises 11 and 15.

Graphs of the basic functions studied in college algebra can be sketched by careful point plotting or generated by a graphing calculator. As you become more familiar with these graphs, you should be able to provide quick rough sketches of them.

The Identity, Squaring, and Cubing Functions The **identity function** $f(x) = x$ pairs every real number with itself. See **Figure 54.**

Teaching Tip Help your students memorize the shapes of these basic functions by using your arms, hands, and so on.

$f(x) = x^2$ (Touchdown!)
$f(x) = x^3$ (John Travolta dancing)
$f(x) = \sqrt{x}$ (Right turn)
$f(x) = \sqrt[3]{x}$ (Swimmer)
$f(x) = |x|$ (Absolute victory)
$f(x) = [\![x]\!]$ (Step function)

Identity Function $f(x) = x$

Domain: $(-\infty, \infty)$ Range: $(-\infty, \infty)$

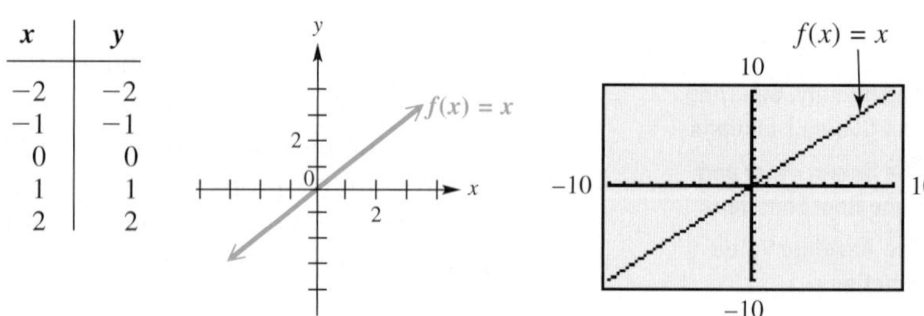

x	y
−2	−2
−1	−1
0	0
1	1
2	2

Figure 54

- $f(x) = x$ is increasing on its entire domain, $(-\infty, \infty)$.
- It is continuous on its entire domain, $(-\infty, \infty)$.

LOOKING AHEAD TO CALCULUS
Many calculus theorems apply only to continuous functions.

The **squaring function** $f(x) = x^2$ pairs each real number with its square. Its graph is called a **parabola.** The point $(0, 0)$ at which the graph changes from decreasing to increasing is called the **vertex** of the parabola. See **Figure 55.** (For a parabola that opens downward, the vertex is the point at which the graph changes from increasing to decreasing.)

Squaring Function $f(x) = x^2$

Domain: $(-\infty, \infty)$ Range: $[0, \infty)$

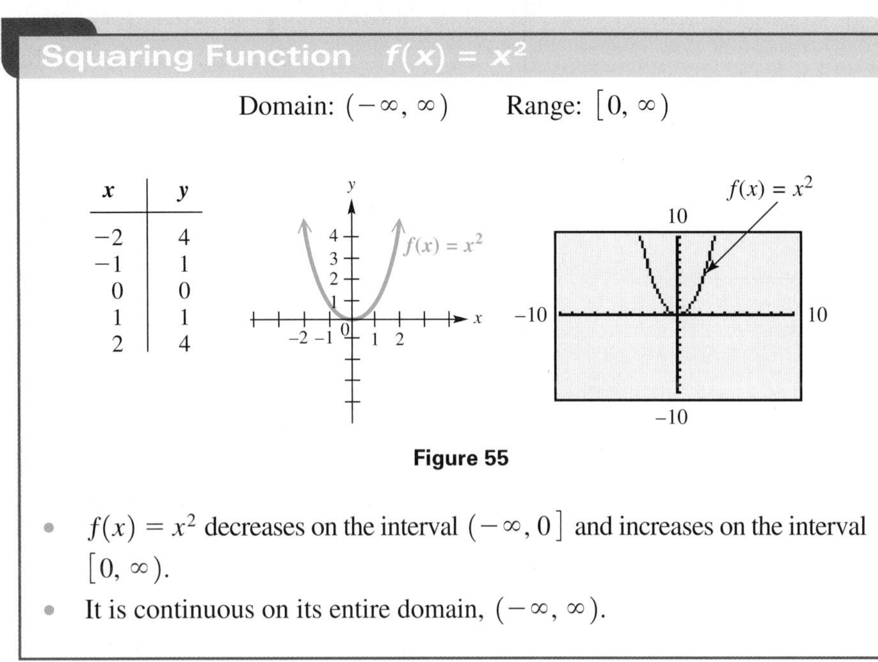

x	y
−2	4
−1	1
0	0
1	1
2	4

Figure 55

- $f(x) = x^2$ decreases on the interval $(-\infty, 0]$ and increases on the interval $[0, \infty)$.
- It is continuous on its entire domain, $(-\infty, \infty)$.

The function $f(x) = x^3$ is called the **cubing function.** It pairs each real number with the cube of the number. See **Figure 56.** The point $(0, 0)$ at which the graph changes from "opening downward" to "opening upward" is an **inflection point.**

Cubing Function $f(x) = x^3$

Domain: $(-\infty, \infty)$ Range: $(-\infty, \infty)$

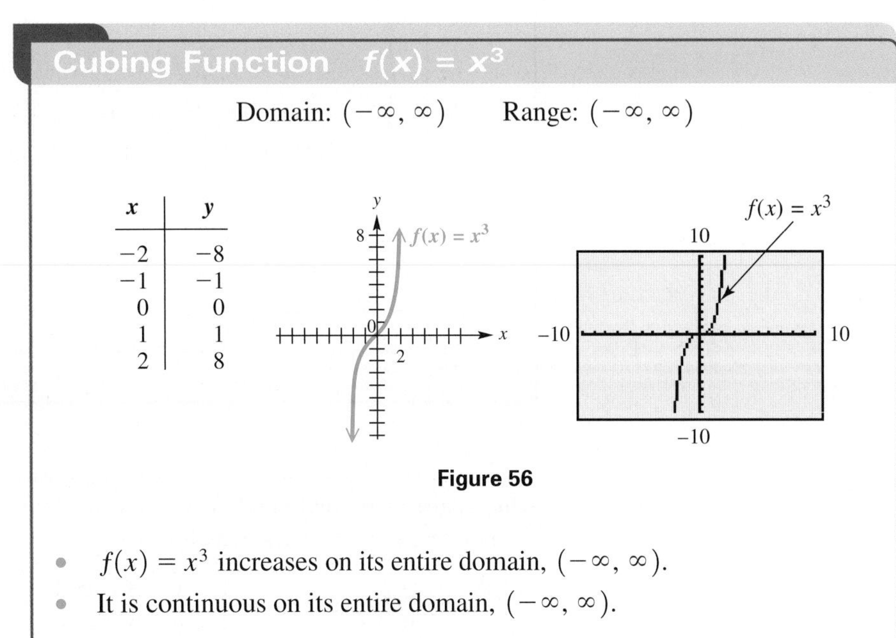

x	y
−2	−8
−1	−1
0	0
1	1
2	8

Figure 56

- $f(x) = x^3$ increases on its entire domain, $(-\infty, \infty)$.
- It is continuous on its entire domain, $(-\infty, \infty)$.

The Square Root and Cube Root Functions The function $f(x) = \sqrt{x}$ is called the **square root function.** It pairs each real number with its principal square root. See **Figure 57.** For the function value to be a real number, the domain must be restricted to $[0, \infty)$.

Square Root Function $f(x) = \sqrt{x}$

Domain: $[0, \infty)$ Range: $[0, \infty)$

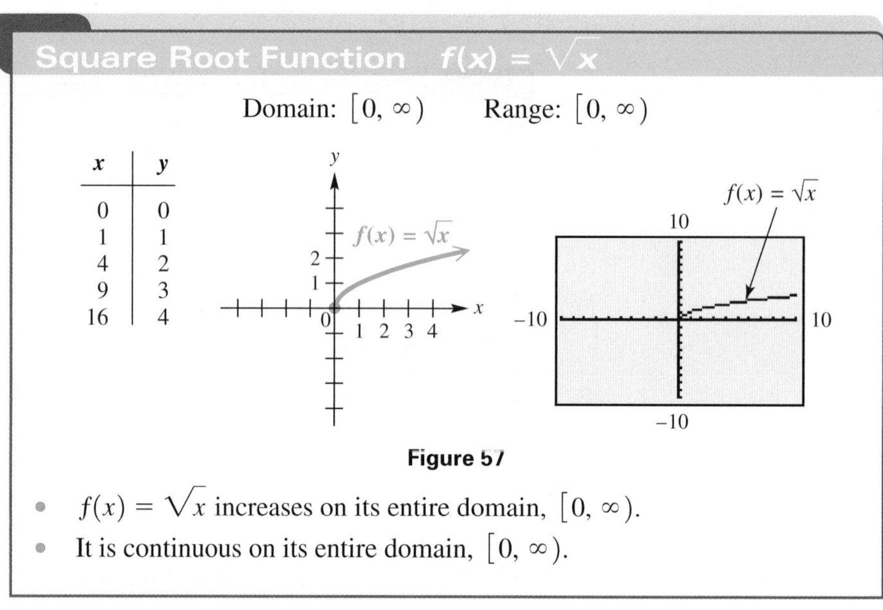

x	y
0	0
1	1
4	2
9	3
16	4

Figure 57

- $f(x) = \sqrt{x}$ increases on its entire domain, $[0, \infty)$.
- It is continuous on its entire domain, $[0, \infty)$.

The **cube root function** $f(x) = \sqrt[3]{x}$ pairs each real number with its cube root. See **Figure 58.** The cube root function differs from the square root function in that *any* real number has a real number cube root. Thus, the domain is $(-\infty, \infty)$.

Cube Root Function $f(x) = \sqrt[3]{x}$

Domain: $(-\infty, \infty)$ Range: $(-\infty, \infty)$

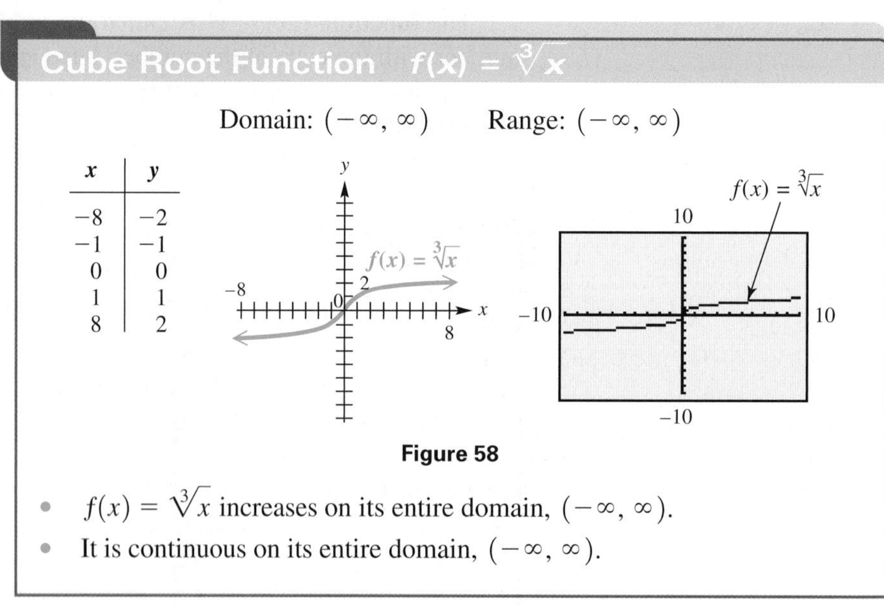

x	y
−8	−2
−1	−1
0	0
1	1
8	2

Figure 58

- $f(x) = \sqrt[3]{x}$ increases on its entire domain, $(-\infty, \infty)$.
- It is continuous on its entire domain, $(-\infty, \infty)$.

The Absolute Value Function The **absolute value function,** $f(x) = |x|$, which pairs every real number with its absolute value, is graphed in **Figure 59** on the next page and is defined as follows.

$$f(x) = |x| = \begin{cases} x & \text{if } x \geq 0 \\ -x & \text{if } x < 0 \end{cases} \quad \text{Absolute value function}$$

That is, we use $|x| = x$ if x is positive or 0, and we use $|x| = -x$ if x is negative.

Classroom Example 2
Graph each function.

(a) $f(x) = \begin{cases} 2x + 4 & \text{if } x < 1 \\ 4 - x & \text{if } x \geq 1 \end{cases}$

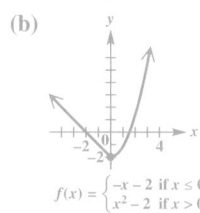

(b) $f(x) = \begin{cases} -x - 2 & \text{if } x \leq 0 \\ x^2 - 2 & \text{if } x > 0 \end{cases}$

Answers:
(a)

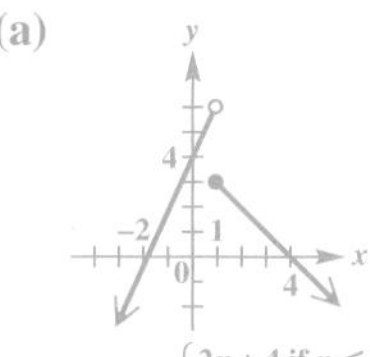

(b)

Teaching Tip Give some examples that show how piecewise-defined functions arise in everyday life. Some examples are income taxes, telephone calls (rates drop in the evening), and postage fees.

Absolute Value Function $f(x) = |x|$

Domain: $(-\infty, \infty)$ Range: $[0, \infty)$

x	y
−2	2
−1	1
0	0
1	1
2	2

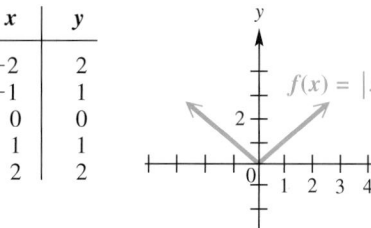

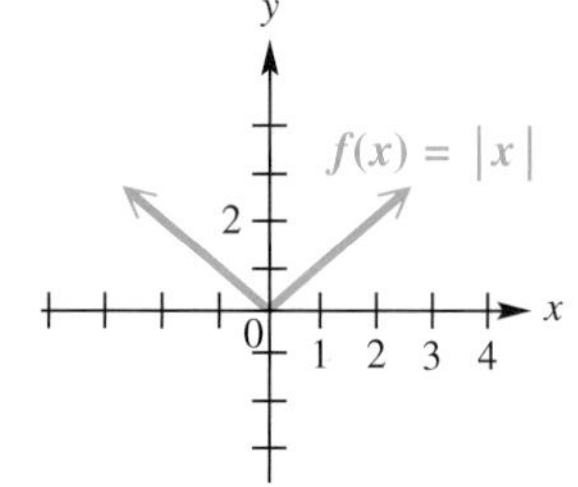

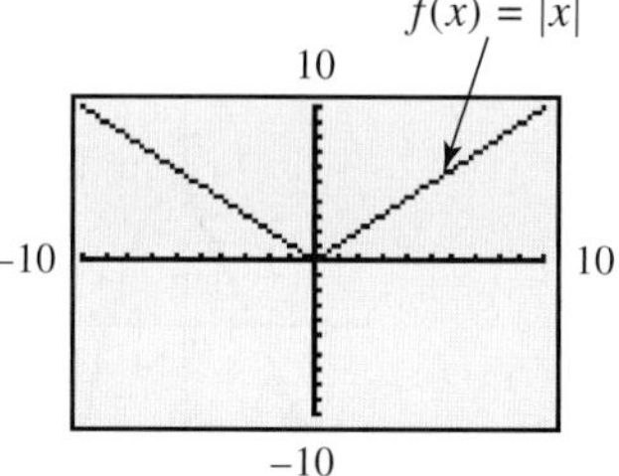

Figure 59

- $f(x) = |x|$ decreases on the interval $(-\infty, 0]$ and increases on $[0, \infty)$.
- It is continuous on its entire domain, $(-\infty, \infty)$.

Piecewise-Defined Functions The absolute value function is defined by different rules over different intervals of its domain. Such functions are called **piecewise-defined functions.**

EXAMPLE 2 Graphing Piecewise-Defined Functions

Graph each function.

(a) $f(x) = \begin{cases} -2x + 5 & \text{if } x \leq 2 \\ x + 1 & \text{if } x > 2 \end{cases}$ **(b)** $f(x) = \begin{cases} 2x + 3 & \text{if } x \leq 0 \\ -x^2 + 3 & \text{if } x > 0 \end{cases}$

ALGEBRAIC SOLUTION

(a) We graph each interval of the domain separately. If $x \leq 2$, the graph of $f(x) = -2x + 5$ has an endpoint at $x = 2$. We find the corresponding y-value by substituting 2 for x in $-2x + 5$ to get $y = 1$. To get another point on this part of the graph, we choose $x = 0$, so $y = 5$. We draw the graph through $(2, 1)$ and $(0, 5)$ as a partial line with endpoint $(2, 1)$.

We graph the function for $x > 2$ similarly, using $f(x) = x + 1$. This partial line has an open endpoint at $(2, 3)$. We use $y = x + 1$ to find another point with x-value greater than 2 to complete the graph. See **Figure 60.**

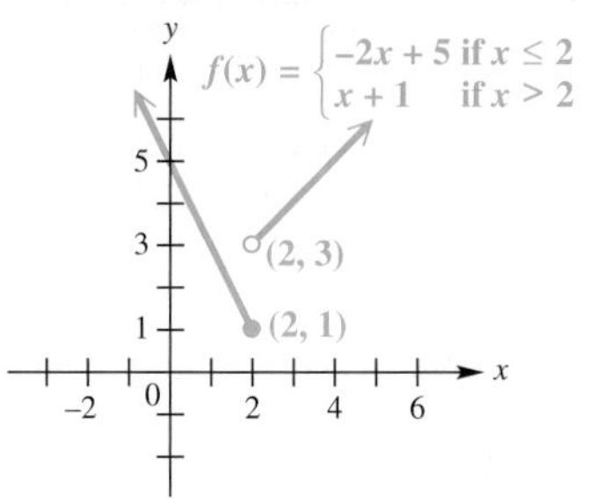

Figure 60

GRAPHING CALCULATOR SOLUTION

(a) By defining

$$Y_1 = (-2X + 5)(X \leq 2)$$

and

$$Y_2 = (X + 1)(X > 2)$$

we obtain the graph of f shown in **Figure 61.** Remember that inclusion or exclusion of endpoints is not readily apparent when we are observing a calculator graph. We must make this decision using our knowledge of inequalities.

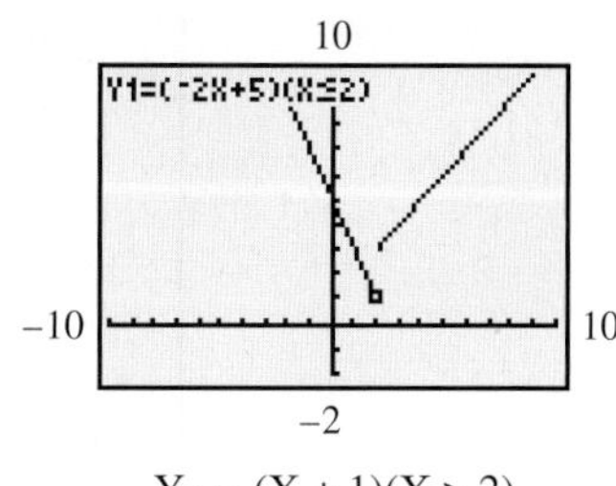

$Y_2 = (X + 1)(X > 2)$

Figure 61

(b) First graph $f(x) = 2x + 3$ for $x \leq 0$. Then for $x > 0$, graph $f(x) = -x^2 + 3$. The two graphs meet at the point $(0, 3)$. See **Figure 62.**

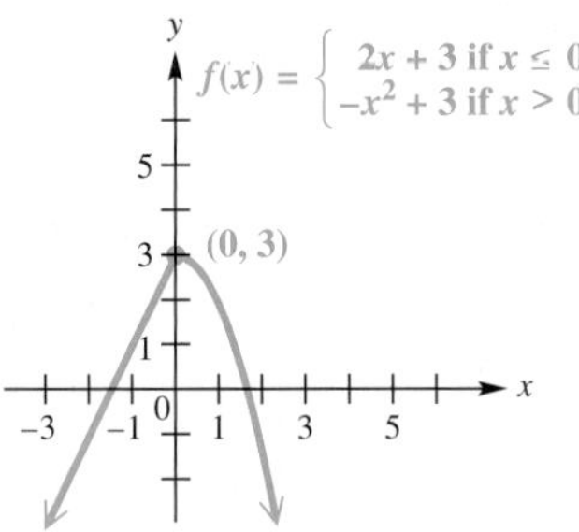

Figure 62

(b) **Figure 63** shows an alternative method that can be used to obtain a calculator graph of this function. Here we entered the function as one expression.

$$Y_1 = (2X + 3)(X \leq 0) + (-X^2 + 3)(X > 0)$$

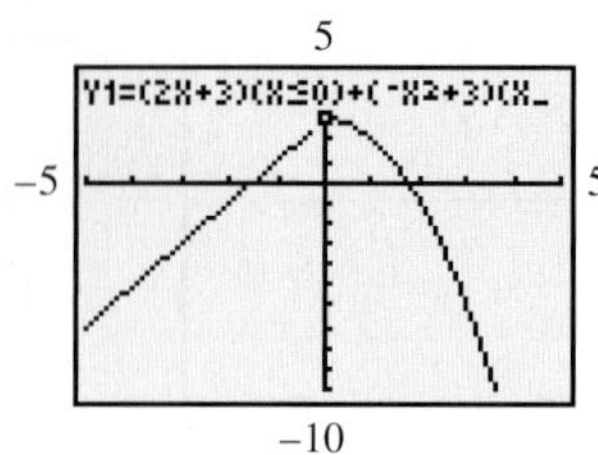

Figure 63

✔ *Now Try Exercises 23 and 29.*

Another piecewise-defined function is the *greatest integer function.*

LOOKING AHEAD TO CALCULUS

The **greatest integer function** is used in calculus as a classic example of how the limit of a function may not exist at a particular value in its domain. For a limit to exist, the left- and right-hand limits must be equal. We can see from the graph of the greatest integer function that for an integer value such as 3, as x approaches 3 from the left, function values are all 2, while as x approaches 3 from the right, function values are all 3. Since the left- and right-hand limits are different, the limit as x approaches 3 does not exist.

$f(x) = [\![x]\!]$

The **greatest integer function** $f(x) = [\![x]\!]$ pairs every real number x with the greatest integer less than or equal to x.

For example, $[\![8.4]\!] = 8$, $[\![-5]\!] = -5$, $[\![\pi]\!] = 3$, and $[\![-6.9]\!] = -7$. In general, if $f(x) = [\![x]\!]$, then

$$\begin{aligned}
&\text{for } -2 \leq x < -1, && f(x) = -2,\\
&\text{for } -1 \leq x < 0, && f(x) = -1,\\
&\text{for } 0 \leq x < 1, && f(x) = 0,\\
&\text{for } 1 \leq x < 2, && f(x) = 1,\\
&\text{for } 2 \leq x < 3, && f(x) = 2, \quad \text{and so on.}
\end{aligned}$$

The graph of the greatest integer function is shown in **Figure 64.**

Greatest Integer Function $f(x) = [\![x]\!]$

Domain: $(-\infty, \infty)$

Range: $\{y \mid y \text{ is an integer}\} = \{\dots, -3, -2, -1, 0, 1, 2, 3, \dots\}$

x	y
-2	-2
-1.5	-2
-0.99	-1
0	0
0.001	0
3	3
3.99	3

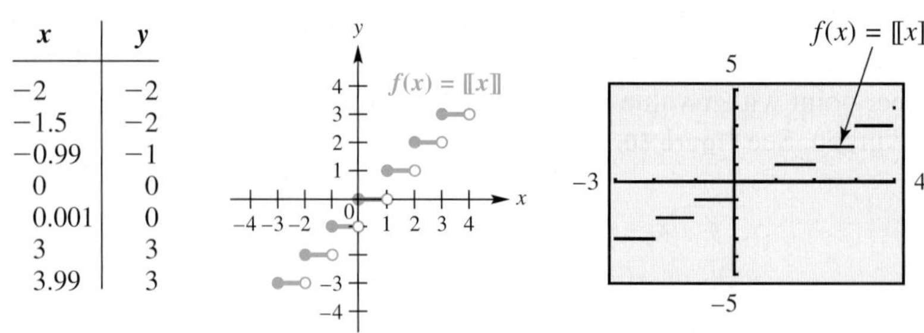

Figure 64

- $f(x) = [\![x]\!]$ is constant on the intervals $\dots, [-2, -1), [-1, 0), [0, 1), [1, 2), [2, 3), \dots$.
- It is discontinuous at all integer values in its domain, $(-\infty, \infty)$.

Classroom Example 3
Graph $f(x) = [\![\frac{1}{3}x - 2]\!]$.

Answer:

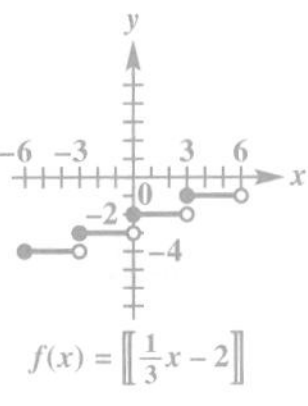

EXAMPLE 3 Graphing a Greatest Integer Function

Graph $f(x) = [\![\frac{1}{2}x + 1]\!]$.

SOLUTION If x is in the interval $[0, 2)$, then $y = 1$. For x in $[2, 4)$, $y = 2$, and so on. Some sample ordered pairs are given here.

x	0	$\frac{1}{2}$	1	$\frac{3}{2}$	2	3	4	−1	−2	−3
y	1	1	1	1	2	2	3	0	0	−1

The ordered pairs in the table suggest the graph shown in **Figure 65.** The domain is $(-\infty, \infty)$. The range is $\{\ldots, -2, -1, 0, 1, 2, \ldots\}$.

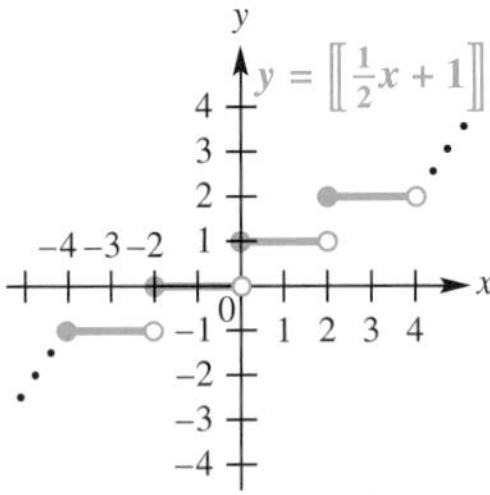

The dots indicate that the graph continues indefinitely in the same pattern.

Figure 65

✔ *Now Try Exercise 45.*

The greatest integer function is an example of a **step function,** a function with a graph that looks like a series of steps.

Classroom Example 4
Suppose that the express mail company in **Example 4** decreases its charges to \$20 for a package weighing up to 2 lb and \$2 for each additional pound or fraction of a pound. Let $y = C(x)$ represent the cost to send a package weighing x pounds. Graph $y = C(x)$ for x in the interval $(0, 6]$.

Answer:

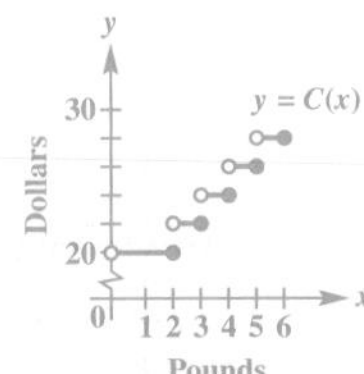

EXAMPLE 4 Applying a Greatest Integer Function

An express mail company charges \$25 for a package weighing up to 2 lb. For each additional pound or fraction of a pound, there is an additional charge of \$3. Let $y = D(x)$ represent the cost to send a package weighing x pounds. Graph $y = D(x)$ for x in the interval $(0, 6]$.

SOLUTION For x in the interval $(0, 2]$, $y = 25$. For x in $(2, 3]$, $y = 25 + 3 = 28$. For x in $(3, 4]$, $y = 28 + 3 = 31$, and so on. The graph, which is that of a step function, is shown in **Figure 66.** In this case, the first step has a different width.

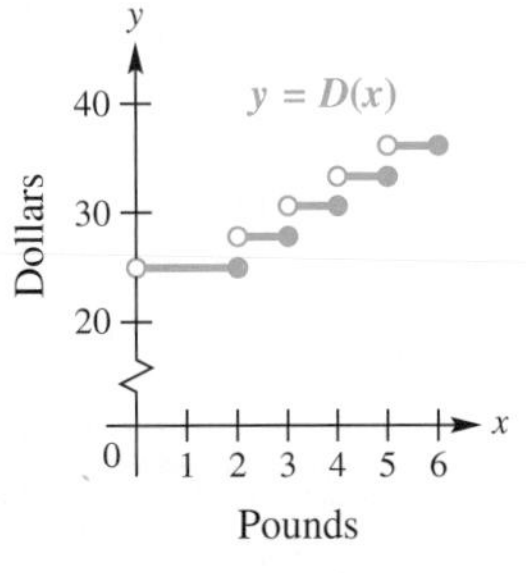

Figure 66

✔ *Now Try Exercise 47.*

The Relation $x = y^2$ ***Recall that a function is a relation where every domain value is paired with one and only one range value.*** Consider the relation defined by the equation $x = y^2$. Notice from the table of selected ordered pairs on the next page that this relation has two different y-values for each positive value of x.

If we plot the points from the table and join them with a smooth curve, we find that the graph of $x = y^2$ is a parabola opening to the right with vertex $(0, 0)$. See **Figure 67(a).** The domain is $[0, \infty)$ and the range is $(-\infty, \infty)$.

Selected Ordered Pairs for $x = y^2$

x	y
0	0
1	± 1
4	± 2
9	± 3

There are two different y-values for the same x-value.

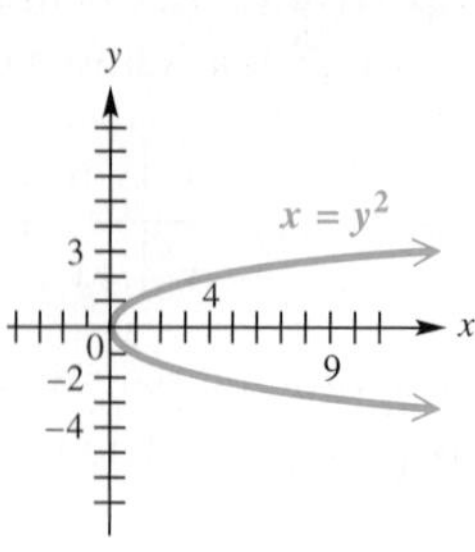

(a)

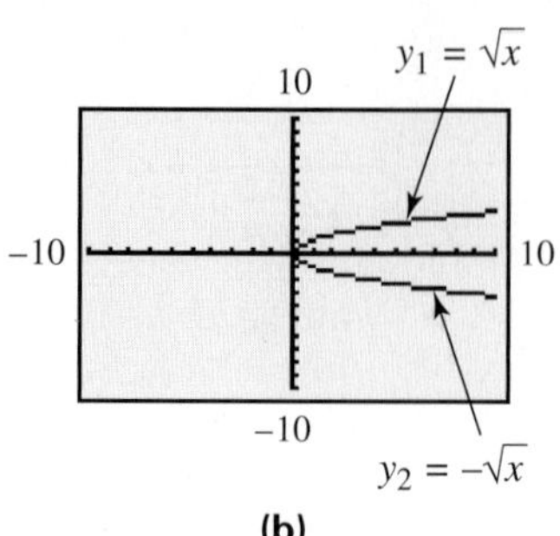

(b)

Figure 67

To use a calculator in function mode to graph the relation $x = y^2$, we graph the two functions $y_1 = \sqrt{x}$ (to generate the top half of the parabola) and $y_2 = -\sqrt{x}$ (to generate the bottom half). See **Figure 67(b).** ■

1. E; $(-\infty, \infty)$ **2.** G; $[0, \infty)$ **3.** A; $(-\infty, \infty)$ **4.** C; $x = y^2$ **5.** F; $y = x$ **6.** B; 1 **7.** H; no **8.** D; $[0, \infty)$
9. B; $\{\dots, -3, -2, -1, 0, 1, 2, 3, \dots\}$ **10.** E and G; $[0, \infty)$; $(-\infty, 0]$

2.6 Exercises

11. $(-\infty, \infty)$ **12.** $(-\infty, \infty)$
13. $[0, \infty)$ **14.** $(-\infty, 0]$
15. $(-\infty, 3)$; $(3, \infty)$
16. $(-\infty, 1)$; $(1, \infty)$
17. (a) -10 (b) -2 (c) -1 (d) 2
18. (a) -7 (b) -3 (c) -2 (d) 2
19. (a) -3 (b) 1 (c) 0 (d) 9
20. (a) 10 (b) -4 (c) -1 (d) -12

21.

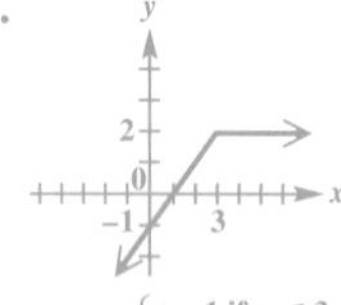

22.

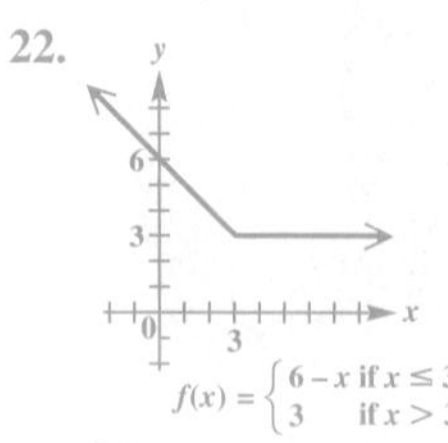

23. **24.**

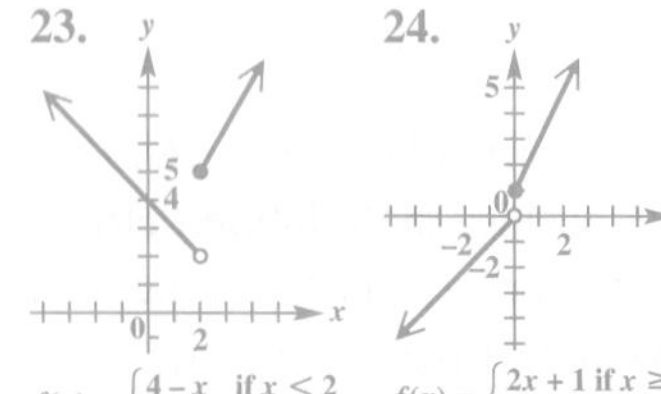

f(x) = {4 − x if x < 2; 1 + 2x if x ≥ 2}
f(x) = {2x + 1 if x ≥ 0; x if x < 0}

Concept Check For Exercises 1–10, refer to the following basic graphs.

A.

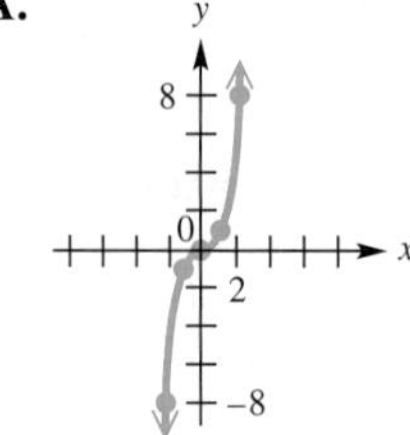

B.

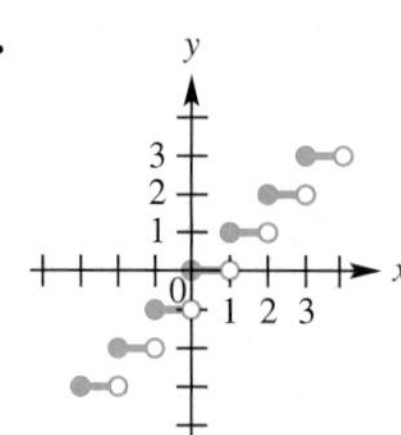

C.

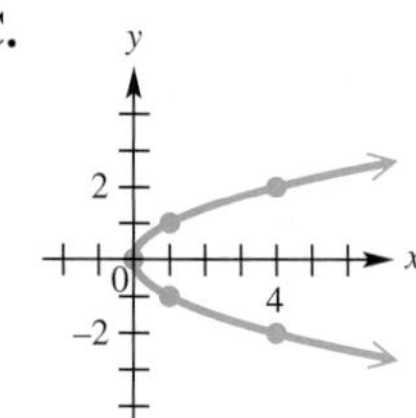

D.

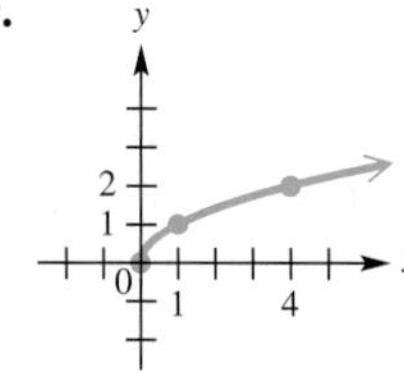

E.

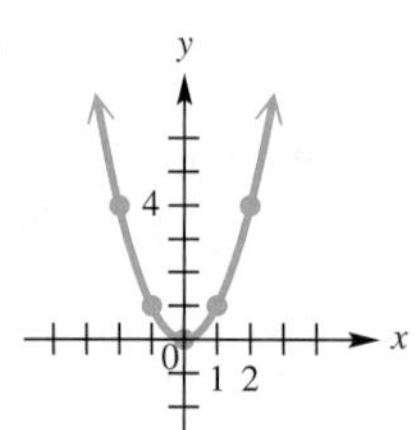

F.

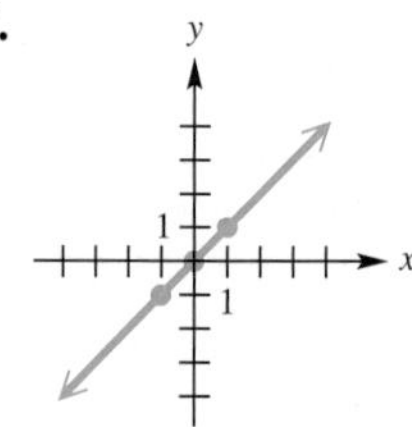

G.

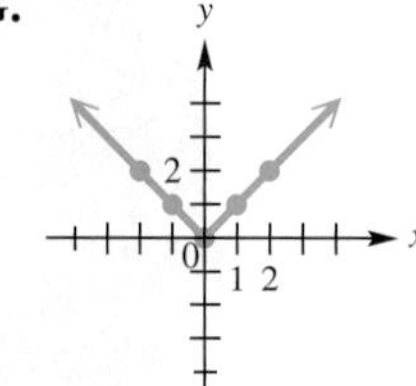

H.

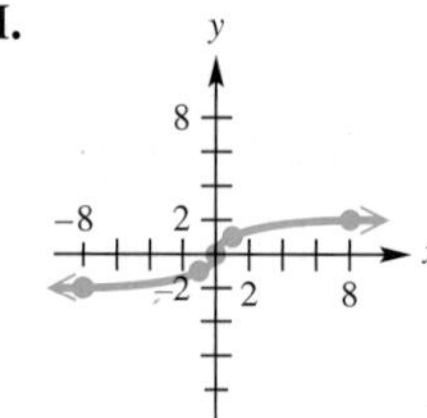

I.

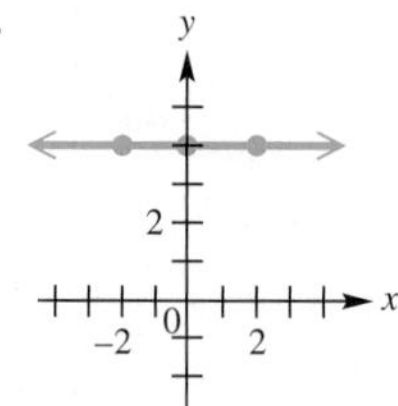

1. Which one is the graph of $y = x^2$? What is its domain?
2. Which one is the graph of $y = |x|$? On what interval is it increasing?
3. Which one is the graph of $y = x^3$? What is its range?
4. Which one is not the graph of a function? What is its equation?
5. Which one is the identity function? What is its equation?

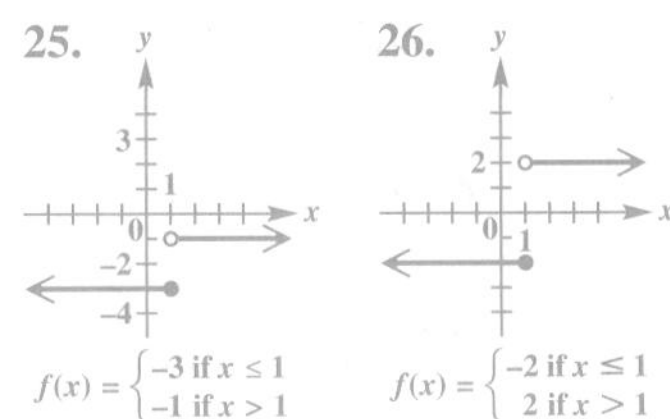

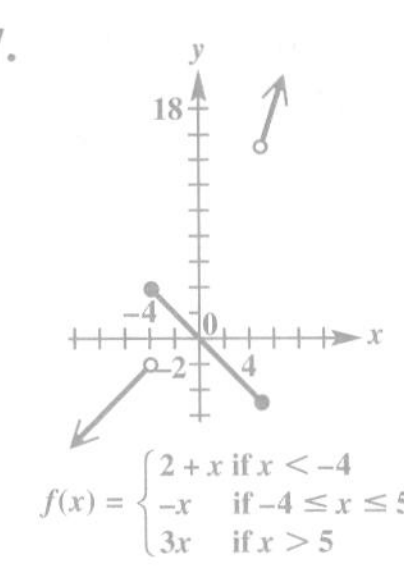

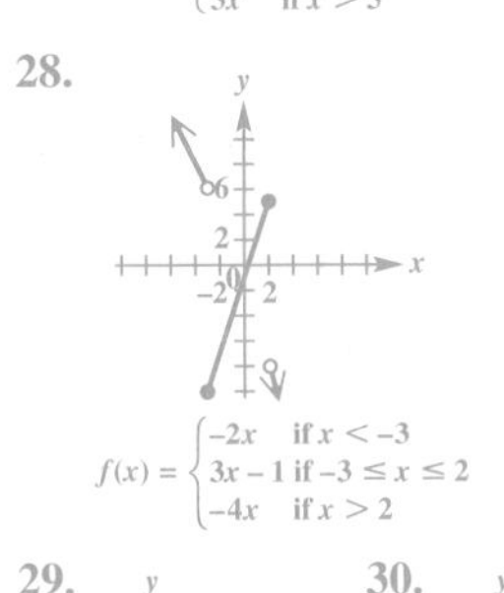

28.
$f(x) = \begin{cases} -2x \quad \text{if } x < -3 \\ 3x - 1 \text{ if } -3 \le x \le 2 \\ -4x \quad \text{if } x > 2 \end{cases}$

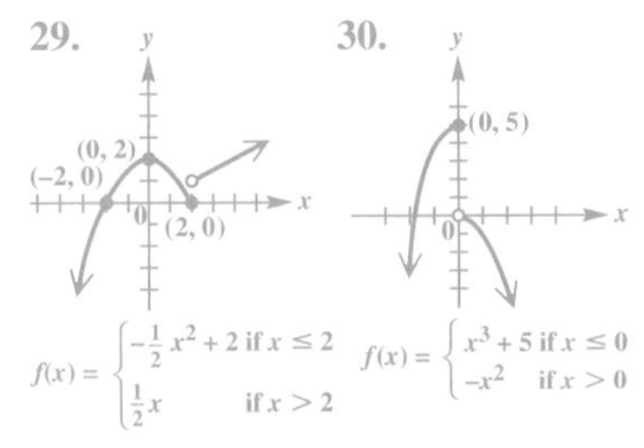

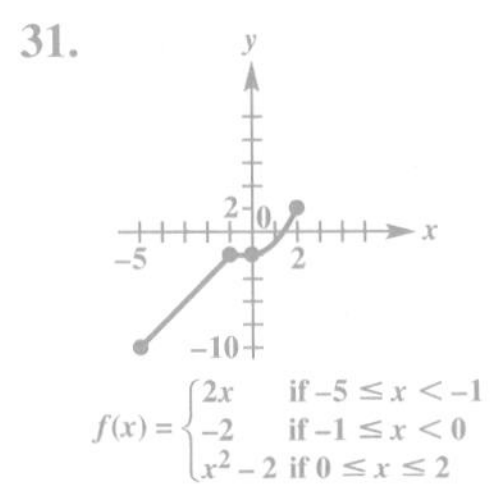

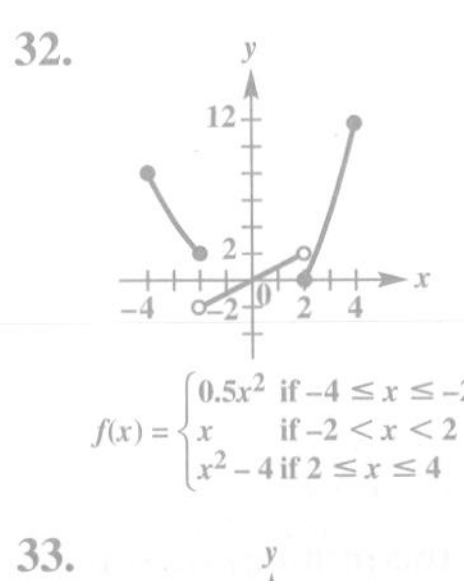

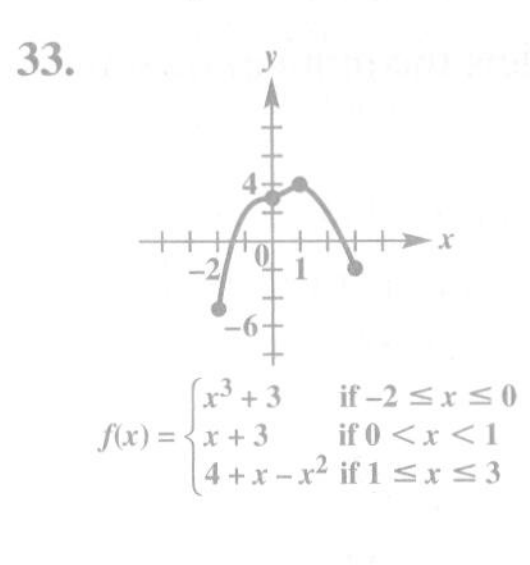

6. Which one is the graph of $y = [\![x]\!]$? What is the value of y when $x = 1.5$?

7. Which one is the graph of $y = \sqrt[3]{x}$? Is there any interval over which the function is decreasing?

8. Which one is the graph of $y = \sqrt{x}$? What is its domain?

9. Which one is discontinuous at many points? What is its range?

10. Which graphs of functions decrease over part of the domain and increase over the rest of the domain? On what intervals do they increase? decrease?

Determine the intervals of the domain over which each function is continuous. ***See Example 1.***

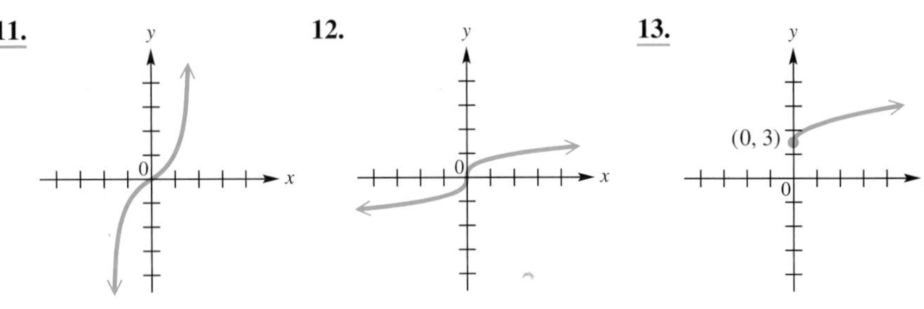

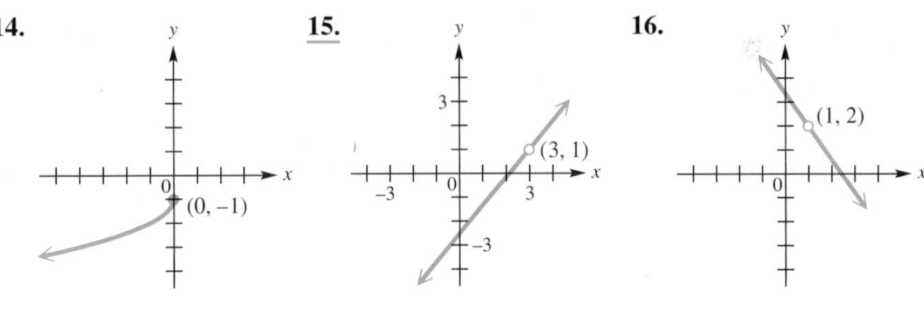

For each piecewise-defined function, find ***(a)*** $f(-5)$, ***(b)*** $f(-1)$, ***(c)*** $f(0)$, *and* ***(d)*** $f(3)$. ***See Example 2.***

17. $f(x) = \begin{cases} 2x & \text{if } x \le -1 \\ x - 1 & \text{if } x > -1 \end{cases}$

18. $f(x) = \begin{cases} x - 2 & \text{if } x < 3 \\ 5 - x & \text{if } x \ge 3 \end{cases}$

19. $f(x) = \begin{cases} 2 + x & \text{if } x < -4 \\ -x & \text{if } -4 \le x \le 2 \\ 3x & \text{if } x > 2 \end{cases}$

20. $f(x) = \begin{cases} -2x & \text{if } x < -3 \\ 3x - 1 & \text{if } -3 \le x \le 2 \\ -4x & \text{if } x > 2 \end{cases}$

Graph each piecewise-defined function. ***See Example 2.***

21. $f(x) = \begin{cases} x - 1 & \text{if } x \le 3 \\ 2 & \text{if } x > 3 \end{cases}$

22. $f(x) = \begin{cases} 6 - x & \text{if } x \le 3 \\ 3 & \text{if } x > 3 \end{cases}$

23. $f(x) = \begin{cases} 4 - x & \text{if } x < 2 \\ 1 + 2x & \text{if } x \ge 2 \end{cases}$

24. $f(x) = \begin{cases} 2x + 1 & \text{if } x \ge 0 \\ x & \text{if } x < 0 \end{cases}$

25. $f(x) = \begin{cases} -3 & \text{if } x \le 1 \\ -1 & \text{if } x > 1 \end{cases}$

26. $f(x) = \begin{cases} -2 & \text{if } x \le 1 \\ 2 & \text{if } x > 1 \end{cases}$

27. $f(x) = \begin{cases} 2 + x & \text{if } x < -4 \\ -x & \text{if } -4 \le x \le 5 \\ 3x & \text{if } x > 5 \end{cases}$

28. $f(x) = \begin{cases} -2x & \text{if } x < -3 \\ 3x - 1 & \text{if } -3 \le x \le 2 \\ -4x & \text{if } x > 2 \end{cases}$

29. $f(x) = \begin{cases} -\frac{1}{2}x^2 + 2 & \text{if } x \le 2 \\ \frac{1}{2}x & \text{if } x > 2 \end{cases}$

30. $f(x) = \begin{cases} x^3 + 5 & \text{if } x \le 0 \\ -x^2 & \text{if } x > 0 \end{cases}$

31. $f(x) = \begin{cases} 2x & \text{if } -5 \le x < -1 \\ -2 & \text{if } -1 \le x < 0 \\ x^2 - 2 & \text{if } 0 \le x \le 2 \end{cases}$

32. $f(x) = \begin{cases} 0.5x^2 & \text{if } -4 \le x \le -2 \\ x & \text{if } -2 < x < 2 \\ x^2 - 4 & \text{if } 2 \le x \le 4 \end{cases}$

33. $f(x) = \begin{cases} x^3 + 3 & \text{if } -2 \le x \le 0 \\ x + 3 & \text{if } 0 < x < 1 \\ 4 + x - x^2 & \text{if } 1 \le x \le 3 \end{cases}$

34. $f(x) = \begin{cases} -2x & \text{if } -3 \le x < -1 \\ x^2 + 1 & \text{if } -1 \le x \le 2 \\ \frac{1}{2}x^3 + 1 & \text{if } 2 < x \le 3 \end{cases}$

Connecting Graphs with Equations *Give a rule for each piecewise-defined function. Also give the domain and range.*

35.

36.

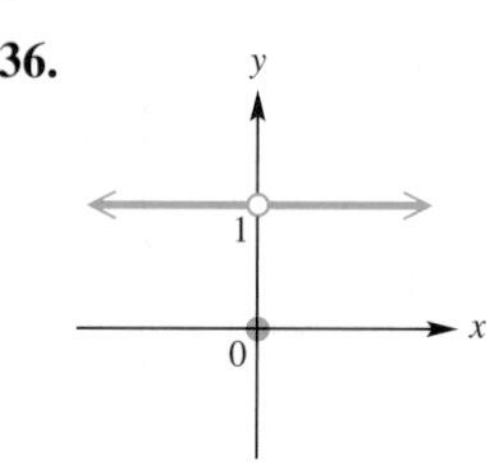

37.

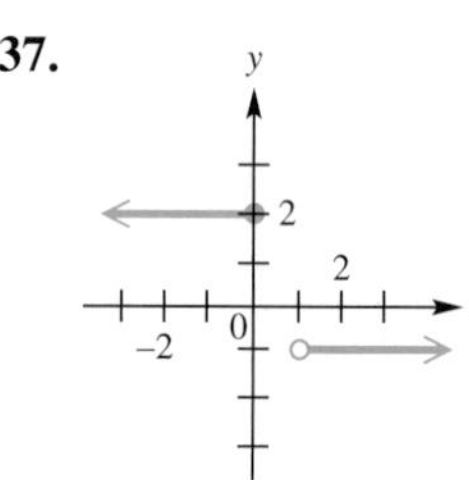

38.

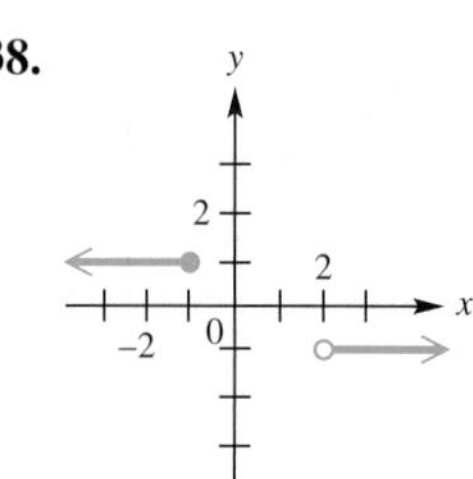

39.

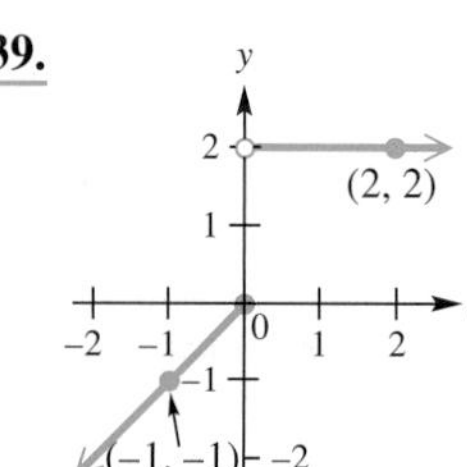

40.

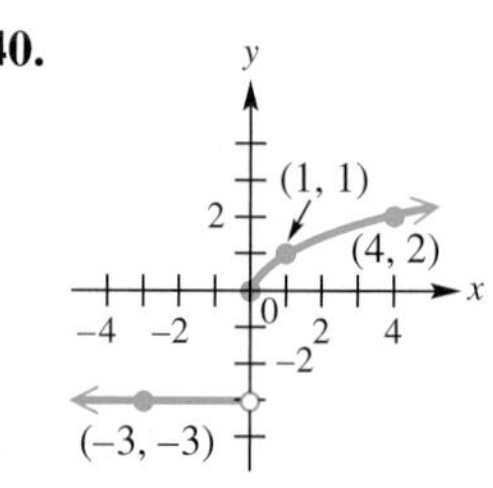

41.

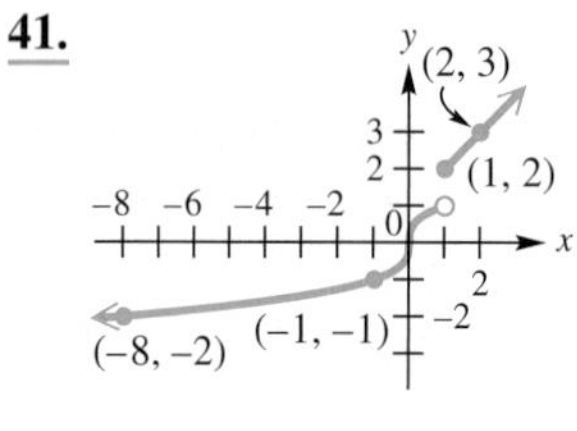

42.

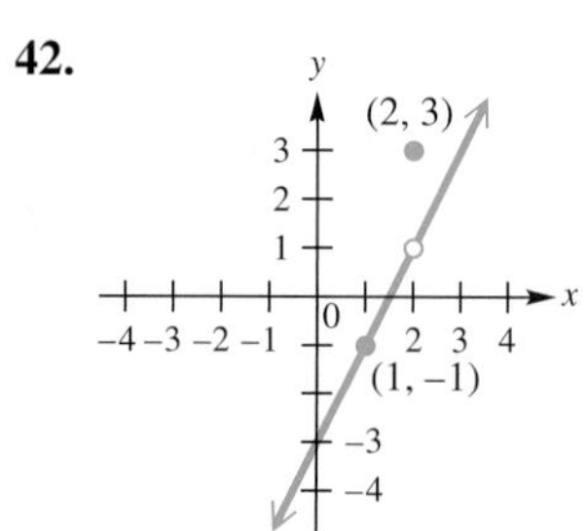

Graph each function. Give the domain and range. ***See Example 3.***

43. $f(x) = [\![-x]\!]$

44. $f(x) = -[\![x]\!]$

45. $f(x) = [\![2x]\!]$

46. $g(x) = [\![2x - 1]\!]$

(Modeling) *Solve each problem.* ***See Example 4.***

47. *Postage Charges* Assume that postage rates are \$0.44 for the first ounce, plus \$0.20 for each additional ounce, and that each letter carries one \$0.44 stamp and as many \$0.20 stamps as necessary. Graph the function f that models the number of stamps on a letter weighing x ounces over the interval $(0, 5]$.

48. *Airport Parking Charges* The cost of parking a car at an airport hourly parking lot is \$3 for the first half-hour and \$2 for each additional half-hour or fraction of a half-hour. Graph the function f that models the cost of parking a car for x hours over the interval $(0, 2]$.

49. *Water in a Tank* Sketch a graph that depicts the amount of water in a 100-gal tank. The tank is initially empty and then filled at a rate of 5 gal per minute. Immediately after it is full, a pump is used to empty the tank at 2 gal per minute.

---

34. $f(x) = \begin{cases} -2x & \text{if } -3 \le x < -1 \\ x^2 + 1 & \text{if } -1 \le x \le 2 \\ \frac{1}{2}x^3 + 1 & \text{if } 2 < x \le 3 \end{cases}$

In Exercises 35–42, we give one of the possible rules, then the domain, and then the range.

35. $f(x) = \begin{cases} -1 & \text{if } x \le 0 \\ 1 & \text{if } x > 0 \end{cases}$; $(-\infty, \infty)$; $\{-1, 1\}$

36. $f(x) = \begin{cases} 1 & \text{if } x \ne 0 \\ 0 & \text{if } x = 0 \end{cases}$; $(-\infty, \infty)$; $\{0, 1\}$

37. $f(x) = \begin{cases} 2 & \text{if } x \le 0 \\ -1 & \text{if } x > 1 \end{cases}$; $(-\infty, 0] \cup (1, \infty)$; $\{-1, 2\}$

38. $f(x) = \begin{cases} 1 & \text{if } x \le -1 \\ -1 & \text{if } x > 2 \end{cases}$; $(-\infty, -1] \cup (2, \infty)$; $\{-1, 1\}$

39. $f(x) = \begin{cases} x & \text{if } x \le 0 \\ 2 & \text{if } x > 0 \end{cases}$; $(-\infty, \infty)$; $(-\infty, 0] \cup \{2\}$

40. $f(x) = \begin{cases} -3 & \text{if } x < 0 \\ \sqrt{x} & \text{if } x \ge 0 \end{cases}$; $(-\infty, \infty)$; $\{-3\} \cup [0, \infty)$

41. $f(x) = \begin{cases} \sqrt[3]{x} & \text{if } x < 1 \\ x + 1 & \text{if } x \ge 1 \end{cases}$; $(-\infty, \infty)$; $(-\infty, 1) \cup [2, \infty)$

42. $f(x) = \begin{cases} 3 & \text{if } x = 2 \\ 2x - 3 & \text{if } x \ne 2 \end{cases}$; $(-\infty, \infty)$; $(-\infty, 1) \cup (1, \infty)$

43. $(-\infty, \infty)$; $\{\ldots, -2, -1, 0, 1, 2, \ldots\}$

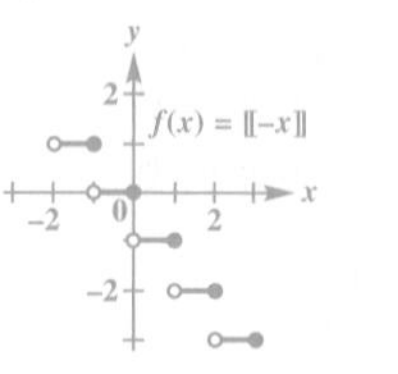

44. $(-\infty, \infty)$; $\{\ldots, -2, -1, 0, 1, 2, \ldots\}$

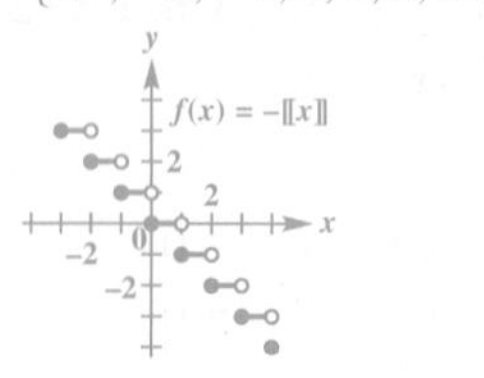

45. $(-\infty, \infty)$; $\{\ldots, -2, -1, 0, 1, 2, \ldots\}$

$f(x) = [\![2x]\!]$

46. $(-\infty, \infty)$; $\{\ldots, -2, -1, 0, 1, 2, \ldots\}$

$g(x) = [\![2x - 1]\!]$

47. Number of Stamps; Weight (in ounces)

48. Dollars; Hours

49. Water (in gallons); Time (in minutes); (20, 100); (70, 0)

50. Distance (in miles); Time (in hours); (0.5, 20); (2.5, 20)

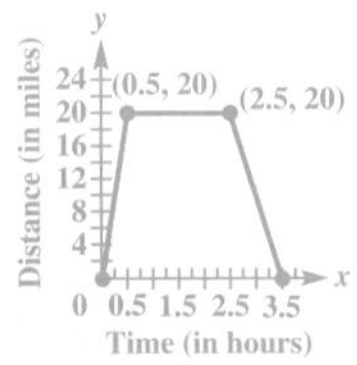

51. (a) for $[0, 4]$: $y = -0.9x + 42.8$; for $(4, 8]$: $y = -1.625x + 45.7$

(b) $f(x) = \begin{cases} -0.9x + 42.8 & \text{if } 0 \le x \le 4 \\ -1.625x + 45.7 & \text{if } 4 < x \le 8 \end{cases}$

52. When $0 \le x \le 3$, the slope is 5, which means the inlet pipe is open and the outlet pipe is closed. When $3 < x \le 5$, the slope is 2, which means both pipes are open. When $5 < x \le 8$, the slope is 0, which means both pipes are closed. When $8 < x \le 10$, the slope is -3, which means the inlet pipe is closed and the outlet pipe is open.

50. *Distance from Home* Sketch a graph showing the distance a person is from home after x hours if he or she drives on a straight road at 40 mph to a park 20 mi away, remains at the park for 2 hr, and then returns home at a speed of 20 mph.

51. *Pickup Truck Market Share* The light vehicle market share (in percent) in the United States for pickup trucks is shown in the graph. Let $x = 0$ represent 1995, $x = 4$ represent 1999, and so on.

(a) Use the points on the graph to write equations for the line segments in the intervals $[0, 4]$ and $(4, 8]$.

(b) Define this graph as a piecewise-defined function f.

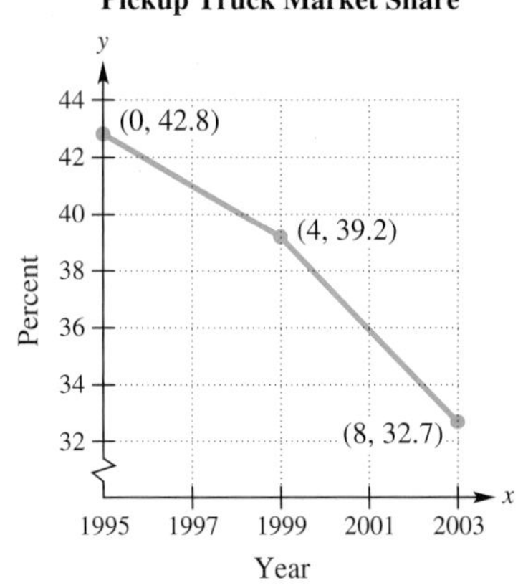

Source: Bureau of Transportation Statistics.

52. *Flow Rates* A water tank has an inlet pipe with a flow rate of 5 gal per minute and an outlet pipe with a flow rate of 3 gal per minute. A pipe can be either closed or completely open. The graph shows the number of gallons of water in the tank after x minutes. Use the concept of slope to interpret each piece of this graph.

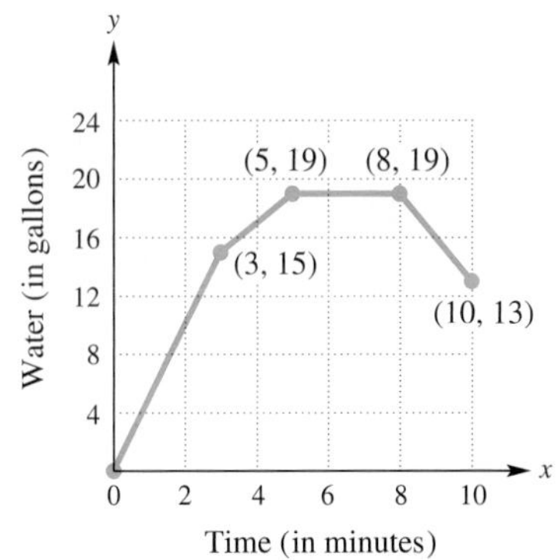

53. *Swimming Pool Levels* The graph of $y = f(x)$ represents the amount of water in thousands of gallons remaining in a swimming pool after x days.

(a) Estimate the initial and final amounts of water contained in the pool.

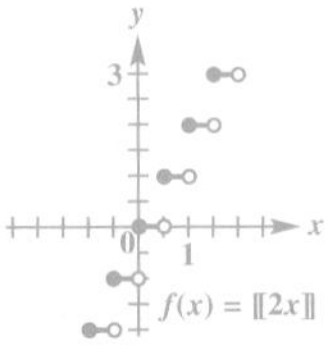

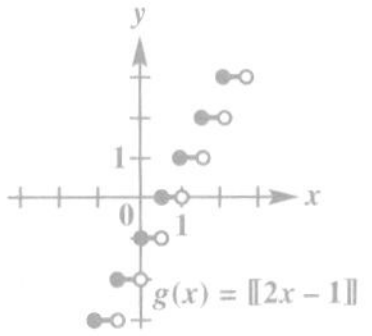

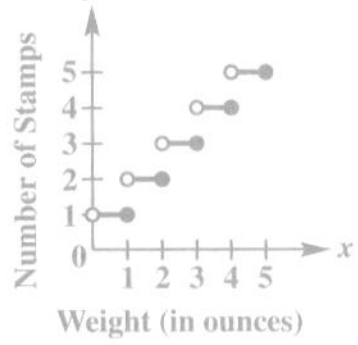

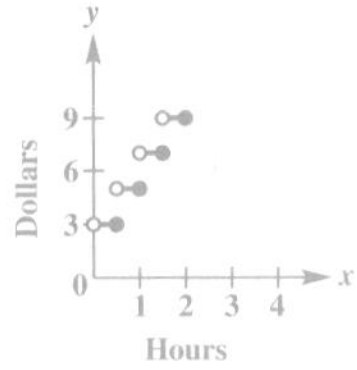

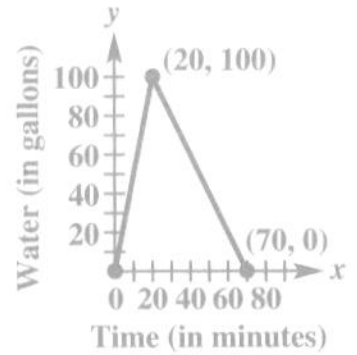

(b) When did the amount of water in the pool remain constant?

(c) Approximate $f(2)$ and $f(4)$.

(d) At what rate was water being drained from the pool when $1 \le x \le 3$?

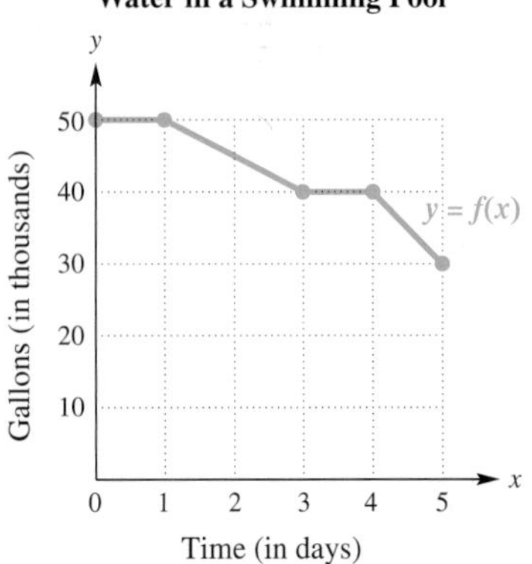

54. *Gasoline Usage* The graph shows the gallons of gasoline y in the gas tank of a car after x hours.

(a) Estimate how much gasoline was in the gas tank when $x = 3$.

(b) When did the car burn gasoline at the greatest rate?

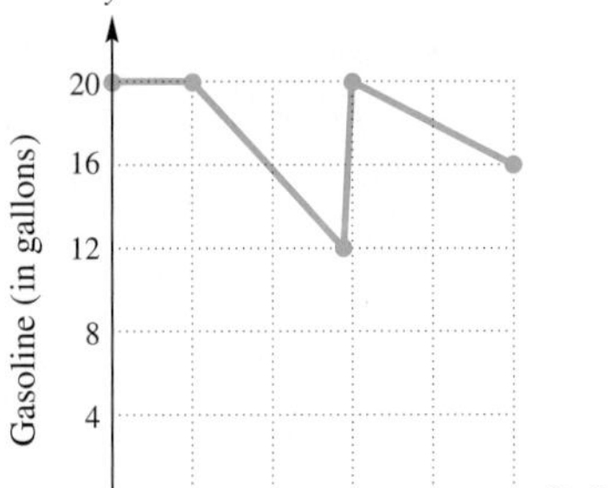

53. (a) 50,000 gal; 30,000 gal
(b) during the first and fourth days
(c) 45,000; 40,000
(d) 5000 gal per day

54. (a) 20 gal
(b) between 1 and approximately 2.9 hr

55. (a) $f(x) = 0.80[\![\frac{x}{2}]\!]$ if $6 \leq x \leq 18$
(b) \$3.20; \$5.60

56. (a)

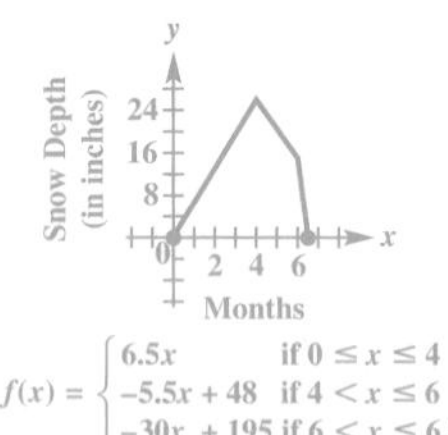

$$f(x) = \begin{cases} 6.5x & \text{if } 0 \leq x \leq 4 \\ -5.5x + 48 & \text{if } 4 < x \leq 6 \\ -30x + 195 & \text{if } 6 < x \leq 6.5 \end{cases}$$

(b) at the beginning of February; 26 in.
(c) begins at the beginning of October; ends in the middle of April

55. *Lumber Costs* Lumber that is used to frame walls of houses is frequently sold in multiples of 2 ft. If the length of a board is not exactly a multiple of 2 ft, there is often no charge for the additional length. For example, if a board measures at least 8 ft, but less than 10 ft, then the consumer is charged for only 8 ft.

(a) Suppose that the cost of lumber is \$0.80 every 2 ft. Find a formula for a function f that computes the cost of a board x feet long for $6 \leq x \leq 18$.
(b) Determine the costs of boards with lengths of 8.5 ft and 15.2 ft.

56. *Snow Depth* The snow depth in Michigan's Isle Royale National Park varies throughout the winter. In a typical winter, the snow depth in inches is approximated by the following function.

$$f(x) = \begin{cases} 6.5x & \text{if } 0 \leq x \leq 4 \\ -5.5x + 48 & \text{if } 4 < x \leq 6 \\ -30x + 195 & \text{if } 6 < x \leq 6.5 \end{cases}$$

Here, x represents the time in months with $x = 0$ representing the beginning of October, $x = 1$ representing the beginning of November, and so on.

(a) Graph $y = f(x)$.
(b) In what month is the snow deepest? What is the deepest snow depth?
(c) In what months does the snow begin and end?

2.7 Graphing Techniques

- Stretching and Shrinking
- Reflecting
- Symmetry
- Even and Odd Functions
- Translations

Graphing techniques presented in this section show how to graph functions that are defined by altering the equation of a basic function.

Stretching and Shrinking

We begin by considering how the graphs of $y = af(x)$ and $y = f(ax)$ compare to the graph of $y = f(x)$, where $a > 0$.

Classroom Example 1
Graph each function.

(a) $g(x) = 2x^2$ (b) $h(x) = \frac{1}{2}x^2$

(c) $k(x) = \left(\frac{1}{2}x\right)^2$

Answers:

(a)

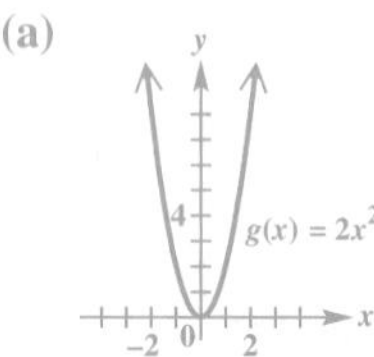

(b)

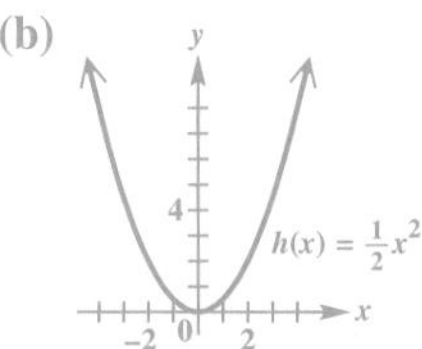

(c)

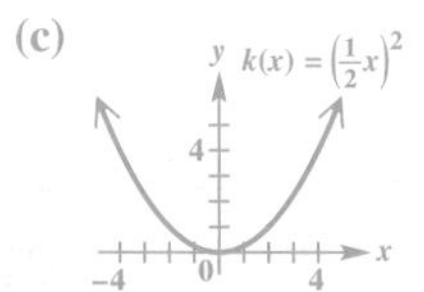

EXAMPLE 1 Stretching or Shrinking a Graph

Graph each function.

(a) $g(x) = 2|x|$ **(b)** $h(x) = \frac{1}{2}|x|$ **(c)** $k(x) = |2x|$

SOLUTION

(a) Comparing the tables of values for $f(x) = |x|$ and $g(x) = 2|x|$ in **Figure 68,** we see that for corresponding x-values, the y-values of g are each twice those of f. The graph of $f(x) = |x|$ is *vertically stretched.* The graph of $g(x)$, shown in blue in **Figure 68,** is narrower than that of $f(x)$, shown in red for comparison.

x	$f(x) = \|x\|$	$g(x) = 2\|x\|$
−2	2	4
−1	1	2
0	0	0
1	1	2
2	2	4

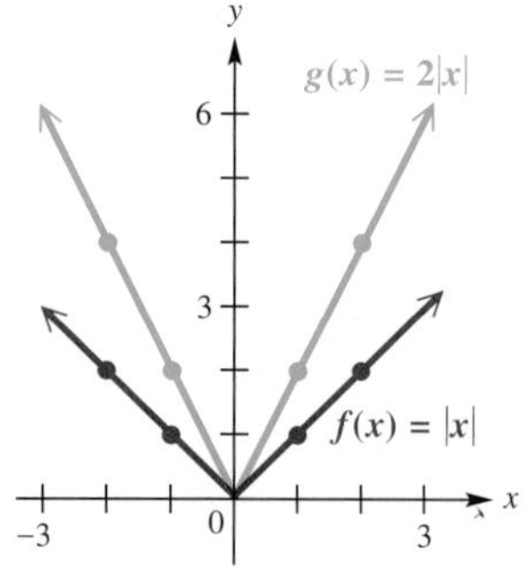

Figure 68

(b) The graph of $h(x) = \frac{1}{2}|x|$ is also the same general shape as that of $f(x)$, but here the coefficient $\frac{1}{2}$ is between 0 and 1 and causes a *vertical shrink*. The graph of $h(x)$ is wider than the graph of $f(x)$, as we see by comparing the tables of values. See **Figure 69.**

x	$f(x) = \lvert x \rvert$	$h(x) = \frac{1}{2}\lvert x \rvert$
-2	2	1
-1	1	$\frac{1}{2}$
0	0	0
1	1	$\frac{1}{2}$
2	2	1

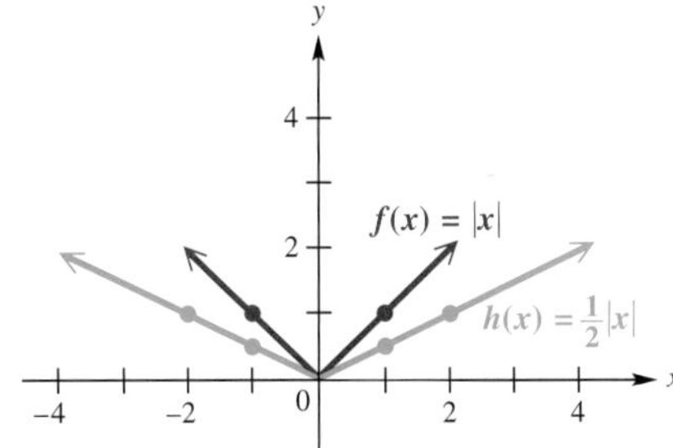

Figure 69

(c) Use Property 2 of absolute value ($|ab| = |a| \cdot |b|$) to rewrite $|2x|$.

$$k(x) = |2x| = |2| \cdot |x| = 2|x| \quad \text{Property 2 (Section R.2)}$$

Therefore, the graph of $k(x) = |2x|$ is the same as the graph of $g(x) = 2|x|$ in part (a). This is a *horizontal shrink* of the graph of $f(x) = |x|$. See **Figure 68** on the previous page.

✔ *Now Try Exercises 7 and 9.*

Vertical Stretching or Shrinking of the Graph of a Function

Suppose that $a > 0$. If a point (x, y) lies on the graph of $y = f(x)$, then the point (x, ay) lies on the graph of $\boldsymbol{y = af(x)}$.

(a) If $a > 1$, then the graph of $y = af(x)$ is a **vertical stretching** of the graph of $y = f(x)$.

(b) If $0 < a < 1$, then the graph of $y = af(x)$ is a **vertical shrinking** of the graph of $y = f(x)$.

Vertical stretching
$a > 1$

Vertical shrinking
$0 < a < 1$

Figure 70

Figure 70 shows graphical interpretations of vertical stretching and shrinking. Notice that in both cases, the x-intercepts of the graph remain the same but the y-intercepts *are* affected.

Graphs of functions can also be stretched and shrunk horizontally.

Horizontal Stretching or Shrinking of the Graph of a Function

Suppose that $a > 0$. If a point (x, y) lies on the graph of $y = f(x)$, then the point $\left(\frac{x}{a}, y\right)$ lies on the graph of $\boldsymbol{y = f(ax)}$.

(a) If $0 < a < 1$, then the graph of $y = f(ax)$ is a **horizontal stretching** of the graph of $y = f(x)$.

(b) If $a > 1$, then the graph of $y = f(ax)$ is a **horizontal shrinking** of the graph of $y = f(x)$.

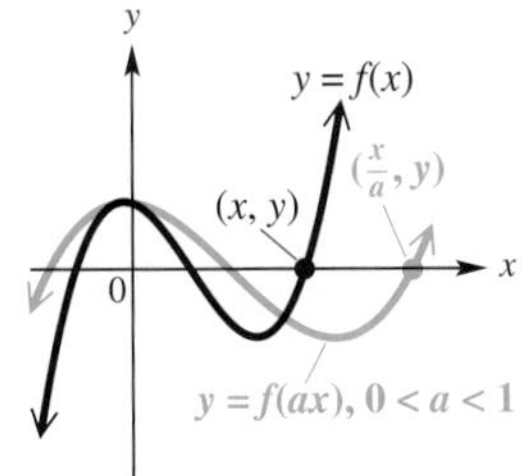

Horizontal stretching
$0 < a < 1$

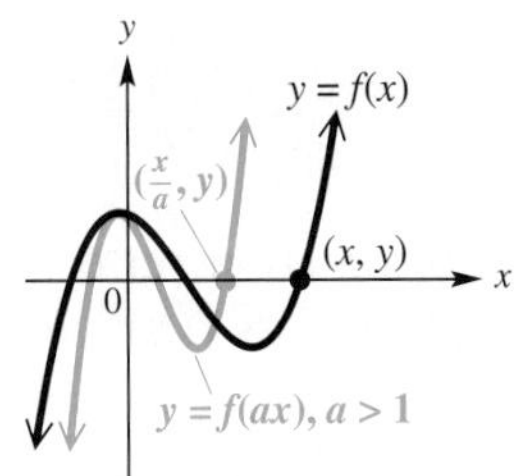

Horizontal shrinking
$a > 1$

Figure 71

See **Figure 71** for graphical interpretations of horizontal stretching and shrinking. Notice that in both cases, the y-intercept remains the same but the x-intercepts *are* affected.

Reflecting Forming the mirror image of a graph across a line is called **reflecting the graph across the line.**

Classroom Example 2
Graph each function.
(a) $g(x) = -|x|$
(b) $h(x) = |-x|$

Answers:

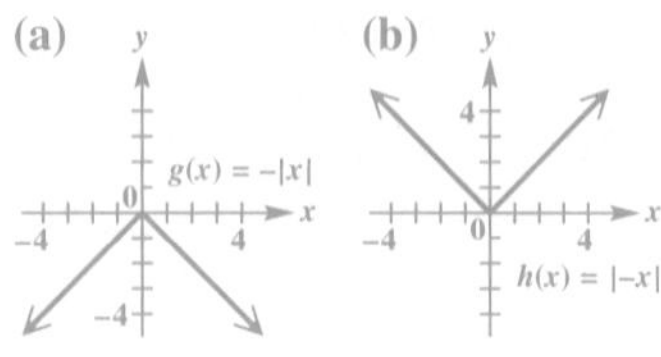

EXAMPLE 2 Reflecting a Graph across an Axis

Graph each function.

(a) $g(x) = -\sqrt{x}$ **(b)** $h(x) = \sqrt{-x}$

SOLUTION

(a) The tables of values for $g(x) = -\sqrt{x}$ and $f(x) = \sqrt{x}$ are shown with their graphs in **Figure 72.** As the tables suggest, every y-value of the graph of $g(x) = -\sqrt{x}$ is the negative of the corresponding y-value of $f(x) = \sqrt{x}$. This has the effect of reflecting the graph across the x-axis.

x	$f(x) = \sqrt{x}$	$g(x) = -\sqrt{x}$
0	0	0
1	1	−1
4	2	−2

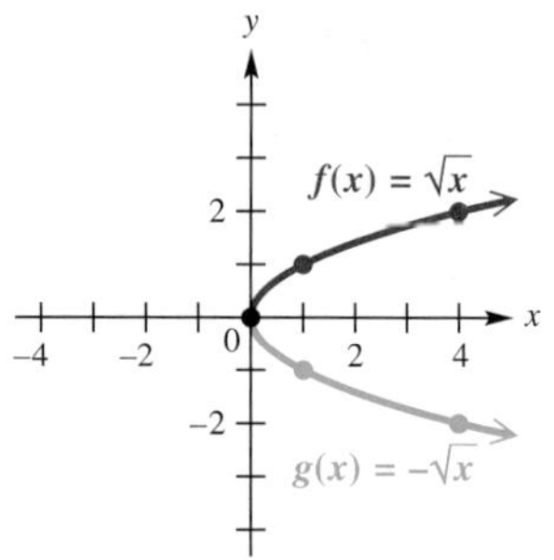

Figure 72

(b) The domain of $h(x) = \sqrt{-x}$ is $x \leq 0$, while the domain of $f(x) = \sqrt{x}$ is $x \geq 0$. Choosing x-values for $h(x)$ that are negatives of those used for $f(x)$, we see that corresponding y-values are the same. The graph of h is a reflection of the graph of f across the y-axis. See **Figure 73.**

x	$f(x) = \sqrt{x}$	$h(x) = \sqrt{-x}$
−4	undefined	2
−1	undefined	1
0	0	0
1	1	undefined
4	2	undefined

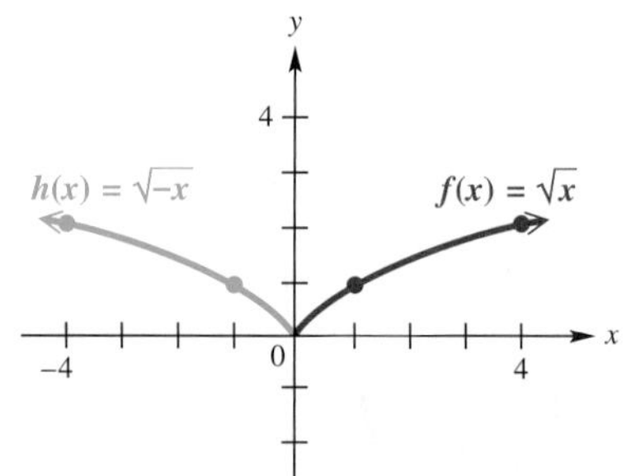

Figure 73

Now Try Exercises 17 and 23.

The graphs in **Example 2** suggest the following generalizations.

Reflecting across an Axis

The graph of $\boldsymbol{y = -f(x)}$ is the same as the graph of $y = f(x)$ reflected across the x-axis. (If a point (x, y) lies on the graph of $y = f(x)$, then $(x, -y)$ lies on this reflection.)

The graph of $\boldsymbol{y = f(-x)}$ is the same as the graph of $y = f(x)$ reflected across the y-axis. (If a point (x, y) lies on the graph of $y = f(x)$, then $(-x, y)$ lies on this reflection.)

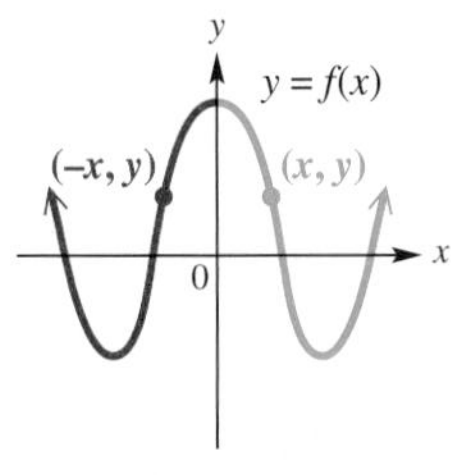

y-axis symmetry

(a)

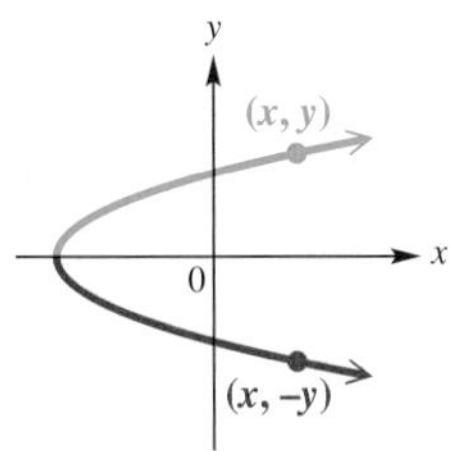

x-axis symmetry

(b)

Figure 74

Symmetry The graph of f shown in **Figure 74(a)** is cut in half by the y-axis with each half the mirror image of the other half. Such a graph is *symmetric with respect to the y-axis.* ***The point*** $(-x, y)$ ***is on the graph whenever the point*** (x, y) ***is on the graph.***

Similarly, if the graph in **Figure 74(b)** were folded in half along the x-axis, the portion at the top would exactly match the portion at the bottom. Such a graph is *symmetric with respect to the x-axis.* ***The point*** $(x, -y)$ ***is on the graph whenever the point*** (x, y) ***is on the graph.***

Symmetry with Respect to an Axis

The graph of an equation is **symmetric with respect to the y-axis** if the replacement of x with $-x$ results in an equivalent equation.

The graph of an equation is **symmetric with respect to the x-axis** if the replacement of y with $-y$ results in an equivalent equation.

In **Section 2.6** we introduced graphs of basic functions. The squaring function and the absolute value function are examples of functions that are symmetric with respect to the y-axis.

Teaching Tip Use numerical coordinates to demonstrate the properties of symmetry.

EXAMPLE 3 Testing for Symmetry with Respect to an Axis

Test for symmetry with respect to the x-axis and the y-axis.

(a) $y = x^2 + 4$ **(b)** $x = y^2 - 3$ **(c)** $x^2 + y^2 = 16$ **(d)** $2x + y = 4$

Classroom Example 3
Test for symmetry with respect to the x-axis and the y-axis.
(a) $x = |y|$
(b) $y = |x| - 3$
(c) $2x - y = 6$
(d) $x^2 + y^2 = 25$

Answers:
(a) x-axis
(b) y-axis
(c) neither axis
(d) x-axis, y-axis

SOLUTION

(a) In $y = x^2 + 4$, replace x with $-x$.

$$y = x^2 + 4$$
$$y = (-x)^2 + 4 \quad \text{Equivalent}$$
$$y = x^2 + 4$$

Use parentheses around $-x$.

The result is the same as the original equation, so the graph, shown in **Figure 75,** is symmetric with respect to the y-axis. Substituting $-y$ for y does not result in an equivalent equation, and thus the graph is *not* symmetric with respect to the x-axis.

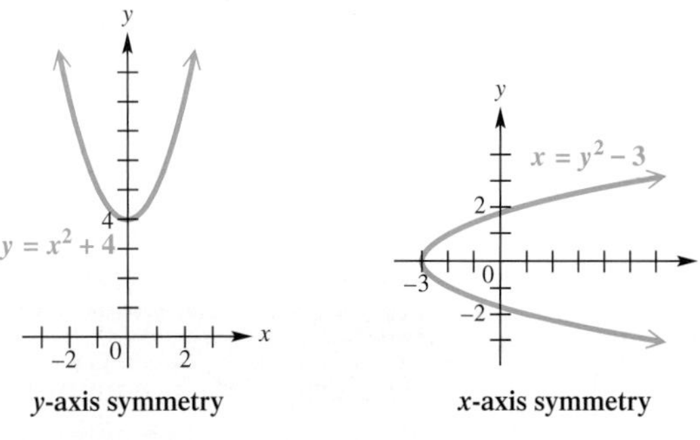

y-axis symmetry

Figure 75

x-axis symmetry

Figure 76

(b) In $x = y^2 - 3$, replace y with $-y$.

$$x = (-y)^2 - 3 = y^2 - 3 \quad \text{Same as the original equation}$$

The graph is symmetric with respect to the x-axis, as shown in **Figure 76.** It is *not* symmetric with respect to the y-axis.

LOOKING AHEAD TO CALCULUS

The tools of calculus enable us to find areas of regions in the plane. To find the area of the region below the graph of $y = x^2$, above the x-axis, bounded on the left by the line $x = -2$ and on the right by $x = 2$, draw a sketch of this region. Notice that due to the symmetry of the graph of $y = x^2$, the desired area is twice the area to the right of the y-axis. Thus, symmetry can be used to reduce the original problem to an easier one by simply finding the area to the right of the y-axis and then doubling the answer.

(c) Substitute $-x$ for x and then $-y$ for y in $x^2 + y^2 = 16$.

$$(-x)^2 + y^2 = 16 \quad \text{and} \quad x^2 + (-y)^2 = 16$$

Both simplify to the original equation,

$$x^2 + y^2 = 16.$$

The graph, a circle of radius 4 centered at the origin, is symmetric with respect to *both* axes. See **Figure 77.**

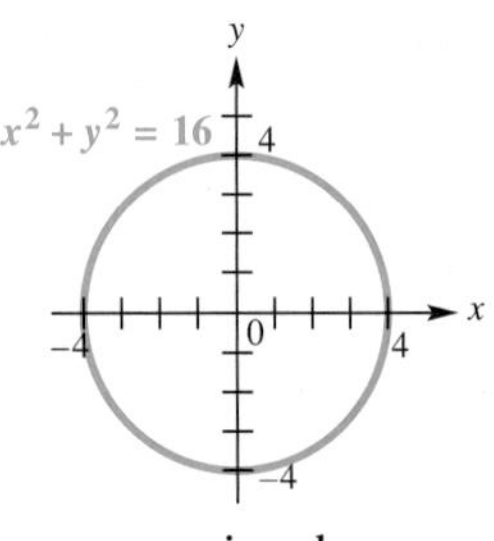

x-axis and y-axis symmetry

Figure 77

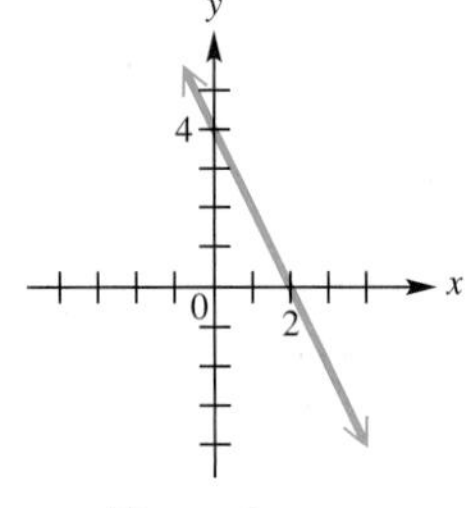

No x-axis or y-axis symmetry

Figure 78

(d) In $2x + y = 4$, replace x with $-x$ to get $-2x + y = 4$. Then replace y with $-y$ to get $2x - y = 4$. Neither case produces an equivalent equation, so this graph is not symmetric with respect to either axis. See **Figure 78.**

Now Try Exercise 35.

Another kind of symmetry occurs when a graph can be rotated 180° about the origin, with the result coinciding exactly with the original graph. Symmetry of this type is called *symmetry with respect to the origin.* ***The point $(-x, -y)$ is on the graph whenever the point (x, y) is on the graph.*** See **Figure 79.**

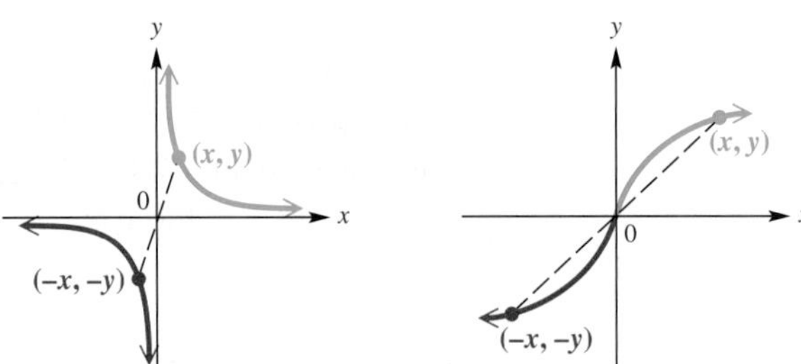

Origin symmetry

Figure 79

Symmetry with Respect to the Origin

The graph of an equation is **symmetric with respect to the origin** if the replacement of both x with $-x$ and y with $-y$ at the same time results in an equivalent equation.

The cubing and cube root functions are examples of functions whose graphs are symmetric with respect to the origin.

Classroom Example 4
Are the following graphs symmetric with respect to the origin?
(a) $y = -2x^3$
(b) $y = -2x^2$

Answers:
(a) yes (b) no

EXAMPLE 4 Testing for Symmetry with Respect to the Origin

Are the following graphs symmetric with respect to the origin?

(a) $x^2 + y^2 = 16$ **(b)** $y = x^3$

SOLUTION

(a) Replace x with $-x$ and y with $-y$.

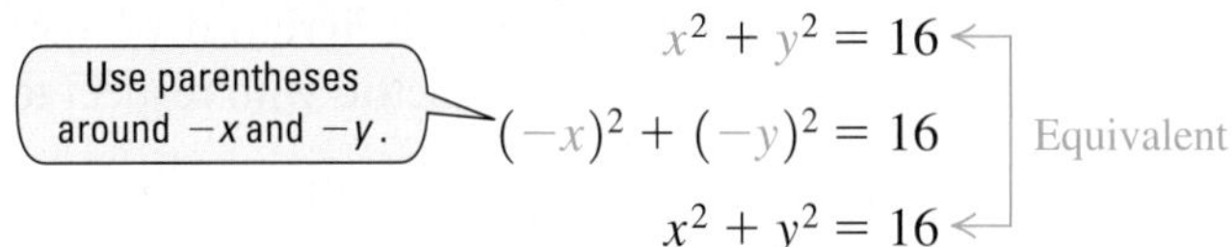

$$x^2 + y^2 = 16$$
$$(-x)^2 + (-y)^2 = 16$$
$$x^2 + y^2 = 16$$

The graph, which is the circle shown in **Figure 77** in **Example 3(c),** is symmetric with respect to the origin.

(b) Replace x with $-x$ and y with $-y$.

$$y = x^3$$
$$-y = (-x)^3$$
$$-y = -x^3$$
$$y = x^3$$

Equivalent

The graph, which is that of the cubing function, is symmetric with respect to the origin and is shown in **Figure 80.**

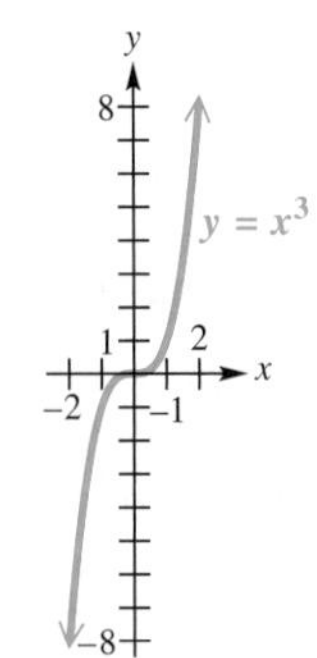

Origin symmetry

Figure 80

Now Try Exercise 39.

Notice the following important concepts regarding symmetry:

- A graph symmetric with respect to both the x- and y-axes is automatically symmetric with respect to the origin. (See **Figure 77.**)
- A graph symmetric with respect to the origin need *not* be symmetric with respect to either axis. (See **Figure 80.**)
- Of the three types of symmetry—with respect to the x-axis, with respect to the y-axis, and with respect to the origin—a graph possessing any two types must also exhibit the third type of symmetry.

The various tests for symmetry are summarized below.

Tests for Symmetry

	Symmetry with Respect to:		
	***x*-axis**	***y*-axis**	**Origin**
Equation is unchanged if:	y is replaced with $-y$	x is replaced with $-x$	x is replaced with $-x$ and y is replaced with $-y$
Example:			

Even and Odd Functions The concepts of symmetry with respect to the y-axis and symmetry with respect to the origin are closely associated with the concepts of *even* and *odd functions.*

Even and Odd Functions

A function f is an **even function** if $f(-x) = f(x)$ for all x in the domain of f. (Its graph is symmetric with respect to the y-axis.)

A function f is an **odd function** if $f(-x) = -f(x)$ for all x in the domain of f. (Its graph is symmetric with respect to the origin.)

Classroom Example 5
Determine whether each function defined is *even, odd,* or *neither.*

(a) $g(x) = x^5 + 2x^3 - 3x$

(b) $h(x) = 2x^2 - 3$

(c) $k(x) = x^2 + 6x + 9$

Answers:
(a) odd (b) even
(c) neither

EXAMPLE 5 Determining Whether Functions Are Even, Odd, or Neither

Determine whether each function defined is *even, odd,* or *neither.*

(a) $f(x) = 8x^4 - 3x^2$ **(b)** $f(x) = 6x^3 - 9x$ **(c)** $f(x) = 3x^2 + 5x$

SOLUTION

(a) Replacing x with $-x$ gives the following.

$$f(x) = 8x^4 - 3x^2$$

$$f(-x) = 8(-x)^4 - 3(-x)^2 \quad \text{Replace } x \text{ with } -x.$$

$$f(-x) = 8x^4 - 3x^2 \quad \text{Apply the exponents.}$$

$$f(-x) = f(x) \quad 8x^4 - 3x^2 = f(x)$$

Since $f(-x) = f(x)$ for each x in the domain of the function, f is even.

(b) $$f(x) = 6x^3 - 9x$$

$$f(-x) = 6(-x)^3 - 9(-x) \quad \text{Replace } x \text{ with } -x.$$

$$f(-x) = -6x^3 + 9x$$ Be careful with signs.

$$f(-x) = -f(x) \quad -6x^3 + 9x = -(6x^3 - 9x) = -f(x)$$

The function f is odd because $f(-x) = -f(x)$.

(c) $$f(x) = 3x^2 + 5x$$

$$f(-x) = 3(-x)^2 + 5(-x) \quad \text{Replace } x \text{ with } -x.$$

$$f(-x) = 3x^2 - 5x \quad \text{Simplify.}$$

Since $f(-x) \neq f(x)$ and $f(-x) \neq -f(x)$, the function f is neither even nor odd.

✔ *Now Try Exercises 43, 45, and 47.*

NOTE Consider a function defined by a polynomial in x.

- If the function has only *even* exponents on x (including the case of a constant where x^0 is understood to have the even exponent 0), it will *always* be an even function.
- Similarly, if only *odd* exponents appear on x, the function will be an odd function.

Translations The next examples show the results of horizontal and vertical shifts, or **translations,** of the graph of $f(x) = |x|$.

Classroom Example 6
Graph $f(x) = x^2 + 2$.

Answer:

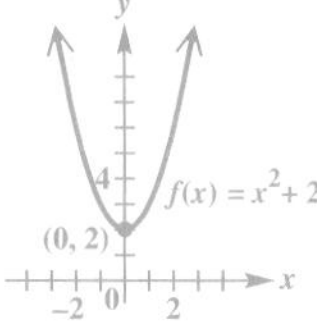

EXAMPLE 6 Translating a Graph Vertically

Graph $g(x) = |x| - 4$.

SOLUTION By comparing the table of values for $g(x) = |x| - 4$ and $f(x) = |x|$ shown with **Figure 81,** we see that for corresponding x-values, the y-values of g are each 4 *less* than those for f. Thus, the graph of $g(x) = |x| - 4$ is the same as that of $f(x) = |x|$, but translated 4 units down. See **Figure 81.** The lowest point is at $(0, -4)$. The graph is symmetric with respect to the y-axis and is therefore the graph of an even function.

x	$f(x) = \|x\|$	$g(x) = \|x\| - 4$
−4	4	0
−1	1	−3
0	0	−4
1	1	−3
4	4	0

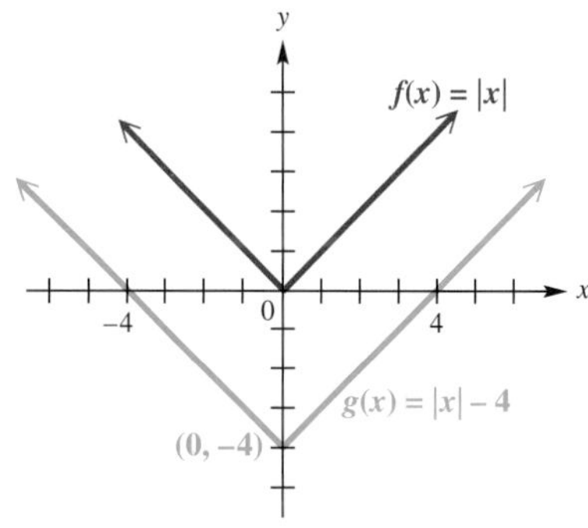

Figure 81

✔ *Now Try Exercise 57.*

The graphs in **Example 6** suggest the following generalization.

Vertical Translations

If a function g is defined by $\mathbf{g(x) = f(x) + c}$, where c is a real number, then for every point (x, y) on the graph of f, there will be a corresponding point $(x, y + c)$ on the graph of g.

The graph of g will be the same as the graph of f, but translated c units up if c is positive or $|c|$ units down if c is negative. The graph of g is called a **vertical translation** of the graph of f.

Figure 82 shows a graph of a function f and two vertical translations of f. **Figure 83** shows two vertical translations of $y_1 = x^2$ on a TI-83/84 Plus calculator screen.

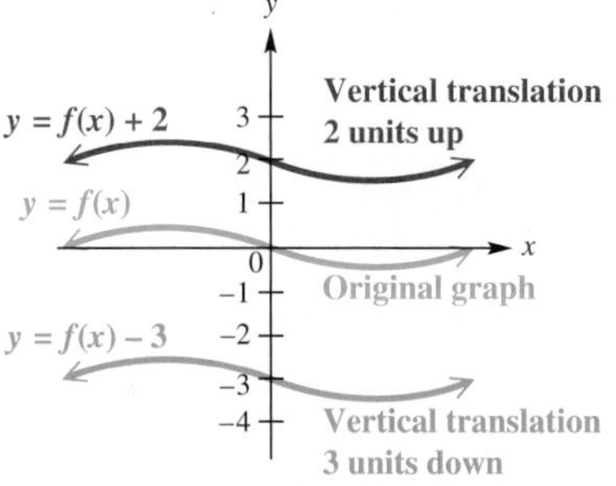

Figure 82

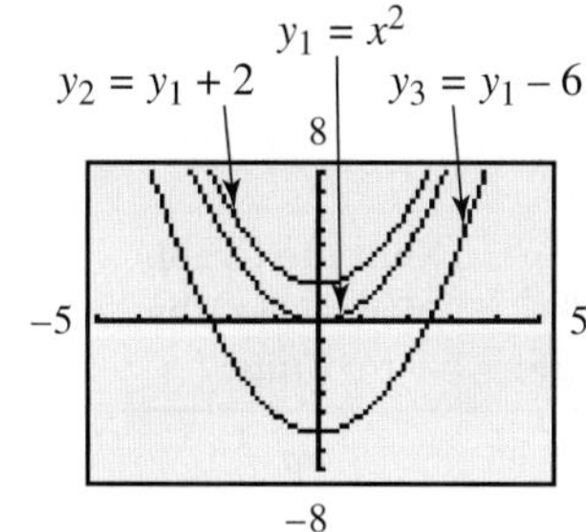

y_2 is the graph of y_1 translated 2 units *up*. y_3 is that of y_1 translated 6 units *down*.

Figure 83

Classroom Example 7
Graph $f(x) = (x + 4)^2$.

Answer:

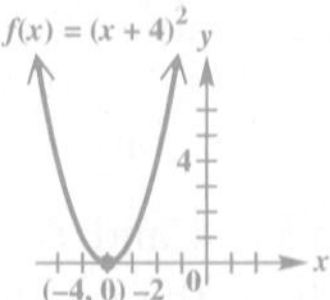

EXAMPLE 7 Translating a Graph Horizontally

Graph $g(x) = |x - 4|$.

SOLUTION Comparing the tables of values given with **Figure 84** shows that for corresponding y-values, the x-values of g are each 4 *more* than those for f. The graph of $g(x) = |x - 4|$ is the same as that of $f(x) = |x|$, but translated 4 units to the right. The lowest point is at $(4, 0)$. As suggested by the graphs in **Figure 84,** this graph is symmetric with respect to the line $x = 4$.

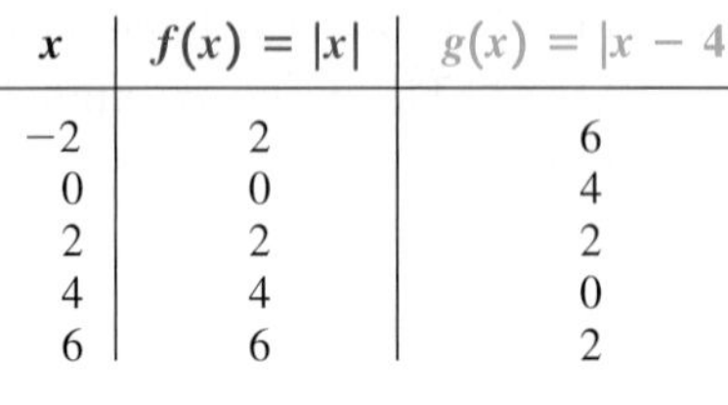

x	$f(x) = \|x\|$	$g(x) = \|x - 4\|$
−2	2	6
0	0	4
2	2	2
4	4	0
6	6	2

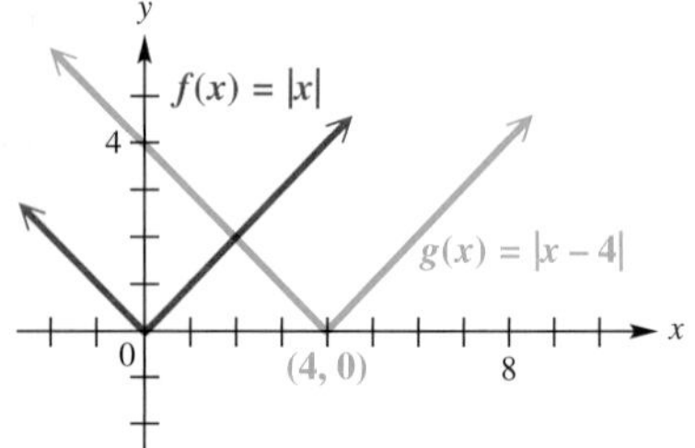

Figure 84

✔ *Now Try Exercise 55.*

The graphs in **Example 7** suggest the following generalization.

Horizontal Translations

If a function g is defined by $\mathbf{g(x) = f(x - c)}$, where c is a real number, then for every point (x, y) on the graph of f, there will be a corresponding point $(x + c, y)$ on the graph of g.

The graph of g will be the same as the graph of f, but translated c units to the right if c is positive or $|c|$ units to the left if c is negative. The graph of g is called a **horizontal translation** of the graph of f.

Figure 85 shows a graph of a function f and two horizontal translations of f. **Figure 86** shows two horizontal translations of $y_1 = x^2$ on a TI-83/84 Plus calculator screen.

$(c > 0)$ To Graph:	Shift the Graph of $y = f(x)$ by c Units:
$y = f(x) + c$	up
$y = f(x) - c$	down
$y = f(x + c)$	left
$y = f(x - c)$	right

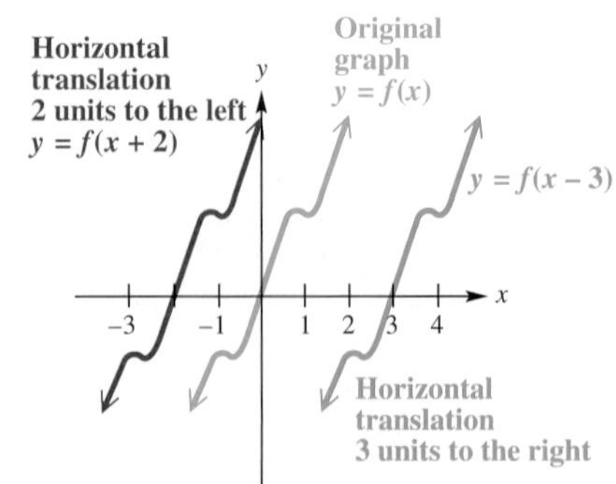

Figure 85

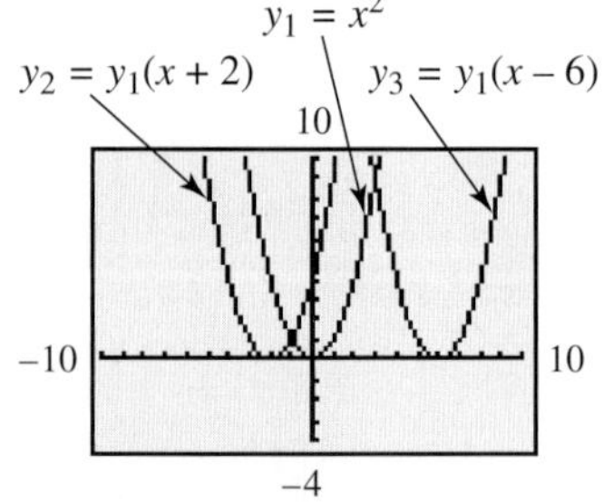

y_2 is the graph of y_1 translated 2 units to the *left*. y_3 is that of y_1 translated 6 units to the *right*.

Figure 86

Vertical and horizontal translations are summarized in the table in the margin, where f is a function, and c is a positive number.

CAUTION ***Be careful when translating graphs horizontally.*** To determine the direction and magnitude of horizontal translations, find the value that would cause the expression in parentheses to equal 0.

For example, the graph of $y = (x - 5)^2$ would be translated 5 units to the *right* of $y = x^2$, because $x = +5$ would cause $x - 5$ to equal 0. On the other hand, the graph of $y = (x + 5)^2$ would be translated 5 units to the *left* of $y = x^2$, because $x = -5$ would cause $x + 5$ to equal 0.

Classroom Example 8
Graph each function.

(a) $f(x) = -(x-1)^2 + 4$

(b) $g(x) = -2|x+3|$

(c) $h(x) = \frac{1}{2}\sqrt{x+2} - 3$

Answers:

(a)

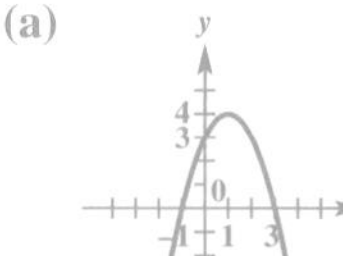

(b)

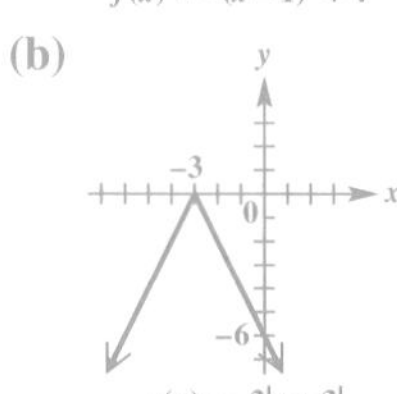

(c)

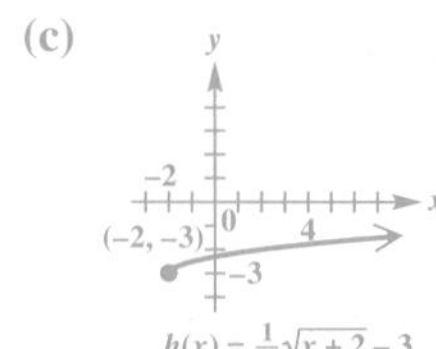

EXAMPLE 8 Using More Than One Transformation

Graph each function.

(a) $f(x) = -|x+3| + 1$ **(b)** $h(x) = |2x - 4|$ **(c)** $g(x) = -\frac{1}{2}x^2 + 4$

SOLUTION

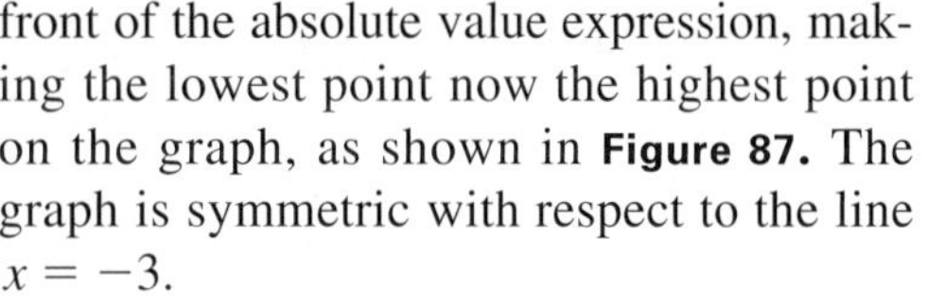

(a) To graph $f(x) = -|x+3| + 1$, the *lowest* point on the graph of $y = |x|$ is translated 3 units to the left and 1 unit up. The graph opens down because of the negative sign in front of the absolute value expression, making the lowest point now the highest point on the graph, as shown in **Figure 87.** The graph is symmetric with respect to the line $x = -3$.

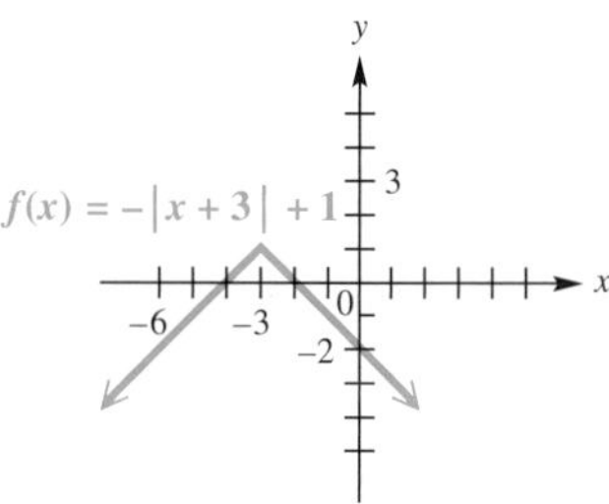

Figure 87

(b) To determine the horizontal translation, factor out 2.

$$\begin{aligned} h(x) &= |2x - 4| \\ &= |2(x-2)| && \text{Factor out 2.} \\ &= |2| \cdot |x-2| && |ab| = |a| \cdot |b| \text{ (Section R.2)} \\ &= 2|x-2| && |2| = 2 \end{aligned}$$

The graph of h is the graph of $y = |x|$ translated 2 units to the right, and vertically stretched by a factor of 2. Horizontal shrinking gives the same appearance as vertical stretching for this function. See **Figure 88.**

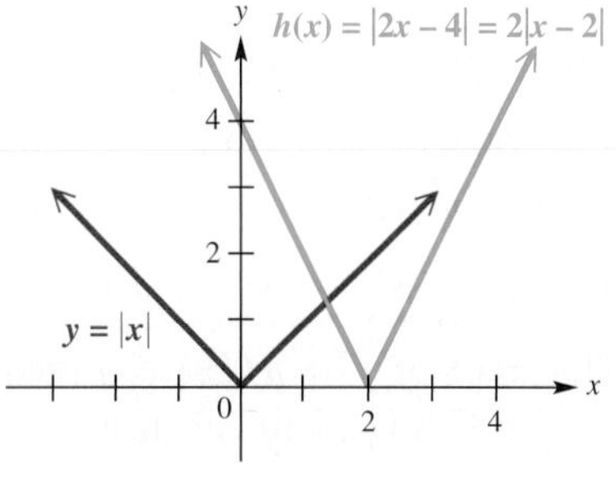

Figure 88

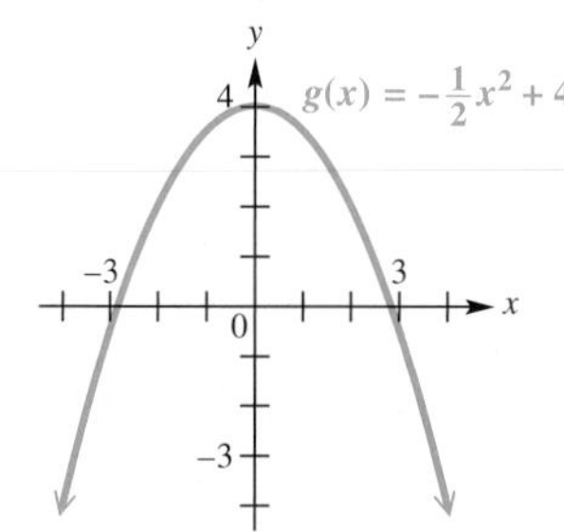

Figure 89

(c) The graph of $g(x) = -\frac{1}{2}x^2 + 4$ has the same shape as that of $y = x^2$, but it is wider (that is, shrunken vertically), reflected across the x-axis because the coefficient $-\frac{1}{2}$ is negative, and then translated 4 units up. See **Figure 89.**

Now Try Exercises 61, 63, and 71.

Classroom Example 9
Use the graph of $y = f(x)$ shown in **Figure 90** to sketch each of the following graphs.

(a) $g(x) = f(x) - 2$
(b) $h(x) = f(x - 2)$
(c) $k(x) = f(x + 1) + 2$

Answers:

(a)

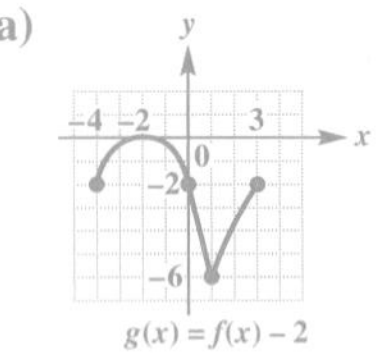

(b)

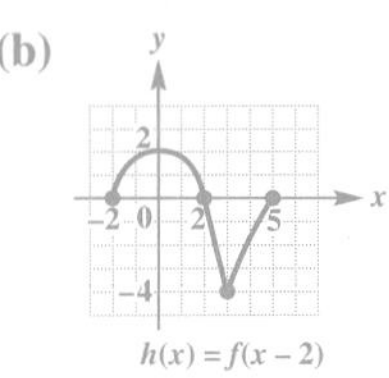

(c)

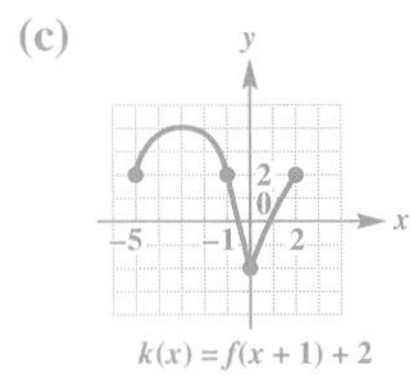

EXAMPLE 9 Graphing Translations of a Given Graph

A graph of a function defined by $y = f(x)$ is shown in **Figure 90.** Use this graph to sketch each of the following graphs.

(a) $g(x) = f(x) + 3$ **(b)** $h(x) = f(x + 3)$

(c) $k(x) = f(x - 2) + 3$

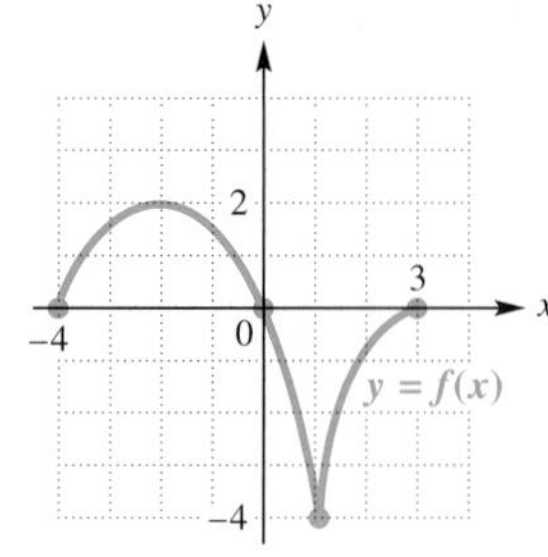

Figure 90

SOLUTION In each part, pay close attention to how the plotted points in **Figure 90** are translated.

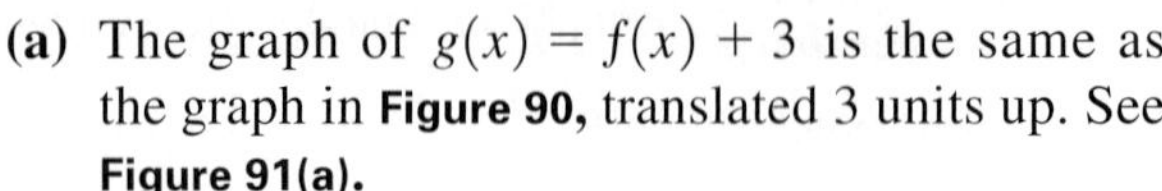

(a) The graph of $g(x) = f(x) + 3$ is the same as the graph in **Figure 90,** translated 3 units up. See **Figure 91(a).**

(b) To get the graph of $h(x) = f(x + 3)$, the graph of $y = f(x)$ must be translated 3 units to the left since $x + 3 = 0$ if $x = -3$. See **Figure 91(b).**

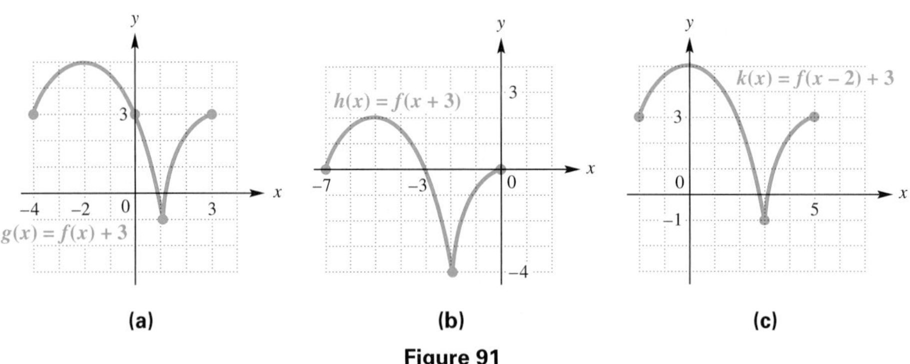

(a) **(b)** **(c)**

Figure 91

(c) The graph of $k(x) = f(x - 2) + 3$ will look like the graph of $f(x)$ translated 2 units to the right and 3 units up, as shown in **Figure 91(c).**

✔ Now Try Exercise 77.

Summary of Graphing Techniques

In the descriptions that follow, assume that $a > 0$, $h > 0$, and $k > 0$. In comparison with the graph of $y = f(x)$:

1. The graph of $\boldsymbol{y = f(x) + k}$ is translated k units up.
2. The graph of $\boldsymbol{y = f(x) - k}$ is translated k units down.
3. The graph of $\boldsymbol{y = f(x + h)}$ is translated h units to the left.
4. The graph of $\boldsymbol{y = f(x - h)}$ is translated h units to the right.
5. The graph of $\boldsymbol{y = af(x)}$ is a vertical stretching of the graph of $y = f(x)$ if $a > 1$. It is a vertical shrinking if $0 < a < 1$.
6. The graph of $\boldsymbol{y = f(ax)}$ is a horizontal stretching of the graph of $y = f(x)$ if $0 < a < 1$. It is a horizontal shrinking if $a > 1$.
7. The graph of $\boldsymbol{y = -f(x)}$ is reflected across the x-axis.
8. The graph of $\boldsymbol{y = f(-x)}$ is reflected across the y-axis.

Teaching Tip Show students graphs of basic functions that have been transformed, and ask them to write the corresponding equations.

2.7 Exercises

1. *Concept Check* Match each equation in Column I with a description of its graph from Column II as it relates to the graph of $y = x^2$.

I	II
(a) $y = (x - 7)^2$	**A.** a translation 7 units to the left
(b) $y = x^2 - 7$	**B.** a translation 7 units to the right
(c) $y = 7x^2$	**C.** a translation 7 units up
(d) $y = (x + 7)^2$	**D.** a translation 7 units down
(e) $y = x^2 + 7$	**E.** a vertical stretching by a factor of 7

2. *Concept Check* Match each equation in Column I with a description of its graph from Column II as it relates to the graph of $y = \sqrt[3]{x}$.

I	II
(a) $y = 4\sqrt[3]{x}$	**A.** a translation 4 units to the right
(b) $y = -\sqrt[3]{x}$	**B.** a translation 4 units down
(c) $y = \sqrt[3]{-x}$	**C.** a reflection across the x-axis
(d) $y = \sqrt[3]{x - 4}$	**D.** a reflection across the y-axis
(e) $y = \sqrt[3]{x} - 4$	**E.** a vertical stretching by a factor of 4

3. *Concept Check* Match each equation in parts (a)–(i) with the sketch of its graph.

(a) $y = x^2 + 2$ **(b)** $y = x^2 - 2$ **(c)** $y = (x + 2)^2$

(d) $y = (x - 2)^2$ **(e)** $y = 2x^2$ **(f)** $y = -x^2$

(g) $y = (x - 2)^2 + 1$ **(h)** $y = (x + 2)^2 + 1$ **(i)** $y = (x + 2)^2 - 1$

A.

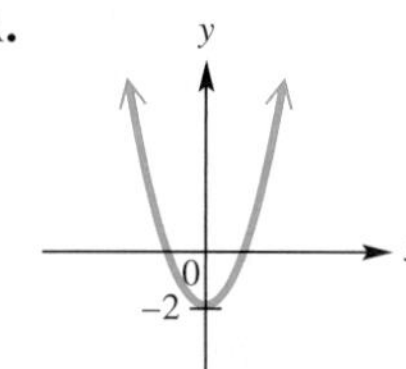

B.

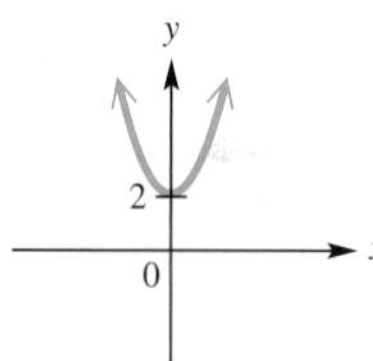

C.

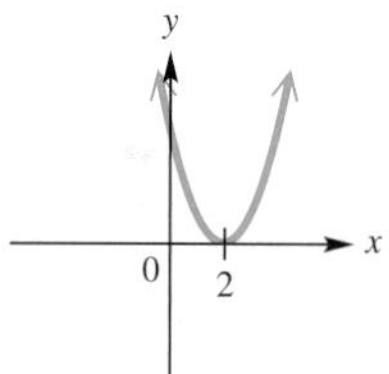

D.

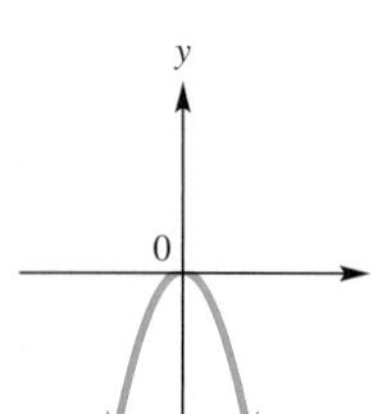

E.

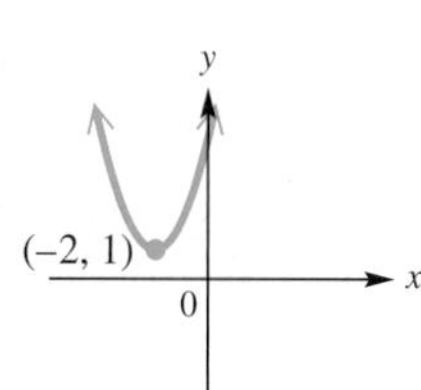

F.

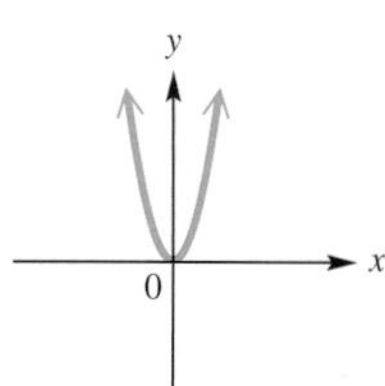

G.

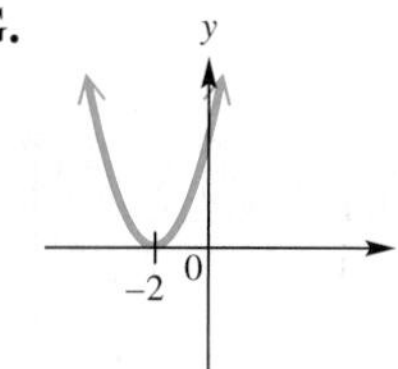

H.

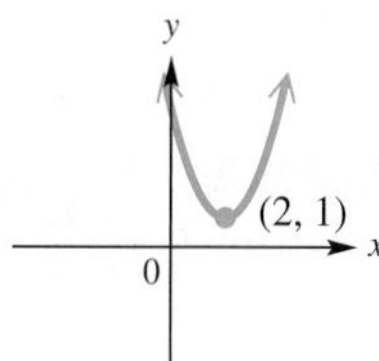

I.

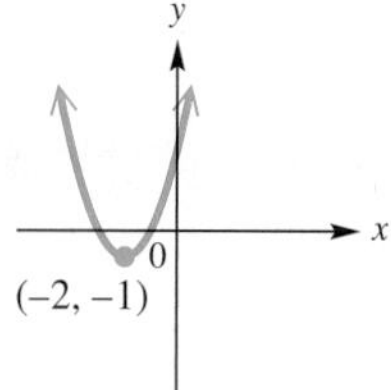

1. (a) B (b) D (c) E (d) A (e) C
2. (a) E (b) C (c) D (d) A (e) B
3. (a) B (b) A (c) G (d) C (e) F (f) D (g) H (h) E (i) I
4. (a) G (b) D (c) E (d) B (e) C (f) A (g) H (h) F (i) I
5. (a) F (b) C (c) H (d) D (e) G (f) A (g) E (h) I (i) B

7. $y = 3|x|$ 8. $y = 4|x|$

9. $y = \frac{2}{3}|x|$ 10. $y = \frac{3}{4}|x|$

11. $y = 2x^2$ 12. $y = 3x^2$

13. $y = \frac{1}{2}x^2$ 14. $y = \frac{1}{3}x^2$

15. $y = -\frac{1}{2}x^2$ 16. $y = -\frac{1}{3}x^2$

17.

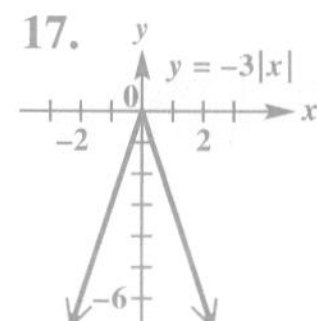

18. $y = -2|x|$

19. 20.

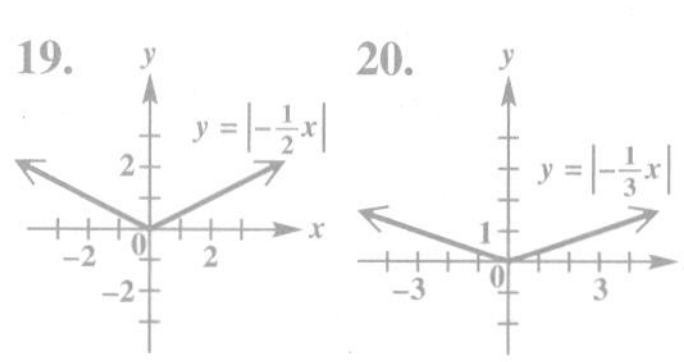

21. 22.

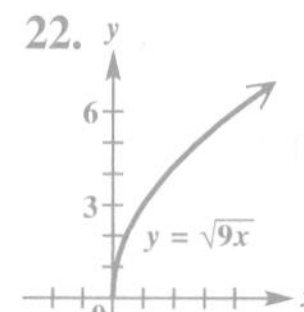

23. 24.

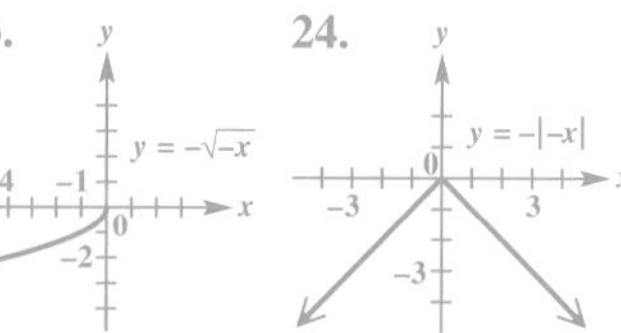

25. (a) $(4, 12)$ (b) $(8, 16)$
26. (a) $(8, 3)$ (b) $(8, 48)$
27. (a) $(2, 12)$ (b) $(32, 12)$
28. (a) $(8, -12)$ (b) $(-8, 12)$

29.

30.

31.

32.

33. $x = 2$ 34. $x = -1$

35. y-axis 36. y-axis
37. x-axis, y-axis, origin
38. x-axis, y-axis, origin
39. origin 40. origin
41. none of these
42. none of these

43. odd 44. odd
45. even 46. even
47. neither 48. neither

49. 50.

51. 52.

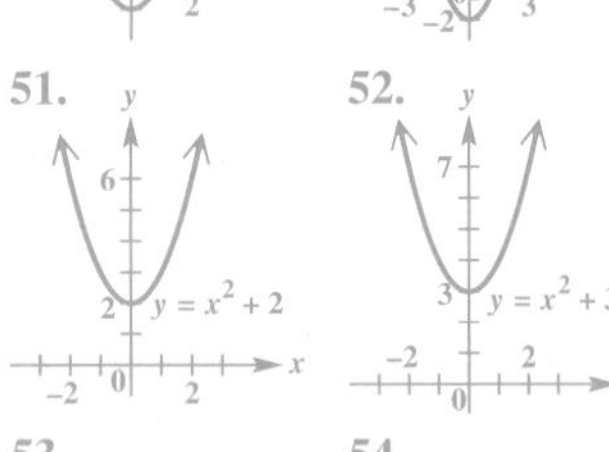

53. 54.

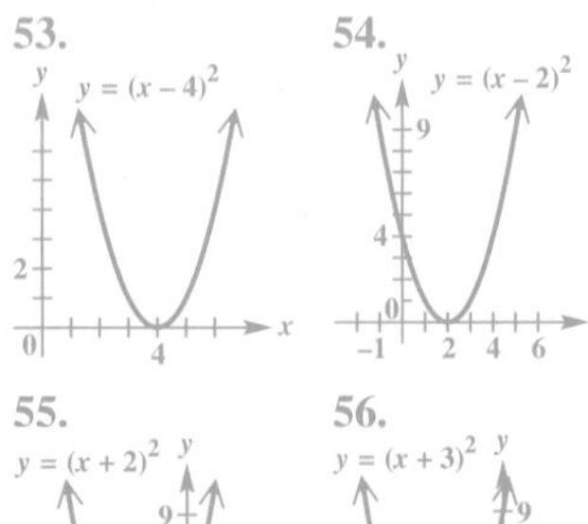

55. 56.

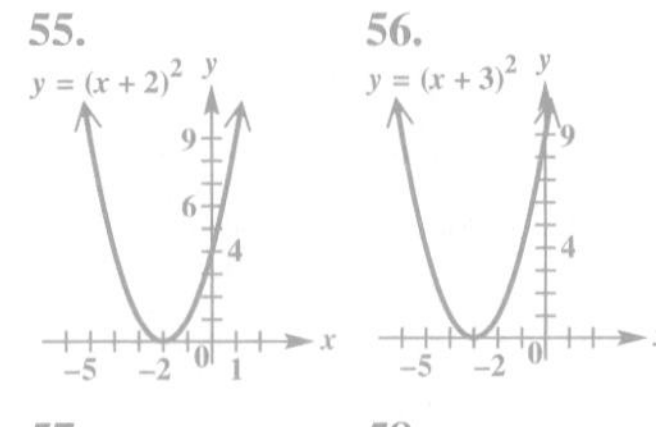

57. 58.

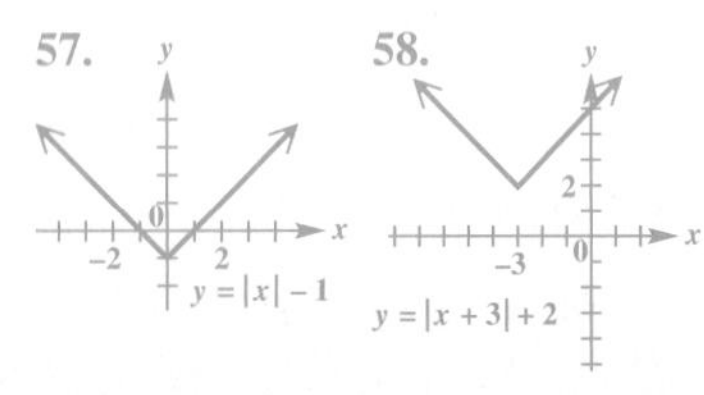

4. *Concept Check* Match each equation in parts (a)–(i) with the sketch of its graph.

(a) $y = \sqrt{x + 3}$ (b) $y = \sqrt{x} - 3$ (c) $y = \sqrt{x} + 3$
(d) $y = 3\sqrt{x}$ (e) $y = -\sqrt{x}$ (f) $y = \sqrt{x - 3}$
(g) $y = \sqrt{x - 3} + 2$ (h) $y = \sqrt{x + 3} + 2$ (i) $y = \sqrt{x - 3} - 2$

A.

B.

C.

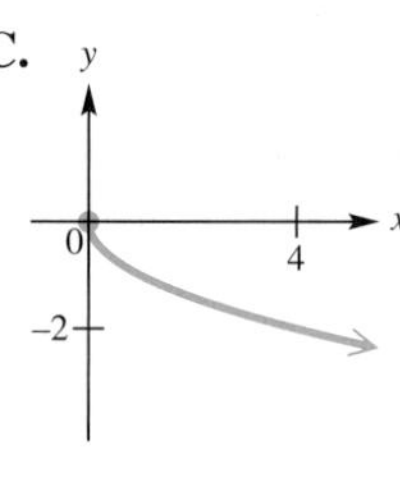

D.

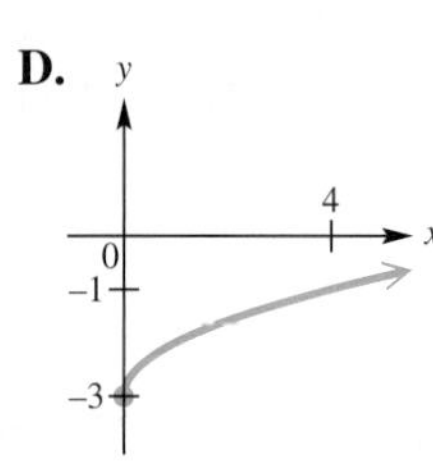

E.

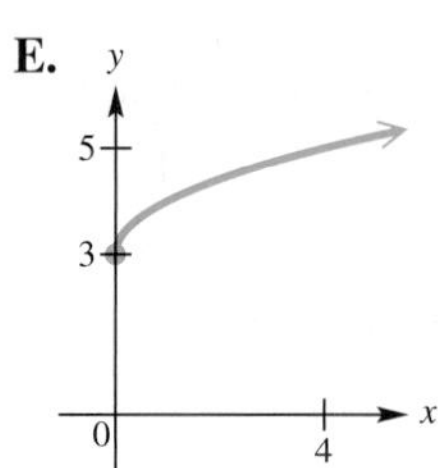

F.

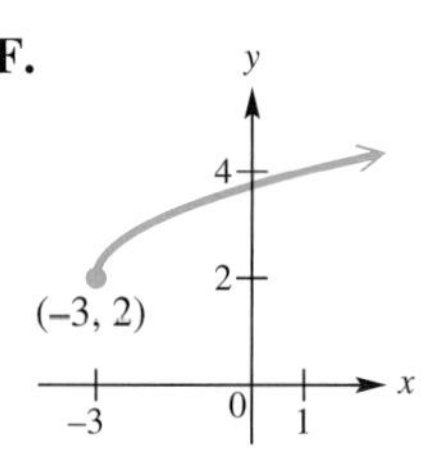

G.

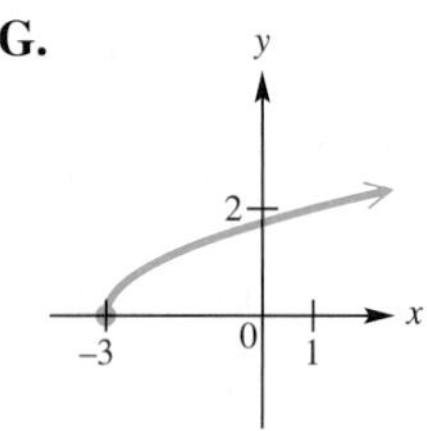

H.

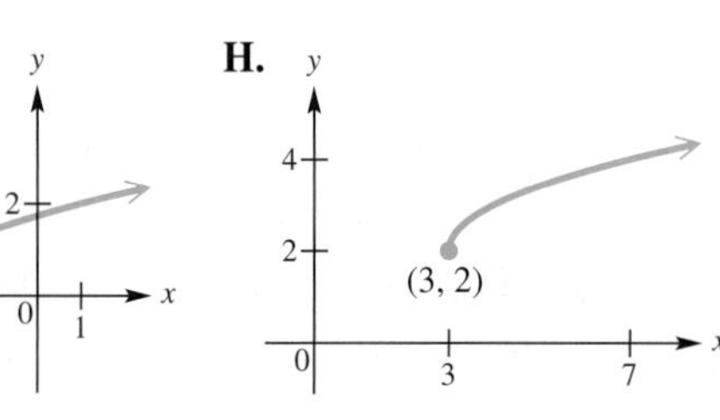

I.

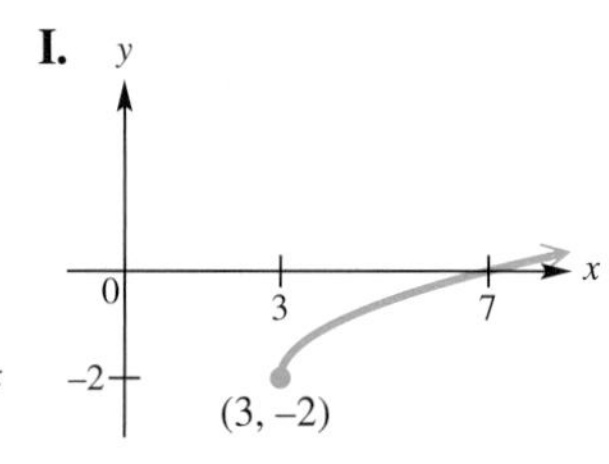

5. *Concept Check* Match each equation in parts (a)–(i) with the sketch of its graph.

(a) $y = |x - 2|$ (b) $y = |x| - 2$ (c) $y = |x| + 2$
(d) $y = 2|x|$ (e) $y = -|x|$ (f) $y = |-x|$
(g) $y = -2|x|$ (h) $y = |x - 2| + 2$ (i) $y = |x + 2| - 2$

A.

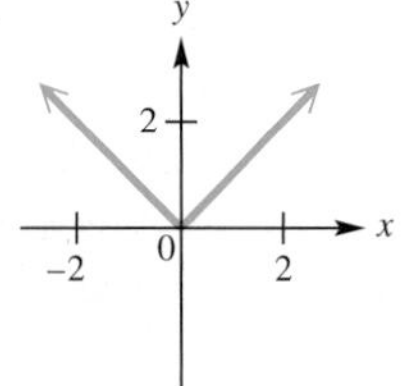

B.

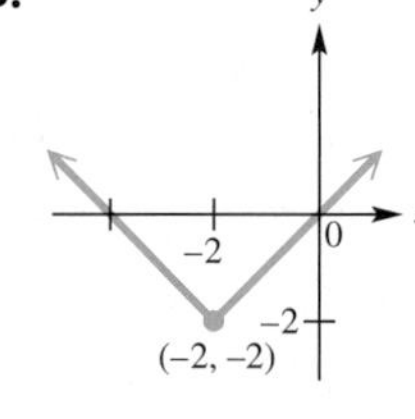

C.

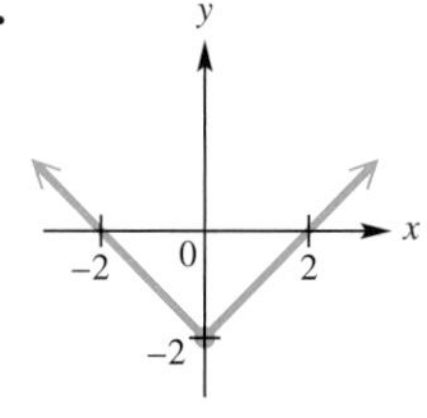

D.

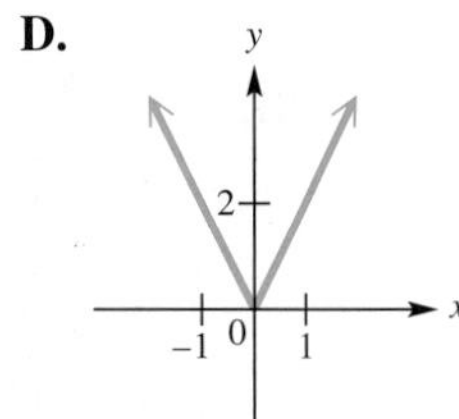

E.

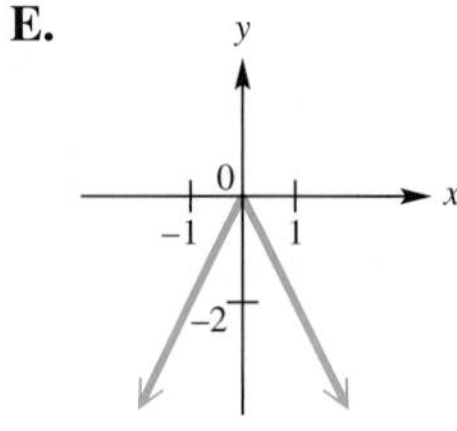

F.

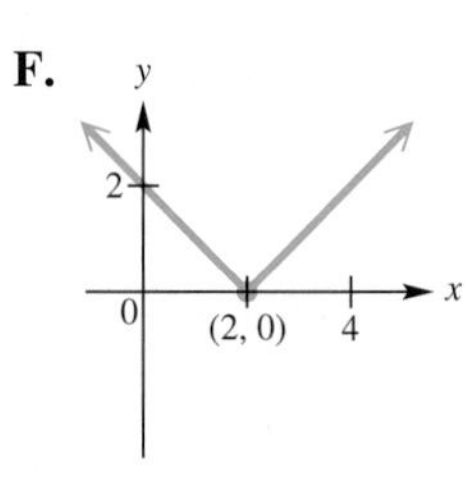

G.

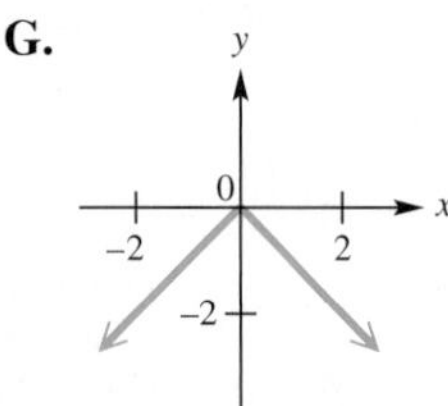

H.

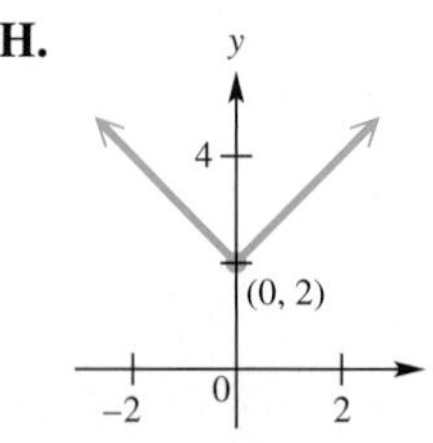

I.

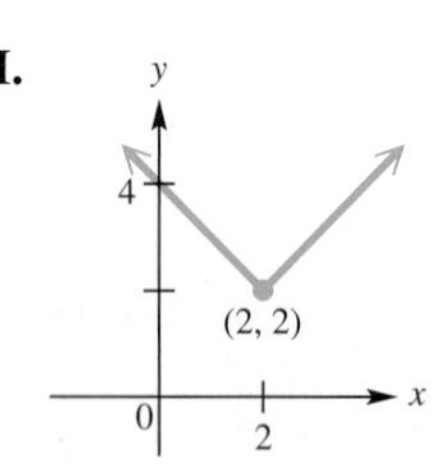

59. 60.

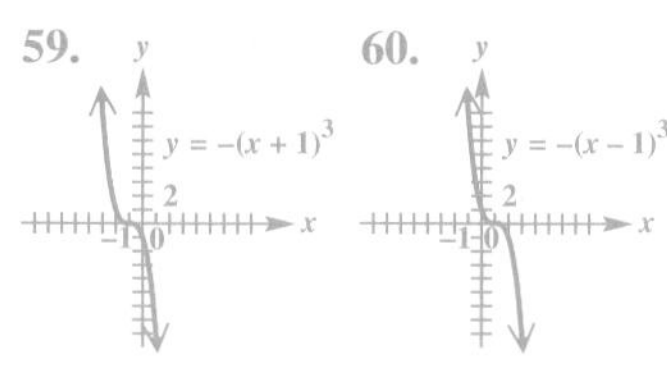

61. 62.

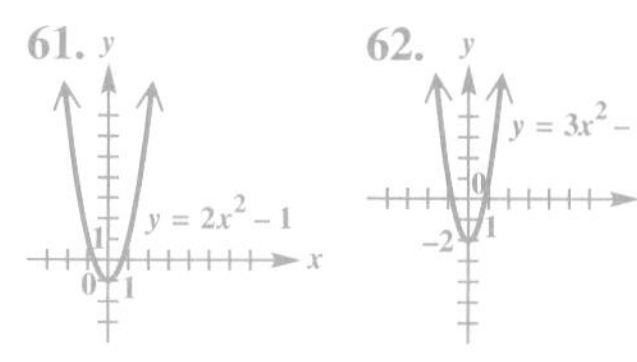

63. 64.

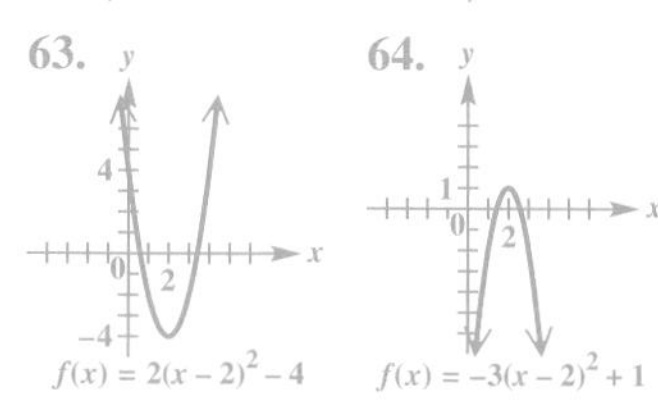

65. 66.

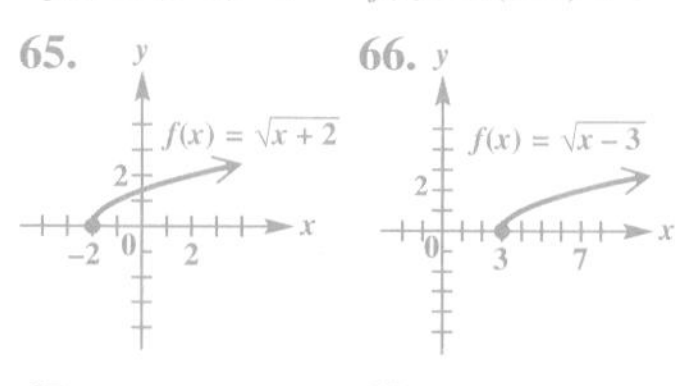

67. 68.

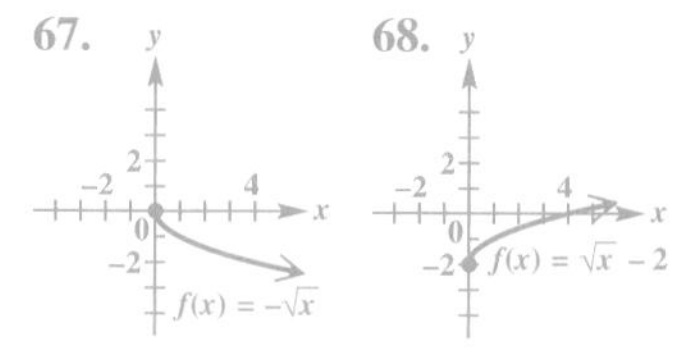

69. 70.

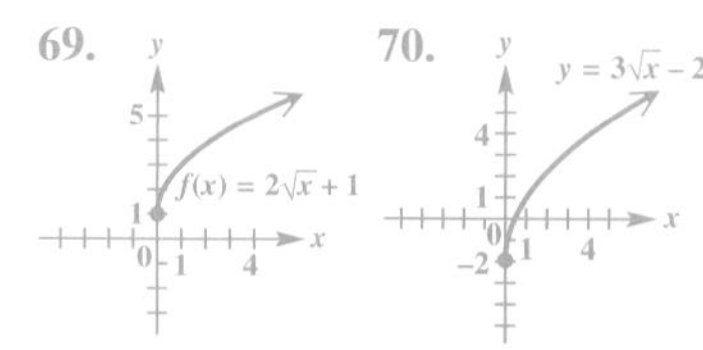

71. 72.

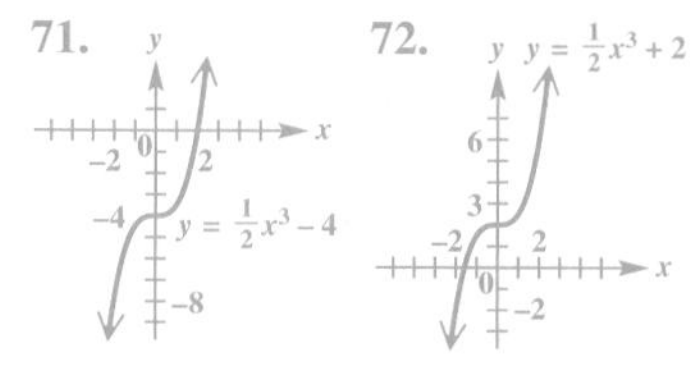

73. 74.

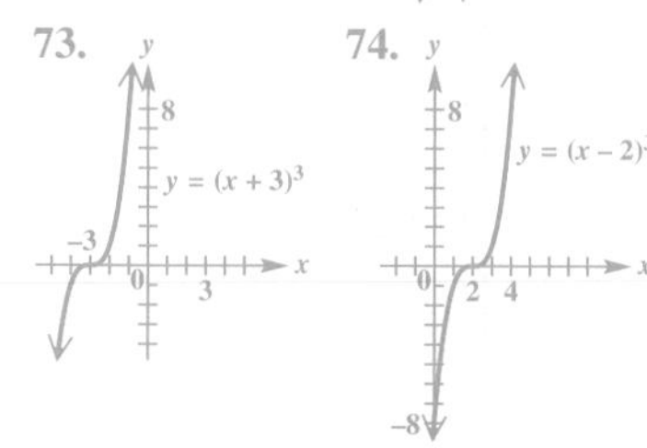

75.

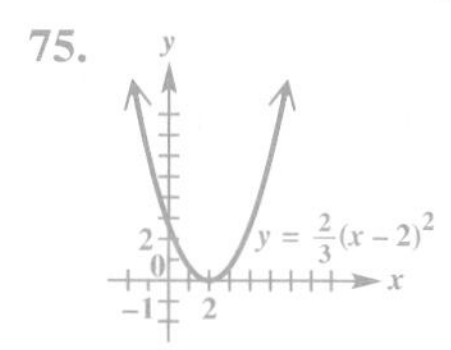

76. They have the same graph.

6. **(a)** Suppose the equation $y = F(x)$ is changed to $y = c \cdot F(x)$, for some constant c. What is the effect on the graph of $y = F(x)$? Discuss the effect depending on whether $c > 0$ or $c < 0$, and on whether $|c| > 1$ or $|c| < 1$.

   (b) Suppose $y = F(x)$ is changed to $y = F(x + h)$. How are the graphs of these equations related? Is the graph of $y = F(x) + h$ the same as the graph of $y = F(x + h)$? If not, how do they differ?

Graph each function. ***See Examples 1 and 2.***

7. $y = 3|x|$ **8.** $y = 4|x|$ **9.** $y = \frac{2}{3}|x|$ **10.** $y = \frac{3}{4}|x|$

11. $y = 2x^2$ **12.** $y = 3x^2$ **13.** $y = \frac{1}{2}x^2$ **14.** $y = \frac{1}{3}x^2$

15. $y = -\frac{1}{2}x^2$ **16.** $y = -\frac{1}{3}x^2$ **17.** $y = -3|x|$ **18.** $y = -2|x|$

19. $y = \left|-\frac{1}{2}x\right|$ **20.** $y = \left|-\frac{1}{3}x\right|$ **21.** $y = \sqrt{4x}$

22. $y = \sqrt{9x}$ **23.** $y = -\sqrt{-x}$ **24.** $y = -|-x|$

Concept Check In Exercises 25–28, suppose the point (8, 12) is on the graph of $y = f(x)$. Find a point on the graph of each function.

25. **(a)** $y = f(x + 4)$ **(b)** $y = f(x) + 4$

26. **(a)** $y = \frac{1}{4}f(x)$ **(b)** $y = 4f(x)$

27. **(a)** $y = f(4x)$ **(b)** $y = f\left(\frac{1}{4}x\right)$

28. **(a)** the reflection of the graph of $y = f(x)$ across the x-axis

(b) the reflection of the graph of $y = f(x)$ across the y-axis

Concept Check Plot each point, and then plot the points that are symmetric to the given point with respect to the ***(a)*** *x-axis,* ***(b)*** *y-axis, and* ***(c)*** *origin.*

29. $(5, -3)$ **30.** $(-6, 1)$ **31.** $(-4, -2)$ **32.** $(-8, 0)$

33. *Concept Check* The graph of $y = |x - 2|$ is symmetric with respect to a vertical line. What is the equation of that line?

34. *Concept Check* Repeat **Exercise 33** for $y = -|x + 1|$.

Without graphing, determine whether each equation has a graph that is symmetric with respect to the x-axis, the y-axis, the origin, or none of these. ***See Examples 3 and 4.***

35. $y = x^2 + 5$ **36.** $y = 2x^4 - 3$ **37.** $x^2 + y^2 = 12$ **38.** $y^2 - x^2 = -6$

39. $y = -4x^3 + x$ **40.** $y = x^3 - x$ **41.** $y = x^2 - x + 8$ **42.** $y = x + 15$

Determine whether each function is even, odd, or neither. ***See Example 5.***

43. $f(x) = -x^3 + 2x$ **44.** $f(x) = x^5 - 2x^3$

45. $f(x) = 0.5x^4 - 2x^2 + 6$ **46.** $f(x) = 0.75x^2 + |x| + 4$

47. $f(x) = x^3 - x + 9$ **48.** $f(x) = x^4 - 5x + 8$

Graph each function. ***See Examples 6–8.***

49. $y = x^2 - 1$ **50.** $y = x^2 - 2$ **51.** $y = x^2 + 2$

52. $y = x^2 + 3$ **53.** $y = (x - 4)^2$ **54.** $y = (x - 2)^2$

55. $y = (x + 2)^2$ **56.** $y = (x + 3)^2$ **57.** $y = |x| - 1$

58. $y = |x + 3| + 2$ **59.** $y = -(x + 1)^3$ **60.** $y = -(x - 1)^3$

61. $y = 2x^2 - 1$ **62.** $y = 3x^2 - 2$ **63.** $f(x) = 2(x - 2)^2 - 4$

64. $f(x) = -3(x - 2)^2 + 1$ **65.** $f(x) = \sqrt{x + 2}$ **66.** $f(x) = \sqrt{x - 3}$

67. $f(x) = -\sqrt{x}$ **68.** $f(x) = \sqrt{x} - 2$ **69.** $f(x) = 2\sqrt{x} + 1$

70. $y = 3\sqrt{x} - 2$ **71.** $y = \frac{1}{2}x^3 - 4$ **72.** $y = \frac{1}{2}x^3 + 2$

73. $y = (x + 3)^3$ **74.** $y = (x - 2)^3$ **75.** $y = \frac{2}{3}(x - 2)^2$

76. *Concept Check* What is the relationship between the graphs of $f(x) = |x|$ and $g(x) = |-x|$?

For Exercises 77 and 78, ***see Example 9.***

77. Given the graph of $y = g(x)$ in the figure, sketch the graph of each function, and explain how it is obtained from the graph of $y = g(x)$.

(a) $y = g(-x)$
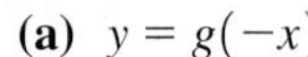

(b) $y = g(x - 2)$

(c) $y = -g(x) + 2$

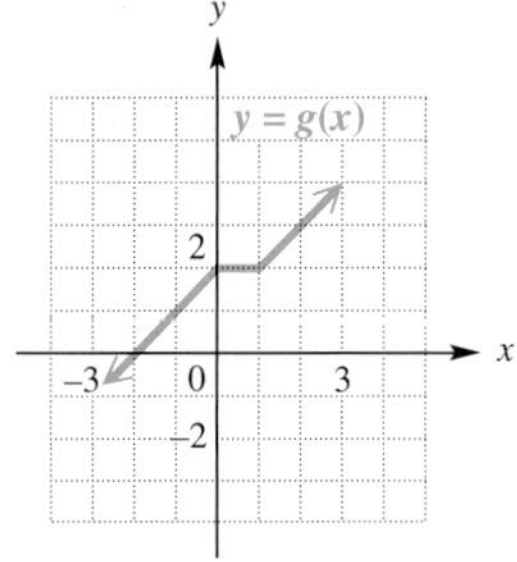

78. Given the graph of $y = f(x)$ in the figure, sketch the graph of each function, and explain how it is obtained from the graph of $y = f(x)$.

(a) $y = -f(x)$

(b) $y = 2f(x)$

(c) $y = f(-x)$

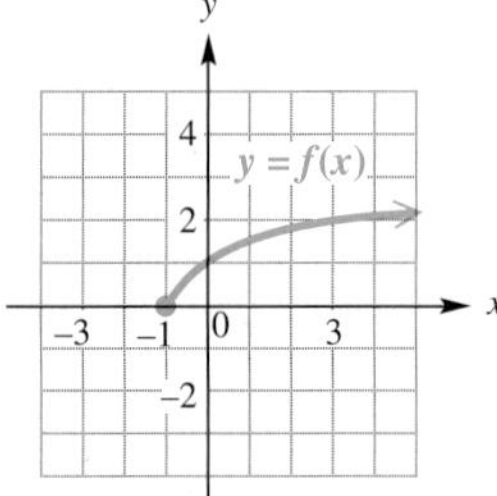

Connecting Graphs with Equations Each of the following graphs is obtained from the graph of $f(x) = |x|$ or $g(x) = \sqrt{x}$ by applying several of the transformations discussed in this section. Describe the transformations and give the equation for the graph.

79.

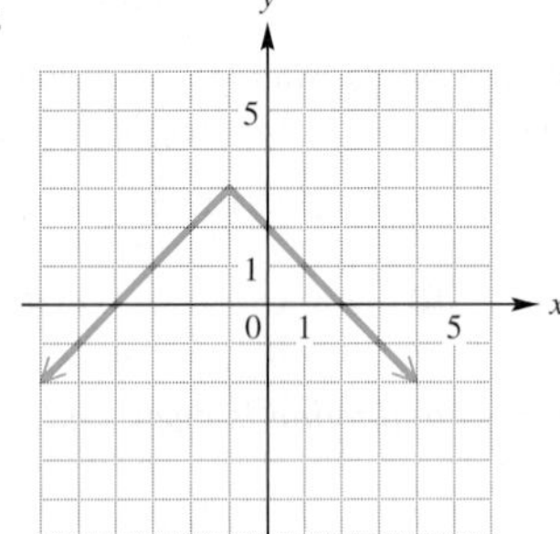

80.

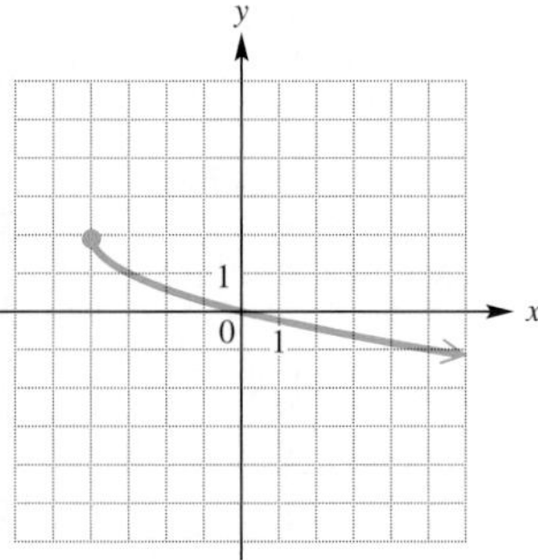

81.

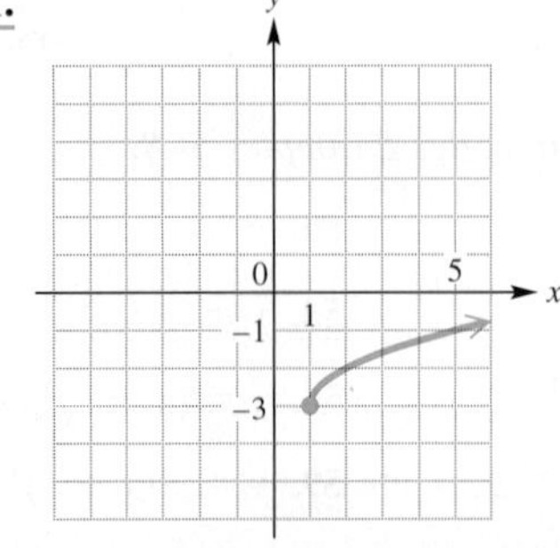

82.

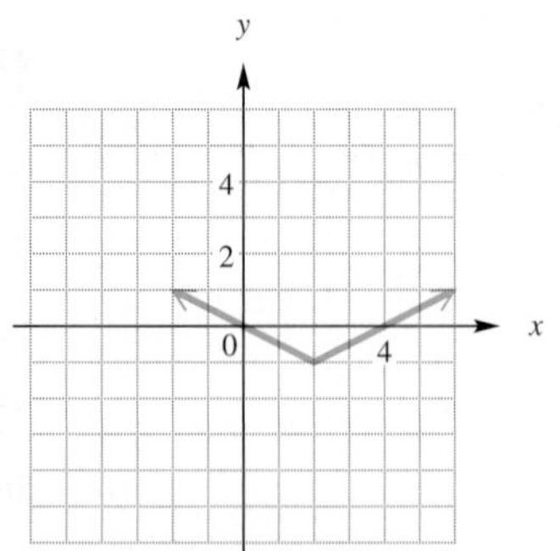

77. (a)

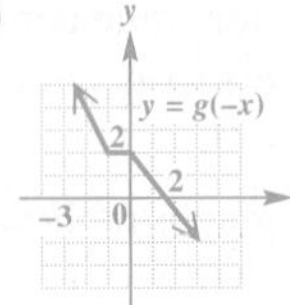

The graph of $g(x)$ is reflected across the y-axis

(b)

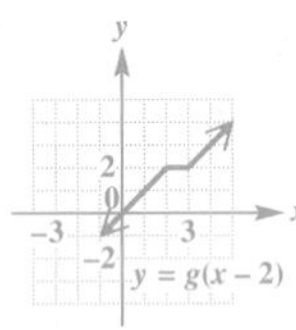

The graph of $g(x)$ is translated 2 units to the right.

(c)

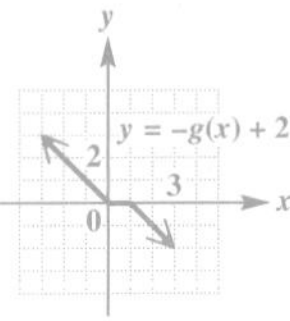

The graph of $g(x)$ is reflected across the x-axis and translated 2 units up.

78. (a)

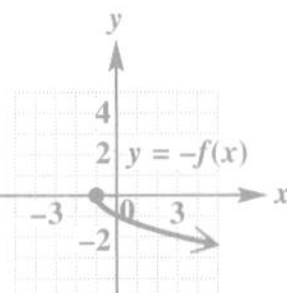

The graph of $f(x)$ is reflected across the x-axis.

(b)

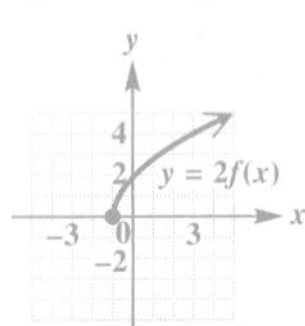

The graph is the same shape as that of $f(x)$ but is stretched vertically by a factor of 2.

(c)

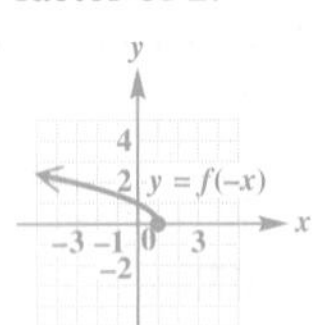

The graph of $f(x)$ is reflected across the y-axis.

79. It is the graph of $f(x) = |x|$ translated 1 unit to the left, reflected across the x-axis, and translated 3 units up. The equation is $y = -|x + 1| + 3$.

80. It is the graph of $g(x) = \sqrt{x}$ translated 4 units to the left, reflected across the x-axis, and translated 2 units up. The equation is $y = -\sqrt{x + 4} + 2$.

81. It is the graph of $g(x) = \sqrt{x}$ translated 1 unit to the right and translated 3 units down. The equation is $y = \sqrt{x - 1} - 3$.

82. It is the graph of $f(x) = |x|$ translated 2 units to the right, shrunken vertically by a factor of $\frac{1}{2}$, and translated 1 unit down. The equation is $y = \frac{1}{2}|x - 2| - 1$.

83. It is the graph of $g(x) = \sqrt{x}$ translated 4 units to the left, stretched vertically by a factor of 2, and translated 4 units down. The equation is $y = 2\sqrt{x + 4} - 4$.

84. It is the graph of $f(x) = |x|$ reflected across the x-axis and translated 2 units down. The equation is $y = -|x| - 2$.

85. $f(-3) = -6$ **86.** $f(-3) = 6$

87. $f(9) = 6$ **88.** $f(-3) = 6$

89. $f(-3) = -6$

90. $f(-3) = -6$

91. $g(x) = 2x + 13$

92. $g(x) = -x + 4$

93.
(a) (b)

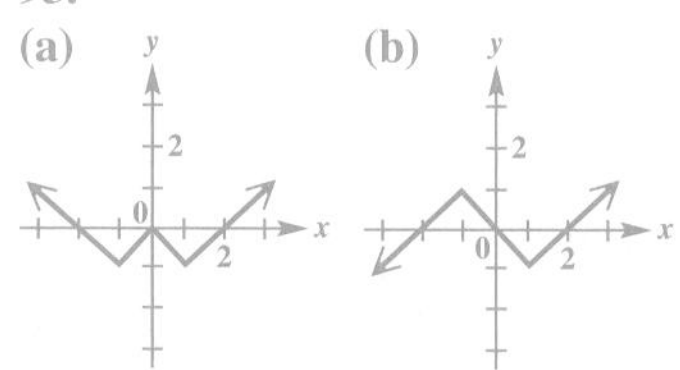

94.
(a) (b)

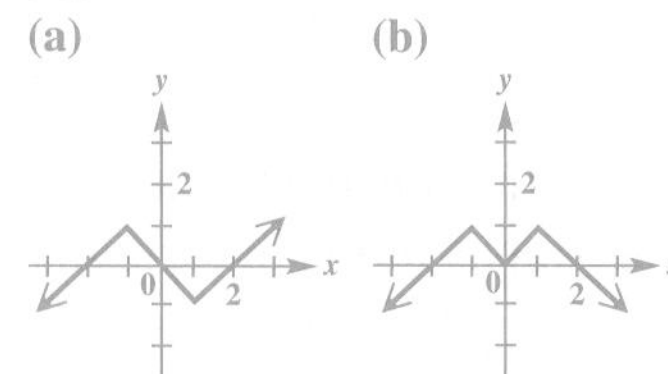

83.

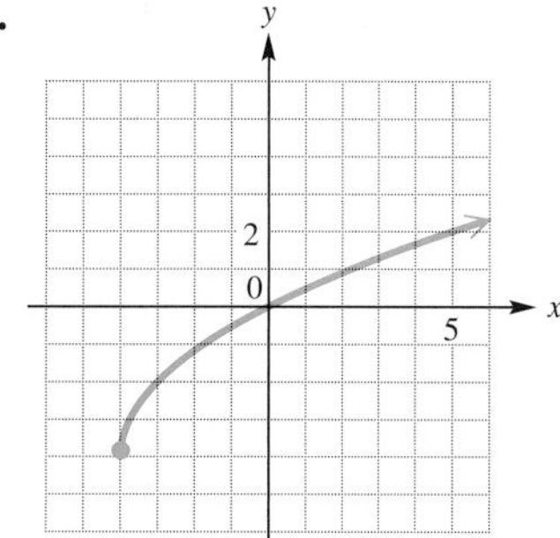

84.

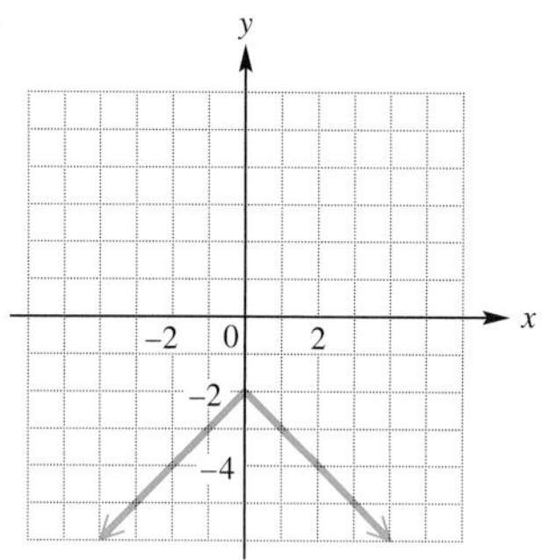

Concept Check Suppose that for a function f, $f(3) = 6$. For the given assumptions in Exercises 85–90, find another function value.

85. The graph of $y = f(x)$ is symmetric with respect to the origin.

86. The graph of $y = f(x)$ is symmetric with respect to the y-axis.

87. The graph of $y = f(x)$ is symmetric with respect to the line $x = 6$.

88. For all x, $f(-x) = f(x)$.

89. For all x, $f(-x) = -f(x)$.

90. f is an odd function.

91. Find the function $g(x) = ax + b$ whose graph can be obtained by translating the graph of $f(x) = 2x + 5$ up 2 units and to the left 3 units.

92. Find the function $g(x) = ax + b$ whose graph can be obtained by translating the graph of $f(x) = 3 - x$ down 2 units and to the right 3 units.

93. *Concept Check* Complete the left half of the graph of $y = f(x)$ in the figure for each condition.

(a) $f(-x) = f(x)$ **(b)** $f(-x) = -f(x)$

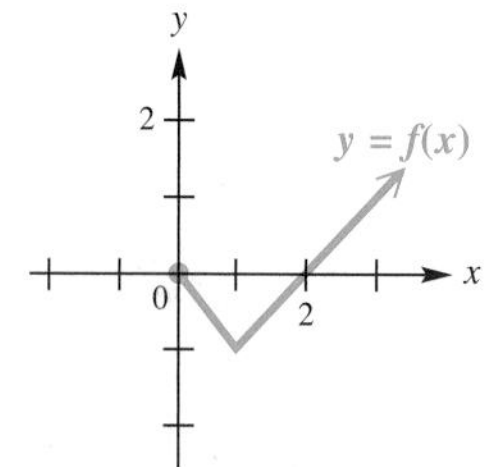

94. *Concept Check* Complete the right half of the graph of $y = f(x)$ in the figure for each condition.

(a) f is odd. **(b)** f is even.

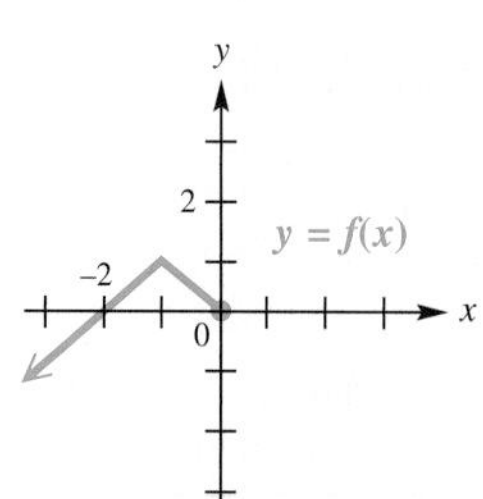

Chapter 2 Quiz (Sections 2.5–2.7)

[2.5]

1. (a) $y = 2x + 11$ **(b)** $-\frac{11}{2}$

2. $y = -\frac{2}{3}x$

3. (a) $x = -8$ **(b)** $y = 5$

1. For the line passing through the points $(-3, 5)$ and $(-1, 9)$, find the following:

(a) the slope-intercept form of its equation **(b)** its x-intercept.

2. Find the slope-intercept form of the equation of the line passing through the point $(-6, 4)$ and perpendicular to the graph of $3x - 2y = 6$.

3. Suppose that P has coordinates $(-8, 5)$. Find the equation of the line through P that is

(a) vertical **(b)** horizontal.

4. For each basic function graphed, give the name of the function, the domain, the range, and intervals over which it is decreasing, increasing, or constant.

(a)

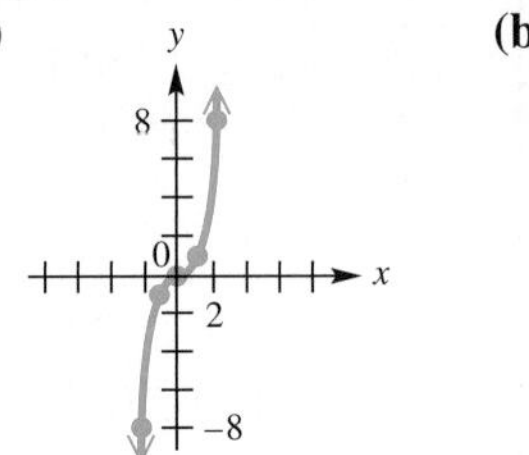

(b)

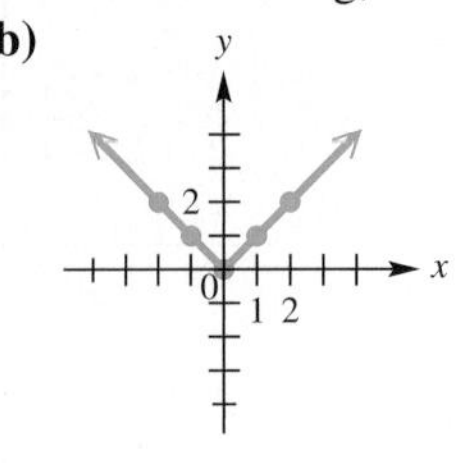

(c)

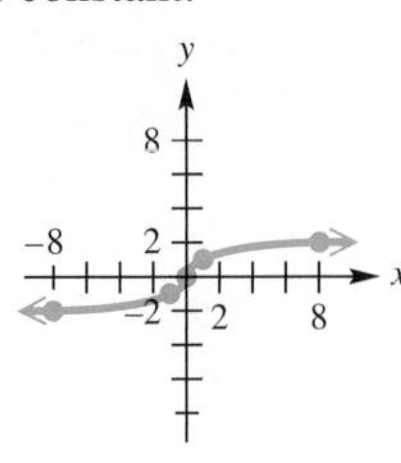

5. *(Modeling) Long-Distance Call Charges* A certain long-distance carrier provides service between Podunk and Nowheresville. If x represents the number of minutes for the call, where $x > 0$, then the function

$$f(x) = 0.40[\![x]\!] + 0.75$$

gives the total cost of the call in dollars. Find the cost of a 5.5-min call.

Graph each function.

6. $f(x) = \begin{cases} \sqrt{x} & \text{if } x \geq 0 \\ 2x + 3 & \text{if } x < 0 \end{cases}$ **7.** $f(x) = -x^3 + 1$ **8.** $f(x) = 2|x - 1| + 3$

9. *Connecting Graphs with Equations* The function graphed here is obtained by stretching, shrinking, reflecting, and/or translating the graph of $f(x) = \sqrt{x}$. Give the equation that defines this function.

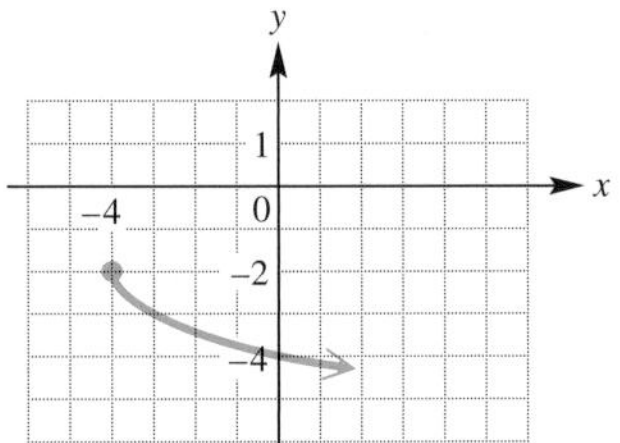

10. Determine whether each function is *even*, *odd*, or *neither*.

(a) $f(x) = x^2 - 7$ **(b)** $f(x) = x^3 - x - 1$ **(c)** $f(x) = x^{101} - x^{99}$

[2.6]

4. (a) cubing function; domain: $(-\infty, \infty)$; range: $(-\infty, \infty)$; increasing over $(-\infty, \infty)$

(b) absolute value function; domain: $(-\infty, \infty)$; range: $[0, \infty)$; decreasing over $(-\infty, 0]$; increasing over $[0, \infty)$

(c) cube root function; domain: $(-\infty, \infty)$; range: $(-\infty, \infty)$; increasing over $(-\infty, \infty)$

5. \$2.75

6.

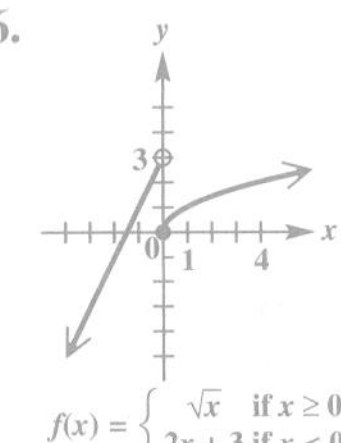

[2.7]

7. **8.** $f(x) = 2|x - 1| + 3$

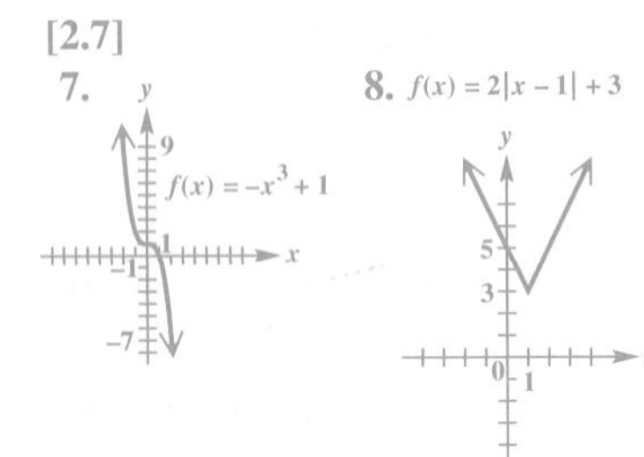

9. $y = -\sqrt{x + 4} - 2$ **10. (a)** even **(b)** neither **(c)** odd

2.8 Function Operations and Composition

- Arithmetic Operations on Functions
- The Difference Quotient
- Composition of Functions and Domain

Arithmetic Operations on Functions **Figure 92** shows the situation for a company that manufactures DVDs. The two lines are the graphs of the linear functions for revenue $R(x) = 168x$ and cost $C(x) = 118x + 800$, where x is the number of DVDs produced and sold, and x, $R(x)$, and $C(x)$ are given in thousands. When 30,000 (that is, 30 thousand) DVDs are produced and sold, profit is found as follows.

$$
\begin{aligned}
P(x) &= R(x) - C(x) && \text{Profit function (Section 2.4)} \\
P(30) &= R(30) - C(30) && \text{Let } x = 30. \\
&= 5040 - 4340 && R(30) = 168(30);\ C(30) = 118(30) + 800 \\
P(30) &= 700 && \text{Subtract.}
\end{aligned}
$$

Thus, the profit from the sale of 30,000 DVDs is \$700,000.

The profit function is found by *subtracting* the cost function from the revenue function. New functions can be formed by using other operations as well.

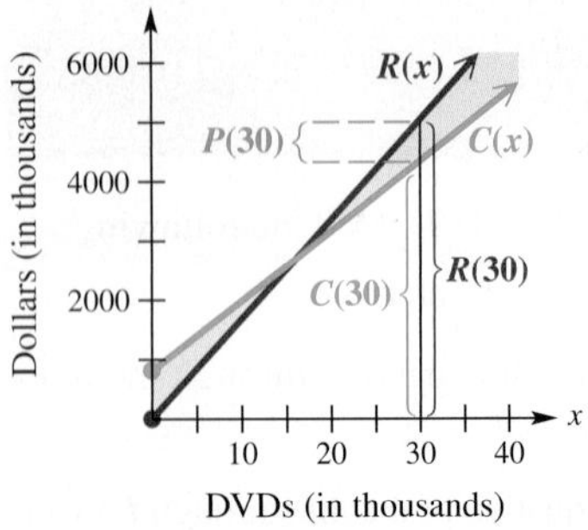

Figure 92

Operations on Functions and Domains

Given two functions f and g, then for all values of x for which both $f(x)$ and $g(x)$ are defined, the functions $f+g$, $f-g$, fg, and $\frac{f}{g}$ are defined as follows.

$$(f+g)(x) = f(x) + g(x) \quad \text{Sum}$$
$$(f-g)(x) = f(x) - g(x) \quad \text{Difference}$$
$$(fg)(x) = f(x) \cdot g(x) \quad \text{Product}$$
$$\left(\frac{f}{g}\right)(x) = \frac{f(x)}{g(x)}, \quad g(x) \neq 0 \quad \text{Quotient}$$

The **domains of $f+g$, $f-g$,** and **fg** include all real numbers in the intersection of the domains of f and g, while the **domain of $\frac{f}{g}$** includes those real numbers in the intersection of the domains of f and g for which $g(x) \neq 0$.

Teaching Tip In preparation for determining domains, ask students to determine the intersection of the following interval pairs:

(a) $[0, \infty)$ and $(-20, 20)$
(b) $(-4, 2) \cup (3, 10)$ and $(-\infty, \infty)$
(c) $(-\infty, 0]$ and $[0, \infty)$
(d) $(-\infty, 5]$ and $(1, \infty)$.

NOTE The condition $g(x) \neq 0$ in the definition of the quotient means that the domain of $\left(\frac{f}{g}\right)(x)$ is restricted to all values of x for which $g(x)$ is not 0. The condition does not mean that $g(x)$ is a function that is never 0.

Classroom Example 1
Let $f(x) = 3x - 4$ and $g(x) = 2x^2 - 1$. Find each of the following.

(a) $(f+g)(0)$
(b) $(f-g)(4)$
(c) $(fg)(-2)$
(d) $\left(\frac{f}{g}\right)(3)$

Answers:
(a) -5 (b) -23
(c) -70 (d) $\frac{5}{17}$

EXAMPLE 1 Using Operations on Functions

Let $f(x) = x^2 + 1$ and $g(x) = 3x + 5$. Find each of the following.

(a) $(f+g)(1)$ **(b)** $(f-g)(-3)$ **(c)** $(fg)(5)$ **(d)** $\left(\frac{f}{g}\right)(0)$

SOLUTION

(a) First determine $f(1) = 2$ and $g(1) = 8$. Then use the definition.

$$\begin{aligned}(f+g)(1) &= f(1) + g(1) && (f+g)(x) = f(x) + g(x)\\ &= 2 + 8 && f(1) = 1^2 + 1;\ g(1) = 3(1) + 5\\ &= 10 && \text{Add.}\end{aligned}$$

(b)
$$\begin{aligned}(f-g)(-3) &= f(-3) - g(-3) && (f-g)(x) = f(x) - g(x)\\ &= 10 - (-4) && f(-3) = (-3)^2 + 1;\ g(-3) = 3(-3) + 5\\ &= 14 && \text{Subtract.}\end{aligned}$$

(c)
$$\begin{aligned}(fg)(5) &= f(5) \cdot g(5)\\ &= (5^2 + 1)(3 \cdot 5 + 5)\\ &= 26 \cdot 20\\ &= 520\end{aligned}$$

(d)
$$\begin{aligned}\left(\frac{f}{g}\right)(0) &= \frac{f(0)}{g(0)} && \left(\frac{f}{g}\right)(x) = \frac{f(x)}{g(x)}\\ &= \frac{0^2 + 1}{3(0) + 5} && f(x) = x^2 + 1,\ g(x) = 3x + 5\\ &= \frac{1}{5} && \text{Simplify.}\end{aligned}$$

✔ Now Try Exercises 1, 3, 5, and 7.

Classroom Example 2
Let $f(x) = x^2 - 3x$ and $g(x) = 4x + 5$. Find each of the following.
(a) $(f+g)(x)$ (b) $(f-g)(x)$
(c) $(fg)(x)$ (d) $\left(\frac{f}{g}\right)(x)$
(e) Give the domains of the functions in parts (a)–(d).

Answers:
(a) $x^2 + x + 5$
(b) $x^2 - 7x - 5$
(c) $4x^3 - 7x^2 - 15x$
(d) $\frac{x^2 - 3x}{4x + 5}$
(e) domains of $f+g$, $f-g$, and fg: $(-\infty, \infty)$; domain of $\frac{f}{g}$: $\left(-\infty, -\frac{5}{4}\right) \cup \left(-\frac{5}{4}, \infty\right)$

EXAMPLE 2 Using Operations on Functions and Determining Domains

Let $f(x) = 8x - 9$ and $g(x) = \sqrt{2x - 1}$. Find each function in (a)–(d).

(a) $(f+g)(x)$ **(b)** $(f-g)(x)$ **(c)** $(fg)(x)$ **(d)** $\left(\frac{f}{g}\right)(x)$

(e) Give the domains of the functions in parts (a)–(d).

SOLUTION

(a) $(f+g)(x) = f(x) + g(x)$
$$= 8x - 9 + \sqrt{2x - 1}$$

(b) $(f-g)(x) = f(x) - g(x)$
$$= 8x - 9 - \sqrt{2x - 1}$$

(c) $(fg)(x) = f(x) \cdot g(x)$
$$= (8x - 9)\sqrt{2x - 1}$$

(d) $\left(\frac{f}{g}\right)(x) = \frac{f(x)}{g(x)}$
$$= \frac{8x - 9}{\sqrt{2x - 1}}$$

(e) To find the domains of the functions in parts (a)–(d), we first find the domains of f and g. The domain of f is the set of all real numbers $(-\infty, \infty)$. Because g is defined by a square root radical, the radicand must be nonnegative (that is, greater than or equal to 0).

$$g(x) = \sqrt{2x - 1} \quad \text{Rule for } g(x)$$
$$2x - 1 \geq 0 \quad 2x - 1 \text{ must be nonnegative.}$$
$$2x \geq 1 \quad \text{Add 1. (Section 1.7)}$$
$$x \geq \frac{1}{2} \quad \text{Divide by 2.}$$

Thus, the domain of g is $\left[\frac{1}{2}, \infty\right)$.

The domains of $f + g$, $f - g$, and fg are the intersection of the domains of f and g, which is

$$(-\infty, \infty) \cap \left[\frac{1}{2}, \infty\right) = \left[\frac{1}{2}, \infty\right). \quad \text{(Section R.1)}$$

The domain of $\frac{f}{g}$ includes those real numbers in the intersection above for which $g(x) = \sqrt{2x - 1} \neq 0$. That is, the domain of $\frac{f}{g}$ is $\left(\frac{1}{2}, \infty\right)$.

✔ Now Try Exercises 9 and 13.

Classroom Example 3
If possible, use the given representations of functions f and g to evaluate $(f+g)(1)$, $(f-g)(0)$, $(fg)(-1)$, and $\left(\frac{f}{g}\right)(-2)$.

(a)

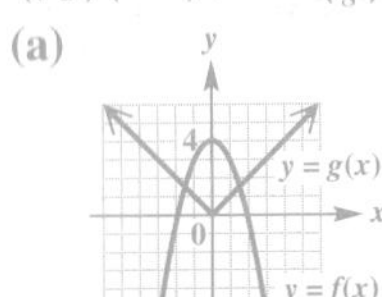

(b)

x	$f(x)$	$g(x)$
−2	−5	0
−1	−3	2
0	−1	4
1	1	6

(c) $f(x) = 3x + 4$, $g(x) = -|x|$

Answers:
(a) $(f+g)(1) = 4$;
$(f-g)(0) = 4$;
$(fg)(-1) = 3$;
$\left(\frac{f}{g}\right)(-2) = 0$
(b) $(f+g)(1) = 7$;
$(f-g)(0) = -5$;
$(fg)(-1) = -6$;
$\left(\frac{f}{g}\right)(-2)$ is undefined.
(c) $(f+g)(1) = 6$;
$(f-g)(0) = 4$;
$(fg)(-1) = -1$;
$\left(\frac{f}{g}\right)(-2) = 1$

EXAMPLE 3 Evaluating Combinations of Functions

If possible, use the given representations of functions f and g to evaluate

$$(f+g)(4), \quad (f-g)(-2), \quad (fg)(1), \quad \text{and} \quad \left(\frac{f}{g}\right)(0).$$

(a)

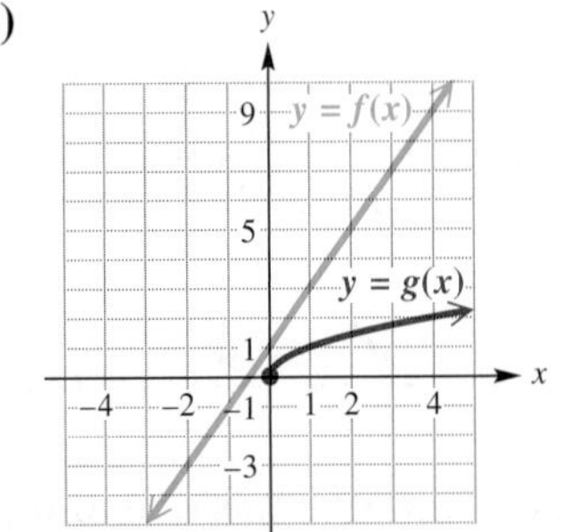

(b)

x	$f(x)$	$g(x)$
−2	−3	undefined
0	1	0
1	3	1
4	9	2

(c) $f(x) = 2x + 1$, $g(x) = \sqrt{x}$

SOLUTION

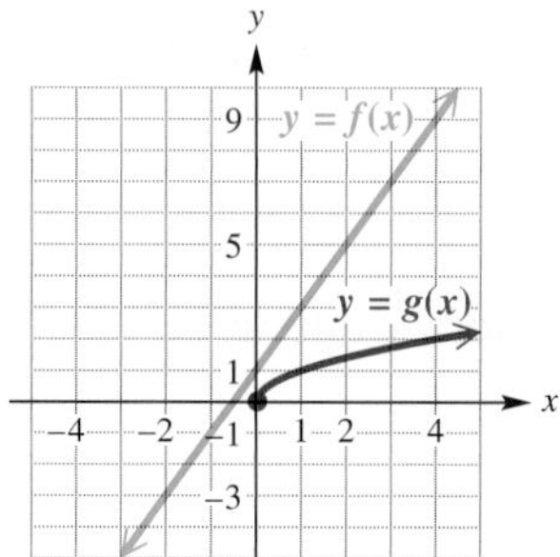

(a) From the figure, repeated in the margin, $f(4) = 9$ and $g(4) = 2$.

$$(f+g)(4) = f(4) + g(4) \quad (f+g)(x) = f(x) + g(x)$$
$$= 9 + 2 \quad \text{Substitute.}$$
$$= 11 \quad \text{Add.}$$

For $(f-g)(-2)$, although $f(-2) = -3$, $g(-2)$ is undefined because -2 is not in the domain of g. Thus $(f-g)(-2)$ is undefined.

The domains of f and g include 1.

$$(fg)(1) = f(1) \cdot g(1) \quad (fg)(x) = f(x) \cdot g(x)$$
$$= 3 \cdot 1 \quad \text{Substitute.}$$
$$= 3 \quad \text{Multiply.}$$

The graph of g includes the origin, so $g(0) = 0$. Thus $\left(\frac{f}{g}\right)(0)$ is undefined.

x	$f(x)$	$g(x)$
-2	-3	undefined
0	1	0
1	3	1
4	9	2

(b) From the table, repeated in the margin, $f(4) = 9$ and $g(4) = 2$.

$$(f+g)(4) = f(4) + g(4) \quad (f+g)(x) = f(x) + g(x)$$
$$= 9 + 2 \quad \text{Substitute.}$$
$$= 11 \quad \text{Add.}$$

In the table, $g(-2)$ is undefined, and thus $(f-g)(-2)$ is also undefined.

$$(fg)(1) = f(1) \cdot g(1) \quad (fg)(x) = f(x) \cdot g(x)$$
$$= 3 \cdot 1 \quad f(1) = 3 \text{ and } g(1) = 1$$
$$= 3 \quad \text{Multiply.}$$

The quotient function value $\left(\frac{f}{g}\right)(0)$ is undefined since the denominator, $g(0)$, equals 0.

Teaching Tip Point out that you cannot determine the domain by looking only at the resulting function. For instance, let

$$f(x) = 1 + \sqrt{x}$$

and $\quad g(x) = 1 - \sqrt{x},$

and consider $f + g$.

(c) Using $f(x) = 2x + 1$ and $g(x) = \sqrt{x}$, we can find $(f+g)(4)$ and $(fg)(1)$. Since -2 is not in the domain of g, $(f-g)(-2)$ is not defined.

$$(f+g)(4) = f(4) + g(4) = (2 \cdot 4 + 1) + \sqrt{4} = 9 + 2 = 11$$

$$(fg)(1) = f(1) \cdot g(1) = (2 \cdot 1 + 1) \cdot \sqrt{1} = 3(1) = 3$$

$\left(\frac{f}{g}\right)(0)$ is undefined since $g(0) = 0$.

✔ *Now Try Exercises 23 and 27.*

The Difference Quotient Suppose the point P lies on the graph of $y = f(x)$ as in **Figure 93,** and suppose h is a positive number. If we let $(x, f(x))$ denote the coordinates of P and let $(x+h, f(x+h))$ denote the coordinates of Q, then the line joining P and Q has slope as follows.

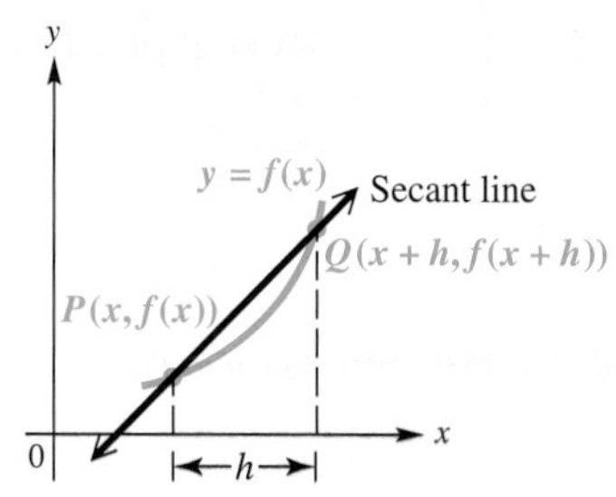

Figure 93

$$m = \frac{f(x+h) - f(x)}{(x+h) - x} \quad \text{Slope formula (Section 2.4)}$$

$$= \frac{\boldsymbol{f(x+h) - f(x)}}{\boldsymbol{h}}, \quad h \neq 0 \quad \text{Difference quotient}$$

This boldface expression is called the **difference quotient.**

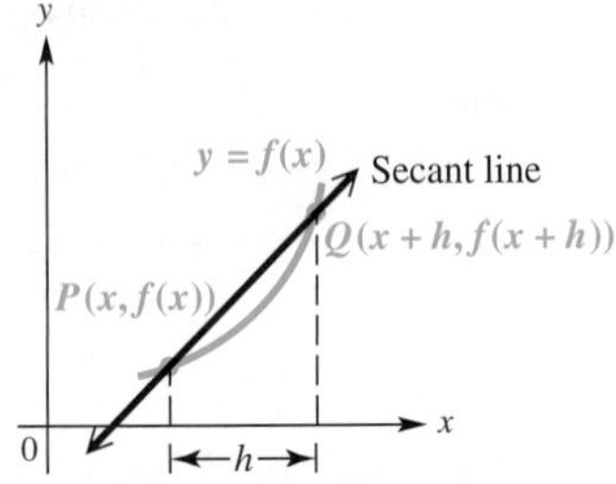

Figure 93 (repeated)

Classroom Example 4
Let $f(x) = 3x^2 - 2x + 4$. Find the difference quotient and simplify the expression.

Answer: $6x + 3h - 2$

Teaching Tip Refer to **Exercise 39** and its selected solution for a more involved difference quotient example $\left(f(x) = \frac{1}{x}\right)$.

Figure 93 shows the graph of the line PQ (called a **secant line**). As h approaches 0, the slope of this secant line approaches the slope of the line tangent to the curve at P. Important applications of this idea are developed in calculus.

EXAMPLE 4 Finding the Difference Quotient

Let $f(x) = 2x^2 - 3x$. Find and simplify the expression for the difference quotient,

$$\frac{f(x+h) - f(x)}{h}.$$

SOLUTION We use a three-step process.

Step 1 Find the first term in the numerator, $f(x+h)$. Replace x in $f(x)$ with $x + h$.

$$f(x+h) = 2(x+h)^2 - 3(x+h)$$

Step 2 Find the entire numerator, $f(x+h) - f(x)$. (From Step 1)

$$f(x+h) - f(x) = \left[2(x+h)^2 - 3(x+h)\right] - (2x^2 - 3x) \quad \text{Substitute.}$$

$$= 2(x^2 + 2xh + h^2) - 3(x+h) - (2x^2 - 3x) \quad \text{Square } x + h. \text{ (Section R.3)}$$

(Remember this term when squaring $x + h$.)

$$= 2x^2 + 4xh + 2h^2 - 3x - 3h - 2x^2 + 3x \quad \text{Distributive property (Section R.2)}$$

$$= 4xh + 2h^2 - 3h \quad \text{Combine like terms.}$$

Step 3 Find the difference quotient by dividing by h.

$$\frac{f(x+h) - f(x)}{h} = \frac{4xh + 2h^2 - 3h}{h} \quad \text{Substitute.}$$

$$= \frac{h(4x + 2h - 3)}{h} \quad \text{Factor out } h. \text{ (Section R.4)}$$

$$= 4x + 2h - 3 \quad \text{Divide. (Section R.5)}$$

✔ ***Now Try Exercises 35 and 45.***

CAUTION In **Example 4,** notice that the expression $f(x+h)$ is not equivalent to $f(x) + f(h)$.

$$f(x+h) = 2(x+h)^2 - 3(x+h) = 2x^2 + 4xh + 2h^2 - 3x - 3h$$

$$f(x) + f(h) = (2x^2 - 3x) + (2h^2 - 3h) = 2x^2 - 3x + 2h^2 - 3h$$

These expressions differ by $4xh$. In general, for a function f, $f(x+h)$ is *not* equivalent to $f(x) + f(h)$.

LOOKING AHEAD TO CALCULUS

The difference quotient is essential in the definition of the **derivative of a function** in calculus. The derivative provides a formula, in function form, for finding the slope of the tangent line to the graph of the function at a given point.

To illustrate, it is shown in calculus that the derivative of $f(x) = x^2 + 3$ is given by the function $f'(x) = 2x$. Now, $f'(0) = 2(0) = 0$, meaning that the slope of the tangent line to $f(x) = x^2 + 3$ at $x = 0$ is 0, which implies that the tangent line is horizontal. If you draw this tangent line, you will see that it is the line $y = 3$, which is indeed a horizontal line.

Composition of Functions and Domain The diagram in **Figure 94** shows a function f that assigns to each x in its domain a value $f(x)$. Then another function g assigns to each $f(x)$ in its domain a value $g(f(x))$. This two-step process takes an element x and produces a corresponding element $g(f(x))$.

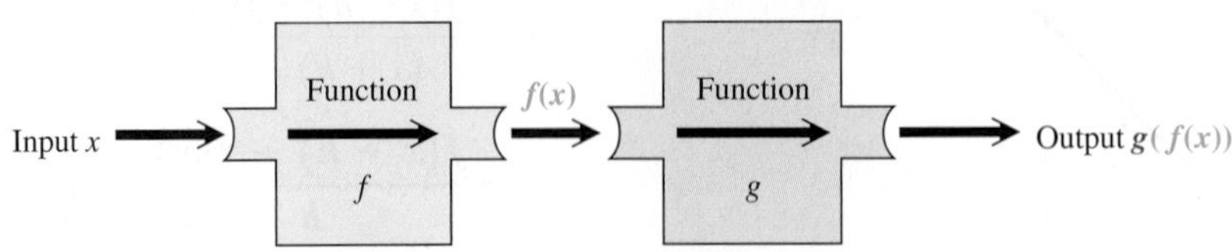

Figure 94

The function with y-values $g(f(x))$ is called the *composition* of functions g and f, which is written $\boldsymbol{g \circ f}$ and read "$\boldsymbol{g}$ **of** $\boldsymbol{f}$."

Teaching Tip Stress that with $f \circ g$, apply g first, and that $g \circ f$ is usually a different function from $f \circ g$. Also, students frequently misinterpret $f \circ g$ as a product.

Composition of Functions and Domain

If f and g are functions, then the **composite function,** or **composition,** of g and f is defined by

$$(g \circ f)(x) = g(f(x)).$$

The **domain of $g \circ f$** is the set of all numbers x in the domain of f such that $f(x)$ is in the domain of g.

As a real-life example of how composite functions occur, consider the following retail situation:

A \$40 pair of blue jeans is on sale for 25% off. If you purchase the jeans before noon, the retailer offers an additional 10% off. What is the final sale price of the blue jeans?

You might be tempted to say that the jeans are 35% off and calculate $\$40(0.35) = \$14$, giving a final sale price of

$$\$40 - \$14 = \$26$$

for the jeans. ***This is not correct.*** To find the final sale price, we must first find the price after taking 25% off and then take an additional 10% off *that* price. See **Figure 95.**

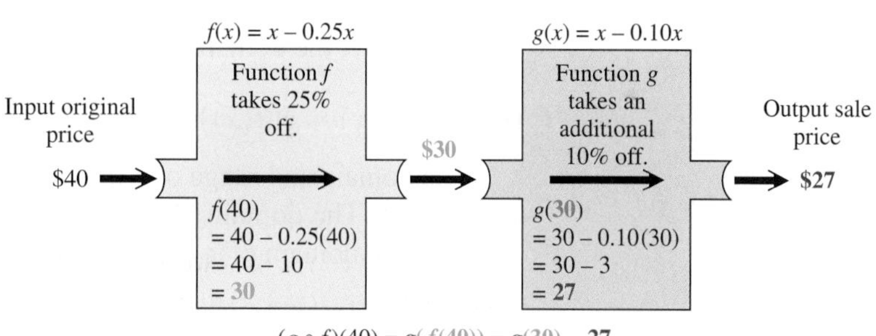

Figure 95

Classroom Example 5

Let $f(x) = \sqrt{x+4}$ and $g(x) = \frac{2}{x}$.

(a) Find $(f \circ g)(2)$.

(b) Find $(g \circ f)(5)$.

Answers:

(a) $\sqrt{5}$ **(b)** $\frac{2}{3}$

EXAMPLE 5 Evaluating Composite Functions

Let $f(x) = 2x - 1$ and $g(x) = \frac{4}{x-1}$.

(a) Find $(f \circ g)(2)$. **(b)** Find $(g \circ f)(-3)$.

SOLUTION

(a) First find $g(2)$: $g(2) = \frac{4}{2-1} = \frac{4}{1} = 4.$

Now find $(f \circ g)(2)$.

$$\begin{aligned} (f \circ g)(2) &= f(g(2)) && \text{Definition of composition} \\ &= f(4) && \text{See above.} \\ &= 2(4) - 1 && \text{Definition of } f \\ &= 7 && \text{Simplify.} \end{aligned}$$

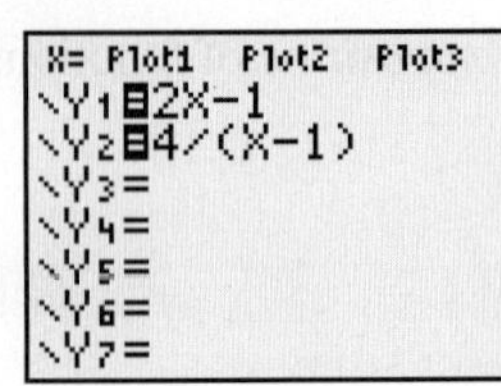

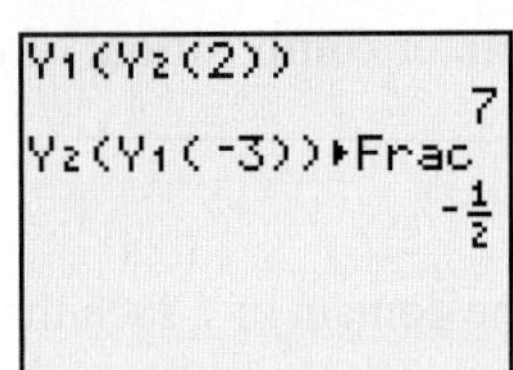

The screens show how a graphing calculator evaluates the expressions in **Example 5.**

(b) $(g \circ f)(-3) = g(f(-3))$ Definition of composition

$= g[2(-3) - 1]$ $f(-3) = 2(-3) - 1$

$= g(-7)$ $2(-3) - 1 = -7$

$= \dfrac{4}{-7-1}$ $g(x) = \frac{4}{x+1}$

$= \dfrac{4}{-8}$, or $-\dfrac{1}{2}$ Simplify.

Now Try Exercise 47.

Classroom Example 6
Given that $f(x) = \sqrt{x-1}$ and $g(x) = 2x + 5$, find each of the following.
(a) $(f \circ g)(x)$ and its domain
(b) $(g \circ f)(x)$ and its domain

Answers:
(a) $\sqrt{2x+4}$; $[-2, \infty)$
(b) $2\sqrt{x-1} + 5$; $[1, \infty)$

EXAMPLE 6 Determining Composite Functions and Their Domains

Given that $f(x) = \sqrt{x}$ and $g(x) = 4x + 2$, find each of the following.

(a) $(f \circ g)(x)$ and its domain **(b)** $(g \circ f)(x)$ and its domain

SOLUTION

(a) $(f \circ g)(x) = f(g(x)) = f(4x+2) = \sqrt{4x+2}$

The domain and range of g are both the set of all real numbers, $(-\infty, \infty)$. The domain of f is the set of all nonnegative real numbers, $[0, \infty)$. Thus, $g(x)$, which is defined as $4x + 2$, must be greater than or equal to zero.

The radicand must be nonnegative.

$4x + 2 \geq 0$ Solve the inequality.

$4x \geq -2$ Subtract 2.

$x \geq -\dfrac{1}{2}$ Divide by 4.

Therefore, the domain of $f \circ g$ is $\left[-\frac{1}{2}, \infty\right)$.

(b) $(g \circ f)(x) = g(f(x)) = g(\sqrt{x}) = 4\sqrt{x} + 2$

The domain and range of f are both the set of all nonnegative real numbers, $[0, \infty)$. The domain of g is the set of all real numbers, $(-\infty, \infty)$. Therefore, the domain of $g \circ f$ is $[0, \infty)$.

Now Try Exercise 65.

Classroom Example 7
Given that

$f(x) = \dfrac{5}{x+4}$ and $g(x) = \dfrac{2}{x}$,

find each of the following.
(a) $(f \circ g)(x)$ and its domain
(b) $(g \circ f)(x)$ and its domain

Answers:
(a) $\frac{5x}{2+4x}$; $\left(-\infty, -\frac{1}{2}\right) \cup \left(-\frac{1}{2}, 0\right) \cup (0, \infty)$
(b) $\frac{2x+8}{5}$; $(-\infty, -4) \cup (-4, \infty)$

EXAMPLE 7 Determining Composite Functions and Their Domains

Given that $f(x) = \frac{6}{x-3}$ and $g(x) = \frac{1}{x}$, find each of the following.

(a) $(f \circ g)(x)$ and its domain **(b)** $(g \circ f)(x)$ and its domain

SOLUTION

(a) $(f \circ g)(x) = f(g(x)) = f\left(\dfrac{1}{x}\right)$ $g(x) = \frac{1}{x}$

$= \dfrac{6}{\frac{1}{x} - 3}$ $f(x) = \frac{6}{x-3}$

$= \dfrac{6x}{1-3x}$ Multiply the numerator and denominator by x. (Section R.5)

The domain of g is all real numbers *except* 0, which makes $g(x)$ undefined. The domain of f is all real numbers *except* 3. The expression for $g(x)$, therefore, cannot equal 3. We determine the value that makes $g(x) = 3$ and *exclude* it from the domain of $f \circ g$.

LOOKING AHEAD TO CALCULUS

Finding the derivative of a function in calculus is called **differentiation.** To differentiate a composite function such as $h(x) = (3x + 2)^4$, we interpret $h(x)$ as $(f \circ g)(x)$, where $g(x) = 3x + 2$ and $f(x) = x^4$. The **chain rule** allows us to differentiate composite functions. Notice the use of the composition symbol and function notation in the following, which comes from the chain rule.

If $h(x) = (f \circ g)(x)$, then
$h'(x) = f'(g(x)) \cdot g'(x)$.

$$\frac{1}{x} = 3 \quad \text{The solution must be excluded.}$$

$$1 = 3x \quad \text{Multiply by } x.$$

$$x = \frac{1}{3} \quad \text{Divide by 3.}$$

Therefore, the domain of $f \circ g$ is the set of all real numbers *except* 0 and $\frac{1}{3}$, written in interval notation as

$$(-\infty, 0) \cup \left(0, \frac{1}{3}\right) \cup \left(\frac{1}{3}, \infty\right).$$

(b) $(g \circ f)(x) = g(f(x)) = g\left(\frac{6}{x-3}\right)$

$$= \frac{1}{\frac{6}{x-3}} \quad \text{Note that this is meaningless if } x = 3.$$

$$= \frac{x-3}{6} \quad \frac{1}{\frac{a}{b}} = 1 \div \frac{a}{b} = 1 \cdot \frac{b}{a} = \frac{b}{a}$$

The domain of f is all real numbers *except* 3, and the domain of g is all real numbers *except* 0. The expression for $f(x)$, which is $\frac{6}{x-3}$, is never zero, since the numerator is the nonzero number 6. Therefore, the domain of $g \circ f$ is the set of all real numbers *except* 3, written

$$(-\infty, 3) \cup (3, \infty).$$

Now Try Exercise 77.

NOTE In a situation like **Example 7(b),** it often helps to consider the *unsimplified* form of the composition expression when determining the domain.

Classroom Example 8
Let $f(x) = 2x - 5$ and $g(x) = 3x^2 + x$. Show that

$$(g \circ f)(x) \neq (f \circ g)(x).$$

Answer: $12x^2 - 58x + 70 \neq 6x^2 + 2x - 5$

EXAMPLE 8 Showing That $(g \circ f)(x)$ Is Not Equivalent to $(f \circ g)(x)$

Let $f(x) = 4x + 1$ and $g(x) = 2x^2 + 5x$. Show that $(g \circ f)(x) \neq (f \circ g)(x)$. (This is sufficient to prove that this inequality is true in general.)

SOLUTION First, find $(g \circ f)(x)$.

$$\begin{aligned}
(g \circ f)(x) &= g(f(x)) = g(4x + 1) && f(x) = 4x + 1 \\
&= 2(4x + 1)^2 + 5(4x + 1) && g(x) = 2x^2 + 5x \\
&= 2(16x^2 + 8x + 1) + 20x + 5 && \text{Square } 4x + 1 \text{ and apply the distributive property.} \\
&= 32x^2 + 16x + 2 + 20x + 5 && \text{Distributive property} \\
(g \circ f)(x) &= 32x^2 + 36x + 7 && \text{Combine like terms.}
\end{aligned}$$

Now, find $(f \circ g)(x)$.

$$\begin{aligned}
(f \circ g)(x) &= f(g(x)) && \text{By definition} \\
&= f(2x^2 + 5x) && g(x) = 2x^2 + 5x \\
&= 4(2x^2 + 5x) + 1 && f(x) = 4x + 1 \\
(f \circ g)(x) &= 8x^2 + 20x + 1 && \text{Distributive property}
\end{aligned}$$

Thus, $(g \circ f)(x) \neq (f \circ g)(x)$.

Now Try Exercise 81.

As **Example 8** shows, ***it is not always true that $f \circ g = g \circ f$.*** In fact, the composite functions $f \circ g$ and $g \circ f$ are equal only for a special class of functions, discussed in **Section 4.1.**

In calculus it is sometimes necessary to treat a function as a composition of two functions. The next example shows how this can be done.

Classroom Example 9
Find functions f and g such that

$(f \circ g)(x)$
$= 4(3x + 2)^2 - 5(3x + 2) - 8.$

Answer:
$f(x) = 4x^2 - 5x - 8;$
$g(x) = 3x + 2$
(Other answers are possible.)

EXAMPLE 9 Finding Functions That Form a Given Composite

Find functions f and g such that

$$(f \circ g)(x) = (x^2 - 5)^3 - 4(x^2 - 5) + 3.$$

SOLUTION Note the repeated quantity $x^2 - 5$. If we choose $g(x) = x^2 - 5$ and $f(x) = x^3 - 4x + 3$, then we have the following.

$$\begin{aligned}(f \circ g)(x) &= f(g(x)) && \text{By definition}\\ &= f(x^2 - 5) && g(x) = x^2 - 5\\ &= (x^2 - 5)^3 - 4(x^2 - 5) + 3 && \text{Use the rule for } f.\end{aligned}$$

There are other pairs of functions f and g that also satisfy these conditions. Here is another such pair.

$$f(x) = (x - 5)^3 - 4(x - 5) + 3 \quad \text{and} \quad g(x) = x^2$$

Now Try Exercise 89.

2.8 Exercises

1. 12 2. 44
3. −4 4. 21
5. −38 6. 144
7. $\frac{1}{2}$ 8. −7
9. $5x - 1$; $x + 9$; $6x^2 - 7x - 20$; $\frac{3x + 4}{2x - 5}$; All domains are $(-\infty, \infty)$ except for that of $\frac{f}{g}$, which is $\left(-\infty, \frac{5}{2}\right) \cup \left(\frac{5}{2}, \infty\right)$.
10. $-7x + 7$; $x + 5$; $12x^2 - 27x + 6$; $\frac{6 - 3x}{-4x + 1}$; All domains are $(-\infty, \infty)$ except for that of $\frac{f}{g}$, which is $\left(-\infty, \frac{1}{4}\right) \cup \left(\frac{1}{4}, \infty\right)$.
11. $3x^2 - 4x + 3$; $x^2 - 2x - 3$; $2x^4 - 5x^3 + 9x^2 - 9x$; $\frac{2x^2 - 3x}{x^2 - x + 3}$; All domains are $(-\infty, \infty)$.
12. $5x^2 - x + 2$; $3x^2 + 5x - 2$; $4x^4 - 10x^3 + 2x^2 + 4x$; $\frac{4x^2 + 2x}{x^2 - 3x + 2}$; All domains are $(-\infty, \infty)$ except for that of $\frac{f}{g}$, which is $(-\infty, 1) \cup (1, 2) \cup (2, \infty)$.

Let $f(x) = x^2 + 3$ and $g(x) = -2x + 6$. Find each of the following. ***See Example 1.***

1. $(f + g)(3)$
2. $(f + g)(-5)$
3. $(f - g)(-1)$
4. $(f - g)(4)$
5. $(fg)(4)$
6. $(fg)(-3)$
7. $\left(\frac{f}{g}\right)(-1)$
8. $\left(\frac{f}{g}\right)(5)$

For the pair of functions defined, find $(f + g)(x)$, $(f - g)(x)$, $(fg)(x)$, and $\left(\frac{f}{g}\right)(x)$. Give the domain of each. ***See Example 2.***

9. $f(x) = 3x + 4$, $g(x) = 2x - 5$
10. $f(x) = 6 - 3x$, $g(x) = -4x + 1$
11. $f(x) = 2x^2 - 3x$, $g(x) = x^2 - x + 3$
12. $f(x) = 4x^2 + 2x$, $g(x) = x^2 - 3x + 2$
13. $f(x) = \sqrt{4x - 1}$, $g(x) = \frac{1}{x}$
14. $f(x) = \sqrt{5x - 4}$, $g(x) = -\frac{1}{x}$

Associate's Degrees Earned *The graph shows the number of associate's degrees earned (in thousands) in the United States from 2000 through 2008. $M(x)$ gives the number of degrees earned by males, $F(x)$ gives the number earned by females, and $T(x)$ gives the total number for both groups. Use the graph in Exercises 15–18.*

15. Estimate $M(2004)$ and $F(2004)$, and use your results to estimate $T(2004)$.
16. Estimate $M(2008)$ and $F(2008)$, and use your results to estimate $T(2008)$.

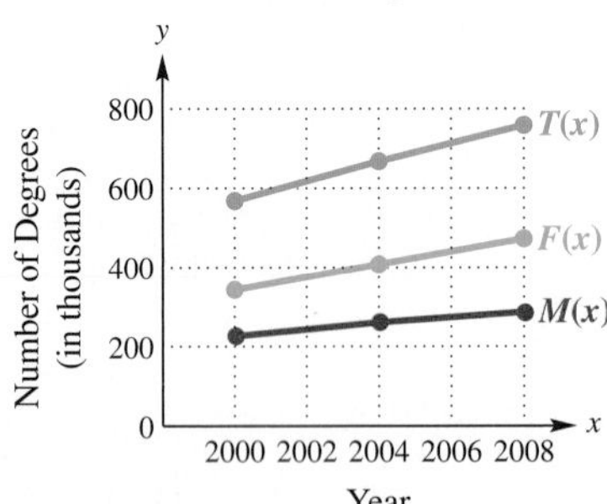

Source: U.S. National Center for Education Statistics.

13. $\sqrt{4x-1}+\frac{1}{x}$; $\sqrt{4x-1}-\frac{1}{x}$; $\frac{\sqrt{4x-1}}{x}$; $x\sqrt{4x-1}$; All domains are $\left[\frac{1}{4}, \infty\right)$.
14. $\sqrt{5x-4}-\frac{1}{x}$; $\sqrt{5x-4}+\frac{1}{x}$; $-\frac{\sqrt{5x-4}}{x}$; $-x\sqrt{5x-4}$; All domains are $\left[\frac{4}{5}, \infty\right)$.
15. 260; 400; 660 (all in thousands)
16. 290; 470; 760 (all in thousands)
17. 2000–2004
18. $r - s$
19. 6; It represents the dollars (in billions) spent for general science in 2000.
20. about 18; It represents the dollars (in billions) spent for space and other technologies in 2010.
21. space and other technologies; 1995–2000
22. space and other technologies in 2005–2010
23. (a) 2 (b) 4 (c) 0 (d) $-\frac{1}{3}$
24. (a) 2 (b) -3 (c) 2 (d) -2
25. (a) 3 (b) -5 (c) 2 (d) undefined
26. (a) -2 (b) -2 (c) 3 (d) -3

17. Use the slopes of the line segments to decide in which period (2000–2004 or 2004–2008) the total number of associate's degrees earned increased more rapidly.

18. *Concept Check* Refer to the graph of Associate's Degrees Earned on the previous page. If $2000 \le k \le 2008$, $T(k) = r$, and $F(k) = s$, then $M(k) =$ ______.

Science and Space/Technology Spending *The graph shows dollars (in billions) spent for general science and for space/other technologies in selected years. $G(x)$ represents the dollars spent for general science, and $S(x)$ represents the dollars spent for space and other technologies. $T(x)$ represents the total expenditures for these two categories. Use the graph in Exercises 19–22.*

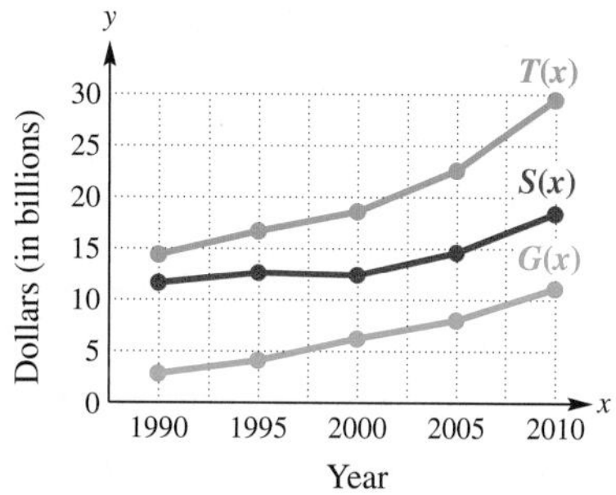

Source: U.S. Office of Management and Budget.

19. Estimate $(T - S)(2000)$. What does this function represent?

20. Estimate $(T - G)(2010)$. What does this function represent?

21. In which of the categories was spending almost static for several years? In which years did this occur?

22. In which period and which category does spending for $G(x)$ or $S(x)$ increase most?

Use the graph to evaluate each expression. ***See Example 3(a).***

23. (a) $(f+g)(2)$ (b) $(f-g)(1)$ (c) $(fg)(0)$ (d) $\left(\frac{f}{g}\right)(1)$

24. (a) $(f+g)(0)$ (b) $(f-g)(-1)$ (c) $(fg)(1)$ (d) $\left(\frac{f}{g}\right)(2)$

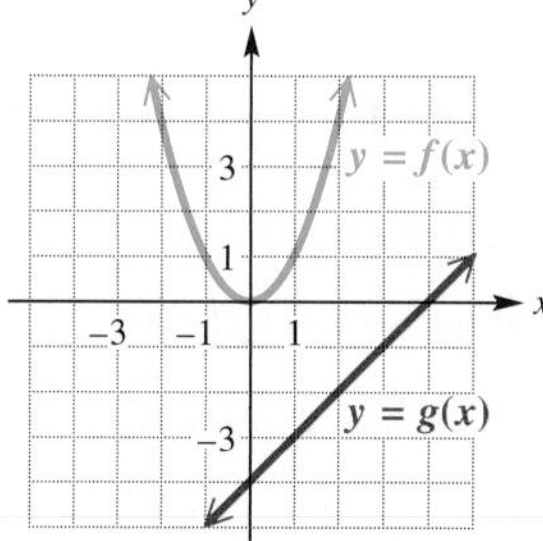

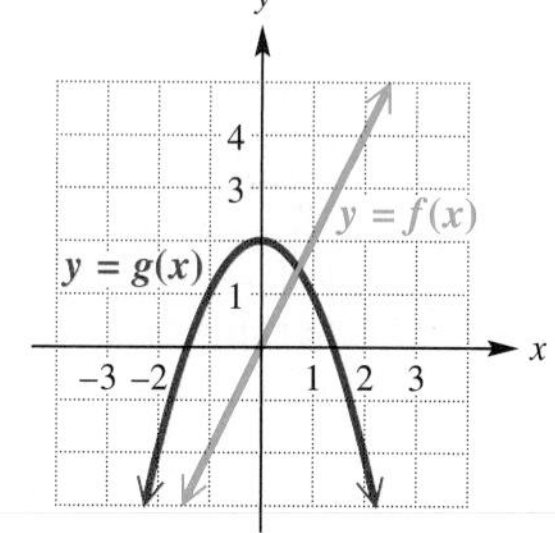

25. (a) $(f+g)(-1)$ (b) $(f-g)(-2)$ (c) $(fg)(0)$ (d) $\left(\frac{f}{g}\right)(2)$

26. (a) $(f+g)(1)$ (b) $(f-g)(0)$ (c) $(fg)(-1)$ (d) $\left(\frac{f}{g}\right)(1)$

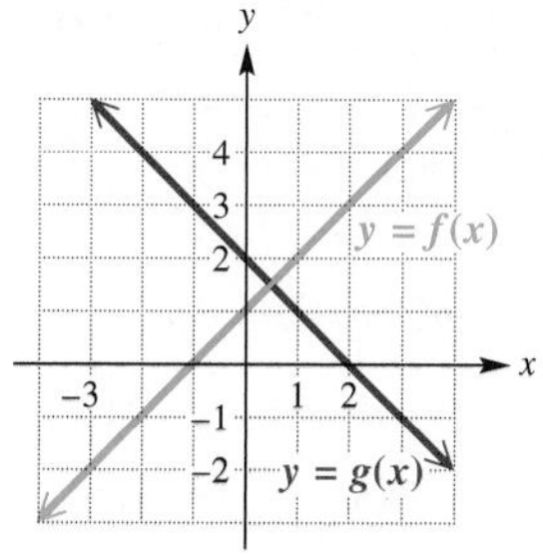

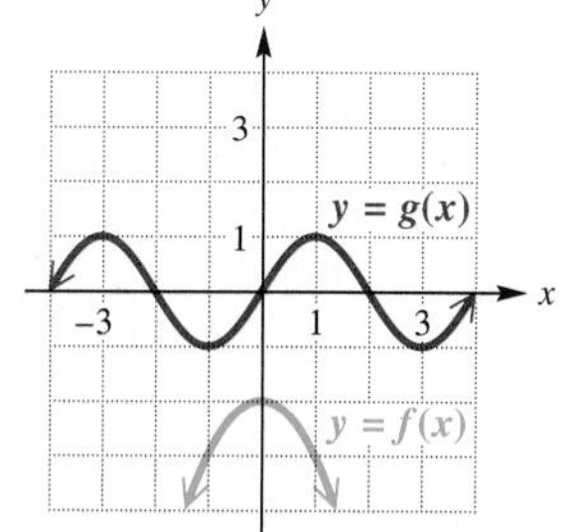

In Exercises 27 and 28, use the table to evaluate each expression in parts (a)–(d), if possible. **See Example 3(b).**

(a) $(f+g)(2)$ **(b)** $(f-g)(4)$ **(c)** $(fg)(-2)$ **(d)** $\left(\frac{f}{g}\right)(0)$

27.

x	$f(x)$	$g(x)$
−2	0	6
0	5	0
2	7	−2
4	10	5

28.

x	$f(x)$	$g(x)$
−2	−4	2
0	8	−1
2	5	4
4	0	0

29. Use the table in **Exercise 27** to complete the following table.

x	$(f+g)(x)$	$(f-g)(x)$	$(fg)(x)$	$\left(\frac{f}{g}\right)(x)$
−2				
0				
2				
4				

30. Use the table in **Exercise 28** to complete the following table.

x	$(f+g)(x)$	$(f-g)(x)$	$(fg)(x)$	$\left(\frac{f}{g}\right)(x)$
−2				
0				
2				
4				

31. How is the difference quotient related to slope?

32. Refer to **Figure 93.** How is the secant line PQ related to the tangent line to a curve at point P?

For each of the functions in Exercises 33–46, find **(a)** $f(x+h)$, **(b)** $f(x+h)-f(x)$, *and* **(c)** $\frac{f(x+h)-f(x)}{h}$. **See Example 4.**

33. $f(x)=2-x$ **34.** $f(x)=1-x$ **35.** $f(x)=6x+2$

36. $f(x)=4x+11$ **37.** $f(x)=-2x+5$ **38.** $f(x)=-4x+2$

39. $f(x)=\frac{1}{x}$ **40.** $f(x)=\frac{1}{x^2}$ **41.** $f(x)=x^2$

42. $f(x)=-x^2$ **43.** $f(x)=1-x^2$ **44.** $f(x)=1+2x^2$

45. $f(x)=x^2+3x+1$ **46.** $f(x)=x^2-4x+2$

Let $f(x)=2x-3$ *and* $g(x)=-x+3$. *Find each function value.* **See Example 5.**

47. $(f\circ g)(4)$ **48.** $(f\circ g)(2)$ **49.** $(f\circ g)(-2)$ **50.** $(g\circ f)(3)$

51. $(g\circ f)(0)$ **52.** $(g\circ f)(-2)$ **53.** $(f\circ f)(2)$ **54.** $(g\circ g)(-2)$

27. (a) 5 (b) 5 (c) 0 (d) undefined
28. (a) 9 (b) 0 (c) −8 (d) −8
29.

x	$(f+g)(x)$	$(f-g)(x)$
−2	6	−6
0	5	5
2	5	9
4	15	5

x	$(fg)(x)$	$\left(\frac{f}{g}\right)(x)$
−2	0	0
0	0	undefined
2	−14	−3.5
4	50	2

30.

x	$(f+g)(x)$	$(f-g)(x)$
−2	−2	−6
0	7	9
2	9	1
4	0	0

x	$(fg)(x)$	$\left(\frac{f}{g}\right)(x)$
−2	−8	−2
0	−8	−8
2	20	1.25
4	0	undefined

33. (a) $2-x-h$ (b) $-h$ (c) -1
34. (a) $1-x-h$ (b) $-h$ (c) -1
35. (a) $6x+6h+2$ (b) $6h$ (c) 6
36. (a) $4x+4h+11$ (b) $4h$ (c) 4
37. (a) $-2x-2h+5$ (b) $-2h$ (c) -2
38. (a) $-4x-4h+2$ (b) $-4h$ (c) -4
39. (a) $\frac{1}{x+h}$ (b) $\frac{-h}{x(x+h)}$ (c) $\frac{-1}{x(x+h)}$
40. (a) $\frac{1}{(x+h)^2}$ (b) $\frac{-2xh-h^2}{x^2(x+h)^2}$ (c) $\frac{-2x-h}{x^2(x+h)^2}$
41. (a) $x^2+2xh+h^2$ (b) $2xh+h^2$ (c) $2x+h$
42. (a) $-x^2-2xh-h^2$ (b) $-2xh-h^2$ (c) $-2x-h$
43. (a) $1-x^2-2xh-h^2$ (b) $-2xh-h^2$ (c) $-2x-h$
44. (a) $1+2x^2+4xh+2h^2$ (b) $4xh+2h^2$ (c) $4x+2h$

45. (a) $x^2 + 2xh + h^2 + 3x + 3h + 1$
(b) $2xh + h^2 + 3h$
(c) $2x + h + 3$
46. (a) $x^2 + 2xh + h^2 - 4x - 4h + 2$
(b) $2xh + h^2 - 4h$
(c) $2x + h - 4$

47. -5 48. -1
49. 7 50. 0
51. 6 52. 10
53. -1 54. -2

55. 1 56. 9
57. 9 58. 12
59. 1 60. 12
61. $g(1) = 9$, and $f(9)$ cannot be determined from the table given.
62. 12

63. (a) $-30x - 33$; $(-\infty, \infty)$
(b) $-30x + 52$; $(-\infty, \infty)$
64. (a) $24x + 4$; $(-\infty, \infty)$
(b) $24x + 35$; $(-\infty, \infty)$
65. (a) $\sqrt{x+3}$; $[-3, \infty)$
(b) $\sqrt{x} + 3$; $[0, \infty)$
66. (a) $\sqrt{x-1}$; $[1, \infty)$
(b) $\sqrt{x} - 1$; $[0, \infty)$
67. (a) $(x^2 + 3x - 1)^3$; $(-\infty, \infty)$
(b) $x^6 + 3x^3 - 1$; $(-\infty, \infty)$
68. (a) $x^4 + x^2 - 2$; $(-\infty, \infty)$
(b) $(x+2)^4 + (x+2)^2 - 4$; $(-\infty, \infty)$
69. (a) $\sqrt{3x-1}$; $\left[\frac{1}{3}, \infty\right)$
(b) $3\sqrt{x-1}$; $[1, \infty)$
70. (a) $\sqrt{2x-2}$; $[1, \infty)$
(b) $2\sqrt{x-2}$; $[2, \infty)$
71. (a) $\frac{2}{x+1}$; $(-\infty, -1) \cup (-1, \infty)$
(b) $\frac{2}{x} + 1$; $(-\infty, 0) \cup (0, \infty)$
72. (a) $\frac{4}{x+4}$; $(-\infty, -4) \cup (-4, \infty)$
(b) $\frac{4}{x} + 4$; $(-\infty, 0) \cup (0, \infty)$
73. (a) $\sqrt{-\frac{1}{x} + 2}$; $(-\infty, 0) \cup \left[\frac{1}{2}, \infty\right)$
(b) $-\frac{1}{\sqrt{x+2}}$; $(-2, \infty)$
74. (a) $\sqrt{-\frac{2}{x} + 4}$; $(-\infty, 0) \cup \left[\frac{1}{2}, \infty\right)$
(b) $\frac{-2}{\sqrt{x+4}}$; $(-4, \infty)$

Concept Check The tables give some selected ordered pairs for functions f and g.

x	3	4	6
$f(x)$	1	3	9

x	2	7	1	9
$g(x)$	3	6	9	12

Find each of the following.

55. $(f \circ g)(2)$ 56. $(f \circ g)(7)$ 57. $(g \circ f)(3)$

58. $(g \circ f)(6)$ 59. $(f \circ f)(4)$ 60. $(g \circ g)(1)$

61. *Concept Check* Why can you not determine $(f \circ g)(1)$ given the information in the tables for **Exercises 55–60?**

62. *Concept Check* Extend the concept of composition of functions to evaluate $(g \circ (f \circ g))(7)$ using the tables for **Exercises 55–60.**

Given functions f and g, find ***(a)*** *$(f \circ g)(x)$ and its domain, and* ***(b)*** *$(g \circ f)(x)$ and its domain.* ***See Examples 6 and 7.***

63. $f(x) = -6x + 9$, $g(x) = 5x + 7$
64. $f(x) = 8x + 12$, $g(x) = 3x - 1$
65. $f(x) = \sqrt{x}$, $g(x) = x + 3$
66. $f(x) = \sqrt{x}$, $g(x) = x - 1$
67. $f(x) = x^3$, $g(x) = x^2 + 3x - 1$
68. $f(x) = x + 2$, $g(x) = x^4 + x^2 - 4$
69. $f(x) = \sqrt{x-1}$, $g(x) = 3x$
70. $f(x) = \sqrt{x-2}$, $g(x) = 2x$
71. $f(x) = \frac{2}{x}$, $g(x) = x + 1$
72. $f(x) = \frac{4}{x}$, $g(x) = x + 4$
73. $f(x) = \sqrt{x+2}$, $g(x) = -\frac{1}{x}$
74. $f(x) = \sqrt{x+4}$, $g(x) = -\frac{2}{x}$
75. $f(x) = \sqrt{x}$, $g(x) = \frac{1}{x+5}$
76. $f(x) = \sqrt{x}$, $g(x) = \frac{3}{x+6}$
77. $f(x) = \frac{1}{x-2}$, $g(x) = \frac{1}{x}$
78. $f(x) = \frac{1}{x+4}$, $g(x) = -\frac{1}{x}$

79. *Concept Check* Fill in the missing entries in the table.

x	$f(x)$	$g(x)$	$g(f(x))$
1	3	2	7
2	1	5	
3	2		

80. *Concept Check* Suppose $f(x)$ is an odd function and $g(x)$ is an even function. Fill in the missing entries in the table.

x	-2	-1	0	1	2
$f(x)$			0	-2	
$g(x)$	0	2	1		
$(f \circ g)(x)$		1	-2		

81. Show that $(f \circ g)(x)$ is not equivalent to $(g \circ f)(x)$ for

$$f(x) = 3x - 2 \quad \text{and} \quad g(x) = 2x - 3.$$

82. Describe the steps required to find the composite function $f \circ g$, given

$$f(x) = 2x - 5 \quad \text{and} \quad g(x) = x^2 + 3.$$

75. (a) $\sqrt{\frac{1}{x+5}}$; $(-5, \infty)$
(b) $\frac{1}{\sqrt{x}+5}$; $[0, \infty)$

76. (a) $\sqrt{\frac{3}{x+6}}$; $(-6, \infty)$
(b) $\frac{3}{\sqrt{x}+6}$; $[0, \infty)$

77. (a) $\frac{x}{1-2x}$; $(-\infty, 0) \cup \left(0, \frac{1}{2}\right) \cup \left(\frac{1}{2}, \infty\right)$
(b) $x-2$; $(-\infty, 2) \cup (2, \infty)$

78. (a) $\frac{x}{-1+4x}$; $(-\infty, 0) \cup \left(0, \frac{1}{4}\right) \cup \left(\frac{1}{4}, \infty\right)$
(b) $-x-4$; $(-\infty, -4) \cup (-4, \infty)$

79.

x	$f(x)$	$g(x)$	$g(f(x))$
1	3	2	7
2	1	5	2
3	2	7	5

80.

x	−2	−1	0	1	2
$f(x)$	−1	2	0	−2	1
$g(x)$	0	2	1	2	0
$(f \circ g)(x)$	0	1	−2	1	0

81. $(f \circ g)(x) = 6x - 11$ and $(g \circ f)(x) = 6x - 7$, so they are not equivalent.

In Exercises 87–92, we give only one of the many possible ways.

87. $g(x) = 6x - 2$, $f(x) = x^2$

88. $g(x) = 11x^2 + 12x$, $f(x) = x^2$

89. $g(x) = x^2 - 1$, $f(x) = \sqrt{x}$

90. $g(x) = 2x - 3$, $f(x) = x^3$

91. $g(x) = 6x$, $f(x) = \sqrt{x} + 12$

92. $g(x) = 2x + 3$, $f(x) = \sqrt[3]{x} - 4$

93. $(f \circ g)(x) = 63{,}360x$ computes the number of inches in x miles.

94. (a) $s = \frac{x}{4}$ (b) $y = \frac{x^2}{16}$
(c) 2.25 square units

95. (a) $\mathcal{A}(2x) = \sqrt{3}x^2$
(b) $64\sqrt{3}$ square units

96. (a) $y_1 = 0.04x$
(b) $y_2 = 0.025(x + 500)$
(c) total annual interest
(d) $28.75

97. (a) $(\mathcal{A} \circ r)(t) = 16\pi t^2$
(b) It defines the area of the leak in terms of the time t, in minutes.
(c) 144π ft^2

For certain pairs of functions f and g, $(f \circ g)(x) = x$ and $(g \circ f)(x) = x$. Show that this is true for each pair in ***Exercises 83–86.***

83. $f(x) = 4x + 2$, $g(x) = \frac{1}{4}(x - 2)$

84. $f(x) = -3x$, $g(x) = -\frac{1}{3}x$

85. $f(x) = \sqrt[3]{5x + 4}$, $g(x) = \frac{1}{5}x^3 - \frac{4}{5}$

86. $f(x) = \sqrt[3]{x + 1}$, $g(x) = x^3 - 1$

Find functions f and g such that $(f \circ g)(x) = h(x)$. (There are many possible ways to do this.) ***See Example 9.***

87. $h(x) = (6x - 2)^2$

88. $h(x) = (11x^2 + 12x)^2$

89. $h(x) = \sqrt{x^2 - 1}$

90. $h(x) = (2x - 3)^3$

91. $h(x) = \sqrt{6x} + 12$

92. $h(x) = \sqrt[3]{2x + 3} - 4$

Solve each problem.

93. *Relationship of Measurement Units* The function defined by $f(x) = 12x$ computes the number of inches in x feet, and the function defined by $g(x) = 5280x$ computes the number of feet in x miles. What does $(f \circ g)(x)$ compute?

94. *Perimeter of a Square* The perimeter x of a square with side of length s is given by the formula $x = 4s$.

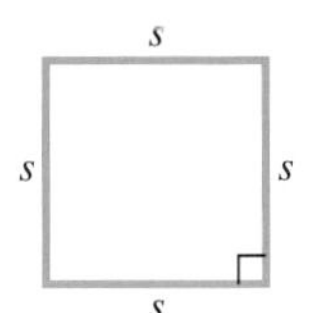

(a) Solve for s in terms of x.
(b) If y represents the area of this square, write y as a function of the perimeter x.
(c) Use the composite function of part (b) to find the area of a square with perimeter 6.

95. *Area of an Equilateral Triangle* The area of an equilateral triangle with sides of length x is given by the function defined by $\mathcal{A}(x) = \frac{\sqrt{3}}{4}x^2$.

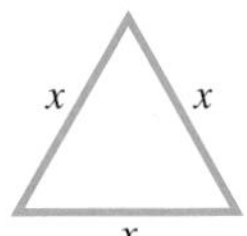

(a) Find $\mathcal{A}(2x)$, the function representing the area of an equilateral triangle with sides of length twice the original length.
(b) Find the area of an equilateral triangle with side length 16. Use the formula $\mathcal{A}(2x)$ found in part (a).

96. *Software Author Royalties* A software author invests his royalties in two accounts for 1 yr.

(a) The first account pays 4% simple interest. If he invests x dollars in this account, write an expression for y_1 in terms of x, where y_1 represents the amount of interest earned.
(b) He invests in a second account $500 more than he invested in the first account. This second account pays 2.5% simple interest. Write an expression for y_2, where y_2 represents the amount of interest earned.
(c) What does $y_1 + y_2$ represent?
(d) How much interest will he receive if $250 is invested in the first account?

97. *Oil Leak* An oil well off the Gulf Coast is leaking, with the leak spreading oil over the water's surface as a circle. At any time t, in minutes, after the beginning of the leak, the radius of the circular oil slick on the surface is $r(t) = 4t$ feet. Let $\mathcal{A}(r) = \pi r^2$ represent the area of a circle of radius r.

(a) Find $(\mathcal{A} \circ r)(t)$.
(b) Interpret $(\mathcal{A} \circ r)(t)$.
(c) What is the area of the oil slick after 3 min?

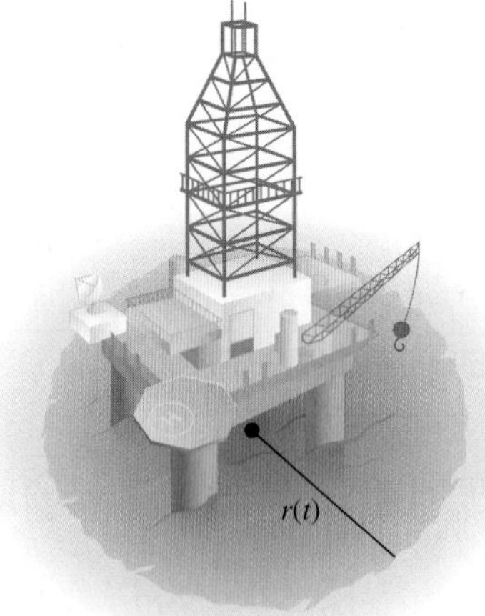

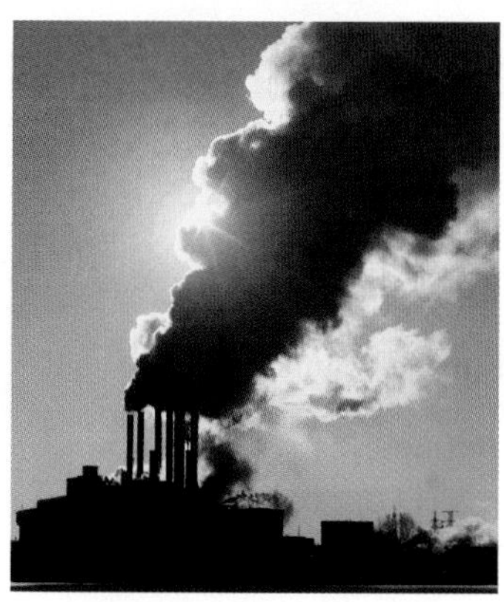

98. *Emission of Pollutants* When a thermal inversion layer is over a city (as happens in Los Angeles), pollutants cannot rise vertically but are trapped below the layer and must disperse horizontally. Assume that a factory smokestack begins emitting a pollutant at 8 A.M. Assume that the pollutant disperses horizontally over a circular area. If t represents the time, in hours, since the factory began emitting pollutants ($t = 0$ represents 8 A.M.), assume that the radius of the circle of pollutants at time t is $r(t) = 2t$ miles. Let $\mathcal{A}(r) = \pi r^2$ represent the area of a circle of radius r.

(a) Find $(\mathcal{A} \circ r)(t)$. **(b)** Interpret $(\mathcal{A} \circ r)(t)$.
(c) What is the area of the circular region covered by the layer at noon?

99. *(Modeling) Catering Cost* The cost to hire a caterer for a party depends on the number of guests attending. If 100 people attend, the cost per person will be \$20. For each person less than 100, the cost will increase by \$5. Assume that no more than 100 people will attend. Let x represent the number less than 100 who do not attend. For example, if 95 attend, $x = 5$.

(a) Write a function defined by $N(x)$ giving the number of guests.
(b) Write a function defined by $G(x)$ giving the cost per guest.
(c) Write a function defined by $N(x) \cdot G(x)$ for the total cost, $C(x)$.
(d) What is the total cost if 80 people attend?

100. *Area of a Square* The area of a square is x^2 square inches. Suppose that 3 in. is added to one dimension and 1 in. is subtracted from the other dimension. Express the area $\mathcal{A}(x)$ of the resulting rectangle as a product of two functions.

98. (a) $(\mathcal{A} \circ r)(t) = 4\pi t^2$
(b) It defines the area of the circular layer in terms of the time t, in hours.
(c) 64π mi^2
99. (a) $N(x) = 100 - x$
(b) $G(x) = 20 + 5x$
(c) $C(x) = (100 - x)(20 + 5x)$
(d) \$9600
100. $\mathcal{A}(x) = (x + 3)(x - 1)$

Chapter 2 Test Prep

Key Terms

2.1 ordered pair; origin; x-axis; y-axis; rectangular (Cartesian) coordinate system; coordinate plane (xy-plane); quadrants; coordinates; collinear; graph of an equation; x-intercept; y-intercept

2.2 circle; radius; center of a circle

2.3 dependent variable; independent variable; relation; function; domain; range; function notation; increasing function; decreasing function; constant function

2.4 linear function; standard form; relatively prime; change in x; change in y; slope; average rate of change; linear cost function; cost; fixed cost; revenue function; profit function

2.5 point-slope form; slope-intercept form; scatter diagram

2.6 continuous function; parabola; vertex; piecewise-defined function; step function

2.7 symmetry; even function; odd function; vertical translation; horizontal translation

2.8 difference quotient; composite function (composition)

New Symbols

(a, b)	ordered pair
$f(x)$	function f evaluated at x (read "f of x")
Δx	change in x
Δy	change in y
m	slope
$[\![x]\!]$	the greatest integer less than or equal to x
$g \circ f$	composite function

(continued)

Quick Review

Concepts | **Examples**

2.1 Rectangular Coordinates and Graphs

Distance Formula

Suppose $P(x_1, y_1)$ and $Q(x_2, y_2)$ are two points in a coordinate plane. Then the distance between P and Q, written $d(P, Q)$, is given by the following.

$$d(P, Q) = \sqrt{(x_2 - x_1)^2 + (y_2 - y_1)^2}$$

Find the distance between the points $P(-1, 4)$ and $Q(6, -3)$.

$$\begin{aligned} d(P, Q) &= \sqrt{[6 - (-1)]^2 + (-3 - 4)^2} \\ &= \sqrt{49 + 49} \\ &= \sqrt{98}, \text{ or } 7\sqrt{2} \end{aligned}$$

Midpoint Formula

The midpoint of the line segment with endpoints (x_1, y_1) and (x_2, y_2) is given by the following.

$$\left(\frac{x_1 + x_2}{2}, \frac{y_1 + y_2}{2}\right)$$

Find the coordinates of the midpoint of the line segment with endpoints $(-1, 4)$ and $(6, -3)$.

$$\left(\frac{-1 + 6}{2}, \frac{4 + (-3)}{2}\right) = \left(\frac{5}{2}, \frac{1}{2}\right)$$

2.2 Circles

Center-Radius Form of the Equation of a Circle

The equation of a circle with center at (h, k) and radius r is given by the following.

$$(x - h)^2 + (y - k)^2 = r^2$$

General Form of the Equation of a Circle

$$x^2 + y^2 + cx + dy + e = 0$$

Find the center-radius form of the equation of the circle with center at $(-2, 3)$ and radius 4.

$$[x - (-2)]^2 + (y - 3)^2 = 4^2$$

or

$$(x + 2)^2 + (y - 3)^2 = 16$$

The general form of the equation of the preceding circle is

$$x^2 + y^2 + 4x - 6y - 3 = 0.$$

2.3 Functions

A **relation** is a set of ordered pairs. A **function** is a relation in which, for each value of the first component of the ordered pairs, there is *exactly one* value of the second component. The set of first components is the **domain,** and the set of second components is the **range.**

The relation $y = x^2$ defines a function, because each choice of a number for x corresponds to one and only one number for y. The domain is $(-\infty, \infty)$, and the range is $[0, \infty)$.

The relation $x = y^2$ does *not* define a function because a number x may correspond to two numbers for y. The domain is $[0, \infty)$, and the range is $(-\infty, \infty)$.

Vertical Line Test

If every vertical line intersects the graph of a relation in no more than one point, then the relation is a function.

Determine whether the graphs are functions.

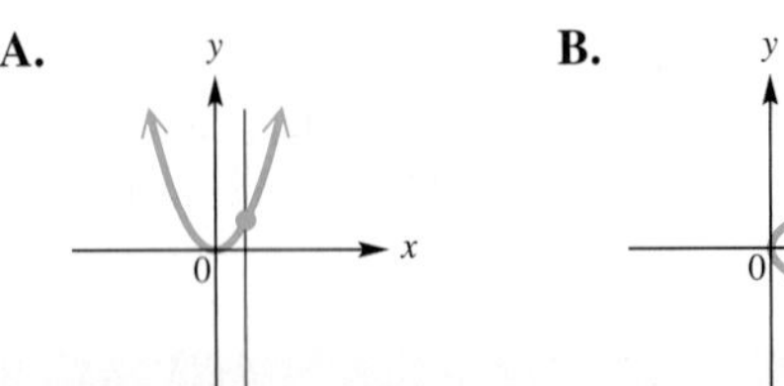

By the vertical line test, graph A is the graph of a function, but graph B is not.

Concepts	Examples

Increasing, Decreasing, and Constant Functions

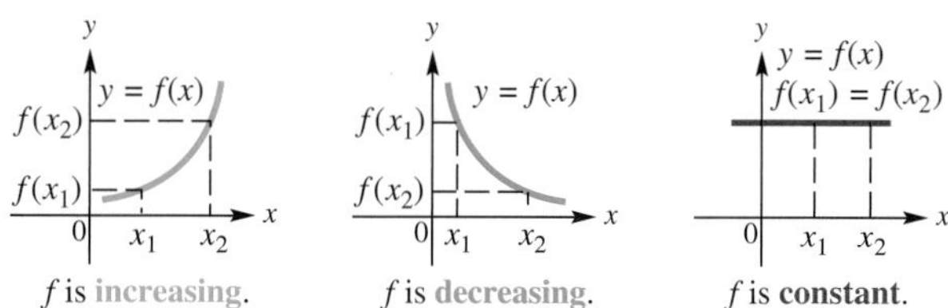

f is **increasing**. *f* is **decreasing**. *f* is **constant**.

Discuss the function in graph A on the preceding page in terms of whether it is increasing, decreasing, or constant.

The function in graph A is decreasing on the interval $(-\infty, 0]$ and increasing on the interval $[0, \infty)$.

2.4 Linear Functions

A function f is a **linear function** if, for real numbers a and b,

$$f(x) = ax + b.$$

The graph of a linear function is a line.

The equation

$$y = \frac{1}{2}x - 4$$

defines y as a linear function of x.

Definition of Slope

The slope m of the line through the points (x_1, y_1) and (x_2, y_2) is given by the following.

$$m = \frac{\text{rise}}{\text{run}} = \frac{\Delta y}{\Delta x} = \frac{y_2 - y_1}{x_2 - x_1}, \quad \text{where } \Delta x \neq 0$$

Find the slope of the line through the points $(2, 4)$ and $(-1, 7)$.

$$m = \frac{7 - 4}{-1 - 2} = \frac{3}{-3} = -1$$

2.5 Equations of Lines and Linear Models

Forms of Linear Equations

Equation	Description
$y = mx + b$	**Slope-Intercept Form** Slope is m. y-intercept is b.
$y - y_1 = m(x - x_1)$	**Point-Slope Form** Slope is m. Line passes through (x_1, y_1).
$Ax + By = C$	**Standard Form** (A, B, and C integers, $A \geq 0$) Slope is $-\frac{A}{B}$ $(B \neq 0)$. x-intercept is $\frac{C}{A}$ $(A \neq 0)$. y-intercept is $\frac{C}{B}$ $(B \neq 0)$.
$y = b$	**Horizontal Line** Slope is 0. y-intercept is b.
$x = a$	**Vertical Line** Slope is undefined. x-intercept is a.

Consider the following equation.

$$y = 3x + \frac{2}{3} \quad \text{Slope-intercept form}$$

The slope of the graph is $m = 3$ and the y-intercept is $b = \frac{2}{3}$.

Consider the following equation.

$$y - 3 = -2(x - 4) \quad \text{Point-slope form}$$

The slope is $m = -2$. The line passes through the point $(4, 3)$.

Consider the following equation.

$$4x + 5y = 7 \quad \text{Standard form with } A = 4, B = 5, C = 7$$

The slope is $m = -\frac{A}{B} = -\frac{4}{5}$.
The x-intercept is $\frac{C}{A} = \frac{7}{4}$.
The y-intercept is $\frac{C}{B} = \frac{7}{5}$.

Consider the following equations.

$$y = -6 \quad \text{Horizontal line}$$

The slope is 0. The y-intercept is -6.

$$x = 3 \quad \text{Vertical line}$$

The slope is undefined. The x-intercept is 3.

(continued)

Concepts	Examples

2.6 Graphs of Basic Functions

Basic Functions
Identity Function $f(x) = x$
Squaring Function $f(x) = x^2$
Cubing Function $f(x) = x^3$
Square Root Function $f(x) = \sqrt{x}$
Cube Root Function $f(x) = \sqrt[3]{x}$
Absolute Value Function $f(x) = |x|$
Greatest Integer Function $f(x) = [\![x]\!]$

Refer to the function boxes in **Section 2.6.** Graphs of the basic functions are also shown on the back inside cover.

2.7 Graphing Techniques

Stretching and Shrinking
If $a > 1$, then the graph of $y = af(x)$ is a **vertical stretching** of the graph of $y = f(x)$. If $0 < a < 1$, then the graph of $y = af(x)$ is a **vertical shrinking** of the graph of $y = f(x)$.

If $0 < a < 1$, then the graph of $y = f(ax)$ is a **horizontal stretching** of the graph of $y = f(x)$. If $a > 1$, then the graph of $y = f(ax)$ is a **horizontal shrinking** of the graph of $y = f(x)$.

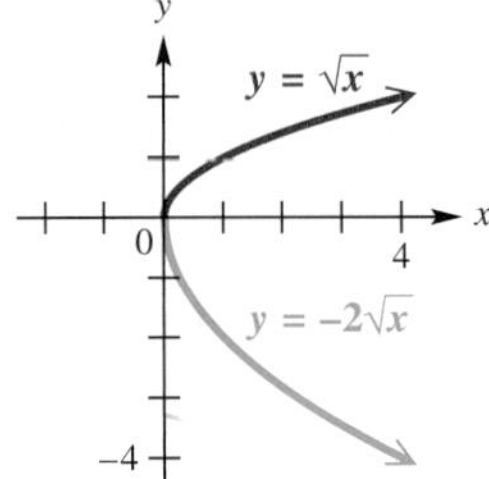

The graph of $y = -2\sqrt{x}$ is the graph of $y = \sqrt{x}$ stretched vertically by a factor of 2 and reflected across the x-axis.

Reflection Across an Axis
The graph of $y = -f(x)$ is the same as the graph of $y = f(x)$ reflected across the x-axis.

The graph of $y = f(-x)$ is the same as the graph of $y = f(x)$ reflected across the y-axis.

Symmetry
The graph of an equation is **symmetric with respect to the y-axis** if the replacement of x with $-x$ results in an equivalent equation.

The graph of an equation is **symmetric with respect to the x-axis** if the replacement of y with $-y$ results in an equivalent equation.

The graph of an equation is **symmetric with respect to the origin** if the replacement of both x with $-x$ and y with $-y$ at the same time results in an equivalent equation.

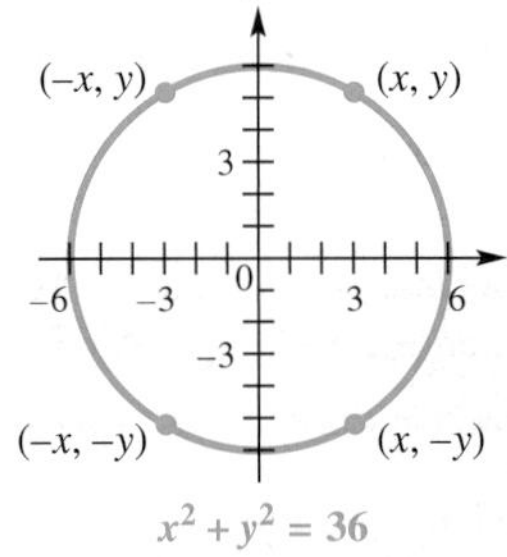

The graph of $x^2 + y^2 = 36$ is symmetric with respect to the y-axis, the x-axis, and the origin.

Translations
Let f be a function and c be a positive number.

To Graph:	Shift the Graph of $y = f(x)$ by c Units:
$y = f(x) + c$	up
$y = f(x) - c$	down
$y = f(x + c)$	left
$y = f(x - c)$	right

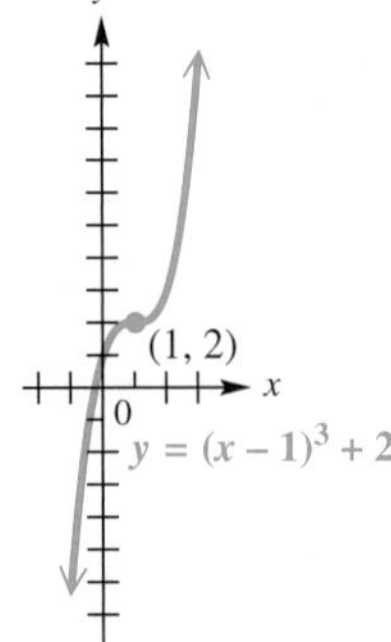

The graph of

$$y = (x - 1)^3 + 2$$

is the graph of $y = x^3$ translated 1 unit to the right and 2 units up.

Concepts	Examples

2.8 Function Operations and Composition

Operations on Functions

Given two functions f and g, then for all values of x for which both $f(x)$ and $g(x)$ are defined, the following operations are defined.

$$(f+g)(x) = f(x) + g(x) \quad \text{Sum}$$

$$(f-g)(x) = f(x) - g(x) \quad \text{Difference}$$

$$(fg)(x) = f(x) \cdot g(x) \quad \text{Product}$$

$$\left(\frac{f}{g}\right)(x) = \frac{f(x)}{g(x)}, \quad g(x) \neq 0 \quad \text{Quotient}$$

Let $f(x) = 2x - 4$ and $g(x) = \sqrt{x}$.

$$(f+g)(x) = 2x - 4 + \sqrt{x}$$

$$(f-g)(x) = 2x - 4 - \sqrt{x}$$

$$(fg)(x) = (2x-4)\sqrt{x}$$

The domain is $[0, \infty)$.

$$\left(\frac{f}{g}\right)(x) = \frac{2x-4}{\sqrt{x}}$$

The domain is $(0, \infty)$.

Difference Quotient

The line joining $P(x, f(x))$ and $Q(x+h, f(x+h))$ has slope

$$m = \frac{f(x+h) - f(x)}{h}, \quad h \neq 0.$$

Refer to **Example 4** in **Section 2.8.**

Composition of Functions

If f and g are functions, then the composite function, or composition, of g and f is defined by

$$(g \circ f)(x) = g(f(x)).$$

The domain of $g \circ f$ is the set of all x in the domain of f such that $f(x)$ is in the domain of g.

Using f and g as defined above,

$$(g \circ f)(x) = \sqrt{2x-4}.$$

The domain is all x such that $2x - 4 \geq 0$. Solving this inequality gives the interval $[2, \infty)$.

Chapter 2 Review Exercises

1. $\sqrt{85}$; $\left(-\frac{1}{2}, 2\right)$
2. $\sqrt{202}$; $\left(-\frac{5}{2}, -\frac{5}{2}\right)$
3. 5; $\left(-6, \frac{11}{2}\right)$
4. yes; $(5, 7)$
5. $(22, -6)$
6. yes
7. $(x+2)^2 + (y-3)^2 = 225$
8. $\left(x - \sqrt{5}\right)^2 + \left(y + \sqrt{7}\right)^2 = 3$
9. $(x+8)^2 + (y-1)^2 = 289$
10. $(x-3)^2 + (y+6)^2 = 36$

Find the distance between each pair of points, and give the coordinates of the midpoint of the segment joining them.

1. $P(3, -1)$, $Q(-4, 5)$ **2.** $M(-8, 2)$, $N(3, -7)$ **3.** $A(-6, 3)$, $B(-6, 8)$

4. Are the points $(5, 7)$, $(3, 9)$, and $(6, 8)$ the vertices of a right triangle? If so, at what point is the right angle?

5. Determine the coordinates of B for segment AB, given that A has coordinates $(-6, 10)$ and the coordinates of its midpoint M are $(8, 2)$.

6. Use the distance formula to determine whether the points $(-2, -5)$, $(1, 7)$, and $(3, 15)$ are collinear.

Find the center-radius form of the equation for each circle described.

7. center $(-2, 3)$, radius 15 **8.** center $\left(\sqrt{5}, -\sqrt{7}\right)$, radius $\sqrt{3}$

9. center $(-8, 1)$, passing through $(0, 16)$ **10.** center $(3, -6)$, tangent to the x-axis

11. $x^2 + y^2 = 34$
12. $x^2 + y^2 = 13$
13. $x^2 + (y - 3)^2 = 13$
14. $(x - 5)^2 + (y - 6)^2 = 10$
15. $(2, -3)$; 1
16. $(3, 5)$; 2
17. $\left(-\frac{7}{2}, -\frac{3}{2}\right)$; $\frac{3\sqrt{6}}{2}$
18. $\left(-\frac{11}{2}, \frac{5}{2}\right)$; $\frac{\sqrt{146}}{2}$
19. $3 + 2\sqrt{5}$; $3 - 2\sqrt{5}$
20. no; $(-\infty, \infty)$; $[0, \infty)$
21. no; $[-6, 6]$; $[-6, 6]$
22. yes; $(-\infty, \infty)$; $[0, \infty)$
23. no; $(-\infty, \infty)$; $(-\infty, -1] \cup [1, \infty)$
24. yes; $(-\infty, \infty)$; $(-\infty, \infty)$
25. no; $[0, \infty)$; $(-\infty, \infty)$
26. not a function of x
27. function of x
28. function of x
29. not a function of x
30. $(-\infty, \infty)$
31. $(-\infty, 8) \cup (8, \infty)$
32. $(-\infty, 2]$
33. (a) $[2, \infty)$ (b) $(-\infty, -2]$ (c) $[-2, 2]$
34. 0
35. -15
36. -8
37. $-2k^2 + 3k - 6$
38.

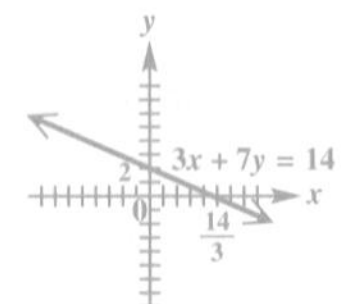

39. 40.

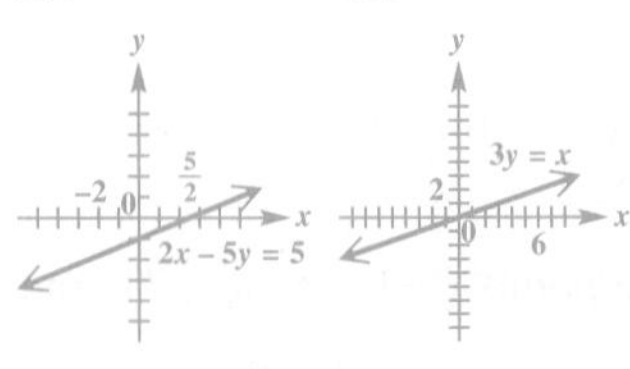

41. 42.

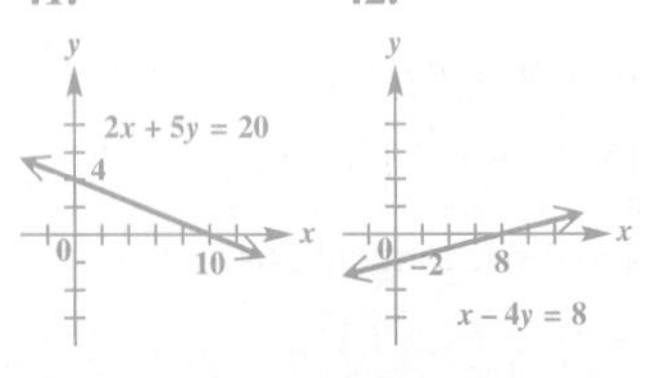

Connecting Graphs with Equations Use each graph to determine the equation of the circle. Express in center-radius form.

11.

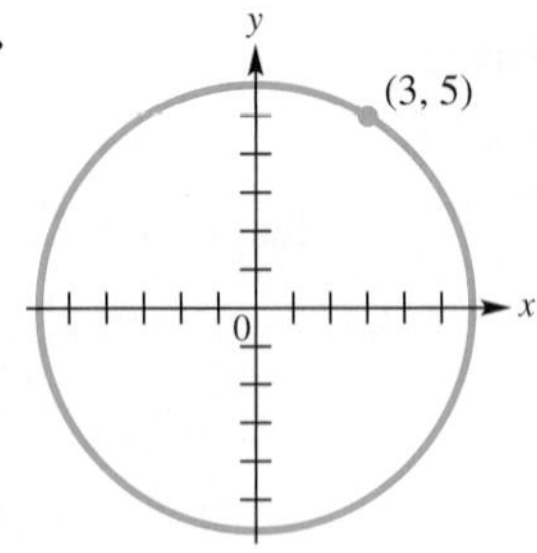

12.

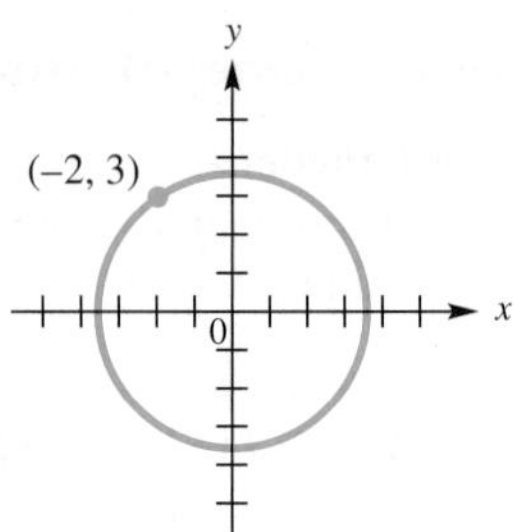

13.

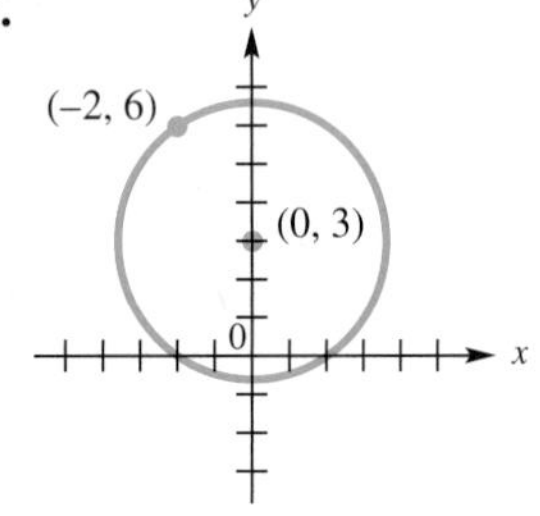

14.

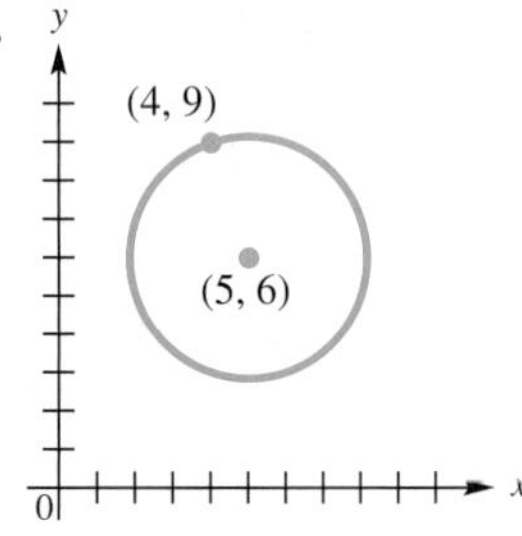

Find the center and radius of each circle.

15. $x^2 + y^2 - 4x + 6y + 12 = 0$

16. $x^2 + y^2 - 6x - 10y + 30 = 0$

17. $2x^2 + 2y^2 + 14x + 6y + 2 = 0$

18. $3x^2 + 3y^2 + 33x - 15y = 0$

19. Find all possible values of x so that the distance between $(x, -9)$ and $(3, -5)$ is 6.

For each graph, determine whether y is a function of x. Give the domain and range of each relation.

20.

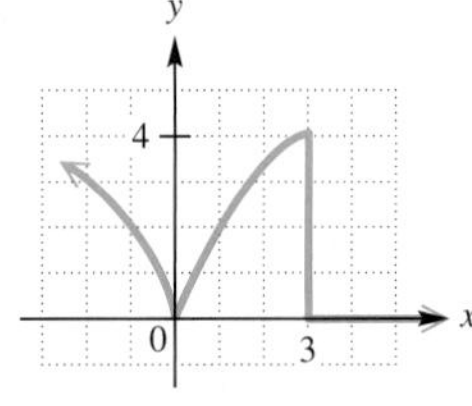

21.

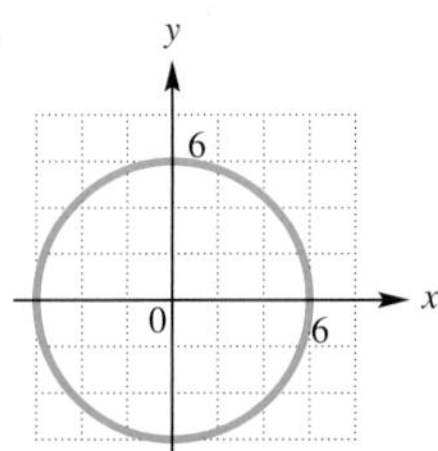

22.

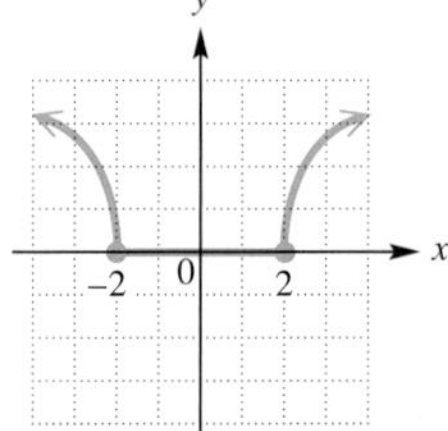

23.

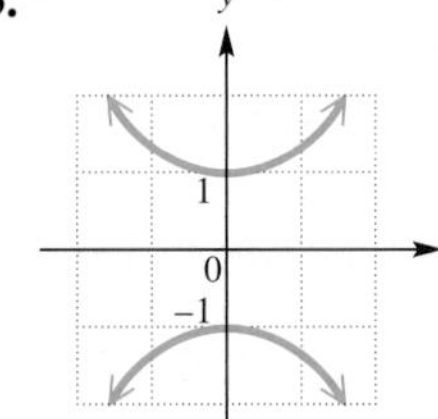

24.

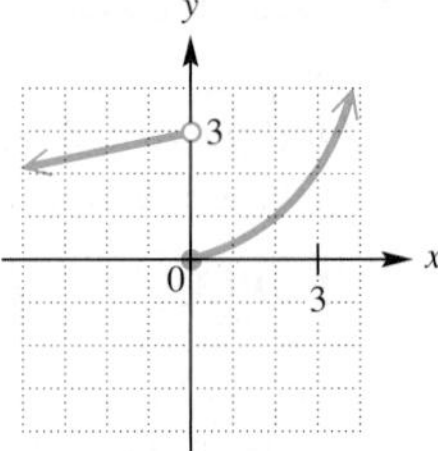

25.

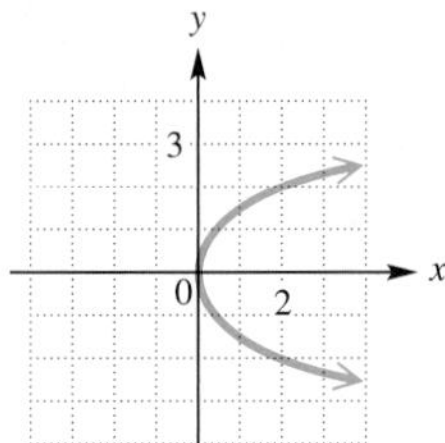

Determine whether each equation defines y as a function of x.

26. $x = \frac{1}{3}y^2$

27. $y = 6 - x^2$

28. $y = -\frac{4}{x}$

29. $y = \pm\sqrt{x - 2}$

Give the domain of each function.

30. $f(x) = -4 + |x|$

31. $f(x) = \frac{8 + x}{8 - x}$

32. $f(x) = \sqrt{6 - 3x}$

33. For the function graphed in **Exercise 22,** give the largest interval over which it is **(a)** increasing, **(b)** decreasing, and **(c)** constant.

34. Suppose the graph in **Exercise 24** is that of $y = f(x)$. What is $f(0)$?

Given $f(x) = -2x^2 + 3x - 6$, *find each function value or expression.*

35. $f(3)$ **36.** $f(-0.5)$ **37.** $f(k)$

Graph each equation.

38. $3x + 7y = 14$ **39.** $2x - 5y = 5$ **40.** $3y = x$

41. $2x + 5y = 20$ **42.** $x - 4y = 8$ **43.** $f(x) = x$

44. $f(x) = 3$ **45.** $x = -5$ **46.** $y + 2 = 0$

47. *Concept Check* The equation of the line that lies along the x-axis is ____.

Graph the line satisfying the given conditions.

48. through $(2, -4)$, $m = \frac{3}{4}$ **49.** through $(0, 5)$, $m = -\frac{2}{3}$

Find the slope for each line, provided that it has a slope.

50. through $(8, 7)$ and $\left(\frac{1}{2}, -2\right)$ **51.** through $(2, -2)$ and $(3, -4)$

52. through $(5, 6)$ and $(5, -2)$ **53.** through $(0, -7)$ and $(3, -7)$

54. $9x - 4y = 2$ **55.** $11x + 2y = 3$

56. $x - 5y = 0$ **57.** $x - 2 = 0$

58. *(Modeling) Job Market* The figure shows the number of jobs gained or lost in a recent period from September to May.

(a) Is this the graph of a function?

(b) In what month were the most jobs lost? the most gained?

(c) What was the largest number of jobs lost? of jobs gained?

(d) Do these data show an upward or a downward trend? If so, which is it?

Job Market Trends

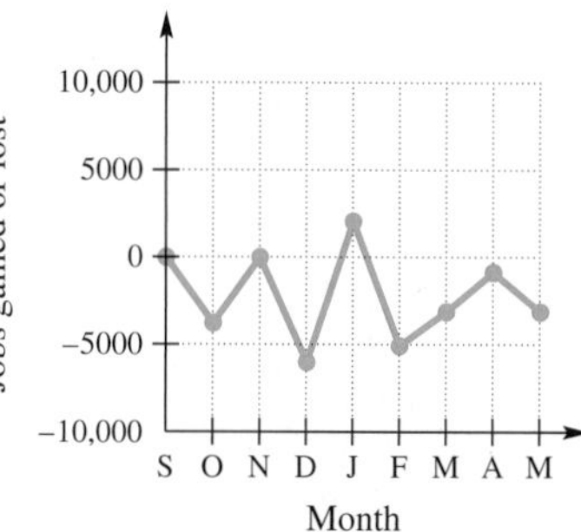

59. *(Modeling) Distance from Home* The graph depicts the distance y that a person driving a car on a straight road is from home after x hours. Interpret the graph. At what speeds did the car travel?

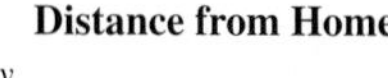
Distance from Home

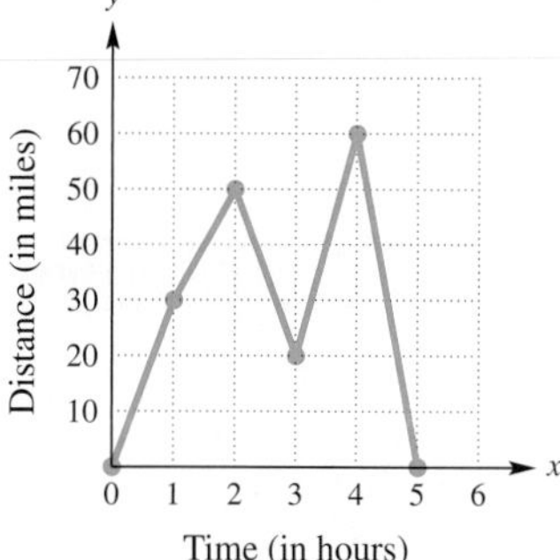

60. *Family Income* Family income in the United States has steadily increased for many years. In 1980 the median family income was about \$21,000 per year. In 2008 it was about \$61,500 per year. Find the average annual rate of change of median family income to the nearest dollar over that period. (*Source:* U.S. Census Bureau.)

43. **44.**

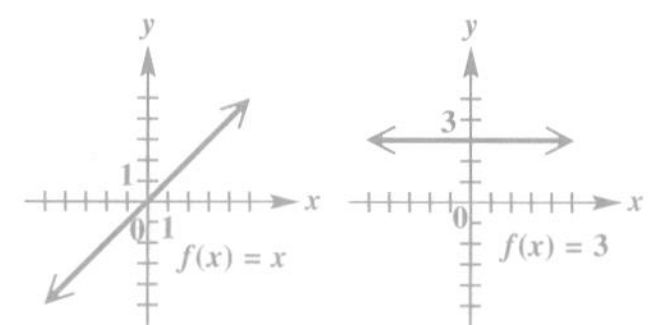

45. **46.**

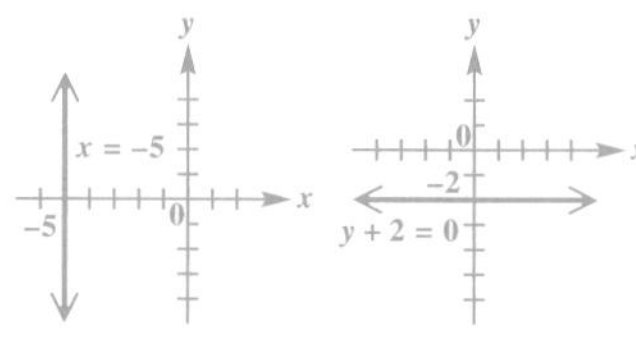

47. $y = 0$

48.

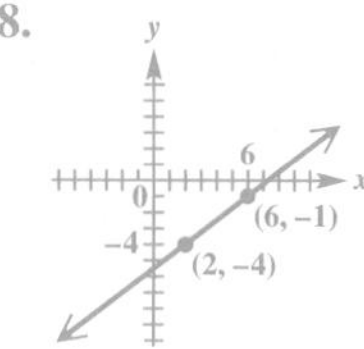

49.

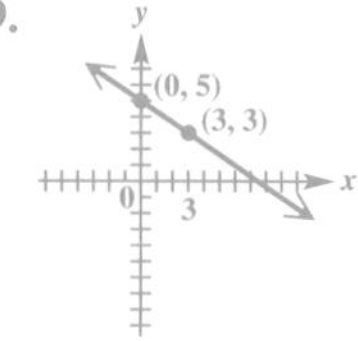

50. $\frac{6}{5}$

51. -2 **52.** undefined

53. 0 **54.** $\frac{9}{4}$

55. $-\frac{11}{2}$ **56.** $\frac{1}{5}$

57. undefined

58. **(a)** yes
(b) December; January
(c) about 6000; about 2000
(d) It shows a slight downward trend.

59. Initially, the car is at home. After traveling 30 mph for 1 hr, the car is 30 mi away from home. During the second hour, the car travels 20 mph until it is 50 mi away. During the third hour, the car travels toward home at 30 mph until it is 20 mi away. During the fourth hour, the car travels away from home at 40 mph until it is 60 mi away from home. During the last hour, the car travels 60 mi at 60 mph until it arrives home.

60. \$1446 per year

61. **(a)** $y = 4.56x + 30.7$; The slope 4.56 indicates that the percent of returns filed electronically rose an average of 4.56% per year during this period.
(b) 58.1%

62. (a) $y = -\frac{1}{3}x + \frac{10}{3}$
(b) $x + 3y = 10$

63. (a) $y = -2x + 1$
(b) $2x + y = 1$

64. (a) $y = \frac{5}{3}x + 5$
(b) $5x - 3y = -15$

65. (a) $y = 3x - 7$
(b) $3x - y = 7$

66. (a) $y = \frac{5}{8}x + 5$
(b) $5x - 8y = -40$

67. (a) $y = -10$
(b) $y = -10$

68. (a) $y = -5$
(b) $y = -5$

69. (a) not possible
(b) $x = -7$

70. $f(x) = -|x|$

71. $f(x) = |x| - 3$

72. $f(x) = -\sqrt{x} - 2$

73. $f(x) = -(x + 1)^2 + 3$

74.

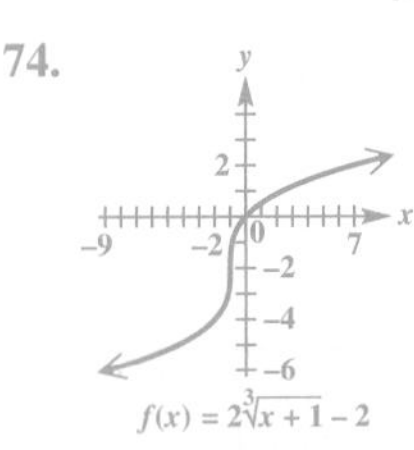

75. 76.

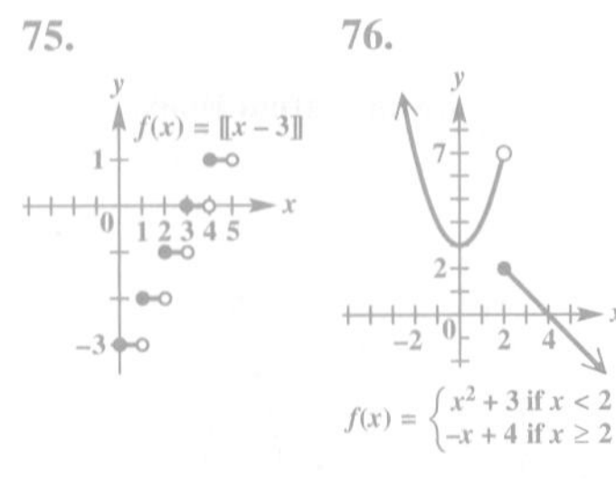

77.

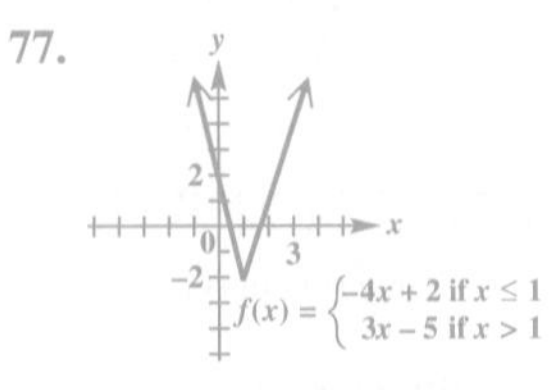

61. *(Modeling) E-Filing Tax Returns* The percent of tax returns filed electronically in 2001 was 30.7%. In 2009, the figure was 67.2%. (*Source:* Internal Revenue Service.)

(a) Use the information given for the years 2001 and 2009, letting $x = 0$ represent 2001, $x = 8$ represent 2009, and y represent the percent of returns filed electronically, to find a linear equation that models the data. Write the equation in slope-intercept form. Interpret the slope of the graph of this equation.

(b) Use your equation from part (a) to approximate the percent of tax returns that were filed electronically in 2007.

*For each line described, write the equation in **(a)** slope-intercept form, if possible, and **(b)** standard form.*

62. through $(-2, 4)$ and $(1, 3)$

63. through $(3, -5)$ with slope -2

64. x-intercept -3, y-intercept 5

65. through $(2, -1)$, parallel to $3x - y = 1$

66. through $(0, 5)$, perpendicular to $8x + 5y = 3$

67. through $(2, -10)$, perpendicular to a line with undefined slope

68. through $(3, -5)$, parallel to $y = 4$

69. through $(-7, 4)$, perpendicular to $y = 8$

Graph each function.

70. $f(x) = -|x|$

71. $f(x) = |x| - 3$

72. $f(x) = -\sqrt{x} - 2$

73. $f(x) = -(x + 1)^2 + 3$

74. $f(x) = 2\sqrt[3]{x + 1} - 2$

75. $f(x) = [\![x - 3]\!]$

76. $f(x) = \begin{cases} x^2 + 3 & \text{if } x < 2 \\ -x + 4 & \text{if } x \geq 2 \end{cases}$

77. $f(x) = \begin{cases} -4x + 2 & \text{if } x \leq 1 \\ 3x - 5 & \text{if } x > 1 \end{cases}$

78. $f(x) = \begin{cases} |x| & \text{if } x < 3 \\ 6 - x & \text{if } x \geq 3 \end{cases}$

79. *Concept Check* If x represents an integer, then what is the simplest form of the expression

$$[\![x]\!] + x?$$

Concept Check Decide whether each statement is true *or* false. *If false, tell why.*

80. The graph of a nonzero function cannot be symmetric with respect to the x-axis.

81. The graph of an even function is symmetric with respect to the y-axis.

82. The graph of an odd function is symmetric with respect to the origin.

83. If (a, b) is on the graph of an even function, so is $(a, -b)$.

84. If (a, b) is on the graph of an odd function, so is $(-a, b)$.

85. The constant function $f(x) = 0$ is both even and odd.

Decide whether each equation has a graph that is symmetric with respect to the x-axis, the y-axis, the origin, or none of these.

86. $5y^2 + 5x^2 = 30$

87. $x + y^2 = 10$

88. $y^3 = x + 4$

89. $x^2 = y^3$

90. $|y| = -x$

91. $6x + y = 4$

92. $|x| = |y|$

93. $y = 1$

94. $x^2 - y^2 = 0$

Describe how the graph of each function can be obtained from the graph of $f(x) = |x|$.

95. $g(x) = -|x|$

96. $h(x) = |x| - 2$

97. $k(x) = 2|x - 4|$

78. 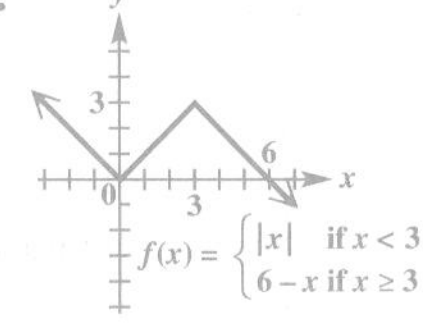

79. $2x$

80. true
81. true
82. true
83. false; For example, $f(x) = x^2$ is even, and $(2, 4)$ is on the graph but $(2, -4)$ is not.
84. false; For example, $f(x) = x^3$ is odd, and $(2, 8)$ is on the graph but $(-2, 8)$ is not.
85. true

86. x-axis, y-axis, origin
87. x-axis
88. none of these
89. y-axis
90. x-axis
91. none of these
92. x-axis, y-axis, origin
93. y-axis
94. x-axis, y-axis, origin

95. Reflect the graph of $f(x) = |x|$ across the x-axis.
96. Translate the graph of $f(x) = |x|$ down 2 units.
97. Translate the graph of $f(x) = |x|$ to the right 4 units and stretch it vertically by a factor of 2.

98. $y = -3x + 4$
99. $y = -3x - 4$
100. $y = 3x + 4$

101. (a)

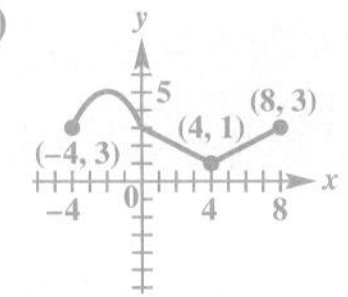

(b)

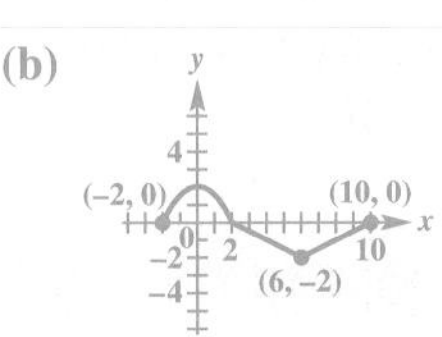

(c)

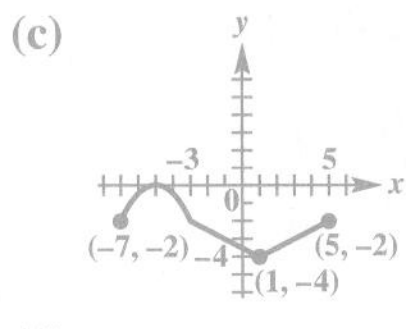

(d)

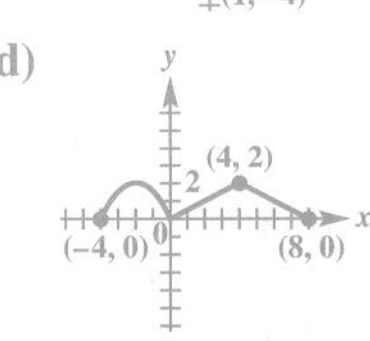

Let $f(x) = 3x - 4$. Find an equation for each reflection of the graph of $f(x)$.

98. across the x-axis **99.** across the y-axis **100.** across the origin

101. *Concept Check* The graph of a function f is shown in the figure. Sketch the graph of each function defined as follows.

(a) $y = f(x) + 3$
(b) $y = f(x - 2)$
(c) $y = f(x + 3) - 2$
(d) $y = |f(x)|$

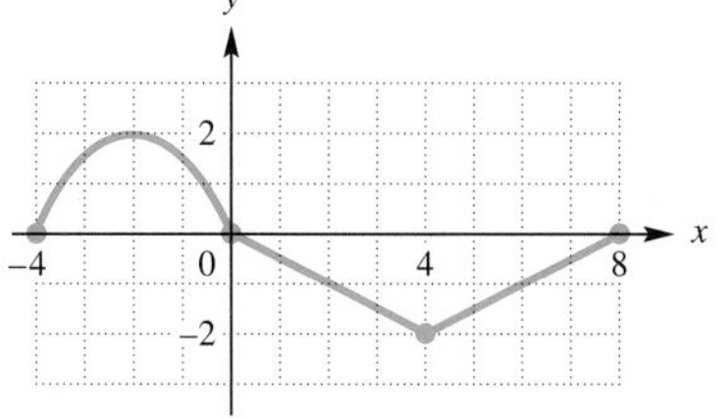

Let $f(x) = 3x^2 - 4$ and $g(x) = x^2 - 3x - 4$. Find each of the following.

102. $(f + g)(x)$ **103.** $(fg)(x)$ **104.** $(f - g)(4)$

105. $(f + g)(-4)$ **106.** $(f + g)(2k)$ **107.** $\left(\frac{f}{g}\right)(3)$

108. $\left(\frac{f}{g}\right)(-1)$ **109.** the domain of $(fg)(x)$ **110.** the domain of $\left(\frac{f}{g}\right)(x)$

111. Which of the following is *not* equivalent to $(f \circ g)(x)$ for $f(x) = \frac{1}{x}$ and $g(x) = x^2 + 1$? (*Hint:* There may be more than one.)

A. $f(g(x))$ **B.** $\frac{1}{x^2 + 1}$ **C.** $\frac{1}{x^2}$ **D.** $(g \circ f)(x)$

For each function, find and simplify $\frac{f(x + h) - f(x)}{h}$, $h \neq 0$.

112. $f(x) = 2x + 9$ **113.** $f(x) = x^2 - 5x + 3$

Let $f(x) = \sqrt{x - 2}$ and $g(x) = x^2$. Find each of the following, if possible.

114. $(f \circ g)(x)$ **115.** $(g \circ f)(x)$ **116.** $(f \circ g)(-6)$

117. $(g \circ f)(3)$ **118.** $(g \circ f)(-1)$ **119.** the domain of $f \circ g$

Use the table to evaluate each expression, if possible.

120. $(f + g)(1)$
121. $(f - g)(3)$
122. $(fg)(-1)$
123. $\left(\frac{f}{g}\right)(0)$
124. $\left(\frac{f}{g}\right)(3)$

x	$f(x)$	$g(x)$
−1	3	−2
0	5	0
1	7	1
3	9	9

Tables for f and g are given. Use them to evaluate the expressions in Exercises 125 and 126.

125. $(g \circ f)(-2)$
126. $(f \circ g)(3)$

x	$f(x)$
−2	1
0	4
2	3
4	2

x	$g(x)$
1	2
2	4
3	−2
4	0

Concept Check *The graphs of two functions f and g are shown in the figures.*

127. Find $(f \circ g)(2)$.

128. Find $(g \circ f)(3)$.

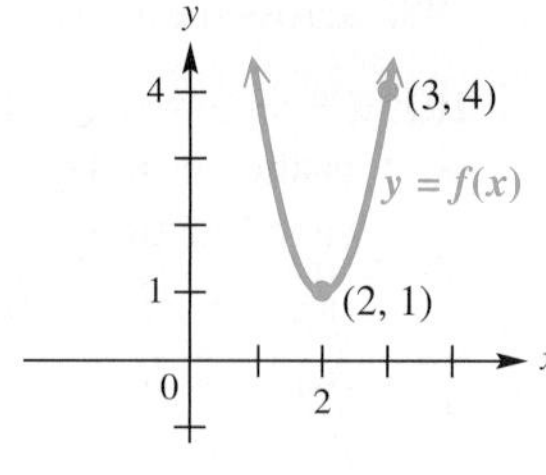

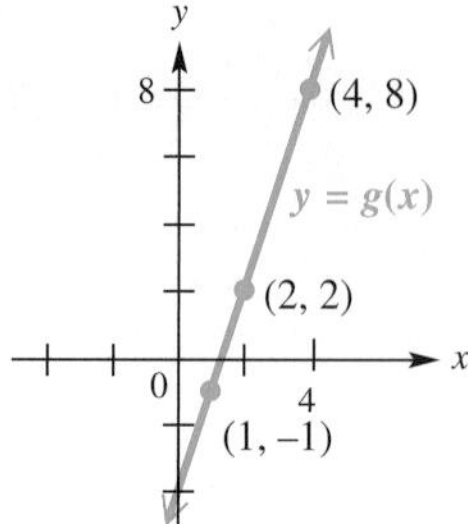

Solve each problem.

129. *Relationship of Measurement Units* There are 36 in. in 1 yd, and there are 1760 yd in 1 mi. Express the number of inches in x miles by forming two functions and then considering their composition.

130. *(Modeling) Perimeter of a Rectangle* Suppose the length of a rectangle is twice its width. Let x represent the width of the rectangle. Write a formula for the perimeter P of the rectangle in terms of x alone. Then use $P(x)$ notation to describe it as a function. What type of function is this?

131. *(Modeling) Volume of a Sphere* The formula for the volume of a sphere is $V(r) = \frac{4}{3}\pi r^3$, where r represents the radius of the sphere. Construct a model function V representing the amount of volume gained when the radius r (in inches) of a sphere is increased by 3 in.

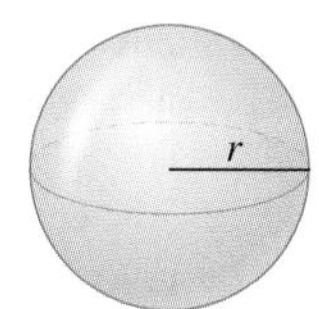

132. *(Modeling) Dimensions of a Cylinder* A cylindrical can makes the most efficient use of materials when its height is the same as the diameter of its top.

(a) Express the volume V of such a can as a function of the diameter d of its top.

(b) Express the surface area S of such a can as a function of the diameter d of its top. (*Hint:* The curved side is made from a rectangle whose length is the circumference of the top of the can.)

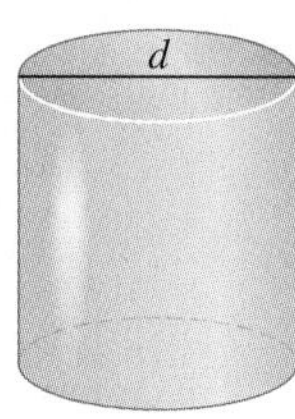

102. $4x^2 - 3x - 8$
103. $3x^4 - 9x^3 - 16x^2 + 12x + 16$
104. 44
105. 68
106. $16k^2 - 6k - 8$
107. $-\frac{23}{4}$
108. undefined
109. $(-\infty, \infty)$
110. $(-\infty, -1) \cup (-1, 4) \cup (4, \infty)$
111. C and D
112. 2
113. $2x + h - 5$
114. $\sqrt{x^2 - 2}$
115. $x - 2$ **116.** $\sqrt{34}$
117. 1 **118.** undefined
119. $(-\infty, -\sqrt{2}] \cup [\sqrt{2}, \infty)$
120. 8
121. 0 **122.** -6
123. undefined **124.** 1
125. 2 **126.** 1
127. 1 **128.** 8
129. $f(x) = 36x$; $g(x) = 1760x$; $(f \circ g)(x) = f(g(x)) = f(1760x) = 36(1760x) = 63{,}360x$
130. $P = 2x + x + 2x + x$; $P(x) = 6x$; linear function
131. $V(r) = \frac{4}{3}\pi(r+3)^3 - \frac{4}{3}\pi r^3$
132. **(a)** $V(d) = \frac{\pi d^3}{4}$ **(b)** $S(d) = \frac{3\pi d^2}{2}$

Chapter 2 Test

1. Match the set described in Column I with the correct interval notation from Column II. Choices in Column II may be used once, more than once, or not at all.

I	II
(a) Domain of $f(x) = \sqrt{x} + 3$	**A.** $[-3, \infty)$
(b) Range of $f(x) = \sqrt{x - 3}$	**B.** $[3, \infty)$
(c) Domain of $f(x) = x^2 - 3$	**C.** $(-\infty, \infty)$
(d) Range of $f(x) = x^2 + 3$	**D.** $[0, \infty)$
(e) Domain of $f(x) = \sqrt[3]{x - 3}$	**E.** $(-\infty, 3)$
(f) Range of $f(x) = \sqrt[3]{x} + 3$	**F.** $(-\infty, 3]$
(g) Domain of $f(x) = \lvert x \rvert - 3$	**G.** $(3, \infty)$
(h) Range of $f(x) = \lvert x + 3 \rvert$	**H.** $(-\infty, 0]$
(i) Domain of $x = y^2$	
(j) Range of $x = y^2$	

[2.3]
1. (a) D **(b)** D **(c)** C **(d)** B **(e)** C **(f)** C **(g)** C **(h)** D **(i)** D **(j)** C

[2.4]
2. $\frac{3}{5}$

[2.1]
3. $\sqrt{34}$ 4. $\left(\frac{1}{2}, \frac{5}{2}\right)$

[2.5]
5. $3x - 5y = -11$
6. $f(x) = \frac{3}{5}x + \frac{11}{5}$

[2.2]
7. (a) $x^2 + y^2 = 4$
(b) $(x - 1)^2 + (y - 4)^2 = 1$
8.

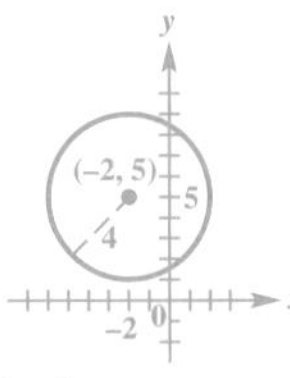

$x^2 + y^2 + 4x - 10y + 13 = 0$

[2.3]
9. (a) not a function; domain: $[0, 4]$; range: $[-4, 4]$
(b) function; domain: $(-\infty, -1) \cup (-1, \infty)$; range: $(-\infty, 0) \cup (0, \infty)$; decreasing on $(-\infty, -1)$ and on $(-1, \infty)$

[2.5]
10. (a) $x = 5$ (b) $y = -3$
11. (a) $y = -3x + 9$
(b) $y = \frac{1}{3}x + \frac{7}{3}$

[2.3]
12. (a) $(-\infty, -3)$
(b) $(4, \infty)$
(c) $[-3, 4]$
(d) $(-\infty, -3)$; $[-3, 4]$; $(4, \infty)$
(e) $(-\infty, \infty)$
(f) $(-\infty, 2)$

[2.6, 2.7]
13.

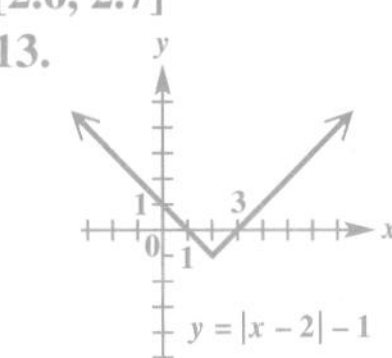

14.

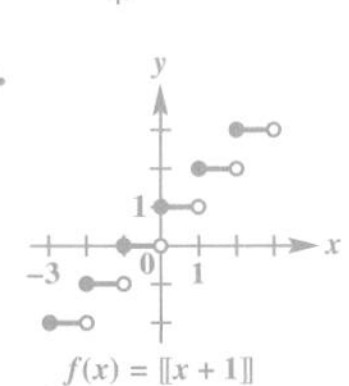

The graph shows the line that passes through the points $(-2, 1)$ and $(3, 4)$. Refer to it to answer Exercises 2–6.

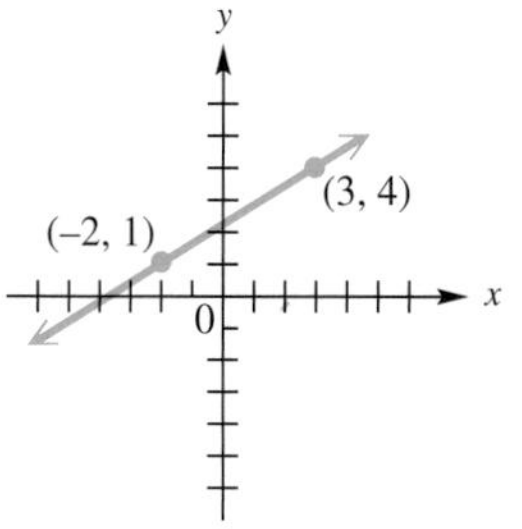

2. What is the slope of the line?

3. What is the distance between the two points shown?

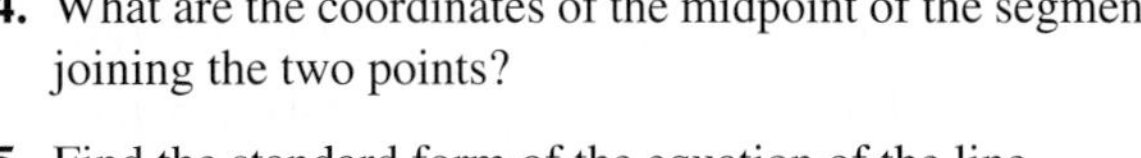

4. What are the coordinates of the midpoint of the segment joining the two points?

5. Find the standard form of the equation of the line.

6. Write the linear function $f(x) = ax + b$ that has this line as its graph.

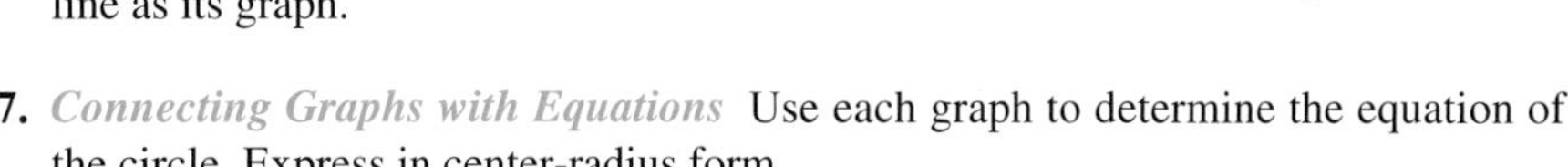

7. *Connecting Graphs with Equations* Use each graph to determine the equation of the circle. Express in center-radius form.

(a)

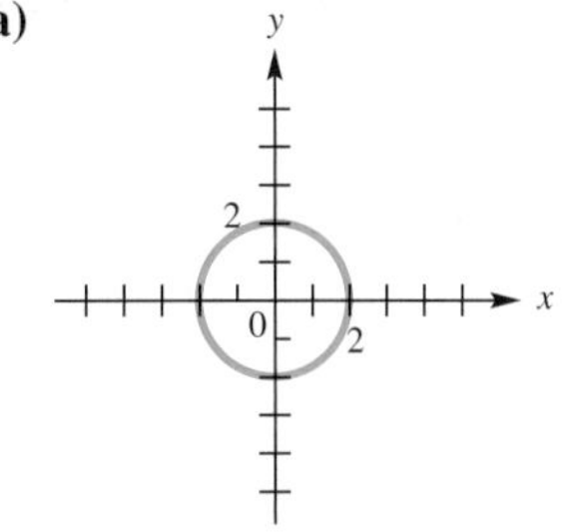

(b)

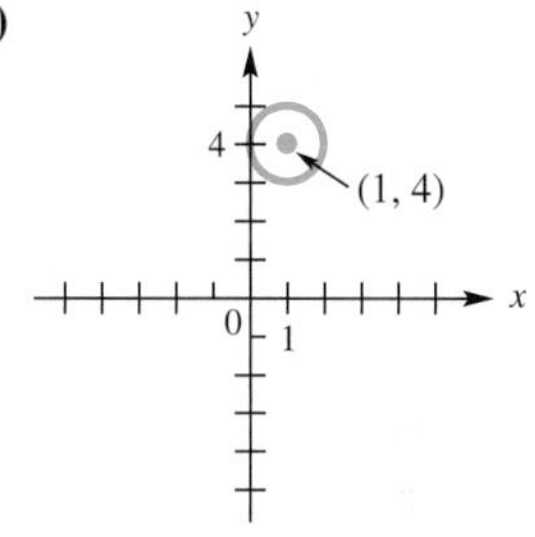

8. Graph the circle with equation $x^2 + y^2 + 4x - 10y + 13 = 0$.

9. In each case, determine whether y is a function of x. Give the domain and range. If it is a function, give the intervals where it is increasing, decreasing, or constant.

(a)

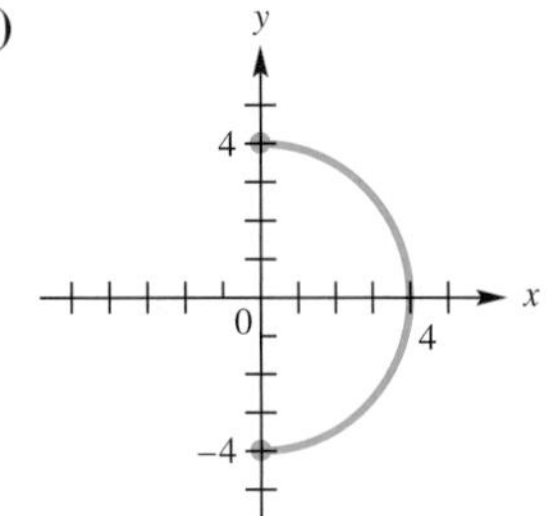

(b)

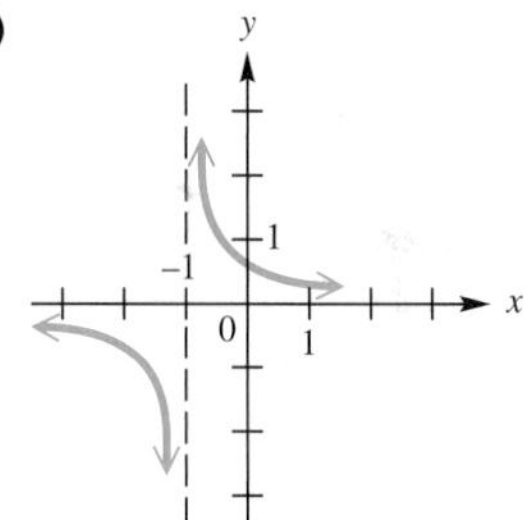

10. Suppose point A has coordinates $(5, -3)$.

(a) What is the equation of the vertical line through A?

(b) What is the equation of the horizontal line through A?

11. Find the slope-intercept form of the equation of the line passing through $(2, 3)$ and

(a) parallel to the graph of $y = -3x + 2$;

(b) perpendicular to the graph of $y = -3x + 2$.

12. Consider the graph of the function shown here. Give the interval(s) over which the function is

(a) increasing,

(b) decreasing,

(c) constant,

(d) continuous.

(e) What is the domain of this function?

(f) What is the range of this function?

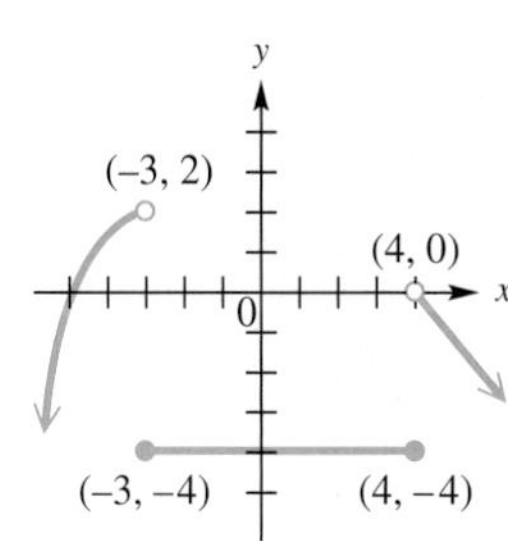

Graph each function.

13. $y = |x - 2| - 1$

14. $f(x) = [\![x + 1]\!]$

15. $f(x) = \begin{cases} 3 & \text{if } x < -2 \\ 2 - \frac{1}{2}x & \text{if } x \geq -2 \end{cases}$

16. The graph of $y = f(x)$ is shown here. Sketch the graph of each of the following. Use ordered pairs to indicate three points on the graph.

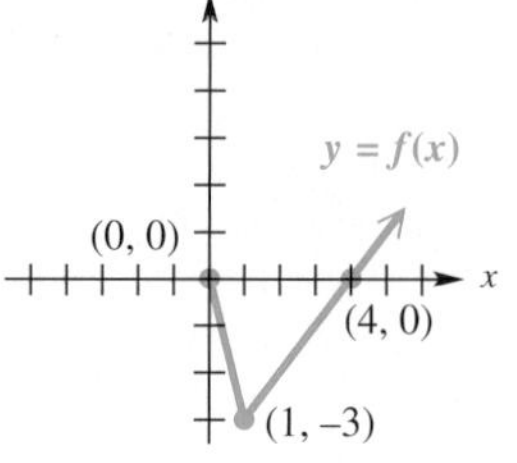

(a) $y = f(x) + 2$ **(b)** $y = f(x + 2)$

(c) $y = -f(x)$ **(d)** $y = f(-x)$

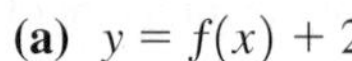

(e) $y = 2f(x)$

17. Describe how the graph of $y = -2\sqrt{x + 2} - 3$ can be obtained from the graph of $y = \sqrt{x}$.

18. Determine whether the graph of $3x^2 - 2y^2 = 3$ is symmetric with respect to

(a) the x-axis, **(b)** the y-axis, and **(c)** the origin.

In Exercises 19 and 20, given $f(x) = 2x^2 - 3x + 2$ and $g(x) = -2x + 1$, find each of the following. Simplify the expressions when possible.

19. (a) $(f - g)(x)$ **(b)** $\left(\frac{f}{g}\right)(x)$

(c) the domain of $\frac{f}{g}$ **(d)** $\frac{f(x + h) - f(x)}{h}$ $(h \neq 0)$

20. (a) $(f + g)(1)$ **(b)** $(fg)(2)$ **(c)** $(f \circ g)(0)$

In Exercises 21 and 22, let $f(x) = \sqrt{x + 1}$ and $g(x) = 2x - 7$.

21. Find $(f \circ g)(x)$ and give its domain.

22. Find $(g \circ f)(x)$ and give its domain.

23. ***(Modeling) Cell Phone Charges*** A cell phone service provides communication between two cities. If x represents the number of minutes for the call, where $x > 0$, then the function

$$f(x) = 0.75[\![x]\!] + 1.50$$

gives the total cost of the call in dollars. Find the cost of a 7.5-min call.

24. ***(Modeling) Cost, Revenue, and Profit Analysis*** Dotty McGinnis starts up a small business manufacturing bobble-head figures of famous soccer players. Her initial cost is \$3300. Each figure costs \$4.50 to manufacture.

(a) Write a cost function C, where x represents the number of figures manufactured.
(b) Find the revenue function R if each figure in part (a) sells for \$10.50.
(c) Give the profit function P.
(d) How many figures must be produced and sold for Dotty to earn a profit?

15.

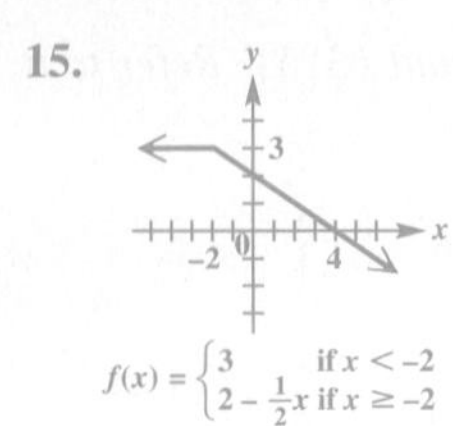

$f(x) = \begin{cases} 3 & \text{if } x < -2 \\ 2 - \frac{1}{2}x & \text{if } x \geq -2 \end{cases}$

16. (a)

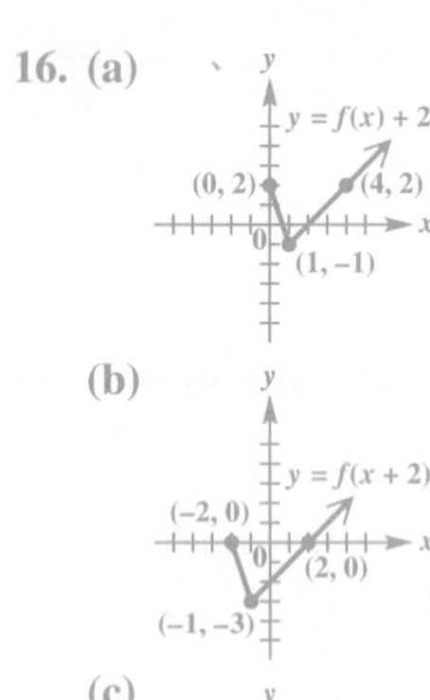

(b)

(c)

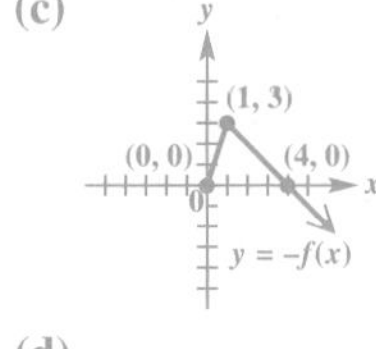

(d)

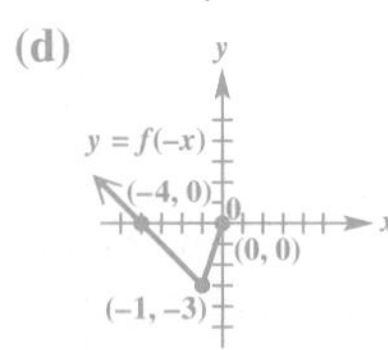

(e)

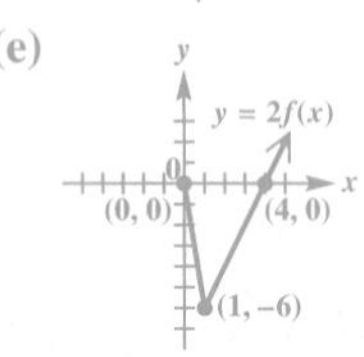

17. It is translated 2 units to the left, stretched vertically by a factor of 2, reflected across the x-axis, and translated 3 units down.

18. (a) yes **(b)** yes **(c)** yes

[2.8]

19. (a) $2x^2 - x + 1$ **(b)** $\frac{2x^2 - 3x + 2}{-2x + 1}$ **(c)** $\left(-\infty, \frac{1}{2}\right) \cup \left(\frac{1}{2}, \infty\right)$ **(d)** $4x + 2h - 3$

20. (a) 0 **(b)** -12 **(c)** 1

21. $\sqrt{2x - 6}$; $[3, \infty)$

22. $2\sqrt{x + 1} - 7$; $[-1, \infty)$

[2.6]

23. \$6.75

[2.4]

24. (a) $C(x) = 3300 + 4.50x$ **(b)** $R(x) = 10.50x$ **(c)** $R(x) - C(x) = 6.00x - 3300$ **(d)** 551

3 Polynomial and Rational Functions

Polynomial functions are used as models in many practical applications including the height of a thrown ball, the volume of a box, and, as seen in the photo here, the trajectories of water spouts.

3.1 Quadratic Functions and Models

- Quadratic Functions
- Graphing Techniques
- Completing the Square
- The Vertex Formula
- Quadratic Models

A *polynomial function* is defined as follows.

Polynomial Function

A **polynomial function** f of degree n, where n is a nonnegative integer, is given by

$$f(x) = a_n x^n + a_{n-1}x^{n-1} + \cdots + a_1 x + a_0,$$

where $a_n, a_{n-1}, \ldots, a_1$, and a_0 are real numbers, with $a_n \neq 0$.

LOOKING AHEAD TO CALCULUS
In calculus, polynomial functions are used to approximate more complicated functions. For example, the trigonometric function $\sin x$ is approximated by the polynomial

$$x - \frac{x^3}{6} + \frac{x^5}{120} - \frac{x^7}{5040}.$$

When we are analyzing a polynomial function, the degree n and the **leading coefficient** a_n play an important role. These are both given in the **leading term** $a_n x^n$. The table provides examples.

Polynomial Function	Function Name	Degree n	Leading Coefficient a_n
$f(x) = 2$	Constant	0	2
$f(x) = 5x - 1$	Linear	1	5
$f(x) = 4x^2 - x + 1$	Quadratic	2	4
$f(x) = 2x^3 - \frac{1}{2}x + 5$	Cubic	3	2
$f(x) = x^4 + \sqrt{2}x^3 - 3x^2$	Quartic	4	1

The function $f(x) = 0$ is the **zero polynomial** and has no degree.

Teaching Tip Ask students to explain why the following are *not* polynomial functions as defined on this page.
(a) $f(x) = 3x^3 - 2x^{-1}$
(b) $f(x) = 4x^{2/3} - 7$
(c) $f(x) = \frac{3x^2 - 7x + 4}{4x}$
(d) $f(x) = -5x^4 + 2x^3 - (3 + 2i)x - 8$

Quadratic Functions Earlier we discussed constant and linear polynomial functions. Polynomial functions of degree 2 are *quadratic functions*.

Quadratic Function

A function f is a **quadratic function** if

$$f(x) = ax^2 + bx + c,$$

where a, b, and c are real numbers, with $a \neq 0$.

The simplest quadratic function is

$$f(x) = x^2, \quad \text{Squaring function (Section 2.6)}$$

x	$f(x)$
-2	4
-1	1
0	0
1	1
2	4

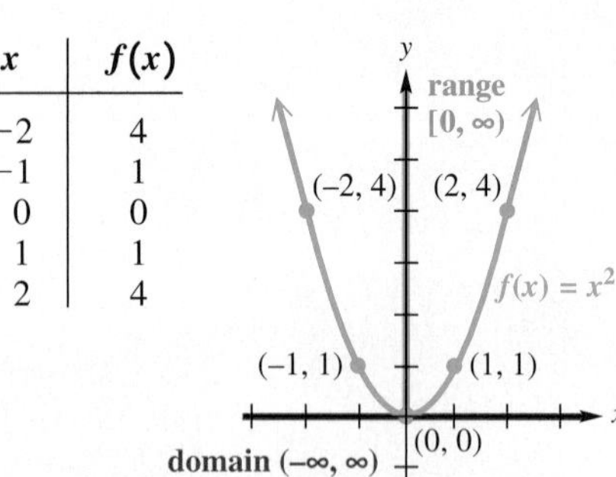

Figure 1

as shown in **Figure 1.** This graph is a **parabola.** Every quadratic function defined over the real numbers has a graph that is a parabola.

The domain of $f(x) = x^2$ is $(-\infty, \infty)$, and the range is $[0, \infty)$. The lowest point on the graph occurs at the origin $(0, 0)$. Thus, the function decreases on the interval $(-\infty, 0]$ and increases on the interval $[0, \infty)$. (Remember that these intervals indicate x-values.)

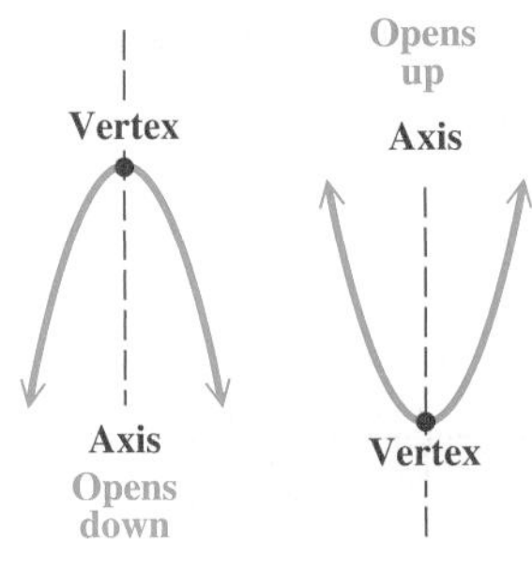

Figure 2

Parabolas are symmetric with respect to a line (the y-axis in **Figure 1**). This line is the **axis of symmetry,** or **axis,** of the parabola. The point where the axis intersects the parabola is the **vertex** of the parabola. As **Figure 2** shows, the vertex of a parabola that opens down is the highest point of the graph, and the vertex of a parabola that opens up is the lowest point of the graph.

Graphing Techniques The graphing techniques of **Section 2.7** may be applied to the graph of $f(x) = x^2$ to give the graph of *any* quadratic function. Compared to the basic graph of $f(x) = x^2$, the graph of $F(x) = a(x - h)^2 + k$ has the following characteristics.

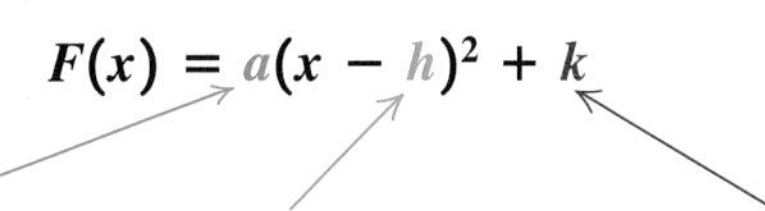

- Opens up if $a > 0$
- Opens down if $a < 0$
- Vertically stretched (narrower) if $|a| > 1$
- Vertically shrunk (wider) if $0 < |a| < 1$

Horizontal shift:

- h units right if $h > 0$
- $|h|$ units left if $h < 0$

Vertical shift:

- k units up if $k > 0$
- $|k|$ units down if $k < 0$

Classroom Example 1

Graph each function. Give the domain and range.

(a) $f(x) = x^2 + 4x - 4$ (by plotting points)

(b) $g(x) = -2x^2$ (and compare to $y = x^2$ and $y = 2x^2$)

(c) $F(x) = -2(x + 3)^2 + 5$ (and compare to the graph in part (b))

Answers:

(a)

$f(x) = x^2 + 4x - 4$

domain: $(-\infty, \infty)$; range: $[-8, \infty)$

(b)

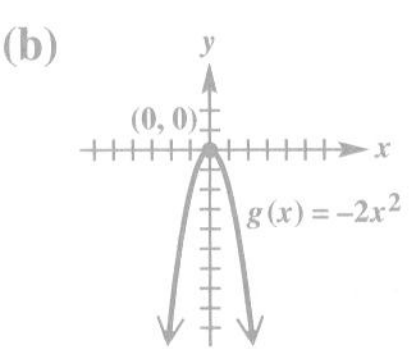

domain: $(-\infty, \infty)$; range: $(-\infty, 0]$; The graph of $y = 2x^2$ is a narrower version of the graph of $y = x^2$. The graph of $y = -2x^2$ is a reflection of the graph of $y = 2x^2$ across the x-axis.

(c)

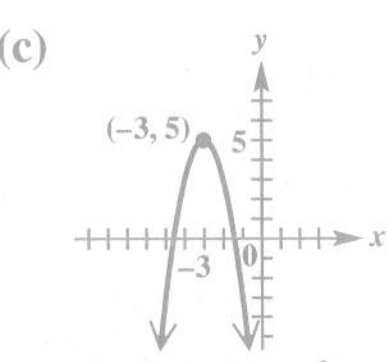

domain: $(-\infty, \infty)$; range: $(-\infty, 5]$; The graph of $F(x)$ is the graph of $g(x)$ translated 3 units to the left and 5 units up.

EXAMPLE 1 Graphing Quadratic Functions

Graph each function. Give the domain and range.

(a) $f(x) = x^2 - 4x - 2$ (by plotting points as in **Section 2.1**)

(b) $g(x) = -\frac{1}{2}x^2$ (and compare to $y = x^2$ and $y = \frac{1}{2}x^2$)

(c) $F(x) = -\frac{1}{2}(x - 4)^2 + 3$ (and compare to the graph in part (b))

SOLUTION

(a) See the table with **Figure 3.** The domain of $f(x) = x^2 - 4x - 2$ is $(-\infty, \infty)$, the range is $[-6, \infty)$, the vertex is $(2, -6)$, and the axis has equation $x = 2$. **Figure 4** shows how a graphing calculator displays this graph.

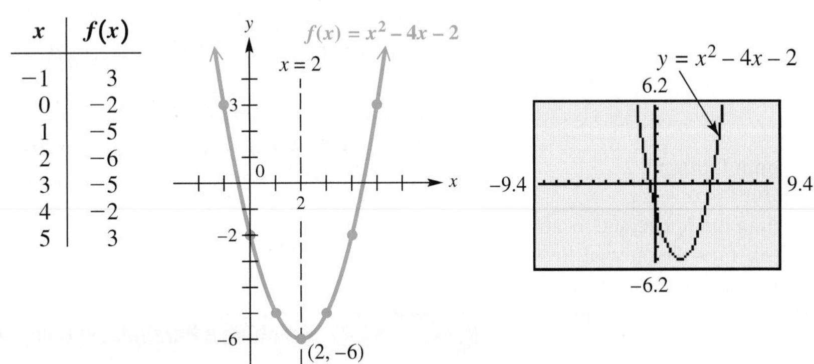

x	$f(x)$
−1	3
0	−2
1	−5
2	−6
3	−5
4	−2
5	3

Figure 3 Figure 4

(b) Think of $g(x) = -\frac{1}{2}x^2$ as $g(x) = -\left(\frac{1}{2}x^2\right)$. The graph of $y = \frac{1}{2}x^2$ is a wider version of the graph of $y = x^2$, and the graph of $g(x) = -\left(\frac{1}{2}x^2\right)$ is a reflection of the graph of $y = \frac{1}{2}x^2$ across the x-axis. See **Figure 5** on the next page. The vertex is $(0, 0)$, and the axis of the parabola is the line $x = 0$ (the y-axis). The domain is $(-\infty, \infty)$, and the range is $(-\infty, 0]$.

Calculator graphs are shown in **Figure 6.** These may be used to verify the accuracy of the graphs drawn in **Figure 5.**

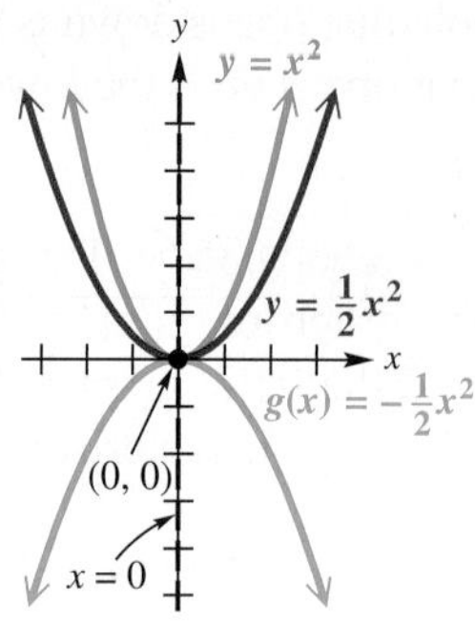

Figure 5

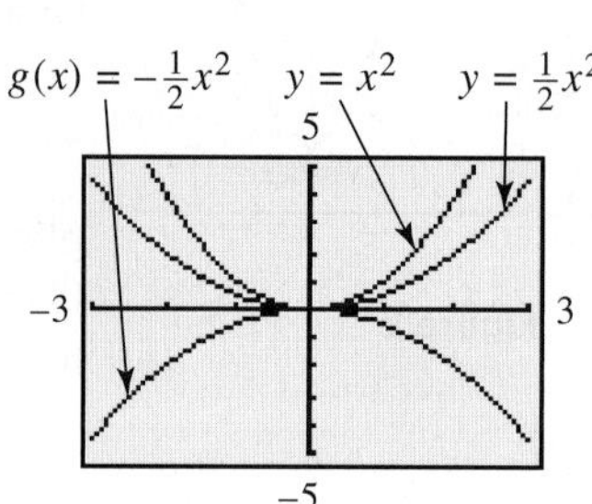

Figure 6

(c) Notice that $F(x) = -\frac{1}{2}(x-4)^2 + 3$ is related to $g(x) = -\frac{1}{2}x^2$ from part (b). The graph of $F(x)$ is the graph of $g(x)$ translated 4 units to the right and 3 units up. See **Figure 7.** The vertex is $(4, 3)$, which is also shown in the calculator graph in **Figure 8,** and the axis of the parabola is the line $x = 4$. The domain is $(-\infty, \infty)$, and the range is $(-\infty, 3]$.

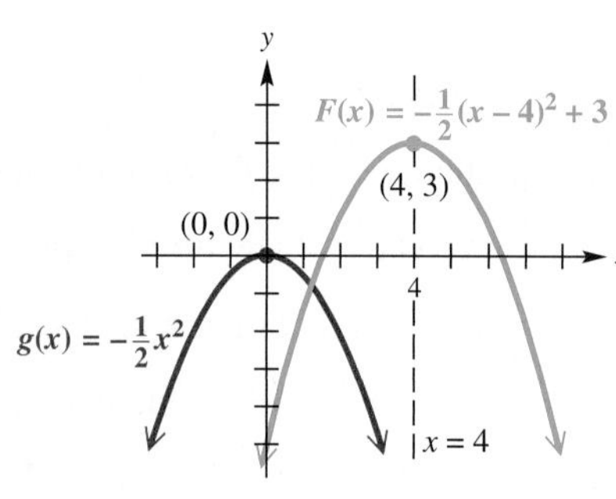

Figure 7

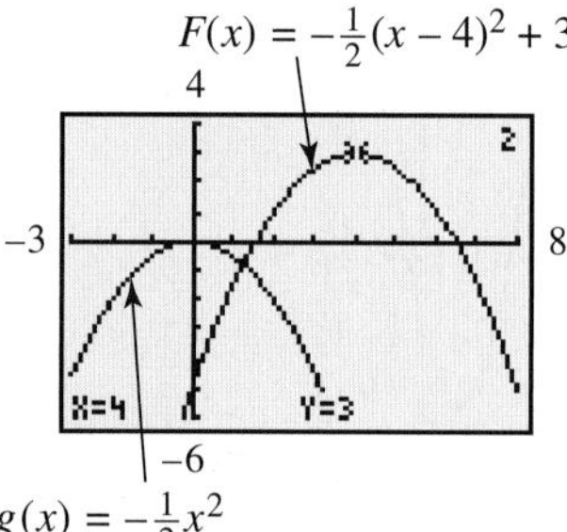

Figure 8

Now Try Exercises 9 and 11.

Completing the Square In general, the graph of the quadratic function

$$f(x) = a(x-h)^2 + k$$

is a parabola with ***vertex*** (h, k) and ***axis*** $x = h$. The parabola opens up if a is positive and down if a is negative. With these facts in mind, we *complete the square* to graph the general quadratic function

$$f(x) = ax^2 + bx + c.$$

Classroom Example 2
Graph $f(x) = x^2 + 2x - 5$ by completing the square and locating the vertex. Find the intervals over which the function is increasing or decreasing.

Answer:

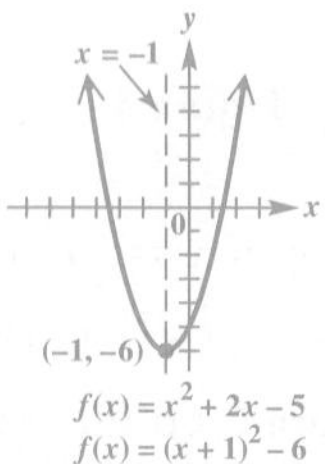

decreasing on $(-\infty, -1]$;
increasing on $[-1, \infty)$

EXAMPLE 2 Graphing a Parabola by Completing the Square ($a = 1$)

Graph $f(x) = x^2 - 6x + 7$ by completing the square and locating the vertex. Find the intervals over which the function is increasing or decreasing.

SOLUTION We express $x^2 - 6x + 7$ in the form $(x-h)^2 + k$ by completing the square. In preparation for this, we first write

$$f(x) = (x^2 - 6x \qquad) + 7. \quad \text{Prepare to complete the square. (Section 1.4)}$$

We must add a number inside the parentheses to get a perfect square trinomial. Find this number by taking half the coefficient of x and squaring the result.

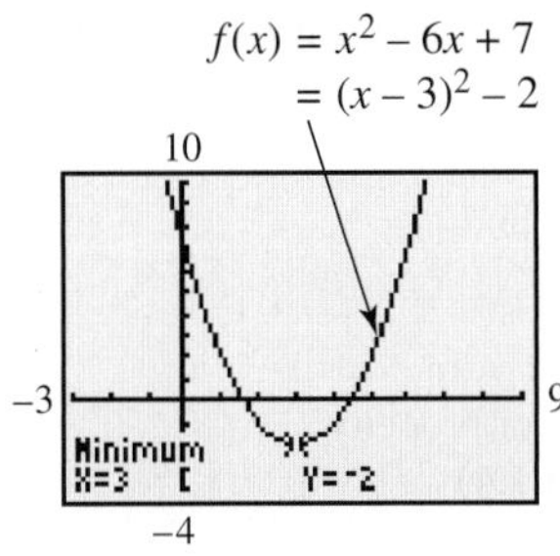

This screen shows that the vertex of the graph in **Figure 9** is (3, −2). Because it is the *lowest* point on the graph, we direct the calculator to find the *minimum*.

$$\left[\frac{1}{2}(-6)\right]^2 = (-3)^2 = 9 \quad \text{Take half the coefficient of } x. \text{ Square the result.}$$

$$f(x) = (x^2 - 6x + 9 - 9) + 7 \quad \text{Add and subtract 9.}$$

This is the same as adding 0.

$$f(x) = (x^2 - 6x + 9) - 9 + 7 \quad \text{Regroup terms.}$$

$$f(x) = (x - 3)^2 - 2 \quad \text{Factor and simplify. (Section R.4)}$$

The vertex of the parabola is $(3, -2)$, and the axis is the line $x = 3$. Now we find additional ordered pairs that satisfy the equation, as shown in the table, and plot and connect these points to obtain the graph in **Figure 9.**

x	y	
0	7	← y-intercept
1	2	
5	2	Find by using symmetry about the axis.
6	7	

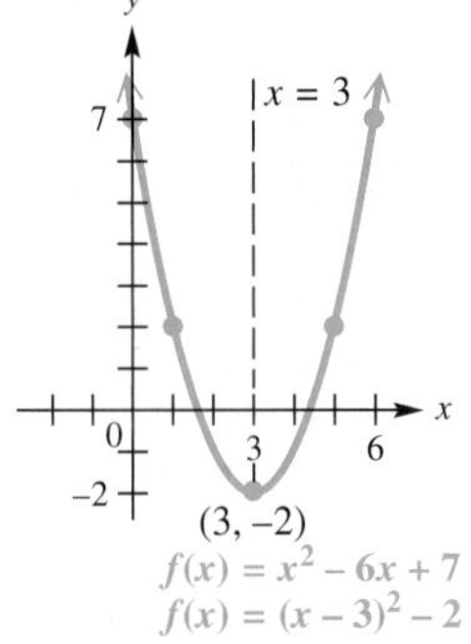

Figure 9

The domain of this function is $(-\infty, \infty)$, and the range is $[-2, \infty)$. Since the lowest point on the graph is the vertex $(3, -2)$, the function is decreasing on $(-\infty, 3]$ and increasing on $[3, \infty)$.

Now Try Exercise 21.

NOTE In **Example 2** we added and subtracted 9 *on the same side* of the equation to complete the square. This differs from adding the same number to *each side of the equation,* as when we completed the square in **Section 1.4.** Since we want $f(x)$ (or y) alone on one side of the equation, we adjusted that step in the process of completing the square.

Classroom Example 3
Graph $f(x) = 2x^2 + x - 6$ by completing the square and locating the vertex. Identify the intercepts of the graph.

Answer:

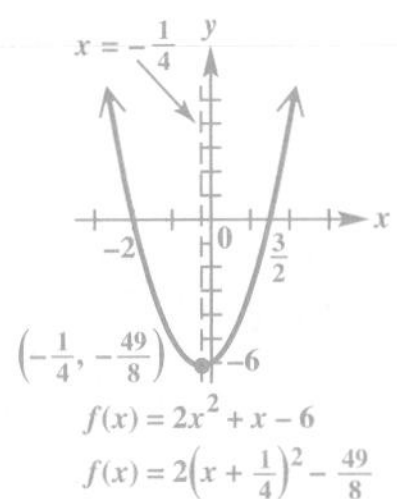

EXAMPLE 3 Graphing a Parabola by Completing the Square ($a \neq 1$)

Graph $f(x) = -3x^2 - 2x + 1$ by completing the square and locating the vertex. Identify the intercepts of the graph.

SOLUTION To complete the square, the coefficient of x^2 must be 1.

$$f(x) = -3\left(x^2 + \frac{2}{3}x \qquad\right) + 1 \quad \text{Factor } -3 \text{ from the first two terms.}$$

$$f(x) = -3\left(x^2 + \frac{2}{3}x + \frac{1}{9} - \frac{1}{9}\right) + 1 \quad \left[\frac{1}{2}\left(\frac{2}{3}\right)\right]^2 = \left(\frac{1}{3}\right)^2 = \frac{1}{9}, \text{ so add and subtract } \frac{1}{9}.$$

$$f(x) = -3\left(x^2 + \frac{2}{3}x + \frac{1}{9}\right) - 3\left(-\frac{1}{9}\right) + 1 \quad \text{Distributive property (Section R.2)}$$

Be careful here.

$$f(x) = -3\left(x + \frac{1}{3}\right)^2 + \frac{4}{3} \quad \text{Factor and simplify.}$$

The vertex is $\left(-\frac{1}{3}, \frac{4}{3}\right)$. The intercepts are good additional points to find. The y-intercept is found by evaluating $f(0)$.

$$f(0) = -3(0)^2 - 2(0) + 1 = 1 \leftarrow \text{The } y\text{-intercept is 1.}$$

Teaching Tip In **Example 3**, encourage students to verify that the vertex form is correct by squaring $\left(x+\frac{1}{3}\right)$, multiplying by -3, adding $\frac{4}{3}$, and combining like terms. This result should match the original expression.

The x-intercepts are found by setting $f(x)$ equal to 0 and solving for x.

$$0 = -3x^2 - 2x + 1 \qquad \text{Set } f(x) = 0.$$

$$0 = 3x^2 + 2x - 1 \qquad \text{Multiply by } -1.$$

$$0 = (3x-1)(x+1) \qquad \text{Factor.}$$

$$x = \frac{1}{3} \quad \text{or} \quad x = -1 \qquad \text{Zero-factor property (Section 1.4)}$$

Therefore, the x-intercepts are $\frac{1}{3}$ and -1. The graph is shown in **Figure 10.**

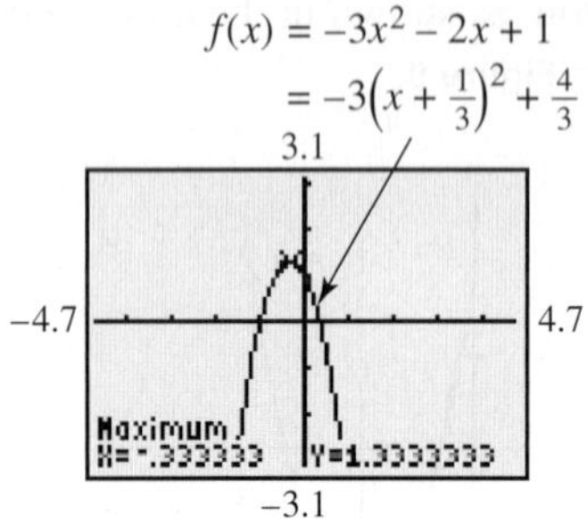

This screen gives the vertex of the graph in **Figure 10** as $(-0.\bar{3}, 1.\bar{3}) = \left(-\frac{1}{3}, \frac{4}{3}\right)$. We want the highest point on the graph, so we direct the calculator to find the *maximum*.

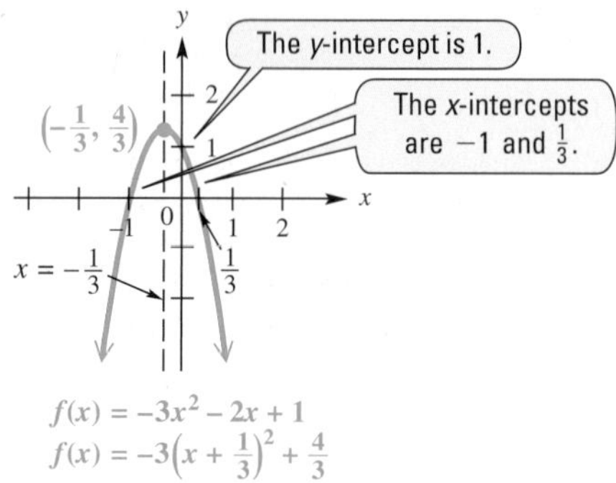

Figure 10

✔ *Now Try Exercise 23.*

NOTE It is possible to reverse the process of **Example 3** and write the quadratic function from its graph if the vertex and any other point on the graph are known. Since quadratic functions take the form

$$f(x) = a(x-h)^2 + k,$$

substitute the x- and y-values of the vertex, $\left(-\frac{1}{3}, \frac{4}{3}\right)$, for h and k, respectively.

$$f(x) = a\left[x - \left(-\frac{1}{3}\right)\right]^2 + \frac{4}{3} \qquad \text{Let } h = -\tfrac{1}{3} \text{ and } k = \tfrac{4}{3}.$$

$$f(x) = a\left(x + \frac{1}{3}\right)^2 + \frac{4}{3} \qquad \text{Simplify.}$$

Now find the value of a by substituting the x- and y-coordinates of any other point on the graph, say $(0, 1)$, into this equation and solving for a.

$$1 = a\left(0 + \frac{1}{3}\right)^2 + \frac{4}{3} \qquad \text{Let } x = 0 \text{ and } y = 1.$$

$$1 = a\left(\frac{1}{9}\right) + \frac{4}{3} \qquad \text{Square.}$$

$$-\frac{1}{3} = \frac{1}{9}a \qquad \text{Subtract } \tfrac{4}{3}.$$

$$a = -3 \qquad \text{Multiply by 9. Interchange sides.}$$

Verify in **Example 3** that the vertex form of the function is

$$f(x) = -3\left(x + \frac{1}{3}\right)^2 + \frac{4}{3}.$$

In the Exercise set, problems of this type are labeled *Connecting Graphs with Equations*.

The Vertex Formula We can generalize the earlier work to obtain a formula for the vertex of a parabola.

$$f(x) = ax^2 + bx + c \quad \text{General quadratic form}$$

$$= a\left(x^2 + \frac{b}{a}x \qquad\right) + c \quad \text{Factor } a \text{ from the first two terms.}$$

$$= a\left(x^2 + \frac{b}{a}x + \frac{b^2}{4a^2}\right) + c - a\left(\frac{b^2}{4a^2}\right) \quad \text{Add } \left[\tfrac{1}{2}\left(\tfrac{b}{a}\right)\right]^2 = \tfrac{b^2}{4a^2} \text{ inside the parentheses. Subtract } a\left(\tfrac{b^2}{4a^2}\right) \text{ outside the parentheses.}$$

$$= a\left(x + \frac{b}{2a}\right)^2 + c - \frac{b^2}{4a} \quad \text{Factor and simplify.}$$

$$f(x) = a\Big[x - \underbrace{\left(-\frac{b}{2a}\right)}_{h}\Big]^2 + \underbrace{\frac{4ac - b^2}{4a}}_{k} \quad \text{Vertex form of } f(x) = a(x-h)^2 + k$$

LOOKING AHEAD TO CALCULUS

An important concept in calculus is the **definite integral.** If the graph of f lies above the x-axis, the symbol

$$\int_a^b f(x)\,dx$$

represents the area of the region above the x-axis and below the graph of f from $x = a$ to $x = b$. For example, in **Figure 10** with

$$f(x) = -3x^2 - 2x + 1,$$

$a = -1$, and $b = \frac{1}{3}$, calculus provides the tools for determining that the area enclosed by the parabola and the x-axis is $\frac{32}{27}$ (square units).

Thus, the vertex (h, k) can be expressed in terms of a, b, and c. ***It is not necessary to memorize the expression for k, since it is equal to $f(h) = f\left(-\frac{b}{2a}\right)$.***

The following statements summarize this discussion.

Graph of a Quadratic Function

The quadratic function defined by $f(x) = ax^2 + bx + c$ can be written as

$$y = f(x) = a(x - h)^2 + k, \quad a \neq 0,$$

where $\quad h = -\dfrac{b}{2a} \quad$ and $\quad k = f(h)$. Vertex formula

The graph of f has the following characteristics.

1. It is a parabola with vertex (h, k) and the vertical line $x = h$ as axis.
2. It opens up if $a > 0$ and down if $a < 0$.
3. It is wider than the graph of $y = x^2$ if $|a| < 1$ and narrower if $|a| > 1$.
4. The y-intercept is $f(0) = c$.
5. The x-intercepts are found by solving the equation $ax^2 + bx + c = 0$.
   - If $b^2 - 4ac > 0$, the x-intercepts are $\frac{-b \pm \sqrt{b^2 - 4ac}}{2a}$.
   - If $b^2 - 4ac = 0$, the x-intercept is $-\frac{b}{2a}$.
   - If $b^2 - 4ac < 0$, there are no x-intercepts.

EXAMPLE 4 Using the Vertex Formula

Find the axis and vertex of the parabola having equation $f(x) = 2x^2 + 4x + 5$.

SOLUTION The axis of the parabola is the vertical line

$$x = h = -\frac{b}{2a} = -\frac{4}{2(2)} = -1. \quad \text{Use the vertex formula. Here } a = 2 \text{ and } b = 4.$$

The vertex is $(-1, f(-1))$. Since

$$f(-1) = 2(-1)^2 + 4(-1) + 5 = 3,$$

the vertex is $(-1, 3)$.

Now Try Exercise 21(a).

Classroom Example 4
Find the axis and vertex of the parabola having equation

$$f(x) = -3x^2 + 12x - 8$$

using the vertex formula.

Answer: axis: $x = 2$; vertex: $(2, 4)$

Quadratic Models Since the vertex of a vertical parabola is the highest or lowest point on the graph, equations of the form

$$y = ax^2 + bx + c$$

are important in certain problems where we must find the maximum or minimum value of some quantity.

- When $a < 0$, the y-coordinate of the vertex gives the maximum value of y.
- When $a > 0$, the y-coordinate of the vertex gives the minimum value of y.

The x-coordinate of the vertex tells *where* the maximum or minimum value occurs.

If air resistance is neglected, the height s (in feet) of an object projected directly upward from an initial height s_0 feet with initial velocity v_0 feet per second is

$$s(t) = -16t^2 + v_0t + s_0, \quad \text{(Section 1.5)}$$

where t is the number of seconds after the object is projected. The coefficient of t^2 (that is, -16) is a constant based on the gravitational force of Earth. This constant is different on other surfaces, such as the moon and the other planets.

Classroom Example 5
A ball is projected directly upward from an initial height of 75 ft with an initial velocity of 112 ft per sec.
(a) Give the function that describes the height of the ball in terms of time t.
(b) After how many seconds does the ball reach its maximum height? What is the maximum height?
(c) For what interval of time is the height of the ball greater than 200 ft?
(d) After how many seconds will the ball hit the ground?

Answers:
(a) $s(t) = -16t^2 + 112t + 75$
(b) 3.5 sec; 271 ft
(c) between 1.39 sec and 5.61 sec
(d) about 7.62 sec

EXAMPLE 5 Solving a Problem Involving Projectile Motion

A ball is projected directly upward from an initial height of 100 ft with an initial velocity of 80 ft per sec.

(a) Give the function that describes the height of the ball in terms of time t.

(b) After how many seconds does the ball reach its maximum height? What is this maximum height?

(c) For what interval of time is the height of the ball greater than 160 ft?

(d) After how many seconds will the ball hit the ground?

ALGEBRAIC SOLUTION

(a) Use the projectile height function with $v_0 = 80$ and $s_0 = 100$.

$$s(t) = -16t^2 + v_0t + s_0$$
$$s(t) = -16t^2 + 80t + 100$$

(b) Since the coefficient of t^2 is -16, the graph of the projectile function is a parabola that opens downward. Find the coordinates of the vertex to determine the maximum height and when it occurs. Let $a = -16$ and $b = 80$ in the vertex formula.

$$t = -\frac{b}{2a} = -\frac{80}{2(-16)} = 2.5$$
$$s(t) = -16t^2 + 80t + 100$$
$$s(2.5) = -16(2.5)^2 + 80(2.5) + 100$$
$$s(2.5) = 200$$

Therefore, after 2.5 sec the ball reaches its maximum height of 200 ft.

GRAPHING CALCULATOR SOLUTION

(a) Use the projectile height function as in the algebraic solution.

$$s(t) = -16t^2 + 80t + 100$$

(b) Using the capabilities of the calculator, we see in **Figure 11** that the vertex coordinates are indeed (2.5, 200).

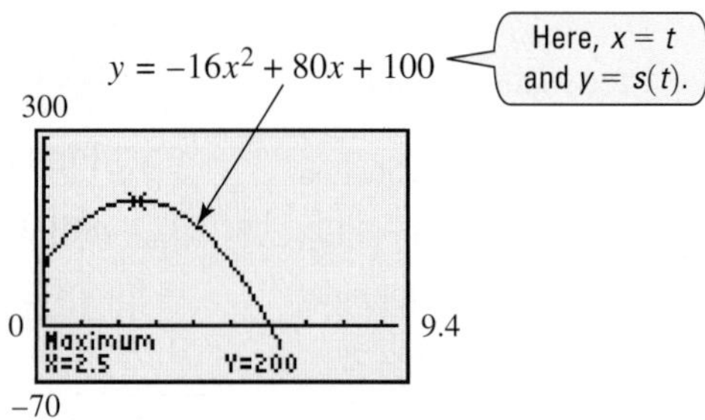

Figure 11

Do not misinterpret the graph in* Figure 11. *It does not show the path followed by the ball. It defines height as a function of time.

(c) We must solve the quadratic *inequality*

$$-16t^2 + 80t + 100 > 160.$$

$$-16t^2 + 80t - 60 > 0 \quad \text{Subtract 160.}$$

$$4t^2 - 20t + 15 < 0 \quad \text{(Section 1.7)}$$

Divide by -4; reverse the inequality symbol.

Use the quadratic formula to find the solutions of $4t^2 - 20t + 15 = 0$.

$$t = \frac{-(-20) \pm \sqrt{(-20)^2 - 4(4)(15)}}{2(4)} \quad \text{(Section 1.4)}$$

$$t = \frac{5 - \sqrt{10}}{2} \approx 0.92 \text{ or } t = \frac{5 + \sqrt{10}}{2} \approx 4.08$$

These numbers divide the number line into three intervals: $(-\infty, 0.92)$, $(0.92, 4.08)$, and $(4.08, \infty)$. Using a test value from each interval shows that $(0.92, 4.08)$ satisfies the *inequality.* The ball is more than 160 ft above the ground between 0.92 sec and 4.08 sec.

(d) The height is 0 when the ball hits the ground. We use the quadratic formula to find the *positive* solution of

$$-16t^2 + 80t + 100 = 0.$$

Here, $a = -16$, $b = 80$, and $c = 100$.

$$t = \frac{-80 \pm \sqrt{80^2 - 4(-16)(100)}}{2(-16)}$$

$t \approx$ ~~-1.04~~ (Reject) or $t \approx 6.04$

The ball hits the ground after about 6.04 sec.

(c) If we graph

$$y_1 = -16x^2 + 80x + 100 \quad \text{and} \quad y_2 = 160,$$

as shown in **Figures 12 and 13,** and locate the two points of intersection, we find that the x-coordinates for these points are approximately 0.92 and 4.08. Therefore, between 0.92 sec and 4.08 sec, y_1 is greater than y_2, and the ball is more than 160 ft above the ground.

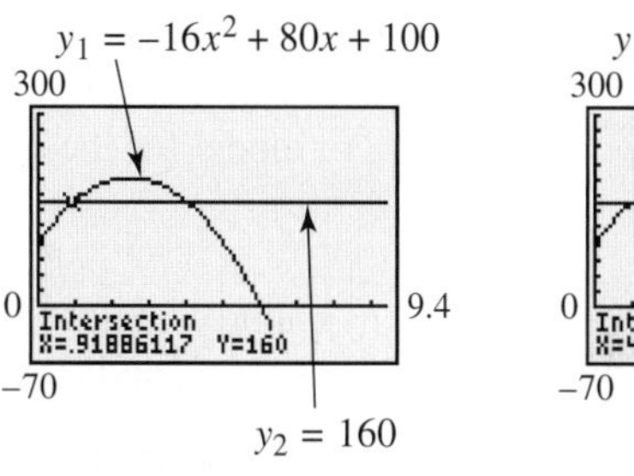

Figure 12

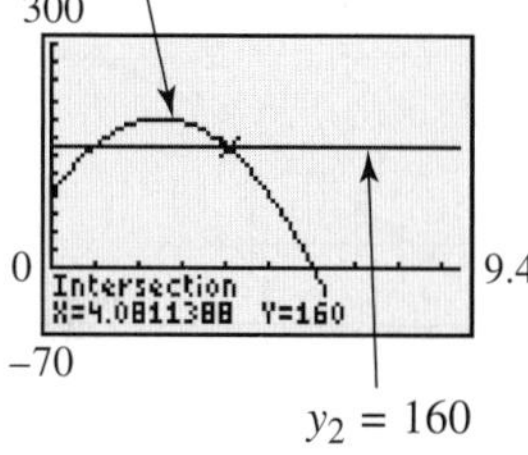

Figure 13

(d) **Figure 14** shows that the positive x-intercept of the graph of $y = -16x^2 + 80x + 100$ is approximately 6.04, which means that the ball hits the ground after about 6.04 sec.

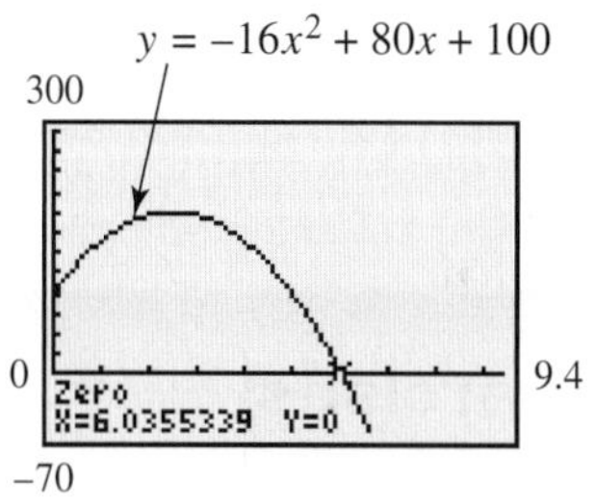

Figure 14

✔ *Now Try Exercise 53.*

Classroom Example 6
Consider the data for hospital outpatient visits for the years 2000–2008 only.
(a) Determine a quadratic model for these data.
(b) Use the model from part (a) to predict the number of visits in 2010.

Answers:
(a) $y = -0.8162x^2 + 183.8x - 9618$
(b) about 724 million visits

EXAMPLE 6 Modeling the Number of Hospital Outpatient Visits

The number of hospital outpatient visits (in millions) for selected years is shown in the table.

Year	Visits	Year	Visits
95	483.2	102	640.5
96	505.5	103	648.6
97	520.6	104	662.1
98	545.5	105	673.7
99	573.5	106	690.4
100	592.7	107	693.5
101	612.0	108	710.0

Source: American Hospital Association.

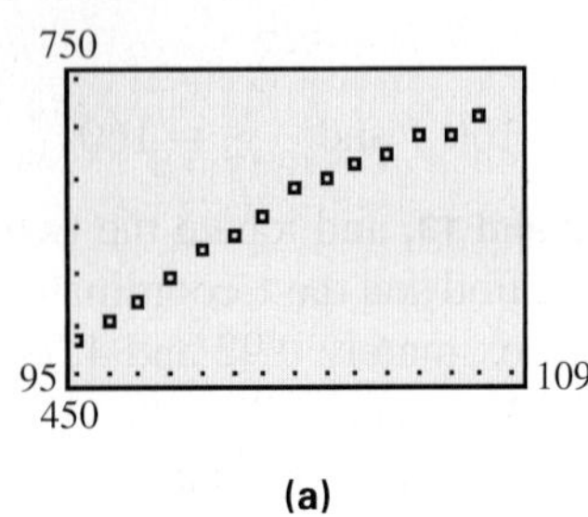

(a)

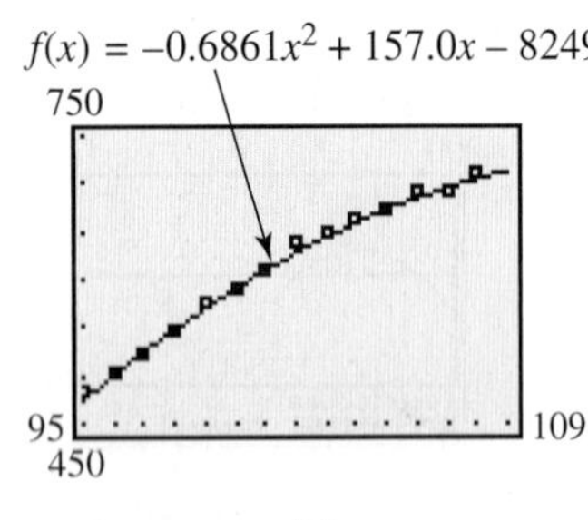

(b)

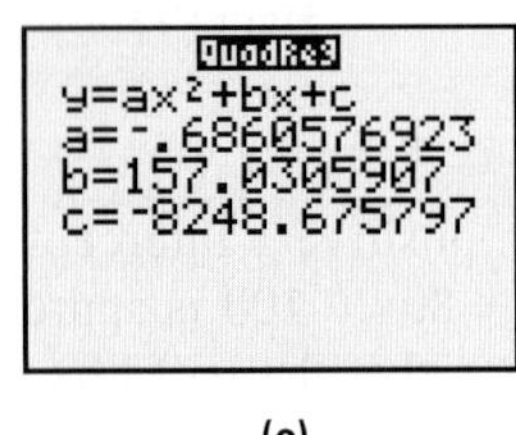

(c)

Figure 15

In the table on the preceding page, 95 represents 1995, 100 represents 2000, and so on, and the number of outpatient visits is given in millions.

(a) Prepare a scatter diagram, and determine a quadratic model for these data.

(b) Use the model from part (a) to predict the number of visits in 2012.

SOLUTION

(a) In **Section 2.5** we used linear regression to determine linear equations that modeled data. With a graphing calculator, we can use **quadratic regression** to find quadratic equations that model data.

The scatter diagram in **Figure 15(a)** suggests that a quadratic function with a negative value of a (so the graph opens down) would be a reasonable model for the data. Using quadratic regression, the quadratic function

$$f(x) = -0.6861x^2 + 157.0x - 8249$$

approximates the data well. See **Figure 15(b).** The quadratic regression values of a, b, and c are displayed in **Figure 15(c).**

(b) The year 2012 corresponds to $x = 112$. The model predicts that there will be 729 million visits in 2012.

$$f(x) = -0.6861x^2 + 157.0x - 8249$$
$$f(112) = -0.6861(112)^2 + 157.0(112) - 8249$$
$$\approx 729 \text{ million}$$

✔ *Now Try Exercise 65.*

3.1 Exercises

1. (a) domain: $(-\infty, \infty)$; range: $[-4, \infty)$ (b) $(-3, -4)$ (c) $x = -3$ (d) 5 (e) $-5, -1$
2. (a) domain: $(-\infty, \infty)$; range: $[-4, \infty)$ (b) $(5, -4)$ (c) $x = 5$ (d) 21 (e) 3, 7
3. (a) domain: $(-\infty, \infty)$; range: $(-\infty, 2]$ (b) $(-3, 2)$ (c) $x = -3$ (d) -16 (e) $-4, -2$
4. (a) domain: $(-\infty, \infty)$; range: $(-\infty, 1]$ (b) $(2, 1)$ (c) $x = 2$ (d) -11 (e) $\frac{6-\sqrt{3}}{3}, \frac{6+\sqrt{3}}{3}$
5. B 6. A
7. D 8. C

In Exercises 1–4, you are given an equation and the graph of a quadratic function. Do each of the following. ***See Examples 1–4.***

(a) *Give the domain and range.* ***(b)*** *Give the coordinates of the vertex.*
(c) *Give the equation of the axis.* ***(d)*** *Find the y-intercept.*
(e) *Find the x-intercepts.*

1.

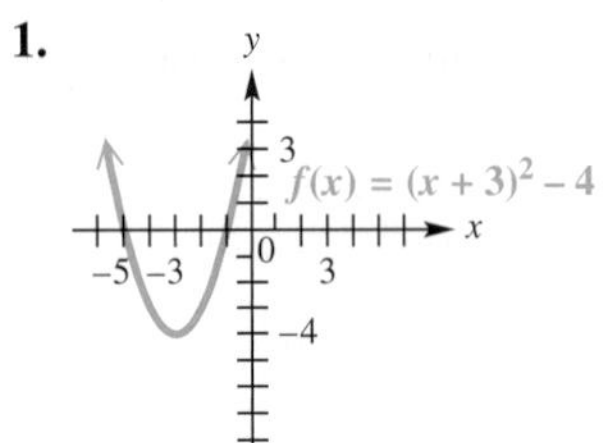
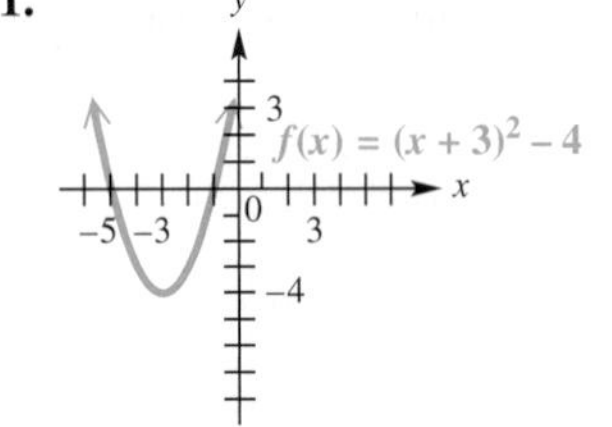

2.

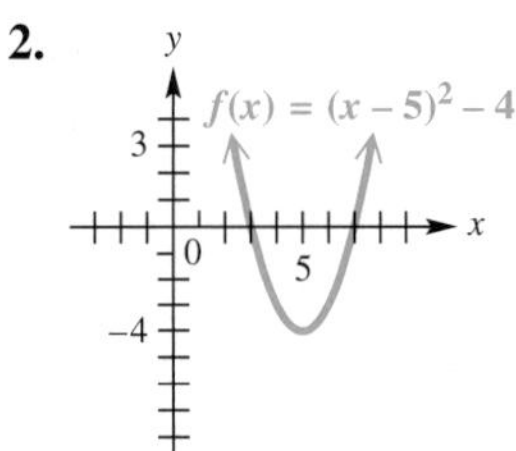

3.

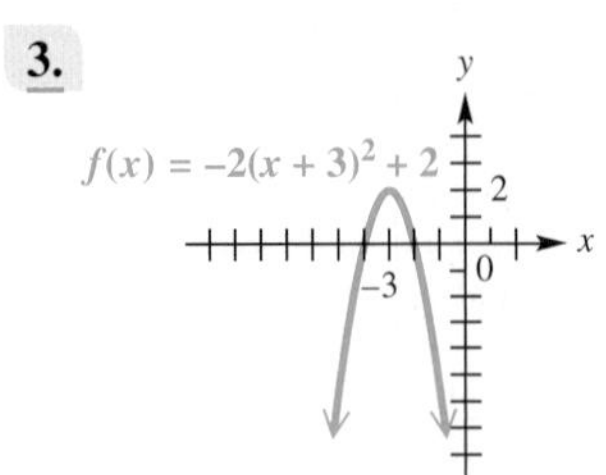

4.

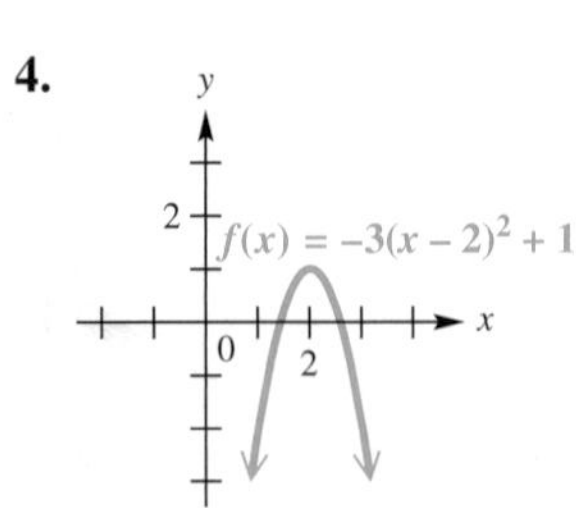

9. 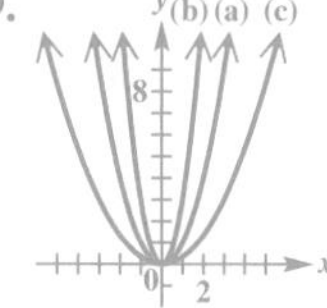

(d) The greater $|a|$ is, the narrower the parabola will be. The smaller $|a|$ is, the wider the parabola will be.

10. 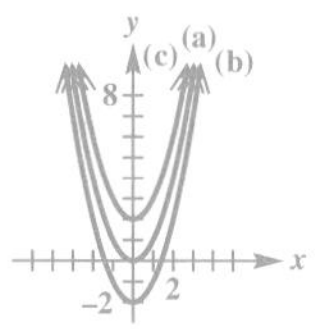

(d) The graph of $y = x^2 + k$ is translated k units up if k is positive and $|k|$ units down if k is negative.

11. 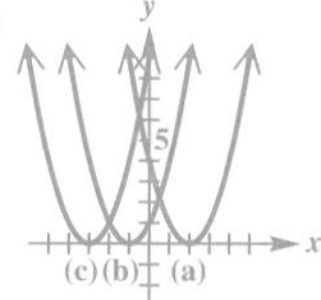

(d) The graph of $y = (x - h)^2$ is translated h units to the right if h is positive and $|h|$ units to the left if h is negative.

12. (a) C (b) A
(c) D (d) B

13. (a) $(2, 0)$ (b) $x = 2$
(c) $(-\infty, \infty)$ (d) $[0, \infty)$
(e) $[2, \infty)$ (f) $(-\infty, -2]$

$f(x) = (x - 2)^2$

14. (a) $(-4, 0)$ (b) $x = -4$
(c) $(-\infty, \infty)$ (d) $[0, \infty)$
(e) $[-4, \infty)$ (f) $(-\infty, -4]$

$f(x) = (x + 4)^2$

15. (a) $(-3, -4)$ (b) $x = -3$
(c) $(-\infty, \infty)$ (d) $[-4, \infty)$
(e) $[-3, \infty)$ (f) $(-\infty, -3]$

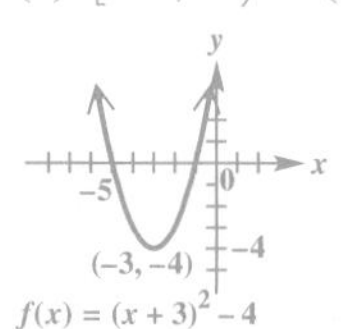

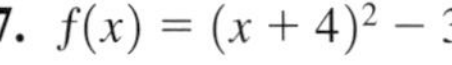

Concept Check *Calculator graphs of the functions in Exercises 5–8 are shown in Figures A–D. Match each function with its graph without actually entering it into your calculator. Then, after you have completed the exercises, check your answers with your calculator. Use the standard viewing window.*

5. $f(x) = (x - 4)^2 - 3$

6. $f(x) = -(x - 4)^2 + 3$

7. $f(x) = (x + 4)^2 - 3$

8. $f(x) = -(x + 4)^2 + 3$

A.

B.

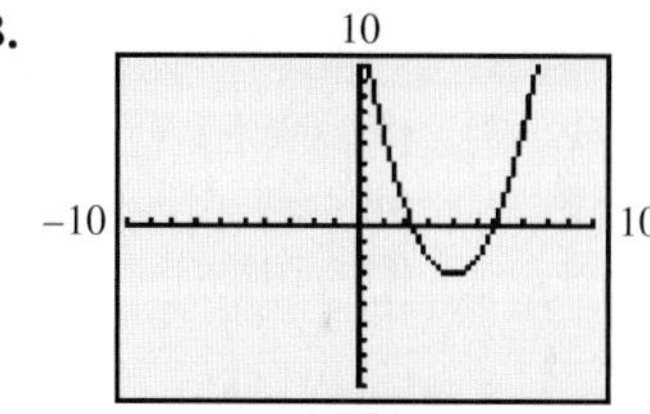

C.

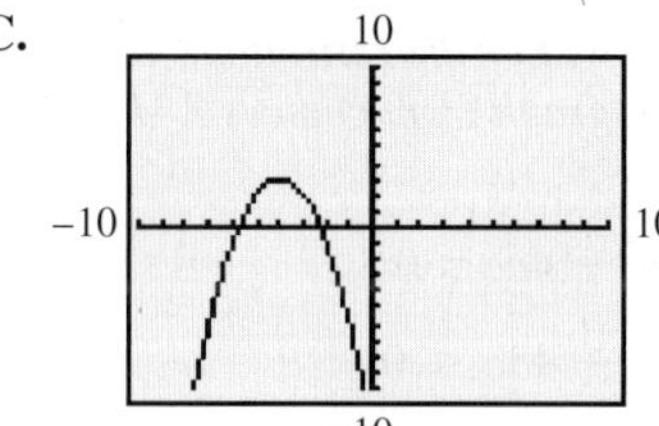

D.

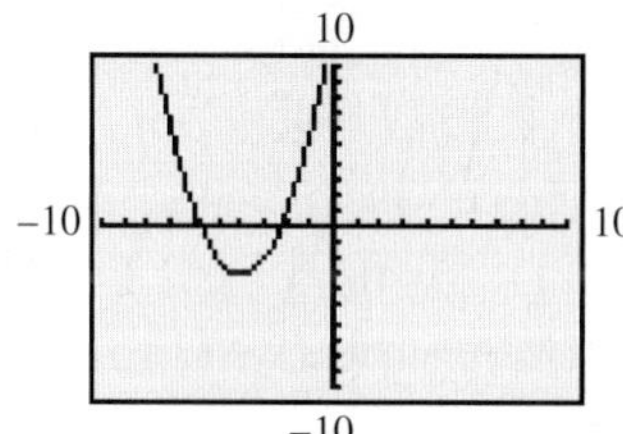

9. Graph the following on the same coordinate system.

(a) $y = x^2$ (b) $y = 3x^2$ (c) $y = \frac{1}{3}x^2$

(d) How does the coefficient of x^2 affect the shape of the graph?

10. Graph the following on the same coordinate system.

(a) $y = x^2$ (b) $y = x^2 - 2$ (c) $y = x^2 + 2$

(d) How do the graphs in parts (b) and (c) differ from the graph of $y = x^2$?

11. Graph the following on the same coordinate system.

(a) $y = (x - 2)^2$ (b) $y = (x + 1)^2$ (c) $y = (x + 3)^2$

(d) How do these graphs differ from the graph of $y = x^2$?

12. *Concept Check* Match each equation with the description of the parabola that is its graph.

(a) $y = (x + 4)^2 + 2$	A. vertex $(-2, 4)$, opens up
(b) $y = (x + 2)^2 + 4$	B. vertex $(-2, 4)$, opens down
(c) $y = -(x + 4)^2 + 2$	C. vertex $(-4, 2)$, opens up
(d) $y = -(x + 2)^2 + 4$	D. vertex $(-4, 2)$, opens down

Graph each quadratic function. Give the **(a)** *vertex,* **(b)** *axis,* **(c)** *domain, and* **(d)** *range. Then determine* **(e)** *the interval of the domain for which the function is increasing and* **(f)** *the interval for which the function is decreasing.* **See Examples 1–4.**

13. $f(x) = (x - 2)^2$

14. $f(x) = (x + 4)^2$

15. $f(x) = (x + 3)^2 - 4$

16. $f(x) = (x - 5)^2 - 4$

17. $f(x) = -\frac{1}{2}(x + 1)^2 - 3$

18. $f(x) = -3(x - 2)^2 + 1$

19. $f(x) = x^2 - 2x + 3$

20. $f(x) = x^2 + 6x + 5$

21. $f(x) = x^2 - 10x + 21$

22. $f(x) = 2x^2 - 4x + 5$

23. $f(x) = -2x^2 - 12x - 16$

24. $f(x) = -3x^2 + 24x - 46$

25. $f(x) = -\frac{1}{2}x^2 - 3x - \frac{1}{2}$

26. $f(x) = \frac{2}{3}x^2 - \frac{8}{3}x + \frac{5}{3}$

16. (a) $(5, -4)$ (b) $x = 5$
(c) $(-\infty, \infty)$ (d) $[-4, \infty)$
(e) $[5, \infty)$ (f) $(-\infty, 5]$

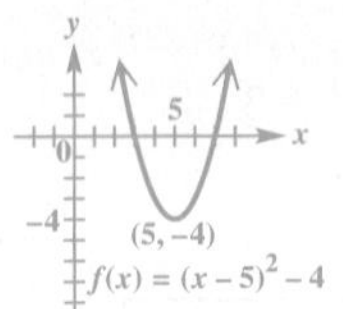

17. (a) $(-1, -3)$ (b) $x = -1$
(c) $(-\infty, \infty)$ (d) $(-\infty, -3]$
(e) $(-\infty, -1]$ (f) $[-1, \infty)$

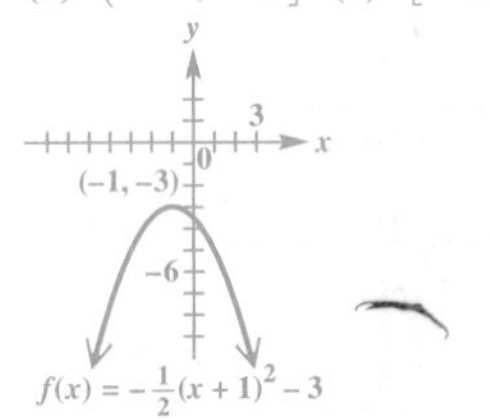

18. (a) $(2, 1)$ (b) $x = 2$
(c) $(-\infty, \infty)$ (d) $(-\infty, 1]$
(e) $(-\infty, 2]$ (f) $[2, \infty)$

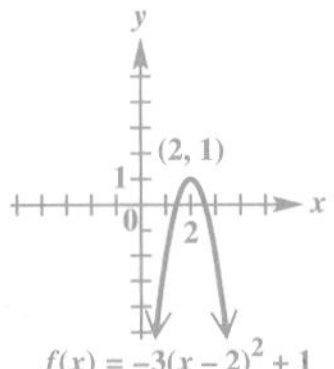

19. (a) $(1, 2)$ (b) $x = 1$
(c) $(-\infty, \infty)$ (d) $[2, \infty)$
(e) $[1, \infty)$ (f) $(-\infty, 1]$

20. (a) $(-3, -4)$ (b) $x = -3$
(c) $(-\infty, \infty)$ (d) $[-4, \infty)$
(e) $[-3, \infty)$ (f) $(-\infty, -3]$

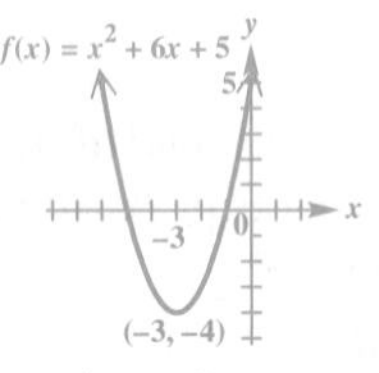

21. (a) $(5, -4)$ (b) $x = 5$
(c) $(-\infty, \infty)$ (d) $[-4, \infty)$
(e) $[5, \infty)$ (f) $(-\infty, 5]$

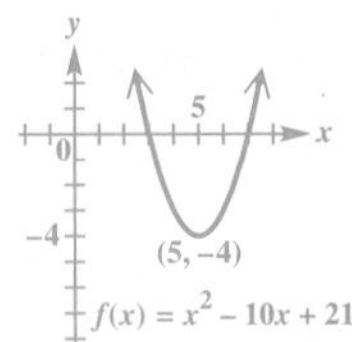

Concept Check The figure shows the graph of a quadratic function $y = f(x)$. Use it to work Exercises 27–30.

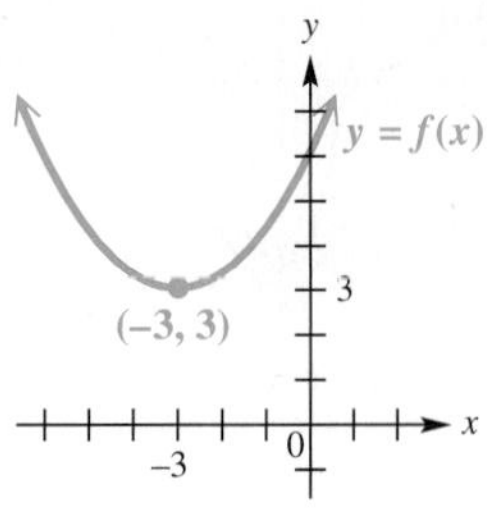

27. What is the minimum value of $f(x)$?

28. For what value of x is $f(x)$ as small as possible?

29. How many real solutions are there to the equation $f(x) = 1$?

30. How many real solutions are there to the equation $f(x) = 4$?

Concept Check Several possible graphs of the quadratic function

$$f(x) = ax^2 + bx + c$$

*are shown below. For the restrictions on a, b, and c given in Exercises 31–36, select the corresponding graph from choices A–F. (Hint: Use the discriminant. See **Section 1.4** if necessary.)*

31. $a < 0;\ b^2 - 4ac = 0$

32. $a > 0;\ b^2 - 4ac < 0$

33. $a < 0;\ b^2 - 4ac < 0$

34. $a < 0;\ b^2 - 4ac > 0$

35. $a > 0;\ b^2 - 4ac > 0$

36. $a > 0;\ b^2 - 4ac = 0$

A.

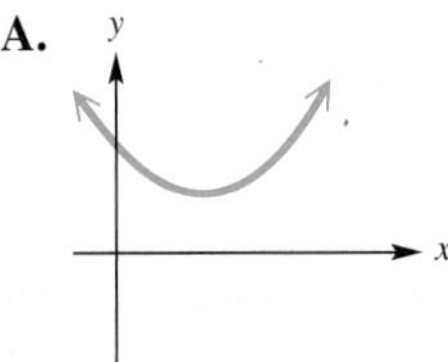

B.

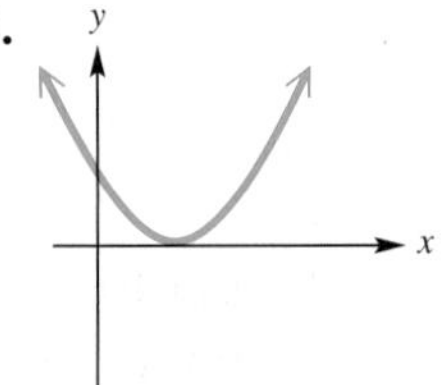

C.

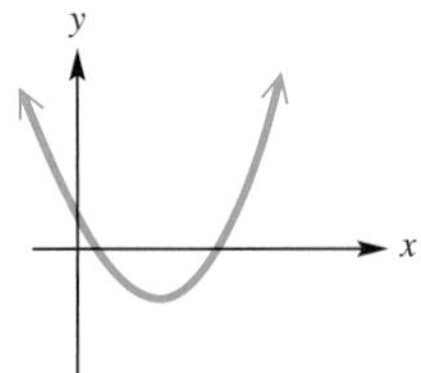

D.

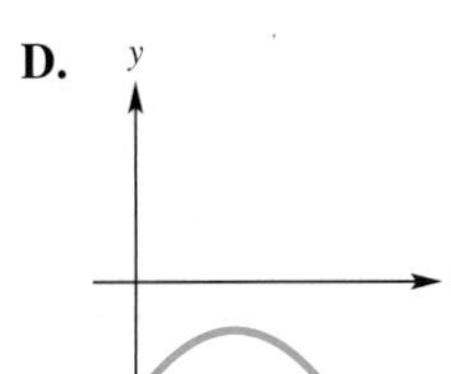

E.

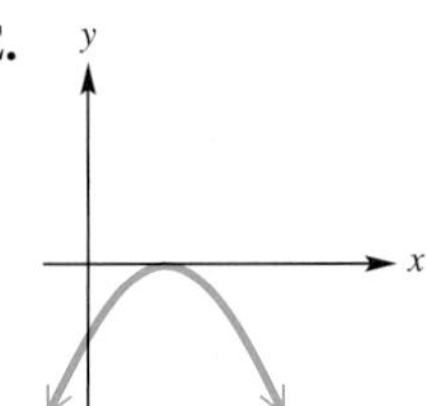

F.

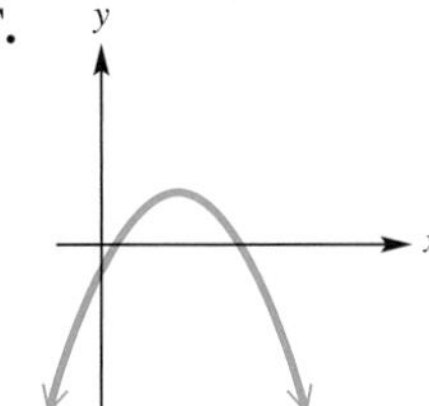

*Connecting Graphs with Equations In Exercises 37–40, find a quadratic function f whose graph matches the one in the figure. (Hint: See the Note following **Example 3.**)*

37.

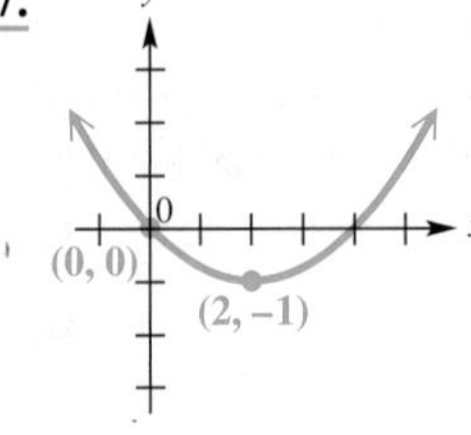

38.

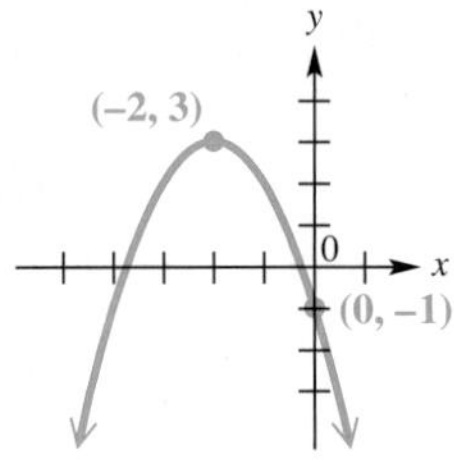

39.

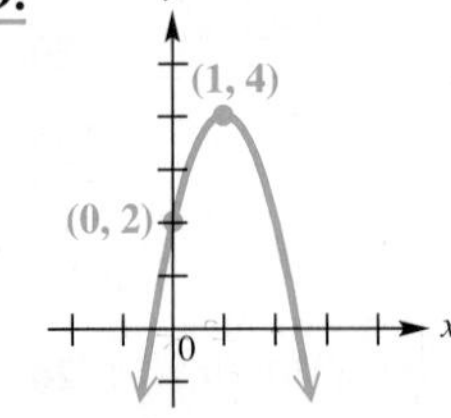

40.

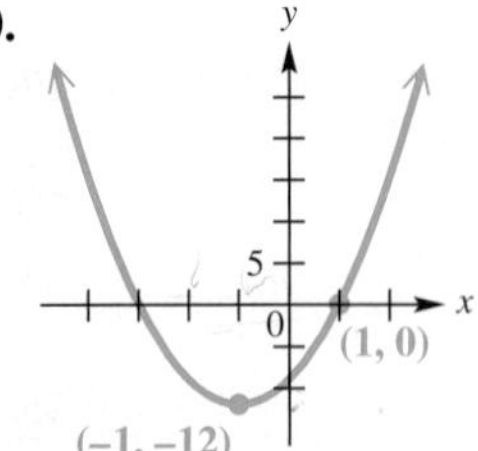

22. (a) $(1, 3)$ (b) $x = 1$
(c) $(-\infty, \infty)$ (d) $[3, \infty)$
(e) $[1, \infty)$ (f) $(-\infty, 1]$

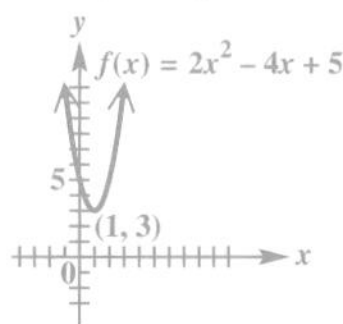

23. (a) $(-3, 2)$ (b) $x = -3$
(c) $(-\infty, \infty)$ (d) $(-\infty, 2]$
(e) $(-\infty, -3]$ (f) $[-3, \infty)$

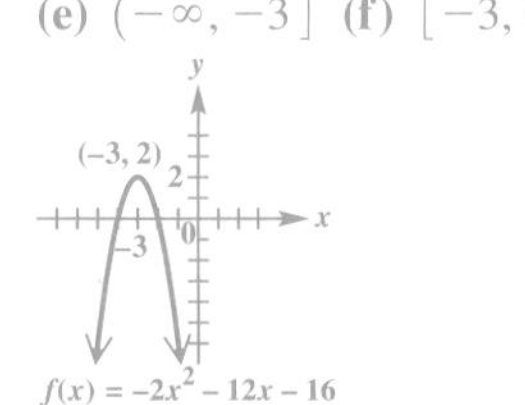

24. (a) $(4, 2)$ (b) $x = 4$
(c) $(-\infty, \infty)$ (d) $(-\infty, 2]$
(e) $(-\infty, 4]$ (f) $[4, \infty)$

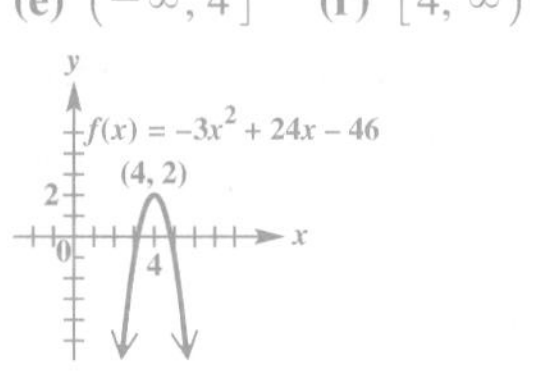

25. (a) $(-3, 4)$ (b) $x = -3$
(c) $(-\infty, \infty)$ (d) $(-\infty, 4]$
(e) $(-\infty, -3]$ (f) $[-3, \infty)$

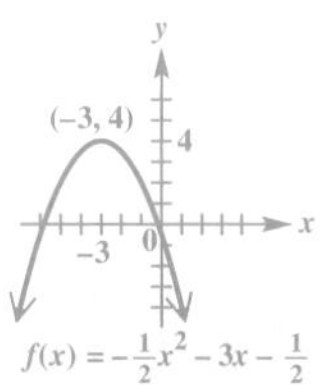

26. (a) $(2, -1)$ (b) $x = 2$
(c) $(-\infty, \infty)$ (d) $[-1, \infty)$
(e) $[2, \infty)$ (f) $(-\infty, 2]$

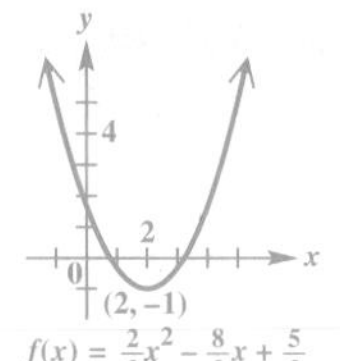

27. 3 28. -3
29. none 30. two

31. E 32. A
33. D 34. F
35. C 36. B

37. $f(x) = \frac{1}{4}(x - 2)^2 - 1$, or $f(x) = \frac{1}{4}x^2 - x$
38. $f(x) = -(x + 2)^2 + 3$, or $f(x) = -x^2 - 4x - 1$
39. $f(x) = -2(x - 1)^2 + 4$, or $f(x) = -2x^2 + 4x + 2$

(Modeling) In each scatter diagram, tell whether a linear or a quadratic model is appropriate for the data. If linear, tell whether the slope should be positive or negative. If quadratic, tell whether the leading coefficient of x^2 should be positive or negative.

41. number of shopping centers as a function of time

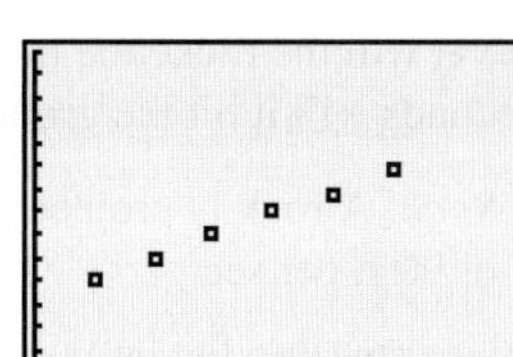

42. growth in science centers/museums as a function of time

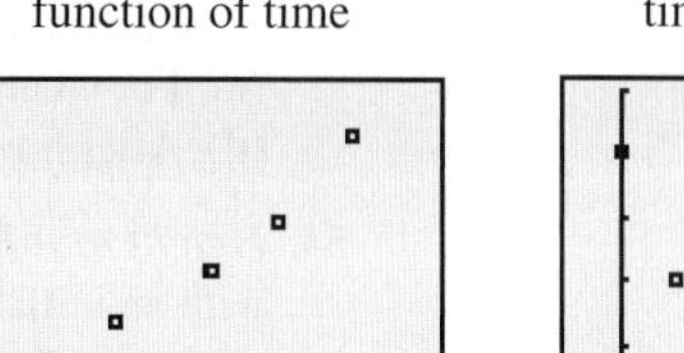

43. value of U.S. salmon catch as a function of time

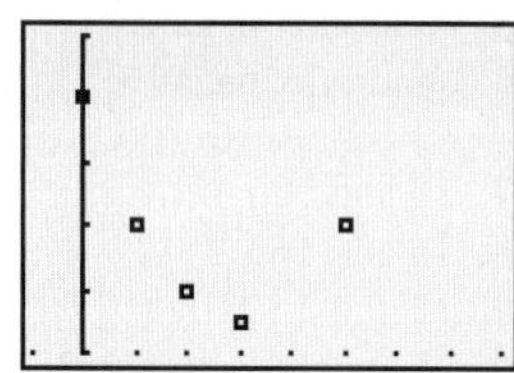

44. height of an object projected upward as a function of time

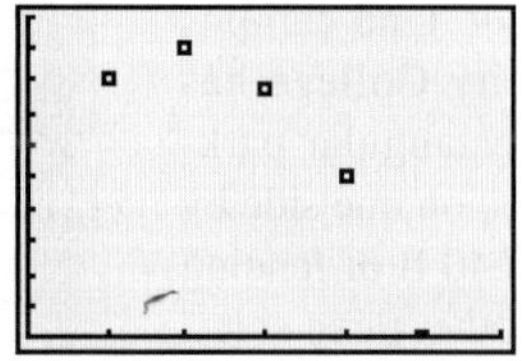

45. Social Security assets as a function of time

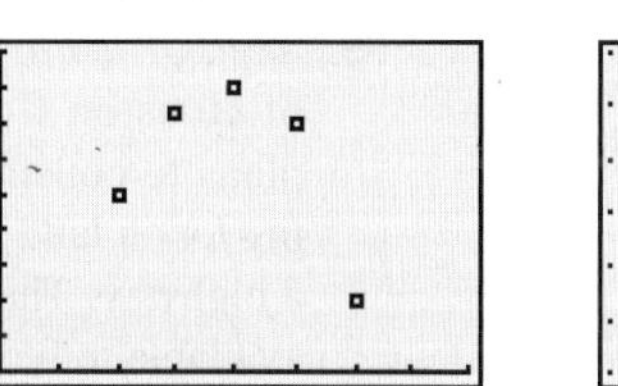

46. newborns with AIDS as a function of time

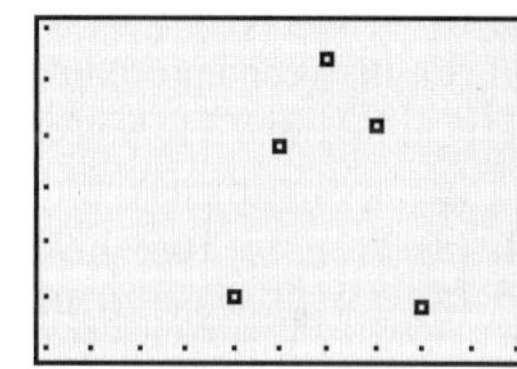

Solve each problem.

47. *Sum and Product of Two Numbers* Find two numbers whose sum is 20 and whose product is the maximum possible value. (*Hint: Let x be one number. Then* $20 - x$ *is the other number. Form a quadratic function by multiplying them, and then find the maximum value of the function.*)

48. *Sum and Product of Two Numbers* Find two numbers whose sum is 32 and whose product is the maximum possible value.

49. *Minimum Cost* Brigette Cole has a taco stand. She has found that her daily costs are approximated by

$$C(x) = x^2 - 40x + 610,$$

where $C(x)$ is the cost, in dollars, to sell x units of tacos. Find the number of units of tacos she should sell to minimize her costs. What is the minimum cost?

50. *Maximum Revenue* The revenue of a charter bus company depends on the number of unsold seats. If the revenue in dollars, $R(x)$, is given by

$$R(x) = -x^2 + 50x + 5000,$$

where x is the number of unsold seats, find the number of unsold seats that produce maximum revenue. What is the maximum revenue?

51. *Maximum Number of Mosquitos* The number of mosquitos, $M(x)$, in millions, in a certain area of Florida depends on the June rainfall, x, in inches. The function

$$M(x) = 10x - x^2$$

models this phenomenon. Find the amount of rainfall that will maximize the number of mosquitos. What is the maximum number of mosquitos?

52. *Height of an Object* If an object is projected upward from ground level with an initial velocity of 32 ft per sec, then its height in feet after t seconds is given by

$$s(t) = -16t^2 + 32t.$$

Find the number of seconds it will take to reach its maximum height. What is this maximum height?

40. $f(x) = 3(x + 1)^2 - 12$, or $f(x) = 3x^2 + 6x - 9$

41. linear; positive
42. linear; positive
43. quadratic; positive
44. quadratic; negative
45. quadratic; negative
46. quadratic; negative

47. 10 and 10
48. 16 and 16
49. 20 units; \$210
50. 25 unsold seats; \$5625
51. 5 in.; 25 million mosquitos
52. 1 sec; 16 ft

53. (a) $f(t) = -16t^2 + 200t + 50$
(b) 6.25 sec; 675 ft
(c) between approximately 1.41 and 11.09 sec
(d) approximately 12.75 sec
54. (a) $f(t) = -16t^2 + 90t$
(b) 2.8125 sec; 126.5625 ft
(c) between approximately 2.17 and 3.45 sec
(d) 5.625 sec
55. (a) $640 - 2x$
(b) $0 < x < 320$
(c) $\mathscr{A}(x) = -2x^2 + 640x$
(d) between approximately 57.04 ft and 85.17 ft or 234.83 ft and 262.96 ft
(e) 160 ft by 320 ft; The maximum area is 51,200 ft^2.
56. (a) $600 - 3x$
(b) $\mathscr{A}(x) = -3x^2 + 600x$; $0 < x < 200$
(c) 50 ft by 450 ft or 150 ft by 150 ft
(d) 30,000 ft^2
57. (a) $2x$
(b) length: $2x - 4$; width: $x - 4$; $x > 4$
(c) $V(x) = 4x^2 - 24x + 32$
(d) 8 in. by 20 in.
(e) 13.0 in. to 14.2 in.

(Modeling) Solve each problem. **See Example 5.**

53. *Height of a Toy Rocket* A toy rocket (not internally powered) is launched straight up from the top of a building 50 ft tall at an initial velocity of 200 ft per sec.

(a) Give the function that describes the height of the rocket in terms of time t.

(b) Determine the time at which the rocket reaches its maximum height and the maximum height in feet.

(c) For what time interval will the rocket be more than 300 ft above ground level?

(d) After how many seconds will it hit the ground?

54. *Height of a Projected Rock* A rock is projected directly upward from ground level with an initial velocity of 90 ft per sec.

(a) Give the function that describes the height of the rock in terms of time t.

(b) Determine the time at which the rock reaches its maximum height and the maximum height in feet.

(c) For what time interval will the rock be more than 120 ft above ground level?

(d) After how many seconds will it return to the ground?

55. *Area of a Parking Lot* One campus of Houston Community College has plans to construct a rectangular parking lot on land bordered on one side by a highway. There are 640 ft of fencing available to fence the other three sides. Let x represent the length of each of the two parallel sides of fencing.

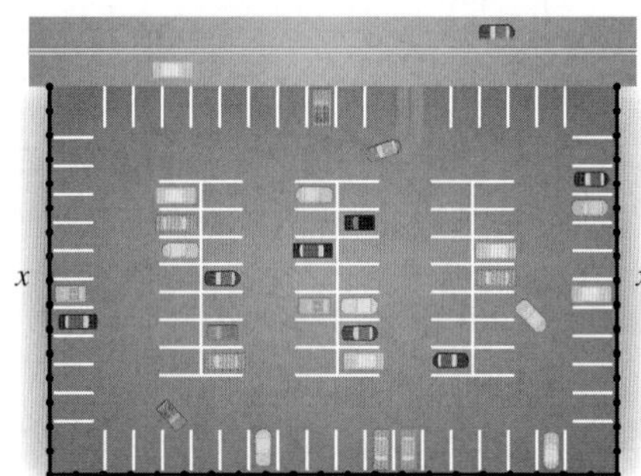

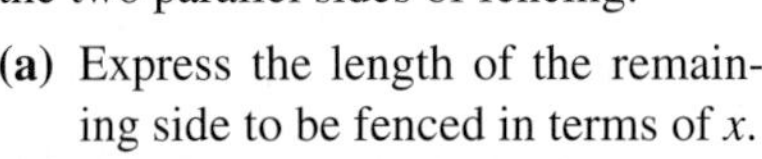

(a) Express the length of the remaining side to be fenced in terms of x.

(b) What are the restrictions on x?

(c) Determine a function $\mathscr{A}$ that represents the area of the parking lot in terms of x.

(d) Determine the values of x that will give an area between 30,000 and 40,000 ft^2.

(e) What dimensions will give a maximum area, and what will this area be?

56. *Area of a Rectangular Region* A farmer wishes to enclose a rectangular region bordering a river with fencing, as shown in the diagram. Suppose that x represents the length of each of the three parallel pieces of fencing. She has 600 ft of fencing available.

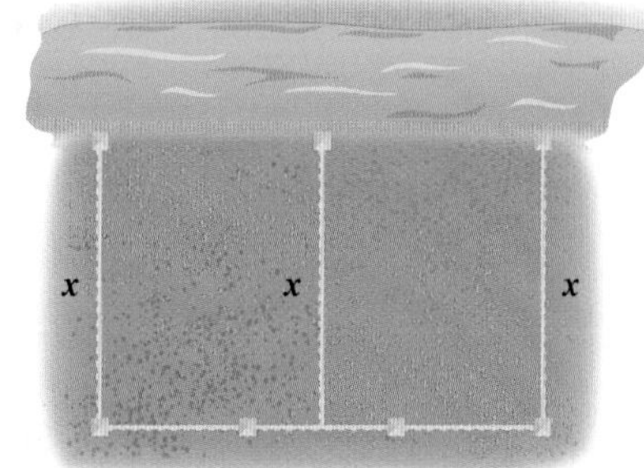

(a) What is the length of the remaining piece of fencing in terms of x?

(b) Determine a function $\mathscr{A}$ that represents the total area of the enclosed region. Give any restrictions on x.

(c) What dimensions for the total enclosed region would give an area of 22,500 ft^2?

(d) What is the maximum area that can be enclosed?

57. *Volume of a Box* A piece of cardboard is twice as long as it is wide. It is to be made into a box with an open top by cutting 2-in. squares from each corner and folding up the sides. Let x represent the width (in inches) of the original piece of cardboard.

(a) Represent the length of the original piece of cardboard in terms of x.

(b) What will be the dimensions of the bottom rectangular base of the box? Give the restrictions on x.

(c) Determine a function V that represents the volume of the box in terms of x.

(d) For what dimensions of the bottom of the box will the volume be 320 in.^{3}?

(e) Find the values of x (to the nearest tenth of an inch) if such a box is to have a volume between 400 and 500 in.^{3}.

58. (a) $2.5x$ (b) $x > 6$
(c) $V(x) = 7.5x^2 - 63x + 108$
(d) 13.3 to 14.7 in.
59. (a) approximately 23.32 ft per sec
(b) approximately 12.88 ft
60. (a) approximately 29.97 ft per sec
(b) approximately 15.42 ft; The underhand shot produces a higher arc.

61. 49.0 (percent)

58. *Volume of a Box* A piece of sheet metal is 2.5 times as long as it is wide. It is to be made into a box with an open top by cutting 3-in. squares from each corner and folding up the sides. Let x represent the width (in inches) of the original piece.

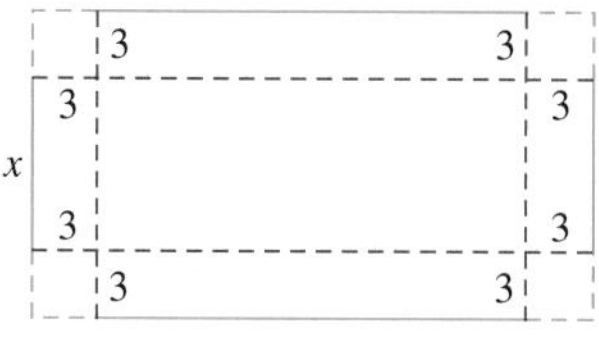

(a) Represent the length of the original piece of sheet metal in terms of x.

(b) What are the restrictions on x?

(c) Determine a function V that represents the volume of the box in terms of x.

(d) For what values of x (that is, original widths) will the volume of the box be between 600 and 800 in.^{3}? Give values to the nearest tenth of an inch.

59. *Shooting a Free Throw* If a person shoots a free throw from a position 8 ft above the floor, then the path of the ball may be modeled by the parabola

$$y = \frac{-16x^2}{0.434v^2} + 1.15x + 8,$$

where v is the initial velocity of the ball in feet per second, as illustrated in the figure. (*Source:* Rist, C., "The Physics of Foul Shots," *Discover.*)

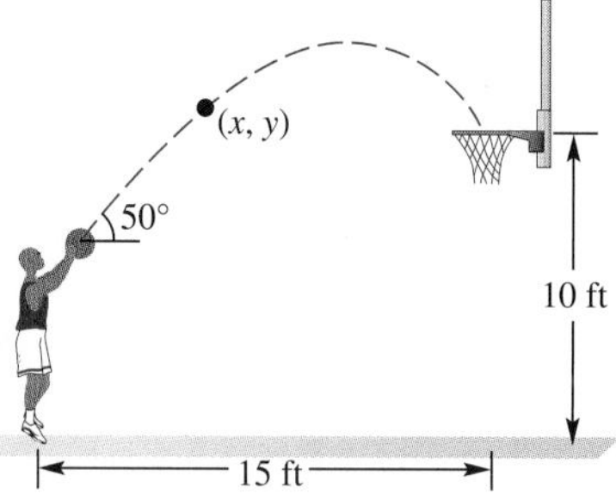

(a) If the basketball hoop is 10 ft high and located 15 ft away, what initial velocity v should the basketball have?

(b) What is the maximum height of the basketball?

60. *Shooting a Free Throw* See **Exercise 59.** If a person shoots a free throw from an underhand position 3 ft above the floor, the path of the ball may be modeled by

$$y = \frac{-16x^2}{0.117v^2} + 2.75x + 3.$$

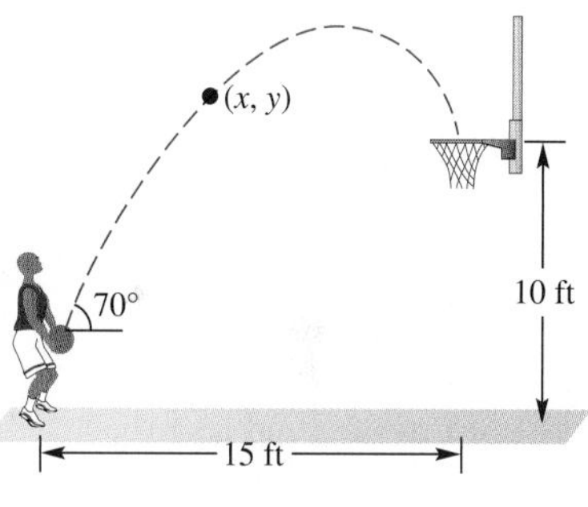

Repeat parts (a) and (b) from **Exercise 59.** Then compare the paths for the overhand shot and the underhand shot.

(Modeling) Solve each problem. ***See Example 6.***

61. *Births to Unmarried Women* The percent of births to unmarried women in the United States from 1996 to 2007 are shown in the table. The data are modeled by the quadratic function

$$f(x) = 0.0773x^2 - 0.2115x + 32.64,$$

where $x = 0$ corresponds to 1996 and $f(x)$ is the percent. If this model continues to apply, what would it predict for the percent of these births in 2012?

Year	Percent	Year	Percent
1996	32.4	2002	34.0
1997	32.4	2003	34.6
1998	32.8	2004	35.8
1999	33.0	2005	36.9
2000	33.2	2006	38.5
2001	33.5	2007	39.7

Source: National Center for Health Statistics.

62. about 406 ppm
63. 2007
64. 49 yr; 3.98
65. (a) 1,100,000

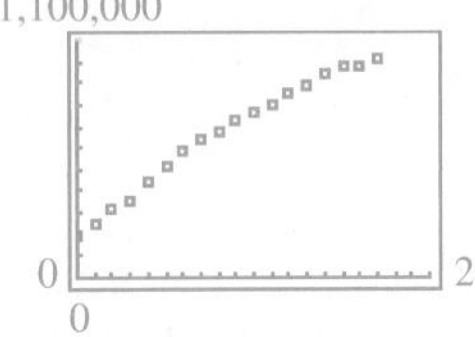

(b) quadratic; The rate at which the number of cases is increasing appears to be slowing down.
(c) $f(x) = -1328x^2 + 71{,}882x + 181{,}195$
(d) 1,100,000

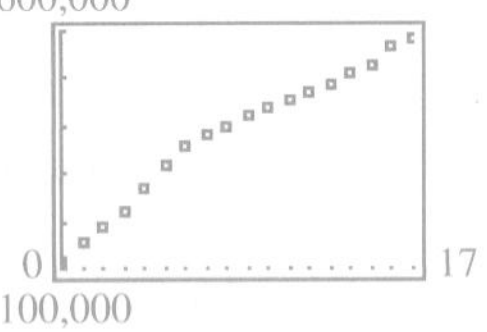

The graph of f models the data very well.
(e) 2009: 1,067,545; 2010: 1,087,635
(f) 20,090

66. (a) 600,000

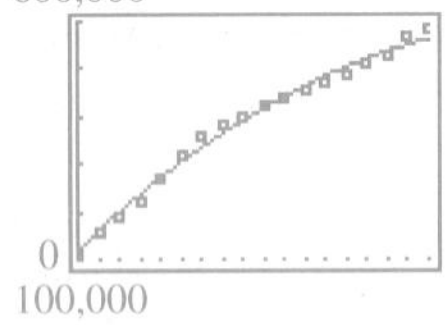

(b) quadratic; The rate at which the cumulative number of deaths is increasing appears to be slowing down.
(c) $g(x) = -884x^2 + 41{,}047x + 120{,}469$
(d) 600,000

The graph of g models the data closely.

62. *Concentration of Atmospheric CO_2* The quadratic function

$$f(x) = 0.0098x^2 + 0.9010x + 316.8$$

models the worldwide atmospheric concentration of carbon dioxide in parts per million (ppm) over the period 1960–2009, where $x = 0$ represents the year 1960. If this model continues to hold, what will be the atmospheric CO_2 concentration in 2020? (*Source*: U.S. Department of Energy.)

63. *Spending on Shoes and Clothing* The total amount spent by Americans on shoes and clothing from 2000 to 2009 can be modeled by

$$f(x) = -4.979x^2 + 71.73x + 97.29,$$

where $x = 0$ represents 2000 and $f(x)$ is in billions of dollars. Based on this model, in what year did spending on shoes and clothing reach a maximum? (*Source*: Bureau of Economic Analysis.)

64. *Accident Rate* According to data from the National Highway Traffic Safety Administration, the accident rate as a function of the age of the driver in years x can be approximated by the function

$$f(x) = 0.0232x^2 - 2.28x + 60.0,$$

for $16 \le x \le 85$. Find both the age at which the accident rate is a minimum and the minimum rate.

65. *AIDS Cases in the United States* The table* lists the total (cumulative) number of AIDS cases diagnosed in the United States through 2007. For example, a total of 361,509 AIDS cases were diagnosed through 1993.

Year	AIDS Cases	Year	AIDS Cases
1990	193,245	1999	718,676
1991	248,023	2000	759,434
1992	315,329	2001	801,302
1993	361,509	2002	844,047
1994	441,406	2003	888,279
1995	515,586	2004	932,387
1996	584,394	2005	978,056
1997	632,249	2006	982,498
1998	673,572	2007	1,018,428

(a) Plot the data. Let $x = 0$ correspond to the year 1990.
(b) Would a linear or a quadratic function model the data better? Explain.
(c) Find a quadratic function defined by $f(x) = ax^2 + bx + c$ that models the data.
(d) Plot the data together with f on the same coordinate plane. How well does f model the number of AIDS cases?
(e) Use f to estimate the total number of AIDS cases diagnosed in the years 2009 and 2010.
(f) According to the model, how many new cases were diagnosed in the year 2010?

66. *AIDS Deaths in the United States* The table* lists the total (cumulative) number of known deaths caused by AIDS in the United States up to 2007.

Year	AIDS Deaths	Year	AIDS Deaths
1990	119,821	1999	419,234
1991	154,567	2000	436,373
1992	191,508	2001	454,099
1993	220,592	2002	471,417
1994	269,992	2003	489,437
1995	320,692	2004	507,536
1996	359,892	2005	524,547
1997	381,738	2006	565,927
1998	400,743	2007	583,298

(a) Plot the data. Let $x = 0$ correspond to the year 1990.
(b) Would a linear or a quadratic function model the data better? Explain.
(c) Find a quadratic function defined by $g(x) = ax^2 + bx + c$ that models the data.
(d) Plot the data together with g on the same coordinate plane. How well does g model the number of AIDS cases?

*Sources: "Facts and Figures," Joint United Nations Programme on HIV/AIDS (UNAIDS), February, 1999; U.S. Census Bureau, 2007; and Centers for Disease Control and Prevention.

For instance, we saw in **Example 11** of **Section R.3** that

$$\frac{3x^3 - 2x^2 - 150}{x^2 - 4} = 3x - 2 + \frac{12x - 158}{x^2 - 4}. \quad \text{(Section R.3)}$$

We can express this result using the division algorithm.

$$\underbrace{3x^3 - 2x^2 - 150}_{\substack{f(x) \\ \text{Dividend} \\ \text{(original polynomial)}}} = \underbrace{(x^2 - 4)}_{\substack{g(x) \\ \text{Divisor}}} \cdot \underbrace{(3x - 2)}_{\substack{q(x) \\ \text{Quotient}}} + \underbrace{12x - 158}_{\substack{r(x) \\ \text{Remainder}}}$$

Synthetic Division When a given polynomial in x is divided by a first-degree binomial of the form $x - k$, a shortcut method called **synthetic division** may be used. The example on the left below is simplified by omitting all variables and writing only coefficients, with 0 used to represent the coefficient of any missing terms. Since the coefficient of x in the divisor is always 1 in these divisions, it too can be omitted. These omissions simplify the problem, as shown on the right.

$$\begin{array}{r|rrrr}
 & 3x^2 & +10x & +40 & \\ \hline
x-4) & 3x^3 & -2x^2 & +0x & -150 \\
 & 3x^3 & -12x^2 & & \\ \hline
 & & 10x^2 & +0x & \\
 & & 10x^2 & -40x & \\ \hline
 & & & 40x & -150 \\
 & & & 40x & -160 \\ \hline
 & & & & 10
\end{array}
\qquad
\begin{array}{r|rrrr}
 & 3 & 10 & 40 & \\ \hline
-4) & 3 & -2 & 0 & -150 \\
 & 3 & -12 & & \\ \hline
 & & 10 & 0 & \\
 & & 10 & -40 & \\ \hline
 & & & 40 & -150 \\
 & & & 40 & -160 \\ \hline
 & & & & 10
\end{array}$$

The numbers in color that are repetitions of the numbers directly above them can also be omitted, as shown on the left below.

$$\begin{array}{r|rrrr}
 & 3 & 10 & 40 & \\ \hline
-4) & 3 & -2 & 0 & -150 \\
 & & -12 & & \\ \hline
 & & 10 & 0 & \\
 & & & -40 & \\ \hline
 & & & 40 & -150 \\
 & & & & -160 \\ \hline
 & & & & 10
\end{array}
\qquad
\begin{array}{r|rrrr}
 & 3 & 10 & 40 & \\ \hline
-4) & 3 & -2 & 0 & -150 \\
 & & -12 & & \\ \hline
 & & 10 & & \\
 & & & -40 & \\ \hline
 & & & 40 & \\
 & & & & -160 \\ \hline
 & & & & 10
\end{array}$$

The numbers in color are again repetitions of those directly above them. They may be omitted, as shown on the right above.

The entire process can now be condensed vertically. The top row of numbers can be omitted since it duplicates the bottom row if the 3 is brought down.

$$\begin{array}{r|rrrr}
-4) & 3 & -2 & 0 & -150 \\
 & & -12 & -40 & -160 \\ \hline
 & 3 & 10 & 40 & 10
\end{array}$$

The rest of the bottom row is obtained by subtracting -12, -40, and -160 from the corresponding terms above them.

To simplify the arithmetic, we replace subtraction in the second row by addition and compensate by changing the -4 at the upper left to its additive inverse, 4.

$$\begin{array}{l r|rrrr l}
\text{Additive inverse} \longrightarrow & 4) & 3 & -2 & 0 & -150 & \\
 & & & 12 & 40 & 160 & \longleftarrow \text{Signs changed} \\ \hline
 & & 3 & 10 & 40 & 10 & \\
 & & \downarrow & \downarrow & \downarrow & \downarrow & \\
\text{Quotient} \longrightarrow & & \multicolumn{4}{l}{3x^2 + 10x + 40 + \dfrac{10}{x-4}} & \longleftarrow \text{Remainder}
\end{array}$$

Teaching Tip Remind students when using synthetic division to
(a) verify that the divisor is of the form $x - k$,
(b) include zeros as placeholders, and
(c) avoid accidentally using an incorrect value for k.

Synthetic division provides an efficient process for dividing a polynomial by a binomial of the form $x - k$. Begin by writing the coefficients of the polynomial in decreasing powers of the variable, using 0 as the coefficient of any missing powers. The number k is written to the left in the same row. In the example above, $x - k$ is $x - 4$, so k is 4. The answer is found on the bottom row with the remainder farthest to the right and the coefficients of the quotient on the left when written in order of decreasing degree.

CAUTION ***To avoid errors, use 0 as the coefficient for any missing terms, including a missing constant, when setting up the division.***

Classroom Example 1
Use synthetic division to divide.

$$\frac{4x^3 - 15x^2 + 11x - 10}{x - 3}$$

Answer: $4x^2 - 3x + 2 + \frac{-4}{x-3}$

EXAMPLE 1 Using Synthetic Division

Use synthetic division to divide.

$$\frac{5x^3 - 6x^2 - 28x - 2}{x + 2}$$

SOLUTION Express $x + 2$ in the form $x - k$ by writing it as $x - (-2)$.

$x + 2$ leads to -2. → $-2\overline{)5 \quad -6 \quad -28 \quad -2}$ ← Coefficients of the polynomial

Bring down the 5, and multiply: $-2(5) = -10$.

$$\begin{array}{r|rrrr} -2 & 5 & -6 & -28 & -2 \\ & \downarrow & -10 & & \\ \hline & 5 & & & \end{array}$$

Add -6 and -10 to obtain -16. Multiply: $-2(-16) = 32$.

$$\begin{array}{r|rrrr} -2 & 5 & -6 & -28 & -2 \\ & & -10 & 32 & \\ \hline & 5 & -16 & & \end{array}$$

Teaching Tip Remind students to use 0 as a constant in the following.

$$\frac{5x^3 - 6x^2 + 3x}{x - 1}$$

$$1\overline{)5 \quad -6 \quad 3 \quad 0}$$

Add -28 and 32, obtaining 4. Finally, $-2(4) = -8$.

$$\begin{array}{r|rrrr} -2 & 5 & -6 & -28 & -2 \\ & & -10 & 32 & -8 \\ \hline & 5 & -16 & 4 & \end{array}$$

Add columns. Be careful with signs.

Add -2 and -8 to obtain -10.

$$\begin{array}{r|rrrr} -2 & 5 & -6 & -28 & -2 \\ & & -10 & 32 & -8 \\ \hline & 5 & -16 & 4 & -10 \end{array}$$

← Remainder; 5, −16, 4: Quotient

Since the divisor $x - k$ has degree 1, the degree of the quotient will always be one less than the degree of the polynomial to be divided.

$$\frac{5x^3 - 6x^2 - 28x - 2}{x + 2} = 5x^2 - 16x + 4 + \frac{-10}{x + 2}$$

Remember to add $\frac{\text{remainder}}{\text{divisor}}$.

Now Try Exercise 9.

The result of the division in **Example 1** can be written as

$$5x^3 - 6x^2 - 28x - 2 = (x + 2)(5x^2 - 16x + 4) + (-10)$$

by multiplying each side by the denominator $x + 2$. The theorem that follows is a generalization of this product form.

Special Case of the Division Algorithm

For any polynomial $f(x)$ and any complex number k, there exists a unique polynomial $q(x)$ and number r such that the following holds.

$$f(x) = (x - k)q(x) + r$$

The mathematical statement

$$\underbrace{5x^3 - 6x^2 - 28x - 2}_{f(x)} = \underbrace{(x + 2)}_{(x-k)}\cdot\underbrace{(5x^2 - 16x + 4)}_{q(x)} + \underbrace{(-10)}_{r}$$

illustrates this connection. This form of the division algorithm is useful in developing the *remainder theorem.*

Evaluating Polynomial Functions Using the Remainder Theorem Suppose that $f(x)$ is written as $f(x) = (x - k)q(x) + r$. This equality is true for all complex values of x, so it is true for $x = k$. Replace x with k.

$$f(k) = (k - k)q(k) + r, \quad \text{or} \quad f(k) = r$$

This proves the following **remainder theorem,** which gives a new method of evaluating polynomial functions.

Remainder Theorem

If the polynomial $f(x)$ is divided by $x - k$, then the remainder is equal to $f(k)$.

In **Example 1,** when $f(x) = 5x^3 - 6x^2 - 28x - 2$ was divided by $x + 2$, or $x - (-2)$, the remainder was -10. Substitute -2 for x in $f(x)$.

$$\begin{aligned} f(-2) &= 5(-2)^3 - 6(-2)^2 - 28(-2) - 2 \\ &= -40 - 24 + 56 - 2 \\ &= -10 \end{aligned}$$

Use parentheses around substituted values to avoid errors.

As shown below, an alternative way to find the value of a polynomial is to use synthetic division. By the remainder theorem, instead of replacing x by -2 to find $f(-2)$, divide $f(x)$ by $x + 2$ using synthetic division as in **Example 1.** Then $f(-2)$ is the remainder, -10.

$$\begin{array}{r|rrrr} -2 & 5 & -6 & -28 & -2 \\ & & -10 & 32 & -8 \\ \hline & 5 & -16 & 4 & -10 \end{array} \leftarrow f(-2)$$

Classroom Example 2
Let
$f(x) = -3x^4 + 15x^3 - 50x + 25.$
Use the remainder theorem to find $f(4)$.
Answer: 17

EXAMPLE 2 Applying the Remainder Theorem

Let $f(x) = -x^4 + 3x^2 - 4x - 5$. Use the remainder theorem to find $f(-3)$.

SOLUTION Use synthetic division with $k = -3$.

$$\begin{array}{r|rrrrr} -3 & -1 & 0 & 3 & -4 & -5 \\ & & 3 & -9 & 18 & -42 \\ \hline & -1 & 3 & -6 & 14 & -47 \end{array} \leftarrow \text{Remainder}$$

$f(-3)$ is equal to the remainder when dividing by $x + 3$.

By this result, $f(-3) = -47$.

Now Try Exercise 31.

Testing Potential Zeros A **zero** of a polynomial function $f(x)$ is a number k such that $f(k) = 0$. ***The real number zeros are the x-intercepts of the graph of the function.***

The remainder theorem gives a quick way to decide whether a number k is a zero of a polynomial function defined by $f(x)$, as follows.

1. Use synthetic division to find $f(k)$.
2. If the remainder is 0, then $f(k) = 0$ and k is a zero of $f(x)$. If the remainder is not 0, then k is not a zero of $f(x)$.

A zero of $f(x)$ is a **root,** or **solution,** of the equation $f(x) = 0$.

Classroom Example 3
Decide whether the given number k is a zero of $f(x)$.
(a) $f(x) = 2x^3 - 3x^2 - 18;\ k = 2$
(b) $f(x) = x^4 - 4x^3 - 14x^2 + 36x + 45;\ k = -3$
(c) $f(x) = x^4 - x^3 + 6x^2 + 14x - 20;\ k = 1 + 3i$

Answers:
(a) no (b) yes (c) yes

EXAMPLE 3 Deciding Whether a Number Is a Zero

Decide whether the given number k is a zero of $f(x)$.

(a) $f(x) = x^3 - 4x^2 + 9x - 6;\quad k = 1$

(b) $f(x) = x^4 + x^2 - 3x + 1;\quad k = -1$

(c) $f(x) = x^4 - 2x^3 + 4x^2 + 2x - 5;\quad k = 1 + 2i$

SOLUTION

(a) Use synthetic division to decide whether 1 is a zero of $f(x) = x^3 - 4x^2 + 9x - 6$.

Proposed zero → 1)	1	−4	9	−6	← $f(x) = x^3 - 4x^2 + 9x - 6$
		1	−3	6	
	1	−3	6	0	← Remainder

$f(x) = x^3 - 4x^2 + 9x - 6$

8, −5, 5, −8

Zero X=1 Y=0

Figure 16

Since the remainder is 0, $f(1) = 0$, and 1 is a zero of the given polynomial function. An x-intercept of the graph of $f(x) = x^3 - 4x^2 + 9x - 6$ is 1, so the graph includes the point $(1, 0)$. The graph in **Figure 16** supports this.

(b) For $f(x) = x^4 + x^2 - 3x + 1$, remember to use 0 as coefficient for the missing x^3-term in the synthetic division.

Proposed zero → −1)	1	0	1	−3	1	
		−1	1	−2	5	
	1	−1	2	−5	6	← Remainder

Y1=X^4+X2-3X+1

(−1, 6)

8, −4, 4, −2

X=-1 Y=6

Figure 17

The remainder is not 0, so -1 is not a zero of $f(x) = x^4 + x^2 - 3x + 1$. In fact, $f(-1) = 6$, indicating that $(-1, 6)$ is on the graph of $f(x)$. The graph in **Figure 17** supports this.

(c) Use synthetic division and operations with complex numbers to determine whether $1 + 2i$ is a zero of $f(x) = x^4 - 2x^3 + 4x^2 + 2x - 5$.

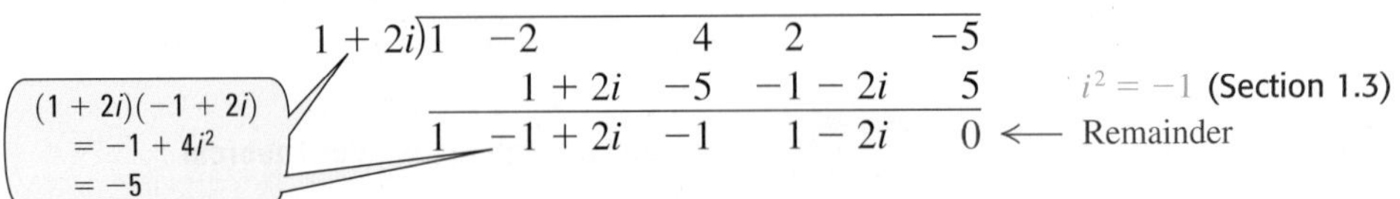

1 + 2i)	1	−2	4	2	−5	
		1 + 2i	−5	−1 − 2i	5	$i^2 = -1$ (Section 1.3)
	1	−1 + 2i	−1	1 − 2i	0	← Remainder

$(1 + 2i)(-1 + 2i)$
$= -1 + 4i^2$
$= -5$

Since the remainder is 0, $1 + 2i$ is a zero of the given polynomial function. Notice that $1 + 2i$ is *not* a real number zero. Therefore, it cannot appear as an x-intercept on the graph of $f(x)$.

Now Try Exercises 43 and 55.

3.2 Exercises

1. $x^2 + 2x + 9$
2. $x^2 + 5x + 3$
3. $5x^3 + 2x - 3$
4. $2x^3 + 3x^2 - x + 5$
5. $x^3 + 2x + 1$
6. $x^3 + 2x^2 - 2x + 3$
7. $x^4 + x^3 + 2x - 1 + \frac{3}{x+2}$
8. $x^5 - 2x^4 + x^3 - 6x + 7 + \frac{-11}{x+2}$
9. $-9x^2 - 10x - 27 + \frac{-52}{x-2}$
10. $-11x^3 + 13x^2 - 21x + 21 + \frac{-25}{x+1}$
11. $\frac{1}{3}x^2 - \frac{1}{9}x + \frac{1}{27}$
12. $x^2 + \frac{1}{2}x + \frac{1}{4}$
13. $x^3 - x^2 - 6x$
14. $x^3 - 2x^2 - 3x$
15. $x^2 + x + 1$
16. $x^3 + x^2 + x + 1$
17. $x^4 - x^3 + x^2 - x + 1$
18. $x^6 - x^5 + x^4 - x^3 + x^2 - x + 1$
19. $f(x) = (x+1) \cdot (2x^2 - x + 2) - 10$
20. $f(x) = (x+4) \cdot (2x^2 - 5x + 4) - 6$
21. $f(x) = (x+2) \cdot (x^2 + 2x + 1) + 0$
22. $f(x) = (x-2) \cdot (-x^2 - x + 1) + 0$
23. $f(x) = (x-3) \cdot (4x^3 + 9x^2 + 7x + 20) + 60$
24. $f(x) = (x+3) \cdot (2x^3 - 5x^2 + 3) - 9$
25. $f(x) = (x+1) \cdot (3x^3 + x^2 - 11x + 11) + 4$
26. $f(x) = (x-1) \cdot (-5x^3 - 4x^2 - 2x + 1) + 2$

27. 0
28. 0
29. −1
30. 127
31. −6
32. −2
33. −5
34. 793
35. 7
36. −1
37. $-6 - i$
38. $5 - 10i$
39. 0
40. 0

Use synthetic division to perform each division. ***See Example 1.***

1. $\dfrac{x^3 + 3x^2 + 11x + 9}{x + 1}$
2. $\dfrac{x^3 + 7x^2 + 13x + 6}{x + 2}$
3. $\dfrac{5x^4 + 5x^3 + 2x^2 - x - 3}{x + 1}$
4. $\dfrac{2x^4 - x^3 - 7x^2 + 7x - 10}{x - 2}$
5. $\dfrac{x^4 + 4x^3 + 2x^2 + 9x + 4}{x + 4}$
6. $\dfrac{x^4 + 5x^3 + 4x^2 - 3x + 9}{x + 3}$
7. $\dfrac{x^5 + 3x^4 + 2x^3 + 2x^2 + 3x + 1}{x + 2}$
8. $\dfrac{x^6 - 3x^4 + 2x^3 - 6x^2 - 5x + 3}{x + 2}$
9. $\dfrac{-9x^3 + 8x^2 - 7x + 2}{x - 2}$
10. $\dfrac{-11x^4 + 2x^3 - 8x^2 - 4}{x + 1}$
11. $\dfrac{\frac{1}{3}x^3 - \frac{2}{9}x^2 + \frac{2}{27}x - \frac{1}{81}}{x - \frac{1}{3}}$
12. $\dfrac{x^3 + x^2 + \frac{1}{2}x + \frac{1}{8}}{x + \frac{1}{2}}$
13. $\dfrac{x^4 - 3x^3 - 4x^2 + 12x}{x - 2}$
14. $\dfrac{x^4 - x^3 - 5x^2 - 3x}{x + 1}$
15. $\dfrac{x^3 - 1}{x - 1}$
16. $\dfrac{x^4 - 1}{x - 1}$
17. $\dfrac{x^5 + 1}{x + 1}$
18. $\dfrac{x^7 + 1}{x + 1}$

Express $f(x)$ in the form $f(x) = (x - k)q(x) + r$ for the given value of k.

19. $f(x) = 2x^3 + x^2 + x - 8;\ k = -1$
20. $f(x) = 2x^3 + 3x^2 - 16x + 10;\ k = -4$
21. $f(x) = x^3 + 4x^2 + 5x + 2;\ k = -2$
22. $f(x) = -x^3 + x^2 + 3x - 2;\ k = 2$
23. $f(x) = 4x^4 - 3x^3 - 20x^2 - x;\ k = 3$
24. $f(x) = 2x^4 + x^3 - 15x^2 + 3x;\ k = -3$
25. $f(x) = 3x^4 + 4x^3 - 10x^2 + 15;\ k = -1$
26. $f(x) = -5x^4 + x^3 + 2x^2 + 3x + 1;\ k = 1$

For each polynomial function, use the remainder theorem and synthetic division to find $f(k)$. ***See Example 2.***

27. $f(x) = x^2 + 5x + 6;\ k = -2$
28. $f(x) = x^2 - 4x - 5;\ k = 5$
29. $f(x) = 2x^2 - 3x - 3;\ k = 2$
30. $f(x) = -x^3 + 8x^2 + 63;\ k = 4$
31. $f(x) = x^3 - 4x^2 + 2x + 1;\ k = -1$
32. $f(x) = 2x^3 - 3x^2 - 5x + 4;\ k = 2$
33. $f(x) = 2x^5 - 10x^3 - 19x^2 - 50;\ k = 3$
34. $f(x) = x^4 + 6x^3 + 9x^2 + 3x - 3;\ k = 4$
35. $f(x) = 6x^4 + x^3 - 8x^2 + 5x + 6;\ k = \frac{1}{2}$
36. $f(x) = 6x^3 - 31x^2 - 15x;\ k = -\frac{1}{2}$
37. $f(x) = x^2 - 5x + 1;\ k = 2 + i$
38. $f(x) = x^2 - x + 3;\ k = 3 - 2i$
39. $f(x) = x^2 + 4;\ k = 2i$
40. $f(x) = 2x^2 + 10;\ k = i\sqrt{5}$

41. yes 42. yes
43. yes 44. yes
45. no; -9 46. no; 7
47. yes 48. yes
49. yes 50. yes
51. no; $\frac{357}{125}$ 52. no; $-\frac{1}{4}$
53. yes 54. yes
55. no; $13+7i$
56. no; $-1+2i$
57. no; $-2+7i$
58. no; $2-21i$

59. -12; $(-2,-12)$
60. 0; $(-1,0)$
61. $\frac{15}{8}$; $\left(-\frac{1}{2},\frac{15}{8}\right)$
62. 2; $(0,2)$
63. 0; $(1,0)$
64. $-\frac{5}{8}$; $\left(\frac{3}{2},-\frac{5}{8}\right)$
65. 0; $(2,0)$
66. 8; $(3,8)$
67.

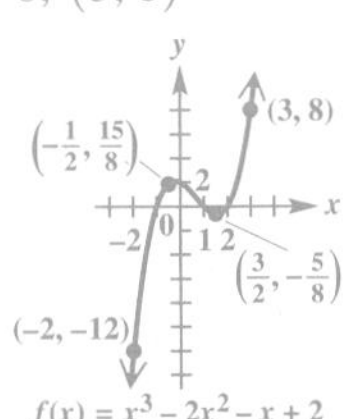

68.

(−3, 12)
$\left(\frac{1}{2},\frac{5}{8}\right)$
(−1, −2)
(2, −8)
$f(x)=-x^3-x^2+2x$

Use synthetic division to decide whether the given number k is a zero of the given polynomial function. If it is not, give the value of $f(k)$. ***See Examples 2 and 3.***

41. $f(x)=x^2+2x-8;\ k=2$
42. $f(x)=x^2+4x-5;\ k=-5$
43. $f(x)=x^3-3x^2+4x-4;\ k=2$
44. $f(x)=x^3+2x^2-x+6;\ k=-3$
45. $f(x)=2x^3-6x^2-9x+4;\ k=1$
46. $f(x)=2x^3+9x^2-16x+12;\ k=1$
47. $f(x)=x^3+7x^2+10x;\ k=0$
48. $f(x)=2x^3-3x^2-5x;\ k=0$
49. $f(x)=4x^4+x^2+17x+3;\ k=-\frac{3}{2}$
50. $f(x)=3x^4+13x^3-10x+8;\ k=-\frac{4}{3}$
51. $f(x)=5x^4+2x^3-x+3;\ k=\frac{2}{5}$
52. $f(x)=16x^4+3x^2-2;\ k=\frac{1}{2}$
53. $f(x)=x^2-2x+2;\ k=1-i$
54. $f(x)=x^2-4x+5;\ k=2-i$
55. $f(x)=x^2+3x+4;\ k=2+i$
56. $f(x)=x^2-3x+5;\ k=1-2i$
57. $f(x)=x^3+3x^2-x+1;\ k=1+i$
58. $f(x)=2x^3-x^2+3x-5;\ k=2-i$

Relating Concepts

For individual or collaborative investigation *(Exercises 59–68)*

The remainder theorem indicates that when a polynomial $f(x)$ is divided by $x-k$, the remainder is equal to $f(k)$. For

$$f(x)=x^3-2x^2-x+2,$$

use the remainder theorem to find each of the following. Then determine the coordinates of the corresponding point on the graph of $f(x)$.

59. $f(-2)$ 60. $f(-1)$ 61. $f\left(-\frac{1}{2}\right)$ 62. $f(0)$

63. $f(1)$ 64. $f\left(\frac{3}{2}\right)$ 65. $f(2)$ 66. $f(3)$

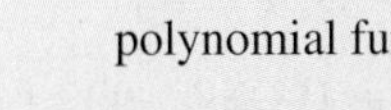

67. Use the results from **Exercises 59–66** to plot eight points on the graph of $f(x)$. Connect these points with a smooth curve. Describe a method for graphing polynomial functions using the remainder theorem.

68. Use the method described in **Exercises 59–67** to sketch the graph of $f(x)=-x^3-x^2+2x$.

3.3 Zeros of Polynomial Functions

- Factor Theorem
- Rational Zeros Theorem
- Number of Zeros
- Conjugate Zeros Theorem
- Finding Zeros of a Polynomial Function
- Descartes' Rule of Signs

Factor Theorem Consider the polynomial function

$$f(x)=x^2+x-2,$$

which is written in factored form as

$$f(x)=(x-1)(x+2).\quad \text{(Section R.4)}$$

For this function, $f(1)=0$ and $f(-2)=0$, and thus 1 and -2 are zeros of $f(x)$. Notice the special relationship between each linear factor and its corresponding zero. The **factor theorem** summarizes this relationship.

Factor Theorem

For any polynomial function $f(x)$, $x - k$ is a factor of the polynomial if and only if $f(k) = 0$.

Classroom Example 1
Determine whether $x + 4$ is a factor of each polynomial.
(a) $f(x) = 3x^4 - 48x^2 + 8x + 32$
(b) $f(x) = x^5 + 6x^4 + 11x^3 + 12x^2 + 5x - 20$

Answers:
(a) yes (b) no

EXAMPLE 1 Deciding Whether $x - k$ Is a Factor

Determine whether $x - 1$ is a factor of each polynomial.

(a) $f(x) = 2x^4 + 3x^2 - 5x + 7$

(b) $f(x) = 3x^5 - 2x^4 + x^3 - 8x^2 + 5x + 1$

SOLUTION

(a) By the factor theorem, $x - 1$ will be a factor if $f(1) = 0$. Use synthetic division and the remainder theorem to decide.

$$\begin{array}{r|rrrrr} 1 & 2 & 0 & 3 & -5 & 7 \\ & & 2 & 2 & 5 & 0 \\ \hline & 2 & 2 & 5 & 0 & 7 \end{array} \leftarrow f(1) = 7 \quad \text{(Section 3.2)}$$

Use a zero coefficient for the missing term.

The remainder is 7 and not 0, so $x - 1$ is not a factor of $2x^4 + 3x^2 - 5x + 7$.

Teaching Tip Point out that **Example 1** also can be solved by calculating $f(1)$ directly. Explain that for any polynomial function f, $f(1)$ can be found quickly by summing the coefficients and constant term of $f(x)$.

(b)
$$\begin{array}{r|rrrrrr} 1 & 3 & -2 & 1 & -8 & 5 & 1 \\ & & 3 & 1 & 2 & -6 & -1 \\ \hline & 3 & 1 & 2 & -6 & -1 & 0 \end{array} \leftarrow f(1) = 0$$

Because the remainder is 0, $x - 1$ is a factor. Additionally, we can determine from the coefficients in the bottom row that the other factor is

$$3x^4 + x^3 + 2x^2 - 6x - 1.$$

Thus, we can express the polynomial in factored form.

$$f(x) = (x - 1)(3x^4 + x^3 + 2x^2 - 6x - 1)$$

✔ *Now Try Exercises 5 and 7.*

We can use the factor theorem to factor a polynomial of greater degree into linear factors of the form $ax - b$.

Classroom Example 2
Factor
$f(x) = 6x^3 - 37x^2 + 32x + 15$
into linear factors if 5 is a zero of f.

Answer:
$f(x) = (x - 5)(2x - 3)(3x + 1)$

EXAMPLE 2 Factoring a Polynomial Given a Zero

Factor $f(x) = 6x^3 + 19x^2 + 2x - 3$ into linear factors if -3 is a zero of f.

SOLUTION Since -3 is a zero of f, $x - (-3) = x + 3$ is a factor.

$$\begin{array}{r|rrrr} -3 & 6 & 19 & 2 & -3 \\ & & -18 & -3 & 3 \\ \hline & 6 & 1 & -1 & 0 \end{array} \quad \text{Use synthetic division to divide } f(x) \text{ by } x + 3.$$

The quotient is $6x^2 + x - 1$, which is the factor that accompanies $x + 3$.

$$f(x) = (x + 3)(6x^2 + x - 1)$$

$$f(x) = (x + 3)(2x + 1)(3x - 1) \quad \text{Factor } 6x^2 + x - 1.$$

These factors are all linear.

✔ *Now Try Exercise 17.*

LOOKING AHEAD TO CALCULUS

Finding the derivative of a polynomial function is one of the basic skills required in a first calculus course. For the functions

$$f(x) = x^4 - x^2 + 5x - 4,$$
$$g(x) = -x^6 + x^2 - 3x + 4,$$
$$\text{and } h(x) = 3x^3 - x^2 + 2x - 4,$$

the derivatives are

$$f'(x) = 4x^3 - 2x + 5,$$
$$g'(x) = -6x^5 + 2x - 3,$$
$$\text{and } h'(x) = 9x^2 - 2x + 2.$$

Notice the use of the "prime" notation. For example, the derivative of $f(x)$ is denoted $f'(x)$.

Look for the pattern among the exponents and the coefficients. Using this pattern, what is the derivative of

$$F(x) = 4x^4 - 3x^3 + 6x - 4?$$

The answer is at the bottom of the page.

Rational Zeros Theorem The **rational zeros theorem** gives a method to determine all possible candidates for rational zeros of a polynomial function with integer coefficients.

Rational Zeros Theorem

If $\frac{p}{q}$ is a rational number written in lowest terms, and if $\frac{p}{q}$ is a zero of f, a polynomial function with integer coefficients, then p is a factor of the constant term, and q is a factor of the leading coefficient.

Proof $f\left(\frac{p}{q}\right) = 0$ since $\frac{p}{q}$ is a zero of $f(x)$.

$$a_n\left(\frac{p}{q}\right)^n + a_{n-1}\left(\frac{p}{q}\right)^{n-1} + \cdots + a_1\left(\frac{p}{q}\right) + a_0 = 0 \qquad \text{Definition of zero of } f$$

$$a_n\left(\frac{p^n}{q^n}\right) + a_{n-1}\left(\frac{p^{n-1}}{q^{n-1}}\right) + \cdots + a_1\left(\frac{p}{q}\right) + a_0 = 0 \qquad \text{Power rule for exponents (Section R.3)}$$

$$a_np^n + a_{n-1}p^{n-1}q + \cdots + a_1pq^{n-1} = -a_0q^n \qquad \text{Multiply by } q^n \text{ and subtract } a_0q^n.$$

$$p(a_np^{n-1} + a_{n-1}p^{n-2}q + \cdots + a_1q^{n-1}) = -a_0q^n \qquad \text{Factor out } p.$$

This result shows that $-a_0q^n$ equals the product of the two factors p and $(a_np^{n-1} + \cdots + a_1q^{n-1})$. For this reason, p must be a factor of $-a_0q^n$. Since it was assumed that $\frac{p}{q}$ is written in lowest terms, p and q have no common factor other than 1, so p is not a factor of q^n. Thus, p must be a factor of a_0. In a similar way, it can be shown that q is a factor of a_n.

Classroom Example 3
Repeat **Example 3** for

$$f(x) = 8x^4 - 26x^3 - 27x^2 + 11x + 4.$$

Answers:
(a) $\pm 1, \pm 2, \pm 4, \pm\frac{1}{2}, \pm\frac{1}{4}, \pm\frac{1}{8}$
(b) $-1, 4, \frac{1}{2}, -\frac{1}{4}$; $f(x) = (x+1)(x-4) \cdot (2x-1)(4x+1)$

EXAMPLE 3 Using the Rational Zeros Theorem

Consider the polynomial function.

$$f(x) = 6x^4 + 7x^3 - 12x^2 - 3x + 2$$

(a) List all possible rational zeros.

(b) Find all rational zeros and factor $f(x)$ into linear factors.

SOLUTION

(a) For a rational number $\frac{p}{q}$ to be a zero, p must be a factor of $a_0 = 2$, and q must be a factor of $a_4 = 6$. Thus, p can be ± 1 or ± 2, and q can be ± 1, ± 2, ± 3, or ± 6. The possible rational zeros, $\frac{p}{q}$, are ± 1, ± 2, $\pm\frac{1}{2}$, $\pm\frac{1}{3}$, $\pm\frac{1}{6}$, and $\pm\frac{2}{3}$.

(b) Use the remainder theorem to show that 1 is a zero.

Use "trial and error" to find zeros.

$$\begin{array}{r|rrrrr} 1 & 6 & 7 & -12 & -3 & 2 \\ & & 6 & 13 & 1 & -2 \\ \hline & 6 & 13 & 1 & -2 & 0 \end{array} \leftarrow f(1) = 0$$

The 0 remainder shows that 1 is a zero. The quotient is $6x^3 + 13x^2 + x - 2$.

$$f(x) = (x-1)(6x^3 + 13x^2 + x - 2) \qquad \text{Begin factoring } f(x).$$

Now, use the quotient polynomial and synthetic division to find that -2 is a zero.

$$\begin{array}{r|rrrr} -2 & 6 & 13 & 1 & -2 \\ & & -12 & -2 & 2 \\ \hline & 6 & 1 & -1 & 0 \end{array} \leftarrow f(-2) = 0$$

Teaching Tip Point out that looking at the graph of the function on a graphing calculator can be useful in deciding which potential rational zeros are the best candidates. Help students link x-intercepts with zeros and linear factors of a polynomial function.

Answer to Looking Ahead to Calculus:
$F'(x) = 16x^3 - 9x^2 + 6$

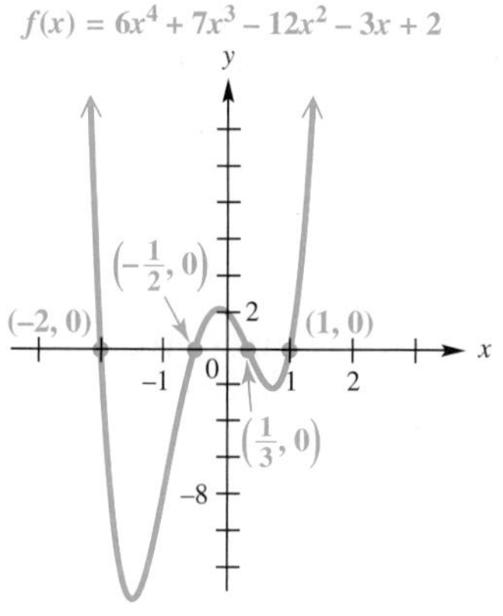

Figure 18

The new quotient polynomial is $6x^2 + x - 1$. Therefore, $f(x)$ can now be completely factored as follows.

$$f(x) = (x - 1)(x + 2)(6x^2 + x - 1)$$
$$f(x) = (x - 1)(x + 2)(3x - 1)(2x + 1)$$

Setting $3x - 1 = 0$ and $2x + 1 = 0$ yields the zeros $\frac{1}{3}$ and $-\frac{1}{2}$. In summary, the rational zeros are 1, −2, $\frac{1}{3}$, and $-\frac{1}{2}$, and they can be seen as x-intercepts on the graph of $f(x)$ in **Figure 18.** The linear factorization of $f(x)$ is

$$f(x) = 6x^4 + 7x^3 - 12x^2 - 3x + 2$$
$$f(x) = (x - 1)(x + 2)(3x - 1)(2x + 1).$$

Check by multiplying these factors.

✔ *Now Try Exercise 35.*

NOTE In **Example 3,** once we obtained the quadratic factor

$$6x^2 + x - 1,$$

we were able to complete the work by factoring it directly. Had it not been easily factorable, we could have used the quadratic formula to find the other two zeros (and factors).

Teaching Tip Stress that common factors in coefficients should be eliminated before applying the rational zeros theorem. For instance, since each coefficient of

$$f(x) = 6x^3 + 33x^2 - 21x - 18$$

is divisible by 3, $f(x)$ has the same zeros as

$$g(x) = 2x^3 + 11x^2 - 7x - 6.$$

CAUTION ***The rational zeros theorem gives only possible rational zeros. It does not tell us whether these rational numbers are actual zeros.*** We must rely on other methods to determine whether or not they are indeed zeros. Furthermore, the polynomial must have integer coefficients.

To apply the rational zeros theorem to a polynomial with fractional coefficients, multiply through by the least common denominator of all the fractions. For example, any rational zeros of $p(x)$ defined below will also be rational zeros of $q(x)$.

$$p(x) = x^4 - \frac{1}{6}x^3 + \frac{2}{3}x^2 - \frac{1}{6}x - \frac{1}{3}$$
$$q(x) = 6x^4 - x^3 + 4x^2 - x - 2 \quad \text{Multiply the terms of } p(x) \text{ by 6.}$$

Carl Friedrich Gauss (1777–1855)

The **fundamental theorem of algebra** was first proved by Carl Friedrich Gauss in his doctoral thesis in 1799, when he was 22 years old.

Number of Zeros The **fundamental theorem of algebra** says that every function defined by a polynomial of degree 1 or more has a zero, which means that every such polynomial can be factored.

Fundamental Theorem of Algebra

Every function defined by a polynomial of degree 1 or more has at least one complex zero.

From the fundamental theorem, if $f(x)$ is of degree 1 or more, then there is some number k_1 such that $f(k_1) = 0$. By the factor theorem,

$$f(x) = (x - k_1)q_1(x), \quad \text{for some polynomial } q_1(x).$$

If $q_1(x)$ is of degree 1 or more, the fundamental theorem and the factor theorem can be used to factor $q_1(x)$ in the same way. There is some number k_2 such that $q_1(k_2) = 0$, so

$$q_1(x) = (x - k_2)q_2(x)$$

and
$$f(x) = (x - k_1)(x - k_2)q_2(x).$$

Assuming that $f(x)$ has degree n and repeating this process n times gives

$$f(x) = a(x - k_1)(x - k_2)\cdots(x - k_n). \quad a \text{ is the leading coefficient.}$$

Each of these factors leads to a zero of $f(x)$, so $f(x)$ has the n zeros k_1, k_2, $k_3, \ldots, k_n$. This result suggests the **number of zeros theorem.**

Number of Zeros Theorem

A function defined by a polynomial of degree n has *at most n* distinct zeros.

For example, a polynomial function of degree 3 has *at most* three distinct zeros but can have as few as one zero. Consider the following polynomial.

$$f(x) = x^3 + 3x^2 + 3x + 1$$

$$f(x) = (x + 1)^3$$

The function f is of degree 3 but has only one zero, -1. Actually, the zero -1 occurs *three* times, since there are three factors of $x + 1$. The number of times a zero occurs is referred to as the **multiplicity of the zero.**

Classroom Example 4
Find a function f defined by a polynomial of degree 3 that satisfies the given conditions.
(a) Zeros of -3, -2, and 5; $f(-1) = 6$
(b) 4 is a zero of multiplicity 3; $f(2) = -24$

Answers:
(a) $f(x) = -\frac{1}{2}x^3 + \frac{19}{2}x + 15$
(b) $f(x) = 3x^3 - 36x^2 + 144x - 192$

EXAMPLE 4 Finding a Polynomial Function That Satisfies Given Conditions (Real Zeros)

Find a function f defined by a polynomial of degree 3 that satisfies the given conditions.

(a) Zeros of -1, 2, and 4; $f(1) = 3$

(b) -2 is a zero of multiplicity 3; $f(-1) = 4$

SOLUTION

(a) These three zeros give $x - (-1) = x + 1$, $x - 2$, and $x - 4$ as factors of $f(x)$. Since $f(x)$ is to be of degree 3, these are the only possible factors by the number of zeros theorem. Therefore, $f(x)$ has the form

$$f(x) = a(x + 1)(x - 2)(x - 4), \quad \text{for some real number } a.$$

To find a, use the fact that $f(1) = 3$.

$$f(1) = a(1 + 1)(1 - 2)(1 - 4) \quad \text{Let } x = 1.$$

$$3 = a(2)(-1)(-3) \quad f(1) = 3$$

$$3 = 6a \quad \text{Multiply.}$$

$$a = \frac{1}{2} \quad \text{Divide by 6.}$$

Thus, $$f(x) = \frac{1}{2}(x + 1)(x - 2)(x - 4),$$

or, $$f(x) = \frac{1}{2}x^3 - \frac{5}{2}x^2 + x + 4. \quad \text{Multiply.}$$

(b) The polynomial function $f(x)$ has the following form.

$$f(x) = a(x+2)(x+2)(x+2) \quad \text{Factor theorem}$$

$$f(x) = a(x+2)^3$$

To find a, use the fact that $f(-1) = 4$.

$$f(-1) = a(-1+2)^3 \quad \text{Let } x = -1.$$

$$4 = a(1)^3 \quad f(-1) = 4$$

$$a = 4 \quad \text{Solve for } a.$$

Remember: $(x+2)^3 \neq x^3 + 2^3$

Thus, $f(x) = 4(x+2)^3$,

or, $f(x) = 4x^3 + 24x^2 + 48x + 32$. Multiply.

Now Try Exercises 49 and 53.

NOTE In **Example 4(a),** we cannot clear the denominators in $f(x)$ by multiplying each side by 2 because the result would equal $2 \cdot f(x)$, not $f(x)$.

Conjugate Zeros Theorem The following properties of complex conjugates are needed to prove the **conjugate zeros theorem.** We use a simplified notation for conjugates here. If $z = a + bi$, then the conjugate of z is written $\bar{z}$, where $\bar{z} = a - bi$. For example, if $z = -5 + 2i$, then $\bar{z} = -5 - 2i$. The proofs of the first two of these properties are left for **Exercises 113 and 114.**

Properties of Conjugates

For any complex numbers c and d, the following properties hold.

$$\overline{c+d} = \bar{c} + \bar{d}, \quad \overline{c \cdot d} = \bar{c} \cdot \bar{d}, \quad \text{and} \quad \overline{c^n} = (\bar{c})^n$$

The remainder theorem can be used to show that both $2 + i$ and $2 - i$ are zeros of $f(x) = x^3 - x^2 - 7x + 15$. In general, if z is a zero of a polynomial function with *real* coefficients, then so is $\bar{z}$.

Conjugate Zeros Theorem

If $f(x)$ defines a polynomial function ***having only real coefficients*** and if $z = a + bi$ is a zero of $f(x)$, where a and b are real numbers, then

$\bar{z} = a - bi$ is also a zero of $f(x)$.

Proof Start with the polynomial function

$$f(x) = a_n x^n + a_{n-1}x^{n-1} + \cdots + a_1 x + a_0,$$

where all coefficients are real numbers. If the complex number z is a zero of $f(x)$, then we have the following.

$$f(z) = a_n z^n + a_{n-1}z^{n-1} + \cdots + a_1 z + a_0 = 0$$

$$a_n z^n + a_{n-1} z^{n-1} + \cdots + a_1 z + a_0 = 0$$ From the preceding discussion

Take the conjugate of both sides of this equation.

$$\overline{a_n z^n + a_{n-1} z^{n-1} + \cdots + a_1 z + a_0} = \overline{0}$$

$$\overline{a_n z^n} + \overline{a_{n-1} z^{n-1}} + \cdots + \overline{a_1 z} + \overline{a_0} = \overline{0}$$ Use generalizations of the properties $\overline{c+d} = \overline{c} + \overline{d}$ and $\overline{c \cdot d} = \overline{c} \cdot \overline{d}$.

$$\overline{a_n}\, \overline{z^n} + \overline{a_{n-1}}\, \overline{z^{n-1}} + \cdots + \overline{a_1}\, \overline{z} + \overline{a_0} = \overline{0}$$

$$a_n(\overline{z})^n + a_{n-1}(\overline{z})^{n-1} + \cdots + a_1(\overline{z}) + a_0 = 0$$ Use the property $\overline{c^n} = (\overline{c})^n$ and the fact that for any real number a, $\overline{a} = a$.

$$f(\overline{z}) = 0$$

Hence $\overline{z}$ is also a zero of $f(x)$, which completes the proof.

CAUTION ***When the conjugate zeros theorem is applied, it is essential that the polynomial have only real coefficients.*** For example,

$$f(x) = x - (1 + i)$$

has $1 + i$ as a zero, but the conjugate $1 - i$ is not a zero.

Classroom Example 5
Find a polynomial function of least degree having only real coefficients and zeros -4 and $3 - i$.

Answer:
$f(x) = x^3 - 2x^2 - 14x + 40$

EXAMPLE 5 Finding a Polynomial Function That Satisfies Given Conditions (Complex Zeros)

Find a polynomial function of least degree having only real coefficients and zeros 3 and $2 + i$.

SOLUTION The complex number $2 - i$ must also be a zero, so the polynomial has at least three zeros: 3, $2 + i$, and $2 - i$. For the polynomial to be of least degree, these must be the only zeros. By the factor theorem there must be three factors: $x - 3$, $x - (2 + i)$, and $x - (2 - i)$.

$$f(x) = (x - 3)[x - (2 + i)][x - (2 - i)]$$ Factor theorem

$$f(x) = (x - 3)(x - 2 - i)(x - 2 + i)$$ Distribute negative signs.

$$f(x) = (x - 3)(x^2 - 4x + 5)$$ Multiply and combine like terms; $i^2 = -1$. **(Section 1.3)**

$$f(x) = x^3 - 7x^2 + 17x - 15$$ Multiply again.

Any nonzero multiple of $x^3 - 7x^2 + 17x - 15$ also satisfies the given conditions on zeros. The information on zeros given in the problem is not sufficient to give a specific value for the leading coefficient.

Now Try Exercise 65.

Finding Zeros of a Polynomial Function The theorem on conjugate zeros helps predict the number of real zeros of polynomial functions with real coefficients.

- A polynomial function with real coefficients of odd degree n, where $n \geq 1$, must have at least one real zero (since zeros of the form $a + bi$, where $b \neq 0$, occur in conjugate pairs).
- A polynomial function with real coefficients of even degree n may have no real zeros.

Classroom Example 6
Find all zeros of

$$f(x) = x^4 - x^3 - 17x^2 + 55x - 50,$$

given that $2 + i$ is a zero.

Answer: $2 + i, 2 - i, -5, 2$

EXAMPLE 6 Finding All Zeros Given One Zero

Find all zeros of $f(x) = x^4 - 7x^3 + 18x^2 - 22x + 12$, given that $1 - i$ is a zero.

SOLUTION Since the polynomial function has only real coefficients and since $1 - i$ is a zero, by the conjugate zeros theorem $1 + i$ is also a zero. To find the remaining zeros, first use synthetic division to divide the original polynomial by $x - (1 - i)$.

$(1 - i)(-6 - i) = -6 - i + 6i + i^2 = -7 + 5i$

$$\begin{array}{r|rrrrr} 1-i & 1 & -7 & 18 & -22 & 12 \\ & & 1-i & -7+5i & 16-6i & -12 \\ \hline & 1 & -6-i & 11+5i & -6-6i & 0 \end{array}$$

By the factor theorem, since $x = 1 - i$ is a zero of $f(x)$, $x - (1 - i)$ is a factor, and $f(x)$ can be written as follows.

$$f(x) = [x - (1 - i)][x^3 + (-6 - i)x^2 + (11 + 5i)x + (-6 - 6i)]$$

We know that $x = 1 + i$ is also a zero of $f(x)$. Continue to use synthetic division and divide the quotient polynomial above by $x - (1 + i)$.

Teaching Tip You might have students multiply out the pair of factors that contain nonreal complex zeros and express $f(x)$ as the product of real factors $(x^2 - 2x + 2)$, $(x - 2)$, and $(x - 3)$. Also, to avoid division with complex numbers, you can divide $f(x)$ by

$$x^2 - 2x + 2$$

using long division to get

$$x^2 - 5x + 6.$$

$$\begin{array}{r|rrrr} 1+i & 1 & -6-i & 11+5i & -6-6i \\ & & 1+i & -5-5i & 6+6i \\ \hline & 1 & -5 & 6 & 0 \end{array}$$

Now $f(x)$ can be written in the following factored form.

$$f(x) = [x - (1 - i)][x - (1 + i)](x^2 - 5x + 6)$$
$$f(x) = [x - (1 - i)][x - (1 + i)](x - 2)(x - 3)$$

The remaining zeros are 2 and 3. The four zeros are $1 - i$, $1 + i$, 2, and 3.

✔ *Now Try Exercise 31.*

NOTE In **Example 6,** if we had been unable to factor $x^2 - 5x + 6$ into linear factors, we would have used the quadratic formula to solve the equation $x^2 - 5x + 6 = 0$ to find the remaining two zeros of the function.

Descartes' Rule of Signs

The following rule helps to determine the number of positive and negative real zeros of a polynomial function. A **variation in sign** is a change from positive to negative or from negative to positive in successive terms of the polynomial when they are written in order of descending powers of the variable. ***Missing terms (those with 0 coefficients) are counted as no change in sign and can be ignored.***

Teaching Tip Point out that Descartes' rule of signs can be used in conjunction with the rational zeros theorem to narrow the list of potential rational zeros.

Descartes' Rule of Signs

Let $f(x)$ define a polynomial function with real coefficients and a nonzero constant term, with terms in descending powers of x.

(a) The number of positive real zeros of f either equals the number of variations in sign occurring in the coefficients of $f(x)$, or is less than the number of variations by a positive even integer.

(b) The number of negative real zeros of f either equals the number of variations in sign occurring in the coefficients of $f(-x)$, or is less than the number of variations by a positive even integer.

Classroom Example 7
Determine the different possibilities for the numbers of positive, negative, and nonreal complex zeros of

$$f(x) = -2x^4 + 3x^3 - 5x^2 + 4x - 1.$$

Answer: four, two, or zero positive; zero negative

Positive	Negative	Nonreal Complex
4	0	0
2	0	2
0	0	4

EXAMPLE 7 Applying Descartes' Rule of Signs

Determine the different possibilities for the numbers of positive, negative, and nonreal complex zeros of

$$f(x) = x^4 - 6x^3 + 8x^2 + 2x - 1.$$

SOLUTION We first consider the possible number of positive zeros by observing that $f(x)$ has three variations in signs.

$$f(x) = +x^4 - 6x^3 + 8x^2 + 2x - 1$$

1 2 3

Thus, by Descartes' rule of signs, $f(x)$ has either three or one (since $3 - 2 = 1$) positive real zeros.

For negative zeros, consider the variations in signs for $f(-x)$.

$$f(-x) = (-x)^4 - 6(-x)^3 + 8(-x)^2 + 2(-x) - 1$$
$$= x^4 + 6x^3 + 8x^2 - 2x - 1$$

1

Since there is only one variation in sign, $f(x)$ has exactly one negative real zero.

Because $f(x)$ is a fourth-degree polynomial function, it must have four complex zeros, some of which may be repeated. Descartes' rule of signs has indicated that exactly one of these zeros is a negative real number.

- One possible combination of the zeros is one negative real zero, three positive real zeros, and no nonreal complex zeros.
- Another possible combination of the zeros is one negative real zero, one positive real zero, and two nonreal complex zeros.

By the conjugate zeros theorem, any possible nonreal complex zeros must occur in conjugate pairs since $f(x)$ has real coefficients. The table below summarizes these possibilities.

Possible Number of Zeros		
Positive	**Negative**	**Nonreal Complex**
3	1	0
1	1	2

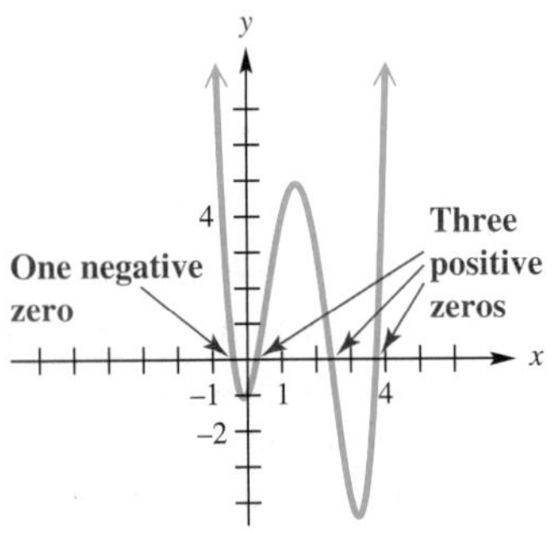

$f(x) = x^4 - 6x^3 + 8x^2 + 2x - 1$

Figure 19

The graph of $f(x)$ in **Figure 19** verifies the correct combination of three positive real zeros with one negative real zero, as seen in the first row of the table.*

Now Try Exercise 75.

NOTE ***Descartes' rule of signs does not identify the multiplicity of the zeros of a function.*** For example, if it indicates that a function $f(x)$ has exactly two positive real zeros, then $f(x)$ may have two distinct positive real zeros or one positive real zero of multiplicity 2.

* The authors would like to thank Mary Hill of College of DuPage for her input into **Example 7.**

3.3 Exercises

Concept Check *Decide whether each statement is* true *or* false. *If false, tell why.*

1. Since $x - 1$ is a factor of $f(x) = x^6 - x^4 + 2x^2 - 2$, we can conclude that $f(1) = 0$.
2. Since $f(1) = 0$ for $f(x) = x^6 - x^4 + 2x^2 - 2$, we can conclude that $x - 1$ is a factor of $f(x)$.
3. For $f(x) = (x + 2)^4(x - 3)$, 2 is a zero of multiplicity 4.
4. Since $2 + 3i$ is a zero of $f(x) = x^2 - 4x + 13$, we can conclude that $2 - 3i$ is also a zero.

Use the factor theorem and synthetic division to decide whether the second polynomial is a factor of the first. **See Example 1.**

5. $x^3 - 5x^2 + 3x + 1;\ x - 1$
6. $x^3 + 6x^2 - 2x - 7;\ x + 1$
7. $2x^4 + 5x^3 - 8x^2 + 3x + 13;\ x + 1$
8. $-3x^4 + x^3 - 5x^2 + 2x + 4;\ x - 1$
9. $-x^3 + 3x - 2;\ x + 2$
10. $-2x^3 + x^2 - 63;\ x + 3$
11. $4x^2 + 2x + 54;\ x - 4$
12. $5x^2 - 14x + 10;\ x + 2$
13. $x^3 + 2x^2 + 3;\ x - 1$
14. $2x^3 + x + 2;\ x + 1$
15. $2x^4 + 5x^3 - 2x^2 + 5x + 6;\ x + 3$
16. $5x^4 + 16x^3 - 15x^2 + 8x + 16;\ x + 4$

Factor $f(x)$ *into linear factors given that k is a zero of* $f(x)$. **See Example 2.**

17. $f(x) = 2x^3 - 3x^2 - 17x + 30;\ k = 2$
18. $f(x) = 2x^3 - 3x^2 - 5x + 6;\ k = 1$
19. $f(x) = 6x^3 + 13x^2 - 14x + 3;\ k = -3$
20. $f(x) = 6x^3 + 17x^2 - 63x + 10;\ k = -5$
21. $f(x) = 6x^3 + 25x^2 + 3x - 4;\ k = -4$
22. $f(x) = 8x^3 + 50x^2 + 47x - 15;\ k = -5$
23. $f(x) = x^3 + (7 - 3i)x^2 + (12 - 21i)x - 36i;\ k = 3i$
24. $f(x) = 2x^3 + (3 + 2i)x^2 + (1 + 3i)x + i;\ k = -i$
25. $f(x) = 2x^3 + (3 - 2i)x^2 + (-8 - 5i)x + (3 + 3i);\ k = 1 + i$
26. $f(x) = 6x^3 + (19 - 6i)x^2 + (16 - 7i)x + (4 - 2i);\ k = -2 + i$
27. $f(x) = x^4 + 2x^3 - 7x^2 - 20x - 12;\ k = -2$ (multiplicity 2)
28. $f(x) = 2x^4 + x^3 - 9x^2 - 13x - 5;\ k = -1$ (multiplicity 3)

For each polynomial function, one zero is given. Find all others. **See Examples 2 and 6.**

29. $f(x) = x^3 - x^2 - 4x - 6;\ 3$
30. $f(x) = x^3 + 4x^2 - 5;\ 1$
31. $f(x) = x^3 - 7x^2 + 17x - 15;\ 2 - i$
32. $f(x) = 4x^3 + 6x^2 - 2x - 1;\ \frac{1}{2}$
33. $f(x) = x^4 + 5x^2 + 4;\ -i$
34. $f(x) = x^4 + 26x^2 + 25;\ i$

For each polynomial function, **(a)** *list all possible rational zeros,* **(b)** *find all rational zeros, and* **(c)** *factor* $f(x)$. **See Example 3.**

35. $f(x) = x^3 - 2x^2 - 13x - 10$
36. $f(x) = x^3 + 5x^2 + 2x - 8$
37. $f(x) = x^3 + 6x^2 - x - 30$
38. $f(x) = x^3 - x^2 - 10x - 8$

Answers (margin)

1. true 2. true
3. false; -2 is a zero of multiplicity 4.
4. true
5. yes 6. yes
7. no 8. no
9. yes 10. yes
11. no 12. no
13. no 14. no
15. yes 16. yes
17. $f(x) = (x - 2)(2x - 5)\cdot(x + 3)$
18. $f(x) = (x - 1)(2x + 3)\cdot(x - 2)$
19. $f(x) = (x + 3)(3x - 1)\cdot(2x - 1)$
20. $f(x) = (x + 5)(6x - 1)\cdot(x - 2)$
21. $f(x) = (x + 4)(3x - 1)\cdot(2x + 1)$
22. $f(x) = (x + 5)(4x - 1)\cdot(2x + 3)$
23. $f(x) = (x - 3i)(x + 4)\cdot(x + 3)$
24. $f(x) = (x + i)(2x + 1)\cdot(x + 1)$
25. $f(x) = [x - (1 + i)]\cdot(2x - 1)(x + 3)$
26. $f(x) = [x - (-2 + i)]\cdot(3x + 2)(2x + 1)$
27. $f(x) = (x + 2)^2(x + 1)(x - 3)$
28. $f(x) = (x + 1)^3(2x - 5)$
29. $-1 \pm i$ 30. $\frac{-5 \pm \sqrt{5}}{2}$
31. $3, 2 + i$ 32. $\frac{-2 \pm \sqrt{2}}{2}$
33. $i, \pm 2i$ 34. $-i, \pm 5i$
35. (a) $\pm 1, \pm 2, \pm 5, \pm 10$ (b) $-1, -2, 5$ (c) $f(x) = (x + 1)(x + 2)\cdot(x - 5)$
36. (a) $\pm 1, \pm 2, \pm 4, \pm 8$ (b) $-4, -2, 1$ (c) $f(x) = (x + 4)(x + 2)\cdot(x - 1)$
37. (a) $\pm 1, \pm 2, \pm 3, \pm 5, \pm 6, \pm 10, \pm 15, \pm 30$ (b) $-5, -3, 2$ (c) $f(x) = (x + 5)(x + 3)(x - 2)$
38. (a) $\pm 1, \pm 2, \pm 4, \pm 8$ (b) $-2, -1, 4$ (c) $f(x) = (x + 2)(x + 1)\cdot(x - 4)$

39. (a) $\pm 1, \pm 2, \pm 3, \pm 4, \pm 6, \pm 12, \pm\frac{1}{2}, \pm\frac{3}{2}, \pm\frac{1}{3}, \pm\frac{2}{3}, \pm\frac{4}{3}, \pm\frac{1}{6}$
(b) $-4, -\frac{1}{3}, \frac{3}{2}$
(c) $f(x) = (x+4)(3x+1) \cdot (2x-3)$

40. (a) $\pm 1, \pm 2, \pm 4, \pm 8, \pm\frac{1}{3}, \pm\frac{2}{3}, \pm\frac{4}{3}, \pm\frac{8}{3}, \pm\frac{1}{5}, \pm\frac{2}{5}, \pm\frac{4}{5}, \pm\frac{8}{5}, \pm\frac{1}{15}, \pm\frac{2}{15}, \pm\frac{4}{15}, \pm\frac{8}{15}$
(b) $-4, -\frac{2}{5}, \frac{1}{3}$
(c) $f(x) = (x+4)(5x+2) \cdot (3x-1)$

41. (a) $\pm 1, \pm 2, \pm 3, \pm 4, \pm 6, \pm 12, \pm\frac{1}{2}, \pm\frac{3}{2}, \pm\frac{1}{3}, \pm\frac{2}{3}, \pm\frac{4}{3}, \pm\frac{1}{4}, +\frac{3}{4}, +\frac{1}{6}, +\frac{1}{8}, \pm\frac{3}{8}, \pm\frac{1}{12}, \pm\frac{1}{24}$
(b) $-\frac{3}{2}, -\frac{2}{3}, \frac{1}{2}$
(c) $f(x) = 2(2x+3) \cdot (3x+2)(2x-1)$

42. (a) $\pm 1, \pm 2, \pm 3, \pm 4, \pm 6, \pm 8, \pm 12, \pm 24, \pm\frac{1}{2}, \pm\frac{1}{3}, \pm\frac{1}{4}, \pm\frac{1}{6}, \pm\frac{1}{8}, \pm\frac{1}{12}, \pm\frac{1}{24}, \pm\frac{2}{3}, \pm\frac{3}{2}, \pm\frac{3}{4}, \pm\frac{3}{8}, \pm\frac{4}{3}, \pm\frac{8}{3}$
(b) $-\frac{3}{2}, -\frac{4}{3}, -\frac{1}{2}$
(c) $f(x) = 2(2x+3) \cdot (3x+4)(2x+1)$

43. 2 (multiplicity 3), $\pm\sqrt{7}$
44. -1 (multiplicity 2), 1 (multiplicity 3), $\pm\sqrt{10}$
45. $0, 2, -3, 1, -1$
46. 0 (multiplicity 2), $\pm 4, -5$
47. -2 (multiplicity 5), 1 (multiplicity 5), $1-\sqrt{3}$ (multiplicity 2)
48. 3 (multiplicity 3), $\frac{1}{2}$ (multiplicity 3), $2+\sqrt{5}$

49. $f(x) = -3x^3 + 6x^2 + 33x - 36$
50. $f(x) = \frac{1}{2}x^3 - \frac{1}{2}x$
51. $f(x) = -\frac{1}{2}x^3 - \frac{1}{2}x^2 + x$
52. $f(x) = -\frac{1}{2}x^3 + 2x^2 + \frac{11}{2}x - 15$
53. $f(x) = \frac{1}{6}x^3 + \frac{3}{2}x^2 + \frac{9}{2}x + \frac{9}{2}$
54. $f(x) = 2x^3 - 20x^2 + 64x - 64$
55. $f(x) = 5x^3 - 10x^2 + 5x$
56. $f(x) = -2x^3 - 8x^2$

39. $f(x) = 6x^3 + 17x^2 - 31x - 12$ **40.** $f(x) = 15x^3 + 61x^2 + 2x - 8$

41. $f(x) = 24x^3 + 40x^2 - 2x - 12$ **42.** $f(x) = 24x^3 + 80x^2 + 82x + 24$

For each polynomial function, find all zeros and their multiplicities.

43. $f(x) = (x-2)^3(x^2-7)$ **44.** $f(x) = (x+1)^2(x-1)^3(x^2-10)$

45. $f(x) = 3x(x-2)(x+3)(x^2-1)$ **46.** $f(x) = 5x^2(x^2-16)(x+5)$

47. $f(x) = (x^2+x-2)^5\left(x-1+\sqrt{3}\right)^2$

48. $f(x) = (2x^2-7x+3)^3\left(x-2-\sqrt{5}\right)$

Find a polynomial function $f(x)$ of degree 3 with real coefficients that satisfies the given conditions. ***See Example 4.***

49. Zeros of -3, 1, and 4; $f(2) = 30$ **50.** Zeros of 1, -1, and 0; $f(2) = 3$

51. Zeros of -2, 1, and 0; $f(-1) = -1$ **52.** Zeros of 2, -3, and 5; $f(3) = 6$

53. Zero of -3 having multiplicity 3; $f(3) = 36$

54. Zero of 2 and zero of 4 having multiplicity 2; $f(1) = -18$

55. Zero of 0 and zero of 1 having multiplicity 2; $f(2) = 10$

56. Zero of -4 and zero of 0 having multiplicity 2; $f(-1) = -6$

Find a polynomial function $f(x)$ of least degree having only real coefficients with zeros as given. ***See Examples 4–6.***

57. $5+i$ and $5-i$ **58.** $7-2i$ and $7+2i$

59. 0, i, and $1+i$ **60.** 0, $-i$, and $2+i$

61. $1+\sqrt{2}$, $1-\sqrt{2}$, and 1 **62.** $1-\sqrt{3}$, $1+\sqrt{3}$, and 1

63. $2-i$, 3, and -1 **64.** $3+2i$, -1, and 2

65. 2 and $3+i$ **66.** -1 and $4-2i$

67. $1-\sqrt{2}$, $1+\sqrt{2}$, and $1-i$ **68.** $2+\sqrt{3}$, $2-\sqrt{3}$, and $2+3i$

69. $2-i$ and $6-3i$ **70.** $5+i$ and $4-i$

71. 4, $1-2i$, and $3+4i$ **72.** -1, $5-i$, and $1+4i$

73. $1+2i$ and 2 (multiplicity 2) **74.** $2+i$ and -3 (multiplicity 2)

Use Descartes' rule of signs to determine the different possibilities for the numbers of positive, negative, and nonreal complex zeros for each function. ***See Example 7.***

75. $f(x) = 2x^3 - 4x^2 + 2x + 7$ **76.** $f(x) = x^3 + 2x^2 + x - 10$

77. $f(x) = 4x^3 - x^2 + 2x - 7$ **78.** $f(x) = 3x^3 + 6x^2 + x + 7$

79. $f(x) = 5x^4 + 3x^2 + 2x - 9$ **80.** $f(x) = 3x^4 + 2x^3 - 8x^2 - 10x - 1$

81. $f(x) = -8x^4 + 3x^3 - 6x^2 + 5x - 7$ **82.** $f(x) = 6x^4 + 2x^3 + 9x^2 + x + 5$

83. $f(x) = x^5 + 3x^4 - x^3 + 2x + 3$ **84.** $f(x) = 2x^5 - x^4 + x^3 - x^2 + x + 5$

85. $f(x) = 7x^5 + 6x^4 + 2x^3 + 9x^2 + x + 5$

86. $f(x) = -2x^5 + 10x^4 - 6x^3 + 8x^2 - x + 1$

87. $f(x) = 2x^5 - 7x^3 + 6x + 8$ **88.** $f(x) = 11x^5 - x^3 + 7x - 5$

89. $f(x) = 5x^6 - 6x^5 + 7x^3 - 4x^2 + x + 2$ **90.** $f(x) = 9x^6 - 7x^4 + 8x^2 + x + 6$

In Exercises 57–74, we give only one possible answer.

57. $f(x) = x^2 - 10x + 26$
58. $f(x) = x^2 - 14x + 53$
59. $f(x) = x^5 - 2x^4 + 3x^3 - 2x^2 + 2x$
60. $f(x) = x^5 - 4x^4 + 6x^3 - 4x^2 + 5x$
61. $f(x) = x^3 - 3x^2 + x + 1$
62. $f(x) = x^3 - 3x^2 + 2$
63. $f(x) = x^4 - 6x^3 + 10x^2 + 2x - 15$
64. $f(x) = x^4 - 7x^3 + 17x^2 - x - 26$
65. $f(x) = x^3 - 8x^2 + 22x - 20$
66. $f(x) = x^3 - 7x^2 + 12x + 20$
67. $f(x) = x^4 - 4x^3 + 5x^2 - 2x - 2$
68. $f(x) = x^4 - 8x^3 + 30x^2 - 56x + 13$
69. $f(x) = x^4 - 16x^3 + 98x^2 - 240x + 225$
70. $f(x) = x^4 - 18x^3 + 123x^2 - 378x + 442$
71. $f(x) = x^5 - 12x^4 + 74x^3 - 248x^2 + 445x - 500$
72. $f(x) = x^5 - 11x^4 + 51x^3 - 159x^2 + 220x + 442$
73. $f(x) = x^4 - 6x^3 + 17x^2 - 28x + 20$
74. $f(x) = x^4 + 2x^3 - 10x^2 - 6x + 45$

75.

Positive	Negative	Nonreal Complex
2	1	0
0	1	2

76.

Positive	Negative	Nonreal Complex
1	2	0
1	0	2

77.

Positive	Negative	Nonreal Complex
3	0	0
1	0	2

78.

Positive	Negative	Nonreal Complex
0	3	0
0	1	2

Answers for the remaining exercises are given in the answer section at the back of the text.

*Find all complex zeros of each polynomial function. Give exact values. List multiple zeros as necessary.**

91. $f(x) = x^4 + 2x^3 - 3x^2 + 24x - 180$

92. $f(x) = x^3 - x^2 - 8x + 12$

93. $f(x) = x^4 + x^3 - 9x^2 + 11x - 4$

94. $f(x) = x^3 - 14x + 8$

95. $f(x) = 2x^5 + 11x^4 + 16x^3 + 15x^2 + 36x$

96. $f(x) = 3x^3 - 9x^2 - 31x + 5$

97. $f(x) = x^5 - 6x^4 + 14x^3 - 20x^2 + 24x - 16$

98. $f(x) = 9x^4 + 30x^3 + 241x^2 + 720x + 600$

99. $f(x) = 2x^4 - x^3 + 7x^2 - 4x - 4$

100. $f(x) = 32x^4 - 188x^3 + 261x^2 + 54x - 27$

101. $f(x) = 5x^3 - 9x^2 + 28x + 6$

102. $f(x) = 4x^3 + 3x^2 + 8x + 6$

103. $f(x) = x^4 + 29x^2 + 100$

104. $f(x) = x^4 + 4x^3 + 6x^2 + 4x + 1$

105. $f(x) = x^4 + 2x^2 + 1$

106. $f(x) = x^4 - 8x^3 + 24x^2 - 32x + 16$

107. $f(x) = x^4 - 6x^3 + 7x^2$

108. $f(x) = 4x^4 - 65x^2 + 16$

109. $f(x) = x^4 - 8x^3 + 29x^2 - 66x + 72$

110. $f(x) = 12x^4 - 43x^3 + 50x^2 + 38x - 12$

111. $f(x) = x^6 - 9x^4 - 16x^2 + 144$

112. $f(x) = x^6 - x^5 - 26x^4 + 44x^3 + 91x^2 - 139x + 30$

If c and d are complex numbers, prove each statement. (Hint: Let $c = a + bi$ and $d = m + ni$ and form all the conjugates, the sums, and the products.)

113. $\overline{c + d} = \overline{c} + \overline{d}$

114. $\overline{c \cdot d} = \overline{c} \cdot \overline{d}$

115. $\overline{a} = a$ for any real number a

116. $\overline{c^2} = (\overline{c})^2$

In 1545, a method of solving a cubic equation of the form

$$x^3 + mx = n,$$

developed by Niccolo Tartaglia, was published in the Ars Magna, *a work by Girolamo Cardano. The formula for finding the one real solution of the equation is*

$$x = \sqrt[3]{\frac{n}{2} + \sqrt{\left(\frac{n}{2}\right)^2 + \left(\frac{m}{3}\right)^3}} - \sqrt[3]{\frac{-n}{2} + \sqrt{\left(\frac{n}{2}\right)^2 + \left(\frac{m}{3}\right)^3}}.$$

(Source: Gullberg, J., *Mathematics from the Birth of Numbers,* W.W. Norton & Company.) *Use the formula to solve each equation for the one real solution.*

117. $x^3 + 9x = 26$

118. $x^3 + 15x = 124$

* The authors would like to thank Aileen Solomon of Trident Technical College for preparing and suggesting the inclusion of **Exercises 91–104.**

3.4 Polynomial Functions: Graphs, Applications, and Models

- Graphs of $f(x) = ax^n$
- Graphs of General Polynomial Functions
- Behavior at Zeros
- Turning Points and End Behavior
- Graphing Techniques
- Intermediate Value and Boundedness Theorems
- Approximating Real Zeros
- Polynomial Models

Graphs of $f(x) = ax^n$

We can now graph polynomial functions of degree 3 or more with real number domains (since we will be graphing in the real number plane). We begin by inspecting the graphs of several functions of the form

$$f(x) = ax^n, \quad \text{with } a = 1,$$

and noticing their behavior. The identity function $f(x) = x$, the squaring function $f(x) = x^2$, and the cubing function $f(x) = x^3$ were first graphed in **Section 2.6** using a general point-plotting method of graphing.

Each function in **Figure 20** has odd degree and is an odd function exhibiting symmetry about the origin. Each has domain $(-\infty, \infty)$ and range $(-\infty, \infty)$ and is continuous on its entire domain $(-\infty, \infty)$. Additionally, these odd functions are increasing on their entire domain $(-\infty, \infty)$, appearing as though they fall to the left and rise to the right.

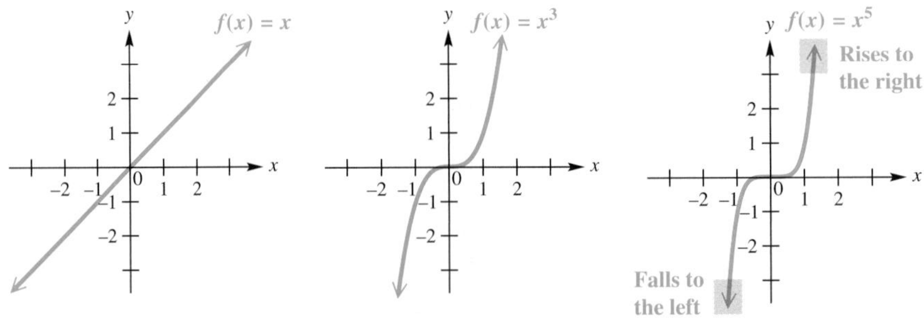

Figure 20

Each function in **Figure 21** has even degree and is an even function exhibiting symmetry about the y-axis. Each has domain $(-\infty, \infty)$ but restricted range $[0, \infty)$. These even functions are also continuous on their entire domain $(-\infty, \infty)$. However, they are decreasing on $(-\infty, 0]$ and increasing on $[0, \infty)$, appearing as though they rise both to the left and to the right.

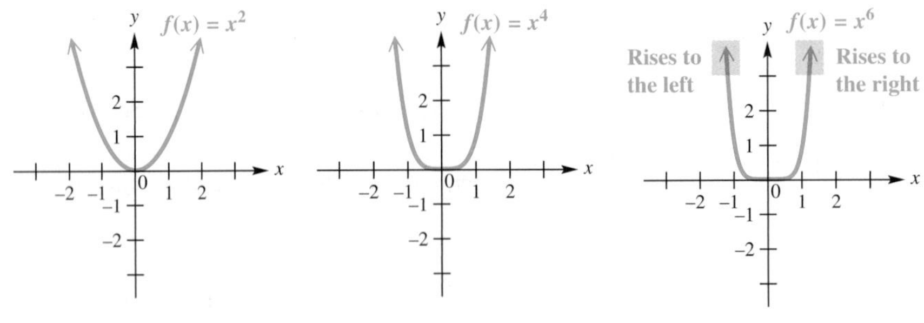

Figure 21

The behaviors that we see in the graphs of these basic polynomial functions also apply to more complicated polynomial functions.

Graphs of General Polynomial Functions

As with quadratic functions, the absolute value of a in $f(x) = ax^n$ determines the width of the graph.

- When $|a| > 1$, the graph is stretched vertically, making it narrower.
- When $0 < |a| < 1$, the graph is shrunk or compressed vertically, making it wider.

Compared to the graph of $f(x) = ax^n$, the following also hold true.

- The graph of $f(x) = -ax^n$ is reflected across the x-axis.
- The graph of $f(x) = ax^n + k$ is translated (shifted) k units up if $k > 0$ and $|k|$ units down if $k < 0$.
- The graph of $f(x) = a(x - h)^n$ is translated h units to the right if $h > 0$ and $|h|$ units to the left if $h < 0$.
- The graph of $f(x) = a(x - h)^n + k$ shows a combination of these translations.

Classroom Example 1
Graph each function. Determine the intervals of the domain for which each function is increasing or decreasing.

(a) $f(x) = x^3 + 1$
(b) $f(x) = (x - 2)^5$
(c) $f(x) = -\frac{1}{2}(x + 3)^4 + 5$

Answers:

(a)

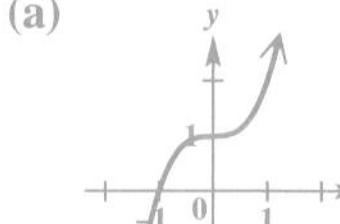

increasing on $(-\infty, \infty)$

(b)

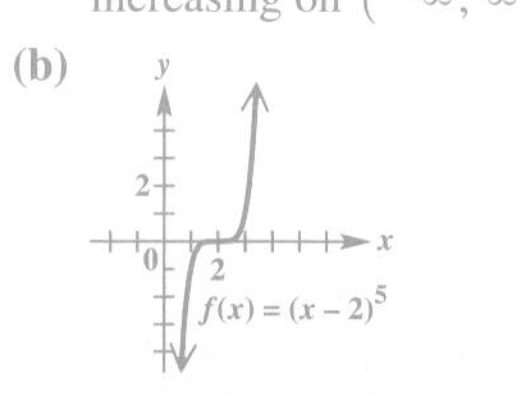

increasing on $(-\infty, \infty)$

(c)

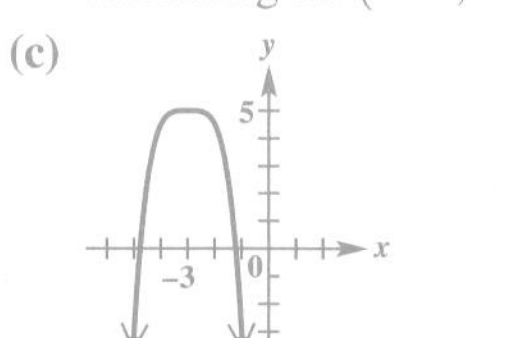

increasing on $(-\infty, -3]$;
decreasing on $[-3, \infty)$

EXAMPLE 1 Examining Vertical and Horizontal Translations

Graph each function. Determine the intervals of the domain for which each function is increasing or decreasing.

(a) $f(x) = x^5 - 2$ **(b)** $f(x) = (x + 1)^6$ **(c)** $f(x) = -2(x - 1)^3 + 3$

SOLUTION

(a) The graph of $f(x) = x^5 - 2$ will be the same as that of $f(x) = x^5$, but translated 2 units down. See **Figure 22.** This function is increasing on its entire domain $(-\infty, \infty)$.

(b) In $f(x) = (x + 1)^6$, function f has a graph like that of $f(x) = x^6$, but since $x + 1 = x - (-1)$, it is translated 1 unit to the left. See **Figure 23.** This function is decreasing on $(-\infty, -1]$ and increasing on $[-1, \infty)$.

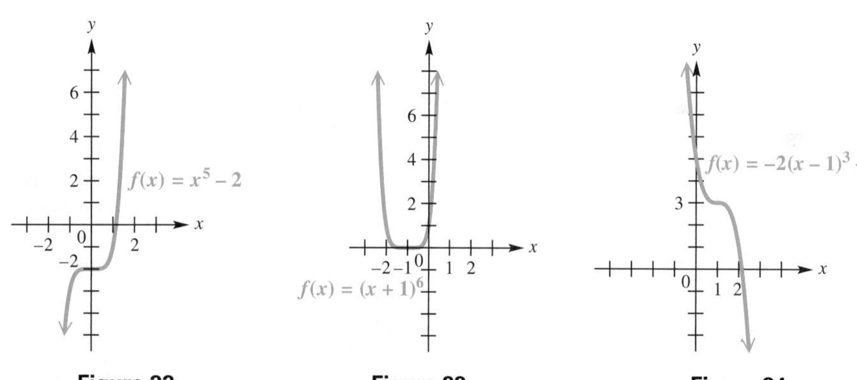

Figure 22 **Figure 23** **Figure 24**

(c) The negative sign in -2 causes the graph of

$$f(x) = -2(x - 1)^3 + 3$$

to be reflected across the x-axis when compared with the graph of $f(x) = x^3$. Because $|-2| > 1$, the graph is stretched vertically when compared to the graph of $f(x) = x^3$. As shown in **Figure 24,** the graph is also translated 1 unit to the right and 3 units up. This function is decreasing on its entire domain $(-\infty, \infty)$.

✔ *Now Try Exercises 13, 15, and 19.*

Unless otherwise restricted, the domain of a polynomial function is the set of all real numbers. Polynomial functions are smooth, continuous curves on the interval $(-\infty, \infty)$. The range of a polynomial function of odd degree is also the set of all real numbers.

Typical graphs of polynomial functions of odd degree are shown in **Figure 25.** These graphs suggest that for every polynomial function f of odd degree there is at least one real value of x that makes $f(x) = 0$. The real zeros are the x-intercepts of the graph and can be determined by inspecting the factored form of each polynomial.

Odd Degree

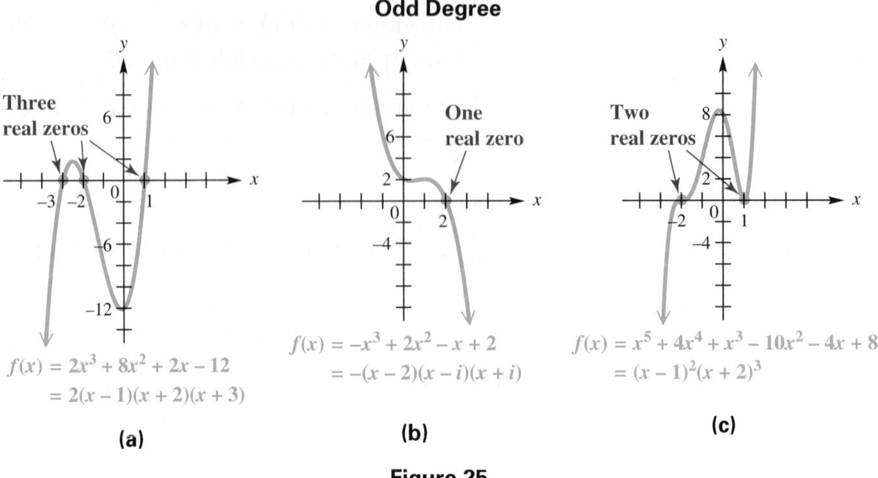

Figure 25

A polynomial function of even degree has a range of the form $(-\infty, k]$ or $[k, \infty)$, for some real number k. **Figure 26** shows two typical graphs.

Even Degree

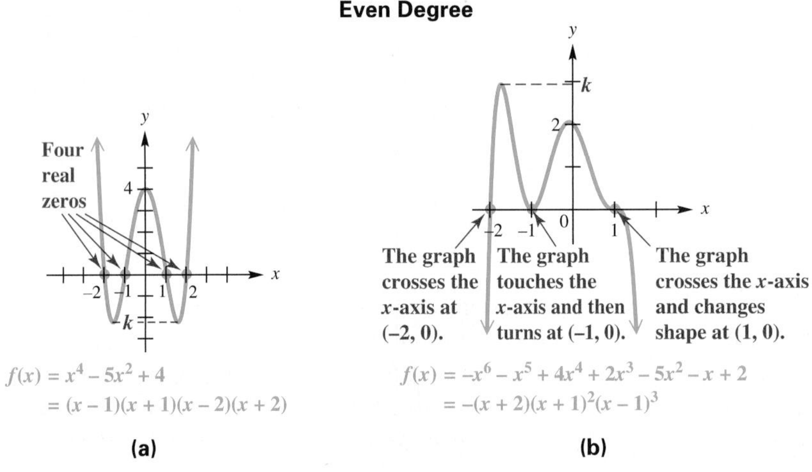

Figure 26

Behavior at Zeros Take a close look at the graph in **Figure 26(b).** Here we see a sixth-degree polynomial function with three distinct zeros, yet the behavior of the graph at each zero is different. This behavior depends on the multiplicity of the zero as determined by the exponent on the corresponding factor. The factored form of the polynomial in **Figure 26(b)** is $f(x) = -(x + 2)^1(x + 1)^2(x - 1)^3$.

- $(x + 2)$ is a factor of multiplicity 1. Therefore, the graph crosses the x-axis at $(-2, 0)$.
- $(x + 1)$ is a factor of multiplicity 2. Therefore, the graph is tangent to the x-axis at $(-1, 0)$. This means that it touches the x-axis, then turns and changes behavior from decreasing to increasing similar to that of the squaring function $f(x) = x^2$ at its zero.
- $(x - 1)$ is a factor of multiplicity 3. Therefore, the graph crosses the x-axis ***and*** is tangent to the x-axis at $(1, 0)$. This causes a change in concavity, or shape, at this x-intercept with behavior similar to that of the cubing function $f(x) = x^3$ at its zero.

See **Figure 27** for a generalization of the behavior of the graphs of polynomial functions at their zeros.

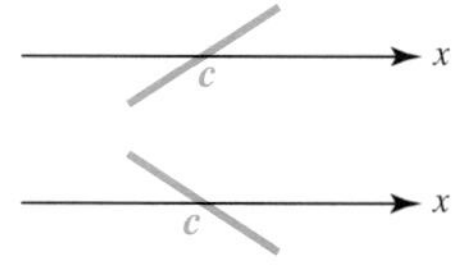

The graph crosses the x-axis at $(c, 0)$ if c is a zero of multiplicity 1.

(a)

The graph is tangent to the x-axis at $(c, 0)$ if c is a zero of even multiplicity. The graph bounces, or turns, at c.

(b)

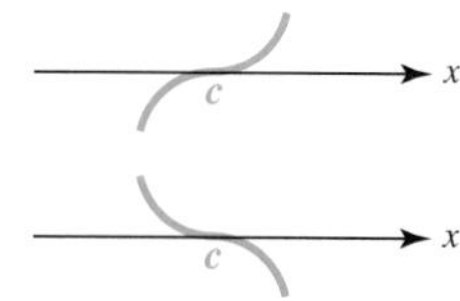

The graph crosses ***and*** is tangent to the x-axis at $(c, 0)$ if c is a zero of odd multiplicity greater than 1. The graph wiggles at c.

(c)

Figure 27

Turning Points and End Behavior The graphs in **Figures 25 and 26** show that polynomial functions often have **turning points** where the function changes from increasing to decreasing or from decreasing to increasing.

Turning Points

A polynomial function of degree n has at most $n - 1$ turning points, with at least one turning point between each pair of successive zeros.

The **end behavior** of a polynomial graph is determined by the **dominating term**—that is, the term of greatest degree. A polynomial of the form

$$f(x) = a_n x^n + a_{n-1}x^{n-1} + \cdots + a_0$$

has the same end behavior as $f(x) = a_n x^n$. For example,

$$f(x) = 2x^3 + 8x^2 + 2x - 12$$

has the same end behavior as $f(x) = 2x^3$. It is large and positive for large positive values of x and large and negative for negative values of x with large absolute value. That is, it rises to the right and falls to the left.

LOOKING AHEAD TO CALCULUS

To find the x-coordinates of the two turning points of the graph of

$$f(x) = 2x^3 + 8x^2 + 2x - 12$$

we can use the "maximum" and "minimum" capabilities of a graphing calculator and determine that, to the nearest thousandth, they are -0.131 and -2.535. In calculus, their exact values can be found by determining the zeros of the derivative function of $f(x)$,

$$f'(x) = 6x^2 + 16x + 2,$$

because the turning points occur precisely where the tangent line has 0 slope. Using the quadratic formula would show that the zeros are

$$\frac{-4 \pm \sqrt{13}}{3},$$

which agree with the calculator approximations.

Figure 25(a) shows that as x increases without bound, y does also. For the same graph, as x decreases without bound, y does also.

$$\text{As } x \to \infty, \quad y \to \infty \quad \text{and as} \quad x \to -\infty, \quad y \to -\infty.$$

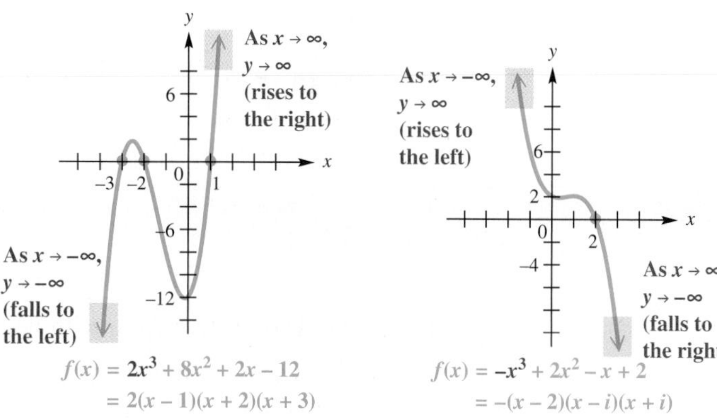

Figure 25(a) (repeated) **Figure 25(b) (repeated)**

The graph in **Figure 25(b)** has the same end behavior as $f(x) = -x^3$.

$$\text{As } x \to \infty, \quad y \to -\infty \quad \text{and as} \quad x \to -\infty, \quad y \to \infty.$$

The graph of a polynomial function with a dominating term of even degree will show end behavior in the same direction. See **Figure 26.**

Teaching Tip Students will benefit from seeing several examples of polynomial functions and an analysis of their corresponding end behaviors. Perhaps stand in front of the class and hold your arms up or down to simulate these end behaviors. (Students usually enjoy this sort of explanation.)

End Behavior of Graphs of Polynomial Functions

Suppose that ax^n is the dominating term of a polynomial function f of ***odd degree.***

1. If $a > 0$, then as $x \rightarrow \infty$, $f(x) \rightarrow \infty$, and as $x \rightarrow -\infty$, $f(x) \rightarrow -\infty$. Therefore, the end behavior of the graph is of the type shown in **Figure 28(a).** We symbolize it as ↙↗.

2. If $a < 0$, then as $x \rightarrow \infty$, $f(x) \rightarrow -\infty$, and as $x \rightarrow -\infty$, $f(x) \rightarrow \infty$. Therefore, the end behavior of the graph is of the type shown in **Figure 28(b).** We symbolize it as ↖↘.

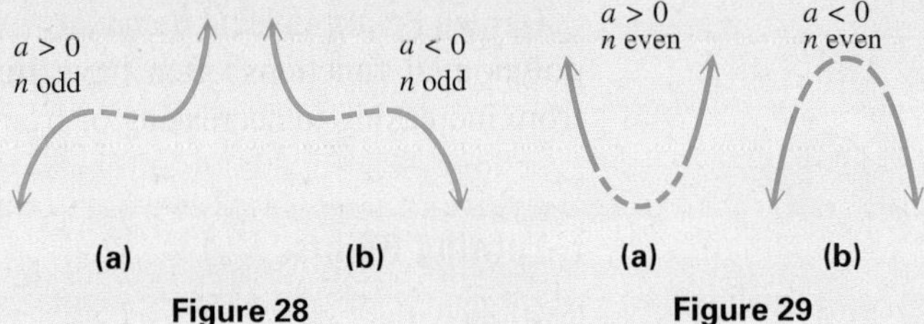

(a) (b) **Figure 28** (a) (b) **Figure 29**

Suppose that ax^n is the dominating term of a polynomial function f of ***even degree.***

1. If $a > 0$, then as $|x| \rightarrow \infty$, $f(x) \rightarrow \infty$. Therefore, the end behavior of the graph is of the type shown in **Figure 29(a).** We symbolize it as ↖↗.

2. If $a < 0$, then as $|x| \rightarrow \infty$, $f(x) \rightarrow -\infty$. Therefore, the end behavior of the graph is of the type shown in **Figure 29(b).** We symbolize it as ↙↘.

Classroom Example 2
Use the symbols for end behavior to describe the end behavior of the graph of each function.

(a) $f(x) = -x^4 + 2x^2 + x - 8$
(b) $g(x) = x^3 + 3x^2 - x + 5$
(c) $h(x) = -x^5 + x^4 - 3x^2 + 2$
(d) $k(x) = x^6 + x^3 + 1$

Answers:

(a) ↙↘ (b) ↙↗ 

(c) ↖↘ (d) ↖↗

EXAMPLE 2 **Determining End Behavior**

The graphs of the functions defined as follows are shown in A–D.

$$f(x) = x^4 - x^2 + 5x - 4, \qquad g(x) = -x^6 + x^2 - 3x - 4,$$

$$h(x) = 3x^3 - x^2 + 2x - 4, \quad \text{and} \quad k(x) = -x^7 + x - 4$$

Based on the previous discussion of end behavior, match each function with its graph.

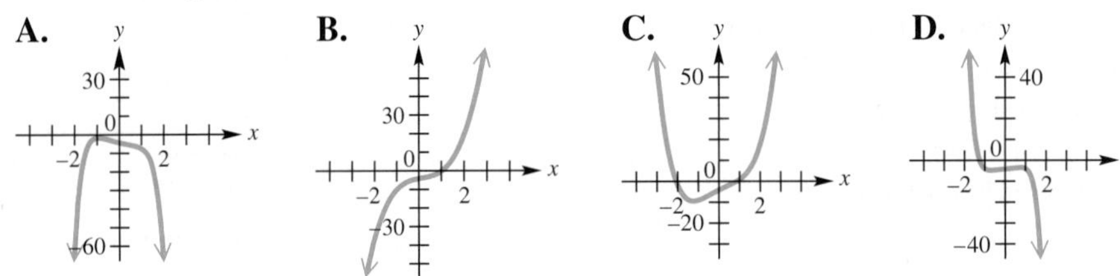

SOLUTION

Function f is of even degree with positive leading coefficient. Its graph is in C.

Function g is of even degree with negative leading coefficient. Its graph is in A.

Function h has odd degree and a dominating term with positive coefficient. Its graph is in B.

Function k has odd degree and a dominating term with negative coefficient. Its graph is in D.

✔ ***Now Try Exercises 21, 23, 25, and 27.***

Graphing Techniques We have discussed several characteristics of the graphs of polynomial functions that are useful for graphing the function by hand. A **comprehensive graph** of a polynomial function $f(x)$ will show the following characteristics.

- all x-intercepts (zeros) and the behavior of the graph at these zeros
- the y-intercept
- the sign of $f(x)$ within the intervals formed by the x-intercepts
- enough of the domain to show the end behavior

In **Examples 3 and 4,** we sketch the graphs of two polynomial functions by hand. While there are several ways to approach this process, we use the following general guidelines.

Graphing a Polynomial Function

Let $f(x) = a_n x^n + a_{n-1}x^{n-1} + \cdots + a_1 x + a_0$, with $a_n \neq 0$, be a polynomial function of degree n. To sketch its graph, follow these steps.

Step 1 Find the real zeros of f. Plot them as x-intercepts.

Step 2 Find $f(0) = a_0$. Plot this as the y-intercept.

Step 3 Use end behavior, whether the graph crosses, bounces on, or wiggles through the x-axis at the x-intercepts, and selected points as necessary to complete the graph.

Classroom Example 3
Graph $f(x) = 2x^3 + 3x^2 - 11x - 6$.

Answer:

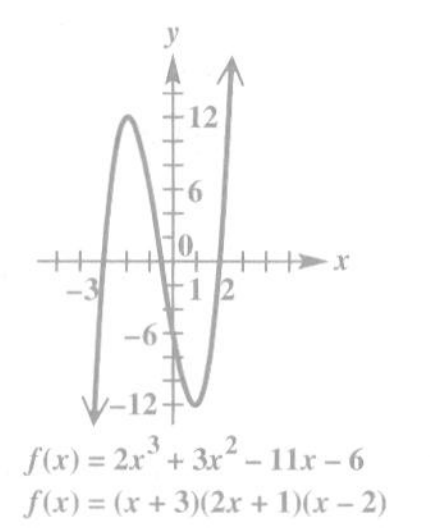

EXAMPLE 3 Graphing a Polynomial Function

Graph $f(x) = 2x^3 + 5x^2 - x - 6$.

SOLUTION

Step 1 The possible rational zeros are ± 1, ± 2, ± 3, ± 6, $\pm \frac{1}{2}$, and $\pm \frac{3}{2}$. Use synthetic division to show that 1 is a zero.

$$\begin{array}{r|rrrr} 1 & 2 & 5 & -1 & -6 \\ & & 2 & 7 & 6 \\ \hline & 2 & 7 & 6 & 0 \end{array} \leftarrow f(1) = 0$$

(Section 3.2)

Thus,

$$f(x) = (x - 1)(2x^2 + 7x + 6)$$

$$f(x) = (x - 1)(2x + 3)(x + 2). \quad \text{Factor. (Section R.4)}$$

Set each linear factor equal to 0, and then solve for x to find zeros. The three zeros of f are 1, $-\frac{3}{2}$, and -2. See **Figure 30.**

Step 2 $f(0) = -6$, so plot $(0, -6)$. See **Figure 30.**

Step 3 Since the dominating term of $f(x)$ is $2x^3$, the graph will have end behavior similar to that of $f(x) = x^3$. It will rise to the right and fall to the left as ↙↗. See **Figure 30.** Each zero of $f(x)$ occurs with multiplicity 1, meaning that the graph of $f(x)$ will cross the x-axis at each of its zeros. Because the graph of a polynomial function has no breaks, gaps, or sudden jumps, we now have sufficient information to sketch the graph of $f(x)$.

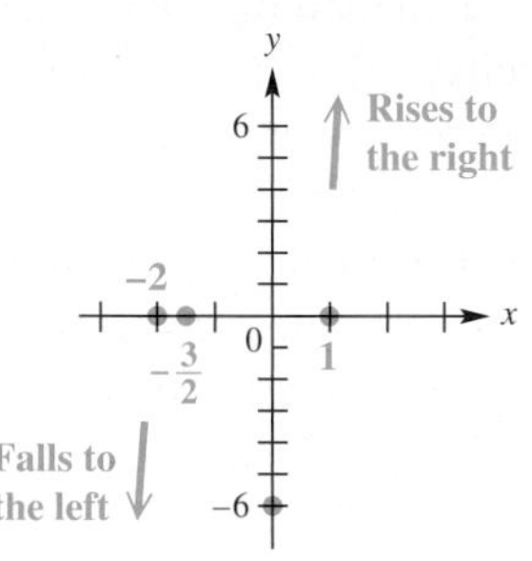

Figure 30

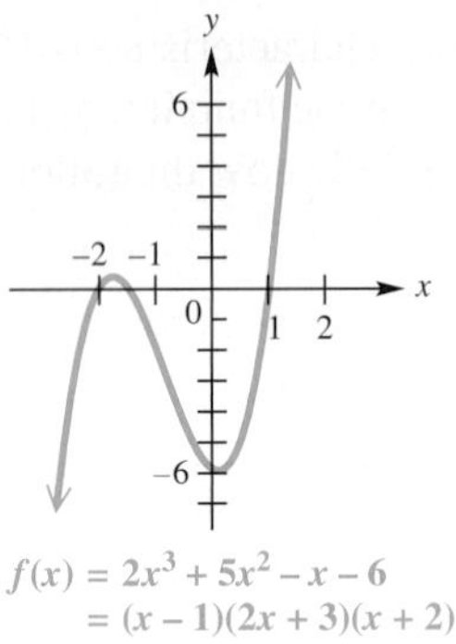

$f(x) = 2x^3 + 5x^2 - x - 6$
$= (x - 1)(2x + 3)(x + 2)$

Figure 31

Begin sketching at either end of the graph with the appropriate end behavior, and draw a smooth curve that crosses the x-axis at each zero, has a turning point between successive zeros, and passes through the y-intercept as shown in **Figure 31.**

Additional points may be used to verify whether the graph is above or below the x-axis between the zeros and to add detail to the sketch of the graph. The zeros divide the x-axis into four intervals:

$$(-\infty, -2), \quad \left(-2, -\frac{3}{2}\right), \quad \left(-\frac{3}{2}, 1\right) \quad \text{and} \quad (1, \infty).$$

Select an x-value as a test point in each interval, and substitute it into the equation for $f(x)$ to determine additional points on the graph. A typical selection of test points and the results of the tests are shown in the table.

Interval	Test Point	Value of $f(x)$	Sign of $f(x)$	Graph Above or Below x-Axis
$(-\infty, -2)$	-3	-12	Negative	Below
$\left(-2, -\frac{3}{2}\right)$	$-\frac{7}{4}$	$\frac{11}{32}$	Positive	Above
$\left(-\frac{3}{2}, 1\right)$	0	-6	Negative	Below
$(1, \infty)$	2	28	Positive	Above

The sketch could be improved by plotting the points found in each interval in the table.

✔ *Now Try Exercise 29.*

Classroom Example 4
Graph $f(x) = (x + 2)^3(x - 1)^2$.

Answer:

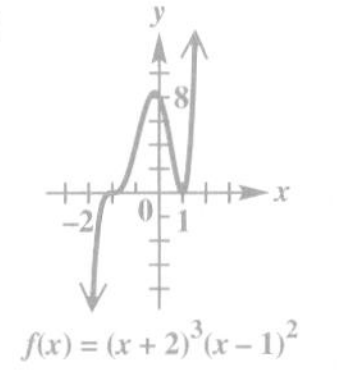

EXAMPLE 4 Graphing a Polynomial Function

Graph $f(x) = -(x - 1)(x - 3)(x + 2)^2$.

SOLUTION

Step 1 Since the polynomial is given in factored form, the zeros can be determined by inspection. They are 1, 3, and -2. Plot these as x-intercepts of the graph of $f(x)$. See **Figure 32.**

Step 2

$$f(0) = -(0 - 1)(0 - 3)(0 + 2)^2 \qquad \text{Find } f(0).$$
$$= -(-1)(-3)(2)^2, \text{ or } -12 \leftarrow y\text{-intercept}$$

Plot $(0, -12)$ on the y-axis. See **Figure 32.**

Step 3 The dominating term of $f(x)$ can be found by multiplying the factors and identifying the term of greatest degree. Here it is $-(x)(x)(x)^2 = -x^4$, indicating that the end behavior of the graph is ↙ ↘. Because 1 and 3 are zeros of multiplicity 1, the graph will cross the x-axis at these zeros. The graph of $f(x)$ will touch the x-axis at -2 and then turn and change direction because it is a zero of even multiplicity.

Begin at either end of the graph with the appropriate end behavior and draw a smooth curve that crosses the x-axis at 1 and 3 and that touches the x-axis at -2, then turns and changes direction. The graph will also pass through the y-intercept -12. See **Figure 32.** Using test points within intervals formed by the x-intercepts is a good way to add detail to the graph and verify the accuracy of the sketch. A typical selection of test points is $(-3, -24)$, $(-1, -8)$, $(2, 16)$, and $(4, -108)$.

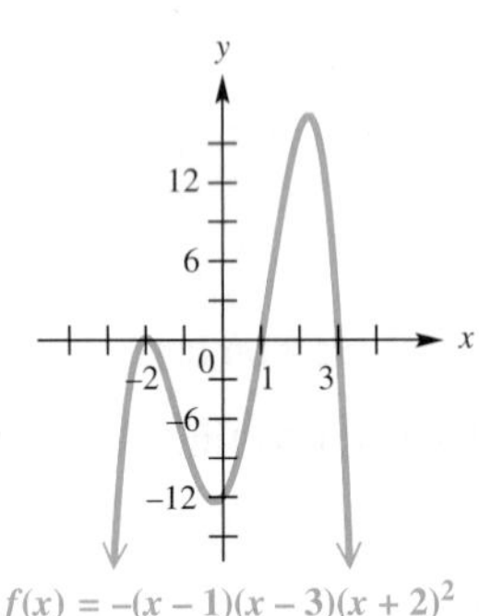

$f(x) = -(x - 1)(x - 3)(x + 2)^2$

Figure 32

✔ *Now Try Exercise 33.*

NOTE It is possible to reverse the process of **Example 4** and write the polynomial function from its graph if the zeros and any other point on the graph are known. Suppose that you are asked to find a polynomial function of least possible degree having the graph shown in **Figure 32.** Because the graph crosses the x-axis at 1 and 3 and bounces at -2, we know that the factored form of the function is as follows.

Multiplicity one — Multiplicity two

$$f(x) = a(x-1)^1(x-3)^1(x+2)^2$$

Now find the value of a by substituting the x- and y-values of any other point on the graph, say $(0, -12)$, into this function and solving for a.

$$f(x) = a(x-1)(x-3)(x+2)^2$$

$$-12 = a(0-1)(0-3)(0+2)^2 \quad \text{Let } x = 0 \text{ and } y = -12.$$

$$-12 = a(12) \quad \text{Simplify.}$$

$$a = -1 \quad \text{Divide by 12. Interchange sides.}$$

Verify in **Example 4** that the polynomial function is

$$f(x) = -(x-1)(x-3)(x+2)^2.$$

In the Exercise set, problems of this type are labeled ***Connecting Graphs with Equations***.

We emphasize the important relationships among the following concepts.

- the x-intercepts of the graph of $y = f(x)$
- the zeros of the function f
- the solutions of the equation $f(x) = 0$
- the factors of $f(x)$

For example, the graph of the function in **Example 3,**

$$f(x) = 2x^3 + 5x^2 - x - 6$$

$$f(x) = (x-1)(2x+3)(x+2), \quad \text{Factored form}$$

has x-intercepts 1, $-\frac{3}{2}$, and -2, as shown in **Figure 31** on the previous page. Since 1, $-\frac{3}{2}$, and -2 are the x-values where the function is 0, they are the zeros of f. Also, 1, $-\frac{3}{2}$, and -2 are the solutions of the polynomial equation

$$2x^3 + 5x^2 - x - 6 = 0.$$

This discussion is summarized as follows.

x-Intercepts, Zeros, Solutions, and Factors

If f is a polynomial function and a is an x-intercept of the graph of $y = f(x)$, then

$$a \text{ is a zero of } f, \quad a \text{ is a solution of } f(x) = 0,$$

and $\quad x - a$ is a factor of $f(x)$.

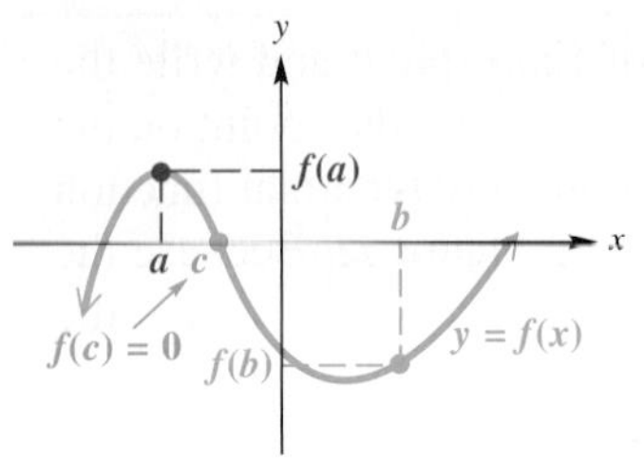

Figure 33

Intermediate Value and Boundedness Theorems As **Examples 3 and 4** show, one key to graphing a polynomial function is locating its zeros. In the special case where the potential zeros are rational numbers, the zeros are found by the rational zeros theorem (**Section 3.3**).

Occasionally, irrational zeros can be found by inspection. For instance, $f(x) = x^3 - 2$ has the irrational zero $\sqrt[3]{2}$. The next two theorems presented in this section apply to the zeros of every polynomial function with real coefficients. The first theorem uses the fact that graphs of polynomial functions are continuous curves. The proof requires advanced methods, so it is not given here. **Figure 33** illustrates the theorem.

Classroom Example 5
Use synthetic division and a graph to show that

$$f(x) = -x^3 + 2x^2 - 4x + 5$$

has a real zero between 1 and 2.

Answer:
$f(1) = 2 > 0; \quad f(2) = -3 < 0$

$f(x) = -x^3 + 2x^2 - 4x + 5$

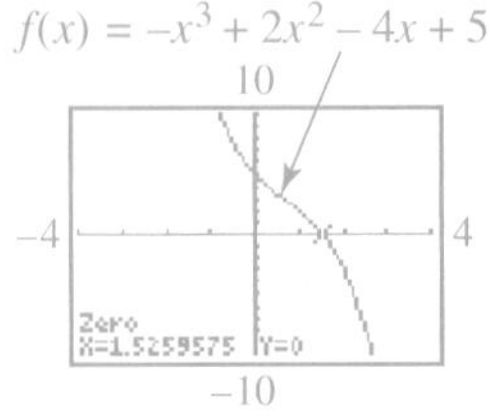

Intermediate Value Theorem for Polynomials

If $f(x)$ is a polynomial function with only real coefficients, and if for real numbers a and b, the values $f(a)$ and $f(b)$ are opposite in sign, then there exists at least one real zero between a and b.

This theorem helps identify intervals where zeros of polynomial functions are located. If $f(a)$ and $f(b)$ are opposite in sign, then 0 is between $f(a)$ and $f(b)$, and so there must be a number c between a and b where $f(c) = 0$.

EXAMPLE 5 Locating a Zero

Use synthetic division and a graph to show that $f(x) = x^3 - 2x^2 - x + 1$ has a real zero between 2 and 3.

ALGEBRAIC SOLUTION

Use synthetic division to find $f(2)$ and $f(3)$.

$$\begin{array}{r|rrrr} 2) & 1 & -2 & -1 & 1 \\ & & 2 & 0 & -2 \\ \hline & 1 & 0 & -1 & -1 = f(2) \end{array}$$

$$\begin{array}{r|rrrr} 3) & 1 & -2 & -1 & 1 \\ & & 3 & 3 & 6 \\ \hline & 1 & 1 & 2 & 7 = f(3) \end{array}$$

Since $f(2)$ is negative and $f(3)$ is positive, by the intermediate value theorem there must be a real zero between 2 and 3.

GRAPHING CALCULATOR SOLUTION

The graphing calculator screen in **Figure 34** indicates that this zero is approximately 2.2469796. (Notice that there are two other zeros as well.)

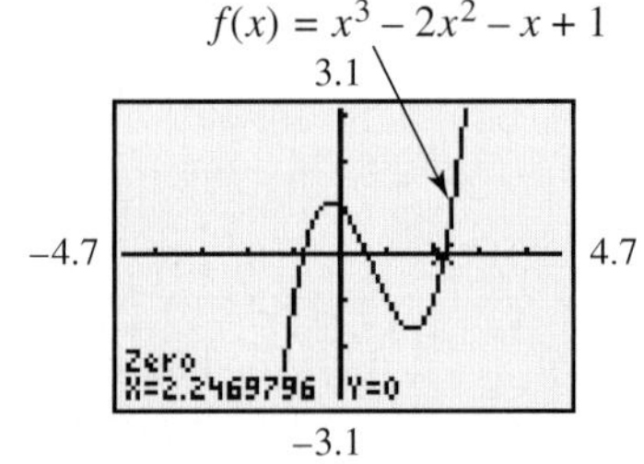

Figure 34

Now Try Exercise 49.

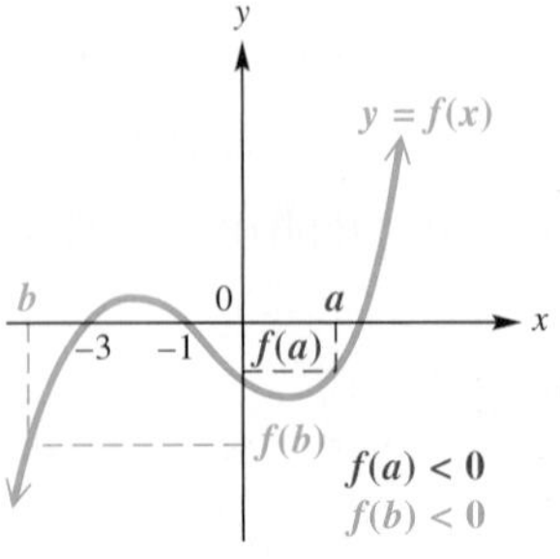

Figure 35

CAUTION ***Be careful how you interpret the intermediate value theorem.*** If $f(a)$ and $f(b)$ are ***not*** opposite in sign, it does not necessarily mean that there is no zero between a and b. In **Figure 35,** $f(a)$ and $f(b)$ are both negative, but -3 and -1, which are between a and b, are zeros of $f(x)$.

The intermediate value theorem for polynomials helps limit the search for real zeros to smaller and smaller intervals. In **Example 5,** we used the theorem to verify that there is a real zero between 2 and 3. We can use the theorem repeatedly to locate the zero more accurately.

Teaching Tip A common error is misinterpretation of the intermediate value theorem. Show students the example that for

$$f(x) = x^2 - 2x - 1,$$

$f(-1) = 2 > 0$ and $f(3) = 2 > 0$, and f has two zeros between -1 and 3.

The **boundedness theorem** shows how the bottom row of a synthetic division is used to place upper and lower bounds on possible real zeros of a polynomial function.

Boundedness Theorem

Let $f(x)$ be a polynomial function of degree $n \geq 1$ with real coefficients and with a positive leading coefficient. Suppose $f(x)$ is divided synthetically by $x - c$.

(a) If $c > 0$ and all numbers in the bottom row of the synthetic division are nonnegative, then $f(x)$ has no zero greater than c.

(b) If $c < 0$ and the numbers in the bottom row of the synthetic division alternate in sign (with 0 considered positive or negative, as needed), then $f(x)$ has no zero less than c.

Proof We outline the proof of part (a). The proof for part (b) is similar. By the division algorithm, if $f(x)$ is divided by $x - c$, then for some $q(x)$ and r,

$$f(x) = (x - c)q(x) + r,$$

where all coefficients of $q(x)$ are nonnegative, $r \geq 0$, and $c > 0$. If $x > c$, then $x - c > 0$. Since $q(x) > 0$ and $r \geq 0$,

$$f(x) = (x - c)q(x) + r > 0.$$

This means that $f(x)$ will never be 0 for $x > c$.

Classroom Example 6
Show that the real zeros of

$$f(x) = x^4 + 5x^2 + 3x - 7$$

satisfy the following conditions.

(a) No real zero is greater than 1.

(b) No real zero is less than -2.

Answers:

(a)

$$\begin{array}{r|rrrrr} 1 & 1 & 0 & 5 & 3 & -7 \\ & & 1 & 1 & 6 & 9 \\ \hline & 1 & 1 & 6 & 9 & 2 \end{array}$$

All numbers in the last row of the synthetic division are nonnegative.

(b)

$$\begin{array}{r|rrrrr} -2 & 1 & 0 & 5 & 3 & -7 \\ & & -2 & 4 & -18 & 30 \\ \hline & 1 & -2 & 9 & -15 & 23 \end{array}$$

The numbers in the last row of the synthetic division alternate in sign.

EXAMPLE 6 Using the Boundedness Theorem

Show that the real zeros of $f(x) = 2x^4 - 5x^3 + 3x + 1$ satisfy these conditions.

(a) No real zero is greater than 3. **(b)** No real zero is less than -1.

SOLUTION

(a) Since $f(x)$ has real coefficients and the leading coefficient, 2, is positive, use the boundedness theorem. Divide $f(x)$ synthetically by $x - 3$.

$$\begin{array}{r|rrrrr} 3 & 2 & -5 & 0 & 3 & 1 \\ & & 6 & 3 & 9 & 36 \\ \hline & 2 & 1 & 3 & 12 & 37 \end{array} \leftarrow \text{All are nonnegative.}$$

Since $3 > 0$ and all numbers in the last row of the synthetic division are nonnegative, $f(x)$ has no real zero greater than 3.

(b) We use the boundedness theorem again and divide $f(x)$ synthetically by $x - (-1)$, or $x + 1$.

$$\begin{array}{r|rrrrr} -1 & 2 & -5 & 0 & 3 & 1 \\ & & -2 & 7 & -7 & 4 \\ \hline & 2 & -7 & 7 & -4 & 5 \end{array} \leftarrow \text{These numbers alternate in sign.}$$

Here $-1 < 0$ and the numbers in the last row alternate in sign, so $f(x)$ has no real zero less than -1.

Now Try Exercises 57 and 59.

Approximating Real Zeros We can approximate the irrational real zeros of a polynomial function using a graphing calculator.

Classroom Example 7
Approximate the real zeros of

$$f(x) = -x^3 - 8x^2 + 4x + 10.$$

Answer: $-8.33594, -0.9401088, 1.2760488$

EXAMPLE 7 Approximating Real Zeros of a Polynomial Function

Approximate the real zeros of $f(x) = x^4 - 6x^3 + 8x^2 + 2x - 1$.

SOLUTION The dominating term is x^4, so the graph will have end behavior similar to the graph of $f(x) = x^4$, which is positive for all values of x with large absolute values. That is, the end behavior is up at the left and the right, ↖ ↗. There are at most four real zeros, since the polynomial is fourth-degree.

Since $f(0) = -1$, the y-intercept is -1. Because the end behavior is positive on the left and the right, by the intermediate value theorem f has at least one real zero on either side of $x = 0$. To approximate the zeros, we use a graphing calculator. The graph in **Figure 36** shows that there are four real zeros, and the table indicates that they are between

$$-1 \text{ and } 0, \quad 0 \text{ and } 1, \quad 2 \text{ and } 3, \quad \text{and} \quad 3 \text{ and } 4$$

because there is a sign change in $f(x) = \text{Y}_1$ in each case.

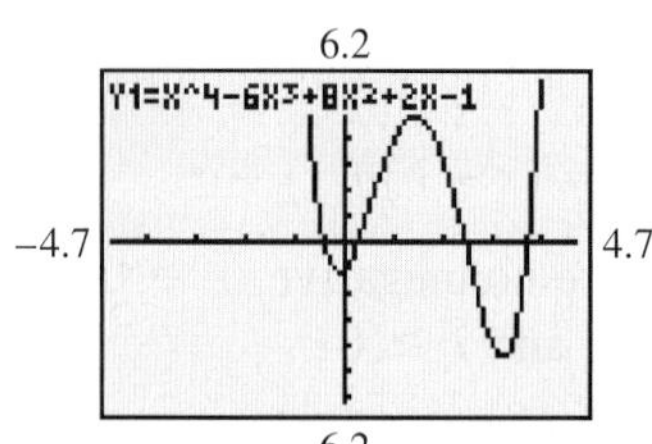

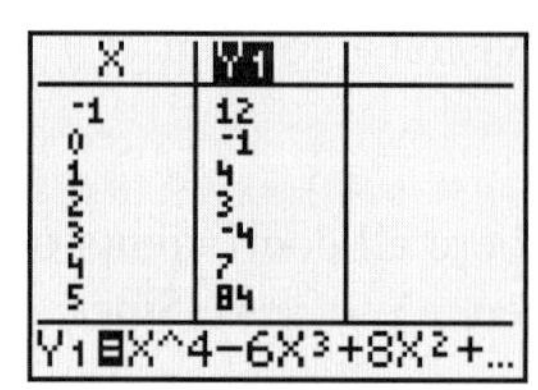

Figure 36

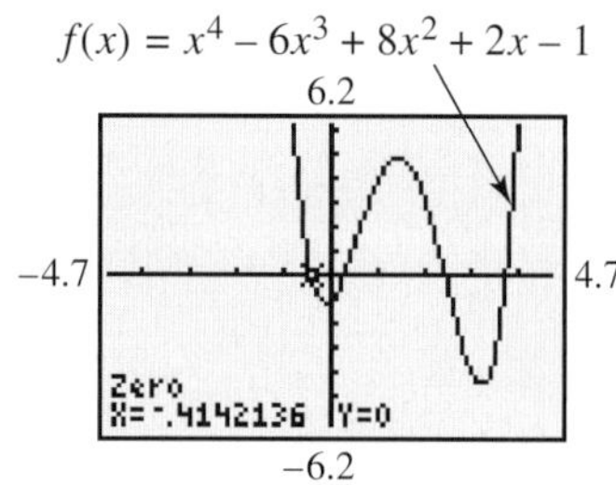

Figure 37

Using the capability of the calculator, we can find the zeros to a great degree of accuracy. **Figure 37** shows that the negative zero is approximately -0.4142136. Similarly, we find that the other three zeros are approximately

$$0.26794919, \quad 2.4142136, \quad \text{and} \quad 3.7320508.$$

✔ *Now Try Exercise 77.*

Polynomial Models

EXAMPLE 8 Examining a Polynomial Model

The table shows the number of transactions, in millions, by users of bank debit cards for selected years.

(a) Using $x = 0$ to represent 1995, $x = 3$ to represent 1998, and so on, use the regression feature of a calculator to determine the quadratic function that best fits the data. Plot the data and the graph.

(b) Repeat part (a) for a cubic function (degree 3).

(c) Repeat part (a) for a quartic function (degree 4).

(d) The **correlation coefficient,** R, is a measure of the strength of the relationship between two variables. The values of R and R^2 are used to determine how well a regression model fits a set of data. The closer the value of R^2 is to 1, the better the fit. Compare R^2 for the three functions found in parts (a)–(c) to decide which function best fits the data.

Year	Transactions (in millions)
1995	829
1998	3765
2000	5290
2004	14,106
2008	28,464
2011*	39,049

Source: Statistical Abstract of the United States.
*Projected

Classroom Example 8
There were 127 million bank debit card transactions in 1990 and 204 million in 1992. Repeat **Example 8** after adding these data to the table, using $x = 0$ to represent 1990 rather than 1995.

Answer:
(a) $y = 117.9x^2 - 630.3x + 622.8$
(b) $y = 0.7473x^3 + 94.64x^2 - 453.5x + 428.8$
(c) $y = -0.3890x^4 + 17.13x^3 - 120.1x^2 + 438.2x - 42.08$
(d) quadratic: $R^2 = 0.9977956177$
cubic: $R^2 = 0.9979380452$
quartic: $R^2 = 0.9990930205$;
The quartic function provides the best fit.

SOLUTION

(a) The best-fitting quadratic function for the data is

$$y = 127.7x^2 + 377.3x + 868.7.$$

The regression coordinates screen and the graph are shown in **Figure 38.**

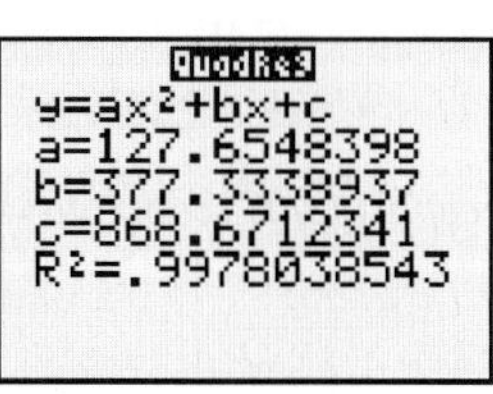

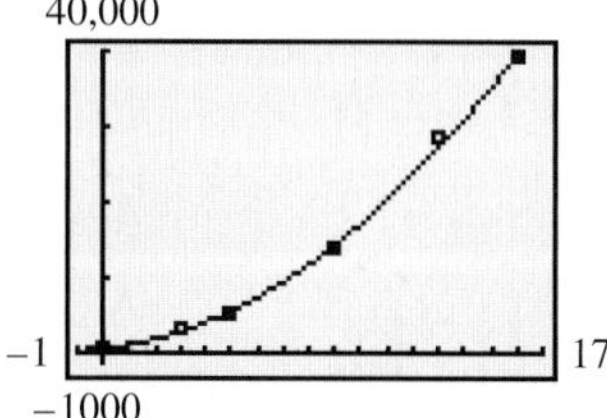

Figure 38

(b) The best-fitting cubic function is shown in **Figure 39** and is

$$y = -2.412x^3 + 184.7x^2 + 50.58x + 1125.$$

Teaching Tip Point out to students that R^2 is available on the TI-83/84 Plus from the Diagnostics On option.

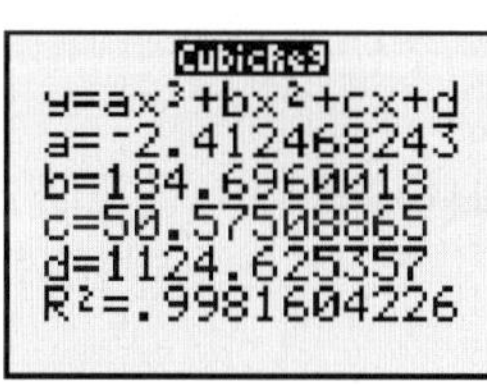

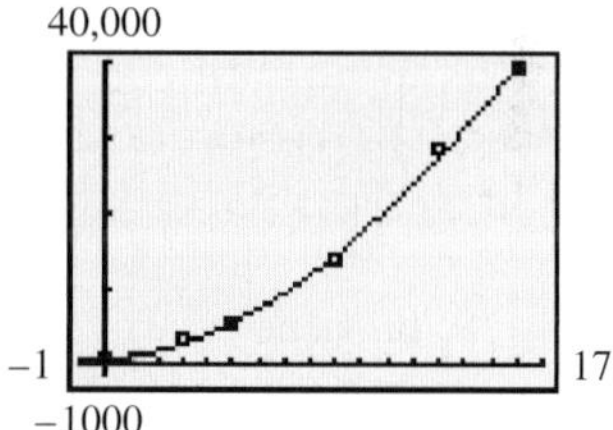

Figure 39

(c) The best-fitting quartic function is shown in **Figure 40** and is

$$y = -1.496x^4 + 45.07x^3 - 274.5x^2 + 1367x + 859.0.$$

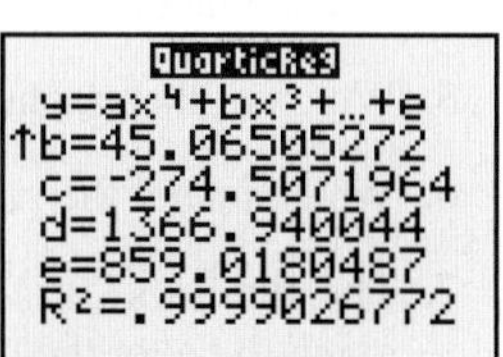

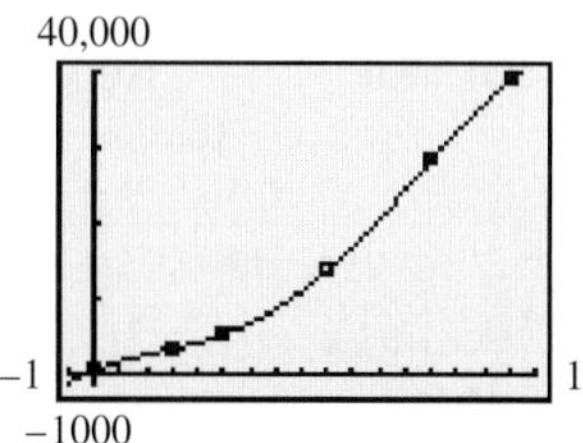

Figure 40

(d) With the statistical diagnostics turned on, the value of R^2 is displayed with the regression results on the TI 83/84 Plus each time that a regression model is executed. By inspecting the R^2 value for each model above, we see that the quartic function provides the best fit since it has the largest R^2 value of 0.9999026772.

✔ *Now Try Exercise 105.*

NOTE In **Example 8(d),** we selected the quartic function as the best model based on the comparison of R^2 values of the models. In practice, however, the best choice of a model should also depend on the set of data being analyzed as well as analysis of its trends and attributes.

3.4 Exercises

1. A 2. B
3. one 4. C
5. B and D 6. B
7. $f(x) = x(x+5)^2(x-3)$
8. $f(x) = -(x+6) \cdot (x+1)^2(x-3)(x-5)$

9.

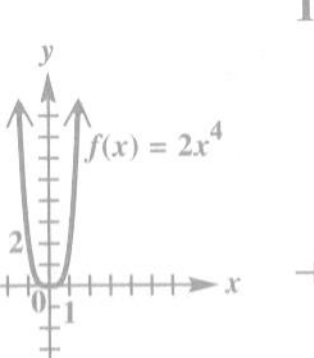

(a) $[0, \infty)$
(b) $(-\infty, 0]$

10.

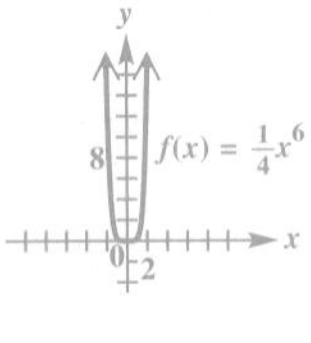

$f(x) = \frac{1}{4}x^6$

(a) $[0, \infty)$
(b) $(-\infty, 0]$

11.

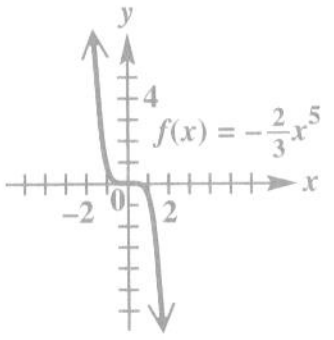

$f(x) = -\frac{2}{3}x^5$

(a) none
(b) $(-\infty, \infty)$

12.

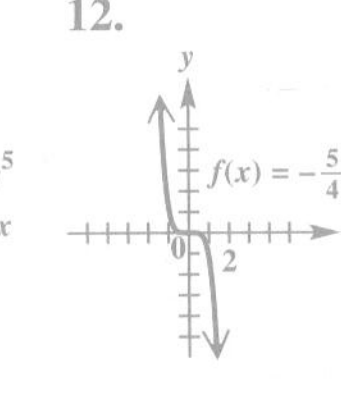

(a) none
(b) $(-\infty, \infty)$

13.

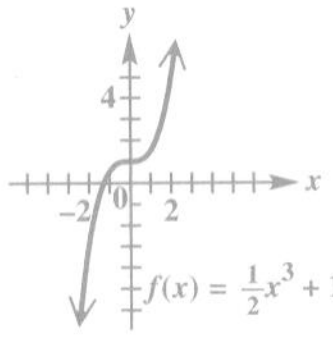

$f(x) = \frac{1}{2}x^3 + 1$

(a) $(-\infty, \infty)$
(b) none

14.

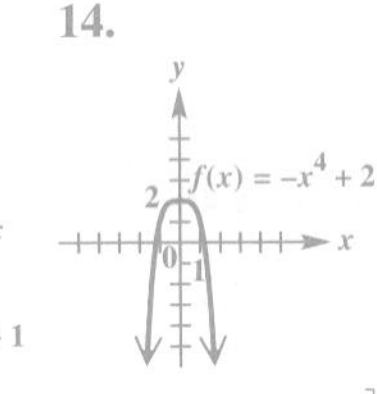

(a) $(-\infty, 0]$
(b) $[0, \infty)$

15. 16.

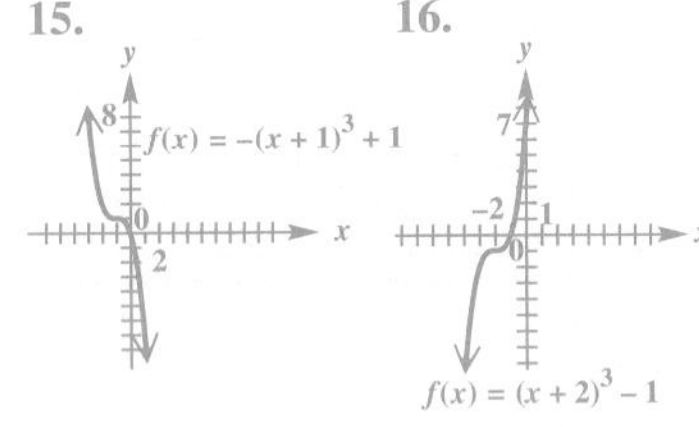

15. (a) none
(b) $(-\infty, \infty)$

16. (a) $(-\infty, \infty)$
(b) none

17.

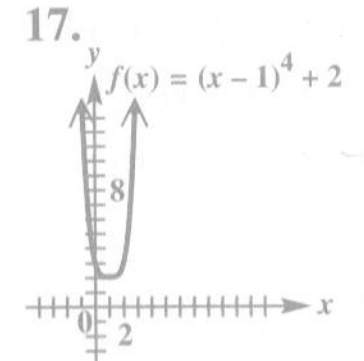

(a) $[1, \infty)$
(b) $(-\infty, 1]$

18.

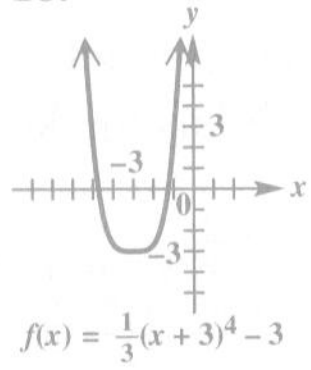

$f(x) = \frac{1}{3}(x+3)^4 - 3$

(a) $[-3, \infty)$
(b) $(-\infty, -3]$

Concept Check Comprehensive graphs of four polynomial functions are shown in A–D. They represent the graphs of functions defined by these four equations, but not necessarily in the order listed.

$$y = x^3 - 3x^2 - 6x + 8 \qquad y = x^4 + 7x^3 - 5x^2 - 75x$$

$$y = -x^3 + 9x^2 - 27x + 17 \qquad y = -x^5 + 36x^3 - 22x^2 - 147x - 90$$

Apply the concepts of this section to work Exercises 1–8.

A.

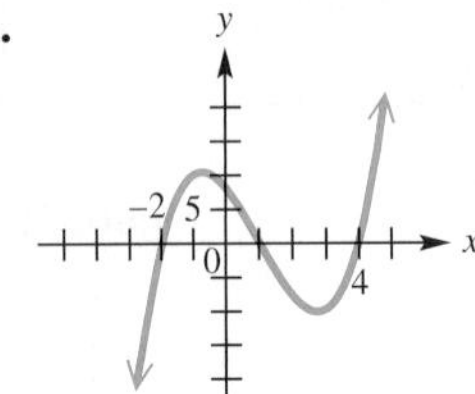

B.

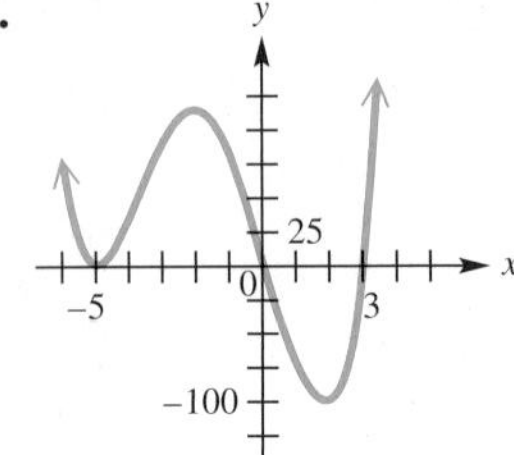

C.

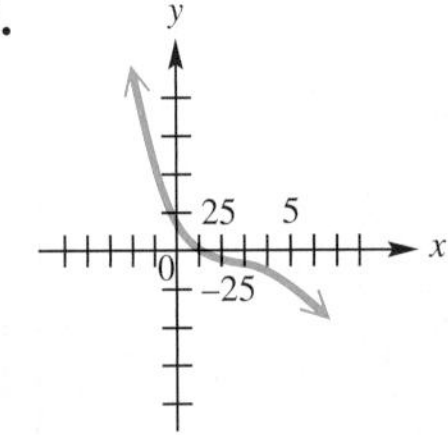

D.

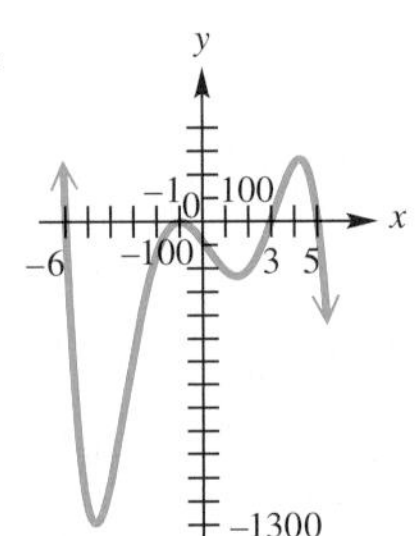

1. Which one of the graphs is that of $y = x^3 - 3x^2 - 6x + 8$?

2. Which one of the graphs is that of $y = x^4 + 7x^3 - 5x^2 - 75x$?

3. How many real zeros does the function graphed in C have?

4. Which one of C and D is the graph of $y = -x^3 + 9x^2 - 27x + 17$?

5. Which of the graphs cannot be that of a cubic polynomial function?

6. Which one of the graphs is that of a function whose range is *not* $(-\infty, \infty)$?

7. The function $f(x) = x^4 + 7x^3 - 5x^2 - 75x$ has the graph shown in B. Use the graph to factor the polynomial.

8. The function $f(x) = -x^5 + 36x^3 - 22x^2 - 147x - 90$ has the graph shown in D. Use the graph to factor the polynomial.

*Sketch the graph of each polynomial function. Determine the intervals of the domain for which each function is **(a)** increasing or **(b)** decreasing.* ***See Example 1.***

9. $f(x) = 2x^4$

10. $f(x) = \frac{1}{4}x^6$

11. $f(x) = -\frac{2}{3}x^5$

12. $f(x) = -\frac{5}{4}x^5$

13. $f(x) = \frac{1}{2}x^3 + 1$

14. $f(x) = -x^4 + 2$

15. $f(x) = -(x+1)^3 + 1$

16. $f(x) = (x+2)^3 - 1$

19. 20.

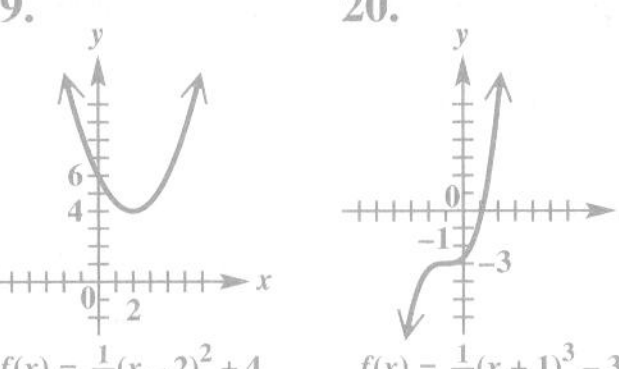

$f(x)=\frac{1}{2}(x-2)^2+4$ $f(x)=\frac{1}{3}(x+1)^3-3$

(a) $[2, \infty)$ (a) $(-\infty, \infty)$

(b) $(-\infty, 2]$ (b) none

21. 22.

23. 24.

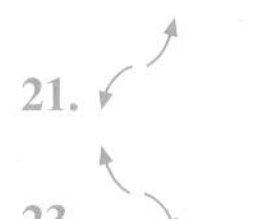

25. 26.

27. 28.

29. 30.

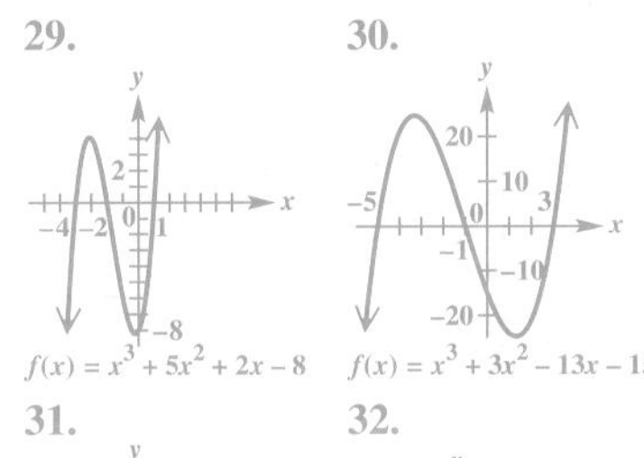

$f(x)=x^3+5x^2+2x-8$ $f(x)=x^3+3x^2-13x-15$

31. 32.

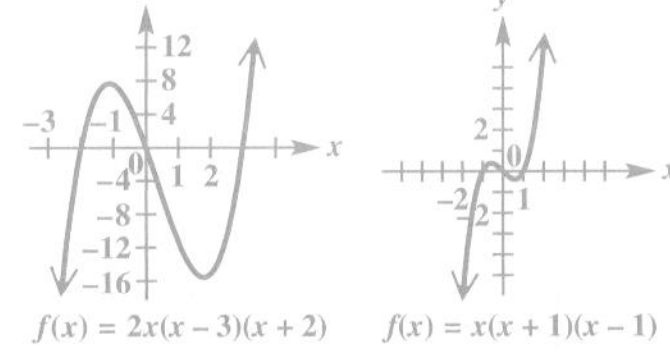

$f(x)=2x(x-3)(x+2)$ $f(x)=x(x+1)(x-1)$

33.

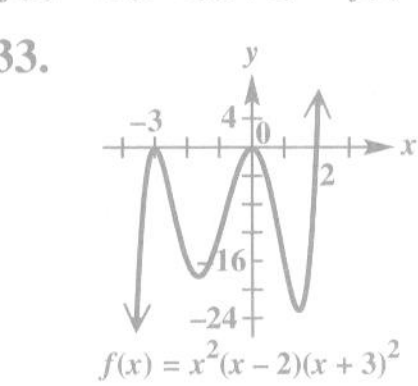

$f(x)=x^2(x-2)(x+3)^2$

34.

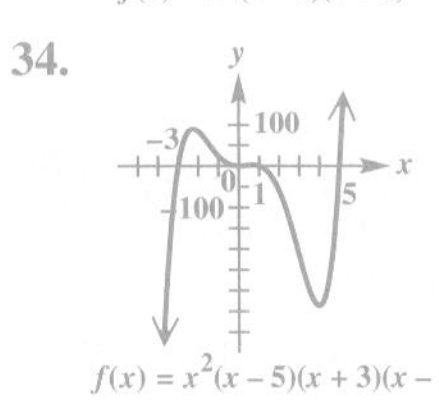

$f(x)=x^2(x-5)(x+3)(x-1)$

35. 36.

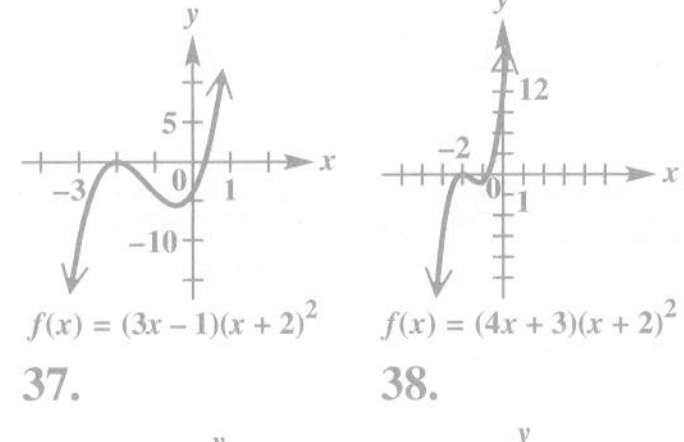

$f(x)=(3x-1)(x+2)^2$ $f(x)=(4x+3)(x+2)^2$

37. 38.

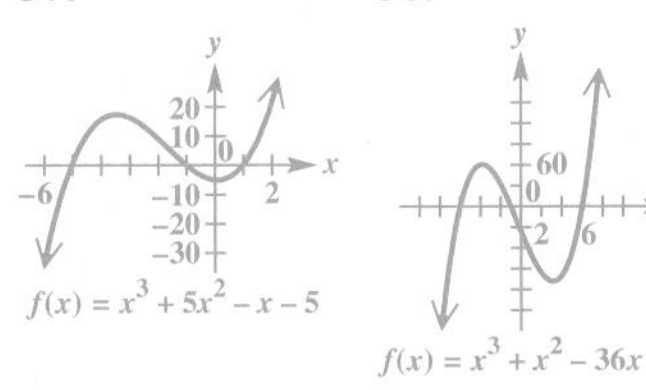

$f(x)=x^3+5x^2-x-5$ $f(x)=x^3+x^2-36x-36$

17. $f(x)=(x-1)^4+2$

18. $f(x)=\frac{1}{3}(x+3)^4-3$

19. $f(x)=\frac{1}{2}(x-2)^2+4$

20. $f(x)=\frac{1}{3}(x+1)^3-3$

Use an end behavior diagram, ↖↗ (both up), ↙↘ (both down), ↖↘, *or* ↙↗, *to describe the end behavior of the graph of each polynomial function.* ***See Example 2.***

21. $f(x)=5x^5+2x^3-3x+4$

22. $f(x)=-x^3-4x^2+2x-1$

23. $f(x)=-4x^3+3x^2-1$

24. $f(x)=4x^7-x^5+x^3-1$

25. $f(x)=9x^6-3x^4+x^2-2$

26. $f(x)=10x^6-x^5+2x-2$

27. $f(x)=3+2x-4x^2-5x^{10}$

28. $f(x)=7+2x-5x^2-10x^4$

Graph each polynomial function. Factor first if the expression is not in factored form. ***See Examples 3 and 4.***

29. $f(x)=x^3+5x^2+2x-8$

30. $f(x)=x^3+3x^2-13x-15$

31. $f(x)=2x(x-3)(x+2)$

32. $f(x)=x(x+1)(x-1)$

33. $f(x)=x^2(x-2)(x+3)^2$

34. $f(x)=x^2(x-5)(x+3)(x-1)$

35. $f(x)=(3x-1)(x+2)^2$

36. $f(x)=(4x+3)(x+2)^2$

37. $f(x)=x^3+5x^2-x-5$

38. $f(x)=x^3+x^2-36x-36$

39. $f(x)=x^3-x^2-2x$

40. $f(x)=3x^4+5x^3-2x^2$

41. $f(x)=2x^3(x^2-4)(x-1)$

42. $f(x)=x^2(x-3)^3(x+1)$

43. $f(x)=2x^3-5x^2-x+6$

44. $f(x)=2x^4+x^3-6x^2-7x-2$

45. $f(x)=3x^4-7x^3-6x^2+12x+8$

46. $f(x)=x^4+3x^3-3x^2-11x-6$

Use the intermediate value theorem for polynomials to show that each polynomial function has a real zero between the numbers given. ***See Example 5.***

47. $f(x)=2x^2-7x+4$; 2 and 3

48. $f(x)=3x^2-x-4$; 1 and 2

49. $f(x)=2x^3-5x^2-5x+7$; 0 and 1

50. $f(x)=2x^3-9x^2+x+20$; 2 and 2.5

51. $f(x)=2x^4-4x^2+4x-8$; 1 and 2

52. $f(x)=x^4-4x^3-x+3$; 0.5 and 1

53. $f(x)=x^4+x^3-6x^2-20x-16$; 3.2 and 3.3

54. $f(x)=x^4-2x^3-2x^2-18x+5$; 3.7 and 3.8

55. $f(x)=x^4-4x^3-20x^2+32x+12$; -1 and 0

56. $f(x)=x^5+2x^4+x^3+3$; -1.8 and -1.7

Show that the real zeros of each polynomial function satisfy the given conditions. ***See Example 6.***

57. $f(x)=x^4-x^3+3x^2-8x+8$; no real zero greater than 2

58. $f(x)=2x^5-x^4+2x^3-2x^2+4x-4$; no real zero greater than 1

59. $f(x)=x^4+x^3-x^2+3$; no real zero less than -2

39.

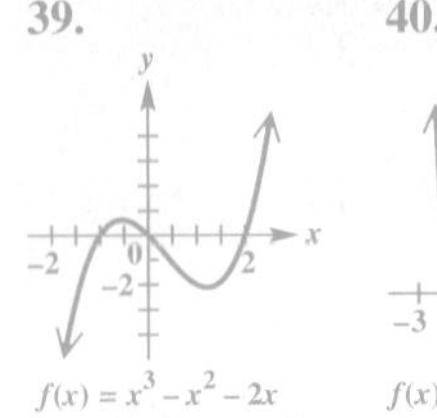

$f(x) = x^3 - x^2 - 2x$

40.

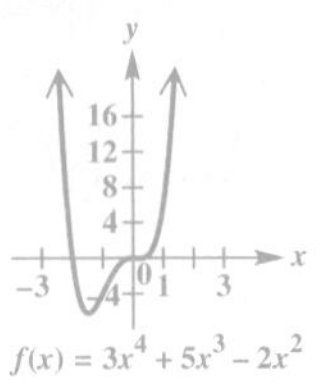

$f(x) = 3x^4 + 5x^3 - 2x^2$

41.

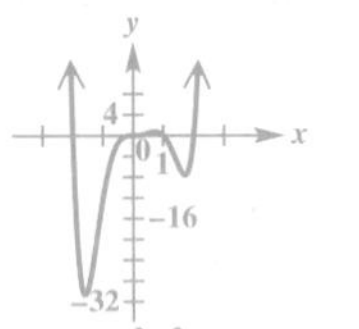

$f(x) = 2x^3(x^2 - 4)(x - 1)$

42.

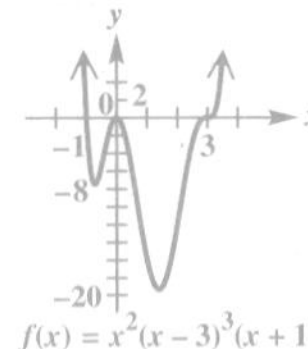

$f(x) = x^2(x - 3)^3(x + 1)$

43.

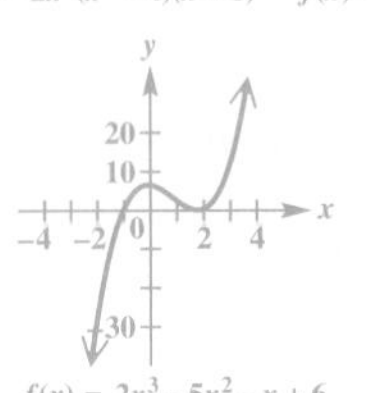

$f(x) = 2x^3 - 5x^2 - x + 6$

44.

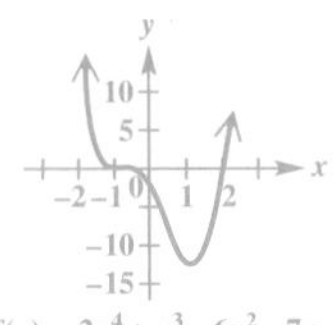

$f(x) = 2x^4 + x^3 - 6x^2 - 7x - 2$

45.

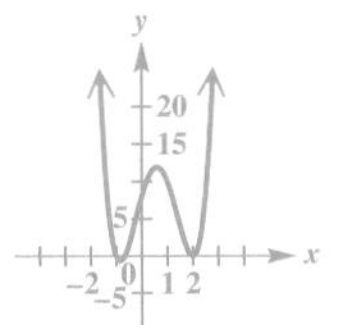

$f(x) = 3x^4 - 7x^3 - 6x^2 + 12x + 8$

46.

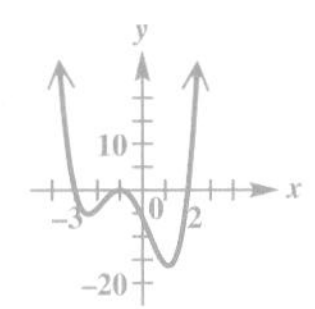

$f(x) = x^4 + 3x^3 - 3x^2 - 11x - 6$

47. $f(2) = -2 < 0$; $f(3) = 1 > 0$

48. $f(1) = -2 < 0$; $f(2) = 6 > 0$

49. $f(0) = 7 > 0$; $f(1) = -1 < 0$

50. $f(2) = 2 > 0$; $f(2.5) = -2.5 < 0$

51. $f(1) = -6 < 0$; $f(2) = 16 > 0$

52. $f(0.5) = 2.0625 > 0$; $f(1) = -1 < 0$

53. $f(3.2) = -3.8144 < 0$; $f(3.3) = 7.1891 > 0$

54. $f(3.7) = -2.8699 < 0$; $f(3.8) = 6.4896 > 0$

55. $f(-1) = -35 < 0$; $f(0) = 12 > 0$

60. $f(x) = x^5 + 2x^3 - 2x^2 + 5x + 5$; no real zero less than -1

61. $f(x) = 3x^4 + 2x^3 - 4x^2 + x - 1$; no real zero greater than 1

62. $f(x) = 3x^4 + 2x^3 - 4x^2 + x - 1$; no real zero less than -2

63. $f(x) = x^5 - 3x^3 + x + 2$; no real zero greater than 2

64. $f(x) = x^5 - 3x^3 + x + 2$; no real zero less than -3

Connecting Graphs with Equations In Exercises 65–70, find a polynomial function $f(x)$ of least possible degree having the graph shown. (Hint: See the Note following ***Example 4.****)*

65.

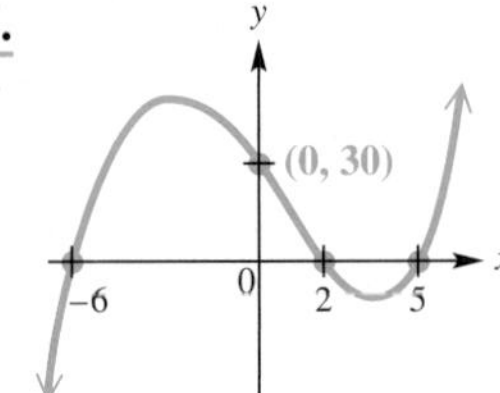

66.

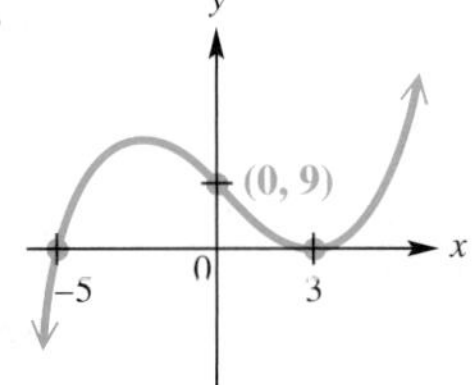

67.

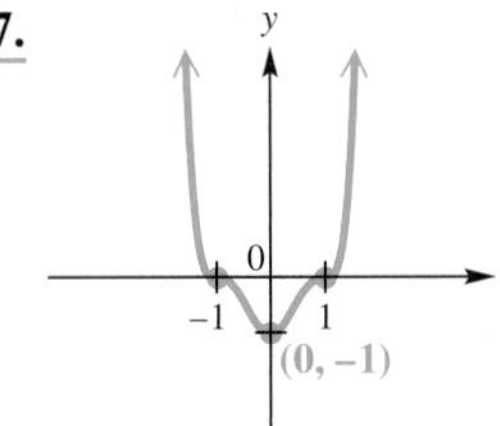

68.

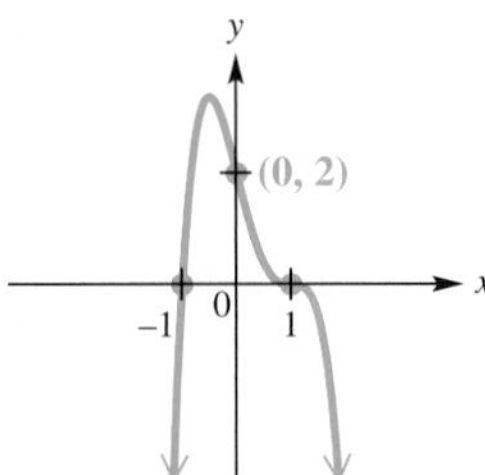

69.

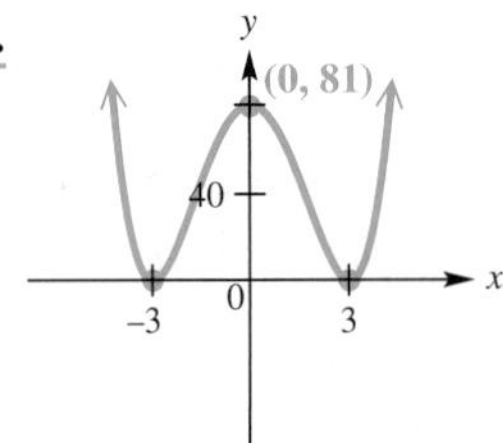

70.

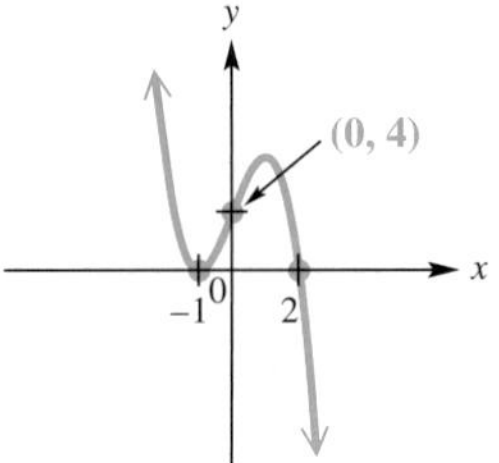

Use a graphing calculator to graph the function in the viewing window specified. Compare the graph to the one shown in the answer section of this text. Then use the graph to find $f(1.25)$.

71. $f(x) = 2x(x - 3)(x + 2)$; window: $[-3, 4]$ by $[-20, 12]$
Compare to **Exercise 31.**

72. $f(x) = x^2(x - 2)(x + 3)^2$; window: $[-4, 3]$ by $[-24, 4]$
Compare to **Exercise 33.**

73. $f(x) = (3x - 1)(x + 2)^2$; window: $[-4, 2]$ by $[-15, 15]$
Compare to **Exercise 35.**

74. $f(x) = x^3 + 5x^2 - x - 5$; window: $[-6, 2]$ by $[-30, 30]$
Compare to **Exercise 37.**

Use a graphing calculator to approximate the real zero discussed in each specified exercise. See ***Example 7.***

75. Exercise 47

76. Exercise 49

77. Exercise 51

78. Exercise 50

56. $f(-1.8) = -0.73248 < 0$; $f(-1.7) = 0.59263 > 0$

65. $f(x) = \frac{1}{2}(x+6)(x-2) \cdot (x-5)$, or $f(x) = \frac{1}{2}x^3 - \frac{1}{2}x^2 - 16x + 30$

66. $f(x) = \frac{1}{5}(x-3)^2(x+5)$, or $f(x) = \frac{1}{5}x^3 - \frac{1}{5}x^2 - \frac{21}{5}x + 9$

67. $f(x) = (x-1)^3(x+1)^3$, or $f(x) = x^6 - 3x^4 + 3x^2 - 1$

68. $f(x) = -2(x-1)^3(x+1)$, or $f(x) = -2x^4 + 4x^3 - 4x + 2$

69. $f(x) = (x-3)^2(x+3)^2$, or $f(x) = x^4 - 18x^2 + 81$

70. $f(x) = -2(x-2)(x+1)^2$, or $f(x) = -2x^3 + 6x + 4$

71. $f(1.25) = -14.21875$

$f(x) = 2x(x-3)(x+2)$

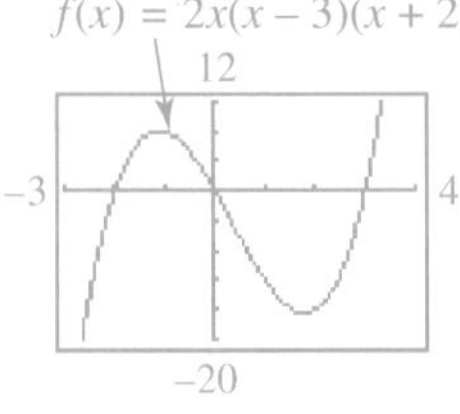

72. $f(1.25) \approx -21.16699$

$f(x) = x^2(x-2)(x+3)^2$

4, −4, 3, −24

73. $f(1.25) = 29.046875$

$f(x) = (3x-1)(x+2)^2$

15, −4, 2, −15

74. $f(1.25) = 3.515625$

$f(x) = x^3 + 5x^2 - x - 5$

30, −6, 2, −30

75. 2.7807764 76. 0.88993856
77. 1.543689 78. 2.193325

79. −3.0, −1.4, 1.4
80. −2.1
81. −1.1, 1.2 82. −1.5, 3.1

For the given polynomial function, approximate each zero as a decimal to the nearest tenth. ***See Example 7.***

79. $f(x) = x^3 + 3x^2 - 2x - 6$

80. $f(x) = x^3 - 3x + 3$

81. $f(x) = -2x^4 - x^2 + x + 5$

82. $f(x) = -x^4 + 2x^3 + 3x^2 + 6$

Use a graphing calculator to find the coordinates of the turning points of the graph of each polynomial function in the given domain interval. Give answers to the nearest hundredth.

83. $f(x) = x^3 + 4x^2 - 8x - 8$; $[-3.8, -3]$

84. $f(x) = x^3 + 4x^2 - 8x - 8$; $[0.3, 1]$

85. $f(x) = 2x^3 - 5x^2 - x + 1$; $[-1, 0]$

86. $f(x) = 2x^3 - 5x^2 - x + 1$; $[1.4, 2]$

87. $f(x) = x^4 - 7x^3 + 13x^2 + 6x - 28$; $[-1, 0]$

88. $f(x) = x^3 - x + 3$; $[-1, 0]$

Solve each problem.

89. ***(Modeling) Social Security Numbers*** Your Social Security number (SSN) is unique, and with it you can construct your own personal Social Security polynomial. Let the polynomial function be defined as follows, where a_i represents the ith digit in your SSN:

$$\text{SSN}(x) = (x - a_1)(x + a_2)(x - a_3)(x + a_4)(x - a_5) \cdot (x + a_6)(x - a_7)(x + a_8)(x - a_9).$$

For example, if the SSN is 539-58-0954, the polynomial function is

$$\text{SSN}(x) = (x-5)(x+3)(x-9)(x+5)(x-8)(x+0)(x-9)(x+5)(x-4).$$

A comprehensive graph of this function is shown in **Figure A.** In **Figure B,** we show a screen obtained by zooming in on the positive zeros, as the comprehensive graph does not show the local behavior well in this region. Use a graphing calculator to graph your own "personal polynomial."

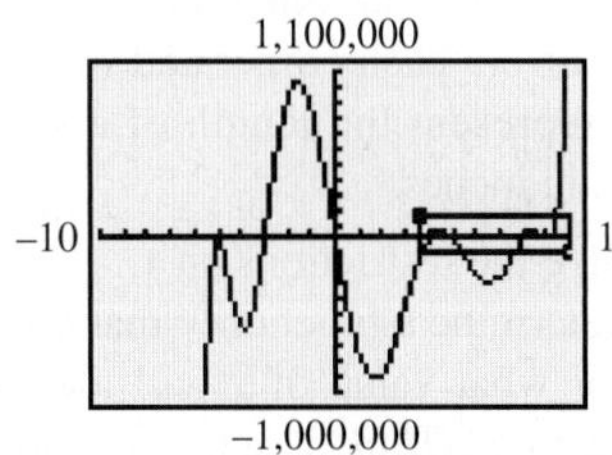

Figure A

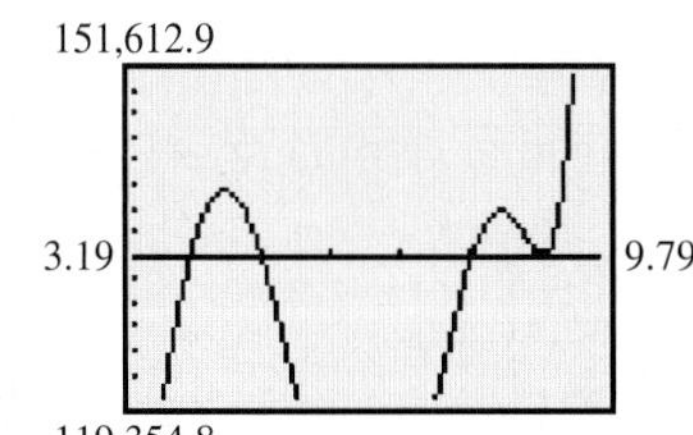

Figure B

90. A comprehensive graph of $f(x) = x^4 - 7x^3 + 18x^2 - 22x + 12$ is shown in the two screens, along with displays of the two real zeros. Find the two remaining nonreal complex zeros.

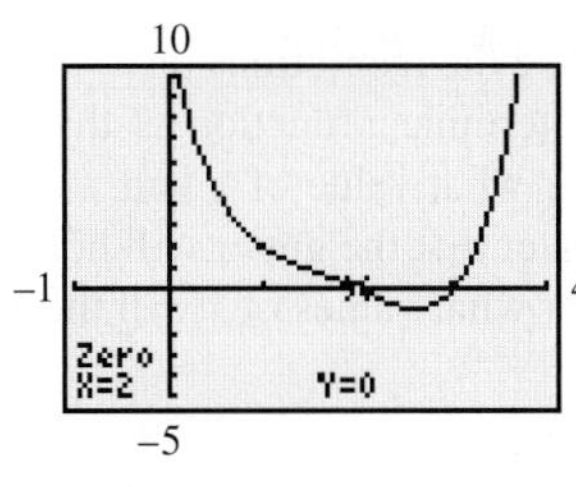

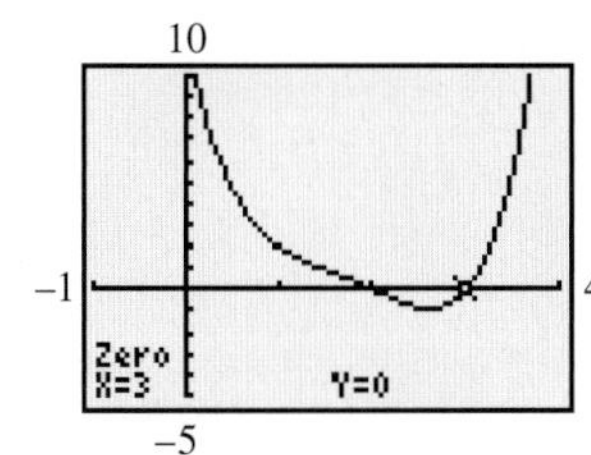

83. $(-3.44, 26.15)$
84. $(0.77, -11.33)$
85. $(-0.09, 1.05)$
86. $(1.76, -5.34)$
87. $(-0.20, -28.62)$
88. $(-0.58, 3.38)$

89. Answers will vary.
90. $1 - i$ and $1 + i$

91.

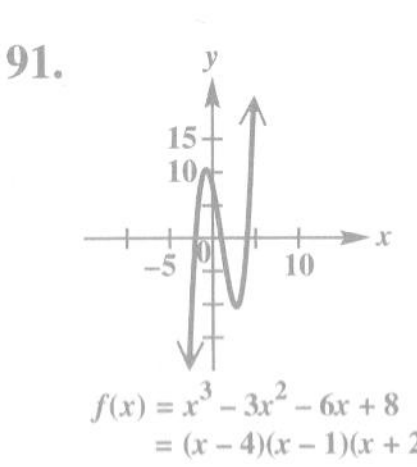

$f(x) = x^3 - 3x^2 - 6x + 8 = (x - 4)(x - 1)(x + 2)$

(a) $\{-2, 1, 4\}$
(b) $(-\infty, -2) \cup (1, 4)$
(c) $(-2, 1) \cup (4, \infty)$

92.

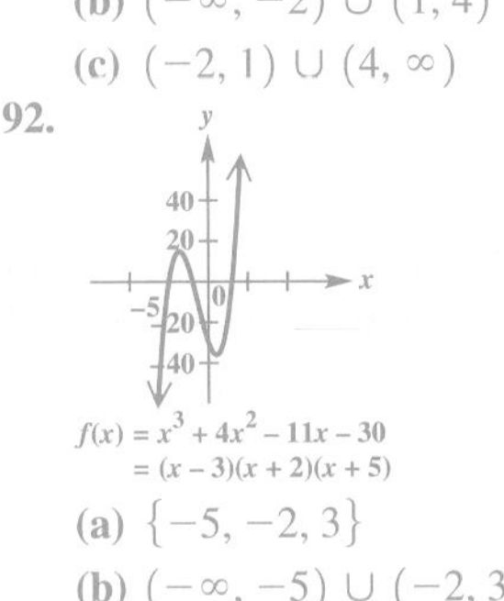

$f(x) = x^3 + 4x^2 - 11x - 30 = (x - 3)(x + 2)(x + 5)$

(a) $\{-5, -2, 3\}$
(b) $(-\infty, -5) \cup (-2, 3)$
(c) $(-5, -2) \cup (3, \infty)$

93.

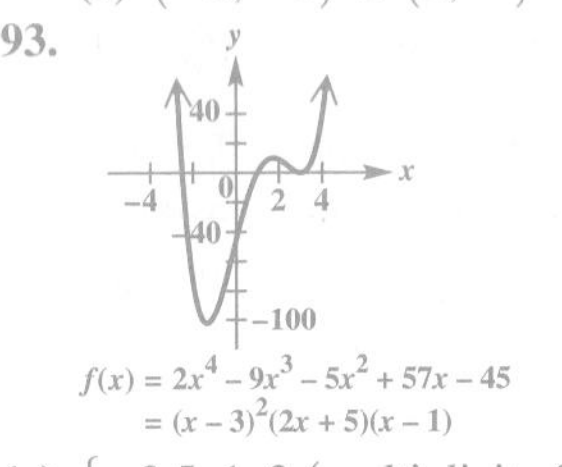

$f(x) = 2x^4 - 9x^3 - 5x^2 + 57x - 45 = (x - 3)^2(2x + 5)(x - 1)$

(a) $\{-2.5, 1, 3 \text{ (multiplicity 2)}\}$
(b) $(-2.5, 1)$
(c) $(-\infty, -2.5) \cup (1, 3) \cup (3, \infty)$

94.

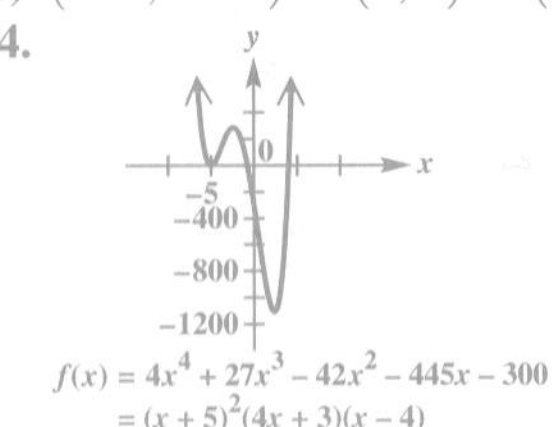

$f(x) = 4x^4 + 27x^3 - 42x^2 - 445x - 300 = (x + 5)^2(4x + 3)(x - 4)$

(a) $\{-5 \text{ (multiplicity 2)}, -0.75, 4\}$
(b) $(-0.75, 4)$
(c) $(-\infty, -5) \cup (-5, -0.75) \cup (4, \infty)$

Relating Concepts

For individual or collaborative investigation *(Exercises 91–96)*

For any function $y = f(x)$, the following hold true.

(a) The real solutions of $f(x) = 0$ are the x-intercepts of the graph.

(b) The real solutions of $f(x) < 0$ are the x-values for which the graph lies *below* the x-axis.

(c) The real solutions of $f(x) > 0$ are the x-values for which the graph lies *above* the x-axis.

In Exercises 91–96, a polynomial function $f(x)$ is given in both expanded and factored forms. Graph the function, and solve the equations and inequalities. Give multiplicities of solutions when applicable.

91. $f(x) = x^3 - 3x^2 - 6x + 8 = (x - 4)(x - 1)(x + 2)$

(a) $f(x) = 0$ **(b)** $f(x) < 0$ **(c)** $f(x) > 0$

92. $f(x) = x^3 + 4x^2 - 11x - 30 = (x - 3)(x + 2)(x + 5)$

(a) $f(x) = 0$ **(b)** $f(x) < 0$ **(c)** $f(x) > 0$

93. $f(x) = 2x^4 - 9x^3 - 5x^2 + 57x - 45 = (x - 3)^2(2x + 5)(x - 1)$

(a) $f(x) = 0$ **(b)** $f(x) < 0$ **(c)** $f(x) > 0$

94. $f(x) = 4x^4 + 27x^3 - 42x^2 - 445x - 300 = (x + 5)^2(4x + 3)(x - 4)$

(a) $f(x) = 0$ **(b)** $f(x) < 0$ **(c)** $f(x) > 0$

95. $f(x) = -x^4 - 4x^3 + 3x^2 + 18x = x(2 - x)(x + 3)^2$

(a) $f(x) = 0$ **(b)** $f(x) \geq 0$ **(c)** $f(x) \leq 0$

96. $f(x) = -x^4 + 2x^3 + 8x^2 = x^2(4 - x)(x + 2)$

(a) $f(x) = 0$ **(b)** $f(x) \geq 0$ **(c)** $f(x) \leq 0$

(Modeling) Exercises 97–104 are geometric in nature and lead to polynomial models. Solve each problem.

97. ***Volume of a Box*** A rectangular piece of cardboard measuring 12 in. by 18 in. is to be made into a box with an open top by cutting equal-size squares from each corner and folding up the sides. Let x represent the length of a side of each such square in inches.

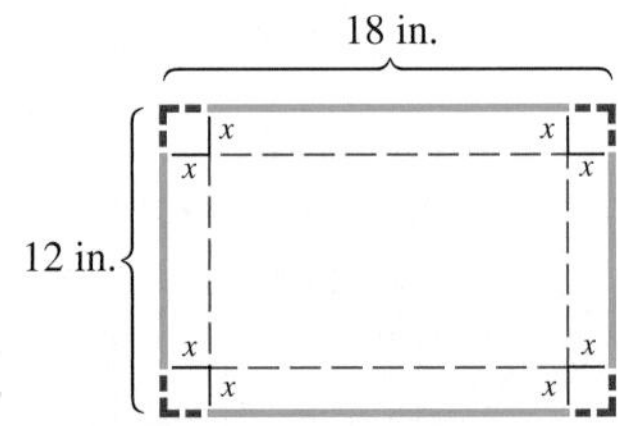

(a) Give the restrictions on x.

(b) Determine a function V that gives the volume of the box as a function of x.

(c) For what value of x will the volume be a maximum? What is this maximum volume? (*Hint:* Use the function of a graphing calculator that allows you to determine a maximum point within a given interval.)

(d) For what values of x will the volume be greater than 80 in.^3?

98. ***Construction of a Rain Gutter*** A piece of rectangular sheet metal is 20 in. wide. It is to be made into a rain gutter by turning up the edges to form parallel sides. Let x represent the length of each of the parallel sides.

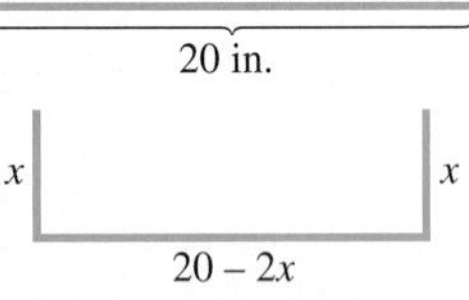

(a) Give the restrictions on x.

(b) Determine a function $\mathcal{A}$ that gives the area of a cross section of the gutter.

(c) For what value of x will $\mathcal{A}$ be a maximum (and thus maximize the amount of water that the gutter will hold)? What is this maximum area?

(d) For what values of x will the area of a cross section be less than 40 in.^2?

95.

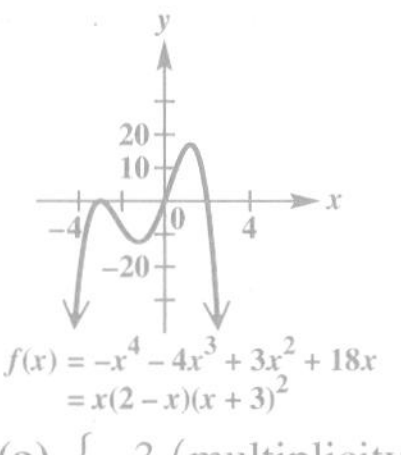

$f(x) = -x^4 - 4x^3 + 3x^2 + 18x$
$= x(2 - x)(x + 3)^2$

(a) $\{-3 \text{ (multiplicity 2)}, 0, 2\}$
(b) $\{-3\} \cup [0, 2]$
(c) $(-\infty, 0] \cup [2, \infty)$

96.

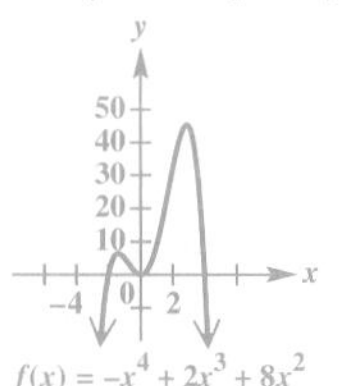

$f(x) = -x^4 + 2x^3 + 8x^2$
$= x^2(4 - x)(x + 2)$

(a) $\{-2, 0 \text{ (multiplicity 2)}, 4\}$
(b) $[-2, 4]$
(c) $(-\infty, -2] \cup \{0\} \cup [4, \infty)$

97. (a) $0 < x < 6$
(b) $V(x) = x(18 - 2x) \cdot (12 - 2x)$, or $V(x) = 4x^3 - 60x^2 + 216x$
(c) $x \approx 2.35$; about 228.16 in.^3
(d) $0.42 < x < 5$

98. (a) $0 < x < 10$
(b) $\mathcal{A}(x) = x(20 - 2x) = -2x^2 + 20x$
(c) $x = 5$; maximum cross section area: 50 in.^2
(d) between 0 and 2.76 or between 7.24 and 10

99. (a) $x - 1$; $(1, \infty)$
(b) $\sqrt{x^2 - (x - 1)^2}$
(c) $2x^3 - 5x^2 + 4x - 28{,}225 = 0$
(d) hypotenuse: 25 in.; legs: 24 in. and 7 in.

100. approximately 1.732

101. 3 ft

102. (a) about 66.15 in.^3
(b) $0.54 \text{ in.} < x < 2.92 \text{ in.}$

103. (a) about 7.13 cm; The ball floats partly above the surface.
(b) The sphere is more dense than water and sinks below the surface.
(c) 10 cm; The balloon is submerged with its top even with the surface.

104. about 11.34 cm

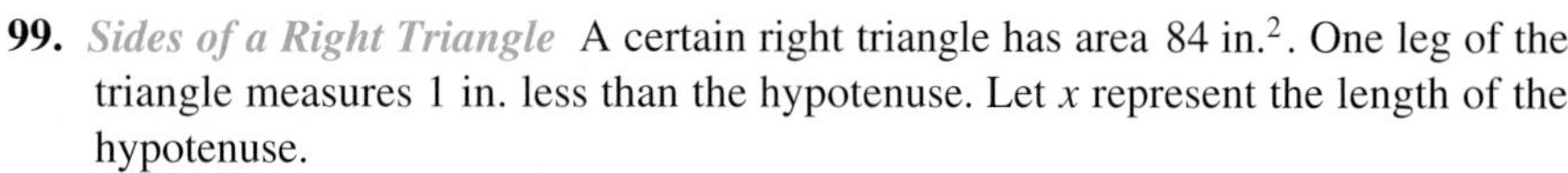

99. *Sides of a Right Triangle* A certain right triangle has area 84 in.^2. One leg of the triangle measures 1 in. less than the hypotenuse. Let x represent the length of the hypotenuse.

(a) Express the length of the leg mentioned above in terms of x. Give the domain of x.

(b) Express the length of the other leg in terms of x.

(c) Write an equation based on the information determined thus far. Square both sides and then write the equation with one side as a polynomial with integer coefficients, in descending powers, and the other side equal to 0.

(d) Solve the equation in part (c) graphically. Find the lengths of the three sides of the triangle.

100. *Area of a Rectangle* Find the value of x in the figure that will maximize the area of rectangle $ABCD$.

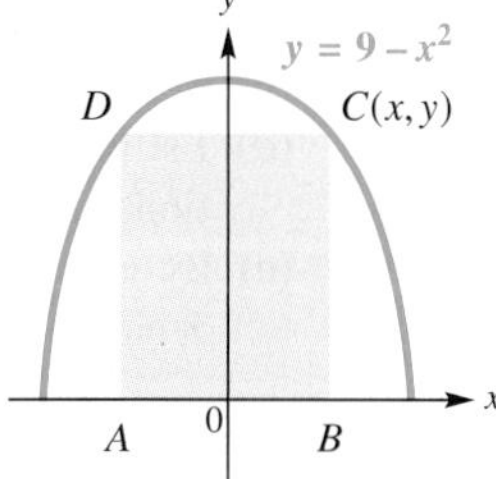

101. *Butane Gas Storage* A storage tank for butane gas is to be built in the shape of a right circular cylinder of altitude 12 ft, with a half sphere attached to each end. If x represents the radius of each half sphere, what radius should be used to cause the volume of the tank to be $144\pi \text{ ft}^3$?

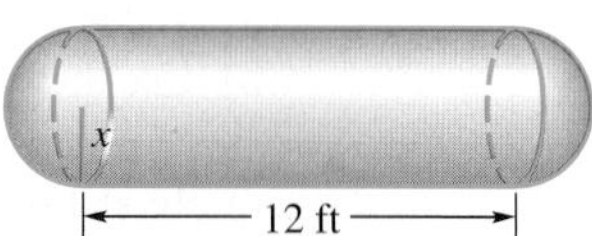

102. *Volume of a Box* A standard piece of notebook paper measuring 8.5 in. by 11 in. is to be made into a box with an open top by cutting equal-size squares from each corner and folding up the sides. Let x represent the length of a side of each such square in inches. Use the table feature of your graphing calculator to do the following.

(a) Find the maximum volume of the box.

(b) Determine when the volume of the box will be greater than 40 in.^3.

103. *Floating Ball* The polynomial function

$$f(x) = \frac{\pi}{3}x^3 - 5\pi x^2 + \frac{500\pi d}{3}$$

can be used to find the depth that a ball 10 cm in diameter sinks in water. The constant d is the density of the ball, where the density of water is 1. The smallest *positive* zero of $f(x)$ equals the depth that the ball sinks. Approximate this depth for each material and interpret the results.

(a) A wooden ball with $d = 0.8$

(b) A solid aluminum ball with $d = 2.7$

(c) A spherical water balloon with $d = 1$

104. *Floating Ball* Refer to **Exercise 103.** If a ball has a 20-cm diameter, then the function becomes

$$f(x) = \frac{\pi}{3}x^3 - 10\pi x^2 + \frac{4000\pi d}{3}.$$

This function can be used to determine the depth that the ball sinks in water. Find the depth that this size ball sinks when $d = 0.6$.

(Modeling) Solve each problem involving a polynomial function model. ***See Example 8.***

105. *Highway Design* To allow enough distance for cars to pass on two-lane highways, engineers calculate minimum sight distances between curves and hills. The table shows the minimum sight distance y in feet for a car traveling at x miles per hour.

x (in mph)	20	30	40	50	60	65	70
y (in feet)	810	1090	1480	1840	2140	2310	2490

Source: Haefner, L., *Introduction to Transportation Systems,* Holt, Rinehart and Winston.

(a) Make a scatter diagram of the data.

(b) Use the regression feature of a calculator to find the best-fitting linear function for the data. Graph the function with the data.

(c) Repeat part (b) for a cubic function.

(d) Estimate the minimum sight distance for a car traveling 43 mph using the functions from parts (b) and (c).

(e) By comparing the graphs of the functions in parts (b) and (c) with the data, decide which function best fits the given data.

106. *Water Pollution* Copper in high doses can be lethal to aquatic life. The table lists copper concentrations in freshwater mussels after 45 days at various distances downstream from an electroplating plant. The concentration C is measured in micrograms of copper per gram of mussel x kilometers downstream.

x	5	21	37	53	59
C	20	13	9	6	5

Source: Foster, R., and J. Bates, "Use of mussels to monitor point source industrial discharges," *Environ. Sci. Technol.*; Mason, C., *Biology of Freshwater Pollution,* John Wiley & Sons.

(a) Make a scatter diagram of the data.

(b) Use the regression feature of a calculator to find the best-fitting quadratic function for the data. Graph the function with the data.

(c) Repeat part (b) for a cubic function.

(d) By comparing graphs of the functions in parts (b) and (c) with the data, decide which function best fits the given data.

(e) Concentrations above 10 are lethal to mussels. Find the values of x (using the cubic function) for which this is the case.

107. *Government Spending on Health Research and Training* The table lists the annual amount (in billions of dollars) spent by the federal government on health research and training programs over an 8-yr period.

Year	Amount	Year	Amount
2002	21.4	2006	28.8
2003	24.0	2007	29.3
2004	27.1	2008	29.9
2005	28.1	2009	30.6

Source: U.S. Office of Management and Budget.

Which one of the following provides the best model for these data, where x represents the year?

A. $f(x) = 0.5(x - 2002)^2 + 21.4$

B. $g(x) = (x - 2002) + 21.4$

C. $h(x) = 3.5\sqrt{x - 2002} + 21.4$

D. $k(x) = (x - 2002)^3 + 21.4$

105. (a)

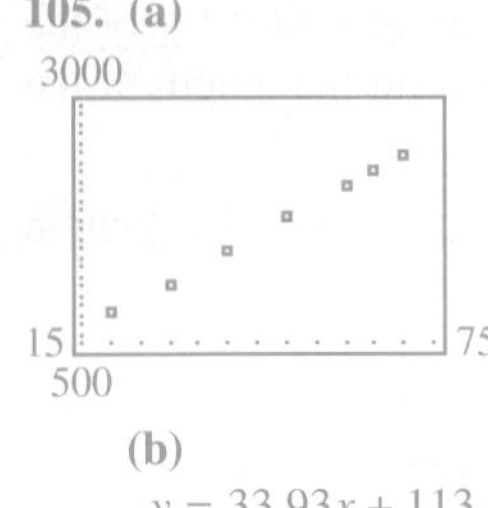

(b) $y = 33.93x + 113.4$

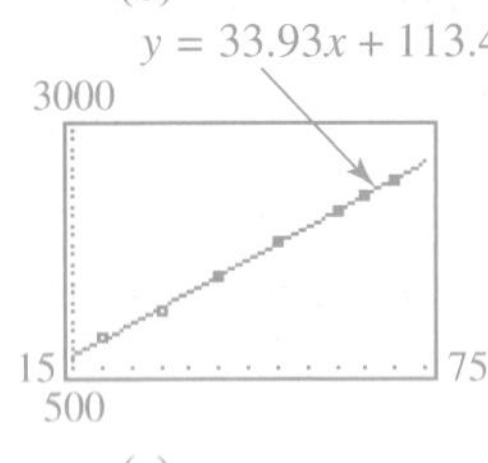

(c) $y = -0.0032x^3 + 0.4245x^2 + 16.64x + 323.1$

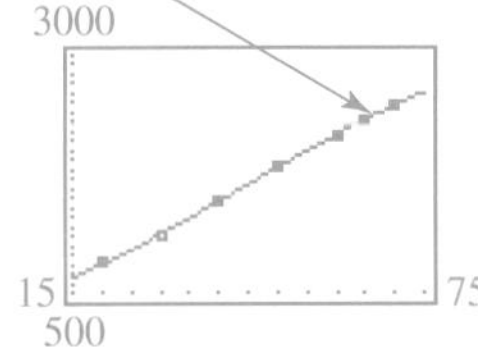

(d) linear: 1572 ft; cubic: 1569 ft

(e) The cubic function appears slightly better because only one data point is not on the curve.

106. (a)

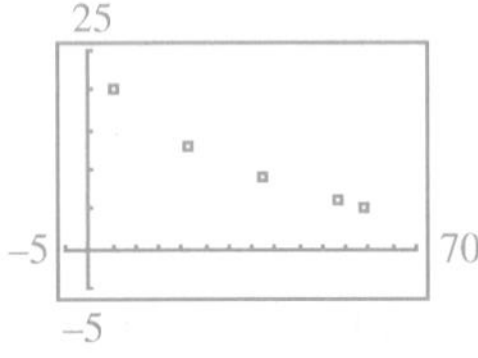

(b) $C(x) = 0.0035x^2 - 0.49x + 22$

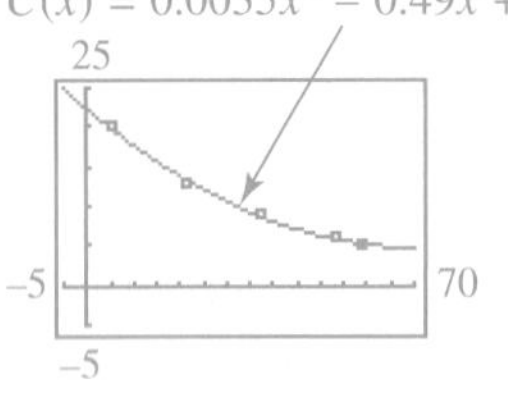

(c) $C(x) = -0.000068x^3 + 0.00987x^2 - 0.653x + 23$

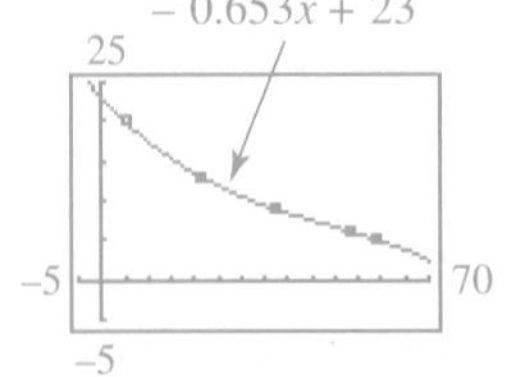

(d) The cubic function is a slightly better fit.

(e) $0 \le x < 31.9$

107. C

108. (a) If the length of the pendulum increases, so does the period of oscillation, T.
(c) $k \approx 0.81$; $n = 2$
(d) about 2.48 sec
(e) T increases by a factor of $\sqrt{2} \approx 1.414$.

108. *Swing of a Pendulum* Grandfather clocks use pendulums to keep accurate time. The relationship between the length of a pendulum L and the time T for one complete oscillation can be expressed by the equation $L = kT^n$, where k is a constant and n is a positive integer to be determined. The data in the table were taken for different lengths of pendulums.

L (ft)	T (sec)	L (ft)	T (sec)
1.0	1.11	3.0	1.92
1.5	1.36	3.5	2.08
2.0	1.57	4.0	2.22
2.5	1.76		

(a) As the length of the pendulum increases, what happens to T?
(b) Discuss how n and k could be found.
(c) Use the data to approximate k and determine the best value for n.
(d) Using the values of k and n from part (c), predict T for a pendulum having length 5 ft.
(e) If the length L of a pendulum doubles, what happens to the period T?

Summary Exercises on Polynomial Functions, Zeros, and Graphs

Classroom Example
Complete parts (a)–(d) of the directions for the **Summary Exercises 1–10** for

$f(x) = x^4 - x^3 + 3x^2 - 5x - 10.$

Answers:

(a)

Positive	Negative	Nonreal Complex
3	1	0
1	1	2

(b) $\pm 1, \pm 2, \pm 5, \pm 10$
(c) $-1, 2$
(d) $-i\sqrt{5}, i\sqrt{5}$

Answers for Summary Exercises

1. (a)

Positive	Negative	Nonreal Complex
2	1	0
0	1	2

(b) $\pm 1, \pm 2, \pm 3, \pm 4, \pm 6, \pm 8, \pm 12, \pm 24, \pm \frac{1}{2}, \pm \frac{3}{2}, \pm \frac{1}{3}, \pm \frac{2}{3}, \pm \frac{4}{3}, \pm \frac{8}{3}, \pm \frac{1}{6}$
(c) $-\frac{1}{2}, \frac{4}{3}, 6$
(d) no other complex zeros

2. (a)

Positive	Negative	Nonreal Complex
2	1	0
0	1	2

(b) $\pm 1, \pm 3, \pm \frac{1}{2}, \pm \frac{3}{2}$
(c) $-1, \frac{1}{2}, 3$
(d) no other complex zeros

We use all of the theorems for finding complex zeros of polynomial functions in the next example.

EXAMPLE **Finding All Zeros of a Polynomial Function**

Find all zeros of $f(x) = x^4 - 3x^3 + 6x^2 - 12x + 8$.

SOLUTION We consider the number of positive zeros by observing the variations in signs for $f(x)$.

$$f(x) = +x^4 - 3x^3 + 6x^2 - 12x + 8 \quad \text{(Section 3.3)}$$

1 2 3 4

Since $f(x)$ has four sign changes, we can use Descartes' rule of signs to determine that there are four, two, or zero positive real zeros. For negative zeros, we consider the variations in signs for $f(-x)$.

$$f(-x) = (-x)^4 - 3(-x)^3 + 6(-x)^2 - 12(-x) + 8$$

$$f(-x) = x^4 + 3x^3 + 6x^2 + 12x + 8$$

Since $f(-x)$ has no sign changes, there are no negative real zeros. Because the function is of degree 4, it has a maximum of four zeros with possibilities summarized as follows.

Positive	Negative	Nonreal Complex
4	0	0
2	0	2
0	0	4

We can now use the rational zeros theorem to determine that the possible rational zeros are ± 1, ± 2, ± 4, and ± 8. Based on Descartes' rule of signs, we discard the negative rational zeros from this list and try to find a positive rational zero. Suppose we start by using synthetic division to check 4.

$$\text{Proposed zero} \longrightarrow 4\overline{)\begin{array}{rrrrr} 1 & -3 & 6 & -12 & 8 \\ & 4 & 4 & 40 & 112 \\ \hline 1 & 1 & 10 & 28 & 120 \end{array}} \longleftarrow f(4) = 120$$

(Section 3.2)

We find that 4 is not a zero. However, $4 > 0$, and the numbers in the bottom row of the synthetic division are nonnegative. Thus, the boundedness theorem indicates that there are no zeros greater than 4. We can discard 8 as a possible rational zero and use synthetic division to show that 1 and 2 are zeros.

$$\begin{array}{r|rrrrr} 1) & 1 & -3 & 6 & -12 & 8 \\ & & 1 & -2 & 4 & -8 \\ \hline 2) & 1 & -2 & 4 & -8 & 0 \leftarrow f(1) = 0 \\ & & 2 & 0 & 8 & \\ \hline & 1 & 0 & 4 & 0 & \leftarrow f(2) = 0 \end{array}$$

The polynomial now factors as

$$f(x) = (x - 1)(x - 2)(x^2 + 4).$$

We find the remaining two zeros using algebra to solve for x in the quadratic factor of the following equation.

$$(x - 1)(x - 2)(x^2 + 4) = 0$$

$$x - 1 = 0 \quad \text{or} \quad x - 2 = 0 \quad \text{or} \quad x^2 + 4 = 0$$ Zero-factor property (Section 1.4)

$$x = 1 \quad \text{or} \quad x = 2 \quad \text{or} \quad x^2 = -4$$

$$x = \pm 2i$$ Square root property (Section 1.4)

The linear factored form of the polynomial is

$$f(x) = (x - 1)(x - 2)(x - 2i)(x + 2i),$$

and the corresponding zeros are 1, 2, $2i$, and $-2i$.

Now Try Exercise 3.

For each polynomial function, do the following in order.

(a) Use Descartes' rule of signs to determine the different possibilities for the numbers of positive, negative, and nonreal complex zeros.
(b) Use the rational zeros theorem to determine the possible rational zeros.
(c) Use synthetic division with the boundedness theorem where appropriate and/or factoring to find the rational zeros, if any.
(d) Find all other complex zeros (both real and nonreal), if any.

1. $f(x) = 6x^3 - 41x^2 + 26x + 24$

2. $f(x) = 2x^3 - 5x^2 - 4x + 3$

3. $f(x) = 3x^4 - 5x^3 + 14x^2 - 20x + 8$

4. $f(x) = 2x^4 - 3x^3 + 16x^2 - 27x - 18$

5. $f(x) = 6x^4 - 5x^3 - 11x^2 + 10x - 2$

6. $f(x) = 5x^4 + 8x^3 - 19x^2 - 24x + 12$

7. $f(x) = x^5 - 6x^4 + 16x^3 - 24x^2 + 16x$ (*Hint:* Factor out x first.)

8. $f(x) = 2x^4 + 8x^3 - 7x^2 - 42x - 9$

9. $f(x) = 8x^4 + 8x^3 - x - 1$ (*Hint:* Factor the polynomial.)

10. $f(x) = 2x^5 + 5x^4 - 9x^3 - 11x^2 + 19x - 6$

3. (a)

Positive	Negative	Nonreal Complex
4	0	0
2	0	2
0	0	4

(b) $\pm 1, \pm 2, \pm 4, \pm 8, \pm\frac{1}{3}, \pm\frac{2}{3}, \pm\frac{4}{3}, \pm\frac{8}{3}$

(c) $\frac{2}{3}, 1$ **(d)** $-2i, 2i$

4. (a)

Positive	Negative	Nonreal Complex
3	1	0
1	1	2

(b) $\pm 1, \pm 2, \pm 3, \pm 6, \pm 9, \pm 18, \pm\frac{1}{2}, \pm\frac{3}{2}, \pm\frac{9}{2}$

(c) $-\frac{1}{2}, 2$ **(d)** $-3i, 3i$

5. (a)

Positive	Negative	Nonreal Complex
3	1	0
1	1	2

(b) $\pm 1, \pm 2, \pm\frac{1}{2}, \pm\frac{1}{3}, \pm\frac{2}{3}, \pm\frac{1}{6}$

(c) $\frac{1}{3}, \frac{1}{2}$ **(d)** $-\sqrt{2}, \sqrt{2}$

6. (a)

Positive	Negative	Nonreal Complex
2	2	0
2	0	2
0	2	2
0	0	4

(b) $\pm 1, \pm 2, \pm 3, \pm 4, \pm 6, \pm 12, \pm\frac{1}{5}, \pm\frac{2}{5}, \pm\frac{3}{5}, \pm\frac{4}{5}, \pm\frac{6}{5}, \pm\frac{12}{5}$

(c) $-2, \frac{2}{5}$ **(d)** $-\sqrt{3}, \sqrt{3}$

7. (a)

Positive	Negative	Nonreal Complex
4	0	0
2	0	2
0	0	4

(b) $0, \pm 1, \pm 2, \pm 4, \pm 8, \pm 16$

(c) 0, 2 (multiplicity 2)

(d) $1 - i\sqrt{3}, 1 + i\sqrt{3}$

8. (a)

Positive	Negative	Nonreal Complex
1	3	0
1	1	2

(b) $\pm 1, \pm 3, \pm 9, \pm\frac{1}{2}, \pm\frac{3}{2}, \pm\frac{9}{2}$

(c) -3 (multiplicity 2)

(d) $\frac{2 - \sqrt{6}}{2}, \frac{2 + \sqrt{6}}{2}$

9. (a)

Positive	Negative	Nonreal Complex
1	3	0
1	1	2

(b) $\pm 1, \pm \frac{1}{2}, \pm \frac{1}{4}, \pm \frac{1}{8}$

(c) $-1, \frac{1}{2}$

(d) $-\frac{1}{4} - \frac{\sqrt{3}}{4}i, -\frac{1}{4} + \frac{\sqrt{3}}{4}i$

10. (a)

Positive	Negative	Nonreal Complex
3	2	0
3	0	2
1	2	2
1	0	4

(b) $\pm 1, \pm 2, \pm 3, \pm 6, \pm \frac{1}{2}, \pm \frac{3}{2}$

(c) $-3, -2, \frac{1}{2}$, 1 (multiplicity 2)

(d) no other complex zeros

Answers for the remaining exercises are given in the answer section at the back of the text.

For each polynomial function, do the following in order.

(a) Use Descartes' rule of signs to determine the different possibilities for the numbers of positive, negative, and nonreal complex zeros.
(b) Use the rational zeros theorem to determine the possible rational zeros.
(c) Find the rational zeros, if any.
(d) Find all other real zeros, if any.
(e) Find any other complex zeros (that is, zeros that are not real), if any.
(f) Find the x-intercepts of the graph, if any.
(g) Find the y-intercept of the graph.
(h) Use synthetic division to find $f(4)$, and give the coordinates of the corresponding point on the graph.
(i) Determine the end behavior of the graph.
(j) Sketch the graph.

11. $f(x) = x^4 + 3x^3 - 3x^2 - 11x - 6$

12. $f(x) = -2x^5 + 5x^4 + 34x^3 - 30x^2 - 84x + 45$

13. $f(x) = 2x^5 - 10x^4 + x^3 - 5x^2 - x + 5$

14. $f(x) = 3x^4 - 4x^3 - 22x^2 + 15x + 18$

15. $f(x) = -2x^4 - x^3 + x + 2$

16. $f(x) = 4x^5 + 8x^4 + 9x^3 + 27x^2 + 27x$ (*Hint:* Factor out x first.)

17. $f(x) = 3x^4 - 14x^2 - 5$ (*Hint:* Factor the polynomial.)

18. $f(x) = -x^5 - x^4 + 10x^3 + 10x^2 - 9x - 9$

19. $f(x) = -3x^4 + 22x^3 - 55x^2 + 52x - 12$

20. For the polynomial functions in **Exercises 11–19** that have irrational zeros, find approximations to the nearest thousandth.

3.5 Rational Functions: Graphs, Applications, and Models

- The Reciprocal Function $f(x) = \frac{1}{x}$
- The Function $f(x) = \frac{1}{x^2}$
- Asymptotes
- Steps for Graphing Rational Functions
- Rational Function Models

A rational expression is a fraction that is the quotient of two polynomials. A function defined by a rational expression is a *rational function.*

Rational Function

A function f of the form

$$f(x) = \frac{p(x)}{q(x)},$$

where $p(x)$ and $q(x)$ are polynomials, with $q(x) \neq 0$, is a **rational function.**

Some examples of rational functions are

$$f(x) = \frac{1}{x}, \quad f(x) = \frac{x+1}{2x^2 + 5x - 3}, \quad \text{and} \quad f(x) = \frac{3x^2 - 3x - 6}{x^2 + 8x + 16}.$$

Rational functions

Since any values of x such that $q(x) = 0$ are excluded from the domain of a rational function, this type of function often has a **discontinuous graph**—that is, a graph that has one or more breaks in it. (See **Chapter 2.**)

Teaching Tip Explain that the graphs of rational functions differ from those of polynomial functions in that their graphs are typically discontinuous at one or more values of x and may include asymptotes.

The Reciprocal Function $f(x) = \frac{1}{x}$ The simplest rational function with a variable denominator is the **reciprocal function.**

$$f(x) = \frac{1}{x} \quad \text{Reciprocal function}$$

The domain of this function is the set of all real numbers except 0. The number 0 cannot be used as a value of x, but it is helpful to find values of $f(x)$ for some values of x very close to 0. We use the table feature of a graphing calculator to do this. The tables in **Figure 41** suggest that $|f(x)|$ increases without bound as x gets closer and closer to 0, which is written in symbols as

$$|f(x)| \rightarrow \infty \quad \text{as} \quad x \rightarrow 0.$$

(The symbol $x \rightarrow 0$ means that x approaches 0, without necessarily ever being equal to 0.) Since x cannot equal 0, the graph of $f(x) = \frac{1}{x}$ will never intersect the vertical line $x = 0$. This line is a **vertical asymptote.**

X	Y1
-1	-1
-.1	-10
-.01	-100
-.001	-1000
-1E-4	-10000
-1E-5	-1E5
-1E-6	-1E6

Y1=1/X

As X approaches 0 from the left, $Y_1 = \frac{1}{X}$ approaches $-\infty$. ($-1E-6$ means -1×10^{-6}.)

X	Y1
1	1
.1	10
.01	100
.001	1000
1E-4	10000
1E-5	100000
1E-6	1E6

Y1=1/X

As X approaches 0 from the right, $Y_1 = \frac{1}{X}$ approaches ∞.

Figure 41

X	Y1
1	1
10	.1
100	.01
1000	.001
10000	1E-4
100000	1E-5
1E6	1E-6

Y1=1/X

As X approaches ∞, $Y_1 = \frac{1}{X}$ approaches 0 through positive values.

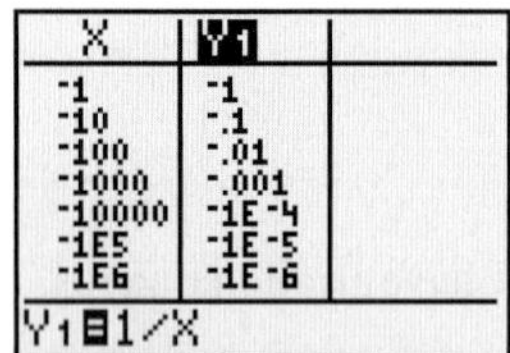

X	Y1
-1	-1
-10	-.1
-100	-.01
-1000	-.001
-10000	-1E-4
-1E5	-1E-5
-1E6	-1E-6

Y1=1/X

As X approaches $-\infty$, $Y_1 = \frac{1}{X}$ approaches 0 through negative values.

Figure 42

As $|x|$ increases without bound, the values of $f(x) = \frac{1}{x}$ get closer and closer to 0, as shown in the tables in **Figure 42.** Letting $|x|$ increase without bound (written $|x| \rightarrow \infty$) causes the graph of $f(x) = \frac{1}{x}$ to move closer and closer to the horizontal line $y = 0$. This line is a **horizontal asymptote.**

The graph and important features of $f(x) = \frac{1}{x}$ are summarized in the following box and shown in **Figure 43.**

Reciprocal Function $f(x) = \frac{1}{x}$

Domain: $(-\infty, 0) \cup (0, \infty)$ Range: $(-\infty, 0) \cup (0, \infty)$

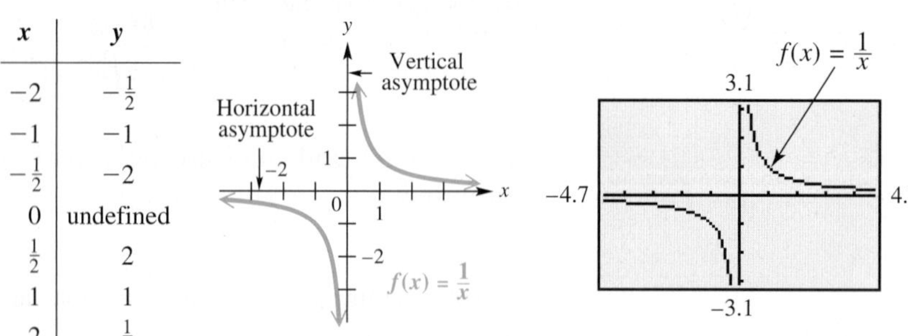

x	y
-2	$-\frac{1}{2}$
-1	-1
$-\frac{1}{2}$	-2
0	undefined
$\frac{1}{2}$	2
1	1
2	$\frac{1}{2}$

Figure 43

- $f(x) = \frac{1}{x}$ decreases on the intervals $(-\infty, 0)$ and $(0, \infty)$.
- It is discontinuous at $x = 0$.
- The y-axis is a vertical asymptote, and the x-axis is a horizontal asymptote.
- It is an odd function, and its graph is symmetric with respect to the origin.

Classroom Example 1
Graph $y = \frac{4}{x}$. Give the domain and range and the intervals of the domain for which the function is increasing or decreasing.

Answer:

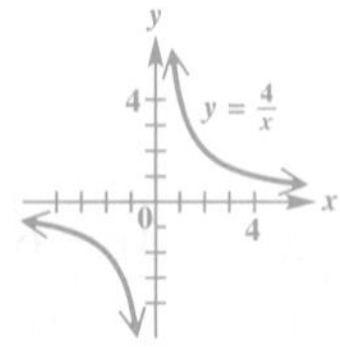

domain: $(-\infty, 0) \cup (0, \infty)$; range: $(-\infty, 0) \cup (0, \infty)$; decreasing on $(-\infty, 0)$ and on $(0, \infty)$

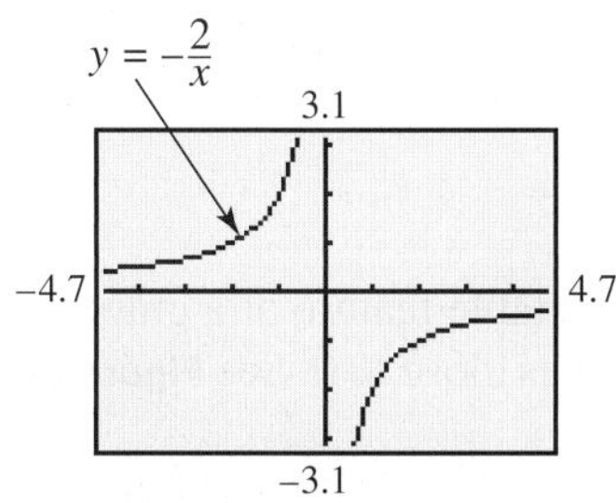

The graph in **Figure 44** is shown here using a **decimal window.** In older models of graphing calculators, using a nondecimal window *may* produce an extraneous vertical line that is not part of the graph.

The graph of $y = \frac{1}{x}$ can be translated and reflected in the same way as other basic graphs in **Chapter 2.**

EXAMPLE 1 Graphing a Rational Function

Graph $y = -\dfrac{2}{x}$. Give the domain and range and the intervals of the domain for which the function is increasing or decreasing.

SOLUTION The expression $-\frac{2}{x}$ can be written as $-2\left(\frac{1}{x}\right)$ or $2\left(\frac{1}{-x}\right)$, indicating that the graph may be obtained by stretching the graph of $y = \frac{1}{x}$ vertically by a factor of 2 and reflecting it across either the x-axis or y-axis. The x- and y-axes remain the horizontal and vertical asymptotes. The domain and range are both still $(-\infty, 0) \cup (0, \infty)$. See **Figure 44.**

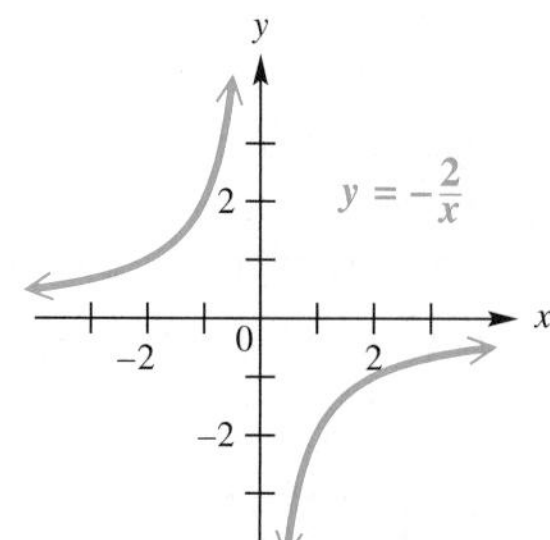

Figure 44

The graph shows that $f(x)$ is increasing on both sides of its vertical asymptote. Thus, it is increasing on

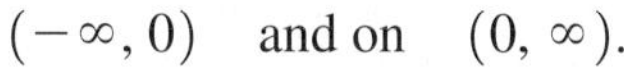

$$(-\infty, 0) \quad \text{and on} \quad (0, \infty).$$

Now Try Exercise 17.

Classroom Example 2
Graph $f(x) = -\frac{1}{x-3}$. Give the domain and range and the intervals of the domain for which the function is increasing or decreasing.

Answer:

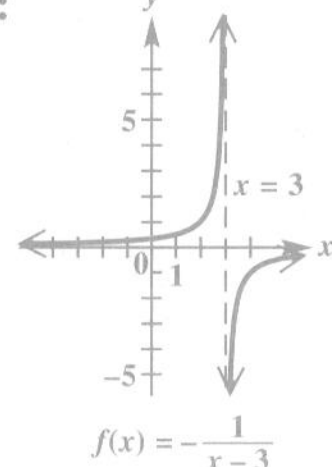

domain: $(-\infty, 3) \cup (3, \infty)$;
range: $(-\infty, 0) \cup (0, \infty)$;
increasing on $(-\infty, 3)$ and on $(3, \infty)$

EXAMPLE 2 Graphing a Rational Function

Graph $f(x) = \dfrac{2}{x+1}$. Give the domain and range and the intervals of the domain for which the function is increasing or decreasing.

ALGEBRAIC SOLUTION

The expression $\frac{2}{x+1}$ can be written as $2\left(\frac{1}{x+1}\right)$, indicating that the graph may be obtained by shifting the graph of $y = \frac{1}{x}$ to the left 1 unit and stretching it vertically by a factor of 2. The graph is shown in **Figure 45.**

The horizontal shift affects the domain, which is now $(-\infty, -1) \cup (-1, \infty)$. The line $x = -1$ is the vertical asymptote, and the line $y = 0$ (the x-axis) remains the horizontal asymptote. The range is still $(-\infty, 0) \cup (0, \infty)$. The graph shows that $f(x)$ is decreasing on both sides of its vertical asymptote. Thus, it is decreasing on $(-\infty, -1)$ and on $(-1, \infty)$.

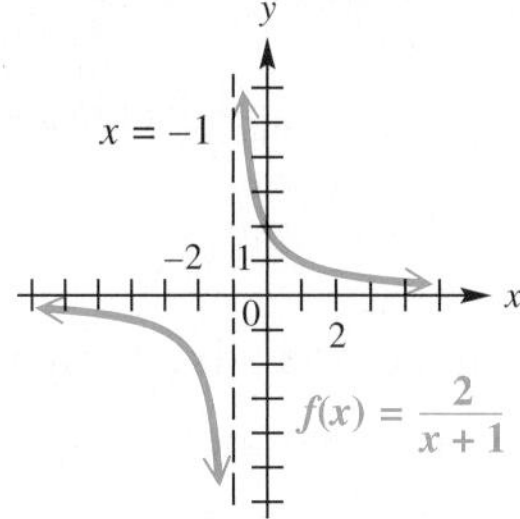

Figure 45

GRAPHING CALCULATOR SOLUTION

When entering this rational function into the function editor, make sure that the numerator is 2 and the denominator is the entire expression $(x + 1)$. The graph of this function has a vertical asymptote at $x = -1$ and a horizontal asymptote at $y = 0$, so it is reasonable to choose a viewing window that contains both asymptotes as well as enough of the graph to determine its basic characteristics. Notice in **Figure 46** that the asymptotes do not actually appear on the calculator screen.

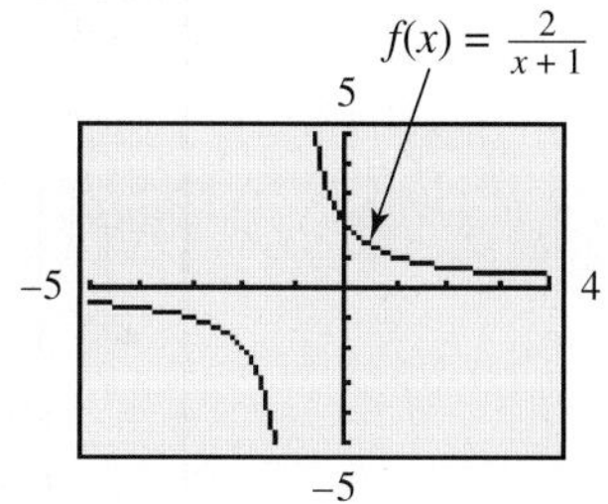

Figure 46

We may need to experiment with various modes and windows to obtain an accurate calculator graph of a rational function.

Now Try Exercise 19.

The Function $f(x) = \frac{1}{x^2}$ The rational function

$$f(x) = \frac{1}{x^2} \quad \text{Rational function}$$

also has domain $(-\infty, 0) \cup (0, \infty)$. We can use the table feature of a graphing calculator to examine values of $f(x)$ for some x-values close to 0. See **Figure 47.**

X	Y_1
-1	1
-.1	100
-.01	10000
-.001	1E6
-1E-4	1E8
-1E-5	1E10
-1E-6	1E12

Y_1=1/X²

As X approaches 0 from the left, $Y_1 = \frac{1}{X^2}$ approaches ∞.

X	Y_1
1	1
.1	100
.01	10000
.001	1E6
1E-4	1E8
1E-5	1E10
1E-6	1E12

Y_1=1/X²

As X approaches 0 from the right, $Y_1 = \frac{1}{X^2}$ approaches ∞.

Figure 47

X	Y_1
1	1
10	.01
100	1E-4
1000	1E-6
10000	1E-8
100000	1E-10
1E6	1E-12

Y_1=1/X²

As X approaches ∞, $Y_1 = \frac{1}{X^2}$ approaches 0 through positive values.

X	Y_1
-1	1
-10	.01
-100	1E-4
-1000	1E-6
-10000	1E-8
-1E5	1E-10
-1E6	1E-12

Y_1=1/X²

As X approaches $-\infty$, $Y_1 = \frac{1}{X^2}$ approaches 0 through positive values.

Figure 48

The tables suggest that $f(x)$ increases without bound as x gets closer and closer to 0. Notice that as x approaches 0 from *either* side, function values are all positive and there is symmetry with respect to the y-axis. Thus, $f(x) \rightarrow \infty$ as $x \rightarrow 0$. The y-axis $(x = 0)$ is the vertical asymptote.

As $|x|$ increases without bound, $f(x)$ approaches 0, as suggested by the tables in **Figure 48.** Again, function values are all positive. The x-axis is the horizontal asymptote of the graph.

The graph and important features of $f(x) = \frac{1}{x^2}$ are summarized in the following box and shown in **Figure 49.**

Rational Function $f(x) = \frac{1}{x^2}$

Domain: $(-\infty, 0) \cup (0, \infty)$ Range: $(0, \infty)$

x	y
± 3	$\frac{1}{9}$
± 2	$\frac{1}{4}$
± 1	1
$\pm \frac{1}{2}$	4
$\pm \frac{1}{4}$	16
0	undefined

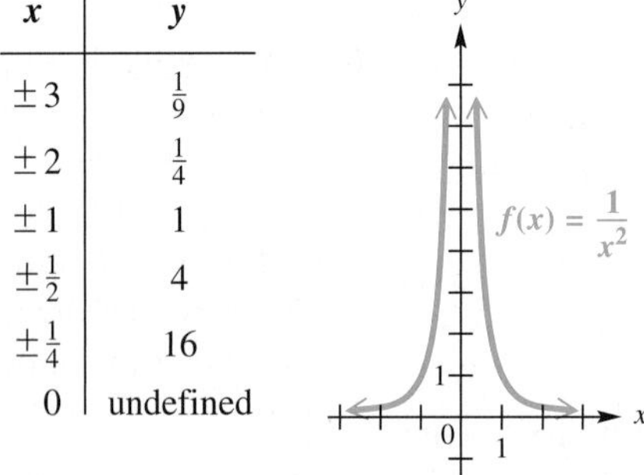

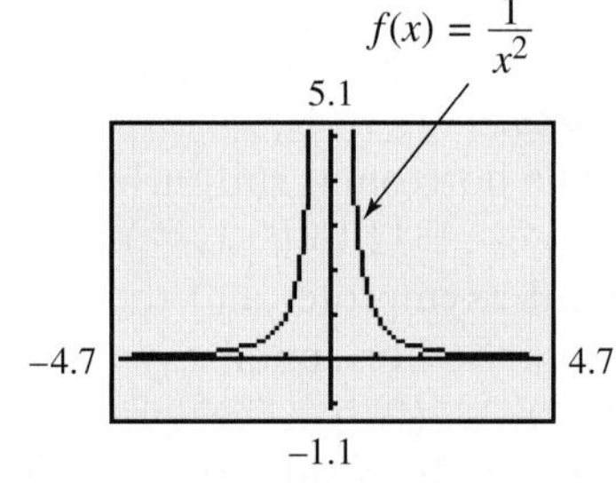

Figure 49

- $f(x) = \frac{1}{x^2}$ increases on the interval $(-\infty, 0)$ and decreases on the interval $(0, \infty)$.
- It is discontinuous at $x = 0$.
- The y-axis is a vertical asymptote, and the x-axis is a horizontal asymptote.
- It is an even function, and its graph is symmetric with respect to the y-axis.

Classroom Example 3
Graph $y = \frac{2}{(x-2)^2} - 4$. Give the domain and range and the intervals of the domain for which the function is increasing or decreasing.

Answer:

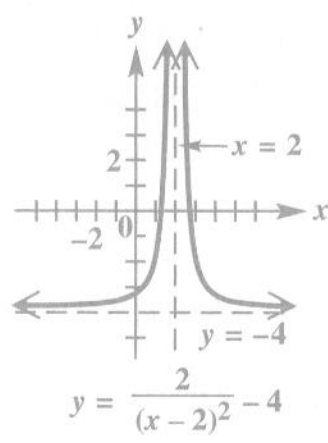

domain: $(-\infty, 2) \cup (2, \infty)$;
range: $(-4, \infty)$;
increasing on $(-\infty, 2)$;
decreasing on $(2, \infty)$

EXAMPLE 3 Graphing a Rational Function

Graph $y = \frac{1}{(x+2)^2} - 1$. Give the domain and range and the intervals of the domain for which the function is increasing or decreasing.

SOLUTION The equation $y = \frac{1}{(x+2)^2} - 1$ is equivalent to

$$y = f(x+2) - 1, \quad \text{where} \quad f(x) = \frac{1}{x^2}.$$

This indicates that the graph will be shifted 2 units to the left and 1 unit down. The horizontal shift affects the domain, now $(-\infty, -2) \cup (-2, \infty)$, while the vertical shift affects the range, now $(-1, \infty)$.

The vertical asymptote has equation $x = -2$, and the horizontal asymptote has equation $y = -1$. A traditional graph is shown in **Figure 50,** with a calculator graph in **Figure 51.** Both graphs show that this function is increasing on $(-\infty, -2)$ and decreasing on $(-2, \infty)$.

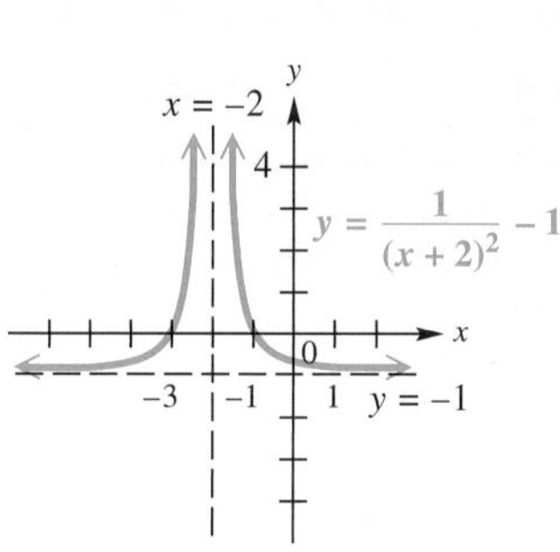

This is the graph of $y = \frac{1}{x^2}$ shifted 2 units to the left and 1 unit down.

Figure 50

$y = \frac{1}{(x+2)^2} - 1$

3.1

-6.7 2.7

-3.1

With older model calculators, a choice of window other than the one here *may* produce an undesired vertical line at $x = -2$.

Figure 51

Now Try Exercise 27.

Asymptotes The preceding examples suggest the following definitions of vertical and horizontal asymptotes.

Asymptotes

Let $p(x)$ and $q(x)$ define polynomials. Consider the rational function $f(x) = \frac{p(x)}{q(x)}$, written in lowest terms, and real numbers a and b.

1. If $|f(x)| \rightarrow \infty$ as $x \rightarrow a$, then the line $x = a$ is a **vertical asymptote.**

2. If $f(x) \rightarrow b$ as $|x| \rightarrow \infty$, then the line $y = b$ is a **horizontal asymptote.**

Locating asymptotes is important when graphing rational functions. We find *vertical asymptotes* by determining the values of x that make the denominator equal to 0. To find *horizontal asymptotes* (and, in some cases, *oblique asymptotes*), we must consider what happens to $f(x)$ as $|x| \rightarrow \infty$. These asymptotes determine the end behavior of the graph.

LOOKING AHEAD TO CALCULUS

The rational function defined by

$$f(x) = \frac{2}{x+1}$$

in **Example 2** has a vertical asymptote at $x = -1$. In calculus, the behavior of the graph of this function for values close to -1 is described using **one-sided limits.** As x approaches -1 from the *left,* the function values decrease without bound. This is written

$$\lim_{x \to -1^-} f(x) = -\infty.$$

As x approaches -1 from the *right,* the function values increase without bound. This is written

$$\lim_{x \to -1^+} f(x) = \infty.$$

Determining Asymptotes

To find the asymptotes of a rational function defined by a rational expression in *lowest terms,* use the following procedures.

1. Vertical Asymptotes

Find any vertical asymptotes by setting the denominator equal to 0 and solving for x. If a is a zero of the denominator, then **the line $x = a$ is a vertical asymptote.**

2. Other Asymptotes

Determine any other asymptotes by considering three possibilities:

(a) If the numerator has lesser degree than the denominator, then there is a **horizontal asymptote $y = 0$** (the x-axis).

(b) If the numerator and denominator have the same degree, and the function is of the form

$$f(x) = \frac{a_n x^n + \cdots + a_0}{b_n x^n + \cdots + b_0}, \quad \text{where} \quad a_n, b_n \neq 0,$$

then the **horizontal asymptote has equation $y = \frac{a_n}{b_n}$.**

(c) If the numerator is of degree exactly one more than the denominator, then there will be an **oblique (slanted) asymptote.** To find it, divide the numerator by the denominator and disregard the remainder. Set the rest of the quotient equal to y to obtain the equation of the asymptote.

NOTE The graph of a rational function may have more than one vertical asymptote, or it may have none at all. ***The graph cannot intersect any vertical asymptote. There can be at most one other (nonvertical) asymptote, and the graph may intersect that asymptote,*** as we shall see in **Example 7.**

Classroom Example 4
For each rational function f, find all asymptotes. (*Note:* We will use these results in **Classroom Examples 5, 6, and 8.**)

(a) $f(x) = \frac{x-2}{x^2-x-6}$

(b) $f(x) = \frac{4x-2}{x+1}$

(c) $f(x) = \frac{x^2+x}{x-1}$

Answers:

(a) vertical: $x = -2$, $x = 3$; horizontal: $y = 0$

(b) vertical: $x = -1$; horizontal: $y = 4$

(c) vertical: $x = 1$; oblique: $y = x + 2$

EXAMPLE 4 Finding Asymptotes of Rational Functions

For each rational function f, find all asymptotes.

(a) $f(x) = \frac{x+1}{(2x-1)(x+3)}$ **(b)** $f(x) = \frac{2x+1}{x-3}$ **(c)** $f(x) = \frac{x^2+1}{x-2}$

SOLUTION

(a) ***To find the vertical asymptotes, set the denominator equal to 0 and solve.***

$$(2x-1)(x+3) = 0$$

$$2x - 1 = 0 \quad \text{or} \quad x + 3 = 0 \qquad \text{Zero-factor property (Section 1.4)}$$

$$x = \frac{1}{2} \quad \text{or} \quad x = -3 \qquad \text{Solve each equation. (Section 1.1)}$$

The equations of the vertical asymptotes are $x = \frac{1}{2}$ and $x = -3$.

To find the equation of the horizontal asymptote, divide each term by the greatest power of x in the expression. First, multiply the factors in the denominator.

$$f(x) = \frac{x+1}{(2x-1)(x+3)} = \frac{x+1}{2x^2+5x-3}$$

Now divide each term in the numerator and denominator by x^2 since 2 is the greatest power of x.

$$f(x) = \frac{\frac{x}{x^2} + \frac{1}{x^2}}{\frac{2x^2}{x^2} + \frac{5x}{x^2} - \frac{3}{x^2}} = \frac{\frac{1}{x} + \frac{1}{x^2}}{2 + \frac{5}{x} - \frac{3}{x^2}}$$

Stop here. Leave the expression in complex form.

Teaching Tip In **Example 4** encourage students to verify their analysis of the behavior of $f(x)$ as $|x| \to \infty$ using large (positive and negative) values of x in the original function.

As $|x|$ increases without bound, the quotients $\frac{1}{x}$, $\frac{1}{x^2}$, $\frac{5}{x}$, and $\frac{3}{x^2}$ all approach 0, and the value of $f(x)$ approaches

$$\frac{0+0}{2+0-0} = \frac{0}{2}, \quad \text{or} \quad 0.$$

The line $y = 0$ (that is, the x-axis) is therefore the horizontal asymptote. This supports procedure 2(a) of determining asymptotes on the previous page.

(b) Set the denominator $x - 3$ equal to 0 to find that the vertical asymptote has equation $x = 3$. To find the horizontal asymptote, divide each term in the rational expression by x since the greatest power of x in the expression is 1.

$$f(x) = \frac{2x+1}{x-3} = \frac{\frac{2x}{x} + \frac{1}{x}}{\frac{x}{x} - \frac{3}{x}} = \frac{2 + \frac{1}{x}}{1 - \frac{3}{x}}$$

As $|x|$ increases without bound, both $\frac{1}{x}$ and $\frac{3}{x}$ approach 0, and $f(x)$ approaches

$$\frac{2+0}{1-0} = \frac{2}{1}, \quad \text{or} \quad 2,$$

so the line $y = 2$ is the horizontal asymptote. This supports procedure 2(b) of determining asymptotes on the previous page.

(c) Setting the denominator $x - 2$ equal to 0 shows that the vertical asymptote has equation $x = 2$. If we divide by the greatest power of x as before (x^2 in this case), we see that there is no horizontal asymptote because

$$f(x) = \frac{x^2+1}{x-2} = \frac{\frac{x^2}{x^2} + \frac{1}{x^2}}{\frac{x}{x^2} - \frac{2}{x^2}} = \frac{1 + \frac{1}{x^2}}{\frac{1}{x} - \frac{2}{x^2}}$$

does not approach any real number as $|x| \to \infty$, since $\frac{1+0}{0-0} = \frac{1}{0}$ is undefined. This happens whenever the degree of the numerator is greater than the degree of the denominator.

In such cases, divide the denominator into the numerator to write the expression in another form. We use synthetic division, as shown in the margin. The result enables us to write the function as follows.

$$\begin{array}{r|rrr} 2 & 1 & 0 & 1 \\ & & 2 & 4 \\ \hline & 1 & 2 & 5 \end{array}$$

(Section 3.2)

$$f(x) = x + 2 + \frac{5}{x-2}$$

For very large values of $|x|$, $\frac{5}{x-2}$ is close to 0, and the graph approaches the line $y = x + 2$. This line is an **oblique asymptote** (slanted, neither vertical nor horizontal) for the graph of the function. This supports procedure 2(c) of determining asymptotes.

Now Try Exercises 37, 39, and 41.

Steps for Graphing Rational Functions A comprehensive graph of a rational function will show the following characteristics.

- all x- and y-intercepts
- all asymptotes: vertical, horizontal, and/or oblique
- the point at which the graph intersects its nonvertical asymptote (if there is any such point)
- the behavior of the function on each domain interval determined by the vertical asymptotes and x-intercepts

Graphing a Rational Function

Let $f(x) = \frac{p(x)}{q(x)}$ define a function where $p(x)$ and $q(x)$ are polynomials and the rational expression is written in lowest terms. To sketch its graph, follow these steps.

Step 1 Find any vertical asymptotes.

Step 2 Find any horizontal or oblique asymptotes.

Step 3 Find the y-intercept by evaluating $f(0)$.

Step 4 Find the x-intercepts, if any, by solving $f(x) = 0$. (These will be the zeros of the numerator, $p(x)$.)

Step 5 Determine whether the graph will intersect its nonvertical asymptote $y = b$ or $y = mx + b$ by solving $f(x) = b$ or $f(x) = mx + b$.

Step 6 Plot selected points, as necessary. Choose an x-value in each domain interval determined by the vertical asymptotes and x-intercepts.

Step 7 Complete the sketch.

Classroom Example 5

Graph $f(x) = \frac{x-2}{x^2 - x - 6}$.

Answer:

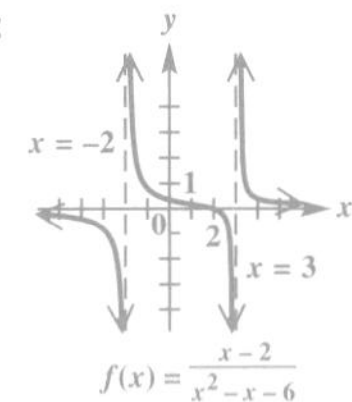

EXAMPLE 5 **Graphing a Rational Function with the *x*-Axis as Horizontal Asymptote**

Graph $f(x) = \frac{x+1}{2x^2 + 5x - 3}$.

SOLUTION

Step 1 Since $2x^2 + 5x - 3 = (2x - 1)(x + 3)$, from **Example 4(a),** the vertical asymptotes have equations $x = \frac{1}{2}$ and $x = -3$.

Step 2 As shown in **Example 4(a),** the horizontal asymptote is the x-axis.

Step 3 The y-intercept is $-\frac{1}{3}$, as justified below.

$$f(0) = \frac{0+1}{2(0)^2 + 5(0) - 3} = -\frac{1}{3}$$ The y-intercept is the ratio of the constant terms.

Step 4 The x-intercept is found by solving $f(x) = 0$.

$$\frac{x+1}{2x^2 + 5x - 3} = 0$$ Set $f(x) = 0$.

$$x + 1 = 0$$ If a rational expression is equal to 0, then its numerator must equal 0.

$$x = -1$$ The x-intercept is -1.

Step 5 To determine whether the graph intersects its horizontal asymptote, solve this equation.

$$f(x) = 0 \longleftarrow$$ y-value of horizontal asymptote

Since the horizontal asymptote is the x-axis, the solution of the equation $f(x) = 0$ was found in Step 4. The graph intersects its horizontal asymptote at $(-1, 0)$.

Step 6 Plot a point in each of the intervals determined by the x-intercepts and vertical asymptotes, $(-\infty, -3)$, $(-3, -1)$, $\left(-1, \frac{1}{2}\right)$ and $\left(\frac{1}{2}, \infty\right)$, to get an idea of how the graph behaves in each interval.

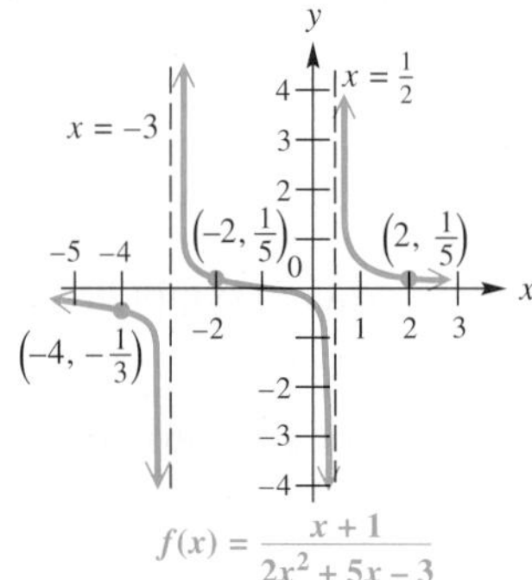

Figure 52

Interval	Test Point	Value of $f(x)$	Sign of $f(x)$	Graph Above or Below x-Axis
$(-\infty, -3)$	-4	$-\frac{1}{3}$	Negative	Below
$(-3, -1)$	-2	$\frac{1}{5}$	Positive	Above
$\left(-1, \frac{1}{2}\right)$	0	$-\frac{1}{3}$	Negative	Below
$\left(\frac{1}{2}, \infty\right)$	2	$\frac{1}{5}$	Positive	Above

Step 7 Complete the sketch as shown in **Figure 52.** This function is decreasing on each interval of its domain—that is, on $(-\infty, -3)$, $\left(-3, \frac{1}{2}\right)$ and $\left(\frac{1}{2}, \infty\right)$.

Now Try Exercise 67.

Classroom Example 6

Graph $f(x) = \frac{4x-2}{x+1}$.

Answer:

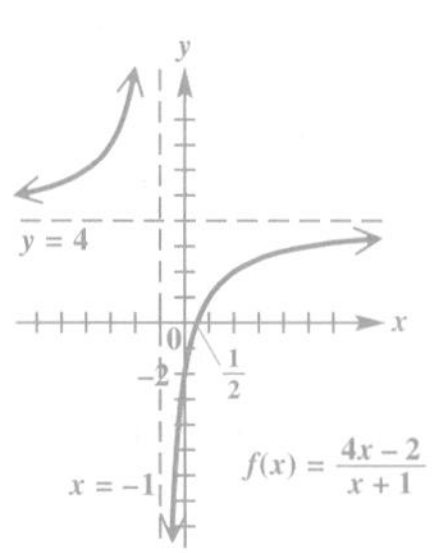

EXAMPLE 6 Graphing a Rational Function That Does Not Intersect Its Horizontal Asymptote

Graph $f(x) = \frac{2x+1}{x-3}$.

SOLUTION

Steps 1 and 2 As determined in **Example 4(b),** the equation of the vertical asymptote is $x = 3$. The horizontal asymptote has equation $y = 2$.

Step 3 $f(0) = -\frac{1}{3}$, so the y-intercept is $-\frac{1}{3}$.

Step 4 Solve $f(x) = 0$ to find any x-intercepts.

$$\frac{2x+1}{x-3} = 0 \quad \text{Set } f(x) = 0.$$

$$2x + 1 = 0 \quad \text{If a rational expression is equal to 0, then its numerator must equal 0.}$$

$$x = -\frac{1}{2} \quad \text{The } x\text{-intercept is } -\tfrac{1}{2}.$$

Step 5 The graph does not intersect its horizontal asymptote since $f(x) = 2$ has no solution.

$$\frac{2x+1}{x-3} = 2 \quad \text{Set } f(x) = 2.$$

$$2x + 1 = 2x - 6 \quad \text{Multiply each side by } x - 3.$$

$$1 = -6 \quad \text{Subtract } 2x.$$

A false statement results. The solution set is ∅.

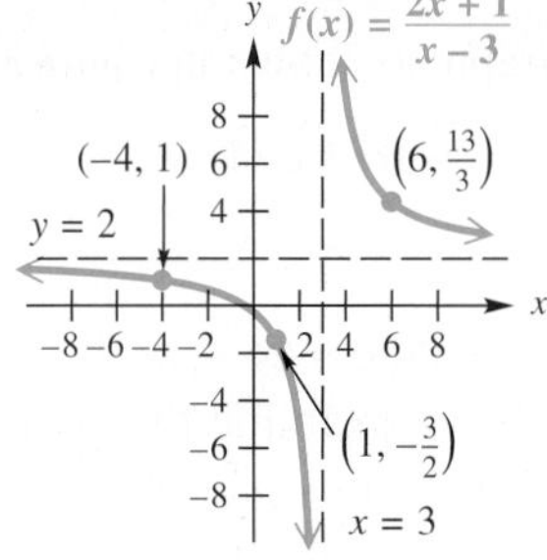

Figure 53

Steps 6 and 7 The points $(-4, 1)$, $\left(1, -\frac{3}{2}\right)$, and $\left(6, \frac{13}{3}\right)$ are on the graph and can be used to complete the sketch of this function, which decreases on every interval of its domain. See **Figure 53.**

Now Try Exercise 63.

LOOKING AHEAD TO CALCULUS

The rational function

$$f(x) = \frac{2x+1}{x-3},$$

seen in **Example 6,** has horizontal asymptote $y = 2$. In calculus, the behavior of the graph of this function as x approaches $-\infty$ and as x approaches ∞ is described using **limits at infinity.** As x approaches $-\infty$, $f(x)$ approaches 2. This is written

$$\lim_{x \to -\infty} f(x) = 2.$$

As x approaches ∞, $f(x)$ approaches 2. This is written

$$\lim_{x \to \infty} f(x) = 2.$$

EXAMPLE 7 Graphing a Rational Function That Intersects Its Horizontal Asymptote

Graph $f(x) = \dfrac{3x^2 - 3x - 6}{x^2 + 8x + 16}$.

SOLUTION

Step 1 To find the vertical asymptote(s), solve $x^2 + 8x + 16 = 0$.

$$x^2 + 8x + 16 = 0 \quad \text{Set the denominator equal to 0.}$$

$$(x+4)^2 = 0 \quad \text{Factor. (Section R.4)}$$

$$x = -4 \quad \text{Zero-factor property}$$

Since the numerator is not 0 when $x = -4$, the vertical asymptote has equation $x = -4$.

Step 2 We divide all terms by x^2 and consider the behavior of each term as $|x|$ increases without bound to get the equation of the horizontal asymptote,

$$y = \frac{3}{1}, \quad \begin{matrix} \leftarrow \text{Leading coefficient of numerator} \\ \leftarrow \text{Leading coefficient of denominator} \end{matrix} \quad \text{or} \quad y = 3.$$

Step 3 The y-intercept is $f(0) = -\frac{3}{8}$.

Step 4 To find the x-intercept(s), if any, we solve $f(x) = 0$.

$$\frac{3x^2 - 3x - 6}{x^2 + 8x + 16} = 0 \quad \text{Set } f(x) = 0.$$

$$3x^2 - 3x - 6 = 0 \quad \text{Set the numerator equal to 0.}$$

$$x^2 - x - 2 = 0 \quad \text{Divide by 3.}$$

$$(x-2)(x+1) = 0 \quad \text{Factor. (Section R.4)}$$

$$x = 2 \quad \text{or} \quad x = -1 \quad \text{Zero-factor property}$$

The x-intercepts are -1 and 2.

Step 5 We set $f(x) = 3$ and solve to locate the point where the graph intersects the horizontal asymptote.

$$\frac{3x^2 - 3x - 6}{x^2 + 8x + 16} = 3 \quad \text{Set } f(x) = 3.$$

$$3x^2 - 3x - 6 = 3x^2 + 24x + 48 \quad \text{Multiply each side by } x^2 + 8x + 16.$$

$$-3x - 6 = 24x + 48 \quad \text{Subtract } 3x^2.$$

$$-27x = 54 \quad \text{Subtract } 24x \text{ and add 6.}$$

$$x = -2 \quad \text{Divide by } -27.$$

The graph intersects its horizontal asymptote at $(-2, 3)$.

Steps 6 and 7 Some other points that lie on the graph are $(-10, 9)$, $\left(-8, 13\frac{1}{8}\right)$, and $\left(5, \frac{2}{3}\right)$. These are used to complete the graph, as shown in **Figure 54.**

Now Try Exercise 83.

Classroom Example 7

Graph $f(x) = \dfrac{2x^2 + 3x - 4}{x^2 + 6x + 9}$.

Answer:

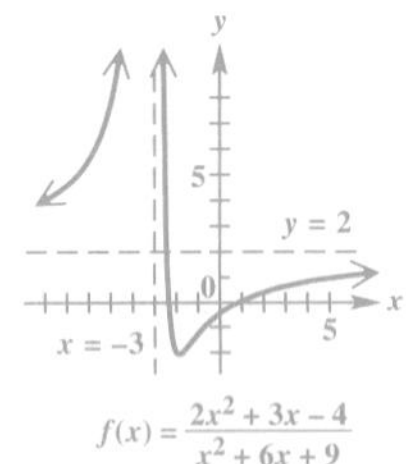

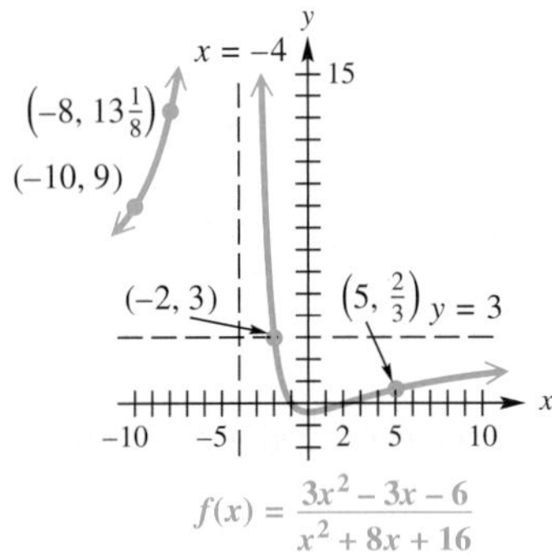

Figure 54

Notice the behavior of the graph of the function in **Figure 54** near the line $x = -4$. As $x \to -4$ from either side, $f(x) \to \infty$. If we examine the behavior of the graph of the function in **Figure 53** (on the previous page) near the line $x = 3$, we find that $f(x) \to -\infty$ as x approaches 3 from the left, while $f(x) \to \infty$ as x approaches 3 from the right.

LOOKING AHEAD TO CALCULUS

Different types of discontinuity are discussed in calculus. The function in **Example 9,**

$$f(x) = \frac{x^2 - 4}{x - 2},$$

is said to have a **removable discontinuity** at $x = 2$, since the discontinuity can be removed by redefining f at 2. The function in **Example 8,**

$$f(x) = \frac{x^2 + 1}{x - 2},$$

has **infinite discontinuity** at $x = 2$, as indicated by the vertical asymptote there. The greatest integer function, discussed in **Section 2.6,** has **jump discontinuities** because the function values "jump" from one value to another for integer domain values.

The behavior of the graph of a rational function near a vertical asymptote $x = a$ partially depends on the exponent on $x - a$ in the denominator.

Behavior of Graphs of Rational Functions near Vertical Asymptotes

Suppose that $f(x)$ is a rational expression in lowest terms. If n is the largest positive integer such that $(x - a)^n$ is a factor of the denominator of $f(x)$, the graph will behave in the manner illustrated.

If n is even:

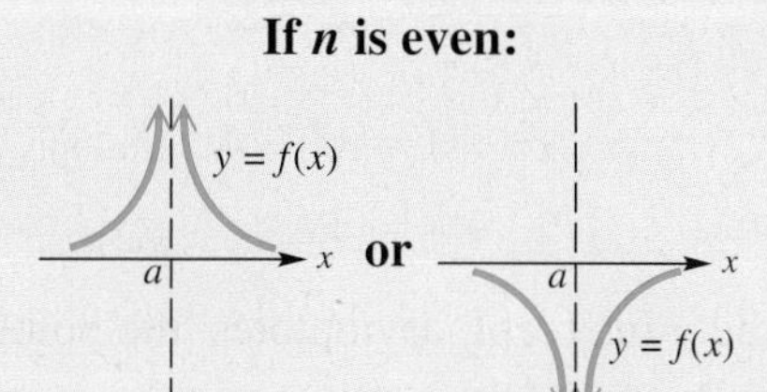

If n is odd:

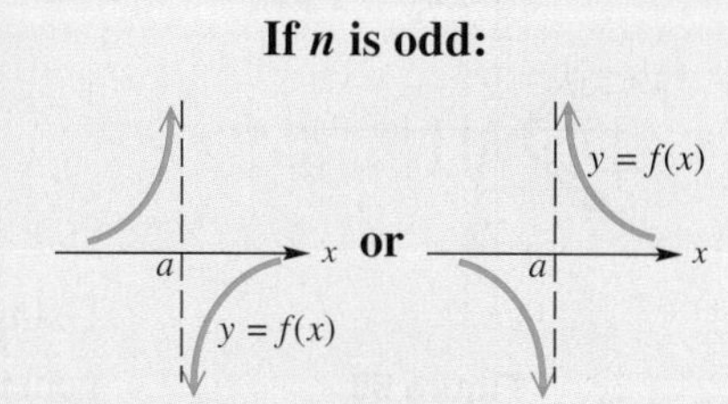

In **Section 3.4** we observed that the behavior of the graph of a polynomial function near its zeros is dependent on the multiplicity of the zero. The same statement can be made for rational functions. Suppose that $f(x)$ is defined by a rational expression in lowest terms. If n is the greatest positive integer such that $(x - c)^n$ is a factor of the numerator of $f(x)$, the graph will behave in the manner illustrated.

If $n = 1$:

x or x

If n is even:

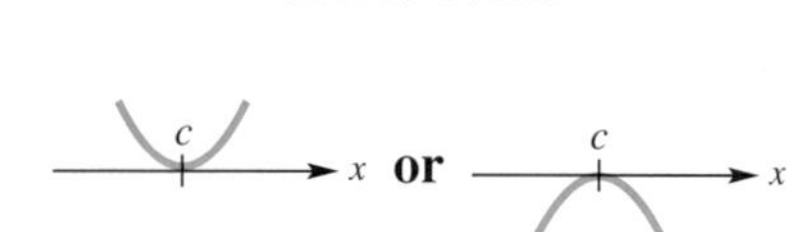

If n is an odd integer greater than 1:

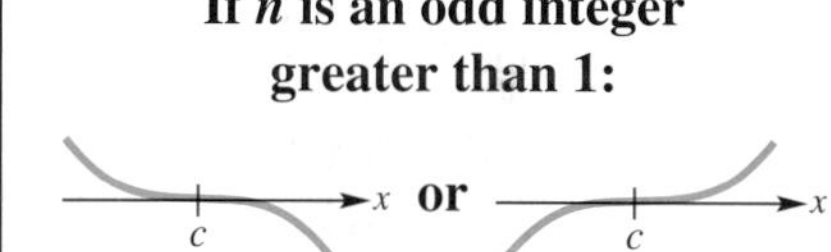

NOTE Suppose that we are asked to reverse the process of **Example 7** and find the equation of a rational function having the graph shown in **Figure 54** on the previous page. Since the graph crosses the x-axis at its x-intercepts -1 and 2, the numerator must have factors $(x + 1)$ and $(x - 2)$, each of degree 1.

The graph's behavior at its vertical asymptote $x = -4$ suggests that there is a factor of $(x + 4)$ of even degree in the denominator. The horizontal asymptote at $y = 3$ indicates that the numerator and denominator have the same degree (both 2) and that the ratio of leading coefficients is 3.

Verify in **Example 7** that the rational function is as follows.

$$f(x) = \frac{3(x + 1)(x - 2)}{(x + 4)^2},$$

or $$f(x) = \frac{3x^2 - 3x - 6}{x^2 + 8x + 16}$$ Multiply the factors in the numerator. Square in the denominator.

In the Exercise set, problems of this type are labeled *Connecting Graphs with Equations*.

EXAMPLE 8 Graphing a Rational Function with an Oblique Asymptote

Graph $f(x)=\dfrac{x^2+1}{x-2}$.

SOLUTION As shown in **Example 4,** the vertical asymptote has equation $x = 2$, and the graph has an oblique asymptote with equation $y = x + 2$. The y-intercept is $-\frac{1}{2}$, and the graph has no x-intercepts since the numerator, $x^2 + 1$, has no real zeros. The graph does not intersect its oblique asymptote because the following has no solution.

$$\frac{x^2+1}{x-2}=x+2$$

$$x^2+1=x^2-4 \quad \text{Multiply each side by } x-2.$$

$$1=-4 \quad \text{False}$$

Using the y intercept, asymptotes, the points $\left(4, \frac{17}{2}\right)$ and $\left(-1, -\frac{2}{3}\right)$, and the general behavior of the graph near its asymptotes leads to the graph in **Figure 55.**

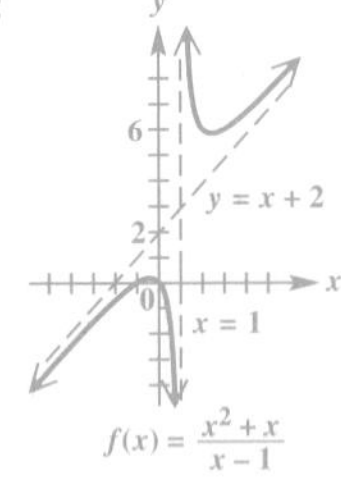

Figure 55

Now Try Exercise 87.

Classroom Example 8

Graph $f(x)=\dfrac{x^2+x}{x-1}$.

Answer:

A rational function that is not in lowest terms often has a "hole," or **point of discontinuity,** in its graph.

EXAMPLE 9 Graphing a Rational Function Defined by an Expression That Is Not in Lowest Terms

Graph $f(x)=\dfrac{x^2-4}{x-2}$.

ALGEBRAIC SOLUTION

The domain of this function cannot include 2. The expression $\frac{x^2-4}{x-2}$ should be written in lowest terms.

$$f(x)=\frac{x^2-4}{x-2} \quad \text{Factor and then divide.}$$

$$=\frac{(x+2)(x-2)}{x-2} \quad \text{Factor.}$$

$$=x+2, \quad x\neq 2$$

Therefore, the graph of this function will be the same as the graph of $y = x + 2$ (a straight line), with the exception of the point with x-value 2. A "hole" appears in the graph at $(2, 4)$. See **Figure 56.**

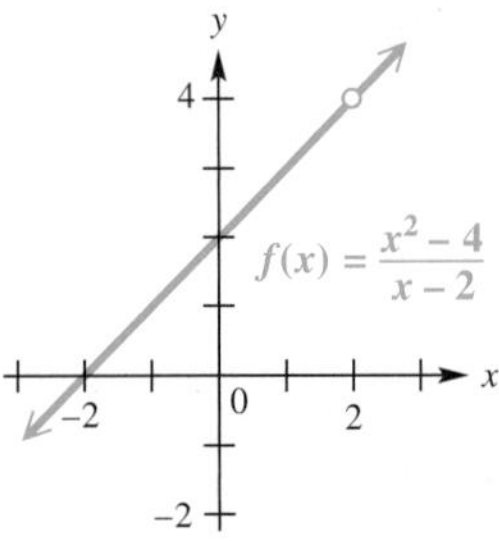

Figure 56

GRAPHING CALCULATOR SOLUTION

If we set the window of a graphing calculator so that an x-value of 2 is displayed, then we can see that the calculator cannot determine a value for y. We define

$$Y_1=\frac{X^2-4}{X-2}$$

and graph it in such a window, as in **Figure 57.** The error message in the table further supports the existence of a discontinuity at $X = 2$. (For the table, $Y_2 = X + 2$.)

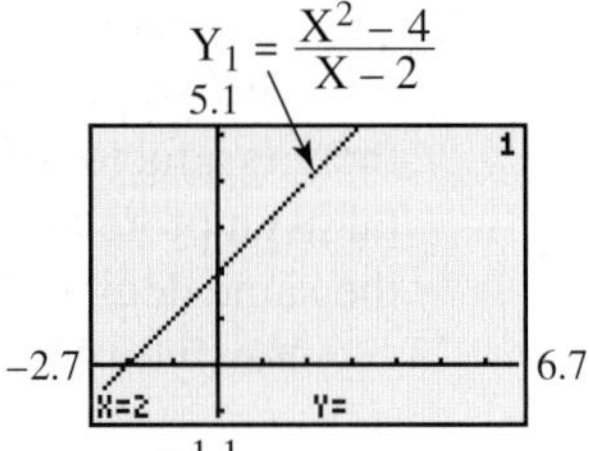

Figure 57

Notice the visible discontinuity at $X = 2$ in the graph. The window was chosen so the "hole" would be visible. This requires a decimal viewing window or a window with x-values centered at 2. Other window choices may not show this discontinuity.

Now Try Exercise 91.

Classroom Example 9

Graph $f(x) = \frac{x^2 - x - 6}{x - 3}$.

Answer:

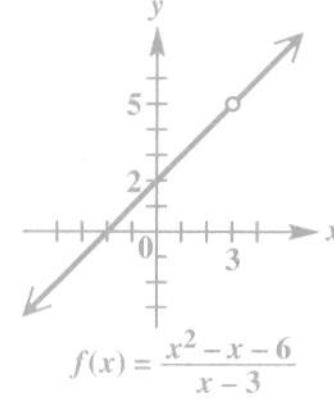

Classroom Example 10

Refer to the function in **Example 10.** Suppose that vehicles arrive randomly at a parking ramp at an average rate of 3.6 vehicles per minute and the parking attendant can admit 4.0 vehicles per minute.

(a) Determine the traffic intensity x for this parking ramp.

(b) Compute the average number of vehicles waiting in line to enter the ramp.

Answers:

(a) $x = 0.9$ **(b)** 4.05 vehicles

Rational Function Models

Rational functions have a variety of applications.

EXAMPLE 10 Modeling Traffic Intensity with a Rational Function

Vehicles arrive randomly at a parking ramp at an average rate of 2.6 vehicles per minute. The parking attendant can admit 3.2 vehicles per minute. However, since arrivals are random, lines form at various times. (*Source:* Mannering, F. and W. Kilareski, *Principles of Highway Engineering and Traffic Analysis,* 2nd ed., John Wiley & Sons.)

(a) The **traffic intensity** x is defined as the ratio of the average arrival rate to the average admittance rate. Determine x for this parking ramp.

(b) The average number of vehicles waiting in line to enter the ramp is given by

$$f(x) = \frac{x^2}{2(1 - x)},$$

where $0 \le x < 1$ is the traffic intensity. Graph $f(x)$ and compute $f(0.8125)$ for this parking ramp.

(c) What happens to the number of vehicles waiting as the traffic intensity approaches 1?

SOLUTION

(a) The average arrival rate is 2.6 vehicles and the average admittance rate is 3.2 vehicles, so

$$x = \frac{2.6}{3.2} = 0.8125.$$

(b) A calculator graph of f is shown in **Figure 58.**

$$f(0.8125) = \frac{0.8125^2}{2(1 - 0.8125)} \approx 1.76 \text{ vehicles}$$

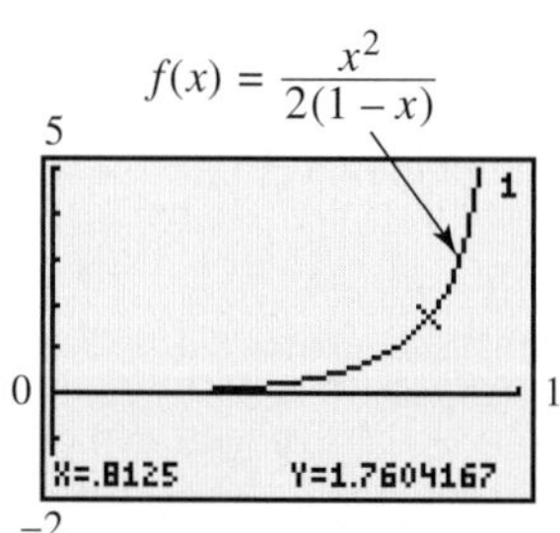

Figure 58

(c) From the graph we see that as x approaches 1, $y = f(x)$ gets very large. Thus, the average number of waiting vehicles gets very large. This is what we would expect.

Now Try Exercise 113.

3.5 Exercises

Concept Check Provide a short answer to each question.

1. What is the domain of $f(x) = \frac{1}{x}$? What is its range?

2. What is the domain of $f(x) = \frac{1}{x^2}$? What is its range?

3. What is the interval over which $f(x) = \frac{1}{x}$ increases? decreases? is constant?

1. $(-\infty, 0) \cup (0, \infty)$; $(-\infty, 0) \cup (0, \infty)$
2. $(-\infty, 0) \cup (0, \infty)$; $(0, \infty)$
3. none; $(-\infty, 0)$ and $(0, \infty)$; none
4. $(-\infty, 0)$; $(0, \infty)$; none
5. $x = 3$; $y = 2$
6. $x = -2$; $y = -4$
7. even; symmetry with respect to the y-axis

8. odd; symmetry with respect to the origin

9. A, B, C **10.** B
11. A **12.** C, D
13. A **14.** D
15. A, C, D **16.** C

17. To obtain the graph of f, stretch the graph of $y = \frac{1}{x}$ vertically by a factor of 2.

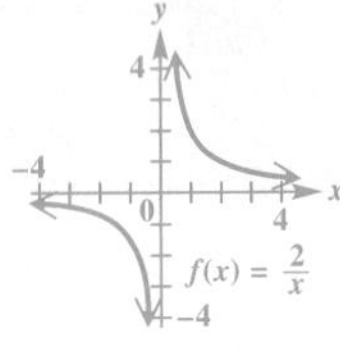

(a) $(-\infty, 0) \cup (0, \infty)$
(b) $(-\infty, 0) \cup (0, \infty)$
(c) none
(d) $(-\infty, 0)$ and $(0, \infty)$

18. To obtain the graph of f, stretch the graph of $y = \frac{1}{x}$ vertically by a factor of 3 and reflect across the x-axis *or* the y-axis.

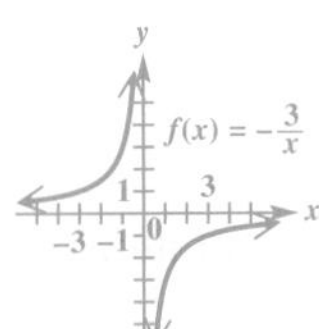

(a) $(-\infty, 0) \cup (0, \infty)$
(b) $(-\infty, 0) \cup (0, \infty)$
(c) $(-\infty, 0)$ and $(0, \infty)$
(d) none

19. To obtain the graph of f, shift the graph of $y = \frac{1}{x}$ to the left 2 units.

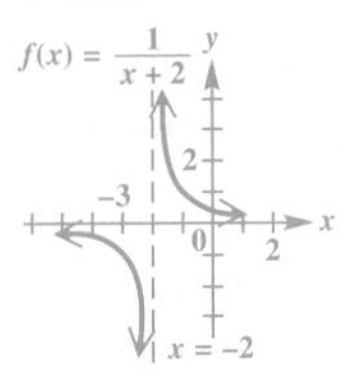

(a) $(-\infty, -2) \cup (-2, \infty)$
(b) $(-\infty, 0) \cup (0, \infty)$
(c) none
(d) $(-\infty, -2)$ and $(-2, \infty)$

20. To obtain the graph of f, shift the graph of $y = \frac{1}{x}$ to the right 3 units.

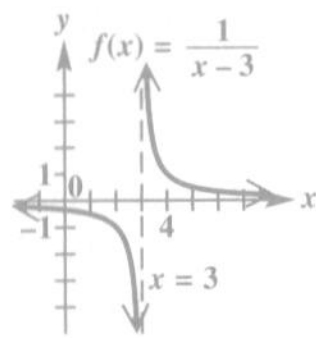

(a) $(-\infty, 3) \cup (3, \infty)$
(b) $(-\infty, 0) \cup (0, \infty)$
(c) none
(d) $(-\infty, 3)$ and $(3, \infty)$

4. What is the interval over which $f(x) = \frac{1}{x^2}$ increases? decreases? is constant?

5. What is the equation of the vertical asymptote of the graph of $y = \frac{1}{x-3} + 2$? of the horizontal asymptote?

6. What is the equation of the vertical asymptote of the graph of $y = \frac{1}{(x+2)^2} - 4$? of the horizontal asymptote?

7. Is $f(x) = \frac{1}{x^2}$ an even or an odd function? What symmetry does its graph exhibit?

8. Is $f(x) = \frac{1}{x}$ an even or an odd function? What symmetry does its graph exhibit?

Concept Check *Use the graphs of the rational functions in choices A–D to answer each question in Exercises 9–16. There may be more than one correct choice.*

A.

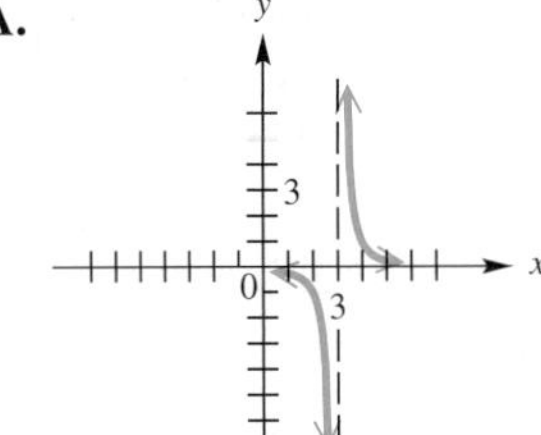

B.

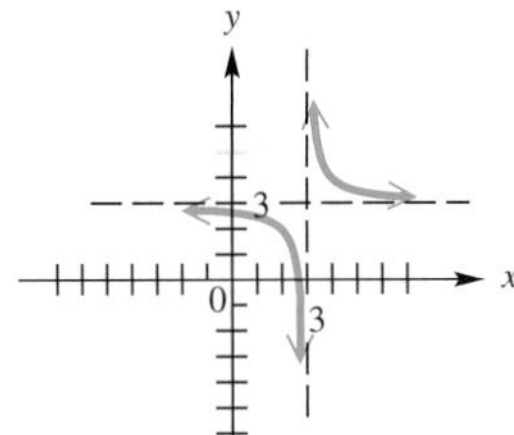

C.

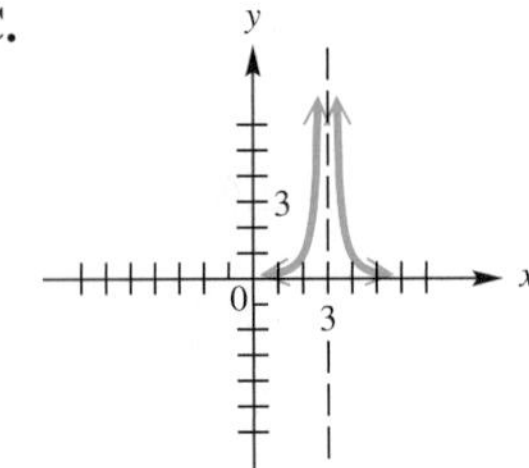

D.

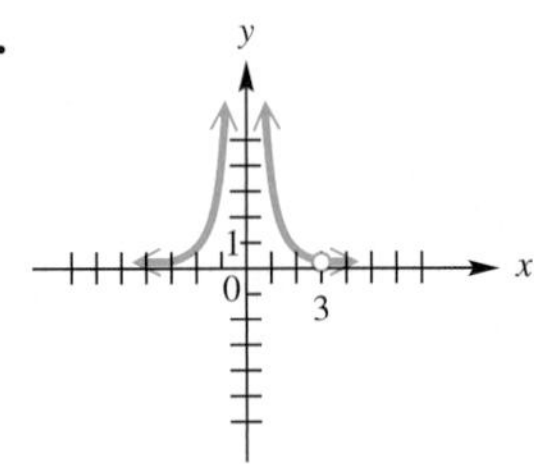

9. Which choices have domain $(-\infty, 3) \cup (3, \infty)$?

10. Which choices have range $(-\infty, 3) \cup (3, \infty)$?

11. Which choices have range $(-\infty, 0) \cup (0, \infty)$?

12. Which choices have range $(0, \infty)$?

13. If f represents the function, only one choice has a single solution to the equation $f(x) = 3$. Which one is it?

14. Which choices have domain $(-\infty, 0) \cup (0, 3) \cup (3, \infty)$?

15. Which choices have the x-axis as a horizontal asymptote?

16. Which choices are symmetric with respect to a vertical line?

*Explain how the graph of each function can be obtained from the graph of $y = \frac{1}{x}$ or $y = \frac{1}{x^2}$. Then graph f and give the **(a)** domain and **(b)** range. Determine the intervals of the domain for which the function is **(c)** increasing or **(d)** decreasing.* ***See Examples 1–3.***

17. $f(x) = \frac{2}{x}$

18. $f(x) = -\frac{3}{x}$

19. $f(x) = \frac{1}{x+2}$

20. $f(x) = \frac{1}{x-3}$

21. $f(x) = \frac{1}{x} + 1$

22. $f(x) = \frac{1}{x} - 2$

23. $f(x) = -\frac{2}{x^2}$

24. $f(x) = \frac{1}{x^2} + 3$

25. $f(x) = \frac{1}{(x-3)^2}$

21. To obtain the graph of f, shift the graph of $y = \frac{1}{x}$ up 1 unit.

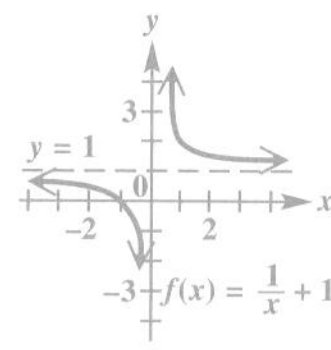

(a) $(-\infty, 0) \cup (0, \infty)$
(b) $(-\infty, 1) \cup (1, \infty)$
(c) none
(d) $(-\infty, 0)$ and $(0, \infty)$

22. To obtain the graph of f, shift the graph of $y = \frac{1}{x}$ down 2 units.

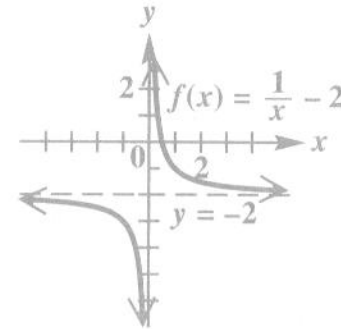

(a) $(-\infty, 0) \cup (0, \infty)$
(b) $(-\infty, -2) \cup (-2, \infty)$
(c) none
(d) $(-\infty, 0)$ and $(0, \infty)$

23. To obtain the graph of f, stretch the graph of $y = \frac{1}{x^2}$ vertically by a factor of 2 and reflect across the x-axis.

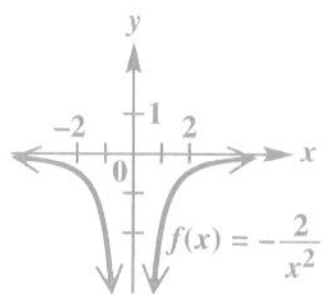

(a) $(-\infty, 0) \cup (0, \infty)$
(b) $(-\infty, 0)$ (c) $(0, \infty)$
(d) $(-\infty, 0)$

24. To obtain the graph of f, shift the graph of $y = \frac{1}{x^2}$ up 3 units.

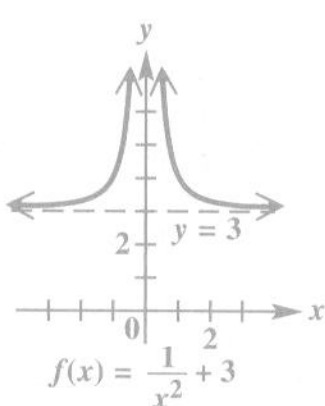

(a) $(-\infty, 0) \cup (0, \infty)$
(b) $(3, \infty)$ (c) $(-\infty, 0)$
(d) $(0, \infty)$

25. To obtain the graph of f, shift the graph of $y = \frac{1}{x^2}$ to the right 3 units.

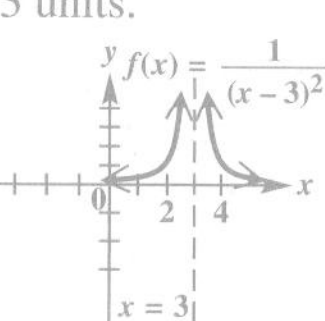

(a) $(-\infty, 3) \cup (3, \infty)$
(b) $(0, \infty)$ (c) $(-\infty, 3)$
(d) $(3, \infty)$

26. $f(x) = \dfrac{-2}{(x-3)^2}$ 27. $f(x) = \dfrac{-1}{(x+2)^2} - 3$ 28. $f(x) = \dfrac{-1}{(x-4)^2} + 2$

Concept Check Match the rational function in Column I with the appropriate description in Column II. Choices in Column II can be used only once.

I	II
29. $f(x) = \dfrac{x+7}{x+1}$	A. The x-intercept is -3.
30. $f(x) = \dfrac{x+10}{x+2}$	B. The y-intercept is 5.
31. $f(x) = \dfrac{1}{x+4}$	C. The horizontal asymptote is $y = 4$.
32. $f(x) = \dfrac{-3}{x^2}$	D. The vertical asymptote is $x = -1$.
33. $f(x) = \dfrac{x^2-16}{x+4}$	E. There is a "hole" in its graph at $x = -4$.
34. $f(x) = \dfrac{4x+3}{x-7}$	F. The graph has an oblique asymptote.
35. $f(x) = \dfrac{x^2+3x+4}{x-5}$	G. The x-axis is its horizontal asymptote, and the y-axis is not its vertical asymptote.
36. $f(x) = \dfrac{x+3}{x-6}$	H. The x-axis is its horizontal asymptote, and the y-axis is its vertical asymptote.

Give the equations of any vertical, horizontal, or oblique asymptotes for the graph of each rational function. ***See Example 4.***

37. $f(x) = \dfrac{3}{x-5}$ 38. $f(x) = \dfrac{-6}{x+9}$ 39. $f(x) = \dfrac{4-3x}{2x+1}$

40. $f(x) = \dfrac{2x+6}{x-4}$ 41. $f(x) = \dfrac{x^2-1}{x+3}$ 42. $f(x) = \dfrac{x^2+4}{x-1}$

43. $f(x) = \dfrac{x^2-2x-3}{2x^2-x-10}$ 44. $f(x) = \dfrac{3x^2-6x-24}{5x^2-26x+5}$

45. $f(x) = \dfrac{x^2+1}{x^2+9}$ 46. $f(x) = \dfrac{4x^2+25}{x^2+9}$

47. *Concept Check* Let f be the function whose graph is obtained by translating the graph of $y = \frac{1}{x}$ to the right 3 units and up 2 units.
(a) Write an equation for $f(x)$ as a quotient of two polynomials.
(b) Determine the zero(s) of f.
(c) Identify the asymptotes of the graph of $f(x)$.

48. *Concept Check* Repeat **Exercise 47** with f the function whose graph is obtained by translating the graph of $y = -\frac{1}{x^2}$ to the left 3 units and up 1 unit.

49. *Concept Check* After the numerator is divided by the denominator,

$$f(x) = \frac{x^5 + x^4 + x^2 + 1}{x^4 + 1} \quad \text{becomes} \quad f(x) = x + 1 + \frac{x^2 - x}{x^4 + 1}.$$

(a) What is the oblique asymptote of the graph of the function?
(b) Where does the graph of the function intersect its asymptote?
(c) As $x \rightarrow \infty$, does the graph of the function approach its asymptote from above or below?

26. To obtain the graph of f, shift the graph of $y = \frac{1}{x^2}$ to the right 3 units, stretch by a factor of 2, and reflect across the x-axis.

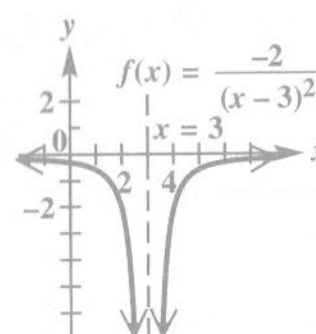

(a) $(-\infty, 3) \cup (3, \infty)$
(b) $(-\infty, 0)$ (c) $(3, \infty)$
(d) $(-\infty, 3)$

27. To obtain the graph of f, shift the graph of $y = \frac{1}{x^2}$ to the left 2 units, reflect across the x-axis, and shift 3 units down.

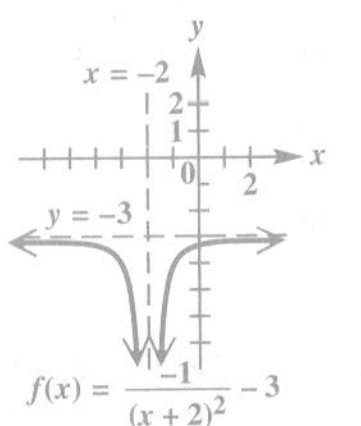

(a) $(-\infty, -2) \cup (-2, \infty)$
(b) $(-\infty, -3)$ (c) $(-2, \infty)$
(d) $(-\infty, -2)$

28. To obtain the graph of f, shift the graph of $y = \frac{1}{x^2}$ to the right 4 units, reflect across the x-axis, and shift 2 units up.

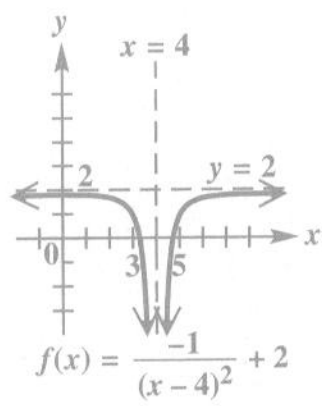

(a) $(-\infty, 4) \cup (4, \infty)$
(b) $(-\infty, 2)$ (c) $(4, \infty)$
(d) $(-\infty, 4)$

29. D 30. B
31. G 32. H
33. E 34. C
35. F 36. A

In selected Exercises 37–60, V.A. represents vertical asymptote, H.A. represents horizontal asymptote, and O.A. represents oblique asymptote.

37. V.A.: $x = 5$; H.A.: $y = 0$
38. V.A.: $x = -9$; H.A.: $y = 0$
39. V.A.: $x = -\frac{1}{2}$; H.A.: $y = -\frac{3}{2}$
40. V.A.: $x = 4$; H.A.: $y = 2$

50. *Concept Check* Choices A–D below show the four ways in which the graph of a rational function can approach the vertical line $x = 2$ as an asymptote. Identify the graph of each rational function defined in parts (a)–(d).

(a) $f(x) = \frac{1}{(x-2)^2}$ **(b)** $f(x) = \frac{1}{x-2}$ **(c)** $f(x) = \frac{-1}{x-2}$ **(d)** $f(x) = \frac{-1}{(x-2)^2}$

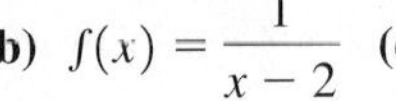

A.

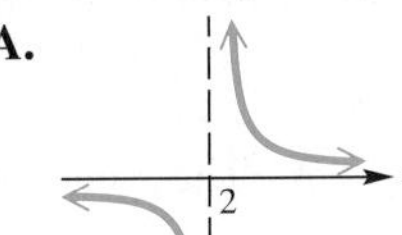

B.

C.

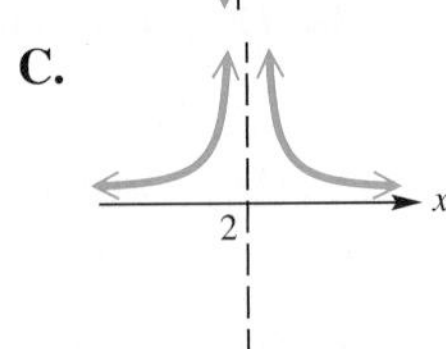

D.

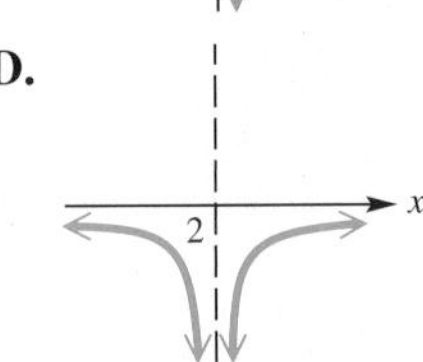

51. *Concept Check* Which function has a graph that does not have a vertical asymptote?

A. $f(x) = \frac{1}{x^2 + 2}$ **B.** $f(x) = \frac{1}{x^2 - 2}$ **C.** $f(x) = \frac{3}{x^2}$ **D.** $f(x) = \frac{2x + 1}{x - 8}$

52. *Concept Check* Which function has a graph that does not have a horizontal asymptote?

A. $f(x) = \frac{2x - 7}{x + 3}$ **B.** $f(x) = \frac{3x}{x^2 - 9}$

C. $f(x) = \frac{x^2 - 9}{x + 3}$ **D.** $f(x) = \frac{x + 5}{(x + 2)(x - 3)}$

Identify any vertical, horizontal, or oblique asymptotes in the graph of $y = f(x)$. State the domain of f.

53.

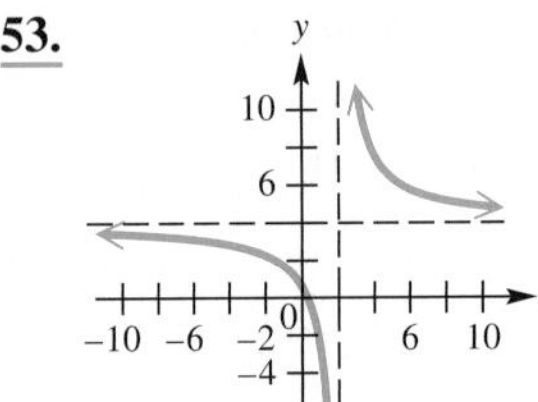

54.

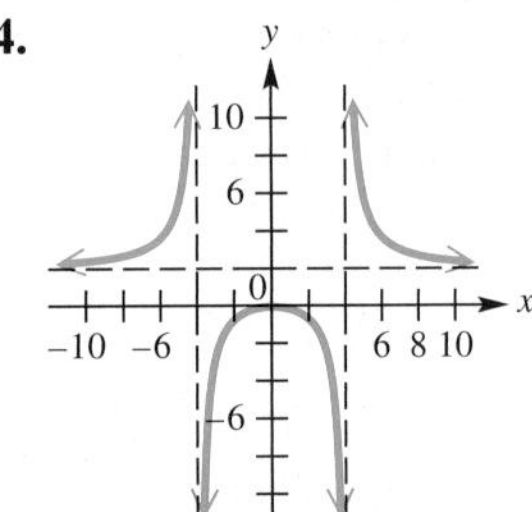

55.

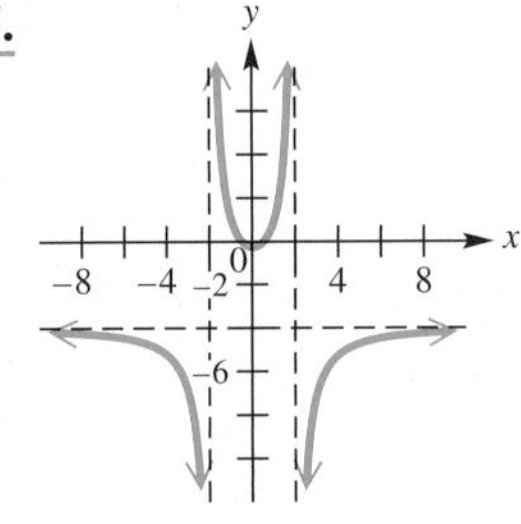

56.

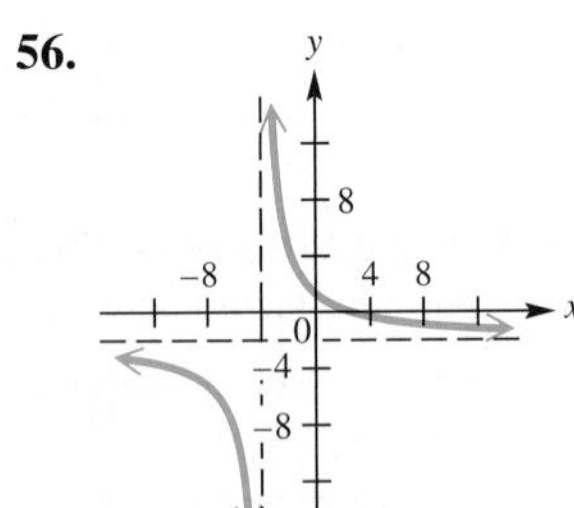

57.

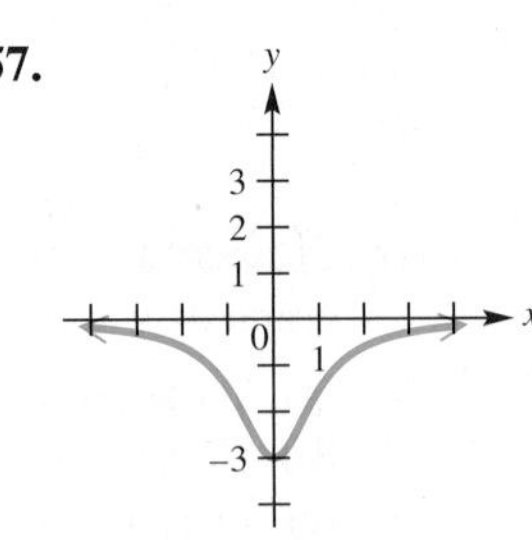

58.

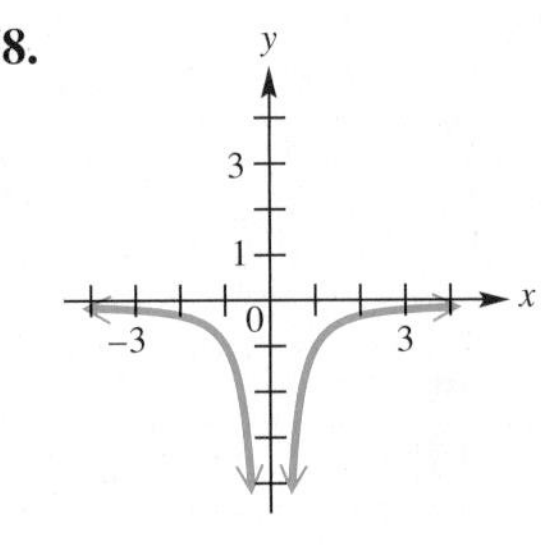

59.

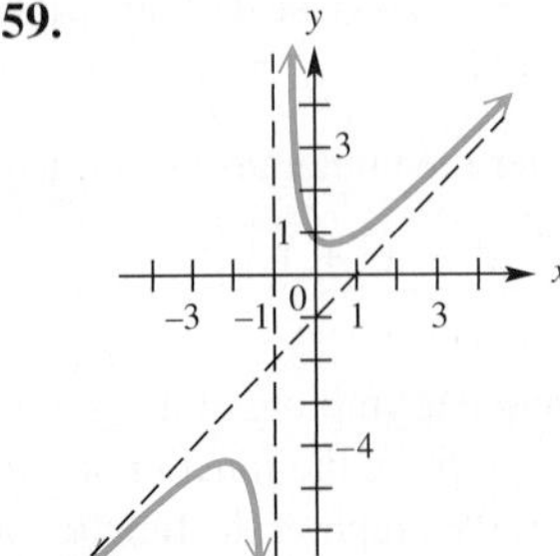

60.

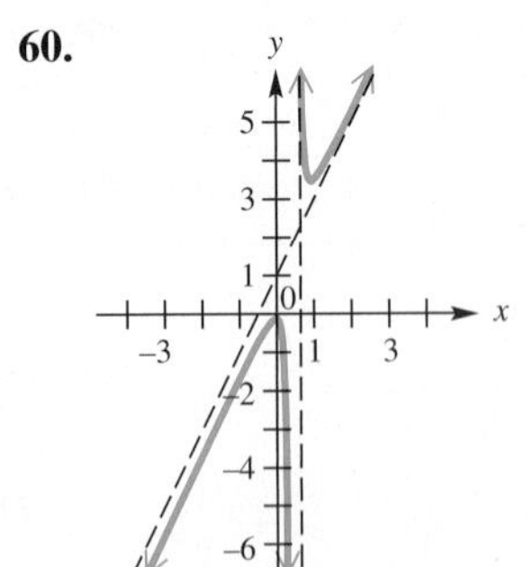

41. V.A.: $x = -3$; O.A.: $y = x - 3$
42. V.A.: $x = 1$; O.A.: $y = x + 1$
43. V.A.: $x = -2$, $x = \frac{5}{2}$; H.A.: $y = \frac{1}{2}$
44. V.A.: $x = \frac{1}{5}$, $x = 5$; H.A.: $y = \frac{3}{5}$
45. V.A.: none; H.A.: $y = 1$
46. V.A.: none; H.A.: $y = 4$
47. (a) $f(x) = \frac{2x - 5}{x - 3}$
 (b) $\frac{5}{2}$
 (c) H.A.: $y = 2$; V.A.: $x = 3$
48. (a) $f(x) = \frac{x^2 + 6x + 8}{x^2 + 6x + 9}$
 (b) $-2, -4$
 (c) H.A.: $y = 1$; V.A.: $x = -3$
49. (a) $y = x + 1$
 (b) at $x = 0$ and $x = 1$
 (c) above
50. (a) C (b) A (c) B (d) D
51. A
52. C
53. V.A.: $x = 2$; H.A.: $y = 4$; $(-\infty, 2) \cup (2, \infty)$
54. V.A.: $x = \pm 4$; H.A.: $y = 2$; $(-\infty, -4) \cup (-4, 4) \cup (4, \infty)$
55. V.A.: $x = \pm 2$; H.A.: $y = -4$; $(-\infty, -2) \cup (-2, 2) \cup (2, \infty)$
56. V.A.: $x = -4$; H.A.: $y = -2$; $(-\infty, -4) \cup (-4, \infty)$
57. V.A.: none; H.A.: $y = 0$; $(-\infty, \infty)$
58. V.A.: $x = 0$; H.A.: $y = 0$; $(-\infty, 0) \cup (0, \infty)$
59. V.A.: $x = -1$; O.A.: $y = x - 1$; $(-\infty, -1) \cup (-1, \infty)$
60. V.A.: $x = \frac{1}{2}$; O.A.: $y = 2x + 1$; $\left(-\infty, \frac{1}{2}\right) \cup \left(\frac{1}{2}, \infty\right)$

61.

$f(x) = \frac{x+1}{x-4}$

62.

$f(x) = \frac{x-5}{x+3}$

Sketch the graph of each rational function. ***See Examples 5–9.***

61. $f(x) = \frac{x + 1}{x - 4}$
62. $f(x) = \frac{x - 5}{x + 3}$
63. $f(x) = \frac{x + 2}{x - 3}$
64. $f(x) = \frac{x - 3}{x + 4}$
65. $f(x) = \frac{4 - 2x}{8 - x}$
66. $f(x) = \frac{6 - 3x}{4 - x}$
67. $f(x) = \frac{3x}{x^2 - x - 2}$
68. $f(x) = \frac{2x + 1}{x^2 + 6x + 8}$
69. $f(x) = \frac{5x}{x^2 - 1}$
70. $f(x) = \frac{x}{4 - x^2}$
71. $f(x) = \frac{(x + 6)(x - 2)}{(x + 3)(x - 4)}$
72. $f(x) = \frac{(x + 3)(x - 5)}{(x + 1)(x - 4)}$
73. $f(x) = \frac{3x^2 + 3x - 6}{x^2 - x - 12}$
74. $f(x) = \frac{4x^2 + 4x - 24}{x^2 - 3x - 10}$
75. $f(x) = \frac{9x^2 - 1}{x^2 - 4}$
76. $f(x) = \frac{16x^2 - 9}{x^2 - 9}$
77. $f(x) = \frac{(x - 3)(x + 1)}{(x - 1)^2}$
78. $f(x) = \frac{x(x - 2)}{(x + 3)^2}$
79. $f(x) = \frac{x}{x^2 - 9}$
80. $f(x) = \frac{-5}{2x + 4}$
81. $f(x) = \frac{1}{x^2 + 1}$
82. $f(x) = \frac{(x - 5)(x - 2)}{x^2 + 9}$
83. $f(x) = \frac{(x + 4)^2}{(x - 1)(x + 5)}$
84. $f(x) = \frac{(x + 1)^2}{(x + 2)(x - 3)}$
85. $f(x) = \frac{20 + 6x - 2x^2}{8 + 6x - 2x^2}$
86. $f(x) = \frac{18 + 6x - 4x^2}{4 + 6x + 2x^2}$
87. $f(x) = \frac{x^2 + 1}{x + 3}$
88. $f(x) = \frac{2x^2 + 3}{x - 4}$
89. $f(x) = \frac{x^2 + 2x}{2x - 1}$
90. $f(x) = \frac{x^2 - x}{x + 2}$
91. $f(x) = \frac{x^2 - 9}{x + 3}$
92. $f(x) = \frac{x^2 - 16}{x + 4}$
93. $f(x) = \frac{2x^2 - 5x - 2}{x - 2}$
94. $f(x) = \frac{x^2 - 5}{x - 3}$
95. $f(x) = \frac{x^2 - 1}{x^2 - 4x + 3}$
96. $f(x) = \frac{x^2 - 4}{x^2 + 3x + 2}$
97. $f(x) = \frac{(x^2 - 9)(2 + x)}{(x^2 - 4)(3 + x)}$
98. $f(x) = \frac{(x^2 - 16)(3 + x)}{(x^2 - 9)(4 + x)}$
99. $f(x) = \frac{x^4 - 20x^2 + 64}{x^4 - 10x^2 + 9}$
100. $f(x) = \frac{x^4 - 5x^2 + 4}{x^4 - 24x^2 + 108}$

63.

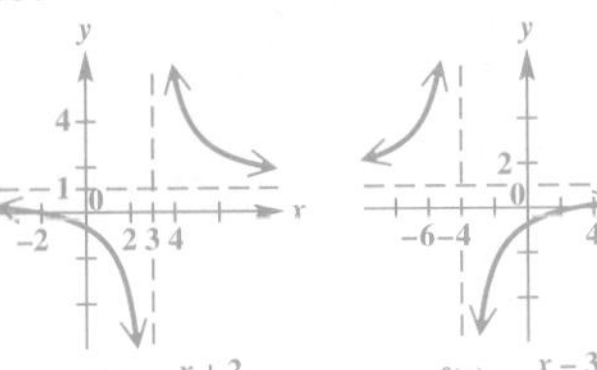

$f(x) = \frac{x+2}{x-3}$

64. $f(x) = \frac{x-3}{x+4}$

65.

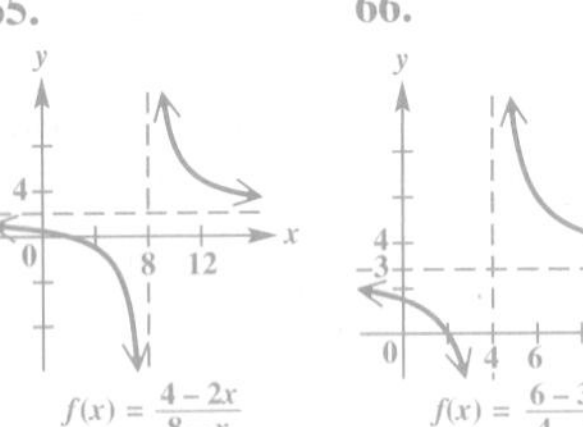

$f(x) = \frac{4-2x}{8-x}$

66. $f(x) = \frac{6-3x}{4-x}$

67.

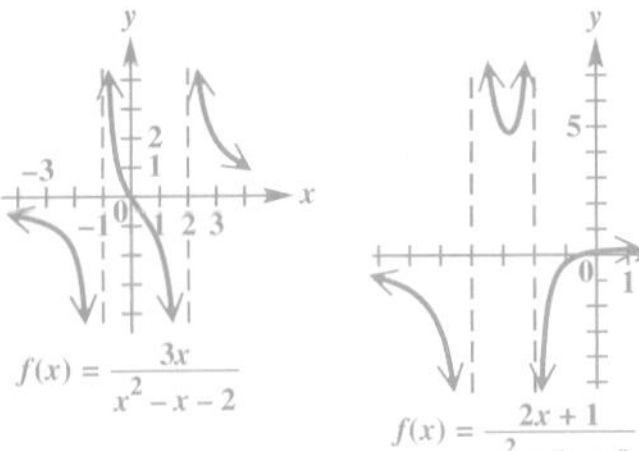

$f(x) = \frac{3x}{x^2-x-2}$

68. $f(x) = \frac{2x+1}{x^2+6x+8}$

69.

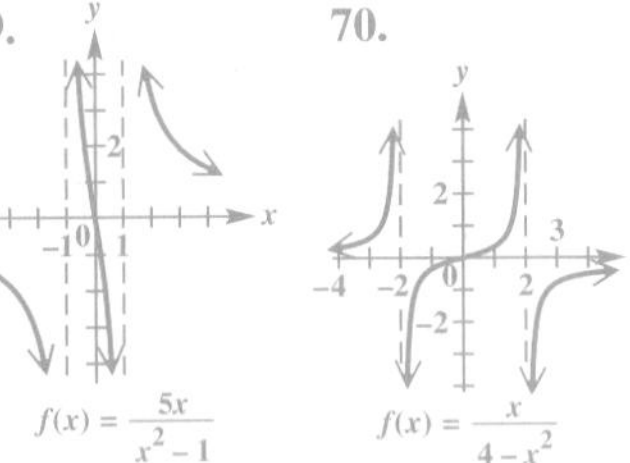

$f(x) = \frac{5x}{x^2-1}$

70. $f(x) = \frac{x}{4-x^2}$

71.

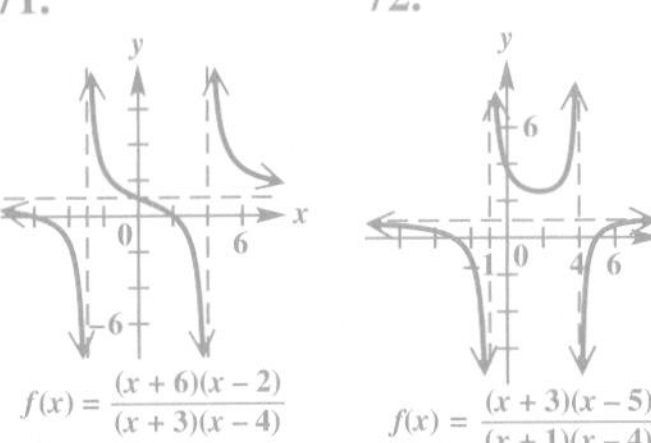

$f(x) = \frac{(x+6)(x-2)}{(x+3)(x-4)}$

72. $f(x) = \frac{(x+3)(x-5)}{(x+1)(x-4)}$

73.

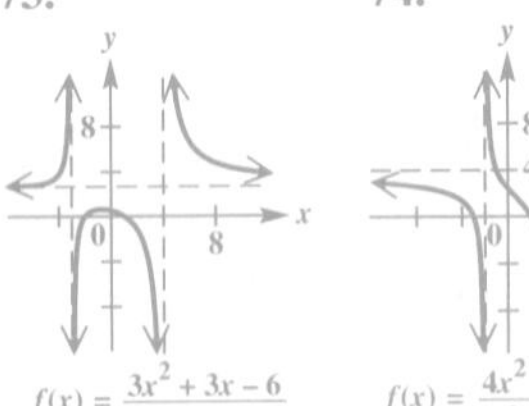

$f(x) = \frac{3x^2+3x-6}{x^2-x-12}$

74. $f(x) = \frac{4x^2+4x-24}{x^2-3x-10}$

75.

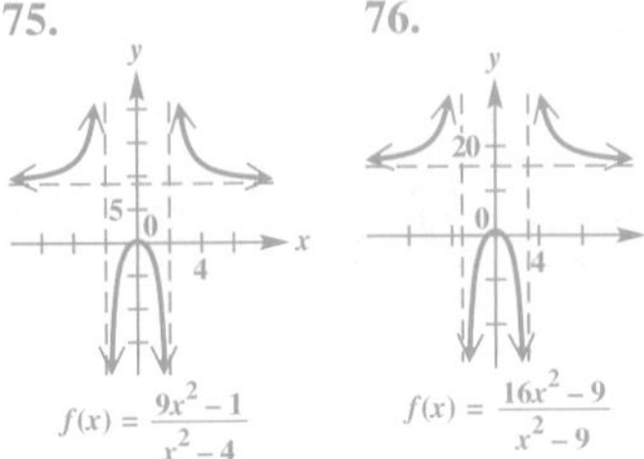

$f(x) = \frac{9x^2-1}{x^2-4}$

76. $f(x) = \frac{16x^2-9}{x^2-9}$

Connecting Graphs with Equations *Find an equation for each rational function graph. (Hint: See the note preceding* ***Example 8.****)*

101. **102.**

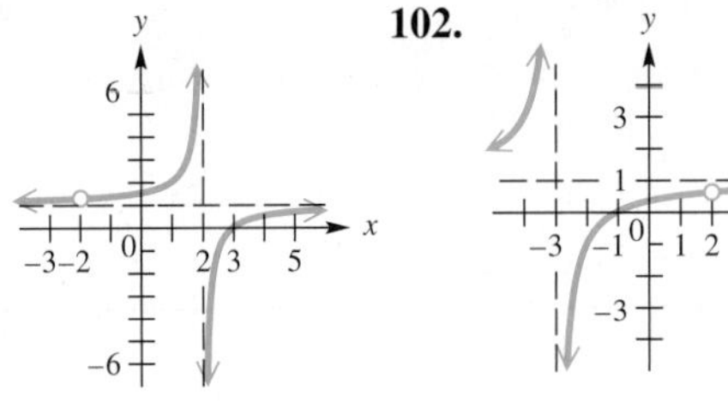

103.

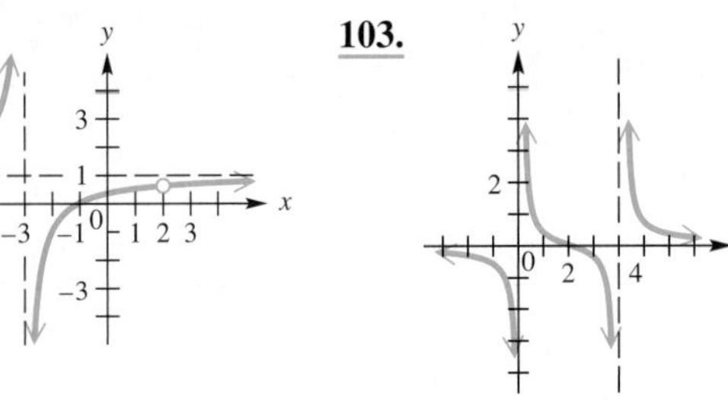

104.

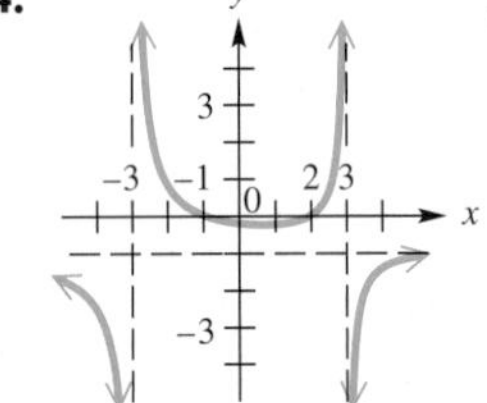

105.

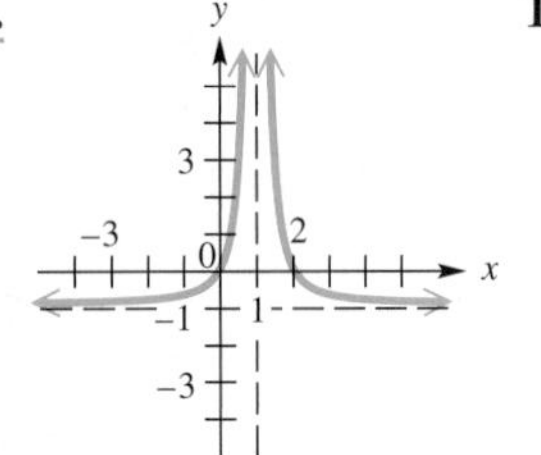

106.

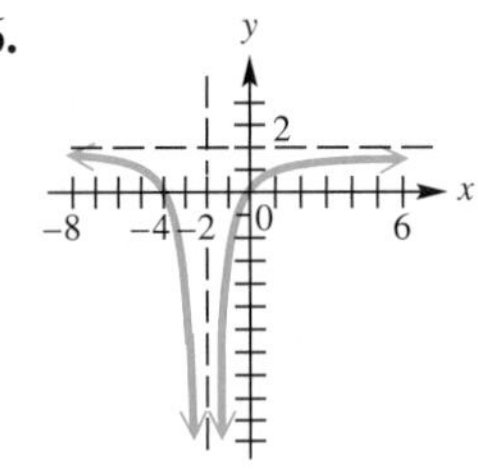

Concept Check *In Exercises 107 and 108, find a possible equation for the function with a graph having the given features.*

107. x-intercepts: -1 and 3
y-intercept: -3
vertical asymptote: $x = 1$
horizontal asymptote: $y = 1$

108. x-intercepts: 1 and 3
y-intercept: none
vertical asymptotes: $x = 0$ and $x = 2$
horizontal asymptote: $y = 1$

Use a graphing calculator to graph the rational function in the specified exercise. Then use the graph to find $f(1.25)$.

109. Exercise 61 **110. Exercise 67** **111. Exercise 89** **112. Exercise 91**

(Modeling) *Solve each problem.* ***See Example 10.***

113. *Traffic Intensity* Let the average number of vehicles arriving at the gate of an amusement park per minute be equal to k, and let the average number of vehicles admitted by the park attendants be equal to r. Then, the average waiting time T (in minutes) for each vehicle arriving at the park is given by the rational function

$$T(r) = \frac{2r-k}{2r^2-2kr},$$

where $r > k$. (*Source:* Mannering, F., and W. Kilareski, *Principles of Highway Engineering and Traffic Analysis,* 2nd ed., John Wiley & Sons.)

(a) It is known from experience that on Saturday afternoon $k = 25$. Use graphing to estimate the admittance rate r that is necessary to keep the average waiting time T for each vehicle to 30 sec.

(b) If one park attendant can serve 5.3 vehicles per minute, how many park attendants will be needed to keep the average wait to 30 sec?

114. *Waiting in Line* **Queuing theory** (also known as **waiting-line theory**) investigates the problem of providing adequate service economically to customers waiting in line. Suppose customers arrive at a fast-food service window at the rate of 9 people per hour. With reasonable assumptions, the average time (in hours) that a customer will wait in line before being served is modeled by

$$f(x) = \frac{9}{x(x-9)},$$

where x is the average number of people served per hour. A graph of $f(x)$ for $x > 9$ is shown in the figure on the next page.

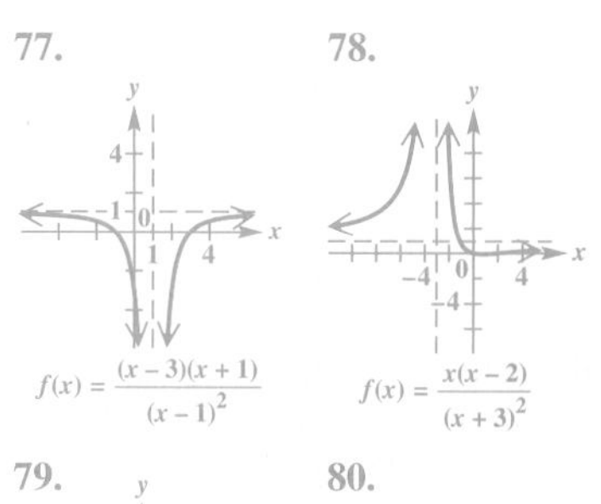

77. $f(x) = \frac{(x-3)(x+1)}{(x-1)^2}$

78. $f(x) = \frac{x(x-2)}{(x+3)^2}$

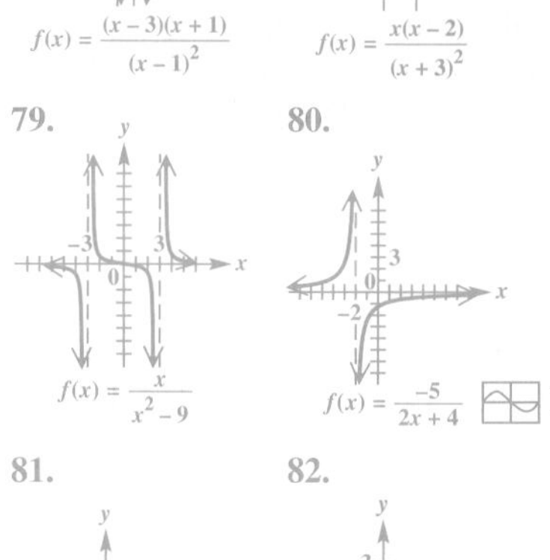

79. $f(x) = \frac{x}{x^2-9}$

80. $f(x) = \frac{-5}{2x+4}$

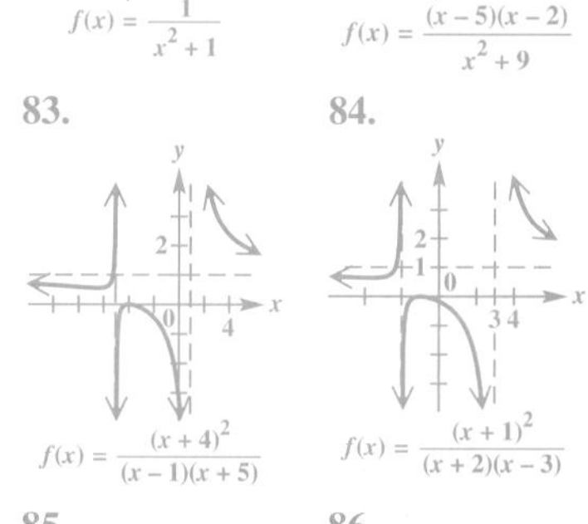

81. $f(x) = \frac{1}{x^2+1}$

82. $f(x) = \frac{(x-5)(x-2)}{x^2+9}$

83. $f(x) = \frac{(x+4)^2}{(x-1)(x+5)}$

84. $f(x) = \frac{(x+1)^2}{(x+2)(x-3)}$

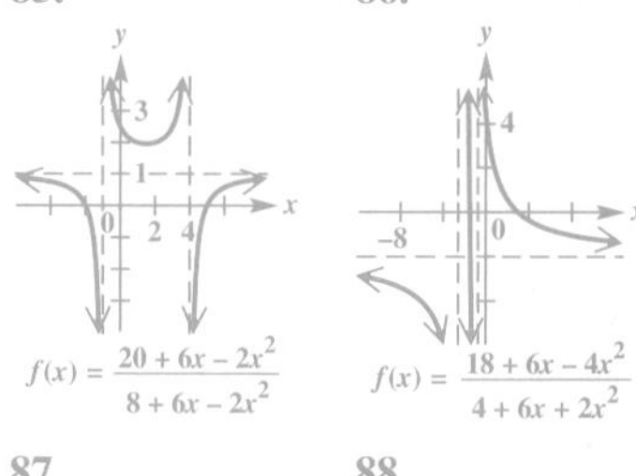

85. $f(x) = \frac{20+6x-2x^2}{8+6x-2x^2}$

86. $f(x) = \frac{18+6x-4x^2}{4+6x+2x^2}$

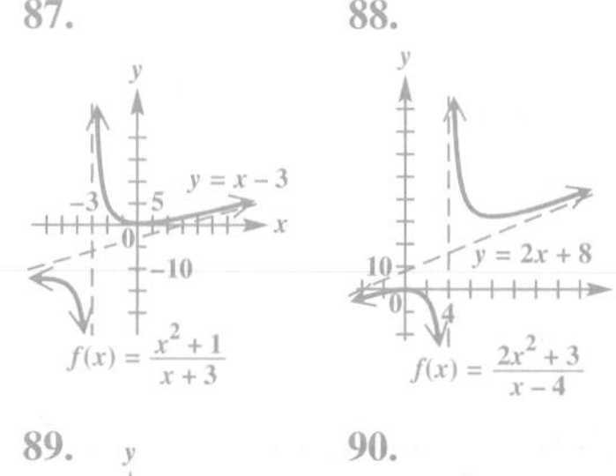

87. $f(x) = \frac{x^2+1}{x+3}$

88. $f(x) = \frac{2x^2+3}{x-4}$

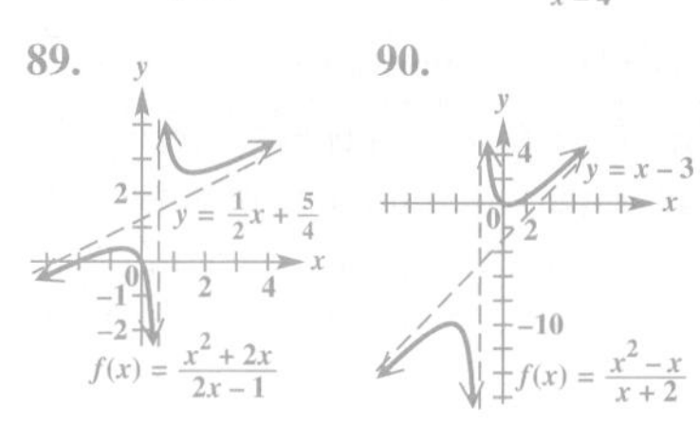

89. $f(x) = \frac{x^2+2x}{2x-1}$

90. $f(x) = \frac{x^2-x}{x+2}$

(a) Why is the function meaningless if the average number of people served per hour is less than 9?

Suppose the average time to serve a customer is 5 min.

(b) How many customers can be served in an hour?

(c) How many minutes will a customer have to wait in line (on the average)?

Average Waiting Time

Hours; People served; $x = 9$

(d) Suppose we want to halve the average waiting time to 7.5 min $\left(\frac{1}{8}\text{ hr}\right)$. How fast must an employee work to serve a customer (on the average)? (*Hint:* Let $f(x) = \frac{1}{8}$ and solve the equation for x. Convert your answer to minutes.) How might this reduction in serving time be accomplished?

115. *Braking Distance* The rational function

$$d(x) = \frac{8710x^2 - 69{,}400x + 470{,}000}{1.08x^2 - 324x + 82{,}200}$$

can be used to accurately model the braking distance for automobiles traveling at x miles per hour, where $20 \le x \le 70$. (*Source:* Mannering, F., and W. Kilareski, *Principles of Highway Engineering and Traffic Analysis,* 2nd ed., John Wiley & Sons.)

(a) Use graphing to estimate x when $d(x) = 300$.

(b) Complete the table for each value of x.

(c) If a car doubles its speed, does the braking distance double or more than double? Explain.

(d) Suppose that the automobile braking distance doubled whenever the speed doubled. What type of relationship would exist between the braking distance and the speed?

x	$d(x)$	x	$d(x)$
20		50	
25		55	
30		60	
35		65	
40		70	
45			

116. *Braking Distance* The **grade** x of a hill is a measure of its steepness. For example, if a road rises 10 ft for every 100 ft of horizontal distance, then it has an uphill grade of

$$x = \frac{10}{100}, \quad \text{or} \quad 10\%.$$

Grades are typically kept quite small—usually less than 10%. The braking distance D for a car traveling at 50 mph on a wet, uphill grade is given by

$$D(x) = \frac{2500}{30(0.3 + x)}.$$

(*Source:* Haefner, L., *Introduction to Transportation Systems,* Holt, Rinehart and Winston.)

(a) Evaluate $D(0.05)$ and interpret the result.

(b) Describe what happens to braking distance as the hill becomes steeper. Does this agree with your driving experience?

(c) Estimate the grade associated with a braking distance of 220 ft.

91. **92.**

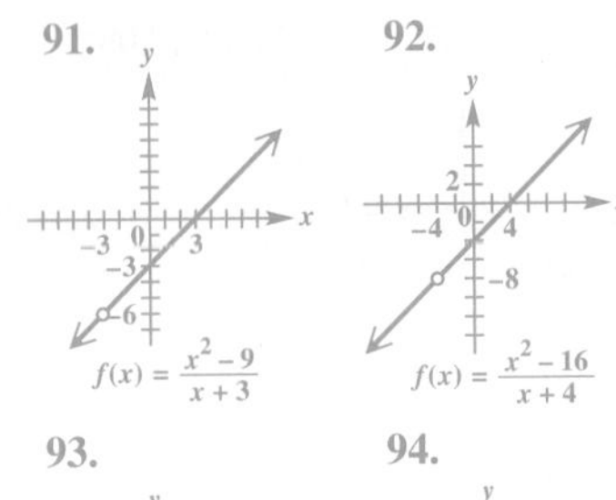

$f(x) = \frac{x^2 - 9}{x + 3}$ $f(x) = \frac{x^2 - 16}{x + 4}$

93. **94.**

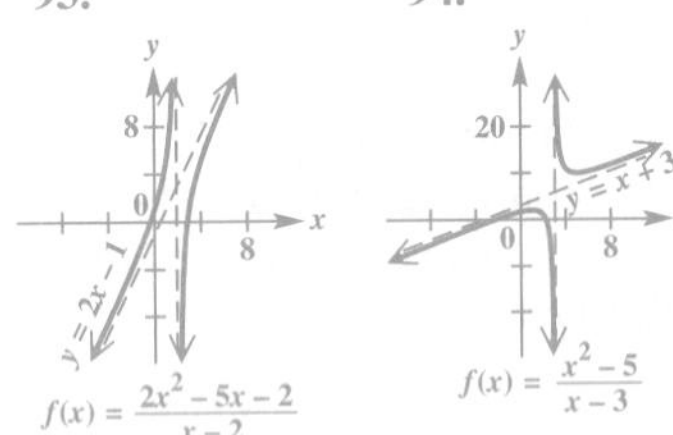

$f(x) = \frac{2x^2 - 5x - 2}{x - 2}$ $f(x) = \frac{x^2 - 5}{x - 3}$

95. **96.**

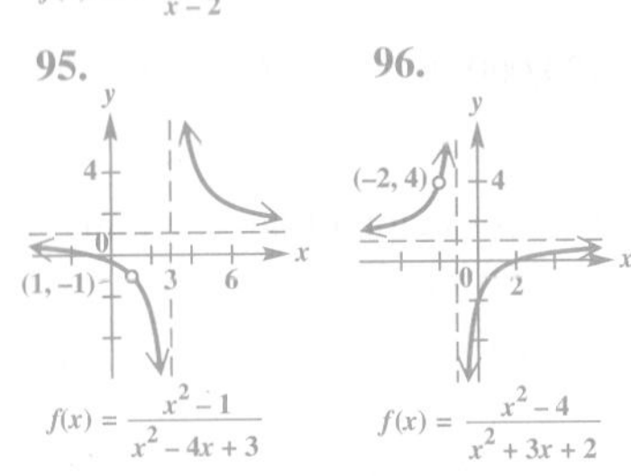

$f(x) = \frac{x^2 - 1}{x^2 - 4x + 3}$ $f(x) = \frac{x^2 - 4}{x^2 + 3x + 2}$

97. **98.**

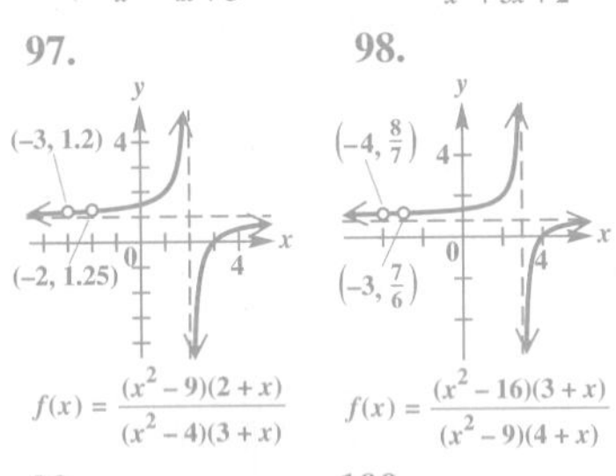

$f(x) = \frac{(x^2 - 9)(2 + x)}{(x^2 - 4)(3 + x)}$ $f(x) = \frac{(x^2 - 16)(3 + x)}{(x^2 - 9)(4 + x)}$

99. **100.**

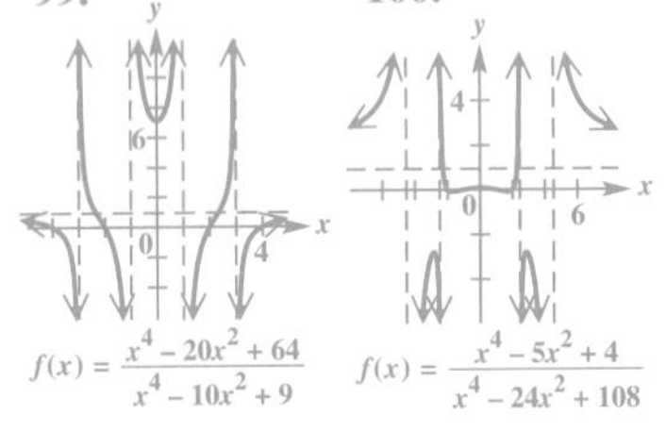

$f(x) = \frac{x^4 - 20x^2 + 64}{x^4 - 10x^2 + 9}$ $f(x) = \frac{x^4 - 5x^2 + 4}{x^4 - 24x^2 + 108}$

101. $f(x) = \frac{(x - 3)(x + 2)}{(x - 2)(x + 2)}$, or
$f(x) = \frac{x^2 - x - 6}{x^2 - 4}$

102. $f(x) = \frac{(x - 2)(x + 1)}{(x - 2)(x + 3)}$, or
$f(x) = \frac{x^2 - x - 2}{x^2 + x - 6}$

103. $f(x) = \frac{x - 2}{x(x - 4)}$, or
$f(x) = \frac{x - 2}{x^2 - 4x}$

104. $f(x) = \frac{-(x + 1)(x - 2)}{(x + 3)(x - 3)}$, or
$f(x) = \frac{-x^2 + x + 2}{x^2 - 9}$

105. $f(x) = \frac{-x(x - 2)}{(x - 1)^2}$, or
$f(x) = \frac{-x^2 + 2x}{x^2 - 2x + 1}$

106. $f(x) = \frac{2x(x + 4)}{(x + 2)^2}$, or
$f(x) = \frac{2x^2 + 8x}{x^2 + 4x + 4}$

117. *Tax Revenue* Economist Arthur Laffer has been a center of controversy because of his **Laffer curve,** an idealized version of which is shown here. According to this curve, increasing a tax rate, say from x_1 percent to x_2 percent on the graph, can actually lead to a decrease in government revenue. All economists agree on the endpoints, 0 revenue at tax rates of both 0% and 100%, but there is much disagreement on the location of the rate x_1 that produces maximum revenue. Suppose an economist studying the Laffer curve produces the rational function

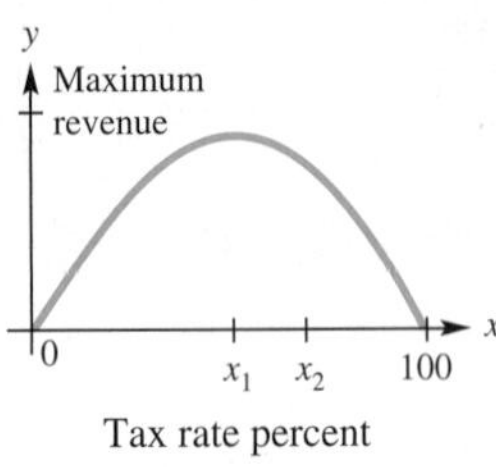

$$R(x) = \frac{80x - 8000}{x - 110},$$

where $R(x)$ is government revenue in tens of millions of dollars for a tax rate of x percent, with the function valid for $55 \le x \le 100$. Find the revenue for the following tax rates.

(a) 55% **(b)** 60% **(c)** 70% **(d)** 90% **(e)** 100%

(f) Graph R in the window $[0, 100]$ by $[0, 80]$.

118. *Tax Revenue* **See Exercise 117.** Suppose an economist determines that

$$R(x) = \frac{60x - 6000}{x - 120},$$

where $y = R(x)$ is government revenue in tens of millions of dollars for a tax rate of x percent, with $y = R(x)$ valid for $50 \le x \le 100$. Find the revenue for each tax rate.

(a) 50% **(b)** 60% **(c)** 80% **(d)** 100%

(e) Graph R in the window $[0, 100]$ by $[0, 50]$.

Relating Concepts

For individual or collaborative investigation *(Exercises 119–128)*

Consider the following "monster" rational function.

$$f(x) = \frac{x^4 - 3x^3 - 21x^2 + 43x + 60}{x^4 - 6x^3 + x^2 + 24x - 20}$$

Analyzing this function will synthesize many of the concepts of this and earlier chapters. ***Work Exercises 119–128 in order.***

119. Find the equation of the horizontal asymptote.

120. Given that -4 and -1 are zeros of the numerator, factor the numerator completely.

121. **(a)** Given that 1 and 2 are zeros of the denominator, factor the denominator completely.
(b) Write the entire quotient for f so that the numerator and the denominator are in factored form.

122. **(a)** What is the common factor in the numerator and the denominator?
(b) For what value of x will there be a point of discontinuity (i.e., a "hole")?

123. What are the x-intercepts of the graph of f?

124. What is the y-intercept of the graph of f?

125. Find the equations of the vertical asymptotes.

(continued)

Several answers are possible in Exercises 107 and 108.

107. $f(x) = \frac{(x-3)(x+1)}{(x-1)^2}$

108. $f(x) = \frac{(x-1)(x-3)}{x(x-2)}$

Answers for the remaining exercises are given in the answer section at the back of the text.

126. Determine the point or points of intersection of the graph of f with its horizontal asymptote.

127. Sketch the graph of f.

128. Use the graph of f to solve each inequality.

(a) $f(x) < 0$ **(b)** $f(x) > 0$

Chapter 3 Quiz (Sections 3.1–3.5)

1. Graph each quadratic function. Give the vertex, axis, domain, range, and intervals of the domain over which each function is increasing or decreasing.

(a) $f(x) = -2(x+3)^2 - 1$ **(b)** $f(x) = 2x^2 - 8x + 3$

2. *Height of a Projected Object* A ball is projected directly upward from an initial height of 200 ft with an initial velocity of 64 ft per sec.

(a) Use the projectile height function $s(t) = -16t^2 + v_0 t + s_0$ to describe the height of the ball in terms of time t.

(b) For what interval of time is the height of the ball greater than 240 ft?

Use synthetic division to decide whether the given number k is a zero of the polynomial function. If it is not, give the value of f(k).

3. $f(x) = 2x^4 + x^3 - 3x + 4;\ k = 2$ **4.** $f(x) = x^2 - 4x + 5;\ k = 2 + i$

5. Find a polynomial of least degree having only real coefficients with zeros -2, 3, and $3 - i$.

Graph each polynomial function. Factor first if the function is not in factored form.

6. $f(x) = x(x-2)^3(x+2)^2$

7. $f(x) = 2x^4 - 9x^3 - 5x^2 + 57x - 45$

8. $f(x) = -4x^5 + 16x^4 + 13x^3 - 76x^2 - 3x + 18$

Sketch the graph of each rational function.

9. $f(x) = \frac{3x+1}{x^2+7x+10}$ **10.** $f(x) = \frac{x^2+2x+1}{x-1}$

[3.1]
1. (a) vertex: $(-3, -1)$; axis: $x = -3$; domain: $(-\infty, \infty)$; range: $(-\infty, -1]$; increasing: $(-\infty, -3]$; decreasing: $[-3, \infty)$

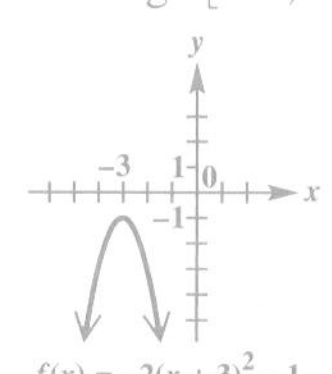

(b) vertex: $(2, -5)$; axis: $x = 2$; domain: $(-\infty, \infty)$; range: $[-5, \infty)$; increasing: $[2, \infty)$; deceasing: $(-\infty, 2]$

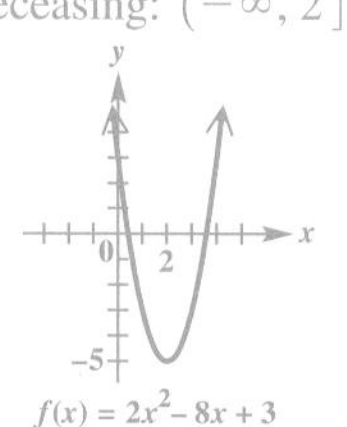

2. (a) $s(t) = -16t^2 + 64t + 200$ (b) between approximately 0.78 sec and 3.22 sec

[3.2]
3. no; 38 4. yes

[3.3]
5. $f(x) = x^4 - 7x^3 + 10x^2 + 26x - 60$

[3.4–3.5]
6.–10. See the answer section at the back of the text.

Summary Exercises on Solving Equations and Inequalities

In **Section 1.7** we solved rational inequalities by rewriting them so that 0 is on one side. We then determined the values that cause either the numerator or denominator to equal 0, and by using a test value from each interval determined by these values, we identified the solution set.

We can now solve rational inequalities by inspecting the graph of a related function. The graphs can be obtained by using technology or by using the steps for graphing a rational function given in **Section 3.5.**

Classroom Example
Solve the inequality.

$$\frac{x+3}{x-2} > 0$$

Answer: $(-\infty, -3) \cup (2, \infty)$

EXAMPLE **Solving a Rational Inequality**

Solve the inequality.

$$\frac{1-x}{x+4} \geq 0$$

SOLUTION Graph the related rational function

$$y = \frac{1-x}{x+4}.$$

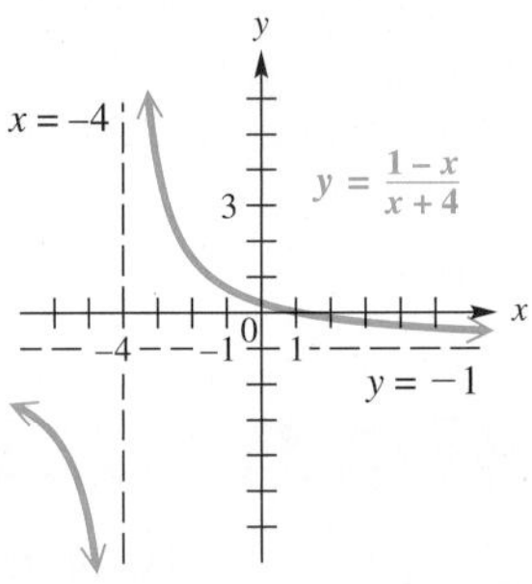

The real solutions of $\frac{1-x}{x+4} \geq 0$ are the x-values for which the graph lies above or on the x-axis. This is true for all x to the right of the vertical asymptote at $x = -4$, up to and including the x-intercept 1. Therefore, the solution set of the inequality is $(-4, 1]$.

By inspecting the graph of the related function, we can also determine that the solution set of $\frac{1-x}{x+4} < 0$ is $(-\infty, -4) \cup (1, \infty)$ and that the solution set of the equation $\frac{1-x}{x+4} = 0$ is $\{1\}$, the value of the x-intercept. This graphical method may be used to solve other equations and inequalities including those defined by polynomials.

Now Try Exercise 19.

1. $\{\frac{4}{3}\}$ 2. $\{20\}$
3. $\{\frac{9}{2}, \frac{14}{3}, \frac{16}{3}, \frac{11}{2}\}$ 4. $\{25, 64\}$
5. $\{3, 7\}$ 6. $\{2\}$
7. $\{\pm\frac{1}{8}\}$ 8. $\{13\}$
9. (a) $(-1, 1) \cup (1, \infty)$ (b) $(-\infty, -1] \cup \{1\}$
10. (a) $(-\infty, -1) \cup (-1, 1) \cup (1, \infty)$ (b) $\emptyset$
11. (a) $\{0\}$ (b) $(-3, 0) \cup (3, \infty)$
12. (a) $\{1, 3\}$ (b) $[1, 2) \cup (2, 3]$

In Exercises 1–8, solve each equation.

1. $\frac{5x+8}{-2} = 2x - 10$

2. $\frac{1}{5}x + 0.25x = \frac{1}{2}x - 1$

3. $(x-5)^{-4} - 13(x-5)^{-2} = -36$

4. $x = 13\sqrt{x} - 40$

5. $\sqrt{2x-5} - \sqrt{x-3} = 1$

6. $3 = \sqrt{x+2} + \sqrt{x-1}$

7. $x^{2/3} + \frac{1}{2} = \frac{3}{4}$

8. $27 - (x-4)^{3/2} = 0$

In Exercises 9–12, use the graph of the function to solve each equation or inequality.

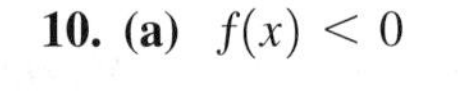

9. (a) $f(x) > 0$ **(b)** $f(x) \leq 0$

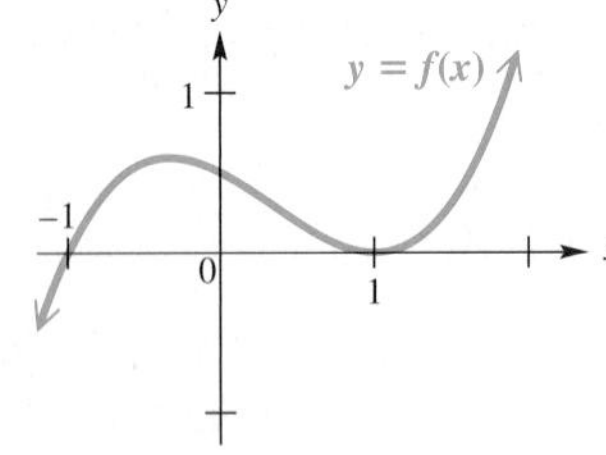

10. (a) $f(x) < 0$ **(b)** $f(x) > 0$

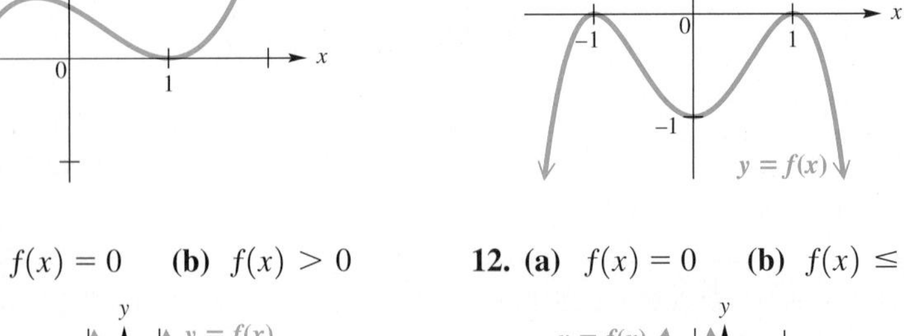

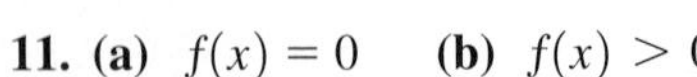

11. (a) $f(x) = 0$ **(b)** $f(x) > 0$

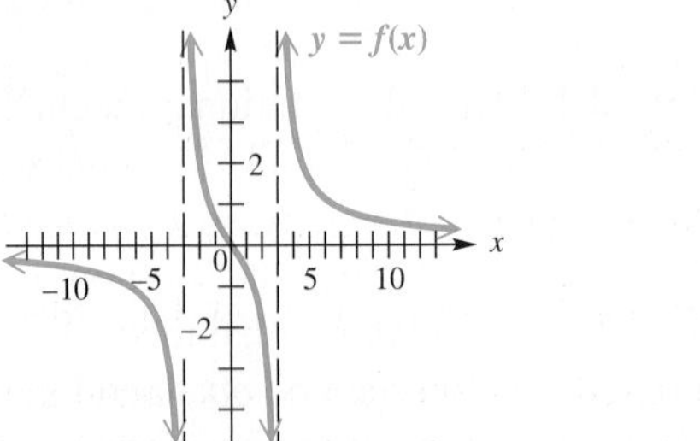

12. (a) $f(x) = 0$ **(b)** $f(x) \leq 0$

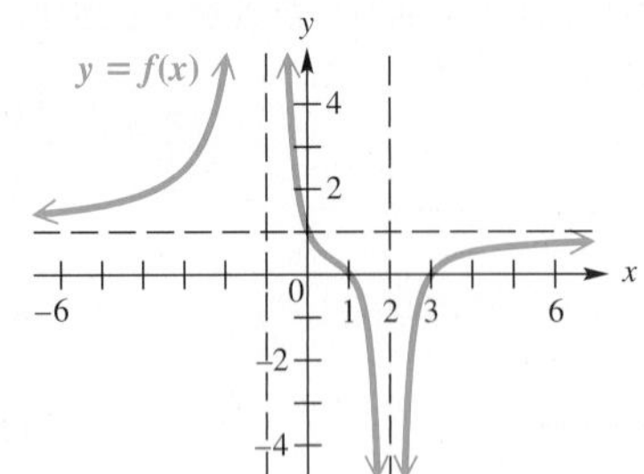

13. inequality; $\left(-\infty, \frac{2}{5}\right) \cup \left(\frac{2}{5}, \infty\right)$
14. inequality; $(-\infty, -1] \cup [4, \infty)$
15. equation; $\{-1, 2\}$
16. inequality; $\{-3\} \cup [1, \infty)$
17. equation; $\{-4, 0, 3\}$
18. inequality; $\left(-\sqrt{3}, -\frac{1}{2}\right) \cup \left(\frac{1}{2}, \sqrt{3}\right)$
19. inequality; $(-\infty, 0) \cup \left[\frac{3}{2}, 3\right) \cup [5, \infty)$
20. inequality; $(-2, -1) \cup (2, \infty)$
21. equation; $\{1\}$
22. inequality; $(-2, 0) \cup (2, \infty)$

In Exercises 13–22, sketch the graph of an appropriate function and then use the graph to solve each equation or inequality. (Note: First identify the expression as an equation or an inequality. When appropriate, use the steps for graphing polynomial functions from **Section 3.4** *or rational functions from* **Section 3.5.**)

13. $25x^2 - 20x + 4 > 0$

14. $3x + 4 \leq x^2$

15. $x^4 - 2x^3 - 3x^2 + 4x + 4 = 0$

16. $x^3 + 5x^2 + 3x - 9 \geq 0$

17. $-x^4 - x^3 + 12x^2 = 0$

18. $-4x^4 + 13x^2 - 3 > 0$

19. $\dfrac{2x^2 - 13x + 15}{x^2 - 3x} \geq 0$

20. $\dfrac{x^2 + 3x - 1}{x + 1} > 3$

21. $\dfrac{x - 1}{(x - 3)^2} = 0$

22. $\dfrac{x}{x^2 - 4} > 0$

3.6 Variation

- Direct Variation
- Inverse Variation
- Combined and Joint Variation

To apply mathematics we often need to express relationships between quantities. For example,

- In chemistry, the ideal gas law describes how temperature, pressure, and volume are related.
- In physics, various formulas in optics describe the relationship between the focal length of a lens and the size of an image.

This section introduces some special applications of polynomial and rational functions.

Direct Variation When one quantity is a constant multiple of another quantity, the two quantities are said to *vary directly.* For example, if you work for an hourly wage of \$10, then

$$[\text{pay}] = 10[\text{hours worked}].$$

Doubling the hours doubles the pay. Tripling the hours triples the pay, and so on. This is stated more precisely as follows.

Direct Variation

y **varies directly** as x, or y is **directly proportional** to x, if there exists a nonzero real number k, called the **constant of variation,** such that for all x,

$$y = kx.$$

The direct variation equation $y = kx$ defines a linear function, where the constant of variation k is the slope of the line. For $k > 0$,

- As the value of x *increases,* the value of y *increases.*
- As the value of x *decreases,* the value of y *decreases.*

When used to describe a direct variation relationship, the phrase "directly proportional" is sometimes abbreviated to just "proportional."

The steps involved in solving a variation problem are summarized here.

Solving Variation Problems

Step 1 Write the general relationship among the variables as an equation. Use the constant k.

Step 2 Substitute given values of the variables and find the value of k.

Step 3 Substitute this value of k into the equation from Step 1, obtaining a specific formula.

Step 4 Substitute the remaining values and solve for the required unknown.

Classroom Example 1
At a given average speed, the distance traveled by a vehicle varies directly as the time. If a vehicle travels 156 mi in 3 hr, find the distance it will travel in 5 hr at the same average speed.

Answer: 260 mi

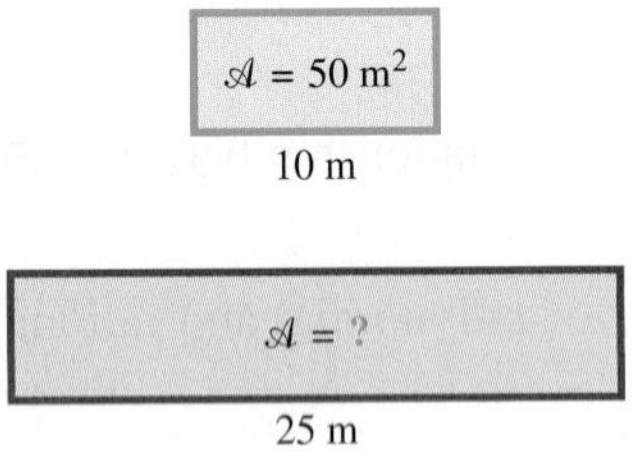

$\mathcal{A} = 50 \text{ m}^2$

10 m

$\mathcal{A} = ?$

25 m

Figure 59

EXAMPLE 1 Solving a Direct Variation Problem

The area of a rectangle varies directly as its length. If the area is 50 m^2 when the length is 10 m, find the area when the length is 25 m. (See **Figure 59.**)

SOLUTION

Step 1 Since the area varies directly as the length,

$$\mathcal{A} = kL,$$

where $\mathcal{A}$ represents the area of the rectangle, L is the length, and k is a nonzero constant.

Step 2 Since $\mathcal{A} = 50$ when $L = 10$, the equation $\mathcal{A} = kL$ becomes

$$50 = 10k \quad \text{Substitute for } \mathcal{A} \text{ and } L.$$

$$k = 5. \quad \text{Divide by 10. Interchange sides.}$$

Step 3 Using this value of k, we can express the relationship between the area and the length as

$$\mathcal{A} = 5L. \quad \text{Direct variation equation}$$

Step 4 To find the area when the length is 25, we replace L with 25.

$$\mathcal{A} = 5L$$

$$\mathcal{A} = 5(25) \quad \text{Substitute for } L.$$

$$\mathcal{A} = 125 \quad \text{Multiply.}$$

The area of the rectangle is 125 m^2 when the length is 25 m.

✔ *Now Try Exercise 21.*

Sometimes y varies as a power of x. If n is a positive integer greater than or equal to 2, then y is a greater-power polynomial function of x.

Direct Variation as *n*th Power

Let n be a positive real number. Then y **varies directly as the *n*th power** of x, or y is **directly proportional to the *n*th power** of x, if for all x there exists a nonzero real number k such that

$$y = kx^n.$$

For example, the area of a square of side x is given by the formula $\mathcal{A} = x^2$, so the area varies directly as the square of the length of a side. Here $k = 1$.

Teaching Tip Give students everyday situations, and have them tell whether direct or inverse variation is involved.

An example of direct variation might be "the amount of gasoline you pump and the amount of money you will pay."

An example of inverse variation might be "the duration of a certain trip and the average speed."

Inverse Variation Another type of variation is *inverse variation.* With inverse variation, where $k > 0$, as the value of one variable increases, the value of the other decreases. This relationship can be expressed as a rational function.

Inverse Variation as *n*th Power

Let n be a positive real number. Then y **varies inversely as the *n*th power** of x, or y is **inversely proportional to the *n*th power** of x, if for all x there exists a nonzero real number k such that

$$y = \frac{k}{x^n}.$$

If $n = 1$, then $y = \frac{k}{x}$, and y **varies inversely** as x.

Classroom Example 2
Refer to **Example 2**. Suppose that manufacturing costs have decreased so that if 100 items are produced, each costs \$1.50. Find the cost per item if 250 items are produced.

Answer: \$0.24, or 24 cents

EXAMPLE 2 Solving an Inverse Variation Problem

In a certain manufacturing process, the cost of producing a single item varies inversely as the square of the number of items produced. If 100 items are produced, each costs \$2. Find the cost per item if 400 items are produced.

SOLUTION

Step 1 Let x represent the number of items produced and y represent the cost per item. Then, for some nonzero constant k,

$$y = \frac{k}{x^2}. \qquad y \text{ varies inversely as the square of } x.$$

Step 2

$$2 = \frac{k}{100^2} \qquad \text{Substitute; } y = 2 \text{ when } x = 100.$$

$$k = 20{,}000 \qquad \text{Solve for } k. \text{ (Section 1.1)}$$

Step 3 The relationship between x and y is $y = \frac{20{,}000}{x^2}$.

Step 4 When 400 items are produced, the cost per item is

$$y = \frac{20{,}000}{x^2} = \frac{20{,}000}{400^2} = 0.125, \quad \text{or} \quad 12.5 \text{ cents.}$$

Now Try Exercise 31.

Teaching Tip Suggest that students break combined variation equations into pieces. For example, the statement "y varies inversely as the square of x and directly as the fourth power of z" can be broken into

$$y = \frac{k}{x^2} \quad \text{and} \quad y = kz^4,$$

and then combined as

$$y = \frac{kz^4}{x^2}.$$

Combined and Joint Variation In **combined variation,** one variable depends on more than one other variable. Specifically, when a variable depends on the *product* of two or more other variables, it is referred to as *joint variation.*

Joint Variation

Let m and n be real numbers. Then y **varies jointly** as the nth power of x and the mth power of z if for all x and z, there exists a nonzero real number k such that

$$y = kx^nz^m.$$

CAUTION Note that *and* in the expression "y varies jointly as x and z" translates as the product $y = kxz$. ***The word "and" does not indicate addition here.***

Classroom Example 3
The volume of a cylinder varies directly as the height and the square of the radius. A cylinder with radius 5 cm and height 10 cm has volume 785 cm^3. Find the volume of a cylinder with radius 10 cm and height 15 cm.

Answer: 4710 cm^3

EXAMPLE 3 Solving a Joint Variation Problem

The area of a triangle varies jointly as the lengths of the base and the height. A triangle with base 10 ft and height 4 ft has area 20 ft^2. Find the area of a triangle with base 3 ft and height 8 ft. (See **Figure 60.**)

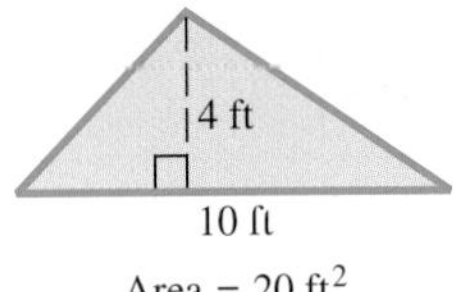

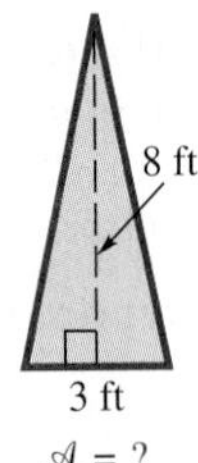

Figure 60

SOLUTION

Step 1 Let $\mathcal{A}$ represent the area, b the base, and h the height of the triangle. Then, for some number k,

$$\mathcal{A} = kbh. \quad \mathcal{A} \text{ varies jointly as } b \text{ and } h.$$

Step 2 Since $\mathcal{A}$ is 20 when b is 10 and h is 4, substitute and solve for k.

$$20 = k(10)(4)$$

$$\frac{1}{2} = k$$

Step 3 The relationship among the variables is the familiar formula for the area of a triangle,

$$\mathcal{A} = \frac{1}{2}bh.$$

Step 4 To find $\mathcal{A}$ when $b = 3$ ft and $h = 8$ ft, substitute into the formula.

$$\mathcal{A} = \frac{1}{2}(3)(8) = 12 \text{ ft}^2$$

✔ *Now Try Exercise 33.*

Classroom Example 4
Vanessa Jones has determined that the number of children enrolled in her day care center varies directly with the amount she spends on advertising per year and inversely with her weekly charge for each child. This year she spent \$900 on advertising, charged \$150 per child per week, and had 51 children enrolled. Next year, if she spends \$1000 on advertising and charges \$170 per week, how many children should she expect to enroll?

Answer: 50 children

EXAMPLE 4 Solving a Combined Variation Problem

The number of vibrations per second (the pitch) of a steel guitar string varies directly as the square root of the tension and inversely as the length of the string. If the number of vibrations per second is 50 when the tension is 225 newtons and the length is 0.60 m, find the number of vibrations per second when the tension is 196 newtons and the length is 0.65 m.

SOLUTION Let n represent the number of vibrations per second, T represent the tension, and L represent the length of the string. Then, from the information in the problem, write the variation equation (Step 1).

$$n = \frac{k\sqrt{T}}{L} \quad n \text{ varies directly as the square root of } T \text{ and inversely as } L.$$

Substitute the given values for n, T, and L to find k (Step 2).

$$50 = \frac{k\sqrt{225}}{0.60} \quad \text{Let } n = 50,\ T = 225,\ L = 0.60.$$

$$30 = k\sqrt{225} \quad \text{Multiply by 0.60.}$$

$$30 = 15k \quad \sqrt{225} = 15$$

$$k = 2 \quad \text{Divide by 15. Interchange sides.}$$

Substitute for k to find the relationship among the variables (Step 3).

$$n = \frac{2\sqrt{T}}{L}$$

Now use the second set of values for T and L to find n (Step 4).

$$n = \frac{2\sqrt{196}}{0.65} \approx 43 \quad \text{Let } T = 196,\ L = 0.65.$$

The number of vibrations per second is approximately 43.

Now Try Exercise 37.

3.6 Exercises

Concept Check *Write each formula as an English phrase using the word* varies *or* proportional.

1. $C = 2\pi r$, where C is the circumference of a circle of radius r
2. $d = \frac{1}{5}s$, where d is the approximate distance (in miles) from a storm, and s is the number of seconds between seeing lightning and hearing thunder
3. $r = \frac{d}{t}$, where r is the speed when traveling d miles in t hours
4. $d = \dfrac{1}{4\pi nr^2}$, where d is the distance a gas atom of radius r travels between collisions, and n is the number of atoms per unit volume
5. $s = kx^3$, where s is the strength of a muscle that has length x
6. $f = \dfrac{mv^2}{r}$, where f is the centripetal force of an object of mass m moving along a circle of radius r at velocity v

Concept Check *Match each statement in Exercises 7–10 with its corresponding graph in choices A–D. In each case,* $k > 0$.

7. y varies directly as x. $(y = kx)$
8. y varies inversely as x. $\left(y = \frac{k}{x}\right)$
9. y varies directly as the second power of x. $(y = kx^2)$
10. x varies directly as the second power of y. $(x = ky^2)$

A.
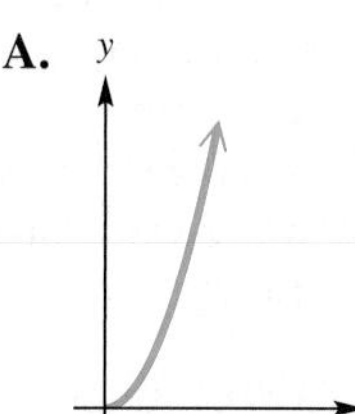

B.
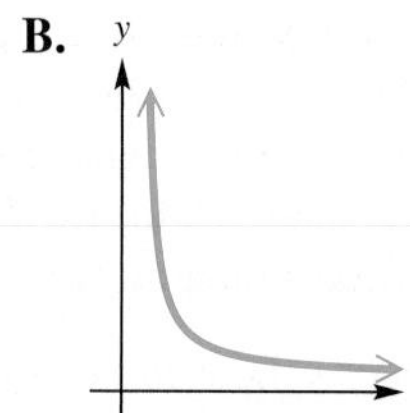

C.
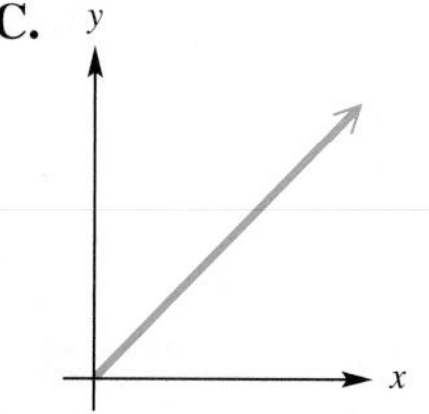

D.
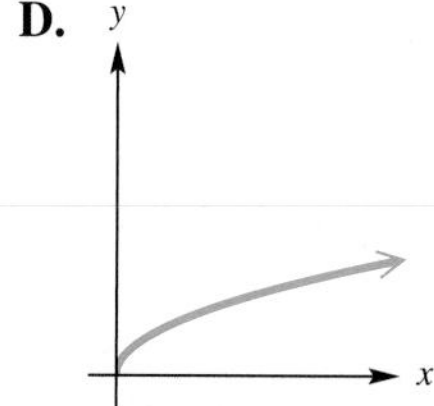

Solve each variation problem. ***See Examples 1–4.***

11. If y varies directly as x, and $y = 20$ when $x = 4$, find y when $x = -6$.
12. If y varies directly as x, and $y = 9$ when $x = 30$, find y when $x = 40$.
13. If m varies jointly as x and y, and $m = 10$ when $x = 2$ and $y = 14$, find m when $x = 21$ and $y = 8$.
14. If m varies jointly as z and p, and $m = 10$ when $z = 2$ and $p = 7.5$, find m when $z = 6$ and $p = 9$.

1. The circumference of a circle varies directly as (or is proportional to) its radius.
2. The distance from a storm varies directly as (or is proportional to) the number of seconds between seeing lightning and hearing thunder.
3. The speed varies directly as (or is proportional to) the distance traveled and inversely as the time.
4. The distance a gas atom travels between collisions varies inversely as the square of its radius and the number of atoms per unit volume.
5. The strength of a muscle varies directly as (or is proportional to) the cube of its length.
6. The centripetal force of an object varies directly as (or is proportional to) its mass and the square of its velocity and inversely as the radius of the circle it moves along.

7. C 8. B
9. A 10. D

11. -30 12. 12
13. 60 14. 36

15. $\frac{3}{2}$ 16. $\frac{1}{3}$
17. $\frac{12}{5}$ 18. $\frac{9}{8}$
19. $\frac{18}{125}$ 20. $\frac{45}{16}$
21. 69.08 in. 22. 75 psi
23. 850 ohms 24. 8 lb
25. 12 km 26. 800 gal
27. 16 in.

15. If y varies inversely as x, and $y = 10$ when $x = 3$, find y when $x = 20$.

16. If y varies inversely as x, and $y = 20$ when $x = \frac{1}{4}$, find y when $x = 15$.

17. Suppose r varies directly as the square of m, and inversely as s. If $r = 12$ when $m = 6$ and $s = 4$, find r when $m = 6$ and $s = 20$.

18. Suppose p varies directly as the square of z, and inversely as r. If $p = \frac{32}{5}$ when $z = 4$ and $r = 10$, find p when $z = 3$ and $r = 32$.

19. Let a be directly proportional to m and n^2, and inversely proportional to y^3. If $a = 9$ when $m = 4$, $n = 9$, and $y = 3$, find a when $m = 6$, $n = 2$, and $y = 5$.

20. If y varies directly as x, and inversely as m^2 and r^2, and $y = \frac{5}{3}$ when $x = 1$, $m = 2$, and $r = 3$, find y when $x = 3$, $m = 1$, and $r = 8$.

Solve each problem. **See Examples 1–4.**

21. *Circumference of a Circle* The circumference of a circle varies directly as the radius. A circle with radius 7 in. has circumference 43.96 in. Find the circumference of the circle if the radius changes to 11 in.

22. *Pressure Exerted by a Liquid* The pressure exerted by a certain liquid at a given point varies directly as the depth of the point beneath the surface of the liquid. The pressure at 10 ft is 50 pounds per square inch (psi). What is the pressure at 15 ft?

23. *Resistance of a Wire* The resistance in ohms of a platinum wire temperature sensor varies directly as the temperature in kelvins (K). If the resistance is 646 ohms at a temperature of 190 K, find the resistance at a temperature of 250 K.

24. *Weight on the Moon* The weight of an object on Earth is directly proportional to the weight of that same object on the moon. A 200-lb astronaut would weigh 32 lb on the moon. How much would a 50-lb dog weigh on the moon?

25. *Distance to the Horizon* The distance that a person can see to the horizon on a clear day from a point above the surface of Earth varies directly as the square root of the height at that point. If a person 144 m above the surface of Earth can see 18 km to the horizon, how far can a person see to the horizon from a point 64 m above the surface?

26. *Water Emptied by a Pipe* The amount of water emptied by a pipe varies directly as the square of the diameter of the pipe. For a certain constant water flow, a pipe emptying into a canal will allow 200 gal of water to escape in an hour. The diameter of the pipe is 6 in. How much water would a 12-in. pipe empty into the canal in an hour, assuming the same water flow?

27. *Hooke's Law for a Spring* Hooke's law for an elastic spring states that the distance a spring stretches varies directly as the force applied. If a force of 15 lb stretches a certain spring 8 in., how much will a force of 30 lb stretch the spring?

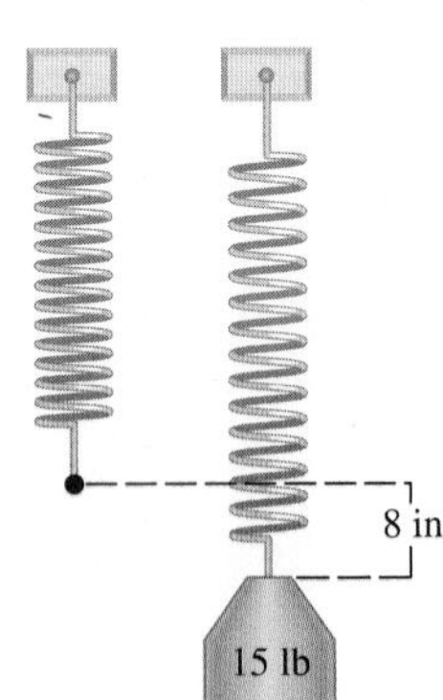

28. 100 amps
29. 90 revolutions per minute
30. 40 lb
31. 0.0444 ohm
32. $\frac{875}{72}$ candela
33. \$875
34. 1.105 L
35. 800 lb
36. 799.5 cm^3

28. *Current in a Circuit* The current in a simple electrical circuit varies inversely as the resistance. If the current is 50 amps when the resistance is 10 ohms, find the current if the resistance is 5 ohms.

29. *Speed of a Pulley* The speed of a pulley varies inversely as its diameter. One kind of pulley, with diameter 3 in., turns at 150 revolutions per minute. Find the speed of a similar pulley with diameter 5 in.

30. *Weight of an Object* The weight of an object varies inversely as the square of its distance from the center of Earth. If an object 8000 mi from the center of Earth weighs 90 lb, find its weight when it is 12,000 mi from the center of Earth.

31. *Current Flow* In electric current flow, it is found that the resistance offered by a fixed length of wire of a given material varies inversely as the square of the diameter of the wire. If a wire 0.01 in. in diameter has a resistance of 0.4 ohm, what is the resistance of a wire of the same length and material with diameter 0.03 in., to the nearest ten-thousandth of an ohm?

32. *Illumination* The illumination produced by a light source varies inversely as the square of the distance from the source. The illumination of a light source at 5 m is 70 candela. What is the illumination 12 m from the source?

33. *Simple Interest* Simple interest varies jointly as principal and time. If \$1000 invested for 2 yr earned \$70, find the amount of interest earned by \$5000 for 5 yr.

34. *Volume of a Gas* The volume of a gas varies inversely as the pressure and directly as the temperature in kelvins (K). If a certain gas occupies a volume of 1.3 L at 300 K and a pressure of 18 newtons, find the volume at 340 K and a pressure of 24 newtons.

35. *Force of Wind* The force of the wind blowing on a vertical surface varies jointly as the area of the surface and the square of the velocity. If a wind of 40 mph exerts a force of 50 lb on a surface of $\frac{1}{2}$ ft^2, how much force will a wind of 80 mph place on a surface of 2 ft^2?

36. *Volume of a Cylinder* The volume of a right circular cylinder is jointly proportional to the square of the radius of the circular base and to the height. If the volume is 300 cm^3 when the height is 10.62 cm and the radius is 3 cm, find the volume to the nearest tenth of a cylinder with radius 4 cm and height 15.92 cm.

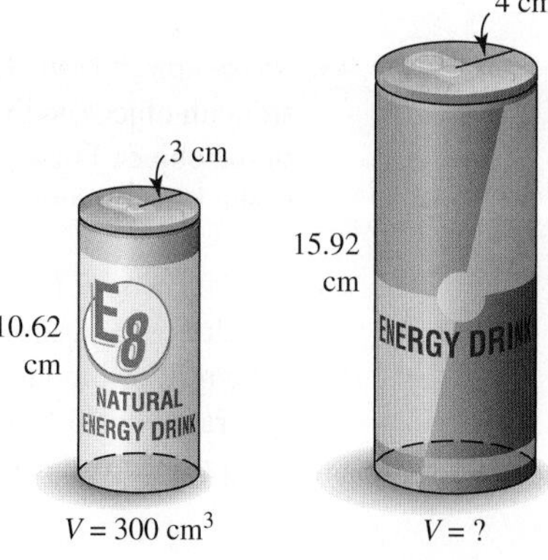

37. $\frac{8}{9}$ metric ton
38. $\frac{1024}{9}$ kg
39. $\frac{66\pi}{17}$ sec
40. 1600 calls
41. 21
42. 4.94
43. 365.24
44. 7.4 km

37. *Sports Arena Construction* The roof of a new sports arena rests on round concrete pillars. The maximum load a cylindrical column of circular cross section can hold varies directly as the fourth power of the diameter and inversely as the square of the height. The arena has 9-m-tall columns that are 1 m in diameter and will support a load of 8 metric tons. How many metric tons will be supported by a column 12 m high and $\frac{2}{3}$ m in diameter?

Load = 8 metric tons

38. *Sports Arena Construction* The sports arena in **Exercise 37** requires a horizontal beam 16 m long, 24 cm wide, and 8 cm high. The maximum load of a horizontal beam that is supported at both ends varies directly as the width of the beam and the square of its height and inversely as the length between supports. If a beam of the same material 8 m long, 12 cm wide, and 15 cm high can support a maximum of 400 kg, what is the maximum load the beam in the arena will support?

39. *Period of a Pendulum* The period of a pendulum varies directly as the square root of the length of the pendulum and inversely as the square root of the acceleration due to gravity. Find the period when the length is 121 cm and the acceleration due to gravity is 980 cm per second squared, if the period is 6π seconds when the length is 289 cm and the acceleration due to gravity is 980 cm per second squared.

40. *Long-Distance Phone Calls* The number of long-distance phone calls between two cities in a certain time period varies directly as the populations p_1 and p_2 of the cities and inversely as the distance between them. If 10,000 calls are made between two cities 500 mi apart, having populations of 50,000 and 125,000, find the number of calls between two cities 800 mi apart, having populations of 20,000 and 80,000.

41. *Body Mass Index* The federal government has developed the **body mass index** (BMI) to determine ideal weights. A person's BMI is directly proportional to his or her weight in pounds and inversely proportional to the square of his or her height in inches. (A BMI of 19 to 25 corresponds to a healthy weight.) A 6-foot-tall person weighing 177 lb has BMI 24. Find the BMI (to the nearest whole number) of a person whose weight is 130 lb and whose height is 66 in.

42. *Poiseuille's Law* According to Poiseuille's law, the resistance to flow of a blood vessel, R, is directly proportional to the length, l, and inversely proportional to the fourth power of the radius, r. If $R = 25$ when $l = 12$ and $r = 0.2$, find R to the nearest hundredth as r increases to 0.3, while l is unchanged.

43. *Stefan-Boltzmann Law* The Stefan-Boltzmann law says that the radiation of heat R from an object is directly proportional to the fourth power of the kelvin temperature of the object. For a certain object, $R = 213.73$ at room temperature (293 K). Find R to the nearest hundredth if the temperature increases to 335 K.

44. *Nuclear Bomb Detonation* Suppose the effects of detonating a nuclear bomb will be felt over a distance from the point of detonation that is directly proportional to the cube root of the yield of the bomb. Suppose a 100-kiloton bomb has certain effects to a radius of 3 km from the point of detonation. Find the distance to the nearest tenth that the effects would be felt for a 1500-kiloton bomb.

45. 92; undernourished
46. (a) $F = 4$
(b) $t = \frac{1}{250}$ sec
47. increases; decreases
48. decreases; increases
49. y is half as large as before.
50. y is half as large as before.
51. y is one-third as large as before.
52. y is one-third as large as before.
53. p is $\frac{1}{32}$ as large as before.
54. m is 324 times as large as before.

45. *Malnutrition Measure* A measure of malnutrition, called the **pelidisi,** varies directly as the cube root of a person's weight in grams and inversely as the person's sitting height in centimeters. A person with a pelidisi below 100 is considered to be undernourished, while a pelidisi greater than 100 indicates overfeeding. A person who weighs 48,820 g with a sitting height of 78.7 cm has a pelidisi of 100. Find the pelidisi (to the nearest whole number) of a person whose weight is 54,430 g and whose sitting height is 88.9 cm. Is this individual undernourished or overfed?

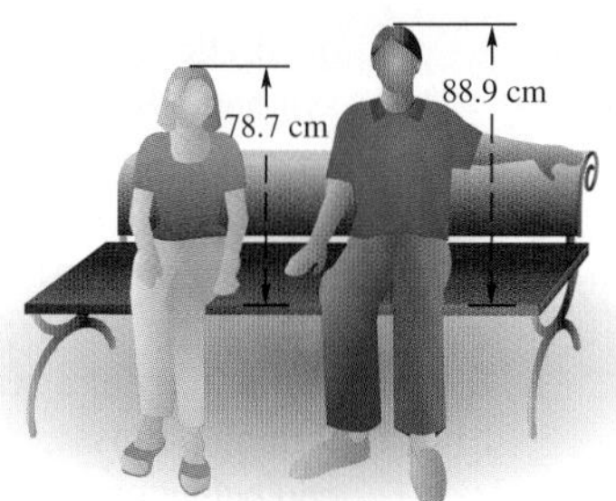

Weight: 48,820 g   Weight: 54,430 g

46. *Photography* Variation occurs in a formula from photography. In

$$L = \frac{25F^2}{st},$$

the luminance, L, varies directly as the square of the F-stop, F, and inversely as the product of the film ASA number, s, and the shutter speed, t.

(a) What would an appropriate F-stop be for 200 ASA film and a shutter speed of $\frac{1}{250}$ sec when 500 footcandles of light is available?
(b) If 125 footcandles of light is available and an F-stop of 2 is used with 200 ASA film, what shutter speed should be used?

Concept Check Work each problem.

47. For $k > 0$, if y varies directly as x, then when x increases, y ________, and when x decreases, y ________.

48. For $k > 0$, if y varies inversely as x, then when x increases, y ________, and when x decreases, y ________.

49. What happens to y if y varies inversely as x, and x is doubled?

50. What happens to y if y varies directly as x, and x is halved?

51. Suppose y is directly proportional to x, and x is replaced by $\frac{1}{3}x$. What happens to y?

52. Suppose y is inversely proportional to x, and x is tripled. What happens to y?

53. Suppose p varies directly as r^3 and inversely as t^2. If r is halved and t is doubled, what happens to p?

54. Suppose m varies directly as p^2 and q^4. If p doubles and q triples, what happens to m?

Chapter 3 Test Prep

Key Terms

3.1 polynomial function, leading coefficient, leading term, zero polynomial, quadratic function, parabola, axis (axis of symmetry), vertex, quadratic regression

3.2 synthetic division, zero of a polynomial function, root (or solution) of an equation

3.3 multiplicity of a zero

3.4 turning points, end behavior, dominating term

3.5 rational function, discontinuous graph, vertical asymptote, horizontal asymptote, oblique asymptote, point of discontinuity

3.6 varies directly (directly proportional to), constant of variation, varies inversely (inversely proportional to), combined variation, varies jointly

New Symbols

$\bar{z}$	conjugate of $z = a + bi$
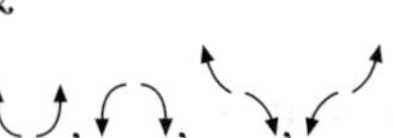	end behavior diagrams
$\lvert f(x) \rvert \to \infty$	absolute value of $f(x)$ increases without bound
$x \to a$	x approaches a

Quick Review

Concepts	Examples

3.1 Quadratic Functions and Models

Concepts

1. The graph of

$$f(x) = a(x - h)^2 + k, \quad a \neq 0,$$

is a parabola with vertex at (h, k) and the vertical line $x = h$ as axis.

2. The graph opens up if a is positive and down if a is negative.

3. The graph is wider than the graph of $f(x) = x^2$ if $|a| < 1$ and narrower if $|a| > 1$.

Examples

Graph $f(x) = -(x + 3)^2 + 1$.

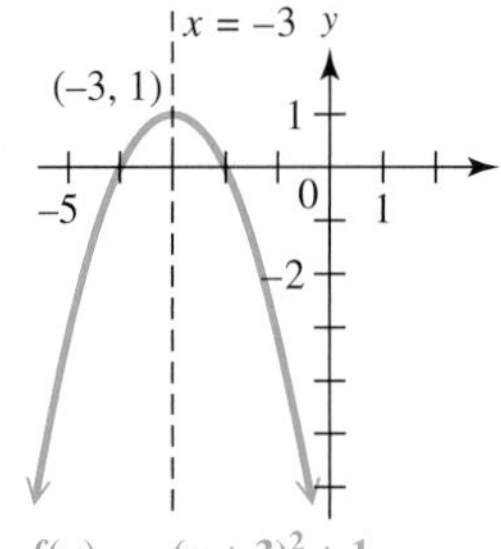

$f(x) = -(x + 3)^2 + 1$

The graph opens down since $a < 0$. It is the graph of $y = -x^2$ shifted 3 units left and 1 unit up, so the vertex is $(-3, 1)$, with axis $x = -3$. The domain is $(-\infty, \infty)$, and the range is $(-\infty, 1]$. The function is increasing on $(-\infty, -3]$ and decreasing on $[-3, \infty)$.

Vertex Formula

The vertex of the graph of $f(x) = ax^2 + bx + c$, $a \neq 0$, may be found by completing the square. The vertex has coordinates

$$\left(-\frac{b}{2a}, f\left(-\frac{b}{2a}\right)\right).$$

Graphing a Quadratic Function

Step 1 Find the vertex either by using the vertex formula or by completing the square.

Step 2 Find the y-intercept by evaluating $f(0)$.

Step 3 Find any x-intercepts by solving $f(x) = 0$.

Step 4 Find and plot any additional points as needed, using symmetry about the axis.

The graph opens up if $a > 0$ and down if $a < 0$.

Graph $f(x) = x^2 + 4x + 3$. The vertex of the graph is

$$\left(-\frac{b}{2a}, f\left(-\frac{b}{2a}\right)\right) = (-2, -1). \quad a = 1, b = 4, c = 3$$

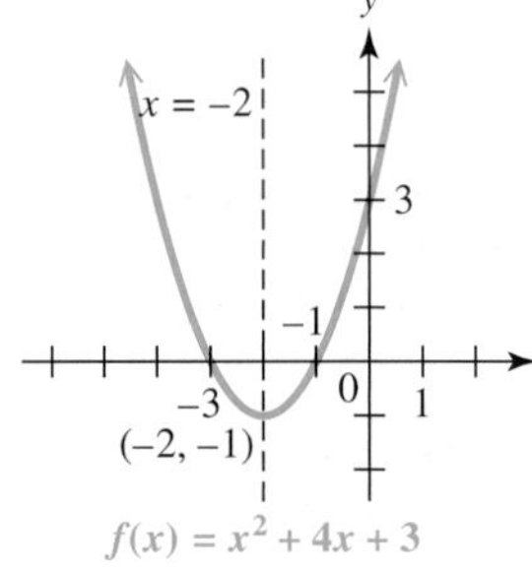

$f(x) = x^2 + 4x + 3$

The graph opens up since $a > 0$. Since $f(0) = 3$, the y-intercept is 3. The solutions of $x^2 + 4x + 3 = 0$ are -1 and -3, which are the x-intercepts. The domain is $(-\infty, \infty)$, and the range is $[-1, \infty)$. The function is decreasing on $(-\infty, -2]$ and increasing on $[-2, \infty)$.

Concepts	Examples
3.2 Synthetic Division	
Synthetic division is a shortcut method for dividing a polynomial by a binomial of the form $x - k$.	Use synthetic division to divide $$f(x) = 2x^3 - 3x + 2 \quad \text{by} \quad x - 1,$$ and write the result as $f(x) = g(x) \cdot q(x) + r(x)$. $$\begin{array}{r\|rrrr} 1 & 2 & 0 & -3 & 2 \\ & & 2 & 2 & -1 \\ \hline & 2 & 2 & -1 & 1 \end{array}$$ Coefficients of the quotient (2, 2, −1); Remainder (1) $$2x^3 - 3x + 2 = \underbrace{(x-1)}_{g(x)}\underbrace{(2x^2 + 2x - 1)}_{q(x)} + \underbrace{1}_{r(x)}$$
Remainder Theorem If the polynomial $f(x)$ is divided by $x - k$, the remainder is $f(k)$.	By the result above, for $f(x) = 2x^3 - 3x + 2$, $$f(1) = 1.$$
3.3 Zeros of Polynomial Functions	
Factor Theorem For any polynomial function $f(x)$, $x - k$ is a factor of the polynomial if and only if $f(k) = 0$.	For the polynomial functions $$f(x) = x^3 + x + 2 \quad \text{and} \quad g(x) = x^3 - 1,$$ $f(-1) = 0$. Therefore, $x - (-1)$, or $x + 1$, is a factor of $f(x)$. Also, since $x - 1$ is a factor of $g(x)$, $g(1) = 0$.
Rational Zeros Theorem If $\frac{p}{q}$ is a rational number written in lowest terms, and if $\frac{p}{q}$ is a zero of f, a polynomial function with integer coefficients, then p is a factor of the constant term, and q is a factor of the leading coefficient.	The only rational numbers that can possibly be zeros of $$f(x) = 2x^3 - 9x^2 - 4x - 5$$ are ± 1, ± 5, $\pm \frac{1}{2}$, and $\pm \frac{5}{2}$. By synthetic division, it can be shown that the only rational zero of $f(x)$ is 5. $$\begin{array}{r\|rrrr} 5 & 2 & -9 & -4 & -5 \\ & & 10 & 5 & 5 \\ \hline & 2 & 1 & 1 & 0 \leftarrow f(5) \end{array}$$
Fundamental Theorem of Algebra Every function defined by a polynomial of degree 1 or more has at least one complex zero.	$f(x) = x^3 + x + 2$ has at least one and at most three distinct zeros.
Number of Zeros Theorem A function defined by a polynomial of degree n has at most n distinct zeros.	
Conjugate Zeros Theorem If $f(x)$ defines a polynomial function ***having only real coefficients*** and if $a + bi$ is a zero of $f(x)$, where a and b are real numbers, then the conjugate $a - bi$ is also a zero of $f(x)$.	Since $1 + 2i$ is a zero of $$f(x) = x^3 - 5x^2 + 11x - 15,$$ its conjugate $1 - 2i$ is a zero as well.

(continued)

Concepts	Examples

Descartes' Rule of Signs
See the discussion in **Section 3.3.**

For $f(x) = 3x^3 - 2x^2 + x - 4$, there are three sign changes, so there will be three or one positive real zeros. Since $f(-x) = -3x^3 - 2x^2 - x - 4$ has no sign changes, there will be no negative real zeros. The table shows the possibilities for the numbers of positive, negative, and nonreal complex zeros.

Positive	Negative	Nonreal Complex
3	0	0
1	0	2

3.4 Polynomial Functions: Graphs, Applications, and Models

Graphing Using Translations
The graph of the function

$$f(x) = a(x - h)^n + k$$

can be found by considering the effects of the constants a, h, and k on the graph of $y = x^n$.

Graph $f(x) = -(x + 2)^4 + 1$.

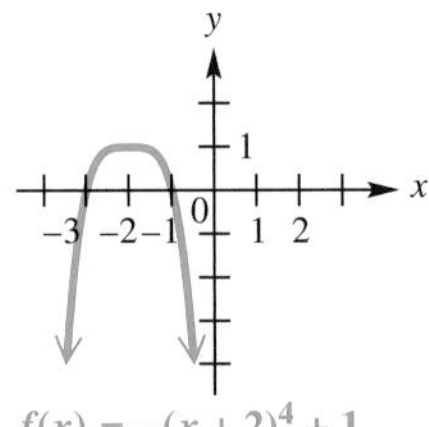

$f(x) = -(x + 2)^4 + 1$

The negative sign causes the graph to be reflected across the x-axis compared to the graph of $y = x^4$. The graph is translated 2 units to the left and 1 unit up. The graph shows that this function is increasing on $(-\infty, -2]$ and decreasing on $[-2, \infty)$.

Multiplicity of a Zero
The behavior of the graph of a polynomial function $f(x)$ near a zero depends on the multiplicity of the zero. If $(x - c)^n$ is a factor of $f(x)$, the graph will behave in the following manner.

- For $n = 1$, the graph will cross the x-axis at $(c, 0)$.
- For n even, the graph will bounce, or turn, at $(c, 0)$.
- For n an odd integer greater than 1, the graph will wiggle through the x-axis at $(c, 0)$.

Determine the behavior of

$$f(x) = (x - 1)(x - 3)^2(x + 1)^3$$

near its zeros, and graph.

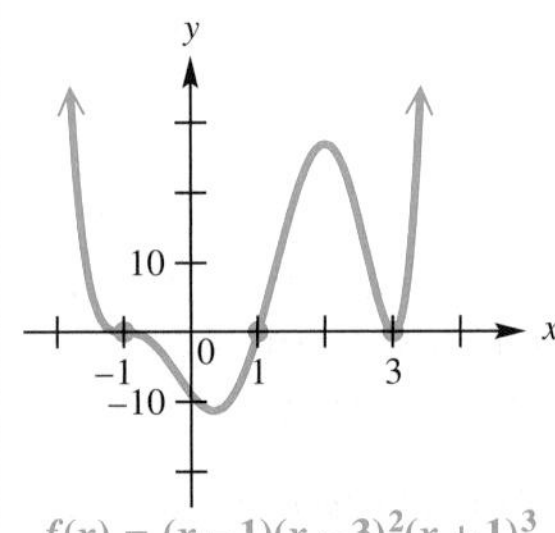

$f(x) = (x - 1)(x - 3)^2(x + 1)^3$

The graph will cross the x-axis at $x = 1$, bounce at $x = 3$, and wiggle through the x-axis at $x = -1$. Since the dominating term is x^6, the end behavior is ↖ ↗. The y-intercept is $f(0) = -9$.

Turning Points
A polynomial function of degree n has at most $n - 1$ turning points.

The graph of

$$f(x) = 4x^5 - 2x^3 + 3x^2 + x - 10$$

has at most four turning points (since $5 - 1 = 4$).

Concepts	Examples

End Behavior

The end behavior of the graph of a polynomial function $f(x)$ is determined by the dominating term, or term of greatest degree. If ax^n is the dominating term of $f(x)$, then the end behavior is as follows.

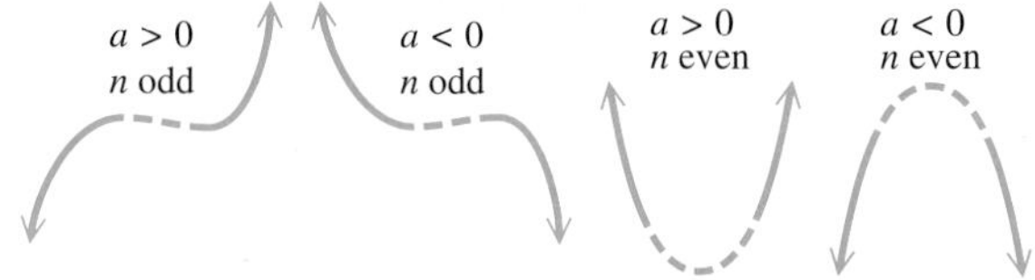

The end behavior of

$$f(x) = 3x^5 + 2x^2 + 7$$

is .

The end behavior of

$$f(x) = -x^4 - 3x^3 + 2x - 9$$

is .

Graphing Polynomial Functions

To graph a polynomial function f, first find the x-intercepts and y-intercepts.

Then use end behavior, whether the graph crosses, bounces on, or wiggles through the x-axis at the x-intercepts, and selected points as necessary to complete the graph.

Graph $f(x) = (x + 2)(x - 1)(x + 3)$.

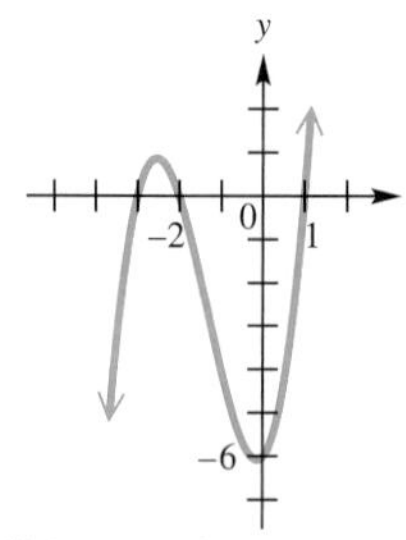

$f(x) = (x + 2)(x - 1)(x + 3)$

The zeros of f are -2, 1, and -3. Since $f(0) = 2(-1)(3) = -6$, the y-intercept is -6. The dominating term is $x(x)(x)$ or x^3, so the end behavior is . Begin at either end of the graph with the correct end behavior, and draw a smooth curve that crosses the x-axis at each zero, has a turning point between successive zeros, and passes through the y-intercept.

Intermediate Value Theorem for Polynomials

If $f(x)$ is a polynomial function with only *real coefficients,* and if for real numbers a and b, the values of $f(a)$ and $f(b)$ are opposite in sign, then there exists at least one real zero between a and b.

For the polynomial function

$$f(x) = -x^4 + 2x^3 + 3x^2 + 6,$$

$$f(3.1) = 2.0599 \text{ and } f(3.2) = -2.6016.$$

Since $f(3.1) > 0$ and $f(3.2) < 0$, there exists at least one real zero between 3.1 and 3.2.

Boundedness Theorem

Let $f(x)$ be a polynomial function of degree $n \geq 1$ with *real coefficients* and with a *positive* leading coefficient. Suppose $f(x)$ is divided synthetically by $x - c$.

(a) If $c > 0$ and all numbers in the bottom row of the synthetic division are nonnegative, then $f(x)$ has no zero greater than c.

(b) If $c < 0$ and the numbers in the bottom row of the synthetic division alternate in sign (with 0 considered positive or negative, as needed), then $f(x)$ has no zero less than c.

Show that $f(x) = x^3 - x^2 - 8x + 12$ has no zero greater than 4 and no zero less than -4.

4)	1	−1	−8	12	
		4	12	16	
	1	3	4	28	⟵ All signs positive

−4)	1	−1	−8	12	
		−4	20	−48	
	1	−5	12	−36	⟵ Alternating signs

(continued)

Concepts	Examples

3.5 Rational Functions: Graphs, Applications, and Models

Graphing Rational Functions

To graph a rational function in lowest terms, find the asymptotes and intercepts. Determine whether the graph intersects a nonvertical asymptote. Plot a few points, as necessary, to complete the sketch.

Graph $f(x) = \dfrac{x^2 - 1}{(x + 3)(x - 2)}$.

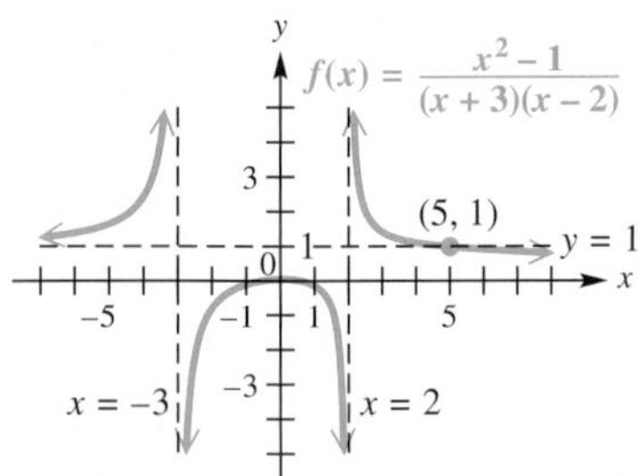

Point of Discontinuity

If a rational function is not written in lowest terms, there may be a "hole" in the graph instead of an asymptote.

Graph $f(x) = \dfrac{x^2 - 1}{x + 1}$.

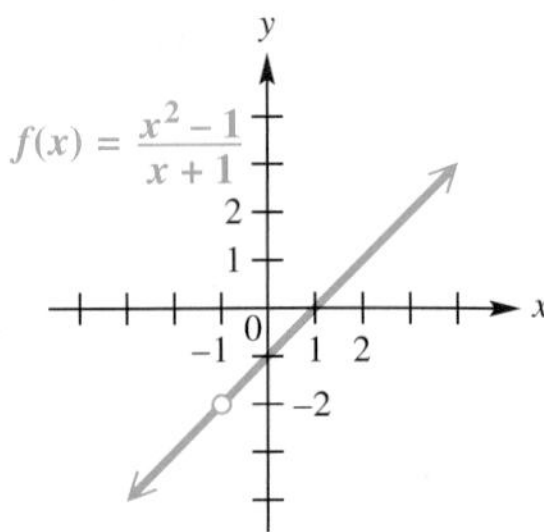

$$\begin{aligned} f(x) &= \frac{x^2 - 1}{x + 1} \\ &= \frac{(x + 1)(x - 1)}{x + 1} \\ &= x - 1, \quad x \neq -1 \end{aligned}$$

3.6 Variation

Direct Variation

y varies directly as the nth power of x if for all x there exists a nonzero real number k such that

$$y = kx^n.$$

The area of a circle varies directly as the square of the radius.

$$\mathcal{A} = kr^2 \quad (k = \pi)$$

Inverse Variation

y varies inversely as the nth power of x if for all x there exists a nonzero real number k such that

$$y = \frac{k}{x^n}.$$

Pressure of a gas varies inversely as volume.

$$P = \frac{k}{V}$$

Joint Variation

For real numbers m and n, y varies jointly as the nth power of x and the mth power of z if for all x and z there exists a nonzero real number k such that

$$y = kx^n z^m.$$

The area of a triangle varies jointly as its base and its height.

$$\mathcal{A} = kbh \quad \left(k = \tfrac{1}{2}\right)$$

Chapter 3 Review Exercises

1. vertex: $(-4, -5)$;
axis: $x = -4$;
x-intercepts: $\frac{-12 \pm \sqrt{15}}{3}$;
y-intercept: 43;
domain: $(-\infty, \infty)$;
range: $[-5, \infty)$;
increasing on $[-4, \infty)$;
decreasing on $(-\infty, -4]$

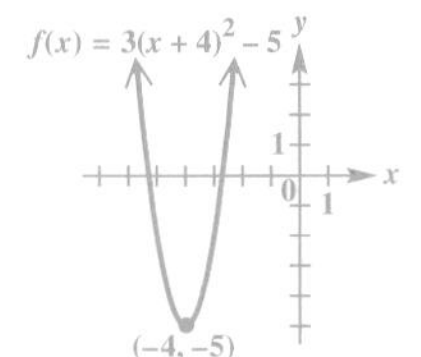

2. vertex: $(6, 7)$;
axis: $x = 6$;
x-intercepts: $\frac{12 \pm \sqrt{42}}{2}$;
y-intercept: -17;
domain: $(-\infty, \infty)$;
range: $(-\infty, 7]$;
increasing on $(-\infty, 6]$;
decreasing on $[6, \infty)$

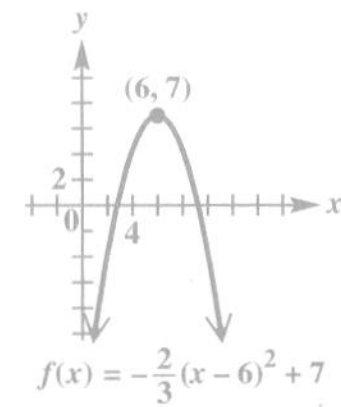

3. vertex: $(-2, 11)$;
axis: $x = -2$;
x-intercepts: $\frac{-6 \pm \sqrt{33}}{3}$;
y-intercept: -1;
domain: $(-\infty, \infty)$;
range: $(-\infty, 11]$;
increasing on $(-\infty, -2]$;
decreasing on $[-2, \infty)$

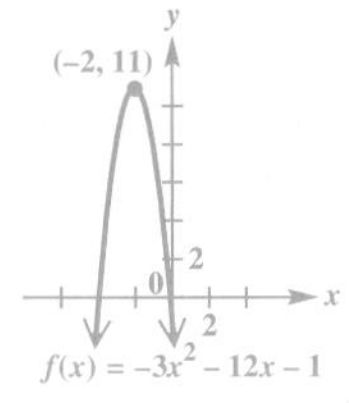

Graph each quadratic function. Give the vertex, axis, x-intercepts, y-intercept, domain, range, and the intervals of the domain for which each function is increasing or decreasing.

1. $f(x) = 3(x + 4)^2 - 5$

2. $f(x) = -\frac{2}{3}(x - 6)^2 + 7$

3. $f(x) = -3x^2 - 12x - 1$

4. $f(x) = 4x^2 - 4x + 3$

Concept Check *In Exercises 5–8, consider the function*

$$f(x) = a(x - h)^2 + k, \quad \text{for } a > 0.$$

5. What are the coordinates of the lowest point of its graph?

6. What is the y-intercept of its graph?

7. Under what conditions will its graph have one or more x-intercepts? For these conditions, express the x-intercept(s) in terms of a, h, and k.

8. If a is positive, what is the least value of $ax^2 + bx + c$ in terms of a, b, and c?

(Modeling) *Solve each problem.*

9. *Area of a Rectangle* Use a quadratic function to find the dimensions of the rectangular region of maximum area that can be enclosed with 180 m of fencing, if no fencing is needed along one side of the region.

10. *Height of a Projectile* A projectile is fired vertically upward, and its height $s(t)$ in feet after t seconds is given by the function

$$s(t) = -16t^2 + 800t + 600.$$

(a) From what height was the projectile fired?
(b) After how many seconds will it reach its maximum height?
(c) What is the maximum height it will reach?
(d) Between what two times (in seconds, to the nearest tenth) will it be more than 5000 ft above the ground?
(e) After how many seconds, to the nearest tenth, will the projectile hit the ground?

11. *Food Bank Volunteers* During the course of a year, the number of volunteers available to run a food bank each month is modeled by $V(x)$, where

$$V(x) = 2x^2 - 32x + 150$$

between the months of January and August. Here x is time in months, with $x = 1$ representing January. From August to December, $V(x)$ is modeled by

$$V(x) = 31x - 226.$$

Find the number of volunteers in each of the following months.

(a) January **(b)** May **(c)** August **(d)** October **(e)** December
(f) Sketch a graph of $y = V(x)$ for January through December. In what month are the fewest volunteers available?

12. *Concentration of Atmospheric CO_2* In 1990, the International Panel on Climate Change (IPCC) stated that if current trends of burning fossil fuel and deforestation were to continue, then future amounts of atmospheric carbon dioxide in parts per million (ppm) would increase, as shown in the table.

Year	Carbon Dioxide
1990	353
2000	375
2075	590
2175	1090
2275	2000

Source: IPCC.

4. vertex: $\left(\frac{1}{2}, 2\right)$;
axis: $x = \frac{1}{2}$;
x-intercepts: none;
y-intercept: 3;
domain: $(-\infty, \infty)$;
range: $[2, \infty)$;
increasing on $\left[\frac{1}{2}, \infty\right)$;
decreasing on $\left(-\infty, \frac{1}{2}\right]$

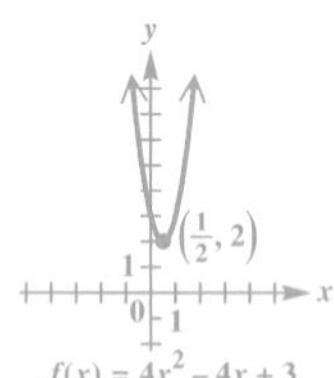

5. (h, k) 6. $ah^2 + k$
7. $k \le 0$; $h \pm \sqrt{\frac{-k}{a}}$
8. $c - \frac{b^2}{4a}$
9. 90 m by 45 m
10. (a) 600 ft (b) 25 sec
(c) 10,600 ft
(d) 6.3 sec and 43.7 sec
(e) 50.7 sec
11. (a) 120 (b) 40
(c) 22 (d) 84
(e) 146
(f)

y
(8, 22)
10
0
8
x
August

The minimum occurs in August when $x = 8$.
12. (a) $f(x) = 0.0203x^2 + 353$
(b) about 2304 ppm
13. Because the discriminant is 67.3033, a positive number, there are two x-intercepts.
14. $\{-0.52, 2.59\}$
15. (a) the open interval $(-0.52, 2.59)$
(b) $(-\infty, -0.52) \cup (2.59, \infty)$
16. $(1.04, 6.37)$
17. $x^2 + 4x + 1 + \frac{-7}{x-3}$
18. $3x^2 + 2x + 1 + \frac{8}{x+2}$
19. $2x^2 - 8x + 31 + \frac{-118}{x+4}$
20. $3x^2 - 3x + 1$
21. $(x-2)(5x^2 + 7x + 16) + 26$
22. $(x+1)(-3x^2 + 3x + 2) - 8$

(a) Let $x = 0$ represent 1990, $x = 10$ represent 2000, and so on. Find a function of the form

$$f(x) = a(x-h)^2 + k$$

that models the data. Use $(0, 353)$ as the vertex and $(285, 2000)$ as another point to determine a.

(b) Use the function to predict the amount of carbon dioxide in 2300.

Consider the function $f(x) = -2.64x^2 + 5.47x + 3.54$ for Exercises 13–16.

13. Use the discriminant to explain how you can determine the number of x-intercepts the graph of $f(x)$ will have even before graphing it on your calculator. (See **Section 1.4.**)

14. Graph the function in the standard viewing window of your calculator, and use the calculator to solve the equation $f(x) = 0$. Express solutions as approximations to the nearest hundredth.

15. Use your answer to **Exercise 14** and the graph of f to solve

(a) $f(x) > 0$, and **(b)** $f(x) < 0$.

16. Use the capabilities of your calculator to find the coordinates of the vertex of the graph. Express coordinates to the nearest hundredth.

Use synthetic division to perform each division.

17. $\dfrac{x^3 + x^2 - 11x - 10}{x - 3}$ **18.** $\dfrac{3x^3 + 8x^2 + 5x + 10}{x + 2}$

19. $\dfrac{2x^3 - x + 6}{x + 4}$ **20.** $\dfrac{3x^3 + 6x^2 - 8x + 3}{x + 3}$

Express $f(x)$ in the form $f(x) = (x-k)q(x) + r$ for the given value of k.

21. $5x^3 - 3x^2 + 2x - 6$; $k = 2$ **22.** $-3x^3 + 5x - 6$; $k = -1$

Use synthetic division to find $f(2)$.

23. $f(x) = -x^3 + 5x^2 - 7x + 1$ **24.** $f(x) = 2x^3 - 3x^2 + 7x - 12$

25. $f(x) = 5x^4 - 12x^2 + 2x - 8$ **26.** $f(x) = x^5 + 4x^2 - 2x - 4$

Use synthetic division to determine whether k is a zero of the function.

27. $f(x) = x^3 + 2x^2 + 3x + 2$; $k = -1$ **28.** $f(x) = 2x^3 + 5x^2 + 30$; $k = -4$

29. *Concept Check* If $f(x)$ is a polynomial function with real coefficients, and if $7 + 2i$ is a zero of the function, then what other complex number must also be a zero?

30. *Concept Check* Suppose the polynomial function f has a zero at $x = -3$. Which of the following statements *must* be true?

A. 3 is an x-intercept of the graph of f. **B.** 3 is a y-intercept of the graph of f.
C. $x - 3$ is a factor of $f(x)$. **D.** $f(-3) = 0$

Find a polynomial function with real coefficients and least degree having the given zeros.

31. $-1, 4, 7$ **32.** $8, 2, 3$

33. $\sqrt{3}, -\sqrt{3}, 2, 3$ **34.** $-2 + \sqrt{5}, -2 - \sqrt{5}, -2, 1$

35. $2, 4, -i$ **36.** $0, 5, 1 + 2i$

Find all rational zeros of each function.

37. $f(x) = 2x^3 - 9x^2 - 6x + 5$ **38.** $f(x) = 8x^4 - 14x^3 - 29x^2 - 4x + 3$

23. -1 24. 6
25. 28 26. 40

27. yes 28. no
29. $7 - 2i$ 30. D

In Exercises 31–36, other answers are possible.

31. $f(x) = x^3 - 10x^2 + 17x + 28$
32. $f(x) = x^3 - 13x^2 + 46x - 48$
33. $f(x) = x^4 - 5x^3 + 3x^2 + 15x - 18$
34. $f(x) = x^4 + 5x^3 + x^2 - 9x + 2$
35. $f(x) = x^4 - 6x^3 + 9x^2 - 6x + 8$
36. $f(x) = x^4 - 7x^3 + 15x^2 - 25x$

37. $\frac{1}{2}, -1, 5$
38. $3, -1, \frac{1}{4}, -\frac{1}{2}$

39. (a) $f(-1) = -10 < 0$; $f(0) = 2 > 0$
(b) $f(2) = -4 < 0$; $f(3) = 14 > 0$
(c) 2.414

40. (a) $f(2) = 44 > 0$; $f(3) = -15 < 0$
(b) $f(7) = -31 < 0$; $f(8) = 140 > 0$
(c) 7.236

42.

Positive	Negative	Nonreal Complex
1	2	0
1	0	2

43. yes
44. $f(x) = 2x^4 - 4x^3 + 10x^2 - 68x + 60$
45. $f(x) = -2x^3 + 6x^2 + 12x - 16$
46. $1 - i, 1 + i, 4, -3$
47. $1, -\frac{1}{2}, \pm 2i$
48. -9
49. $\frac{13}{2}$
50. Any polynomial that can be factored into $a(x - b)^2(x - c)^2$ satisfies the conditions. One example is $f(x) = 2(x - 1)^2(x - 3)^2$.

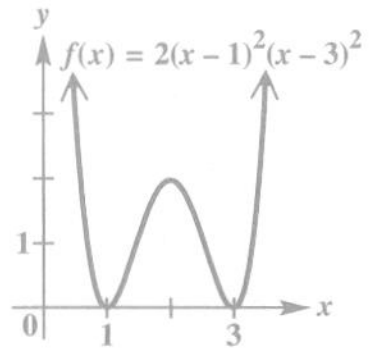

In Exercises 39–41, show that the polynomial function has a real zero as described in parts (a) and (b). In Exercises 39 and 40, then work part (c).

39. $f(x) = 3x^3 - 8x^2 + x + 2$
(a) between -1 and 0
(b) between 2 and 3
(c) Find the zero in part (b) to three decimal places.

40. $f(x) = 4x^3 - 37x^2 + 50x + 60$
(a) between 2 and 3
(b) between 7 and 8
(c) Find the zero in part (b) to three decimal places.

41. $f(x) = 6x^4 + 13x^3 - 11x^2 - 3x + 5$
(a) no zero greater than 1
(b) no zero less than -3

Solve each problem.

42. Use Descartes' rule of signs to determine the different possibilities for the numbers of positive, negative, and nonreal complex zeros of

$$f(x) = x^3 + 3x^2 - 4x - 2.$$

43. Is $x + 1$ a factor of $f(x) = x^3 + 2x^2 + 3x + 2$?

44. Find a polynomial function f with real coefficients of degree 4 with 3, 1, and $-1 + 3i$ as zeros, and $f(2) = -36$.

45. Find a polynomial function f of degree 3 with -2, 1, and 4 as zeros, and $f(2) = 16$.

46. Find all zeros of $f(x) = x^4 - 3x^3 - 8x^2 + 22x - 24$, given that $1 + i$ is a zero.

47. Find all zeros of $f(x) = 2x^4 - x^3 + 7x^2 - 4x - 4$, given that 1 and $-2i$ are zeros.

48. Find a value of k such that $x - 4$ is a factor of

$$f(x) = x^3 - 2x^2 + kx + 4.$$

49. Find a value of k such that when the polynomial $x^3 - 3x^2 + kx - 4$ is divided by $x - 2$, the remainder is 5.

50. *Concept Check* Give an example of a fourth-degree polynomial function having exactly two distinct real zeros, and then sketch its graph.

51. *Concept Check* Give an example of a cubic polynomial function having exactly one real zero, and then sketch its graph.

52. Give the maximum number of turning points of the graph of each function.
(a) $f(x) = x^5 - 9x^2$ (b) $f(x) = 4x^3 - 6x^2 + 2$

53. *Concept Check* Suppose the leading term of a polynomial function is $10x^7$. What can you conclude about each of the following features of the graph of the function?
(a) domain (b) range (c) end behavior (d) number of zeros
(e) number of turning points

54. *Concept Check* Repeat **Exercise 53** for a polynomial function with leading term $-9x^6$.

Sketch the graph of each polynomial function.

55. $f(x) = (x - 2)^2(x + 3)$ **56.** $f(x) = -2x^3 + 7x^2 - 2x - 3$

57. $f(x) = 2x^3 + x^2 - x$ **58.** $f(x) = x^4 - 3x^2 + 2$

59. $f(x) = x^4 + x^3 - 3x^2 - 4x - 4$ **60.** $f(x) = -2x^4 + 7x^3 - 4x^2 - 4x$

Concept Check For each polynomial function, identify its graph from choices A–F.

61. $f(x) = (x-2)^2(x-5)$ **62.** $f(x) = -(x-2)^2(x-5)$

63. $f(x) = (x-2)^2(x-5)^2$ **64.** $f(x) = (x-2)(x-5)$

65. $f(x) = -(x-2)(x-5)$ **66.** $f(x) = -(x-2)^2(x-5)^2$

A.

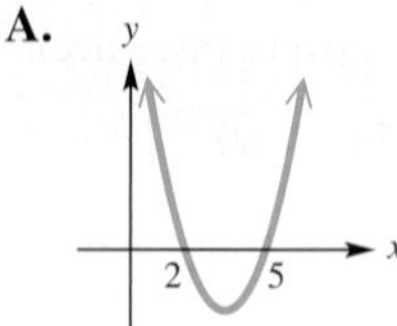

B.

C.

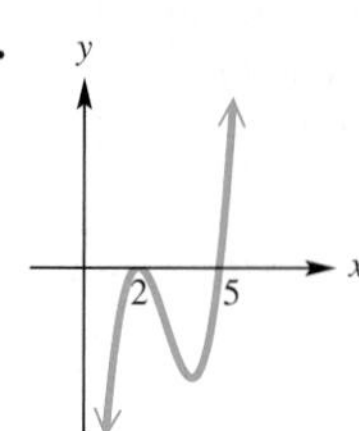

D.

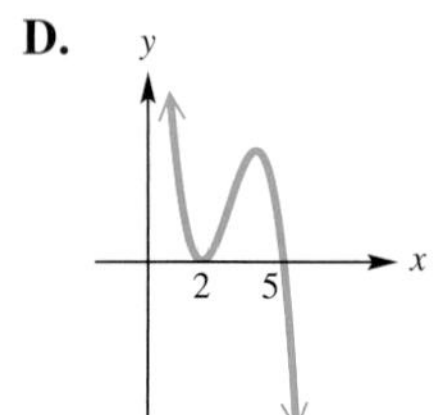

E.

F.

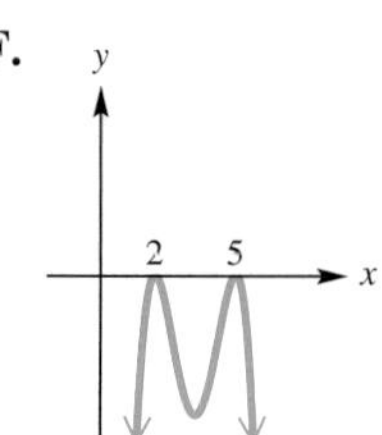

Use a graphing calculator to graph each polynomial function in the viewing window specified. Then determine the real zeros to as many decimal places as the calculator will provide.

67. $f(x) = x^3 - 8x^2 + 2x + 5$; window: $[-10, 10]$ by $[-60, 60]$

68. $f(x) = x^4 - 4x^3 - 5x^2 + 14x - 15$; window: $[-10, 10]$ by $[-60, 60]$

Solve each problem.

69. *(Modeling) Medicare Beneficiary Spending* Out-of-pocket spending projections for a typical Medicare beneficiary as a share of his or her income are given in the table. Let $x = 0$ represent 1990, so $x = 8$ represents 1998. Use a graphing calculator to do the following.

Year	Percent of Income
1998	18.6
2000	19.3
2005	21.7
2010	24.7
2015	27.5
2020	28.3
2025	28.6

Source: Urban Institute's Analysis of 1998 Medicare Trustees' Report.

(a) Graph the data points.
(b) Find a quadratic function to model the data.
(c) Find a cubic function to model the data.
(d) Graph each function in the same viewing window as the data points.
(e) Compare the two functions. Which is a better fit for the data?

70. *Dimensions of a Cube* After a 2-in. slice is cut off the top of a cube, the resulting solid has a volume of 32 in.^{3}. Find the dimensions of the original cube.

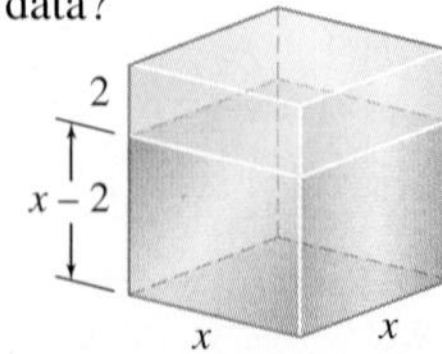

51. Any polynomial that can be factored into $a(x-b)^3$ satisfies the conditions. One example is $f(x) = 2(x-1)^3$.

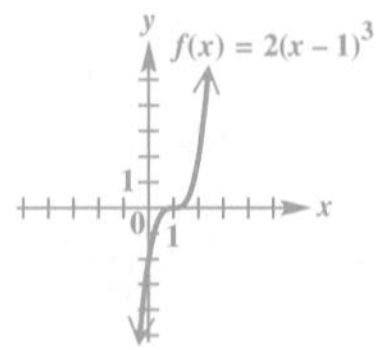

52. (a) four **(b)** two
53. (a) $(-\infty, \infty)$ **(b)** $(-\infty, \infty)$
(c) $f(x) \to \infty$ as $x \to \infty$, $f(x) \to -\infty$ as $x \to -\infty$;
(d) at most seven
(e) at most six
54. (a) $(-\infty, \infty)$
(b) $(-\infty, M]$, where M is the greatest value assumed by the function
(c) $f(x) \to -\infty$ as $x \to \infty$, $f(x) \to -\infty$ as $x \to -\infty$;
(d) at most six
(e) at most five

55. $f(x) = (x-2)^2(x+3)$ **56.** $f(x) = -2x^3 + 7x^2 - 2x - 3$

57. $f(x) = 2x^3 + x^2 - x$ **58.** $f(x) = x^4 - 3x^2 + 2$

59. $f(x) = x^4 + x^3 - 3x^2 - 4x - 4$ **60.** $f(x) = -2x^4 + 7x^3 - 4x^2 - 4x$

61. C **62.** D **63.** E
64. A **65.** B **66.** F

67. 7.6533119, 1, −0.6533119
68. 4.5803997, −2.258838
69. (a)

69. (b) $f(x) = -0.0109x^2 + 0.8693x + 11.85$
(c) $f(x) = -0.00087x^3 + 0.0456x^2 - 0.2191x + 17.83$
(d)

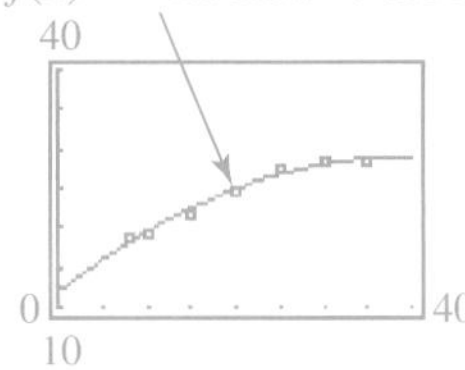

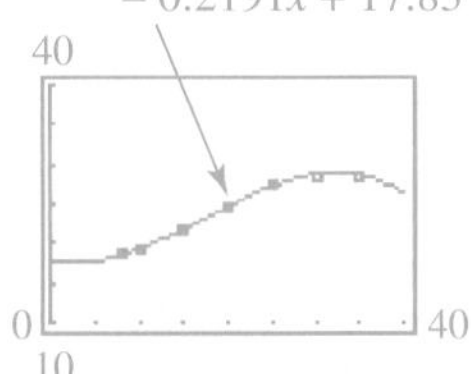

(e) Both functions approximate the data well. The quadratic function is probably better for prediction, because it is unlikely that the percent of out-of-pocket spending would decrease after 2025 (as the cubic function shows) unless changes were made in Medicare law.

70. 4 in. × 4 in. × 4 in.
71. 12 in. × 4 in. × 15 in.

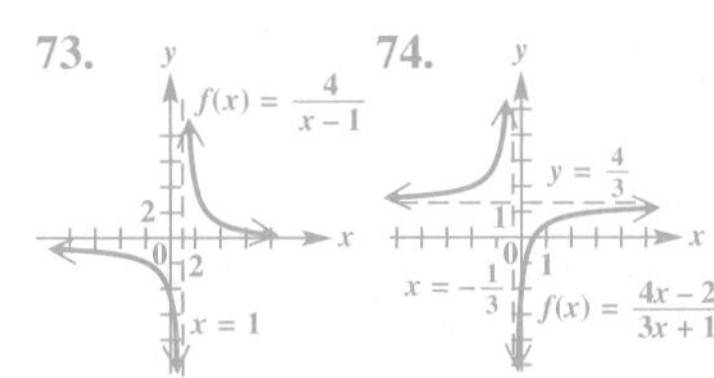

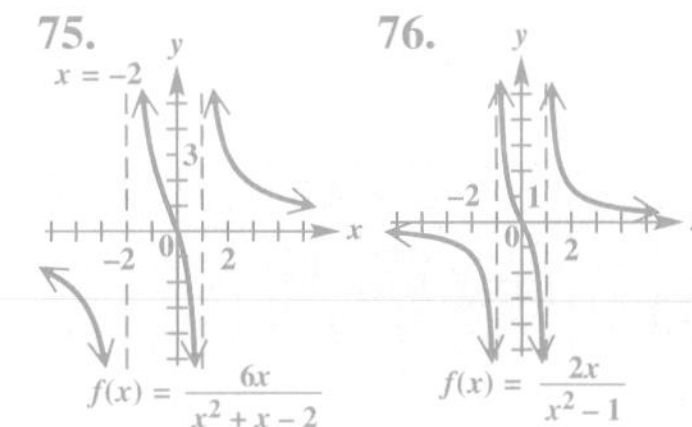

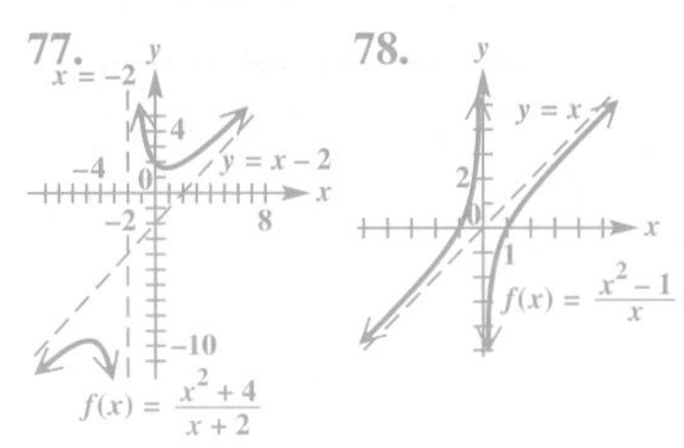

71. *Dimensions of a Box* The width of a rectangular box is three times its height, and its length is 11 in. more than its height. Find the dimensions of the box if its volume is 720 in.^{3}.

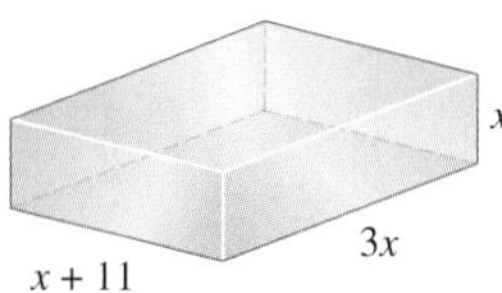

72. The function $f(x) = \frac{1}{x}$ is negative at $x = -1$ and positive at $x = 1$ but has no zero between -1 and 1. Explain why this does not contradict the intermediate value theorem.

Graph each rational function.

73. $f(x) = \dfrac{4}{x-1}$ **74.** $f(x) = \dfrac{4x-2}{3x+1}$ **75.** $f(x) = \dfrac{6x}{x^2+x-2}$

76. $f(x) = \dfrac{2x}{x^2-1}$ **77.** $f(x) = \dfrac{x^2+4}{x+2}$ **78.** $f(x) = \dfrac{x^2-1}{x}$

79. $f(x) = \dfrac{-2}{x^2+1}$ **80.** $f(x) = \dfrac{4x^2-9}{2x+3}$

81. *Concept Check*
(a) Sketch the graph of a function that does not intersect its horizontal asymptote $y = 1$, has the line $x = 3$ as a vertical asymptote, and has x-intercepts 2 and 4.
(b) Find an equation for a possible corresponding rational function.

82. *Concept Check*
(a) Sketch the graph of a function that is never negative and has the lines $x = -1$ and $x = 1$ as vertical asymptotes, the x-axis as a horizontal asymptote, and 0 as an x-intercept.
(b) Find an equation for a possible corresponding rational function.

83. *Connecting Graphs with Equations* Find an equation for the rational function graphed here.

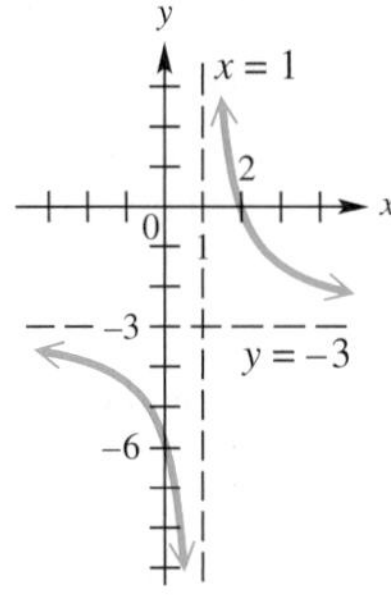

(Modeling) Solve each problem.

84. *Antique-Car Competition* Antique-car owners often enter their cars in a **concours d'elegance** in which a maximum of 100 points can be awarded to a particular car. Points are awarded for the general attractiveness of the car. The function

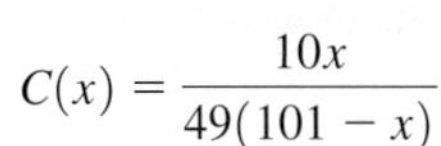

$$C(x) = \frac{10x}{49(101-x)}$$

models the cost, in thousands of dollars, of restoring a car so that it will win x points.

(a) Graph the function in the window $[0, 101]$ by $[0, 10]$.
(b) How much would an owner expect to pay to restore a car in order to earn 95 points?

79. **80.**

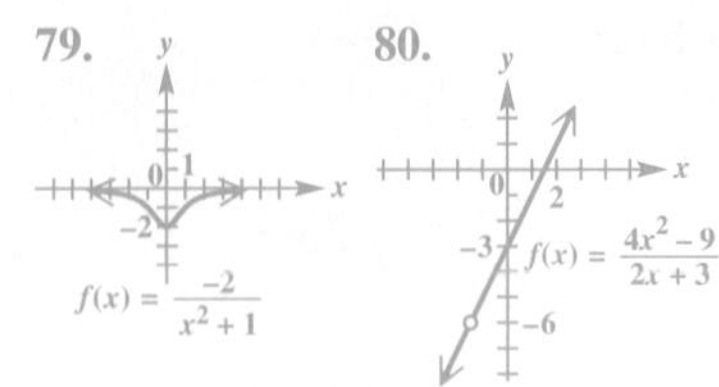

81. (a)

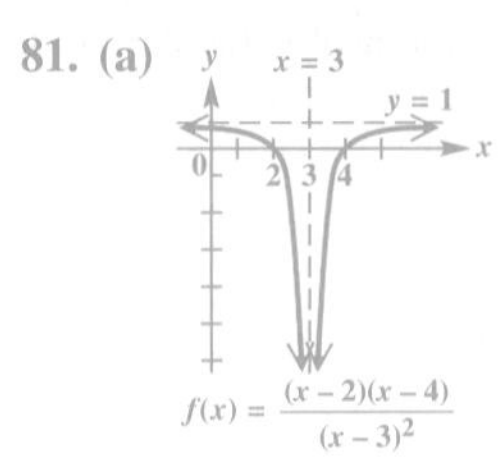

(b) One possibility is $f(x) = \frac{(x - 2)(x - 4)}{(x - 3)^2}$.

82. (a)

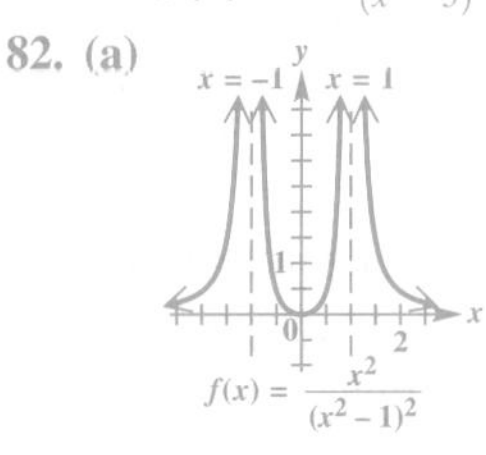

(b) One possibility is $f(x) = \frac{x^2}{(x^2 - 1)^2}$.

83. $f(x) = \frac{-3x + 6}{x - 1}$

Calculator graphs are not included for Exercises 84(a) and 85(a).

84. (b) approximately \$3.23 thousand

85. (b) approximately \$127.3 thousand

86. 36
87. 35 **88.** 0.5
89. 0.75 **90.** $111\frac{1}{9}$
91. 27 **92.** 150 kg per m^2
93. 7500 lb **94.** 33,750 units

85. *Environmental Pollution* In situations involving environmental pollution, a cost-benefit model expresses cost as a function of the percentage of pollutant removed from the environment. Suppose a cost-benefit model is expressed as

$$C(x) = \frac{6.7x}{100 - x},$$

where $C(x)$ is cost in thousands of dollars of removing x percent of a pollutant.

(a) Graph the function in the window $[0, 100]$ by $[0, 100]$.

(b) How much would it cost to remove 95% of the pollutant?

Solve each variation problem.

86. If x varies directly as y, and $x = 12$ when $y = 4$, find x when $y = 12$.

87. If x varies directly as y, and $x = 20$ when $y = 14$, find y when $x = 50$.

88. If z varies inversely as w, and $z = 10$ when $w = \frac{1}{2}$, find z when $w = 10$.

89. If t varies inversely as s, and $t = 3$ when $s = 5$, find s when $t = 20$.

90. p varies jointly as q and r^2, and $p = 100$ when $q = 2$ and $r = 3$. Find p when $q = 5$ and $r = 2$.

91. f varies jointly as g^2 and h, and $f = 50$ when $g = 5$ and $h = 4$. Find f when $g = 3$ and $h = 6$.

92. *Pressure in a Liquid* The pressure on a point in a liquid is directly proportional to the distance from the surface to the point. In a certain liquid, the pressure at a depth of 4 m is 60 kg per m^2. Find the pressure at a depth of 10 m.

93. *Skidding Car* The force needed to keep a car from skidding on a curve varies inversely as the radius r of the curve and jointly as the weight of the car and the square of the speed. It takes 3000 lb of force to keep a 2000-lb car from skidding on a curve of radius 500 ft at 30 mph. What force will keep the same car from skidding on a curve of radius 800 ft at 60 mph?

94. *Power of a Windmill* The power a windmill obtains from the wind varies directly as the cube of the wind velocity. If a wind of 10 km per hr produces 10,000 units of power, how much power is produced by a wind of 15 km per hr?

Chapter 3 Test

1. Sketch the graph of the quadratic function $f(x) = -2x^2 + 6x - 3$. Give the intercepts, vertex, axis, domain, range, and the intervals of the domain for which the function is increasing or decreasing.

2. *(Modeling) Height of a Projectile* A small rocket is fired directly upward, and its height s in feet after t seconds is given by the function

$$s(t) = -16t^2 + 88t + 48.$$

(a) Determine the time at which the rocket reaches its maximum height.

(b) Determine the maximum height.

(c) Between what two times (in seconds, to the nearest tenth) will the rocket be more than 100 ft above ground level?

(d) After how many seconds will the rocket return to the ground?

[3.1]

1.

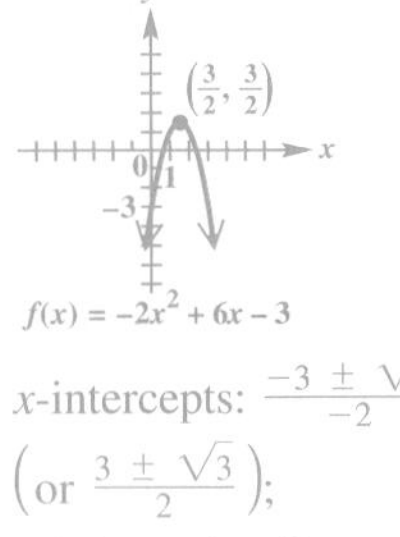

x-intercepts: $\frac{-3 \pm \sqrt{3}}{-2}$ $\left(\text{or } \frac{3 \pm \sqrt{3}}{2}\right)$;
y-intercept: -3;
vertex: $\left(\frac{3}{2}, \frac{3}{2}\right)$;
axis: $x = \frac{3}{2}$;
domain: $(-\infty, \infty)$;
range: $\left(-\infty, \frac{3}{2}\right]$;
increasing on $\left(-\infty, \frac{3}{2}\right]$;
decreasing on $\left[\frac{3}{2}, \infty\right)$

2. (a) 2.75 sec
(b) 169 ft
(c) 0.7 sec and 4.8 sec
(d) 6 sec

[3.2]

3. $3x^2 - 2x - 5 + \frac{16}{x+2}$
4. $2x^2 - x - 5$
5. 53

[3.3]

6. It is a factor. The other factor is $6x^3 + 7x^2 - 14x - 8$.
7. $-2, -3 - 2i, -3 + 2i$
8. $f(x) = 2x^4 - 2x^3 - 2x^2 - 2x - 4$
9. Because $f(x) > 0$ for all x, the graph never intersects or touches the x-axis. Therefore, $f(x)$ has no real zeros.

[3.3, 3.4]

10. (a) $f(1) = 5 > 0$; $f(2) = -1 < 0$
(b)

Positive	Negative	Nonreal Complex
2	1	0
0	1	2

(c) 4.0937635, 1.8370381, −0.9308016

Use synthetic division to perform each division.

3. $\dfrac{3x^3 + 4x^2 - 9x + 6}{x + 2}$ **4.** $\dfrac{2x^3 - 11x^2 + 25}{x - 5}$

5. Use synthetic division to determine $f(5)$, if $f(x) = 2x^3 - 9x^2 + 4x + 8$.

6. Use the factor theorem to determine whether the polynomial $x - 3$ is a factor of

$$6x^4 - 11x^3 - 35x^2 + 34x + 24.$$

If it is, what is the other factor? If it is not, explain why.

7. Find all zeros of $f(x)$, given that $f(x) = x^3 + 8x^2 + 25x + 26$ and -2 is one zero.

8. Find a fourth degree polynomial function $f(x)$ having only real coefficients, -1, 2, and i as zeros, and $f(3) = 80$.

9. Why can't the polynomial function $f(x) = x^4 + 8x^2 + 12$ have any real zeros?

10. Consider the function defined by $f(x) = x^3 - 5x^2 + 2x + 7$.

(a) Use the intermediate value theorem to show that f has a zero between 1 and 2.
(b) Use Descartes' rule of signs to determine the different possibilities for the numbers of positive, negative, and nonreal complex zeros.
(c) Use a graphing calculator to find all real zeros to as many decimal places as the calculator will give.

11. Graph the functions $f(x) = x^4$ and $g(x) = -2(x + 5)^4 + 3$ on the same axes. How can the graph of g be obtained by a transformation of the graph of f?

12. Use end behavior to determine which one of the following graphs is that of $f(x) = -x^7 + x - 4$.

A.

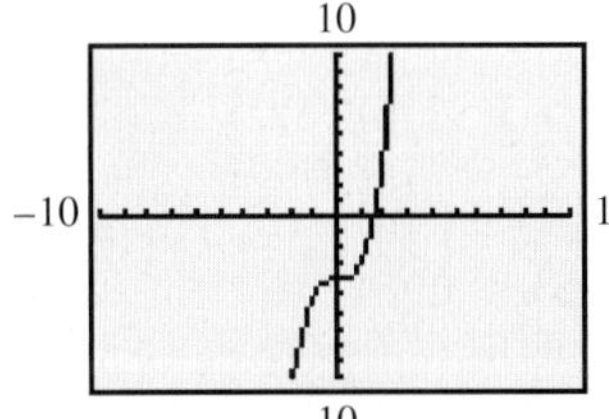

B.

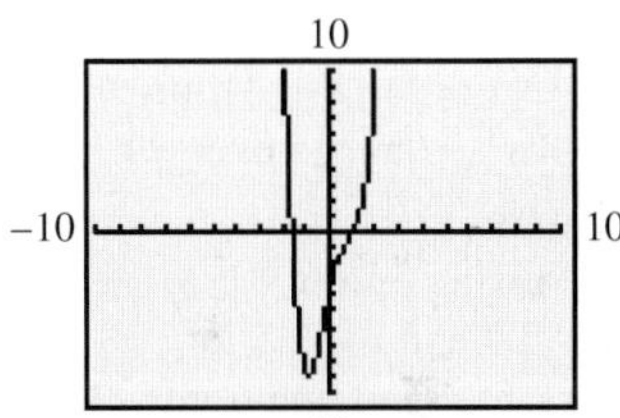

C.

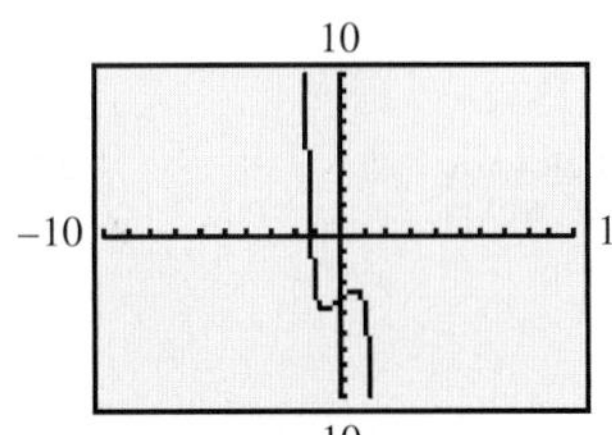

D.

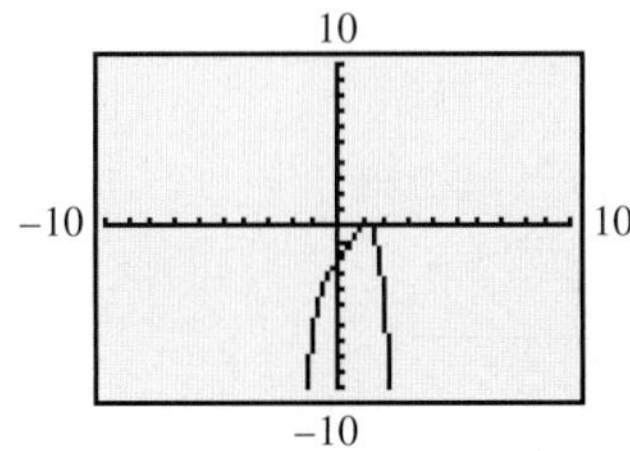

Graph each polynomial function.

13. $f(x) = x^3 - 5x^2 + 3x + 9$ **14.** $f(x) = 2x^2(x - 2)^2$

15. $f(x) = -x^3 - 4x^2 + 11x + 30$

16. *Connecting Graphs with Equations* Find a cubic polynomial function having the graph shown.

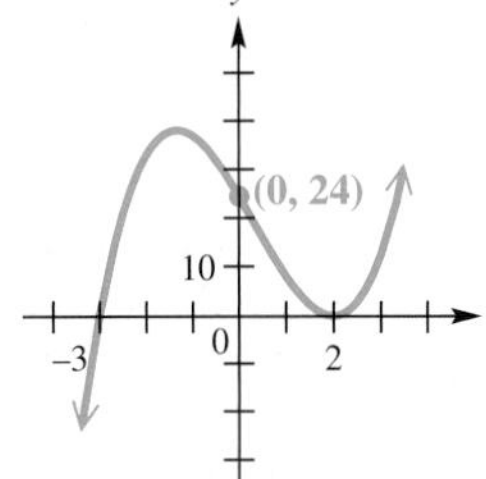

[3.4]

11.

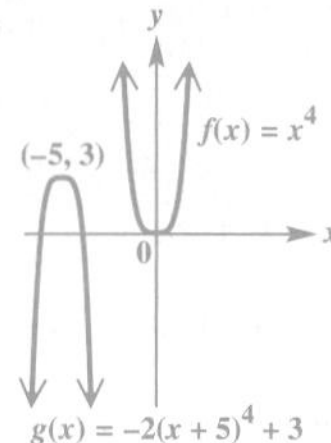

To obtain the graph of g, translate the graph of f 5 units to the left, stretch by a factor of 2, reflect across the x-axis, and translate 3 units up.

12. C

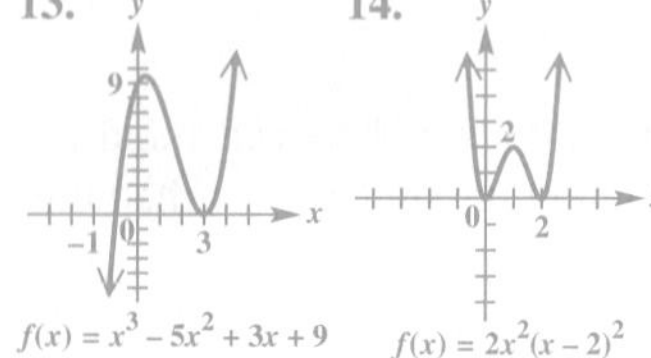

15.

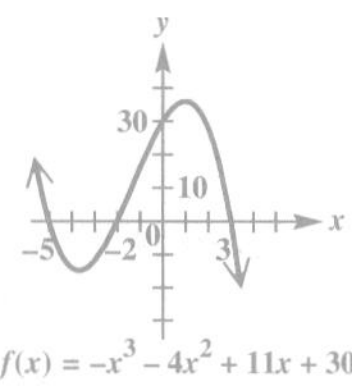

16. $f(x) = 2(x-2)^2(x+3)$, or $f(x) = 2x^3 - 2x^2 - 16x + 24$

17. (a) 270.08
(b) increasing from $t = 0$ to $t = 5.9$ and $t = 9.5$ to $t = 15$; decreasing from $t = 5.9$ to $t = 9.5$

[3.5]

18.

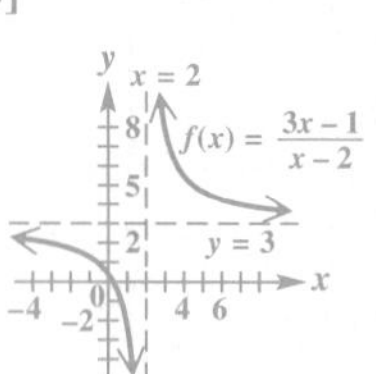

19.

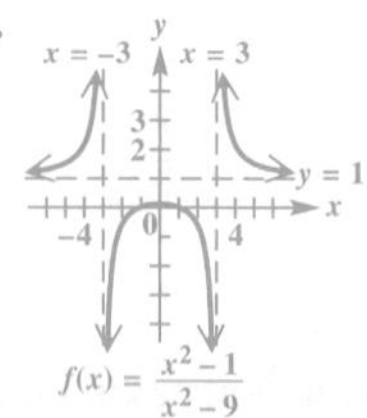

17. *(Modeling) Oil Pressure* The pressure of oil in a reservoir tends to drop with time. Engineers found that the change in pressure is modeled by

$$f(t) = 1.06t^3 - 24.6t^2 + 180t,$$

for t (in years) in the interval $[0, 15]$.

(a) What was the change after 2 yr?

(b) For what time periods, to the nearest tenth of a year, is the amount of change in pressure increasing? decreasing? Use a graph to decide.

Graph each rational function.

18. $f(x) = \dfrac{3x - 1}{x - 2}$

19. $f(x) = \dfrac{x^2 - 1}{x^2 - 9}$

20. Consider the rational function $f(x) = \dfrac{2x^2 + x - 6}{x - 1}$.

(a) Determine the equation of the oblique asymptote.

(b) Determine the x-intercepts.

(c) Determine the y-intercept.

(d) Determine the equation of the vertical asymptote.

(e) Sketch the graph.

21. If y varies directly as the square root of x, and $y = 12$ when $x = 4$, find y when $x = 100$.

22. *Weight On and Above Earth* The weight w of an object varies inversely as the square of the distance d between the object and the center of Earth. If a man weighs 90 kg on the surface of Earth, how much would he weigh 800 km above the surface? (*Hint:* The radius of Earth is about 6400 km.)

20. (a) $y = 2x + 3$ (b) $-2, \frac{3}{2}$ (c) 6 (d) $x = 1$
(e)

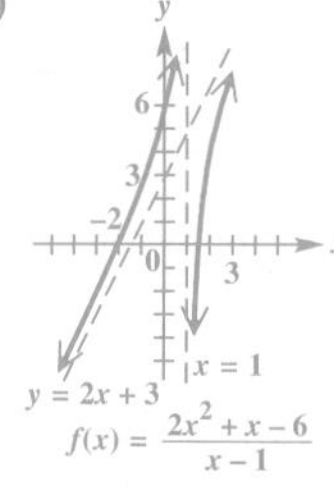

[3.6] 21. 60

22. $\frac{640}{9}$ kg

384 CHAPTER 4 Inverse, Exponential, and Logarithmic Functions

4 Inverse, Exponential, and Logarithmic Functions

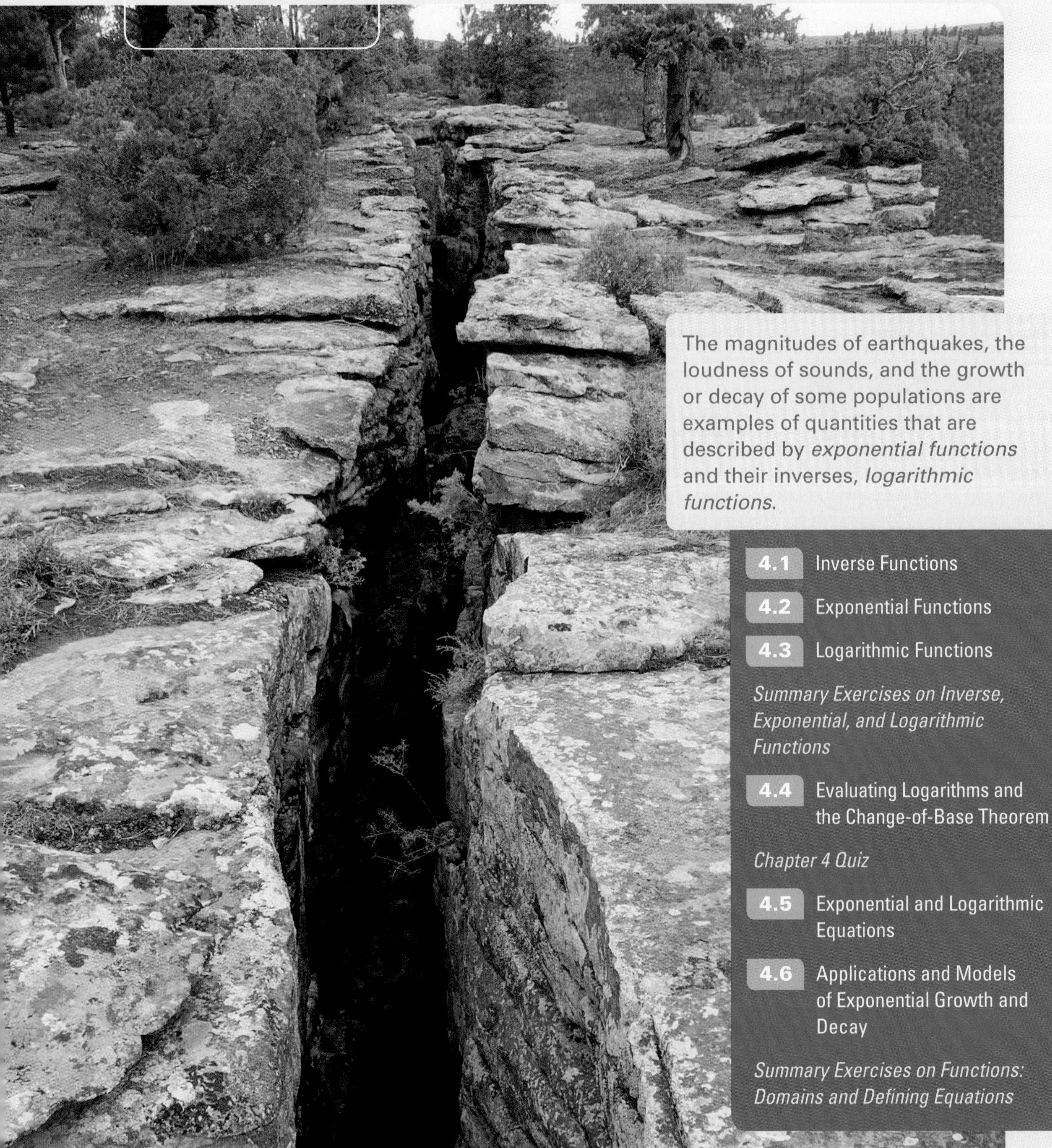

The magnitudes of earthquakes, the loudness of sounds, and the growth or decay of some populations are examples of quantities that are described by *exponential functions* and their inverses, *logarithmic functions*.

4.1 Inverse Functions

- One-to-One Functions
- Inverse Functions
- Equations of Inverses
- An Application of Inverse Functions to Cryptography

One-to-One Functions Suppose we define the function

$$F = \{(-2, 2), (-1, 1), (0, 0), (1, 3), (2, 5)\}.$$

(Notice that we have defined F so that each *second* component is used only once.) We can form another set of ordered pairs from F by interchanging the x- and y-values of each pair in F. We call this set G, so

$$G = \{(2, -2), (1, -1), (0, 0), (3, 1), (5, 2)\}.$$

To show that these two sets are related, G is called the *inverse* of F. For a function f to have an inverse, f must be a *one-to-one function*.

In a one-to-one function, each x-value corresponds to only one y-value, and each y-value corresponds to only one x-value.

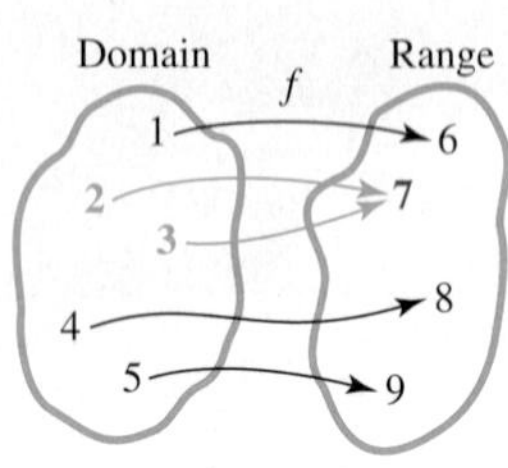

Not One-to-One

Figure 1

The function f shown in **Figure 1** is not one-to-one because the y-value 7 corresponds to *two* x-values, 2 and 3. That is, the ordered pairs $(2, 7)$ and $(3, 7)$ both belong to the function. The function f in **Figure 2** is one-to-one.

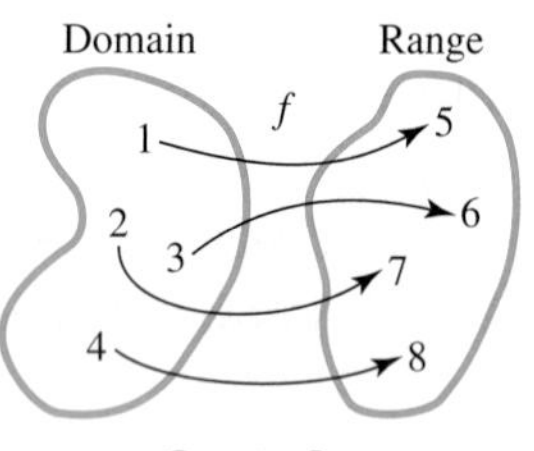

One-to-One

Figure 2

One-to-One Function

A function f is a **one-to-one function** if, for elements a and b in the domain of f,

$$a \neq b \quad \text{implies} \quad f(a) \neq f(b).$$

Using the concept of the *contrapositive* from the study of logic, the last line in the preceding box is equivalent to

$$f(a) = f(b) \quad \text{implies} \quad a = b.$$

We use this statement to decide whether a function f is one-to-one in the next example.

Classroom Example 1
Decide whether each function is one-to-one.
(a) $f(x) = -3x + 7$
(b) $f(x) = \sqrt{49 - x^2}$

Answers:
(a) It is one-to-one.
(b) It is not one-to-one.

EXAMPLE 1 Deciding Whether Functions Are One-to-One

Decide whether each function is one-to-one.

(a) $f(x) = -4x + 12$ **(b)** $f(x) = \sqrt{25 - x^2}$

SOLUTION

(a) We must show that $f(a) = f(b)$ leads to the result $a = b$.

$$f(a) = f(b)$$

$$-4a + 12 = -4b + 12 \qquad f(x) = -4x + 12$$

$$-4a = -4b \qquad \text{Subtract 12. (Section 1.1)}$$

$$a = b \qquad \text{Divide by } -4.$$

By the definition, $f(x) = -4x + 12$ is one-to-one.

(b) For the function $f(x) = \sqrt{25 - x^2}$, if we choose $a = 3$ and $b = -3$, then $3 \neq -3$, but

$$f(3) = \sqrt{25 - 3^2} = \sqrt{25 - 9} = \sqrt{16} = 4$$

and $$f(-3) = \sqrt{25 - (-3)^2} = \sqrt{25 - 9} = 4.$$

Here, even though $3 \neq -3$, $f(3) = f(-3) = 4$. By the definition, f is *not* a one-to-one function.

Now Try Exercises 9 and 11.

As illustrated in **Example 1(b),** a way to show that a function is *not* one-to-one is to produce a pair of different domain elements that lead to the same function value. There is also a useful graphical test, the **horizontal line test,** that tells whether or not a function is one-to-one.

Horizontal Line Test

A function is one-to-one if every horizontal line intersects the graph of the function at most once.

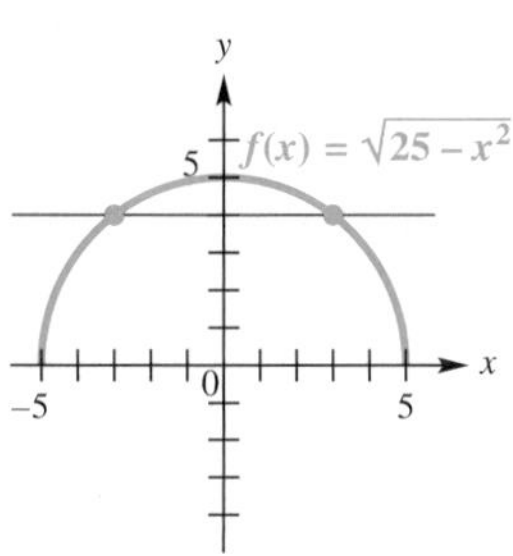

Figure 3

NOTE In **Example 1(b),** the graph of the function is a semicircle, as shown in **Figure 3.** Because there is at least one horizontal line that intersects the graph in more than one point, this function is not one-to-one.

Classroom Example 2
Determine whether each graph is the graph of a one-to-one function.

(a)

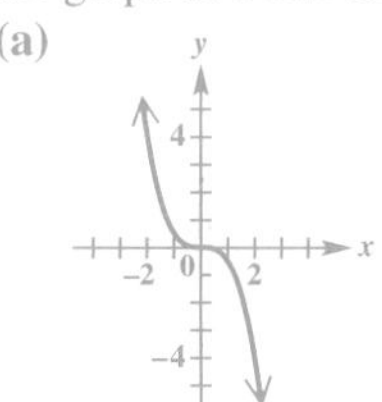

(b)

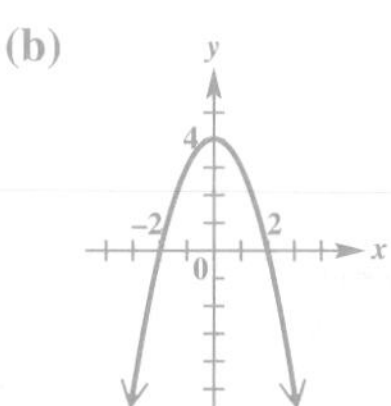

Answers:
(a) It is one-to-one.
(b) It is not one-to-one.

EXAMPLE 2 **Using the Horizontal Line Test**

Determine whether each graph is the graph of a one-to-one function.

(a)

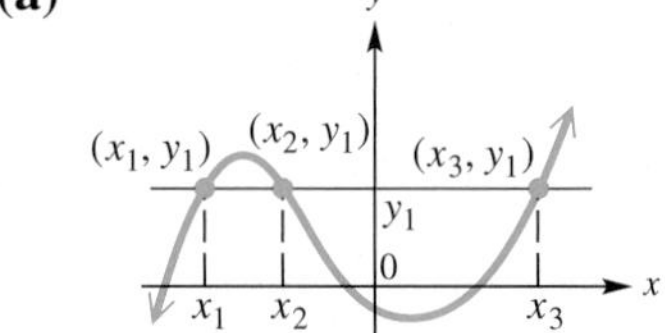

(b)

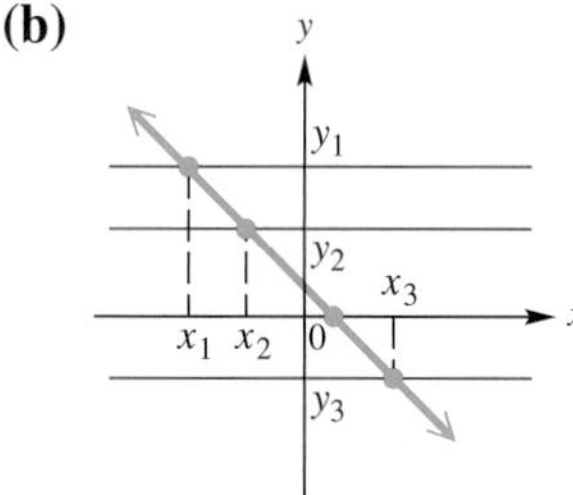

SOLUTION

(a) Each point where the horizontal line intersects the graph has the same value of y but a different value of x. Since more than one (here three) different values of x lead to the same value of y, the function is not one-to-one.

(b) Since every horizontal line will intersect the graph at exactly one point, this function is one-to-one.

Now Try Exercises 3 and 5.

Notice that the function graphed in **Example 2(b)** decreases on its entire domain. ***In general, a function that is either increasing or decreasing on its entire domain, such as*** $f(x) = -x$, $g(x) = x^3$, ***and*** $h(x) = \sqrt{x}$, ***must be one-to-one.***

In summary, there are four ways to decide whether a function is one-to-one.

Tests to Determine Whether a Function Is One-to-One

1. Show that $f(a) = f(b)$ implies $a = b$. This means that f is one-to-one. **(Example 1(a))**
2. In a one-to-one function, every y-value corresponds to no more than one x-value. To show that a function is not one-to-one, find at least two x-values that produce the same y-value. **(Example 1(b))**
3. Sketch the graph and use the horizontal line test. **(Example 2)**
4. If the function either increases or decreases on its entire domain, then it is one-to-one. A sketch is helpful here, too. **(Example 2(b))**

Teaching Tip This is a good place to review composition of functions, first introduced in **Section 2.8.**

Inverse Functions Consider the functions

$$f(x) = 8x + 5 \quad \text{and} \quad g(x) = \frac{1}{8}x - \frac{5}{8}.$$

Let us choose an arbitrary element from the domain of f, say 10. Evaluate $f(10)$.

$$f(10) = 8 \cdot 10 + 5 \qquad \text{Let } x = 10.$$

$$f(10) = 85 \qquad \text{Multiply and then add.}$$

Now, we evaluate $g(85)$.

$$g(85) = \frac{1}{8}(85) - \frac{5}{8} \qquad \text{Let } x = 85.$$

$$= \frac{85}{8} - \frac{5}{8} \qquad \text{Multiply.}$$

$$= \frac{80}{8} \qquad \text{Subtract.}$$

$$g(85) = 10 \qquad \text{Divide.}$$

Starting with 10, we "applied" function f and then "applied" function g to the result, which returned the number 10. See **Figure 4.**

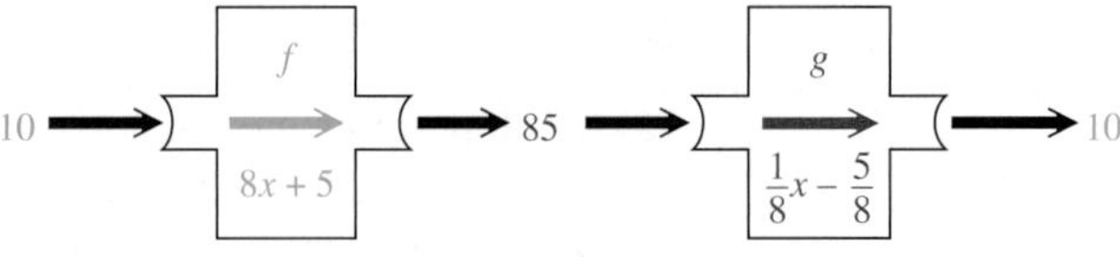

Figure 4

As further examples, check that

$$f(3) = 29 \quad \text{and} \quad g(29) = 3,$$

$$f(-5) = -35 \quad \text{and} \quad g(-35) = -5,$$

$$g(2) = -\frac{3}{8} \quad \text{and} \quad f\left(-\frac{3}{8}\right) = 2.$$

In particular, for this pair of functions,

$$f(g(2)) = 2 \quad \text{and} \quad g(f(2)) = 2.$$

In fact, for *any* value of x,

$$f(g(x)) = x \quad \text{and} \quad g(f(x)) = x.$$

Using the notation for composition introduced in **Section 2.8,** these two equations can be written as follows.

$$(f \circ g)(x) = x \quad \text{and} \quad (g \circ f)(x) = x \quad \text{The result is the identity function.}$$

Because the compositions of f and g yield the *identity* function, they are *inverses* of each other.

Inverse Function

Let f be a one-to-one function. Then g is the **inverse function** of f if

$$(f \circ g)(x) = x \quad \text{for every } x \text{ in the domain of } g,$$

and

$$(g \circ f)(x) = x \quad \text{for every } x \text{ in the domain of } f.$$

The condition that f is one-to-one in the definition of inverse function is essential. Otherwise, g will not define a function.

Classroom Example 3
Let functions f and g be defined by

$$f(x) = 2x + 5$$

and

$$g(x) = \frac{1}{2}x - 5,$$

respectively. Is g the inverse function of f?

Answer: no

EXAMPLE 3 Deciding Whether Two Functions Are Inverses

Let functions f and g be defined by

$$f(x) = x^3 - 1 \quad \text{and} \quad g(x) = \sqrt[3]{x+1},$$

respectively. Is g the inverse function of f?

SOLUTION As shown in **Figure 5,** the horizontal line test applied to the graph indicates that f is one-to-one, so the function does have an inverse. Since it is one-to-one, we now find $(f \circ g)(x)$ and $(g \circ f)(x)$.

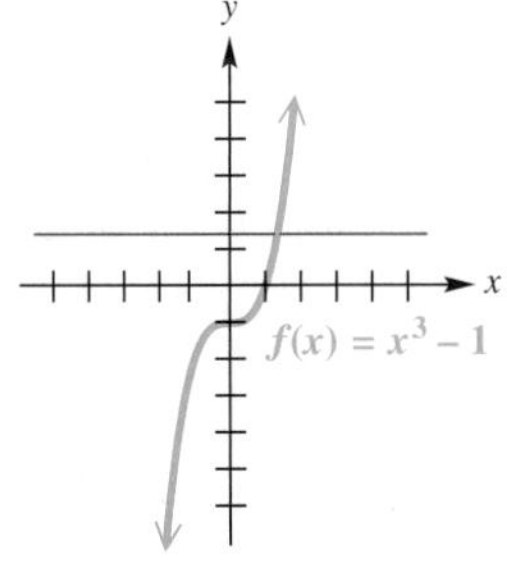

Figure 5

$$\begin{aligned}(f \circ g)(x) = f(g(x)) &= \left(\sqrt[3]{x+1}\right)^3 - 1 \quad \text{(Section 2.8)}\\ &= x + 1 - 1\\ &= x\end{aligned}$$

$$\begin{aligned}(g \circ f)(x) = g(f(x)) &= \sqrt[3]{(x^3 - 1) + 1}\\ &= \sqrt[3]{x^3}\\ &= x\end{aligned}$$

Since $(f \circ g)(x) = x$ and $(g \circ f)(x) = x$, function g is the inverse of function f.

Now Try Exercise 41.

A special notation is used for inverse functions: If g is the inverse of a function f, then g is written as f^{-1} (read "**f-inverse**"). In **Example 3,**

$$f(x) = x^3 - 1 \quad \text{has inverse} \quad f^{-1}(x) = \sqrt[3]{x+1}.$$

CAUTION ***Do not confuse the −1 in f^{-1} with a negative exponent.*** The symbol $f^{-1}(x)$ does not represent $\frac{1}{f(x)}$. It represents the inverse function of f.

Teaching Tip Remind students that composition of functions is not commutative. Therefore, $(f \circ g)(x)$ and $(g \circ f)(x)$ must both be checked. Use an example such as

$$f(x) = x^2 \quad \text{and} \quad g(x) = \sqrt{x}$$

to illustrate the *necessity* of checking both compositions if you do not verify the functions are one-to-one.

By the definition of inverse function, the domain of f is the range of f^{-1}, and the range of f is the domain of f^{-1}. See **Figure 6.**

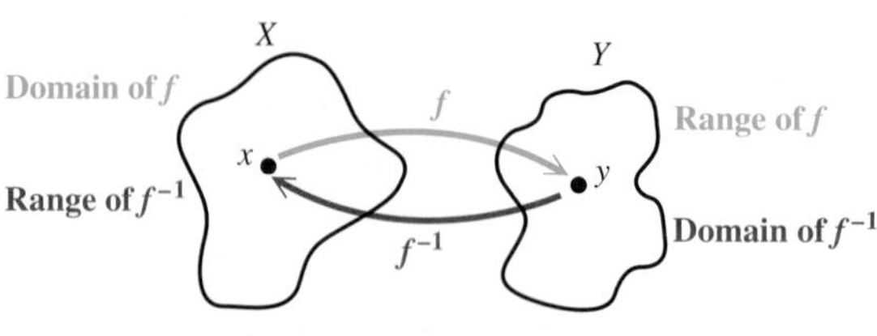

Figure 6

Year	Number of Unhealthy Days
2004	7
2005	32
2006	8
2007	24
2008	14
2009	13

Source: Illinois Environmental Protection Agency.

EXAMPLE 4 Finding Inverses of One-to-One Functions

Find the inverse of each function that is one-to-one.

(a) $F = \{(-2, 1), (-1, 0), (0, 1), (1, 2), (2, 2)\}$

(b) $G = \{(3, 1), (0, 2), (2, 3), (4, 0)\}$

(c) The table in the margin shows the number of days in Illinois that were unhealthy for sensitive groups for selected years using the Air Quality Index (AQI). Let f be the function defined in the table, with the years forming the domain and the numbers of unhealthy days forming the range.

SOLUTION

(a) Each x-value in F corresponds to just one y-value. However, the y-value 2 corresponds to two x-values, 1 and 2. Also, the y-value 1 corresponds to both -2 and 0. Because at least one y-value corresponds to more than one x-value, F is not one-to-one and does not have an inverse.

(b) Every x-value in G corresponds to only one y-value, and every y-value corresponds to only one x-value, so G is a one-to-one function. The inverse function is found by interchanging the x- and y-values in each ordered pair.

$$G^{-1} = \{(1, 3), (2, 0), (3, 2), (0, 4)\}$$

Notice how the domain and range of G become the range and domain, respectively, of G^{-1}.

(c) Each x-value in f corresponds to only one y-value, and each y-value corresponds to only one x-value, so f is a one-to-one function. The inverse function is found by interchanging the x- and y-values in the table.

$$f^{-1}(x) = \{(7, 2004), (32, 2005), (8, 2006), (24, 2007), (14, 2008), (13, 2009)\}$$

The domain and range of f become the range and domain of f^{-1}.

Now Try Exercises 37, 51, and 53.

Classroom Example 4
Find the inverse of each function that is one-to-one.

(a) $F = \{(-2, -8), (-1, -1), (0, 0), (1, 1), (2, 8)\}$

(b) $G = \{(-2, 5), (-1, 2), (0, 1), (1, 2), (2, 5)\}$

(c) Let h be the function defined by the table in **Example 4(c)** if the number of unhealthy days for 2008 is decreased by 1.

Answers:

(a) $F^{-1} = \{(-8, -2), (-1, -1), (0, 0), (1, 1), (8, 2)\}$

(b) G is not one-to-one.

(c) h is not one-to-one.

Equations of Inverses The inverse of a one-to-one function is found by interchanging the x- and y-values of each of its ordered pairs. The equation of the inverse of a function defined by $y = f(x)$ is found in the same way, as given in the box on the next page.

Finding the Equation of the Inverse of $y = f(x)$

For a one-to-one function f defined by an equation $y = f(x)$, find the defining equation of the inverse as follows. (If necessary, replace $f(x)$ with y first. Any restrictions on x and y should be considered.)

Step 1 Interchange x and y.

Step 2 Solve for y.

Step 3 Replace y with $f^{-1}(x)$.

Classroom Example 5
Decide whether each function defines a one-to-one function. If so, find the equation of the inverse.

(a) $f(x) = |x|$

(b) $g(x) = 4x - 7$

(c) $h(x) = x^3 + 5$

Answers:

(a) f is not one-to-one.

(b) g is one-to-one; $g^{-1}(x) = \frac{1}{4}x + \frac{7}{4}$.

(c) h is one-to-one; $h^{-1}(x) = \sqrt[3]{x - 5}$.

EXAMPLE 5 Finding Equations of Inverses

Decide whether each equation defines a one-to-one function. If so, find the equation of the inverse.

(a) $f(x) = 2x + 5$ **(b)** $y = x^2 + 2$ **(c)** $f(x) = (x - 2)^3$

SOLUTION

(a) The graph of $y = 2x + 5$ is a nonhorizontal line, so by the horizontal line test, f is a one-to-one function. To find the equation of the inverse, follow the steps in the preceding box, first replacing $f(x)$ with y.

$$f(x) = 2x + 5$$

$$y = 2x + 5 \quad \text{Let } y = f(x).$$

$$x = 2y + 5 \quad \text{Interchange } x \text{ and } y. \text{ (Step 1)}$$

$$x - 5 = 2y \quad \text{Subtract 5.}$$

$$y = \frac{x - 5}{2} \quad \text{Divide by 2. Rewrite.}$$

(Solve for y. (Step 2))

$$f^{-1}(x) = \frac{1}{2}x - \frac{5}{2} \quad \text{Replace } y \text{ with } f^{-1}(x). \text{ (Step 3)} \quad \frac{a-b}{c} = \left(\frac{1}{c}\right)a - \frac{b}{c}$$

Thus, $f^{-1}(x) = \frac{x-5}{2} = \frac{1}{2}x - \frac{5}{2}$ is a linear function. In the function defined by $y = 2x + 5$, the value of y is found by starting with a value of x, multiplying by 2, and adding 5.

The form $f^{-1}(x) = \frac{x-5}{2}$ for the equation of the inverse has us *subtract* 5 and then *divide* by 2. This shows how an inverse is used to "undo" what a function does to the variable x.

Teaching Tip To emphasize the relationship between inverse functions, for $f(x) = 2x + 5$ and $x = 3$, find

$$y = 2 \cdot 3 + 5 = 11.$$

Now, for $f^{-1}(x) = \frac{1}{2}x - \frac{5}{2}$ and $x = 11$, find

$$y = \frac{1}{2}(11) - \frac{5}{2} = 3.$$

This shows numerically that when we substitute the original y-value in $f^{-1}(x)$, we obtain the original x-value.

(b) The equation $y = x^2 + 2$ has a parabola opening up as its graph, so some horizontal lines will intersect the graph at two points. For example, both $x = 3$ and $x = -3$ correspond to $y = 11$. Because of the presence of the x^2-term, there are many pairs of x-values that correspond to the same y-value. This means that the function defined by $y = x^2 + 2$ is not one-to-one and does not have an inverse.

The steps for finding the equation of an inverse lead to the following.

$$y = x^2 + 2$$

$$x = y^2 + 2 \quad \text{Interchange } x \text{ and } y.$$

$$x - 2 = y^2 \quad \text{Solve for } y.$$

Remember *both* roots.

$$\pm\sqrt{x - 2} = y \quad \text{Square root property } \textbf{(Section 1.4)}$$

The last step shows that there are two y-values for each choice of x greater than 2, so the given function is not one-to-one and cannot have an inverse.

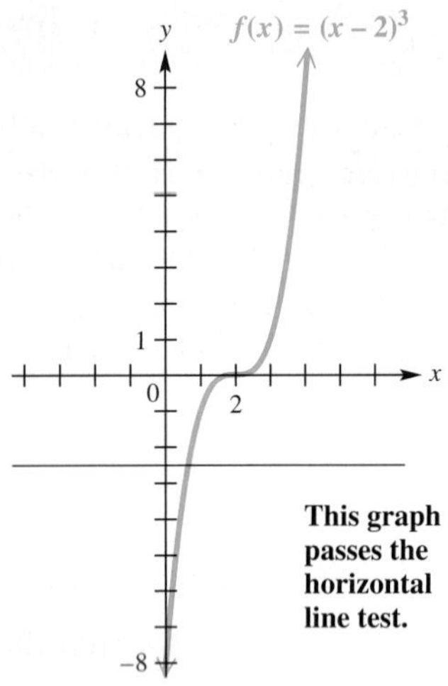

Figure 7

(c) **Figure 7** shows that the horizontal line test assures us that this horizontal translation of the graph of the cubing function is one-to-one.

$$f(x) = (x-2)^3 \quad \text{Given function}$$
$$y = (x-2)^3 \quad \text{Replace } f(x) \text{ with } y.$$
$$x = (y-2)^3 \quad \text{Interchange } x \text{ and } y.$$
$$\sqrt[3]{x} = \sqrt[3]{(y-2)^3} \quad \text{Take the cube root on each side. (Section 1.6)}$$
$$\sqrt[3]{x} = y - 2 \quad \sqrt[3]{a^3} = a \text{ (Section R.7)}$$
$$\sqrt[3]{x} + 2 = y \quad \text{Solve for } y \text{ by adding 2.}$$
$$f^{-1}(x) = \sqrt[3]{x} + 2 \quad \text{Replace } y \text{ with } f^{-1}(x). \text{ Rewrite.}$$

✔ ***Now Try Exercises 59(a), 63(a), and 65(a).***

Classroom Example 6

The rational function

$$f(x) = \frac{-3x+1}{x-5}, \quad x \neq 5$$

is a one-to-one function. Find its inverse.

Answer: $f^{-1}(x) = \frac{5x+1}{x+3}$, $x \neq -3$

EXAMPLE 6 Finding the Equation of the Inverse of a Rational Function

The rational function

$$f(x) = \frac{2x+3}{x-4}, \quad x \neq 4 \quad \text{(Section 3.5)}$$

is a one-to-one function. Find its inverse.

SOLUTION

$$f(x) = \frac{2x+3}{x-4}, \quad x \neq 4$$

$$y = \frac{2x+3}{x-4} \quad \text{Replace } f(x) \text{ with } y.$$

$$x = \frac{2y+3}{y-4}, \quad y \neq 4 \quad \text{Interchange } x \text{ and } y. \text{ (Step 1)}$$

$$x(y-4) = 2y+3 \quad \text{Multiply by } y-4.$$

Pay close attention here.

$$xy - 4x = 2y + 3 \quad \text{Distributive property}$$

$$xy - 2y = 4x + 3 \quad \text{Add } 4x \text{ and } -2y.$$

$$y(x-2) = 4x+3 \quad \text{Factor out } y. \text{ (Section R.4)}$$

$$y = \frac{4x+3}{x-2}, \quad x \neq 2 \quad \text{Divide by } x-2.$$

(Solve for y. (Step 2))

In the final line, we give the condition $x \neq 2$. (Note that 2 was not in the *range* of f, so it is not in the domain of f^{-1}.)

$$f^{-1}(x) = \frac{4x+3}{x-2}, \quad x \neq 2 \quad \text{Replace } y \text{ with } f^{-1}(x). \text{ (Step 3)}$$

✔ ***Now Try Exercise 71(a).***

One way to graph the inverse of a function f whose equation is known follows.

Step 1 Find some ordered pairs that are on the graph of f.

Step 2 Interchange x and y to get ordered pairs that are on the graph of f^{-1}.

Step 3 Plot those points, and sketch the graph of f^{-1} through them.

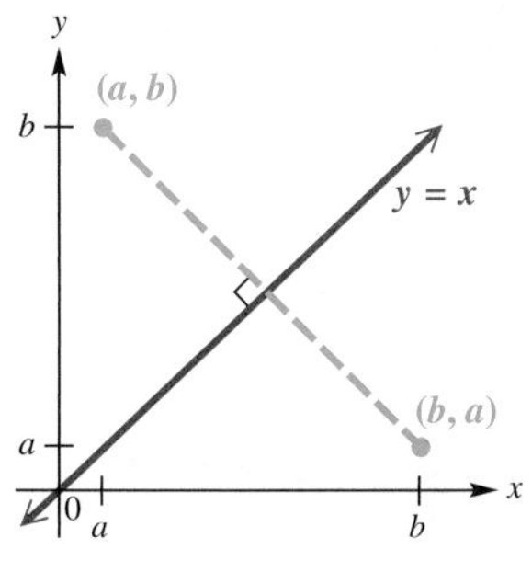

Figure 8

Another way is to select points on the graph of f and use symmetry to find corresponding points on the graph of f^{-1}.

For example, suppose the point (a, b) shown in **Figure 8** is on the graph of a one-to-one function f. Then the point (b, a) is on the graph of f^{-1}. The line segment connecting (a, b) and (b, a) is perpendicular to, and cut in half by, the line $y = x$. The points (a, b) and (b, a) are "mirror images" of each other with respect to $y = x$. ***Thus, we can find the graph of f^{-1} from the graph of f by locating the mirror image of each point in f with respect to the line $y = x$.***

Classroom Example 7
Determine whether functions f and g graphed are inverses of each other.

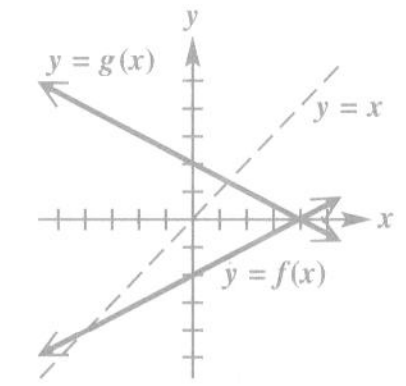

Answer: They are not inverses.

EXAMPLE 7 Graphing f^{-1} Given the Graph of f

In each set of axes in **Figure 9**, the graph of a one-to-one function f is shown in blue. Graph f^{-1} in red.

SOLUTION In **Figure 9,** the graphs of two functions f shown in blue are given with their inverses shown in red. In each case, the graph of f^{-1} is a reflection of the graph of f with respect to the line $y = x$.

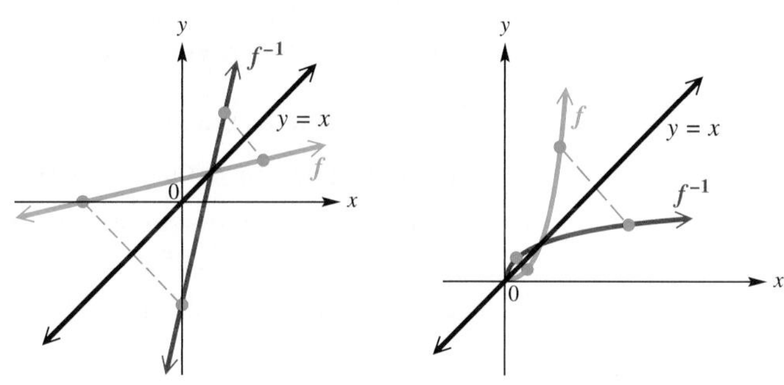

Figure 9

Now Try Exercises 77 and 81.

Classroom Example 8
Let $f(x) = x^2 + 4,\ x \le 0$. Find $f^{-1}(x)$.

Answer:
$f^{-1}(x) = -\sqrt{x - 4},\ x \ge 4$

EXAMPLE 8 Finding the Inverse of a Function with a Restricted Domain

Let $f(x) = \sqrt{x + 5},\quad x \ge -5$. Find $f^{-1}(x)$.

SOLUTION First, notice that the domain of f is restricted to the interval $[-5, \infty)$. Function f is one-to-one because it is increasing on its entire domain and, thus, has an inverse function. Now we find the equation of the inverse.

$$f(x) = \sqrt{x + 5},\quad x \ge -5$$

$$y = \sqrt{x + 5},\quad x \ge -5 \qquad \text{Replace } f(x) \text{ with } y.$$

$$x = \sqrt{y + 5},\quad y \ge -5 \qquad \text{Interchange } x \text{ and } y.$$

$$x^2 = \left(\sqrt{y + 5}\right)^2 \qquad \text{Square each side. (Section 1.6)}$$

$$x^2 = y + 5 \qquad \left(\sqrt{a}\right)^2 = a \text{ for } a \ge 0 \text{ (Section R.7)}$$

$$y = x^2 - 5 \qquad \text{Solve for } y.$$

However, we cannot define $f^{-1}(x)$ as $x^2 - 5$. The domain of f is $[-5, \infty)$, and its range is $[0, \infty)$. The range of f is the domain of f^{-1}, so f^{-1} must be defined as

$$f^{-1}(x) = x^2 - 5,\quad x \ge 0.$$

As a check, the range of f^{-1}, $[-5, \infty)$, is the domain of f.

Graphs of f and f^{-1} are shown in **Figures 10** and **11.** The line $y = x$ is included on the graphs to show that the graphs of f and f^{-1} are mirror images with respect to this line.

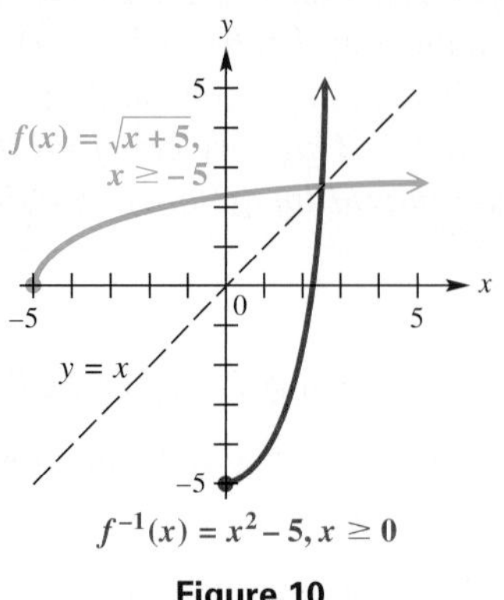

Figure 10

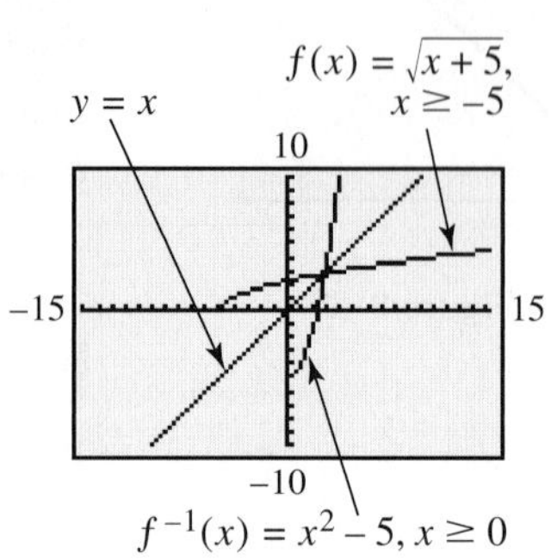

Figure 11

Now Try Exercise 75.

Important Facts about Inverses

1. If f is one-to-one, then f^{-1} exists.
2. The domain of f is the range of f^{-1}, and the range of f is the domain of f^{-1}.
3. If the point (a, b) lies on the graph of f, then (b, a) lies on the graph of f^{-1}. The graphs of f and f^{-1} are reflections of each other across the line $y = x$.
4. To find the equation for f^{-1}, replace $f(x)$ with y, interchange x and y, and solve for y. This gives $f^{-1}(x)$.

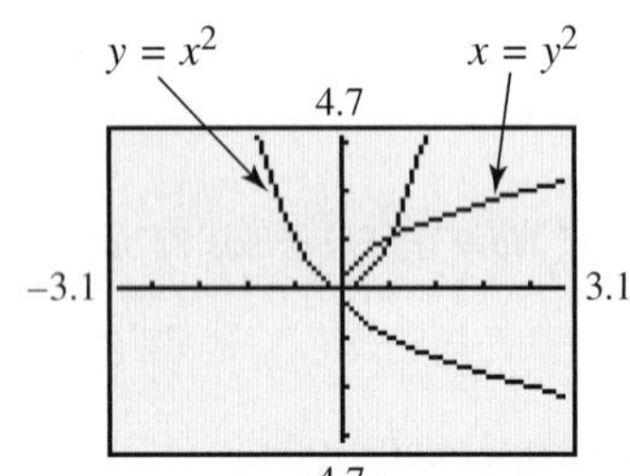

Despite the fact that $y = x^2$ is not one-to-one, the calculator will draw its "inverse," $x = y^2$.

Figure 12

Some graphing calculators have the capability of "drawing" the reflection of a graph across the line $y = x$. This feature does not require that the function be one-to-one, however, so the resulting figure may not be the graph of a function. See **Figure 12.** ***Again, it is necessary to understand the mathematics to interpret results correctly.*** ■

An Application of Inverse Functions to Cryptography A one-to-one function and its inverse can be used to make information secure. The function is used to encode a message, and its inverse is used to decode the coded message. In practice, complicated functions are used. We illustrate the process with a simple function in **Example 9.**

Classroom Example 9
The function

$$f(x) = 3x - 1$$

was used to encode a message as

11 44 74 44 62
59 68 14 14 59

Find the inverse function and decode the message. (Use the same numerical values for the letters as in **Example 9.**)

Answer: $f^{-1}(x) = \frac{1}{3}x + \frac{1}{3}$; DO YOU TWEET

EXAMPLE 9 Using Functions to Encode and Decode a Message

Use the one-to-one function $f(x) = 3x + 1$ and the following numerical values assigned to each letter of the alphabet to encode and decode the message BE MY FACEBOOK FRIEND.

A	1	**H**	8	**O**	15	**V**	22
B	2	**I**	9	**P**	16	**W**	23
C	3	**J**	10	**Q**	17	**X**	24
D	4	**K**	11	**R**	18	**Y**	25
E	5	**L**	12	**S**	19	**Z**	26
F	6	**M**	13	**T**	20		
G	7	**N**	14	**U**	21		

A	1	N	14
B	2	O	15
C	3	P	16
D	4	Q	17
E	5	R	18
F	6	S	19
G	7	T	20
H	8	U	21
I	9	V	22
J	10	W	23
K	11	X	24
L	12	Y	25
M	13	Z	26

SOLUTION The message BE MY FACEBOOK FRIEND would be encoded as

7 16 40 76 19 4 10 16 7
46 46 34 19 55 28 16 43 13

because

B corresponds to 2 and $f(2) = 3(2) + 1 = 7,$

E corresponds to 5 and $f(5) = 3(5) + 1 = 16,$ and so on.

Using the inverse $f^{-1}(x) = \frac{1}{3}x - \frac{1}{3}$ to decode yields

$$f^{-1}(7) = \frac{1}{3}(7) - \frac{1}{3} = 2,$$ which corresponds to B,

$$f^{-1}(16) = \frac{1}{3}(16) - \frac{1}{3} = 5,$$ which corresponds to E, and so on.

✔ Now Try Exercise 97.

1. Yes, it is one-to-one, because every number in the list of registered passenger cars is used only once. **2.** It is not one-to-one because both Illinois and Texas are paired with the same range element, 49. **3.** one-to-one **4.** one-to-one **5.** not one-to-one **6.** not one-to-one **7.** one-to-one **8.** one-to-one **9.** one-to-one **10.** one-to-one **11.** not one-to-one **12.** not one-to-one **13.** one-to-one **14.** one-to-one **15.** one-to-one **16.** one-to-one **17.** not one-to-one **18.** not one-to-one

4.1 Exercises

19. one-to-one
20. one-to-one

23. one-to-one
24. x; $(g \circ f)(x)$
25. range; domain
26. (b, a)
27. false
28. $y = x$
29. -3
30. does not; It is not one-to-one.

31. untying your shoelaces
32. stopping a car
33. leaving a room
34. descending the stairs
35. unscrewing a light bulb
36. emptying a cup

37. inverses **38.** not inverses
39. not inverses **40.** inverses

41. inverses **42.** inverses
43. not inverses **44.** not inverses
45. inverses **46.** inverses
47. not inverses **48.** not inverses
49. inverses **50.** inverses

51. $\{(6, -3), (1, 2), (8, 5)\}$
52. $\{(-1, 3), (0, 5), (5, 0), (\frac{2}{3}, 4)\}$
53. not one-to-one
54. not one-to-one

Concept Check In Exercises 1 and 2, answer the question and then write a short explanation.

1. The table shows the number of registered passenger cars in the United States for the years 2004–2008.

If this correspondence is considered to be a function that pairs each year with the number of registered passenger cars, is it one-to-one? If not, explain why.

Year	Registered Passenger Cars (in thousands)
2004	136,431
2005	136,568
2006	135,400
2007	135,933
2008	137,080

Source: U.S. Federal Highway Administration.

2. The table shows the number of hazardous waste sites on the National Priority List for the Superfund program in 2008 for seven states in the top ten.

If this correspondence is considered to be a function that pairs each state with its number of hazardous waste sites, is it one-to-one? If not, explain why.

State	Number of Sites
New Jersey	116
California	97
Pennsylvania	96
New York	86
Florida	52
Illinois	49
Texas	49

Source: U.S. Environmental Protection Agency.

55. inverses **56.** inverses
57. not inverses **58.** not inverses

59. (a) $f^{-1}(x) = \frac{1}{3}x + \frac{4}{3}$
(b)

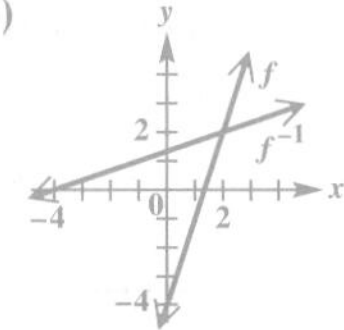

(c) Domains and ranges of both f and f^{-1} are $(-\infty, \infty)$.

60. (a) $f^{-1}(x) = \frac{1}{4}x + \frac{5}{4}$
(b)

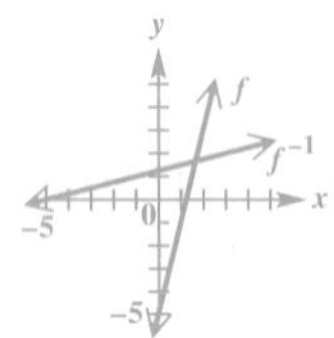

(c) Domains and ranges of both f and f^{-1} are $(-\infty, \infty)$.

61. (a) $f^{-1}(x) = -\frac{1}{4}x + \frac{3}{4}$
(b)

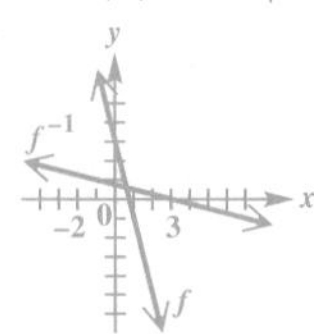

(c) Domains and ranges of both f and f^{-1} are $(-\infty, \infty)$.

62. (a) $f^{-1}(x) = -\frac{1}{6}x - \frac{4}{3}$
(b)

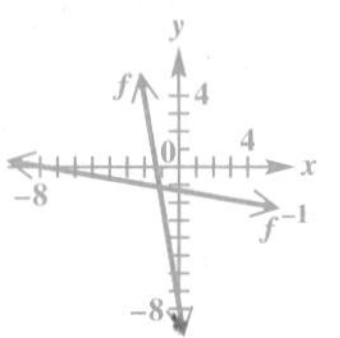

(c) Domains and ranges of both f and f^{-1} are $(-\infty, \infty)$.

63. (a) $f^{-1}(x) = \sqrt[3]{x - 1}$
(b)

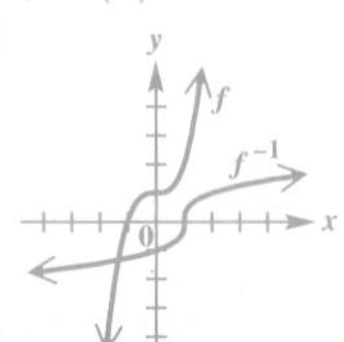

(c) Domains and ranges of both f and f^{-1} are $(-\infty, \infty)$.

64. (a) $f^{-1}(x) = \sqrt[3]{-x - 2}$
(b)

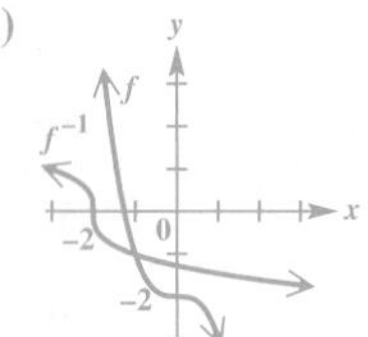

(c) Domains and ranges of both f and f^{-1} are $(-\infty, \infty)$.

65. not one-to-one
66. not one-to-one

Decide whether each function as graphed or defined is one-to-one. ***See Examples 1 and 2.***

3.

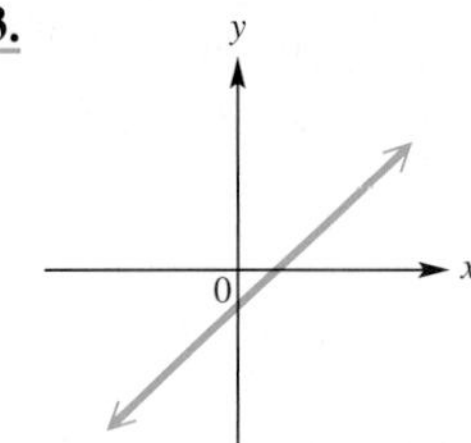

4.

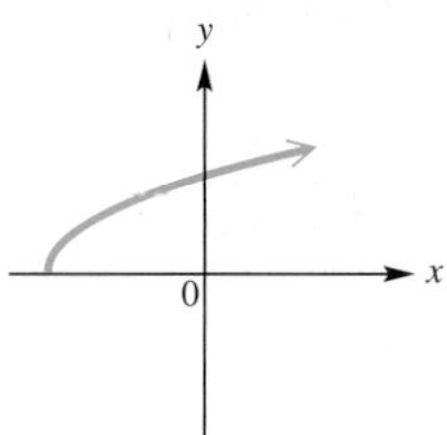

5.

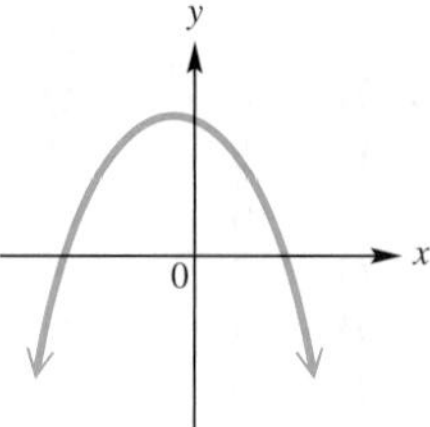

6.

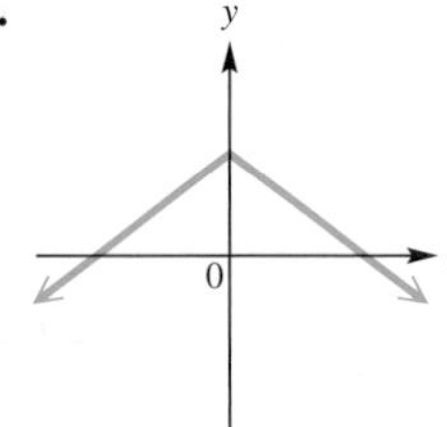

7.

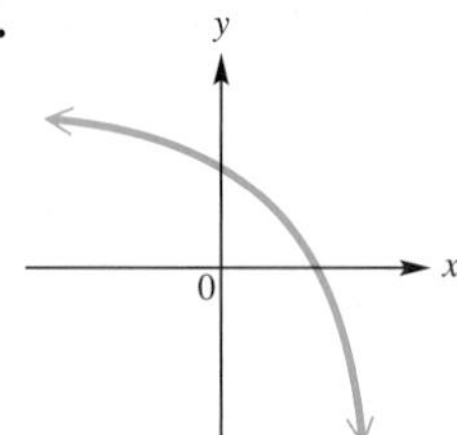

8.

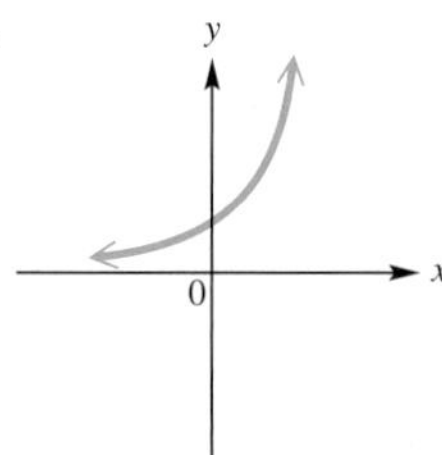

9. $y = 2x - 8$

10. $y = 4x + 20$

11. $y = \sqrt{36 - x^2}$

12. $y = -\sqrt{100 - x^2}$

13. $y = 2x^3 - 1$

14. $y = 3x^3 - 6$

15. $y = \dfrac{-1}{x + 2}$

16. $y = \dfrac{4}{x - 8}$

17. $y = 2(x + 1)^2 - 6$

18. $y = -3(x - 6)^2 + 8$

19. $y = \sqrt[3]{x + 1} - 3$

20. $y = -\sqrt[3]{x + 2} - 8$

21. Explain why a constant function, such as $f(x) = 3$, defined over the set of real numbers, cannot be one-to-one.

22. (a) Explain why a polynomial function of even degree cannot have an inverse.
(b) Explain why a polynomial function of odd degree *may* not be one-to-one.

Concept Check Answer each of the following.

23. For a function to have an inverse, it must be ______________.

24. If two functions f and g are inverses, then $(f \circ g)(x) =$ ______________, and ______________ $= x$.

25. The domain of f is equal to the ______________ of f^{-1}, and the range of f is equal to the ______________ of f^{-1}.

26. If the point (a, b) lies on the graph of f, and f has an inverse, then the point ______________ lies on the graph of f^{-1}.

27. *True* or *false*: If $f(x) = x^2$, then $f^{-1}(x) = \sqrt{x}$.

28. If a function f has an inverse, then the graph of f^{-1} may be obtained by reflecting the graph of f across the line with equation ______________.

29. If a function f has an inverse and $f(-3) = 6$, then $f^{-1}(6) =$ ______________.

30. If $f(-4) = 16$ and $f(4) = 16$, then f ______________ (does/does not) have an inverse because ______________.

Concept Check In Exercises 31–36, an everyday activity is described. Keeping in mind that an inverse operation "undoes" what an operation does, describe each inverse activity.

31. tying your shoelaces

32. starting a car

33. entering a room

34. climbing the stairs

35. screwing in a light bulb

36. filling a cup

67. (a) $f^{-1}(x) = \frac{1}{x}, x \neq 0$
(b)

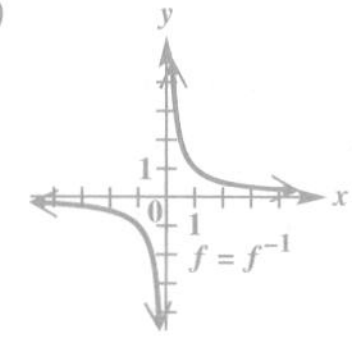

(c) Domains and ranges of both f and f^{-1} are $(-\infty, 0) \cup (0, \infty)$.

68. (a) $f^{-1}(x) = \frac{4}{x}, x \neq 0$
(b)

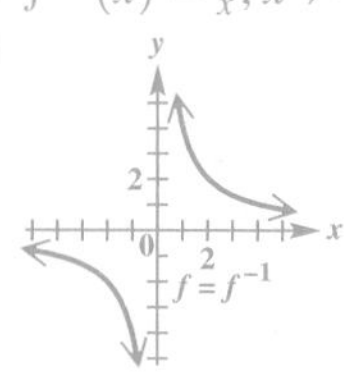

(c) Domains and ranges of both f and f^{-1} are $(-\infty, 0) \cup (0, \infty)$.

69. (a) $f^{-1}(x) = \frac{1 + 3x}{x}, x \neq 0$
(b)

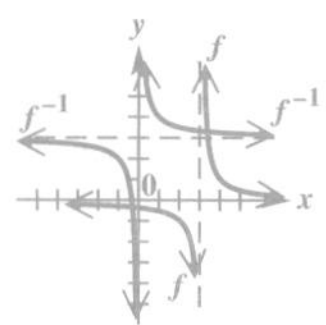

(c) Domain of f = range of $f^{-1} = (-\infty, 3) \cup (3, \infty)$. Domain of f^{-1} = range of $f = (-\infty, 0) \cup (0, \infty)$.

70. (a) $f^{-1}(x) = \frac{1 - 2x}{x}, x \neq 0$
(b)

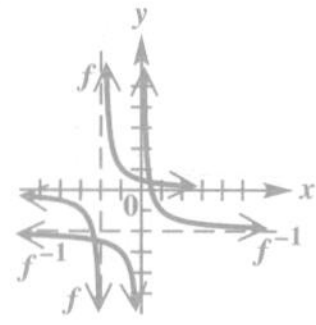

(c) Domain of f = range of $f^{-1} = (-\infty, -2) \cup (-2, \infty)$.
Domain of f^{-1} = range of $f = (-\infty, 0) \cup (0, \infty)$.

71. (a) $f^{-1}(x) = \frac{3x + 1}{x - 1}, x \neq 1$
(b)

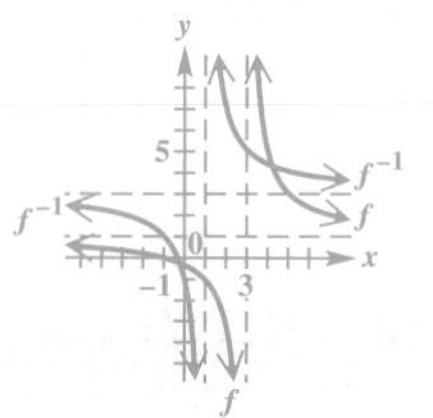

(c) Domain of f = range of $f^{-1} = (-\infty, 3) \cup (3, \infty)$. Domain of f^{-1} = range of $f = (-\infty, 1) \cup (1, \infty)$.

Decide whether the given functions are inverses. ***See Example 4.***

37.

x	$f(x)$
3	−4
2	−6
5	8
1	9
4	3

x	$g(x)$
−4	3
−6	2
8	5
9	1
3	4

38.

x	$f(x)$
−2	−8
−1	−1
0	0
1	1
2	8

x	$g(x)$
8	−2
1	−1
0	0
−1	1
−8	2

39. $f = \{(2, 5), (3, 5), (4, 5)\}; \quad g = \{(5, 2)\}$

40. $f = \{(1, 1), (3, 3), (5, 5)\}; \quad g = \{(1, 1), (3, 3), (5, 5)\}$

Use the definition of inverses to determine whether f and g are inverses. ***See Example 3.***

41. $f(x) = 2x + 4, \quad g(x) = \frac{1}{2}x - 2$

42. $f(x) = 3x + 9, \quad g(x) = \frac{1}{3}x - 3$

43. $f(x) = -3x + 12, \quad g(x) = -\frac{1}{3}x - 12$

44. $f(x) = -4x + 2, \quad g(x) = -\frac{1}{4}x - 2$

45. $f(x) = \frac{x + 1}{x - 2}, \quad g(x) = \frac{2x + 1}{x - 1}$

46. $f(x) = \frac{x - 3}{x + 4}, \quad g(x) = \frac{4x + 3}{1 - x}$

47. $f(x) = \frac{2}{x + 6}, \quad g(x) = \frac{6x + 2}{x}$

48. $f(x) = \frac{-1}{x + 1}, \quad g(x) = \frac{1 - x}{x}$

49. $f(x) = x^2 + 3, \quad x \geq 0; \quad g(x) = \sqrt{x - 3}, \quad x \geq 3$

50. $f(x) = \sqrt{x + 8}, \quad x \geq -8; \quad g(x) = x^2 - 8, \quad x \geq 0$

If the function is one-to-one, find its inverse. ***See Example 4.***

51. $\{(-3, 6), (2, 1), (5, 8)\}$

52. $\left\{(3, -1), (5, 0), (0, 5), \left(4, \frac{2}{3}\right)\right\}$

53. $\{(1, -3), (2, -7), (4, -3), (5, -5)\}$

54. $\{(6, -8), (3, -4), (0, -8), (5, -4)\}$

Decide whether each pair of functions graphed are inverses. ***See Example 7.***

55.

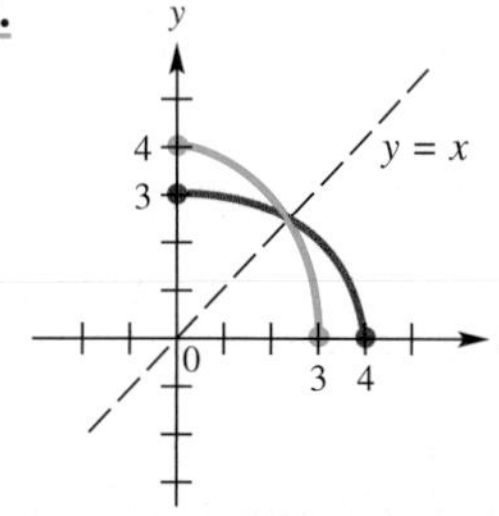

56.

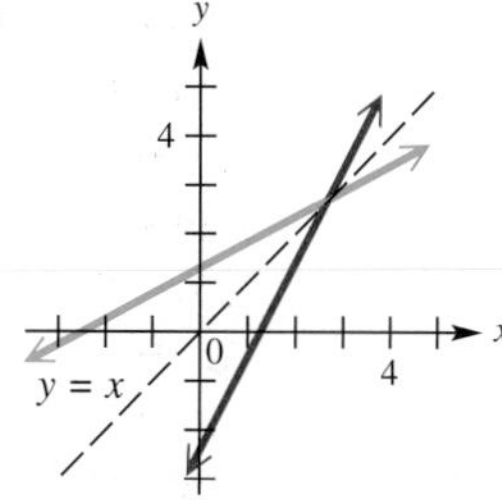

57.

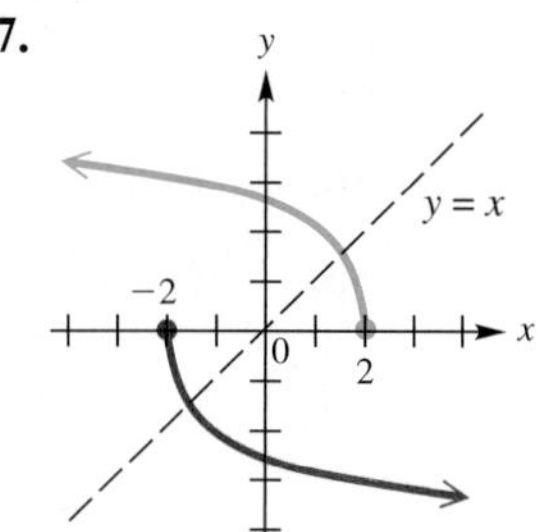

58.

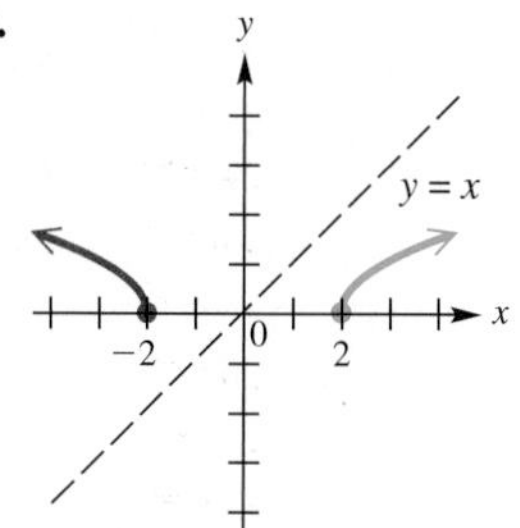

For each function as defined that is one-to-one, **(a)** *write an equation for the inverse function in the form* $y = f^{-1}(x)$, **(b)** *graph* f *and* f^{-1} *on the same axes, and* **(c)** *give the domain and the range of* f *and* f^{-1}. *If the function is not one-to-one, say so.* ***See Examples 5–8.***

59. $y = 3x - 4$ **60.** $y = 4x - 5$ **61.** $f(x) = -4x + 3$

62. $f(x) = -6x - 8$ **63.** $f(x) = x^3 + 1$ **64.** $f(x) = -x^3 - 2$

65. $y = x^2 + 8$ **66.** $y = -x^2 + 2$ **67.** $y = \frac{1}{x}, \quad x \neq 0$

68. $y = \frac{4}{x}, \quad x \neq 0$ **69.** $f(x) = \frac{1}{x-3}, \quad x \neq 3$ **70.** $f(x) = \frac{1}{x+2}, \quad x \neq -2$

71. $f(x) = \frac{x+1}{x-3}, \quad x \neq 3$ **72.** $f(x) = \frac{x+2}{x-1}, \quad x \neq 1$

73. $f(x) = \frac{2x+6}{x-3}, \quad x \neq 3$ **74.** $f(x) = \frac{-3x+12}{x-6}, \quad x \neq 6$

75. $f(x) = \sqrt{6+x}, \quad x \geq -6$ **76.** $f(x) = -\sqrt{x^2 - 16}, \quad x \geq 4$

Graph the inverse of each one-to-one function. ***See Example 7.***

77.

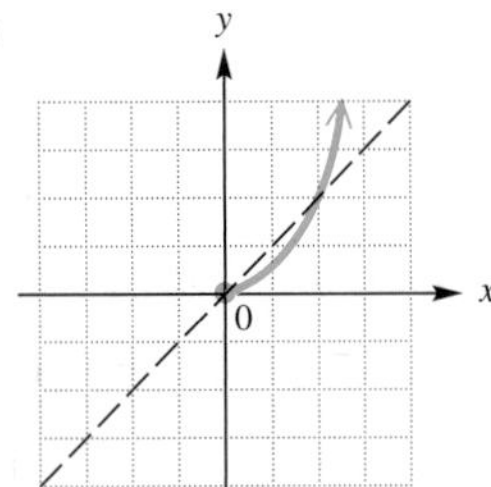

78.

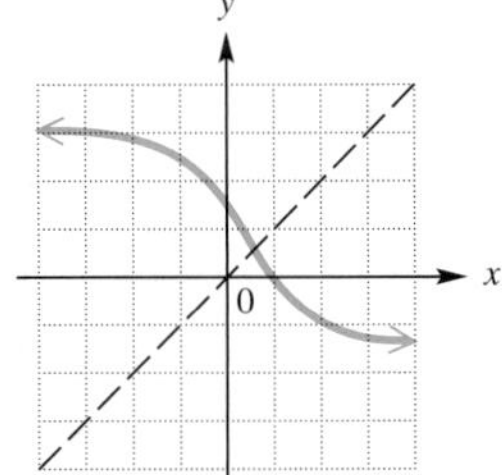

79.

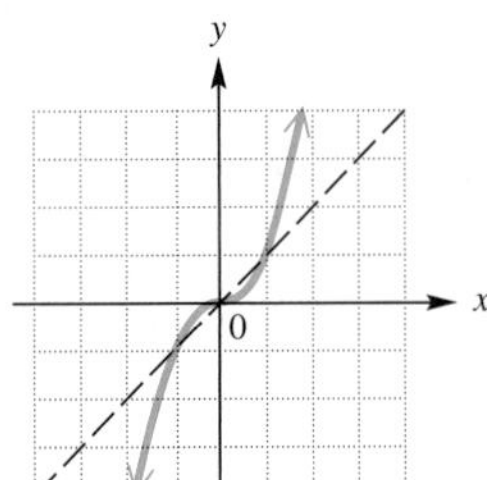

80.

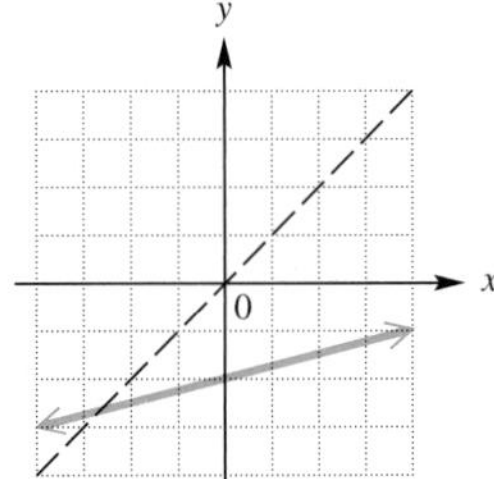

81.

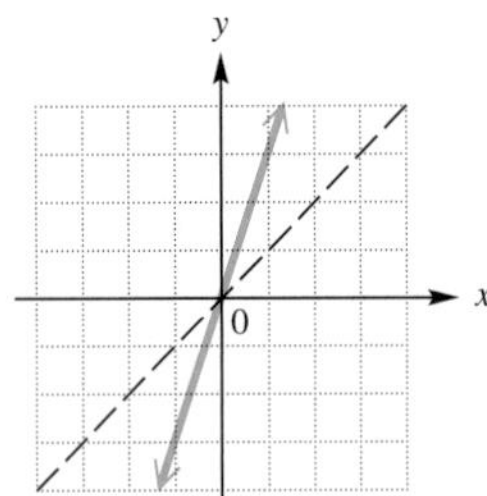

82.

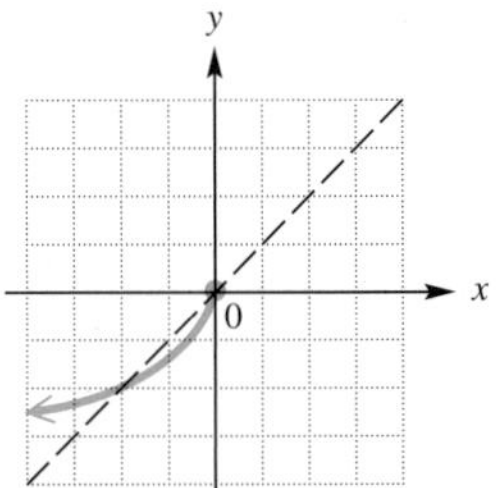

Concept Check The graph of a function f is shown in the figure. Use the graph to find each value.

83. $f^{-1}(4)$ **84.** $f^{-1}(2)$

85. $f^{-1}(0)$ **86.** $f^{-1}(-2)$

87. $f^{-1}(-3)$ **88.** $f^{-1}(-4)$

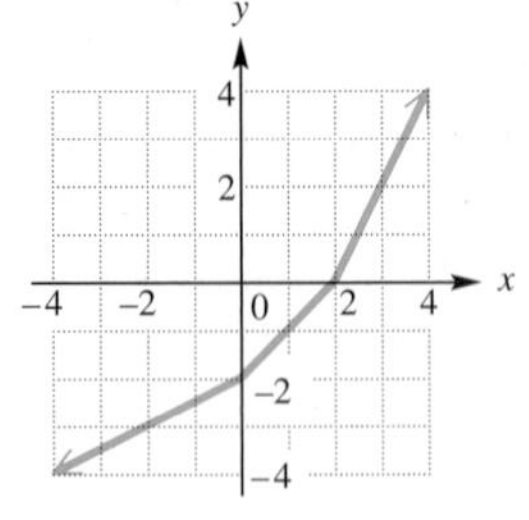

72. (a) $f^{-1}(x) = \frac{x+2}{x-1}, x \neq 1$
(b)
(c) Domain of f = range of $f^{-1} = (-\infty, 1) \cup (1, \infty)$. Domain of f^{-1} = range of $f = (-\infty, 1) \cup (1, \infty)$.

73. (a) $f^{-1}(x) = \frac{3x+6}{x-2}, x \neq 2$
(b)
(c) Domain of f = range of $f^{-1} = (-\infty, 3) \cup (3, \infty)$. Domain of f^{-1} = range of $f = (-\infty, 2) \cup (2, \infty)$.

74. (a) $f^{-1}(x) = \frac{6x+12}{x+3}, x \neq -3$
(b)
(c) Domain of f = range of $f^{-1} = (-\infty, 6) \cup (6, \infty)$. Domain of f^{-1} = range of $f = (-\infty, -3) \cup (-3, \infty)$.

75. (a) $f^{-1}(x) = x^2 - 6, x \geq 0$
(b)
(c) Domain of f = range of $f^{-1} = [-6, \infty)$. Domain of f^{-1} = range of $f = [0, \infty)$.

76. (a) $f^{-1}(x) = \sqrt{x^2 + 16}, x \leq 0$
(b)
(c) Domain of f = range of $f^{-1} = [4, \infty)$. Domain of f^{-1} = range of $f = (-\infty, 0]$.

77.

78.

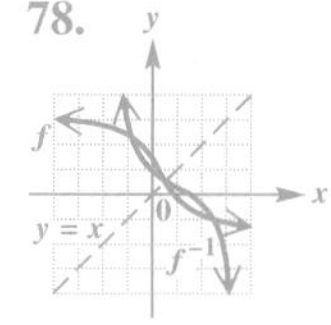

79.

80.

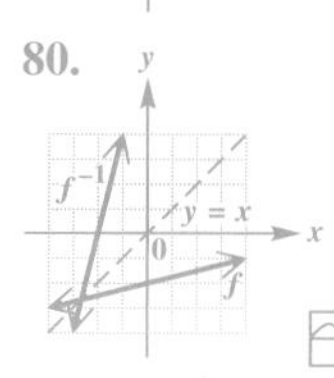

81.

82.

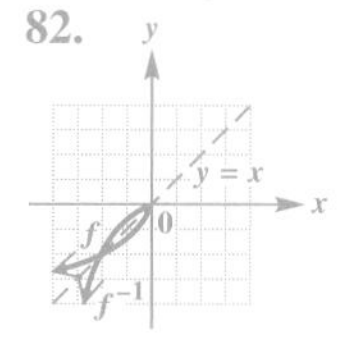

83. 4 84. 3
85. 2 86. 0
87. −2 88. −4

89. It represents the cost, in dollars, of building 1000 cars.
90. It represents the radius of a sphere with volume 5 in.^{3}.
91. $\frac{1}{a}$ 92. 2

93. not one-to-one
94. not one-to-one
95. one-to-one; $f^{-1}(x) = \frac{-5-3x}{x-1}$, $x \neq 1$
96. one-to-one; $f^{-1}(x) = \frac{4x}{x+1}$, $x \neq -1$
97. $f^{-1}(x) = \frac{1}{3}x + \frac{2}{3}$; MIGUEL HAS ARRIVED
98. $f^{-1}(x) = \frac{1}{2}x + \frac{9}{2}$; BIG GIRLS DONT CRY
99. 6858 124 2743 63 511 124 1727 4095; $f^{-1}(x) = \sqrt[3]{x+1}$
100. 8000 8 1000 2197 4096 6859 27 216 13824 8 6859 216; $f^{-1}(x) = \sqrt[3]{x} - 1$

Concept Check Answer each of the following.

89. Suppose $f(x)$ is the number of cars that can be built for x dollars. What does $f^{-1}(1000)$ represent?

90. Suppose $f(r)$ is the volume (in cubic inches) of a sphere of radius r inches. What does $f^{-1}(5)$ represent?

91. If a line has slope a, what is the slope of its reflection across the line $y = x$?

92. For a one-to-one function f, find $(f^{-1} \circ f)(2)$, where $f(2) = 3$.

Use a graphing calculator to graph each function defined as follows, using the given viewing window. Use the graph to decide which functions are one-to-one. If a function is one-to-one, give the equation of its inverse.

93. $f(x) = 6x^3 + 11x^2 - 6$; $[-3, 2]$ by $[-10, 10]$

94. $f(x) = x^4 - 5x^2$; $[-3, 3]$ by $[-8, 8]$

95. $f(x) = \frac{x-5}{x+3}$, $x \neq -3$; $[-8, 8]$ by $[-6, 8]$

96. $f(x) = \frac{-x}{x-4}$, $x \neq 4$; $[-1, 8]$ by $[-6, 6]$

Use the alphabet coding assignment given in ***Example 9*** *for Exercises 97–100.*

97. The function $f(x) = 3x - 2$ was used to encode a message as

37 25 19 61 13 34 22 1 55 1 52 52 25 64 13 10.

Find the inverse function and determine the message.

98. The function $f(x) = 2x - 9$ was used to encode a message as

−5 9 5 5 9 27 15 29 −1 21 19 31 −3 27 41.

Find the inverse function and determine the message.

99. Encode the message SEND HELP, using the one-to-one function $f(x) = x^3 - 1$. Give the inverse function that the decoder will need when the message is received.

100. Encode the message SAILOR BEWARE, using the one-to-one function $f(x) = (x+1)^3$. Give the inverse function that the decoder will need when the message is received.

4.2 Exponential Functions

- Exponents and Properties
- Exponential Functions
- Exponential Equations
- Compound Interest
- The Number *e* and Continuous Compounding
- Exponential Models

Exponents and Properties Recall the definition of $a^{m/n}$: If a is a real number, m is an integer, n is a positive integer, and $\sqrt[n]{a}$ is a real number, then

$$a^{m/n} = \left(\sqrt[n]{a}\right)^m. \quad \text{(Section R.7)}$$

For example,

$$16^{3/4} = \left(\sqrt[4]{16}\right)^3 = 2^3 = 8,$$

$$27^{-1/3} = \frac{1}{27^{1/3}} = \frac{1}{\sqrt[3]{27}} = \frac{1}{3}, \quad \text{and} \quad 64^{-1/2} = \frac{1}{64^{1/2}} = \frac{1}{\sqrt{64}} = \frac{1}{8}.$$

Teaching Tip Use this opportunity to emphasize the distinction between *exact* and *approximate* values.

In this section we extend the definition of a^r to include all real (not just rational) values of the exponent r. For example, $2^{\sqrt{3}}$ might be evaluated by *approximating* the exponent $\sqrt{3}$ with the rational numbers 1.7, 1.73, 1.732, and so on. Since these decimals approach the value of $\sqrt{3}$ more and more closely, it seems reasonable that $2^{\sqrt{3}}$ should be approximated more and more closely by the numbers $2^{1.7}$, $2^{1.73}$, $2^{1.732}$, and so on. (Recall, for example, that $2^{1.7} = 2^{17/10} = (\sqrt[10]{2})^{17}$.) To show that this assumption is reasonable, **Figure 13** gives graphs of the function $f(x) = 2^x$ with three different domains.

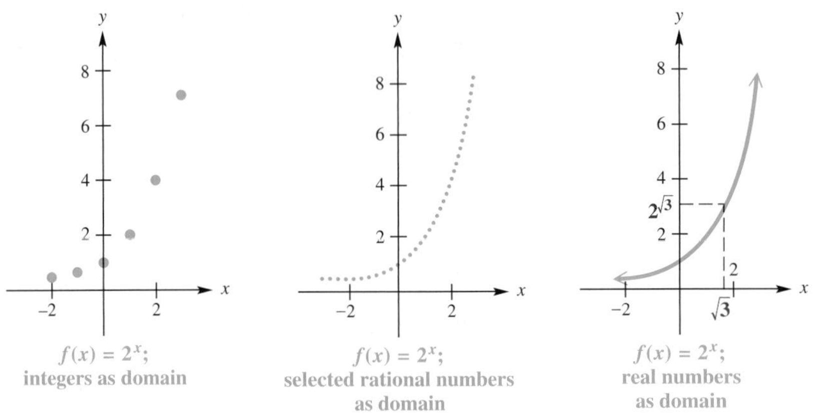

Figure 13

Using this interpretation of real exponents, all rules and theorems for exponents are valid for all real number exponents, not just rational ones. In addition to the rules for exponents presented earlier, we use several new properties in this chapter. These properties are generalized below. Proofs of the properties are not given here, because they require more advanced mathematics.

Teaching Tip Give interpretations of the properties. That is, property (a) says that

$$f(x) = a^x$$

is a function with domain $(-\infty, \infty)$, property (b) says that f is one-to one, property (c) says that f is increasing if $a > 1$, and property (d) says that f is decreasing if $0 < a < 1$.

Additional Properties of Exponents

For any real number $a > 0$, $a \neq 1$, the following statements are true.

(a) a^x **is a unique real number for all real numbers** x**.**

(b) $a^b = a^c$ **if and only if** $b = c$**.**

(c) If $a > 1$ **and** $m < n$**, then** $a^m < a^n$**.**

(d) If $0 < a < 1$ **and** $m < n$**, then** $a^m > a^n$**.**

Properties (a) and (b) require $a > 0$ so that a^x is always defined. For example, $(-6)^x$ is not a real number if $x = \frac{1}{2}$. This means that a^x will always be positive, since a must be positive. In property (a), a cannot equal 1 because $1^x = 1$ for every real number value of x, so each value of x leads to the same real number, 1. For property (b) to hold, a must not equal 1 since, for example, $1^4 = 1^5$, even though $4 \neq 5$.

Properties (c) and (d) say that ***when*** $a > 1$***, increasing the exponent on "a" leads to a greater number, but when*** $0 < a < 1$***, increasing the exponent on "a" leads to a lesser number.***

Classroom Example 1
If $f(x) = 4^x$, find each of the following.

(a) $f(-2)$ (b) $f(5)$

(c) $f\left(\frac{2}{3}\right)$ (d) $f(2.15)$

Answers:
(a) $\frac{1}{16}$ (b) 1024 (c) $2\sqrt[3]{2}$
(d) 19.69831061

EXAMPLE 1 Evaluating an Exponential Expression

If $f(x) = 2^x$, find each of the following.

(a) $f(-1)$ **(b)** $f(3)$ **(c)** $f\left(\frac{5}{2}\right)$ **(d)** $f(4.92)$

SOLUTION

(a) $f(-1) = 2^{-1} = \frac{1}{2}$ Replace x with -1.

(b) $f(3) = 2^3 = 8$

(c) $f\left(\frac{5}{2}\right) = 2^{5/2} = (2^5)^{1/2} = 32^{1/2} = \sqrt{32} = \sqrt{16 \cdot 2} = 4\sqrt{2}$ (Section R.7)

(d) $f(4.92) = 2^{4.92} \approx 30.2738447$ Use a calculator.

Now Try Exercises 3, 9, and 13.

Exponential Functions We now define a function $f(x) = a^x$ whose domain is the set of all real numbers. Notice how the independent variable x appears in the exponent in this function. In earlier chapters the independent variable did not appear in exponents.

Exponential Function

If $a > 0$ and $a \neq 1$, then

$$f(x) = a^x$$

defines the **exponential function with base a.**

NOTE ***We do not allow* 1 *as the base for an exponential function.*** If $a = 1$, the function becomes the constant function defined by $f(x) = 1$, which is not an exponential function.

Figure 13 showed the graph of $f(x) = 2^x$ with three different domains. We repeat the final graph (with real numbers as domain) here.

- The y-intercept is $y = 2^0 = 1$.
- Since $2^x > 0$ for all x and $2^x \rightarrow 0$ as $x \rightarrow -\infty$, the x-axis is a horizontal asymptote.
- As the graph suggests, the domain of the function is $(-\infty, \infty)$ and the range is $(0, \infty)$.
- The function is increasing on its entire domain, and it therefore is one-to-one.

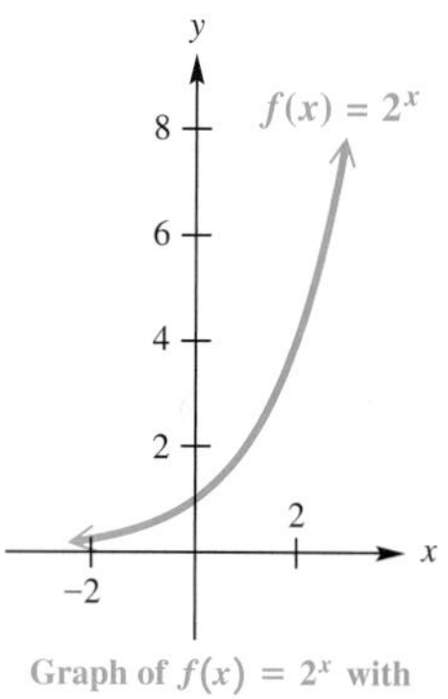

Graph of $f(x) = 2^x$ with domain $(-\infty, \infty)$

Figure 13 (repeated)

These observations from **Figure 13** lead to the following generalizations about the graphs of exponential functions.

Exponential Function $f(x) = a^x$

Domain: $(-\infty, \infty)$ Range: $(0, \infty)$

For $f(x) = 2^x$:

x	$f(x)$
-2	$\frac{1}{4}$
-1	$\frac{1}{2}$
0	1
1	2
2	4
3	8

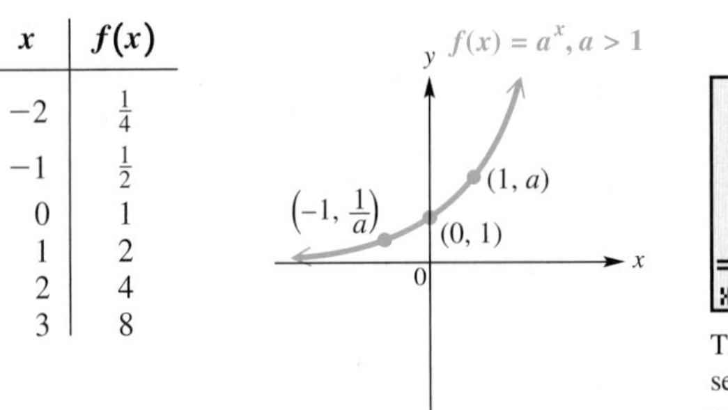

f(x) = a^x, a > 1

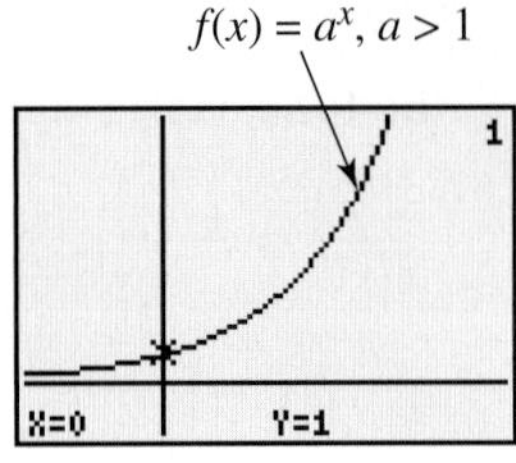

This is the general behavior seen on a calculator graph for **any base *a*, for *a* > 1.**

Figure 14

- **$f(x) = a^x$, for $a > 1$,** is increasing and continuous on its entire domain, $(-\infty, \infty)$.
- The x-axis is a horizontal asymptote as $x \to -\infty$.
- The graph passes through the points $\left(-1, \frac{1}{a}\right)$, $(0, 1)$, and $(1, a)$.

For $f(x) = \left(\frac{1}{2}\right)^x$:

x	$f(x)$
-3	8
-2	4
-1	2
0	1
1	$\frac{1}{2}$
2	$\frac{1}{4}$

f(x) = a^x, 0 < a < 1

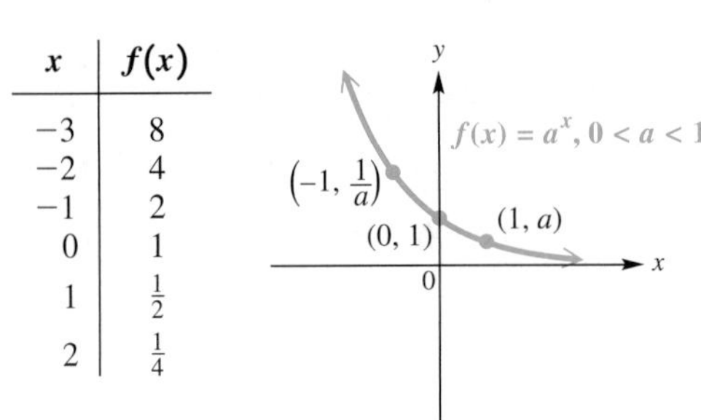

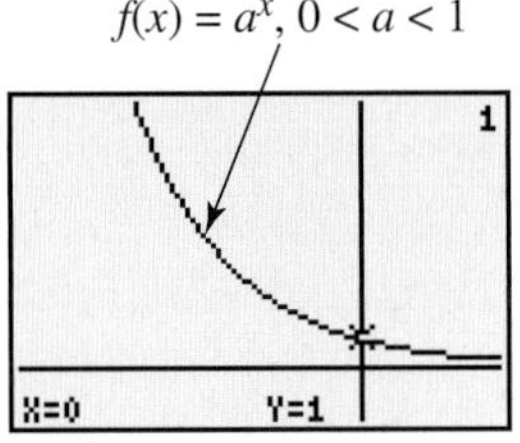

This is the general behavior seen on a calculator graph for **any base *a*, for 0 < *a* < 1.**

Figure 15

- **$f(x) = a^x$, for $0 < a < 1$,** is decreasing and continuous on its entire domain, $(-\infty, \infty)$.
- The x-axis is a horizontal asymptote as $x \to \infty$.
- The graph passes through the points $\left(-1, \frac{1}{a}\right)$, $(0, 1)$, and $(1, a)$.

From **Section 2.7,** the graph of $y = f(-x)$ is the graph of $y = f(x)$ reflected across the y-axis. Thus, we have the following.

$$\text{If } f(x) = 2^x, \quad \text{then} \quad f(-x) = 2^{-x} = 2^{-1 \cdot x} = (2^{-1})^x = \left(\frac{1}{2}\right)^x.$$

This is supported by the graphs in **Figures 14 and 15.**

The graph of $f(x) = 2^x$ is typical of graphs of $f(x) = a^x$ where $a > 1$. For larger values of a, the graphs rise more steeply, but the general shape is similar to the graph in **Figure 14.** When $0 < a < 1$, the graph decreases in a manner similar to the graph of $f(x) = \left(\frac{1}{2}\right)^x$. In **Figure 16** on the next page, the graphs of several typical exponential functions illustrate these facts.

Teaching Tip The point (0, 1) lies on the graph of

$$f(x) = a^x$$

for all a. We can informally consider it "home plate" when graphing translations and reflections, as seen later in **Example 3.**

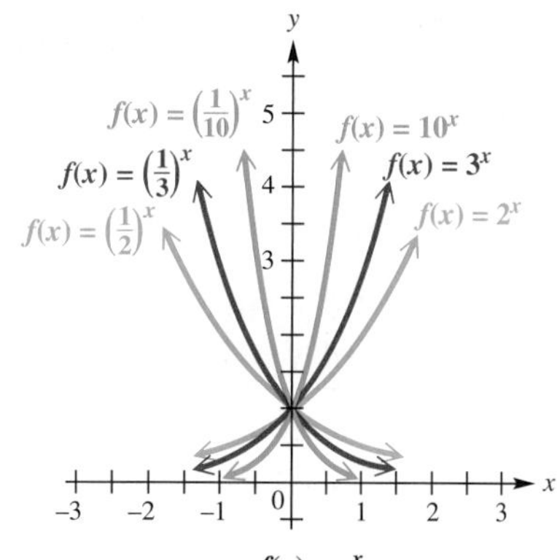

$f(x) = a^x$

Domain: $(-\infty, \infty)$; Range: $(0, \infty)$

- When $a > 1$, the function is increasing.
- When $0 < a < 1$, the function is decreasing.
- In every case, the x-axis is a horizontal asymptote.

Figure 16

In summary, the graph of a function of the form $f(x) = a^x$ has the following features.

Teaching Tip Encourage students to learn the characteristics of the graph of

$$f(x) = a^x.$$

Ask them how the coordinates $(0, 1)$ and $(1, a)$ are affected by different variations of the graph of

$$f(x) = -a^{x-h} + k.$$

Characteristics of the Graph of $f(x) = a^x$

1. The points $\left(-1, \frac{1}{a}\right)$, $(0, 1)$, and $(1, a)$ are on the graph.
2. If $a > 1$, then f is an increasing function. If $0 < a < 1$, then f is a decreasing function.
3. The x-axis is a horizontal asymptote.
4. The domain is $(-\infty, \infty)$, and the range is $(0, \infty)$.

Classroom Example 2

Graph $f(x) = \left(\frac{1}{2}\right)^x$.

Answer:

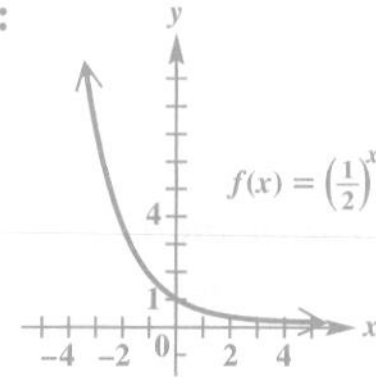

EXAMPLE 2 Graphing an Exponential Function

Graph $f(x) = 5^x$. Give the domain and range.

SOLUTION The y-intercept is 1, and the x-axis is a horizontal asymptote. Plot a few ordered pairs, and draw a smooth curve through them as shown in **Figure 17.** Like the function $f(x) = 2^x$, this function also has domain $(-\infty, \infty)$ and range $(0, \infty)$ and is one-to-one. The function is increasing on its entire domain.

x	$f(x)$
-1	0.2
0	1
0.5	≈ 2.2
1	5
1.5	≈ 11.2
2	25

$f(x) = 5^x$

Figure 17

Now Try Exercise 17.

Classroom Example 3
Graph each function. Give the domain and range.
(a) $f(x) = -3^x$ (b) $f(x) = 3^{x-2}$
(c) $f(x) = 3^{x+2} - 2$

Answers:
(a) $(-\infty, \infty)$; $(-\infty, 0)$

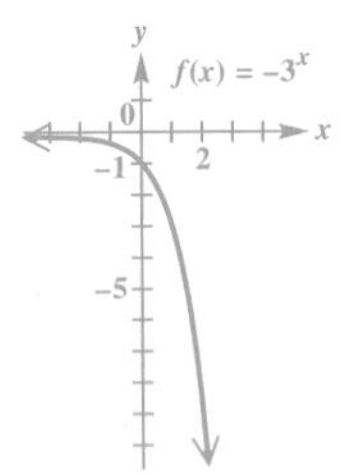

(b) $(-\infty, \infty)$; $(0, \infty)$

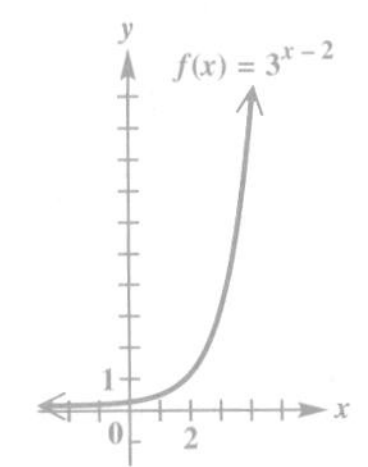

(c) $(-\infty, \infty)$; $(-2, \infty)$

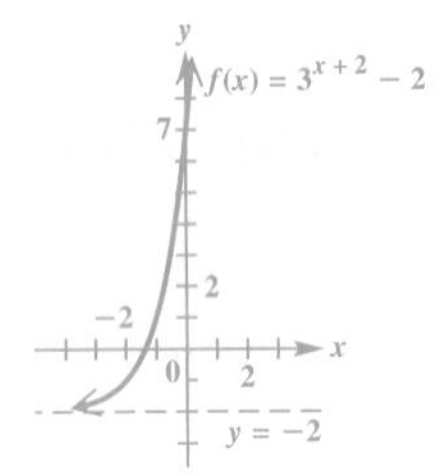

EXAMPLE 3 Graphing Reflections and Translations

Graph each function. Show the graph of $y = 2^x$ for comparison. Give the domain and range.

(a) $f(x) = -2^x$ **(b)** $f(x) = 2^{x+3}$ **(c)** $f(x) = 2^{x-2} - 1$

SOLUTION In each graph, we show in particular how the point $(0, 1)$ on the graph of $y = 2^x$ has been translated.

(a) The graph of $f(x) = -2^x$ is that of $f(x) = 2^x$ reflected across the x-axis. The domain is $(-\infty, \infty)$, and the range is $(-\infty, 0)$. See **Figure 18.**

(b) The graph of $f(x) = 2^{x+3}$ is the graph of $f(x) = 2^x$ translated 3 units to the left, as shown in **Figure 19.** The domain is $(-\infty, \infty)$, and the range is $(0, \infty)$.

(c) The graph of $f(x) = 2^{x-2} - 1$ is that of $f(x) = 2^x$ translated 2 units to the right and 1 unit down. See **Figure 20.** The domain is $(-\infty, \infty)$, and the range is $(-1, \infty)$.

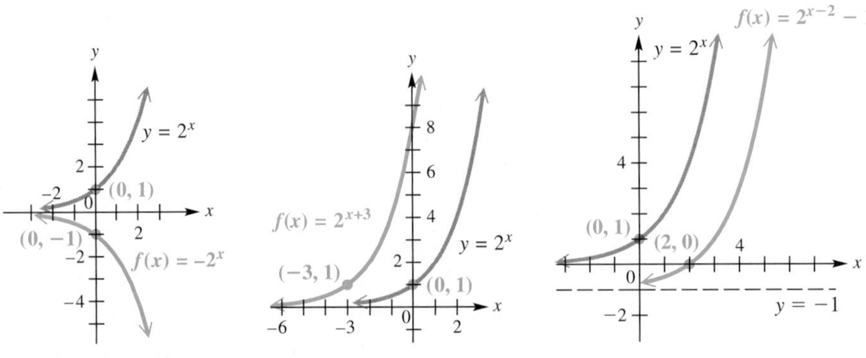

Figure 18 **Figure 19** **Figure 20**

✔ *Now Try Exercises 29, 31, and 37.*

Exponential Equations Because the graph of $y = a^x$ is that of a one-to-one function, to solve $a^{x_1} = a^{x_2}$, we need only show that $x_1 = x_2$. Property (b) given earlier in this section is used to solve **exponential equations,** which are equations with variables as exponents.

Classroom Example 4
Solve $5^x = \frac{1}{125}$.

Answer: $\{-3\}$

EXAMPLE 4 Solving an Exponential Equation

Solve $\left(\frac{1}{3}\right)^x = 81$.

SOLUTION ***Write each side of the equation using a common base.***

$$\left(\frac{1}{3}\right)^x = 81$$

$$(3^{-1})^x = 81 \quad \text{Definition of negative exponent (Section R.6)}$$

$$3^{-x} = 81 \quad (a^m)^n = a^{mn} \text{ (Section R.3)}$$

$$3^{-x} = 3^4 \quad \text{Write 81 as a power of 3. (Section R.2)}$$

$$-x = 4 \quad \text{Set exponents equal (Property (b)).}$$

$$x = -4 \quad \text{Multiply by } -1. \text{ (Section 1.1)}$$

The solution set of the original equation is $\{-4\}$.

✔ *Now Try Exercise 63.*

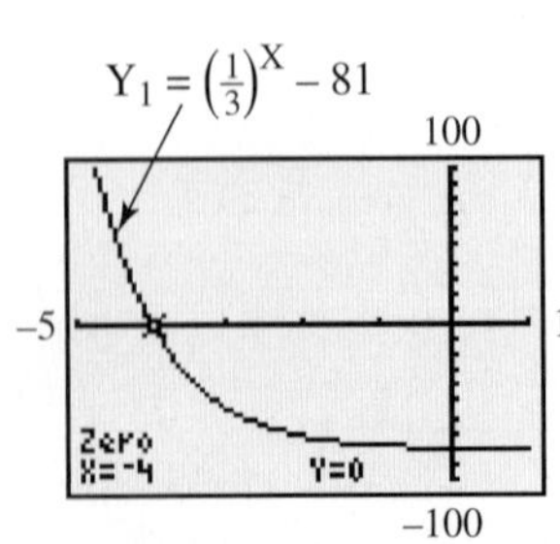

By extending the method of solving linear equations graphically from **Section 2.5,** we can support the result in **Example 4.**

Classroom Example 5
Solve $3^{x+1} = 9^{x-3}$.

Answer: $\{7\}$

EXAMPLE 5 Solving an Exponential Equation

Solve $2^{x+4} = 8^{x-6}$.

SOLUTION Write each side of the equation using a common base.

$$2^{x+4} = 8^{x-6}$$

$$2^{x+4} = (2^3)^{x-6} \quad \text{Write 8 as a power of 2.}$$

$$2^{x+4} = 2^{3x-18} \quad (a^m)^n = a^{mn}$$

$$x + 4 = 3x - 18 \quad \text{Set exponents equal (Property (b)).}$$

$$-2x = -22 \quad \text{Subtract } 3x \text{ and 4. (Section 1.1)}$$

$$x = 11 \quad \text{Divide by } -2.$$

Check by substituting 11 for x in the original equation. The solution set is $\{11\}$.

✔ *Now Try Exercise 71.*

Teaching Tip Review the properties of exponents from **Section R.6.**

Teaching Tip Give an example such as

$$16^{x-2} = 9^{1-x}$$

that cannot be solved using the methods of this section. Show how a graphing calculator can be used to find the approximate solution 1.558. (Mention that **Section 4.5** presents algebraic techniques for finding the solution.)

Later in this chapter, we describe a general method for solving exponential equations where the approach used in **Examples 4 and 5** is not possible. For instance, the above method could not be used to solve an equation like

$$7^x = 12,$$

since it is not easy to express both sides as exponential expressions with the same base.

In **Example 6,** we review a solution process first seen in **Section 1.6.**

Classroom Example 6
Solve $x^{5/2} = 243$.

Answer: $\{9\}$

EXAMPLE 6 Solving an Equation with a Fractional Exponent

Solve $x^{4/3} = 81$.

SOLUTION Notice that the variable is in the base rather than in the exponent.

$$x^{4/3} = 81$$

$$(\sqrt[3]{x})^4 = 81 \quad \text{Radical notation for } a^{m/n} \text{ (Section R.7)}$$

$$\sqrt[3]{x} = \pm 3 \quad \text{Take fourth roots on each side. Remember to use } \pm. \text{ (Section 1.6)}$$

$$x = \pm 27 \quad \text{Cube each side.}$$

Check *both* solutions in the original equation. Both check, so the solution set is $\{\pm 27\}$.

Teaching Tip Caution students that in solving equations like the one in **Example 6,** they must consider *both* even roots.

Alternative Method There may be more than one way to solve an exponential equation, as shown here.

$$x^{4/3} = 81$$

$$(x^{4/3})^3 = 81^3 \quad \text{Cube each side.}$$

$$x^4 = (3^4)^3 \quad \text{Write 81 as } 3^4.$$

$$x^4 = 3^{12} \quad (a^m)^n = a^{mn}$$

$$x = \pm\sqrt[4]{3^{12}} \quad \text{Take fourth roots on each side.}$$

$$x = \pm 3^3 \quad \text{Simplify the radical.}$$

$$x = \pm 27 \quad \text{Apply the exponent.}$$

The same solution set, $\{\pm 27\}$, results.

✔ *Now Try Exercise 79.*

Compound Interest Recall the formula for simple interest, $I = Prt$, where P is principal (amount deposited), r is annual rate of interest expressed as a decimal, and t is time in years that the principal earns interest. Suppose $t = 1$ yr. Then at the end of the year, the amount has grown to

$$P + Pr = P(1 + r),$$

the original principal plus interest. If this balance earns interest at the same interest rate for another year, the balance at the end of *that* year will be

$$[P(1 + r)] + [P(1 + r)]r = [P(1 + r)](1 + r) \quad \text{Factor.}$$
$$= P(1 + r)^2. \quad a \cdot a = a^2$$

After the third year, this will grow to

$$[P(1 + r)^2] + [P(1 + r)^2]r = [P(1 + r)^2](1 + r) \quad \text{Factor.}$$
$$= P(1 + r)^3. \quad a^2 \cdot a = a^3$$

Continuing in this way produces a formula for interest compounded annually.

$$A = P(1 + r)^t$$

The general formula for compound interest can be derived in the same way.

Compound Interest

If P dollars are deposited in an account paying an annual rate of interest r compounded (paid) n times per year, then after t years the account will contain A dollars, according to the following formula.

$$A = P\left(1 + \frac{r}{n}\right)^{tn}$$

Classroom Example 7
Repeat **Example 7** if \$2500 is deposited in an account paying 6% per year compounded semiannually (twice per year).

Answers:
(a) \$4515.28 (b) \$2015.28

EXAMPLE 7 Using the Compound Interest Formula

Suppose \$1000 is deposited in an account paying 4% interest per year compounded quarterly (four times per year).

(a) Find the amount in the account after 10 yr with no withdrawals.

(b) How much interest is earned over the 10-yr period?

SOLUTION

(a) $A = P\left(1 + \frac{r}{n}\right)^{tn}$ Compound interest formula

$A = 1000\left(1 + \frac{0.04}{4}\right)^{10(4)}$ Let $P = 1000$, $r = 0.04$, $n = 4$, and $t = 10$.

$A = 1000(1 + 0.01)^{40}$ Simplify.

$A = 1488.86$ Round to the nearest cent.

Thus, \$1488.86 is in the account after 10 yr.

(b) The interest earned for that period is

$$\$1488.86 - \$1000 = \$488.86.$$

✔ *Now Try Exercise 87(a).*

In the formula for compound interest

$$A = P\left(1 + \frac{r}{n}\right)^{tn},$$

A is sometimes called the **future value** and P the **present value.** A is also called the **compound amount** and is the balance *after* interest has been earned.

Classroom Example 8
Brendan Berger must pay a lump sum of \$15,000 in 8 yr.

(a) What amount deposited today (present value) at 4.8% compounded annually will give \$15,000 in 8 yr?

(b) If only \$10,000 is available to deposit now, what annual interest rate is necessary for the money to increase to \$15,000?

Answers:
(a) \$10,308.63 (b) 5.20%

EXAMPLE 8 Finding Present Value

Becky Anderson must pay a lump sum of \$6000 in 5 yr.

(a) What amount deposited today (present value) at 3.1% compounded annually will grow to \$6000 in 5 yr?

(b) If only \$5000 is available to deposit now, what annual interest rate is necessary for the money to increase to \$6000 in 5 yr?

SOLUTION

(a)

$$A = P\left(1 + \frac{r}{n}\right)^{tn} \quad \text{Compound interest formula}$$

$$6000 = P\left(1 + \frac{0.031}{1}\right)^{5(1)} \quad \text{Let } A = 6000,\ r = 0.031,\ n = 1, \text{ and } t = 5.$$

$$6000 = P(1.031)^5 \quad \text{Simplify.}$$

$$P = \frac{6000}{(1.031)^5} \quad \text{Divide by } (1.031)^5 \text{ to solve for } P. \textbf{ (Section 1.1)}$$

$$P \approx 5150.60 \quad \text{Use a calculator.}$$

If Becky leaves \$5150.60 for 5 yr in an account paying 3.1% compounded annually, she will have \$6000 when she needs it. Thus, \$5150.60 is the present value of \$6000 if interest of 3.1% is compounded annually for 5 yr.

(b)

$$A = P\left(1 + \frac{r}{n}\right)^{tn} \quad \text{Compound interest formula}$$

$$6000 = 5000(1 + r)^5 \quad \text{Let } A = 6000,\ P = 5000,\ n = 1, \text{ and } t = 5.$$

$$\frac{6}{5} = (1 + r)^5 \quad \text{Divide by 5000.}$$

$$\left(\frac{6}{5}\right)^{1/5} = 1 + r \quad \text{Take the fifth root on each side.}$$

$$\left(\frac{6}{5}\right)^{1/5} - 1 = r \quad \text{Subtract 1.}$$

$$r \approx 0.0371 \quad \text{Use a calculator.}$$

An interest rate of 3.71% will produce enough interest to increase the \$5000 to \$6000 by the end of 5 yr.

Now Try Exercises 89 and 93.

CAUTION When performing the computations in problems like those in **Examples 7 and 8,** do not round off during intermediate steps. Keep all calculator digits and round at the end of the process.

---

n	$\left(1+\frac{1}{n}\right)^n$ (rounded)
1	2
2	2.25
5	2.48832
10	2.59374
100	2.70481
1000	2.71692
10,000	2.71815
1,000,000	2.71828

The Number *e* and Continuous Compounding The more often interest is compounded within a given time period, the more interest will be earned. Surprisingly, however, there is a limit on the amount of interest, no matter how often it is compounded.

Suppose that \$1 is invested at 100% interest per year, compounded n times per year. Then the interest rate (in decimal form) is 1.00, and the interest rate per period is $\frac{1}{n}$. According to the formula (with $P = 1$), the compound amount at the end of 1 yr will be

$$A = \left(1 + \frac{1}{n}\right)^n.$$

A calculator gives the results in the margin for various values of n. The table suggests that as n increases, the value of $\left(1 + \frac{1}{n}\right)^n$ gets closer and closer to some fixed number. This is indeed the case. This fixed number is called e. ***(Note that in mathematics, e is a real number and not a variable.)***

Value of *e*

$$e \approx 2.718281828459045$$

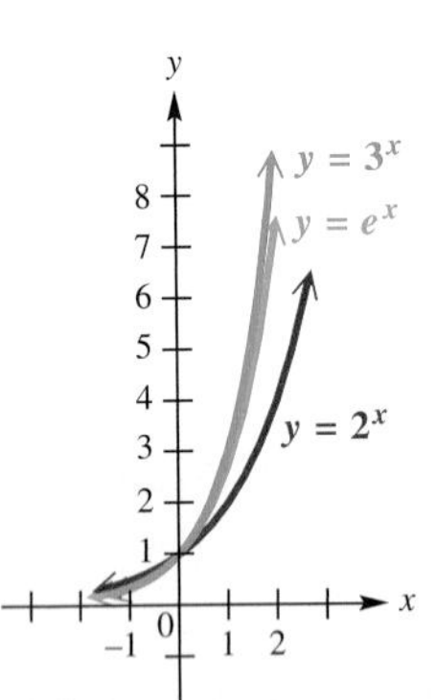

Figure 21

Figure 21 shows graphs of the functions

$$y = 2^x, \quad y = 3^x, \quad \text{and} \quad y = e^x.$$

Because $2 < e < 3$, the graph of $y = e^x$ lies "between" the other two graphs.

As mentioned above, the amount of interest earned increases with the frequency of compounding, but the value of the expression $\left(1 + \frac{1}{n}\right)^n$ approaches e as n gets larger. Consequently, the formula for compound interest approaches a limit as well, called the compound amount from **continuous compounding.**

Teaching Tip Tell students that as n increases without bound, $\left(1 + \frac{1}{n}\right)^n$ approaches e.

Continuous Compounding

If P dollars are deposited at a rate of interest r compounded continuously for t years, the compound amount A in dollars on deposit is given by the following formula.

$$A = Pe^{rt}$$

Classroom Example 9
Suppose \$8000 is deposited in an account paying 5% interest compounded continuously for 6 yr. Find the total amount on deposit at the end of 6 yr.

Answer: \$10,798.87

EXAMPLE 9 Solving a Continuous Compounding Problem

Suppose \$5000 is deposited in an account paying 3% interest compounded continuously for 5 yr. Find the total amount on deposit at the end of 5 yr.

SOLUTION

$A = Pe^{rt}$ — Continuous compounding formula

$= 5000e^{0.03(5)}$ — Let $P = 5000$, $r = 0.03$, and $t = 5$.

$= 5000e^{0.15}$ — Multiply exponents.

$A \approx 5809.17$ or \$5809.17 — Use a calculator.

Check that daily compounding would have produced a compound amount about \$0.03 less.

Now Try Exercise 87(b).

Classroom Example 10
In **Classroom Example 7,** we found the total amount on deposit after 10 yr in an account paying 6% semiannually in which \$2500 was invested. Find the amounts from the same investment for interest compounded quarterly, monthly, daily, and continuously.

Answer:
quarterly: \$4535.05;
monthly: \$4548.49;
daily: \$4555.07;
continuously: \$4555.30

EXAMPLE 10 Comparing Interest Earned as Compounding Is More Frequent

In **Example 7,** we found that \$1000 invested at 4% compounded quarterly for 10 yr grew to \$1488.86. Compare this same investment compounded annually, semiannually, monthly, daily, and continuously.

SOLUTION Substitute 0.04 for r, 10 for t, and the appropriate number of compounding periods for n into

$$A = P\left(1 + \frac{r}{n}\right)^{tn} \quad \text{Compound interest formula}$$

and also into

$$A = Pe^{rt}. \quad \text{Continuous compounding formula}$$

The results for amounts of \$1 and \$1000 are given in the table.

Compounded	\$1	\$1000
Annually	$(1 + 0.04)^{10} \approx 1.48024$	\$1480.24
Semiannually	$\left(1 + \frac{0.04}{2}\right)^{10(2)} \approx 1.48595$	\$1485.95
Quarterly	$\left(1 + \frac{0.04}{4}\right)^{10(4)} \approx 1.48886$	\$1488.86
Monthly	$\left(1 + \frac{0.04}{12}\right)^{10(12)} \approx 1.49083$	\$1490.83
Daily	$\left(1 + \frac{0.04}{365}\right)^{10(365)} \approx 1.49179$	\$1491.79
Continuously	$e^{10(0.04)} \approx 1.49182$	\$1491.82

Comparing the results, we notice the following.

- Compounding semiannually rather than annually increases the value of the account after 10 yr by \$5.71.
- Quarterly compounding grows to \$2.91 more than semiannual compounding after 10 yr.
- Daily compounding yields only \$0.96 more than monthly compounding.
- Continuous compounding yields only \$0.03 more than daily compounding.

Each increase in compounding frequency earns less additional interest.

Now Try Exercise 95.

LOOKING AHEAD TO CALCULUS
In calculus, the derivative allows us to determine the slope of a tangent line to the graph of a function. For the function $f(x) = e^x$, the derivative is the function f itself: $f'(x) = e^x$. Therefore, in calculus the exponential function with base e is much easier to work with than exponential functions having other bases.

Exponential Models The number e is important as the base of an exponential function in many practical applications. In situations involving growth or decay of a quantity, the amount or number present at time t often can be closely modeled by a function of the form

$$y = y_0e^{kt},$$

where y_0 is the amount or number present at time $t = 0$ and k is a constant.

Example 11 on the next page illustrates exponential growth. Further examples of exponential growth and decay are given in **Section 4.6.**

Classroom Example 11
Refer to the model given in **Example 11.**

(a) What will be the atmospheric carbon dioxide level in parts per million in 2015?

(b) Use a graph of this model to estimate when the carbon dioxide level will be double the level that it was in 2000.

Answers:
(a) 415 ppm (b) by 2112

EXAMPLE 11 Using Data to Model Exponential Growth

Data from recent past years indicate that future amounts of carbon dioxide in the atmosphere may grow according to the table. Amounts are given in parts per million.

Year	Carbon Dioxide (ppm)
1990	353
2000	375
2075	590
2175	1090
2275	2000

Source: International Panel on Climate Change (IPCC).

(a) Make a scatter diagram of the data. Do the carbon dioxide levels appear to grow exponentially?

(b) One model for the data is the function

$$y = 0.001942e^{0.00609x},$$

where x is the year and $1990 \leq x \leq 2275$. Use a graph of this model to estimate when future levels of carbon dioxide will double and triple over the preindustrial level of 280 ppm.

SOLUTION

(a) We show a calculator graph for the data in **Figure 22(a).** The data appear to resemble the graph of an increasing exponential function.

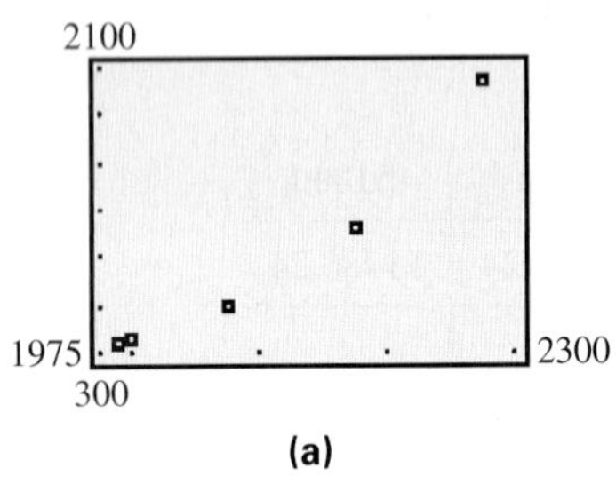

(a)

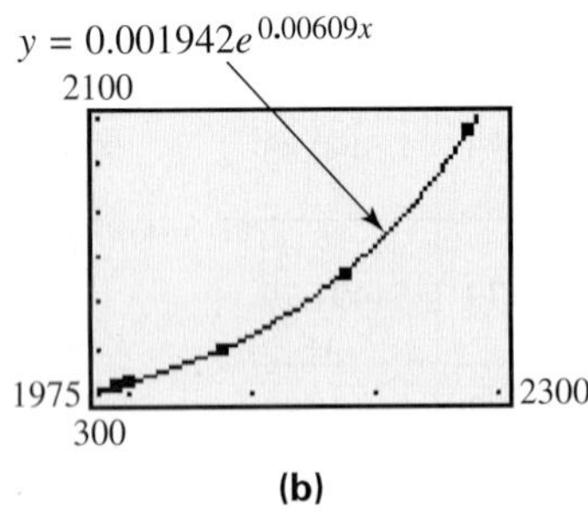

(b)

Figure 22

(b) A graph of $y = 0.001942e^{0.00609x}$ in **Figure 22(b)** shows that it is very close to the data points. We graph $y = 2 \cdot 280 = 560$ in **Figure 23(a)** and $y = 3 \cdot 280 = 840$ in **Figure 23(b)** on the same coordinate axes as the given function, and we use the calculator to find the intersection points.

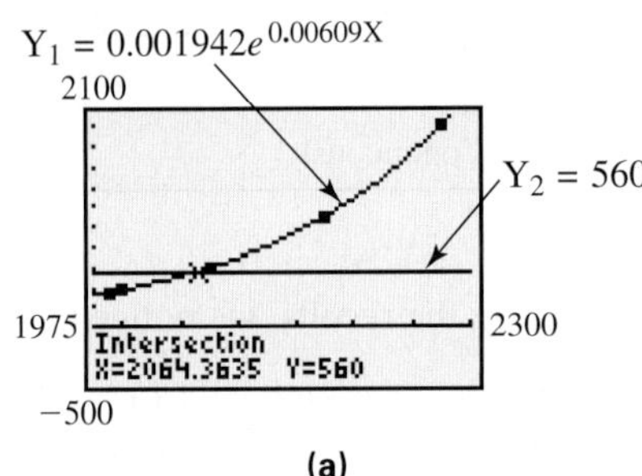

(a)

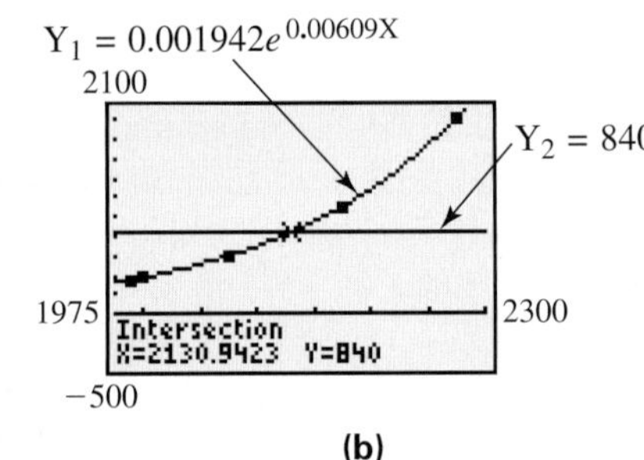

(b)

Figure 23

The graph of the function intersects the horizontal lines at approximately 2064.4 and 2130.9. According to this model, carbon dioxide levels will have doubled by 2064 and tripled by 2131.

✔ *Now Try Exercise 97.*

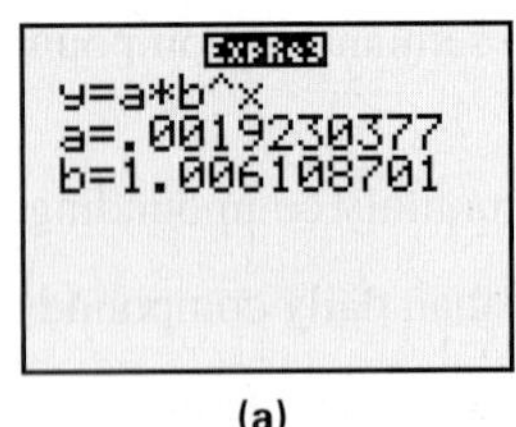

(a)

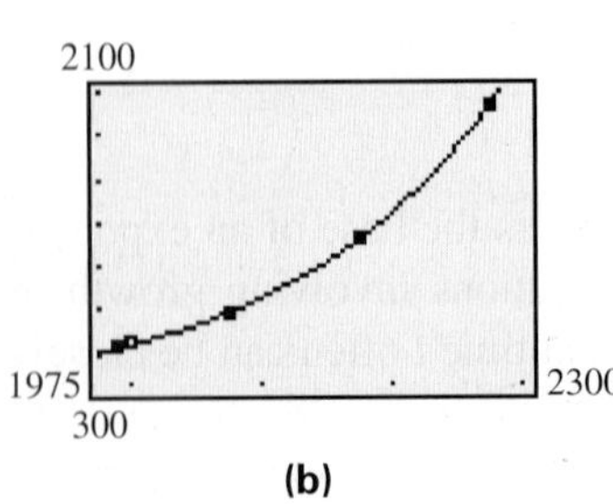

(b)

Figure 24

Graphing calculators are capable of fitting exponential curves to scatter diagrams like the one found in **Example 11. Figure 24(a)** shows how the TI-83/84 Plus displays another (different) equation for the atmospheric carbon dioxide example, approximated as follows.

$$y = 0.001923(1.006109)^x$$

Notice that this calculator form differs from the model in **Example 11. Figure 24(b)** shows the data points and the graph of this exponential regression equation. ■

4.2 Exercises

1. 9 2. 27 3. $\frac{1}{9}$ 4. $\frac{1}{27}$
5. $\frac{1}{16}$ 6. $\frac{1}{64}$ 7. 16 8. 64
9. $3\sqrt{3}$ 10. $\frac{\sqrt{3}}{27}$
11. $\frac{1}{8}$ 12. 32
13. 13.076 14. 0.158
15. 10.267 16. 0.039

17. $f(x) = 3^x$ 18. $f(x) = 4^x$

19. 20. $f(x) = \left(\frac{1}{4}\right)^x$

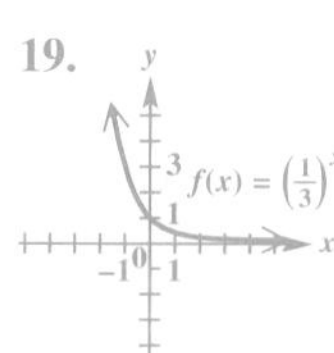

21. 22. $f(x) = \left(\frac{5}{3}\right)^x$

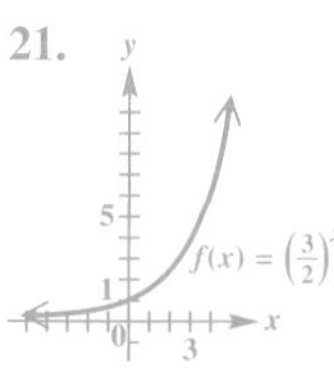

23. 24.

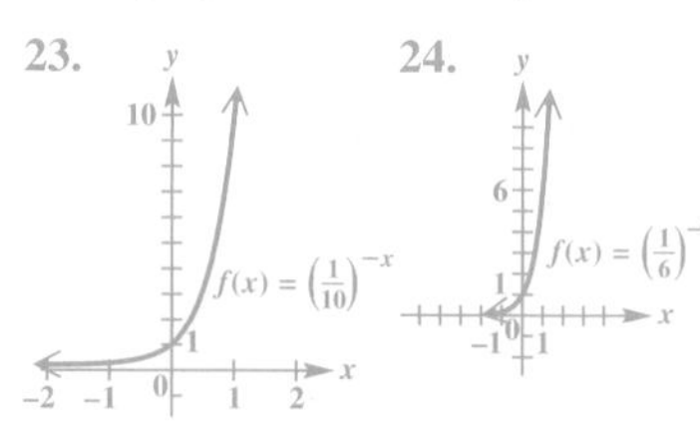

25. 26. $f(x) = 10^{-x}$

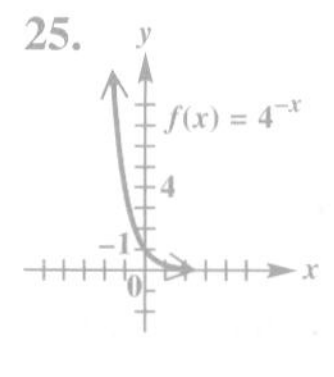

27. 28.

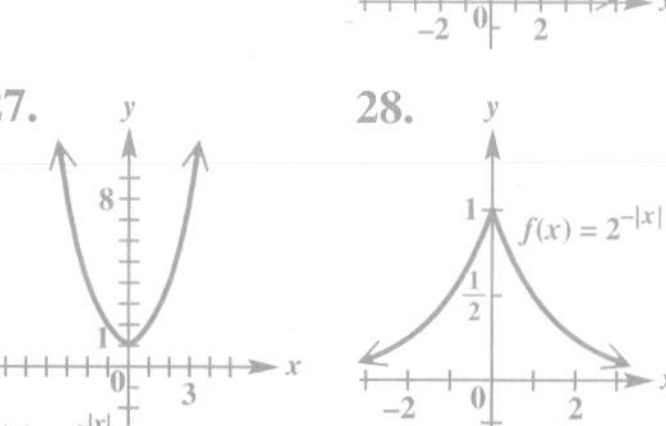

29. 30.

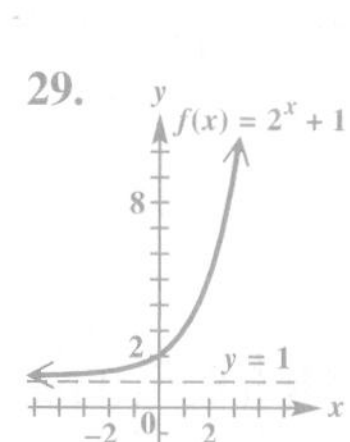

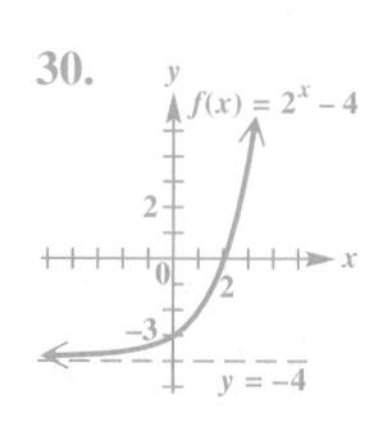

For $f(x) = 3^x$ and $g(x) = \left(\frac{1}{4}\right)^x$, find each of the following. In Exercises 13–16, round the answer to the nearest thousandth. ***See Example 1.***

1. $f(2)$
2. $f(3)$
3. $f(-2)$
4. $f(-3)$
5. $g(2)$
6. $g(3)$
7. $g(-2)$
8. $g(-3)$
9. $f\left(\frac{3}{2}\right)$
10. $f\left(-\frac{5}{2}\right)$
11. $g\left(\frac{3}{2}\right)$
12. $g\left(-\frac{5}{2}\right)$
13. $f(2.34)$
14. $f(-1.68)$
15. $g(-1.68)$
16. $g(2.34)$

Graph each function. ***See Example 2.***

17. $f(x) = 3^x$
18. $f(x) = 4^x$
19. $f(x) = \left(\frac{1}{3}\right)^x$
20. $f(x) = \left(\frac{1}{4}\right)^x$
21. $f(x) = \left(\frac{3}{2}\right)^x$
22. $f(x) = \left(\frac{5}{3}\right)^x$
23. $f(x) = \left(\frac{1}{10}\right)^{-x}$
24. $f(x) = \left(\frac{1}{6}\right)^{-x}$
25. $f(x) = 4^{-x}$
26. $f(x) = 10^{-x}$
27. $f(x) = 2^{|x|}$
28. $f(x) = 2^{-|x|}$

Sketch the graph of $f(x) = 2^x$. Then refer to it and use the techniques of ***Chapter 2*** *to graph each function as defined.* ***See Example 3.***

29. $f(x) = 2^x + 1$
30. $f(x) = 2^x - 4$
31. $f(x) = 2^{x+1}$
32. $f(x) = 2^{x-4}$
33. $f(x) = -2^{x+2}$
34. $f(x) = -2^{x-3}$
35. $f(x) = 2^{-x}$
36. $f(x) = -2^{-x}$
37. $f(x) = 2^{x-1} + 2$
38. $f(x) = 2^{x+3} + 1$
39. $f(x) = 2^{x+2} - 4$
40. $f(x) = 2^{x-3} - 1$

Sketch the graph of $f(x) = \left(\frac{1}{3}\right)^x$. Then refer to it and use the techniques of ***Chapter 2*** *to graph each function as defined.* ***See Example 3.***

41. $f(x) = \left(\frac{1}{3}\right)^x - 2$
42. $f(x) = \left(\frac{1}{3}\right)^x + 4$
43. $f(x) = \left(\frac{1}{3}\right)^{x+2}$
44. $f(x) = \left(\frac{1}{3}\right)^{x-4}$
45. $f(x) = \left(\frac{1}{3}\right)^{-x+1}$
46. $f(x) = \left(\frac{1}{3}\right)^{-x-2}$
47. $f(x) = \left(\frac{1}{3}\right)^{-x}$
48. $f(x) = -\left(\frac{1}{3}\right)^{-x}$
49. $f(x) = \left(\frac{1}{3}\right)^{x-2} + 2$
50. $f(x) = \left(\frac{1}{3}\right)^{x-1} + 3$
51. $f(x) = \left(\frac{1}{3}\right)^{x+2} - 1$
52. $f(x) = \left(\frac{1}{3}\right)^{x+3} - 2$

53. *Concept Check* Fill in the blank: The graph of $f(x) = a^{-x}$ is the same as that of $g(x) = (\_\_\_\_\_\_)^x$.

54. *Concept Check* Fill in the blanks: If $a > 1$, then the graph of $f(x) = a^x$ __________ (rises/falls) from left to right. If $0 < a < 1$, then the graph of $g(x) = a^x$ __________ (rises/falls) from left to right.

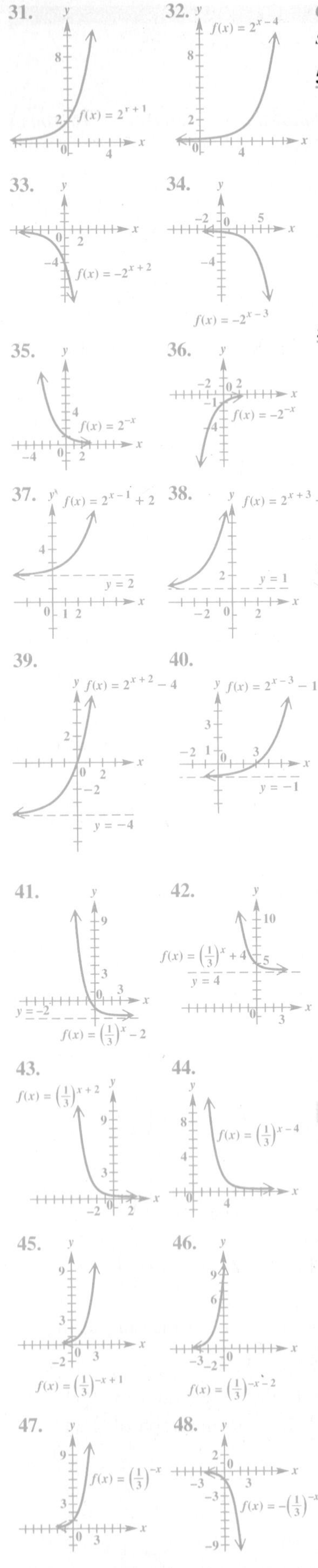

31. 32.

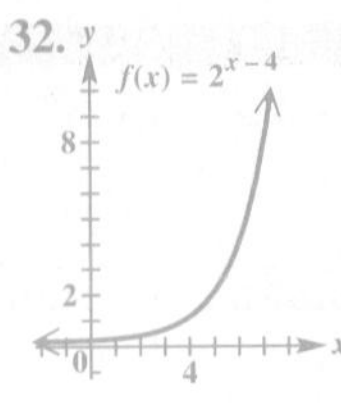
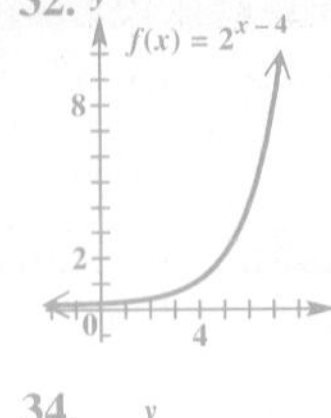

33. 34.

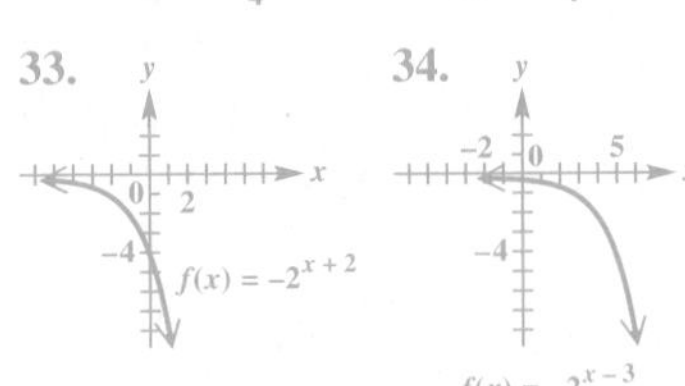

35. 36.

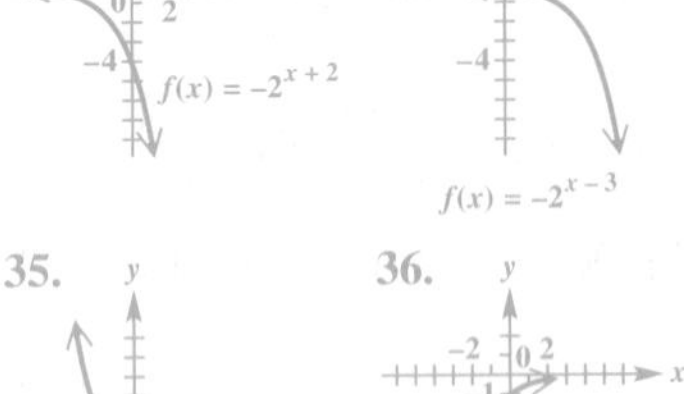

37. 38.

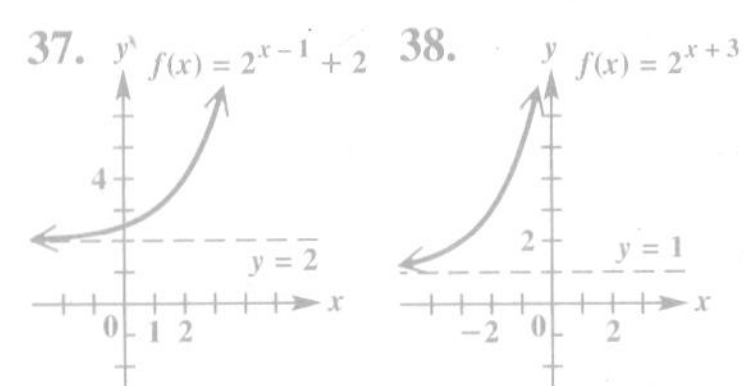

39. 40.

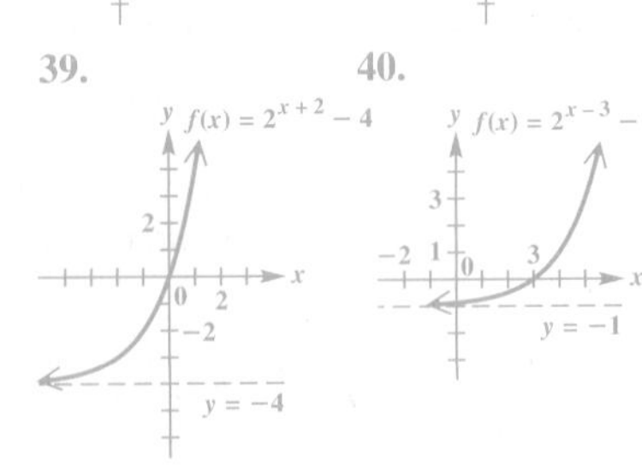

41. 42.

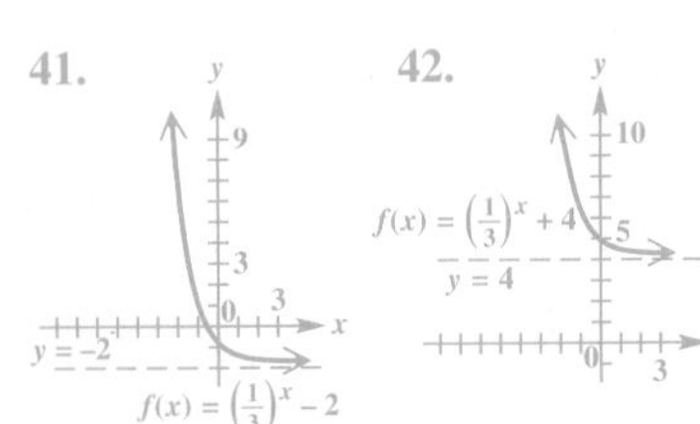
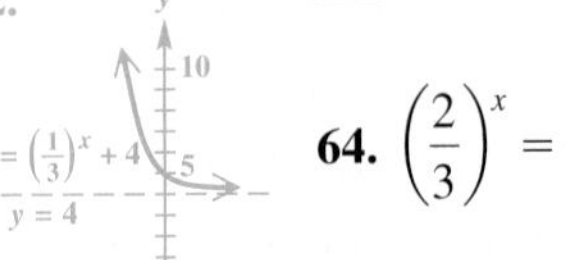

43. $f(x) = \left(\frac{1}{3}\right)^{x+2}$

44. $f(x) = \left(\frac{1}{3}\right)^{x-4}$

45. $f(x) = \left(\frac{1}{3}\right)^{-x+1}$

46. $f(x) = \left(\frac{1}{3}\right)^{-x-2}$

47. $f(x) = \left(\frac{1}{3}\right)^{-x}$

48. $f(x) = -\left(\frac{1}{3}\right)^{-x}$

Connecting Graphs with Equations *Write an equation for the graph given. Each represents an exponential function f, with base 2 or 3, translated and/or reflected.*

55.

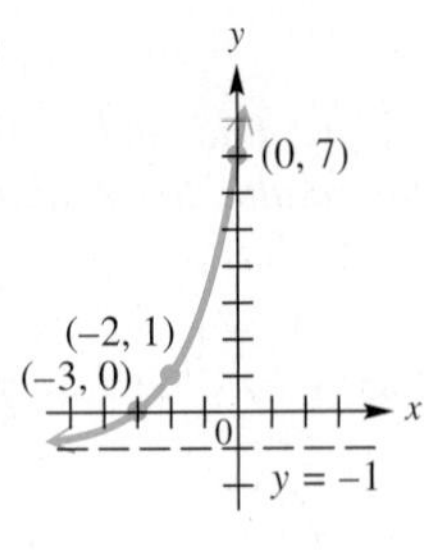

56.

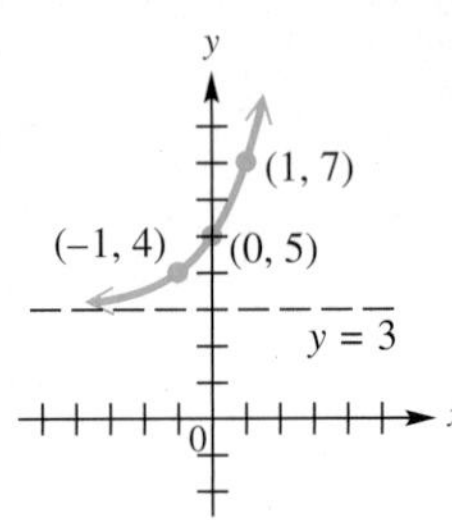

57.

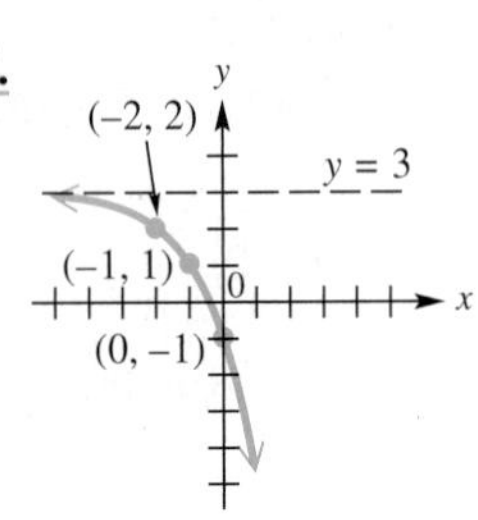

58.

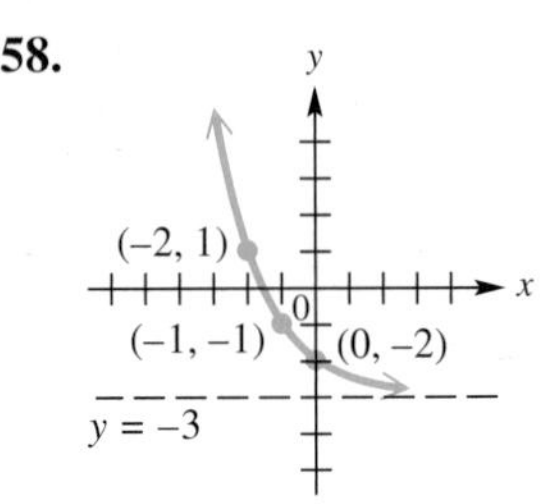

59.

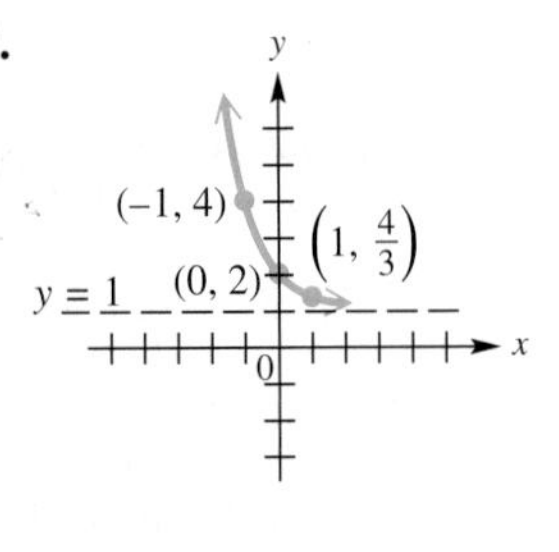

60.

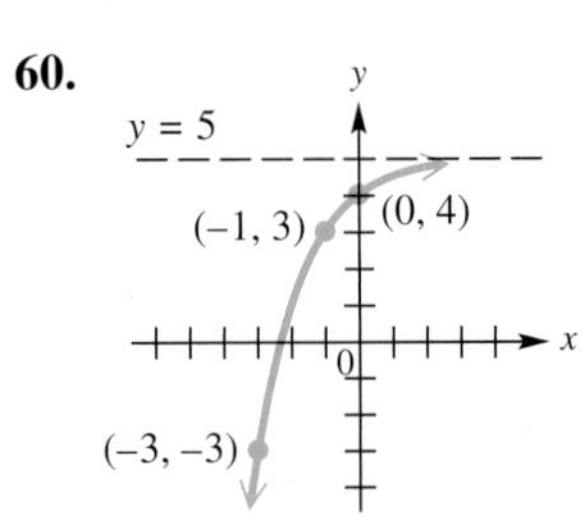

Solve each equation. ***See Examples 4–6.***

61. $4^x = 2$

62. $125^x = 5$

63. $\left(\frac{5}{2}\right)^x = \frac{4}{25}$

64. $\left(\frac{2}{3}\right)^x = \frac{9}{4}$

65. $2^{3-2x} = 8$

66. $5^{2+2x} = 25$

67. $e^{4x-1} = (e^2)^x$

68. $e^{3-x} = (e^3)^{-x}$

69. $27^{4x} = 9^{x+1}$

70. $32^{2x} = 16^{x-1}$

71. $4^{x-2} = 2^{3x+3}$

72. $2^{6-3x} = 8^{x+1}$

73. $\left(\frac{1}{e}\right)^{-x} = \left(\frac{1}{e^2}\right)^{x+1}$

74. $e^{x-1} = \left(\frac{1}{e^4}\right)^{x+1}$

75. $(\sqrt{2})^{x+4} = 4^x$

76. $(\sqrt[3]{5})^{-x} = \left(\frac{1}{5}\right)^{x+2}$

77. $\frac{1}{27} = x^{-3}$

78. $\frac{1}{32} = x^{-5}$

79. $x^{2/3} = 4$

80. $x^{2/5} = 16$

81. $x^{5/2} = 32$

82. $x^{3/2} = 27$

83. $x^{-6} = \frac{1}{64}$

84. $x^{-4} = \frac{1}{256}$

85. $x^{5/3} = -243$

86. $x^{7/5} = -128$

Solve each problem involving compound interest. ***See Examples 7–9.***

87. *Future Value* Find the future value and interest earned if \$8906.54 is invested for 9 yr at 5% compounded

(a) semiannually **(b)** continuously.

49. $f(x) = \left(\frac{1}{3}\right)^{x-2} + 2$

50. $f(x) = \left(\frac{1}{3}\right)^{x-1} + 3$

51.

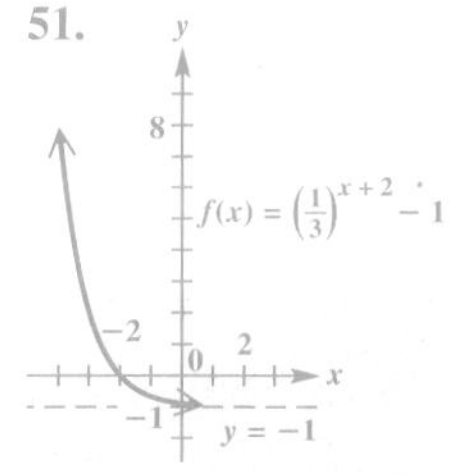

52.

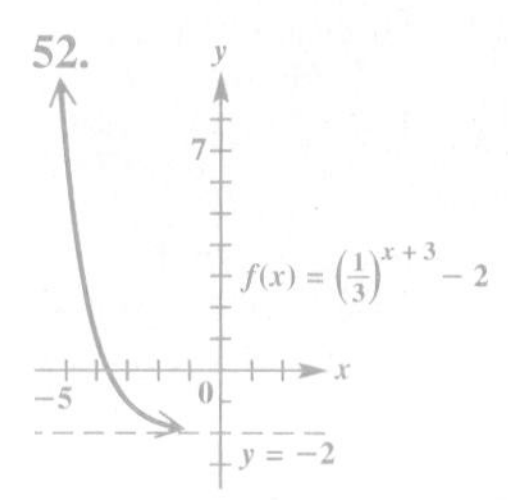

53. $\frac{1}{a}$ 54. rises; falls

55. $f(x) = 2^{x+3} - 1$
56. $f(x) = 2^{x+1} + 3$
57. $f(x) = -2^{x+2} + 3$
58. $f(x) = 2^{-x} - 3$
59. $f(x) = 3^{-x} + 1$
60. $f(x) = -2^{-x} + 5$

61. $\{\frac{1}{2}\}$ 62. $\{\frac{1}{3}\}$
63. $\{-2\}$ 64. $\{-2\}$
65. $\{0\}$ 66. $\{0\}$
67. $\{\frac{1}{2}\}$ 68. $\{-\frac{3}{2}\}$
69. $\{\frac{1}{5}\}$ 70. $\{-\frac{2}{3}\}$
71. $\{-7\}$ 72. $\{\frac{1}{2}\}$
73. $\{-\frac{2}{3}\}$ 74. $\{-\frac{3}{5}\}$
75. $\{\frac{4}{3}\}$ 76. $\{-3\}$
77. $\{3\}$ 78. $\{2\}$
79. $\{\pm 8\}$ 80. $\{\pm 1024\}$
81. $\{4\}$ 82. $\{9\}$
83. $\{\pm 2\}$ 84. $\{\pm 4\}$
85. $\{-27\}$ 86. $\{-32\}$

87. (a) \$13,891.16; \$4984.62 (b) \$13,968.24; \$5061.70
88. (a) \$76,855.95; \$20,075.95 (b) \$125,735.96; \$68,955.96
89. \$21,223.33 90. \$43,411.15
91. \$3528.81 92. 1.0%
93. 4.5% 94. 6.5%

95. Bank A (even though it has the greatest stated rate)

88. *Future Value* Find the future value and interest earned if \$56,780 is invested at 5.3% compounded

(a) quarterly for 23 quarters **(b)** continuously for 15 yr.

89. *Present Value* Find the present value that will grow to \$25,000 if interest is 6% compounded quarterly for 11 quarters.

90. *Present Value* Find the present value that will grow to \$45,000 if interest is 3.6% compounded monthly for 1 yr.

91. *Present Value* Find the present value that will grow to \$5000 if interest is 3.5% compounded quarterly for 10 yr.

92. *Interest Rate* Find the required annual interest rate to the nearest tenth of a percent for \$65,000 to grow to \$65,325 if interest is compounded monthly for 6 months.

93. *Interest Rate* Find the required annual interest rate to the nearest tenth of a percent for \$1200 to grow to \$1500 if interest is compounded quarterly for 5 yr.

94. *Interest Rate* Find the required annual interest rate to the nearest tenth of a percent for \$5000 to grow to \$8400 if interest is compounded quarterly for 8 yr.

*Solve each problem. **See Example 10.***

95. *Comparing Loans* Bank A is lending money at 6.4% interest compounded annually. The rate at Bank B is 6.3% compounded monthly, and the rate at Bank C is 6.35% compounded quarterly. At which bank will you pay the *least* interest?

96. *Future Value* Suppose \$10,000 is invested at an annual rate of 5% for 10 yr. Find the future value if interest is compounded as follows.

(a) annually **(b)** quarterly **(c)** monthly **(d)** daily (365 days)

*(Modeling) Solve each problem. **See Example 11.***

97. *Atmospheric Pressure* The atmospheric pressure (in millibars) at a given altitude (in meters) is shown in the table.

Altitude	Pressure	Altitude	Pressure
0	1013	6000	472
1000	899	7000	411
2000	795	8000	357
3000	701	9000	308
4000	617	10,000	265
5000	541		

Source: Miller, A. and J. Thompson, *Elements of Meteorology,* Fourth Edition, Charles E. Merrill Publishing Company, Columbus, Ohio.

(a) Use a graphing calculator to make a scatter diagram of the data for atmospheric pressure P at altitude x.

(b) Would a linear or an exponential function fit the data better?

(c) The function

$$P(x) = 1013e^{-0.0001341x}$$

approximates the data. Use a graphing calculator to graph P and the data on the same coordinate axes.

(d) Use P to predict the pressures at 1500 m and 11,000 m, and compare them to the actual values of 846 millibars and 227 millibars, respectively.

96. **(a)** \$16,288.95
(b) \$16,436.19
(c) \$16,470.09
(d) \$16,486.65

97. **(a)**

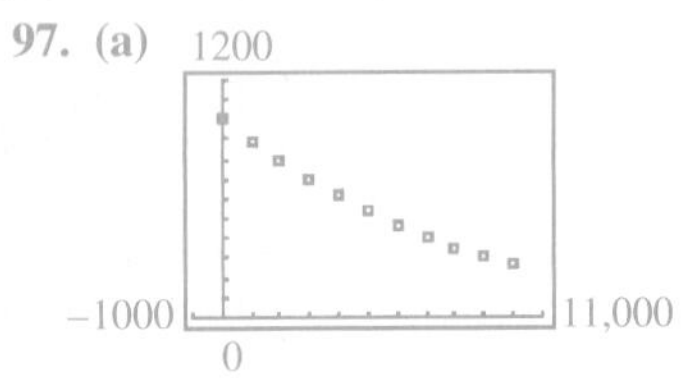

(b) exponential

(c)

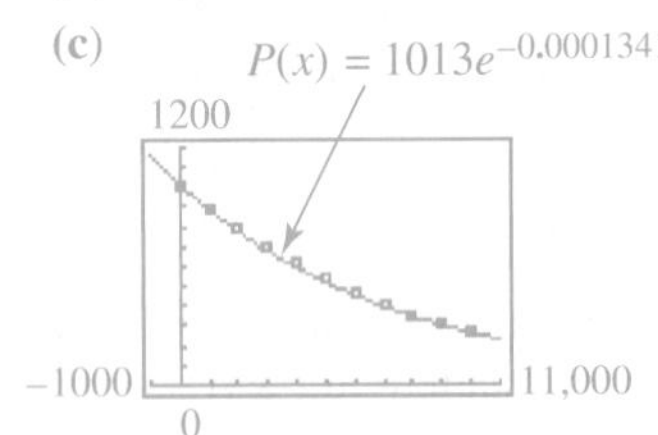

(d) $P(1500) \approx 828$ mb; $P(11{,}000) \approx 232$ mb

98. **(a)** The function gives approximately 6860 million, which differs by 7 million from the actual value.
(b) 7734 million
(c) 8720 million

99. **(a)** about 63,000
(b) about 42,000
(c) about 21,000

100. **(a)** about 207
(b) about 235
(c) about 249
(d) The number of symbols approaches 250.

101. $\{0.9\}$ **102.** $\{0, 0.7\}$
103. $\{-0.5, 1.3\}$ **104.** $\emptyset$

106. **(a)** 3 **(b)** $\frac{1}{3}$
(c) 9 **(d)** 1

107. $f(x) = 2^x$
108. $f(x) = 5^x$
109. $f(x) = \left(\frac{1}{4}\right)^x$
110. $f(x) = \left(\frac{1}{6}\right)^x$

111. $f(t) = 27 \cdot 9^t$
112. $f(t) = 4 \cdot 8^t$
113. $f(t) = \left(\frac{1}{3}\right)9^t$
114. $f(t) = \left(\frac{1}{2}\right)4^t$

---

98. *World Population Growth* Since 2000, world population in millions closely fits the exponential function

$$y = 6084e^{0.0120x},$$

where x is the number of years since 2000. (*Source:* U.S. Census Bureau.)

(a) The world population was about 6853 million in 2010. How closely does the function approximate this value?
(b) Use this model to predict the population in 2020.
(c) Use this model to predict the population in 2030.
(d) Explain why this model may not be accurate for 2030.

99. *Deer Population* The exponential growth of the deer population in Massachusetts can be calculated using the model

$$f(x) = 50{,}000(1 + 0.06)^x,$$

where 50,000 is the initial deer population and 0.06 is the rate of growth. $f(x)$ is the total population after x years have passed.

(a) Predict the total population after 4 yr.
(b) If the initial population was 30,000 and the growth rate was 0.12, approximately how many deer would be present after 3 yr?
(c) How many additional deer can we expect in 5 yr if the initial population is 45,000 and the current growth rate is 0.08?

100. *Employee Training* A person learning certain skills involving repetition tends to learn quickly at first. Then learning tapers off and skill acquisition approaches some upper limit. Suppose the number of symbols per minute that a person using a keyboard can type is given by

$$f(t) = 250 - 120(2.8)^{-0.5t},$$

where t is the number of months the operator has been in training. Find each value.

(a) $f(2)$ **(b)** $f(4)$ **(c)** $f(10)$
(d) What happens to the number of symbols per minute after several months of training?

Use a graphing calculator to find the solution set of each equation. Approximate the solution(s) to the nearest tenth.

101. $5e^{3x} = 75$ **102.** $6^{-x} = 1 - x$ **103.** $3x + 2 = 4^x$ **104.** $x = 2^x$

105. A function of the form $f(x) = x^r$, where r is a constant, is called a **power function.** Discuss the difference between an exponential function and a power function.

106. *Concept Check* If $f(x) = a^x$ and $f(3) = 27$, find each value of $f(x)$.

(a) $f(1)$ **(b)** $f(-1)$ **(c)** $f(2)$ **(d)** $f(0)$

Concept Check Give an equation of the form $f(x) = a^x$ to define the exponential function whose graph contains the given point.

107. $(3, 8)$ **108.** $(3, 125)$ **109.** $(-3, 64)$ **110.** $(-2, 36)$

Concept Check Use properties of exponents to write each function in the form $f(t) = ka^t$, where k is a constant. (Hint: Recall that $a^{x+y} = a^x \cdot a^y$.)

111. $f(t) = 3^{2t+3}$ **112.** $f(t) = 2^{3t+2}$ **113.** $f(t) = \left(\frac{1}{3}\right)^{1-2t}$ **114.** $f(t) = \left(\frac{1}{2}\right)^{1-2t}$

115. 2.717 (A calculator gives 2.718.)
116. 0.9512 (A calculator gives 0.9512.)

117. yes; an inverse function
118.

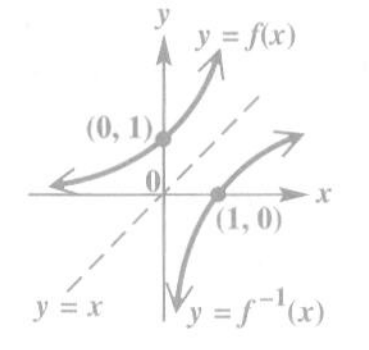

119. $x = a^y$ 120. $x = 10^y$
121. $x = e^y$ 122. (q, p)

In calculus, it is shown that

$$e^x = 1 + x + \frac{x^2}{2 \cdot 1} + \frac{x^3}{3 \cdot 2 \cdot 1} + \frac{x^4}{4 \cdot 3 \cdot 2 \cdot 1} + \frac{x^5}{5 \cdot 4 \cdot 3 \cdot 2 \cdot 1} + \cdots.$$

By using more terms, one can obtain a more accurate approximation for e^x.

115. Use the terms shown, and replace x with 1 to approximate $e^1 = e$ to three decimal places. Check your result with a calculator.

116. Use the terms shown, and replace x with -0.05 to approximate $e^{-0.05}$ to four decimal places. Check your result with a calculator.

Relating Concepts

For individual or collaborative investigation *(Exercises 117–122)*

Assume $f(x) = a^x$, *where* $a > 1$. ***Work these exercises in order.***

117. Is f a one-to-one function? If so, based on **Section 4.1,** what kind of related function exists for f?

118. If f has an inverse function f^{-1}, sketch f and f^{-1} on the same set of axes.

119. If f^{-1} exists, find an equation for $y = f^{-1}(x)$ using the method described in **Section 4.1.** You need not solve for y.

120. If $a = 10$, what is the equation for $y = f^{-1}(x)$? (You need not solve for y.)

121. If $a = e$, what is the equation for $y = f^{-1}(x)$? (You need not solve for y.)

122. If the point (p, q) is on the graph of f, then the point ________ is on the graph of f^{-1}.

4.3 Logarithmic Functions

- Logarithms
- Logarithmic Equations
- Logarithmic Functions
- Properties of Logarithms

Logarithms The previous section dealt with exponential functions of the form $y = a^x$ for all positive values of a, where $a \neq 1$. The horizontal line test shows that exponential functions are one-to-one and thus have inverse functions. The equation defining the inverse of a function is found by interchanging x and y in the equation that defines the function. Starting with $y = a^x$ and interchanging x and y yields

$$x = a^y.$$

Here y is the exponent to which a must be raised in order to obtain x. We call this exponent a **logarithm,** symbolized by the abbreviation **"log."** The expression $\log_a x$ represents the logarithm in this discussion. The number a is called the **base** of the logarithm, and x is called the **argument** of the expression. It is read **"logarithm with base *a* of *x*,"** or **"logarithm of *x* with base *a*,"** or **"base *a* logarithm of *x*."**

Logarithm

For all real numbers y and all positive numbers a and x, where $a \neq 1$,

$$y = \log_a x \quad \text{is equivalent to} \quad x = a^y.$$

The expression $\log_a x$ ***represents the exponent to which the base a must be raised in order to obtain x.***

Classroom Example 1
Complete the missing equivalent forms in the table.

	Logarithmic Form	Exponential Form
(a)		$3^4 = 81$
(b)	$\log_{1/2} 8 = -3$	
(c)		$10^3 = 1000$
(d)	$\log_5 \frac{1}{125} = -3$	
(e)		$12^1 = 12$
(f)	$\log_6 1 = 0$	

Answers:
(a) $\log_3 81 = 4$ (b) $\left(\frac{1}{2}\right)^{-3} = 8$
(c) $\log_{10} 1000 = 3$ (d) $5^{-3} = \frac{1}{125}$
(e) $\log_{12} 12 = 1$ (f) $6^0 = 1$

Teaching Tip Have students verbalize the statements in the left column of the table in **Example 1.** For instance, "$\log_2 8$ is the exponent to which 2 must be raised in order to obtain 8."

EXAMPLE 1 Writing Equivalent Logarithmic and Exponential Forms

The table shows several pairs of equivalent statements, written in both logarithmic and exponential forms.

SOLUTION

Logarithmic Form	Exponential Form
$\log_2 8 = 3$	$2^3 = 8$
$\log_{1/2} 16 = -4$	$\left(\frac{1}{2}\right)^{-4} = 16$
$\log_{10} 100{,}000 = 5$	$10^5 = 100{,}000$
$\log_3 \frac{1}{81} = -4$	$3^{-4} = \frac{1}{81}$
$\log_5 5 = 1$	$5^1 = 5$
$\log_{3/4} 1 = 0$	$\left(\frac{3}{4}\right)^0 = 1$

To remember the relationships among a, x, and y in the two equivalent forms $y = \log_a x$ and $x = a^y$, refer to these diagrams.

A logarithm is an exponent.

Exponent ↓
Logarithmic form: $y = \log_a x$
↑ Base

Exponent ↓
Exponential form: $a^y = x$
↑ Base

✔ *Now Try Exercises 3, 5, 7, and 9.*

Logarithmic Equations The definition of logarithm can be used to solve a **logarithmic equation,** which is an equation with a logarithm in at least one term. Many logarithmic equations can be solved by first writing the equation in exponential form.

Classroom Example 2
Solve each equation.

(a) $\log_x \frac{16}{9} = -2$

(b) $\log_{16} x = \frac{3}{4}$

(c) $\log_{36} \sqrt{6} = x$

Answers:
(a) $\left\{\frac{3}{4}\right\}$ (b) $\{8\}$ (c) $\left\{\frac{1}{4}\right\}$

Teaching Tip Point out that **Example 2** illustrates three different locations for the variable: the base, the argument, and the exponent. Remind students of the restrictions on x in parts (a) and (b), and that there is no restriction on x in part (c).

EXAMPLE 2 Solving Logarithmic Equations

Solve each equation.

(a) $\log_x \frac{8}{27} = 3$ **(b)** $\log_4 x = \frac{5}{2}$ **(c)** $\log_{49} \sqrt[3]{7} = x$

SOLUTION

(a)
$$\log_x \frac{8}{27} = 3$$

$$x^3 = \frac{8}{27} \quad \text{Write in exponential form.}$$

$$x^3 = \left(\frac{2}{3}\right)^3 \quad \frac{8}{27} = \left(\frac{2}{3}\right)^3 \text{ (Section R.2)}$$

$$x = \frac{2}{3} \quad \text{Take cube roots. (Section 1.6)}$$

CHECK

$$\log_x \frac{8}{27} = 3 \quad \text{Original equation}$$

$$\log_{2/3} \frac{8}{27} \stackrel{?}{=} 3 \quad \text{Let } x = \tfrac{2}{3}.$$

$$\left(\frac{2}{3}\right)^3 \stackrel{?}{=} \frac{8}{27} \quad \text{Write in exponential form.}$$

$$\frac{8}{27} = \frac{8}{27} \ ✓ \quad \text{True}$$

The solution set is $\left\{\frac{2}{3}\right\}$.

(b)

$$\log_4 x = \frac{5}{2}$$

$$4^{5/2} = x \quad \text{Write in exponential form.}$$

$$(4^{1/2})^5 = x \quad a^{mn} = (a^m)^n \text{ (Section R.3)}$$

$$2^5 = x \quad 4^{1/2} = (2^2)^{1/2} = 2$$

$$32 = x \quad \text{Apply the exponent.}$$

A check shows that the solution set is $\{32\}$.

(c)

$$\log_{49} \sqrt[3]{7} = x$$

$$49^x = \sqrt[3]{7} \quad \text{Write in exponential form.}$$

$$(7^2)^x = 7^{1/3} \quad \text{Write with the same base.}$$

$$7^{2x} = 7^{1/3} \quad \text{Power rule for exponents}$$

$$2x = \frac{1}{3} \quad \text{Set exponents equal.}$$

$$x = \frac{1}{6} \quad \text{Divide by 2.}$$

A check shows that the solution set is $\{\frac{1}{6}\}$.

✔ *Now Try Exercises 15, 27, and 29.*

Logarithmic Functions We define the logarithmic function with base a as follows.

Logarithmic Function

If $a > 0$, $a \neq 1$, and $x > 0$, then

$$f(x) = \log_a x$$

defines the **logarithmic function with base *a*.**

Exponential and logarithmic functions are inverses of each other. The graph of $y = 2^x$ is shown in red in **Figure 25(a).** The graph of its inverse is found by reflecting the graph of $y = 2^x$ across the line $y = x$. The graph of the inverse function, defined by $y = \log_2 x$, shown in blue, has the y-axis as a vertical asymptote. **Figure 25(b)** shows a calculator graph of the two functions.

x	2^x
-2	0.25
-1	0.5
0	1
1	2
2	4

x	$\log_2 x$
0.25	-2
0.5	-1
1	0
2	1
4	2

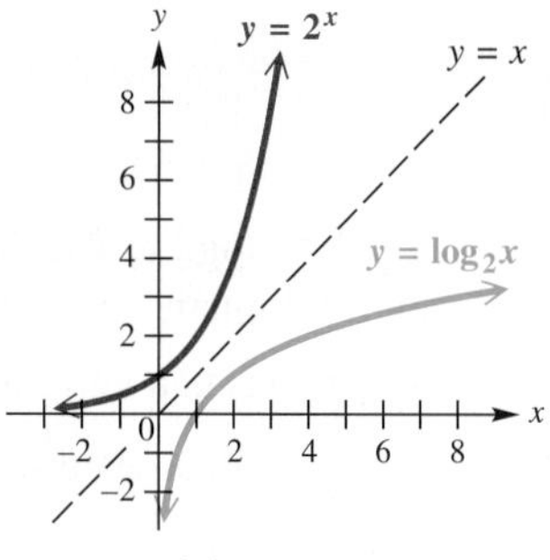

(a)

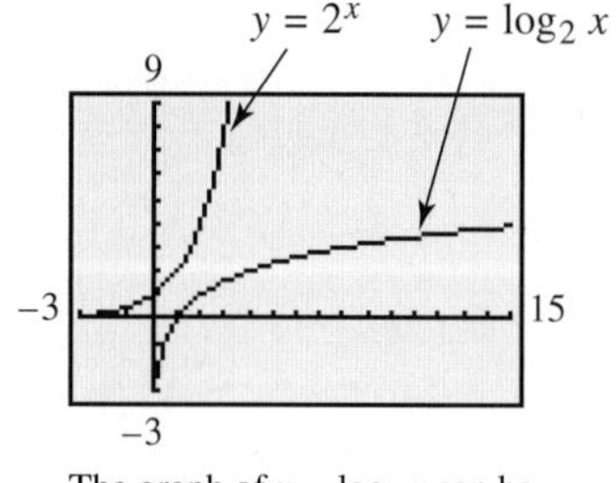

The graph of $y = \log_2 x$ can be obtained by drawing the inverse of $y = 2^x$.

(b)

Figure 25

Since the domain of an exponential function is the set of all real numbers, the range of a logarithmic function also will be the set of all real numbers. In the same way, both the range of an exponential function and the domain of a logarithmic function are the set of all positive real numbers.

Thus, logarithms can be found for positive numbers only.

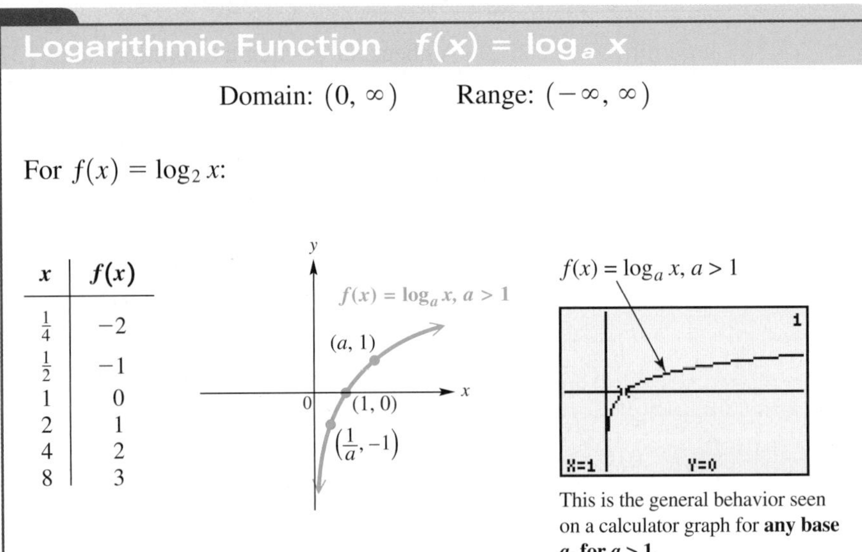

x	$f(x)$
$\frac{1}{4}$	-2
$\frac{1}{2}$	-1
1	0
2	1
4	2
8	3

This is the general behavior seen on a calculator graph for **any base *a*, for *a* > 1.**

Figure 26

- $f(x) = \log_a x$**, for** $a > 1$**,** is increasing and continuous on its entire domain, $(0, \infty)$.
- The y-axis is a vertical asymptote as $x \to 0$ from the right.
- The graph passes through the points $\left(\frac{1}{a}, -1\right)$, $(1, 0)$, and $(a, 1)$.

For $f(x) = \log_{1/2} x$:

x	$f(x)$
$\frac{1}{4}$	2
$\frac{1}{2}$	1
1	0
2	-1
4	-2
8	-3

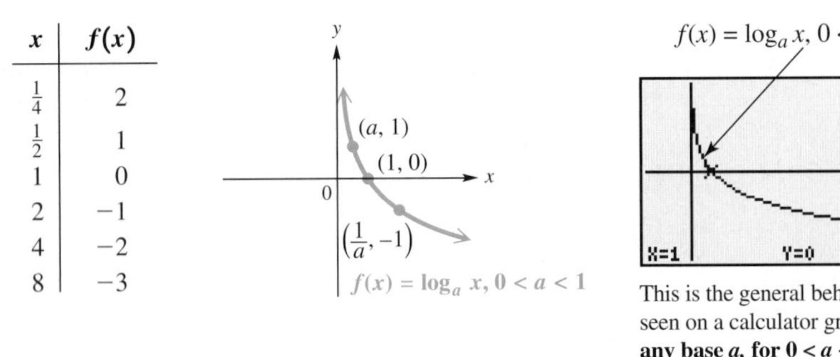

This is the general behavior seen on a calculator graph for **any base *a*, for 0 < *a* < 1.**

Figure 27

- $f(x) = \log_a x$**, for** $0 < a < 1$**,** is decreasing and continuous on its entire domain, $(0, \infty)$.
- The y-axis is a vertical asymptote as $x \to 0$ from the right.
- The graph passes through the points $\left(\frac{1}{a}, -1\right)$, $(1, 0)$, and $(a, 1)$.

Calculator graphs of logarithmic functions do not, in general, give an accurate picture of the behavior of the graphs near the vertical asymptotes. While it may seem as if the graph has an endpoint, this is not the case. The resolution of the calculator screen is not precise enough to indicate that the graph approaches the vertical asymptote as the value of x gets closer to it. Do not draw incorrect conclusions just because the calculator does not show this behavior. ■

The graphs in **Figures 26 and 27** and the information with them suggest the following generalizations about the graphs of logarithmic functions of the form $f(x) = \log_a x$.

Teaching Tip Encourage students to learn the characteristics of the graph of

$$f(x) = \log_a x.$$

Ask them how the coordinates $\left(\frac{1}{a}, -1\right)$, $(1, 0)$, and $(a, 1)$ are affected by different translations of the graph. Also, emphasize that horizontal shifts affect the domains of logarithmic functions.

Characteristics of the Graph of $f(x) = \log_a x$

1. The points $\left(\frac{1}{a}, -1\right)$, $(1, 0)$, and $(a, 1)$ are on the graph.
2. If $a > 1$, then f is an increasing function. If $0 < a < 1$, then f is a decreasing function.
3. The y-axis is a vertical asymptote.
4. The domain is $(0, \infty)$, and the range is $(-\infty, \infty)$.

Classroom Example 3
Graph each function.

(a) $f(x) = \log_{1/3} x$

(b) $f(x) = \log_5 x$

Answers:

(a) (b)

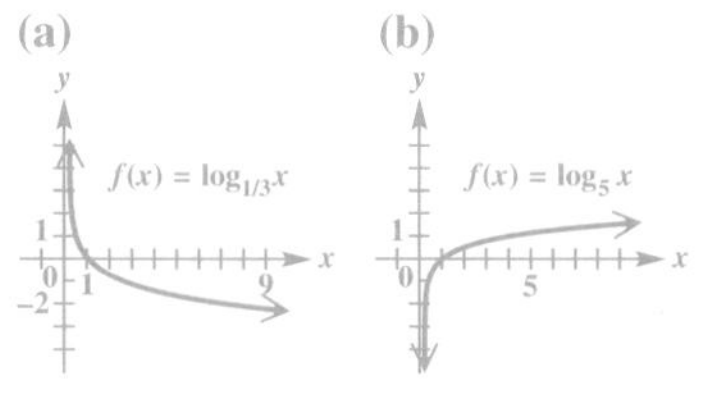

EXAMPLE 3 Graphing Logarithmic Functions

Graph each function.

(a) $f(x) = \log_{1/2} x$ **(b)** $f(x) = \log_3 x$

SOLUTION

(a) One approach is to first graph $y = \left(\frac{1}{2}\right)^x$, which defines the inverse function of f, by plotting points. Some ordered pairs are given in the table with the graph shown in red in **Figure 28.** The graph of $f(x) = \log_{1/2} x$ is the reflection of the graph of $y = \left(\frac{1}{2}\right)^x$ across the line $y = x$. The ordered pairs for $y = \log_{1/2} x$ are found by interchanging the x- and y-values in the ordered pairs for $y = \left(\frac{1}{2}\right)^x$. See the graph in blue in **Figure 28.**

x	$y = \left(\frac{1}{2}\right)^x$
-2	4
-1	2
0	1
1	$\frac{1}{2}$
2	$\frac{1}{4}$
4	$\frac{1}{16}$

x	$f(x) = \log_{1/2} x$
4	-2
2	-1
1	0
$\frac{1}{2}$	1
$\frac{1}{4}$	2
$\frac{1}{16}$	4

Figure 28

x	$f(x) = \log_3 x$
$\frac{1}{3}$	-1
1	0
3	1
9	2

Think: $x = 3^y$

Figure 29

(b) Another way to graph a logarithmic function is to write $f(x) = y = \log_3 x$ in exponential form as $x = 3^y$, and then select y-values and calculate corresponding x-values. Several selected ordered pairs are shown in the table for the graph in **Figure 29.**

Now Try Exercise 51.

CAUTION If you write a logarithmic function in exponential form to graph, as in **Example 3(b),** start *first* with y-values to calculate corresponding x-values. ***Be careful to write the values in the ordered pairs in the correct order.***

More general logarithmic functions can be obtained by forming the composition of $f(x) = \log_a x$ with a function $g(x)$. For example, if $f(x) = \log_2 x$ and $g(x) = x - 1$, then

$$(f \circ g)(x) = f(g(x)) = \log_2(x - 1). \quad \text{(Section 2.8)}$$

The next example shows how to graph such functions.

Classroom Example 4
Graph each function. Give the domain and range.
(a) $f(x) = \log_3(x + 2)$
(b) $f(x) = (\log_4 x) + 2$
(c) $f(x) = \log_2(x - 3) + 2$

Answers:
(a) $(-2, \infty)$; $(-\infty, \infty)$

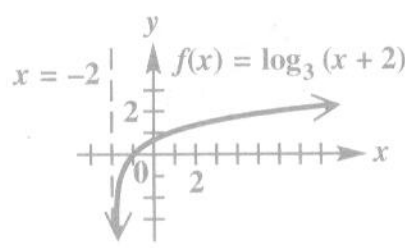

(b) $(0, \infty)$; $(-\infty, \infty)$

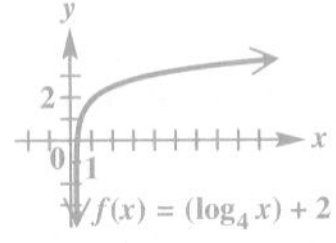

(c) $(3, \infty)$; $(-\infty, \infty)$

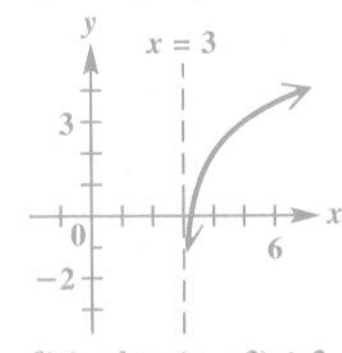

EXAMPLE 4 Graphing Translated Logarithmic Functions

Graph each function. Give the domain and range.

(a) $f(x) = \log_2(x - 1)$ **(b)** $f(x) = (\log_3 x) - 1$

(c) $f(x) = \log_4(x + 2) + 1$

SOLUTION

(a) The graph of $f(x) = \log_2(x - 1)$ is the graph of $f(x) = \log_2 x$ translated 1 unit to the right. The vertical asymptote has equation $x = 1$. Since logarithms can be found only for positive numbers, we solve $x - 1 > 0$ to find the domain, $(1, \infty)$. To determine ordered pairs to plot, use the equivalent exponential form of the equation $y = \log_2(x - 1)$.

$$y = \log_2(x - 1)$$

$$x - 1 = 2^y \quad \text{Write in exponential form.}$$

$$x = 2^y + 1 \quad \text{Add 1.}$$

We first choose values for y and then calculate each of the corresponding x-values. The range is $(-\infty, \infty)$. See **Figure 30.**

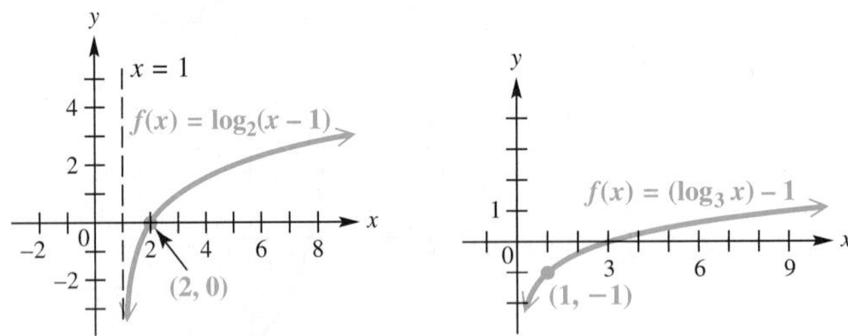

Figure 30 **Figure 31**

(b) The function $f(x) = (\log_3 x) - 1$ has the same graph as $g(x) = \log_3 x$ translated 1 unit down. We find ordered pairs to plot by writing the equation $y = (\log_3 x) - 1$ in exponential form.

$$y = (\log_3 x) - 1$$

$$y + 1 = \log_3 x \quad \text{Add 1.}$$

$$x = 3^{y+1} \quad \text{Write in exponential form.}$$

Again, choose y-values and calculate the corresponding x-values. The graph is shown in **Figure 31.** The domain is $(0, \infty)$, and the range is $(-\infty, \infty)$.

(c) The graph of $f(x) = \log_4(x + 2) + 1$ is obtained by shifting the graph of $y = \log_4 x$ to the left 2 units and up 1 unit. The domain is found by solving $x + 2 > 0$, which yields $(-2, \infty)$. The vertical asymptote has been shifted to the left 2 units as well, and it has equation $x = -2$. The range is unaffected by the vertical shift and remains $(-\infty, \infty)$. See **Figure 32.**

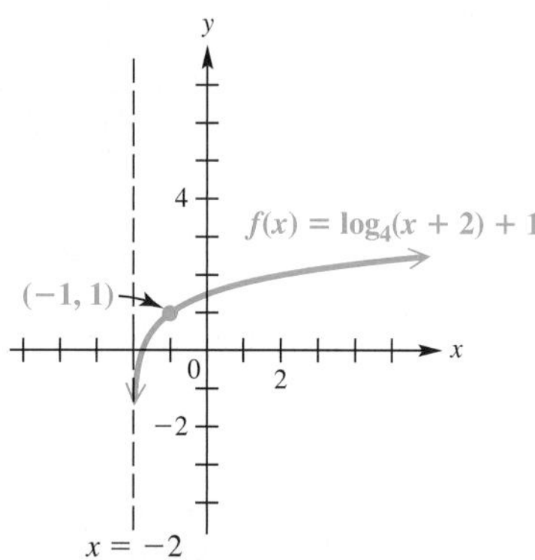

Figure 32

✔ *Now Try Exercises 39, 43, and 55.*

NOTE If we are given a graph such as the one in **Figure 31** and are asked to find its equation, we could reason as follows: The point $(1, 0)$ on the basic logarithmic graph has been shifted *down* 1 unit, and the point $(3, 0)$ on the given graph is 1 unit lower than $(3, 1)$, which is on the graph of $y = \log_3 x$. Thus, the equation will be

$$y = (\log_3 x) - 1.$$

Teaching Tip Encourage students to memorize the properties of logarithms.

Properties of Logarithms The properties of logarithms enable us to change the form of logarithmic statements so that products can be converted to sums, quotients can be converted to differences, and powers can be converted to products.

Teaching Tip Point out that the phrase "quotients can be converted to differences" refers to the quotient property. It does **NOT** say that

$$\frac{\log_a x}{\log_a y} = \log_a x - \log_a y$$

or

$$\frac{\log_a x}{\log_a y} = \log_a (x - y),$$

which are **COMMON ERRORS** that students make. A quotient or product of logarithms cannot be simplified by using the properties of logarithms given in the box.

Properties of Logarithms

For $x > 0$, $y > 0$, $a > 0$, $a \neq 1$, and any real number r, the following properties hold.

Property	Description
Product Property $\log_a xy = \log_a x + \log_a y$	The logarithm of the product of two numbers is equal to the sum of the logarithms of the numbers.
Quotient Property $\log_a \frac{x}{y} = \log_a x - \log_a y$	The logarithm of the quotient of two numbers is equal to the difference between the logarithms of the numbers.
Power Property $\log_a x^r = r \log_a x$	The logarithm of a number raised to a power is equal to the exponent multiplied by the logarithm of the number.
Logarithm of 1 $\log_a 1 = 0$	The base a logarithm of 1 is 0.
Base *a* Logarithm of *a* $\log_a a = 1$	The base a logarithm of a is 1.

LOOKING AHEAD TO CALCULUS
A technique called **logarithmic differentiation,** which uses the properties of logarithms, can often be used to differentiate complicated functions.

Proof To prove the product property, let $m = \log_a x$ and $n = \log_a y$.

$$\log_a x = m \quad \text{means} \quad a^m = x$$
$$\log_a y = n \quad \text{means} \quad a^n = y$$

Now consider the product xy.

$$xy = a^m \cdot a^n \qquad \text{Substitute.}$$
$$xy = a^{m+n} \qquad \text{Product rule for exponents (Section R.3)}$$
$$\log_a xy = m + n \qquad \text{Write in logarithmic form.}$$
$$\log_a xy = \log_a x + \log_a y \qquad \text{Substitute.}$$

The last statement is the result we wished to prove. The quotient and power properties are proved similarly. (See **Exercises 103 and 104.**)

Classroom Example 5
Rewrite each expression. Assume all variables represent positive real numbers, with $a \neq 1$ and $b \neq 1$.

(a) $\log_7(8 \cdot 6)$ (b) $\log_6 \frac{12}{5}$

(c) $\log_2 \sqrt[3]{9}$ (d) $\log_b \frac{rs^2t}{u^3v^5}$

(e) $\log_a \sqrt[5]{r^3}$ (f) $\log_a \sqrt[m]{\frac{r^3s^2}{t^4}}$

Answers:
(a) $\log_7 8 + \log_7 6$
(b) $\log_6 12 - \log_6 5$
(c) $\frac{1}{3} \log_2 9$
(d) $\log_b r + 2 \log_b s + \log_b t - 3 \log_b u - 5 \log_b v$
(e) $\frac{3}{5} \log_a r$
(f) $\frac{3}{m} \log_a r + \frac{2}{m} \log_a s - \frac{4}{m} \log_a t$

EXAMPLE 5 Using the Properties of Logarithms

Rewrite each expression. Assume all variables represent positive real numbers, with $a \neq 1$ and $b \neq 1$.

(a) $\log_6(7 \cdot 9)$ **(b)** $\log_9 \frac{15}{7}$ **(c)** $\log_5 \sqrt{8}$

(d) $\log_a \frac{mnq}{p^2t^4}$ **(e)** $\log_a \sqrt[3]{m^2}$ **(f)** $\log_b \sqrt[n]{\frac{x^3y^5}{z^m}}$

SOLUTION

(a) $\log_6(7 \cdot 9) = \log_6 7 + \log_6 9$ Product property

(b) $\log_9 \frac{15}{7} = \log_9 15 - \log_9 7$ Quotient property

(c) $\log_5 \sqrt{8} = \log_5(8^{1/2}) = \frac{1}{2} \log_5 8$ Power property

(d) $\log_a \frac{mnq}{p^2t^4} = \log_a m + \log_a n + \log_a q - (\log_a p^2 + \log_a t^4)$ Product and quotient properties

Use parentheses to avoid errors.

$= \log_a m + \log_a n + \log_a q - (2 \log_a p + 4 \log_a t)$ Power property

$= \log_a m + \log_a n + \log_a q - 2 \log_a p - 4 \log_a t$

Be careful with signs.

(e) $\log_a \sqrt[3]{m^2} = \log_a m^{2/3} = \frac{2}{3} \log_a m$

(f) $\log_b \sqrt[n]{\frac{x^3y^5}{z^m}} = \log_b \left(\frac{x^3y^5}{z^m}\right)^{1/n}$ $\sqrt[n]{a} = a^{1/n}$ (Section R.7)

$= \frac{1}{n} \log_b \frac{x^3y^5}{z^m}$ Power property

$= \frac{1}{n}(\log_b x^3 + \log_b y^5 - \log_b z^m)$ Product and quotient properties

$= \frac{1}{n}(3 \log_b x + 5 \log_b y - m \log_b z)$ Power property

$= \frac{3}{n} \log_b x + \frac{5}{n} \log_b y - \frac{m}{n} \log_b z$ Distributive property (Section R.2)

✔ Now Try Exercises 63, 65, 69, and 73.

Teaching Tip Point out that in **Example 5(d),** the product and quotient properties of logarithms were applied before the power property. In **Example 5(f),** the power property was applied first.

In general, properties of logarithms should follow the order of operations with respect to exponents, multiplication, and division.

Classroom Example 6
Write each expression as a single logarithm with coefficient 1. Assume all variables represent positive real numbers, with $a \neq 1$ and $b \neq 1$.

(a) $\log_4 x - \log_4 y + \log_4 z$

(b) $4 \log_b r - 5 \log_b s$

(c) $\frac{1}{3}\log_a x + \frac{2}{3}\log_a y - \log_a xy$

Answers:

(a) $\log_4 \frac{xz}{y}$ (b) $\log_b \frac{r^4}{s^5}$

(c) $\log_a \frac{\sqrt[3]{xy^2}}{xy}$

EXAMPLE 6 Using the Properties of Logarithms

Write each expression as a single logarithm with coefficient 1. Assume all variables represent positive real numbers, with $a \neq 1$ and $b \neq 1$.

(a) $\log_3(x+2) + \log_3 x - \log_3 2$ **(b)** $2 \log_a m - 3 \log_a n$

(c) $\frac{1}{2}\log_b m + \frac{3}{2}\log_b 2n - \log_b m^2 n$

SOLUTION

(a) $\log_3(x+2) + \log_3 x - \log_3 2 = \log_3 \frac{(x+2)x}{2}$ Product and quotient properties

(b) $2 \log_a m - 3 \log_a n = \log_a m^2 - \log_a n^3$ Power property

$= \log_a \frac{m^2}{n^3}$ Quotient property

(c) $\frac{1}{2}\log_b m + \frac{3}{2}\log_b 2n - \log_b m^2 n$

$= \log_b m^{1/2} + \log_b (2n)^{3/2} - \log_b m^2 n$ Power property

$= \log_b \frac{m^{1/2}(2n)^{3/2}}{m^2 n}$ Product and quotient properties

Use parentheses on $2n$.

$= \log_b \frac{2^{3/2} n^{1/2}}{m^{3/2}}$ Rules for exponents (Sections R.3 and R.6)

$= \log_b \left(\frac{2^3 n}{m^3}\right)^{1/2}$ Rules for exponents

$= \log_b \sqrt{\frac{8n}{m^3}}$ Definition of $a^{1/n}$

Now Try Exercises 75, 79, and 81.

Classroom Example 7
Assume that $\log_{10} 7 = 0.8451$. Find each logarithm.

(a) $\log_{10} 49$ (b) $\log_{10} 70$

Answers:

(a) 1.6902 (b) 1.8451

CAUTION ***There is no property of logarithms to rewrite a logarithm of a sum or difference.*** That is why, in **Example 6(a),** $\log_3(x+2)$ was not written as $\log_3 x + \log_3 2$. The distributive property does not apply in a situation like this because $\log_3(x + y)$ is one term. The abbreviation "log" is a function name, *not* a factor.

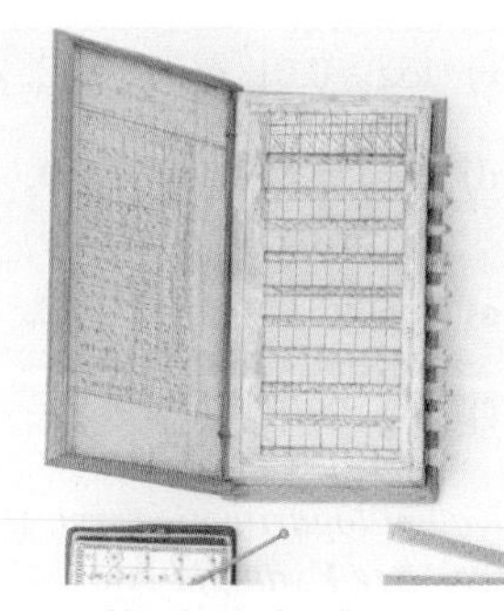

Napier's Rods

The search for ways to make calculations easier has been a long, ongoing process. Machines built by Charles Babbage and Blaise Pascal, a system of "rods" used by John Napier, and slide rules were the forerunners of today's calculators and computers. The invention of logarithms by John Napier in the 16th century was a great breakthrough in the search for easier calculation methods.

Source: IBM Corporate Archives.

The next example uses = symbols for values of logarithms. These are actually approximations.

EXAMPLE 7 Using the Properties of Logarithms with Numerical Values

Assume that $\log_{10} 2 = 0.3010$. Find each logarithm.

(a) $\log_{10} 4$ **(b)** $\log_{10} 5$

SOLUTION

(a) $\log_{10} 4 = \log_{10} 2^2$

$= 2 \log_{10} 2$

$= 2(0.3010)$

$= 0.6020$

(b) $\log_{10} 5 = \log_{10} \frac{10}{2}$

$= \log_{10} 10 - \log_{10} 2$

$= 1 - 0.3010$

$= 0.6990$

Now Try Exercises 85 and 87.

Teaching Tip Point out that the theorem on inverses really follows from the definition of logarithm. Since $\log_a x$ is the power to which a must be raised to obtain x, a raised to that power will be x. In practice, x will often be an algebraic expression.

Recall that for inverse functions f and g, $(f \circ g)(x) = (g \circ f)(x) = x$. We can use this property with exponential and logarithmic functions to state two more properties. If $f(x) = a^x$ and $g(x) = \log_a x$, then

$$(f \circ g)(x) = a^{\log_a x} \quad \text{and} \quad (g \circ f)(x) = \log_a(a^x).$$

Theorem on Inverses

For $a > 0$, $a \neq 1$, the following properties hold.

$$a^{\log_a x} = x \ (\text{for } x > 0) \quad \text{and} \quad \log_a a^x = x$$

The following are examples of applications of this theorem.

$$7^{\log_7 10} = 10, \quad \log_5 5^3 = 3, \quad \text{and} \quad \log_r r^{k+1} = k + 1$$

The second statement in the theorem will be useful in **Sections 4.5 and 4.6** when we solve other logarithmic and exponential equations.

4.3 Exercises

1. (a) C (b) A (c) E (d) B (e) F (f) D
2. (a) F (b) B (c) A (d) D (e) C (f) E
3. $\log_3 81 = 4$ 4. $\log_2 32 = 5$
5. $\log_{2/3} \frac{27}{8} = -3$
6. $\log_{10} 0.0001 = -4$
7. $6^2 = 36$ 8. $5^1 = 5$
9. $(\sqrt{3})^8 = 81$ 10. $4^{-3} = \frac{1}{64}$
12. $a^0 = 1$, $(a \neq 0)$ for all real numbers a.
13. $\{-4\}$ 14. $\{-4\}$
15. $\{\frac{1}{2}\}$ 16. $\{\frac{3}{4}\}$
17. $\{\frac{1}{4}\}$ 18. $\{\frac{1}{5}\}$
19. $\{8\}$ 20. $\{5\}$
21. $\{9\}$ 22. $\{11\}$

Concept Check In Exercises 1 and 2, match the logarithm in Column I with its value in Column II. Remember that $\log_a x$ is the exponent to which a must be raised in order to obtain x.

	I	II
1.	**(a)** $\log_2 16$	**A.** 0
	(b) $\log_3 1$	**B.** $\frac{1}{2}$
	(c) $\log_{10} 0.1$	**C.** 4
	(d) $\log_2 \sqrt{2}$	**D.** -3
	(e) $\log_e \frac{1}{e^2}$	**E.** -1
	(f) $\log_{1/2} 8$	**F.** -2

	I	II
2.	**(a)** $\log_3 81$	**A.** -2
	(b) $\log_3 \frac{1}{3}$	**B.** -1
	(c) $\log_{10} 0.01$	**C.** 0
	(d) $\log_6 \sqrt{6}$	**D.** $\frac{1}{2}$
	(e) $\log_e 1$	**E.** $\frac{9}{2}$
	(f) $\log_3 27^{3/2}$	**F.** 4

If the statement is in exponential form, write it in an equivalent logarithmic form. If the statement is in logarithmic form, write it in exponential form. ***See Example 1.***

3. $3^4 = 81$ **4.** $2^5 = 32$ **5.** $\left(\frac{2}{3}\right)^{-3} = \frac{27}{8}$ **6.** $10^{-4} = 0.0001$

7. $\log_6 36 = 2$ **8.** $\log_5 5 = 1$ **9.** $\log_{\sqrt{3}} 81 = 8$ **10.** $\log_4 \frac{1}{64} = -3$

11. Explain why logarithms of negative numbers are not defined.

12. *Concept Check* Why is $\log_a 1$ always equal to 0 for any valid base a?

Solve each logarithmic equation. ***See Example 2.***

13. $x = \log_5 \frac{1}{625}$ **14.** $x = \log_3 \frac{1}{81}$ **15.** $\log_x \frac{1}{32} = 5$

23. $\left\{\frac{1}{5}\right\}$ 24. $\left\{\frac{1}{4}\right\}$
25. $\{64\}$ 26. $\{8\}$
27. $\left\{\frac{2}{3}\right\}$ 28. $\left\{\frac{1}{2}\right\}$
29. $\{243\}$ 30. $\{128\}$
31. $\{13\}$ 32. $\{3\}$
33. $\{3\}$ 34. $\{23\}$
35. $\{5\}$ 36. $\{6\}$

38. Y(2.3219281) represents the exponent to which 2 must be raised in order to obtain X(5).

39. $(0, \infty)$; $(-\infty, \infty)$

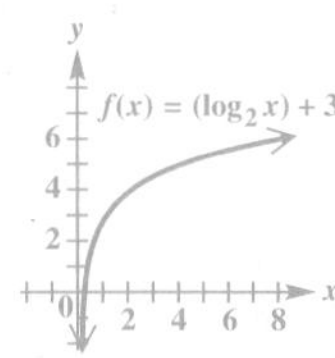

40. $(-3, \infty)$; $(-\infty, \infty)$

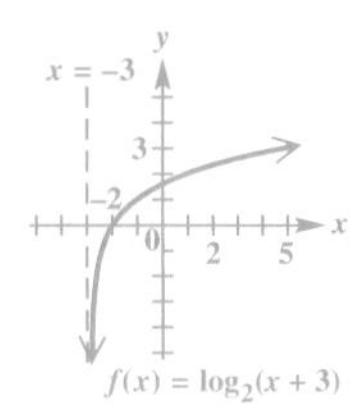

41. $(-3, \infty)$; $[0, \infty)$

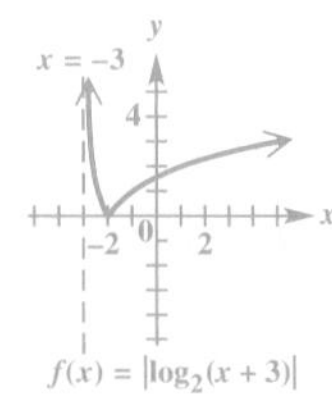

42. $(0, \infty)$; $(-\infty, \infty)$

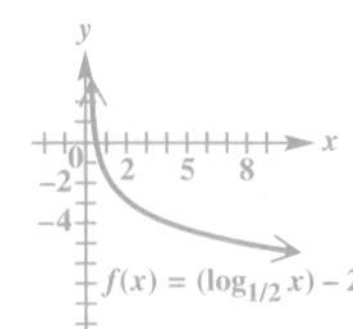

43. $(2, \infty)$; $(-\infty, \infty)$

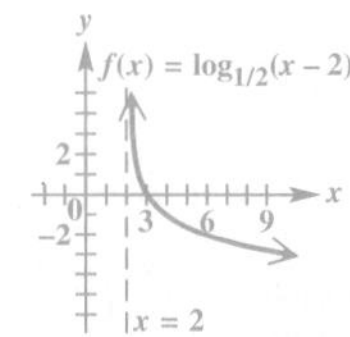

44. $(2, \infty)$; $[0, \infty)$

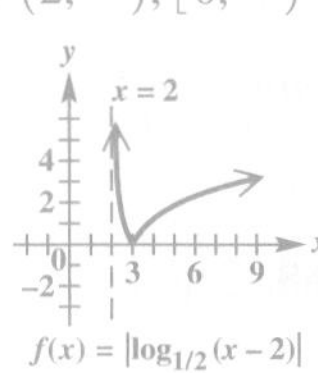

45. E 46. D 47. B
48. C 49. F 50. A

16. $\log_x \frac{27}{64} = 3$ **17.** $x = \log_8 \sqrt[4]{8}$ **18.** $x = \log_7 \sqrt[5]{7}$

19. $x = 3^{\log_3 8}$ **20.** $x = 12^{\log_{12} 5}$ **21.** $x = 2^{\log_2 9}$

22. $x = 8^{\log_8 11}$ **23.** $\log_x 25 = -2$ **24.** $\log_x 16 = -2$

25. $\log_4 x = 3$ **26.** $\log_2 x = 3$ **27.** $x = \log_4 \sqrt[3]{16}$

28. $x = \log_5 \sqrt[4]{25}$ **29.** $\log_9 x = \frac{5}{2}$ **30.** $\log_4 x = \frac{7}{2}$

31. $\log_{1/2}(x + 3) = -4$ **32.** $\log_{1/3}(x + 6) = -2$

33. $\log_{(x+3)} 6 = 1$ **34.** $\log_{(x-4)} 19 = 1$

35. $3x - 15 = \log_x 1 \quad (x > 0, x \neq 1)$ **36.** $4x - 24 = \log_x 1 \quad (x > 0, x \neq 1)$

37. Compare the summary of characteristics of the graph of $f(x) = \log_a x$ with the similar summary about the graph of $f(x) = a^x$ in **Section 4.2.** Make a list of characteristics that reinforce the idea that these are inverse functions.

38. *Concept Check* The calculator graph of $\text{Y} = \log_2 \text{X}$ shows the values of the ordered pair with X = 5. What does the value of Y represent?

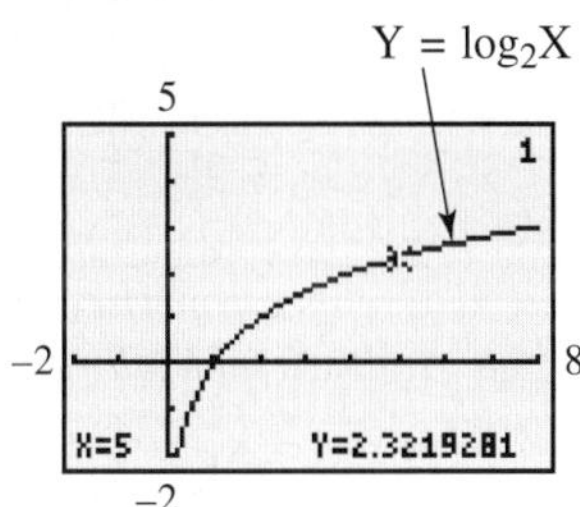

*Sketch the graph of $f(x) = \log_2 x$. Then refer to it and use the techniques of **Chapter 2** to graph each function. Give the domain and range. **See Example 4.***

39. $f(x) = (\log_2 x) + 3$ **40.** $f(x) = \log_2(x + 3)$ **41.** $f(x) = |\log_2(x + 3)|$

*Sketch the graph of $f(x) = \log_{1/2} x$. Then refer to it and use the techniques of **Chapter 2** to graph each function. Give the domain and range. **See Example 4.***

42. $f(x) = (\log_{1/2} x) - 2$ **43.** $f(x) = \log_{1/2}(x - 2)$ **44.** $f(x) = |\log_{1/2}(x - 2)|$

Concept Check In Exercises 45–50, match the function with its graph from choices A–F.

45. $f(x) = \log_2 x$ **46.** $f(x) = \log_2 2x$ **47.** $f(x) = \log_2 \frac{1}{x}$

48. $f(x) = \log_2\left(\frac{1}{2}x\right)$ **49.** $f(x) = \log_2(x - 1)$ **50.** $f(x) = \log_2(-x)$

A.

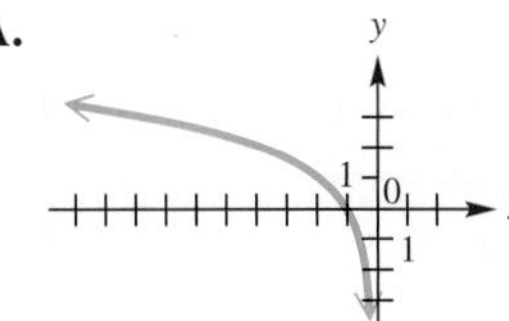

B.

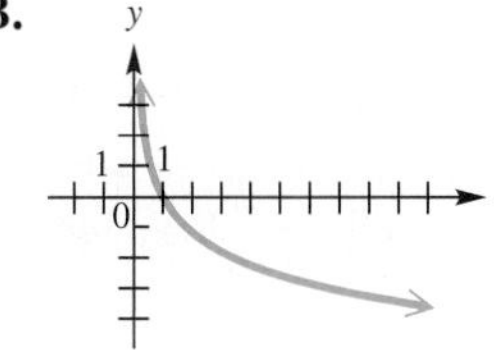

C.

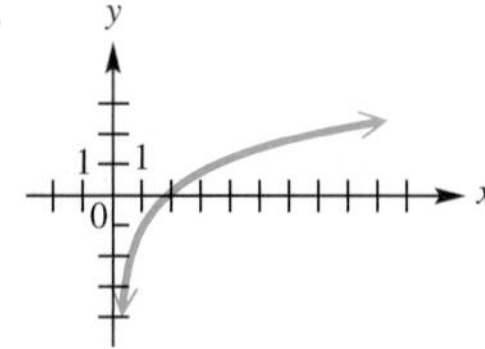

D.

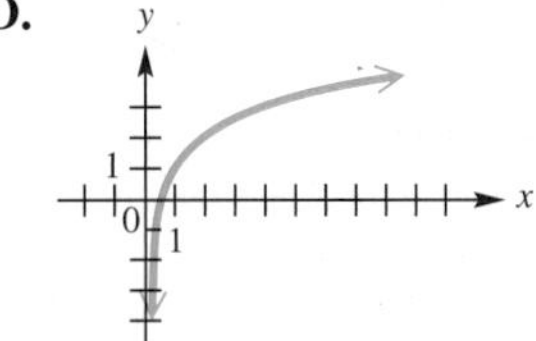

E.

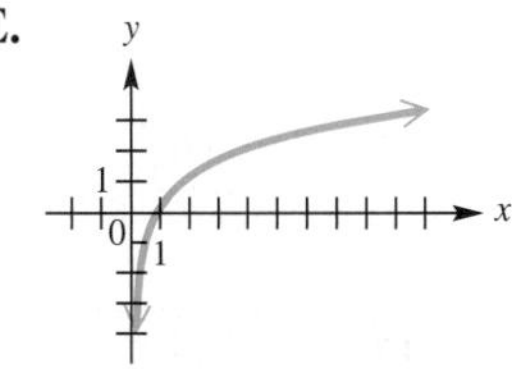

F.

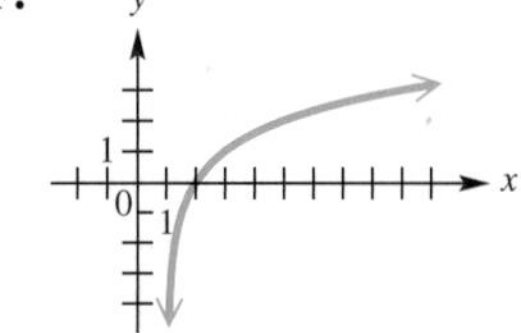

Graph each function. ***See Examples 3 and 4.***

51. $f(x) = \log_5 x$
52. $f(x) = \log_{10} x$
53. $f(x) = \log_{1/2}(1 - x)$
54. $f(x) = \log_{1/3}(3 - x)$
55. $f(x) = \log_3(x - 1) + 2$
56. $f(x) = \log_2(x + 2) - 3$

Connecting Graphs with Equations *In Exercises 57–62, write an equation for the graph given. Each is a logarithmic function f with base 2 or 3, translated and/or reflected.* ***See the Note following Example 4.***

57.

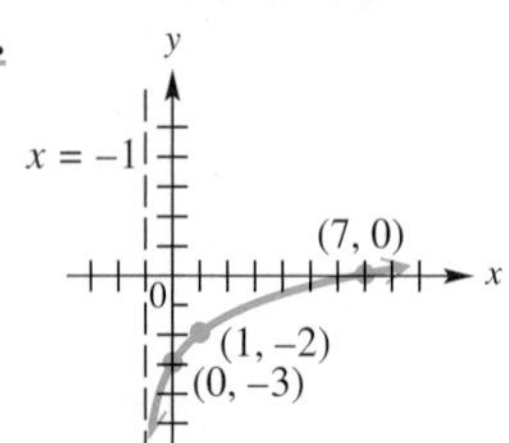

58.

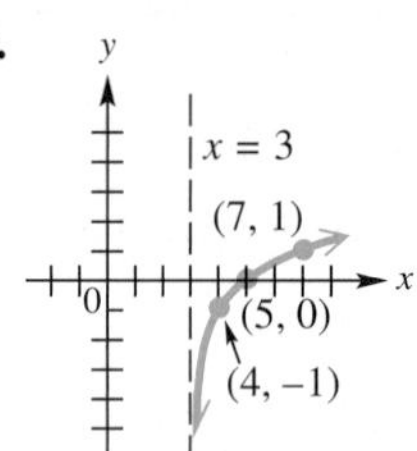

59.

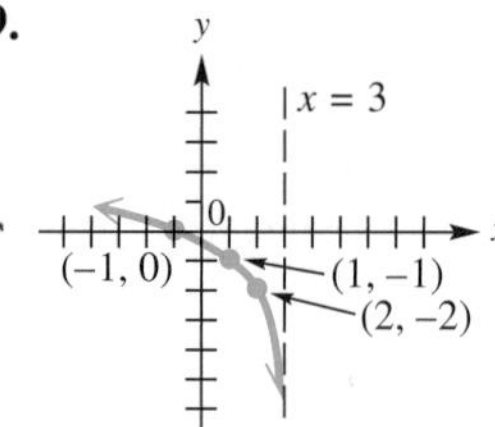

60.

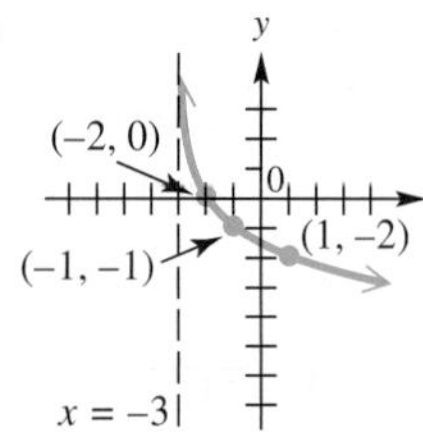

61.

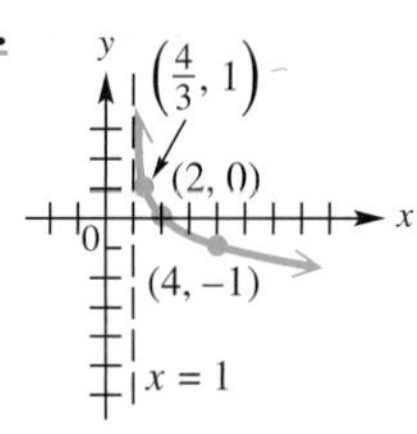

62.

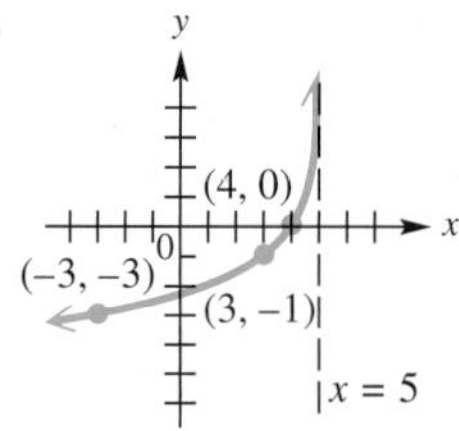

Use the properties of logarithms to rewrite each expression. Simplify the result if possible. Assume all variables represent positive real numbers. ***See Example 5.***

63. $\log_2 \frac{6x}{y}$
64. $\log_3 \frac{4p}{q}$
65. $\log_5 \frac{5\sqrt{7}}{3}$
66. $\log_2 \frac{2\sqrt{3}}{5}$
67. $\log_4(2x + 5y)$
68. $\log_6(7m + 3q)$
69. $\log_m \sqrt{\frac{5r^3}{z^5}}$
70. $\log_p \sqrt[3]{\frac{m^5 n^4}{t^2}}$
71. $\log_2 \frac{ab}{cd}$
72. $\log_2 \frac{xy}{tqr}$
73. $\log_3 \frac{\sqrt{x} \cdot \sqrt[3]{y}}{w^2\sqrt{z}}$
74. $\log_4 \frac{\sqrt[3]{a} \cdot \sqrt[4]{b}}{\sqrt{c} \cdot \sqrt[3]{d^2}}$

Write each expression as a single logarithm with coefficient 1. Assume all variables represent positive real numbers. ***See Example 6.***

75. $\log_a x + \log_a y - \log_a m$
76. $\log_b k + \log_b m - \log_b a$
77. $\log_a m - \log_a n - \log_a t$
78. $\log_b p - \log_b q - \log_b r$
79. $\frac{1}{3}\log_b x^4y^5 - \frac{3}{4}\log_b x^2y$
80. $\frac{1}{2}\log_y p^3q^4 - \frac{2}{3}\log_y p^4q^3$
81. $2\log_a(z + 1) + \log_a(3z + 2)$
82. $5\log_a(z + 7) + \log_a(2z + 9)$
83. $-\frac{2}{3}\log_5 5m^2 + \frac{1}{2}\log_5 25m^2$
84. $-\frac{3}{4}\log_3 16p^4 - \frac{2}{3}\log_3 8p^3$

Given the approximations $\log_{10} 2 = 0.3010$ *and* $\log_{10} 3 = 0.4771$, *find each logarithm without using a calculator.* ***See Example 7.***

85. $\log_{10} 6$
86. $\log_{10} 12$
87. $\log_{10} \frac{3}{2}$
88. $\log_{10} \frac{2}{9}$
89. $\log_{10} \frac{9}{4}$
90. $\log_{10} \frac{20}{27}$
91. $\log_{10} \sqrt{30}$
92. $\log_{10} 36^{1/3}$

51. 52.

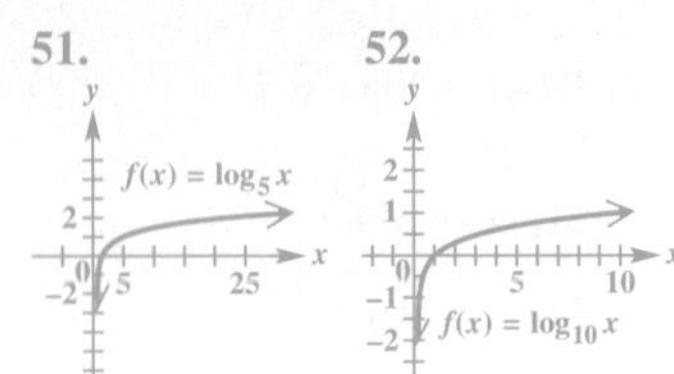

53. 54.

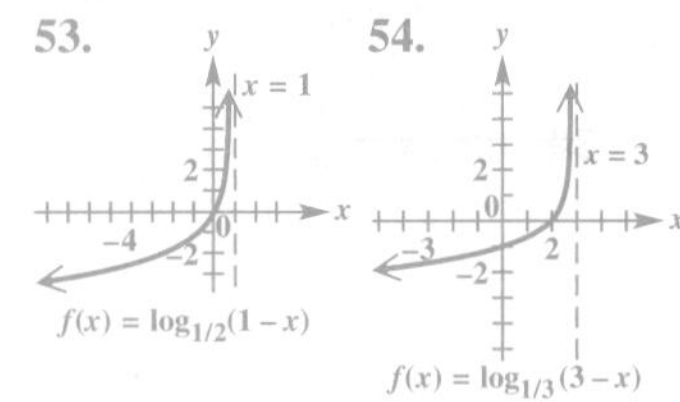

55. 56.

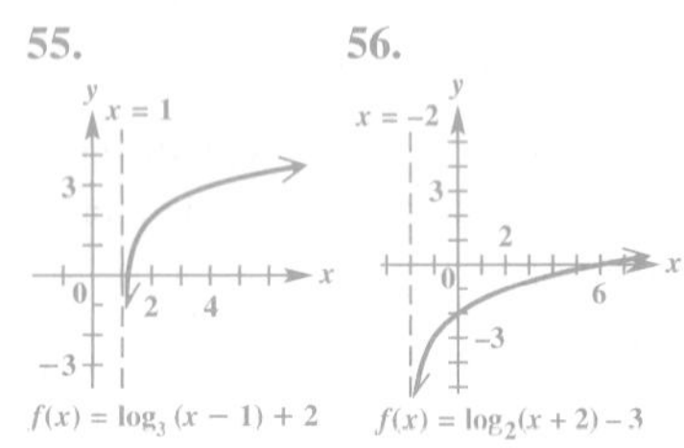

57. $f(x) = \log_2(x + 1) - 3$
58. $f(x) = \log_2(x - 3) - 1$
59. $f(x) = \log_2(-x + 3) - 2$
60. $f(x) = -\log_2(x + 3)$
61. $f(x) = -\log_3(x - 1)$
62. $f(x) = -\log_2(-x + 5)$

63. $\log_2 6 + \log_2 x - \log_2 y$
64. $\log_3 4 + \log_3 p - \log_3 q$
65. $1 + \frac{1}{2}\log_5 7 - \log_5 3$
66. $1 + \frac{1}{2}\log_2 3 - \log_2 5$
67. This cannot be simplified.
68. This cannot be simplified.
69. $\frac{1}{2}(\log_m 5 + 3\log_m r - 5\log_m z)$
70. $\frac{1}{3}(5\log_p m + 4\log_p n - 2\log_p t)$
71. $\log_2 a + \log_2 b - \log_2 c - \log_2 d$
72. $\log_2 x + \log_2 y - \log_2 t - \log_2 q - \log_2 r$
73. $\frac{1}{2}\log_3 x + \frac{1}{3}\log_3 y - 2\log_3 w - \frac{1}{2}\log_3 z$
74. $\frac{1}{3}\log_4 a + \frac{1}{4}\log_4 b - \frac{1}{2}\log_4 c - \frac{2}{3}\log_4 d$

75. $\log_a \frac{xy}{m}$
76. $\log_b \frac{km}{a}$
77. $\log_a \frac{m}{nt}$
78. $\log_b \frac{p}{qr}$
79. $\log_b(x^{-1/6}y^{11/12})$
80. $\log_y(p^{-7/6})$
81. $\log_a[(z + 1)^2(3z + 2)]$
82. $\log_a[(z + 7)^5(2z + 9)]$
83. $\log_5 \frac{5^{1/3}}{m^{1/3}}$, or $\log_5 \sqrt[3]{\frac{5}{m}}$
84. $\log_3 \frac{1}{32p^5}$

85. 0.7781 **86.** 1.0791
87. 0.1761 **88.** −0.6532
89. 0.3522 **90.** −0.1303
91. 0.7386 **92.** 0.5187

93. (a)

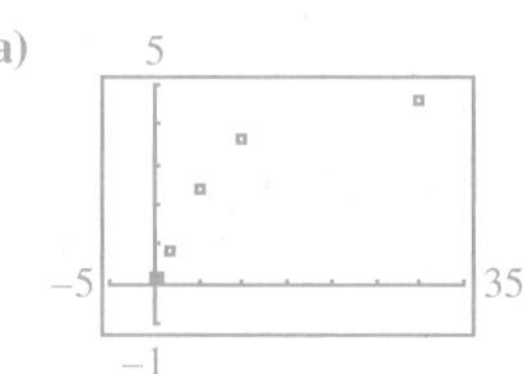

(b) logarithmic

94. (a) −1.1 **(b)** −0.2

95. (a) −4 **(b)** 6 **(c)** 4 **(d)** −1

96. (a) 9 **(b)** −6 **(c)** −20 **(d)** 125

98. $4 = \log_a 5$

99.

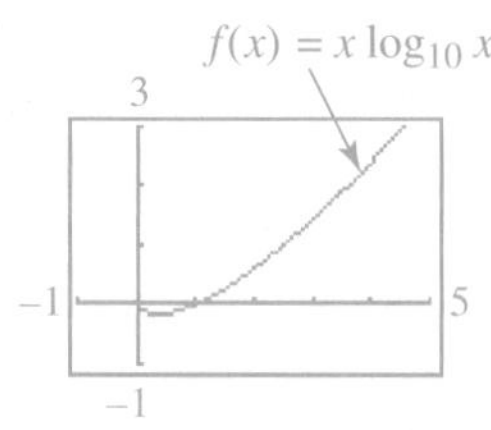

100.

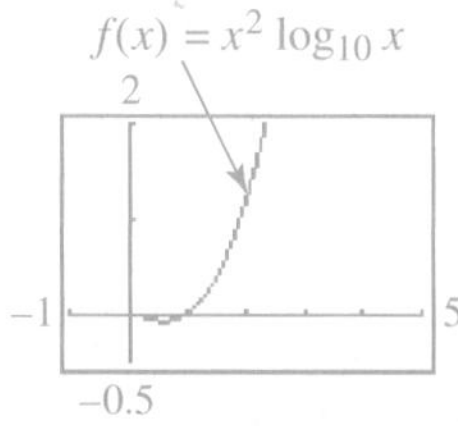

101. $\{0.01, 2.38\}$ **102.** $\{1.87\}$

Solve each problem.

93. *(Modeling) Interest Rates of Treasury Securities* The table gives interest rates for various U.S. Treasury Securities on February 15, 2011.

(a) Make a scatter diagram of the data.

(b) Which type of function will model these data best: linear, exponential, or logarithmic?

Time	Yield
3-month	0.13%
6-month	0.17%
2-year	0.84%
5-year	2.35%
10-year	3.61%
30-year	4.66%

Source: www.federalreserve.gov

94. *Concept Check* Use the graph to estimate each logarithm.

(a) $\log_3 0.3$ **(b)** $\log_3 0.8$

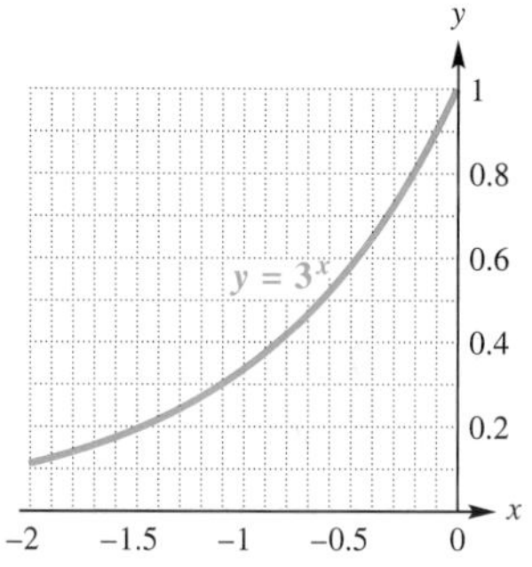

95. *Concept Check* Suppose $f(x) = \log_a x$ and $f(3) = 2$. Determine each function value.

(a) $f\left(\frac{1}{9}\right)$ **(b)** $f(27)$ **(c)** $f(9)$ **(d)** $f\left(\frac{\sqrt{3}}{3}\right)$

96. Use properties of logarithms to evaluate each expression.

(a) $100^{\log_{10} 3}$ **(b)** $\log_{10}(0.01)^3$ **(c)** $\log_{10}(0.0001)^5$ **(d)** $1000^{\log_{10} 5}$

97. Refer to the compound interest formula from **Section 4.2.** Show that the amount of time required for a deposit to double is

$$\frac{1}{\log_2\left(1 + \frac{r}{n}\right)^n}.$$

98. *Concept Check* If $(5, 4)$ is on the graph of the logarithmic function with base a, which of the following statements is true:

$$5 = \log_a 4 \quad \text{or} \quad 4 = \log_a 5?$$

Standard mathematical notation uses $\log x$ *as an abbreviation for* $\log_{10} x$. *Use the* log *key on your graphing calculator to graph each function.*

99. $f(x) = x \log_{10} x$ **100.** $f(x) = x^2 \log_{10} x$

Use a graphing calculator to find the solution set of each equation. Give solutions to the nearest hundredth.

101. $\log_{10} x = x - 2$ **102.** $2^{-x} = \log_{10} x$

103. Prove the quotient property of logarithms: $\log_a \frac{x}{y} = \log_a x - \log_a y$.

104. Prove the power property of logarithms: $\log_a x^r = r \log_a x$.

Summary Exercises on Inverse, Exponential, and Logarithmic Functions

1. They are inverses.
2. They are not inverses.
3. They are inverses.
4. They are inverses.

5. 6.

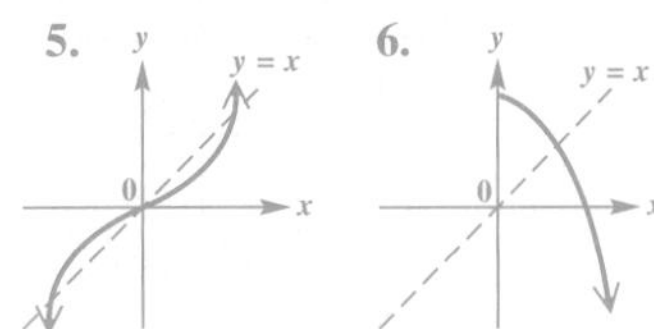

7. It is not one-to-one.
8.

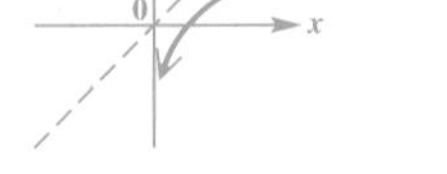

9. B 10. D 11. C 12. A

13. The functions in Exercises 9 and 12 are inverses of one another. The functions in Exercises 10 and 11 are inverses of one another.
14. $f^{-1}(x) = 5^x$

The following exercises are designed to help solidify your understanding of inverse, exponential, and logarithmic functions from **Sections 4.1–4.3.**

Determine whether the functions in each pair are inverses of each other.

1. $f(x) = 3x - 4, \quad g(x) = \frac{1}{3}x + \frac{4}{3}$

2. $f(x) = 8 - 5x, \quad g(x) = 8 + \frac{1}{5}x$

3. $f(x) = 1 + \log_2 x, \quad g(x) = 2^{x-1}$

4. $f(x) = 3^{x/5} - 2, \quad g(x) = 5 \log_3(x + 2)$

Determine whether each function is one-to-one. If it is, then sketch the graph of its inverse function.

5.

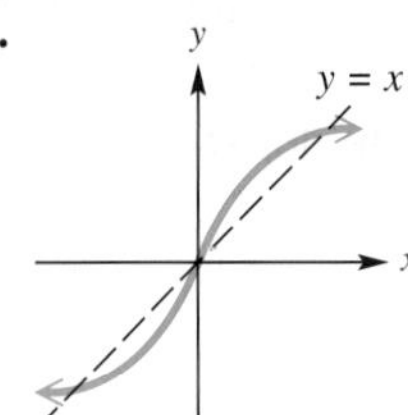

6.

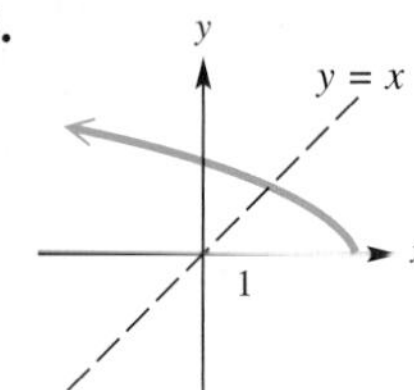

7.

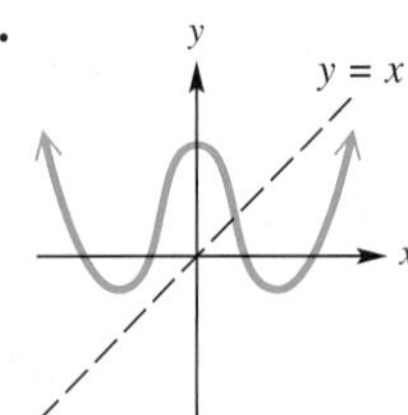

8.

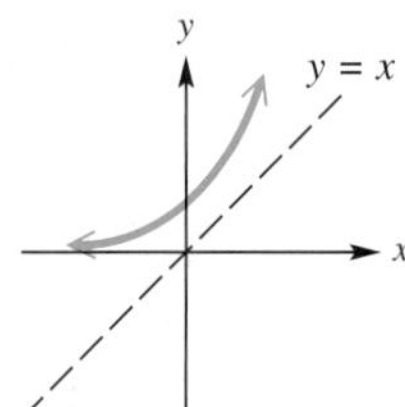

In Exercises 9–12, match the function with its graph from choices A–D.

9. $y = \log_3(x + 2)$

10. $y = 5 - 2^x$

11. $y = \log_2(5 - x)$

12. $y = 3^x - 2$

A.

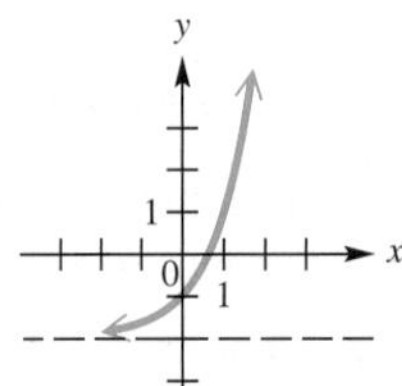

B.

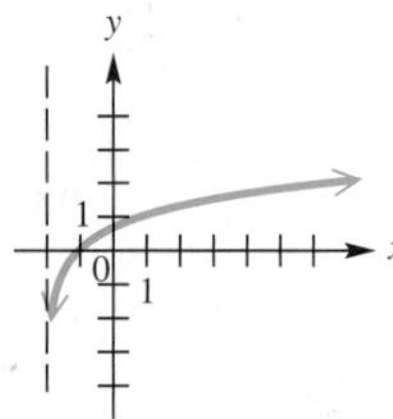

C.

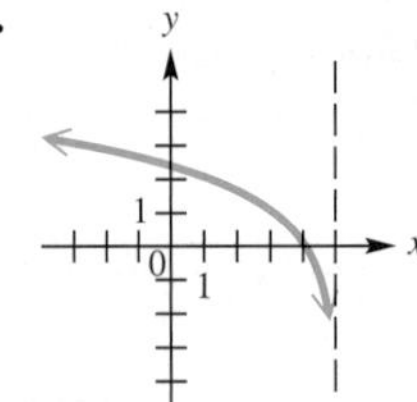

D.

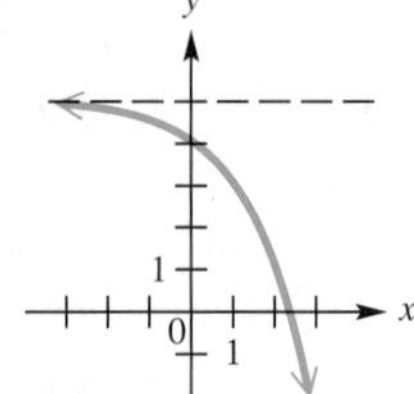

13. The functions in **Exercises 9–12** form two pairs of inverse functions. Determine which functions are inverses of each other.

14. Determine the inverse of the function $f(x) = \log_5 x$. (*Hint:* Replace $f(x)$ with y, and write in exponential form.)

15. $f^{-1}(x) = \frac{1}{3}x + 2$; Domains and ranges of both f and f^{-1} are $(-\infty, \infty)$.
16. $f^{-1}(x) = \sqrt[3]{\frac{x}{2}} - 1$; Domains and ranges of both f and f^{-1} are $(-\infty, \infty)$.
17. f is not one-to-one.
18. $f^{-1}(x) = \frac{5x+1}{2+3x}$; Domain of f = range of $f^{-1} = \left(-\infty, \frac{5}{3}\right) \cup \left(\frac{5}{3}, \infty\right)$. Domain of f^{-1} = range of $f = \left(-\infty, -\frac{2}{3}\right) \cup \left(-\frac{2}{3}, \infty\right)$.
19. f is not one-to-one.
20. $f^{-1}(x) = \sqrt{x^2 + 9}$, $x \geq 0$; Domain of f = range of $f^{-1} = [3, \infty)$. Domain of f^{-1} = range of $f = [0, \infty)$.

21. $\log_{1/10} 1000 = -3$
22. $\log_a c = b$
23. $\log_{\sqrt{3}} 9 = 4$
24. $\log_4 \frac{1}{8} = -\frac{3}{2}$
25. $\log_2 32 = x$
26. $\log_{27} 81 = \frac{4}{3}$

27. $\{2\}$ 28. $\{-3\}$
29. $\{-3\}$ 30. $\{25\}$
31. $\{-2\}$ 32. $\{\frac{1}{3}\}$
33. $(0, 1) \cup (1, \infty)$ 34. $\{\frac{3}{2}\}$
35. $\{5\}$ 36. $\{243\}$
37. $\{1\}$ 38. $\{-2\}$
39. $\{1\}$ 40. $\{2\}$
41. $\{2\}$ 42. $\{\frac{1}{9}\}$
43. $\{-\frac{1}{3}\}$ 44. $(-\infty, \infty)$

For each function that is one-to-one, write an equation for its inverse function. Give the domain and range of both f and f^{-1}. If the function is not one-to-one, say so.

15. $f(x) = 3x - 6$

16. $f(x) = 2(x + 1)^3$

17. $f(x) = 3x^2$

18. $f(x) = \dfrac{2x - 1}{5 - 3x}$

19. $f(x) = \sqrt[3]{5 - x^4}$

20. $f(x) = \sqrt{x^2 - 9},\ x \geq 3$

Write an equivalent statement in logarithmic form.

21. $\left(\dfrac{1}{10}\right)^{-3} = 1000$

22. $a^b = c$

23. $\left(\sqrt{3}\right)^4 = 9$

24. $4^{-3/2} = \dfrac{1}{8}$

25. $2^x = 32$

26. $27^{4/3} = 81$

Solve each equation.

27. $3x = 7^{\log_7 6}$

28. $x = \log_{10} 0.001$

29. $x = \log_6 \dfrac{1}{216}$

30. $\log_x 5 = \dfrac{1}{2}$

31. $\log_{10} 0.01 = x$

32. $\log_x 3 = -1$

33. $\log_x 1 = 0$

34. $x = \log_2 \sqrt{8}$

35. $\log_x \sqrt[3]{5} = \dfrac{1}{3}$

36. $\log_{1/3} x = -5$

37. $\log_{10}(\log_2 2^{10}) = x$

38. $x = \log_{4/5} \dfrac{25}{16}$

39. $2x - 1 = \log_6 6^x$

40. $x = \sqrt{\log_{1/2} \dfrac{1}{16}}$

41. $2^x = \log_2 16$

42. $\log_3 x = -2$

43. $\left(\dfrac{1}{3}\right)^{x+1} = 9^x$

44. $5^{2x-6} = 25^{x-3}$

4.4 Evaluating Logarithms and the Change-of-Base Theorem

- Common Logarithms
- Applications and Models with Common Logarithms
- Natural Logarithms
- Applications and Models with Natural Logarithms
- Logarithms with Other Bases

Common Logarithms Two of the most important bases for logarithms are 10 and e. Base 10 logarithms are called **common logarithms.** The common logarithm of x is written $\log x$, where the base is understood to be 10.

Common Logarithm

For all positive numbers x,

$$\log x = \log_{10} x.$$

A calculator with a log key can be used to find the base 10 logarithm of any positive number.

Classroom Example 1
Use a calculator to find the values of log 10,000, log 341, and log 0.06894.

Answers: 4; 2.532754379; −1.161528721

EXAMPLE 1 Evaluating Common Logarithms with a Calculator

Use a calculator to find the values of

$$\log 1000, \quad \log 142, \quad \text{and} \quad \log 0.005832.$$

SOLUTION **Figure 33** shows that the exact value of log 1000 is 3 (because $10^3 = 1000$), and that

$$\log 142 \approx 2.152288344$$

and $$\log 0.005832 \approx -2.234182485.$$

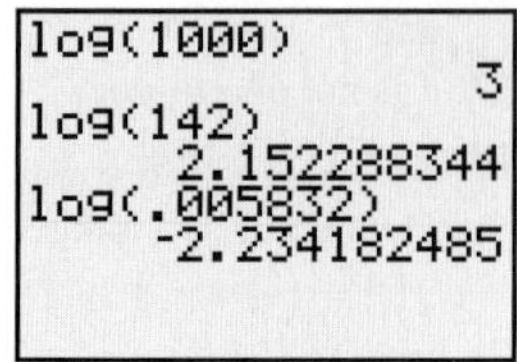

Figure 33

Most common logarithms that appear in calculations are approximations, as seen in the second and third displays.

✔ *Now Try Exercises 11, 15, and 17.*

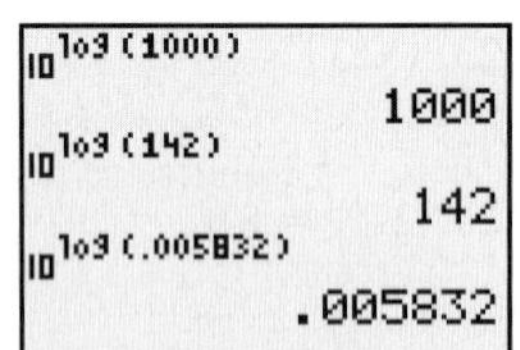

Figure 34

Figure 34 reinforces the concept presented in the previous section: **log *x* *is the exponent to which* 10 *must be raised to obtain x.***

NOTE ***Base a logarithms of numbers between* 0 *and* 1, *where a* > 1, *are always negative,*** as suggested by the graphs in **Section 4.3.**

Applications and Models with Common Logarithms In chemistry, the **pH** of a solution is defined as

$$\mathbf{pH = -\log[H_3O^+],}$$

where $[H_3O^+]$ is the hydronium ion concentration in moles* per liter. The pH value is a measure of the acidity or alkalinity of a solution. Pure water has pH 7.0, substances with pH values greater than 7.0 are alkaline, and substances with pH values less than 7.0 are acidic. It is customary to round pH values to the nearest tenth.

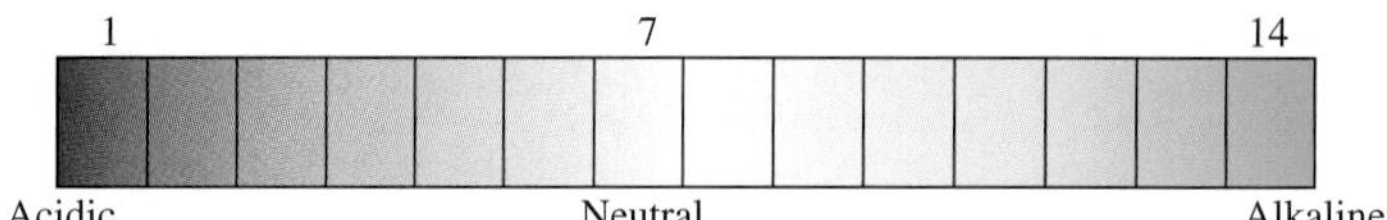

Classroom Example 2
(a) Find the pH of a solution with $[H_3O^+] = 6.8 \times 10^{-8}$.
(b) Find the hydronium ion concentration of a solution with pH = 4.3.

Answers:
(a) 7.2 (b) 5.0×10^{-5}

EXAMPLE 2 Finding pH

(a) Find the pH of a solution with $[H_3O^+] = 2.5 \times 10^{-4}$.

(b) Find the hydronium ion concentration of a solution with pH = 7.1.

SOLUTION

(a)
$$\begin{aligned} \text{pH} &= -\log[H_3O^+] \\ &= -\log(2.5 \times 10^{-4}) && \text{Substitute.} \\ &= -(\log 2.5 + \log 10^{-4}) && \text{Product property (Section 4.3)} \\ &= -(0.3979 - 4) && \log 10^{-4} = -4 \text{ (Section 4.3)} \\ &= -0.3979 + 4 && \text{Distributive property (Section R.2)} \\ \text{pH} &\approx 3.6 && \text{Add.} \end{aligned}$$

*A *mole* is the amount of a substance that contains the same number of molecules as the number of atoms in exactly 12 grams of carbon-12.

(b)

$$\text{pH} = -\log[\text{H}_3\text{O}^+]$$

$$7.1 = -\log[\text{H}_3\text{O}^+] \quad \text{Substitute.}$$

$$-7.1 = \log[\text{H}_3\text{O}^+] \quad \text{Multiply by } -1. \text{ (Section 1.1)}$$

$$[\text{H}_3\text{O}^+] = 10^{-7.1} \quad \text{Write in exponential form. (Section 4.3)}$$

$$[\text{H}_3\text{O}^+] \approx 7.9 \times 10^{-8} \quad \text{Evaluate } 10^{-7.1} \text{ with a calculator.}$$

Now Try Exercises 29 and 33.

NOTE In the fourth line of the solution in **Example 2(a),** we use the equality symbol, =, rather than the approximate equality symbol, ≈, when replacing log 2.5 with 0.3979. This is often done for convenience, despite the fact that most logarithms used in applications are indeed approximations.

Classroom Example 3
Refer to **Example 3.** The hydronium ion concentration of a water sample from a wetland is 4.5×10^{-3}. Classify this wetland.

Answer: bog

EXAMPLE 3 Using pH in an Application

Wetlands are classified as *bogs, fens, marshes,* and *swamps* based on pH values. A pH value between 6.0 and 7.5 indicates that the wetland is a "rich fen." When the pH is between 3.0 and 6.0, it is a "poor fen," and if the pH falls to 3.0 or less, the wetland is a "bog." (*Source:* R. Mohlenbrock, "Summerby Swamp, Michigan," *Natural History*.)

Suppose that the hydronium ion concentration of a sample of water from a wetland is 6.3×10^{-5}. How would this wetland be classified?

SOLUTION

$$\text{pH} = -\log[\text{H}_3\text{O}^+] \quad \text{Definition of pH}$$

$$= -\log(6.3 \times 10^{-5}) \quad \text{Substitute.}$$

$$= -(\log 6.3 + \log 10^{-5}) \quad \text{Product property}$$

$$= -\log 6.3 - (-5) \quad \text{Distributive property; } \log 10^n = n$$

$$= -\log 6.3 + 5 \quad \text{Definition of subtraction}$$

$$\text{pH} \approx 4.2 \quad \text{Use a calculator.}$$

Since the pH is between 3.0 and 6.0, the wetland is a poor fen.

Now Try Exercise 37.

Classroom Example 4
Find the decibel rating of a sound with intensity $10{,}000{,}000 I_0$.

Answer: 70

EXAMPLE 4 Measuring the Loudness of Sound

The loudness of sounds is measured in **decibels.** We first assign an intensity of I_0 to a very faint **threshold sound.** If a particular sound has intensity I, then the decibel rating d of this louder sound is given by the following formula.

$$d = 10 \log \frac{I}{I_0}$$

Find the decibel rating d of a sound with intensity $10{,}000 I_0$.

SOLUTION

$$d = 10 \log \frac{10{,}000 I_0}{I_0} \quad \text{Let } I = 10{,}000 I_0.$$

$$= 10 \log 10{,}000 \quad \frac{I_0}{I_0} = 1$$

$$= 10(4) \quad \log 10{,}000 = \log 10^4 = 4 \text{ (Section 4.3)}$$

$$= 40 \quad \text{Multiply.}$$

The sound has a decibel rating of 40.

Now Try Exercise 63.

LOOKING AHEAD TO CALCULUS

The natural logarithmic function $f(x) = \ln x$ and the reciprocal function $g(x) = \frac{1}{x}$ have an important relationship in calculus. The derivative of the natural logarithmic function is the reciprocal function. Using **Leibniz notation** (named after one of the co-inventors of calculus), we write this fact as $\frac{d}{dx}(\ln x) = \frac{1}{x}$.

Natural Logarithms In **Section 4.2,** we introduced the irrational number e. In most practical applications of logarithms, e is used as base. Logarithms with base e are called **natural logarithms,** since they occur in the life sciences and economics in natural situations that involve growth and decay. The base e logarithm of x is written $\ln x$ (read **"el-en x"**). ***The expression* $\ln x$ *represents the exponent to which e must be raised to obtain x.***

Natural Logarithm

For all positive numbers x,

$$\ln x = \log_e x.$$

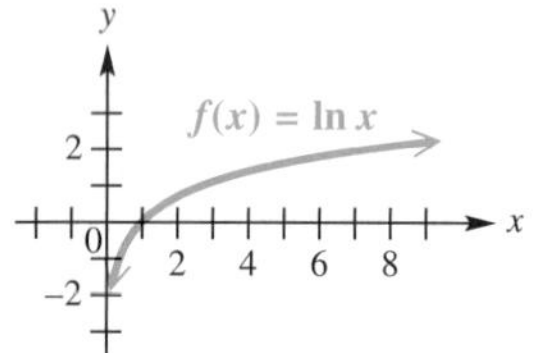

Figure 35

A graph of the natural logarithmic function $f(x) = \ln x$ is given in **Figure 35.**

EXAMPLE 5 Evaluating Natural Logarithms with a Calculator

Use a calculator to find the values of

$$\ln e^3, \quad \ln 142, \quad \text{and} \quad \ln 0.005832.$$

Classroom Example 5
Use a calculator to find the values of $\ln e^4$, $\ln 341$, and $\ln 0.06894$.

Answers: 4; 5.831882477; -2.674518718

SOLUTION **Figure 36** shows that the exact value of $\ln e^3$ is 3, and that

$$\ln 142 \approx 4.955827058$$

and $$\ln 0.005832 \approx -5.144395284.$$

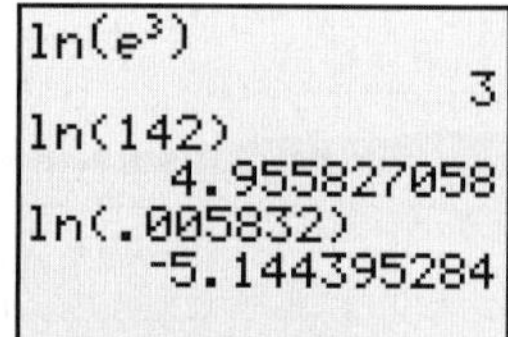

Figure 36

✔ *Now Try Exercises 45, 49, and 51.*

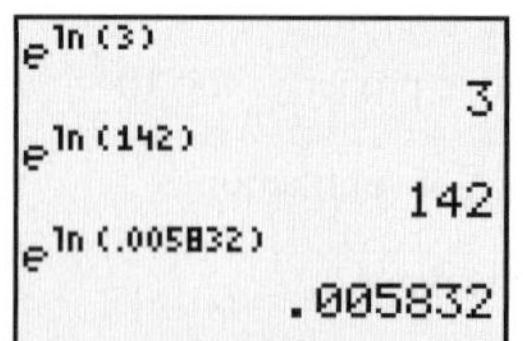

Figure 37

Figure 37 illustrates that $\ln x$ ***is the exponent to which e must be raised to obtain x.***

Applications and Models with Natural Logarithms We now consider two applications of natural logarithms.

EXAMPLE 6 Measuring the Age of Rocks

Classroom Example 6
Refer to the model given in **Example 6.**
(a) How old is a rock in which $A = K$?
(b) How old is a rock in which $\frac{A}{K} = 0.325$?

Answers:
(a) about 4.06 billion yr old
(b) about 2.38 billion yr old

Geologists sometimes measure the age of rocks by using "atomic clocks." By measuring the amounts of potassium-40 and argon-40 in a rock, it is possible to find the age t of the specimen in years with the formula

$$t = (1.26 \times 10^9)\frac{\ln\left(1 + 8.33\left(\frac{A}{K}\right)\right)}{\ln 2},$$

where A and K are the numbers of atoms of argon-40 and potassium-40, respectively, in the specimen.

(a) How old is a rock in which $A = 0$ and $K > 0$?

(b) The ratio $\frac{A}{K}$ for a sample of granite from New Hampshire is 0.212. How old is the sample?

SOLUTION

(a) If $A = 0$, then $\frac{A}{K} = 0$ and the equation is as follows.

$$t = (1.26 \times 10^9)\frac{\ln\left(1 + 8.33\left(\frac{A}{K}\right)\right)}{\ln 2} \quad \text{Given formula}$$

$$= (1.26 \times 10^9)\frac{\ln 1}{\ln 2} \quad \frac{A}{K} = 0, \text{ so } \ln(1 + 0) = \ln 1$$

$$= (1.26 \times 10^9)(0) \quad \ln 1 = 0$$

$$t = 0$$

The rock is new (0 yr old).

(b) Since $\frac{A}{K} = 0.212$, we have the following.

$$t = (1.26 \times 10^9)\frac{\ln(1 + 8.33(0.212))}{\ln 2} \quad \text{Substitute.}$$

$$t \approx 1.85 \times 10^9 \quad \text{Use a calculator.}$$

The granite is about 1.85 billion yr old.

✔ *Now Try Exercise 77.*

Classroom Example 7
Refer to the equation given in **Example 7.** Suppose that $C = 2C_0$ and $k = 14$.
(a) Find the radiative forcing given these conditions.
(b) Find the average global temperature increase to the nearest degree Fahrenheit under the same conditions.

Answers:
(a) about 9.7 w/m^2 (b) 10°F

EXAMPLE 7 Modeling Global Temperature Increase

Carbon dioxide in the atmosphere traps heat from the sun. The additional solar radiation trapped by carbon dioxide is called **radiative forcing.** It is measured in watts per square meter (w/m^2). In 1896 the Swedish scientist Svante Arrhenius modeled radiative forcing R caused by additional atmospheric carbon dioxide, using the logarithmic equation

$$R = k \ln \frac{C}{C_0},$$

where C_0 is the preindustrial amount of carbon dioxide, C is the current carbon dioxide level, and k is a constant. Arrhenius determined that $10 \le k \le 16$ when $C = 2C_0$. (*Source:* Clime, W., *The Economics of Global Warming,* Institute for International Economics, Washington, D.C.)

(a) Let $C = 2C_0$. Is the relationship between R and k linear or logarithmic?

(b) The average global temperature increase T (in °F) is given by $T(R) = 1.03R$. Write T as a function of k.

SOLUTION

(a) If $C = 2C_0$, then $\frac{C}{C_0} = 2$, so $R = k \ln 2$ is a linear relation, because $\ln 2$ is a constant.

(b)

$$T(R) = 1.03R$$

$$T(k) = 1.03k \ln \frac{C}{C_0} \quad \text{Use the given expression for } R.$$

✔ *Now Try Exercise 75.*

Logarithms with Other Bases We can use a calculator to find the values of either natural logarithms (base e) or common logarithms (base 10). However, sometimes we must use logarithms with other bases. The change-of-base theorem can be used to convert logarithms from one base to another.

LOOKING AHEAD TO CALCULUS

In calculus, natural logarithms are more convenient to work with than logarithms with other bases. The change-of-base theorem enables us to convert any logarithmic function to a *natural* logarithmic function.

Change-of-Base Theorem

For any positive real numbers x, a, and b, where $a \neq 1$ and $b \neq 1$, the following holds.

$$\log_a x = \frac{\log_b x}{\log_b a}$$

Proof Let $y = \log_a x$.

$$a^y = x \quad \text{Change to exponential form.}$$

$$\log_b a^y = \log_b x \quad \text{Take the base } b \text{ logarithm on each side.}$$

$$y \log_b a = \log_b x \quad \text{Power property (Section 4.3)}$$

$$y = \frac{\log_b x}{\log_b a} \quad \text{Divide each side by } \log_b a.$$

$$\log_a x = \frac{\log_b x}{\log_b a} \quad \text{Substitute } \log_a x \text{ for } y.$$

Any positive number other than 1 can be used for base b in the change-of-base theorem, but usually the only practical bases are e and 10 since calculators give logarithms for these two bases.

Teaching Tip Show students that $\log_2 10$ can be evaluated using $\frac{\log 10}{\log 2}$ or $\frac{\ln 10}{\ln 2}$.

For example, with the change-of-base theorem, we can now graph the equation $y = \log_2 x$ by directing the calculator to graph $y = \frac{\log x}{\log 2}$, or, equivalently, $y = \frac{\ln x}{\ln 2}$. ■

Classroom Example 8
Use the change-of-base theorem to find an approximation to four decimal places for each logarithm.
(a) $\log_4 20$ (b) $\log_2 0.7$
Answers:
(a) 2.1610 (b) −0.5146

EXAMPLE 8 **Using the Change-of-Base Theorem**

Use the change-of-base theorem to find an approximation to four decimal places for each logarithm.

(a) $\log_5 17$ **(b)** $\log_2 0.1$

SOLUTION

(a) We will arbitrarily use natural logarithms.

There is no need to actually write this step.

$$\log_5 17 = \frac{\ln 17}{\ln 5} \approx \frac{2.8332}{1.6094} \approx 1.7604$$

(b) Here, we use common logarithms.

$$\log_2 0.1 = \frac{\log 0.1}{\log 2} \approx -3.3219$$

✔ *Now Try Exercises 79 and 81.*

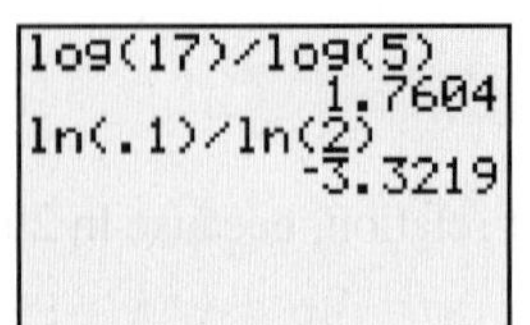

The screen shows how the result of **Example 8(a)** can be found using *common* logarithms, and how the result of **Example 8(b)** can be found using *natural* logarithms. The results are the same as those in **Example 8.**

NOTE In **Example 8,** logarithms evaluated in the intermediate steps, such as ln 17 and ln 5, were shown to four decimal places. However, the final answers were obtained *without* rounding these intermediate values, using all the digits obtained with the calculator. ***In general, it is best to wait until the final step to round off the answer; otherwise, a build-up of round-off errors may cause the final answer to have an incorrect digit in the final decimal place.***

EXAMPLE 9 Modeling Diversity of Species

One measure of the diversity of the species in an ecological community is modeled by the formula

$$H = -\left[P_1 \log_2 P_1 + P_2 \log_2 P_2 + \cdots + P_n \log_2 P_n\right],$$

where $P_1, P_2, \ldots, P_n$ are the proportions of a sample that belong to each of n species found in the sample. (*Source:* Ludwig, J., and J. Reynolds, *Statistical Ecology: A Primer on Methods and Computing,* New York, Wiley.)

Find the measure of diversity in a community with two species where there are 90 of one species and 10 of the other.

SOLUTION Since there are 100 members in the community, $P_1 = \frac{90}{100} = 0.9$ and $P_2 = \frac{10}{100} = 0.1$, so

$$H = -\left[0.9 \log_2 0.9 + 0.1 \log_2 0.1\right]. \quad \text{Substitute for } P_1 \text{ and } P_2.$$

Classroom Example 9
Refer to the formula in **Example 9.** Find the measure of diversity in a community with two species where there are 60 of one species and 140 of the other.

Answer: 0.881

In **Example 8(b),** we found that $\log_2 0.1 \approx -3.32$. Now we find $\log_2 0.9$.

$$\log_2 0.9 = \frac{\log 0.9}{\log 2} \approx -0.152 \quad \text{Change-of-base theorem}$$

Therefore,

$$H = -\left[0.9 \log_2 0.9 + 0.1 \log_2 0.1\right]$$
$$H \approx -\left[0.9(-0.152) + 0.1(-3.32)\right] \quad \text{Substitute approximate values.}$$
$$H \approx 0.469. \quad \text{Simplify.}$$

Verify that $H \approx 0.971$ if there are 60 of one species and 40 of the other. As the proportions of n species get closer to $\frac{1}{n}$ each, the measure of diversity increases to a maximum of $\log_2 n$.

Now Try Exercise 73.

At the end of **Section 4.2,** we saw that graphing calculators are capable of fitting exponential curves to data that suggest such behavior. The same is true for logarithmic curves. For example, during the early 2000s on one particular day, interest rates for various U.S. Treasury Securities were as shown in the table.

Time	3-mo	6-mo	2-yr	5-yr	10-yr	30-yr
Yield	0.83%	0.91%	1.35%	2.46%	3.54%	4.58%

Source: U.S. Treasury.

Figure 38 shows how a calculator gives the best-fitting natural logarithmic curve for the data, as well as the data points and the graph of this curve.

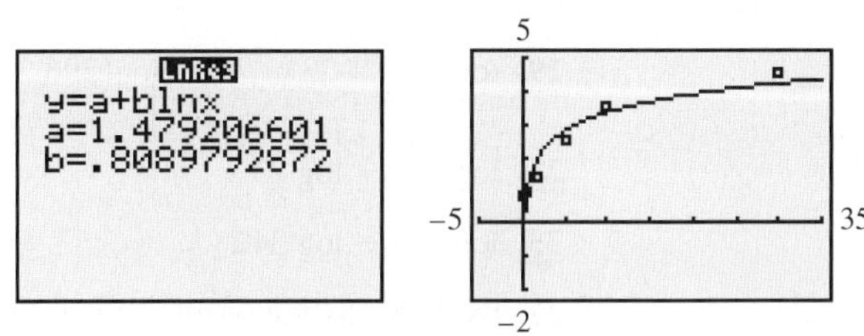

Figure 38

4.4 Exercises

1. increasing
2. increasing
3. $f^{-1}(x) = \log_5 x$
4. $\log_4 11$
5. natural; common
6. $\frac{\ln 12}{\ln 3}$
7. There is no power of 2 that yields a result of 0.
8. 3 and 4
9. $\log 8 = 0.90308999$
10. $\ln 2.75 = 1.0116009$

11. 12 12. 7
13. −1 14. −2
15. 1.7993 16. 1.9731
17. −2.6576 18. −2.2596
19. 3.9494 20. 3.5505
21. 0.1803 22. 0.3503
23. 3.9494 24. 3.5505
25. 0.1803 26. 0.3503

Concept Check Answer each of the following.

1. For the exponential function $f(x) = a^x$, where $a > 1$, is the function increasing or decreasing over its entire domain?

2. For the logarithmic function $g(x) = \log_a x$, where $a > 1$, is the function increasing or decreasing over its entire domain?

3. If $f(x) = 5^x$, what is the rule for $f^{-1}(x)$?

4. What is the name given to the exponent to which 4 must be raised to obtain 11?

5. A base e logarithm is called a(n) ________ logarithm, and a base 10 logarithm is called a(n) ________ logarithm.

6. How is $\log_3 12$ written in terms of natural logarithms?

7. Why is $\log_2 0$ undefined?

8. Between what two consecutive integers must $\log_2 12$ lie?

9. The graph of $y = \log x$ shows a point on the graph. Write the logarithmic equation associated with that point.

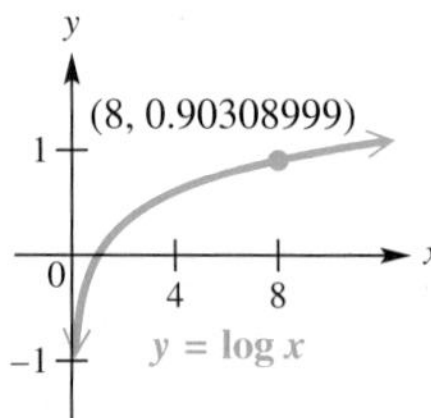

10. The graph of $y = \ln x$ shows a point on the graph. Write the logarithmic equation associated with that point.

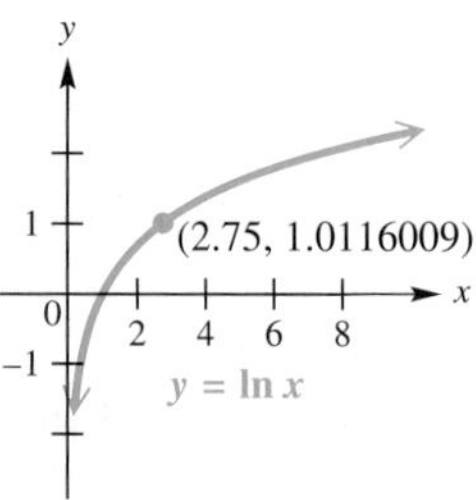

Find each value. If applicable, give an approximation to four decimal places. ***See Example 1.***

11. $\log 10^{12}$ **12.** $\log 10^{7}$ **13.** $\log 0.1$ **14.** $\log 0.01$

15. $\log 63$ **16.** $\log 94$ **17.** $\log 0.0022$ **18.** $\log 0.0055$

19. $\log(387 \times 23)$ **20.** $\log(296 \times 12)$ **21.** $\log\left(\frac{518}{342}\right)$ **22.** $\log\left(\frac{643}{287}\right)$

23. $\log 387 + \log 23$ **24.** $\log 296 + \log 12$

25. $\log 518 - \log 342$ **26.** $\log 643 - \log 287$

27. Explain why the result in **Exercise 23** is the same as that in **Exercise 19.**

28. Explain why the result in **Exercise 25** is the same as that in **Exercise 21.**

29. 3.2 30. 1.8
31. 8.4 32. 13.5

33. 2.0×10^{-3} 34. 4.0×10^{-4}
35. 1.6×10^{-5} 36. 3.2×10^{-7}

37. poor fen 38. poor fen
39. bog 40. bog
41. rich fen 42. rich fen

43. (a) 2.60031933
(b) 1.60031933
(c) 0.6003193298
(d) The whole number parts will vary, but the decimal parts will be the same.

45. 1.6 46. 5.8
47. −2 48. −4
49. 3.3322 50. 3.6636
51. −8.9480 52. −4.8665
53. 10.1449 54. 9.8386
55. 2.0200 56. 1.5976
57. 10.1449 58. 9.8386
59. 2.0200 60. 1.5976

63. (a) 20 (b) 30 (c) 50
(d) 60 (e) about 3 decibels
64. (a) 21 (b) 70 (c) 91
(d) 120 (e) 140

For each substance, find the pH from the given hydronium ion concentration. ***See Example 2(a).***

29. grapefruit, 6.3×10^{-4}

30. limes, 1.6×10^{-2}

31. crackers, 3.9×10^{-9}

32. sodium hydroxide (lye), 3.2×10^{-14}

Find the $[H_3O^+]$ *for each substance with the given pH.* ***See Example 2(b).***

33. soda pop, 2.7

34. wine, 3.4

35. beer, 4.8

36. drinking water, 6.5

In Exercises 37–42, suppose that water from a wetland area is sampled and found to have the given hydronium ion concentration. Determine whether the wetland is a rich fen, a poor fen, or a bog. ***See Example 3.***

37. 2.49×10^{-5}

38. 6.22×10^{-5}

39. 2.49×10^{-2}

40. 3.14×10^{-2}

41. 2.49×10^{-7}

42. 5.86×10^{-7}

43. Use your calculator to find an approximation for each logarithm.
(a) log 398.4 **(b)** log 39.84 **(c)** log 3.984
(d) From your answers to parts (a)–(c), make a conjecture concerning the decimal values in the approximations of common logarithms of numbers greater than 1 that have the same digits.

44. Given that $\log 25 \approx 1.3979$, $\log 250 \approx 2.3979$, and $\log 2500 \approx 3.3979$, make a conjecture for an approximation of log 25,000. Then explain why this pattern continues.

Find each value. If applicable, give an approximation to four decimal places. ***See Example 5.***

45. $\ln e^{1.6}$

46. $\ln e^{5.8}$

47. $\ln\left(\frac{1}{e^2}\right)$

48. $\ln\left(\frac{1}{e^4}\right)$

49. ln 28

50. ln 39

51. ln 0.00013

52. ln 0.0077

53. $\ln(27 \times 943)$

54. $\ln(33 \times 568)$

55. $\ln\left(\frac{98}{13}\right)$

56. $\ln\left(\frac{84}{17}\right)$

57. ln 27 + ln 943

58. ln 33 + ln 568

59. ln 98 − ln 13

60. ln 84 − ln 17

61. Explain why the result in **Exercise 57** is the same as that in **Exercise 53.**

62. Explain why the result in **Exercise 59** is the same as that in **Exercise 55.**

Solve each application of logarithms. ***See Examples 4, 6, 7, and 9.***

63. *Decibel Levels* Find the decibel ratings of sounds having the following intensities.
(a) $100I_0$ **(b)** $1000I_0$ **(c)** $100{,}000I_0$ **(d)** $1{,}000{,}000I_0$
(e) If the intensity of a sound is doubled, by how much is the decibel rating increased?

64. *Decibel Levels* Find the decibel ratings of the following sounds, having intensities as given. Round each answer to the nearest whole number.
(a) whisper, $115I_0$ **(b)** busy street, $9{,}500{,}000I_0$
(c) heavy truck, 20 m away, $1{,}200{,}000{,}000I_0$
(d) rock music, $895{,}000{,}000{,}000I_0$
(e) jetliner at takeoff, $109{,}000{,}000{,}000{,}000I_0$

65. (a) 3 (b) 6 (c) 8
66. $1{,}258{,}925{,}412\,I_0$
67. $630{,}957{,}345\,I_0$
68. 1.995 times greater
69. 94.6 thousand; We must assume that the model continues to be logarithmic.

65. *Earthquake Intensity* The magnitude of an earthquake, measured on the Richter scale, is $\log_{10}\frac{I}{I_0}$, where I is the amplitude registered on a seismograph 100 km from the epicenter of the earthquake, and I_0 is the amplitude of an earthquake of a certain (small) size. Find the Richter scale ratings for earthquakes having the following amplitudes.

(a) $1000I_0$ **(b)** $1{,}000{,}000I_0$ **(c)** $100{,}000{,}000I_0$

66. *Earthquake Intensity* On December 26, 2004, an earthquake struck in the Indian Ocean with a magnitude of 9.1 on the Richter scale. The resulting tsunami killed an estimated 229,900 people in several countries. Express this reading in terms of I_0.

67. *Earthquake Intensity* On February 27, 2010, a massive earthquake struck Chile with a magnitude of 8.8 on the Richter scale. Express this reading in terms of I_0.

68. *Earthquake Intensity Comparison* Compare your answers to **Exercises 66 and 67.** How many times greater was the force of the 2004 earthquake than that of the 2010 earthquake?

69. *(Modeling) Bachelor's Degrees in Psychology* The table gives the number of bachelor's degrees in psychology (in thousands) earned at U.S. colleges and universities for selected years from 1980 through 2008. Suppose x represents the number of years since 1950. Thus, 1980 is represented by 30, 1990 is represented by 40, and so on.

Year	Degrees Earned (in thousands)
1980	42.1
1990	54.0
2000	74.2
2005	85.6
2007	90.0
2008	92.6

Source: U.S. National Center for Education Statistics.

The logarithmic function

$$f(x) = -228.1 + 78.19 \ln x$$

is the best-fitting logarithmic model for the data. Use this function to estimate the number of bachelor's degrees in psychology earned in the year 2012. What assumption must we make to estimate the number of degrees in years beyond 2012?

70. *(Modeling) Domestic Leisure Travel* The bar graph shows numbers of leisure trips within the United States (in millions of person-trips of 50 or more miles one-way) over the years 2003–2008. The function

$$f(t) = 1393 + 69.49 \ln t, \quad t \geq 1,$$

where t represents the number of years since 2002 and $f(t)$ is the number of person-trips, in millions, approximates the curve reasonably well.

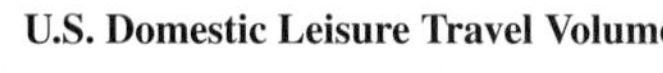

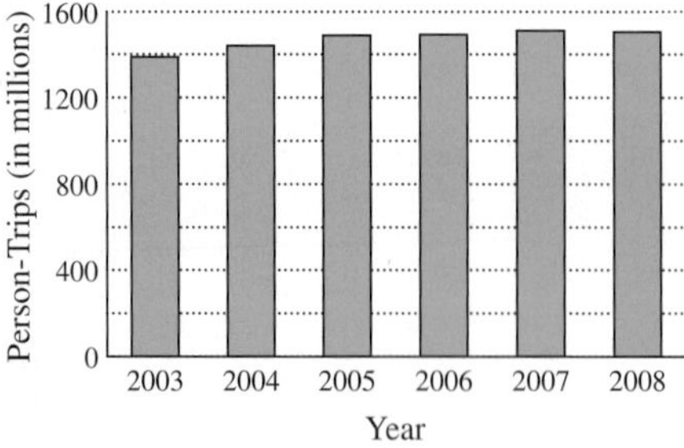

Source: U.S. Travel Association.

70. **(a)** approximately 1489 million person-trips; This is very close to the actual number.
71. **(a)** 2 **(b)** 2 **(c)** 2 **(d)** 1
72. **(a)** 4 **(b)** 4 **(c)** 5
73. 1
74. 1.59
75. between 7°F and 11°F
76. **(a)** $T(x) =$

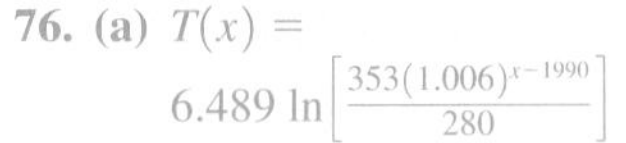

$6.489 \ln\left[\frac{353(1.006)^{x-1990}}{280}\right]$

(b) $C(x) = 353(1.006)^{x-1990}$

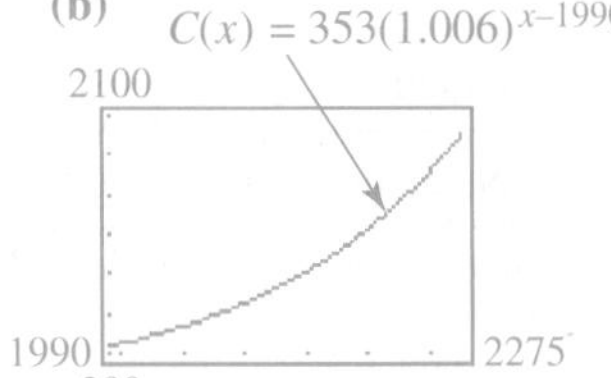

$T(x) = 6.489 \ln\left(\frac{C}{280}\right)$, where $C(x)$ is defined as given earlier

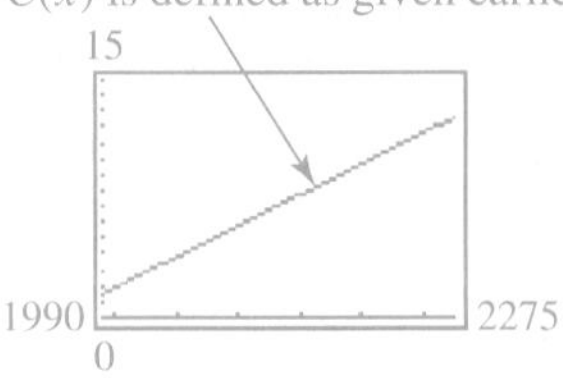

C is an exponential function, and T is a linear function over the same time period. While the carbon dioxide levels in the atmosphere increase at an exponential rate, the average global temperature will rise at a linear rate.

(c) The slope is 0.0388. This means that the temperature is expected to rise at an average rate of 0.04°F per year from 1990 to 2275.

(d) $x \approx 2208.9$; $C \approx 1307$ ppm (about 4.67 times above preindustrial carbon dioxide levels)

77. 1.126 billion yr

(a) Use the function to approximate the number of person-trips in 2006. How does this approximation compare to the actual number of 1492 million?

(b) Explain why an exponential function would *not* provide a good model for these data.

71. *(Modeling) Diversity of Species* The number of species $S(n)$ in a sample is given by

$$S(n) = a \ln\left(1 + \frac{n}{a}\right),$$

where n is the number of individuals in the sample, and a is a constant that indicates the diversity of species in the community. If $a = 0.36$, find $S(n)$ for each value of n. (*Hint:* $S(n)$ must be a whole number.)

(a) 100 **(b)** 200 **(c)** 150 **(d)** 10

72. *(Modeling) Diversity of Species* In **Exercise 71,** find $S(n)$ if a changes to 0.88. Use the following values of n.

(a) 50 **(b)** 100 **(c)** 250

73. *(Modeling) Diversity of Species* Suppose a sample of a small community shows two species with 50 individuals each. Find the measure of diversity H.

74. *(Modeling) Diversity of Species* A virgin forest in northwestern Pennsylvania has 4 species of large trees with the following proportions of each: hemlock, 0.521; beech, 0.324; birch, 0.081; maple, 0.074. Find the measure of diversity H.

75. *(Modeling) Global Temperature Increase* In **Example 7,** we expressed the average global temperature increase T (in °F) as

$$T(k) = 1.03k \ln \frac{C}{C_0},$$

where C_0 is the preindustrial amount of carbon dioxide, C is the current carbon dioxide level, and k is a constant. Arrhenius determined that $10 \leq k \leq 16$ when C was double the value C_0. Use $T(k)$ to find the range of the rise in global temperature T (rounded to the nearest degree) that Arrhenius predicted. (*Source:* Clime, W., *The Economics of Global Warming,* Institute for International Economics, Washington, D.C.)

76. *(Modeling) Global Temperature Increase* (Refer to **Exercise 75.**) According to one study by the IPCC, future increases in average global temperatures (in °F) can be modeled by

$$T(C) = 6.489 \ln \frac{C}{280},$$

where C is the concentration of atmospheric carbon dioxide (in ppm). C can be modeled by the function

$$C(x) = 353(1.006)^{x-1990},$$

where x is the year. (*Source:* International Panel on Climate Change (IPCC).)

(a) Write T as a function of x.

(b) Using a graphing calculator, graph $C(x)$ and $T(x)$ on the interval [1990, 2275] using different coordinate axes. Describe the graph of each function. How are C and T related?

(c) Approximate the slope of the graph of T. What does this slope represent?

(d) Use graphing to estimate x and $C(x)$ when $T(x) = 10$°F.

77. *Age of Rocks* Use the formula of **Example 6** to estimate the age of a rock sample having $\frac{A}{K} = 0.103$.

78. (a)

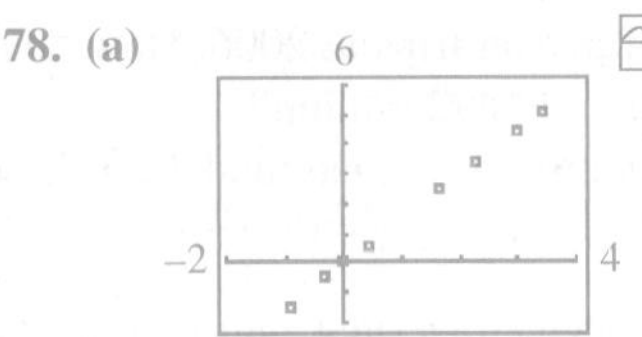

Let $x = \ln D$ and $y = \ln P$ for each planet. From the graph, the data appear to be linear.

(b)

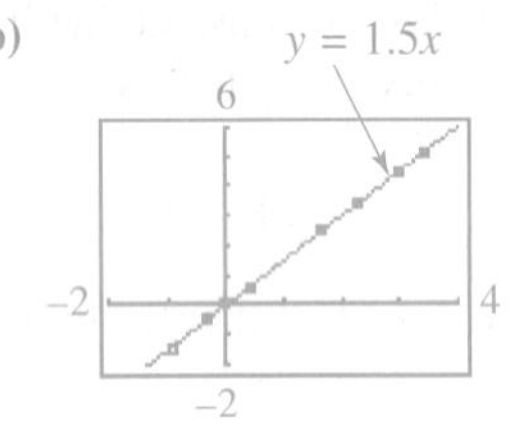

The points $(0, 0)$ and $(3.40, 5.10)$ determine the line $y = 1.5x$ or $\ln P = 1.5 \ln D$. (Answers will vary.)

(c) $P \approx 248.3$ yr

79. 2.3219 **80.** 3.1699
81. −0.2537 **82.** −0.1647
83. −1.5850 **84.** −0.6309
85. 0.8736 **86.** 0.3028
87. 1.9376 **88.** 1.0932
89. −1.4125 **90.** −22.0488

91. $4v + \frac{1}{2}u$ **92.** $3u - 2v$
93. $\frac{3}{2}u - \frac{5}{2}v$ **94.** $\frac{1}{3}u + 4v$

95. (a) 4 (b) 5^2, or 25 (c) $\frac{1}{e}$
96. (a) 2 (b) ln 3 (c) 2 ln 3, or ln 9
97. (a) 6 (b) ln 3 (c) 2 ln 3, or ln 9
98. (a) 7 (b) 1 (c) 2

99. D **100.** D

101. domain: $(-\infty, 0) \cup (0, \infty)$; range: $(-\infty, \infty)$; symmetric with respect to the y-axis

102. (a) $(-\infty, 0) \cup (0, \infty)$

(b)

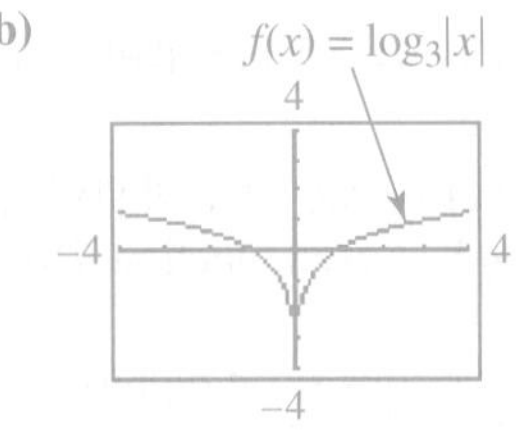

(c) The graph appears to show a point with x-value of 0, which does not exist on this graph.

---

78. *(Modeling) Planets' Distances from the Sun and Periods of Revolution* The table contains the planets' average distances D from the sun and their periods P of revolution around the sun in years. The distances have been normalized so that Earth is one unit away from the sun. For example, since Jupiter's distance is 5.2, its distance from the sun is 5.2 times farther than Earth's.

(a) Using a graphing calculator, make a scatter diagram by plotting the point $(\ln D, \ln P)$ for each planet on the xy-coordinate axes. Do the data points appear to be linear?

(b) Determine a linear equation that models the data points. Graph your line and the data on the same coordinate axes.

(c) Use this linear model to predict the period of Pluto if its distance is 39.5. Compare your answer to the actual value of 248.5 yr.

Planet	D	P
Mercury	0.39	0.24
Venus	0.72	0.62
Earth	1	1
Mars	1.52	1.89
Jupiter	5.2	11.9
Saturn	9.54	29.5
Uranus	19.2	84.0
Neptune	30.1	164.8

Source: Ronan, C., *The Natural History of the Universe,* MacMillan Publishing Co., New York.

Use the change-of-base theorem to find an approximation to four decimal places for each logarithm. ***See Example 8.***

79. $\log_2 5$ **80.** $\log_2 9$ **81.** $\log_8 0.59$ **82.** $\log_8 0.71$

83. $\log_{1/2} 3$ **84.** $\log_{1/3} 2$ **85.** $\log_\pi e$ **86.** $\log_\pi \sqrt{2}$

87. $\log_{\sqrt{13}} 12$ **88.** $\log_{\sqrt{19}} 5$ **89.** $\log_{0.32} 5$ **90.** $\log_{0.91} 8$

Let $u = \ln a$ and $v = \ln b$. Write each expression in terms of u and v without using the ln *function.*

91. $\ln(b^4\sqrt{a})$ **92.** $\ln \frac{a^3}{b^2}$ **93.** $\ln \sqrt{\frac{a^3}{b^5}}$ **94.** $\ln(\sqrt[3]{a} \cdot b^4)$

Concept Check In Exercises 95–98, use the various properties of exponential and logarithmic functions to evaluate the expressions in parts (a)–(c).

95. Given $g(x) = e^x$, find **(a)** $g(\ln 4)$ **(b)** $g(\ln(5^2))$ **(c)** $g(\ln(\frac{1}{e}))$.

96. Given $f(x) = 3^x$, find **(a)** $f(\log_3 2)$ **(b)** $f(\log_3(\ln 3))$ **(c)** $f(\log_3(2 \ln 3))$.

97. Given $f(x) = \ln x$, find **(a)** $f(e^6)$ **(b)** $f(e^{\ln 3})$ **(c)** $f(e^{2 \ln 3})$.

98. Given $f(x) = \log_2 x$, find **(a)** $f(2^7)$ **(b)** $f(2^{\log_2 2})$ **(c)** $f(2^{2 \log_2 2})$.

Work each problem.

99. *Concept Check* Which of the following is equivalent to $2 \ln(3x)$ for $x > 0$?

A. $\ln 9 + \ln x$ **B.** $\ln(6x)$ **C.** $\ln 6 + \ln x$ **D.** $\ln(9x^2)$

100. *Concept Check* Which of the following is equivalent to $\ln(4x) - \ln(2x)$ for $x > 0$?

A. $2 \ln x$ **B.** $\ln(2x)$ **C.** $\frac{\ln(4x)}{\ln(2x)}$ **D.** $\ln 2$

101. The function $f(x) = \ln |x|$ plays a prominent role in calculus. Find its domain, its range, and the symmetries of its graph.

102. Consider the function $f(x) = \log_3 |x|$.

(a) What is the domain of this function?

(b) Use a graphing calculator to graph $f(x) = \log_3 |x|$ in the window $[-4, 4]$ by $[-4, 4]$.

(c) How might one easily misinterpret the domain of the function by merely observing the calculator graph?

Use the properties of logarithms and the terminology of **Chapter 2** to describe how the graph of the given function compares to the graph of $g(x) = \ln x$.

103. $f(x) = \ln(e^2x)$ **104.** $f(x) = \ln \frac{x}{e}$ **105.** $f(x) = \ln \frac{x}{e^2}$

103. $f(x) = 2 + \ln x$, so it is the graph of $g(x) = \ln x$ translated 2 units up.
104. $f(x) = \ln x - 1$, so it is the graph of $g(x) = \ln x$ translated 1 unit down.
105. $f(x) = \ln x - 2$, so it is the graph of $g(x) = \ln x$ translated 2 units down.

Chapter 4 Quiz (Sections 4.1–4.4)

1. For the one-to-one function $f(x) = \sqrt[3]{3x - 6}$, find $f^{-1}(x)$.

2. Solve $4^{2x+1} = 8^{3x-6}$.

3. Graph $f(x) = -3^x$. Give the domain and range.

4. Graph $f(x) = \log_4(x + 2)$. Give the domain and range.

5. *Future Value* Suppose that \$15,000 is deposited in a bank certificate of deposit at an annual rate of 3.5% for 8 yr. Find the future value if interest is compounded as follows.

(a) annually **(b)** quarterly **(c)** monthly **(d)** daily (365 days)

6. Use a calculator to evaluate each logarithm to four decimal places.

(a) log 34.56 **(b)** ln 34.56

7. What is the meaning of the expression $\log_6 25$?

8. Solve.

(a) $x = 3^{\log_3 4}$ **(b)** $\log_x 25 = 2$ **(c)** $\log_4 x = -2$

9. Assuming all variables represent positive real numbers, use properties of logarithms to rewrite

$$\log_3 \frac{\sqrt{x} \cdot y}{pq^4}.$$

10. Given $\log_b 9 = 3.1699$ and $\log_b 5 = 2.3219$, find the value of $\log_b 225$.

11. Find the value of $\log_3 40$ to four decimal places.

12. If $f(x) = 4^x$, what is the value of $f(\log_4 12)$?

[4.1]
1. $f^{-1}(x) = \frac{x^3 + 6}{3}$

[4.2]
2. {4}
3.

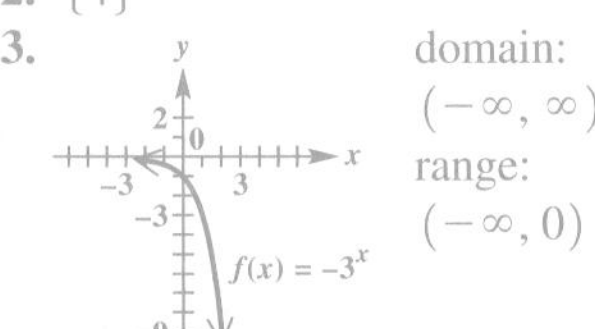

domain: $(-\infty, \infty)$; range: $(-\infty, 0)$

[4.3]
4.

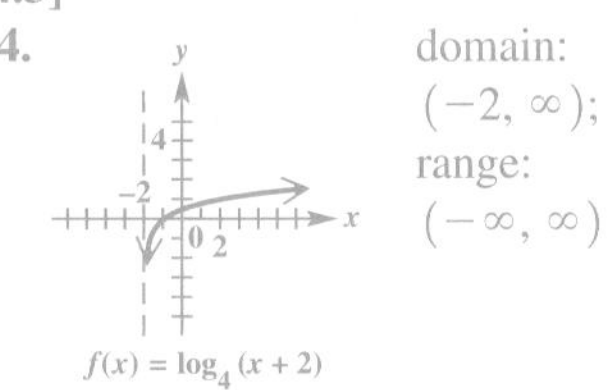

domain: $(-2, \infty)$; range: $(-\infty, \infty)$

[4.2]
5. (a) \$19,752.14 (b) \$19,822.79 (c) \$19,838.86 (d) \$19,846.68

[4.4]
6. (a) 1.5386 (b) 3.5427

[4.3]
7. The expression $\log_6 25$ represents the exponent to which 6 must be raised to obtain 25.
8. (a) {4} (b) {5} (c) $\{\frac{1}{16}\}$

[4.4]
9. $\frac{1}{2}\log_3 x + \log_3 y - \log_3 p - 4\log_3 q$ 10. 7.8137 11. 3.3578 12. 12

4.5 Exponential and Logarithmic Equations

- Exponential Equations
- Logarithmic Equations
- Applications and Models

Exponential Equations We solved exponential equations in earlier sections. General methods for solving these equations depend on the property below, which follows from the fact that logarithmic functions are one-to-one.

Property of Logarithms

If $x > 0$, $y > 0$, $a > 0$, and $a \neq 1$, then the following holds.

$\boldsymbol{x = y}$ is equivalent to $\boldsymbol{\log_a x = \log_a y}$.

Classroom Example 1
Solve $8^x = 21$. Give the solution to the nearest thousandth.

Answer: $\{1.464\}$

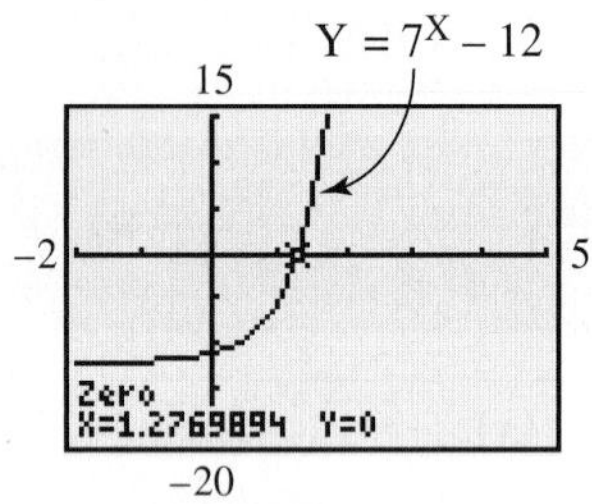

As seen in the display at the bottom of the screen, when rounded to three decimal places, the solution agrees with that found in **Example 1.**

EXAMPLE 1 Solving an Exponential Equation

Solve $7^x = 12$. Give the solution to the nearest thousandth.

SOLUTION The properties of exponents given in **Section 4.2** cannot be used to solve this equation, so we apply the preceding property of logarithms. While any appropriate base b can be used, the best practical base is base 10 or base e. We choose base e (natural) logarithms here.

$$7^x = 12$$

$$\ln 7^x = \ln 12 \quad \text{Property of logarithms}$$

$$x \ln 7 = \ln 12 \quad \text{Power property (Section 4.3)}$$

$$x = \frac{\ln 12}{\ln 7} \quad \text{Divide by } \ln 7. \quad \text{(This is exact.)}$$

$$x \approx 1.277 \quad \text{Use a calculator. (Section 4.4)}$$

The solution set is $\{1.277\}$. (This is *approximate*.)

Now Try Exercise 5.

CAUTION When evaluating a quotient like $\frac{\ln 12}{\ln 7}$ in **Example 1,** do not confuse this quotient with $\ln \frac{12}{7}$, which can be written as $\ln 12 - \ln 7$. ***We cannot change the quotient of two logarithms to a difference of logarithms.***

$$\frac{\ln 12}{\ln 7} \neq \ln \frac{12}{7}$$

Classroom Example 2
Solve $5^{2x-3} = 8^{x+1}$. Give the solution to the nearest thousandth.

Answer: $\{6.062\}$

EXAMPLE 2 Solving an Exponential Equation

Solve $3^{2x-1} = 0.4^{x+2}$. Give the solution to the nearest thousandth.

SOLUTION

$$3^{2x-1} = 0.4^{x+2}$$

$$\ln 3^{2x-1} = \ln 0.4^{x+2} \quad \text{Take the natural logarithm on each side.}$$

$$(2x - 1) \ln 3 = (x + 2) \ln 0.4 \quad \text{Power property}$$

$$2x \ln 3 - \ln 3 = x \ln 0.4 + 2 \ln 0.4 \quad \text{Distributive property (Section R.2)}$$

$$2x \ln 3 - x \ln 0.4 = 2 \ln 0.4 + \ln 3 \quad \text{Write the terms with } x \text{ on one side.}$$

$$x(2 \ln 3 - \ln 0.4) = 2 \ln 0.4 + \ln 3 \quad \text{Factor out } x. \text{ (Section R.4)}$$

$$x = \frac{2 \ln 0.4 + \ln 3}{2 \ln 3 - \ln 0.4} \quad \text{Divide by } 2 \ln 3 - \ln 0.4.$$

$$x = \frac{\ln 0.4^2 + \ln 3}{\ln 3^2 - \ln 0.4} \quad \text{Power property}$$

$$x = \frac{\ln 0.16 + \ln 3}{\ln 9 - \ln 0.4} \quad \text{Apply the exponents.}$$

$$x = \frac{\ln 0.48}{\ln 22.5} \quad \text{Product and quotient properties (Section 4.3)} \quad \text{(This is exact.)}$$

$$x \approx -0.236 \quad \text{Use a calculator.}$$

The solution set is $\{-0.236\}$. (This is *approximate*.)

Now Try Exercise 13.

Teaching Tip Caution students against applying the quotient property of logarithms incorrectly in **Example 2.**

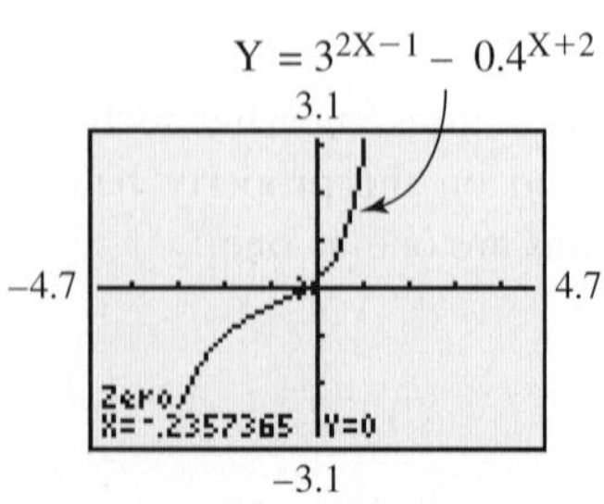

This screen supports the solution found in **Example 2.**

Classroom Example 3
Solve each equation. Give solutions to the nearest thousandth.
(a) $e^{|x|} = 50$
(b) $e^{4x} \cdot e^{x-1} = 5e$

Answers:
(a) $\{\pm 3.912\}$ (b) $\{0.722\}$

EXAMPLE 3 Solving Base *e* Exponential Equations

Solve each equation. Give solutions to the nearest thousandth.

(a) $e^{x^2} = 200$ **(b)** $e^{2x+1} \cdot e^{-4x} = 3e$

SOLUTION

(a)

$$e^{x^2} = 200$$

$$\ln e^{x^2} = \ln 200 \quad \text{Take the natural logarithm on each side.}$$

$$x^2 = \ln 200 \quad \ln e^{x^2} = x^2 \text{ (Section 4.3)}$$

Remember both roots.

$$x = \pm\sqrt{\ln 200} \quad \text{Square root property (Section 1.4)}$$

$$x \approx \pm 2.302 \quad \text{Use a calculator.}$$

The solution set is $\{\pm 2.302\}$.

(b)

$$e^{2x+1} \cdot e^{-4x} = 3e$$

$$e^{-2x+1} = 3e \quad a^m \cdot a^n = a^{m+n} \text{ (Section R.3)}$$

$$e^{-2x} = 3 \quad \text{Divide by } e;\ \frac{a^m}{a^n} = a^{m-n}. \text{ (Section R.3)}$$

$$\ln e^{-2x} = \ln 3 \quad \text{Take the natural logarithm on each side.}$$

$$-2x \ln e = \ln 3 \quad \text{Power property}$$

$$-2x = \ln 3 \quad \ln e = 1$$

$$x = -\frac{1}{2}\ln 3 \quad \text{Multiply by } -\tfrac{1}{2}.$$

$$x \approx -0.549 \quad \text{Use a calculator.}$$

The solution set is $\{-0.549\}$.

Now Try Exercises 15 and 17.

Classroom Example 4
Solve $e^{2x} - 6e^x + 5 = 0$. Give exact value(s) for x.

Answer: $\{0, \ln 5\}$

EXAMPLE 4 Solving an Exponential Equation Quadratic in Form

Solve $e^{2x} - 4e^x + 3 = 0$. Give exact value(s) for x.

SOLUTION This is an equation that is quadratic in form, because it can be rewritten with $u = e^x$.

$$e^{2x} - 4e^x + 3 = 0$$

$$(e^x)^2 - 4e^x + 3 = 0 \quad a^{mn} = (a^n)^m \text{ (Section R.3)}$$

$$u^2 - 4u + 3 = 0 \quad \text{Let } u = e^x. \text{ (Section 1.6)}$$

$$(u-1)(u-3) = 0 \quad \text{Factor. (Section R.4)}$$

$$u - 1 = 0 \quad \text{or} \quad u - 3 = 0 \quad \text{Zero-factor property (Section 1.4)}$$

$$u = 1 \quad \text{or} \quad u = 3 \quad \text{Solve for } u.$$

$$e^x = 1 \quad \text{or} \quad e^x = 3 \quad \text{Substitute } e^x \text{ for } u.$$

$$\ln e^x = \ln 1 \quad \text{or} \quad \ln e^x = \ln 3 \quad \text{Take the natural logarithm on each side.}$$

$$x = 0 \quad \text{or} \quad x = \ln 3 \quad \ln e^x = x \text{ (Section 4.3)}$$

Both values check, so the solution set is $\{0, \ln 3\}$.

Now Try Exercise 29.

Logarithmic Equations The following equations involve logarithms of variable expressions.

Classroom Example 5
Solve each equation. Give exact values.
(a) $4 \ln x = 36$
(b) $\log_3(x^3 - 5) = 1$

Answers:
(a) $\{e^9\}$ (b) $\{2\}$

EXAMPLE 5 Solving Logarithmic Equations

Solve each equation. Give exact values.

(a) $7 \ln x = 28$ **(b)** $\log_2(x^3 - 19) = 3$

SOLUTION

(a)

$$7 \ln x = 28$$
$$\ln x = 4 \quad \text{Divide by 7.}$$
$$x = e^4 \quad \text{Write the natural logarithm in exponential form.}$$

The solution set is $\{e^4\}$.

(b)

$$\log_2(x^3 - 19) = 3$$
$$x^3 - 19 = 2^3 \quad \text{Write in exponential form.}$$
$$x^3 - 19 = 8 \quad \text{Apply the exponent.}$$
$$x^3 = 27 \quad \text{Add 19.}$$
$$x = \sqrt[3]{27} \quad \text{Take cube roots. (Section 1.6)}$$
$$x = 3 \quad \sqrt[3]{27} = 3$$

The solution set is $\{3\}$.

Now Try Exercises 35 and 43.

Classroom Example 6
Solve
$\log(2x + 1) + \log x = \log(x + 8)$.
Give exact value(s).
Answer: $\{2\}$

EXAMPLE 6 Solving a Logarithmic Equation

Solve $\log(x + 6) - \log(x + 2) = \log x$. Give exact value(s).

SOLUTION Keep in mind that logarithms are defined only for nonnegative numbers.

$$\log(x + 6) - \log(x + 2) = \log x$$
$$\log \frac{x + 6}{x + 2} = \log x \quad \text{Quotient property}$$
$$\frac{x + 6}{x + 2} = x \quad \text{Property of logarithms}$$
$$x + 6 = x(x + 2) \quad \text{Multiply by } x + 2. \text{ (Section 1.6)}$$
$$x + 6 = x^2 + 2x \quad \text{Distributive property}$$
$$x^2 + x - 6 = 0 \quad \text{Standard form (Section 1.4)}$$
$$(x + 3)(x - 2) = 0 \quad \text{Factor.}$$
$$x + 3 = 0 \quad \text{or} \quad x - 2 = 0 \quad \text{Zero-factor property}$$
$$x = -3 \quad \text{or} \quad x = 2 \quad \text{Solve for } x. \text{ (Section 1.1)}$$

Teaching Tip Tell students that a logarithmic equation cannot be solved with two logarithms on the same side of the equation. The equation must be transformed so that each side contains an expression with at most one logarithm.

The proposed negative solution (-3) is not in the domain of $\log x$ in the original equation, so the only valid solution is the positive number 2. The solution set is $\{2\}$.

Now Try Exercise 63.

CAUTION Recall that the domain of $y = \log_a x$ is $(0, \infty)$. For this reason, ***it is always necessary to check that proposed solutions of a logarithmic equation result in logarithms of positive numbers in the original equation.***

Classroom Example 7
Solve $\log_3[(4x + 1)(x + 1)] = 3$.
Give exact value(s).

Answer: $\left\{-\frac{13}{4}, 2\right\}$

EXAMPLE 7 **Solving a Logarithmic Equation**

Solve $\log_2[(3x - 7)(x - 4)] = 3$. Give exact value(s).

SOLUTION $\log_2[(3x - 7)(x - 4)] = 3$

$$(3x - 7)(x - 4) = 2^3 \quad \text{Write in exponential form.}$$

$$3x^2 - 19x + 28 = 8 \quad \text{Multiply. (Section R.3)}$$

$$3x^2 - 19x + 20 = 0 \quad \text{Standard form}$$

$$(3x - 4)(x - 5) = 0 \quad \text{Factor.}$$

$$3x - 4 = 0 \quad \text{or} \quad x - 5 = 0 \quad \text{Zero-factor property}$$

$$x = \frac{4}{3} \quad \text{or} \quad x = 5 \quad \text{Solve for } x.$$

A check is necessary to be sure that the argument of the logarithm in the given equation is positive. In both cases, the product $(3x - 7)(x - 4)$ leads to 8, and $\log_2 8 = 3$ is true. The solution set is $\left\{\frac{4}{3}, 5\right\}$.

✔ *Now Try Exercise 47.*

Classroom Example 8
Solve

$\log_2(2x - 5) + \log_2(x - 3) = 3$.

Give exact value(s).

Answer: $\left\{\frac{11 + \sqrt{65}}{4}\right\}$

EXAMPLE 8 **Solving a Logarithmic Equation**

Solve $\log(3x + 2) + \log(x - 1) = 1$. Give exact value(s).

SOLUTION The notation $\log x$ is an abbreviation for $\log_{10} x$, and $1 = \log_{10} 10$.

$$\log(3x + 2) + \log(x - 1) = 1$$

$$\log(3x + 2) + \log(x - 1) = \log 10 \quad \text{Substitute.}$$

$$\log[(3x + 2)(x - 1)] = \log 10 \quad \text{Product property}$$

$$(3x + 2)(x - 1) = 10 \quad \text{Property of logarithms}$$

$$3x^2 - x - 2 = 10 \quad \text{Multiply.}$$

$$3x^2 - x - 12 = 0 \quad \text{Subtract 10.}$$

$$x = \frac{1 \pm \sqrt{1 + 144}}{6} \quad \text{Quadratic formula (Section 1.4)}$$

The two proposed solutions are

$$\frac{1 - \sqrt{145}}{6} \quad \text{and} \quad \frac{1 + \sqrt{145}}{6}.$$

The first of these proposed solutions, $\frac{1 - \sqrt{145}}{6}$, is negative and when substituted for x in $\log(x - 1)$ results in a negative argument, which is not allowed. Therefore, this solution must be rejected.

The second proposed solution, $\frac{1 + \sqrt{145}}{6}$, is positive. Substituting it for x in $\log(3x + 2)$ results in a positive argument, and substituting it for x in $\log(x + 1)$ also results in a positive argument, both of which are necessary conditions. Therefore, the solution set is $\left\{\frac{1 + \sqrt{145}}{6.}\right\}$.

✔ *Now Try Exercise 71.*

NOTE We could have used the definition of logarithm in **Example 8** by first writing

$$\log(3x + 2) + \log(x - 1) = 1 \quad \text{Equation from Example 8}$$
$$\log_{10}[(3x + 2)(x - 1)] = 1 \quad \text{Product property}$$
$$(3x + 2)(x - 1) = 10^1, \quad \text{Definition of logarithm (Section 4.3)}$$

and then continuing as shown on the preceding page.

Classroom Example 9
Solve $\ln e^{\ln x} - \ln(x - 4) = \ln 5$. Give exact value(s).

Answer: $\{5\}$

EXAMPLE 9 Solving a Base *e* Logarithmic Equation

Solve $\ln e^{\ln x} - \ln(x - 3) = \ln 2$. Give exact value(s).

SOLUTION This logarithmic equation differs from those in **Examples 7 and 8** because the expression on the right side involves a logarithm.

$$\ln e^{\ln x} - \ln(x - 3) = \ln 2$$
$$\ln x - \ln(x - 3) = \ln 2 \quad e^{\ln x} = x \text{ (Section 4.3)}$$
$$\ln \frac{x}{x - 3} = \ln 2 \quad \text{Quotient property}$$
$$\frac{x}{x - 3} = 2 \quad \text{Property of logarithms}$$
$$x = 2(x - 3) \quad \text{Multiply by } x - 3.$$
$$x = 2x - 6 \quad \text{Distributive property}$$
$$6 = x \quad \text{Solve for } x.$$

Check that the solution set is $\{6\}$.

✔ *Now Try Exercise 73.*

Solving Exponential or Logarithmic Equations

To solve an exponential or logarithmic equation, change the given equation into one of the following forms, where a and b are real numbers, $a > 0$ and $a \neq 1$, and follow the guidelines.

1. $a^{f(x)} = b$

   Solve by taking logarithms on both sides.
2. $\log_a f(x) = b$

   Solve by changing to exponential form $a^b = f(x)$.
3. $\log_a f(x) = \log_a g(x)$

   The given equation is equivalent to the equation $f(x) = g(x)$. Solve algebraically.
4. In a more complicated equation, such as

   $$e^{2x+1} \cdot e^{-4x} = 3e$$

   in **Example 3(b),** it may be necessary to first solve for $a^{f(x)}$ or $\log_a f(x)$ and then solve the resulting equation using one of the methods given above.
5. Check that each proposed solution is in the domain.

Applications and Models

Classroom Example 10
The equation $T = T_0 + Ce^{-kt}$ can be used to describe Newton's law of cooling, which is discussed in **Example 6** of **Section 4.6.** Solve this equation for k.

Answer: $k = -\frac{1}{t}\ln\left(\frac{T - T_0}{C}\right)$

EXAMPLE 10 Applying an Exponential Equation to the Strength of a Habit

The strength of a habit is a function of the number of times the habit is repeated. If N is the number of repetitions and H is the strength of the habit, then, according to psychologist C.L. Hull,

$$H = 1000(1 - e^{-kN}),$$

where k is a constant. Solve this equation for k.

SOLUTION

$$H = 1000(1 - e^{-kN}) \quad \text{First solve for } e^{-kN}.$$

$$\frac{H}{1000} = 1 - e^{-kN} \quad \text{Divide by 1000.}$$

$$\frac{H}{1000} - 1 = -e^{-kN} \quad \text{Subtract 1.}$$

$$e^{-kN} = 1 - \frac{H}{1000} \quad \text{Multiply by } -1 \text{ and rewrite.}$$

Now solve for k.

$$\ln e^{-kN} = \ln\left(1 - \frac{H}{1000}\right) \quad \text{Take the natural logarithm on each side.}$$

$$-kN = \ln\left(1 - \frac{H}{1000}\right) \quad \ln e^x = x$$

$$k = -\frac{1}{N}\ln\left(1 - \frac{H}{1000}\right) \quad \text{Multiply by } -\tfrac{1}{N}.$$

With the final equation, if one pair of values for H and N is known, k can be found, and the equation can then be used to find either H or N for given values of the other variable.

Now Try Exercise 85.

Classroom Example 11
Refer to the model in **Example 11.**

(a) Approximately what amount of coal was consumed in the United States in 1998? How does this figure compare to the actual figure of 21.66 quads?

(b) If this trend continues, approximately when will annual consumption reach 28 quads?

Answers:

(a) 20.95 quads; This is fairly close to the actual figure.

(b) It will reach 28 quads in the year 2030.

EXAMPLE 11 Modeling Coal Consumption in the U.S.

The table gives U.S. coal consumption (in quadrillions of British thermal units, or *quads*) for several years. The data can be modeled by the function

$$f(t) = 24.92 \ln t - 93.31, \quad t \geq 80,$$

where t is the number of years after 1900, and $f(t)$ is in quads.

Year	Coal Consumption (in quads)
1980	15.42
1985	17.48
1990	19.17
1995	20.09
2000	22.58
2005	22.80
2008	22.39

Source: U.S. Energy Information Administration.

(a) Approximately what amount of coal was consumed in the United States in 2003? How does this figure compare to the actual figure of 22.32 quads?

(b) If this trend continues, approximately when will annual consumption reach 25 quads?

SOLUTION

(a) The year 2003 is represented by $t = 2003 - 1900 = 103$.

$$f(103) = 24.92 \ln 103 - 93.31 \quad \text{Let } t = 103.$$

$$\approx 22.19 \quad \text{Use a calculator.}$$

Based on this model, 22.19 quads were used in 2003. This figure is very close to the actual amount of 22.32 quads.

(b) Replace $f(t)$ with 25, and solve for t.

$$25 = 24.92 \ln t - 93.31 \qquad f(t) = 25 \text{ in the given model.}$$

$$118.31 = 24.92 \ln t \qquad \text{Add 93.31.}$$

$$\ln t = \frac{118.31}{24.92} \qquad \text{Divide by 24.92. Rewrite.}$$

$$t = e^{118.31/24.92} \qquad \text{Write in exponential form.}$$

$$t \approx 115.3 \qquad \text{Use a calculator.}$$

Add 115 to 1900 to get 2015. Based on this model, annual consumption will reach 25 quads in 2015.

✔ ***Now Try Exercise 105.***

4.5 Exercises

1. $\log_7 19$; $\frac{\log 19}{\log 7}$; $\frac{\ln 19}{\ln 7}$
2. $\log_3 10$; $\frac{\log 10}{\log 3}$, or $\frac{1}{\log 3}$; $\frac{\ln 10}{\ln 3}$
3. $\log_{1/2} 12$; $\frac{\log 12}{\log\left(\frac{1}{2}\right)}$; $\frac{\ln 12}{\ln\left(\frac{1}{2}\right)}$
4. $\log_{1/3} 4$; $\frac{\log 4}{\log\left(\frac{1}{3}\right)}$; $\frac{\ln 4}{\ln\left(\frac{1}{3}\right)}$
5. $\{1.771\}$ 6. $\{1.594\}$
7. $\{-2.322\}$ 8. $\{-1.631\}$
9. $\{-6.213\}$ 10. $\{-2.151\}$
11. $\{-1.710\}$ 12. $\{0.823\}$
13. $\{3.240\}$ 14. $\{-5.057\}$
15. $\{\pm 2.146\}$ 16. $\{\pm 1.621\}$
17. $\{9.386\}$ 18. $\{0.347\}$
19. $\emptyset$ 20. $\emptyset$
21. $\{32.950\}$ 22. $\{6.579\}$
23. $\{7.044\}$ 24. $\{1\}$
25. $\{25.677\}$ 26. $\{9.095\}$
27. $\{2011.568\}$ 28. $\{1917.096\}$
29. $\{\ln 2, \ln 4\}$ 30. $\{\ln 3, \ln 5\}$
31. $\{\ln \frac{3}{2}\}$ 32. $\{\ln \frac{1}{3}\}$
33. $\{\log_5 4\}$ 34. $\{\log_3 5, \log_3 7\}$

Concept Check *An exponential equation such as*

$$5^x = 9$$

can be solved for its exact solution using the meaning of logarithm and the change-of-base theorem. Since x is the exponent to which 5 must be raised in order to obtain 9, the exact solution is

$$\log_5 9, \quad or \quad \frac{\log 9}{\log 5}, \quad or \quad \frac{\ln 9}{\ln 5}.$$

For each equation, give the exact solution in three forms similar to the forms explained above.

1. $7^x = 19$ **2.** $3^x = 10$ **3.** $\left(\frac{1}{2}\right)^x = 12$ **4.** $\left(\frac{1}{3}\right)^x = 4$

Solve each exponential equation. In Exercises 5–28, express irrational solutions as decimals correct to the nearest thousandth. In Exercises 29–34, express solutions in exact form. ***See Examples 1–4.***

5. $3^x = 7$ **6.** $5^x = 13$ **7.** $\left(\frac{1}{2}\right)^x = 5$

8. $\left(\frac{1}{3}\right)^x = 6$ **9.** $0.8^x = 4$ **10.** $0.6^x = 3$

11. $4^{x-1} = 3^{2x}$ **12.** $2^{x+3} = 5^{2x}$ **13.** $6^{x+1} = 4^{2x-1}$

14. $3^{x-4} = 7^{2x+5}$ **15.** $e^{x^2} = 100$ **16.** $e^{x^4} = 1000$

17. $e^{3x-7} \cdot e^{-2x} = 4e$ **18.** $e^{1-3x} \cdot e^{5x} = 2e$ **19.** $\left(\frac{1}{3}\right)^x = -3$

20. $\left(\frac{1}{9}\right)^x = -9$ **21.** $0.05(1.15)^x = 5$ **22.** $1.2(0.9)^x = 0.6$

23. $3(2)^{x-2} + 1 = 100$ **24.** $5(1.2)^{3x-2} + 1 = 7$ **25.** $2(1.05)^x + 3 = 10$

26. $3(1.4)^x - 4 = 60$ **27.** $5(1.015)^{x-1980} = 8$ **28.** $6(1.024)^{x-1900} = 9$

29. $e^{2x} - 6e^x + 8 = 0$ **30.** $e^{2x} - 8e^x + 15 = 0$ **31.** $2e^{2x} + e^x = 6$

32. $3e^{2x} + 2e^x = 1$ **33.** $5^{2x} + 3(5^x) = 28$ **34.** $3^{2x} - 12(3^x) = -35$

35. $\{e^2\}$ 36. $\{e^3\}$
37. $\left\{\frac{e^{1.5}}{4}\right\}$ 38. $\left\{\frac{e^5}{2}\right\}$
39. $\{2-\sqrt{10}\}$ 40. $\{3-\sqrt[4]{1000}\}$
41. $\{16\}$ 42. $\{-39\}$
43. $\{3\}$ 44. $\{-4\}$
45. $\{e\}$ 46. $\{10\}$
47. $\{-6, 4\}$ 48. $\left\{-\frac{14}{3}, 8\right\}$
49. $\{-8, 0\}$ 50. $\left\{-\frac{8}{3}, 0\right\}$
51. $\{5\}$ 52. $\{2\}$
53. $\{-5\}$ 54. $\{-1\}$
55. $\emptyset$ 56. $\emptyset$
57. $\{-8\}$ 58. $\{-13\}$
59. $\{0\}$ 60. $\left\{\frac{9}{2}\right\}$
61. $\{12\}$ 62. $\{3\}$
63. $\{25\}$ 64. $\{5\}$
65. $\emptyset$ 66. $\emptyset$
67. $\left\{\frac{5}{2}\right\}$ 68. $\{4\}$
69. $\{3\}$ 70. $\{8\}$
71. $\left\{\frac{1+\sqrt{41}}{4}\right\}$ 72. $\left\{\frac{1+\sqrt{85}}{6}\right\}$
73. $\{6\}$ 74. $\{6\}$
75. $\{4\}$ 76. $\{1, 10\}$
77. $\{1, 100\}$ 78. $\{-2, 2\}$

79. Proposed solutions that cause *any argument of a logarithm* to be negative or zero must be rejected. The statement is not correct. For example, the solution set of $\log(-x+99)=2$ is $\{-1\}$.

80. any real numbers less than or equal to $\frac{7}{4}$

81. $x = e^{k/(p-a)}$
82. $t = e^{(p-r)/k}$
83. $t = -\frac{1}{k}\log\left(\frac{T-T_0}{T_1-T_0}\right)$
84. $n = -\dfrac{\log\left(\frac{A-Pr}{A}\right)}{\log(1+r)}$

Solve each logarithmic equation. Express all solutions in exact form. ***See Examples 5–9.***

35. $5 \ln x = 10$
36. $3 \ln x = 9$
37. $\ln(4x) = 1.5$
38. $\ln(2x) = 5$
39. $\log(2-x) = 0.5$
40. $\log(3-x) = 0.75$
41. $\log_6(2x+4) = 2$
42. $\log_5(8-3x) = 3$
43. $\log_4(x^3+37) = 3$
44. $\log_7(x^3+65) = 0$
45. $\ln x + \ln x^2 = 3$
46. $\log x + \log x^2 = 3$
47. $\log_3[(x+5)(x-3)] = 2$
48. $\log_4[(3x+8)(x-6)] = 3$
49. $\log_2[(2x+8)(x+4)] = 5$
50. $\log_5[(3x+5)(x+1)] = 1$
51. $\log x + \log(x+15) = 2$
52. $\log x + \log(2x+1) = 1$
53. $\log(x+25) = \log(x+10) + \log 4$
54. $\log(3x+5) - \log(2x+4) = 0$
55. $\log(x-10) - \log(x-6) = \log 2$
56. $\log(x^2+10x-39) - \log(x-3) = \log 10$
57. $\ln(5-x) + \ln(-3-x) = \ln(1-8x)$
58. $\ln(10-x) + \ln(-6-x) = \ln(-34-15x)$
59. $\log_8(x+2) + \log_8(x+4) = \log_8 8$
60. $\log_2(5x-6) - \log_2(x+1) = \log_2 3$
61. $\log_2(x^2-100) - \log_2(x+10) = 1$
62. $\log_2(x-2) + \log_2(x-1) = 1$
63. $\log x + \log(x-21) = \log 100$
64. $\log x + \log(3x-13) = \log 10$
65. $\log(9x+5) = 3 + \log(x+2)$
66. $\log(11x+9) = 3 + \log(x+3)$
67. $\ln(4x-2) - \ln 4 = -\ln(x-2)$
68. $\ln(5+4x) - \ln(3+x) = \ln 3$
69. $\log_5(x+2) + \log_5(x-2) = 1$
70. $\log_2(x-7) + \log_2 x = 3$
71. $\log_2(2x-3) + \log_2(x+1) = 1$
72. $\log_5(3x+2) + \log_5(x-1) = 1$
73. $\ln e^x - 2\ln e = \ln e^4$
74. $\ln e^x - \ln e^3 = \ln e^3$
75. $\log_2(\log_2 x) = 1$
76. $\log x = \sqrt{\log x}$
77. $\log x^2 = (\log x)^2$
78. $\log_2 \sqrt{2x^2} = \dfrac{3}{2}$

79. *Concept Check* Suppose you overhear the following statement: "I must reject any negative proposed solution when I solve an equation involving logarithms." Is this correct? Why or why not?

80. *Concept Check* What values of x could not possibly be solutions of the following equation?

$$\log_a(4x-7) + \log_a(x^2+4) = 0$$

Solve each equation for the indicated variable. Use logarithms with the appropriate bases. ***See Example 10.***

81. $p = a + \dfrac{k}{\ln x}$, for x
82. $r = p - k \ln t$, for t
83. $T = T_0 + (T_1 - T_0)10^{-kt}$, for t
84. $A = \dfrac{Pr}{1-(1+r)^{-n}}$, for n

85. $t = -\frac{2}{R}\ln\left(1 - \frac{RI}{E}\right)$
86. $b = \frac{\ln\left(\frac{K-y}{ay}\right)}{-x}$
87. $x = \frac{\ln\left(\frac{A+B-y}{B}\right)}{-C}$
88. $M = M_0 \cdot 10^{(6-m)/2.5}$
89. $A = \frac{B}{x^C}$
90. $I = I_0 \cdot 10^{d/10}$
91. $t = \frac{\log\left(\frac{A}{P}\right)}{n\log\left(1 + \frac{r}{n}\right)}$
92. $x = 10^{D/10-16}$

93. \$11,611.84 94. \$6885.64
95. 2.6 yr 96. 5.55 yr
97. 5.02% 98. 4.27%

99. (a) 10.9% (b) 35.8% (c) 84.1%
100. (a) approximately \$3891 (b) approximately \$6990 (c) approximately \$8495

85. $I = \frac{E}{R}(1 - e^{-Rt/2})$, for t

86. $y = \frac{K}{1 + ae^{-bx}}$, for b

87. $y = A + B(1 - e^{-Cx})$, for x

88. $m = 6 - 2.5\log\left(\frac{M}{M_0}\right)$, for M

89. $\log A = \log B - C\log x$, for A

90. $d = 10\log\left(\frac{I}{I_0}\right)$, for I

91. $A = P\left(1 + \frac{r}{n}\right)^{tn}$, for t

92. $D = 160 + 10\log x$, for x

For Exercises 93–98, refer to the formulas for compound interest.

$$A = P\left(1 + \frac{r}{n}\right)^{tn} \quad \text{and} \quad A = Pe^{rt} \quad \text{(Section 4.2)}$$

93. *Compound Amount* If \$10,000 is invested in an account at 3% annual interest compounded quarterly, how much will be in the account in 5 yr if no money is withdrawn?

94. *Compound Amount* If \$5000 is invested in an account at 4% annual interest compounded continuously, how much will be in the account in 8 yr if no money is withdrawn?

95. *Investment Time* Kurt Daniels wants to buy a \$30,000 car. He has saved \$27,000. Find the number of years (to the nearest tenth) it will take for his \$27,000 to grow to \$30,000 at 4% interest compounded quarterly.

96. *Investment Time* Find t to the nearest hundredth of a year if \$1786 becomes \$2063 at 2.6%, with interest compounded monthly.

97. *Interest Rate* Find the interest rate to the nearest hundredth of a percent that will produce \$2500, if \$2000 is left at interest compounded semiannually for 4.5 yr.

98. *Interest Rate* At what interest rate, to the nearest hundredth of a percent, will \$16,000 grow to \$20,000 if invested for 5.25 yr and interest is compounded quarterly?

(Modeling) Solve each application. ***See Example 11.***

99. In the central Sierra Nevada (a mountain range in California), the percent of moisture that falls as snow rather than rain is approximated reasonably well by

$$f(x) = 86.3\ln x - 680,$$

where x is the altitude in feet and $f(x)$ is the percent of moisture that falls as snow. Find the percent of moisture that falls as snow at each altitude.

(a) 3000 ft **(b)** 4000 ft **(c)** 7000 ft

100. C. Horne Creations finds that its total sales in dollars, $T(x)$, from the distribution of x thousand catalogues is approximated by

$$T(x) = 5000\log(x + 1).$$

Find the total sales resulting from the distribution of each number of catalogues.

(a) 5000 **(b)** 24,000 **(c)** 49,000

101. during 2013
102. (a) 11.6451 m per sec
(b) 2.4823 sec
103. (a) about 61%
(b) 1989
104. (b) 984 ft (c) 39 ft

101. *Average Annual Public University Costs* The table shows the cost of a year's tuition, room and board, and fees at 4-year public colleges for the years 2002–2010. Letting y represent the cost and x the number of years since 2002, we find that the function

$$f(x) = 9318(1.06)^x$$

models the data quite well. According to this function, when will the cost in 2002 be doubled?

Year	Average Annual Cost
2002	$ 9,119
2003	$ 9,951
2004	$10,648
2005	$11,322
2006	$11,929
2007	$12,698
2008	$13,467
2009	$14,265
2010	$15,067

Source: The College Board, *Annual Survey of Colleges.*

102. *Race Speed* At the World Championship races held at Rome's Olympic Stadium in 1987, American sprinter Carl Lewis ran the 100-m race in 9.86 sec. His speed in meters per second after t seconds is closely modeled by the function

$$f(t) = 11.65(1 - e^{-t/1.27}).$$

(*Source:* Banks, Robert B., *Towing Icebergs, Falling Dominoes, and Other Adventures in Applied Mathematics,* Princeton University Press.)

(a) How fast was he running as he crossed the finish line?
(b) After how many seconds was he running at the rate of 10 m per sec?

103. *Women in Labor Force* The percent of women in the United States who were in the civilian labor force increased rapidly for several decades and then stabilized. If x represents the number of years since 1950, the function

$$f(x) = \frac{67.21}{1 + 1.081e^{-x/24.71}}$$

models the percent fairly well. (*Source: Monthly Labor Review,* U.S. Bureau of Labor Statistics.)

(a) What percent of U.S. women were in the civilian labor force in 2008?
(b) In what year were 55% of U.S. women in the civilian labor force?

104. *Height of the Eiffel Tower* One side of the Eiffel Tower in Paris has a shape that can be approximated by the graph of the function

$$f(x) = -301 \ln \frac{x}{207}.$$

See the figure. (*Source:* Banks, Robert B., *Towing Icebergs, Falling Dominoes, and Other Adventures in Applied Mathematics,* Princeton University Press.)

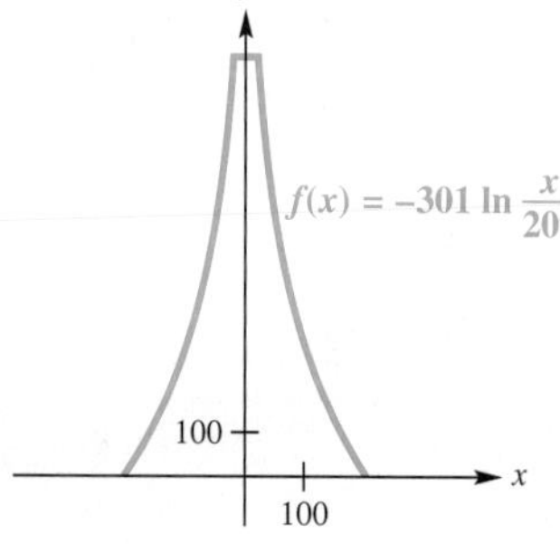

(a) Explain why the shape of the left side of the Eiffel Tower has the formula given by $f(-x)$.
(b) The short horizontal line at the top of the figure has length 15.7488 ft. Approximately how tall is the Eiffel Tower?
(c) Approximately how far from the center of the tower is the point on the right side that is 500 ft above the ground?

105. (a) $P(T) = 1 - e^{-0.0034-0.0053T}$
(b) For $T = x$, $P(x) = 1 - e^{-0.0034-0.0053x}$

1

0 1000
0

(c) $P(60) \approx 0.275$, or 27.5%; The reduction in carbon emissions from a tax of \$60 per ton of carbon is 27.5%.
(d) $T = \$130.14$

106. (a) $R \approx 4.4$ w/m^2
(b) $T \approx 4.5°$F; While this is less than that predicted by Arrhenius in 1896, his values are still consistent with some current computer models.

107. $f^{-1}(x) = \ln(x + 4) - 1$; domain: $(-4, \infty)$; range: $(-\infty, \infty)$

108. $f^{-1}(x) = \frac{1}{3}e^{x/2}$; domain: $(-\infty, \infty)$; range: $(0, \infty)$

109. $\{1.52\}$ 110. $\{-0.93, 1.35\}$

111. $\{0\}$ 112. $\{0.69, 1.10\}$

113. $\{2.45, 5.66\}$

114. $\{0.23\}$

105. *CO_2 Emissions Tax* One action that government could take to reduce carbon emissions into the atmosphere is to levy a tax on fossil fuel. This tax would be based on the amount of carbon dioxide emitted into the air when the fuel is burned. The **cost-benefit equation**

$$\ln(1 - P) = -0.0034 - 0.0053T$$

models the approximate relationship between a tax of T dollars per ton of carbon and the corresponding percent reduction P (in decimal form) of emissions of carbon dioxide. (*Source:* Nordhause, W., "To Slow or Not to Slow: The Economics of the Greenhouse Effect," Yale University, New Haven, Connecticut.)

(a) Write P as a function of T.

(b) Graph P for $0 \le T \le 1000$. Discuss the benefit of continuing to raise taxes on carbon.

(c) Determine P when $T = \$60$, and interpret this result.

(d) What value of T will give a 50% reduction in carbon emissions?

106. *Radiative Forcing* (Refer to **Example 7** in **Section 4.4.**) Using computer models, the International Panel on Climate Change (IPCC) in 1990 estimated k to be 6.3 in the radiative forcing equation

$$R = k \ln \frac{C}{C_0},$$

where C_0 is the preindustrial amount of carbon dioxide and C is the current level. (*Source:* Clime, W., *The Economics of Global Warming,* Institute for International Economics, Washington, D.C.)

(a) Use the equation $R = 6.3 \ln \frac{C}{C_0}$ to determine the radiative forcing R (in watts per square meter) expected by the IPCC if the carbon dioxide level in the atmosphere doubles from its preindustrial level.

(b) Determine the global temperature increase T that the IPCC predicted would occur if atmospheric carbon dioxide levels were to double. (*Hint:* $T(R) = 1.03R$.)

Find $f^{-1}(x)$, and give the domain and range.

107. $f(x) = e^{x+1} - 4$

108. $f(x) = 2 \ln 3x$

Use a graphing calculator to solve each equation. Give irrational solutions correct to the nearest hundredth.

109. $e^x + \ln x = 5$

110. $e^x - \ln(x + 1) = 3$

111. $2e^x + 1 = 3e^{-x}$

112. $e^x + 6e^{-x} = 5$

113. $\log x = x^2 - 8x + 14$

114. $\ln x = -\sqrt[3]{x + 3}$

115. Explain the **error** in the following **"proof"** that $2 < 1$.

$\frac{1}{9} < \frac{1}{3}$ True statement

$\left(\frac{1}{3}\right)^2 < \frac{1}{3}$ Rewrite the left side.

$\log\left(\frac{1}{3}\right)^2 < \log \frac{1}{3}$ Take the logarithm on each side.

$2 \log \frac{1}{3} < 1 \log \frac{1}{3}$ Property of logarithms; identity property

$2 < 1$ Divide each side by $\log \frac{1}{3}$.

4.6 Applications and Models of Exponential Growth and Decay

- The Exponential Growth or Decay Function
- Growth Function Models
- Decay Function Models

The Exponential Growth or Decay Function In many situations that occur in ecology, biology, economics, and the social sciences, a quantity changes at a rate proportional to the amount present. In such cases the amount present at time t is a special function of t called an **exponential growth or decay function.**

LOOKING AHEAD TO CALCULUS
The exponential growth and decay function formulas are studied in calculus in conjunction with the topic known as **differential equations.**

Exponential Growth or Decay Function

Let y_0 be the amount or number present at time $t = 0$. Then, under certain conditions, the amount present at any time t is modeled by

$$y = y_0 e^{kt}, \quad \text{where } k \text{ is a constant.}$$

When $k > 0$, the function describes growth. In **Section 4.2,** we saw examples of exponential growth: compound interest and atmospheric carbon dioxide, for example. When $k < 0$, the function describes decay; one example of exponential decay is radioactivity.

Growth Function Models The amount of time it takes for a quantity that grows exponentially to become twice its initial amount is its **doubling time.**

EXAMPLE 1 Determining an Exponential Function to Model the Increase of Carbon Dioxide

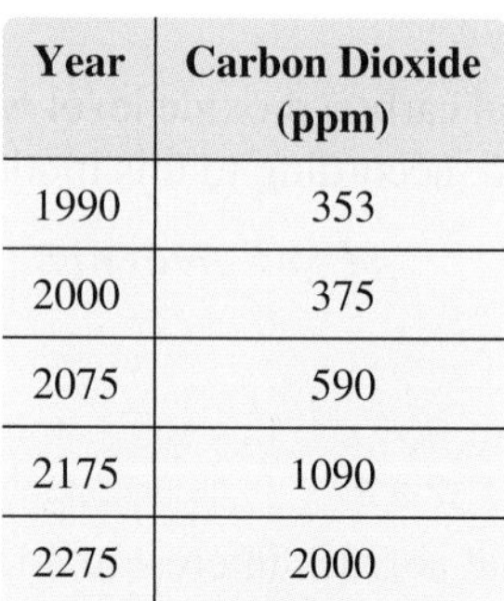

Year	Carbon Dioxide (ppm)
1990	353
2000	375
2075	590
2175	1090
2275	2000

Source: International Panel on Climate Change (IPCC).

In **Example 11, Section 4.2,** we discussed the growth of atmospheric carbon dioxide over time. A function based on the data from the table was given in that example. Now we can see how to determine such a function from the data.

(a) Find an exponential function that gives the amount of carbon dioxide y in year x.

(b) Estimate the year when future levels of carbon dioxide will be double the preindustrial level of 280 ppm.

SOLUTION

(a) Recall that the graph of the data points showed exponential growth, so the equation will take the form

$$y = y_0 e^{kx}.$$

We must find the values of y_0 and k. The data begin with the year 1990, so to simplify our work we let 1990 correspond to $x = 0$, 1991 correspond to $x = 1$, and so on. Since y_0 is the initial amount, $y_0 = 353$ in 1990 when $x = 0$. Thus the equation is

$$y = 353e^{kx}. \quad \text{Let } y_0 = 353.$$

From the last pair of values in the table, we know that in 2275 the carbon dioxide level is expected to be 2000 ppm. The year 2275 corresponds to $2275 - 1990 = 285$. Substitute 2000 for y and 285 for x, and solve for k.

Classroom Example 1
Refer to the table given with **Example 1.**

(a) Find a different exponential model using the data for 2000 and 2175. Let the year 2000 correspond to $x = 0$.

(b) Use your model from part (a) to estimate when future levels of carbon dioxide will triple from the preindustrial level of 280 ppm.

Answers:

(a) $y = 375e^{0.00610x}$

(b) It will triple from the preindustrial level during 2132.

$$y = 353e^{kx}$$

$$2000 = 353e^{k(285)} \quad \text{Substitute 2000 for } y \text{ and 285 for } x.$$

$$\frac{2000}{353} = e^{285k} \quad \text{Divide by 353.}$$

$$\ln\left(\frac{2000}{353}\right) = \ln(e^{285k}) \quad \text{Take the logarithm on each side. (Section 4.5)}$$

$$\ln\left(\frac{2000}{353}\right) = 285k \quad \ln e^x = x, \text{ for all } x. \text{ (Section 4.3)}$$

$$k = \frac{1}{285} \cdot \ln\left(\frac{2000}{353}\right) \quad \text{Multiply by } \tfrac{1}{285} \text{ and rewrite.}$$

$$k \approx 0.00609 \quad \text{Use a calculator.}$$

A function that models the data is

$$y = 353e^{0.00609x}.$$

(b) Let $y = 2(280) = 560$ in $y = 353e^{0.00609x}$, and find x.

$$560 = 353e^{0.00609x} \quad \text{Let } y = 560.$$

$$\frac{560}{353} = e^{0.00609x} \quad \text{Divide by 353.}$$

$$\ln\left(\frac{560}{353}\right) = \ln e^{0.00609x} \quad \text{Take the logarithm on each side.}$$

$$\ln\left(\frac{560}{353}\right) = 0.00609x \quad \ln e^x = x, \text{ for all } x.$$

$$x = \frac{1}{0.00609} \cdot \ln\left(\frac{560}{353}\right) \quad \text{Multiply by } \tfrac{1}{0.00609} \text{ and rewrite.}$$

$$x \approx 75.8 \quad \text{Use a calculator.}$$

Since $x = 0$ corresponds to 1990, the preindustrial carbon dioxide level will double in the 75th year after 1990, or during 2065, according to this model.

Now Try Exercise 21.

Teaching Tip In **Example 1(b)**, stress that 75.8 does not represent the time for carbon dioxide to double $e^{0.00609x}$. It represents the time since 1990 that the preindustrial level of 280 ppm took to double. Mention that the solution to

$$2 = e^{0.00609x}$$

yields a doubling time of 114 yr.

Classroom Example 2
How long will it take for the money in an account that accrues interest at a rate of 4.25%, compounded continuously, to double?

Answer: about 16 yr

Teaching Tip Have students solve **Example 2** using interest rates of 5%, 10%, and 15%. Explain that the **Rule of 72,**

$$t = \frac{72}{r},$$

is often used to approximate doubling times. Have students check their answers against this rule.

EXAMPLE 2 Finding Doubling Time for Money

How long will it take for the money in an account that accrues interest at a rate of 3%, compounded continuously, to double?

SOLUTION

$$A = Pe^{rt} \quad \text{Continuous compounding formula (Section 4.2)}$$

$$2P = Pe^{0.03t} \quad \text{Let } A = 2P \text{ and } r = 0.03.$$

$$2 = e^{0.03t} \quad \text{Divide by } P.$$

$$\ln 2 = \ln e^{0.03t} \quad \text{Take the logarithm on each side.}$$

$$\ln 2 = 0.03t \quad \ln e^x = x$$

$$\frac{\ln 2}{0.03} = t \quad \text{Divide by 0.03.}$$

$$23.10 \approx t \quad \text{Use a calculator.}$$

It will take about 23 yr for the amount to double.

Now Try Exercise 27.

EXAMPLE 3 Determining an Exponential Function to Model Population Growth

According to the U.S. Census Bureau, the world population reached 6 billion people during 1999 and was growing exponentially. By the end of 2010, the population had grown to 6.947 billion. The projected world population (in billions of people) t years after 2010 is given by the function

$$f(t) = 6.947e^{0.00745t}.$$

(a) Based on this model, what will the world population be in 2020?

(b) In what year will the world population reach 9 billion?

Classroom Example 3
Refer to the model in **Example 3.**
(a) What will the world population be at the end of 2015?
(b) In what year will the world population reach 8 billion?

Answers:
(a) 7.211 billion
(b) It will reach 8 billion during the year 2028.

SOLUTION

(a) Since $t = 0$ represents the year 2010, in 2020, t would be $2020 - 2010 = 10$ yr. We must find $f(t)$ when t is 10.

$$f(t) = 6.947e^{0.00745t} \quad \text{Given formula}$$

$$f(10) = 6.947e^{0.00745(10)} \quad \text{Let } t = 10.$$

$$f(10) \approx 7.484 \quad \text{Use a calculator.}$$

The population will be 7.484 billion at the end of 2020.

(b)

$$f(t) = 6.947e^{0.00745t} \quad \text{Given formula}$$

$$9 = 6.947e^{0.00745t} \quad \text{Let } f(t) = 9.$$

$$\frac{9}{6.947} = e^{0.00745t} \quad \text{Divide by 6.947.}$$

$$\ln \frac{9}{6.947} = \ln e^{0.00745t} \quad \text{Take the logarithm on each side.}$$

$$\ln \frac{9}{6.947} = 0.00745t \quad \ln e^x = x, \text{ for all } x.$$

$$t = \frac{\ln \frac{9}{6.947}}{0.00745} \quad \text{Divide by 0.00745 and rewrite.}$$

$$t \approx 34.8 \quad \text{Use a calculator.}$$

Thus, 34.8 yr after 2010, during the year 2044, world population will reach 9 billion.

Now Try Exercise 35.

Decay Function Models **Half-life** is the amount of time it takes for a quantity that decays exponentially to become half its initial amount.

EXAMPLE 4 Determining an Exponential Function to Model Radioactive Decay

Suppose 600 g of a radioactive substance are present initially and 3 yr later only 300 g remain.

(a) Determine the exponential equation that models this decay.

(b) How much of the substance will be present after 6 yr?

Classroom Example 4
Suppose 800 g of a radioactive substance are present initially and 2.5 yr later only 400 g remain.
(a) Determine the exponential equation that models this decay.
(b) How much of the substance will be present after 4 yr?

Answers:
(a) $y = 800e^{-0.277t}$ (b) 264 g

SOLUTION

(a) To express the situation as an exponential equation

$$y = y_0e^{kt},$$

we use the given values first to find y_0 and then to find k.

$$y = y_0 e^{kt}$$

$$600 = y_0 e^{k(0)} \quad \text{Let } y = 600 \text{ and } t = 0.$$

$$600 = y_0 \quad e^0 = 1 \text{ (Section R.3)}$$

Thus, $y_0 = 600$, which gives $y = 600e^{kt}$. Because the initial amount (600 g) decays to half that amount (300 g) in 3 yr, its half-life is 3 yr. Now we solve this exponential equation for k.

$$y = 600e^{kt} \quad \text{Let } y_0 = 600.$$

$$300 = 600e^{3k} \quad \text{Let } y = 300 \text{ and } t = 3.$$

$$0.5 = e^{3k} \quad \text{Divide by 600.}$$

$$\ln 0.5 = \ln e^{3k} \quad \text{Take logarithms on each side.}$$

$$\ln 0.5 = 3k \quad \ln e^x = x, \text{ for all } x.$$

$$\frac{\ln 0.5}{3} = k \quad \text{Divide by 3.}$$

$$k \approx -0.231 \quad \text{Use a calculator.}$$

Thus, the exponential decay equation is

$$y = 600e^{-0.231t}.$$

(b) To find the amount present after 6 yr, let $t = 6$.

$$y = 600e^{-0.231(6)} \quad \text{Let } t = 6.$$

$$y = 600e^{-1.386} \quad \text{Multiply.}$$

$$y \approx 150 \quad \text{Use a calculator.}$$

After 6 yr, 150 g of the substance will remain.

Now Try Exercise 15.

NOTE In **Example 4** the initial amount of the substance was given as 600 g. Notice that it could have been *any* amount y_0, and substituting $\frac{1}{2}y_0$ for y on the left would allow the common factors of y to be divided out, leaving the equation $0.5 = e^{3k}$. The rest of the work would be the same.

Classroom Example 5
Suppose that the skeleton of a woman who lived in the Classical Greek period was discovered in 2005. Carbon-14 testing at that time determined that the skeleton contained $\frac{3}{4}$ of the carbon-14 of a living woman of the same size. Estimate the year in which the Greek woman died.

Answer: 361 B.C.

EXAMPLE 5 Solving a Carbon Dating Problem

Carbon-14, also known as radiocarbon, is a radioactive form of carbon that is found in all living plants and animals. After a plant or animal dies, the radiocarbon disintegrates. Scientists can determine the age of the remains by comparing the amount of radiocarbon with the amount present in living plants and animals. This technique is called **carbon dating.** The amount of radiocarbon present after t years is given by

$$y = y_0 e^{-0.0001216t},$$

where y_0 is the amount present in living plants and animals.

(a) Find the half-life of carbon-14.

(b) Charcoal from an ancient fire pit on Java contained $\frac{1}{4}$ the carbon-14 of a living sample of the same size. Estimate the age of the charcoal.

SOLUTION

(a) If y_0 is the amount of radiocarbon present in a living thing, then $\frac{1}{2}y_0$ is half this initial amount. Thus, we substitute and solve the given equation for t.

$$y = y_0e^{-0.0001216t} \quad \text{Given equation}$$

$$\frac{1}{2}y_0 = y_0e^{-0.0001216t} \quad \text{Let } y = \tfrac{1}{2}y_0.$$

$$\frac{1}{2} = e^{-0.0001216t} \quad \text{Divide by } y_0.$$

$$\ln\frac{1}{2} = \ln e^{-0.0001216t} \quad \text{Take the logarithm on each side.}$$

$$\ln\frac{1}{2} = -0.0001216t \quad \ln e^x = x, \text{ for all } x.$$

$$\frac{\ln\frac{1}{2}}{-0.0001216} = t \quad \text{Divide by } -0.0001216.$$

$$5700 \approx t \quad \text{Use a calculator.}$$

The half-life is about 5700 yr.

(b) Solve again for t, this time letting the amount $y = \frac{1}{4}y_0$.

$$y = y_0e^{-0.0001216t} \quad \text{Given equation}$$

$$\frac{1}{4}y_0 = y_0e^{-0.0001216t} \quad \text{Let } y = \tfrac{1}{4}y_0.$$

$$\frac{1}{4} = e^{-0.0001216t} \quad \text{Divide by } y_0.$$

$$\ln\frac{1}{4} = \ln e^{-0.0001216t} \quad \text{Take the logarithm on each side.}$$

$$\frac{\ln\frac{1}{4}}{-0.0001216} = t \quad \ln e^x = x; \text{ Divide by } -0.0001216.$$

$$t \approx 11{,}400 \quad \text{Use a calculator.}$$

The charcoal is about 11,400 yr old.

Now Try Exercise 17.

EXAMPLE 6 Modeling Newton's Law of Cooling

Newton's law of cooling says that the rate at which a body cools is proportional to the difference in temperature between the body and the environment around it. The temperature $f(t)$ of the body at time t in appropriate units after being introduced into an environment having constant temperature T_0 is

$$f(t) = T_0 + Ce^{-kt}, \quad \text{where } C \text{ and } k \text{ are constants.}$$

A pot of coffee with a temperature of 100°C is set down in a room with a temperature of 20°C. The coffee cools to 60°C after 1 hr.

(a) Write an equation to model the data.

(b) Find the temperature after half an hour.

(c) How long will it take for the coffee to cool to 50°C?

Classroom Example 6
Newton's law of cooling also applies to warming. Mary Hill took a leg of lamb out of her refrigerator, which is set at 34°F, and placed it in her oven, which she had preheated to 350°F. After 1 hr, her meat thermometer registered 70°F.

(a) Write a function to model the data.

(b) Find the temperature 90 min after the leg of lamb was placed in the oven.

(c) Mary wants to serve the leg of lamb medium rare, which requires an internal temperature of 145°F. What is the total amount of time it will take to cook the leg of lamb?

Answers:

(a) $f(t) = 350 - 316e^{-0.121t}$

(b) 86.5°F

(c) 3.576 hr, or about 3 hr, 35 min

SOLUTION

(a) We must find values for C and k in the given formula. From the given information, when $t = 0$, $T_0 = 20$, and the temperature of the coffee is $f(0) = 100$. Also, when $t = 1$, $f(1) = 60$. Substitute the first pair of values into the function along with $T_0 = 20$.

$$f(t) = T_0 + Ce^{-kt} \quad \text{Given formula}$$

$$100 = 20 + Ce^{-0k} \quad \text{Let } t = 0,\ f(0) = 100, \text{ and } T_0 = 20.$$

$$100 = 20 + C \quad e^0 = 1$$

$$80 = C \quad \text{Subtract 20.}$$

The function that models the data can now be given.

$$f(t) = 20 + 80e^{-kt} \quad \text{Let } T_0 = 20 \text{ and } C = 80.$$

Now use the remaining pair of values in this function to find k.

$$f(t) = 20 + 80e^{-kt} \quad \text{Formula with } T_0 = 20 \text{ and } C = 80$$

$$60 = 20 + 80e^{-1k} \quad \text{Let } t = 1 \text{ and } f(1) = 60.$$

$$40 = 80e^{-k} \quad \text{Subtract 20.}$$

$$\frac{1}{2} = e^{-k} \quad \text{Divide by 80.}$$

$$\ln\frac{1}{2} = \ln e^{-k} \quad \text{Take the logarithm on each side.}$$

$$\ln\frac{1}{2} = -k \quad \ln e^x = x, \text{ for all } x.$$

$$k = -\ln\frac{1}{2} \quad \text{Multiply by } -1 \text{ and rewrite.}$$

$$k \approx 0.693 \quad \text{Use a calculator.}$$

Thus, the model is $f(t) = 20 + 80e^{-0.693t}$.

(b) To find the temperature after $\frac{1}{2}$ hr, let $t = \frac{1}{2}$ in the model from part (a).

$$f(t) = 20 + 80e^{-0.693t} \quad \text{Model from part (a)}$$

$$f\left(\frac{1}{2}\right) = 20 + 80e^{(-0.693)(1/2)} \quad \text{Let } t = \tfrac{1}{2}.$$

$$f\left(\frac{1}{2}\right) \approx 76.6^\circ\text{C} \quad \text{Use a calculator.}$$

(c) To find how long it will take for the coffee to cool to 50°C, let $f(t) = 50$.

$$50 = 20 + 80e^{-0.693t} \quad \text{Let } f(t) = 50 \text{ in model from part (a).}$$

$$30 = 80e^{-0.693t} \quad \text{Subtract 20.}$$

$$\frac{3}{8} = e^{-0.693t} \quad \text{Divide by 80.}$$

$$\ln\frac{3}{8} = \ln e^{-0.693t} \quad \text{Take the logarithm on each side.}$$

$$\ln\frac{3}{8} = -0.693t \quad \ln e^x = x, \text{ for all } x.$$

$$t = \frac{\ln\frac{3}{8}}{-0.693} \quad \text{Divide by } -0.693 \text{ and rewrite.}$$

$$t \approx 1.415 \text{ hr}, \quad \text{or} \quad \text{about 1 hr, 25 min}$$

✔ *Now Try Exercise 23.*

4.6 Exercises

1. B 2. D 3. C 4. A

5. $\frac{1}{3}\ln\frac{1}{3}$ 6. $\frac{1}{6}\ln\frac{1}{3}$

7. $\frac{1}{100}\ln\frac{1}{2}$ 8. $\frac{1}{200}\ln\frac{1}{2}$

9. $\frac{1}{2}\ln\frac{1}{4}$ 10. $\frac{1}{4}\ln\frac{1}{9}$

11. (a) 440 g (b) 387 g (c) 264 g (d) 21.66 yr

12. (a) 404 g (b) 327 g (c) 173 g (d) 13.08 yr

13. 1611.97 yr 14. 15.93 days

15. 3.57 g

16. Magnitude 1 is about 6.3 times as great as magnitude 3.

Population Growth A population is increasing according to the exponential function $y = 2e^{0.02x}$, where y is in millions and x is the number of years. Match each question in Column I with the correct procedure in Column II to answer the question.

I	II
1. How long will it take for the population to triple?	**A.** Evaluate $y = 2e^{0.02(1/3)}$.
2. When will the population reach 3 million?	**B.** Solve $2e^{0.02x} = 6$.
3. How large will the population be in 3 yr?	**C.** Evaluate $y = 2e^{0.02(3)}$.
4. How large will the population be in 4 months?	**D.** Solve $2e^{0.02x} = 3$.

(Modeling) The exercises in this set are grouped according to discipline. They involve exponential or logarithmic models. ***See Examples 1–6.***

Physical Sciences (Exercises 5–24)

In Exercises 5–10, an initial amount of a radioactive substance y_0 is given, along with information about the amount remaining after a given time t in appropriate units. For an equation of the form $y = y_0e^{kt}$ that models the situation, give the exact value of k in terms of natural logarithms.

5. $y_0 = 60$ g; After 3 hr, 20 g remain.

6. $y_0 = 30$ g; After 6 hr, 10 g remain.

7. $y_0 = 10$ mg; The half-life is 100 days.

8. $y_0 = 20$ mg; The half-life is 200 days.

9. $y_0 = 2.56$ lb; After 2 yr, 0.64 lb remains.

10. $y_0 = 8.1$ kg; After 4 yr, 0.9 kg remains.

11. *Decay of Lead* A sample of 500 g of radioactive lead-210 decays to polonium-210 according to the function

$$A(t) = 500e^{-0.032t},$$

where t is time in years. Find the amount of radioactive lead remaining after

(a) 4 yr, **(b)** 8 yr, **(c)** 20 yr. **(d)** Find the half-life.

12. *Decay of Plutonium* Repeat **Exercise 11** for 500 g of plutonium-241, which decays according to the function $A(t) = A_0e^{-0.053t}$, where t is time in years.

13. *Decay of Radium* Find the half-life of radium-226, which decays according to the function $A(t) = A_0e^{-0.00043t}$, where t is time in years.

14. *Decay of Iodine* How long will it take any quantity of iodine-131 to decay to 25% of its initial amount, knowing that it decays according to the exponential function $A(t) = A_0e^{-0.087t}$, where t is time in days?

15. *Radioactive Decay* If 12 g of a radioactive substance are present initially and 4 yr later only 6 g remain, how much of the substance will be present after 7 yr?

16. *Magnitude of a Star* The magnitude M of a star is modeled by

$$M = 6 - \frac{5}{2}\log\frac{I}{I_0},$$

where I_0 is the intensity of a just-visible star and I is the actual intensity of the star being measured. The dimmest stars are of magnitude 6, and the brightest are of magnitude 1. Determine the ratio of light intensities between a star of magnitude 1 and a star of magnitude 3.

17. about 9000 yr
18. about 4200 yr
19. about 15,600 yr
20. 43°C
21. (a) possible answer: $f(x) = 31,000(0.97)^{x-1970}$
(b) 3%
22. about 1.635 billion yr
23. 6.25°C 24. 81.25°C
25. (a) 5% compounded quarterly
(b) $837.73
26. about 6.06 yr
27. about 27.73 yr

17. *Carbon-14 Dating* Suppose an Egyptian mummy is discovered in which the amount of carbon-14 present is only about one-third the amount found in living human beings. About how long ago did the Egyptian die?

18. *Carbon-14 Dating* A sample from a refuse deposit near the Strait of Magellan had 60% of the carbon-14 of a contemporary sample. How old was the sample?

19. *Carbon-14 Dating* Paint from the Lascaux caves of France contains 15% of the normal amount of carbon-14. Estimate the age of the paintings.

20. *Dissolving a Chemical* The amount of a chemical that will dissolve in a solution increases exponentially as the (Celsius) temperature t is increased according to the model

$$A(t) = 10e^{0.0095t}.$$

At what temperature will 15 g dissolve?

21. *Sulfur Dioxide Levels* Sulfur dioxide (SO_2), a gas that is emitted by burning fossil fuels, is a source of air pollution and one of the major contributors to acid rain. Reductions in emissions of sulfur dioxide in the United States were mandated by the 1990 revision of the Clean Air Act. The graph displays approximate U.S. emissions of sulfur dioxide (in thousands of tons) over the period 1970–2008.

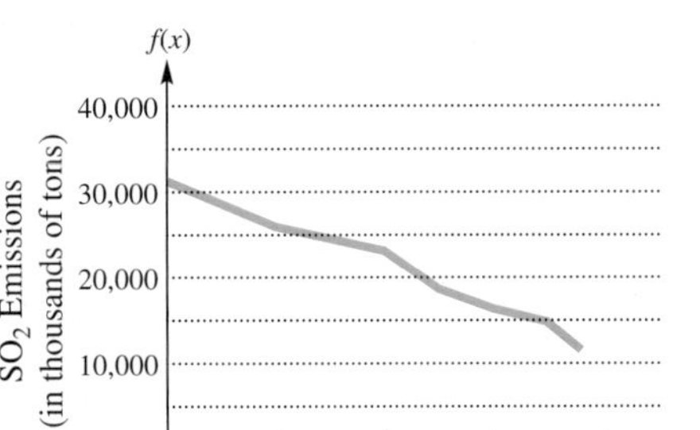

Source: U.S. Environmental Protection Agency.

(a) Use the graph to find an exponential function

$$f(x) = A_0 a^{x-1970}$$

that models U.S. emissions of sulfur dioxide over the years 1970–2008.

(b) Approximate the average annual percent decrease in sulfur dioxide emissions over this time period.

22. *Rock Sample Age* Estimate the age of a rock sample using the function

$$t = T\frac{\ln\left(1 + 8.33\left(\frac{A}{K}\right)\right)}{\ln 2}.$$

Tests show that $\frac{A}{K}$ is 0.175 for the sample. Let $T = 1.26 \times 10^9$.

23. *Newton's Law of Cooling* Boiling water, at 100°C, is placed in a freezer at 0°C. The temperature of the water is 50°C after 24 min. Find the temperature of the water after 96 min. (*Hint:* Change minutes to hours.)

24. *Newton's Law of Cooling* A piece of metal is heated to 300°C and then placed in a cooling liquid at 50°C. After 4 min, the metal has cooled to 175°C. Find its temperature after 12 min. (*Hint:* Change minutes to hours.)

Finance (Exercises 25–30)

25. *Comparing Investments* Russ McClelland, who is self-employed, wants to invest $60,000 in a pension plan. One investment offers 5% compounded quarterly. Another offers 4.75% compounded continuously.

(a) Which investment will earn more interest in 5 yr?
(b) How much more will the better plan earn?

26. *Growth of an Account* If Russ (see **Exercise 25**) chooses the plan with continuous compounding, how long will it take for his $60,000 to grow to $80,000?

27. *Doubling Time* Find the doubling time of an investment earning 2.5% interest if interest is compounded continuously.

28. The time will be divided by 3.
29. about 21.97 yr
30. 19 yr, 39 days

31. (a) 315 (b) 229 (c) 142
32. 1968
33. (a) $P = 1$; $a \approx 1.01355$
(b) 1.3 billion
(c) 2030
34. (a) 961,000
(b) 7.2 yr
(c) 17.3 yr
35. (a) \$2969
(b) 2006

28. *Doubling Time* If interest is compounded continuously and the interest rate is tripled, what effect will this have on the time required for an investment to double?

29. *Growth of an Account* How long will it take an investment to triple if interest is compounded continuously at 5%?

30. *Growth of an Account* Use the Table feature of a graphing calculator to find how long it will take \$1500 invested at 5.75% compounded daily to triple in value. Zoom in on the solution by systematically decreasing the increment for x. Find the answer to the nearest day. (Find the answer to the nearest day by eventually letting the increment of x equal $\frac{1}{365}$. The decimal part of the solution can be multiplied by 365 to determine the number of days greater than the nearest year. For example, if the solution is determined to be 16.2027 yr, then multiply 0.2027 by 365 to get 73.9855. The solution is then, to the nearest day, 16 yr, 74 days.) Confirm the answer algebraically.

Social Sciences (Exercises 31–38)

31. *Legislative Turnover* The turnover of legislators is a problem of interest to political scientists. In a study during the 1970s, it was found that one model of legislative turnover in the U.S. House of Representatives was

$$M(t) = 434e^{-0.08t},$$

where $M(t)$ represents the number of continuously serving members at time t. Here, $t = 0$ represents 1965, $t = 1$ represents 1966, and so on. Use this model to approximate the number of continuously serving members in each year.

(a) 1969 (b) 1973 (c) 1979

32. *Legislative Turnover* Use the model in **Exercise 31** to determine the year in which the number of continuously serving members was 338.

33. *Population Growth* In 2000 India's population reached 1 billion, and it is projected to be 1.4 billion in 2025. (*Source:* U.S. Census Bureau.)

(a) Find values for P_0 and a so that $f(x) = P_0 a^{x-2000}$ models the population of India in year x.
(b) Predict India's population in 2020 to the nearest tenth of a billion.
(c) Use f to determine the year when India's population might reach 1.5 billion.

34. *Population Decline* A midwestern city finds its residents moving to the suburbs. Its population is declining according to the function defined by

$$P(t) = P_0 e^{-0.04t},$$

where t is time measured in years and P_0 is the population at time $t = 0$. Assume that $P_0 = 1{,}000{,}000$.

(a) Find the population at time $t = 1$.
(b) Estimate the time it will take for the population to decline to 750,000.
(c) How long will it take for the population to decline to half the initial number?

35. *Health Care Spending* Out-of-pocket spending in the United States for health care increased between 2004 and 2008. The function

$$f(x) = 2572e^{0.0359x}$$

models average annual expenditures per household, in dollars. In this model, x represents the year, where $x = 0$ corresponds to 2004. (*Source*: U.S. Bureau of Labor Statistics.)

(a) Estimate out-of-pocket household spending on health care in 2008.
(b) Determine the year when spending reached \$2775 per household.

36. $941 billion
37. (a) $12,879
(b) $15,590
(c) $17,483
38. (a) about 350 yr
(b) about 4000 yr
(c) about 2300 yr
39. (a) 15,000
(b) 9098
(c) 5249
40. about 14 days
41. (a) 611 million
(b) 746 million
(c) 1007 million
42. 6.9 days

36. *Recreational Expenditures* Personal consumption expenditures for recreation in billions of dollars in the United States during the years 2004–2008 can be approximated by the function

$$A(t) = 769.5e^{0.0503t},$$

where $t = 0$ corresponds to the year 2004. Based on this model, how much were personal consumption expenditures in 2008? (*Source:* U.S. Bureau of Economic Analysis.)

37. *Housing Costs* Average annual per-household spending on housing over the years 1990–2008 is approximated by

$$H = 8790e^{0.0382t},$$

where t is the number of years since 1990. Find H to the nearest dollar for each year. (*Source:* U.S. Bureau of Labor Statistics.)

(a) 2000 **(b)** 2005 **(c)** 2008

38. *Evolution of Language* The number of years, n, since two independently evolving languages split off from a common ancestral language is approximated by

$$n \approx -7600 \log r,$$

where r is the proportion of words from the ancestral language common to both languages.

(a) Find n if $r = 0.9$. **(b)** Find n if $r = 0.3$.

(c) How many years have elapsed since the split if half of the words of the ancestral language are common to both languages?

Life Sciences (Exercises 39–44)

39. *Spread of Disease* During an epidemic, the number of people who have never had the disease and who are not immune (they are *susceptible*) decreases exponentially according to the function

$$f(t) = 15{,}000e^{-0.05t},$$

where t is time in days. Find the number of susceptible people at each time.

(a) at the beginning of the epidemic **(b)** after 10 days **(c)** after 3 weeks

40. *Spread of Disease* Refer to **Exercise 39** and determine how long it will take for the initial number of people susceptible to decrease to half its amount.

41. *Growth of Bacteria* The growth of bacteria makes it necessary to time-date some food products so that they will be sold and consumed before the bacteria count is too high. Suppose for a certain product the number of bacteria present is given by

$$f(t) = 500e^{0.1t},$$

where t is time in days and the value of $f(t)$ is in millions. Find the number of bacteria present at each time.

(a) 2 days **(b)** 4 days **(c)** 1 week

42. *Growth of Bacteria* How long will it take the bacteria population in **Exercise 41** to double?

43. *Medication Effectiveness* Drug effectiveness decreases over time. If, each hour, a drug is only 90% as effective as the previous hour, at some point the patient will not be receiving enough medication and must receive another dose. If the initial dose was 200 mg and the drug was administered 3 hr ago, the expression $200(0.90)^3$, which equals 145.8, represents the amount of effective medication still in the system. (The exponent is equal to the number of hours since the drug was administered.)

43. about 13.2 hr

44. (a) $G(x) = \frac{250{,}000}{100 + 2400e^{-x}}$

2500

0 8

0

(b) 590; 589

(c) 2.8; 2.7726

45. 2016

46. (a) $S(1) \approx 45{,}200$; $S(3) \approx 37{,}000$

(b) $S(2) \approx 72{,}400$; $S(10) \approx 48{,}500$

47. 6.9 yr

The amount of medication still available in the system is given by the function

$$f(t) = 200(0.90)^t.$$

In this model, t is in hours and $f(t)$ is in milligrams. How long will it take for this initial dose to reach the dangerously low level of 50 mg?

*Population Size Many environmental situations place effective limits on the growth of the number of an organism in an area. Many such limited-growth situations are described by the **logistic function***

$$G(x) = \frac{MG_0}{G_0 + (M - G_0)e^{-kMx}},$$

where G_0 is the initial number present, M is the maximum possible size of the population, and k is a positive constant. The screens illustrate a typical logistic function calculation and graph.

L1	L2	L3
10	14	
20	25	
30	42	
40	50	
50	58	
60	60	

L1 ={10, 20, 30, 40...

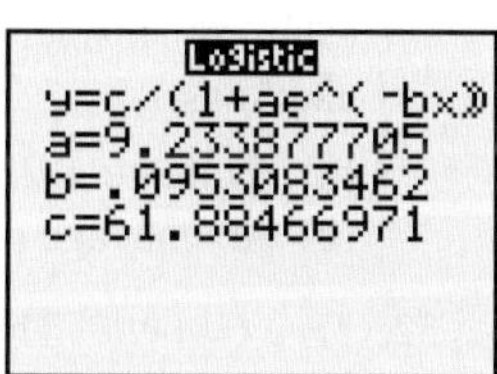

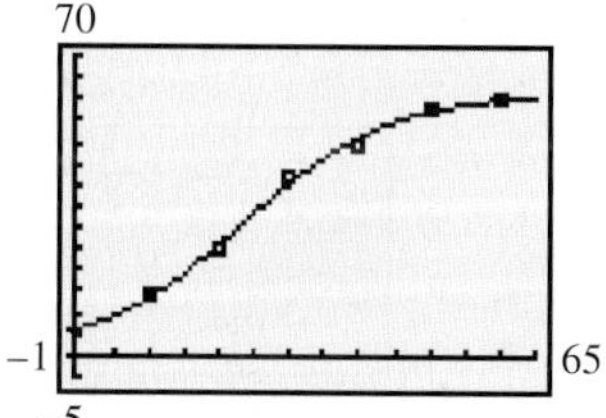

44. Assume that $G_0 = 100$, $M = 2500$, $k = 0.0004$, and $x =$ time in decades (10-yr periods).

(a) Use a calculator to graph the function, using $0 \le x \le 8$, $0 \le y \le 2500$.

(b) Estimate the value of $G(2)$ from the graph. Then evaluate $G(2)$ algebraically to find the population after 20 yr.

(c) Find the x-coordinate of the intersection of the curve with the horizontal line $y = 1000$ to estimate the number of decades required for the population to reach 1000. Then solve $G(x) = 1000$ algebraically to obtain the exact value of x.

Economics (Exercises 45–50)

45. *Consumer Price Index* The U.S. Consumer Price Index for the years 1990–2009 is approximated by

$$A(t) = 100e^{0.026t},$$

where t represents the number of years after 1990. (Since $A(16)$ is about 152, the amount of goods that could be purchased for \$100 in 1990 cost about \$152 in 2006.) Use the function to determine the year in which costs will be 100% higher than in 1990. (*Source:* U.S. Bureau of Labor Statistics.)

46. *Product Sales* Sales of a product, under relatively stable market conditions but in the absence of promotional activities such as advertising, tend to decline at a constant yearly rate. This rate of sales decline varies considerably from product to product, but it seems to remain the same for any particular product. The sales decline can be expressed by the function

$$S(t) = S_0 e^{-at},$$

where $S(t)$ is the rate of sales at time t measured in years, S_0 is the rate of sales at time $t = 0$, and a is the sales decay constant.

(a) Suppose the sales decay constant for a particular product is $a = 0.10$. Let $S_0 = 50{,}000$ and find $S(1)$ and $S(3)$.

(b) Find $S(2)$ and $S(10)$ if $S_0 = 80{,}000$ and $a = 0.05$.

47. *Product Sales* Use the sales decline function given in **Exercise 46.** If $a = 0.10$, $S_0 = 50{,}000$, and t is time measured in years, find the number of years it will take for sales to fall to half the initial sales.

48. 27.5 yr
49. 11.6 yr
50. 34.7 yr

51. **(a)** 0.065; 0.82; Among people age 25, 6.5% have some CHD, while among people age 65, 82% have some CHD.
(b) about 48 yr

52. **(a)**

X	Y_1
10	7.9832
20	31.582
30	46.965
40	49.645
50	49.96
60	49.996
70	50

X=10

The maximum height appears to be 50 ft.

(b)

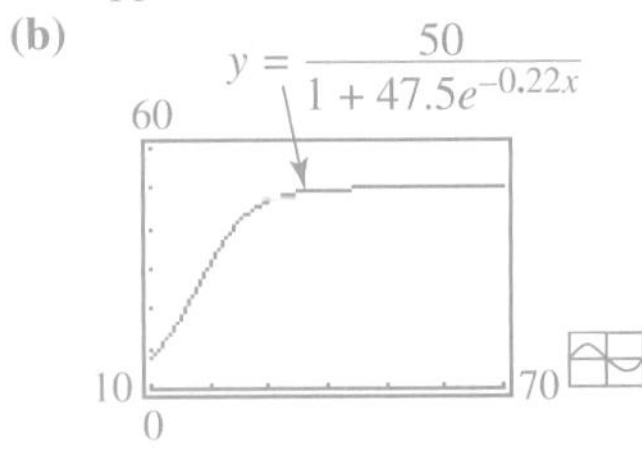

The horizontal asymptote is $y = 50$. It tells us that this tree cannot grow taller than 50 ft.
(c) after about 19.4 yr

48. *Cost of Bread* Assume the cost of a loaf of bread is $4. With continuous compounding, find the time it would take for the cost to triple at an annual inflation rate of 4%.

49. *Electricity Consumption* Suppose that in a certain area the consumption of electricity has increased at a continuous rate of 6% per year. If it continued to increase at this rate, find the number of years before twice as much electricity would be needed.

50. *Electricity Consumption* Suppose a conservation campaign, together with higher rates, caused demand for electricity to increase at only 2% per year. (See **Exercise 49.**) Find the number of years before twice as much electricity would be needed.

(Modeling) Exercises 51 and 52 use another type of logistic function.

51. *Heart Disease* As age increases, so does the likelihood of coronary heart disease (CHD). The fraction of people x years old with some CHD is modeled by

$$f(x) = \frac{0.9}{1 + 271e^{-0.122x}}.$$

(*Source:* Hosmer, D., and S. Lemeshow, *Applied Logistic Regression,* John Wiley and Sons.)

(a) Evaluate $f(25)$ and $f(65)$. Interpret the results.
(b) At what age does this likelihood equal 50%?

52. *Tree Growth* The height of a certain tree in feet after x years is modeled by

$$f(x) = \frac{50}{1 + 47.5e^{-0.22x}}.$$

(a) Make a table for f starting at $x = 10$, and incrementing by 10. What appears to be the maximum height of the tree?
(b) Graph f and identify the horizontal asymptote. Explain its significance.
(c) After how long was the tree 30 ft tall?

Summary Exercises on Functions: Domains and Defining Equations

Finding the Domain of a Function: A Summary To find the domain of a function, given the equation that defines the function, remember that the value of x input into the equation must yield a real number for y when the function is evaluated. For the functions studied so far in this book, there are three cases to consider when determining domains.

Guidelines for Domain Restrictions

1. No input value can lead to 0 in a denominator, because division by 0 is undefined.
2. No input value can lead to an even root of a negative number, because this situation does not yield a real number.
3. No input value can lead to the logarithm of a negative number or 0, because this situation does not yield a real number.

Unless domains are otherwise specified, we can determine domains as follows.

- The domain of a **polynomial function** is the set of all real numbers.
- The domain of an **absolute value function** is the set of all real numbers for which the expression inside the absolute value bars (the argument) is defined.
- If a **function is defined by a rational expression,** the domain is the set of all real numbers for which the denominator is not zero.
- The domain of a **function defined by a radical with *even* root index** is the set of all real numbers that make the radicand greater than or equal to zero. **If the root index is *odd*,** the domain is the set of all real numbers for which the radicand is itself a real number.
- For an **exponential function** with constant base, the domain is the set of all real numbers for which the exponent is a real number.
- For a **logarithmic function,** the domain is the set of all real numbers that make the argument of the logarithm greater than zero.

Determining Whether an Equation Defines *y* as a Function of *x* ***For y to be a function of x, it is necessary that every input value of x in the domain leads to one and only one value of y.*** To determine whether an equation such as

$$x - y^3 = 0 \quad \text{or} \quad x - y^2 = 0$$

represents a function, solve the equation for y. In the first equation above, doing so leads to

$$y = \sqrt[3]{x}.$$

Notice that every value of x in the domain (that is, all real numbers) leads to one and only one value of y. So in the first equation, we can write y as a function of x. However, in the second equation above, solving for y leads to

$$y = \pm\sqrt{x}.$$

If we let $x = 4$, for example, we get two values of y: -2 and 2. Thus, in the second equation, we cannot write y as a function of x.

Find the domain of each function. Write answers using interval notation.

1. $f(x) = 3x - 6$ **2.** $f(x) = \sqrt{2x - 7}$ **3.** $f(x) = |x + 4|$

4. $f(x) = \dfrac{x + 2}{x - 6}$ **5.** $f(x) = \dfrac{-2}{x^2 + 7}$ **6.** $f(x) = \sqrt{x^2 - 9}$

7. $f(x) = \dfrac{x^2 + 7}{x^2 - 9}$ **8.** $f(x) = \sqrt[3]{x^3 + 7x - 4}$ **9.** $f(x) = \log_5(16 - x^2)$

10. $f(x) = \log \dfrac{x + 7}{x - 3}$ **11.** $f(x) = \sqrt{x^2 - 7x - 8}$ **12.** $f(x) = 2^{1/x}$

13. $f(x) = \dfrac{1}{2x^2 - x + 7}$ **14.** $f(x) = \dfrac{x^2 - 25}{x + 5}$ **15.** $f(x) = \sqrt{x^3 - 1}$

16. $f(x) = \ln|x^2 - 5|$ **17.** $f(x) = e^{x^2+x+4}$ **18.** $f(x) = \dfrac{x^3 - 1}{x^2 - 1}$

1. $(-\infty, \infty)$
2. $\left[\frac{7}{2}, \infty\right)$
3. $(-\infty, \infty)$
4. $(-\infty, 6) \cup (6, \infty)$
5. $(-\infty, \infty)$
6. $(-\infty, -3] \cup [3, \infty)$
7. $(-\infty, -3) \cup (-3, 3) \cup (3, \infty)$
8. $(-\infty, \infty)$
9. $(-4, 4)$
10. $(-\infty, -7) \cup (3, \infty)$
11. $(-\infty, -1] \cup [8, \infty)$
12. $(-\infty, 0) \cup (0, \infty)$
13. $(-\infty, \infty)$
14. $(-\infty, -5) \cup (-5, \infty)$
15. $[1, \infty)$
16. $(-\infty, -\sqrt{5}) \cup (-\sqrt{5}, \sqrt{5}) \cup (\sqrt{5}, \infty)$
17. $(-\infty, \infty)$
18. $(-\infty, -1) \cup (-1, 1) \cup (1, \infty)$

19. $(-\infty, 1)$
20. $(-\infty, 2) \cup (2, \infty)$
21. $(-\infty, \infty)$
22. $[-2, 3] \cup [4, \infty)$
23. $(-\infty, -2) \cup (-2, 3) \cup (3, \infty)$
24. $[-3, \infty)$
25. $(-\infty, 0) \cup (0, \infty)$
26. $(-\infty, -\sqrt{7}) \cup (-\sqrt{7}, \sqrt{7}) \cup (\sqrt{7}, \infty)$
27. $(-\infty, \infty)$
28. $\varnothing$
29. $[-2, 2]$
30. $(-\infty, \infty)$
31. $(-\infty, -7] \cup (-4, 3) \cup [9, \infty)$
32. $(-\infty, \infty)$
33. $(-\infty, 5]$
34. $(-\infty, 3)$
35. $(-\infty, 4) \cup (4, \infty)$
36. $(-\infty, \infty)$
37. $(-\infty, -5] \cup [5, \infty)$
38. $(-\infty, \infty)$
39. $(-2, 6)$
40. $(0, 1) \cup (1, \infty)$

41. A 42. B
43. C 44. D
45. A 46. B
47. D 48. C
49. C 50. B

19. $f(x) = \sqrt{\dfrac{-1}{x^3 - 1}}$ **20.** $f(x) = \sqrt[3]{\dfrac{1}{x^3 - 8}}$

21. $f(x) = \ln(x^2 + 1)$ **22.** $f(x) = \sqrt{(x - 3)(x + 2)(x - 4)}$

23. $f(x) = \log\left(\dfrac{x + 2}{x - 3}\right)^2$ **24.** $f(x) = \sqrt[12]{(4 - x)^2(x + 3)}$

25. $f(x) = e^{|1/x|}$ **26.** $f(x) = \dfrac{1}{|x^2 - 7|}$

27. $f(x) = x^{100} - x^{50} + x^2 + 5$ **28.** $f(x) = \sqrt{-x^2 - 9}$

29. $f(x) = \sqrt[4]{16 - x^4}$ **30.** $f(x) = \sqrt[3]{16 - x^4}$

31. $f(x) = \sqrt{\dfrac{x^2 - 2x - 63}{x^2 + x - 12}}$ **32.** $f(x) = \sqrt[5]{5 - x}$

33. $f(x) = |\sqrt{5 - x}|$ **34.** $f(x) = \sqrt{\dfrac{-1}{x - 3}}$

35. $f(x) = \log\left|\dfrac{1}{4 - x}\right|$ **36.** $f(x) = 6^{x^2 - 9}$

37. $f(x) = 6^{\sqrt{x^2 - 25}}$ **38.** $f(x) = 6^{\sqrt[3]{x^2 - 25}}$

39. $f(x) = \ln\left(\dfrac{-3}{(x + 2)(x - 6)}\right)$ **40.** $f(x) = \dfrac{-2}{\log x}$

Determine which one of the choices (A, B, C, or D) is an equation in which y can be written as a function of x.

41. A. $3x + 2y = 6$ **B.** $x = \sqrt{|y|}$ **C.** $x = |y + 3|$ **D.** $x^2 + y^2 = 9$

42. A. $3x^2 + 2y^2 = 36$ **B.** $x^2 + y - 2 = 0$ **C.** $x - |y| = 0$ **D.** $x = y^2 - 4$

43. A. $x = \sqrt{y^2}$ **B.** $x = \log y^2$ **C.** $x^3 + y^3 = 5$ **D.** $x = \dfrac{1}{y^2 + 3}$

44. A. $\dfrac{x^2}{4} + \dfrac{y^2}{4} = 1$ **B.** $x = 5y^2 - 3$ **C.** $\dfrac{x^2}{4} - \dfrac{y^2}{9} = 1$ **D.** $x = 10^y$

45. A. $x = \dfrac{2 - y}{y + 3}$ **B.** $x = \ln(y + 1)^2$ **C.** $\sqrt{x} = |y + 1|$ **D.** $\sqrt[4]{x} = y^2$

46. A. $e^{y^2} = x$ **B.** $e^{y+2} = x$ **C.** $e^{|y|} = x$ **D.** $10^{|y+2|} = x$

47. A. $x^2 = \dfrac{1}{y^2}$ **B.** $x + 2 = \dfrac{1}{y^2}$ **C.** $3x = \dfrac{1}{y^4}$ **D.** $2x = \dfrac{1}{y^3}$

48. A. $|x| = |y|$ **B.** $x = |y^2|$ **C.** $x = \dfrac{1}{y}$ **D.** $x^4 + y^4 = 81$

49. A. $\dfrac{x^2}{4} - \dfrac{y^2}{9} = 1$ **B.** $\dfrac{y^2}{4} - \dfrac{x^2}{9} = 1$ **C.** $\dfrac{x}{4} - \dfrac{y}{9} = 0$ **D.** $\dfrac{x^2}{4} - \dfrac{y^2}{9} = 0$

50. A. $y^2 - \sqrt{(x + 2)^2} = 0$ **B.** $y - \sqrt{(x + 2)^2} = 0$
C. $y^6 - \sqrt{(x + 1)^2} = 0$ **D.** $y^4 - \sqrt{x^2} = 0$

Chapter 4 Test Prep

Key Terms

4.1 one-to-one function
inverse function

4.2 exponential function
exponential equation
compound interest
future value
present value
compound amount
continuous compounding

4.3 logarithm
base
argument
logarithmic equation
logarithmic function

4.4 common logarithm
pH
natural logarithm

4.6 doubling time
half-life

New Symbols

$f^{-1}(x)$	the inverse of $f(x)$
e	a constant, approximately 2.718281828459045
$\log_a x$	the logarithm of x with the base a
$\log x$	common (base 10) logarithm of x
$\ln x$	natural (base e) logarithm of x

Quick Review

Concepts / **Examples**

4.1 Inverse Functions

One-to-One Function

In a one-to-one function, each x-value corresponds to only one y-value, and each y-value corresponds to only one x-value.

A function f is one-to-one if, for elements a and b in the domain of f, $a \neq b$ implies $f(a) \neq f(b)$.

Example: The function $y = x^2$ is not one-to-one, because $y = 16$, for example, corresponds to both $x = 4$ and $x = -4$.

Horizontal Line Test

A function is one-to-one if every horizontal line intersects the graph of the function at most once.

Example: The graph of $f(x) = 2x - 1$ is a straight line with slope 2. f is a one-to-one function by the horizontal line test.

Inverse Functions

Let f be a one-to-one function. Then g is the inverse function of f if

$$(f \circ g)(x) = x \quad \text{for every } x \text{ in the domain of } g$$

and

$$(g \circ f)(x) = x \quad \text{for every } x \text{ in the domain of } f.$$

To find $g(x)$, interchange x and y in $y = f(x)$, solve for y, and replace y with $g(x)$, which is $f^{-1}(x)$.

Example: Find the inverse of $y = f(x) = 2x - 1$.

$$x = 2y - 1 \quad \text{Interchange } x \text{ and } y.$$

$$y = \frac{x+1}{2} \quad \text{Solve for } y.$$

$$f^{-1}(x) = \frac{x+1}{2} \quad \text{Replace } y \text{ with } f^{-1}(x).$$

$$f^{-1}(x) = \frac{1}{2}x + \frac{1}{2} \quad \frac{x+1}{2} = \frac{x}{2} + \frac{1}{2} = \frac{1}{2}x + \frac{1}{2}$$

(continued)

Concepts	Examples

4.2 Exponential Functions

Additional Properties of Exponents

For any real number $a > 0$, $a \neq 1$, the following are true.

(a) a^x is a unique real number for all real numbers x.
(b) $a^b = a^c$ if and only if $b = c$.
(c) If $a > 1$ and $m < n$, then $a^m < a^n$.
(d) If $0 < a < 1$ and $m < n$, then $a^m > a^n$.

Examples:

(a) 2^x is a unique real number for all real numbers x.
(b) $2^x = 2^3$ if and only if $x = 3$.
(c) $2^5 < 2^{10}$, because $2 > 1$ and $5 < 10$.
(d) $\left(\frac{1}{2}\right)^5 > \left(\frac{1}{2}\right)^{10}$ because $0 < \frac{1}{2} < 1$ and $5 < 10$.

Exponential Function

If $a > 0$ and $a \neq 1$, then $\boldsymbol{f(x) = a^x}$ defines the exponential function with base a.

Example: $f(x) = 3^x$ is the exponential function with base 3.

Graph of $f(x) = a^x$

1. The graph contains the points $\left(-1, \frac{1}{a}\right)$, $(0, 1)$, and $(1, a)$.
2. If $a > 1$, then f is an increasing function. If $0 < a < 1$, then f is a decreasing function.
3. The x-axis is a horizontal asymptote.
4. The domain is $(-\infty, \infty)$, and the range is $(0, \infty)$.

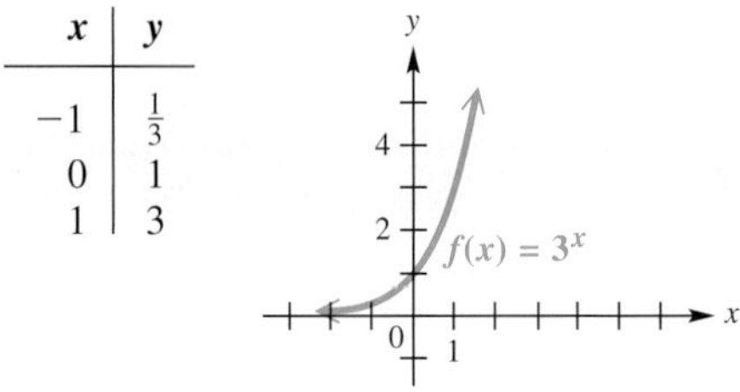

x	y
-1	$\frac{1}{3}$
0	1
1	3

4.3 Logarithmic Functions

Logarithmic Function

If $a > 0$, $a \neq 1$, and $x > 0$, then $\boldsymbol{f(x) = \log_a x}$ defines the logarithmic function with base a.

Example: $f(x) = \log_3 x$ is the logarithmic function with base 3.

Graph of $f(x) = \log_a x$

1. The graph contains the points $\left(\frac{1}{a}, -1\right)$, $(1, 0)$, and $(a, 1)$.
2. If $a > 1$, then f is an increasing function. If $0 < a < 1$, then f is a decreasing function.
3. The y-axis is a vertical asymptote.
4. The domain is $(0, \infty)$, and the range is $(-\infty, \infty)$.

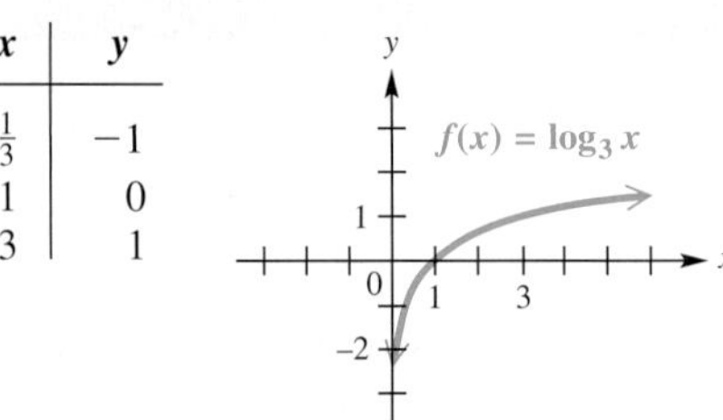

x	y
$\frac{1}{3}$	-1
1	0
3	1

Properties of Logarithms

For $x > 0$, $y > 0$, $a > 0$, $a \neq 1$, and any real number r, the following properties hold.

Property	Name	Example
$\log_a xy = \log_a x + \log_a y$	Product property	$\log_2(3 \cdot 5) = \log_2 3 + \log_2 5$
$\log_a \frac{x}{y} = \log_a x - \log_a y$	Quotient property	$\log_2 \frac{3}{5} = \log_2 3 - \log_2 5$
$\log_a x^r = r \log_a x$	Power property	$\log_6 3^5 = 5 \log_6 3$
$\log_a 1 = 0$	Logarithm of 1	$\log_{10} 1 = 0$
$\log_a a = 1$	Base a logarithm of a	$\log_{10} 10 = 1$

Theorem on Inverses

For $a > 0$ and $a \neq 1$, the following properties hold.

$$a^{\log_a x} = x \ (x > 0) \quad \text{and} \quad \log_a a^x = x$$

Example: $2^{\log_2 5} = 5$ and $\log_2 2^5 = 5$

Concepts	Examples

4.4 Evaluating Logarithms and the Change-of-Base Theorem

Common and Natural Logarithms
For all positive numbers x, base 10 logarithms and base e logarithms are written as follows.

$$\log x = \log_{10} x \quad \text{Common logarithm}$$
$$\ln x = \log_e x \quad \text{Natural logarithm}$$

Approximate log 0.045 and ln 247.1.

$$\log 0.045 \approx -1.3468$$
$$\ln 247.1 \approx 5.5098$$
Use a calculator.

Change-of-Base Theorem
For any positive real numbers x, a, and b, where $a \neq 1$ and $b \neq 1$, the following holds.

$$\log_a x = \frac{\log_b x}{\log_b a}$$

Approximate $\log_8 7$.

$$\log_8 7 = \frac{\log 7}{\log 8} = \frac{\ln 7}{\ln 8} \approx 0.9358 \quad \text{Use a calculator.}$$

4.5 Exponential and Logarithmic Equations

Property of Logarithms
If $x > 0$, $y > 0$, $a > 0$, and $a \neq 1$, then the following holds.

$$x = y \quad \text{is equivalent to} \quad \log_a x = \log_a y.$$

Solve $e^{5x} = 10$.

$$\ln e^{5x} = \ln 10 \quad \text{Take logarithms.}$$
$$5x = \ln 10 \quad \ln e^x = x, \text{ for all } x.$$
$$x = \frac{\ln 10}{5} \approx 0.461 \quad \text{Use a calculator.}$$

The solution set can be written with the exact value, $\left\{\frac{\ln 10}{5}\right\}$, or with the approximate value, $\{0.461\}$.

Solve $\log_2(x^2 - 3) = \log_2 6$.

$$x^2 - 3 = 6 \quad \text{Property of logarithms}$$
$$x^2 = 9 \quad \text{Add 3.}$$
$$x = \pm 3 \quad \text{Take square roots.}$$

Both values check, so the solution set is $\{\pm 3\}$.

4.6 Applications and Models of Exponential Growth and Decay

Exponential Growth or Decay Function
The exponential growth or decay function is

$$y = y_0 e^{kt},$$

for constant k, where y_0 is the amount or number present at time $t = 0$.

The formula for continuous compounding,

$$A = Pe^{rt},$$

is an example of exponential growth. Here, A is the compound amount if P is invested at an annual interest rate r for t years.

If $P = \$200$, $r = 3\%$, and $t = 5$ yr, find A.

$$A = 200e^{0.03(5)} \quad \text{Substitute.}$$
$$A \approx \$232.37 \quad \text{Use a calculator.}$$

Chapter 4 Review Exercises

1. not one-to-one
2. one-to-one
3. one-to-one
4. one-to-one
5. not one-to-one
6. not one-to-one
7. $f^{-1}(x) = \sqrt[3]{x+3}$
8. not possible
9. It represents the number of years after 2004 for the investment to reach \$50,000.
10. yes
11. one-to-one
12. yes
13. B 14. A
15. C 16. D
17. $\log_2 32 = 5$
18. $\log_{100} 10 = \frac{1}{2}$
19. $\log_{3/4} \frac{4}{3} = -1$
20.

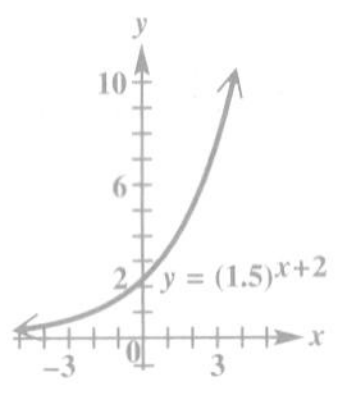

21. $10^3 = 1000$ 22. $9^{3/2} = 27$
23. $e^{1/2} = \sqrt{e}$
24. 3
25. 2

Decide whether each function as defined or graphed is one-to-one.

1.

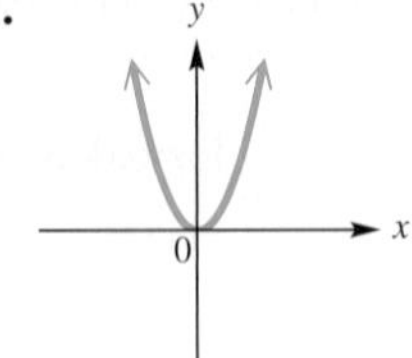

2.

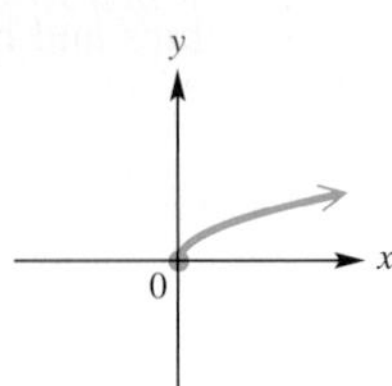

3. $y = 5x - 4$

4. $y = x^3 + 1$ **5.** $y = (x+3)^2$ **6.** $y = \sqrt{3x^2 + 2}$

Write an equation, if possible, for each inverse function in the form $y = f^{-1}(x)$.

7. $f(x) = x^3 - 3$ **8.** $f(x) = \sqrt{25 - x^2}$

9. *Concept Check* Suppose $f(t)$ is the amount an investment will grow to t years after 2004. What does $f^{-1}(\$50{,}000)$ represent?

10. *Concept Check* The graphs of two functions are shown. Based on their graphs, are these functions inverses?

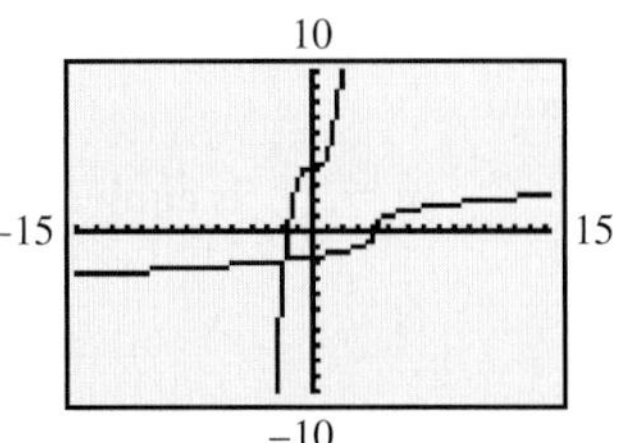

11. *Concept Check* To have an inverse, a function must be a(n) ________ function.

12. *Concept Check* Assuming that f has an inverse, is it true that every x-intercept of the graph of $y = f(x)$ is a y-intercept of the graph of $y = f^{-1}(x)$?

Match each equation with the figure that most closely resembles its graph.

13. $y = \log_{0.3} x$ **14.** $y = e^x$ **15.** $y = \ln x$ **16.** $y = 0.3^x$

A.

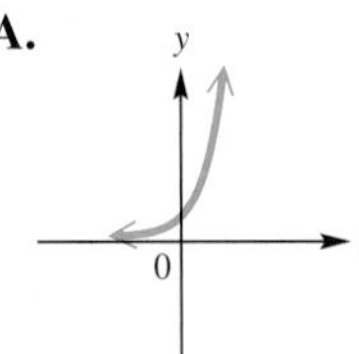

B.

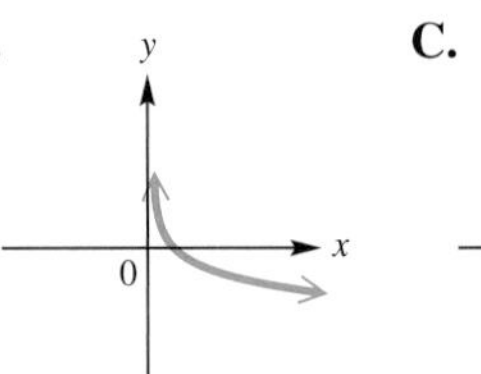

C. **D.**

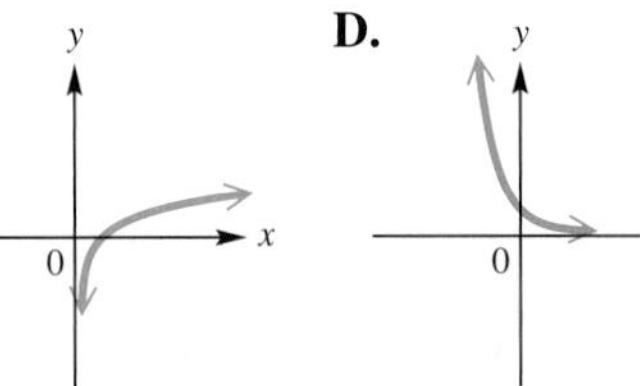

Write each equation in logarithmic form.

17. $2^5 = 32$ **18.** $100^{1/2} = 10$ **19.** $\left(\frac{3}{4}\right)^{-1} = \frac{4}{3}$

20. Graph $y = (1.5)^{x+2}$.

Write each equation in exponential form.

21. $\log 1000 = 3$ **22.** $\log_9 27 = \frac{3}{2}$ **23.** $\ln \sqrt{e} = \frac{1}{2}$

24. *Concept Check* What is the base of the logarithmic function whose graph contains the point $(81, 4)$?

25. *Concept Check* What is the base of the exponential function whose graph contains the point $\left(-4, \frac{1}{16}\right)$?

26. $2 \log_5 x + 4 \log_5 y + \frac{1}{5}(3 \log_5 m + \log_5 p)$
27. $\log_3 m + \log_3 n - \log_3 5 - \log_3 r$
28. This cannot be simplified.

29. -1.3862 30. 1.6590
31. 11.8776 32. 6.1527
33. 1.1592 34. 6.0486

35. $\{\frac{22}{5}\}$ 36. $\{1.792\}$
37. $\{3.667\}$ 38. $\{2.269\}$
39. $\{-13.257\}$ 40. $\{2.386\}$
41. $\{-0.485\}$ 42. $\{-0.123\}$
43. $\{2.102\}$ 44. $\{-17.531\}$
45. $\{-2.487\}$ 46. $\{2\}$
47. $\{3\}$ 48. $\{140.011\}$
49. $\{\ln 3\}$ 50. $\emptyset$
51. $\{6.959\}$

52. A, C, D

53. $\{e^{13/3}\}$ 54. $\{\frac{e^{16}}{5}\}$
55. $\{\frac{\sqrt[4]{10}-7}{2}\}$ 56. $\{e^3\}$
57. $\{3\}$ 58. $\{\pm 6\}$
59. $\{-\frac{4}{3}, 5\}$ 60. $\{6\}$
61. $\{1, \frac{10}{3}\}$ 62. $\{4\}$
63. $\{2\}$ 64. $\{3\}$
65. $\{-3\}$
66. $n = a(e^{S/a} - 1)$
67. $I_0 = \frac{I}{10^{d/10}}$
68. $x = 10^{D/100-2}$
69. $\{1.315\}$

70. (a) about $31{,}600{,}000 I_0$
(b) about $79{,}400{,}000 I_0$
(c) about 2.51 times as great

Use properties of logarithms to rewrite each expression. Simplify the result, if possible. Assume all variables represent positive numbers.

26. $\log_5\left(x^2 y^4 \sqrt[5]{m^3 p}\right)$ **27.** $\log_3 \frac{mn}{5r}$ **28.** $\log_7(7k + 5r^2)$

Find each logarithm. Round to four decimal places.

29. $\log 0.0411$ **30.** $\log 45.6$ **31.** $\ln 144{,}000$

32. $\ln 470$ **33.** $\log_{2/3} \frac{5}{8}$ **34.** $\log_3 769$

Solve each exponential equation. Unless otherwise specified, express irrational solutions as decimals correct to the nearest thousandth.

35. $16^{x+4} = 8^{3x-2}$ **36.** $4^x = 12$ **37.** $3^{2x-5} = 13$

38. $2^{x+3} = 5^x$ **39.** $6^{x+3} = 4^x$ **40.** $e^{x-1} = 4$

41. $e^{2-x} = 12$ **42.** $2e^{5x+2} = 8$ **43.** $10e^{3x-7} = 5$

44. $5^{x+2} = 2^{2x-1}$ **45.** $6^{x-3} = 3^{4x+1}$ **46.** $e^{8x} \cdot e^{2x} = e^{20}$

47. $e^{6x} \cdot e^x = e^{21}$ **48.** $100(1.02)^{x/4} = 200$ **49.** $2e^{2x} - 5e^x - 3 = 0$ (Give exact form.)

50. $\left(\frac{1}{2}\right)^x + 2 = 0$ **51.** $4(1.06)^x + 2 = 8$

52. *Concept Check* Which one or more of the following choices is the solution set of $5^x = 9$?

A. $\{\log_5 9\}$ **B.** $\{\log_9 5\}$ **C.** $\left\{\frac{\log 9}{\log 5}\right\}$ **D.** $\left\{\frac{\ln 9}{\ln 5}\right\}$

Solve each logarithmic equation. Express all solutions in exact form.

53. $3 \ln x = 13$ **54.** $\ln(5x) = 16$

55. $\log(2x + 7) = 0.25$ **56.** $\ln x + \ln x^3 = 12$

57. $\log_2(x^3 + 5) = 5$ **58.** $\log_3(x^2 - 9) = 3$

59. $\log_4[(3x + 1)(x - 4)] = 2$ **60.** $\ln e^{\ln x} - \ln(x - 4) = \ln 3$

61. $\log x + \log(13 - 3x) = 1$ **62.** $\log_7(3x + 2) - \log_7(x - 2) = 1$

63. $\ln(6x) - \ln(x + 1) = \ln 4$ **64.** $\log_{16}\sqrt{x + 1} = \frac{1}{4}$

65. $\ln[\ln(e^{-x})] = \ln 3$ **66.** $S = a \ln\left(1 + \frac{n}{a}\right)$, for n

67. $d = 10 \log\left(\frac{I}{I_0}\right)$, for I_0 **68.** $D = 200 + 100 \log x$, for x

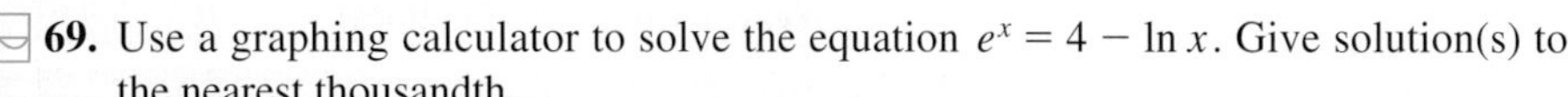
69. Use a graphing calculator to solve the equation $e^x = 4 - \ln x$. Give solution(s) to the nearest thousandth.

Solve each problem.

70. *Earthquake Intensity* On September 30, 2009, the region of Indonesia known as Sumatra was shaken by an earthquake that measured 7.5 on the Richter scale.

(a) Express this reading in terms of I_0. (See **Section 4.4** Exercises.)
(b) On May 12, 2008, a quake measuring 7.9 on the Richter scale killed about 88,000 people in East Sichuan Province, China. Express the magnitude of a 7.9 reading in terms of I_0.
(c) How much greater was the force of the earthquake that measured 7.9?

71. **(a)** about $200{,}000{,}000 I_0$
(b) about $13{,}000{,}000 I_0$
(c) The 1906 earthquake had a magnitude almost 16 times greater than the 1989 earthquake.
72. 89 decibels is about twice as loud as 86 decibels. This is a 100% increase.
73. 5.1%
74. 4.0 yr
75. \$24,970.64
76. \$25,149.59
77. 17.3 yr
78. **(a)** For $t = x$, $A(x) = x^2 - x + 350$

500, 0, 10, 0

(b) For $t = x$, $A(x) = 350 \log(x + 1)$

500, 0, 10, 0

(c) For $t = x$, $A(x) = 350(0.75)^x$

500, 0, 10, 0

(d) For $t = x$, $A(x) = 100(0.95)^x$

500, 0, 10, 0

Function (c) best describes $A(t)$.

71. *Earthquake Intensity*

(a) The San Francisco earthquake of 1906 had a Richter scale rating of 8.3. Express the magnitude of this earthquake as a multiple of I_0.

(b) In 1989, the San Francisco region experienced an earthquake with a Richter scale rating of 7.1. Express the magnitude of this earthquake as a multiple of I_0.

(c) Compare the magnitudes of the two San Francisco earthquakes discussed in parts (a) and (b).

72. *(Modeling) Decibel Levels* In **Section 4.4** we found that the model for the decibel rating of the loudness of a sound is

$$d = 10 \log \frac{I}{I_0}.$$

A few years ago, there was a controversy about a proposed government limit on factory noise. One group wanted a maximum of 89 decibels, while another group wanted 86. This difference seemed very small to many people. Find the percent by which the 89-decibel intensity exceeds that for 86 decibels.

73. *Interest Rate* What annual interest rate, to the nearest tenth, will produce \$5760 if \$3500 is left at interest compounded annually for 10 yr?

74. *Growth of an Account* Find the number of years (to the nearest tenth) needed for \$48,000 to become \$58,344 at 5% interest compounded semiannually.

75. *Growth of an Account* Manuel deposits \$10,000 for 12 yr in an account paying 4% compounded annually. He then puts this total amount on deposit in another account paying 5% compounded semiannually for another 9 yr. Find the total amount on deposit after the entire 21-yr period.

76. *Growth of an Account* Anne Kelly deposits \$12,000 for 8 yr in an account paying 5% compounded annually. She then leaves the money alone with no further deposits at 6% compounded annually for an additional 6 yr. Find the total amount on deposit after the entire 14-yr period.

77. *Cost from Inflation* Suppose the inflation rate is 4%. Use the formula for continuous compounding to find the number of years, to the nearest tenth, for a \$1 item to cost \$2.

78. *(Modeling) Drug Level in the Bloodstream* After a medical drug is injected directly into the bloodstream, it is gradually eliminated from the body. Graph the following functions on the interval $[0, 10]$. Use $[0, 500]$ for the range of $A(t)$. Determine the function that best models the amount $A(t)$ (in milligrams) of a drug remaining in the body after t hours if 350 mg were initially injected.

(a) $A(t) = t^2 - t + 350$ **(b)** $A(t) = 350 \log(t + 1)$

(c) $A(t) = 350(0.75)^t$ **(d)** $A(t) = 100(0.95)^t$

79. *(Modeling) Chicago Cubs' Payroll* The table shows the total payroll (in millions of dollars) of the Chicago Cubs baseball team for the years 2007–2010.

Year	Total Payroll (millions of dollars)
2007	115.9
2008	118.3
2009	134.8
2010	146.6

Source: www.baseball.about.com; www.espn.go.com

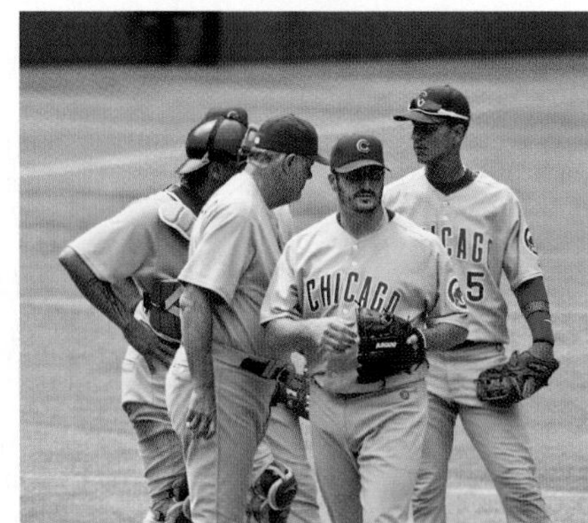

79. 2018

80. (a) See the graph in part (c).
(b) An exponential function describes the data best.
(c) Answers will vary. One answer is $f(x) = 220{,}034(1.402)^x$.

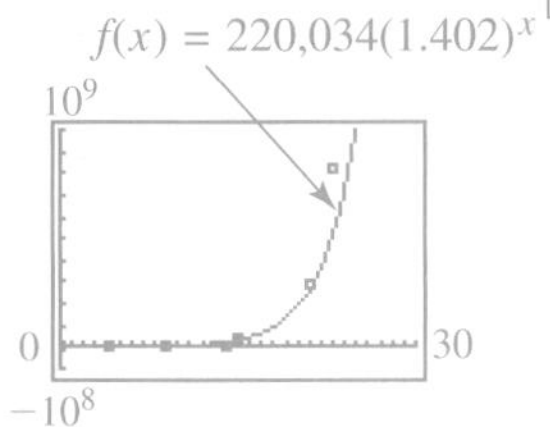

(d) 2,016,900,000

81. (a) \$15,207
(b) \$10,716
(c) \$4491
(d) They are the same.

82. (a) $\log_4(2x^2 - x) = \frac{\ln(2x^2 - x)}{\ln 4}$
(b)

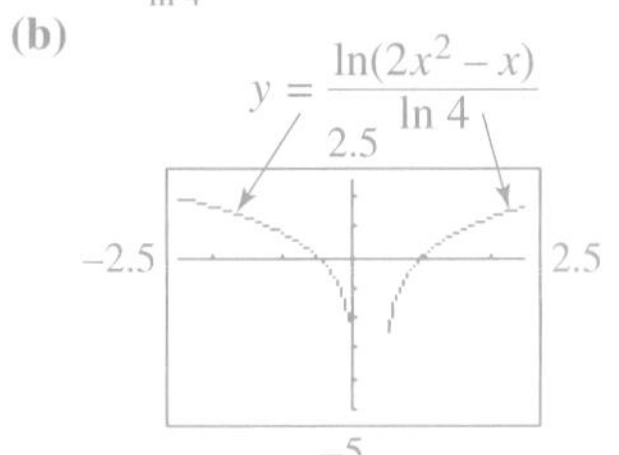

(c) $-\frac{1}{2}, 1$
(d) $x = 0,\ x = \frac{1}{2}$

Letting y represent the total payroll and x represent the number of years since 2007, we find that the function

$$f(x) = 113.2e^{0.0836x}$$

models the data quite well. According to this function, when would the total payroll double its 2010 value?

80. *(Modeling) Transistors on Computer Chips* Computing power of personal computers has increased dramatically as a result of the ability to place an increasing number of transistors on a single processor chip. The table lists the number of transistors on some popular computer chips made by Intel.

Year	Chip	Transistors
1985	386DX	275,000
1989	486DX	1,200,000
1994	Pentium	3,300,000
1999	Pentium 3	9,500,000
2000	Pentium 4	42,000,000
2006	Core 2 Duo	291,000,000
2008	Core 2 Quad	820,000,000

Source: Intel.

(a) Make a scatter diagram of the data. Let the x-axis represent the year, where $x = 0$ corresponds to 1985, and let the y-axis represent the number of transistors.

(b) Decide whether a linear, a logarithmic, or an exponential function describes the data best.

(c) Determine a function f that approximates these data. Plot f and the data on the same coordinate axes.

(d) Assuming that this trend continues, use f to predict the number of transistors on a chip in the year 2012.

81. *Financial Planning* The traditional IRA (individual retirement account) is a common tax-deferred saving plan in the United States. Earned income deposited into an IRA is not taxed in the current year, and no taxes are incurred on the interest paid in subsequent years. However, when you withdraw the money from the account after age $59\frac{1}{2}$, you pay taxes on the entire amount you withdraw.

Suppose you deposited \$5000 of earned income into an IRA, you can earn an annual interest rate of 4%, and you are in a 25% tax bracket. (*Note:* Interest rates and tax brackets are subject to change over time, but some assumptions must be made to evaluate the investment.) Also, suppose that you deposit the \$5000 at age 25 and withdraw it at age 60, and that interest is compounded continuously.

(a) How much money will remain after you pay the taxes at age 60?

(b) Suppose that instead of depositing the money into an IRA, you pay taxes on the money and the annual interest. How much money will you have at age 60? (*Note:* You effectively start with \$3750 (75% of \$5000), and the money earns 3% (75% of 4%) interest after taxes.)

(c) To the nearest dollar, how much additional money will you earn with the IRA?

(d) Suppose you pay taxes on the original \$5000 but are then able to earn 4% in a tax-free investment. Compare your balance at age 60 with the IRA balance.

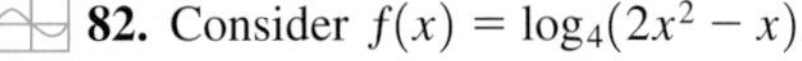

82. Consider $f(x) = \log_4(2x^2 - x)$.

(a) Use the change-of-base theorem with base e to write $\log_4(2x^2 - x)$ in a suitable form to graph with a calculator.

(b) Graph the function using a graphing calculator. Use the window $[-2.5, 2.5]$ by $[-5, 2.5]$.

(c) What are the x-intercepts?

(d) Give the equations of the vertical asymptotes.

(e) Explain why there is no y-intercept.

Chapter 4 Test

1. Consider the function $f(x) = \sqrt[3]{2x - 7}$.
   (a) What are the domain and range of f?
   (b) Explain why f^{-1} exists.
   (c) Find the rule for $f^{-1}(x)$.
   (d) What are the domain and range of f^{-1}?
   (e) Graph both f and f^{-1}. How are the two graphs related to the line $y = x$?

2. Match each equation with its graph.

   (a) $y = \log_{1/3} x$ (b) $y = e^x$ (c) $y = \ln x$ (d) $y = \left(\frac{1}{3}\right)^x$

A.
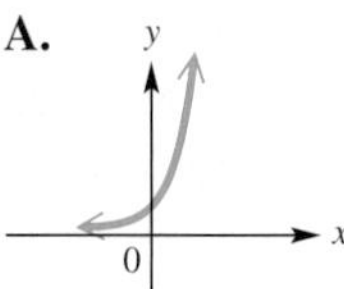

B.
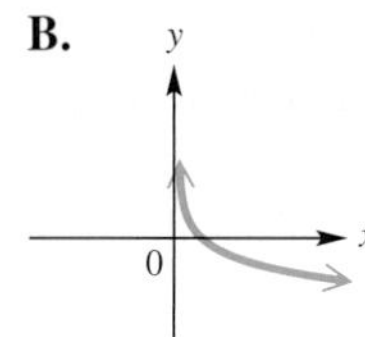

C.
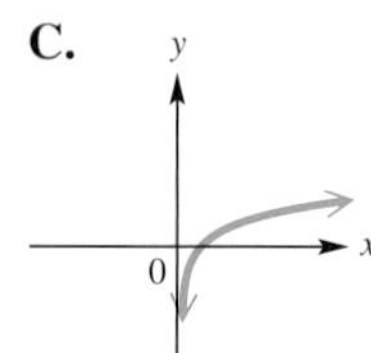

D.
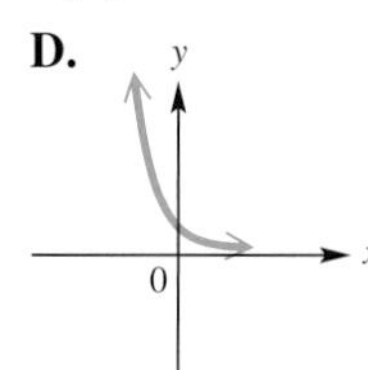

3. Solve $\left(\frac{1}{8}\right)^{2x-3} = 16^{x+1}$.

4. (a) Write $4^{3/2} = 8$ in logarithmic form.
   (b) Write $\log_8 4 = \frac{2}{3}$ in exponential form.

5. Graph $f(x) = \left(\frac{1}{2}\right)^x$ and $g(x) = \log_{1/2} x$ on the same axes. What is their relationship?

6. Use properties of logarithms to write

$$\log_7 \frac{x^2 \sqrt[4]{y}}{z^3}$$

as a sum, difference, or product of logarithms. Assume all variables represent positive numbers.

Use a calculator to find an approximation for each logarithm. Express answers to four decimal places.

7. $\log 2388$ 8. $\ln 2388$ 9. $\log_9 13$

10. Solve the following.

$$x^{2/3} = 25$$

Solve each exponential equation. If applicable, express solutions to the nearest thousandth.

11. $12^x = 1$ 12. $9^x = 4$ 13. $16^{2x+1} = 8^{3x}$

14. $2^{x+1} = 3^{x-4}$ 15. $e^{0.4x} = 4^{x-2}$

16. $2e^{2x} - 5e^x + 3 = 0$ (Give both exact and approximate values.)

Solve each logarithmic equation. Express all solutions in exact form.

17. $\log_x \frac{9}{16} = 2$ 18. $\log_2(x - 4)(x - 2) = 3$

19. $\log_2 x + \log_2(x + 2) = 3$ 20. $\ln x - 4 \ln 3 = \ln\left(\frac{1}{5}x\right)$

21. $\log_3(x + 1) - \log_3(x - 3) = 2$

[4.1]
1. (a) $(-\infty, \infty)$; $(-\infty, \infty)$
   (b) The graph is a stretched translation of $y = \sqrt[3]{x}$, which passes the horizontal line test and is thus a one-to-one function.
   (c) $f^{-1}(x) = \frac{x^3 + 7}{2}$
   (d) $(-\infty, \infty)$; $(-\infty, \infty)$
   (e)

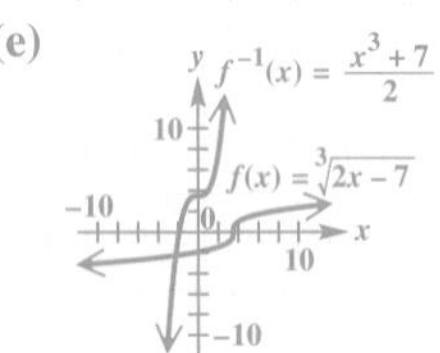

The graphs are reflections of each other across the line $y = x$.

[4.2, 4.3]
2. (a) B (b) A
   (c) C (d) D

[4.2]
3. $\left\{\frac{1}{2}\right\}$

[4.3]
4. (a) $\log_4 8 = \frac{3}{2}$
   (b) $8^{2/3} = 4$

[4.1–4.3]
5. They are inverses.

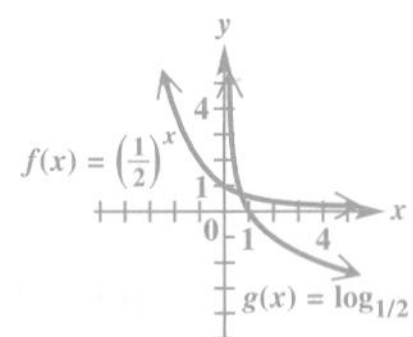

[4.3]
6. $2 \log_7 x + \frac{1}{4} \log_7 y - 3 \log_7 z$

[4.4]
7. 3.3780
8. 7.7782
9. 1.1674

[4.2]
10. $\{\pm 125\}$
11. $\{0\}$

[4.5]
12. $\{0.631\}$

[4.2]
13. $\{4\}$

[4.5]
14. $\{12.548\}$
15. $\{2.811\}$
16. $\{0, \ln \frac{3}{2}\}$; $\{0, 0.405\}$

[4.3]
17. $\{\frac{3}{4}\}$

[4.5]
18. $\{0, 6\}$
19. $\{2\}$
20. $\emptyset$
21. $\{\frac{7}{2}\}$

[4.6]
23. 10 sec
24. **(a)** 33.8 yr **(b)** 33.7 yr
25. 28.9 yr
26. **(a)** 329.3 g **(b)** 13.9 days

22. One of your friends is taking another mathematics course and tells you, "I have no idea what an expression like $\log_5 27$ really means." Write an explanation of what it means, and tell how you can find an approximation for it with a calculator.

Solve each problem.

23. *(Modeling) Skydiver Fall Speed* A skydiver in free fall travels at a speed modeled by

$$v(t) = 176(1 - e^{-0.18t})$$

feet per second after t seconds. How long will it take for the skydiver to attain a speed of 147 ft per sec (100 mph)?

24. *Growth of an Account* How many years, to the nearest tenth, will be needed for \$5000 to increase to \$18,000 at 3.8% annual interest compounded **(a)** monthly **(b)** continuously?

25. *Tripling Time* For *any* amount of money invested at 3.8% annual interest compounded continuously, how long will it take to triple?

26. *(Modeling) Radioactive Decay* The amount of radioactive material, in grams, present after t days is modeled by

$$A(t) = 600e^{-0.05t}.$$

(a) Find the amount present after 12 days.
(b) Find the half-life of the material.

5 Trigonometric Functions

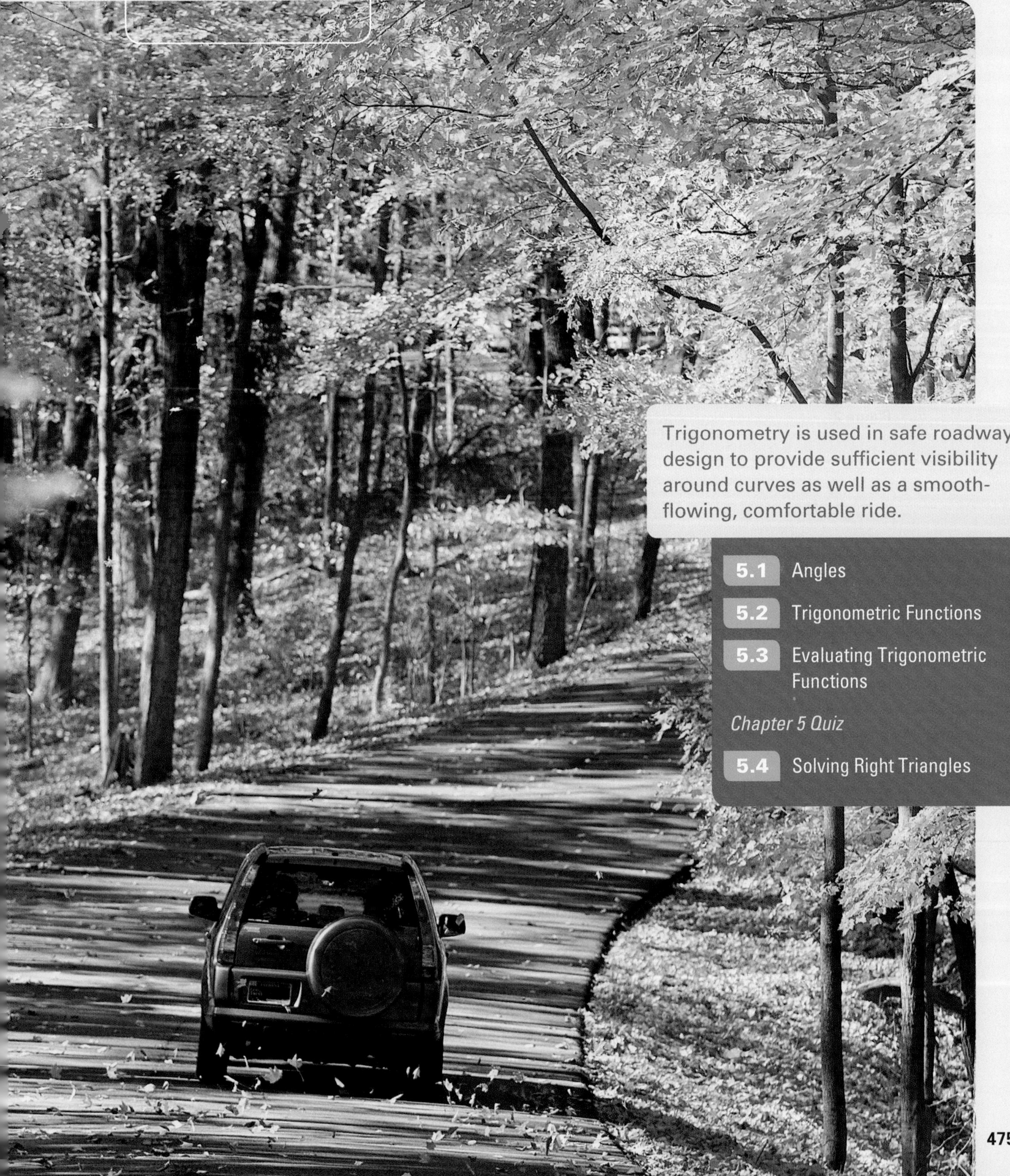

Trigonometry is used in safe roadway design to provide sufficient visibility around curves as well as a smooth-flowing, comfortable ride.

5.1 Angles

- Basic Terminology
- Degree Measure
- Standard Position
- Coterminal Angles

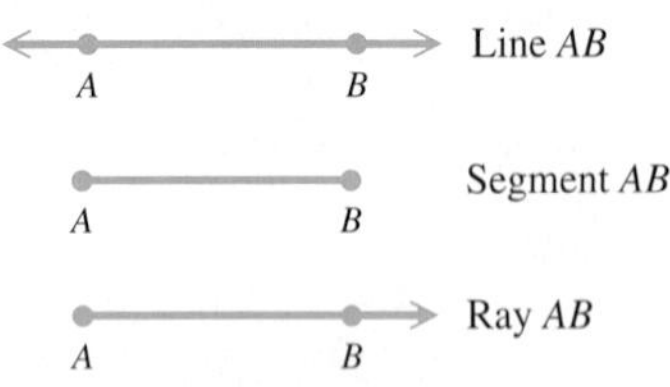

Figure 1

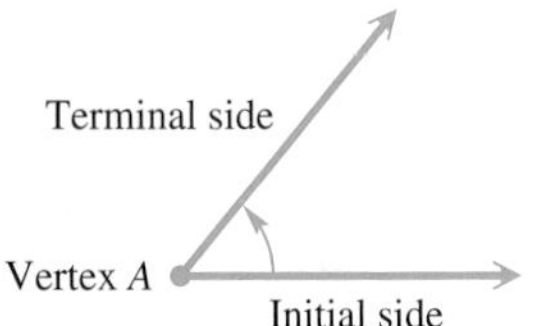

Figure 2

Basic Terminology

Two distinct points A and B determine a line called **line AB.** The portion of the line between A and B, including points A and B themselves, is **line segment AB,** or simply **segment AB.** The portion of line AB that starts at A and continues through B, and on past B, is the **ray AB.** Point A is the **endpoint of the ray.** See **Figure 1.**

In trigonometry, an **angle** consists of two rays in a plane with a common endpoint, or two line segments with a common endpoint. These two rays (or segments) are the **sides** of the angle, and the common endpoint is the **vertex** of the angle. Associated with an angle is its measure, generated by a rotation about the vertex. See **Figure 2.** This measure is determined by rotating a ray starting at one side of the angle, the **initial side,** to the position of the other side, the **terminal side.** ***A counterclockwise rotation generates a positive measure, and a clockwise rotation generates a negative measure.*** The rotation can consist of more than one complete revolution.

Figure 3 shows two angles, one **positive** and one **negative.**

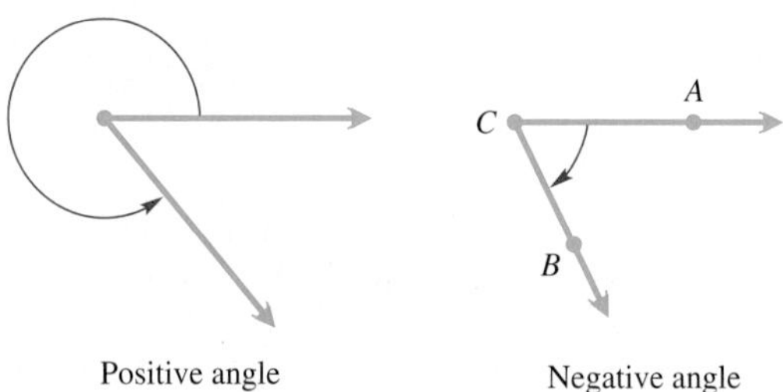

Figure 3

An angle can be named by using the name of its vertex. For example, the angle on the right in **Figure 3** can be named angle C. Alternatively, an angle can be named using three letters, with the vertex letter in the middle. Thus, the angle on the right also could be named angle ACB or angle BCA.

Teaching Tip Explain to students that we often say "angle" when we are actually referring to "angle measure."

Degree Measure

The most common unit for measuring angles is the **degree.** Degree measure was developed by the Babylonians 4000 yr ago. To use degree measure, we assign 360 degrees to a complete rotation of a ray.* In **Figure 4,** notice that the terminal side of the angle corresponds to its initial side when it makes a complete rotation.

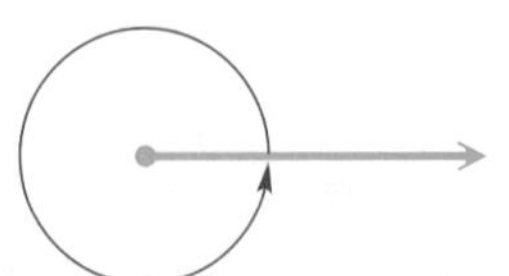

A complete rotation of a ray gives an angle whose measure is $360°$. $\frac{1}{360}$ of a complete rotation gives an angle whose measure is $1°$.

Figure 4

$$\text{One degree, written } 1°, \text{ represents } \frac{1}{360} \text{ of a rotation.}$$

Therefore, $90°$ represents $\frac{90}{360} = \frac{1}{4}$ of a complete rotation, and $180°$ represents $\frac{180}{360} = \frac{1}{2}$ of a complete rotation.

An angle measuring between $0°$ and $90°$ is an **acute angle.** An angle measuring exactly $90°$ is a **right angle.** The symbol ⌉ is often used at the vertex of a right angle to denote the $90°$ measure. An angle measuring more than $90°$ but less than $180°$ is an **obtuse angle,** and an angle of exactly $180°$ is a **straight angle.**

*The Babylonians were the first to subdivide the circumference of a circle into 360 parts. There are various theories about why the number 360 was chosen. One is that it is approximately the number of days in a year, and it has many divisors, which makes it convenient to work with.

In **Figure 5,** we use the **Greek letter θ (theta)*** to name each angle.

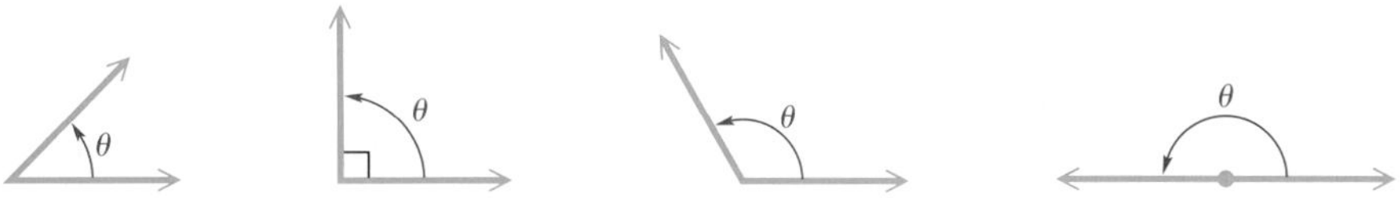

Acute angle $0° < \theta < 90°$ | Right angle $\theta = 90°$ | Obtuse angle $90° < \theta < 180°$ | Straight angle $\theta = 180°$

Figure 5

If the sum of the measures of two positive angles is 90°, the angles are **complementary** and the angles are **complements** of each other. Two positive angles with measures whose sum is 180° are **supplementary,** and the angles are **supplements.**

Classroom Example 1
For an angle measuring 55°, find the measure of **(a)** its complement and **(b)** its supplement.

Answers:
(a) 35° (b) 125°

EXAMPLE 1 Finding the Complement and the Supplement of an Angle

For an angle measuring 40°, find the measure of **(a)** its complement and **(b)** its supplement.

SOLUTION

(a) To find the measure of its complement, subtract the measure of the angle from 90°.

$$90° - 40° = 50° \quad \text{Complement of } 40°$$

(b) To find the measure of its supplement, subtract the measure of the angle from 180°.

$$180° - 40° = 140° \quad \text{Supplement of } 40°$$

✔ *Now Try Exercise 1.*

(6x)° (3x)°

(a)

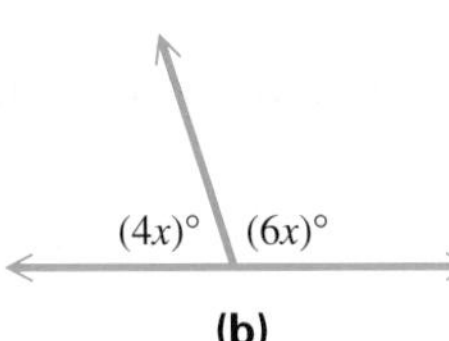

(b)

Figure 6

EXAMPLE 2 Finding Measures of Complementary and Supplementary Angles

Find the measure of each marked angle in **Figure 6.**

SOLUTION

(a) Since the two angles in **Figure 6(a)** form a right angle, they are complementary angles.

$$6x + 3x = 90 \quad \text{Complementary angles sum to } 90°.$$

$$9x = 90 \quad \text{Combine like terms.}$$

Don't stop here. $x = 10$ Divide by 9. **(Section 1.1)**

Be sure to determine the measure of each angle by substituting 10 for x. The two angles have measures of $6(10) = 60°$ and $3(10) = 30°$.

(b) The angles in **Figure 6(b)** are supplementary, so their sum must be 180°.

$$4x + 6x = 180 \quad \text{Supplementary angles sum to } 180°.$$

$$10x = 180 \quad \text{Combine like terms.}$$

$$x = 18 \quad \text{Divide by 10.}$$

These angle measures are $4(18) = 72°$ and $6(18) = 108°$.

✔ *Now Try Exercises 13 and 15.*

Classroom Example 2
Find the measure of each angle.

(a) (2x)° (3x)°

(b) (7x)° (2x)°

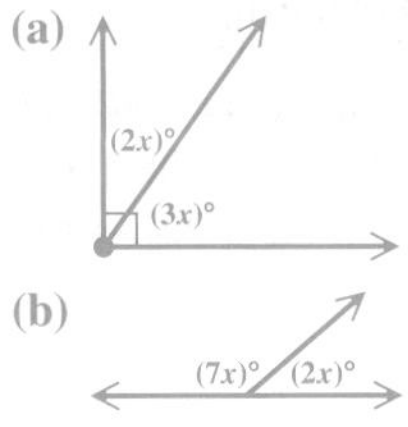

Answers:
(a) 36° and 54° (b) 140° and 40°

*In addition to θ (theta), other Greek letters such as α (alpha) and β (beta) are often used.

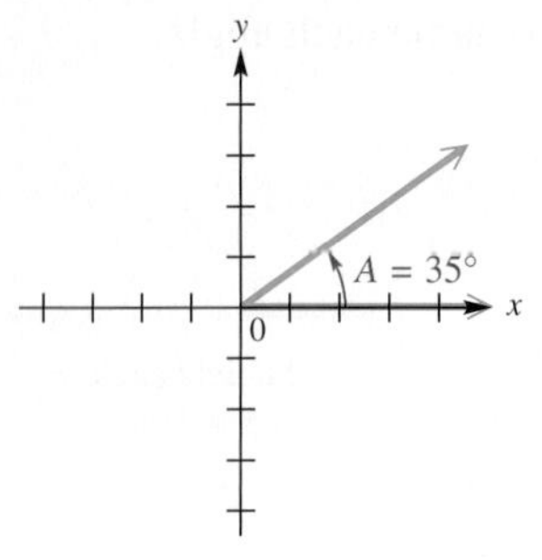

Figure 7

The measure of angle A in **Figure 7** is 35°. This measure is often expressed by saying that $m(\text{angle } A)$ is 35°, where $m(\text{angle } A)$ is read **"the measure of angle A."** It is convenient, however, to abbreviate the symbolism $m(\text{angle } A) = 35°$ as $A = 35°$.

Traditionally, portions of a degree have been measured with minutes and seconds. One **minute,** written **1′**, is $\frac{1}{60}$ of a degree.

$$1' = \left(\frac{1}{60}\right)^\circ \quad \text{or} \quad 60' = 1^\circ$$

One **second, 1″**, is $\frac{1}{60}$ of a minute.

$$1'' = \left(\frac{1}{60}\right)' = \left(\frac{1}{3600}\right)^\circ \quad \text{or} \quad 60'' = 1'$$

The measure 12° 42′ 38″ represents 12 degrees, 42 minutes, 38 seconds.

Classroom Example 3
Perform each calculation.
(a) 28° 35′ + 63° 52′
(b) 180° − 117° 29′

Answers:
(a) 92° 27′ (b) 62° 31′

EXAMPLE 3 Calculating with Degrees, Minutes, and Seconds

Perform each calculation.

(a) 51° 29′ + 32° 46′ **(b)** 90° − 73° 12′

SOLUTION

(a)
$$\begin{array}{r} 51^\circ\ 29' \\ +\ 32^\circ\ 46' \\ \hline 83^\circ\ 75' \end{array} \quad \text{Add degrees and minutes separately.}$$

The sum 83° 75′ can be rewritten as follows.

$$\begin{aligned} 83^\circ\ 75' &= 83^\circ + 1^\circ\ 15' && 75' = 60' + 15' = 1^\circ\ 15' \\ &= 84^\circ\ 15' && \text{Add.} \end{aligned}$$

(b)
$$\begin{array}{r} 89^\circ\ 60' \\ -\ 73^\circ\ 12' \\ \hline 16^\circ\ 48' \end{array} \quad \text{Write } 90^\circ \text{ as } 89^\circ\ 60'.$$

✔ *Now Try Exercises 37 and 41.*

Because calculators are so prevalent, angles are commonly measured in decimal degrees. For example, 12.4238° represents

$$12.4238^\circ = 12\frac{4238}{10{,}000}^\circ.$$

Classroom Example 4
(a) Convert 105° 20′ 32″ to decimal degrees to the nearest thousandth.
(b) Convert 85.263° to degrees, minutes, and seconds to the nearest second.

Answers:
(a) 105.342° (b) 85° 15′ 47″

EXAMPLE 4 Converting between Decimal Degrees and Degrees, Minutes, and Seconds

(a) Convert 74° 08′ 14″ to decimal degrees to the nearest thousandth.

(b) Convert 34.817° to degrees, minutes, and seconds to the nearest second.

SOLUTION

(a)
$$\begin{aligned} 74^\circ\ 08'\ 14'' &= 74^\circ + \left(\frac{8}{60}\right)^\circ + \left(\frac{14}{3600}\right)^\circ && 1' = \left(\frac{1}{60}\right)^\circ \text{ and } 1'' = \left(\frac{1}{3600}\right)^\circ \\ &\approx 74^\circ + 0.1333^\circ + 0.0039^\circ \\ &\approx 74.137^\circ && \text{Add and round to the nearest thousandth.} \end{aligned}$$

```
74°8'14"
      74.13722222
74°8'14"
           74.137
34.817▸DMS
      34°49'1.2"
```

A graphing calculator performs the conversions in **Example 4** as shown above. The ▸DMS option is found in the ANGLE Menu of the TI-83/84 Plus calculator.

(b)

$34.817° = 34° + 0.817°$	Write as a sum.
$= 34° + 0.817(60')$	$1° = 60'$
$= 34° + 49.02'$	Multiply.
$= 34° + 49' + 0.02'$	Write as a sum.
$= 34° + 49' + 0.02(60'')$	$1' = 60''$
$= 34° + 49' + 1.2''$	Write as a sum.
$\approx 34° \, 49' \, 01''$	Approximate to the nearest second.

✔ *Now Try Exercises 53 and 63.*

Standard Position An angle is in **standard position** if its vertex is at the origin and its initial side lies on the positive x-axis. The angles in **Figures 8(a) and 8(b)** are in standard position. An angle in standard position is said to lie in the quadrant in which its terminal side lies. An acute angle is in quadrant I (**Figure 8(a)**) and an obtuse angle is in quadrant II (**Figure 8(b)**). **Figure 8(c)** shows ranges of angle measures for each quadrant when $0° < \theta < 360°$.

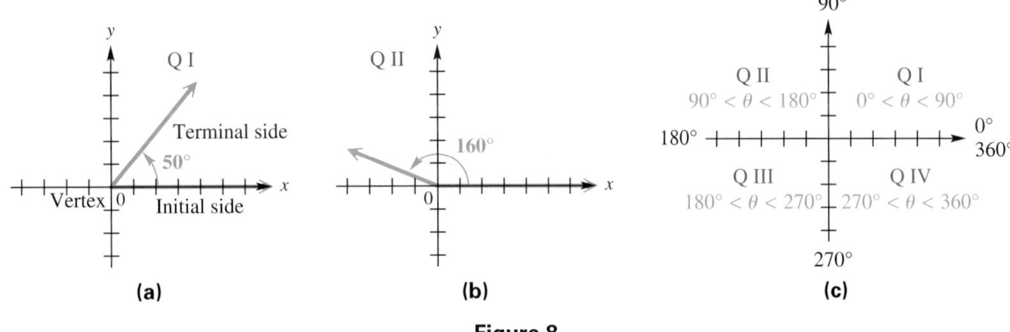

Figure 8

Quadrantal Angles

Angles in standard position whose terminal sides lie on the x-axis or y-axis, such as angles with measures 90°, 180°, 270°, and so on, are **quadrantal angles.**

Coterminal Angles A complete rotation of a ray results in an angle measuring 360°. By continuing the rotation, angles of measure larger than 360° can be produced. The angles in **Figure 9** with measures 60° and 420° have the same initial side and the same terminal side, but different amounts of rotation. Such angles are **coterminal angles.** ***Their measures differ by a multiple of*** **360°.** As shown in **Figure 10,** angles with measures 110° and 830° are coterminal.

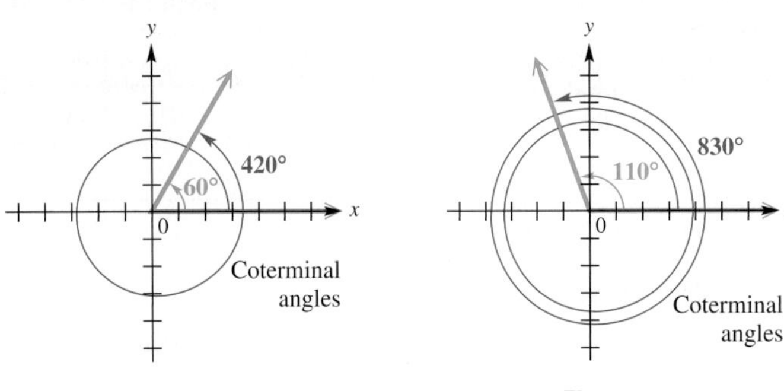

Figure 9 **Figure 10**

Classroom Example 5
Find the angles of least positive measure that are coterminal with each angle.
(a) 1106° (b) −150°
(c) −603°

Answers:
(a) 26° (b) 210° (c) 117°

EXAMPLE 5 Finding Measures of Coterminal Angles

Find the angles of least positive measure that are coterminal with each angle.

(a) 908° **(b)** −75° **(c)** −800°

SOLUTION

(a) Subtract 360° as many times as needed to obtain an angle with measure greater than 0° but less than 360°. Since

$$908^\circ - 2 \cdot 360^\circ = 188^\circ,$$

an angle of 188° is coterminal with an angle of 908°. See **Figure 11.**

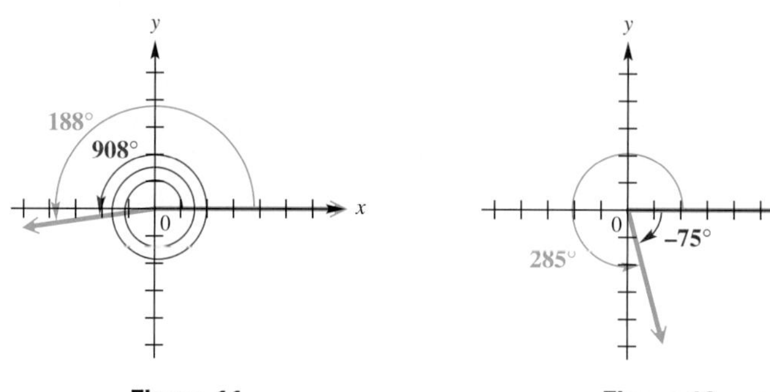

Figure 11 **Figure 12**

(b) See **Figure 12.** Use a rotation of

$$360^\circ + (-75^\circ) = 285^\circ.$$

(c) The least integer multiple of 360° greater than 800° is

$$360^\circ \cdot 3 = 1080^\circ.$$

Add 1080° to −800° to obtain

$$1080^\circ + (-800^\circ) = 280^\circ.$$

✔ *Now Try Exercises 77, 87, and 91.*

Teaching Tip Remind students that the set of integers is $\{\ldots, -3, -2, -1, 0, 1, 2, 3, \ldots\}$.

Sometimes it is necessary to find an expression that will generate all angles coterminal with a given angle. For example, we can obtain any angle coterminal with 60° by adding an integer multiple of 360° to 60°. Let n represent any integer. Then the following expression represents all such coterminal angles.

$$60^\circ + \boldsymbol{n} \cdot \mathbf{360^\circ} \quad \text{Angles coterminal with } 60^\circ$$

The table below shows a few possibilities.

Value of n	Angle Coterminal with 60°
2	$60^\circ + 2 \cdot 360^\circ = 780^\circ$
1	$60^\circ + 1 \cdot 360^\circ = 420^\circ$
0	$60^\circ + 0 \cdot 360^\circ = 60^\circ$ (the angle itself)
−1	$60^\circ + (-1) \cdot 360^\circ = -300^\circ$

Examples of Coterminal Quadrantal Angles

Quadrantal Angle θ	Coterminal with θ
0°	±360°, ±720°
90°	−630°, −270°, 450°
180°	−180°, 540°, 900°
270°	−450°, −90°, 630°

The table in the margin shows some examples of coterminal quadrantal angles.

Classroom Example 6
A wheel makes 270 revolutions per min. Through how many degrees will a point on the edge of the wheel move in 5 sec?

Answer: 8100°

EXAMPLE 6 Analyzing the Revolutions of a CD Player

CD players always spin at the same speed. Suppose a player makes 480 revolutions per min. Through how many degrees will a point on the edge of a CD move in 2 sec?

SOLUTION The player revolves 480 times in 1 min, or $\frac{480}{60}$ times = 8 times per sec (since 60 sec = 1 min). In 2 sec, the player will revolve $2 \cdot 8 = 16$ times. Each revolution is 360°, so in 2 sec a point on the edge of the CD will revolve

$$16 \cdot 360^\circ = 5760^\circ.$$

A unit analysis expression can also be used.

$$\frac{480 \text{ rev}}{1 \text{ min}} \times \frac{1 \text{ min}}{60 \text{ sec}} \times \frac{360^\circ}{1 \text{ rev}} \times 2 \text{ sec} = 5760^\circ \quad \text{Divide out common units.}$$

Now Try Exercise 131.

5.1 Exercises

1. (a) 60° (b) 150°
2. (a) 30° (b) 120°
3. (a) 45° (b) 135°
4. (a) 72° (b) 162°
5. (a) 36° (b) 126°
6. (a) 1° (b) 91°
7. (a) 89° (b) 179°
8. (a) 80° (b) 170°
9. (a) 75° 40′ (b) 165° 40′
10. (a) 50° 10′ (b) 140° 10′
11. (a) 69° 49′ 30″ (b) 159° 49′ 30″
12. (a) 39° 19′ 10″ (b) 129° 19′ 10″

13. 70°; 110° 14. 150°; 30°
15. 30°; 60° 16. 55°; 35°
17. 40°; 140° 18. 90°; 90°
19. 107°; 73° 20. 80°; 100°
21. 69°; 21° 22. 40°; 50°

23. 45° 24. 90°

25. 150° 26. 142° 30′
27. 7° 30′ 28. 22° 30′
29. 130° 30. 125°

31. $(90 - x)^\circ$ 32. $(180 - x)^\circ$
33. $(x - 360)^\circ$ 34. $(360 + x)^\circ$

35. 83° 59′ 36. 158° 47′
37. 179° 19′ 38. 143° 20′
39. 23° 49′ 40. −26° 25′
41. 38° 32′ 42. 72° 47′
43. 60° 34′ 44. 55° 09′
45. 30° 27′ 46. 59° 59′
47. 17° 01′ 49″ 48. 53° 41′ 13″

*Find **(a)** the complement and **(b)** the supplement of an angle with the given measure.* ***See Examples 1 and 3.***

1. 30° 2. 60° 3. 45° 4. 18°
5. 54° 6. 89° 7. 1° 8. 10°
9. 14° 20′ 10. 39° 50′ 11. 20° 10′ 30″ 12. 50° 40′ 50″

Find the measure of each unknown angle in Exercises 13–22. ***See Example 2.***

13.
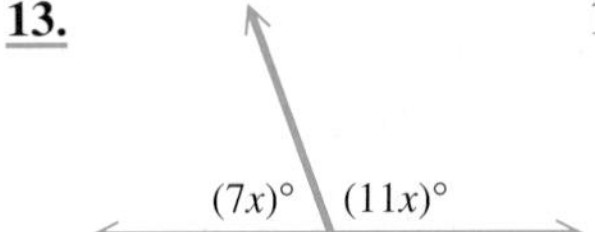

14.
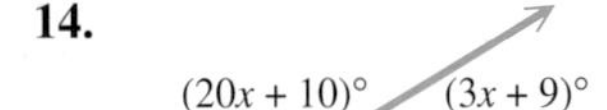

15.
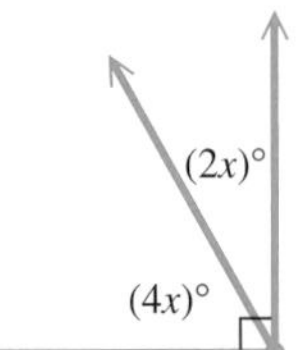

16.
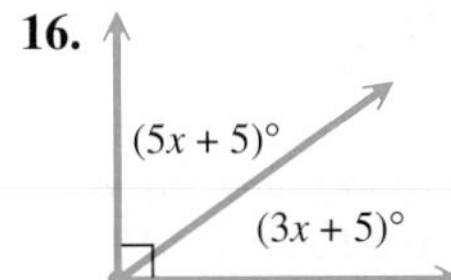

17.
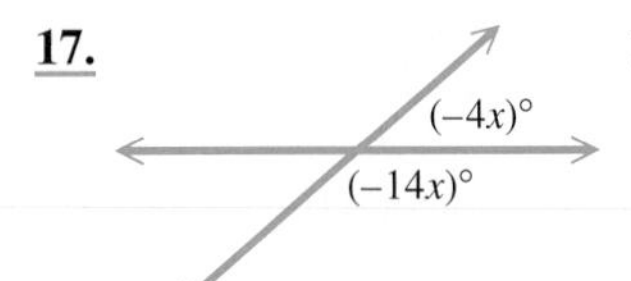

18.
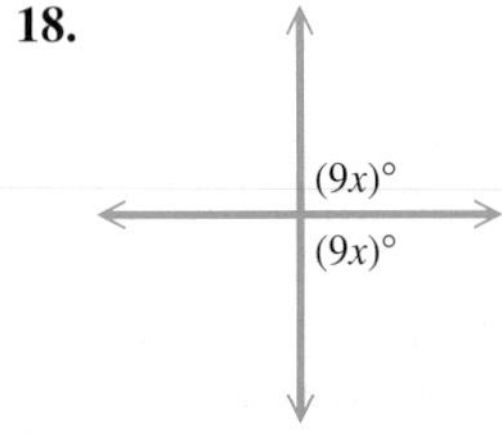

19. supplementary angles with measures $10x + 7$ and $7x + 3$ degrees
20. supplementary angles with measures $6x - 4$ and $8x - 12$ degrees
21. complementary angles with measures $9x + 6$ and $3x$ degrees
22. complementary angles with measures $3x - 5$ and $6x - 40$ degrees
23. *Concept Check* What is the measure of an angle that is its own complement?
24. *Concept Check* What is the measure of an angle that is its own supplement?

Find the measure of the smaller angle formed by the hands of a clock at the following times.

25.

26.

27. 3:15 **28.** 9:45 **29.** 8:20 **30.** 6:10

Concept Check *Answer each question.*

31. If an angle measures $x°$, how can we represent its complement?

32. If an angle measures $x°$, how can we represent its supplement?

33. If a positive angle has measure $x°$ between 0° and 60°, how can we represent the first negative angle coterminal with it?

34. If a negative angle has measure $x°$ between 0° and −60°, how can we represent the first positive angle coterminal with it?

Perform each calculation. ***See Example 3.***

35. 62° 18′ + 21° 41′	**36.** 75° 15′ + 83° 32′	**37.** 97° 42′ + 81° 37′
38. 110° 25′ + 32° 55′	**39.** 71° 18′ − 47° 29′	**40.** 47° 23′ − 73° 48′

41. 90° − 51° 28′ **42.** 90° − 17° 13′

43. 180° − 119° 26′ **44.** 180° − 124° 51′

45. 26° 20′ + 18° 17′ − 14° 10′ **46.** 55° 30′ + 12° 44′ − 8° 15′

47. 90° − 72° 58′ 11″ **48.** 90° − 36° 18′ 47″

Convert each angle measure to decimal degrees. If applicable, round to the nearest thousandth of a degree. ***See Example 4(a).***

49. 35° 30′	**50.** 82° 30′	**51.** 112° 15′
52. 133° 45′	**53.** −60° 12′	**54.** −70° 48′
55. 20° 54′ 00″	**56.** 38° 42′ 00″	**57.** 91° 35′ 54″
58. 34° 51′ 35″	**59.** 274° 18′ 59″	**60.** 165° 51′ 09″

Convert each angle measure to degrees, minutes, and seconds. Round answers to the nearest second, if applicable. ***See Example 4(b).***

61. 39.25°	**62.** 46.75°	**63.** 126.76°	**64.** 174.255°
65. −18.515°	**66.** −25.485°	**67.** 31.4296°	**68.** 59.0854°
69. 89.9004°	**70.** 102.3771°	**71.** 178.5994°	**72.** 122.6853°

Find the angle of least positive measure (not equal to the given measure) that is coterminal with each angle. ***See Example 5.***

73. 32°	**74.** 86°	**75.** 26° 30′	**76.** 58° 40′
77. −40°	**78.** −98°	**79.** −125°	**80.** −203°
81. 361°	**82.** 541°	**83.** −361°	**84.** −541°

---

49. 35.5° **50.** 82.5°
51. 112.25° **52.** 133.75°
53. −60.2° **54.** −70.8°
55. 20.9° **56.** 38.7°
57. 91.598° **58.** 34.860°
59. 274.316° **60.** 165.853°

61. 39° 15′ 00″ **62.** 46° 45′ 00″
63. 126° 45′ 36″ **64.** 174° 15′ 18″
65. −18° 30′ 54″ **66.** −25° 29′ 06″
67. 31° 25′ 47″ **68.** 59° 05′ 07″
69. 89° 54′ 01″ **70.** 102° 22′ 38″
71. 178° 35′ 58″ **72.** 122° 41′ 07″

73. 392° **74.** 446°
75. 386° 30′ **76.** 418° 40′
77. 320° **78.** 262°
79. 235° **80.** 157°
81. 1° **82.** 181°
83. 359° **84.** 179°
85. 179° **86.** 339°
87. 130° **88.** 280°
89. 240° **90.** 160°
91. 120° **92.** 200°

In Exercises 93–96, answers may vary.
93. 450°, 810°; −270°, −630°
94. 540°, 900°; −180°, −540°
95. 360°, 720°; −360°, −720°
96. 630°, 990°; −90°, −450°

97. $30° + n \cdot 360°$
98. $45° + n \cdot 360°$
99. $135° + n \cdot 360°$
100. $225° + n \cdot 360°$
101. $-90° + n \cdot 360°$
102. $-180° + n \cdot 360°$
103. $0° + n \cdot 360°$, or $n \cdot 360°$
104. $360° + n \cdot 360°$, or $n \cdot 360°$

106. C and D

Angles other than those given are possible in Exercises 107–118.

107. **108.**

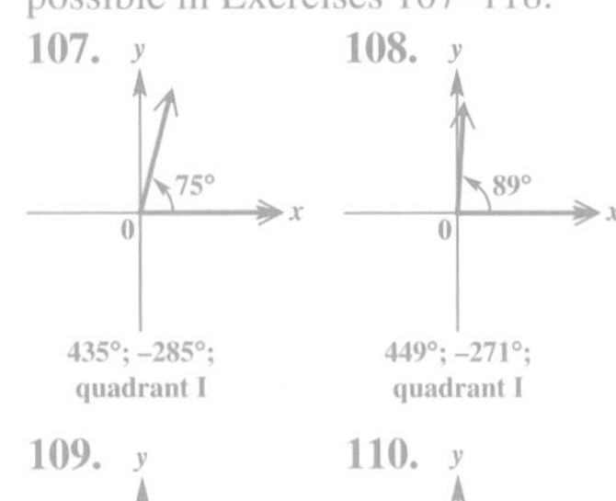

109. **110.**

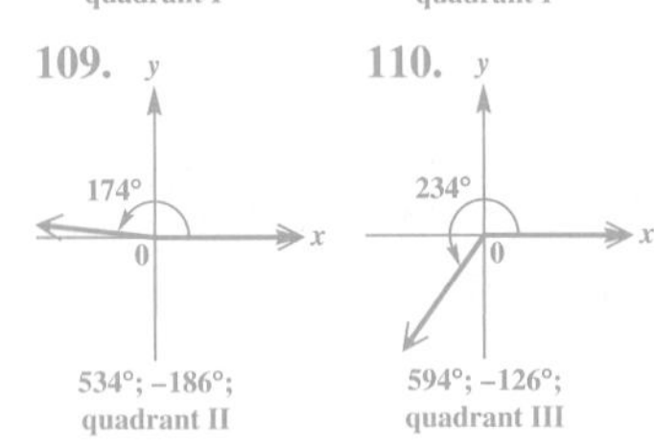

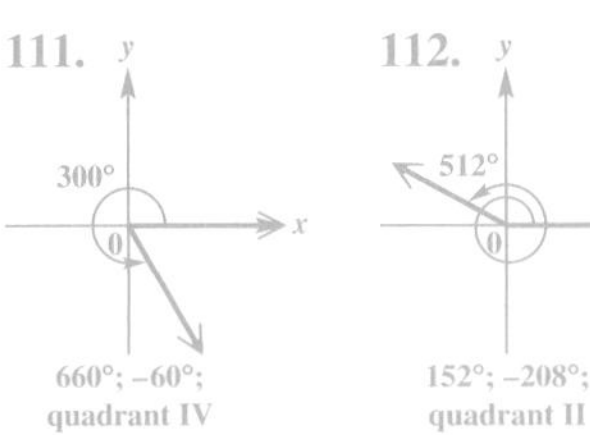

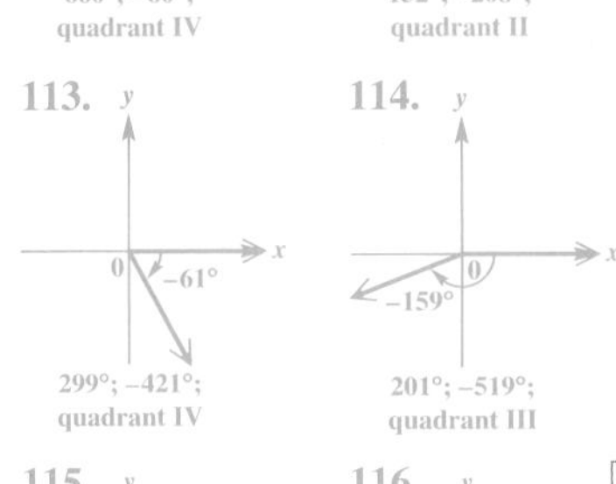

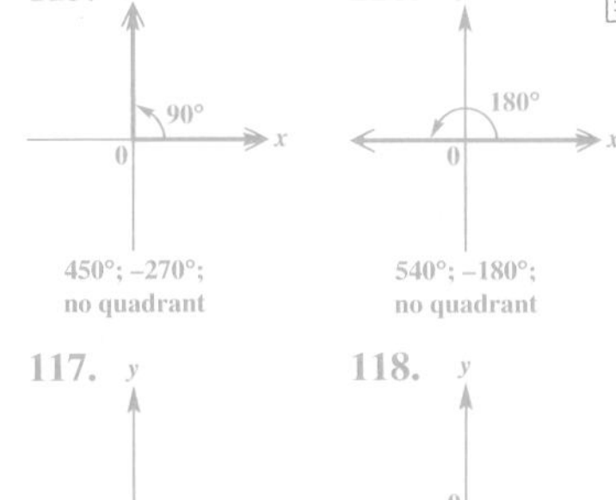

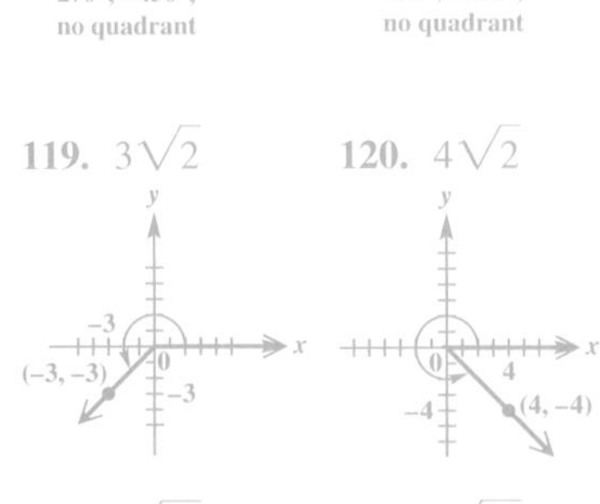

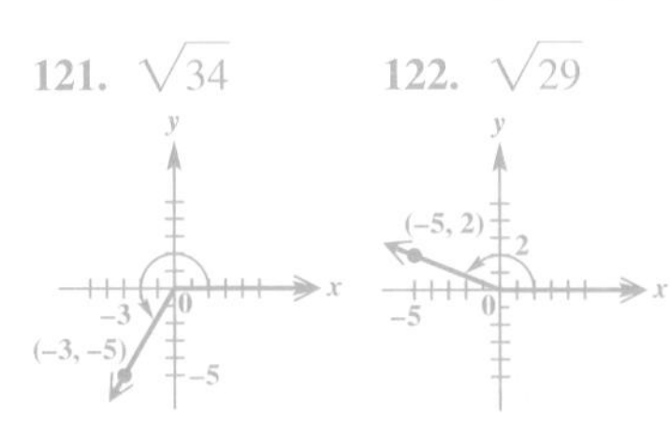

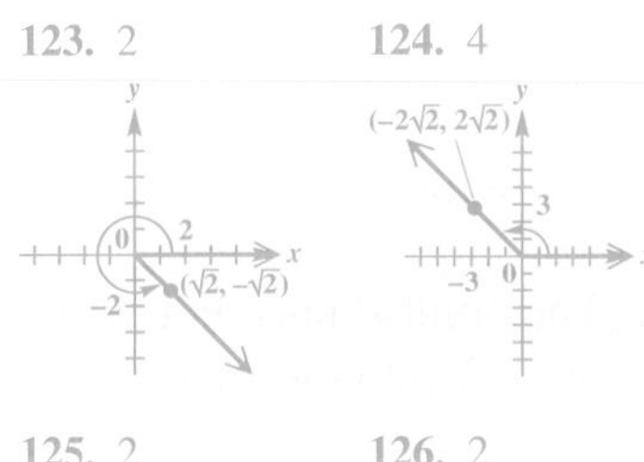

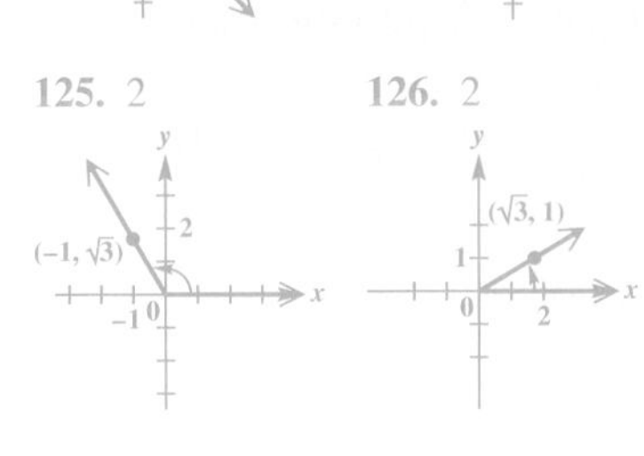

85. 539° **86.** 699° **87.** 850° **88.** 1000°

89. 5280° **90.** 8440° **91.** −5280° **92.** −8440°

Give two positive and two negative angles that are coterminal with the given quadrantal angle.

93. 90° **94.** 180° **95.** 0° **96.** 270°

Give an expression that generates all angles coterminal with each angle. Let n represent any integer.

97. 30° **98.** 45° **99.** 135° **100.** 225°

101. −90° **102.** −180° **103.** 0° **104.** 360°

105. Explain why the answers to **Exercises 103 and 104** give the same set of angles.

106. *Concept Check* Which two of the following are not coterminal with $r°$?

A. $360° + r°$ **B.** $r° - 360°$ **C.** $360° - r°$ **D.** $r° + 180°$

Concept Check Sketch each angle in standard position. Draw an arrow representing the correct amount of rotation. Find the measure of two other angles, one positive and one negative, that are coterminal with the given angle. Give the quadrant of each angle, if applicable.

107. 75° **108.** 89° **109.** 174° **110.** 234°

111. 300° **112.** 512° **113.** −61° **114.** −159°

115. 90° **116.** 180° **117.** −90° **118.** −180°

Concept Check Locate each point in a coordinate system. Draw a ray from the origin through the given point. Indicate with an arrow the angle in standard position having least positive measure. Then find the distance r from the origin to the point, using the distance formula of ***Section 2.1.***

119. $(-3, -3)$ **120.** $(4, -4)$ **121.** $(-3, -5)$ **122.** $(-5, 2)$

123. $(\sqrt{2}, -\sqrt{2})$ **124.** $(-2\sqrt{2}, 2\sqrt{2})$ **125.** $(-1, \sqrt{3})$ **126.** $(\sqrt{3}, 1)$

127. $(-2, 2\sqrt{3})$ **128.** $(4\sqrt{3}, -4)$ **129.** $(0, -4)$ **130.** $(0, 2)$

Solve each problem. ***See Example 6.***

131. *Revolutions of a Turntable* A turntable in a shop makes 45 revolutions per min. How many revolutions does it make per second?

132. *Revolutions of a Windmill* A windmill makes 90 revolutions per min. How many revolutions does it make per second?

133. *Rotating Tire* A tire is rotating 600 times per min. Through how many degrees does a point on the edge of the tire move in $\frac{1}{2}$ sec?

134. *Rotating Airplane Propeller* An airplane propeller rotates 1000 times per min. Find the number of degrees that a point on the edge of the propeller will rotate in 1 sec.

135. *Rotating Pulley* A pulley rotates through 75° in 1 min. How many rotations does the pulley make in an hour?

127. 4 128. 8

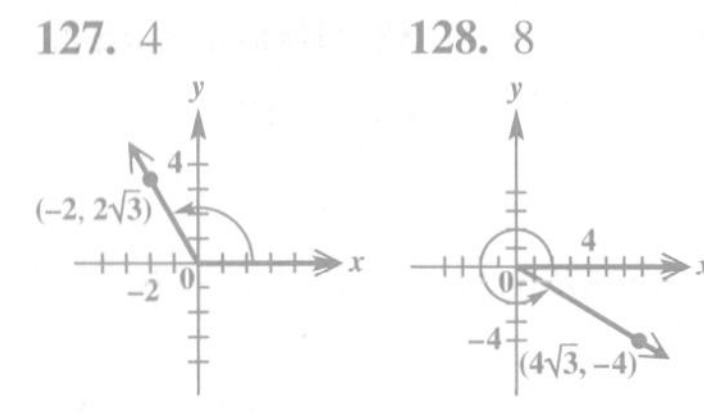

129. 4 130. 2

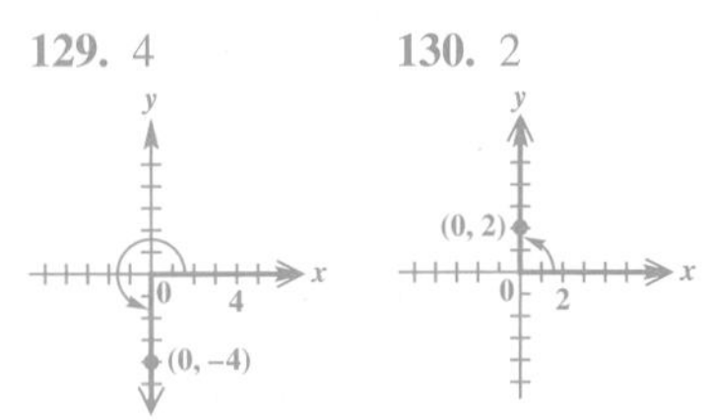

131. $\frac{3}{4}$ 132. 1.5
133. 1800° 134. 6000°
135. 12.5 rotations per hr
136. 5′, or 0.08°
137. 4 sec 138. 36°

136. *Surveying* One student in a surveying class measures an angle as 74.25°, while another student measures the same angle as 74° 20′. Find the difference between these measurements, both to the nearest minute and to the nearest hundredth of a degree.

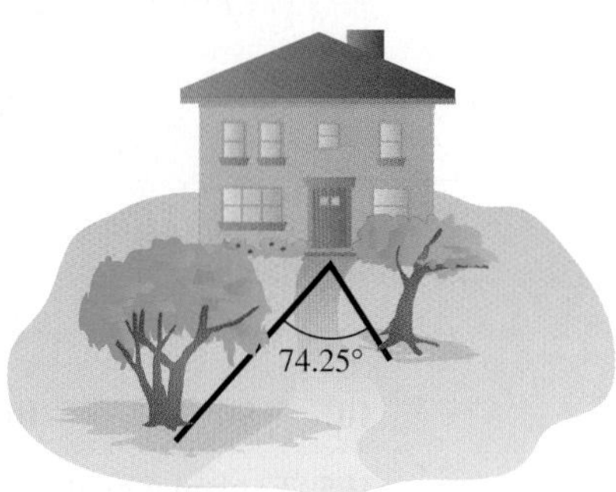

137. *Viewing Field of a Telescope* As a consequence of Earth's rotation, celestial objects such as the moon and the stars appear to move across the sky, rising in the east and setting in the west. As a result, if a telescope on Earth remains stationary while viewing a celestial object, the object will slowly move outside the viewing field of the telescope. For this reason, a motor is often attached to telescopes so that the telescope rotates at the same rate as Earth. Determine how long it should take the motor to turn the telescope through an angle of 1 min in a direction perpendicular to Earth's axis.

138. *Angle Measure of a Star on the American Flag* Determine the measure of the angle in each point of the five-pointed star appearing on the American flag. (*Hint:* Inscribe the star in a circle, and use the following theorem from geometry: *An angle whose vertex lies on the circumference of a circle is equal to half the central angle that cuts off the same arc.* See the figure.)

5.2 Trigonometric Functions

- Trigonometric Functions
- Quadrantal Angles
- Reciprocal Identities
- Signs and Ranges of Function Values
- Pythagorean Identities
- Quotient Identities

Trigonometric Functions To define the six **trigonometric functions,** we start with an angle θ in standard position and choose any point P having coordinates (x, y) on the terminal side of angle θ. (The point P must not be the vertex of the angle.) See **Figure 13** on the next page. A perpendicular from point P to the x-axis at point Q determines a right triangle, having vertices at O, P, and Q. We find the distance r from $P(x, y)$ to the origin, $(0, 0)$, using the distance formula.

$$r = \sqrt{(x - 0)^2 + (y - 0)^2} \quad \text{(Section 2.1)}$$

$$r = \sqrt{x^2 + y^2}$$

Notice that $r > 0$ since this is the undirected distance.

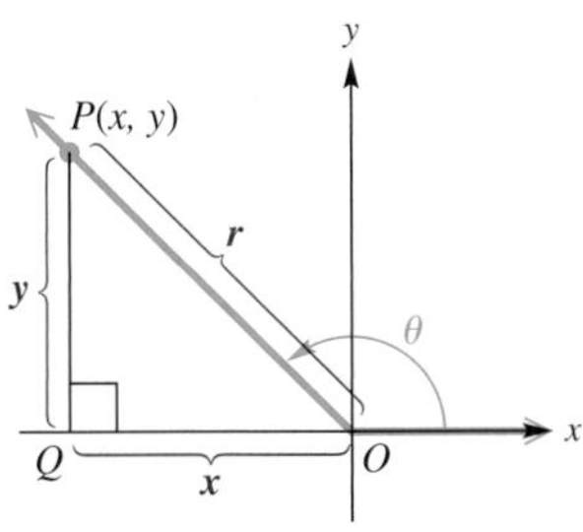

Figure 13

The six trigonometric functions of angle θ are **sine, cosine, tangent, cotangent, secant,** and **cosecant,** abbreviated **sin, cos, tan, cot, sec,** and **csc.**

Trigonometric Functions

Let (x, y) be a point other than the origin on the terminal side of an angle θ in standard position. The distance from the point to the origin is $r = \sqrt{x^2 + y^2}$. The six trigonometric functions of θ are defined as follows.

$$\sin\theta = \frac{y}{r} \qquad \cos\theta = \frac{x}{r} \qquad \tan\theta = \frac{y}{x} \quad (x \neq 0)$$

$$\csc\theta = \frac{r}{y} \quad (y \neq 0) \qquad \sec\theta = \frac{r}{x} \quad (x \neq 0) \qquad \cot\theta = \frac{x}{y} \quad (y \neq 0)$$

Classroom Example 1
The terminal side of an angle θ in standard position passes through the point $(12, 5)$. Find the values of the six trigonometric functions of angle θ.

Answer:
$\sin\theta = \frac{5}{13}$; $\cos\theta = \frac{12}{13}$;
$\tan\theta = \frac{5}{12}$; $\cot\theta = \frac{12}{5}$;
$\sec\theta = \frac{13}{12}$; $\csc\theta = \frac{13}{5}$

EXAMPLE 1 Finding Function Values of an Angle

The terminal side of an angle θ in standard position passes through the point $(8, 15)$. Find the values of the six trigonometric functions of angle θ.

SOLUTION **Figure 14** shows angle θ and the triangle formed by dropping a perpendicular from the point $(8, 15)$ to the x-axis. The point $(8, 15)$ is 8 units to the right of the y-axis and 15 units above the x-axis, so $x = 8$ and $y = 15$. Now use $r = \sqrt{x^2 + y^2}$.

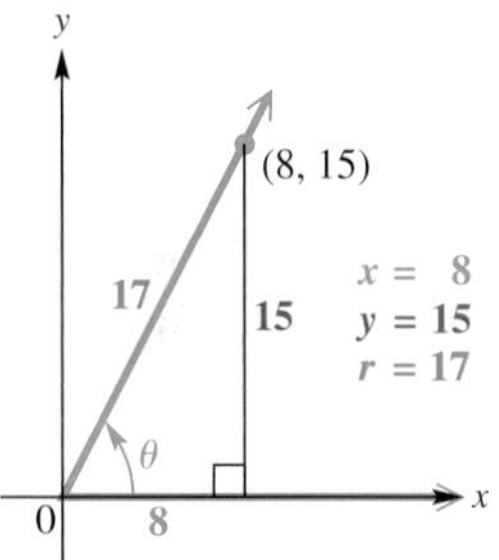

Figure 14

$$r = \sqrt{8^2 + 15^2} = \sqrt{64 + 225} = \sqrt{289} = 17$$

We can now find the values of the six trigonometric functions of angle θ.

$$\sin\theta = \frac{y}{r} = \frac{15}{17} \qquad \cos\theta = \frac{x}{r} = \frac{8}{17} \qquad \tan\theta = \frac{y}{x} = \frac{15}{8}$$

$$\csc\theta = \frac{r}{y} = \frac{17}{15} \qquad \sec\theta = \frac{r}{x} = \frac{17}{8} \qquad \cot\theta = \frac{x}{y} = \frac{8}{15}$$

✔ *Now Try Exercise 5.*

Classroom Example 2
The terminal side of an angle θ in standard position passes through the point $(8, -6)$. Find the values of the six trigonometric functions of angle θ.

Answer:
$\sin\theta = -\frac{3}{5}$; $\cos\theta = \frac{4}{5}$;
$\tan\theta = -\frac{3}{4}$; $\cot\theta = -\frac{4}{3}$;
$\sec\theta = \frac{5}{4}$; $\csc\theta = -\frac{5}{3}$

EXAMPLE 2 Finding Function Values of an Angle

The terminal side of an angle θ in standard position passes through the point $(-3, -4)$. Find the values of the six trigonometric functions of angle θ.

SOLUTION As shown in **Figure 15,** $x = -3$ and $y = -4$.

$r = \sqrt{(-3)^2 + (-4)^2}$ $r = \sqrt{x^2 + y^2}$

$r = \sqrt{25}$ Simplify the radicand.

$r = 5$ $r > 0$

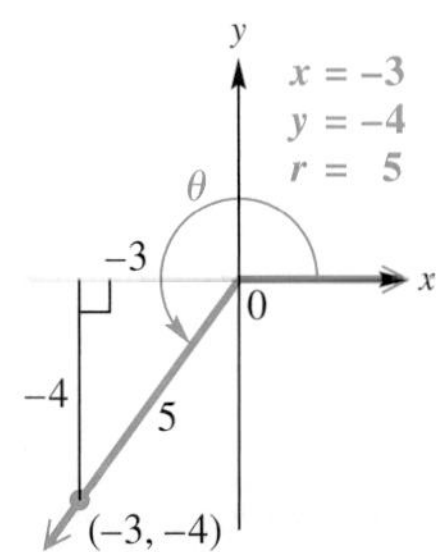

Figure 15

Now use the definitions of the trigonometric functions.

$$\sin\theta = \frac{-4}{5} = -\frac{4}{5} \qquad \cos\theta = \frac{-3}{5} = -\frac{3}{5} \qquad \tan\theta = \frac{-4}{-3} = \frac{4}{3}$$

$$\csc\theta = \frac{5}{-4} = -\frac{5}{4} \qquad \sec\theta = \frac{5}{-3} = -\frac{5}{3} \qquad \cot\theta = \frac{-3}{-4} = \frac{3}{4}$$

✔ *Now Try Exercise 7.*

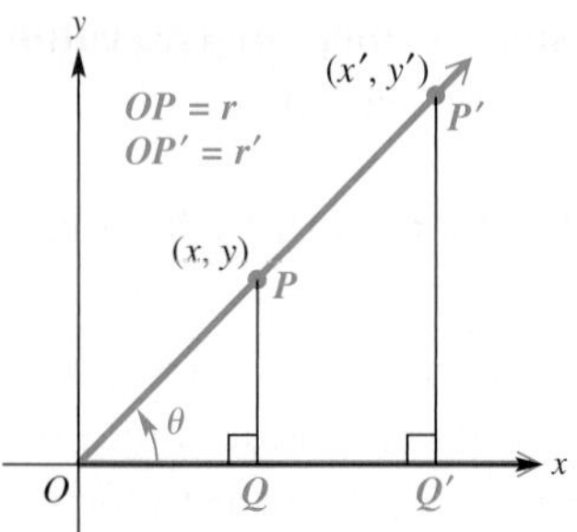

Figure 16

We can find the six trigonometric functions using *any* point other than the origin on the terminal side of an angle. To see why any point can be used, refer to **Figure 16,** which shows an angle θ and two distinct points on its terminal side. Point P has coordinates (x, y), and point P' (read **"P-prime"**) has coordinates (x', y'). Let r be the length of the hypotenuse of triangle OPQ, and let r' be the length of the hypotenuse of triangle $OP'Q'$. Since corresponding sides of similar triangles are proportional,

$$\frac{y}{r} = \frac{y'}{r'},$$

so $\sin \theta = \frac{y}{r}$ is the same no matter which point is used to find it. A similar result holds for the other five trigonometric functions.

We can also find the trigonometric function values of an angle if we know the equation of the line coinciding with the terminal ray. Recall from algebra that the graph of the equation

$$Ax + By = 0 \quad \text{(Section 2.4)}$$

is a line that passes through the origin. If we restrict x to have only nonpositive or only nonnegative values, we obtain as the graph a ray with endpoint at the origin. For example, the graph of $x + 2y = 0,\ x \geq 0$, shown in **Figure 17,** is a ray that can serve as the terminal side of an angle θ in standard position. By choosing a point on the ray, we can find the trigonometric function values of the angle.

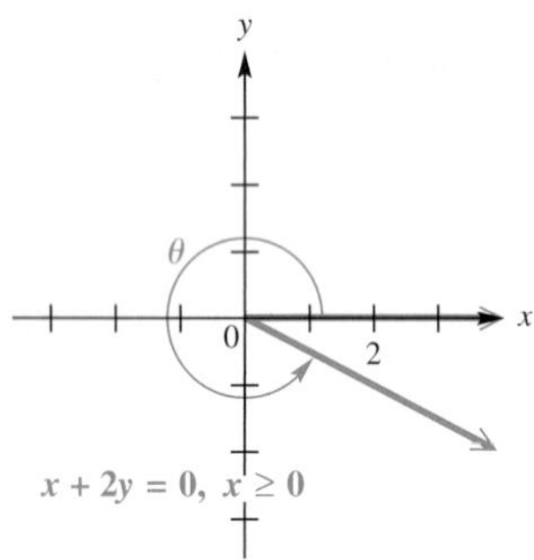

Figure 17

Classroom Example 3
Find the six trigonometric function values of the angle θ in standard position, if the terminal side of θ is defined by $3x - 2y = 0, x \leq 0$.

Answer:

$\sin \theta = -\frac{3\sqrt{13}}{13}$;

$\cos \theta = -\frac{2\sqrt{13}}{13}$;

$\tan \theta = \frac{3}{2}$; $\cot \theta = \frac{2}{3}$;

$\sec \theta = -\frac{\sqrt{13}}{2}$; $\csc \theta = -\frac{\sqrt{13}}{3}$

EXAMPLE 3 Finding Function Values of an Angle

Find the six trigonometric function values of the angle θ in standard position, if the terminal side of θ is defined by $x + 2y = 0,\ x \geq 0$.

SOLUTION The angle is shown in **Figure 18.** We can use *any* point except $(0, 0)$ on the terminal side of θ to find the trigonometric function values. We choose $x = 2$ and find the corresponding y-value.

$x + 2y = 0, \quad x \geq 0$

$2 + 2y = 0$ Let $x = 2$.

$2y = -2$ Subtract 2. (Section 1.1)

$y = -1$ Divide by 2.

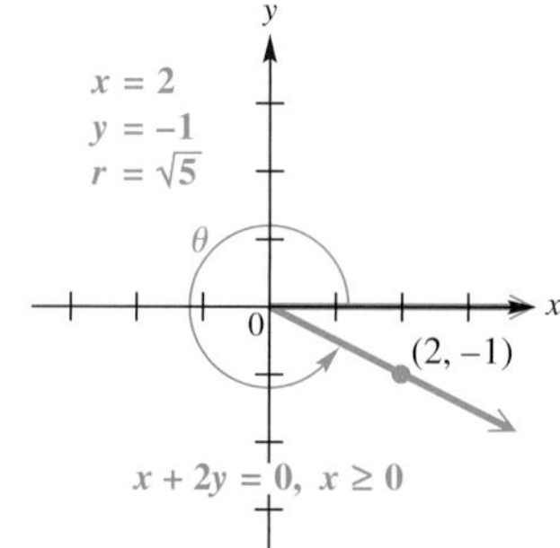

Figure 18

The point $(2, -1)$ lies on the terminal side, and the corresponding value of r is $r = \sqrt{2^2 + (-1)^2} = \sqrt{5}$. Now we use the definitions of the trigonometric functions.

$$\sin \theta = \frac{y}{r} = \frac{-1}{\sqrt{5}} = \frac{-1}{\sqrt{5}} \cdot \frac{\sqrt{5}}{\sqrt{5}} = -\frac{\sqrt{5}}{5}$$

$$\cos \theta = \frac{x}{r} = \frac{2}{\sqrt{5}} = \frac{2}{\sqrt{5}} \cdot \frac{\sqrt{5}}{\sqrt{5}} = \frac{2\sqrt{5}}{5}$$

Multiply by $\frac{\sqrt{5}}{\sqrt{5}}$, which equals 1, to rationalize the denominators. (Section R.7)

$$\tan \theta = \frac{y}{x} = \frac{-1}{2} = -\frac{1}{2}$$

$$\csc \theta = \frac{r}{y} = \frac{\sqrt{5}}{-1} = -\sqrt{5} \qquad \sec \theta = \frac{r}{x} = \frac{\sqrt{5}}{2} \qquad \cot \theta = \frac{x}{y} = \frac{2}{-1} = -2$$

✔ *Now Try Exercise 35.*

Recall that when the equation of a line is written in slope-intercept form

$$y = mx + b, \quad \text{(Section 2.5)}$$

the coefficient m of x is the slope of the line. In **Example 3,** the equation $x + 2y = 0$ can be written as $y = -\frac{1}{2}x$, so the slope is $-\frac{1}{2}$. Notice that $\tan \theta = -\frac{1}{2}$.

In general, it is true that m = tan θ.

NOTE The trigonometric function values we found in **Examples 1–3** are *exact.* If we were to use a calculator to approximate these values, the decimal results would not be acceptable if exact values were required.

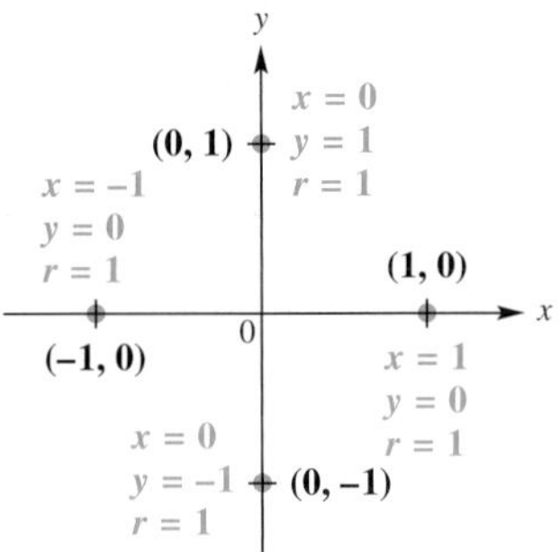

Figure 19

Quadrantal Angles If the terminal side of an angle in standard position lies along the y-axis, any point on this terminal side has x-coordinate 0. Similarly, an angle with terminal side on the x-axis has y-coordinate 0 for any point on the terminal side. Since the values of x and y appear in the denominators of some trigonometric functions, and since a fraction is undefined if its denominator is 0, some trigonometric function values of quadrantal angles (i.e., those with terminal side on an axis) are undefined.

When determining trigonometric function values of quadrantal angles, **Figure 19** can help find the ratios. Because *any* point on the terminal side can be used, it is convenient to choose the point one unit from the origin, with $r = 1$. (In **Chapter 6** we extend this idea to the *unit circle.*)

To find the function values of a quadrantal angle, determine the position of the terminal side, choose the one of these four points that lies on this terminal side, and then use the definitions involving x, y, and r.

Classroom Example 4 is on the next page.

EXAMPLE 4 Finding Function Values of Quadrantal Angles

Find the values of the six trigonometric functions for each angle.

(a) an angle of 90°

(b) an angle θ in standard position with terminal side through $(-3, 0)$

SOLUTION

(a) **Figure 20** shows that the terminal side passes through $(0, 1)$. So $x = 0$, $y = 1$, and $r = 1$. Thus, we have the following.

$$\sin 90° = \frac{1}{1} = 1 \qquad \cos 90° = \frac{0}{1} = 0 \qquad \tan 90° = \frac{1}{0} \text{ (undefined)}$$

$$\csc 90° = \frac{1}{1} = 1 \qquad \sec 90° = \frac{1}{0} \text{ (undefined)} \qquad \cot 90° = \frac{0}{1} = 0$$

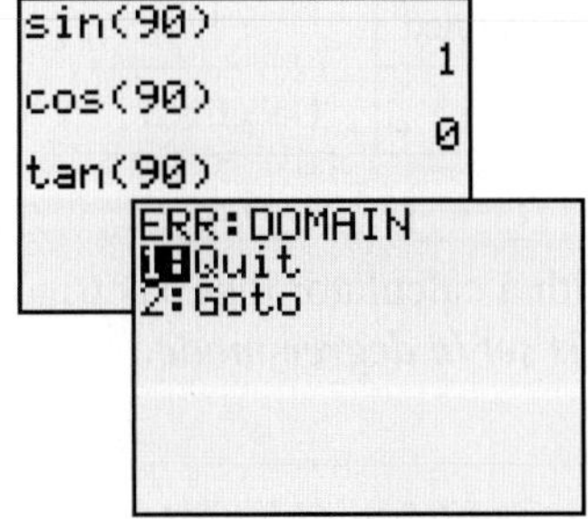

A calculator in degree mode returns the correct values for sin 90° and cos 90°. The second screen shows an ERROR message for tan 90°, because 90° is not in the domain of the tangent function.

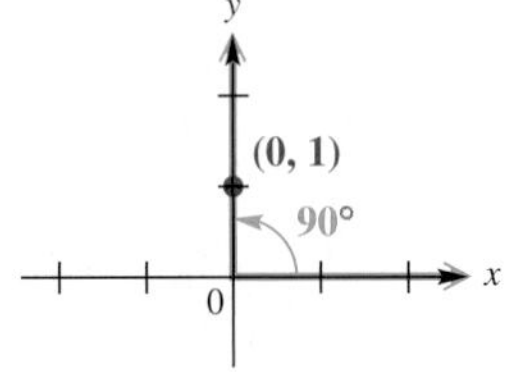

Figure 20

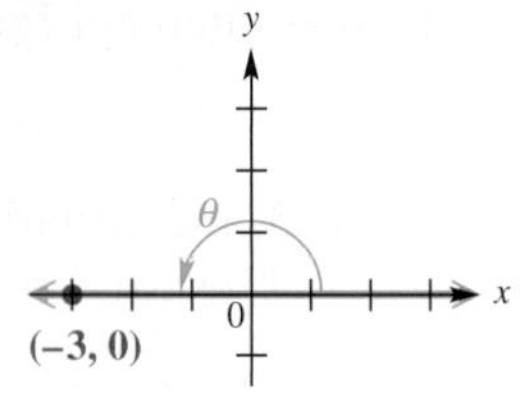

Figure 21

(b) **Figure 21** shows the angle. Here, $x = -3$, $y = 0$, and $r = 3$, so the trigonometric functions have the following values.

$$\sin\theta = \frac{0}{3} = 0 \qquad \cos\theta = \frac{-3}{3} = -1 \qquad \tan\theta = \frac{0}{-3} = 0$$

$$\csc\theta = \frac{3}{0} \text{ (undefined)} \qquad \sec\theta = \frac{3}{-3} = -1 \qquad \cot\theta = \frac{-3}{0} \text{ (undefined)}$$

Verify that these values can also be found by using the point $(-1, 0)$.

✔ *Now Try Exercises 9, 45, 47, 51, and 53.*

Classroom Example 4
Find the values of the six trigonometric functions for each angle.
(a) an angle of 360°
(b) an angle θ in standard position with terminal side through $(0, -5)$

Answers:
(a) $\sin\theta = 0$; $\cos\theta = 1$; $\tan\theta = 0$; $\cot\theta$: undefined; $\sec\theta = 1$; $\csc\theta$: undefined
(b) $\sin\theta = -1$; $\cos\theta = 0$; $\tan\theta$: undefined; $\cot\theta = 0$; $\sec\theta$: undefined; $\csc\theta = -1$

The conditions under which the trigonometric function values of quadrantal angles are undefined are summarized here.

Conditions for Undefined Function Values

Identify the terminal side of a quadrantal angle.

- If the terminal side of the quadrantal angle lies along the *y*-axis, then the tangent and secant functions are undefined.
- If the terminal side of the quadrantal angle lies along the *x*-axis, then the cotangent and cosecant functions are undefined.

The function values of some commonly used quadrantal angles, 0°, 90°, 180°, 270°, and 360°, are summarized in the table. They can be determined when needed by using **Figure 19** and the method of **Example 4(a).**

For other quadrantal angles such as −90°, −270°, and 450°, first determine the coterminal angle that lies between 0° and 360°, and then refer to the table entries for that particular angle. For example, the function values of a −90° angle would correspond to those of a 270° angle.

Function Values of Quadrantal Angles

θ	$\sin\theta$	$\cos\theta$	$\tan\theta$	$\cot\theta$	$\sec\theta$	$\csc\theta$
0°	0	1	0	Undefined	1	Undefined
90°	1	0	Undefined	0	Undefined	1
180°	0	−1	0	Undefined	−1	Undefined
270°	−1	0	Undefined	0	Undefined	−1
360°	0	1	0	Undefined	1	Undefined

The values given in this table can be found with a calculator that has trigonometric function keys. ***Make sure the calculator is set in degree mode.***

TI-83 Plus

NORMAL SCI ENG
FLOAT 0123456789
RADIAN DEGREE
FUNC PAR POL SEQ
CONNECTED DOT
SEQUENTIAL SIMUL
REAL a+bi re^θi
FULL HORIZ G-T
↓NEXT↓

TI-84 Plus

Figure 22

CAUTION ***One of the most common errors involving calculators in trigonometry occurs when the calculator is set for radian measure, rather than degree measure.*** (Radian measure of angles is discussed in **Chapter 6.**) Be sure you know how to set your calculator in degree mode. See **Figure 22,** which illustrates degree mode for TI-83/84 Plus calculators.

Reciprocal Identities Identities are equations that are true for all values of the variables for which all expressions are defined. Identities are studied in more detail in **Chapter 7.**

$$(x + y)^2 = x^2 + 2xy + y^2 \qquad 2(x + 3) = 2x + 6$$ Identities (Section 1.1)

Recall the definition of a reciprocal: the **reciprocal** of the nonzero number x is $\frac{1}{x}$. For example, the reciprocal of 2 is $\frac{1}{2}$, and the reciprocal of $\frac{8}{11}$ is $\frac{11}{8}$. There is no reciprocal for 0.

The definitions of the trigonometric functions earlier in this section were written so that functions in the same column were reciprocals of each other. Since $\sin\theta = \frac{y}{r}$ and $\csc\theta = \frac{r}{y}$,

$$\sin\theta = \frac{1}{\csc\theta} \quad \text{and} \quad \csc\theta = \frac{1}{\sin\theta}, \quad \text{provided } \sin\theta \neq 0.$$

Teaching Tip Students may be tempted to associate secant with sine and cosecant with cosine. Note this common misconception.

Also, $\cos\theta$ and $\sec\theta$ are reciprocals, as are $\tan\theta$ and $\cot\theta$. The **reciprocal identities** hold for any angle θ that does not lead to a 0 denominator.

Reciprocal Identities

For all angles θ for which both functions are defined, the following identities hold.

$$\sin\theta = \frac{1}{\csc\theta} \qquad \cos\theta = \frac{1}{\sec\theta} \qquad \tan\theta = \frac{1}{\cot\theta}$$

$$\csc\theta = \frac{1}{\sin\theta} \qquad \sec\theta = \frac{1}{\cos\theta} \qquad \cot\theta = \frac{1}{\tan\theta}$$

The screens in **Figures 23(a) and (b)** show how to find

$$\csc 90^\circ, \quad \sec 180^\circ, \quad \text{and} \quad \csc(-270^\circ),$$

using the appropriate reciprocal identities and the reciprocal key of a graphing calculator in degree mode. Attempting to find sec 90° by entering $\frac{1}{\cos 90^\circ}$ produces an ERROR message, indicating that the reciprocal is undefined. See **Figure 23(c).** Compare these results with the ones found in the table of quadrantal angle function values.

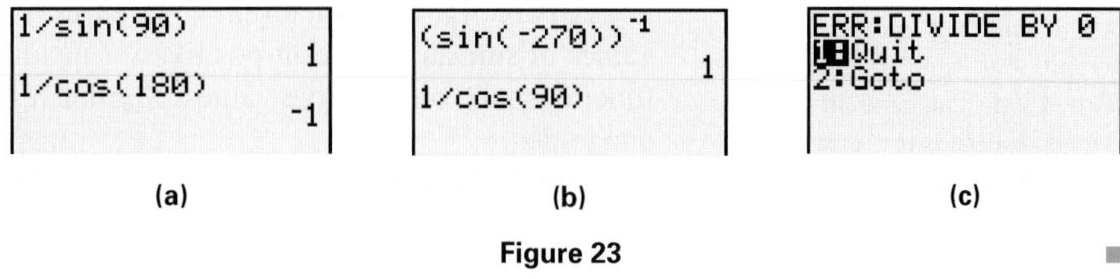

(a) (b) (c)

Figure 23

CAUTION ***Be sure not to use the inverse trigonometric function keys to find reciprocal function values.*** For example,

$$\sin^{-1}(90^\circ) \neq \frac{1}{\sin(90^\circ)}.$$

Inverse trigonometric functions are covered in **Section 5.3.**

The reciprocal identities can be written in different forms. For example,

$$\sin\theta = \frac{1}{\csc\theta} \quad \text{can be written} \quad \csc\theta = \frac{1}{\sin\theta}, \quad \text{or} \quad (\sin\theta)(\csc\theta) = 1.$$

Classroom Example 5
Find each function value.
(a) $\tan\theta$, given that $\cot\theta = 4$
(b) $\sec\theta$, given that
$\cos\theta = -\frac{2}{\sqrt{20}}$

Answers:
(a) $\frac{1}{4}$ (b) $-\sqrt{5}$

EXAMPLE 5 Using the Reciprocal Identities

Find each function value.

(a) $\cos\theta$, given that $\sec\theta = \frac{5}{3}$ **(b)** $\sin\theta$, given that $\csc\theta = -\frac{\sqrt{12}}{2}$

SOLUTION

(a) Since $\cos\theta$ is the reciprocal of $\sec\theta$,

$$\cos\theta = \frac{1}{\sec\theta} = \frac{1}{\frac{5}{3}} = 1 \div \frac{5}{3} = 1 \cdot \frac{3}{5} = \frac{3}{5}.$$ Simplify the complex fraction. (Section R.5)

(b) $\sin\theta = \dfrac{1}{-\frac{\sqrt{12}}{2}}$ $\quad$ $\sin\theta = \frac{1}{\csc\theta}$ and $\csc\theta = -\frac{\sqrt{12}}{2}$

$= -\dfrac{2}{\sqrt{12}}$ $\quad$ Simplify the complex fraction as in part (a).

$= -\dfrac{2}{2\sqrt{3}}$ $\quad$ $\sqrt{12} = \sqrt{4 \cdot 3} = 2\sqrt{3}$ (Section R.7)

$= -\dfrac{1}{\sqrt{3}}$ $\quad$ Divide out the common factor 2.

$= -\dfrac{1}{\sqrt{3}} \cdot \dfrac{\sqrt{3}}{\sqrt{3}}$ $\quad$ Rationalize the denominator.

$= -\dfrac{\sqrt{3}}{3}$ $\quad$ Multiply.

✔ *Now Try Exercises 77 and 83.*

Teaching Tip Some students use the sentence "**A**ll **S**tudents **T**ake **C**alculus" to remember which of the three basic functions are positive in each quadrant. **A** indicates "all" in quadrant I, **S** represents "sine" in quadrant II, **T** represents "tangent" in quadrant III, and **C** stands for "cosine" in quadrant IV.

Signs and Ranges of Function Values In the definitions of the trigonometric functions, r is the distance from the origin to the point (x, y). This distance is undirected, so $r > 0$. If we choose a point (x, y) in quadrant I, then both x and y will be positive, and the values of all six functions will be positive.

A point (x, y) in quadrant II satisfies $x < 0$ and $y > 0$. This makes the values of sine and cosecant positive for quadrant II angles, while the other four functions take on negative values. Similar results can be obtained for the other quadrants.

This important information is summarized here.

Signs of Function Values

θ in Quadrant	$\sin\theta$	$\cos\theta$	$\tan\theta$	$\cot\theta$	$\sec\theta$	$\csc\theta$
I	+	+	+	+	+	+
II	+	−	−	−	−	+
III	−	−	+	+	−	−
IV	−	+	−	−	+	−

Classroom Example 6
Determine the signs of the trigonometric functions of an angle in standard position with the given measure.

(a) 54° (b) 260° (c) -60°

Answers:

(a) All values are positive.
(b) Tangent and cotangent are positive. All others are negative.
(c) Cosine and secant are positive. All others are negative.

EXAMPLE 6 Determining Signs of Functions of Nonquadrantal Angles

Determine the signs of the trigonometric functions of an angle in standard position with the given measure.

(a) 87° **(b)** 300° **(c)** -200°

SOLUTION

(a) An angle of 87° is in the first quadrant, with x, y, and r all positive, so all of its trigonometric function values are positive.

(b) A 300° angle is in quadrant IV, so the cosine and secant are positive, while the sine, cosecant, tangent, and cotangent are negative.

(c) A -200° angle is in quadrant II. The sine and cosecant are positive, and all other function values are negative.

Now Try Exercises 87, 89, and 93.

NOTE Because numbers that are reciprocals always have the same sign, the sign of a function value automatically determines the sign of the reciprocal function value.

Classroom Example 7
Identify the quadrant (or possible quadrants) of an angle θ that satisfies the given conditions.

(a) $\tan\theta > 0$, $\csc\theta < 0$
(b) $\sin\theta > 0$, $\csc\theta > 0$

Answers:

(a) III (b) I or II

EXAMPLE 7 Identifying the Quadrant of an Angle

Identify the quadrant (or possible quadrants) of an angle θ that satisfies the given conditions.

(a) $\sin\theta > 0$, $\tan\theta < 0$ **(b)** $\cos\theta < 0$, $\sec\theta < 0$

SOLUTION

(a) Since $\sin\theta > 0$ in quadrants I and II and $\tan\theta < 0$ in quadrants II and IV, both conditions are met only in quadrant II.

(b) The cosine and secant functions are both negative in quadrants II and III, so in this case θ could be in either of these two quadrants.

Now Try Exercises 101 and 107.

Figure 24(a) shows an angle θ as it increases in measure from near 0° toward 90°. In each case, the value of r is the same. As the measure of the angle increases, y increases but never exceeds r, so $y \le r$. Dividing both sides by the positive number r gives $\frac{y}{r} \le 1$.

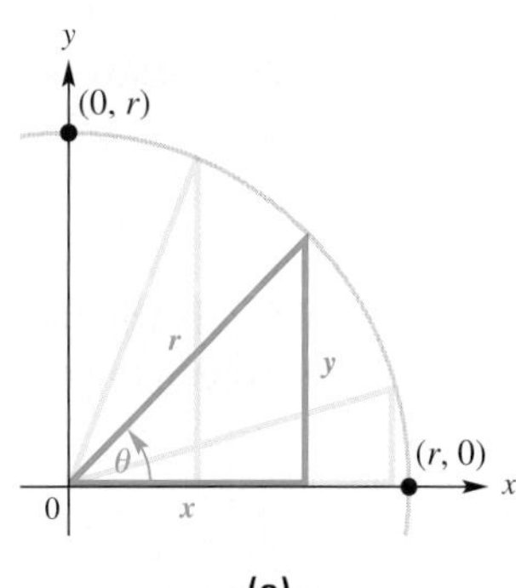

(a)

Figure 24

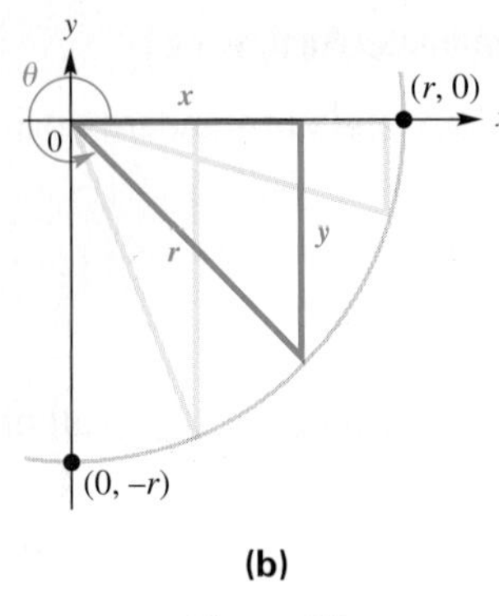

(b)

Figure 24

In a similar way, angles in quadrant IV as in **Figure 24(b)** suggest that

$$-1 \le \frac{y}{r},$$

so $$-1 \le \frac{y}{r} \le 1$$

and $$\mathbf{-1 \le \sin\theta \le 1.} \quad \frac{y}{r} = \sin\theta \text{ for any angle } \theta.$$

Similarly, $$\mathbf{-1 \le \cos\theta \le 1.}$$

The tangent of an angle is defined as $\frac{y}{x}$. It is possible that $x < y$, $x = y$, or $x > y$. Thus, $\frac{y}{x}$ can take any value, so **tan θ can be any real number, as can cot θ.**

The functions $\sec\theta$ and $\csc\theta$ are reciprocals of the functions $\cos\theta$ and $\sin\theta$, respectively, making

$$\mathbf{\sec\theta \le -1} \quad \text{or} \quad \mathbf{\sec\theta \ge 1} \quad \text{and} \quad \mathbf{\csc\theta \le -1} \quad \text{or} \quad \mathbf{\csc\theta \ge 1.}$$

In summary, the ranges of the trigonometric functions are as follows.

Ranges of Trigonometric Functions

Trigonometric Function of θ	Range (Set-Builder Notation)	Range (Interval Notation)
$\sin\theta$, $\cos\theta$	$\{y \mid \lvert y \rvert \le 1\}$	$[-1, 1]$
$\tan\theta$, $\cot\theta$	$\{y \mid y \text{ is a real number}\}$	$(-\infty, \infty)$
$\sec\theta$, $\csc\theta$	$\{y \mid \lvert y \rvert \ge 1\}$	$(-\infty, -1] \cup [1, \infty)$

Classroom Example 8

Decide whether each statement is *possible* or *impossible*.

(a) $\cot\theta = -0.999$

(b) $\cos\theta = -1.7$

(c) $\csc\theta = 0$

Answers:

(a) possible

(b) impossible

(c) impossible

EXAMPLE 8 Deciding Whether a Value Is in the Range of a Trigonometric Function

Decide whether each statement is *possible* or *impossible*.

(a) $\sin\theta = 2.5$ **(b)** $\tan\theta = 110.47$ **(c)** $\sec\theta = 0.6$

SOLUTION

(a) For any value of θ, we know that

$$-1 \le \sin\theta \le 1.$$

Since $2.5 > 1$, it is impossible to find a value of θ that satisfies $\sin\theta = 2.5$.

(b) The tangent function can take on any real number value. Thus, $\tan\theta = 110.47$ is possible.

(c) Since $\lvert \sec\theta \rvert \ge 1$ for all θ for which the secant is defined, the statement $\sec\theta = 0.6$ is impossible.

✔ *Now Try Exercises 109, 113, and 115.*

The six trigonometric functions are defined in terms of x, y, and r, where the Pythagorean theorem shows that

$$r^2 = x^2 + y^2 \quad \text{and} \quad r > 0.$$

With these relationships, knowing the value of only one function and the quadrant in which the angle lies makes it possible to find the values of the other trigonometric functions.

Classroom Example 9
Suppose that angle θ is in quadrant III and $\tan\theta = \frac{8}{5}$. Find the values of the other five trigonometric functions.

Answer:
$\sin\theta = -\frac{8\sqrt{89}}{89}$; $\cos\theta = -\frac{5\sqrt{89}}{89}$;
$\csc\theta = -\frac{\sqrt{89}}{8}$; $\sec\theta = -\frac{\sqrt{89}}{5}$;
$\cot\theta = \frac{5}{8}$

EXAMPLE 9 Finding All Function Values Given One Value and the Quadrant

Suppose that angle θ is in quadrant II and $\sin\theta = \frac{2}{3}$. Find the values of the other five trigonometric functions.

SOLUTION Choose any point on the terminal side of angle θ. For simplicity, since $\sin\theta = \frac{y}{r}$, choose the point with $r = 3$.

$$\sin\theta = \frac{2}{3} \quad \text{Given value}$$

$$\frac{y}{r} = \frac{2}{3} \quad \text{Substitute } \tfrac{y}{r} \text{ for } \sin\theta.$$

Since $\frac{y}{r} = \frac{2}{3}$ and $r = 3$, then $y = 2$. To find x, use the equation $x^2 + y^2 = r^2$.

$$x^2 + y^2 = r^2$$

$$x^2 + 2^2 = 3^2 \quad \text{Substitute.}$$

$$x^2 + 4 = 9 \quad \text{Apply exponents.}$$

$$x^2 = 5 \quad \text{Subtract 4.}$$

Remember *both* roots.

$$x = \sqrt{5} \quad \text{or} \quad x = -\sqrt{5} \quad \text{Square root property (Section 1.4)}$$

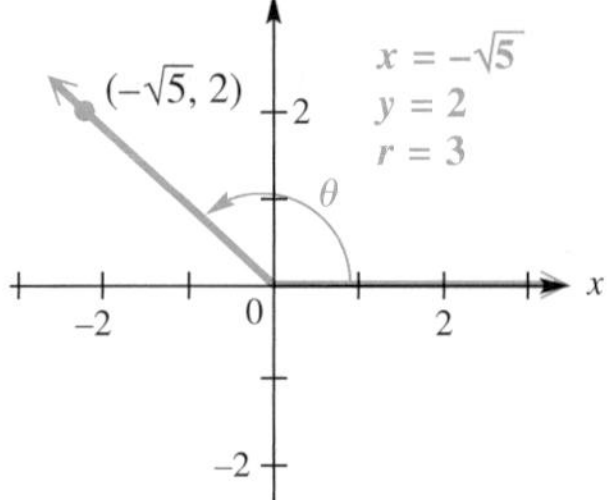

Figure 25

Since θ is in quadrant II, x must be negative. Choose $x = -\sqrt{5}$ so that the point $\left(-\sqrt{5}, 2\right)$ is on the terminal side of θ. See **Figure 25.** Now we can find the values of the remaining trigonometric functions.

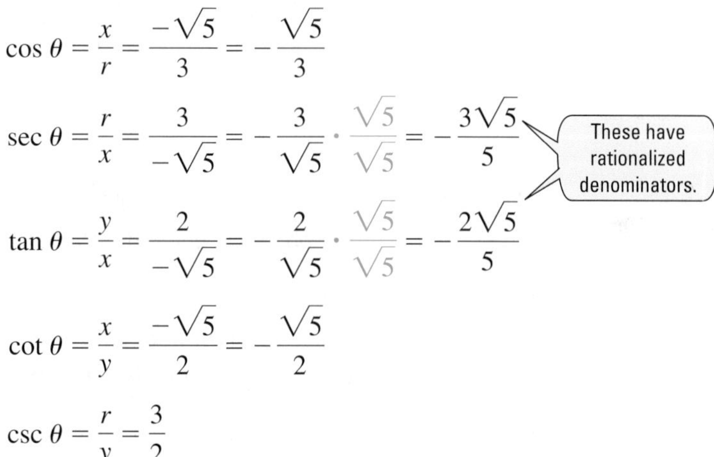

$$\cos\theta = \frac{x}{r} = \frac{-\sqrt{5}}{3} = -\frac{\sqrt{5}}{3}$$

$$\sec\theta = \frac{r}{x} = \frac{3}{-\sqrt{5}} = -\frac{3}{\sqrt{5}} \cdot \frac{\sqrt{5}}{\sqrt{5}} = -\frac{3\sqrt{5}}{5}$$

$$\tan\theta = \frac{y}{x} = \frac{2}{-\sqrt{5}} = -\frac{2}{\sqrt{5}} \cdot \frac{\sqrt{5}}{\sqrt{5}} = -\frac{2\sqrt{5}}{5}$$

These have rationalized denominators.

$$\cot\theta = \frac{x}{y} = \frac{-\sqrt{5}}{2} = -\frac{\sqrt{5}}{2}$$

$$\csc\theta = \frac{r}{y} = \frac{3}{2}$$

Now Try Exercise 129.

Pythagorean Identities We derive three new identities from the relationship $x^2 + y^2 = r^2$.

$$x^2 + y^2 = r^2 \quad \text{Equation of the Pythagorean theorem (Section 1.5)}$$

$$\frac{x^2}{r^2} + \frac{y^2}{r^2} = \frac{r^2}{r^2} \quad \text{Divide by } r^2.$$

$$\left(\frac{x}{r}\right)^2 + \left(\frac{y}{r}\right)^2 = 1 \quad \text{Power rule for exponents; } \frac{a^m}{b^m} = \left(\frac{a}{b}\right)^m \text{ (Section R.3)}$$

$$(\cos\theta)^2 + (\sin\theta)^2 = 1 \quad \cos\theta = \tfrac{x}{r},\ \sin\theta = \tfrac{y}{r}$$

$$\mathbf{\sin^2\theta + \cos^2\theta = 1} \quad \text{Apply exponents; commutative property}$$

Starting again with $x^2 + y^2 = r^2$ and dividing through by x^2 gives the following.

$$\frac{x^2}{x^2} + \frac{y^2}{x^2} = \frac{r^2}{x^2} \qquad \text{Divide by } x^2.$$

$$1 + \left(\frac{y}{x}\right)^2 = \left(\frac{r}{x}\right)^2 \qquad \text{Power rule for exponents}$$

$$1 + (\tan\theta)^2 = (\sec\theta)^2 \qquad \tan\theta = \tfrac{y}{x},\ \sec\theta = \tfrac{r}{x}$$

$$\boldsymbol{\tan^2\theta + 1 = \sec^2\theta} \qquad \text{Apply exponents; commutative property}$$

Similarly, dividing through by y^2 leads to another identity.

$$\boldsymbol{1 + \cot^2\theta = \csc^2\theta}$$

These three identities are the **Pythagorean identities** since the original equation that led to them, $x^2 + y^2 = r^2$, comes from the Pythagorean theorem.

Teaching Tip Point out the importance of the exponents in the Pythagorean identities. Emphasize that without the exponents, the resulting equations are not identities.

Pythagorean Identities

For all angles θ for which the function values are defined, the following identities hold.

$$\boldsymbol{\sin^2\theta + \cos^2\theta = 1 \qquad \tan^2\theta + 1 = \sec^2\theta \qquad 1 + \cot^2\theta = \csc^2\theta}$$

As before, we have given only one form of each identity. However, algebraic transformations produce equivalent identities. For example, by subtracting $\sin^2\theta$ from both sides of

$$\sin^2\theta + \cos^2\theta = 1,$$

we obtain an equivalent identity.

$$\cos^2\theta = 1 - \sin^2\theta \qquad \text{Alternative form}$$

It is important to be able to transform these identities quickly and also to recognize their equivalent forms.

LOOKING AHEAD TO CALCULUS
The reciprocal, Pythagorean, and quotient identities are used in calculus to find derivatives and integrals of trigonometric functions. A standard technique of integration called **trigonometric substitution** relies on the Pythagorean identities.

Quotient Identities Consider the quotient of $\sin\theta$ and $\cos\theta$, for $\cos\theta \neq 0$.

$$\frac{\sin\theta}{\cos\theta} = \frac{\frac{y}{r}}{\frac{x}{r}} = \frac{y}{r} \div \frac{x}{r} = \frac{y}{r} \cdot \frac{r}{x} = \frac{y}{x} = \tan\theta$$

Similarly, $\frac{\cos\theta}{\sin\theta} = \cot\theta$, for $\sin\theta \neq 0$. Thus, we have the **quotient identities.**

Quotient Identities

For all angles θ for which the denominators are not zero, the following identities hold.

$$\boldsymbol{\frac{\sin\theta}{\cos\theta} = \tan\theta \qquad \frac{\cos\theta}{\sin\theta} = \cot\theta}$$

Classroom Example 10
Find $\cos\theta$ and $\tan\theta$, given that $\sin\theta = -\frac{\sqrt{2}}{3}$ and $\cos\theta > 0$.

Answer:
$\cos\theta = \frac{\sqrt{7}}{3}$; $\tan\theta = -\frac{\sqrt{14}}{7}$

EXAMPLE 10 Using Identities to Find Function Values

Find $\sin\theta$ and $\tan\theta$, given that $\cos\theta = -\frac{\sqrt{3}}{4}$ and $\sin\theta > 0$.

SOLUTION Start with $\sin^2\theta + \cos^2\theta = 1$.

$$\sin^2\theta + \left(-\frac{\sqrt{3}}{4}\right)^2 = 1 \quad \text{Replace } \cos\theta \text{ with } -\tfrac{\sqrt{3}}{4}.$$

$$\sin^2\theta + \frac{3}{16} = 1 \quad \text{Square } -\tfrac{\sqrt{3}}{4}.$$

$$\sin^2\theta = \frac{13}{16} \quad \text{Subtract } \tfrac{3}{16}.$$

$$\sin\theta = \pm\frac{\sqrt{13}}{4} \quad \text{Take square roots. (Section 1.4)}$$

$$\sin\theta = \frac{\sqrt{13}}{4} \quad \text{Choose the positive square root since } \sin\theta \text{ is positive.}$$

Choose the correct sign here.

Teaching Tip In **Example 10,** the answer can be checked by solving $\left(-\sqrt{3}\right)^2 + y^2 = 4^2$ for y and then applying the trigonometric function definitions.

To find $\tan\theta$, use the quotient identity $\tan\theta = \frac{\sin\theta}{\cos\theta}$.

$$\tan\theta = \frac{\sin\theta}{\cos\theta} = \frac{\frac{\sqrt{13}}{4}}{-\frac{\sqrt{3}}{4}} = \frac{\sqrt{13}}{4}\left(-\frac{4}{\sqrt{3}}\right) = -\frac{\sqrt{13}}{\sqrt{3}}$$

$$= -\frac{\sqrt{13}}{\sqrt{3}} \cdot \frac{\sqrt{3}}{\sqrt{3}} = -\frac{\sqrt{39}}{3} \quad \text{Rationalize the denominator.}$$

✔ *Now Try Exercise 133.*

CAUTION ***Be careful to choose the correct sign when taking square roots.***

Classroom Example 11
Find $\sin\theta$ and $\cos\theta$, given that $\cot\theta = -\frac{7}{24}$ and θ is in quadrant II.

Answer: $\sin\theta = \frac{24}{25}$; $\cos\theta = -\frac{7}{25}$

EXAMPLE 11 Using Identities to Find Function Values

Find $\sin\theta$ and $\cos\theta$, given that $\tan\theta = \frac{4}{3}$ and θ is in quadrant III.

SOLUTION Since θ is in quadrant III, $\sin\theta$ and $\cos\theta$ will both be negative. It is tempting to say that since $\tan\theta = \frac{\sin\theta}{\cos\theta}$ and $\tan\theta = \frac{4}{3}$, then $\sin\theta = -4$ and $\cos\theta = -3$. This is *incorrect,* because $\sin\theta$ and $\cos\theta$ must be in the interval $[-1, 1]$.

We use the Pythagorean identity $\tan^2\theta + 1 = \sec^2\theta$ to find $\sec\theta$, and then the reciprocal identity $\cos\theta = \frac{1}{\sec\theta}$ to find $\cos\theta$.

$$\tan^2\theta + 1 = \sec^2\theta \quad \text{Pythagorean identity}$$

$$\left(\frac{4}{3}\right)^2 + 1 = \sec^2\theta \quad \tan\theta = \tfrac{4}{3}$$

$$\frac{16}{9} + 1 = \sec^2\theta \quad \text{Square } \tfrac{4}{3}.$$

$$\frac{25}{9} = \sec^2\theta \quad \text{Add.}$$

$$-\frac{5}{3} = \sec\theta \quad \text{Choose the negative square root since } \sec\theta \text{ is negative when } \theta \text{ is in quadrant III.}$$

Be careful to choose the correct sign here.

$$-\frac{3}{5} = \cos\theta \quad \text{Secant and cosine are reciprocals.}$$

Since $\sin^2 \theta = 1 - \cos^2 \theta$,

$$\sin^2 \theta = 1 - \left(-\frac{3}{5}\right)^2 \qquad \cos \theta = -\tfrac{3}{5}$$

$$\sin^2 \theta = 1 - \frac{9}{25} \qquad \text{Square } -\tfrac{3}{5}.$$

$$\sin^2 \theta = \frac{16}{25} \qquad \text{Subtract.}$$

Again, be careful.

$$\sin \theta = -\frac{4}{5}. \qquad \text{Choose the negative square root.}$$

Teaching Tip When students use the alternative method described in the Note, they often make errors involving quadrant placement and signs. Warn them about this.

Now Try Exercise 131.

NOTE **Example 11** can also be worked by sketching θ in standard position in quadrant III, finding r to be 5, and then using the definitions of $\sin \theta$ and $\cos \theta$ in terms of x, y, and r. See **Figure 26.**

When using this method, be sure to choose the correct signs for x and y as determined by the quadrant in which the terminal side of θ lies. This is analogous to choosing the correct signs after applying the Pythagorean identities.

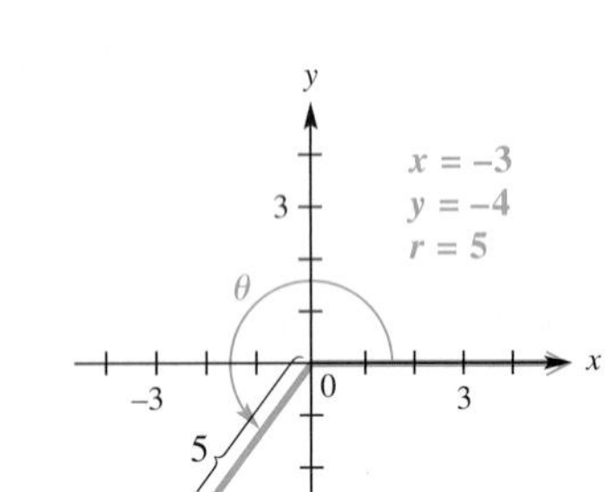

Figure 26

5.2 Exercises

In Exercises 1–16 and 35–42, we give, in order, sine, cosine, tangent, cotangent, secant, and cosecant.

1.

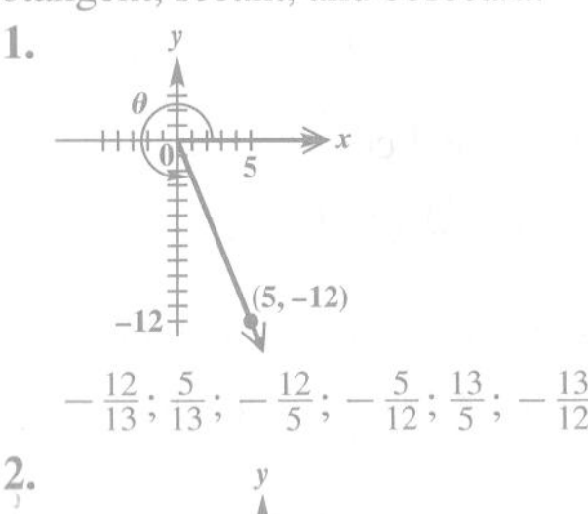

$-\frac{12}{13}$; $\frac{5}{13}$; $-\frac{12}{5}$; $-\frac{5}{12}$; $\frac{13}{5}$; $-\frac{13}{12}$

2.

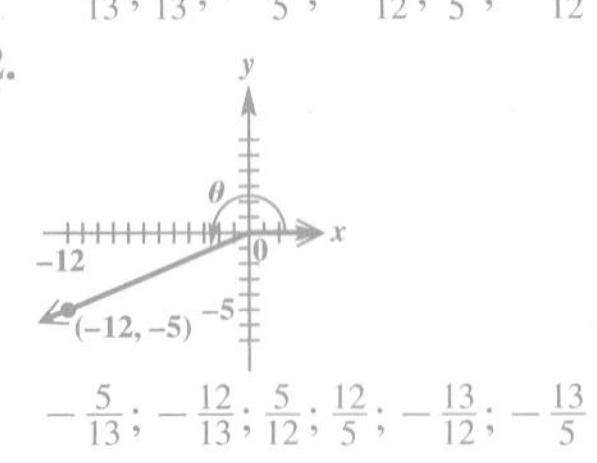

$-\frac{5}{13}$; $-\frac{12}{13}$; $\frac{5}{12}$; $\frac{12}{5}$; $-\frac{13}{12}$; $-\frac{13}{5}$

Answer graphs are not included for Exercises 3–16. They are similar to the ones in Exercises 1 and 2.

3. $\frac{4}{5}$; $-\frac{3}{5}$; $-\frac{4}{3}$; $-\frac{3}{4}$; $-\frac{5}{3}$; $\frac{5}{4}$

4. $-\frac{3}{5}$; $-\frac{4}{5}$; $\frac{3}{4}$; $\frac{4}{3}$; $-\frac{5}{4}$; $-\frac{5}{3}$

5. $\frac{15}{17}$; $-\frac{8}{17}$; $-\frac{15}{8}$; $-\frac{8}{15}$; $-\frac{17}{8}$; $\frac{17}{15}$

6. $-\frac{8}{17}$; $\frac{15}{17}$; $-\frac{8}{15}$; $-\frac{15}{8}$; $\frac{17}{15}$; $-\frac{17}{8}$

7. $-\frac{24}{25}$; $\frac{7}{25}$; $-\frac{24}{7}$; $-\frac{7}{24}$; $\frac{25}{7}$; $-\frac{25}{24}$

Concept Check *Sketch an angle θ in standard position such that θ has the least positive measure, and the given point is on the terminal side of θ. Then find the values of the six trigonometric functions for each angle. Rationalize denominators when applicable.* ***See Examples 1, 2, and 4.***

1. $(5, -12)$	**2.** $(-12, -5)$	**3.** $(-3, 4)$	**4.** $(-4, -3)$
5. $(-8, 15)$	**6.** $(15, -8)$	**7.** $(7, -24)$	**8.** $(-24, -7)$
9. $(0, 2)$	**10.** $(0, 5)$	**11.** $(-4, 0)$	**12.** $(-5, 0)$
13. $(1, \sqrt{3})$	**14.** $(-1, \sqrt{3})$	**15.** $(-2\sqrt{3}, -2)$	**16.** $(-2\sqrt{3}, 2)$

17. For any nonquadrantal angle θ, $\sin \theta$ and $\csc \theta$ will have the same sign. Explain why.

18. *Concept Check* How is the value of r interpreted geometrically in the definitions of the sine, cosine, secant, and cosecant functions?

Concept Check *Suppose that the point (x, y) is in the indicated quadrant. Decide whether the given ratio is positive or negative. Recall that $r = \sqrt{x^2 + y^2}$. (Hint: Drawing a sketch may help.)*

19. II, $\frac{x}{r}$	**20.** III, $\frac{y}{r}$	**21.** IV, $\frac{y}{x}$	**22.** IV, $\frac{x}{y}$
23. II, $\frac{y}{r}$	**24.** III, $\frac{x}{r}$	**25.** IV, $\frac{x}{r}$	**26.** IV, $\frac{y}{r}$
27. II, $\frac{x}{y}$	**28.** II, $\frac{y}{x}$	**29.** III, $\frac{y}{x}$	**30.** III, $\frac{x}{y}$
31. III, $\frac{r}{x}$	**32.** III, $\frac{r}{y}$	**33.** I, $\frac{x}{y}$	**34.** I, $\frac{y}{x}$

8. $-\frac{7}{25}; -\frac{24}{25}; \frac{7}{24}; \frac{24}{7}; -\frac{25}{24}; -\frac{25}{7}$

9. 1; 0; undefined; 0; undefined; 1

10. 1; 0; undefined; 0; undefined; 1

11. 0; -1; 0; undefined; -1; undefined

12. 0; -1; 0; undefined; -1; undefined

13. $\frac{\sqrt{3}}{2}; \frac{1}{2}; \sqrt{3}; \frac{\sqrt{3}}{3}; 2; \frac{2\sqrt{3}}{3}$

14. $\frac{\sqrt{3}}{2}; -\frac{1}{2}; -\sqrt{3}; -\frac{\sqrt{3}}{3}; -2; \frac{2\sqrt{3}}{3}$

15. $-\frac{1}{2}; -\frac{\sqrt{3}}{2}; \frac{\sqrt{3}}{3}; \sqrt{3}; -\frac{2\sqrt{3}}{3}; -2$

16. $\frac{1}{2}; -\frac{\sqrt{3}}{2}; -\frac{\sqrt{3}}{3}; -\sqrt{3}; -\frac{2\sqrt{3}}{3}; 2$

18. r is the distance from (x, y) to the origin.

19. negative	**20.** negative
21. negative	**22.** negative
23. positive	**24.** negative
25. positive	**26.** negative
27. negative	**28.** negative
29. positive	**30.** positive
31. negative	**32.** negative
33. positive	**34.** positive

Answer graphs are not included for Exercises 35–42.

35. $-\frac{2\sqrt{5}}{5}; \frac{\sqrt{5}}{5}; -2; -\frac{1}{2}; \sqrt{5}; -\frac{\sqrt{5}}{2}$

36. $-\frac{3\sqrt{34}}{34}; \frac{5\sqrt{34}}{34}; -\frac{3}{5}; -\frac{5}{3}; \frac{\sqrt{34}}{5}; -\frac{\sqrt{34}}{3}$

37. $\frac{6\sqrt{37}}{37}; -\frac{\sqrt{37}}{37}; -6; -\frac{1}{6}; -\sqrt{37}; \frac{\sqrt{37}}{6}$

38. $\frac{5\sqrt{34}}{34}; -\frac{3\sqrt{34}}{34}; -\frac{5}{3}; -\frac{3}{5}; -\frac{\sqrt{34}}{3}; \frac{\sqrt{34}}{5}$

39. $-\frac{\sqrt{2}}{2}; \frac{\sqrt{2}}{2}; -1; -1; \sqrt{2}; -\sqrt{2}$

40. $\frac{\sqrt{2}}{2}; \frac{\sqrt{2}}{2}; 1; 1; \sqrt{2}; \sqrt{2}$

41. $-\frac{\sqrt{3}}{2}; -\frac{1}{2}; \sqrt{3}; \frac{\sqrt{3}}{3}; -2; -\frac{2\sqrt{3}}{3}$

42. $\frac{\sqrt{3}}{2}; -\frac{1}{2}; -\sqrt{3}; -\frac{\sqrt{3}}{3}; -2; \frac{2\sqrt{3}}{3}$

43. 0	**44.** 1
45. 0	**46.** 0

In Exercises 35–42, an equation of the terminal side of an angle θ in standard position is given with a restriction on x. Sketch the least positive such angle θ, and find the values of the six trigonometric functions of θ. ***See Example 3.***

35. $2x + y = 0, x \geq 0$

36. $3x + 5y = 0, x \geq 0$

37. $-6x - y = 0, x \leq 0$

38. $-5x - 3y = 0, x \leq 0$

39. $x + y = 0, x \geq 0$

40. $x - y = 0, x \geq 0$

41. $-\sqrt{3}x + y = 0, x \leq 0$

42. $\sqrt{3}x + y = 0, x \leq 0$

To work Exercises 43–60, begin by reproducing the graph in **Figure 19.** *Keep in mind that for each of the four points labeled in the figure, $r = 1$. For each quadrantal angle, identify the appropriate values of x, y, and r to find the indicated function value. If it is undefined, say so.* ***See Example 4.***

43. cos 90°	**44.** sin 90°	**45.** tan 180°
46. cot 90°	**47.** sec 180°	**48.** csc 270°
49. sin(−270°)	**50.** cos(−90°)	**51.** cot 540°
52. tan 450°	**53.** csc(−450°)	**54.** sec(−540°)
55. sin 1800°	**56.** cos 1800°	**57.** csc 1800°
58. cot 1800°	**59.** sec 1800°	**60.** tan 1800°

Use the trigonometric function values of quadrantal angles given in this section to evaluate each expression. An expression such as $\cot^2 90°$ *means* $(\cot 90°)^2$, *which is equal to* $0^2 = 0$.

61. $\cos 90° + 3 \sin 270°$

62. $\tan 0° - 6 \sin 90°$

63. $3 \sec 180° - 5 \tan 360°$

64. $4 \csc 270° + 3 \cos 180°$

65. $\tan 360° + 4 \sin 180° + 5 \cos^2 180°$

66. $2 \sec 0° + 4 \cot^2 90° + \cos 360°$

67. $-2 \sin^4 0° + 3 \tan^2 0°$

68. $-3 \sin^4 90° + 4 \cos^3 180°$

69. $\sin^2(-90°) + \cos^2(-90°)$

70. $\cos^2(-180°) + \sin^2(-180°)$

If n is an integer, $n \cdot 180°$ represents an integer multiple of 180°, $(2n + 1) \cdot 90°$ represents an odd integer multiple of 90°, and so on. Decide whether each expression is equal to 0, 1, or −1 or is undefined.

71. $\cos[(2n + 1) \cdot 90°]$

72. $\sin[n \cdot 180°]$

73. $\tan[n \cdot 180°]$

74. $\tan[(2n + 1) \cdot 90°]$

75. $\sin[270° + n \cdot 360°]$

76. $\cot[n \cdot 180°]$

Use the appropriate reciprocal identity to find each function value. Rationalize denominators when applicable. ***See Example 5.***

77. $\sec \theta$, given that $\cos \theta = \frac{2}{3}$

78. $\sec \theta$, given that $\cos \theta = \frac{5}{8}$

79. $\csc \theta$, given that $\sin \theta = -\frac{3}{7}$

80. $\csc \theta$, given that $\sin \theta = -\frac{8}{43}$

81. $\cot \theta$, given that $\tan \theta = 5$

82. $\cot \theta$, given that $\tan \theta = 18$

83. $\sin \theta$, given that $\csc \theta = \frac{\sqrt{8}}{2}$

84. $\sin \theta$, given that $\csc \theta = \frac{\sqrt{24}}{3}$

85. $\sin \theta$, given that $\csc \theta = 1.42716321$

86. $\cos \theta$, given that $\sec \theta = 9.80425133$

47. -1 48. -1
49. 1 50. 0
51. undefined 52. undefined
53. -1 54. -1
55. 0 56. 1
57. undefined 58. undefined
59. 1 60. 0

61. -3 62. -6
63. -3 64. -7
65. 5 66. 3
67. 0 68. -7
69. 1 70. 1

71. 0 72. 0
73. 0 74. undefined
75. -1 76. undefined

77. $\frac{3}{2}$ 78. $\frac{8}{5}$
79. $-\frac{7}{3}$ 80. $-\frac{43}{8}$
81. $\frac{1}{5}$ 82. $\frac{1}{18}$
83. $\frac{\sqrt{2}}{2}$ 84. $\frac{\sqrt{6}}{4}$
85. 0.70069071 86. 0.10199657

87. All are positive.
88. All are positive.
89. Tangent and cotangent are positive. All others are negative.
90. Tangent and cotangent are positive. All others are negative.
91. Sine and cosecant are positive. All others are negative.
92. Sine and cosecant are positive. All others are negative.
93. Cosine and secant are positive. All others are negative.
94. Cosine and secant are positive. All others are negative.
95. Sine and cosecant are positive. All others are negative.
96. Cosine and secant are positive. All others are negative.

97. I, II 98. I, IV
99. I 100. I
101. II 102. III
103. I 104. I
105. III 106. II
107. III, IV 108. II, IV

109. impossible 110. impossible
111. possible 112. possible
113. possible 114. possible
115. impossible 116. impossible
117. possible 118. possible

119. $-\frac{4}{5}$ 120. $-\frac{3}{5}$
121. $-\frac{\sqrt{5}}{2}$ 122. $-\frac{4}{3}$

Determine the signs of the trigonometric functions of an angle in standard position with the given measure. ***See Example 6.***

87. 74° **88.** 84° **89.** 218° **90.** 195° **91.** 178°
92. 125° **93.** −80° **94.** −15° **95.** 855° **96.** 1005°

Identify the quadrant (or possible quadrants) of an angle θ that satisfies the given conditions. ***See Example 7.***

97. $\sin\theta > 0, \csc\theta > 0$ **98.** $\cos\theta > 0, \sec\theta > 0$ **99.** $\cos\theta > 0, \sin\theta > 0$

100. $\sin\theta > 0, \tan\theta > 0$ **101.** $\tan\theta < 0, \cos\theta < 0$ **102.** $\cos\theta < 0, \sin\theta < 0$

103. $\sec\theta > 0, \csc\theta > 0$ **104.** $\csc\theta > 0, \cot\theta > 0$ **105.** $\sec\theta < 0, \csc\theta < 0$

106. $\cot\theta < 0, \sec\theta < 0$ **107.** $\sin\theta < 0, \csc\theta < 0$ **108.** $\tan\theta < 0, \cot\theta < 0$

Decide whether each statement is possible *or* impossible *for some angle θ.* ***See Example 8.***

109. $\sin\theta = 2$ **110.** $\sin\theta = 3$ **111.** $\cos\theta = -0.96$

112. $\cos\theta = -0.56$ **113.** $\tan\theta = 0.93$ **114.** $\cot\theta = 0.93$

115. $\sec\theta = -0.3$ **116.** $\sec\theta = -0.9$ **117.** $\csc\theta = 100$ **118.** $\csc\theta = -100$

Use identities to solve each of the following. ***See Examples 9–11.***

119. Find $\cos\theta$, given that $\sin\theta = \frac{3}{5}$ and θ is in quadrant II.

120. Find $\sin\theta$, given that $\cos\theta = \frac{4}{5}$ and θ is in quadrant IV.

121. Find $\csc\theta$, given that $\cot\theta = -\frac{1}{2}$ and θ is in quadrant IV.

122. Find $\sec\theta$, given that $\tan\theta = \frac{\sqrt{7}}{3}$ and θ is in quadrant III.

123. Find $\tan\theta$, given that $\sin\theta = \frac{1}{2}$ and θ is in quadrant II.

124. Find $\cot\theta$, given that $\csc\theta = -2$ and θ is in quadrant III.

125. Find $\cot\theta$, given that $\csc\theta = -3.5891420$ and θ is in quadrant III.

126. Find $\tan\theta$, given that $\sin\theta = 0.49268329$ and θ is in quadrant II.

Find the five remaining trigonometric function values for each angle θ. ***See Examples 9–11.***

127. $\tan\theta = -\frac{15}{8}$, and θ is in quadrant II **128.** $\cos\theta = -\frac{3}{5}$, and θ is in quadrant III

129. $\sin\theta = \frac{\sqrt{5}}{7}$, and θ is in quadrant I **130.** $\tan\theta = \sqrt{3}$, and θ is in quadrant III

131. $\cot\theta = \frac{\sqrt{3}}{8}$, and θ is in quadrant I **132.** $\csc\theta = 2$, and θ is in quadrant II

133. $\sin\theta = \frac{\sqrt{2}}{6}$, and $\cos\theta < 0$ **134.** $\cos\theta = \frac{\sqrt{5}}{8}$, and $\tan\theta < 0$

135. $\sec\theta = -4$, and $\sin\theta > 0$ **136.** $\csc\theta = -3$, and $\cos\theta > 0$

137. $\sin\theta = 0.164215$, and θ is in quadrant II

138. $\cot\theta = -1.49586$, and θ is in quadrant IV

Work each problem.

139. Derive the identity $1 + \cot^2\theta = \csc^2\theta$ by dividing $x^2 + y^2 = r^2$ by y^2.

140. Using a method similar to the one given in this section showing that $\frac{\sin\theta}{\cos\theta} = \tan\theta$, show that $\frac{\cos\theta}{\sin\theta} = \cot\theta$.

123. $-\frac{\sqrt{3}}{3}$ 124. $\sqrt{3}$
125. 3.44701905
126. −0.56616682

In Exercises 127–138, we give, in order, sine, cosine, tangent, cotangent, secant, and cosecant.

127. $\frac{15}{17}$; $-\frac{8}{17}$; $-\frac{15}{8}$; $-\frac{8}{15}$; $-\frac{17}{8}$; $\frac{17}{15}$
128. $-\frac{4}{5}$; $-\frac{3}{5}$; $\frac{4}{3}$; $\frac{3}{4}$; $-\frac{5}{3}$; $-\frac{5}{4}$
129. $\frac{\sqrt{5}}{7}$; $\frac{2\sqrt{11}}{7}$; $\frac{\sqrt{55}}{22}$; $\frac{2\sqrt{55}}{5}$; $\frac{7\sqrt{11}}{22}$; $\frac{7\sqrt{5}}{5}$

141. *Concept Check* *True* or *false*: For all angles θ, $\sin\theta + \cos\theta = 1$. If the statement is false, give an example showing why.

142. *Concept Check* *True* or *false*: Since $\cot\theta = \frac{\cos\theta}{\sin\theta}$, if $\cot\theta = \frac{1}{2}$ with θ in quadrant I, then $\cos\theta = 1$ and $\sin\theta = 2$. If the statement is false, give an explanation showing why.

Concept Check Suppose that $90° < \theta < 180°$. *Find the sign of each function value.*

143. $\sin 2\theta$ **144.** $\csc 2\theta$ **145.** $\tan\frac{\theta}{2}$ **146.** $\cot\frac{\theta}{2}$

Concept Check Suppose that $-90° < \theta < 90°$. *Find the sign of each function value.*

147. $\cos\frac{\theta}{2}$ **148.** $\sec\frac{\theta}{2}$ **149.** $\sec(-\theta)$ **150.** $\cos(-\theta)$

Answers for the remaining exercises are given in the answer section at the back of the text.

5.3 Evaluating Trigonometric Functions

- Right-Triangle-Based Definitions of the Trigonometric Functions
- Cofunctions
- Trigonometric Function Values of Special Angles
- Reference Angles
- Special Angles as Reference Angles
- Finding Function Values Using a Calculator
- Finding Angle Measures

Right-Triangle-Based Definitions of the Trigonometric Functions We used angles in standard position to define the trigonometric functions in **Section 5.2.** There is another way to approach them: As ratios of the lengths of the sides of right triangles.

Figure 27 shows an acute angle A in standard position. The definitions of the trigonometric function values of angle A require x, y, and r. As drawn in **Figure 27,** x and y are the lengths of the two legs of the right triangle ABC, and r is the length of the hypotenuse.

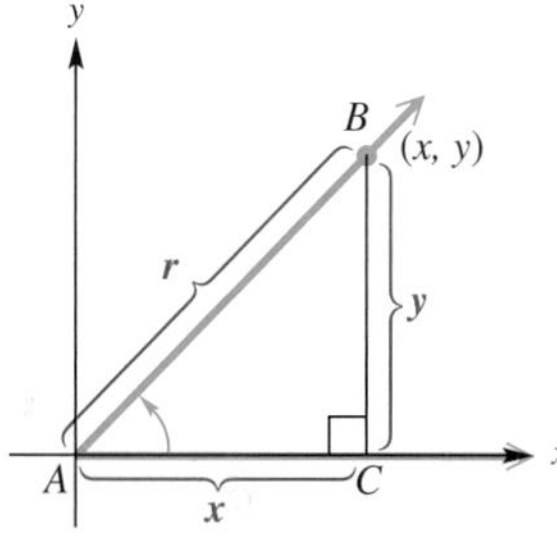

Figure 27

The side of length y is called the **side opposite** angle A, and the side of length x is called the **side adjacent** to angle A. We use the lengths of these sides to replace x and y in the definitions of the trigonometric functions, and the length of the hypotenuse to replace r, to get the following right-triangle-based definitions.

Teaching Tip Introduce the mnemonic **sohcahtoa** to help students remember that "**s**ine is **o**pposite over **h**ypotenuse, **c**osine is **a**djacent over **h**ypotenuse, and **t**angent is **o**pposite over **a**djacent."

Right-Triangle-Based Definitions of Trigonometric Functions

Let A represent any acute angle in standard position.

$$\sin A = \frac{y}{r} = \frac{\text{side opposite } A}{\text{hypotenuse}} \qquad \csc A = \frac{r}{y} = \frac{\text{hypotenuse}}{\text{side opposite } A}$$

$$\cos A = \frac{x}{r} = \frac{\text{side adjacent to } A}{\text{hypotenuse}} \qquad \sec A = \frac{r}{x} = \frac{\text{hypotenuse}}{\text{side adjacent to } A}$$

$$\tan A = \frac{y}{x} = \frac{\text{side opposite } A}{\text{side adjacent to } A} \qquad \cot A = \frac{x}{y} = \frac{\text{side adjacent to } A}{\text{side opposite } A}$$

NOTE We will sometimes shorten wording like "side opposite A" to just "side opposite" when the meaning is obvious.

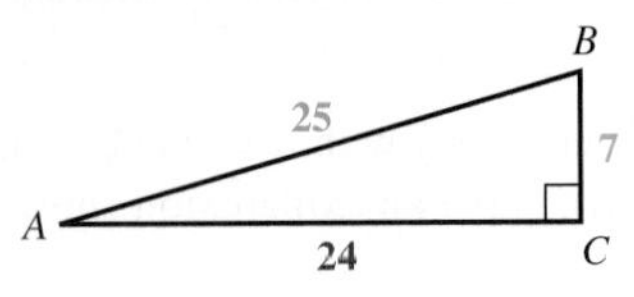

Figure 28

EXAMPLE 1 Finding Trigonometric Function Values of an Acute Angle

Find the sine, cosine, and tangent values for angles A and B in the right triangle in **Figure 28.**

SOLUTION The length of the side opposite angle A is 7, the length of the side adjacent to angle A is 24, and the length of the hypotenuse is 25.

$$\sin A = \frac{\text{side opposite}}{\text{hypotenuse}} = \frac{7}{25} \qquad \cos A = \frac{\text{side adjacent}}{\text{hypotenuse}} = \frac{24}{25} \qquad \tan A = \frac{\text{side opposite}}{\text{side adjacent}} = \frac{7}{24}$$

The length of the side opposite angle B is 24, and the length of the side adjacent to B is 7.

$$\sin B = \frac{24}{25} \qquad \cos B = \frac{7}{25} \qquad \tan B = \frac{24}{7} \qquad \text{Use the relationships given in the box.}$$

✔ *Now Try Exercise 1.*

Classroom Example 1
Find the sine, cosine, and tangent values for angles A and B in the figure.

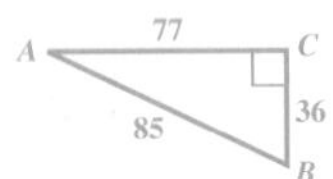

Answer:
$\sin A = \frac{36}{85}$; $\cos A = \frac{77}{85}$; $\tan A = \frac{36}{77}$; $\sin B = \frac{77}{85}$; $\cos B = \frac{36}{85}$; $\tan B = \frac{77}{36}$

NOTE Because the cosecant, secant, and cotangent ratios are the reciprocals of the sine, cosine, and tangent values, respectively, in **Example 1,**

$$\csc A = \frac{25}{7}, \quad \sec A = \frac{25}{24}, \quad \cot A = \frac{24}{7}, \quad \csc B = \frac{25}{24},$$

$$\sec B = \frac{25}{7}, \quad \text{and} \quad \cot B = \frac{7}{24}.$$

Cofunctions In **Example 1,** notice that $\sin A = \cos B$ and $\cos A = \sin B$. Such relationships are always true for the two acute angles of a right triangle.

Figure 29 shows a right triangle with acute angles A and B and a right angle at C. The length of the side opposite angle A is a, and the length of the side opposite angle B is b. The length of the hypotenuse is c.

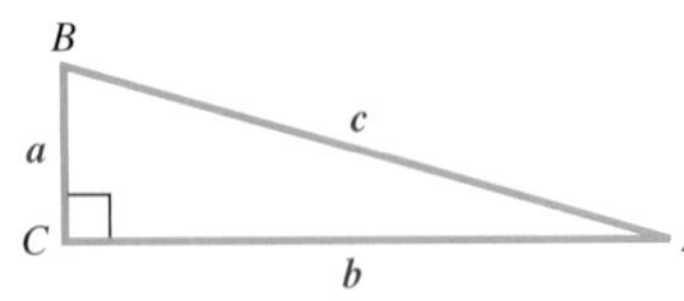

Whenever we use A, B, and C to name angles in a right triangle, C will be the right angle.

Figure 29

By the preceding definitions, $\sin A = \frac{a}{c}$. Also, $\cos B = \frac{a}{c}$. Thus,

$$\mathbf{\sin A = \frac{a}{c} = \cos B.}$$

Similarly, $\mathbf{\tan A = \frac{a}{b} = \cot B}$ and $\mathbf{\sec A = \frac{c}{b} = \csc B.}$

Since the sum of the three angles in any triangle is 180° and angle C equals 90°, angles A and B must have a sum of $180° - 90° = 90°$. As mentioned in **Section 5.1,** angles with a sum of 90° are complementary angles. Since angles A and B are complementary and $\sin A = \cos B$, the functions sine and cosine are **cofunctions.** Tangent and cotangent are also cofunctions, as are secant and cosecant. And since the angles A and B are complementary, $A + B = 90°$, or $B = 90° - A$, giving the following.

$$\sin A = \cos B = \cos(90° - A)$$

Similar **cofunction identities** are true for the other trigonometric functions.

Cofunction Identities

For any acute angle A, cofunction values of complementary angles are equal.

$$\mathbf{\sin A = \cos(90° - A) \quad \sec A = \csc(90° - A) \quad \tan A = \cot(90° - A)}$$

$$\mathbf{\cos A = \sin(90° - A) \quad \csc A = \sec(90° - A) \quad \cot A = \tan(90° - A)}$$

Classroom Example 2
Write each function in terms of its cofunction.

(a) sin 9° **(b)** cot 76°
(c) csc 45°

Answers:
(a) cos 81° **(b)** tan 14°
(c) sec 45°

EXAMPLE 2 Writing Functions in Terms of Cofunctions

Write each function in terms of its cofunction.

(a) cos 52° **(b)** tan 71° **(c)** sec 24°

SOLUTION

(a)

Cofunctions

$$\cos 52° = \sin(90° - 52°) = \sin 38° \qquad \cos A = \sin(90° - A)$$

Complementary angles

(b) $\tan 71° = \cot(90° - 71°) = \cot 19°$ **(c)** $\sec 24° = \csc 66°$

✔ ***Now Try Exercises 25 and 27.***

Equilateral triangle

(a)

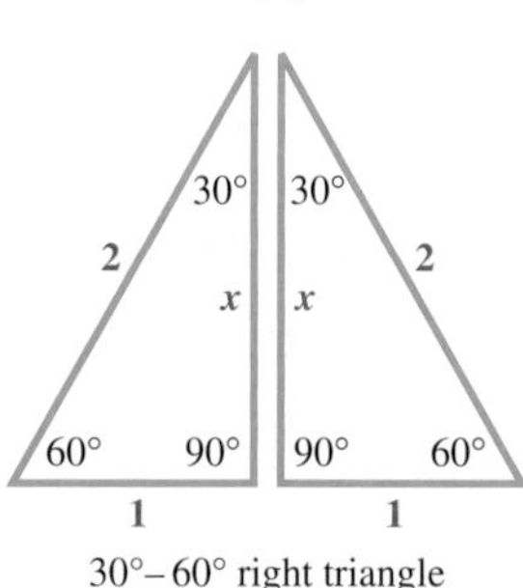

30°–60° right triangle

(b)

Figure 30

Trigonometric Function Values of Special Angles Certain special angles, such as 30°, 45°, and 60°, occur so often in trigonometry and in more advanced mathematics that they deserve special study. We start with an equilateral triangle, a triangle with all sides of equal length. Each angle of such a triangle measures 60°. Although the results we will obtain are independent of the length, for convenience we choose the length of each side to be 2 units. See **Figure 30(a).**

Bisecting one angle of this equilateral triangle leads to two right triangles, each of which has angles of 30°, 60°, and 90°, as shown in **Figure 30(b).** An angle bisector of an equilateral triangle also bisects the opposite side; therefore, the shorter leg has length 1. Let x represent the length of the longer leg.

$$2^2 = 1^2 + x^2 \quad \text{Pythagorean theorem (Section 1.5)}$$

$$4 = 1 + x^2 \quad \text{Apply the exponents.}$$

$$3 = x^2 \quad \text{Subtract 1 from each side. (Section 1.1)}$$

$$\sqrt{3} = x \quad \text{Square root property (Section 1.4); choose the positive root.}$$

Figure 31 summarizes our results using a 30°–60° right triangle. As shown in the figure, the side opposite the 30° angle has length 1; that is, for the 30° angle,

$$\text{hypotenuse} = 2, \quad \text{side opposite} = 1, \quad \text{side adjacent} = \sqrt{3}.$$

Now we use the definitions of the trigonometric functions.

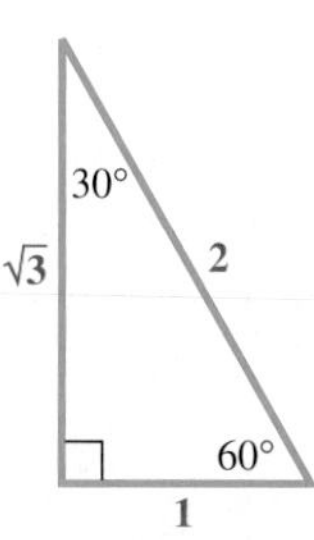

Figure 31

$$\sin 30° = \frac{\text{side opposite}}{\text{hypotenuse}} = \frac{1}{2}$$

$$\cos 30° = \frac{\text{side adjacent}}{\text{hypotenuse}} = \frac{\sqrt{3}}{2}$$

$$\tan 30° = \frac{\text{side opposite}}{\text{side adjacent}} = \frac{1}{\sqrt{3}} = \frac{1}{\sqrt{3}} \cdot \frac{\sqrt{3}}{\sqrt{3}} = \frac{\sqrt{3}}{3}$$

$$\csc 30° = \frac{2}{1} = 2$$

Rationalize the denominator.

$$\sec 30° = \frac{2}{\sqrt{3}} = \frac{2}{\sqrt{3}} \cdot \frac{\sqrt{3}}{\sqrt{3}} = \frac{2\sqrt{3}}{3}$$

$$\cot 30° = \frac{\sqrt{3}}{1} = \sqrt{3}$$

Teaching Tip Tell students that a good way to obtain one of these function values is to draw the appropriate "famous" right triangle as shown in **Figure 31 or 32** and then use the right-triangle-based definition of the function value.

Figure 31 (repeated)

EXAMPLE 3 Finding Trigonometric Function Values for 60°

Find the six trigonometric function values for a 60° angle.

SOLUTION Refer to **Figure 31** to find the following ratios.

$$\sin 60° = \frac{\sqrt{3}}{2} \qquad \cos 60° = \frac{1}{2} \qquad \tan 60° = \frac{\sqrt{3}}{1} = \sqrt{3}$$

$$\csc 60° = \frac{2}{\sqrt{3}} = \frac{2\sqrt{3}}{3} \qquad \sec 60° = \frac{2}{1} = 2 \qquad \cot 60° = \frac{1}{\sqrt{3}} = \frac{\sqrt{3}}{3}$$

✔ *Now Try Exercises 29, 31, and 33.*

NOTE The results in **Example 3** can also be found using the fact that cofunction values of complementary angles are equal.

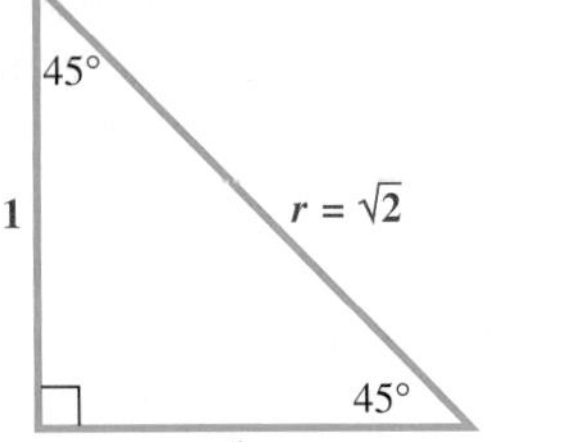

45°–45° right triangle

Figure 32

We find the values of the trigonometric functions for 45° by starting with a 45°–45° right triangle, as shown in **Figure 32.** This triangle is isosceles. For simplicity, we choose the lengths of the equal sides to be 1 unit. (As before, the results are independent of the length of the equal sides.) If r represents the length of the hypotenuse, then we can find its value using the Pythagorean theorem.

$$1^2 + 1^2 = r^2 \quad \text{Pythagorean theorem (Section 1.5)}$$

$$2 = r^2 \quad \text{Simplify.}$$

$$\sqrt{2} = r \quad \text{Choose the positive root.}$$

Classroom Example 3
Find the six trigonometric function values for a 30° angle.

Answer:
$\sin 30° = \frac{1}{2}$; $\cos 30° = \frac{\sqrt{3}}{2}$; $\tan 30° = \frac{\sqrt{3}}{3}$; $\csc 30° = 2$; $\sec 30° = \frac{2\sqrt{3}}{3}$; $\cot 30° = \sqrt{3}$

Now we use the measures indicated on the 45°–45° right triangle in **Figure 32.**

$$\sin 45° = \frac{1}{\sqrt{2}} = \frac{\sqrt{2}}{2} \qquad \cos 45° = \frac{1}{\sqrt{2}} = \frac{\sqrt{2}}{2} \qquad \tan 45° = \frac{1}{1} = 1$$

$$\csc 45° = \frac{\sqrt{2}}{1} = \sqrt{2} \qquad \sec 45° = \frac{\sqrt{2}}{1} = \sqrt{2} \qquad \cot 45° = \frac{1}{1} = 1$$

Function values for 30°, 45°, and 60° are summarized in the table that follows.

Function Values of Special Angles

θ	$\sin\theta$	$\cos\theta$	$\tan\theta$	$\cot\theta$	$\sec\theta$	$\csc\theta$
30°	$\frac{1}{2}$	$\frac{\sqrt{3}}{2}$	$\frac{\sqrt{3}}{3}$	$\sqrt{3}$	$\frac{2\sqrt{3}}{3}$	2
45°	$\frac{\sqrt{2}}{2}$	$\frac{\sqrt{2}}{2}$	1	1	$\sqrt{2}$	$\sqrt{2}$
60°	$\frac{\sqrt{3}}{2}$	$\frac{1}{2}$	$\sqrt{3}$	$\frac{\sqrt{3}}{3}$	2	$\frac{2\sqrt{3}}{3}$

NOTE You will be able to reproduce this table quickly if you learn the values of sin 30°, sin 45°, and sin 60°. Then you can complete the rest of the table using the reciprocal, cofunction, and quotient identities.

Reference Angles Associated with every nonquadrantal angle in standard position is a positive acute angle called its *reference angle.* A **reference angle** for an angle θ, written θ', is the positive acute angle made by the terminal side of angle θ and the x-axis.

Figure 33 illustrates several angles θ (each less than one complete counterclockwise revolution) in quadrants II, III, and IV, respectively, with the reference angle θ' also shown. In quadrant I, θ and θ' are the same. If an angle θ is negative or has measure greater than 360°, its reference angle is found by first finding its coterminal angle that is between 0° and 360°, and then using the diagrams in **Figure 33.**

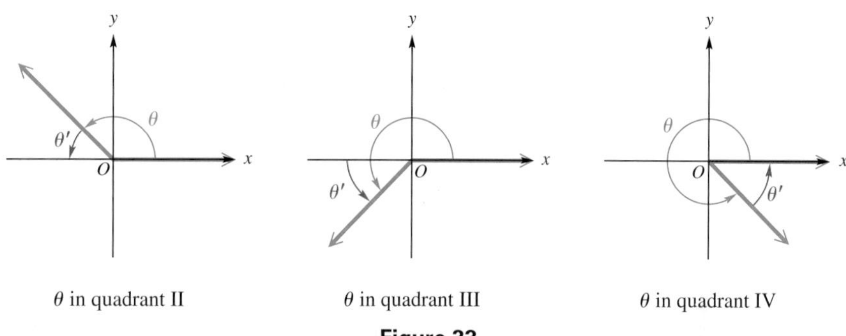

θ in quadrant II θ in quadrant III θ in quadrant IV

Figure 33

Classroom Example 4
Find the reference angle for each angle.
(a) 294° (b) 883°
Answers:
(a) 66° (b) 17°

CAUTION A common error is to find the reference angle by using the terminal side of θ and the y-axis. ***The reference angle is always found with reference to the x-axis.***

EXAMPLE 4 Finding Reference Angles

Find the reference angle for each angle.

(a) 218° **(b)** 1387°

SOLUTION

(a) As shown in **Figure 34(a),** the positive acute angle made by the terminal side of this angle and the x-axis is

$$218° - 180° = 38°.$$

For $\theta = 218°$, the reference angle $\theta' = 38°$.

y
218°
38°
0
x

218° – 180° = 38°

(a)

(b) First find a coterminal angle between 0° and 360°. Divide 1387° by 360° to get a quotient of about 3.9. Begin by subtracting 360° three times (because of the whole number 3 in 3.9).

$$1387° - 3 \cdot 360° = 1387° - 1080° \quad \text{Multiply. (Section 5.1)}$$
$$= 307° \quad \text{Subtract.}$$

The reference angle for 307° (and thus for 1387°) is

$$360° - 307° = 53°.$$

See **Figure 34(b).**

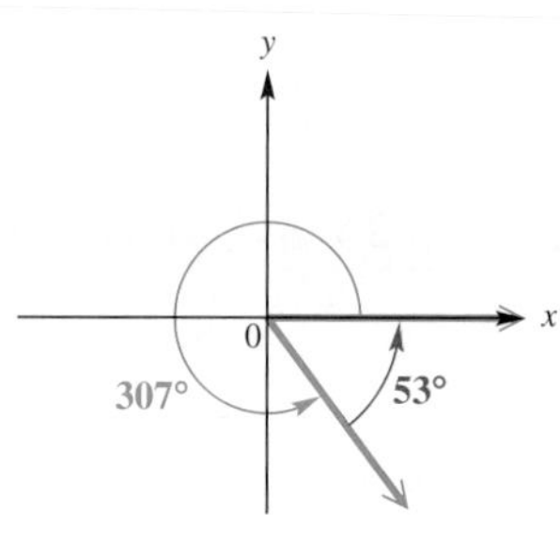

360° – 307° = 53°

(b)

Figure 34

✔ *Now Try Exercises 59 and 63.*

The preceding example suggests the following table for finding the reference angle θ' for any angle θ between 0° and 360°.

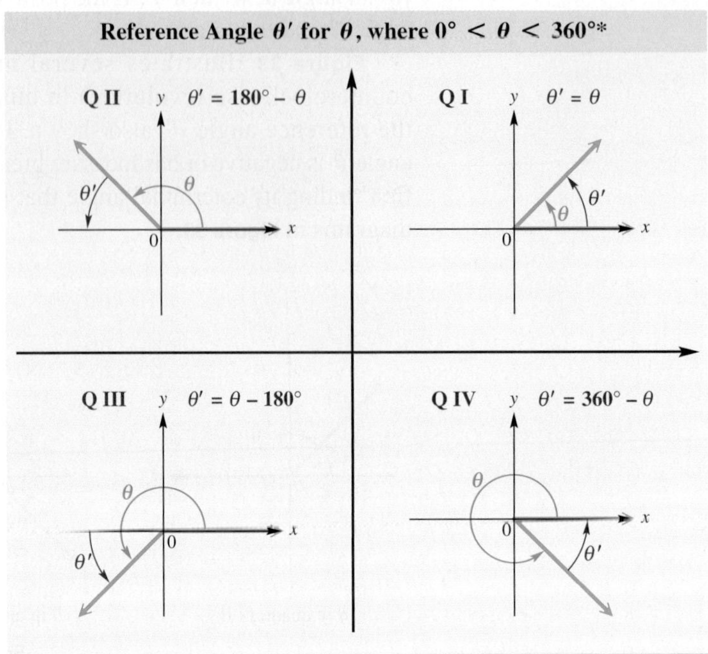

Special Angles as Reference Angles We can now find exact trigonometric function values of angles with reference angles of 30°, 45°, or 60°.

EXAMPLE 5 Finding Trigonometric Function Values of a Quadrant III Angle

Find the values of the six trigonometric functions for 210°.

SOLUTION An angle of 210° is shown in **Figure 35.** The reference angle is

$$210° - 180° = 30°.$$

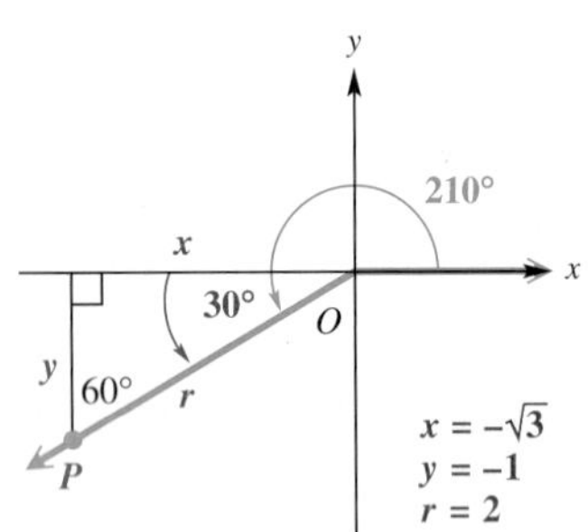

Figure 35

To find the trigonometric function values of 210°, choose point P on the terminal side of the angle so that the distance from the origin O to P is 2. By the results from 30°–60° right triangles, the coordinates of point P become $(-\sqrt{3}, -1)$, with

$$x = -\sqrt{3}, \quad y = -1, \quad \text{and} \quad r = 2.$$

Then, by the definitions of the trigonometric functions in **Section 5.2,** we obtain the following.

$$\sin 210° = \frac{-1}{2} = -\frac{1}{2} \qquad \csc 210° = \frac{2}{-1} = -2$$

$$\cos 210° = \frac{-\sqrt{3}}{2} = -\frac{\sqrt{3}}{2} \qquad \sec 210° = \frac{2}{-\sqrt{3}} = -\frac{2\sqrt{3}}{3}$$

Rationalize denominators as needed. **(Section R.7)**

$$\tan 210° = \frac{-1}{-\sqrt{3}} = \frac{\sqrt{3}}{3} \qquad \cot 210° = \frac{-\sqrt{3}}{-1} = \sqrt{3}$$

Now Try Exercise 73.

Classroom Example 5
Find the values of the six trigonometric functions for 225°.

Answer:
$\sin 225° = -\frac{\sqrt{2}}{2}$;
$\cos 225° = -\frac{\sqrt{2}}{2}$;
$\tan 225° = 1$;
$\csc 225° = -\sqrt{2}$;
$\sec 225° = -\sqrt{2}$;
$\cot 225° = 1$

*The authors would like to thank Bethany Vaughn and Theresa Matick, of Vincennes Lincoln High School, for their suggestions concerning this table.

Teaching Tip Review the signs of each trigonometric function by quadrant. As mentioned previously, the sentence "**A**ll **S**tudents **T**ake **C**alculus" may help some students remember signs of the basic functions by quadrant. (You could tell them that you graduated from **A**ll **S**tate **T**eachers' **C**ollege.)

Notice in **Example 5** that the trigonometric function values of 210° correspond in absolute value to those of its reference angle 30°. The signs are different for the sine, cosine, secant, and cosecant functions because 210° is a quadrant III angle. These results suggest a shortcut for finding the trigonometric function values of a non-acute angle, using the reference angle. In **Example 5,** the reference angle for 210° is 30°. Using the trigonometric function values of 30°, and choosing the correct signs for a quadrant III angle, we obtain the same results.

We determine the values of the trigonometric functions for any nonquadrantal angle θ as follows.

Finding Trigonometric Function Values for Any Nonquadrantal Angle θ

Step 1 If $\theta > 360°$, or if $\theta < 0°$, then find a coterminal angle by adding or subtracting 360° as many times as needed to get an angle greater than 0° but less than 360°.

Step 2 Find the reference angle θ'.

Step 3 Find the trigonometric function values for reference angle θ'.

Step 4 Determine the correct signs for the values found in Step 3. (Use the table of signs in **Section 5.2,** if necessary.) This gives the values of the trigonometric functions for angle θ.

NOTE To avoid sign errors when finding the trigonometric function values of an angle, sketch it in standard position. Include a reference triangle complete with appropriate values for x, y, and r as done in **Figure 35.**

Classroom Example 6
Find the exact value of each expression.
(a) $\sin(-150°)$ **(b)** $\cot 780°$

Answers:
(a) $-\frac{1}{2}$ **(b)** $\frac{\sqrt{3}}{3}$

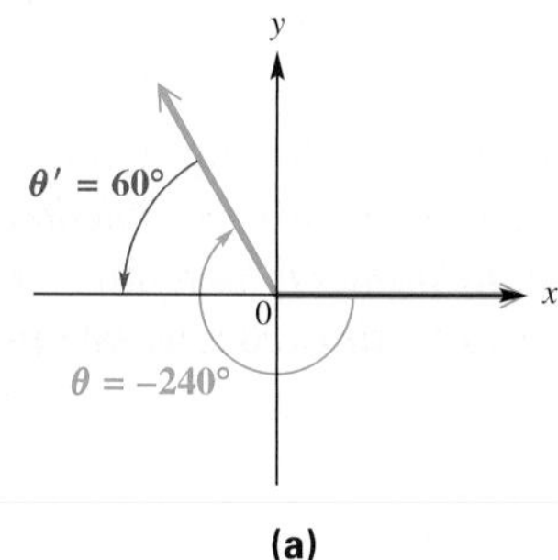

(a)

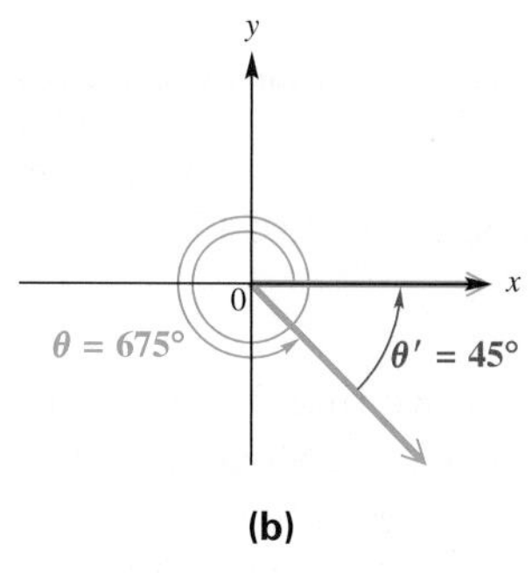

(b)

Figure 36

EXAMPLE 6 **Finding Trigonometric Function Values Using Reference Angles**

Find the exact value of each expression.

(a) $\cos(-240°)$ **(b)** $\tan 675°$

SOLUTION

(a) Since an angle of $-240°$ is coterminal with an angle of

$$-240° + 360° = 120°, \quad \text{(Section 5.1)}$$

the reference angle is $180° - 120° = 60°$, as shown in **Figure 36(a).** Since the cosine is negative in quadrant II,

$$\cos(-240°) = \underset{\substack{\uparrow \\ \text{Coterminal angle}}}{\cos 120°} = \underset{\substack{\uparrow \\ \text{Reference angle}}}{-\cos 60°} = -\frac{1}{2}.$$

(b) Begin by subtracting 360° to get a coterminal angle between 0° and 360°.

$$675° - 360° = 315°$$

As shown in **Figure 36(b),** the reference angle is $360° - 315° = 45°$. An angle of 315° is in quadrant IV, so the tangent will be negative.

$$\begin{aligned} \tan 675° &= \tan 315° && \text{Coterminal angle} \\ &= -\tan 45° && \text{Reference angle; quadrant-based sign choice} \\ &= -1 && \text{Evaluate.} \end{aligned}$$

Now Try Exercises 91 and 93.

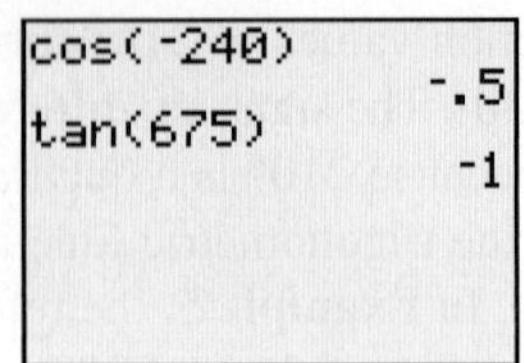

Degree mode

Figure 37

Finding Function Values Using a Calculator Calculators are capable of finding trigonometric function values. For example, the values of $\cos(-240°)$ and $\tan 675°$ in **Example 6** are found with a calculator as shown in **Figure 37.**

CAUTION ***When evaluating trigonometric functions of angles given in degrees, remember that the calculator must be set in degree mode.*** Get in the habit of always starting work by entering sin 90. If the displayed answer is 1, then the calculator is set for degree measure. Remember that most calculator values of trigonometric functions are *approximations.*

Teaching Tip Caution students *not* to use the $\sin^{-1}$, $\cos^{-1}$, and $\tan^{-1}$ keys when evaluating reciprocal functions with a calculator.

EXAMPLE 7 Finding Function Values with a Calculator

Approximate the value of each expression.

(a) $\sin 49° 12'$ **(b)** $\sec 97.977°$ **(c)** $\dfrac{1}{\cot 51.4283°}$ **(d)** $\sin(-246°)$

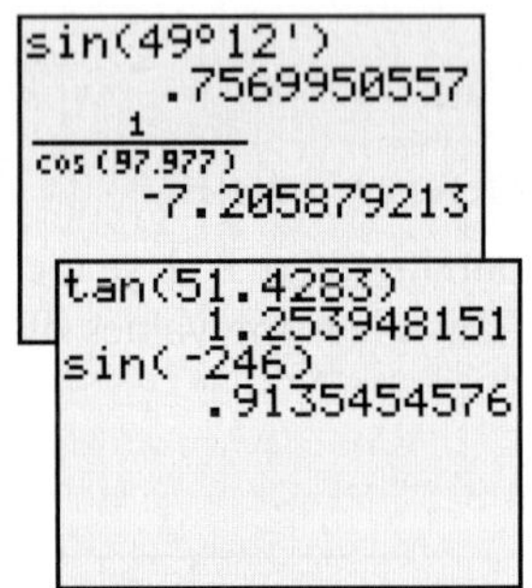

These screens support the results of **Example 7.** We entered the angle measure in degrees and minutes for part (a).

SOLUTION

(a) $49° 12' = 49\frac{12}{60}° = 49.2°$ Convert 49° 12′ to decimal degrees. (Section 5.1)

$\sin 49° 12' = \sin 49.2° \approx 0.75699506$ To eight decimal places

(b) Calculators do not have secant keys. However, $\sec\theta = \frac{1}{\cos\theta}$ for all angles θ where $\cos\theta \neq 0$. Therefore, we use the reciprocal of the cosine function to evaluate the secant function.

$$\sec 97.977° = \frac{1}{\cos 97.977°} \approx -7.20587921$$

(c) Use the reciprocal identity $\tan\theta = \frac{1}{\cot\theta}$ to simplify the expression first.

$$\frac{1}{\cot 51.4283°} = \tan 51.4283° \approx 1.25394815$$

(d) $\sin(-246°) \approx 0.91354546$

Now Try Exercises 101, 103, 107, and 111.

Classroom Example 7
Approximate the value of each expression.
(a) $\tan 68° 43'$
(b) $\cos 193.622°$
(c) $\csc 35.8471°$
(d) $\sec(-287°)$

Answers:
(a) 2.56707352
(b) −0.97187064
(c) 1.70757967
(d) 3.42030362

Finding Angle Measures To find the measure of an angle having a certain trigonometric function value, graphing calculators have three *inverse functions* (denoted $\sin^{-1}$, $\cos^{-1}$, and $\tan^{-1}$). ***If x is an appropriate number, then $\sin^{-1} x$, $\cos^{-1} x$, or $\tan^{-1} x$ gives the measure of an angle whose sine, cosine, or tangent, respectively, is x.*** For applications in this chapter, these functions will return angles in quadrant I.

EXAMPLE 8 Using Inverse Trigonometric Functions to Find Angles

Use a calculator to find an angle θ in the interval $[0°, 90°]$ that satisfies each condition.

(a) $\sin\theta \approx 0.96770915$ **(b)** $\sec\theta \approx 1.0545829$

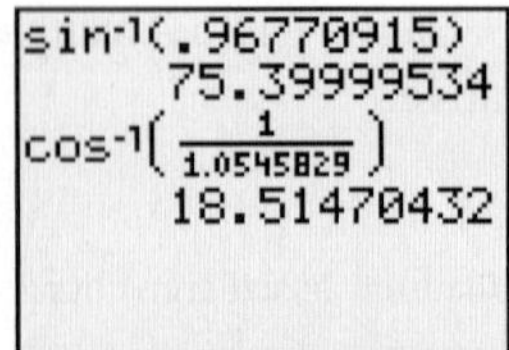

Degree mode

Figure 38

SOLUTION

(a) Using degree mode and the inverse sine function, we find that an angle θ having sine value 0.96770915 is about 75.399995°. See **Figure 38.** (There are infinitely many such angles, but the calculator gives only this one.)

$$\theta \approx \sin^{-1} 0.96770915 \approx 75.399995°$$

Classroom Example 8
Use a calculator to find an angle θ in the interval $[0°, 90°]$ that satisfies each condition.

(a) $\cos \theta \approx 0.92118541$

(b) $\cot \theta \approx 1.4466474$

Answers:

(a) $22.899999°$

(b) $34.654300°$

(b) Use the identity $\cos \theta = \frac{1}{\sec \theta}$. If $\sec \theta \approx 1.0545829$, then

$$\cos \theta \approx \frac{1}{1.0545829}.$$

Now, find θ using the inverse cosine function. See **Figure 38** at the bottom of the preceding page.

$$\theta \approx \cos^{-1}\left(\frac{1}{1.0545829}\right) \approx 18.514704°$$

Now Try Exercises 117 and 121.

CAUTION **Compare Examples 7(b) and 8(b).** To determine the secant of an angle, as in **Example 7(b),** we find the *reciprocal of the cosine* of the angle. To determine an angle with a given secant value, as in **Example 8(b),** we find the *inverse cosine of the reciprocal* of the value.

Classroom Example 9
Find all values of θ, if θ is in the interval $[0°, 360°)$ and $\sin \theta = -\frac{\sqrt{3}}{2}$.

Answer: $240°$; $300°$

EXAMPLE 9 Finding Angle Measures Given an Interval and a Function Value

Find all values of θ, if θ is in the interval $[0°, 360°)$ and $\cos \theta = -\frac{\sqrt{2}}{2}$.

SOLUTION Since $\cos \theta$ is negative, θ must lie in quadrant II or III. Since the absolute value of $\cos \theta$ is $\frac{\sqrt{2}}{2}$, the reference angle θ' must be $45°$. The two possible angles θ are sketched in **Figure 39.**

$$180° - 45° = 135° \quad \text{Quadrant II angle } \theta$$

$$180° + 45° = 225° \quad \text{Quadrant III angle } \theta$$

Teaching Tip Relate cosine to the value of the x-coordinate, mentioning that the value of x does not change if a point is reflected from quadrant II to quadrant III.

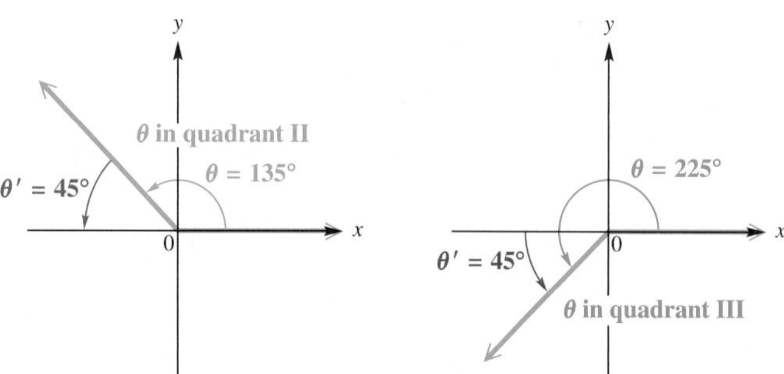

Figure 39

Now Try Exercise 127.

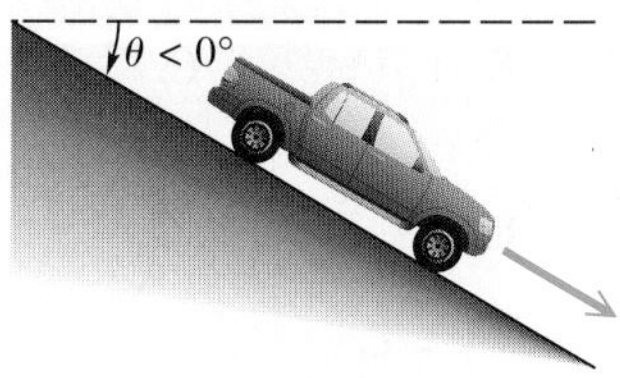

Figure 40

EXAMPLE 10 Finding Grade Resistance

When an automobile travels uphill or downhill on a highway, it experiences a force due to gravity. This force F in pounds is the **grade resistance** and is modeled by the equation

$$F = W \sin \theta,$$

where θ is the grade and W is the weight of the automobile. If the automobile is moving uphill, then $\theta > 0°$; if downhill, then $\theta < 0°$. See **Figure 40.** (*Source:* Mannering, F. and W. Kilareski, *Principles of Highway Engineering and Traffic Analysis,* Second Edition, John Wiley and Sons.)

Classroom Example 10
Refer to **Example 10.**

(a) Calculate F to the nearest 10 lb for a 5500-lb car traveling an uphill grade with $\theta = 3.9°$.

(b) Calculate F to the nearest 10 lb for a 2800-lb car traveling a downhill grade with $\theta = -4.8°$.

(c) A 2400-lb car traveling uphill has a grade resistance of 288 lb. What is the angle of the grade?

Answers:
(a) 370 lb **(b)** -230 lb
(c) 6.89°

(a) Calculate F to the nearest 10 lb for a 2500-lb car traveling an uphill grade with $\theta = 2.5°$.

(b) Calculate F to the nearest 10 lb for a 5000-lb truck traveling a downhill grade with $\theta = -6.1°$.

(c) Calculate F for $\theta = 0°$ and $\theta = 90°$. Do these answers agree with your intuition?

SOLUTION

(a) $F = W \sin \theta = 2500 \sin 2.5° \approx 110$ lb

(b) $F = W \sin \theta = 5000 \sin(-6.1°) \approx -530$ lb
F is negative because the truck is moving downhill.

(c) $F = W \sin \theta = W \sin 0° = W(0) = 0$ lb

$F = W \sin \theta = W \sin 90° = W(1) = W$ lb

This agrees with intuition because if $\theta = 0°$, then there is level ground and gravity does not cause the vehicle to roll. If θ were 90°, the road would be vertical and the full weight of the vehicle would be pulled downward by gravity, so $F = W$.

Now Try Exercises 141 and 143.

5.3 Exercises

In Exercises 1–4, we give, in order, sine, cosine, and tangent.

1. $\frac{21}{29}$; $\frac{20}{29}$; $\frac{21}{20}$ **2.** $\frac{45}{53}$; $\frac{28}{53}$; $\frac{45}{28}$

3. $\frac{n}{p}$; $\frac{m}{p}$; $\frac{n}{m}$ **4.** $\frac{k}{z}$; $\frac{y}{z}$; $\frac{k}{y}$

5. C **6.** H **7.** B
8. G **9.** E **10.** A

In Exercises 11–19, we give, in order, the unknown side, sine, cosine, tangent, cotangent, secant, and cosecant.

11. $c = 13$; $\frac{12}{13}$; $\frac{5}{13}$; $\frac{12}{5}$; $\frac{5}{12}$; $\frac{13}{5}$; $\frac{13}{12}$

12. $c = 5$; $\frac{4}{5}$; $\frac{3}{5}$; $\frac{4}{3}$; $\frac{3}{4}$; $\frac{5}{3}$; $\frac{5}{4}$

13. $b = \sqrt{13}$; $\frac{\sqrt{13}}{7}$; $\frac{6}{7}$; $\frac{\sqrt{13}}{6}$; $\frac{6\sqrt{13}}{13}$; $\frac{7}{6}$; $\frac{7\sqrt{13}}{13}$

14. $a = \sqrt{95}$; $\frac{7}{12}$; $\frac{\sqrt{95}}{12}$; $\frac{7\sqrt{95}}{95}$; $\frac{\sqrt{95}}{7}$; $\frac{12\sqrt{95}}{95}$; $\frac{12}{7}$

15. $b = \sqrt{91}$; $\frac{\sqrt{91}}{10}$; $\frac{3}{10}$; $\frac{\sqrt{91}}{3}$; $\frac{3\sqrt{91}}{91}$; $\frac{10}{3}$; $\frac{10\sqrt{91}}{91}$

Find exact values or expressions for sin *A*, cos *A*, *and* tan *A*. ***See Example 1.***

1.

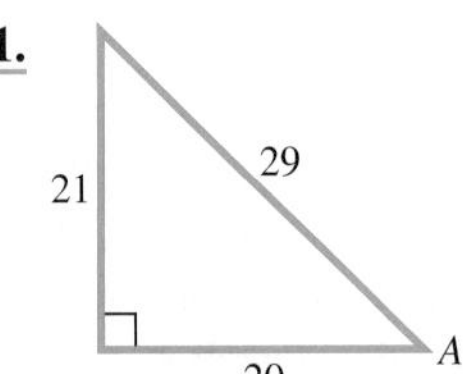

2.

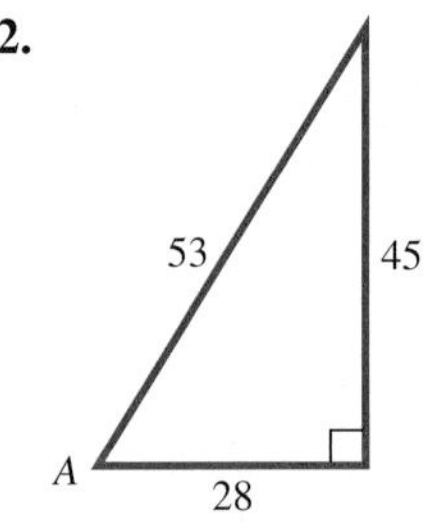

3.

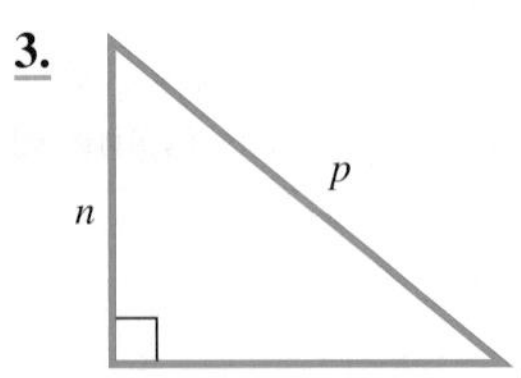

4.

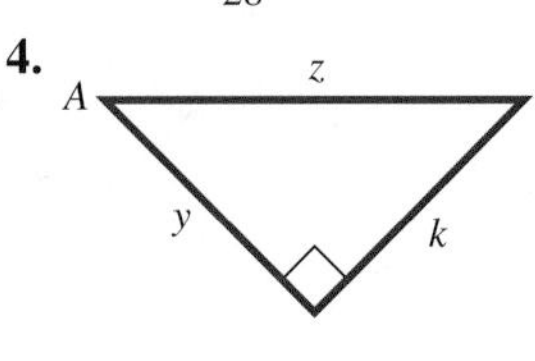

Concept Check For each trigonometric function in Column I, choose its value from Column II.

I		II		
5. sin 30°	**6.** cos 45°	**A.** $\sqrt{3}$	**B.** 1	**C.** $\frac{1}{2}$
7. tan 45°	**8.** sec 60°	**D.** $\frac{\sqrt{3}}{2}$	**E.** $\frac{2\sqrt{3}}{3}$	**F.** $\frac{\sqrt{3}}{3}$
9. csc 60°	**10.** cot 30°	**G.** 2	**H.** $\frac{\sqrt{2}}{2}$	**I.** $\sqrt{2}$

Suppose ABC is a right triangle with sides of lengths a, b, and c and right angle at C. (See **Figure 29.**) *Find the unknown side length using the Pythagorean theorem (**Section 1.5**), and then find the values of the six trigonometric functions for angle B. Rationalize denominators when applicable.*

11. $a = 5,\ b = 12$ **12.** $a = 3,\ b = 4$ **13.** $a = 6,\ c = 7$

14. $b = 7,\ c = 12$ **15.** $a = 3,\ c = 10$ **16.** $b = 8,\ c = 11$

17. $a = 1,\ c = 2$ **18.** $a = \sqrt{2},\ c = 2$ **19.** $b = 2,\ c = 5$

20. *Concept Check* Give a summary of the six cofunction relationships.

Write each function in terms of its cofunction. Assume that all angles in which an unknown appears are acute angles. ***See Example 2.***

21. cos 30° **22.** sin 45° **23.** csc 60° **24.** cot 73°

25. sec 39° **26.** tan 25.4° **27.** sin 38.7° **28.** csc 49.9°

For each expression, give the exact value. ***See Example 3.***

29. tan 30° **30.** cot 30° **31.** sin 30° **32.** cos 30°

33. sec 30° **34.** csc 30° **35.** csc 45° **36.** sec 45°

37. cos 45° **38.** cot 45° **39.** tan 45° **40.** sin 45°

41. sin 60° **42.** cos 60° **43.** tan 60° **44.** csc 60°

Relating Concepts

For individual or collaborative investigation *(Exercises 45–48)*

The figure shows a 45° central angle in a circle with radius 4 units. To find the coordinates of point P on the circle, ***work Exercises 45–48 in order.***

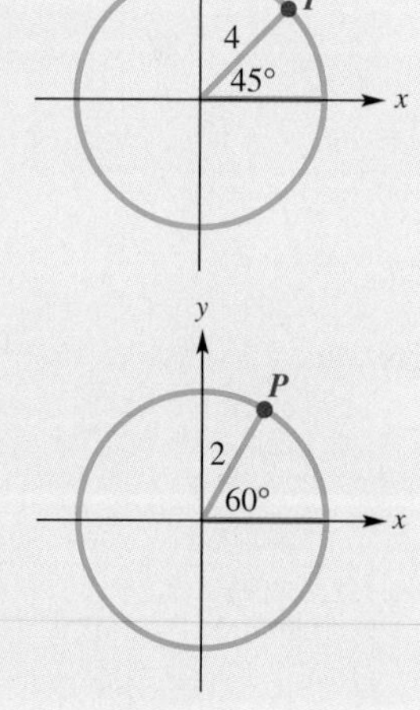

45. Sketch a line segment from P perpendicular to the x-axis.

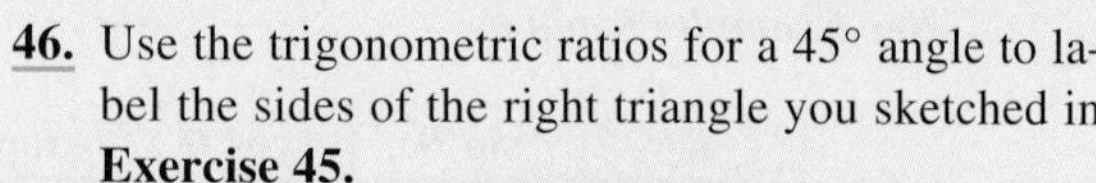

46. Use the trigonometric ratios for a 45° angle to label the sides of the right triangle you sketched in **Exercise 45.**

47. Which sides of the right triangle give the coordinates of point P? What are the coordinates of P?

48. The figure at the right shows a 60° central angle in a circle of radius 2 units. Follow the same procedure as in **Exercises 45–47** to find the coordinates of P in the figure.

Concept Check *Work each problem.*

49. Find the equation of the line that passes through the origin and makes a 30° angle with the x-axis.

50. Find the equation of the line that passes through the origin and makes a 60° angle with the x-axis.

51. What angle does the line $y = \sqrt{3}x$ make with the positive x-axis?

52. What angle does the line $y = \frac{\sqrt{3}}{3}x$ make with the positive x-axis?

---

16. $a = \sqrt{57}$; $\frac{8}{11}$; $\frac{\sqrt{57}}{11}$; $\frac{8\sqrt{57}}{57}$; $\frac{\sqrt{57}}{8}$; $\frac{11\sqrt{57}}{57}$; $\frac{11}{8}$

17. $b = \sqrt{3}$; $\frac{\sqrt{3}}{2}$; $\frac{1}{2}$; $\sqrt{3}$; $\frac{\sqrt{3}}{3}$; 2; $\frac{2\sqrt{3}}{3}$

18. $b = \sqrt{2}$; $\frac{\sqrt{2}}{2}$; $\frac{\sqrt{2}}{2}$; 1; 1; $\sqrt{2}$; $\sqrt{2}$

19. $a = \sqrt{21}$; $\frac{2}{5}$; $\frac{\sqrt{21}}{5}$; $\frac{2\sqrt{21}}{21}$; $\frac{\sqrt{21}}{2}$; $\frac{5\sqrt{21}}{21}$; $\frac{5}{2}$

20. $\sin A = \cos(90° - A)$; $\cos A = \sin(90° - A)$; $\tan A = \cot(90° - A)$; $\cot A = \tan(90° - A)$; $\sec A = \csc(90° - A)$; $\csc A = \sec(90° - A)$

21. sin 60° 22. cos 45°
23. sec 30° 24. tan 17°
25. csc 51° 26. cot 64.6°
27. cos 51.3° 28. sec 40.1°

29. $\frac{\sqrt{3}}{3}$ 30. $\sqrt{3}$

31. $\frac{1}{2}$ 32. $\frac{\sqrt{3}}{2}$

33. $\frac{2\sqrt{3}}{3}$ 34. 2

35. $\sqrt{2}$ 36. $\sqrt{2}$

37. $\frac{\sqrt{2}}{2}$ 38. 1

39. 1 40. $\frac{\sqrt{2}}{2}$

41. $\frac{\sqrt{3}}{2}$ 42. $\frac{1}{2}$

43. $\sqrt{3}$ 44. $\frac{2\sqrt{3}}{3}$

45. 46.

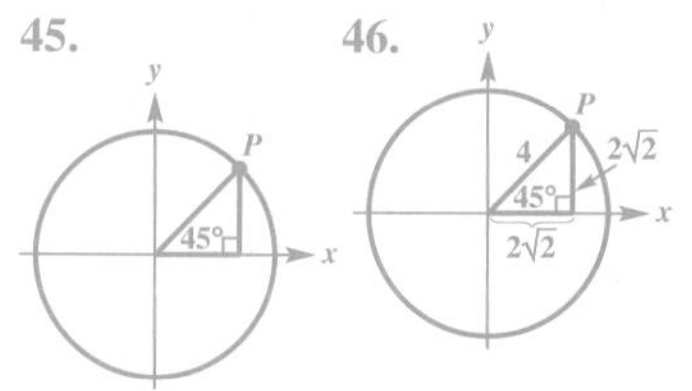

47. the legs; $(2\sqrt{2}, 2\sqrt{2})$

48. $(1, \sqrt{3})$

49. $y = \frac{\sqrt{3}}{3}x$

50. $y = \sqrt{3}x$

51. 60°

52. 30°

53. $x = \frac{9\sqrt{3}}{2}$; $y = \frac{9}{2}$; $z = \frac{3\sqrt{3}}{2}$; $w = 3\sqrt{3}$

54. $a = 12$; $b = 12\sqrt{3}$; $d = 12\sqrt{3}$; $c = 12\sqrt{6}$

55. $p = 15$; $r = 15\sqrt{2}$; $q = 5\sqrt{6}$; $t = 10\sqrt{6}$

56. $m = \frac{7\sqrt{3}}{3}$; $a = \frac{14\sqrt{3}}{3}$; $n = \frac{14\sqrt{3}}{3}$; $q = \frac{14\sqrt{6}}{3}$

57. $\mathcal{A} = \frac{s^2}{2}$ 58. $\mathcal{A} = \frac{s^2\sqrt{3}}{4}$

59. C 60. F
61. A 62. B
63. D 64. B

65. $\frac{\sqrt{3}}{3}$; $\sqrt{3}$

66. $\frac{\sqrt{2}}{2}$; $\frac{\sqrt{2}}{2}$; $\sqrt{2}$; $\sqrt{2}$

67. $\frac{\sqrt{3}}{2}$; $\frac{\sqrt{3}}{3}$; $\frac{2\sqrt{3}}{3}$

68. $-\frac{1}{2}$; $-\frac{\sqrt{3}}{3}$; -2

69. -1; -1

70. $\frac{1}{2}$; $-\sqrt{3}$; $-\frac{2\sqrt{3}}{3}$

71. $-\frac{\sqrt{3}}{2}$; $-\frac{2\sqrt{3}}{3}$

72. $\sqrt{3}$; $\frac{\sqrt{3}}{3}$

In Exercises 73–90, we give, in order, sine, cosine, tangent, cotangent, secant, and cosecant.

73. $-\frac{\sqrt{3}}{2}$; $\frac{1}{2}$; $-\sqrt{3}$; $-\frac{\sqrt{3}}{3}$; 2; $-\frac{2\sqrt{3}}{3}$

74. $-\frac{\sqrt{2}}{2}$; $\frac{\sqrt{2}}{2}$; -1; -1; $\sqrt{2}$; $-\sqrt{2}$

75. $\frac{\sqrt{2}}{2}$; $\frac{\sqrt{2}}{2}$; 1; 1; $\sqrt{2}$; $\sqrt{2}$

76. $\frac{\sqrt{3}}{2}$; $\frac{1}{2}$; $\sqrt{3}$; $\frac{\sqrt{3}}{3}$; 2; $\frac{2\sqrt{3}}{3}$

77. $\frac{\sqrt{3}}{2}$; $-\frac{1}{2}$; $-\sqrt{3}$; $-\frac{\sqrt{3}}{3}$; -2; $\frac{2\sqrt{3}}{3}$

78. $\frac{\sqrt{2}}{2}$; $-\frac{\sqrt{2}}{2}$; -1; -1; $-\sqrt{2}$; $\sqrt{2}$

79. $-\frac{1}{2}$; $-\frac{\sqrt{3}}{2}$; $\frac{\sqrt{3}}{3}$; $\sqrt{3}$; $-\frac{2\sqrt{3}}{3}$; -2

80. $\frac{1}{2}$; $\frac{\sqrt{3}}{2}$; $\frac{\sqrt{3}}{3}$; $\sqrt{3}$; $\frac{2\sqrt{3}}{3}$; 2

81. $-\frac{\sqrt{2}}{2}$; $-\frac{\sqrt{2}}{2}$; 1; 1; $-\sqrt{2}$; $-\sqrt{2}$

82. $\frac{\sqrt{3}}{2}$; $\frac{1}{2}$; $\sqrt{3}$; $\frac{\sqrt{3}}{3}$; 2; $\frac{2\sqrt{3}}{3}$

83. $\frac{\sqrt{3}}{2}$; $\frac{1}{2}$; $\sqrt{3}$; $\frac{\sqrt{3}}{3}$; 2; $\frac{2\sqrt{3}}{3}$

Find the exact value of each part labeled with a variable in each figure.

53.

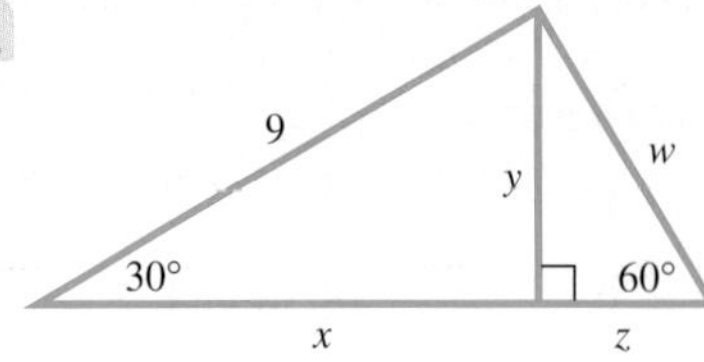

54.

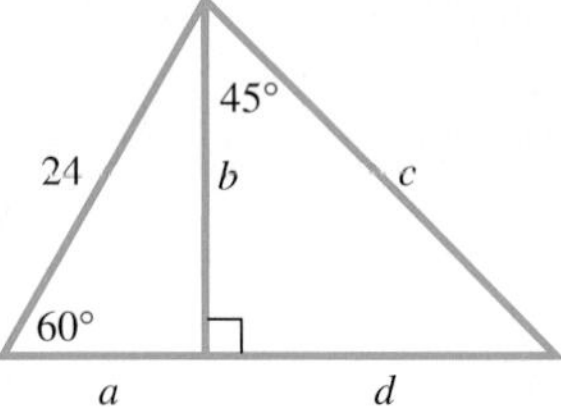

55.

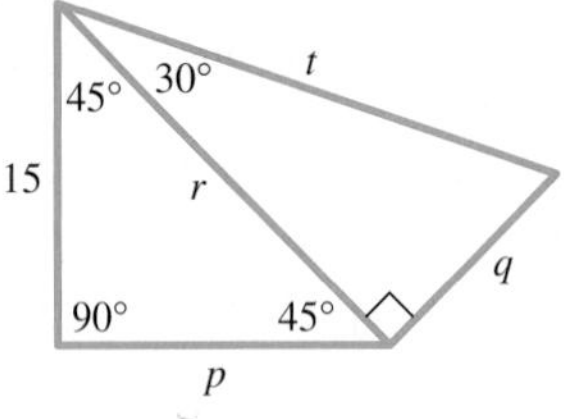

56.

Find a formula for the area of each figure in terms of s.

57.

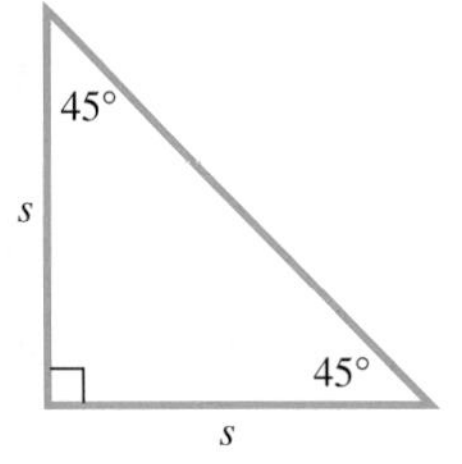

58.

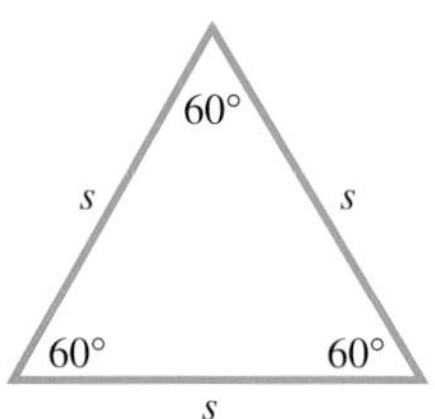

Match each angle in Column I with its reference angle in Column II. Choices may be used once, more than once, or not at all. ***See Example 4.***

I		II	
59. 98°	**60.** 212°	**A.** 45°	**B.** 60°
61. −135°	**62.** −60°	**C.** 82°	**D.** 30°
63. 750°	**64.** 480°	**E.** 38°	**F.** 32°

Complete the table with exact trigonometric function values. Do not use a calculator. ***See Examples 5 and 6.***

	θ	$\sin\theta$	$\cos\theta$	$\tan\theta$	$\cot\theta$	$\sec\theta$	$\csc\theta$
65.	30°	$\frac{1}{2}$	$\frac{\sqrt{3}}{2}$			$\frac{2\sqrt{3}}{3}$	2
66.	45°			1	1		
67.	60°		$\frac{1}{2}$	$\sqrt{3}$		2	
68.	120°	$\frac{\sqrt{3}}{2}$		$-\sqrt{3}$			$\frac{2\sqrt{3}}{3}$
69.	135°	$\frac{\sqrt{2}}{2}$	$-\frac{\sqrt{2}}{2}$			$-\sqrt{2}$	$\sqrt{2}$
70.	150°		$-\frac{\sqrt{3}}{2}$	$-\frac{\sqrt{3}}{3}$			2
71.	210°	$-\frac{1}{2}$		$\frac{\sqrt{3}}{3}$	$\sqrt{3}$		-2
72.	240°	$-\frac{\sqrt{3}}{2}$	$-\frac{1}{2}$			-2	$-\frac{2\sqrt{3}}{3}$

Find exact values of the six trigonometric functions for each angle. Rationalize denominators when applicable. ***See Examples 5 and 6.***

73. 300° **74.** 315° **75.** 405° **76.** 420° **77.** 480° **78.** 495°

79. 570° **80.** 750° **81.** 1305° **82.** 1500° **83.** −300° **84.** −390°

85. −510° **86.** −1020° **87.** −1290° **88.** −855° **89.** −1860° **90.** −2205°

Find the exact value of each expression. ***See Example 6.***

91. sin 1305° **92.** sin 1500° **93.** cos(−510°) **94.** tan(−1020°)

95. csc(−855°) **96.** sec(−495°) **97.** tan 3015° **98.** cot 2280°

Concept Check Fill in the blanks to complete each statement.

99. The CAUTION preceding **Example 7** suggests verifying that a calculator is in degree mode by finding ________ 90°. If the calculator is in degree mode,
(sin/cos/tan)
then the display should be ________.

100. To find values of the cotangent, secant, and cosecant functions with a calculator, it is necessary to find the ________ of the ________ function value.

Use a calculator to find a decimal approximation for each value. Give as many digits as your calculator displays. In Exercises 111–114, simplify the expression before using the calculator. ***See Example 7.***

101. sin 38° 42′ **102.** cos 41° 24′ **103.** sec 13° 15′

104. csc 145° 45′ **105.** cot 183° 48′ **106.** tan 421° 30′

107. sin(−312° 12′) **108.** tan(−80° 06′) **109.** csc(−317° 36′)

110. cot(−512° 20′) **111.** $\frac{1}{\cot 23.4°}$ **112.** $\frac{1}{\sec 14.8°}$

113. $\frac{\cos 77°}{\sin 77°}$ **114.** $\frac{\sin 33°}{\cos 33°}$

Find a value of θ in the interval [0°, 90°] that satisfies each statement. Write each answer in decimal degrees to six decimal places as needed. ***See Example 8.***

115. $\tan\theta = 1.4739716$ **116.** $\tan\theta = 6.4358841$ **117.** $\sin\theta = 0.27843196$

118. $\sin\theta = 0.84802194$ **119.** $\cot\theta = 1.2575516$ **120.** $\csc\theta = 1.3861147$

121. $\sec\theta = 2.7496222$ **122.** $\sec\theta = 1.1606249$ **123.** $\cos\theta = 0.70058013$

124. $\cos\theta = 0.85536428$ **125.** $\csc\theta = 4.7216543$ **126.** $\cot\theta = 0.21563481$

Find all values of θ, if θ is in the interval [0°, 360°) and has the given function value. ***See Example 9.***

127. $\sin\theta = \frac{1}{2}$ **128.** $\cos\theta = \frac{\sqrt{3}}{2}$ **129.** $\tan\theta = -\sqrt{3}$

130. $\sec\theta = -\sqrt{2}$ **131.** $\cos\theta = \frac{\sqrt{2}}{2}$ **132.** $\cot\theta = -\frac{\sqrt{3}}{3}$

133. $\csc\theta = -2$ **134.** $\sin\theta = -\frac{\sqrt{3}}{2}$ **135.** $\tan\theta = \frac{\sqrt{3}}{3}$

136. $\cos\theta = -\frac{1}{2}$ **137.** $\csc\theta = -\sqrt{2}$ **138.** $\cot\theta = -1$

---

84. $-\frac{1}{2}$; $\frac{\sqrt{3}}{2}$; $-\frac{\sqrt{3}}{3}$; $-\sqrt{3}$; $\frac{2\sqrt{3}}{3}$; -2

85. $-\frac{1}{2}$; $-\frac{\sqrt{3}}{2}$; $\frac{\sqrt{3}}{3}$; $\sqrt{3}$; $-\frac{2\sqrt{3}}{3}$; -2

86. $\frac{\sqrt{3}}{2}$; $\frac{1}{2}$; $\sqrt{3}$; $\frac{\sqrt{3}}{3}$; 2; $\frac{2\sqrt{3}}{3}$

87. $\frac{1}{2}$; $-\frac{\sqrt{3}}{2}$; $-\frac{\sqrt{3}}{3}$; $-\sqrt{3}$; $-\frac{2\sqrt{3}}{3}$; 2

88. $-\frac{\sqrt{2}}{2}$; $-\frac{\sqrt{2}}{2}$; 1; 1; $-\sqrt{2}$; $-\sqrt{2}$

89. $-\frac{\sqrt{3}}{2}$; $\frac{1}{2}$; $-\sqrt{3}$; $-\frac{\sqrt{3}}{3}$; 2; $-\frac{2\sqrt{3}}{3}$

90. $-\frac{\sqrt{2}}{2}$; $\frac{\sqrt{2}}{2}$; −1; −1; $\sqrt{2}$; $-\sqrt{2}$

91. $-\frac{\sqrt{2}}{2}$ **92.** $\frac{\sqrt{3}}{2}$

93. $-\frac{\sqrt{3}}{2}$ **94.** $\sqrt{3}$

95. $-\sqrt{2}$ **96.** $-\sqrt{2}$

97. −1 **98.** $-\frac{\sqrt{3}}{3}$

99. sin; 1

100. reciprocal; reciprocal

In Exercises 101–114, the number of decimal places may vary depending on the calculator used.

101. 0.62524266

102. 0.75011107

103. 1.0273488

104. 1.7768146

105. 15.055723

106. 1.8417709

107. 0.74080460

108. −5.7297416

109. 1.4830142

110. 1.9074147

111. tan 23.4° ≈ 0.43273864

112. cos 14.8° ≈ 0.96682339

113. cot 77° ≈ 0.23086819

114. tan 33° ≈ 0.64940759

115. 55.845496° **116.** 81.168073°

117. 16.166641° **118.** 57.997172°

119. 38.491580° **120.** 46.173582°

121. 68.673241° **122.** 30.502748°

123. 45.526434° **124.** 31.199998°

125. 12.227282° **126.** 77.831359°

127. 30°; 150° **128.** 30°; 330°

129. 120°; 300° **130.** 135°; 225°

131. 45°; 315° 132. 120°; 300°
133. 210°; 330° 134. 240°; 300°
135. 30°; 210° 136. 120°; 240°
137. 225°; 315° 138. 135°; 315°

139. A: 68.94 mph; B: 65.78 mph

141. 70 lb 142. −100 lb
143. −2.9° 144. 2.9°
145. 2500 lb 146. 2800 lb

Work each problem.

139. *Measuring Speed by Radar* Any offset between a stationary radar gun and a moving target creates a "cosine effect" that reduces the radar reading by the cosine of the angle between the gun and the vehicle. That is, the radar speed reading is the product of the actual speed and the cosine of the angle. Find the radar readings, to the nearest hundredth, for Auto A and Auto B shown in the figure. (*Source:* Fischetti, M., "Working Knowledge," *Scientific American.*)

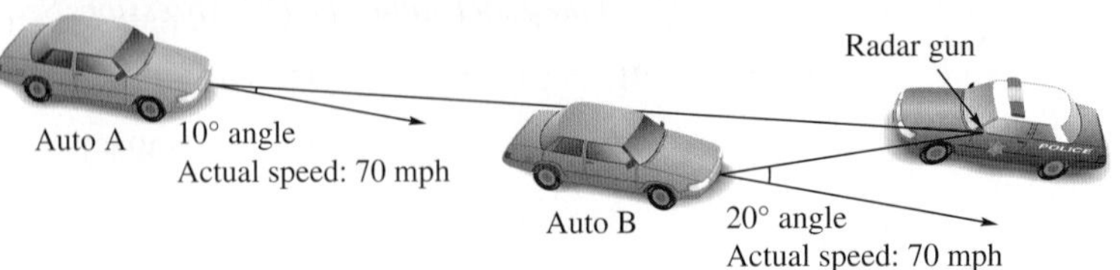

140. *Measuring Speed by Radar* In **Exercise 139,** we saw that the speed reported by a radar gun is reduced by the cosine of angle θ, shown in the figure. In the figure, r represents reduced speed and a represents the actual speed. Use the figure to show why this "cosine effect" occurs.

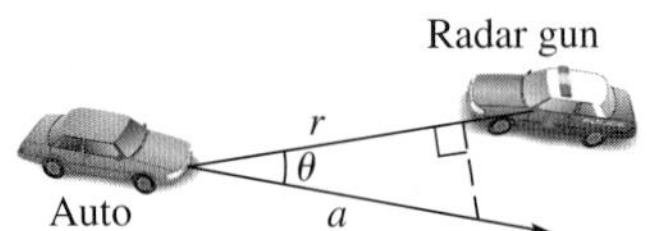

(Modeling) Grade Resistance See **Example 10** *to work Exercises 141–146.*

141. Find the grade resistance, to the nearest ten pounds, for a 2100-lb car traveling on a 1.8° uphill grade.

142. Find the grade resistance, to the nearest ten pounds, for a 2400-lb car traveling on a −2.4° downhill grade.

143. A 2600-lb car traveling downhill has a grade resistance of −130 lb. Find the angle of the grade to the nearest tenth of a degree.

144. A 3000-lb car traveling uphill has a grade resistance of 150 lb. Find the angle of the grade to the nearest tenth of a degree.

145. A car traveling on a 2.7° uphill grade has a grade resistance of 120 lb. Determine the weight of the car to the nearest hundred pounds.

146. A car traveling on a −3° downhill grade has a grade resistance of −145 lb. Determine the weight of the car to the nearest hundred pounds.

(Modeling) Solve each problem.

147. *Design of Highway Curves* When highway curves are designed, the outside of the curve is often slightly elevated or inclined above the inside of the curve. See the figure. This inclination is the **superelevation.** For safety reasons, it is important that both the curve's radius and superelevation be correct for a given speed limit. If an automobile is traveling at velocity V (in feet per second), the safe radius R for a curve with superelevation θ is modeled by the formula

$$R = \frac{V^2}{g(f + \tan \theta)},$$

where f and g are constants. (*Source:* Mannering, F. and W. Kilareski, *Principles of Highway Engineering and Traffic Analysis,* Second Edition, John Wiley and Sons.)

147. (a) 703 ft (b) 1701 ft
(c) R would decrease; 644 ft, 1559 ft

148. 55 mph

(a) A roadway is being designed for automobiles traveling at 45 mph. If $\theta = 3°$, $g = 32.2$, and $f = 0.14$, calculate R to the nearest foot. (*Hint:* 45 mph = 66 ft per sec)

(b) Determine the radius of the curve, to the nearest foot, if the speed in part (a) is increased to 70 mph.

(c) How would increasing the angle θ affect the results? Verify your answer by repeating parts (a) and (b) with $\theta = 4°$.

148. *Speed Limit on a Curve* Refer to **Exercise 147** and use the same values for f and g. A highway curve has radius $R = 1150$ ft and a superelevation of $\theta = 2.1°$. What should the speed limit (in miles per hour) be for this curve?

Chapter 5 Quiz (Sections 5.1–5.3)

[5.1]
1. (a) 71° (b) 161°
2. 65°; 115°
3. 26°; 64°
4. (a) 77.2025° (b) 22° 01′ 30″
5. (a) 50° (b) 300° (c) 170° (d) 417°
6. 1800°

[5.2]
7. $\sin\theta = \frac{7}{25}$; $\cos\theta = -\frac{24}{25}$; $\tan\theta = -\frac{7}{24}$; $\cot\theta = -\frac{24}{7}$; $\sec\theta = -\frac{25}{24}$; $\csc\theta = \frac{25}{7}$

[5.3]
8. $\sin A = \frac{3}{5}$; $\cos A = \frac{4}{5}$; $\tan A = \frac{3}{4}$; $\cot A = \frac{4}{3}$; $\sec A = \frac{5}{4}$; $\csc A = \frac{5}{3}$

9.

θ	$\sin\theta$	$\cos\theta$	$\tan\theta$
30°	$\frac{1}{2}$	$\frac{\sqrt{3}}{2}$	$\frac{\sqrt{3}}{3}$
45°	$\frac{\sqrt{2}}{2}$	$\frac{\sqrt{2}}{2}$	1
60°	$\frac{\sqrt{3}}{2}$	$\frac{1}{2}$	$\sqrt{3}$

θ	$\cot\theta$	$\sec\theta$	$\csc\theta$
30°	$\sqrt{3}$	$\frac{2\sqrt{3}}{3}$	2
45°	1	$\sqrt{2}$	$\sqrt{2}$
60°	$\frac{\sqrt{3}}{3}$	2	$\frac{2\sqrt{3}}{3}$

1. For an angle measuring 19°, give the measure of **(a)** its complement and **(b)** its supplement.

Find the measure of each unknown angle.

2.

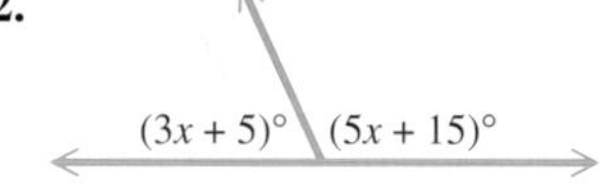

3.

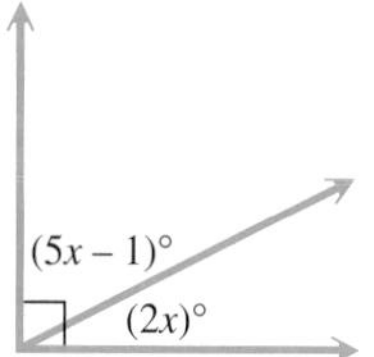

4. Perform each indicated conversion.

(a) 77° 12′ 09″ to decimal degrees **(b)** 22.0250° to degrees, minutes, seconds

5. Find the angle of least positive measure (not equal to the given angle) coterminal with each angle.

(a) 410° **(b)** −60° **(c)** 890° **(d)** 57°

6. *Rotating Flywheel* A flywheel rotates 300 times per min. Through how many degrees does a point on the edge of the flywheel move in 1 sec?

7. The terminal side of an angle θ in standard position passes through the point $(-24, 7)$. Find the values of the six trigonometric functions of θ.

8. Find the exact values of the six trigonometric functions for angle A in the figure.

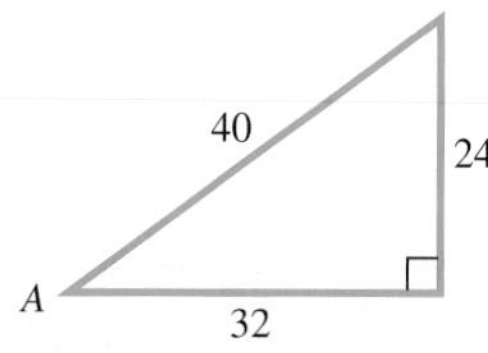

9. Find exact values of the trigonometric functions to complete the table.

θ	$\sin\theta$	$\cos\theta$	$\tan\theta$	$\cot\theta$	$\sec\theta$	$\csc\theta$
30°						
45°						
60°						

10. $w = 18$; $x = 18\sqrt{3}$; $y = 18$; $z = 18\sqrt{2}$

11. $\mathcal{A} = 3x^2 \sin \theta$

In Exercises 12–14, we give, in order, sine, cosine, tangent, cotangent, secant, and cosecant.

12. $\frac{\sqrt{2}}{2}$; $-\frac{\sqrt{2}}{2}$; -1; -1; $-\sqrt{2}$; $\sqrt{2}$

13. $-\frac{1}{2}$; $-\frac{\sqrt{3}}{2}$; $\frac{\sqrt{3}}{3}$; $\sqrt{3}$; $-\frac{2\sqrt{3}}{3}$; -2

14. $-\frac{\sqrt{3}}{2}$; $\frac{1}{2}$; $-\sqrt{3}$; $-\frac{\sqrt{3}}{3}$; 2; $-\frac{2\sqrt{3}}{3}$

15. 60°; 120°

16. 135°; 225°

17. 0.67301251

18. −1.1817633

19. 69.497888°

20. 24.777233°

10. Find the exact value of each variable in the figure.

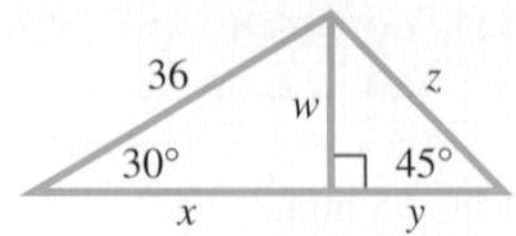

11. *Area of a Solar Cell* A solar cell converts the energy of sunlight directly into electrical energy. The amount of energy a cell produces depends on its area. Suppose a solar cell is hexagonal, as shown in the figure on the left below. Express its area $\mathcal{A}$ in terms of $\sin \theta$ and any side x. (*Hint:* Consider one of the six equilateral triangles from the hexagon. See the figure on the right below.) (*Source:* Kastner, B., *Space Mathematics,* NASA.)

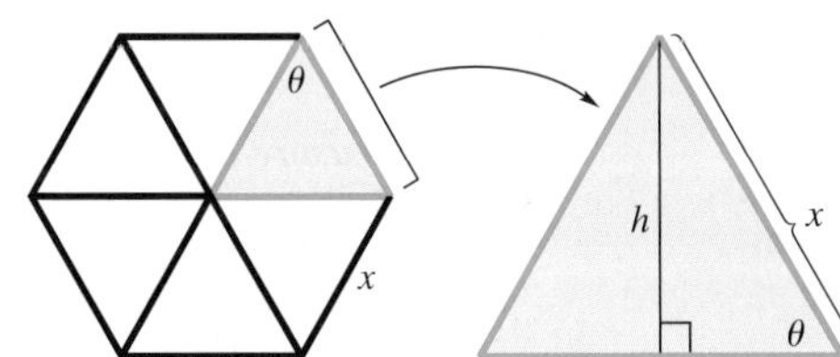

Find exact values of the six trigonometric functions for each angle. Rationalize denominators when applicable.

12. 135° **13.** −150° **14.** 1020°

Find all values of θ in the interval $[0°, 360°)$ *that have the given function value.*

15. $\sin \theta = \frac{\sqrt{3}}{2}$ **16.** $\sec \theta = -\sqrt{2}$

Use a calculator to approximate each value. Give as many digits as your calculator displays.

17. $\sin 42° \, 18'$ **18.** $\sec(-212° \, 12')$

Use a calculator to find the value of θ in the interval $[0°, 90°]$ *that satisfies each statement. Write each answer in decimal degrees to six decimal places as needed.*

19. $\tan \theta = 2.6743210$ **20.** $\csc \theta = 2.3861147$

5.4 Solving Right Triangles

- Significant Digits
- Solving Triangles
- Angles of Elevation or Depression
- Bearing
- Further Applications

Significant Digits A number that represents the result of counting, or a number that results from theoretical work and is not the result of measurement, is an **exact number.** There are 50 states in the United States. In this statement, 50 is an exact number.

Most values obtained for trigonometric applications are measured values that are *not* exact. Suppose we quickly measure a room as 15 ft by 18 ft. See **Figure 41.** We can calculate the length of a diagonal of the room as follows.

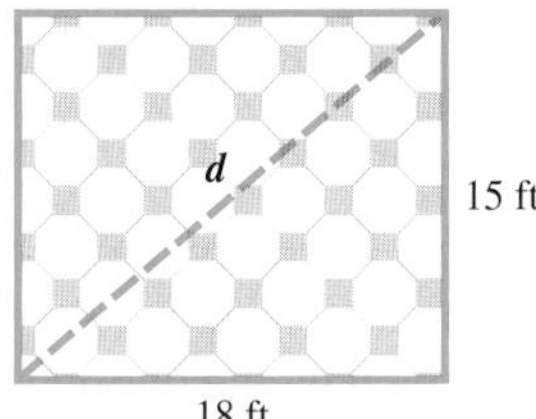

Figure 41

Teaching Tip Make the analogy that this discussion is similar to the idea that a chain is only as strong as its weakest link.

Teaching Tip To help students identify significant digits, write numbers using scientific notation. For example,

$$408 = 4.08 \times 10^2$$
$$6.700 = 6.700 \times 10^0$$
$$0.0025 = 2.5 \times 10^{-3}$$
$$7300 = 7.3 \times 10^3.$$

The number of digits in the first factor of a number written in scientific notation represents the number of significant digits.

$d^2 = 15^2 + 18^2$ Pythagorean theorem (Section 1.5)

$d^2 = 549$ Apply the exponents and add.

$d = \sqrt{549}$ Square root property (Section 1.4); choose the positive root.

$d \approx 23.430749$

Should this answer be given as the length of the diagonal of the room? Of course not. The number 23.430749 contains six decimal places, while the original data of 15 ft and 18 ft are accurate only to the nearest foot. In practice, the results of a calculation can be no more accurate than the least accurate number in the calculation. Thus, we should indicate that the diagonal of the 15-by-18-ft room is approximately 23 ft.

If a wall measured to the nearest foot is 18 ft long, this actually means that the wall has length between 17.5 ft and 18.5 ft. If the wall is measured more accurately as 18.3 ft long, then its length is really between 18.25 ft and 18.35 ft. The results of physical measurement are only approximately accurate and depend on the precision of the measuring instrument as well as the aptness of the observer. The digits obtained by actual measurement are called **significant digits.** The measurement 18 ft is said to have two significant digits; 18.3 ft has three significant digits.

In the following numbers, the significant digits are identified in color.

408 21.5 18.00 6.700 0.0025 0.09810 7300

Notice that 18.00 has four significant digits. The zeros in this number represent measured digits accurate to the nearest hundredth. The number 0.0025 has only two significant digits, 2 and 5, because the zeros here are used only to locate the decimal point. The number 7300 causes some confusion because it is impossible to determine whether the zeros are measured values. The number 7300 may have two, three, or four significant digits. When presented with this situation, we assume that the zeros are not significant, unless the context of the problem indicates otherwise.

To determine the number of significant digits for answers in applications of angle measure, use the following table.

Angle Measure to Nearest	Examples	Answer to Number of Significant Digits
Degree	62°, 36°	two
Ten minutes, or nearest tenth of a degree	52° 30′, 60.4°	three
Minute, or nearest hundredth of a degree	81° 48′, 71.25°	four
Ten seconds, or nearest thousandth of a degree	10° 52′ 20″, 21.264°	five

To perform calculations with measured numbers, start by identifying the number with the least number of significant digits. Round your final answer to the same number of significant digits as this number. ***Remember that your answer is no more accurate than the least accurate number in your calculation.***

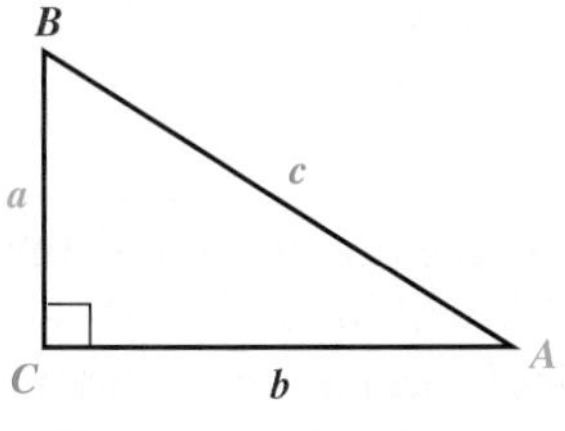

When we are solving triangles, a labeled sketch is an important aid.

Figure 42

Solving Triangles To *solve a triangle* means to find the measures of all the angles and sides of the triangle. As shown in **Figure 42,** we use a to represent the length of the side opposite angle A, b for the length of the side opposite angle B, and so on. In a right triangle, the letter c is reserved for the hypotenuse.

EXAMPLE 1 Solving a Right Triangle Given an Angle and a Side

Solve right triangle ABC, if $A = 34° 30'$ and $c = 12.7$ in. See **Figure 43.**

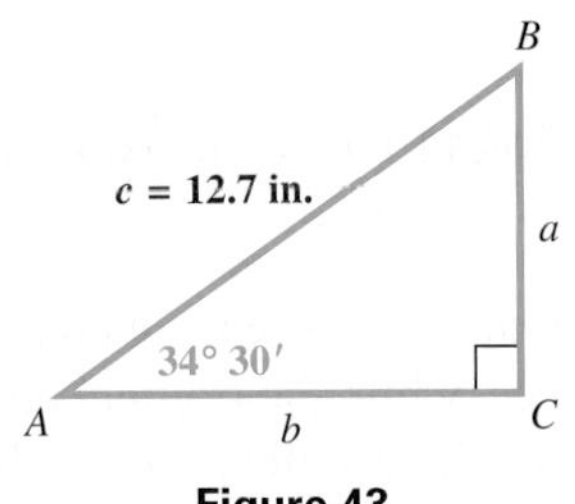

Figure 43

Classroom Example 1
Solve right triangle ABC, if $B = 28° 40'$ and $a = 25.3$ cm.

Answer:
$b = 13.8$ cm; $c = 28.8$ cm; $A = 61° 20'$

SOLUTION To solve the triangle, find the measures of the remaining sides and angles. To find the value of a, use a trigonometric function involving the known values of angle A and side c. Since the sine of angle A is given by the quotient of the side opposite A and the hypotenuse, use sin A.

$$\sin A = \frac{a}{c} \qquad \sin A = \frac{\text{side opposite}}{\text{hypotenuse}} \quad \text{(Section 5.3)}$$

$$\sin 34° 30' = \frac{a}{12.7} \qquad A = 34° 30',\ c = 12.7$$

$$a = 12.7 \sin 34° 30' \qquad \text{Multiply by 12.7 and rewrite.}$$

$$a = 12.7 \sin 34.5° \qquad \text{Convert to decimal degrees. (Section 5.1)}$$

$$a \approx 12.7(0.56640624) \qquad \text{Use a calculator. (Section 5.3)}$$

$$a \approx 7.19 \text{ in.} \qquad \text{Three significant digits}$$

Assuming that $34° 30'$ is given to the nearest ten minutes, we rounded the answer to three significant digits.

To find the value of b, we could substitute the value of a just calculated and the given value of c in the Pythagorean theorem. It is better, however, to use the information given in the problem rather than a result just calculated. If an error is made in finding a, then b also would be incorrect. And, rounding more than once may cause the result to be less accurate. To find b, use cos A.

$$\cos A = \frac{b}{c} \qquad \cos A = \frac{\text{side adjacent}}{\text{hypotenuse}} \quad \text{(Section 5.3)}$$

$$\cos 34° 30' = \frac{b}{12.7} \qquad A = 34° 30',\ c = 12.7$$

$$b = 12.7 \cos 34° 30' \qquad \text{Multiply by 12.7 and rewrite.}$$

$$b \approx 10.5 \text{ in.} \qquad \text{Three significant digits}$$

Teaching Tip When solving triangles, encourage students to refer to a sketch of the problem to see if their answer is reasonable.

Once b is found, the Pythagorean theorem can be used to verify the results.

All that remains to solve triangle ABC is to find the measure of angle B.

$$A + B = 90° \qquad \text{(Section 5.1)}$$

$$B = 90° - A \qquad \text{Solve for } B.$$

$$B = 89° 60' - 34° 30' \qquad \text{Rewrite 90°. Substitute } 34° 30' \text{ for } A.$$

$$B = 55° 30' \qquad \text{Subtract degrees and minutes separately.}$$

✔ *Now Try Exercise 13.*

LOOKING AHEAD TO CALCULUS
The derivatives of the **parametric equations** $x = f(t)$ and $y = g(t)$ often represent the rate of change of physical quantities, such as velocities. When x and y are related by an equation, the derivatives are **related rates** because a change in one causes a related change in the other. Determining these rates in calculus often requires solving a right triangle.

NOTE In **Example 1,** we could have found the measure of angle B first and then used the trigonometric function values of B to find the unknown sides. A right triangle can usually be solved in several ways, each producing the correct answer. ***To maintain accuracy, always use given information as much as possible, and avoid rounding in intermediate steps.***

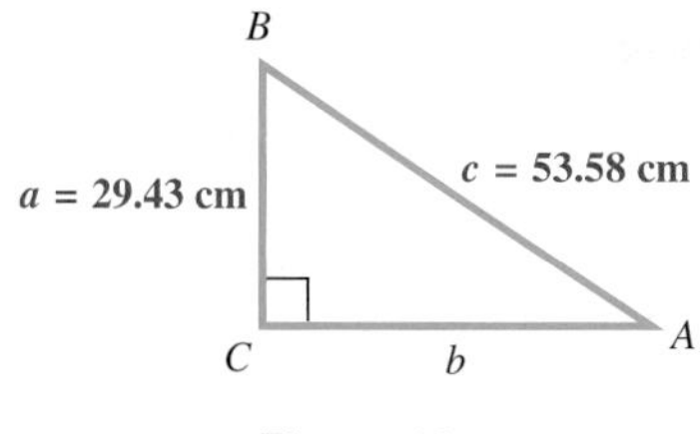

Figure 44

Classroom Example 2
Solve right triangle ABC, if $a = 44.25$ cm and $b = 55.87$ cm.

Answer:
$A = 38° 23'$; $B = 51° 37'$; $c = 71.27$ cm

EXAMPLE 2 Solving a Right Triangle Given Two Sides

Solve right triangle ABC, if $a = 29.43$ cm and $c = 53.58$ cm.

SOLUTION We draw a sketch showing the given information, as in **Figure 44.** One way to begin is to find angle A by using the sine function.

$$\sin A = \frac{a}{c} \qquad \sin A = \frac{\text{side opposite}}{\text{hypotenuse}}$$

$$\sin A = \frac{29.43}{53.58} \qquad a = 29.43,\ c = 53.58$$

$$\sin A \approx 0.5492721165 \qquad \text{Use a calculator.}$$

$$A \approx \sin^{-1}(0.5492721165) \qquad \text{Use the inverse sine function. (Section 5.3)}$$

$$A \approx 33.32°, \quad \text{or} \quad 33° 19' \qquad \text{Four significant digits}$$

The measure of B is approximately

$$90° - 33° 19' = 56° 41'. \qquad 90° = 89° 60' \text{ (Section 5.1)}$$

We now find b from the Pythagorean theorem.

$$b^2 = c^2 - a^2 \qquad \text{Pythagorean theorem solved for } b^2 \text{ (Section 1.5)}$$

$$b^2 = 53.58^2 - 29.43^2 \qquad c = 53.58,\ a = 29.43$$

$$b = \sqrt{2004.6915} \qquad \text{Simplify on the right; square root property}$$

$$b \approx 44.77 \text{ cm}$$

Choose the positive square root.

✔ *Now Try Exercise 23.*

Teaching Tip Remind students to use the appropriate **trigonometric function** key to find a **length** and to use the appropriate **inverse trigonometric function** key to find an **angle.**

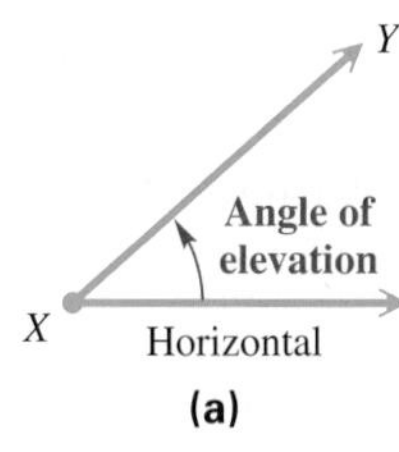

(a)

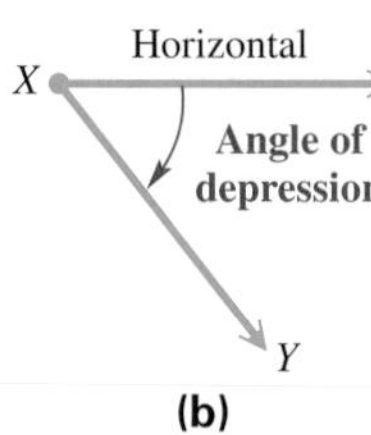

(b)

Figure 45

Angles of Elevation or Depression In applications of right triangles, the **angle of elevation** from point X to point Y (above X) is the acute angle formed by ray XY and a horizontal ray with endpoint at X. See **Figure 45(a).** The **angle of depression** from point X to point Y (below X) is the acute angle formed by ray XY and a horizontal ray with endpoint X. See **Figure 45(b).**

CAUTION Be careful when interpreting the angle of depression. ***Both the angle of elevation and the angle of depression are measured between the line of sight and a horizontal line.***

To solve applied trigonometry problems, follow the same procedure as solving a triangle. ***Drawing a sketch and labeling it correctly in Step 1 is crucial.***

Solving an Applied Trigonometry Problem

Step 1 Draw a sketch, and label it with the given information. Label the quantity to be found with a variable.

Step 2 Use the sketch to write an equation relating the given quantities to the variable.

Step 3 Solve the equation, and check that your answer makes sense.

Classroom Example 3
The angle of elevation from a point on the ground 15.5 m from the base of a tree to the top of the tree is 60.4°. Find the height of the tree.

Answer: 27.3 m

EXAMPLE 3 Finding a Length Given the Angle of Elevation

Pat Porterfield knows that when she stands 123 ft from the base of a flagpole, the angle of elevation to the top of the flagpole is $26° 40'$. If her eyes are 5.30 ft above the ground, find the height of the flagpole.

SOLUTION

Step 1 The length of the side adjacent to Pat is known, and the length of the side opposite her must be found. See **Figure 46.**

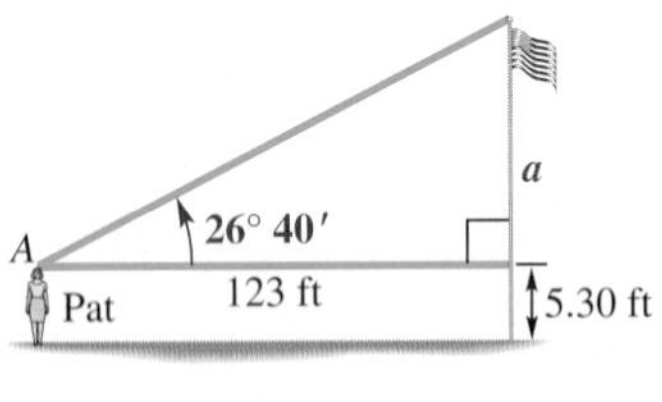

Figure 46

Step 2 The tangent ratio involves the given values. Write an equation.

$$\tan A = \frac{\text{side opposite}}{\text{side adjacent}} \qquad \text{Tangent ratio (Section 5.3)}$$

$$\tan 26° 40' = \frac{a}{123} \qquad A = 26° 40',\ \text{side adjacent} = 123$$

Step 3

$$a = 123 \tan 26° 40' \qquad \text{Multiply by 123 and rewrite.}$$

$$a \approx 123(0.50221888) \qquad \text{Use a calculator.}$$

$$a \approx 61.8 \text{ ft} \qquad \text{Three significant digits}$$

The height of the flagpole is

$$61.8 + 5.30 = 67.1 \text{ ft.} \qquad \text{Pat's eyes are 5.30 ft above the ground.}$$

✔ *Now Try Exercise 41.*

EXAMPLE 4 Finding an Angle of Depression

Classroom Example 4
Find the angle of depression from the top of a 97-ft building to the base of a building across the street located 42 ft away.

Answer: 67°

From the top of a 210-ft cliff, David observes a lighthouse that is 430 ft offshore. Find the angle of depression from the top of the cliff to the base of the lighthouse.

SOLUTION As shown in **Figure 47,** the angle of depression is measured from a horizontal line down to point B. The angle of depression and angle B, in the right triangle shown, are alternate interior angles whose measures are equal.

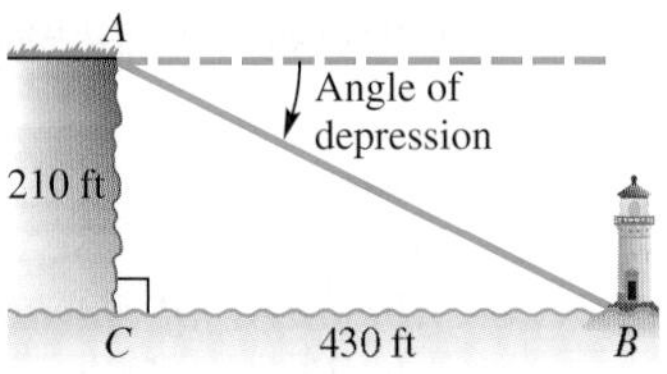

Figure 47

We use the tangent ratio to solve for angle B.

$$\tan B = \frac{210}{430}, \quad \text{so} \quad B = \tan^{-1}\frac{210}{430} \approx 26° \qquad \text{Angle of depression}$$

✔ *Now Try Exercise 43.*

Bearing Other applications of right triangles involve **bearing,** an important concept in navigation. There are two methods for expressing bearing.

Method 1 When a single angle is given, such as 164°, it is understood that the bearing is measured in a clockwise direction from due north.

Several sample bearings using Method 1 are shown in **Figure 48.**

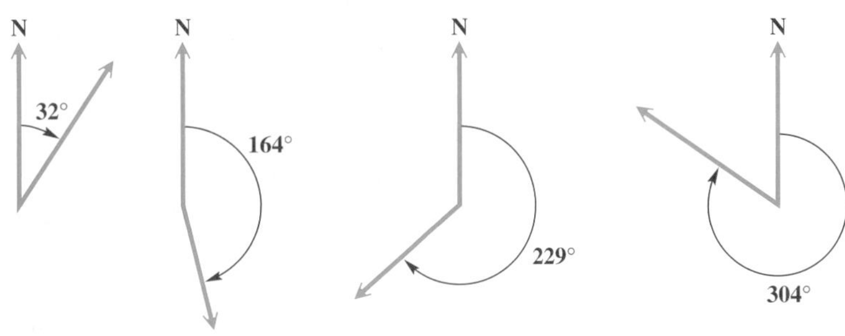

Figure 48

Classroom Example 5
Radar stations A and B are on an east-west line, 8.6 km apart. Station A detects a plane at C, on a bearing of 53°. Station B simultaneously detects the same plane, on a bearing of 323°. Find the distance from B to C.

Answer: 5.2 km

EXAMPLE 5 Solving a Problem Involving Bearing (Method 1)

Radar stations A and B are on an east-west line, 3.7 km apart. Station A detects a plane at C, on a bearing of 61°. Station B simultaneously detects the same plane, on a bearing of 331°. Find the distance from A to C.

SOLUTION Draw a sketch showing the given information, as in **Figure 49.** Since a line drawn due north is perpendicular to an east-west line, right angles are formed at A and B, so angles CAB and CBA can be found as shown in **Figure 49.** Angle C is a right angle because angles CAB and CBA are complementary.

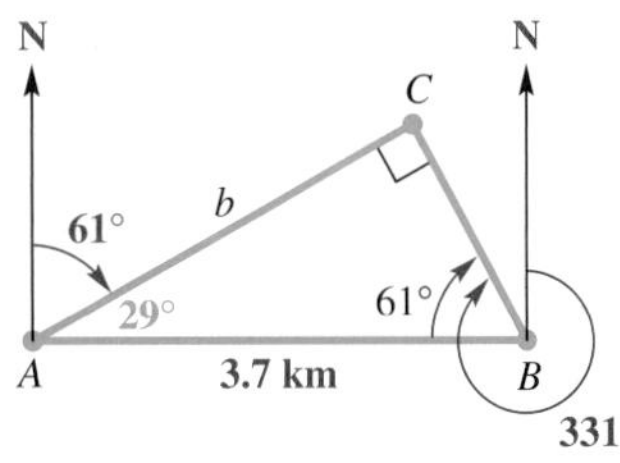

Figure 49

Find distance b by using the cosine function for angle A.

$$\cos 29° = \frac{b}{3.7} \quad \text{Cosine ratio (Section 5.1)}$$

$$3.7 \cos 29° = b \quad \text{Multiply by 3.7.}$$

$$b \approx 3.2 \text{ km} \quad \text{Use a calculator and round to the nearest tenth.}$$

The distance from A to C is approximately 3.2 km.

✔ *Now Try Exercise 61.*

Teaching Tip Emphasize the need to draw accurate sketches describing each application problem. Encourage students to use their sketches to check that their answers are reasonable.

CAUTION ***A correctly labeled sketch is crucial*** when solving applications like that in **Example 5.** Some of the necessary information is often not directly stated in the problem and can be determined only from the sketch.

Method 2 The second method for expressing bearing starts with a north-south line and uses an acute angle to show the direction, either east or west, from this line.

Figure 50 shows several sample bearings using this method. Either N or S always comes first, followed by an acute angle, and then E or W.

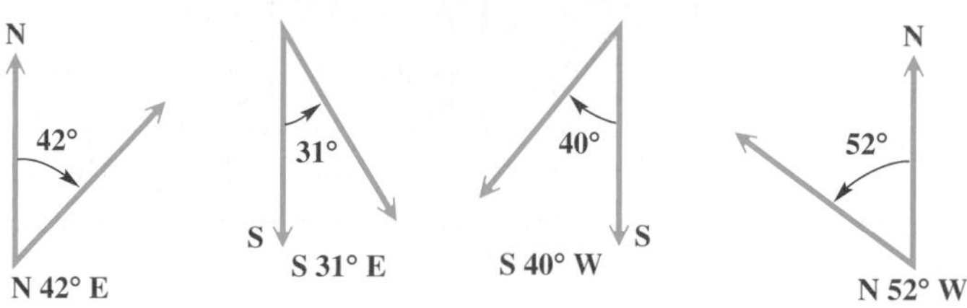

Figure 50

Classroom Example 6
A ship leaves port and sails on a bearing of S 34° W for 2.5 hr. It then turns and sails on a bearing of N 56° W for 3.0 hr. If the ship's rate of speed is 18 knots (nautical miles per hour), find the distance that the ship is from port.

Answer: 70 nautical mi

EXAMPLE 6 Solving a Problem Involving Bearing (Method 2)

A ship leaves port and sails on a bearing of N 47° E for 3.5 hr. It then turns and sails on a bearing of S 43° E for 4.0 hr. If the ship's rate of speed is 22 knots (nautical miles per hour), find the distance that the ship is from port.

SOLUTION Draw a sketch as in **Figure 51.** Choose a point C on a bearing of N 47° E from port at point A. Then choose a point B on a bearing of S 43° E from point C. North-south lines are parallel. Therefore, angle ACD measures 47° because it is an alternate interior angle to the angle at A measured clockwise from due north. The measure of angle ACB is

$$47^\circ + 43^\circ = 90^\circ,$$

making triangle ABC a right triangle.

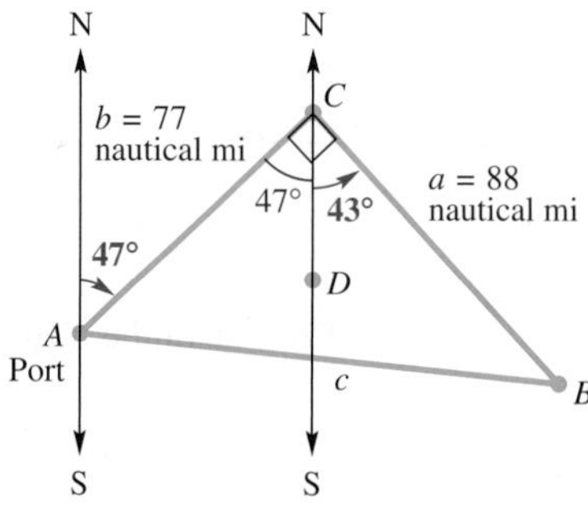

Figure 51

Next, use the formula relating distance, rate, and time to find the distances from A to C and from C to B.

$$b = 22 \times 3.5 = 77 \text{ nautical mi}$$
$$a = 22 \times 4.0 = 88 \text{ nautical mi}$$

distance = rate × time (Section 1.2)

Now find c, the distance from port at point A to the ship at point B.

$$a^2 + b^2 = c^2 \quad \text{Pythagorean theorem}$$
$$88^2 + 77^2 = c^2 \quad a = 88,\ b = 77$$
$$c = \sqrt{88^2 + 77^2} \quad \text{Use the square root property.}$$
$$c \approx 120 \text{ nautical mi} \quad \text{Two significant digits}$$

The ship is about 120 nautical mi from port.

✔ *Now Try Exercise 67.*

Further Applications

Classroom Example 7
Repeat **Example 7** for $\theta = 2^\circ 41' 38''$ and $b = 3.5000$ cm.

Answers:
(a) 74.427 cm
(b) 0.0076765 cm

EXAMPLE 7 Using Trigonometry to Measure a Distance

The **subtense bar method** is a method that surveyors use to determine a small distance d between two points P and Q. The subtense bar with length b is centered at Q and situated perpendicular to the line of sight between P and Q. See **Figure 52.** Angle θ is measured, and then the distance d can be determined.

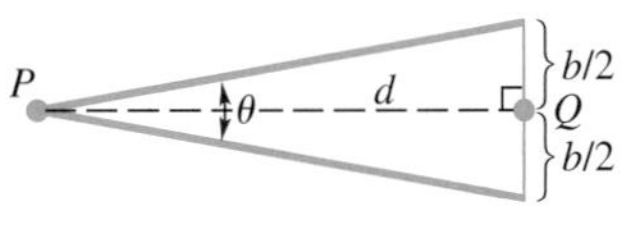

Figure 52

(a) Find d when $\theta = 1^\circ 23' 12''$ and $b = 2.0000$ cm.

(b) How much change would there be in the value of d if θ measured $1''$ larger?

SOLUTION

(a) From **Figure 52,** we obtain the following.

$$\cot \frac{\theta}{2} = \frac{d}{\frac{b}{2}} \qquad \text{Cotangent ratio}$$

$$d = \frac{b}{2} \cot \frac{\theta}{2} \qquad \text{Multiply and rewrite.}$$

Let $b = 2$. To evaluate $\frac{\theta}{2}$, we change θ to decimal degrees.

$$1^\circ 23' 12'' \approx 1.386666667^\circ$$

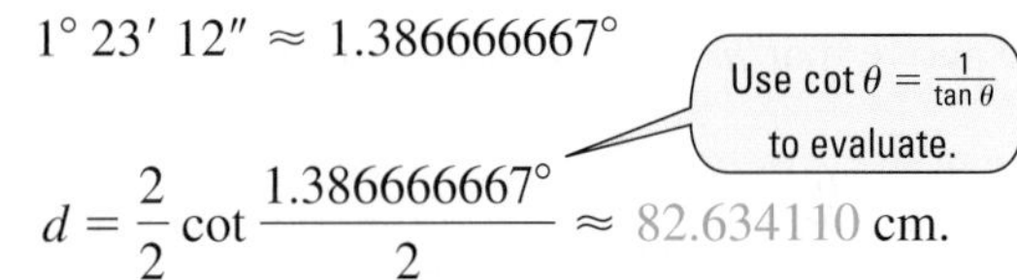

Then $$d = \frac{2}{2} \cot \frac{1.386666667^\circ}{2} \approx 82.634110 \text{ cm.}$$

(b) Since θ is $1''$ larger, use $\theta = 1^\circ 23' 13'' \approx 1.386944444^\circ$.

$$d = \frac{2}{2} \cot \frac{1.386944444^\circ}{2} \approx 82.617558 \text{ cm}$$

The difference is $82.634110 - 82.617558 = 0.016552$ cm.

Now Try Exercise 75.

Classroom Example 8
Marla needs to find the height of a building. From a given point on the ground, she finds that the angle of elevation to the top of the building is 74.2°. She then walks back 35 ft. From the second point, the angle of elevation to the top of the building is 51.8°. Find the height of the building.

Answer: approximately 69 ft

EXAMPLE 8 Solving a Problem Involving Angles of Elevation

Francisco needs to know the height of a tree. From a given point on the ground, he finds that the angle of elevation to the top of the tree is 36.7°. He then moves back 50 ft. From the second point, the angle of elevation to the top of the tree is 22.2°. See **Figure 53.** Find the height of the tree to the nearest foot.

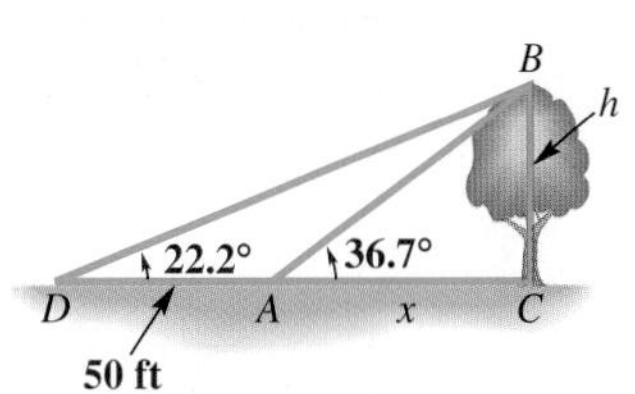

Figure 53

ALGEBRAIC SOLUTION

Figure 53 on the preceding page shows two unknowns: x, the distance from the center of the trunk of the tree to the point where the first observation was made, and h, the height of the tree. See **Figure 54** in the Graphing Calculator Solution. Since nothing is given about the length of the hypotenuse of either triangle ABC or triangle BCD, use a ratio that does not involve the hypotenuse—namely, the tangent.

In triangle ABC, $\tan 36.7° = \dfrac{h}{x}$ or $h = x \tan 36.7°$.

In triangle BCD, $\tan 22.2° = \dfrac{h}{50 + x}$ or $h = (50 + x) \tan 22.2°$.

Each expression equals h, so the expressions must be equal.

$$x \tan 36.7° = (50 + x) \tan 22.2°$$

Equate expressions for h.

$$x \tan 36.7° = 50 \tan 22.2° + x \tan 22.2°$$

Distributive property (Section R.2)

$$x \tan 36.7° - x \tan 22.2° = 50 \tan 22.2°$$

Write the x-terms on one side. (Section 1.1)

$$x(\tan 36.7° - \tan 22.2°) = 50 \tan 22.2°$$

Factor out x. (Section R.4)

$$x = \frac{50 \tan 22.2°}{\tan 36.7° - \tan 22.2°}$$

Divide by the coefficient of x.

We saw above that $h = x \tan 36.7°$. Substitute for x.

$$h = \left(\frac{50 \tan 22.2°}{\tan 36.7° - \tan 22.2°}\right) \tan 36.7°$$

Use a calculator.

$$\tan 36.7° = 0.74537703 \quad \text{and} \quad \tan 22.2° = 0.40809244$$

Thus,

$$\tan 36.7° - \tan 22.2° = 0.74537703 - 0.40809244 = 0.33728459$$

and

$$h = \left(\frac{50(0.40809244)}{0.33728459}\right) 0.74537703 \approx 45.$$

The height of the tree is approximately 45 ft.

GRAPHING CALCULATOR SOLUTION*

In **Figure 54,** we have superimposed **Figure 53** on coordinate axes with the origin at D. By definition, the tangent of the angle between the x-axis and the graph of a line with equation $y = mx + b$ is the slope of the line, m. For line DB, $m = \tan 22.2°$. Since b equals 0, the equation of line DB is

$$y_1 = (\tan 22.2°)x.$$

The equation of line AB is

$$y_2 = (\tan 36.7°)x + b.$$

Since $b \neq 0$ here, we use the point $A(50, 0)$ and the point-slope form to find the equation.

$$y_2 - y_1 = m(x - x_1) \quad \text{(Section 2.5)}$$

$$y_2 - 0 = m(x - 50) \quad x_1 = 50,\ y_1 = 0$$

$$y_2 = \tan 36.7°(x - 50)$$

Lines y_1 and y_2 are graphed in **Figure 55.** The y-coordinate of the point of intersection of the graphs gives the length of BC, or h. Thus, $h \approx 45$.

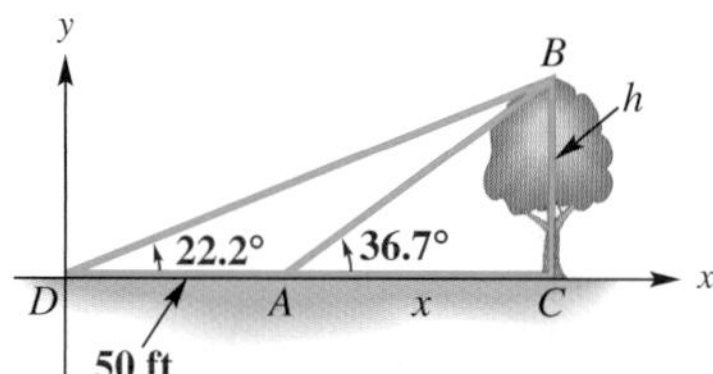

Figure 54

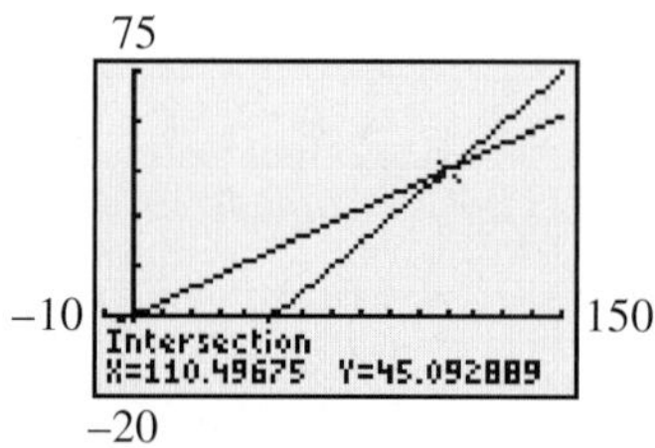

Figure 55

✔ *Now Try Exercise 69.*

NOTE In practice, we usually do not write down intermediate calculator approximation steps. We did in **Example 8** so that you could follow the steps more easily.

---

*_Source:_ Adapted from "Letter to the Editor," by Robert Ruzich (_Mathematics Teacher,_ Volume 88, Number 1). Copyright © 1995 by the National Council of Teachers of Mathematics.

5.4 Exercises

1. 22,894.5 to 22,895.5
2. 28,999.5 to 29,000.5
3. 8958.5 to 8959.5

Note to instructor: While most of the measures resulting from solving triangles in this chapter are approximations, for convenience we use = rather than ≈ in the answers.

5. $B = 53° 40'$; $a = 571$ m; $b = 777$ m
6. $Y = 42.2°$; $x = 66.4$ cm; $y = 60.2$ cm
7. $M = 38.8°$; $n = 154$ m; $p = 198$ m
8. $B = 58° 20'$; $c = 68.4$ km; $b = 58.2$ km
9. $A = 47.9108°$; $c = 84.816$ cm; $a = 62.942$ cm
10. $A = 21.4858°$; $b = 3330.68$ m; $a = 1311.04$ m
11. $A = 37° 40'$; $B = 52° 20'$; $c = 20.5$ ft
12. $A = 18° 20'$; $B = 71° 40'$; $b = 14.5$ m

Concept Check Refer to the discussion of accuracy and significant digits in this section to work Exercises 1–4.

1. *Leading NFL Receiver* As of the end of the 2009 National Football League season, Jerry Rice was the leading career receiver with 22,895 yd. State the range represented by this number. (*Source:* www.nfl.com)

2. *Height of Mt. Everest* When Mt. Everest was first surveyed, the surveyors obtained a height of 29,000 ft to the nearest foot. State the range represented by this number. (The surveyors thought no one would believe a measurement of 29,000 ft, so they reported it as 29,002.) (*Source:* Dunham, W., *The Mathematical Universe,* John Wiley and Sons.)

3. *Longest Vehicular Tunnel* The E. Johnson Memorial Tunnel in Colorado, which measures 8959 ft, is one of the longest land vehicular tunnels in the United States. What is the range of this number? (*Source: World Almanac and Book of Facts.*)

4. *Top WNBA Scorer* Women's National Basketball Association player Cappie Pondexter of the New York Liberty received the 2010 award for most points scored, 729. Is it appropriate to consider this number as between 728.5 and 729.5? Why or why not? (*Source:* www.wnba.com)

Solve each right triangle. When two sides are given, give angles in degrees and minutes. ***See Examples 1 and 2.***

5.
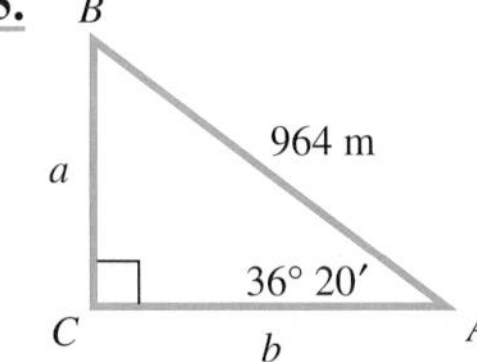

6.
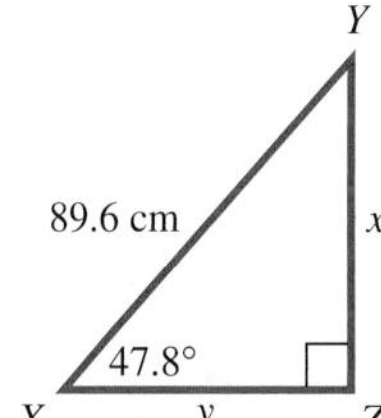

7.
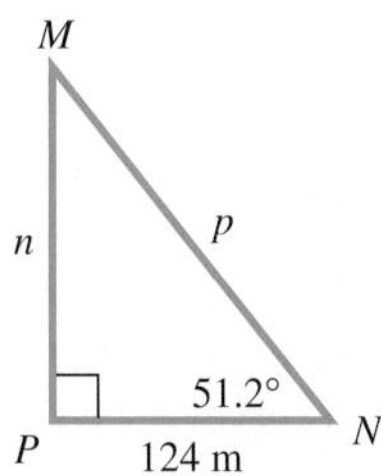

8.
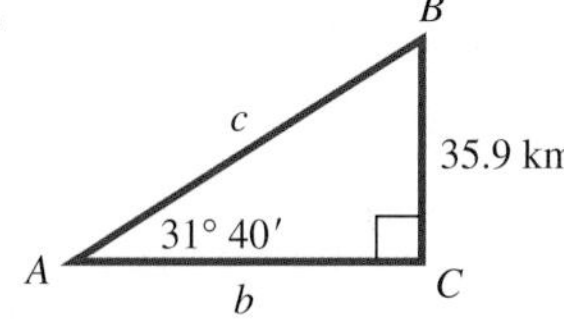

9.
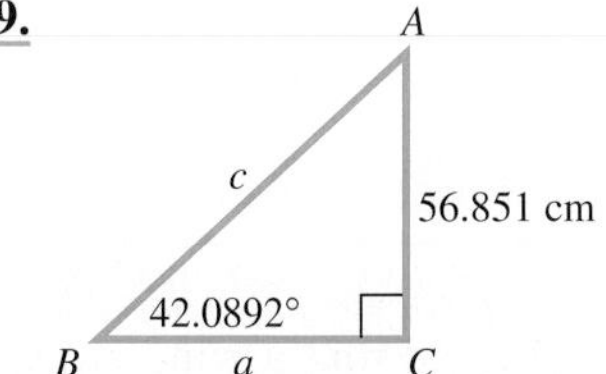

10.
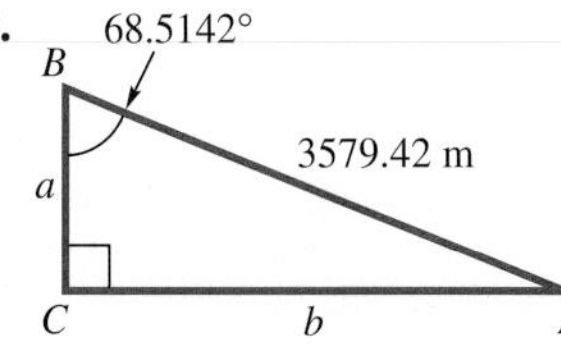

11.
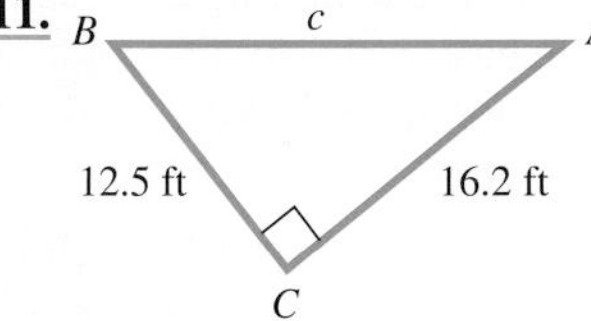

12.
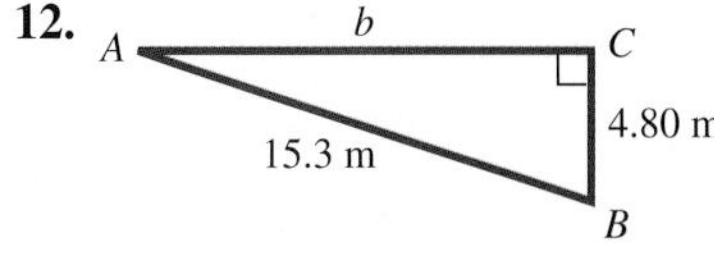

13. $B = 62.0°$; $a = 8.17$ ft; $b = 15.4$ ft
14. $A = 44.0°$; $a = 20.6$ m; $b = 21.4$ m
15. $A = 17.0°$; $a = 39.1$ in.; $c = 134$ in.
16. $B = 27.5°$; $b = 6.61$ m; $c = 14.3$ m
17. $B = 29.0°$; $a = 70.7$ cm; $c = 80.9$ cm
18. $A = 38.3°$; $b = 35.6$ ft; $c = 45.3$ ft
19. $A = 36°$; $B = 54°$; $b = 18$ m
20. $A = 51°$; $B = 39°$; $a = 40$ ft
21. $c = 85.9$ yd; $A = 62° 50'$; $B = 27° 10'$
22. $c = 1080$ m; $A = 63° 00'$; $B = 27° 00'$
23. $b = 42.3$ cm; $A = 24° 10'$; $B = 65° 50'$
24. $a = 609$ m; $A = 70° 10'$; $B = 19° 50'$
25. $B = 36° 36'$; $a = 310.8$ ft; $b = 230.8$ ft
26. $B = 76° 13'$; $a = 306.2$ m; $b = 1248$ m
27. $A = 50° 51'$; $a = 0.4832$ m; $b = 0.3934$ m
28. $A = 7° 09'$; $b = 4.787$ cm; $a = 0.6006$ cm

31. 9.35 m 32. 33.4 m
33. 128 ft

Solve each right triangle. In each case, $C = 90°$. If angle information is given in degrees and minutes, give answers in the same way. If angle information is given in decimal degrees, do likewise in answers. When two sides are given, give angles in degrees and minutes. ***See Examples 1 and 2.***

13. $A = 28.0°$, $c = 17.4$ ft

14. $B = 46.0°$, $c = 29.7$ m

15. $B = 73.0°$, $b = 128$ in.

16. $A = 62.5°$, $a = 12.7$ m

17. $A = 61.0°$, $b = 39.2$ cm

18. $B = 51.7°$, $a = 28.1$ ft

19. $a = 13$ m, $c = 22$ m

20. $b = 32$ ft, $c = 51$ ft

21. $a = 76.4$ yd, $b = 39.3$ yd

22. $a = 958$ m, $b = 489$ m

23. $a = 18.9$ cm, $c = 46.3$ cm

24. $b = 219$ m, $c = 647$ m

25. $A = 53° 24'$, $c = 387.1$ ft

26. $A = 13° 47'$, $c = 1285$ m

27. $B = 39° 09'$, $c = 0.6231$ m

28. $B = 82° 51'$, $c = 4.825$ cm

29. Explain why the angle of depression *DAB* has the same measure as the angle of elevation *ABC* in the figure.

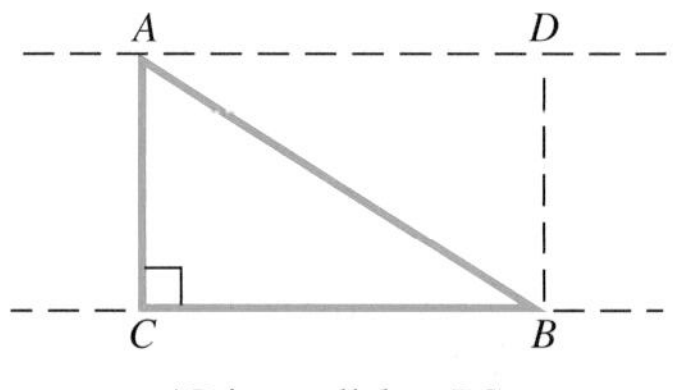

AD is parallel to *BC*.

30. Why is angle *CAB* *not* an angle of depression in the figure for **Exercise 29?**

Solve each problem involving triangles. ***See Examples 1–4.***

31. *Height of a Ladder on a Wall* A 13.5-m fire truck ladder is leaning against a wall. Find the distance *d* the ladder goes up the wall (above the top of the fire truck) if the ladder makes an angle of $43° 50'$ with the horizontal.

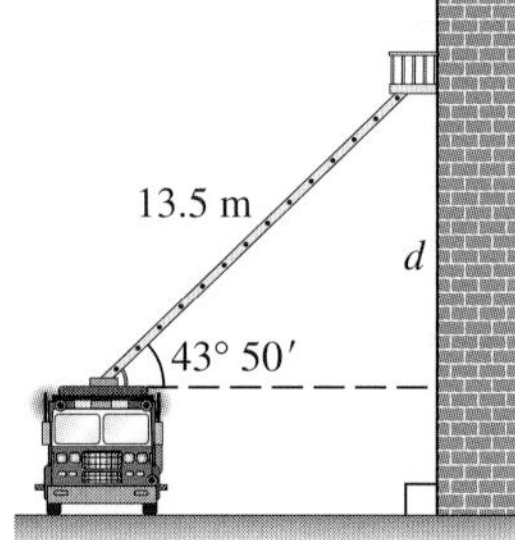

32. *Distance across a Lake* To find the distance *RS* across a lake, a surveyor lays off length $RT = 53.1$ m, so that angle $T = 32° 10'$ and angle $S = 57° 50'$. Find length *RS*.

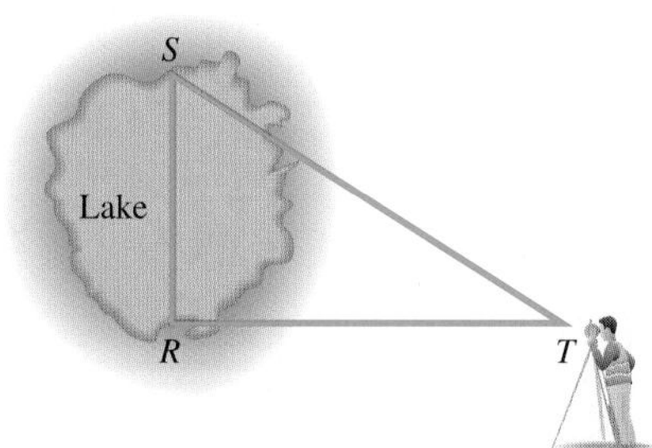

33. *Height of a Building* From a window 30.0 ft above the street, the angle of elevation to the top of the building across the street is 50.0° and the angle of depression to the base of this building is 20.0°. Find the height of the building across the street.

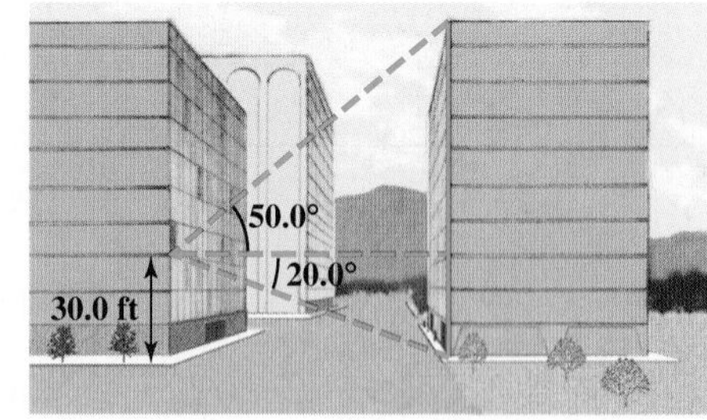

34. 864,900 mi
35. 26.92 in.
36. 134.7 cm
37. 22°
38. 583 ft
39. 28.0 m
40. 469 m

34. *Diameter of the Sun* To determine the diameter of the sun, an astronomer might sight with a **transit** (a device used by surveyors for measuring angles) first to one edge of the sun and then to the other, estimating that the included angle equals 32′. Assuming that the distance d from Earth to the sun is 92,919,800 mi, approximate the diameter of the sun.

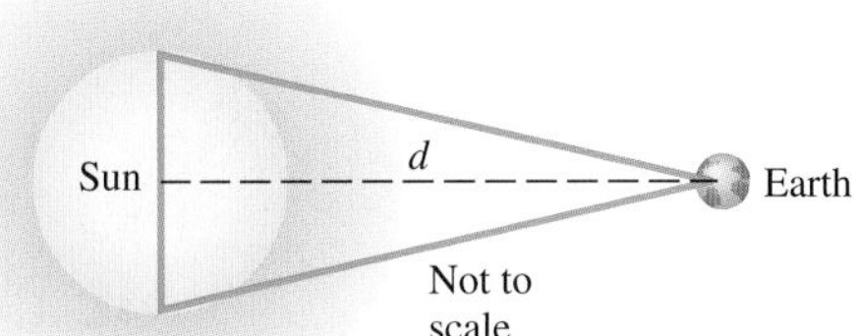

35. *Side Lengths of a Triangle* The length of the base of an isosceles triangle is 42.36 in. Each base angle is 38.12°. Find the length of each of the two equal sides of the triangle. (*Hint:* Divide the triangle into two right triangles.)

36. *Altitude of a Triangle* Find the altitude of an isosceles triangle having base 184.2 cm if the angle opposite the base is 68° 44′.

Solve each problem involving an angle of elevation or depression. ***See Examples 3 and 4.***

37. *Angle of Elevation of the Pyramid of the Sun* The Pyramid of the Sun in the ancient Mexican city of Teotihuacan was the largest and most important structure in the city. The base is a square with sides about 700 ft long, and the height of the pyramid is about 200 ft. Find the angle of elevation of the edge indicated in the figure to two significant digits. (*Hint:* The base of the triangle in the figure is half the diagonal of the square base of the pyramid.) (*Source*: www.britannica.com)

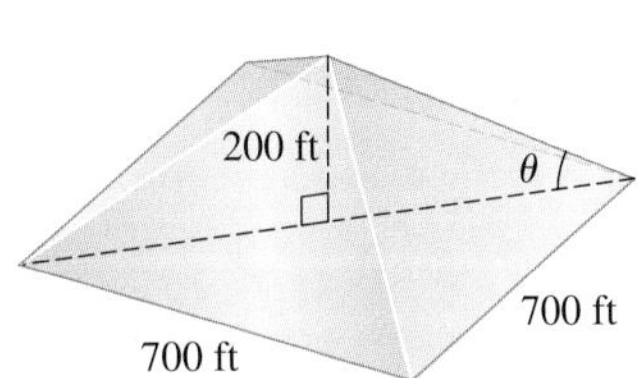

38. *Cloud Ceiling* The U.S. Weather Bureau defines a **cloud ceiling** as the altitude of the lowest clouds that cover more than half the sky. To determine a cloud ceiling, a powerful searchlight projects a circle of light vertically on the bottom of the cloud. An observer sights the circle of light in the crosshairs of a tube called a **clinometer.** A pendant hanging vertically from the tube and resting on a protractor gives the angle of elevation. Find the cloud ceiling if the searchlight is located 1000 ft from the observer and the angle of elevation is 30.0° as measured with a clinometer at eye-height 6 ft. (Assume three significant digits.)

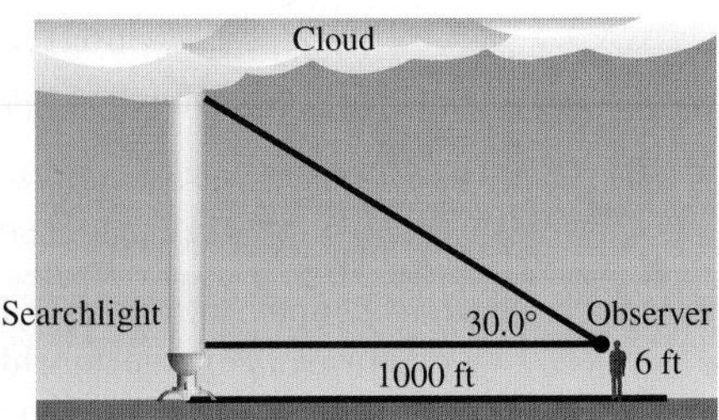

39. *Height of a Tower* The shadow of a vertical tower is 40.6 m long when the angle of elevation of the sun is 34.6°. Find the height of the tower.

40. *Distance from the Ground to the Top of a Building* The angle of depression from the top of a building to a point on the ground is 32° 30′. How far is the point on the ground from the top of the building if the building is 252 m high?

41. 13.3 ft 42. 42,600 ft
43. 37° 35′ 44. 146 m
45. 42.18° 46. 63.39°

41. *Length of a Shadow* Suppose that the angle of elevation of the sun is 23.4°. Find the length of the shadow cast by Dot Peterson, who is 5.75 ft tall.

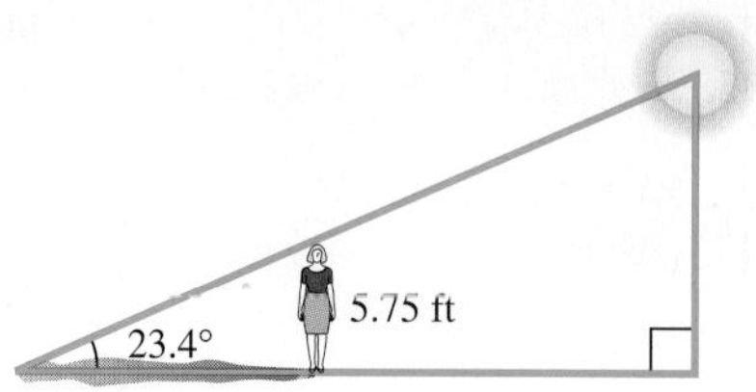

42. *Airplane Distance* An airplane is flying 10,500 ft above level ground. The angle of depression from the plane to the base of a tree is 13° 50′. How far horizontally must the plane fly to be directly over the tree?

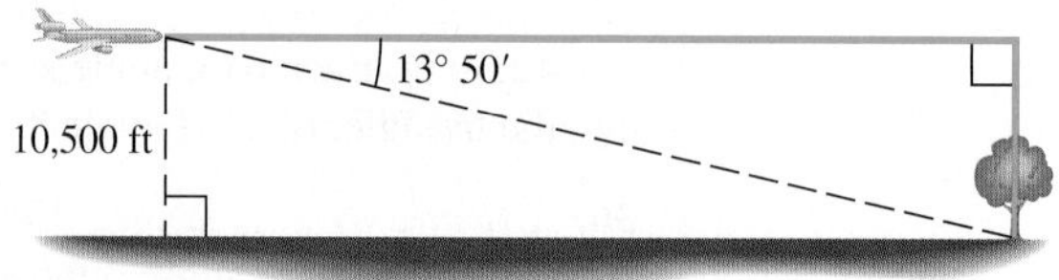

43. *Angle of Depression of a Light* A company safety committee has recommended that a floodlight be mounted in a parking lot so as to illuminate the employee exit. Find the angle of depression of the light to the nearest minute.

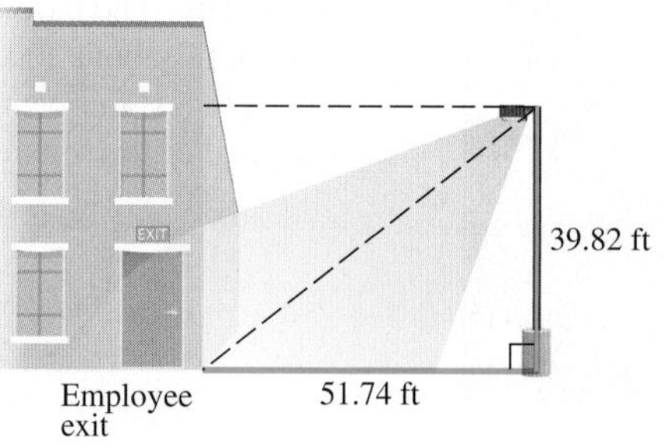

44. *Height of a Building* The angle of elevation from the top of a small building to the top of a nearby taller building is 46° 40′, and the angle of depression to the bottom is 14° 10′. If the shorter building is 28.0 m high, find the height of the taller building.

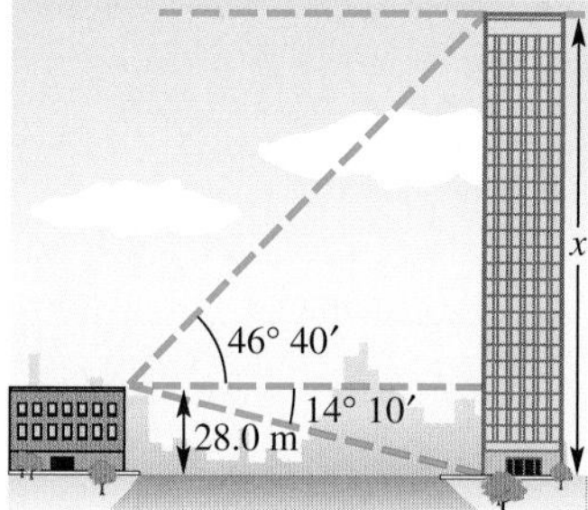

45. *Angle of Elevation of the Sun* The length of the shadow of a building 34.09 m tall is 37.62 m. Find the angle of elevation of the sun to the nearest hundredth of a degree.

46. *Angle of Elevation of the Sun* The length of the shadow of a flagpole 55.20 ft tall is 27.65 ft. Find the angle of elevation of the sun to the nearest hundredth of a degree.

47. *Height of Mt. Everest* The highest mountain peak in the world is Mt. Everest, located in the Himalayas. The height of this enormous mountain was determined in 1856 by surveyors using trigonometry long before it was first climbed in 1953. This difficult measurement had to be done from a great distance. At an altitude of 14,545 ft on a different mountain, the straight-line distance to the peak of Mt. Everest is 27.0134 mi and its angle of elevation is $\theta = 5.82°$. (*Source:* Dunham, W., *The Mathematical Universe,* John Wiley and Sons.)

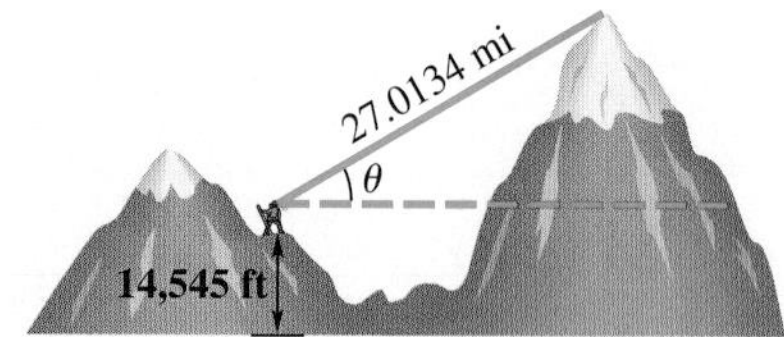

47. (a) 29,000 ft (b) shorter
48. 34.0 mi
49. 270°; N 90° W, or S 90° W
50. 90°; N 90° E, or S 90° E
51. 0°; N 0° E, or N 0° W
52. 180°; S 0° E, or S 0° W
53. 315°; N 45° W
54. 225°; S 45° W
55. 135°; S 45° E
56. 45°; N 45° E
57. 220 mi
58. 150 km
59. 47 nautical mi
60. 70 nautical mi
61. 2203 ft
62. 5856 m

(a) Approximate the height (in feet) of Mt. Everest.

(b) In the actual measurement, Mt. Everest was over 100 mi away and the curvature of Earth had to be taken into account. Would the curvature of Earth make the peak appear taller or shorter than it actually is?

48. *Error in Measurement* A degree may seem like a very small unit, but an error of one degree in measuring an angle may be very significant. For example, suppose a laser beam directed toward the visible center of the moon misses its assigned target by 30 sec. How far is it (in miles) from its assigned target? Take the distance from the surface of Earth to that of the moon to be 234,000 mi. (*Source: A Sourcebook of Applications of School Mathematics* by Donald Bushaw et al.)

Concept Check An observer for a radar station is located at the origin of a coordinate system. For each of the points in Exercises 49–56, find the bearing of an airplane located at that point. Express the bearing using both methods.

49. $(-4, 0)$ **50.** $(5, 0)$ **51.** $(0, 4)$ **52.** $(0, -2)$

53. $(-5, 5)$ **54.** $(-3, -3)$ **55.** $(2, -2)$ **56.** $(2, 2)$

Work each problem. In these exercises, assume the course of a plane or ship is on the indicated bearing. ***See Examples 5 and 6.***

57. *Distance Flown by a Plane* A plane flies 1.3 hr at 110 mph on a bearing of 38°. It then turns and flies 1.5 hr at the same speed on a bearing of 128°. How far is the plane from its starting point?

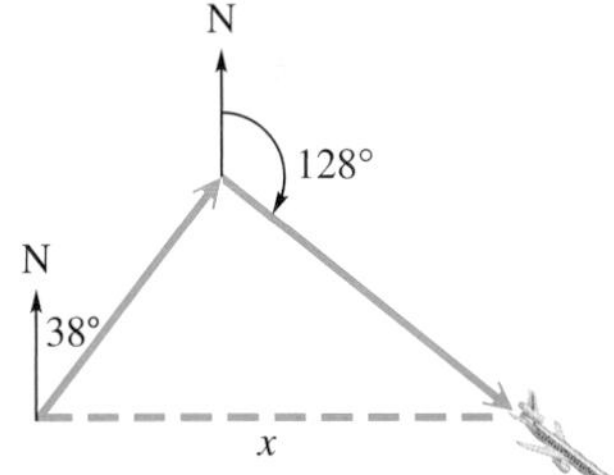

58. *Distance Traveled by a Ship* A ship travels 55 km on a bearing of 27° and then travels on a bearing of 117° for 140 km. Find the distance from the starting point to the ending point.

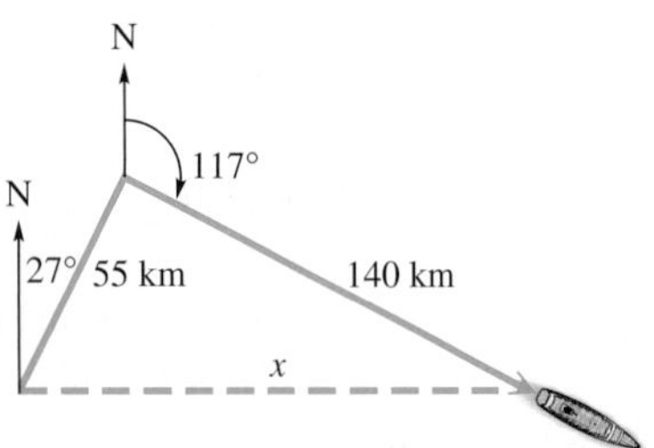

59. *Distance between Two Ships* Two ships leave a port at the same time. The first ship sails on a bearing of 40° at 18 knots (nautical miles per hour) and the second on a bearing of 130° at 26 knots. How far apart are they after 1.5 hr?

60. *Distance between Two Ships* Two ships leave a port at the same time. The first ship sails on a bearing of 52° at 17 knots and the second on a bearing of 322° at 22 knots. How far apart are they after 2.5 hr?

61. *Distance between Two Docks* Two docks are located on an east-west line 2587 ft apart. From dock A, the bearing of a coral reef is 58° 22′. From dock B, the bearing of the coral reef is 328° 22′. Find the distance from dock A to the coral reef.

62. *Distance between Two Lighthouses* Two lighthouses are located on a north-south line. From lighthouse A, the bearing of a ship 3742 m away is 129° 43′. From lighthouse B, the bearing of the ship is 39° 43′. Find the distance between the lighthouses.

63. 148 mi 64. 2.01 mi
65. 430 mi 66. 350 mi
67. 140 mi 68. 130 mi

69. 433 ft 70. 448 m

63. *Distance between Two Ships* A ship leaves its home port and sails on a bearing of S 61° 50′ E. Another ship leaves the same port at the same time and sails on a bearing of N 28° 10′ E. If the first ship sails at 24.0 mph and the second sails at 28.0 mph, find the distance between the two ships after 4 hr.

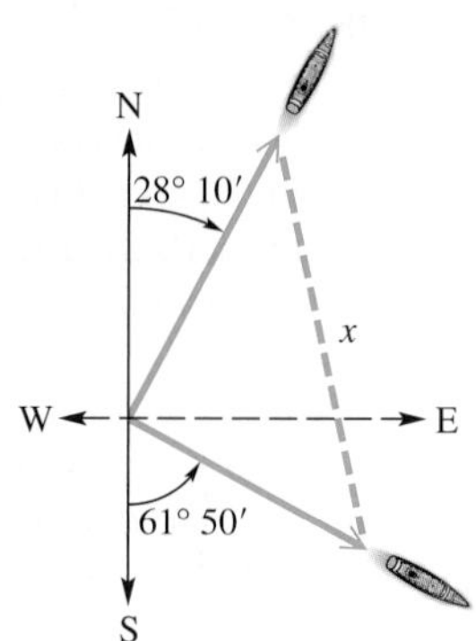

64. *Distance between Transmitters* Radio direction finders are set up at two points A and B, which are 2.50 mi apart on an east-west line. From A, it is found that the bearing of a signal from a radio transmitter is N 36° 20′ E, and from B the bearing of the same signal is N 53° 40′ W. Find the distance of the transmitter from B.

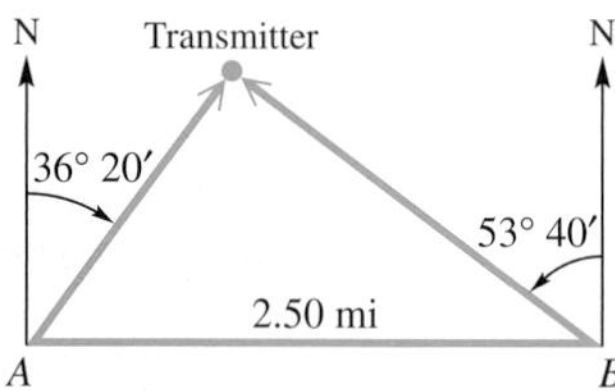

65. *Flying Distance* The bearing from A to C is S 52° E. The bearing from A to B is N 84° E. The bearing from B to C is S 38° W. A plane flying at 250 mph takes 2.4 hr to go from A to B. Find the distance from A to C.

66. *Flying Distance* The bearing from A to C is N 64° W. The bearing from A to B is S 82° W. The bearing from B to C is N 26° E. A plane flying at 350 mph takes 1.8 hr to go from A to B. Find the distance from B to C.

67. *Distance between Two Cities* The bearing from Winston-Salem, North Carolina, to Danville, Virginia, is N 42° E. The bearing from Danville to Goldsboro, North Carolina, is S 48° E. A car driven by Ellen Winchell, traveling at 65 mph, takes 1.1 hr to go from Winston-Salem to Danville and 1.8 hr to go from Danville to Goldsboro. Find the distance from Winston-Salem to Goldsboro.

68. *Distance between Two Cities* The bearing from Atlanta to Macon is S 27° E, and the bearing from Macon to Augusta is N 63° E. An automobile traveling at 62 mph needs $1\frac{1}{4}$ hr to go from Atlanta to Macon and $1\frac{3}{4}$ hr to go from Macon to Augusta. Find the distance from Atlanta to Augusta.

In Exercises 69–74, use the method of ***Example 8.***

69. Find h as indicated in the figure.

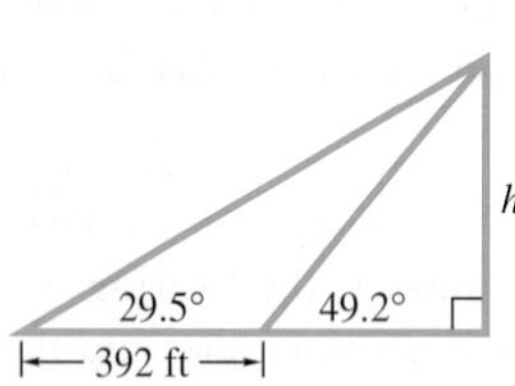

70. Find h as indicated in the figure.

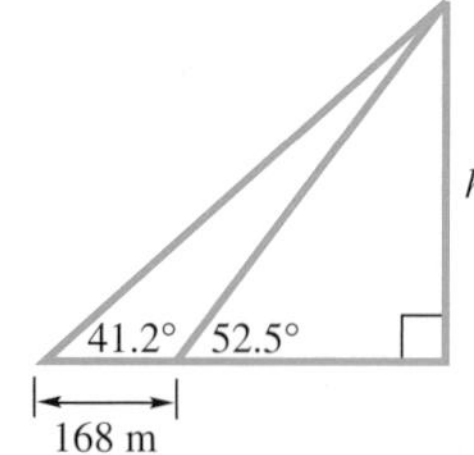

71. 114 ft   72. 147 m
73. 5.18 m   74. 2.47 km

75. (a) $d = \frac{b}{2}\left(\cot \frac{\alpha}{2} + \cot \frac{\beta}{2}\right)$
(b) 345.4 cm
76. 1.95 mi

71. *Height of a Pyramid* The angle of elevation from a point on the ground to the top of a pyramid is 35° 30′. The angle of elevation from a point 135 ft farther back to the top of the pyramid is 21° 10′. Find the height of the pyramid.

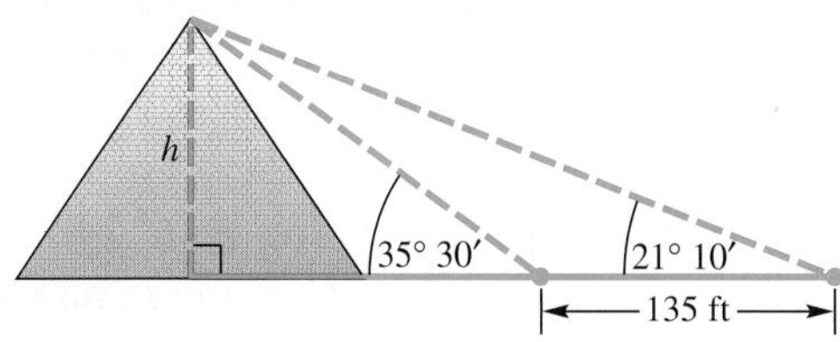

72. *Distance between a Whale and a Lighthouse* Debbie Glockner-Ferrari, a whale researcher, is watching a whale approach directly toward a lighthouse as she observes from the top of this lighthouse. When she first begins watching the whale, the angle of depression to the whale is 15° 50′. Just as the whale turns away from the lighthouse, the angle of depression is 35° 40′. If the height of the lighthouse is 68.7 m, find the distance traveled by the whale as it approached the lighthouse.

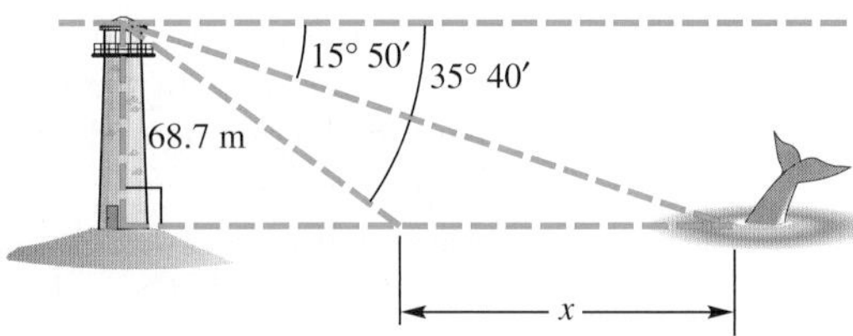

73. *Height of an Antenna* A scanner antenna is on top of the center of a house. The angle of elevation from a point 28.0 m from the center of the house to the top of the antenna is 27° 10′, and the angle of elevation to the bottom of the antenna is 18° 10′. Find the height of the antenna.

74. *Height of Mt. Whitney* The angle of elevation from Lone Pine to the top of Mt. Whitney is 10° 50′. Van Dong Le, traveling 7.00 km from Lone Pine along a straight, level road toward Mt. Whitney, finds the angle of elevation to be 22° 40′. Find the height of the top of Mt. Whitney above the level of the road.

Solve each problem.

75. *(Modeling) Distance between Two Points* Refer to **Example 7.** A variation of the subtense bar method that surveyors use to determine larger distances d between two points P and Q is shown in the figure. In this case the subtense bar with length b is placed between the points P and Q so that the bar is centered on and perpendicular to the line of sight connecting P and Q. The angles α and β are measured from points P and Q, respectively. (*Source:* Mueller, I. and K. Ramsayer, *Introduction to Surveying,* Frederick Ungar Publishing Co.)

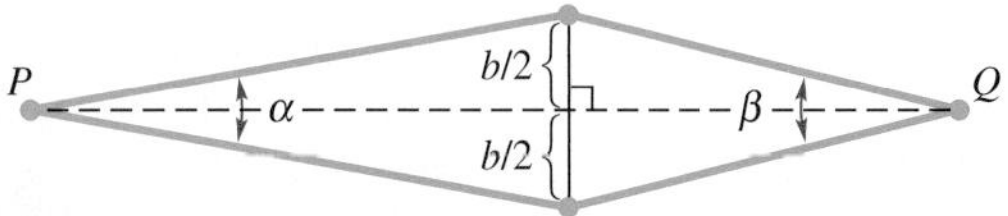

(a) Find a formula for d involving α, β, and b.
(b) Use your formula to determine d if $\alpha = 37' \, 48''$, $\beta = 42' \, 03''$, and $b = 2.000$ cm.

76. *Height of a Plane above Earth* Find the minimum height h above the surface of Earth so that a pilot at point A in the figure can see an object on the horizon at C, 125 mi away. Assume that the radius of Earth is 4.00×10^3 mi.

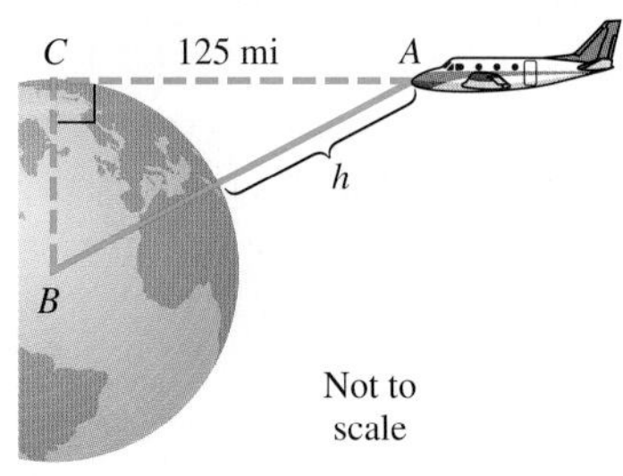

77. 10.8 ft
78. $A = 35.987°$, or $35° 59' 10''$; $B = 54.013°$, or $54° 00' 50''$
79. (a) 320 ft
(b) $R\left(1 - \cos \frac{\theta}{2}\right)$
80. (a) 23 ft (b) 48 ft
(c) As the speed limit increases, more land needs to be cleared inside the curve.

77. *Distance of a Plant from a Fence* In one area, the lowest angle of elevation of the sun in winter is $23° 20'$. Find the minimum distance x that a plant needing full sun can be placed from a fence 4.65 ft high.

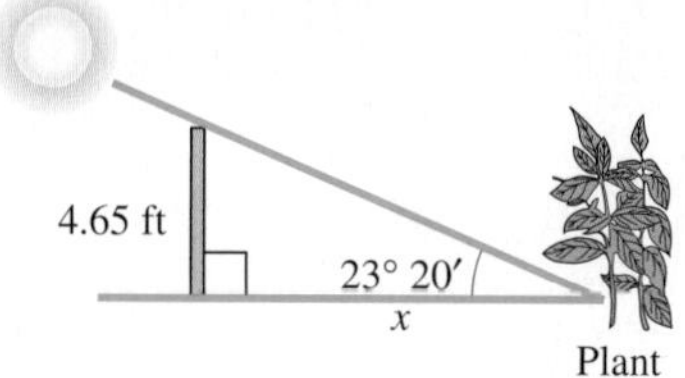

78. *Distance through a Tunnel* A tunnel is to be built from A to B. Both A and B are visible from C. If AC is 1.4923 mi and BC is 1.0837 mi, and if C is 90°, find the measures of angles A and B.

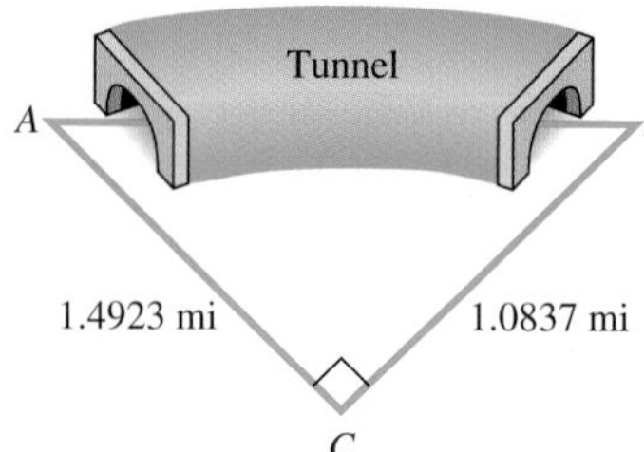

79. *(Modeling) Highway Curves* A basic highway curve connecting two straight sections of road is often circular. In the figure, the points P and S mark the beginning and end of the curve. Let Q be the point of intersection where the two straight sections of highway leading into the curve would meet if extended. The radius of the curve is R, and the central angle θ denotes how many degrees the curve turns. (*Source:* Mannering, F. and W. Kilareski, *Principles of Highway Engineering and Traffic Analysis,* Second Edition, John Wiley and Sons.)

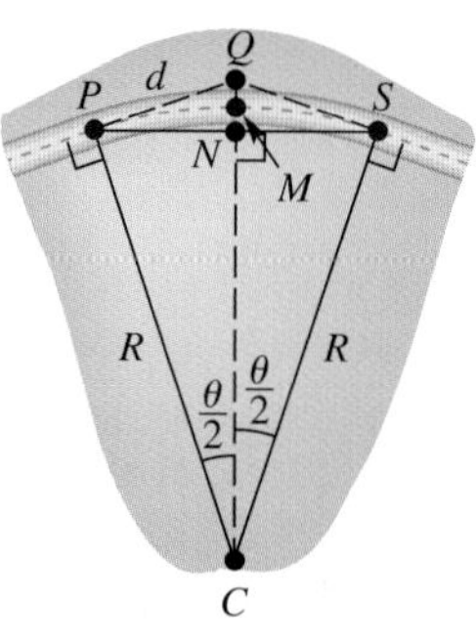

(a) If $R = 965$ ft and $\theta = 37°$, find the distance d between P and Q.

(b) Find an expression in terms of R and θ for the distance between points M and N.

80. *(Modeling) Stopping Distance on a Curve* Refer to **Exercise 79.** When an automobile travels along a circular curve, objects like trees and buildings situated on the inside of the curve can obstruct the driver's vision. These obstructions prevent the driver from seeing sufficiently far down the highway to ensure a safe stopping distance. In the figure, the *minimum* distance d that should be cleared on the inside of the highway is modeled by the equation

$$d = R\left(1 - \cos \frac{\theta}{2}\right).$$

(*Source:* Mannering, F. and W. Kilareski, *Principles of Highway Engineering and Traffic Analysis,* Second Edition, John Wiley and Sons.)

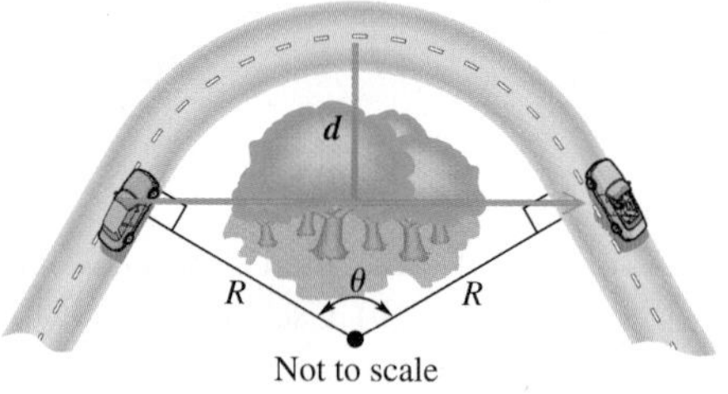

Not to scale

(a) It can be shown that if θ is measured in degrees, then $\theta \approx \frac{57.3S}{R}$, where S is the safe stopping distance for the given speed limit. Compute d to the nearest foot for a 55 mph speed limit if $S = 336$ ft and $R = 600$ ft.

(b) Compute d to the nearest foot for a 65 mph speed limit given $S = 485$ ft and $R = 600$ ft.

(c) How does the speed limit affect the amount of land that should be cleared on the inside of the curve?

Chapter 5 Test Prep

Key Terms

5.1 line
line segment (or segment)
ray
endpoint of a ray
angle
side of an angle
vertex of an angle
initial side
terminal side
positive angle
negative angle
degree
acute angle
right angle
obtuse angle
straight angle
complementary angles (complements)
supplementary angles (supplements)
minute
second
angle in standard position
quadrantal angle
coterminal angles

5.2 sine (sin)
cosine (cos)
tangent (tan)
cotangent (cot)
secant (sec)
cosecant (csc)
degree mode
reciprocal

5.3 side opposite
side adjacent
cofunctions
reference angle

5.4 exact number
significant digits
angle of elevation
angle of depression
bearing

New Symbols

⌉	right angle symbol (for a right triangle)
θ	Greek letter theta
°	degree
′	minute
″	second

Quick Review

Concepts	Examples
5.1 Angles	
Types of Angles Two angles with a sum of 90° are complementary angles, and two angles with a sum of 180° are supplementary angles. $1 \text{ degree} = 60 \text{ minutes} \quad (1° = 60')$ $1 \text{ minute} = 60 \text{ seconds} \quad (1' = 60'')$	70° and $90° - 70° = 20°$ are complementary. 70° and $180° - 70° = 110°$ are supplementary. $15°\,30'\,45'' = 15° + \frac{30°}{60} + \frac{45°}{3600}$ $= 15.5125°$ Decimal degrees
Coterminal angles have measures that differ by a multiple of 360°. Their terminal sides coincide when in standard position.	The acute angle θ in the figure is in standard position. If θ measures 46°, find the measure of a negative coterminal angle. $46° - 360° = -314°$ 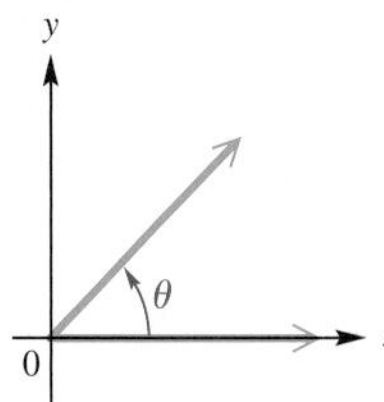

(continued)

Concepts	Examples

5.2 Trigonometric Functions

Definitions of the Trigonometric Functions

Let (x, y) be a point other than the origin on the terminal side of an angle θ in standard position. Let $r = \sqrt{x^2 + y^2}$ represent the distance from the origin to (x, y). Then

$$\sin\theta = \frac{y}{r} \qquad \cos\theta = \frac{x}{r} \qquad \tan\theta = \frac{y}{x}\ (x \neq 0)$$

$$\csc\theta = \frac{r}{y}\ (y \neq 0) \quad \sec\theta = \frac{r}{x}\ (x \neq 0) \quad \cot\theta = \frac{x}{y}\ (y \neq 0).$$

See the summary table of trigonometric function values for quadrantal angles in **Section 5.2.**

If the point $(-2, 3)$ is on the terminal side of angle θ in standard position, then $x = -2,\ y = 3$, and

$$r = \sqrt{(-2)^2 + 3^2} = \sqrt{4 + 9} = \sqrt{13}.$$

Then

$$\sin\theta = \frac{3\sqrt{13}}{13}, \quad \cos\theta = -\frac{2\sqrt{13}}{13}, \quad \tan\theta = -\frac{3}{2},$$

$$\csc\theta = \frac{\sqrt{13}}{3}, \quad \sec\theta = -\frac{\sqrt{13}}{2}, \quad \cot\theta = -\frac{2}{3}.$$

Reciprocal Identities

$$\sin\theta = \frac{1}{\csc\theta} \qquad \cos\theta = \frac{1}{\sec\theta} \qquad \tan\theta = \frac{1}{\cot\theta}$$

$$\csc\theta = \frac{1}{\sin\theta} \qquad \sec\theta = \frac{1}{\cos\theta} \qquad \cot\theta = \frac{1}{\tan\theta}$$

If $\cot\theta = -\frac{2}{3}$, find $\tan\theta$.

$$\tan\theta = \frac{1}{\cot\theta} = \frac{1}{-\frac{2}{3}} = -\frac{3}{2}$$

Pythagorean Identities

$$\sin^2\theta + \cos^2\theta = 1 \qquad \tan^2\theta + 1 = \sec^2\theta$$

$$1 + \cot^2\theta = \csc^2\theta$$

Use the function values for the example directly above to illustrate the Pythagorean identities.

$$\sin^2\theta + \cos^2\theta = \left(\frac{3\sqrt{13}}{13}\right)^2 + \left(-\frac{2\sqrt{13}}{13}\right)^2 = \frac{9}{13} + \frac{4}{13} = 1$$

$$\tan^2\theta + 1 = \left(-\frac{3}{2}\right)^2 + 1 = \frac{13}{4} = \left(-\frac{\sqrt{13}}{2}\right)^2 = \sec^2\theta$$

$$1 + \cot^2\theta = 1 + \left(-\frac{2}{3}\right)^2 = \frac{13}{9} = \left(\frac{\sqrt{13}}{3}\right)^2 = \csc^2\theta$$

Quotient Identities

$$\frac{\sin\theta}{\cos\theta} = \tan\theta \qquad \frac{\cos\theta}{\sin\theta} = \cot\theta$$

Use the function values for the example directly above to illustrate $\frac{\sin\theta}{\cos\theta} = \tan\theta$.

$$\frac{\sin\theta}{\cos\theta} = \frac{\frac{3\sqrt{13}}{13}}{-\frac{2\sqrt{13}}{13}} = \frac{3\sqrt{13}}{13}\left(-\frac{13}{2\sqrt{13}}\right) = -\frac{3}{2} = \tan\theta$$

Signs of the Trigonometric Functions

$x < 0, y > 0, r > 0$ II Sine and cosecant positive	$x > 0, y > 0, r > 0$ I All functions positive
$x < 0, y < 0, r > 0$ III Tangent and cotangent positive	$x > 0, y < 0, r > 0$ IV Cosine and secant positive

Identify the quadrant(s) of any angle θ that satisfies $\sin\theta < 0,\ \tan\theta > 0$.

Since $\sin\theta < 0$ in quadrants III and IV, and $\tan\theta > 0$ in quadrants I and III, both conditions are met only in quadrant III.

Concepts	Examples

5.3 Evaluating Trigonometric Functions

Right-Triangle-Based Definitions of the Trigonometric Functions

Let A represent any acute angle in standard position.

$$\sin A = \frac{y}{r} = \frac{\text{side opposite}}{\text{hypotenuse}} \qquad \csc A = \frac{r}{y} = \frac{\text{hypotenuse}}{\text{side opposite}}$$

$$\cos A = \frac{x}{r} = \frac{\text{side adjacent}}{\text{hypotenuse}} \qquad \sec A = \frac{r}{x} = \frac{\text{hypotenuse}}{\text{side adjacent}}$$

$$\tan A = \frac{y}{x} = \frac{\text{side opposite}}{\text{side adjacent}} \qquad \cot A = \frac{x}{y} = \frac{\text{side adjacent}}{\text{side opposite}}$$

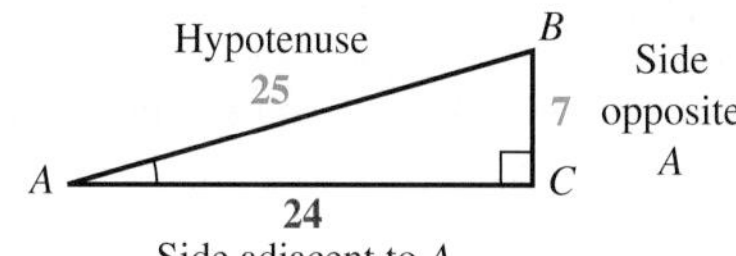

$$\sin A = \frac{7}{25} \qquad \cos A = \frac{24}{25} \qquad \tan A = \frac{7}{24}$$

$$\csc A = \frac{25}{7} \qquad \sec A = \frac{25}{24} \qquad \cot A = \frac{24}{7}$$

Cofunction Identities

For any acute angle A, cofunction values of complementary angles are equal.

$$\sin A = \cos(90° - A) \qquad \cos A = \sin(90° - A)$$

$$\sec A = \csc(90° - A) \qquad \csc A = \sec(90° - A)$$

$$\tan A = \cot(90° - A) \qquad \cot A = \tan(90° - A)$$

$$\sin 55° = \cos(90° - 55°) = \cos 35°$$

$$\sec 48° = \csc(90° - 48°) = \csc 42°$$

$$\tan 72° = \cot(90° - 72°) = \cot 18°$$

Function Values of Special Angles

θ	$\sin\theta$	$\cos\theta$	$\tan\theta$	$\cot\theta$	$\sec\theta$	$\csc\theta$
30°	$\frac{1}{2}$	$\frac{\sqrt{3}}{2}$	$\frac{\sqrt{3}}{3}$	$\sqrt{3}$	$\frac{2\sqrt{3}}{3}$	2
45°	$\frac{\sqrt{2}}{2}$	$\frac{\sqrt{2}}{2}$	1	1	$\sqrt{2}$	$\sqrt{2}$
60°	$\frac{\sqrt{3}}{2}$	$\frac{1}{2}$	$\sqrt{3}$	$\frac{\sqrt{3}}{3}$	2	$\frac{2\sqrt{3}}{3}$

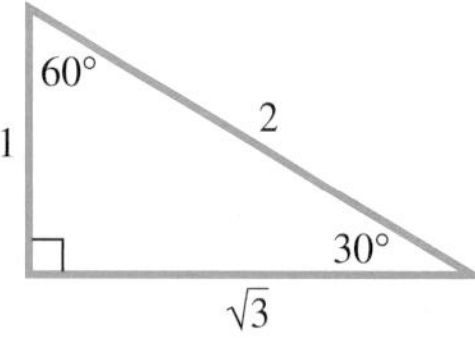

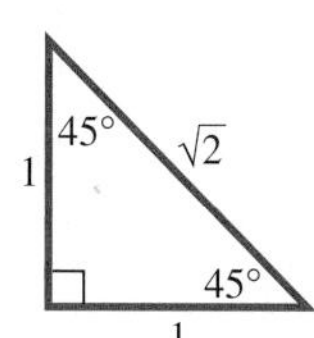

Reference Angle θ' for θ in $(0°, 360°)$

θ in Quadrant	I	II	III	IV
θ' is	θ	$180° - \theta$	$\theta - 180°$	$360° - \theta$

See the figure in **Section 5.3** for illustrations of reference angles.

Quadrant I: For $\theta = 25°$, $\theta' = 25°$
Quadrant II: For $\theta = 152°$, $\theta' = 28°$
Quadrant III: For $\theta = 200°$, $\theta' = 20°$
Quadrant IV: For $\theta = 320°$, $\theta' = 40°$

Finding Trigonometric Function Values for Any Nonquadrantal Angle

Step 1 Add or subtract 360° as many times as needed to get an angle greater than 0° but less than 360°.

Step 2 Find the reference angle θ'.

Step 3 Find the trigonometric function values for θ'.

Step 4 Determine the correct signs for the values found in Step 3.

Find sin 1050°.

$$1050° - 2(360°) = 330° \quad \text{Coterminal angle in quadrant IV}$$

Thus, $\theta' = 30°$.

$$\sin 1050° = -\sin 30° \quad \text{Reference angle}$$

$$= -\frac{1}{2}$$

(continued)

Concepts	Examples
To approximate a trigonometric function value of an angle in degrees, make sure your calculator is in degree mode.	Approximate each value. $\cos 50° 15' = \cos 50.25° \approx 0.63943900$ $\csc 32.5° = \frac{1}{\sin 32.5°} \approx 1.86115900$ $\quad \csc \theta = \frac{1}{\sin \theta}$
To find the corresponding angle measure given a trigonometric function value, use an appropriate inverse function.	Find an angle θ in the interval $[0°, 90°]$ that satisfies each condition in color. $\cos \theta \approx 0.73677482$ $\theta \approx \cos^{-1}(0.73677482)$ $\theta \approx 42.542600°$ $\csc \theta \approx 1.04766792$ $\sin \theta \approx \frac{1}{1.04766792}$ $\quad \sin \theta = \frac{1}{\csc \theta}$ $\theta \approx \sin^{-1}\left(\frac{1}{1.04766792}\right)$ $\theta \approx 72.65°$

5.4 Solving Right Triangles

Concepts	Examples
Solving an Applied Trigonometry Problem *Step 1* Draw a sketch, and label it with the given information. Label the quantity to be found with a variable.	Find the angle of elevation of the sun if a 48.6-ft flagpole casts a shadow 63.1 ft long. *Step 1* See the sketch. We must find θ. 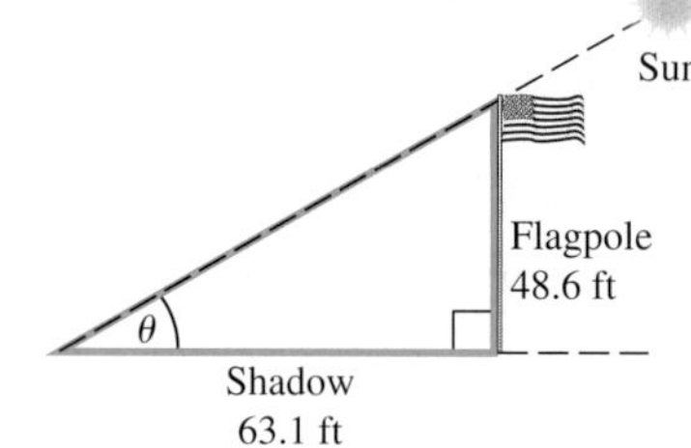
Step 2 Use the sketch to write an equation relating the given quantities to the variable.	*Step 2* $\tan \theta = \frac{48.6}{63.1} \approx 0.770206$
Step 3 Solve the equation, and check that your answer makes sense.	*Step 3* $\theta = \tan^{-1} 0.770206 \approx 37.6°$ The angle of elevation rounded to three significant digits is 37.6°, or 37° 40′.
Expressing Bearing *Method 1* When a single angle is given, such as 220°, this bearing is measured in a clockwise direction from due north. *Method 2* Start with a north-south line and use an acute angle to show direction, either east or west, from this line.	*Example:* 220° 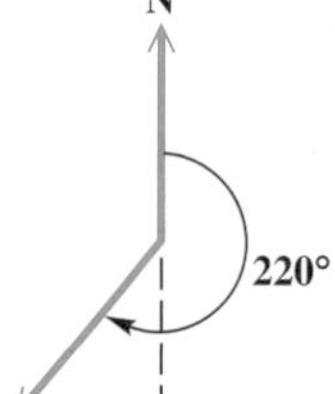  Method 1 *Example:* S 40° W 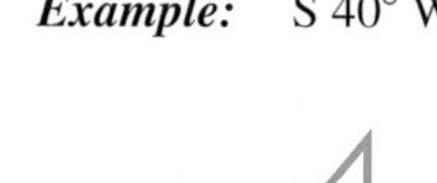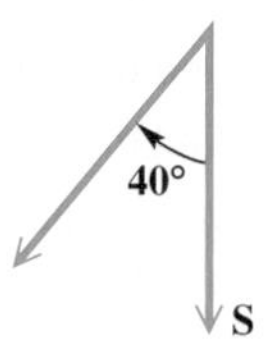 Method 2

Chapter 5 Review Exercises

1. Give the measures of the complement and the supplement of an angle measuring 35°.

Find the angle of least positive measure that is coterminal with each angle.

2. $-51°$
3. $-174°$
4. $792°$

5. Find the measure of each marked angle.

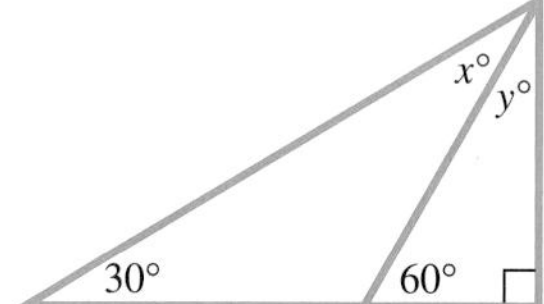

Work each problem.

6. **Rotating Pulley** A pulley is rotating 320 times per min. Through how many degrees does a point on the edge of the pulley move in $\frac{2}{3}$ sec?

7. **Rotating Propeller** The propeller of a speedboat rotates 650 times per min. Through how many degrees does a point on the edge of the propeller rotate in 2.4 sec?

Convert decimal degrees to degrees, minutes, seconds, and convert degrees, minutes, seconds to decimal degrees. Round to the nearest second or the nearest thousandth of a degree, as appropriate. Use a calculator as necessary.

8. $47° \, 25' \, 11''$
9. $119° \, 08' \, 03''$
10. $-61.5034°$
11. $275.1005°$

Find the six trigonometric function values for each angle. If a value is undefined, say so.

12.

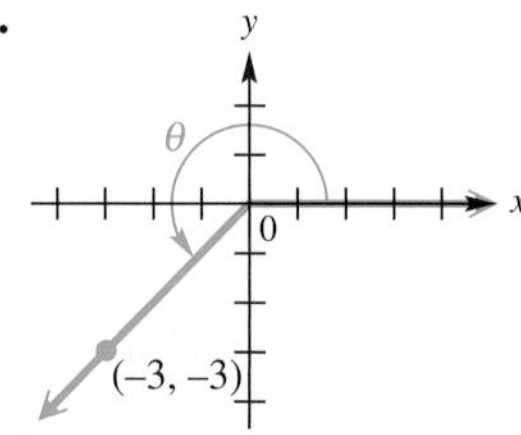

13.

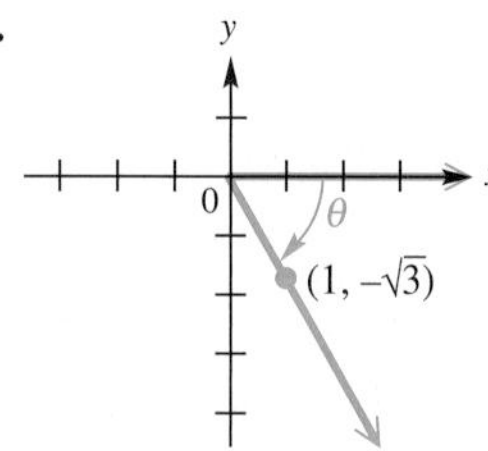

14.

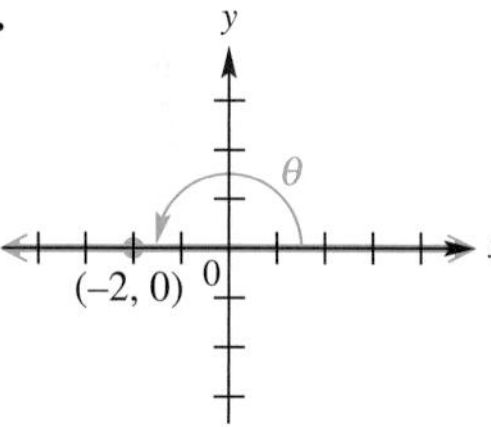

Find the values of the six trigonometric functions for an angle in standard position having each given point on its terminal side.

15. $(3, -4)$
16. $(9, -2)$
17. $(-8, 15)$
18. $(1, -5)$
19. $(6\sqrt{3}, -6)$
20. $(-2\sqrt{2}, 2\sqrt{2})$

21. **Concept Check** If the terminal side of a quadrantal angle lies along the y-axis, which of its trigonometric functions are undefined?

22. Find the values of all six trigonometric functions for an angle in standard position having its terminal side defined by the equation $5x - 3y = 0$, with $x \geq 0$.

In Exercises 23 and 24, consider an angle θ in standard position whose terminal side has the equation $y = -5x$, with $x \leq 0$.

23. Sketch θ and use an arrow to show the rotation if $0° \leq \theta < 360°$.

24. Find the exact values of $\sin \theta$, $\cos \theta$, $\tan \theta$, $\cot \theta$, $\sec \theta$, and $\csc \theta$.

---

1. complement: 55°; supplement: 145°
2. 309°
3. 186°
4. 72°
5. $x = 30$; $y = 30$
6. 1280°
7. 9360°
8. 47.420°
9. 119.134°
10. $-61° \, 30' \, 12''$
11. $275° \, 06' \, 02''$

In Exercises 12–20 and 22, we give, in order, sine, cosine, tangent, cotangent, secant, and cosecant.

12. $-\frac{\sqrt{2}}{2}$; $-\frac{\sqrt{2}}{2}$; 1; 1; $-\sqrt{2}$; $-\sqrt{2}$
13. $-\frac{\sqrt{3}}{2}$; $\frac{1}{2}$; $-\sqrt{3}$; $-\frac{\sqrt{3}}{3}$; 2; $-\frac{2\sqrt{3}}{3}$
14. 0; -1; 0; undefined; -1; undefined
15. $-\frac{4}{5}$; $\frac{3}{5}$; $-\frac{4}{3}$; $-\frac{3}{4}$; $\frac{5}{3}$; $-\frac{5}{4}$
16. $-\frac{2\sqrt{85}}{85}$; $\frac{9\sqrt{85}}{85}$; $-\frac{2}{9}$; $-\frac{9}{2}$; $\frac{\sqrt{85}}{9}$; $-\frac{\sqrt{85}}{2}$
17. $\frac{15}{17}$; $-\frac{8}{17}$; $-\frac{15}{8}$; $-\frac{8}{15}$; $-\frac{17}{8}$; $\frac{17}{15}$
18. $-\frac{5\sqrt{26}}{26}$; $\frac{\sqrt{26}}{26}$; -5; $-\frac{1}{5}$; $\sqrt{26}$; $-\frac{\sqrt{26}}{5}$
19. $-\frac{1}{2}$; $\frac{\sqrt{3}}{2}$; $-\frac{\sqrt{3}}{3}$; $-\sqrt{3}$; $\frac{2\sqrt{3}}{3}$; -2
20. $\frac{\sqrt{2}}{2}$; $-\frac{\sqrt{2}}{2}$; -1; -1; $-\sqrt{2}$; $\sqrt{2}$
21. tangent and secant
22. $\frac{5\sqrt{34}}{34}$; $\frac{3\sqrt{34}}{34}$; $\frac{5}{3}$; $\frac{3}{5}$; $\frac{\sqrt{34}}{3}$; $\frac{\sqrt{34}}{5}$
23.

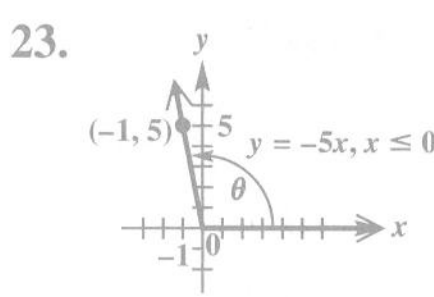

24. $\sin \theta = \frac{5\sqrt{26}}{26}$; $\cos \theta = -\frac{\sqrt{26}}{26}$; $\tan \theta = -5$; $\cot \theta = -\frac{1}{5}$; $\sec \theta = -\sqrt{26}$; $\csc \theta = \frac{\sqrt{26}}{5}$

25. 0; −1; 0; undefined; −1; undefined

26. −1; 0; undefined; 0; undefined; −1

In Exercises 27–38, we give, in order, sine, cosine, tangent, cotangent, secant, and cosecant.

27. $-\frac{\sqrt{39}}{8}$; $-\frac{5}{8}$; $\frac{\sqrt{39}}{5}$; $\frac{5\sqrt{39}}{39}$; $-\frac{8}{5}$; $-\frac{8\sqrt{39}}{39}$

28. $\frac{\sqrt{3}}{5}$; $-\frac{\sqrt{22}}{5}$; $-\frac{\sqrt{66}}{22}$; $-\frac{\sqrt{66}}{3}$; $-\frac{5\sqrt{22}}{22}$; $\frac{5\sqrt{3}}{3}$

29. $\frac{2\sqrt{5}}{5}$; $-\frac{\sqrt{5}}{5}$; -2; $-\frac{1}{2}$; $-\sqrt{5}$; $\frac{\sqrt{5}}{2}$

30. $-\frac{2\sqrt{5}}{5}$; $-\frac{\sqrt{5}}{5}$; 2; $\frac{1}{2}$; $-\sqrt{5}$; $-\frac{\sqrt{5}}{2}$

31. $-\frac{3}{5}$; $\frac{4}{5}$; $-\frac{3}{4}$; $-\frac{4}{3}$; $\frac{5}{4}$; $-\frac{5}{3}$

32. $-\frac{2}{5}$; $-\frac{\sqrt{21}}{5}$; $\frac{2\sqrt{21}}{21}$; $\frac{\sqrt{21}}{2}$; $-\frac{5\sqrt{21}}{21}$; $-\frac{5}{2}$

33. $\frac{60}{61}$; $\frac{11}{61}$; $\frac{60}{11}$; $\frac{11}{60}$; $\frac{61}{11}$; $\frac{61}{60}$

34. $\frac{20}{29}$; $\frac{21}{29}$; $\frac{20}{21}$; $\frac{21}{20}$; $\frac{29}{21}$; $\frac{29}{20}$

35. $-\frac{\sqrt{3}}{2}$; $\frac{1}{2}$; $-\sqrt{3}$; $-\frac{\sqrt{3}}{3}$; 2; $-\frac{2\sqrt{3}}{3}$

36. $\frac{\sqrt{3}}{2}$; $-\frac{1}{2}$; $-\sqrt{3}$; $-\frac{\sqrt{3}}{3}$; -2; $\frac{2\sqrt{3}}{3}$

37. $-\frac{1}{2}$; $\frac{\sqrt{3}}{2}$; $-\frac{\sqrt{3}}{3}$; $-\sqrt{3}$; $\frac{2\sqrt{3}}{3}$; -2

38. $\frac{\sqrt{2}}{2}$; $-\frac{\sqrt{2}}{2}$; -1; -1; $-\sqrt{2}$; $\sqrt{2}$

39. 120°; 240°
40. 210°; 330°
41. 150°; 210°
42. 135°; 315°

43. $-\frac{\sqrt{2}}{2}$; $-\frac{\sqrt{2}}{2}$; 1

44. $-\frac{\sqrt{3}}{2}$; $\frac{1}{2}$; $-\sqrt{3}$

45. −1.3563417
46. 0.95371695
47. 1.0210339
48. −0.71592968
49. 0.20834446
50. 1.9362132

Complete the table with the appropriate function values of the given quadrantal angles. If the value is undefined, say so.

	θ	$\sin\theta$	$\cos\theta$	$\tan\theta$	$\cot\theta$	$\sec\theta$	$\csc\theta$
25.	180°						
26.	−90°						

Find all six trigonometric function values for each angle θ. Rationalize denominators when applicable.

27. $\cos\theta = -\frac{5}{8}$, and θ is in quadrant III

28. $\sin\theta = \frac{\sqrt{3}}{5}$, and $\cos\theta < 0$

29. $\sec\theta = -\sqrt{5}$, and θ is in quadrant II

30. $\tan\theta = 2$, and θ is in quadrant III

31. $\sec\theta = \frac{5}{4}$, and θ is in quadrant IV

32. $\sin\theta = -\frac{2}{5}$, and θ is in quadrant III

Find the values of the six trigonometric functions for each angle A.

33.

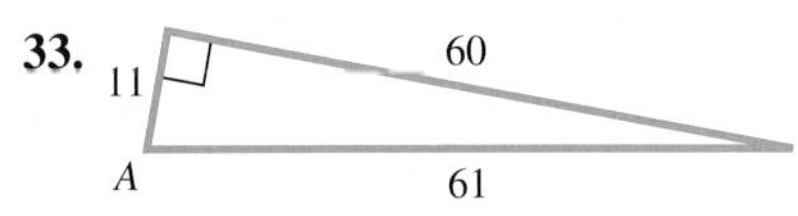

34.

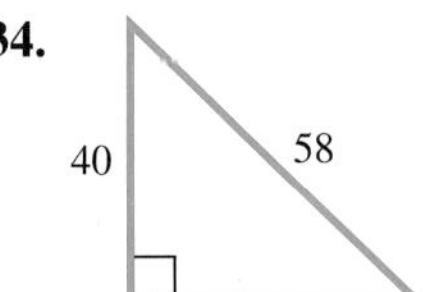

Find exact values of the six trigonometric functions for each angle. Do not use a calculator. Rationalize denominators when applicable.

35. 1020° **36.** 120° **37.** −1470° **38.** −225°

Find all values of θ, if θ is in the interval $[0°, 360°)$ and θ has the given function value.

39. $\cos\theta = -\frac{1}{2}$ **40.** $\sin\theta = -\frac{1}{2}$ **41.** $\sec\theta = -\frac{2\sqrt{3}}{3}$ **42.** $\cot\theta = -1$

Find the sine, cosine, and tangent function values for each angle.

43.

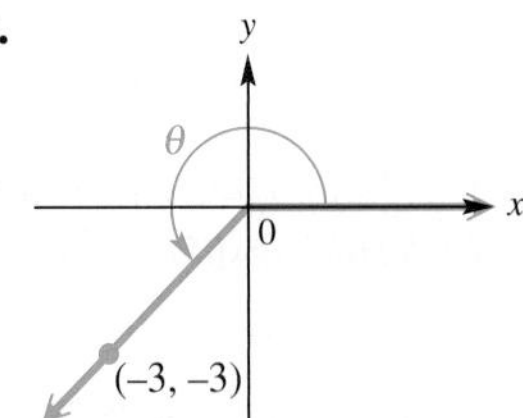

44.

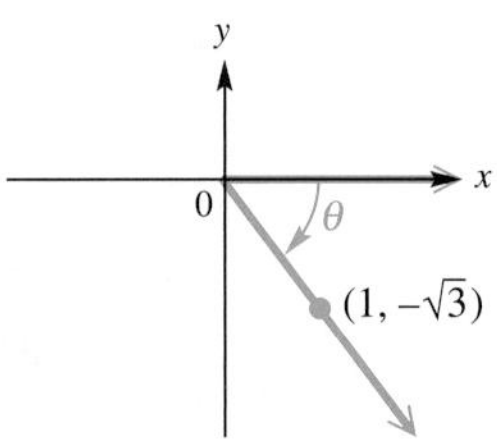

Use a calculator to find each value.

45. sec 222° 30′ **46.** sin 72° 30′ **47.** csc 78° 21′

48. cot 305.6° **49.** tan 11.7689° **50.** sec 58.9041°

Use a calculator to find each value of θ, where θ is in the interval $[0°, 90°)$. Give answers in decimal degrees.

51. $\sin\theta = 0.82584121$ **52.** $\cot\theta = 1.1249386$ **53.** $\cos\theta = 0.97540415$

54. $\sec\theta = 1.2637891$ **55.** $\tan\theta = 1.9633124$ **56.** $\csc\theta = 9.5670466$

51. 55.673870°
52. 41.635092°
53. 12.733938°
54. 37.695528°
55. 63.008286°
56. 5.9998273°

57. 47.1°; 132.9°
58. 54.2°; 234.2°

61. $B = 31° 30'$; $a = 638$; $b = 391$
62. $A = 19° 25'$; $B = 70° 35'$; $c = 390.3$
63. $B = 50.28°$; $a = 32.38$ m; $c = 50.66$ m
64. $A = 42° 07'$; $c = 402.5$ m; $a = 270.0$ m

65. 73.7 ft **66.** 20.4 m
67. 18.75 cm **68.** 50.24 m
69. 1200 m **70.** 110 km
71. 140 mi

Find two angles in the interval $[0°, 360°)$ that satisfy each of the following. Leave answers in decimal degrees rounded to the nearest tenth.

57. $\sin \theta = 0.73254290$

58. $\tan \theta = 1.3865342$

59. A student wants to use a calculator to find the value of $\cot 25°$. However, instead of entering $\frac{1}{\tan 25}$, he enters $\tan^{-1} 25$. Assuming the calculator is in degree mode, will this produce the correct answer? Explain.

60. Explain the process for using a calculator to find $\sec^{-1} 10$.

Solve each right triangle. In Exercise 62, give angles to the nearest minute. In Exercises 63 and 64, label the triangle ABC as in Exercises 61 and 62.

61. B, a, $c = 748$, $58° 30'$, C, b, A

62.

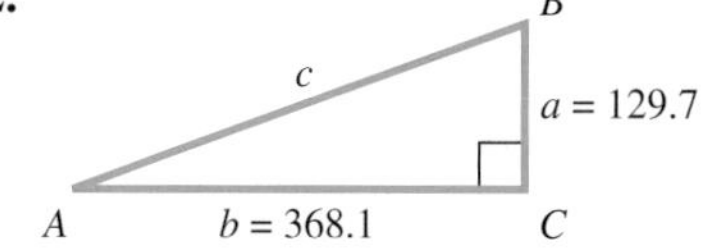

63. $A = 39.72°$, $b = 38.97$ m

64. $B = 47° 53'$, $b = 298.6$ m

Solve each problem.

65. *Height of a Tower* The angle of elevation from a point 93.2 ft from the base of a tower to the top of the tower is $38° 20'$. Find the height of the tower.

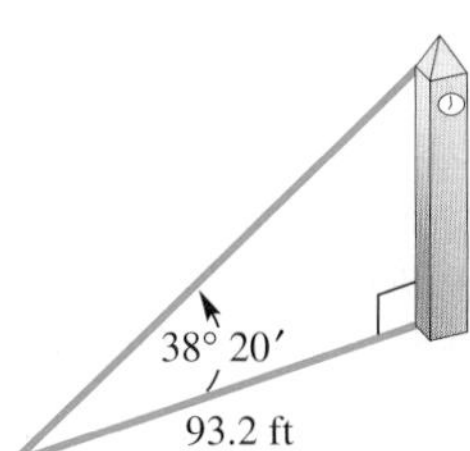

66. *Height of a Tower* The angle of depression from a television tower to a point on the ground 36.0 m from the bottom of the tower is 29.5°. Find the height of the tower.

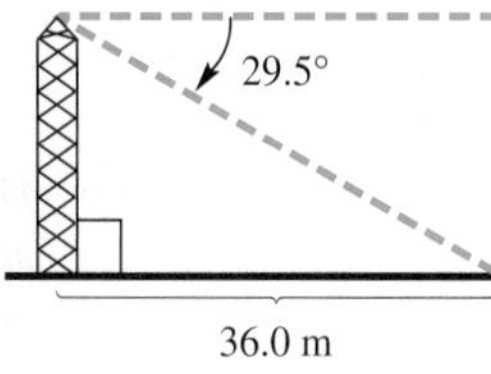

67. *Length of a Diagonal* One side of a rectangle measures 15.24 cm. The angle between the diagonal and that side is 35.65°. Find the length of the diagonal.

68. *Length of Sides of an Isosceles Triangle* An isosceles triangle has a base of length 49.28 m. The angle opposite the base is 58.746°. Find the length of each of the two equal sides.

69. *Distance between Two Points* The bearing of point B from point C is 254°. The bearing of point A from point C is 344°. The bearing of point A from point B is 32°. If the distance from A to C is 780 m, find the distance from A to B.

70. *Distance a Ship Sails* The bearing from point A to point B is S 55° E, and the bearing from point B to point C is N 35° E. If a ship sails from A to B, a distance of 81 km, and then from B to C, a distance of 74 km, how far is it from A to C?

71. *Distance between Two Points* Two cars leave an intersection at the same time. One heads due south at 55 mph. The other travels due west. After 2 hr, the bearing of the car headed west from the car headed south is 324°. How far apart are they at that time?

72. (a) 716 mi (b) 1104 mi

72. *(Modeling) Height of a Satellite* Artificial satellites that orbit Earth often use VHF signals to communicate with the ground. VHF signals travel in straight lines. The height h of the satellite above Earth and the time T that the satellite can communicate with a fixed location on the ground are related by the model

$$h = R\left(\frac{1}{\cos \frac{180T}{P}} - 1\right),$$

where $R = 3955$ mi is the radius of Earth and P is the period for the satellite to orbit Earth. (*Source:* Schlosser, W., T. Schmidt-Kaler, and E. Milone, *Challenges of Astronomy,* Springer-Verlag.)

(a) Find h to the nearest mile when $T = 25$ min and $P = 140$ min. (Evaluate the cosine function in degree mode.)

(b) What is the value of h to the nearest mile if T is increased to 30 min?

Chapter 5 Test

[5.1]
1. (a) 23° (b) 113°
2. 74.31°
3. 45° 12′ 09″
4. (a) 30° (b) 280° (c) 90°
5. 2700°

[5.2]
6.

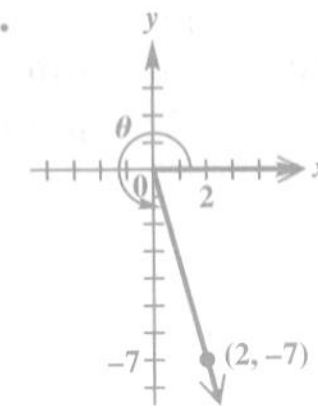

$\sin\theta = -\frac{7\sqrt{53}}{53}$; $\cos\theta = \frac{2\sqrt{53}}{53}$; $\tan\theta = -\frac{7}{2}$; $\cot\theta = -\frac{2}{7}$; $\sec\theta = \frac{\sqrt{53}}{2}$; $\csc\theta = -\frac{\sqrt{53}}{7}$

7.

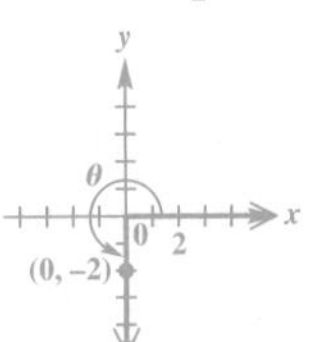

$\sin\theta = -1$; $\cos\theta = 0$; $\tan\theta$ is undefined; $\cot\theta = 0$; $\sec\theta$ is undefined; $\csc\theta = -1$

8.

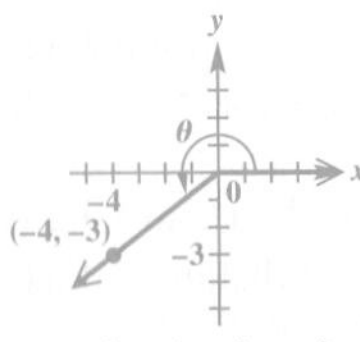

$\sin\theta = -\frac{3}{5}$; $\cos\theta = -\frac{4}{5}$; $\tan\theta = \frac{3}{4}$; $\cot\theta = \frac{4}{3}$; $\sec\theta = -\frac{5}{4}$; $\csc\theta = -\frac{5}{3}$

1. For an angle measuring 67°, give the measure of

(a) its complement **(b)** its supplement.

Perform each conversion.

2. 74° 18′ 36″ to decimal degrees

3. 45.2025° to degrees, minutes, seconds

4. Find the least positive measure of an angle that is coterminal with an angle of the given measure.

(a) 390° **(b)** −80° **(c)** 810°

5. *Rotating Tire* A tire rotates 450 times per min. Through how many degrees does a point on the edge of the tire move in 1 sec?

Draw a sketch of an angle in standard position having the given point on its terminal side. Indicate the angle of least positive measure θ, *and give the values of* sin θ, cos θ, tan θ, cot θ, sec θ, *and* csc θ. *If any of these are undefined, say so.*

6. $(2, -7)$

7. $(0, -2)$

8. Draw a sketch of an angle in standard position having the equation $3x - 4y = 0$, $x \leq 0$, as its terminal side. Indicate the angle of least positive measure θ, and give the values of $\sin\theta$, $\cos\theta$, $\tan\theta$, $\cot\theta$, $\sec\theta$, and $\csc\theta$.

9. Complete the table with the appropriate function values of the given quadrantal angles. If the value is undefined, say so.

θ	sin θ	cos θ	tan θ	cot θ	sec θ	csc θ
90°						
−360°						
630°						

10. If the terminal side of a quadrantal angle lies along the negative x-axis, which two of its trigonometric function values are undefined?

9. row 1: 1, 0, undefined, 0, undefined, 1;
row 2: 0, 1, 0, undefined, 1, undefined;
row 3: -1, 0, undefined, 0, undefined, -1

10. cosecant and cotangent

11. **(a)** I **(b)** III, IV **(c)** III

12. $\cos\theta = -\frac{2\sqrt{10}}{7}$; $\tan\theta = -\frac{3\sqrt{10}}{20}$; $\cot\theta = -\frac{2\sqrt{10}}{3}$; $\sec\theta = -\frac{7\sqrt{10}}{20}$; $\csc\theta = \frac{7}{3}$

[5.3]

13. $\sin A = \frac{12}{13}$; $\cos A = \frac{5}{13}$; $\tan A = \frac{12}{5}$; $\cot A = \frac{5}{12}$; $\sec A = \frac{13}{5}$; $\csc A = \frac{13}{12}$

14. $x = 4$; $y = 4\sqrt{3}$; $z = 4\sqrt{2}$; $w = 8$

In Exercises 15–17, we give, in order, sine, cosine, tangent, cotangent, secant, and cosecant.

15. $-\frac{\sqrt{3}}{2}$; $-\frac{1}{2}$; $\sqrt{3}$; $\frac{\sqrt{3}}{3}$; -2; $-\frac{2\sqrt{3}}{3}$

16. $-\frac{\sqrt{2}}{2}$; $-\frac{\sqrt{2}}{2}$; 1; 1; $-\sqrt{2}$; $-\sqrt{2}$

17. -1; 0; undefined; 0; undefined; -1

18. 135°; 225°

19. 240°; 300°

20. 45°; 225°

21. Take the reciprocal of $\tan\theta$ to get $\cot\theta = 0.59600119$.

22. **(a)** 0.97939940 **(b)** -1.9056082 **(c)** 1.9362132

23. 16.166641°

[5.4]

24. $B = 31°\ 30'$; $c = 877$; $b = 458$

25. 67.1°, or 67° 10′

26. 15.5 ft

27. 8800 ft

---

11. Identify the possible quadrant(s) in which θ must lie under the given conditions.

(a) $\cos\theta > 0$, $\tan\theta > 0$ **(b)** $\sin\theta < 0$, $\csc\theta < 0$ **(c)** $\cot\theta > 0$, $\cos\theta < 0$

12. Find the five remaining trigonometric function values of θ if $\sin\theta = \frac{3}{7}$ and θ is in quadrant II.

13. Give the six trigonometric function values of angle A.

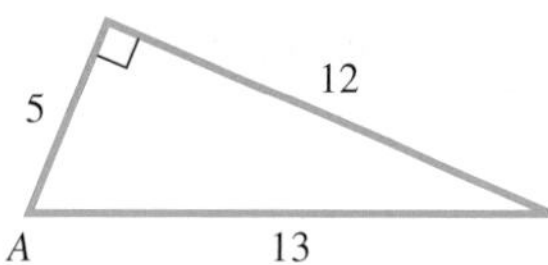

14. Find the exact value of each part labeled with a letter.

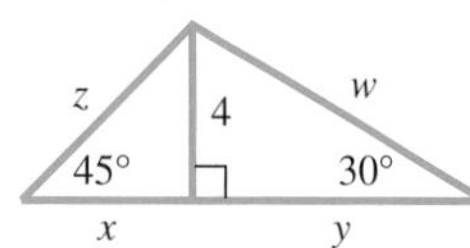

Find the exact values of the six trigonometric functions for each angle. Rationalize denominators when applicable.

15. 240° **16.** $-135°$ **17.** 990°

Find all values of θ in the interval $[0°, 360°)$ that have the given function value.

18. $\cos\theta = -\frac{\sqrt{2}}{2}$ **19.** $\csc\theta = -\frac{2\sqrt{3}}{3}$ **20.** $\tan\theta = 1$

21. How would you find $\cot\theta$ using a calculator, if $\tan\theta = 1.6778490$? Give $\cot\theta$.

22. Use a calculator to approximate each value.

(a) $\sin 78°\ 21'$ **(b)** $\tan 117.689°$ **(c)** $\sec 58.9041°$

23. Find a value of θ in the interval $[0°, 90°)$ in decimal degrees, if

$$\sin\theta = 0.27843196.$$

24. Solve the triangle.

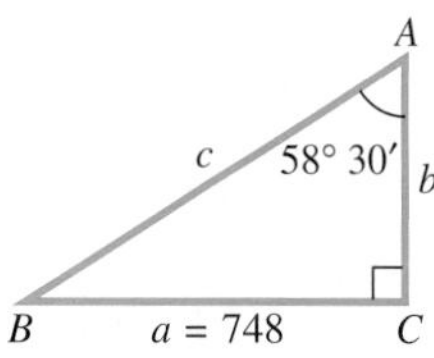

Solve each problem.

25. *Antenna Mast Guy Wire* A guy wire 77.4 m long is attached to the top of an antenna mast that is 71.3 m high. Find the angle that the wire makes with the ground.

26. *Height of a Flagpole* To measure the height of a flagpole, Amado Carillo found that the angle of elevation from a point 24.7 ft from the base to the top is 32° 10′. What is the height of the flagpole?

27. *Altitude of a Mountain* The highest point in Texas is Guadalupe Peak. The angle of depression from the top of this peak to a small miner's cabin at an approximate elevation of 2000 ft is 26°. The cabin is located 14,000 ft horizontally from a point directly under the top of the mountain. Find the altitude of the top of the mountain to the nearest hundred feet.

28. 72 nautical mi
29. 92 km
30. 448 m

28. *Distance between Two Points* Two ships leave a port at the same time. The first ship sails on a bearing of 32° at 16 knots (nautical miles per hour) and the second on a bearing of 122° at 24 knots. How far apart are they after 2.5 hr?

29. *Distance of a Ship from a Pier* A ship leaves a pier on a bearing of S 62° E and travels for 75 km. It then turns and continues on a bearing of N 28° E for 53 km. How far is the ship from the pier?

30. Find h as indicated in the figure.

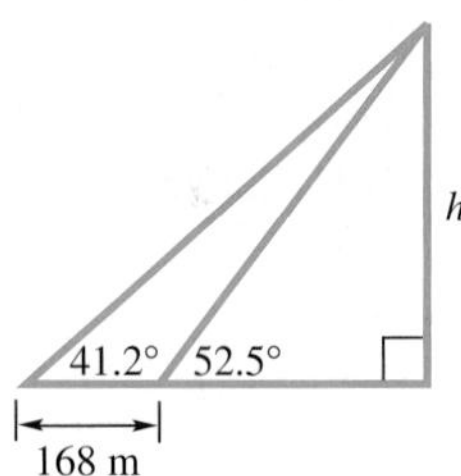

6 The Circular Functions and Their Graphs

Phenomena that repeat in a regular pattern, such as average monthly temperature, rotation of a planet on its axis, and high and low tides, can be modeled by *periodic functions*.

6.1 Radian Measure

- Radian Measure
- Converting between Degrees and Radians
- Arc Length on a Circle
- Area of a Sector of a Circle

Radian Measure We have seen that angles can be measured in degrees. In more theoretical work in mathematics, *radian measure* of angles is preferred. Radian measure enables us to treat the trigonometric functions as functions with domains of *real numbers,* rather than angles.

Figure 1 shows an angle θ in standard position, along with a circle of radius r. The vertex of θ is at the center of the circle. Because angle θ intercepts an arc on the circle equal in length to the radius of the circle, we say that angle θ has a measure of *1 radian*.

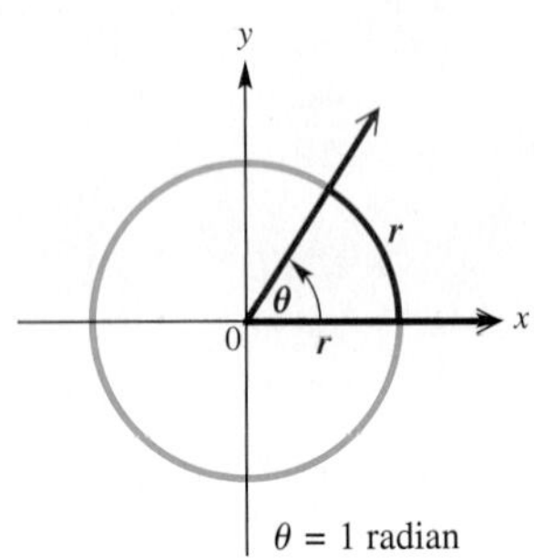

Figure 1

Radian

An angle with its vertex at the center of a circle that intercepts an arc on the circle equal in length to the radius of the circle has a measure of **1 radian.**

Teaching Tip Students may resist the idea of using radian measure. Explain that radians have wide applications in angular motion problems (seen in **Section 6.2**), as well as engineering and science.

It follows that an angle of measure 2 radians intercepts an arc equal in length to twice the radius of the circle, an angle of measure $\frac{1}{2}$ radian intercepts an arc equal in length to half the radius of the circle, and so on. ***In general, if θ is a central angle of a circle of radius r, and θ intercepts an arc of length s, then the radian measure of θ is $\frac{s}{r}$. See*** **Figure 2.**

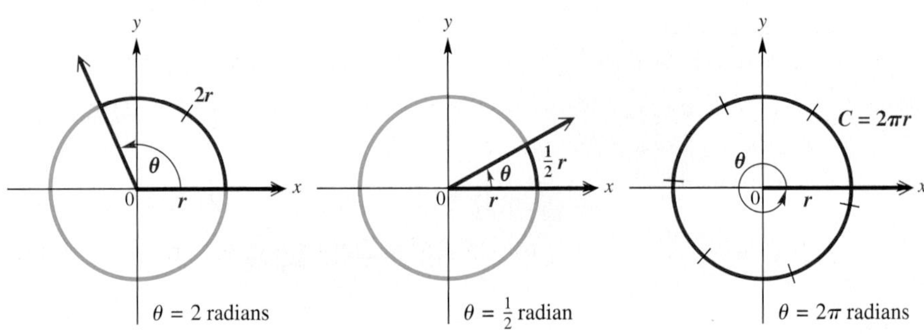

Figure 2

The ratio $\frac{s}{r}$ is a pure number, where s and r are expressed in the same units. ***Thus, "radians" is not a unit of measure like feet or centimeters.***

Converting between Degrees and Radians The **circumference** of a circle—the distance around the circle—is given by $C = 2\pi r$, where r is the radius of the circle. The formula $C = 2\pi r$ shows that the radius can be measured off 2π times around a circle. Therefore, an angle of 360°, which corresponds to a complete circle, intercepts an arc equal in length to 2π times the radius of the circle. Thus, an angle of 360° has a measure of 2π radians.

$$360^\circ = 2\pi \text{ radians}$$

Teaching Tip Emphasize the difference between an angle of a degrees (written a°) and an angle of a radians (sometimes written a^r). Tell students that although radian measures are often given exactly in terms of π, they can also be approximated using decimals.

An angle of 180° is half the size of an angle of 360°, so an angle of 180° has half the radian measure of an angle of 360°.

$$180^\circ = \frac{1}{2}(2\pi) \text{ radians} = \pi \text{ radians} \quad \text{Degree/radian relationship}$$

We can use the relationship $180° = \pi$ radians to develop a method for converting between degrees and radians as follows.

$$180° = \pi \text{ radians}$$

$$1° = \frac{\pi}{180} \text{ radian} \quad \text{Divide by 180.} \qquad \text{or} \qquad 1 \text{ radian} = \frac{180°}{\pi} \quad \text{Divide by } \pi.$$

Converting between Degrees and Radians

1. Multiply a degree measure by $\frac{\pi}{180}$ radian and simplify to convert to radians.
2. Multiply a radian measure by $\frac{180°}{\pi}$ and simplify to convert to degrees.

Classroom Example 1
Convert each degree measure to radians.
(a) 108° (b) −135°
(c) 325.7°

Answers:
(a) $\frac{3\pi}{5}$ radians
(b) $-\frac{3\pi}{4}$ radians
(c) 5.685 radians

EXAMPLE 1 Converting Degrees to Radians

Convert each degree measure to radians.

(a) 45° **(b)** −270° **(c)** 249.8°

SOLUTION

(a) $45° = 45\left(\frac{\pi}{180} \text{ radian}\right) = \frac{\pi}{4} \text{ radian}$ Multiply by $\frac{\pi}{180}$ radian.

(b) $-270° = -270\left(\frac{\pi}{180} \text{ radian}\right)$ Multiply by $\frac{\pi}{180}$ radian.

$$= -\frac{270\pi}{180} \text{ radians}$$

$$= -\frac{3\pi}{2} \text{ radians} \quad \text{Write in lowest terms.}$$

(c) $249.8° = 249.8\left(\frac{\pi}{180} \text{ radian}\right) \approx 4.360 \text{ radians}$ Nearest thousandth

```
45°
          .7853981634
-270°
          -4.71238898
249.8°
          4.359832471
```

This radian mode screen shows TI-83/84 Plus conversions for **Example 1.** Verify that the first two results are *approximations* for the *exact* values of $\frac{\pi}{4}$ and $-\frac{3\pi}{2}$.

Now Try Exercises 7, 13, and 47.

Classroom Example 2
Convert each radian measure to degrees.
(a) $\frac{11\pi}{12}$ (b) $-\frac{7\pi}{6}$
(c) −2.92

Answers:
(a) 165° (b) −210°
(c) −167.3°, or −167° 18′

EXAMPLE 2 Converting Radians to Degrees

Convert each radian measure to degrees.

(a) $\frac{9\pi}{4}$ **(b)** $-\frac{5\pi}{6}$ **(c)** 4.25

SOLUTION

(a) $\frac{9\pi}{4} \text{ radians} = \frac{9\pi}{4}\left(\frac{180°}{\pi}\right) = 405°$ Multiply by $\frac{180°}{\pi}$.

(b) $-\frac{5\pi}{6} \text{ radians} = -\frac{5\pi}{6}\left(\frac{180°}{\pi}\right) = -150°$ Multiply by $\frac{180°}{\pi}$.

(c) $4.25 \text{ radians} = 4.25\left(\frac{180°}{\pi}\right)$

$$\approx 243.5°, \text{ or } 243° 30' \quad 0.50706(60') \approx 30'$$

```
(9π/4)r
                 405
(-5π/6)r
                -150
4.25r►DMS
  243°30'25.427"
```

This degree mode screen shows how a TI-83/84 Plus calculator converts the radian measures in **Example 2** to degree measures.

Now Try Exercises 31, 35, and 59.

NOTE Another way to convert a radian measure that is a rational multiple of π, such as $\frac{9\pi}{4}$, to degrees is to just substitute $180°$ for π. In **Example 2(a),** this would be

$$\frac{9(180°)}{4} = 405°.$$

One of the most important facts to remember when working with angles and their measures is summarized in the following statement.

Agreement on Angle Measurement Units

If no unit of angle measure is specified, then the angle is understood to be measured in radians.

For example, **Figure 3(a)** shows an angle of 30°, and **Figure 3(b)** shows an angle of 30 (which means 30 radians).

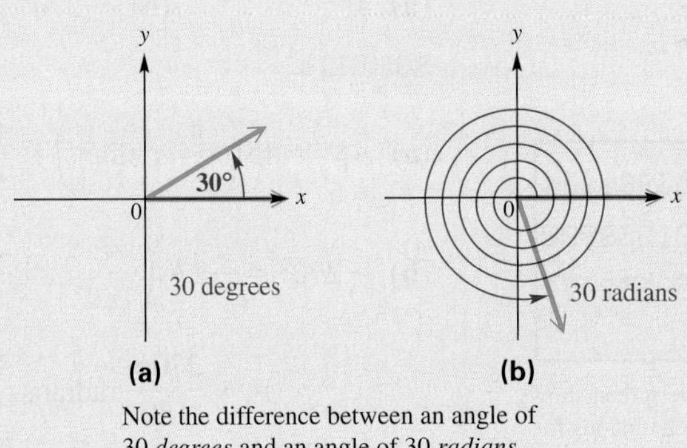

Note the difference between an angle of 30 *degrees* and an angle of 30 *radians.*

Figure 3

The following table and **Figure 4** on the next page give some equivalent angle measures in degrees and radians. Keep in mind that

$$\mathbf{180° = \pi \text{ radians.}}$$

Degrees	Radians		Degrees	Radians	
	Exact	Approximate		Exact	Approximate
0°	0	0	90°	$\frac{\pi}{2}$	1.57
30°	$\frac{\pi}{6}$	0.52	180°	π	3.14
45°	$\frac{\pi}{4}$	0.79	270°	$\frac{3\pi}{2}$	4.71
60°	$\frac{\pi}{3}$	1.05	360°	2π	6.28

These exact values are *rational multiples of* π.

LOOKING AHEAD TO CALCULUS

In calculus, radian measure is much easier to work with than degree measure. If x is measured in radians, then the derivative of $f(x) = \sin x$ is

$$f'(x) = \cos x.$$

However, if x is measured in degrees, then the derivative of $f(x) = \sin x$ is

$$f'(x) = \frac{\pi}{180}\cos x.$$

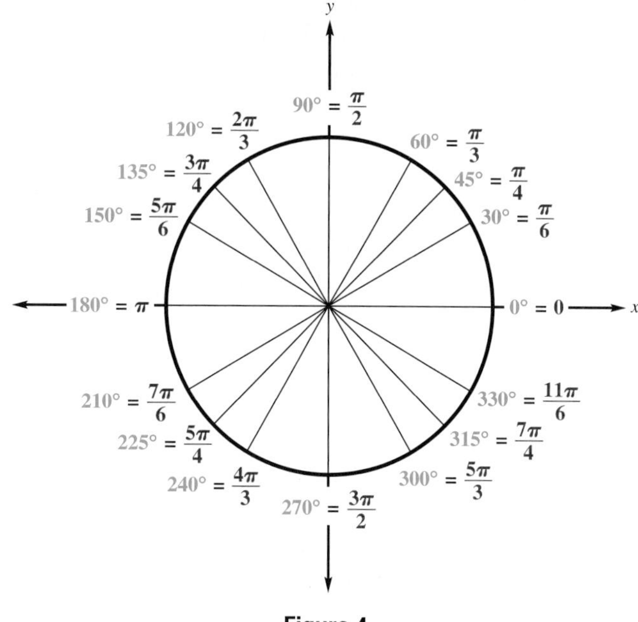

Figure 4

The angles marked in **Figure 4** are extremely important in the study of trigonometry. ***You should learn these equivalences. They will appear often in the chapters to follow.***

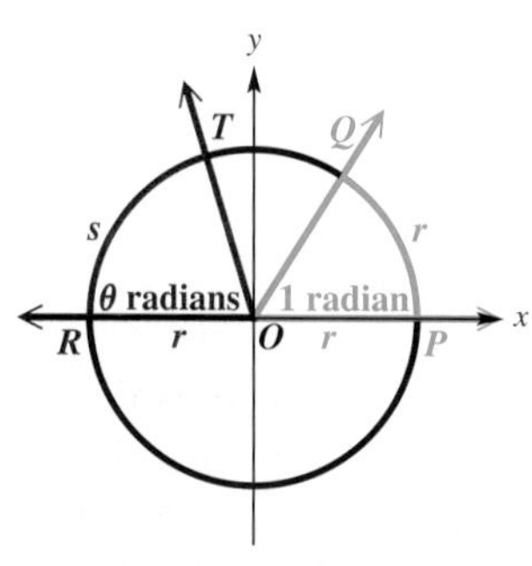

Figure 5

Arc Length on a Circle The formula for finding the length of an arc of a circle follows directly from the definition of an angle θ in radians, where $\theta = \frac{s}{r}$.

In **Figure 5,** we see that angle QOP has measure 1 radian and intercepts an arc of length r on the circle. Angle ROT has measure θ radians and intercepts an arc of length s on the circle. From plane geometry, we know that the lengths of the arcs are proportional to the measures of their central angles.

$$\frac{s}{r} = \frac{\theta}{1} \quad \text{Set up a proportion.}$$

Multiplying each side by r gives

$$s = r\theta. \quad \text{Solve for } s.$$

Arc Length

The length s of the arc intercepted on a circle of radius r by a central angle of measure θ radians is given by the product of the radius and the radian measure of the angle.

$$s = r\theta, \quad \textbf{where } \theta \textbf{ is in radians}$$

CAUTION ***When the formula $s = r\theta$ is applied, the value of θ* MUST *be expressed in radians, not degrees.***

Classroom Example 3
A circle has radius 25.60 cm. Find the length of the arc intercepted by a central angle having each of the following measures.

(a) $\frac{7\pi}{8}$ radians (b) 54°

Answers:
(a) 22.4π cm $\approx$ 70.37 cm
(b) 7.68π cm $\approx$ 24.13 cm

EXAMPLE 3 Finding Arc Length Using $s = r\theta$

A circle has radius 18.20 cm. Find the length of the arc intercepted by a central angle having each of the following measures.

(a) $\frac{3\pi}{8}$ radians (b) 144°

SOLUTION

(a) As shown in **Figure 6**, $r = 18.20$ cm and $\theta = \frac{3\pi}{8}$.

$s = r\theta$ Arc length formula

$s = 18.20\left(\frac{3\pi}{8}\right)$ cm Substitute for r and θ.

$s \approx 21.44$ cm Use a calculator.

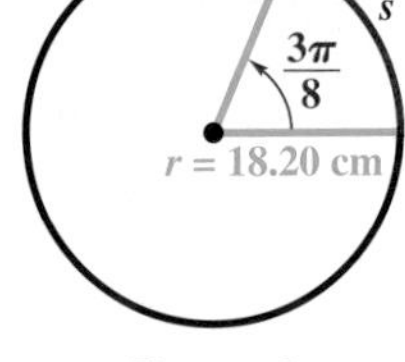

Figure 6

Teaching Tip To help students appreciate radian measure, show them that in degrees, the formula for arc length is

$$s = \frac{\theta}{360}(2\pi r) = \frac{\theta\pi r}{180}.$$

(See **Exercise 86.**)

(b) The formula $s = r\theta$ requires that θ be measured in radians. First, convert θ to radians by multiplying 144° by $\frac{\pi}{180}$ radian.

$144° = 144\left(\frac{\pi}{180}\right) = \frac{4\pi}{5}$ radians Convert from degrees to radians.

The length s is found by using the arc length formula.

$s = r\theta$ Be sure to use radians for θ in $s = r\theta$.

$s = 18.20\left(\frac{4\pi}{5}\right)$ Let $r = 18.20$ cm and $\theta = \frac{4\pi}{5}$.

$s \approx 45.74$ cm Use a calculator.

Now Try Exercises 67, 77, and 81.

Classroom Example 4
Erie, Pennsylvania, is approximately due north of Columbia, South Carolina. The latitude of Erie is 42° N, and that of Columbia is 34° N. The radius of Earth is 6400 km. Find the north-south distance between the two cities.

Answer: 890 km

EXAMPLE 4 Finding the Distance between Two Cities

Latitude gives the measure of a central angle with vertex at Earth's center whose initial side goes through the equator and whose terminal side goes through the given location. Reno, Nevada, is approximately due north of Los Angeles. The latitude of Reno is 40° N, and that of Los Angeles is 34° N. (The N in 34° N means *north* of the equator.) The radius of Earth is 6400 km. Find the north-south distance between the two cities.

SOLUTION As shown in **Figure 7**, the central angle between Reno and Los Angeles is

$$40° - 34° = 6°.$$

The distance between the two cities can be found by the formula $s = r\theta$, after 6° is converted to radians.

$$6° = 6\left(\frac{\pi}{180}\right) = \frac{\pi}{30} \text{ radian}$$

The distance between the two cities is given by s.

$s = r\theta = 6400\left(\frac{\pi}{30}\right) \approx 670$ km Let $r = 6400$ and $\theta = \frac{\pi}{30}$.

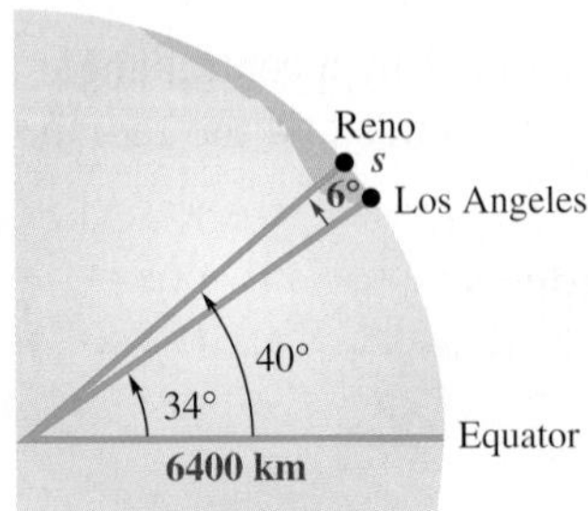

Figure 7

Now Try Exercise 87.

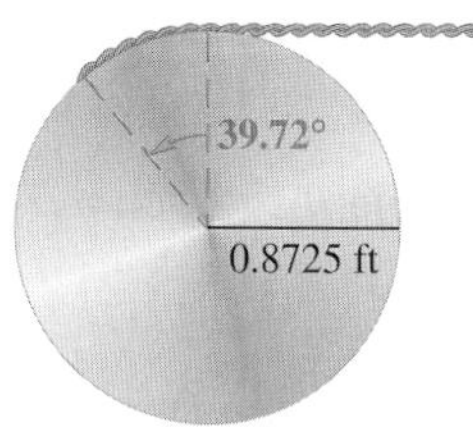

Figure 8

Classroom Example 5
Repeat **Example 5** if the radius of the drum is 0.327 m and the drum is rotated through an angle of 132.6°.

Answer: 0.757 m

EXAMPLE 5 Finding a Length Using $s = r\theta$

A rope is being wound around a drum with radius 0.8725 ft. (See **Figure 8.**) How much rope will be wound around the drum if the drum is rotated through an angle of 39.72°?

SOLUTION The length of rope wound around the drum is the arc length for a circle of radius 0.8725 ft and a central angle of 39.72°. Use the formula $s = r\theta$, with the angle converted to radian measure. The length of the rope wound around the drum is approximated by s.

$$s = r\theta = 0.8725\left[39.72\left(\frac{\pi}{180}\right)\right] \approx 0.6049 \text{ ft}$$

Now Try Exercise 99(a).

Figure 9

Classroom Example 6
Repeat **Example 6** if the radii of the gears are 3.6 in. and 5.4 in. and the smaller gear rotates through 150°.

Answer: 100°

EXAMPLE 6 Finding an Angle Measure Using $s = r\theta$

Two gears are adjusted so that the smaller gear drives the larger one, as shown in **Figure 9.** If the smaller gear rotates through an angle of 225°, through how many degrees will the larger gear rotate?

SOLUTION First find the radian measure of the angle of rotation for the smaller gear, and then find the arc length on the smaller gear. This arc length will correspond to the arc length of the motion of the larger gear. Since $225° = \frac{5\pi}{4}$ radians, for the smaller gear,

$$s = r\theta = 2.5\left(\frac{5\pi}{4}\right) = \frac{12.5\pi}{4} = \frac{25\pi}{8} \text{ cm.}$$

The tips of the two mating gear teeth must move at the same linear speed, or the teeth will break. So we must have "equal arc lengths in equal times." An arc with this length s on the larger gear corresponds to an angle measure θ, in radians, where $s = r\theta$.

$$s = r\theta$$

$$\frac{25\pi}{8} = 4.8\theta \quad \text{Substitute } \tfrac{25\pi}{8} \text{ for } s \text{ and 4.8 for } r \text{ (for the larger gear).}$$

$$\frac{125\pi}{192} = \theta \quad 4.8 = \tfrac{48}{10} = \tfrac{24}{5}. \text{ Multiply by } \tfrac{5}{24} \text{ to solve for } \theta.$$

Converting θ back to degrees shows that the larger gear rotates through

$$\frac{125\pi}{192}\left(\frac{180°}{\pi}\right) \approx 117°. \quad \text{Convert } \theta = \tfrac{125\pi}{192} \text{ to degrees.}$$

Now Try Exercise 93.

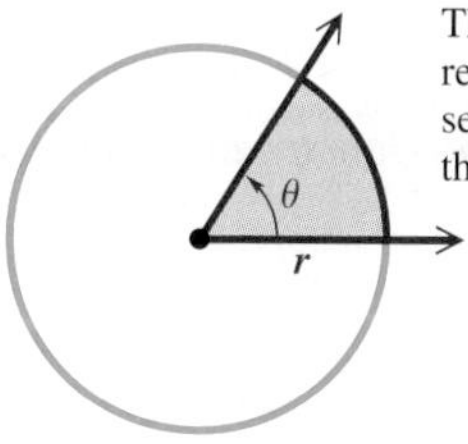

The shaded region is a sector of the circle.

Figure 10

Area of a Sector of a Circle A **sector of a circle** is the portion of the interior of a circle intercepted by a central angle. Think of it as a "piece of pie." See **Figure 10.** A complete circle can be thought of as an angle with measure 2π radians. If a central angle for a sector has measure θ radians, then the sector makes up the fraction $\frac{\theta}{2\pi}$ of a complete circle. The area $\mathcal{A}$ of a complete circle with radius r is $\mathcal{A} = \pi r^2$. Therefore, we have the following.

$$\text{Area } \mathcal{A} \text{ of a sector} = \frac{\theta}{2\pi}(\pi r^2) = \frac{1}{2}r^2\theta, \quad \text{where } \theta \text{ is in radians.}$$

This discussion can be summarized.

Teaching Tip To find the arc length or area of a sector, you may first find the ratio

$$\frac{\theta}{2\pi}, \quad \text{or} \quad \frac{\theta}{360^\circ}.$$

Then, the arc length is this fraction of the circumference $(2\pi r)$, and the sector area is this fraction of the total area (πr^2).

Area of a Sector

The area $\mathscr{A}$ of a sector of a circle of radius r and central angle θ is given by the following formula.

$$\mathscr{A} = \frac{1}{2}r^2\theta, \quad \text{where } \theta \text{ is in radians}$$

CAUTION ***As in the formula for arc length, the value of θ must be in radians when this formula is used for the area of a sector.***

EXAMPLE 7 Finding the Area of a Sector-Shaped Field

Center-pivot irrigation system

A center-pivot irrigation system provides water to a sector-shaped field with the measures shown in **Figure 11.** Find the area of the field.

Figure 11

SOLUTION First, convert 15° to radians.

$$15^\circ = 15\left(\frac{\pi}{180}\right) = \frac{\pi}{12} \text{ radian} \qquad \text{Convert to radians.}$$

Now use the formula to find the area of a sector of a circle with radius $r = 321$.

$$\mathscr{A} = \frac{1}{2}r^2\theta$$

$$\mathscr{A} = \frac{1}{2}(321)^2\left(\frac{\pi}{12}\right) \qquad \text{Substitute for } r \text{ and } \theta.$$

$$\mathscr{A} \approx 13{,}500 \text{ m}^2 \qquad \text{Multiply.}$$

Classroom Example 7
Find the area of a sector of a circle having radius 15.20 ft and central angle 108.0°.

Answer: 217.8 ft^2

✔ *Now Try Exercise 123.*

6.1 Exercises

1. 1 2. 2 3. 3

Concept Check In Exercises 1–6, each angle θ is an integer (e.g., 0, ±1, ±2, ...) when measured in radians. Give the radian measure of the angle. (It helps to remember that $\pi \approx 3$.)

1.

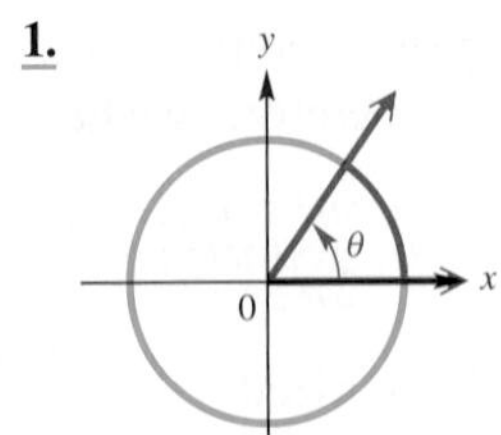

2.

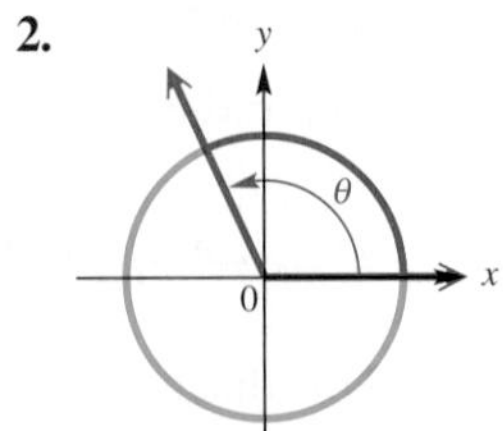

3.

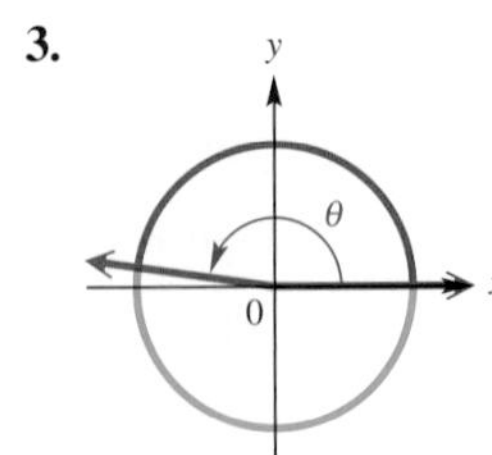

4. -1 5. -3 6. -2

7. $\frac{\pi}{3}$ 8. $\frac{\pi}{6}$
9. $\frac{\pi}{2}$ 10. $\frac{2\pi}{3}$
11. $\frac{5\pi}{6}$ 12. $\frac{3\pi}{2}$
13. $-\frac{5\pi}{3}$ 14. $-\frac{7\pi}{4}$
15. $\frac{5\pi}{2}$ 16. $\frac{8\pi}{3}$
17. 10π 18. 20π
19. 0 20. π
21. -5π 22. -10π

29. $60°$ 30. $480°$
31. $315°$ 32. $120°$
33. $330°$ 34. $675°$
35. $-30°$ 36. $-288°$
37. $126°$ 38. $132°$
39. $-48°$ 40. $-63°$
41. $153°$ 42. $66°$
43. $-900°$ 44. $2700°$

45. 0.68 46. 1.3
47. 0.742 48. 4.623
49. 2.43 50. 3.05
51. 1.122 52. 1.484
53. 0.9847 54. 2.140
55. -0.832391 56. -0.401675

57. $114° 35'$ 58. $286° 29'$
59. $99° 42'$ 60. $175° 20'$
61. $19° 35'$ 62. $564° 14'$
63. $-287° 06'$ 64. $-198° 55'$

4.

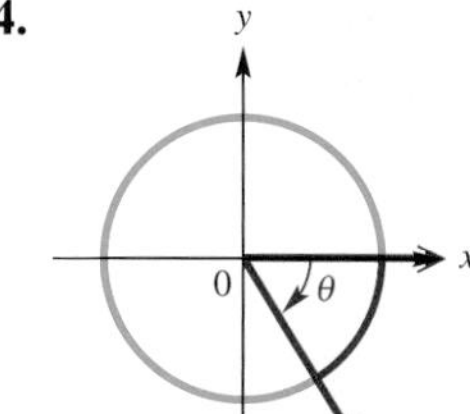

5.

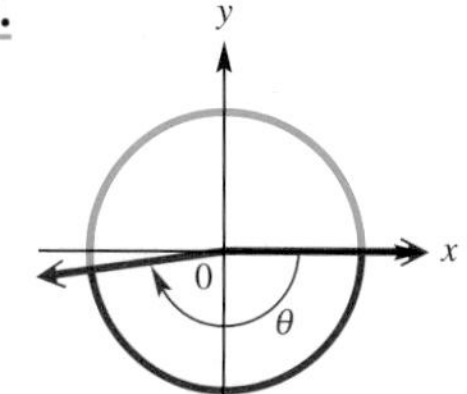

6.

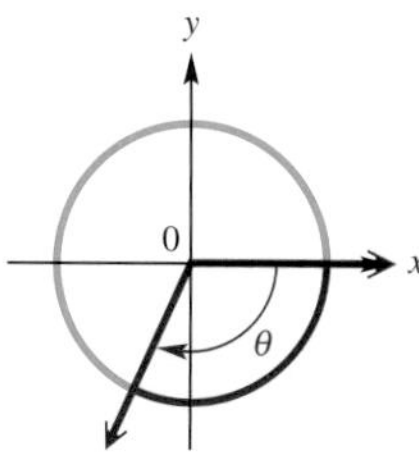

Convert each degree measure to radians. Leave answers as multiples of π. ***See Examples 1(a) and 1(b).***

7. $60°$ **8.** $30°$ **9.** $90°$ **10.** $120°$

11. $150°$ **12.** $270°$ **13.** $-300°$ **14.** $-315°$

15. $450°$ **16.** $480°$ **17.** $1800°$ **18.** $3600°$

19. $0°$ **20.** $180°$ **21.** $-900°$ **22.** $-1800°$

Give a short explanation in Exercises 23–28.

23. Explain how to convert degree measure to radian measure.

24. Explain how to convert radian measure to degree measure.

25. Explain the meaning of radian measure.

26. Explain the difference between degree measure and radian measure.

27. Use an example to show that you can convert from radian measure to degree measure by multiplying by $\frac{180°}{\pi}$.

28. Explain why an angle of radian measure t in standard position intercepts an arc of length t on a circle of radius 1.

Convert each radian measure to degrees. ***See Examples 2(a) and 2(b).***

29. $\frac{\pi}{3}$ **30.** $\frac{8\pi}{3}$ **31.** $\frac{7\pi}{4}$ **32.** $\frac{2\pi}{3}$

33. $\frac{11\pi}{6}$ **34.** $\frac{15\pi}{4}$ **35.** $-\frac{\pi}{6}$ **36.** $-\frac{8\pi}{5}$

37. $\frac{7\pi}{10}$ **38.** $\frac{11\pi}{15}$ **39.** $-\frac{4\pi}{15}$ **40.** $-\frac{7\pi}{20}$

41. $\frac{17\pi}{20}$ **42.** $\frac{11\pi}{30}$ **43.** -5π **44.** 15π

Convert each degree measure to radians. ***See Example 1(c).***

45. $39°$ **46.** $74°$ **47.** $42.5°$ **48.** $264.9°$

49. $139° 10'$ **50.** $174° 50'$ **51.** $64.29°$ **52.** $85.04°$

53. $56° 25'$ **54.** $122° 37'$ **55.** $-47.6925°$ **56.** $-23.0143°$

Convert each radian measure to degrees. Write answers to the nearest minute. ***See Example 2(c).***

57. 2 **58.** 5 **59.** 1.74 **60.** 3.06

61. 0.3417 **62.** 9.84763 **63.** -5.01095 **64.** -3.47189

65. We begin the answers with the blank next to 30°, and then proceed counterclockwise from there: $\frac{\pi}{6}$; 45; $\frac{\pi}{3}$; 120; 135; $\frac{5\pi}{6}$; π; $\frac{7\pi}{6}$; $\frac{5\pi}{4}$; 240; 300; $\frac{7\pi}{4}$; $\frac{11\pi}{6}$.

66. $\frac{\pi^2}{180}$ radian

67. 2π 68. 4π 69. 20π
70. 8 71. 6 72. 8
73. 1 74. 1.5 75. 2

65. *Concept Check* The figure shows the same angles measured in both degrees and radians. Complete the missing measures.

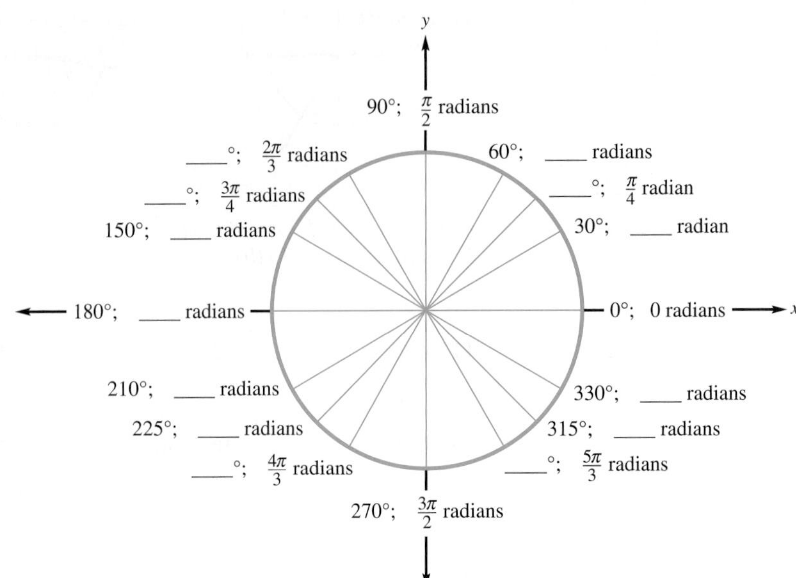

66. *Concept Check* What would be the exact radian measure of an angle that measures π degrees?

Concept Check Find the exact length of each arc intercepted by the given central angle.

67.

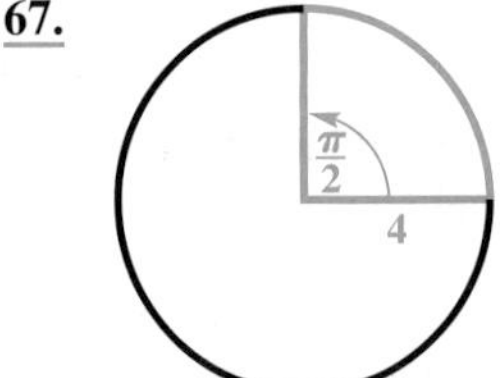

68.

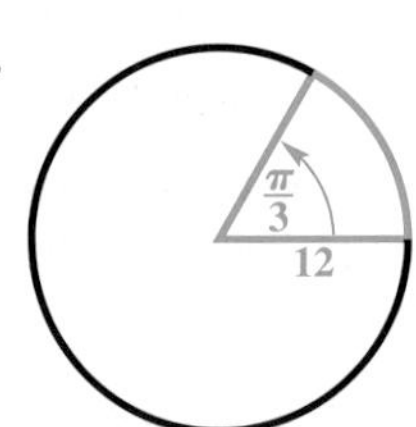

69.

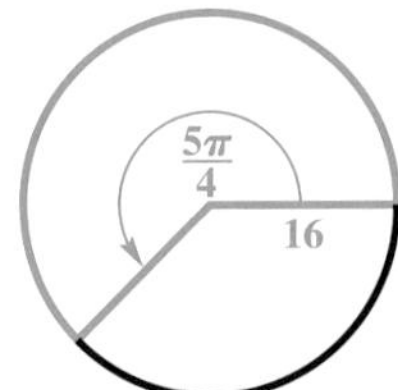

Concept Check Find the radius of each circle.

70.

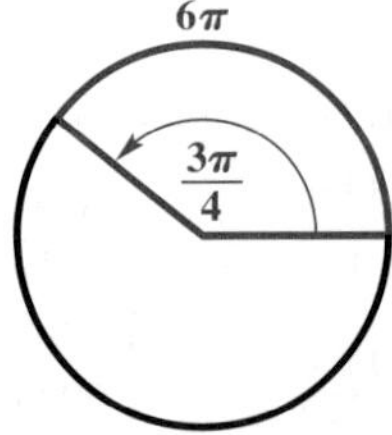

71.

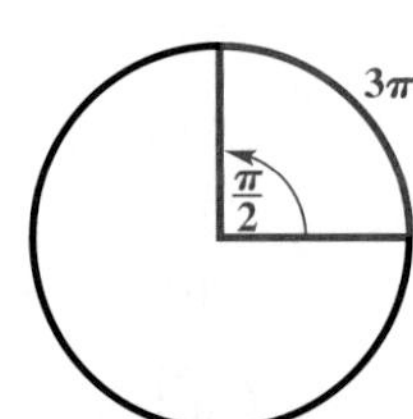

72.

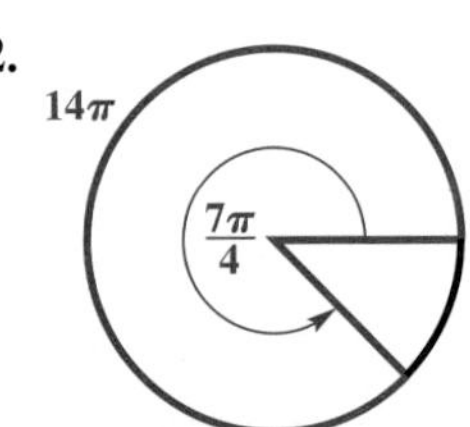

Concept Check Find the measure of each central angle (in radians).

73.

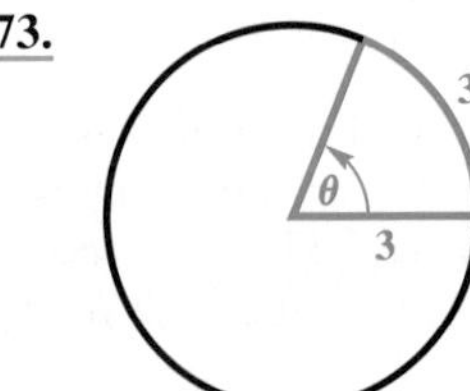

74.

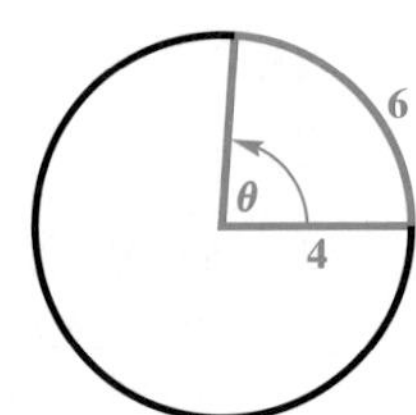

75.

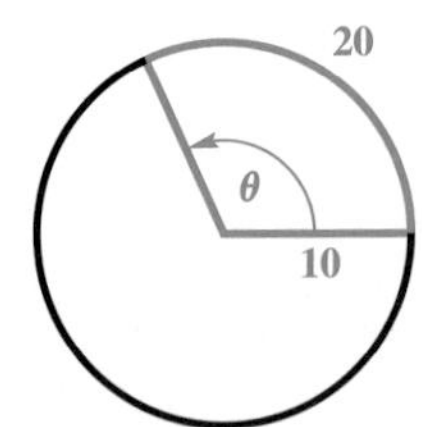

76. Explain how to find the *degree* measure of a central angle in a circle if both the radius and the length of the intercepted arc are known.

77. 25.8 cm 78. 3.08 cm
79. 3.61 ft 80. 11.9 mi
81. 5.05 m 82. 169 cm
83. 55.3 in. 84. 71.4 ft

85. The length is doubled.
86. $s = \frac{\pi r\theta}{180}$

87. 3500 km 88. 1500 km
89. 5900 km 90. 8800 km
91. 44° N 92. 43° N

93. 156° 94. 213°
95. 38.5° 96. 82.3°

Unless otherwise directed, give calculator approximations in your answers in the rest of this exercise set.

Find the length to three significant digits of each arc intercepted by a central angle θ in a circle of radius r. **See Example 3.**

77. $r = 12.3$ cm, $\theta = \frac{2\pi}{3}$ radians

78. $r = 0.892$ cm, $\theta = \frac{11\pi}{10}$ radians

79. $r = 1.38$ ft, $\theta = \frac{5\pi}{6}$ radians

80. $r = 3.24$ mi, $\theta = \frac{7\pi}{6}$ radians

81. $r = 4.82$ m, $\theta = 60°$

82. $r = 71.9$ cm, $\theta = 135°$

83. $r = 15.1$ in., $\theta = 210°$

84. $r = 12.4$ ft, $\theta = 330°$

85. *Concept Check* If the radius of a circle is doubled, how is the length of the arc intercepted by a fixed central angle changed?

86. *Concept Check* Radian measure simplifies many formulas, such as the formula for arc length, $s = r\theta$. Give the corresponding formula when θ is measured in degrees instead of radians.

Distance between Cities Find the distance in kilometers between each pair of cities, assuming they lie on the same north-south line. Use $r = 6400$ km for the radius of Earth. **See Example 4.**

87. Panama City, Panama, 9° N, and Pittsburgh, Pennsylvania, 40° N

88. Farmersville, California, 36° N, and Penticton, British Columbia, 49° N

89. New York City, New York, 41° N, and Lima, Peru, 12° S

90. Halifax, Nova Scotia, 45° N, and Buenos Aires, Argentina, 34° S

91. *Latitude of Madison* Madison, South Dakota, and Dallas, Texas, are 1200 km apart and lie on the same north-south line. The latitude of Dallas is 33° N. What is the latitude of Madison?

92. *Latitude of Toronto* Charleston, South Carolina, and Toronto, Canada, are 1100 km apart and lie on the same north-south line. The latitude of Charleston is 33° N. What is the latitude of Toronto?

Work each problem. **See Examples 5 and 6.**

93. *Gear Movement* Two gears are adjusted so that the smaller gear drives the larger one, as shown in the figure. If the smaller gear rotates through an angle of 300°, through how many degrees does the larger gear rotate?

94. *Gear Movement* Repeat **Exercise 93** for gear radii of 4.8 in. and 7.1 in. and for an angle of 315° for the smaller gear.

95. *Rotating Wheels* The rotation of the smaller wheel in the figure causes the larger wheel to rotate. Through how many degrees does the larger wheel rotate if the smaller one rotates through 60.0°?

96. *Rotating Wheels* Repeat **Exercise 95** for wheel radii of 6.84 in. and 12.46 in. and an angle of 150° for the smaller wheel.

97. 18.7 cm 98. 29.2 in.
99. (a) 11.6 in. (b) 37° 05′
100. 12.7 cm
101. 146 in.
102. (a) 39,616 rotations
(b) 62.9 mi; yes

103. 3π in. 104. 4π in.
105. 27π in. 106. 39π in.

107. 0.20 km

97. *Rotating Wheels* Find the radius of the larger wheel in the figure if the smaller wheel rotates 80.0° when the larger wheel rotates 50.0°.

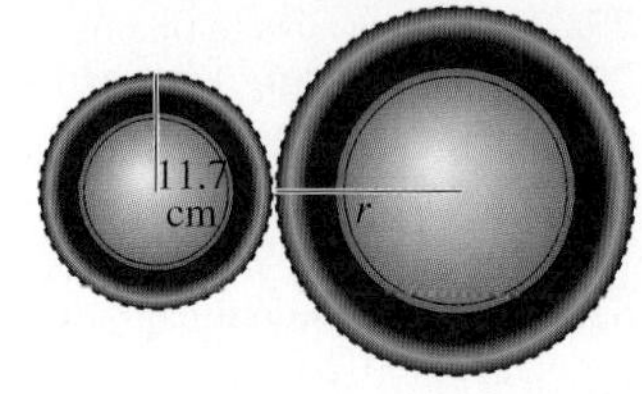

98. *Rotating Wheels* Repeat **Exercise 97** if the smaller wheel of radius 14.6 in. rotates 120° when the larger wheel rotates 60°.

99. *Pulley Raising a Weight* Refer to the figure.

(a) How many inches will the weight in the figure rise if the pulley is rotated through an angle of 71° 50′?

(b) Through what angle, to the nearest minute, must the pulley be rotated to raise the weight 6 in.?

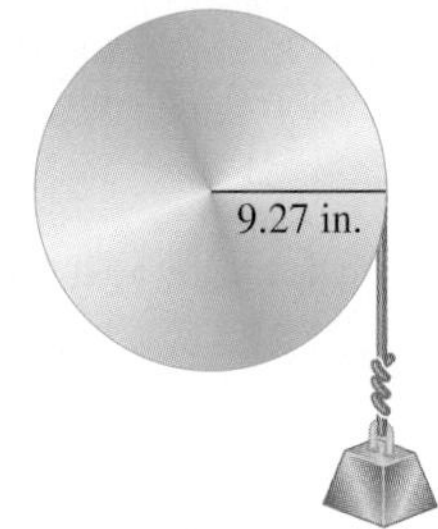

100. *Pulley Raising a Weight* Find the radius of the pulley in the figure if a rotation of 51.6° raises the weight 11.4 cm.

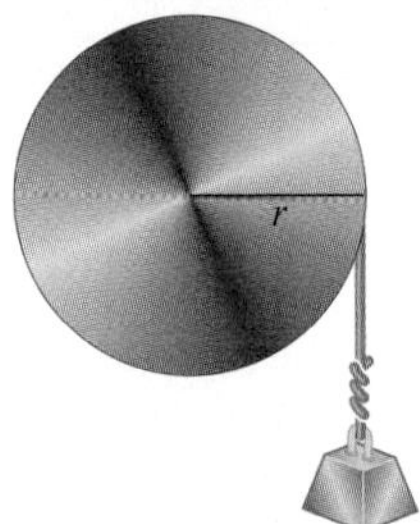

101. *Bicycle Chain Drive* The figure shows the chain drive of a bicycle. How far will the bicycle move if the pedals are rotated through 180°? Assume the radius of the bicycle wheel is 13.6 in.

102. *Car Speedometer* The speedometer of Terry's Honda CR-V is designed to be accurate with tires of radius 14 in.

(a) Find the number of rotations of a tire in 1 hr if the car is driven at 55 mph.

(b) Suppose that oversize tires of radius 16 in. are placed on the car. If the car is now driven for 1 hr with the speedometer reading 55 mph, how far has the car gone? If the speed limit is 55 mph, does Terry deserve a speeding ticket?

Suppose the tip of the minute hand of a clock is 3 in. *from the center of the clock. For each duration, determine the distance traveled by the tip of the minute hand.*

103. 30 min
104. 40 min
105. 4.5 hr
106. $6\frac{1}{2}$ hr

*If a central angle is very small, there is little difference in length between an arc and the inscribed chord. See the figure. Approximate each of the following lengths by finding the necessary arc length. (Note: When a central angle intercepts an arc, the arc is said to **subtend** the angle.)*

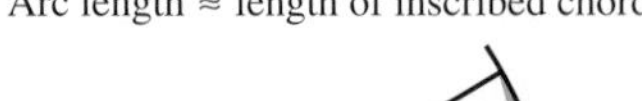

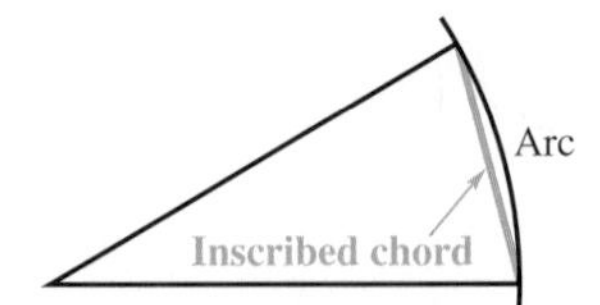

107. *Length of a Train* A railroad track in the desert is 3.5 km away. A train on the track subtends (horizontally) an angle of 3° 20′. Find the length of the train.

108. 850 ft

109. 6π 110. 16π
111. 72π 112. 75π

113. 60° 114. 240°

115. 1.5 116. 1

117. 1116.1 m^2
118. 3744.8 km^2
119. 706.9 ft^2
120. 10,602.9 yd^2
121. 114.0 cm^2
122. 365.3 m^2
123. 1885.0 mi^2
124. 19,085.2 km^2

125. 3.6 126. 16 m

108. *Distance to a Boat* The mast of Brent Simon's boat is 32 ft high. If it subtends an angle of 2° 10′, how far away is it?

Concept Check Find the area of each sector.

109.

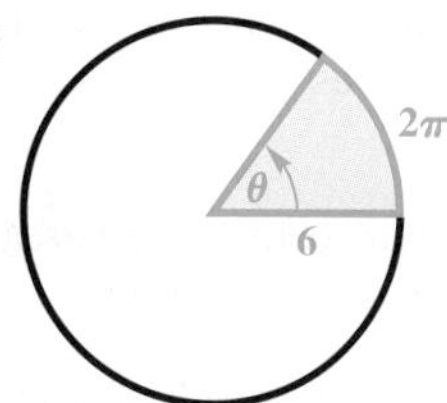

110.

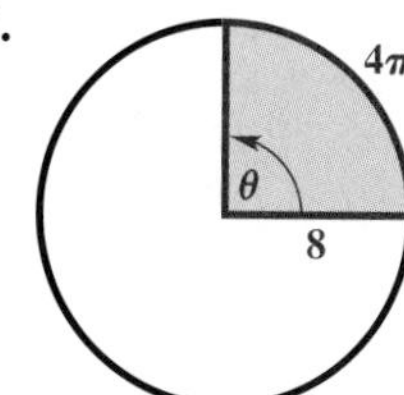

111.

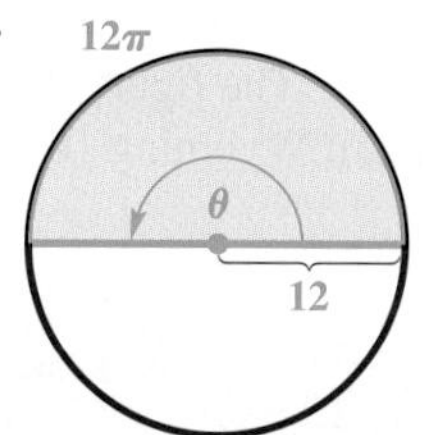

112.

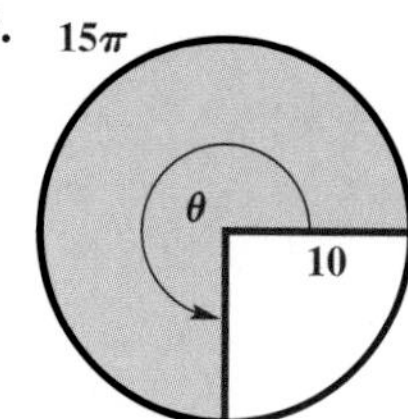

Concept Check Find the measure (in degrees) of each central angle. The number inside the sector is the area.

113.

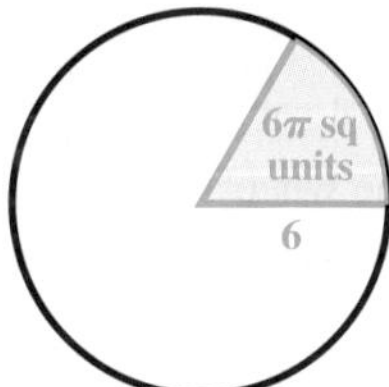

114.

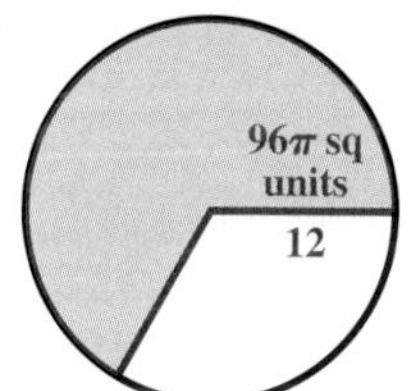

Concept Check Find the measure (in radians) of each central angle. The number inside the sector is the area.

115.

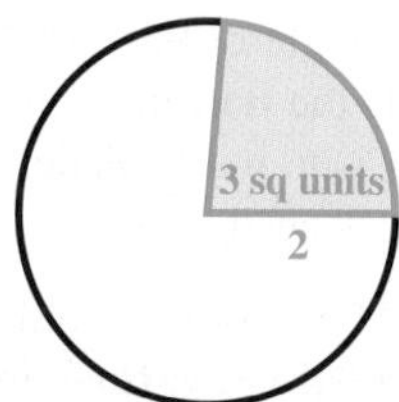

116.

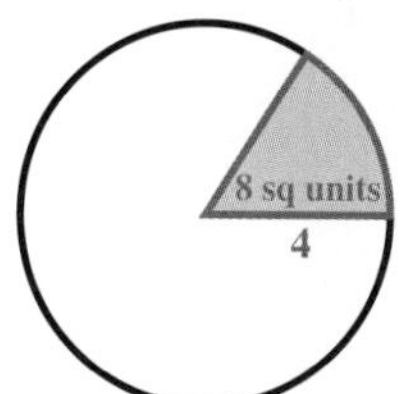

Find the area of a sector of a circle having radius r and central angle θ. Express answers to the nearest tenth. ***See Example 7.***

117. $r = 29.2$ m, $\theta = \dfrac{5\pi}{6}$ radians

118. $r = 59.8$ km, $\theta = \dfrac{2\pi}{3}$ radians

119. $r = 30.0$ ft, $\theta = \dfrac{\pi}{2}$ radians

120. $r = 90.0$ yd, $\theta = \dfrac{5\pi}{6}$ radians

121. $r = 12.7$ cm, $\theta = 81°$

122. $r = 18.3$ m, $\theta = 125°$

123. $r = 40.0$ mi, $\theta = 135°$

124. $r = 90.0$ km, $\theta = 270°$

Work each problem. ***See Example 7.***

125. *Angle Measure* Find the measure (in radians) of a central angle of a sector of area 16 in.^{2} in a circle of radius 3.0 in.

126. *Radius Length* Find the radius of a circle in which a central angle of $\frac{\pi}{6}$ radian determines a sector of area 64 m^2.

127. approximately 8060 yd^2
128. The new area is double the original area; yes
129. 20 in.
130. $\frac{405°}{2\pi}$, or approximately 64°
131. (a) $13\frac{1}{3}°$; $\frac{2\pi}{27}$ (b) 478 ft (c) 17.7 ft (d) approximately 672 ft^2
132. 75.4 $in.^2$
133. (a) 140 ft (b) 102 ft (c) 622 ft^2
134. 550 m; 1800 m
135. 1900 yd^2

127. *Irrigation Area* A center-pivot irrigation system provides water to a sector-shaped field as shown in the figure. Find the area of the field if $\theta = 40.0°$ and $r = 152$ yd.

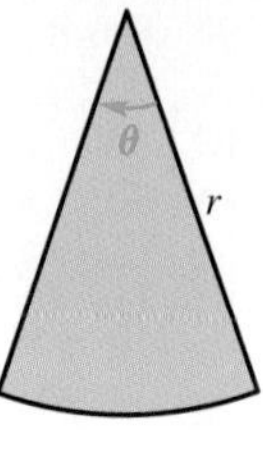

128. *Irrigation Area* Suppose that in **Exercise 127** the angle is halved and the radius length is doubled. How does the new area compare to the original area? Does this result hold in general for any values of θ and r?

129. *Arc Length* A circular sector has an area of 50 $in.^2$. The radius of the circle is 5 in. What is the arc length of the sector?

130. *Angle Measure* In a circle, a sector has an area of 16 cm^2 and an arc length of 6.0 cm. What is the measure of the central angle in degrees?

131. *Measures of a Structure* The figure illustrates Medicine Wheel, a Native American structure in northern Wyoming. There are 27 aboriginal spokes in the wheel, all equally spaced.

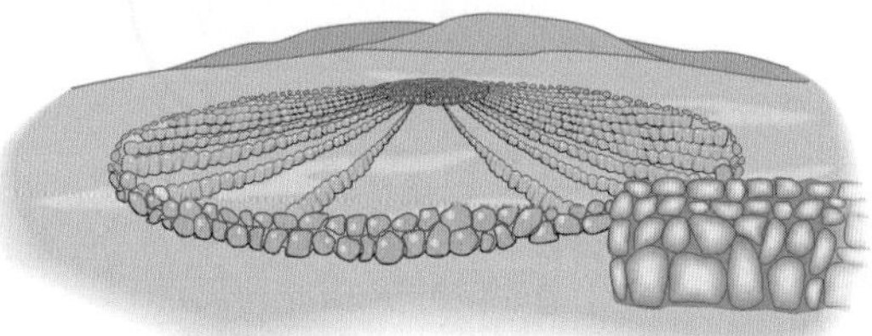

(a) Find the measure of each central angle in degrees and in radians.

(b) If the radius of the wheel is 76.0 ft, find the circumference.

(c) Find the length of each arc intercepted by consecutive pairs of spokes.

(d) Find the area of each sector formed by consecutive spokes.

132. *Area Cleaned by a Windshield Wiper* The Ford Model A, built from 1928 to 1931, had a single windshield wiper on the driver's side. The total arm and blade was 10 in. long and rotated back and forth through an angle of 95°. The shaded region in the figure is the portion of the windshield cleaned by the 7-in. wiper blade. What is the area of the region cleaned?

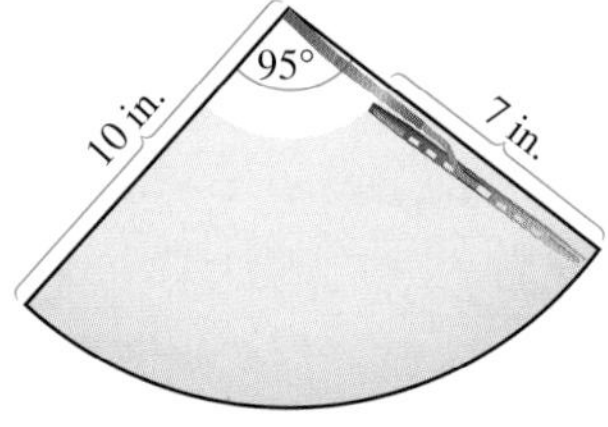

133. *Circular Railroad Curves* In the United States, circular railroad curves are designated by the **degree of curvature**, the central angle subtended by a chord of 100 ft. Suppose a portion of track has curvature 42.0°. (*Source:* Hay, W., *Railroad Engineering,* John Wiley and Sons.)

(a) What is the radius of the curve?

(b) What is the length of the arc determined by the 100-ft chord?

(c) What is the area of the portion of the circle bounded by the arc and the 100-ft chord?

134. *Land Required for a Solar-Power Plant* A 300-megawatt solar-power plant requires approximately 950,000 m^2 of land area to collect the required amount of energy from sunlight. If this land area is circular, what is its radius? If this land area is a 35° sector of a circle, what is its radius?

135. *Area of a Lot* A frequent problem in surveying city lots and rural lands adjacent to curves of highways and railways is that of finding the area when one or more of the boundary lines is the arc of a circle. Find the area (to two significant digits) of the lot shown in the figure. (*Source:* Anderson, J. and E. Michael, *Introduction to Surveying,* McGraw-Hill.)

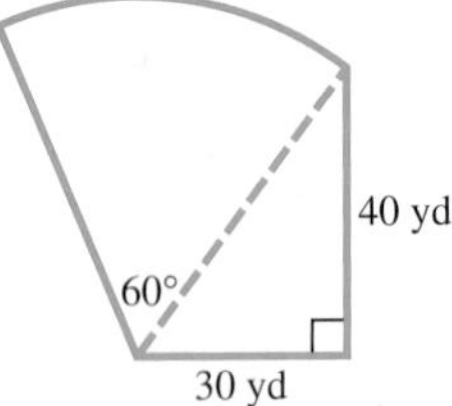

136. 1.15 mi

136. *Nautical Miles* **Nautical miles** are used by ships and airplanes. They are different from **statute miles,** which equal 5280 ft. A nautical mile is defined to be the arc length along the equator intercepted by a central angle AOB of 1 min, as illustrated in the figure. If the equatorial radius of Earth is 3963 mi, use the arc length formula to approximate the number of statute miles in 1 nautical mile. Round your answer to two decimal places.

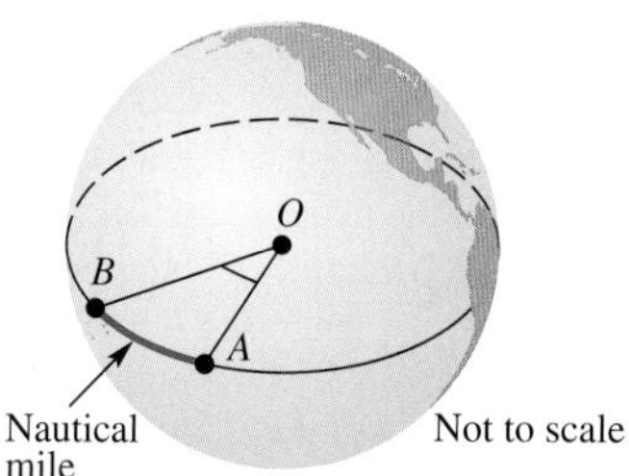

6.2 The Unit Circle and Circular Functions

- Circular Functions
- Finding Values of Circular Functions
- Determining a Number with a Given Circular Function Value
- Expressing Function Values as Lengths of Line Segments
- Linear and Angular Speed

In **Section 5.2,** we defined the six trigonometric functions in such a way that the domain of each function was a set of *angles* in standard position. These angles can be measured in degrees or in radians. In advanced courses, such as calculus, it is necessary to modify the trigonometric functions so that their domains consist of *real numbers* rather than angles. We do this by using the relationship between an angle θ and an arc of length s on a circle.

Circular Functions In **Figure 12,** we start at the point $(1, 0)$ and measure an arc of length s along the circle. If $s > 0$, then the arc is measured in a counterclockwise direction, and if $s < 0$, then the direction is clockwise. (If $s = 0$, then no arc is measured.) Let the endpoint of this arc be at the point (x, y). The circle in **Figure 12** is the **unit circle**—it has center at the origin and radius 1 unit (hence the name *unit circle*). Recall from algebra that the equation of this circle is

$$x^2 + y^2 = 1. \qquad \text{(Section 2.2)}$$

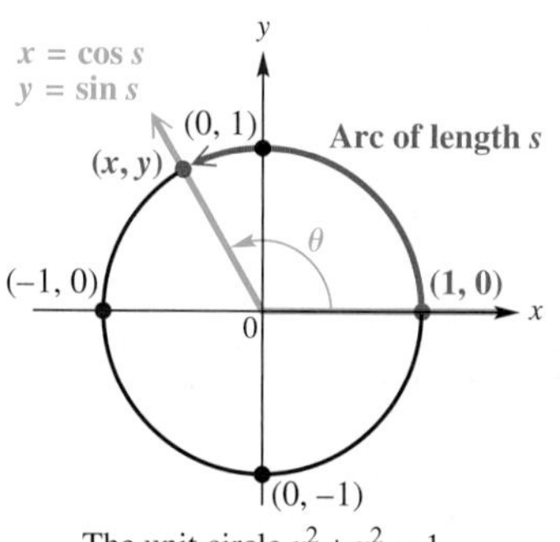

The unit circle $x^2 + y^2 = 1$

Figure 12

The radian measure of θ is related to the arc length s. For θ measured in radians, we know that $s = r\theta$. Here $r = 1$, so s, which is measured in linear units such as inches or centimeters, is equal to θ, measured in radians. Thus, the trigonometric functions of angle θ in radians found by choosing a point (x, y) on the unit circle can be rewritten as functions of the arc length s, a real number. When interpreted this way, they are called **circular functions.**

Circular Functions

For any real number s represented by a directed arc on the unit circle,

$$\sin s = y \qquad \cos s = x \qquad \tan s = \frac{y}{x} \quad (x \neq 0)$$

$$\csc s = \frac{1}{y} \quad (y \neq 0) \qquad \sec s = \frac{1}{x} \quad (x \neq 0) \qquad \cot s = \frac{x}{y} \quad (y \neq 0).$$

Since x represents the cosine of s and y represents the sine of s, and because of the discussion in **Section 6.1** on converting between degrees and radians, we can summarize a great deal of information in a concise manner, as seen in **Figure 13** on the next page.

Teaching Tip This is a good opportunity to review the concept of symmetry of a graph with respect to the axes and to the origin. (See **Section 2.7.**) Draw the unit circle and discuss the relative positions of the points (a, b), $(-a, b)$, $(-a, -b)$, and $(a, -b)$.

The unit circle is symmetric with respect to the x-axis, the y-axis, and the origin. (See **Section 2.7.**) Thus, if a point (a, b) lies on the unit circle, so do $(a, -b)$, $(-a, b)$, and $(-a, -b)$. Furthermore, each of these points has a *reference arc* of equal magnitude. For a point on the unit circle, its **reference arc** is the shortest arc from the point itself to the nearest point on the x-axis. (This concept is analogous to the reference angle concept introduced in **Chapter 5.**) Using the concept of symmetry makes determining sines and cosines of the real numbers identified in **Figure 13*** a relatively simple procedure if we know the coordinates of the points labeled in quadrant I.

Teaching Tip Draw the unit circle on the chalkboard and trace a point slowly counterclockwise, starting at $(1, 0)$. Students should see that the y-value (sine function)

- starts at 0
- increases to 1 (at $\frac{\pi}{2}$ radians)
- decreases to 0 (at π radians)
- decreases again to -1 (at $\frac{3\pi}{2}$ radians)
- returns to 0 after one full revolution (2π radians).

This analysis will help illustrate the range associated with the sine and cosine functions.

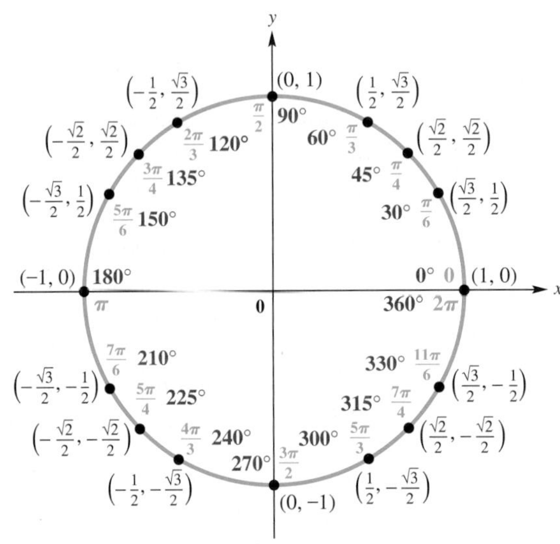

The unit circle $x^2 + y^2 = 1$

Figure 13

For example, the quadrant I real number $\frac{\pi}{3}$ is associated with the point $\left(\frac{1}{2}, \frac{\sqrt{3}}{2}\right)$ on the unit circle. Therefore, we can use symmetry to identify the coordinates of the points associated with

$$\underset{\text{Quadrant II}}{\pi - \frac{\pi}{3} = \frac{2\pi}{3}}, \quad \underset{\text{Quadrant III}}{\pi + \frac{\pi}{3} = \frac{4\pi}{3}}, \quad \text{and} \quad \underset{\text{Quadrant IV}}{2\pi - \frac{\pi}{3} = \frac{5\pi}{3}}.$$

The following chart summarizes this information.

s	Quadrant of s	Symmetry Type and Corresponding Point	$\cos s$	$\sin s$
$\frac{\pi}{3}$	I	not applicable; $\left(\frac{1}{2}, \frac{\sqrt{3}}{2}\right)$	$\frac{1}{2}$	$\frac{\sqrt{3}}{2}$
$\pi - \frac{\pi}{3} = \frac{2\pi}{3}$	II	y-axis; $\left(-\frac{1}{2}, \frac{\sqrt{3}}{2}\right)$	$-\frac{1}{2}$	$\frac{\sqrt{3}}{2}$
$\pi + \frac{\pi}{3} = \frac{4\pi}{3}$	III	origin; $\left(-\frac{1}{2}, -\frac{\sqrt{3}}{2}\right)$	$-\frac{1}{2}$	$-\frac{\sqrt{3}}{2}$
$2\pi - \frac{\pi}{3} = \frac{5\pi}{3}$	IV	x-axis; $\left(\frac{1}{2}, -\frac{\sqrt{3}}{2}\right)$	$\frac{1}{2}$	$-\frac{\sqrt{3}}{2}$

LOOKING AHEAD TO CALCULUS

If you plan to study calculus, you must become very familiar with radian measure. In calculus, the trigonometric or circular functions are always understood to have real number domains.

*The authors thank Professor Marvel Townsend of the University of Florida for her suggestion to include **Figure 13.**

NOTE Because $\cos s = x$ and $\sin s = y$, we can replace x and y in the equation of the unit circle $x^2 + y^2 = 1$ and obtain the following.

$$\cos^2 s + \sin^2 s = 1 \quad \text{Pythagorean identity (Section 5.2)}$$

The ordered pair (x, y) represents a point on the unit circle, and therefore

$$-1 \le x \le 1 \quad \text{and} \quad -1 \le y \le 1,$$

$$\mathbf{-1 \le \cos s \le 1} \quad \text{and} \quad \mathbf{-1 \le \sin s \le 1.}$$

For any value of s, both $\sin s$ and $\cos s$ exist, so the domain of these functions is the set of all real numbers.

For $\tan s$, defined as $\frac{y}{x}$, x must not equal 0. The only way x can equal 0 is when the arc length s is $\frac{\pi}{2}, -\frac{\pi}{2}, \frac{3\pi}{2}, -\frac{3\pi}{2}$, and so on. To avoid a 0 denominator, the domain of the tangent function must be restricted to those values of s that satisfy

$$s \neq (2n + 1)\frac{\pi}{2}, \quad \text{where } n \text{ is any integer.}$$

The definition of secant also has x in the denominator, so the domain of secant is the same as the domain of tangent. Both cotangent and cosecant are defined with a denominator of y. To guarantee that $y \neq 0$, the domain of these functions must be the set of all values of s that satisfy

$$s \neq n\pi, \quad \text{where } n \text{ is any integer.}$$

Domains of the Circular Functions

The domains of the circular functions are as follows.

Sine and Cosine Functions: $(-\infty, \infty)$

Tangent and Secant Functions:

$$\left\{s \,\middle|\, s \neq (2n + 1)\frac{\pi}{2}, \text{ where } n \text{ is any integer}\right\}$$

Cotangent and Cosecant Functions:

$$\{s \,|\, s \neq n\pi, \text{ where } n \text{ is any integer}\}$$

Finding Values of Circular Functions The circular functions of real numbers correspond to the trigonometric functions of angles measured in radians. Let us assume that angle θ is in standard position, superimposed on the unit circle. See **Figure 14**. Suppose that θ is the *radian* measure of this angle. Using the arc length formula

$$s = r\theta \quad \text{with } r = 1, \quad \text{we have} \quad s = \theta.$$

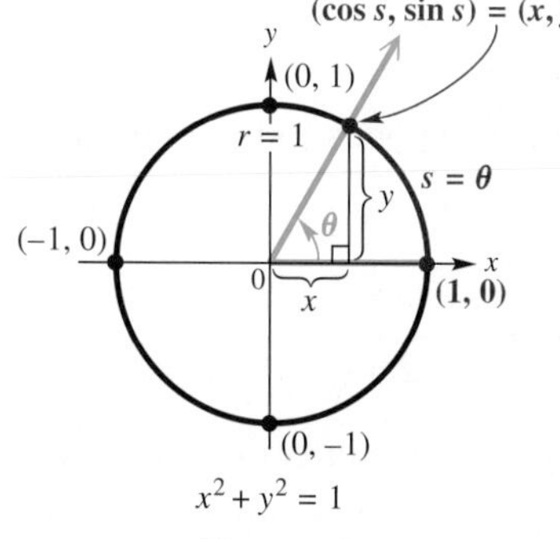

Figure 14

Thus, the length of the intercepted arc is the real number that corresponds to the radian measure of θ. Use the trigonometric function definitions from **Section 5.2** to obtain the following.

$$\sin\theta = \frac{y}{r} = \frac{y}{1} = y = \sin s, \quad \cos\theta = \frac{x}{r} = \frac{x}{1} = x = \cos s, \quad \text{and so on.}$$

As shown here, the trigonometric functions and the circular functions lead to the same function values, provided that we think of the angles as being in radian measure. This leads to the following important result.

Evaluating a Circular Function

Circular function values of real numbers are obtained in the same manner as trigonometric function values of angles measured in radians. This applies both to methods of finding exact values (such as reference angle analysis) and to calculator approximations. ***Calculators must be in radian mode when finding circular function values.***

Classroom Example 1
Find the exact values of $\sin(-3\pi)$, $\cos(-3\pi)$, and $\cot(-3\pi)$.

Answer:
$\sin(-3\pi) = 0$;
$\cos(-3\pi) = -1$;
$\cot(-3\pi)$ is undefined.

Teaching Tip Point out the importance of being able to sketch the unit circle as in **Figure 15.** This will enable students to more easily determine function values of multiples of $\frac{\pi}{2}$.

EXAMPLE 1 Finding Exact Circular Function Values

Find the exact values of $\sin \frac{3\pi}{2}$, $\cos \frac{3\pi}{2}$, and $\tan \frac{3\pi}{2}$.

SOLUTION Evaluating a circular function at the real number $\frac{3\pi}{2}$ is equivalent to evaluating it at $\frac{3\pi}{2}$ radians. An angle of $\frac{3\pi}{2}$ radians intersects the unit circle at the point $(0, -1)$, as shown in **Figure 15.** Since

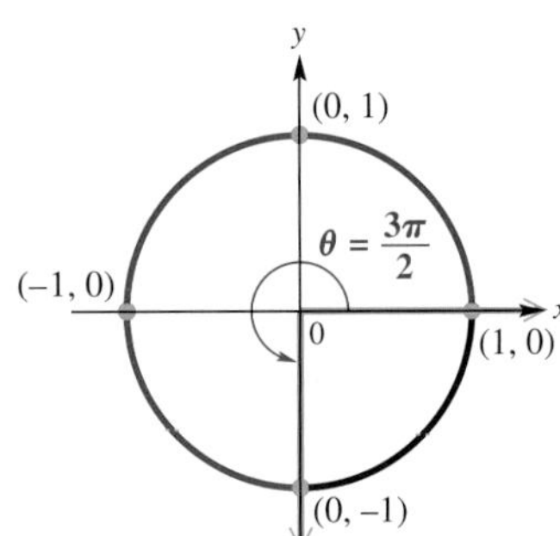

Figure 15

$$\sin s = y, \quad \cos s = x, \quad \text{and} \quad \tan s = \frac{y}{x},$$

it follows that

$$\sin \frac{3\pi}{2} = -1, \quad \cos \frac{3\pi}{2} = 0, \quad \text{and} \quad \tan \frac{3\pi}{2} \text{ is undefined.}$$

Now Try Exercises 1 and 3.

Classroom Example 2
Find each exact value using the specified method.
(a) Use **Figure 13** to find the exact values of $\sin \frac{4\pi}{3}$ and $\cos \frac{4\pi}{3}$.
(b) Use **Figure 13** and the definition of the tangent to find the exact value of $\tan(-\frac{9\pi}{4})$.
(c) Use reference angles and radian-to-degree conversion to find the exact value of $\sin \frac{11\pi}{6}$.

Answers:
(a) $\sin \frac{4\pi}{3} = -\frac{\sqrt{3}}{2}$; $\cos \frac{4\pi}{3} = -\frac{1}{2}$
(b) -1
(c) $-\frac{1}{2}$

EXAMPLE 2 Finding Exact Circular Function Values

Find each exact value using the specified method.

(a) Use **Figure 13** to find the exact values of $\cos \frac{7\pi}{4}$ and $\sin \frac{7\pi}{4}$.

(b) Use **Figure 13** and the definition of the tangent to find the exact value of $\tan(-\frac{5\pi}{3})$.

(c) Use reference angles and radian-to-degree conversion to find the exact value of $\cos \frac{2\pi}{3}$.

SOLUTION

(a) In **Figure 13,** we see that the real number $\frac{7\pi}{4}$ corresponds to the unit circle point $\left(\frac{\sqrt{2}}{2}, -\frac{\sqrt{2}}{2}\right)$.

$$\cos \frac{7\pi}{4} = \frac{\sqrt{2}}{2} \quad \text{and} \quad \sin \frac{7\pi}{4} = -\frac{\sqrt{2}}{2}$$

(b) Moving around the unit circle $\frac{5\pi}{3}$ units in the *negative* direction yields the same ending point as moving around $\frac{\pi}{3}$ units in the positive direction. Thus, $-\frac{5\pi}{3}$ corresponds to $\left(\frac{1}{2}, \frac{\sqrt{3}}{2}\right)$.

$\tan s = \frac{y}{x}$

$$\tan\left(-\frac{5\pi}{3}\right) = \tan \frac{\pi}{3} = \frac{\frac{\sqrt{3}}{2}}{\frac{1}{2}} = \frac{\sqrt{3}}{2} \div \frac{1}{2} = \frac{\sqrt{3}}{2} \cdot \frac{2}{1} = \sqrt{3}$$

(c) An angle of $\frac{2\pi}{3}$ radians corresponds to an angle of 120°. In standard position, 120° lies in quadrant II with a reference angle of 60°.

Cosine is negative in quadrant II.

$$\cos\frac{2\pi}{3} = \cos 120° = -\cos 60° = -\frac{1}{2}$$

Reference angle (Section 5.3)

✔ *Now Try Exercises 7, 13, 17, and 21.*

EXAMPLE 3 Approximating Circular Function Values

Find a calculator approximation for each circular function value.

(a) $\cos 1.85$ **(b)** $\cos 0.5149$ **(c)** $\cot 1.3209$ **(d)** $\sec(-2.9234)$

SOLUTION

(a) $\cos 1.85 \approx -0.2756$ Use a calculator in radian mode.

```
cos(1.85)
        -.2756
```

Radian mode

This is how the TI-83/84 Plus calculator displays the result of **Example 3(a),** fixed to four decimal digits.

(b) $\cos 0.5149 \approx 0.8703$ Use a calculator in radian mode.

(c) As before, to find cotangent, secant, and cosecant function values, we must use the appropriate reciprocal functions. To find cot 1.3209, first find tan 1.3209 and then find the reciprocal.

$$\cot 1.3209 = \frac{1}{\tan 1.3209} \approx 0.2552$$ Tangent and cotangent are reciprocals.

We may also find cot 1.3209 by determining the ratio of cos 1.3209 to sin 1.3209. This will give the same result as using the reciprocal function.

(d) $$\sec(-2.9234) = \frac{1}{\cos(-2.9234)} \approx -1.0243$$ Cosine and secant are reciprocals.

✔ *Now Try Exercises 23, 29, and 33.*

Classroom Example 3
Find a calculator approximation for each circular function value.
(a) sin 3.42
(b) tan 0.8234
(c) sec 5.6041
(d) csc(−2.7335)

Answers:
(a) −0.2748 (b) 1.0790
(c) 1.2851 (d) −2.5198

CAUTION A common error is using a calculator in degree mode when radian mode should be used. ***Remember, when finding a circular function value of a real number, the calculator must be in radian mode.***

Determining a Number with a Given Circular Function Value Recall from **Section 5.3** how we used a calculator to determine an angle measure, given a trigonometric function value of the angle.

Remember that the keys marked $\sin^{-1}$, $\cos^{-1}$, and $\tan^{-1}$ do not represent reciprocal functions. They enable us to find inverse function values.

For reasons explained in **Chapter 7,** the following statements are true.

- For all x in $[-1, 1]$, a calculator in radian mode returns a single value in $\left[-\frac{\pi}{2}, \frac{\pi}{2}\right]$ for $\sin^{-1} x$.
- For all x in $[-1, 1]$, a calculator in radian mode returns a single value in $[0, \pi]$ for $\cos^{-1} x$.
- For all real numbers x, a calculator in radian mode returns a single value in $\left(-\frac{\pi}{2}, \frac{\pi}{2}\right)$ for $\tan^{-1} x$.

Classroom Example 4
Find each value as specified.

(a) Approximate the value of s in the interval $\left[0, \frac{\pi}{2}\right]$ if $\sin s = 0.3210$.

(b) Find the exact value of s in the interval $\left[\frac{3\pi}{2}, 2\pi\right]$ if $\tan s = -\frac{\sqrt{3}}{3}$.

Answers:
(a) 0.3268 (b) $\frac{11\pi}{6}$

EXAMPLE 4 Finding a Number Given Its Circular Function Value

Find each value as specified.

(a) Approximate the value of s in the interval $\left[0, \frac{\pi}{2}\right]$ if $\cos s = 0.9685$.

(b) Find the exact value of s in the interval $\left[\pi, \frac{3\pi}{2}\right]$ if $\tan s = 1$.

SOLUTION

(a) Since we are given a cosine value and want to determine the real number in $\left[0, \frac{\pi}{2}\right]$ that has this cosine value, we use the *inverse cosine* function of a calculator in radian mode. See **Figure 16.**

$$\cos^{-1}(0.9685) \approx 0.2517 \quad \text{(Section 5.3)}$$

Radian mode

Figure 16

(b) Recall that $\tan \frac{\pi}{4} = 1$, and in quadrant III $\tan s$ is positive.

$$\tan\left(\pi + \frac{\pi}{4}\right) = \tan \frac{5\pi}{4} = 1$$

Thus, $s = \frac{5\pi}{4}$. See **Figure 17.**

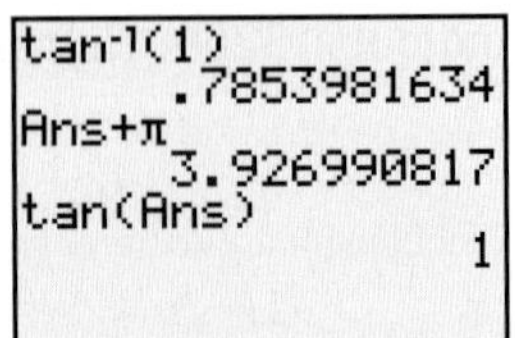

This screen supports the result in **Example 4(b)** with calculator approximations.

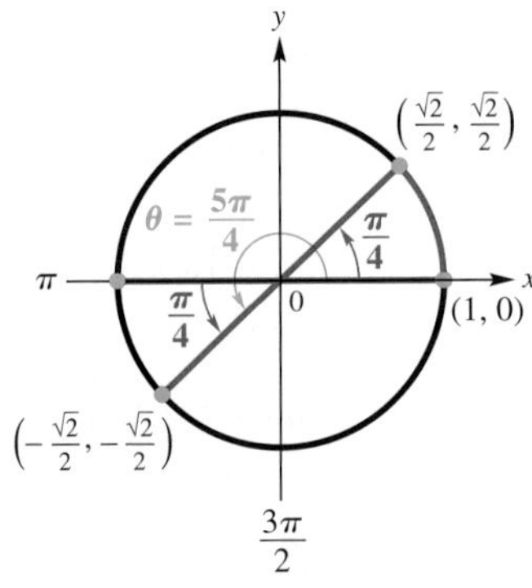

Figure 17

Now Try Exercises 55 and 65.

Expressing Function Values as Lengths of Line Segments The diagram shown in **Figure 18** illustrates a correspondence that ties together the right triangle ratio definitions of the trigonometric functions introduced in **Chapter 5** and the unit circle interpretation. The arc SR is the first-quadrant portion of the unit circle, and the standard-position angle POQ is designated θ. By definition, the coordinates of P are $(\cos \theta, \sin \theta)$. The six trigonometric functions of θ can be interpreted as lengths of line segments found in **Figure 18.**

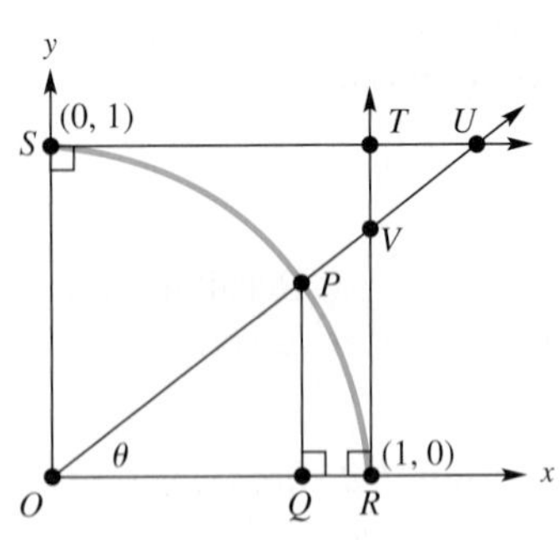

Figure 18

For $\cos \theta$ and $\sin \theta$, use right triangle POQ and right triangle ratios.

$$\cos \theta = \frac{\text{side adjacent to } \theta}{\text{hypotenuse}} = \frac{OQ}{OP} = \frac{OQ}{1} = OQ$$

$$\sin \theta = \frac{\text{side opposite } \theta}{\text{hypotenuse}} = \frac{PQ}{OP} = \frac{PQ}{1} = PQ$$

For $\tan \theta$ and $\sec \theta$, use right triangle VOR in **Figure 18** and right triangle ratios.

$$\tan \theta = \frac{\text{side opposite } \theta}{\text{side adjacent to } \theta} = \frac{VR}{OR} = \frac{VR}{1} = VR$$

$$\sec \theta = \frac{\text{hypotenuse}}{\text{side adjacent to } \theta} = \frac{OV}{OR} = \frac{OV}{1} = OV$$

For csc θ and cot θ, first note that US and OR are parallel. Thus angle SUO is equal to θ because it is an alternate interior angle to angle POQ, which is equal to θ. Use right triangle USO and right triangle ratios.

$$\csc SUO = \mathbf{\csc \theta} = \frac{\text{hypotenuse}}{\text{side opposite } \theta} = \frac{OU}{OS} = \frac{OU}{1} = \boldsymbol{OU}$$

$$\cot SUO = \mathbf{\cot \theta} = \frac{\text{side adjacent to } \theta}{\text{side opposite } \theta} = \frac{US}{OS} = \frac{US}{1} = \boldsymbol{US}$$

Figure 19 uses color to illustrate the results found above.

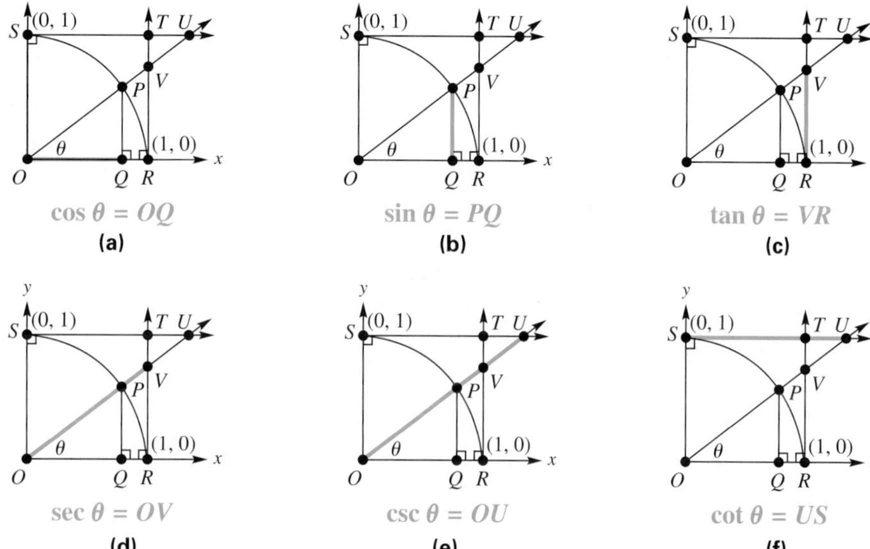

Figure 19

Classroom Example 5
Repeat **Example 5,** but suppose that angle TVU measures 45°.

Answer:

$OQ = \frac{\sqrt{2}}{2}$ $\quad PQ = \frac{\sqrt{2}}{2}$

$VR = 1$ $\quad OV = \sqrt{2}$

$OU = \sqrt{2}$ $\quad US = 1$

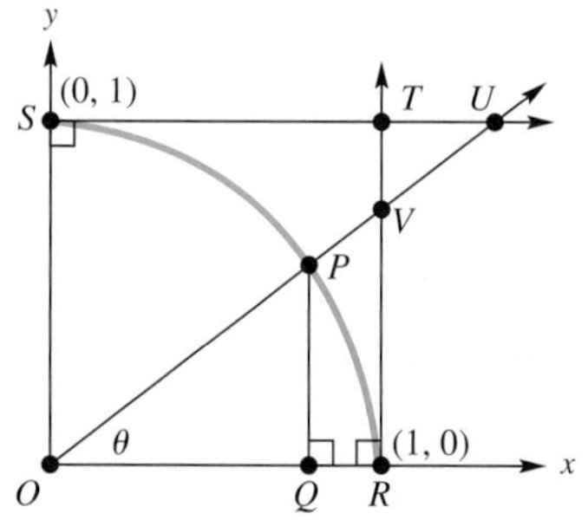

Figure 18 (repeated)

EXAMPLE 5 Finding Lengths of Line Segments

Figure 18 is repeated in the margin. Suppose that angle TVU measures 60°. Find the exact lengths of segments OQ, PQ, VR, OV, OU, and US.

SOLUTION Angle TVU has the same measure as angle OVR because they are vertical angles. Therefore, angle OVR measures 60°. Because it is one of the acute angles in right triangle VOR, θ must be its complement, measuring 30°. Now use the equations found in **Figure 19,** with $\theta = 30°$.

$$OQ = \cos 30° = \frac{\sqrt{3}}{2} \qquad OV = \sec 30° = \frac{2\sqrt{3}}{3}$$

$$PQ = \sin 30° = \frac{1}{2} \qquad OU = \csc 30° = 2$$

$$VR = \tan 30° = \frac{\sqrt{3}}{3} \qquad US = \cot 30° = \sqrt{3}$$

✔ *Now Try Exercise 73.*

Linear and Angular Speed There are situations when we need to know how fast a point on a circular disk is moving or how fast the central angle of such a disk is changing. Some examples occur with machinery involving gears or pulleys or the speed of a car around a curved portion of highway.

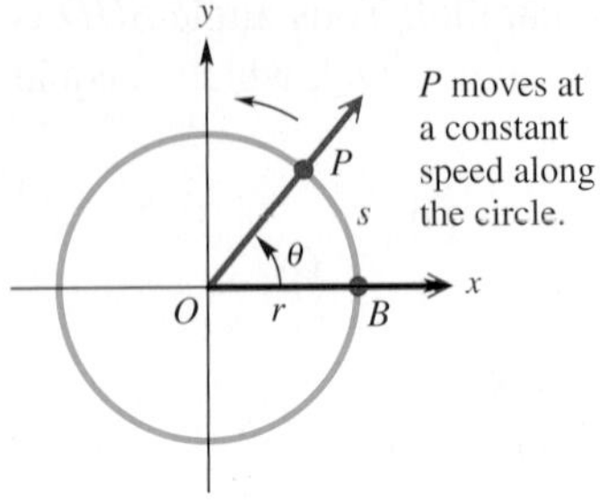

Figure 20

Suppose that point P moves at a constant speed along a circle of radius r and center O. See **Figure 20.** The measure of how fast the position of P is changing is the **linear speed.** If v represents linear speed, then

$$\text{speed} = \frac{\text{distance}}{\text{time}}, \quad \text{or} \quad v = \frac{s}{t},$$

where s is the length of the arc traced by point P at time t. (This formula is just a restatement of $r = \frac{d}{t}$ with s as distance, v as rate (speed), and t as time.)

Refer to **Figure 20.** As point P in the figure moves along the circle, ray OP rotates around the origin. Since ray OP is the terminal side of angle POB, the measure of the angle changes as P moves along the circle. The measure of how fast angle POB is changing is its **angular speed.** Angular speed, symbolized ω, is given as

$$\boldsymbol{\omega = \frac{\theta}{t},} \quad \textbf{where } \boldsymbol{\theta} \textbf{ is in radians.}$$

Here θ is the measure of angle POB at time t. ***As with earlier formulas in this chapter, θ must be measured in radians, with ω expressed in radians per unit of time.***

In **Section 6.1,** the length s of the arc intercepted on a circle of radius r by a central angle of measure θ radians was found to be $s = r\theta$. Using this formula, the formula for linear speed, $v = \frac{s}{t}$, becomes

$$v = \frac{s}{t} \qquad \text{Formula for linear speed}$$

$$v = \frac{r\theta}{t} \qquad s = r\theta$$

$$v = r \cdot \frac{\theta}{t}$$

$$v = r\omega. \qquad \omega = \frac{\theta}{t}$$

The formulas for angular and linear speed are summarized in the table.

Angular Speed ω	Linear Speed v
$\omega = \frac{\theta}{t}$ (ω in radians per unit time t, θ in radians)	$v = \frac{s}{t}$ $v = \frac{r\theta}{t}$ $v = r\omega$

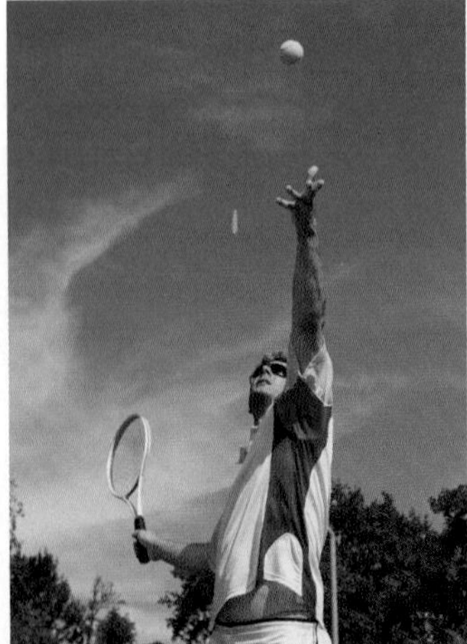

As an example of linear and angular speeds, consider the following. The human joint that can be flexed the fastest is the wrist, which can rotate through 90°, or $\frac{\pi}{2}$ radians, in 0.045 sec while holding a tennis racket. The angular speed of a human wrist swinging a tennis racket is

$$\omega = \frac{\theta}{t} \qquad \text{Formula for angular speed}$$

$$\omega = \frac{\frac{\pi}{2}}{0.045} \qquad \text{Substitute.}$$

$$\omega \approx 35 \text{ radians per sec.} \qquad \text{Use a calculator.}$$

If the radius (distance) from the tip of the racket to the wrist joint is 2 ft, then the speed at the tip of the racket is

$v = r\omega$ Formula for linear speed

$v \approx 2(35)$ Substitute.

$v = 70$ ft per sec, or about 48 mph. Use a calculator.

In a tennis serve the arm rotates at the shoulder, so the final speed of the racket is considerably greater. (*Source:* Cooper, J. and R. Glassow, *Kinesiology,* Second Edition, C.V. Mosby.)

Classroom Example 6
Suppose that P is on a circle with radius 15 in., and ray OP is rotating with angular speed $\frac{\pi}{12}$ radian per sec.

(a) Find the angle generated by P in 10 sec.

(b) Find the distance traveled by P along the circle in 10 sec.

(c) Find the linear speed of P in inches per second.

Answers:
(a) $\frac{5\pi}{6}$ radians **(b)** $\frac{25\pi}{2}$ in.
(c) $\frac{5\pi}{4}$ in. per sec

Teaching Tip Mention that, while problems involving linear and angular speed can be solved using the formula for circumference, the formulas presented in this section are much more efficient.

EXAMPLE 6 Using Linear and Angular Speed Formulas

Suppose that point P is on a circle with radius 10 cm, and ray OP is rotating with angular speed $\frac{\pi}{18}$ radian per sec.

(a) Find the angle generated by P in 6 sec.

(b) Find the distance traveled by P along the circle in 6 sec.

(c) Find the linear speed of P in centimeters per second.

SOLUTION

(a) The speed of ray OP is $\omega = \frac{\pi}{18}$ radian per sec. Use $\omega = \frac{\theta}{t}$ and $t = 6$ sec.

$$\frac{\pi}{18} = \frac{\theta}{6}$$ Let $\omega = \frac{\pi}{18}$ and $t = 6$ in the angular speed formula.

$$\theta = \frac{6\pi}{18}, \quad \text{or} \quad \frac{\pi}{3} \text{ radians}$$ Solve for θ.

(b) From part (a), P generates an angle of $\frac{\pi}{3}$ radians in 6 sec. The distance traveled by P along the circle is found as follows.

$$s = r\theta = 10\left(\frac{\pi}{3}\right) = \frac{10\pi}{3} \text{ cm} \quad \text{(Section 6.1)}$$

(c) From part (b), $s = \frac{10\pi}{3}$ cm for 6 sec, so for 1 sec we divide by 6.

Be careful simplifying this complex fraction.

$$v = \frac{s}{t} = \frac{\frac{10\pi}{3}}{6} = \frac{10\pi}{3} \div 6 = \frac{10\pi}{3} \cdot \frac{1}{6} = \frac{5\pi}{9} \text{ cm per sec}$$

✔ Now Try Exercise 77.

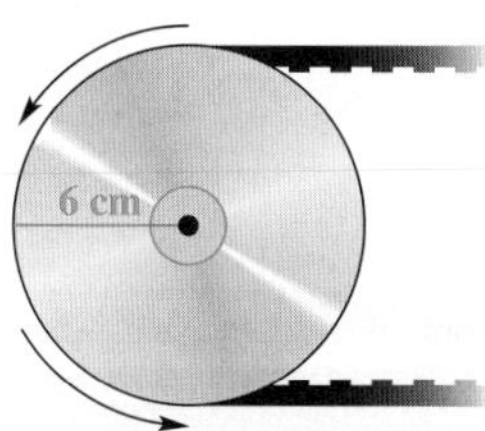

Figure 21

EXAMPLE 7 Finding Angular Speed of a Pulley and Linear Speed of a Belt

A belt runs a pulley of radius 6 cm at 80 revolutions per min. See **Figure 21.**

(a) Find the angular speed of the pulley in radians per second.

(b) Find the linear speed of the belt in centimeters per second.

SOLUTION

(a) In 1 min, the pulley makes 80 revolutions. Each revolution is 2π radians.

$$80(2\pi) = 160\pi \text{ radians per min}$$

Classroom Example 7 is on the next page.

Since there are 60 sec in 1 min, we find ω, the angular speed in radians per second, by dividing 160π by 60.

$$\omega = \frac{160\pi}{60} = \frac{8\pi}{3} \text{ radians per sec}$$

Classroom Example 7
Repeat **Example 7** for a belt that runs a pulley of radius 5 in. at 120 revolutions per min.

Answers:
(a) 4π radians per sec
(b) $20\pi \approx 63$ in. per sec

(b) The linear speed v of the belt will be the same as that of a point on the circumference of the pulley.

$$v = r\omega = 6\left(\frac{8\pi}{3}\right) = 16\pi \approx 50 \text{ cm per sec} \qquad \text{Let } r = 6 \text{ and } \omega = \frac{8\pi}{3}.$$

✔ *Now Try Exercise 115.*

6.2 Exercises

1. (a) 1 (b) 0 (c) undefined
2. (a) 0 (b) −1 (c) 0
3. (a) 0 (b) 1 (c) 0
4. (a) 0 (b) −1 (c) 0
5. (a) 0 (b) −1 (c) 0
6. (a) 1 (b) 0 (c) undefined
7. $-\frac{1}{2}$ 8. $\frac{1}{2}$
9. -1 10. -2
11. -2 12. $-\sqrt{3}$
13. $-\frac{1}{2}$ 14. $\sqrt{3}$
15. $\frac{\sqrt{2}}{2}$ 16. $-\sqrt{2}$
17. $\frac{\sqrt{3}}{2}$ 18. $-\frac{1}{2}$
19. $\frac{2\sqrt{3}}{3}$ 20. $\frac{2\sqrt{3}}{3}$
21. $-\frac{\sqrt{3}}{3}$ 22. $-\frac{\sqrt{2}}{2}$
23. 0.5736 24. 0.7314
25. 0.4068 26. 0.5397
27. 1.2065 28. 0.1944
29. 14.3338 30. 1.0170
31. −1.0460 32. −2.1291
33. −3.8665 34. 1.1848
35. 0.7 36. 0.8
37. 0.9 38. −0.75
39. −0.6 40. −1.0
41. 2.3 or 4.0 42. 4.4 or 5.0
43. 0.8 or 2.4 44. 1.3 or 5.0

For each value of the real number s, find **(a)** $\sin s$, **(b)** $\cos s$, *and* **(c)** $\tan s$. **See Example 1.**

1. $s = \frac{\pi}{2}$ **2.** $s = \pi$ **3.** $s = 2\pi$

4. $s = 3\pi$ **5.** $s = -\pi$ **6.** $s = -\frac{3\pi}{2}$

Find the exact circular function value for each of the following. **See Example 2.**

7. $\sin \frac{7\pi}{6}$ **8.** $\cos \frac{5\pi}{3}$ **9.** $\tan \frac{3\pi}{4}$ **10.** $\sec \frac{2\pi}{3}$

11. $\csc \frac{11\pi}{6}$ **12.** $\cot \frac{5\pi}{6}$ **13.** $\cos\left(-\frac{4\pi}{3}\right)$ **14.** $\tan\left(-\frac{17\pi}{3}\right)$

15. $\cos \frac{7\pi}{4}$ **16.** $\sec \frac{5\pi}{4}$ **17.** $\sin\left(-\frac{4\pi}{3}\right)$ **18.** $\sin\left(-\frac{5\pi}{6}\right)$

19. $\sec \frac{23\pi}{6}$ **20.** $\csc \frac{13\pi}{3}$ **21.** $\tan \frac{5\pi}{6}$ **22.** $\cos \frac{3\pi}{4}$

Find a calculator approximation for each circular function value. **See Example 3.**

23. $\sin 0.6109$ **24.** $\sin 0.8203$ **25.** $\cos(-1.1519)$

26. $\cos(-5.2825)$ **27.** $\tan 4.0203$ **28.** $\tan 6.4752$

29. $\csc(-9.4946)$ **30.** $\csc 1.3875$ **31.** $\sec 2.8440$

32. $\sec(-8.3429)$ **33.** $\cot 6.0301$ **34.** $\cot 3.8426$

Concept Check *The figure displays a unit circle and an angle of* 1 *radian. The tick marks on the circle are spaced at every two-tenths radian. Use the figure to estimate each value.*

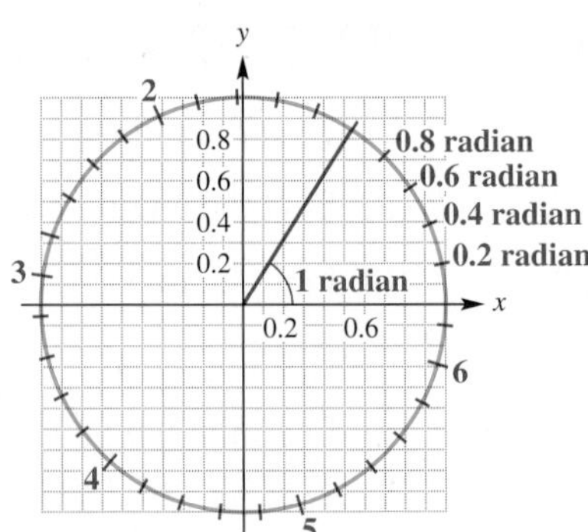

35. $\cos 0.8$ **36.** $\cos 0.6$ **37.** $\sin 2$ **38.** $\sin 4$ **39.** $\sin 3.8$ **40.** $\cos 3.2$

41. a positive angle whose cosine is -0.65 **42.** a positive angle whose sine is -0.95

43. a positive angle whose sine is 0.7 **44.** a positive angle whose cosine is 0.3

45. negative **46.** negative
47. negative **48.** positive
49. positive **50.** negative

51. $\sin\theta = \frac{\sqrt{2}}{2}$; $\cos\theta = \frac{\sqrt{2}}{2}$; $\tan\theta = 1$; $\cot\theta = 1$; $\sec\theta = \sqrt{2}$; $\csc\theta = \sqrt{2}$

52. $\sin\theta = \frac{8}{17}$; $\cos\theta = -\frac{15}{17}$; $\tan\theta = -\frac{8}{15}$; $\cot\theta = -\frac{15}{8}$; $\sec\theta = -\frac{17}{15}$; $\csc\theta = \frac{17}{8}$

53. $\sin\theta = -\frac{12}{13}$; $\cos\theta = \frac{5}{13}$; $\tan\theta = -\frac{12}{5}$; $\cot\theta = -\frac{5}{12}$; $\sec\theta = \frac{13}{5}$; $\csc\theta = -\frac{13}{12}$

54. $\sin\theta = -\frac{1}{2}$; $\cos\theta = -\frac{\sqrt{3}}{2}$; $\tan\theta = \frac{\sqrt{3}}{3}$; $\cot\theta = \sqrt{3}$; $\sec\theta = -\frac{2\sqrt{3}}{3}$; $\csc\theta = -2$

55. 0.2095 **56.** 0.6720
57. 1.4426 **58.** 1.2799
59. 0.3887 **60.** 1.3634

61. $\frac{5\pi}{6}$ **62.** $\frac{2\pi}{3}$
63. $\frac{4\pi}{3}$ **64.** $\frac{7\pi}{6}$
65. $\frac{7\pi}{4}$ **66.** $\frac{11\pi}{6}$

67. $\frac{4\pi}{3}, \frac{5\pi}{3}$ **68.** $\frac{2\pi}{3}, \frac{4\pi}{3}$
69. $\frac{\pi}{4}, \frac{3\pi}{4}, \frac{5\pi}{4}, \frac{7\pi}{4}$
70. $\frac{\pi}{3}, \frac{2\pi}{3}, \frac{4\pi}{3}, \frac{5\pi}{3}$
71. $-\frac{11\pi}{6}, -\frac{7\pi}{6}, -\frac{5\pi}{6}, -\frac{\pi}{6}, \frac{\pi}{6}, \frac{5\pi}{6}$
72. $-\frac{3\pi}{4}, -\frac{\pi}{4}, \frac{\pi}{4}, \frac{3\pi}{4}$

Concept Check *Without using a calculator, decide whether each function value is positive or negative. (Hint: Consider the radian measures of the quadrantal angles, and remember that $\pi \approx 3.14$.)*

45. $\cos 2$ **46.** $\sin(-1)$ **47.** $\sin 5$ **48.** $\cos 6$ **49.** $\tan 6.29$ **50.** $\tan(-6.29)$

Concept Check *Each figure in Exercises 51–54 shows an angle θ in standard position with its terminal side intersecting the unit circle. Evaluate the six circular function values of θ.*

51.

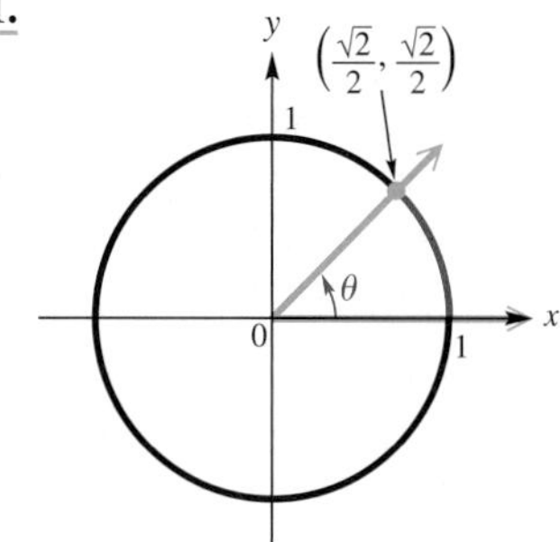

52.

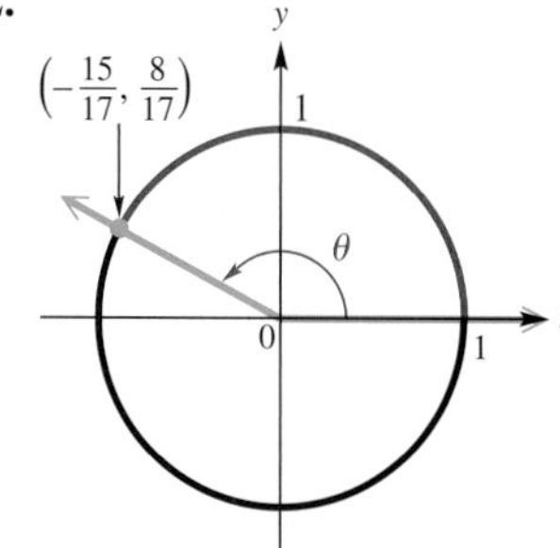

53.

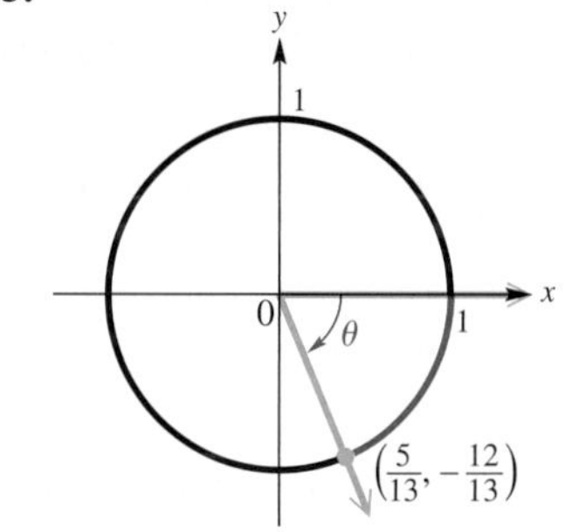

54.

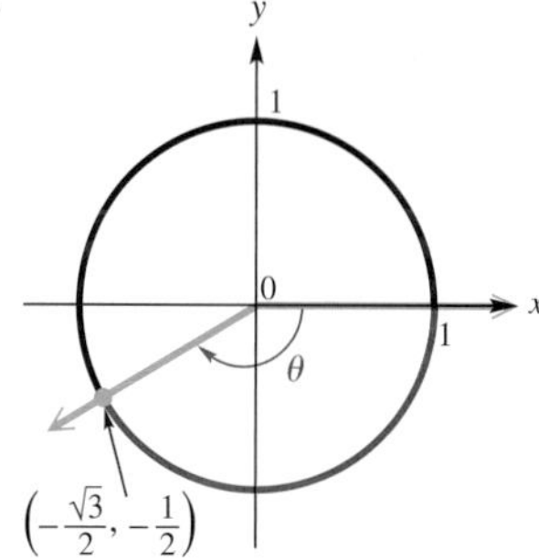

Find the value of s in the interval $\left[0, \frac{\pi}{2}\right]$ that makes each statement true. ***See Example 4(a).***

55. $\tan s = 0.2126$ **56.** $\cos s = 0.7826$ **57.** $\sin s = 0.9918$

58. $\cot s = 0.2994$ **59.** $\sec s = 1.0806$ **60.** $\csc s = 1.0219$

Find the exact value of s in the given interval that has the given circular function value. Do not use a calculator. ***See Example 4(b).***

61. $\left[\frac{\pi}{2}, \pi\right]$; $\sin s = \frac{1}{2}$ **62.** $\left[\frac{\pi}{2}, \pi\right]$; $\cos s = -\frac{1}{2}$

63. $\left[\pi, \frac{3\pi}{2}\right]$; $\tan s = \sqrt{3}$ **64.** $\left[\pi, \frac{3\pi}{2}\right]$; $\sin s = -\frac{1}{2}$

65. $\left[\frac{3\pi}{2}, 2\pi\right]$; $\tan s = -1$ **66.** $\left[\frac{3\pi}{2}, 2\pi\right]$; $\cos s = \frac{\sqrt{3}}{2}$

Find the exact values of s in the given interval that satisfy the given condition.

67. $[0, 2\pi)$; $\sin s = -\frac{\sqrt{3}}{2}$ **68.** $[0, 2\pi)$; $\cos s = -\frac{1}{2}$

69. $[0, 2\pi)$; $\cos^2 s = \frac{1}{2}$ **70.** $[0, 2\pi)$; $\tan^2 s = 3$

71. $[-2\pi, \pi)$; $3\tan^2 s = 1$ **72.** $[-\pi, \pi)$; $\sin^2 s = \frac{1}{2}$

73. (a) $\frac{1}{2}$ (b) $\frac{\sqrt{3}}{2}$
(c) $\sqrt{3}$ (d) 2
(e) $\frac{2\sqrt{3}}{3}$ (f) $\frac{\sqrt{3}}{3}$
74. (a) 0.7880 (b) 0.6157
(c) 0.7813 (d) 1.269
(e) 1.624 (f) 1.280
75. 2π sec 76. 2π sec
77. (a) $\frac{\pi}{2}$ radians (b) 10π cm
(c) $\frac{5\pi}{3}$ cm per sec
78. (a) $\frac{2\pi}{5}$ radians (b) 12π cm
(c) 3π cm per sec
79. 2π radians
80. $\frac{5\pi}{4}$ radians
81. $\frac{3\pi}{32}$ radian per sec
82. $\frac{\pi}{25}$ radian per sec
83. $\frac{6}{5}$ min
84. 9 min
85. 0.1803 radian per sec
86. 2.078 radians per sec
87. 10.77 radians
88. 20.51 radians
89. 8π m per sec
90. $\frac{72\pi}{5}$ cm per sec
91. $\frac{9}{5}$ radians per sec
92. 6 radians per sec
93. 1.834 radians per sec
94. 9.296 cm per sec

In Exercises 73 and 74, see ***Example 5.***

73. Refer to **Figures 18 and 19.** Suppose that angle θ measures 60°. Find the exact length of each segment.

(a) OQ (b) PQ (c) VR (d) OV (e) OU (f) US

74. Refer to **Figures 18 and 19.** Repeat **Exercise 73** for $\theta = 38°$, but give lengths as approximations to four significant digits.

Concept Check Refer to the figure and answer Exercises 75 and 76.

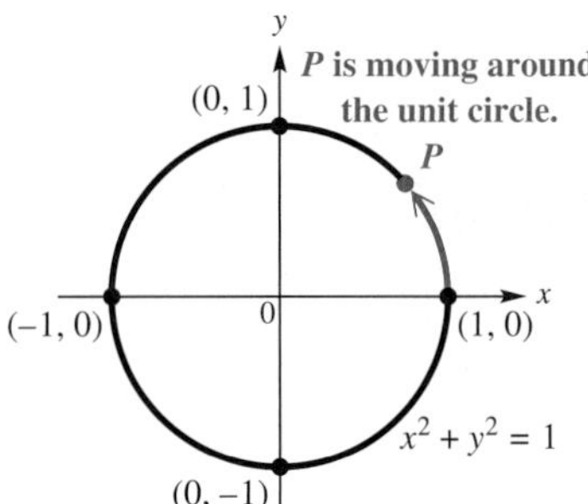

75. If the point P moves around the circumference of the unit circle at an angular velocity of 1 radian per sec, how long will it take for P to move around the entire circle?

76. If the point P moves around the circumference of the unit circle at a speed of 1 unit per sec, how long will it take for P to move around the entire circle?

Suppose that point P is on a circle with radius r, and ray OP is rotating with angular speed ω. For the given values of r, ω, and t, find each of the following. ***See Example 6.***

(a) the angle generated by P in time t
(b) the distance traveled by P along the circle in time t
(c) the linear speed of P

77. $r = 20$ cm, $\omega = \frac{\pi}{12}$ radian per sec, $t = 6$ sec

78. $r = 30$ cm, $\omega = \frac{\pi}{10}$ radian per sec, $t = 4$ sec

Use the formula $\omega = \frac{\theta}{t}$ to find the value of the missing variable.

79. $\omega = \frac{2\pi}{3}$ radians per sec, $t = 3$ sec

80. $\omega = \frac{\pi}{4}$ radian per min, $t = 5$ min

81. $\theta = \frac{3\pi}{4}$ radians, $t = 8$ sec

82. $\theta = \frac{2\pi}{5}$ radians, $t = 10$ sec

83. $\theta = \frac{2\pi}{9}$ radian, $\omega = \frac{5\pi}{27}$ radian per min

84. $\theta = \frac{3\pi}{8}$ radians, $\omega = \frac{\pi}{24}$ radian per min

85. $\theta = 3.871$ radians, $t = 21.47$ sec

86. $\theta = 5.225$ radians, $t = 2.515$ sec

87. $\omega = 0.9067$ radian per min, $t = 11.88$ min

88. $\omega = 4.316$ radians per min, $t = 4.752$ min

Use the formula $v = r\omega$ to find the value of the missing variable.

89. $r = 12$ m, $\omega = \frac{2\pi}{3}$ radians per sec

90. $r = 8$ cm, $\omega = \frac{9\pi}{5}$ radians per sec

91. $v = 9$ m per sec, $r = 5$ m

92. $v = 18$ ft per sec, $r = 3$ ft

93. $v = 107.7$ m per sec, $r = 58.74$ m

94. $r = 24.93$ cm, $\omega = 0.3729$ radian per sec

95. 18π cm
96. $\frac{216\pi}{5}$ yd
97. 12 sec
98. 4 sec
99. $\frac{3\pi}{32}$ radian per sec
100. $\frac{\pi}{18}$ radian per sec

101. $\frac{\pi}{6}$ radian per hr
102. $\frac{\pi}{30}$ radian per sec
103. $\frac{\pi}{30}$ radian per min
104. 600π radians per min

105. $\frac{7\pi}{30}$ cm per min
106. $\frac{14\pi}{15}$ mm per sec
107. 168π m per min
108. 1260π cm per min
109. 1500π m per min
110. $112{,}880\pi$ cm per min

111. 16.6 mph
112. 24.62 hr

The formula $\omega = \frac{\theta}{t}$ can be rewritten as $\theta = \omega t$. Substituting ωt for θ converts $s = r\theta$ to $s = r\omega t$. Use the formula $s = r\omega t$ to find the value of the missing variable.

95. $r = 6$ cm, $\omega = \frac{\pi}{3}$ radians per sec, $t = 9$ sec

96. $r = 9$ yd, $\omega = \frac{2\pi}{5}$ radians per sec, $t = 12$ sec

97. $s = 6\pi$ cm, $r = 2$ cm, $\omega = \frac{\pi}{4}$ radian per sec

98. $s = \frac{12\pi}{5}$ m, $r = \frac{3}{2}$ m, $\omega = \frac{2\pi}{5}$ radians per sec

99. $s = \frac{3\pi}{4}$ km, $r = 2$ km, $t = 4$ sec

100. $s = \frac{8\pi}{9}$ m, $r = \frac{4}{3}$ m, $t = 12$ sec

Find the angular speed ω for each of the following.

101. the hour hand of a clock

102. the second hand of a clock

103. the minute hand of a clock

104. a line from the center to the edge of a CD revolving 300 times per min

Find the linear speed v for each of the following.

105. the tip of the minute hand of a clock, if the hand is 7 cm long

106. the tip of the second hand of a clock, if the hand is 28 mm long

107. a point on the edge of a flywheel of radius 2 m, rotating 42 times per min

108. a point on the tread of a tire of radius 18 cm, rotating 35 times per min

109. the tip of a propeller 3 m long, rotating 500 times per min (*Hint:* $r = 1.5$ m)

110. a point on the edge of a gyroscope of radius 83 cm, rotating 680 times per min

Solve each problem. ***See Examples 6 and 7.***

111. *Speed of a Bicycle* The tires of a bicycle have radius 13.0 in. and are turning at the rate of 215 revolutions per min. See the figure. How fast is the bicycle traveling in miles per hour? (*Hint:* 5280 ft = 1 mi)

112. *Hours in a Martian Day* Mars rotates on its axis at the rate of about 0.2552 radian per hr. Approximately how many hours are in a Martian day (or *sol*)? (*Source: World Almanac and Book of Facts.*)

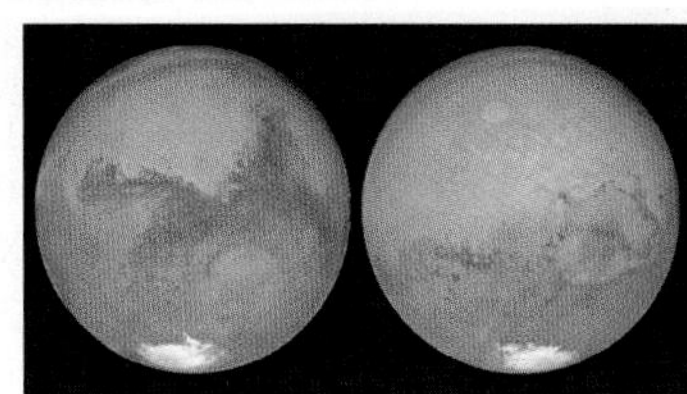

Opposite sides of Mars

113. (a) $\frac{2\pi}{365}$ radian
(b) $\frac{\pi}{4380}$ radian per hr
(c) about 67,000 mph

114. (a) 2π radians per day; $\frac{\pi}{12}$ radian per hr
(b) 0
(c) $12{,}800\pi$ km per day or about 533π km per hr
(d) about 28,000 km per day or about 1200 km per hr

115. (a) 3.1 cm per sec
(b) 0.24 radian per sec

116. larger pulley: $\frac{25\pi}{18}$ radians per sec; smaller pulley: $\frac{125\pi}{48}$ radians per sec

117. 3.73 cm

118. about 29 sec

119. 523.6 radians per sec

120. 125 ft per sec

113. *Angular and Linear Speeds of Earth* The orbit of Earth about the sun is almost circular. Assume that the orbit is a circle with radius 93,000,000 mi. Its angular and linear speeds are used in designing solar-power facilities.

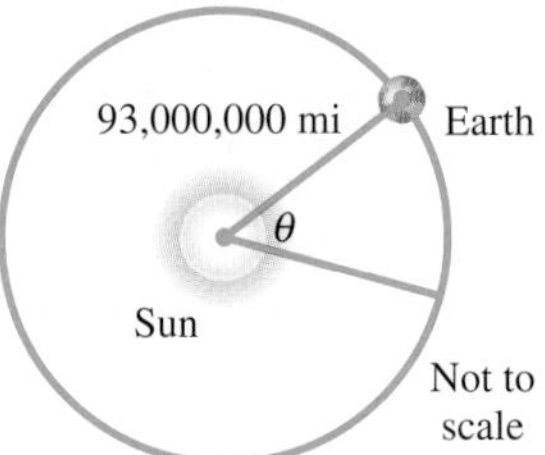

(a) Assume that a year is 365 days, and find the angle formed by Earth's movement in one day.
(b) Give the angular speed in radians per hour.
(c) Find the linear speed of Earth in miles per hour.

114. *Angular and Linear Speeds of Earth* Earth revolves on its axis once every 24 hr. Assuming that Earth's radius is 6400 km, find the following.

(a) angular speed of Earth in radians per day and radians per hour
(b) linear speed at the North Pole or South Pole
(c) linear speed at Quito, Ecuador, a city on the equator
(d) linear speed at Salem, Oregon (halfway from the equator to the North Pole)

115. *Speeds of a Pulley and a Belt* The pulley shown has a radius of 12.96 cm. Suppose it takes 18 sec for 56 cm of belt to go around the pulley.

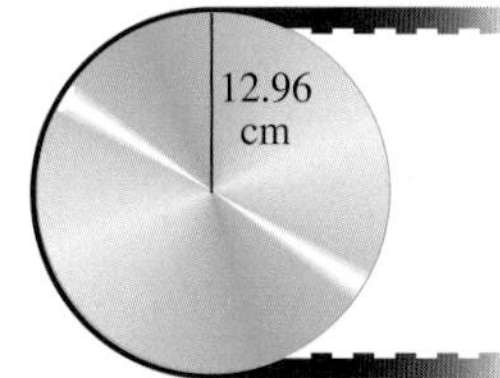

(a) Find the linear speed of the belt in centimeters per second.
(b) Find the angular speed of the pulley in radians per second.

116. *Angular Speeds of Pulleys* The two pulleys in the figure have radii of 15 cm and 8 cm, respectively. The larger pulley rotates 25 times in 36 sec. Find the angular speed of each pulley in radians per second.

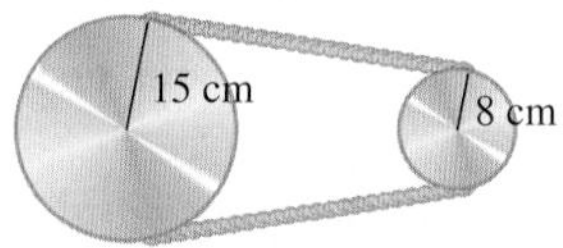

117. *Radius of a Spool of Thread* A thread is being pulled off a spool at the rate of 59.4 cm per sec. Find the radius of the spool if it makes 152 revolutions per min.

118. *Time to Move along a Railroad Track* A railroad track is laid along the arc of a circle of radius 1800 ft. The circular part of the track subtends a central angle of 40°. How long (in seconds) will it take a point on the front of a train traveling 30.0 mph to go around this portion of the track?

119. *Angular Speed of a Motor Propeller* The propeller of a 90-horsepower outboard motor at full throttle rotates at exactly 5000 revolutions per min. Find the angular speed of the propeller in radians per second.

120. *Linear Speed of a Golf Club* The shoulder joint can rotate at 25.0 radians per sec. If a golfer's arm is straight and the distance from the shoulder to the club head is 5.00 ft, find the linear speed of the club head from shoulder rotation. (*Source:* Cooper, J. and R. Glassow, *Kinesiology,* Second Edition, C.V. Mosby.)

6.3 Graphs of the Sine and Cosine Functions

- Periodic Functions
- Graph of the Sine Function
- Graph of the Cosine Function
- Graphing Techniques, Amplitude, and Period
- Connecting Graphs with Equations
- Using a Trigonometric Model

Teaching Tip In preparation for the material presented in this chapter, have students review **Section 2.3** on Functions and also **Section 2.7** on Graphing Techniques.

Periodic Functions Many things in daily life repeat with a predictable pattern, such as weather, tides, and hours of daylight. Because the sine and cosine functions repeat their values in a regular pattern, they are *periodic functions.* **Figure 22** shows a periodic graph that represents a normal heartbeat.

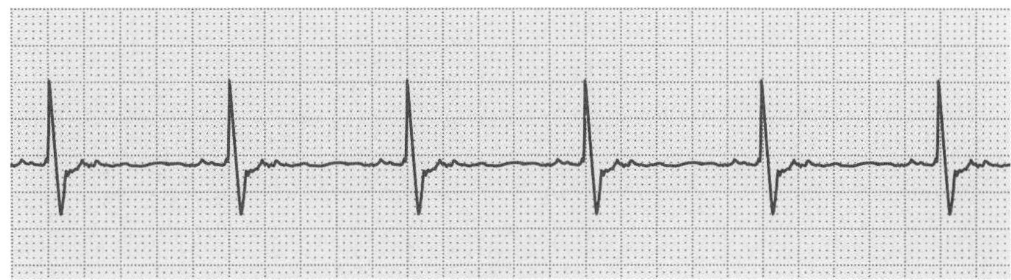

Figure 22

Periodic Function

A **periodic function** is a function f such that

$$f(x) = f(x + np),$$

for every real number x in the domain of f, every integer n, and some positive real number p. The least possible positive value of p is the **period** of the function.

LOOKING AHEAD TO CALCULUS
Periodic functions are used throughout calculus, so you will need to know their characteristics. One use of these functions is to describe the location of a point in the plane using **polar coordinates,** an alternative to rectangular coordinates. (See **Chapter 8.**)

The circumference of the unit circle is 2π, so the least value of p for which the sine and cosine functions repeat is 2π. ***Therefore, the sine and cosine functions are periodic functions with period 2π,*** and the following statements are true for every integer n.

$$\sin x = \sin(x + n \cdot 2\pi) \quad \text{and} \quad \cos x = \cos(x + n \cdot 2\pi)$$

Graph of the Sine Function In **Section 6.2** we saw that for a real number s, the point on the unit circle corresponding to s has coordinates $(\cos s, \sin s)$. See **Figure 23.** Trace along the circle to verify the results shown in the table.

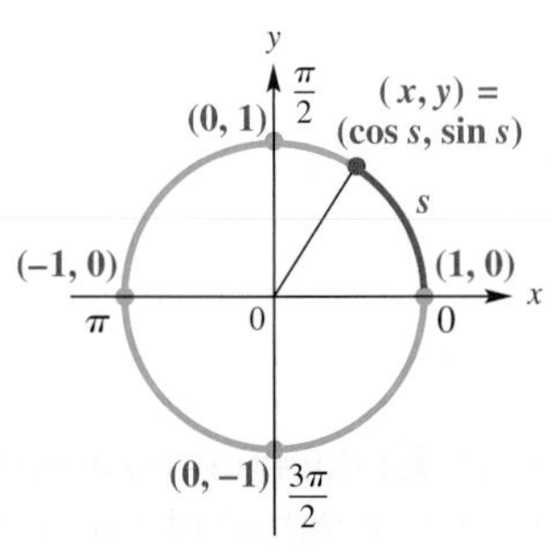

Figure 23

As s Increases from	$\sin s$	$\cos s$
0 to $\frac{\pi}{2}$	Increases from 0 to 1	Decreases from 1 to 0
$\frac{\pi}{2}$ to π	Decreases from 1 to 0	Decreases from 0 to -1
π to $\frac{3\pi}{2}$	Decreases from 0 to -1	Increases from -1 to 0
$\frac{3\pi}{2}$ to 2π	Increases from -1 to 0	Increases from 0 to 1

To avoid confusion when graphing the sine function, we use x rather than s; this corresponds to the letters in the xy-coordinate system. Selecting key values of x and finding the corresponding values of $\sin x$ leads to the table in **Figure 24** on the next page.

To obtain the traditional graph in **Figure 24,** we plot the points from the table, use symmetry, and join them with a smooth curve. Since $y = \sin x$ is periodic with period 2π and has domain $(-\infty, \infty)$, the graph continues in the same pattern in both directions. This graph is called a **sine wave,** or **sinusoid.**

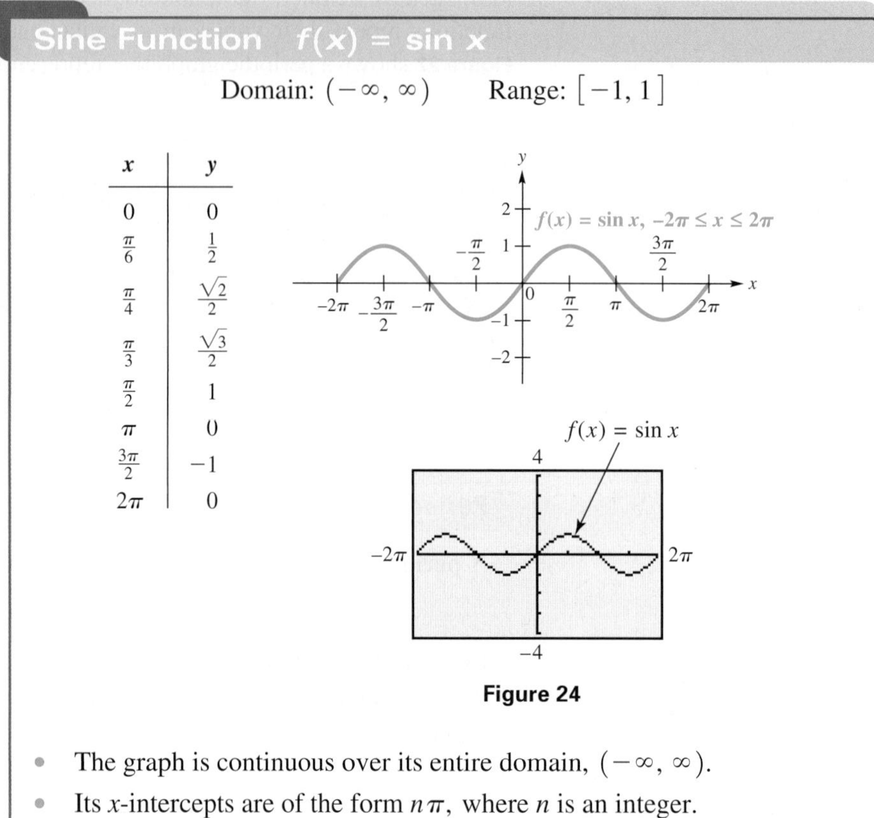

Sine Function $f(x) = \sin x$

Domain: $(-\infty, \infty)$ Range: $[-1, 1]$

x	y
0	0
$\frac{\pi}{6}$	$\frac{1}{2}$
$\frac{\pi}{4}$	$\frac{\sqrt{2}}{2}$
$\frac{\pi}{3}$	$\frac{\sqrt{3}}{2}$
$\frac{\pi}{2}$	1
π	0
$\frac{3\pi}{2}$	-1
2π	0

Figure 24

- The graph is continuous over its entire domain, $(-\infty, \infty)$.
- Its x-intercepts are of the form $n\pi$, where n is an integer.
- Its period is 2π.
- The graph is symmetric with respect to the origin, so the function is an odd function. For all x in the domain, $\sin(-x) = -\sin x$.

NOTE A function f is an **odd function** if for all x in the domain of f,

$$f(-x) = -f(x). \quad \text{(Section 2.7)}$$

The graph of an odd function is symmetric with respect to the origin. This means that if (x, y) belongs to the function, then $(-x, -y)$ also belongs to the function. For example, $\left(\frac{\pi}{2}, 1\right)$ and $\left(-\frac{\pi}{2}, -1\right)$ are points on the graph of $y = \sin x$, illustrating the property $\sin(-x) = -\sin x$.

The sine function is closely related to the unit circle. **Its domain consists of real numbers corresponding to angle measures (or arc lengths) of the unit circle, and its range corresponds to the y-coordinates (or sine values) of the unit circle.**

Consider the unit circle in **Figure 23** and assume that the line from the origin to some point on the circle is part of the pedal of a bicycle, with a foot placed on the circle itself. As the pedal is rotated from 0 radians on the horizontal axis through various angles, the angle (or arc length) giving the pedal's location and its corresponding height from the horizontal axis given by $\sin x$ are used to create points on the sine graph. See **Figure 25** on the next page.

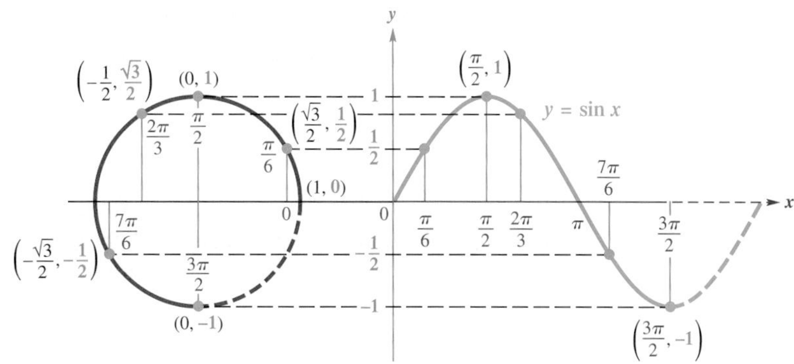

Figure 25

> **LOOKING AHEAD TO CALCULUS**
> The discussion of the derivative of a function in calculus shows that for the sine function, the slope of the tangent line at any point x is given by $\cos x$. For example, look at the graph of $y = \sin x$ and notice that a tangent line at $x = \pm\frac{\pi}{2}, \pm\frac{3\pi}{2}, \pm\frac{5\pi}{2}, \ldots$ will be horizontal and thus have slope 0. Now look at the graph of $y = \cos x$ and see that for these values, $\cos x = 0$.

Graph of the Cosine Function The graph of $y = \cos x$ in **Figure 26** has the same shape as the graph of $y = \sin x$. ***The graph of the cosine function is, in fact, the graph of the sine function shifted, or translated,*** $\frac{\pi}{2}$ ***units to the left.***

Cosine Function $f(x) = \cos x$

Domain: $(-\infty, \infty)$ Range: $[-1, 1]$

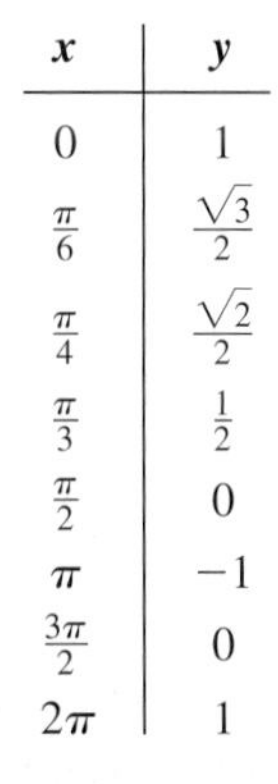

x	y
0	1
$\frac{\pi}{6}$	$\frac{\sqrt{3}}{2}$
$\frac{\pi}{4}$	$\frac{\sqrt{2}}{2}$
$\frac{\pi}{3}$	$\frac{1}{2}$
$\frac{\pi}{2}$	0
π	-1
$\frac{3\pi}{2}$	0
2π	1

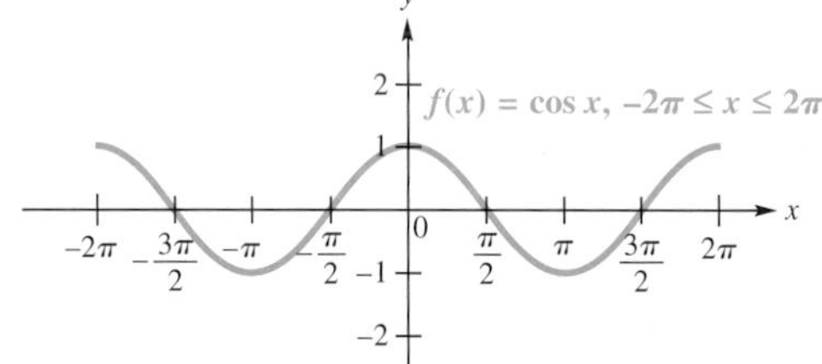

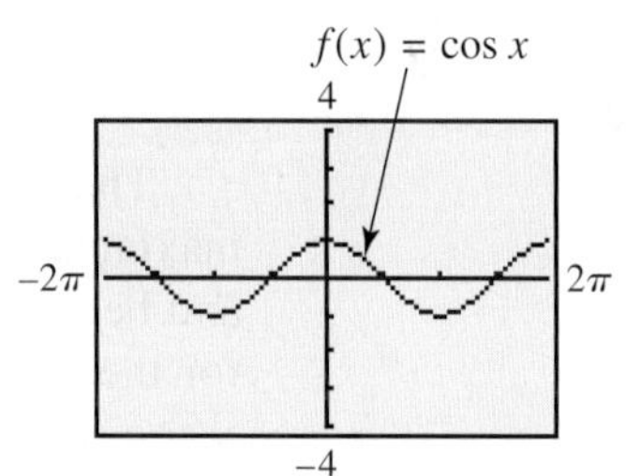

Figure 26

- The graph is continuous over its entire domain, $(-\infty, \infty)$.
- Its x-intercepts are of the form $(2n + 1)\frac{\pi}{2}$, where n is an integer.
- Its period is 2π.
- The graph is symmetric with respect to the y-axis, so the function is an even function. For all x in the domain, $\cos(-x) = \cos x$.

NOTE A function f is an **even function** if for all x in the domain of f,

$$f(-x) = f(x). \quad \text{(Section 2.7)}$$

The graph of an even function is symmetric with respect to the y-axis. This means that if (x, y) belongs to the function, then $(-x, y)$ also belongs to the function. For example, $\left(\frac{\pi}{2}, 0\right)$ and $\left(-\frac{\pi}{2}, 0\right)$ are points on the graph of $y = \cos x$, illustrating the property $\cos(-x) = \cos x$.

The calculator graphs of $f(x) = \sin x$ in **Figure 24** and $f(x) = \cos x$ in **Figure 26** are graphed in the window approximately $[-2\pi, 2\pi]$ by $[-4, 4]$, with Xscl $= \frac{\pi}{2}$ and Yscl $= 1$. This is the **trig viewing window.** (Your model may use a different "standard" trig window. Consult your owner's manual.) ■

Graphing Techniques, Amplitude, and Period The examples that follow show graphs that are "stretched" or "compressed" (shrunk) either vertically, horizontally, or both when compared with the graphs of $y = \sin x$ or $y = \cos x$.

Classroom Example 1

Graph $y = \frac{1}{2} \sin x$, and compare to the graph of $y = \sin x$.

Answer:

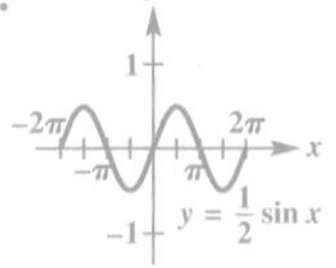

The only change in the graph is the range, which becomes $\left[-\frac{1}{2}, \frac{1}{2}\right]$.

EXAMPLE 1 Graphing $y = a \sin x$

Graph $y = 2 \sin x$, and compare to the graph of $y = \sin x$.

SOLUTION For a given value of x, the value of y is twice what it would be for $y = \sin x$, as shown in the table of values. The only change in the graph is the range, which becomes $[-2, 2]$. See **Figure 27,** which includes a graph of $y = \sin x$ for comparison.

x	0	$\frac{\pi}{2}$	π	$\frac{3\pi}{2}$	2π
$\sin x$	0	1	0	-1	0
$2 \sin x$	0	2	0	-2	0

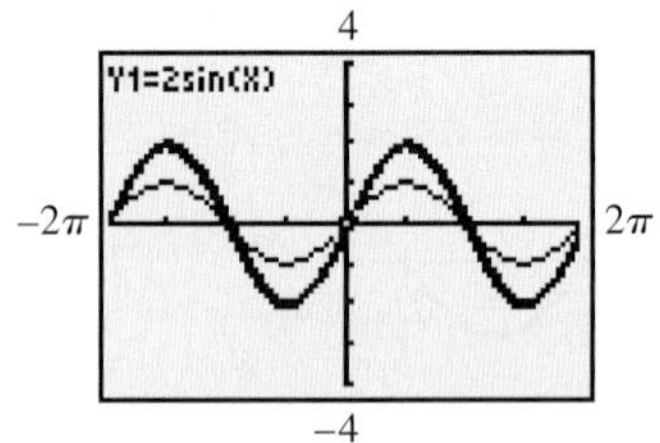

The thick graph style represents the function $y = 2 \sin x$ in **Example 1.**

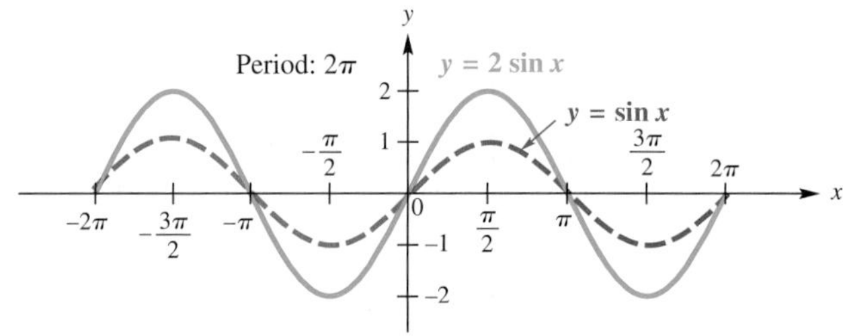

Figure 27

The **amplitude** of a periodic function is half the difference between the maximum and minimum values. It describes the height of the graph both above and below a horizontal line passing through the "middle" of the graph. Thus, for the basic sine function $y = \sin x$ (and also for the basic cosine function $y = \cos x$) the amplitude is computed as follows.

$$\frac{1}{2}[1 - (-1)] = \frac{1}{2}(2) = 1 \quad \text{Amplitude of } y = \sin x$$

For $y = 2 \sin x$, the amplitude is

$$\frac{1}{2}[2 - (-2)] = \frac{1}{2}(4) = 2. \quad \text{Amplitude of } y = 2 \sin x$$

We can think of the graph of $y = a \sin x$ as a vertical stretching of the graph of $y = \sin x$ when $a > 1$ and a vertical shrinking when $0 < a < 1$.

✔ *Now Try Exercise 15.*

Generalizing from **Example 1** gives the following.

Amplitude

The graph of $y = a \sin x$ or $y = a \cos x$, with $a \neq 0$, will have the same shape as the graph of $y = \sin x$ or $y = \cos x$, respectively, except with range $[-|a|, |a|]$. The amplitude is $|a|$.

Teaching Tip Students may think that the period of $y = \sin 2x$ is 4π. Explain that a factor of 2 causes values of the argument to *increase twice as fast,* thereby shortening the length of the period to half its original value.

While the coefficient a in $y = a \sin x$ or $y = a \cos x$ affects the amplitude of the graph, the coefficient of x in the argument affects the period. Consider $y = \sin 2x$. We can complete a table of values for the interval $[0, 2\pi]$.

x	0	$\frac{\pi}{4}$	$\frac{\pi}{2}$	$\frac{3\pi}{4}$	π	$\frac{5\pi}{4}$	$\frac{3\pi}{2}$	$\frac{7\pi}{4}$	2π
sin 2*x*	0	1	0	−1	0	1	0	−1	0

Note that one complete cycle occurs in π units, not 2π units. Therefore, the period here is π, which equals $\frac{2\pi}{2}$. Now consider $y = \sin 4x$. Look at the next table.

x	0	$\frac{\pi}{8}$	$\frac{\pi}{4}$	$\frac{3\pi}{8}$	$\frac{\pi}{2}$	$\frac{5\pi}{8}$	$\frac{3\pi}{4}$	$\frac{7\pi}{8}$	π
sin 4*x*	0	1	0	−1	0	1	0	−1	0

These values suggest that one complete cycle is achieved in $\frac{\pi}{2}$ or $\frac{2\pi}{4}$ units, which is reasonable since

$$\sin\left(4 \cdot \frac{\pi}{2}\right) = \sin 2\pi = 0.$$

***In general, the graph of a function of the form y* = sin *bx or y* = cos *bx*, *for b* > 0, *will have a period different from* 2π *when b* ≠ 1.** To see why this is so, remember that the values of $\sin bx$ or $\cos bx$ will take on all possible values as bx ranges from 0 to 2π. Therefore, to find the period of either of these functions, we must solve the following three-part inequality.

$$0 \le bx \le 2\pi \quad \text{(Section 1.7)}$$

$$0 \le x \le \frac{2\pi}{b} \quad \text{Divide each part by the positive number } b.$$

Thus, the period is $\boldsymbol{\frac{2\pi}{b}}$**.** By dividing the interval $\left[0, \frac{2\pi}{b}\right]$ into four equal parts, we obtain the values for which $\sin bx$ or $\cos bx$ is $-1, 0,$ or 1. These values will give minimum points, x-intercepts, and maximum points on the graph. Once these points are determined, we can sketch the graph by joining the points with a smooth sinusoidal curve. (If a function has $b < 0$, then the identities of the next chapter can be used to rewrite the function so that $b > 0$.)

NOTE One method to divide an interval into four equal parts is as follows.

Step 1 Find the midpoint of the interval by adding the x-values of the endpoints and dividing by 2. (See **Section 2.1.**)

Step 2 Find the quarter points (the midpoints of the two intervals found in Step 1) using the same procedure.

Classroom Example 2

Graph $y = \sin \frac{1}{3}x$, and compare to the graph of $y = \sin x$.

Answer:

The period changes from 2π to 6π.

EXAMPLE 2 Graphing $y = \sin bx$

Graph $y = \sin 2x$, and compare to the graph of $y = \sin x$.

SOLUTION In this function the coefficient of x is 2, so $b = 2$ and the period is $\frac{2\pi}{2} = \pi$. Therefore, the graph will complete one period over the interval $[0, \pi]$.

We can divide the interval $[0, \pi]$ into four equal parts by first finding its midpoint: $\frac{1}{2}(0 + \pi) = \frac{\pi}{2}$. The quarter points are found next by determining the midpoints of the two intervals $\left[0, \frac{\pi}{2}\right]$ and $\left[\frac{\pi}{2}, \pi\right]$.

$\frac{1}{2}\left(\frac{\pi}{2} + \pi\right) = \frac{1}{2}\left(\frac{3\pi}{2}\right) = \frac{3\pi}{4}$

$$\frac{1}{2}\left(0 + \frac{\pi}{2}\right) = \frac{\pi}{4} \quad \text{and} \quad \frac{1}{2}\left(\frac{\pi}{2} + \pi\right) = \frac{3\pi}{4} \quad \text{Quarter points}$$

The interval $[0, \pi]$ is divided into four equal parts using these x-values.

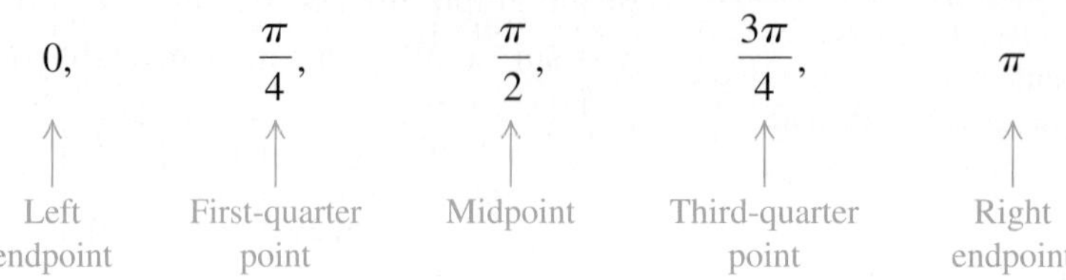

We plot the points from the table of values given at the top of the previous page, and join them with a smooth sinusoidal curve. More of the graph can be sketched by repeating this cycle, as shown in **Figure 28.** The amplitude is not changed.

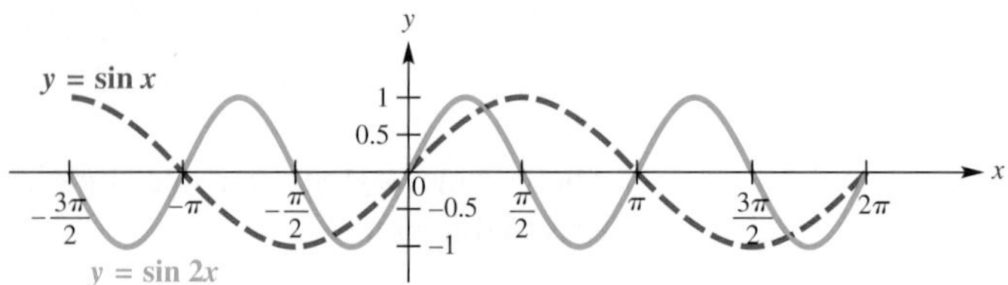

Figure 28

We can think of the graph of $y = \sin bx$ as a horizontal stretching of the graph of $y = \sin x$ when $0 < b < 1$ and as a horizontal shrinking when $b > 1$.

Now Try Exercise 27.

Period

For $b > 0$, the graph of $y = \sin bx$ will resemble that of $y = \sin x$, but with period $\frac{2\pi}{b}$. Also, the graph of $y = \cos bx$ will resemble that of $y = \cos x$, but with period $\frac{2\pi}{b}$.

Classroom Example 3

Graph $y = \cos \frac{1}{2}x$ over one period.

Answer:

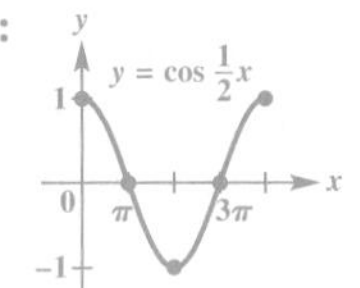

EXAMPLE 3 Graphing $y = \cos bx$

Graph $y = \cos \frac{2}{3}x$ over one period.

SOLUTION The period is

$$\frac{2\pi}{\frac{2}{3}} = 2\pi \div \frac{2}{3} = 2\pi \cdot \frac{3}{2} = 3\pi.$$

To divide by a number, multiply by its reciprocal.

We divide the interval $[0, 3\pi]$ into four equal parts to get the x-values 0, $\frac{3\pi}{4}$, $\frac{3\pi}{2}$, $\frac{9\pi}{4}$, and 3π that yield minimum points, maximum points, and x-intercepts. We use these values to obtain a table of key points for one period.

x	0	$\frac{3\pi}{4}$	$\frac{3\pi}{2}$	$\frac{9\pi}{4}$	3π
$\frac{2}{3}x$	0	$\frac{\pi}{2}$	π	$\frac{3\pi}{2}$	2π
$\cos \frac{2}{3}x$	1	0	−1	0	1

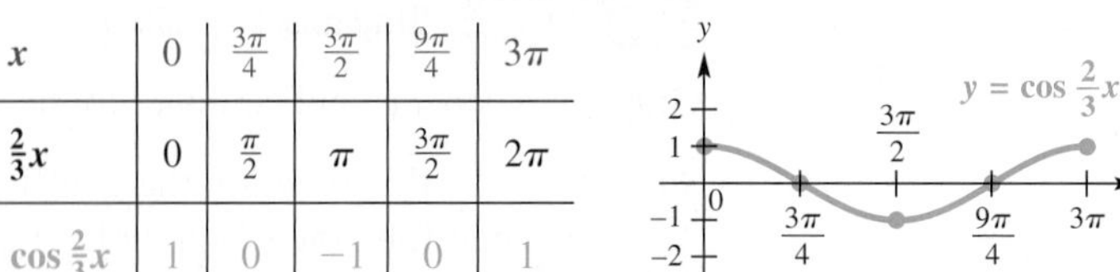

Figure 29

2

Y1=cos((2/3)X)

0 3π

−2

This screen shows a graph of the function in **Example 3.** By our choosing Xscl $= \frac{3\pi}{4}$, the x-intercepts, maxima, and minima coincide with tick marks on the x-axis.

The amplitude is 1 because the maximum value is 1, the minimum value is -1, and $\frac{1}{2}[1 - (-1)] = \frac{1}{2}(2) = 1$. We plot these points and join them with a smooth curve. The graph is shown in **Figure 29.**

Now Try Exercise 25.

NOTE Look back at the middle row of the table in **Example 3.** Dividing the interval $\left[0, \frac{2\pi}{b}\right]$ into four equal parts will always give the values 0, $\frac{\pi}{2}$, π, $\frac{3\pi}{2}$, and 2π for this row, in this case resulting in values of -1, 0, or 1. These values lead to key points on the graph, which can then be easily sketched.

Guidelines for Sketching Graphs of Sine and Cosine Functions

To graph $y = a \sin bx$ or $y = a \cos bx$, with $b > 0$, follow these steps.

Step 1 Find the period, $\frac{2\pi}{b}$. Start at 0 on the x-axis, and lay off a distance of $\frac{2\pi}{b}$.

Step 2 Divide the interval into four equal parts. (See the Note preceding **Example 2.**)

Step 3 Evaluate the function for each of the five x-values resulting from Step 2. The points will be maximum points, minimum points, and x-intercepts.

Step 4 Plot the points found in Step 3, and join them with a sinusoidal curve having amplitude $|a|$.

Step 5 Draw the graph over additional periods as needed.

Classroom Example 4
Graph $y = -3 \sin 2x$ over one period using the guidelines preceding **Example 4.**

Answer:

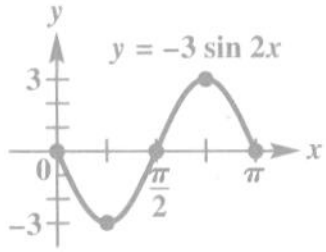

EXAMPLE 4 Graphing $y = a \sin bx$

Graph $y = -2 \sin 3x$ over one period using the preceding guidelines.

SOLUTION

Step 1 For this function, $b = 3$, so the period is $\frac{2\pi}{3}$. The function will be graphed over the interval $\left[0, \frac{2\pi}{3}\right]$.

Step 2 Divide the interval $\left[0, \frac{2\pi}{3}\right]$ into four equal parts to get the x-values 0, $\frac{\pi}{6}$, $\frac{\pi}{3}$, $\frac{\pi}{2}$, and $\frac{2\pi}{3}$.

Step 3 Make a table of values determined by the x-values from Step 2.

x	0	$\frac{\pi}{6}$	$\frac{\pi}{3}$	$\frac{\pi}{2}$	$\frac{2\pi}{3}$
$3x$	0	$\frac{\pi}{2}$	π	$\frac{3\pi}{2}$	2π
$\sin 3x$	0	1	0	-1	0
$-2 \sin 3x$	0	-2	0	2	0

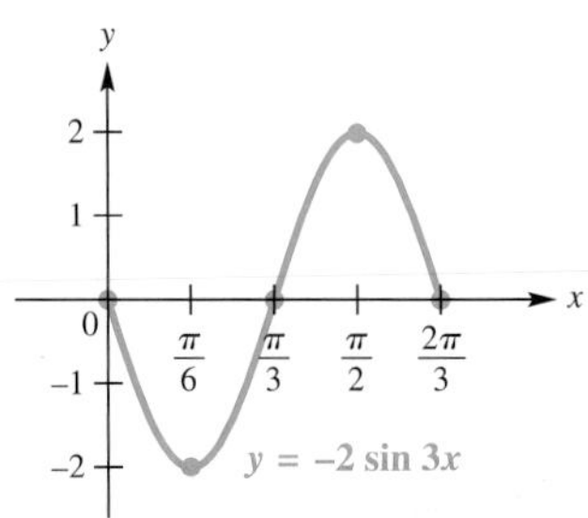

Figure 30

Step 4 Plot the points $(0, 0)$, $\left(\frac{\pi}{6}, -2\right)$, $\left(\frac{\pi}{3}, 0\right)$, $\left(\frac{\pi}{2}, 2\right)$, and $\left(\frac{2\pi}{3}, 0\right)$, and join them with a sinusoidal curve with amplitude 2. See **Figure 30.**

Step 5 The graph can be extended by repeating the cycle.

Notice that when a is negative, the graph of $y = a \sin bx$ is the reflection across the x-axis of the graph of $y = |a| \sin bx$.

✔ *Now Try Exercise 29.*

Classroom Example 5
Graph $y = 2\cos\frac{\pi}{2}x$ over one period.

Answer:

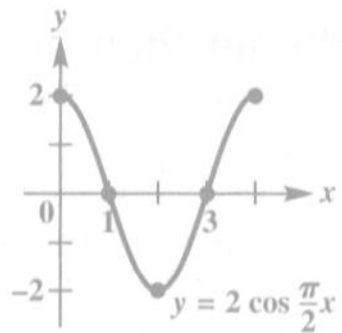

EXAMPLE 5 Graphing $y = a\cos bx$ for b That Is a Multiple of π

Graph $y = -3\cos\pi x$ over one period.

SOLUTION

Step 1 Since $b = \pi$, the period is $\frac{2\pi}{\pi} = 2$, so we will graph the function over the interval $[0, 2]$.

Step 2 Dividing $[0, 2]$ into four equal parts yields the x-values $0, \frac{1}{2}, 1, \frac{3}{2}$, and 2.

Step 3 Make a table using these x-values.

x	0	$\frac{1}{2}$	1	$\frac{3}{2}$	2
πx	0	$\frac{\pi}{2}$	π	$\frac{3\pi}{2}$	2π
$\cos \pi x$	1	0	−1	0	1
$-3\cos\pi x$	−3	0	3	0	−3

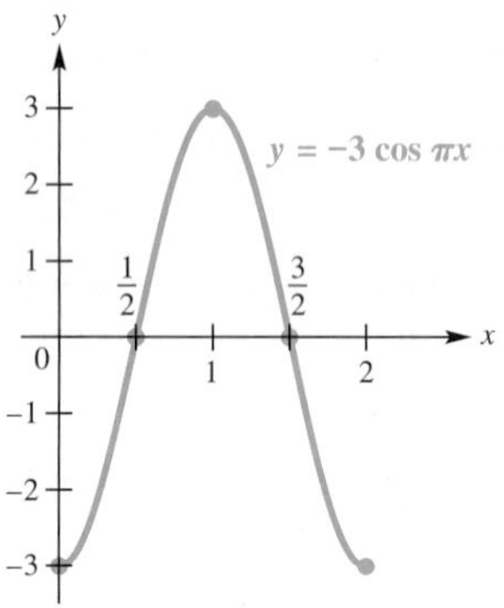

Figure 31

Step 4 Plot the points $(0, -3)$, $\left(\frac{1}{2}, 0\right)$, $(1, 3)$, $\left(\frac{3}{2}, 0\right)$, and $(2, -3)$, and join them with a sinusoidal curve having amplitude $|-3| = 3$. See **Figure 31.**

Step 5 The graph can be extended by repeating the cycle.

Notice that when b is an integer multiple of π, the x-intercepts of the graph are rational numbers.

✔ *Now Try Exercise 37.*

Connecting Graphs with Equations

Classroom Example 6
Determine an equation of the form $y = a\cos bx$ or $y = a\sin bx$, where $b > 0$, for the given graph.

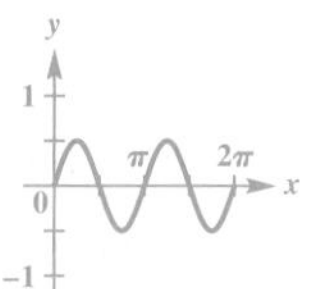

Answer: $y = \frac{1}{2}\sin 2x$

EXAMPLE 6 Determining an Equation for a Graph

Determine an equation of the form $y = a\cos bx$ or $y = a\sin bx$, where $b > 0$, for the given graph.

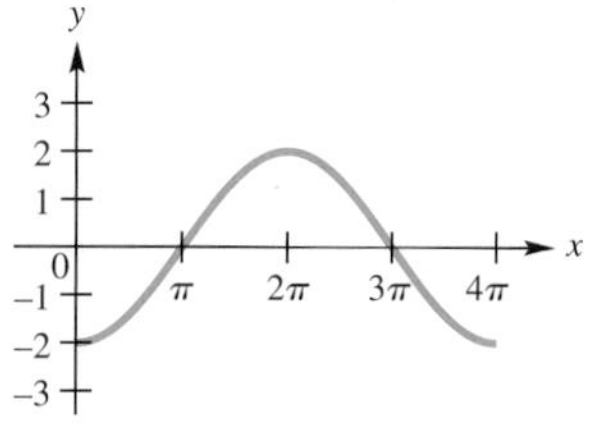

SOLUTION This graph is that of a cosine function that is reflected across its horizontal axis, the x-axis. The amplitude is half the distance between the maximum and minimum values.

$$\frac{1}{2}[2 - (-2)] = \frac{1}{2}(4) = 2 \quad \text{The amplitude } |a| \text{ is 2.}$$

Because the graph completes a cycle on the interval $[0, 4\pi]$, the period is 4π. We use this fact to solve for b.

$$4\pi = \frac{2\pi}{b} \quad \text{Period} = \frac{2\pi}{b}$$

$$4\pi b = 2\pi \quad \text{Multiply each side by } b. \text{ (Section 1.1)}$$

$$b = \frac{1}{2} \quad \text{Divide each side by } 4\pi.$$

An equation for the graph is

$$y = -2\cos\frac{1}{2}x.$$

x-axis reflection (points to −2); Horizontal stretch (points to $\frac{1}{2}$)

✔ *Now Try Exercise 41.*

Using a Trigonometric Model Sine and cosine functions may be used to model many real-life phenomena that repeat their values in a cyclical, or periodic, manner. Average temperature in a certain geographic location is one such example.

Classroom Example 7
The average temperature (in °F) in Phoenix can be approximated by the function

$$f(x) = 19\sin\left[\frac{\pi}{6}(x - 4.3)\right] + 74,$$

where x is the month and $x = 1$ corresponds to January, $x = 2$ to February, and so on.

(a) To observe the graph over a two-year interval, graph f in the window $[0, 25]$ by $[0, 125]$.

(b) According to this model, what is the average temperature during the month of October?

(c) What would be an approximation for the average *yearly* temperature in Phoenix?

Answers:

(a) 125

Y1=19sin((π/6)(X-4.3))+74

X=10 Y=76.972255

0 25

0

(b) 77°F (c) 74°F

EXAMPLE 7 Interpreting a Sine Function Model

The average temperature (in °F) at Mould Bay, Canada, can be approximated by the function

$$f(x) = 34\sin\left[\frac{\pi}{6}(x - 4.3)\right],$$

where x is the month and $x = 1$ corresponds to January, $x = 2$ to February, and so on.

(a) To observe the graph over a two-year interval and to see the maximum and minimum points, graph f in the window $[0, 25]$ by $[-45, 45]$.

(b) According to this model, what is the average temperature during the month of May?

(c) What would be an approximation for the average *yearly* temperature at Mould Bay?

SOLUTION

(a) The graph of $f(x) = 34\sin\left[\frac{\pi}{6}(x - 4.3)\right]$ is shown in **Figure 32.** Its amplitude is 34, and the period is

$$\frac{2\pi}{\frac{\pi}{6}} = 2\pi \div \frac{\pi}{6} = 2\pi \cdot \frac{6}{\pi} = 12.$$ Simplify the complex fraction. (Section R.5)

The function f has a period of 12 months, or 1 year, which agrees with the changing of the seasons.

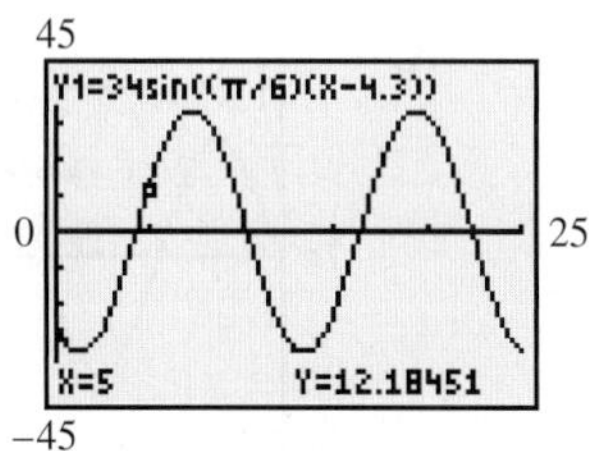

Figure 32

(b) May is the fifth month, so the average temperature during May is

$$f(5) = 34\sin\left[\frac{\pi}{6}(5 - 4.3)\right] \approx 12°\text{F}.$$ Let $x = 5$. (Section 2.3)

See the display at the bottom of the screen in **Figure 32.**

(c) From the graph, it appears that the average yearly temperature is about 0°F since the graph is centered vertically about the line $y = 0$.

Now Try Exercise 59.

6.3 Exercises

1. G 2. A
3. E 4. D
5. B 6. H
7. F 8. C

9. D 10. B
11. C 12. A

13. 2 14. 3

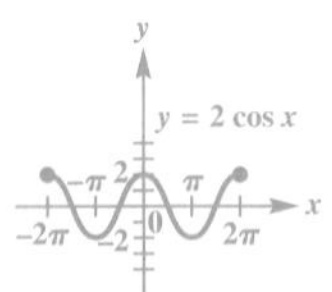

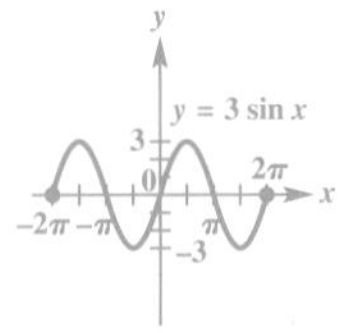

15. $\frac{2}{3}$ 16. $\frac{3}{4}$

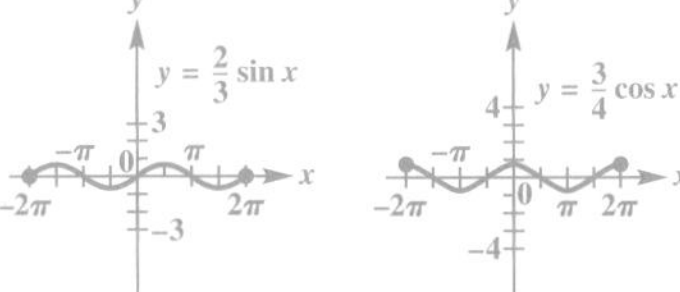

17. 1 18. 1

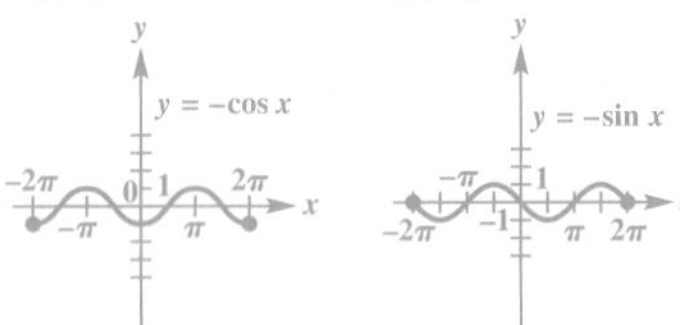

19. 2 20. 3

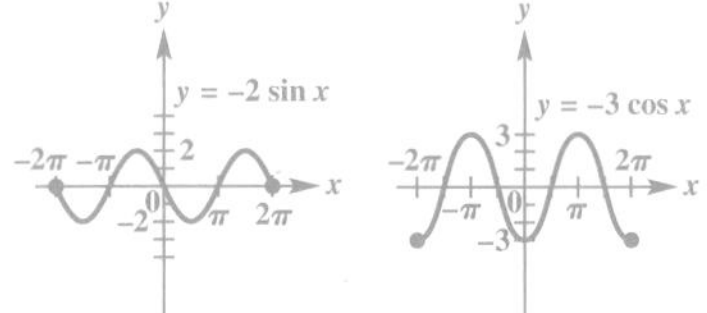

21. 1

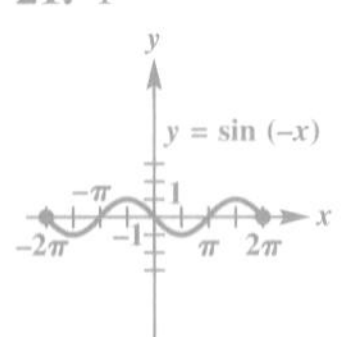

22. Because $\sin(-x) = -\sin x$ for all x, the graphs are the same.

23. 4π; 1 24. 3π; 1

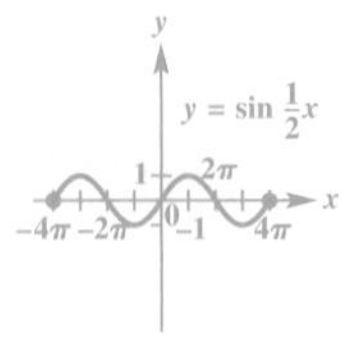

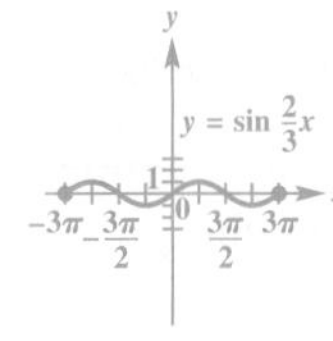

Concept Check In Exercises 1–8, match each function with its graph in choices A–I. (One choice will not be used.)

1. $y = \sin x$
2. $y = \cos x$
3. $y = -\sin x$
4. $y = -\cos x$
5. $y = \sin 2x$
6. $y = \cos 2x$
7. $y = 2 \sin x$
8. $y = 2 \cos x$

A.

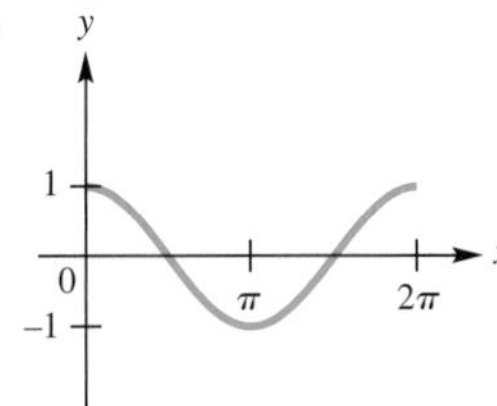

B.

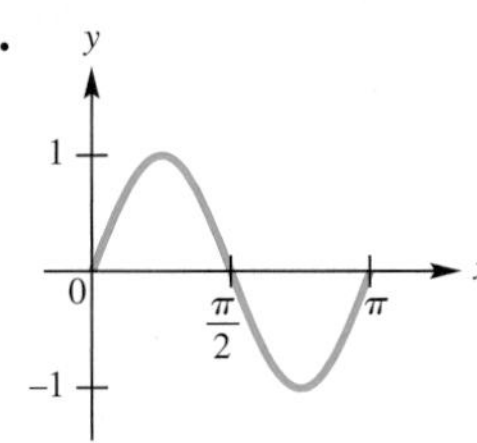

C.

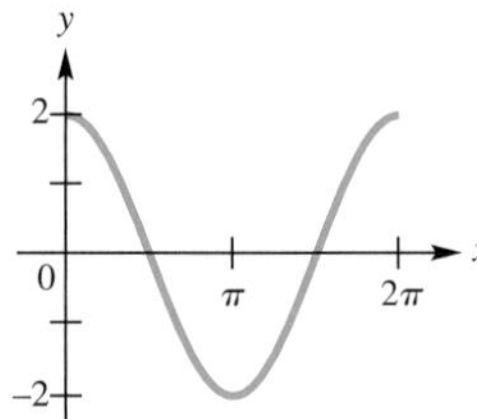

D.

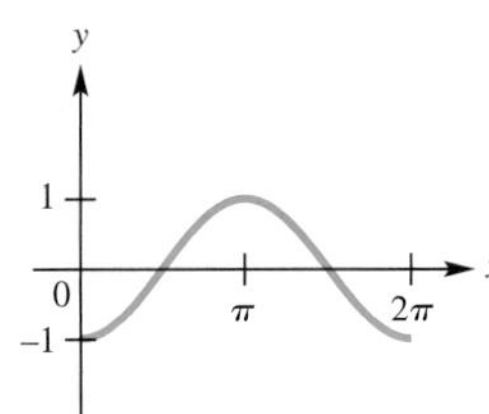

E.

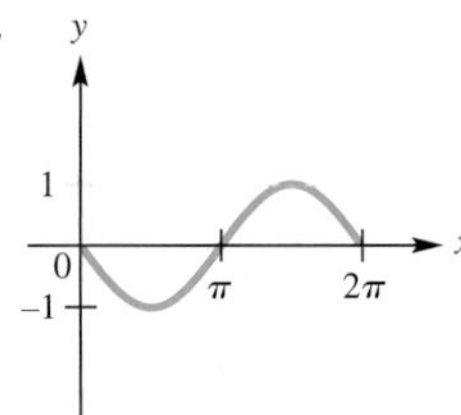

F.

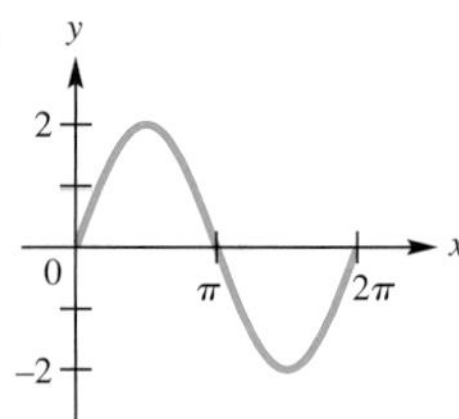

G.

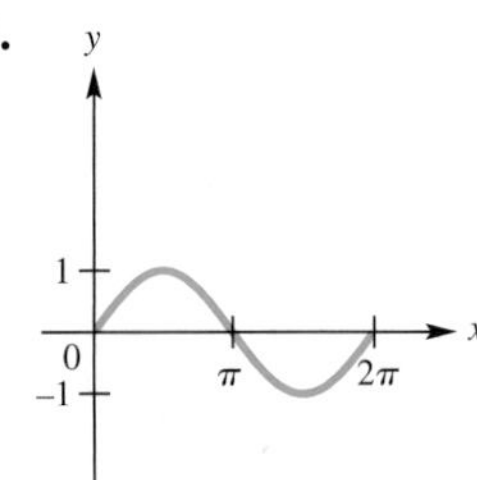

H.

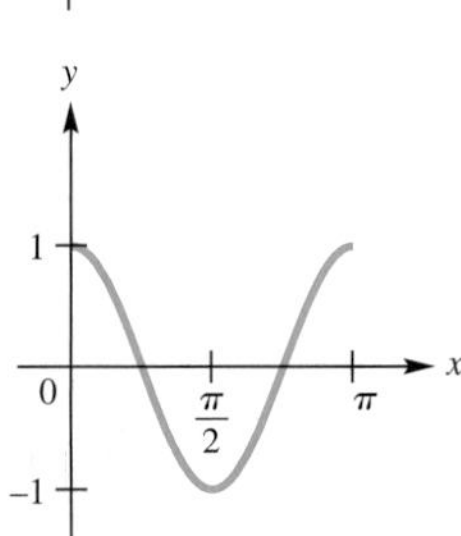

I.

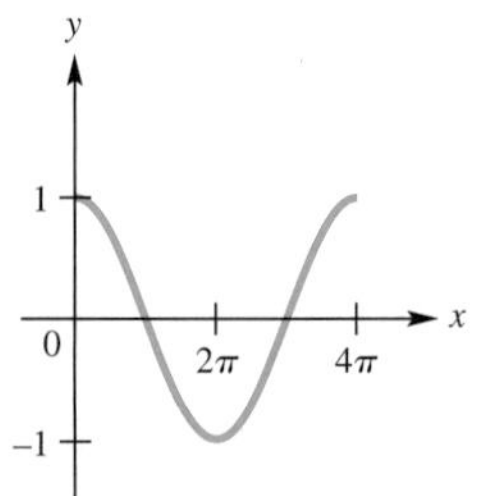

Concept Check In Exercises 9–12, match each function with its calculator graph.

9. $y = \sin 3x$
10. $y = \cos 3x$
11. $y = 3 \cos x$
12. $y = 3 \sin x$

A.

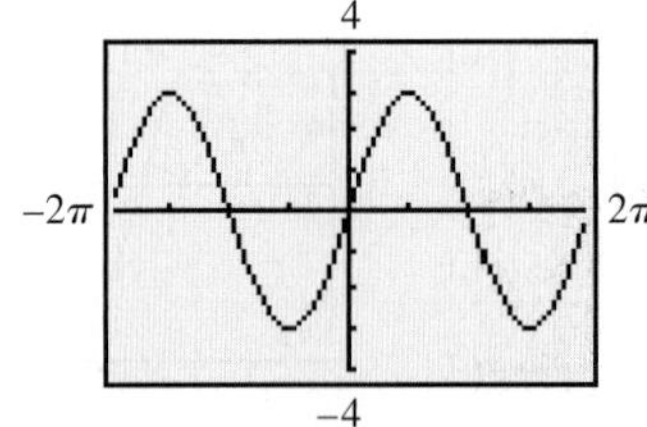

B.

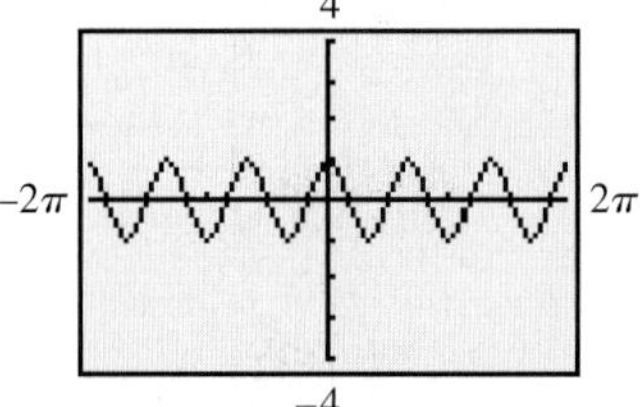

C.

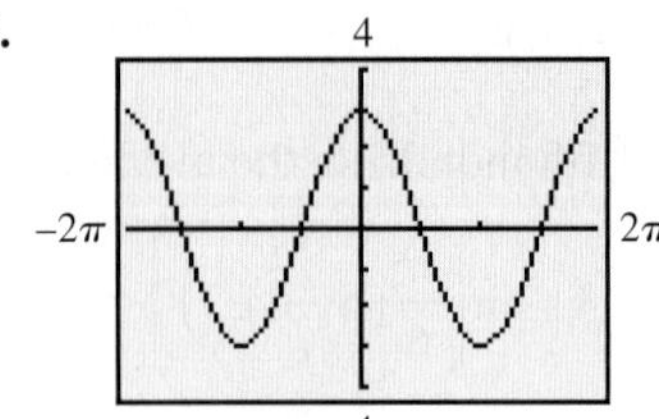

D.

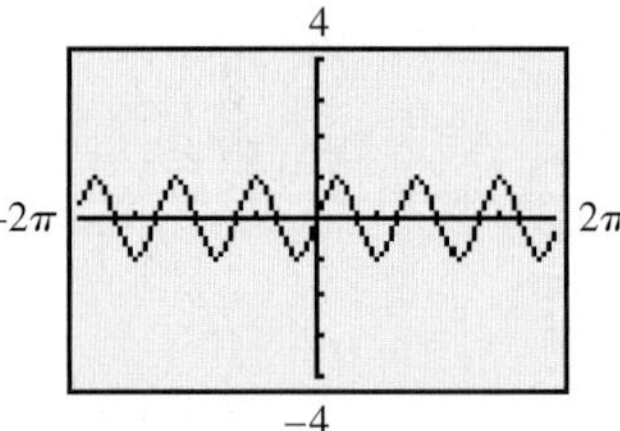

Graph each function over the interval $[-2\pi, 2\pi]$. Give the amplitude. ***See Example 1.***

13. $y = 2 \cos x$
14. $y = 3 \sin x$
15. $y = \frac{2}{3} \sin x$
16. $y = \frac{3}{4} \cos x$
17. $y = -\cos x$
18. $y = -\sin x$
19. $y = -2 \sin x$
20. $y = -3 \cos x$
21. $y = \sin(-x)$

25. $\frac{8\pi}{3}$; 1 26. 6π; 1

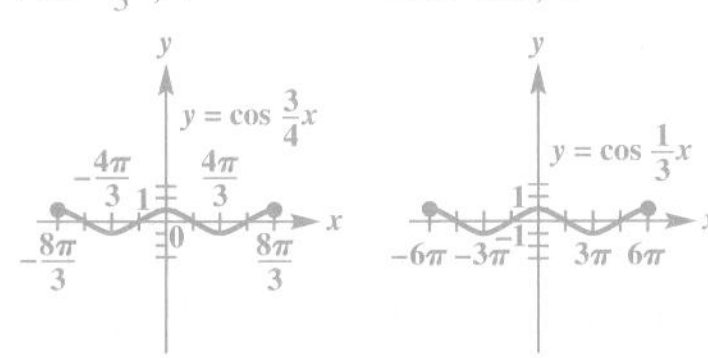

27. $\frac{2\pi}{3}$; 1 28. π; 1

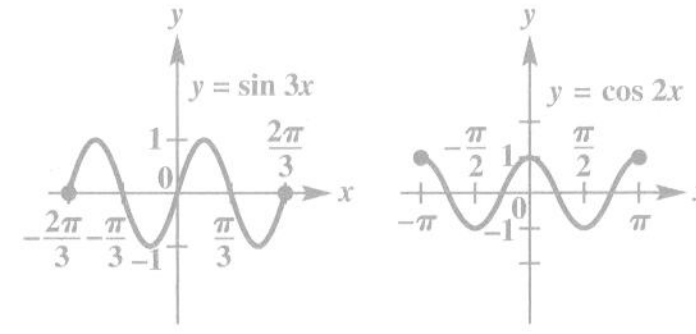

29. 8π; 2 30. π; 3

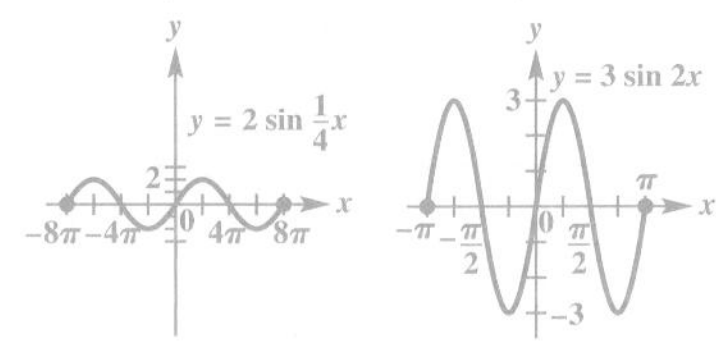

31. $\frac{2\pi}{3}$; 2 32. π; 5

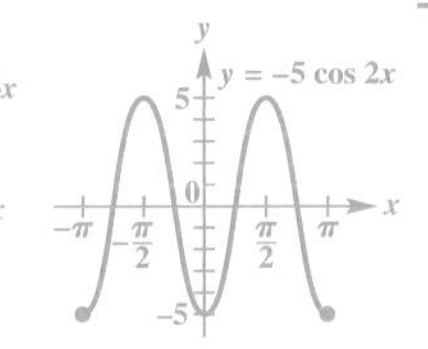

33. 2; 1 34. 2; 1

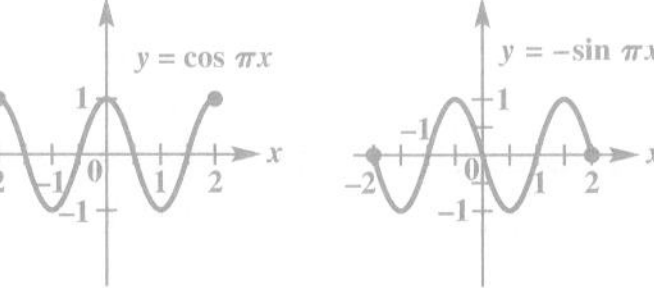

35. 1; 2 36. 1; 3

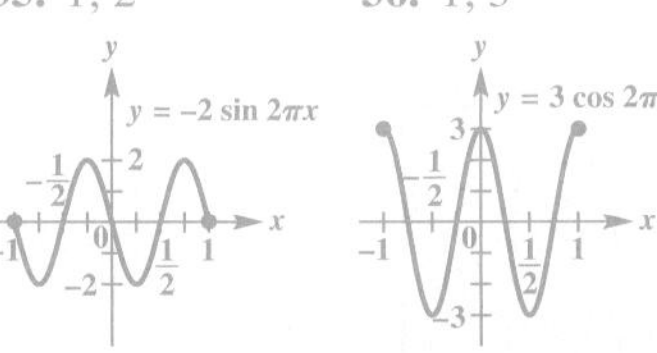

37. 4; $\frac{1}{2}$ 38. 8; $\frac{2}{3}$

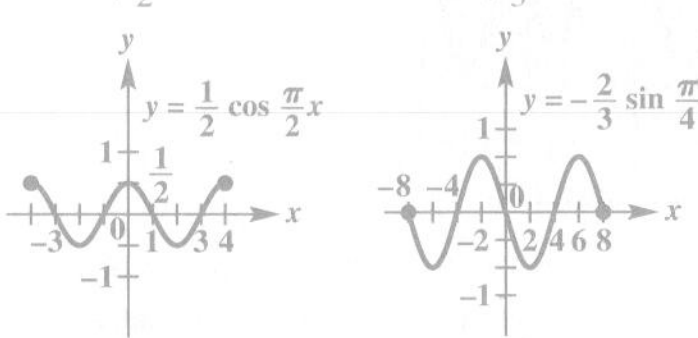

39. 2; π 40. 2; π

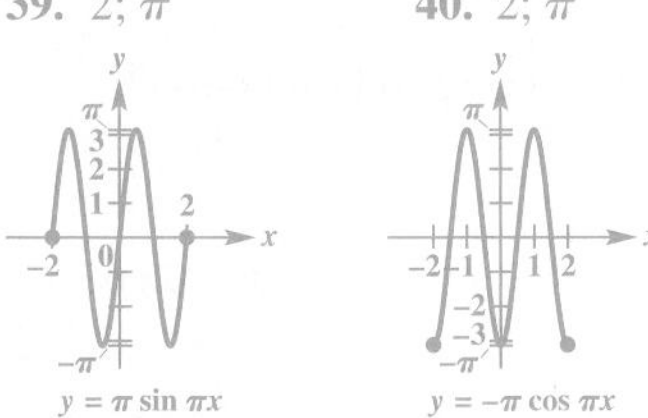

22. *Concept Check* In **Exercise 21,** why is the graph the same as that of $y = -\sin x$?

Graph each function over a two-period interval. Give the period and amplitude. ***See Examples 2–5.***

23. $y = \sin \frac{1}{2}x$ **24.** $y = \sin \frac{2}{3}x$ **25.** $y = \cos \frac{3}{4}x$

26. $y = \cos \frac{1}{3}x$ **27.** $y = \sin 3x$ **28.** $y = \cos 2x$

29. $y = 2 \sin \frac{1}{4}x$ **30.** $y = 3 \sin 2x$ **31.** $y = -2 \cos 3x$

32. $y = -5 \cos 2x$ **33.** $y = \cos \pi x$ **34.** $y = -\sin \pi x$

35. $y = -2 \sin 2\pi x$ **36.** $y = 3 \cos 2\pi x$ **37.** $y = \frac{1}{2} \cos \frac{\pi}{2}x$

38. $y = -\frac{2}{3} \sin \frac{\pi}{4}x$ **39.** $y = \pi \sin \pi x$ **40.** $y = -\pi \cos \pi x$

Connecting Graphs with Equations *Each function graphed is of the form* $y = a \sin bx$ *or* $y = a \cos bx$*, where* $b > 0$*. Determine the equation of the graph.* ***See Example 6.***

41.

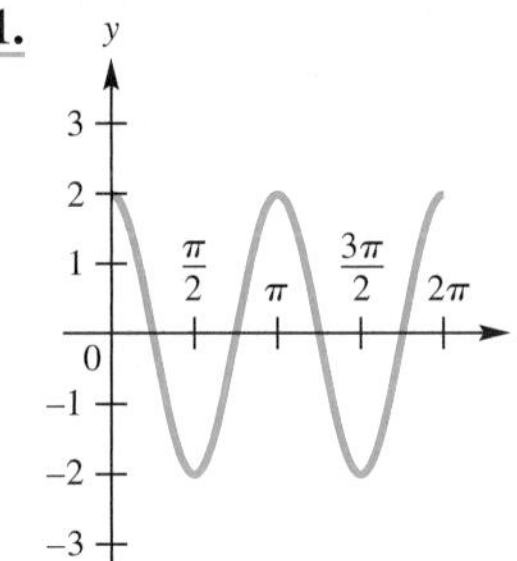

42.

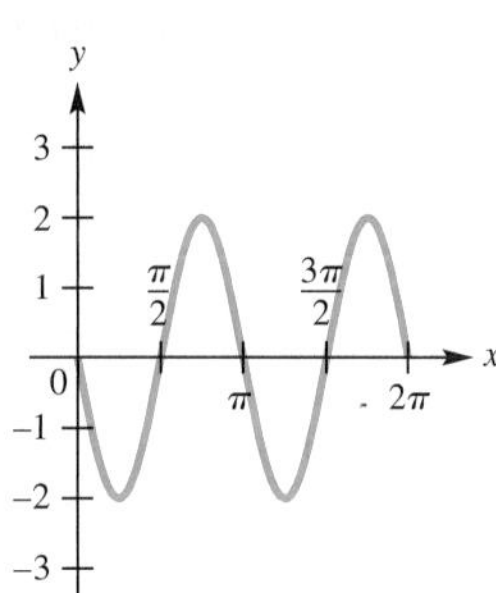

43.

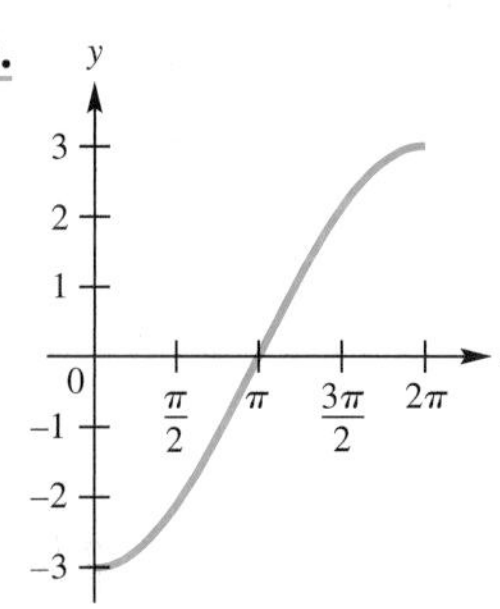

44.

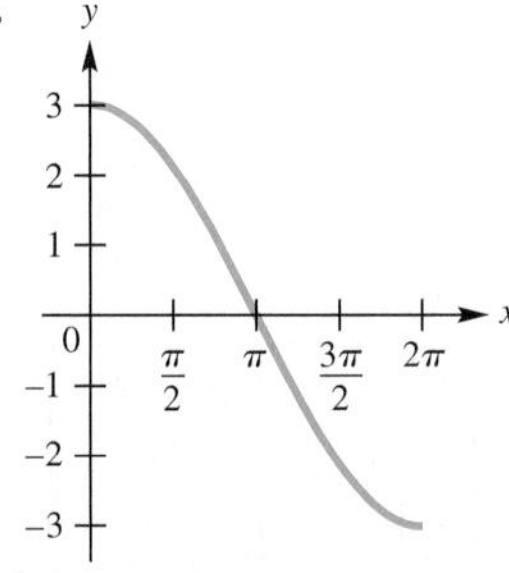

45.

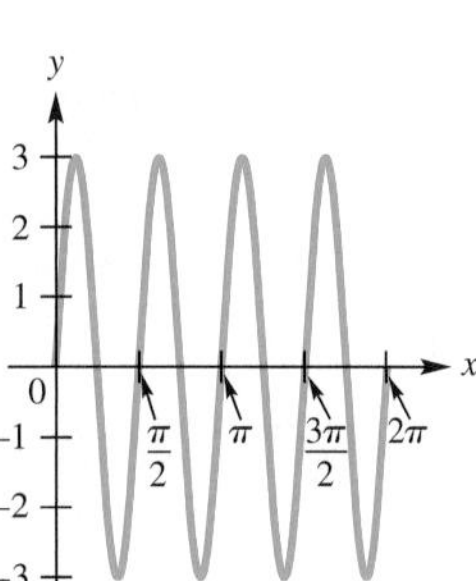

46.

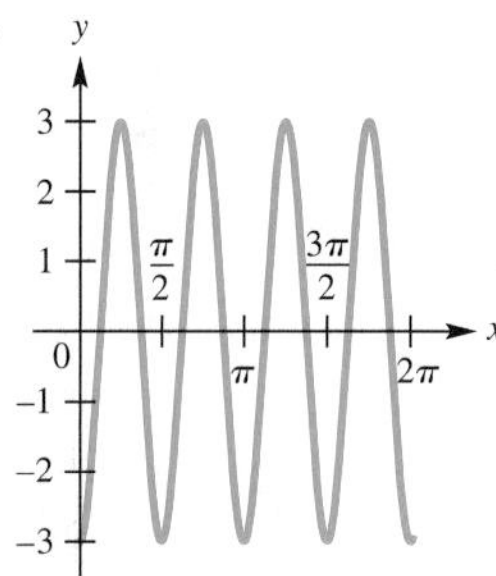

(Modeling) Solve each problem.

47. *Average Annual Temperature* Scientists believe that the average annual temperature in a given location is periodic. The average temperature at a given place during a given season fluctuates as time goes on, from colder to warmer, and back to colder. The graph shows an idealized description of the temperature (in °F) for approximately the last 150 thousand years of a location at the same latitude as Anchorage, Alaska.

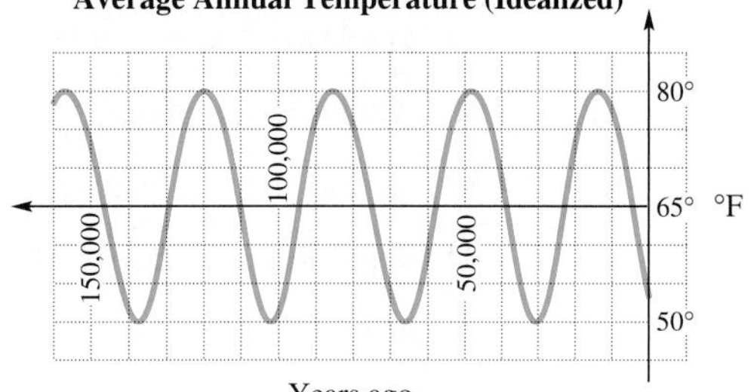

41. $y = 2 \cos 2x$
42. $y = -2 \sin 2x$
43. $y = -3 \cos \frac{1}{2}x$
44. $y = 3 \cos \frac{1}{2}x$
45. $y = 3 \sin 4x$
46. $y = -3 \cos 4x$

47. **(a)** 80°F; 50°F **(b)** 15 **(c)** about 35,000 yr **(d)** downward
48. **(a)** 120; 80 **(b)** 20 **(c)** 75

49. 24 hr
50. approximately $\frac{2.6 - 0.2}{2} = 1.2$
51. approximately 6:00 P.M.; approximately 0.2 ft
52. approximately 7:19 P.M.; approximately 0 ft
53. approximately 3:18 A.M.; approximately 2.4 ft

(a) Find the highest and lowest temperatures recorded.
(b) Use these two numbers to find the amplitude.
(c) Find the period of the function.
(d) What is the trend of the temperature now?

48. *Blood Pressure Variation* The graph gives the variation in blood pressure for a typical person. **Systolic** and **diastolic pressures** are the upper and lower limits of the periodic changes in pressure that produce the pulse. The length of time between peaks is called the period of the pulse.

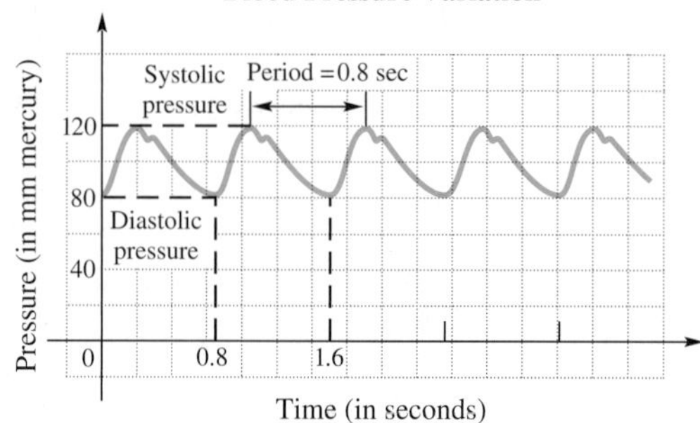

(a) Find the systolic and diastolic pressures.
(b) Find the amplitude of the graph.
(c) Find the pulse rate (the number of pulse beats in 1 min) for this person.

Tides for Kahului Harbor The chart shows the tides for Kahului Harbor (on the island of Maui, Hawaii). To identify high and low tides and times for other Maui areas, the following adjustments must be made.

Hana:	High, +40 min, +0.1 ft; Low, +18 min, −0.2 ft	Makena:	High, +1:21, −0.5 ft; Low, +1:09, −0.2 ft
Maalaea:	High, +1:52, −0.1 ft; Low, +1:19, −0.2 ft	Lahaina:	High, +1:18, −0.2 ft; Low, +1:01, −0.1 ft

JANUARY

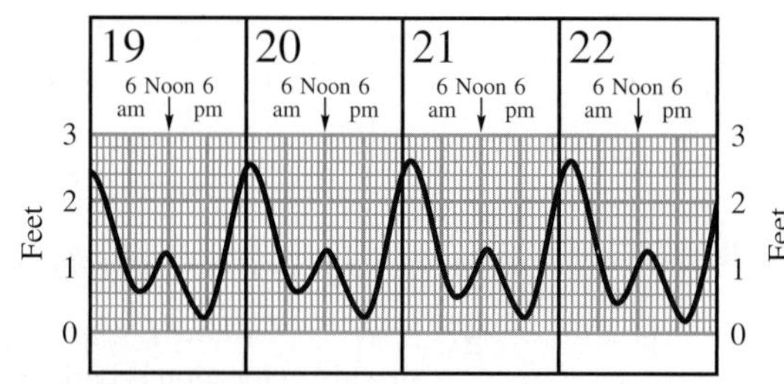

Source: Maui News. Original chart prepared by Edward K. Noda and Associates.

Use the graph to work Exercises 49–53.

49. The graph is an example of a periodic function. What is the period (in hours)?

50. What is the amplitude?

51. At what time on January 20 was low tide at Kahului? What was the height then?

52. Repeat **Exercise 51** for Maalaea.

53. At what time on January 22 was high tide at Lahaina? What was the height then?

54. **(a)** about 2 hr **(b)** 1 yr

55. **(a)** 5; $\frac{1}{60}$ **(b)** 60

(c) 5; 1.545; -4.045; -4.045; 1.545

(d)

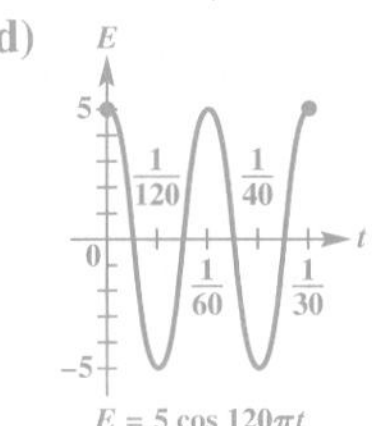

$E = 5 \cos 120\pi t$

56. **(a)** 3.8; $\frac{1}{20}$ **(b)** 20

(c) -3.074; 1.174; -3.074; -3.074; 1.174

(d)

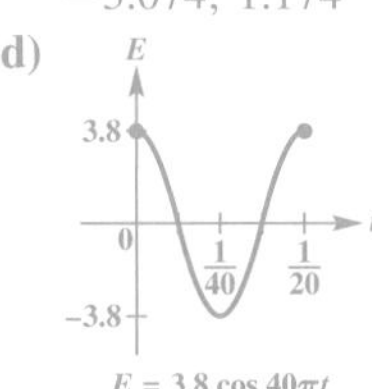

$E = 3.8 \cos 40\pi t$

57. **(a)** $L(x) = 0.022x^2 + 0.55x + 316 + 3.5 \sin 2\pi x$

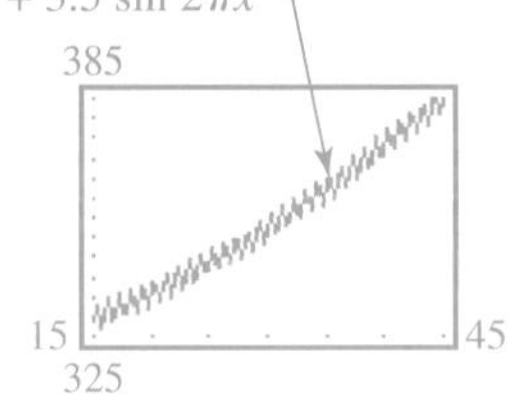

(b) maxima: $x = \frac{1}{4}, \frac{5}{4}, \frac{9}{4}, \ldots$; minima: $x = \frac{3}{4}, \frac{7}{4}, \frac{11}{4}, \ldots$

(Modeling) Solve each problem.

54. *Activity of a Nocturnal Animal* Many of the activities of living organisms are periodic. For example, the graph at the right shows the time that a certain nocturnal animal begins its evening activity.

(a) Find the amplitude of this graph.

(b) Find the period.

Activity of a Nocturnal Animal

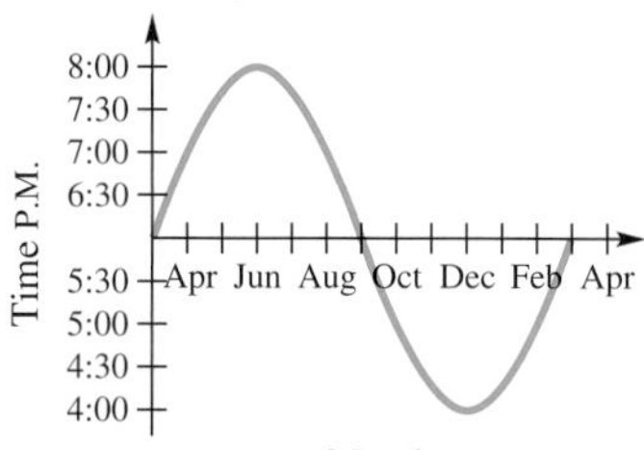

55. *Voltage of an Electrical Circuit* The voltage E in an electrical circuit is modeled by

$$E = 5 \cos 120\pi t,$$

where t is time measured in seconds.

(a) Find the amplitude and the period.

(b) How many cycles are completed in 1 sec? (The number of cycles, or periods, completed in 1 sec is the **frequency** of the function.)

(c) Find E when $t = 0, 0.03, 0.06, 0.09, 0.12$.

(d) Graph E for $0 \le t \le \frac{1}{30}$.

56. *Voltage of an Electrical Circuit* For another electrical circuit, the voltage E is modeled by

$$E = 3.8 \cos 40\pi t,$$

where t is time measured in seconds.

(a) Find the amplitude and the period.

(b) Find the frequency. See **Exercise 55(b).**

(c) Find E when $t = 0.02, 0.04, 0.08, 0.12, 0.14$.

(d) Graph one period of E.

57. *Atmospheric Carbon Dioxide* At Mauna Loa, Hawaii, atmospheric carbon dioxide levels in parts per million (ppm) were measured regularly from 1958 to 2004. The function

$$L(x) = 0.022x^2 + 0.55x + 316 + 3.5 \sin 2\pi x$$

can be used to model these levels, where x is in years and $x = 0$ corresponds to 1960. (*Source:* Nilsson, A., *Greenhouse Earth,* John Wiley and Sons.)

(a) Graph L in the window $[15, 45]$ by $[325, 385]$.

(b) When do the seasonal maximum and minimum carbon dioxide levels occur?

(c) L is the sum of a quadratic function and a sine function. What is the significance of each of these functions? Discuss what physical phenomena may be responsible for each function.

58. (a)

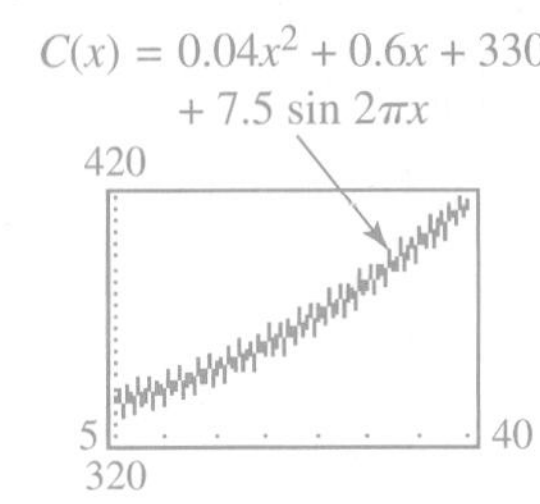

(c) $C(x) = 0.04(x - 1970)^2 + 0.6(x - 1970) + 330 + 7.5 \sin[2\pi(x - 1970)]$

59. (a) 31°F (b) 38°F (c) 57°F (d) 58°F (e) 37°F (f) 16°F

60. (a) 1.998 watts per m^2
(b) -46.461 watts per m^2
(c) 46.478 watts per m^2
(d) Answers may vary. A possible answer is $N = 82.5$. (Since N represents a day number, which should be a natural number, we might interpret day 82.5 as noon on the 82nd day.)

61. 1; 240°, or $\frac{4\pi}{3}$

62. 1; 120°, or $\frac{2\pi}{3}$

58. *Atmospheric Carbon Dioxide* Refer to **Exercise 57.** The carbon dioxide content in the atmosphere at Barrow, Alaska, in parts per million (ppm) can be modeled using the function

$$C(x) = 0.04x^2 + 0.6x + 330 + 7.5 \sin 2\pi x,$$

where $x = 0$ corresponds to 1970. (*Source:* Zeilik, M. and S. Gregory, *Introductory Astronomy and Astrophysics,* Brooks/Cole.)

(a) Graph C in the window $[5, 40]$ by $[320, 420]$.

(b) Discuss possible reasons why the amplitude of the oscillations in the graph of C is larger than the amplitude of the oscillations in the graph of L in **Exercise 57,** which models Hawaii.

(c) Define a new function C that is valid if x represents the actual year, where $1970 \le x \le 2010$. (See horizontal translations in **Section 2.7.**)

59. *Average Daily Temperature* The temperature in Anchorage, Alaska, is modeled by

$$T(x) = 37 + 21 \sin\left[\frac{2\pi}{365}(x - 91)\right],$$

where $T(x)$ is the temperature in degrees Fahrenheit on day x, with $x = 1$ corresponding to January 1 and $x = 365$ corresponding to December 31. Use a calculator to estimate the temperature on the following days. (*Source: World Almanac and Book of Facts.*)

(a) March 15 (day 74) **(b)** April 5 (day 95) **(c)** Day 200
(d) June 25 **(e)** October 1 **(f)** December 31

60. *Fluctuation in the Solar Constant* The **solar constant** S is the amount of energy per unit area that reaches Earth's atmosphere from the sun. It is equal to 1367 watts per m^2 but varies slightly throughout the seasons. This fluctuation ΔS in S can be calculated using the formula

$$\Delta S = 0.034S \sin\left[\frac{2\pi(82.5 - N)}{365.25}\right].$$

In this formula, N is the day number covering a four-year period, where $N = 1$ corresponds to January 1 of a leap year and $N = 1461$ corresponds to December 31 of the fourth year. (*Source:* Winter, C., R. Sizmann, and L. L.Vant-Hull, Editors, *Solar Power Plants,* Springer-Verlag.)

(a) Calculate ΔS for $N = 80$, which is the spring equinox in the first year.
(b) Calculate ΔS for $N = 1268$, which is the summer solstice in the fourth year.
(c) What is the maximum value of ΔS?
(d) Find a value for N where ΔS is equal to 0.

Musical Sound Waves Pure sounds produce single sine waves on an oscilloscope. Find the amplitude and period of each sine wave graph in Exercises 61 and 62. On the vertical scale, each square represents 0.5; *on the horizontal scale, each square represents* 30° *or* $\frac{\pi}{6}$.

61.

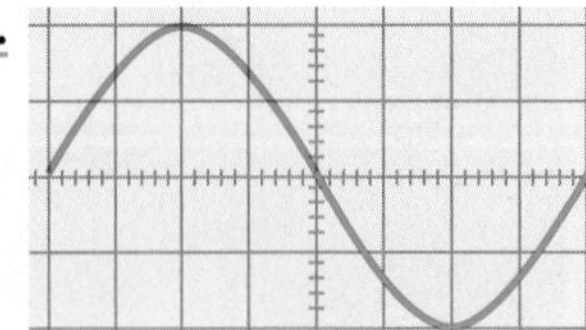

62.

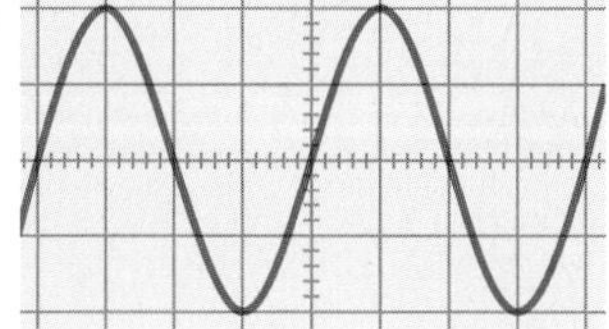

63. Over the interval $[0, 2\pi]$, compare the graphs of

$$y = \sin 2x \quad \text{and} \quad y = 2 \sin x.$$

Can we say that, in general, $\sin bx = b \sin x$? Explain.

64. Over the interval $[0, 2\pi]$, compare the graphs of

$$y = \cos 3x \quad \text{and} \quad y = 3 \cos x.$$

Can we say that, in general, $\cos bx = b \cos x$? Explain.

65. X = −0.4161468, Y = 0.90929743; X is cos 2 and Y is sin 2.
66. X = 2, Y = 0.90929743; sin 2 = 0.90929743
67. X = 2, Y = −0.4161468; cos 2 = −0.4161468

Relating Concepts

For individual or collaborative investigation *(Exercises 65–68)*

Connecting the Unit Circle and Sine Graph *Using a TI-83/84 Plus calculator, adjust the settings to correspond to the following screens.*

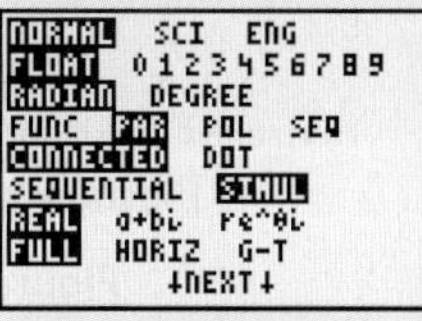

MODE

FORMAT

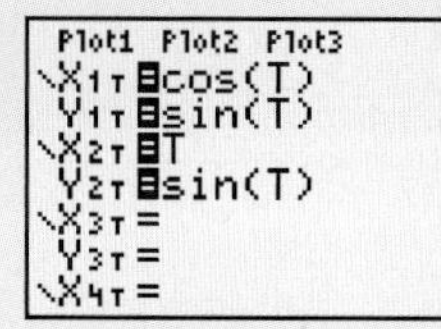

Y = editor

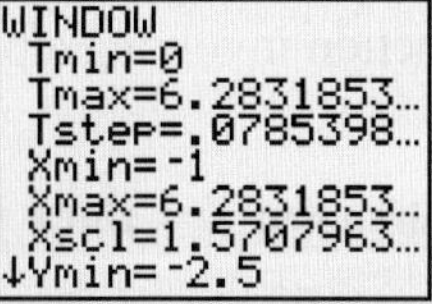

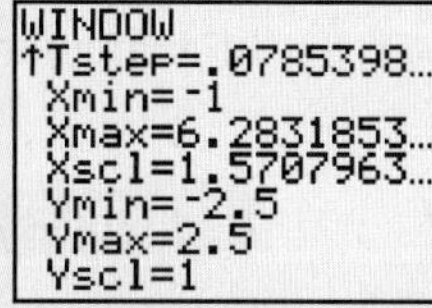

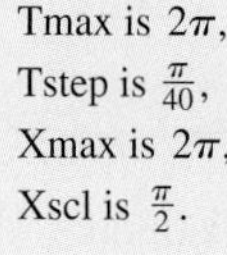

Tmax is 2π,
Tstep is $\frac{\pi}{40}$,
Xmax is 2π,
Xscl is $\frac{\pi}{2}$.

Graph the two equations (which are in ***parametric form****), and watch as the unit circle and the sine function are graphed simultaneously. Press the* TRACE *key once to get the screen shown on the left below, and then press the up-arrow key to get the screen shown on the right below. The screen on the left gives a unit circle interpretation of* cos 0 = 1 *and* sin 0 = 0. *The screen on the right gives a rectangular coordinate graph interpretation of* sin 0 = 0.

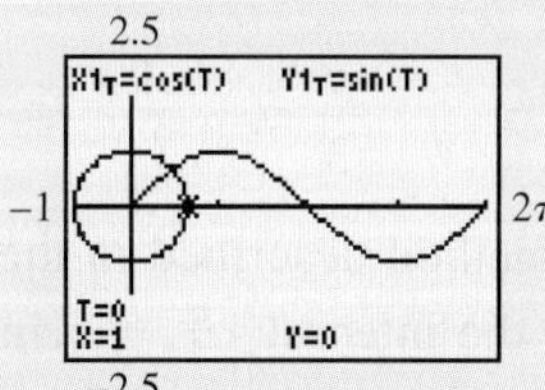

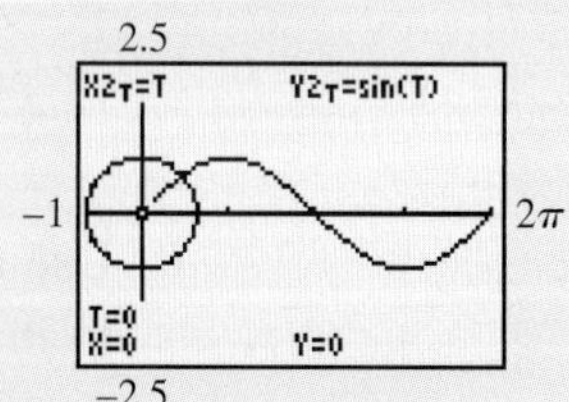

65. On the unit circle graph, let T = 2. Find X and Y, and interpret their values.

66. On the sine graph, let T = 2. What values of X and Y are displayed? Interpret these values with an equation in X and Y.

67. Now go back and redefine Y_{2T} as cos(T). Graph both equations. On the cosine graph, let T = 2. What values of X and Y are displayed? Interpret these values with an equation in X and Y.

68. Explain the relationship between the coordinates of the unit circle and the coordinates of the sine and cosine graphs.

6.4 Translations of the Graphs of the Sine and Cosine Functions

- Horizontal Translations
- Vertical Translations
- Combinations of Translations
- Determining a Trigonometric Model

Horizontal Translations

The graph of the function

$$y = f(x - d)$$

is translated *horizontally* compared to the graph of $y = f(x)$. The translation is d units to the right if $d > 0$ and is $|d|$ units to the left if $d < 0$. See **Figure 33** on the next page.

With circular functions, a horizontal translation is called a **phase shift.** In the function $y = f(x - d)$, the expression $x - d$ is the **argument.**

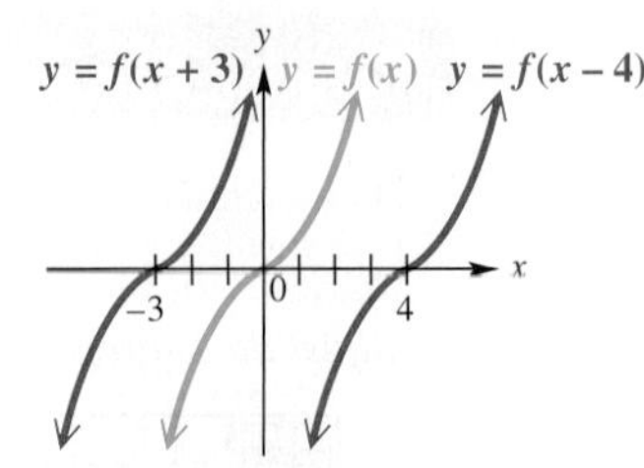

Horizontal translations of $y = f(x)$
(Section 2.7)

Figure 33

In **Examples 1–3,** we give two methods that can be used to sketch the graph of a circular function involving a phase shift.

Classroom Example 1

Graph $y = \sin\left(x + \frac{3\pi}{4}\right)$ over one period.

Answer:

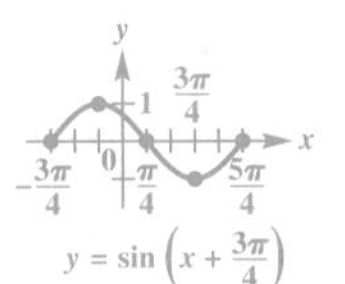

EXAMPLE 1 Graphing $y = \sin(x - d)$

Graph $y = \sin\left(x - \frac{\pi}{3}\right)$ over one period.

SOLUTION ***Method 1*** For the argument $x - \frac{\pi}{3}$ to result in all possible values throughout one period, it must take on all values between 0 and 2π, inclusive. To find an interval of one period, we solve the following three-part inequality.

$$0 \leq x - \frac{\pi}{3} \leq 2\pi \quad \text{(Section 1.7)}$$

$$\frac{\pi}{3} \leq x \leq \frac{7\pi}{3} \quad \text{Add } \tfrac{\pi}{3} \text{ to each part.}$$

Use the method described in the Note preceding **Example 2** in **Section 6.3** to divide the interval $\left[\frac{\pi}{3}, \frac{7\pi}{3}\right]$ into four equal parts, obtaining the following x-values.

$$\frac{\pi}{3}, \frac{5\pi}{6}, \frac{4\pi}{3}, \frac{11\pi}{6}, \frac{7\pi}{3}$$

These are *key* x-values.

A table of values using these x-values follows.

x	$\frac{\pi}{3}$	$\frac{5\pi}{6}$	$\frac{4\pi}{3}$	$\frac{11\pi}{6}$	$\frac{7\pi}{3}$
$x - \frac{\pi}{3}$	0	$\frac{\pi}{2}$	π	$\frac{3\pi}{2}$	2π
$\sin\left(x - \frac{\pi}{3}\right)$	0	1	0	−1	0

Figure 34

We join the corresponding points with a smooth curve to get the solid blue graph shown in **Figure 34.** The period is 2π, and the amplitude is 1.

Method 2 We can also graph $y = \sin\left(x - \frac{\pi}{3}\right)$ by using a horizontal translation of the graph of $y = \sin x$. The argument $x - \frac{\pi}{3}$ indicates that the graph will be translated $\frac{\pi}{3}$ units to the *right* (the phase shift) compared to the graph of $y = \sin x$. See **Figure 34.**

Therefore, to graph a function using this method, first graph the basic circular function, and then graph the desired function by using the appropriate translation.

✔ *Now Try Exercise 35.*

NOTE The graph in **Figure 34** of **Example 1** can be extended through additional periods by repeating the given portion of the graph, as necessary.

Classroom Example 2

Graph $y = -2\cos\left(x - \frac{\pi}{3}\right)$ over one period.

Answer:

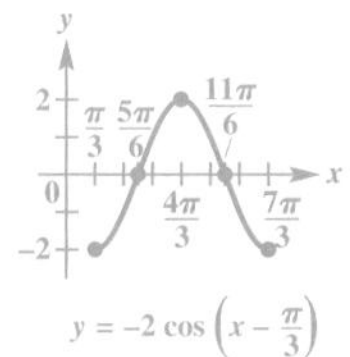

EXAMPLE 2 Graphing $y = a\cos(x - d)$

Graph $y = 3\cos\left(x + \frac{\pi}{4}\right)$ over one period.

SOLUTION *Method 1* First solve the following three-part inequality.

$$0 \le x + \frac{\pi}{4} \le 2\pi$$

$$-\frac{\pi}{4} \le x \le \frac{7\pi}{4} \quad \text{Subtract } \tfrac{\pi}{4} \text{ from each part.}$$

Dividing this interval into four equal parts gives these x-values.

$$-\frac{\pi}{4}, \quad \frac{\pi}{4}, \quad \frac{3\pi}{4}, \quad \frac{5\pi}{4}, \quad \frac{7\pi}{4} \quad \text{Key } x\text{-values}$$

Use these x-values to make a table of points.

x	$-\frac{\pi}{4}$	$\frac{\pi}{4}$	$\frac{3\pi}{4}$	$\frac{5\pi}{4}$	$\frac{7\pi}{4}$
$x + \frac{\pi}{4}$	0	$\frac{\pi}{2}$	π	$\frac{3\pi}{2}$	2π
$\cos\left(x + \frac{\pi}{4}\right)$	1	0	-1	0	1
$3\cos\left(x + \frac{\pi}{4}\right)$	3	0	-3	0	3

These x-values lead to maximum points, minimum points, and x-intercepts.

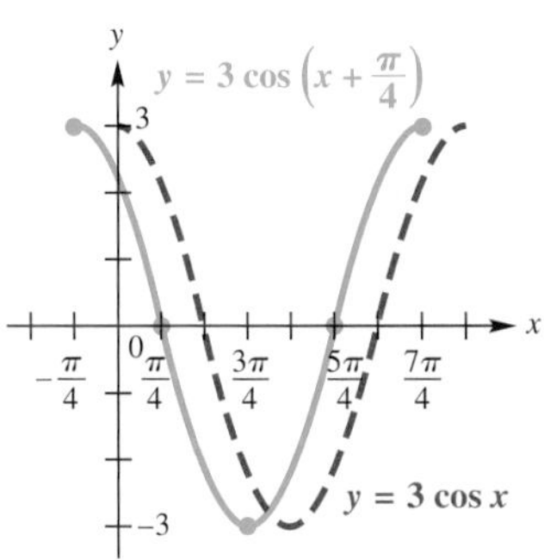

Figure 35

We join the corresponding points with a smooth curve to get the solid blue graph shown in **Figure 35.** The period is 2π, and the amplitude is 3.

Method 2 Write $y = 3\cos\left(x + \frac{\pi}{4}\right)$ in the form $y = a\cos(x - d)$.

$$y = 3\cos\left(x + \frac{\pi}{4}\right), \quad \text{or} \quad y = 3\cos\left[x - \left(-\frac{\pi}{4}\right)\right] \quad \text{Rewrite to subtract } -\tfrac{\pi}{4}.$$

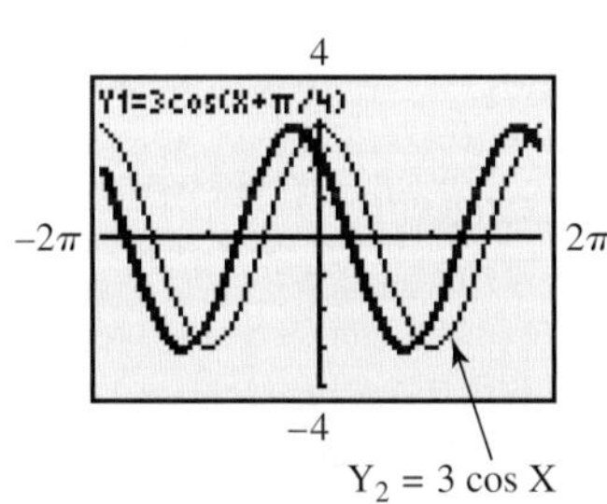

This result shows that $d = -\frac{\pi}{4}$. Since $-\frac{\pi}{4}$ is negative, the phase shift is $\left|-\frac{\pi}{4}\right| = \frac{\pi}{4}$ unit to the left. The graph is the same as that of $y = 3\cos x$ (the thin-lined graph in the margin calculator screen), except that it is translated $\frac{\pi}{4}$ unit to the left (the thick-lined graph).

Now Try Exercise 37.

Classroom Example 3

Graph $y = \frac{3}{2}\cos(2x - \pi)$ over two periods.

Answer:

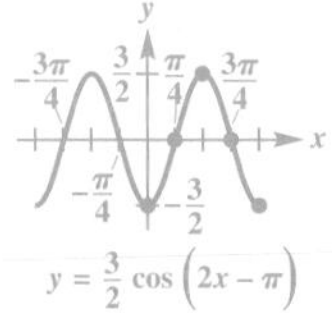

EXAMPLE 3 Graphing $y = a\cos[b(x - d)]$

Graph $y = -2\cos(3x + \pi)$ over two periods.

SOLUTION *Method 1* The function can be sketched over one period by solving the three-part inequality

$$0 \le 3x + \pi \le 2\pi$$

to find the interval $\left[-\frac{\pi}{3}, \frac{\pi}{3}\right]$. Divide this interval into four equal parts to find the points $\left(-\frac{\pi}{3}, -2\right)$, $\left(-\frac{\pi}{6}, 0\right)$, $(0, 2)$, $\left(\frac{\pi}{6}, 0\right)$, and $\left(\frac{\pi}{3}, -2\right)$. Plot these points and join them with a smooth curve. By graphing an additional half period to the left and to the right, we obtain the graph shown in **Figure 36.**

Method 2 First write the equation in the form $y = a\cos[b(x - d)]$.

$$y = -2\cos(3x + \pi), \quad \text{or} \quad y = -2\cos\left[3\left(x + \frac{\pi}{3}\right)\right] \quad \text{Rewrite by factoring out 3.}$$

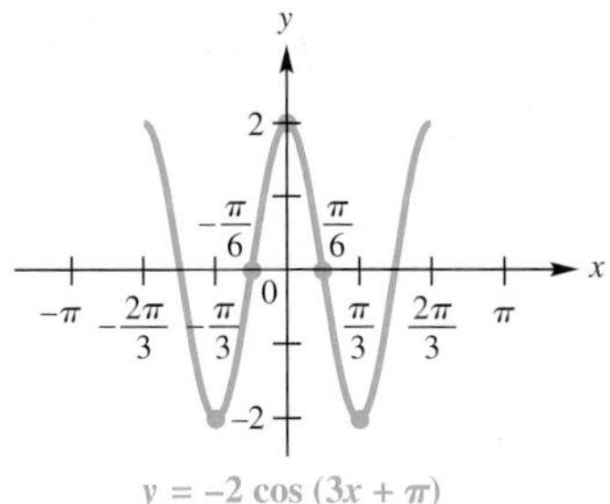

Figure 36

Then $a = -2$, $b = 3$, and $d = -\frac{\pi}{3}$. The amplitude is $|-2| = 2$, and the period is $\frac{2\pi}{3}$ (since the value of b is 3). The phase shift is $\left|-\frac{\pi}{3}\right| = \frac{\pi}{3}$ units to the left compared to the graph of $y = -2\cos 3x$. Again, see **Figure 36.**

Now Try Exercise 43.

Vertical Translations The graph of a function of the form

$$y = c + f(x)$$

is translated *vertically* compared to the graph of $y = f(x)$. See **Figure 37.** The translation is c units up if $c > 0$ and is $|c|$ units down if $c < 0$.

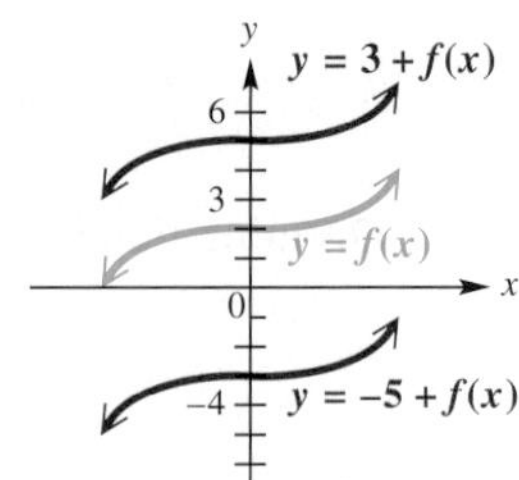

Vertical translations of $y = f(x)$
(Section 2.7)

Figure 37

Classroom Example 4
Graph $y = -2 + 3 \cos 2x$ over two periods.

Answer:

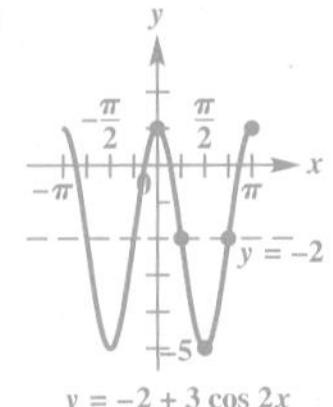

EXAMPLE 4 Graphing $y = c + a \cos bx$

Graph $y = 3 - 2 \cos 3x$ over two periods.

SOLUTION The values of y will be 3 greater than the corresponding values of y in $y = -2 \cos 3x$. This means that the graph of $y = 3 - 2 \cos 3x$ is the same as the graph of $y = -2 \cos 3x$, vertically translated 3 units up. Since the period of $y = -2 \cos 3x$ is $\frac{2\pi}{3}$, the key points have these x-values.

$$0, \quad \frac{\pi}{6}, \quad \frac{\pi}{3}, \quad \frac{\pi}{2}, \quad \frac{2\pi}{3} \qquad \text{Key } x\text{-values}$$

Use these x-values to make a table of points.

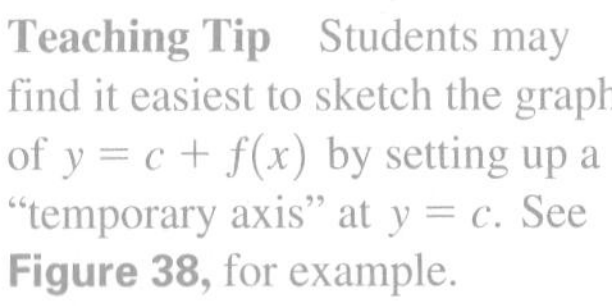

Teaching Tip Students may find it easiest to sketch the graph of $y = c + f(x)$ by setting up a "temporary axis" at $y = c$. See **Figure 38,** for example.

x	0	$\frac{\pi}{6}$	$\frac{\pi}{3}$	$\frac{\pi}{2}$	$\frac{2\pi}{3}$
$\cos 3x$	1	0	−1	0	1
$2 \cos 3x$	2	0	−2	0	2
$3 - 2 \cos 3x$	1	3	5	3	1

The key points are shown on the graph in **Figure 38,** along with more of the graph, which is sketched using the fact that the function is periodic.

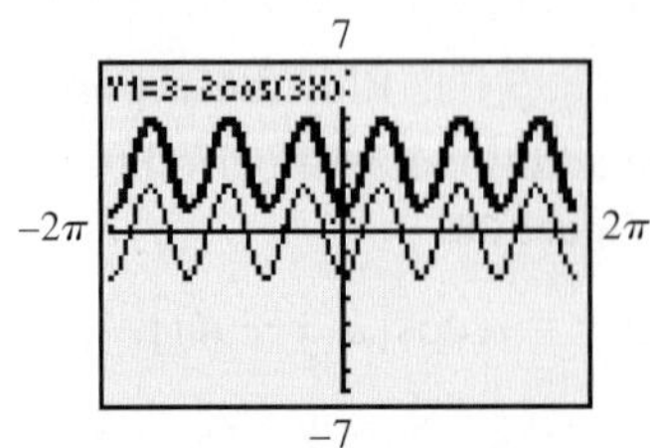

The function in **Example 4** is shown using the thick graph style. Notice also the thin graph style for $y = -2 \cos 3x$.

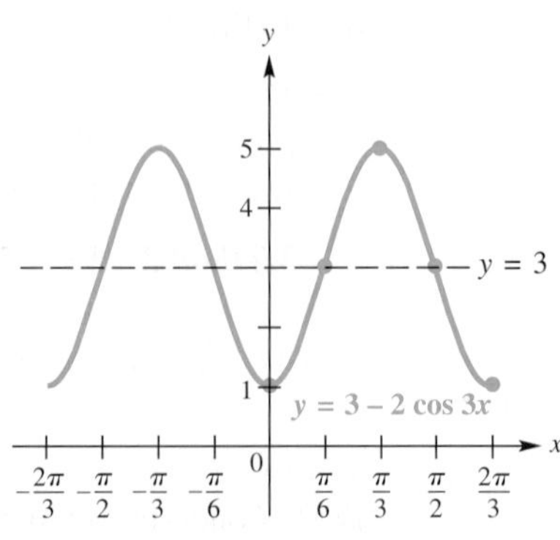

Figure 38

Now Try Exercise 47.

Combinations of Translations A function of the form

$$y = c + a \sin[b(x - d)] \quad \text{or} \quad y = c + a \cos[b(x - d)], \quad \text{where } b > 0,$$

which involves stretching, shrinking, and translating, can be graphed according to the following guidelines.

Further Guidelines for Sketching Graphs of Sine and Cosine Functions

Method 1 Follow these steps.

Step 1 Find an interval whose length is one period $\frac{2\pi}{b}$ by solving the three-part inequality $0 \leq b(x - d) \leq 2\pi$. (See **Section 1.7.**)

Step 2 Divide the interval into four equal parts. (See the Note preceding **Example 2** in **Section 6.3.**)

Step 3 Evaluate the function for each of the five x-values resulting from Step 2. The points will be maximum points, minimum points, and points that intersect the line $y = c$ ("middle" points of the wave).

Step 4 Plot the points found in Step 3, and join them with a sinusoidal curve having amplitude $|a|$.

Step 5 Draw the graph over additional periods, as needed.

Method 2 Follow these steps.

Step 1 Graph $y = a \sin bx$ or $y = a \cos bx$. The amplitude of the function is $|a|$, and the period is $\frac{2\pi}{b}$.

Step 2 Use translations to graph the desired function. The vertical translation is c units up if $c > 0$ and is $|c|$ units down if $c < 0$. The horizontal translation (phase shift) is d units to the right if $d > 0$ and is $|d|$ units to the left if $d < 0$.

Teaching Tip Give several examples of sinusoidal graphs, and ask students to determine their corresponding equations. Point out that any given graph may have several different corresponding equations, all of which are valid.

Many computer graphing utilities have animation features that enable you to incrementally change the values of a, b, c, and d in the sinusoidal graphs of

$$y = c + a \sin b(x - d)$$
and
$$y = c + a \cos b(x - d).$$

This provides an excellent visual image of how these parameters affect sinusoidal graphs.

Classroom Example 5
Graph $y = 4 - 2\sin(3x - \pi)$ over two periods.

Answer:

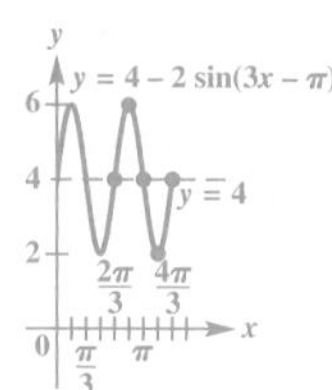

EXAMPLE 5 **Graphing $y = c + a \sin[b(x - d)]$**

Graph $y = -1 + 2\sin(4x + \pi)$ over two periods.

SOLUTION We use Method 1. First write the expression on the right side of the equation in the form $c + a \sin[b(x - d)]$.

$$y = -1 + 2\sin(4x + \pi), \quad \text{or} \quad y = -1 + 2\sin\left[4\left(x + \frac{\pi}{4}\right)\right] \quad \text{Rewrite by factoring out 4.}$$

Step 1 Find an interval whose length is one period.

$$0 \leq 4\left(x + \frac{\pi}{4}\right) \leq 2\pi$$

$$0 \leq x + \frac{\pi}{4} \leq \frac{\pi}{2} \quad \text{Divide each part by 4.}$$

$$-\frac{\pi}{4} \leq x \leq \frac{\pi}{4} \quad \text{Subtract } \tfrac{\pi}{4} \text{ from each part.}$$

Step 2 Divide the interval $\left[-\frac{\pi}{4}, \frac{\pi}{4}\right]$ into four equal parts to get these x-values.

$$-\frac{\pi}{4}, \quad -\frac{\pi}{8}, \quad 0, \quad \frac{\pi}{8}, \quad \frac{\pi}{4} \quad \text{Key } x\text{-values}$$

Step 3 Make a table of values.

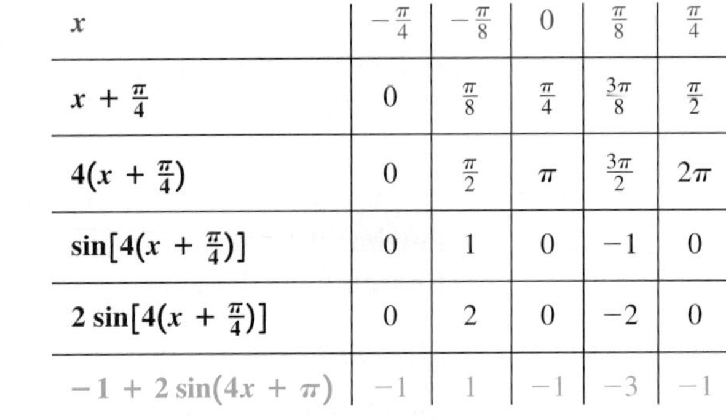

x	$-\frac{\pi}{4}$	$-\frac{\pi}{8}$	0	$\frac{\pi}{8}$	$\frac{\pi}{4}$
$x + \frac{\pi}{4}$	0	$\frac{\pi}{8}$	$\frac{\pi}{4}$	$\frac{3\pi}{8}$	$\frac{\pi}{2}$
$4(x + \frac{\pi}{4})$	0	$\frac{\pi}{2}$	π	$\frac{3\pi}{2}$	2π
$\sin[4(x + \frac{\pi}{4})]$	0	1	0	-1	0
$2\sin[4(x + \frac{\pi}{4})]$	0	2	0	-2	0
$-1 + 2\sin(4x + \pi)$	-1	1	-1	-3	-1

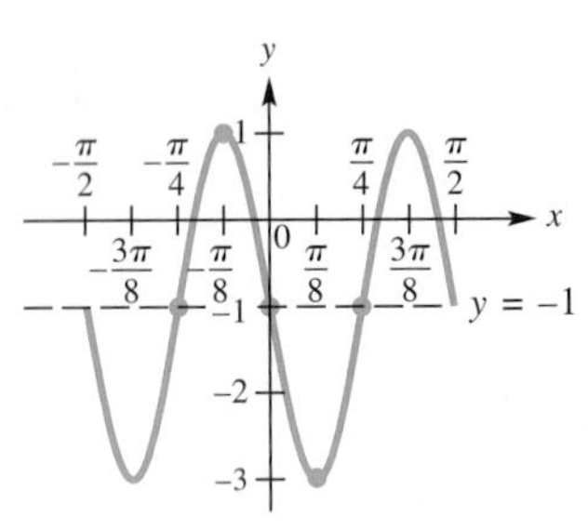

$y = -1 + 2\sin(4x + \pi)$

Figure 39

Steps 4 and 5 Plot the points found in the table and join them with a sinusoidal curve. **Figure 39** shows the graph, extended to the right and left to include two full periods.

✔ ***Now Try Exercise 53.***

Determining a Trigonometric Model A sinusoidal function is often a good approximation of a set of real data points.

EXAMPLE 6 Modeling Temperature with a Sine Function

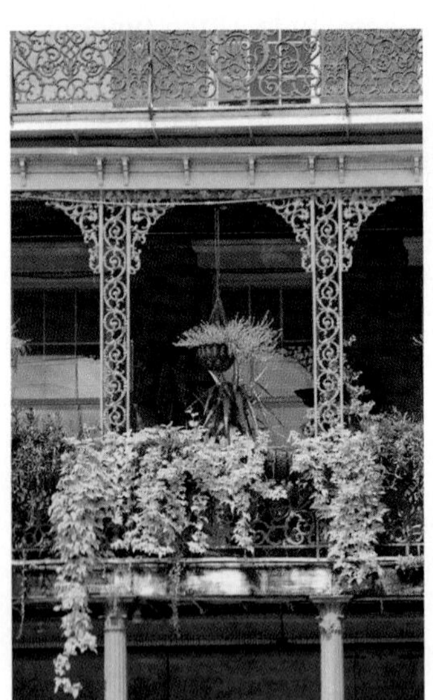

The maximum average monthly temperature in New Orleans is 83°F, and the minimum is 53°F. The table shows the average monthly temperatures. The scatter diagram for a two-year interval in **Figure 40** strongly suggests that the temperatures can be modeled with a sine curve.

Month	°F	Month	°F
Jan	53	July	83
Feb	56	Aug	83
Mar	62	Sept	79
Apr	68	Oct	70
May	76	Nov	61
June	81	Dec	55

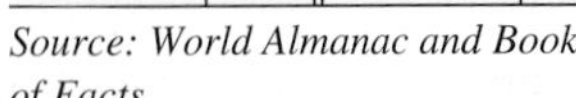

Source: World Almanac and Book of Facts.

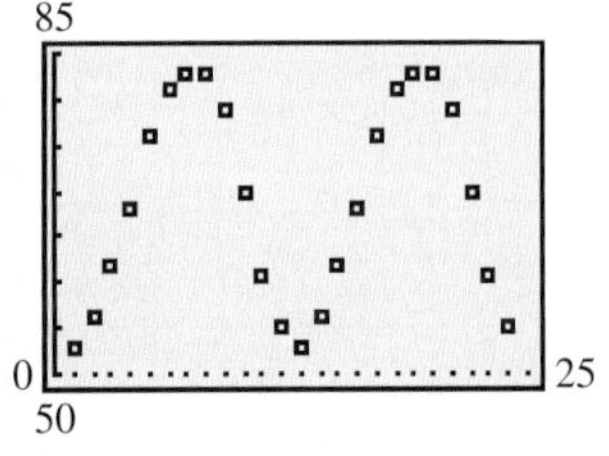

Figure 40

(a) Using only the maximum and minimum temperatures, determine a function of the form

$$f(x) = a\sin[b(x - d)] + c, \quad \text{where } a, b, c, \text{ and } d \text{ are constants,}$$

that models the average monthly temperature in New Orleans. Let x represent the month, with January corresponding to $x = 1$.

(b) On the same coordinate axes, graph f for a two-year period together with the actual data values found in the table.

(c) Use the **sine regression** feature of a graphing calculator to determine a second model for these data.

Classroom Example 6
In Minneapolis, the average monthly temperatures include 13°F in January, 47°F in April, 73°F in July, and 49°F in October. (*Source: World Almanac and Book of Facts.*) On average, January is the coldest month of the year, and July is the warmest.

(a) Using this information, repeat part (a) of **Example 6** for Minneapolis.

(b) Repeat part (c) of **Example 6** for Minneapolis. (Use the given data for a two-year period, and round all constants to the nearest hundredth.)

(c) Use your models from parts (a) and (b) to predict the average temperature for September in Minneapolis. Compare the results.

Answers:

(a) $f(x) = 30 \sin\left[\frac{\pi}{6}(x-4)\right] + 43$

(b) $f(x) = 30.03 \sin(0.52x - 2.05) + 45.23$

(c) 58°F; 59.9°F; They are quite close.

SOLUTION

(a) We use the maximum and minimum average monthly temperatures to find the amplitude a.

$$a = \frac{83 - 53}{2} = 15 \quad \text{Amplitude}$$

The average of the maximum and minimum temperatures is a good choice for c. The average is

$$\frac{83 + 53}{2} = 68. \quad \text{Vertical translation}$$

Since temperatures repeat every 12 months, b can be found as follows.

$$12 = \frac{2\pi}{b} \quad \text{Period} = \frac{2\pi}{b}$$

$$b = \frac{\pi}{6} \quad \text{Solve for } b. \text{ (Section 1.1)}$$

The coldest month is January, when $x = 1$, and the hottest month is July, when $x = 7$. A good choice for d is 4 because April, when $x = 4$, is located at the midpoint between January and July. Also, notice that the average monthly temperature in April is 68°F, which is the value of the vertical translation, c. The average monthly temperature in New Orleans is modeled closely by the following equation.

$$f(x) = a \sin[b(x - d)] + c$$

$$f(x) = 15 \sin\left[\frac{\pi}{6}(x - 4)\right] + 68 \quad \text{Substitute.}$$

(b) **Figure 41** shows the data points from the table, along with the graph of $y = 15 \sin\left[\frac{\pi}{6}(x-4)\right] + 68$ and the graph of $y = 15 \sin \frac{\pi}{6}x + 68$ for comparison.

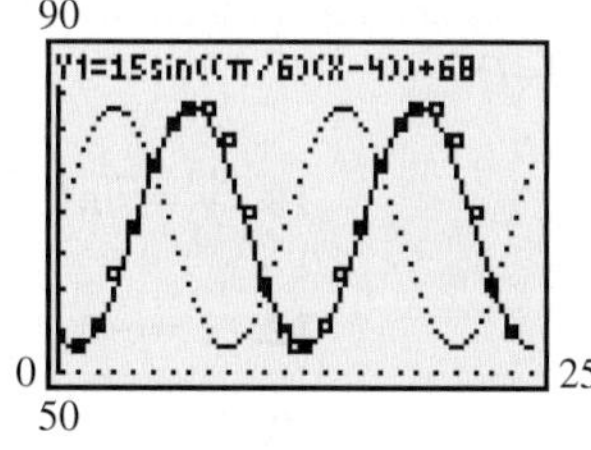

Figure 41

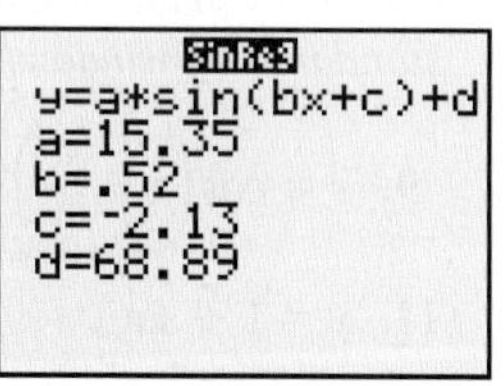

Values are rounded to the nearest hundredth.

(a)

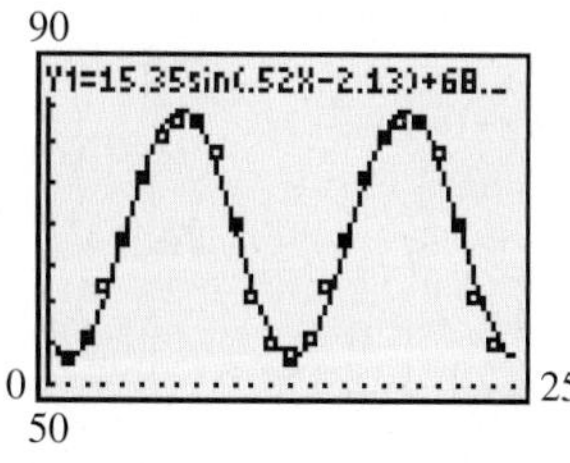

(b)

Figure 42

(c) We used the given data for a two-year period and the sine regression capability of a graphing calculator to produce the model

$$f(x) = 15.35 \sin(0.52x - 2.13) + 68.89$$

described in **Figure 42(a)**. Its graph along with the data points is shown in **Figure 42(b)**.

Now Try Exercise 57.

6.4 Exercises

1. D 2. G 3. H 4. A
5. B 6. E 7. I 8. C

9. C 10. B
11. A 12. D

14. $y = \sin x + 1$

15. B 16. D
17. C 18. A

19. right 20. left

21. $y = -1 + \sin x$
22. $y = 2 + \cos x$
23. $y = \cos\left(x - \frac{\pi}{3}\right)$
24. $y = \cos\left(x - \frac{\pi}{6}\right)$

25. 2; 2π; none; π to the left
26. 3; 2π; none; $\frac{\pi}{2}$ to the left
27. $\frac{1}{4}$; 4π; none; π to the left
28. $\frac{1}{2}$; 4π; none; 2π to the left
29. 3; 4; none; $\frac{1}{2}$ to the right
30. 1; 2; none; $\frac{1}{3}$ to the right
31. 1; $\frac{2\pi}{3}$; up 2; $\frac{\pi}{15}$ to the right
32. $\frac{1}{2}$; π; down 1; $\frac{3\pi}{2}$ to the right

33.

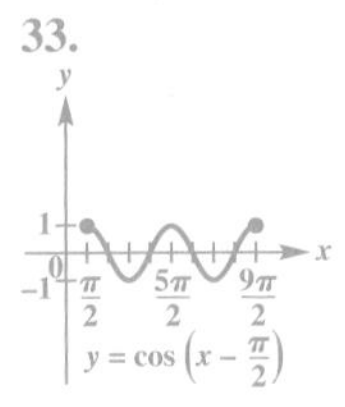

34.

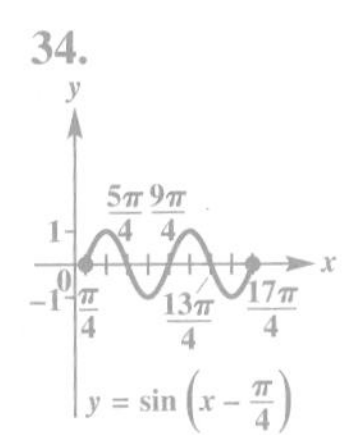

35.

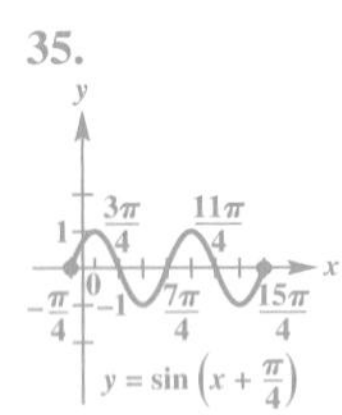

36.

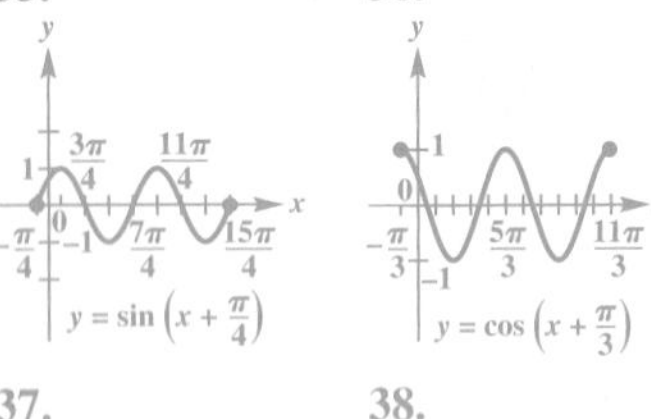

37.

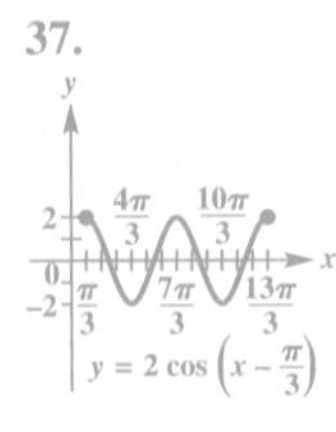

38.

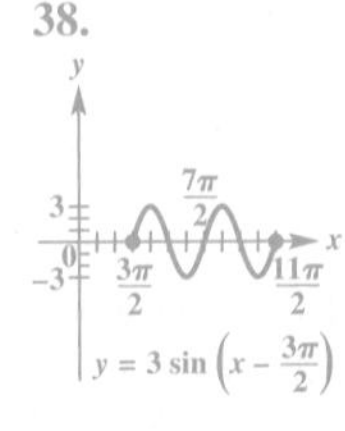

39. 40.

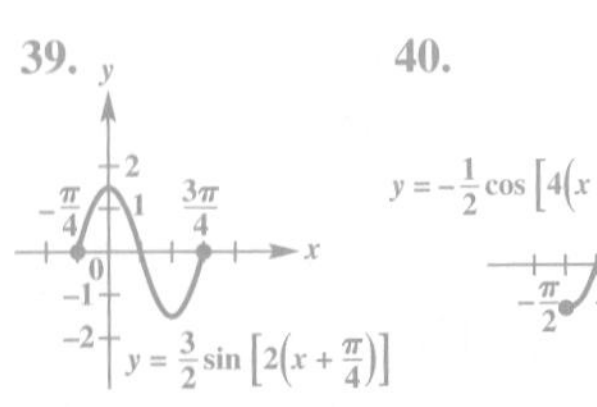

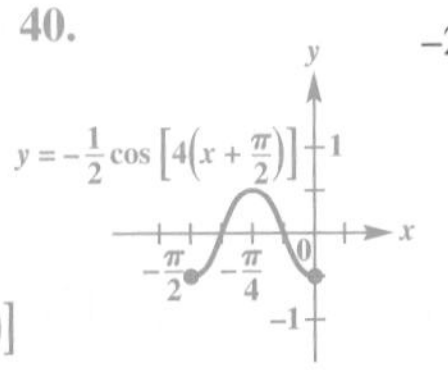

Concept Check *In Exercises 1–8, match each function with its graph in choices A–I. (One choice will not be used.)*

1. $y = \sin\left(x - \frac{\pi}{4}\right)$
2. $y = \sin\left(x + \frac{\pi}{4}\right)$
3. $y = \cos\left(x - \frac{\pi}{4}\right)$
4. $y = \cos\left(x + \frac{\pi}{4}\right)$
5. $y = 1 + \sin x$
6. $y = -1 + \sin x$
7. $y = 1 + \cos x$
8. $y = -1 + \cos x$

A.

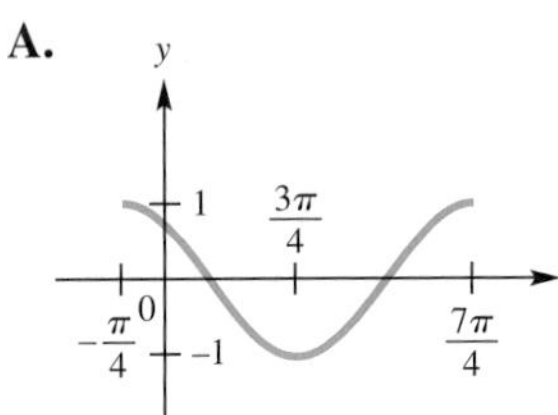

B.

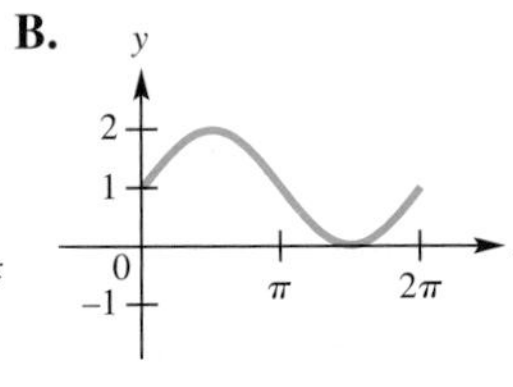

C.

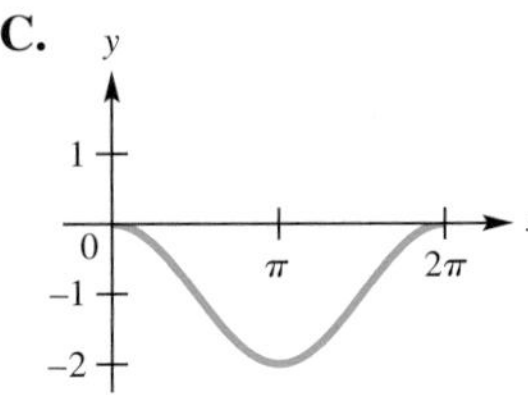

D.

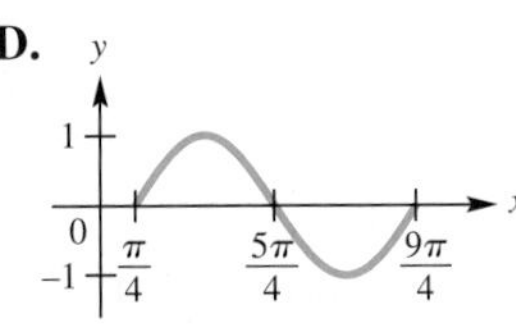

E.

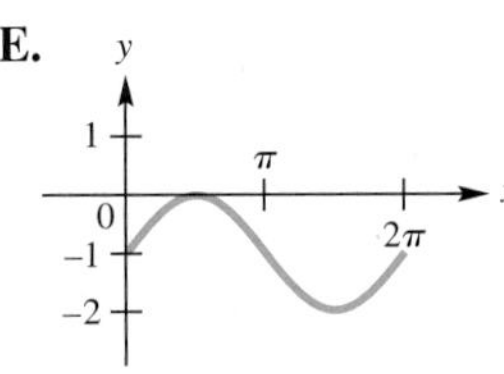

F.

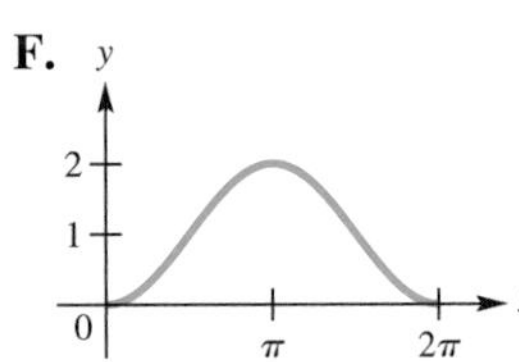

G.

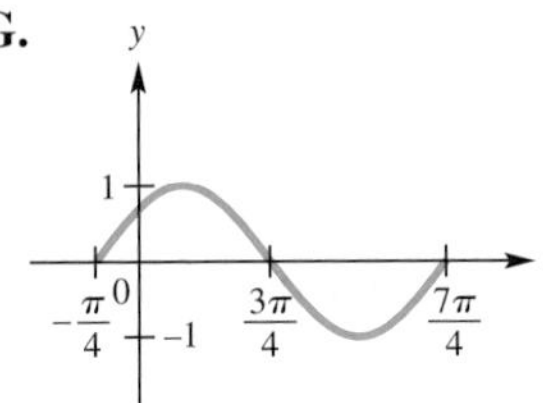

H.

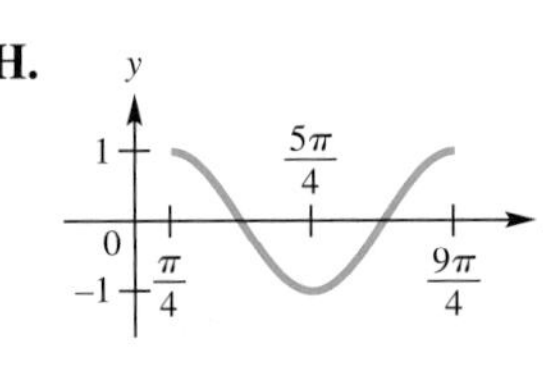

I.

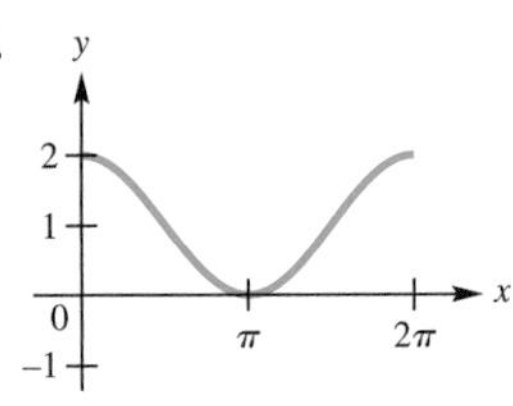

Concept Check *In Exercises 9–12, match each function with its calculator graph in the standard trig window in choices A–D.*

9. $y = \cos\left(x - \frac{\pi}{4}\right)$
10. $y = \sin\left(x - \frac{\pi}{4}\right)$
11. $y = 1 + \sin x$
12. $y = -1 + \cos x$

A.

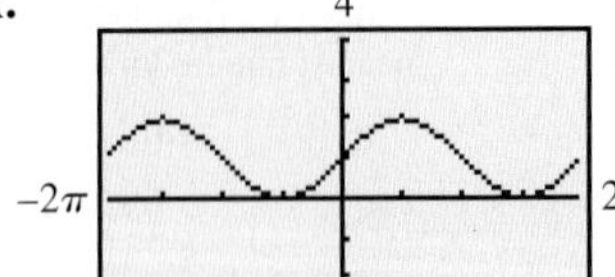

B.

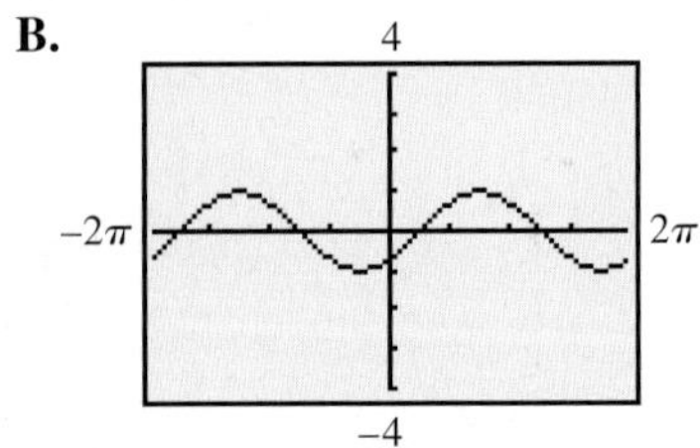

C.

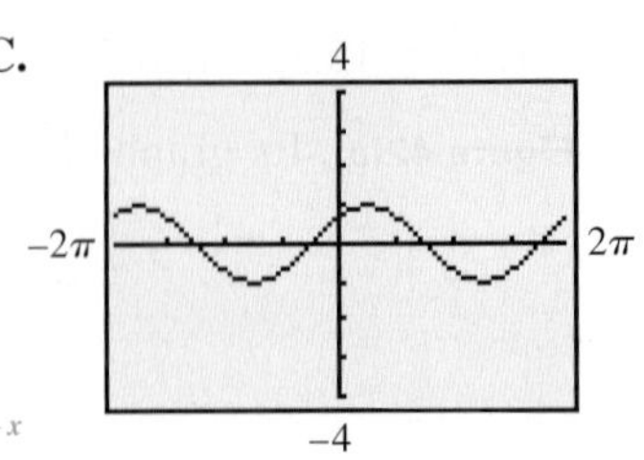

D.

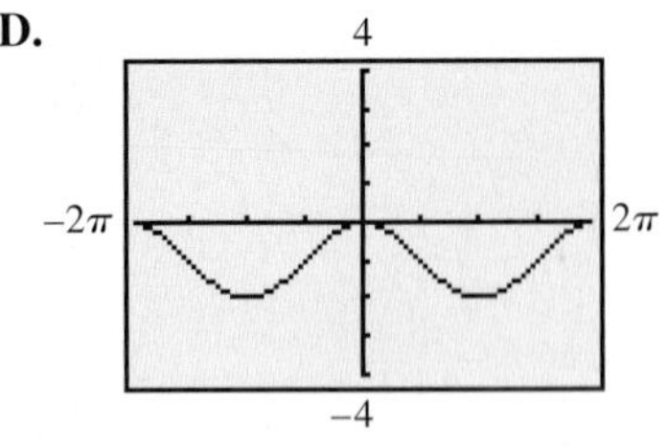

41. 42.

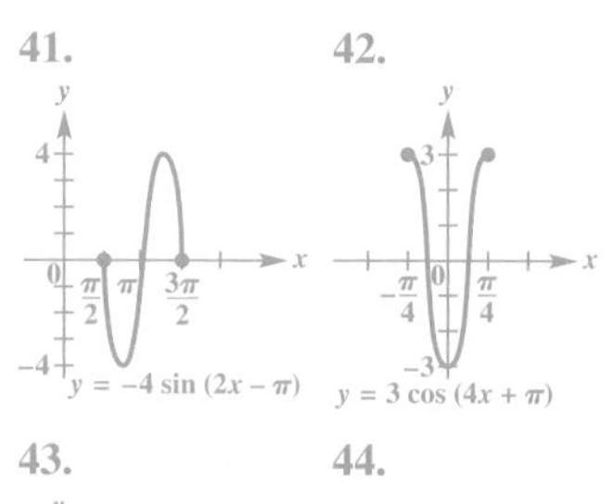

43. 44.

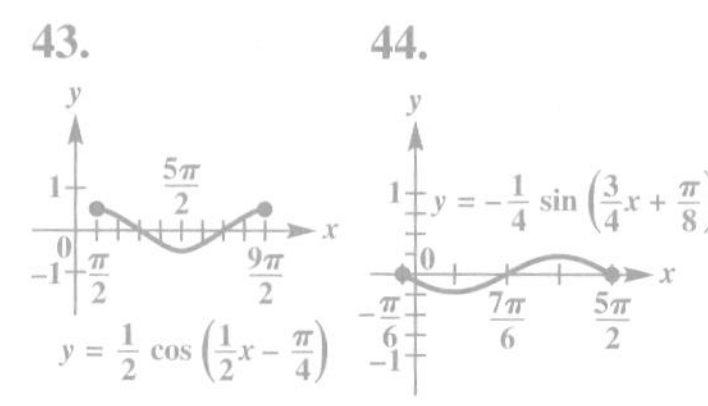

45.

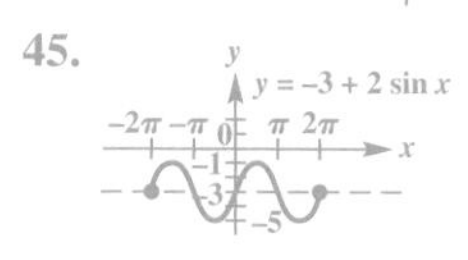

46.

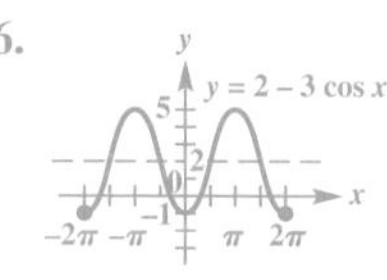

47.

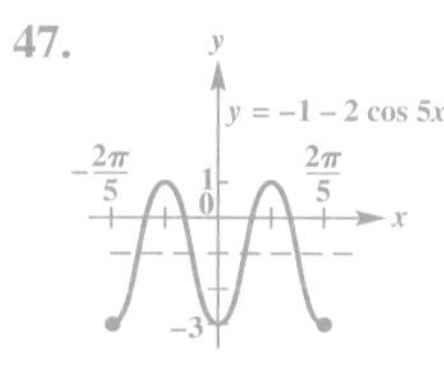

48.

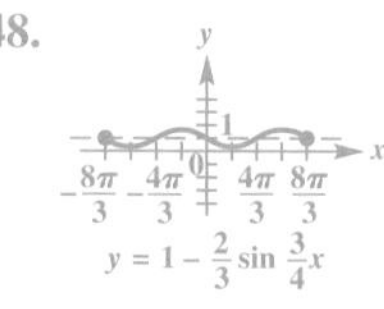

49.

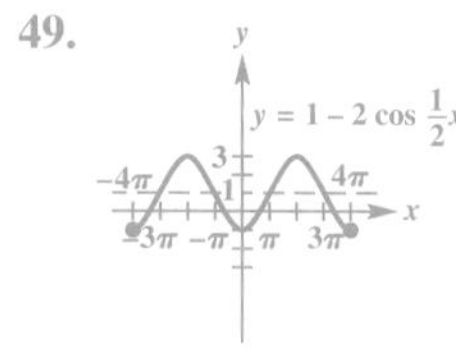

50.

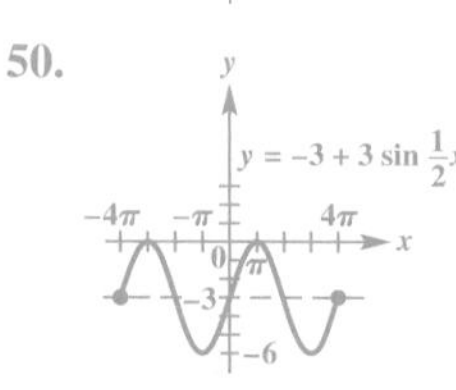

51.

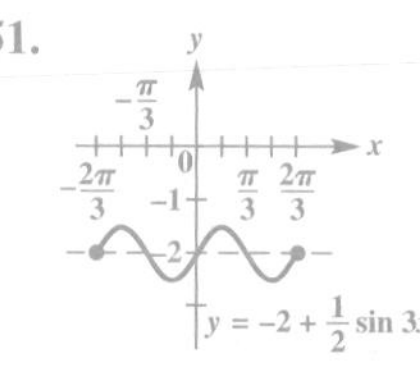

52.

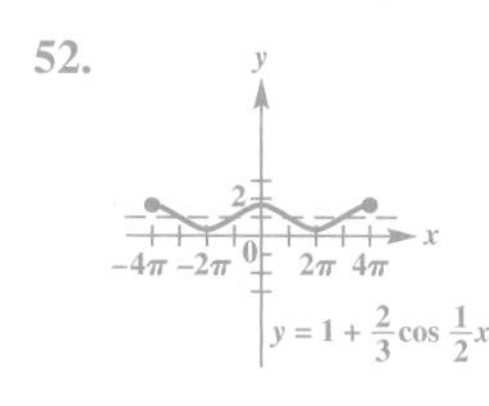

13. The graphs of $y = \sin x + 1$ and $y = \sin(x + 1)$ are **NOT** the same. Explain why this is so.

14. *Concept Check* Refer to **Exercise 13.** Which one of the two graphs is the same as that of $y = 1 + \sin x$?

Concept Check Match each function in Column I with the appropriate description in Column II.

I	II
15. $y = 3 \sin(2x - 4)$	**A.** amplitude = 2, period = $\frac{\pi}{2}$, phase shift = $\frac{3}{4}$
16. $y = 2 \sin(3x - 4)$	**B.** amplitude = 3, period = π, phase shift = 2
17. $y = -4 \sin(3x - 2)$	**C.** amplitude = 4, period = $\frac{2\pi}{3}$, phase shift = $\frac{2}{3}$
18. $y = -2 \sin(4x - 3)$	**D.** amplitude = 2, period = $\frac{2\pi}{3}$, phase shift = $\frac{4}{3}$

Concept Check In Exercises 19 and 20, fill in the blanks with the word right *or the word* left.

19. If the graph of $y = \cos x$ is translated $\frac{\pi}{2}$ units horizontally to the __________, it will coincide with the graph of $y = \sin x$.

20. If the graph of $y = \sin x$ is translated $\frac{\pi}{2}$ units horizontally to the __________, it will coincide with the graph of $y = \cos x$.

Connecting Graphs with Equations Each function graphed in Exercises 21–24 is of the form $y = c + \cos x$, $y = c + \sin x$, $y = \cos(x - d)$, or $y = \sin(x - d)$, where d is the least possible positive value. Determine the equation of the graph.

21.

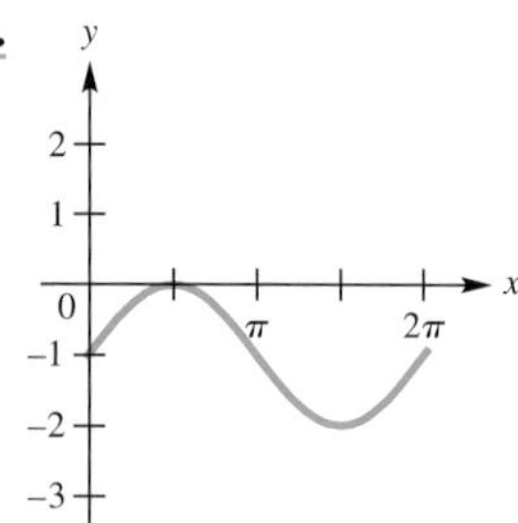

22.

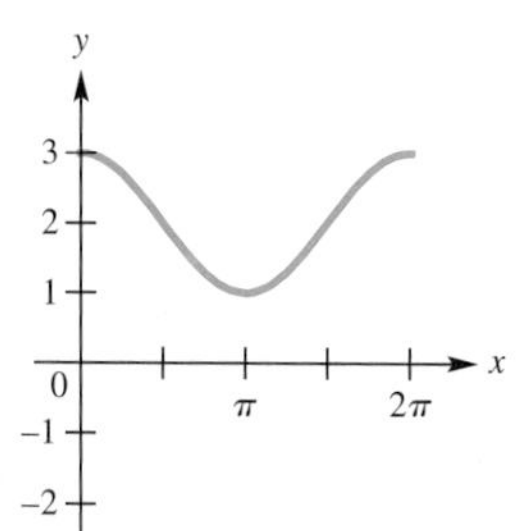

23.

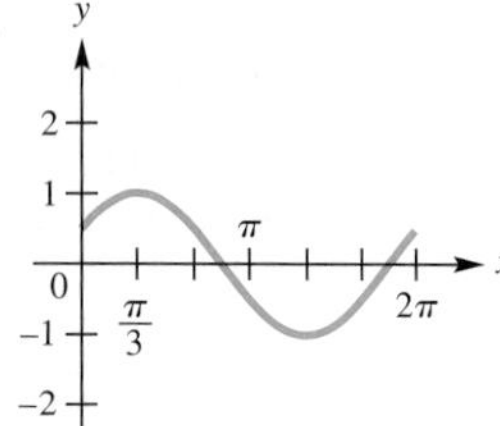

24.

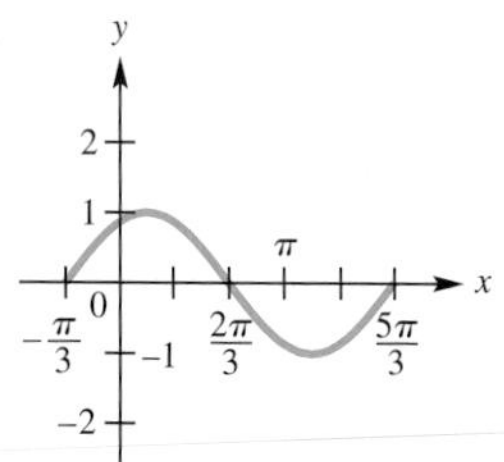

Find the amplitude, the period, any vertical translation, and any phase shift of the graph of each function. ***See Examples 1–5.***

25. $y = 2 \sin(x + \pi)$

26. $y = 3 \sin\left(x + \frac{\pi}{2}\right)$

27. $y = -\frac{1}{4} \cos\left(\frac{1}{2}x + \frac{\pi}{2}\right)$

28. $y = -\frac{1}{2} \sin\left(\frac{1}{2}x + \pi\right)$

29. $y = 3 \cos\left[\frac{\pi}{2}\left(x - \frac{1}{2}\right)\right]$

30. $y = -\cos\left[\pi\left(x - \frac{1}{3}\right)\right]$

31. $y = 2 - \sin\left(3x - \frac{\pi}{5}\right)$

32. $y = -1 + \frac{1}{2} \cos(2x - 3\pi)$

Graph each function over a two-period interval. **See Examples 1 and 2.**

33. $y = \cos\left(x - \frac{\pi}{2}\right)$

34. $y = \sin\left(x - \frac{\pi}{4}\right)$

35. $y = \sin\left(x + \frac{\pi}{4}\right)$

36. $y = \cos\left(x + \frac{\pi}{3}\right)$

37. $y = 2\cos\left(x - \frac{\pi}{3}\right)$

38. $y = 3\sin\left(x - \frac{3\pi}{2}\right)$

Graph each function over a one-period interval. **See Example 3.**

39. $y = \frac{3}{2}\sin\left[2\left(x + \frac{\pi}{4}\right)\right]$

40. $y = -\frac{1}{2}\cos\left[4\left(x + \frac{\pi}{2}\right)\right]$

41. $y = -4\sin(2x - \pi)$

42. $y = 3\cos(4x + \pi)$

43. $y = \frac{1}{2}\cos\left(\frac{1}{2}x - \frac{\pi}{4}\right)$

44. $y = -\frac{1}{4}\sin\left(\frac{3}{4}x + \frac{\pi}{8}\right)$

Graph each function over a two-period interval. **See Example 4.**

45. $y = -3 + 2\sin x$

46. $y = 2 - 3\cos x$

47. $y = -1 - 2\cos 5x$

48. $y = 1 - \frac{2}{3}\sin\frac{3}{4}x$

49. $y = 1 - 2\cos\frac{1}{2}x$

50. $y = -3 + 3\sin\frac{1}{2}x$

51. $y = -2 + \frac{1}{2}\sin 3x$

52. $y = 1 + \frac{2}{3}\cos\frac{1}{2}x$

Graph each function over a one-period interval. **See Example 5.**

53. $y = -3 + 2\sin\left(x + \frac{\pi}{2}\right)$

54. $y = 4 - 3\cos(x - \pi)$

55. $y = \frac{1}{2} + \sin\left[2\left(x + \frac{\pi}{4}\right)\right]$

56. $y = -\frac{5}{2} + \cos\left[3\left(x - \frac{\pi}{6}\right)\right]$

(Modeling) Solve each problem. **See Example 6.**

57. ***Average Monthly Temperature*** The average monthly temperature (in °F) in Seattle, Washington, is shown in the table.

Month	°F	Month	°F
Jan	41	July	65
Feb	43	Aug	66
Mar	46	Sept	61
Apr	50	Oct	53
May	56	Nov	45
June	61	Dec	41

Source: World Almanac and Book of Facts.

(a) Plot the average monthly temperature over a two-year period, letting $x = 1$ correspond to January during the first year. Do the data seem to indicate a translated sine graph?

(b) The highest average monthly temperature is 66°F in August, and the lowest average monthly temperature is 41°F in January. Their average is 53.5°F. Graph the data together with the line $y = 53.5$. What does this line represent with regard to temperature in Seattle?

(c) Approximate the amplitude, period, and phase shift of the translated sine wave.

(d) Determine a function of the form $f(x) = a\sin[b(x - d)] + c$, where a, b, c, and d are constants, that models the data.

(e) Graph f together with the data on the same coordinate axes. How well does f model the given data?

(f) Use the sine regression capability of a graphing calculator to find the equation of a sine curve that fits these data.

53.

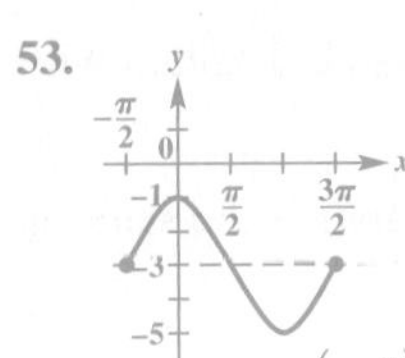

54.

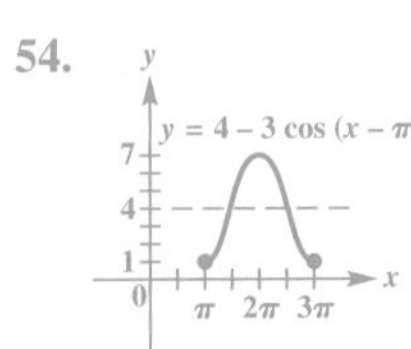

55.

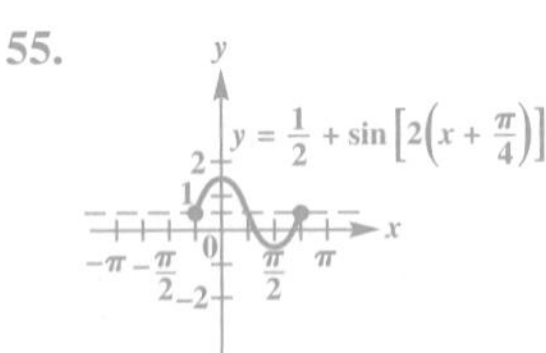

56.

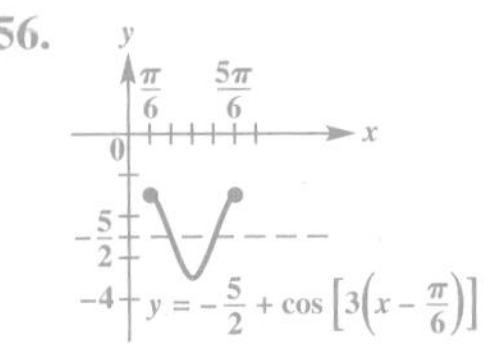

57. (a) yes

(b) It represents the average yearly temperature.

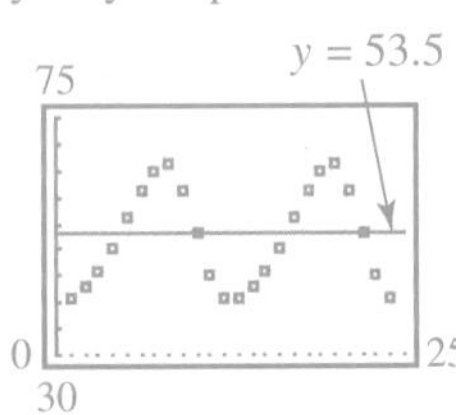

(c) 12.5; 12; 4.5

(d) $f(x) = 12.5\sin\left[\frac{\pi}{6}(x - 4.5)\right] + 53.5$

(e) The function gives a good model for the data.

$f(x) = 12.5\sin\left[\frac{\pi}{6}(x - 4.5)\right] + 53.5$

(f)

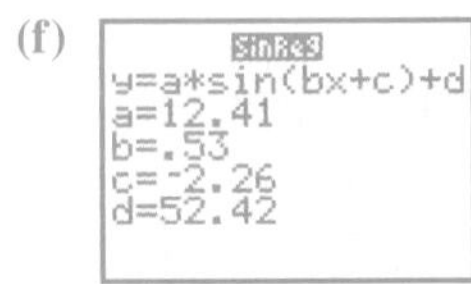

TI-83/84 Plus fixed to the nearest hundredth

58. (a) 73.5°F
(b) See the graph in part (d).
(c) $f(x) = 19.5 \cos\left[\frac{\pi}{6}(x - 7)\right] + 73.5$
(d) The function gives a good model for the data.

$f(x) = 19.5 \cos\left[\frac{\pi}{6}(x - 7)\right] + 73.5$

100, 0, 25, 45

(e)

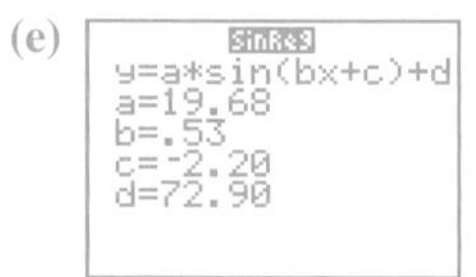

TI-83/84 Plus fixed to the nearest hundredth

59.

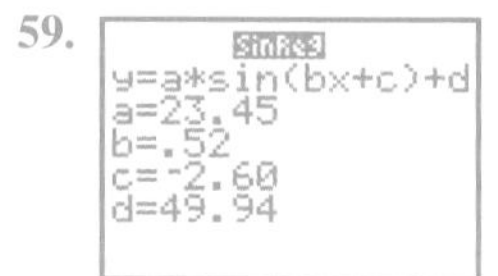

TI-83/84 Plus fixed to the nearest hundredth

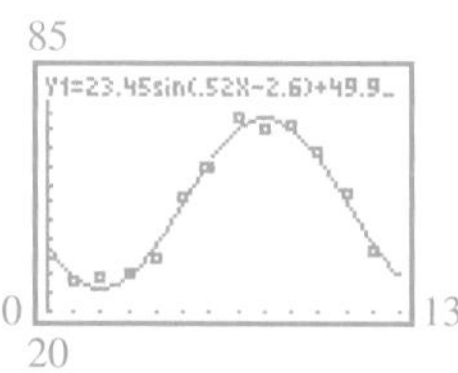

60.

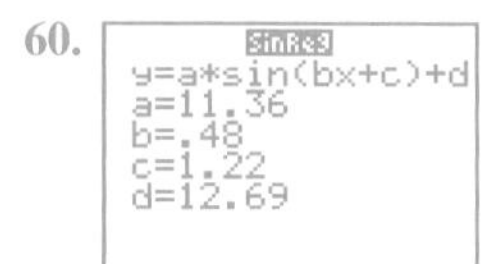

TI-83/84 Plus fixed to the nearest hundredth

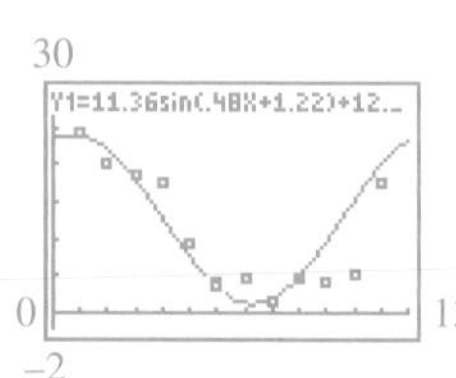

58. *Average Monthly Temperature* The average monthly temperature (in °F) in Phoenix, Arizona, is shown in the table.

(a) Predict the average yearly temperature.
(b) Plot the average monthly temperature over a two-year period, letting $x = 1$ correspond to January of the first year.
(c) Determine a function of the form $f(x) = a \cos\left[b(x - d)\right] + c$, where a, b, c, and d are constants, that models the data.
(d) Graph f together with the data on the same coordinate axes. How well does f model the data?
(e) Use the sine regression capability of a graphing calculator to find the equation of a sine curve that fits these data (two years).

Month	°F	Month	°F
Jan	54	July	93
Feb	58	Aug	91
Mar	63	Sept	86
Apr	70	Oct	75
May	79	Nov	62
June	89	Dec	54

Source: World Almanac and Book of Facts.

(Modeling) Utility Bills In an article entitled "I Found Sinusoids in My Gas Bill" (Mathematics Teacher, *January 2000), Cathy G. Schloemer presents the following graph* that accompanied her gas bill.*

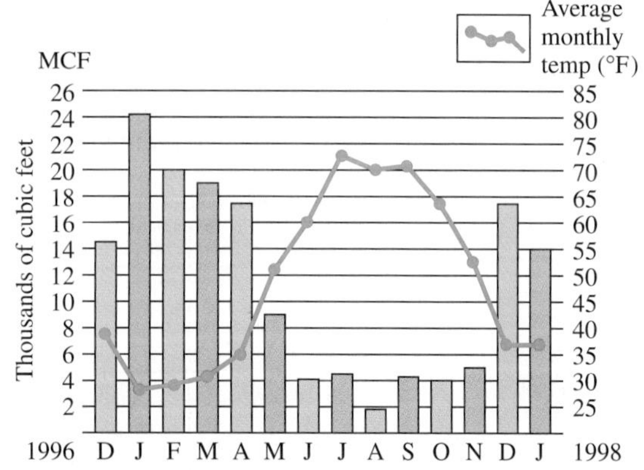

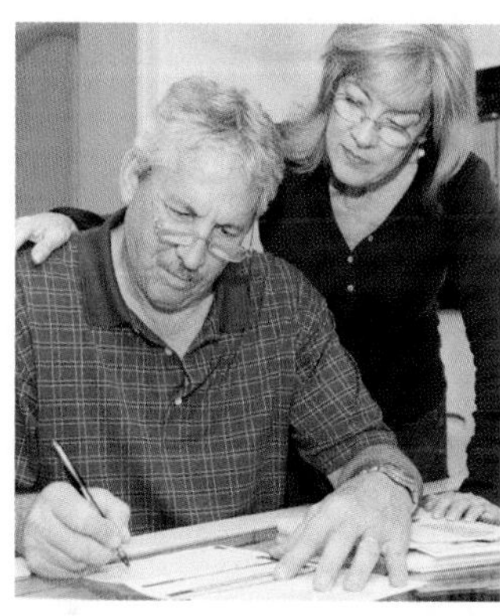

Notice that two sinusoids are suggested here: one for the behavior of the average monthly temperature and another for gas use in MCF (thousands of cubic feet). Use this information in Exercises 59 and 60.

59. If January 1997 is represented by $x = 1$, the data of estimated ordered pairs (month, temperature) are given in the list shown on the two graphing calculator screens below.

L1	L2	L3
1	28	------
2	29	
3	31	
4	35	
5	51	
6	60	
7	73	

L1(1) = 1

L1	L2	L3
7	73	
8	70	
9	71	
10	64	
11	53	
12	37	

L1(13) =

Use the sine regression feature of a graphing calculator to find a sine function that fits these data points. Then make a scatter diagram, and graph the function.

60. If January 1997 is again represented by $x = 1$, the data of estimated ordered pairs (month, gas use in thousands of cubic feet (MCF)) are given in the list shown on the two graphing calculator screens below.

L1	L2	L3
1	24.2	------
2	20	
3	18.8	
4	17.5	
5	9.2	
6	4.2	
7	4.8	

L1(1) = 1

L1	L2	L3
7	4.8	
8	1.8	
9	4.8	
10	4	
11	5	
12	17.5	

L1(13) =

Use the sine regression feature of a graphing calculator to find a sine function that fits these data points. Then make a scatter diagram, and graph the function.

Chapter 6 Quiz (Sections 6.1–6.4)

[6.1]

1. $\frac{5\pi}{4}$ 2. $-210°$
3. 1.5 4. 67,500 in.^{2}

[6.2]

5. (a) $\frac{\sqrt{2}}{2}$ (b) $-\frac{1}{2}$ (c) 0
6. $\frac{2\pi}{3}$

[6.3]

7. 2π; 4 8. π; $\frac{1}{2}$

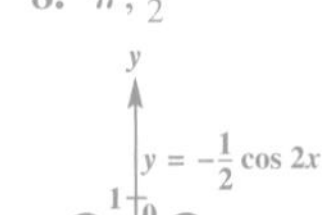

[6.4]

9. 2π; 2

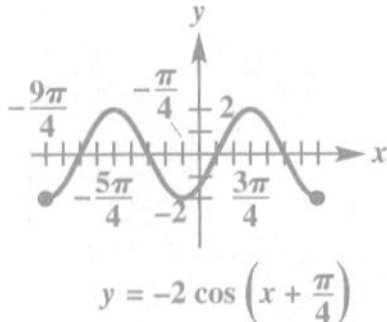

$y = -2\cos\left(x + \frac{\pi}{4}\right)$

10. π; 1

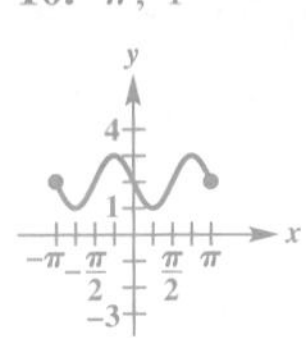

$y = 2 + \sin(2x - \pi)$

[6.3]

11. $y = 2 \sin x$ 12. $y = \cos 2x$
13. $y = -\sin x$

[6.3, 6.4]

14. 73°F 15. 60°F; 84°F

1. Convert 225° to radians. **2.** Convert $\frac{7\pi}{6}$ radians to degrees.

A central angle of a circle with radius 300 in. *intercepts an arc of* 450 in. *(These measures are accurate to the nearest inch.) Find each measure.*

3. the radian measure of the angle **4.** the area of the sector

5. Find each circular function value. Give exact values.

(a) $\cos \frac{7\pi}{4}$ **(b)** $\sin\left(-\frac{5\pi}{6}\right)$ **(c)** $\tan 3\pi$

6. Find the exact value of s in the interval $\left[\frac{\pi}{2}, \pi\right]$ if $\sin s = \frac{\sqrt{3}}{2}$.

Graph each function over a two-period interval. Give the period and amplitude.

7. $y = -4 \sin x$ **8.** $y = -\frac{1}{2}\cos 2x$

9. $y = -2\cos\left(x + \frac{\pi}{4}\right)$ **10.** $y = 2 + \sin(2x - \pi)$

Connecting Graphs with Equations *Each function graphed is of the form $y = a \cos bx$ or $y = a \sin bx$, where $b > 0$. Determine the equation of the graph.*

11.

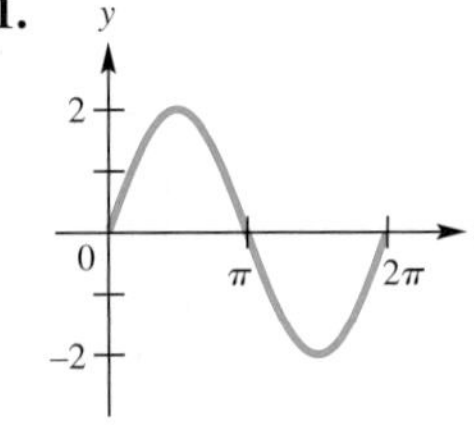

12.

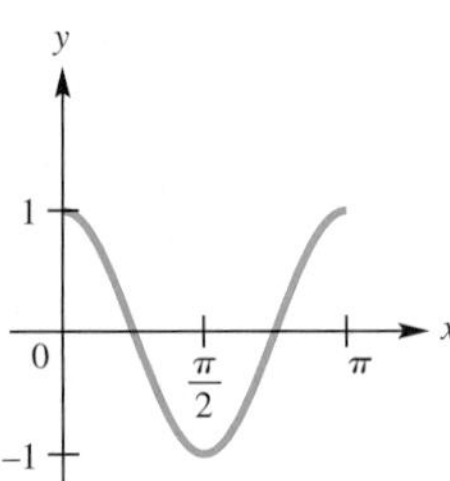

13.

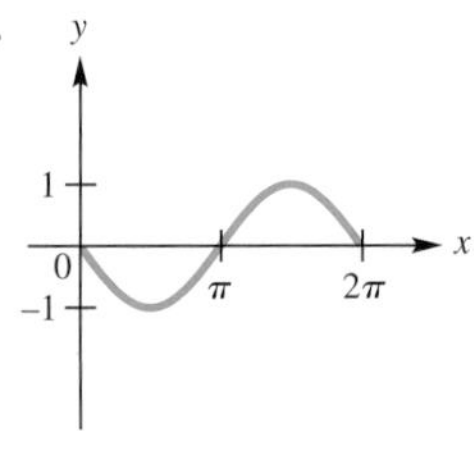

(Modeling) Average Monthly Temperature *The average temperature (in °F) at a certain location can be approximated by the function*

$$f(x) = 12 \sin\left[\frac{\pi}{6}(x - 3.9)\right] + 72,$$

where $x = 1$ represents January, $x = 2$ represents February, and so on.

14. What is the average temperature in April?

15. What is the lowest average monthly temperature? What is the highest?

6.5 Graphs of the Tangent and Cotangent Functions

- Graph of the Tangent Function
- Graph of the Cotangent Function
- Graphing Techniques
- Connecting Graphs with Equations

Graph of the Tangent Function Consider the table of selected points accompanying the graph of the tangent function in **Figure 43** on the next page. These points include special values between $-\frac{\pi}{2}$ and $\frac{\pi}{2}$. The tangent function is undefined for odd multiples of $\frac{\pi}{2}$ and, thus, has *vertical asymptotes* for such values. A **vertical asymptote** is a vertical line that the graph approaches but does not intersect. As the x-values get closer and closer to the line, the function values increase or decrease without bound. Furthermore, since

$$\tan(-x) = -\tan x, \quad \text{(See Exercise 45.)}$$

the graph of the tangent function is symmetric with respect to the origin.

x	$y = \tan x$
$-\frac{\pi}{3}$	$-\sqrt{3} \approx -1.7$
$-\frac{\pi}{4}$	-1
$-\frac{\pi}{6}$	$-\frac{\sqrt{3}}{3} \approx -0.6$
0	0
$\frac{\pi}{6}$	$\frac{\sqrt{3}}{3} \approx 0.6$
$\frac{\pi}{4}$	1
$\frac{\pi}{3}$	$\sqrt{3} \approx 1.7$

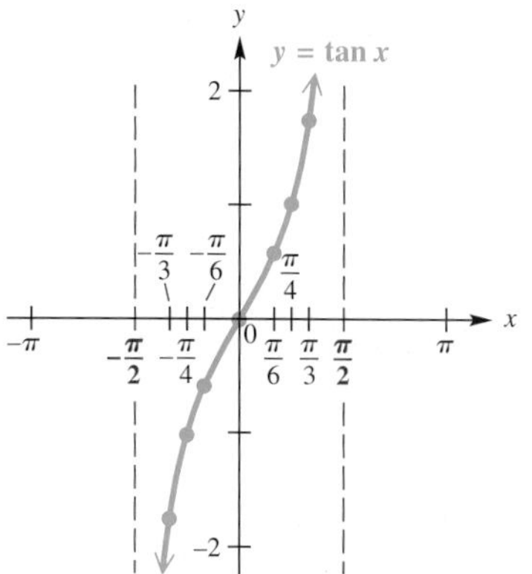

Figure 43

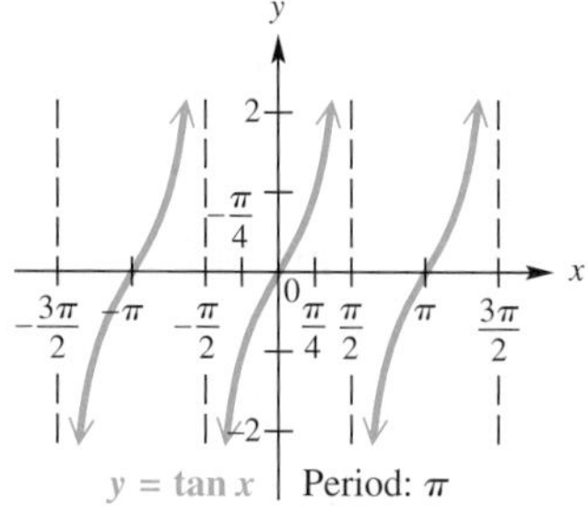

Figure 44

The tangent function has period π. Because $\tan x = \frac{\sin x}{\cos x}$, tangent values are 0 when sine values are 0, and are undefined when cosine values are 0. As x-values increase from $-\frac{\pi}{2}$ to $\frac{\pi}{2}$, tangent values range from $-\infty$ to ∞ and increase throughout the interval. Those same values are repeated as x increases from $\frac{\pi}{2}$ to $\frac{3\pi}{2}$, from $\frac{3\pi}{2}$ to $\frac{5\pi}{2}$, and so on. The graph of $y = \tan x$ from $-\frac{3\pi}{2}$ to $\frac{3\pi}{2}$ is shown in **Figure 44.** The graph continues in this pattern.

Tangent Function $f(x) = \tan x$

Domain: $\{x \mid x \neq (2n+1)\frac{\pi}{2}$, where n is any integer$\}$ Range: $(-\infty, \infty)$

x	y
$-\frac{\pi}{2}$	undefined
$-\frac{\pi}{4}$	-1
0	0
$\frac{\pi}{4}$	1
$\frac{\pi}{2}$	undefined

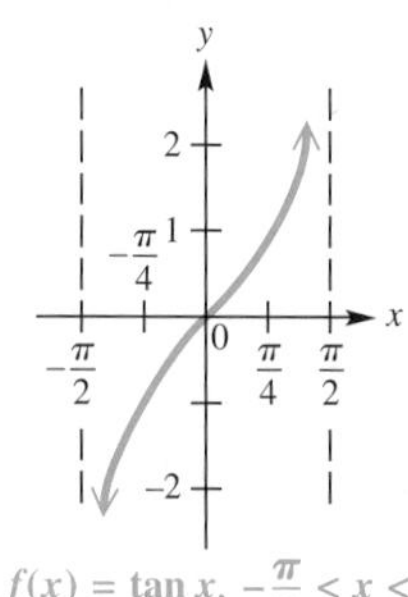

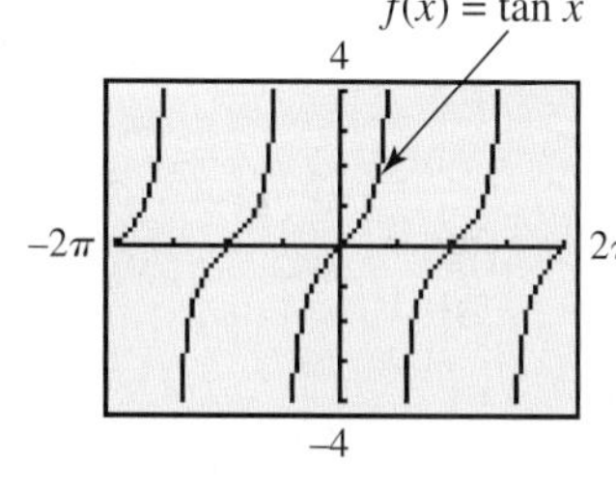

Figure 45

- The graph is discontinuous at values of x of the form $x = (2n+1)\frac{\pi}{2}$ and has vertical asymptotes at these values.
- Its x-intercepts are of the form $x = n\pi$.
- Its period is π.
- Its graph has no amplitude, since there are no minimum or maximum values.
- The graph is symmetric with respect to the origin, so the function is an odd function. For all x in the domain, $\tan(-x) = -\tan x$.

Graph of the Cotangent Function A similar analysis for selected points between 0 and π for the graph of the cotangent function yields the graph in **Figure 46** on the next page. Here the vertical asymptotes are at x-values that are integer multiples of π. Because

$$\cot(-x) = -\cot x, \quad \text{(See Exercise 46.)}$$

this graph is also symmetric with respect to the origin. (This can be seen when more of the graph is plotted.)

x	$y = \cot x$
$\frac{\pi}{6}$	$\sqrt{3} \approx 1.7$
$\frac{\pi}{4}$	1
$\frac{\pi}{3}$	$\frac{\sqrt{3}}{3} \approx 0.6$
$\frac{\pi}{2}$	0
$\frac{2\pi}{3}$	$-\frac{\sqrt{3}}{3} \approx -0.6$
$\frac{3\pi}{4}$	-1
$\frac{5\pi}{6}$	$-\sqrt{3} \approx -1.7$

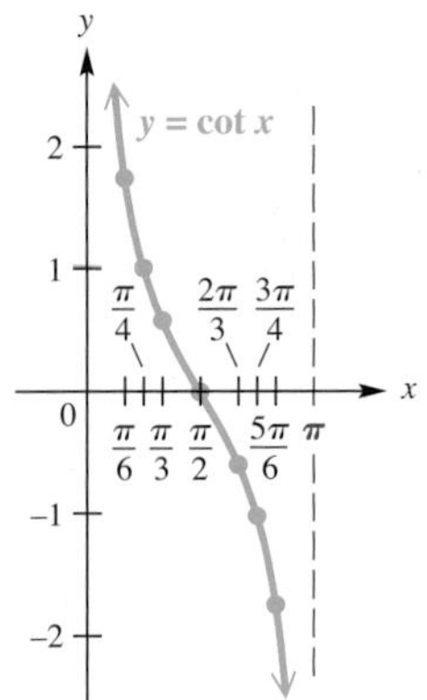

Figure 46

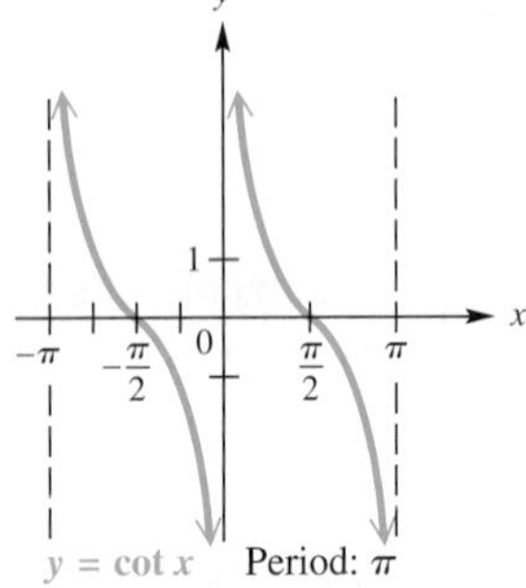

Figure 47

The cotangent function also has period π. Cotangent values are 0 when cosine values are 0, and are undefined when sine values are 0. As x-values increase from 0 to π, cotangent values range from ∞ to $-\infty$ and decrease throughout the interval. Those same values are repeated as x increases from π to 2π, from 2π to 3π, and so on. The graph of $y = \cot x$ from $-\pi$ to π is shown in **Figure 47.** The graph continues in this pattern.

Cotangent Function $f(x) = \cot x$

Domain: $\{x \mid x \neq n\pi$, where n is any integer$\}$ Range: $(-\infty, \infty)$

x	y
0	undefined
$\frac{\pi}{4}$	1
$\frac{\pi}{2}$	0
$\frac{3\pi}{4}$	-1
π	undefined

$f(x) = \cot x,\ 0 < x < \pi$

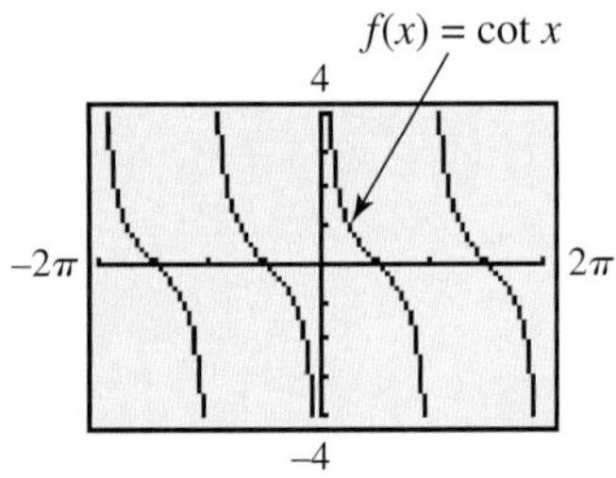

Figure 48

- The graph is discontinuous at values of x of the form $x = n\pi$ and has vertical asymptotes at these values.
- Its x-intercepts are of the form $x = (2n + 1)\frac{\pi}{2}$.
- Its period is π.
- Its graph has no amplitude, since there are no minimum or maximum values.
- The graph is symmetric with respect to the origin, so the function is an odd function. For all x in the domain, $\cot(-x) = -\cot x$.

The tangent function can be graphed directly with a graphing calculator, using the tangent key. To graph the cotangent function, however, we must use one of the identities

$$\cot x = \frac{1}{\tan x} \quad \text{or} \quad \cot x = \frac{\cos x}{\sin x}, \qquad \text{(Section 5.2)}$$

because graphing calculators generally do not have cotangent keys. ■

Graphing Techniques

Guidelines for Sketching Graphs of Tangent and Cotangent Functions

To graph $\boldsymbol{y = a \tan bx}$ or $\boldsymbol{y = a \cot bx}$, with $b > 0$, follow these steps.

Step 1 Determine the period, $\frac{\pi}{b}$. To locate two adjacent vertical asymptotes, solve the following equations for x:

For $y = a \tan bx$: $\quad bx = -\dfrac{\pi}{2}$ and $bx = \dfrac{\pi}{2}$.

For $y = a \cot bx$: $\quad bx = 0$ and $bx = \pi$.

Step 2 Sketch the two vertical asymptotes found in Step 1.

Step 3 Divide the interval formed by the vertical asymptotes into four equal parts.

Step 4 Evaluate the function for the first-quarter point, midpoint, and third-quarter point, using the x-values found in Step 3.

Step 5 Join the points with a smooth curve, approaching the vertical asymptotes. Indicate additional asymptotes and periods of the graph as necessary.

Teaching Tip Emphasize to students that the period of the tangent and cotangent functions is $\frac{\pi}{b}$, not $\frac{2\pi}{b}$.

Classroom Example 1
Graph $y = \tan \frac{2}{3}x$.

Answer:

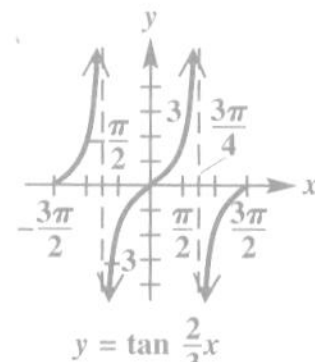

EXAMPLE 1 **Graphing $y = \tan bx$**

Graph $y = \tan 2x$.

SOLUTION

Step 1 The period of this function is $\frac{\pi}{2}$. To locate two adjacent vertical asymptotes, solve $2x = -\frac{\pi}{2}$ and $2x = \frac{\pi}{2}$ (because this is a tangent function). The two asymptotes have equations $x = -\frac{\pi}{4}$ and $x = \frac{\pi}{4}$.

Step 2 Sketch the two vertical asymptotes $x = \pm\frac{\pi}{4}$, as shown in **Figure 49.**

Step 3 Divide the interval $\left(-\frac{\pi}{4}, \frac{\pi}{4}\right)$ into four equal parts. This gives the following key x-values.

first-quarter value: $-\dfrac{\pi}{8}$, middle value: 0, third-quarter value: $\dfrac{\pi}{8}$ Key x-values

Step 4 Evaluate the function for the x-values found in Step 3.

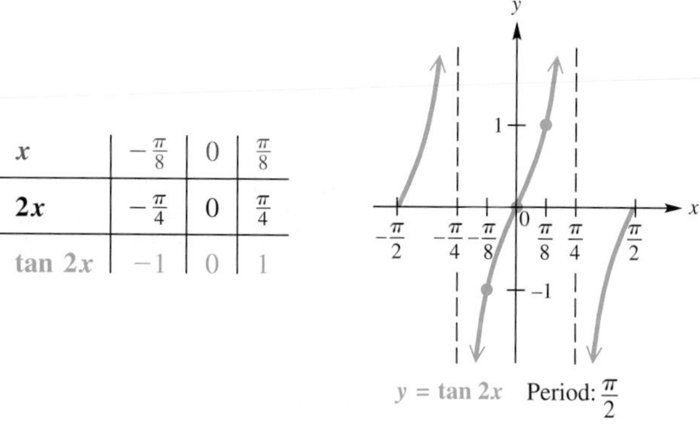

x	$-\frac{\pi}{8}$	0	$\frac{\pi}{8}$
$2x$	$-\frac{\pi}{4}$	0	$\frac{\pi}{4}$
$\tan 2x$	-1	0	1

Figure 49

Step 5 Join these points with a smooth curve, approaching the vertical asymptotes. See **Figure 49.** Another period has been graphed, one half period to the left and one half period to the right.

✔ *Now Try Exercise 7.*

EXAMPLE 2 Graphing $y = a \tan bx$

Graph $y = -3 \tan \frac{1}{2}x$.

SOLUTION The period is $\frac{\pi}{\frac{1}{2}} = \pi \div \frac{1}{2} = \pi \cdot \frac{2}{1} = 2\pi$. Adjacent asymptotes are at $x = -\pi$ and $x = \pi$. Dividing the interval $(-\pi, \pi)$ into four equal parts gives key x-values of $-\frac{\pi}{2}$, 0, and $\frac{\pi}{2}$. Evaluating the function at these x-values gives the following key points.

$$\left(-\frac{\pi}{2}, 3\right), \quad (0, 0), \quad \left(\frac{\pi}{2}, -3\right) \qquad \text{Key points}$$

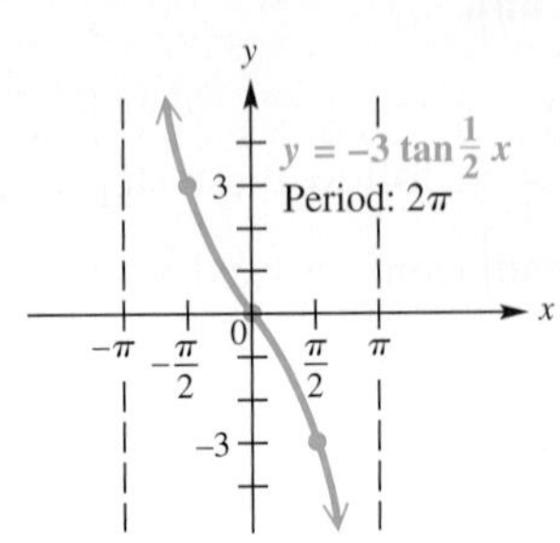

Figure 50

By plotting these points and joining them with a smooth curve, we obtain the graph shown in **Figure 50.** Because the coefficient -3 is negative, the graph is reflected across the x-axis compared to the graph of $y = 3 \tan \frac{1}{2}x$.

Now Try Exercise 15.

Classroom Example 2
Graph $y = -\frac{1}{2} \tan 2x$.

Answer:

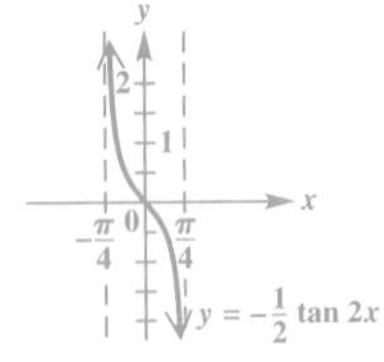

NOTE The function $y = -3 \tan \frac{1}{2}x$ in **Example 2,** graphed in **Figure 50,** has a graph that compares to the graph of $y = \tan x$ as follows.

1. The period is larger because $b = \frac{1}{2}$, and $\frac{1}{2} < 1$.
2. The graph is "stretched" vertically because $a = -3$, and $|-3| > 1$.
3. Each branch of the graph falls from left to right (that is, the function decreases) between each pair of adjacent asymptotes because $a = -3$, and $-3 < 0$. When $a < 0$, the graph is reflected across the x-axis compared to the graph of $y = |a| \tan bx$.

Classroom Example 3
Graph $y = 3 \cot \frac{1}{2}x$.

Answer:

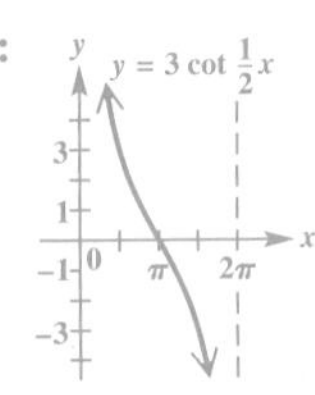

EXAMPLE 3 Graphing $y = a \cot bx$

Graph $y = \frac{1}{2} \cot 2x$.

SOLUTION Because this function involves the cotangent, we can locate two adjacent asymptotes by solving the equations $2x = 0$ and $2x = \pi$. The lines $x = 0$ (the y-axis) and $x = \frac{\pi}{2}$ are two such asymptotes. We divide the interval $\left(0, \frac{\pi}{2}\right)$ into four equal parts, getting key x-values of $\frac{\pi}{8}$, $\frac{\pi}{4}$, and $\frac{3\pi}{8}$. Evaluating the function at these x-values gives the key points $\left(\frac{\pi}{8}, \frac{1}{2}\right), \left(\frac{\pi}{4}, 0\right), \left(\frac{3\pi}{8}, -\frac{1}{2}\right)$. We plot these points and join them with a smooth curve approaching the asymptotes to obtain the graph shown in **Figure 51.**

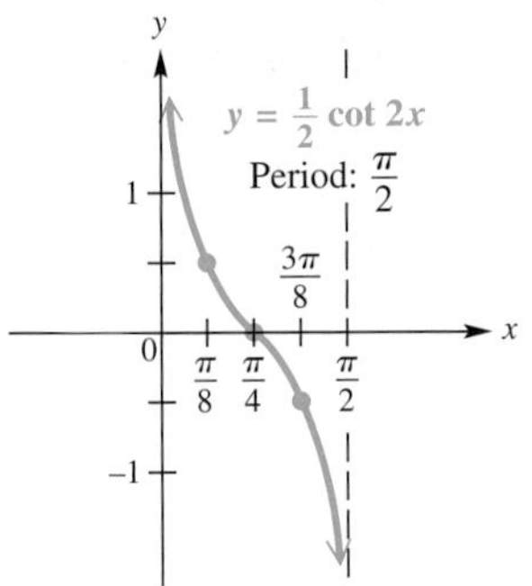

Figure 51

Now Try Exercise 17.

Like the other circular functions, the graphs of the tangent and cotangent functions may be translated horizontally and vertically.

EXAMPLE 4 Graphing $y = c + \tan x$

Graph $y = 2 + \tan x$.

ANALYTIC SOLUTION

Every value of y for this function will be 2 units more than the corresponding value of y in $y = \tan x$, causing the graph of $y = 2 + \tan x$ to be translated 2 units up compared to the graph of $y = \tan x$. See **Figure 52.**

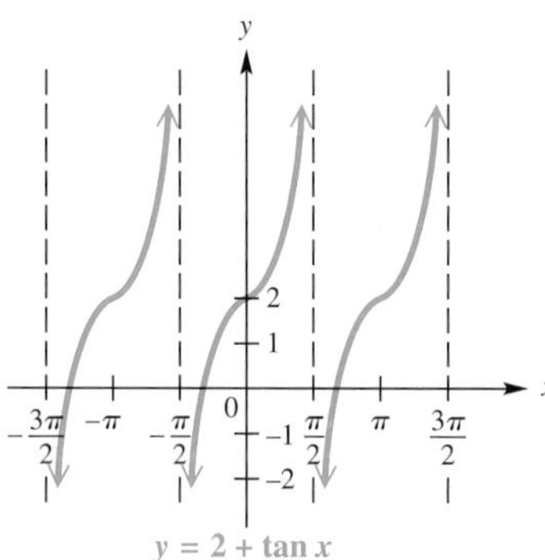

Figure 52

Three periods of the function are shown in **Figure 52.** Because the period of $y = 2 + \tan x$ is π, additional asymptotes and periods of the function can be drawn by repeating the basic graph every π units on the x-axis to the left or to the right of the graph shown.

GRAPHING CALCULATOR SOLUTION

To see the vertical translation, observe the coordinates displayed at the bottoms of the screens in **Figures 53 and 54.** For $X = \frac{\pi}{4} \approx 0.78539816$,

$$Y_1 = \tan X = 1,$$

while for the same X-value,

$$Y_2 = 2 + \tan X = 2 + 1 = 3.$$

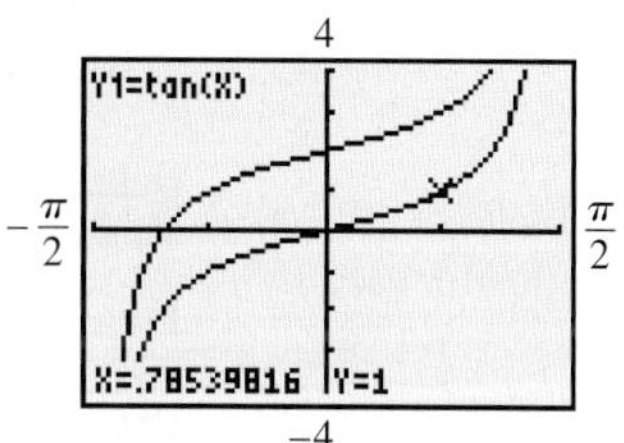

Figure 53

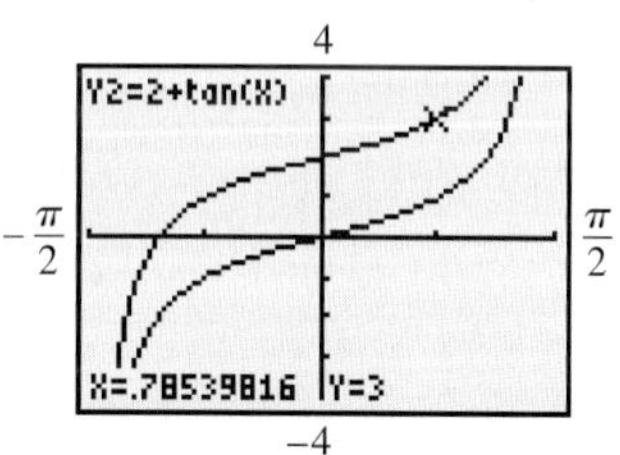

Figure 54

Now Try Exercise 23.

Classroom Example 4
Graph $y = -3 + \tan x$.

Answer:

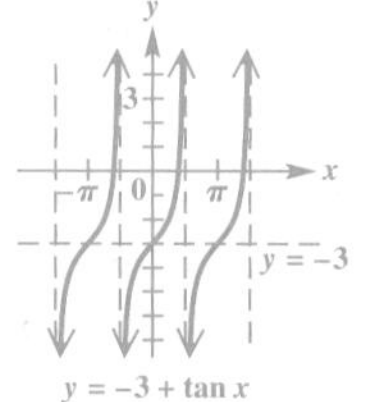

Classroom Example 5
Graph $y = 3 + \cot\left(x + \frac{\pi}{2}\right)$.

Answer:

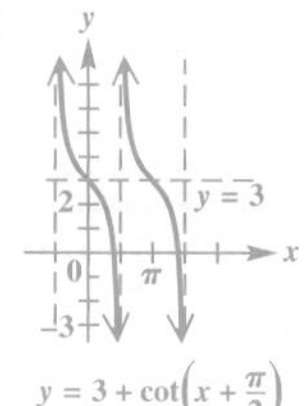

EXAMPLE 5 Graphing $y = c + a\cot(x - d)$

Graph $y = -2 - \cot\left(x - \frac{\pi}{4}\right)$.

SOLUTION Here $b = 1$, so the period is π. The negative sign in front of the cotangent will cause the graph to be reflected across the x-axis, and the argument $\left(x - \frac{\pi}{4}\right)$ indicates a phase shift (horizontal shift) $\frac{\pi}{4}$ unit to the right. Because $c = -2$, the graph will then be translated down 2 units. To locate adjacent asymptotes, since this function involves the cotangent, we solve the following equations.

$$x - \frac{\pi}{4} = 0 \quad \text{and} \quad x - \frac{\pi}{4} = \pi$$

$$x = \frac{\pi}{4} \quad \text{and} \quad x = \frac{5\pi}{4} \qquad \text{Add } \tfrac{\pi}{4}. \text{ (Section 1.1)}$$

Dividing the interval $\left(\frac{\pi}{4}, \frac{5\pi}{4}\right)$ into four equal parts and evaluating the function at the three key x-values within the interval give these points.

$$\left(\frac{\pi}{2}, -3\right), \quad \left(\frac{3\pi}{4}, -2\right), \quad (\pi, -1) \qquad \text{Key points}$$

We join these points with a smooth curve. This period of the graph, along with the one in the domain interval $\left(-\frac{3\pi}{4}, \frac{\pi}{4}\right)$, is shown in **Figure 55** on the next page.

Figure 55

✔ Now Try Exercise 31.

Connecting Graphs with Equations

Classroom Example 6
Determine an equation for each graph.

(a)

(b)

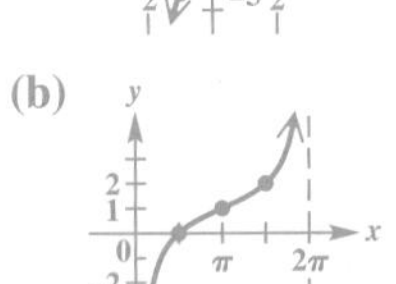

Answers:
(a) $y = 3 \tan x$
(b) $y = 1 - \cot \frac{1}{2}x$

EXAMPLE 6 Determining an Equation for a Graph

Determine an equation for each graph.

(a)

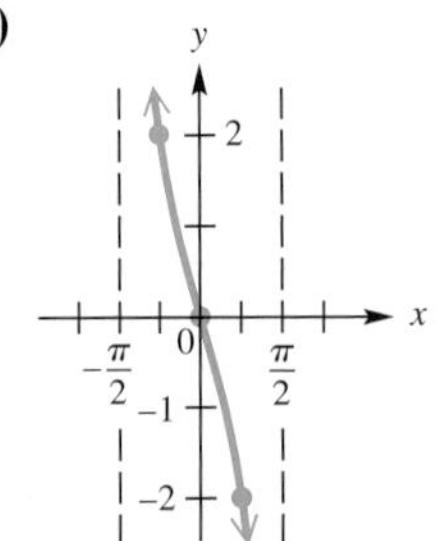

(b)

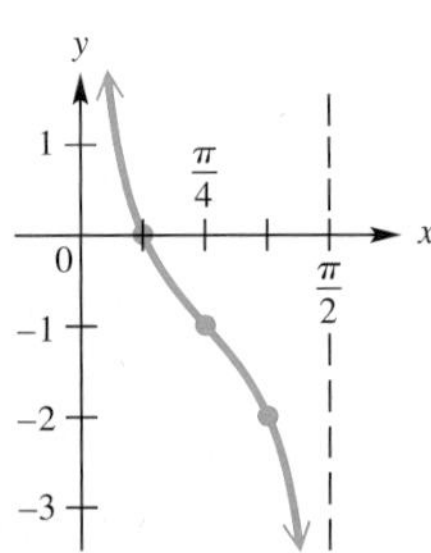

SOLUTION

(a) This graph is that of $y = \tan x$ but reflected across the x-axis and stretched vertically by a factor of 2. Therefore, an equation for this graph is

$$y = -2 \tan x.$$

Vertical stretch
x-axis reflection

(b) This is the graph of a cotangent function, but the period is $\frac{\pi}{2}$ rather than π. Therefore, the coefficient of x is 2. This graph is vertically translated 1 unit down compared to the graph of $y = \cot 2x$. An equation for this graph is

$$y = -1 + \cot 2x.$$

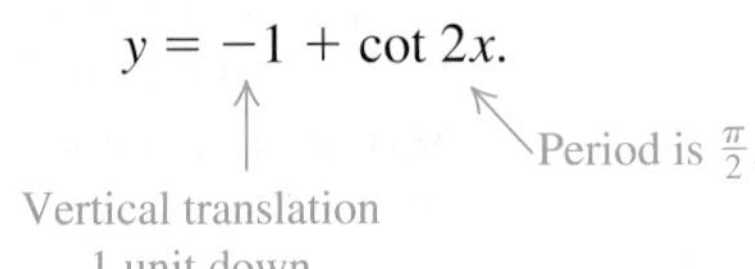

✔ Now Try Exercises 33 and 37.

NOTE Because the circular functions are periodic, there are infinitely many equations that correspond to each graph in **Example 6.** Confirm that both

$$y = -1 - \cot(-2x) \quad \text{and} \quad y = -1 - \tan\left(2x - \frac{\pi}{2}\right)$$

are equations for the graph in **Example 6(b).** When writing the equation from a graph, it is practical to write the simplest form. Therefore, we choose values of b where $b > 0$ and write the function without a phase shift when possible.

6.5 Exercises

1. C 2. A
3. B 4. D
5. F 6. E

7. 8.

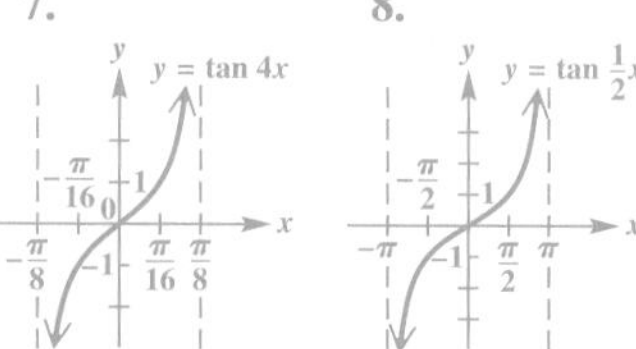

9. 10.

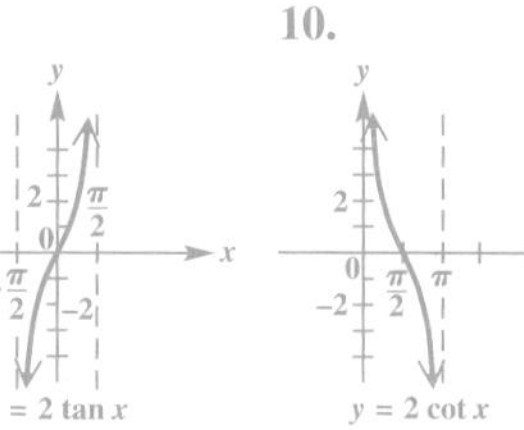

11. 12.

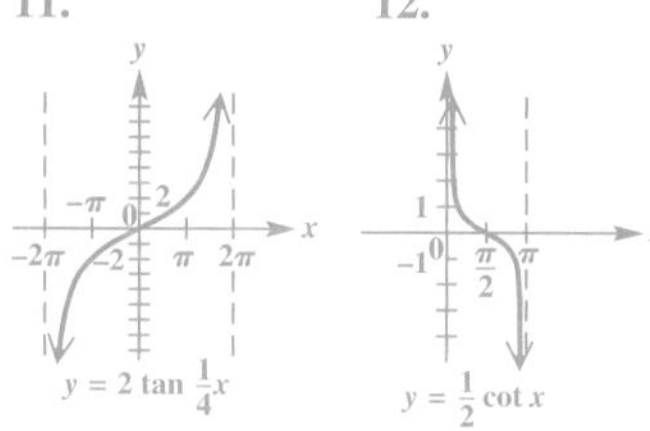

13. 14.

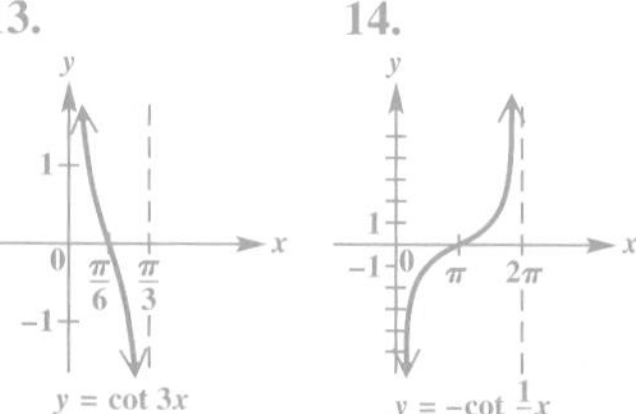

15. 16.

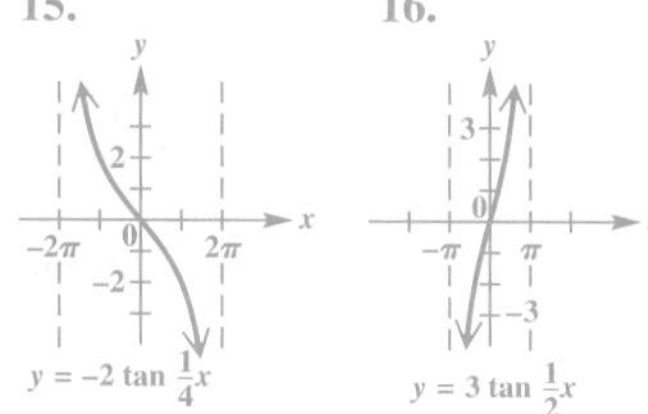

17. 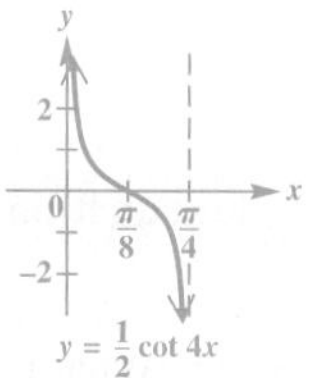 18.

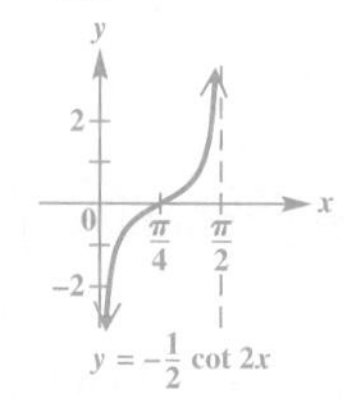

Concept Check In Exercises 1–6, match each function with its graph from choices A–F.

1. $y = -\tan x$
2. $y = -\cot x$
3. $y = \tan\left(x - \dfrac{\pi}{4}\right)$
4. $y = \cot\left(x - \dfrac{\pi}{4}\right)$
5. $y = \cot\left(x + \dfrac{\pi}{4}\right)$
6. $y = \tan\left(x + \dfrac{\pi}{4}\right)$

A.

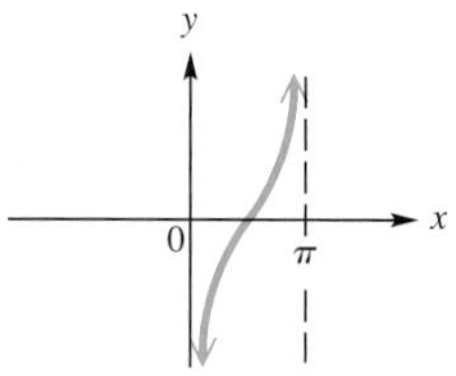

B.

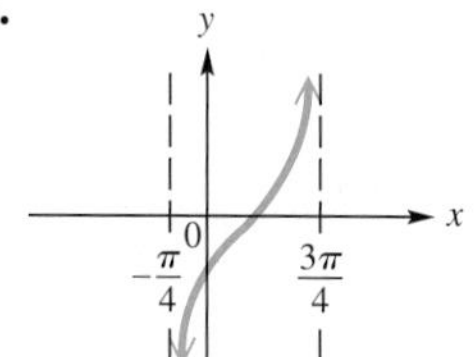

C.

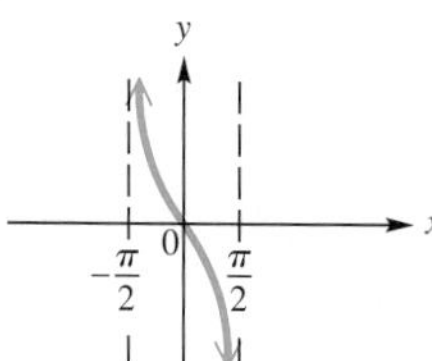

D.

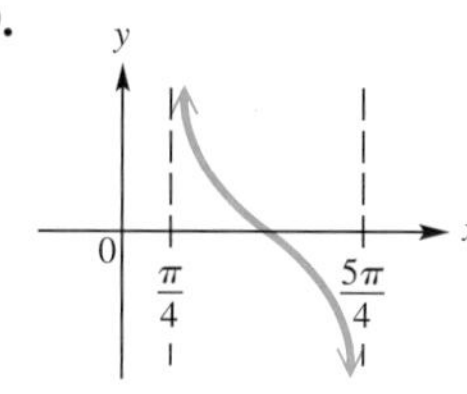

E.

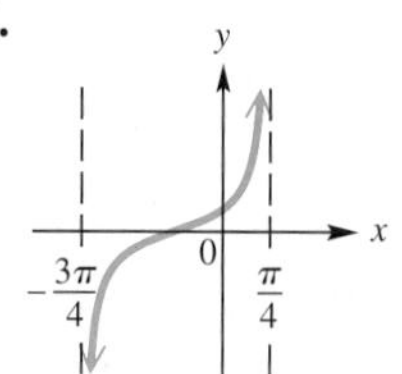

F.

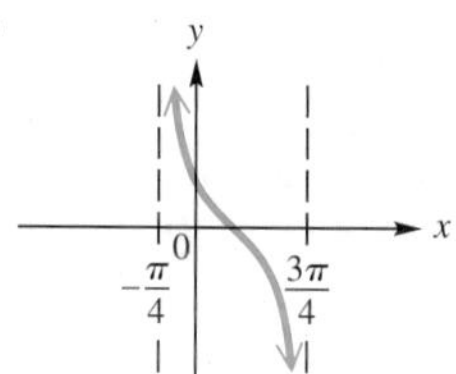

Graph each function over a one-period interval. **See Examples 1–3.**

7. $y = \tan 4x$
8. $y = \tan \dfrac{1}{2}x$
9. $y = 2 \tan x$
10. $y = 2 \cot x$
11. $y = 2 \tan \dfrac{1}{4}x$
12. $y = \dfrac{1}{2} \cot x$
13. $y = \cot 3x$
14. $y = -\cot \dfrac{1}{2}x$
15. $y = -2 \tan \dfrac{1}{4}x$
16. $y = 3 \tan \dfrac{1}{2}x$
17. $y = \dfrac{1}{2} \cot 4x$
18. $y = -\dfrac{1}{2} \cot 2x$

Graph each function over a two-period interval. **See Examples 4 and 5.**

19. $y = \tan(2x - \pi)$
20. $y = \tan\left(\dfrac{x}{2} + \pi\right)$
21. $y = \cot\left(3x + \dfrac{\pi}{4}\right)$
22. $y = \cot\left(2x - \dfrac{3\pi}{2}\right)$
23. $y = 1 + \tan x$
24. $y = 1 - \tan x$
25. $y = 1 - \cot x$
26. $y = -2 - \cot x$
27. $y = -1 + 2 \tan x$
28. $y = 3 + \dfrac{1}{2} \tan x$
29. $y = -1 + \dfrac{1}{2} \cot(2x - 3\pi)$
30. $y = -2 + 3 \tan(4x + \pi)$
31. $y = 1 - 2 \cot\left[2\left(x + \dfrac{\pi}{2}\right)\right]$
32. $y = -2 + \dfrac{2}{3} \tan\left(\dfrac{3}{4}x - \pi\right)$

19. $y = \tan(2x - \pi)$

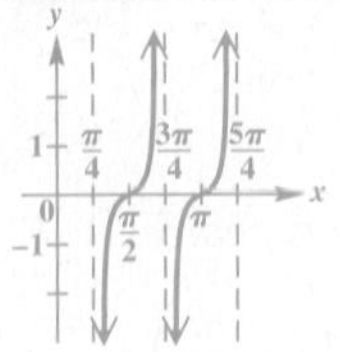

20. $y = \tan\left(\frac{x}{2} + \pi\right)$

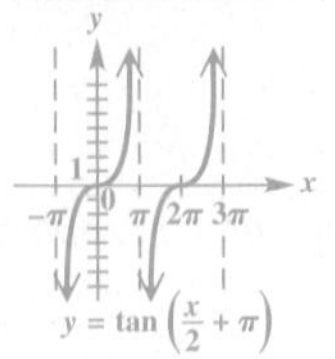

21. $y = \cot\left(3x + \frac{\pi}{4}\right)$

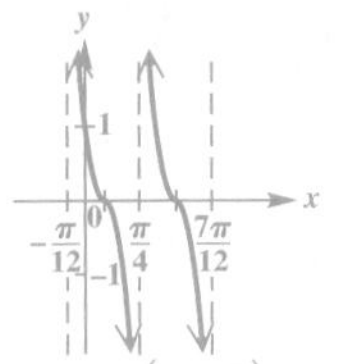

22. $y = \cot\left(2x - \frac{3\pi}{2}\right)$

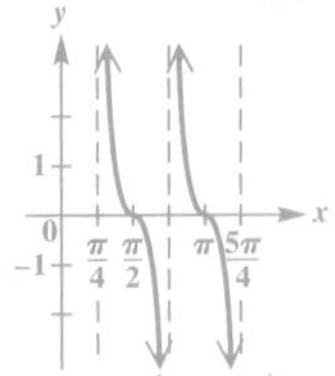

23. $y = 1 + \tan x$

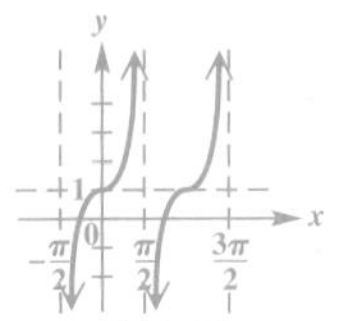

24. $y = 1 - \tan x$

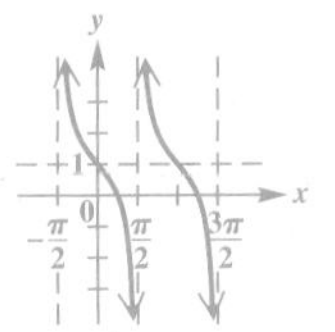

25. $y = 1 - \cot x$

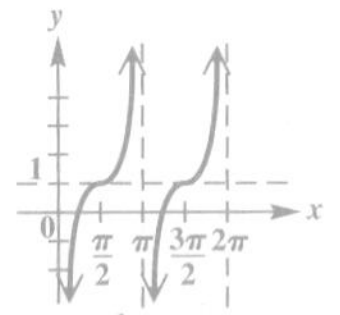

26. $y = -2 - \cot x$

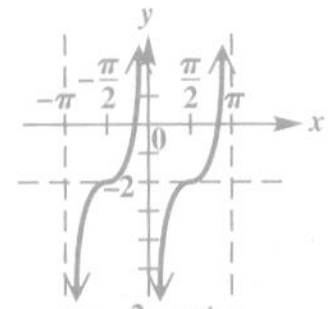

27. $y = -1 + 2\tan x$

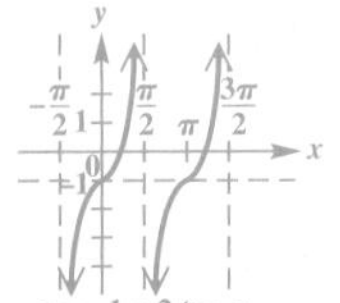

28. $y = 3 + \frac{1}{2}\tan x$

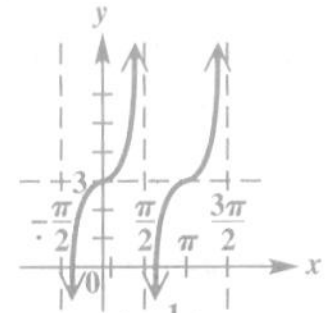

29. $y = -1 + \frac{1}{2}\cot(2x - 3\pi)$

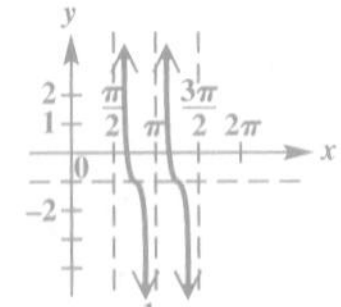

30. $y = -2 + 3\tan(4x + \pi)$

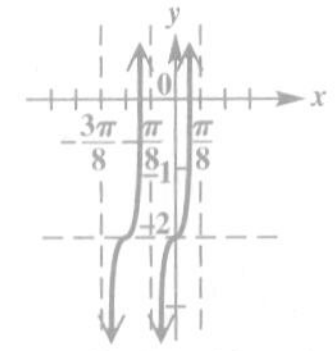

31. $y = 1 - 2\cot\left[2\left(x + \frac{\pi}{2}\right)\right]$

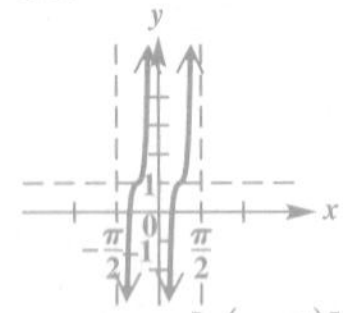

32. $y = -2 + \frac{2}{3}\tan\left(\frac{3}{4}x - \pi\right)$

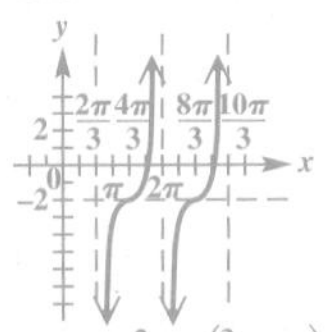

Connecting Graphs with Equations *Determine the simplest form of an equation for each graph. Choose $b > 0$, and include no phase shifts. (Midpoints and quarter-points are identified by dots.) See* ***Example 6.***

33.

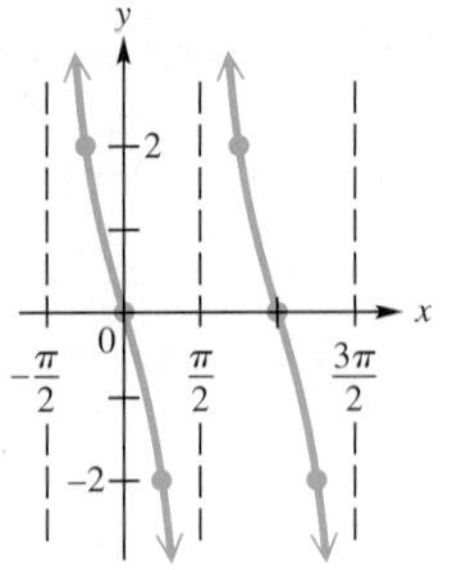

34.

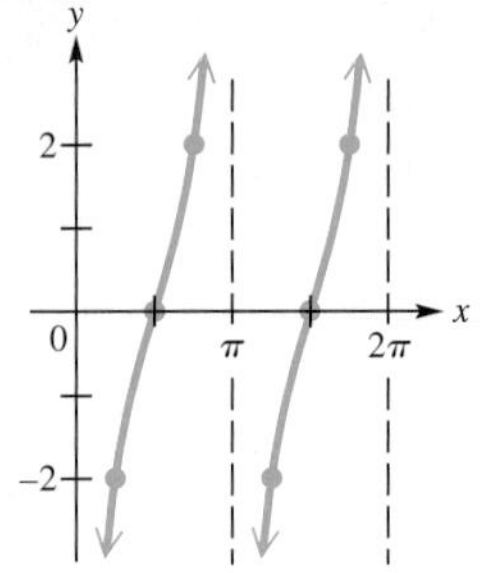

35.

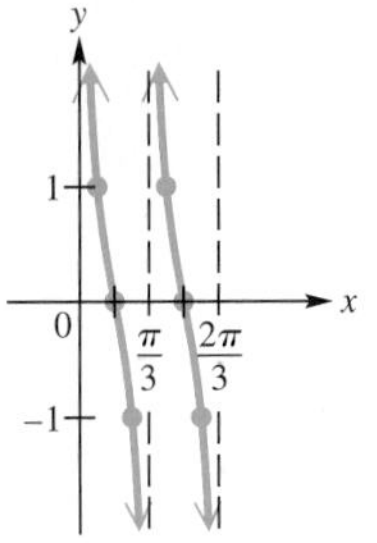

36.

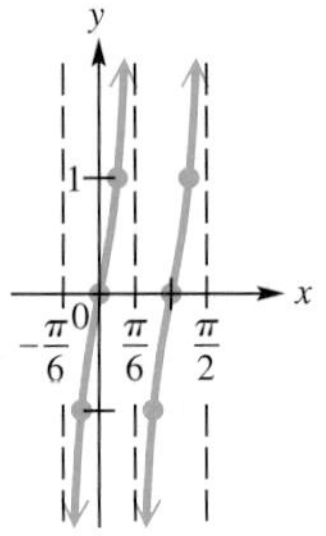

37.

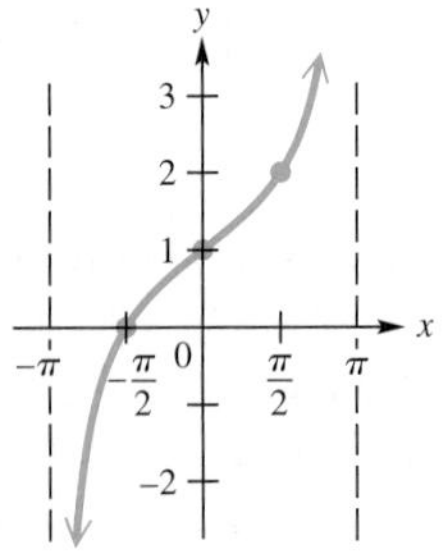

38.

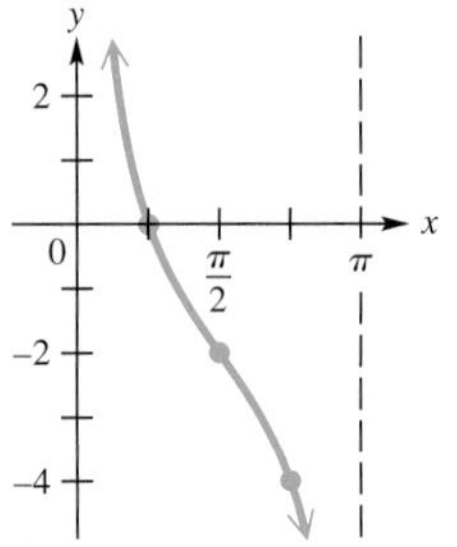

Concept Check *In Exercises 39–42, tell whether each statement is* true *or* false. *If* false, *tell why.*

39. The least positive number k for which $x = k$ is an asymptote for the tangent function is $\frac{\pi}{2}$.

40. The least positive number k for which $x = k$ is an asymptote for the cotangent function is $\frac{\pi}{2}$.

41. The graph of $y = \tan x$ in **Figure 44** suggests that $\tan(-x) = \tan x$ for all x in the domain of $\tan x$.

42. The graph of $y = \cot x$ in **Figure 47** suggests that $\cot(-x) = -\cot x$ for all x in the domain of $\cot x$.

Work each exercise.

43. *Concept Check* If c is any number, then how many solutions does the equation $c = \tan x$ have in the interval $(-2\pi, 2\pi]$?

44. *Concept Check* Consider the function defined by $f(x) = -4\tan(2x + \pi)$. What is the domain of f? What is its range?

45. Show that $\tan(-x) = -\tan x$ by writing $\tan(-x)$ as $\frac{\sin(-x)}{\cos(-x)}$ and then using the relationships for $\sin(-x)$ and $\cos(-x)$.

46. Show that $\cot(-x) = -\cot x$ by writing $\cot(-x)$ as $\frac{\cos(-x)}{\sin(-x)}$ and then using the relationships for $\cos(-x)$ and $\sin(-x)$.

33. $y = -2\tan x$
34. $y = -2\cot x$
35. $y = \cot 3x$
36. $y = \tan 3x$
37. $y = 1 + \tan\frac{1}{2}x$
38. $y = -2 + 2\cot x$

39. true
40. false; The least such k is π.
41. false; $\tan(-x) = -\tan x$ for all x in the domain.
42. true

43. four
44. domain: $\{x \mid x \neq (2n+1)\frac{\pi}{4}$, where n is any integer$\}$; range: $(-\infty, \infty)$

47. (a) 0 m (b) −2.9 m (c) −12.3 m (d) 12.3 m (e) It leads to $\tan \frac{\pi}{2}$, which is undefined.

49. π
50. $\frac{5\pi}{4}$
51. $x = \frac{5\pi}{4} + n\pi$
52. approximately 0.3217505544
53. approximately 3.463343208
54. $\{x \mid x = 0.3217505544 + n\pi\}$

47. *(Modeling) Distance of a Rotating Beacon* A rotating beacon is located at point A next to a long wall. The beacon is 4 m from the wall. The distance d is given by

$$d = 4 \tan 2\pi t,$$

where t is time measured in seconds since the beacon started rotating. (When $t = 0$, the beacon is aimed at point R. When the beacon is aimed to the right of R, the value of d is positive; d is negative when the beacon is aimed to the left of R.) Find d for each time.

(a) $t = 0$
(b) $t = 0.4$
(c) $t = 0.8$
(d) $t = 1.2$
(e) Why is 0.25 a meaningless value for t?

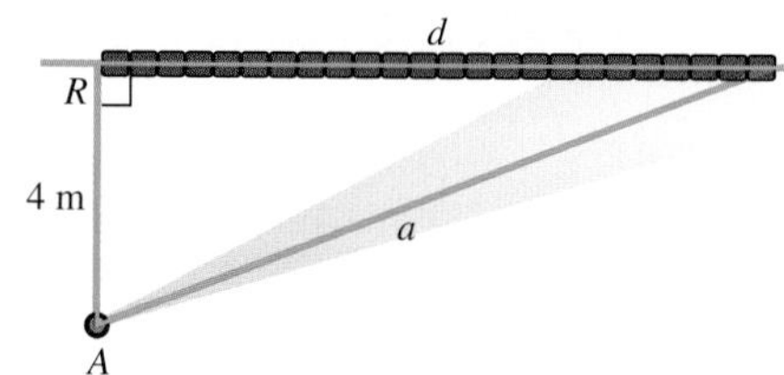

48. Simultaneously graph $y = \tan x$ and $y = x$ in the window $[-1, 1]$ by $[-1, 1]$ with a graphing calculator. Write a short description of the relationship between $\tan x$ and x for small x-values.

Relating Concepts

For individual or collaborative investigation *(Exercises 49–54)*

Consider the following function from ***Example 5. Work these exercises in order.***

$$y = -2 - \cot\left(x - \frac{\pi}{4}\right)$$

49. What is the least positive number for which $y = \cot x$ is undefined?

50. Let k represent the number you found in **Exercise 49.** Set $x - \frac{\pi}{4}$ equal to k, and solve to find a positive number for which $\cot\left(x - \frac{\pi}{4}\right)$ is undefined.

51. Based on your answer in **Exercise 50** and the fact that the cotangent function has period π, give the general form of the equations of the asymptotes of the graph of $y = -2 - \cot\left(x - \frac{\pi}{4}\right)$. Let n represent any integer.

52. Use the capabilities of your calculator to find the least positive x-intercept of the graph of this function.

53. Use the fact that the period of this function is π to find the next positive x-intercept.

54. Give the solution set of the equation $-2 - \cot\left(x - \frac{\pi}{4}\right) = 0$ over all real numbers. Let n represent any integer.

6.6 Graphs of the Secant and Cosecant Functions

- Graph of the Secant Function
- Graph of the Cosecant Function
- Graphing Techniques
- Connecting Graphs with Equations
- Addition of Ordinates

Graph of the Secant Function Consider the table of selected points accompanying the graph of the secant function in **Figure 56** on the next page. These points include special values from $-\pi$ to π. The secant function is undefined for odd multiples of $\frac{\pi}{2}$ and thus, like the tangent function, has vertical asymptotes for such values. Furthermore, since

$$\sec(-x) = \sec x, \quad \text{(See Exercise 31.)}$$

the graph of the secant function is symmetric with respect to the y-axis.

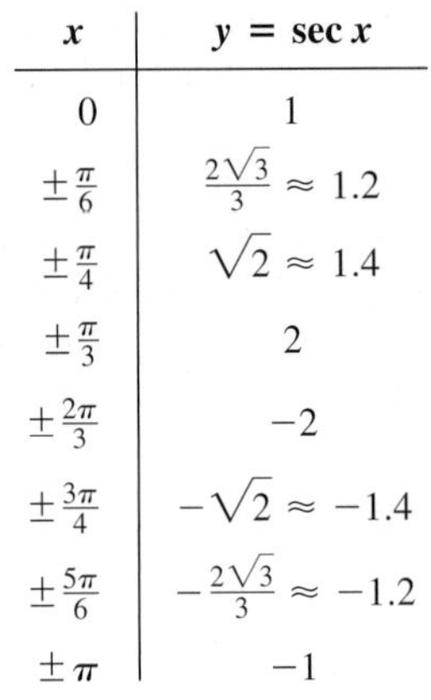

x	$y = \sec x$
0	1
$\pm\frac{\pi}{6}$	$\frac{2\sqrt{3}}{3} \approx 1.2$
$\pm\frac{\pi}{4}$	$\sqrt{2} \approx 1.4$
$\pm\frac{\pi}{3}$	2
$\pm\frac{2\pi}{3}$	-2
$\pm\frac{3\pi}{4}$	$-\sqrt{2} \approx -1.4$
$\pm\frac{5\pi}{6}$	$-\frac{2\sqrt{3}}{3} \approx -1.2$
$\pm\pi$	-1

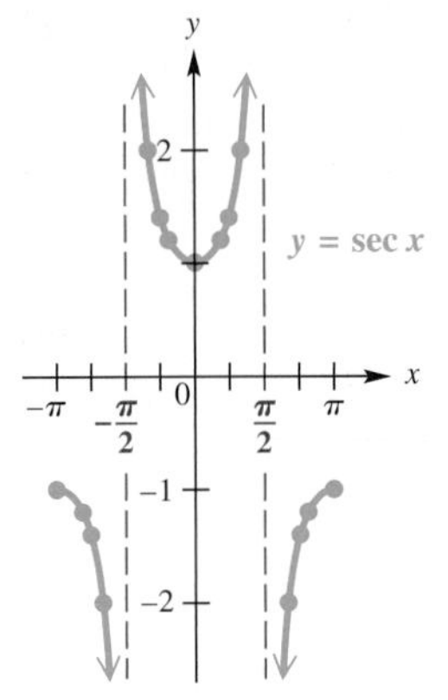

Figure 56

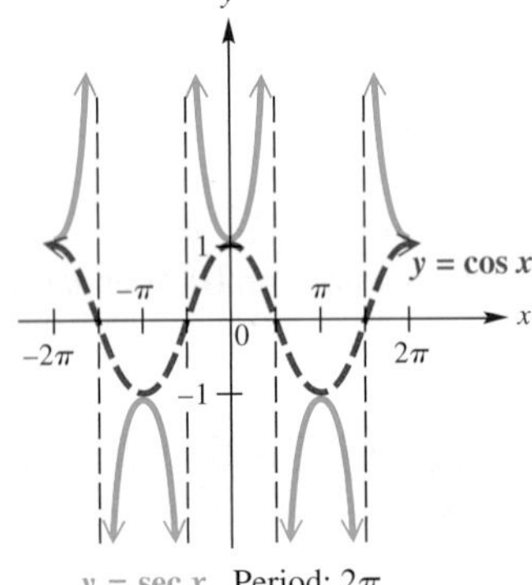

Figure 57

Because secant values are reciprocals of corresponding cosine values, the period of the secant function is 2π, the same as for $y = \cos x$. When $\cos x = 1$, the value of $\sec x$ is also 1. Likewise, when $\cos x = -1$, $\sec x = -1$. For all x, $-1 \leq \cos x \leq 1$, and thus,

$$|\sec x| \geq 1 \quad \text{for all } x \text{ in its domain.}$$

Figure 57 shows how the graphs of $y = \cos x$ and $y = \sec x$ are related.

Secant Function $f(x) = \sec x$

Domain: $\{x \mid x \neq (2n + 1)\frac{\pi}{2}$, where n is any integer$\}$ Range: $(-\infty, -1] \cup [1, \infty)$

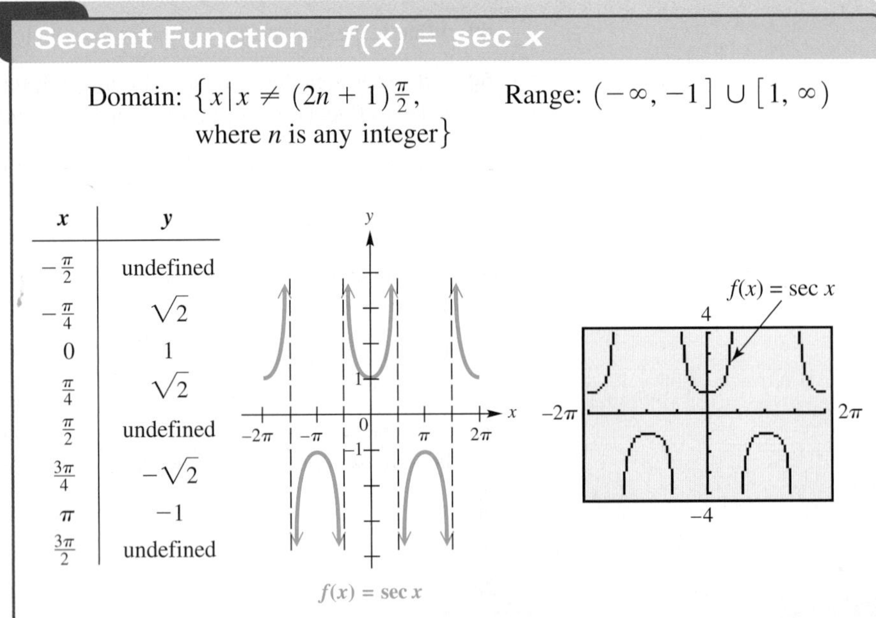

x	y
$-\frac{\pi}{2}$	undefined
$-\frac{\pi}{4}$	$\sqrt{2}$
0	1
$\frac{\pi}{4}$	$\sqrt{2}$
$\frac{\pi}{2}$	undefined
$\frac{3\pi}{4}$	$-\sqrt{2}$
π	-1
$\frac{3\pi}{2}$	undefined

Figure 58

- The graph is discontinuous at values of x of the form $x = (2n + 1)\frac{\pi}{2}$ and has vertical asymptotes at these values.
- There are no x-intercepts.
- Its period is 2π.
- Its graph has no amplitude, since there are no minimum or maximum values.
- The graph is symmetric with respect to the y-axis, so the function is an even function. For all x in the domain, $\sec(-x) = \sec x$.

Graph of the Cosecant Function A similar analysis for selected points between $-\pi$ and π for the graph of the cosecant function yields the graph in **Figure 59.** The vertical asymptotes are at x-values that are integer multiples of π. Because

$$\csc(-x) = -\csc x, \quad \text{(See Exercise 32.)}$$

this graph is symmetric with respect to the origin.

x	$y = \csc x$	x	$y = \csc x$
$\frac{\pi}{6}$	2	$-\frac{\pi}{6}$	-2
$\frac{\pi}{4}$	$\sqrt{2} \approx 1.4$	$-\frac{\pi}{4}$	$-\sqrt{2} \approx -1.4$
$\frac{\pi}{3}$	$\frac{2\sqrt{3}}{3} \approx 1.2$	$-\frac{\pi}{3}$	$-\frac{2\sqrt{3}}{3} \approx -1.2$
$\frac{\pi}{2}$	1	$-\frac{\pi}{2}$	-1
$\frac{2\pi}{3}$	$\frac{2\sqrt{3}}{3} \approx 1.2$	$-\frac{2\pi}{3}$	$-\frac{2\sqrt{3}}{3} \approx -1.2$
$\frac{3\pi}{4}$	$\sqrt{2} \approx 1.4$	$-\frac{3\pi}{4}$	$-\sqrt{2} \approx -1.4$
$\frac{5\pi}{6}$	2	$-\frac{5\pi}{6}$	-2

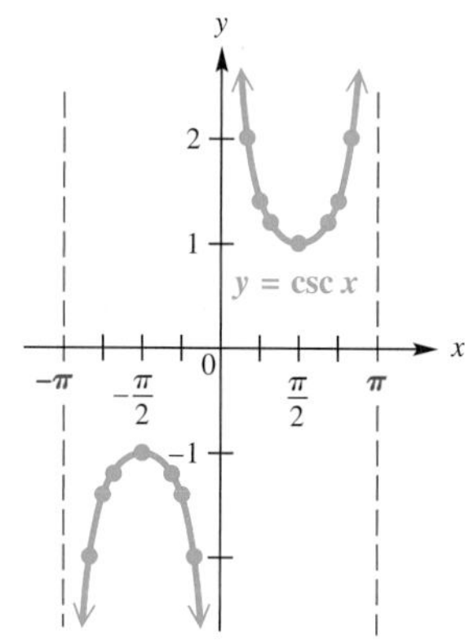

Figure 59

y = csc x Period: 2π

Figure 60

Because cosecant values are reciprocals of corresponding sine values, the period of the cosecant function is 2π, the same as for $y = \sin x$. When $\sin x = 1$, the value of $\csc x$ is also 1. Likewise, when $\sin x = -1$, $\csc x = -1$. For all x, $-1 \le \sin x \le 1$, and thus $|\csc x| \ge 1$ for all x in its domain. **Figure 60** shows how the graphs of $y = \sin x$ and $y = \csc x$ are related.

Cosecant Function $f(x) = \csc x$

Domain: $\{x \mid x \neq n\pi,$ where n is any integer$\}$ Range: $(-\infty, -1] \cup [1, \infty)$

x	y
0	undefined
$\frac{\pi}{6}$	2
$\frac{\pi}{3}$	$\frac{2\sqrt{3}}{3}$
$\frac{\pi}{2}$	1
$\frac{2\pi}{3}$	$\frac{2\sqrt{3}}{3}$
π	undefined
$\frac{3\pi}{2}$	-1
2π	undefined

$f(x) = \csc x$

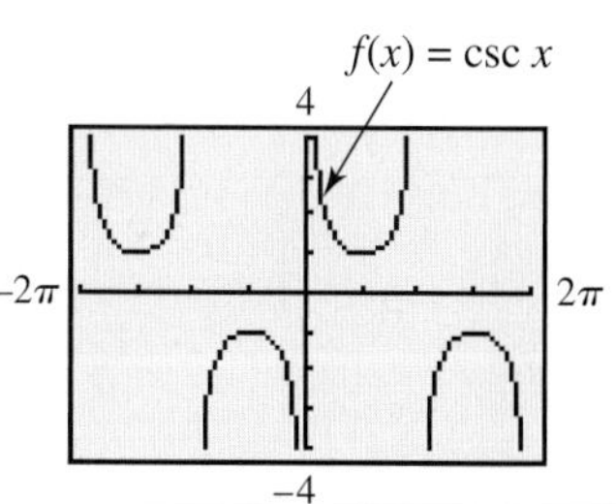

Figure 61

- The graph is discontinuous at values of x of the form $x = n\pi$ and has vertical asymptotes at these values.
- There are no x-intercepts.
- Its period is 2π.
- Its graph has no amplitude, since there are no minimum or maximum values.
- The graph is symmetric with respect to the origin, so the function is an odd function. For all x in the domain, $\csc(-x) = -\csc x$.

Typically, calculators do not have keys for the cosecant and secant functions. To graph $y = \csc x$ with a graphing calculator, use

$$\csc x = \frac{1}{\sin x}.$$ Reciprocal identity (Section 5.2)

Figure 62 shows the graph of $Y_1 = \sin X$ as a thin graph and that of $Y_2 = \csc X$ as a thick graph. Although this calculator screen does not show the vertical asymptotes, they occur at each x-intercept of the guide function $Y_1 = \sin X$. **Figure 63** shows the graph of the secant function, graphed in a similar manner, using the identity

$$\sec x = \frac{1}{\cos x}.$$ Reciprocal identity (Section 5.2)

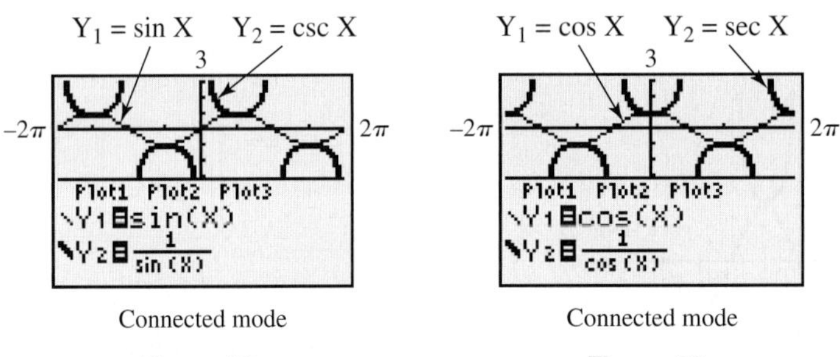

Connected mode

Figure 62

Connected mode

Figure 63

Graphing Techniques In the previous section, we gave guidelines for sketching graphs of tangent and cotangent functions. We now present similar guidelines for graphing cosecant and secant functions.

Guidelines for Sketching Graphs of Cosecant and Secant Functions

To graph $y = a \csc bx$ or $y = a \sec bx$, with $b > 0$, follow these steps.

Step 1 Graph the corresponding reciprocal function as a guide, using a dashed curve.

To Graph	Use as a Guide
$y = a \csc bx$	$y = a \sin bx$
$y = a \sec bx$	$y = a \cos bx$

Step 2 Sketch the vertical asymptotes. They will have equations of the form $x = k$, where k is an x-intercept of the graph of the guide function.

Step 3 Sketch the graph of the desired function by drawing the typical U-shaped branches between the adjacent asymptotes. The branches will be above the graph of the guide function when the guide function values are positive and below the graph of the guide function when the guide function values are negative. The graph will resemble those in **Figures 58 and 61** in the function boxes given earlier in this section.

Like graphs of the sine and cosine functions, graphs of the secant and cosecant functions may be translated vertically and horizontally. The period of both basic functions is 2π.

Classroom Example 1
Graph $y = 3 \sec 2x$.

Answer:

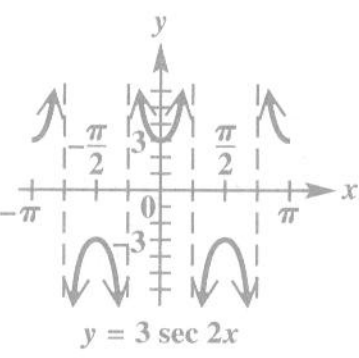

EXAMPLE 1 Graphing $y = a \sec bx$

Graph $y = 2 \sec \frac{1}{2}x$.

SOLUTION

Step 1 This function involves the secant, so the corresponding reciprocal function will involve the cosine. The guide function to graph is

$$y = 2 \cos \frac{1}{2}x.$$

Using the guidelines of **Section 6.3,** we find that this guide function has amplitude 2 and that one period of the graph lies along the interval that satisfies the following inequality.

$$0 \le \frac{1}{2}x \le 2\pi$$

$$0 \le x \le 4\pi, \quad \text{or} \quad [0, 4\pi] \quad \text{(Section 1.7)}$$

Dividing this interval into four equal parts gives these key points.

$$(0, 2), \quad (\pi, 0), \quad (2\pi, -2), \quad (3\pi, 0), \quad (4\pi, 2) \quad \text{Key points}$$

These points are plotted and joined with a dashed red curve to indicate that this graph is only a guide. An additional period is graphed as shown in **Figure 64(a).**

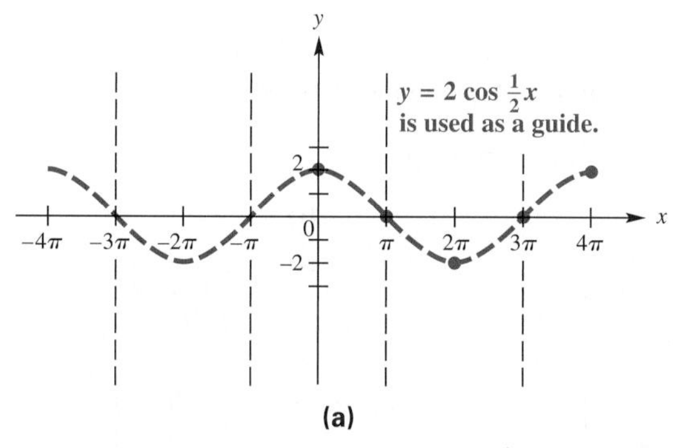

(a)

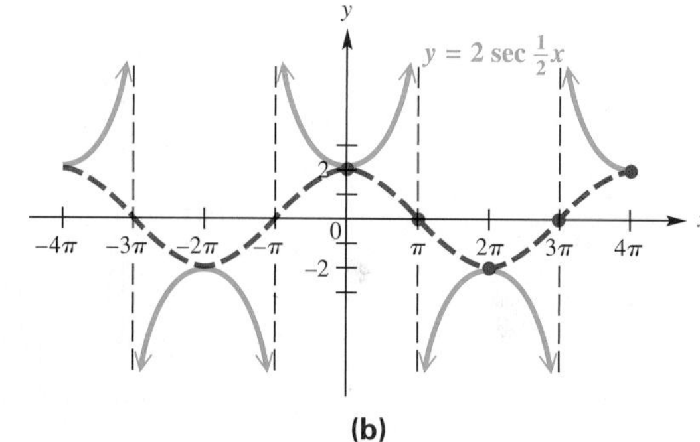

(b)

Figure 64

Y1=2/cos((1/2)X)

This is a calculator graph of the function in **Example 1.**

Step 2 Sketch the vertical asymptotes as shown in **Figure 64(a).** These occur at x-values for which the guide function equals 0, such as

$$x = -3\pi, \quad x = -\pi, \quad x = \pi, \quad x = 3\pi.$$

Step 3 Sketch the graph of $y = 2 \sec \frac{1}{2}x$ by drawing the typical U-shaped branches, approaching the asymptotes. See the solid blue graph in **Figure 64(b).**

Now Try Exercise 5.

Classroom Example 2
Graph $y = \frac{1}{2} \csc\left(x + \frac{\pi}{4}\right)$.

Answer:

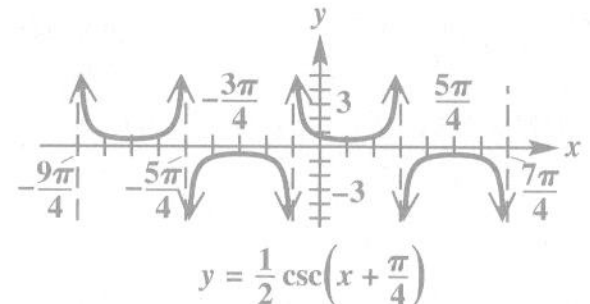

EXAMPLE 2 Graphing $y = a \csc(x - d)$

Graph $y = \frac{3}{2} \csc\left(x - \frac{\pi}{2}\right)$.

SOLUTION

Step 1 Use the guidelines of **Section 6.4** to graph the corresponding reciprocal function defined by

$$y = \frac{3}{2} \sin\left(x - \frac{\pi}{2}\right),$$

shown as a red dashed curve in **Figure 65** on the next page.

Step 2 Sketch the vertical asymptotes through the x-intercepts of the graph of $y = \frac{3}{2}\sin\left(x - \frac{\pi}{2}\right)$. These have the form $x = (2n + 1)\frac{\pi}{2}$, where n is any integer. See the black dashed lines in **Figure 65.**

Step 3 Sketch the graph of $y = \frac{3}{2}\csc\left(x - \frac{\pi}{2}\right)$ by drawing the typical U-shaped branches between adjacent asymptotes. See the solid blue graph in **Figure 65.**

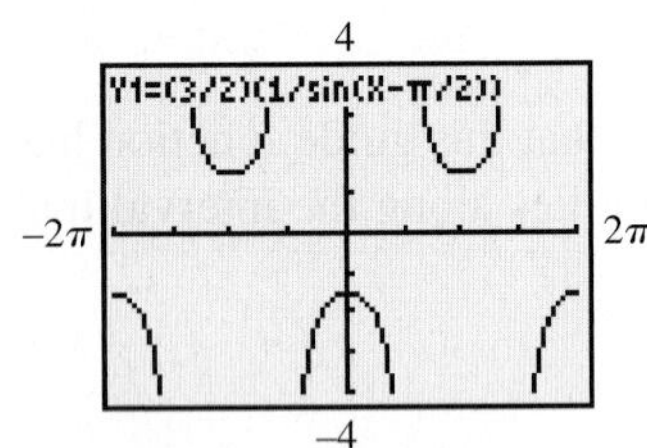

This is a calculator graph of the function in **Example 2.**

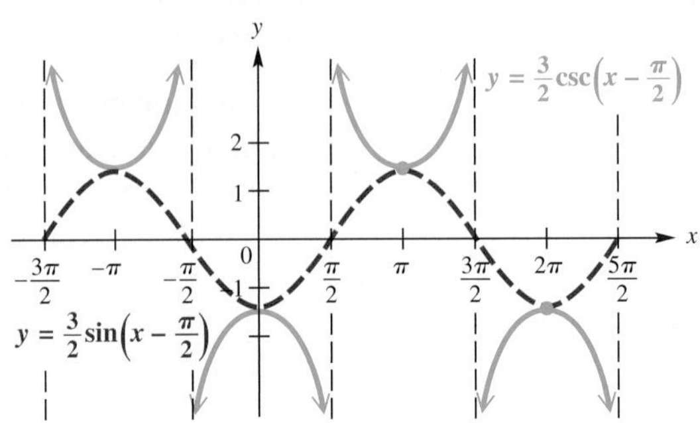

Figure 65

✔ *Now Try Exercise 7.*

Connecting Graphs with Equations

Classroom Example 3
Determine an equation for each graph.

(a)

(b)

Answers:
(a) $y = \sec \frac{1}{2}x$
(b) $y = -1 + \csc x$

EXAMPLE 3 Determining an Equation for a Graph

Determine an equation for each graph.

(a)

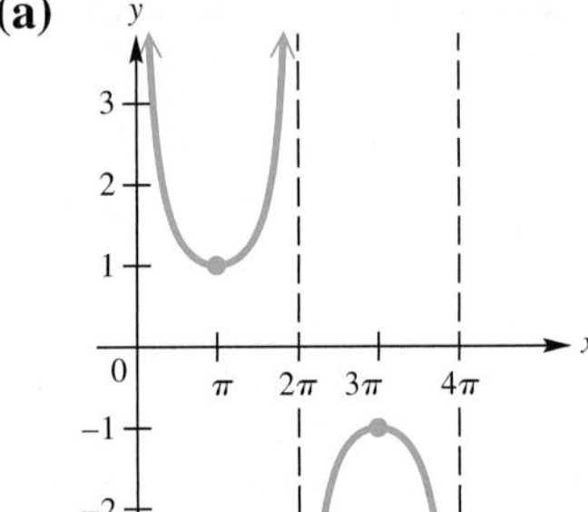

(b)

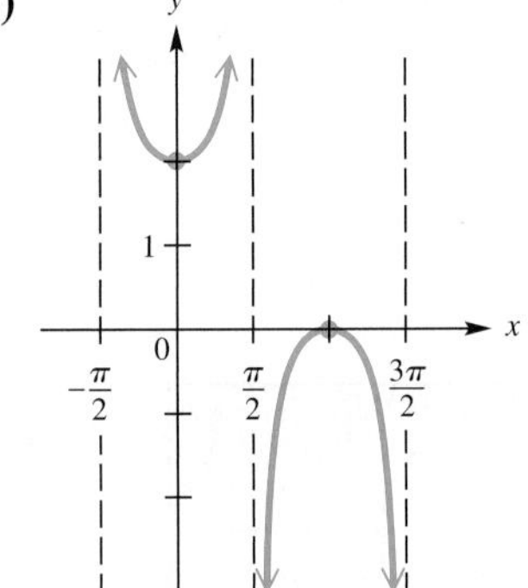

SOLUTION

(a) This graph is that of a cosecant function that is stretched horizontally having period 4π. Therefore, if $y = \csc bx$, where $b > 0$, we must have $b = \frac{1}{2}$. An equation for this graph is

$$y = \csc \frac{1}{2}x.$$

Horizontal stretch

(b) This is the graph of $y = \sec x$, translated 1 unit upward. An equation is

$$y = 1 + \sec x.$$

Vertical translation

✔ *Now Try Exercises 19 and 21.*

Addition of Ordinates New functions can be formed by adding or subtracting other functions. A function formed by combining two other functions, such as

$$y = \cos x + \sin x,$$

has historically been graphed using a method known as **addition of ordinates.** (The x-value of a point is sometimes called its **abscissa,** while its y-value is called its **ordinate.**)

To apply this method to this function, we graph the functions $y = \cos x$ and $y = \sin x$. Then, for selected values of x, we add $\cos x$ and $\sin x$, and plot the points $(x, \cos x + \sin x)$. Joining the resulting points with a sinusoidal curve gives the graph of the desired function. Although this method illustrates some valuable concepts involving the arithmetic of functions, it is time-consuming.

This technique is easily illustrated with graphing calculators. Consider $Y_1 = \cos X$, $Y_2 = \sin X$, and $Y_3 = Y_1 + Y_2$. **Figure 66** shows the result when Y_1 and Y_2 are graphed in thin graph style, and $Y_3 = \cos X + \sin X$ is graphed in thick graph style. Notice that for $X = \frac{\pi}{6} \approx 0.52359878$, $Y_1 + Y_2 = Y_3$.

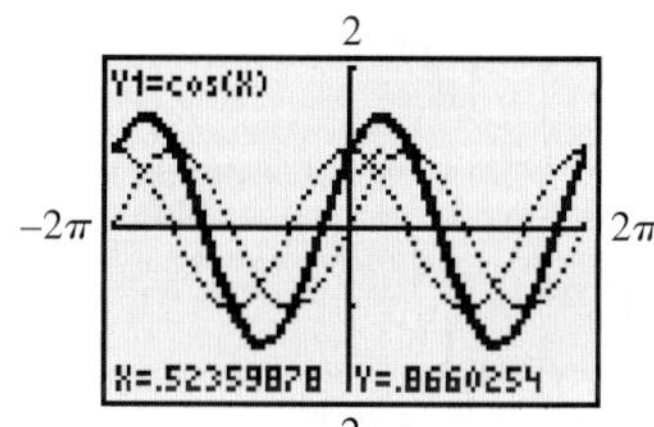

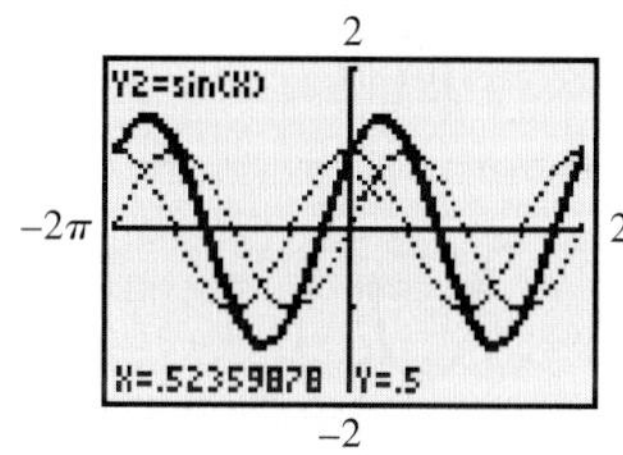

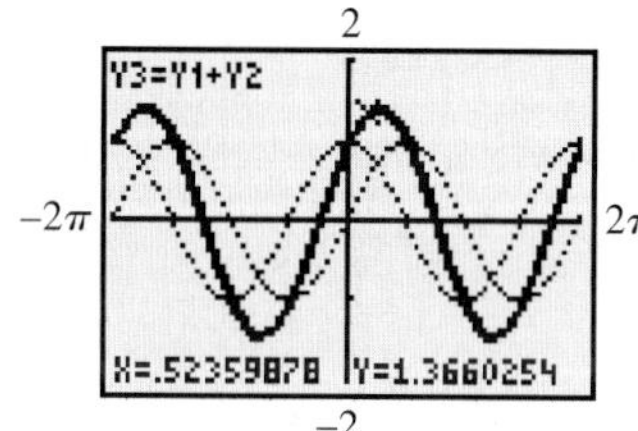

Figure 66 ■

6.6 Exercises

1. B 2. C 3. D 4. A

Concept Check In Exercises 1–4, match each function with its graph from choices A–D.

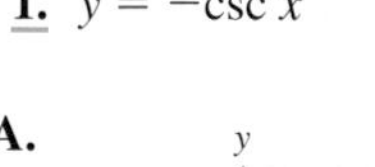
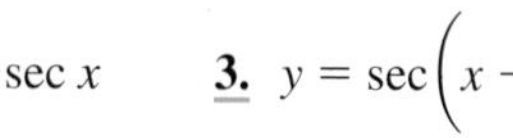

1. $y = -\csc x$ 2. $y = -\sec x$ 3. $y = \sec\left(x - \frac{\pi}{2}\right)$ 4. $y = \csc\left(x + \frac{\pi}{2}\right)$

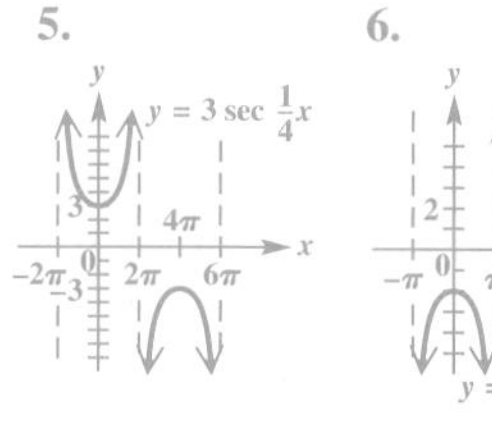
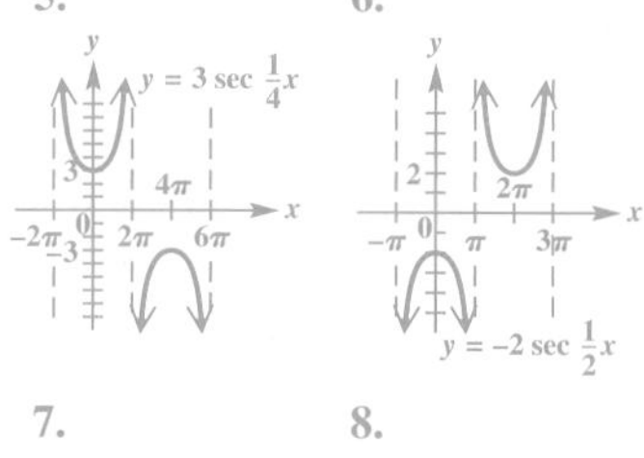

A.

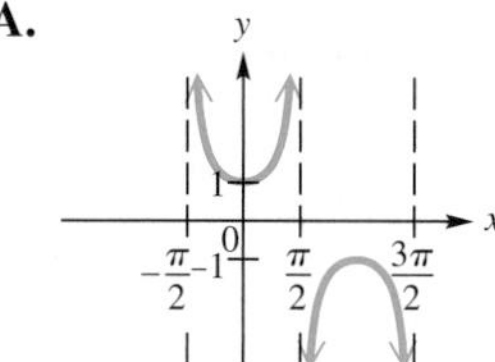

B.

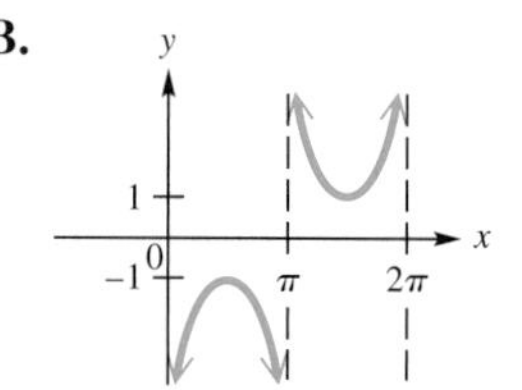

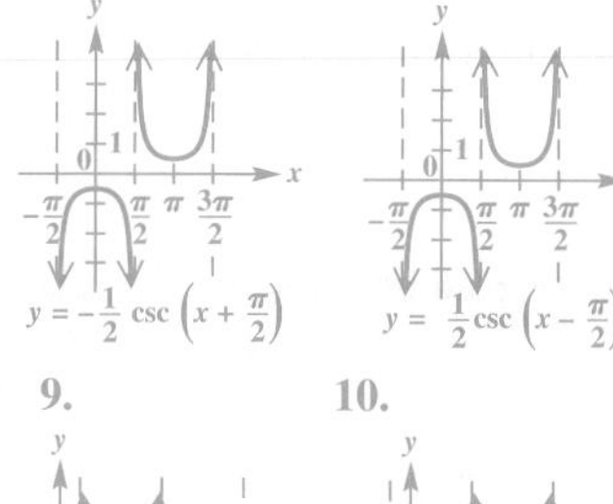

C.

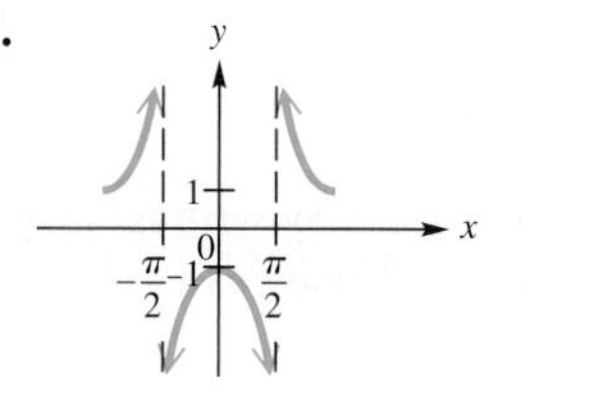

D.

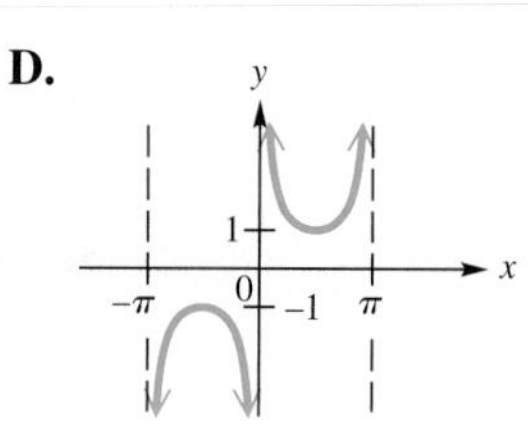

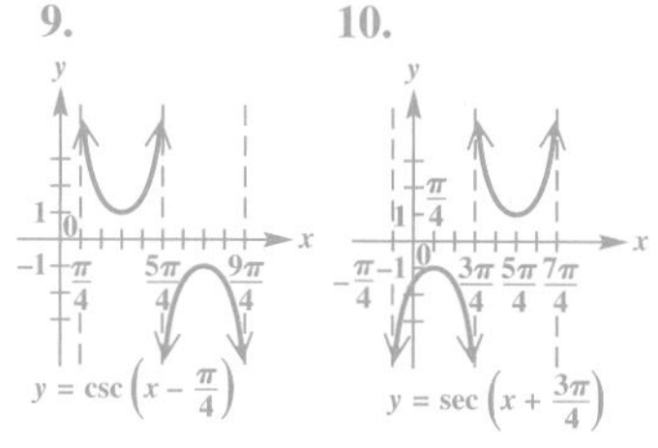

Graph each function over a one-period interval. ***See Examples 1 and 2.***

5. $y = 3 \sec \frac{1}{4}x$ 6. $y = -2 \sec \frac{1}{2}x$ 7. $y = -\frac{1}{2} \csc\left(x + \frac{\pi}{2}\right)$

8. $y = \frac{1}{2} \csc\left(x - \frac{\pi}{2}\right)$ 9. $y = \csc\left(x - \frac{\pi}{4}\right)$ 10. $y = \sec\left(x + \frac{3\pi}{4}\right)$

11. 12.

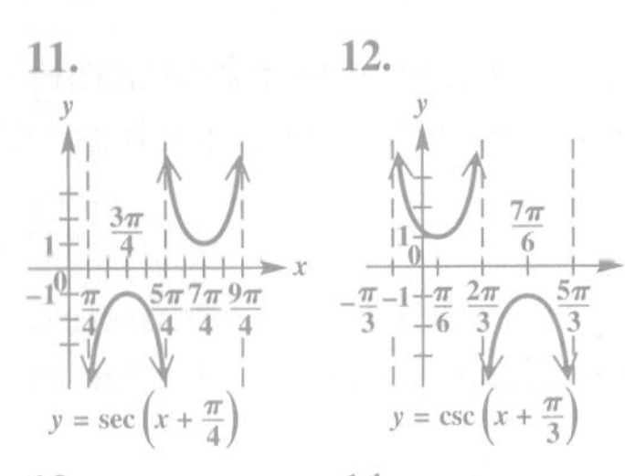

13. 14.

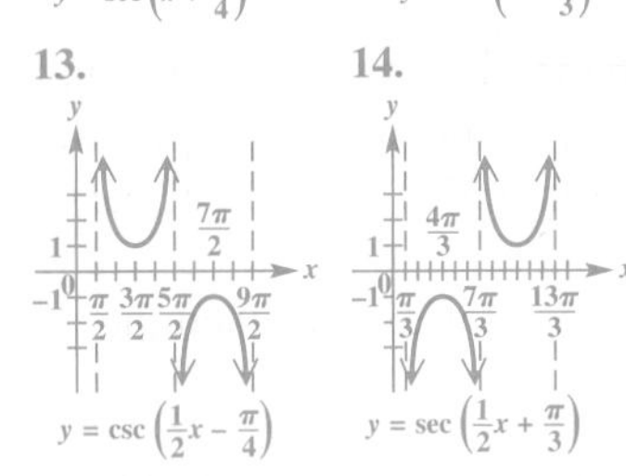

15. 16.

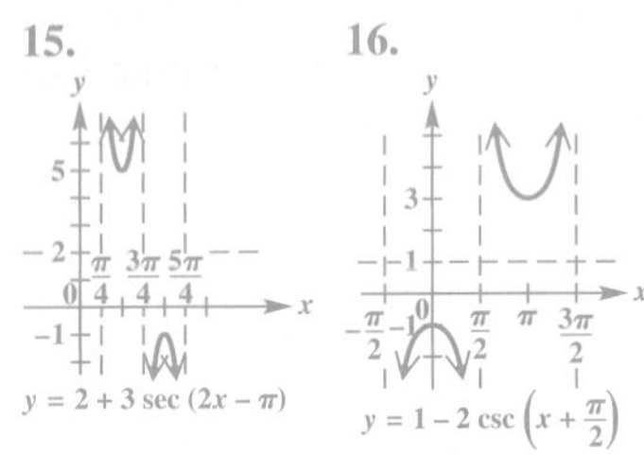

17. 18.

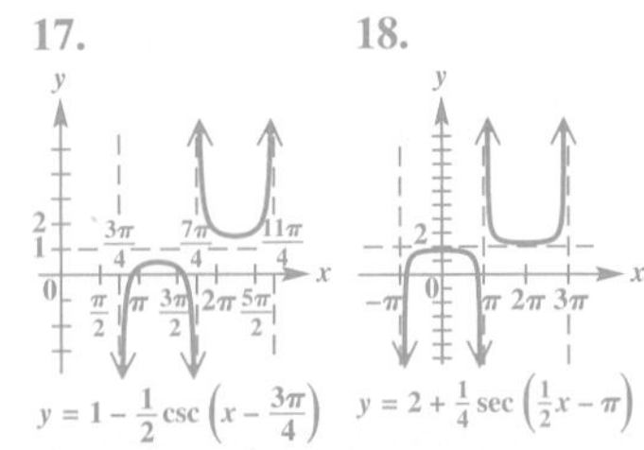

19. $y = \sec 4x$ 20. $y = \sec 2x$
21. $y = -2 + \csc x$
22. $y = 1 + \csc x$
23. $y = -1 - \sec x$
24. $y = 1 - \csc \frac{1}{2}x$

25. true
26. false; Secant values are undefined when $x = (2n + 1)\frac{\pi}{2}$, while cosecant values are undefined when $x = n\pi$.
27. true
28. true

29. none
30. domain: $\left\{x \mid x \neq \frac{n\pi}{4}\right.$, where n is any integer$\}$; range: $(-\infty, -2] \cup [2, \infty)$
33. (a) 4 m (b) 6.3 m (c) 63.7 m

In Exercises 35 and 36, we show the display for $Y_1 + Y_2$ at $X = \frac{\pi}{6}$.

35.

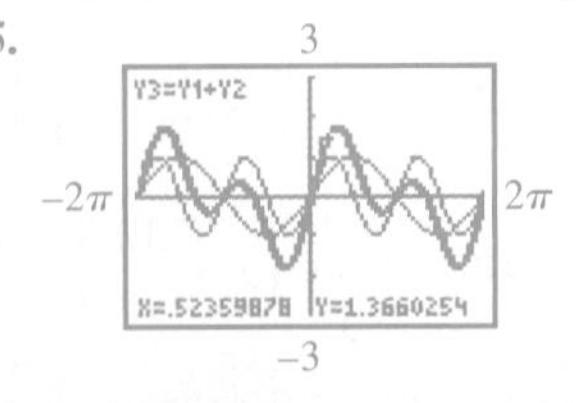

11. $y = \sec\left(x + \frac{\pi}{4}\right)$

12. $y = \csc\left(x + \frac{\pi}{3}\right)$

13. $y = \csc\left(\frac{1}{2}x - \frac{\pi}{4}\right)$

14. $y = \sec\left(\frac{1}{2}x + \frac{\pi}{3}\right)$

15. $y = 2 + 3\sec(2x - \pi)$

16. $y = 1 - 2\csc\left(x + \frac{\pi}{2}\right)$

17. $y = 1 - \frac{1}{2}\csc\left(x - \frac{3\pi}{4}\right)$

18. $y = 2 + \frac{1}{4}\sec\left(\frac{1}{2}x - \pi\right)$

Connecting Graphs with Equations Determine an equation for each graph. ***See Example 3.***

19.

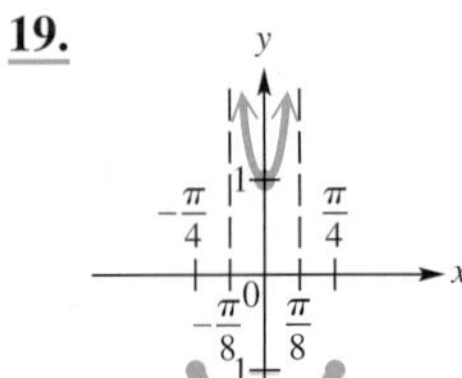

20.

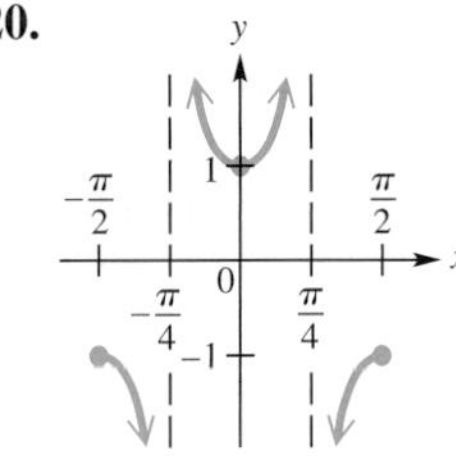

21.

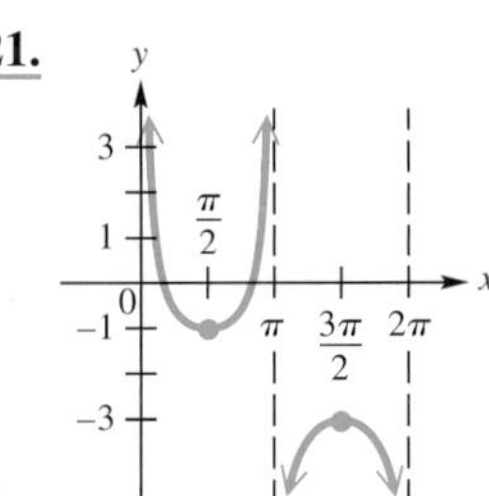

22.

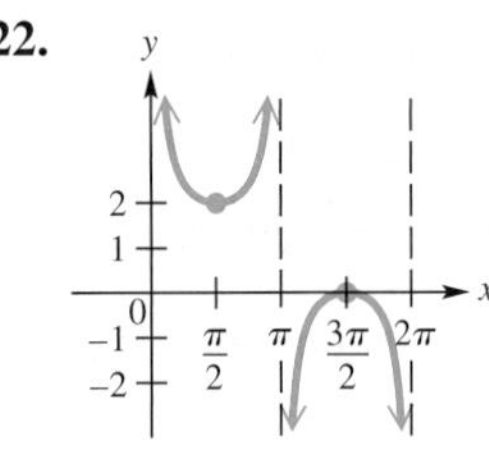

23.

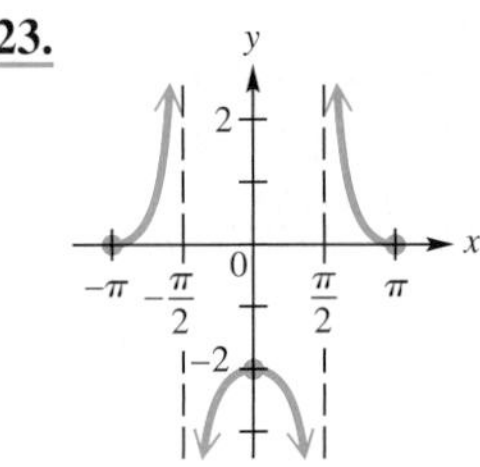

24.

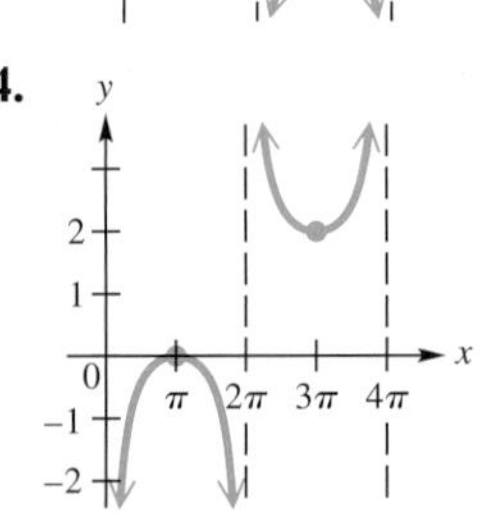

Concept Check In Exercises 25–28, tell whether each statement is true *or* false. *If* false, *tell why.*

25. The tangent and secant functions are undefined for the same values.

26. The secant and cosecant functions are undefined for the same values.

27. The graph of $y = \sec x$ in **Figure 58** suggests that $\sec(-x) = \sec x$ for all x in the domain of $\sec x$.

28. The graph of $y = \csc x$ in **Figure 61** suggests that $\csc(-x) = -\csc x$ for all x in the domain of $\csc x$.

Work each exercise.

29. *Concept Check* If c is any number such that $-1 < c < 1$, then how many solutions does the equation $c = \sec x$ have over the entire domain of the secant function?

30. *Concept Check* Consider the function $g(x) = -2\csc(4x + \pi)$. What is the domain of g? What is its range?

31. Show that $\sec(-x) = \sec x$ by writing $\sec(-x)$ as $\frac{1}{\cos(-x)}$ and then using the relationship between $\cos(-x)$ and $\cos x$.

32. Show that $\csc(-x) = -\csc x$ by writing $\csc(-x)$ as $\frac{1}{\sin(-x)}$ and then using the relationship between $\sin(-x)$ and $\sin x$.

33. *(Modeling) Distance of a Rotating Beacon* In the figure for **Exercise 47** in **Section 6.5**, the distance a is given by

$$a = 4|\sec 2\pi t|.$$

Find a for each time.

(a) $t = 0$ (b) $t = 0.86$ (c) $t = 1.24$

36.

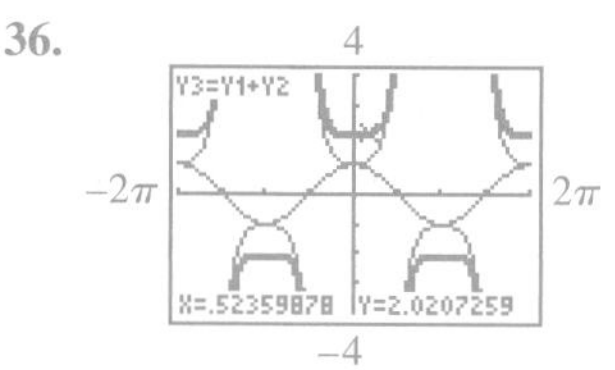

34. Between each pair of successive asymptotes, a portion of the graph of $y = \sec x$ or $y = \csc x$ resembles a parabola. Can each of these portions actually be a parabola? Explain.

Use a graphing calculator to graph Y_1, Y_2, and $Y_1 + Y_2$ on the same screen. Evaluate each of the three functions at $X = \frac{\pi}{6}$, and verify that $Y_1\left(\frac{\pi}{6}\right) + Y_2\left(\frac{\pi}{6}\right) = (Y_1 + Y_2)\left(\frac{\pi}{6}\right)$. See the discussion on addition of ordinates.

35. $Y_1 = \sin X, \quad Y_2 = \sin 2X$

36. $Y_1 = \cos X, \quad Y_2 = \sec X$

Summary Exercises on Graphing Circular Functions

Answer graphs for the Summary Exercises are given in the answer section at the back of the text.

These summary exercises provide practice with the various graphing techniques presented in this chapter. Graph each function over a one-period interval.

1. $y = 2 \sin \pi x$

2. $y = 4 \cos \frac{3}{2}x$

3. $y = -2 + \frac{1}{2} \cos \frac{\pi}{4}x$

4. $y = 3 \sec \frac{\pi}{2}x$

5. $y = -4 \csc \frac{1}{2}x$

6. $y = 3 \tan\left(\frac{\pi}{2}x + \pi\right)$

Graph each function over a two-period interval.

7. $y = -5 \sin \frac{x}{3}$

8. $y = 10 \cos\left(\frac{x}{4} + \frac{\pi}{2}\right)$

9. $y = 3 - 4 \sin\left(\frac{5}{2}x + \pi\right)$

10. $y = 2 - \sec[\pi(x - 3)]$

6.7 Harmonic Motion

- Simple Harmonic Motion
- Damped Oscillatory Motion

Simple Harmonic Motion In part A of **Figure 67,** a spring with a weight attached to its free end is in equilibrium (or rest) position. If the weight is pulled down *a* units and released (part B of the figure), the spring's elasticity causes the weight to rise *a* units $(a > 0)$ above the equilibrium position, as seen in part C, and then to oscillate about the equilibrium position.

If friction is neglected, this oscillatory motion is described mathematically by a sinusoid. Other applications of this type of motion include sound, electric current, and electromagnetic waves.

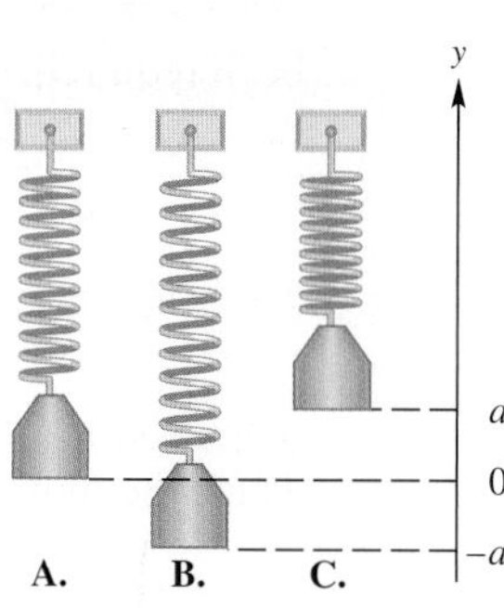

Figure 67

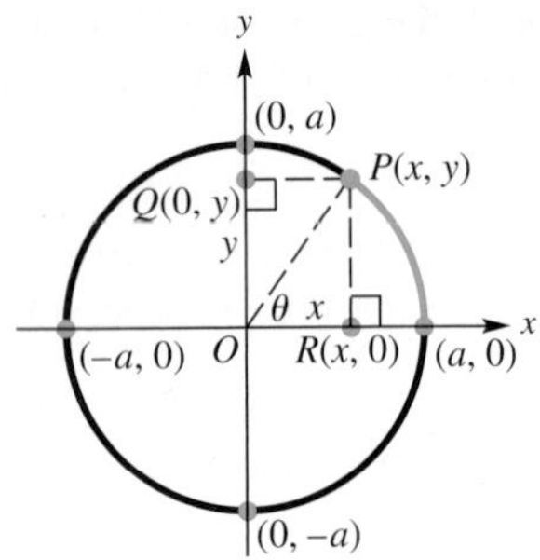

Figure 68

To develop a general equation for such motion, consider **Figure 68.** Suppose the point $P(x, y)$ moves around the circle counterclockwise at a uniform angular speed ω. Assume that at time $t = 0$, P is at $(a, 0)$. The angle swept out by ray OP at time t is given by $\theta = \omega t$. The coordinates of point P at time t are

$$x = a \cos \theta = a \cos \omega t \quad \text{and} \quad y = a \sin \theta = a \sin \omega t.$$

As P moves around the circle from the point $(a, 0)$, the point $Q(0, y)$ oscillates back and forth along the y-axis between the points $(0, a)$ and $(0, -a)$. Similarly, the point $R(x, 0)$ oscillates back and forth between $(a, 0)$ and $(-a, 0)$. This oscillatory motion is called **simple harmonic motion.**

The amplitude of the motion is $|a|$, and the period is $\frac{2\pi}{\omega}$. The moving points P and Q or P and R complete one oscillation or cycle per period. The number of cycles per unit of time, called the **frequency,** is the reciprocal of the period, $\frac{\omega}{2\pi}$, where $\omega > 0$.

Teaching Tip Mention that simple harmonic motion is an important topic for students majoring in engineering or physics.

Simple Harmonic Motion

The position of a point oscillating about an equilibrium position at time t is modeled by either

$$s(t) = a \cos \omega t \quad \text{or} \quad s(t) = a \sin \omega t,$$

where a and ω are constants, with $\omega > 0$. The amplitude of the motion is $|a|$, the period is $\frac{2\pi}{\omega}$, and the frequency is $\frac{\omega}{2\pi}$ oscillations per time unit.

Classroom Example 1
Repeat **Example 1** if the object is pulled down a distance of 16 cm from its equilibrium position and then released, and the time for one complete oscillation is 6 sec.

Answers:
(a) $s(t) = -16 \cos \frac{\pi}{3} t$
(b) $s(1.5) = 0$; The object is at the equilibrium position.
(c) $\frac{1}{6}$ oscillation per sec

EXAMPLE 1 Modeling the Motion of a Spring

Suppose that an object is attached to a coiled spring such as the one in **Figure 67** on the preceding page. It is pulled down a distance of 5 in. from its equilibrium position and then released. The time for one complete oscillation is 4 sec.

(a) Give an equation that models the position of the object at time t.

(b) Determine the position at $t = 1.5$ sec.

(c) Find the frequency.

SOLUTION

(a) When the object is released at $t = 0$, the distance of the object from the equilibrium position is 5 in. below equilibrium. If $s(t)$ is to model the motion, then $s(0)$ must equal -5. We use

$$s(t) = a \cos \omega t, \quad \text{with } a = -5.$$

We choose the cosine function because $\cos \omega(0) = \cos 0 = 1$, and $-5 \cdot 1 = -5$. (Had we chosen the sine function, a phase shift would have been required.) Use the fact that the period is 4 to solve for ω.

$$\frac{2\pi}{\omega} = 4 \qquad \text{The period is } \frac{2\pi}{\omega}.$$

$$\omega = \frac{\pi}{2} \qquad \text{Solve for } \omega. \text{ (Section 1.1)}$$

Thus, the motion is modeled by

$$s(t) = -5 \cos \frac{\pi}{2} t.$$

(b) $s(1.5) = -5 \cos\left[\frac{\pi}{2}(1.5)\right]$ Let $t = 1.5$ in the equation from part (a). (Section 2.3)

≈ 3.54 in.

Because $3.54 > 0$, the object is above the equilibrium position.

(c) The frequency is the reciprocal of the period, or $\frac{1}{4}$ oscillation per sec.

✔ *Now Try Exercise 9.*

Classroom Example 2
Suppose that an object oscillates according to the model

$$s(t) = 2.5 \sin 5t,$$

where t is in seconds and $s(t)$ is in meters. Analyze the motion.

Answer: The motion is harmonic. The object oscillates 2.5 m from its starting point. The period is $\frac{2\pi}{5} \approx 1.3$ sec, and the frequency is $\frac{5}{2\pi} \approx 0.80$ oscillation per sec.

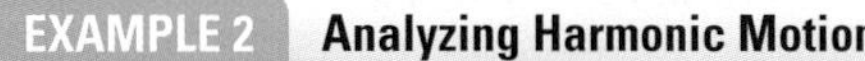

EXAMPLE 2 Analyzing Harmonic Motion

Suppose that an object oscillates according to the model

$$s(t) = 8 \sin 3t,$$

where t is in seconds and $s(t)$ is in feet. Analyze the motion.

SOLUTION The motion is harmonic because the model is $s(t) = a \sin \omega t$. Because $a = 8$, the object oscillates 8 ft in either direction from its starting point. The period $\frac{2\pi}{3} \approx 2.1$ is the time, in seconds, it takes for one complete oscillation. The frequency is the reciprocal of the period, so the object completes $\frac{3}{2\pi} \approx 0.48$ oscillation per sec.

✔ *Now Try Exercise 17.*

Damped Oscillatory Motion In the example of the stretched spring, we disregard the effect of friction. Friction causes the amplitude of the motion to diminish gradually until the weight comes to rest. In this situation, we say that the motion has been *damped* by the force of friction. Most oscillatory motions are damped, and the decrease in amplitude follows the pattern of exponential decay. An example of **damped oscillatory motion** is provided by the function

$$s(t) = e^{-t} \sin t.$$

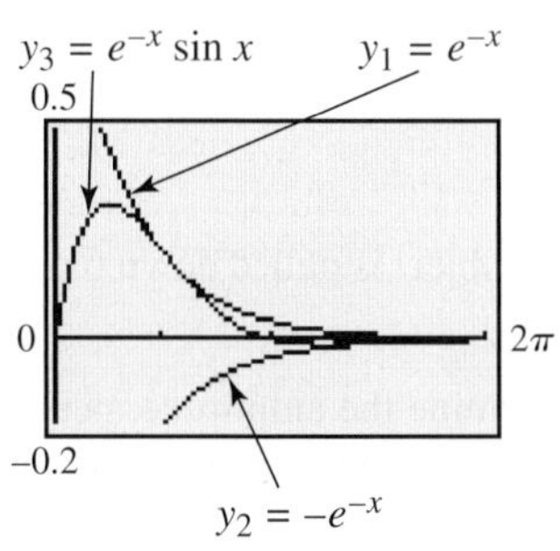

Figure 69

(The number $e \approx 2.718$ is the base of the natural logarithmic function, first introduced in **Chapter 4.**) **Figure 69** shows how the graph of $y_3 = e^{-x} \sin x$ is bounded above by the graph of $y_1 = e^{-x}$ and below by the graph of $y_2 = -e^{-x}$. The damped motion curve dips below the x-axis at $x = \pi$ but stays above the graph of y_2. **Figure 70** shows a traditional graph of $s(t) = e^{-t} \sin t$, along with the graph of $y = \sin t$.

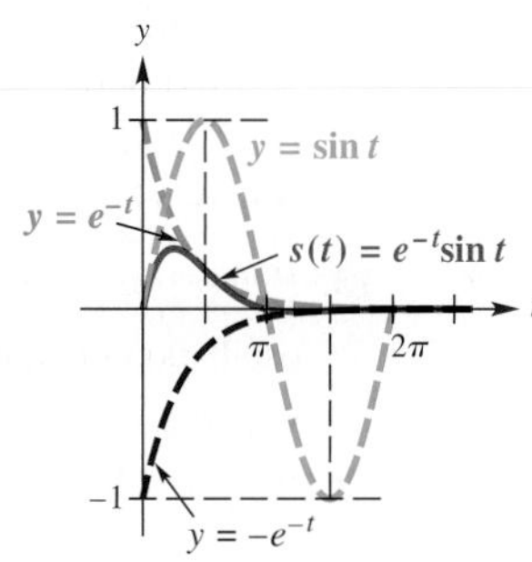

Figure 70

Shock absorbers are put on an automobile in order to damp oscillatory motion. Instead of the car oscillating up and down for a long while after hitting a bump or pothole, the oscillations of the car are quickly damped out for a smoother ride.

6.7 Exercises

(Modeling) Springs *A weight on a spring has initial position $s(0)$ and period P.*

(a) Find a function s given by $s(t) = a \cos \omega t$ that models the displacement of the weight.

(b) Evaluate $s(1)$. Is the weight moving upward, downward, or neither when $t = 1$? Support your results graphically or numerically.

1. $s(0) = 2$ in.; $P = 0.5$ sec
2. $s(0) = 5$ in.; $P = 1.5$ sec
3. $s(0) = -3$ in.; $P = 0.8$ sec
4. $s(0) = -4$ in.; $P = 1.2$ sec

(Modeling) Music *A note on the piano has given frequency F. Suppose the maximum displacement at the center of the piano wire is given by $s(0)$. Find constants a and ω so that the equation*

$$s(t) = a \cos \omega t$$

models this displacement. Graph s in the viewing window $[0, 0.05]$ by $[-0.3, 0.3]$.

5. $F = 27.5$; $s(0) = 0.21$
6. $F = 110$; $s(0) = 0.11$
7. $F = 55$; $s(0) = 0.14$
8. $F = 220$; $s(0) = 0.06$

(Modeling) Solve each problem. ***See Examples 1 and 2.***

9. *Spring Motion* An object is attached to a coiled spring, as in **Figure 67.** It is pulled down a distance of 4 units from its equilibrium position and then released. The time for one complete oscillation is 3 sec.
   (a) Give an equation that models the position of the object at time t.
   (b) Determine the position at $t = 1.25$ sec.
   (c) Find the frequency.

10. *Spring Motion* Repeat **Exercise 9,** but assume that the object is pulled down 6 units and that the time for one complete oscillation is 4 sec.

11. *Particle Movement* Write the equation and then determine the amplitude, period, and frequency of the simple harmonic motion of a particle moving uniformly around a circle of radius 2 units, with the given angular speed.
   (a) 2 radians per sec **(b)** 4 radians per sec

12. *Spring Motion* The height attained by a weight attached to a spring set in motion is

$$s(t) = -4 \cos 8\pi t \text{ inches after } t \text{ seconds.}$$

   (a) Find the maximum height that the weight rises above the equilibrium position of $s(t) = 0$.
   (b) When does the weight first reach its maximum height if $t \geq 0$?
   (c) What are the frequency and the period?

13. *Pendulum Motion* What are the period P and frequency T of oscillation of a pendulum of length $\frac{1}{2}$ ft? $\left(Hint: P = 2\pi\sqrt{\frac{L}{32}}\right.$, where L is the length of the pendulum in feet and the period P is in seconds.$\left.\right)$

14. *Pendulum Motion* In **Exercise 13,** how long should the pendulum be to have a period of 1 sec?

---

1. **(a)** $s(t) = 2 \cos 4\pi t$ **(b)** $s(1) = 2$; The weight is moving neither upward nor downward. At $t = 1$, the motion of the weight is changing from up to down.
2. **(a)** $s(t) = 5 \cos \frac{4\pi}{3} t$ **(b)** $s(1) = -2.5$; upward
3. **(a)** $s(t) = -3 \cos 2.5\pi t$ **(b)** $s(1) = 0$; upward
4. **(a)** $s(t) = -4 \cos \frac{5\pi}{3} t$ **(b)** $s(1) = -2$; downward
5. $s(t) = 0.21 \cos 55\pi t$

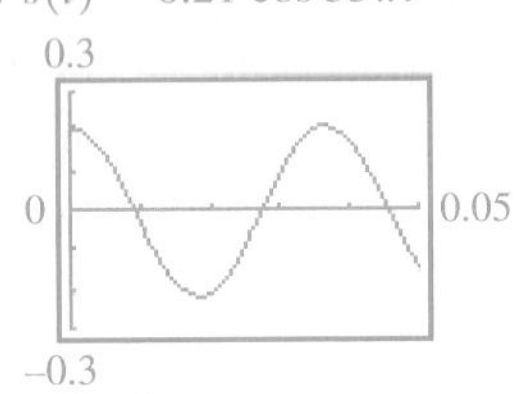

6. $s(t) = 0.11 \cos 220\pi t$

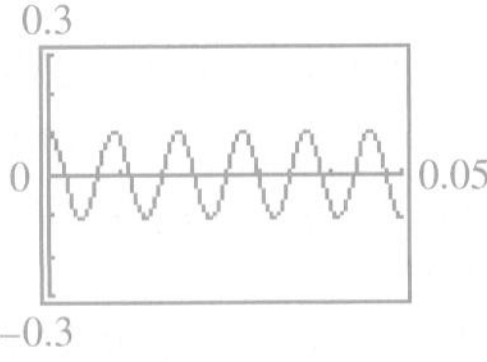

7. $s(t) = 0.14 \cos 110\pi t$

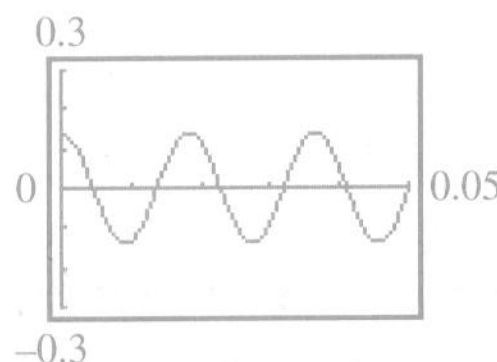

8. $s(t) = 0.06 \cos 440\pi t$

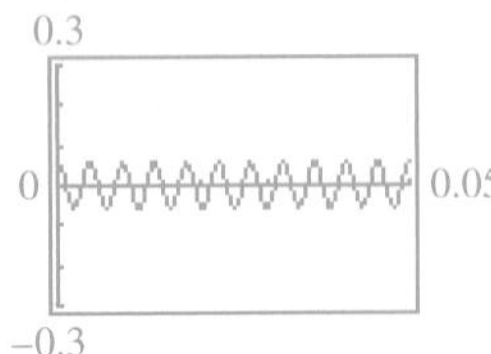

9. **(a)** $s(t) = -4 \cos \frac{2\pi}{3} t$ **(b)** 3.46 units **(c)** $\frac{1}{3}$ oscillation per sec
10. **(a)** $s(t) = -6 \cos \frac{\pi}{2} t$ **(b)** 2.30 units **(c)** $\frac{1}{4}$ oscillation per sec

11. **(a)** $s(t) = 2 \sin 2t$; amplitude: 2; period: π; frequency: $\frac{1}{\pi}$ rotation per sec
(b) $s(t) = 2 \sin 4t$; amplitude: 2; period: $\frac{\pi}{2}$; frequency: $\frac{2}{\pi}$ rotation per sec
12. **(a)** 4 in. **(b)** after $\frac{1}{8}$ sec **(c)** 4 cycles per sec; $\frac{1}{4}$ sec
13. period: $\frac{\pi}{4}$; frequency: $\frac{4}{\pi}$ oscillations per sec
14. $\frac{8}{\pi^2}$ ft
15. $\frac{1}{\pi^2}$
16. **(a)** amplitude: $\frac{1}{2}$; period: $\sqrt{2}\pi$; frequency: $\frac{\sqrt{2}}{2\pi}$ oscillation per sec **(b)** $s(t) = \frac{1}{2} \sin \sqrt{2}t$
17. **(a)** 5 in. **(b)** 2 cycles per sec; $\frac{1}{2}$ sec **(c)** after $\frac{1}{4}$ sec **(d)** approximately 4; After 1.3 sec, the weight is about 4 in. above the equilibrium position.
18. **(a)** 4 in. **(b)** $\frac{5}{\pi}$ cycles per sec; $\frac{\pi}{5}$ sec **(c)** after $\frac{\pi}{10}$ sec **(d)** approximately 2; After 1.466 sec, the weight is about 2 in. above the equilibrium position.
19. **(a)** $s(t) = -3 \cos 12t$ **(b)** $\frac{\pi}{6}$ sec
20. **(a)** $s(t) = -2 \cos 6\pi t$ **(b)** 3 cycles per sec
21. 0; π; They are the same.
22. for y_1 and y_2: $\left(\frac{\pi}{2}, e^{-\pi/2}\right)$; for y_1 and y_3: none in $[0, \pi]$; Because $\sin \frac{\pi}{2} = 1$, $e^{-\pi/2} \sin \frac{\pi}{2} = e^{-\pi/2}$.

15. *Spring Motion* The formula for the up and down motion of a weight on a spring is given by

$$s(t) = a \sin \sqrt{\frac{k}{m}}\, t.$$

If the spring constant k is 4, what mass m must be used to produce a period of 1 sec?

16. *Spring Motion* (See **Exercise 15.**) A spring with spring constant $k = 2$ and a 1-unit mass m attached to it is stretched and then allowed to come to rest.
(a) If the spring is stretched $\frac{1}{2}$ ft and released, what are the amplitude, period, and frequency of the resulting oscillatory motion?
(b) What is the equation of the motion?

17. *Spring Motion* The position of a weight attached to a spring is

$$s(t) = -5 \cos 4\pi t \text{ inches after } t \text{ seconds.}$$

(a) What is the maximum height that the weight rises above the equilibrium position?
(b) What are the frequency and period?
(c) When does the weight first reach its maximum height?
(d) Calculate and interpret $s(1.3)$.

18. *Spring Motion* The position of a weight attached to a spring is

$$s(t) = -4 \cos 10t \text{ inches after } t \text{ seconds.}$$

(a) What is the maximum height that the weight rises above the equilibrium position?
(b) What are the frequency and period?
(c) When does the weight first reach its maximum height?
(d) Calculate and interpret $s(1.466)$.

19. *Spring Motion* A weight attached to a spring is pulled down 3 in. below the equilibrium position.
(a) Assuming that the frequency is $\frac{6}{\pi}$ cycles per sec, determine a model that gives the position of the weight at time t seconds.
(b) What is the period?

20. *Spring Motion* A weight attached to a spring is pulled down 2 in. below the equilibrium position.
(a) Assuming that the period is $\frac{1}{3}$ sec, determine a model that gives the position of the weight at time t seconds.
(b) What is the frequency?

Damped Oscillatory Motion Use a graphing calculator to graph

$$y_1 = e^{-t} \sin t, \quad y_2 = e^{-t}, \quad \textit{and} \quad y_3 = -e^{-t}$$

in the viewing window $[0, \pi]$ *by* $[-0.5, 0.5]$.

21. Find the t-intercepts of the graph of y_1. Explain the relationship of these intercepts to the x-intercepts of the graph of $y = \sin x$.

22. Find any points of intersection of y_1 and y_2 or y_1 and y_3. How are these points related to the graph of $y = \sin x$?

Chapter 6 Test Prep

Key Terms

6.1 radian circumference latitude sector of a circle longitude subtend degree of curvature	nautical mile statute mile **6.2** unit circle circular functions reference arc linear speed v angular speed ω	**6.3** periodic function period sine wave (sinusoid) amplitude **6.4** phase shift argument **6.5** vertical asymptote	**6.6** addition of ordinates **6.7** simple harmonic motion frequency damped oscillatory motion

Quick Review

Concepts | **Examples**

6.1 Radian Measure

An angle with its vertex at the center of a circle that intercepts an arc on the circle equal in length to the radius of the circle has a measure of **1 radian.**

Degree/Radian Relationship $\quad 180° = \pi$ **radians**

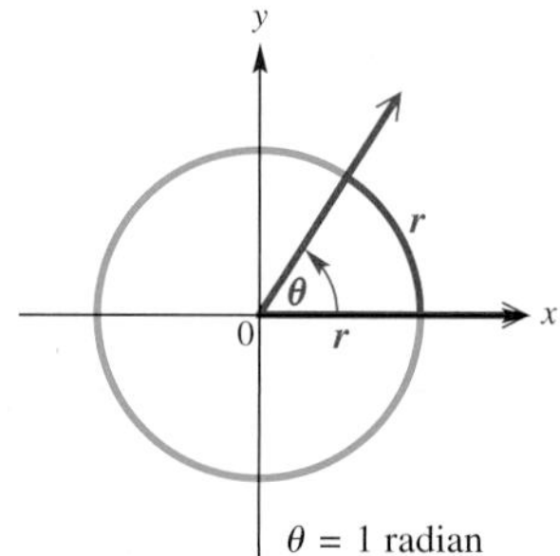

Converting between Degrees and Radians

1. Multiply a degree measure by $\frac{\pi}{180}$ radian and simplify to convert to radians.

Convert 135° to radians.

$$135° = 135\left(\frac{\pi}{180}\text{ radian}\right) = \frac{3\pi}{4}\text{ radians}$$

2. Multiply a radian measure by $\frac{180°}{\pi}$ and simplify to convert to degrees.

Convert $-\frac{5\pi}{3}$ radians to degrees.

$$-\frac{5\pi}{3}\text{ radians} = -\frac{5\pi}{3}\left(\frac{180°}{\pi}\right) = -300°$$

Arc Length

The length s of the arc intercepted on a circle of radius r by a central angle of measure θ radians is given by the product of the radius and the radian measure of the angle.

$$s = r\theta, \quad \textbf{where } \theta \textbf{ is in radians}$$

Find the central angle θ in the figure.

$$\theta = \frac{s}{r} = \frac{3}{4}\text{ radian}$$

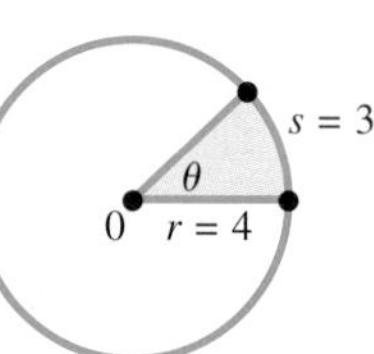

Area of a Sector

The area $\mathcal{A}$ of a sector of a circle of radius r and central angle θ is given by the following formula.

$$\mathcal{A} = \frac{1}{2}r^2\theta, \quad \textbf{where } \theta \textbf{ is in radians}$$

Find the area $\mathcal{A}$ of the sector in the figure above.

$$\mathcal{A} = \frac{1}{2}(4)^2\left(\frac{3}{4}\right) = 6\text{ sq units}$$

Concepts	Examples

6.2 The Unit Circle and Circular Functions

Circular Functions

Start at the point $(1, 0)$ on the unit circle $x^2 + y^2 = 1$ and measure off an arc of length $|s|$ along the circle, going counterclockwise if s is positive and clockwise if s is negative. Let the endpoint of the arc be at the point (x, y). The six circular functions of s are defined as follows. (Assume that no denominators are 0.)

$$\sin s = y \qquad \cos s = x \qquad \tan s = \frac{y}{x}$$

$$\csc s = \frac{1}{y} \qquad \sec s = \frac{1}{x} \qquad \cot s = \frac{x}{y}$$

Use the unit circle to find each value.

$$\sin \frac{5\pi}{6} = \frac{1}{2}$$

$$\cos \frac{3\pi}{2} = 0$$

$$\tan \frac{\pi}{4} = \frac{\frac{\sqrt{2}}{2}}{\frac{\sqrt{2}}{2}} = 1$$

$$\csc \frac{7\pi}{4} = \frac{1}{-\frac{\sqrt{2}}{2}} = -\sqrt{2}$$

$$\sec \frac{7\pi}{6} = \frac{1}{-\frac{\sqrt{3}}{2}} = -\frac{2\sqrt{3}}{3}$$

$$\cot \frac{\pi}{3} = \frac{\frac{1}{2}}{\frac{\sqrt{3}}{2}} = \frac{\sqrt{3}}{3}$$

$$\sin 0 = 0$$

$$\cos \frac{\pi}{2} = 0$$

The Unit Circle

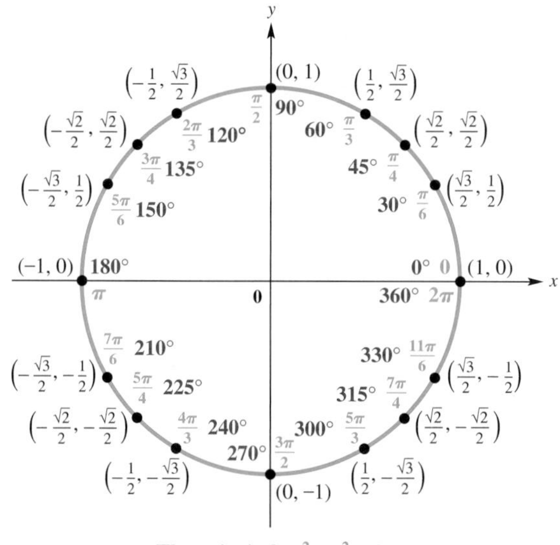

The unit circle $x^2 + y^2 = 1$

Find the value of s in $\left[0, \frac{\pi}{2}\right]$ that makes $\cos s = \frac{\sqrt{3}}{2}$ true.

In $\left[0, \frac{\pi}{2}\right]$, the arc length $s = \frac{\pi}{6}$ is associated with the point $\left(\frac{\sqrt{3}}{2}, \frac{1}{2}\right)$. The first coordinate is

$$\cos s = \cos \frac{\pi}{6} = \frac{\sqrt{3}}{2},$$

so $s = \frac{\pi}{6}$ makes the statement true.

Formulas for Angular and Linear Speed

Angular Speed ω	Linear Speed v
$\omega = \frac{\theta}{t}$ (ω in radians per unit time t, θ in radians)	$v = \frac{s}{t}$ $v = \frac{r\theta}{t}$ $v = r\omega$

A belt runs a machine pulley of radius 8 in. at 60 revolutions per min. Find each of the following.

(a) the angular speed ω in radians per minute

$$\omega = 60(2\pi)$$
$$= 120\pi \text{ radians per min}$$

(b) the linear speed v of the belt in inches per minute

$$v = r\omega$$
$$= 8(120\pi)$$
$$= 960\pi \text{ in. per min}$$

(continued)

Concepts	Examples

6.3 Graphs of the Sine and Cosine Functions

6.4 Translations of the Graphs of the Sine and Cosine Functions

Sine and Cosine Functions

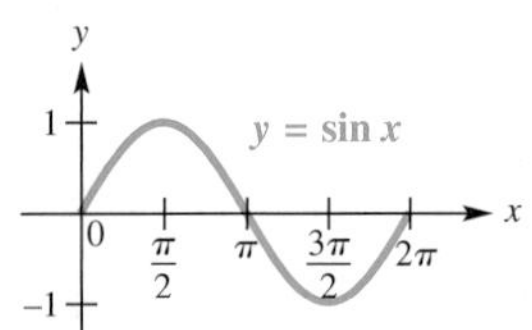

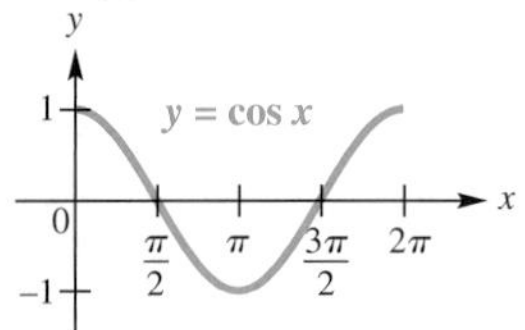

Domain: $(-\infty, \infty)$
Range: $[-1, 1]$
Amplitude: 1
Period: 2π

Domain: $(-\infty, \infty)$
Range: $[-1, 1]$
Amplitude: 1
Period: 2π

The graph of

$$y = c + a \sin[b(x - d)] \text{ or } y = c + a \cos[b(x - d)],$$

with $b > 0$, has the following characteristics.

1. amplitude $|a|$
2. period $\frac{2\pi}{b}$
3. vertical translation c units up if $c > 0$ or $|c|$ units down if $c < 0$
4. phase shift d units to the right if $d > 0$ or $|d|$ units to the left if $d < 0$

See **Sections 6.3 and 6.4** for a summary of graphing techniques.

Graph $y = 1 + \sin 3x$.

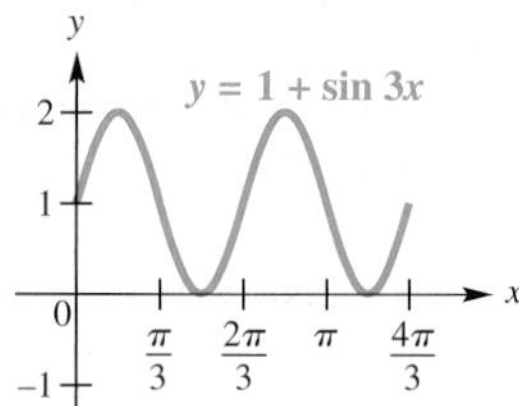

amplitude: 1
period: $\frac{2\pi}{3}$
vertical translation: 1 unit up

domain: $(-\infty, \infty)$
range: $[0, 2]$

Graph $y = -2 \cos\left(x + \frac{\pi}{2}\right)$.

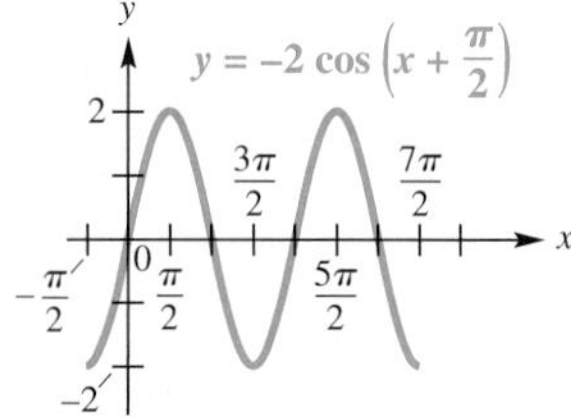

amplitude: 2
period: 2π
phase shift: $\frac{\pi}{2}$ left

domain: $(-\infty, \infty)$
range: $[-2, 2]$

6.5 Graphs of the Tangent and Cotangent Functions

Tangent and Cotangent Functions

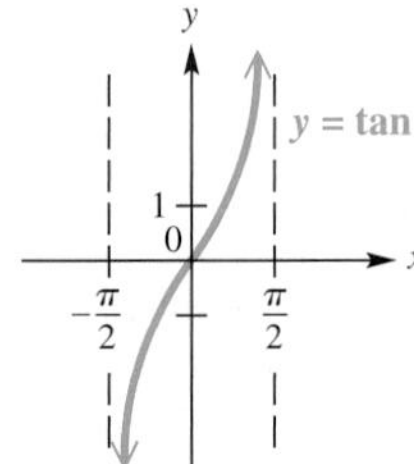

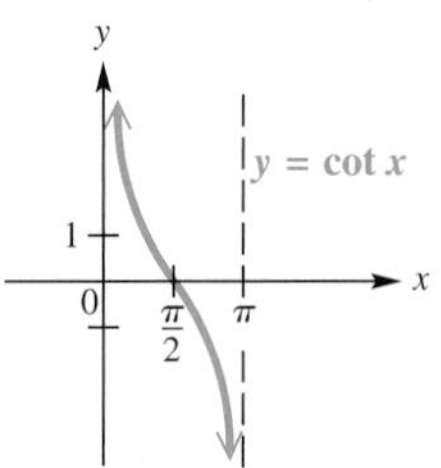

Domain: $\left\{x \mid x \neq (2n + 1)\frac{\pi}{2},\right.$ where n is any integer$\left.\right\}$
Range: $(-\infty, \infty)$
Period: π

Domain: $\{x \mid x \neq n\pi,$ where n is any integer$\}$
Range: $(-\infty, \infty)$
Period: π

See **Section 6.5** for a summary of graphing techniques.

Graph one period of $y = 2 \tan x$.

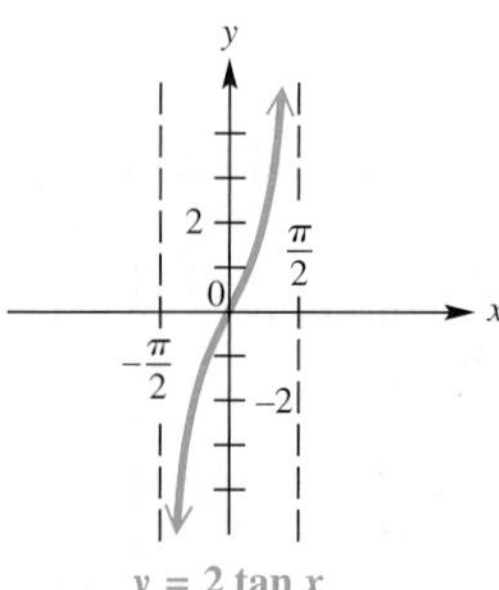

period: π
domain: $\left\{x \mid x \neq (2n + 1)\frac{\pi}{2},\right.$ where n is any integer$\left.\right\}$
range: $(-\infty, \infty)$

Concepts	Examples

6.6 Graphs of the Secant and Cosecant Functions

Secant and Cosecant Functions

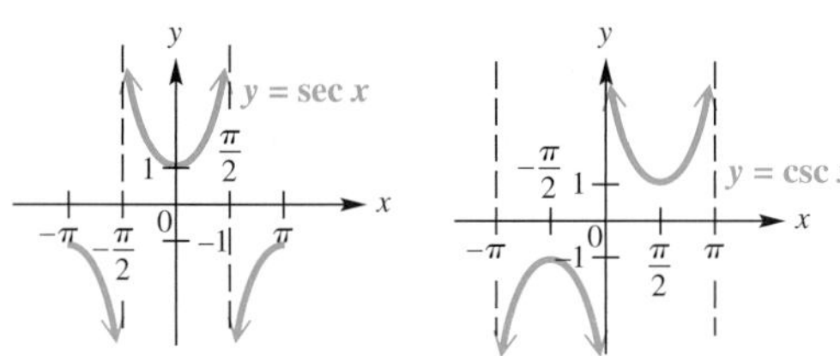

Domain: $\{x \mid x \neq (2n+1)\frac{\pi}{2}$, where n is any integer$\}$
Range: $(-\infty, -1] \cup [1, \infty)$
Period: 2π

Domain: $\{x \mid x \neq n\pi$, where n is any integer$\}$
Range: $(-\infty, -1] \cup [1, \infty)$
Period: 2π

See **Section 6.6** for a summary of graphing techniques.

Graph one period of $y = \sec\left(x + \frac{\pi}{4}\right)$.

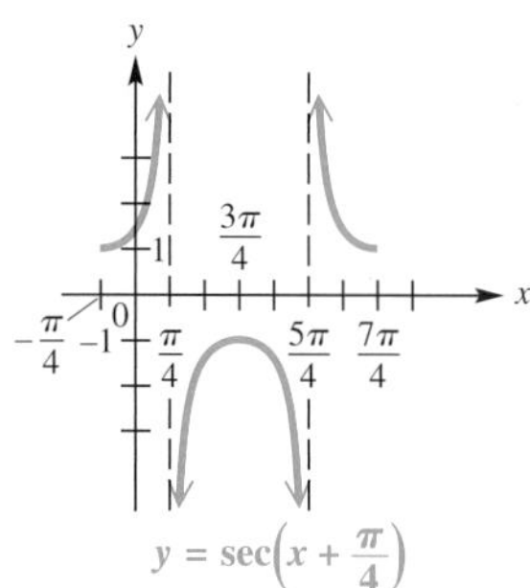

period: 2π
phase shift: $\frac{\pi}{4}$ left
domain: $\{x \mid x \neq \frac{\pi}{4} + n\pi$, where n is any integer$\}$
range: $(-\infty, -1] \cup [1, \infty)$

6.7 Harmonic Motion

Simple Harmonic Motion

The position of a point oscillating about an equilibrium position at time t is modeled by either

$$s(t) = a \cos \omega t \quad \text{or} \quad s(t) = a \sin \omega t,$$

where a and ω are constants, with $\omega > 0$. The amplitude of the motion is $|a|$, the period is $\frac{2\pi}{\omega}$, and the frequency is $\frac{\omega}{2\pi}$ oscillations per time unit.

A spring oscillates according to

$$s(t) = -5 \cos 6t,$$

where t is in seconds and $s(t)$ is in inches. Find the amplitude, period, and frequency.

$$\text{amplitude} = |-5| = 5 \text{ in.} \quad \text{period} = \frac{2\pi}{6} = \frac{\pi}{3} \text{ sec}$$

$$\text{frequency} = \frac{3}{\pi} \text{ oscillation per sec}$$

Chapter 6 Review Exercises

1. *Concept Check* What is the meaning of "an angle with measure 2 radians"?

2. *Concept Check* Consider each angle in standard position having the given radian measure. In what quadrant does the terminal side lie?

   (a) 3 **(b)** 4 **(c)** -2 **(d)** 7

3. Find three angles coterminal with an angle of 1 radian.

4. Give an expression that generates all angles coterminal with an angle of $\frac{\pi}{6}$ radian. Let n represent any integer.

Answers:

1. A central angle of a circle that intercepts an arc of length 2 times the radius of the circle has a measure of 2 radians.
2. (a) II (b) III (c) III (d) I
3. Three of many possible answers are $1 + 2\pi$, $1 + 4\pi$, and $1 + 6\pi$.
4. $\frac{\pi}{6} + 2n\pi$

5. $\frac{\pi}{4}$ 6. $\frac{2\pi}{3}$
7. $\frac{35\pi}{36}$ 8. $\frac{11\pi}{6}$
9. $\frac{40\pi}{9}$ 10. $\frac{17\pi}{3}$

Convert each degree measure to radians. Leave answers as multiples of π.

5. 45° **6.** 120° **7.** 175°

8. 330° **9.** 800° **10.** 1020°

11. 225° 12. 162°
13. 480° 14. 216°
15. −110° 16. −756°

Convert each radian measure to degrees.

11. $\frac{5\pi}{4}$ **12.** $\frac{9\pi}{10}$ **13.** $\frac{8\pi}{3}$

14. $\frac{6\pi}{5}$ **15.** $-\frac{11\pi}{18}$ **16.** $-\frac{21\pi}{5}$

17. π in. 18. $\frac{4\pi}{3}$ in.
19. 12π in.

Suppose the tip of the minute hand of a clock is 2 in. *from the center of the clock. For each duration, determine the distance traveled by the tip of the minute hand.*

17. 15 min **18.** 20 min **19.** 3 hr

20. 8.77 cm
21. 35.8 cm 22. 2263 in.^{2}
23. 49.06° 24. 273 m^2
25. 4500 km 26. 12,000 km

Solve each problem. Use a calculator as necessary.

20. *Arc Length* Find the length of an arc intercepted by a central angle of 0.769 radian on a circle with radius 11.4 cm.

21. *Arc Length* The radius of a circle is 15.2 cm. Find the length of an arc of the circle intercepted by a central angle of $\frac{3\pi}{4}$ radians.

22. *Area of a Sector* A central angle of $\frac{7\pi}{4}$ radians forms a sector of a circle. Find the area of the sector if the radius of the circle is 28.69 in.

23. *Angle Measure* Find the measure (in degrees) of a central angle that intercepts an arc of length 7.683 cm in a circle of radius 8.973 cm.

24. *Area of a Sector* Find the area of a sector of a circle having a central angle of $21° 40'$ in a circle of radius 38.0 m.

Distance between Cities *Assume that the radius of Earth is* 6400 km.

25. Find the distance in kilometers between cities on a north-south line that are on latitudes 28° N and 12° S, respectively.

26. Two cities on the equator have longitudes of 72° E and 35° W, respectively. Find the distance between the cities.

27. $\frac{3}{4}$; 1.5 sq units
28. $\frac{1}{2}$; 16 sq units

Concept Check *Find the measure of the central angle θ (in radians) and the area of the sector.*

27.

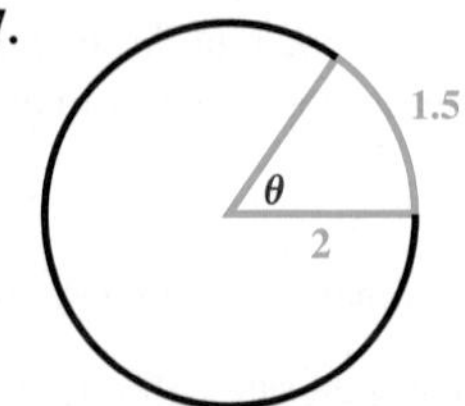

28.

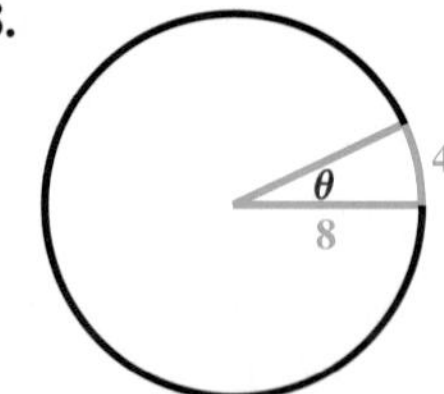

29. $\sqrt{3}$ 30. $-\frac{1}{2}$
31. $-\frac{1}{2}$ 32. $-\sqrt{3}$
33. 2 34. undefined
35. 0.8660 36. 2.7976
37. 0.9703 38. −11.4266
39. 1.9513 40. −1.0080
41. 0.3898 42. 1.3265
43. 0.5148 44. 0.9424
45. 1.1054 46. 1.3497
47. $\frac{\pi}{4}$ 48. $\frac{2\pi}{3}$
49. $\frac{7\pi}{6}$ 50. $\frac{11\pi}{6}$
51. $\frac{15}{32}$ sec
52. 108 radians
53. $\frac{\pi}{20}$ radian per sec
54. 285.3 cm
55. 1260π cm per sec
56. $\frac{\pi}{36}$ radian per sec

Find each exact function value. Do not use a calculator.

29. $\tan \frac{\pi}{3}$ **30.** $\cos \frac{2\pi}{3}$ **31.** $\sin\left(-\frac{5\pi}{6}\right)$

32. $\tan\left(-\frac{7\pi}{3}\right)$ **33.** $\csc\left(-\frac{11\pi}{6}\right)$ **34.** $\cot(-13\pi)$

Use a calculator to find an approximation for each circular function value. Be sure your calculator is set in radian mode.

35. $\sin 1.0472$ **36.** $\tan 1.2275$ **37.** $\cos(-0.2443)$

38. $\cot 3.0543$ **39.** $\sec 7.3159$ **40.** $\csc 4.8386$

Find the value of s in the interval $\left[0, \frac{\pi}{2}\right]$ *that makes each statement true.*

41. $\cos s = 0.9250$ **42.** $\tan s = 4.0112$

43. $\sin s = 0.4924$ **44.** $\csc s = 1.2361$

45. $\cot s = 0.5022$ **46.** $\sec s = 4.5600$

Find the exact value of s in the given interval that has the given circular function value. Do not use a calculator.

47. $\left[0, \frac{\pi}{2}\right]$; $\cos s = \frac{\sqrt{2}}{2}$ **48.** $\left[\frac{\pi}{2}, \pi\right]$; $\tan s = -\sqrt{3}$

49. $\left[\pi, \frac{3\pi}{2}\right]$; $\sec s = -\frac{2\sqrt{3}}{3}$ **50.** $\left[\frac{3\pi}{2}, 2\pi\right]$; $\sin s = -\frac{1}{2}$

Solve each problem, where t, ω, θ, and s are as defined in ***Section 6.2.***

51. Find t if $\theta = \frac{5\pi}{12}$ radians and $\omega = \frac{8\pi}{9}$ radians per sec.

52. Find θ if $t = 12$ sec and $\omega = 9$ radians per sec.

53. Find ω if $t = 8$ sec and $\theta = \frac{2\pi}{5}$ radians.

54. Find s if $r = 11.46$ cm, $\omega = 4.283$ radians per sec, and $t = 5.813$ sec.

Solve each problem.

55. *Linear Speed of a Flywheel* Find the linear speed of a point on the edge of a flywheel of radius 7 cm if the flywheel is rotating 90 times per sec.

56. *Angular Speed of a Ferris Wheel* A Ferris wheel has radius 25 ft. If it takes 30 sec for the wheel to turn $\frac{5\pi}{6}$ radians, what is the angular speed of the wheel?

57. 5 in.

58. (a) 0; The face of the moon is not visible.
(b) $\frac{1}{2}$; Half the face of the moon is visible.
(c) 1; The face of the moon is completely visible.
(d) $\frac{1}{2}$; Half the face of the moon is visible.

59. B 60. D

61. 2; 2π; none; none
62. not applicable; $\frac{\pi}{3}$; none; none
63. $\frac{1}{2}$; $\frac{2\pi}{3}$; none; none
64. 2; $\frac{2\pi}{5}$; none; none
65. 2; 8π; 1 up; none
66. $\frac{1}{4}$; 3π; 3 up; none
67. 3; 2π; none; $\frac{\pi}{2}$ to the left
68. 1; 2π; none; $\frac{3\pi}{4}$ to the right
69. not applicable; π; none; $\frac{\pi}{8}$ to the right
70. not applicable; 2; none; 2 to the right
71. not applicable; $\frac{\pi}{3}$; none; $\frac{\pi}{9}$ to the right
72. not applicable; 2π; none; $\frac{3\pi}{2}$ to the left

73. tangent 74. sine
75. cosine 76. cosecant
77. cotangent 78. secant

79.

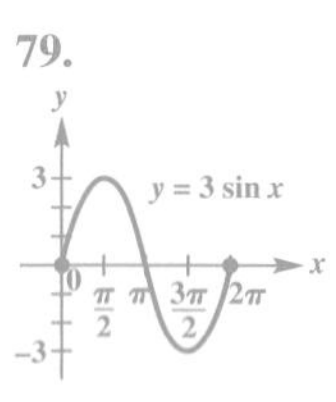

80.

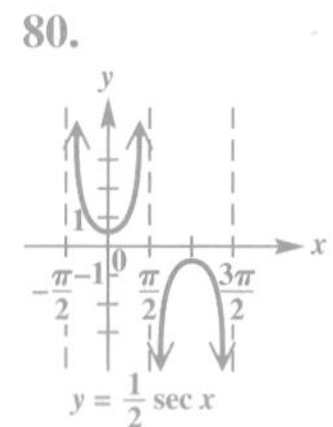

81. 82.

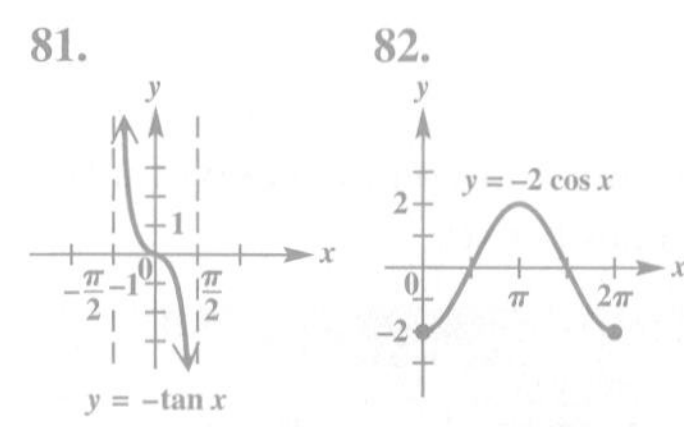

57. *(Modeling) Archaeology* An archaeology professor believes that an unearthed fragment is a piece of the edge of a circular ceremonial plate and uses a formula that will give the radius of the original plate using measurements from the fragment, shown in **Figure A.** Measurements are in inches.

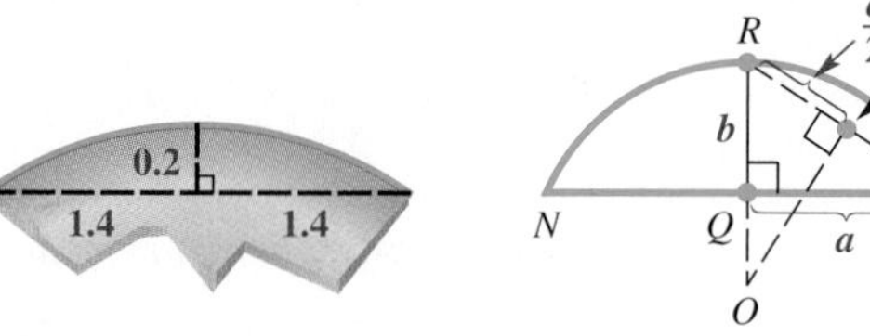

Figure A **Figure B**

In **Figure B,** a is $\frac{1}{2}$ the length of chord NP, and b is the distance from the midpoint of chord NP to the circle. According to the formula, the radius r of the circle, OR, is given by

$$r = \frac{a^2 + b^2}{2b}.$$

What is the radius of the original plate from which the fragment came?

58. *(Modeling) Phase Angle of the Moon* Because the moon orbits Earth, we observe different phases of the moon during the period of a month. In the figure, t is the **phase angle.**

The **phase** F of the moon is modeled by

$$F(t) = \frac{1}{2}(1 - \cos t)$$

and gives the fraction of the moon's face that is illuminated by the sun. (*Source:* Duffet-Smith, P., *Practical Astronomy with Your Calculator,* Cambridge University Press.) Evaluate each expression and interpret the result.

(a) $F(0)$ (b) $F\left(\frac{\pi}{2}\right)$ (c) $F(\pi)$ (d) $F\left(\frac{3\pi}{2}\right)$

59. *Concept Check* Which one of the following is true about the graph of $y = 4 \sin 2x$?

A. It has amplitude 2 and period $\frac{\pi}{2}$.
B. It has amplitude 4 and period π.
C. Its range is $[0, 4]$.
D. Its range is $[-4, 0]$.

60. *Concept Check* Which one of the following is false about the graph of $y = -3 \cos \frac{1}{2}x$?

A. Its range is $[-3, 3]$.
B. Its domain is $(-\infty, \infty)$.
C. Its amplitude is 3, and its period is 4π.
D. Its amplitude is -3, and its period is π.

For each function, give the amplitude, period, vertical translation, and phase shift, as applicable.

61. $y = 2 \sin x$
62. $y = \tan 3x$
63. $y = -\frac{1}{2} \cos 3x$
64. $y = 2 \sin 5x$

83. 84.

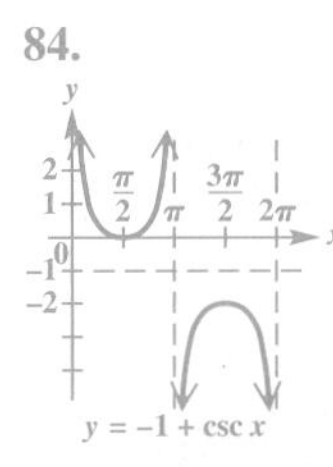

85. 86.

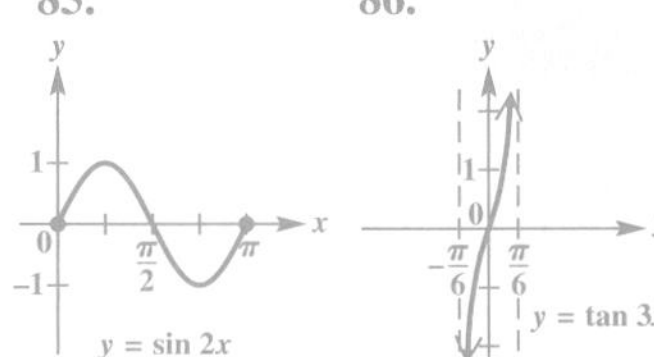

87. 88.

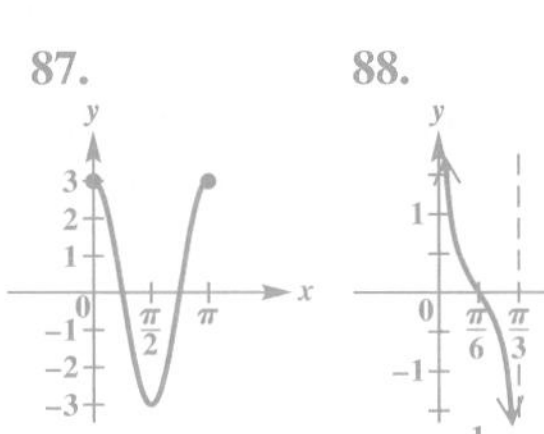

89. 90.

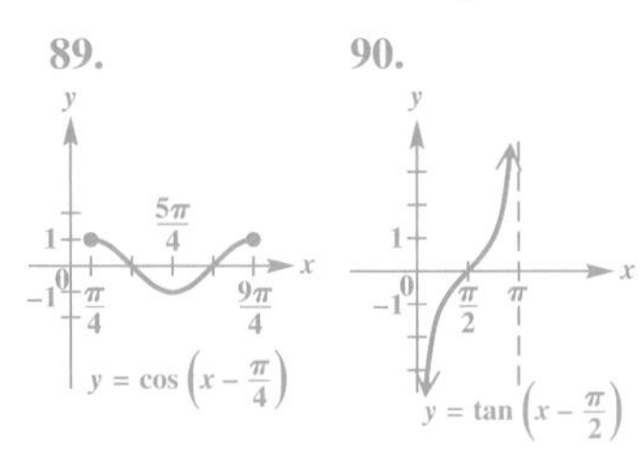

91. 92.

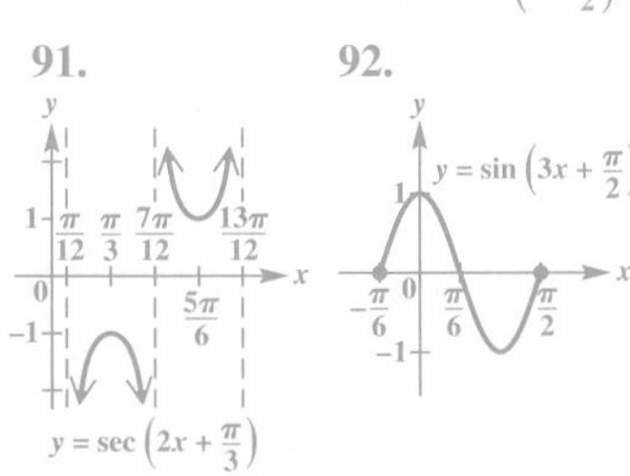

93. 94.

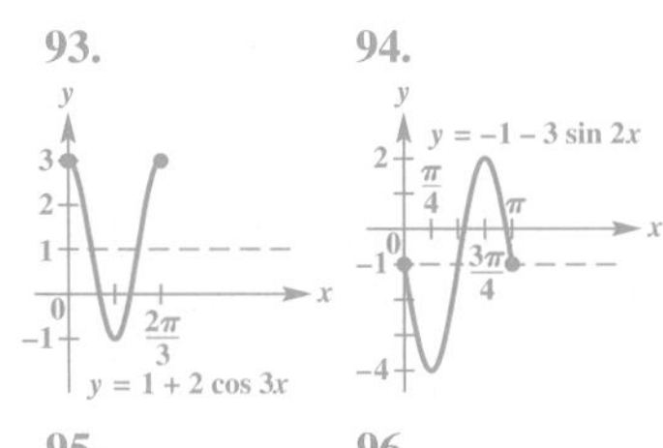

95. 96.

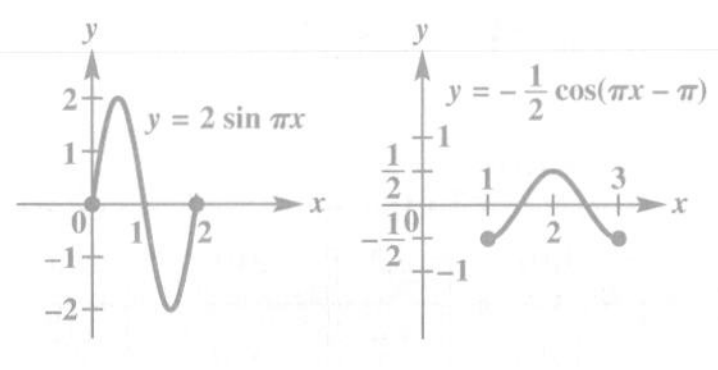

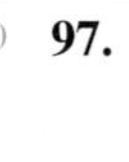

97. $y = 1 - \sin x$

98. $y = \frac{1}{2}\cos 2x$

99. $y = 2\tan\frac{1}{2}x$

100. $y = -1 + 2\csc x$

65. $y = 1 + 2\sin\frac{1}{4}x$

66. $y = 3 - \frac{1}{4}\cos\frac{2}{3}x$

67. $y = 3\cos\left(x + \frac{\pi}{2}\right)$

68. $y = -\sin\left(x - \frac{3\pi}{4}\right)$

69. $y = \frac{1}{2}\csc\left(2x - \frac{\pi}{4}\right)$

70. $y = 2\sec(\pi x - 2\pi)$

71. $y = \frac{1}{3}\tan\left(3x - \frac{\pi}{3}\right)$

72. $y = \cot\left(\frac{x}{2} + \frac{3\pi}{4}\right)$

Concept Check *Identify the circular function that satisfies each description.*

73. period is π, x-intercepts are of the form $n\pi$, where n is any integer

74. period is 2π, graph passes through the origin

75. period is 2π, graph passes through the point $\left(\frac{\pi}{2}, 0\right)$

76. period is 2π, domain is $\{x \mid x \neq n\pi$, where n is any integer$\}$

77. period is π, function is decreasing on the interval $(0, \pi)$

78. period is 2π, has vertical asymptotes of the form $x = (2n + 1)\frac{\pi}{2}$, where n is any integer

Graph each function over a one-period interval.

79. $y = 3\sin x$

80. $y = \frac{1}{2}\sec x$

81. $y = -\tan x$

82. $y = -2\cos x$

83. $y = 2 + \cot x$

84. $y = -1 + \csc x$

85. $y = \sin 2x$

86. $y = \tan 3x$

87. $y = 3\cos 2x$

88. $y = \frac{1}{2}\cot 3x$

89. $y = \cos\left(x - \frac{\pi}{4}\right)$

90. $y = \tan\left(x - \frac{\pi}{2}\right)$

91. $y = \sec\left(2x + \frac{\pi}{3}\right)$

92. $y = \sin\left(3x + \frac{\pi}{2}\right)$

93. $y = 1 + 2\cos 3x$

94. $y = -1 - 3\sin 2x$

95. $y = 2\sin \pi x$

96. $y = -\frac{1}{2}\cos(\pi x - \pi)$

Connecting Graphs with Equations *Determine the simplest form of an equation for each graph. Choose $b > 0$, and include no phase shifts.*

97.

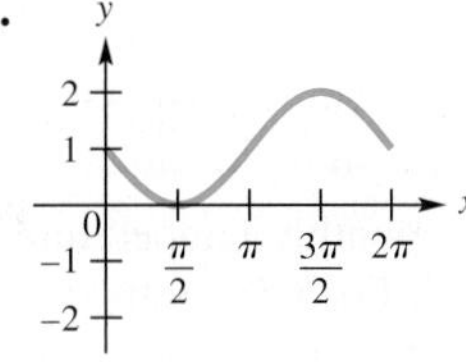

98.

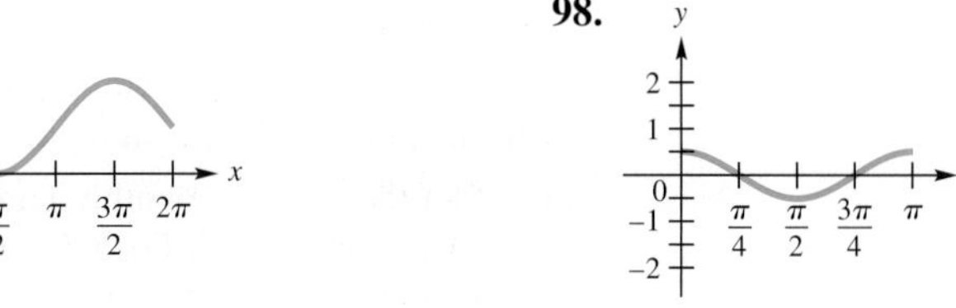

99.

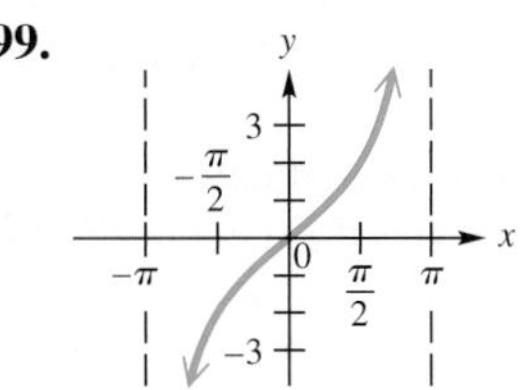

100.

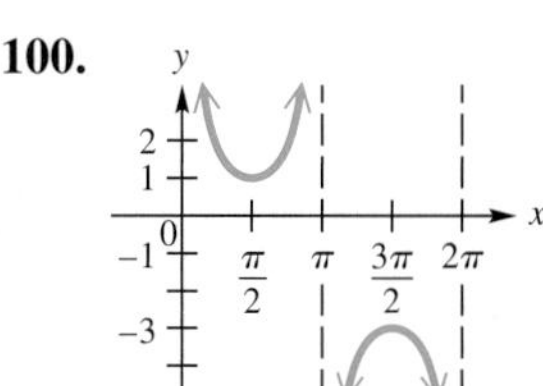

101. (b)

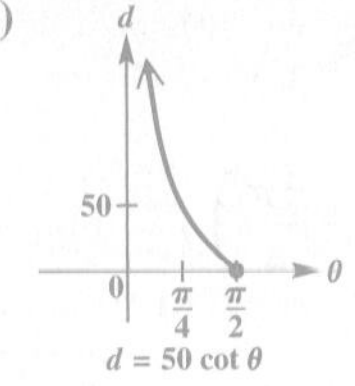

102. (a) 12.3 hr (b) 1.2 ft
(c) 1.56 ft

103. (a) 30°F (b) 60°F
(c) 75°F (d) 86°F
(e) 86°F (f) 60°F

104. (a) See the graph in part (d).
(b) $f(x) = 25.5 \sin\left[\frac{\pi}{6}(x-4)\right] + 47.5$
(d) The function gives an excellent model for the data.

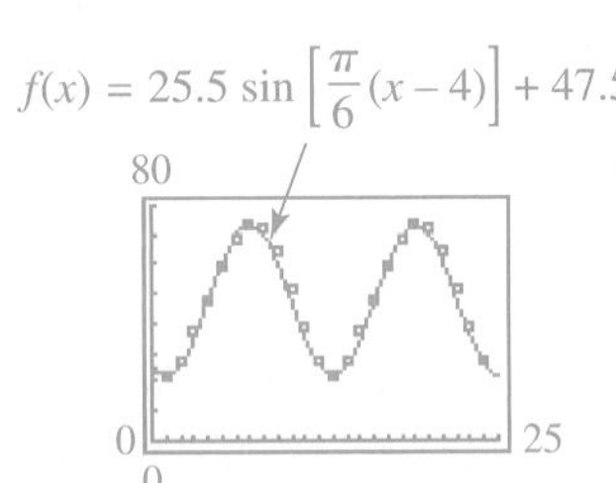

Solve each problem.

101. *Viewing Angle to an Object* Let a person whose eyes are h_1 feet from the ground stand d feet from an object h_2 feet tall, where $h_2 > h_1$. Let θ be the angle of elevation to the top of the object. See the figure.

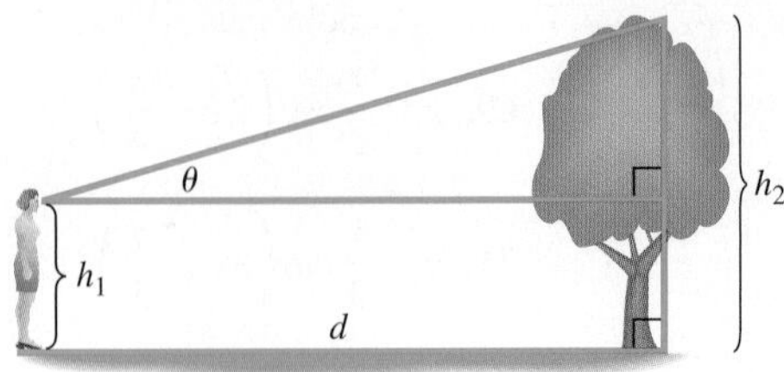

(a) Show that $d = (h_2 - h_1) \cot \theta$.

(b) Let $h_2 = 55$ and $h_1 = 5$. Graph d for the interval $0 < \theta \leq \frac{\pi}{2}$.

102. *(Modeling) Tides* The figure shows a function f that models the tides in feet at Clearwater Beach, Florida, x hours after midnight. (*Source:* Pentcheff, D., *WWW Tide and Current Predictor.*)

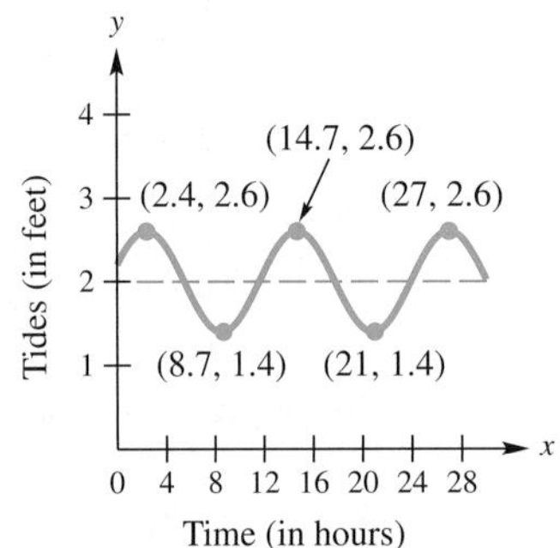

(a) Find the time between high tides.

(b) What is the difference in water levels between high tide and low tide?

(c) The tides can be modeled by

$$f(x) = 0.6 \cos[0.511(x - 2.4)] + 2.$$

Estimate the tides when $x = 10$.

103. *(Modeling) Maximum Temperatures* The maximum afternoon temperature (in °F) in a given city might be modeled by

$$t = 60 - 30 \cos \frac{x\pi}{6},$$

where t represents the maximum afternoon temperature in month x, with $x = 0$ representing January, $x = 1$ representing February, and so on. Find the maximum afternoon temperature to the nearest degree for each month.

(a) January **(b)** April **(c)** May
(d) June **(e)** August **(f)** October

104. *(Modeling) Average Monthly Temperature* The average monthly temperature (in °F) in Chicago, Illinois, is shown in the table.

Month	°F	Month	°F
Jan	22	July	73
Feb	27	Aug	72
Mar	37	Sept	64
Apr	48	Oct	52
May	59	Nov	39
June	68	Dec	27

Source: World Almanac and Book of Facts.

(a) Plot the average monthly temperature over a two-year period. Let $x = 1$ correspond to January of the first year.

(b) Determine a model function of the form $f(x) = a \sin[b(x - d)] + c$, where a, b, c, and d are constants.

(c) Explain the significance of each constant.

(d) Graph f together with the data on the same coordinate axes. How well does f model the data?

104. (e)
```
SinReg
y=a*sin(bx+c)+d
a=24.96
b=.52
c=-2.15
d=48.90
```
TI-83/84 Plus fixed to the nearest hundredth

105. (a) 100 (b) 258 (c) 122 (d) 296

106. (a) about 20 yr (b) maximum: about 150,000; minimum: about 5000

107. amplitude: 4; period: 2; frequency: $\frac{1}{2}$ cycle per sec

108. amplitude: 3; period: π; frequency: $\frac{1}{\pi}$ cycle per sec

109. The frequency is the number of cycles in one unit of time; -4; 0; $-2\sqrt{2}$

110. The period is the time to complete one cycle. The amplitude is the maximum distance (on either side) from the equilibrium point.

(e) Use the sine regression capability of a graphing calculator to find the equation of a sine curve that fits these data.

105. *(Modeling) Pollution Trends* The amount of pollution in the air is lower after heavy spring rains and higher after periods of little rain. In addition to this seasonal fluctuation, the long-term trend is upward. An idealized graph of this situation is shown in the figure.

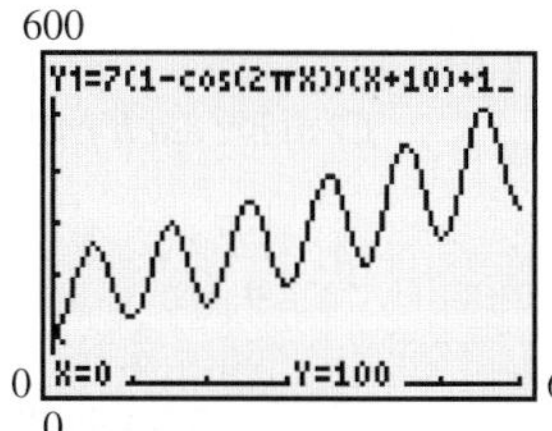

Circular functions can be used to model the fluctuating part of the pollution levels. Powers of the number e (e is the base of the natural logarithm; $e \approx 2.718282$) can be used to model long-term growth. The pollution level in a certain area might be given by

$$y = 7(1 - \cos 2\pi x)(x + 10) + 100e^{0.2x},$$

where x is the time in years, with $x = 0$ representing January 1 of the base year. July 1 of the same year would be represented by $x = 0.5$, October 1 of the following year would be represented by $x = 1.75$, and so on. Find the pollution levels on each date.

(a) January 1, base year
(b) July 1, base year
(c) January 1, following year
(d) July 1, following year

106. *(Modeling) Lynx and Hare Populations* The figure shows the populations of lynx and hares in Canada for the years 1847–1903. The hares are food for the lynx. An increase in hare population causes an increase in lynx population some time later. The increasing lynx population then causes a decline in hare population. The two graphs have the same period.

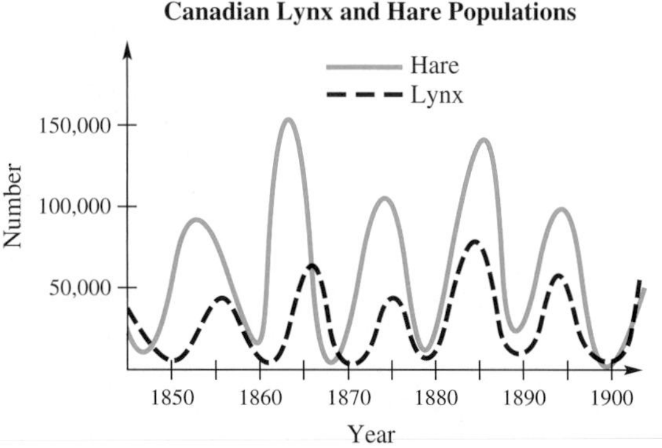

(a) Estimate the length of one period.
(b) Estimate the maximum and minimum hare populations.

An object in simple harmonic motion has position function $s(t)$ inches from an equilibrium point, where t is the time in seconds. Find the amplitude, period, and frequency.

107. $s(t) = 4 \sin \pi t$

108. $s(t) = 3 \cos 2t$

109. In **Exercise 107,** what does the frequency represent? Find the position of the object relative to the equilibrium point at 1.5 sec, 2 sec, and 3.25 sec.

110. Refer to **Exercise 108.** What does the period represent? What does the amplitude represent?

Chapter 6 Test

[6.1]
1. $\frac{2\pi}{3}$ 2. $-\frac{\pi}{4}$
3. 0.09 4. 135°
5. −210° 6. 229.18°
7. (a) $\frac{4}{3}$ (b) 15,000 cm^2
8. 2 radians

[6.2]
9. $\frac{\sqrt{2}}{2}$ 10. $-\frac{\sqrt{3}}{2}$
11. undefined 12. −2
13. 0 14. 0
15. $\sin \frac{7\pi}{6} = -\frac{1}{2}$; $\cos \frac{7\pi}{6} = -\frac{\sqrt{3}}{2}$; $\tan \frac{7\pi}{6} = \frac{\sqrt{3}}{3}$; $\csc \frac{7\pi}{6} = -2$; $\sec \frac{7\pi}{6} = -\frac{2\sqrt{3}}{3}$; $\cot \frac{7\pi}{6} = \sqrt{3}$
16. (a) 0.9716 (b) $\frac{\pi}{3}$
17. (a) $\frac{2\pi}{3}$ radians (b) 40π cm (c) 5π cm per sec
18. (a) 75 ft (b) $\frac{\pi}{45}$ radian per sec

Convert each degree measure to radians.

1. 120° **2.** −45° **3.** 5° (to the nearest hundredth)

Convert each radian measure to degrees.

4. $\frac{3\pi}{4}$ **5.** $-\frac{7\pi}{6}$ **6.** 4 (to the nearest hundredth)

7. A central angle of a circle with radius 150 cm intercepts an arc of 200 cm. Find each measure.

(a) the radian measure of the angle

(b) the area of a sector with that central angle

8. *Rotation of Gas Gauge Arrow* The arrow on a car's gasoline gauge is $\frac{1}{2}$ in. long. See the figure. Through what angle does the arrow rotate when it moves 1 in. on the gauge?

Find each circular function value.

9. $\sin \frac{3\pi}{4}$ **10.** $\cos\left(-\frac{7\pi}{6}\right)$ **11.** $\tan \frac{3\pi}{2}$

12. $\sec \frac{8\pi}{3}$ **13.** $\tan \pi$ **14.** $\cos \frac{3\pi}{2}$

15. Determine the six exact circular function values of s in the figure.

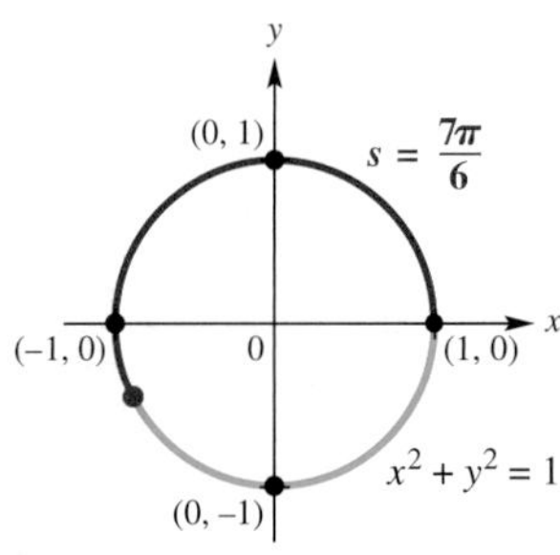

16. (a) Use a calculator to approximate s in the interval $\left[0, \frac{\pi}{2}\right]$ if $\sin s = 0.8258$.

(b) Find the exact value of s in the interval $\left[0, \frac{\pi}{2}\right]$ if $\cos s = \frac{1}{2}$.

17. *Angular and Linear Speed of a Point* Suppose that point P is on a circle with radius 60 cm, and ray OP is rotating with angular speed $\frac{\pi}{12}$ radian per sec.

(a) Find the angle generated by P in 8 sec.

(b) Find the distance traveled by P along the circle in 8 sec.

(c) Find the linear speed of P.

18. *Ferris Wheel* A Ferris wheel has radius 50.0 ft. A person takes a seat and then the wheel turns $\frac{2\pi}{3}$ radians.

(a) How far is the person above the ground?

(b) If it takes 30 sec for the wheel to turn $\frac{2\pi}{3}$ radians, what is the angular speed of the wheel?

[6.3–6.6]
19. (a) $y = \sec x$ (b) $y = \sin x$
(c) $y = \cos x$ (d) $y = \tan x$
(e) $y = \csc x$ (f) $y = \cot x$

20. (a) $y = 1 + \cos \frac{1}{2}x$
(b) $y = -\frac{1}{2} \cot x$

[6.3, 6.5, 6.6]
21. (a) $(-\infty, \infty)$ (b) $[-1, 1]$
(c) $\frac{\pi}{2}$
(d) $(-\infty, -1] \cup [1, \infty)$

[6.4]
22. (a) π (b) 6
(c) $[-3, 9]$ (d) -3
(e) $\frac{\pi}{4}$ to the left $\left(\text{that is, } -\frac{\pi}{4}\right)$

23.

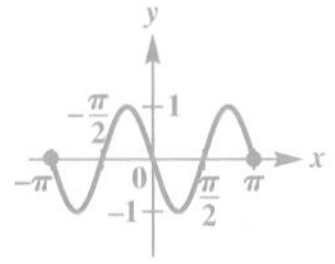

$y = \sin(2x + \pi)$

24.

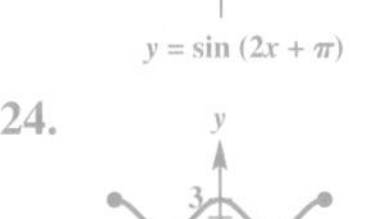

$y = 2 + \cos x$

25.

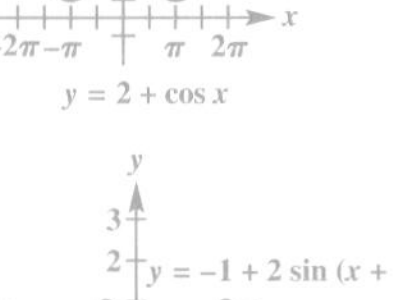

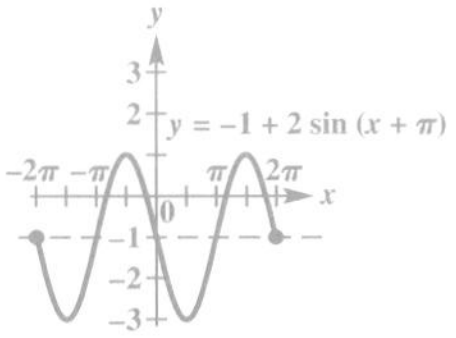

[6.5]
26.

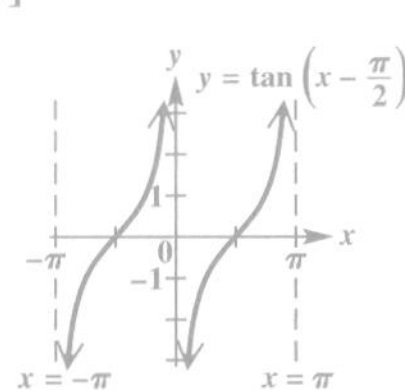

27.

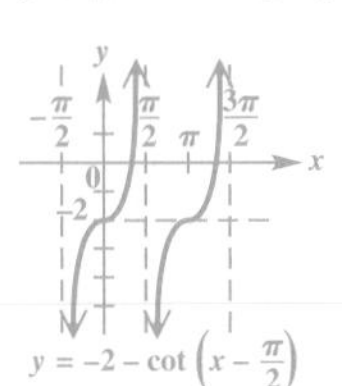

[6.6]
28.

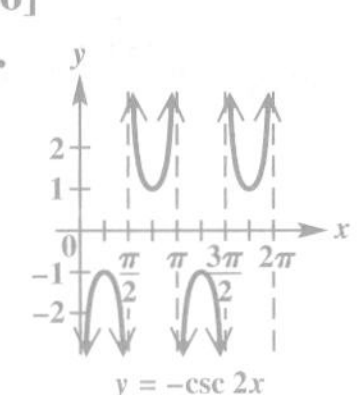

19. Identify each of the following basic circular function graphs.

(a)

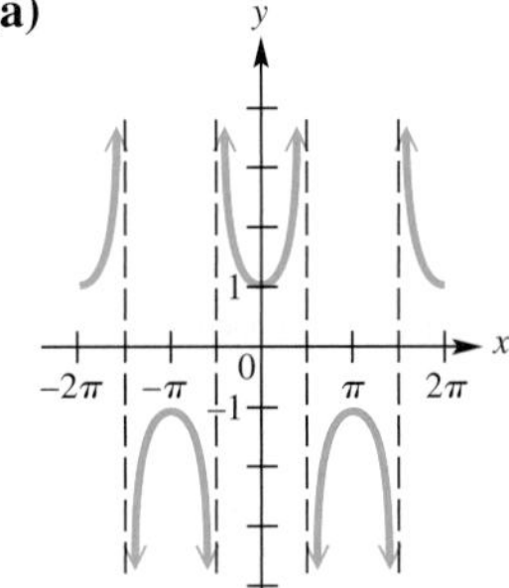

(b)

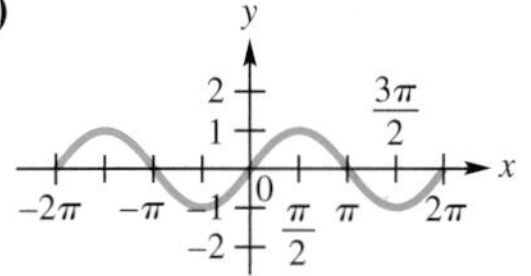

(c)

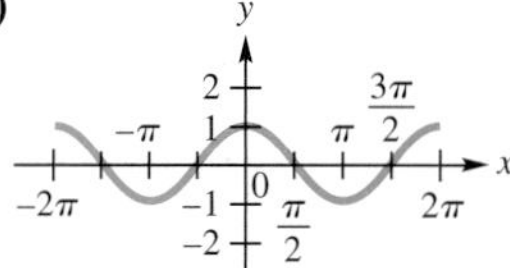

(d)

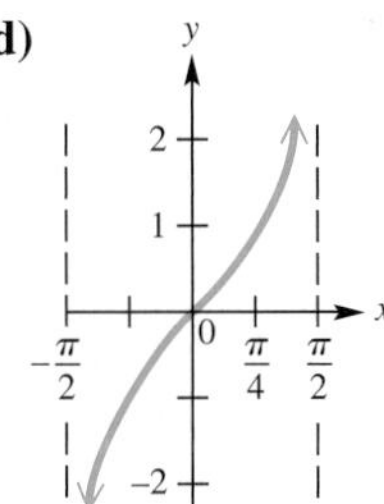

(e)

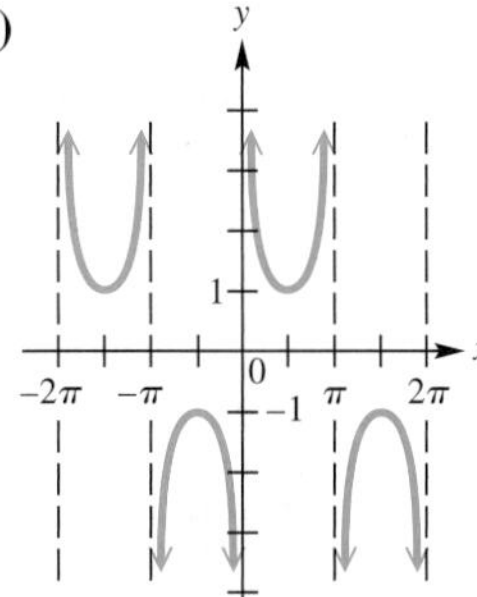

(f)

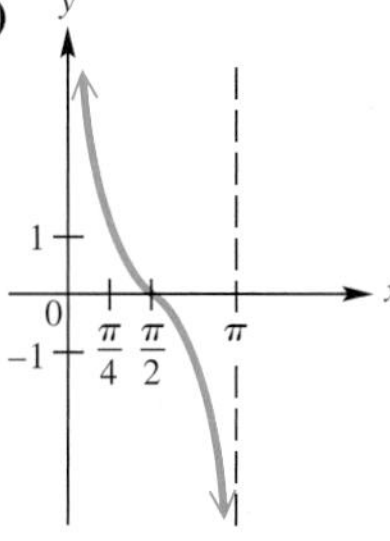

20. *Connecting Graphs with Equations* Determine the simplest form of an equation for each graph. Choose $b > 0$, and include no phase shifts.

(a)

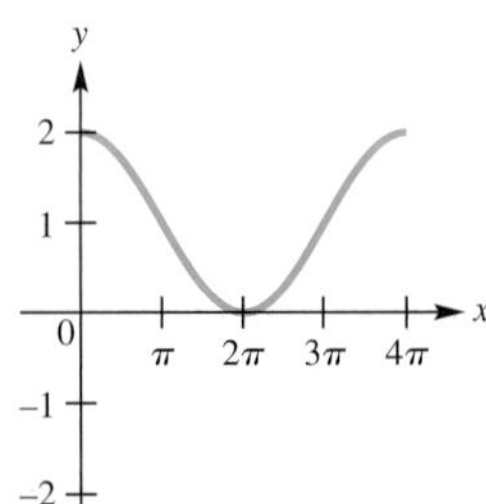

(b)

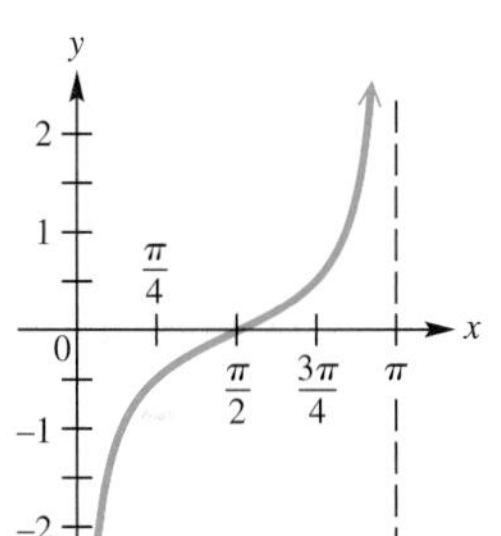

21. Give a short answer to each of the following.

(a) What is the domain of the cosine function?
(b) What is the range of the sine function?
(c) What is the least positive value for which the tangent function is undefined?
(d) What is the range of the secant function?

22. Consider the function $y = 3 - 6 \sin\left(2x + \frac{\pi}{2}\right)$.

(a) What is its period?
(b) What is the amplitude of its graph?
(c) What is its range?
(d) What is the y-intercept of its graph?
(e) What is its phase shift?

Graph each function over a two-period interval. Identify asymptotes when applicable.

23. $y = \sin(2x + \pi)$

24. $y = 2 + \cos x$

25. $y = -1 + 2 \sin(x + \pi)$

26. $y = \tan\left(x - \frac{\pi}{2}\right)$

27. $y = -2 - \cot\left(x - \frac{\pi}{2}\right)$

28. $y = -\csc 2x$

(Modeling) *Solve each problem.*

29. *Average Monthly Temperature* The average monthly temperature (in °F) in San Antonio, Texas, can be modeled using the circular function

$$f(x) = 16.5 \sin\left[\frac{\pi}{6}(x - 4)\right] + 67.5,$$

where x is the month and $x = 1$ corresponds to January. (*Source: World Almanac and Book of Facts.*)

(a) Graph f in the window $[0, 25]$ by $[40, 90]$.

(b) Determine the amplitude, period, phase shift, and vertical translation of f.

(c) What is the average monthly temperature for the month of December?

(d) Determine the minimum and maximum average monthly temperatures and the months when they occur.

(e) What would be an approximation for the average *yearly* temperature in San Antonio? How is this related to the vertical translation of the sine function in the formula for f?

30. *Spring Motion* The height of a weight attached to a spring is

$$s(t) = -4 \cos 8\pi t \text{ inches after } t \text{ seconds.}$$

(a) Find the maximum height that the weight rises above the equilibrium position of $s(t) = 0$.

(b) When does the weight first reach its maximum height if $t \geq 0$?

(c) What are the frequency and period?

[6.3, 6.4]

29. (a)

$f(x) = 16.5 \sin\left[\frac{\pi}{6}(x-4)\right] + 67.5$

90

0 25

40

(b) 16.5; 12; 4 to the right; 67.5 up

(c) approximately 53°F

(d) 51°F in January; 84°F in July

(e) approximately 67.5°F; This is the vertical translation.

[6.7]

30. (a) 4 in. **(b)** after $\frac{1}{8}$ sec

(c) 4 cycles per sec; $\frac{1}{4}$ sec

7 Trigonometric Identities and Equations

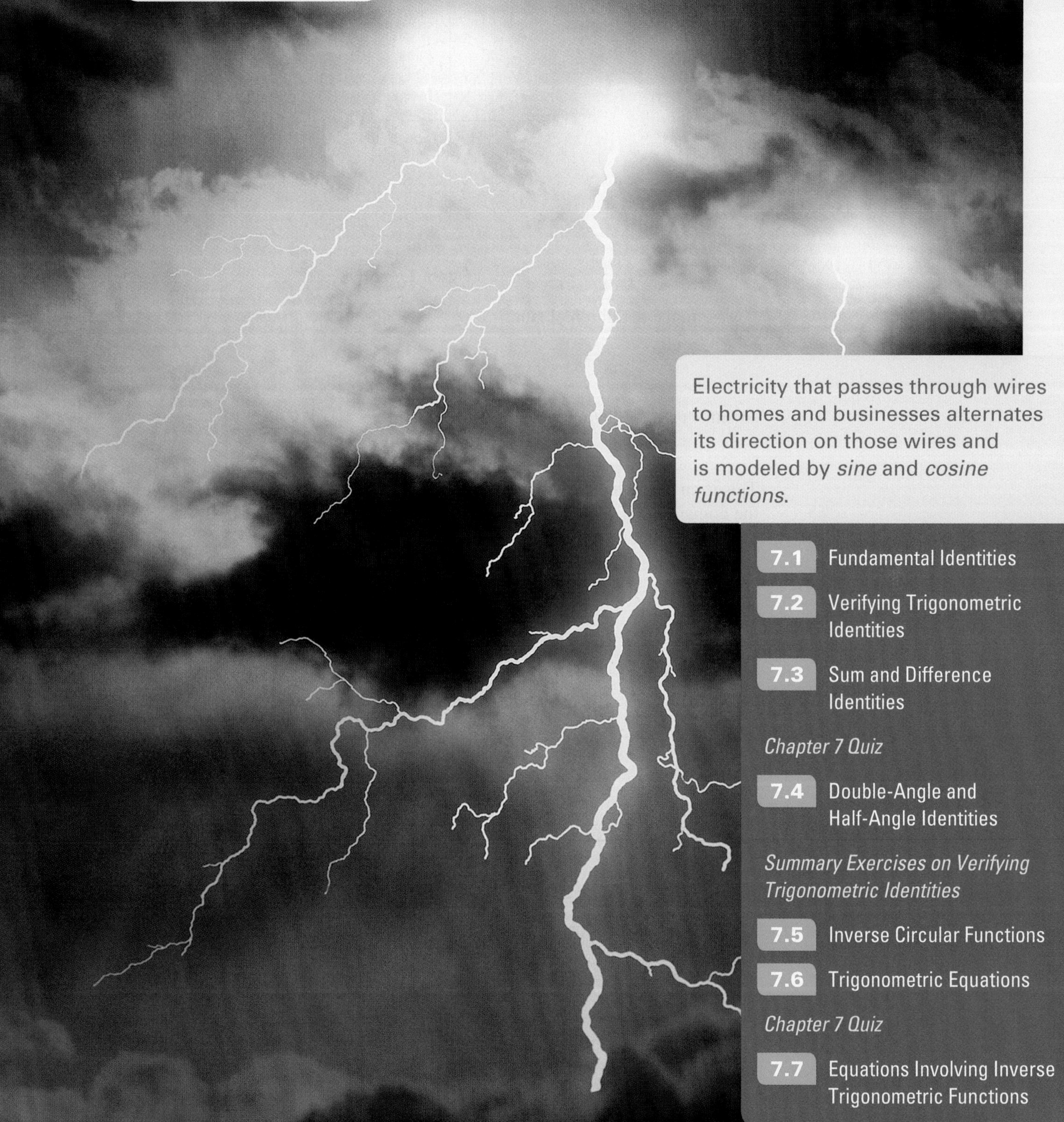

Electricity that passes through wires to homes and businesses alternates its direction on those wires and is modeled by *sine* and *cosine functions*.

7.1 Fundamental Identities

- Fundamental Identities
- Using the Fundamental Identities

Fundamental Identities As suggested by the circle shown in **Figure 1,** an angle θ having the point (x, y) on its terminal side has a corresponding angle $-\theta$ with the point $(x, -y)$ on its terminal side.

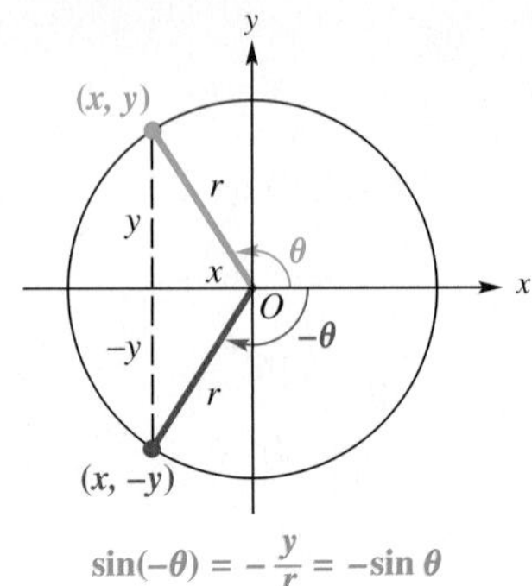

Figure 1

From the definition of sine,

$$\sin(-\theta) = \frac{-y}{r} \quad \text{and} \quad \sin\theta = \frac{y}{r}, \qquad \text{(Section 5.2)}$$

so $\sin(-\theta)$ and $\sin\theta$ are negatives of each other.

$$\mathbf{\sin(-\theta) = -\sin\theta}$$

This is an example of an **identity,** an equation that is satisfied by *every* value in the domain of its variable. (See **Section 1.1.**) Some examples of identities from algebra are

$$x^2 - y^2 = (x + y)(x - y),$$

$$x(x + y) = x^2 + xy, \qquad \text{(Sections R.2 and R.4)}$$

and

$$x^2 + 2xy + y^2 = (x + y)^2.$$

Figure 1 shows an angle θ in quadrant II, but the same result holds for θ in any quadrant. The figure also suggests the following identity.

$$\cos(-\theta) = \frac{x}{r} \quad \text{and} \quad \cos\theta = \frac{x}{r} \qquad \text{(Section 5.2)}$$

$$\mathbf{\cos(-\theta) = \cos\theta}$$

We use the identities for $\sin(-\theta)$ and $\cos(-\theta)$ to find $\tan(-\theta)$ in terms of $\tan\theta$.

$$\tan(-\theta) = \frac{\sin(-\theta)}{\cos(-\theta)} = \frac{-\sin\theta}{\cos\theta} = -\frac{\sin\theta}{\cos\theta}$$

$$\mathbf{\tan(-\theta) = -\tan\theta}$$

Similar reasoning gives the remaining three **negative-angle** or **negative-number identities,** which, together with the reciprocal, quotient, and Pythagorean identities from **Chapter 5,** make up the **fundamental identities.** For reference, we summarize these identities in the box at the top of the next page.

NOTE In trigonometric identities, θ can be an angle in degrees, a real number, or a variable.

Teaching Tip Encourage students to memorize the identities presented in this section as well as subsequent sections. Point out that numerical values can be used to help check whether or not an identity was recalled correctly.

Fundamental Identities

Reciprocal Identities

$$\cot\theta = \frac{1}{\tan\theta} \qquad \sec\theta = \frac{1}{\cos\theta} \qquad \csc\theta = \frac{1}{\sin\theta}$$

Quotient Identities

$$\tan\theta = \frac{\sin\theta}{\cos\theta} \qquad \cot\theta = \frac{\cos\theta}{\sin\theta}$$

Pythagorean Identities

$$\sin^2\theta + \cos^2\theta = 1 \qquad \tan^2\theta + 1 = \sec^2\theta \qquad 1 + \cot^2\theta = \csc^2\theta$$

Negative-Angle Identities

$$\sin(-\theta) = -\sin\theta \qquad \cos(-\theta) = \cos\theta \qquad \tan(-\theta) = -\tan\theta$$

$$\csc(-\theta) = -\csc\theta \qquad \sec(-\theta) = \sec\theta \qquad \cot(-\theta) = -\cot\theta$$

NOTE We will also use alternative forms of the fundamental identities. *For example, two other forms of* $\sin^2\theta + \cos^2\theta = 1$ *are*

$$\sin^2\theta = 1 - \cos^2\theta \quad \textit{and} \quad \cos^2\theta = 1 - \sin^2\theta.$$

Using the Fundamental Identities We can use these identities to find the values of other trigonometric functions from the value of a given trigonometric function.

Classroom Example 1

If $\cos\theta = \frac{5}{8}$ and θ is in quadrant IV, find each function value.

(a) $\sin\theta$ **(b)** $\tan\theta$

(c) $\sec(-\theta)$

Answers:

(a) $-\frac{\sqrt{39}}{8}$ **(b)** $-\frac{\sqrt{39}}{5}$

(c) $\frac{8}{5}$

EXAMPLE 1 Finding Trigonometric Function Values Given One Value and the Quadrant

If $\tan\theta = -\frac{5}{3}$ and θ is in quadrant II, find each function value.

(a) $\sec\theta$ **(b)** $\sin\theta$ **(c)** $\cot(-\theta)$

SOLUTION

(a) We use an identity that relates the tangent and secant functions. Remember that $\sec\theta$ will be negative because θ is in quadrant II.

$$\tan^2\theta + 1 = \sec^2\theta \quad \text{Pythagorean identity}$$

$$\left(-\frac{5}{3}\right)^2 + 1 = \sec^2\theta \quad \tan\theta = -\tfrac{5}{3}$$

$$\frac{25}{9} + 1 = \sec^2\theta \quad \text{Square } -\tfrac{5}{3}.$$

$$\frac{34}{9} = \sec^2\theta \quad \text{Add; } 1 = \tfrac{25}{25}$$

Teaching Tip Warn students that the given information in **Example 1,** $\tan\theta = -\frac{5}{3} = \frac{-5}{3}$, does not mean that $\sin\theta = -5$ and $\cos\theta = 3$. Ask them why these values cannot be correct.

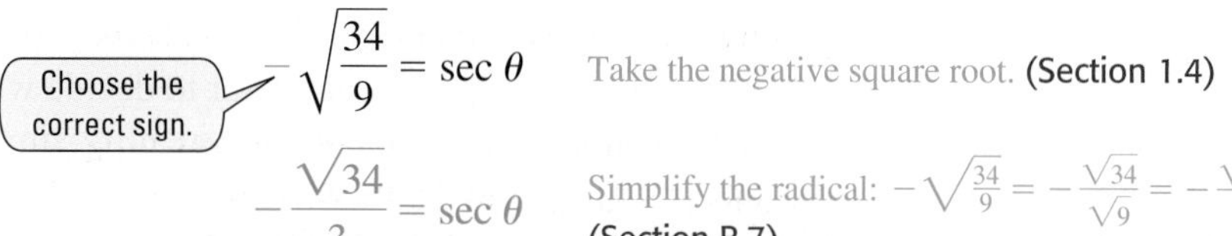
Choose the correct sign.

$$-\sqrt{\frac{34}{9}} = \sec\theta \quad \text{Take the negative square root. (Section 1.4)}$$

$$-\frac{\sqrt{34}}{3} = \sec\theta \quad \text{Simplify the radical: } -\sqrt{\tfrac{34}{9}} = -\tfrac{\sqrt{34}}{\sqrt{9}} = -\tfrac{\sqrt{34}}{3} \text{ (Section R.7)}$$

(b)

$$\tan\theta = \frac{\sin\theta}{\cos\theta} \quad \text{Quotient identity}$$

$$\cos\theta\tan\theta = \sin\theta \quad \text{Multiply each side by } \cos\theta.$$

$$\left(\frac{1}{\sec\theta}\right)\tan\theta = \sin\theta \quad \text{Reciprocal identity}$$

$$\left(-\frac{3\sqrt{34}}{34}\right)\left(-\frac{5}{3}\right) = \sin\theta$$

$\frac{1}{\sec\theta} = \frac{1}{-\frac{\sqrt{34}}{3}} = -\frac{3}{\sqrt{34}} = -\frac{3}{\sqrt{34}}\cdot\frac{\sqrt{34}}{\sqrt{34}} = -\frac{3\sqrt{34}}{34}$ (Section R.5); $\tan\theta = -\frac{5}{3}$, as given

$$\sin\theta = \frac{5\sqrt{34}}{34} \quad \text{Multiply and rewrite.}$$

(c)

$$\cot(-\theta) = \frac{1}{\tan(-\theta)} \quad \text{Reciprocal identity}$$

$$\cot(-\theta) = \frac{1}{-\tan\theta} \quad \text{Negative-angle identity}$$

$$\cot(-\theta) = \frac{1}{-\left(-\frac{5}{3}\right)} = \frac{3}{5}$$

Use $\tan\theta = -\frac{5}{3}$, and simplify the complex fraction.

✔ *Now Try Exercises 7, 15, and 31.*

CAUTION ***To avoid a common error, when taking the square root, be sure to choose the sign based on the quadrant of θ and the function being evaluated.***

Classroom Example 2
Write $\tan\theta$ in terms of $\cos\theta$.

Answer:

$\tan\theta = \frac{\pm\sqrt{1-\cos^2\theta}}{\cos\theta}$

EXAMPLE 2 Writing One Trigonometric Function in Terms of Another

Write $\cos x$ in terms of $\tan x$.

SOLUTION Since $\sec x$ is related to both $\cos x$ and $\tan x$ by identities, we start with $1 + \tan^2 x = \sec^2 x$.

$$1 + \tan^2 x = \sec^2 x \quad \text{Pythagorean identity}$$

$$\frac{1}{1+\tan^2 x} = \frac{1}{\sec^2 x} \quad \text{Take reciprocals.}$$

$$\frac{1}{1+\tan^2 x} = \cos^2 x \quad \text{The reciprocal of } \sec^2 x \text{ is } \cos^2 x.$$

Remember both the positive and negative roots.

$$\pm\sqrt{\frac{1}{1+\tan^2 x}} = \cos x \quad \text{Take the square root of each side.}$$

$$\cos x = \frac{\pm 1}{\sqrt{1+\tan^2 x}}$$

Quotient rule for radicals: $\sqrt[n]{\frac{a}{b}} = \frac{\sqrt[n]{a}}{\sqrt[n]{b}}$ (Section R.7); rewrite.

$$\cos x = \frac{\pm\sqrt{1+\tan^2 x}}{1+\tan^2 x}$$

Rationalize the denominator. (Section R.7)

The choice of the + sign or the − sign is made depending on the quadrant of x.

✔ *Now Try Exercise 53.*

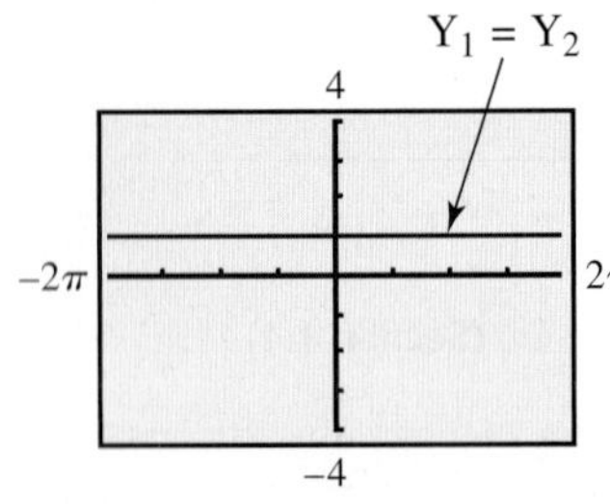

Figure 2

We can use a graphing calculator to decide whether two functions are identical. See **Figure 2,** which supports the identity $\sin^2 x + \cos^2 x = 1$. Y_1 is defined as $\sin^2 x + \cos^2 x$, and Y_2 is defined as 1. With an identity, you should see no difference between the two graphs. ■

Each of the functions $\tan\theta$, $\cot\theta$, $\sec\theta$, and $\csc\theta$ can easily be expressed in terms of $\sin\theta$, $\cos\theta$, or both. We often make such substitutions in an expression to simplify it.

Classroom Example 3

Write $\frac{1+\tan^2\theta}{1-\sec^2\theta}$ in terms of $\sin\theta$ and $\cos\theta$, and then simplify the expression so that no quotients appear.

Answer: $-\csc^2\theta$

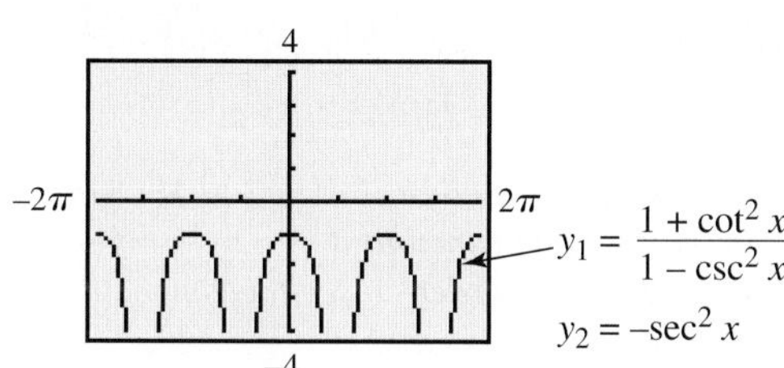

The graph supports the result in **Example 3.** The graphs of y_1 and y_2 coincide.

EXAMPLE 3 Rewriting an Expression in Terms of Sine and Cosine

Write $\frac{1+\cot^2\theta}{1-\csc^2\theta}$ in terms of $\sin\theta$ and $\cos\theta$, and then simplify the expression so that no quotients appear.

SOLUTION

$$\frac{1+\cot^2\theta}{1-\csc^2\theta}$$

$$= \frac{1+\frac{\cos^2\theta}{\sin^2\theta}}{1-\frac{1}{\sin^2\theta}} \quad \text{Quotient identities}$$

$$= \frac{\left(1+\frac{\cos^2\theta}{\sin^2\theta}\right)\sin^2\theta}{\left(1-\frac{1}{\sin^2\theta}\right)\sin^2\theta} \quad \text{Simplify the complex fraction by multiplying both numerator and denominator by the LCD. (Section R.5)}$$

$$= \frac{\sin^2\theta+\cos^2\theta}{\sin^2\theta-1} \quad \text{Distributive property: } a(b+c)=ab+ac$$

$$= \frac{1}{-\cos^2\theta} \quad \text{Pythagorean identities}$$

$$= -\sec^2\theta \quad \text{Reciprocal identity}$$

Now Try Exercise 65.

CAUTION ***When working with trigonometric expressions and identities, be sure to write the argument of the function.*** For example, we would *not* write $\sin^2 + \cos^2 = 1$. An argument such as θ is necessary in this identity.

7.1 Exercises

1. -2.6
2. -0.65
3. 0.625
4. -0.75
5. $\frac{2}{3}$
6. $\frac{1}{5}$
7. $\frac{\sqrt{7}}{4}$
8. $\frac{\sqrt{11}}{6}$
9. $-\frac{5\sqrt{26}}{26}$
10. $-\frac{3\sqrt{10}}{10}$

Concept Check In Exercises 1–6, use identities to fill in the blanks.

1. If $\tan\theta = 2.6$, then $\tan(-\theta) =$ ______.
2. If $\cos\theta = -0.65$, then $\cos(-\theta) =$ ______.
3. If $\tan\theta = 1.6$, then $\cot\theta =$ ______.
4. If $\cos\theta = 0.8$ and $\sin\theta = 0.6$, then $\tan(-\theta) =$ ______.
5. If $\sin\theta = \frac{2}{3}$, then $-\sin(-\theta) =$ ______.
6. If $\cos\theta = -\frac{1}{5}$, then $-\cos(-\theta) =$ ______.

Find $\sin\theta$. ***See Example 1.***

7. $\cos\theta = \frac{3}{4}$, θ in quadrant I
8. $\cos\theta = \frac{5}{6}$, θ in quadrant I
9. $\cot\theta = -\frac{1}{5}$, θ in quadrant IV
10. $\cot\theta = -\frac{1}{3}$, θ in quadrant IV

11. $-\frac{2\sqrt{5}}{5}$ 12. $-\frac{\sqrt{33}}{6}$
13. $-\frac{\sqrt{15}}{5}$ 14. $-\frac{\sqrt{77}}{11}$
15. $-\frac{\sqrt{105}}{11}$ 16. $-\frac{3\sqrt{5}}{7}$
17. $-\frac{4}{9}$ 18. $-\frac{5}{8}$

20. There is no number or angle whose cosine is 3.

21. $-\sin x$ 22. odd
23. $\cos x$ 24. even
25. $-\tan x$ 26. odd

27. $f(-x) = f(x)$
28. $f(-x) = -f(x)$
29. $f(-x) = -f(x)$
30. $f(-x) = -f(x)$

31. $\cos\theta = -\frac{\sqrt{5}}{3}$; $\tan\theta = -\frac{2\sqrt{5}}{5}$; $\cot\theta = -\frac{\sqrt{5}}{2}$; $\sec\theta = -\frac{3\sqrt{5}}{5}$; $\csc\theta = \frac{3}{2}$

32. $\sin\theta = \frac{2\sqrt{6}}{5}$; $\tan\theta = 2\sqrt{6}$; $\cot\theta = \frac{\sqrt{6}}{12}$; $\sec\theta = 5$; $\csc\theta = \frac{5\sqrt{6}}{12}$

33. $\sin\theta = -\frac{\sqrt{17}}{17}$; $\cos\theta = \frac{4\sqrt{17}}{17}$; $\cot\theta = -4$; $\sec\theta = \frac{\sqrt{17}}{4}$; $\csc\theta = -\sqrt{17}$

34. $\sin\theta = -\frac{2}{5}$; $\cos\theta = -\frac{\sqrt{21}}{5}$; $\tan\theta = \frac{2\sqrt{21}}{21}$; $\cot\theta = \frac{\sqrt{21}}{2}$; $\sec\theta = -\frac{5\sqrt{21}}{21}$

35. $\sin\theta = \frac{3}{5}$; $\cos\theta = \frac{4}{5}$; $\tan\theta = \frac{3}{4}$; $\sec\theta = \frac{5}{4}$; $\csc\theta = \frac{5}{3}$

36. $\cos\theta = -\frac{3}{5}$; $\tan\theta = \frac{4}{3}$; $\cot\theta = \frac{3}{4}$; $\sec\theta = -\frac{5}{3}$; $\csc\theta = -\frac{5}{4}$

37. $\sin\theta = -\frac{\sqrt{7}}{4}$; $\cos\theta = \frac{3}{4}$; $\tan\theta = -\frac{\sqrt{7}}{3}$; $\cot\theta = -\frac{3\sqrt{7}}{7}$; $\csc\theta = -\frac{4\sqrt{7}}{7}$

38. $\sin\theta = \frac{\sqrt{15}}{4}$; $\tan\theta = -\sqrt{15}$; $\cot\theta = -\frac{\sqrt{15}}{15}$; $\sec\theta = -4$; $\csc\theta = \frac{4\sqrt{15}}{15}$

11. $\cos(-\theta) = \frac{\sqrt{5}}{5}$, $\tan\theta < 0$

12. $\cos(-\theta) = \frac{\sqrt{3}}{6}$, $\cot\theta < 0$

13. $\tan\theta = -\frac{\sqrt{6}}{2}$, $\cos\theta > 0$

14. $\tan\theta = -\frac{\sqrt{7}}{2}$, $\sec\theta > 0$

15. $\sec\theta = \frac{11}{4}$, $\cot\theta < 0$

16. $\sec\theta = \frac{7}{2}$, $\tan\theta < 0$

17. $\csc\theta = -\frac{9}{4}$

18. $\csc\theta = -\frac{8}{5}$

19. Why is it unnecessary to give the quadrant of θ in **Exercises 17 and 18?**

20. *Concept Check* What is **WRONG** with the statement of this problem?

Find $\cos(-\theta)$ *if* $\cos\theta = 3$.

Relating Concepts

For individual or collaborative investigation *(Exercises 21–26)*

Recall from **Section 2.7** *that a function is an* **even function** *if* $f(-x) = f(x)$ *for all x in the domain of f. Similarly, a function is an* **odd function** *if* $f(-x) = -f(x)$ *for all x in the domain of f.* **Work Exercises 21–26 in order,** *to see the connection between the negative-angle identities and even and odd functions.*

21. Complete the statement: $\sin(-x) =$ ______.

22. Is the function $f(x) = \sin x$ *even* or *odd?*

23. Complete the statement: $\cos(-x) =$ ______.

24. Is the function $f(x) = \cos x$ *even* or *odd?*

25. Complete the statement: $\tan(-x) =$ ______.

26. Is the function $f(x) = \tan x$ *even* or *odd?*

Concept Check For each graph of a circular function $y = f(x)$, *determine whether* $f(-x) = f(x)$ *or* $f(-x) = -f(x)$ *is true.*

27.

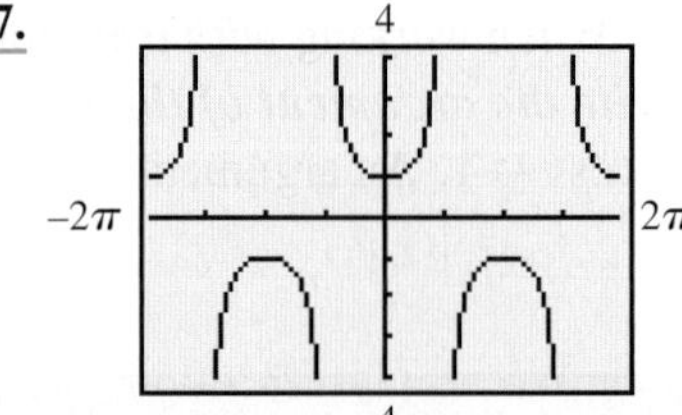

28.

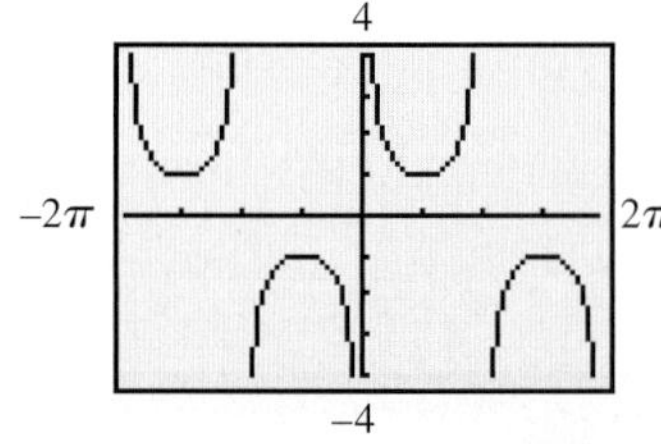

29.

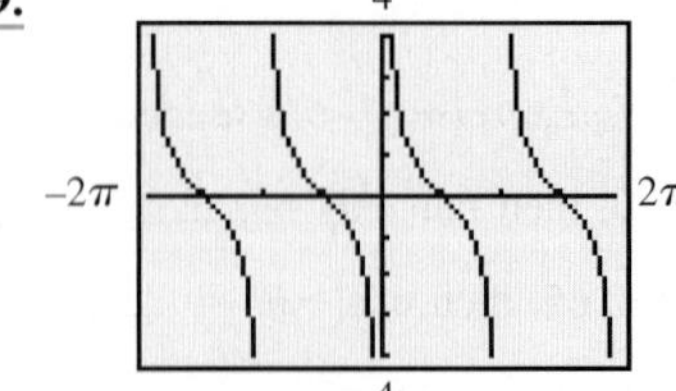

30.

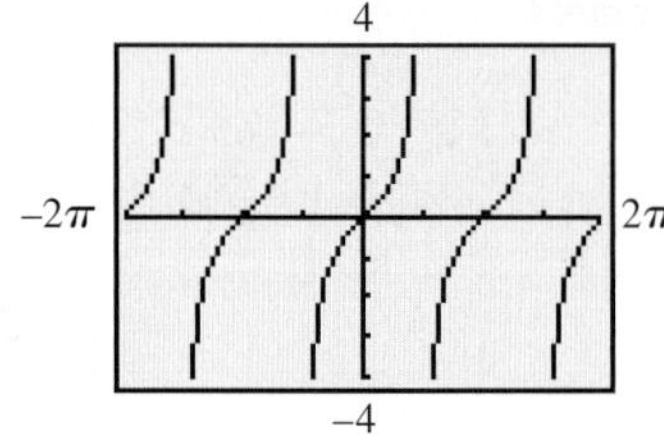

Find the remaining five trigonometric functions of θ. **See Example 1.**

31. $\sin\theta = \frac{2}{3}$, θ in quadrant II

32. $\cos\theta = \frac{1}{5}$, θ in quadrant I

33. $\tan\theta = -\frac{1}{4}$, θ in quadrant IV

34. $\csc\theta = -\frac{5}{2}$, θ in quadrant III

35. $\cot\theta = \frac{4}{3}$, $\sin\theta > 0$

36. $\sin\theta = -\frac{4}{5}$, $\cos\theta < 0$

37. $\sec\theta = \frac{4}{3}$, $\sin\theta < 0$

38. $\cos\theta = -\frac{1}{4}$, $\sin\theta > 0$

39. B 40. D
41. E 42. C
43. A

44. C
45. A 46. E
47. D 48. B

51. $\sin\theta = \frac{\pm\sqrt{2x+1}}{x+1}$

52. $\tan\theta = \frac{\pm 2\sqrt{2x+4}}{x}$

53. $\sin x = \pm\sqrt{1-\cos^2 x}$

54. $\cot x = \frac{\pm\sqrt{1-\sin^2 x}}{\sin x}$

55. $\tan x = \pm\sqrt{\sec^2 x - 1}$

56. $\cot x = \pm\sqrt{\csc^2 x - 1}$

57. $\csc x = \frac{\pm\sqrt{1-\cos^2 x}}{1-\cos^2 x}$

58. $\sec x = \frac{\pm\sqrt{1-\sin^2 x}}{1-\sin^2 x}$

In Exercises 59–84, there may be more than one possible answer.

59. $\cos\theta$ 60. $\sin\theta$
61. 1 62. 1
63. $\cot\theta$ 64. $\tan\theta$
65. $\cos^2\theta$ 66. $\csc^2\theta$
67. $\sec\theta - \cos\theta$ 68. $\tan^2\theta$
69. $-\cot\theta + 1$ 70. $\tan\theta + 1$
71. $\sin^2\theta\cos^2\theta$
72. $\sin^2\theta\cos^2\theta$
73. $\tan\theta\sin\theta$
74. $\cot\theta\cos\theta$
75. $\cot\theta - \tan\theta$
76. $\tan\theta - \cot\theta$

Concept Check For each expression in Column I, choose the expression from Column II that completes an identity.

I	II
39. $\frac{\cos x}{\sin x} =$ ______	**A.** $\sin^2 x + \cos^2 x$
40. $\tan x =$ ______	**B.** $\cot x$
41. $\cos(-x) =$ ______	**C.** $\sec^2 x$
42. $\tan^2 x + 1 =$ ______	**D.** $\frac{\sin x}{\cos x}$
43. $1 =$ ______	**E.** $\cos x$

Concept Check For each expression in Column I, choose the expression from Column II that completes an identity. You may have to rewrite one or both expressions.

I	II
44. $-\tan x \cos x =$ ______	**A.** $\frac{\sin^2 x}{\cos^2 x}$
45. $\sec^2 x - 1 =$ ______	**B.** $\frac{1}{\sec^2 x}$
46. $\frac{\sec x}{\csc x} =$ ______	**C.** $\sin(-x)$
47. $1 + \sin^2 x =$ ______	**D.** $\csc^2 x - \cot^2 x + \sin^2 x$
48. $\cos^2 x =$ ______	**E.** $\tan x$

49. A student writes "$1 + \cot^2 = \csc^2$." Comment on this student's work.

50. A student makes the following claim: "Since $\sin^2\theta + \cos^2\theta = 1$, I should be able to also say that $\sin\theta + \cos\theta = 1$ if I take the square root of each side." Comment on this student's statement.

51. *Concept Check* Suppose that $\cos\theta = \frac{x}{x+1}$. Find an expression in x for $\sin\theta$.

52. *Concept Check* Suppose that $\sec\theta = \frac{x+4}{x}$. Find an expression in x for $\tan\theta$.

Perform each transformation. ***See Example 2.***

53. Write $\sin x$ in terms of $\cos x$.

54. Write $\cot x$ in terms of $\sin x$.

55. Write $\tan x$ in terms of $\sec x$.

56. Write $\cot x$ in terms of $\csc x$.

57. Write $\csc x$ in terms of $\cos x$.

58. Write $\sec x$ in terms of $\sin x$.

Write each expression in terms of sine and cosine, and simplify so that no quotients appear in the final expression and all functions are of θ only. ***See Example 3.***

59. $\cot\theta\sin\theta$

60. $\tan\theta\cos\theta$

61. $\sec\theta\cot\theta\sin\theta$

62. $\csc\theta\cos\theta\tan\theta$

63. $\cos\theta\csc\theta$

64. $\sin\theta\sec\theta$

65. $\sin^2\theta(\csc^2\theta - 1)$

66. $\cot^2\theta(1 + \tan^2\theta)$

67. $(1 - \cos\theta)(1 + \sec\theta)$

68. $(\sec\theta - 1)(\sec\theta + 1)$

69. $\frac{1 + \tan(-\theta)}{\tan(-\theta)}$

70. $\frac{1 + \cot\theta}{\cot\theta}$

71. $\frac{1 - \cos^2(-\theta)}{1 + \tan^2(-\theta)}$

72. $\frac{1 - \sin^2(-\theta)}{1 + \cot^2(-\theta)}$

73. $\sec\theta - \cos\theta$

74. $\csc\theta - \sin\theta$

75. $(\sec\theta + \csc\theta)(\cos\theta - \sin\theta)$

76. $(\sin\theta - \cos\theta)(\csc\theta + \sec\theta)$

77. $\cos^2\theta$ 78. $-\sin^2\theta$
79. $\tan^2\theta$ 80. $\tan^4\theta$
81. $\sec^2\theta$ 82. $-\tan^2\theta$
83. $-\sec\theta$ 84. $-\sin\theta$

85. $\frac{25\sqrt{6}-60}{12}$; $\frac{-25\sqrt{6}-60}{12}$

86. $\frac{2\sqrt{2}+8}{9}$; $\frac{-2\sqrt{2}+8}{9}$

87. $y = -\sin(2x)$
88. It is the negative of $y = \sin(2x)$.
89. $y = \cos(4x)$
90. It is the same function.
91. (a) $y = -\sin(4x)$
(b) $y = \cos(2x)$
(c) $y = 5\sin(3x)$

93. identity
94. not an identity
95. not an identity
96. identity

77. $\sin\theta(\csc\theta - \sin\theta)$

78. $\cos\theta(\cos\theta - \sec\theta)$

79. $\dfrac{1 + \tan^2\theta}{1 + \cot^2\theta}$

80. $\dfrac{\sec^2\theta - 1}{\csc^2\theta - 1}$

81. $\sin^2(-\theta) + \tan^2(-\theta) + \cos^2(-\theta)$

82. $-\sec^2(-\theta) + \sin^2(-\theta) + \cos^2(-\theta)$

83. $\dfrac{\csc\theta}{\cot(-\theta)}$

84. $\dfrac{\tan(-\theta)}{\sec\theta}$

Work each problem.

85. Let $\cos x = \frac{1}{5}$. Find all possible values of $\frac{\sec x - \tan x}{\sin x}$.

86. Let $\csc x = -3$. Find all possible values of $\frac{\sin x + \cos x}{\sec x}$.

Relating Concepts

For individual or collaborative investigation *(Exercises 87–92)*

In ***Chapter 6*** *we graphed functions defined by*

$$y = c + a \cdot f[b(x - d)]$$

with the assumption that $b > 0$. *To see what happens when* $b < 0$, ***work Exercises 87–92 in order.***

87. Use a negative-angle identity to write $y = \sin(-2x)$ as a function of $2x$.

88. How is your answer to **Exercise 87** related to $y = \sin(2x)$?

89. Use a negative-angle identity to write $y = \cos(-4x)$ as a function of $4x$.

90. How is your answer to **Exercise 89** related to $y = \cos(4x)$?

91. Use your results from **Exercises 87–90** to rewrite the following with a positive value of b.

(a) $y = \sin(-4x)$ **(b)** $y = \cos(-2x)$ **(c)** $y = -5\sin(-3x)$

92. Write a short response to this statement, which is often used by one of the authors of this text in trigonometry classes: *Students who tend to ignore negative signs should enjoy graphing functions involving the cosine and the secant.*

Use a graphing calculator to make a conjecture about whether each equation is an identity.

93. $\cos 2x = 1 - 2\sin^2 x$

94. $2\sin x = \sin 2x$

95. $\sin x = \sqrt{1 - \cos^2 x}$

96. $\cos 2x = \cos^2 x - \sin^2 x$

7.2 Verifying Trigonometric Identities

- Strategies
- Verifying Identities by Working with One Side
- Verifying Identities by Working with Both Sides

Strategies One of the skills required for more advanced work in mathematics, especially in calculus, is the ability to use identities to write expressions in alternative forms. We develop this skill by using the fundamental identities to verify that a trigonometric equation is an identity (for those values of the variable for which it is defined). Here are some helpful hints.

LOOKING AHEAD TO CALCULUS

Trigonometric identities are used in calculus to simplify trigonometric expressions, determine derivatives of trigonometric functions, and change the form of some integrals.

Teaching Tip There is no substitute for experience when it comes to verifying identities. Guide students through several examples, giving hints such as "Apply a reciprocal identity" or "Use a different form of the Pythagorean identity $\sin^2\theta + \cos^2\theta = 1$."

Hints for Verifying Identities

1. ***Learn the fundamental identities given in Section 7.1.*** Whenever you see either side of a fundamental identity, the other side should come to mind. ***Also, be aware of equivalent forms of the fundamental identities.*** For example,

$$\sin^2\theta = 1 - \cos^2\theta \quad \text{is an alternative form of} \quad \sin^2\theta + \cos^2\theta = 1.$$

2. ***Try to rewrite the more complicated side*** of the equation so that it is identical to the simpler side.
3. ***It is sometimes helpful to express all trigonometric functions in the equation in terms of sine and cosine*** and then simplify the result.
4. ***Usually, any factoring or indicated algebraic operations should be performed.*** These *algebraic* identities are often used in verifying trigonometric identities.

$$(a+b)^2 = a^2 + 2ab + b^2 \qquad (a-b)^2 = a^2 - 2ab + b^2$$

$$a^3 - b^3 = (a-b)(a^2+ab+b^2) \qquad a^3 + b^3 = (a+b)(a^2-ab+b^2)$$

$$a^2 - b^2 = (a+b)(a-b)$$

(Sections R.3 and R.4)

For example, the expression

$$\sin^2 x + 2\sin x + 1 \quad \text{can be factored as} \quad (\sin x + 1)^2.$$

The sum or difference of two trigonometric expressions can be found in the same way as any other rational expression. For example,

$$\frac{1}{\sin\theta} + \frac{1}{\cos\theta} = \frac{1\cdot\cos\theta}{\sin\theta\cos\theta} + \frac{1\cdot\sin\theta}{\cos\theta\sin\theta} \qquad \text{Write with the LCD.}$$

$$= \frac{\cos\theta + \sin\theta}{\sin\theta\cos\theta}. \qquad \frac{a}{c} + \frac{b}{c} = \frac{a+b}{c} \text{ (Section R.5)}$$

5. ***As you select substitutions, keep in mind the side you are not changing, because it represents your goal.*** For example, to verify the identity

$$\tan^2 x + 1 = \frac{1}{\cos^2 x},$$

try to think of an identity that relates $\tan x$ to $\cos x$. In this case, since $\sec x = \frac{1}{\cos x}$ and $\sec^2 x = \tan^2 x + 1$, the secant function is the best link between the two sides.

6. If an expression contains $1 + \sin x$, ***multiplying both numerator and denominator*** by $1 - \sin x$ would give $1 - \sin^2 x$, which could be replaced with $\cos^2 x$. Similar procedures apply for $1 - \sin x$, $1 + \cos x$, and $1 - \cos x$.

CAUTION ***The procedure for verifying identities is not the same as that of solving equations.*** Techniques used in solving equations, such as adding the same term to each side, and multiplying each side by the same term, should not be used when working with identities.

Verifying Identities by Working with One Side To avoid the temptation to use algebraic properties of equations to verify identities, ***one strategy is to work with only one side and rewrite it to match the other side.***

Classroom Example 1
Verify that the following equation is an identity.

$$\sec\theta(\sin\theta + \cos\theta) = 1 + \tan\theta$$

Answer: The verification appears in the *Solutions to Classroom Examples.*

EXAMPLE 1 Verifying an Identity (Working with One Side)

Verify that the following equation is an identity.

$$\cot\theta + 1 = \csc\theta(\cos\theta + \sin\theta)$$

SOLUTION We use the fundamental identities from **Section 7.1** to rewrite one side of the equation so that it is identical to the other side. Since the right side is more complicated, we work with it, as suggested in Hint 2, and use Hint 3 to change all functions to expressions involving sine or cosine.

Steps	Reasons
$\underbrace{\csc\theta(\cos\theta + \sin\theta)}_{\text{Right side of given equation}} = \frac{1}{\sin\theta}(\cos\theta + \sin\theta)$	$\csc\theta = \frac{1}{\sin\theta}$ (Section 7.1)
$= \frac{\cos\theta}{\sin\theta} + \frac{\sin\theta}{\sin\theta}$	Distributive property: $a(b + c) = ab + ac$
$= \underbrace{\cot\theta + 1}_{\text{Left side of given equation}}$	$\frac{\cos\theta}{\sin\theta} = \cot\theta$; $\frac{\sin\theta}{\sin\theta} = 1$

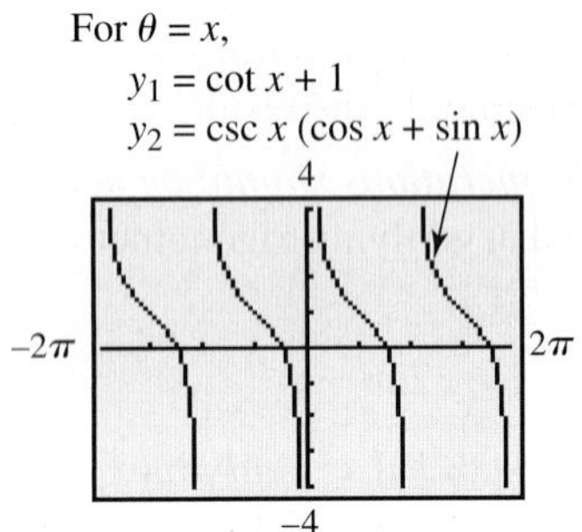

The graphs coincide, which supports the conclusion in **Example 1.**

The given equation is an identity. The right side of the equation is identical to the left side.

Now Try Exercise 35.

Classroom Example 2
Verify that the following equation is an identity.

$$\cot^2 x(\tan^2 x + 1) = \csc^2 x$$

Answer: The verification appears in the *Solutions to Classroom Examples.*

EXAMPLE 2 Verifying an Identity (Working with One Side)

Verify that the following equation is an identity.

$$\tan^2 x(1 + \cot^2 x) = \frac{1}{1 - \sin^2 x}$$

SOLUTION We work with the more complicated left side, as suggested in Hint 2. Again, we use the fundamental identities from **Section 7.1.**

Steps	Reasons
$\underbrace{\tan^2 x(1 + \cot^2 x)}_{\text{Left side of given equation}} = \tan^2 x + \tan^2 x\cot^2 x$	Distributive property (Section R.2)
$= \tan^2 x + \tan^2 x \cdot \frac{1}{\tan^2 x}$	$\cot^2 x = \frac{1}{\tan^2 x}$
$= \tan^2 x + 1$	$\tan^2 x \cdot \frac{1}{\tan^2 x} = 1$
$= \sec^2 x$	Pythagorean identity (Section 7.1)
$= \frac{1}{\cos^2 x}$	$\sec^2 x = \frac{1}{\cos^2 x}$
$= \underbrace{\frac{1}{1 - \sin^2 x}}_{\text{Right side of given equation}}$	Pythagorean identity (Section 7.1)

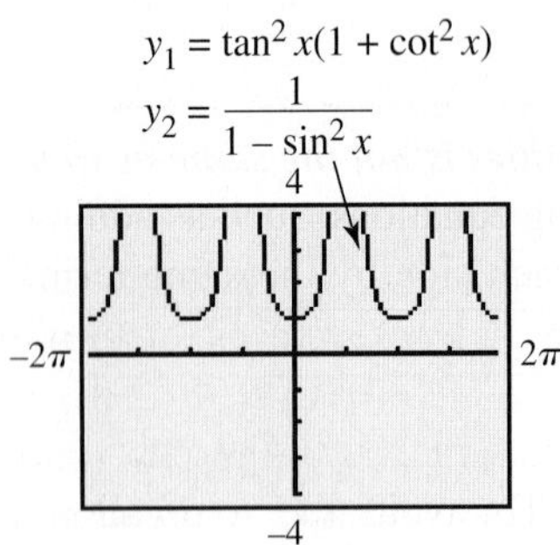

The screen supports the conclusion in **Example 2.**

Since the left side of the equation is identical to the right side, the given equation is an identity.

Now Try Exercise 39.

Classroom Example 3
Verify that the following equation is an identity.

$$\frac{\tan^2 t}{\sec^2 t} = (1 + \cos t)(1 - \cos t)$$

Answer: The verification appears in the *Solutions to Classroom Examples.*

EXAMPLE 3 Verifying an Identity (Working with One Side)

Verify that the following equation is an identity.

$$\frac{\tan t - \cot t}{\sin t \cos t} = \sec^2 t - \csc^2 t$$

SOLUTION We transform the more complicated left side to match the right side.

$$\frac{\tan t - \cot t}{\sin t \cos t} = \frac{\tan t}{\sin t \cos t} - \frac{\cot t}{\sin t \cos t} \qquad \frac{a-b}{c} = \frac{a}{c} - \frac{b}{c} \text{ (Section R.5)}$$

$$= \tan t \cdot \frac{1}{\sin t \cos t} - \cot t \cdot \frac{1}{\sin t \cos t} \qquad \frac{a}{b} = a \cdot \frac{1}{b}$$

$$= \frac{\sin t}{\cos t} \cdot \frac{1}{\sin t \cos t} - \frac{\cos t}{\sin t} \cdot \frac{1}{\sin t \cos t} \qquad \tan t = \frac{\sin t}{\cos t};\ \cot t = \frac{\cos t}{\sin t}$$

$$= \frac{1}{\cos^2 t} - \frac{1}{\sin^2 t} \qquad \text{Multiply.}$$

$$= \sec^2 t - \csc^2 t \qquad \frac{1}{\cos^2 t} = \sec^2 t;\ \frac{1}{\sin^2 t} = \csc^2 t$$

Hint 3 about writing all trigonometric functions in terms of sine and cosine was used in the third line of the solution.

✔ *Now Try Exercise 43.*

Classroom Example 4
Verify that the following equation is an identity.

$$\frac{\sec x + \tan x}{\sin x} = \frac{\csc x}{\sec x - \tan x}$$

Answer: The verification appears in the *Solutions to Classroom Examples.*

EXAMPLE 4 Verifying an Identity (Working with One Side)

Verify that the following equation is an identity.

$$\frac{\cos x}{1 - \sin x} = \frac{1 + \sin x}{\cos x}$$

SOLUTION We work on the right side, using Hint 6 in the list given earlier to multiply the numerator and denominator on the right by $1 - \sin x$.

$$\frac{1 + \sin x}{\cos x} = \frac{(1 + \sin x)(1 - \sin x)}{\cos x(1 - \sin x)} \qquad \text{Multiply by 1 in the form } \frac{1 - \sin x}{1 - \sin x}.$$

$$= \frac{1 - \sin^2 x}{\cos x(1 - \sin x)} \qquad (x + y)(x - y) = x^2 - y^2 \text{ (Section R.3)}$$

$$= \frac{\cos^2 x}{\cos x(1 - \sin x)} \qquad 1 - \sin^2 x = \cos^2 x$$

$$= \frac{\cos x \cdot \cos x}{\cos x(1 - \sin x)} \qquad a^2 = a \cdot a$$

$$= \frac{\cos x}{1 - \sin x} \qquad \text{Write in lowest terms. (Section R.5)}$$

✔ *Now Try Exercise 49.*

Teaching Tip Show that the identity in **Example 4** can also be verified by multiplying the numerator and denominator on the left side by $1 + \sin x$.

Verifying Identities by Working with Both Sides If both sides of an identity appear to be equally complex, the identity can be verified by working independently on the left side and on the right side, until each side is changed into some common third result. ***Each step, on each side, must be reversible.*** With all steps reversible, the procedure is as shown in the margin. The left side leads to a common third expression, which leads back to the right side.

left = right

common third expression

NOTE Working with both sides is often a good alternative for identities that are difficult. In practice, if working with one side does not seem to be effective, switch to the other side. Somewhere along the way it may happen that the same expression occurs on both sides.

Classroom Example 5
Verify that the following equation is an identity.

$$\frac{\cot \alpha - \csc \alpha}{\cot \alpha + \csc \alpha} = \frac{1 - 2\cos \alpha + \cos^2 \alpha}{-\sin^2 \alpha}$$

Answer: The verification appears in the *Solutions to Classroom Examples.*

EXAMPLE 5 Verifying an Identity (Working with Both Sides)

Verify that the following equation is an identity.

$$\frac{\sec \alpha + \tan \alpha}{\sec \alpha - \tan \alpha} = \frac{1 + 2\sin \alpha + \sin^2 \alpha}{\cos^2 \alpha}$$

SOLUTION Both sides appear equally complex, so we verify the identity by changing each side into a common third expression. We work first on the left, multiplying the numerator and denominator by $\cos \alpha$.

$$\underbrace{\frac{\sec \alpha + \tan \alpha}{\sec \alpha - \tan \alpha}}_{\text{Left side of given equation}} = \frac{(\sec \alpha + \tan \alpha)\cos \alpha}{(\sec \alpha - \tan \alpha)\cos \alpha} \quad \text{Multiply by 1 in the form } \frac{\cos \alpha}{\cos \alpha}.$$

$$= \frac{\sec \alpha \cos \alpha + \tan \alpha \cos \alpha}{\sec \alpha \cos \alpha - \tan \alpha \cos \alpha} \quad \text{Distributive property}$$

$$= \frac{1 + \tan \alpha \cos \alpha}{1 - \tan \alpha \cos \alpha} \quad \sec \alpha \cos \alpha = 1$$

$$= \frac{1 + \frac{\sin \alpha}{\cos \alpha} \cdot \cos \alpha}{1 - \frac{\sin \alpha}{\cos \alpha} \cdot \cos \alpha} \quad \tan \alpha = \frac{\sin \alpha}{\cos \alpha}$$

$$= \frac{1 + \sin \alpha}{1 - \sin \alpha} \quad \text{Simplify.}$$

On the right side of the original equation, begin by factoring.

$$\underbrace{\frac{1 + 2\sin \alpha + \sin^2 \alpha}{\cos^2 \alpha}}_{\text{Right side of given equation}} = \frac{(1 + \sin \alpha)^2}{\cos^2 \alpha} \quad \text{Factor the numerator; } x^2 + 2xy + y^2 = (x + y)^2 \text{ (Section R.4)}$$

$$= \frac{(1 + \sin \alpha)^2}{1 - \sin^2 \alpha} \quad \cos^2 \alpha = 1 - \sin^2 \alpha$$

$$= \frac{(1 + \sin \alpha)^2}{(1 + \sin \alpha)(1 - \sin \alpha)} \quad \text{Factor the denominator; } x^2 - y^2 = (x + y)(x - y)$$

$$= \frac{1 + \sin \alpha}{1 - \sin \alpha} \quad \text{Write in lowest terms.}$$

We have shown that

$$\overbrace{\frac{\sec \alpha + \tan \alpha}{\sec \alpha - \tan \alpha}}^{\text{Left side of given equation}} = \overbrace{\frac{1 + \sin \alpha}{1 - \sin \alpha}}^{\text{Common third expression}} = \overbrace{\frac{1 + 2\sin \alpha + \sin^2 \alpha}{\cos^2 \alpha}}^{\text{Right side of given equation}},$$

and thus have verified that the given equation is an identity.

✔ *Now Try Exercise 65.*

CAUTION Use the method of **Example 5** *only* if the steps are reversible.

There are usually several ways to verify a given identity. For instance, another way to begin verifying the identity in **Example 5** is to work on the left as follows.

$$\underbrace{\frac{\sec \alpha + \tan \alpha}{\sec \alpha - \tan \alpha}}_{\text{Left side of given equation in Example 5}} = \frac{\frac{1}{\cos \alpha} + \frac{\sin \alpha}{\cos \alpha}}{\frac{1}{\cos \alpha} - \frac{\sin \alpha}{\cos \alpha}} \quad \text{Fundamental identities (Section 7.1)}$$

$$= \frac{\frac{1 + \sin \alpha}{\cos \alpha}}{\frac{1 - \sin \alpha}{\cos \alpha}} \quad \text{Add and subtract fractions. (Section R.5)}$$

$$= \frac{1 + \sin \alpha}{\cos \alpha} \div \frac{1 - \sin \alpha}{\cos \alpha} \quad \text{Simplify the complex fraction. Use the definition of division.}$$

$$= \frac{1 + \sin \alpha}{\cos \alpha} \cdot \frac{\cos \alpha}{1 - \sin \alpha} \quad \text{Multiply by the reciprocal.}$$

$$= \frac{1 + \sin \alpha}{1 - \sin \alpha} \quad \text{Multiply and write in lowest terms. (Section R.5)}$$

Compare this with the result shown in **Example 5** for the right side to see that the two sides indeed agree.

Classroom Example 6
FM radio stations broadcast at higher frequencies than AM stations. For a certain classical music FM station, the energy stored in the inductor is given by

$$L(t) = 5 \cos^2(620{,}000{,}000t)$$

and the energy in the capacitor is given by

$$C(t) = 5 \sin^2(620{,}000{,}000t).$$

Write a simplified expression for $E(t)$, the total energy in the circuit.

Answer: 5

EXAMPLE 6 Applying a Pythagorean Identity to Electronics

Tuners in radios select a radio station by adjusting the frequency. A tuner may contain an inductor L and a capacitor C, as illustrated in **Figure 3.** The energy stored in the inductor at time t is given by

$$L(t) = k \sin^2(2\pi F t)$$

and the energy stored in the capacitor is given by

$$C(t) = k \cos^2(2\pi F t),$$

where F is the frequency of the radio station and k is a constant. The total energy E in the circuit is given by

$$E(t) = L(t) + C(t).$$

Show that E is a constant function. (*Source:* Weidner, R. and R. Sells, *Elementary Classical Physics,* Vol. 2, Allyn & Bacon.)

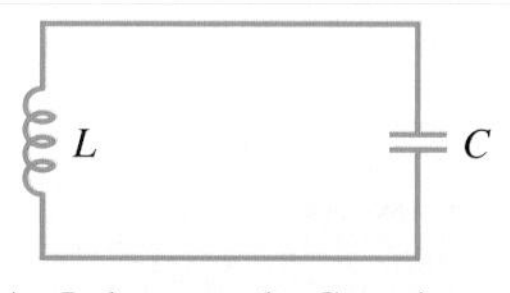

An Inductor and a Capacitor

Figure 3

SOLUTION

$$\begin{aligned} E(t) &= L(t) + C(t) && \text{Given equation} \\ &= k \sin^2(2\pi F t) + k \cos^2(2\pi F t) && \text{Substitute.} \\ &= k[\sin^2(2\pi F t) + \cos^2(2\pi F t)] && \text{Factor out } k. \text{ (Section R.4)} \\ &= k(1) && \sin^2\theta + \cos^2\theta = 1 \text{ (Here } \theta = 2\pi F t.) \\ &= k && \text{Identity property} \end{aligned}$$

Since k is a constant, $E(t)$ is a constant function.

Now Try Exercise 95.

7.2 Exercises

1. $\csc \theta \sec \theta$
2. $\csc x \sec x$
3. $1 + \sec x$
4. $1 + \cot \beta$
5. 1
6. 1
7. $1 - 2 \sin \alpha \cos \alpha$
8. $\sec^2 x + \csc^2 x$
9. $2 + 2 \sin t$
10. $\sec^2 \theta$
11. $-2 \cot x \csc x$
12. $-2 \sec^2 \alpha$
13. $(\sin \theta + 1)(\sin \theta - 1)$
14. $(\sec \theta + 1)(\sec \theta - 1)$
15. $4 \sin x$
16. 4
17. $(2 \sin x + 1)(\sin x + 1)$
18. $(4 \tan \beta - 3)(\tan \beta + 1)$
19. $(\cos^2 x + 1)^2$
20. $(\cot^2 x + 2)(\cot^2 x + 1)$, or $\csc^2 x(\cot^2 x + 2)$
21. $(\sin x - \cos x)(1 + \sin x \cos x)$
22. $(\sin \alpha + \cos \alpha)(1 - \sin \alpha \cos \alpha)$
23. $\sin \theta$
24. $\cos \alpha$
25. 1
26. 1
27. $\tan^2 \beta$
28. $\sec^2 \theta$
29. $\tan^2 x$
30. $\cot^2 t$
31. $\sec^2 x$
32. $\csc^2 \alpha$
33. $\cos^2 x$
34. $\sin^2 x$

To the student: **Exercises 1–34** are designed for practice in applying algebraic techniques to trigonometric expressions. These techniques are essential in verifying the identities that follow.

Perform each indicated operation and simplify the result so that there are no quotients.

1. $\cot \theta + \frac{1}{\cot \theta}$ **2.** $\frac{\sec x}{\csc x} + \frac{\csc x}{\sec x}$ **3.** $\tan x(\cot x + \csc x)$

4. $\cos \beta(\sec \beta + \csc \beta)$ **5.** $\frac{1}{\csc^2 \theta} + \frac{1}{\sec^2 \theta}$ **6.** $\frac{\cos x}{\sec x} + \frac{\sin x}{\csc x}$

7. $(\sin \alpha - \cos \alpha)^2$ **8.** $(\tan x + \cot x)^2$ **9.** $(1 + \sin t)^2 + \cos^2 t$

10. $(1 + \tan \theta)^2 - 2 \tan \theta$ **11.** $\frac{1}{1 + \cos x} - \frac{1}{1 - \cos x}$ **12.** $\frac{1}{\sin \alpha - 1} - \frac{1}{\sin \alpha + 1}$

Factor each trigonometric expression.

13. $\sin^2 \theta - 1$ **14.** $\sec^2 \theta - 1$

15. $(\sin x + 1)^2 - (\sin x - 1)^2$ **16.** $(\tan x + \cot x)^2 - (\tan x - \cot x)^2$

17. $2 \sin^2 x + 3 \sin x + 1$ **18.** $4 \tan^2 \beta + \tan \beta - 3$

19. $\cos^4 x + 2 \cos^2 x + 1$ **20.** $\cot^4 x + 3 \cot^2 x + 2$

21. $\sin^3 x - \cos^3 x$ **22.** $\sin^3 \alpha + \cos^3 \alpha$

Each expression simplifies to a constant, a single function, or a power of a function. Use fundamental identities to simplify each expression.

23. $\tan \theta \cos \theta$ **24.** $\cot \alpha \sin \alpha$ **25.** $\sec r \cos r$

26. $\cot t \tan t$ **27.** $\frac{\sin \beta \tan \beta}{\cos \beta}$ **28.** $\frac{\csc \theta \sec \theta}{\cot \theta}$

29. $\sec^2 x - 1$ **30.** $\csc^2 t - 1$ **31.** $\frac{\sin^2 x}{\cos^2 x} + \sin x \csc x$

32. $\frac{1}{\tan^2 \alpha} + \cot \alpha \tan \alpha$ **33.** $1 - \frac{1}{\csc^2 x}$ **34.** $1 - \frac{1}{\sec^2 x}$

In Exercises 35–78, verify that each trigonometric equation is an identity. **See Examples 1–5.**

35. $\frac{\cot \theta}{\csc \theta} = \cos \theta$ **36.** $\frac{\tan \alpha}{\sec \alpha} = \sin \alpha$

37. $\frac{1 - \sin^2 \beta}{\cos \beta} = \cos \beta$ **38.** $\frac{\tan^2 \alpha + 1}{\sec \alpha} = \sec \alpha$

39. $\cos^2 \theta(\tan^2 \theta + 1) = 1$ **40.** $\sin^2 \beta(1 + \cot^2 \beta) = 1$

41. $\cot \theta + \tan \theta = \sec \theta \csc \theta$ **42.** $\sin^2 \alpha + \tan^2 \alpha + \cos^2 \alpha = \sec^2 \alpha$

43. $\frac{\cos \alpha}{\sec \alpha} + \frac{\sin \alpha}{\csc \alpha} = \sec^2 \alpha - \tan^2 \alpha$ **44.** $\frac{\sin^2 \theta}{\cos \theta} = \sec \theta - \cos \theta$

45. $\sin^4 \theta - \cos^4 \theta = 2 \sin^2 \theta - 1$ **46.** $\sec^4 x - \sec^2 x = \tan^4 x + \tan^2 x$

47. $\frac{1 - \cos x}{1 + \cos x} = (\cot x - \csc x)^2$ **48.** $(\sec \alpha - \tan \alpha)^2 = \frac{1 - \sin \alpha}{1 + \sin \alpha}$

49. $\dfrac{\cos\theta + 1}{\tan^2\theta} = \dfrac{\cos\theta}{\sec\theta - 1}$

50. $\dfrac{(\sec\theta - \tan\theta)^2 + 1}{\sec\theta\csc\theta - \tan\theta\csc\theta} = 2\tan\theta$

51. $\dfrac{1}{1 - \sin\theta} + \dfrac{1}{1 + \sin\theta} = 2\sec^2\theta$

52. $\dfrac{1}{\sec\alpha - \tan\alpha} = \sec\alpha + \tan\alpha$

53. $\dfrac{\cot\alpha + 1}{\cot\alpha - 1} = \dfrac{1 + \tan\alpha}{1 - \tan\alpha}$

54. $\dfrac{\csc\theta + \cot\theta}{\tan\theta + \sin\theta} = \cot\theta\csc\theta$

55. $\dfrac{\cos\theta}{\sin\theta\cot\theta} = 1$

56. $\sin^2\theta(1 + \cot^2\theta) - 1 = 0$

57. $\dfrac{\sec^4\theta - \tan^4\theta}{\sec^2\theta + \tan^2\theta} = \sec^2\theta - \tan^2\theta$

58. $\dfrac{\sin^4\alpha - \cos^4\alpha}{\sin^2\alpha - \cos^2\alpha} = 1$

59. $\dfrac{\tan^2 t - 1}{\sec^2 t} = \dfrac{\tan t - \cot t}{\tan t + \cot t}$

60. $\dfrac{\cot^2 t - 1}{1 + \cot^2 t} = 1 - 2\sin^2 t$

61. $\sin^2\alpha\sec^2\alpha + \sin^2\alpha\csc^2\alpha = \sec^2\alpha$

62. $\tan^2\alpha\sin^2\alpha = \tan^2\alpha + \cos^2\alpha - 1$

63. $\dfrac{\tan x}{1 + \cos x} + \dfrac{\sin x}{1 - \cos x} = \cot x + \sec x\csc x$

64. $\dfrac{\sin\theta}{1 - \cos\theta} - \dfrac{\sin\theta\cos\theta}{1 + \cos\theta} = \csc\theta(1 + \cos^2\theta)$

65. $\dfrac{1 + \cos x}{1 - \cos x} - \dfrac{1 - \cos x}{1 + \cos x} = 4\cot x\csc x$

66. $\dfrac{1 + \sin\theta}{1 - \sin\theta} - \dfrac{1 - \sin\theta}{1 + \sin\theta} = 4\tan\theta\sec\theta$

67. $\dfrac{1 - \sin\theta}{1 + \sin\theta} = \sec^2\theta - 2\sec\theta\tan\theta + \tan^2\theta$

68. $\sin\theta + \cos\theta = \dfrac{\sin\theta}{1 - \cot\theta} + \dfrac{\cos\theta}{1 - \tan\theta}$

69. $\dfrac{-1}{\tan\alpha - \sec\alpha} + \dfrac{-1}{\tan\alpha + \sec\alpha} = 2\tan\alpha$

70. $(1 + \sin x + \cos x)^2 = 2(1 + \sin x)(1 + \cos x)$

71. $(1 - \cos^2\alpha)(1 + \cos^2\alpha) = 2\sin^2\alpha - \sin^4\alpha$

72. $(\sec\alpha + \csc\alpha)(\cos\alpha - \sin\alpha) = \cot\alpha - \tan\alpha$

73. $\dfrac{1 - \cos x}{1 + \cos x} = \csc^2 x - 2\csc x\cot x + \cot^2 x$

74. $\dfrac{1 - \cos\theta}{1 + \cos\theta} = 2\csc^2\theta - 2\csc\theta\cot\theta - 1$

75. $(2\sin x + \cos x)^2 + (2\cos x - \sin x)^2 = 5$

76. $\sin^2 x(1 + \cot x) + \cos^2 x(1 - \tan x) + \cot^2 x = \csc^2 x$

77. $\sec x - \cos x + \csc x - \sin x - \sin x\tan x = \cos x\cot x$

78. $\sin^3\theta + \cos^3\theta = (\cos\theta + \sin\theta)(1 - \cos\theta\sin\theta)$

Graph each expression and use the graph to make a conjecture, predicting what might be an identity. Then verify your conjecture algebraically.

79. $(\sec\theta + \tan\theta)(1 - \sin\theta)$

80. $(\csc\theta + \cot\theta)(\sec\theta - 1)$

81. $\dfrac{\cos\theta + 1}{\sin\theta + \tan\theta}$

82. $\tan\theta\sin\theta + \cos\theta$

83. identity
84. identity
85. not an identity
86. not an identity

91. It is true when $\sin x \le 0$.
92. It is true when $\cos x \le 0$.

93. (a) $I = k(1 - \sin^2 \theta)$
(b) For $\theta = 2\pi n$ and all integers n, $\cos^2 \theta = 1$, its maximum value, and I attains a maximum value of k.
94. (a) $P = 16k \cos^2(2\pi t)$
(b) $P = 16k[1 - \sin^2(2\pi t)]$
95. (a) The sum of L and C equals 3.

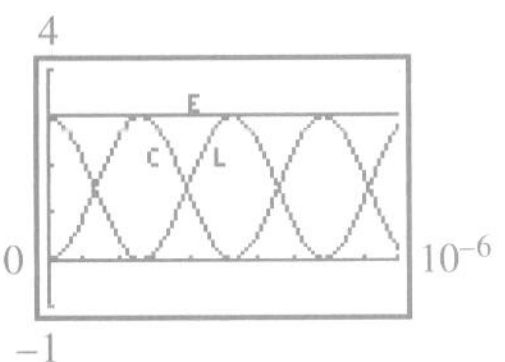

(b) Let $Y_1 = L(t)$, $Y_2 = C(t)$, and $Y_3 = E(t)$. $Y_3 = 3$ for all inputs.

X	Y2	Y3
0	0	3
1E-7	.95646	3
2E-7	2.6061	3
3E-7	2.8451	3
4E-7	1.3688	3
5E-7	.05974	3
6E-7	.58747	3

Y3 = Y1+Y2

(c) $E(t) = 3$

Graph the expressions on each side of the equals symbol to determine whether the equation might be an identity. (Note: Use a domain whose length is at least 2π.) If the equation looks like an identity, verify it algebraically. ***See Example 1.***

83. $\dfrac{2 + 5 \cos x}{\sin x} = 2 \csc x + 5 \cot x$

84. $1 + \cot^2 x = \dfrac{\sec^2 x}{\sec^2 x - 1}$

85. $\dfrac{\tan x - \cot x}{\tan x + \cot x} = 2 \sin^2 x$

86. $\dfrac{1}{1 + \sin x} + \dfrac{1}{1 - \sin x} = \sec^2 x$

By substituting a number for t, show that the equation is not an identity.

87. $\sin(\csc t) = 1$

88. $\sqrt{\cos^2 t} = \cos t$

89. $\csc t = \sqrt{1 + \cot^2 t}$

90. $\cos t = \sqrt{1 - \sin^2 t}$

91. *Concept Check* When is $\sin x = -\sqrt{1 - \cos^2 x}$ a true statement?

92. *Concept Check* When is $\cos x = -\sqrt{1 - \sin^2 x}$ a true statement?

(Modeling) Work each problem.

93. *Intensity of a Lamp* According to **Lambert's law,** the intensity of light from a single source on a flat surface at point P is given by

$$I = k \cos^2 \theta,$$

where k is a constant. (*Source:* Winter, C., *Solar Power Plants,* Springer-Verlag.)

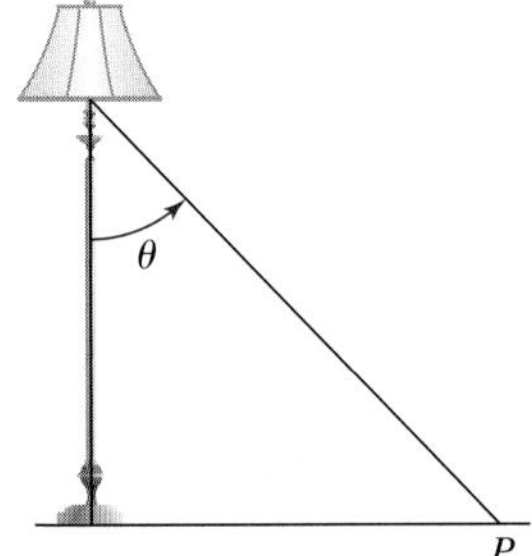

(a) Write I in terms of the sine function.
(b) Why does the maximum value of I occur when $\theta = 0$?

94. *Oscillating Spring* The distance or displacement y of a weight attached to an oscillating spring from its natural position is modeled by

$$y = 4 \cos(2\pi t),$$

where t is time in seconds. Potential energy is the energy of position and is given by

$$P = ky^2,$$

where k is a constant. The weight has the greatest potential energy when the spring is stretched the most. (*Source:* Weidner, R. and R. Sells, *Elementary Classical Physics,* Vol. 2, Allyn & Bacon.)

(a) Write an expression for P that involves the cosine function.
(b) Use a fundamental identity to write P in terms of $\sin(2\pi t)$.

95. *Radio Tuners* Refer to **Example 6.** Let the energy stored in the inductor be given by

$$L(t) = 3 \cos^2(6{,}000{,}000t)$$

and let the energy in the capacitor be given by

$$C(t) = 3 \sin^2(6{,}000{,}000t),$$

where t is time in seconds. The total energy E in the circuit is given by $E(t) = L(t) + C(t)$.

(a) Graph L, C, and E in the window $[0, 10^{-6}]$ by $[-1, 4]$, with Xscl $= 10^{-7}$ and Yscl $= 1$. Interpret the graph.
(b) Make a table of values for L, C, and E starting at $t = 0$, incrementing by 10^{-7}. Interpret your results.
(c) Use a fundamental identity to derive a simplified expression for $E(t)$.

7.3 Sum and Difference Identities

Cosine Sum and Difference Identities Several examples presented earlier should have convinced you by now that

$$\cos(A - B) \textit{ does not equal } \cos A - \cos B.$$

For example, if $A = \frac{\pi}{2}$ and $B = 0$, then

$$\cos(A - B) = \cos\left(\frac{\pi}{2} - 0\right) = \cos\frac{\pi}{2} = 0,$$

while $$\cos A - \cos B = \cos\frac{\pi}{2} - \cos 0 = 0 - 1 = -1.$$

To derive a formula for $\cos(A - B)$, we start by locating angles A and B in standard position on a unit circle, with $B < A$. Let S and Q be the points where the terminal sides of angles A and B, respectively, intersect the circle. Let P be the point $(1, 0)$, and locate point R on the unit circle so that angle POR equals the difference $A - B$. See **Figure 4.**

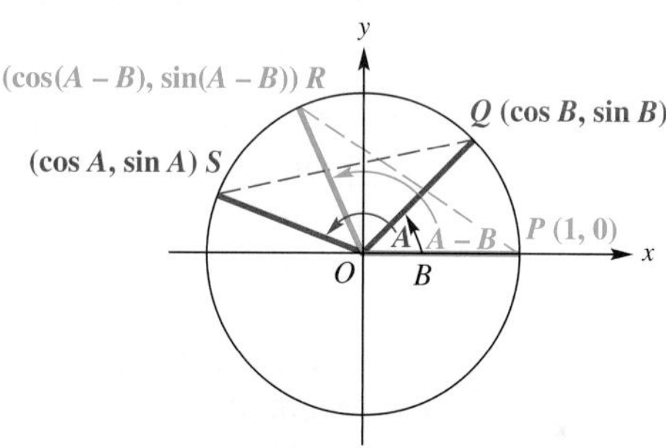

Figure 4

Because point Q is on the unit circle, the x-coordinate of Q is the cosine of angle B, while the y-coordinate of Q is the sine of angle B.

$$Q \text{ has coordinates } (\cos B, \sin B).$$

In the same way,

$$S \text{ has coordinates } (\cos A, \sin A),$$

and $$R \text{ has coordinates } (\cos(A - B), \sin(A - B)).$$

Angle SOQ also equals $A - B$. Since the central angles SOQ and POR are equal, chords PR and SQ are equal. By the distance formula, since $PR = SQ$,

$$\sqrt{[\cos(A - B) - 1]^2 + [\sin(A - B) - 0]^2}$$
$$= \sqrt{(\cos A - \cos B)^2 + (\sin A - \sin B)^2}. \quad \text{(Section 2.1)}$$

Square each side and clear parentheses.

$$\cos^2(A - B) - 2\cos(A - B) + 1 + \sin^2(A - B)$$
$$= \cos^2 A - 2\cos A\cos B + \cos^2 B + \sin^2 A - 2\sin A\sin B + \sin^2 B$$

Since $\sin^2 x + \cos^2 x = 1$ for any value of x, we can rewrite the equation, as shown on the next page.

$$2 - 2\cos(A - B) = 2 - 2\cos A \cos B - 2\sin A \sin B$$ Use $\sin^2 x + \cos^2 x = 1$ three times and add like terms.

$$\mathbf{\cos(A - B) = \cos A \cos B + \sin A \sin B}$$ Subtract 2 and divide by -2.

This is the identity for $\cos(A - B)$. Although **Figure 4** shows angles A and B in the second and first quadrants, respectively, this result is the same for any values of these angles.

To find a similar expression for $\cos(A + B)$, rewrite $A + B$ as $A - (-B)$ and use the identity for $\cos(A - B)$.

$$\cos(A + B) = \cos[A - (-B)]$$ Definition of subtraction

$$= \cos A \cos(-B) + \sin A \sin(-B)$$ Cosine difference identity

$$= \cos A \cos B + \sin A(-\sin B)$$ Negative-angle identities (Section 7.1)

$$\mathbf{\cos(A + B) = \cos A \cos B - \sin A \sin B}$$ Multiply.

Cosine of a Sum or Difference

$$\mathbf{\cos(A + B) = \cos A \cos B - \sin A \sin B}$$
$$\mathbf{\cos(A - B) = \cos A \cos B + \sin A \sin B}$$

These identities are important in calculus and useful in certain applications. For example, the method shown in **Example 1** can be applied to find an exact value for $\cos 15°$.

Classroom Example 1
Find the *exact* value of each expression.

(a) $\cos(-75°)$ **(b)** $\cos \frac{17\pi}{12}$

(c) $\cos 173° \cos 83° + \sin 173° \sin 83°$

Answers:
(a) $\frac{\sqrt{6} - \sqrt{2}}{4}$ **(b)** $\frac{\sqrt{2} - \sqrt{6}}{4}$
(c) 0

EXAMPLE 1 Finding Exact Cosine Function Values

Find the *exact* value of each expression.

(a) $\cos 15°$ **(b)** $\cos \frac{5\pi}{12}$ **(c)** $\cos 87° \cos 93° - \sin 87° \sin 93°$

SOLUTION

(a) To find $\cos 15°$, we write $15°$ as the sum or difference of two angles with known function values, such as $45°$ and $30°$, since

$$15° = 45° - 30°.$$

(We could also use $60° - 45°$.) Then we use the cosine difference identity.

$\cos 15°$

$$= \cos(45° - 30°)$$ $15° = 45° - 30°$

$$= \cos 45° \cos 30° + \sin 45° \sin 30°$$ Cosine difference identity

$$= \frac{\sqrt{2}}{2} \cdot \frac{\sqrt{3}}{2} + \frac{\sqrt{2}}{2} \cdot \frac{1}{2}$$ Substitute known values. (Section 5.3)

$$= \frac{\sqrt{6} + \sqrt{2}}{4}$$ Multiply and then add fractions. (Sections R.5 and R.7)

(b) $\cos \frac{5\pi}{12}$

```
cos(5π/12)
      .2588190451
(√6-√2)/4
      .2588190451
```

This screen supports the solution in **Example 1(b)** by showing that the decimal approximations for $\cos \frac{5\pi}{12}$ and $\frac{\sqrt{6}-\sqrt{2}}{4}$ agree.

$= \cos\left(\frac{\pi}{6} + \frac{\pi}{4}\right)$ $\quad \frac{\pi}{6} = \frac{2\pi}{12}$ and $\frac{\pi}{4} = \frac{3\pi}{12}$

$= \cos \frac{\pi}{6} \cos \frac{\pi}{4} - \sin \frac{\pi}{6} \sin \frac{\pi}{4}$ Cosine sum identity

$= \frac{\sqrt{3}}{2} \cdot \frac{\sqrt{2}}{2} - \frac{1}{2} \cdot \frac{\sqrt{2}}{2}$ Substitute known values. (Section 6.2)

$= \frac{\sqrt{6} - \sqrt{2}}{4}$ Multiply and then subtract fractions.

(c) $\cos 87° \cos 93° - \sin 87° \sin 93°$

$= \cos(87° + 93°)$ Cosine sum identity

$= \cos 180°$ Add.

$= -1$ (Section 5.2)

Teaching Tip Write the following equivalences so that students can see how to express multiples of $\frac{\pi}{12}$, like the one in **Example 1(b).**

$$\frac{\pi}{6} = \frac{2\pi}{12}, \quad \frac{\pi}{4} = \frac{3\pi}{12}, \quad \frac{\pi}{3} = \frac{4\pi}{12}$$

✔ ***Now Try Exercises 7, 11, and 15.***

Cofunction Identities We can use the identity for the cosine of the difference of two angles and the fundamental identities to derive *cofunction identities,* presented originally in **Section 5.3** for values of θ in the interval $[0°, 90°]$.

Teaching Tip Mention that these identities state that the trigonometric function of an acute angle is the same as the cofunction of its complement. Verify the cofunction identities for acute angles using complementary angles in a right triangle along with the right-triangle-based definitions of the trigonometric functions. Emphasize that these identities apply to *any* angle θ, not just to acute angles.

Cofunction Identities

The following identities hold for any angle θ for which the functions are defined.

$$\cos(90° - \theta) = \sin \theta \qquad \cot(90° - \theta) = \tan \theta$$
$$\sin(90° - \theta) = \cos \theta \qquad \sec(90° - \theta) = \csc \theta$$
$$\tan(90° - \theta) = \cot \theta \qquad \csc(90° - \theta) = \sec \theta$$

The same identities can be obtained for a real number domain by replacing 90° with $\frac{\pi}{2}$.

Substituting 90° for A and θ for B in the identity for $\cos(A - B)$ gives the following.

$\cos(90° - \theta) = \cos 90° \cos \theta + \sin 90° \sin \theta$ Cosine difference identity

$= 0 \cdot \cos \theta + 1 \cdot \sin \theta$ Substitute.

$= \sin \theta$ Simplify.

This result is true for *any* value of θ since the identity for $\cos(A - B)$ is true for any values of A and B.

NOTE Because trigonometric (circular) functions are periodic, the solutions that follow in **Example 2** on the next page are not unique. We give only one of infinitely many possibilities.

Classroom Example 2
Find one value of θ or x that satisfies each of the following.
(a) $\sec \theta = \csc 62°$
(b) $\tan \theta = \cot(-54°)$
(c) $\cos x = \sin \frac{7\pi}{6}$

Answers:
(a) $\theta = 28°$ (b) $\theta = 144°$
(c) $x = -\frac{2\pi}{3}$

Teaching Tip Support the results from **Example 2** using a calculator.

EXAMPLE 2 Using Cofunction Identities to Find θ

Find one value of θ or x that satisfies each of the following.

(a) $\cot \theta = \tan 25°$ **(b)** $\sin \theta = \cos(-30°)$ **(c)** $\csc \frac{3\pi}{4} = \sec x$

SOLUTION

(a) Since tangent and cotangent are cofunctions, $\tan(90° - \theta) = \cot \theta$.

$$\cot \theta = \tan 25°$$
$$\tan(90° - \theta) = \tan 25° \quad \text{Cofunction identity}$$
$$90° - \theta = 25° \quad \text{Set angle measures equal.}$$
$$\theta = 65° \quad \text{Solve for } \theta.$$

(b)
$$\sin \theta = \cos(-30°)$$
$$\cos(90° - \theta) = \cos(-30°) \quad \text{Cofunction identity}$$
$$90° - \theta = -30° \quad \text{Set angle measures equal.}$$
$$\theta = 120° \quad \text{Solve for } \theta.$$

(c)
$$\csc \frac{3\pi}{4} = \sec x$$
$$\csc \frac{3\pi}{4} = \csc\left(\frac{\pi}{2} - x\right) \quad \text{Cofunction identity}$$
$$\frac{3\pi}{4} = \frac{\pi}{2} - x \quad \text{Set angle measures equal.}$$
$$x = -\frac{\pi}{4} \quad \text{Solve for } x;\ \frac{\pi}{2} - \frac{3\pi}{4} = \frac{2\pi}{4} - \frac{3\pi}{4} = -\frac{\pi}{4}$$

✔ *Now Try Exercises 33 and 37.*

Sine and Tangent Sum and Difference Identities We can use the cosine sum and difference identities to derive similar identities for sine and tangent. In $\sin \theta = \cos(90° - \theta)$, replace θ with $A + B$.

$$\sin(A + B) = \cos[90° - (A + B)] \quad \text{Cofunction identity}$$
$$= \cos[(90° - A) - B] \quad \text{Distribute negative sign and regroup.}$$
$$= \cos(90° - A)\cos B + \sin(90° - A)\sin B \quad \text{Cosine difference identity}$$
$$\mathbf{\sin(A + B) = \sin A \cos B + \cos A \sin B} \quad \text{Cofunction identities}$$

Now we write $\sin(A - B)$ as $\sin[A + (-B)]$ and use the identity just found for $\sin(A + B)$.

$$\sin(A - B) = \sin[A + (-B)] \quad \text{Definition of subtraction}$$
$$= \sin A \cos(-B) + \cos A \sin(-B) \quad \text{Sine sum identity}$$
$$\mathbf{\sin(A - B) = \sin A \cos B - \cos A \sin B} \quad \text{Negative-angle identities}$$

Sine of a Sum or Difference

$$\mathbf{\sin(A + B) = \sin A \cos B + \cos A \sin B}$$
$$\mathbf{\sin(A - B) = \sin A \cos B - \cos A \sin B}$$

We can now derive the identity for $\tan(A + B)$ as follows.

$$\tan(A+B) = \frac{\sin(A+B)}{\cos(A+B)}$$ Fundamental identity (Section 7.1)

We express this result in terms of the tangent function.

$$= \frac{\sin A \cos B + \cos A \sin B}{\cos A \cos B - \sin A \sin B}$$ Sum identities

$$= \frac{\dfrac{\sin A \cos B + \cos A \sin B}{1}}{\dfrac{\cos A \cos B - \sin A \sin B}{1}} \cdot \frac{\dfrac{1}{\cos A \cos B}}{\dfrac{1}{\cos A \cos B}}$$ Multiply by 1, where $1 = \frac{\frac{1}{\cos A \cos B}}{\frac{1}{\cos A \cos B}}$.

$$= \frac{\dfrac{\sin A \cos B}{\cos A \cos B} + \dfrac{\cos A \sin B}{\cos A \cos B}}{\dfrac{\cos A \cos B}{\cos A \cos B} - \dfrac{\sin A \sin B}{\cos A \cos B}}$$ Multiply numerators and multiply denominators.

$$= \frac{\dfrac{\sin A}{\cos A} + \dfrac{\sin B}{\cos B}}{1 - \dfrac{\sin A}{\cos A} \cdot \dfrac{\sin B}{\cos B}}$$ Simplify.

$$\mathbf{\tan(A+B) = \frac{\tan A + \tan B}{1 - \tan A \tan B}}$$ $\frac{\sin \theta}{\cos \theta} = \tan \theta$

We can replace B with $-B$ and use the fact that $\tan(-B) = -\tan B$ to obtain the identity for the tangent of the difference of two angles, as seen below.

Tangent of a Sum or Difference

$$\tan(A+B) = \frac{\tan A + \tan B}{1 - \tan A \tan B} \qquad \tan(A-B) = \frac{\tan A - \tan B}{1 + \tan A \tan B}$$

Applying the Sum and Difference Identities

Classroom Example 3
Find the *exact* value of each expression.
(a) $\sin(-15°)$
(b) $\tan \frac{13\pi}{12}$
(c) $\frac{\tan 100° - \tan 70°}{1 + \tan 100° \tan 70°}$

Answers:
(a) $\frac{\sqrt{2} - \sqrt{6}}{4}$ (b) $2 - \sqrt{3}$
(c) $\frac{\sqrt{3}}{3}$

EXAMPLE 3 Finding Exact Sine and Tangent Function Values

Find the *exact* value of each expression.

(a) $\sin 75°$ **(b)** $\tan \frac{7\pi}{12}$ **(c)** $\sin 40° \cos 160° - \cos 40° \sin 160°$

SOLUTION

(a) $\sin 75°$

$= \sin(45° + 30°)$ $\quad 75° = 45° + 30°$

$= \sin 45° \cos 30° + \cos 45° \sin 30°$ $\quad$ Sine sum identity

$= \frac{\sqrt{2}}{2} \cdot \frac{\sqrt{3}}{2} + \frac{\sqrt{2}}{2} \cdot \frac{1}{2}$ $\quad$ Substitute known values. (Section 5.3)

$= \frac{\sqrt{6} + \sqrt{2}}{4}$ $\quad$ Multiply and then add fractions.

(b) $\tan \frac{7\pi}{12}$

$= \tan\left(\frac{\pi}{3} + \frac{\pi}{4}\right)$ $\quad$ $\frac{\pi}{3} = \frac{4\pi}{12}$ and $\frac{\pi}{4} = \frac{3\pi}{12}$

$= \frac{\tan \frac{\pi}{3} + \tan \frac{\pi}{4}}{1 - \tan \frac{\pi}{3} \tan \frac{\pi}{4}}$ $\quad$ Tangent sum identity

$= \frac{\sqrt{3} + 1}{1 - \sqrt{3} \cdot 1}$ $\quad$ Substitute known values. (Section 6.2)

$= \frac{\sqrt{3} + 1}{1 - \sqrt{3}} \cdot \frac{1 + \sqrt{3}}{1 + \sqrt{3}}$ $\quad$ Rationalize the denominator. (Section R.7)

$= \frac{\sqrt{3} + 3 + 1 + \sqrt{3}}{1 - 3}$ $\quad$ $(a + b)(c + d) = ac + ad + bc + bd$
$(a - b)(a + b) = a^2 - b^2$ (Section R.3)

$= \frac{4 + 2\sqrt{3}}{-2}$ $\quad$ Combine like terms.

Factor first. Then divide out the common factor.

$= \frac{2(2 + \sqrt{3})}{2(-1)}$ $\quad$ Factor out 2.

$= -2 - \sqrt{3}$ $\quad$ Write in lowest terms. (Section R.5)

(c) $\sin 40° \cos 160° - \cos 40° \sin 160°$

$= \sin(40° - 160°)$ $\quad$ Sine difference identity

$= \sin(-120°)$ $\quad$ Subtract.

$= -\sin 120°$ $\quad$ Negative-angle identity

$= -\frac{\sqrt{3}}{2}$ $\quad$ (Section 5.3)

✔ ***Now Try Exercises 43, 51, and 57.***

Classroom Example 4
Write each function as an expression involving functions of θ.
(a) $\cos(\theta - 270°)$
(b) $\tan(\theta + 3\pi)$
(c) $\sin(120° + \theta)$

Answers:
(a) $-\sin \theta$ $\quad$ (b) $\tan \theta$
(c) $\frac{\sqrt{3} \cos \theta - \sin \theta}{2}$

EXAMPLE 4 Writing Functions as Expressions Involving Functions of θ

Write each function as an expression involving functions of θ.

(a) $\cos(30° + \theta)$ $\quad$ **(b)** $\tan(45° - \theta)$ $\quad$ **(c)** $\sin(180° - \theta)$

SOLUTION

(a) $\cos(30° + \theta)$

$= \cos 30° \cos \theta - \sin 30° \sin \theta$ $\quad$ Cosine sum identity

$= \frac{\sqrt{3}}{2} \cos \theta - \frac{1}{2} \sin \theta$ $\quad$ $\cos 30° = \frac{\sqrt{3}}{2}$ and $\sin 30° = \frac{1}{2}$.

$= \frac{\sqrt{3} \cos \theta - \sin \theta}{2}$ $\quad$ Subtract.

(b) $\tan(45° - \theta)$

$= \frac{\tan 45° - \tan \theta}{1 + \tan 45° \tan \theta}$ $\quad$ Tangent difference identity

$= \frac{1 - \tan \theta}{1 + \tan \theta}$ $\quad$ $\tan 45° = 1$

(c) $\sin(180° - \theta)$

$= \sin 180° \cos\theta - \cos 180° \sin\theta$ Sine difference identity

$= 0 \cdot \cos\theta - (-1)\sin\theta$ $\sin 180° = 0$ and $\cos 180° = -1$

$= \sin\theta$ Simplify.

Now Try Exercises 71, 75, and 79.

Classroom Example 5
Suppose that A and B are angles in standard position, with $\cos A = -\frac{7}{25}$, $\pi < A < \frac{3\pi}{2}$, and $\sin B = -\frac{3}{5}$, $\frac{3\pi}{2} < B < 2\pi$.
Find each of the following.
(a) $\sin(A - B)$ (b) $\tan(A - B)$
(c) the quadrant of $A - B$

Answers:
(a) $-\frac{117}{125}$ (b) $-\frac{117}{44}$
(c) quadrant IV

EXAMPLE 5 Finding Function Values and the Quadrant of $A + B$

Suppose that A and B are angles in standard position, with $\sin A = \frac{4}{5}$, $\frac{\pi}{2} < A < \pi$, and $\cos B = -\frac{5}{13}$, $\pi < B < \frac{3\pi}{2}$. Find each of the following.

(a) $\sin(A + B)$ **(b)** $\tan(A + B)$ **(c)** the quadrant of $A + B$

SOLUTION

(a) The identity for $\sin(A + B)$ involves $\sin A$, $\cos A$, $\sin B$, and $\cos B$. We are given values of $\sin A$ and $\cos B$. We must find values of $\cos A$ and $\sin B$.

$\sin^2 A + \cos^2 A = 1$ Fundamental identity **(Section 7.1)**

$\left(\frac{4}{5}\right)^2 + \cos^2 A = 1$ $\sin A = \frac{4}{5}$

$\frac{16}{25} + \cos^2 A = 1$ Square.

$\cos^2 A = \frac{9}{25}$ Subtract $\frac{16}{25}$.

Pay attention to signs.

$\cos A = -\frac{3}{5}$ Take square roots **(Section 1.4)**. Since A is in quadrant II, $\cos A < 0$.

In the same way, $\sin B = -\frac{12}{13}$. Now find $\sin(A + B)$.

$\sin(A + B) = \sin A \cos B + \cos A \sin B$ Sine sum identity

$= \frac{4}{5}\left(-\frac{5}{13}\right) + \left(-\frac{3}{5}\right)\left(-\frac{12}{13}\right)$ Substitute the given values for $\sin A$ and $\cos B$ and the values found for $\cos A$ and $\sin B$.

$= -\frac{20}{65} + \frac{36}{65}$ Multiply.

$\sin(A + B) = \frac{16}{65}$ Add.

(b) To find $\tan(A + B)$, use the values of sine and cosine from part (a), $\sin A = \frac{4}{5}$, $\cos A = -\frac{3}{5}$, $\sin B = -\frac{12}{13}$, and $\cos B = -\frac{5}{13}$, to get $\tan A$ and $\tan B$.

$$\tan A = \frac{\sin A}{\cos A} = \frac{\frac{4}{5}}{-\frac{3}{5}} = \frac{4}{5} \div \left(-\frac{3}{5}\right)$$

$$\tan A = \frac{4}{5} \cdot \left(-\frac{5}{3}\right), \text{ or } -\frac{4}{3}$$

$$\tan B = \frac{\sin B}{\cos B} = \frac{-\frac{12}{13}}{-\frac{5}{13}} = -\frac{12}{13} \div \left(-\frac{5}{13}\right)$$

$$\tan B = -\frac{12}{13} \cdot \left(-\frac{13}{5}\right), \text{ or } \frac{12}{5}$$

Now use the identity for $\tan(A + B)$.

$$\tan(A+B) = \frac{\tan A + \tan B}{1 - \tan A \tan B} \quad \text{Tangent sum identity}$$

$$= \frac{\left(-\frac{4}{3}\right) + \frac{12}{5}}{1 - \left(-\frac{4}{3}\right)\left(\frac{12}{5}\right)} \quad \text{Substitute.}$$

$$= \frac{\frac{16}{15}}{1 + \frac{48}{15}} \quad \text{Perform the indicated operations.}$$

$$= \frac{\frac{16}{15}}{\frac{63}{15}} \quad \text{Add terms in the denominator.}$$

$$= \frac{16}{15} \div \frac{63}{15} \quad \text{Simplify the complex fraction.}$$

$$= \frac{16}{15} \cdot \frac{15}{63} \quad \text{Definition of division}$$

$$\tan(A+B) = \frac{16}{63} \quad \text{Multiply.}$$

(c) From parts (a) and (b),

$$\sin(A+B) = \frac{16}{65} \quad \text{and} \quad \tan(A+B) = \frac{16}{63},$$

and thus both are positive. Therefore, $A + B$ must be in quadrant I, since it is the only quadrant in which both sine and tangent are positive.

✔ *Now Try Exercise 91.*

Classroom Example 6

See **Example 6.** Because household current is supplied at different voltages in different countries, international travelers often carry electrical adapters to connect items they have brought from home to a power source. The voltage V in a typical European 220-volt outlet can be expressed by the function

$$V(t) = 311 \sin \omega t.$$

(a) European generators rotate at precisely 50 cycles per sec. Determine ω for these electric generators.

(b) What is the maximum voltage in the outlet?

(c) Determine the least positive value of ϕ in radians so that the graph of

$$V(t) = 311 \cos(\omega t + \phi)$$

is the same as the graph of

$$V(t) = 311 \sin \omega t.$$

Answers:

(a) 100π radians per sec

(b) 311 volts

(c) $\phi = \frac{3\pi}{2}$

EXAMPLE 6 Applying the Cosine Difference Identity to Voltage

Common household electric current is called **alternating current** because the current alternates direction within the wires. The voltage V in a typical 115-volt outlet can be expressed by the function

$$V(t) = 163 \sin \omega t,$$

where ω is the angular speed (in radians per second) of the rotating generator at the electrical plant and t is time measured in seconds. (*Source:* Bell, D., *Fundamentals of Electric Circuits,* Fourth Edition, Prentice-Hall.)

(a) It is essential for electric generators to rotate at precisely 60 cycles per sec so household appliances and computers will function properly. Determine ω for these electric generators.

(b) Graph V in the window $[0, 0.05]$ by $[-200, 200]$.

(c) Determine a value of ϕ so that the graph of

$$V(t) = 163 \cos(\omega t - \phi)$$

is the same as the graph of $V(t) = 163 \sin \omega t$.

SOLUTION

(a) Each cycle is 2π radians at 60 cycles per sec, so the angular speed is

$$\omega = 60(2\pi) = 120\pi \text{ radians per sec.}$$

(b) $V(t) = 163 \sin \omega t$

$V(t) = 163 \sin 120\pi t$ From part (a), $\omega = 120\pi$ radians per sec.

Because the amplitude of the function $V(t)$ is 163 (from **Section 6.3**), $[-200, 200]$ is an appropriate interval for the range, as shown in the graph in **Figure 5.**

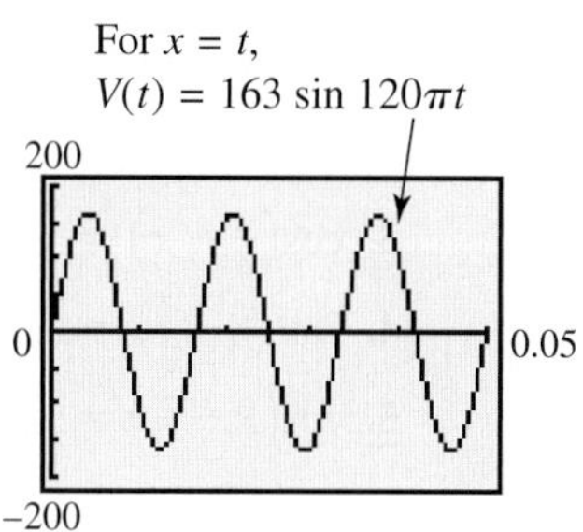

Figure 5

(c) Using the negative-angle identity for cosine and a cofunction identity gives

$$\cos\left(x - \frac{\pi}{2}\right) = \cos\left[-\left(\frac{\pi}{2} - x\right)\right] = \cos\left(\frac{\pi}{2} - x\right) = \sin x.$$

Therefore, if $\phi = \frac{\pi}{2}$, then

$$V(t) = 163 \cos(\omega t - \phi) = 163 \cos\left(\omega t - \frac{\pi}{2}\right) = 163 \sin \omega t.$$

✔ *Now Try Exercise 111.*

Verifying an Identity

Classroom Example 7
Verify that the following equation is an identity.

$$\tan\left(\frac{\pi}{4} + t\right) + \tan\left(\frac{\pi}{4} - t\right) = \frac{2 \sec^2 t}{1 - \tan^2 t}$$

Answer: The verification appears in the *Solutions to Classroom Examples.*

EXAMPLE 7 Verifying an Identity Using Sum and Difference Identities

Verify that the equation is an identity.

$$\sin\left(\frac{\pi}{6} + \theta\right) + \cos\left(\frac{\pi}{3} + \theta\right) = \cos \theta$$

SOLUTION Work on the more complicated left side.

$$\sin\left(\frac{\pi}{6} + \theta\right) + \cos\left(\frac{\pi}{3} + \theta\right)$$

$$= \left(\sin \frac{\pi}{6} \cos \theta + \cos \frac{\pi}{6} \sin \theta\right) + \left(\cos \frac{\pi}{3} \cos \theta - \sin \frac{\pi}{3} \sin \theta\right)$$

Sine sum identity; cosine sum identity

$$= \left(\frac{1}{2} \cos \theta + \frac{\sqrt{3}}{2} \sin \theta\right) + \left(\frac{1}{2} \cos \theta - \frac{\sqrt{3}}{2} \sin \theta\right)$$

$\sin \frac{\pi}{6} = \frac{1}{2}$; $\cos \frac{\pi}{6} = \frac{\sqrt{3}}{2}$; $\cos \frac{\pi}{3} = \frac{1}{2}$; $\sin \frac{\pi}{3} = \frac{\sqrt{3}}{2}$

$$= \frac{1}{2} \cos \theta + \frac{1}{2} \cos \theta$$ Simplify.

$$= \cos \theta$$ Add.

The left side is identical to the right side, so the given equation is an identity.

✔ *Now Try Exercise 103.*

7.3 Exercises

Concept Check *Match each expression in Column I with the correct expression in Column II to form an identity. Choices may be used once, more than once, or not at all.*

I	II
1. $\cos(x+y) =$ ______	**A.** $\cos x \cos y + \sin x \sin y$
2. $\cos(x-y) =$ ______	**B.** $\cos x$
3. $\cos\left(\frac{\pi}{2} - x\right) =$ ______	**C.** $-\cos x$
4. $\sin\left(\frac{\pi}{2} - x\right) =$ ______	**D.** $-\sin x$
5. $\cos\left(x - \frac{\pi}{2}\right) =$ ______	**E.** $\sin x$
6. $\sin\left(x - \frac{\pi}{2}\right) =$ ______	**F.** $\cos x \cos y - \sin x \sin y$

Use identities to find each exact value. (Do not use a calculator.) **See Example 1.**

7. $\cos 75°$

8. $\cos(-15°)$

9. $\cos(-105°)$
(*Hint:* $-105° = -60° + (-45°)$)

10. $\cos 105°$
(*Hint:* $105° = 60° + 45°$)

11. $\cos \frac{7\pi}{12}$

12. $\cos\left(\frac{\pi}{12}\right)$

13. $\cos\left(-\frac{\pi}{12}\right)$

14. $\cos\left(-\frac{7\pi}{12}\right)$

15. $\cos 40° \cos 50° - \sin 40° \sin 50°$

16. $\cos \frac{7\pi}{9} \cos \frac{2\pi}{9} - \sin \frac{7\pi}{9} \sin \frac{2\pi}{9}$

Use a graphing or scientific calculator to support your answer for each of the following. **See Example 1.**

17. Exercise 15

18. Exercise 16

Write each function value in terms of the cofunction of a complementary angle. **See Example 2.**

19. $\tan 87°$

20. $\sin 15°$

21. $\cos \frac{\pi}{12}$

22. $\sin \frac{2\pi}{5}$

23. $\csc(14° 24')$

24. $\sin 142° 14'$

25. $\sin \frac{5\pi}{8}$

26. $\cot \frac{9\pi}{10}$

Use identities to fill in each blank with the appropriate trigonometric function name. **See Example 2.**

27. $\cot \frac{\pi}{3} =$ ______ $\frac{\pi}{6}$

28. $\sin \frac{2\pi}{3} =$ ______ $\left(-\frac{\pi}{6}\right)$

29. ______ $33° = \sin 57°$

30. ______ $72° = \cot 18°$

31. $\cos 70° = \frac{1}{\text{\_\_\_\_\_\_ } 20°}$

32. $\tan 24° = \frac{1}{\text{\_\_\_\_\_\_ } 66°}$

Find one angle θ *that satisfies each of the following.* **See Example 2.**

33. $\tan \theta = \cot(45° + 2\theta)$

34. $\sin \theta = \cos(2\theta + 30°)$

Answers (margin)

1. F 2. A
3. E 4. B
5. E 6. C

7. $\frac{\sqrt{6}-\sqrt{2}}{4}$ 8. $\frac{\sqrt{6}+\sqrt{2}}{4}$
9. $\frac{\sqrt{2}-\sqrt{6}}{4}$ 10. $\frac{\sqrt{2}-\sqrt{6}}{4}$
11. $\frac{\sqrt{2}-\sqrt{6}}{4}$ 12. $\frac{\sqrt{6}+\sqrt{2}}{4}$
13. $\frac{\sqrt{6}+\sqrt{2}}{4}$ 14. $\frac{\sqrt{2}-\sqrt{6}}{4}$
15. 0 16. -1

17. The calculator gives a value of 0 for the expression.
18. The calculator gives a value of -1 for the expression.

19. $\cot 3°$ 20. $\cos 75°$
21. $\sin \frac{5\pi}{12}$ 22. $\cos \frac{\pi}{10}$
23. $\sec 75° 36'$
24. $\cos(-52° 14')$
25. $\cos\left(-\frac{\pi}{8}\right)$
26. $\tan\left(-\frac{2\pi}{5}\right)$

27. tan 28. cos
29. cos 30. tan
31. csc 32. tan

For Exercises 33–38, other answers are possible. We give the most obvious one.
33. 15° 34. 20°

35. $\frac{140}{3}°$ 36. $\frac{348}{5}°$
37. 20° 38. 40°
40. $\frac{-\sqrt{6}-\sqrt{2}}{4}$
41. $\frac{-\sqrt{6}-\sqrt{2}}{4}$
42. (a) $\frac{\sqrt{2}-\sqrt{6}}{4}$
(b) $\frac{-\sqrt{6}-\sqrt{2}}{4}$
43. C 44. A
45. E 46. F
47. B 48. D
49. $\frac{\sqrt{6}+\sqrt{2}}{4}$ 50. $\frac{\sqrt{2}-\sqrt{6}}{4}$
51. $2-\sqrt{3}$ 52. $2+\sqrt{3}$
53. $\frac{-\sqrt{6}-\sqrt{2}}{4}$ 54. $\frac{-\sqrt{2}-\sqrt{6}}{4}$
55. $-2-\sqrt{3}$ 56. $2+\sqrt{3}$
57. $\frac{\sqrt{2}}{2}$ 58. 1
59. -1 60. -1
61. 1 62. 1
63. $\frac{\sqrt{6}-\sqrt{2}}{4}$ 64. $\frac{-\sqrt{6}-\sqrt{2}}{4}$
65. $-2+\sqrt{3}$ 66. $\frac{\sqrt{6}-\sqrt{2}}{4}$
67. $-\cos\theta$ 68. $-\sin\theta$
69. $-\cos\theta$ 70. $\sin\theta$
71. $\frac{\cos\theta-\sqrt{3}\sin\theta}{2}$
72. $\frac{\sqrt{3}\cos\theta+\sin\theta}{2}$
73. $\frac{\sqrt{2}(\sin x-\cos x)}{2}$
74. $\frac{\sqrt{2}(\cos\theta+\sin\theta)}{2}$

35. $\sec\theta = \csc\left(\frac{\theta}{2} + 20°\right)$ **36.** $\cos\theta = \sin\left(\frac{\theta}{4} + 3°\right)$

37. $\sin(3\theta - 15°) = \cos(\theta + 25°)$ **38.** $\cot(\theta - 10°) = \tan(2\theta - 20°)$

Relating Concepts

For individual or collaborative investigation *(Exercises 39–42)*

The identities for $\cos(A + B)$ *and* $\cos(A - B)$ *can be used to find exact values of expressions like* $\cos 195°$ *and* $\cos 255°$, *where the angle is not in the first quadrant.* ***Work Exercises 39–42 in order,*** *to see how this is done.*

39. By writing 195° as $180° + 15°$, use the identity for $\cos(A + B)$ to express $\cos 195°$ as $-\cos 15°$.

40. Use the identity for $\cos(A - B)$ to find $-\cos 15°$.

41. By the results of **Exercises 39 and 40,** $\cos 195° =$ ______.

42. Find each exact value using the method shown in **Exercises 39–41.**

(a) $\cos 255°$ **(b)** $\cos\frac{11\pi}{12}$

Concept Check Match each expression in Column I with its value in Column II. ***See Example 3.***

I		II	
43. $\sin 15°$	**44.** $\sin 105°$	**A.** $\frac{\sqrt{6}+\sqrt{2}}{4}$	**B.** $\frac{-\sqrt{6}-\sqrt{2}}{4}$
45. $\tan 15°$	**46.** $\tan 105°$	**C.** $\frac{\sqrt{6}-\sqrt{2}}{4}$	**D.** $2+\sqrt{3}$
47. $\sin(-105°)$	**48.** $\tan(-105°)$	**E.** $2-\sqrt{3}$	**F.** $-2-\sqrt{3}$

Use identities to find each exact value. ***See Example 3.***

49. $\sin\frac{5\pi}{12}$ **50.** $\sin\frac{13\pi}{12}$ **51.** $\tan\frac{\pi}{12}$ **52.** $\tan\frac{5\pi}{12}$

53. $\sin\left(-\frac{7\pi}{12}\right)$ **54.** $\sin\left(-\frac{5\pi}{12}\right)$ **55.** $\tan\left(-\frac{5\pi}{12}\right)$ **56.** $\tan\left(-\frac{7\pi}{12}\right)$

57. $\sin 76°\cos 31° - \cos 76°\sin 31°$ **58.** $\sin 40°\cos 50° + \cos 40°\sin 50°$

59. $\frac{\tan 80° + \tan 55°}{1 - \tan 80°\tan 55°}$ **60.** $\frac{\tan 80° - \tan(-55°)}{1 + \tan 80°\tan(-55°)}$

61. $\sin\frac{\pi}{5}\cos\frac{3\pi}{10} + \cos\frac{\pi}{5}\sin\frac{3\pi}{10}$ **62.** $\sin 100°\cos 10° - \cos 100°\sin 10°$

Find each exact value. Use an appropriate sum or difference identity and the technique developed in ***Relating Concepts Exercises 39–42.***

63. $\sin 165°$ **64.** $\sin 255°$ **65.** $\tan\frac{11\pi}{12}$ **66.** $\sin\left(-\frac{13\pi}{12}\right)$

Use identities to write each expression as a single function of x or θ. ***See Example 4.***

67. $\cos(\theta - 180°)$ **68.** $\cos(\theta - 270°)$ **69.** $\cos(180° + \theta)$ **70.** $\cos(270° + \theta)$

71. $\cos(60° + \theta)$ **72.** $\cos(\theta - 30°)$ **73.** $\cos\left(\frac{3\pi}{4} - x\right)$ **74.** $\sin(45° + \theta)$

75. $\frac{\sqrt{3}\tan\theta + 1}{\sqrt{3} - \tan\theta}$
76. $\frac{1 + \tan x}{1 - \tan x}$
77. $\frac{\sqrt{2}(\cos x + \sin x)}{2}$
78. $\frac{\sqrt{2}(\cos x + \sin x)}{2}$
79. $-\cos\theta$ 80. $\tan\theta$
81. $-\tan x$ 82. $-\sin x$

83. $\cos(90° + \theta) = -\sin\theta$
84. $\cos(270° - \theta) = -\sin\theta$
85. $\sin(180° + \theta) = -\sin\theta$
86. $\tan(270° - \theta) = \cot\theta$

75. $\tan(\theta + 30°)$ **76.** $\tan\left(\frac{\pi}{4} + x\right)$ **77.** $\sin\left(\frac{\pi}{4} + x\right)$ **78.** $\sin\left(\frac{3\pi}{4} - x\right)$

79. $\sin(270° - \theta)$ **80.** $\tan(180° + \theta)$ **81.** $\tan(2\pi - x)$ **82.** $\sin(\pi + x)$

Relating Concepts

For individual or collaborative investigation *(Exercises 83–86)*

(This discussion applies to functions of both angles and real numbers.) Consider the following.

$$\cos(180° - \theta)$$

$$= \cos 180° \cos\theta + \sin 180° \sin\theta \quad \text{Cosine difference identity}$$

$$= (-1)\cos\theta + (0)\sin\theta \quad \text{(Section 5.2)}$$

$$= -\cos\theta \quad \text{Simplify.}$$

$\mathbf{\cos(180° - \theta) = -\cos\theta}$ is an example of a **reduction formula,** which is an identity that *reduces* a function of a quadrantal angle plus or minus θ to a function of θ alone. Another example of a reduction formula is $\mathbf{\cos(270° + \theta) = \sin\theta}$.

Here is an interesting method for quickly determining a reduction formula for a trigonometric function f of the form $f(Q \pm \theta)$, where Q is a quadrantal angle. ***There are two cases to consider, and in each case, think of θ as a small positive angle*** in order to determine the quadrant in which $Q \pm \theta$ will lie.

Case 1 **Suppose that Q is a quadrantal angle whose terminal side lies along the x-axis.** Determine the quadrant in which $Q \pm \theta$ will lie for a small positive angle θ. If the given function f is positive in that quadrant, use a + sign on the reduced form. If f is negative in that quadrant, use a − sign. The reduced form will have that sign, f as the function, and θ as the argument. For example:

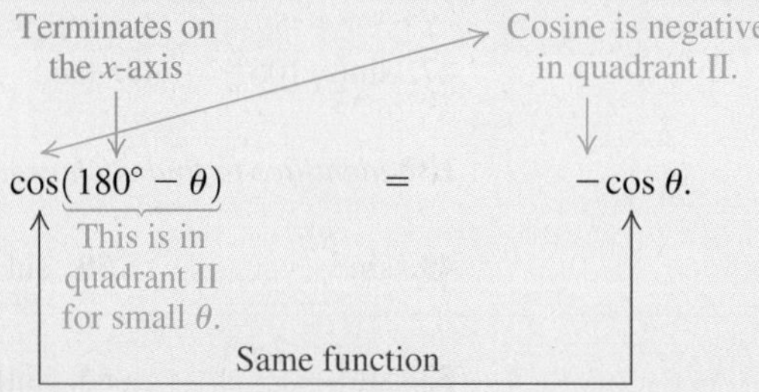

Case 2 **Suppose that Q is a quadrantal angle whose terminal side lies along the y-axis.** Determine the quadrant in which $Q \pm \theta$ will lie for a small positive angle θ. If the given function f is positive in that quadrant, use a + sign on the reduced form. If f is negative in that quadrant, use a − sign. The reduced form will have that sign, the ***cofunction of f*** as the function, and θ as the argument. For example:

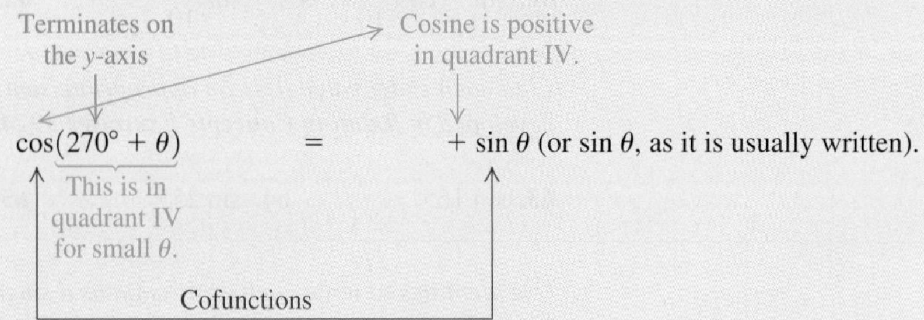

Use these ideas to write reduction formulas for each of the following.

83. $\cos(90° + \theta)$ **84.** $\cos(270° - \theta)$ **85.** $\sin(180° + \theta)$ **86.** $\tan(270° - \theta)$

87. $\frac{16}{65}$; $-\frac{56}{65}$ 88. $-\frac{36}{85}$; $\frac{84}{85}$

89. $\frac{4-6\sqrt{6}}{25}$; $\frac{4+6\sqrt{6}}{25}$

90. $\frac{-2\sqrt{10}+2}{9}$; $\frac{-2\sqrt{10}-2}{9}$

91. (a) $\frac{63}{65}$ (b) $\frac{63}{16}$ (c) I

92. (a) $-\frac{63}{65}$ (b) $-\frac{63}{16}$ (c) IV

93. (a) $\frac{77}{85}$ (b) $-\frac{77}{36}$ (c) II

94. (a) $-\frac{36}{85}$ (b) $\frac{36}{77}$ (c) III

95. (a) $\frac{4\sqrt{2}+\sqrt{5}}{9}$
(b) $\frac{-\sqrt{5}-\sqrt{2}}{2}$
(c) II

96. (a) $\frac{-8\sqrt{6}-3}{25}$
(b) $\frac{\sqrt{6}+6}{4}$
(c) III

97. $\sin\left(\frac{\pi}{2}+\theta\right)=\cos\theta$

98. $\sin\left(\frac{3\pi}{2}+\theta\right)=-\cos\theta$

99. $\tan\left(\frac{\pi}{2}+\theta\right)=-\cot\theta$

100. $\tan\left(\frac{\pi}{2}-\theta\right)=\cot\theta$

111. (a) 3
(b) 163 and -163; no

Use the given information to find $\cos(s+t)$ *and* $\cos(s-t)$. ***See Example 5.***

87. $\sin s=\frac{3}{5}$ and $\sin t=-\frac{12}{13}$, s in quadrant I and t in quadrant III

88. $\cos s=-\frac{8}{17}$ and $\cos t=-\frac{3}{5}$, s and t in quadrant III

89. $\cos s=-\frac{1}{5}$ and $\sin t=\frac{3}{5}$, s and t in quadrant II

90. $\sin s=\frac{2}{3}$ and $\sin t=-\frac{1}{3}$, s in quadrant II and t in quadrant IV

Use the given information to find ***(a)*** $\sin(s+t)$, ***(b)*** $\tan(s+t)$, *and* ***(c)*** *the quadrant of* $s+t$. ***See Example 5.***

91. $\cos s=\frac{3}{5}$ and $\sin t=\frac{5}{13}$, s and t in quadrant I

92. $\sin s=\frac{3}{5}$ and $\sin t=-\frac{12}{13}$, s in quadrant I and t in quadrant III

93. $\cos s=-\frac{8}{17}$ and $\cos t=-\frac{3}{5}$, s and t in quadrant III

94. $\cos s=-\frac{15}{17}$ and $\sin t=\frac{4}{5}$, s in quadrant II and t in quadrant I

95. $\sin s=\frac{2}{3}$ and $\sin t=-\frac{1}{3}$, s in quadrant II and t in quadrant IV

96. $\cos s=-\frac{1}{5}$ and $\sin t=\frac{3}{5}$, s and t in quadrant II

Graph each expression and use the graph to make a conjecture, predicting what might be an identity. Then verify your conjecture algebraically.

97. $\sin\left(\frac{\pi}{2}+\theta\right)$ **98.** $\sin\left(\frac{3\pi}{2}+\theta\right)$

99. $\tan\left(\frac{\pi}{2}+\theta\right)$ **100.** $\tan\left(\frac{\pi}{2}-\theta\right)$

Verify that each equation is an identity. ***See Example 7.***

101. $\sin 2x=2\sin x\cos x$ (*Hint:* $\sin 2x=\sin(x+x)$)

102. $\sin(x+y)+\sin(x-y)=2\sin x\cos y$

103. $\sin\left(\frac{7\pi}{6}+x\right)-\cos\left(\frac{2\pi}{3}+x\right)=0$

104. $\tan(x-y)-\tan(y-x)=\frac{2(\tan x-\tan y)}{1+\tan x\tan y}$

105. $\frac{\cos(\alpha-\beta)}{\cos\alpha\sin\beta}=\tan\alpha+\cot\beta$ **106.** $\frac{\sin(s+t)}{\cos s\cos t}=\tan s+\tan t$

107. $\frac{\sin(x-y)}{\sin(x+y)}=\frac{\tan x-\tan y}{\tan x+\tan y}$ **108.** $\frac{\sin(x+y)}{\cos(x-y)}=\frac{\cot x+\cot y}{1+\cot x\cot y}$

109. $\frac{\sin(s-t)}{\sin t}+\frac{\cos(s-t)}{\cos t}=\frac{\sin s}{\sin t\cos t}$

110. $\frac{\tan(\alpha+\beta)-\tan\beta}{1+\tan(\alpha+\beta)\tan\beta}=\tan\alpha$

(Modeling) Solve each problem.

111. *Electric Current* Refer to **Example 6.**

(a) How many times does the current oscillate in 0.05 sec?

(b) What are the maximum and minimum voltages in this outlet? Is the voltage always equal to 115 volts?

112. (a) The pressure P is oscillating.

For $x = t$,

$P(t) = \frac{0.4}{10}\cos\left[\frac{20\pi}{4.9} - 1026t\right]$

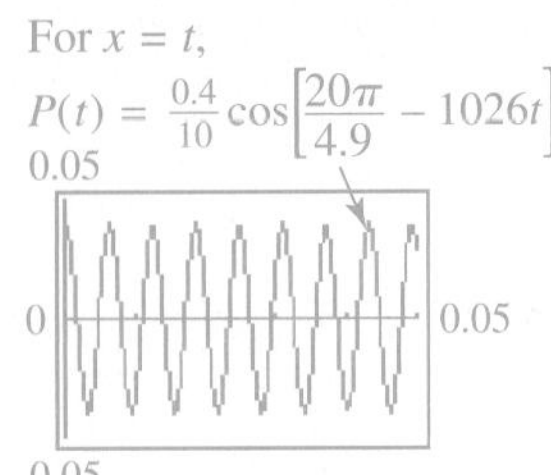

(b) The pressure oscillates and amplitude decreases as r increases.

For $x = r$,

$P(r) = \frac{3}{r}\cos\left[\frac{2\pi r}{4.9} - 10{,}260\right]$

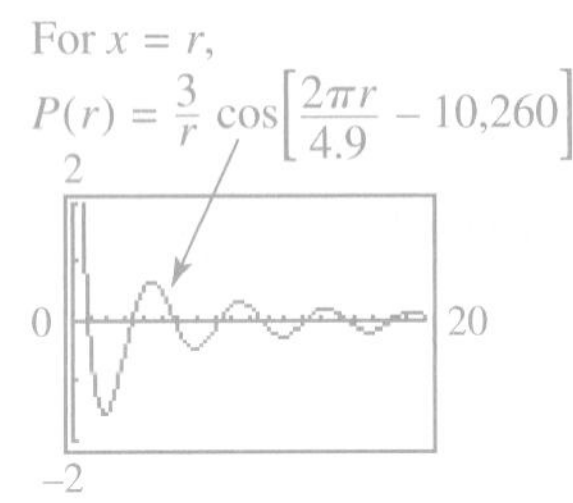

(c) $P = \frac{a}{n\lambda}\cos ct$

113. $-20\cos\frac{\pi t}{4}$

114. (a)

For $x = t$,

$V = V_1 + V_2$
$= 30\sin 120\pi t + 40\cos 120\pi t$

60 / 0 / 0.05 / −60

(b) $a = 50$; $\phi \approx -5.356$

115. (a) 425 lb (c) 0°

116. (a) 408 lb (b) 46.1°

112. *Sound Waves* Sound is a result of waves applying pressure to a person's eardrum. For a pure sound wave radiating outward in a spherical shape, the trigonometric function

$$P = \frac{a}{r}\cos\left(\frac{2\pi r}{\lambda} - ct\right)$$

can be used to model the sound pressure at a radius of r feet from the source, where t is time in seconds, λ is length of the sound wave in feet, c is speed of sound in feet per second, and a is maximum sound pressure at the source measured in pounds per square foot. (*Source:* Beranek, L., *Noise and Vibration Control,* Institute of Noise Control Engineering, Washington, D.C.) Let $\lambda = 4.9$ ft and $c = 1026$ ft per sec.

(a) Let $a = 0.4$ lb per ft^2. Graph the sound pressure at distance $r = 10$ ft from its source in the window $[0, 0.05]$ by $[-0.05, 0.05]$. Describe P at this distance.

(b) Now let $a = 3$ and $t = 10$. Graph the sound pressure in the window $[0, 20]$ by $[-2, 2]$. What happens to pressure P as radius r increases?

(c) Suppose a person stands at a radius r so that $r = n\lambda$, where n is a positive integer. Use the difference identity for cosine to simplify P in this situation.

113. *Voltage* A coil of wire rotating in a magnetic field induces a voltage

$$E = 20\sin\left(\frac{\pi t}{4} - \frac{\pi}{2}\right).$$

Use an identity from this section to express this in terms of $\cos\frac{\pi t}{4}$.

114. *Voltage of a Circuit* When the two voltages

$$V_1 = 30\sin 120\pi t \quad \text{and} \quad V_2 = 40\cos 120\pi t$$

are applied to the same circuit, the resulting voltage V will be equal to their sum. (*Source:* Bell, D., *Fundamentals of Electric Circuits,* Second Edition, Reston Publishing Company.)

(a) Graph the sum in the window $[0, 0.05]$ by $[-60, 60]$.

(b) Use the graph to estimate values for a and ϕ so that $V = a\sin(120\pi t + \phi)$.

(c) Use identities to verify that your expression for V is valid.

115. *Back Stress* If a person bends at the waist with a straight back making an angle of θ degrees with the horizontal, then the force F exerted on the back muscles can be modeled by the equation

$$F = \frac{0.6W\sin(\theta + 90°)}{\sin 12°},$$

where W is the weight of the person. (*Source:* Metcalf, H., *Topics in Classical Biophysics,* Prentice-Hall.)

(a) Calculate force F for $W = 170$ lb and $\theta = 30°$.

(b) Use an identity to show that F is approximately equal to $2.9W\cos\theta$.

(c) For what value of θ is F maximum?

116. *Back Stress* Refer to **Exercise 115.**

(a) Suppose a 200-lb person bends at the waist so that $\theta = 45°$. Estimate the force exerted on the person's back muscles.

(b) Approximate graphically the value of θ that results in the back muscles exerting a force of 400 lb.

Chapter 7 Quiz (Sections 7.1–7.3)

[7.1]

1. $\cos\theta = \frac{24}{25}$; $\tan\theta = -\frac{7}{24}$; $\cot\theta = -\frac{24}{7}$; $\sec\theta = \frac{25}{24}$; $\csc\theta = -\frac{25}{7}$
2. $\frac{\cos^2 x + 1}{\sin^2 x}$

[7.3]

3. $\frac{-\sqrt{6}-\sqrt{2}}{4}$
4. $-\cos\theta$
5. (a) $-\frac{16}{65}$ (b) $-\frac{63}{65}$ (c) III

[7.1–7.3]

6. $\frac{-1+\tan x}{1+\tan x}$

1. If $\sin\theta = -\frac{7}{25}$ and θ is in quadrant IV, find the remaining five trigonometric function values of θ.
2. Express $\cot^2 x + \csc^2 x$ in terms of $\sin x$ and $\cos x$, and simplify.
3. Find the exact value of $\sin\left(-\frac{7\pi}{12}\right)$.
4. Express $\cos(180° - \theta)$ as a function of θ alone.
5. If $\cos A = \frac{3}{5}$, $\sin B = -\frac{5}{13}$, $0 < A < \frac{\pi}{2}$, and $\pi < B < \frac{3\pi}{2}$, find each of the following.
   (a) $\cos(A + B)$ (b) $\sin(A + B)$ (c) the quadrant of $A + B$
6. Express $\tan\left(\frac{3\pi}{4} + x\right)$ as a function of x alone.

Verify each identity.

7. $\frac{1+\sin\theta}{\cot^2\theta} = \frac{\sin\theta}{\csc\theta - 1}$
8. $\sin\left(\frac{\pi}{3}+\theta\right) - \sin\left(\frac{\pi}{3}-\theta\right) = \sin\theta$
9. $\frac{\sin^2\theta - \cos^2\theta}{\sin^4\theta - \cos^4\theta} = 1$
10. $\frac{\cos(x+y)+\cos(x-y)}{\sin(x-y)+\sin(x+y)} = \cot x$

7.4 Double-Angle and Half-Angle Identities

- Double-Angle Identities
- An Application
- Product-to-Sum and Sum-to-Product Identities
- Half-Angle Identities
- Verifying an Identity

Double-Angle Identities When $A = B$ in the identities for the sum of two angles, the **double-angle identities** result. To derive an expression for $\cos 2A$, we let $B = A$ in the identity $\cos(A + B) = \cos A\cos B - \sin A\sin B$.

$$\cos 2A = \cos(A + A)$$

$$= \cos A\cos A - \sin A\sin A \quad \text{Cosine sum identity (Section 7.3)}$$

$$\mathbf{\cos 2A = \cos^2 A - \sin^2 A} \quad a \cdot a = a^2$$

Two other useful forms of this identity can be obtained by substituting either $\cos^2 A = 1 - \sin^2 A$ or $\sin^2 A = 1 - \cos^2 A$. Replacing $\cos^2 A$ with the expression $1 - \sin^2 A$ gives the following.

$$\cos 2A = \cos^2 A - \sin^2 A \quad \text{From above}$$

$$= (1 - \sin^2 A) - \sin^2 A \quad \text{Fundamental identity (Section 7.1)}$$

$$\mathbf{\cos 2A = 1 - 2\sin^2 A} \quad \text{Subtract.}$$

Replacing $\sin^2 A$ with $1 - \cos^2 A$ gives a third form.

$$\cos 2A = \cos^2 A - \sin^2 A$$

$$= \cos^2 A - (1 - \cos^2 A) \quad \text{Fundamental identity}$$

$$= \cos^2 A - 1 + \cos^2 A \quad \text{Distributive property (Section R.2)}$$

$$\mathbf{\cos 2A = 2\cos^2 A - 1} \quad \text{Add.}$$

Teaching Tip Students might find it helpful to see each formula illustrated with a concrete example that they can check. For instance, you might show that

$$\cos 60° = \cos(2 \cdot 30°)$$
$$= \cos^2 30° - \sin^2 30°.$$

LOOKING AHEAD TO CALCULUS

The identities

$$\cos 2A = 1 - 2\sin^2 A$$

and $\cos 2A = 2\cos^2 A - 1$

can be rewritten as

$$\sin^2 A = \frac{1}{2}(1 - \cos 2A)$$

and $\cos^2 A = \frac{1}{2}(1 + \cos 2A)$.

These identities are used to integrate the functions $f(A) = \sin^2 A$ and $g(A) = \cos^2 A$.

We find sin 2*A* using $\sin(A + B) = \sin A \cos B + \cos A \sin B$, with $B = A$.

$$\sin 2A = \sin(A + A)$$

$$= \sin A \cos A + \cos A \sin A \quad \text{Sine sum identity (Section 7.3)}$$

$$\mathbf{\sin 2A = 2 \sin A \cos A} \quad \text{Add.}$$

Using the identity for $\tan(A + B)$, we find tan 2*A*.

$$\tan 2A = \tan(A + A)$$

$$= \frac{\tan A + \tan A}{1 - \tan A \tan A} \quad \text{Tangent sum identity (Section 7.3)}$$

$$\mathbf{\tan 2A = \frac{2 \tan A}{1 - \tan^2 A}} \quad \text{Simplify.}$$

NOTE In general, for a trigonometric function f,

$$f(2A) \neq 2f(A).$$

Teaching Tip Students often make the error of writing cos 2*A* as 2 cos *A*. Refer them to the Note above.

Double-Angle Identities

$$\cos 2A = \cos^2 A - \sin^2 A \qquad \cos 2A = 1 - 2\sin^2 A$$

$$\cos 2A = 2\cos^2 A - 1 \qquad \sin 2A = 2 \sin A \cos A$$

$$\tan 2A = \frac{2 \tan A}{1 - \tan^2 A}$$

Classroom Example 1
Given $\sin\theta = \frac{8}{17}$ and $\cos\theta < 0$, find $\sin 2\theta$, $\cos 2\theta$, and $\tan 2\theta$.

Answer: $\sin 2\theta = -\frac{240}{289}$; $\cos 2\theta = \frac{161}{289}$; $\tan 2\theta = -\frac{240}{161}$

EXAMPLE 1 Finding Function Values of 2θ Given Information about θ

Given $\cos\theta = \frac{3}{5}$ and $\sin\theta < 0$, find $\sin 2\theta$, $\cos 2\theta$, and $\tan 2\theta$.

SOLUTION To find $\sin 2\theta$, we must first find the value of $\sin\theta$.

$$\sin^2\theta + \left(\frac{3}{5}\right)^2 = 1 \quad \sin^2\theta + \cos^2\theta = 1 \text{ and } \cos\theta = \tfrac{3}{5}$$

$$\sin^2\theta = \frac{16}{25} \quad \left(\tfrac{3}{5}\right)^2 = \tfrac{9}{25}\text{; subtract } \tfrac{9}{25}.$$

Pay attention to signs here.

$$\sin\theta = -\frac{4}{5} \quad \text{Take square roots (Section 1.4). Choose the negative square root since } \sin\theta < 0.$$

Now use the double-angle identity for sine.

$$\sin 2\theta = 2\sin\theta\cos\theta = 2\left(-\frac{4}{5}\right)\left(\frac{3}{5}\right) = -\frac{24}{25} \quad \sin\theta = -\tfrac{4}{5} \text{ and } \cos\theta = \tfrac{3}{5}$$

Now we find $\cos 2\theta$, using the first of the double-angle identities for cosine.

Any of the three forms may be used.

$$\cos 2\theta = \cos^2\theta - \sin^2\theta = \frac{9}{25} - \frac{16}{25} = -\frac{7}{25} \quad \cos\theta = \tfrac{3}{5} \text{ and } \sin\theta = -\tfrac{4}{5}$$

The value of $\tan 2\theta$ can be found in either of two ways. We can use the double-angle identity and the fact that $\tan\theta = \frac{\sin\theta}{\cos\theta} = \frac{-\frac{4}{5}}{\frac{3}{5}} = -\frac{4}{5} \div \frac{3}{5} = -\frac{4}{5}\cdot\frac{5}{3} = -\frac{4}{3}$.

(Section R.5)

$$\tan 2\theta = \frac{2\tan\theta}{1-\tan^2\theta} = \frac{2\left(-\frac{4}{3}\right)}{1-\left(-\frac{4}{3}\right)^2} = \frac{-\frac{8}{3}}{-\frac{7}{9}} = \frac{24}{7}$$

Teaching Tip Just as $\cos(A+B) \neq \cos A + \cos B$, caution students not to simplify $\cos 2A = \frac{4}{5}$ in **Example 2** as $\cos A = \frac{2}{5}$. Show that $\cos 2A = 2\cos^2 A - 1$ can also be used to work **Example 2.**

Alternatively, we can find $\tan 2\theta$ by finding the quotient of $\sin 2\theta$ and $\cos 2\theta$.

$$\tan 2\theta = \frac{\sin 2\theta}{\cos 2\theta} = \frac{-\frac{24}{25}}{-\frac{7}{25}} = \frac{24}{7} \quad \text{Same result as above}$$

Now Try Exercise 11.

In **Example 1** we found function values of 2θ given information about θ. In **Example 2** we do the opposite.

Classroom Example 2
Find the values of the six trigonometric functions of θ if $\cos 2\theta = -\frac{12}{13}$ and $180° < \theta < 270°$.

Answer:

$\sin\theta = -\frac{5\sqrt{26}}{26}$; $\cos\theta = -\frac{\sqrt{26}}{26}$;

$\tan\theta = 5$; $\cot\theta = \frac{1}{5}$;

$\sec\theta = -\sqrt{26}$; $\csc\theta = -\frac{\sqrt{26}}{5}$

EXAMPLE 2 Finding Function Values of θ Given Information about 2θ

Find the values of the six trigonometric functions of θ if $\cos 2\theta = \frac{4}{5}$ and $90° < \theta < 180°$.

SOLUTION We must obtain a trigonometric function value of θ alone.

$$\cos 2\theta = 1 - 2\sin^2\theta \quad \text{Double-angle identity}$$

$$\frac{4}{5} = 1 - 2\sin^2\theta \quad \cos 2\theta = \tfrac{4}{5}$$

$$-\frac{1}{5} = -2\sin^2\theta \quad \text{Subtract 1 from each side.}$$

$$\frac{1}{10} = \sin^2\theta \quad \text{Multiply by } -\tfrac{1}{2}.$$

$$\sin\theta = \sqrt{\frac{1}{10}} \quad \text{Take square roots and choose the positive square root since } \theta \text{ terminates in quadrant II.}$$

$$\sin\theta = \frac{1}{\sqrt{10}}\cdot\frac{\sqrt{10}}{\sqrt{10}} \quad \text{Use the quotient rule and rationalize the denominator. (Section R.7)}$$

$$\sin\theta = \frac{\sqrt{10}}{10} \quad \sqrt{a}\cdot\sqrt{a} = a$$

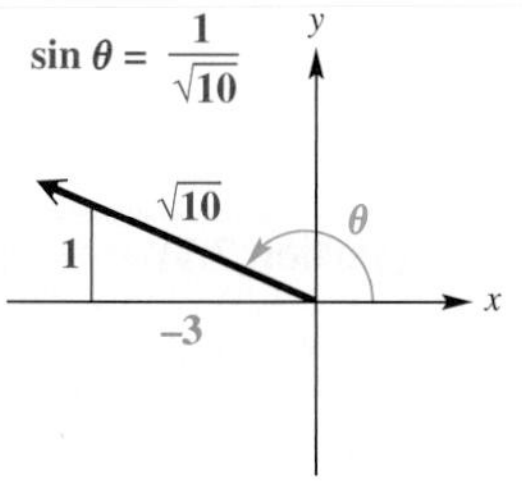

Figure 6

Now find values of $\cos\theta$ and $\tan\theta$ by sketching and labeling a right triangle in quadrant II. Since $\sin\theta = \frac{1}{\sqrt{10}}$, the triangle in **Figure 6** is labeled accordingly. The Pythagorean theorem is used to find the remaining leg. Now,

$$\cos\theta = \frac{-3}{\sqrt{10}} = -\frac{3\sqrt{10}}{10}, \quad \text{and} \quad \tan\theta = \frac{1}{-3} = -\frac{1}{3}. \quad \text{(Section 5.2)}$$

Find the other three functions using reciprocals.

$$\csc\theta = \frac{1}{\sin\theta} = \sqrt{10}, \quad \sec\theta = \frac{1}{\cos\theta} = -\frac{\sqrt{10}}{3}, \quad \cot\theta = \frac{1}{\tan\theta} = -3$$

Now Try Exercise 15.

Classroom Example 3
Simplify each expression.
(a) $2\cos^2 5x - 1$
(b) $\sin 165° \cos 165°$

Answers:
(a) $\cos 10x$ (b) $-\frac{1}{4}$

EXAMPLE 3 Simplifying Expressions Using Double-Angle Identities

Simplify each expression.

(a) $\cos^2 7x - \sin^2 7x$ **(b)** $\sin 15° \cos 15°$

SOLUTION

(a) This expression suggests one of the double-angle identities for cosine: $\cos 2A = \cos^2 A - \sin^2 A$. Substitute $7x$ for A.

$$\cos^2 7x - \sin^2 7x = \cos 2(7x) = \cos 14x$$

(b) If the expression $\sin 15° \cos 15°$ were

$$2 \sin 15° \cos 15°,$$

we could apply the identity for $\sin 2A$ directly because $\sin 2A = 2 \sin A \cos A$.

$\sin 15° \cos 15°$

$= \frac{1}{2}(2) \sin 15° \cos 15°$ Multiply by 1 in the form $\frac{1}{2}(2)$.

This is not an obvious way to begin, but it is indeed valid.

$= \frac{1}{2}(2 \sin 15° \cos 15°)$ Associative property (Section R.2)

$= \frac{1}{2}\sin(2 \cdot 15°)$ $2 \sin A \cos A = \sin 2A$, with $A = 15°$

$= \frac{1}{2}\sin 30°$ Multiply.

$= \frac{1}{2} \cdot \frac{1}{2}$ $\sin 30° = \frac{1}{2}$ (Section 5.3)

$= \frac{1}{4}$ Multiply.

✔ *Now Try Exercises 17 and 19.*

Identities involving larger multiples of the variable can be derived by repeated use of the double-angle identities and other identities.

Classroom Example 4
Write $\cos 3x$ in terms of $\cos x$.

Answer:
$\cos 3x = 4\cos^3 x - 3\cos x$

EXAMPLE 4 Deriving a Multiple-Angle Identity

Write $\sin 3x$ in terms of $\sin x$.

SOLUTION

$\sin 3x$

$= \sin(2x + x)$ Use the simple fact that $3 = 2 + 1$ here.

$= \sin 2x \cos x + \cos 2x \sin x$ Sine sum identity (Section 7.3)

$= (2 \sin x \cos x)\cos x + (\cos^2 x - \sin^2 x)\sin x$ Double-angle identities

$= 2 \sin x \cos^2 x + \cos^2 x \sin x - \sin^3 x$ Multiply.

$= 2 \sin x(1 - \sin^2 x) + (1 - \sin^2 x)\sin x - \sin^3 x$ $\cos^2 x = 1 - \sin^2 x$

$= 2 \sin x - 2 \sin^3 x + \sin x - \sin^3 x - \sin^3 x$ Distributive property

$= 3 \sin x - 4 \sin^3 x$ Combine like terms.

✔ *Now Try Exercise 29.*

An Application

Classroom Example 5
Refer to **Example 5.** The voltage V in a typical European outlet can be expressed by the function

$$V(t) = 311 \sin 100\pi t.$$

Find the maximum wattage of a European light bulb with $R = 686$.

Answer:
approximately 141 watts

EXAMPLE 5 Determining Wattage Consumption

If a toaster is plugged into a common household outlet, the wattage consumed is not constant. Instead, it varies at a high frequency according to the model

$$W = \frac{V^2}{R},$$

where V is the voltage and R is a constant that measures the resistance of the toaster in ohms. (*Source:* Bell, D., *Fundamentals of Electric Circuits,* Fourth Edition, Prentice-Hall.) Graph the wattage W consumed by a typical toaster with $R = 15$ and $V = 163 \sin 120\pi t$ in the window $[0, 0.05]$ by $[-500, 2000]$. How many oscillations are there?

SOLUTION Substituting the given values into the wattage equation gives

$$W = \frac{V^2}{R} = \frac{(163 \sin 120\pi t)^2}{15}.$$

To determine the range of W, we note that $\sin 120\pi t$ has maximum value 1, so the expression for W has maximum value $\frac{163^2}{15} \approx 1771$. The minimum value is 0. The graph in **Figure 7** shows that there are six oscillations.

Now Try Exercise 107.

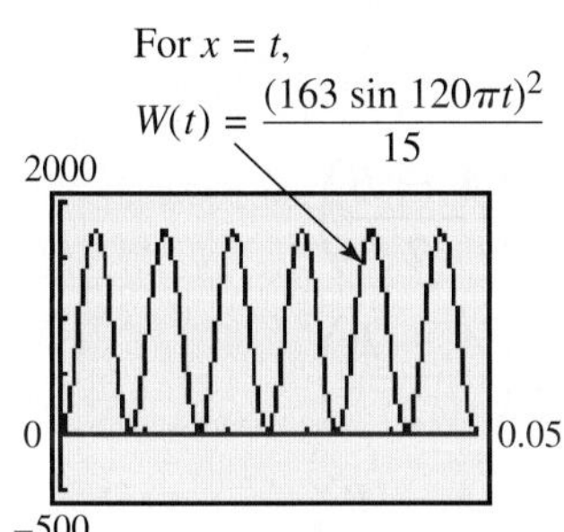

Figure 7

Product-to-Sum and Sum-to-Product Identities We can add the identities for $\cos(A + B)$ and $\cos(A - B)$ to derive an identity useful in calculus.

$$\begin{aligned} \cos(A + B) &= \cos A \cos B - \sin A \sin B \\ \cos(A - B) &= \cos A \cos B + \sin A \sin B \\ \hline \cos(A + B) + \cos(A - B) &= 2 \cos A \cos B \quad \text{Add.} \end{aligned}$$

or $$\mathbf{\cos A \cos B = \frac{1}{2}[\cos(A + B) + \cos(A - B)]}$$

Similarly, subtracting $\cos(A + B)$ from $\cos(A - B)$ gives

$$\mathbf{\sin A \sin B = \frac{1}{2}[\cos(A - B) - \cos(A + B)].}$$

Using the identities for $\sin(A + B)$ and $\sin(A - B)$ in the same way, we obtain two more identities. Those and the previous ones are now summarized.

LOOKING AHEAD TO CALCULUS
The product-to-sum identities are used in calculus to find **integrals** of functions that are products of trigonometric functions. The classic calculus text by Earl Swokowski includes the following example:

Evaluate $\int \cos 5x \cos 3x \, dx$.

The first solution line reads:
"We may write

$\cos 5x \cos 3x = \frac{1}{2}[\cos 8x + \cos 2x]$."

Product-to-Sum Identities

$$\cos A \cos B = \frac{1}{2}[\cos(A + B) + \cos(A - B)]$$

$$\sin A \sin B = \frac{1}{2}[\cos(A - B) - \cos(A + B)]$$

$$\sin A \cos B = \frac{1}{2}[\sin(A + B) + \sin(A - B)]$$

$$\cos A \sin B = \frac{1}{2}[\sin(A + B) - \sin(A - B)]$$

Classroom Example 6
Write $6 \sin 40° \sin 15°$ as the sum or difference of two functions.

Answer: $3 \cos 25° - 3 \cos 55°$

EXAMPLE 6 Using a Product-To-Sum Identity

Write $4 \cos 75° \sin 25°$ as the sum or difference of two functions.

SOLUTION

$$4 \cos 75° \sin 25°$$

$$= 4\left[\frac{1}{2}\left(\sin (75° + 25°) - \sin (75° - 25°)\right)\right] \quad \text{Use the identity for } \cos A \sin B, \text{ with } A = 75° \text{ and } B = 25°.$$

$$= 2 \sin 100° - 2 \sin 50° \quad \text{Simplify.}$$

✔ *Now Try Exercise 33.*

We can convert the product-to-sum identities into equivalent useful forms that enable us to write sums as products.

Sum-to-Product Identities

$$\sin A + \sin B = 2 \sin\left(\frac{A + B}{2}\right) \cos\left(\frac{A - B}{2}\right)$$

$$\sin A - \sin B = 2 \cos\left(\frac{A + B}{2}\right) \sin\left(\frac{A - B}{2}\right)$$

$$\cos A + \cos B = 2 \cos\left(\frac{A + B}{2}\right) \cos\left(\frac{A - B}{2}\right)$$

$$\cos A - \cos B = -2 \sin\left(\frac{A + B}{2}\right) \sin\left(\frac{A - B}{2}\right)$$

Classroom Example 7
Write $\cos 3\theta + \cos 7\theta$ as a product of two functions.

Answer: $2 \cos 5\theta \cos 2\theta$

EXAMPLE 7 Using a Sum-to-Product Identity

Write $\sin 2\theta - \sin 4\theta$ as a product of two functions.

SOLUTION

$$\sin 2\theta - \sin 4\theta$$

$$= 2 \cos\left(\frac{2\theta + 4\theta}{2}\right) \sin\left(\frac{2\theta - 4\theta}{2}\right) \quad \text{Use the identity for } \sin A - \sin B, \text{ with } A = 2\theta \text{ and } B = 4\theta.$$

$$= 2 \cos \frac{6\theta}{2} \sin\left(\frac{-2\theta}{2}\right) \quad \text{Simplify the numerators.}$$

$$= 2 \cos 3\theta \sin(-\theta) \quad \text{Divide.}$$

$$= -2 \cos 3\theta \sin \theta \quad \sin(-\theta) = -\sin \theta \text{ (Section 7.1)}$$

✔ *Now Try Exercise 39.*

Half-Angle Identities From the alternative forms of the identity for $\cos 2A$, we derive identities for $\sin \frac{A}{2}$, $\cos \frac{A}{2}$, and $\tan \frac{A}{2}$. These are known as **half-angle identities.**

We derive the identity for $\sin \frac{A}{2}$ as follows.

$$\cos 2x = 1 - 2 \sin^2 x \quad \text{Cosine double-angle identity}$$

$$2 \sin^2 x = 1 - \cos 2x \quad \text{Add } 2 \sin^2 x \text{ and subtract } \cos 2x.$$

$$\sin x = \pm\sqrt{\frac{1 - \cos 2x}{2}}$$ Divide by 2 and take square roots.

Remember both the positive and negative square roots.

$$\sin \frac{A}{2} = \pm\sqrt{\frac{1 - \cos A}{2}}$$ Let $2x = A$, so $x = \frac{A}{2}$. Substitute.

The $\pm$ sign in this identity indicates that the appropriate sign is chosen depending on the quadrant of $\frac{A}{2}$. For example, if $\frac{A}{2}$ is a quadrant III angle, we choose the negative sign because the sine function is negative in quadrant III.

We derive the identity for $\cos \frac{A}{2}$ using another double-angle identity.

$$\cos 2x = 2\cos^2 x - 1$$ Cosine double-angle identity

$$1 + \cos 2x = 2\cos^2 x$$ Add 1.

$$\cos^2 x = \frac{1 + \cos 2x}{2}$$ Rewrite and divide by 2.

$$\cos x = \pm\sqrt{\frac{1 + \cos 2x}{2}}$$ Take square roots.

$$\cos \frac{A}{2} = \pm\sqrt{\frac{1 + \cos A}{2}}$$ Replace x with $\frac{A}{2}$.

An identity for $\tan \frac{A}{2}$ comes from the identities for $\sin \frac{A}{2}$ and $\cos \frac{A}{2}$.

$$\tan \frac{A}{2} = \frac{\sin \frac{A}{2}}{\cos \frac{A}{2}} = \frac{\pm\sqrt{\frac{1 - \cos A}{2}}}{\pm\sqrt{\frac{1 + \cos A}{2}}} = \pm\sqrt{\frac{1 - \cos A}{1 + \cos A}}$$

We derive an alternative identity for $\tan \frac{A}{2}$ using double-angle identities.

$$\tan \frac{A}{2} = \frac{\sin \frac{A}{2}}{\cos \frac{A}{2}} = \frac{2\sin \frac{A}{2}\cos \frac{A}{2}}{2\cos^2 \frac{A}{2}}$$ Multiply by $2\cos \frac{A}{2}$ in numerator and denominator.

$$= \frac{\sin 2\left(\frac{A}{2}\right)}{1 + \cos 2\left(\frac{A}{2}\right)}$$ Double-angle identities

$$\tan \frac{A}{2} = \frac{\sin A}{1 + \cos A}$$ Simplify.

From the identity $\tan \frac{A}{2} = \frac{\sin A}{1 + \cos A}$, we can also derive an equivalent identity.

$$\tan \frac{A}{2} = \frac{1 - \cos A}{\sin A}$$

Teaching Tip Point out that the first identity for $\tan \frac{A}{2}$ follows directly from the cosine and sine half-angle identities. However, the other two identities for $\tan \frac{A}{2}$ have the advantage of requiring no sign choice.

Half-Angle Identities

In the following identities, the symbol $\pm$ indicates that the sign is chosen based on the function under consideration and the quadrant of $\frac{A}{2}$.

$$\cos \frac{A}{2} = \pm\sqrt{\frac{1 + \cos A}{2}} \qquad \sin \frac{A}{2} = \pm\sqrt{\frac{1 - \cos A}{2}}$$

$$\tan \frac{A}{2} = \pm\sqrt{\frac{1 - \cos A}{1 + \cos A}} \qquad \tan \frac{A}{2} = \frac{\sin A}{1 + \cos A} \qquad \tan \frac{A}{2} = \frac{1 - \cos A}{\sin A}$$

The final two identities for $\tan \frac{A}{2}$ do not require a sign choice. When using the other half-angle identities, select the plus or minus sign according to the quadrant in which $\frac{A}{2}$ terminates. For example, if an angle $A = 324°$, then $\frac{A}{2} = 162°$, which lies in quadrant II. So when $A = 324°$, $\cos \frac{A}{2}$ and $\tan \frac{A}{2}$ are negative, and $\sin \frac{A}{2}$ is positive.

Classroom Example 8
Find the exact value of sin 22.5° using the half-angle identity for sine.

Answer: $\frac{\sqrt{2-\sqrt{2}}}{2}$

EXAMPLE 8 Using a Half-Angle Identity to Find an Exact Value

Find the exact value of cos 15° using the half-angle identity for cosine.

SOLUTION

$$\cos 15° = \cos \frac{1}{2}(30°) = \sqrt{\frac{1 + \cos 30°}{2}}$$

Choose the positive square root.

$$= \sqrt{\frac{1 + \frac{\sqrt{3}}{2}}{2}} = \sqrt{\frac{\left(1 + \frac{\sqrt{3}}{2}\right) \cdot 2}{2 \cdot 2}} = \frac{\sqrt{2 + \sqrt{3}}}{2}$$

Simplify the radicals. **(Section R.7)**

Now Try Exercise 51.

Teaching Tip Have students compare the value of cos 15° in **Example 1** to the value in **Example 1(a)** of **Section 7.3,** where we used the identity for the cosine of the difference of two angles. Although the expressions look completely different, they are equal, as suggested by a calculator approximation for both, 0.96592583.

EXAMPLE 9 Using a Half-Angle Identity to Find an Exact Value

Find the exact value of tan 22.5° using the identity $\tan \frac{A}{2} = \frac{\sin A}{1 + \cos A}$.

SOLUTION Since $22.5° = \frac{1}{2}(45°)$, replace A with 45°.

$$\tan 22.5° = \tan \frac{45°}{2} = \frac{\sin 45°}{1 + \cos 45°} = \frac{\frac{\sqrt{2}}{2}}{1 + \frac{\sqrt{2}}{2}} = \frac{\frac{\sqrt{2}}{2}}{1 + \frac{\sqrt{2}}{2}} \cdot \frac{2}{2}$$

$$= \frac{\sqrt{2}}{2 + \sqrt{2}} = \frac{\sqrt{2}}{2 + \sqrt{2}} \cdot \frac{2 - \sqrt{2}}{2 - \sqrt{2}} = \frac{2\sqrt{2} - 2}{2}$$

Rationalize the denominator.

$$= \frac{2(\sqrt{2} - 1)}{2} = \sqrt{2} - 1$$

Factor out 2. **(Section R.5)**

Factor first, and then divide out the common factor.

Now Try Exercise 53.

Classroom Example 9
Find the exact value of tan 75° using the identity $\tan \frac{A}{2} = \frac{\sin A}{1 + \cos A}$.

Answer: $2 + \sqrt{3}$

Classroom Example 10
Given $\cos s = -\frac{3}{7}$, with $\pi < s < \frac{3\pi}{2}$, find $\sin \frac{s}{2}$, $\cos \frac{s}{2}$, and $\tan \frac{s}{2}$.

Answer:
$\sin \frac{s}{2} = \frac{\sqrt{35}}{7}$; $\cos \frac{s}{2} = -\frac{\sqrt{14}}{7}$;
$\tan \frac{s}{2} = -\frac{\sqrt{10}}{2}$

EXAMPLE 10 Finding Function Values of $\frac{s}{2}$ Given Information about s

Given $\cos s = \frac{2}{3}$, with $\frac{3\pi}{2} < s < 2\pi$, find $\sin \frac{s}{2}$, $\cos \frac{s}{2}$, and $\tan \frac{s}{2}$.

SOLUTION The angle associated with $\frac{s}{2}$ terminates in quadrant II, since

$$\frac{3\pi}{2} < s < 2\pi \quad \text{and} \quad \frac{3\pi}{4} < \frac{s}{2} < \pi.$$

Divide by 2. **(Section 1.7)**

See **Figure 8.** In quadrant II, the values of $\cos \frac{s}{2}$ and $\tan \frac{s}{2}$ are negative and the value of $\sin \frac{s}{2}$ is positive. We use the appropriate half-angle identities and simplify, as shown on the next page.

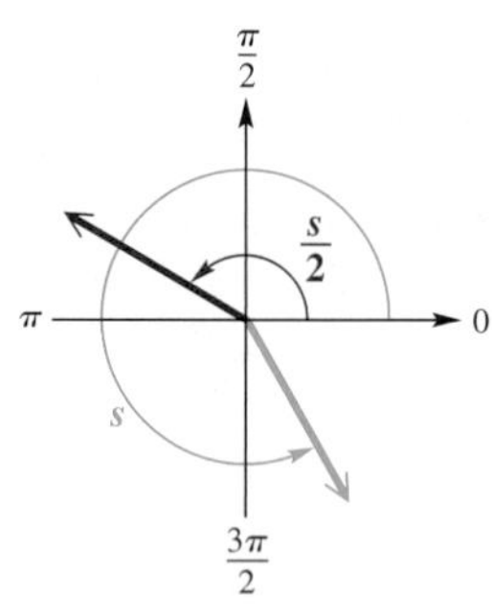

Figure 8

$$\sin\frac{s}{2}=\sqrt{\frac{1-\frac{2}{3}}{2}}=\sqrt{\frac{1}{6}}=\frac{\sqrt{1}}{\sqrt{6}}\cdot\frac{\sqrt{6}}{\sqrt{6}}=\frac{\sqrt{6}}{6}$$

Rationalize all denominators.

$$\cos\frac{s}{2}=-\sqrt{\frac{1+\frac{2}{3}}{2}}=-\sqrt{\frac{5}{6}}=-\frac{\sqrt{5}}{\sqrt{6}}\cdot\frac{\sqrt{6}}{\sqrt{6}}=-\frac{\sqrt{30}}{6}$$

$$\tan\frac{s}{2}=\frac{\sin\frac{s}{2}}{\cos\frac{s}{2}}=\frac{\frac{\sqrt{6}}{6}}{-\frac{\sqrt{30}}{6}}=\frac{\sqrt{6}}{-\sqrt{30}}=-\frac{\sqrt{6}}{\sqrt{30}}\cdot\frac{\sqrt{30}}{\sqrt{30}}=-\frac{\sqrt{180}}{30}=-\frac{6\sqrt{5}}{6\cdot 5}=-\frac{\sqrt{5}}{5}$$

Notice that it is not necessary to use a half-angle identity for $\tan\frac{s}{2}$ once we find $\sin\frac{s}{2}$ and $\cos\frac{s}{2}$. However, using this identity provides an excellent check.

Now Try Exercise 57.

Classroom Example 11
Simplify each expression.

(a) $\pm\sqrt{\frac{1-\cos 8x}{2}}$

(b) $\pm\sqrt{\frac{1-\cos 9\beta}{1+\cos 9\beta}}$

Answers:
(a) $\sin 4x$ (b) $\tan\frac{9\beta}{2}$

EXAMPLE 11 Simplifying Expressions Using the Half-Angle Identities

Simplify each expression.

(a) $\pm\sqrt{\frac{1+\cos 12x}{2}}$ **(b)** $\frac{1-\cos 5\alpha}{\sin 5\alpha}$

SOLUTION

(a) This matches part of the identity for $\cos\frac{A}{2}$. Replace A with $12x$ to get

$$\cos\frac{A}{2}=\pm\sqrt{\frac{1+\cos A}{2}}=\pm\sqrt{\frac{1+\cos 12x}{2}}=\cos\frac{12x}{2}=\cos 6x.$$

(b) Use the third identity for $\tan\frac{A}{2}$ given earlier with $A=5\alpha$ to get

$$\frac{1-\cos 5\alpha}{\sin 5\alpha}=\tan\frac{5\alpha}{2}.$$

Now Try Exercises 73 and 75.

Verifying an Identity

Classroom Example 12
Verify that the following equation is an identity.

$$\tan^2\left(\frac{x}{2}\right)=\frac{\sec x+\cos x-2}{\sec x-\cos x}$$

Answer: The verification appears in the *Solutions to Classroom Examples.*

EXAMPLE 12 Verifying an Identity

Verify that the following equation is an identity.

$$\left(\sin\frac{x}{2}+\cos\frac{x}{2}\right)^2=1+\sin x$$

SOLUTION We work on the more complicated left side.

$$\left(\sin\frac{x}{2}+\cos\frac{x}{2}\right)^2$$

Remember the term $2ab$ when squaring a binomial.

$$=\sin^2\frac{x}{2}+2\sin\frac{x}{2}\cos\frac{x}{2}+\cos^2\frac{x}{2}$$ $(a+b)^2=a^2+2ab+b^2$ (Section R.3)

$$=1+2\sin\frac{x}{2}\cos\frac{x}{2}$$ $\sin^2\frac{x}{2}+\cos^2\frac{x}{2}=1$

$$=1+\sin 2\left(\frac{x}{2}\right)$$ $2\sin\frac{x}{2}\cos\frac{x}{2}=\sin 2\left(\frac{x}{2}\right)$

$$=1+\sin x$$ Multiply.

Now Try Exercise 95.

7.4 Exercises

Concept Check *Match each expression in Column I with its value in Column II.*

I		II	
1. $2\cos^2 15° - 1$	**2.** $\dfrac{2\tan 15°}{1-\tan^2 15°}$	**A.** $\dfrac{1}{2}$	**B.** $\dfrac{\sqrt{2}}{2}$
3. $2\sin 22.5° \cos 22.5°$	**4.** $\cos^2\dfrac{\pi}{6} - \sin^2\dfrac{\pi}{6}$	**C.** $\dfrac{\sqrt{3}}{2}$	**D.** $-\sqrt{3}$
5. $4\sin\dfrac{\pi}{3}\cos\dfrac{\pi}{3}$	**6.** $\dfrac{2\tan\frac{\pi}{3}}{1-\tan^2\frac{\pi}{3}}$	**E.** $\dfrac{\sqrt{3}}{3}$	**F.** $\sqrt{3}$

Use identities to find values of the sine and cosine functions for each angle measure. **See Examples 1 and 2.**

7. 2θ, given $\sin\theta = \frac{2}{5}$ and $\cos\theta < 0$

8. 2θ, given $\cos\theta = -\frac{12}{13}$ and $\sin\theta > 0$

9. $2x$, given $\tan x = 2$ and $\cos x > 0$

10. $2x$, given $\tan x = \frac{5}{3}$ and $\sin x < 0$

11. 2θ, given $\sin\theta = -\frac{\sqrt{5}}{7}$ and $\cos\theta > 0$

12. 2θ, given $\cos\theta = \frac{\sqrt{3}}{5}$ and $\sin\theta > 0$

13. θ, given $\cos 2\theta = \frac{3}{5}$ and θ terminates in quadrant I

14. θ, given $\cos 2\theta = \frac{3}{4}$ and θ terminates in quadrant III

15. θ, given $\cos 2\theta = -\frac{5}{12}$ and $90° < \theta < 180°$

16. θ, given $\cos 2\theta = \frac{2}{3}$ and $90° < \theta < 180°$

Use an identity to write each expression as a single trigonometric function value or as a single number. **See Example 3.**

17. $\cos^2 15° - \sin^2 15°$ **18.** $\dfrac{2\tan 15°}{1-\tan^2 15°}$ **19.** $1 - 2\sin^2 15°$

20. $1 - 2\sin^2 22\dfrac{1°}{2}$ **21.** $2\cos^2 67\dfrac{1°}{2} - 1$ **22.** $\cos^2\dfrac{\pi}{8} - \dfrac{1}{2}$

23. $\dfrac{\tan 51°}{1-\tan^2 51°}$ **24.** $\dfrac{\tan 34°}{2(1-\tan^2 34°)}$ **25.** $\dfrac{1}{4} - \dfrac{1}{2}\sin^2 47.1°$

26. $\dfrac{1}{8}\sin 29.5° \cos 29.5°$ **27.** $\sin^2\dfrac{2\pi}{5} - \cos^2\dfrac{2\pi}{5}$ **28.** $\cos^2 2x - \sin^2 2x$

Express each function as a trigonometric function of x. **See Example 4.**

29. $\sin 4x$ **30.** $\cos 3x$ **31.** $\tan 3x$ **32.** $\cos 4x$

Write each expression as a sum or difference of trigonometric functions. **See Example 6.**

33. $2\sin 58° \cos 102°$ **34.** $2\cos 85° \sin 140°$ **35.** $2\sin\dfrac{\pi}{6}\cos\dfrac{\pi}{3}$

36. $5\cos 3x \cos 2x$ **37.** $6\sin 4x \sin 5x$ **38.** $8\sin 7x \sin 9x$

Write each expression as a product of trigonometric functions. **See Example 7.**

39. $\cos 4x - \cos 2x$ **40.** $\cos 5x + \cos 8x$ **41.** $\sin 25° + \sin(-48°)$

42. $\sin 102° - \sin 95°$ **43.** $\cos 4x + \cos 8x$ **44.** $\sin 9x - \sin 3x$

Answers (margin)

1. C 2. E
3. B 4. A
5. F 6. D

7. $\cos 2\theta = \frac{17}{25}$; $\sin 2\theta = -\frac{4\sqrt{21}}{25}$
8. $\cos 2\theta = \frac{119}{169}$; $\sin 2\theta = -\frac{120}{169}$
9. $\cos 2x = -\frac{3}{5}$; $\sin 2x = \frac{4}{5}$
10. $\cos 2x = -\frac{8}{17}$; $\sin 2x = \frac{15}{17}$
11. $\cos 2\theta = \frac{39}{49}$; $\sin 2\theta = -\frac{4\sqrt{55}}{49}$
12. $\cos 2\theta = -\frac{19}{25}$; $\sin 2\theta = \frac{2\sqrt{66}}{25}$
13. $\cos\theta = \frac{2\sqrt{5}}{5}$; $\sin\theta = \frac{\sqrt{5}}{5}$
14. $\cos\theta = -\frac{\sqrt{14}}{4}$; $\sin\theta = -\frac{\sqrt{2}}{4}$
15. $\cos\theta = -\frac{\sqrt{42}}{12}$; $\sin\theta = \frac{\sqrt{102}}{12}$
16. $\cos\theta = -\frac{\sqrt{30}}{6}$; $\sin\theta = \frac{\sqrt{6}}{6}$

17. $\frac{\sqrt{3}}{2}$ 18. $\frac{\sqrt{3}}{3}$
19. $\frac{\sqrt{3}}{2}$ 20. $\frac{\sqrt{2}}{2}$
21. $-\frac{\sqrt{2}}{2}$ 22. $\frac{\sqrt{2}}{4}$
23. $\frac{1}{2}\tan 102°$ 24. $\frac{1}{4}\tan 68°$
25. $\frac{1}{4}\cos 94.2°$ 26. $\frac{1}{16}\sin 59°$
27. $-\cos\frac{4\pi}{5}$ 28. $\cos 4x$

29. $\sin 4x = 4\sin x\cos^3 x - 4\sin^3 x\cos x$
30. $\cos 3x = 4\cos^3 x - 3\cos x$
31. $\tan 3x = \frac{3\tan x - \tan^3 x}{1 - 3\tan^2 x}$
32. $\cos 4x = 8\cos^4 x - 8\cos^2 x + 1$

33. $\sin 160° - \sin 44°$
34. $\sin 225° + \sin 55°$
35. $\sin\frac{\pi}{2} - \sin\frac{\pi}{6}$
36. $\frac{5}{2}\cos 5x + \frac{5}{2}\cos x$
37. $3\cos x - 3\cos 9x$
38. $4\cos 2x - 4\cos 16x$

39. $-2\sin 3x\sin x$
40. $2\cos 6.5x\cos 1.5x$
41. $-2\sin 11.5°\cos 36.5°$
42. $2\cos 98.5°\sin 3.5°$
43. $2\cos 6x\cos 2x$
44. $2\cos 6x\sin 3x$

45. C 46. A
47. D 48. E
49. F 50. B

51. $\frac{\sqrt{2+\sqrt{2}}}{2}$ 52. $-\frac{\sqrt{2-\sqrt{3}}}{2}$
53. $2-\sqrt{3}$ 54. $-\frac{\sqrt{2+\sqrt{3}}}{2}$
55. $-\frac{\sqrt{2+\sqrt{3}}}{2}$ 56. $\frac{\sqrt{2-\sqrt{3}}}{2}$

57. $\frac{\sqrt{10}}{4}$ 58. $\frac{\sqrt{13}}{4}$
59. 3 60. $-\frac{\sqrt{5}}{5}$
61. $\frac{\sqrt{50-10\sqrt{5}}}{10}$ 62. $\frac{\sqrt{50-15\sqrt{10}}}{10}$
63. $-\sqrt{7}$ 64. $\frac{\sqrt{5}}{5}$
65. $\frac{\sqrt{5}}{5}$ 66. $-\frac{\sqrt{3}}{2}$
67. $-\frac{\sqrt{42}}{12}$ 68. $-\frac{\sqrt{6}}{6}$

69. $\sin 20°$ 70. $\cos 38°$
71. $\tan 73.5°$ 72. $\cot 82.5°$
73. $\tan 29.87°$ 74. $\tan 79.1°$
75. $\cos 9x$ 76. $\cos 10\alpha$
77. $\tan 4\theta$ 78. $\tan \frac{5A}{2}$
79. $\cos \frac{x}{8}$ 80. $\sin \frac{3\theta}{10}$

Match each expression in Column I with its value in Column II. ***See Examples 8 and 9.***

I		II	
45. $\sin 15°$	**46.** $\tan 15°$	**A.** $2-\sqrt{3}$	**B.** $\frac{\sqrt{2-\sqrt{2}}}{2}$
47. $\cos \frac{\pi}{8}$	**48.** $\tan\left(-\frac{\pi}{8}\right)$	**C.** $\frac{\sqrt{2-\sqrt{3}}}{2}$	**D.** $\frac{\sqrt{2+\sqrt{2}}}{2}$
49. $\tan 67.5°$	**50.** $\cos 67.5°$	**E.** $1-\sqrt{2}$	**F.** $1+\sqrt{2}$

Use a half-angle identity to find each exact value. ***See Examples 8 and 9.***

51. $\sin 67.5°$ **52.** $\sin 195°$ **53.** $\tan 195°$

54. $\cos 195°$ **55.** $\cos 165°$ **56.** $\sin 165°$

Find each of the following. ***See Example 10.***

57. $\cos \frac{x}{2}$, given $\cos x = \frac{1}{4}$, with $0 < x < \frac{\pi}{2}$

58. $\sin \frac{x}{2}$, given $\cos x = -\frac{5}{8}$, with $\frac{\pi}{2} < x < \pi$

59. $\tan \frac{\theta}{2}$, given $\sin \theta = \frac{3}{5}$, with $90° < \theta < 180°$

60. $\cos \frac{\theta}{2}$, given $\sin \theta = -\frac{4}{5}$, with $180° < \theta < 270°$

61. $\sin \frac{x}{2}$, given $\tan x = 2$, with $0 < x < \frac{\pi}{2}$

62. $\cos \frac{x}{2}$, given $\cot x = -3$, with $\frac{\pi}{2} < x < \pi$

63. $\tan \frac{\theta}{2}$, given $\tan \theta = \frac{\sqrt{7}}{3}$, with $180° < \theta < 270°$

64. $\cot \frac{\theta}{2}$, given $\tan \theta = -\frac{\sqrt{5}}{2}$, with $90° < \theta < 180°$

65. $\sin \theta$, given $\cos 2\theta = \frac{3}{5}$ and θ terminates in quadrant I

66. $\cos \theta$, given $\cos 2\theta = \frac{1}{2}$ and θ terminates in quadrant II

67. $\cos x$, given $\cos 2x = -\frac{5}{12}$, with $\frac{\pi}{2} < x < \pi$

68. $\sin x$, given $\cos 2x = \frac{2}{3}$, with $\pi < x < \frac{3\pi}{2}$

Use an identity to write each expression as a single trigonometric function. ***See Example 11.***

69. $\sqrt{\frac{1-\cos 40°}{2}}$ **70.** $\sqrt{\frac{1+\cos 76°}{2}}$ **71.** $\sqrt{\frac{1-\cos 147°}{1+\cos 147°}}$

72. $\sqrt{\frac{1+\cos 165°}{1-\cos 165°}}$ **73.** $\frac{1-\cos 59.74°}{\sin 59.74°}$ **74.** $\frac{\sin 158.2°}{1+\cos 158.2°}$

75. $\pm\sqrt{\frac{1+\cos 18x}{2}}$ **76.** $\pm\sqrt{\frac{1+\cos 20\alpha}{2}}$ **77.** $\pm\sqrt{\frac{1-\cos 8\theta}{1+\cos 8\theta}}$

78. $\pm\sqrt{\frac{1-\cos 5A}{1+\cos 5A}}$ **79.** $\pm\sqrt{\frac{1+\cos \frac{x}{4}}{2}}$ **80.** $\pm\sqrt{\frac{1-\cos \frac{3\theta}{5}}{2}}$

Verify that each equation is an identity. ***See Example 12.***

81. $(\sin x + \cos x)^2 = \sin 2x + 1$ **82.** $\sec 2x = \frac{\sec^2 x + \sec^4 x}{2 + \sec^2 x - \sec^4 x}$

83. $(\cos 2x + \sin 2x)^2 = 1 + \sin 4x$ **84.** $(\cos 2x - \sin 2x)^2 = 1 - \sin 4x$

85. $\tan 8\theta - \tan 8\theta \tan^2 4\theta = 2 \tan 4\theta$ **86.** $\sin 2x = \frac{2 \tan x}{1 + \tan^2 x}$

101. $\cos^4 x - \sin^4 x = \cos 2x$

102. $\frac{4 \tan x \cos^2 x - 2 \tan x}{1 - \tan^2 x} = \sin 2x$

103. $\frac{2 \tan x}{2 - \sec^2 x} = \tan 2x$

104. $\frac{\cot^2 x - 1}{2 \cot x} = \cot 2x$

105. $\frac{\sin x}{1 + \cos x} = \tan \frac{x}{2}$

106. $\frac{1 - \cos x}{\sin x} = \tan \frac{x}{2}$

107. $\frac{\tan \frac{x}{2} + \cot \frac{x}{2}}{\cot \frac{x}{2} - \tan \frac{x}{2}} = \sec x$

108. $1 - 8 \sin^2 \frac{x}{2} \cos^2 \frac{x}{2} = \cos 2x$

109. $a = -885.6$; $c = 885.6$; $\omega = 240\pi$

87. $\cos 2\theta = \dfrac{2 - \sec^2 \theta}{\sec^2 \theta}$

88. $\tan 2\theta = \dfrac{-2 \tan \theta}{\sec^2 \theta - 2}$

89. $\sin 4x = 4 \sin x \cos x \cos 2x$

90. $\dfrac{1 + \cos 2x}{\sin 2x} = \cot x$

91. $\dfrac{2 \cos 2\theta}{\sin 2\theta} = \cot \theta - \tan \theta$

92. $\cot 4\theta = \dfrac{1 - \tan^2 2\theta}{2 \tan 2\theta}$

93. $\sec^2 \dfrac{x}{2} = \dfrac{2}{1 + \cos x}$

94. $\cot^2 \dfrac{x}{2} = \dfrac{(1 + \cos x)^2}{\sin^2 x}$

95. $\sin^2 \dfrac{x}{2} = \dfrac{\tan x - \sin x}{2 \tan x}$

96. $\dfrac{\sin 2x}{2 \sin x} = \cos^2 \dfrac{x}{2} - \sin^2 \dfrac{x}{2}$

97. $\dfrac{2}{1 + \cos x} - \tan^2 \dfrac{x}{2} = 1$

98. $\tan \dfrac{\theta}{2} = \csc \theta - \cot \theta$

99. $1 - \tan^2 \dfrac{\theta}{2} = \dfrac{2 \cos \theta}{1 + \cos \theta}$

100. $\cos x = \dfrac{1 - \tan^2 \frac{x}{2}}{1 + \tan^2 \frac{x}{2}}$

Graph each expression and use the graph to make a conjecture, predicting what might be an identity. Then verify your conjecture algebraically.

101. $\cos^4 x - \sin^4 x$

102. $\dfrac{4 \tan x \cos^2 x - 2 \tan x}{1 - \tan^2 x}$

103. $\dfrac{2 \tan x}{2 - \sec^2 x}$

104. $\dfrac{\cot^2 x - 1}{2 \cot x}$

105. $\dfrac{\sin x}{1 + \cos x}$

106. $\dfrac{1 - \cos x}{\sin x}$

107. $\dfrac{\tan \frac{x}{2} + \cot \frac{x}{2}}{\cot \frac{x}{2} - \tan \frac{x}{2}}$

108. $1 - 8 \sin^2 \dfrac{x}{2} \cos^2 \dfrac{x}{2}$

(Modeling) Solve each problem. ***See Example 5.***

109. *Wattage Consumption* Use an identity to determine values of a, c, and ω in **Example 5** so that

$$W = a \cos(\omega t) + c.$$

Check your answer by graphing both expressions for W on the same coordinate axes.

110. *Amperage, Wattage, and Voltage* Amperage is a measure of the amount of electricity that is moving through a circuit, whereas voltage is a measure of the force pushing the electricity. The wattage W consumed by an electrical device can be determined by calculating the product of the amperage I and voltage V. (*Source:* Wilcox, G. and C. Hesselberth, *Electricity for Engineering Technology,* Allyn & Bacon.)

110. (a)

For $x = t$, $W = VI =$ $(163 \sin 120\pi t)(1.23 \sin 120\pi t)$

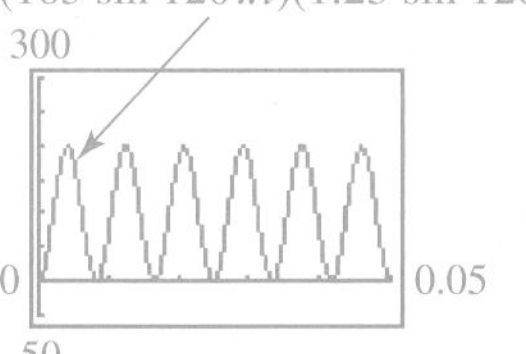

(b) maximum: 200.49 watts; minimum: 0 watts

(c) $a = -100.245$; $\omega = 240\pi$; $c = 100.245$

(e) 100 watts

111. 106°
112. 84°
113. 2
114. 3.9

115. (a) $\cos \frac{\theta}{2} = \frac{R-b}{R}$

(b) $\tan \frac{\theta}{4} = \frac{b}{50}$

116. 54°

(a) A household circuit has voltage

$$V = 163 \sin 120\pi t$$

when an incandescent light bulb is turned on with amperage

$$I = 1.23 \sin 120\pi t.$$

Graph the wattage $W = VI$ consumed by the light bulb in the window $[0, 0.05]$ by $[-50, 300]$.

(b) Determine the maximum and minimum wattages used by the light bulb.

(c) Use identities to determine values for a, c, and ω so that $W = a \cos(\omega t) + c$.

(d) Check your answer by graphing both expressions for W on the same coordinate axes.

(e) Use the graph to estimate the average wattage used by the light. For how many watts (to the nearest integer) do you think this incandescent light bulb is rated?

(Modeling) Mach Number An airplane flying faster than sound sends out sound waves that form a cone, as shown in the figure. The cone intersects the ground to form a ***hyperbola.*** *As this hyperbola passes over a particular point on the ground, a sonic boom is heard at that point. If θ is the angle at the vertex of the cone, then*

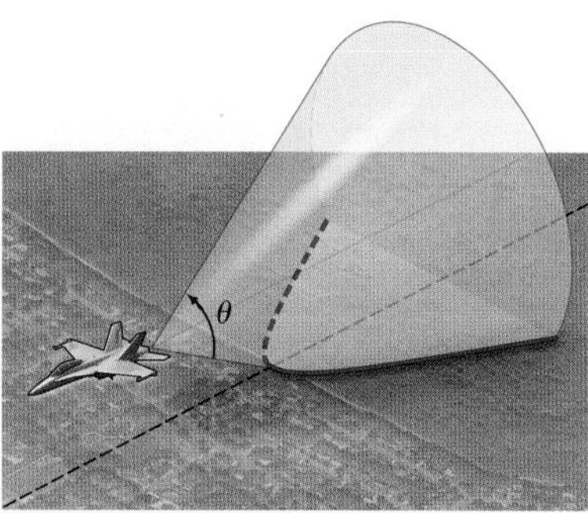

$$\sin \frac{\theta}{2} = \frac{1}{m},$$

where m is the Mach number for the speed of the plane. (We assume $m > 1$.) The Mach number is the ratio of the speed of the plane to the speed of sound. Thus, a speed of Mach 1.4 means that the plane is flying at 1.4 times the speed of sound. In Exercises 111–114, one of the values θ or m is given. Find the other value.

111. $m = \frac{5}{4}$

112. $m = \frac{3}{2}$

113. $\theta = 60°$

114. $\theta = 30°$

(Modeling) Solve each problem.

115. *Railroad Curves* In the United States, circular railroad curves are designated by the **degree of curvature,** the central angle subtended by a chord of 100 ft. See the figure. (*Source:* Hay, W. W., *Railroad Engineering,* John Wiley and Sons.)

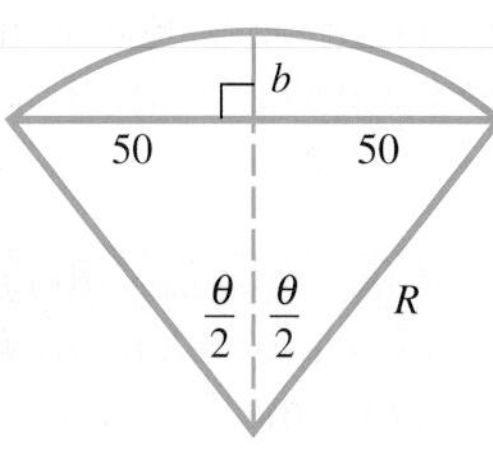

(a) Use the figure to write an expression for $\cos \frac{\theta}{2}$.

(b) Use the result of part (a) and the third half-angle identity for tangent to write an expression for $\tan \frac{\theta}{4}$.

116. In **Exercise 115,** if $b = 12$, what is the measure of angle θ to the nearest degree?

Summary Exercises on Verifying Trigonometric Identities

Teaching Tip Remind students that if they get stuck while trying to verify an identity, they should move on to the next one. Their brains may continue to work on the original one subconsciously, and they may be able to come back and finish it very quickly at a later time.

These summary exercises provide practice with the various types of trigonometric identities presented in this chapter. Verify that each equation is an identity.

1. $\tan \theta + \cot \theta = \sec \theta \csc \theta$

2. $\csc \theta \cos^2 \theta + \sin \theta = \csc \theta$

3. $\tan \frac{x}{2} = \csc x - \cot x$

4. $\sec(\pi - x) = -\sec x$

5. $\frac{\sin t}{1 + \cos t} = \frac{1 - \cos t}{\sin t}$

6. $\frac{1 - \sin t}{\cos t} = \frac{1}{\sec t + \tan t}$

7. $\sin 2\theta = \frac{2 \tan \theta}{1 + \tan^2 \theta}$

8. $\frac{2}{1 + \cos x} - \tan^2 \frac{x}{2} = 1$

9. $\cot \theta - \tan \theta = \frac{2 \cos^2 \theta - 1}{\sin \theta \cos \theta}$

10. $\frac{1}{\sec t - 1} + \frac{1}{\sec t + 1} = 2 \cot t \csc t$

11. $\frac{\sin(x + y)}{\cos(x - y)} = \frac{\cot x + \cot y}{1 + \cot x \cot y}$

12. $1 - \tan^2 \frac{\theta}{2} = \frac{2 \cos \theta}{1 + \cos \theta}$

13. $\frac{\sin \theta + \tan \theta}{1 + \cos \theta} = \tan \theta$

14. $\csc^4 x - \cot^4 x = \frac{1 + \cos^2 x}{1 - \cos^2 x}$

15. $\cos x = \frac{1 - \tan^2 \frac{x}{2}}{1 + \tan^2 \frac{x}{2}}$

16. $\cos 2x = \frac{2 - \sec^2 x}{\sec^2 x}$

17. $\frac{\tan^2 t + 1}{\tan t \csc^2 t} = \tan t$

18. $\frac{\sin s}{1 + \cos s} + \frac{1 + \cos s}{\sin s} = 2 \csc s$

19. $\tan 4\theta = \frac{2 \tan 2\theta}{2 - \sec^2 2\theta}$

20. $\tan\left(\frac{x}{2} + \frac{\pi}{4}\right) = \sec x + \tan x$

21. $\frac{\cot s - \tan s}{\cos s + \sin s} = \frac{\cos s - \sin s}{\sin s \cos s}$

22. $\frac{\tan \theta - \cot \theta}{\tan \theta + \cot \theta} = 1 - 2 \cos^2 \theta$

23. $\frac{\tan(x + y) - \tan y}{1 + \tan(x + y) \tan y} = \tan x$

24. $2 \cos^2 \frac{x}{2} \tan x = \tan x + \sin x$

25. $\frac{\cos^4 x - \sin^4 x}{\cos^2 x} = 1 - \tan^2 x$

26. $\frac{\csc t + 1}{\csc t - 1} = (\sec t + \tan t)^2$

27. $\frac{2(\sin x - \sin^3 x)}{\cos x} = \sin 2x$

28. $\frac{1}{2} \cot \frac{x}{2} - \frac{1}{2} \tan \frac{x}{2} = \cot x$

7.5 Inverse Circular Functions

- Review of Inverse Functions
- Inverse Sine Function
- Inverse Cosine Function
- Inverse Tangent Function
- Remaining Inverse Circular Functions
- Inverse Function Values

Review of Inverse Functions

Recall from **Section 4.1** that if a function is defined so that ***each range element is used only once,*** then it is a **one-to-one function.** For example, the function

$$f(x) = x^3 \text{ is a one-to-one function}$$

because every real number has exactly one real cube root. However,

$$g(x) = x^2 \text{ is not a one-to-one function}$$

because $g(2) = 4$ and $g(-2) = 4$. There are two domain elements, 2 and -2, that correspond to the range element 4.

By interchanging the components of the ordered pairs of a one-to-one function f, we obtain a new set of ordered pairs that satisfies the definition of a function. Recall that the **inverse function** of the one-to-one function f is defined as follows.

$$f^{-1} = \{(y, x) \mid (x, y) \text{ belongs to } f\}$$

The special notation used for inverse functions is f^{-1} (read **"f-inverse"**). In simple terms, it represents the function created by interchanging the input (domain) and the output (range) of a one-to-one function.

CAUTION ***Do not confuse the −1 in f^{-1} with a negative exponent.*** The symbol $f^{-1}(x)$ does *not* represent $\frac{1}{f(x)}$. It represents the inverse function of f.

The following statements review the concepts of inverse functions.

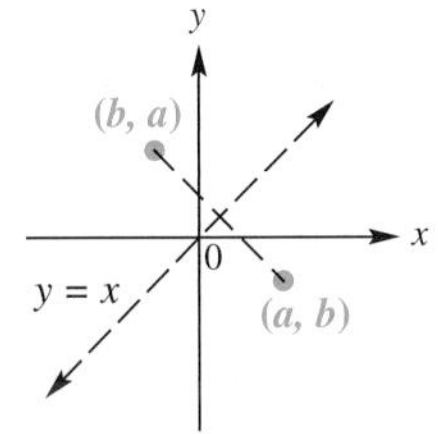

(b, a) is the reflection of (a, b) across the line y = x.

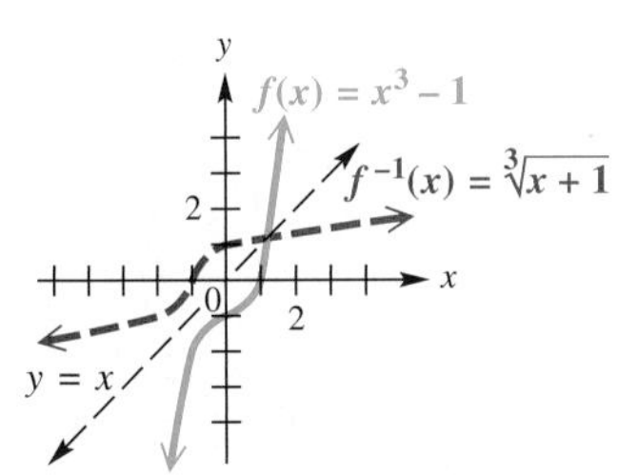

The graph of f^{-1} is the reflection of the graph of f across the line $y = x$.

Figure 9

Review of Inverse Functions

1. In a one-to-one function, each x-value corresponds to only one y-value and each y-value corresponds to only one x-value.
2. If a function f is one-to-one, then f has an inverse function f^{-1}.
3. The domain of f is the range of f^{-1}, and the range of f is the domain of f^{-1}. That is, if the point (a, b) is on the graph of f, then (b, a) is on the graph of f^{-1}.
4. The graphs of f and f^{-1} are reflections of each other across the line $y = x$.
5. To find $f^{-1}(x)$ from $f(x)$, follow these steps.

   Step 1 Replace $f(x)$ with y and interchange x and y.

   Step 2 Solve for y.

   Step 3 Replace y with $f^{-1}(x)$.

Figure 9 illustrates some of these concepts.

NOTE Recall that we often restrict the domain of a function that is not one-to-one to make it one-to-one, without changing the range. For example, the function $g(x) = x^2$, with its natural domain $(-\infty, \infty)$, is not one-to-one. However, if we restrict its domain to the set of nonnegative numbers $[0, \infty)$, we obtain a new function f that is one-to-one and has the same range as g, $[0, \infty)$. See the figure in the margin.

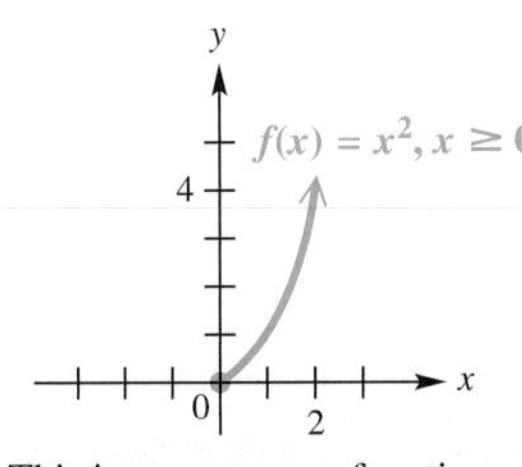

This is a one-to-one function.

Inverse Sine Function Refer to the graph of the sine function in **Figure 10** on the next page. Applying the horizontal line test, we see that $y = \sin x$ does not define a one-to-one function. If we restrict the domain to the interval $\left[-\frac{\pi}{2}, \frac{\pi}{2}\right]$, which is the part of the graph in **Figure 10** shown in color, this restricted function is one-to-one and has an inverse function. The range of $y = \sin x$ is $[-1, 1]$, so the domain of the inverse function will be $[-1, 1]$, and its range will be $\left[-\frac{\pi}{2}, \frac{\pi}{2}\right]$.

Teaching Tip Mention that the interval $\left[-\frac{\pi}{2}, \frac{\pi}{2}\right]$ contains enough of the graph of the sine function to include all possible values of y. Other intervals could also be used, but this interval is an accepted convention that is adopted by mathematicians and used by scientific calculators and graphing calculators.

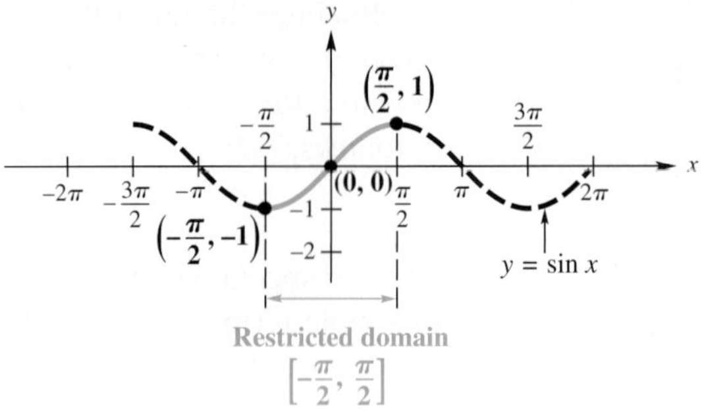

Figure 10

Reflecting the graph of $y = \sin x$ on the restricted domain, shown in **Figure 11(a),** across the line $y = x$ gives the graph of the inverse function, shown in **Figure 11(b).** Some key points are labeled on the graph. The equation of the inverse of $y = \sin x$ is found by interchanging x and y to get

$$x = \sin y.$$

This equation is solved for y by writing

$$\boldsymbol{y = \sin^{-1} x} \quad \text{(read \textbf{“inverse sine of } \textit{x}\textbf{”})}.$$

As **Figure 11(b)** shows, the domain of $y = \sin^{-1} x$ is $[-1, 1]$, while the restricted domain of $y = \sin x$, $\left[-\frac{\pi}{2}, \frac{\pi}{2}\right]$, is the range of $y = \sin^{-1} x$. An alternative notation for $\sin^{-1} x$ is $\arcsin x$.

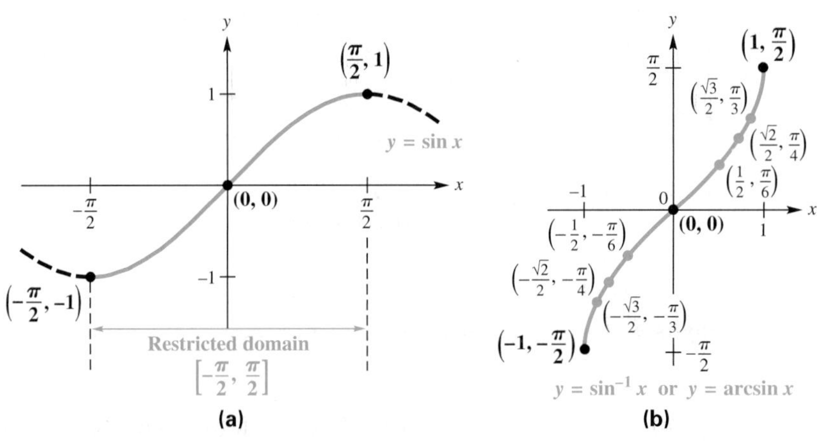

Figure 11

Teaching Tip In problems like **Example 1(a),** use phrases such as

“y is a value *in radians* between $-\frac{\pi}{2}$ and $\frac{\pi}{2}$ whose sine is equal to $\frac{1}{2}$”

to help students understand the meaning of inverse sine. Illustrate the answer graphically by graphing $y = \sin x$ and $y = \frac{1}{2}$ in the interval $\left[-\frac{\pi}{2}, \frac{\pi}{2}\right]$, and then determine the point of intersection.

Inverse Sine Function

$\boldsymbol{y = \sin^{-1} x}$ or $\boldsymbol{y = \arcsin x}$ means that $x = \sin y$, for $-\frac{\pi}{2} \le y \le \frac{\pi}{2}$.

We can think of $\boldsymbol{y = \sin^{-1} x}$ ***or*** $\boldsymbol{y = \arcsin x}$ ***as***

“y is the number (angle) in the interval $\left[-\frac{\pi}{2}, \frac{\pi}{2}\right]$ ***whose sine is x.”***

Thus, we can write $y = \sin^{-1} x$ as $\sin y = x$ to evaluate it. We must pay close attention to the domain and range intervals.

EXAMPLE 1 Finding Inverse Sine Values

Find y in each equation.

(a) $y = \arcsin \frac{1}{2}$ **(b)** $y = \sin^{-1}(-1)$ **(c)** $y = \sin^{-1}(-2)$

ALGEBRAIC SOLUTION

(a) The graph of the function defined by $y = \arcsin x$ (**Figure 11(b)**) includes the point $\left(\frac{1}{2}, \frac{\pi}{6}\right)$. Therefore, $\arcsin \frac{1}{2} = \frac{\pi}{6}$.

Alternatively, we can think of $y = \arcsin \frac{1}{2}$ as "y is the number in $\left[-\frac{\pi}{2}, \frac{\pi}{2}\right]$ whose sine is $\frac{1}{2}$." Then we can write the given equation as $\sin y = \frac{1}{2}$. Since $\sin \frac{\pi}{6} = \frac{1}{2}$ and $\frac{\pi}{6}$ is in the range of the arcsine function, $y = \frac{\pi}{6}$.

(b) Writing the equation $y = \sin^{-1}(-1)$ in the form $\sin y = -1$ shows that $y = -\frac{\pi}{2}$. Notice that the point $\left(-1, -\frac{\pi}{2}\right)$ is on the graph of $y = \sin^{-1} x$.

(c) Because -2 is not in the domain of the inverse sine function, $\sin^{-1}(-2)$ does not exist.

GRAPHING CALCULATOR SOLUTION

We graph the equation $Y_1 = \sin^{-1} X$ and find the points with X-values $\frac{1}{2} = 0.5$ and -1. For these two X-values, **Figure 12** indicates that $Y = \frac{\pi}{6} \approx 0.52359878$ and $Y = -\frac{\pi}{2} \approx -1.570796$.

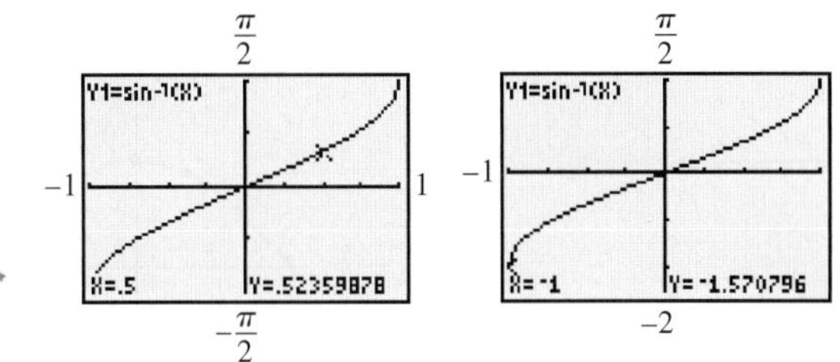

Figure 12

Since $\sin^{-1}(-2)$ does not exist, a calculator will give an error message for this input.

✔ *Now Try Exercises 13, 21, and 25.*

Classroom Example 1
Find y in each equation.

(a) $y = \arcsin \frac{\sqrt{3}}{2}$

(b) $y = \sin^{-1}\left(-\frac{1}{2}\right)$

(c) $y = \sin^{-1} \sqrt{2}$

Answers:

(a) $\frac{\pi}{3}$ (b) $-\frac{\pi}{6}$

(c) $\sin^{-1}\sqrt{2}$ does not exist.

CAUTION In **Example 1(b),** it is tempting to give the value of $\sin^{-1}(-1)$ as $\frac{3\pi}{2}$, since $\sin \frac{3\pi}{2} = -1$. Notice, however, that $\frac{3\pi}{2}$ is not in the range of the inverse sine function. ***Be certain that the number given for an inverse function value is in the range of the particular inverse function being considered.***

We summarize this discussion about the inverse sine function as follows.

Inverse Sine Function $y = \sin^{-1} x$ or $y = \arcsin x$

Domain: $[-1, 1]$ Range: $\left[-\frac{\pi}{2}, \frac{\pi}{2}\right]$

x	y
-1	$-\frac{\pi}{2}$
$-\frac{\sqrt{2}}{2}$	$-\frac{\pi}{4}$
0	0
$\frac{\sqrt{2}}{2}$	$\frac{\pi}{4}$
1	$\frac{\pi}{2}$

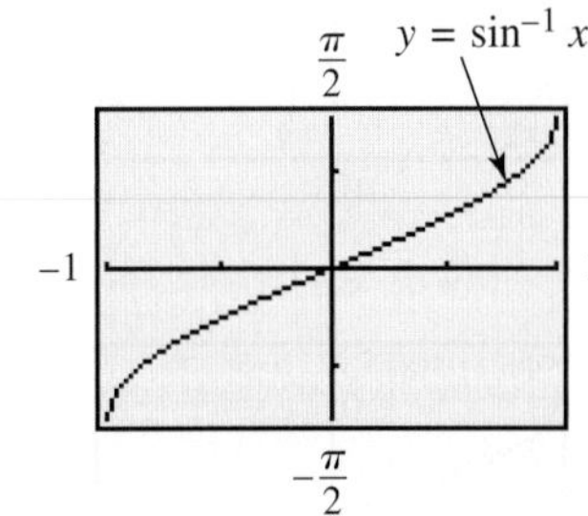

Figure 13

- The inverse sine function is increasing and continuous on its domain $[-1, 1]$.
- Its x-intercept is 0, and its y-intercept is 0.
- Its graph is symmetric with respect to the origin, so the function is an odd function. For all x in the domain, $\sin^{-1}(-x) = -\sin^{-1} x$.

LOOKING AHEAD TO CALCULUS

The **inverse circular functions** are used in calculus to solve certain types of related-rates problems and to integrate certain rational functions.

Inverse Cosine Function The function

$$y = \cos^{-1} x \quad (\text{or } y = \arccos x)$$

is defined by restricting the domain of the function $y = \cos x$ to the interval $[0, \pi]$ as in **Figure 14.** This restricted function, which is the part of the graph in **Figure 14** shown in color, is one-to-one and has an inverse function. The inverse function, $y = \cos^{-1} x$, is found by interchanging the roles of x and y. Reflecting the graph of $y = \cos x$ across the line $y = x$ gives the graph of the inverse function shown in **Figure 15.** Some key points are shown on the graph.

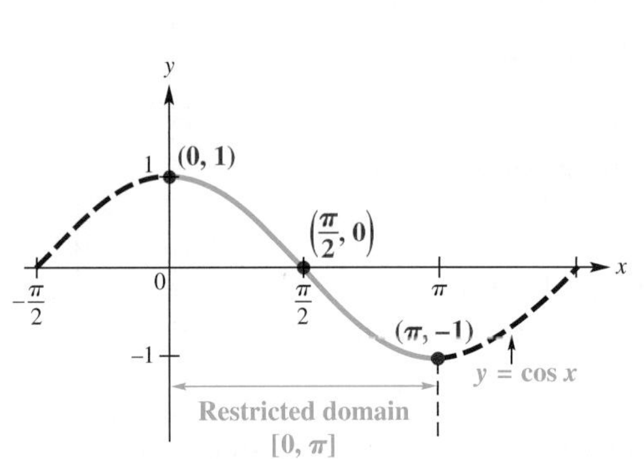

Figure 14

Figure 15

Classroom Example 2
Find y in each equation.
(a) $y = \arccos 0$
(b) $y = \cos^{-1} \frac{1}{2}$

Answers:
(a) $\frac{\pi}{2}$ (b) $\frac{\pi}{3}$

Inverse Cosine Function

$y = \cos^{-1} x$ or $y = \arccos x$ means that $x = \cos y$, for $0 \leq y \leq \pi$.

We can think of $y = \cos^{-1} x$ or $y = \arccos x$ as

"y is the number (angle) in the interval $[0, \pi]$ whose cosine is x."

EXAMPLE 2 Finding Inverse Cosine Values

Find y in each equation.

(a) $y = \arccos 1$ **(b)** $y = \cos^{-1}\left(-\frac{\sqrt{2}}{2}\right)$

SOLUTION

(a) Since the point $(1, 0)$ lies on the graph of $y = \arccos x$ in **Figure 15,** the value of y, or arccos 1, is 0. Alternatively, we can think of $y = \arccos 1$ as

"y is the number in $[0, \pi]$ whose cosine is 1," or $\cos y = 1$.

Thus, $y = 0$, since $\cos 0 = 1$ and 0 is in the range of the arccosine function.

(b) We must find the value of y that satisfies

$$\cos y = -\frac{\sqrt{2}}{2}, \quad \text{where } y \text{ is in the interval } [0, \pi],$$

which is the range of the function $y = \cos^{-1} x$. The only value for y that satisfies these conditions is $\frac{3\pi}{4}$. Again, this can be verified from the graph in **Figure 15.**

Now Try Exercises 15 and 23.

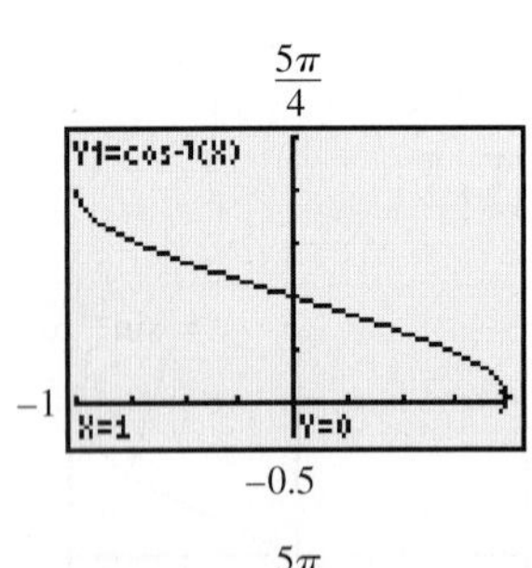

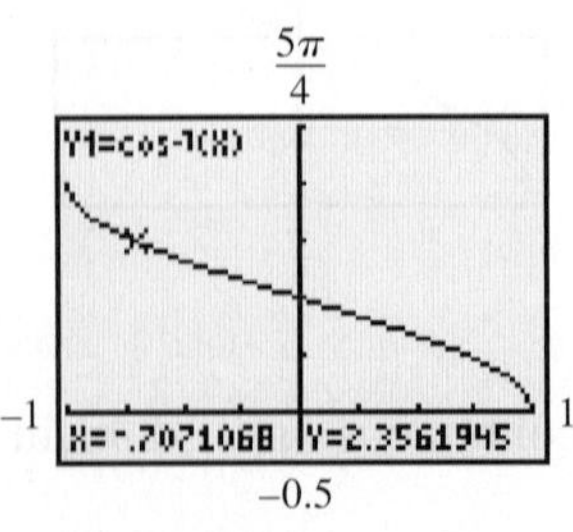

These screens support the results of **Example 2** because $-\frac{\sqrt{2}}{2} \approx -0.7071068$ and $\frac{3\pi}{4} \approx 2.3561945$.

Our observations about the inverse cosine function lead to the following generalizations.

Inverse Cosine Function $y = \cos^{-1} x$ or $y = \arccos x$

Domain: $[-1, 1]$ Range: $[0, \pi]$

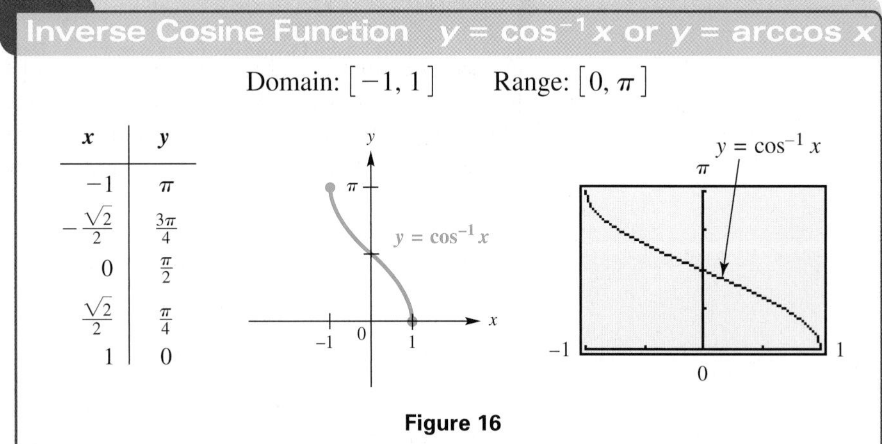

x	y
-1	π
$-\frac{\sqrt{2}}{2}$	$\frac{3\pi}{4}$
0	$\frac{\pi}{2}$
$\frac{\sqrt{2}}{2}$	$\frac{\pi}{4}$
1	0

Figure 16

- The inverse cosine function is decreasing and continuous on its domain $[-1, 1]$.
- Its x-intercept is 1, and its y-intercept is $\frac{\pi}{2}$.
- Its graph is not symmetric with respect to either the y-axis or the origin.

Inverse Tangent Function Restricting the domain of the function $y = \tan x$ to the open interval $\left(-\frac{\pi}{2}, \frac{\pi}{2}\right)$ yields a one-to-one function. By interchanging the roles of x and y, we obtain the inverse tangent function given by $y = \tan^{-1} x$ or $y = \arctan x$. **Figure 17** shows the graph of the restricted tangent function. **Figure 18** gives the graph of $y = \tan^{-1} x$.

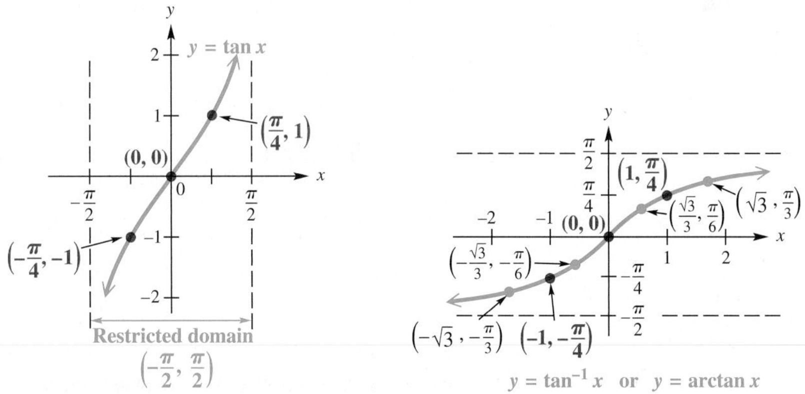

Figure 17 **Figure 18**

Inverse Tangent Function

$\mathbf{y = \tan^{-1} x}$ or $\mathbf{y = \arctan x}$ means that $x = \tan y$, for $-\frac{\pi}{2} < y < \frac{\pi}{2}$.

We can think of $y = \tan^{-1} x$ or $y = \arctan x$ as

"y is the number (angle) in the interval $\left(-\frac{\pi}{2}, \frac{\pi}{2}\right)$ whose tangent is x."

We summarize this discussion about the inverse tangent function as follows.

Inverse Tangent Function $y = \tan^{-1} x$ or $y = \arctan x$

Domain: $(-\infty, \infty)$ Range: $\left(-\frac{\pi}{2}, \frac{\pi}{2}\right)$

x	y
-1	$-\frac{\pi}{4}$
$-\frac{\sqrt{3}}{3}$	$-\frac{\pi}{6}$
0	0
$\frac{\sqrt{3}}{3}$	$\frac{\pi}{6}$
1	$\frac{\pi}{4}$

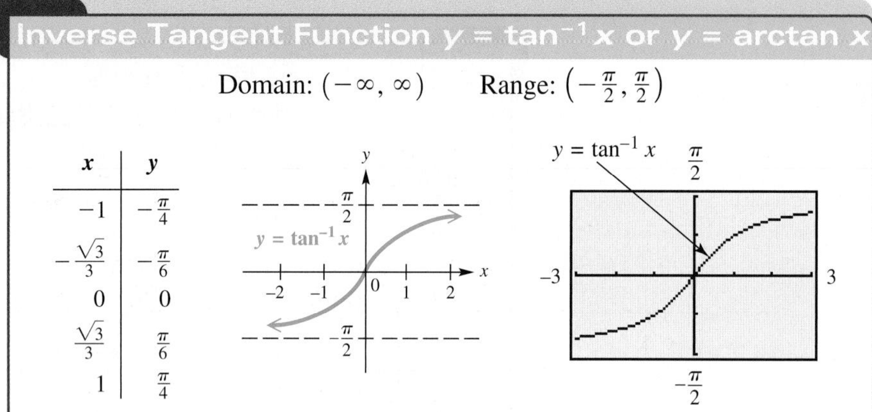

Figure 19

- The inverse tangent function is increasing and continuous on its domain $(-\infty, \infty)$.
- Its x-intercept is 0, and its y-intercept is 0.
- Its graph is symmetric with respect to the origin so the function is an odd function. For all x in the domain, $\tan^{-1}(-x) = -\tan^{-1} x$.
- The lines $y = \frac{\pi}{2}$ and $y = -\frac{\pi}{2}$ are horizontal asymptotes.

Remaining Inverse Circular Functions The remaining three inverse trigonometric functions are defined similarly. Their graphs are shown in **Figure 20.**

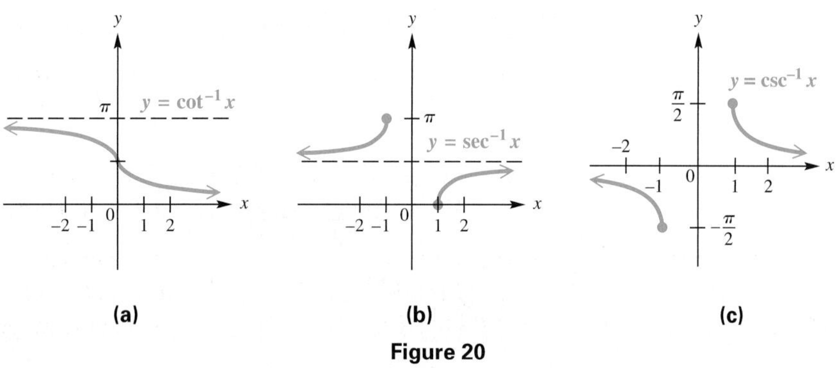

Figure 20

Inverse Cotangent, Secant, and Cosecant Functions*

$y = \cot^{-1} x$ or $y = \operatorname{arccot} x$ means that $x = \cot y$, for $0 < y < \pi$.

$y = \sec^{-1} x$ or $y = \operatorname{arcsec} x$ means that $x = \sec y$, for $0 \le y \le \pi$, $y \ne \frac{\pi}{2}$.

$y = \csc^{-1} x$ or $y = \operatorname{arccsc} x$ means that $x = \csc y$, for $-\frac{\pi}{2} \le y \le \frac{\pi}{2}$, $y \ne 0$.

*The inverse secant and inverse cosecant functions are sometimes defined with different ranges. We use intervals that match those of the inverse cosine and inverse sine functions, respectively (except for one missing point).

The table gives all six inverse trigonometric functions with their domains and ranges.

Inverse Function	Domain	Range	
		Interval	Quadrants of the Unit Circle
$y = \sin^{-1} x$	$[-1, 1]$	$\left[-\frac{\pi}{2}, \frac{\pi}{2}\right]$	I and IV
$y = \cos^{-1} x$	$[-1, 1]$	$[0, \pi]$	I and II
$y = \tan^{-1} x$	$(-\infty, \infty)$	$\left(-\frac{\pi}{2}, \frac{\pi}{2}\right)$	I and IV
$y = \cot^{-1} x$	$(-\infty, \infty)$	$(0, \pi)$	I and II
$y = \sec^{-1} x$	$(-\infty, -1] \cup [1, \infty)$	$\left[0, \frac{\pi}{2}\right) \cup \left(\frac{\pi}{2}, \pi\right]$	I and II
$y = \csc^{-1} x$	$(-\infty, -1] \cup [1, \infty)$	$\left[-\frac{\pi}{2}, 0\right) \cup \left(0, \frac{\pi}{2}\right]$	I and IV

Inverse Function Values The inverse circular functions are formally defined with real number ranges. However, there are times when it may be convenient to find degree-measured angles equivalent to these real number values. It is also often convenient to think in terms of the unit circle and choose the inverse function values on the basis of the quadrants given in the preceding table.

Classroom Example 3
Find the *degree measure* of θ in the following.
(a) $\theta = \arctan \sqrt{3}$
(b) $\theta = \csc^{-1}(-\sqrt{2})$

Answers:
(a) $60°$ (b) $-45°$

EXAMPLE 3 Finding Inverse Function Values (Degree-Measured Angles)

Find the *degree measure* of θ in the following.

(a) $\theta = \arctan 1$ **(b)** $\theta = \sec^{-1} 2$

SOLUTION

(a) Here θ must be in $(-90°, 90°)$, but since 1 is positive, θ must be in quadrant I. The alternative statement, $\tan \theta = 1$, leads to $\theta = 45°$.

(b) Write the equation as $\sec \theta = 2$. For $\sec^{-1} x$, θ is in quadrant I or II. Because 2 is positive, θ is in quadrant I and $\theta = 60°$, since $\sec 60° = 2$. Note that $60°$ (the degree equivalent of $\frac{\pi}{3}$) is in the range of the inverse secant function.

Now Try Exercises 37 and 45.

The inverse trigonometric function keys on a calculator give correct results for the inverse sine, inverse cosine, and inverse tangent functions.

$$\sin^{-1} 0.5 = 30°, \quad \sin^{-1}(-0.5) = -30°,$$
$$\tan^{-1}(-1) = -45°, \quad \text{and} \quad \cos^{-1}(-0.5) = 120° \qquad \text{Degree mode}$$

However, finding $\cot^{-1} x$, $\sec^{-1} x$, and $\csc^{-1} x$ with a calculator is not as straightforward, because these functions must first be expressed in terms of $\tan^{-1} x$, $\cos^{-1} x$, and $\sin^{-1} x$, respectively. If $y = \sec^{-1} x$, for example, then $\sec y = x$, which must be written in terms of cosine as follows.

$$\text{If } \sec y = x, \quad \text{then} \quad \frac{1}{\cos y} = x, \quad \text{or} \quad \cos y = \frac{1}{x}, \quad \text{and} \quad y = \cos^{-1} \frac{1}{x}.$$

Use the following to evaluate these inverse trigonometric functions on a calculator.

$$\sec^{-1} x \textit{ can be evaluated as } \cos^{-1}\frac{1}{x}; \quad \csc^{-1} x \textit{ can be evaluated as } \sin^{-1}\frac{1}{x};$$

$$\cot^{-1} x \textit{ can be evaluated as } \begin{cases} \tan^{-1}\frac{1}{x} & \text{if } x > 0 \\ 180° + \tan^{-1}\frac{1}{x} & \text{if } x < 0. \end{cases}$$

Degree mode

Classroom Example 4
Use a calculator to give each value.
(a) Find y in radians if $y = \sec^{-1}(-4)$.
(b) Find θ in degrees if $\theta = \text{arccot}(-0.2528)$.

Answers:
(a) 1.823476582
(b) 104.1871349°

EXAMPLE 4 Finding Inverse Function Values with a Calculator

Use a calculator to give each value.

(a) Find y in radians if $y = \csc^{-1}(-3)$.

(b) Find θ in degrees if $\theta = \text{arccot}(-0.3541)$.

SOLUTION

(a) With the calculator in radian mode, enter $\csc^{-1}(-3)$ as $\sin^{-1}\left(\frac{1}{-3}\right)$ to get $y \approx -0.3398369095$. See **Figure 21.**

(b) Now set the calculator to degree mode. A calculator gives the inverse tangent value of a negative number as a quadrant IV angle. The restriction on the range of arccotangent implies that θ must be in quadrant II, so enter

$$\text{arccot}(-0.3541) \quad \text{as} \quad \tan^{-1}\left(\frac{1}{-0.3541}\right) + 180°.$$

As shown in **Figure 21,**

$$\theta \approx 109.4990544°.$$

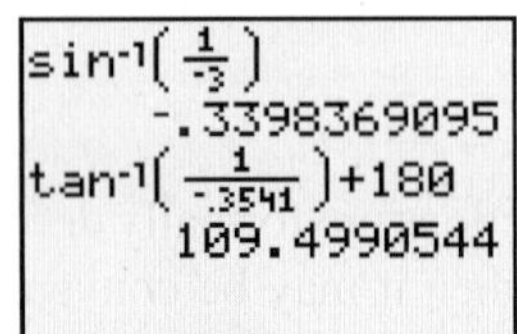

Figure 21

✔ *Now Try Exercises 53 and 65.*

CAUTION ***Be careful when using your calculator to evaluate the inverse cotangent of a negative quantity.*** To do this, we must enter the inverse tangent of the *reciprocal* of the negative quantity, which returns an angle in quadrant IV. Since inverse cotangent is negative in quadrant II, adjust your calculator result by adding 180° or π accordingly. Note that $\cot^{-1} 0 = \frac{\pi}{2}$.

Classroom Example 5
Evaluate each expression without using a calculator.
(a) $\cos\left(\sin^{-1}\frac{2}{3}\right)$
(b) $\sec\left(\cot^{-1}\left(-\frac{15}{8}\right)\right)$

Answers:
(a) $\frac{\sqrt{5}}{3}$ (b) $-\frac{17}{15}$

EXAMPLE 5 Finding Function Values Using Definitions of the Trigonometric Functions

Evaluate each expression without using a calculator.

(a) $\sin\left(\tan^{-1}\frac{3}{2}\right)$ **(b)** $\tan\left(\cos^{-1}\left(-\frac{5}{13}\right)\right)$

SOLUTION

(a) Let $\theta = \tan^{-1}\frac{3}{2}$, so $\tan\theta = \frac{3}{2}$. The inverse tangent function yields values only in quadrants I and IV, and since $\frac{3}{2}$ is positive, θ is in quadrant I. Sketch θ in quadrant I, and label a triangle, as shown in **Figure 22** on the next page. By the Pythagorean theorem, the hypotenuse is $\sqrt{13}$. The value of sine is the quotient of the side opposite and the hypotenuse.

$$\sin\left(\tan^{-1}\frac{3}{2}\right) = \sin\theta = \frac{3}{\sqrt{13}} = \frac{3}{\sqrt{13}} \cdot \frac{\sqrt{13}}{\sqrt{13}} = \frac{3\sqrt{13}}{13} \quad \text{(Section 5.3)}$$

Rationalize the denominator. (Section R.7)

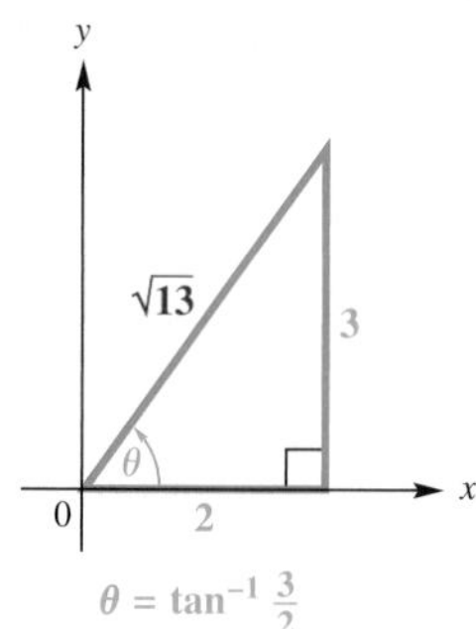

Figure 22

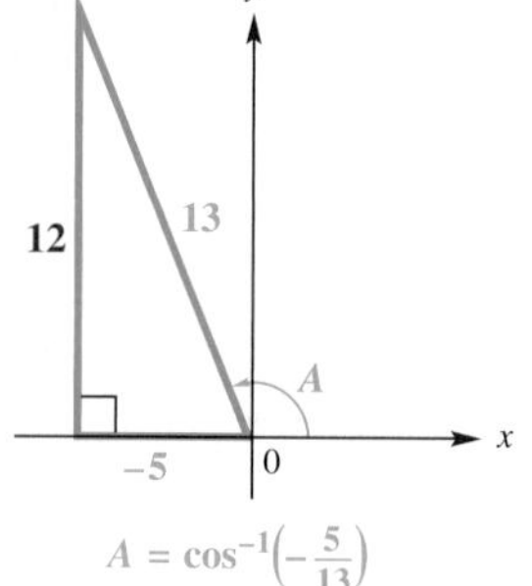

Figure 23

(b) Let $A = \cos^{-1}\left(-\frac{5}{13}\right)$. Then, $\cos A = -\frac{5}{13}$. Since $\cos^{-1} x$ for a negative value of x is in quadrant II, sketch A in quadrant II, as shown in **Figure 23.**

$$\tan\left(\cos^{-1}\left(-\frac{5}{13}\right)\right) = \tan A = -\frac{12}{5}$$

✔ ***Now Try Exercises 79 and 81.***

Classroom Example 6
Evaluate each expression without using a calculator.

(a) $\sin\left(\arctan\frac{4}{3} - \arccos\frac{12}{13}\right)$

(b) $\sin(2\,\text{arccot}(-5))$

Answers:
(a) $\frac{33}{65}$ (b) $-\frac{5}{13}$

EXAMPLE 6 Finding Function Values Using Identities

Evaluate each expression without using a calculator.

(a) $\cos\left(\arctan\sqrt{3} + \arcsin\frac{1}{3}\right)$ **(b)** $\tan\left(2\arcsin\frac{2}{5}\right)$

SOLUTION

(a) Let $A = \arctan\sqrt{3}$ and $B = \arcsin\frac{1}{3}$, so $\tan A = \sqrt{3}$ and $\sin B = \frac{1}{3}$. Sketch both A and B in quadrant I, as shown in **Figure 24,** and use the Pythagorean theorem to find the unknown side in each triangle. Then, use the cosine sum identity.

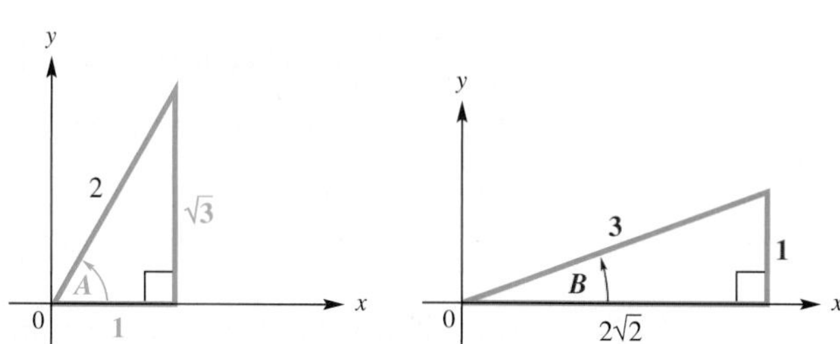

Figure 24

$$\cos\left(\arctan\sqrt{3} + \arcsin\frac{1}{3}\right)$$

$= \cos(A + B)$ Let $A = \arctan\sqrt{3}$ and $B = \arcsin\frac{1}{3}$.

$= \cos A \cos B - \sin A \sin B$ Cosine sum identity **(Section 7.3)**

$= \frac{1}{2} \cdot \frac{2\sqrt{2}}{3} - \frac{\sqrt{3}}{2} \cdot \frac{1}{3}$ Substitute values using **Figure 24.**

$= \frac{2\sqrt{2} - \sqrt{3}}{6}$ Multiply and write as a single fraction. **(Section R.5)**

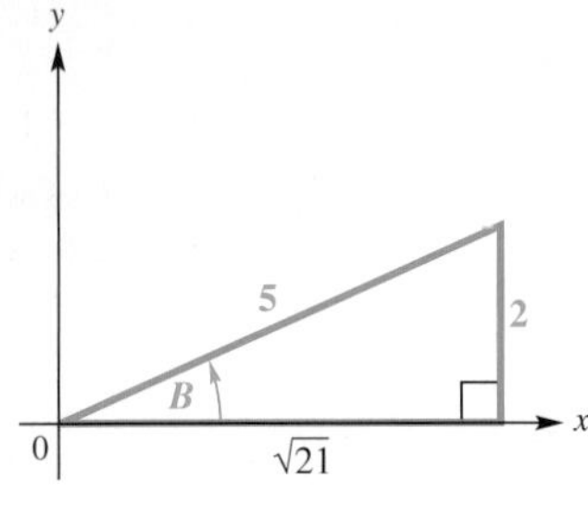

Figure 25

(b) Let $B = \arcsin \frac{2}{5}$, so that $\sin B = \frac{2}{5}$. Sketch angle B in quadrant I, and find the length of the third side of the triangle. Then, use the double-angle tangent identity $\tan 2B = \frac{2 \tan B}{1 - \tan^2 B}$.

$$\tan\left(2 \arcsin \tfrac{2}{5}\right)$$

$$= \frac{2\left(\frac{2}{\sqrt{21}}\right)}{1 - \left(\frac{2}{\sqrt{21}}\right)^2} \quad \tan B = \frac{2}{\sqrt{21}} \text{ from Figure 25. (Section 7.4)}$$

$$= \frac{\frac{4}{\sqrt{21}}}{1 - \frac{4}{21}} \quad \text{Multiply and apply the exponent.}$$

$$= \frac{\frac{4}{\sqrt{21}} \cdot \frac{\sqrt{21}}{\sqrt{21}}}{\frac{17}{21}} \quad \text{Rationalize in the numerator. Subtract in the denominator.}$$

$$= \frac{\frac{4\sqrt{21}}{21}}{\frac{17}{21}} \quad \text{Multiply in the numerator.}$$

$$= \frac{4\sqrt{21}}{17} \quad \text{Divide; } \frac{\frac{a}{b}}{\frac{c}{d}} = \frac{a}{b} \div \frac{c}{d} = \frac{a}{b} \cdot \frac{d}{c}. \text{ (Section R.5)}$$

✔ *Now Try Exercises 83 and 91.*

Teaching Tip Point out that the equations $\sin(\sin^{-1} x) = x$, $\cos(\cos^{-1} x) = x$, and $\tan(\tan^{-1} x) = x$ are true wherever they are defined. However, $\sin^{-1}(\sin x) = x$, $\cos^{-1}(\cos x) = x$, and $\tan^{-1}(\tan x) = x$ are true only for values of x in the *restricted* domains of the sine, cosine, and tangent functions.

While the work shown in **Examples 5 and 6** does not rely on a calculator, we can support our algebraic work with one. By entering $\cos\left(\arctan \sqrt{3} + \arcsin \frac{1}{3}\right)$ from **Example 6(a)** into a calculator, we get the approximation 0.1827293862, the same approximation as when we enter $\frac{2\sqrt{2} - \sqrt{3}}{6}$ (the exact value we obtained algebraically). Similarly, we obtain the same approximation when we evaluate $\tan\left(2 \arcsin \frac{2}{5}\right)$ and $\frac{4\sqrt{21}}{17}$, supporting our answer in **Example 6(b).**

EXAMPLE 7 Writing Function Values in Terms of *u*

Write each trigonometric expression as an algebraic expression in u.

(a) $\sin(\tan^{-1} u)$ **(b)** $\cos(2 \sin^{-1} u)$

SOLUTION

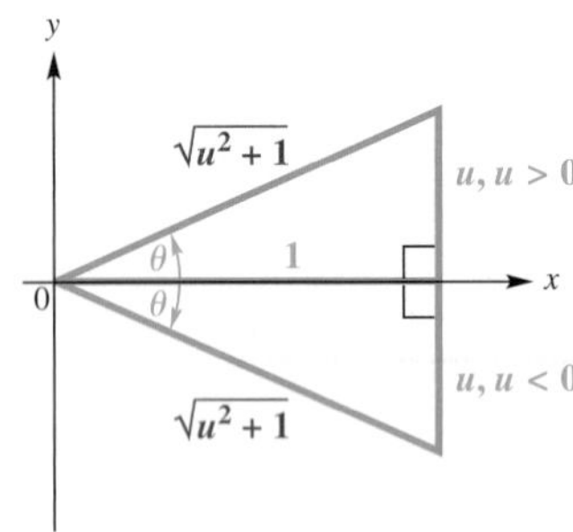

Figure 26

(a) Let $\theta = \tan^{-1} u$, so $\tan \theta = u$. Here, u may be positive or negative. Since $-\frac{\pi}{2} < \tan^{-1} u < \frac{\pi}{2}$, sketch θ in quadrants I and IV and label two triangles, as shown in **Figure 26.** Because sine is given by the quotient of the side opposite and the hypotenuse, we have the following.

$$\sin(\tan^{-1} u) = \sin \theta = \frac{u}{\sqrt{u^2+1}} = \frac{u}{\sqrt{u^2+1}} \cdot \frac{\sqrt{u^2+1}}{\sqrt{u^2+1}} = \frac{u\sqrt{u^2+1}}{u^2+1}$$

Rationalize the denominator.

The result is positive when u is positive and negative when u is negative.

(b) Let $\theta = \sin^{-1} u$, so $\sin \theta = u$. To find $\cos 2\theta$, use the double-angle identity $\cos 2\theta = 1 - 2\sin^2 \theta$.

$$\cos(2 \sin^{-1} u) = \cos 2\theta = 1 - 2\sin^2 \theta = 1 - 2u^2 \quad \text{(Section 7.4)}$$

✔ *Now Try Exercises 99 and 103.*

Classroom Example 7
Write each trigonometric expression as an algebraic expression in u.
(a) $\cot(\sec^{-1} u)$, $u > 0$
(b) $\sin(2 \cos^{-1} u)$

Answers:
(a) $\frac{\sqrt{u^2-1}}{u^2-1}$ (b) $2u\sqrt{1-u^2}$

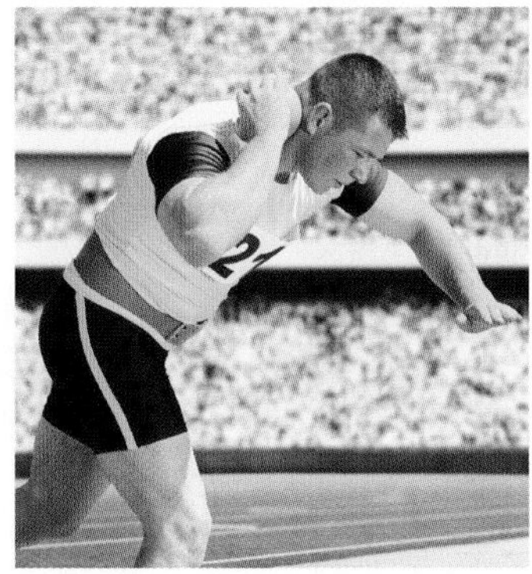

Classroom Example 8
Repeat **Example 8** if $h = 7.5$ ft and $v = 50$ ft per sec.

Answer: 42°

EXAMPLE 8 Finding the Optimal Angle of Elevation of a Shot Put

The optimal angle of elevation θ that a shot-putter should aim for in order to throw the greatest distance depends on the velocity v of the throw and the initial height h of the shot. See **Figure 27.** One model for θ that achieves this greatest distance is

$$\theta = \arcsin\left(\sqrt{\frac{v^2}{2v^2 + 64h}}\right).$$

(*Source:* Townend, M. S., *Mathematics in Sport,* Chichester, Ellis Horwood Limited.)

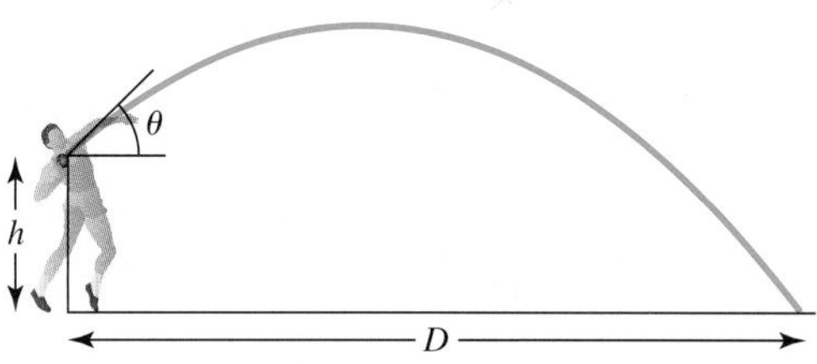

Figure 27

Suppose a shot-putter can consistently throw the steel ball with $h = 6.6$ ft and $v = 42$ ft per sec. At what angle should he release the ball to maximize distance?

SOLUTION To find this angle, substitute and use a calculator in degree mode.

$$\theta = \arcsin\left(\sqrt{\frac{42^2}{2(42^2) + 64(6.6)}}\right) \approx 42° \quad h = 6.6,\ v = 42$$

Now Try Exercise 109.

7.5 Exercises

1. one-to-one
2. range
3. $\cos y$
4. $\left(1, \frac{\pi}{4}\right)$; $y = \tan^{-1} x$, or $y = \arctan x$
5. π
6. Sketch the reflection of the graph of f across the line $y = x$.
7. (a) $[-1, 1]$ (b) $\left[-\frac{\pi}{2}, \frac{\pi}{2}\right]$ (c) increasing (d) -2 is not in the domain.

Concept Check Complete each statement, or answer the question.

1. For a function to have an inverse, it must be ______.
2. The domain of $y = \arcsin x$ equals the ______ of $y = \sin x$.
3. $y = \cos^{-1} x$ means that $x =$ ______, for $0 \le y \le \pi$.
4. The point $\left(\frac{\pi}{4}, 1\right)$ lies on the graph of $y = \tan x$. Therefore, the point ______ lies on the graph of ______.
5. If a function f has an inverse and $f(\pi) = -1$, then $f^{-1}(-1) =$ ______.
6. How can the graph of f^{-1} be sketched if the graph of f is known?

Concept Check In Exercises 7–10, write short answers.

7. Consider the inverse sine function, defined by $y = \sin^{-1} x$ or $y = \arcsin x$.
   (a) What is its domain?
   (b) What is its range?
   (c) Is this function increasing or decreasing?
   (d) Why is $\arcsin(-2)$ not defined?

8. (a) $[-1, 1]$ (b) $[0, \pi]$
(c) decreasing
(d) $-\frac{4\pi}{3}$ is not in the range.
9. (a) $(-\infty, \infty)$ (b) $\left(-\frac{\pi}{2}, \frac{\pi}{2}\right)$
(c) increasing (d) no
10. (a) $(-\infty, -1] \cup [1, \infty)$; $\left[-\frac{\pi}{2}, 0\right) \cup \left(0, \frac{\pi}{2}\right]$
(b) $(-\infty, -1] \cup [1, \infty)$; $\left[0, \frac{\pi}{2}\right) \cup \left(\frac{\pi}{2}, \pi\right]$
(c) $(-\infty, \infty)$; $(0, \pi)$

11. $\cos^{-1}\frac{1}{a}$
12. Find $\tan^{-1}\frac{1}{a} + \pi$, or $\tan^{-1}\frac{1}{a} + 180°$.

13. 0 14. $-\frac{\pi}{2}$
15. π 16. $\frac{\pi}{2}$
17. $\frac{\pi}{4}$ 18. $-\frac{\pi}{4}$
19. 0 20. $-\frac{\pi}{4}$
21. $-\frac{\pi}{3}$ 22. $\frac{\pi}{4}$
23. $\frac{5\pi}{6}$ 24. $\frac{2\pi}{3}$
25. $\sin^{-1}\sqrt{3}$ does not exist.
26. $\arcsin(-\sqrt{2})$ does not exist.
27. $\frac{3\pi}{4}$ 28. $\frac{5\pi}{6}$
29. $-\frac{\pi}{6}$ 30. $\frac{\pi}{4}$
31. $\frac{\pi}{6}$ 32. $\frac{3\pi}{4}$
33. 0
34. $\sec^{-1} 0$ does not exist.
35. $\csc^{-1}\frac{\sqrt{2}}{2}$ does not exist.
36. $\text{arccsc}\left(-\frac{1}{2}\right)$ does not exist.

37. $-45°$ 38. $60°$
39. $-60°$ 40. $-45°$
41. $120°$ 42. $120°$
43. $120°$ 44. $60°$
45. $-30°$ 46. $-90°$
47. $\sin^{-1} 2$ does not exist.
48. $\cos^{-1}(-2)$ does not exist.

49. $-7.6713835°$
50. $51.1691219°$
51. $113.500970°$
52. $97.671207°$
53. $30.987961°$
54. $29.506181°$
55. $121.267893°$
56. $160.172137°$
57. $-82.678329°$
58. $101.267354°$

59. 1.1900238
60. 0.96012698
61. 1.9033723

8. Consider the inverse cosine function, defined by $y = \cos^{-1} x$, or $y = \arccos x$.

(a) What is its domain?
(b) What is its range?
(c) Is this function increasing or decreasing?
(d) $\arccos\left(-\frac{1}{2}\right) = \frac{2\pi}{3}$. Why is $\arccos\left(-\frac{1}{2}\right)$ not equal to $-\frac{4\pi}{3}$?

9. Consider the inverse tangent function, defined by $y = \tan^{-1} x$, or $y = \arctan x$.

(a) What is its domain?
(b) What is its range?
(c) Is this function increasing or decreasing?
(d) Is there any real number x for which $\arctan x$ is not defined? If so, what is it (or what are they)?

10. Give the domain and range of each inverse trigonometric function, as defined in this section.

(a) inverse cosecant function
(b) inverse secant function
(c) inverse cotangent function

11. *Concept Check* Is $\sec^{-1} a$ calculated as $\cos^{-1}\frac{1}{a}$ or as $\frac{1}{\cos^{-1} a}$?

12. *Concept Check* For positive values of a, $\cot^{-1} a$ is calculated as $\tan^{-1}\frac{1}{a}$. How is $\cot^{-1} a$ calculated for negative values of a?

Find the exact value of each real number y if it exists. Do not use a calculator. ***See Examples 1 and 2.***

13. $y = \sin^{-1} 0$ **14.** $y = \sin^{-1}(-1)$ **15.** $y = \cos^{-1}(-1)$

16. $y = \arccos 0$ **17.** $y = \tan^{-1} 1$ **18.** $y = \arctan(-1)$

19. $y = \arctan 0$ **20.** $y = \tan^{-1}(-1)$ **21.** $y = \arcsin\left(-\frac{\sqrt{3}}{2}\right)$

22. $y = \sin^{-1}\frac{\sqrt{2}}{2}$ **23.** $y = \arccos\left(-\frac{\sqrt{3}}{2}\right)$ **24.** $y = \cos^{-1}\left(-\frac{1}{2}\right)$

25. $y = \sin^{-1}\sqrt{3}$ **26.** $y = \arcsin(-\sqrt{2})$ **27.** $y = \cot^{-1}(-1)$

28. $y = \text{arccot}(-\sqrt{3})$ **29.** $y = \csc^{-1}(-2)$ **30.** $y = \csc^{-1}\sqrt{2}$

31. $y = \text{arcsec}\,\frac{2\sqrt{3}}{3}$ **32.** $y = \sec^{-1}(-\sqrt{2})$ **33.** $y = \sec^{-1} 1$

34. $y = \sec^{-1} 0$ **35.** $y = \csc^{-1}\frac{\sqrt{2}}{2}$ **36.** $y = \text{arccsc}\left(-\frac{1}{2}\right)$

Give the degree measure of θ if it exists. Do not use a calculator. ***See Example 3.***

37. $\theta = \arctan(-1)$ **38.** $\theta = \tan^{-1}\sqrt{3}$ **39.** $\theta = \arcsin\left(-\frac{\sqrt{3}}{2}\right)$

40. $\theta = \arcsin\left(-\frac{\sqrt{2}}{2}\right)$ **41.** $\theta = \arccos\left(-\frac{1}{2}\right)$ **42.** $\theta = \sec^{-1}(-2)$

43. $\theta = \cot^{-1}\left(-\frac{\sqrt{3}}{3}\right)$ **44.** $\theta = \cot^{-1}\frac{\sqrt{3}}{3}$ **45.** $\theta = \csc^{-1}(-2)$

46. $\theta = \csc^{-1}(-1)$ **47.** $\theta = \sin^{-1} 2$ **48.** $\theta = \cos^{-1}(-2)$

Use a calculator to give each value in decimal degrees. ***See Example 4.***

49. $\theta = \sin^{-1}(-0.13349122)$ **50.** $\theta = \arcsin 0.77900016$

51. $\theta = \arccos(-0.39876459)$ **52.** $\theta = \cos^{-1}(-0.13348816)$

53. $\theta = \csc^{-1} 1.9422833$

54. $\theta = \cot^{-1} 1.7670492$

55. $\theta = \cot^{-1}(-0.60724226)$

56. $\theta = \cot^{-1}(-2.7733744)$

57. $\theta = \tan^{-1}(-7.7828641)$

58. $\theta = \sec^{-1}(-5.1180378)$

Use a calculator to give each real number value. (Be sure the calculator is in radian mode.) ***See Example 4.***

59. $y = \arcsin 0.92837781$

60. $y = \arcsin 0.81926439$

61. $y = \cos^{-1}(-0.32647891)$

62. $y = \arccos 0.44624593$

63. $y = \arctan 1.1111111$

64. $y = \cot^{-1} 1.0036571$

65. $y = \cot^{-1}(-0.92170128)$

66. $y = \cot^{-1}(-36.874610)$

67. $y = \sec^{-1}(-1.2871684)$

68. $y = \sec^{-1} 4.7963825$

The screen here shows how to define the inverse secant, cosecant, and cotangent functions in order to graph them using a TI-83/84 Plus graphing calculator.

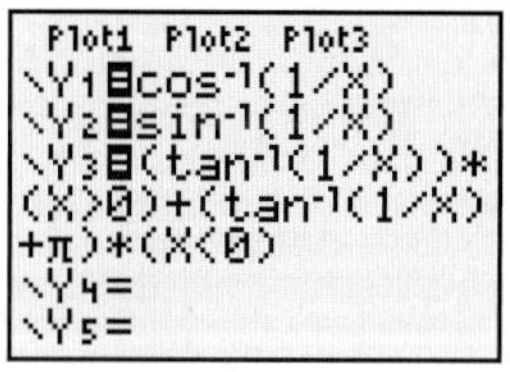

Use this information to graph each inverse circular function and compare your graphs to those in **Figure 20.**

69. $y = \sec^{-1} x$

70. $y = \csc^{-1} x$

71. $y = \cot^{-1} x$

Graph each inverse circular function by hand.

72. $y = \text{arccsc } 2x$

73. $y = \text{arcsec } \frac{1}{2}x$

74. $y = 2 \cot^{-1} x$

75. *Concept Check* Explain why attempting to find $\sin^{-1} 1.003$ on your calculator will result in an error message.

Relating Concepts

For individual or collaborative investigation *(Exercises 76–78)**

76. Consider the function

$$f(x) = 3x - 2 \quad \text{and its inverse} \quad f^{-1}(x) = \frac{1}{3}x + \frac{2}{3}.$$

Simplify $f(f^{-1}(x))$ and $f^{-1}(f(x))$. What do you notice in each case? What would the graph look like in each case?

77. Use a graphing calculator to graph $y = \tan(\tan^{-1} x)$ in the standard viewing window, using radian mode. How does this compare to the graph you described in **Exercise 76?**

78. Use a graphing calculator to graph $y = \tan^{-1}(\tan x)$ in the standard viewing window, using radian and dot modes. Why does this graph not agree with the graph you found in **Exercise 77?**

*The authors wish to thank Carol Walker of Hinds Community College for making a suggestion on which these exercises are based.

62. 1.1082303
63. 0.83798122
64. 0.78357295
65. 2.3154725
66. 3.1144804
67. 2.4605221
68. 1.3607651

69.

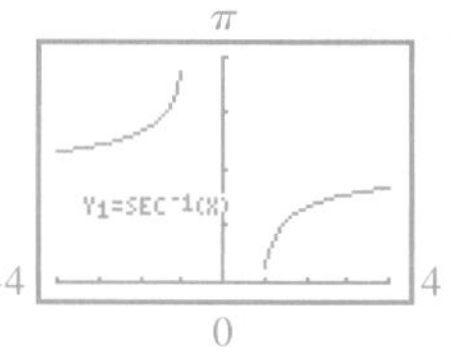

70.

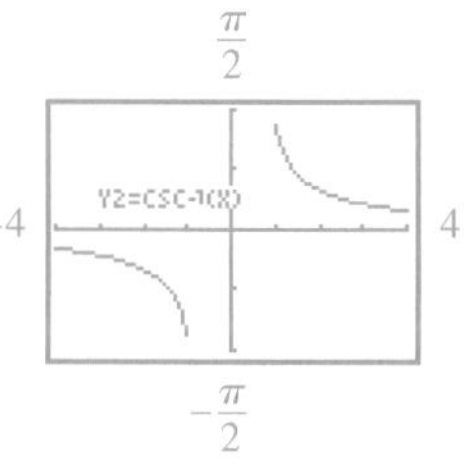

71.

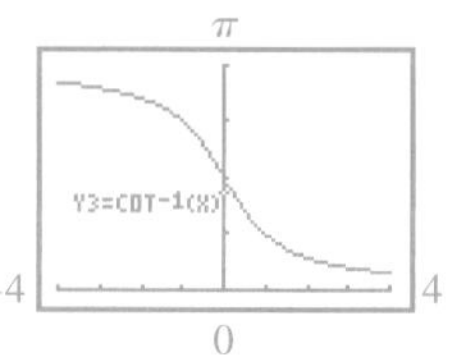

72.

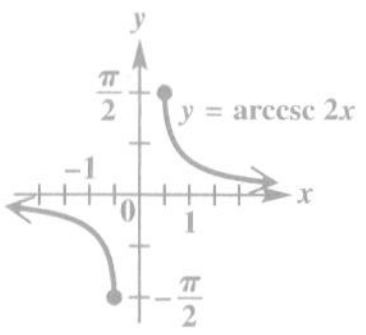

73.

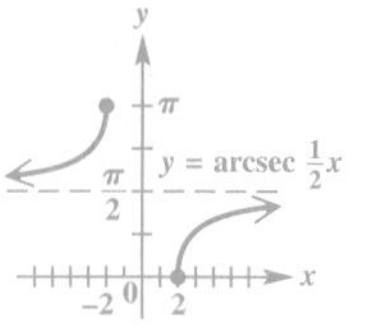

74.

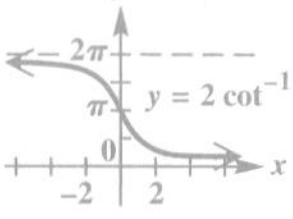

75. 1.003 is not in the domain of $y = \sin^{-1} x$.

76. In both cases, the result is x. In each case, the graph is a straight line bisecting quadrants I and III (i.e., the line $y = x$).

77. It is the graph of $y = x$.

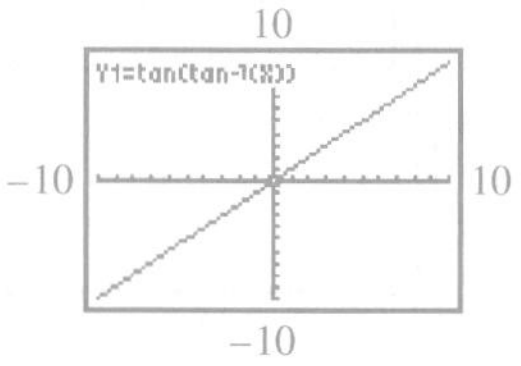

78. It does not agree because the range of the inverse tangent function is $\left(-\frac{\pi}{2}, \frac{\pi}{2}\right)$, not $(-\infty, \infty)$, as was the case in Exercise 77.

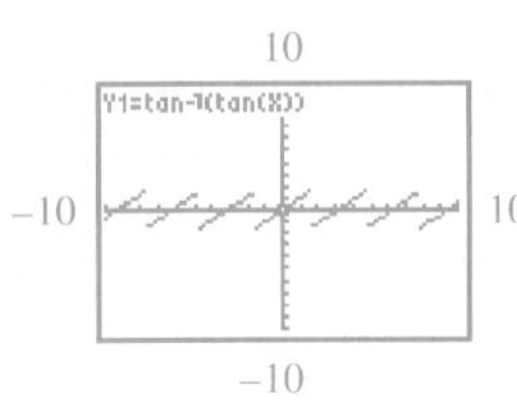

79. $\frac{\sqrt{7}}{3}$ **80.** $\frac{\sqrt{15}}{4}$

81. $\frac{\sqrt{5}}{5}$ **82.** $\frac{5\sqrt{6}}{12}$

83. $\frac{120}{169}$ **84.** $\frac{7}{8}$

85. $-\frac{7}{25}$ **86.** $-\frac{\sqrt{15}}{7}$

87. $\frac{4\sqrt{6}}{25}$ **88.** $-\frac{3}{5}$

89. 2 **90.** $\sqrt{2}$

91. $\frac{63}{65}$ **92.** $-\frac{16}{65}$

93. $\frac{\sqrt{10} - 3\sqrt{30}}{20}$ **94.** $\frac{48 + 25\sqrt{3}}{39}$

95. 0.894427191
96. 0.9682458366
97. 0.1234399811
98. 0.716386406

99. $\sqrt{1 - u^2}$ **100.** $\frac{\sqrt{1 - u^2}}{u}$

101. $\sqrt{1 - u^2}$ **102.** $\frac{\sqrt{1 - u^2}}{u}$

103. $\frac{4\sqrt{u^2 - 4}}{u^2}$ **104.** $\frac{u^2 - 9}{u^2 + 9}$

105. $\frac{u\sqrt{2}}{2}$ **106.** $\frac{\sqrt{u^2 + 5}}{u}$

107. $\frac{2\sqrt{4 - u^2}}{4 - u^2}$ **108.** $\frac{3\sqrt{9 - u^2}}{9 - u^2}$

109. 41°
110. **(a)** 45° **(b)** $\theta = 45°$
111. **(a)** 18° **(b)** 18°
(c) 15°
(e) 1.4142151 m
(Note: Due to the computational routine, there may be a discrepancy in the last few decimal places.)

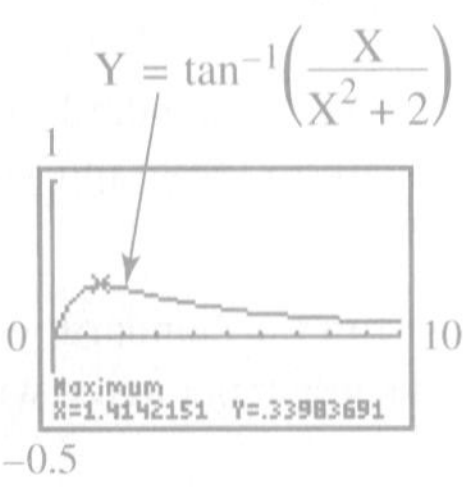

Radian mode

Give the exact value of each expression without using a calculator. **See Examples 5 and 6.**

79. $\tan\left(\arccos \frac{3}{4}\right)$ **80.** $\sin\left(\arccos \frac{1}{4}\right)$ **81.** $\cos(\tan^{-1}(-2))$

82. $\sec\left(\sin^{-1}\left(-\frac{1}{5}\right)\right)$ **83.** $\sin\left(2 \tan^{-1} \frac{12}{5}\right)$ **84.** $\cos\left(2 \sin^{-1} \frac{1}{4}\right)$

85. $\cos\left(2 \arctan \frac{4}{3}\right)$ **86.** $\tan\left(2 \cos^{-1} \frac{1}{4}\right)$ **87.** $\sin\left(2 \cos^{-1} \frac{1}{5}\right)$

88. $\cos(2 \tan^{-1}(-2))$ **89.** $\sec(\sec^{-1} 2)$ **90.** $\csc(\csc^{-1}\sqrt{2})$

91. $\cos\left(\tan^{-1} \frac{5}{12} - \tan^{-1} \frac{3}{4}\right)$ **92.** $\cos\left(\sin^{-1} \frac{3}{5} + \cos^{-1} \frac{5}{13}\right)$

93. $\sin\left(\sin^{-1} \frac{1}{2} + \tan^{-1}(-3)\right)$ **94.** $\tan\left(\cos^{-1} \frac{\sqrt{3}}{2} - \sin^{-1}\left(-\frac{3}{5}\right)\right)$

Use a calculator to find each value. Give answers as real numbers.

95. $\cos(\tan^{-1} 0.5)$ **96.** $\sin(\cos^{-1} 0.25)$

97. $\tan(\arcsin 0.12251014)$ **98.** $\cot(\arccos 0.58236841)$

Write each expression as an algebraic (nontrigonometric) expression in u, for $u > 0$. **See Example 7.**

99. $\sin(\arccos u)$ **100.** $\tan(\arccos u)$ **101.** $\cos(\arcsin u)$

102. $\cot(\arcsin u)$ **103.** $\sin\left(2 \sec^{-1} \frac{u}{2}\right)$ **104.** $\cos\left(2 \tan^{-1} \frac{3}{u}\right)$

105. $\tan\left(\sin^{-1} \frac{u}{\sqrt{u^2 + 2}}\right)$ **106.** $\sec\left(\cos^{-1} \frac{u}{\sqrt{u^2 + 5}}\right)$

107. $\sec\left(\operatorname{arccot} \frac{\sqrt{4 - u^2}}{u}\right)$ **108.** $\csc\left(\arctan \frac{\sqrt{9 - u^2}}{u}\right)$

(Modeling) Solve each problem.

109. *Angle of Elevation of a Shot Put* Refer to **Example 8.** Suppose a shot-putter can consistently release the steel ball with velocity v of 32 ft per sec from an initial height h of 5.0 ft. What angle, to the nearest degree, will maximize the distance?

110. *Angle of Elevation of a Shot Put* Refer to **Example 8.**

(a) What is the optimal angle, to the nearest degree, when $h = 0$?

(b) Fix h at 6 ft and regard θ as a function of v. As v increases without bound, the graph approaches an asymptote. Find the equation of that asymptote.

111. *Observation of a Painting* A painting 1 m high and 3 m from the floor will cut off an angle θ to an observer, where

$$\theta = \tan^{-1}\left(\frac{x}{x^2 + 2}\right),$$

assuming that the observer is x meters from the wall where the painting is displayed and that the eyes of the observer are 2 m above the ground. (See the figure.) Find the value of θ for the following values of x. Round to the nearest degree.

(a) 1 **(b)** 2 **(c)** 3

(d) Derive the formula given above. (*Hint:* Use the identity for $\tan(\theta + \alpha)$. Use right triangles.)

(e) Graph the function for θ with a graphing calculator, and determine the distance that maximizes the angle.

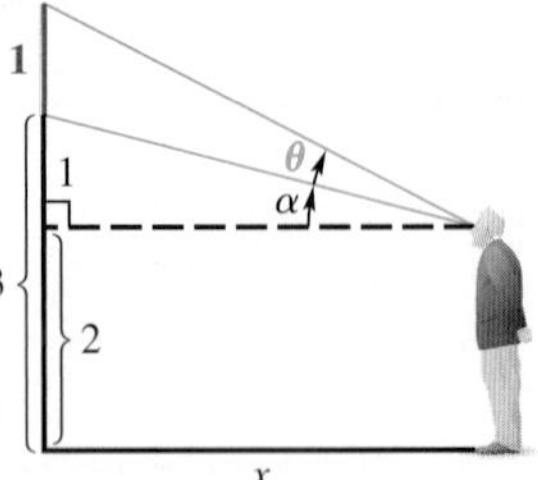

111. (f) $\sqrt{2}$
113. 44.7%
114. $20\left[9 \arctan\left(\sqrt{8}\right) - \sqrt{8}\right] \approx 165 \text{ ft}^3$

(f) The concept in part (e) was first investigated in 1471 by the astronomer Regiomontanus. (*Source:* Maor, E., *Trigonometric Delights,* Princeton University Press.) If the bottom of the picture is a meters above eye level and the top of the picture is b meters above eye level, then the optimum value of x is $\sqrt{ab}$ meters. Use this result to find the exact answer to part (e).

112. *Landscaping Formula* A shrub is planted in a 100-ft-wide space between buildings measuring 75 ft and 150 ft tall. The location of the shrub determines how much sun it receives each day. Show that if θ is the angle in the figure and x is the distance of the shrub from the taller building, then the value of θ (in radians) is given by

$$\theta = \pi - \arctan\left(\frac{75}{100 - x}\right) - \arctan\left(\frac{150}{x}\right).$$

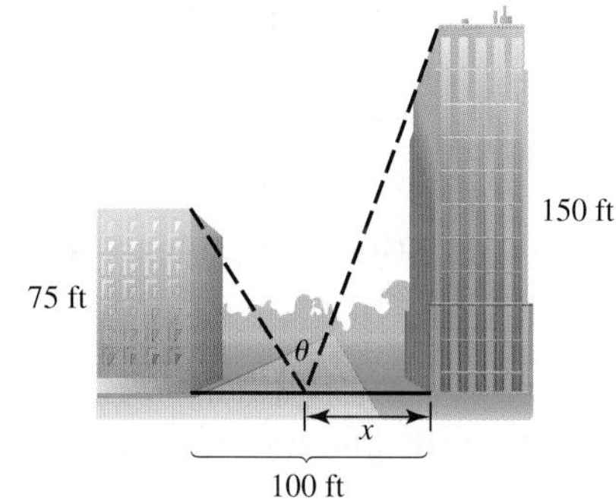

113. *Communications Satellite Coverage* The figure shows a stationary communications satellite positioned 20,000 mi above the equator. What percent, to the nearest tenth, of the equator can be seen from the satellite? The diameter of Earth is 7927 mi at the equator.

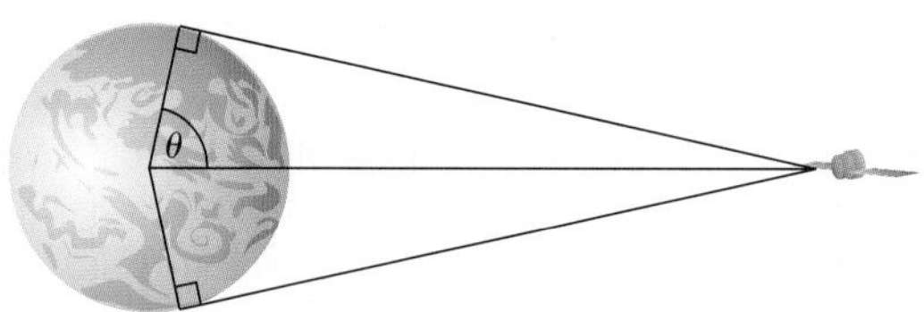

114. *Oil in a Storage Tank* The level of oil in a storage tank buried in the ground can be found in much the same way as a dipstick is used to determine the oil level in an automobile crankcase. Suppose the ends of the cylindrical storage tank in the figure are circles of radius 3 ft and the cylinder is 20 ft long. Determine the volume of oil in the tank to the nearest cubic foot if the rod shows a depth of 2 ft. (*Hint:* The volume will be 20 times the area of the shaded segment of the circle shown in the figure on the right.)

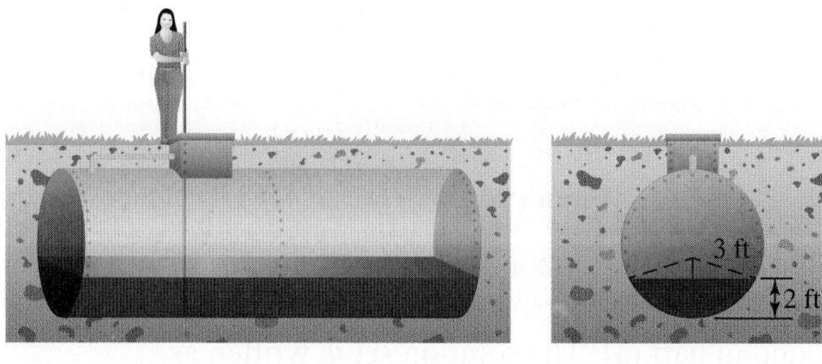

7.6 Trigonometric Equations

Earlier in this chapter, we studied trigonometric equations that were identities. We now consider trigonometric equations that are *conditional.* These equations are satisfied by some values but not others. (See **Section 1.1.**)

Solving by Linear Methods The most basic trigonometric equations are solved by first using properties of equality to isolate a trigonometric expression on one side of the equation.

EXAMPLE 1 Solving a Trigonometric Equation by Linear Methods

Solve the equation $2 \sin \theta + 1 = 0$

(a) over the interval $[0°, 360°)$, and

(b) for all solutions.

ALGEBRAIC SOLUTION

(a) Because $\sin \theta$ is to the first power, we use the same method as we would to solve the linear equation $2x + 1 = 0$.

$2 \sin \theta + 1 = 0$ Original equation

$2 \sin \theta = -1$ Subtract 1. (Section 1.1)

$\sin \theta = -\frac{1}{2}$ Divide by 2.

To find values of θ that satisfy $\sin \theta = -\frac{1}{2}$, we observe that θ must be in either quadrant III or quadrant IV because the sine function is negative only in these two quadrants. Furthermore, the reference angle must be 30°. The graph of the unit circle in **Figure 28** shows the two possible values of θ. The solution set is $\{210°, 330°\}$.

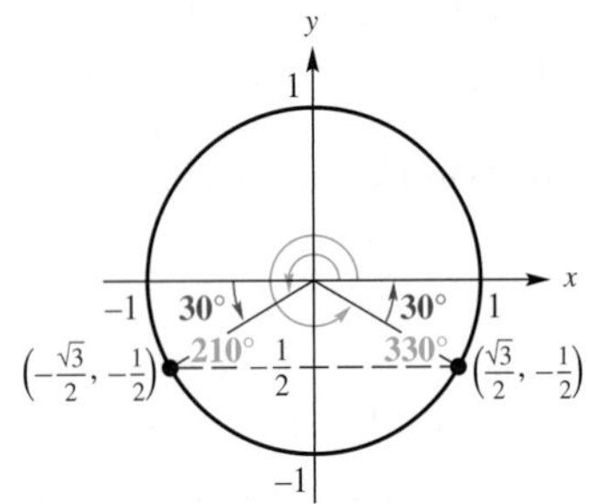

Figure 28

(b) To find all solutions, we add integer multiples of the period of the sine function, 360°, to each solution found in part (a). The solution set is written as follows.

$$\{210° + 360°n, 330° + 360°n, \text{ where } n \text{ is any integer}\}$$

GRAPHING CALCULATOR SOLUTION

(a) Consider the original equation.

$$2 \sin \theta + 1 = 0$$

We can find the solution set of this equation by graphing the function

$$Y_1 = 2 \sin X + 1$$

and then determining its x-intercepts, or zeros. Since we are finding solutions over the interval $[0°, 360°)$, we use degree mode and choose this interval of values for the input X on the graph.

The screen in **Figure 29(a)** indicates that one solution is 210°, and the screen in **Figure 29(b)** indicates that the other solution is 330°. The solution set is $\{210°, 330°\}$, which agrees with the algebraic solution.

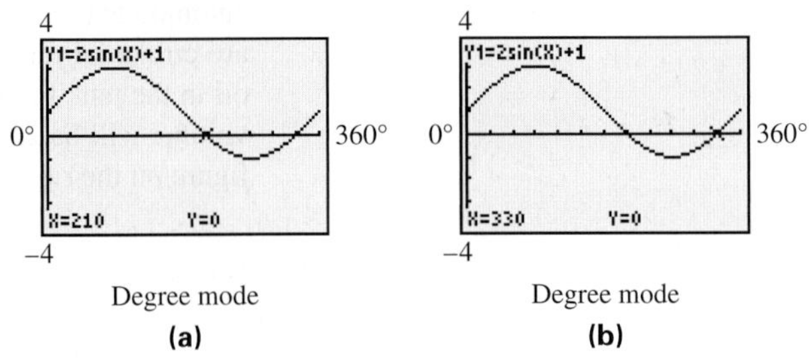

Figure 29

(b) Because the graph of

$$Y_1 = 2 \sin X + 1$$

repeats the same y-values every 360°, all solutions are found by adding integer multiples of 360° to the solutions found in part (a). See the algebraic solution.

Now Try Exercises 11 and 43.

Classroom Example 1
Solve $3 \tan \theta - \sqrt{3} = 0$
(a) over the interval $[0°, 360°)$, and
(b) for all solutions.

Answers:
(a) $\{30°, 210°\}$
(b) $\{30° + 180°n$, where n is any integer$\}$

Solving by Factoring

EXAMPLE 2 Solving a Trigonometric Equation by Factoring

Solve $\sin \theta \tan \theta = \sin \theta$ over the interval $[0°, 360°)$.

SOLUTION

$$\sin \theta \tan \theta = \sin \theta \quad \text{Original equation}$$

$$\sin \theta \tan \theta - \sin \theta = 0 \quad \text{Subtract } \sin \theta.$$

$$\sin \theta (\tan \theta - 1) = 0 \quad \text{Factor out } \sin \theta. \text{ (Section R.4)}$$

$$\sin \theta = 0 \quad \text{or} \quad \tan \theta - 1 = 0 \quad \text{Zero-factor property (Section 1.4)}$$

$$\tan \theta = 1$$

$$\theta = 0° \quad \text{or} \quad \theta = 180° \qquad \theta = 45° \quad \text{or} \quad \theta = 225° \quad \text{Apply the inverse function. (Section 7.5)}$$

See **Figure 30.** The solution set is $\{0°, 45°, 180°, 225°\}$.

✔ *Now Try Exercise 31.*

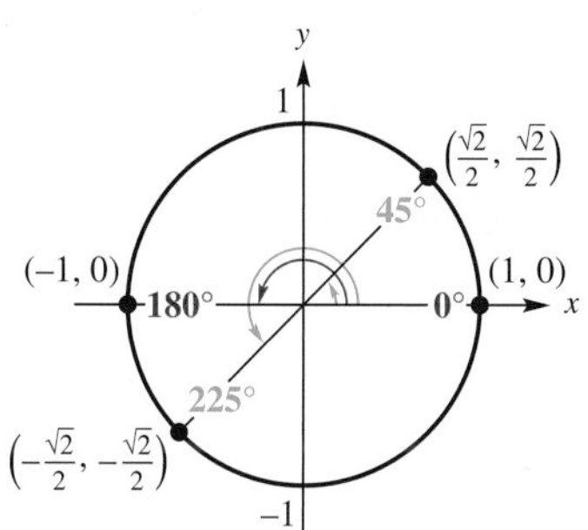

Figure 30

CAUTION Trying to solve the equation in **Example 2** by dividing each side by $\sin \theta$ would lead to $\tan \theta = 1$, which would give $\theta = 45°$ or $\theta = 225°$. The missing two solutions are the ones that make the divisor, $\sin \theta$, equal 0. ***For this reason, we avoid dividing by a variable expression.***

Classroom Example 2
Solve $\cos \theta \cot \theta = -\cos \theta$ over the interval $[0°, 360°)$.

Answer: $\{90°, 135°, 270°, 315°\}$

Solving by Quadratic Methods The equation $au^2 + bu + c = 0$, where u is an algebraic expression, is solved by quadratic methods. The expression u may be a trigonometric function, as in the next example.

Classroom Example 3
Solve $3 \sin^2 x - \sin x - 2 = 0$ over the interval $[0, 2\pi)$.

Answer: $\left\{\frac{\pi}{2}, 3.8713, 5.5535\right\}$

EXAMPLE 3 Solving a Trigonometric Equation by Factoring

Solve $\tan^2 x + \tan x - 2 = 0$ over the interval $[0, 2\pi)$.

SOLUTION

$$\tan^2 x + \tan x - 2 = 0 \quad \text{This equation is quadratic in form.}$$

$$(\tan x - 1)(\tan x + 2) = 0 \quad \text{Factor. (Section R.4)}$$

$$\tan x - 1 = 0 \quad \text{or} \quad \tan x + 2 = 0 \quad \text{Zero-factor property}$$

$$\tan x = 1 \quad \text{or} \quad \tan x = -2 \quad \text{Solve each equation.}$$

The solutions for $\tan x = 1$ over the interval $[0, 2\pi)$ are $x = \frac{\pi}{4}$ and $x = \frac{5\pi}{4}$.

To solve $\tan x = -2$ over that interval, we use a scientific calculator set in *radian* mode. We find that $\tan^{-1}(-2) \approx -1.1071487$. This is a quadrant IV number, based on the range of the inverse tangent function. However, since we want solutions over the interval $[0, 2\pi)$, we must first add π to -1.1071487, and then add 2π. See **Figure 31.**

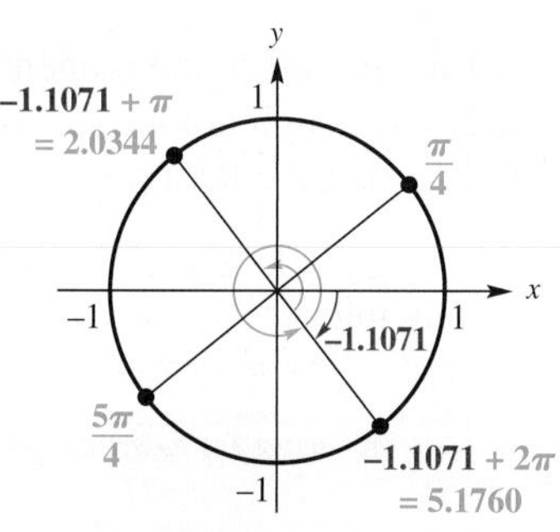

The solutions shown in blue represent angle measures, in radians, ***and*** their intercepted arc lengths on the unit circle.

Figure 31

$$x \approx -1.1071487 + \pi \approx 2.0344439$$

$$x \approx -1.1071487 + 2\pi \approx 5.1760366$$

The solutions over the required interval form the following solution set.

$$\left\{\underbrace{\frac{\pi}{4}, \frac{5\pi}{4}}_{\text{Exact values}}, \underbrace{2.0344, 5.1760}_{\text{Approximate values to four decimal places}}\right\}$$

✔ *Now Try Exercise 21.*

Classroom Example 4
Find all solutions of

$$\cos x(\cos x + 2) = 1.$$

Write the solution set.

Answer:
$\{1.1437 + 2n\pi, 5.1395 + 2n\pi,$ where n is any integer$\}$

Teaching Tip Equations of the form $y = \tan x$ will not contain a second value for x between $-\frac{\pi}{2}$ and $\frac{\pi}{2}$. Remind students that other solutions to $y = \tan x$ are found using period π radians.

EXAMPLE 4 Solving a Trigonometric Equation Using the Quadratic Formula

Find all solutions of $\cot x(\cot x + 3) = 1$. Write the solution set.

SOLUTION We multiply the factors on the left and subtract 1 to write the equation in standard quadratic form.

$$\cot x(\cot x + 3) = 1 \quad \text{Original equation}$$

$$\cot^2 x + 3\cot x - 1 = 0 \quad \text{(Section 1.4)}$$

This equation is quadratic in form, but cannot be solved by factoring. Therefore, we use the quadratic formula, with $a = 1$, $b = 3$, $c = -1$, and $\cot x$ as the variable.

$$\cot x = \frac{-b \pm \sqrt{b^2 - 4ac}}{2a} \quad \text{Quadratic formula (Section 1.4)}$$

$$= \frac{-3 \pm \sqrt{3^2 - 4(1)(-1)}}{2(1)} \quad a = 1, b = 3, c = -1$$

Be careful with signs.

$$= \frac{-3 \pm \sqrt{9 + 4}}{2} \quad \text{Simplify.}$$

$$= \frac{-3 \pm \sqrt{13}}{2} \quad \text{Add.}$$

$$\cot x \approx -3.302775638 \quad \text{or} \quad \cot x \approx 0.3027756377 \quad \text{Use a calculator.}$$

$$x \approx \cot^{-1}(-3.302775638) \quad \text{or} \quad x \approx \cot^{-1}(0.3027756377) \quad \text{Definition of inverse cotangent}$$

$$x \approx \tan^{-1}\left(\frac{1}{-3.302775638}\right) + \pi \quad \text{or} \quad x \approx \tan^{-1}\left(\frac{1}{0.3027756377}\right)$$

Reciprocal identity: $\tan x = \frac{1}{\cot x}$ (Section 7.1)

$$x \approx -0.2940013018 + \pi \quad \text{or} \quad x \approx 1.276795025 \quad \text{Use a calculator in radian mode.}$$

$$x \approx 2.847591352$$

To find *all* solutions, we add integer multiples of the period of the tangent function, which is π, to each solution found previously. Although not unique, a common form of the solution set of the equation, written using the least possible nonnegative angle measures, is given as follows.

$$\{2.8476 + n\pi, 1.2768 + n\pi, \text{ where } n \text{ is any integer}\} \quad \text{Round to four decimal places.}$$

Now Try Exercise 53.

Solving by Using Trigonometric Identities Recall that squaring each side of an equation, such as

$$\sqrt{x + 4} = x + 2,$$

will yield all solutions but may also give extraneous solutions—solutions that satisfy the final equation but *not* the original equation. As a result, all proposed solutions *must* be checked in the original equation as shown in **Example 5.**

Classroom Example 5
Solve $\cot x - \sqrt{3} = \csc x$ over the interval $[0, 2\pi)$.

Answer: $\{\frac{4\pi}{3}\}$

EXAMPLE 5 **Solving a Trigonometric Equation by Squaring**

Solve $\tan x + \sqrt{3} = \sec x$ over the interval $[0, 2\pi)$.

SOLUTION Our first goal is to rewrite the equation in terms of a single trigonometric function. Since the tangent and secant functions are related by the identity $1 + \tan^2 x = \sec^2 x$, square each side and express $\sec^2 x$ in terms of $\tan^2 x$.

$$\left(\tan x + \sqrt{3}\right)^2 = (\sec x)^2$$ Square each side. (Section 1.6)

Don't forget the middle term.

$$\tan^2 x + 2\sqrt{3}\tan x + 3 = \sec^2 x$$ $(x+y)^2 = x^2 + 2xy + y^2$ (Section R.3)

$$\tan^2 x + 2\sqrt{3}\tan x + 3 = 1 + \tan^2 x$$ Pythagorean identity (Section 7.1)

$$2\sqrt{3}\tan x = -2$$ Subtract $3 + \tan^2 x$.

$$\tan x = -\frac{1}{\sqrt{3}}, \quad \text{or} \quad -\frac{\sqrt{3}}{3}$$ Divide by $2\sqrt{3}$. Rationalize the denominator. (Section R.7)

Teaching Tip Explain that whenever possible, answers should be given in exact form, such as $\frac{11\pi}{6}$, rather than as decimal approximations.

Solutions of $\tan x = -\frac{\sqrt{3}}{3}$ over $[0, 2\pi)$ are $\frac{5\pi}{6}$ and $\frac{11\pi}{6}$. These possible, or proposed, solutions must be checked to determine whether they are also solutions of the original equation.

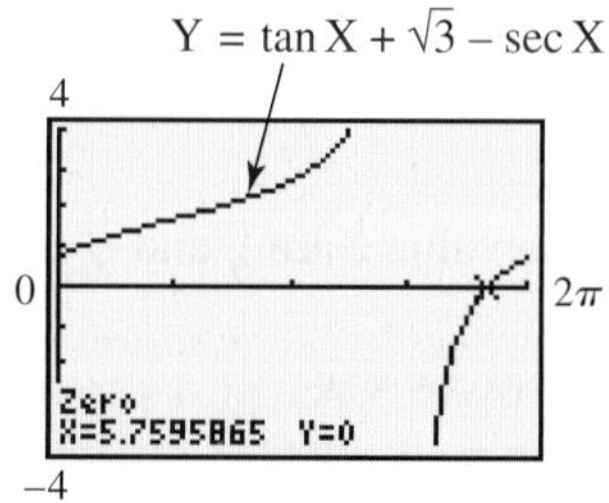

Radian mode

The graph shows that on the interval $[0, 2\pi)$, the only x-intercept of the graph of $Y = \tan X + \sqrt{3} - \sec X$ is 5.7595865, which is an approximation for $\frac{11\pi}{6}$, the solution found in **Example 5.**

CHECK $\quad \tan x + \sqrt{3} = \sec x$ Original equation

$\tan\left(\frac{5\pi}{6}\right) + \sqrt{3} \stackrel{?}{=} \sec\left(\frac{5\pi}{6}\right)$ Let $x = \frac{5\pi}{6}$.	$\tan\left(\frac{11\pi}{6}\right) + \sqrt{3} \stackrel{?}{=} \sec\left(\frac{11\pi}{6}\right)$ Let $x = \frac{11\pi}{6}$.
$-\frac{\sqrt{3}}{3} + \frac{3\sqrt{3}}{3} \stackrel{?}{=} -\frac{2\sqrt{3}}{3}$	$-\frac{\sqrt{3}}{3} + \frac{3\sqrt{3}}{3} \stackrel{?}{=} \frac{2\sqrt{3}}{3}$
$\frac{2\sqrt{3}}{3} = -\frac{2\sqrt{3}}{3}$ False	$\frac{2\sqrt{3}}{3} = \frac{2\sqrt{3}}{3}$ ✓ True

As the check shows, only $\frac{11\pi}{6}$ is a solution, so the solution set is $\left\{\frac{11\pi}{6}\right\}$.

✔ *Now Try Exercise 41.*

Methods for solving trigonometric equations can be summarized as follows.

Solving a Trigonometric Equation

1. Decide whether the equation is linear or quadratic in form, so that you can determine the solution method.
2. If only one trigonometric function is present, solve the equation for that function.
3. If more than one trigonometric function is present, rearrange the equation so that one side equals 0. Then try to factor and set each factor equal to 0 to solve.
4. If the equation is quadratic in form, but not factorable, use the quadratic formula. Check that solutions are in the desired interval.
5. Try using identities to change the form of the equation. It may be helpful to square each side of the equation first. In this case, check for extraneous solutions.

LOOKING AHEAD TO CALCULUS

There are many instances in calculus where it is necessary to solve trigonometric equations. Examples include solving related-rates problems and optimization problems.

Equations with Half-Angles

Teaching Tip As a slight variation of the problem in **Example 6,** replace $\frac{x}{2}$ with u, solve $2 \sin u = 1$ for u, and then multiply the solutions by 2 to find x.

EXAMPLE 6 Solving an Equation with a Half-Angle

Solve the equation $2 \sin \frac{x}{2} = 1$

(a) over the interval $[0, 2\pi)$, and **(b)** for all solutions.

Classroom Example 6

Solve $2 \cos \frac{x}{2} - \sqrt{2} = 0$

(a) over the interval $[0, 2\pi)$, and
(b) for all solutions.

Answers:
(a) $\left\{\frac{\pi}{2}\right\}$
(b) $\left\{\frac{\pi}{2} + 4n\pi, \frac{7\pi}{2} + 4n\pi\right.$, where n is any integer$\left.\right\}$

SOLUTION

(a) Write the interval $[0, 2\pi)$ as the inequality

$$0 \leq x < 2\pi.$$

The corresponding interval for $\frac{x}{2}$ is

$$0 \leq \frac{x}{2} < \pi. \quad \text{Divide by 2. (Section 1.7)}$$

To find all values of $\frac{x}{2}$ over the interval $[0, \pi)$ that satisfy the given equation, first solve for $\sin \frac{x}{2}$.

$$2 \sin \frac{x}{2} = 1 \quad \text{Original equation}$$

$$\sin \frac{x}{2} = \frac{1}{2} \quad \text{Divide by 2.}$$

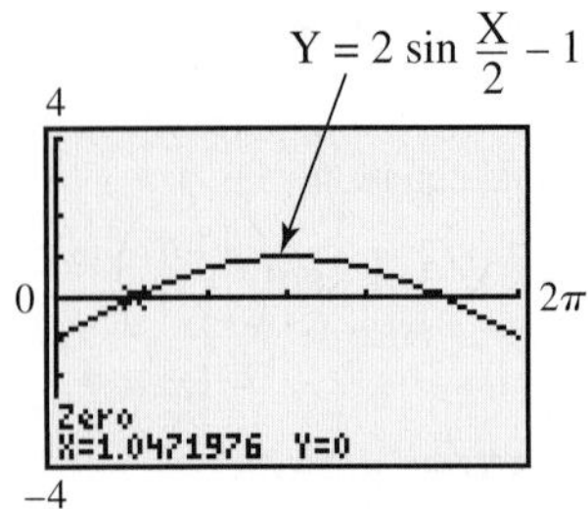

The x-intercepts are the solutions found in **Example 6(a).** Using Xscl = $\frac{\pi}{3}$ makes it possible to support the exact solutions by counting the tick marks from 0 on the graph.

The two numbers over the interval $[0, \pi)$ with sine value $\frac{1}{2}$ are $\frac{\pi}{6}$ and $\frac{5\pi}{6}$, so

$$\frac{x}{2} = \frac{\pi}{6} \quad \text{or} \quad \frac{x}{2} = \frac{5\pi}{6} \quad \text{Definition of inverse sine (Section 7.5)}$$

$$x = \frac{\pi}{3} \quad \text{or} \quad x = \frac{5\pi}{3}. \quad \text{Multiply by 2.}$$

The solution set over the given interval is $\left\{\frac{\pi}{3}, \frac{5\pi}{3}\right\}$.

(b) Because this is a sine function with period 4π, all solutions are found by adding integer multiples of 4π.

$$\left\{\frac{\pi}{3} + 4n\pi, \frac{5\pi}{3} + 4n\pi, \text{ where } n \text{ is any integer}\right\}$$

Now Try Exercises 73 and 87.

Equations with Multiple Angles

Classroom Example 7
Solve $\cos 2x = \sin x$ over the interval $[0, 2\pi)$.

Answer: $\left\{\frac{\pi}{6}, \frac{5\pi}{6}, \frac{3\pi}{2}\right\}$

EXAMPLE 7 Solving an Equation Using a Double-Angle Identity

Solve $\cos 2x = \cos x$ over the interval $[0, 2\pi)$.

SOLUTION First change $\cos 2x$ to a trigonometric function of x. Use the identity $\cos 2x = 2 \cos^2 x - 1$ so that the equation involves only $\cos x$. Then factor.

$$\cos 2x = \cos x \quad \text{Original equation}$$

$$2 \cos^2 x - 1 = \cos x \quad \text{Cosine double-angle identity (Section 7.4)}$$

$$2 \cos^2 x - \cos x - 1 = 0 \quad \text{Subtract } \cos x.$$

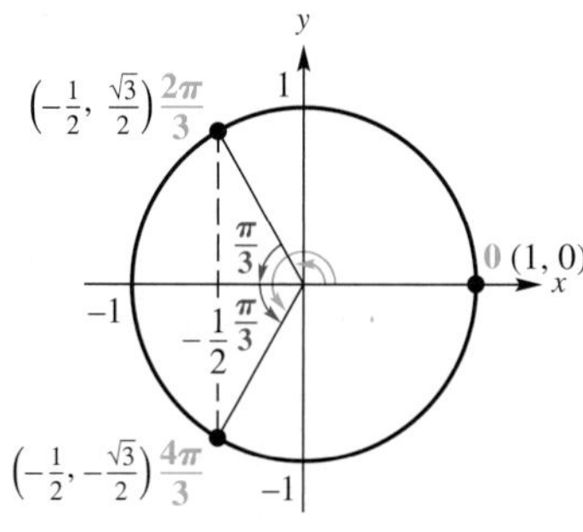

Figure 32

$$(2\cos x + 1)(\cos x - 1) = 0 \qquad \text{Factor. (Section R.4)}$$

$$2\cos x + 1 = 0 \quad \text{or} \quad \cos x - 1 = 0 \qquad \text{Zero-factor property}$$

$$\cos x = -\frac{1}{2} \quad \text{or} \quad \cos x = 1 \qquad \text{Solve each equation for } \cos x.$$

Cosine is $-\frac{1}{2}$ in quadrants II and III with reference angle $\frac{\pi}{3}$, and it has a value of 1 at 0 radians. We can use **Figure 32** to determine that solutions over the required interval are $x = \frac{2\pi}{3}$ or $x = \frac{4\pi}{3}$ or $x = 0$. The solution set is $\left\{0, \frac{2\pi}{3}, \frac{4\pi}{3}\right\}$.

✔ *Now Try Exercise 75.*

CAUTION In **Example 7,** because 2 is not a factor of $\cos 2x$, $\frac{\cos 2x}{2} \neq \cos x$. The only way to change $\cos 2x$ to a trigonometric function of x is by using one of the identities for $\cos 2x$.

Classroom Example 8
Solve $2\cos^2\theta - 2\sin^2\theta + 1 = 0$
(a) over the interval $[0, 360°)$, and
(b) for all solutions.

Answers:
(a) $\{60°, 120°, 240°, 300°\}$
(b) $\{60° + 180°n, 120° + 180°n$, where n is any integer$\}$

EXAMPLE 8 Solving an Equation Using a Multiple-Angle Identity

Solve the equation $4\sin\theta\cos\theta = \sqrt{3}$

(a) over the interval $[0°, 360°)$, and **(b)** for all solutions.

SOLUTION

(a) The identity $2\sin\theta\cos\theta = \sin 2\theta$ is useful here.

$$4\sin\theta\cos\theta = \sqrt{3} \qquad \text{Original equation}$$

$$2(2\sin\theta\cos\theta) = \sqrt{3} \qquad 4 = 2 \cdot 2$$

$$2\sin 2\theta = \sqrt{3} \qquad 2\sin\theta\cos\theta = \sin 2\theta \text{ (Section 7.4)}$$

$$\sin 2\theta = \frac{\sqrt{3}}{2} \qquad \text{Divide by 2.}$$

From the given interval $0° \le \theta < 360°$, the corresponding interval for 2θ is $0° \le 2\theta < 720°$. Because the sine is positive in quadrants I and II, solutions over this interval are as follows.

$$2\theta = 60°, 120°, 420°, 480°, \qquad \text{Reference angle is } 60°. \text{ (Section 5.3)}$$

or $$\theta = 30°, 60°, 210°, 240° \qquad \text{Divide by 2.}$$

Teaching Tip Students should be aware that

$$y = \sin ax$$

may have as many as $2a$ solutions from 0° to 360°.

The final two solutions for 2θ were found by adding 360° to 60° and 120°, respectively, which gives the solution set $\{30°, 60°, 210°, 240°\}$.

(b) All angles 2θ that are solutions of the equation $\sin 2\theta = \frac{\sqrt{3}}{2}$ are found by adding integer multiples of 360° to the basic solution angles, 60° and 120°.

$$2\theta = 60° + 360°n \quad \text{and} \quad 2\theta = 120° + 360°n \qquad \text{Add integer multiples of } 360°.$$

$$\theta = 30° + 180°n \quad \text{and} \quad \theta = 60° + 180°n \qquad \text{Divide by 2.}$$

All solutions are given by the following set, where 180° represents the period of $\sin 2\theta$.

$$\{30° + 180°n, 60° + 180°n, \text{ where } n \text{ is any integer}\}$$

✔ *Now Try Exercises 71 and 95.*

Applications

Classroom Example 9

(a) Repeat **Example 9(a)** if the equation for the graph is changed to

$$y = 0.003 \sin 440\pi x,$$

and the function is graphed over the interval $[0, 0.0182]$.

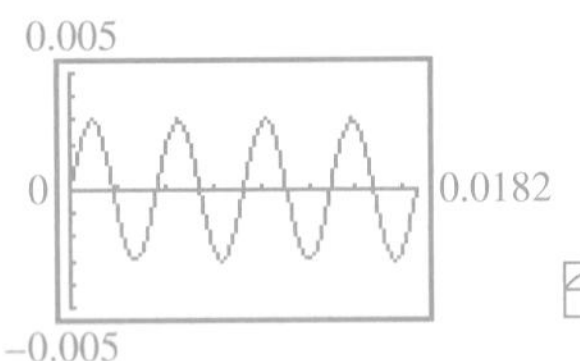

(b) Repeat **Example 9(b)** for the new function.

(c) Use a calculator to estimate the first solution to the equation that makes $y = 0.002$ over the interval $[0, 0.0182]$.

Answers:

(a) 220 Hz **(b)** 0.0045 sec

(c) 0.00053 sec

EXAMPLE 9 Describing a Musical Tone from a Graph

A basic component of music is a pure tone. The graph in **Figure 33** models the sinusoidal pressure $y = P$ in pounds per square foot from a pure tone at time $x = t$ in seconds.

(a) The frequency of a pure tone is often measured in hertz. One hertz is equal to one cycle per second and is abbreviated Hz. What is the frequency f, in hertz, of the pure tone shown in the graph?

(b) The time for the tone to produce one complete cycle is the **period.** Approximate the period T, in seconds, of the pure tone.

(c) An equation for the graph is $y = 0.004 \sin 300\pi x$. Use a calculator to estimate all solutions to the equation that make $y = 0.004$ over the interval $[0, 0.02]$.

SOLUTION

(a) From the graph in **Figure 33,** we see that there are 6 cycles in 0.04 sec. This is equivalent to $\frac{6}{0.04} = 150$ cycles per sec. The pure tone has a frequency of $f = 150$ Hz.

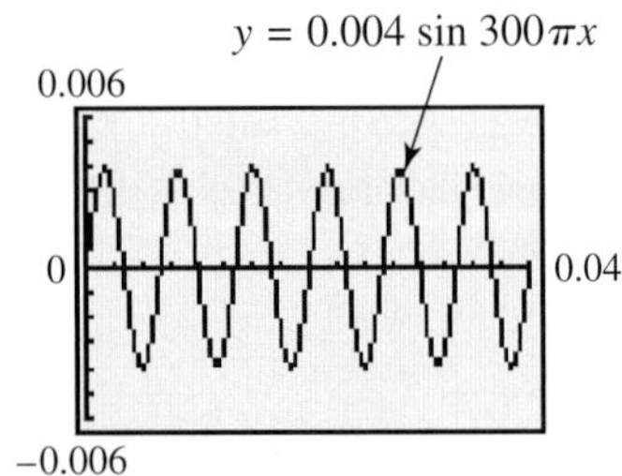

Figure 33

(b) Six periods cover a time interval of 0.04 sec. One period would be equal to $T = \frac{0.04}{6} = \frac{1}{150}$, or $0.00\overline{6}$ sec.

(c) If we reproduce the graph in **Figure 33** on a calculator as Y_1 and also graph a second function as $Y_2 = 0.004$, we can determine that the approximate values of x at the points of intersection of the graphs over the interval $[0, 0.02]$ are

$$0.0017, \quad 0.0083, \quad \text{and} \quad 0.015.$$

The first value is shown in **Figure 34.** These values represent time in seconds.

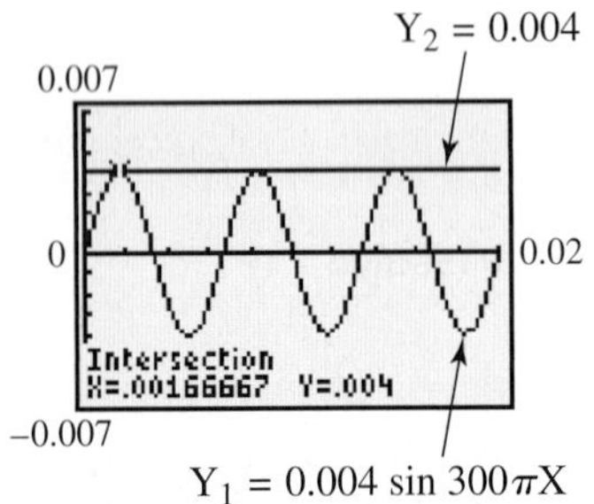

Figure 34

✔ *Now Try Exercise 101.*

A piano string can vibrate at more than one frequency when it is struck. It produces a complex wave that can mathematically be modeled by a sum of several pure tones. When a piano key with a frequency of f_1 is played, the corresponding string vibrates not only at f_1 but also at the higher frequencies of $2f_1, 3f_1, 4f_1, \ldots, nf_1$. f_1 is the **fundamental frequency** of the string, and higher frequencies are the **upper harmonics.** The human ear will hear the sum of these frequencies as one complex tone. (*Source:* Roederer, J., *Introduction to the Physics and Psychophysics of Music,* Second Edition, Springer-Verlag.)

EXAMPLE 10 Analyzing Pressures of Upper Harmonics

Suppose that the A key above middle C is played on a piano. Its fundamental frequency is $f_1 = 440$ Hz, and its associated pressure is expressed as

$$P_1 = 0.002 \sin 880\pi t.$$

The string will also vibrate at

$$f_2 = 880, \; f_3 = 1320, \; f_4 = 1760, \; f_5 = 2200, \ldots \text{Hz}.$$

Classroom Example 10
Suppose that the E key above middle C is played on a piano. Its fundamental frequency is

$$f_1 = 330\,\text{Hz},$$

and its associated pressure is expressed as

$$P_1 = 0.002 \sin 660\pi t.$$

(a) What are the next four frequencies at which the string will vibrate?

(b) What are the pressures corresponding to these four upper harmonics?

Answers:

(a) $f_2 = 660$, $f_3 = 990$, $f_4 = 1320$, $f_5 = 1650$

(b) $P_2 = \frac{0.002}{2} \sin 1320\pi t$, $P_3 = \frac{0.002}{3} \sin 1980\pi t$, $P_4 = \frac{0.002}{4} \sin 2640\pi t$, $P_5 = \frac{0.002}{5} \sin 3300\pi t$

The corresponding pressures of these upper harmonics are as follows.

$$P_2 = \frac{0.002}{2} \sin 1760\pi t, \qquad P_3 = \frac{0.002}{3} \sin 2640\pi t,$$

$$P_4 = \frac{0.002}{4} \sin 3520\pi t, \qquad \text{and} \qquad P_5 = \frac{0.002}{5} \sin 4400\pi t$$

The graph of

$$P = P_1 + P_2 + P_3 + P_4 + P_5,$$

shown in **Figure 35,** is "saw-toothed."

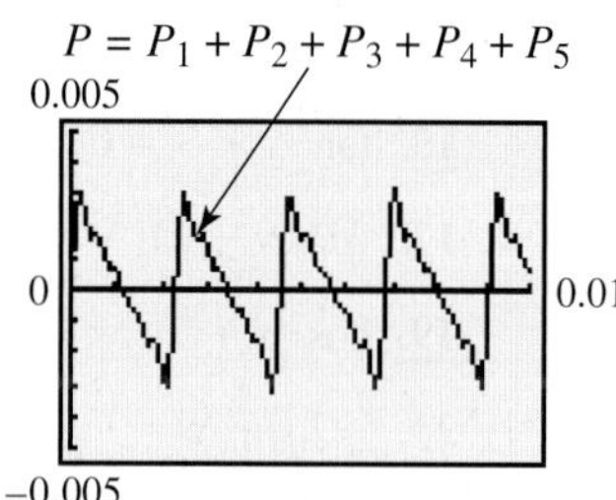

Figure 35

(a) What is the maximum value of P?

(b) At what values of $t = x$ does this maximum occur over the interval $[0, 0.01]$?

SOLUTION

(a) A graphing calculator shows that the maximum value of P is approximately 0.00317. See **Figure 36.**

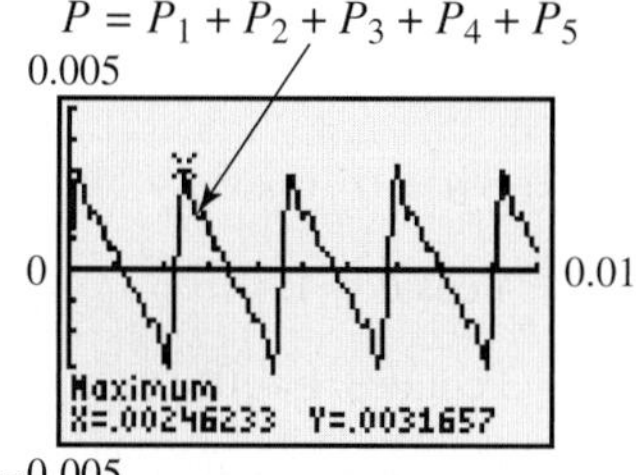

Figure 36

(b) The maximum occurs at

$$t = x \approx 0.000191, 0.00246, 0.00474, 0.00701, \text{ and } 0.00928.$$

Figure 36 shows how the second value is found. The other values are found similarly.

Now Try Exercise 103.

7.6 Exercises

Concept Check Refer to the summary box on solving a trigonometric equation following ***Example 5.*** *Decide on the appropriate technique to begin the solution of each equation. Do not solve the equation.*

1. $2 \cot x + 1 = -1$

2. $\sin x + 2 = 3$

3. $5 \sec^2 x = 6 \sec x$

4. $2 \cos^2 x - \cos x = 1$

5. $9 \sin^2 x - 5 \sin x = 1$

6. $\tan^2 x - 4 \tan x + 2 = 0$

7. $\tan x - \cot x = 0$

8. $\cos^2 x = \sin^2 x + 1$

1. Solve the linear equation for $\cot x$.
2. Solve the linear equation for $\sin x$.
3. Solve the quadratic equation for $\sec x$ by factoring.
4. Solve the quadratic equation for $\cos x$ by factoring.

5. Solve the quadratic equation for $\sin x$ using the quadratic formula.
6. Solve the quadratic equation for $\tan x$ using the quadratic formula.
7. Use an identity to rewrite as an equation with one trigonometric function.
8. Use an identity to rewrite as an equation with one trigonometric function.

9. Suppose that in solving an equation over the interval $[0°, 360°)$, you reach the step $\sin \theta = -\frac{1}{2}$. Why is $-30°$ not a correct answer?

10. Lindsay solved the equation $\sin x = 1 - \cos x$ by squaring each side to get

$$\sin^2 x = 1 - 2 \cos x + \cos^2 x.$$

Several steps later, using correct algebra, she concluded that the solution set for solutions over the interval $[0, 2\pi)$ is $\{0, \frac{\pi}{2}, \frac{3\pi}{2}\}$. Explain why this is not the correct solution set.

11. $\{\frac{3\pi}{4}, \frac{7\pi}{4}\}$ 12. $\{\frac{\pi}{2}\}$
13. $\{\frac{\pi}{6}, \frac{5\pi}{6}\}$ 14. $\{\frac{\pi}{3}, \frac{5\pi}{3}\}$
15. $\emptyset$ 16. $\emptyset$
17. $\{\frac{\pi}{4}, \frac{2\pi}{3}, \frac{5\pi}{4}, \frac{5\pi}{3}\}$
18. $\{\frac{\pi}{4}, \frac{3\pi}{4}, \frac{7\pi}{6}, \frac{11\pi}{6}\}$
19. $\{\pi\}$
20. $\{\frac{\pi}{6}, \frac{\pi}{2}, \frac{3\pi}{2}, \frac{11\pi}{6}\}$
21. $\{\frac{7\pi}{6}, \frac{3\pi}{2}, \frac{11\pi}{6}\}$
22. $\{0, \frac{2\pi}{3}, \frac{4\pi}{3}\}$

Solve each equation for exact solutions over the interval $[0, 2\pi)$. ***See Examples 1–3.***

11. $2 \cot x + 1 = -1$
12. $\sin x + 2 = 3$
13. $2 \sin x + 3 = 4$
14. $2 \sec x + 1 = \sec x + 3$
15. $\tan^2 x + 3 = 0$
16. $\sec^2 x + 2 = -1$
17. $(\cot x - 1)(\sqrt{3} \cot x + 1) = 0$
18. $(\csc x + 2)(\csc x - \sqrt{2}) = 0$
19. $\cos^2 x + 2 \cos x + 1 = 0$
20. $2 \cos^2 x - \sqrt{3} \cos x = 0$
21. $-2 \sin^2 x = 3 \sin x + 1$
22. $2 \cos^2 x - \cos x = 1$

23. $\{30°, 210°, 240°, 300°\}$
24. $\{0°, 45°, 225°\}$
25. $\{90°, 210°, 330°\}$
26. $\{60°, 135°, 240°, 315°\}$
27. $\{45°, 135°, 225°, 315°\}$
28. $\{0°, 180°\}$
29. $\{45°, 225°\}$
30. $\{90°, 270°\}$
31. $\{0°, 30°, 150°, 180°\}$
32. $\{0°, 90°, 180°, 270°\}$
33. $\{0°, 45°, 135°, 180°, 225°, 315°\}$
34. $\{45°, 135°, 225°, 315°\}$
35. $\{53.6°, 126.4°, 187.9°, 352.1°\}$
36. $\{78.0°, 282.0°\}$
37. $\{149.6°, 329.6°, 106.3°, 286.3°\}$
38. $\{38.4°, 218.4°, 104.8°, 284.8°\}$
39. $\emptyset$
40. $\emptyset$
41. $\{57.7°, 159.2°\}$
42. $\{114.3°, 335.7°\}$

Solve each equation for solutions over the interval $[0°, 360°)$. *Give solutions to the nearest tenth as appropriate.* ***See Examples 2–5.***

23. $(\cot \theta - \sqrt{3})(2 \sin \theta + \sqrt{3}) = 0$
24. $(\tan \theta - 1)(\cos \theta - 1) = 0$
25. $2 \sin \theta - 1 = \csc \theta$
26. $\tan \theta + 1 = \sqrt{3} + \sqrt{3} \cot \theta$
27. $\tan \theta - \cot \theta = 0$
28. $\cos^2 \theta = \sin^2 \theta + 1$
29. $\csc^2 \theta - 2 \cot \theta = 0$
30. $\sin^2 \theta \cos \theta = \cos \theta$
31. $2 \tan^2 \theta \sin \theta - \tan^2 \theta = 0$
32. $\sin^2 \theta \cos^2 \theta = 0$
33. $\sec^2 \theta \tan \theta = 2 \tan \theta$
34. $\cos^2 \theta - \sin^2 \theta = 0$
35. $9 \sin^2 \theta - 6 \sin \theta = 1$
36. $4 \cos^2 \theta + 4 \cos \theta = 1$
37. $\tan^2 \theta + 4 \tan \theta + 2 = 0$
38. $3 \cot^2 \theta - 3 \cot \theta - 1 = 0$
39. $\sin^2 \theta - 2 \sin \theta + 3 = 0$
40. $2 \cos^2 \theta + 2 \cos \theta + 1 = 0$
41. $\cot \theta + 2 \csc \theta = 3$
42. $2 \sin \theta = 1 - 2 \cos \theta$

43. $\{180° + 360°n$, where n is any integer$\}$
44. $\{135° + 180°n$, where n is any integer$\}$
45. $\{\frac{\pi}{3} + 2n\pi, \frac{2\pi}{3} + 2n\pi$, where n is any integer$\}$

Solve each equation (x in radians and θ in degrees) for all exact solutions where appropriate. Round approximate answers in radians to four decimal places and approximate answers in degrees to the nearest tenth. Write answers using the least possible nonnegative angle measures. ***See Examples 1–5.***

43. $\cos \theta + 1 = 0$
44. $\tan \theta + 1 = 0$
45. $3 \csc x - 2\sqrt{3} = 0$
46. $\cot x + \sqrt{3} = 0$
47. $6 \sin^2 \theta + \sin \theta = 1$
48. $3 \sin^2 \theta - \sin \theta = 2$
49. $2 \cos^2 x + \cos x - 1 = 0$
50. $4 \cos^2 x - 1 = 0$
51. $\sin \theta \cos \theta - \sin \theta = 0$
52. $\tan \theta \csc \theta - \sqrt{3} \csc \theta = 0$
53. $\sin x (3 \sin x - 1) = 1$
54. $\tan x (\tan x - 2) = 5$
55. $5 + 5 \tan^2 \theta = 6 \sec \theta$
56. $\sec^2 \theta = 2 \tan \theta + 4$
57. $\dfrac{2 \tan \theta}{3 - \tan^2 \theta} = 1$
58. $\dfrac{2 \cot^2 \theta}{\cot \theta + 3} = 1$

46. $\left\{\frac{5\pi}{6} + n\pi\text{, where } n \text{ is any integer}\right\}$

47. $\{19.5° + 360°n, 160.5° + 360°n, 210° + 360°n, 330° + 360°n$, where n is any integer$\}$

48. $\{90° + 360°n, 221.8° + 360°n, 318.2° + 360°n$, where n is any integer$\}$

49. $\left\{\frac{\pi}{3} + 2n\pi, \pi + 2n\pi, \frac{5\pi}{3} + 2n\pi\right.$, where n is any integer$\}$

50. $\left\{\frac{\pi}{3} + n\pi, \frac{2\pi}{3} + n\pi\right.$, where n is any integer$\}$

51. $\{180°n$, where n is any integer$\}$

52. $\{60° + 180°n$, where n is any integer$\}$

53. $\{0.8751 + 2n\pi, 2.2665 + 2n\pi, 3.5908 + 2n\pi, 5.8340 + 2n\pi$, where n is any integer$\}$

54. $\{1.2886 + n\pi, 2.1747 + n\pi$, where n is any integer$\}$

55. $\{33.6° + 360°n, 326.4° + 360°n$, where n is any integer$\}$

56. $\{71.6° + 180°n, 135° + 180°n$, where n is any integer$\}$

57. $\{45° + 180°n, 108.4° + 180°n$, where n is any integer$\}$

58. $\{33.7° + 180°n, 135° + 180°n$, where n is any integer$\}$

59. $\{0.6806, 1.4159\}$

60. $\{0, 0.3760\}$

61. $\left\{\frac{\pi}{3}, \pi, \frac{4\pi}{3}\right\}$

62. $\left\{\frac{\pi}{8}, \frac{5\pi}{6}, \frac{5\pi}{4}\right\}$

63. $\{60°, 210°, 240°, 310°\}$

64. $\{135°, 180°, 225°, 270°\}$

65. $\left\{\frac{\pi}{12}, \frac{11\pi}{12}, \frac{13\pi}{12}, \frac{23\pi}{12}\right\}$

66. $\left\{\frac{\pi}{3}, \frac{2\pi}{3}, \frac{4\pi}{3}, \frac{5\pi}{3}\right\}$

67. $\{90°, 210°, 330°\}$

68. $\{0°, 60°, 120°, 180°, 240°, 300°\}$

69. $\left\{\frac{\pi}{18}, \frac{7\pi}{18}, \frac{13\pi}{18}, \frac{19\pi}{18}, \frac{25\pi}{18}, \frac{31\pi}{18}\right\}$

70. $\left\{\frac{\pi}{18}, \frac{7\pi}{18}, \frac{13\pi}{18}, \frac{19\pi}{18}, \frac{25\pi}{18}, \frac{31\pi}{18}\right\}$

71. $\{67.5°, 112.5°, 247.5°, 292.5°\}$

72. $\{15°, 75°, 195°, 255°\}$

73. $\left\{\frac{\pi}{2}, \frac{3\pi}{2}\right\}$

74. $\left\{0, \frac{\pi}{4}, \frac{\pi}{2}, \frac{3\pi}{4}, \pi, \frac{5\pi}{4}, \frac{3\pi}{2}, \frac{7\pi}{4}\right\}$

75. $\left\{0, \frac{\pi}{3}, \pi, \frac{5\pi}{3}\right\}$ 76. $\left\{0, \frac{2\pi}{3}, \frac{4\pi}{3}\right\}$

77. $\emptyset$ 78. $\emptyset$

The following equations cannot be solved by algebraic methods. Use a graphing calculator to find all solutions over the interval $[0, 2\pi)$. Express solutions to four decimal places.

59. $x^2 + \sin x - x^3 - \cos x = 0$

60. $x^3 - \cos^2 x = \frac{1}{2}x - 1$

Concept Check *Answer each question.*

61. Suppose you are solving a trigonometric equation for solutions over the interval $[0, 2\pi)$, and your work leads to $2x = \frac{2\pi}{3}, 2\pi, \frac{8\pi}{3}$. What are the corresponding values of x?

62. Suppose you are solving a trigonometric equation for solutions over the interval $[0, 2\pi)$, and your work leads to $\frac{1}{2}x = \frac{\pi}{16}, \frac{5\pi}{12}, \frac{5\pi}{8}$. What are the corresponding values of x?

63. Suppose you are solving a trigonometric equation for solutions over the interval $[0°, 360°)$, and your work leads to $3\theta = 180°, 630°, 720°, 930°$. What are the corresponding values of θ?

64. Suppose you are solving a trigonometric equation for solutions over the interval $[0°, 360°)$, and your work leads to $\frac{1}{3}\theta = 45°, 60°, 75°, 90°$. What are the corresponding values of θ?

Solve each equation in x for exact solutions over the interval $[0, 2\pi)$ and each equation in θ for exact solutions over the interval $[0°, 360°)$. ***See Examples 6–8.***

65. $\cos 2x = \frac{\sqrt{3}}{2}$

66. $\cos 2x = -\frac{1}{2}$

67. $\sin 3\theta = -1$

68. $\sin 3\theta = 0$

69. $3 \tan 3x = \sqrt{3}$

70. $\cot 3x = \sqrt{3}$

71. $\sqrt{2} \cos 2\theta = -1$

72. $2\sqrt{3} \sin 2\theta = \sqrt{3}$

73. $\sin \frac{x}{2} = \sqrt{2} - \sin \frac{x}{2}$

74. $\tan 4x = 0$

75. $\sin x = \sin 2x$

76. $\cos 2x - \cos x = 0$

77. $8 \sec^2 \frac{x}{2} = 4$

78. $\sin^2 \frac{x}{2} - 2 = 0$

79. $\sin \frac{\theta}{2} = \csc \frac{\theta}{2}$

80. $\sec \frac{\theta}{2} = \cos \frac{\theta}{2}$

81. $\cos 2x + \cos x = 0$

82. $\sin x \cos x = \frac{1}{4}$

Solve each equation (x in radians and θ in degrees) for all exact solutions where appropriate. Round approximate answers in radians to four decimal places and approximate answers in degrees to the nearest tenth. Write answers using the least possible nonnegative angle measures. ***See Examples 6–8.***

83. $\sqrt{2} \sin 3x - 1 = 0$

84. $-2 \cos 2x = \sqrt{3}$

85. $\cos \frac{\theta}{2} = 1$

86. $\sin \frac{\theta}{2} = 1$

87. $2\sqrt{3} \sin \frac{x}{2} = 3$

88. $2\sqrt{3} \cos \frac{x}{2} = -3$

89. $2 \sin \theta = 2 \cos 2\theta$

90. $\cos \theta - 1 = \cos 2\theta$

91. $1 - \sin x = \cos 2x$

92. $\sin 2x = 2 \cos^2 x$

93. $3 \csc^2 \frac{x}{2} = 2 \sec x$

94. $\cos x = \sin^2 \frac{x}{2}$

95. $2 - \sin 2\theta = 4 \sin 2\theta$

96. $4 \cos 2\theta = 8 \sin \theta \cos \theta$

97. $2 \cos^2 2\theta = 1 - \cos 2\theta$

98. $\sin \theta - \sin 2\theta = 0$

The following equations cannot be solved by algebraic methods. Use a graphing calculator to find all solutions over the interval $[0, 2\pi)$. Express solutions to four decimal places.

99. $2 \sin 2x - x^3 + 1 = 0$

100. $3 \cos \frac{x}{2} + \sqrt{x} - 2 = -\frac{1}{2}x + 2$

79. $\{180°\}$ **80.** $\{0°\}$

81. $\left\{\frac{\pi}{3}, \pi, \frac{5\pi}{3}\right\}$

82. $\left\{\frac{\pi}{12}, \frac{5\pi}{12}, \frac{13\pi}{12}, \frac{17\pi}{12}\right\}$

83. $\left\{\frac{\pi}{12} + \frac{2n\pi}{3}, \frac{\pi}{4} + \frac{2n\pi}{3},\right.$ where n is any integer$\left.\right\}$

84. $\left\{\frac{5\pi}{12} + n\pi, \frac{7\pi}{12} + n\pi,\right.$ where n is any integer$\left.\right\}$

85. $\{720°n$, where n is any integer$\}$

86. $\{180° + 720°n$, where n is any integer$\}$

87. $\left\{\frac{2\pi}{3} + 4n\pi, \frac{4\pi}{3} + 4n\pi,\right.$ where n is any integer$\left.\right\}$

88. $\left\{\frac{5\pi}{3} + 4n\pi, \frac{7\pi}{3} + 4n\pi,\right.$ where n is any integer$\left.\right\}$

89. $\{30° + 360°n, 150° + 360°n, 270° + 360°n$, where n is any integer$\}$

90. $\{60° + 360°n, 90° + 180°n, 300° + 360°n$, where n is any integer$\}$

91. $\left\{n\pi, \frac{\pi}{6} + 2n\pi, \frac{5\pi}{6} + 2n\pi,\right.$ where n is any integer$\left.\right\}$

92. $\left\{\frac{\pi}{2} + n\pi, \frac{\pi}{4} + n\pi,\right.$ where n is any integer$\left.\right\}$

93. $\{1.3181 + 2n\pi, 4.9651 + 2n\pi$, where n is any integer$\}$

94. $\{1.2310 + 2n\pi, 5.0522 + 2n\pi$, where n is any integer$\}$

95. $\{11.8° + 180°n, 78.2° + 180°n$, where n is any integer$\}$

96. $\{22.5° + 180°n, 112.5° + 180°n$, where n is any integer$\}$

97. $\{30° + 180°n, 90° + 180°n, 150° + 180°n$, where n is any integer$\}$

98. $\{180°n, 60° + 360°n, 300° + 360°n$, where n is any integer$\}$

99. $\{1.2802\}$

100. $\{0.6919, 2.0820\}$

101. **(a)** 0.00164 and 0.00355 **(b)** $[0.00164, 0.00355]$ **(c)** outward

(Modeling) Solve each problem. ***See Examples 9 and 10.***

101. *Pressure on the Eardrum* No musical instrument can generate a true pure tone. A pure tone has a unique, constant frequency and amplitude that sounds rather dull and uninteresting. The pressures caused by pure tones on the eardrum are sinusoidal. The change in pressure P in pounds per square foot on a person's eardrum from a pure tone at time t in seconds can be modeled using the equation

$$P = A \sin(2\pi ft + \phi),$$

where f is the frequency in cycles per second, and ϕ is the phase angle. When P is positive, there is an increase in pressure and the eardrum is pushed inward. When P is negative, there is a decrease in pressure and the eardrum is pushed outward. (*Source:* Roederer, J., *Introduction to the Physics and Psychophysics of Music,* Second Edition, Springer-Verlag.) A graph of the tone middle C is shown in the figure.

(a) Determine algebraically the values of t for which $P = 0$ over $[0, 0.005]$.

(b) From the graph and your answer in part (a), determine the interval for which $P \leq 0$ over $[0, 0.005]$.

(c) Would an eardrum hearing this tone be vibrating outward or inward when $P < 0$?

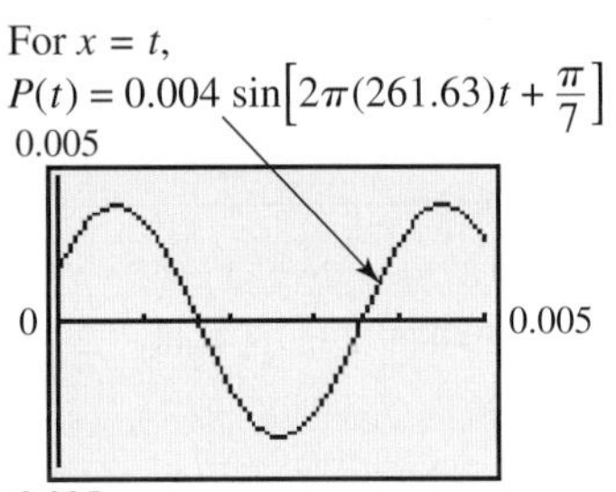

102. *Hearing Beats in Music* Musicians sometimes tune instruments by playing the same tone on two different instruments and listening for a phenomenon known as **beats.** Beats occur when two tones vary in frequency by only a few hertz. When the two instruments are in tune, the beats disappear. The ear hears beats because the pressure slowly rises and falls as a result of this slight variation in the frequency. This phenomenon can be seen using a graphing calculator. (*Source:* Pierce, J., *The Science of Musical Sound,* Scientific American Books.)

(a) Consider the two tones with frequencies of 220 Hz and 223 Hz and pressures $P_1 = 0.005 \sin 440\pi t$ and $P_2 = 0.005 \sin 446\pi t$, respectively. Graph the pressure $P = P_1 + P_2$ felt by an eardrum over the 1-sec interval $[0.15, 1.15]$. How many beats are there in 1 sec?

(b) Repeat part (a) with frequencies of 220 and 216 Hz.

(c) Determine a simple way to find the number of beats per second if the frequency of each tone is given.

103. *Pressure of a Plucked String* If a string with a fundamental frequency of 110 Hz is plucked in the middle, it will vibrate at the odd harmonics of 110, 330, 550, . . . Hz but not at the even harmonics of 220, 440, 660, . . . Hz. The resulting pressure P caused by the string can be modeled by the equation

$$P = 0.003 \sin 220\pi t + \frac{0.003}{3} \sin 660\pi t + \frac{0.003}{5} \sin 1100\pi t + \frac{0.003}{7} \sin 1540\pi t.$$

(*Source:* Benade, A., *Fundamentals of Musical Acoustics,* Dover Publications. Roederer, J., *Introduction to the Physics and Psychophysics of Music,* Second Edition, Springer-Verlag.)

(a) Graph P in the viewing window $[0, 0.03]$ by $[-0.005, 0.005]$.

(b) Use the graph to describe the shape of the sound wave that is produced.

(c) Refer to **Exercise 101.** At lower frequencies, the inner ear will hear a tone only when the eardrum is moving outward. Determine the times over the interval $[0, 0.03]$ when this will occur.

102. (a) 3 beats per sec
For $x = t$,
$P(t) = 0.005 \sin 440\pi t + 0.005 \sin 446\pi t$

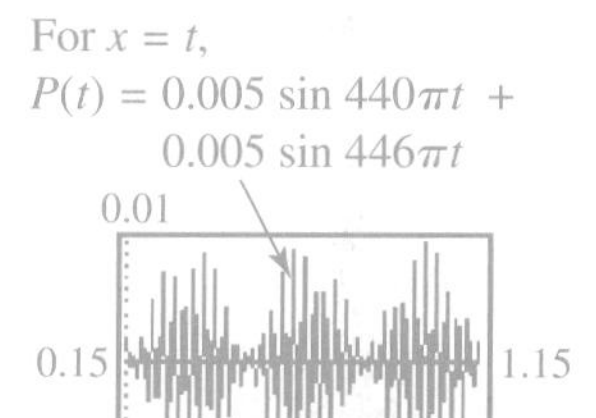

(b) 4 beats per sec
For $x = t$,
$P(t) = 0.005 \sin 440\pi t + 0.005 \sin 432\pi t$

0.01, 0.15, 1.15, −0.01

(c) The number of beats is equal to the absolute value of the difference in the frequencies of the two tones.

103. (a)
For $x = t$,
$P(t) = 0.003 \sin 220\pi t + \frac{0.003}{3} \sin 660\pi t + \frac{0.003}{5} \sin 1100\pi t + \frac{0.003}{7} \sin 1540\pi t$

0.005, 0, 0.03, −0.005

(b) The graph is periodic, and the wave has "jagged square" tops and bottoms.
(c) This will occur when t is in one of these intervals: $(0.0045, 0.0091)$, $(0.0136, 0.0182)$, $(0.0227, 0.0273)$.

104. (a) For $x = t$,
$P(t) = \frac{1}{2} \sin[2\pi(220)t] + \frac{1}{3} \sin[2\pi(330)t] + \frac{1}{4} \sin[2\pi(440)t]$

1, 0, 0.03, −1

104. *Hearing Difference Tones* When a musical instrument creates a tone of 110 Hz, it also creates tones at 220, 330, 440, 550, 660, . . . Hz. A small speaker cannot reproduce the 110-Hz vibration but it can reproduce the higher frequencies, which are the **upper harmonics.** The low tones can still be heard because the speaker produces **difference tones** of the upper harmonics. The difference between consecutive frequencies is 110 Hz, and this difference tone will be heard by a listener. (*Source:* Benade, A., *Fundamentals of Musical Acoustics,* Dover Publications.)

(a) We can model this phenomenon using a graphing calculator. In the window $[0, 0.03]$ by $[-1, 1]$, graph the upper harmonics represented by the pressure

$$P = \frac{1}{2}\sin[2\pi(220)t] + \frac{1}{3}\sin[2\pi(330)t] + \frac{1}{4}\sin[2\pi(440)t].$$

(b) Estimate all t-coordinates where P is maximum.

(c) What does a person hear in addition to the frequencies of 220, 330, and 440 Hz?

(d) Graph the pressure produced by a speaker that can vibrate at 110 Hz and above.

(Modeling) Solve each problem.

105. *Electromotive Force* In an electric circuit, suppose that

$$V = \cos 2\pi t$$

models the electromotive force in volts at t seconds. Find the least value of t where $0 \le t \le \frac{1}{2}$ for each value of V.

(a) $V = 0$ **(b)** $V = 0.5$ **(c)** $V = 0.25$

106. *Accident Reconstruction* The model

$$0.342D \cos\theta + h\cos^2\theta = \frac{16D^2}{V_0{}^2}$$

is used to reconstruct accidents in which a vehicle vaults into the air after hitting an obstruction. V_0 is velocity in feet per second of the vehicle when it hits the obstruction, D is distance (in feet) from the obstruction to the landing point, and h is the difference in height (in feet) between landing point and takeoff point. Angle θ is the takeoff angle, the angle between the horizontal and the path of the vehicle. Find θ to the nearest degree if $V_0 = 60$, $D = 80$, and $h = 2$.

107. *Movement of a Particle* A particle moves along a straight line. The distance of the particle from the origin at time t is modeled by

$$s(t) = \sin t + 2\cos t.$$

Find a value of t that satisfies each equation.

(a) $s(t) = \frac{2 + \sqrt{3}}{2}$ **(b)** $s(t) = \frac{3\sqrt{2}}{2}$

108. *Voltage Induced by a Coil of Wire* A coil of wire rotating in a magnetic field induces a voltage modeled by

$$E = 20\sin\left(\frac{\pi t}{4} - \frac{\pi}{2}\right),$$

where t is time in seconds. Find the least positive time to produce each voltage.

(a) 0 **(b)** $10\sqrt{3}$

109. *Average Monthly Temperature in Vancouver* The following function approximates the average monthly temperature y (in °F) in Vancouver, Canada. Here x represents the month, where $x = 1$ corresponds to January, $x = 2$ corresponds to February, and so on. (*Source:* www.weather.com)

$$f(x) = 14\sin\left[\frac{\pi}{6}(x - 4)\right] + 50$$

When is the average monthly temperature **(a)** 64°F **(b)** 39°F?

104. (b) 0.0007576, 0.009847, 0.01894, 0.02803
(c) 110 Hz
(d) For $x = t$,
$P(t) = \sin[2\pi(110)t] + \frac{1}{2}\sin[2\pi(220)t] + \frac{1}{3}\sin[2\pi(330)t] + \frac{1}{4}\sin[2\pi(440)t]$

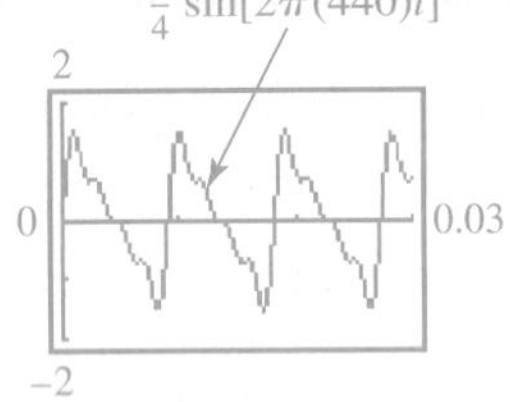

105. (a) $\frac{1}{4}$ sec (b) $\frac{1}{6}$ sec (c) 0.21 sec
106. 14°
107. (a) One such value is $\frac{\pi}{3}$. (b) One such value is $\frac{\pi}{4}$.
108. (a) 2 sec (b) $3\frac{1}{3}$ sec

110. *Average Monthly Temperature in Phoenix* The following function approximates the average monthly temperature y (in °F) in Phoenix, Arizona. Here x represents the month, where $x = 1$ corresponds to January, $x = 2$ corresponds to February, and so on. (*Source:* www.weather.com)

$$f(x) = 19.5\cos\left[\frac{\pi}{6}(x - 7)\right] + 70.5$$

When is the average monthly temperature **(a)** 70.5°F **(b)** 55°F?

(Modeling) Alternating Electric Current The study of alternating electric current requires the solutions of equations of the form

$$i = I_{max}\sin 2\pi ft,$$

for time t in seconds, where i is instantaneous current in amperes, I_{max} is maximum current in amperes, and f is the number of cycles per second. (Source: Hannon, R. H., *Basic Technical Mathematics with Calculus,* W. B. Saunders Company.*) Find the least positive value of t, given the following data.*

111. $i = 40$, $I_{max} = 100$, $f = 60$ **112.** $i = 50$, $I_{max} = 100$, $f = 120$

113. $i = I_{max}$, $f = 60$ **114.** $i = \frac{1}{2}I_{max}$, $f = 60$

109. (a) when $x = 7$ (during July)
(b) when $x = 2.3$ (during February) and when $x = 11.7$ (during November)
110. (a) when $x = 4$ (during April) and when $x = 10$ (during October)
(b) when $x = 2.2$ (during February) and when $x = 11.8$ (during November)
111. 0.001 sec 112. 0.0007 sec 113. 0.004 sec 114. 0.0014 sec

Chapter 7 Quiz (Sections 7.5–7.6)

1. Graph $y = \cos^{-1} x$, and indicate the coordinates of three points on the graph. Give the domain and range.

2. Find the exact value of each real number y.

(a) $y = \sin^{-1}\left(-\frac{\sqrt{2}}{2}\right)$ **(b)** $y = \tan^{-1}\sqrt{3}$ **(c)** $y = \sec^{-1}\left(-\frac{2\sqrt{3}}{3}\right)$

3. Use a calculator to give each value in decimal degrees.

(a) $\theta = \arccos 0.92341853$ **(b)** $\theta = \cot^{-1}(-1.0886767)$

4. Give the exact value of each expression without using a calculator.

(a) $\cos\left(\tan^{-1}\frac{4}{5}\right)$ **(b)** $\sin\left(\cos^{-1}\left(-\frac{1}{2}\right) + \tan^{-1}(-\sqrt{3})\right)$

Solve each equation for exact solutions over the interval $[0°, 360°)$.

5. $2\sin\theta - \sqrt{3} = 0$ **6.** $\cos\theta + 1 = 2\sin^2\theta$

Solve each equation for solutions over the interval $[0, 2\pi)$.

7. $\tan^2 x - 5\tan x + 3 = 0$ **8.** $3\cot 2x - \sqrt{3} = 0$

9. Solve $\cos\frac{x}{2} + \sqrt{3} = -\cos\frac{x}{2}$, giving all solutions in radians.

10. *(Modeling) Electromotive Force* In an electric circuit, suppose that

$$V = \cos 2\pi t$$

models the electromotive force in volts at t seconds. Find the least value of t where $0 \le t \le \frac{1}{2}$ for each value of V.

(a) $V = 1$ **(b)** $V = 0.30$

[7.5]
1. $[-1, 1]$; $[0, \pi]$

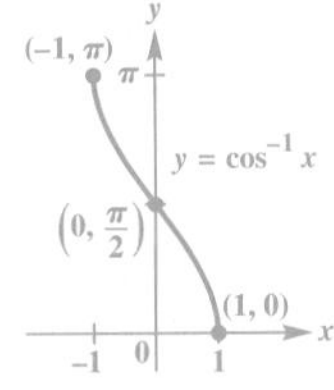

2. (a) $-\frac{\pi}{4}$ (b) $\frac{\pi}{3}$ (c) $\frac{5\pi}{6}$
3. (a) 22.568922° (b) 137.431085°
4. (a) $\frac{5\sqrt{41}}{41}$ (b) $\frac{\sqrt{3}}{2}$

[7.6]
5. {60°, 120°}
6. {60°, 180°, 300°}
7. {0.6089, 1.3424, 3.7505, 4.4840}
8. $\left\{\frac{\pi}{6}, \frac{2\pi}{3}, \frac{7\pi}{6}, \frac{5\pi}{3}\right\}$
9. $\left\{\frac{5\pi}{3} + 4n\pi, \frac{7\pi}{3} + 4n\pi\right.$, where n is any integer$\}$
10. (a) 0 sec (b) 0.20 sec

7.7 Equations Involving Inverse Trigonometric Functions

- Solving for x in Terms of y Using Inverse Functions
- Solving Inverse Trigonometric Equations

Solving for x in Terms of y Using Inverse Functions

Classroom Example 1
Solve $y = 4 \tan 3x$ for x, where x is restricted to the interval $\left(-\frac{\pi}{6}, \frac{\pi}{6}\right)$.

Answer: $x = \frac{1}{3} \arctan \frac{y}{4}$

EXAMPLE 1 Solving an Equation for a Specified Variable

Solve $y = 3 \cos 2x$ for x, where x is restricted to the interval $\left[0, \frac{\pi}{2}\right]$.

SOLUTION We want cos $2x$ alone on one side of the equation so that we can solve for $2x$, and then for x.

$$y = 3 \cos 2x$$ Our goal is to isolate x.

$$\frac{y}{3} = \cos 2x$$ Divide by 3. (Section 1.1)

$$2x = \arccos \frac{y}{3}$$ Definition of arccosine (Section 7.5)

$$x = \frac{1}{2} \arccos \frac{y}{3}$$ Multiply by $\frac{1}{2}$.

An equivalent form of this answer is $x = \frac{1}{2} \cos^{-1} \frac{y}{3}$.

Because the function $y = 3 \cos 2x$ is periodic, with period π, there are infinitely many domain values (x-values) that will result in a given range value (y-value). For example, the x-values 0 and π both correspond to the y-value 3. See **Figure 37.** The restriction $0 \le x \le \frac{\pi}{2}$ given in the original problem ensures that this function is one-to-one, and, correspondingly, that

$$x = \frac{1}{2} \arccos \frac{y}{3}$$

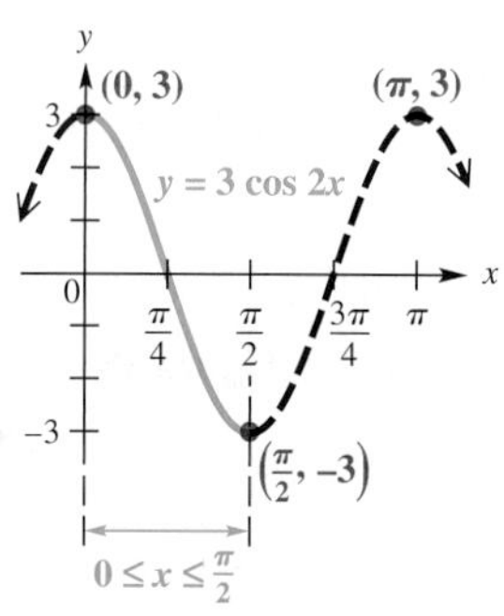

Figure 37

has a one-to-one relationship. Thus, each y-value in $[-3, 3]$ substituted into this equation will lead to a single x-value.

Now Try Exercise 9.

Solving Inverse Trigonometric Equations

Classroom Example 2
Solve $3 \arctan x = \pi$.

Answer: $\{\sqrt{3}\}$

EXAMPLE 2 Solving an Equation Involving an Inverse Trigonometric Function

Solve $2 \arcsin x = \pi$.

SOLUTION First solve for arcsin x, and then for x.

$$2 \arcsin x = \pi$$ Original equation

$$\arcsin x = \frac{\pi}{2}$$ Divide by 2.

$$x = \sin \frac{\pi}{2}$$ Definition of arcsine (Section 7.5)

$$x = 1$$ (Section 6.2)

CHECK

$$2 \arcsin x = \pi \quad \text{Original equation}$$

$$2 \arcsin 1 \stackrel{?}{=} \pi \quad \text{Let } x = 1.$$

$$2\left(\frac{\pi}{2}\right) \stackrel{?}{=} \pi \quad \text{Substitute the inverse value.}$$

$$\pi = \pi \ \checkmark \ \text{True}$$

The solution set is $\{1\}$.

✔ *Now Try Exercise 25.*

Classroom Example 3
Solve $\sec^{-1} x = \csc^{-1} 2$.

Answer: $\left\{\frac{2\sqrt{3}}{3}\right\}$

EXAMPLE 3 Solving an Equation Involving Inverse Trigonometric Functions

Solve $\cos^{-1} x = \sin^{-1} \frac{1}{2}$.

SOLUTION Let $\sin^{-1} \frac{1}{2} = u$. Then $\sin u = \frac{1}{2}$, and for u in quadrant I we have the following.

$$\cos^{-1} x = \sin^{-1} \frac{1}{2} \quad \text{Original equation}$$

$$\cos^{-1} x = u \quad \text{Substitute.}$$

$$\cos u = x \quad \text{Alternative form (Section 7.5)}$$

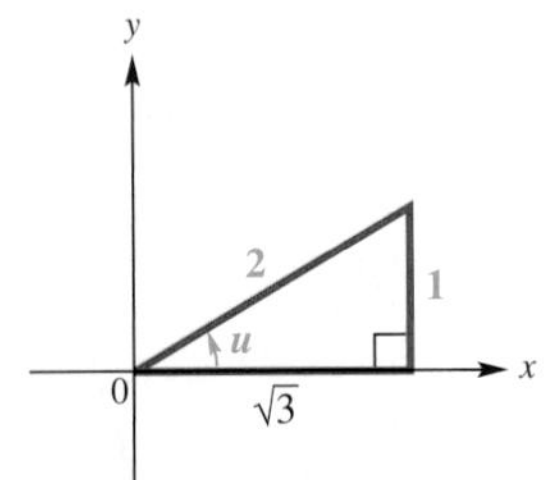

Figure 38

Sketch a triangle and label it using the facts that u is in quadrant I and $\sin u = \frac{1}{2}$. See **Figure 38.** Since $x = \cos u$, $x = \frac{\sqrt{3}}{2}$, and the solution set is $\left\{\frac{\sqrt{3}}{2}\right\}$. A check indicates that this is correct.

✔ *Now Try Exercise 33.*

Classroom Example 4
Solve $\arccos x - \arcsin x = \pi$.

Answer: $\left\{-\frac{\sqrt{2}}{2}\right\}$

EXAMPLE 4 Solving an Inverse Trigonometric Equation Using an Identity

Solve $\arcsin x - \arccos x = \frac{\pi}{6}$.

SOLUTION Isolate one inverse function on one side of the equation.

$$\arcsin x - \arccos x = \frac{\pi}{6} \quad \text{Original equation}$$

$$\arcsin x = \arccos x + \frac{\pi}{6} \quad \text{Add } \arccos x. \quad (1)$$

$$x = \sin\left(\arccos x + \frac{\pi}{6}\right) \quad \text{Definition of arcsine}$$

Let $u = \arccos x$. The arccosine function yields angles in quadrants I and II, so $0 \le u \le \pi$ by definition.

$$x = \sin\left(u + \frac{\pi}{6}\right) \quad \text{Substitute.}$$

$$x = \sin u \cos \frac{\pi}{6} + \cos u \sin \frac{\pi}{6} \quad \text{Sine sum identity (Section 7.3)} \quad (2)$$

From equation (1) and by the definition of the arcsine function,

$$-\frac{\pi}{2} \leq \arccos x + \frac{\pi}{6} \leq \frac{\pi}{2} \qquad \text{Range of arcsine is } \left[-\frac{\pi}{2}, \frac{\pi}{2}\right]. \text{ (Section 7.5)}$$

$$-\frac{2\pi}{3} \leq \arccos x \leq \frac{\pi}{3}. \qquad \text{Subtract } \frac{\pi}{6} \text{ from each part. (Section 1.7)}$$

Since $0 \leq \arccos x \leq \pi$ **and** $-\frac{2\pi}{3} \leq \arccos x \leq \frac{\pi}{3}$, the intersection yields $0 \leq \arccos x \leq \frac{\pi}{3}$. This places u in quadrant I, and we can sketch the triangle in **Figure 39.** From this triangle we find that $\sin u = \sqrt{1 - x^2}$. Now substitute into equation (2) using $\sin u = \sqrt{1 - x^2}$, $\sin \frac{\pi}{6} = \frac{1}{2}$, $\cos \frac{\pi}{6} = \frac{\sqrt{3}}{2}$, and $\cos u = x$.

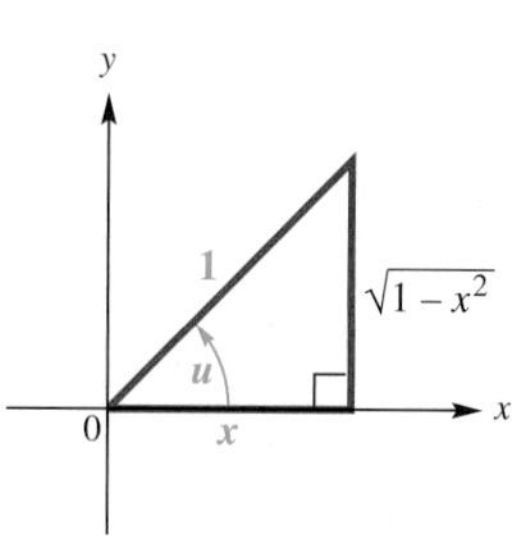

Figure 39

$$x = \sin u \cos \frac{\pi}{6} + \cos u \sin \frac{\pi}{6} \qquad (2)$$

$$x = \left(\sqrt{1 - x^2}\right)\frac{\sqrt{3}}{2} + x \cdot \frac{1}{2} \qquad \text{Substitute.}$$

$$2x = \left(\sqrt{1 - x^2}\right)\sqrt{3} + x \qquad \text{Multiply by 2.}$$

$$x = \left(\sqrt{3}\right)\sqrt{1 - x^2} \qquad \text{Subtract } x\text{; commutative property (Section R.2)}$$

Square *each* factor.

$$x^2 = 3(1 - x^2) \qquad \text{Square each side. (Section 1.6)}$$

$$x^2 = 3 - 3x^2 \qquad \text{Distributive property}$$

$$4x^2 = 3 \qquad \text{Add } 3x^2.$$

$$x^2 = \frac{3}{4} \qquad \text{Divide by 4.}$$

Choose the positive square root, $x > 0$.

$$x = \sqrt{\frac{3}{4}} \qquad \text{Take the square root on each side. (Section 1.4)}$$

$$x = \frac{\sqrt{3}}{2} \qquad \text{Quotient rule: } \sqrt[n]{\frac{a}{b}} = \frac{\sqrt[n]{a}}{\sqrt[n]{b}} \text{ (Section R.7)}$$

CHECK

$$\arcsin x - \arccos x = \frac{\pi}{6} \qquad \text{Original equation}$$

$$\arcsin \frac{\sqrt{3}}{2} - \arccos \frac{\sqrt{3}}{2} \stackrel{?}{=} \frac{\pi}{6} \qquad \text{Let } x = \frac{\sqrt{3}}{2}.$$

$$\frac{\pi}{3} - \frac{\pi}{6} \stackrel{?}{=} \frac{\pi}{6} \qquad \text{Substitute inverse values.}$$

$$\frac{\pi}{6} = \frac{\pi}{6} \qquad ✓ \text{ True}$$

The solution set is $\left\{\frac{\sqrt{3}}{2}\right\}$.

Now Try Exercise 35.

7.7 Exercises

1. C
2. A
3. C
4. C
5. $x = \arccos \frac{y}{5}$
6. $x = \arcsin 4y$
7. $x = \frac{1}{3} \text{arccot}\ 2y$
8. $x = \text{arcsec}\ 12y$
9. $x = \frac{1}{2} \arctan \frac{y}{3}$
10. $x = 2 \arcsin \frac{y}{3}$
11. $x = 4 \arccos \frac{y}{6}$
12. $x = 3 \arcsin(-y)$
13. $x = \frac{1}{5} \arccos\left(-\frac{y}{2}\right)$
14. $x = \frac{1}{5} \text{arccot} \frac{y}{3}$
15. $x = -3 + \arccos y$
16. $x = \frac{1}{2}(1 + \arctan y)$
17. $x = \arcsin(y + 2)$
18. $x = \text{arccot}(y - 1)$
19. $x = \arcsin\left(\frac{y + 4}{2}\right)$
20. $x = \arccos\left(\frac{y - 4}{3}\right)$
21. $x = \frac{1}{2} \sec^{-1}\left(\frac{y - \sqrt{2}}{3}\right)$
22. $x = 2 \csc^{-1}\left(\frac{y + \sqrt{3}}{2}\right)$

Concept Check *Answer each question.*

1. Which one of the following equations has solution 0?

A. $\arctan 1 = x$ **B.** $\arccos 0 = x$ **C.** $\arcsin 0 = x$

2. Which one of the following equations has solution $\frac{\pi}{4}$?

A. $\arcsin \frac{\sqrt{2}}{2} = x$ **B.** $\arccos\left(-\frac{\sqrt{2}}{2}\right) = x$ **C.** $\arctan \frac{\sqrt{3}}{3} = x$

3. Which one of the following equations has solution $\frac{3\pi}{4}$?

A. $\arctan 1 = x$ **B.** $\arcsin \frac{\sqrt{2}}{2} = x$ **C.** $\arccos\left(-\frac{\sqrt{2}}{2}\right) = x$

4. Which one of the following equations has solution $-\frac{\pi}{6}$?

A. $\arctan \frac{\sqrt{3}}{3} = x$ **B.** $\arccos\left(-\frac{1}{2}\right) = x$ **C.** $\arcsin\left(-\frac{1}{2}\right) = x$

Solve each equation for x, where x is restricted to the given interval. **See Example 1.**

5. $y = 5 \cos x$, for x in $[0, \pi]$

6. $y = \frac{1}{4} \sin x$, for x in $\left[-\frac{\pi}{2}, \frac{\pi}{2}\right]$

7. $y = \frac{1}{2} \cot 3x$, for x in $\left(0, \frac{\pi}{3}\right)$

8. $y = \frac{1}{12} \sec x$, for x in $\left[0, \frac{\pi}{2}\right) \cup \left(\frac{\pi}{2}, \pi\right]$

9. $y = 3 \tan 2x$, for x in $\left(-\frac{\pi}{4}, \frac{\pi}{4}\right)$

10. $y = 3 \sin \frac{x}{2}$, for x in $[-\pi, \pi]$

11. $y = 6 \cos \frac{x}{4}$, for x in $[0, 4\pi]$

12. $y = -\sin \frac{x}{3}$, for x in $\left[-\frac{3\pi}{2}, \frac{3\pi}{2}\right]$

13. $y = -2 \cos 5x$, for x in $\left[0, \frac{\pi}{5}\right]$

14. $y = 3 \cot 5x$, for x in $\left(0, \frac{\pi}{5}\right)$

15. $y = \cos(x + 3)$, for x in $[-3, \pi - 3]$

16. $y = \tan(2x - 1)$, for x in $\left(\frac{1}{2} - \frac{\pi}{4}, \frac{1}{2} + \frac{\pi}{4}\right)$

17. $y = \sin x - 2$, for x in $\left[-\frac{\pi}{2}, \frac{\pi}{2}\right]$

18. $y = \cot x + 1$, for x in $(0, \pi)$

19. $y = -4 + 2 \sin x$, for x in $\left[-\frac{\pi}{2}, \frac{\pi}{2}\right]$

20. $y = 4 + 3 \cos x$, for x in $[0, \pi]$

21. $y = \sqrt{2} + 3 \sec 2x$, for x in $\left[0, \frac{\pi}{4}\right) \cup \left(\frac{\pi}{4}, \frac{\pi}{2}\right]$

22. $y = -\sqrt{3} + 2 \csc \frac{x}{2}$, for x in $[-\pi, 0) \cup (0, \pi]$

23. Refer to **Exercise 17.** A student attempting to solve this equation wrote as the first step $y = \sin(x - 2)$, inserting parentheses as shown. Explain why this is incorrect.

24. Explain why the equation $\sin^{-1} x = \cos^{-1} 2$ cannot have a solution. (No work is required.)

Solve each equation for exact solutions. ***See Examples 2 and 3.***

25. $-4 \arcsin x = \pi$

26. $6 \arccos x = 5\pi$

27. $\frac{4}{3} \cos^{-1} \frac{x}{4} = \pi$

28. $4\pi + 4 \tan^{-1} x = \pi$

29. $2 \arccos\left(\frac{x - \pi}{3}\right) = 2\pi$

30. $\arccos\left(x - \frac{\pi}{3}\right) = \frac{\pi}{6}$

31. $\arcsin x = \arctan \frac{3}{4}$

32. $\arctan x = \arccos \frac{5}{13}$

33. $\cos^{-1} x = \sin^{-1} \frac{3}{5}$

34. $\cot^{-1} x = \tan^{-1} \frac{4}{3}$

Solve each equation for exact solutions. ***See Example 4.***

35. $\sin^{-1} x - \tan^{-1} 1 = -\frac{\pi}{4}$

36. $\sin^{-1} x + \tan^{-1} \sqrt{3} = \frac{2\pi}{3}$

37. $\arccos x + 2 \arcsin \frac{\sqrt{3}}{2} = \pi$

38. $\arccos x + 2 \arcsin \frac{\sqrt{3}}{2} = \frac{\pi}{3}$

39. $\arcsin 2x + \arccos x = \frac{\pi}{6}$

40. $\arcsin 2x + \arcsin x = \frac{\pi}{2}$

41. $\cos^{-1} x + \tan^{-1} x = \frac{\pi}{2}$

42. $\sin^{-1} x + \tan^{-1} x = 0$

43. Provide graphical support for the solution in **Example 4** by showing that the graph of

$$y = \arcsin x - \arccos x - \frac{\pi}{6} \quad \text{has } x\text{-intercept} \quad \frac{\sqrt{3}}{2} \approx 0.8660254.$$

44. Provide graphical support for the solution in **Example 4** by showing that the x-coordinate of the point of intersection of the graphs of

$$\text{Y}_1 = \arcsin \text{X} - \arccos \text{X} \quad \text{and} \quad \text{Y}_2 = \frac{\pi}{6} \quad \text{is} \quad \frac{\sqrt{3}}{2} \approx 0.8660254.$$

The following equations cannot be solved by algebraic methods. Use a graphing calculator to find all solutions over the interval $[0, 6]$. *Express solutions to four decimal places.*

45. $(\arctan x)^3 - x + 2 = 0$

46. $\pi \sin^{-1}(0.2x) - 3 = -\sqrt{x}$

(Modeling) Solve each problem.

47. ***Tone Heard by a Listener*** When two sources located at different positions produce the same pure tone, the human ear will often hear one sound that is equal to the sum of the individual tones. Since the sources are at different locations, they will have different phase angles ϕ. If two speakers located at different positions produce pure tones $P_1 = A_1 \sin(2\pi f t + \phi_1)$ and $P_2 = A_2 \sin(2\pi f t + \phi_2)$, where $-\frac{\pi}{4} \le \phi_1, \phi_2 \le \frac{\pi}{4}$, then the resulting tone heard by a listener can be written as $P = A \sin(2\pi f t + \phi)$, where

$$A = \sqrt{(A_1 \cos \phi_1 + A_2 \cos \phi_2)^2 + (A_1 \sin \phi_1 + A_2 \sin \phi_2)^2}$$

and

$$\phi = \arctan\left(\frac{A_1 \sin \phi_1 + A_2 \sin \phi_2}{A_1 \cos \phi_1 + A_2 \cos \phi_2}\right).$$

(*Source:* Fletcher, N. and T. Rossing, *The Physics of Musical Instruments,* Second Edition, Springer-Verlag.)

(a) Calculate A and ϕ if $A_1 = 0.0012$, $\phi_1 = 0.052$, $A_2 = 0.004$, and $\phi_2 = 0.61$. Also find an expression for $P = A \sin(2\pi f t + \phi)$ if $f = 220$.

(b) Graph $\text{Y}_1 = P$ and $\text{Y}_2 = P_1 + P_2$ on the same coordinate axes over the interval $[0, 0.01]$. Are the two graphs the same?

---

25. $\left\{-\frac{\sqrt{2}}{2}\right\}$ 26. $\left\{-\frac{\sqrt{3}}{2}\right\}$

27. $\{-2\sqrt{2}\}$ 28. $\emptyset$

29. $\{\pi - 3\}$ 30. $\left\{\frac{3\sqrt{3} + 2\pi}{6}\right\}$

31. $\left\{\frac{3}{5}\right\}$ 32. $\left\{\frac{12}{5}\right\}$

33. $\left\{\frac{4}{5}\right\}$ 34. $\left\{\frac{3}{4}\right\}$

35. $\{0\}$ 36. $\left\{\frac{\sqrt{3}}{2}\right\}$

37. $\left\{\frac{1}{2}\right\}$ 38. $\emptyset$

39. $\left\{-\frac{1}{2}\right\}$ 40. $\left\{\frac{\sqrt{5}}{5}\right\}$

41. $\{0\}$ 42. $\{0\}$

43. $\text{Y} = \arcsin \text{X} - \arccos \text{X} - \frac{\pi}{6}$

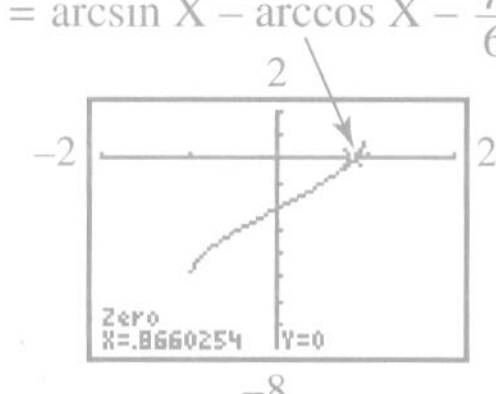

44.

$\text{Y}_1 = \arcsin \text{X} - \arccos \text{X} \quad \text{Y}_2 = \frac{\pi}{6}$

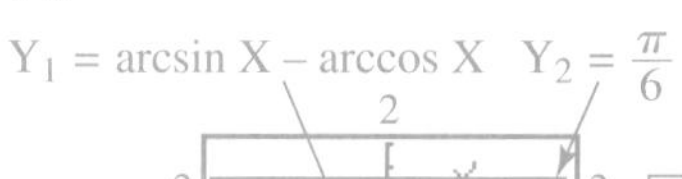

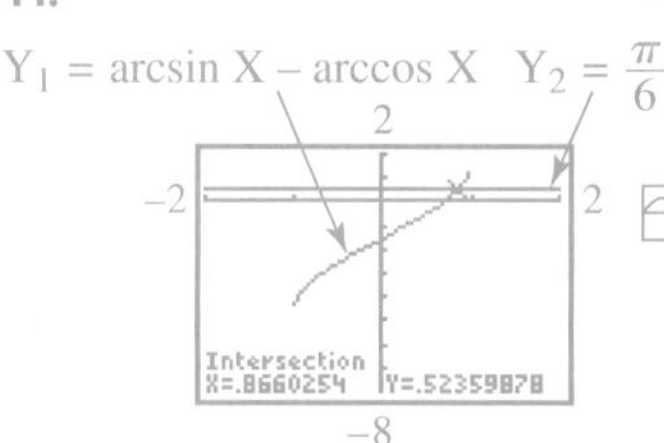

45. $\{4.4622\}$ 46. $\{2.2824\}$

47. (a) $A \approx 0.00506$, $\phi \approx 0.484$; $P = 0.00506 \sin(440\pi t + 0.484)$ (b) The two graphs are the same.

For $x = t$,
$P(t) = 0.00506 \sin(440\pi t + 0.484)$
$P_1(t) + P_2(t) = 0.0012 \sin(440\pi t + 0.052) + 0.004 \sin(440\pi t + 0.61)$

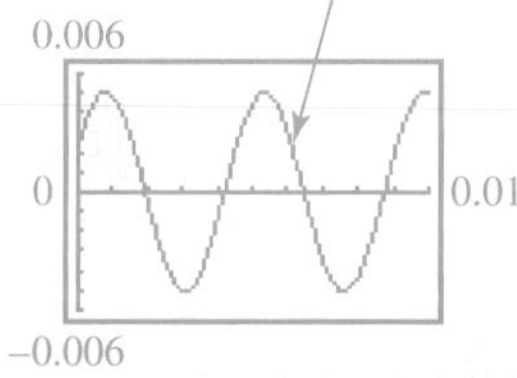

48. *Tone Heard by a Listener* Repeat **Exercise 47.** Use $A_1 = 0.0025$, $\phi_1 = \frac{\pi}{7}$, $A_2 = 0.001$, $\phi_2 = \frac{\pi}{6}$, and $f = 300$.

49. *Depth of Field* When a large-view camera is used to take a picture of an object that is not parallel to the film, the lens board should be tilted so that the planes containing the subject, the lens board, and the film intersect in a line. This gives the best "depth of field." See the figure. (*Source:* Bushaw, D., et al., *A Sourcebook of Applications of School Mathematics,* Mathematical Association of America.)

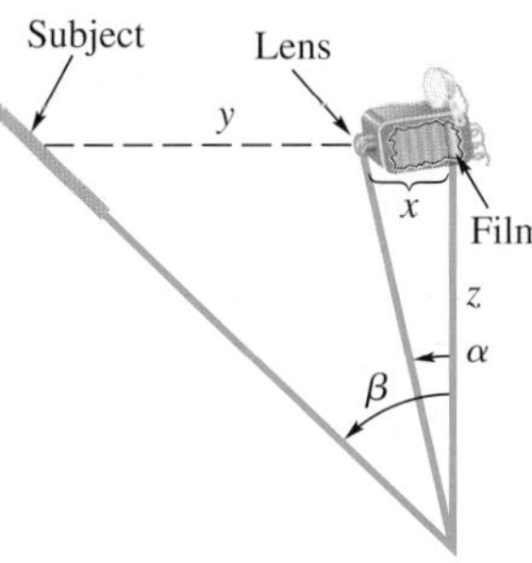

(a) Write two equations, one relating α, x, and z, and the other relating β, x, y, and z.

(b) Eliminate z from the equations in part (a) to get one equation relating α, β, x, and y.

(c) Solve the equation from part (b) for α.

(d) Solve the equation from part (b) for β.

50. *Programming Language for Inverse Functions* In Visual Basic, a widely used programming language for PCs, the only inverse trigonometric function available is arctangent. The other inverse trigonometric functions can be expressed in terms of arctangent as follows.

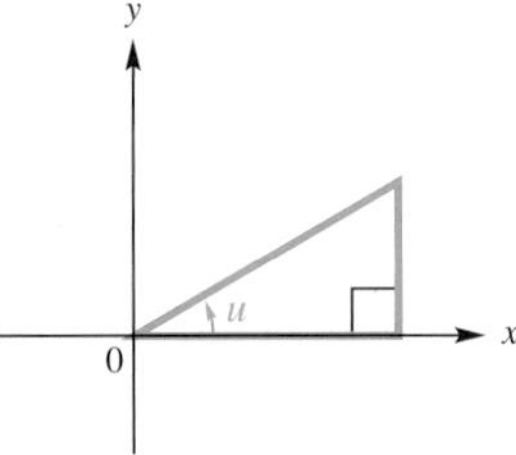

(a) Let $u = \arcsin x$. Solve the equation for x in terms of u.

(b) Use the result of part (a) to label the three sides of the triangle in the figure in terms of x.

(c) Use the triangle from part (b) to write an equation for $\tan u$ in terms of x.

(d) Solve the equation from part (c) for u.

51. *Alternating Electric Current* In the study of alternating electric current, instantaneous voltage is modeled by

$$E = E_{\max} \sin 2\pi f t,$$

where f is the number of cycles per second, $E_{\max}$ is the maximum voltage, and t is time in seconds.

(a) Solve the equation for t.

(b) Find the least positive value of t if $E_{\max} = 12$, $E = 5$, and $f = 100$. Use a calculator.

52. *Viewing Angle of an Observer* While visiting a museum, Marsha Langlois views a painting that is 3 ft high and hangs 6 ft above the ground. See the figure. Assume her eyes are 5 ft above the ground, and let x be the distance from the spot where she is standing to the wall displaying the painting.

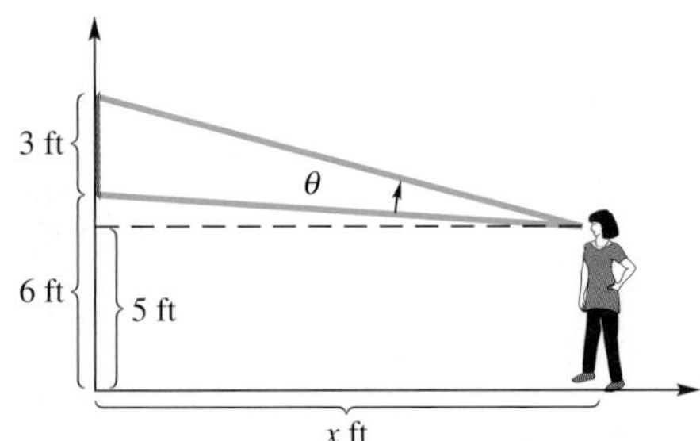

(a) Show that θ, the viewing angle subtended by the painting, is given by

$$\theta = \tan^{-1}\left(\frac{4}{x}\right) - \tan^{-1}\left(\frac{1}{x}\right).$$

(b) Find the value of x to the nearest hundredth for each value of θ.

(i) $\theta = \frac{\pi}{6}$ **(ii)** $\theta = \frac{\pi}{8}$

(c) Find the value of θ to the nearest hundredth for each value of x.

(i) $x = 4$ **(ii)** $x = 3$

48. (a) $A \approx 0.0035$, $\phi \approx 0.470$; $P = 0.0035 \sin(600\pi t + 0.47)$
(b) The two graphs are the same.

For $x = t$,
$P(t) = 0.0035 \sin(600\pi t + 0.47)$
$P_1(t) + P_2(t) = 0.0025 \sin\left(600\pi t + \frac{\pi}{7}\right) + 0.001 \sin\left(600\pi t + \frac{\pi}{6}\right)$

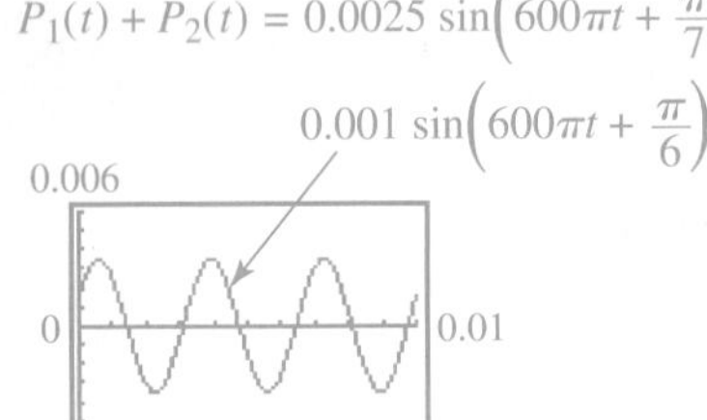

49. (a) $\tan \alpha = \frac{x}{z}$; $\tan \beta = \frac{x+y}{z}$
(b) $\frac{x}{\tan \alpha} = \frac{x+y}{\tan \beta}$
(c) $\alpha = \arctan\left(\frac{x \tan \beta}{x+y}\right)$
(d) $\beta = \arctan\left(\frac{(x+y)\tan \alpha}{x}\right)$

50. (a) $x = \sin u$, $-\frac{\pi}{2} \le u \le \frac{\pi}{2}$
(b)

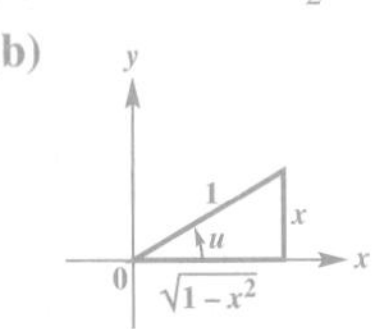

(c) $\tan u = \frac{x\sqrt{1-x^2}}{1-x^2}$
(d) $u = \arctan \frac{x\sqrt{1-x^2}}{1-x^2}$

51. (a) $t = \frac{1}{2\pi f} \arcsin \frac{E}{E_{\max}}$
(b) 0.00068 sec

52. (b) (i) 0.94 or 4.26
(ii) 0.60 or 6.64
(c) (i) 0.54
(ii) 0.61

Chapter 7 Test Prep

New Symbols

$\sin^{-1} x$ $(\arcsin x)$	inverse sine of x	$\cot^{-1} x$ $(\text{arccot } x)$	inverse cotangent of x
$\cos^{-1} x$ $(\arccos x)$	inverse cosine of x	$\sec^{-1} x$ $(\text{arcsec } x)$	inverse secant of x
$\tan^{-1} x$ $(\arctan x)$	inverse tangent of x	$\csc^{-1} x$ $(\text{arccsc } x)$	inverse cosecant of x

Quick Review

Concepts	Examples

7.1 Fundamental Identities

Reciprocal Identities

$$\cot\theta = \frac{1}{\tan\theta} \qquad \sec\theta = \frac{1}{\cos\theta} \qquad \csc\theta = \frac{1}{\sin\theta}$$

Quotient Identities

$$\tan\theta = \frac{\sin\theta}{\cos\theta} \qquad \cot\theta = \frac{\cos\theta}{\sin\theta}$$

Pythagorean Identities

$$\sin^2\theta + \cos^2\theta = 1 \qquad \tan^2\theta + 1 = \sec^2\theta$$

$$1 + \cot^2\theta = \csc^2\theta$$

Negative-Angle Identities

$$\sin(-\theta) = -\sin\theta \quad \cos(-\theta) = \cos\theta \quad \tan(-\theta) = -\tan\theta$$

$$\csc(-\theta) = -\csc\theta \quad \sec(-\theta) = \sec\theta \quad \cot(-\theta) = -\cot\theta$$

Example:

If θ is in quadrant IV and $\sin\theta = -\frac{3}{5}$, find $\csc\theta$, $\cos\theta$, and $\sin(-\theta)$.

$$\csc\theta = \frac{1}{\sin\theta} = \frac{1}{-\frac{3}{5}} = -\frac{5}{3} \quad \text{Reciprocal identity}$$

$$\sin^2\theta + \cos^2\theta = 1 \quad \text{Pythagorean identity}$$

$$\left(-\frac{3}{5}\right)^2 + \cos^2\theta = 1 \quad \text{Substitute.}$$

$$\cos^2\theta = 1 - \frac{9}{25} = \frac{16}{25} \quad \text{Subtract } \tfrac{9}{25}.$$

$$\cos\theta = +\sqrt{\frac{16}{25}} = \frac{4}{5} \quad \cos\theta \text{ is positive in quadrant IV.}$$

$$\sin(-\theta) = -\sin\theta = -\left(-\frac{3}{5}\right) = \frac{3}{5}$$

Negative-angle identity

7.2 Verifying Trigonometric Identities

See the box titled Hints for Verifying Identities in **Section 7.2.**

7.3 Sum and Difference Identities

Cofunction Identities

$$\cos(90° - \theta) = \sin\theta \qquad \cot(90° - \theta) = \tan\theta$$

$$\sin(90° - \theta) = \cos\theta \qquad \sec(90° - \theta) = \csc\theta$$

$$\tan(90° - \theta) = \cot\theta \qquad \csc(90° - \theta) = \sec\theta$$

Example:

Find one value of θ such that $\tan\theta = \cot 78°$.

$$\tan\theta = \cot 78°$$

$$\cot(90° - \theta) = \cot 78° \quad \text{Cofunction identity}$$

$$90° - \theta = 78° \quad \text{Set angles equal.}$$

$$\theta = 12° \quad \text{Solve for } \theta.$$

(continued)

Concepts	Examples

Sum and Difference Identities

$$\cos(A - B) = \cos A \cos B + \sin A \sin B$$

$$\cos(A + B) = \cos A \cos B - \sin A \sin B$$

$$\sin(A + B) = \sin A \cos B + \cos A \sin B$$

$$\sin(A - B) = \sin A \cos B - \cos A \sin B$$

$$\tan(A + B) = \frac{\tan A + \tan B}{1 - \tan A \tan B}$$

$$\tan(A - B) = \frac{\tan A - \tan B}{1 + \tan A \tan B}$$

Find the exact value of $\cos(-15°)$.

$\cos(-15°)$

$= \cos(30° - 45°)$

$= \cos 30° \cos 45° + \sin 30° \sin 45°$ Cosine difference identity

$= \frac{\sqrt{3}}{2} \cdot \frac{\sqrt{2}}{2} + \frac{1}{2} \cdot \frac{\sqrt{2}}{2}$ Substitute values.

$= \frac{\sqrt{6} + \sqrt{2}}{4}$ Simplify.

Write $\tan\left(\frac{\pi}{4} + \theta\right)$ in terms of $\tan \theta$.

$$\tan\left(\frac{\pi}{4} + \theta\right) = \frac{\tan \frac{\pi}{4} + \tan \theta}{1 - \tan \frac{\pi}{4} \tan \theta} = \frac{1 + \tan \theta}{1 - \tan \theta} \qquad \tan \frac{\pi}{4} = 1$$

7.4 Double-Angle and Half-Angle Identities

Double-Angle Identities

$$\cos 2A = \cos^2 A - \sin^2 A \qquad \cos 2A = 1 - 2\sin^2 A$$

$$\cos 2A = 2\cos^2 A - 1 \qquad \sin 2A = 2 \sin A \cos A$$

$$\tan 2A = \frac{2 \tan A}{1 - \tan^2 A}$$

Given $\cos \theta = -\frac{5}{13}$ and $\sin \theta > 0$, find $\sin 2\theta$.

Sketch a triangle in quadrant II since $\cos \theta < 0$ and $\sin \theta > 0$. Use it to find that $\sin \theta = \frac{12}{13}$.

$\sin 2\theta = 2 \sin \theta \cos \theta$

$= 2\left(\frac{12}{13}\right)\left(-\frac{5}{13}\right)$

$= -\frac{120}{169}$

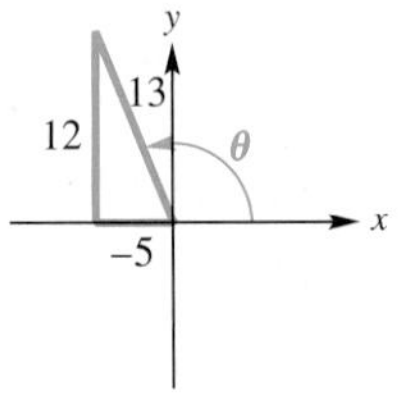

Product-to-Sum Identities

$$\cos A \cos B = \frac{1}{2}[\cos(A + B) + \cos(A - B)]$$

$$\sin A \sin B = \frac{1}{2}[\cos(A - B) - \cos(A + B)]$$

$$\sin A \cos B = \frac{1}{2}[\sin(A + B) + \sin(A - B)]$$

$$\cos A \sin B = \frac{1}{2}[\sin(A + B) - \sin(A - B)]$$

Write $\sin(-\theta) \sin 2\theta$ as the difference of two functions.

$\sin(-\theta) \sin 2\theta$

$= \frac{1}{2}[\cos(-\theta - 2\theta) - \cos(-\theta + 2\theta)]$

$= \frac{1}{2}[\cos(-3\theta) - \cos \theta]$

$= \frac{1}{2}\cos(-3\theta) - \frac{1}{2}\cos \theta$

$= \frac{1}{2}\cos 3\theta - \frac{1}{2}\cos \theta$

Sum-to-Product Identities

$$\sin A + \sin B = 2 \sin\left(\frac{A + B}{2}\right) \cos\left(\frac{A - B}{2}\right)$$

$$\sin A - \sin B = 2 \cos\left(\frac{A + B}{2}\right) \sin\left(\frac{A - B}{2}\right)$$

$$\cos A + \cos B = 2 \cos\left(\frac{A + B}{2}\right) \cos\left(\frac{A - B}{2}\right)$$

$$\cos A - \cos B = -2 \sin\left(\frac{A + B}{2}\right) \sin\left(\frac{A - B}{2}\right)$$

Write $\cos \theta + \cos 3\theta$ as a product of two functions.

$\cos \theta + \cos 3\theta$

$= 2 \cos\left(\frac{\theta + 3\theta}{2}\right) \cos\left(\frac{\theta - 3\theta}{2}\right)$

$= 2 \cos\left(\frac{4\theta}{2}\right) \cos\left(\frac{-2\theta}{2}\right)$

$= 2 \cos 2\theta \cos(-\theta)$

$= 2 \cos 2\theta \cos \theta$

Concepts	Examples

Half-Angle Identities

$$\cos\frac{A}{2} = \pm\sqrt{\frac{1+\cos A}{2}} \qquad \sin\frac{A}{2} = \pm\sqrt{\frac{1-\cos A}{2}}$$

$$\tan\frac{A}{2} = \pm\sqrt{\frac{1-\cos A}{1+\cos A}} \qquad \tan\frac{A}{2} = \frac{\sin A}{1+\cos A}$$

$$\tan\frac{A}{2} = \frac{1-\cos A}{\sin A}$$

(In the identities involving radicals, the sign is chosen on the basis of the function under consideration and the quadrant of $\frac{A}{2}$.)

Find the exact value of tan 67.5°.

We choose the last form with $A = 135°$.

$$\tan 67.5° = \tan\frac{135°}{2} = \frac{1-\cos 135°}{\sin 135°} = \frac{1-\left(-\frac{\sqrt{2}}{2}\right)}{\frac{\sqrt{2}}{2}}$$

$$= \frac{1+\frac{\sqrt{2}}{2}}{\frac{\sqrt{2}}{2}} \cdot \frac{2}{2} = \frac{2+\sqrt{2}}{\sqrt{2}}, \quad \text{or} \quad \sqrt{2}+1$$

Rationalize the denominator and simplify.

7.5 Inverse Circular Functions

Inverse Function	Domain	Range: Interval	Range: Quadrants of the Unit Circle
$y = \sin^{-1} x$	$[-1, 1]$	$\left[-\frac{\pi}{2}, \frac{\pi}{2}\right]$	I and IV
$y = \cos^{-1} x$	$[-1, 1]$	$[0, \pi]$	I and II
$y = \tan^{-1} x$	$(-\infty, \infty)$	$\left(-\frac{\pi}{2}, \frac{\pi}{2}\right)$	I and IV
$y = \cot^{-1} x$	$(-\infty, \infty)$	$(0, \pi)$	I and II
$y = \sec^{-1} x$	$(-\infty, -1] \cup [1, \infty)$	$\left[0, \frac{\pi}{2}\right) \cup \left(\frac{\pi}{2}, \pi\right]$	I and II
$y = \csc^{-1} x$	$(-\infty, -1] \cup [1, \infty)$	$\left[-\frac{\pi}{2}, 0\right) \cup \left(0, \frac{\pi}{2}\right]$	I and IV

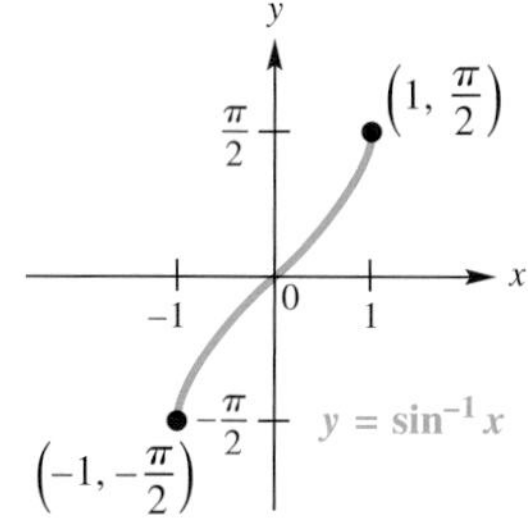

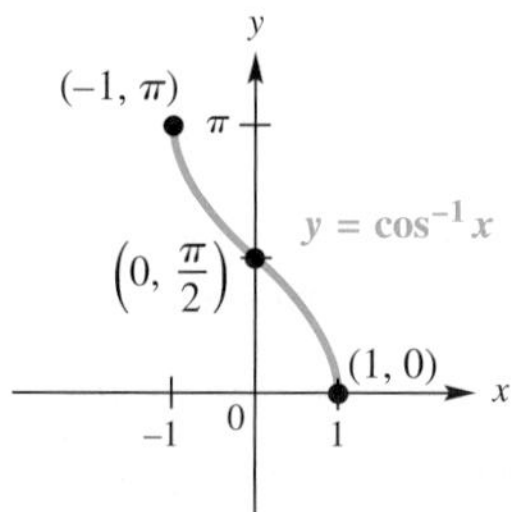

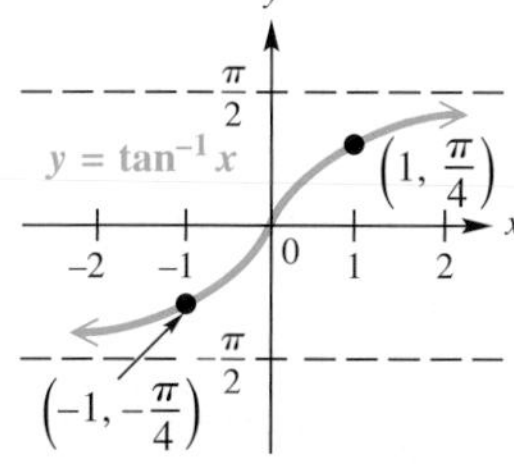

See **Section 7.5** for graphs of the other inverse circular (trigonometric) functions.

Evaluate $y = \cos^{-1} 0$.

Write $y = \cos^{-1} 0$ as $\cos y = 0$. Then

$$y = \frac{\pi}{2},$$

because $\cos\frac{\pi}{2} = 0$ and $\frac{\pi}{2}$ is in the range of $\cos^{-1} x$.

Use a calculator to find y in radians if $y = \sec^{-1}(-3)$.

With the calculator in radian mode, enter $\sec^{-1}(-3)$ as $\cos^{-1}\left(\frac{1}{-3}\right)$ to get

$$y \approx 1.9106332.$$

Evaluate $\sin\left(\tan^{-1}\left(-\frac{3}{4}\right)\right)$.

Let $u = \tan^{-1}\left(-\frac{3}{4}\right)$. Then $\tan u = -\frac{3}{4}$. Since $\tan u$ is negative when u is in quadrant IV, sketch a triangle as shown.

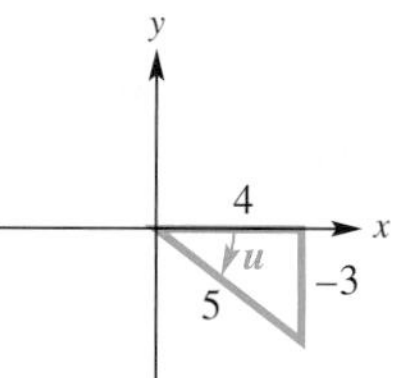

We want $\sin\left(\tan^{-1}\left(-\frac{3}{4}\right)\right) = \sin u$. From the triangle,

$$\sin u = -\frac{3}{5}.$$

(continued)

Concepts	Examples

7.6 Trigonometric Equations

Solving a Trigonometric Equation

1. Decide whether the equation is linear or quadratic in form, so that you can determine the solution method.
2. If only one trigonometric function is present, solve the equation for that function.
3. If more than one trigonometric function is present, rearrange the equation so that one side equals 0. Then try to factor and set each factor equal to 0 to solve.
4. If the equation is quadratic in form, but not factorable, use the quadratic formula. Check that solutions are in the desired interval.
5. Try using identities to change the form of the equation. It may be helpful to square each side of the equation first. In this case, check for extraneous solutions.

Solve $\tan \theta + \sqrt{3} = 2\sqrt{3}$ over the interval $[0°, 360°)$. Use a linear method.

$$\tan \theta + \sqrt{3} = 2\sqrt{3} \quad \text{Original equation}$$

$$\tan \theta = \sqrt{3} \quad \text{Subtract } \sqrt{3}.$$

$$\theta = 60° \quad \text{Definition of inverse tangent}$$

Another solution over $[0°, 360°)$ is

$$\theta = 60° + 180° = 240°.$$

The solution set is $\{60°, 240°\}$.

Solve $2 \cos^2 x = 1$ for all solutions, using a double-angle identity.

$$2 \cos^2 x = 1 \quad \text{Original equation}$$

$$2 \cos^2 x - 1 = 0 \quad \text{Subtract 1.}$$

$$\cos 2x = 0 \quad \text{Cosine double-angle identity}$$

$$2x = \frac{\pi}{2} + 2n\pi \quad \text{and} \quad 2x = \frac{3\pi}{2} + 2n\pi$$

Add integer multiples of 2π.

$$x = \frac{\pi}{4} + n\pi \quad \text{and} \quad x = \frac{3\pi}{4} + n\pi$$

Divide by 2.

All solutions are given by the following set, where π represents the period of $\cos 2x$.

$$\left\{\frac{\pi}{4} + n\pi, \frac{3\pi}{4} + n\pi, \text{ where } n \text{ is any integer}\right\}$$

7.7 Equations Involving Inverse Trigonometric Functions

We can solve equations of the form $y = f(x)$, where $f(x)$ is a trigonometric function, using inverse trigonometric functions.

Solve $y = 2 \sin 3x$ for x, where x is restricted to the interval $\left[-\frac{\pi}{6}, \frac{\pi}{6}\right]$.

$$y = 2 \sin 3x \quad \text{Original equation}$$

$$\frac{y}{2} = \sin 3x \quad \text{Divide by 2.}$$

$$3x = \arcsin \frac{y}{2} \quad \text{Definition of arcsine}$$

$$x = \frac{1}{3} \arcsin \frac{y}{2} \quad \text{Multiply by } \tfrac{1}{3}.$$

Techniques introduced in this section also show how to solve equations that involve inverse functions.

Solve.

$$4 \tan^{-1} x = \pi \quad \text{Original equation}$$

$$\tan^{-1} x = \frac{\pi}{4} \quad \text{Divide by 4.}$$

$$x = \tan \frac{\pi}{4} = 1 \quad \text{Definition of arctangent}$$

The solution set is $\{1\}$.

Chapter 7 Review Exercises

Concept Check For each expression in Column I, choose the expression from Column II that completes an identity.

I		II	
1. $\sec x =$ ______	**2.** $\csc x =$ ______	**A.** $\frac{1}{\sin x}$	**B.** $\frac{1}{\cos x}$
3. $\tan x =$ ______	**4.** $\cot x =$ ______	**C.** $\frac{\sin x}{\cos x}$	**D.** $\frac{1}{\cot^2 x}$
5. $\tan^2 x =$ ______	**6.** $\sec^2 x =$ ______	**E.** $\frac{1}{\cos^2 x}$	**F.** $\frac{\cos x}{\sin x}$

Use identities to write each expression in terms of $\sin\theta$ *and* $\cos\theta$, *and simplify.*

7. $\sec^2\theta - \tan^2\theta$ **8.** $\frac{\cot\theta}{\sec\theta}$ **9.** $\tan^2\theta(1 + \cot^2\theta)$

10. $\csc\theta + \cot\theta$ **11.** $\tan\theta - \sec\theta\csc\theta$ **12.** $\csc^2\theta + \sec^2\theta$

13. Use the trigonometric identities to find $\sin x$, $\tan x$, and $\cot(-x)$, given $\cos x = \frac{3}{5}$ and x in quadrant IV.

14. Given $\tan x = -\frac{5}{4}$, where $\frac{\pi}{2} < x < \pi$, use the trigonometric identities to find $\cot x$, $\csc x$, and $\sec x$.

15. Find the exact values of the six trigonometric functions of 165°.

16. Find the exact values of $\sin x$, $\cos x$, and $\tan x$, for $x = \frac{\pi}{12}$, using

(a) difference identities **(b)** half-angle identities.

Concept Check For each expression in Column I, use an identity to choose an expression from Column II with the same value. Choices may be used once, more than once, or not at all.

I		II	
17. $\cos 210°$	**18.** $\sin 35°$	**A.** $\sin(-35°)$	**B.** $\cos 55°$
19. $\tan(-35°)$	**20.** $-\sin 35°$	**C.** $\sqrt{\frac{1 + \cos 150°}{2}}$	**D.** $2\sin 150°\cos 150°$
21. $\cos 35°$	**22.** $\cos 75°$	**E.** $\cot(-35°)$	**F.** $\cos^2 150° - \sin^2 150°$
23. $\sin 75°$	**24.** $\sin 300°$	**G.** $\cos(-35°)$	**H.** $\cot 125°$
25. $\cos 300°$	**26.** $\cos(-55°)$	**I.** $\cos 150°\cos 60° - \sin 150°\sin 60°$	
		J. $\sin 15°\cos 60° + \cos 15°\sin 60°$	

For each of the following, find $\sin(x + y)$, $\cos(x - y)$, $\tan(x + y)$, *and the quadrant of* $x + y$.

27. $\sin x = -\frac{3}{5}$, $\cos y = -\frac{7}{25}$, x and y in quadrant III

28. $\sin x = \frac{3}{5}$, $\cos y = \frac{24}{25}$, x in quadrant I, y in quadrant IV

29. $\sin x = -\frac{1}{2}$, $\cos y = -\frac{2}{5}$, x and y in quadrant III

30. $\sin y = -\frac{2}{3}$, $\cos x = -\frac{1}{5}$, x in quadrant II, y in quadrant III

31. $\sin x = \frac{1}{10}$, $\cos y = \frac{4}{5}$, x in quadrant I, y in quadrant IV

32. $\cos x = \frac{2}{9}$, $\sin y = -\frac{1}{2}$, x in quadrant IV, y in quadrant III

1. B 2. A
3. C 4. F
5. D 6. E
7. 1 8. $\frac{\cos^2\theta}{\sin\theta}$
9. $\frac{1}{\cos^2\theta}$ 10. $\frac{1 + \cos\theta}{\sin\theta}$
11. $-\frac{\cos\theta}{\sin\theta}$ 12. $\frac{1}{\sin^2\theta\cos^2\theta}$
13. $\sin x = -\frac{4}{5}$; $\tan x = -\frac{4}{3}$; $\cot(-x) = \frac{3}{4}$
14. $\cot x = -\frac{4}{5}$; $\csc x = \frac{\sqrt{41}}{5}$; $\sec x = -\frac{\sqrt{41}}{4}$
15. $\sin 165° = \frac{\sqrt{6} - \sqrt{2}}{4}$; $\cos 165° = \frac{-\sqrt{6} - \sqrt{2}}{4}$; $\tan 165° = -2 + \sqrt{3}$; $\csc 165° = \sqrt{6} + \sqrt{2}$; $\sec 165° = -\sqrt{6} + \sqrt{2}$; $\cot 165° = -2 - \sqrt{3}$
16. (a) $\sin\frac{\pi}{12} = \frac{\sqrt{6} - \sqrt{2}}{4}$; $\cos\frac{\pi}{12} = \frac{\sqrt{6} + \sqrt{2}}{4}$; $\tan\frac{\pi}{12} = 2 - \sqrt{3}$
(b) $\sin\frac{\pi}{12} = \frac{\sqrt{2 - \sqrt{3}}}{2}$; $\cos\frac{\pi}{12} = \frac{\sqrt{2 + \sqrt{3}}}{2}$; $\tan\frac{\pi}{12} = 2 - \sqrt{3}$
17. I 18. B
19. H 20. A
21. G 22. C
23. J 24. D
25. F 26. B
27. $\frac{117}{125}$; $\frac{4}{5}$; $-\frac{117}{44}$; II
28. $\frac{44}{125}$; $\frac{3}{5}$; $\frac{44}{117}$; I
29. $\frac{2 + 3\sqrt{7}}{10}$; $\frac{2\sqrt{3} + \sqrt{21}}{10}$; $\frac{-25\sqrt{3} - 8\sqrt{21}}{9}$; II
30. $\frac{2 - 2\sqrt{30}}{15}$; $\frac{\sqrt{5} - 4\sqrt{6}}{15}$; $\frac{18\sqrt{6} - 50\sqrt{5}}{91}$; IV
31. $\frac{4 - 9\sqrt{11}}{50}$; $\frac{12\sqrt{11} - 3}{50}$; $\frac{\sqrt{11} - 16}{21}$; IV
32. $\frac{\sqrt{231} - 2}{18}$; $\frac{\sqrt{77} - 2\sqrt{3}}{18}$; $\frac{8\sqrt{77} - 81\sqrt{3}}{65}$; II

33. $\sin\theta = \frac{\sqrt{14}}{4}$; $\cos\theta = \frac{\sqrt{2}}{4}$

34. $\sin B = -\frac{\sqrt{7}}{4}$; $\cos B = \frac{3}{4}$

35. $\sin 2x = \frac{3}{5}$; $\cos 2x = -\frac{4}{5}$

36. $\sin 2y = -\frac{24}{25}$; $\cos 2y = -\frac{7}{25}$

37. $\frac{1}{2}$ 38. $\frac{\sqrt{14}}{4}$

39. $\frac{\sqrt{5}-1}{2}$ 40. $\frac{\sqrt{6}}{3}$

41. 0.5 42. -0.96

43. $-\frac{\sin 2x + \sin x}{\cos 2x - \cos x} = \cot\frac{x}{2}$

44. $\frac{1-\cos 2x}{\sin 2x} = \tan x$

45. $\frac{\sin x}{1-\cos x} = \cot\frac{x}{2}$

46. $\frac{\cos x \sin 2x}{1+\cos 2x} = \sin x$

47. $\frac{2(\sin x - \sin^3 x)}{\cos x} = \sin 2x$

48. $\csc x - \cot x = \tan\frac{x}{2}$

67. $[-1, 1]$; $\left[-\frac{\pi}{2}, \frac{\pi}{2}\right]$

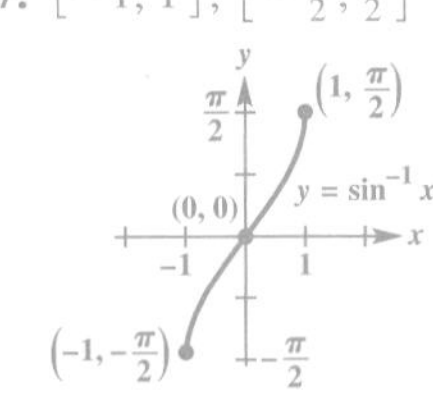

$[-1, 1]$; $[0, \pi]$

(−1, π) $y = \cos^{-1} x$ $(0, \frac{\pi}{2})$ (1, 0)

$(-\infty, \infty)$; $\left(-\frac{\pi}{2}, \frac{\pi}{2}\right)$

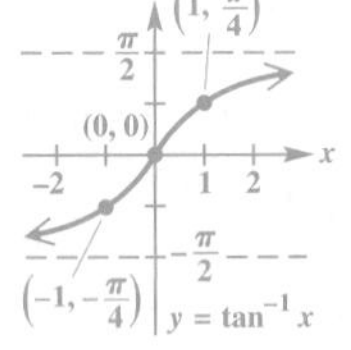

Find sine and cosine of each of the following.

33. θ, given $\cos 2\theta = -\frac{3}{4}$, $90° < 2\theta < 180°$

34. B, given $\cos 2B = \frac{1}{8}$, $540° < 2B < 720°$

35. $2x$, given $\tan x = 3$, $\sin x < 0$

36. $2y$, given $\sec y = -\frac{5}{3}$, $\sin y > 0$

Find each of the following.

37. $\cos\frac{\theta}{2}$, given $\cos\theta = -\frac{1}{2}$, $90° < \theta < 180°$

38. $\sin\frac{A}{2}$, given $\cos A = -\frac{3}{4}$, $90° < A < 180°$

39. $\tan x$, given $\tan 2x = 2$, $\pi < x < \frac{3\pi}{2}$

40. $\sin y$, given $\cos 2y = -\frac{1}{3}$, $\frac{\pi}{2} < y < \pi$

41. $\tan\frac{x}{2}$, given $\sin x = 0.8$, $0 < x < \frac{\pi}{2}$

42. $\sin 2x$, given $\sin x = 0.6$, $\frac{\pi}{2} < x < \pi$

Graph each expression and use the graph to make a conjecture, predicting what might be an identity. Then verify your conjecture algebraically.

43. $-\frac{\sin 2x + \sin x}{\cos 2x - \cos x}$

44. $\frac{1 - \cos 2x}{\sin 2x}$

45. $\frac{\sin x}{1 - \cos x}$

46. $\frac{\cos x \sin 2x}{1 + \cos 2x}$

47. $\frac{2(\sin x - \sin^3 x)}{\cos x}$

48. $\csc x - \cot x$

Verify that each equation is an identity.

49. $\sin^2 x - \sin^2 y = \cos^2 y - \cos^2 x$

50. $2\cos^3 x - \cos x = \frac{\cos^2 x - \sin^2 x}{\sec x}$

51. $\frac{\sin^2 x}{2 - 2\cos x} = \cos^2\frac{x}{2}$

52. $\frac{\sin 2x}{\sin x} = \frac{2}{\sec x}$

53. $2\cos A - \sec A = \cos A - \frac{\tan A}{\csc A}$

54. $\frac{2\tan B}{\sin 2B} = \sec^2 B$

55. $1 + \tan^2\alpha = 2\tan\alpha\csc 2\alpha$

56. $\frac{2\cot x}{\tan 2x} = \csc^2 x - 2$

57. $\tan\theta\sin 2\theta = 2 - 2\cos^2\theta$

58. $\csc A\sin 2A - \sec A = \cos 2A\sec A$

59. $2\tan x\csc 2x - \tan^2 x = 1$

60. $2\cos^2\theta - 1 = \frac{1 - \tan^2\theta}{1 + \tan^2\theta}$

61. $\tan\theta\cos^2\theta = \frac{2\tan\theta\cos^2\theta - \tan\theta}{1 - \tan^2\theta}$

62. $\sec^2\alpha - 1 = \frac{\sec 2\alpha - 1}{\sec 2\alpha + 1}$

63. $\frac{\sin^2 x - \cos^2 x}{\csc x} = 2\sin^3 x - \sin x$

64. $\sin^3\theta = \sin\theta - \cos^2\theta\sin\theta$

65. $\tan 4\theta = \frac{2\tan 2\theta}{2 - \sec^2 2\theta}$

66. $2\cos^2\frac{x}{2}\tan x = \tan x + \sin x$

67. Graph the inverse sine, cosine, and tangent functions, indicating three points on each graph. Give the domain and range for each.

68. false; The range of the inverse tangent function is $\left(-\frac{\pi}{2}, \frac{\pi}{2}\right)$, while that of the inverse cotangent is $(0, \pi)$.
69. false; $\arcsin\left(-\frac{1}{2}\right) = -\frac{\pi}{6}$, not $\frac{11\pi}{6}$.
70. true

71. $\frac{\pi}{4}$ 72. $\frac{2\pi}{3}$ 73. $-\frac{\pi}{3}$
74. $-\frac{\pi}{2}$ 75. $\frac{3\pi}{4}$ 76. $\frac{\pi}{6}$
77. $\frac{2\pi}{3}$ 78. $\frac{\pi}{3}$ 79. $\frac{3\pi}{4}$
80. $60°$ 81. $-60°$ 82. $0°$

83. $60.68°$ 84. $-41.33°$
85. $36.49°$ 86. $12.52°$
87. $73.26°$ 88. $7.67°$

89. -1 90. $-\frac{\sqrt{3}}{2}$ 91. $\frac{3\pi}{4}$
92. π 93. $\frac{\pi}{4}$ 94. 0
95. $\frac{\sqrt{7}}{4}$ 96. $\frac{\sqrt{10}}{10}$ 97. $\frac{\sqrt{3}}{2}$
98. $\frac{9}{7}$ 99. $\frac{294 + 125\sqrt{6}}{92}$
100. $\sqrt{1 - u^2}$ 101. $\frac{1}{u}$

102. $\left\{\frac{\pi}{2}, \frac{3\pi}{2}\right\}$
103. $\{0.4636, 3.6052\}$
104. $\left\{0.7297, \frac{\pi}{2}, 2.4119\right\}$
105. $\left\{\frac{\pi}{4}, \frac{3\pi}{4}, \frac{5\pi}{4}, \frac{7\pi}{4}\right\}$
106. $\left\{\frac{\pi}{8}, \frac{3\pi}{8}, \frac{5\pi}{8}, \frac{7\pi}{8}, \frac{9\pi}{8}, \frac{11\pi}{8}, \frac{13\pi}{8}, \frac{15\pi}{8}\right\}$
107. $\left\{\frac{\pi}{8}, \frac{3\pi}{8}, \frac{5\pi}{8}, \frac{7\pi}{8}, \frac{9\pi}{8}, \frac{11\pi}{8}, \frac{13\pi}{8}, \frac{15\pi}{8}\right\}$
108. $\{2n\pi$, where n is any integer$\}$
109. $\left\{\frac{\pi}{3} + 2n\pi, \pi + 2n\pi, \frac{5\pi}{3} + 2n\pi\right.$, where n is any integer$\}$
110. $\left\{\frac{\pi}{6} + n\pi, \frac{\pi}{3} + n\pi\right.$, where n is any integer$\}$

Concept Check Determine whether each statement is true *or* false. *If false, tell why.*

68. The ranges of the inverse tangent and inverse cotangent functions are the same.

69. It is true that $\sin \frac{11\pi}{6} = -\frac{1}{2}$, and therefore $\arcsin\left(-\frac{1}{2}\right) = \frac{11\pi}{6}$.

70. For all x, $\tan(\tan^{-1} x) = x$.

Give the exact real number value of y. Do not use a calculator.

71. $y = \sin^{-1} \frac{\sqrt{2}}{2}$ **72.** $y = \arccos\left(-\frac{1}{2}\right)$ **73.** $y = \tan^{-1}\left(-\sqrt{3}\right)$

74. $y = \arcsin(-1)$ **75.** $y = \cos^{-1}\left(-\frac{\sqrt{2}}{2}\right)$ **76.** $y = \arctan \frac{\sqrt{3}}{3}$

77. $y = \sec^{-1}(-2)$ **78.** $y = \text{arccsc} \frac{2\sqrt{3}}{3}$ **79.** $y = \text{arccot}(-1)$

Give the degree measure of θ. Do not use a calculator.

80. $\theta = \arccos \frac{1}{2}$ **81.** $\theta = \arcsin\left(-\frac{\sqrt{3}}{2}\right)$ **82.** $\theta = \tan^{-1} 0$

Use a calculator to give the degree measure of θ to the nearest hundredth.

83. $\theta = \arctan 1.7804675$ **84.** $\theta = \sin^{-1}(-0.66045320)$ **85.** $\theta = \cos^{-1} 0.80396577$

86. $\theta = \cot^{-1} 4.5046388$ **87.** $\theta = \text{arcsec } 3.4723155$ **88.** $\theta = \csc^{-1} 7.4890096$

Evaluate the following without using a calculator.

89. $\cos(\arccos(-1))$ **90.** $\sin\left(\arcsin\left(-\frac{\sqrt{3}}{2}\right)\right)$ **91.** $\arccos\left(\cos \frac{3\pi}{4}\right)$

92. $\text{arcsec}(\sec \pi)$ **93.** $\tan^{-1}\left(\tan \frac{\pi}{4}\right)$ **94.** $\cos^{-1}(\cos 0)$

95. $\sin\left(\arccos \frac{3}{4}\right)$ **96.** $\cos(\arctan 3)$ **97.** $\cos(\csc^{-1}(-2))$

98. $\sec\left(2 \sin^{-1}\left(-\frac{1}{3}\right)\right)$ **99.** $\tan\left(\arcsin \frac{3}{5} + \arccos \frac{5}{7}\right)$

Write each of the following as an algebraic (nontrigonometric) expression in u, u > 0.

100. $\cos\left(\arctan \frac{u}{\sqrt{1 - u^2}}\right)$ **101.** $\tan\left(\text{arcsec} \frac{\sqrt{u^2 + 1}}{u}\right)$

Solve each equation for exact solutions over the interval $[0, 2\pi)$ where appropriate. Round approximate solutions to four decimal places.

102. $\sin^2 x = 1$ **103.** $2 \tan x - 1 = 0$

104. $3 \sin^2 x - 5 \sin x + 2 = 0$ **105.** $\tan x = \cot x$

106. $\sec^2 2x = 2$ **107.** $\tan^2 2x - 1 = 0$

Give all exact solutions, in radians, for each equation.

108. $\sec \frac{x}{2} = \cos \frac{x}{2}$ **109.** $\cos 2x + \cos x = 0$ **110.** $4 \sin x \cos x = \sqrt{3}$

111. $\{270°\}$
112. $\{45°, 153.4°, 225°, 333.4°\}$
113. $\{45°, 90°, 225°, 270°\}$
114. $\{15°, 75°, 195°, 255°\}$
115. $\{70.5°, 180°, 289.5°\}$
116. $\{53.5°, 118.4°, 233.5°, 298.4°\}$
117. $\{300° + 720°n, 420° + 720°n$, where n is any integer$\}$
118. $\{30° + 360°n, 150° + 360°n, 270° + 360°n$, where n is any integer$\}$
119. $\{180° + 360°n$, where n is any integer$\}$
120. $\{-1\}$ 121. $\emptyset$
122. $\left\{\frac{3\sqrt{5}}{7}\right\}$ 123. $\left\{-\frac{1}{2}\right\}$
124. $x = 2 \arccos \frac{y}{3}$
125. $x = \arcsin 2y$
126. $x = \arcsin\left(\frac{5y+3}{4}\right)$
127. $x = \left(\frac{1}{3} \arctan 2y\right) - \frac{2}{3}$
128. $t = \frac{50}{\pi} \arccos\left(\frac{d-550}{450}\right)$
129. **(b)** 8.6602567 ft; There may be a discrepancy in the final digits.

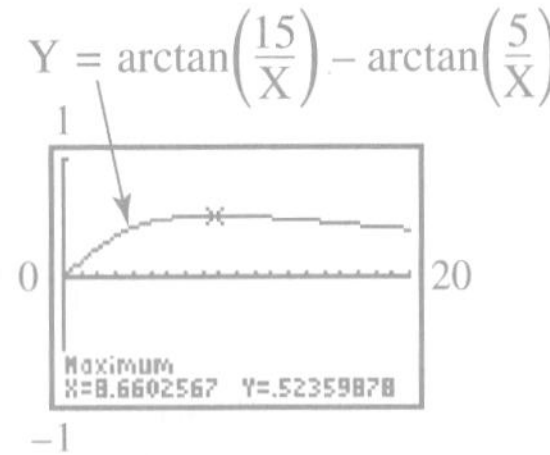

130. 48.8°

Solve each equation for exact solutions over the interval $[0°, 360°)$ where appropriate. Round approximate solutions to the nearest tenth of a degree.

111. $\sin^2 \theta + 3 \sin \theta + 2 = 0$
112. $2 \tan^2 \theta = \tan \theta + 1$
113. $\sin 2\theta = \cos 2\theta + 1$
114. $2 \sin 2\theta = 1$
115. $3 \cos^2 \theta + 2 \cos \theta - 1 = 0$
116. $5 \cot^2 \theta - \cot \theta - 2 = 0$

Give all exact solutions, in degrees, for each equation.

117. $2\sqrt{3} \cos \frac{\theta}{2} = -3$
118. $\sin \theta - \cos 2\theta = 0$
119. $\tan \theta - \sec \theta = 1$

Solve each equation for x. In Exercises 124–127, x is restricted to the given interval.

120. $4\pi - 4 \cot^{-1} x = \pi$
121. $\frac{4}{3} \arctan \frac{x}{2} = \pi$
122. $\arccos x = \arcsin \frac{2}{7}$
123. $\arccos x + \arctan 1 = \frac{11\pi}{12}$
124. $y = 3 \cos \frac{x}{2}$, for x in $[0, 2\pi]$
125. $y = \frac{1}{2} \sin x$, for x in $\left[-\frac{\pi}{2}, \frac{\pi}{2}\right]$
126. $y = \frac{4}{5} \sin x - \frac{3}{5}$, for x in $\left[-\frac{\pi}{2}, \frac{\pi}{2}\right]$
127. $y = \frac{1}{2} \tan(3x + 2)$, for x in $\left(-\frac{2}{3} - \frac{\pi}{6}, -\frac{2}{3} + \frac{\pi}{6}\right)$

128. Solve the equation for t, where t is in the interval $[0, 50]$.

$$d = 550 + 450 \cos\left(\frac{\pi}{50} t\right)$$

(Modeling) Solve each problem.

129. ***Viewing Angle of an Observer*** A 10-ft-wide chalkboard is situated 5 ft from the left wall of a classroom. See the figure. A student sitting next to the wall x feet from the front of the classroom has a viewing angle of θ radians.

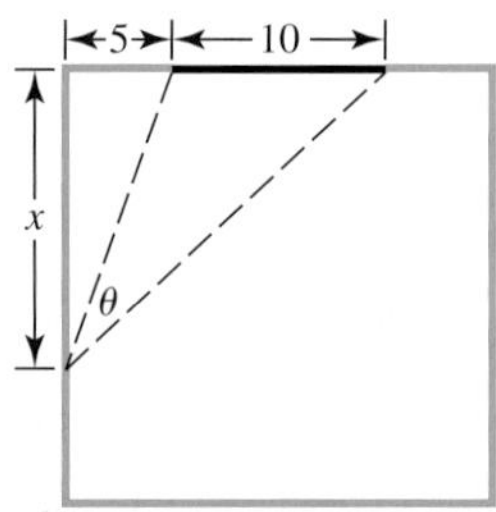

(a) Show that the value of θ is given by the function

$$f(x) = \arctan\left(\frac{15}{x}\right) - \arctan\left(\frac{5}{x}\right).$$

(b) Graph $f(x)$ with a graphing calculator to estimate the value of x that maximizes the viewing angle.

130. ***Snell's Law*** Snell's law says that

$$\frac{c_1}{c_2} = \frac{\sin \theta_1}{\sin \theta_2},$$

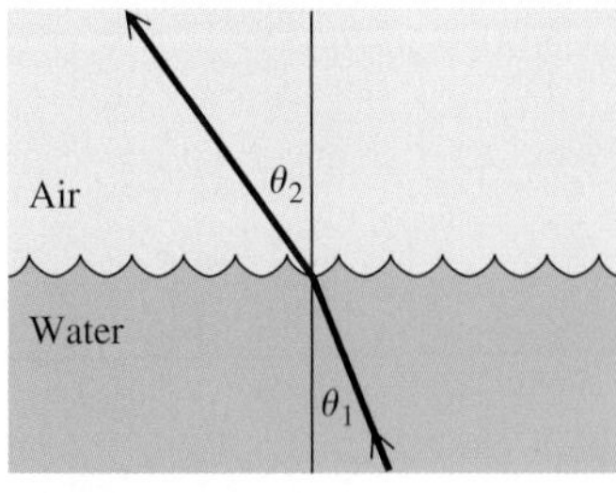

where c_1 is the speed of light in one medium, c_2 is the speed of light in a second medium, and θ_1 and θ_2 are the angles shown in the figure. Suppose a light is shining up through water into the air as in the figure. As θ_1 increases, θ_2 approaches 90°, at which point no light will emerge from the water. Assume the ratio $\frac{c_1}{c_2}$ in this case is 0.752. For what value of θ_1 does $\theta_2 = 90°$? This value of θ_1 is the **critical angle** for water.

131. No light will emerge from the water.

132. (a) 42.2° (b) 90° (c) 48.0°

131. *Snell's Law* Refer to **Exercise 130.** What happens when θ_1 is greater than the critical angle?

132. *British Nautical Mile* The British nautical mile is defined as the length of a minute of arc of a meridian. Since Earth is flat at its poles, the nautical mile, in feet, is given by

$$L = 6077 - 31 \cos 2\theta,$$

where θ is the latitude in degrees. See the figure. (*Source:* Bushaw, D., et al., *A Sourcebook of Applications of School Mathematics,* Mathematical Association of America.)

A nautical mile is the length on any of the meridians cut by a central angle of measure 1 minute.

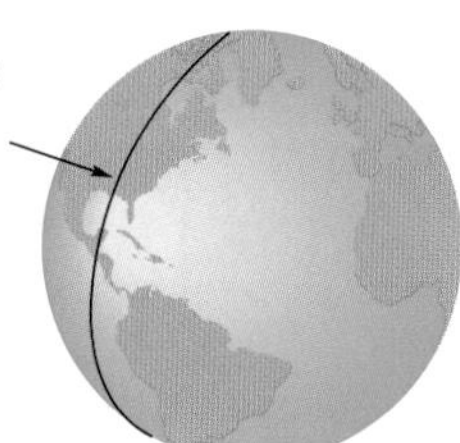

(a) Find the latitude between 0° and 90° at which the nautical mile is 6074 ft.

(b) At what latitude between 0° and 180° is the nautical mile 6108 ft?

(c) In the United States, the nautical mile is defined everywhere as 6080.2 ft. At what latitude between 0° and 90° does this agree with the British nautical mile?

Chapter 7 Test

[7.1]

1. $\sin\theta = -\frac{7}{25}$; $\tan\theta = -\frac{7}{24}$; $\cot\theta = -\frac{24}{7}$; $\sec\theta = \frac{25}{24}$; $\csc\theta = -\frac{25}{7}$

2. $\cos\theta$

3. -1

[7.3]

4. $\frac{\sqrt{6}-\sqrt{2}}{4}$

5. (a) $-\sin x$ (b) $\tan x$

[7.4]

6. $-\frac{\sqrt{2-\sqrt{2}}}{2}$

[7.3]

7. (a) $\frac{33}{65}$ (b) $-\frac{56}{65}$ (c) $\frac{63}{16}$ (d) II

[7.4]

8. (a) $-\frac{7}{25}$ (b) $-\frac{24}{25}$ (c) $\frac{24}{7}$ (d) $\frac{\sqrt{5}}{5}$ (e) 2

[7.3]

13. (a) $V = 163\cos\left(\frac{\pi}{2} - \omega t\right)$ (b) 163 volts; $\frac{1}{240}$ sec

1. If $\cos\theta = \frac{24}{25}$ and θ is in quadrant IV, find the five remaining trigonometric function values of θ.

2. Express $\sec\theta - \sin\theta\tan\theta$ as a single function of θ.

3. Express $\tan^2 x - \sec^2 x$ in terms of $\sin x$ and $\cos x$, and simplify.

4. Find the exact value of $\cos\frac{5\pi}{12}$.

5. Express as a function of x alone.

(a) $\cos(270° - x)$ **(b)** $\tan(\pi + x)$

6. Use a half-angle identity to find the exact value of $\sin(-22.5°)$.

7. Given that $\sin A = \frac{5}{13}$, $\cos B = -\frac{3}{5}$, A is a quadrant I angle, and B is a quadrant II angle, find each of the following.

(a) $\sin(A + B)$ **(b)** $\cos(A + B)$ **(c)** $\tan(A - B)$ **(d)** the quadrant of $A + B$

8. Given that $\cos\theta = -\frac{3}{5}$ and $90° < \theta < 180°$, find each of the following.

(a) $\cos 2\theta$ **(b)** $\sin 2\theta$ **(c)** $\tan 2\theta$ **(d)** $\cos\frac{\theta}{2}$ **(e)** $\tan\frac{\theta}{2}$

Verify each identity.

9. $\sec^2 B = \dfrac{1}{1 - \sin^2 B}$

10. $\cos 2A = \dfrac{\cot A - \tan A}{\csc A \sec A}$

11. $\tan^2 x - \sin^2 x = (\tan x \sin x)^2$

12. $\dfrac{\tan x - \cot x}{\tan x + \cot x} = 2\sin^2 x - 1$

13. *(Modeling) Voltage* The voltage in common household current is expressed as $V = 163 \sin \omega t$, where ω is the angular speed (in radians per second) of the generator at the electrical plant and t is time (in seconds).

(a) Use an identity to express V in terms of cosine.

(b) If $\omega = 120\pi$, what is the maximum voltage? Give the least positive value of t when the maximum voltage occurs.

14. Graph $y = \sin^{-1} x$, and indicate the coordinates of three points on the graph. Give the domain and range.

[7.5]

14. $[-1, 1]$; $\left[-\frac{\pi}{2}, \frac{\pi}{2}\right]$

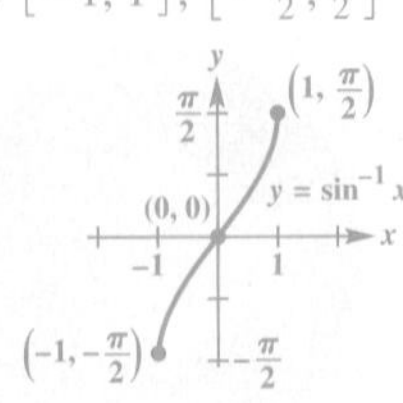

15. (a) $\frac{2\pi}{3}$ (b) $-\frac{\pi}{3}$ (c) 0 (d) $\frac{2\pi}{3}$

16. (a) $30°$ (b) $-45°$ (c) $135°$ (d) $-60°$

17. (a) $42.54°$ (b) $22.72°$ (c) $125.47°$

18. (a) $\frac{\sqrt{5}}{3}$ (b) $\frac{4\sqrt{2}}{9}$

19. $\frac{u\sqrt{1-u^2}}{1-u^2}$

[7.6]

20. $\{30°, 330°\}$

21. $\{90°, 270°\}$

22. $\{18.4°, 135°, 198.4°, 315°\}$

23. $\left\{0, \frac{2\pi}{3}, \frac{4\pi}{3}\right\}$

24. $\left\{\frac{\pi}{12}, \frac{7\pi}{12}, \frac{3\pi}{4}, \frac{5\pi}{4}, \frac{17\pi}{12}, \frac{23\pi}{12}\right\}$

25. $\{0.3649, 1.2059, 3.5065, 4.3475\}$

26. $\{90° + 180°n$, where n is any integer$\}$

27. $\left\{\frac{2\pi}{3} + 4n\pi, \frac{4\pi}{3} + 4n\pi\right.$, where n is any integer$\}$

[7.7]

28. (a) $x = \frac{1}{3}\arccos y$ (b) $x = \operatorname{arccot}\left(\frac{y-4}{3}\right)$

29. (a) $\left\{\frac{4}{5}\right\}$ (b) $\left\{\frac{\sqrt{3}}{3}\right\}$

30. $\frac{5}{6}$ sec, $\frac{11}{6}$ sec, $\frac{17}{6}$ sec

15. Find the exact value of y for each equation.

(a) $y = \arccos\left(-\frac{1}{2}\right)$ **(b)** $y = \sin^{-1}\left(-\frac{\sqrt{3}}{2}\right)$

(c) $y = \tan^{-1} 0$ **(d)** $y = \operatorname{arcsec}(-2)$

16. Give the degree measure of θ.

(a) $\theta = \arccos\frac{\sqrt{3}}{2}$ **(b)** $\theta = \tan^{-1}(-1)$

(c) $\theta = \cot^{-1}(-1)$ **(d)** $\theta = \csc^{-1}\left(-\frac{2\sqrt{3}}{3}\right)$

17. Use a calculator to give each value in decimal degrees to the nearest hundredth.

(a) $\sin^{-1} 0.67610476$ **(b)** $\sec^{-1} 1.0840880$

(c) $\cot^{-1}(-0.7125586)$

18. Find each exact value.

(a) $\cos\left(\arcsin\frac{2}{3}\right)$ **(b)** $\sin\left(2\cos^{-1}\frac{1}{3}\right)$

19. Write $\tan(\arcsin u)$ as an algebraic (nontrigonometric) expression in u, $u > 0$.

Solve each equation for exact solutions over the interval $[0°, 360°)$ where appropriate. Round approximate solutions to the nearest tenth of a degree.

20. $-3\sec\theta + 2\sqrt{3} = 0$ **21.** $\sin^2\theta = \cos^2\theta + 1$ **22.** $\csc^2\theta - 2\cot\theta = 4$

Solve each equation for exact solutions over the interval $[0, 2\pi)$ where appropriate. Round approximate solutions to four decimal places.

23. $\cos x = \cos 2x$ **24.** $\sqrt{2}\cos 3x - 1 = 0$ **25.** $\sin x \cos x = \frac{1}{3}$

Solve each equation for all exact solutions in radians (for x) or in degrees (for θ). Write answers using the least possible nonnegative angle measures.

26. $\sin^2\theta = -\cos 2\theta$ **27.** $2\sqrt{3}\sin\frac{x}{2} = 3$

28. Solve each equation for x, where x is restricted to the given interval.

(a) $y = \cos 3x$, for x in $\left[0, \frac{\pi}{3}\right]$ **(b)** $y = 4 + 3\cot x$, for x in $(0, \pi)$

29. Solve each equation for exact solutions.

(a) $\arcsin x = \arctan\frac{4}{3}$ **(b)** $\operatorname{arccot} x + 2\arcsin\frac{\sqrt{3}}{2} = \pi$

30. ***(Modeling) Movement of a Runner's Arm*** A runner's arm swings rhythmically according to the model

$$y = \frac{\pi}{8}\cos\left[\pi\left(t - \frac{1}{3}\right)\right],$$

where y represents the angle between the actual position of the upper arm and the downward vertical position, and t represents time in seconds. At what times over the interval $[0, 3)$ is the angle y equal to 0?

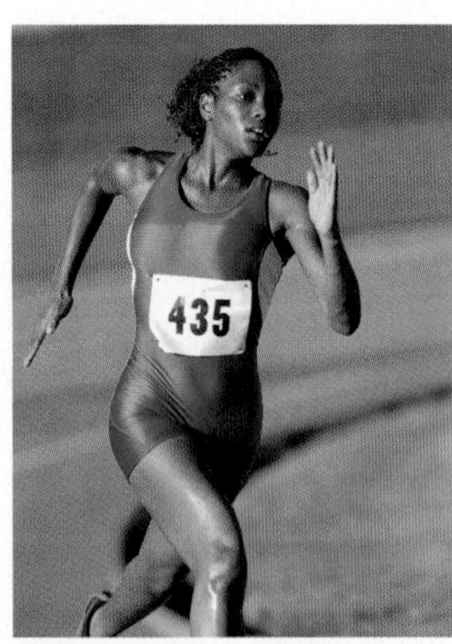

8 Applications of Trigonometry

Surveyors use a method known as *triangulation* to measure distances when direct measurements cannot be made due to obstructions in the line of sight.

8.1 The Law of Sines

- Congruency and Oblique Triangles
- Derivation of the Law of Sines
- Using the Law of Sines
- The Ambiguous Case
- Area of a Triangle

Congruency and Oblique Triangles The concepts of solving triangles developed in **Chapter 5** can be extended to *all* triangles. The following axioms from geometry enable us to prove that two triangles are congruent (that is, their corresponding sides and angles are equal).

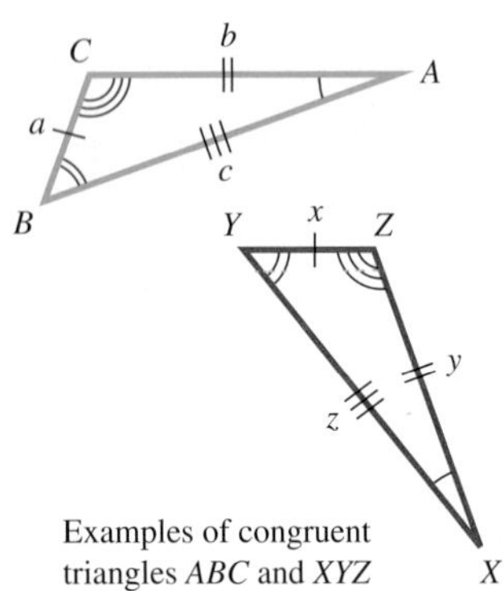

Examples of congruent triangles *ABC* and *XYZ*

Congruence Axioms

Side-Angle-Side (SAS)	If two sides and the included angle of one triangle are equal, respectively, to two sides and the included angle of a second triangle, then the triangles are congruent.
Angle-Side-Angle (ASA)	If two angles and the included side of one triangle are equal, respectively, to two angles and the included side of a second triangle, then the triangles are congruent.
Side-Side-Side (SSS)	If three sides of one triangle are equal, respectively, to three sides of a second triangle, then the triangles are congruent.

If a side and *any* two angles are given **(SAA),** the third angle is easily determined by using

$$A + B + C = 180^\circ, \quad \text{Angle sum formula}$$

and then the ASA axiom can be applied. Keep in mind that whenever SAS, ASA, or SSS is given, the triangle is unique.

A triangle that is not a right triangle is called an **oblique triangle.** ***The measures of the three sides and the three angles of a triangle can be found if at least one side and any other two measures are known.*** There are four possible cases.

Data Required for Solving Oblique Triangles

Case 1	One side and two angles are known (SAA or ASA).
Case 2	Two sides and one angle not included between the two sides are known (SSA). This case may lead to more than one triangle.
Case 3	Two sides and the angle included between the two sides are known (SAS).
Case 4	Three sides are known (SSS).

NOTE ***If we know three angles of a triangle, we cannot find unique side lengths since AAA assures us only of similarity, not congruence.*** For example, there are infinitely many triangles *ABC* of different sizes with $A = 35^\circ$, $B = 65^\circ$, and $C = 80^\circ$.

Cases 1 and 2, discussed in this section, require the *law of sines.* Cases 3 and 4, discussed in **Section 8.2,** require the *law of cosines.*

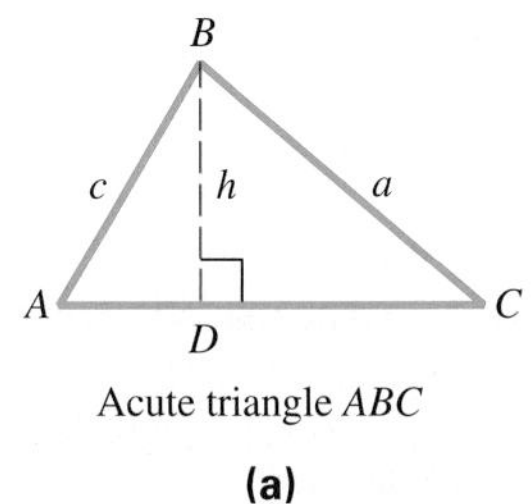

Acute triangle *ABC*

(a)

Obtuse triangle *ABC*

(b)

We label oblique triangles as we did right triangles: side a opposite angle A, side b opposite angle B, and side c opposite angle C.

Figure 1

Derivation of the Law of Sines To derive the law of sines, we start with an oblique triangle, such as the **acute triangle** in **Figure 1(a)** or the **obtuse triangle** in **Figure 1(b).** This discussion applies to both triangles. First, construct the perpendicular from B to side AC (or its extension). Let h be the length of this perpendicular. Then c is the hypotenuse of right triangle ADB, and a is the hypotenuse of right triangle BDC.

$$\text{In triangle } ADB, \quad \sin A = \frac{h}{c}, \quad \text{or} \quad h = c \sin A.$$

(Section 5.3)

$$\text{In triangle } BDC, \quad \sin C = \frac{h}{a}, \quad \text{or} \quad h = a \sin C.$$

Since $h = c \sin A$ and $h = a \sin C$, we set these two expressions equal.

$$a \sin C = c \sin A$$

$$\frac{a}{\sin A} = \frac{c}{\sin C} \quad \text{Divide each side by } \sin A \sin C.$$

In a similar way, by constructing perpendicular lines from the other vertices, we can show that these two equations are also true.

$$\frac{a}{\sin A} = \frac{b}{\sin B} \quad \text{and} \quad \frac{b}{\sin B} = \frac{c}{\sin C}$$

This discussion proves the following theorem.

Teaching Tip The law of sines says that in a triangle, the ratio of the length of any side to the sine of its opposite angle is constant. (This constant is the diameter of the circumscribed circle of the triangle. See **Exercise 85** in this section.)

Law of Sines

In any triangle ABC, with sides a, b, and c,

$$\frac{a}{\sin A} = \frac{b}{\sin B}, \quad \frac{a}{\sin A} = \frac{c}{\sin C}, \quad \text{and} \quad \frac{b}{\sin B} = \frac{c}{\sin C}.$$

This can be written in compact form as follows.

$$\frac{a}{\sin A} = \frac{b}{\sin B} = \frac{c}{\sin C}$$

That is, according to the law of sines, the lengths of the sides in a triangle are proportional to the sines of the measures of the angles opposite them.

In practice we can also use an alternative form of the law of sines.

$$\frac{\sin A}{a} = \frac{\sin B}{b} = \frac{\sin C}{c} \quad \text{Alternative form}$$

NOTE When using the law of sines, a good strategy is to select an equation so that the unknown variable is in the numerator and all other variables are known. This makes computation easier.

Using the Law of Sines If two angles and one side of a triangle are known (Case 1, SAA or ASA), then the law of sines can be used to solve the triangle.

Classroom Example 1
Solve triangle ABC if $A = 28.8°$, $C = 102.6°$, and $c = 25.3$ in.

Answer: $a = 12.5$ in., $B = 48.6°$, $b = 19.4$ in.

EXAMPLE 1 Applying the Law of Sines (SAA)

Solve triangle ABC if $A = 32.0°$, $B = 81.8°$, and $a = 42.9$ cm.

SOLUTION Start by drawing a triangle, roughly to scale, and labeling the given parts as in **Figure 2.** Since the values of A, B, and a are known, use the form of the law of sines that involves these variables, and then solve for b.

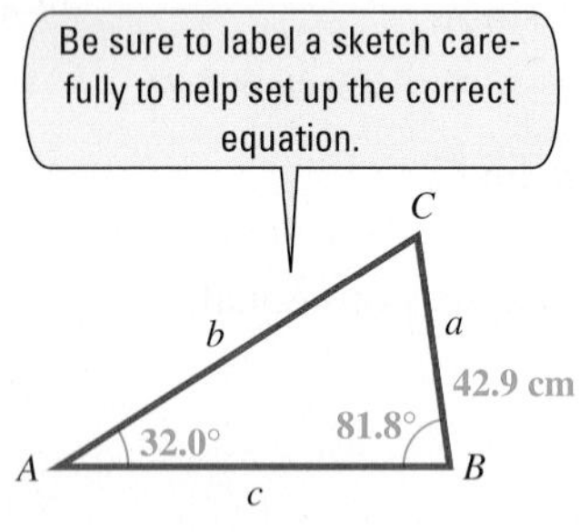

Be sure to label a sketch carefully to help set up the correct equation.

Figure 2

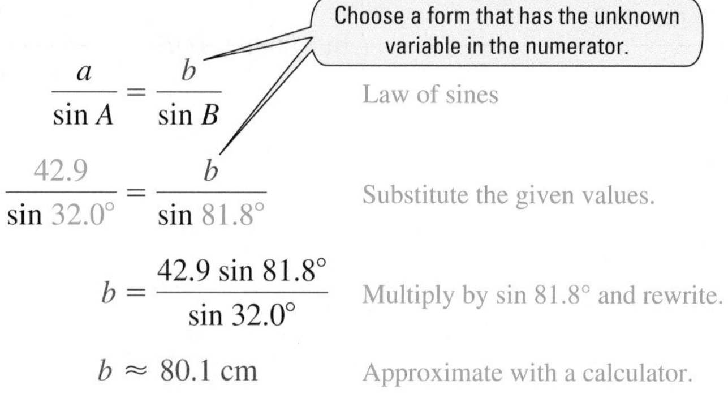

Choose a form that has the unknown variable in the numerator.

$$\frac{a}{\sin A} = \frac{b}{\sin B} \quad \text{Law of sines}$$

$$\frac{42.9}{\sin 32.0°} = \frac{b}{\sin 81.8°} \quad \text{Substitute the given values.}$$

$$b = \frac{42.9 \sin 81.8°}{\sin 32.0°} \quad \text{Multiply by } \sin 81.8° \text{ and rewrite.}$$

$$b \approx 80.1 \text{ cm} \quad \text{Approximate with a calculator.}$$

To find C, use the fact that the sum of the angles of any triangle is 180°.

$$A + B + C = 180° \quad \text{Angle sum formula}$$

$$C = 180° - A - B \quad \text{Solve for } C. \text{ (Section 1.1)}$$

$$C = 180° - 32.0° - 81.8° \quad \text{Substitute.}$$

$$C = 66.2° \quad \text{Subtract.}$$

Teaching Tip Encourage students to draw a scale version of the given triangle to help solve it, and then check answers for reasonableness.

Now use the law of sines to find c. (The Pythagorean theorem does not apply because this is not a right triangle.)

$$\frac{a}{\sin A} = \frac{c}{\sin C} \quad \text{Law of sines}$$

$$\frac{42.9}{\sin 32.0°} = \frac{c}{\sin 66.2°} \quad \text{Substitute known values.}$$

$$c = \frac{42.9 \sin 66.2°}{\sin 32.0°} \quad \text{Multiply by } \sin 66.2° \text{ and rewrite.}$$

$$c \approx 74.1 \text{ cm} \quad \text{Approximate with a calculator.}$$

✔ *Now Try Exercise 9.*

CAUTION Whenever possible, use the given values in solving triangles, rather than values obtained in intermediate steps, to avoid rounding errors.

Classroom Example 2
Repeat **Example 2** if $C = 117.2°$, $A = 28.8°$, and $b = 75.6$ ft.

Answer: 65.1 ft

EXAMPLE 2 Applying the Law of Sines (ASA)

Kurt Daniels wishes to measure the distance across the Gasconade River. See **Figure 3.** He determines that $C = 112.90°$, $A = 31.10°$, and $b = 347.6$ ft. Find the distance a across the river.

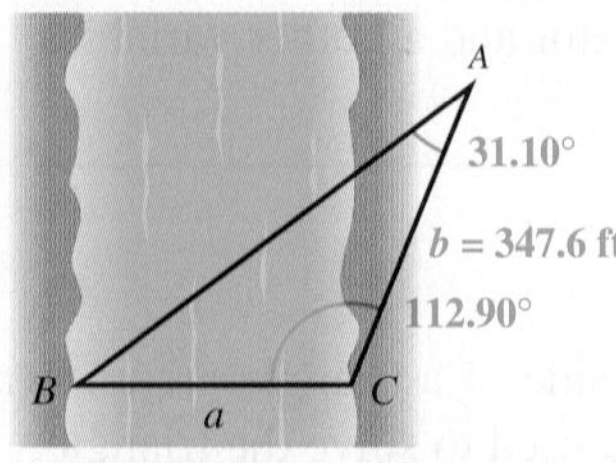

Figure 3

SOLUTION To use the law of sines, one side and the angle opposite it must be known. Since b is the only side whose length is given, angle B must be found before the law of sines can be used.

$$B = 180° - A - C \quad \text{Angle sum formula, solved for } B$$

$$B = 180° - 31.10° - 112.90° \quad \text{Substitute the given values.}$$

$$B = 36.00° \quad \text{Subtract.}$$

Now use the form of the law of sines involving A, B, and b to find a.

Solve for a.

$$\frac{a}{\sin A} = \frac{b}{\sin B} \qquad \text{Law of sines}$$

$$\frac{a}{\sin 31.10°} = \frac{347.6}{\sin 36.00°} \qquad \text{Substitute known values.}$$

$$a = \frac{347.6 \sin 31.10°}{\sin 36.00°} \qquad \text{Multiply by } \sin 31.10°.$$

$$a \approx 305.5 \text{ ft} \qquad \text{Use a calculator.}$$

Now Try Exercise 55.

The next example involves the concept of bearing, first discussed in **Chapter 5.**

Classroom Example 3
The bearing of a lighthouse from a ship was found to be N 52° W. After the ship sailed 5.8 km due south, the new bearing was N 23° W. Find the distance, to the nearest tenth kilometer, between the ship and the lighthouse at each location.

Answer: first location: 4.7 km; second location: 9.4 km

EXAMPLE 3 Applying the Law of Sines (ASA)

Two ranger stations are on an east-west line 110 mi apart. A forest fire is located on a bearing of N 42° E from the western station at A and a bearing of N 15° E from the eastern station at B. To the nearest ten miles, how far is the fire from the western station?

SOLUTION **Figure 4** shows the two stations at points A and B and the fire at point C.

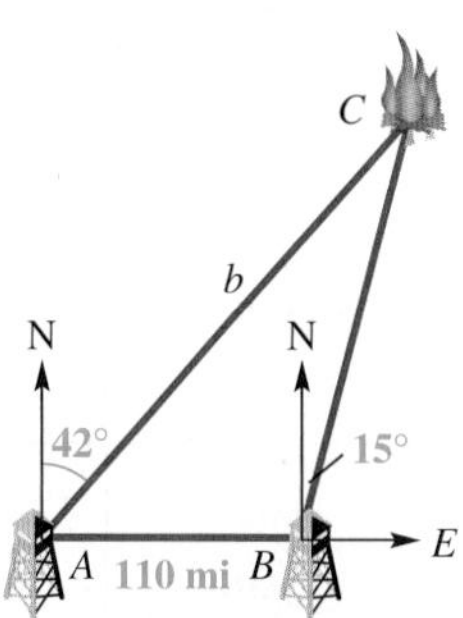

Figure 4

In **Figure 4,** angle $BAC = 90° - 42° = 48°$, the obtuse angle at B measures $90° + 15° = 105°$, and the third angle, C, measures $180° - 105° - 48° = 27°$. We use the law of sines to find side b.

Solve for b.

$$\frac{b}{\sin B} = \frac{c}{\sin C} \qquad \text{Law of sines}$$

$$\frac{b}{\sin 105°} = \frac{110}{\sin 27°} \qquad \text{Substitute known values.}$$

$$b = \frac{110 \sin 105°}{\sin 27°} \qquad \text{Multiply by } \sin 105°.$$

$$b \approx 230 \text{ mi} \qquad \text{Use a calculator and give two significant digits. (Section 5.4)}$$

Now Try Exercise 57.

NOTE There is another method for describing bearing and it was first introduced in Method 1 in **Chapter 5.** It involves measuring clockwise from due north, using a single degree measure between 0° and 360°.

The Ambiguous Case If we are given the lengths of two sides and the angle opposite one of them (Case 2, SSA), then zero, one, or two such triangles may exist. (There is no SSA congruence axiom.)

Suppose we know the measure of acute angle A of triangle ABC, the length of side a, and the length of side b, as shown in **Figure 5.** Now we must draw the side of length a opposite angle A. The table shows possible outcomes. This situation (SSA) is called the **ambiguous case** of the law of sines.

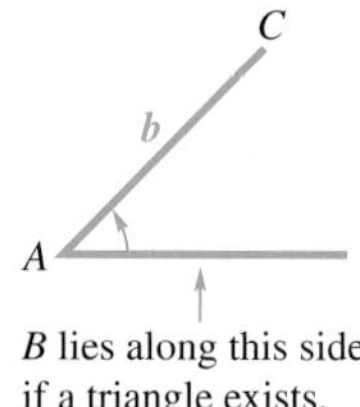

Figure 5

As shown in the table, if angle A is acute, there are four possible outcomes. If A is obtuse, there are two possible outcomes.

Teaching Tip The case leading to two triangles, for acute angle A and known sides a and b, is troublesome for students. It is a good idea to draw the "triangle" below and talk through the possibilities, comparing the lengths of a, b, and h. Use a straightedge anchored at point C, and have it swing an arc. Use different lengths for the straightedge so that students can actually visualize the cases. ***Emphasize making good sense of the situation.*** Note that $\sin A = \frac{h}{b}$, or $h = b \sin A$.

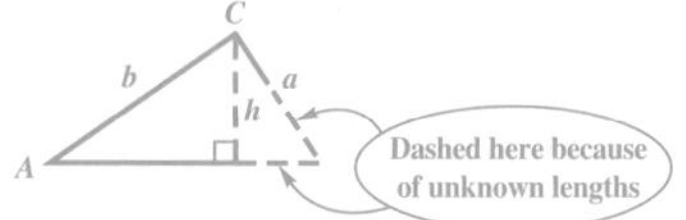

- If $a < h$, no triangle exists.
- If $a = h$, a right triangle exists.
- If $h < a < b$, two triangles exist.
- If $a \geq b$, one triangle exists.

Draw a similar sketch with obtuse angle A to show that a must be greater than b for a single triangle to exist. Ask leading questions to see whether students understand why a must be the longest side of the triangle.

Angle A is	Possible Number of Triangles	Sketch	Applying Law of Sines Leads to
Acute	0		$\sin B > 1,\ a < h < b$
Acute	1		$\sin B = 1,\ a = h < b$
Acute	1		$0 < \sin B < 1,\ a \geq b$
Acute	2		$0 < \sin B_1 < 1,\ h < a < b,$ $A + B_2 < 180°$
Obtuse	0		$\sin B \geq 1,\ a \leq b$
Obtuse	1		$0 < \sin B < 1,\ a > b$

The following basic facts help determine which situation applies.

Applying the Law of Sines

1. For any angle θ of a triangle, $0 < \sin \theta \leq 1$. If $\sin \theta = 1$, then $\theta = 90°$ and the triangle is a right triangle.
2. $\sin \theta = \sin(180° - \theta)$ (Supplementary angles have the same sine value.)
3. The smallest angle is opposite the shortest side, the largest angle is opposite the longest side, and the middle-valued angle is opposite the intermediate side (assuming the triangle has sides that are all of different lengths).

Classroom Example 4
Solve triangle ABC if $a = 17.9$ cm, $c = 13.2$ cm, and $C = 75° 30'$.

Answer: no such triangle

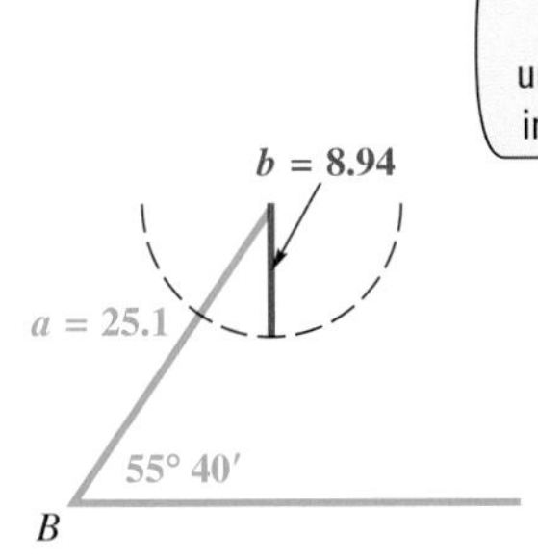

Figure 6

EXAMPLE 4 Solving the Ambiguous Case (No Such Triangle)

Solve triangle ABC if $B = 55° 40'$, $b = 8.94$ m, and $a = 25.1$ m.

SOLUTION We are given B, b, and a, so we can use the law of sines to find A.

Choose a form that has the unknown variable in the numerator.

$$\frac{\sin A}{a} = \frac{\sin B}{b}$$ Law of sines (alternative form)

$$\frac{\sin A}{25.1} = \frac{\sin 55° 40'}{8.94}$$ Substitute the given values.

$$\sin A = \frac{25.1 \sin 55° 40'}{8.94}$$ Multiply by 25.1.

$$\sin A \approx 2.3184379$$ Use a calculator.

Because sin A cannot be greater than 1, there can be no such angle A—and thus no triangle with the given information. An attempt to sketch such a triangle leads to the situation shown in **Figure 6.**

Now Try Exercise 37.

NOTE In the ambiguous case, we are given two sides and an angle opposite one of the sides (SSA). For example, suppose b, c, and angle C are given. This situation represents the ambiguous case because angle C is opposite side c.

Classroom Example 5
Solve triangle ABC if $A = 61.4°$, $a = 35.5$ cm, and $b = 39.2$ cm.

Answer: $B_1 = 75.8°$, $C_1 = 42.8°$, $c_1 = 27.5$ cm; $B_2 = 104.2°$, $C_2 = 14.4°$, $c_2 = 10.1$ cm

EXAMPLE 5 Solving the Ambiguous Case (Two Triangles)

Solve triangle ABC if $A = 55.3°$, $a = 22.8$ ft, and $b = 24.9$ ft.

SOLUTION To begin, use the law of sines to find angle B.

$$\frac{\sin A}{a} = \frac{\sin B}{b}$$ Solve for sin B.

$$\frac{\sin 55.3°}{22.8} = \frac{\sin B}{24.9}$$ Substitute the given values.

$$\sin B = \frac{24.9 \sin 55.3°}{22.8}$$ Multiply by 24.9 and rewrite.

$$\sin B \approx 0.8978678$$ Use a calculator.

There are two angles B between 0° and 180° that satisfy this condition. Since $\sin B \approx 0.8978678$, to the nearest tenth one value of B is

$$B_1 = 63.9°.$$ Use the inverse sine function. **(Section 7.5)**

Supplementary angles have the same sine value, so another *possible* value of B is

$$B_2 = 180° - 63.9° = 116.1°.$$ **(Section 5.1)**

To see whether $B_2 = 116.1°$ is a valid possibility, add 116.1° to the measure of A, 55.3°. Since

$$116.1° + 55.3° = 171.4°,$$

and this sum is less than 180°, it is a valid angle measure for this triangle.

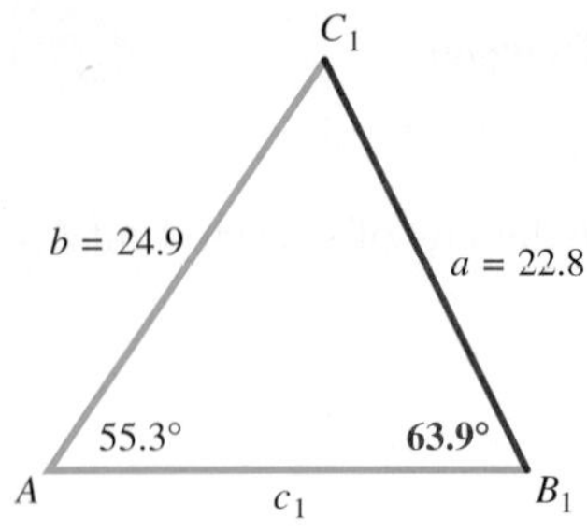

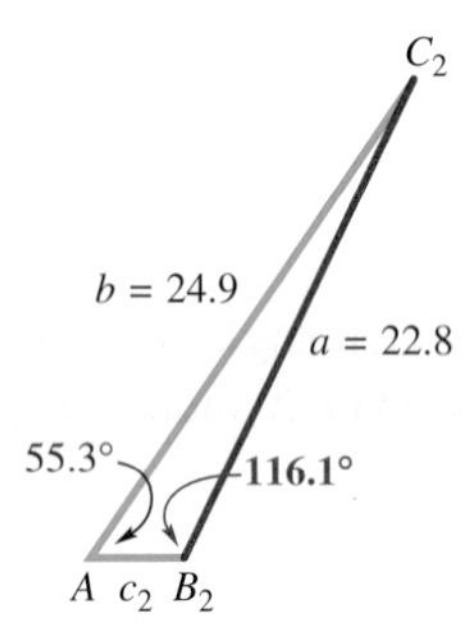

Figure 7

Now separately solve triangles AB_1C_1 and AB_2C_2 shown in **Figure 7.** Begin with AB_1C_1. Find C_1 first.

$$C_1 = 180° - A - B_1 \quad \text{Angle sum formula (Section 5.3)}$$

$$C_1 = 180° - 55.3° - 63.9° \quad \text{Substitute.}$$

$$C_1 = 60.8° \quad \text{Subtract.}$$

Now, use the law of sines to find c_1.

$$\frac{a}{\sin A} = \frac{c_1}{\sin C_1} \quad \text{Solve for } c_1.$$

$$\frac{22.8}{\sin 55.3°} = \frac{c_1}{\sin 60.8°} \quad \text{Substitute.}$$

$$c_1 = \frac{22.8 \sin 60.8°}{\sin 55.3°} \quad \text{Multiply by } \sin 60.8°.$$

$$c_1 \approx 24.2 \text{ ft} \quad \text{Use a calculator.}$$

To solve triangle AB_2C_2, first find C_2.

$$C_2 = 180° - A - B_2 \quad \text{Angle sum formula}$$

$$C_2 = 180° - 55.3° - 116.1° \quad \text{Substitute.}$$

$$C_2 = 8.6° \quad \text{Subtract.}$$

Use the law of sines to find c_2.

$$\frac{a}{\sin A} = \frac{c_2}{\sin C_2} \quad \text{Solve for } c_2.$$

$$\frac{22.8}{\sin 55.3°} = \frac{c_2}{\sin 8.6°} \quad \text{Substitute.}$$

$$c_2 = \frac{22.8 \sin 8.6°}{\sin 55.3°} \quad \text{Multiply by } \sin 8.6°.$$

$$c_2 \approx 4.15 \text{ ft} \quad \text{Use a calculator.}$$

✔ Now Try Exercise 45.

The ambiguous case results in zero, one, or two triangles. The following guidelines can be used to determine how many triangles there are.

Teaching Tip Ask students to outline the steps that they would take to solve each triangle. Do not actually solve the triangle.

1. $B = 85°$, $C = 40°$, $b = 26$
2. $A = 55°$, $B = 65°$, $c = 4$
3. $a = 17$, $b = 22$, $B = 32°$
4. $A = 20°$, $b = 7$, $a = 9$
5. $c = 225$, $A = 103.2°$, $B = 62.5°$

Number of Triangles Satisfying the Ambiguous Case (SSA)

Let sides a and b and angle A be given in triangle ABC. (The law of sines can be used to calculate the value of $\sin B$.)

1. If applying the law of sines results in an equation having $\sin B > 1$, then *no triangle* satisfies the given conditions.
2. If $\sin B = 1$, then *one triangle* satisfies the given conditions and $B = 90°$.
3. If $0 < \sin B < 1$, then either *one or two triangles* satisfy the given conditions.
   (a) If $\sin B = k$, then let $B_1 = \sin^{-1} k$ and use B_1 for B in the first triangle.
   (b) Let $B_2 = 180° - B_1$. If $A + B_2 < 180°$, then a second triangle exists. In this case, use B_2 for B in the second triangle.

Classroom Example 6
Solve triangle ABC, given $B = 68.7°$, $b = 25.4$ in., and $a = 19.6$ in.

Answer: $A = 46.0°$, $C = 65.3°$, $c = 24.8$ in.

EXAMPLE 6 Solving the Ambiguous Case (One Triangle)

Solve triangle ABC, given $A = 43.5°$, $a = 10.7$ in., and $c = 7.2$ in.

SOLUTION To find angle C, use an alternative form of the law of sines.

$$\frac{\sin C}{c} = \frac{\sin A}{a} \quad \text{Law of sines}$$

$$\frac{\sin C}{7.2} = \frac{\sin 43.5°}{10.7} \quad \text{Substitute the given values.}$$

$$\sin C = \frac{7.2 \sin 43.5°}{10.7} \quad \text{Multiply by 7.2.}$$

$$\sin C \approx 0.46319186 \quad \text{Use a calculator.}$$

$$C \approx 27.6° \quad \text{Use the inverse sine function.}$$

There is another angle C that has sine value 0.46319186. It is

$$C = 180° - 27.6° = 152.4°.$$

However, notice in the given information that $c < a$, meaning that in the triangle, angle C must have measure *less than* angle A. Notice also that when we add this obtuse value to the given angle $A = 43.5°$, we obtain

$$152.4° + 43.5° = 195.9°,$$

which is greater than 180°. Thus either of these approaches shows that there can be only one triangle. See **Figure 8.** Then

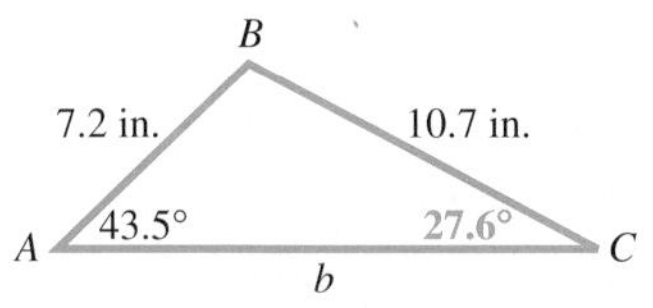

Figure 8

$$B = 180° - 27.6° - 43.5° \quad \text{Substitute.}$$

$$B = 108.9°, \quad \text{Subtract.}$$

and we can find side b with the law of sines.

$$\frac{b}{\sin B} = \frac{a}{\sin A} \quad \text{Law of sines}$$

$$\frac{b}{\sin 108.9°} = \frac{10.7}{\sin 43.5°} \quad \text{Substitute known values.}$$

$$b = \frac{10.7 \sin 108.9°}{\sin 43.5°} \quad \text{Multiply by } \sin 108.9°.$$

$$b \approx 14.7 \text{ in.} \quad \text{Use a calculator.}$$

Now Try Exercise 41.

When solving triangles, it is important to analyze the given information to determine whether it forms a valid triangle.

Classroom Example 7
Without using the law of sines, explain why no triangle ABC exists satisfying $B = 93°$, $b = 42$ cm, and $c = 48$ cm.

Answer: Because B is an obtuse angle, it must be the largest angle of the triangle. Thus b must be the longest side of the triangle. However, we are given that $c > b$, so no such triangle exists.

EXAMPLE 7 Analyzing Data Involving an Obtuse Angle

Without using the law of sines, explain why $A = 104°$, $a = 26.8$ m, and $b = 31.3$ m cannot be valid for a triangle ABC.

SOLUTION Since A is an obtuse angle, it is the largest angle, and so the longest side of the triangle must be a. However, we are given $b > a$.

Thus, $B > A$, which is impossible if A is obtuse.

Therefore, no such triangle ABC exists.

Now Try Exercise 53.

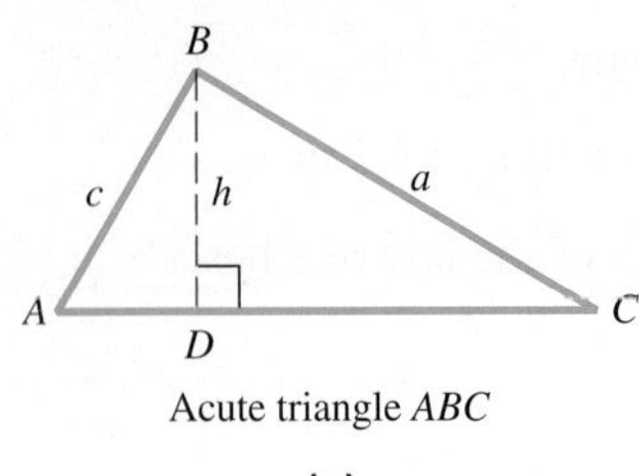

Acute triangle ABC

(a)

Obtuse triangle ABC

(b)

Figure 9

Area of a Triangle The method used to derive the law of sines can also be used to derive a formula to find the area of a triangle. A familiar formula for the area of a triangle is

$$\mathscr{A} = \frac{1}{2}bh, \quad \text{where } \mathscr{A} \text{ represents area, } b \text{ base, and } h \text{ height.}$$

This formula cannot always be used easily because in practice, h is often unknown. To find another formula, refer to acute triangle ABC in **Figure 9(a)** or obtuse triangle ABC in **Figure 9(b).**

A perpendicular has been drawn from B to the base of the triangle (or the extension of the base). Consider right triangle ADB in either figure.

$$\sin A = \frac{h}{c}, \quad \text{or} \quad h = c \sin A$$

Substitute into the formula for the area of a triangle.

$$\mathscr{A} = \frac{1}{2}bh = \frac{1}{2}bc \sin A$$

Any other pair of sides and the angle between them could have been used.

Teaching Tip Emphasize that in each of these formulas, two sides and an included angle are used.

Area of a Triangle (SAS)

In any triangle ABC, the area $\mathscr{A}$ is given by the following formulas.

$$\mathscr{A} = \frac{1}{2}bc \sin A, \quad \mathscr{A} = \frac{1}{2}ab \sin C, \quad \text{and} \quad \mathscr{A} = \frac{1}{2}ac \sin B$$

That is, the area is half the product of the lengths of two sides and the sine of the angle included between them.

Classroom Example 8
Find the area of triangle DEF in the figure.

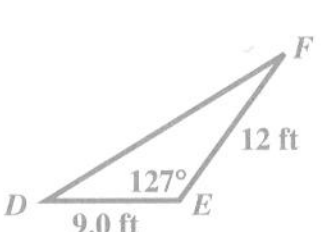

Answer: 43 ft^2

NOTE If the included angle measures 90°, its sine is 1 and the formula becomes the familiar $\mathscr{A} = \frac{1}{2}bh$.

EXAMPLE 8 Finding the Area of a Triangle (SAS)

Find the area of triangle ABC in **Figure 10.**

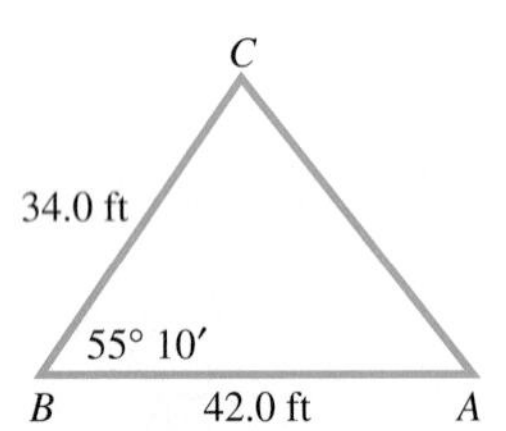

Figure 10

SOLUTION Substitute $B = 55° \, 10'$, $a = 34.0$ ft, and $c = 42.0$ ft into the area formula.

$$\mathscr{A} = \frac{1}{2}ac \sin B = \frac{1}{2}(34.0)(42.0) \sin 55° \, 10' \approx 586 \text{ ft}^2$$

✔ *Now Try Exercise 75.*

EXAMPLE 9 Finding the Area of a Triangle (ASA)

Find the area of triangle ABC in **Figure 11.**

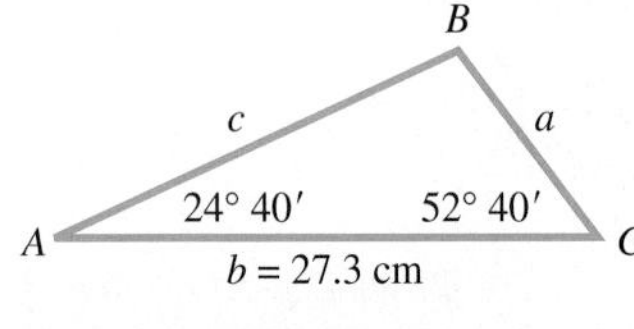

Figure 11

SOLUTION Before the area formula can be used, we must find either a or c. Begin by using the fact that the sum of the measures of the angles of any triangle is 180°.

$180° = A + B + C$	Angle sum formula
$B = 180° - 24° \, 40' - 52° \, 40'$	Substitute and solve for B.
$B = 102° \, 40'$	Subtract.

Classroom Example 9
Find the area of triangle ABC if $B = 58° 10'$, $a = 32.5$ cm, and $C = 73° 30'$.

Answer: 576 cm^2

Next use the law of sines to find a.

$$\frac{a}{\sin A} = \frac{b}{\sin B} \quad \text{Law of sines}$$

Solve for a.

$$\frac{a}{\sin 24° 40'} = \frac{27.3}{\sin 102° 40'} \quad \text{Substitute known values.}$$

$$a = \frac{27.3 \sin 24° 40'}{\sin 102° 40'} \quad \text{Multiply by } \sin 24° 40'.$$

$$a \approx 11.7 \text{ cm} \quad \text{Use a calculator.}$$

Now that we know two sides, a and b, and their included angle C, we find the area.

$$\mathcal{A} = \frac{1}{2}ab \sin C \approx \frac{1}{2}(11.7)(27.3) \sin 52° 40' \approx 127 \text{ cm}^2$$

11.7 is only an approximation. In practice, use your calculator value.

✔ Now Try Exercise 81.

8.1 Exercises

Note to instructor: Although most of the measurements resulting from solving triangles in this chapter are approximations, for convenience we use = rather than ≈ in the answers.

1. C 2. C, D

3. $\sqrt{3}$ 4. $10\sqrt{2}$

5. $C = 95°$, $b = 13$ m, $a = 11$ m
6. $A = 99°$, $b = 34$ cm, $c = 21$ cm

1. *Concept Check* Consider the oblique triangle ABC. Which one of the following proportions is *not* valid?

   A. $\frac{a}{b} = \frac{\sin A}{\sin B}$ **B.** $\frac{a}{\sin A} = \frac{b}{\sin B}$

   C. $\frac{\sin A}{a} = \frac{b}{\sin B}$ **D.** $\frac{\sin A}{a} = \frac{\sin B}{b}$

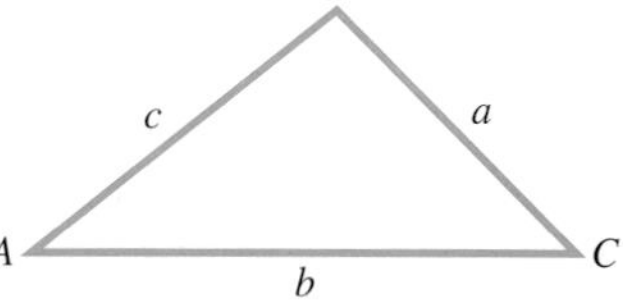

2. *Concept Check* Which two of the following situations do *not* provide sufficient information for solving a triangle by the law of sines?
   A. We are given two angles and the side included between them.
   B. We are given two angles and a side opposite one of them.
   C. We are given two sides and the angle included between them.
   D. We are given three sides.

Find the length of each side a. Do not use a calculator.

3.

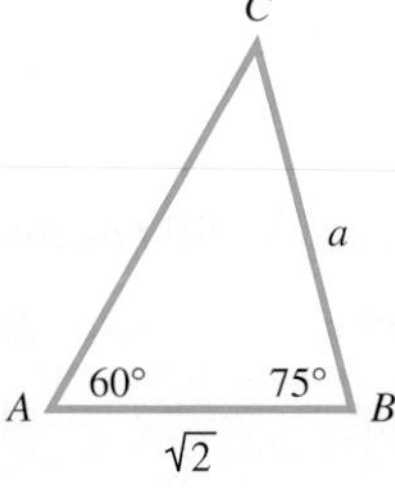

4.

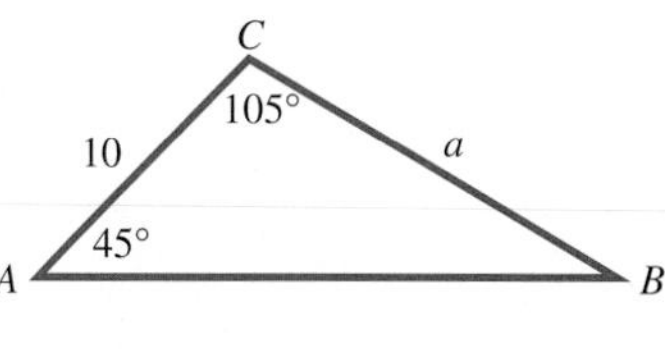

Determine the remaining sides and angles of each triangle ABC. ***See Example 1.***

5.

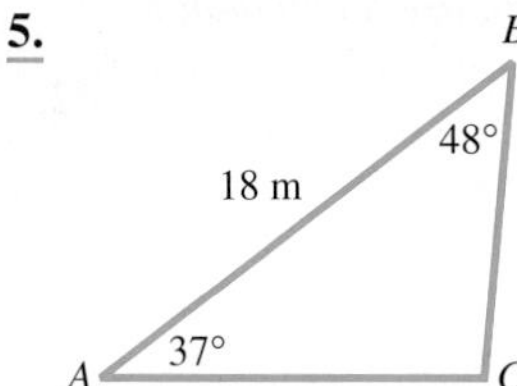

6.

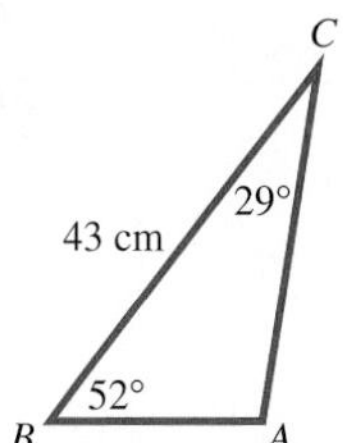

7. $B = 37.3°$, $a = 38.5$ ft, $b = 51.0$ ft
8. $A = 37.2°$, $a = 178$ m, $c = 244$ m
9. $C = 57.36°$, $b = 11.13$ ft, $c = 11.55$ ft
10. $A = 36.54°$, $b = 44.17$ m, $a = 28.10$ m
11. $B = 18.5°$, $a = 239$ yd, $c = 230$ yd
12. $A = 49° 40'$, $b = 16.1$ cm, $c = 25.8$ cm
13. $A = 56° 00'$, $AB = 361$ ft, $BC = 308$ ft
14. $C = 91.9°$, $BC = 490$ ft, $AB = 847$ ft
15. $B = 110.0°$, $a = 27.01$ m, $c = 21.36$ m
16. $A = 65.60°$, $b = 1.942$ cm, $c = 2.727$ cm
17. $A = 34.72°$, $a = 3326$ ft, $c = 5704$ ft
18. $B = 23.75°$, $a = 4663$ yd, $b = 1955$ yd
19. $C = 97° 34'$, $b = 283.2$ m, $c = 415.2$ m
20. $B = 67° 45'$, $b = 37.50$ mm, $a = 22.04$ mm

21. A 22. D

23. (a) $4 < L < 5$ (b) $L = 4$ or $L > 5$ (c) $L < 4$
24. (a) none (b) $L > 5$ (c) $L \leq 5$

25. 1 26. 1
27. 2 28. 2
29. 0 30. 0

31. 45° 32. 30°

7.

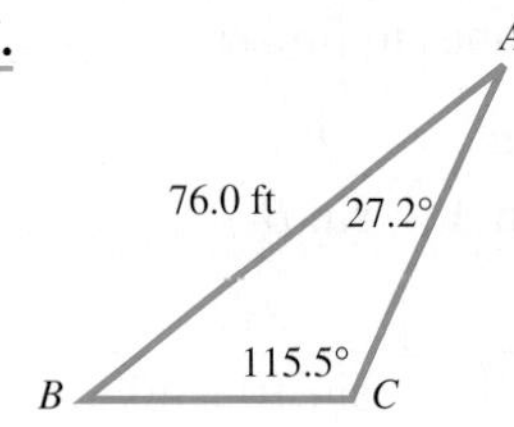

8.

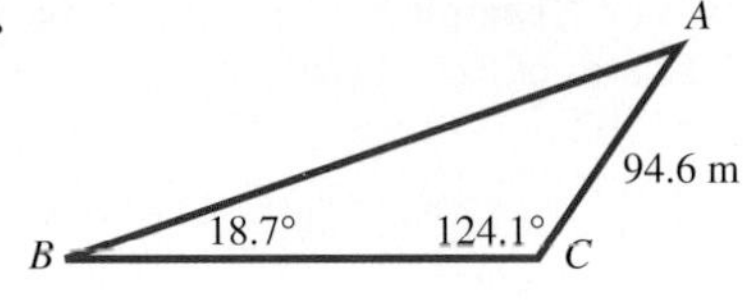

9. $A = 68.41°$, $B = 54.23°$, $a = 12.75$ ft

10. $C = 74.08°$, $B = 69.38°$, $c = 45.38$ m

11. $A = 87.2°$, $b = 75.9$ yd, $C = 74.3°$

12. $B = 38° 40'$, $a = 19.7$ cm, $C = 91° 40'$

13. $B = 20° 50'$, $C = 103° 10'$, $AC = 132$ ft

14. $A = 35.3°$, $B = 52.8°$, $AC = 675$ ft

15. $A = 39.70°$, $C = 30.35°$, $b = 39.74$ m

16. $C = 71.83°$, $B = 42.57°$, $a = 2.614$ cm

17. $B = 42.88°$, $C = 102.40°$, $b = 3974$ ft

18. $C = 50.15°$, $A = 106.1°$, $c = 3726$ yd

19. $A = 39° 54'$, $a = 268.7$ m, $B = 42° 32'$

20. $C = 79° 18'$, $c = 39.81$ mm, $A = 32° 57'$

21. *Concept Check* Which one of the following sets of data does *not* determine a unique triangle?

A. $A = 40°$, $B = 60°$, $C = 80°$ **B.** $a = 5$, $b = 12$, $c = 13$

C. $a = 3$, $b = 7$, $C = 50°$ **D.** $a = 2$, $b = 2$, $c = 2$

22. *Concept Check* Which one of the following sets of data determines a unique triangle?

A. $A = 50°$, $B = 50°$, $C = 80°$ **B.** $a = 3$, $b = 5$, $c = 20$

C. $A = 40°$, $B = 20°$, $C = 30°$ **D.** $a = 7$, $b = 24$, $c = 25$

Concept Check In each figure, a line segment of length L is to be drawn from the given point to the positive x-axis in order to form a triangle. For what value(s) of L can you draw the following?

(a) *two triangles* **(b)** *exactly one triangle* **(c)** *no triangle*

23.

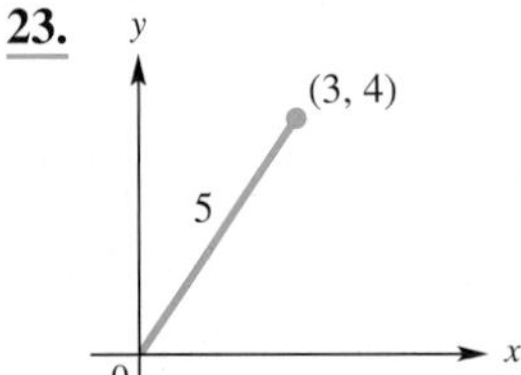

24.

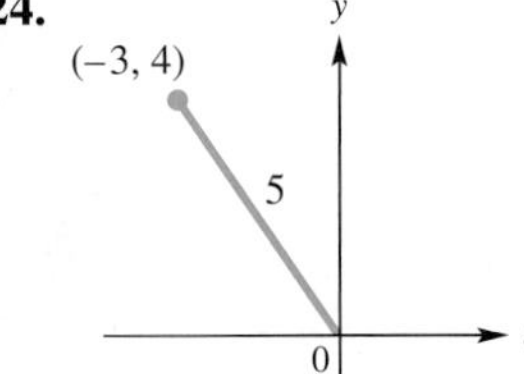

Determine the number of triangles ABC possible with the given parts. **See Examples 4–7.**

25. $a = 50$, $b = 26$, $A = 95°$

26. $a = 35$, $b = 30$, $A = 40°$

27. $a = 31$, $b = 26$, $B = 48°$

28. $B = 54°$, $c = 28$, $b = 23$

29. $a = 50$, $b = 61$, $A = 58°$

30. $b = 60$, $a = 82$, $B = 100°$

Find each angle B. Do not use a calculator.

31.

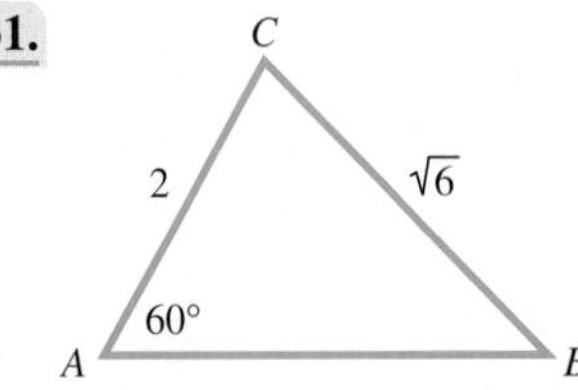

32.

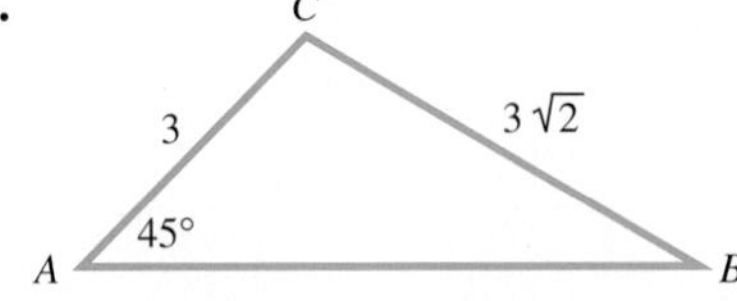

33. $B_1 = 49.1°$, $C_1 = 101.2°$, $B_2 = 130.9°$, $C_2 = 19.4°$
34. $A_1 = 72.2°$, $C_1 = 59.6°$, $A_2 = 107.8°$, $C_2 = 24.0°$
35. $B = 26° 30'$, $A = 112° 10'$
36. $A = 37° 50'$, $C = 93° 20'$
37. no such triangle
38. no such triangle
39. $B = 27.19°$, $C = 10.68°$
40. $A = 45.40°$, $C = 20.88°$
41. $B = 20.6°$, $C = 116.9°$, $c = 20.6$ ft
42. $A = 25.5°$, $B = 102.2°$, $b = 73.9$ yd
43. no such triangle
44. no such triangle
45. $B_1 = 49° 20'$, $C_1 = 92° 00'$, $c_1 = 15.5$ m; $B_2 = 130° 40'$, $C_2 = 10° 40'$, $c_2 = 2.88$ m
46. $A_1 = 55° 20'$, $B_1 = 94° 50'$, $b_1 = 10.4$ m; $A_2 = 124° 40'$, $B_2 = 25° 30'$, $b_2 = 4.51$ m
47. $B = 37.77°$, $C = 45.43°$, $c = 4.174$ ft
48. $B = 30.39°$, $A = 60.91°$, $a = 98.25$ m
49. $A_1 = 53.23°$, $C_1 = 87.09°$, $c_1 = 37.16$ m; $A_2 = 126.77°$, $C_2 = 13.55°$, $c_2 = 8.719$ m
50. $C_1 = 59.71°$, $B_1 = 69.09°$, $b_1 = 8640$ cm; $C_2 = 120.29°$, $B_2 = 8.51°$, $b_2 = 1369$ cm
51. 1; 90°; a right triangle
55. 118 m
56. 448 yd

Find the unknown angles in triangle ABC for each triangle that exists. ***See Examples 4–6.***

33. $A = 29.7°$, $b = 41.5$ ft, $a = 27.2$ ft

34. $B = 48.2°$, $a = 890$ cm, $b = 697$ cm

35. $C = 41° 20'$, $b = 25.9$ m, $c = 38.4$ m

36. $B = 48° 50'$, $a = 3850$ in., $b = 4730$ in.

37. $B = 74.3°$, $a = 859$ m, $b = 783$ m

38. $C = 82.2°$, $a = 10.9$ km, $c = 7.62$ km

39. $A = 142.13°$, $b = 5.432$ ft, $a = 7.297$ ft

40. $B = 113.72°$, $a = 189.6$ yd, $b = 243.8$ yd

Solve each triangle ABC that exists. ***See Examples 4–6.***

41. $A = 42.5°$, $a = 15.6$ ft, $b = 8.14$ ft

42. $C = 52.3°$, $a = 32.5$ yd, $c = 59.8$ yd

43. $B = 72.2°$, $b = 78.3$ m, $c = 145$ m

44. $C = 68.5°$, $c = 258$ cm, $b = 386$ cm

45. $A = 38° 40'$, $a = 9.72$ m, $b = 11.8$ m

46. $C = 29° 50'$, $a = 8.61$ m, $c = 5.21$ m

47. $A = 96.80°$, $b = 3.589$ ft, $a = 5.818$ ft

48. $C = 88.70°$, $b = 56.87$ m, $c = 112.4$ m

49. $B = 39.68°$, $a = 29.81$ m, $b = 23.76$ m

50. $A = 51.20°$, $c = 7986$ cm, $a = 7208$ cm

51. Apply the law of sines to the following: $a = \sqrt{5}$, $c = 2\sqrt{5}$, $A = 30°$. What is the value of sin C? What is the measure of C? Based on its angle measures, what kind of triangle is triangle ABC?

52. Explain the condition that must exist to determine that there is no triangle satisfying the given values of a, b, and B, once the value of sin A is found.

53. Without using the law of sines, explain why no triangle ABC exists satisfying $A = 103° 20'$, $a = 14.6$ ft, $b = 20.4$ ft.

54. Apply the law of sines to the data given in **Example 7.** Describe what happens when you try to find the measure of angle B using a calculator.

Solve each problem. ***See Examples 2 and 3.***

55. *Distance across a River* To find the distance AB across a river, a surveyor laid off a distance $BC = 354$ m on one side of the river. It is found that $B = 112° 10'$ and $C = 15° 20'$. Find AB. See the figure.

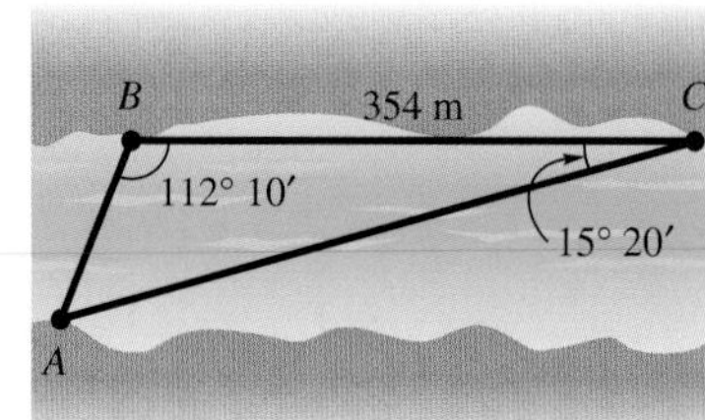

56. *Distance across a Canyon* To determine the distance RS across a deep canyon, Rhonda lays off a distance $TR = 582$ yd. She then finds that $T = 32° 50'$ and $R = 102° 20'$. Find RS.

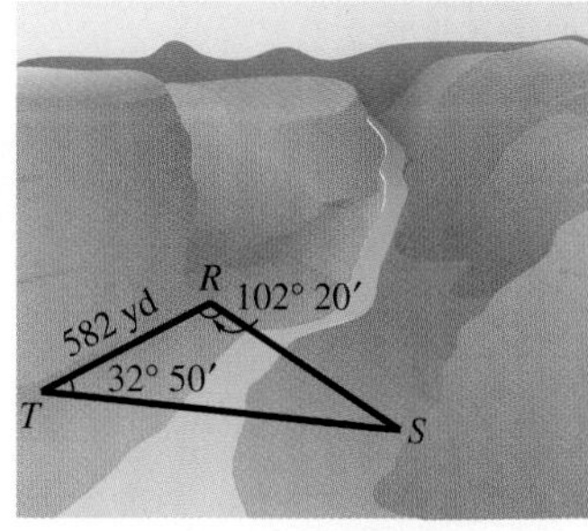

57. 17.8 km
58. 1.93 mi
59. first location: 5.1 mi; second location: 7.2 mi
60. 293.4 m
61. 0.49 mi
62. 10.4 in.
63. 111°
64. 12
65. 664 m

57. *Distance a Ship Travels* A ship is sailing due north. At a certain point the bearing of a lighthouse 12.5 km away is N 38.8° E. Later on, the captain notices that the bearing of the lighthouse has become S 44.2° E. How far did the ship travel between the two observations of the lighthouse?

58. *Distance between Radio Direction Finders* Radio direction finders are placed at points A and B, which are 3.46 mi apart on an east-west line, with A west of B. From A the bearing of a certain radio transmitter is 47.7°, and from B the bearing is 302.5°. Find the distance of the transmitter from A.

59. *Distance between a Ship and a Lighthouse* The bearing of a lighthouse from a ship was found to be N 37° E. After the ship sailed 2.5 mi due south, the new bearing was N 25° E. Find the distance between the ship and the lighthouse at each location.

60. *Distance across a River* Standing on one bank of a river flowing north, Mark notices a tree on the opposite bank at a bearing of 115.45°. Lisa is on the same bank as Mark, but 428.3 m away. She notices that the bearing of the tree is 45.47°. The two banks are parallel. What is the distance across the river?

61. *Height of a Balloon* A balloonist is directly above a straight road 1.5 mi long that joins two villages. She finds that the town closer to her is at an angle of depression of 35°, and the farther town is at an angle of depression of 31°. How high above the ground is the balloon?

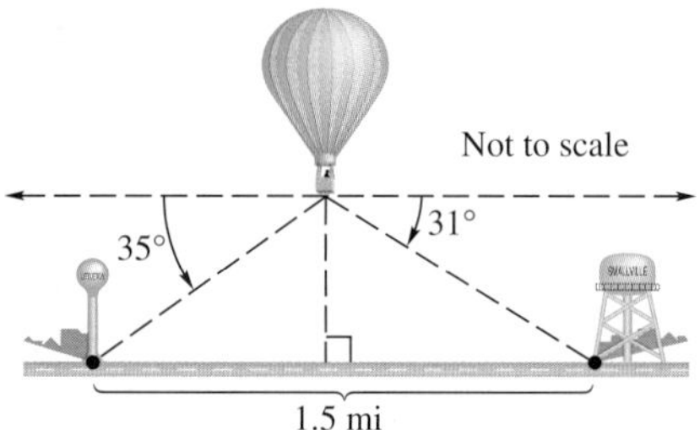

62. *Measurement of a Folding Chair* A folding chair is to have a seat 12.0 in. deep with angles as shown in the figure. How far down from the seat should the crossing legs be joined? (Find length x in the figure.)

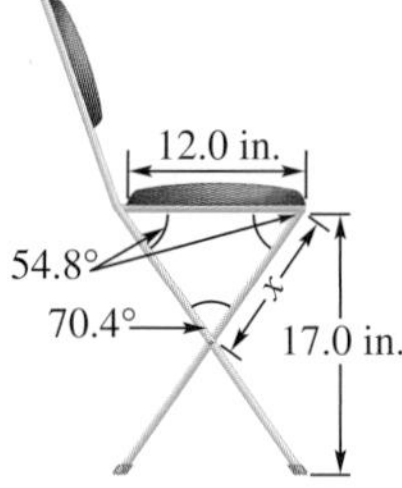

63. *Angle Formed by Radii of Gears* Three gears are arranged as shown in the figure. Find angle θ.

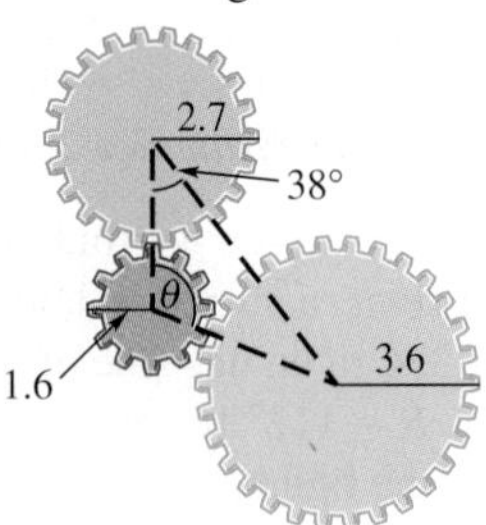

64. *Distance between Atoms* Three atoms with atomic radii of 2.0, 3.0, and 4.5 are arranged as in the figure. Find the distance between the centers of atoms A and C.

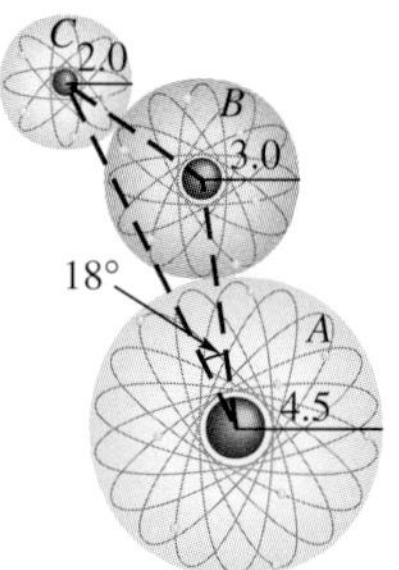

65. *Distance between Inaccessible Points* To find the distance between a point X and an inaccessible point Z, a line segment XY is constructed. It is found that $XY = 960$ m, angle $XYZ = 43° 30'$, and angle $YZX = 95° 30'$. Find the distance between X and Z to the nearest meter.

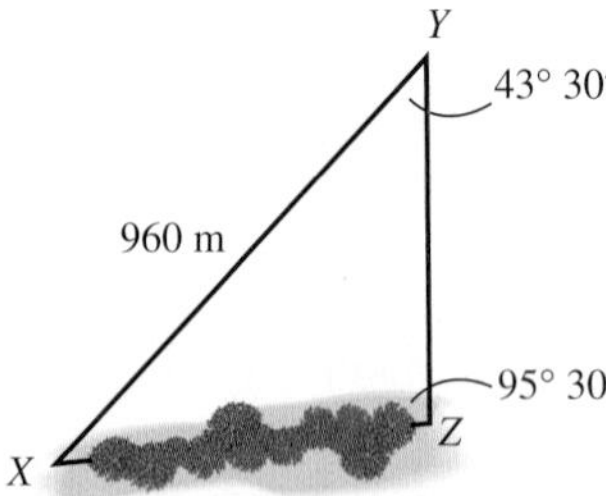

66. 87.3 ft
67. 218 ft
68. 75° 00′; 274 mi
69. The distance is about 419,000 km, which compares favorably to the actual value.
70. 10,285 ft
71. $\frac{\sqrt{3}}{2}$ sq unit
72. $\sqrt{3}$ sq units
73. $\frac{\sqrt{2}}{2}$ sq unit
74. 1 sq unit

66. *Height of an Antenna Tower* The angle of elevation from the top of a building 45.0 ft high to the top of a nearby antenna tower is 15° 20′. From the base of the building, the angle of elevation of the tower is 29° 30′. Find the height of the tower.

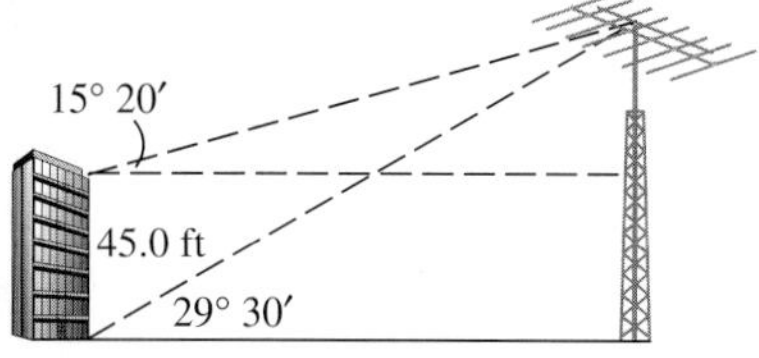

67. *Height of a Building* A flagpole 95.0 ft tall is on the top of a building. From a point on level ground, the angle of elevation of the top of the flagpole is 35.0°, and the angle of elevation of the bottom of the flagpole is 26.0°. Find the height of the building.

68. *Flight Path of a Plane* A pilot flies her plane on a heading of 35° 00′ from point X to point Y, which is 400 mi from X. Then she turns and flies on a heading of 145° 00′ to point Z, which is 400 mi from her starting point X. What is the heading of Z from X, and what is the distance YZ?

69. *Distance to the Moon* Since the moon is a relatively close celestial object, its distance can be measured directly by taking two different photographs at precisely the same time from two different locations. The moon will have a different angle of elevation at each location. On April 29, 1976, at 11:35 A.M., the lunar angles of elevation during a partial solar eclipse at Bochum in upper Germany and at Donaueschingen in lower Germany were measured as 52.6997° and 52.7430°, respectively. The two cities are 398 km apart. Calculate the distance to the moon from Bochum on this day, and compare it with the actual value of 406,000 km. Disregard the curvature of Earth in this calculation. (*Source:* Scholosser, W., T. Schmidt-Kaler, and E. Milone, *Challenges of Astronomy*, Springer-Verlag.)

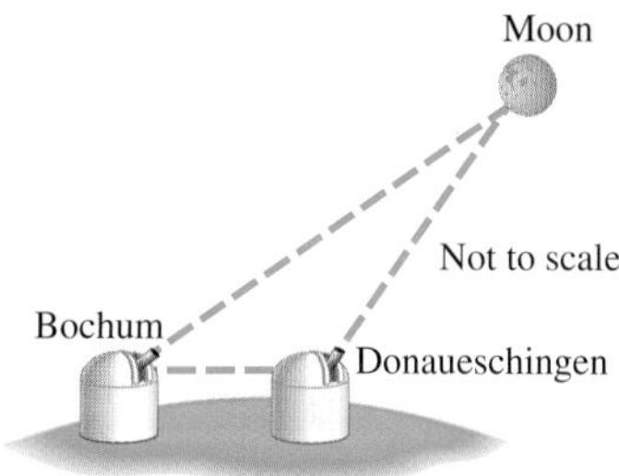

70. *Ground Distances Measured by Aerial Photography* The distance covered by an aerial photograph is determined by both the focal length of the camera and the tilt of the camera from the perpendicular to the ground. A camera lens with a 12-in. focal length will have an angular coverage of 60°. If an aerial photograph is taken with this camera tilted $\theta = 35°$ at an altitude of 5000 ft, calculate to the nearest foot the ground distance d that will be shown in this photograph. (*Source:* Brooks, R. and D. Johannes, *Phytoarchaeology*, Dioscorides Press.)

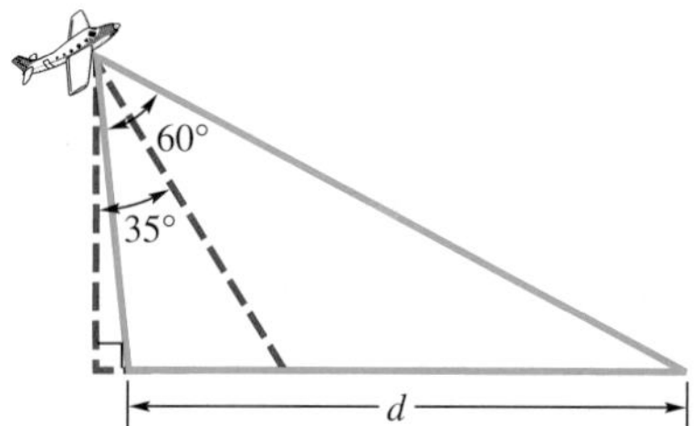

Find the area of each triangle using the formula $\mathcal{A} = \frac{1}{2}bh$, and then verify that the formula $\mathcal{A} = \frac{1}{2}ab \sin C$ gives the same result.

71.

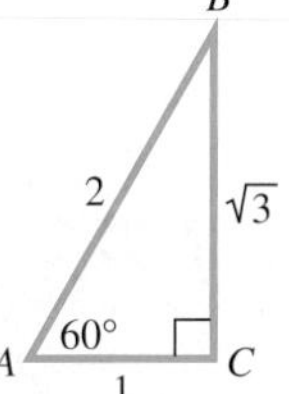

72.

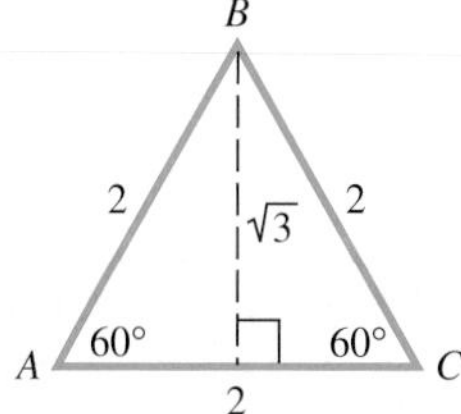

73.

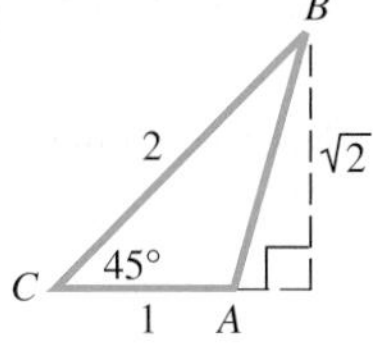

74.

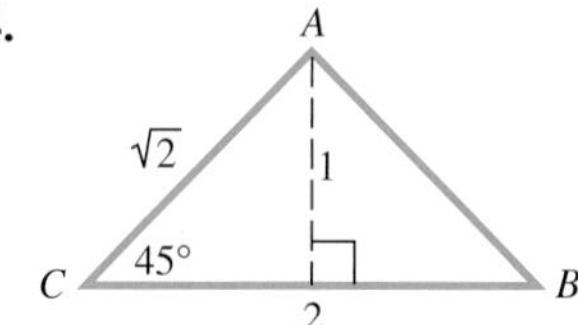

75. 46.4 m^2 76. 732 ft^2
77. 356 cm^2 78. 163 km^2
79. 722.9 $in.^2$ 80. 289.9 m^2
81. 65.94 cm^2 82. 84.41 m^2
83. 100 m^2 84. 373 m^2
85. $a = \sin A$, $b = \sin B$, $c = \sin C$

Find the area of each triangle ABC. ***See Examples 8 and 9.***

75. $A = 42.5°$, $b = 13.6$ m, $c = 10.1$ m

76. $C = 72.2°$, $b = 43.8$ ft, $a = 35.1$ ft

77. $B = 124.5°$, $a = 30.4$ cm, $c = 28.4$ cm

78. $C = 142.7°$, $a = 21.9$ km, $b = 24.6$ km

79. $A = 56.80°$, $b = 32.67$ in., $c = 52.89$ in.

80. $A = 34.97°$, $b = 35.29$ m, $c = 28.67$ m

81. $A = 30.50°$, $b = 13.00$ cm, $C = 112.60°$

82. $A = 59.80°$, $b = 15.00$ m, $C = 53.10°$

Solve each problem.

83. *Area of a Metal Plate* A painter is going to apply a special coating to a triangular metal plate on a new building. Two sides measure 16.1 m and 15.2 m. She knows that the angle between these sides is 125°. What is the area of the surface she plans to cover with the coating?

84. *Area of a Triangular Lot* A real estate agent wants to find the area of a triangular lot. A surveyor takes measurements and finds that two sides are 52.1 m and 21.3 m, and the angle between them is 42.2°. What is the area of the triangular lot?

85. *Triangle Inscribed in a Circle* For a triangle inscribed in a circle of radius r, the law of sines ratios $\frac{a}{\sin A}$, $\frac{b}{\sin B}$, and $\frac{c}{\sin C}$ have value $2r$. The circle in the figure has diameter 1. What are the values of a, b, and c? (*Note:* This result provides an alternative way to define the sine function for angles between 0° and 180°. It was used nearly 2000 years ago by the mathematician Ptolemy to construct one of the earliest trigonometric tables.)

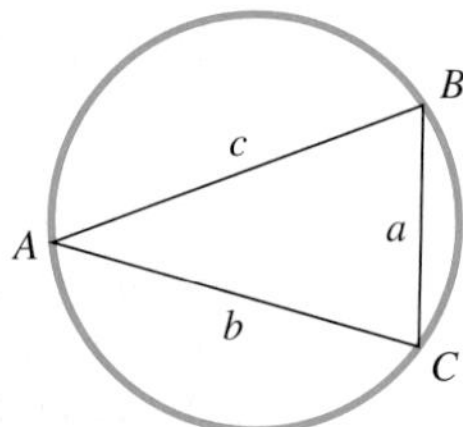

86. *Theorem of Ptolemy* The following theorem is also attributed to Ptolemy: *In a quadrilateral inscribed in a circle, the product of the diagonals is equal to the sum of the products of the opposite sides.* (*Source:* Eves, H., *An Introduction to the History of Mathematics,* Sixth Edition, Saunders College Publishing.) The circle in the figure has diameter 1. Explain why the lengths of the line segments are as shown, and then apply Ptolemy's theorem to derive the formula for the sine of the sum of two angles.

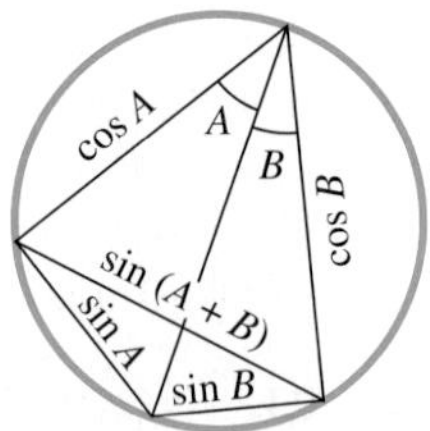

Use the law of sines to prove that each statement is true for any triangle ABC, with corresponding sides a, b, and c.

87. $\frac{a+b}{b} = \frac{\sin A + \sin B}{\sin B}$

88. $\frac{a-b}{a+b} = \frac{\sin A - \sin B}{\sin A + \sin B}$

90. $\mathscr{A} = 1.12257R^2$
91. (a) 8.77 in.^{2} (b) 5.32 in.^{2}
92. red

Relating Concepts

For individual or collaborative investigation. *(Exercises 89–92)*

Colors of the U.S. Flag The flag of the United States includes the colors red, white, and blue. Which color is predominant? Clearly the answer is either red or white. (It can be shown that only 18.73% of the total area is blue.) (Source: *Banks, R.,* Slicing Pizzas, Racing Turtles, and Further Adventures in Applied Mathematics, *Princeton University Press.) Work* ***Exercises 89–92 in order*** *to determine the answer to this question.*

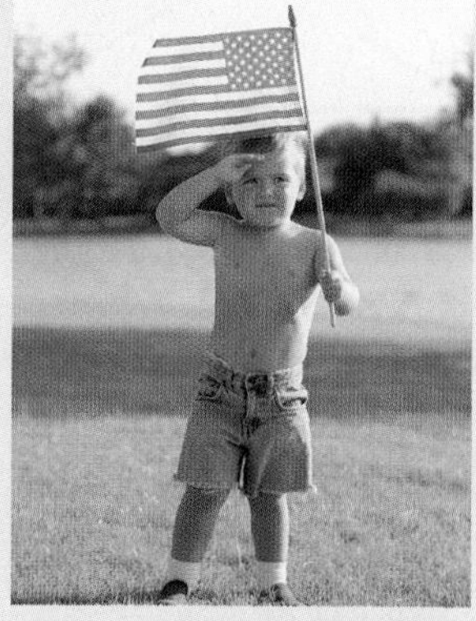

89. Let R denote the radius of the circumscribing circle of a five-pointed star appearing on the American flag. The star can be decomposed into ten congruent triangles. In the figure, r is the radius of the circumscribing circle of the pentagon in the interior of the star. Show that the area of a star is

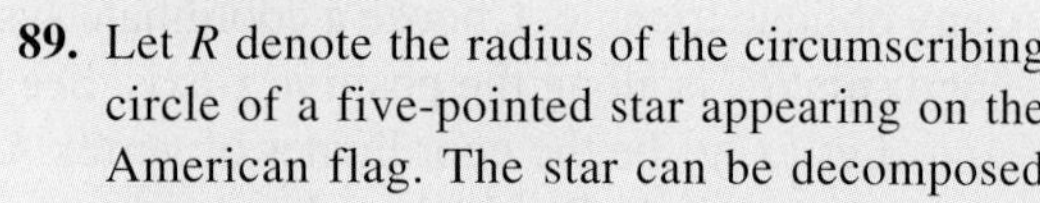

$$\mathscr{A} = \left[5\frac{\sin A \sin B}{\sin(A+B)}\right]R^2.$$

(*Hint:* $\sin C = \sin[180° - (A+B)] = \sin(A+B)$.)

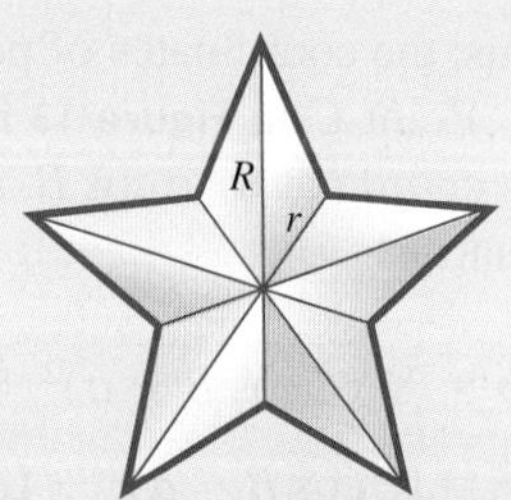

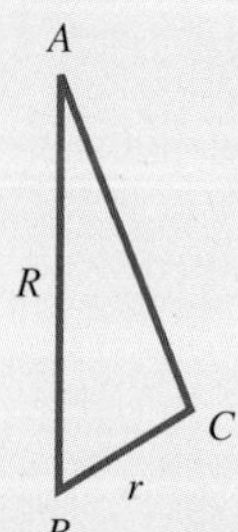

90. Angles A and B have values 18° and 36°, respectively. Express the area $\mathscr{A}$ of a star in terms of its radius, R.

91. To determine whether red or white is predominant, we must know the measurements of the flag. Consider a flag of width 10 in., length 19 in., length of each upper stripe 11.4 in., and radius R of the circumscribing circle of each star 0.308 in. The thirteen stripes consist of six matching pairs of red and white stripes and one additional red, upper stripe. Therefore, we must compare the area of a red, upper stripe with the total area of the 50 white stars.

(a) Compute the area of the red, upper stripe.
(b) Compute the total area of the 50 white stars.

92. Which color occupies the greatest area on the flag?

8.2 The Law of Cosines

- Derivation of the Law of Cosines
- Using the Law of Cosines
- Heron's Formula for the Area of a Triangle
- Derivation of Heron's Formula

As mentioned in **Section 8.1,** if we are given two sides and the included angle (Case 3) or three sides (Case 4) of a triangle, then a unique triangle is determined. These are the SAS and SSS cases, respectively. Both cases require using the *law of cosines.*

The property of triangles given at the top of the next page is important when applying the law of cosines to solve a triangle.

Triangle Side Length Restriction

In any triangle, the sum of the lengths of any two sides must be greater than the length of the remaining side.

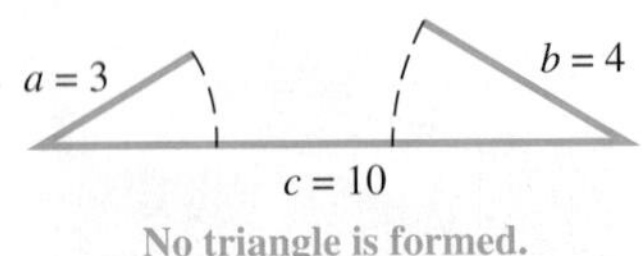

Figure 12

For example, it would be impossible to construct a triangle with sides of lengths 3, 4, and 10. See **Figure 12.**

Derivation of the Law of Cosines To derive the law of cosines, let ABC be any oblique triangle. Choose a coordinate system so that vertex B is at the origin and side BC is along the positive x-axis. See **Figure 13.**

Let (x, y) be the coordinates of vertex A of the triangle. Then the following are true for angle B, whether obtuse or acute.

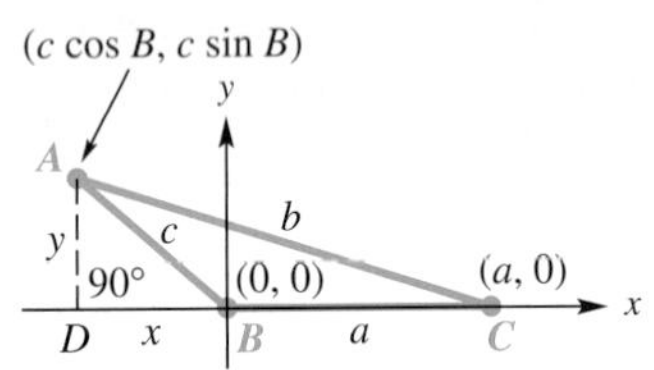

Figure 13

$$\sin B = \frac{y}{c} \quad \text{and} \quad \cos B = \frac{x}{c} \qquad \text{(Section 5.2)}$$

$$y = c \sin B \quad \text{and} \quad x = c \cos B \qquad \text{Here } x \text{ is negative when } B \text{ is obtuse.}$$

Thus, the coordinates of point A become $(c \cos B, c \sin B)$.

Point C in **Figure 13** has coordinates $(a, 0)$, AC has length b, and point A has coordinates $(c \cos B, c \sin B)$. We can use the distance formula to write an equation.

$$b = \sqrt{(c \cos B - a)^2 + (c \sin B - 0)^2} \qquad \text{(Section 2.1)}$$

$$b^2 = (c \cos B - a)^2 + (c \sin B)^2 \qquad \text{Square each side. (Section 1.6)}$$

$$= (c^2 \cos^2 B - 2ac \cos B + a^2) + c^2 \sin^2 B \qquad \text{Multiply (Section R.3); } (x - y)^2 = x^2 - 2xy + y^2$$

Remember the middle term.

$$= a^2 + c^2(\cos^2 B + \sin^2 B) - 2ac \cos B \qquad \text{Properties of real numbers (Section R.2)}$$

$$= a^2 + c^2(1) - 2ac \cos B \qquad \text{Fundamental identity (Section 7.1)}$$

$$b^2 = a^2 + c^2 - 2ac \cos B$$

This result is one form of the law of cosines. In our work, we could just as easily have placed A or C at the origin. This would have given the same result, but with the variables rearranged.

Law of Cosines

In any triangle ABC, with sides a, b, and c, the following hold.

$$a^2 = b^2 + c^2 - 2bc \cos A,$$

$$b^2 = a^2 + c^2 - 2ac \cos B,$$

$$c^2 = a^2 + b^2 - 2ab \cos C$$

That is, according to the law of cosines, the square of a side of a triangle is equal to the sum of the squares of the other two sides, minus twice the product of those two sides and the cosine of the angle included between them.

Classroom Example 1
Two boats leave a harbor at the same time, traveling on courses that make an angle of 82° 20′ between them. When the slower boat has traveled 62.5 km, the faster one has traveled 79.4 km. At that time, what is the distance between the boats?

Answer: 94.3 km

NOTE If we let $C = 90°$ in the third form of the law of cosines, then $\cos C = \cos 90° = 0$, and the formula becomes $c^2 = a^2 + b^2$, the Pythagorean theorem **(Section 1.5).** The Pythagorean theorem is a special case of the law of cosines.

Using the Law of Cosines

EXAMPLE 1 Applying the Law of Cosines (SAS)

A surveyor wishes to find the distance between two inaccessible points A and B on opposite sides of a lake. While standing at point C, she finds that $b = 259$ m, $a = 423$ m, and angle ACB measures 132° 40′. Find the distance c. See **Figure 14.**

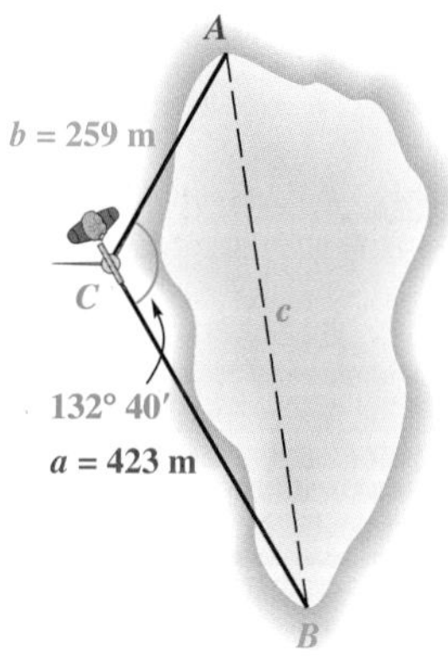

Figure 14

SOLUTION We can use the law of cosines here because we know the lengths of two sides of the triangle and the measure of the included angle.

$$c^2 = a^2 + b^2 - 2ab \cos C \quad \text{Law of cosines}$$

$$c^2 = 423^2 + 259^2 - 2(423)(259) \cos 132° \, 40' \quad \text{Substitute.}$$

$$c^2 \approx 394{,}510.6 \quad \text{Use a calculator.}$$

$$c \approx 628 \quad \text{Take the square root of each side. Choose the positive root. (Section 1.4)}$$

The distance between the points is approximately 628 m.

✔ *Now Try Exercise 39.*

EXAMPLE 2 Applying the Law of Cosines (SAS)

Solve triangle ABC if $A = 42.3°$, $b = 12.9$ m, and $c = 15.4$ m.

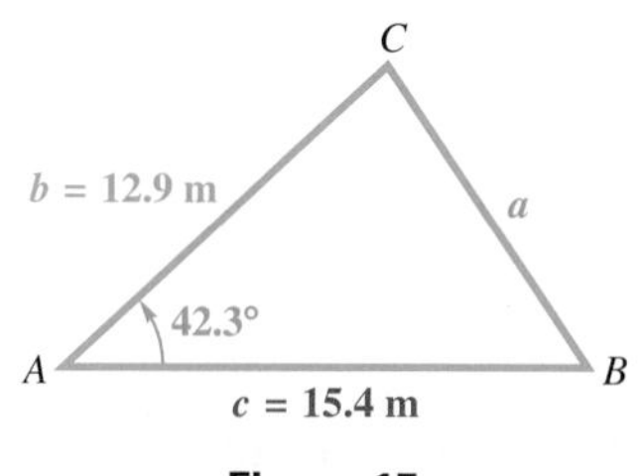

Figure 15

SOLUTION See **Figure 15.** We start by finding a with the law of cosines.

$$a^2 = b^2 + c^2 - 2bc \cos A \quad \text{Law of cosines}$$

$$a^2 = 12.9^2 + 15.4^2 - 2(12.9)(15.4) \cos 42.3° \quad \text{Substitute.}$$

$$a^2 \approx 109.7 \quad \text{Use a calculator.}$$

$$a \approx 10.47 \text{ m} \quad \text{Take square roots and choose the positive root.}$$

Of the two remaining angles B and C, B must be the smaller since it is opposite the shorter of the two sides b and c. Therefore, B cannot be obtuse.

Classroom Example 2
Solve triangle ABC if $B = 73.5°$, $a = 28.2$ ft, and $c = 46.7$ ft.

Answer: $b = 47.2$ ft, $A = 34.9°$, $C = 71.6°$

$$\frac{\sin A}{a} = \frac{\sin B}{b} \quad \text{Law of sines (alternative form) (Section 8.1)}$$

$$\frac{\sin 42.3°}{10.47} = \frac{\sin B}{12.9} \quad \text{Substitute.}$$

$$\sin B = \frac{12.9 \sin 42.3°}{10.47} \quad \text{Multiply by 12.9 and rewrite.}$$

$$B \approx 56.0° \quad \text{Use the inverse sine function. (Section 7.5)}$$

The easiest way to find C is to subtract the measures of A and B from 180°.

$$C = 180° - A - B \quad \text{Angle sum formula, solved for } C$$

$$C \approx 180° - 42.3° - 56.0° \quad \text{Substitute.}$$

$$C \approx 81.7° \quad \text{Subtract.}$$

✔ *Now Try Exercise 19.*

Teaching Tip Emphasize that once the law of cosines has been used, students do not need to be concerned about the ambiguous case of the law of sines.

CAUTION Had we used the law of sines to find C rather than B in **Example 2,** we would not have known whether C was equal to $81.7°$ or its supplement, $98.3°$.

Classroom Example 3
Solve triangle ABC if $a = 25.4$ cm, $b = 42.8$ cm, and $c = 59.3$ cm.

Answer: $A = 22.1°$, $B = 39.3°$, $C = 118.6°$

EXAMPLE 3 Applying the Law of Cosines (SSS)

Solve triangle ABC if $a = 9.47$ ft, $b = 15.9$ ft, and $c = 21.1$ ft.

SOLUTION We can use the law of cosines to solve for any angle of the triangle. We solve for C, the largest angle. We will know that C is obtuse if $\cos C < 0$.

$$c^2 = a^2 + b^2 - 2ab \cos C \quad \text{Law of cosines}$$

$$\cos C = \frac{a^2 + b^2 - c^2}{2ab} \quad \text{Solve for } \cos C.$$

$$\cos C = \frac{9.47^2 + 15.9^2 - 21.1^2}{2(9.47)(15.9)} \quad \text{Substitute.}$$

$$\cos C \approx -0.34109402 \quad \text{Use a calculator.}$$

$$C \approx 109.9° \quad \text{Use the inverse cosine function. (Section 7.5)}$$

Now use the law of sines to find B.

$$\frac{\sin C}{c} = \frac{\sin B}{b} \quad \text{Law of sines (alternative form)}$$

$$\frac{\sin 109.9°}{21.1} = \frac{\sin B}{15.9} \quad \text{Substitute.}$$

$$\sin B = \frac{15.9 \sin 109.9°}{21.1} \quad \text{Multiply by 15.9 and rewrite.}$$

$$B \approx 45.1° \quad \text{Use the inverse sine function.}$$

Since $A = 180° - B - C$, we have $A \approx 180° - 45.1° - 109.9° \approx 25.0°$.

Now Try Exercise 23.

Figure 16

Trusses are frequently used to support roofs on buildings, as illustrated in **Figure 16.** The simplest type of roof truss is a triangle, as shown in **Figure 17.** (*Source:* Riley, W., L. Sturges, and D. Morris, *Statics and Mechanics of Materials,* John Wiley and Sons.)

Classroom Example 4
Find angle C to the nearest degree for the truss shown in **Figure 17.**

Answer: $55°$

EXAMPLE 4 Designing a Roof Truss (SSS)

Find angle B to the nearest degree for the truss shown in **Figure 17.**

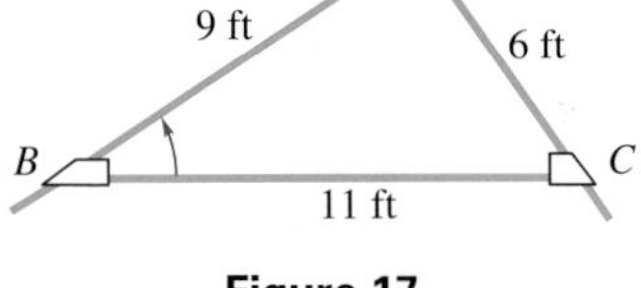

Figure 17

SOLUTION

$$b^2 = a^2 + c^2 - 2ac \cos B \quad \text{Law of cosines}$$

$$\cos B = \frac{a^2 + c^2 - b^2}{2ac} \quad \text{Solve for } \cos B.$$

$$\cos B = \frac{11^2 + 9^2 - 6^2}{2(11)(9)} \quad \text{Let } a = 11,\ b = 6, \text{ and } c = 9.$$

$$\cos B \approx 0.83838384 \quad \text{Use a calculator.}$$

$$B \approx 33° \quad \text{Use the inverse cosine function.}$$

Now Try Exercise 49.

Four possible cases can occur when we solve an oblique triangle. They are summarized in the following table. In all four cases, it is assumed that the given information actually produces a triangle.

Teaching Tip Ask students to outline the steps that they would take to solve each triangle. Do not actually solve the triangle.

1. $A = 32°, B = 67°, a = 20$
2. $b = 22, c = 15, A = 39°$
3. $a = 5, c = 7, C = 22°$
4. $b = 13, c = 15, B = 22°$
5. $a = 5, b = 7, c = 9$

Oblique Triangle	Suggested Procedure for Solving
Case 1: One side and two angles are known. **(SAA or ASA)**	***Step 1*** Find the remaining angle using the angle sum formula $(A + B + C = 180°)$. ***Step 2*** Find the remaining sides using the law of sines.
Case 2: Two sides and one angle (not included between the two sides) are known. **(SSA)**	*This is the ambiguous case. There may be no triangle, one triangle, or two triangles.* ***Step 1*** Find an angle using the law of sines. ***Step 2*** Find the remaining angle using the angle sum formula. ***Step 3*** Find the remaining side using the law of sines. *If two triangles exist, repeat Steps 2 and 3.*
Case 3: Two sides and the included angle are known. **(SAS)**	***Step 1*** Find the third side using the law of cosines. ***Step 2*** Find the smaller of the two remaining angles using the law of sines. ***Step 3*** Find the remaining angle using the angle sum formula.
Case 4: Three sides are known. **(SSS)**	***Step 1*** Find the largest angle using the law of cosines. ***Step 2*** Find either remaining angle using the law of sines. ***Step 3*** Find the remaining angle using the angle sum formula.

Teaching Tip Heron's formula, while intended for use with oblique triangles, can be easily illustrated by finding the area of a right triangle (i.e., apply Heron's formula and $\mathcal{A} = \frac{1}{2}bh$).

Heron's Formula for the Area of a Triangle A formula for finding the area of a triangle given the lengths of the three sides, known as **Heron's formula,** is named after the Greek mathematician Heron of Alexandria, who lived around A.D. 75. It is found in his work *Metrica*. Heron's formula can be used for the case SSS.

Heron's Area Formula (SSS)

If a triangle has sides of lengths a, b, and c, with **semiperimeter**

$$s = \frac{1}{2}(a + b + c),$$

then the area $\mathcal{A}$ of the triangle is given by the following formula.

$$\mathcal{A} = \sqrt{s(s - a)(s - b)(s - c)}$$

That is, according to Heron's formula, the area of a triangle is the square root of the product of four factors: (1) the semiperimeter, (2) the semiperimeter minus the first side, (3) the semiperimeter minus the second side, and (4) the semiperimeter minus the third side.

A derivation of Heron's formula is given at the end of this section.

Classroom Example 5
The distance "as the crow flies" from Chicago to St. Louis is 262 mi, from St. Louis to New Orleans is 599 mi, and from New Orleans to Chicago is 834 mi. What is the area of the triangular region having these three cities as vertices?

Answer: about 40,800 mi^2

EXAMPLE 5 Using Heron's Formula to Find an Area (SSS)

The distance "as the crow flies" from Los Angeles to New York is 2451 mi, from New York to Montreal is 331 mi, and from Montreal to Los Angeles is 2427 mi. What is the area of the triangular region having these three cities as vertices? (Ignore the curvature of Earth.)

SOLUTION In **Figure 18,** we let $a = 2451$, $b = 331$, and $c = 2427$.

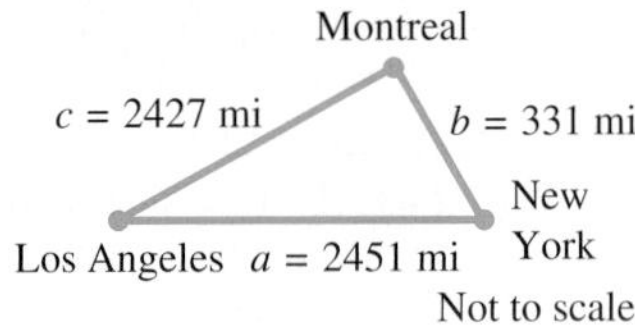

Figure 18

Here,

$$s = \frac{1}{2}(2451 + 331 + 2427) \quad \text{Semiperimeter}$$

$$s = 2604.5. \quad \text{Add, and then multiply.}$$

Now use Heron's formula to find the area $\mathscr{A}$.

$$\mathscr{A} = \sqrt{s(s-a)(s-b)(s-c)}$$

Don't forget the factor s.

$$\mathscr{A} = \sqrt{2604.5(2604.5 - 2451)(2604.5 - 331)(2604.5 - 2427)}$$

$$\mathscr{A} \approx 401{,}700 \text{ mi}^2 \quad \text{Use a calculator.}$$

✔ *Now Try Exercise 73.*

Derivation of Heron's Formula A trigonometric derivation of Heron's formula illustrates some ingenious manipulation involving the law of cosines, algebraic techniques, double-angle identities, and the area formula $\mathscr{A} = \frac{1}{2}bc \sin A$.

Let triangle ABC have sides of lengths a, b, and c. Apply the law of cosines.

$$a^2 = b^2 + c^2 - 2bc \cos A \quad \text{Law of cosines}$$

$$\cos A = \frac{b^2 + c^2 - a^2}{2bc} \quad \text{Solve for } \cos A. \quad (1)$$

The perimeter of the triangle is $a + b + c$, so half of the perimeter (the semi-perimeter) is given by the formula in equation (2) below.

$$s = \frac{1}{2}(a + b + c) \quad (2)$$

$$2s = a + b + c \quad \text{Multiply by 2.} \quad (3)$$

$$b + c - a = 2s - 2a \quad \text{Subtract } 2a \text{ from each side and rewrite.}$$

$$b + c - a = 2(s - a) \quad \text{Factor. } \textbf{(Section R.4)} \quad (4)$$

Subtract $2b$ and $2c$ in a similar way in equation (3) to obtain equations (5) and (6).

$$a - b + c = 2(s - b) \quad (5)$$

$$a + b - c = 2(s - c) \quad (6)$$

Now we obtain an expression for $1 - \cos A$.

$$1 - \cos A = 1 - \underbrace{\frac{b^2 + c^2 - a^2}{2bc}}_{\cos A,\text{ from }(1)}$$

$$= \frac{2bc + a^2 - b^2 - c^2}{2bc} \quad \text{Find a common denominator, and distribute the } - \text{ sign.}$$

$$= \frac{a^2 - (b^2 - 2bc + c^2)}{2bc} \quad \text{Regroup.}$$

Pay attention to signs.

$$= \frac{a^2 - (b - c)^2}{2bc} \quad \text{Factor the perfect square trinomial. (Section R.4)}$$

$$= \frac{[a - (b - c)][a + (b - c)]}{2bc} \quad \text{Factor the difference of squares.}$$

$$= \frac{(a - b + c)(a + b - c)}{2bc} \quad \text{Distributive property}$$

$$= \frac{2(s - b) \cdot 2(s - c)}{2bc} \quad \text{From (5) and (6)}$$

$$1 - \cos A = \frac{2(s - b)(s - c)}{bc} \quad \text{Lowest terms} \quad (7)$$

Similarly, it can be shown that

$$1 + \cos A = \frac{2s(s - a)}{bc}. \quad (8)$$

Recall the double-angle identities for $\cos 2\theta$ from **Section 7.4.**

$$\cos 2\theta = 2\cos^2\theta - 1$$

$$\cos A = 2\cos^2\left(\frac{A}{2}\right) - 1 \quad \text{Let } \theta = \tfrac{A}{2}.$$

$$1 + \cos A = 2\cos^2\left(\frac{A}{2}\right) \quad \text{Add 1.}$$

$$\underbrace{\frac{2s(s - a)}{bc}}_{\text{From (8)}} = 2\cos^2\left(\frac{A}{2}\right) \quad \text{Substitute.}$$

$$\frac{s(s - a)}{bc} = \cos^2\left(\frac{A}{2}\right) \quad \text{Divide by 2.}$$

$$\cos\left(\frac{A}{2}\right) = \sqrt{\frac{s(s - a)}{bc}} \quad (9)$$

$$\cos 2\theta = 1 - 2\sin^2\theta$$

$$\cos A = 1 - 2\sin^2\left(\frac{A}{2}\right) \quad \text{Let } \theta = \tfrac{A}{2}.$$

$$1 - \cos A = 2\sin^2\left(\frac{A}{2}\right) \quad \text{Subtract 1. Multiply by } -1.$$

$$\underbrace{\frac{2(s - b)(s - c)}{bc}}_{\text{From (7)}} = 2\sin^2\left(\frac{A}{2}\right) \quad \text{Substitute.}$$

$$\frac{(s - b)(s - c)}{bc} = \sin^2\left(\frac{A}{2}\right) \quad \text{Divide by 2.}$$

$$\sin\left(\frac{A}{2}\right) = \sqrt{\frac{(s - b)(s - c)}{bc}} \quad (10)$$

The area of triangle ABC can be expressed as follows.

$$\mathcal{A} = \frac{1}{2}bc\sin A \quad \text{(Section 8.1)}$$

$$2\mathcal{A} = bc\sin A \quad \text{Multiply by 2.}$$

$$\frac{2\mathcal{A}}{bc} = \sin A \quad \text{Divide by } bc. \quad (11)$$

Recall the double-angle identity for $\sin 2\theta$.

$$\sin 2\theta = 2 \sin \theta \cos \theta \qquad \text{(Section 7.4)}$$

$$\sin A = 2 \sin\left(\frac{A}{2}\right) \cos\left(\frac{A}{2}\right) \qquad \text{Let } \theta = \tfrac{A}{2}.$$

$$\frac{2\mathcal{A}}{bc} = 2 \sin\left(\frac{A}{2}\right) \cos\left(\frac{A}{2}\right) \qquad \text{Use equation (11).}$$

$$\frac{2\mathcal{A}}{bc} = 2\sqrt{\frac{(s-b)(s-c)}{bc}} \cdot \sqrt{\frac{s(s-a)}{bc}} \qquad \text{Use equations (9) and (10).}$$

$$\frac{2\mathcal{A}}{bc} = 2\sqrt{\frac{s(s-a)(s-b)(s-c)}{b^2c^2}} \qquad \text{Multiply.}$$

$$\frac{2\mathcal{A}}{bc} = \frac{2\sqrt{s(s-a)(s-b)(s-c)}}{bc} \qquad \text{Simplify the denominator.}$$

$$\mathcal{A} = \sqrt{s(s-a)(s-b)(s-c)} \qquad \text{Heron's formula}$$

8.2 Exercises

1. (a) SAS (b) law of cosines
2. (a) SAA (b) law of sines
3. (a) SSA (b) law of sines
4. (a) ASA (b) law of sines
5. (a) ASA (b) law of sines
6. (a) SSA (b) law of sines
7. (a) SSS (b) law of cosines
8. (a) SAS (b) law of cosines

9. 5 10. 7

11. 120° 12. 30°

13. $a = 7.0$, $B = 37.6°$, $C = 21.4°$
14. $a = 5.4$, $B = 40.7°$, $C = 78.3°$

Concept Check Assume a triangle ABC has standard labeling.

(a) Determine whether SAA, ASA, SSA, SAS, *or* SSS *is given.*

(b) Decide whether the law of sines or the law of cosines should be used to begin solving the triangle.

1. a, b, and C **2.** A, C, and c **3.** a, b, and A **4.** a, B, and C

5. A, B, and c **6.** a, c, and A **7.** a, b, and c **8.** b, c, and A

Find the length of the remaining side of each triangle. Do not use a calculator.

9.

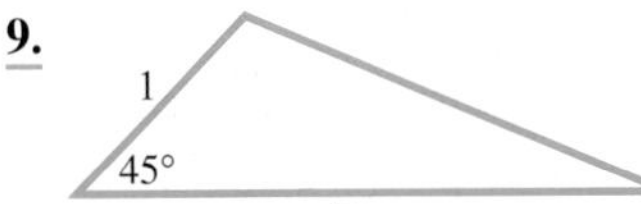

10.

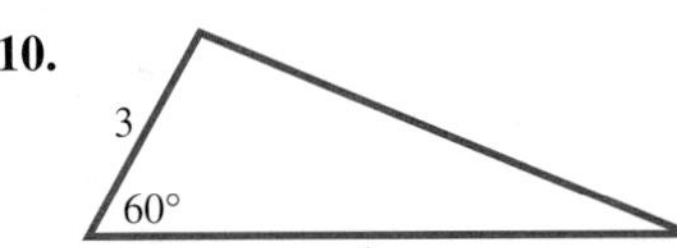

Find the measure of θ in each triangle. Do not use a calculator.

11.

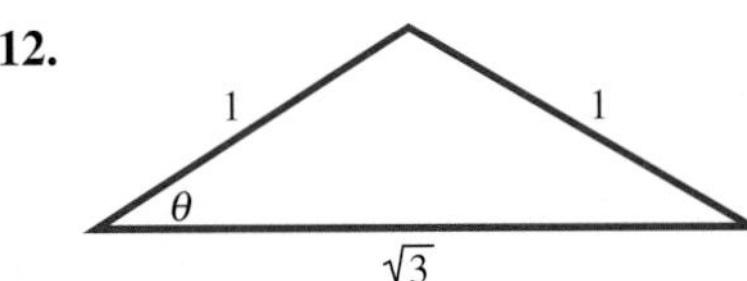

12.

Solve each triangle. Approximate values to the nearest tenth.

13.

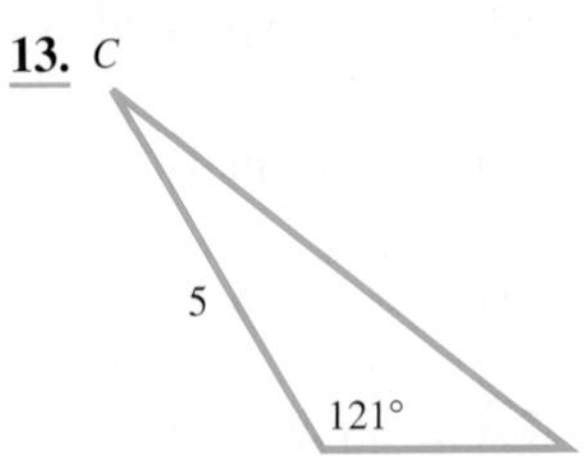

14.

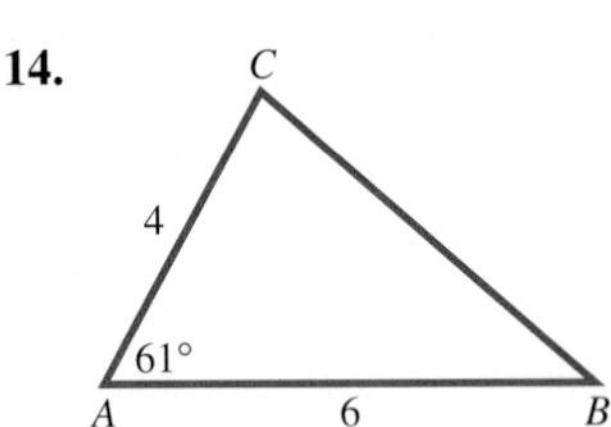

15. $A = 73.7°$, $B = 53.1°$, $C = 53.1°$ (The angles do not sum to 180° due to rounding.)
16. $A = 22.3°$, $B = 108.2°$, $C = 49.5°$
17. $b = 88.2$, $A = 56.7°$, $C = 68.3°$
18. $A = 33.6°$, $B = 50.7°$, $C = 95.7°$

19. $a = 2.60$ yd, $B = 45.1°$, $C = 93.5°$
20. $c = 2.83$ in., $A = 44.9°$, $B = 106.8°$
21. $c = 6.46$ m, $A = 53.1°$, $B = 81.3°$
22. $a = 43.7$ km, $B = 53.2°$, $C = 59.5°$
23. $A = 82°$, $B = 37°$, $C = 61°$
24. $C = 98°$, $A = 29°$, $B = 53°$
25. $C = 102° 10'$, $B = 35° 50'$, $A = 42° 00'$
26. $C = 107° 20'$, $B = 39° 00'$, $A = 33° 40'$
27. $C = 84° 30'$, $B = 44° 40'$, $A = 50° 50'$
28. $B = 85° 10'$, $C = 44° 50'$, $A = 50° 00'$
29. $a = 156$ cm, $B = 64° 50'$, $C = 34° 30'$
30. $c = 348$ ft, $A = 63° 50'$, $B = 43° 30'$
31. $b = 9.53$ in., $A = 64.6°$, $C = 40.6°$
32. $c = 4.28$ mi, $A = 48.8°$, $B = 71.5°$
33. $a = 15.7$ m, $B = 21.6°$, $C = 45.6°$
34. $b = 34.1$ cm, $A = 5.2°$, $C = 6.6°$
35. $A = 30°$, $B = 56°$, $C = 94°$
36. $A = 24°$, $B = 31°$, $C = 125°$

37. The value of $\cos \theta$ will be greater than 1. Your calculator will give you an error message (or a nonreal complex number) when using the inverse cosine function.

39. 257 m

15.

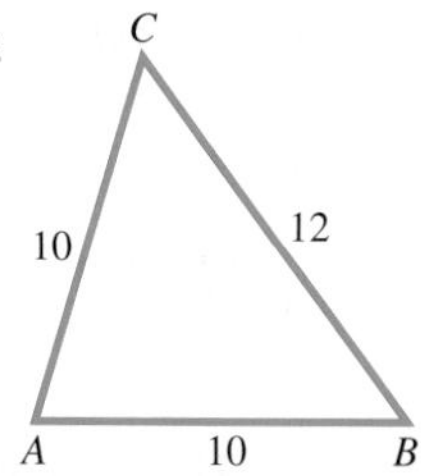

16.

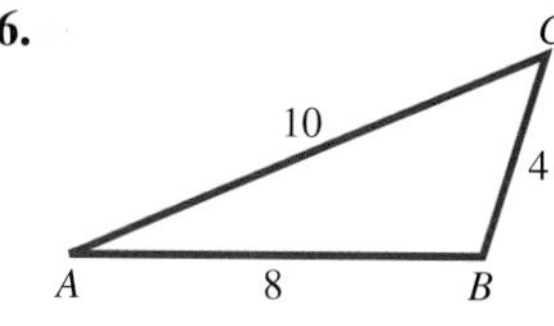

17.

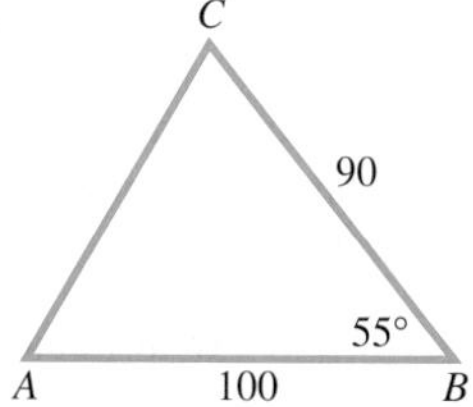

18.

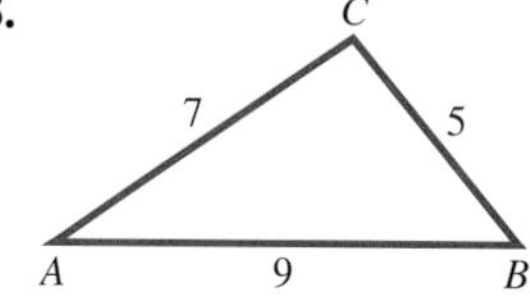

Solve each triangle. ***See Examples 2 and 3.***

19. $A = 41.4°$, $b = 2.78$ yd, $c = 3.92$ yd

20. $C = 28.3°$, $b = 5.71$ in., $a = 4.21$ in.

21. $C = 45.6°$, $b = 8.94$ m, $a = 7.23$ m

22. $A = 67.3°$, $b = 37.9$ km, $c = 40.8$ km

23. $a = 9.3$ cm, $b = 5.7$ cm, $c = 8.2$ cm

24. $a = 28$ ft, $b = 47$ ft, $c = 58$ ft

25. $a = 42.9$ m, $b = 37.6$ m, $c = 62.7$ m

26. $a = 189$ yd, $b = 214$ yd, $c = 325$ yd

27. $a = 965$ ft, $b = 876$ ft, $c = 1240$ ft

28. $a = 324$ m, $b = 421$ m, $c = 298$ m

29. $A = 80° 40'$, $b = 143$ cm, $c = 89.6$ cm

30. $C = 72° 40'$, $a = 327$ ft, $b = 251$ ft

31. $B = 74.8°$, $a = 8.92$ in., $c = 6.43$ in.

32. $C = 59.7°$, $a = 3.73$ mi, $b = 4.70$ mi

33. $A = 112.8°$, $b = 6.28$ m, $c = 12.2$ m

34. $B = 168.2°$, $a = 15.1$ cm, $c = 19.2$ cm

35. $a = 3.0$ ft, $b = 5.0$ ft, $c = 6.0$ ft

36. $a = 4.0$ ft, $b = 5.0$ ft, $c = 8.0$ ft

37. Refer to **Figure 12.** If you attempt to find any angle of a triangle with the values $a = 3$, $b = 4$, and $c = 10$ by using the law of cosines, what happens?

38. "The shortest distance between two points is a straight line." Explain how this is related to the geometric property that states that the sum of the lengths of any two sides of a triangle must be greater than the length of the remaining side.

Solve each problem. ***See Examples 1–4.***

39. *Distance across a River* Points A and B are on opposite sides of False River. From a third point, C, the angle between the lines of sight to A and B is 46.3°. If AC is 350 m long and BC is 286 m long, find AB.

40. 95.3 m
41. 163.5°
42. 5.2 cm, 8.8 cm
43. 281 km
44. 180 mi
45. 438.14 ft
46. 745 mi
47. 10.8 mi

40. *Distance across a Ravine* Points X and Y are on opposite sides of a ravine. From a third point Z, the angle between the lines of sight to X and Y is 37.7°. If XZ is 153 m long and YZ is 103 m long, find XY.

41. *Angle in a Parallelogram* A parallelogram has sides of length 25.9 cm and 32.5 cm. The longer diagonal has length 57.8 cm. Find the measure of the angle opposite the longer diagonal.

42. *Diagonals of a Parallelogram* The sides of a parallelogram are 4.0 cm and 6.0 cm. One angle is 58° while another is 122°. Find the lengths of the diagonals of the parallelogram.

43. *Flight Distance* Airports A and B are 450 km apart, on an east-west line. Tom flies in a northeast direction from A to airport C. From C he flies 359 km on a bearing of 128° 40′ to B. How far is C from A?

44. *Distance Traveled by a Plane* An airplane flies 180 mi from point X at a bearing of 125°, and then turns and flies at a bearing of 230° for 100 mi. How far is the plane from point X?

45. *Distance between Ends of the Vietnam Memorial* The Vietnam Veterans Memorial in Washington, D.C., is V-shaped with equal sides of length 246.75 ft. The angle between these sides measures 125° 12′. Find the distance between the ends of the two sides. (*Source:* Pamphlet obtained at Vietnam Veterans Memorial.)

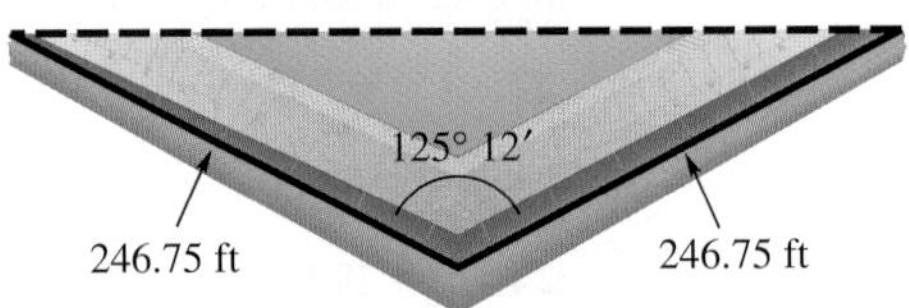

46. *Distance between Two Ships* Two ships leave a harbor together, traveling on courses that have an angle of 135° 40′ between them. If each travels 402 mi, how far apart are they?

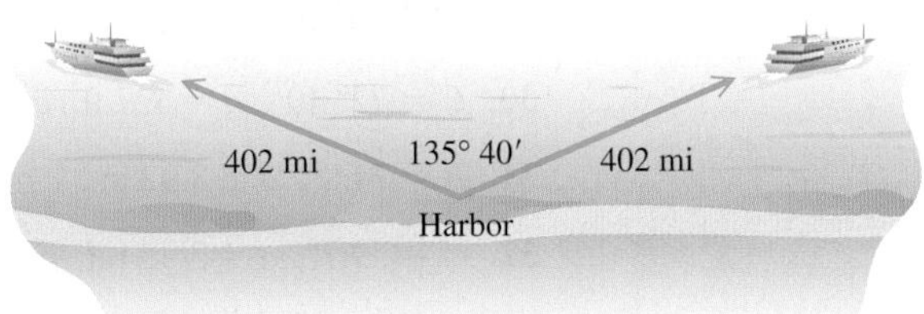

47. *Distance between a Ship and a Rock* A ship is sailing east. At one point, the bearing of a submerged rock is 45° 20′. After the ship has sailed 15.2 mi, the bearing of the rock has become 308° 40′. Find the distance of the ship from the rock at the latter point.

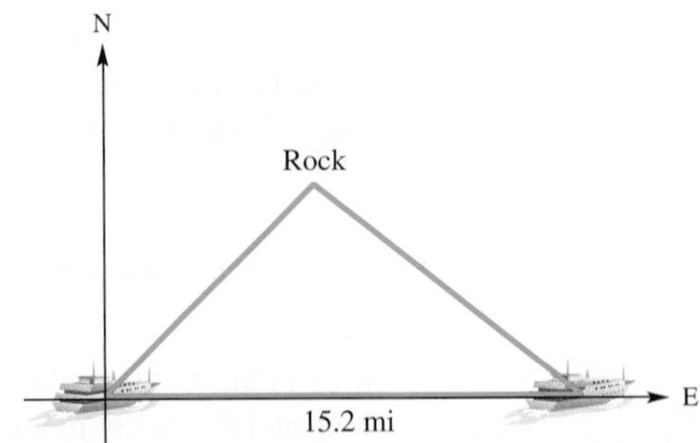

48. 1450 ft
49. 40°
50. 53°
51. 26° and 36°
52. 18 ft
53. second base: 66.8 ft; first and third bases: 63.7 ft
54. second base: 38.9 ft; first and third bases: 42.6 ft

48. *Distance between a Ship and a Submarine* From an airplane flying over the ocean, the angle of depression to a submarine lying under the surface is 24° 10′. At the same moment, the angle of depression from the airplane to a battleship is 17° 30′. See the figure. The distance from the airplane to the battleship is 5120 ft. Find the distance between the battleship and the submarine. (Assume the airplane, submarine, and battleship are in a vertical plane.)

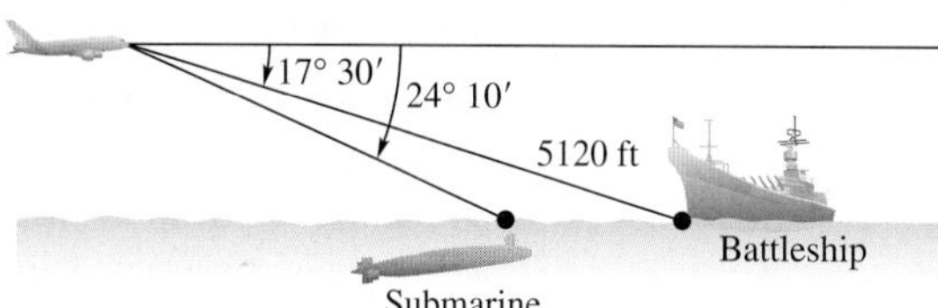

49. *Truss Construction* A triangular truss is shown in the figure. Find angle θ.

50. *Truss Construction* Find angle β in the truss shown in the figure.

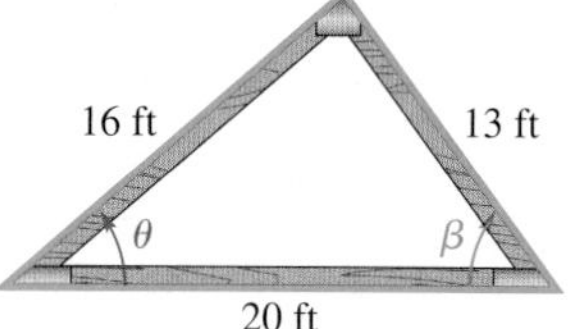

51. *Distance between a Beam and Cables* A weight is supported by cables attached to both ends of a balance beam, as shown in the figure. What angles are formed between the beam and the cables?

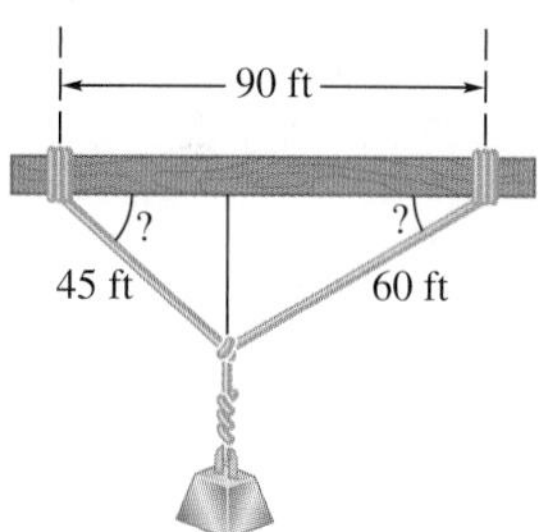

52. *Distance between Points on a Crane* A crane with a counterweight is shown in the figure. Find the horizontal distance between points A and B to the nearest foot.

53. *Distance on a Baseball Diamond* A baseball diamond is a square, 90.0 ft on a side, with home plate and the three bases as vertices. The pitcher's position is 60.5 ft from home plate. Find the distance from the pitcher's position to each of the bases.

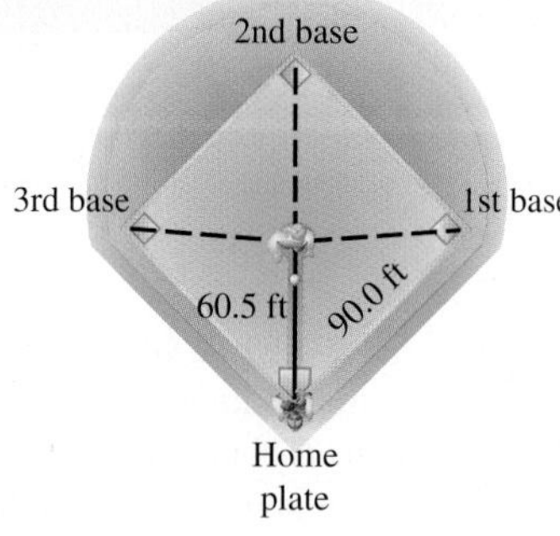

54. *Distance on a Softball Diamond* A softball diamond is a square, 60.0 ft on a side, with home plate and the three bases as vertices. The pitcher's position is 46.0 ft from home plate. Find the distance from the pitcher's position to each of the bases.

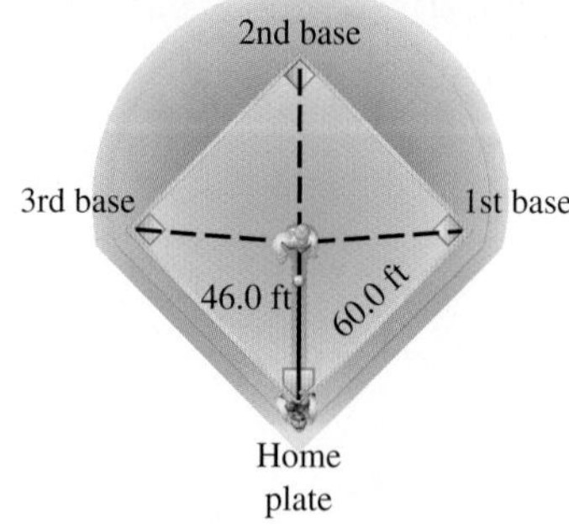

55. 39.2 km
56. 1473 m
57. 47.5 ft
58. 5.9 mi
59. 5500 m
60. 5.99 km

55. *Distance between a Ship and a Point* Starting at point A, a ship sails 18.5 km on a bearing of 189°, then turns and sails 47.8 km on a bearing of 317°. Find the distance of the ship from point A.

56. *Distance between Two Factories* Two factories blow their whistles at exactly 5:00. A man hears the two blasts at 3 sec and 6 sec after 5:00, respectively. The angle between his lines of sight to the two factories is 42.2°. If sound travels 344 m per sec, how far apart are the factories?

57. *Measurement Using Triangulation* Surveyors are often confronted with obstacles, such as trees, when measuring the boundary of a lot. One technique used to obtain an accurate measurement is the so-called **triangulation method.** In this technique, a triangle is constructed around the obstacle and one angle and two sides of the triangle are measured. Use this technique to find the length of the property line (the straight line between the two markers) in the figure. (*Source:* Kavanagh, B., *Surveying Principles and Applications,* Sixth Edition, Prentice-Hall.)

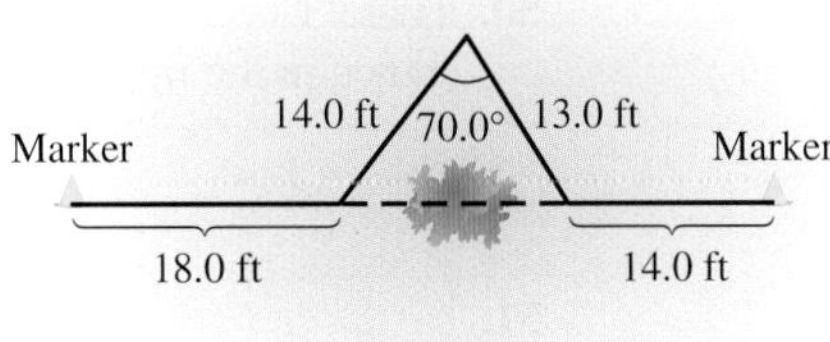

58. *Path of a Ship* A ship sailing due east in the North Atlantic has been warned to change course to avoid icebergs. The captain turns and sails on a bearing of 62°, then changes course again to a bearing of 115° until the ship reaches its original course. See the figure. How much farther did the ship have to travel to avoid the icebergs?

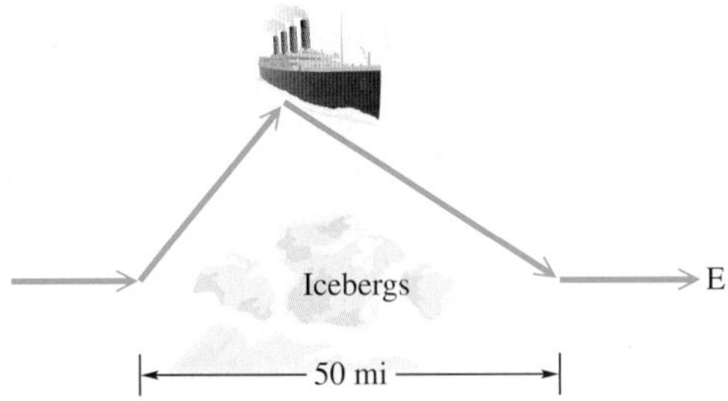

59. *Length of a Tunnel* To measure the distance through a mountain for a proposed tunnel, a point C is chosen that can be reached from each end of the tunnel. See the figure. If $AC = 3800$ m, $BC = 2900$ m, and angle $C = 110°$, find the length of the tunnel.

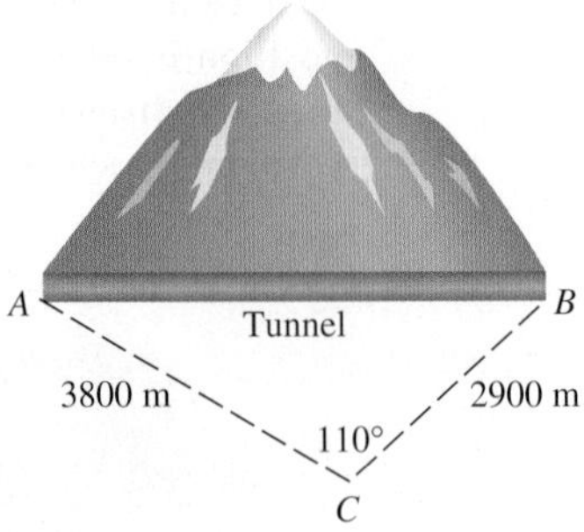

60. *Distance between an Airplane and a Mountain* A person in a plane flying straight north observes a mountain at a bearing of 24.1°. At that time, the plane is 7.92 km from the mountain. A short time later, the bearing to the mountain becomes 32.7°. How far is the airplane from the mountain when the second bearing is taken?

61. 16.26° **62.** 14.25°

63. $24\sqrt{3}$ sq units

64. $15\sqrt{3}$ sq units

65. 78 m^2 **66.** 310 in.^{2}

67. 12,600 cm^2 **68.** 228 yd^2

69. 3650 ft^2 **70.** 83.01 m^2

71. Area and perimeter are both 36.

72. **(a)** Area is the integer 66.
(b) Area is the integer 84.
(c) Area is the integer 42.
(d) Area is the integer 36.

73. 390,000 mi^2

74. 33 cans

75. **(a)** 87.8° and 92.2° both appear possible.
(b) 92.2°
(c) With the law of cosines we are required to find the inverse cosine of a negative number. Therefore, we know that angle C is greater than 90°.

Find the measure of each angle θ to two decimal places.

61.

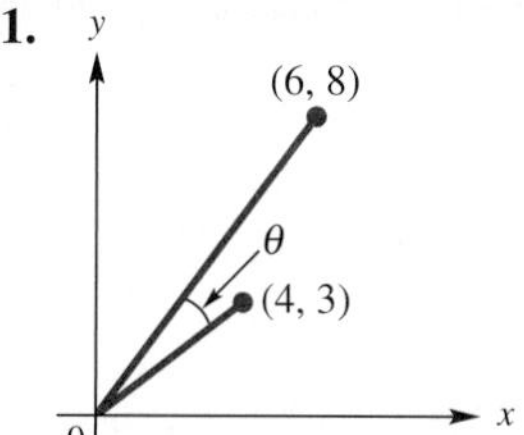

62.

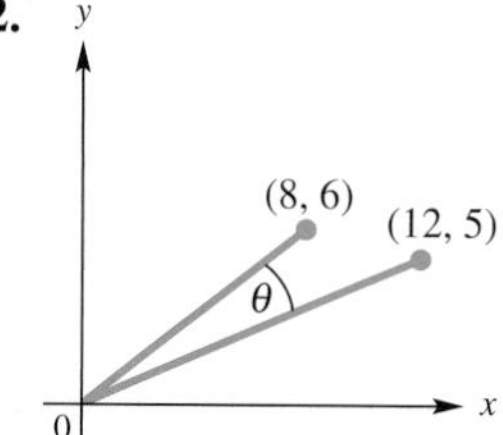

Find the exact area of each triangle using the formula $\mathcal{A} = \frac{1}{2}bh$, and then verify that Heron's formula gives the same result.

63.

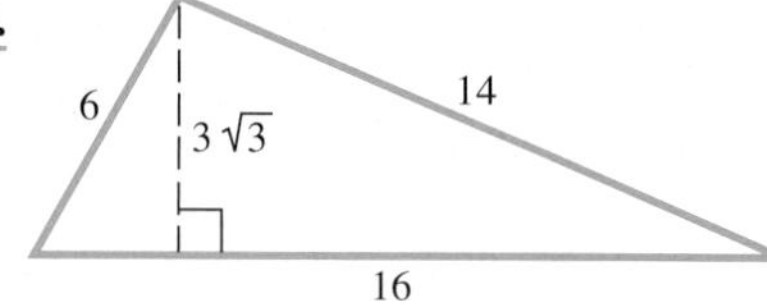

64.

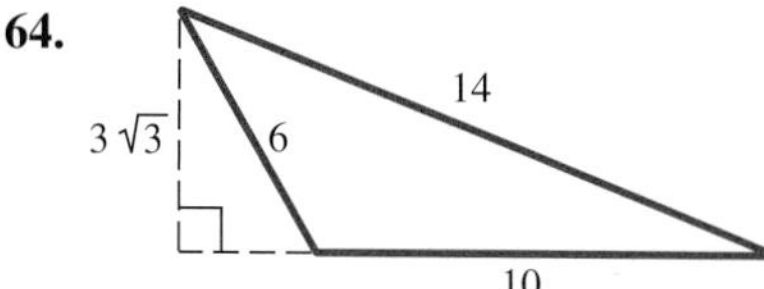

Find the area of each triangle ABC. **See Example 5.**

65. $a = 12$ m, $b = 16$ m, $c = 25$ m

66. $a = 22$ in., $b = 45$ in., $c = 31$ in.

67. $a = 154$ cm, $b = 179$ cm, $c = 183$ cm

68. $a = 25.4$ yd, $b = 38.2$ yd, $c = 19.8$ yd

69. $a = 76.3$ ft, $b = 109$ ft, $c = 98.8$ ft

70. $a = 15.89$ m, $b = 21.74$ m, $c = 10.92$ m

Solve each problem. **See Example 5.**

71. *Perfect Triangles* A **perfect triangle** is a triangle whose sides have whole number lengths and whose area is numerically equal to its perimeter. Show that the triangle with sides of length 9, 10, and 17 is perfect.

72. *Heron Triangles* A **Heron triangle** is a triangle having integer sides and area. Show that each of the following is a Heron triangle.

(a) $a = 11, b = 13, c = 20$ **(b)** $a = 13, b = 14, c = 15$
(c) $a = 7, b = 15, c = 20$ **(d)** $a = 9, b = 10, c = 17$

73. *Area of the Bermuda Triangle* Find the area of the Bermuda Triangle if the sides of the triangle have approximate lengths 850 mi, 925 mi, and 1300 mi.

74. *Required Amount of Paint* A painter needs to cover a triangular region 75 m by 68 m by 85 m. A can of paint covers 75 m^2 of area. How many cans (to the next higher number of cans) will be needed?

75. Consider triangle ABC shown here.

(a) Use the law of sines to find candidates for the value of angle C. Round angle measures to the nearest tenth of a degree.
(b) Rework part (a) using the law of cosines.
(c) Why is the law of cosines a better method in this case?

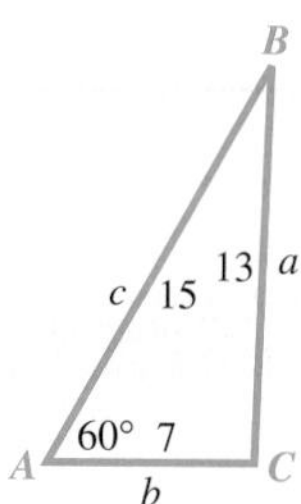

76. Show that the measure of angle A is twice the measure of angle B. (*Hint:* Use the law of cosines to find $\cos A$ and $\cos B$, and then show that $\cos A = 2\cos^2 B - 1$.)

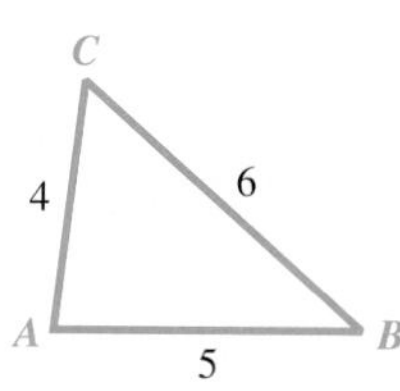

77.

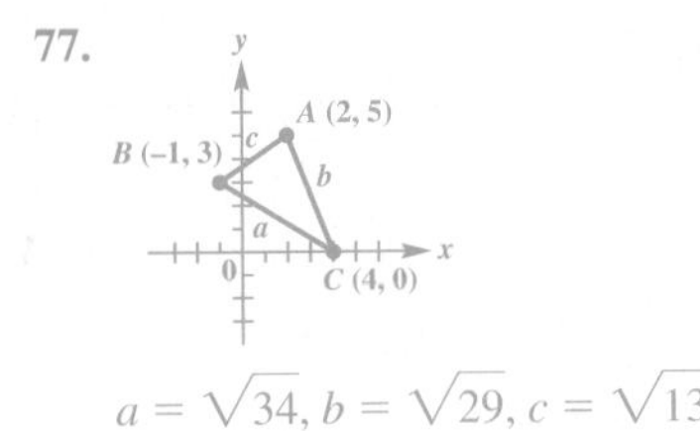

$a = \sqrt{34}, b = \sqrt{29}, c = \sqrt{13}$

78. 9.5 sq units
79. 9.5 sq units
80. 9.5 sq units

Relating Concepts

For individual or collaborative investigation *(Exercises 77–80)*

We have introduced two new formulas for the area of a triangle in this chapter. You should now be able to find the area $\mathcal{A}$ of a triangle using one of three formulas.

(a) $\mathcal{A} = \frac{1}{2}bh$

(b) $\mathcal{A} = \frac{1}{2}ab \sin C \quad \left(\text{or } \mathcal{A} = \frac{1}{2}ac \sin B \text{ or } \mathcal{A} = \frac{1}{2}bc \sin A\right)$

(c) $\mathcal{A} = \sqrt{s(s-a)(s-b)(s-c)}$ (Heron's formula)

Another area formula can be used when the coordinates of the vertices of a triangle are given. If the vertices are the ordered pairs (x_1, y_1), (x_2, y_2), and (x_3, y_3), then the following is valid.

(d) $\mathcal{A} = \frac{1}{2}\left|(x_1y_2 - y_1x_2 + x_2y_3 - y_2x_3 + x_3y_1 - y_3x_1)\right|$

Work Exercises 77–80 in order, *showing that the various formulas all lead to the same area.*

77. Draw a triangle with vertices $A(2, 5)$, $B(-1, 3)$, and $C(4, 0)$, and use the distance formula to find the lengths of the sides a, b, and c.

78. Find the area of triangle ABC using formula (b). (First use the law of cosines to find the measure of an angle.)

79. Find the area of triangle ABC using formula (c)—that is, Heron's formula.

80. Find the area of triangle ABC using new formula (d).

Chapter 8 Quiz (Sections 8.1–8.2)

[8.1]
1. 131°

[8.2]
2. 201 m
3. 48.0°

[8.1]
4. 15.75 sq units

[8.2]
5. 189 km^2

[8.1]
6. 41.6°, 138.4°
7. $a = 648$, $b = 456$, $C = 28°$
8. 3.6 mi

Find the indicated part of each triangle ABC.

1. Find A if $B = 30.6°$, $b = 7.42$ in., and $c = 4.54$ in.

2. Find a if $A = 144°$, $c = 135$ m, and $b = 75.0$ m.

3. Find C if $a = 28.4$ ft, $b = 16.9$ ft, and $c = 21.2$ ft.

4. Find the area of the triangle shown here.

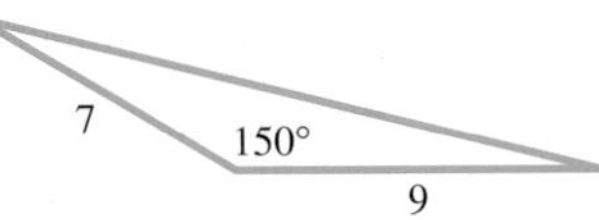

5. Find the area of triangle ABC if $a = 19.5$ km, $b = 21.0$ km, and $c = 22.5$ km.

6. For triangle ABC with $c = 345$, $a = 534$, and $C = 25.4°$, there are two possible values for angle A. What are they?

7. Solve triangle ABC if $c = 326$, $A = 111°$, and $B = 41.0°$.

8. *Height of a Balloon* The angles of elevation of a hot air balloon from two observation points X and Y on level ground are 42° 10′ and 23° 30′, respectively. As shown in the figure, points X, Y, and Z are in the same vertical plane and points X and Y are 12.2 mi apart. Approximate the height of the balloon to the nearest tenth of a mile.

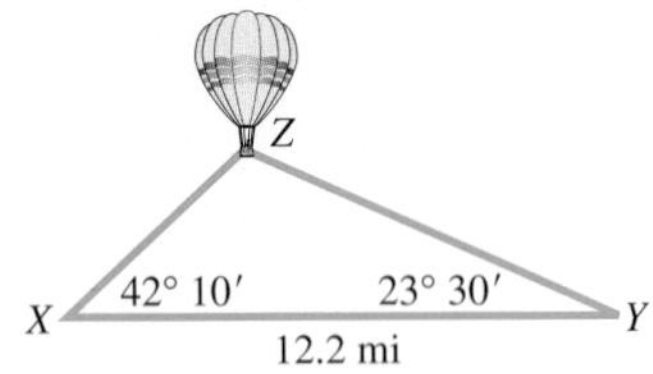

[8.2]
9. 25.24983 mi
10. 3921 m

9. *Volcano Movement* To help predict eruptions from the volcano Mauna Loa on the island of Hawaii, scientists keep track of the volcano's movement by using a "super triangle" with vertices on the three volcanoes shown on the map at the right. Find BC given that $AB = 22.47928$ mi, $AC = 28.14276$ mi, and $A = 58.56989°$.

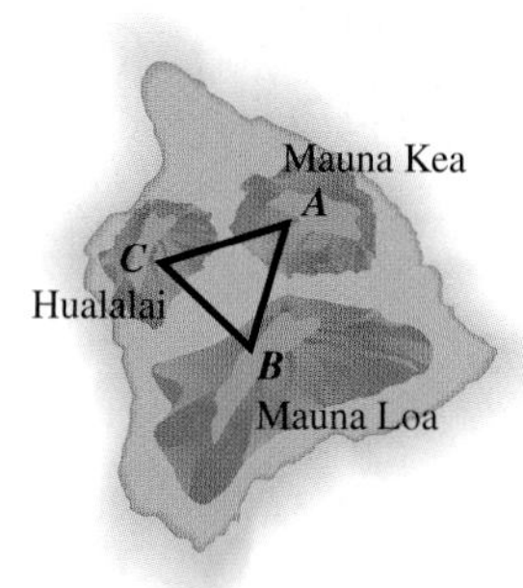

10. *Distance between Two Towns* To find the distance between two small towns, an electronic distance measuring (EDM) instrument is placed on a hill from which both towns are visible. The distance to each town from the EDM and the angle between the two lines of sight are measured. See the figure. Find the distance between the towns.

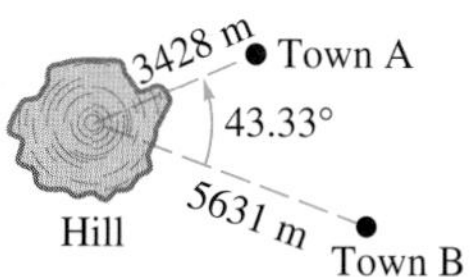

8.3 Vectors, Operations, and the Dot Product

- Basic Terminology
- Algebraic Interpretation of Vectors
- Operations with Vectors
- Dot Product and the Angle between Vectors

Basic Terminology Quantities that involve magnitudes, such as 45 lb or 60 mph, can be represented by real numbers called **scalars.** Other quantities, called **vector quantities,** involve both magnitude *and* direction. Typical vector quantities are velocity, acceleration, and force. For example, traveling 50 mph *east* represents a vector quantity.

A vector quantity can be represented with a directed line segment (a segment that uses an arrowhead to indicate direction) called a **vector.** The *length* of the vector represents the **magnitude** of the vector quantity. The *direction* of the vector, indicated by the arrowhead, represents the direction of the quantity. See **Figure 19.**

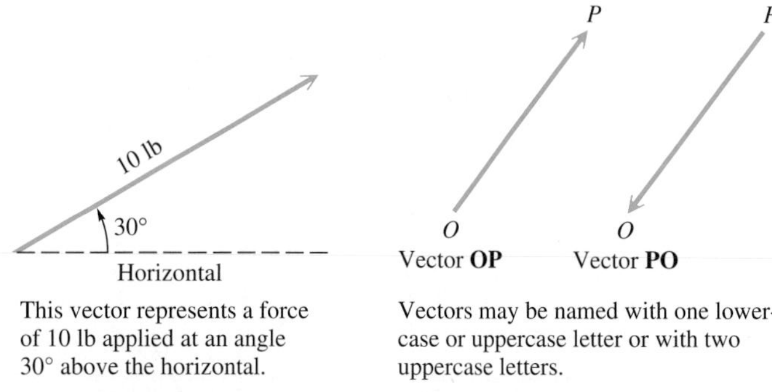

This vector represents a force of 10 lb applied at an angle 30° above the horizontal.

Figure 19

Vectors may be named with one lowercase or uppercase letter or with two uppercase letters.

Figure 20

When we indicate vectors in print, it is customary to use boldface type or an arrow over the letter or letters. Thus, **OP** and $\overrightarrow{OP}$ both represent the vector **OP**. When two letters name a vector, the first indicates the **initial point** and the second indicates the **terminal point** of the vector. Knowing these points gives the direction of the vector. For example, vectors **OP** and **PO** in **Figure 20** are not the same vector. They have the same magnitude but *opposite* directions. The magnitude of vector **OP** is written $|\mathbf{OP}|$.

Two vectors are equal if and only if they have the same direction and the same magnitude. In **Figure 21,** vectors **A** and **B** are equal, as are vectors **C** and **D**. As **Figure 21** shows, equal vectors need not coincide, but they must be parallel and in the same direction. Vectors **A** and **E** are unequal because they do not have the same direction, while $\mathbf{A} \neq \mathbf{F}$ because they have different magnitudes.

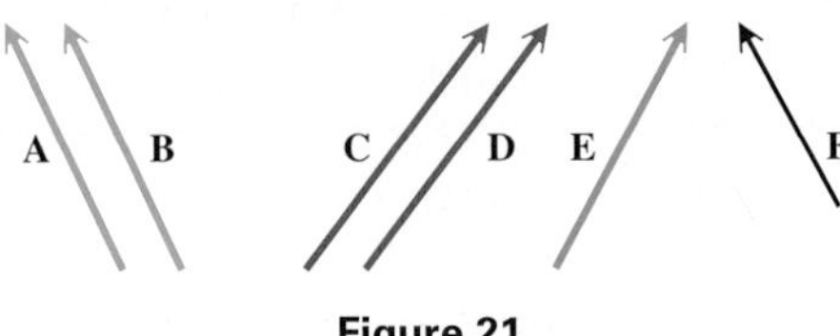

Figure 21

The sum of two vectors is also a vector. There are two ways to find the sum of two vectors **A** and **B** geometrically.

1. Place the initial point of vector **B** at the terminal point of vector **A**, as shown in **Figure 22(a).** The vector with the same initial point as **A** and the same terminal point as **B** is the sum $\mathbf{A} + \mathbf{B}$.

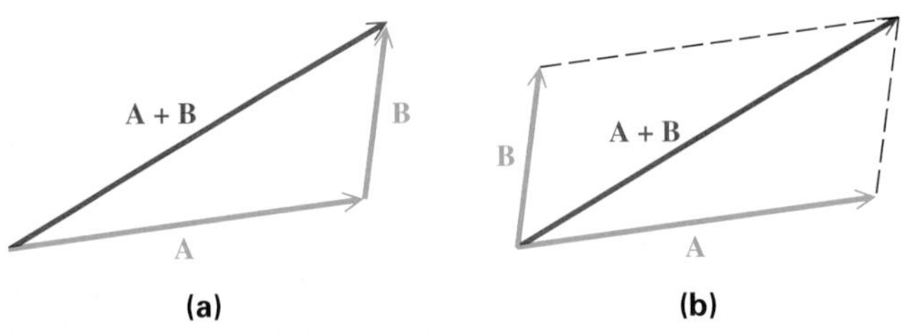

Figure 22

2. Use the **parallelogram rule.** Place vectors **A** and **B** so that their initial points coincide, as in **Figure 22(b).** Then, complete a parallelogram that has **A** and **B** as two sides. The diagonal of the parallelogram with the same initial point as **A** and **B** is the sum $\mathbf{A} + \mathbf{B}$.

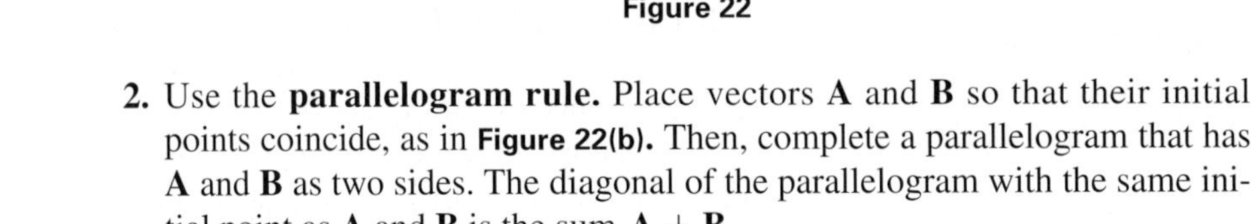

Vectors **v** and **–v** are opposites.

Figure 23

Parallelograms can be used to show that vector $\mathbf{B} + \mathbf{A}$ is the same as vector $\mathbf{A} + \mathbf{B}$, or that $\mathbf{A} + \mathbf{B} = \mathbf{B} + \mathbf{A}$, so ***vector addition is commutative.*** The vector sum $\mathbf{A} + \mathbf{B}$ is the **resultant** of vectors **A** and **B**.

For every vector **v** there is a vector $-\mathbf{v}$ that has the same magnitude as **v** but opposite direction. Vector $-\mathbf{v}$ is the **opposite** of **v**. See **Figure 23.** The sum of **v** and $-\mathbf{v}$ has magnitude 0 and is the **zero vector.** As with real numbers, to subtract vector **B** from vector **A**, find the vector sum $\mathbf{A} + (-\mathbf{B})$. See **Figure 24.**

Teaching Tip Show by geometric analysis that vector addition is commutative. Also, mention that $\mathbf{A} + (-\mathbf{B})$ is the same as $\mathbf{A} - \mathbf{B}$. In **Figure 24,** move vector $-\mathbf{B}$ so that its initial point is the terminal point of **A** to further illustrate the meaning of $\mathbf{A} - \mathbf{B}$.

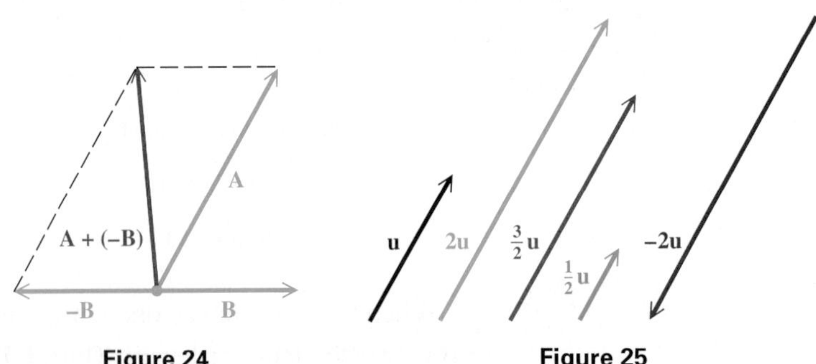

Figure 24 **Figure 25**

The product of a real number (or scalar) k and a vector **u** is the vector $k \cdot \mathbf{u}$, which has magnitude $|k|$ times the magnitude of **u**. As suggested by **Figure 25,** the vector $k \cdot \mathbf{u}$ has the same direction as **u** if $k > 0$, and has the opposite direction if $k < 0$.

LOOKING AHEAD TO CALCULUS

In addition to two-dimensional vectors in a plane, calculus courses introduce three-dimensional vectors in space. The magnitude of the two-dimensional vector $\langle a, b \rangle$ is given by $\sqrt{a^2 + b^2}$. If we extend this to the three-dimensional vector $\langle a, b, c \rangle$, the expression becomes $\sqrt{a^2 + b^2 + c^2}$. Similar extensions are made for other concepts.

Algebraic Interpretation of Vectors A vector with its initial point at the origin in a rectangular coordinate system is called a **position vector.** A position vector **u** with its endpoint at the point (a, b) is written $\langle a, b \rangle$, so

$$\mathbf{u} = \langle a, b \rangle.$$

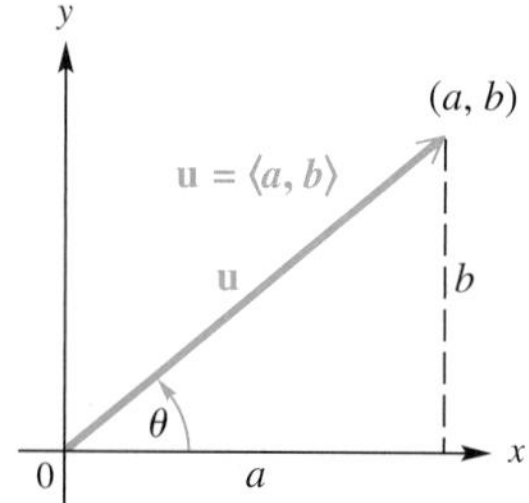

Figure 26

This means that every vector in the real plane corresponds to an ordered pair of real numbers. ***Thus, geometrically a vector is a directed line segment while algebraically it is an ordered pair.*** The numbers a and b are the **horizontal component** and the **vertical component,** respectively, of vector **u**.

Figure 26 shows the vector $\mathbf{u} = \langle a, b \rangle$. The positive angle between the x-axis and a position vector is the **direction angle** for the vector. In **Figure 26,** θ is the direction angle for vector **u**.

From **Figure 26,** we can see that the magnitude and direction of a vector are related to its horizontal and vertical components.

Magnitude and Direction Angle of a Vector $\langle a, b \rangle$

The magnitude (length) of vector $\mathbf{u} = \langle a, b \rangle$ is given by the following.

$$|\mathbf{u}| = \sqrt{a^2 + b^2}$$

The direction angle θ satisfies $\tan \theta = \frac{b}{a}$, where $a \neq 0$.

Classroom Example 1
Find the magnitude and direction angle for $\mathbf{u} = \langle -5, 4 \rangle$.

Answer: $\sqrt{41}$; 141.3°

EXAMPLE 1 **Finding Magnitude and Direction Angle**

Find the magnitude and direction angle for $\mathbf{u} = \langle 3, -2 \rangle$.

ALGEBRAIC SOLUTION

The magnitude is $|\mathbf{u}| = \sqrt{3^2 + (-2)^2} = \sqrt{13}$. To find the direction angle θ, start with $\tan \theta = \frac{b}{a} = \frac{-2}{3} = -\frac{2}{3}$. Vector **u** has a positive horizontal component and a negative vertical component, placing the position vector in quadrant IV. A calculator gives $\tan^{-1}\left(-\frac{2}{3}\right) \approx -33.7°$. Adding 360° yields the direction angle $\theta \approx 326.3°$. See **Figure 27.**

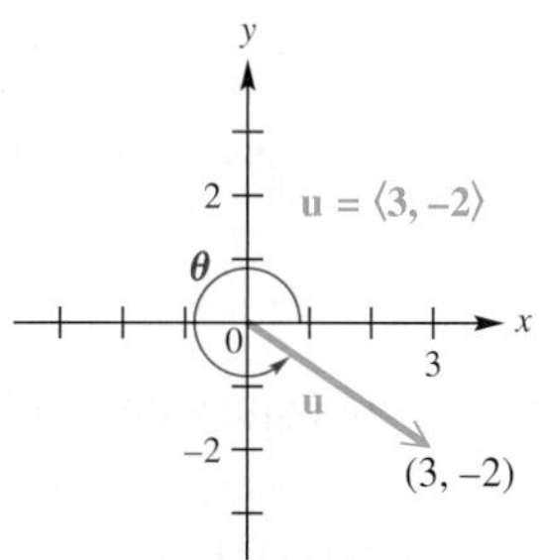

Figure 27

GRAPHING CALCULATOR SOLUTION

A calculator returns the magnitude and direction angle, given the horizontal and vertical components. An approximation for $\sqrt{13}$ is given, and the direction angle has a measure with least possible absolute value. We must add 360° to the value of θ to obtain the positive direction angle. See **Figure 28.**

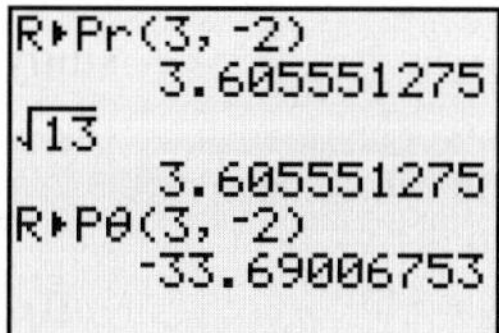

Figure 28

For more information, see your owner's manual or the graphing calculator manual that accompanies this text.

✓ *Now Try Exercise 33.*

Classroom Example 2
Vector **v** has magnitude 14.5 and direction angle 220°. Find the horizontal and vertical components.

Answer: $\langle -11.1, -9.3 \rangle$

Horizontal and Vertical Components

The horizontal and vertical components, respectively, of a vector **u** having magnitude $|\mathbf{u}|$ and direction angle θ are the following.

$$a = |\mathbf{u}| \cos \theta \quad \text{and} \quad b = |\mathbf{u}| \sin \theta$$

That is, $\mathbf{u} = \langle a, b \rangle = \langle |\mathbf{u}| \cos \theta, |\mathbf{u}| \sin \theta \rangle$.

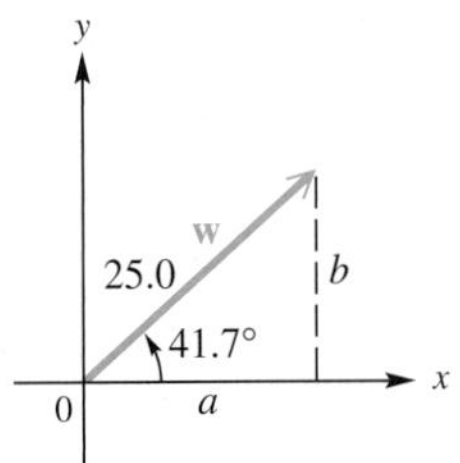

Figure 29

EXAMPLE 2 Finding Horizontal and Vertical Components

Vector **w** in **Figure 29** has magnitude 25.0 and direction angle 41.7°. Find the horizontal and vertical components.

ALGEBRAIC SOLUTION

Use the two formulas in the box, with $|\mathbf{w}| = 25.0$ and $\theta = 41.7°$.

$$a = 25.0 \cos 41.7° \qquad b = 25.0 \sin 41.7°$$

$$a \approx 18.7 \qquad b \approx 16.6$$

Therefore, $\mathbf{w} = \langle 18.7, 16.6 \rangle$. The horizontal component is 18.7, and the vertical component is 16.6 (rounded to the nearest tenth).

GRAPHING CALCULATOR SOLUTION

See **Figure 30.** The results support the algebraic solution.

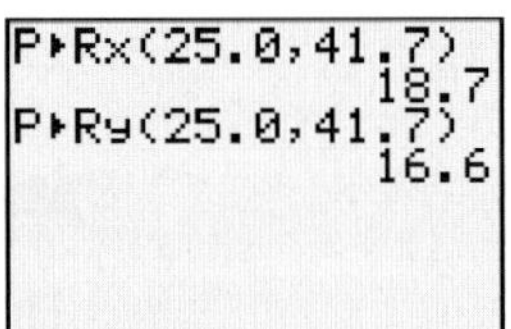

Figure 30

Now Try Exercise 37.

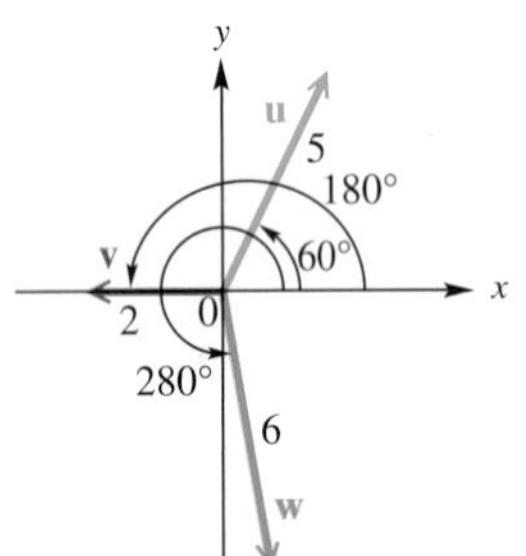

Figure 31

EXAMPLE 3 Writing Vectors in the Form $\langle a, b \rangle$

Write each vector in **Figure 31** in the form $\langle a, b \rangle$.

SOLUTION

$$\mathbf{u} = \langle 5 \cos 60°, 5 \sin 60° \rangle = \left\langle 5 \cdot \frac{1}{2}, 5 \cdot \frac{\sqrt{3}}{2} \right\rangle = \left\langle \frac{5}{2}, \frac{5\sqrt{3}}{2} \right\rangle$$

$$\mathbf{v} = \langle 2 \cos 180°, 2 \sin 180° \rangle = \langle 2(-1), 2(0) \rangle = \langle -2, 0 \rangle$$

$$\mathbf{w} = \langle 6 \cos 280°, 6 \sin 280° \rangle \approx \langle 1.0419, -5.9088 \rangle \quad \text{Use a calculator.}$$

Now Try Exercises 43 and 45.

Classroom Example 3
Write each vector in the form $\langle a, b \rangle$.
u: magnitude 8, direction angle 135°
v: magnitude 4, direction angle 270°
w: magnitude 10, direction angle 340°

Answer:
$\mathbf{u} = \langle -4\sqrt{2}, 4\sqrt{2} \rangle$;
$\mathbf{v} = \langle 0, -4 \rangle$;
$\mathbf{w} \approx \langle 9.3969, -3.4202 \rangle$

The following geometric properties of parallelograms are helpful when studying applications of vectors.

Properties of Parallelograms

1. A parallelogram is a quadrilateral whose opposite sides are parallel.
2. The opposite sides and opposite angles of a parallelogram are equal, and adjacent angles of a parallelogram are supplementary.
3. The diagonals of a parallelogram bisect each other, but they do not necessarily bisect the angles of the parallelogram.

Classroom Example 4
Two forces of 32 and 48 newtons act on a point in the plane. If the angle between the forces is 76°, find the magnitude of the resultant vector.

Answer: 64 newtons

EXAMPLE 4 Finding the Magnitude of a Resultant

Two forces of 15 and 22 newtons act on a point in the plane. (A **newton** is a unit of force that equals 0.225 lb.) If the angle between the forces is 100°, find the magnitude of the resultant force.

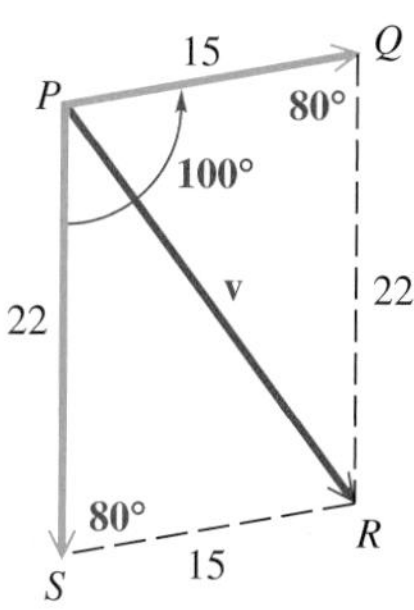

Figure 32

SOLUTION As shown in **Figure 32,** a parallelogram that has the forces as adjacent sides can be formed. The angles of the parallelogram adjacent to angle P measure 80°, since adjacent angles of a parallelogram are supplementary. Opposite sides of the parallelogram are equal in length. The resultant force divides the parallelogram into two triangles. Use the law of cosines with either triangle.

$$|\mathbf{v}|^2 = 15^2 + 22^2 - 2(15)(22)\cos 80^\circ$$ Law of cosines **(Section 8.2)**

$$\approx 225 + 484 - 115$$ Evaluate powers and multiply.

$$|\mathbf{v}|^2 \approx 594$$ Add and subtract.

$$|\mathbf{v}| \approx 24$$ Take the positive square root. **(Section 1.4)**

To the nearest unit, the magnitude of the resultant force is 24 newtons.

Now Try Exercise 49.

Operations with Vectors As shown in **Figure 33,** $\mathbf{m} = \langle a, b\rangle$, $\mathbf{n} = \langle c, d\rangle$, and $\mathbf{p} = \langle a + c, b + d\rangle$. Using geometry, we can show that the endpoints of the three vectors and the origin form a parallelogram. Since a diagonal of this parallelogram gives the resultant of **m** and **n**, we have $\mathbf{p} = \mathbf{m} + \mathbf{n}$ or

$$\langle a + c, b + d\rangle = \langle a, b\rangle + \langle c, d\rangle.$$

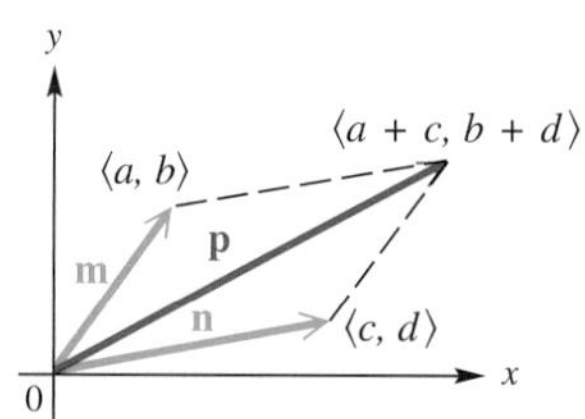

Figure 33

Similarly, we can verify the following vector operations.

Vector Operations

Let a, b, c, d, and k represent real numbers.

$$\langle a, b\rangle + \langle c, d\rangle = \langle a + c, b + d\rangle$$

$$k \cdot \langle a, b\rangle = \langle ka, kb\rangle$$

If $\mathbf{u} = \langle a_1, a_2\rangle$, then $-\mathbf{u} = \langle -a_1, -a_2\rangle$.

$$\langle a, b\rangle - \langle c, d\rangle = \langle a, b\rangle + (-\langle c, d\rangle) = \langle a - c, b - d\rangle$$

Classroom Example 5
Let $\mathbf{u} = \langle 6, -3\rangle$ and $\mathbf{v} = \langle -14, 8\rangle$. Find the following.

(a) $\mathbf{u} + \mathbf{v}$ (b) $-\frac{1}{2}\mathbf{v}$

(c) $5\mathbf{u} + 2\mathbf{v}$

Answers:
(a) $\langle -8, 5\rangle$ (b) $\langle 7, -4\rangle$
(c) $\langle 2, 1\rangle$

EXAMPLE 5 Performing Vector Operations

Let $\mathbf{u} = \langle -2, 1\rangle$ and $\mathbf{v} = \langle 4, 3\rangle$. See **Figure 34.** Find and illustrate each of the following.

(a) $\mathbf{u} + \mathbf{v}$

(b) $-2\mathbf{u}$

(c) $3\mathbf{u} - 2\mathbf{v}$

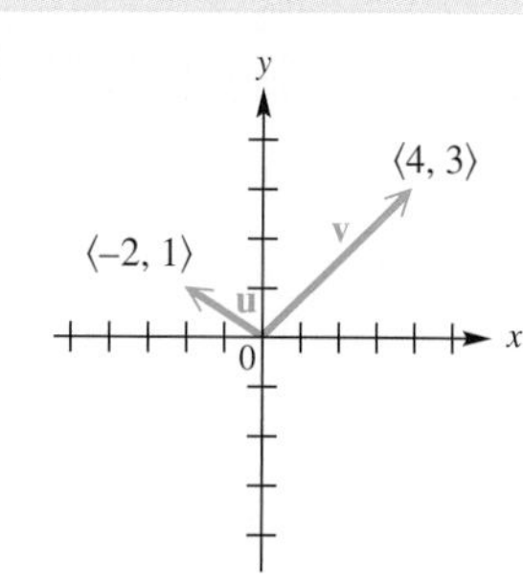

Figure 34

SOLUTION See **Figure 35.**

(a) $\mathbf{u} + \mathbf{v} = \langle -2, 1 \rangle + \langle 4, 3 \rangle$
$= \langle -2 + 4, 1 + 3 \rangle$
$= \langle 2, 4 \rangle$

(b) $-2\mathbf{u} = -2 \cdot \langle -2, 1 \rangle$
$= \langle -2(-2), -2(1) \rangle$
$= \langle 4, -2 \rangle$

(c) $3\mathbf{u} - 2\mathbf{v} = 3 \cdot \langle -2, 1 \rangle - 2 \cdot \langle 4, 3 \rangle$
$= \langle -6, 3 \rangle - \langle 8, 6 \rangle$
$= \langle -6 - 8, 3 - 6 \rangle$
$= \langle -14, -3 \rangle$

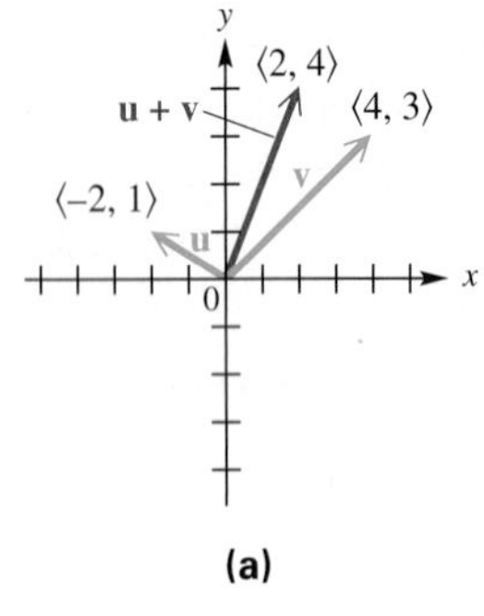

(a)

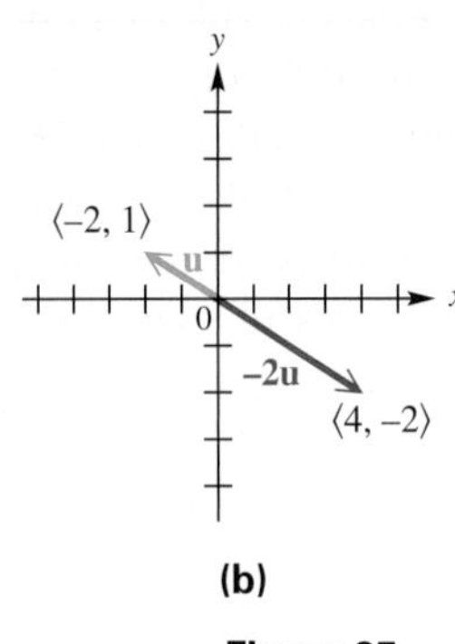

(b)

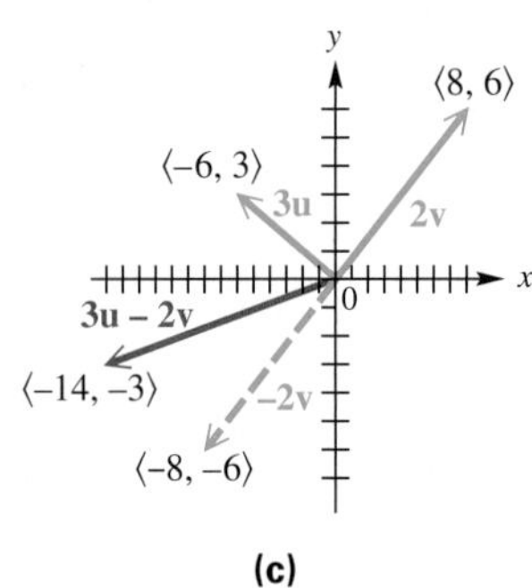

(c)

Figure 35

✔ *Now Try Exercises 59, 61, and 63.*

A **unit vector** is a vector that has magnitude 1. Two very useful unit vectors are defined as follows and shown in **Figure 36(a).**

$$\mathbf{i} = \langle 1, 0 \rangle \qquad \mathbf{j} = \langle 0, 1 \rangle$$

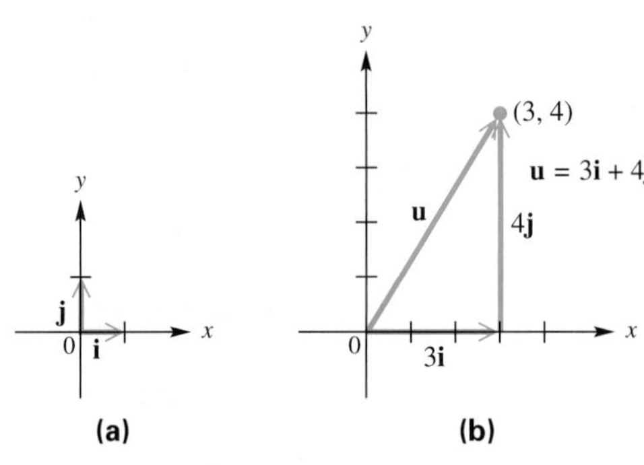

(a) (b)

Figure 36

Teaching Tip Point out that the boldface type for unit vectors helps distinguish the use of **i** and **j** in vector notation from the use of *i* and *j* with complex numbers.

With the unit vectors **i** and **j**, we can express any other vector $\langle a, b \rangle$ in the form $a\mathbf{i} + b\mathbf{j}$, as shown in **Figure 36(b)**, where $\langle 3, 4 \rangle = 3\mathbf{i} + 4\mathbf{j}$. The vector operations previously given can be restated, using $a\mathbf{i} + b\mathbf{j}$ notation.

i, j Form for Vectors

If $\mathbf{v} = \langle a, b \rangle$, then

$$\mathbf{v} = a\mathbf{i} + b\mathbf{j}, \quad \text{where } \mathbf{i} = \langle 1, 0 \rangle \text{ and } \mathbf{j} = \langle 0, 1 \rangle.$$

Dot Product and the Angle between Vectors ***The* dot product *of two vectors is a real number, not a vector.*** It is also known as the *inner product*. Dot products are used to determine the angle between two vectors, to derive geometric theorems, and to solve physics problems.

Dot Product

The **dot product** of the two vectors $\mathbf{u} = \langle a, b \rangle$ and $\mathbf{v} = \langle c, d \rangle$ is denoted $\mathbf{u} \cdot \mathbf{v}$, read "**u** dot **v**," and given by the following.

$$\mathbf{u} \cdot \mathbf{v} = ac + bd$$

That is, the dot product of two vectors is the sum of the product of their first components and the product of their second components.

Classroom Example 6
Find each dot product.
(a) $\langle -5, 2 \rangle \cdot \langle -2, -4 \rangle$
(b) $\langle 10, -1 \rangle \cdot \langle 1, 10 \rangle$

Answers:
(a) 2 (b) 0

EXAMPLE 6 Finding Dot Products

Find each dot product.

(a) $\langle 2, 3 \rangle \cdot \langle 4, -1 \rangle$ **(b)** $\langle 6, 4 \rangle \cdot \langle -2, 3 \rangle$

SOLUTION

(a) $\langle 2, 3 \rangle \cdot \langle 4, -1 \rangle = 2(4) + 3(-1)$
$= 5$

(b) $\langle 6, 4 \rangle \cdot \langle -2, 3 \rangle = 6(-2) + 4(3)$
$= 0$

✔ *Now Try Exercises 71 and 73.*

The following properties of dot products can be verified by using the definitions presented so far.

Properties of the Dot Product

For all vectors **u**, **v**, and **w** and real numbers k, the following hold.

(a) $\mathbf{u} \cdot \mathbf{v} = \mathbf{v} \cdot \mathbf{u}$ **(b)** $\mathbf{u} \cdot (\mathbf{v} + \mathbf{w}) = \mathbf{u} \cdot \mathbf{v} + \mathbf{u} \cdot \mathbf{w}$

(c) $(\mathbf{u} + \mathbf{v}) \cdot \mathbf{w} = \mathbf{u} \cdot \mathbf{w} + \mathbf{v} \cdot \mathbf{w}$ **(d)** $(k\mathbf{u}) \cdot \mathbf{v} = k(\mathbf{u} \cdot \mathbf{v}) = \mathbf{u} \cdot (k\mathbf{v})$

(e) $\mathbf{0} \cdot \mathbf{u} = 0$ **(f)** $\mathbf{u} \cdot \mathbf{u} = |\mathbf{u}|^2$

For example, to prove the first part of (d), we let $\mathbf{u} = \langle a, b \rangle$ and $\mathbf{v} = \langle c, d \rangle$.

$$\begin{aligned}
(k\mathbf{u}) \cdot \mathbf{v} &= (k\langle a, b \rangle) \cdot \langle c, d \rangle && \text{Substitute.} \\
&= \langle ka, kb \rangle \cdot \langle c, d \rangle && \text{Multiply by scalar } k. \\
&= kac + kbd && \text{Dot product} \\
&= k(ac + bd) && \text{Distributive property (Section R.2)} \\
&= k(\langle a, b \rangle \cdot \langle c, d \rangle) && \text{Dot product} \\
&= k(\mathbf{u} \cdot \mathbf{v}) && \text{Substitute.}
\end{aligned}$$

The proofs of the remaining properties are similar.

The dot product of two vectors can be positive, 0, or negative. A geometric interpretation of the dot product explains when each of these cases occurs. This interpretation involves the angle between the two vectors. Consider the vectors $\mathbf{u} = \langle a_1, a_2 \rangle$ and $\mathbf{v} = \langle b_1, b_2 \rangle$, as shown in **Figure 37.** The **angle θ between u and v** is defined to be the angle having the two vectors as its sides for which $0° \le \theta \le 180°$.

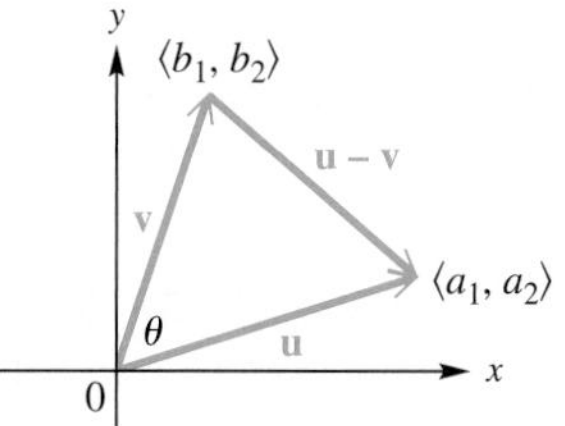

Figure 37

We can use the law of cosines to develop a formula to find angle θ in **Figure 37.**

$$|\mathbf{u}-\mathbf{v}|^2 = |\mathbf{u}|^2 + |\mathbf{v}|^2 - 2|\mathbf{u}||\mathbf{v}|\cos\theta$$

Law of cosines applied to **Figure 37** (Section 8.2)

$$\left(\sqrt{(a_1-b_1)^2+(a_2-b_2)^2}\right)^2 = \left(\sqrt{a_1^2+a_2^2}\right)^2 + \left(\sqrt{b_1^2+b_2^2}\right)^2 - 2|\mathbf{u}||\mathbf{v}|\cos\theta$$

Magnitude of a vector

$$a_1^2 - 2a_1b_1 + b_1^2 + a_2^2 - 2a_2b_2 + b_2^2 = a_1^2 + a_2^2 + b_1^2 + b_2^2 - 2|\mathbf{u}||\mathbf{v}|\cos\theta$$

Square. (Section R.7)

$$-2a_1b_1 - 2a_2b_2 = -2|\mathbf{u}||\mathbf{v}|\cos\theta$$

Subtract like terms from each side.

$$a_1b_1 + a_2b_2 = |\mathbf{u}||\mathbf{v}|\cos\theta$$

Divide by -2.

$$\mathbf{u}\cdot\mathbf{v} = |\mathbf{u}||\mathbf{v}|\cos\theta$$

Definition of dot product

$$\cos\theta = \frac{\mathbf{u}\cdot\mathbf{v}}{|\mathbf{u}||\mathbf{v}|}$$

Divide by $|\mathbf{u}||\mathbf{v}|$ and rewrite.

Geometric Interpretation of Dot Product

If θ is the angle between the two nonzero vectors **u** and **v**, where $0^\circ \le \theta \le 180^\circ$, then the following holds.

$$\cos\theta = \frac{\mathbf{u}\cdot\mathbf{v}}{|\mathbf{u}||\mathbf{v}|}$$

Classroom Example 7
Find the angle θ between the two vectors.

(a) $\mathbf{u} = \langle 5, -12\rangle$ and $\mathbf{v} = \langle 4, 3\rangle$

(b) $\mathbf{u} = \langle 4, -3\rangle$ and $\mathbf{v} = \langle 6, 8\rangle$

Answers:
(a) $\theta \approx 104.25^\circ$ (b) $\theta = 90^\circ$

EXAMPLE 7 Finding the Angle between Two Vectors

Find the angle θ between the two vectors.

(a) $\mathbf{u} = \langle 3, 4\rangle$ and $\mathbf{v} = \langle 2, 1\rangle$ **(b)** $\mathbf{u} = \langle 2, -6\rangle$ and $\mathbf{v} = \langle 6, 2\rangle$

SOLUTION

(a)
$$\cos\theta = \frac{\mathbf{u}\cdot\mathbf{v}}{|\mathbf{u}||\mathbf{v}|} = \frac{\langle 3,4\rangle\cdot\langle 2,1\rangle}{|\langle 3,4\rangle||\langle 2,1\rangle|}$$ Substitute values.

$$= \frac{3(2)+4(1)}{\sqrt{9+16}\cdot\sqrt{4+1}}$$ Use the definitions.

$$= \frac{10}{5\sqrt{5}} \approx 0.894427191$$ Use a calculator.

Therefore, $\theta \approx \cos^{-1} 0.894427191 \approx 26.57^\circ$. Use the inverse cosine function. (Section 7.5)

(b)
$$\cos\theta = \frac{\mathbf{u}\cdot\mathbf{v}}{|\mathbf{u}||\mathbf{v}|} = \frac{\langle 2,-6\rangle\cdot\langle 6,2\rangle}{|\langle 2,-6\rangle||\langle 6,2\rangle|}$$ Substitute values.

$$= \frac{2(6)+(-6)(2)}{\sqrt{4+36}\cdot\sqrt{36+4}}$$ Use the definitions.

$$= \frac{0}{40} = 0$$ Evaluate.

$$\theta = \cos^{-1} 0 = 90^\circ$$ $\cos^{-1} 0 = 90^\circ$

Now Try Exercises 77 and 79.

For angles θ between 0° and 180°, cos θ is positive, 0, or negative when θ is less than, equal to, or greater than 90°, respectively. Therefore, the dot product of nonzero vectors is positive, 0, or negative according to this table.

Dot Product	Angle between Vectors
Positive	Acute
0	Right
Negative	Obtuse

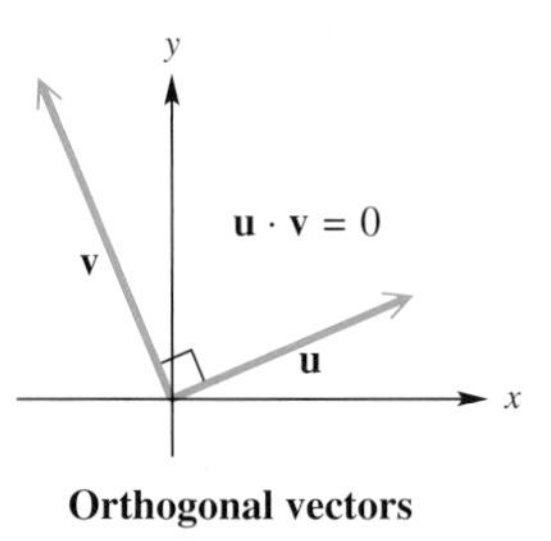

Orthogonal vectors

Figure 38

Thus, in **Example 7** on the preceding page, the vectors in part (a) form an acute angle, and those in part (b) form a right angle. If $\mathbf{u} \cdot \mathbf{v} = 0$ for two nonzero vectors **u** and **v**, then $\cos \theta = 0$ and $\theta = 90°$. Thus, **u** and **v** are perpendicular vectors, also called **orthogonal vectors.** See **Figure 38.**

8.3 Exercises

1. **m** and **p**; **n** and **r**
2. **m** and **q**; **p** and **q**; **r** and **s**; **n** and **s**
3. **m** and **p** equal 2**t**, or **t** equals $\frac{1}{2}$**m** and $\frac{1}{2}$**p**. Also **m** = 1**p** and **n** = 1**r**.
4. **m** = −1**q**; **p** = −1**q**; **r** = −1**s**; **q** = −2**t**; **n** = −1**s**

5.

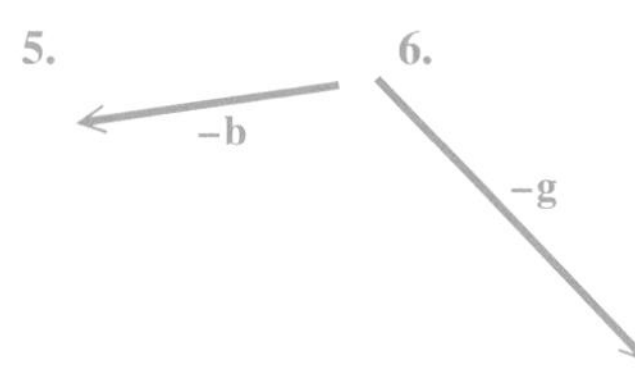

6.

7.

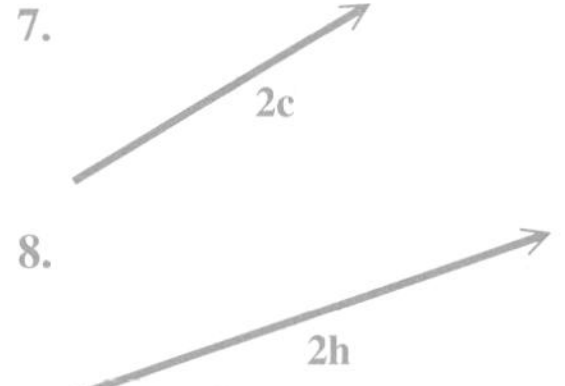

8.

9.

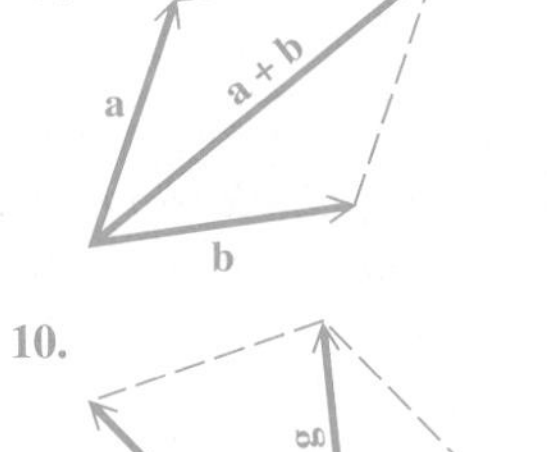

10.

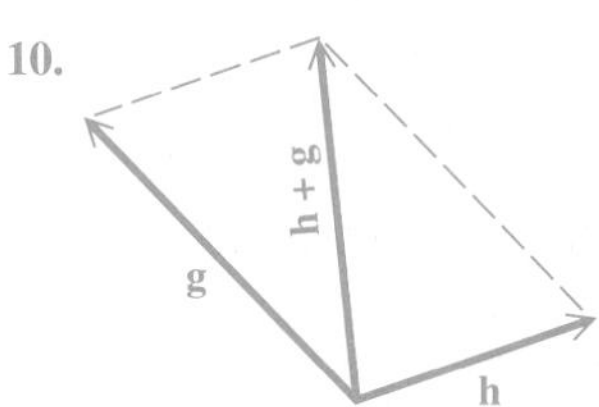

Concept Check *Exercises 1–4 refer to the vectors* **m** *through* **t** *at the right.*

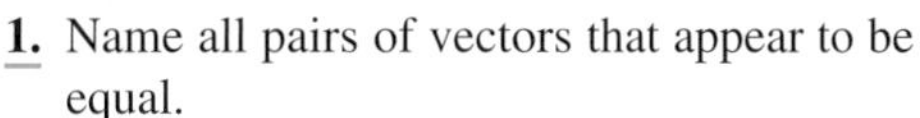

1. Name all pairs of vectors that appear to be equal.

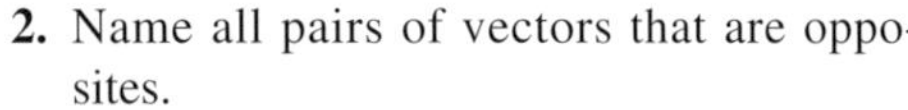

2. Name all pairs of vectors that are opposites.

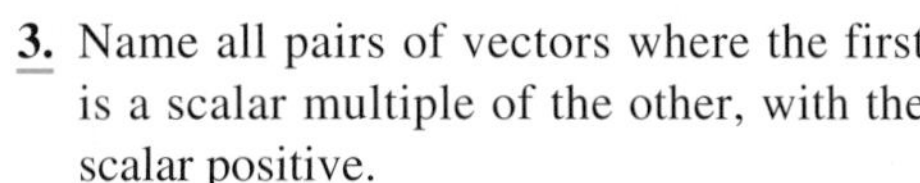

3. Name all pairs of vectors where the first is a scalar multiple of the other, with the scalar positive.

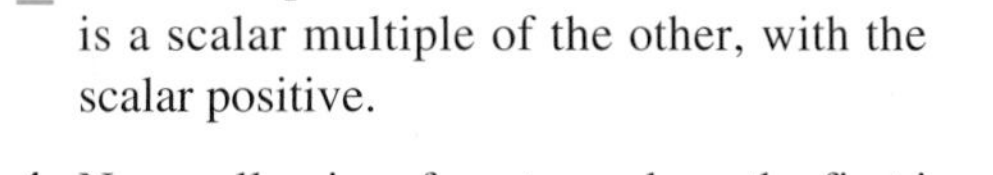

4. Name all pairs of vectors where the first is a scalar multiple of the other, with the scalar negative.

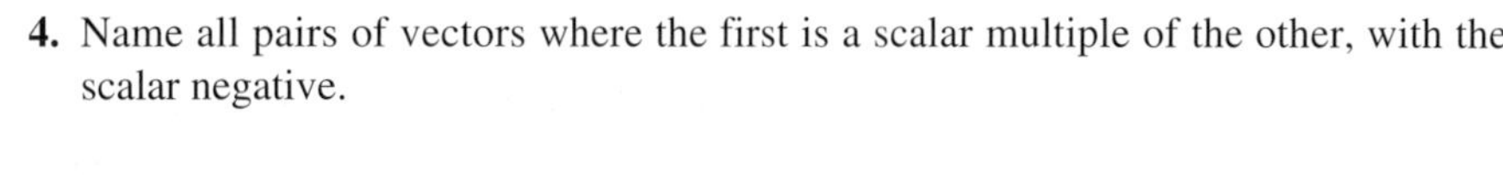

Concept Check *Refer to vectors* **a** *through* **h** *below. Make a copy or a sketch of each vector, and then draw a sketch to represent each vector in Exercises 5–16. For example, find* **a** + **e** *by placing* **a** *and* **e** *so that their initial points coincide. Then use the parallelogram rule to find the resultant, as shown in the figure on the right.*

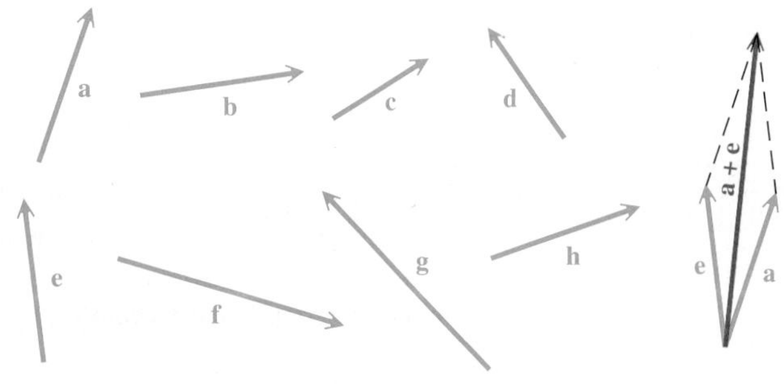

5. −**b**	6. −**g**	7. 2**c**	8. 2**h**
9. **a** + **b**	10. **h** + **g**	11. **a** − **c**	12. **d** − **e**
13. **a** + (**b** + **c**)	14. (**a** + **b**) + **c**	15. **c** + **d**	16. **d** + **c**

11. 12.

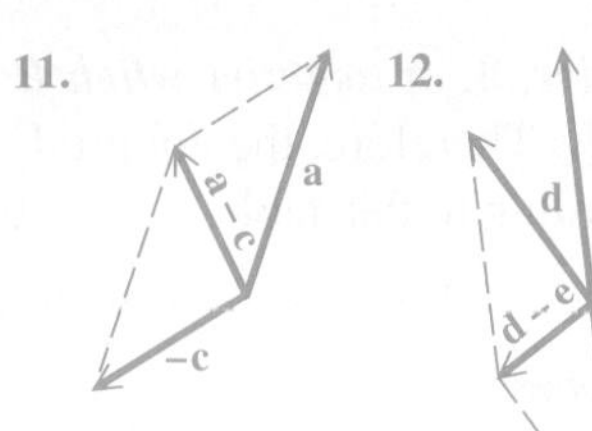

13.

a + (b + c)
a
c
b + c
b

14.

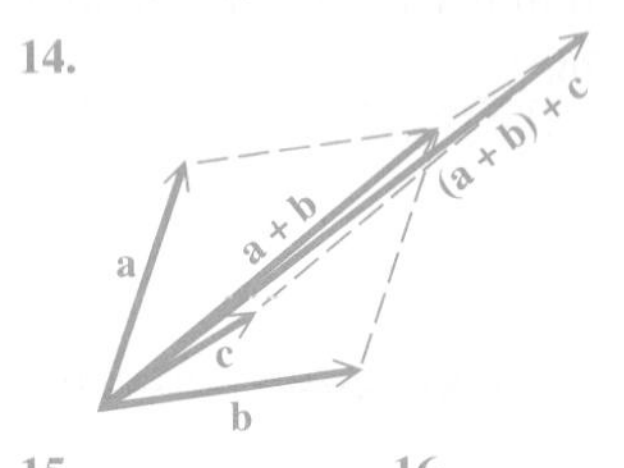

15. 16.

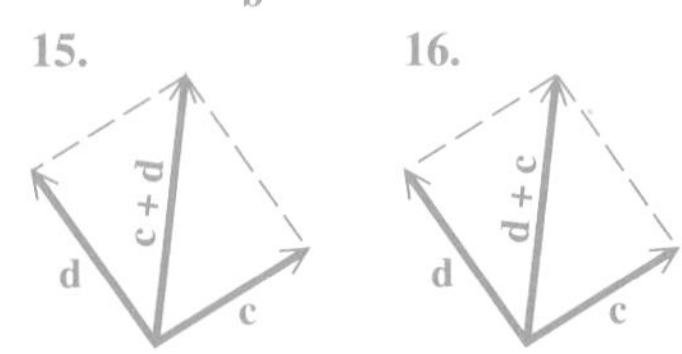

17. Yes, it appears that vector addition is associative (and this is true, in general).

18. Yes, it appears that vector addition is commutative (and this is true, in general).

19. (a) $\langle -4, 16\rangle$ (b) $\langle -12, 0\rangle$ (c) $\langle 8, -8\rangle$

20. (a) $\langle -4, -8\rangle$ (b) $\langle 12, 0\rangle$ (c) $\langle -4, 4\rangle$

21. (a) $\langle 8, 0\rangle$ (b) $\langle 0, 16\rangle$ (c) $\langle -4, -8\rangle$

22. (a) $\langle 4, 0\rangle$ (b) $\langle -12, -8\rangle$ (c) $\langle 4, 4\rangle$

23. (a) $\langle 0, 12\rangle$ (b) $\langle -16, -4\rangle$ (c) $\langle 8, -4\rangle$

24. (a) $\langle 4, 4\rangle$ (b) $\langle 12, -12\rangle$ (c) $\langle -8, 4\rangle$

25. (a) $4\mathbf{i}$ (b) $7\mathbf{i} + 3\mathbf{j}$ (c) $-5\mathbf{i} + \mathbf{j}$

26. (a) $-2\mathbf{i} + 4\mathbf{j}$ (b) $\mathbf{i} + \mathbf{j}$ (c) $4\mathbf{i} - 7\mathbf{j}$

27. (a) $\langle -2, 4\rangle$ (b) $\langle 7, 4\rangle$ (c) $\langle 6, -6\rangle$

17. From the results of **Exercises 13 and 14,** does it appear that vector addition is associative?

18. From the results of **Exercises 15 and 16,** does it appear that vector addition is commutative?

In Exercises 19–24, use the figure to find each vector: (a) $\mathbf{u} + \mathbf{v}$ *(b)* $\mathbf{u} - \mathbf{v}$ *(c)* $-\mathbf{u}$. *Use vector notation as in **Example 3.***

19. **20.** **21.**

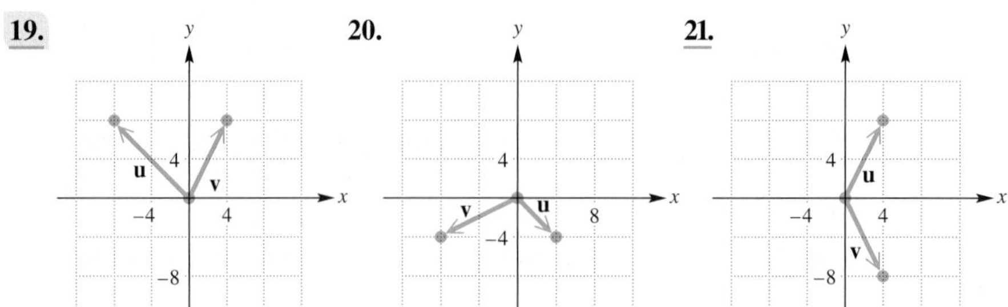

22. **23.** **24.**

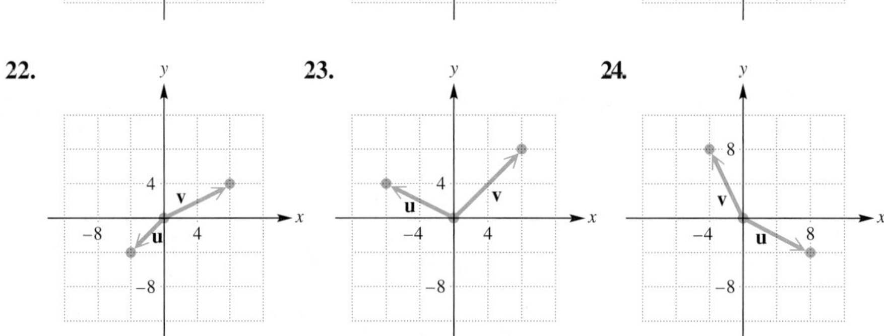

Given vectors **u** *and* **v**, *find: (a)* $2\mathbf{u}$ *(b)* $2\mathbf{u} + 3\mathbf{v}$ *(c)* $\mathbf{v} - 3\mathbf{u}$.

25. $\mathbf{u} = 2\mathbf{i}, \mathbf{v} = \mathbf{i} + \mathbf{j}$

26. $\mathbf{u} = -\mathbf{i} + 2\mathbf{j}, \mathbf{v} = \mathbf{i} - \mathbf{j}$

27. $\mathbf{u} = \langle -1, 2\rangle, \mathbf{v} = \langle 3, 0\rangle$

28. $\mathbf{u} = \langle -2, -1\rangle, \mathbf{v} = \langle -3, 2\rangle$

For each pair of vectors **u** *and* **v** *with angle* θ *between them, sketch the resultant.*

29. $|\mathbf{u}| = 12, |\mathbf{v}| = 20, \theta = 27°$

30. $|\mathbf{u}| = 8, |\mathbf{v}| = 12, \theta = 20°$

31. $|\mathbf{u}| = 20, |\mathbf{v}| = 30, \theta = 30°$

32. $|\mathbf{u}| = 50, |\mathbf{v}| = 70, \theta = 40°$

*Find the magnitude and direction angle for each vector. **See Example 1.***

33. $\langle 15, -8\rangle$

34. $\langle -7, 24\rangle$

35. $\langle -4, 4\sqrt{3}\rangle$

36. $\langle 8\sqrt{2}, -8\sqrt{2}\rangle$

For each of the following, vector **v** *has the given direction and magnitude. Find the magnitudes of the horizontal and vertical components of* **v**, *if* θ *is the direction angle of* **v** *from the horizontal. **See Example 2.***

37. $\theta = 20°, |\mathbf{v}| = 50$

38. $\theta = 50°, |\mathbf{v}| = 26$

39. $\theta = 35° \; 50', |\mathbf{v}| = 47.8$

40. $\theta = 27° \; 30', |\mathbf{v}| = 15.4$

41. $\theta = 128.5°, |\mathbf{v}| = 198$

42. $\theta = 146.3°, |\mathbf{v}| = 238$

28. (a) $\langle -4, -2 \rangle$ (b) $\langle -13, 4 \rangle$ (c) $\langle 3, 5 \rangle$

29. 12, 27°, 20

30. 20°, 8, 12

31. 20, 30°, 30

32. 50, 40°, 70

33. 17; 331.9° 34. 25; 106.3°
35. 8; 120° 36. 16; 315°

37. 47, 17 38. 17, 20
39. 38.8, 28.0 40. 13.7, 7.11
41. 123, 155 42. 198, 132

43. $\left\langle \frac{5\sqrt{3}}{2}, \frac{5}{2} \right\rangle$ 44. $\langle 4, 4\sqrt{3} \rangle$
45. $\langle -3.0642, 2.5712 \rangle$
46. $\langle -1.9284, 2.2981 \rangle$
47. $\langle 4.0958, -2.8679 \rangle$
48. $\langle -1.5321, -1.2856 \rangle$

49. 530 newtons
50. 29 newtons
51. 88.2 lb 52. 76.2 lb

53. 94.2 lb 54. 158.0 lb
55. 24.4 lb 56. 1286.0 lb

57. $\langle a + c, b + d \rangle$

59. $\langle -6, 2 \rangle$ 60. $\langle 6, -2 \rangle$
61. $\langle 8, -20 \rangle$ 62. $\langle -20, -15 \rangle$
63. $\langle -30, -3 \rangle$ 64. $\langle 20, 2 \rangle$
65. $\langle 8, -7 \rangle$ 66. $\langle -24, -5 \rangle$

67. $-5\mathbf{i} + 8\mathbf{j}$ 68. $6\mathbf{i} - 3\mathbf{j}$
69. $2\mathbf{i}$, or $2\mathbf{i} + 0\mathbf{j}$
70. $-4\mathbf{j}$, or $0\mathbf{i} - 4\mathbf{j}$

Write each vector in the form $\langle a, b \rangle$. ***See Example 3.***

43. 44. 45.

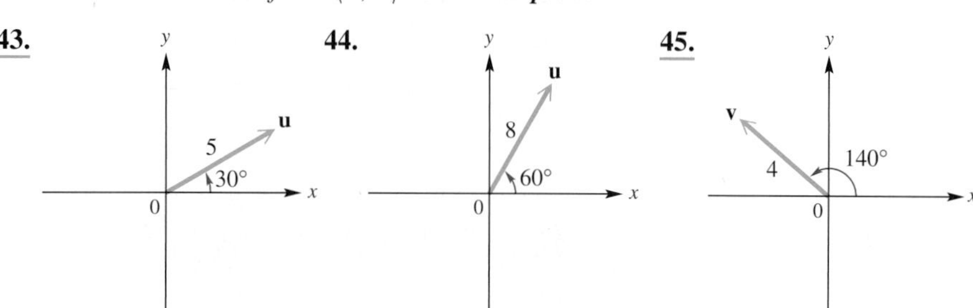

46. 47. 48.

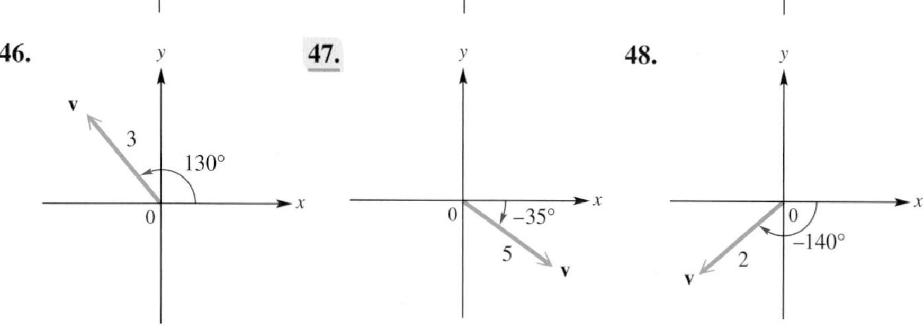

Two forces act at a point in the plane. The angle between the two forces is given. Find the magnitude of the resultant force. ***See Example 4.***

49. forces of 250 and 450 newtons, forming an angle of 85°

50. forces of 19 and 32 newtons, forming an angle of 118°

51. forces of 116 and 139 lb, forming an angle of 140° 50′

52. forces of 37.8 and 53.7 lb, forming an angle of 68.5°

Use the parallelogram rule to find the magnitude of the resultant force for the two forces shown in each figure. Round answers to the nearest tenth.

53.

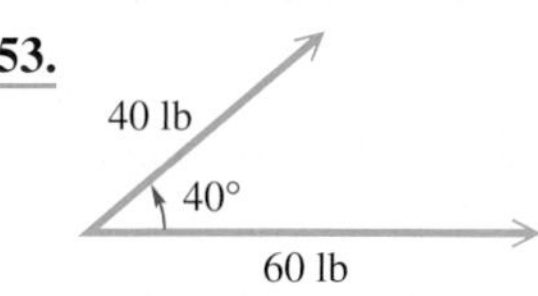

54.

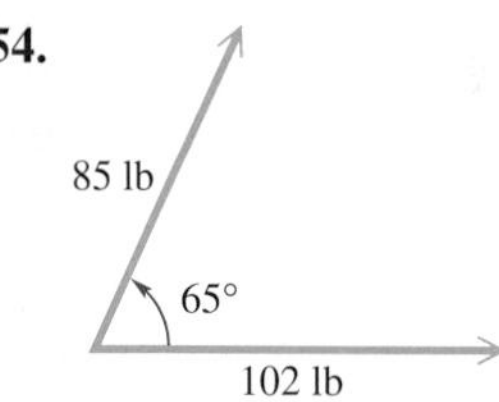

55.

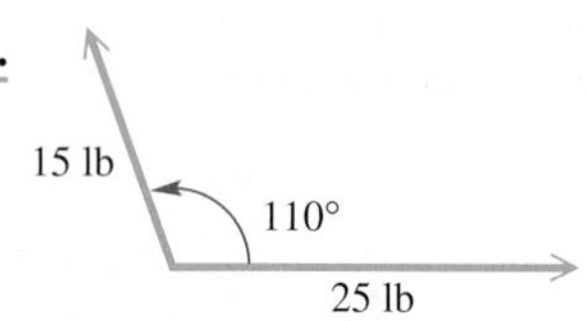

56.

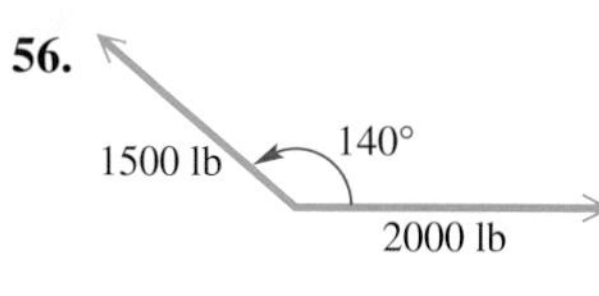

57. *Concept Check* If $\mathbf{u} = \langle a, b \rangle$ and $\mathbf{v} = \langle c, d \rangle$, what is the vector notation for $\mathbf{u} + \mathbf{v}$?

58. Explain how to add vectors.

Given $\mathbf{u} = \langle -2, 5 \rangle$ and $\mathbf{v} = \langle 4, 3 \rangle$, find each of the following. ***See Example 5.***

59. $\mathbf{u} - \mathbf{v}$ 60. $\mathbf{v} - \mathbf{u}$ 61. $-4\mathbf{u}$ 62. $-5\mathbf{v}$

63. $3\mathbf{u} - 6\mathbf{v}$ 64. $-2\mathbf{u} + 4\mathbf{v}$ 65. $\mathbf{u} + \mathbf{v} - 3\mathbf{u}$ 66. $2\mathbf{u} + \mathbf{v} - 6\mathbf{v}$

Write each vector in the form $a\mathbf{i} + b\mathbf{j}$. ***See Figure 36(b).***

67. $\langle -5, 8 \rangle$ 68. $\langle 6, -3 \rangle$ 69. $\langle 2, 0 \rangle$ 70. $\langle 0, -4 \rangle$

71. 7 72. −61
73. 0 74. 0
75. 20 76. −4

77. 135° 78. 36.87°
79. 90° 80. 45°
81. 36.87° 82. 78.93°

83. −6 84. −6
85. −24 86. −24

87. orthogonal
88. orthogonal
89. not orthogonal
90. not orthogonal
91. not orthogonal
92. not orthogonal

In Exercises 93–97, answers may vary due to rounding.

93. magnitude: 9.5208; direction angle: 119.0647°
94. $\langle -4.1042, 11.2763 \rangle$
95. $\langle -0.5209, -2.9544 \rangle$
96. $\langle -4.6252, 8.3219 \rangle$
97. magnitude: 9.5208; direction angle: 119.0647°
98. They are the same. Preference of method is an individual choice.

Find the dot product for each pair of vectors. ***See Example 6.***

71. $\langle 6, -1 \rangle, \langle 2, 5 \rangle$ 72. $\langle -3, 8 \rangle, \langle 7, -5 \rangle$ 73. $\langle 5, 2 \rangle, \langle -4, 10 \rangle$

74. $\langle 7, -2 \rangle, \langle 4, 14 \rangle$ 75. $4\mathbf{i}, 5\mathbf{i} - 9\mathbf{j}$ 76. $2\mathbf{i} + 4\mathbf{j}, -\mathbf{j}$

Find the angle between each pair of vectors. ***See Example 7.***

77. $\langle 2, 1 \rangle, \langle -3, 1 \rangle$ 78. $\langle 1, 7 \rangle, \langle 1, 1 \rangle$ 79. $\langle 1, 2 \rangle, \langle -6, 3 \rangle$

80. $\langle 4, 0 \rangle, \langle 2, 2 \rangle$ 81. $3\mathbf{i} + 4\mathbf{j}, \mathbf{j}$ 82. $-5\mathbf{i} + 12\mathbf{j}, 3\mathbf{i} + 2\mathbf{j}$

Let $\mathbf{u} = \langle -2, 1 \rangle$, $\mathbf{v} = \langle 3, 4 \rangle$, *and* $\mathbf{w} = \langle -5, 12 \rangle$. *Evaluate each expression.*

83. $(3\mathbf{u}) \cdot \mathbf{v}$ 84. $\mathbf{u} \cdot (3\mathbf{v})$ 85. $\mathbf{u} \cdot \mathbf{v} - \mathbf{u} \cdot \mathbf{w}$ 86. $\mathbf{u} \cdot (\mathbf{v} - \mathbf{w})$

Determine whether each pair of vectors is orthogonal. ***See Example 7(b).***

87. $\langle 1, 2 \rangle, \langle -6, 3 \rangle$ 88. $\langle 1, 1 \rangle, \langle 1, -1 \rangle$

89. $\langle 1, 0 \rangle, \langle \sqrt{2}, 0 \rangle$ 90. $\langle 3, 4 \rangle, \langle 6, 8 \rangle$

91. $\sqrt{5}\mathbf{i} - 2\mathbf{j}, -5\mathbf{i} + 2\sqrt{5}\mathbf{j}$ 92. $-4\mathbf{i} + 3\mathbf{j}, 8\mathbf{i} - 6\mathbf{j}$

Relating Concepts

For individual or collaborative investigation *(Exercises 93–98)*

Consider the two vectors **u** *and* **v** *shown. Assume all values are exact.* ***Work Exercises 93–98 in order.***

93. Use trigonometry alone (without using vector notation) to find the magnitude and direction angle of $\mathbf{u} + \mathbf{v}$. Use the law of cosines and the law of sines in your work.

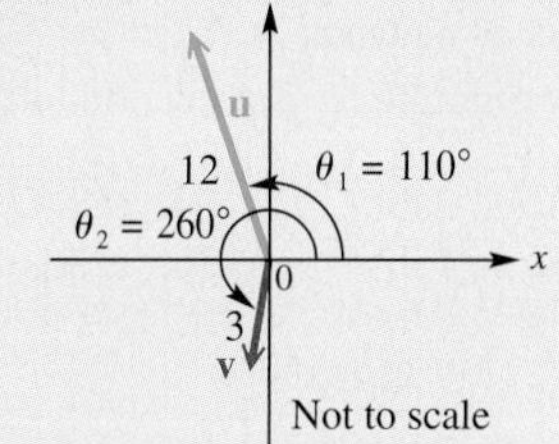

94. Find the horizontal and vertical components of **u**, using your calculator.

95. Find the horizontal and vertical components of **v**, using your calculator.

96. Find the horizontal and vertical components of $\mathbf{u} + \mathbf{v}$ by adding the results you obtained in **Exercises 94 and 95.**

97. Use your calculator to find the magnitude and direction angle of the vector $\mathbf{u} + \mathbf{v}$.

98. Compare your answers in **Exercises 93 and 97.** What do you notice? Which method of solution do you prefer?

8.4 Applications of Vectors

- The Equilibrant
- Incline Applications
- Navigation Applications

The Equilibrant The previous section covered methods for finding the resultant of two vectors. Sometimes it is necessary to find a vector that will counterbalance the resultant. This opposite vector is called the **equilibrant.** That is, the equilibrant of vector **u** is the vector $-\mathbf{u}$.

Classroom Example 1
Find the magnitude of the equilibrant of forces of 54 newtons and 42 newtons acting on a point A, if the angle between the forces is 98°. Then find the angle between the equilibrant and the 42-newton force.

Answer: 64 newtons; 123°

EXAMPLE 1 Finding the Magnitude and Direction of an Equilibrant

Find the magnitude of the equilibrant of forces of 48 newtons and 60 newtons acting on a point A, if the angle between the forces is 50°. Then find the angle between the equilibrant and the 48-newton force.

SOLUTION

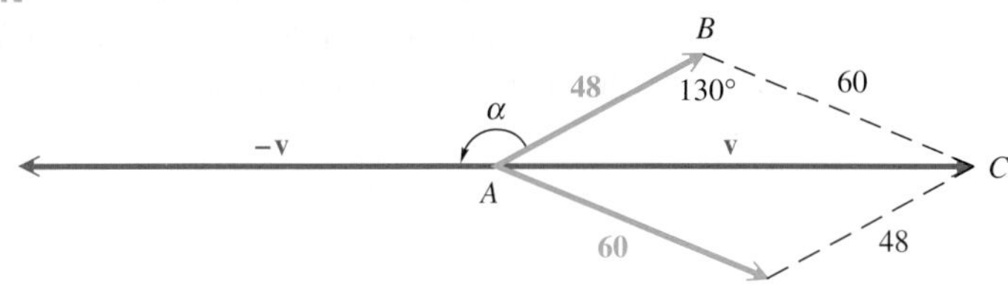

Figure 39

Teaching Tip In **Example 1,** caution students not to assume that $\overrightarrow{AC}$ bisects the angle formed by the two forces. In fact, this happens only if the forces have equal magnitudes.

In **Figure 39,** the equilibrant is $-\mathbf{v}$. The magnitude of $\mathbf{v}$, and hence of $-\mathbf{v}$, is found by using triangle ABC and the law of cosines.

$$|\mathbf{v}|^2 = 48^2 + 60^2 - 2(48)(60)\cos 130° \quad \text{Law of cosines (Section 8.2)}$$

$$|\mathbf{v}|^2 \approx 9606.5 \quad \text{Use a calculator.}$$

$$|\mathbf{v}| \approx 98 \text{ newtons} \quad \text{Two significant digits (Section 5.4)}$$

The required angle, labeled α in **Figure 39,** can be found by subtracting angle CAB from 180°. Use the law of sines to find angle CAB.

$$\frac{\sin CAB}{60} = \frac{\sin 130°}{98} \quad \text{Law of sines (alternative form) (Section 8.1)}$$

$$\sin CAB \approx 0.46900680 \quad \text{Multiply by 60 and use a calculator.}$$

$$CAB \approx 28° \quad \text{Use the inverse sine function. (Section 7.5)}$$

Finally, $\alpha \approx 180° - 28° = 152°$.

✔ *Now Try Exercise 1.*

Classroom Example 2
Find the force required to keep a 2500-lb car parked on a hill that makes a 12° angle with the horizontal.

Answer: 520 lb

Incline Applications

We can use vectors to solve incline problems.

EXAMPLE 2 Finding a Required Force

Find the force required to keep a 50-lb wagon from sliding down a ramp inclined at 20° to the horizontal. (Assume there is no friction.)

SOLUTION In **Figure 40,** the vertical 50-lb force **BA** represents the force of gravity. It is the sum of vectors **BC** and −**AC**. The vector **BC** represents the force with which the weight pushes against the ramp. The vector **BF** represents the force that would pull the weight up the ramp. Since vectors **BF** and **AC** are equal, $|\mathbf{AC}|$ gives the magnitude of the required force.

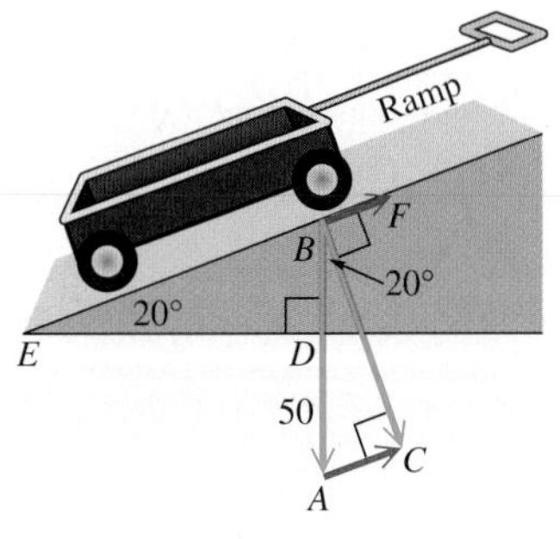

Figure 40

Vectors **BF** and **AC** are parallel, so angle EBD equals angle A. Since angle BDE and angle C are right angles, triangles CBA and DEB have two corresponding angles equal and, thus, are similar triangles. Therefore, angle ABC equals angle E, which is 20°. From right triangle ABC, we have the following.

$$\sin 20° = \frac{|\mathbf{AC}|}{50} \quad \text{(Section 5.3)}$$

$$|\mathbf{AC}| = 50 \sin 20° \quad \text{Multiply by 50 and rewrite.}$$

$$|\mathbf{AC}| \approx 17 \quad \text{Use a calculator.}$$

A force of approximately 17 lb will keep the wagon from sliding down the ramp.

✔ *Now Try Exercise 9.*

Teaching Tip Emphasize that weight is a *force* whose direction is always straight down—toward the center of Earth.

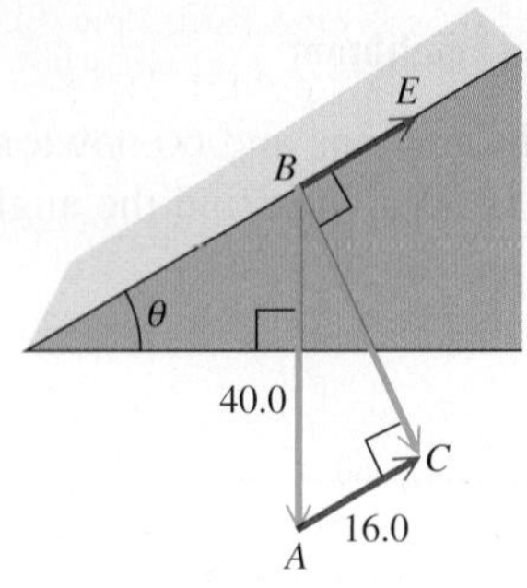

Figure 41

EXAMPLE 3 Finding an Incline Angle

A force of 16.0 lb is required to hold a 40.0-lb lawn mower on an incline. What angle does the incline make with the horizontal?

SOLUTION **Figure 41** illustrates the situation. Consider right triangle ABC. Angle B equals angle θ, the magnitude of vector **BA** represents the weight of the mower, and vector **AC** equals vector **BE**, which represents the force required to hold the mower on the incline.

$$\sin B = \frac{16.0}{40.0} \qquad \sin B = \frac{\text{side opposite } B}{\text{hypotenuse}} \quad \text{(Section 5.3)}$$

$$\sin B = 0.4 \qquad \text{Simplify.}$$

$$B \approx 23.6^\circ \qquad \text{Use the inverse sine function.}$$

Therefore, the hill makes an angle of about 23.6° with the horizontal.

Now Try Exercise 11.

Classroom Example 3
A force of 18.0 lb is required to hold a 74.0-lb crate on a ramp. What angle does the ramp make with the horizontal?

Answer: 14.1°

Navigation Applications Problems involving bearing (defined in **Section 5.4**) can also be worked with vectors.

Classroom Example 4
A ship leaves port on a bearing of 25.0° and travels 61.4 km. The ship then turns due east and travels 84.6 km. How far is the ship from port? What is its bearing from port?

Answer: 124 km; 63.3°

EXAMPLE 4 Applying Vectors to a Navigation Problem

A ship leaves port on a bearing of 28.0° and travels 8.20 mi. The ship then turns due east and travels 4.30 mi. How far is the ship from port? What is its bearing from port?

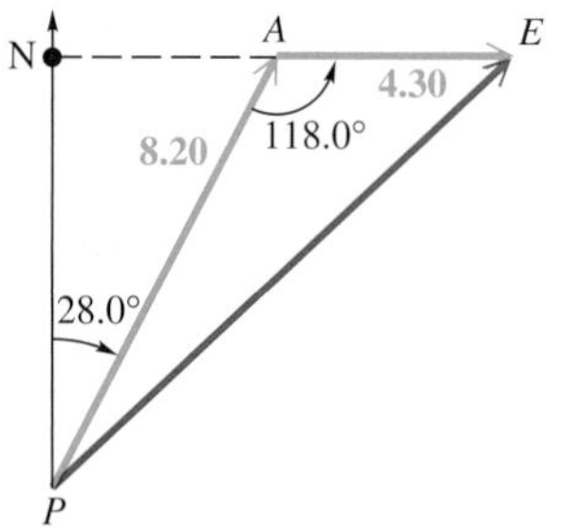

SOLUTION In **Figure 42,** vectors **PA** and **AE** represent the ship's path. The magnitude and bearing of the resultant **PE** can be found as follows. Triangle PNA is a right triangle, so

Figure 42

$$\text{angle } NAP = 90^\circ - 28.0^\circ = 62.0^\circ,$$

and $$\text{angle } PAE = 180^\circ - 62.0^\circ = 118.0^\circ.$$

Use the law of cosines to find $|\mathbf{PE}|$, the magnitude of vector **PE**.

$$|\mathbf{PE}|^2 = 8.20^2 + 4.30^2 - 2(8.20)(4.30)\cos 118.0^\circ \qquad \text{Law of cosines}$$

$$|\mathbf{PE}|^2 \approx 118.84 \qquad \text{Evaluate.}$$

$$|\mathbf{PE}| \approx 10.9 \qquad \text{Square root property (Section 1.4)}$$

The ship is about 10.9 mi from port.

To find the bearing of the ship from port, first find angle APE. Use the law of sines.

$$\frac{\sin APE}{4.30} = \frac{\sin 118.0^\circ}{10.9} \qquad \text{Law of sines}$$

$$\sin APE = \frac{4.30 \sin 118.0^\circ}{10.9} \qquad \text{Multiply by 4.30.}$$

$$APE \approx 20.4^\circ \qquad \text{Use the inverse sine function.}$$

Now add 20.4° to 28.0° to find that the bearing is 48.4°.

Now Try Exercise 15.

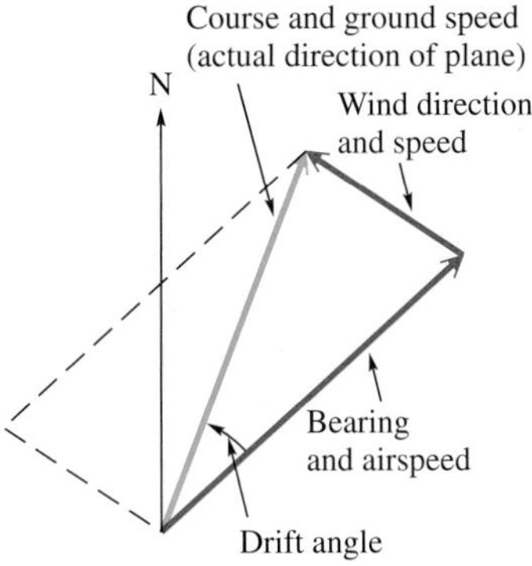

Figure 43

Classroom Example 5

A small plane follows a bearing of 70° at an airspeed of 230 mph and encounters a wind blowing at 25.0 mph from a direction of 290°. Find the resulting bearing and ground speed of the plane.

Answer: bearing: 74°; ground speed: 250 mph

In air navigation, the **airspeed** of a plane is its speed relative to the air, and the **ground speed** is its speed relative to the ground. Because of wind, these two speeds are usually different. The ground speed of the plane is represented by the vector sum of the airspeed and windspeed vectors. See **Figure 43.**

EXAMPLE 5 Applying Vectors to a Navigation Problem

An airplane that is following a bearing of 239° at an airspeed of 425 mph encounters a wind blowing at 36.0 mph from a direction of 115°. Find the resulting bearing and ground speed of the plane.

SOLUTION An accurate sketch is essential to the solution of this problem. We have included two sets of geographical axes which enable us to determine measures of necessary angles. Analyze **Figure 44** carefully.

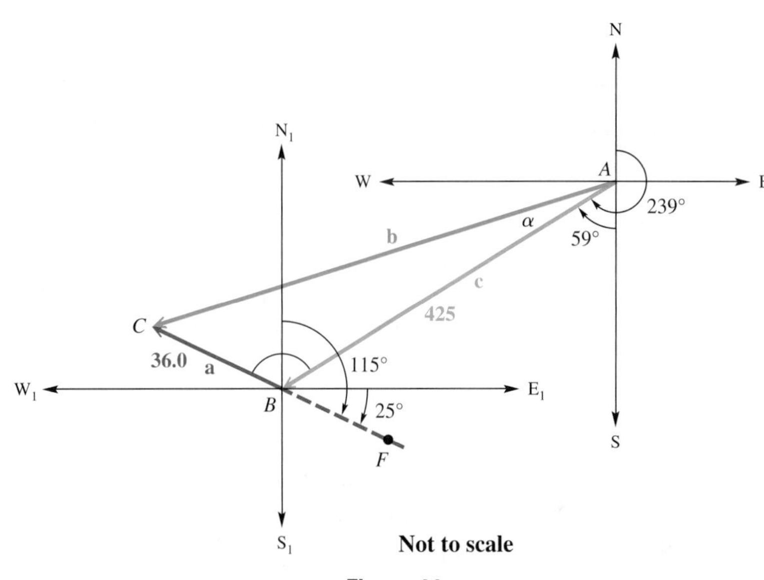

Figure 44

Vector **c** represents the airspeed and bearing of the plane, and vector **a** represents the speed and direction of the wind. Angle ABC has as its measure the sum of angle ABN_1 and angle N_1BC.

- Angle SAB measures $239° - 180° = 59°$. Because angle ABN_1 is an alternate interior angle to it, $ABN_1 = 59°$.
- Angle E_1BF measures $115° - 90° = 25°$. Thus, angle CBW_1 also measures 25° because it is a vertical angle. Angle N_1BC is the complement of 25°, which is

$$90° - 25° = 65°.$$

By these results,

$$\text{angle } ABC = 59° + 65° = 124°.$$

To find $|\mathbf{b}|$, we use the law of cosines.

$$|\mathbf{b}|^2 = |\mathbf{a}|^2 + |\mathbf{c}|^2 - 2|\mathbf{a}||\mathbf{c}| \cos ABC \quad \text{Law of cosines}$$

$$|\mathbf{b}|^2 = 36.0^2 + 425^2 - 2(36.0)(425) \cos 124° \quad \text{Substitute.}$$

$$|\mathbf{b}|^2 \approx 199{,}032 \quad \text{Use a calculator.}$$

$$|\mathbf{b}| \approx 446 \quad \text{Square root property}$$

The ground speed is approximately 446 mph.

To find the resulting bearing of **b**, we must find the measure of angle α in **Figure 44** and then add it to 239°. To find α, we use the law of sines.

$$\frac{\sin \alpha}{36.0} = \frac{\sin 124°}{446}$$ To maintain accuracy, use all the significant digits that your calculator allows.

$$\sin \alpha = \frac{36.0 \sin 124°}{446}$$ Multiply by 36.0.

$$\alpha = \sin^{-1}\left(\frac{36.0 \sin 124°}{446}\right)$$ Use the inverse sine function.

$$\alpha \approx 4°$$ Use a calculator.

Add 4° to 239° to find the resulting bearing of 243°.

Now Try Exercise 21.

8.4 Exercises

1. 2640 lb at an angle of 167.2° with the 1480-lb force
2. 1800 lb at an angle of 167° with the 840-lb force
3. 93.9°
4. 70.1°
5. 190 lb and 283 lb, respectively
6. magnitude of resultant: 117 lb; force: 93.9 lb
7. 18°
8. 800 lb

Solve each problem. ***See Examples 1–3.***

1. *Direction and Magnitude of an Equilibrant* Two tugboats are pulling a disabled speedboat into port with forces of 1240 lb and 1480 lb. The angle between these forces is 28.2°. Find the direction and magnitude of the equilibrant.

2. *Direction and Magnitude of an Equilibrant* Two rescue vessels are pulling a broken-down motorboat toward a boathouse with forces of 840 lb and 960 lb. The angle between these forces is 24.5°. Find the direction and magnitude of the equilibrant.

3. *Angle between Forces* Two forces of 692 newtons and 423 newtons act at a point. The resultant force is 786 newtons. Find the angle between the forces.

4. *Angle between Forces* Two forces of 128 lb and 253 lb act at a point. The resultant force is 320 lb. Find the angle between the forces.

5. *Magnitudes of Forces* A force of 176 lb makes an angle of 78° 50′ with a second force. The resultant of the two forces makes an angle of 41° 10′ with the first force. Find the magnitudes of the second force and of the resultant.

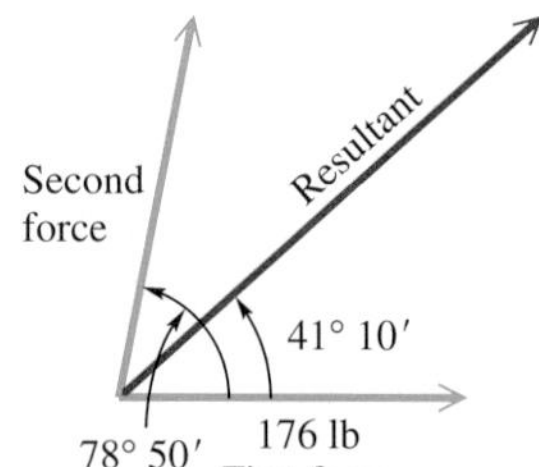

6. *Magnitudes of Forces* A force of 28.7 lb makes an angle of 42° 10′ with a second force. The resultant of the two forces makes an angle of 32° 40′ with the first force. Find the magnitudes of the second force and of the resultant.

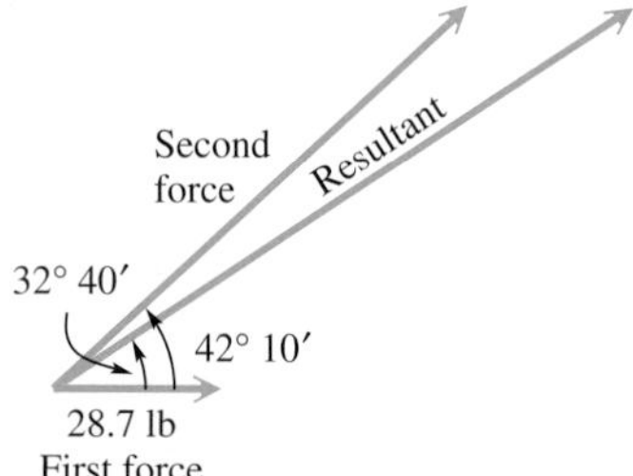

7. *Angle of a Hill Slope* A force of 25 lb is required to hold an 80-lb crate on a hill. What angle does the hill make with the horizontal?

8. *Force Needed to Keep a Car Parked* Find the force required to keep a 3000-lb car parked on a hill that makes an angle of 15° with the horizontal.

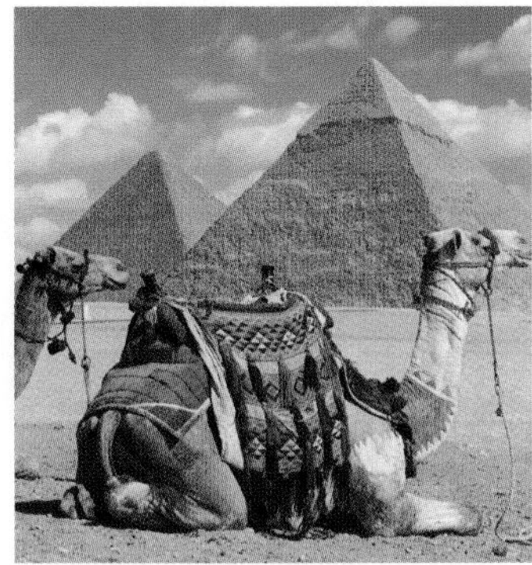

9. *Force Needed for a Monolith* To build the pyramids in Egypt, it is believed that giant causeways were constructed to transport the building materials to the site. One such causeway is said to have been 3000 ft long, with a slope of about 2.3°. How much force would be required to hold a 60-ton monolith on this causeway?

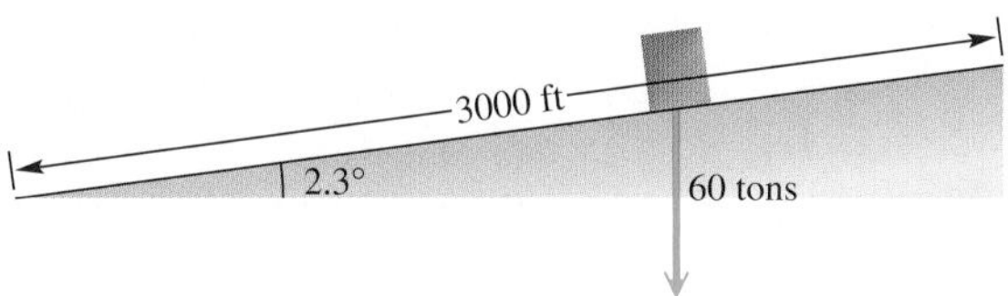

10. *Force Needed for a Monolith* If the causeway in **Exercise 9** were 500 ft longer and the monolith weighed 10 tons more, how much force would be required?

9. 2.4 tons **10.** 2.8 tons
11. 17.5° **12.** 22.0°
13. 226 lb
14. weight: 64.8 lb; tension: 61.9 lb
15. 13.5 mi; 50.4°
16. 6.6 mi; 117.1°
17. 39.2 km
18. 14.5 km
19. current: 3.5 mph; motorboat: 19.7 mph

11. *Incline Angle* A force of 18.0 lb is required to hold a 60.0-lb stump grinder on an incline. What angle does the incline make with the horizontal?

12. *Incline Angle* A force of 30.0 lb is required to hold an 80.0-lb pressure washer on an incline. What angle does the incline make with the horizontal?

13. *Weight of a Box* Two people are carrying a box. One person exerts a force of 150 lb at an angle of 62.4° with the horizontal. The other person exerts a force of 114 lb at an angle of 54.9°. Find the weight of the box.

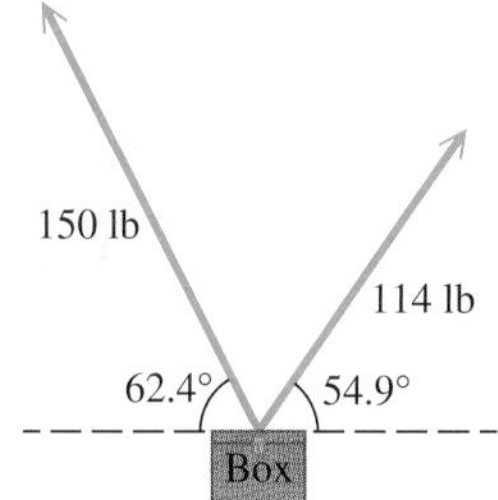

14. *Weight of a Crate and Tension of a Rope* A crate is supported by two ropes. One rope makes an angle of 46° 20′ with the horizontal and has a tension of 89.6 lb on it. The other rope is horizontal. Find the weight of the crate and the tension in the horizontal rope.

Solve each problem. ***See Examples 4 and 5.***

15. *Distance and Bearing of a Ship* A ship leaves port on a bearing of 34.0° and travels 10.4 mi. The ship then turns due east and travels 4.6 mi. How far is the ship from port, and what is its bearing from port?

16. *Distance and Bearing of a Luxury Liner* A luxury liner leaves port on a bearing of 110.0° and travels 8.8 mi. It then turns due west and travels 2.4 mi. How far is the liner from port, and what is its bearing from port?

17. *Distance of a Ship from Its Starting Point* Starting at point A, a ship sails 18.5 km on a bearing of 189°, then turns and sails 47.8 km on a bearing of 317°. Find the distance of the ship from point A.

18. *Distance of a Ship from Its Starting Point* Starting at point X, a ship sails 15.5 km on a bearing of 200°, then turns and sails 2.4 km on a bearing of 320°. Find the distance of the ship from point X.

19. *Distance and Direction of a Motorboat* A motorboat sets out in the direction N 80° 00′ E. The speed of the boat in still water is 20.0 mph. If the current is flowing directly south, and the actual direction of the motorboat is due east, find the speed of the current and the actual speed of the motorboat.

20. (a) 6.3 mph (b) 14 sec (c) 25°
21. bearing: 237°; ground speed: 470 mph
22. 3:21 P.M.
23. ground speed: 161 mph; airspeed: 156 mph
24. 173.1°
25. bearing: 74°; ground speed: 202 mph
26. bearing: 65° 30′; ground speed: 181 mph
27. bearing: 358°; airspeed: 170 mph
28. ground speed: 198 mph; bearing: 186.5°
29. ground speed: 230 km per hr; bearing: 167°

20. *Movement of a Motorboat* Suppose you would like to cross a 132-ft-wide river in a motorboat. Assume that the motorboat can travel at 7.0 mph relative to the water and that the current is flowing west at the rate of 3.0 mph. The bearing θ is chosen so that the motorboat will land at a point exactly across from the starting point.

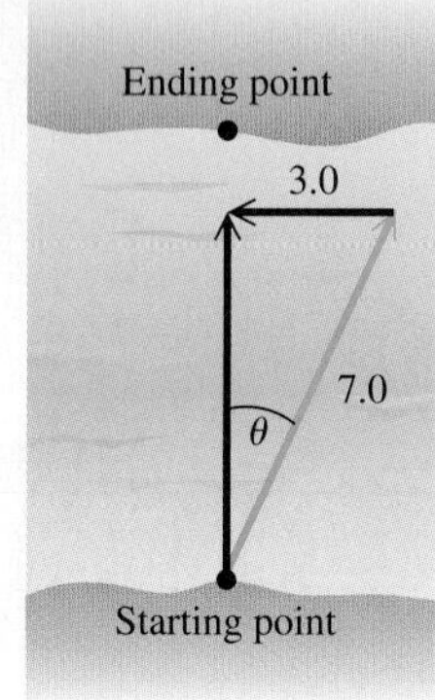

(a) At what speed will the motorboat be traveling relative to the banks?
(b) How long will it take for the motorboat to make the crossing?
(c) What is the measure of angle θ?

21. *Bearing and Ground Speed of a Plane* An airline route from San Francisco to Honolulu is on a bearing of 233.0°. A jet flying at 450 mph on that bearing encounters a wind blowing at 39.0 mph from a direction of 114.0°. Find the resulting bearing and ground speed of the plane.

22. *Path Traveled by a Plane* The aircraft carrier *Tallahassee* is traveling at sea on a steady course with a bearing of 30° at 32 mph. Patrol planes on the carrier have enough fuel for 2.6 hr of flight when traveling at a speed of 520 mph. One of the pilots takes off on a bearing of 338° and then turns and heads in a straight line, so as to be able to catch the carrier and land on the deck at the exact instant that his fuel runs out. If the pilot left at 2 P.M., at what time did he turn to head for the carrier?

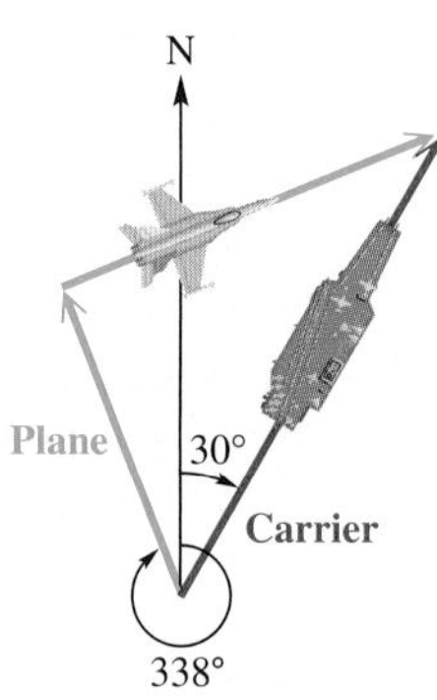

23. *Airspeed and Ground Speed* A pilot wants to fly on a bearing of 74.9°. By flying due east, he finds that a 42.0-mph wind, blowing from the south, puts him on course. Find the airspeed and the ground speed.

24. *Bearing of a Plane* A plane flies 650 mph on a bearing of 175.3°. A 25-mph wind, from a direction of 266.6°, blows against the plane. Find the resulting bearing of the plane.

25. *Bearing and Ground Speed of a Plane* A pilot is flying at 190.0 mph. He wants his flight path to be on a bearing of 64° 30′. A wind is blowing from the south at 35.0 mph. Find the bearing he should fly, and find the plane's ground speed.

26. *Bearing and Ground Speed of a Plane* A pilot is flying at 168 mph. She wants her flight path to be on a bearing of 57° 40′. A wind is blowing from the south at 27.1 mph. Find the bearing the pilot should fly, and find the plane's ground speed.

27. *Bearing and Airspeed of a Plane* What bearing and airspeed are required for a plane to fly 400 mi due north in 2.5 hr if the wind is blowing from a direction of 328° at 11 mph?

28. *Ground Speed and Bearing of a Plane* A plane is headed due south with an airspeed of 192 mph. A wind from a direction of 78.0° is blowing at 23.0 mph. Find the ground speed and resulting bearing of the plane.

29. *Ground Speed and Bearing of a Plane* An airplane is headed on a bearing of 174° at an airspeed of 240 km per hr. A 30-km-per-hr wind is blowing from a direction of 245°. Find the ground speed and resulting bearing of the plane.

30. (a) approximately 56 mi per sec
(b) approximately 87 mi per sec

30. *Velocity of a Star* The space velocity $\mathbf{v}$ of a star relative to the sun can be expressed as the resultant vector of two perpendicular vectors—the radial velocity $\mathbf{v}_r$ and the tangential velocity $\mathbf{v}_t$, where $\mathbf{v} = \mathbf{v}_r + \mathbf{v}_t$. If a star is located near the sun and its space velocity is large, then its motion across the sky will also be large. Barnard's Star is a relatively close star with a distance of 35 trillion mi from the sun. It moves across the sky through an angle of 10.34″ per year, which is the largest motion of any known star. Its radial velocity is $\mathbf{v}_r = 67$ mi per sec toward the sun. (*Sources:* Zeilik, M., S. Gregory, and E. Smith, *Introductory Astronomy and Astrophysics,* Second Edition, Saunders College Publishing; Acker, A. and C. Jaschek, *Astronomical Methods and Calculations,* John Wiley and Sons.)

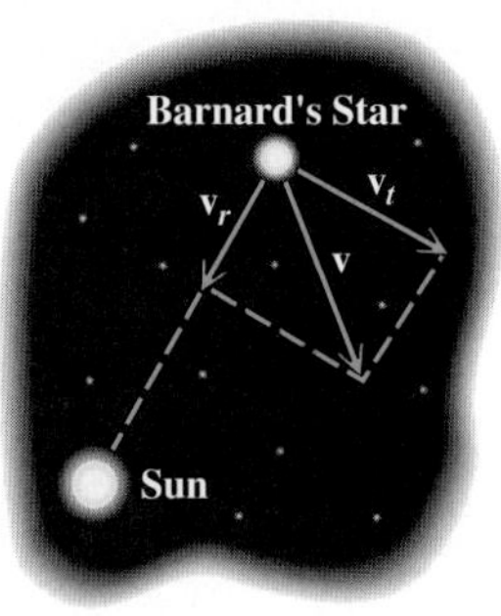

Not to scale

(a) Approximate the tangential velocity $\mathbf{v}_t$ of Barnard's Star. (*Hint:* Use the arc length formula $s = r\theta$ from **Section 6.1.**)

(b) Compute the magnitude of $\mathbf{v}$.

Summary Exercises on Applications of Trigonometry and Vectors

1. 29 ft; 38 ft
2. 38.3 cm
3. 5856 m
4. 15.8 ft per sec; 71.6°
5. 42 lb

These summary exercises provide practice with applications that involve solving triangles and using vectors.

1. *Wires Supporting a Flagpole* A flagpole stands vertically on a hillside that makes an angle of 20° with the horizontal. Two supporting wires are attached as shown in the figure. What are the lengths of the supporting wires?

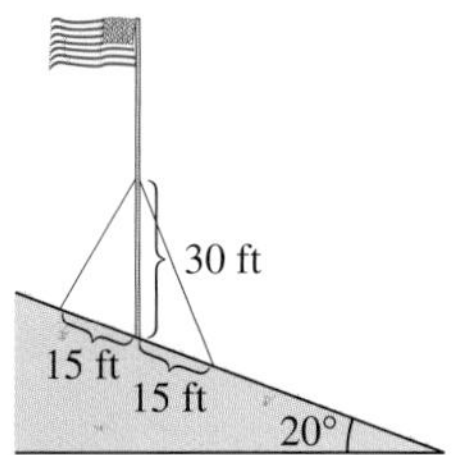

2. *Distance between a Pin and a Rod* A slider crank mechanism is shown in the figure. Find the distance between the wrist pin W and the connecting rod center C.

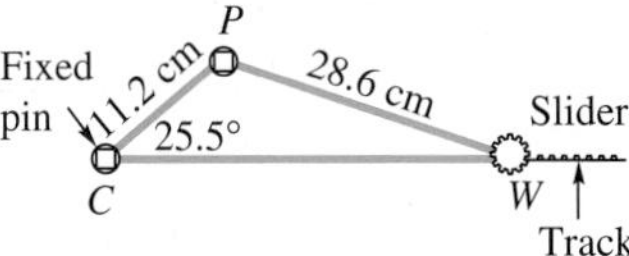

3. *Distance between Two Lighthouses* Two lighthouses are located on a north-south line. From lighthouse A, the bearing of a ship 3742 m away is 129° 43′. From lighthouse B, the bearing of a ship is 39° 43′. Find the distance between the lighthouses.

4. *Hot-Air Balloon* A hot-air balloon is rising straight up at the speed of 15.0 ft per sec. Then a wind starts blowing horizontally at 5.00 ft per sec. What will the new speed of the balloon be and what angle with the horizontal will the balloon's path make?

5. *Playing on a Swing* Mary is playing with her daughter Brittany on a swing. Starting from rest, Mary pulls the swing through an angle of 40° and holds it briefly before releasing the swing. If Brittany weighs 50 lb, what horizontal force, to the nearest pound, must Mary apply while holding the swing?

6. 7200 ft
7. (a) 10 mph
(b) $3\mathbf{v} = 18\mathbf{i} + 24\mathbf{j}$; This represents a 30-mph wind in the direction of **v**.
(c) **u** represents a southeast wind of $\sqrt{128} \approx 11.3$ mph.
8. 380 mph; 64°
9. It cannot exist.
10. Other angles can be 36° 10′, 115° 40′, third side 40.5, or other angles can be 143° 50′, 8° 00′, third side 6.25. (Lengths are in yards.)

6. *Height of an Airplane* Two observation points A and B are 950 ft apart. From these points the angles of elevation of an airplane are 52° and 57°. See the figure. Find the height of the airplane.

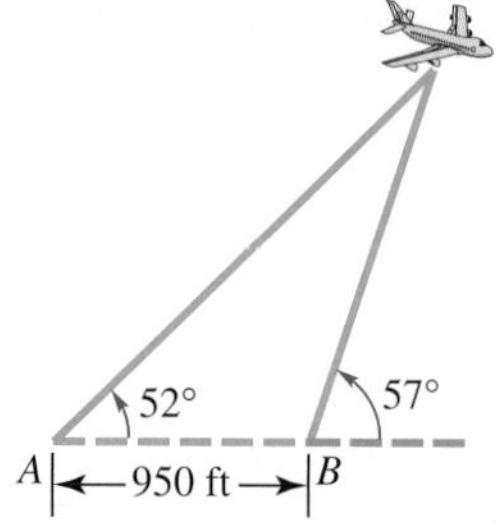

7. *Wind and Vectors* A wind can be described by $\mathbf{v} = 6\mathbf{i} + 8\mathbf{j}$, where vector $\mathbf{j}$ points north and represents a south wind of 1 mph.
(a) What is the speed of the wind? (b) Find $3\mathbf{v}$ and interpret the result.
(c) Interpret the direction and speed of the wind if it changes to $\mathbf{u} = -8\mathbf{i} + 8\mathbf{j}$.

8. *Ground Speed and Bearing* A plane with an airspeed of 355 mph is on a bearing of 62°. A wind is blowing from west to east at 28.5 mph. Find the ground speed and the actual bearing of the plane.

9. *Property Survey* A surveyor reported the following data about a piece of property: "The property is triangular in shape, with dimensions as shown in the figure." Use the law of sines to see whether such a piece of property could exist.

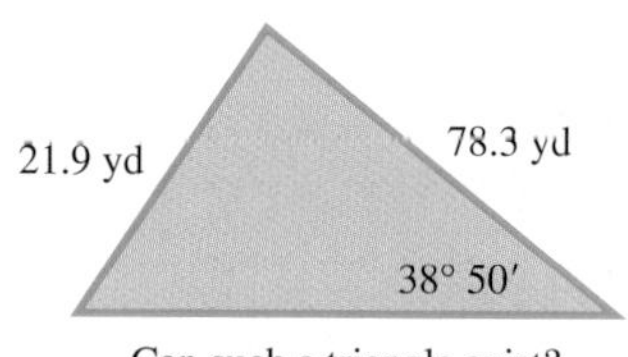

Can such a triangle exist?

10. *Property Survey* A triangular piece of property has the dimensions shown. It turns out that the surveyor did not consider every possible case. Use the law of sines to show why.

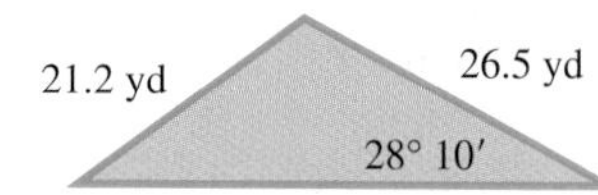

8.5 Trigonometric (Polar) Form of Complex Numbers; Products and Quotients

- The Complex Plane and Vector Representation
- Trigonometric (Polar) Form
- Converting between Rectangular and Trigonometric (Polar) Forms
- An Application of Complex Numbers to Fractals
- Products of Complex Numbers in Trigonometric Form
- Quotients of Complex Numbers in Trigonometric Form

The Complex Plane and Vector Representation Unlike real numbers, complex numbers cannot be ordered. One way to organize and illustrate them is by using a graph.

To graph a complex number such as $2 - 3i$, we modify the familiar coordinate system by calling the horizontal axis the **real axis** and the vertical axis the **imaginary axis.** Then complex numbers can be graphed in this **complex plane,** as shown in **Figure 45.** ***Each complex number $a + bi$ determines a unique position vector with initial point $(0, 0)$ and terminal point (a, b).***

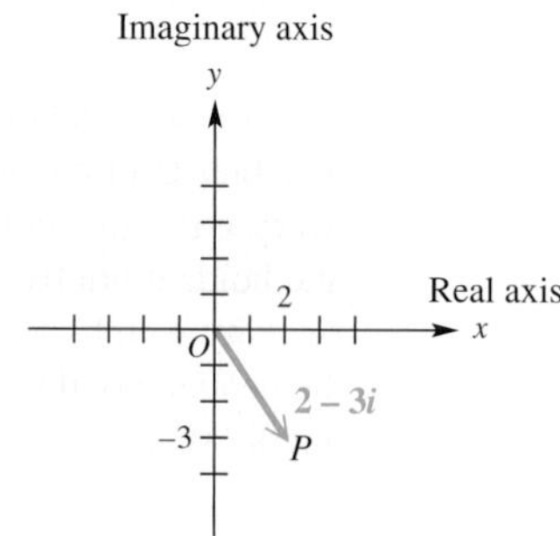

Figure 45

NOTE This geometric representation is the reason that $a + bi$ is called the **rectangular form** (or *standard form*) of a complex number.

Recall that the sum of the two complex numbers $4 + i$ and $1 + 3i$ is

$$(4 + i) + (1 + 3i) = 5 + 4i. \quad \text{(Section 1.3)}$$

Graphically, the sum of two complex numbers is represented by the vector that is the resultant of the vectors corresponding to the two numbers, as shown in **Figure 46.**

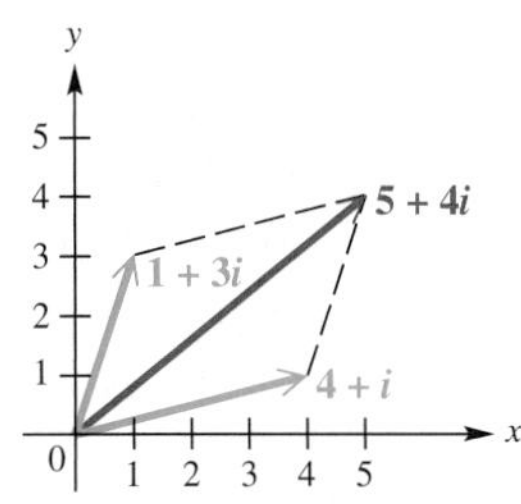

Figure 46

Classroom Example 1
Find the sum of $2 + 3i$ and $-4 + 2i$. Graph both complex numbers and their resultant.

Answer: $-2 + 5i$

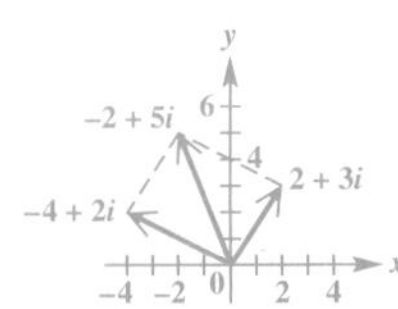

EXAMPLE 1 Expressing the Sum of Complex Numbers Graphically

Find the sum of $6 - 2i$ and $-4 - 3i$. Graph both complex numbers and their resultant.

SOLUTION The sum is found by adding the two numbers.

$$(6 - 2i) + (-4 - 3i) = 2 - 5i \quad \text{Add real parts, and add imaginary parts.}$$

The graphs are shown in **Figure 47.**

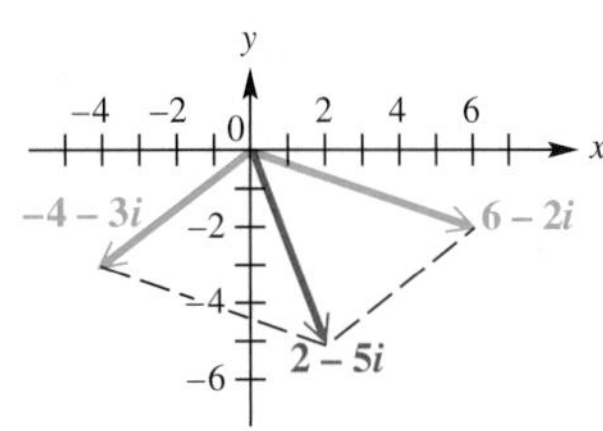

Figure 47

✔ *Now Try Exercise 13.*

Trigonometric (Polar) Form **Figure 48** shows the complex number $x + yi$ that corresponds to a vector **OP** with direction angle θ and magnitude r. The following relationships among x, y, r, and θ can be verified from **Figure 48.**

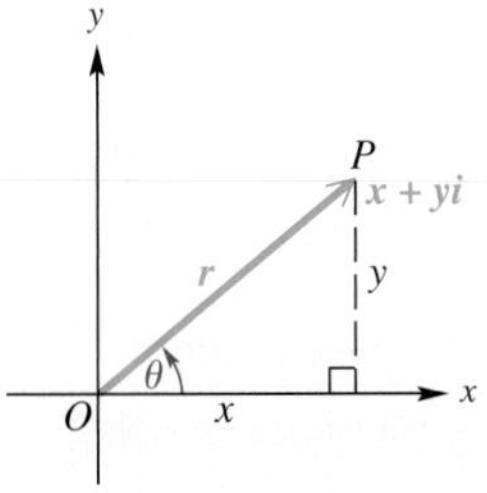

Figure 48

Relationships among x, y, r, and θ

$$x = r\cos\theta \qquad y = r\sin\theta$$

$$r = \sqrt{x^2 + y^2} \qquad \tan\theta = \frac{y}{x}, \quad \text{if } x \neq 0$$

Substituting $x = r\cos\theta$ and $y = r\sin\theta$ into $x + yi$ gives the following.

$$\begin{aligned} x + yi &= r\cos\theta + (r\sin\theta)i && \text{Substitute.} \\ &= r(\cos\theta + i\sin\theta) && \text{Factor out } r. \end{aligned}$$

Trigonometric (Polar) Form of a Complex Number

The expression

$$r(\cos \theta + i \sin \theta)$$

is the **trigonometric form** (or **polar form**) of the complex number $x + yi$. The expression $\cos \theta + i \sin \theta$ is sometimes abbreviated **cis θ.** Using this notation, $r(\cos \theta + i \sin \theta)$ is written **r cis θ.**

The number r is the **absolute value** (or **modulus**) of $x + yi$, and θ is the **argument** of $x + yi$. In this section, we choose the value of θ in the interval $[0°, 360°)$. However, any angle coterminal with θ also could serve as the argument.

Classroom Example 2
Express $10(\cos 135° + i \sin 135°)$ in rectangular form.

Answer: $-5\sqrt{2} + 5i\sqrt{2}$

EXAMPLE 2 **Converting from Trigonometric Form to Rectangular Form**

Express $2(\cos 300° + i \sin 300°)$ in rectangular form.

ALGEBRAIC SOLUTION

$2(\cos 300° + i \sin 300°)$

$= 2\left(\frac{1}{2} - i\frac{\sqrt{3}}{2}\right)$ — $\cos 300° = \frac{1}{2}$; $\sin 300° = -\frac{\sqrt{3}}{2}$ (Section 5.3)

$= 1 - i\sqrt{3}$ — Distributive property (Section R.2)

Note that the real part is positive and the imaginary part is negative. This is consistent with 300° being a quadrant IV angle.

GRAPHING CALCULATOR SOLUTION

We use a calculator in degree mode to confirm the algebraic solution. See **Figure 49.**

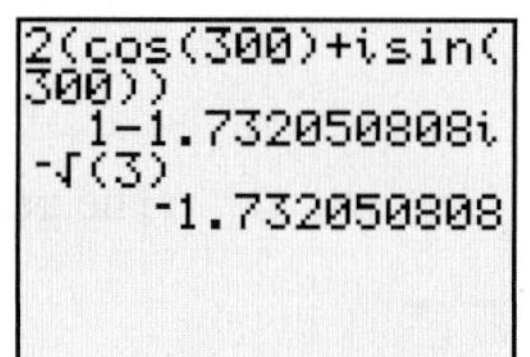

The imaginary part is an approximation for $-\sqrt{3}$.

Figure 49

✔ *Now Try Exercise 29.*

Teaching Tip Relate the definitions from **Chapter 5,** that is,

$$\frac{x}{r} = \cos \theta, \quad \frac{y}{r} = \sin \theta,$$

and $\quad x^2 + y^2 = r^2,$

to the relationships given here. Caution students to be especially careful about determining the correct quadrant of θ in Step 3.

Converting between Rectangular and Trigonometric (Polar) Forms To convert from rectangular form to trigonometric form, we use the following procedure.

Converting from Rectangular to Trigonometric Form

Step 1 Sketch a graph of the number $x + yi$ in the complex plane.

Step 2 Find r by using the equation $r = \sqrt{x^2 + y^2}$.

Step 3 Find θ by using the equation $\tan \theta = \frac{y}{x}$, where $x \neq 0$, choosing the quadrant indicated in Step 1.

CAUTION Errors often occur in Step 3. ***Be sure to choose the correct quadrant for θ by referring to the graph sketched in Step 1.***

Classroom Example 3
Write each complex number in trigonometric form.
(a) $8 - 8i$ (b) -15

Answers:
(a) $8\sqrt{2}$ cis 315°
(b) 15 cis 180°

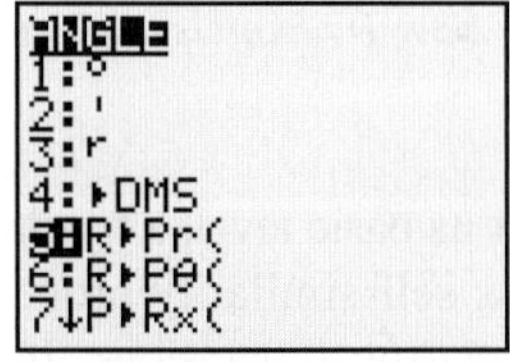

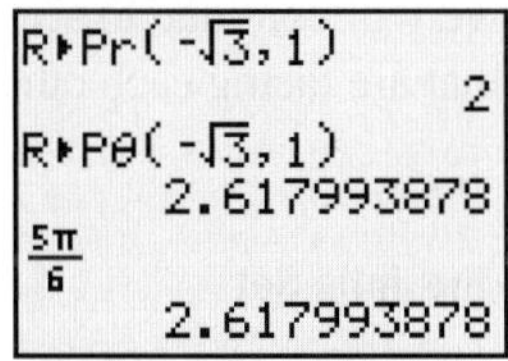

Choices 5 and 6 in the top screen show how to convert from rectangular (x, y) form to trigonometric form. The calculator is in radian mode. The results agree with our algebraic results in **Example 3(a).**

EXAMPLE 3 Converting from Rectangular to Trigonometric Form

Write each complex number in trigonometric form.

(a) $-\sqrt{3} + i$ (Use radian measure.) **(b)** $-3i$ (Use degree measure.)

SOLUTION

(a) We start by sketching the graph of $-\sqrt{3} + i$ in the complex plane, as shown in **Figure 50.** Next, we use $x = -\sqrt{3}$ and $y = 1$ to find r and θ.

$$r = \sqrt{x^2 + y^2} = \sqrt{\left(-\sqrt{3}\right)^2 + 1^2} = \sqrt{3 + 1} = 2$$

and $$\tan\theta = \frac{y}{x} = \frac{1}{-\sqrt{3}} = -\frac{1}{\sqrt{3}} \cdot \frac{\sqrt{3}}{\sqrt{3}} = -\frac{\sqrt{3}}{3}$$

Rationalize the denominator. **(Section R.7)**

Since $\tan\theta = -\frac{\sqrt{3}}{3}$, the reference angle for θ in radians is $\frac{\pi}{6}$. From the graph, we see that θ is in quadrant II, so $\theta = \pi - \frac{\pi}{6} = \frac{5\pi}{6}$. Therefore,

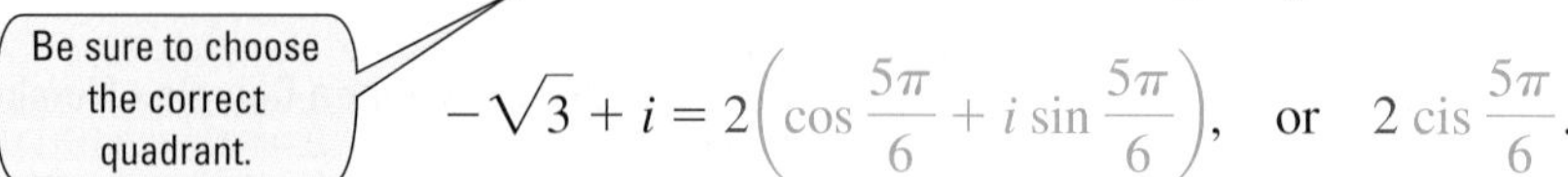

$$-\sqrt{3} + i = 2\left(\cos\frac{5\pi}{6} + i\sin\frac{5\pi}{6}\right), \quad \text{or} \quad 2 \text{ cis } \frac{5\pi}{6}.$$

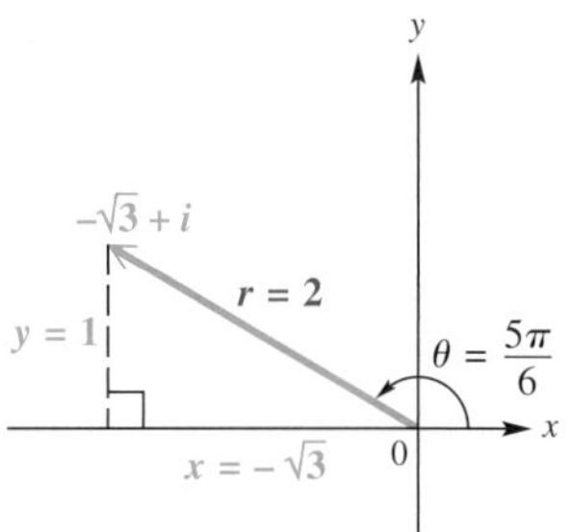

Figure 50

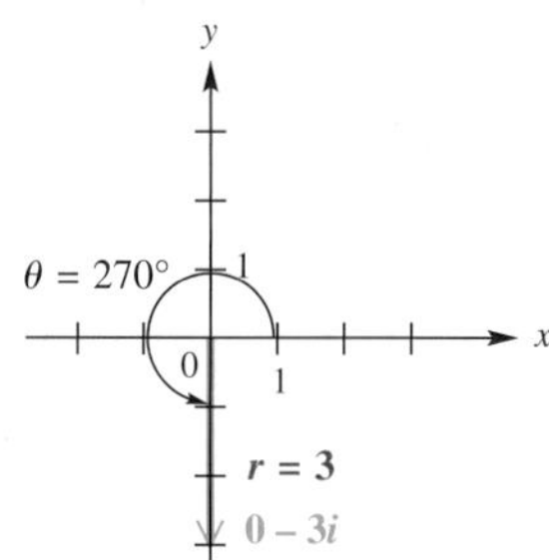

Figure 51

(b) See **Figure 51.** Since $-3i = 0 - 3i$, we have $x = 0$ and $y = -3$.

$$r = \sqrt{0^2 + (-3)^2} = \sqrt{0 + 9} = \sqrt{9} = 3 \quad \text{Substitute.}$$

We cannot find θ by using $\tan\theta = \frac{y}{x}$, because $x = 0$. However, the graph suggests that the value for θ is 270°.

$$-3i = 3(\cos 270° + i\sin 270°), \quad \text{or} \quad 3 \text{ cis } 270° \quad \text{Trigonometric form}$$

Now Try Exercises 41 and 47.

Classroom Example 4
Write each complex number in its alternative form, using calculator approximations as necessary.
(a) $7(\cos 205° + i\sin 205°)$
(b) $-7 + 2i$

Answers:
(a) $-6.3442 - 2.9583i$
(b) $\sqrt{53}$ cis 164.05°

EXAMPLE 4 Converting between Trigonometric and Rectangular Forms Using Calculator Approximations

Write each complex number in its alternative form, using calculator approximations as necessary.

(a) $6(\cos 115° + i\sin 115°)$ **(b)** $5 - 4i$

SOLUTION

(a) Since 115° does not have a special angle as a reference angle, we cannot find exact values for cos 115° and sin 115°. Use a calculator set in degree mode.

$$6(\cos 115° + i\sin 115°)$$
$$\approx 6(-0.4226182617 + 0.906307787i) \quad \text{Use a calculator.}$$
$$\approx -2.5357 + 5.4378i \quad \text{Four decimal places}$$

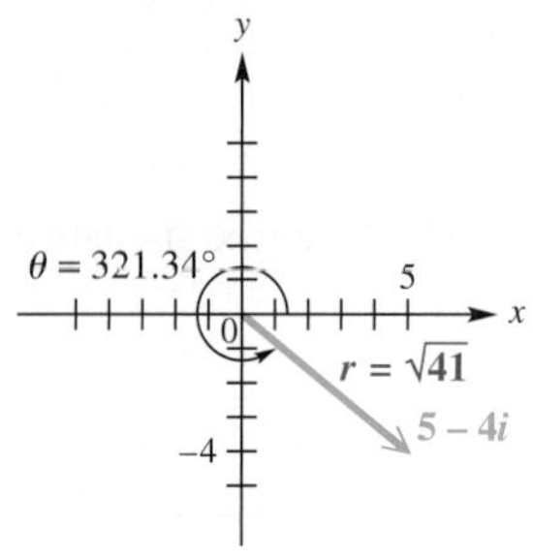

Figure 52

(b) A sketch of $5 - 4i$ shows that θ must be in quadrant IV. See **Figure 52.**

$$r = \sqrt{5^2 + (-4)^2} = \sqrt{41} \quad \text{and} \quad \tan \theta = -\frac{4}{5}$$

Use a calculator to find that one measure of θ is $-38.66°$. In order to express θ in the interval $[0, 360°)$, we find $\theta = 360° - 38.66° = 321.34°$.

$$5 - 4i = \sqrt{41} \text{ cis } 321.34°$$

✔ *Now Try Exercises 53 and 57.*

An Application of Complex Numbers to Fractals At its basic level, a **fractal** is a unique, enchanting geometric figure with an endless self-similarity property. A fractal image repeats itself infinitely with ever-decreasing dimensions. If we look at smaller and smaller portions of a fractal image, we will continue to see the whole—it is much like looking into two parallel mirrors that are facing each other.

Classroom Example 5
Refer to **Example 5.** Determine whether each number belongs to the Julia set.
(a) $z = 1 - 1i$
(b) $z = -1 + 0i$

Answers:
(a) no (b) yes

EXAMPLE 5 Deciding Whether a Complex Number Is in the Julia Set

The fractal called the **Julia set** is shown in **Figure 53.** To determine whether a complex number $z = a + bi$ is in this Julia set, perform the following sequence of calculations.

$$z^2 - 1, \quad (z^2 - 1)^2 - 1, \quad [(z^2 - 1)^2 - 1]^2 - 1, \quad \ldots$$

If the absolute values of any of the resulting complex numbers exceed 2, then the complex number z is not in the Julia set. Otherwise z is part of this set and the point (a, b) should be shaded in the graph.

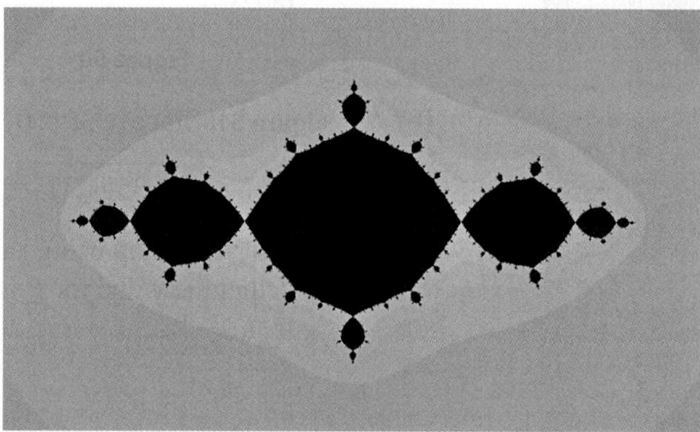

Figure 53

Determine whether each number belongs to the Julia set.

(a) $z = 0 + 0i$ **(b)** $z = 1 + 1i$

SOLUTION

(a) Here

$$z = 0 + 0i = 0,$$
$$z^2 - 1 = 0^2 - 1 = -1,$$
$$(z^2 - 1)^2 - 1 = (-1)^2 - 1 = 0,$$
$$[(z^2 - 1)^2 - 1]^2 - 1 = 0^2 - 1 = -1, \quad \text{and so on.}$$

We see that the calculations repeat as $0, -1, 0, -1$, and so on. The absolute values are either 0 or 1, which do not exceed 2, so $0 + 0i$ is in the Julia set and the point $(0, 0)$ is part of the graph.

(b) For $z = 1 + 1i$, we have the following.

$$z^2 - 1 = (1+i)^2 - 1 \quad \text{Substitute for } z;\ 1 + 1i = 1 + i.$$
$$= (1 + 2i + i^2) - 1 \quad \text{Square the binomial; } (x+y)^2 = x^2 + 2xy + y^2. \text{ (Section R.3)}$$
$$= -1 + 2i \quad i^2 = -1 \text{ (Section 1.3)}$$

The absolute value is

$$\sqrt{(-1)^2 + 2^2} = \sqrt{5}.$$

Since $\sqrt{5}$ is greater than 2, the number $1 + 1i$ is not in the Julia set and $(1, 1)$ is not part of the graph.

✔ *Now Try Exercise 63.*

Products of Complex Numbers in Trigonometric Form Using the FOIL method to multiply complex numbers in rectangular form, we find the product of $1 + i\sqrt{3}$ and $-2\sqrt{3} + 2i$ as follows.

$$(1 + i\sqrt{3})(-2\sqrt{3} + 2i)$$
$$= -2\sqrt{3} + 2i - 2i(3) + 2i^2\sqrt{3} \quad \text{FOIL (Section R.3)}$$
$$= -2\sqrt{3} + 2i - 6i - 2\sqrt{3} \quad i^2 = -1$$
$$= -4\sqrt{3} - 4i \quad \text{Combine like terms.}$$

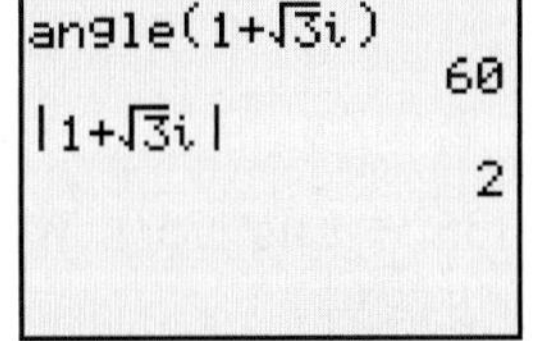

With the TI-83/84 Plus calculator in complex and degree modes, the MATH menu can be used to find the angle and the magnitude (absolute value) of the vector that corresponds to a given complex number.

We can also find this same product by first converting the complex numbers $1 + i\sqrt{3}$ and $-2\sqrt{3} + 2i$ to trigonometric form using the method explained earlier in this section.

$$1 + i\sqrt{3} = 2(\cos 60° + i \sin 60°)$$

and

$$-2\sqrt{3} + 2i = 4(\cos 150° + i \sin 150°)$$

If we multiply the trigonometric forms and use identities for the cosine and the sine of the sum of two angles, then the result is as follows.

$$[2(\cos 60° + i \sin 60°)][4(\cos 150° + i \sin 150°)]$$
$$= 2 \cdot 4(\cos 60° \cdot \cos 150° + i \sin 60° \cdot \cos 150° + i \cos 60° \cdot \sin 150° + i^2 \sin 60° \cdot \sin 150°) \quad \text{Multiply the absolute values and use FOIL.}$$
$$= 8[(\cos 60° \cdot \cos 150° - \sin 60° \cdot \sin 150°) + i(\sin 60° \cdot \cos 150° + \cos 60° \cdot \sin 150°)] \quad i^2 = -1; \text{ Factor out } i. \text{ (Section R.4)}$$
$$= 8[\cos(60° + 150°) + i \sin(60° + 150°)] \quad \text{Use identities for } \cos(A+B) \text{ and } \sin(A+B). \text{ (Section 7.3)}$$
$$= 8(\cos 210° + i \sin 210°) \quad \text{Add.}$$

Notice the following.

- The absolute value of the product, 8, is equal to the product of the absolute values of the factors, $2 \cdot 4$.
- The argument of the product, 210°, is equal to the sum of the arguments of the factors, $60° + 150°$.

The product obtained when multiplying by the first method is the rectangular form of the product obtained when multiplying by the second method.

$$8(\cos 210^\circ + i \sin 210^\circ)$$

$$= 8\left(-\frac{\sqrt{3}}{2} - \frac{1}{2}i\right) \quad \cos 210^\circ = -\frac{\sqrt{3}}{2};\ \sin 210^\circ = -\frac{1}{2} \text{ (Section 5.3)}$$

$$= -4\sqrt{3} - 4i \quad \text{Rectangular form}$$

We can generalize this work in the *product theorem.*

Teaching Tip Use the product theorem to show how to obtain the product of two complex numbers geometrically. Whereas sums of complex numbers involve parallelograms, products involve rotations.

The product theorem often results in a value of $\theta_1 + \theta_2$ greater than 360°. In such cases, students may want to subtract a multiple of 360° to get an angle between 0° and 360°.

Product Theorem

If $r_1(\cos \theta_1 + i \sin \theta_1)$ and $r_2(\cos \theta_2 + i \sin \theta_2)$ are any two complex numbers, then the following holds.

$$[r_1(\cos \theta_1 + i \sin \theta_1)] \cdot [r_2(\cos \theta_2 + i \sin \theta_2)]$$
$$= r_1 r_2[\cos(\theta_1 + \theta_2) + i \sin(\theta_1 + \theta_2)]$$

In compact form, this is written

$$(r_1 \text{ cis } \theta_1)(r_2 \text{ cis } \theta_2) = r_1 r_2 \text{ cis}(\theta_1 + \theta_2).$$

That is, to multiply complex numbers in trigonometric form, multiply their absolute values and add their arguments.

Classroom Example 6
Find the product of

$$4(\cos 120^\circ + i \sin 120^\circ)$$

and

$$5(\cos 30^\circ + i \sin 30^\circ).$$

Write the result in rectangular form.

Answer: $-10\sqrt{3} + 10i$

EXAMPLE 6 Using the Product Theorem

Find the product of $3(\cos 45^\circ + i \sin 45^\circ)$ and $2(\cos 135^\circ + i \sin 135^\circ)$. Write the result in rectangular form.

SOLUTION

$$[3(\cos 45^\circ + i \sin 45^\circ)][2(\cos 135^\circ + i \sin 135^\circ)]$$

$$= 3 \cdot 2[\cos(45^\circ + 135^\circ) + i \sin(45^\circ + 135^\circ)] \quad \text{Product theorem}$$

$$= 6(\cos 180^\circ + i \sin 180^\circ) \quad \text{Multiply and add.}$$

$$= 6(-1 + i \cdot 0) \quad \cos 180^\circ = -1;\ \sin 180^\circ = 0 \text{ (Section 5.3)}$$

$$= 6(-1), \text{ or } -6 \quad \text{Rectangular form}$$

Now Try Exercise 71.

Quotients of Complex Numbers in Trigonometric Form The rectangular form of the quotient of $1 + i\sqrt{3}$ and $-2\sqrt{3} + 2i$ is found as follows.

$$\frac{1 + i\sqrt{3}}{-2\sqrt{3} + 2i}$$

$$= \frac{(1 + i\sqrt{3})(-2\sqrt{3} - 2i)}{(-2\sqrt{3} + 2i)(-2\sqrt{3} - 2i)} \quad \text{Multiply both numerator and denominator by the conjugate of the denominator. (Section 1.3)}$$

$$= \frac{-2\sqrt{3} - 2i - 6i - 2i^2\sqrt{3}}{12 - 4i^2} \quad \text{FOIL; } (x + y)(x - y) = x^2 - y^2 \text{ (Section R.3)}$$

$$= \frac{-8i}{16}, \text{ or } -\frac{1}{2}i \quad \text{Simplify.}$$

Writing $1 + i\sqrt{3}$, $-2\sqrt{3} + 2i$, and $-\frac{1}{2}i$ in trigonometric form gives

$$1 + i\sqrt{3} = 2(\cos 60° + i \sin 60°),$$

$$-2\sqrt{3} + 2i = 4(\cos 150° + i \sin 150°),$$

Use $r = \sqrt{x^2 + y^2}$ and $\tan \theta = \frac{y}{x}$.

and $$-\frac{1}{2}i = \frac{1}{2}[\cos(-90°) + i \sin(-90°)].$$

Here, the absolute value of the quotient, $\frac{1}{2}$, is the quotient of the two absolute values, $\frac{2}{4} = \frac{1}{2}$. The argument of the quotient, $-90°$, is the difference of the two arguments,

$$60° - 150° = -90°.$$

Generalizing this work leads to the *quotient theorem.*

Quotient Theorem

If $r_1(\cos \theta_1 + i \sin \theta_1)$ and $r_2(\cos \theta_2 + i \sin \theta_2)$ are any two complex numbers, where $r_2(\cos \theta_2 + i \sin \theta_2) \neq 0$, then the following holds.

$$\frac{r_1(\cos \theta_1 + i \sin \theta_1)}{r_2(\cos \theta_2 + i \sin \theta_2)} = \frac{r_1}{r_2}[\cos(\theta_1 - \theta_2) + i \sin(\theta_1 - \theta_2)]$$

In compact form, this is written

$$\frac{r_1 \text{ cis } \theta_1}{r_2 \text{ cis } \theta_2} = \frac{r_1}{r_2} \text{cis}(\theta_1 - \theta_2).$$

Teaching Tip The quotient theorem often results in a negative value for $\theta_1 - \theta_2$. In such cases, students may want to add a multiple of 360° to get an angle between 0° and 360°.

That is, to divide complex numbers in trigonometric form, divide their absolute values and subtract their arguments.

Classroom Example 7
Find the quotient

$$\frac{27 \text{ cis } 45°}{9 \text{ cis}(-180°)}.$$

Write the result in rectangular form.

Answer: $-\frac{3\sqrt{2}}{2} - \frac{3\sqrt{2}}{2}i$

EXAMPLE 7 Using the Quotient Theorem

Find the quotient $\frac{10 \text{ cis}(-60°)}{5 \text{ cis } 150°}$. Write the result in rectangular form.

SOLUTION

$$\frac{10 \text{ cis}(-60°)}{5 \text{ cis } 150°}$$

$= \frac{10}{5} \text{cis}(-60° - 150°)$ Quotient theorem

$= 2 \text{ cis}(-210°)$ Divide and subtract.

$= 2[\cos(-210°) + i \sin(-210°)]$ Rewrite.

$= 2\left[-\frac{\sqrt{3}}{2} + i\left(\frac{1}{2}\right)\right]$ $\cos(-210°) = -\frac{\sqrt{3}}{2}$; $\sin(-210°) = \frac{1}{2}$ (Section 5.3)

$= -\sqrt{3} + i$ Rectangular form

Now Try Exercise 81.

8.5 Exercises

1. length (magnitude)
2. It is the angle formed by the vector and the positive x-axis.

3. 4.

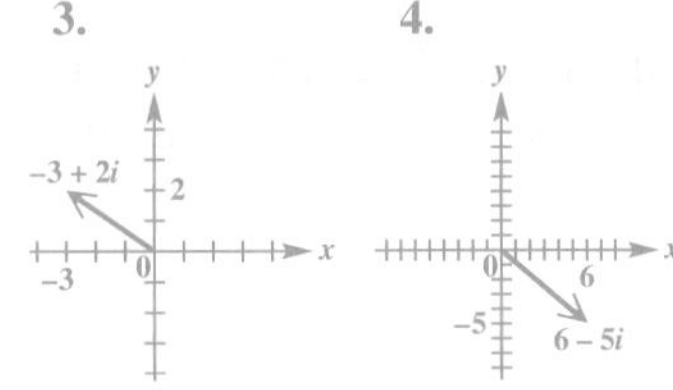

5. 6.

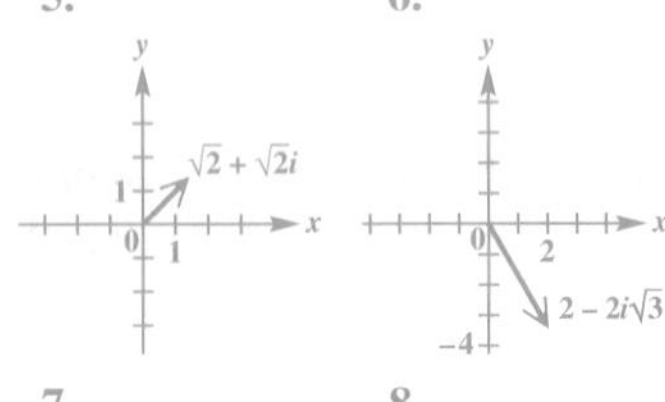

7. 8.

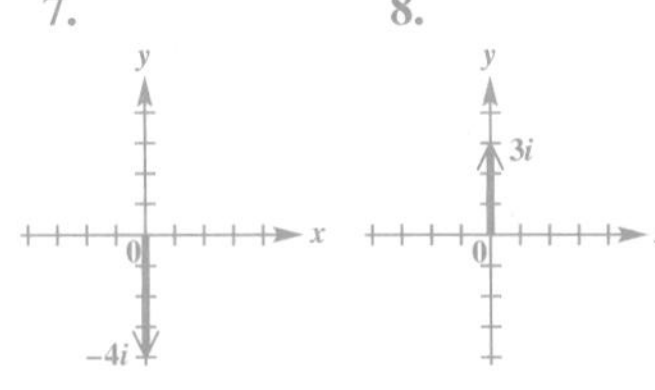

9. 10.

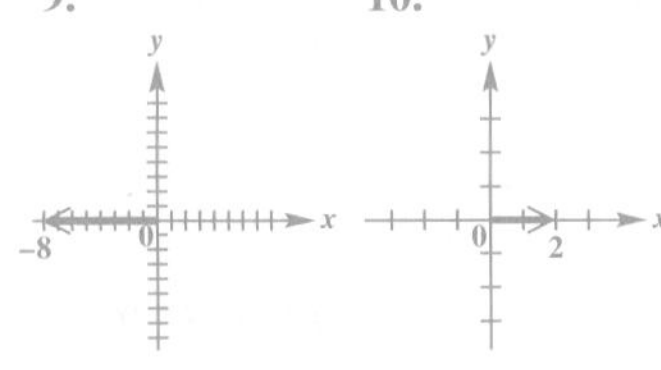

11. $1 - 4i$ 12. $-4 - i$

13. $3 - i$

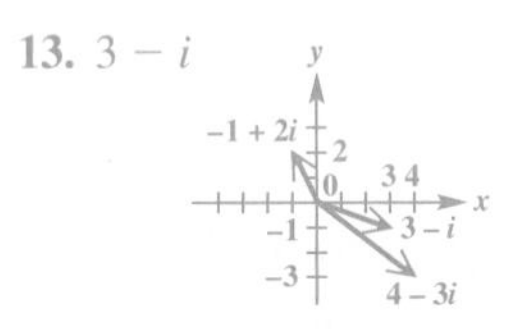

14. $-2 + 2i$

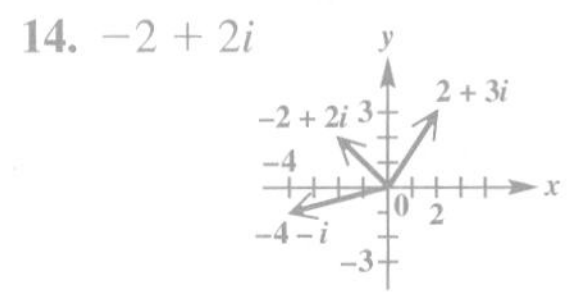

15. $-3i$

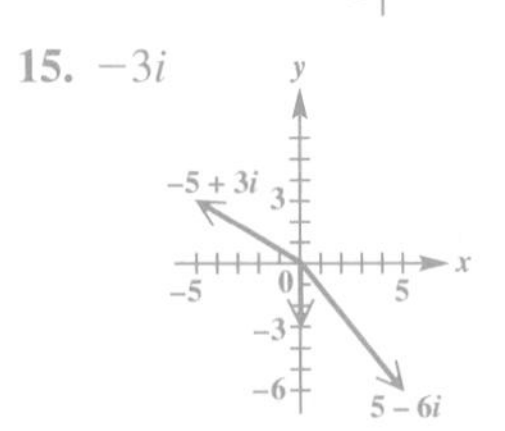

16. 3

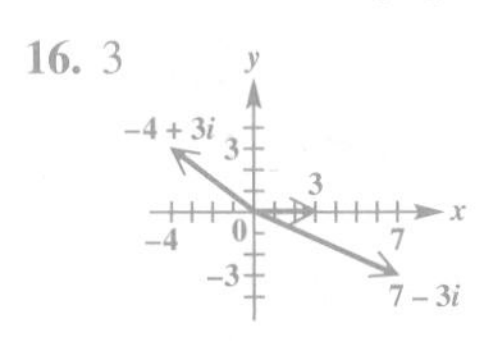

1. *Concept Check* The absolute value (or modulus) of a complex number represents the ________________ of the vector representing it in the complex plane.

2. *Concept Check* What is the geometric interpretation of the argument of a complex number?

Graph each complex number. ***See Example 1.***

3. $-3 + 2i$ 4. $6 - 5i$ 5. $\sqrt{2} + \sqrt{2}i$ 6. $2 - 2i\sqrt{3}$

7. $-4i$ 8. $3i$ 9. -8 10. 2

Concept Check Give the rectangular form of the complex number shown.

11.

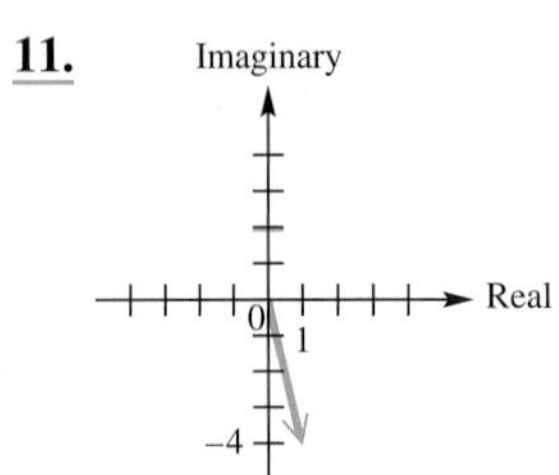

12.

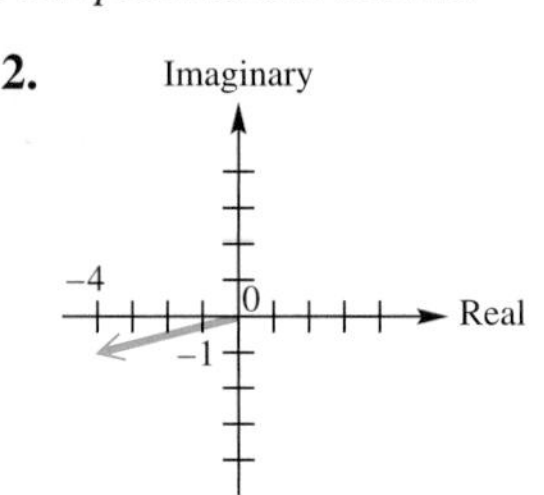

Find the sum of each pair of complex numbers. In Exercises 13–16, graph both complex numbers and their resultant. ***See Example 1.***

13. $4 - 3i, -1 + 2i$ 14. $2 + 3i, -4 - i$ 15. $5 - 6i, -5 + 3i$

16. $7 - 3i, -4 + 3i$ 17. $-3, 3i$ 18. $6, -2i$

19. $-5 - 8i, -1$ 20. $4 - 2i, 5$ 21. $7 + 6i, 3i$

22. $2 + 6i, -2i$ 23. $\frac{1}{2} + \frac{2}{3}i, \frac{2}{3} + \frac{1}{2}i$ 24. $-\frac{1}{5} + \frac{2}{7}i, \frac{3}{7} - \frac{3}{4}i$

Write each complex number in rectangular form. ***See Example 2.***

25. $2(\cos 45° + i \sin 45°)$ 26. $4(\cos 60° + i \sin 60°)$

27. $10(\cos 90° + i \sin 90°)$ 28. $8(\cos 270° + i \sin 270°)$

29. $4(\cos 240° + i \sin 240°)$ 30. $2(\cos 330° + i \sin 330°)$

31. $3 \text{ cis } 150°$ 32. $2 \text{ cis } 30°$

33. $5 \text{ cis } 300°$ 34. $6 \text{ cis } 135°$

35. $\sqrt{2} \text{ cis } 225°$ 36. $\sqrt{3} \text{ cis } 315°$

37. $4(\cos(-30°) + i \sin(-30°))$ 38. $\sqrt{2}(\cos(-60°) + i \sin(-60°))$

Write each complex number in trigonometric form $r(\cos\theta + i\sin\theta)$, with θ in the interval $[0°, 360°)$. ***See Example 3.***

39. $-3 - 3i\sqrt{3}$ 40. $1 + i\sqrt{3}$ 41. $\sqrt{3} - i$

42. $4\sqrt{3} + 4i$ 43. $-5 - 5i$ 44. $-2 + 2i$

45. $2 + 2i$ 46. $4 + 4i$ 47. $5i$

48. $-2i$ 49. -4 50. 7

17. $-3 + 3i$ 18. $6 - 2i$
19. $-6 - 8i$ 20. $9 - 2i$
21. $7 + 9i$ 22. $2 + 4i$
23. $\frac{7}{6} + \frac{7}{6}i$ 24. $\frac{8}{35} - \frac{13}{28}i$
25. $\sqrt{2} + i\sqrt{2}$ 26. $2 + 2i\sqrt{3}$
27. $10i$ 28. $-8i$
29. $-2 - 2i\sqrt{3}$ 30. $\sqrt{3} - i$
31. $-\frac{3\sqrt{3}}{2} + \frac{3}{2}i$ 32. $\sqrt{3} + i$
33. $\frac{5}{2} - \frac{5\sqrt{3}}{2}i$
34. $-3\sqrt{2} + 3i\sqrt{2}$
35. $-1 - i$ 36. $\frac{\sqrt{6}}{2} - \frac{\sqrt{6}}{2}i$
37. $2\sqrt{3} - 2i$ 38. $\frac{\sqrt{2}}{2} - \frac{\sqrt{6}}{2}i$
39. $6(\cos 240° + i \sin 240°)$
40. $2(\cos 60° + i \sin 60°)$
41. $2(\cos 330° + i \sin 330°)$
42. $8(\cos 30° + i \sin 30°)$
43. $5\sqrt{2}(\cos 225° + i \sin 225°)$
44. $2\sqrt{2}(\cos 135° + i \sin 135°)$
45. $2\sqrt{2}(\cos 45° + i \sin 45°)$
46. $4\sqrt{2}(\cos 45° + i \sin 45°)$
47. $5(\cos 90° + i \sin 90°)$
48. $2(\cos 270° + i \sin 270°)$
49. $4(\cos 180° + i \sin 180°)$
50. $7(\cos 0° + i \sin 0°)$
51. $\sqrt{13}(\cos 56.31° + i \sin 56.31°)$
52. $0.8192 + 0.5736i$
53. $-1.0261 - 2.8191i$
54. $\sqrt{17}(\cos 165.96° + i \sin 165.96°)$
55. $12(\cos 90° + i \sin 90°)$
56. -3
57. $\sqrt{34}(\cos 59.04° + i \sin 59.04°)$
58. $-0.3502 + 0.9367i$
59. It is the circle of radius 1 centered at the origin.
60. It is the line $y = x$.
61. It is the vertical line $x = 1$.
62. It is the horizontal line $y = 1$.
63. yes
64. (b) $z_1^2 - 1 = a^2 - b^2 - 1 + 2abi$; $z_2^2 - 1 = a^2 - b^2 - 1 - 2abi$
67. $-3\sqrt{3} + 3i$ 68. $-10\sqrt{3} + 10i$
69. $12\sqrt{3} + 12i$ 70. $20 + 20i\sqrt{3}$
71. 4 72. -16
73. $-3i$ 74. $6i$
75. $-\frac{15\sqrt{2}}{2} + \frac{15\sqrt{2}}{2}i$
76. $-\frac{21\sqrt{3}}{2} - \frac{21}{2}i$

Perform each conversion, using a calculator to approximate answers as necessary. **See Example 4.**

	Rectangular Form	Trigonometric Form
51.	$2 + 3i$	________
52.	________	$\cos 35° + i \sin 35°$
53.	________	$3(\cos 250° + i \sin 250°)$
54.	$-4 + i$	________
55.	$12i$	________
56.	________	$3 \text{ cis } 180°$
57.	$3 + 5i$	________
58.	________	$\text{cis } 110.5°$

Concept Check The complex number z, where $z = x + yi$, can be graphed in the plane as (x, y). Describe the graphs of all complex numbers z satisfying the conditions in Exercises 59–62.

59. The absolute value of z is 1.

60. The real and imaginary parts of z are equal.

61. The real part of z is 1.

62. The imaginary part of z is 1.

Julia Set Refer to ***Example 5*** *to solve Exercises 63 and 64.*

63. Is $z = -0.2i$ in the Julia set?

64. The graph of the Julia set in **Figure 53** appears to be symmetric with respect to both the x-axis and the y-axis. Complete the following to show that this is true.

(a) Show that complex conjugates have the same absolute value.
(b) Compute $z_1^2 - 1$ and $z_2^2 - 1$, where $z_1 = a + bi$ and $z_2 = a - bi$.
(c) Discuss why if (a, b) is in the Julia set, then so is $(a, -b)$.
(d) Conclude that the graph of the Julia set must be symmetric with respect to the x-axis.
(e) Using a similar argument, show that the Julia set must also be symmetric with respect to the y-axis.

In Exercises 65 and 66, suppose $z = r(\cos\theta + i \sin\theta)$.

65. Use vectors to show that the conjugate of z is

$$r[\cos(360° - \theta) + i \sin(360° - \theta)], \quad \text{or} \quad r(\cos\theta - i \sin\theta).$$

66. Use vectors to show that

$$-z = r[\cos(\theta + \pi) + i \sin(\theta + \pi)].$$

Find each product and write it in rectangular form. **See Example 6.**

67. $[3(\cos 60° + i \sin 60°)][2(\cos 90° + i \sin 90°)]$

68. $[4(\cos 30° + i \sin 30°)][5(\cos 120° + i \sin 120°)]$

69. $[4(\cos 60° + i \sin 60°)][6(\cos 330° + i \sin 330°)]$

70. $[8(\cos 300° + i \sin 300°)][5(\cos 120° + i \sin 120°)]$

71. $[2(\cos 135° + i \sin 135°)][2(\cos 225° + i \sin 225°)]$

72. $[8(\cos 210° + i \sin 210°)][2(\cos 330° + i \sin 330°)]$

73. $(\sqrt{3} \text{ cis } 45°)(\sqrt{3} \text{ cis } 225°)$

74. $(\sqrt{6} \text{ cis } 120°)[\sqrt{6} \text{ cis}(-30°)]$

75. $(5 \text{ cis } 90°)(3 \text{ cis } 45°)$

76. $(3 \text{ cis } 300°)(7 \text{ cis } 270°)$

Find each quotient and write it in rectangular form. In Exercises 83–88, first convert the numerator and the denominator to trigonometric form. **See Example 7.**

77. $\dfrac{4(\cos 150° + i \sin 150°)}{2(\cos 120° + i \sin 120°)}$

78. $\dfrac{24(\cos 150° + i \sin 150°)}{2(\cos 30° + i \sin 30°)}$

79. $\dfrac{10(\cos 50° + i \sin 50°)}{5(\cos 230° + i \sin 230°)}$

80. $\dfrac{12(\cos 23° + i \sin 23°)}{6(\cos 293° + i \sin 293°)}$

81. $\dfrac{3 \text{ cis } 305°}{9 \text{ cis } 65°}$

82. $\dfrac{16 \text{ cis } 310°}{8 \text{ cis } 70°}$

83. $\dfrac{8}{\sqrt{3} + i}$

84. $\dfrac{2i}{-1 - i\sqrt{3}}$

85. $\dfrac{-i}{1 + i}$

86. $\dfrac{1}{2 - 2i}$

87. $\dfrac{2\sqrt{6} - 2i\sqrt{2}}{\sqrt{2} - i\sqrt{6}}$

88. $\dfrac{-3\sqrt{2} + 3i\sqrt{6}}{\sqrt{6} + i\sqrt{2}}$

Use a calculator to perform the indicated operations. Give answers in rectangular form, expressing real and imaginary parts to four decimal places.

89. $[2.5(\cos 35° + i \sin 35°)][3.0(\cos 50° + i \sin 50°)]$

90. $[4.6(\cos 12° + i \sin 12°)][2.0(\cos 13° + i \sin 13°)]$

91. $(12 \text{ cis } 18.5°)(3 \text{ cis } 12.5°)$

92. $(4 \text{ cis } 19.25°)(7 \text{ cis } 41.75°)$

93. $\dfrac{45\left(\cos \frac{2\pi}{3} + i \sin \frac{2\pi}{3}\right)}{22.5\left(\cos \frac{3\pi}{5} + i \sin \frac{3\pi}{5}\right)}$

94. $\dfrac{30\left(\cos \frac{2\pi}{5} + i \sin \frac{2\pi}{5}\right)}{10\left(\cos \frac{\pi}{7} + i \sin \frac{\pi}{7}\right)}$

95. $\left[2 \text{ cis } \dfrac{5\pi}{9}\right]^2$

96. $\left[24.3 \text{ cis } \dfrac{7\pi}{12}\right]^2$

Relating Concepts

For individual or collaborative investigation *(Exercises 97–103)*

Consider the following complex numbers, and **work Exercises 97–103 in order.**

$$w = -1 + i \quad \text{and} \quad z = -1 - i$$

97. Multiply w and z using their rectangular forms and the FOIL method. Leave the product in rectangular form.

98. Find the trigonometric forms of w and z.

99. Multiply w and z using their trigonometric forms and the method described in this section.

100. Use the result of **Exercise 99** to find the rectangular form of wz. How does this compare to your result in **Exercise 97?**

101. Find the quotient $\frac{w}{z}$ using their rectangular forms and multiplying both the numerator and the denominator by the conjugate of the denominator. Leave the quotient in rectangular form.

102. Use the trigonometric forms of w and z, found in **Exercise 98,** to divide w by z using the method described in this section.

103. Use the result of **Exercise 102** to find the rectangular form of $\frac{w}{z}$. How does this compare to your result in **Exercise 101?**

77. $\sqrt{3} + i$ 78. $-6 + 6i\sqrt{3}$
79. -2 80. $2i$
81. $-\frac{1}{6} - \frac{\sqrt{3}}{6}i$ 82. $-1 - i\sqrt{3}$
83. $2\sqrt{3} - 2i$ 84. $-\frac{\sqrt{3}}{2} - \frac{1}{2}i$
85. $-\frac{1}{2} - \frac{1}{2}i$ 86. $\frac{1}{4} + \frac{1}{4}i$
87. $\sqrt{3} + i$ 88. $3i$

89. $0.6537 + 7.4715i$
90. $8.3380 + 3.8881i$
91. $30.8580 + 18.5414i$
92. $13.5747 + 24.4894i$
93. $1.9563 + 0.4158i$
94. $2.0732 + 2.1684i$
95. $-3.7588 - 1.3681i$
96. $-511.3793 - 295.2450i$

97. 2
98. $w = \sqrt{2} \text{ cis } 135°$; $z = \sqrt{2} \text{ cis } 225°$
99. $2 \text{ cis } 0°$
100. 2; It is the same.
101. $-i$
102. $\text{cis}(-90°)$
103. $-i$; It is the same.

107. $1.18 - 0.14i$
108. $1.7 + 2.8i$
109. approximately $27.43 + 11.50i$
110. approximately $22.75°$

104. Note that $(r \text{ cis } \theta)^2 = (r \text{ cis } \theta)(r \text{ cis } \theta) = r^2 \text{ cis}(\theta + \theta) = r^2 \text{ cis } 2\theta$. Explain how we can square a complex number in trigonometric form. (In the next section, we will develop this idea more fully.)

105. Without actually performing the operations, state why the following products are the same.

$$[2(\cos 45° + i \sin 45°)] \cdot [5(\cos 90° + i \sin 90°)]$$

and $$[2[\cos(-315°) + i \sin(-315°)]] \cdot [5[\cos(-270°) + i \sin(-270°)]]$$

106. Show that $\frac{1}{z} = \frac{1}{r}(\cos \theta - i \sin \theta)$, where $z = r(\cos \theta + i \sin \theta)$.

(Modeling) Solve each problem.

107. *Electrical Current* The alternating current in an electric inductor is $I = \frac{E}{Z}$ amperes, where E is voltage and $Z = R + X_L i$ is impedance. If $E = 8(\cos 20° + i \sin 20°)$, $R = 6$, and $X_L = 3$, find the current. Give the answer in rectangular form, with real and imaginary parts to the nearest hundredth.

108. *Electrical Current* The current I in a circuit with voltage E, resistance R, capacitive reactance X_c, and inductive reactance X_L is

$$I = \frac{E}{R + (X_L - X_c)i}.$$

Find I if $E = 12(\cos 25° + i \sin 25°)$, $R = 3$, $X_L = 4$, and $X_c = 6$. Give the answer in rectangular form, with real and imaginary parts to the nearest tenth.

(Modeling) Impedance In the parallel electrical circuit shown in the figure, the impedance Z can be calculated using the equation

$$Z = \frac{1}{\frac{1}{Z_1} + \frac{1}{Z_2}},$$

where Z_1 and Z_2 are the impedances for the branches of the circuit.

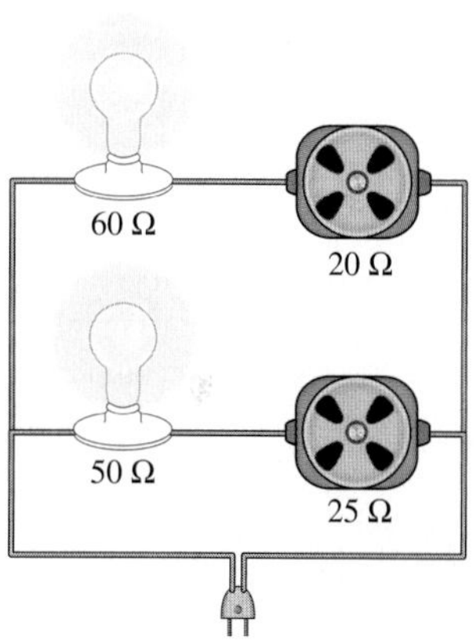

109. If $Z_1 = 50 + 25i$ and $Z_2 = 60 + 20i$, calculate Z.

110. Determine the angle θ for the value of Z found in **Exercise 109.**

8.6 De Moivre's Theorem; Powers and Roots of Complex Numbers

- Powers of Complex Numbers (De Moivre's Theorem)
- Roots of Complex Numbers

Powers of Complex Numbers (De Moivre's Theorem) Because raising a number to a positive integer power is a repeated application of the product rule, it would seem likely that a theorem for finding powers of complex numbers exists. Consider the following.

$$[r(\cos \theta + i \sin \theta)]^2$$

$$= [r(\cos \theta + i \sin \theta)][r(\cos \theta + i \sin \theta)] \quad a^2 = a \cdot a$$

$$= r \cdot r[\cos(\theta + \theta) + i \sin(\theta + \theta)] \quad \text{Product theorem (Section 8.5)}$$

$$= r^2(\cos 2\theta + i \sin 2\theta) \quad \text{Multiply and add.}$$

Abraham De Moivre (1667–1754)

Named after this French expatriate friend of Isaac Newton, De Moivre's theorem relates complex numbers and trigonometry.

In the same way,

$$[r(\cos\theta + i\sin\theta)]^3 \quad \text{is equivalent to} \quad r^3(\cos 3\theta + i\sin 3\theta).$$

These results suggest the following theorem for positive integer values of n. Although the theorem is stated and can be proved for all n, we use it only for positive integer values of n and their reciprocals.

De Moivre's Theorem

If $r(\cos\theta + i\sin\theta)$ is a complex number, and if n is any real number, then the following holds.

$$[r(\cos\theta + i\sin\theta)]^n = r^n(\cos n\theta + i\sin n\theta)$$

In compact form, this is written

$$[r \text{ cis } \theta]^n = r^n(\text{cis } n\theta).$$

Classroom Example 1
Find $(\sqrt{2} + i\sqrt{2})^5$ and express the result in rectangular form.

Answer: $-16\sqrt{2} - 16i\sqrt{2}$

Teaching Tip Mention that De Moivre's theorem greatly simplifies calculations involving powers and roots of complex numbers. Demonstrate this concept by evaluating $(1 + i\sqrt{3})^3$ with and without the theorem.

EXAMPLE 1 Finding a Power of a Complex Number

Find $(1 + i\sqrt{3})^8$ and express the result in rectangular form.

SOLUTION First write $1 + i\sqrt{3}$ in trigonometric form as

$$2(\cos 60° + i\sin 60°). \quad \text{(Section 8.5)}$$

Now, apply De Moivre's theorem.

$$(1 + i\sqrt{3})^8$$

$$= [2(\cos 60° + i\sin 60°)]^8$$

$$= 2^8[\cos(8 \cdot 60°) + i\sin(8 \cdot 60°)] \quad \text{De Moivre's theorem}$$

$$= 256(\cos 480° + i\sin 480°) \quad \text{Apply the exponent and multiply.}$$

$$= 256(\cos 120° + i\sin 120°) \quad \text{480° and 120° are coterminal. (Section 5.1)}$$

$$= 256\left(-\frac{1}{2} + i\frac{\sqrt{3}}{2}\right) \quad \cos 120° = -\tfrac{1}{2};\ \sin 120° = \tfrac{\sqrt{3}}{2} \text{ (Section 5.3)}$$

$$= -128 + 128i\sqrt{3} \quad \text{Rectangular form}$$

✔ *Now Try Exercise 7.*

Roots of Complex Numbers Every nonzero complex number has exactly n distinct complex nth roots. De Moivre's theorem can be extended to find all nth roots of a complex number.

*n*th Root

For a positive integer n, the complex number $a + bi$ is an ***n*th root** of the complex number $x + yi$ if

$$(a + bi)^n = x + yi.$$

To find the three complex cube roots of $8(\cos 135° + i \sin 135°)$, for example, look for a complex number, say $r(\cos \alpha + i \sin \alpha)$, that will satisfy

$$[r(\cos \alpha + i \sin \alpha)]^3 = 8(\cos 135° + i \sin 135°).$$

By De Moivre's theorem, this equation becomes

$$r^3(\cos 3\alpha + i \sin 3\alpha) = 8(\cos 135° + i \sin 135°).$$

Set $r^3 = 8$ and $\cos 3\alpha + i \sin 3\alpha = \cos 135° + i \sin 135°$, to satisfy this equation. The first of these conditions implies that $r = 2$, and the second implies that

$$\cos 3\alpha = \cos 135° \quad \text{and} \quad \sin 3\alpha = \sin 135°.$$

For these equations to be satisfied, 3α must represent an angle that is coterminal with 135°. Therefore, we must have

$$3\alpha = 135° + 360° \cdot k, \quad k \text{ any integer}$$

or

$$\alpha = \frac{135° + 360° \cdot k}{3}, \quad k \text{ any integer.}$$

Now, let k take on the integer values 0, 1, and 2.

$$\text{If } k = 0, \text{ then} \quad \alpha = \frac{135° + 360° \cdot 0}{3} = 45°.$$

$$\text{If } k = 1, \text{ then} \quad \alpha = \frac{135° + 360° \cdot 1}{3} = \frac{495°}{3} = 165°.$$

$$\text{If } k = 2, \text{ then} \quad \alpha = \frac{135° + 360° \cdot 2}{3} = \frac{855°}{3} = 285°.$$

In the same way, $\alpha = 405°$ when $k = 3$. But note that $405° = 45° + 360°$, so $\sin 405° = \sin 45°$ and $\cos 405° = \cos 45°$. Similarly, if $k = 4$, then $\alpha = 525°$, which has the same sine and cosine values as 165°. Continuing with larger values of k would repeat solutions already found. Therefore, all of the cube roots (three of them) can be found by letting $k = 0$, 1, and 2, respectively.

When $k = 0$, the root is $2(\cos 45° + i \sin 45°)$.

When $k = 1$, the root is $2(\cos 165° + i \sin 165°)$.

When $k = 2$, the root is $2(\cos 285° + i \sin 285°)$.

In summary, we see that $2(\cos 45° + i \sin 45°)$, $2(\cos 165° + i \sin 165°)$, and $2(\cos 285° + i \sin 285°)$ are the three cube roots of $8(\cos 135° + i \sin 135°)$.

*n*th Root Theorem

If n is any positive integer, r is a positive real number, and θ is in degrees, then the nonzero complex number $r(\cos \theta + i \sin \theta)$ has exactly n distinct nth roots, given by the following.

$$\sqrt[n]{r}(\cos \alpha + i \sin \alpha) \quad \text{or} \quad \sqrt[n]{r} \operatorname{cis} \alpha,$$

where

$$\alpha = \frac{\theta + 360° \cdot k}{n}, \quad \text{or} \quad \alpha = \frac{\theta}{n} + \frac{360° \cdot k}{n}, \quad k = 0, 1, 2, \ldots, n - 1$$

If θ is in radians, then

$$\alpha = \frac{\theta + 2\pi k}{n}, \quad \text{or} \quad \alpha = \frac{\theta}{n} + \frac{2\pi k}{n}, \quad k = 0, 1, 2, \ldots, n - 1.$$

Classroom Example 2
Find the three cube roots of -8. Write the roots in rectangular form.

Answer: $1 + i\sqrt{3}, -2, 1 - i\sqrt{3}$

EXAMPLE 2 Finding Complex Roots

Find the two square roots of $4i$. Write the roots in rectangular form.

SOLUTION First write $4i$ in trigonometric form.

$$4\left(\cos\frac{\pi}{2} + i\sin\frac{\pi}{2}\right) \quad \text{Trigonometric form}$$

Here $r = 4$ and $\theta = \frac{\pi}{2}$. The square roots have absolute value $\sqrt{4} = 2$ and arguments as follows.

$$\alpha = \frac{\frac{\pi}{2}}{2} + \frac{2\pi k}{2} = \frac{\pi}{4} + \pi k$$

Be careful simplifying here.

Since there are two square roots, let $k = 0$ and 1.

$$\text{If } k = 0, \text{ then } \quad \alpha = \frac{\pi}{4} + \pi \cdot 0 = \frac{\pi}{4}.$$

$$\text{If } k = 1, \text{ then } \quad \alpha = \frac{\pi}{4} + \pi \cdot 1 = \frac{5\pi}{4}.$$

```
(√2+i√2)²
                4i
(-√2-i√2)²
                4i
```

This screen confirms the result of **Example 2.**

Using these values for α, the square roots are $2 \text{ cis } \frac{\pi}{4}$ and $2 \text{ cis } \frac{5\pi}{4}$, which can be written in rectangular form as

$$\sqrt{2} + i\sqrt{2} \quad \text{and} \quad -\sqrt{2} - i\sqrt{2}.$$

✔ *Now Try Exercise 17(a).*

Classroom Example 3
Find all fourth roots of

$$-128 - 128i\sqrt{3}.$$

Write the roots in rectangular form.

Answer:
$2 + 2i\sqrt{3}, -2\sqrt{3} + 2i,$
$-2 - 2i\sqrt{3}, 2\sqrt{3} - 2i$

EXAMPLE 3 Finding Complex Roots

Find all fourth roots of $-8 + 8i\sqrt{3}$. Write the roots in rectangular form.

SOLUTION $\quad -8 + 8i\sqrt{3} = 16 \text{ cis } 120°$ Write in trigonometric form.

Here $r = 16$ and $\theta = 120°$. The fourth roots of this number have absolute value $\sqrt[4]{16} = 2$ and arguments as follows.

$$\alpha = \frac{120°}{4} + \frac{360° \cdot k}{4} = 30° + 90° \cdot k$$

Since there are four fourth roots, let $k = 0, 1, 2,$ and 3.

$$\text{If } k = 0, \text{ then } \quad \alpha = 30° + 90° \cdot 0 = 30°.$$

$$\text{If } k = 1, \text{ then } \quad \alpha = 30° + 90° \cdot 1 = 120°.$$

$$\text{If } k = 2, \text{ then } \quad \alpha = 30° + 90° \cdot 2 = 210°.$$

$$\text{If } k = 3, \text{ then } \quad \alpha = 30° + 90° \cdot 3 = 300°.$$

```
R▸Pr(-8,8√3)
                16
R▸Pθ(-8,8√3)
               120
```

Degree mode

This screen shows how a calculator finds r and θ for the number in **Example 3.**

Using these angles, the fourth roots are

$$2 \text{ cis } 30°, \quad 2 \text{ cis } 120°, \quad 2 \text{ cis } 210°, \quad \text{and} \quad 2 \text{ cis } 300°.$$

These four roots can be written in rectangular form as

$$\sqrt{3} + i, \quad -1 + i\sqrt{3}, \quad -\sqrt{3} - i, \quad \text{and} \quad 1 - i\sqrt{3}.$$

The graphs of these roots lie on a circle with center at the origin and radius 2. See **Figure 54.** The roots are equally spaced about the circle, 90° apart.

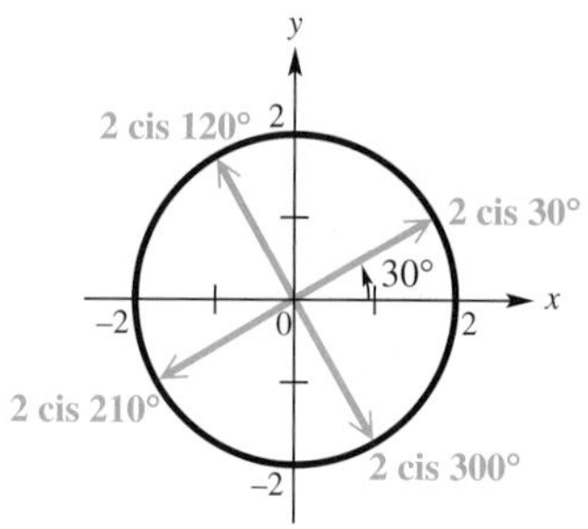

Figure 54

Now Try Exercises 23(a) and (b).

Classroom Example 4
Find all complex number solutions of

$$x^6 + i = 0.$$

Graph them as vectors in the complex plane.

Answer:
$\{$cis 45°, cis 105°, cis 165°, cis 225°, cis 285°, cis 345°$\}$

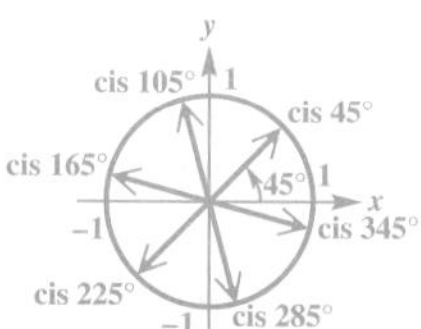

EXAMPLE 4 Solving an Equation (Complex Roots)

Find all complex number solutions of $x^5 - 1 = 0$. Graph them as vectors in the complex plane.

SOLUTION Write the equation as

$$x^5 - 1 = 0, \quad \text{or} \quad x^5 = 1.$$

There is only one real number solution, 1, but there are five complex number solutions. To find these solutions, first write 1 in trigonometric form.

$$1 = 1 + 0i = 1(\cos 0° + i \sin 0°) \quad \text{Trigonometric form}$$

The absolute value of the fifth roots is $\sqrt[5]{1} = 1$. The arguments are given by

$$0° + 72° \cdot k, \quad k = 0, 1, 2, 3, \text{ and } 4.$$

By using these arguments, we find that the fifth roots are as follows.

$$1(\cos 0° + i \sin 0°), \quad k = 0$$

$$1(\cos 72° + i \sin 72°), \quad k = 1$$

$$1(\cos 144° + i \sin 144°), \quad k = 2$$

$$1(\cos 216° + i \sin 216°), \quad k = 3$$

$$1(\cos 288° + i \sin 288°) \quad k = 4$$

The solution set of the equation can be written as

$$\{\text{cis } 0°, \text{cis } 72°, \text{cis } 144°, \text{cis } 216°, \text{cis } 288°\}.$$

The first of these roots equals 1. The others cannot easily be expressed in rectangular form but can be approximated with a calculator.

The tips of the arrows representing the five fifth roots all lie on a unit circle and are equally spaced around it every 72°, as shown in **Figure 55** on the next page.

Teaching Tip Knowing that the roots are evenly spaced on a circle in the complex plane, students may prefer to find the first solution for α and add multiples of $\frac{360°}{n}$ until all n roots are found.

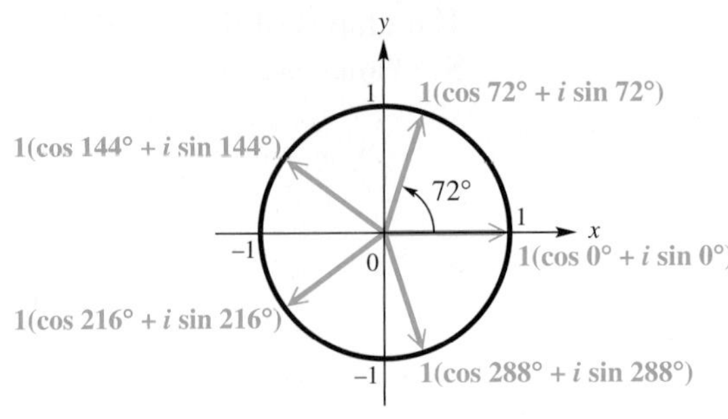

Figure 55

Now Try Exercise 35.

8.6 Exercises

Find each power. Write each answer in rectangular form. **See Example 1.**

1. $[3(\cos 30° + i \sin 30°)]^3$
2. $[2(\cos 135° + i \sin 135°)]^4$
3. $(\cos 45° + i \sin 45°)^8$
4. $[2(\cos 120° + i \sin 120°)]^3$
5. $[3 \text{ cis } 100°]^3$
6. $[3 \text{ cis } 40°]^3$
7. $(\sqrt{3} + i)^5$
8. $(2 - 2i\sqrt{3})^4$
9. $(2\sqrt{2} - 2i\sqrt{2})^6$
10. $\left(\frac{\sqrt{2}}{2} - \frac{\sqrt{2}}{2}i\right)^8$
11. $(-2 - 2i)^5$
12. $(-1 + i)^7$

In Exercises 13–24, **(a)** find all cube roots of each complex number. Leave answers in trigonometric form. **(b)** Graph each cube root as a vector in the complex plane. **See Examples 2 and 3.**

13. $\cos 0° + i \sin 0°$
14. $\cos 90° + i \sin 90°$
15. $8 \text{ cis } 60°$
16. $27 \text{ cis } 300°$
17. $-8i$
18. $27i$
19. -64
20. 27
21. $1 + i\sqrt{3}$
22. $2 - 2i\sqrt{3}$
23. $-2\sqrt{3} + 2i$
24. $\sqrt{3} - i$

Find and graph all specified roots of 1.

25. second (square)
26. fourth
27. sixth

Find and graph all specified roots of *i*.

28. second (square)
29. third (cube)
30. fourth

Find all complex number solutions of each equation. Leave answers in trigonometric form. **See Example 4.**

31. $x^3 - 1 = 0$
32. $x^3 + 1 = 0$
33. $x^3 + i = 0$
34. $x^4 + i = 0$
35. $x^3 - 8 = 0$
36. $x^3 + 27 = 0$
37. $x^4 + 1 = 0$
38. $x^4 + 16 = 0$
39. $x^4 - i = 0$
40. $x^5 - i = 0$
41. $x^3 - (4 + 4i\sqrt{3}) = 0$
42. $x^4 - (8 + 8i\sqrt{3}) = 0$

1. $27i$ 2. -16
3. 1 4. 8
5. $\frac{27}{2} - \frac{27\sqrt{3}}{2}i$
6. $-\frac{27}{2} + \frac{27\sqrt{3}}{2}i$
7. $-16\sqrt{3} + 16i$
8. $-128 + 128i\sqrt{3}$
9. $4096i$ 10. 1
11. $128 + 128i$ 12. $-8 - 8i$

13. (a) $\cos 0° + i \sin 0°$, $\cos 120° + i \sin 120°$, $\cos 240° + i \sin 240°$
(b)

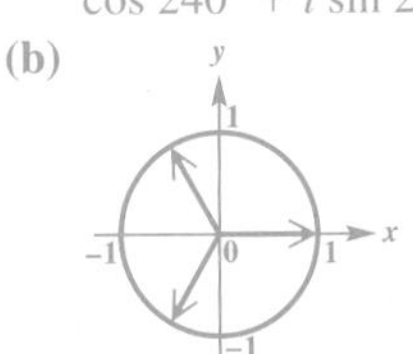

14. (a) $\cos 30° + i \sin 30°$, $\cos 150° + i \sin 150°$, $\cos 270° + i \sin 270°$
(b)

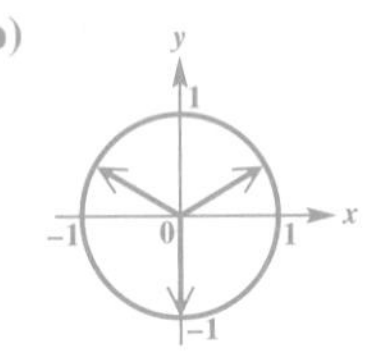

15. (a) $2 \text{ cis } 20°$, $2 \text{ cis } 140°$, $2 \text{ cis } 260°$
(b)

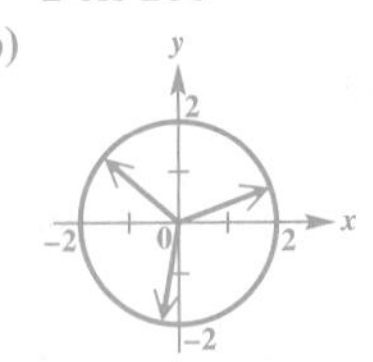

16. (a) 3 cis 100°, 3 cis 220°, 3 cis 340°
(b)

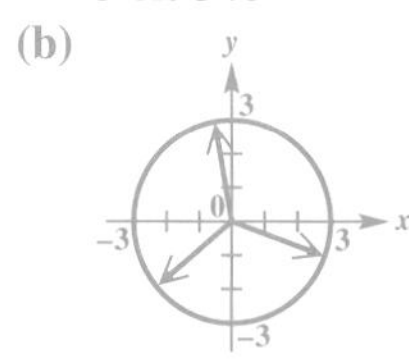

17. (a) $2(\cos 90° + i \sin 90°)$, $2(\cos 210° + i \sin 210°)$, $2(\cos 330° + i \sin 330°)$
(b)

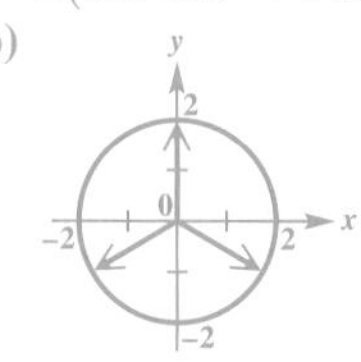

18. (a) $3(\cos 30° + i \sin 30°)$, $3(\cos 150° + i \sin 150°)$, $3(\cos 270° + i \sin 270°)$
(b)

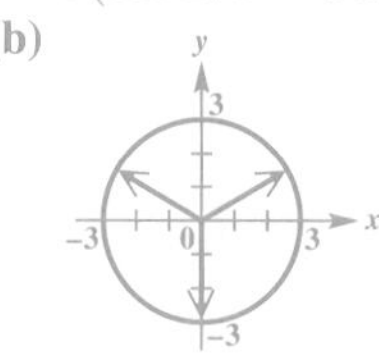

19. (a) $4(\cos 60° + i \sin 60°)$, $4(\cos 180° + i \sin 180°)$, $4(\cos 300° + i \sin 300°)$
(b)

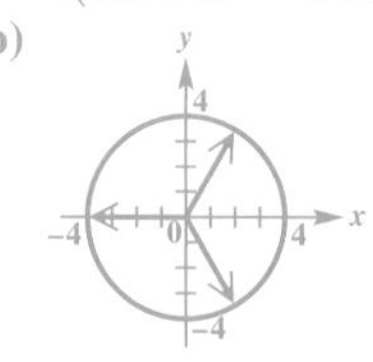

20. (a) $3(\cos 0° + i \sin 0°)$, $3(\cos 120° + i \sin 120°)$, $3(\cos 240° + i \sin 240°)$
(b)

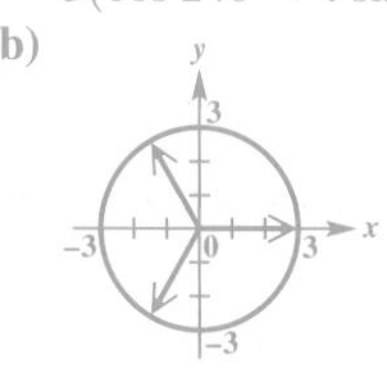

21. (a) $\sqrt[3]{2}(\cos 20° + i \sin 20°)$, $\sqrt[3]{2}(\cos 140° + i \sin 140°)$, $\sqrt[3]{2}(\cos 260° + i \sin 260°)$
(b)

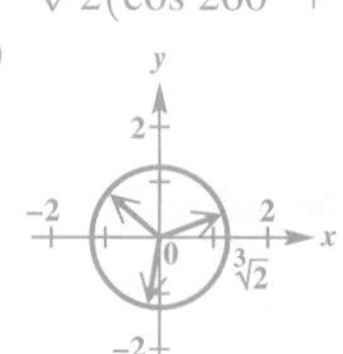

43. Solve the cubic equation

$$x^3 - 1 = 0$$

by factoring the left side as the difference of two cubes and setting each factor equal to 0. Apply the quadratic formula as needed. Then compare your solutions to those of **Exercise 31.**

44. Solve the cubic equation

$$x^3 + 27 = 0$$

by factoring the left side as the sum of two cubes and setting each factor equal to 0. Apply the quadratic formula as needed. Then compare your solutions to those of **Exercise 36.**

Relating Concepts

For individual or collaborative investigation *(Exercises 45–48)*

*In **Chapter 7** we derived identities, or formulas, for* $\cos 2\theta$ *and* $\sin 2\theta$. *These identities can also be derived using De Moivre's theorem.* ***Work Exercises 45–48 in order,*** *to see how this is done.*

45. De Moivre's theorem states that $(\cos \theta + i \sin \theta)^2 =$ ____________.

46. Expand the left side of the equation in **Exercise 45** as a binomial and collect terms to write the left side in the form $a + bi$.

47. Use the result of **Exercise 46** to obtain the double-angle formula for cosine.

48. Repeat **Exercise 47,** but find the double-angle formula for sine.

Solve each problem.

49. *Mandelbrot Set* The fractal known as the **Mandelbrot set** is shown in the figure. To determine if a complex number $z = a + bi$ is in this set, perform the following sequence of calculations. Repeatedly compute

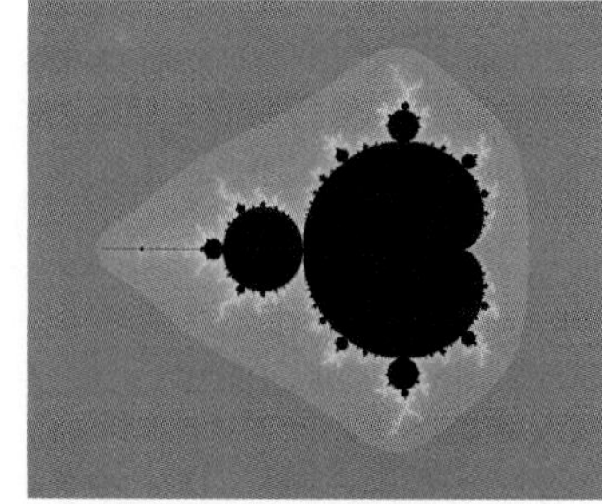

$$z, \quad z^2 + z, \quad (z^2 + z)^2 + z,$$
$$[(z^2 + z)^2 + z]^2 + z, \ldots .$$

In a manner analogous to the Julia set, the complex number z does not belong to the Mandelbrot set if any of the resulting absolute values exceeds 2. Otherwise z is in the set and the point (a, b) should be shaded in the graph. Determine whether or not the following numbers belong to the Mandelbrot set. (*Source:* Lauwerier, H., *Fractals,* Princeton University Press.)

(a) $z = 0 + 0i$ **(b)** $z = 1 - 1i$ **(c)** $z = -0.5i$

50. *Basins of Attraction* The fractal shown in the figure is the solution to Cayley's problem of determining the basins of attraction for the cube roots of unity. The three cube roots of unity are

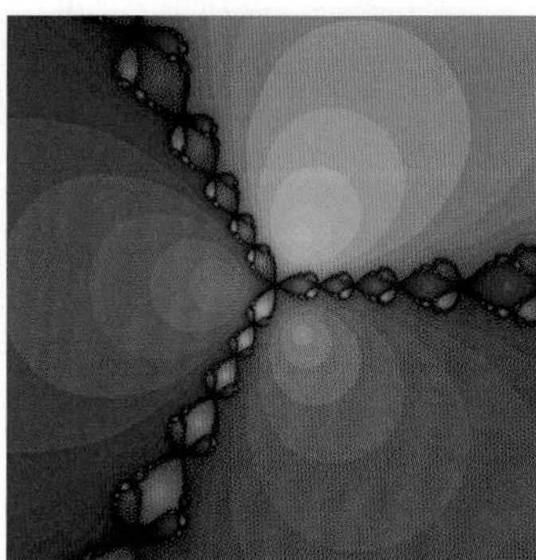

$$w_1 = 1, \quad w_2 = -\frac{1}{2} + \frac{\sqrt{3}}{2}i,$$

and $$w_3 = -\frac{1}{2} - \frac{\sqrt{3}}{2}i.$$

22. (a) $\sqrt[3]{4}(\cos 100° + i \sin 100°)$, $\sqrt[3]{4}(\cos 220° + i \sin 220°)$, $\sqrt[3]{4}(\cos 340° + i \sin 340°)$
(b)

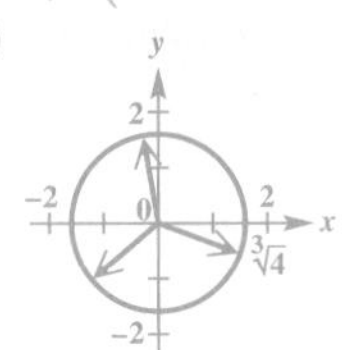

23. (a) $\sqrt[3]{4}(\cos 50° + i \sin 50°)$, $\sqrt[3]{4}(\cos 170° + i \sin 170°)$, $\sqrt[3]{4}(\cos 290° + i \sin 290°)$
(b)

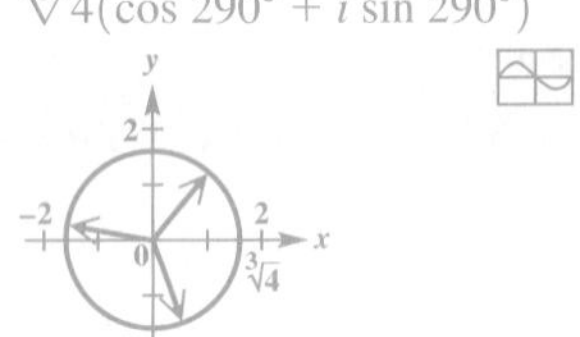

24. (a) $\sqrt[3]{2}(\cos 110° + i \sin 110°)$, $\sqrt[3]{2}(\cos 230° + i \sin 230°)$, $\sqrt[3]{2}(\cos 350° + i \sin 350°)$
(b)

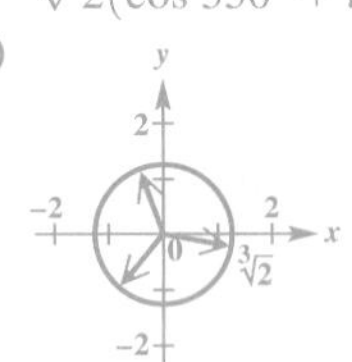

25. $\cos 0° + i \sin 0°$, $\cos 180° + i \sin 180°$

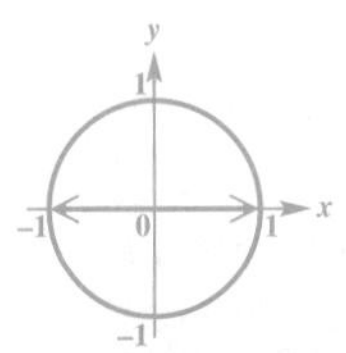

26. $\cos 0° + i \sin 0°$, $\cos 90° + i \sin 90°$, $\cos 180° + i \sin 180°$, $\cos 270° + i \sin 270°$

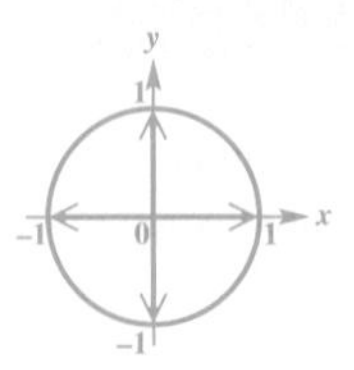

27. $\cos 0° + i \sin 0°$, $\cos 60° + i \sin 60°$, $\cos 120° + i \sin 120°$, $\cos 180° + i \sin 180°$, $\cos 240° + i \sin 240°$, $\cos 300° + i \sin 300°$

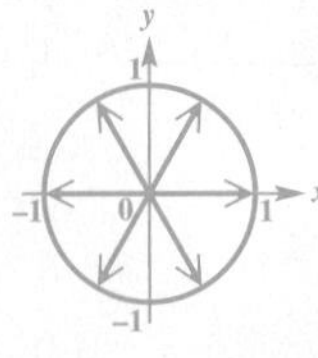

This fractal can be generated by repeatedly evaluating the function

$$f(z) = \frac{2z^3 + 1}{3z^2},$$

where z is a complex number. One begins by picking $z_1 = a + bi$ and then successively computing $z_2 = f(z_1)$, $z_3 = f(z_2)$, $z_4 = f(z_3)$, ... If the resulting values of $f(z)$ approach w_1, color the pixel at (a, b) red. If they approach w_2, color it blue, and if they approach w_3, color it yellow. If this process continues for a large number of different z_1, the fractal in the figure will appear. Determine the appropriate color of the pixel for each value of z_1. (*Source:* Crownover, R., *Introduction to Fractals and Chaos,* Jones and Bartlett Publishers.)

(a) $z_1 = i$ **(b)** $z_1 = 2 + i$ **(c)** $z_1 = -1 - i$

51. The screens here illustrate how a pentagon can be graphed using a graphing calculator. Note that a pentagon has five sides, and the T-step is $\frac{360}{5} = 72$. The display at the bottom of the graph screen indicates that one fifth root of 1 is $1 + 0i = 1$. Use this technique to find all fifth roots of 1, and express the real and imaginary parts in decimal form.

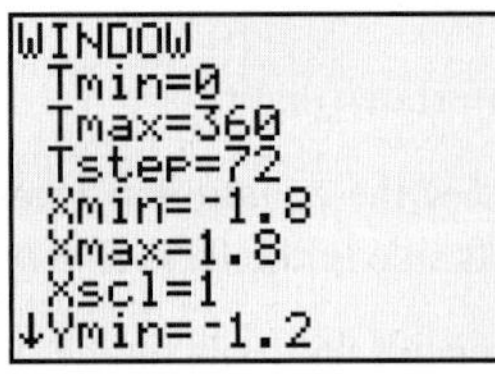

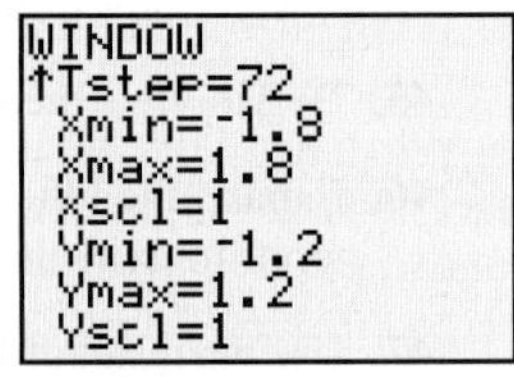

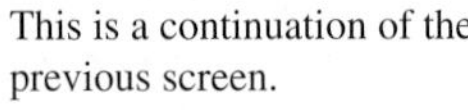
This is a continuation of the previous screen.

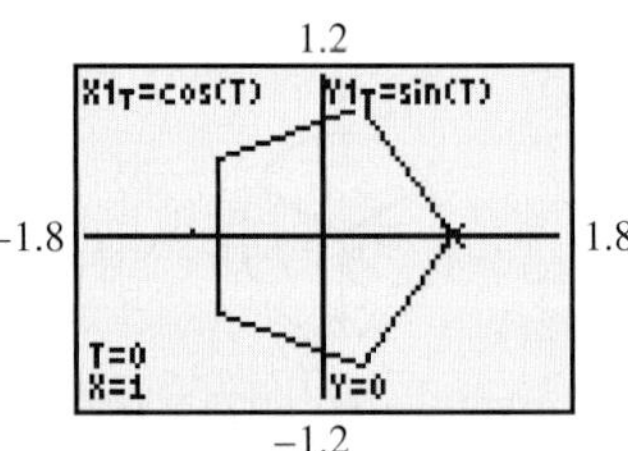

The calculator is in parametric, degree, and connected graph modes.

52. Use the method of **Exercise 51** to find the first three of the ten 10th roots of 1.

53. One of the three cube roots of a complex number is $2 + 2i\sqrt{3}$. Determine the rectangular form of its other two cube roots.

Use a calculator to find all solutions of each equation in rectangular form.

54. $x^2 + 2 - i = 0$

55. $x^2 - 3 + 2i = 0$

56. $x^3 + 4 - 5i = 0$

57. $x^5 + 2 + 3i = 0$

58. *Concept Check* How many complex 64th roots does 1 have? How many are real? How many are not?

59. *Concept Check* *True* or *false*: Every real number must have two distinct real square roots.

60. *Concept Check* *True* or *false*: Some real numbers have three real cube roots.

61. Show that if z is an nth root of 1, then so is $\frac{1}{z}$.

62. Explain why a real number can have only one real cube root.

63. Explain why the n nth roots of 1 are equally spaced around the unit circle.

64. Refer to **Figure 55.** A regular pentagon can be created by joining the tips of the arrows. Explain how you can use this principle to create a regular octagon.

28. $\cos 45° + i \sin 45°$, $\cos 225° + i \sin 225°$

29. $\cos 30° + i \sin 30°$, $\cos 150° + i \sin 150°$, $\cos 270° + i \sin 270°$

30. $\cos 22.5° + i \sin 22.5°$, $\cos 112.5° + i \sin 112.5°$, $\cos 202.5° + i \sin 202.5°$, $\cos 292.5° + i \sin 292.5°$

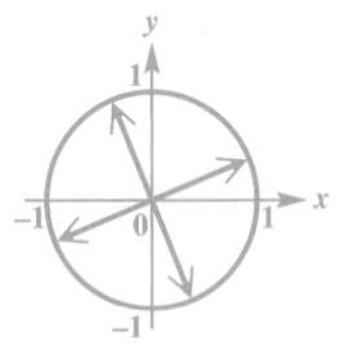

Answers for the remaining exercises are given in the answer section at the back of the text.

Chapter 8 Quiz (Sections 8.3–8.6)

1. Given vectors $\mathbf{a} = \langle -1, 4 \rangle$ and $\mathbf{b} = \langle 5, 2 \rangle$, find each of the following.

   (a) $3\mathbf{a}$ **(b)** $4\mathbf{a} - 2\mathbf{b}$ **(c)** $|\mathbf{a}|$

   (d) $\mathbf{a} \cdot \mathbf{b}$ **(e)** the angle between **a** and **b**

2. *Walking Dogs on Leashes* While Michael is walking his two dogs, Duke and Prince, they reach a corner and must wait for a WALK sign. Michael is holding the two leashes in the same hand, and the dogs are pulling on their leashes at the angles and forces shown in the figure. Find the magnitude of the equilibrant force (to the nearest tenth of a pound) that Michael must apply to restrain the dogs.

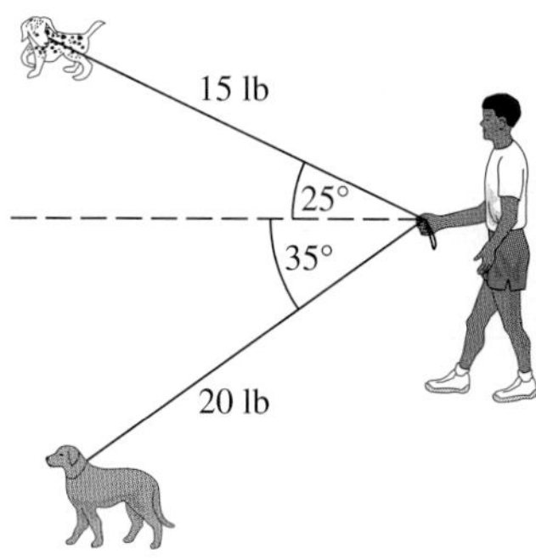

3. For the complex numbers $w = 3 + 5i$ and $z = -4 + i$, find $w + z$ and give a geometric representation of the sum.

4. Express $(1 - i)^3$ in rectangular form.

5. Write each complex number in trigonometric (polar) form, where $0° \le \theta < 360°$.

   (a) $-4i$ **(b)** $1 - i\sqrt{3}$ **(c)** $-3 - i$

6. Write each complex number in rectangular form.

   (a) $4(\cos 60° + i \sin 60°)$ **(b)** $5 \text{ cis } 130°$ **(c)** $7(\cos 270° + i \sin 270°)$

7. Write each of the following in the form specified for the complex numbers

$$w = 12(\cos 80° + i \sin 80°) \quad \text{and} \quad z = 3(\cos 50° + i \sin 50°).$$

   (a) wz (trigonometric form) **(b)** $\frac{w}{z}$ (rectangular form) **(c)** z^3 (rectangular form)

8. Find the four complex fourth roots of -16. Express them in both trigonometric and rectangular forms.

[8.3]
1. (a) $\langle -3, 12 \rangle$ (b) $\langle -14, 12 \rangle$ (c) $\sqrt{17}$ (d) 3 (e) 82.23°

[8.4]
2. 30 lb

[8.5]
3. $-1 + 6i$

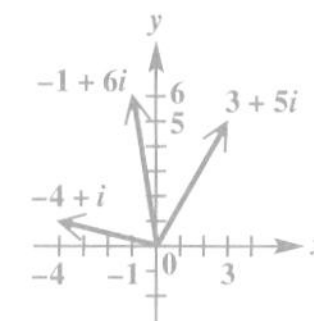

4. $-2 - 2i$
5. (a) $4(\cos 270° + i \sin 270°)$ (b) $2(\cos 300° + i \sin 300°)$ (c) $\sqrt{10}(\cos 198.4° + i \sin 198.4°)$
6. (a) $2 + 2i\sqrt{3}$ (b) $-3.2139 + 3.8302i$ (c) $-7i$, or $0 - 7i$

[8.5, 8.6]
7. (a) $36(\cos 130° + i \sin 130°)$ (b) $2\sqrt{3} + 2i$ (c) $-\frac{27\sqrt{3}}{2} + \frac{27}{2}i$

[8.6]
8. $2(\cos 45° + i \sin 45°)$, $2(\cos 135° + i \sin 135°)$, $2(\cos 225° + i \sin 225°)$, $2(\cos 315° + i \sin 315°)$; $\sqrt{2} + i\sqrt{2}$, $-\sqrt{2} + i\sqrt{2}$, $-\sqrt{2} - i\sqrt{2}$, $\sqrt{2} - i\sqrt{2}$

8.7 Polar Equations and Graphs

- Polar Coordinate System
- Graphs of Polar Equations
- Converting from Polar to Rectangular Equations
- Classifying Polar Equations

Polar Coordinate System Previously we have used the rectangular coordinate system to graph points and equations. In the rectangular coordinate system, each point in the plane is specified by giving two numbers (x, y). These represent the directed distances from a pair of perpendicular axes, the x-axis and the y-axis.

Now we consider the **polar coordinate system** which is based on a point, called the **pole,** and a ray, called the **polar axis.** The polar axis is usually drawn in the direction of the positive x-axis, as shown in **Figure 56.**

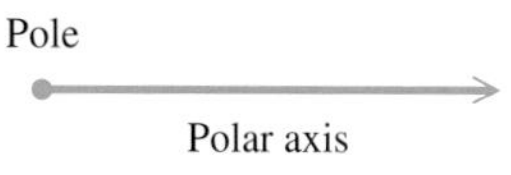

Figure 56

In **Figure 57** the pole has been placed at the origin of a rectangular coordinate system so that the polar axis coincides with the positive x-axis. Point P has rectangular coordinates (x, y). Point P can also be located by giving the directed angle θ from the positive x-axis to ray OP and the *directed* distance r from the pole to point P. The ordered pair (r, θ) gives the **polar coordinates** of point P. If $r > 0$ then point P lies on the terminal side of θ, and if $r < 0$ then point P lies on the ray pointing in the opposite direction of the terminal side of θ, a distance $|r|$ from the pole. **Figure 58** shows rectangular axes superimposed on a polar coordinate grid.

Classroom Example 1
Plot each point by hand in the polar coordinate system. Then determine the rectangular coordinates of each point.
(a) $P(4, 135°)$ (b) $Q(-3, \frac{\pi}{6})$
(c) $R(2, -\frac{\pi}{2})$

Answers:
(a)

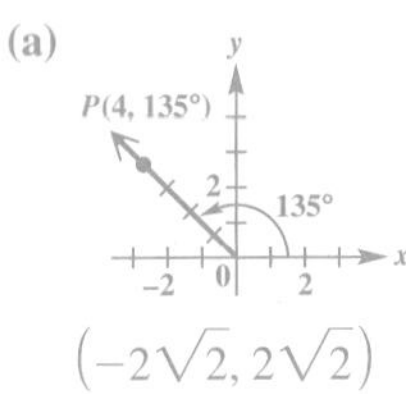

$(-2\sqrt{2}, 2\sqrt{2})$

(b)

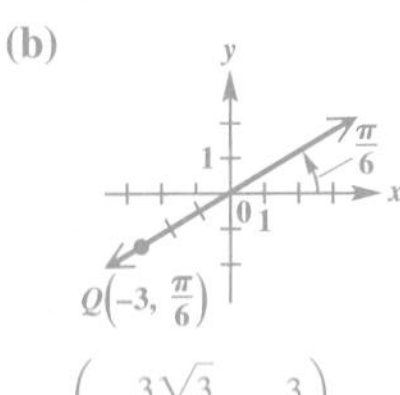

$\left(-\frac{3\sqrt{3}}{2}, -\frac{3}{2}\right)$

(c)

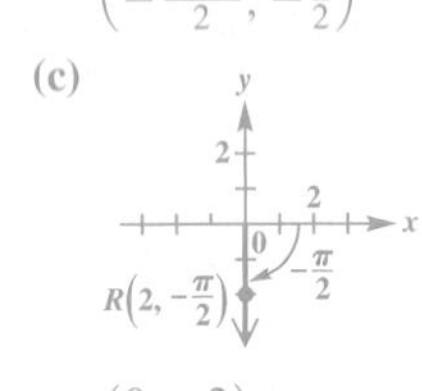

$(0, -2)$

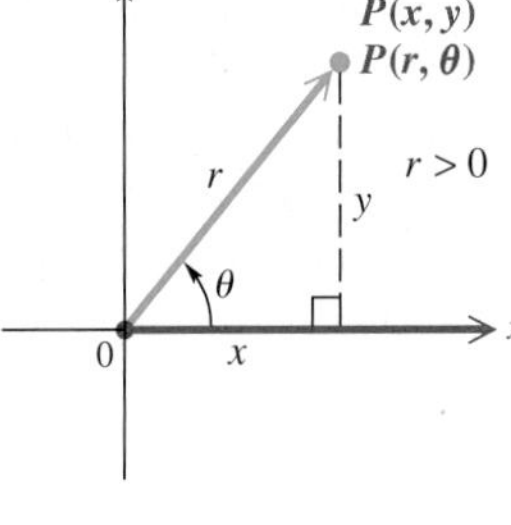

Figure 57

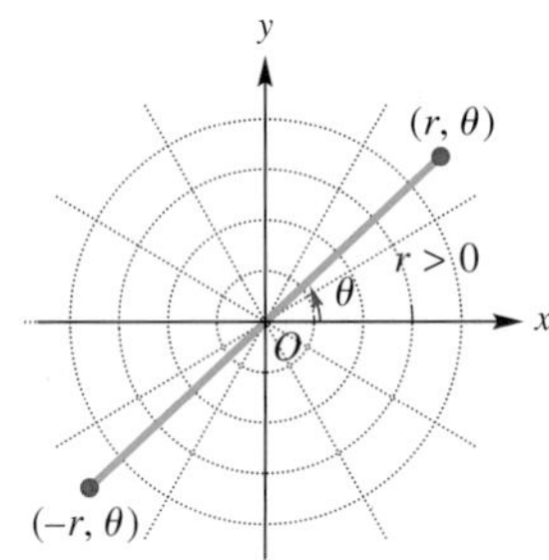

Figure 58

Rectangular and Polar Coordinates

If a point has rectangular coordinates (x, y) and polar coordinates (r, θ), then these coordinates are related as follows.

$$x = r\cos\theta \qquad y = r\sin\theta$$

$$r^2 = x^2 + y^2 \qquad \tan\theta = \frac{y}{x}, \quad \text{if } x \neq 0$$

EXAMPLE 1 **Plotting Points with Polar Coordinates**

Plot each point by hand in the polar coordinate system. Then determine the rectangular coordinates of each point.

(a) $P(2, 30°)$ **(b)** $Q\left(-4, \frac{2\pi}{3}\right)$ **(c)** $R\left(5, -\frac{\pi}{4}\right)$

SOLUTION

(a) In the point $P(2, 30°)$, $r = 2$ and $\theta = 30°$, so P is located 2 units from the origin in the positive direction on a ray making a $30°$ angle with the polar axis, as shown in **Figure 59.**

We find the rectangular coordinates as follows.

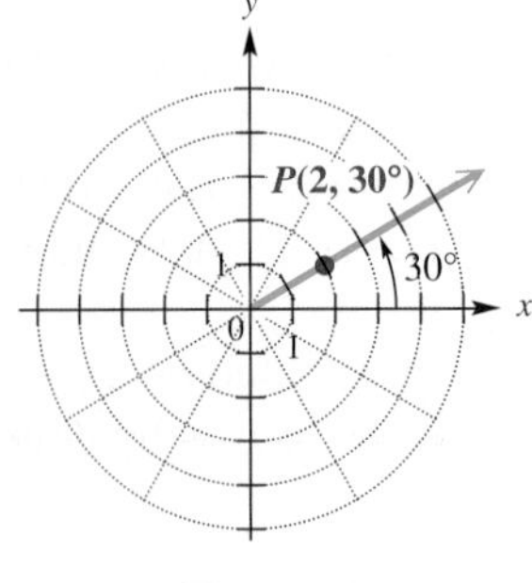

Figure 59

$x = r\cos\theta$	$y = r\sin\theta$	Conversion equations
$x = 2\cos 30°$	$y = 2\sin 30°$	Substitute.
$x = 2\left(\frac{\sqrt{3}}{2}\right)$	$y = 2\left(\frac{1}{2}\right)$	(Section 5.3)
$x = \sqrt{3}$	$y = 1$	Multiply.

The rectangular coordinates are $(\sqrt{3}, 1)$.

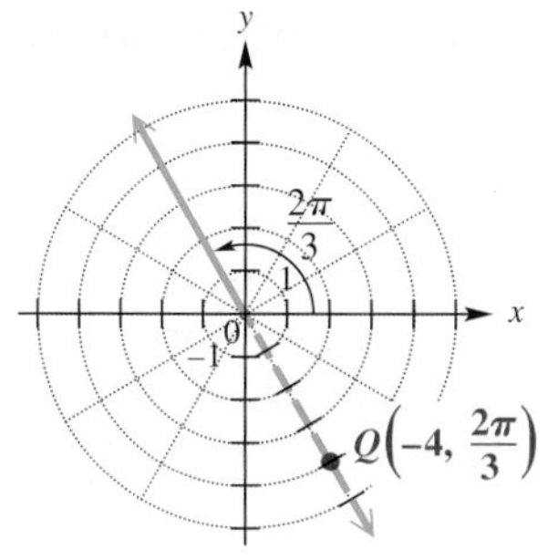

Figure 60

(b) In the point $Q\left(-4, \frac{2\pi}{3}\right)$, r is *negative*, so Q is 4 units in the *opposite* direction from the pole on an extension of the $\frac{2\pi}{3}$ ray. See **Figure 60.** The rectangular coordinates are

$$x = -4 \cos \frac{2\pi}{3} = -4\left(-\frac{1}{2}\right) = 2$$

and $$y = -4 \sin \frac{2\pi}{3} = -4\left(\frac{\sqrt{3}}{2}\right) = -2\sqrt{3}.$$

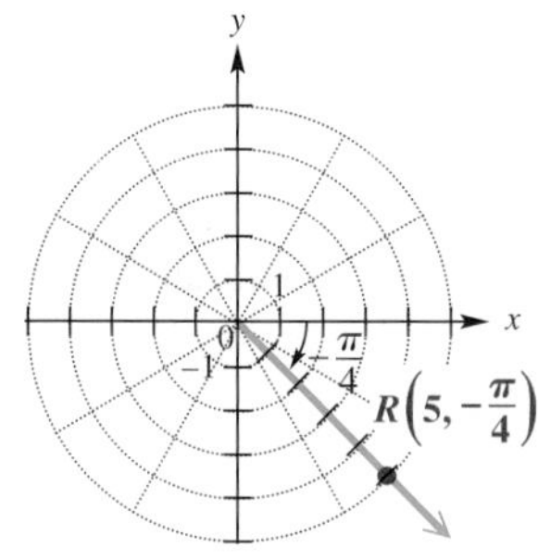

Figure 61

(c) Point $R\left(5, -\frac{\pi}{4}\right)$ is shown in **Figure 61.** Since θ is negative, the angle is measured in the clockwise direction.

$$x = 5 \cos\left(-\frac{\pi}{4}\right) = \frac{5\sqrt{2}}{2} \quad \text{and} \quad y = 5 \sin\left(-\frac{\pi}{4}\right) = -\frac{5\sqrt{2}}{2}$$

✔ *Now Try Exercises 3(a), (c), 5(a), (c), and 11(a), (c).*

While a given point in the plane can have only one pair of rectangular coordinates, this same point can have an infinite number of pairs of polar coordinates. For example, $(2, 30°)$ locates the same point as

$$(2, 390°), \quad (2, -330°), \quad \text{and} \quad (-2, 210°).$$

Classroom Example 2
(a) Give three other pairs of polar coordinates for the point $P(5, -110°)$.
(b) Give two pairs of polar coordinates for the point with rectangular coordinates $(6, -6\sqrt{3})$.

Answers:
(Other answers are possible.)
(a) $(5, 250°)$, $(-5, 70°)$, $(-5, -290°)$
(b) $(12, 300°)$, $(-12, 120°)$

EXAMPLE 2 Giving Alternative Forms for Coordinates of a Point

(a) Give three other pairs of polar coordinates for the point $P(3, 140°)$.

(b) Determine two pairs of polar coordinates for the point with rectangular coordinates $(-1, 1)$.

SOLUTION

(a) Three pairs that could be used for the point are $(3, -220°)$, $(-3, 320°)$, and $(-3, -40°)$. See **Figure 62.**

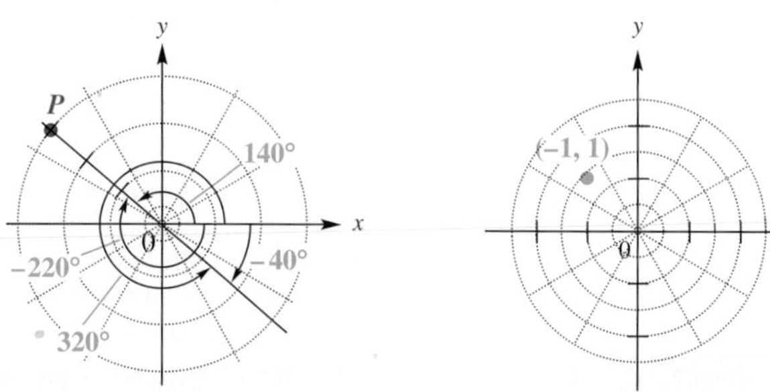

Figure 62 Figure 63

LOOKING AHEAD TO CALCULUS

Techniques studied in calculus associated with derivatives and integrals provide methods of finding slopes of tangent lines to polar curves, areas bounded by such curves, and lengths of their arcs.

(b) As shown in **Figure 63,** the point $(-1, 1)$ lies in the second quadrant. Since $\tan \theta = \frac{1}{-1} = -1$, one possible value for θ is 135°. Also,

$$r = \sqrt{x^2 + y^2} = \sqrt{(-1)^2 + 1^2} = \sqrt{2}.$$

Two pairs of polar coordinates are $\left(\sqrt{2}, 135°\right)$ and $\left(-\sqrt{2}, 315°\right)$.

✔ *Now Try Exercises 3(b), 5(b), 11(b), and 15.*

Classroom Example 3

For each rectangular equation, give the equivalent polar equation and sketch its graph.

(a) $y = 2x - 4$

(b) $x^2 + y^2 = 25$

Answers:

(a) $r = \frac{4}{2 \cos \theta - \sin \theta}$

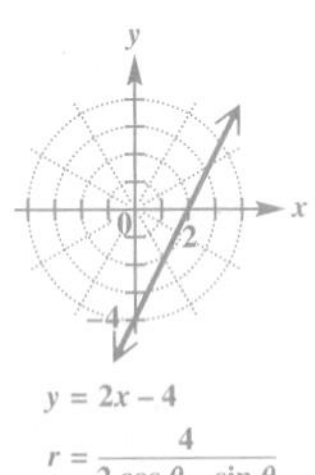

$y = 2x - 4$

$r = \frac{4}{2 \cos \theta - \sin \theta}$

(b) $r = 5$ or $r = -5$

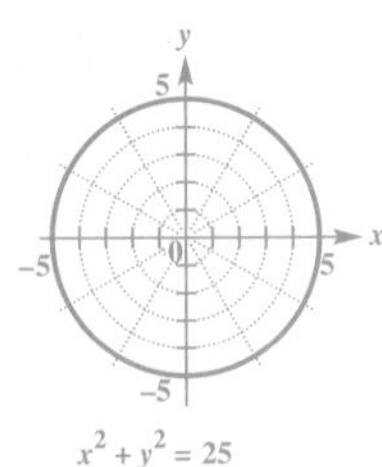

$x^2 + y^2 = 25$

$r = 5$ or $r = -5$

Graphs of Polar Equations Equations in x and y are **rectangular** (or **Cartesian**) **equations.** An equation in which r and θ are the variables instead of x and y is a **polar equation.**

$$r = 3 \sin \theta, \quad r = 2 + \cos \theta, \quad r = \theta \quad \text{Polar equations}$$

Although the rectangular forms of lines and circles are the ones most often encountered, they can also be defined in terms of polar coordinates. The polar equation of the line $ax + by = c$ can be derived as follows.

Line:

$$ax + by = c \quad \text{Rectangular equation of a line}$$

$$a(r \cos \theta) + b(r \sin \theta) = c \quad \text{Convert to polar coordinates.}$$

$$r(a \cos \theta + b \sin \theta) = c \quad \text{Factor out } r. \text{ (Section R.4)}$$

This is the polar equation of $ax + by = c$.

$$\boldsymbol{r = \frac{c}{a \cos \theta + b \sin \theta}} \quad \text{Polar equation of a line}$$

For the circle $x^2 + y^2 = a^2$, the polar equation can be found in a similar manner.

Circle:

$$x^2 + y^2 = a^2 \quad \text{Rectangular equation of a circle (Section 2.2)}$$

$$r^2 = a^2 \quad x^2 + y^2 = r^2$$

These are polar equations of $x^2 + y^2 = a^2$.

$$\boldsymbol{r = \pm a} \quad \text{Polar equation of a circle; } r \text{ can be negative in polar coordinates.}$$

We use these forms in the next example.

EXAMPLE 3 Finding Polar Equations of Lines and Circles

For each rectangular equation, give the equivalent polar equation and sketch its graph.

(a) $y = x - 3$ **(b)** $x^2 + y^2 = 4$

SOLUTION

(a) This is the equation of a line.

$$y = x - 3$$

$$x - y = 3 \quad \text{Write in standard form, } ax + by = c. \text{ (Section 2.4)}$$

$$r \cos \theta - r \sin \theta = 3 \quad \text{Substitute for } x \text{ and } y.$$

$$r(\cos \theta - \sin \theta) = 3 \quad \text{Factor out } r.$$

$$r = \frac{3}{\cos \theta - \sin \theta} \quad \text{Divide by } \cos \theta - \sin \theta.$$

A traditional graph is shown in **Figure 64(a),** and a calculator graph is shown in **Figure 64(b).**

$y = x - 3$ (rectangular)

$r = \frac{3}{\cos \theta - \sin \theta}$ (polar)

(a)

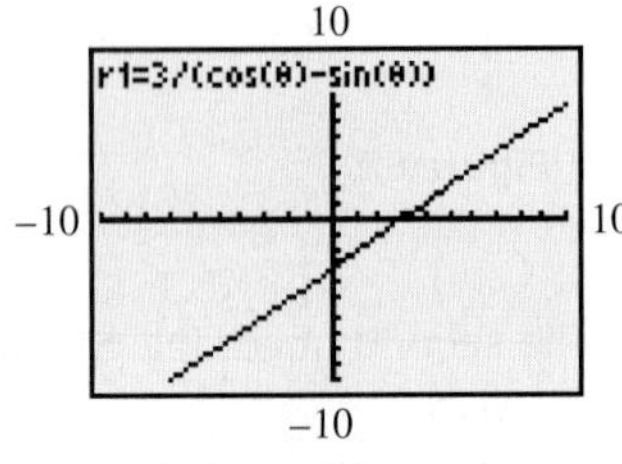

Polar graphing mode

(b)

Figure 64

(b) The graph of $x^2 + y^2 = 4$ is a circle with center at the origin and radius 2.

$$x^2 + y^2 = 4 \quad \text{(Section 2.2)}$$

$$r^2 = 4 \quad x^2 + y^2 = r^2$$

$$r = 2 \quad \text{or} \quad r = -2$$

In polar coordinates, we may have $r < 0$.

The graphs of $r = 2$ and $r = -2$ coincide. See **Figure 65** on the next page.

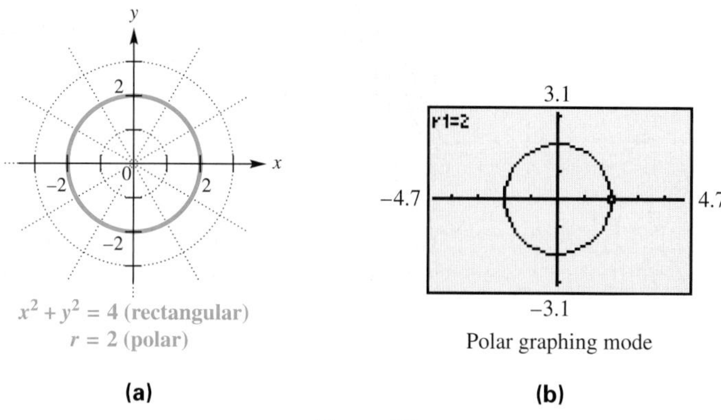

$x^2 + y^2 = 4$ (rectangular)
$r = 2$ (polar)

(a)

Polar graphing mode

(b)

Figure 65

Classroom Example 4
Graph $r = 1 - \sin \theta$.

Answer:

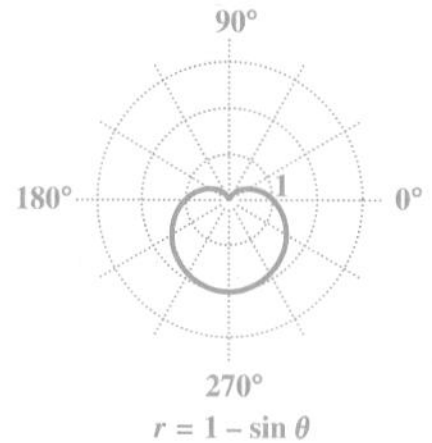

Now Try Exercises 27 and 29.

To graph polar equations, evaluate r for various values of θ until a pattern appears, and then join the points with a smooth curve.

EXAMPLE 4 Graphing a Polar Equation (Cardioid)

Graph $r = 1 + \cos \theta$.

ALGEBRAIC SOLUTION

To graph this equation, find some ordered pairs (as in the table). Once the pattern of values of r becomes clear, it is not necessary to find more ordered pairs. The table includes approximated values for $\cos \theta$ and r.

θ	$\cos \theta$	$r = 1 + \cos \theta$	θ	$\cos \theta$	$r = 1 + \cos \theta$
0°	1	2	135°	−0.7	0.3
30°	0.9	1.9	150°	−0.9	0.1
45°	0.7	1.7	180°	−1	0
60°	0.5	1.5	270°	0	1
90°	0	1	315°	0.7	1.7
120°	−0.5	0.5	330°	0.9	1.9

Connect the points in order—from $(2, 0°)$ to $(1.9, 30°)$ to $(1.7, 45°)$ and so on. See **Figure 66.** This curve is called a **cardioid** because of its heart shape. The curve has been graphed on a **polar grid.**

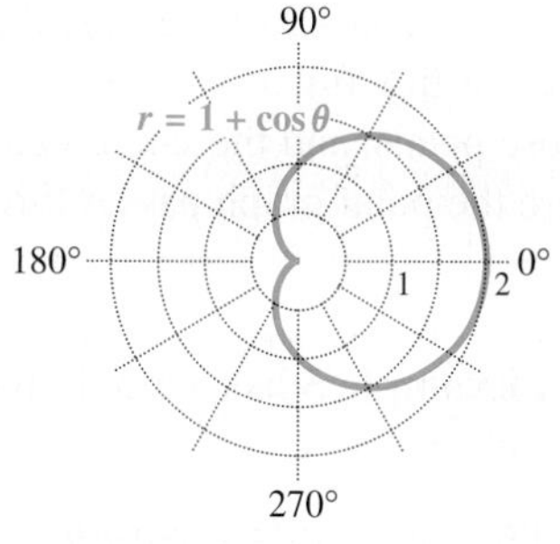

Figure 66

GRAPHING CALCULATOR SOLUTION

We choose degree mode and graph values of θ in the interval $[0°, 360°]$. The screens in **Figure 67(a)** show the choices needed to generate the graph in **Figure 67(b).**

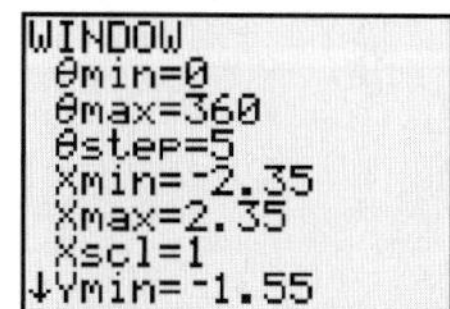

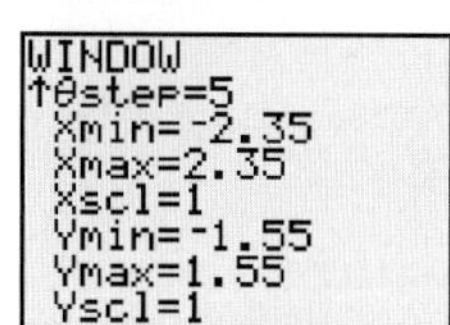

This is a continuation of the previous screen.

(a)

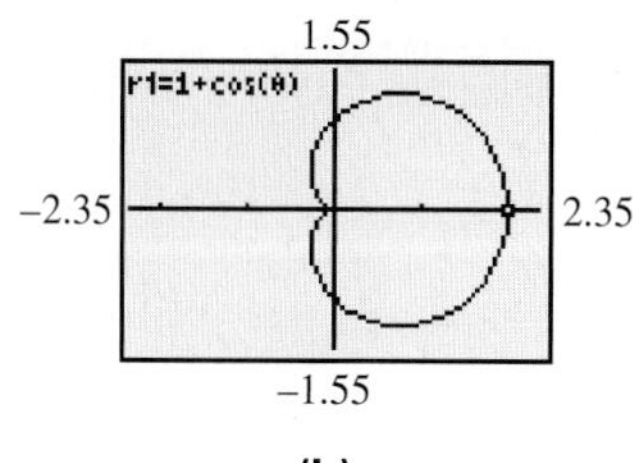

(b)

Figure 67

Now Try Exercise 45.

Classroom Example 5
Graph $r = 4 \cos 3\theta$.

Answer:

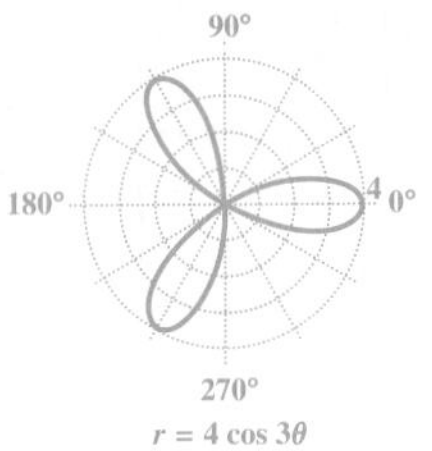

EXAMPLE 5 Graphing a Polar Equation (Rose)

Graph $r = 3 \cos 2\theta$.

SOLUTION Because the argument is 2θ, the graph requires a greater number of points than when the argument is just θ. We complete the table using selected angle measures through 360° in order to see the pattern of the graph. Approximate values in the table have been rounded to the nearest tenth.

θ	2θ	$\cos 2\theta$	$r = 3 \cos 2\theta$	θ	2θ	$\cos 2\theta$	$r = 3 \cos 2\theta$
0°	0°	1	3	120°	240°	−0.5	−1.5
15°	30°	0.9	2.6	135°	270°	0	0
30°	60°	0.5	1.5	180°	360°	1	3
45°	90°	0	0	225°	450°	0	0
60°	120°	−0.5	−1.5	270°	540°	−1	−3
75°	150°	−0.9	−2.6	315°	630°	0	0
90°	180°	−1	−3	360°	720°	1	3

Plotting the points from the table in order gives the graph of a **four-leaved rose.** Note in **Figure 68(a)** how the graph is developed with a continuous curve, beginning with the upper half of the right horizontal leaf and ending with the lower half of that leaf. As the graph is traced, the curve goes through the pole four times. This can actually be seen as a calculator graphs the curve. See **Figure 68(b).**

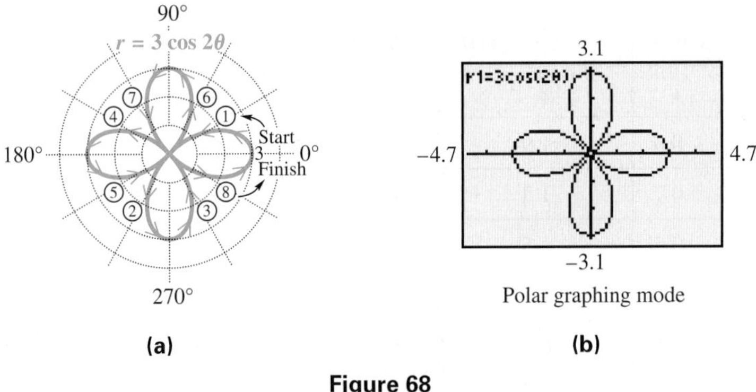

Figure 68

Now Try Exercise 49.

NOTE To sketch the graph of $r = 3 \cos 2\theta$ in polar coordinates, it may be helpful to first sketch the graph of $y = 3 \cos 2x$ in rectangular coordinates. The minimum and maximum values of this function may be used to determine the location of the tips of the rose petals, and the x-intercepts of this function may be used to determine where the polar graph passes through the pole.

Classroom Example 6
Graph $r^2 = 9 \sin 2\theta$.

Answer:

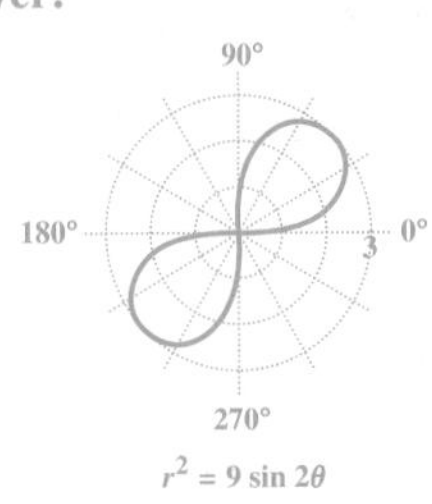

The equation $r = 3 \cos 2\theta$ in **Example 5** has a graph that belongs to a family of curves called **roses.** The graphs of

$$r = a \sin n\theta \quad \text{and} \quad r = a \cos n\theta$$

are roses, with n leaves if n is odd, and $2n$ leaves if n is even. The absolute value of a determines the length of the leaves.

EXAMPLE 6 Graphing a Polar Equation (Lemniscate)

Graph $r^2 = \cos 2\theta$.

ALGEBRAIC SOLUTION

Complete a table of ordered pairs, and sketch the graph, as in **Figure 69.** The point $(-1, 0°)$, with r negative, may be plotted as $(1, 180°)$. Also, $(-0.7, 30°)$ may be plotted as $(0.7, 210°)$, and so on.

Values of θ for $45° < \theta < 135°$ are not included in the table because the corresponding values of $\cos 2\theta$ are negative (quadrants II and III) and so do not have real square roots. Values of θ larger than 180° give 2θ larger than 360° and would repeat the points already found. This curve is called a **lemniscate.**

θ	0°	30°	45°	135°	150°	180°
2θ	0°	60°	90°	270°	300°	360°
$\cos 2\theta$	1	0.5	0	0	0.5	1
$r = \pm\sqrt{\cos 2\theta}$	±1	±0.7	0	0	±0.7	±1

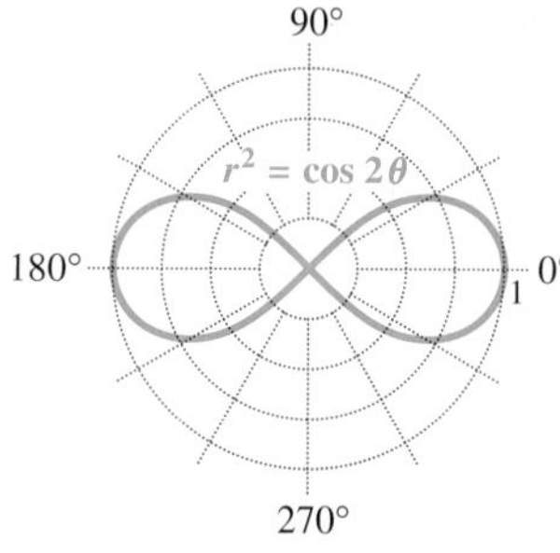

Figure 69

GRAPHING CALCULATOR SOLUTION

To graph $r^2 = \cos 2\theta$ with a graphing calculator, first solve for r by considering both square roots. Enter the two polar equations as

$$r_1 = \sqrt{\cos 2\theta}$$

and $$r_2 = -\sqrt{\cos 2\theta}.$$

See **Figures 70(a) and (b).**

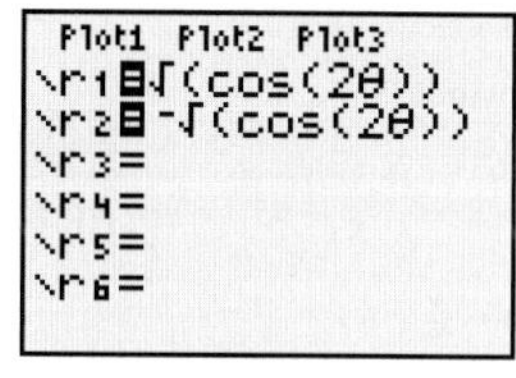

(a)

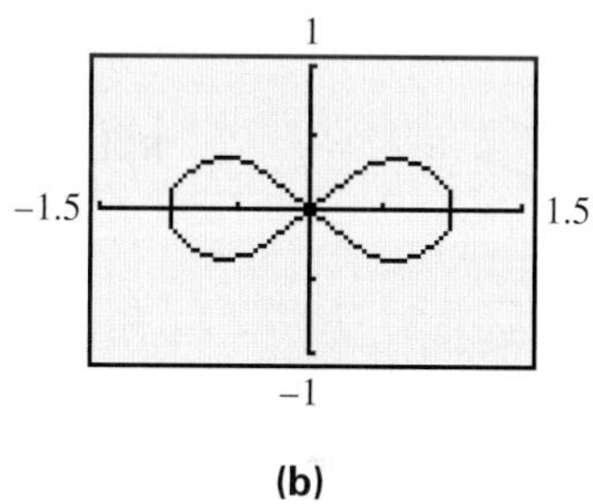

(b)

Figure 70

✓ *Now Try Exercise 51.*

Classroom Example 7
Graph $r = -\theta$ (with θ measured in radians).

Answer:

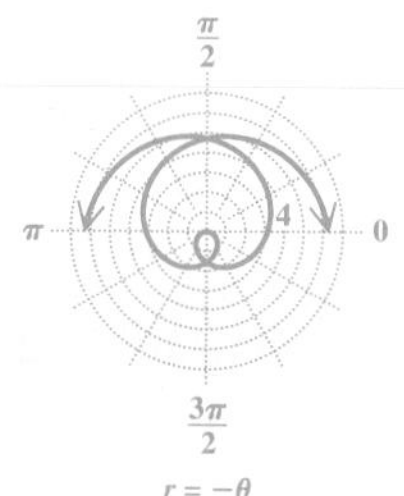

EXAMPLE 7 Graphing a Polar Equation (Spiral of Archimedes)

Graph $r = 2\theta$ (with θ measured in radians).

SOLUTION Some ordered pairs are shown in the table. Since $r = 2\theta$ rather than a trigonometric function of θ, we must also consider negative values of θ. Radian measures have been rounded. The graph in **Figure 71(a)** on the next page is a **spiral of Archimedes. Figure 71(b)** shows a calculator graph of this spiral.

θ (radians)	$r = 2\theta$	θ (radians)	$r = 2\theta$
$-\pi$	−6.3	$\frac{\pi}{3}$	2.1
$-\frac{\pi}{2}$	−3.1	$\frac{\pi}{2}$	3.1
$-\frac{\pi}{4}$	−1.6	π	6.3
0	0	$\frac{3\pi}{2}$	9.4
$\frac{\pi}{6}$	1	2π	12.6

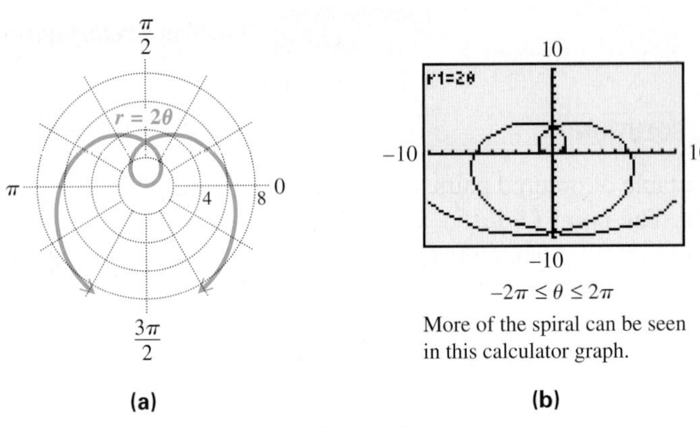

(a)

$-2\pi \le \theta \le 2\pi$

More of the spiral can be seen in this calculator graph.

(b)

Figure 71

Teaching Tip Point out that not all polar equations are easier to graph after being converted to rectangular equations.

✔ *Now Try Exercise 67.*

Classroom Example 8
Convert the equation

$$r = \frac{1}{1 - \cos\theta}$$

to rectangular coordinates, and graph.

Answer: $y^2 = 2\left(x + \frac{1}{2}\right)$

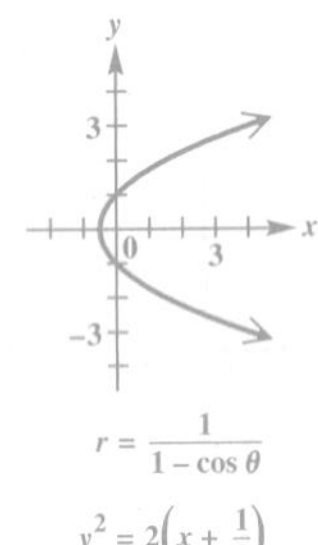

Converting from Polar to Rectangular Equations In **Example 3** we converted rectangular equations to polar equations. We conclude with an example that converts a polar equation to a rectangular one.

EXAMPLE 8 Converting a Polar Equation to a Rectangular Equation

Convert the equation $r = \frac{4}{1 + \sin\theta}$ to rectangular coordinates, and graph.

SOLUTION

$$r = \frac{4}{1 + \sin\theta} \quad \text{Polar equation}$$

$$r(1 + \sin\theta) = 4 \quad \text{Multiply by } 1 + \sin\theta.$$

$$r + r\sin\theta = 4 \quad \text{Distributive property (Section R.2)}$$

$$\sqrt{x^2 + y^2} + y = 4 \quad \text{Let } r = \sqrt{x^2 + y^2} \text{ and } r\sin\theta = y.$$

$$\sqrt{x^2 + y^2} = 4 - y \quad \text{Subtract } y.$$

$$x^2 + y^2 = (4 - y)^2 \quad \text{Square each side. (Section 1.6)}$$

$$x^2 + y^2 = 16 - 8y + y^2 \quad \text{Expand the right side. (Section R.3)}$$

$$x^2 = -8y + 16 \quad \text{Subtract } y^2.$$

$$x^2 = -8(y - 2) \quad \text{Rectangular equation}$$

The final equation represents a parabola and is graphed in **Figure 72.**

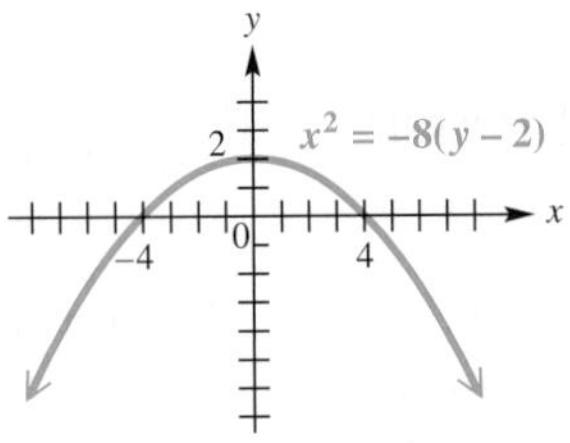

Figure 72

✔ *Now Try Exercise 59.*

The conversion in **Example 8** is not necessary when one is using a graphing calculator. **Figure 73** shows the graph of $r = \frac{4}{1 + \sin\theta}$, graphed directly with the calculator in polar mode. ■

r1=4/(1+sin(θ))

10, −10, 10, −10

$0° \le \theta \le 360°$

Figure 73

Classifying Polar Equations The table on the next page summarizes common polar graphs and forms of their equations. (In addition to circles, lemniscates, and roses, we include **limaçons.** Cardioids are a special case of limaçons, where $\left|\frac{a}{b}\right| = 1$.)

Circles and Lemniscates

Circles		Lemniscates	
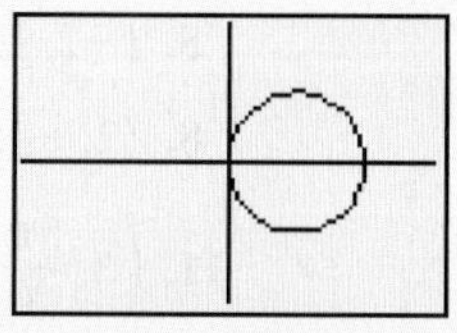	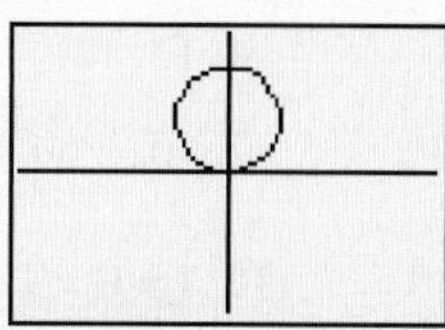	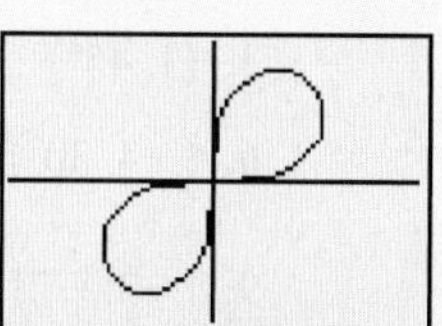	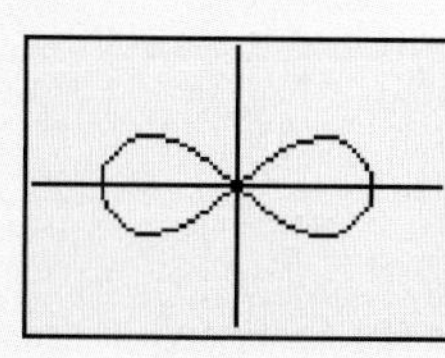
$r = a \cos \theta$	$r = a \sin \theta$	$r^2 = a^2 \sin 2\theta$	$r^2 = a^2 \cos 2\theta$

Limaçons

$r = a \pm b \sin \theta$ or $r = a \pm b \cos \theta$

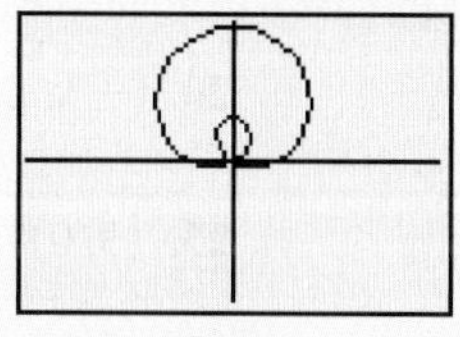	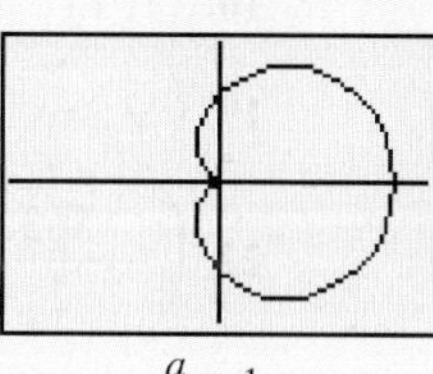	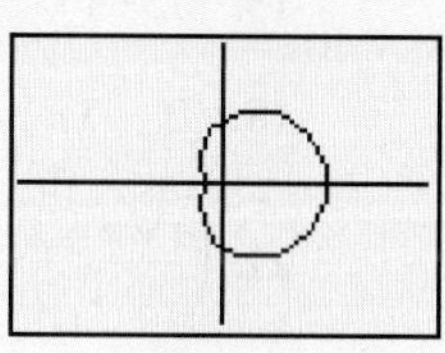	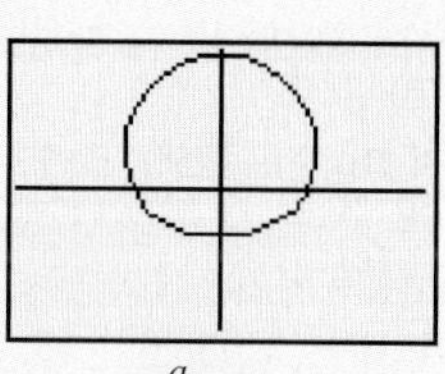
$\frac{a}{b} < 1$	$\frac{a}{b} = 1$	$1 < \frac{a}{b} < 2$	$\frac{a}{b} \geq 2$

Rose Curves

$2n$ leaves if n is even, $n \geq 2$		n leaves if n is odd	
$n = 2$	$n = 4$	$n = 3$	$n = 5$
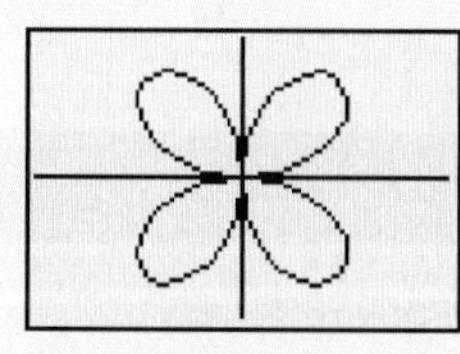	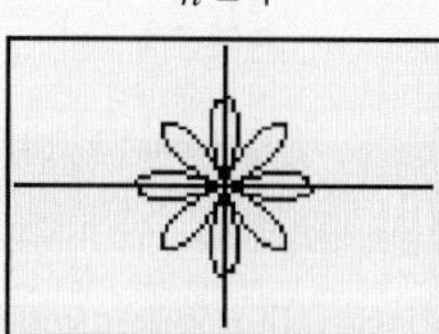	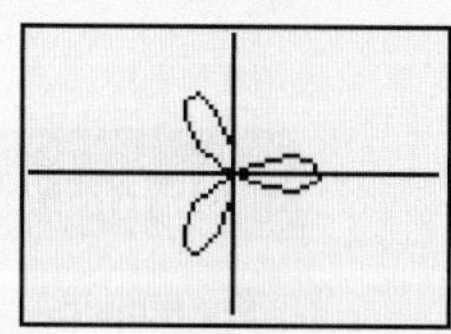	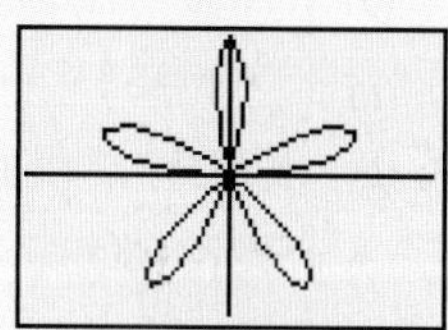
$r = a \sin n\theta$	$r = a \cos n\theta$	$r = a \cos n\theta$	$r = a \sin n\theta$

NOTE Some other polar curves are the **cissoid, kappa curve, conchoid, trisectrix, cruciform, strophoid,** and **lituus.** Refer to older textbooks on analytic geometry or the Internet to investigate them.

8.7 Exercises

1. (a) II (b) I
(c) IV (d) III
2. (a) positive x-axis
(b) negative x-axis
(c) negative y-axis
(d) positive y-axis

1. *Concept Check* For each point given in polar coordinates, state the quadrant in which the point lies if it is graphed in a rectangular coordinate system.

(a) $(5, 135°)$ **(b)** $(2, 60°)$ **(c)** $(6, -30°)$ **(d)** $(4.6, 213°)$

2. *Concept Check* For each point given in polar coordinates, state the axis on which the point lies if it is graphed in a rectangular coordinate system. Also state whether it is on the positive portion or the negative portion of the axis. (For example, $(5, 0°)$ lies on the positive x-axis.)

(a) $(7, 360°)$ **(b)** $(4, 180°)$ **(c)** $(2, -90°)$ **(d)** $(8, 450°)$

Graphs for Exercises 3(a), 5(a), 7(a), 9(a), 11(a), 13(a)

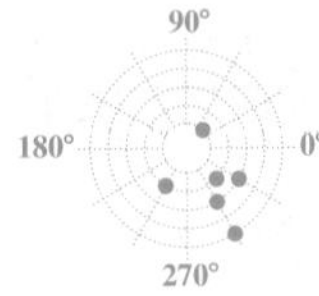

Graphs for Exercises 4(a), 6(a), 8(a), 10(a), 12(a), 14(a)

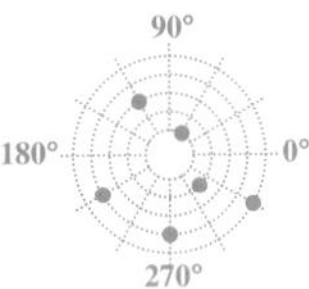

Answers may vary in Exercises 3(b)–14(b).

3. (b) $(1, 405°), (-1, 225°)$
(c) $\left(\frac{\sqrt{2}}{2}, \frac{\sqrt{2}}{2}\right)$

4. (b) $(3, 480°), (-3, 300°)$
(c) $\left(-\frac{3}{2}, \frac{3\sqrt{3}}{2}\right)$

5. (b) $(-2, 495°), (2, 315°)$
(c) $\left(\sqrt{2}, -\sqrt{2}\right)$

6. (b) $(-4, 390°), (4, 210°)$
(c) $\left(-2\sqrt{3}, -2\right)$

7. (b) $(5, 300°), (-5, 120°)$
(c) $\left(\frac{5}{2}, -\frac{5\sqrt{3}}{2}\right)$

8. (b) $(2, 315°), (-2, 135°)$
(c) $\left(\sqrt{2}, -\sqrt{2}\right)$

9. (b) $(-3, 150°), (3, -30°)$
(c) $\left(\frac{3\sqrt{3}}{2}, -\frac{3}{2}\right)$

10. (b) $(-1, 240°), (1, 60°)$
(c) $\left(\frac{1}{2}, \frac{\sqrt{3}}{2}\right)$

11. (b) $\left(3, \frac{11\pi}{3}\right), \left(-3, \frac{2\pi}{3}\right)$
(c) $\left(\frac{3}{2}, -\frac{3\sqrt{3}}{2}\right)$

12. (b) $\left(4, -\frac{\pi}{2}\right), \left(-4, \frac{\pi}{2}\right)$
(c) $(0, -4)$

13. (b) $\left(-2, \frac{7\pi}{3}\right), \left(2, \frac{4\pi}{3}\right)$
(c) $\left(-1, -\sqrt{3}\right)$

14. (b) $\left(-5, \frac{17\pi}{6}\right), \left(5, \frac{11\pi}{6}\right)$
(c) $\left(\frac{5\sqrt{3}}{2}, -\frac{5}{2}\right)$

Graphs for Exercises 15(a), 17(a), 19(a), 21(a), 23(a), 25(a)

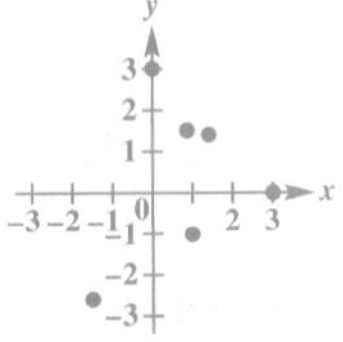

*For each pair of polar coordinates, **(a)** plot the point, **(b)** give two other pairs of polar coordinates for the point, and **(c)** give the rectangular coordinates for the point.* ***See Examples 1 and 2.***

3. $(1, 45°)$ **4.** $(3, 120°)$ **5.** $(-2, 135°)$

6. $(-4, 30°)$ **7.** $(5, -60°)$ **8.** $(2, -45°)$

9. $(-3, -210°)$ **10.** $(-1, -120°)$ **11.** $\left(3, \frac{5\pi}{3}\right)$

12. $\left(4, \frac{3\pi}{2}\right)$ **13.** $\left(-2, \frac{\pi}{3}\right)$ **14.** $\left(-5, \frac{5\pi}{6}\right)$

*For each pair of rectangular coordinates, **(a)** plot the point and **(b)** give two pairs of polar coordinates for the point, where $0° \le \theta < 360°$.* ***See Example 2(b).***

15. $(1, -1)$ **16.** $(1, 1)$ **17.** $(0, 3)$

18. $(0, -3)$ **19.** $\left(\sqrt{2}, \sqrt{2}\right)$ **20.** $\left(-\sqrt{2}, \sqrt{2}\right)$

21. $\left(\frac{\sqrt{3}}{2}, \frac{3}{2}\right)$ **22.** $\left(-\frac{\sqrt{3}}{2}, -\frac{1}{2}\right)$ **23.** $(3, 0)$

24. $(-2, 0)$ **25.** $\left(-\frac{3}{2}, -\frac{3\sqrt{3}}{2}\right)$ **26.** $\left(\frac{1}{2}, -\frac{\sqrt{3}}{2}\right)$

For each rectangular equation, give its equivalent polar equation and sketch its graph. ***See Example 3.***

27. $x - y = 4$ **28.** $x + y = -7$ **29.** $x^2 + y^2 = 16$

30. $x^2 + y^2 = 9$ **31.** $2x + y = 5$ **32.** $3x - 2y = 6$

Relating Concepts

For individual or collaborative investigation *(Exercises 33–40)*

In rectangular coordinates, the graph of

$$ax + by = c$$

is a horizontal line if $a = 0$ or a vertical line if $b = 0$. ***Work Exercises 33–40 in order,*** *to determine the general forms of polar equations for horizontal and vertical lines.*

33. Begin with the equation $y = k$, whose graph is a horizontal line. Make a trigonometric substitution for y using r and θ.

34. Solve the equation in **Exercise 33** for r.

35. Rewrite the equation in **Exercise 34** using the appropriate reciprocal function.

36. Sketch the graph of the equation

$$r = 3 \csc \theta.$$

What is the corresponding rectangular equation?

37. Begin with the equation $x = k$, whose graph is a vertical line. Make a trigonometric substitution for x using r and θ.

38. Solve the equation in **Exercise 37** for r.

39. Rewrite the equation in **Exercise 38** using the appropriate reciprocal function.

40. Sketch the graph of $r = 3 \sec \theta$. What is the corresponding rectangular equation?

Graphs for Exercises 16(a), 18(a), 20(a), 22(a), 24(a), 26(a)

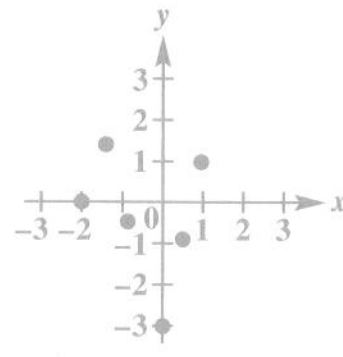

Answers may vary in Exercises 15(b)–26(b).

15. (b) $(\sqrt{2}, 315°), (-\sqrt{2}, 135°)$

16. (b) $(\sqrt{2}, 45°), (-\sqrt{2}, 225°)$

17. (b) $(3, 90°), (-3, 270°)$

18. (b) $(3, 270°), (-3, 90°)$

19. (b) $(2, 45°), (-2, 225°)$

20. (b) $(2, 135°), (-2, 315°)$

21. (b) $(\sqrt{3}, 60°), (-\sqrt{3}, 240°)$

22. (b) $(1, 210°), (-1, 30°)$

23. (b) $(3, 0°), (-3, 180°)$

24. (b) $(-2, 0°), (2, 180°)$

25. (b) $(3, 240°), (-3, 60°)$

26. (b) $(1, 300°), (-1, 120°)$

27. $r = \frac{4}{\cos \theta - \sin \theta}$

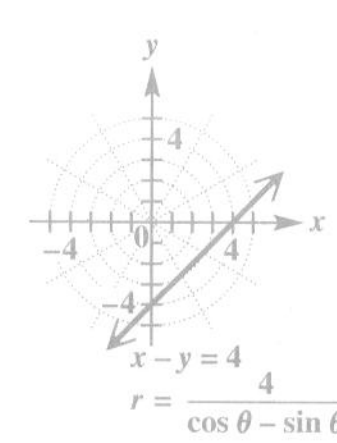

28. $r = \frac{-7}{\cos \theta + \sin \theta}$

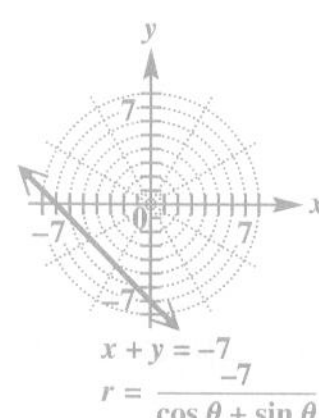

29. $r = 4$ or $r = -4$

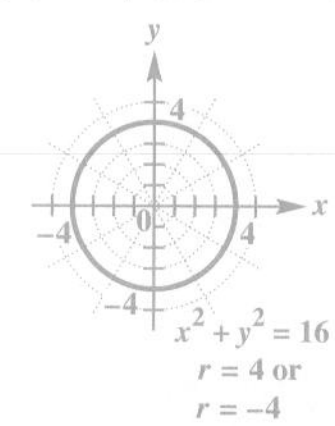

30. $r = 3$ or $r = -3$

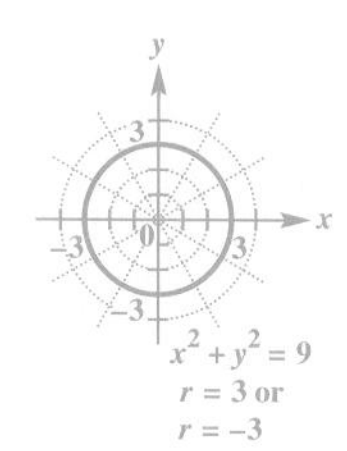

Concept Check In Exercises 41–44, match each equation with its polar graph from choices A–D.

41. $r = 3$ **42.** $r = \cos 3\theta$ **43.** $r = \cos 2\theta$ **44.** $r = \frac{2}{\cos \theta + \sin \theta}$

A.

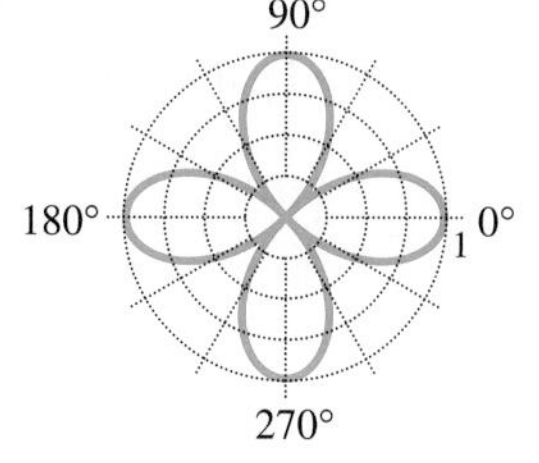

B.

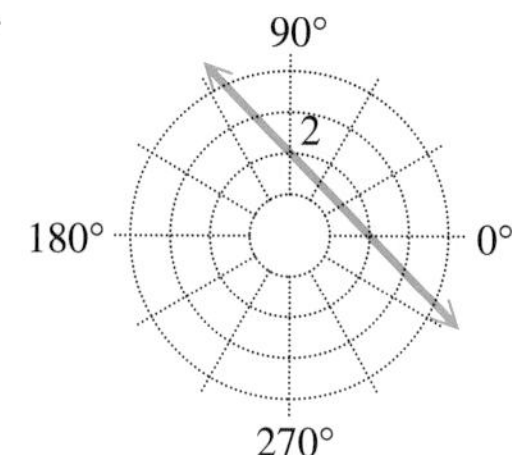

C.

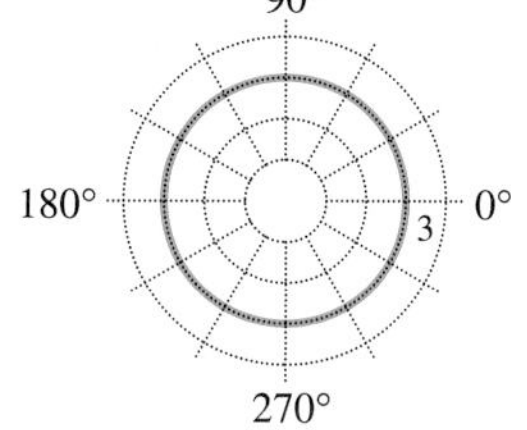

D.

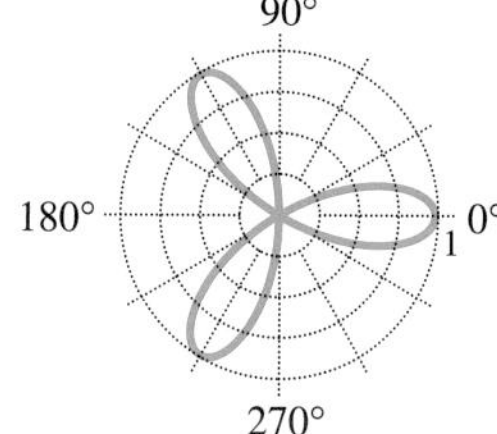

Give a complete graph of each polar equation. In Exercises 45–54, also identify the type of polar graph. ***See Examples 4–6.***

45. $r = 2 + 2 \cos \theta$

46. $r = 8 + 6 \cos \theta$

47. $r = 3 + \cos \theta$

48. $r = 2 - \cos \theta$

49. $r = 4 \cos 2\theta$

50. $r = 3 \cos 5\theta$

51. $r^2 = 4 \cos 2\theta$

52. $r^2 = 4 \sin 2\theta$

53. $r = 4 - 4 \cos \theta$

54. $r = 6 - 3 \cos \theta$

55. $r = 2 \sin \theta \tan \theta$ (This is a **cissoid.**)

56. $r = \frac{\cos 2\theta}{\cos \theta}$ (This is a **cissoid with a loop.**)

For each equation, find an equivalent equation in rectangular coordinates, and graph. ***See Example 8.***

57. $r = 2 \sin \theta$

58. $r = 2 \cos \theta$

59. $r = \frac{2}{1 - \cos \theta}$

60. $r = \frac{3}{1 - \sin \theta}$

61. $r = -2 \cos \theta - 2 \sin \theta$

62. $r = \frac{3}{4 \cos \theta - \sin \theta}$

63. $r = 2 \sec \theta$

64. $r = -5 \csc \theta$

65. $r = \frac{2}{\cos \theta + \sin \theta}$

66. $r = \frac{2}{2 \cos \theta + \sin \theta}$

67. Graph $r = \theta$, a spiral of Archimedes. (**See Example 7.**) Use both positive and nonpositive values for θ.

68. Use a graphing calculator window of $[-1250, 1250]$ by $[-1250, 1250]$, in degree mode, to graph more of

$$r = 2\theta \text{ (a spiral of Archimedes)}$$

than what is shown in **Figure 71.** Use $-1250° \le \theta \le 1250°$.

69. Find the polar equation of the line that passes through the points $(1, 0°)$ and $(2, 90°)$.

70. Explain how to plot a point (r, θ) in polar coordinates, if $r < 0$.

31. 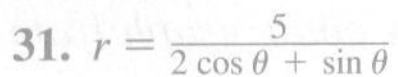

$r = \frac{5}{2 \cos \theta + \sin \theta}$

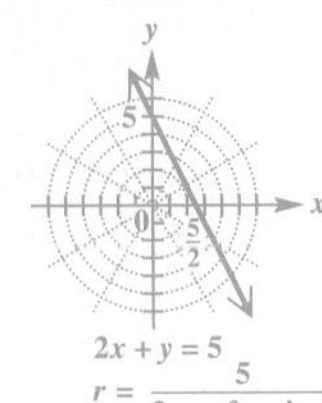

32. $r = \frac{6}{3 \cos \theta - 2 \sin \theta}$

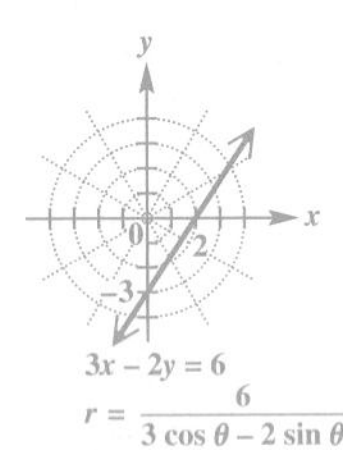

33. $r \sin \theta = k$

34. $r = \frac{k}{\sin \theta}$

35. $r = k \csc \theta$

36. $y = 3$

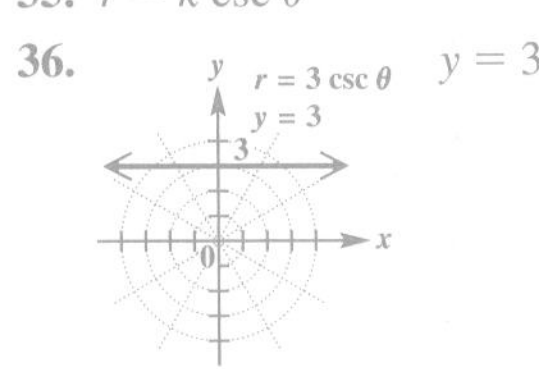

37. $r \cos \theta = k$

38. $r = \frac{k}{\cos \theta}$

39. $r = k \sec \theta$

40. $x = 3$

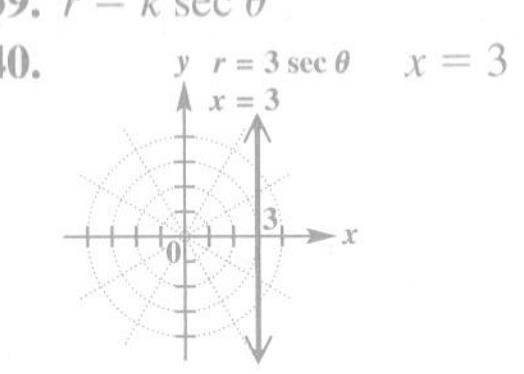

41. C 42. D

43. A 44. B

45. cardioid

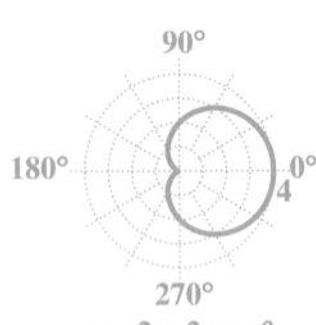

46. limaçon

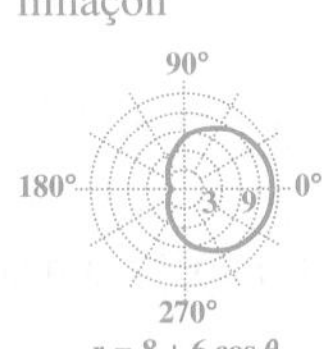

Concept Check The polar graphs in this section exhibit symmetry. ***(See Section 2.7.)*** *Visualize an xy-plane superimposed on the polar coordinate system, with the pole at the origin and the polar axis on the positive x-axis. Then a polar graph may be symmetric with respect to the x-axis (the polar axis), the y-axis (the line $\theta = \frac{\pi}{2}$), or the origin (the pole). Use this information to work Exercises 71 and 72.*

71. Complete the missing ordered pairs in the graphs below.

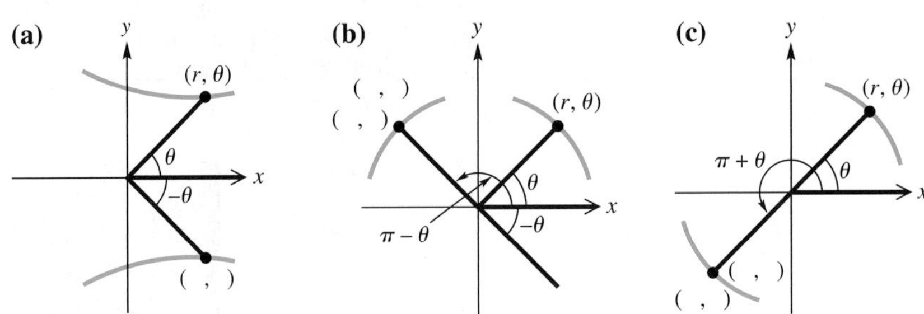

72. Based on your results in **Exercise 71,** fill in the blanks with the correct responses.

(a) The graph of $r = f(\theta)$ is symmetric with respect to the polar axis if substitution of __________ for θ leads to an equivalent equation.

(b) The graph of $r = f(\theta)$ is symmetric with respect to the vertical line $\theta = \frac{\pi}{2}$ if substitution of __________ for θ leads to an equivalent equation.

(c) Alternatively, the graph of $r = f(\theta)$ is symmetric with respect to the vertical line $\theta = \frac{\pi}{2}$ if substitution of __________ for r and __________ for θ leads to an equivalent equation.

(d) The graph of $r = f(\theta)$ is symmetric with respect to the pole if substitution of __________ for r leads to an equivalent equation.

(e) Alternatively, the graph of $r = f(\theta)$ is symmetric with respect to the pole if substitution of __________ for θ leads to an equivalent equation.

(f) In general, the completed statements in parts (a)–(e) mean that the graphs of polar equations of the form $r = a \pm b \cos \theta$ (where a may be 0) are symmetric with respect to __________.

(g) In general, the completed statements in parts (a)–(e) mean that the graphs of polar equations of the form $r = a \pm b \sin \theta$ (where a may be 0) are symmetric with respect to __________.

The graph of $r = a\theta$ in polar coordinates is an example of the spiral of Archimedes. With your calculator set to radian mode, use the given value of a and interval of θ to graph the spiral in the window specified.

73. $a = 1,\ 0 \le \theta \le 4\pi,\ [-15, 15]$ by $[-15, 15]$

74. $a = 2,\ -4\pi \le \theta \le 4\pi,\ [-30, 30]$ by $[-30, 30]$

75. $a = 1.5,\ -4\pi \le \theta \le 4\pi,\ [-20, 20]$ by $[-20, 20]$

76. $a = -1,\ 0 \le \theta \le 12\pi,\ [-40, 40]$ by $[-40, 40]$

Find the polar coordinates of the points of intersection of the given curves for the specified interval of θ.

77. $r = 4 \sin \theta,\ r = 1 + 2 \sin \theta;\ 0 \le \theta < 2\pi$

78. $r = 3,\ r = 2 + 2 \cos \theta;\ 0° \le \theta < 360°$

79. $r = 2 + \sin \theta,\ r = 2 + \cos \theta;\ 0 \le \theta < 2\pi$

80. $r = \sin 2\theta,\ r = \sqrt{2} \cos \theta;\ 0 \le \theta < \pi$

47. limaçon

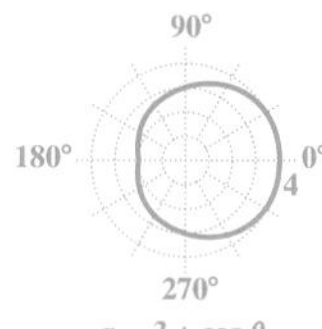

48. limaçon

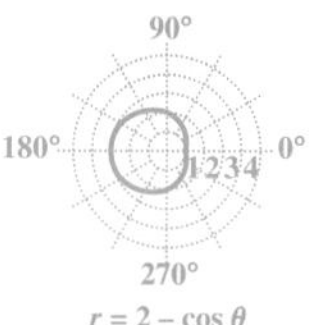

49. four-leaved rose

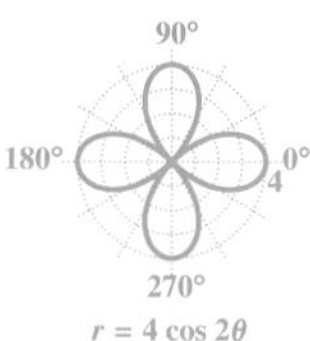

50. five-leaved rose

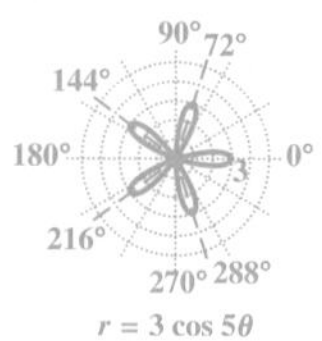

51. lemniscate

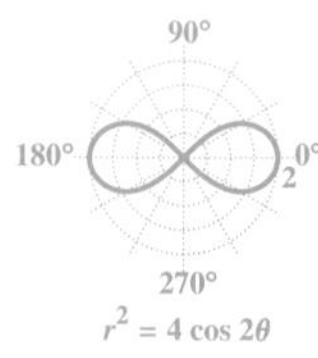

52. lemniscate

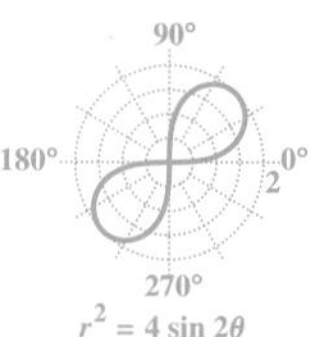

53. cardioid

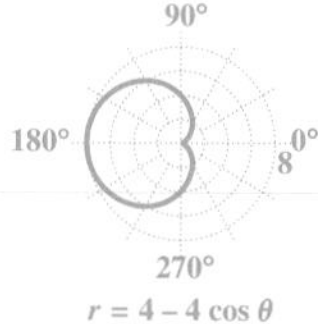

54. limaçon

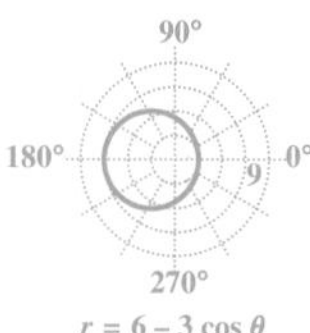

Answers for the remaining exercises are given in the answer section at the back of the text.

(Modeling) Solve each problem.

81. *Orbits of Satellites* The polar equation

$$r = \frac{a(1 - e^2)}{1 + e \cos \theta}$$

can be used to graph the orbits of the satellites of our sun, where a is the average distance in astronomical units from the sun and e is a constant called the **eccentricity**. The sun will be located at the pole. The table lists the values of a and e.

Satellite	a	e
Mercury	0.39	0.206
Venus	0.78	0.007
Earth	1.00	0.017
Mars	1.52	0.093
Jupiter	5.20	0.048
Saturn	9.54	0.056
Uranus	19.20	0.047
Neptune	30.10	0.009
Pluto	39.40	0.249

Source: Karttunen, H., P. Kröger, H. Oja, M. Putannen, and K. Donners (Editors), *Fundamental Astronomy, 4th edition,* Springer-Verlag. Zeilik, M., S. Gregory, and E. Smith, *Introductory Astronomy and Astrophysics,* Saunders College Publishers.

(a) Graph the orbits of the four closest satellites on the same polar grid. Choose a viewing window that results in a graph with nearly circular orbits.

(b) Plot the orbits of Earth, Jupiter, Uranus, and Pluto on the same polar grid. How does Earth's distance from the sun compare to the others' distances from the sun?

(c) Use graphing to determine whether or not Pluto is always farthest from the sun.

82. *Radio Towers and Broadcasting Patterns* Many times radio stations do not broadcast in all directions with the same intensity. To avoid interference with an existing station to the north, a new station may be licensed to broadcast only east and west. To create an east-west signal, two radio towers are sometimes used, as illustrated in the figure. Locations where the radio signal is received correspond to the interior of the curve

$$r^2 = 40{,}000 \cos 2\theta,$$

where the polar axis (or positive x-axis) points east.

(a) Graph $r^2 = 40{,}000 \cos 2\theta$ for $0° \le \theta \le 360°$, where distances are in miles. Assuming the radio towers are located near the pole, use the graph to describe the regions where the signal can be received and where the signal cannot be received.

(b) Suppose a radio signal pattern is given by

$$r^2 = 22{,}500 \sin 2\theta.$$

Graph this pattern and interpret the results.

8.8 Parametric Equations, Graphs, and Applications

- Basic Concepts
- Parametric Graphs and Their Rectangular Equivalents
- The Cycloid
- Applications of Parametric Equations

Basic Concepts Throughout this text, we have graphed sets of ordered pairs of real numbers that correspond to a function of the form $y = f(x)$ or $r = g(\theta)$. Another way to determine a set of ordered pairs involves two functions f and g defined by $x = f(t)$ and $y = g(t)$, where t is a real number in some interval I. Each value of t leads to a corresponding x-value and a corresponding y-value, and thus to an ordered pair (x, y).

Parametric Equations of a Plane Curve

A **plane curve** is a set of points (x, y) such that $x = f(t)$, $y = g(t)$, and f and g are both defined on an interval I. The equations $x = f(t)$ and $y = g(t)$ are **parametric equations** with **parameter *t*.**

Classroom Example 1
Let

$x = t - 2$ and $y = t^2 - 1$,

for t in $[-3, 3]$. Graph the set of ordered pairs (x, y).

Answer:

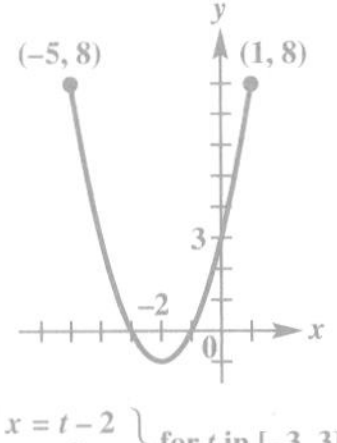

$x = t - 2$, $y = t^2 - 1$ for t in $[-3, 3]$

Graphing calculators are capable of graphing plane curves defined by parametric equations. The calculator must be set in parametric mode, and the window requires intervals for the parameter t, as well as for x and y. ■

Parametric Graphs and Their Rectangular Equivalents

EXAMPLE 1 Graphing a Plane Curve Defined Parametrically

Let $x = t^2$ and $y = 2t + 3$, for t in $[-3, 3]$. Graph the set of ordered pairs (x, y).

ALGEBRAIC SOLUTION

Make a table of corresponding values of t, x, and y over the domain of t. Plot the points as shown in **Figure 74.** The graph is a portion of a parabola with horizontal axis $y = 3$. The arrowheads indicate the direction the curve traces as t increases.

t	x	y
−3	9	−3
−2	4	−1
−1	1	1
0	0	3
1	1	5
2	4	7
3	9	9

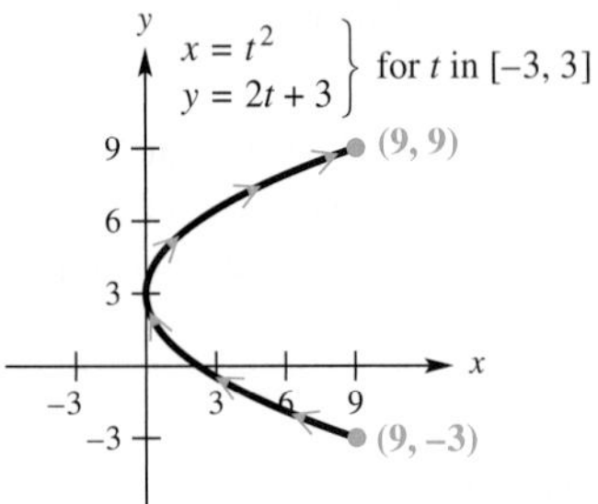

Figure 74

GRAPHING CALCULATOR SOLUTION

We set the parameters of the TI-83/84 Plus as shown in the top two screens to obtain the bottom screen in **Figure 75.**

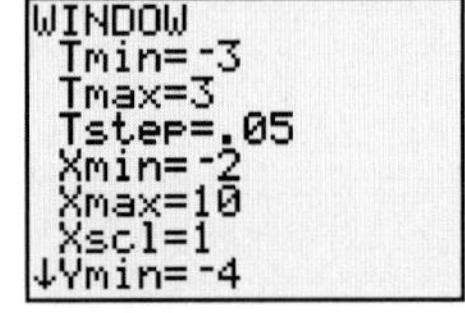

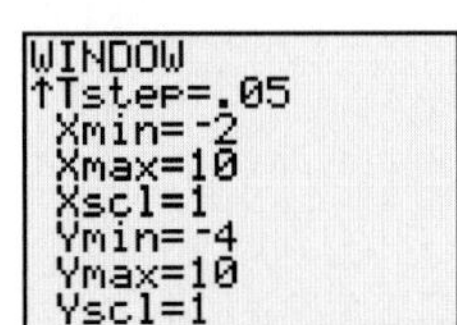

This is a continuation of the previous screen.

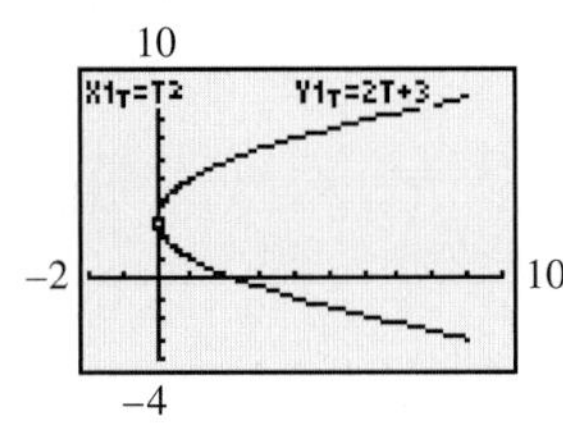

Figure 75

Now Try Exercise 5(a).

Classroom Example 2
Find a rectangular equation for the plane curve in **Classroom Example 1** defined as follows: $x = t - 2$ and $y = t^2 - 1$, for t in $[-3, 3]$.

Answer: $y = (x + 2)^2 - 1$, for x in $[-5, 1]$

EXAMPLE 2 Finding an Equivalent Rectangular Equation

Find a rectangular equation for the plane curve of **Example 1** defined as follows:

$$x = t^2, \quad y = 2t + 3, \qquad \text{for } t \text{ in } [-3, 3].$$

SOLUTION To eliminate the parameter t, first solve either equation for t. Here, only the second equation, $y = 2t + 3$, leads to a unique solution for t, so we choose it.

$$y = 2t + 3 \quad \text{Choose the simpler equation.}$$

$$2t = y - 3 \quad \text{Subtract 3 and rewrite. (Section 1.1)}$$

$$t = \frac{y - 3}{2} \quad \text{Divide by 2.}$$

Now substitute this result into the first equation to eliminate the parameter t.

$$x = t^2$$

$$x = \left(\frac{y - 3}{2}\right)^2 \quad \text{Substitute for } t.$$

$$x = \frac{(y - 3)^2}{4} \quad \left(\frac{a}{b}\right)^2 = \frac{a^2}{b^2} \text{ (Section R.3)}$$

$$4x = (y - 3)^2 \quad \text{Multiply by 4.}$$

This is the equation of a horizontal parabola opening to the right, which agrees with the graph given in **Figure 74.** Because t is in $[-3, 3]$, x is in $[0, 9]$ and y is in $[-3, 9]$. The rectangular equation must be given with its restricted domain as

$$4x = (y - 3)^2, \qquad \text{for } x \text{ in } [0, 9].$$

✔ *Now Try Exercise 5(b).*

Classroom Example 3
Graph the plane curve defined by

$$x = 4 \cos t, \quad y = 2 \sin t,$$

for t in $[0, 2\pi]$.

Answer:

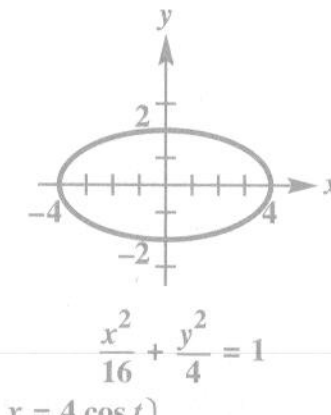

Teaching Tip If students are not sure how to proceed, have them create a table of values for t, x, and y, identify the curve represented by the parametric equations, and then use this information to determine the next step.

EXAMPLE 3 Graphing a Plane Curve Defined Parametrically

Graph the plane curve defined by $x = 2 \sin t$, $y = 3 \cos t$, for t in $[0, 2\pi]$.

SOLUTION To convert to a rectangular equation, it is not productive here to solve either equation for t. Instead, we use the fact that

$$\sin^2 t + \cos^2 t = 1$$

to apply another approach. Square both sides of each equation; solve one for $\sin^2 t$, the other for $\cos^2 t$.

$x = 2 \sin t$	$y = 3 \cos t$	Given equations
$x^2 = 4 \sin^2 t$	$y^2 = 9 \cos^2 t$	Square each side. (Section 1.6)
$\frac{x^2}{4} = \sin^2 t$	$\frac{y^2}{9} = \cos^2 t$	Divide.

Now add corresponding sides of the two equations.

$$\frac{x^2}{4} + \frac{y^2}{9} = \sin^2 t + \cos^2 t$$

$$\frac{x^2}{4} + \frac{y^2}{9} = 1 \qquad \sin^2 t + \cos^2 t = 1 \text{ (Section 5.2)}$$

This is an equation of an **ellipse.** See **Figure 76** on the next page for traditional and calculator graphs. (Ellipses are covered in more detail in **Chapter 10.**)

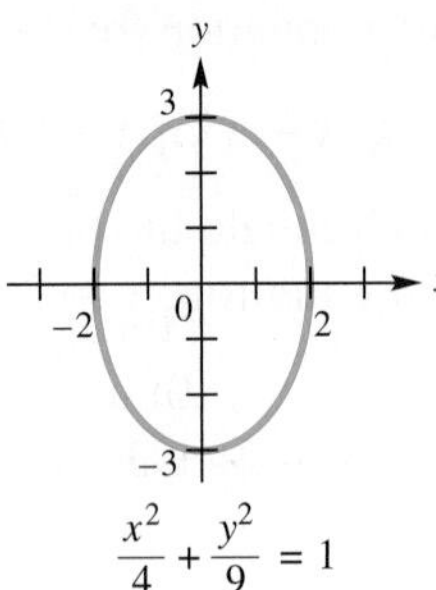

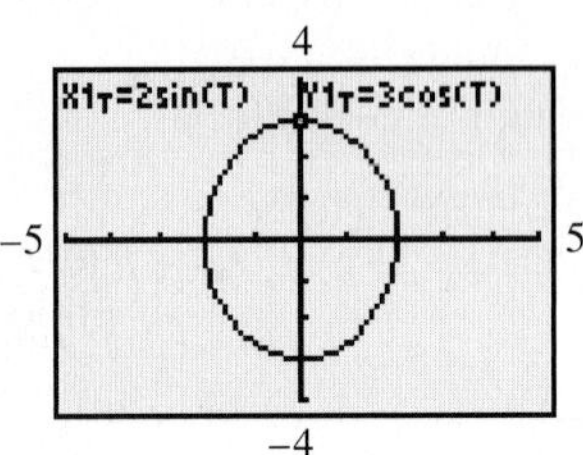

Parametric graphing mode

Figure 76

Now Try Exercise 27.

Parametric representations of a curve are not unique. In fact, there are infinitely many parametric representations of a given curve. If the curve can be described by a rectangular equation $y = f(x)$, with domain X, then one simple parametric representation is

$$x = t, \quad y = f(t), \qquad \text{for } t \text{ in } X.$$

Classroom Example 4
Give two parametric representations for the equation of the parabola.

$$y = (x - 3)^2 - 1$$

Answer:
$x = t,\ y = (t - 3)^2 - 1,$
for t in $(-\infty, \infty)$;

$x = t + 3,\ y = t^2 - 1,$
for t in $(-\infty, \infty)$

(Other answers are possible.)

EXAMPLE 4 Finding Alternative Parametric Equation Forms

Give two parametric representations for the equation of the parabola.

$$y = (x - 2)^2 + 1$$

SOLUTION The simplest choice is to let

$$x = t, \quad y = (t - 2)^2 + 1, \qquad \text{for } t \text{ in } (-\infty, \infty).$$

Another choice, which leads to a simpler equation for y, is

$$x = t + 2, \quad y = t^2 + 1, \qquad \text{for } t \text{ in } (-\infty, \infty).$$

Now Try Exercise 29.

NOTE Sometimes trigonometric functions are desirable. One choice in **Example 4** might be

$$x = 2 + \tan t, \quad y = \sec^2 t, \qquad \text{for } t \text{ in } \left(-\frac{\pi}{2}, \frac{\pi}{2}\right).$$

Classroom Example 5
Graph the cycloid

$$x = 2t - 2\sin t,$$
$$y = 2 - 2\cos t,$$

for t in $[0, 2\pi]$.

Answer:

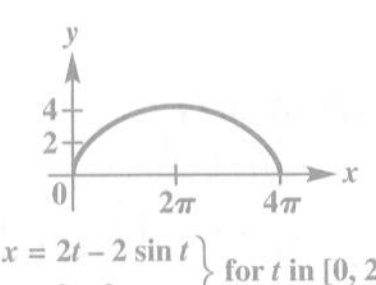

The Cycloid The *cycloid* is a special case of the **trochoid**—a curve traced out by a point at a given distance from the center of a circle as the circle rolls along a straight line. If the given point is on the *circumference* of the circle, then the path traced as the circle rolls along a straight line is a **cycloid,** which is defined parametrically as follows.

$$\boldsymbol{x = at - a\sin t, \quad y = a - a\cos t, \qquad \text{for } t \text{ in } (-\infty, \infty)}$$

Other curves related to trochoids are **hypotrochoids** and **epitrochoids,** which are traced out by a point that is a given distance from the center of a circle that rolls not on a straight line, but on the inside or outside, respectively, of another circle. The classic Spirograph toy can be used to draw these curves.

EXAMPLE 5 Graphing a Cycloid

Graph the cycloid.

$$x = t - \sin t, \quad y = 1 - \cos t, \quad \text{for } t \text{ in } [0, 2\pi]$$

ALGEBRAIC SOLUTION

There is no simple way to find a rectangular equation for the cycloid from its parametric equations. Instead, begin with a table using selected values for t in $[0, 2\pi]$. Approximate values have been rounded as necessary.

t	0	$\frac{\pi}{4}$	$\frac{\pi}{2}$	π	$\frac{3\pi}{2}$	2π
x	0	0.08	0.6	π	5.7	2π
y	0	0.3	1	2	1	0

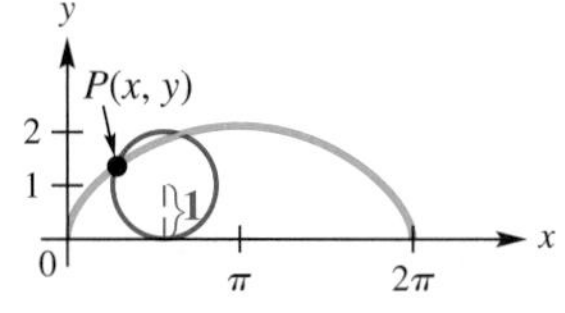

Figure 77

Plotting the ordered pairs (x, y) from the table of values leads to the portion of the graph in **Figure 77** from 0 to 2π.

GRAPHING CALCULATOR SOLUTION

It is easier to graph a cycloid with a graphing calculator in parametric mode than with traditional methods. See **Figure 78.**

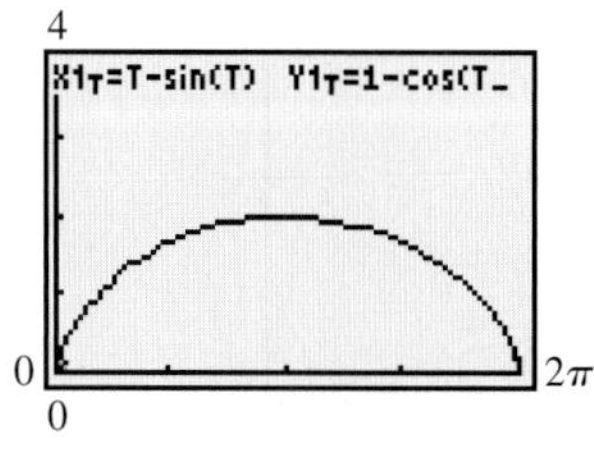

Figure 78

Using a larger interval for t would show that the cycloid repeats the pattern shown here every 2π units.

✔ *Now Try Exercise 33.*

Figure 79

The cycloid has an interesting physical property. If a flexible cord or wire goes through points P and Q as in **Figure 79,** and a bead is allowed to slide due to the force of gravity without friction along this path from P to Q, the path that requires the shortest time takes the shape of the graph of an inverted cycloid.

LOOKING AHEAD TO CALCULUS

At any time t, the velocity of an object is given by the vector $\mathbf{v} = \langle f'(t), g'(t) \rangle$. The object's speed at time t is

$$|\mathbf{v}| = \sqrt{(f'(t))^2 + (g'(t))^2}.$$

Applications of Parametric Equations Parametric equations are used to simulate motion. If a ball is thrown with a velocity of v feet per second at an angle θ with the horizontal, its flight can be modeled by the parametric equations

$$\mathbf{x = (v \cos \theta)t} \quad \text{and} \quad \mathbf{y = (v \sin \theta)t - 16t^2 + h,}$$

where t is in seconds and h is the ball's initial height in feet above the ground. Here, x gives the horizontal position information and y gives the vertical position information. The term $-16t^2$ occurs because gravity is pulling downward. See **Figure 80.** These equations ignore air resistance.

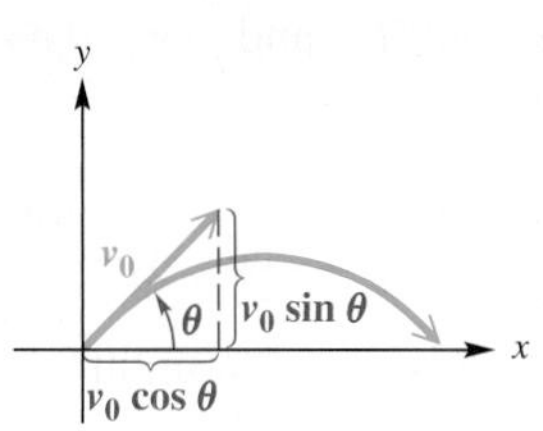

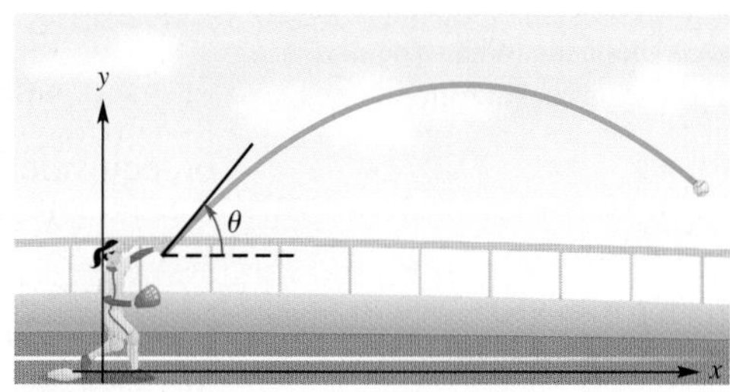

Figure 80

Classroom Example 6
Repeat **Example 6** for three golf balls hit simultaneously into the air at 120 ft per sec at angles of 25°, 45°, and 65° with the horizontal.

Answers:
(a) the ball hit at 45°; 450 ft
(b) the ball hit at 65°; 185 ft

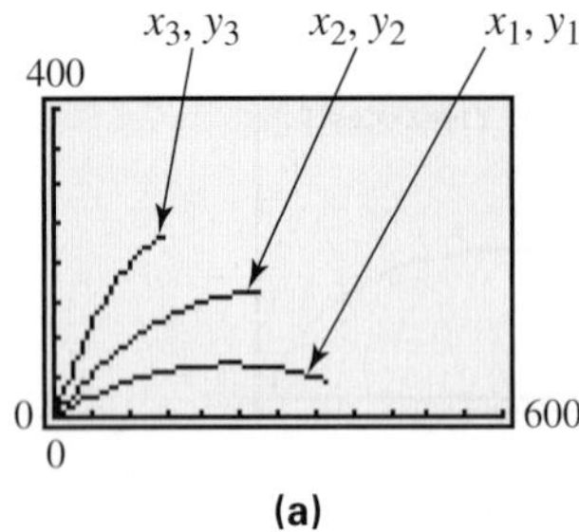

(a)

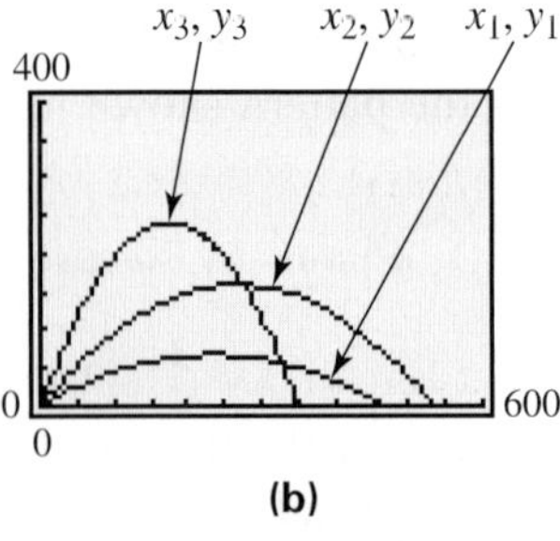

(b)

Figure 81

EXAMPLE 6 Simulating Motion with Parametric Equations

Three golf balls are hit simultaneously into the air at 132 ft per sec (90 mph) at angles of 30°, 50°, and 70° with the horizontal.

(a) Assuming the ground is level, determine graphically which ball travels the greatest distance. Estimate this distance.

(b) Which ball reaches the greatest height? Estimate this height.

SOLUTION

(a) Use the following parametric equations to model the flight of the golf balls.

$$x = (v \cos \theta)t \quad \text{and} \quad y = (v \sin \theta)t - 16t^2 + h$$

Substitute $h = 0$, $v = 132$ ft per sec, and $\theta = 30°, 50°$, and $70°$ to write three sets of parametric equations.

$$x_1 = (132 \cos 30°)t, \quad y_1 = (132 \sin 30°)t - 16t^2$$

$$x_2 = (132 \cos 50°)t, \quad y_2 = (132 \sin 50°)t - 16t^2$$

$$x_3 = (132 \cos 70°)t, \quad y_3 = (132 \sin 70°)t - 16t^2$$

The graphs of the three sets of parametric equations are shown in **Figure 81(a),** where $0 \le t \le 9$. From the graph in **Figure 81(b),** we can see that the ball hit at 50° travels the greatest distance. Using the TRACE feature of the TI-83/84 Plus, we estimate this distance to be about 540 ft.

(b) Again, use the TRACE feature to find that the ball hit at 70° reaches the greatest height, about 240 ft.

✔ *Now Try Exercise 39.*

Teaching Tip Explain that in motion problems, the variable t acts as a control variable (specifying an amount of time appropriate to the situation), while x gives horizontal position information and y gives vertical position information.

NOTE The TI-83/84 Plus graphing calculator allows the user to view the graphing of more than one equation either *sequentially* or *simultaneously*. By choosing the latter, one can view the three golf balls in **Figure 81** in flight at the same time.

EXAMPLE 7 Examining Parametric Equations of Flight

Jack Lukas launches a small rocket from a table that is 3.36 ft above the ground. Its initial velocity is 64 ft per sec, and it is launched at an angle of 30° with respect to the ground. Find the rectangular equation that models its path. What type of path does the rocket follow?

Classroom Example 7
A toy rocket is launched from the ground with velocity 36 ft per sec at an angle of 45° with the ground. Find the rectangular equation that models this path. What type of path does the rocket follow?

Answer:
$y = -\frac{2}{81}x^2 + x$;
It follows a parabolic path.

SOLUTION The path of the rocket is defined by the parametric equations

$$x = (64 \cos 30°)t \quad \text{and} \quad y = (64 \sin 30°)t - 16t^2 + 3.36$$

or, equivalently,

$$x = 32\sqrt{3}t \quad \text{and} \quad y = -16t^2 + 32t + 3.36.$$

From $x = 32\sqrt{3}t$, we solve for t to obtain

$$t = \frac{x}{32\sqrt{3}}. \quad \text{Divide by } 32\sqrt{3}.$$

Substituting for t in the other parametric equation yields the following.

$$y = -16t^2 + 32t + 3.36$$

$$y = -16\left(\frac{x}{32\sqrt{3}}\right)^2 + 32\left(\frac{x}{32\sqrt{3}}\right) + 3.36 \quad \text{Let } t = \frac{x}{32\sqrt{3}}.$$

$$y = -\frac{1}{192}x^2 + \frac{\sqrt{3}}{3}x + 3.36 \quad \text{Simplify.}$$

Because this equation defines a parabola, we can conclude that the rocket follows a parabolic path.

Now Try Exercise 43(a).

Classroom Example 8
Determine the total flight time and the horizontal distance traveled by the rocket in **Classroom Example 7.**

Answer: 1.59 sec; 40.5 ft

EXAMPLE 8 Analyzing the Path of a Projectile

Determine the total flight time and the horizontal distance traveled by the rocket in **Example 7.**

ALGEBRAIC SOLUTION

The equation $y = -16t^2 + 32t + 3.36$ tells the vertical position of the rocket at time t. We need to determine those values of t for which $y = 0$ since these values correspond to the rocket at ground level. This yields

$$0 = -16t^2 + 32t + 3.36.$$

Using the quadratic formula, the solutions are $t = -0.1$ or $t = 2.1$. Since t represents time, $t = -0.1$ is an unacceptable answer. Therefore, the flight time is 2.1 sec.

The rocket was in the air for 2.1 sec, so we can use $t = 2.1$ and the parametric equation that models the horizontal position, $x = 32\sqrt{3}t$, to obtain

$$x = 32\sqrt{3}(2.1) \approx 116.4 \text{ ft.}$$

GRAPHING CALCULATOR SOLUTION

Figure 82 shows that when T = 2.1, the horizontal distance X covered is approximately 116.4 ft, which agrees with the algebraic solution.

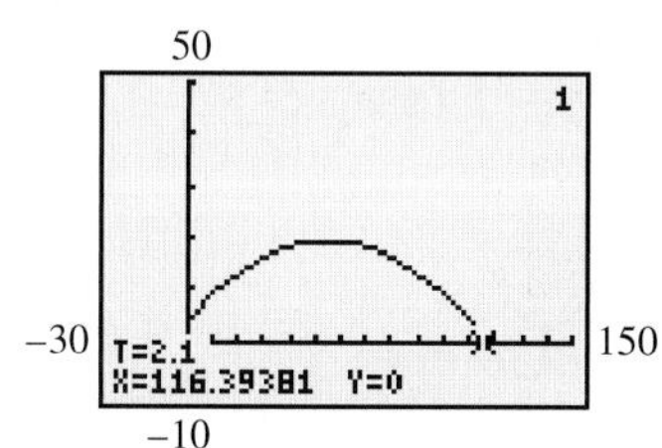

Figure 82

Now Try Exercise 43(b).

8.8 Exercises

1. C 2. D
3. A 4. B

5. (a)

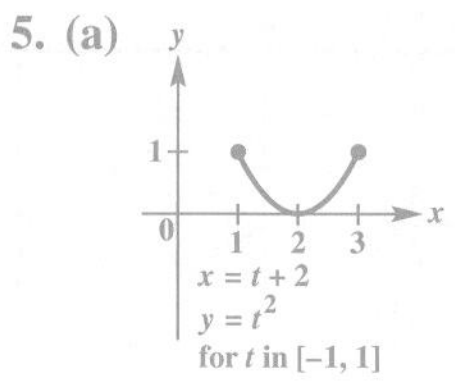

(b) $y = x^2 - 4x + 4$, for x in $[1, 3]$

Concept Check Match the ordered pair from Column II with the pair of parametric equations in Column I on whose graph the point lies. In each case, consider the given value of t.

I	II
1. $x = 3t + 6,\ y = -2t + 4;\quad t = 2$	A. $(5, 25)$
2. $x = \cos t,\ y = \sin t;\quad t = \frac{\pi}{4}$	B. $(7, 2)$
3. $x = t,\ y = t^2;\quad t = 5$	C. $(12, 0)$
4. $x = t^2 + 3,\ y = t^2 - 2;\quad t = 2$	D. $\left(\frac{\sqrt{2}}{2}, \frac{\sqrt{2}}{2}\right)$

6. (a) $x = 2t$, $y = t + 1$, for t in $[-2, 3]$; $(6, 4)$, $(-4, -1)$

(b) $y = \frac{1}{2}x + 1$, for x in $[-4, 6]$

7. (a) $x = \sqrt{t}$, $y = 3t - 4$, for t in $[0, 4]$; $(2, 8)$, $(0, -4)$

(b) $y = 3x^2 - 4$, for x in $[0, 2]$

8. (a) $x = t^2$, $y = \sqrt{t}$, for t in $[0, 4]$

(b) $y = \sqrt[4]{x}$, for x in $[0, 16]$

9. (a) $x = t^3 + 1$, $y = t^3 - 1$, for t in $(-\infty, \infty)$

(b) $y = x - 2$, for x in $(-\infty, \infty)$

10. (a) $(-1, 2)$; $x = 2t - 1$, $y = t^2 + 2$, for t in $(-\infty, \infty)$

(b) $y = \frac{1}{4}(x + 1)^2 + 2$, for x in $(-\infty, \infty)$

11. (a) $x = 2 \sin t$, $y = 2 \cos t$, for t in $[0, 2\pi]$

(b) $x^2 + y^2 = 4$, for x in $[-2, 2]$

12. (a) $(0, \sqrt{3})$, $(\sqrt{5}, 0)$, $(-\sqrt{5}, 0)$, $(0, -\sqrt{3})$; $x = \sqrt{5} \sin t$, $y = \sqrt{3} \cos t$, for t in $[0, 2\pi]$

(b) $\frac{x^2}{5} + \frac{y^2}{3} = 1$, for x in $[-\sqrt{5}, \sqrt{5}]$

For each plane curve, ***(a)*** *graph the curve, and* ***(b)*** *find a rectangular equation for the curve.* ***See Examples 1 and 2.***

5. $x = t + 2,\ y = t^2,$ for t in $[-1, 1]$

6. $x = 2t,\ y = t + 1,$ for t in $[-2, 3]$

7. $x = \sqrt{t},\ y = 3t - 4,$ for t in $[0, 4]$

8. $x = t^2,\ y = \sqrt{t},$ for t in $[0, 4]$

9. $x = t^3 + 1,\ y = t^3 - 1,$ for t in $(-\infty, \infty)$

10. $x = 2t - 1,\ y = t^2 + 2,$ for t in $(-\infty, \infty)$

11. $x = 2 \sin t,\ y = 2 \cos t,$ for t in $[0, 2\pi]$

12. $x = \sqrt{5} \sin t,\ y = \sqrt{3} \cos t,$ for t in $[0, 2\pi]$

13. $x = 3 \tan t,\ y = 2 \sec t,$ for t in $\left(-\frac{\pi}{2}, \frac{\pi}{2}\right)$

14. $x = \cot t,\ y = \csc t,$ for t in $(0, \pi)$

15. $x = \sin t,\ y = \csc t,$ for t in $(0, \pi)$

16. $x = \tan t,\ y = \cot t,$ for t in $\left(0, \frac{\pi}{2}\right)$

17. $x = t,\ y = \sqrt{t^2 + 2},$ for t in $(-\infty, \infty)$

18. $x = \sqrt{t},\ y = t^2 - 1,$ for t in $[0, \infty)$

19. $x = 2 + \sin t,\ y = 1 + \cos t,$ for t in $[0, 2\pi]$

20. $x = 1 + 2 \sin t,\ y = 2 + 3 \cos t,$ for t in $[0, 2\pi]$

21. $x = t + 2,\ y = \dfrac{1}{t + 2},$ for $t \neq -2$

22. $x = t - 3,\ y = \dfrac{2}{t - 3},$ for $t \neq 3$

23. $x = t + 2,\ y = t - 4,$ for t in $(-\infty, \infty)$

24. $x = t^2 + 2,\ y = t^2 - 4,$ for t in $(-\infty, \infty)$

Graph each plane curve defined by the parametric equations for t in $[0, 2\pi]$. Then find a rectangular equation for the plane curve. ***See Example 3.***

25. $x = 3 \cos t,\ y = 3 \sin t$

26. $x = 2 \cos t,\ y = 2 \sin t$

27. $x = 3 \sin t,\ y = 2 \cos t$

28. $x = 4 \sin t,\ y = 3 \cos t$

Give two parametric representations for the equation of each parabola. ***See Example 4.***

29. $y = (x + 3)^2 - 1$

30. $y = (x + 4)^2 + 2$

31. $y = x^2 - 2x + 3$

32. $y = x^2 - 4x + 6$

Graph each cycloid defined by the given equations for t in the specified interval. ***See Example 5.***

33. $x = 2t - 2 \sin t,\ y = 2 - 2 \cos t,$ for t in $[0, 4\pi]$

34. $x = t - \sin t,\ y = 1 - \cos t,$ for t in $[0, 4\pi]$

Lissajous Figures The screen shown here is an example of a ***Lissajous figure.*** *Lissajous figures occur in electronics and may be used to find the frequency of an unknown voltage. Graph each Lissajous figure for t in $[0, 6.5]$ in the window $[-6, 6]$ by $[-4, 4]$.*

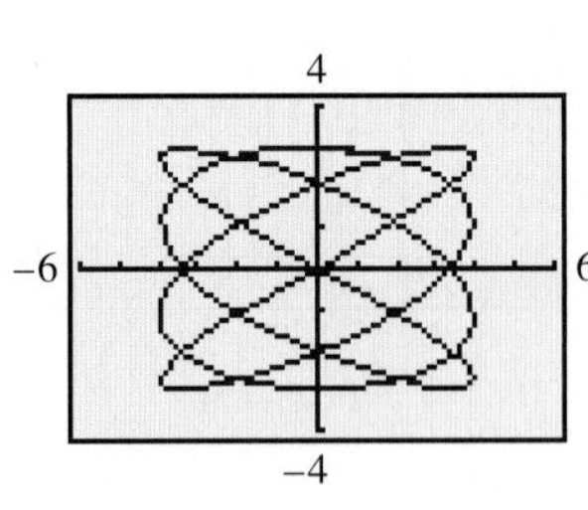

35. $x = 2 \cos t,\ y = 3 \sin 2t$

36. $x = 3 \cos 2t,\ y = 3 \sin 3t$

37. $x = 3 \sin 4t,\ y = 3 \cos 3t$

38. $x = 4 \sin 4t,\ y = 3 \sin 5t$

13. (a)

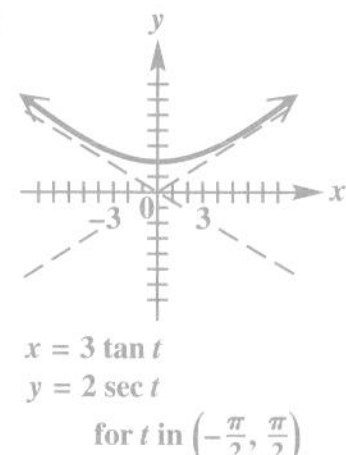

(b) $y = 2\sqrt{1 + \frac{x^2}{9}}$, for x in $(-\infty, \infty)$

14. (a)

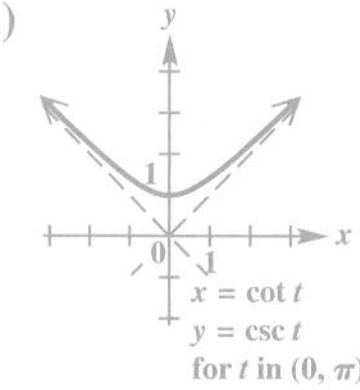

(b) $y = \sqrt{x^2 + 1}$, for x in $(-\infty, \infty)$

15. (a)

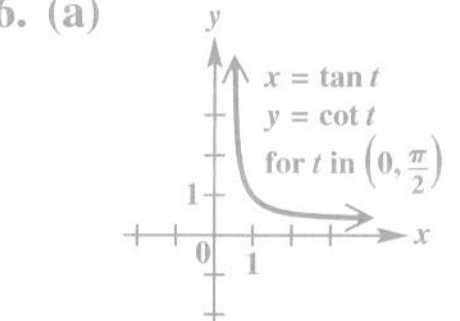

(b) $y = \frac{1}{x}$, for x in $(0, 1]$

16. (a)

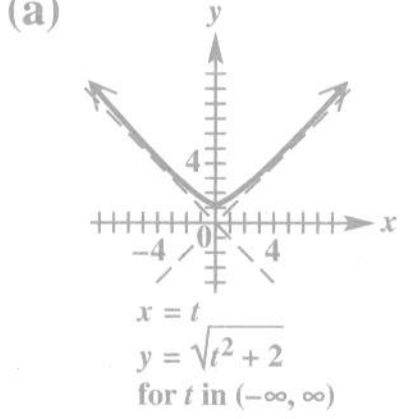

(b) $y = \frac{1}{x}$, for x in $(0, \infty)$

17. (a)

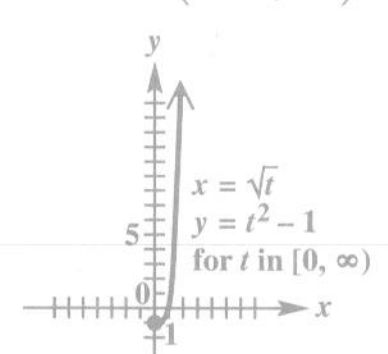

(b) $y = \sqrt{x^2 + 2}$, for x in $(-\infty, \infty)$

18. (a)

$x = \sqrt{t}$
$y = t^2 - 1$
for t in $[0, \infty)$

(b) $y = x^4 - 1$, for x in $[0, \infty)$

19. (a)

$x = 2 + \sin t$
$y = 1 + \cos t$
for t in $[0, 2\pi]$

(b) $(x - 2)^2 + (y - 1)^2 = 1$, for x in $[1, 3]$

(Modeling) In Exercises 39–42, do the following. ***See Examples 6–8.***

(a) Determine the parametric equations that model the path of the projectile.
(b) Determine the rectangular equation that models the path of the projectile.
(c) Determine approximately how long the projectile is in flight and the horizontal distance covered.

39. *Flight of a Model Rocket* A model rocket is launched from the ground with velocity 48 ft per sec at an angle of 60° with respect to the ground.

40. *Flight of a Golf Ball* Tyler is playing golf. He hits a golf ball from the ground at an angle of 60° with respect to the ground at velocity 150 ft per sec.

41. *Flight of a Softball* Sally hits a softball when it is 2 ft above the ground. The ball leaves her bat at an angle of 20° with respect to the ground at velocity 88 ft per sec.

42. *Flight of a Baseball* Carlos hits a baseball when it is 2.5 ft above the ground. The ball leaves his bat at an angle of 29° from the horizontal with velocity 136 ft per sec.

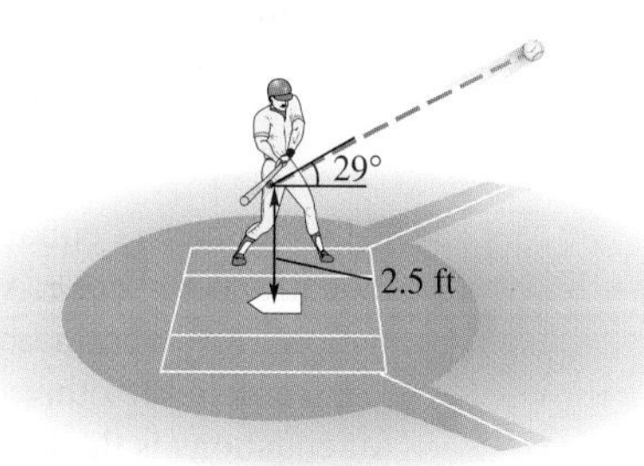

(Modeling) Solve each problem. ***See Examples 7 and 8.***

43. *Path of a Rocket* A rocket is launched from the top of an 8-ft ladder. Its initial velocity is 128 ft per sec, and it is launched at an angle of 60° with respect to the ground.

(a) Find the rectangular equation that models its path. What type of path does the rocket follow?

(b) Determine the total flight time and the horizontal distance the rocket travels.

44. *Simulating Gravity on the Moon* If an object is thrown on the moon, then the parametric equations of flight are

$$x = (v \cos \theta)t \quad \text{and} \quad y = (v \sin \theta)t - 2.66t^2 + h.$$

Estimate the distance that a golf ball hit at 88 ft per sec (60 mph) at an angle of 45° with the horizontal travels on the moon if the moon's surface is level.

45. *Flight of a Baseball* A baseball is hit from a height of 3 ft at a 60° angle above the horizontal. Its initial velocity is 64 ft per sec.

(a) Write parametric equations that model the flight of the baseball.

(b) Determine the horizontal distance traveled by the ball in the air. Assume that the ground is level.

(c) What is the maximum height of the baseball? At that time, how far has the ball traveled horizontally?

(d) Would the ball clear a 5-ft-high fence that is 100 ft from the batter?

(Modeling) Path of a Projectile In Exercises 46 and 47, a projectile has been launched from the ground with initial velocity 88 ft per sec. The parametric equations modeling the path of the projectile are supplied.

(a) Graph the parametric equations.
(b) Approximate θ, the angle the projectile makes with the horizontal at launch, to the nearest tenth of a degree.
(c) Based on your answer to part (b), write parametric equations for the projectile using the cosine and sine functions.

46. $x = 82.69295063t,\ y = -16t^2 + 30.09777261t$

47. $x = 56.56530965t,\ y = -16t^2 + 67.41191099t$

48. Give two parametric representations of the line through the point (x_1, y_1) with slope m.

49. Give two parametric representations of the parabola $y = a(x - h)^2 + k$.

50. Give a parametric representation of the rectangular equation $\frac{x^2}{a^2} - \frac{y^2}{b^2} = 1$.

51. Give a parametric representation of the rectangular equation $\frac{x^2}{a^2} + \frac{y^2}{b^2} = 1$.

52. The spiral of Archimedes has polar equation $r = a\theta$, where $r^2 = x^2 + y^2$. Show that a parametric representation of the spiral of Archimedes is

$$x = a\theta \cos \theta, \quad y = a\theta \sin \theta, \qquad \text{for } \theta \text{ in } (-\infty, \infty).$$

53. Show that the **hyperbolic spiral** $r\theta = a$, where $r^2 = x^2 + y^2$, is given parametrically by

$$x = \frac{a \cos \theta}{\theta}, \quad y = \frac{a \sin \theta}{\theta}, \qquad \text{for } \theta \text{ in } (-\infty, 0) \cup (0, \infty).$$

54. The parametric equations $x = \cos t$, $y = \sin t$, for t in $[0, 2\pi]$ and the parametric equations $x = \cos t$, $y = -\sin t$, for t in $[0, 2\pi]$ both have the unit circle as their graph. However, in one case the circle is traced out clockwise (as t moves from 0 to 2π), and in the other case the circle is traced out counterclockwise. For which pair of equations is the circle traced out in the clockwise direction?

Concept Check Consider the parametric equations $x = f(t)$, $y = g(t)$, for t in $[a, b]$, with $c > 0$, $d > 0$.

55. How is the graph affected if the equation $x = f(t)$ is replaced by $x = c + f(t)$?

56. How is the graph affected if the equation $y = g(t)$ is replaced by $y = d + g(t)$?

20. (a)

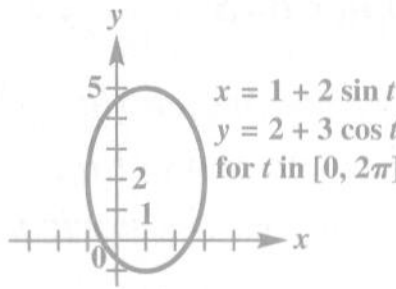

(b) $\frac{(x-1)^2}{4} + \frac{(y-2)^2}{9} = 1$, for x in $[-1, 3]$

21. (a)

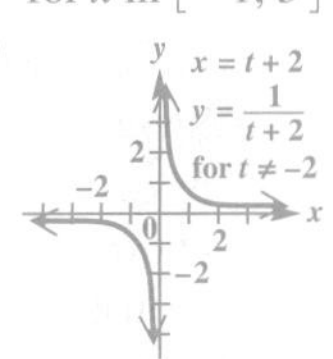

(b) $y = \frac{1}{x}$, for x in $(-\infty, 0) \cup (0, \infty)$

22. (a)

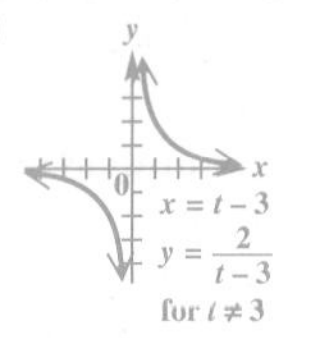

(b) $y = \frac{2}{x}$, for x in $(-\infty, 0) \cup (0, \infty)$

23. (a)

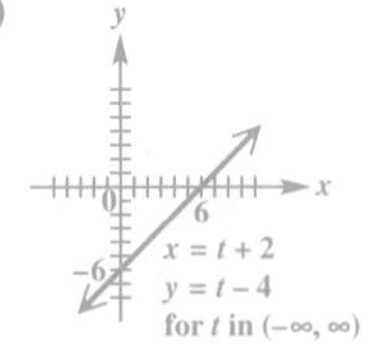

(b) $y = x - 6$, for x in $(-\infty, \infty)$

24. (a)

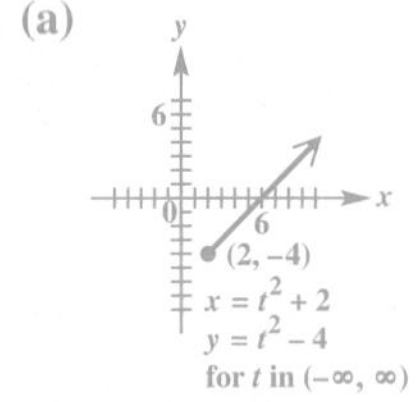

(b) $y = x - 6$, for x in $[2, \infty)$

25.

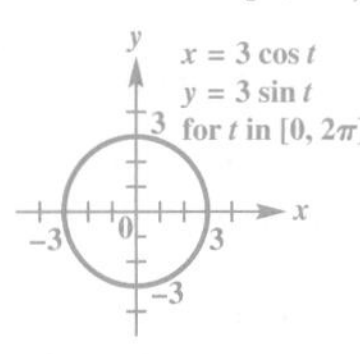

$x^2 + y^2 = 9$

26.

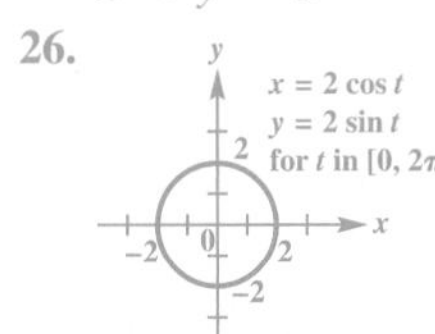

$x^2 + y^2 = 4$

27.

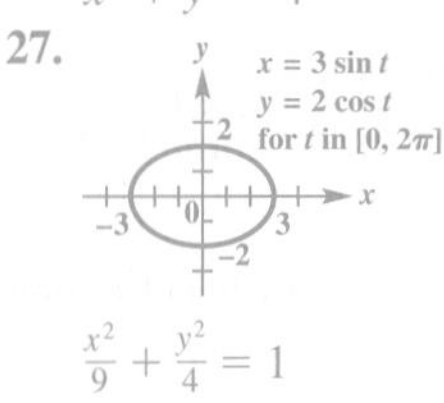

$\frac{x^2}{9} + \frac{y^2}{4} = 1$

28.

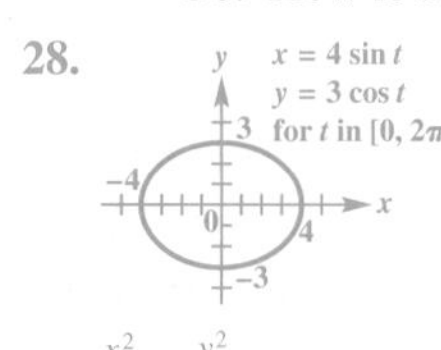

$\frac{x^2}{16} + \frac{y^2}{9} = 1$

Answers may vary for Exercises 29–32.

29. $x = t,\ y = (t + 3)^2 - 1$, for t in $(-\infty, \infty)$; $x = t - 3$, $y = t^2 - 1$, for t in $(-\infty, \infty)$ **30.** $x = t,\ y = (t + 4)^2 + 2$, for t in $(-\infty, \infty)$; $x = t - 4$, $y = t^2 + 2$, for t in $(-\infty, \infty)$

31. $x = t,\ y = t^2 - 2t + 3$, for t in $(-\infty, \infty)$; $x = t + 1$, $y = t^2 + 2$, for t in $(-\infty, \infty)$ **32.** $x = t,\ y = t^2 - 4t + 6$, for t in $(-\infty, \infty)$; $x = t + 2$, $y = t^2 + 2$, for t in $(-\infty, \infty)$

33.

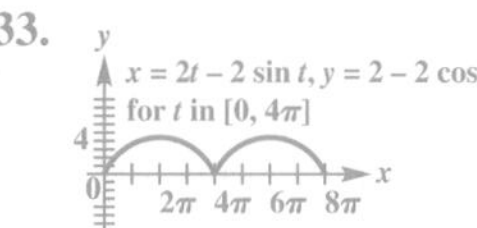

34.

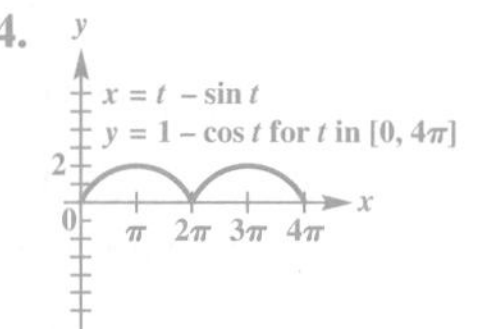

35.

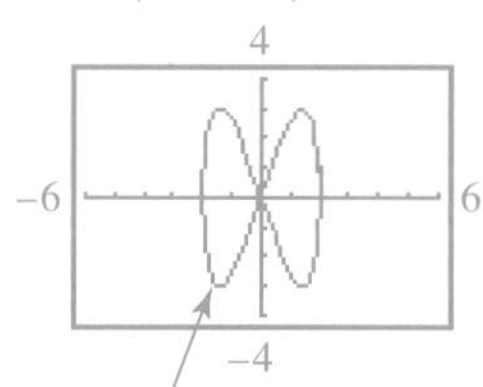

$x = 2 \cos t,\ y = 3 \sin 2t$, for t in $[0, 6.5]$

36.

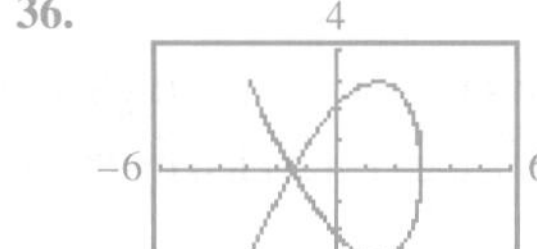

$x = 3 \cos 2t,\ y = 3 \sin 3t$, for t in $[0, 6.5]$

37.

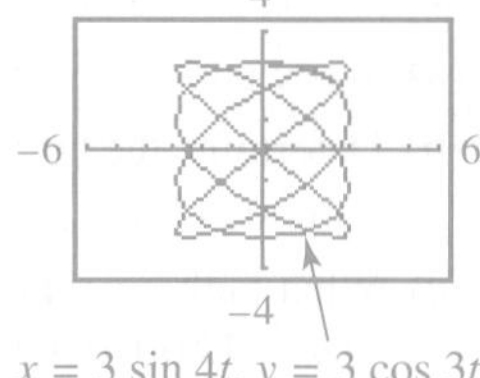

$x = 3 \sin 4t,\ y = 3 \cos 3t$, for t in $[0, 6.5]$

Answers for the remaining exercises are given in the answer section at the back of the text.

Chapter 8 Test Prep

Key Terms

8.1 Side-Angle-Side (SAS)
Angle-Side-Angle (ASA)
Side-Side-Side (SSS)
oblique triangle
Side-Angle-Angle (SAA)
ambiguous case

8.2 semiperimeter

8.3 scalar
vector quantity
vector
magnitude
initial point
terminal point
parallelogram rule
resultant
opposite (of a vector)
zero vector
position vector
horizontal component
vertical component
direction angle
unit vector
dot product
inner product
angle between two vectors
orthogonal vectors

8.4 equilibrant
airspeed
ground speed

8.5 real axis
imaginary axis
complex plane
rectangular form of a complex number
trigonometric (polar) form of a complex number
absolute value (modulus)
argument

8.6 nth root of a complex number

8.7 polar coordinate system
pole
polar axis
polar coordinates
rectangular (Cartesian) equation
polar equation
cardioid
polar grid
rose curve
lemniscate
spiral of Archimedes
limaçon

8.8 plane curve
parametric equations of a plane curve
parameter
cycloid

New Symbols

OP or $\overrightarrow{\mathbf{OP}}$	vector **OP**
$\|\mathbf{OP}\|$	magnitude of vector **OP**
$\langle a, b\rangle$	position vector
i, j	unit vectors

Quick Review

Concepts	Examples

8.1 The Law of Sines

Law of Sines

In any triangle ABC, with sides a, b, and c, the following holds.

$$\frac{a}{\sin A} = \frac{b}{\sin B} = \frac{c}{\sin C}$$

In triangle ABC, find c, to the nearest hundredth, if $A = 44°$, $C = 62°$, and $a = 12.00$ units. Then find its area.

$$\frac{a}{\sin A} = \frac{c}{\sin C} \quad \text{Law of sines}$$

$$\frac{12.00}{\sin 44°} = \frac{c}{\sin 62°} \quad \text{Substitute.}$$

$$c = \frac{12.00 \sin 62°}{\sin 44°} \quad \text{Multiply by } \sin 62° \text{ and rewrite.}$$

$$c \approx 15.25 \text{ units} \quad \text{Use a calculator.}$$

Area of a Triangle

In any triangle ABC, the area is half the product of the lengths of two sides and the sine of the angle between them.

$$\mathcal{A} = \frac{1}{2}bc \sin A, \quad \mathcal{A} = \frac{1}{2}ab \sin C, \quad \mathcal{A} = \frac{1}{2}ac \sin B$$

For triangle ABC above, apply the appropriate area formula. Here, $B = 180° - 44° - 62° = 74°$.

$$\mathcal{A} = \frac{1}{2}ac \sin B = \frac{1}{2}(12.00)(15.25) \sin 74° \approx 87.96 \text{ sq units}$$

(continued)

Concepts	Examples

Ambiguous Case

If we are given the lengths of two sides and the angle opposite one of them (for example, A, a, and b in triangle ABC), then it is possible that zero, one, or two such triangles exist. If A is acute, h is the altitude from C, and

- $a < h < b$, then there is no triangle.
- $a = h$ and $h < b$, then there is one triangle (a right triangle).
- $a \geq b$, then there is one triangle.
- $h < a < b$, then there are two triangles.

If A is obtuse and

- $a \leq b$, then there is no triangle.
- $a > b$, then there is one triangle.

See the guidelines in **Section 8.1** that illustrate the possible outcomes.

Solve triangle ABC, given $A = 44.5°$, $a = 11.0$ in., and $c = 7.0$ in.

Find angle C.

$$\frac{\sin C}{7.0} = \frac{\sin 44.5°}{11.0} \quad \text{Law of sines}$$

$$\sin C \approx 0.4460 \quad \text{Solve for } \sin C.$$

$$C \approx 26.5° \quad \text{Use inverse sine.}$$

Another angle with this sine value is

$$180° - 26.5° \approx 153.5°.$$

However, $153.5° + 44.5° > 180°$, so there is only one triangle.

$$B \approx 180° - 44.5° - 26.5° \quad \text{Angle sum formula}$$

$$B \approx 109° \quad \text{Subtract.}$$

Use the law of sines again to solve for b.

$$b \approx 14.8 \text{ in.}$$

8.2 The Law of Cosines

Law of Cosines

In any triangle ABC, with sides a, b, and c, the following hold.

$$a^2 = b^2 + c^2 - 2bc \cos A$$

$$b^2 = a^2 + c^2 - 2ac \cos B$$

$$c^2 = a^2 + b^2 - 2ab \cos C$$

In triangle ABC, find C if $a = 11$ units, $b = 13$ units, and $c = 20$ units. Then find its area.

$$c^2 = a^2 + b^2 - 2ab \cos C \quad \text{Law of cosines}$$

$$20^2 = 11^2 + 13^2 - 2(11)(13) \cos C \quad \text{Substitute.}$$

$$400 = 121 + 169 - 286 \cos C \quad \text{Square and multiply.}$$

$$\frac{400 - 121 - 169}{-286} = \cos C \quad \text{Solve for } \cos C.$$

$$C = \cos^{-1}\left(\frac{400 - 121 - 169}{-286}\right)$$

$$C \approx 113°$$

Heron's Area Formula

If a triangle has sides of lengths a, b, and c, with semiperimeter

$$s = \frac{1}{2}(a + b + c),$$

then the area $\mathcal{A}$ of the triangle is given by the following.

$$\mathcal{A} = \sqrt{s(s-a)(s-b)(s-c)}$$

The semiperimeter s is

$$s = \frac{1}{2}(11 + 13 + 20) = 22,$$

so

$$\mathcal{A} = \sqrt{22(22-11)(22-13)(22-20)}$$

$$= 66 \text{ sq units.}$$

Concepts	Examples

8.3 Vectors, Operations, and the Dot Product

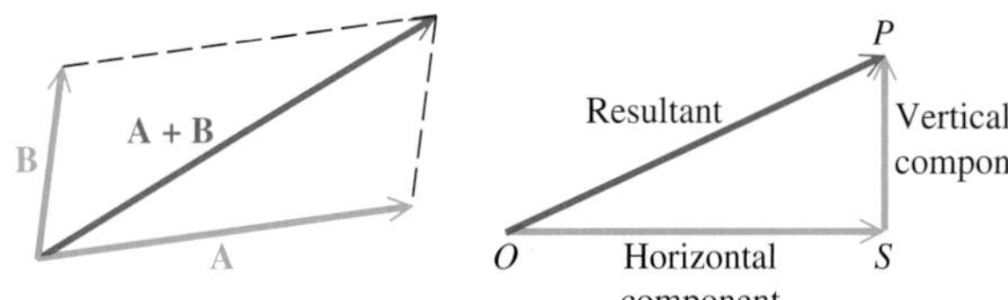

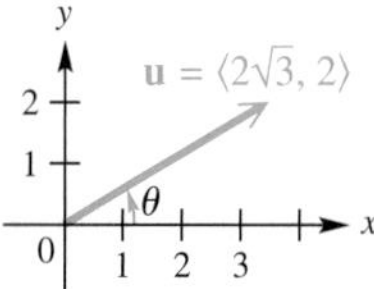

Magnitude and Direction Angle of a Vector

The magnitude (length) of vector $\mathbf{u} = \langle a, b \rangle$ is given by the following.

$$|\mathbf{u}| = \sqrt{a^2 + b^2}$$

The direction angle θ satisfies $\tan\theta = \frac{b}{a}$, where $a \neq 0$.

Example:

$$|\mathbf{u}| = \sqrt{(2\sqrt{3})^2 + 2^2} = \sqrt{16} = 4$$

Since $\tan\theta = \frac{2}{2\sqrt{3}} = \frac{1}{\sqrt{3}} \cdot \frac{\sqrt{3}}{\sqrt{3}} = \frac{\sqrt{3}}{3}$, it follows that $\theta = 30°$.

Vector Operations

Let a, b, c, d, and k represent real numbers.

$$\langle a, b\rangle + \langle c, d\rangle = \langle a + c, b + d\rangle$$

$$k \cdot \langle a, b\rangle = \langle ka, kb\rangle$$

If $\mathbf{u} = \langle a_1, a_2\rangle$, then $-\mathbf{u} = \langle -a_1, -a_2\rangle$.

$$\langle a, b\rangle - \langle c, d\rangle = \langle a, b\rangle + (-\langle c, d\rangle) = \langle a - c, b - d\rangle$$

Examples:

$$\langle 4, 6\rangle + \langle -8, 3\rangle = \langle -4, 9\rangle$$

$$5\langle -2, 1\rangle = \langle -10, 5\rangle$$

$$-\langle -9, 6\rangle = \langle 9, -6\rangle$$

$$\langle 4, 6\rangle - \langle -8, 3\rangle = \langle 12, 3\rangle$$

If $\mathbf{u} = \langle a, b\rangle$ has direction angle θ, then

$$\mathbf{u} = \langle |\mathbf{u}|\cos\theta, |\mathbf{u}|\sin\theta\rangle.$$

For **u** defined at the top of the column,

$$\mathbf{u} = \langle 4\cos 30°, 4\sin 30°\rangle$$

$$= \langle 2\sqrt{3}, 2\rangle \qquad \cos 30° = \frac{\sqrt{3}}{2},\ \sin 30° = \frac{1}{2}$$

and $\mathbf{u} = 2\sqrt{3}\,\mathbf{i} + 2\mathbf{j}$.

i, j Form for Vectors

If $\mathbf{v} = \langle a, b\rangle$, then

$$\mathbf{v} = a\mathbf{i} + b\mathbf{j}, \quad \text{where } \mathbf{i} = \langle 1, 0\rangle \text{ and } \mathbf{j} = \langle 0, 1\rangle.$$

Dot Product

The dot product of the two vectors $\mathbf{u} = \langle a, b\rangle$ and $\mathbf{v} = \langle c, d\rangle$, denoted $\mathbf{u} \cdot \mathbf{v}$, is given by the following.

$$\mathbf{u} \cdot \mathbf{v} = ac + bd$$

$$\langle 2, 1\rangle \cdot \langle 5, -2\rangle = 2 \cdot 5 + 1(-2) = 8$$

If θ is the angle between **u** and **v**, where $0° \leq \theta \leq 180°$, then the following holds.

$$\cos\theta = \frac{\mathbf{u} \cdot \mathbf{v}}{|\mathbf{u}||\mathbf{v}|}$$

Find the angle θ between $\mathbf{u} = \langle 3, 1\rangle$ and $\mathbf{v} = \langle 2, -3\rangle$.

$$\cos\theta = \frac{\langle 3, 1\rangle \cdot \langle 2, -3\rangle}{\sqrt{3^2 + 1^2} \cdot \sqrt{2^2 + (-3)^2}}$$

$$\cos\theta = \frac{6 + (-3)}{\sqrt{10} \cdot \sqrt{13}}$$

$$\cos\theta = \frac{3}{\sqrt{130}}$$

$$\theta = \cos^{-1}\frac{3}{\sqrt{130}}$$

$$\theta \approx 74.7°$$

(continued)

Concepts	Examples

8.5 Trigonometric (Polar) Form of Complex Numbers; Products and Quotients

Trigonometric (Polar) Form of Complex Numbers

Let the complex number $x + yi$ correspond to the vector with direction angle θ and magnitude r.

$$x = r\cos\theta \qquad y = r\sin\theta$$

$$r = \sqrt{x^2 + y^2} \qquad \tan\theta = \frac{y}{x}, \text{ if } x \neq 0$$

The expression

$$r(\cos\theta + i\sin\theta) \quad \text{or} \quad r \text{ cis } \theta$$

is the trigonometric form (or polar form) of $x + yi$.

Write $2(\cos 60° + i \sin 60°)$ in rectangular form.

$$2(\cos 60° + i\sin 60°) = 2\left(\frac{1}{2} + i \cdot \frac{\sqrt{3}}{2}\right)$$
$$= 1 + i\sqrt{3}$$

Write $-\sqrt{2} + i\sqrt{2}$ in trigonometric form.

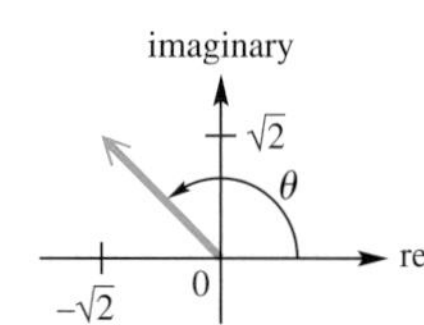

$r = \sqrt{(-\sqrt{2})^2 + (\sqrt{2})^2} = 2$

$\tan\theta = -1$ and θ is in quadrant II, so $\theta = 180° - 45° = 135°$. Therefore,

$$-\sqrt{2} + i\sqrt{2} = 2 \text{ cis } 135°.$$

Product and Quotient Theorems

For any two complex numbers $r_1(\cos\theta_1 + i\sin\theta_1)$ and $r_2(\cos\theta_2 + i\sin\theta_2)$, the following hold.

$$[r_1(\cos\theta_1 + i\sin\theta_1)] \cdot [r_2(\cos\theta_2 + i\sin\theta_2)]$$
$$= r_1r_2[\cos(\theta_1 + \theta_2) + i\sin(\theta_1 + \theta_2)]$$

and

$$\frac{r_1(\cos\theta_1 + i\sin\theta_1)}{r_2(\cos\theta_2 + i\sin\theta_2)}$$
$$= \frac{r_1}{r_2}[\cos(\theta_1 - \theta_2) + i\sin(\theta_1 - \theta_2)],$$

where $r_2 \text{ cis } \theta_2 \neq 0$

Let $\quad z_1 = 4(\cos 135° + i\sin 135°)$

and $\quad z_2 = 2(\cos 45° + i\sin 45°)$.

$$z_1z_2 = 8(\cos 180° + i\sin 180°)$$
$$= 8(-1 + i \cdot 0)$$
$$= -8$$

$$\frac{z_1}{z_2} = 2(\cos 90° + i\sin 90°)$$
$$= 2(0 + i \cdot 1)$$
$$= 2i$$

8.6 De Moivre's Theorem; Powers and Roots of Complex Numbers

De Moivre's Theorem

$$[r(\cos\theta + i\sin\theta)]^n = r^n(\cos n\theta + i\sin n\theta)$$

nth Root Theorem

If n is any positive integer, r is a positive real number, and θ is in degrees, then the nonzero complex number $r(\cos\theta + i\sin\theta)$ has exactly n distinct nth roots, given by the following.

$$\sqrt[n]{r}(\cos\alpha + i\sin\alpha), \quad \text{or} \quad \sqrt[n]{r} \text{ cis } \alpha,$$

where

$$\alpha = \frac{\theta + 360° \cdot k}{n}, \quad k = 0, 1, 2, \ldots, n - 1$$

If θ is in radians, then

$$\alpha = \frac{\theta + 2\pi k}{n}, \quad k = 0, 1, 2, \ldots, n - 1.$$

Let $z = 4(\cos 180° + i\sin 180°)$. Find z^3 and the square roots of z.

$$z^3 = 4^3(\cos 3 \cdot 180° + i\sin 3 \cdot 180°)$$
$$= 64(\cos 540° + i\sin 540°)$$
$$= 64(-1 + i \cdot 0)$$
$$= -64$$

For the given z, $r = 4$ and $\theta = 180°$. Its square roots are

$$\sqrt{4}\left(\cos\frac{180°}{2} + i\sin\frac{180°}{2}\right) = 2(0 + i \cdot 1)$$
$$= 2i$$

and $\sqrt{4}\left(\cos\frac{180° + 360°}{2} + i\sin\frac{180° + 360°}{2}\right)$
$$= 2(0 + i(-1))$$
$$= -2i.$$

Concepts	Examples

8.7 Polar Equations and Graphs

Rectangular and Polar Coordinates

The following relationships hold between the point (x, y) in the rectangular coordinate plane and the same point (r, θ) in the polar coordinate plane.

$$x = r\cos\theta \qquad y = r\sin\theta$$

$$r^2 = x^2 + y^2 \qquad \tan\theta = \frac{y}{x}, \quad \text{if } x \neq 0$$

Find the rectangular coordinates for the point $(5, 60°)$ in polar coordinates.

$$x = 5\cos 60° = 5\left(\frac{1}{2}\right) = \frac{5}{2}$$

$$y = 5\sin 60° = 5\left(\frac{\sqrt{3}}{2}\right) = \frac{5\sqrt{3}}{2}$$

The rectangular coordinates are $\left(\frac{5}{2}, \frac{5\sqrt{3}}{2}\right)$.

Find polar coordinates for $(-1, -1)$ in rectangular coordinates.

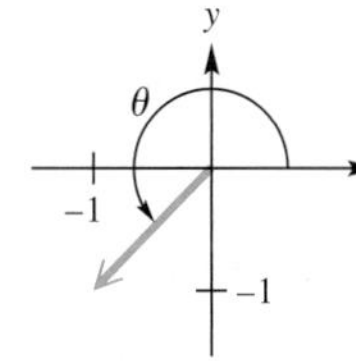

$$r = \sqrt{(-1)^2 + (-1)^2} = \sqrt{2}$$

$\tan\theta = 1$ and θ is in quadrant III, so $\theta = 225°$.

One pair of polar coordinates for $(-1, -1)$ is $(\sqrt{2}, 225°)$.

Polar Equations and Graphs

$r = a\cos\theta$, $r = a\sin\theta$ — Circles

$r^2 = a^2\sin 2\theta$, $r^2 = a^2\cos 2\theta$ — Lemniscates

$r = a \pm b\sin\theta$, $r = a \pm b\cos\theta$ — Limaçons

$r = a\sin n\theta$, $r = a\cos n\theta$ — Rose curves

Graph $r = 4\cos 2\theta$.

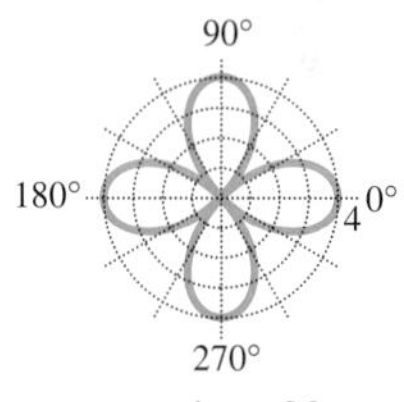

$r = 4\cos 2\theta$

8.8 Parametric Equations, Graphs, and Applications

Plane Curve

A **plane curve** is a set of points (x, y) such that $x = f(t)$, $y = g(t)$, and f and g are both defined on an interval I. The equations

$$x = f(t) \quad \text{and} \quad y = g(t)$$

are **parametric equations** with **parameter t.**

Graph $x = 2 - \sin t$, $y = \cos t - 1$, for $0 \le t \le 2\pi$.

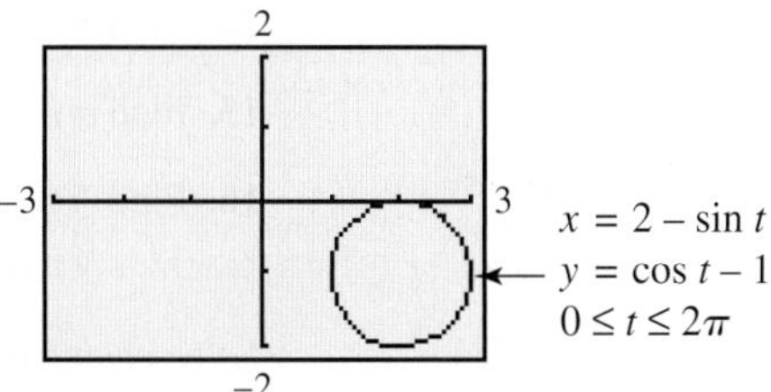

Flight of an Object

If an object has initial velocity v, has initial height h, and travels such that its initial angle of elevation is θ, then its flight after t seconds is modeled by the following parametric equations.

$$x = (v\cos\theta)t \quad \text{and} \quad y = (v\sin\theta)t - 16t^2 + h$$

Joe kicks a football from the ground at an angle of 45° with a velocity of 48 ft per sec. Give the parametric equations that model the path of the football and the distance it travels before hitting the ground.

$$x = (48\cos 45°)t = 24\sqrt{2}t$$
$$y = (48\sin 45°)t - 16t^2 = 24\sqrt{2}t - 16t^2$$

When the ball hits the ground, $y = 0$.

$24\sqrt{2}t - 16t^2 = 0$ — Substitute $y = 0$.

$8t(3\sqrt{2} - 2t) = 0$ — Factor.

$t = 0$ (Reject) or $t = \frac{3\sqrt{2}}{2}$ — Zero-factor property

The distance it travels is $x = 24\sqrt{2}\left(\frac{3\sqrt{2}}{2}\right) = 72$ ft.

Chapter 8 Review Exercises

Use the law of sines to find the indicated part of each triangle ABC.

1. Find b if $C = 74.2°$, $c = 96.3$ m, $B = 39.5°$.
2. Find B if $A = 129.7°$, $a = 127$ ft, $b = 69.8$ ft.
3. Find B if $C = 51.3°$, $c = 68.3$ m, $b = 58.2$ m.
4. Find b if $a = 165$ m, $A = 100.2°$, $B = 25.0°$.
5. Find A if $B = 39° 50'$, $b = 268$ m, $a = 340$ m.
6. Find A if $C = 79° 20'$, $c = 97.4$ mm, $a = 75.3$ mm.
7. If we are given a, A, and C in a triangle ABC, does the possibility of the ambiguous case exist? If not, explain why.
8. Can triangle ABC exist if

$$a = 4.7, \quad b = 2.3, \quad \text{and} \quad c = 7.0?$$

If not, explain why. Answer this question without using trigonometry.

9. Given $a = 10$ and $B = 30°$, determine the values of b for which A has
   (a) exactly one value
   (b) two possible values
   (c) no value.
10. Explain why there can be no triangle ABC satisfying $A = 140°$, $a = 5$, and $b = 7$.

Use the law of cosines to find the indicated part of each triangle ABC.

11. Find A if $a = 86.14$ in., $b = 253.2$ in., $c = 241.9$ in.
12. Find b if $B = 120.7°$, $a = 127$ ft, $c = 69.8$ ft.
13. Find a if $A = 51° 20'$, $c = 68.3$ m, $b = 58.2$ m.
14. Find B if $a = 14.8$ m, $b = 19.7$ m, $c = 31.8$ m.
15. Find a if $A = 60°$, $b = 5.0$ cm, $c = 21$ cm.
16. Find A if $a = 13$ ft, $b = 17$ ft, $c = 8$ ft.

Solve each triangle ABC having the given information.

17. $A = 25.2°$, $a = 6.92$ yd, $b = 4.82$ yd
18. $A = 61.7°$, $a = 78.9$ m, $b = 86.4$ m
19. $a = 27.6$ cm, $b = 19.8$ cm, $C = 42° 30'$
20. $a = 94.6$ yd, $b = 123$ yd, $c = 109$ yd

Find the area of each triangle ABC with the given information.

21. $b = 840.6$ m, $c = 715.9$ m, $A = 149.3°$
22. $a = 6.90$ ft, $b = 10.2$ ft, $C = 35° 10'$
23. $a = 0.913$ km, $b = 0.816$ km, $c = 0.582$ km
24. $a = 43$ m, $b = 32$ m, $c = 51$ m

Answers

1. 63.7 m 2. 25.0°
3. 41.7° 4. 70.9 m
5. 54° 20′ or 125° 40′
6. 49° 30′

9. (a) $b = 5$, $b \geq 10$
(b) $5 < b < 10$
(c) $b < 5$

11. 19.87°, or 19° 52′
12. 173 ft
13. 55.5 m
14. 26.5°, or 26° 30′
15. 19 cm
16. 47°

17. $B = 17.3°$, $C = 137.5°$, $c = 11.0$ yd
18. $B_1 = 74.6°$, $C_1 = 43.7°$, $c_1 = 61.9$ m; $B_2 = 105.4°$, $C_2 = 12.9°$, $c_2 = 20.0$ m
19. $c = 18.7$ cm, $A = 91° 40'$, $B = 45° 50'$
20. $A = 47.7°$, $B = 73.9°$, $C = 58.4°$

21. 153,600 m^2 22. 20.3 ft^2
23. 0.234 km^2 24. 680 m^2

25. 58.6 ft
26. 11 ft
27. 13 m
28. 15.8 ft
29. 53.2 ft

Solve each problem.

25. *Distance across a Canyon* To measure the distance AB across a canyon for a power line, a surveyor measures angles B and C and the distance BC, as shown in the figure. What is the distance from A to B?

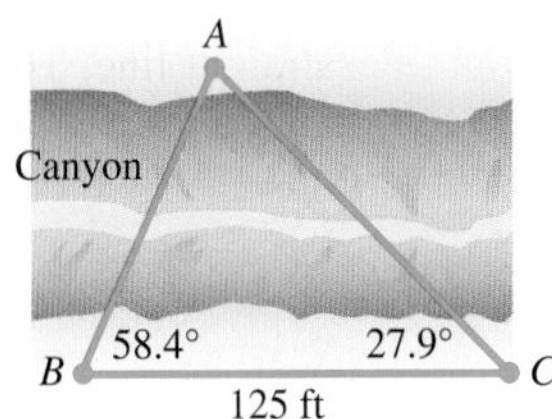

26. *Length of a Brace* A banner on an 8.0-ft pole is to be mounted on a building at an angle of 115°, as shown in the figure. Find the length of the brace.

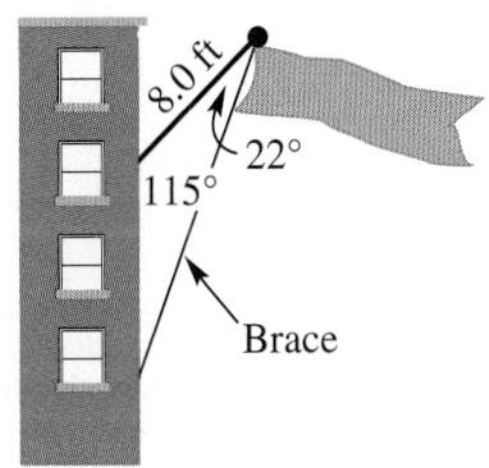

27. *Height of a Tree* A tree leans at an angle of 8.0° from the vertical. From a point 7.0 m from the bottom of the tree, the angle of elevation to the top of the tree is 68°. How tall is the leaning tree?

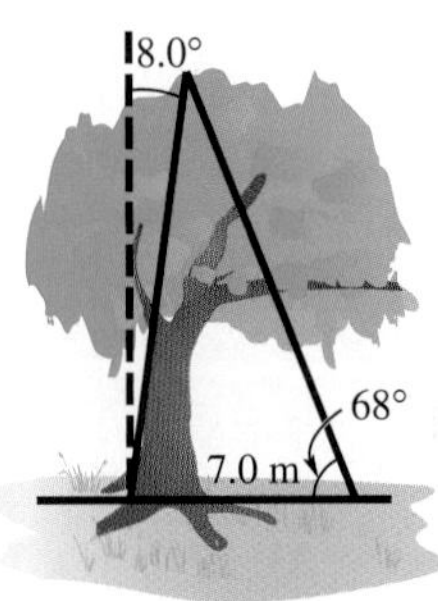

28. *Hanging Sculpture* A hanging sculpture is to be hung in an art gallery with two wires of lengths 15.0 ft and 12.2 ft so that the angle between them is 70.3°. How far apart should the ends of the wire be placed on the ceiling?

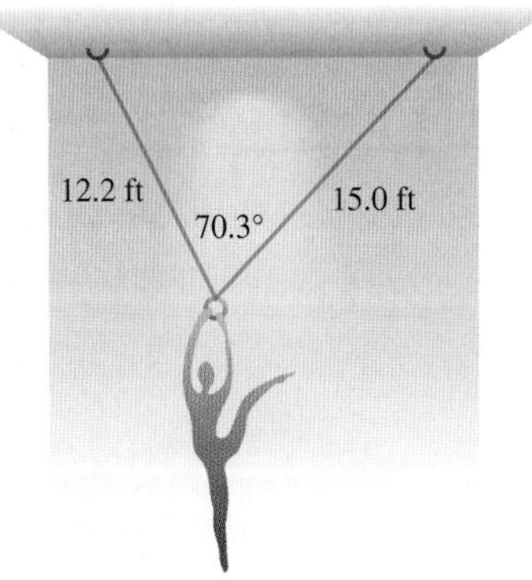

29. *Height of a Tree* A hill makes an angle of 14.3° with the horizontal. From the base of the hill, the angle of elevation to the top of a tree on top of the hill is 27.2°. The distance along the hill from the base to the tree is 212 ft. Find the height of the tree.

30. 77.1°
31. 115 km
32. 2.4 mi
33. 25 sq units
34. The sides measure 5, 10, and $5\sqrt{5}$, which satisfy the converse of the Pythagorean theorem. The area is 25 sq units.

35.

36.

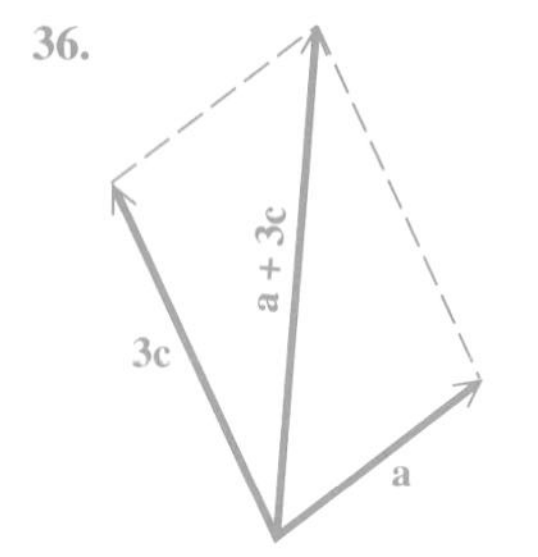

30. *Pipeline Position* A pipeline is to run between points A and B, which are separated by a protected wetlands area. To avoid the wetlands, the pipe will run from point A to C and then to B. The distances involved are $AB = 150$ km, $AC = 102$ km, and $BC = 135$ km. What angle should be used at point C?

31. *Distance between Two Boats* Two boats leave a dock together. Each travels in a straight line. The angle between their courses measures 54° 10′. One boat travels 36.2 km per hr, and the other travels 45.6 km per hr. How far apart will they be after 3 hr?

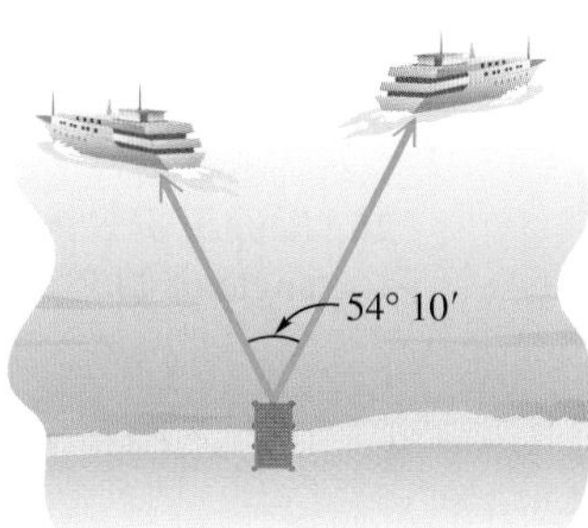

32. *Distance from a Ship to a Lighthouse* A ship sailing parallel to shore sights a lighthouse at an angle of 30° from its direction of travel. After the ship travels 2.0 mi farther, the angle has increased to 55°. At that time, how far is the ship from the lighthouse?

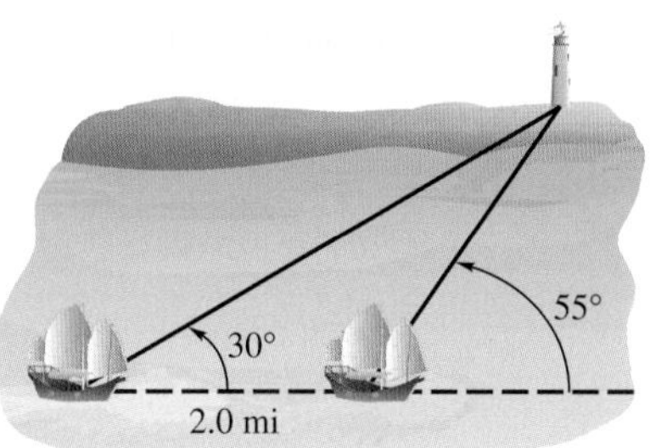

33. *Area of a Triangle* Find the area of the triangle shown in the figure using Heron's area formula.

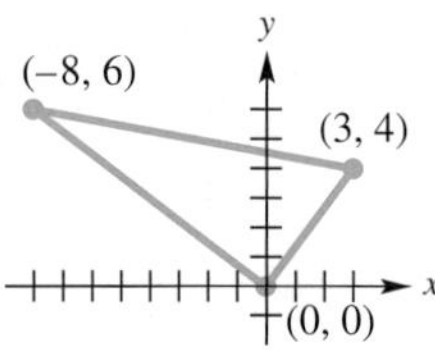

34. Show that the triangle in **Exercise 33** is a right triangle. Then find the area using the formula

$$\mathcal{A} = \frac{1}{2}ac \sin B, \quad \text{with } B = 90°.$$

Use the given vectors to sketch each of the following.

35. $\mathbf{a} - \mathbf{b}$

36. $\mathbf{a} + 3\mathbf{c}$

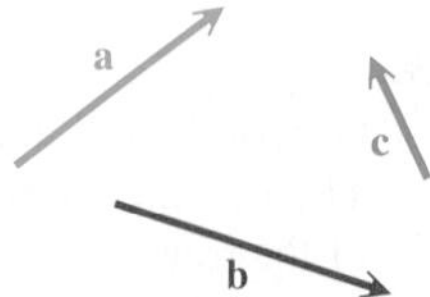

37. 207 lb
38. 209 newtons
39. 869; 418
40. $25\sqrt{2}$; $25\sqrt{2}$
41. 15; 126.9°
42. 29; 316.4°
43. (a) **i** (b) 4**i** − 2**j** (c) 11**i** − 7**j**
44. 142.1°
45. 90°; orthogonal
46. 45°
47. 29 lb
48. 280 newtons; 30.4°
49. bearing: 306°; ground speed: 524 mph
50. 3.8°
51. 34 lb
52. speed: 21 km per hr; bearing: 118°
53. $-30i$
54. $-3 - 3i\sqrt{3}$
55. $-\frac{1}{8} + \frac{\sqrt{3}}{8}i$
56. -2
57. $8i$
58. $-128 + 128i$
59. $-\frac{1}{2} - \frac{\sqrt{3}}{2}i$
60. x
61.
62.
63.
64.
65. $2\sqrt{2}(\cos 135° + i \sin 135°)$
66. $3i$
67. $-\sqrt{2} - i\sqrt{2}$
68. $8(\cos 120° + i \sin 120°)$

Given two forces and the angle between them, find the magnitude of the resultant force.

37. (100 lb, 52°, 130 lb)

38. two forces of 142 and 215 newtons, forming an angle of 112°

Vector **v** *has the given magnitude and direction angle. Find the magnitudes of the horizontal and vertical components of* **v**.

39. $|\mathbf{v}| = 964$, $\theta = 154° \, 20'$

40. $|\mathbf{v}| = 50$, $\theta = 45°$ (Give exact values.)

Find the magnitude and direction angle for **u** *rounded to the nearest tenth.*

41. $\mathbf{u} = \langle -9, 12 \rangle$

42. $\mathbf{u} = \langle 21, -20 \rangle$

43. Let $\mathbf{v} = 2\mathbf{i} - \mathbf{j}$ and $\mathbf{u} = -3\mathbf{i} + 2\mathbf{j}$. Express each in terms of **i** and **j**.

(a) $2\mathbf{v} + \mathbf{u}$ (b) $2\mathbf{v}$ (c) $\mathbf{v} - 3\mathbf{u}$

Find the angle between the vectors. Round to the nearest tenth of a degree. If the vectors are orthogonal, say so.

44. $\langle 3, -2 \rangle, \langle -1, 3 \rangle$

45. $\langle 5, -3 \rangle, \langle 3, 5 \rangle$

46. $\langle 0, 4 \rangle, \langle -4, 4 \rangle$

Solve each problem.

47. *Weight of a Sled and Passenger* Paula and Steve are pulling their daughter Jessie on a sled. Steve pulls with a force of 18 lb at an angle of 10°. Paula pulls with a force of 12 lb at an angle of 15°. Find the magnitude of the resultant force on Jessie and the sled.

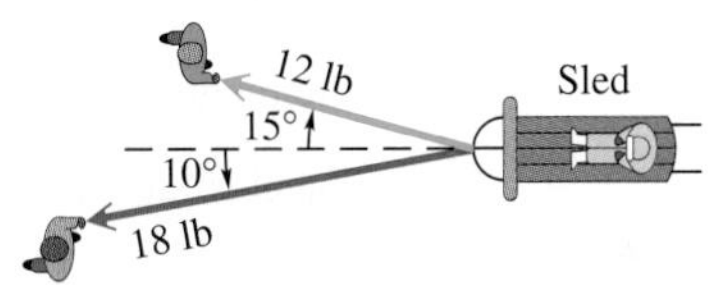

48. *Force Placed on a Barge* One boat pulls a barge with a force of 100 newtons. Another boat pulls the barge at an angle of 45° to the first force, with a force of 200 newtons. Find the resultant force acting on the barge, to the nearest unit, and the angle between the resultant and the first boat, to the nearest tenth.

49. *Direction and Speed of a Plane* A plane has an airspeed of 520 mph. The pilot wishes to fly on a bearing of 310°. A wind of 37 mph is blowing from a bearing of 212°. In what direction should the pilot fly, and what will be her ground speed?

50. *Angle of a Hill* A 186-lb force is required to hold a 2800-lb car on a hill. What angle does the hill make with the horizontal?

51. *Incline Force* Find the force required to keep a 75-lb sled from sliding down an incline that makes an angle of 27° with the horizontal. (Assume there is no friction.)

52. *Speed and Direction of a Boat* A boat travels 15 km per hr in still water. The boat is traveling across a large river, on a bearing of 130°. The current in the river, coming from the west, has a speed of 7 km per hr. Find the resulting speed of the boat and its resulting direction of travel.

Perform each operation. Write answers in rectangular form.

53. $[5(\cos 90° + i \sin 90°)][6(\cos 180° + i \sin 180°)]$

54. $[3 \text{ cis } 135°][2 \text{ cis } 105°]$

55. $\dfrac{2(\cos 60° + i \sin 60°)}{8(\cos 300° + i \sin 300°)}$

69. $\sqrt{2}(\cos 315° + i \sin 315°)$
70. $-2 - 2i\sqrt{3}$
71. $4(\cos 270° + i \sin 270°)$
72. $4.4995 - 5.3623i$

73. It is the line $y = -x$.
74. It is a circle of radius 2 with the origin as center.

75. $\sqrt[6]{2}(\cos 105° + i \sin 105°)$, $\sqrt[6]{2}(\cos 225° + i \sin 225°)$, $\sqrt[6]{2}(\cos 345° + i \sin 345°)$
76. $\sqrt[10]{8}(\cos 27° + i \sin 27°)$, $\sqrt[10]{8}(\cos 99° + i \sin 99°)$, $\sqrt[10]{8}(\cos 171° + i \sin 171°)$, $\sqrt[10]{8}(\cos 243° + i \sin 243°)$, $\sqrt[10]{8}(\cos 315° + i \sin 315°)$

77. none
78. one

79. $\{2(\cos 45° + i \sin 45°)$, $2(\cos 135° + i \sin 135°)$, $2(\cos 225° + i \sin 225°)$, $2(\cos 315° + i \sin 315°)\}$
80. $\{5(\cos 60° + i \sin 60°)$, $5(\cos 180° + i \sin 180°)$, $5(\cos 300° + i \sin 300°)\}$
81. $\{\cos 135° + i \sin 135°$, $\cos 315° + i \sin 315°\}$

82. $\left(\frac{5\sqrt{2}}{2}, -\frac{5\sqrt{2}}{2}\right)$
83. $(2, 120°)$

84. It will be a circle.

85. circle

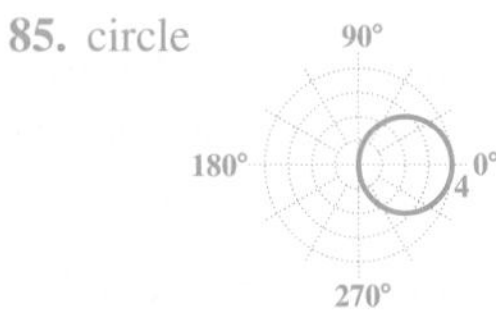

86. cardioid

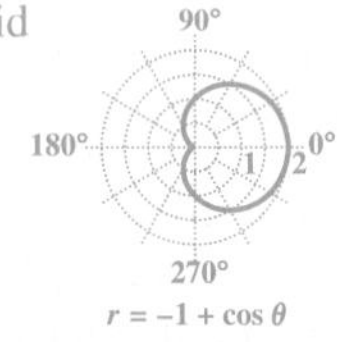

87. eight-leaved rose

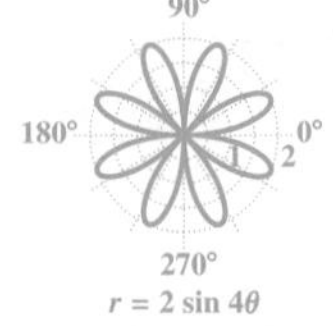

56. $\dfrac{4 \text{ cis } 270°}{2 \text{ cis } 90°}$

57. $(\sqrt{3} + i)^3$

58. $(2 - 2i)^5$

59. $(\cos 100° + i \sin 100°)^6$

60. *Concept Check* The vector representing a real number will lie on the ____-axis in the complex plane.

Graph each complex number as a vector.

61. $5i$ **62.** $-4 + 2i$ **63.** $3 - 3i\sqrt{3}$

64. Find the sum of $7 + 3i$ and $-2 + i$. Graph both complex numbers and their resultant.

Perform each conversion, using a calculator to approximate answers as necessary.

	Rectangular Form	Trigonometric Form
65.	$-2 + 2i$	________
66.	________	$3(\cos 90° + i \sin 90°)$
67.	________	$2(\cos 225° + i \sin 225°)$
68.	$-4 + 4i\sqrt{3}$	________
69.	$1 - i$	________
70.	________	$4 \text{ cis } 240°$
71.	$-4i$	________
72.	________	$7 \text{ cis } 310°$

Concept Check The complex number z, where $z = x + yi$, can be graphed in the plane as (x, y). Describe the graph of all complex numbers z satisfying the given conditions.

73. The imaginary part of z is the negative of the real part of z.

74. The absolute value of z is 2.

Find all roots as indicated. Express them in trigonometric form.

75. the cube roots of $1 - i$ **76.** the fifth roots of $-2 + 2i$

77. *Concept Check* How many real sixth roots does -64 have?

78. *Concept Check* How many real fifth roots does -32 have?

Solve each equation. Leave answers in trigonometric form.

79. $x^4 + 16 = 0$ **80.** $x^3 + 125 = 0$ **81.** $x^2 + i = 0$

82. Convert $(5, 315°)$ to rectangular coordinates.

83. Convert $(-1, \sqrt{3})$ to polar coordinates, with $0° \le \theta < 360°$ and $r > 0$.

84. *Concept Check* What will the graph of $r = k$ be, for $k > 0$?

Identify and graph each polar equation for θ in $[0°, 360°)$.

85. $r = 4 \cos \theta$ **86.** $r = -1 + \cos \theta$

87. $r = 2 \sin 4\theta$ **88.** $r = \dfrac{2}{2 \cos \theta - \sin \theta}$

Find an equivalent equation in rectangular coordinates.

89. $r = \dfrac{3}{1 + \cos \theta}$ **90.** $r = \sin \theta + \cos \theta$ **91.** $r = 2$

88. line

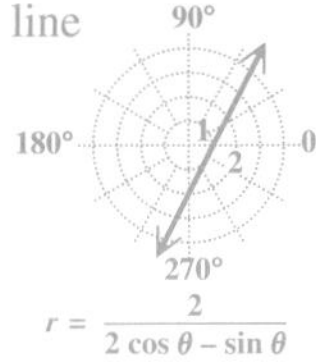

$r = \frac{2}{2\cos\theta - \sin\theta}$

89. $y^2 = -6\left(x - \frac{3}{2}\right)$, or $y^2 + 6x - 9 = 0$

90. $\left(x - \frac{1}{2}\right)^2 + \left(y - \frac{1}{2}\right)^2 = \frac{1}{2}$, or $x^2 + y^2 - x - y = 0$

91. $x^2 + y^2 = 4$

92. $\sin\theta = \cos\theta$, or $\tan\theta = 1$

93. $r = \tan\theta \sec\theta$, or $r = \frac{\tan\theta}{\cos\theta}$

94. $r = 5$ or $r = -5$

95. $r = 2\sec\theta$, or $r = \frac{2}{\cos\theta}$

96. $r = 2\csc\theta$, or $r = \frac{2}{\sin\theta}$

97. $r = \frac{4}{\cos\theta + 2\sin\theta}$

98. $r = 2$ or $r = -2$

99.

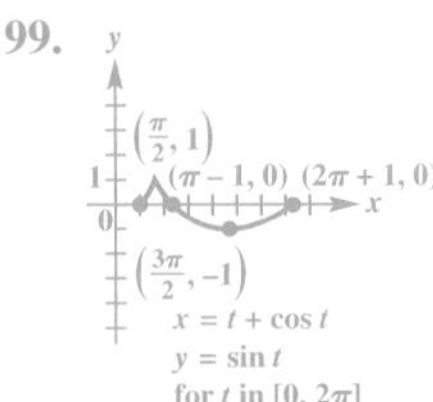

101. $y = \sqrt{x^2 + 1}$, for x in $[0, \infty)$

102. $x - 3y = 5$, for x in $[-13, 17]$

103. $y = 3\sqrt{1 + \frac{x^2}{25}}$, for x in $(-\infty, \infty)$

104. $y = \frac{1}{x - 4}$, for x in $[5, \infty)$

105. $y^2 = -\frac{1}{2}(x - 1)$, or $2y^2 + x - 1 = 0$, for x in $[-1, 1]$

106. $x = 3 + 5\cos t$, $y = 4 + 5\sin t$, for t in $[0, 2\pi]$

107. (a) $x = (118 \cos 27°)t$, $y = 3.2 - 16t^2 + (118 \sin 27°)t$
(b) $y = 3.2 - \frac{4x^2}{3481 \cos^2 27°} + (\tan 27°)x$
(c) 3.4 sec; 358 ft

108. $z^2 + z = 1 + 3i$; $r = \sqrt{10}$ and $\sqrt{10} > 2$

Find an equivalent equation in polar coordinates.

92. $y = x$ **93.** $y = x^2$ **94.** $x^2 + y^2 = 25$

In Exercises 95–98, find a polar equation having the given graph.

95.

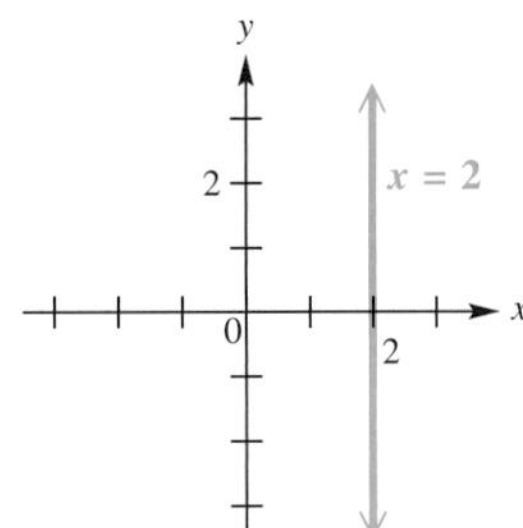

96.

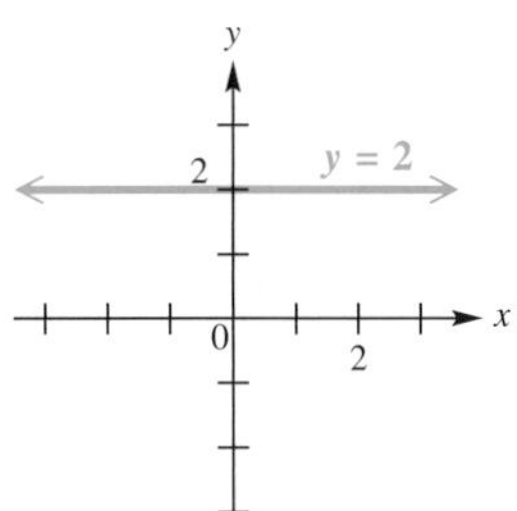

97.

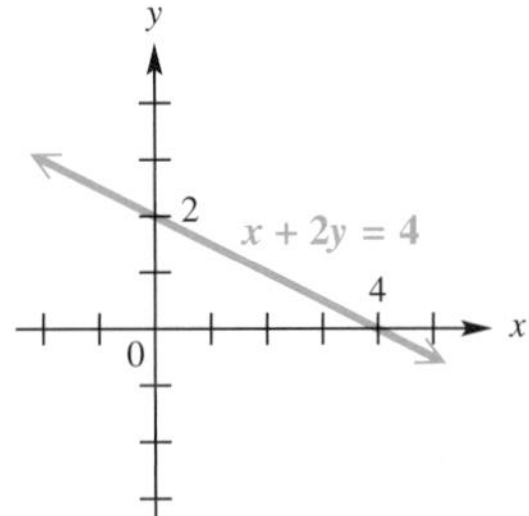

98.

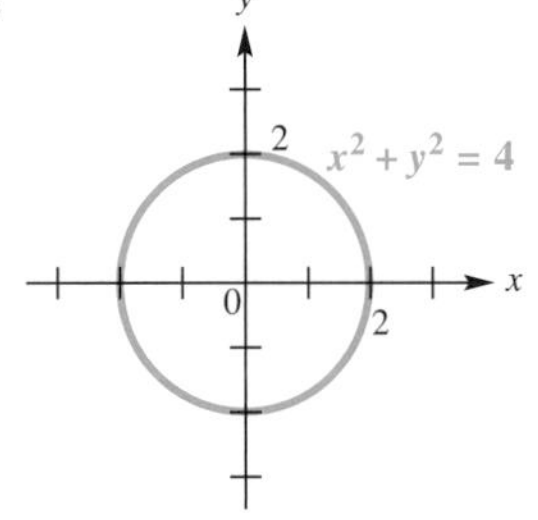

99. Graph the plane curve defined by the parametric equations $x = t + \cos t$, $y = \sin t$, for t in $[0, 2\pi]$.

100. Show that the distance between (r_1, θ_1) and (r_2, θ_2) in polar coordinates is given by

$$d = \sqrt{r_1^2 + r_2^2 - 2r_1r_2\cos(\theta_1 - \theta_2)}.$$

Find a rectangular equation for each plane curve with the given parametric equations.

101. $x = \sqrt{t - 1}$, $y = \sqrt{t}$, for t in $[1, \infty)$

102. $x = 3t + 2$, $y = t - 1$, for t in $[-5, 5]$

103. $x = 5\tan t$, $y = 3\sec t$, for t in $\left(-\frac{\pi}{2}, \frac{\pi}{2}\right)$

104. $x = t^2 + 5$, $y = \frac{1}{t^2 + 1}$, for t in $(-\infty, \infty)$

105. $x = \cos 2t$, $y = \sin t$, for t in $(-\pi, \pi)$

106. Find a pair of parametric equations whose graph is the circle having center $(3, 4)$ and passing through the origin.

107. *Flight of a Baseball* A batter hits a baseball when it is 3.2 ft above the ground. It leaves the bat with velocity 118 ft per sec at an angle of 27° with respect to the ground.

(a) Determine the parametric equations that model the path of the baseball.

(b) Determine the rectangular equation that models the path of the baseball.

(c) Determine approximately how long the projectile is in flight and the horizontal distance it covers.

108. *Mandelbrot Set* Consider the complex number $z = 1 + i$. Compute the value of $z^2 + z$, and show that its absolute value exceeds 2, indicating that $1 + i$ is not in the Mandelbrot set.

Chapter 8 Test

Find the indicated part of each triangle ABC.

1. Find C if $A = 25.2°$, $a = 6.92$ yd, and $b = 4.82$ yd.

2. Find c if $C = 118°$, $a = 75.0$ km, and $b = 131$ km.

3. Find B if $a = 17.3$ ft, $b = 22.6$ ft, $c = 29.8$ ft.

4. Find the area of triangle ABC if $a = 14$, $b = 30$, and $c = 40$.

5. Find the area of triangle XYZ shown here.

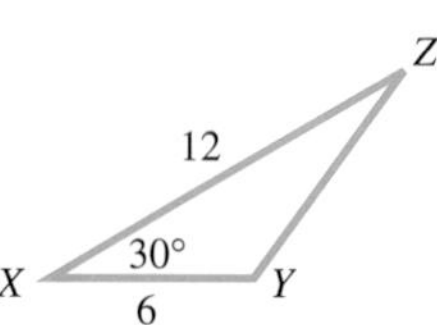

6. Given $a = 10$ and $B = 150°$ in triangle ABC, determine the values of b for which A has

(a) exactly one value **(b)** two possible values **(c)** no value.

Solve each triangle ABC.

7. $A = 60°$, $b = 30$ m, $c = 45$ m

8. $b = 1075$ in., $c = 785$ in., $C = 38° 30'$

9. Find the magnitude and the direction angle for the vector shown in the figure.

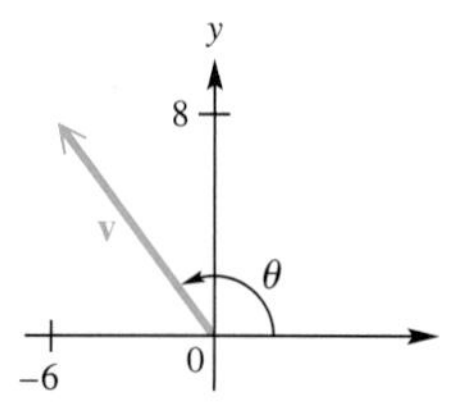

10. For the vectors $\mathbf{u} = \langle -1, 3 \rangle$ and $\mathbf{v} = \langle 2, -6 \rangle$, find each of the following.

(a) $\mathbf{u} + \mathbf{v}$ **(b)** $-3\mathbf{v}$ **(c)** $\mathbf{u} \cdot \mathbf{v}$ **(d)** $|\mathbf{u}|$

11. Find the measure of the angle θ between $\mathbf{u} = \langle 4, 3 \rangle$ and $\mathbf{v} = \langle 1, 5 \rangle$.

Solve each problem.

12. *Height of a Balloon* The angles of elevation of a balloon from two points A and B on level ground are $24° 50'$ and $47° 20'$, respectively. As shown in the figure, points A, B, and C are in the same vertical plane and points A and B are 8.4 mi apart. Approximate the height of the balloon above the ground to the nearest tenth of a mile.

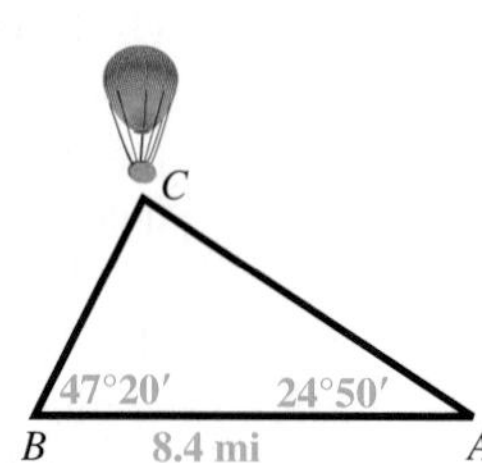

[8.1]
1. 137.5°

[8.2]
2. 179 km
3. 49.0°
4. 168 sq units

[8.1]
5. 18 sq units
6. (a) $b > 10$ (b) none (c) $b \le 10$

[8.1, 8.2]
7. $a = 40$ m, $B = 41°$, $C = 79°$
8. $B_1 = 58° 30'$, $A_1 = 83° 00'$, $a_1 = 1250$ in.; $B_2 = 121° 30'$, $A_2 = 20° 00'$, $a_2 = 431$ in.

[8.3]
9. $|\mathbf{v}| = 10$; $\theta = 126.9°$
10. (a) $\langle 1, -3 \rangle$ (b) $\langle -6, 18 \rangle$ (c) -20 (d) $\sqrt{10}$
11. 41.8°

[8.1]
12. 2.7 mi

[8.3]
13. $\langle -346, 451 \rangle$

[8.4]
14. 1.91 mi

[8.1]
15. 14 m

[8.4]
16. bearing: 357°; airspeed: 220 mph
17. 18.7°

[8.5]
18. $7 - 3i$

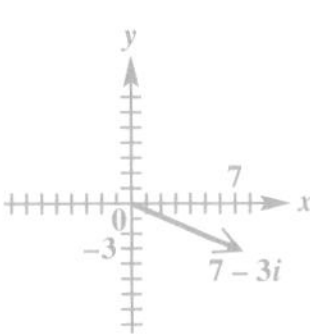

19. (a) $-i$ (b) $2i$
20. (a) $3(\cos 90° + i \sin 90°)$
(b) $\sqrt{5}$ cis 63.43°
(c) $2(\cos 240° + i \sin 240°)$
21. (a) $\frac{3\sqrt{3}}{2} + \frac{3}{2}i$
(b) $3.06 + 2.57i$
(c) $3i$

[8.5, 8.6]
22. (a) $16(\cos 50° + i \sin 50°)$
(b) $2\sqrt{3} + 2i$
(c) $4\sqrt{3} + 4i$

[8.6]
23. 2 cis 67.5°, 2 cis 157.5°, 2 cis 247.5°, 2 cis 337.5°

[8.7]
24. Answers may vary.
(a) (5, 90°), (5, −270°)
(b) $(2\sqrt{2}, 225°)$, $(2\sqrt{2}, -135°)$
25. (a) $\left(\frac{3\sqrt{2}}{2}, -\frac{3\sqrt{2}}{2}\right)$
(b) (0, −4)

13. *Horizontal and Vertical Components* Find the horizontal and vertical components of the vector with magnitude 569 and direction angle 127.5° from the horizontal. Give your answer in the form $\langle a, b \rangle$.

14. *Radio Direction Finders* Radio direction finders are placed at points A and B, which are 3.46 mi apart on an east-west line, with A west of B. From A, the bearing of a certain illegal pirate radio transmitter is 48°, and from B the bearing is 302°. Find the distance between the transmitter and A to the nearest hundredth of a mile.

15. *Height of a Tree* A tree leans at an angle of 8.0° from the vertical, as shown in the figure. From a point 8.0 m from the bottom of the tree, the angle of elevation to the top of the tree is 66°. Find the height of the leaning tree.

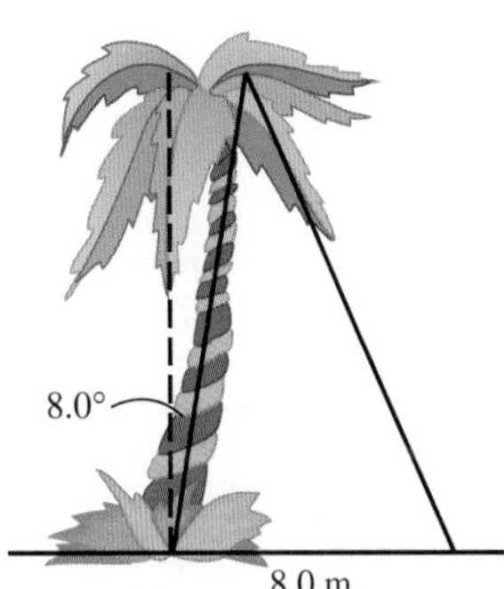

16. *Bearing and Airspeed* Find the bearing and airspeed required for a plane to fly 630 mi due north in 3.0 hr if the wind is blowing from a direction of 318° at 15 mph. Approximate the bearing to the nearest degree and the airspeed to the nearest 10 mph.

17. *Incline Angle* A force of 16.0 lb is required to hold a 50.0-lb wheelbarrow on an incline. What angle does the incline make with the horizontal?

18. For the following complex numbers, find $w + z$ in rectangular form and give a geometric representation of the sum.

$$w = 2 - 4i \quad \text{and} \quad z = 5 + i$$

19. Express each of the following in rectangular form.
(a) i^{15} **(b)** $(1 + i)^2$

20. Write each complex number in trigonometric (polar) form, where $0° \le \theta < 360°$.
(a) $3i$ **(b)** $1 + 2i$ **(c)** $-1 - i\sqrt{3}$

21. Write each complex number in rectangular form.
(a) $3(\cos 30° + i \sin 30°)$ **(b)** 4 cis 40° **(c)** $3(\cos 90° + i \sin 90°)$

22. Find each of the following in the form specified for the complex numbers

$$w = 8(\cos 40° + i \sin 40°) \quad \text{and} \quad z = 2(\cos 10° + i \sin 10°).$$

(a) wz (trigonometric form) **(b)** $\frac{w}{z}$ (rectangular form) **(c)** z^3 (rectangular form)

23. Find the four complex fourth roots of $-16i$. Express them in trigonometric form.

24. Convert the given rectangular coordinates to polar coordinates. Give two pairs of polar coordinates for each point.
(a) (0, 5) **(b)** (−2, −2)

25. Convert the given polar coordinates to rectangular coordinates.
(a) (3, 315°) **(b)** (−4, 90°)

Identify and graph each polar equation for θ in $[0°, 360°)$.

26. $r = 1 - \cos \theta$

27. $r = 3 \cos 3\theta$

28. Convert each polar equation to a rectangular equation, and sketch its graph.

(a) $r = \dfrac{4}{2 \sin \theta - \cos \theta}$

(b) $r = 6$

Graph each pair of parametric equations.

29. $x = 4t - 3,\ y = t^2,\quad$ for t in $[-3, 4]$

30. $x = 2 \cos 2t,\ y = 2 \sin 2t,\quad$ for t in $[0, 2\pi]$

26. cardioid

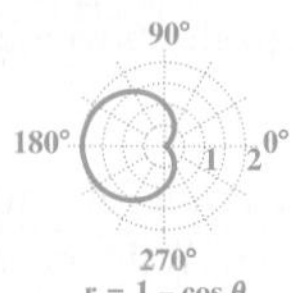

27. three-leaved rose

28. (a) $x - 2y = -4$

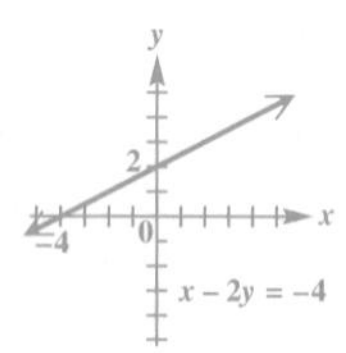

(b) $x^2 + y^2 = 36$

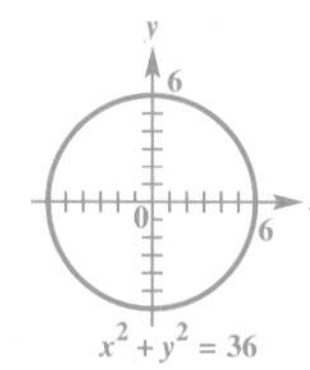

[8.8]

29.

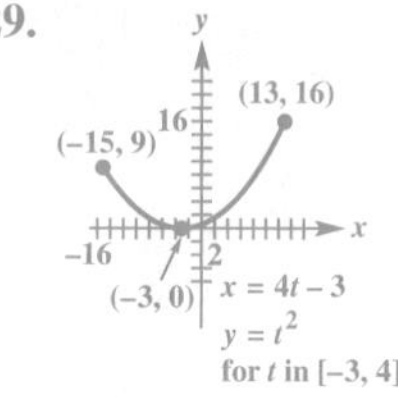

30.

9 Systems and Matrices

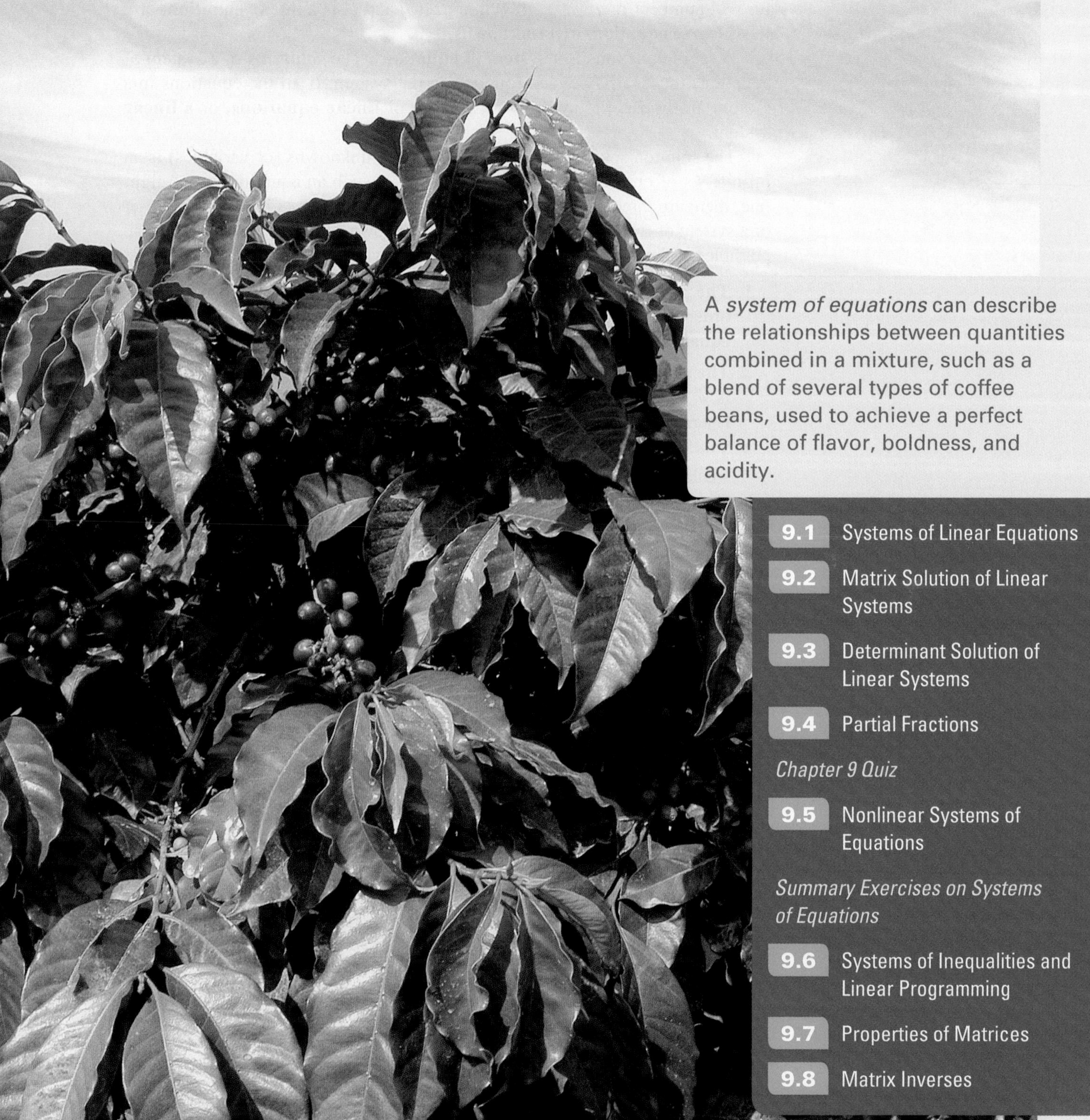

A *system of equations* can describe the relationships between quantities combined in a mixture, such as a blend of several types of coffee beans, used to achieve a perfect balance of flavor, boldness, and acidity.

9.1 Systems of Linear Equations

- Linear Systems
- Substitution Method
- Elimination Method
- Special Systems
- Applying Systems of Equations
- Solving Linear Systems with Three Unknowns (Variables)
- Using Systems of Equations to Model Data

Linear Systems The definition of a linear equation given in **Chapter 1** can be extended to more variables. Any equation of the form

$$a_1x_1 + a_2x_2 + \cdots + a_nx_n = b,$$

for real numbers $a_1, a_2, \ldots, a_n$ (all nonzero) and b, is a **linear equation,** or a **first-degree equation, in *n* unknowns.**

A set of equations is a **system of equations.** The **solutions** of a system of equations must satisfy every equation in the system. If all the equations in a system are linear, the system is a **system of linear equations,** or a **linear system.**

The solution set of a linear equation in two unknowns (or variables) is an infinite set of ordered pairs. Since the graph of such an equation is a straight line, there are three possibilities for the number of elements in the solution set of a system of two linear equations in two unknowns, as shown in **Figure 1.** The possible graphs of a linear system in two unknowns are as follows.

1. **The graphs intersect at exactly one point,** which gives the (single) ordered-pair solution of the system. The **system is consistent** and the **equations are independent.** See **Figure 1(a).**
2. **The graphs are parallel lines,** so there is no solution and the solution set is $\emptyset$. The **system is inconsistent** and the **equations are independent.** See **Figure 1(b).**
3. **The graphs are the same line,** and there are an infinite number of solutions. The **system is consistent** and the **equations are dependent.** See **Figure 1(c).**

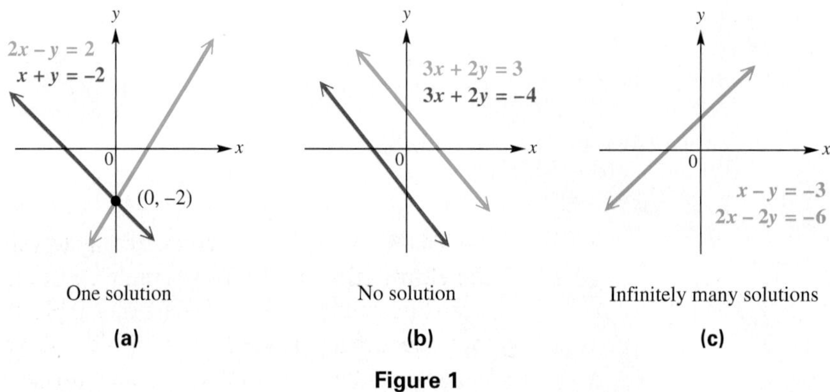

Figure 1

Using graphs to find the solution set of a linear system in two unknowns provides a good visual perspective, but may be inefficient when the solution set contains non-integer values. Thus, we introduce two algebraic methods for solving systems: *substitution* and *elimination.*

Teaching Tip Point out in **Example 1** that we chose to begin with the second equation since it can easily be solved for one of the variables.

Substitution Method In a system of two equations with two variables, the **substitution method** involves using one equation to find an expression for one variable in terms of the other, and then substituting into the other equation of the system.

Classroom Example 1
Solve the system.

$$4x - 3y = 14$$
$$x - 2y = 1$$

Answer: $\{(5, 2)\}$

EXAMPLE 1 Solving a System by Substitution

Solve the system.

$$3x + 2y = 11 \quad (1)$$
$$-x + y = 3 \quad (2)$$

SOLUTION Begin by solving one of the equations for one of the variables. We solve equation (2) for y.

$$-x + y = 3 \quad (2)$$
$$y = x + 3 \quad \text{Add } x. \text{ (Section 1.1)} \quad (3)$$

Now replace y with $x + 3$ in equation (1), and solve for x.

$$3x + 2y = 11 \quad (1)$$
$$3x + 2(x + 3) = 11 \quad \text{Let } y = x + 3 \text{ in (1).}$$
$$3x + 2x + 6 = 11 \quad \text{Distributive property (Section R.2)}$$
$$5x + 6 = 11 \quad \text{Combine like terms.}$$
$$5x = 5 \quad \text{Subtract 6.}$$
$$x = 1 \quad \text{Divide by 5.}$$

Note the careful use of parentheses.

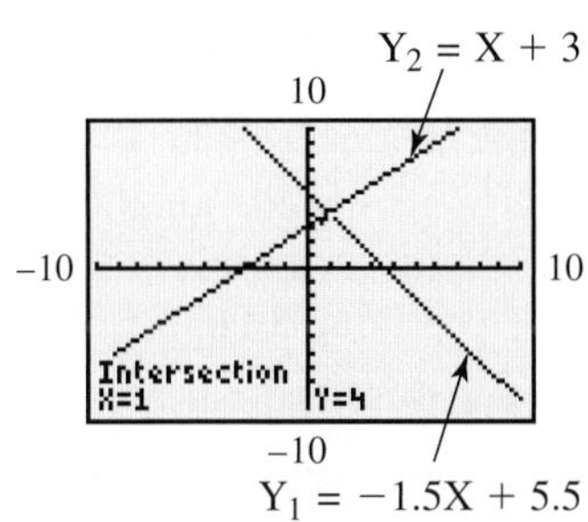

To solve the system in **Example 1** graphically, solve both equations for y:

$3x + 2y = 11$ leads to

$$Y_1 = -1.5X + 5.5.$$

$-x + y = 3$ leads to

$$Y_2 = X + 3.$$

Graph both Y_1 and Y_2 in the standard window to find that their point of intersection is (1, 4).

Replace x with 1 in equation (3) to obtain

$$y = x + 3 = 1 + 3 = 4.$$

The solution of the system is the ordered pair $(1, 4)$. ***Check this solution in both equations (1) and (2).***

CHECK

$3x + 2y = 11 \quad (1)$	$-x + y = 3 \quad (2)$
$3(1) + 2(4) \stackrel{?}{=} 11$	$-1 + 4 \stackrel{?}{=} 3$
$11 = 11$ ✓ True	$3 = 3$ ✓ True

True statements result when the solution is substituted in both equations, confirming that the solution set is $\{(1, 4)\}$.

✔ *Now Try Exercise 7.*

Elimination Method Another way to solve a system of two equations, called the **elimination method,** uses multiplication and addition to eliminate a variable from one equation. To eliminate a variable, the coefficients of that variable in the two equations must be additive inverses. To achieve this, we use properties of algebra to change the system to an **equivalent system,** one with the same solution set.

The three transformations that produce an equivalent system are listed here.

Teaching Tip Point out that the elimination method is used primarily with linear systems of equations. On the other hand, the substitution method is more general. It will be used in **Section 9.5** to find solutions when one or more of the equations are nonlinear.

Transformations of a Linear System

1. Interchange any two equations of the system.
2. Multiply or divide any equation of the system by a nonzero real number.
3. Replace any equation of the system by the sum of that equation and a multiple of another equation in the system.

Classroom Example 2
Solve the system.

$$2x + 3y = -1$$
$$3x - 2y = 18$$

Answer: $\{(4, -3)\}$

EXAMPLE 2 Solving a System by Elimination

Solve the system.

$$3x - 4y = 1 \quad (1)$$
$$2x + 3y = 12 \quad (2)$$

SOLUTION One way to eliminate a variable is to use the second transformation and multiply both sides of equation (2) by -3, to get an equivalent system.

$$3x - 4y = 1 \quad (1)$$
$$-6x - 9y = -36 \quad \text{Multiply (2) by } -3. \quad (3)$$

Now multiply both sides of equation (1) by 2, and use the third transformation to add the result to equation (3), eliminating x. Solve the result for y.

$$6x - 8y = 2 \quad \text{Multiply (1) by 2.}$$
$$\underline{-6x - 9y = -36} \quad (3)$$
$$-17y = -34 \quad \text{Add.}$$
$$y = 2 \quad \text{Solve for } y.$$

Substitute 2 for y in either of the original equations and solve for x.

$$3x - 4y = 1 \quad (1)$$
$$3x - 4(2) = 1 \quad \text{Let } y = 2 \text{ in (1).}$$
$$3x - 8 = 1 \quad \text{Multiply.}$$
$$3x = 9 \quad \text{Add 8.}$$
$$x = 3 \quad \text{Divide by 3.}$$

Write the x-value first.

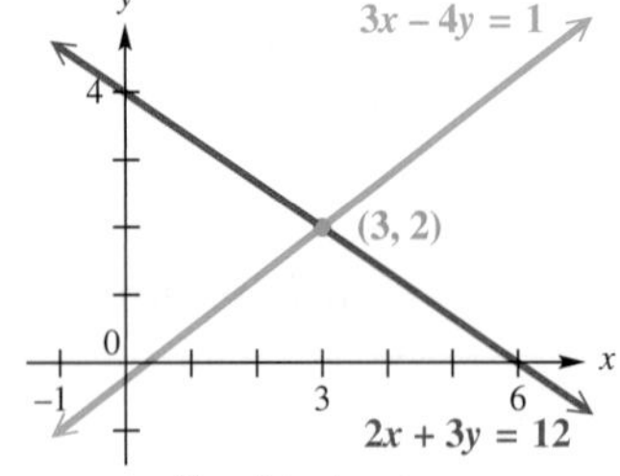

Consistent system

Figure 2

A check shows that $(3, 2)$ satisfies both equations (1) and (2). Therefore, the solution set is $\{(3, 2)\}$. The graph in **Figure 2** confirms this.

✔ *Now Try Exercise 21.*

Special Systems The systems in **Examples 1 and 2** were both consistent, having a single solution. This is not always the case.

Classroom Example 3
Solve the system.

$$7x + 3y = -5$$
$$-14x - 6y = -10$$

Answer: $\emptyset$

EXAMPLE 3 Solving an Inconsistent System

Solve the system.

$$3x - 2y = 4 \quad (1)$$
$$-6x + 4y = 7 \quad (2)$$

SOLUTION To eliminate the variable x, multiply both sides of equation (1) by 2, and add the result to equation (2).

$$6x - 4y = 8 \quad \text{Multiply (1) by 2.}$$
$$\underline{-6x + 4y = 7} \quad (2)$$
$$0 = 15 \quad \text{False}$$

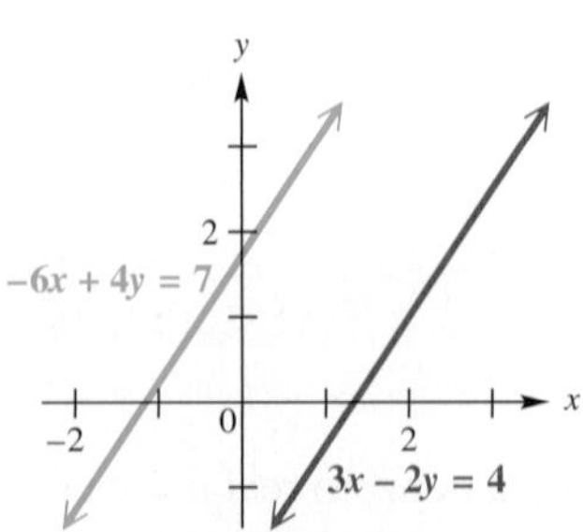

Inconsistent system

Figure 3

Since $0 = 15$ is false, the system is inconsistent and has no solution. As suggested by **Figure 3,** this means that the graphs of the equations of the system never intersect. (The lines are parallel.) The solution set is $\emptyset$.

✔ *Now Try Exercise 31.*

Classroom Example 4
Solve the system.

$$-9x + 3y = -24$$
$$3x - y = 8$$

Answer: $\{(x, 3x - 8)\}$

EXAMPLE 4 Solving a System with Infinitely Many Solutions

Solve the system.

$$8x - 2y = -4 \quad (1)$$
$$-4x + y = 2 \quad (2)$$

ALGEBRAIC SOLUTION

Divide both sides of equation (1) by 2, and add the result to equation (2).

$$4x - y = -2 \quad \text{Divide (1) by 2.}$$
$$-4x + y = 2 \quad (2)$$
$$0 = 0 \quad \text{True}$$

The result, $0 = 0$, is a true statement, which indicates that the equations are equivalent. Any ordered pair (x, y) that satisfies either equation will satisfy the system. Solve for y in equation (2).

$$-4x + y = 2 \quad (2)$$
$$y = 4x + 2$$

The solutions of the system can be written in the form of a set of ordered pairs $(x, 4x + 2)$, for any real number x. Some ordered pairs in the solution set are $(0, 4 \cdot 0 + 2)$, or $(0, 2)$, and $(1, 4 \cdot 1 + 2)$, or $(1, 6)$, as well as $(3, 14)$, and $(-2, -6)$.

As shown in **Figure 4**, the equations of the original system are dependent and lead to the same straight-line graph. Using this method, the solution set can be written $\{(x, 4x + 2)\}$.

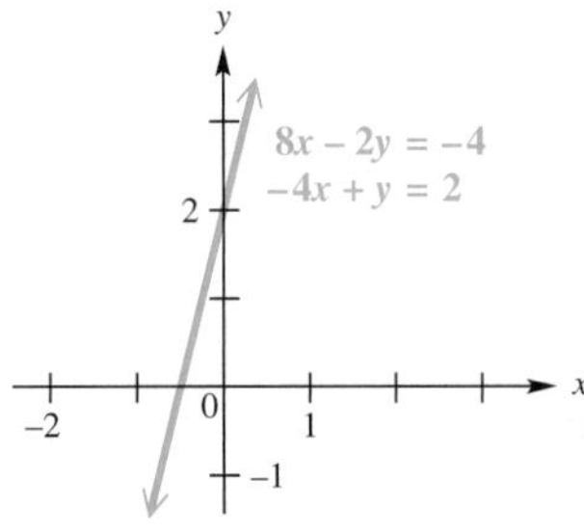

Infinitely many solutions

Figure 4

GRAPHING CALCULATOR SOLUTION

Solving the equations for y gives

$$Y_1 = 4X + 2$$

and

$$Y_2 = 4X + 2.$$

When written in this form, we can immediately determine that the equations are identical. Each has slope 4 and y-intercept 2.

As expected, the graphs coincide. See the top screen in **Figure 5.** The table indicates that

$$Y_1 = Y_2 \text{ for selected values of X,}$$

providing another way to show that the two equations lead to the same graph.

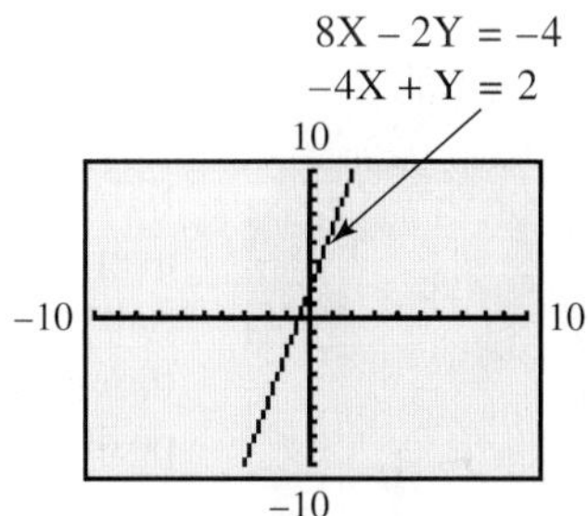

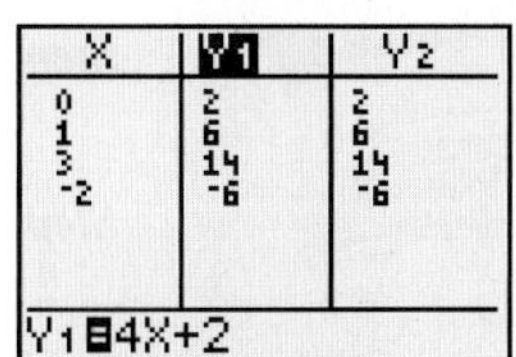

X	Y1	Y2
0	2	2
1	6	6
3	14	14
-2	-6	-6

Y1 = 4X+2

Figure 5

Refer to the algebraic solution to see how the solution set can be written using an arbitrary variable.

Now Try Exercise 33.

NOTE In the algebraic solution for **Example 4,** we wrote the solution set with the variable x arbitrary. We could write the solution set with y arbitrary.

$$\left\{\left(\frac{y-2}{4}, y\right)\right\}$$

By selecting values for y and solving for x in this ordered pair, we can find individual solutions. Verify again that $(0, 2)$ is a solution by letting $y = 2$ and solving for x to obtain $\frac{2-2}{4} = 0$.

Teaching Tip Point out in **Examples 3 and 4** that the graphs of both equations have the same slope. This is the tipoff that the system has either no solution or infinitely many solutions. When both equations in **Example 4** are written in the form $y = mx + b$, the two equations are identical.

Applying Systems of Equations Many applied problems involve more than one unknown quantity. Although some problems with two unknowns can be solved using just one variable, it is often easier to use two variables.

To solve a problem with two unknowns, we must write two equations that relate the unknown quantities. The system formed by the pair of equations can then be solved using the methods of this chapter. The following steps, based on the six-step problem-solving method first introduced in **Section 1.2,** give a strategy for solving such applied problems.

Solving an Applied Problem by Writing a System of Equations

Step 1 **Read** the problem carefully until you understand what is given and what is to be found.

Step 2 **Assign variables** to represent the unknown values, using diagrams or tables as needed. *Write down* what each variable represents.

Step 3 **Write a system of equations** that relates the unknowns.

Step 4 **Solve** the system of equations.

Step 5 **State the answer** to the problem. Does it seem reasonable?

Step 6 **Check** the answer in the words of the original problem.

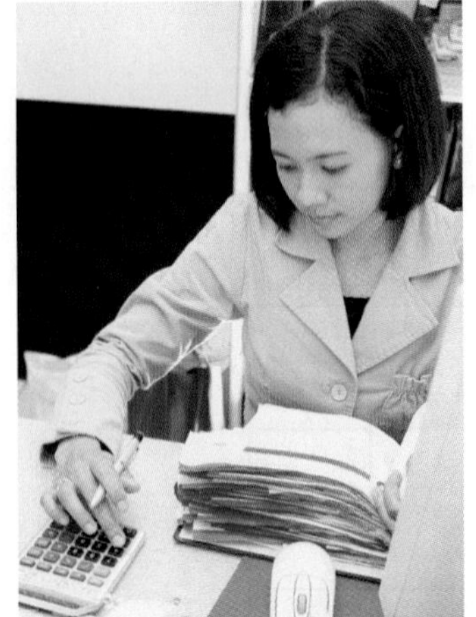

Classroom Example 5
In 2010, the average of the salaries for the position of Environmental Compliance Specialist in Boston, Massachusetts, and Indianapolis, Indiana, was \$67,703.50. The salary in Boston exceeded the salary in Indianapolis by \$9457. Determine the salary for the Environmental Compliance Specialist position in Boston and in Indianapolis.
(*Source:* www.salary.com)

Answer:
Boston: \$72,432;
Indianapolis: \$62,975

EXAMPLE 5 Using a Linear System to Solve an Application

Salaries for the same position can vary depending on the location. In 2010, the average of the salaries for the position of Accountant I in San Diego, California, and Salt Lake City, Utah, was \$45,091.50. The salary in San Diego, however, exceeded the salary in Salt Lake City by \$5231. Determine the salary for the Accountant I position in San Diego and in Salt Lake City. (*Source:* www.salary.com)

SOLUTION

Step 1 **Read** the problem. We must find the salary of the Accountant I position in San Diego and in Salt Lake City.

Step 2 **Assign variables.** Let x represent the salary of the Accountant I position in San Diego and y represent the salary for the same position in Salt Lake City.

Step 3 **Write a system of equations.** Since the average of the salaries for the Accountant I position in San Diego and Salt Lake City was \$45,091.50, one equation is as follows.

$$\frac{x+y}{2} = 45{,}091.50$$

Multiply both sides of this equation by 2 to clear the fraction and get an equivalent equation.

$$x + y = 90{,}183 \quad (1)$$

The salary in San Diego exceeded the salary in Salt Lake City by \$5231. Thus, $x - y = 5231$, which gives the following system of equations.

$$x + y = 90{,}183 \quad (1)$$

$$x - y = 5231 \quad (2)$$

Step 4 **Solve** the system. To eliminate y, add the two equations.

$$\begin{aligned} x + y &= 90{,}183 && (1) \\ x - y &= 5231 && (2) \\ \hline 2x \quad &= 95{,}414 && \text{Add.} \\ x &= 47{,}707 && \text{Solve for } x. \end{aligned}$$

To find y, substitute 47,707 for x in equation (2).

$$\begin{aligned} x - y &= 5231 && (2) \\ 47{,}707 - y &= 5231 && \text{Let } x = 47{,}707. \\ -y &= -42{,}476 && \text{Subtract } 47{,}707. \\ y &= 42{,}476 && \text{Multiply by } -1. \end{aligned}$$

Step 5 **State the answer.** The salary for the position of Accountant I was \$47,707 in San Diego and \$42,476 in Salt Lake City.

Step 6 **Check.** The average of \$47,707 and \$42,476 is

$$\frac{\$47{,}707 + \$42{,}476}{2} = \$45{,}091.50.$$

Also,

$$\$47{,}707 - \$42{,}476 = \$5231, \quad \text{as required.}$$

Now Try Exercise 95.

Teaching Tip Use the walls in the classroom to give students a visual image of how three planes can intersect. Show students, for example, that the intersection of the front wall, left wall, and ceiling represents the solution of a consistent system.

Solving Linear Systems with Three Unknowns (Variables) Earlier, we saw that the graph of a linear equation in two unknowns is a straight line. The graph of a linear equation in three unknowns requires a three-dimensional coordinate system. The three number lines are placed at right angles. The graph of a linear equation in three unknowns is a plane. Some possible intersections of planes representing three equations in three variables are shown in **Figure 6.**

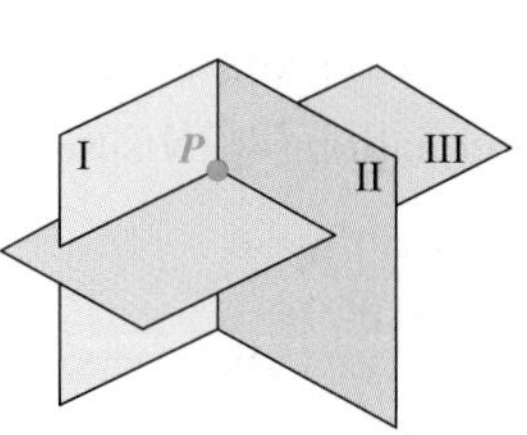

A single solution

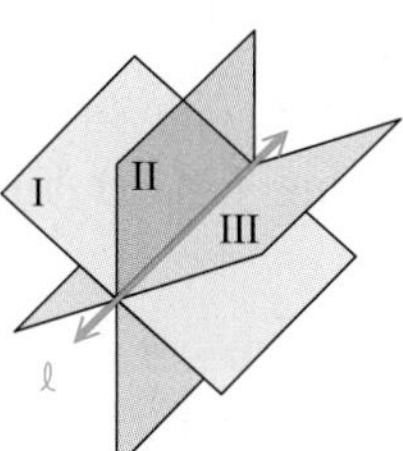

Points of a line in common

All points in common

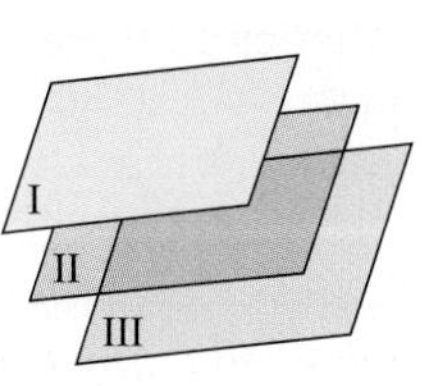

No points in common

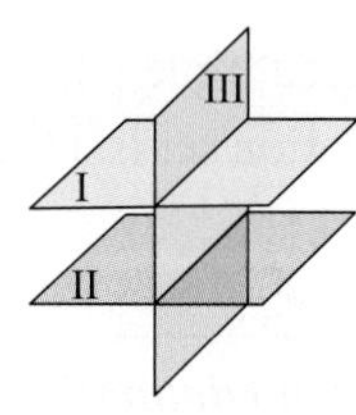

No points in common

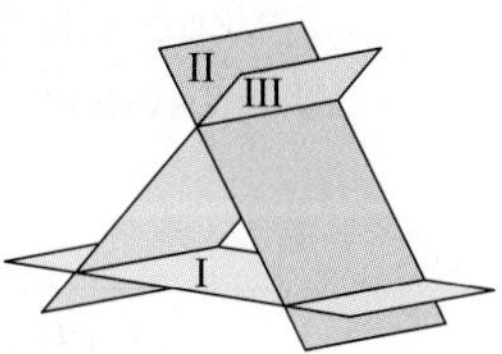

No points in common

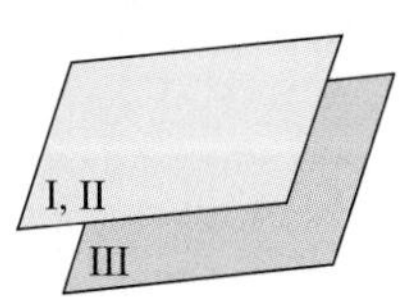

No points in common

Figure 6

Solve a linear system with three unknowns as follows.

Step 1 Eliminate a variable from any two of the equations.

Step 2 Eliminate the ***same variable*** from a different pair of equations.

Step 3 Eliminate a second variable using the resulting two equations in two variables to get an equation with just one variable whose value we can now determine.

Step 4 Find the values of the remaining variables by substitution. Write the solution of the system as an **ordered triple.**

Classroom Example 6
Solve the system.

$$\begin{aligned} 3x + 4y - 2z &= 14 \\ 2x + y + 2z &= -9 \\ x - y + z &= -9 \end{aligned}$$

Answer: $\{(-2, 3, -4)\}$

EXAMPLE 6 Solving a System of Three Equations with Three Variables

Solve the system.

$$\begin{aligned} 3x + 9y + 6z &= 3 \quad (1) \\ 2x + y - z &= 2 \quad (2) \\ x + y + z &= 2 \quad (3) \end{aligned}$$

SOLUTION Eliminate z by adding equations (2) and (3). (Step 1)

$$3x + 2y = 4 \quad (4)$$

To eliminate z from another pair of equations, multiply both sides of equation (2) by 6 and add the result to equation (1). (Step 2)

$$\begin{aligned} 12x + 6y - 6z &= 12 \quad \text{Multiply (2) by 6.} \\ 3x + 9y + 6z &= 3 \quad (1) \\ \hline 15x + 15y \quad\quad &= 15 \quad (5) \end{aligned}$$

Make sure equation (5) has the *same* two variables as equation (4).

To eliminate x from equations (4) and (5), multiply both sides of equation (4) by -5 and add the result to equation (5). Solve the resulting equation for y. (Step 3)

$$\begin{aligned} -15x - 10y &= -20 \quad \text{Multiply (4) by } -5. \\ 15x + 15y &= 15 \quad (5) \\ \hline 5y &= -5 \quad \text{Add.} \\ y &= -1 \quad \text{Divide by 5.} \end{aligned}$$

Using $y = -1$, find x from equation (4) by substitution. (Step 4)

$$\begin{aligned} 3x + 2(-1) &= 4 \quad \text{(4) with } y = -1 \\ x &= 2 \quad \text{Solve for } x. \end{aligned}$$

Substitute 2 for x and -1 for y in equation (3) to find z.

$$\begin{aligned} 2 + (-1) + z &= 2 \quad \text{(3) with } x = 2, y = -1 \\ z &= 1 \quad \text{Solve for } z. \end{aligned}$$

Write the values of *x*, *y*, and *z* in the correct order.

Verify that the ordered triple $(2, -1, 1)$ satisfies all three equations in the *original* system. The solution set is $\{(2, -1, 1)\}$.

✔ *Now Try Exercise 47.*

CAUTION ***When eliminating a variable from any two equations to create equations like (4) and (5) in Example 6, be sure to eliminate the same variable from both equations.*** Otherwise, the result will include two equations that still have three variables, and no progress will have been made.

Classroom Example 7
Solve the system. Write the solution set with z arbitrary.

$$3x + y - 2z = -7$$
$$5x + 2y + z = -6$$

Answer:
$\{(5z - 8, -13z + 17, z)\}$

EXAMPLE 7 Solving a System of Two Equations with Three Variables

Solve the system.

$$x + 2y + z = 4 \quad (1)$$
$$3x - y - 4z = -9 \quad (2)$$

SOLUTION Geometrically, the solution is the intersection of the two planes given by equations (1) and (2). The intersection of two different nonparallel planes is a line. Thus there will be an infinite number of ordered triples in the solution set, representing the points on the line of intersection.

To eliminate x, multiply both sides of equation (1) by -3 and add the result to equation (2). (Either y or z could have been eliminated instead.)

$$-3x - 6y - 3z = -12 \quad \text{Multiply (1) by } -3.$$
$$3x - y - 4z = -9 \quad (2)$$
$$-7y - 7z = -21 \quad (3)$$
$$-7z = 7y - 21 \quad \text{Add } 7y.$$
$$z = -y + 3 \quad \text{Divide } each \text{ term by } -7.$$

Solve this equation for z.

This gives z in terms of y. Express x also in terms of y by solving equation (1) for x and substituting $-y + 3$ for z in the result.

$$x + 2y + z = 4 \quad (1)$$
$$x = -2y - z + 4 \quad \text{Solve for } x.$$
$$x = -2y - (-y + 3) + 4 \quad \text{Substitute } (-y + 3) \text{ for } z.$$
$$x = -y + 1 \quad \text{Simplify.}$$

Use parentheses around $-y + 3$.

The system has an infinite number of solutions. For any value of y, the value of z is $-y + 3$ and the value of x is $-y + 1$. For example, if $y = 1$, then $x = -1 + 1 = 0$ and $z = -1 + 3 = 2$, giving the solution $(0, 1, 2)$. Verify that another solution is $(-1, 2, 1)$.

With y arbitrary, the solution set is of the form $\{(-y + 1, y, -y + 3)\}$.

✔ *Now Try Exercise 59.*

NOTE Had we solved equation (3) in **Example 7** for y instead of z, the solution would have had a different form but would have led to the same set of solutions. In that case we would have z arbitrary, and the solution set would be of the form

$$\{(z - 2, -z + 3, z)\}.$$

By choosing $z = 2$, one solution would be $(0, 1, 2)$, which was found above.

Using Systems of Equations to Model Data Applications with three unknowns usually require solving a system of three equations. If we know three points on the graph, we can find the equation of a parabola in the form

$$y = ax^2 + bx + c \quad \text{(Section 3.1)}$$

by solving a system of three equations with three variables.

Classroom Example 8
Find the equation of the parabola $y = ax^2 + bx + c$ that passes through the points $(-5, 7)$, $(-1, -2)$, and $(3, 5)$.

Answer:

$y = \frac{1}{2}x^2 + \frac{3}{4}x - \frac{7}{4}$, or
$y = 0.5x^2 + 0.75x - 1.75$

EXAMPLE 8 Using Modeling to Find an Equation through Three Points

Find the equation of the parabola $y = ax^2 + bx + c$ that passes through the points $(2, 4)$, $(-1, 1)$, and $(-2, 5)$.

SOLUTION Since the three points lie on the graph of the given equation $y = ax^2 + bx + c$, they must satisfy the equation. Substituting each ordered pair into the equation gives three equations with three unknowns.

$$4 = a(2)^2 + b(2) + c, \quad \text{or} \quad 4 = 4a + 2b + c \quad (1)$$
$$1 = a(-1)^2 + b(-1) + c, \quad \text{or} \quad 1 = a - b + c \quad (2)$$
$$5 = a(-2)^2 + b(-2) + c, \quad \text{or} \quad 5 = 4a - 2b + c \quad (3)$$

To solve this system, first eliminate c using equations (1) and (2).

$$4 = 4a + 2b + c \quad (1)$$
$$-1 = -a + b - c \quad \text{Multiply (2) by } -1.$$
$$3 = 3a + 3b \quad (4)$$

Now, use equations (2) and (3) to eliminate the same unknown, c.

$$1 = a - b + c \quad (2)$$
$$-5 = -4a + 2b - c \quad \text{Multiply (3) by } -1.$$
$$-4 = -3a + b \quad (5)$$

Equation (5) must have the *same* two unknowns as equation (4).

Solve the system of equations (4) and (5) in two unknowns by eliminating a.

$$3 = 3a + 3b \quad (4)$$
$$-4 = -3a + b \quad (5)$$
$$-1 = 4b \quad \text{Add.}$$
$$-\frac{1}{4} = b \quad \text{Divide by 4.}$$

Find a by substituting $-\frac{1}{4}$ for b in equation (4).

$$1 = a + b \quad \text{Equation (4) divided by 3}$$
$$1 = a - \frac{1}{4} \quad \text{Let } b = -\tfrac{1}{4}.$$
$$\frac{5}{4} = a \quad \text{Add } \tfrac{1}{4}.$$

Finally, find c by substituting $a = \frac{5}{4}$ and $b = -\frac{1}{4}$ in equation (2).

$$1 = a - b + c \quad (2)$$
$$1 = \frac{5}{4} - \left(-\frac{1}{4}\right) + c \quad \text{Let } a = \tfrac{5}{4},\ b = -\tfrac{1}{4}.$$
$$1 = \frac{6}{4} + c \quad \text{Add.}$$
$$-\frac{1}{2} = c \quad \text{Subtract } \tfrac{6}{4}.$$

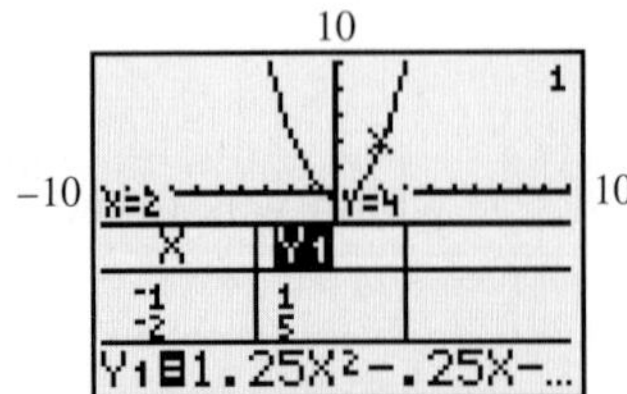

This graph/table screen shows that the points (2, 4), (–1, 1), and (–2, 5) lie on the graph of $Y_1 = 1.25X^2 - 0.25X - 0.5$. This supports the result of **Example 8.**

The required equation is $y = \frac{5}{4}x^2 - \frac{1}{4}x - \frac{1}{2}$, or $y = 1.25x^2 - 0.25x - 0.5$.

Now Try Exercise 79.

Classroom Example 9
Refer to the table in **Example 9**. How many units of each ingredient should be used to make a feed that contains 35 units of protein, 38 units of fat, and 28 units of fiber?

Answer:
60 units of corn;
40 units of soybeans;
20 units of cottonseed

EXAMPLE 9 Solving an Application Using a System of Three Equations

An animal feed is made from three ingredients: corn, soybeans, and cottonseed. One unit of each ingredient provides units of protein, fat, and fiber as shown in the table. How many units of each ingredient should be used to make a feed that contains 22 units of protein, 28 units of fat, and 18 units of fiber?

	Corn	Soybeans	Cottonseed	Total
Protein	0.25	0.4	0.2	22
Fat	0.4	0.2	0.3	28
Fiber	0.3	0.2	0.1	18

SOLUTION

Step 1 **Read** the problem. We must determine the number of units of corn, soybeans, and cottonseed.

Step 2 **Assign variables.** Let x represent the number of units of corn, y the number of units of soybeans, and z the number of units of cottonseed.

Step 3 **Write a system of equations.** The total amount of protein is to be 22 units, so we use the first row of the table to write equation (1).

$$0.25x + 0.4y + 0.2z = 22 \quad (1)$$

We use the second row of the table to obtain 28 units of fat.

$$0.4x + 0.2y + 0.3z = 28 \quad (2)$$

Finally, we use the third row of the table to obtain 18 units of fiber.

$$0.3x + 0.2y + 0.1z = 18 \quad (3)$$

Teaching Tip Point out that it is usually a good idea to eliminate decimal points by multiplying by a power of 10, as done in **Example 9.** Similarly, an equation with fractions can be simplified by multiplying the equation by the least common multiple of the denominators.

Multiply equation (1) on both sides by 100, and equations (2) and (3) by 10, to get an equivalent system.

$$\begin{aligned} 25x + 40y + 20z &= 2200 \quad (4) \\ 4x + 2y + 3z &= 280 \quad (5) \\ 3x + 2y + z &= 180 \quad (6) \end{aligned}$$

Eliminate the decimal points in equations (1), (2), and (3) by multiplying each equation by an appropriate power of 10.

Step 4 **Solve** the system. Using the methods described earlier in this section, we find that

$$x = 40, \quad y = 15, \quad \text{and} \quad z = 30.$$

Step 5 **State the answer.** The feed should contain 40 units of corn, 15 units of soybeans, and 30 units of cottonseed.

Step 6 **Check.** Show that the ordered triple $(40, 15, 30)$ satisfies the system formed by equations (1), (2), and (3).

Now Try Exercise 101.

NOTE Notice how the table in **Example 9** is used to set up the equations of the system. The coefficients in each equation are read from left to right. This idea is extended in the next section, where we introduce the solution of systems by matrices.

9.1 Exercises

1. 2006–2009
2. about 1.26 million
3. approximately (2005.2, 1.26)
4. The population of the New Orleans metropolitan area was constant from 2004 to 2005, decreased from 2005 to 2006, and increased from 2006 to 2009.
5. year; population (in millions)

7. $\{(-4, 1)\}$ 8. $\{(8, -5)\}$
9. $\{(48, 8)\}$ 10. $\{(-1, -11)\}$
11. $\{(1, 3)\}$ 12. $\{(1, 3)\}$
13. $\{(-1, 3)\}$ 14. $\{(-2, 3)\}$
15. $\{(3, -4)\}$ 16. $\{(5, 0)\}$
17. $\{(0, 2)\}$ 18. $\{(6, 2)\}$

19. $\{(0, 4)\}$ 20. $\{(-7, 5)\}$
21. $\{(1, 3)\}$ 22. $\{(-1, 1)\}$
23. $\{(4, -2)\}$ 24. $\{(2, 3)\}$
25. $\{(2, -2)\}$ 26. $\{(2, -3)\}$
27. $\{(4, 6)\}$ 28. $\{(-2, 2)\}$
29. $\{(5, 2)\}$ 30. $\{(66, -34)\}$

Changes in Population Many factors may contribute to population changes in metropolitan areas. The graph shows the populations of the New Orleans, Louisiana, and the Jacksonville, Florida, metropolitan areas over the years 2004–2009.

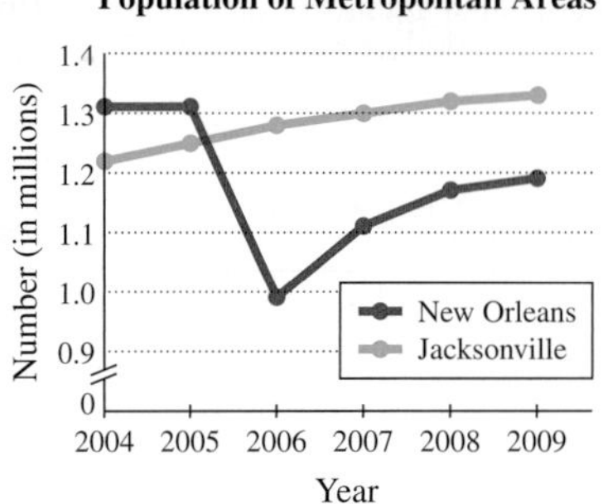

Source: U.S. Census Bureau.

1. In what years was the population of the Jacksonville metropolitan area greater than that of the New Orleans metropolitan area?
2. At the time when the populations of the two metropolitan areas were equal, what was the approximate population of each area?
3. Express the solution of the system as an ordered pair.
4. Use the terms *increasing, decreasing,* and *constant* to describe the trends for the population of the New Orleans metropolitan area.
5. If equations of the form $y = f(t)$ were determined that modeled either of the two graphs, then the variable t would represent ______________ and the variable y would represent ______________.
6. Explain why each graph is that of a function.

Solve each system by substitution. ***See Example 1.***

7. $4x + 3y = -13$, $-x + y = 5$
8. $3x + 4y = 4$, $x - y = 13$
9. $x - 5y = 8$, $x = 6y$
10. $6x - y = 5$, $y = 11x$
11. $8x - 10y = -22$, $3x + y = 6$
12. $4x - 5y = -11$, $2x + y = 5$
13. $7x - y = -10$, $3y - x = 10$
14. $4x + 5y = 7$, $9y = 31 + 2x$
15. $-2x = 6y + 18$, $-29 = 5y - 3x$
16. $3x - 7y = 15$, $3x + 7y = 15$
17. $3y = 5x + 6$, $x + y = 2$
18. $4y = 2x - 4$, $x - y = 4$

Solve each system by elimination. In Exercises 27–30, first clear denominators. ***See Example 2.***

19. $3x - y = -4$, $x + 3y = 12$
20. $4x + y = -23$, $x - 2y = -17$
21. $2x - 3y = -7$, $5x + 4y = 17$
22. $4x + 3y = -1$, $2x + 5y = 3$
23. $5x + 7y = 6$, $10x - 3y = 46$
24. $12x - 5y = 9$, $3x - 8y = -18$
25. $6x + 7y + 2 = 0$, $7x - 6y - 26 = 0$
26. $5x + 4y + 2 = 0$, $4x - 5y - 23 = 0$
27. $\frac{x}{2} + \frac{y}{3} = 4$, $\frac{3x}{2} + \frac{3y}{2} = 15$
28. $\frac{3x}{2} + \frac{y}{2} = -2$, $\frac{x}{2} + \frac{y}{2} = 0$
29. $\frac{2x-1}{3} + \frac{y+2}{4} = 4$, $\frac{x+3}{2} - \frac{x-y}{3} = 3$
30. $\frac{x+6}{5} + \frac{2y-x}{10} = 1$, $\frac{x+2}{4} + \frac{3y+2}{5} = -3$

31. $\emptyset$; inconsistent system
32. $\emptyset$; inconsistent system
33. $\left\{\left(\frac{y+9}{4}, y\right)\right\}$; infinitely many solutions
34. $\left\{\left(\frac{-5y-2}{3}, y\right)\right\}$; infinitely many solutions
35. $\emptyset$; inconsistent system
36. $\emptyset$; inconsistent system
37. $\left\{\left(\frac{-2y+6}{7}, y\right)\right\}$; infinitely many solutions
38. $\{(4y+2, y)\}$; infinitely many solutions
39. $\emptyset$; inconsistent system
40. A
41. $x - 3y = -3$
$3x + 2y = 6$
42. $x + y = 3$
$x - y = 2$
43. $\{(0.820, -2.508)\}$
44. $\{(0.138, -4.762)\}$
45. $\{(0.892, 0.453)\}$
46. $\{(0.236, 0.674)\}$
47. $\{(1, 2, -1)\}$ 48. $\{(2, 1, 4)\}$
49. $\{(2, 0, 3)\}$

Solve each system. State whether it is inconsistent or has infinitely many solutions. If the system has infinitely many solutions, write the solution set with y arbitrary. ***See Examples 3 and 4.***

31. $9x - 5y = 1$
$-18x + 10y = 1$

32. $3x + 2y = 5$
$6x + 4y = 8$

33. $4x - y = 9$
$-8x + 2y = -18$

34. $3x + 5y = -2$
$9x + 15y = -6$

35. $5x - 5y - 3 = 0$
$x - y - 12 = 0$

36. $2x - 3y - 7 = 0$
$-4x + 6y - 14 = 0$

37. $7x + 2y = 6$
$14x + 4y = 12$

38. $2x - 8y = 4$
$x - 4y = 2$

39. $2x - 6y = 0$
$-7x + 21y = 10$

40. *Concept Check* Which screen gives the correct graphical solution of the system? (*Hint:* Solve for y first in each equation and use the slope-intercept forms to help you answer the question.)

$$4x - 5y = -11$$
$$2x + y = 5$$

A.
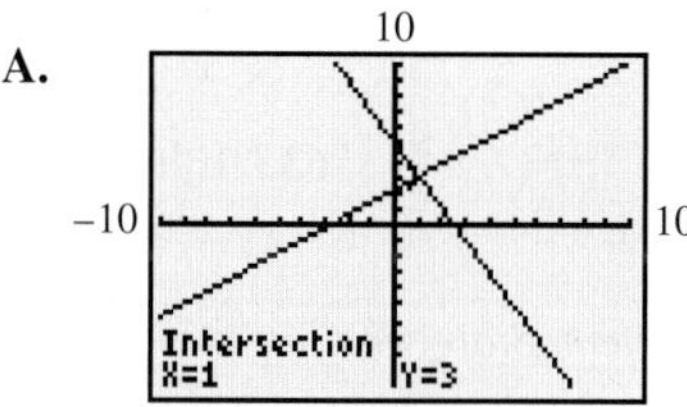

B.
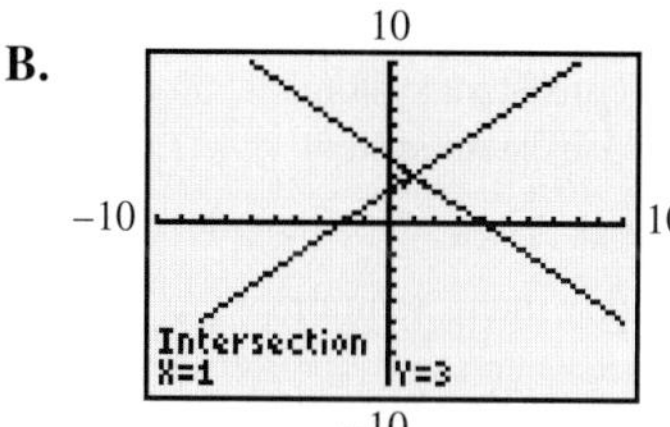

C.
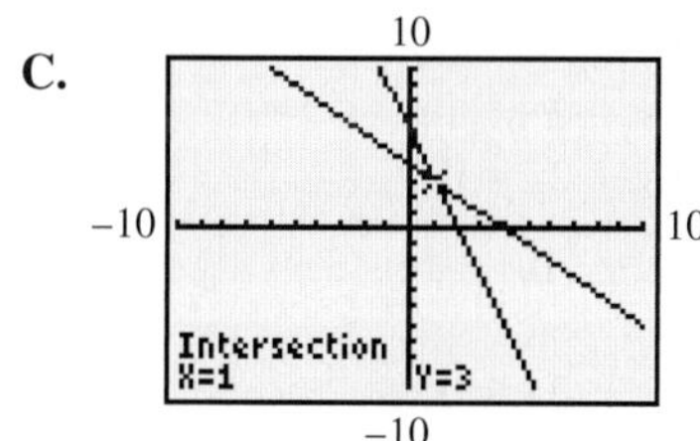

D.
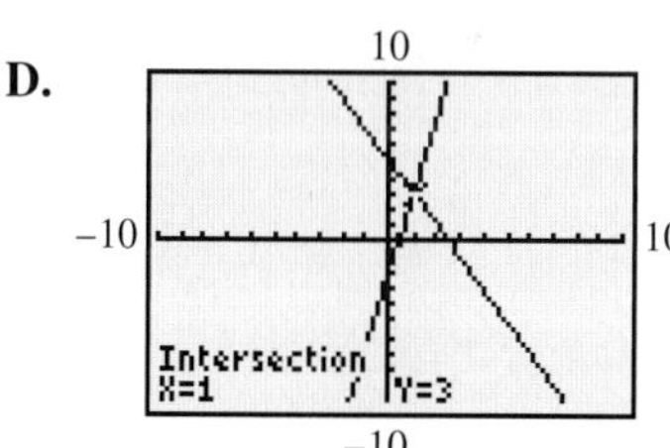

Connecting Graphs with Equations *Determine the system of equations illustrated in each graph. Write equations in standard form.*

41.

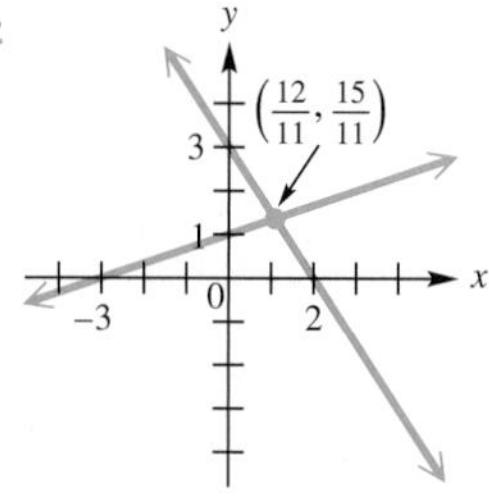

42.

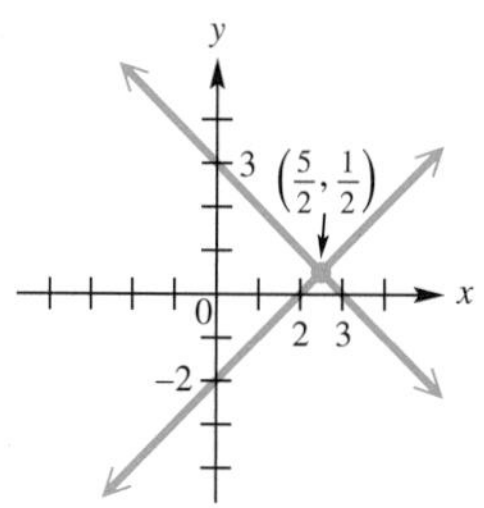

Use a graphing calculator to solve each system. Express solutions with approximations to the nearest thousandth.

43. $\frac{11}{3}x + y = 0.5$
$0.6x - y = 3$

44. $\sqrt{3}x - y = 5$
$100x + y = 9$

45. $\sqrt{7}x + \sqrt{2}y - 3 = 0$
$\sqrt{6}x - y - \sqrt{3} = 0$

46. $0.2x + \sqrt{2}y = 1$
$\sqrt{5}x + 0.7y = 1$

Solve each system. ***See Example 6.***

47. $x + y + z = 2$
$2x + y - z = 5$
$x - y + z = -2$

48. $2x + y + z = 9$
$-x - y + z = 1$
$3x - y + z = 9$

49. $x + 3y + 4z = 14$
$2x - 3y + 2z = 10$
$3x - y + z = 9$

50. $\{(-1, 4, 2)\}$
51. $\{(1, 2, 3)\}$ 52. $\{(2, 0, 1)\}$
53. $\{(4, 1, 2)\}$ 54. $\{(2, -3, 4)\}$
55. $\{(\frac{1}{2}, \frac{2}{3}, -1)\}$
56. $\{(-\frac{3}{4}, \frac{5}{3}, \frac{3}{2})\}$
57. $\{(-3, 1, 6)\}$ 58. $\{(-1, 1, 2)\}$
59. $\{(x, -4x + 3, -3x + 4)\}$
60. $\{(x, -2x + 13, -7x + 41)\}$
61. $\{(x, \frac{5x+6}{5}, \frac{-5x+69}{5})\}$
62. $\{(x, \frac{x-15}{3}, \frac{x+24}{3})\}$
63. $\{(x, \frac{-7x+41}{9}, \frac{-x+47}{9})\}$
64. $\{(x, \frac{-3x+13}{2}, \frac{-5x+1}{2})\}$
65. $\emptyset$; inconsistent system
66. $\emptyset$; inconsistent system
67. $\{(-\frac{z}{9}, \frac{5z}{9}, z)\}$; infinitely many solutions
68. $\{(\frac{2z}{3}, \frac{5z}{3}, z)\}$; infinitely many solutions
69. $\{(2, 2)\}$ 70. $\{(5, 1)\}$
71. $\{(\frac{1}{5}, 1)\}$ 72. $\{(\frac{1}{3}, \frac{1}{4})\}$
73. $\{(4, 6, 1)\}$ 74. $\{(5, 1, -3)\}$
75. $k \neq -6$; $k = -6$
76. Other answers are possible.
(a) One example is
$x + 2y + z = 5$
$2x - y + 3z = 4.$
(b) One example is
$x + y + z = 5$
$2x - y + 3z = 4.$
(c) One example is
$2x + 2y + 2z = 8$
$2x - y + 3z = 4.$
77. $y = -3x - 5$

50. $\begin{aligned} 4x - y + 3z &= -2 \\ 3x + 5y - z &= 15 \\ -2x + y + 4z &= 14 \end{aligned}$

51. $\begin{aligned} x + 4y - z &= 6 \\ 2x - y + z &= 3 \\ 3x + 2y + 3z &= 16 \end{aligned}$

52. $\begin{aligned} 4x - 3y + z &= 9 \\ 3x + 2y - 2z &= 4 \\ x - y + 3z &= 5 \end{aligned}$

53. $\begin{aligned} x - 3y - 2z &= -3 \\ 3x + 2y - z &= 12 \\ -x - y + 4z &= 3 \end{aligned}$

54. $\begin{aligned} x + y + z &= 3 \\ 3x - 3y - 4z &= -1 \\ x + y + 3z &= 11 \end{aligned}$

55. $\begin{aligned} 2x + 6y - z &= 6 \\ 4x - 3y + 5z &= -5 \\ 6x + 9y - 2z &= 11 \end{aligned}$

56. $\begin{aligned} 8x - 3y + 6z &= -2 \\ 4x + 9y + 4z &= 18 \\ 12x - 3y + 8z &= -2 \end{aligned}$

57. $\begin{aligned} 2x - 3y + 2z - 3 &= 0 \\ 4x + 8y + z - 2 &= 0 \\ -x - 7y + 3z - 14 &= 0 \end{aligned}$

58. $\begin{aligned} -x + 2y - z - 1 &= 0 \\ -x - y - z + 2 &= 0 \\ x - y + 2z - 2 &= 0 \end{aligned}$

Solve each system in terms of the arbitrary variable x. ***See Example 7.***

59. $\begin{aligned} x - 2y + 3z &= 6 \\ 2x - y + 2z &= 5 \end{aligned}$

60. $\begin{aligned} 3x - 2y + z &= 15 \\ x + 4y - z &= 11 \end{aligned}$

61. $\begin{aligned} 5x - 4y + z &= 9 \\ y + z &= 15 \end{aligned}$

62. $\begin{aligned} 3x - 5y - 4z &= -7 \\ y - z &= -13 \end{aligned}$

63. $\begin{aligned} 3x + 4y - z &= 13 \\ x + y + 2z &= 15 \end{aligned}$

64. $\begin{aligned} x - y + z &= -6 \\ 4x + y + z &= 7 \end{aligned}$

Solve each system. State whether it is inconsistent or has infinitely many solutions. If the system has infinitely many solutions, write the solution set with z arbitrary. ***See Examples 3, 4, 6, and 7.***

65. $\begin{aligned} 3x + 5y - z &= -2 \\ 4x - y + 2z &= 1 \\ -6x - 10y + 2z &= 0 \end{aligned}$

66. $\begin{aligned} 3x + y + 3z &= 1 \\ x + 2y - z &= 2 \\ 2x - y + 4z &= 4 \end{aligned}$

67. $\begin{aligned} 5x - 4y + z &= 0 \\ x + y &= 0 \\ -10x + 8y - 2z &= 0 \end{aligned}$

68. $\begin{aligned} 2x + y - 3z &= 0 \\ 4x + 2y - 6z &= 0 \\ x - y + z &= 0 \end{aligned}$

Solve each system. (Hint: In Exercises 69–72, let $\frac{1}{x} = t$ and $\frac{1}{y} = u$.)

69. $\begin{aligned} \frac{2}{x} + \frac{1}{y} &= \frac{3}{2} \\ \frac{3}{x} - \frac{1}{y} &= 1 \end{aligned}$

70. $\begin{aligned} \frac{1}{x} + \frac{3}{y} &= \frac{16}{5} \\ \frac{5}{x} + \frac{4}{y} &= 5 \end{aligned}$

71. $\begin{aligned} \frac{2}{x} + \frac{1}{y} &= 11 \\ \frac{3}{x} - \frac{5}{y} &= 10 \end{aligned}$

72. $\begin{aligned} \frac{2}{x} + \frac{3}{y} &= 18 \\ \frac{4}{x} - \frac{5}{y} &= -8 \end{aligned}$

73. $\begin{aligned} \frac{2}{x} + \frac{3}{y} - \frac{2}{z} &= -1 \\ \frac{8}{x} - \frac{12}{y} + \frac{5}{z} &= 5 \\ \frac{6}{x} + \frac{3}{y} - \frac{1}{z} &= 1 \end{aligned}$

74. $\begin{aligned} -\frac{5}{x} + \frac{4}{y} + \frac{3}{z} &= 2 \\ \frac{10}{x} + \frac{3}{y} - \frac{6}{z} &= 7 \\ \frac{5}{x} + \frac{2}{y} - \frac{9}{z} &= 6 \end{aligned}$

75. *Concept Check* For what value(s) of k will the following system of linear equations have no solution? infinitely many solutions?

$$\begin{aligned} x - 2y &= 3 \\ -2x + 4y &= k \end{aligned}$$

76. *Concept Check* Consider the linear equation in three variables $x + y + z = 4$. Find a pair of linear equations in three variables that, when considered together with the given equation, form a system having **(a)** exactly one solution, **(b)** no solution, **(c)** infinitely many solutions.

(Modeling) Use a system of equations to solve each problem. ***See Example 8.***

77. Find the equation of the line $y = ax + b$ that passes through the points $(-2, 1)$ and $(-1, -2)$.

78. $y = -2x + 2$
79. $y = \frac{3}{4}x^2 + \frac{1}{4}x - \frac{1}{2}$
80. $y = \frac{2}{3}x^2 - \frac{1}{2}x + \frac{1}{3}$
81. $y = 3x - 1$
82. $y = x^2 + 2x + 1$
83. $y = -\frac{1}{2}x^2 + x + \frac{1}{4}$
84. $Y_1 = 0.47X^2 - 0.23X + 0.56$
85. $x^2 + y^2 - 4x + 2y - 20 = 0$
86. $x^2 + y^2 - 6x - 4y - 12 = 0$
87. $x^2 + y^2 + x - 7y = 0$
88. $x^2 + y^2 + 2x - 6y - 15 = 0$
89. $x^2 + y^2 - \frac{7}{5}x + \frac{27}{5}y - \frac{126}{5} = 0$
90. $x^2 + y^2 - x - 3y - 12 = 0$

78. Find the equation of the line $y = ax + b$ that passes through the points $(3, -4)$ and $(-1, 4)$.

79. Find the equation of the parabola $y = ax^2 + bx + c$ that passes through the points $(2, 3)$, $(-1, 0)$, and $(-2, 2)$.

80. Find the equation of the parabola $y = ax^2 + bx + c$ that passes through the points $(-2, 4)$, $(2, 2)$, and $(4, 9)$.

81. *Connecting Graphs with Equations* Use a system to find the equation of the line through the given points.

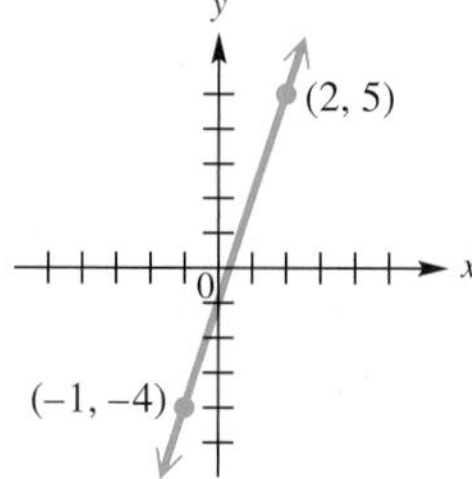

82. *Connecting Graphs with Equations* Use a system to find the equation of the parabola through the given points.

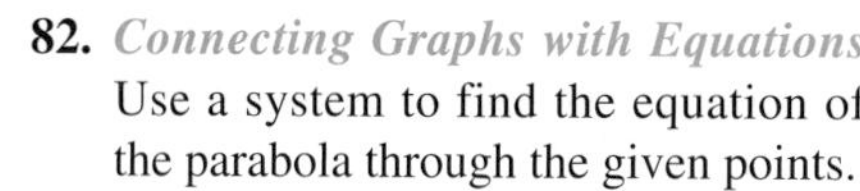

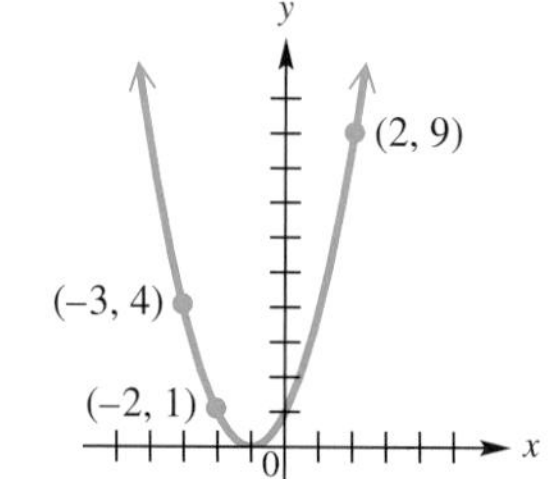

83. *Connecting Graphs with Equations* Find the equation of the parabola. Three views of the same curve are given.

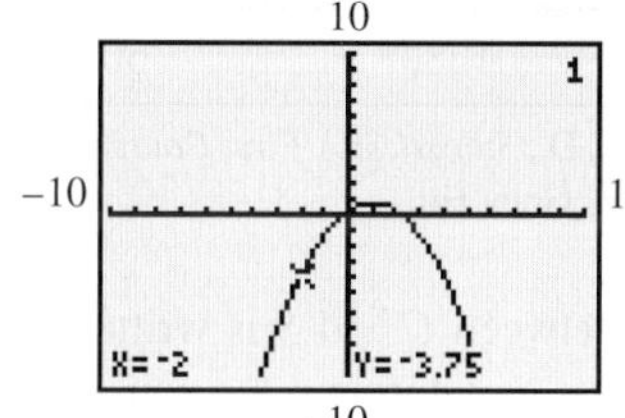

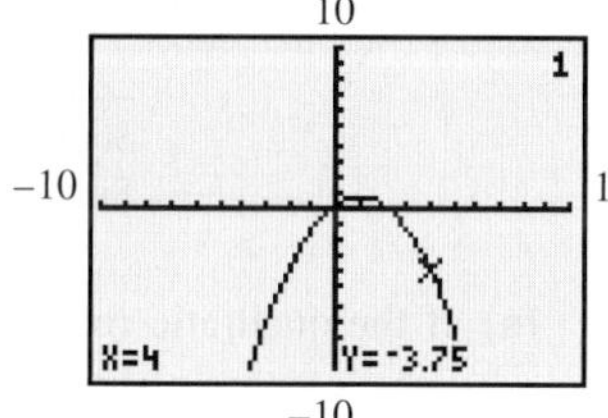

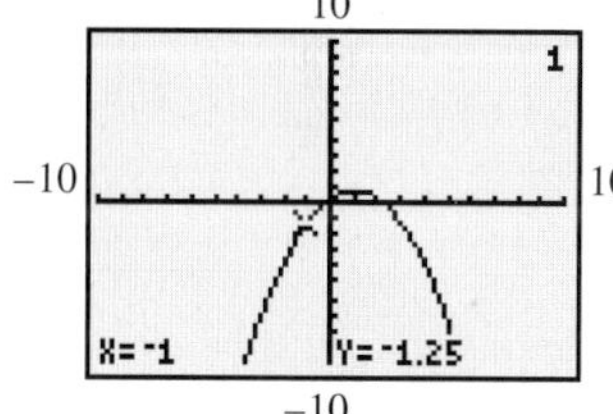

84. *(Modeling)* The table was generated using a function defined by $Y_1 = aX^2 + bX + c$. Use any three points from the table to find the equation that defines the function.

X	Y_1
-3	5.48
-2	2.9
-1	1.26
0	.56
1	.8
2	1.98
3	4.1

X= -3

(Modeling) Given three noncollinear points, there is one and only one circle that passes through them. Knowing that the equation of a circle may be written in the form

$$x^2 + y^2 + ax + by + c = 0, \quad \text{(Section 2.2)}$$

find the equation of the circle passing through the given points.

85. $(-1, 3)$, $(6, 2)$, and $(-2, -4)$

86. $(-1, 5)$, $(6, 6)$, and $(7, -1)$

87. $(2, 1)$, $(-1, 0)$, and $(3, 3)$

88. $(-5, 0)$, $(2, -1)$, and $(4, 3)$

89. *Connecting Graphs with Equations*

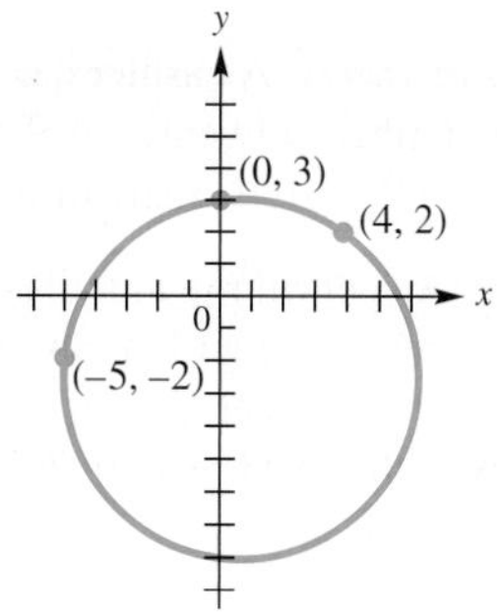

90. *Connecting Graphs with Equations*

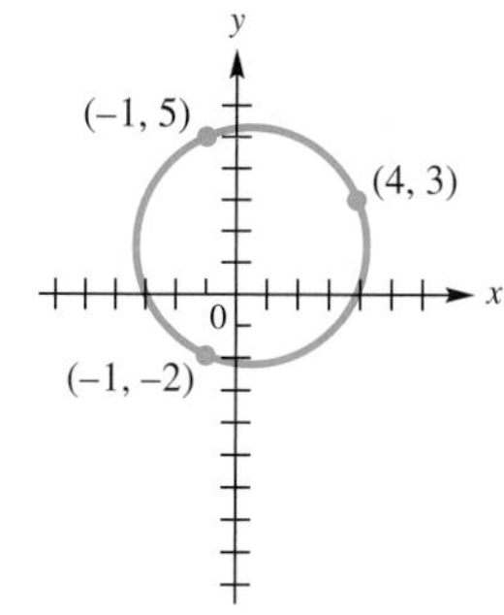

91. (a) $a = 0.009923$, $b = 0.9015$, $c = 317$; $C = 0.009923x^2 + 0.9015x + 317$
(b) near the end of 2098
92. (a) $C = -0.0000107567S^2 + 0.034896S + 22.9347$
(b) approximately 861 knots (2383 knots also satisfies the equation, but speeds between 861 and 2383 knots exceed the ceiling.)
93. 23, 24
94. goat: \$30; sheep: \$25
95. baseball: \$194.98; football: \$420.54
96. 6 ones, 9 fives, 15 twenties
97. 120 gal of \$9.00; 60 gal of \$3.00; 120 gal of \$4.50

(Modeling) Use the method of ***Example 8*** *to work Exercises 91 and 92.*

91. *Atmospheric Carbon Dioxide* Carbon dioxide concentrations (in parts per million) have been measured directly from the atmosphere since 1960. This concentration has increased quadratically. The table lists readings for three years.

Year	CO_2
1960	317
1980	339
2009	385

Source: U.S. Department of Energy; Carbon Dioxide Information Analysis Center.

(a) If the quadratic relationship between the carbon dioxide concentration C and the year t is expressed as $C = at^2 + bt + c$, where $t = 0$ corresponds to 1960, use a system of linear equations to determine the constants a, b, and c, and give the equation.

(b) Predict when the amount of carbon dioxide in the atmosphere will be double its 1960 level.

92. *Aircraft Speed and Altitude* For certain aircraft there exists a quadratic relationship between an airplane's maximum speed S (in knots) and its ceiling C, or highest altitude possible (in thousands of feet). The table lists three airplanes that conform to this relationship.

Airplane	Max Speed (S)	Ceiling (C)
Hawkeye	320	33
Corsair	600	40
Tomcat	1283	50

Source: Sanders, D., *Statistics: A First Course,* Sixth Edition, McGraw Hill.

(a) If the quadratic relationship between C and S is written as $C = aS^2 + bS + c$, use a system of linear equations to determine the constants a, b and c, and give the equation.

(b) A new aircraft of this type has a ceiling of 45,000 ft. Predict its top speed.

Solve each problem. ***See Examples 5 and 9.***

93. *Unknown Numbers* The sum of two numbers is 47, and the difference between the numbers is 1. Find the numbers.

94. *Costs of Goats and Sheep* At the Brendan Berger ranch, 6 goats and 5 sheep sell for \$305, while 2 goats and 9 sheep sell for \$285. Find the cost of a single goat and of a single sheep.

95. *Fan Cost Index* The Fan Cost Index (FCI) is a measure of how much it will cost a family of four to attend a professional sports event. In 2010, the FCI prices for Major League Baseball and the National Football League averaged \$307.76. The FCI for baseball was \$225.56 less than that for football. What were the FCIs for these sports? (*Source:* Team Marketing Report.)

96. *Money Denominations* A cashier has a total of 30 bills, made up of ones, fives, and twenties. The number of twenties is 9 more than the number of ones. The total value of the money is \$351. How many of each denomination of bill are there?

97. *Mixing Water* A sparkling-water distributor wants to make up 300 gal of sparkling water to sell for \$6.00 per gallon. She wishes to mix three grades of water selling for \$9.00, \$3.00, and \$4.50 per gallon, respectively. She must use twice as much of the \$4.50 water as of the \$3.00 water. How many gallons of each should she use?

98. 30 barrels each of \$150 and \$190 glue; 210 barrels of \$120 glue
99. 28 in.; 17 in.; 14 in.
100. 75°; 65°; 40°
101. \$100,000 at 3%; \$40,000 at 2.5%; \$60,000 at 1.5%
102. \$10,000 at 2%; \$20,000 at 2.5%; \$10,000 at 1.25%

98. *Mixing Glue* A glue company needs to make some glue that it can sell for \$120 per barrel. It wants to use 150 barrels of glue worth \$100 per barrel, along with some glue worth \$150 per barrel and some glue worth \$190 per barrel. It must use the same number of barrels of \$150 and \$190 glue. How much of the \$150 and \$190 glue will be needed? How many barrels of \$120 glue will be produced?

99. *Triangle Dimensions* The perimeter of a triangle is 59 in. The longest side is 11 in. longer than the medium side, and the medium side is 3 in. longer than the shortest side. Find the length of each side of the triangle.

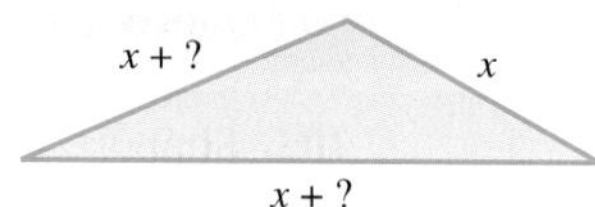

100. *Triangle Dimensions* The sum of the measures of the angles of any triangle is 180°. In a certain triangle, the largest angle measures 55° less than twice the medium angle, and the smallest angle measures 25° less than the medium angle. Find the measures of all three angles.

101. *Investment Decisions* Patrick Summers wins \$200,000 in the Louisiana state lottery. He invests part of the money in real estate with an annual return of 3% and another part in a money market account at 2.5% interest. He invests the rest, which amounts to \$80,000 less than the sum of the other two parts, in certificates of deposit that pay 1.5%. If the total annual interest on the money is \$4900, how much was invested at each rate?

	Amount Invested	Rate (as a decimal)	Annual Interest
Real Estate		0.03	
Money Market		0.025	
CDs		0.015	

102. *Investment Decisions* Jane Hooker invests \$40,000 received as an inheritance in three parts. With one part she buys mutual funds that offer a return of 2% per year. The second part, which amounts to twice the first, is used to buy government bonds paying 2.5% per year. She puts the rest of the money into a savings account that pays 1.25% annual interest. During the first year, the total interest is \$825. How much did she invest at each rate?

Relating Concepts

For individual or collaborative investigation *(Exercises 103–108)*

Supply and Demand In many applications of economics, as the price of an item goes up, demand for the item goes down and supply of the item goes up. The price where supply and demand are equal is the ***equilibrium price,*** *and the resulting supply or demand is the* ***equilibrium supply*** *or* ***equilibrium demand.***

Suppose the supply of a product is related to its price by the equation

$$p = \frac{2}{3}q,$$

where p is in dollars and q is supply in appropriate units. (Here, q stands for quantity.)

(continued)

103. (a) \$16 (b) \$11 (c) \$6
104. (a) 8 (b) 4 (c) 0
105. See the answer to Exercise 107.
106. (a) 0 (b) $\frac{40}{3}$ (c) $\frac{80}{3}$
107.

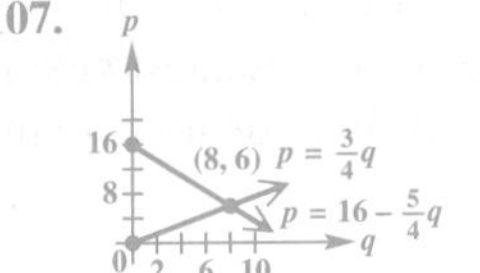

108. price: \$6; demand: 8

109. $\{(40, 15, 30)\}$
111. 11.92 lb of Arabian Mocha Sanani; 14.23 lb of Organic Shade Grown Mexico; 23.85 lb of Guatemala Antigua (Answers are approximations.)
112. −15.5 lb of Arabian Mocha Sanani; 96.5 lb of Organic Shade Grown Mexico; −31 lb of Guatemala Antigua; This is not reasonable, since the mixture cannot include negative amounts of ingredients.

Furthermore, suppose demand and price for the same product are related by

$$p = -\frac{1}{3}q + 18,$$

where p is price and q is demand. The system formed by these two equations has solution (18, 12), as seen in the graph. Use this information to ***work Exercises 103–108 in order.***

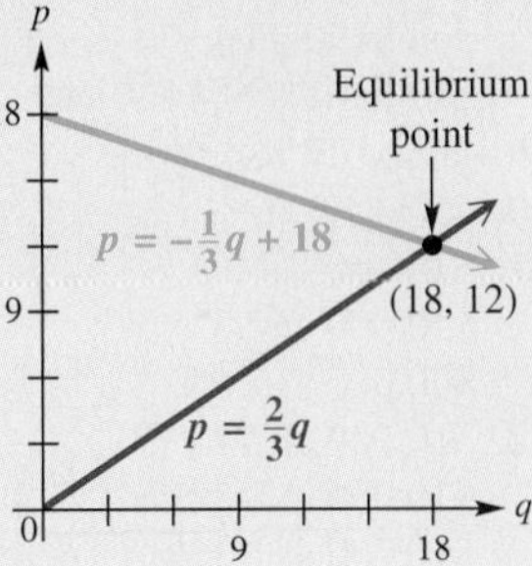

103. Suppose the demand and price for a certain model of electric can opener are related by $p = 16 - \frac{5}{4}q$, where p is price, in dollars, and q is demand, in appropriate units. Find the price when the demand is at each level.

(a) 0 units **(b)** 4 units **(c)** 8 units

104. Find the demand for the electric can opener at each price.

(a) \$6 **(b)** \$11 **(c)** \$16

105. Graph $p = 16 - \frac{5}{4}q$.

106. Suppose the price and supply of the can opener are related by $p = \frac{3}{4}q$, where q represents the supply and p the price. Find the supply at each price.

(a) \$0 **(b)** \$10 **(c)** \$20

107. Graph $p = \frac{3}{4}q$ on the same axes used for **Exercise 105.**

108. Use the result of **Exercise 107** to find the equilibrium price and the equilibrium demand.

109. Solve the system of equations (4), (5), and (6) from **Example 9.**

$$\begin{aligned} 25x + 40y + 20z &= 2200 \quad (4) \\ 4x + 2y + 3z &= 280 \quad (5) \\ 3x + 2y + z &= 180 \quad (6) \end{aligned}$$

110. Check your solution in **Exercise 109,** showing that it satisfies all three equations of the system.

111. *Blending Coffee Beans* Three varieties of coffee—Arabian Mocha Sanani, Organic Shade Grown Mexico, and Guatemala Antigua—are combined and roasted, yielding a 50-lb batch of coffee beans. Twice as many pounds of Guatemala Antigua, which retails for \$10.19 per lb, are needed as of Arabian Mocha Sanani, which retails for \$15.99 per lb. Organic Shade Grown Mexico retails for \$12.99 per lb. How many pounds of each coffee should be used in a blend that sells for \$12.37 per lb?

112. *Blending Coffee Beans* Rework **Exercise 111** if Guatemala Antigua retails for \$12.49 per lb instead of \$10.19 per lb. Does your answer seem reasonable? Explain.

9.2 Matrix Solution of Linear Systems

- The Gauss-Jordan Method
- Special Systems

$$\begin{bmatrix} 2 & 3 & 7 \\ 5 & -1 & 10 \end{bmatrix} \quad \text{Matrix}$$

Since systems of linear equations occur in so many practical situations, computer methods have been developed for efficiently solving linear systems. Computer solutions of linear systems depend on the idea of a **matrix** (plural **matrices**), a rectangular array of numbers enclosed in brackets. Each number is an **element** of the matrix.

The Gauss-Jordan Method In this section, we develop a method for solving linear systems using matrices. We start with a system and write the coefficients of the variables and the constants as an **augmented matrix** of the system.

Linear system of equations / Augmented matrix

$$\begin{aligned} x + 3y + 2z &= 1 \\ 2x + y - z &= 2 \\ x + y + z &= 2 \end{aligned} \quad \text{can be written as} \quad \left[\begin{array}{ccc|c} 1 & 3 & 2 & 1 \\ 2 & 1 & -1 & 2 \\ 1 & 1 & 1 & 2 \end{array}\right] \leftarrow \text{Rows}$$

Columns

Teaching Tip Explain that the numbers in each column of a matrix, except the last column, correspond to coefficients of the same variable. If a variable is missing from one of the equations, a 0 is placed in the corresponding position of the matrix.

The vertical line, which is optional, separates the coefficients from the constants. Because this matrix has 3 rows (horizontal) and 4 columns (vertical), we say its **dimension*** is 3×4 (read "three by four"). ***The number of rows is always given first.*** To refer to a number in the matrix, use its row and column numbers. For example, the number 3 is in the first row, second column.

We can treat the rows of this matrix just like the equations of the corresponding system of linear equations. Since the augmented matrix is nothing more than a shorthand form of the system, any transformation of the matrix that results in an equivalent system of equations can be performed.

Matrix Row Transformations

For any augmented matrix of a system of linear equations, the following row transformations will result in the matrix of an equivalent system.

1. Interchange any two rows.
2. Multiply or divide the elements of any row by a nonzero real number.
3. Replace any row of the matrix by the sum of the elements of that row and a multiple of the elements of another row.

These transformations are restatements in matrix form of the transformations of systems discussed in the previous section. From now on, when referring to the third transformation, we will abbreviate "a multiple of the elements of a row" as "a multiple of a row."

Before matrices can be used to solve a linear system, the system must be arranged in the proper form, with variable terms on the left side of the equation and constant terms on the right. The variable terms must be in the same order in each of the equations.

*Other terms used to describe the dimension of a matrix are *order* and *size*.

The **Gauss-Jordan method** is a systematic technique for applying matrix row transformations in an attempt to reduce a matrix to **diagonal form,** with 1s along the diagonal, from which the solutions are easily obtained.

$$\begin{bmatrix} 1 & 0 & | & a \\ 0 & 1 & | & b \end{bmatrix} \quad \text{or} \quad \begin{bmatrix} 1 & 0 & 0 & | & a \\ 0 & 1 & 0 & | & b \\ 0 & 0 & 1 & | & c \end{bmatrix}$$

Diagonal form, or reduced-row echelon form

This form is also called **reduced-row echelon form.**

Teaching Tip Point out that the algorithmic nature of the Gauss-Jordan method makes it easy for a computer to solve systems of linear equations involving hundreds of variables and hundreds of equations.

Using the Gauss-Jordan Method to Put a Matrix into Diagonal Form

Step 1 Obtain 1 as the first element of the first column.

Step 2 Use the first row to transform the remaining entries in the first column to 0.

Step 3 Obtain 1 as the second entry in the second column.

Step 4 Use the second row to transform the remaining entries in the second column to 0.

Step 5 Continue in this manner as far as possible.

NOTE ***The Gauss-Jordan method proceeds column by column, from left to right.*** In each column, we work to obtain 1 in the appropriate diagonal location, and then use it to transform the remaining elements in that column to 0s. When we are working with a particular column, no row operation should undo the form of a preceding column.

Classroom Example 1
Solve the system.

$$2x + 3y = 7$$
$$3x - 4y = -32$$

Answer: $\{(-4, 5)\}$

EXAMPLE 1 Using the Gauss-Jordan Method

Solve the system.

$$3x - 4y = 1$$
$$5x + 2y = 19$$

SOLUTION Both equations are in the same form, with variable terms in the same order on the left, and constant terms on the right.

$$\begin{bmatrix} 3 & -4 & | & 1 \\ 5 & 2 & | & 19 \end{bmatrix} \quad \text{Write the augmented matrix.}$$

The goal is to transform the augmented matrix into one in which the value of the variables will be easy to see. That is, since each of the first two columns in the matrix represents the coefficients of one variable, the augmented matrix should be transformed so that it is of the following form.

This form is our goal. $$\begin{bmatrix} \mathbf{1} & \mathbf{0} & | & k \\ \mathbf{0} & \mathbf{1} & | & j \end{bmatrix} \quad \text{Here } k \text{ and } j \text{ are real numbers.}$$

Once the augmented matrix is in this form, the matrix can be rewritten as a linear system

$$x = k$$
$$y = j.$$

***It is best to work in columns, beginning in each column with the element that is to become* 1.** In the augmented matrix

$$\left[\begin{array}{cc|c} 3 & -4 & 1 \\ 5 & 2 & 19 \end{array}\right],$$

3 is in the first row, first column position. Use transformation 2, multiplying each entry in the first row by $\frac{1}{3}$ to get 1 in this position. (This step is abbreviated as $\frac{1}{3}$ R1.)

$$\left[\begin{array}{cc|c} 1 & -\frac{4}{3} & \frac{1}{3} \\ 5 & 2 & 19 \end{array}\right] \quad \tfrac{1}{3}\text{R1}$$

Introduce 0 in the second row, first column by multiplying each element of the first row by -5 and adding the result to the corresponding element in the second row, using transformation 3.

$$\left[\begin{array}{cc|c} 1 & -\frac{4}{3} & \frac{1}{3} \\ 0 & \frac{26}{3} & \frac{52}{3} \end{array}\right] \quad -5\text{R1} + \text{R2}$$

Obtain 1 in the second row, second column by multiplying each element of the second row by $\frac{3}{26}$, using transformation 2.

$$\left[\begin{array}{cc|c} 1 & -\frac{4}{3} & \frac{1}{3} \\ 0 & 1 & 2 \end{array}\right] \quad \tfrac{3}{26}\text{R2}$$

Finally, get 0 in the first row, second column by multiplying each element of the second row by $\frac{4}{3}$ and adding the result to the corresponding element in the first row.

$$\left[\begin{array}{cc|c} 1 & 0 & 3 \\ 0 & 1 & 2 \end{array}\right] \quad \tfrac{4}{3}\text{R2} + \text{R1}$$

This last matrix corresponds to the system

$$x = 3$$
$$y = 2,$$

which indicates the solution $(3, 2)$. We can read this solution directly from the third column of the final matrix.

CHECK Substitute the solution in both equations of the *original* system.

$3x - 4y = 1$	$5x + 2y = 19$
$3(3) - 4(2) \stackrel{?}{=} 1$	$5(3) + 2(2) \stackrel{?}{=} 19$
$9 - 8 \stackrel{?}{=} 1$	$15 + 4 \stackrel{?}{=} 19$
$1 = 1$ ✓ True	$19 = 19$ ✓ True

Since true statements result, the solution set is $\{(3, 2)\}$.

✔ *Now Try Exercise 17.*

NOTE Using row operations to write a matrix in diagonal form requires effective use of the inverse properties of addition and multiplication. See **Section R.2.**

A linear system with three equations is solved in a similar way. Row transformations are used to get 1s down the diagonal from left to right and 0s above and below each 1.

Classroom Example 2
Solve the system.

$$\begin{aligned} x + y - 4z &= 10 \\ 2x - 3y + z &= 7 \\ 3x - y - z &= 12 \end{aligned}$$

Answer: $\{(3, -1, -2)\}$

EXAMPLE 2 Using the Gauss-Jordan Method

Solve the system.

$$\begin{aligned} x - y + 5z &= -6 \\ 3x + 3y - z &= 10 \\ x + 3y + 2z &= 5 \end{aligned}$$

SOLUTION

$$\left[\begin{array}{ccc|c} 1 & -1 & 5 & -6 \\ 3 & 3 & -1 & 10 \\ 1 & 3 & 2 & 5 \end{array}\right] \quad \text{Write the augmented matrix.}$$

There is already a 1 in the first row, first column. Introduce 0 in the second row of the first column by multiplying each element in the first row by -3 and adding the result to the corresponding element in the second row.

$$\left[\begin{array}{ccc|c} 1 & -1 & 5 & -6 \\ 0 & 6 & -16 & 28 \\ 1 & 3 & 2 & 5 \end{array}\right] \quad -3\text{R1} + \text{R2}$$

To change the third element in the first column to 0, multiply each element of the first row by -1. Add the result to the corresponding element of the third row.

$$\left[\begin{array}{ccc|c} 1 & -1 & 5 & -6 \\ 0 & 6 & -16 & 28 \\ 0 & 4 & -3 & 11 \end{array}\right] \quad -1\text{R1} + \text{R3}$$

Use the same procedure to transform the second and third columns. For both of these columns, perform the additional step of getting 1 in the appropriate position of each column. Do this by multiplying the elements of the row by the reciprocal of the number in that position.

$$\left[\begin{array}{ccc|c} 1 & -1 & 5 & -6 \\ 0 & 1 & -\frac{8}{3} & \frac{14}{3} \\ 0 & 4 & -3 & 11 \end{array}\right] \quad \tfrac{1}{6}\text{R2}$$

$$\left[\begin{array}{ccc|c} 1 & 0 & \frac{7}{3} & -\frac{4}{3} \\ 0 & 1 & -\frac{8}{3} & \frac{14}{3} \\ 0 & 4 & -3 & 11 \end{array}\right] \quad \text{R2} + \text{R1}$$

$$\left[\begin{array}{ccc|c} 1 & 0 & \frac{7}{3} & -\frac{4}{3} \\ 0 & 1 & -\frac{8}{3} & \frac{14}{3} \\ 0 & 0 & \frac{23}{3} & -\frac{23}{3} \end{array}\right] \quad -4\text{R2} + \text{R3}$$

$$\left[\begin{array}{ccc|c} 1 & 0 & \frac{7}{3} & -\frac{4}{3} \\ 0 & 1 & -\frac{8}{3} & \frac{14}{3} \\ 0 & 0 & 1 & -1 \end{array}\right] \quad \tfrac{3}{23}\text{R3}$$

$$\left[\begin{array}{ccc|c} 1 & 0 & 0 & 1 \\ 0 & 1 & -\frac{8}{3} & \frac{14}{3} \\ 0 & 0 & 1 & -1 \end{array}\right] \quad -\tfrac{7}{3}\text{R3} + \text{R1}$$

$$\left[\begin{array}{ccc|c} 1 & 0 & 0 & 1 \\ 0 & 1 & 0 & 2 \\ 0 & 0 & 1 & -1 \end{array}\right] \quad \tfrac{8}{3}\text{R3} + \text{R2}$$

The linear system associated with this final matrix is

$$x = 1$$
$$y = 2$$
$$z = -1.$$

The solution set is $\{(1, 2, -1)\}$. Check the solution in the original system.

✔ Now Try Exercise 25.

A TI-83/84 Plus graphing calculator with matrix capability is able to perform row operations. See **Figure 7(a).** The screen in **Figure 7(b)** shows typical entries for the matrix in the second step of the solution in **Example 2.** The entire Gauss-Jordan method can be carried out in one step with the rref (reduced-row echelon form) command, as shown in **Figure 7(c).**

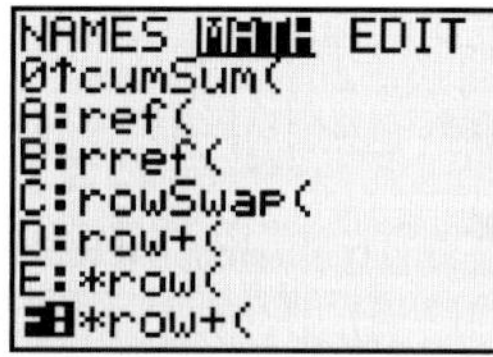

This typical menu shows various options for matrix row transformations in choices C–F.

(a)

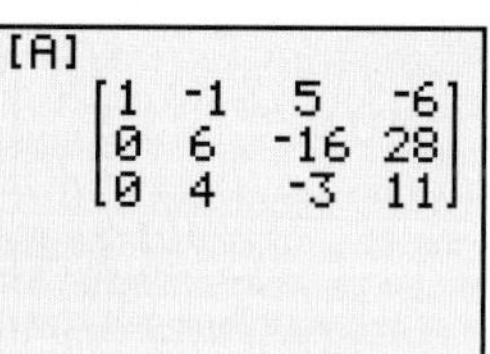

This is the matrix after column 1 has been transformed.

(b)

rref([A])

$$\begin{bmatrix} 1 & 0 & 0 & 1 \\ 0 & 1 & 0 & 2 \\ 0 & 0 & 1 & -1 \end{bmatrix}$$

This screen shows the final matrix in the algebraic solution, found by using the rref command.

(c)

Figure 7

■

Special Systems The next two examples show how to recognize inconsistent systems or systems with infinitely many solutions when solving such systems using row transformations.

Classroom Example 3
Use the Gauss-Jordan method to solve the system.

$$2x - 3y = 7$$
$$-6x + 9y = 0$$

Answer: $\emptyset$

EXAMPLE 3 Solving an Inconsistent System

Use the Gauss-Jordan method to solve the system.

$$x + y = 2$$
$$2x + 2y = 5$$

SOLUTION

$$\left[\begin{array}{cc|c} 1 & 1 & 2 \\ 2 & 2 & 5 \end{array}\right] \quad \text{Write the augmented matrix.}$$

$$\left[\begin{array}{cc|c} 1 & 1 & 2 \\ 0 & 0 & 1 \end{array}\right] \quad -2R1 + R2$$

The next step would be to get 1 in the second row, second column. Because of the 0 there, it is impossible to go further. Since the second row corresponds to the equation

$$0x + 0y = 1,$$

which has no solution, the system is inconsistent and the solution set is $\emptyset$.

✔ Now Try Exercise 21.

Classroom Example 4
Use the Gauss-Jordan method to solve the system. Write the solution set with z arbitrary.

$$2x + y + z = 5$$
$$3x + 2y - z = -8$$

Answer:
$\{(-3z + 18, 5z - 31, z)\}$

EXAMPLE 4 Solving a System with Infinitely Many Solutions

Use the Gauss-Jordan method to solve the system.

$$2x - 5y + 3z = 1$$
$$x - 2y - 2z = 8$$

SOLUTION Recall from the previous section that a system with two equations in three variables usually has an infinite number of solutions. We can use the Gauss-Jordan method to give the solution with z arbitrary.

$$\left[\begin{array}{ccc|c} 2 & -5 & 3 & 1 \\ 1 & -2 & -2 & 8 \end{array}\right] \quad \text{Write the augmented matrix.}$$

$$\left[\begin{array}{ccc|c} 1 & -2 & -2 & 8 \\ 2 & -5 & 3 & 1 \end{array}\right] \quad \text{Interchange rows to get 1 in the first row, first column position.}$$

$$\left[\begin{array}{ccc|c} 1 & -2 & -2 & 8 \\ 0 & -1 & 7 & -15 \end{array}\right] \quad -2R1 + R2$$

$$\left[\begin{array}{ccc|c} 1 & -2 & -2 & 8 \\ 0 & 1 & -7 & 15 \end{array}\right] \quad -1R2$$

$$\left[\begin{array}{ccc|c} 1 & 0 & -16 & 38 \\ 0 & 1 & -7 & 15 \end{array}\right] \quad 2R2 + R1$$

It is not possible to go further with the Gauss-Jordan method. The equations that correspond to the final matrix are

$$x - 16z = 38 \quad \text{and} \quad y - 7z = 15.$$

Solve these equations for x and y, respectively.

$x - 16z = 38$	$y - 7z = 15$
$x = 16z + 38$	$y = 7z + 15$ (Section 1.1)

The solution set, written with z arbitrary, is $\{(16z + 38, 7z + 15, z)\}$.

Now Try Exercise 37.

Summary of Possible Cases

When matrix methods are used to solve a system of linear equations and the resulting matrix is written in diagonal form:

1. If the number of rows with nonzero elements to the left of the vertical line is equal to the number of variables in the system, then the system has a single solution. **See Examples 1 and 2.**
2. If one of the rows has the form $[0 \; 0 \cdots 0 \,|\, a]$ with $a \neq 0$, then the system has no solution. **See Example 3.**
3. If there are fewer rows in the matrix containing nonzero elements than the number of variables, then the system has either no solution or infinitely many solutions. If there are infinitely many solutions, give the solutions in terms of one or more arbitrary variables. **See Example 4.**

9.2 Exercises

Use the given row transformation to change each matrix as indicated. **See Example 1.**

1. $\begin{bmatrix} 1 & 4 \\ 4 & 7 \end{bmatrix}$; -4 times row 1 added to row 2

2. $\begin{bmatrix} 1 & -4 \\ 7 & 0 \end{bmatrix}$; -7 times row 1 added to row 2

3. $\begin{bmatrix} 1 & 5 & 6 \\ -2 & 3 & -1 \\ 4 & 7 & 0 \end{bmatrix}$; 2 times row 1 added to row 2

4. $\begin{bmatrix} 1 & 5 & 6 \\ -4 & -1 & 2 \\ 3 & 7 & 1 \end{bmatrix}$; 4 times row 1 added to row 2

Concept Check Write the augmented matrix for each system and give its dimension. Do not solve.

5. $2x + 3y = 11$
$x + 2y = 8$

6. $3x + 5y = -13$
$2x + 3y = -9$

7. $2x + y + z - 3 = 0$
$3x - 4y + 2z + 7 = 0$
$x + y + z - 2 = 0$

8. $4x - 2y + 3z - 4 = 0$
$3x + 5y + z - 7 = 0$
$5x - y + 4z - 7 = 0$

Concept Check Write the system of equations associated with each augmented matrix. Do not solve.

9. $\left[\begin{array}{ccc|c} 3 & 2 & 1 & 1 \\ 0 & 2 & 4 & 22 \\ -1 & -2 & 3 & 15 \end{array}\right]$

10. $\left[\begin{array}{ccc|c} 2 & 1 & 3 & 12 \\ 4 & -3 & 0 & 10 \\ 5 & 0 & -4 & -11 \end{array}\right]$

11. $\left[\begin{array}{ccc|c} 1 & 0 & 0 & 2 \\ 0 & 1 & 0 & 3 \\ 0 & 0 & 1 & -2 \end{array}\right]$

12. $\left[\begin{array}{ccc|c} 1 & 0 & 0 & 4 \\ 0 & 1 & 0 & 2 \\ 0 & 0 & 1 & 3 \end{array}\right]$

13. [A]
$\begin{bmatrix} 1 & 1 & 0 & 3 \\ 0 & 2 & 1 & -4 \\ 1 & 0 & -1 & 5 \end{bmatrix}$

14. [B]
$\begin{bmatrix} 2 & 0 & 1 & 9 \\ 0 & -1 & -1 & 5 \\ 3 & 1 & 0 & 8 \end{bmatrix}$

Use the Gauss-Jordan method to solve each system of equations. For systems in two variables with infinitely many solutions, give the solution with y arbitrary. For systems in three variables with infinitely many solutions, give the solution with z arbitrary. **See Examples 1–4.**

15. $x + y = 5$
$x - y = -1$

16. $x + 2y = 5$
$2x + y = -2$

17. $3x + 2y = -9$
$2x - 5y = -6$

18. $2x - 5y = 10$
$3x + y = 15$

19. $6x - 3y - 4 = 0$
$3x + 6y - 7 = 0$

20. $2x - 3y - 10 = 0$
$2x + 2y - 5 = 0$

21. $2x - y = 6$
$4x - 2y = 0$

22. $3x - 2y = 1$
$6x - 4y = -1$

23. $\frac{3}{8}x - \frac{1}{2}y = \frac{7}{8}$
$-6x + 8y = -14$

24. $\frac{1}{2}x + \frac{3}{5}y = \frac{1}{4}$
$10x + 12y = 5$

25. $x + y - 5z = -18$
$3x - 3y + z = 6$
$x + 3y - 2z = -13$

26. $-x + 2y + 6z = 2$
$3x + 2y + 6z = 6$
$x + 4y - 3z = 1$

27. $x + y - z = 6$
$2x - y + z = -9$
$x - 2y + 3z = 1$

28. $x + 3y - 6z = 7$
$2x - y + z = 1$
$x + 2y + 2z = -1$

29. $x - z = -3$
$y + z = 9$
$x + z = 7$

---

1. $\begin{bmatrix} 1 & 4 \\ 0 & -9 \end{bmatrix}$ **2.** $\begin{bmatrix} 1 & -4 \\ 0 & 28 \end{bmatrix}$

3. $\begin{bmatrix} 1 & 5 & 6 \\ 0 & 13 & 11 \\ 4 & 7 & 0 \end{bmatrix}$

4. $\begin{bmatrix} 1 & 5 & 6 \\ 0 & 19 & 26 \\ 3 & 7 & 1 \end{bmatrix}$

5. $\left[\begin{array}{cc|c} 2 & 3 & 11 \\ 1 & 2 & 8 \end{array}\right]$; 2×3

6. $\left[\begin{array}{cc|c} 3 & 5 & -13 \\ 2 & 3 & -9 \end{array}\right]$; 2×3

7. $\left[\begin{array}{ccc|c} 2 & 1 & 1 & 3 \\ 3 & -4 & 2 & -7 \\ 1 & 1 & 1 & 2 \end{array}\right]$; 3×4

8. $\left[\begin{array}{ccc|c} 4 & -2 & 3 & 4 \\ 3 & 5 & 1 & 7 \\ 5 & -1 & 4 & 7 \end{array}\right]$; 3×4

9. $3x + 2y + z = 1$
$2y + 4z = 22$
$-x - 2y + 3z = 15$

10. $2x + y + 3z = 12$
$4x - 3y = 10$
$5x - 4z = -11$

11. $x = 2$
$y = 3$
$z = -2$

12. $x = 4$
$y = 2$
$z = 3$

13. $x + y = 3$
$2y + z = -4$
$x - z = 5$

14. $2x + z = 9$
$-y - z = 5$
$3x + y = 8$

15. $\{(2, 3)\}$ **16.** $\{(-3, 4)\}$
17. $\{(-3, 0)\}$ **18.** $\{(5, 0)\}$
19. $\{(1, \frac{2}{3})\}$ **20.** $\{(\frac{7}{2}, -1)\}$
21. $\emptyset$ **22.** $\emptyset$
23. $\{(\frac{4y+7}{3}, y)\}$
24. $\{(\frac{-12y+5}{10}, y)\}$
25. $\{(-1, -2, 3)\}$
26. $\{(1, \frac{3}{10}, \frac{2}{5})\}$
27. $\{(-1, 23, 16)\}$
28. $\{(1, 0, -1)\}$
29. $\{(2, 4, 5)\}$

30. $\{(3, 2, -4)\}$
31. $\{(\frac{1}{2}, 1, -\frac{1}{2})\}$
32. $\{(-1, 2, -2)\}$
33. $\{(2, 1, -1)\}$
34. $\{(2, 2, 2)\}$
35. $\emptyset$
36. $\emptyset$
37. $\{(\frac{-15z-12}{23}, \frac{z-13}{23}, z)\}$
38. $\{(\frac{2z+13}{11}, \frac{7z+18}{11}, z)\}$
39. $\{(1, 1, 2, 0)\}$
40. $\{(2, -1, 1, -2)\}$
41. $\{(0, 2, -2, 1)\}$
42. $\{(-4, 3, -2, 1)\}$
43. $\{(0.571, 7.041, 11.442)\}$
44. $\{(0.407, 9.316, 7.270)\}$
45. none
46. $\{(5, 6)\}$
47. $A = \frac{1}{2}$, $B = -\frac{1}{2}$
48. $A = 1$, $B = 4$
49. $A = \frac{1}{2}$, $B = \frac{1}{2}$
50. $A = \frac{4}{3}$, $B = \frac{2}{3}$
51. day laborer: \$152; concrete finisher: \$160

30. $\begin{aligned} -x + y &= -1 \\ y - z &= 6 \\ x + z &= -1 \end{aligned}$

31. $\begin{aligned} y &= -2x - 2z + 1 \\ x &= -2y - z + 2 \\ z &= x - y \end{aligned}$

32. $\begin{aligned} x &= -y + 1 \\ z &= 2x \\ y &= -2z - 2 \end{aligned}$

33. $\begin{aligned} 2x - y + 3z &= 0 \\ x + 2y - z &= 5 \\ 2y + z &= 1 \end{aligned}$

34. $\begin{aligned} 4x + 2y - 3z &= 6 \\ x - 4y + z &= -4 \\ -x + 2z &= 2 \end{aligned}$

35. $\begin{aligned} 3x + 5y - z + 2 &= 0 \\ 4x - y + 2z - 1 &= 0 \\ -6x - 10y + 2z &= 0 \end{aligned}$

36. $\begin{aligned} 3x + y + 3z - 1 &= 0 \\ x + 2y - z - 2 &= 0 \\ 2x - y + 4z - 4 &= 0 \end{aligned}$

37. $\begin{aligned} x - 8y + z &= 4 \\ 3x - y + 2z &= -1 \end{aligned}$

38. $\begin{aligned} 5x - 3y + z &= 1 \\ 2x + y - z &= 4 \end{aligned}$

39. $\begin{aligned} x - y + 2z + w &= 4 \\ y + z &= 3 \\ z - w &= 2 \\ x - y &= 0 \end{aligned}$

40. $\begin{aligned} x + 2y + z - 3w &= 7 \\ y + z &= 0 \\ x - w &= 4 \\ -x + y &= -3 \end{aligned}$

41. $\begin{aligned} x + 3y - 2z - w &= 9 \\ 4x + y + z + 2w &= 2 \\ -3x - y + z - w &= -5 \\ x - y - 3z - 2w &= 2 \end{aligned}$

42. $\begin{aligned} 2x + y - z + 3w &= 0 \\ 3x - 2y + z - 4w &= -24 \\ x + y - z + w &= 2 \\ x - y + 2z - 5w &= -16 \end{aligned}$

Solve each system using a graphing calculator capable of performing row operations. Give solutions with values correct to the nearest thousandth.

43. $\begin{aligned} 0.3x + 2.7y - \sqrt{2}z &= 3 \\ \sqrt{7}x - 20y + 12z &= -2 \\ 4x + \sqrt{3}y - 1.2z &= \frac{3}{4} \end{aligned}$

44. $\begin{aligned} \sqrt{5}x - 1.2y + z &= -3 \\ \frac{1}{2}x - 3y + 4z &= \frac{4}{3} \\ 4x + 7y - 9z &= \sqrt{2} \end{aligned}$

Graph each system of three equations together on the same axes, and determine the number of solutions (exactly one, none, or infinitely many). If there is exactly one solution, estimate the solution. Then confirm your answer by solving the system with the Gauss-Jordan method.

45. $\begin{aligned} 2x + 3y &= 5 \\ -3x + 5y &= 22 \\ 2x + y &= -1 \end{aligned}$

46. $\begin{aligned} 3x - 2y &= 3 \\ -2x + 4y &= 14 \\ x + y &= 11 \end{aligned}$

For each equation, determine the constants A and B that make the equation an identity. (Hint: Combine terms on the right, and set coefficients of corresponding terms in the numerators equal.)

47. $\dfrac{1}{(x-1)(x+1)} = \dfrac{A}{x-1} + \dfrac{B}{x+1}$

48. $\dfrac{x+4}{x^2} = \dfrac{A}{x} + \dfrac{B}{x^2}$

49. $\dfrac{x}{(x-a)(x+a)} = \dfrac{A}{x-a} + \dfrac{B}{x+a}$

50. $\dfrac{2x}{(x+2)(x-1)} = \dfrac{A}{x+2} + \dfrac{B}{x-1}$

Solve each problem using matrices.

51. **Daily Wages** Dan Abbey is a building contractor. If he hires 7 day laborers and 2 concrete finishers, his payroll for the day is \$1384. If he hires 1 day laborer and 5 concrete finishers, his daily cost is \$952. Find the daily wage for each type of worker.

52. peanuts: \$2.40; cashews: \$3.60
53. 12, 6, 2
54. 10 small, 7 medium, 7 large
55. 9.6 cm^3 of 7%; 30.4 cm^3 of 2%
56. \$10,000 at 4%; \$7000 at 6%; \$8000 at 5%
59. 44.4 g of A; 133.3 g of B; 222.2 g of C
61. (a) 65 or older: $y = 0.0018x + 0.13$; ages 25–34: $y = -0.000175x + 0.135$
(b) $\{(2.5316, 0.1346)\}$; 2012; 13.5%
62. (a) $y = -0.0002x + 0.133$
(b) $\{(1.5, 0.1327)\}$; 2011; 13.3%

52. *Mixing Nuts* At the Everglades Nut Company, 5 lb of peanuts and 6 lb of cashews cost \$33.60, while 3 lb of peanuts and 7 lb of cashews cost \$32.40. Find the cost of a single pound of peanuts and a single pound of cashews.

53. *Unknown Numbers* Find three numbers whose sum is 20, if the first number is three times the difference between the second and the third, and the second number is two more than twice the third.

54. *Car Sales Quota* To meet a sales quota, a car salesperson must sell 24 new cars, consisting of small, medium, and large cars. She must sell 3 more small cars than medium cars, and the same number of medium cars as large cars. How many of each size must she sell?

55. *Mixing Acid Solutions* A chemist has two prepared acid solutions, one of which is 2% acid by volume, the other 7% acid. How many cubic centimeters of each should the chemist mix together to obtain 40 cm^3 of a 3.2% acid solution?

56. *Financing an Expansion* To get the necessary funds for a planned expansion, a small company took out three loans totaling \$25,000. The company was able to borrow some of the money at 4% interest. It borrowed \$2000 more than one-half the amount of the 4% loan at 6%, and the rest at 5%. The total annual interest was \$1220. How much did the company borrow at each rate?

57. *Financing an Expansion* In **Exercise 56,** suppose we drop the condition that the amount borrowed at 6% is \$2000 more than one-half the amount borrowed at 4%. How is the solution changed?

58. *Financing an Expansion* Suppose the company in **Exercise 56** can borrow only \$6000 at 5%. Is a solution possible that still meets the given conditions? Explain.

59. *Planning a Diet* In a special diet for a hospital patient, the total amount per meal of food groups A, B, and C must equal 400 g. The diet should include one-third as much of group A as of group B. The sum of the amounts of group A and group C should equal twice the amount of group B. How many grams of each food group should be included? (Give answers to the nearest tenth.)

60. *Planning a Diet* In **Exercise 59,** suppose that, in addition to the conditions given there, foods A and B cost \$0.02 per gram, food C costs \$0.03 per gram, and a meal must cost \$8. Is a solution possible? Explain.

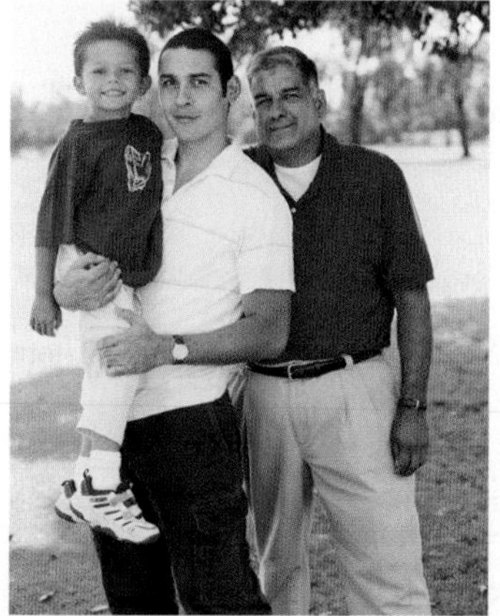

(Modeling) Age Distribution in the United States As people live longer, a larger percentage of the U.S. population is 65 or over and a smaller percentage is in younger age brackets. Use matrices to solve the problems in Exercises 61 and 62. Let $x = 0$ represent 2010 and $x = 40$ represent 2050. Express percents in decimal form.

61. In 2010, 13.0% of the population was 65 or older. By 2050, this percentage is expected to be 20.2%. The percentage of the population aged 25–34 in 2010 was 13.5%. That age group is expected to include 12.8% of the population in 2050. (*Source:* U.S. Census Bureau.)

(a) Assuming these population changes are linear, use the data for the 65-or-older age group to write a linear equation. Then do the same for the 25–34 age group.

(b) Solve the system of linear equations from part (a). In what year will the two age groups include the same percentage of the population? What is that percentage?

(c) Does your answer to part (b) suggest that the *number* of people in the U.S. population aged 25–34 is decreasing? Explain.

62. In 2010, 13.3% of the U.S. population was aged 35–44. This percentage is expected to decrease to 12.5% in 2050. (*Source:* U.S. Census Bureau.)

(a) Write a linear equation representing this population change.

(b) Solve the system containing the equation from part (a) and the equation from **Exercise 61** for the 65-or-older age group. Give the year and percentage when these two age groups will include the same portion of the population.

63. (a) using the first equation, approximately 245 lb; using the second equation, approximately 253 lb
(b) for the first, 7.46 lb; for the second, 7.93 lb
(c) at approximately 66 in. and 118 lb

64. (a) $x_3 + x_4 = 600$
(b) $\begin{bmatrix} 1 & 0 & 0 & 1 & | & 1000 \\ 1 & 1 & 0 & 0 & | & 1100 \\ 0 & 1 & 1 & 0 & | & 700 \\ 0 & 0 & 1 & 1 & | & 600 \end{bmatrix}$; $\begin{bmatrix} 1 & 0 & 0 & 1 & | & 1000 \\ 0 & 1 & 0 & -1 & | & 100 \\ 0 & 0 & 1 & 1 & | & 600 \\ 0 & 0 & 0 & 0 & | & 0 \end{bmatrix}$
(c) $x_4 = 1000 - x_1$, $x_4 = x_2 - 100$, $x_4 = 600 - x_3$
(d) 1000
(e) 100
(f) 600; 600
(g) $x_1 = 1000$, $x_2 = 100$, $x_3 = 600$, $x_4 = 0$; no

63. *(Modeling) Athlete's Weight and Height* The relationship between a professional basketball player's height H (in inches) and weight W (in pounds) was modeled using two different samples of players. The resulting equations that modeled the two samples were

$$W = 7.46H - 374$$

and $$W = 7.93H - 405.$$

(a) Use each equation to predict the weight of a 6 ft 11 in. professional basketball player.

(b) According to each model, what change in weight is associated with a 1-in. increase in height?

(c) Determine the weight and height where the two models agree.

64. *(Modeling) Traffic Congestion* At rush hours, substantial traffic congestion is encountered at the traffic intersections shown in the figure. (All streets are one-way.) The city wishes to improve the signals at these corners to speed the flow of traffic. The traffic engineers first gather data. As the figure shows, 700 cars per hour come down M Street to intersection A, and 300 cars per hour come to intersection A on 10th Street. A total of x_1 of these cars leave A on M Street, while x_4 cars leave A on 10th Street. The number of cars entering A must equal the number leaving, which suggests the following equation.

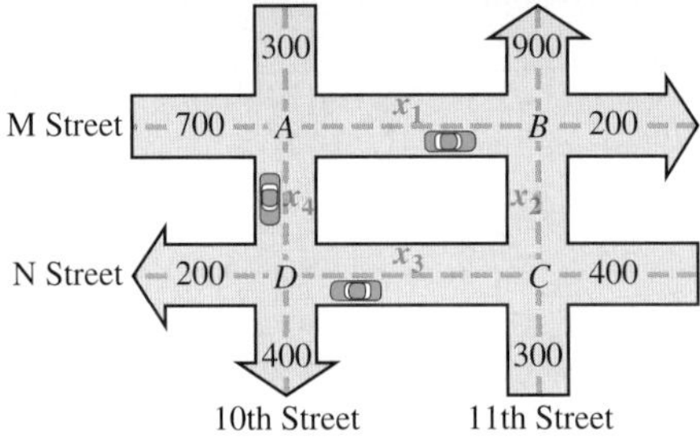

$$x_1 + x_4 = 700 + 300$$
$$x_1 + x_4 = 1000$$

For intersection B, x_1 cars enter B on M Street, and x_2 cars enter B on 11th Street. As the figure shows, 900 cars leave B on 11th, while 200 leave on M, which leads to the following equation.

$$x_1 + x_2 = 900 + 200$$
$$x_1 + x_2 = 1100$$

At intersection C, 400 cars enter on N Street and 300 on 11th Street, while x_2 leave on 11th Street and x_3 leave on N Street.

$$x_2 + x_3 = 400 + 300$$
$$x_2 + x_3 = 700$$

Finally, intersection D has x_3 cars entering on N and x_4 entering on 10th. There are 400 cars leaving D on 10th and 200 leaving on N.

(a) Set up an equation for intersection D.

(b) Use the four equations to write an augmented matrix, and then transform it so that 1s are on the diagonal and 0s are below. This is **triangular form.**

(c) Since you got a row of all 0s, the system of equations does not have a unique solution. Write three equations, corresponding to the three nonzero rows of the matrix. Solve each of the equations for x_4.

(d) One of your equations should have been $x_4 = 1000 - x_1$. What is the greatest possible value of x_1 so that x_4 is not negative?

(e) Another equation should have been $x_4 = x_2 - 100$. Find the least possible value of x_2 so that x_4 is not negative.

(f) Find the greatest possible values of x_3 and x_4 so that neither variable is negative.

(g) Use the results of parts (a)–(f) to give a solution for the problem in which all the equations are satisfied and all variables are nonnegative. Is the solution unique?

65. $a + 871b + 11.5c + 3d = 239$
$a + 847b + 12.2c + 2d = 234$
$a + 685b + 10.6c + 5d = 192$
$a + 969b + 14.2c + 1d = 343$

66. $\left[\begin{array}{cccc|c} 1 & 871 & 11.5 & 3 & 239 \\ 1 & 847 & 12.2 & 2 & 234 \\ 1 & 685 & 10.6 & 5 & 192 \\ 1 & 969 & 14.2 & 1 & 343 \end{array}\right]$

The solution is $a \approx -715.457$, $b \approx 0.34756$, $c \approx 48.6585$, and $d \approx 30.71951$.

67. $F = -715.457 + 0.34756A + 48.6585P + 30.71951W$

68. approximately 323; This estimate is very close to the actual value of 320.

69.

n	T
3	28
6	191
10	805
29	17,487
100	681,550
200	5,393,100
400	42,906,200
1000	668,165,500
5000	8.3×10^{10}
10,000	6.7×10^{11}
100,000	6.7×10^{14}

Relating Concepts

For individual or collaborative investigation *(Exercises 65–68)*

(Modeling) Number of Fawns To model the spring fawn count F from the adult pronghorn population A, the precipitation P, and the severity of the winter W, environmentalists have used the equation

$$F = a + bA + cP + dW,$$

where the coefficients a, b, c, and d are constants that must be determined before using the equation. (Winter severity is scaled between 1 and 5, with 1 being mild and 5 being severe.) ***Work Exercises 65–68 in order.*** (*Source:* Brase, C. and C. Brase, *Understandable Statistics,* D.C. Heath and Company; Bureau of Land Management.)

65. Substitute the values for F, A, P, and W from the table for Years 1–4 into the equation

$$F = a + bA + cP + dW$$

and obtain four linear equations involving a, b, c, and d.

Year	Fawns	Adults	Precip. (in inches)	Winter Severity
1	239	871	11.5	3
2	234	847	12.2	2
3	192	685	10.6	5
4	343	969	14.2	1
5	?	960	12.6	3

66. Write an augmented matrix representing the system in **Exercise 65,** and solve for a, b, c, and d.

67. Write the equation for F using the values found in **Exercise 66** for the coefficients.

68. Use the information in the table to predict the spring fawn count in Year 5. (Compare this with the actual count of 320.)

(Modeling) Number of Calculations When computers are programmed to solve large linear systems involved in applications like designing aircraft or electrical circuits, they frequently use an algorithm that is similar to the Gauss-Jordan method presented in this section. Solving a linear system with n equations and n variables requires the computer to perform a total of

$$T(n) = \frac{2}{3}n^3 + \frac{3}{2}n^2 - \frac{7}{6}n$$

arithmetic calculations, including additions, subtractions, multiplications, and divisions. Use this model to solve the problems in Exercises 69–72. (*Source:* Burden, R. and J. Faires, *Numerical Analysis,* Sixth Edition, Brooks/Cole Publishing Company.)

69. Compute T for the following values of n.

$$n = 3, 6, 10, 29, 100, 200, 400, 1000, 5000, 10,000, 100,000$$

Write the results in a table.

70. 17,487; yes
71. no; It increases by almost a factor of 8.
72. 3.1 hr

70. In 1940, John Atanasoff, a physicist from Iowa State University, wanted to solve a 29×29 linear system of equations. How many arithmetic operations would this have required? Is this too many to do by hand? (Atanasoff's work led to the invention of the first fully electronic digital computer.) (*Source: The Gazette.*)

Atanasoff-Berry Computer

71. If the number of equations and variables is doubled, does the number of arithmetic operations double?

72. A Cray-T90 supercomputer can execute up to 60 billion arithmetic operations per second. How many hours would be required to solve a linear system with 100,000 variables?

9.3 Determinant Solution of Linear Systems

- Determinants
- Cofactors
- Evaluating $n \times n$ Determinants
- Determinant Theorems
- Cramer's Rule

Determinants Every $n \times n$ matrix A is associated with a real number called the **determinant** of A, written $|A|$. The determinant of a 2×2 matrix is defined as follows.

Determinant of a 2 × 2 Matrix

If $A = \begin{bmatrix} a_{11} & a_{12} \\ a_{21} & a_{22} \end{bmatrix}$, then $|A| = \begin{vmatrix} a_{11} & a_{12} \\ a_{21} & a_{22} \end{vmatrix} = a_{11}a_{22} - a_{21}a_{12}.$

NOTE ***Matrices are enclosed with square brackets, while determinants are denoted with vertical bars.*** A matrix is an *array* of numbers, but its determinant is a *single* number.

$$\begin{vmatrix} a_{11} & a_{12} \\ a_{21} & a_{22} \end{vmatrix}$$

The arrows in the diagram in the margin will remind you which products to find when evaluating a 2×2 determinant.

Classroom Example 1

Let $B = \begin{bmatrix} 5 & -8 \\ 2 & -4 \end{bmatrix}$. Find $|B|$.

Answer: -4

EXAMPLE 1 Evaluating a 2 × 2 Determinant

Let $A = \begin{bmatrix} -3 & 4 \\ 6 & 8 \end{bmatrix}$. Find $|A|$.

ALGEBRAIC SOLUTION

Use the definition with

$$a_{11} = -3, \quad a_{12} = 4, \quad a_{21} = 6, \quad a_{22} = 8.$$

$$\begin{aligned} |A| &= \underset{a_{11}}{-3} \cdot \underset{a_{22}}{8} - \underset{a_{21}}{6} \cdot \underset{a_{12}}{4} \\ &= -24 - 24 \\ &= -48 \end{aligned}$$

GRAPHING CALCULATOR SOLUTION

We can define a matrix and then use the capability of a graphing calculator to find the determinant of the matrix. In the screen in **Figure 8,** the symbol det([A]) represents the determinant of [A].

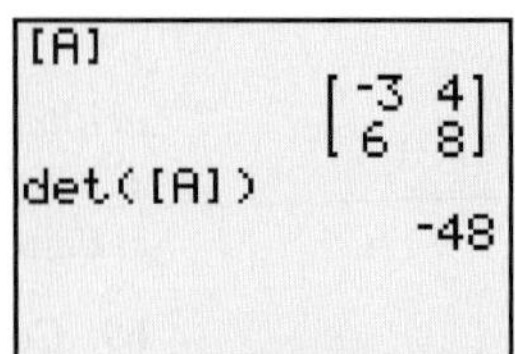

Figure 8

Now Try Exercise 1.

LOOKING AHEAD TO CALCULUS

Determinants are used in calculus to find **vector cross products,** which are used to study the effect of forces in the plane or in space.

Teaching Tip Show students how the products in the definition in the box correspond to "diagonals" of the matrix.

Determinant of a 3 × 3 Matrix

If $A = \begin{bmatrix} a_{11} & a_{12} & a_{13} \\ a_{21} & a_{22} & a_{23} \\ a_{31} & a_{32} & a_{33} \end{bmatrix}$, then the determinant of A, symbolized $|A|$, is

defined as follows.

$$|A| = \begin{vmatrix} a_{11} & a_{12} & a_{13} \\ a_{21} & a_{22} & a_{23} \\ a_{31} & a_{32} & a_{33} \end{vmatrix} = (a_{11}a_{22}a_{33} + a_{12}a_{23}a_{31} + a_{13}a_{21}a_{32}) \\ - (a_{31}a_{22}a_{13} + a_{32}a_{23}a_{11} + a_{33}a_{21}a_{12})$$

The terms on the right side of the equation in the definition of $|A|$ can be rearranged to get the following.

$$\begin{vmatrix} a_{11} & a_{12} & a_{13} \\ a_{21} & a_{22} & a_{23} \\ a_{31} & a_{32} & a_{33} \end{vmatrix} = a_{11}(a_{22}a_{33} - a_{32}a_{23}) - a_{21}(a_{12}a_{33} - a_{32}a_{13}) \\ + a_{31}(a_{12}a_{23} - a_{22}a_{13})$$

Each quantity in parentheses represents the determinant of a 2×2 matrix that is the part of the 3×3 matrix remaining when the row and column of the multiplier are eliminated, as shown below.

$$a_{11}(a_{22}a_{33} - a_{32}a_{23}) \qquad \begin{bmatrix} a_{11} & a_{12} & a_{13} \\ a_{21} & a_{22} & a_{23} \\ a_{31} & a_{32} & a_{33} \end{bmatrix}$$

$$a_{21}(a_{12}a_{33} - a_{32}a_{13}) \qquad \begin{bmatrix} a_{11} & a_{12} & a_{13} \\ a_{21} & a_{22} & a_{23} \\ a_{31} & a_{32} & a_{33} \end{bmatrix}$$

$$a_{31}(a_{12}a_{23} - a_{22}a_{13}) \qquad \begin{bmatrix} a_{11} & a_{12} & a_{13} \\ a_{21} & a_{22} & a_{23} \\ a_{31} & a_{32} & a_{33} \end{bmatrix}$$

Cofactors The determinant of each 2×2 matrix above is the **minor** of the associated element in the 3×3 matrix. The symbol $\boldsymbol{M_{ij}}$ represents the minor that results when row i and column j are eliminated. The following list gives some of the minors from the matrix above.

Element	Minor	Element	Minor
a_{11}	$M_{11} = \begin{vmatrix} a_{22} & a_{23} \\ a_{32} & a_{33} \end{vmatrix}$	a_{22}	$M_{22} = \begin{vmatrix} a_{11} & a_{13} \\ a_{31} & a_{33} \end{vmatrix}$
a_{21}	$M_{21} = \begin{vmatrix} a_{12} & a_{13} \\ a_{32} & a_{33} \end{vmatrix}$	a_{23}	$M_{23} = \begin{vmatrix} a_{11} & a_{12} \\ a_{31} & a_{32} \end{vmatrix}$
a_{31}	$M_{31} = \begin{vmatrix} a_{12} & a_{13} \\ a_{22} & a_{23} \end{vmatrix}$	a_{33}	$M_{33} = \begin{vmatrix} a_{11} & a_{12} \\ a_{21} & a_{22} \end{vmatrix}$

In a 4×4 matrix, the minors are determinants of 3×3 matrices. Similarly, an $n \times n$ matrix has minors that are determinants of $(n-1) \times (n-1)$ matrices.

To find the determinant of a 3×3 or larger matrix, first choose any row or column. Then the minor of each element in that row or column must be multiplied by $+1$ or -1, depending on whether the sum of the row number and column number is even or odd. The product of a minor and the number $+1$ or -1 is called a *cofactor.*

Cofactor

Let M_{ij} be the minor for element a_{ij} in an $n \times n$ matrix. The **cofactor** of a_{ij}, written A_{ij}, is defined as follows.

$$A_{ij} = (-1)^{i+j} \cdot M_{ij}$$

Classroom Example 2
Find the cofactor of each of the following elements of the given matrix.

$$\begin{bmatrix} 5 & -1 & 6 \\ 0 & 2 & -9 \\ 4 & 1 & 3 \end{bmatrix}$$

(a) 5 (b) 0 (c) 3

Answers:
(a) 15 (b) 9 (c) 10

EXAMPLE 2 Finding Cofactors of Elements

Find the cofactor of each of the following elements of the given matrix.

$$\begin{bmatrix} 6 & 2 & 4 \\ 8 & 9 & 3 \\ 1 & 2 & 0 \end{bmatrix}$$

(a) 6 **(b)** 3 **(c)** 8

SOLUTION

(a) Since 6 is in the first row, first column of the matrix, $i = 1$ and $j = 1$, so $M_{11} = \begin{vmatrix} 9 & 3 \\ 2 & 0 \end{vmatrix} = -6$. The cofactor is

$$(-1)^{1+1}(-6) = 1(-6) = -6.$$

(b) Here $i = 2$ and $j = 3$, so $M_{23} = \begin{vmatrix} 6 & 2 \\ 1 & 2 \end{vmatrix} = 10$. The cofactor is

$$(-1)^{2+3}(10) = -1(10) = -10.$$

(c) We have $i = 2$ and $j = 1$, and $M_{21} = \begin{vmatrix} 2 & 4 \\ 2 & 0 \end{vmatrix} = -8$. The cofactor is

$$(-1)^{2+1}(-8) = -1(-8) = 8.$$

Now Try Exercise 11.

Teaching Tip An alternative method for finding 3×3 determinants is described in **Exercises 33–36.**

Evaluating $n \times n$ Determinants The determinant of a 3×3 or larger matrix is found as follows.

Finding the Determinant of a Matrix

Multiply each element in any row or column of the matrix by its cofactor. The sum of these products gives the value of the determinant.

The process of forming this sum of products is called **expansion by a given row or column.**

Classroom Example 3
Evaluate, expanding by the third row.

$$\begin{vmatrix} 2 & -3 & 5 \\ 1 & 4 & -5 \\ 0 & 2 & -6 \end{vmatrix}$$

Answer: -36

EXAMPLE 3 Evaluating a 3 × 3 Determinant

Evaluate $\begin{vmatrix} 2 & -3 & -2 \\ -1 & -4 & -3 \\ -1 & 0 & 2 \end{vmatrix}$, expanding by the second column.

SOLUTION To find this determinant, first find the minor of each element in the second column.

$$M_{12} = \begin{vmatrix} -1 & -3 \\ -1 & 2 \end{vmatrix} = -1(2) - (-1)(-3) = -5$$

$$M_{22} = \begin{vmatrix} 2 & -2 \\ -1 & 2 \end{vmatrix} = 2(2) - (-1)(-2) = 2$$

$$M_{32} = \begin{vmatrix} 2 & -2 \\ -1 & -3 \end{vmatrix} = 2(-3) - (-1)(-2) = -8$$

Use parentheses, and keep track of all negative signs to avoid errors.

Now find the cofactor of each element of these minors.

$$A_{12} = (-1)^{1+2} \cdot M_{12} = (-1)^3 \cdot (-5) = -1(-5) = 5$$

$$A_{22} = (-1)^{2+2} \cdot M_{22} = (-1)^4 \cdot 2 = 1 \cdot 2 = 2$$

$$A_{32} = (-1)^{3+2} \cdot M_{32} = (-1)^5 \cdot (-8) = -1(-8) = 8$$

Find the determinant by multiplying each cofactor by its corresponding element in the matrix and finding the sum of these products.

$$\begin{vmatrix} 2 & -3 & -2 \\ -1 & -4 & -3 \\ -1 & 0 & 2 \end{vmatrix} = a_{12} \cdot A_{12} + a_{22} \cdot A_{22} + a_{32} \cdot A_{32}$$

$$= -3(5) + (-4)2 + 0(8)$$

$$= -15 + (-8) + 0$$

$$= -23$$

```
[B]
        [ 2  -3  -2]
        [-1  -4  -3]
        [-1   0   2]
det([B])
                -23
```

The matrix in **Example 3** can be entered as [B] and its determinant found using a graphing calculator, as shown in the screen above.

✔ *Now Try Exercise 15.*

Teaching Tip Show students how the choice of row or column can potentially reduce the amount of work needed to calculate a determinant.

In **Example 3,** we would have found the same answer using any row or column of the matrix. One reason we used column 2 is that it contains a 0 element, so it was not really necessary to calculate M_{32} and A_{32}.

Instead of calculating $(-1)^{i+j}$ for a given element, we can use the sign checkerboard shown below. The signs alternate for each row and column, beginning with + in the first row, first column position. If we expand a 3 × 3 matrix about row 3, for example, the first minor would have a + sign associated with it, the second minor a − sign, and the third minor a + sign.

Sign array for 3 × 3 matrices

$$\begin{matrix} + & - & + \\ - & + & - \\ + & - & + \end{matrix}$$

This array of signs can be extended for determinants of 4 × 4 and larger matrices.

Determinant Theorems The following theorems are true for square matrices of any dimension and can be used to simplify the process of finding determinants.

Determinant Theorems

1. If every element in a row (or column) of matrix A is 0, then $|A| = 0$.
2. If the rows of matrix A are the corresponding columns of matrix B, then $|B| = |A|$.
3. If any two rows (or columns) of matrix A are interchanged to form matrix B, then $|B| = -|A|$.
4. Suppose matrix B is formed by multiplying every element of a row (or column) of matrix A by the real number k. Then $|B| = k \cdot |A|$.
5. If two rows (or columns) of matrix A are identical, then $|A| = 0$.
6. Changing a row (or column) of a matrix by adding to it a constant times another row (or column) does not change the determinant of the matrix.
7. If matrix A is in triangular form, having zeros either above or below the main diagonal, then $|A|$ is the product of the elements on the main diagonal of A.

Classroom Example 4
Use the determinant theorems to find the value of each determinant.

(a) $\begin{vmatrix} 2 & 1 & 7 \\ -4 & 0 & 8 \\ 6 & 5 & 21 \end{vmatrix}$

(b) $\begin{vmatrix} 1 & 2 & 5 & 8 \\ 0 & 4 & -1 & 7 \\ 0 & 0 & -2 & 5 \\ 0 & 0 & 0 & 9 \end{vmatrix}$

Answers:
(a) -88 (b) -72

EXAMPLE 4 Using the Determinant Theorems

Use the determinant theorems to find the value of each determinant.

(a) $\begin{vmatrix} -2 & 4 & 2 \\ 6 & 7 & 3 \\ 0 & 16 & 8 \end{vmatrix}$ **(b)** $\begin{vmatrix} 3 & -7 & 4 & 10 \\ 0 & 1 & 8 & 3 \\ 0 & 0 & -5 & 2 \\ 0 & 0 & 0 & 6 \end{vmatrix}$

SOLUTION

(a) Use determinant theorem 6 to obtain a 0 in the second row of the first column. Multiply each element in the first row by 3, and add the result to the corresponding element in the second row.

$$\begin{vmatrix} -2 & 4 & 2 \\ 0 & 19 & 9 \\ 0 & 16 & 8 \end{vmatrix} \quad 3R1 + R2$$

Now, find the determinant by expanding by column 1.

$$-2(-1)^{1+1}\begin{vmatrix} 19 & 9 \\ 16 & 8 \end{vmatrix} = -2(1)(8) = -16 \qquad 19(8) - 16(9) = 152 - 144 = 8$$

(b) Use determinant theorem 7 to find the determinant of this triangular matrix by multiplying the elements on the main diagonal.

$$\begin{vmatrix} 3 & -7 & 4 & 10 \\ 0 & 1 & 8 & 3 \\ 0 & 0 & -5 & 2 \\ 0 & 0 & 0 & 6 \end{vmatrix} = 3(1)(-5)(6) = -90$$

✔ *Now Try Exercises 65 and 67.*

Cramer's Rule The elimination method can be used to develop a process for solving a linear system in two unknowns using determinants. Consider the following system.

$$a_1x + b_1y = c_1 \quad (1)$$
$$a_2x + b_2y = c_2 \quad (2)$$

The variable y in this system of equations can be eliminated by using multiplication to create coefficients that are additive inverses and by adding the two equations.

$$a_1b_2x + b_1b_2y = c_1b_2 \qquad \text{Multiply (1) by } b_2.$$
$$-a_2b_1x - b_1b_2y = -c_2b_1 \qquad \text{Multiply (2) by } -b_1.$$
$$(a_1b_2 - a_2b_1)x = c_1b_2 - c_2b_1 \qquad \text{Add.}$$
$$x = \frac{c_1b_2 - c_2b_1}{a_1b_2 - a_2b_1}, \quad \text{if } a_1b_2 - a_2b_1 \neq 0$$

Similarly, the variable x can be eliminated.

$$-a_1a_2x - a_2b_1y = -a_2c_1 \qquad \text{Multiply (1) by } -a_2.$$
$$a_1a_2x + a_1b_2y = a_1c_2 \qquad \text{Multiply (2) by } a_1.$$
$$(a_1b_2 - a_2b_1)y = a_1c_2 - a_2c_1 \qquad \text{Add.}$$
$$y = \frac{a_1c_2 - a_2c_1}{a_1b_2 - a_2b_1}, \quad \text{if } a_1b_2 - a_2b_1 \neq 0$$

Both numerators and the common denominator of these values for x and y can be written as determinants.

$$c_1b_2 - c_2b_1 = \begin{vmatrix} c_1 & b_1 \\ c_2 & b_2 \end{vmatrix}, \quad a_1c_2 - a_2c_1 = \begin{vmatrix} a_1 & c_1 \\ a_2 & c_2 \end{vmatrix}, \quad \text{and } a_1b_2 - a_2b_1 = \begin{vmatrix} a_1 & b_1 \\ a_2 & b_2 \end{vmatrix}$$

The solutions for x and y can be written using these determinants.

$$x = \frac{\begin{vmatrix} c_1 & b_1 \\ c_2 & b_2 \end{vmatrix}}{\begin{vmatrix} a_1 & b_1 \\ a_2 & b_2 \end{vmatrix}} \quad \text{and} \quad y = \frac{\begin{vmatrix} a_1 & c_1 \\ a_2 & c_2 \end{vmatrix}}{\begin{vmatrix} a_1 & b_1 \\ a_2 & b_2 \end{vmatrix}}, \quad \text{if} \quad \begin{vmatrix} a_1 & b_1 \\ a_2 & b_2 \end{vmatrix} \neq 0.$$

We denote the three determinants in the solution as follows.

$$\begin{vmatrix} a_1 & b_1 \\ a_2 & b_2 \end{vmatrix} = D, \quad \begin{vmatrix} c_1 & b_1 \\ c_2 & b_2 \end{vmatrix} = D_x, \quad \text{and} \quad \begin{vmatrix} a_1 & c_1 \\ a_2 & c_2 \end{vmatrix} = D_y$$

NOTE The elements of D are the four coefficients of the variables in the given system. The elements of D_x are obtained by replacing the coefficients of x in D by the respective constants, and the elements of D_y are obtained by replacing the coefficients of y in D by the respective constants.

These results are summarized as **Cramer's rule.**

Teaching Tip Tell students that, as with the Gauss-Jordan method, Cramer's rule offers a systematic approach to solving linear systems of equations. Encourage them to recognize patterns with respect to the sign array and the process by which determinants are evaluated.

Cramer's Rule for Two Equations in Two Variables

Given the system

$$a_1x + b_1y = c_1$$
$$a_2x + b_2y = c_2,$$

if $D \neq 0$, then the system has the unique solution

$$x = \frac{D_x}{D} \quad \text{and} \quad y = \frac{D_y}{D},$$

where $D = \begin{vmatrix} a_1 & b_1 \\ a_2 & b_2 \end{vmatrix}$, $D_x = \begin{vmatrix} c_1 & b_1 \\ c_2 & b_2 \end{vmatrix}$, and $D_y = \begin{vmatrix} a_1 & c_1 \\ a_2 & c_2 \end{vmatrix}$.

CAUTION ***Cramer's rule does not apply if $D = 0$.*** When $D = 0$, the system is inconsistent or has infinitely many solutions. Evaluate D first.

EXAMPLE 5 Applying Cramer's Rule to a 2 × 2 System

Use Cramer's rule to solve the system.

$$5x + 7y = -1$$
$$6x + 8y = 1$$

SOLUTION By Cramer's rule, $x = \frac{D_x}{D}$ and $y = \frac{D_y}{D}$. Find D first, since if $D = 0$, Cramer's rule does not apply. If $D \neq 0$, then find D_x and D_y.

$$D = \begin{vmatrix} 5 & 7 \\ 6 & 8 \end{vmatrix} = 5(8) - 6(7) = -2$$

$$D_x = \begin{vmatrix} -1 & 7 \\ 1 & 8 \end{vmatrix} = -1(8) - 1(7) = -15$$

$$D_y = \begin{vmatrix} 5 & -1 \\ 6 & 1 \end{vmatrix} = 5(1) - 6(-1) = 11$$

$$x = \frac{D_x}{D} = \frac{-15}{-2} = \frac{15}{2} \quad \text{and} \quad y = \frac{D_y}{D} = \frac{11}{-2} = -\frac{11}{2} \qquad \text{Cramer's rule}$$

The solution set is $\left\{\left(\frac{15}{2}, -\frac{11}{2}\right)\right\}$. Verify by substituting in the given system.

✔ *Now Try Exercise 81.*

[A] [5 7; 6 8]
[B] [-1 7; 1 8]

[C] [5 -1; 6 1]
det([B])/det([A]) ▸Frac
15/2

det([C])/det([A]) ▸Frac
-11/2

Because graphing calculators can evaluate determinants, they can also be used to apply Cramer's rule to solve a system of linear equations. The screens above support the result in **Example 5.**

Cramer's rule can be generalized.

Classroom Example 5 Use Cramer's rule to solve the system.

$$4x + 3y = 2$$
$$x - 2y = 5$$

Answer: $\left\{\left(\frac{19}{11}, -\frac{18}{11}\right)\right\}$

General Form of Cramer's Rule

Let an $n \times n$ system have linear equations of the form

$$a_1x_1 + a_2x_2 + a_3x_3 + \cdots + a_nx_n = b.$$

Define D as the determinant of the $n \times n$ matrix of all coefficients of the variables. Define D_{x_1} as the determinant obtained from D by replacing the entries in column 1 of D with the constants of the system. Define D_{x_i} as the determinant obtained from D by replacing the entries in column i with the constants of the system. If $D \neq 0$, the unique solution of the system is

$$x_1 = \frac{D_{x_1}}{D}, \quad x_2 = \frac{D_{x_2}}{D}, \quad x_3 = \frac{D_{x_3}}{D}, \quad \ldots, \quad x_n = \frac{D_{x_n}}{D}.$$

Classroom Example 6
Use Cramer's rule to solve the system.

$$2x - 4y + z - 19 = 0$$
$$4x + 6y - z + 15 = 0$$
$$x + y + 2z - 11 = 0$$

Answer: $\left\{\left(\frac{3}{2}, -\frac{5}{2}, 6\right)\right\}$

EXAMPLE 6 Applying Cramer's Rule to a 3 × 3 System

Use Cramer's rule to solve the system.

$$x + y - z + 2 = 0$$
$$2x - y + z + 5 = 0$$
$$x - 2y + 3z - 4 = 0$$

SOLUTION

$$x + y - z = -2$$
$$2x - y + z = -5$$
$$x - 2y + 3z = 4$$

Rewrite each equation in the form $ax + by + cz + \cdots = k$.

Verify the required determinants.

$$D = \begin{vmatrix} 1 & 1 & -1 \\ 2 & -1 & 1 \\ 1 & -2 & 3 \end{vmatrix} = -3, \quad D_x = \begin{vmatrix} -2 & 1 & -1 \\ -5 & -1 & 1 \\ 4 & -2 & 3 \end{vmatrix} = 7,$$

$$D_y = \begin{vmatrix} 1 & -2 & -1 \\ 2 & -5 & 1 \\ 1 & 4 & 3 \end{vmatrix} = -22, \quad D_z = \begin{vmatrix} 1 & 1 & -2 \\ 2 & -1 & -5 \\ 1 & -2 & 4 \end{vmatrix} = -21$$

$$x = \frac{D_x}{D} = \frac{7}{-3} = -\frac{7}{3}, \quad y = \frac{D_y}{D} = \frac{-22}{-3} = \frac{22}{3}, \quad z = \frac{D_z}{D} = \frac{-21}{-3} = 7$$

The solution set is $\left\{\left(-\frac{7}{3}, \frac{22}{3}, 7\right)\right\}$.

✔ *Now Try Exercise 93.*

CAUTION As shown in **Example 6,** each equation in the system must be written in the form $ax + by + cz + \cdots = k$ before Cramer's rule is used.

Classroom Example 7
Show that Cramer's rule does not apply to the following system.

$$x - 3y + 2z = 5$$
$$3x + y - 4z = 8$$
$$-9x - 3y + 12z = -24$$

Answer: $D = 0$, so Cramer's rule does not apply.

EXAMPLE 7 Showing That Cramer's Rule Does Not Apply

Show that Cramer's rule does not apply to the following system.

$$2x - 3y + 4z = 10$$
$$6x - 9y + 12z = 24$$
$$x + 2y - 3z = 5$$

SOLUTION We need to show that $D = 0$. Expand about column 1.

$$D = \begin{vmatrix} 2 & -3 & 4 \\ 6 & -9 & 12 \\ 1 & 2 & -3 \end{vmatrix} = 2\begin{vmatrix} -9 & 12 \\ 2 & -3 \end{vmatrix} - 6\begin{vmatrix} -3 & 4 \\ 2 & -3 \end{vmatrix} + 1\begin{vmatrix} -3 & 4 \\ -9 & 12 \end{vmatrix}$$
$$= 2(3) - 6(1) + 1(0)$$
$$= 0$$

Since $D = 0$, Cramer's rule does not apply.

✔ *Now Try Exercise 95.*

NOTE ***When D = 0, the system is either inconsistent or has infinitely many solutions.*** Use the elimination method to tell which is the case. Verify that the system in **Example 7** is inconsistent, and thus the solution set is $\emptyset$.

9.3 Exercises

1. −31 2. −3
3. 7 4. −6
5. 0 6. −2
7. −26 8. −68
9. 0

10. The determinant value is 0.

11. 2, −6, 4 12. −5, 1, 3
13. −6, 0, −6 14. 8, 16, 0

15. 186 16. 50
17. 17 18. 10
19. 166 20. −71
21. 0 22. 0
23. 0 24. 0
25. 1 26. −1
27. 2 28. −1
29. $-144-8\sqrt{10}$
30. $2-4\sqrt{21}$
31. −5.5 32. −0.051

Find the value of each determinant. **See Example 1.**

1. $\begin{vmatrix} -5 & 9 \\ 4 & -1 \end{vmatrix}$ **2.** $\begin{vmatrix} -1 & 3 \\ -2 & 9 \end{vmatrix}$ **3.** $\begin{vmatrix} -1 & -2 \\ 5 & 3 \end{vmatrix}$

4. $\begin{vmatrix} 6 & -4 \\ 0 & -1 \end{vmatrix}$ **5.** $\begin{vmatrix} 9 & 3 \\ -3 & -1 \end{vmatrix}$ **6.** $\begin{vmatrix} 0 & 2 \\ 1 & 5 \end{vmatrix}$

7. $\begin{vmatrix} 3 & 4 \\ 5 & -2 \end{vmatrix}$ **8.** $\begin{vmatrix} -9 & 7 \\ 2 & 6 \end{vmatrix}$ **9.** $\begin{vmatrix} -7 & 0 \\ 3 & 0 \end{vmatrix}$

10. *Concept Check* Refer to **Exercise 5.** Make a conjecture about the value of the determinant of a matrix in which one row is a multiple of another row.

Find the cofactor of each element in the second row for each determinant. **See Example 2.**

11. $\begin{vmatrix} -2 & 0 & 1 \\ 1 & 2 & 0 \\ 4 & 2 & 1 \end{vmatrix}$ **12.** $\begin{vmatrix} 1 & -1 & 2 \\ 1 & 0 & 2 \\ 0 & -3 & 1 \end{vmatrix}$ **13.** $\begin{vmatrix} 1 & 2 & -1 \\ 2 & 3 & -2 \\ -1 & 4 & 1 \end{vmatrix}$ **14.** $\begin{vmatrix} 2 & -1 & 4 \\ 3 & 0 & 1 \\ -2 & 1 & 4 \end{vmatrix}$

Find the value of each determinant. **See Example 3.**

15. $\begin{vmatrix} 4 & -7 & 8 \\ 2 & 1 & 3 \\ -6 & 3 & 0 \end{vmatrix}$ **16.** $\begin{vmatrix} 8 & -2 & -4 \\ 7 & 0 & 3 \\ 5 & -1 & 2 \end{vmatrix}$ **17.** $\begin{vmatrix} 1 & 2 & 0 \\ -1 & 2 & -1 \\ 0 & 1 & 4 \end{vmatrix}$

18. $\begin{vmatrix} 2 & 1 & -1 \\ 4 & 7 & -2 \\ 2 & 4 & 0 \end{vmatrix}$ **19.** $\begin{vmatrix} 10 & 2 & 1 \\ -1 & 4 & 3 \\ -3 & 8 & 10 \end{vmatrix}$ **20.** $\begin{vmatrix} 7 & -1 & 1 \\ 1 & -7 & 2 \\ -2 & 1 & 1 \end{vmatrix}$

21. $\begin{vmatrix} 1 & -2 & 3 \\ 0 & 0 & 0 \\ 1 & 10 & -12 \end{vmatrix}$ **22.** $\begin{vmatrix} 2 & 3 & 0 \\ 1 & 9 & 0 \\ -1 & -2 & 0 \end{vmatrix}$ **23.** $\begin{vmatrix} 3 & 3 & -1 \\ 2 & 6 & 0 \\ -6 & -6 & 2 \end{vmatrix}$

24. $\begin{vmatrix} 5 & -3 & 2 \\ -5 & 3 & -2 \\ 1 & 0 & 1 \end{vmatrix}$ **25.** $\begin{vmatrix} 1 & 0 & 0 \\ 0 & 1 & 0 \\ 0 & 0 & 1 \end{vmatrix}$ **26.** $\begin{vmatrix} 1 & 0 & 0 \\ 0 & -1 & 0 \\ 1 & 0 & 1 \end{vmatrix}$

27. $\begin{vmatrix} -2 & 0 & 1 \\ 0 & 1 & 0 \\ 0 & 0 & -1 \end{vmatrix}$ **28.** $\begin{vmatrix} 0 & 0 & -1 \\ -1 & 0 & 1 \\ 0 & -1 & 0 \end{vmatrix}$ **29.** $\begin{vmatrix} \sqrt{2} & 4 & 0 \\ 1 & -\sqrt{5} & 7 \\ -5 & \sqrt{5} & 1 \end{vmatrix}$

30. $\begin{vmatrix} \sqrt{3} & 1 & 0 \\ \sqrt{7} & 4 & -1 \\ 5 & 0 & -\sqrt{7} \end{vmatrix}$ **31.** $\begin{vmatrix} 0.4 & -0.8 & 0.6 \\ 0.3 & 0.9 & 0.7 \\ 3.1 & 4.1 & -2.8 \end{vmatrix}$ **32.** $\begin{vmatrix} -0.3 & -0.1 & 0.9 \\ 2.5 & 4.9 & -3.2 \\ -0.1 & 0.4 & 0.8 \end{vmatrix}$

Relating Concepts

For individual or collaborative investigation *(Exercises 33–36)*

The determinant of a 3×3 *matrix A is defined as follows.*

$$\text{If } A = \begin{bmatrix} a_{11} & a_{12} & a_{13} \\ a_{21} & a_{22} & a_{23} \\ a_{31} & a_{32} & a_{33} \end{bmatrix}, \text{ then } |A| = \begin{vmatrix} a_{11} & a_{12} & a_{13} \\ a_{21} & a_{22} & a_{23} \\ a_{31} & a_{32} & a_{33} \end{vmatrix}$$

$$= (a_{11}a_{22}a_{33} + a_{12}a_{23}a_{31} + a_{13}a_{21}a_{32}) - (a_{31}a_{22}a_{13} + a_{32}a_{23}a_{11} + a_{33}a_{21}a_{12}).$$

34. 102
35. 102; yes
36. no

37. $\left\{-\frac{4}{3}\right\}$ 38. $\{0\}$
39. $\{-1, 4\}$ 40. $\left\{-\frac{1}{2}, 6\right\}$
41. $\{-4\}$ 42. $\{-2\}$
43. $\{13\}$ 44. $\left\{-\frac{2}{3}\right\}$
45. $\left\{-\frac{1}{2}\right\}$

46.
$$\begin{matrix} + & - & + & - \\ - & + & - & + \\ + & - & + & - \\ - & + & - & + \end{matrix}$$

47. 1 unit^2 48. 4.5 units^2
49. 9.5 units^2 50. 7 units^2

51. approximately 19,328.3 ft^2

Work these exercises in order.

33. The determinant of a 3×3 matrix can also be found using the method of "diagonals." To use this method, first rewrite columns 1 and 2 of matrix A to the right of matrix A. Next, identify the diagonals d_1 through d_6 and multiply their elements. Find the sum of the products from d_1, d_2, and d_3. Then subtract the sum of the products from d_4, d_5, and d_6 from that sum: $(d_1 + d_2 + d_3) - (d_4 + d_5 + d_6)$. Verify that this method produces the same results as the method given above.

$$\begin{bmatrix} a_{11} & a_{12} & a_{13} \\ a_{21} & a_{22} & a_{23} \\ a_{31} & a_{32} & a_{33} \end{bmatrix} \begin{matrix} a_{11} & a_{12} \\ a_{21} & a_{22} \\ a_{31} & a_{32} \end{matrix}$$

$d_4 \quad d_5 \quad d_6 \qquad d_1 \quad d_2 \quad d_3$ Each d is a product.

34. Find the determinant $\begin{vmatrix} 1 & 3 & 2 \\ 0 & 2 & 6 \\ 7 & 1 & 5 \end{vmatrix}$ using the method of "diagonals."

35. Refer to **Exercise 34.** Find the determinant by expanding about column 1 and using the method of cofactors. Do these methods give the same determinant for 3×3 matrices?

36. *Concept Check* Does the method of finding a determinant using "diagonals" extend to 4×4 matrices?

Solve each equation for x.

37. $\begin{vmatrix} 5 & x \\ -3 & 2 \end{vmatrix} = 6$

38. $\begin{vmatrix} -0.5 & 2 \\ x & x \end{vmatrix} = 0$

39. $\begin{vmatrix} x & 3 \\ x & x \end{vmatrix} = 4$

40. $\begin{vmatrix} 2x & x \\ 11 & x \end{vmatrix} = 6$

41. $\begin{vmatrix} -2 & 0 & 1 \\ -1 & 3 & x \\ 5 & -2 & 0 \end{vmatrix} = 3$

42. $\begin{vmatrix} 4 & 3 & 0 \\ 2 & 0 & 1 \\ -3 & x & -1 \end{vmatrix} = 5$

43. $\begin{vmatrix} 5 & 3x & -3 \\ 0 & 2 & -1 \\ 4 & -1 & x \end{vmatrix} = -7$

44. $\begin{vmatrix} 2x & 1 & -1 \\ 0 & 4 & x \\ 3 & 0 & 2 \end{vmatrix} = x$

45. $\begin{vmatrix} x & 0 & -1 \\ 2 & -3 & x \\ x & 0 & 7 \end{vmatrix} = 12$

46. *Concept Check* Write the sign array representing $(-1)^{i+j}$ for each element of a 4×4 matrix.

Area of a Triangle *A triangle with vertices at (x_1, y_1), (x_2, y_2), and (x_3, y_3), as shown in the figure, has area equal to the absolute value of D, where*

$$D = \frac{1}{2}\begin{vmatrix} x_1 & y_1 & 1 \\ x_2 & y_2 & 1 \\ x_3 & y_3 & 1 \end{vmatrix}.$$

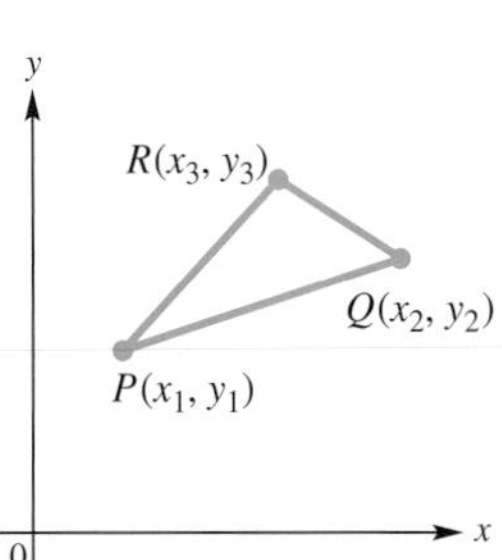

Find the area of each triangle having vertices at P, Q, and R.

47. $P(0, 0), Q(0, 2), R(1, 4)$

48. $P(0, 1), Q(2, 0), R(1, 5)$

49. $P(2, 5), Q(-1, 3), R(4, 0)$

50. $P(2, -2), Q(0, 0), R(-3, -4)$

51. *Area of a Triangle* Find the area of a triangular lot whose vertices have coordinates in feet of

$$(101.3, 52.7), \quad (117.2, 253.9), \quad \text{and} \quad (313.1, 301.6).$$

(*Source:* Al-Khafaji, A. and J. Tooley, *Numerical Methods in Engineering Practice*, Holt, Rinehart, and Winston.)

53. −3 54. −3
55. 15 56. 30
57. 3 58. 3

59. 0 60. 0
61. 0 62. 0
63. 16 64. −49
65. 17 66. −35
67. 54 68. 168
69. 0 70. 0
71. 0 72. 0
73. −88 74. 720
75. 298 76. −311

77. (a) D (b) A
(c) C (d) B

52. Let $A = \begin{bmatrix} a_{11} & a_{12} & a_{13} \\ a_{21} & a_{22} & a_{23} \\ a_{31} & a_{32} & a_{33} \end{bmatrix}$. Find $|A|$ by expansion about row 3 of the matrix. Show that your result is really equal to $|A|$ as given in the definition of the determinant of a 3×3 matrix at the beginning of this section.

Use the determinant theorems and the fact that $\begin{vmatrix} 1 & 2 & 3 \\ 4 & 5 & 6 \\ 7 & 9 & 10 \end{vmatrix} = 3$ *to find the value of each determinant.*

53. $\begin{vmatrix} 4 & 5 & 6 \\ 1 & 2 & 3 \\ 7 & 9 & 10 \end{vmatrix}$

54. $\begin{vmatrix} 3 & 2 & 1 \\ 6 & 5 & 4 \\ 10 & 9 & 7 \end{vmatrix}$

55. $\begin{vmatrix} 5 & 10 & 15 \\ 4 & 5 & 6 \\ 7 & 9 & 10 \end{vmatrix}$

56. $\begin{vmatrix} 1 & 20 & 3 \\ 4 & 50 & 6 \\ 7 & 90 & 10 \end{vmatrix}$

57. $\begin{vmatrix} 1 & 2 & 3 \\ 4 & 5 & 6 \\ 8 & 11 & 13 \end{vmatrix}$

58. $\begin{vmatrix} 1 & 2 & 0 \\ 4 & 5 & -6 \\ 7 & 9 & -11 \end{vmatrix}$

Use the determinant theorems to find the value of each determinant. **See Example 4.**

59. $\begin{vmatrix} 1 & 0 & 0 \\ 1 & 0 & 1 \\ 3 & 0 & 0 \end{vmatrix}$

60. $\begin{vmatrix} -1 & 2 & 4 \\ 4 & -8 & -16 \\ 0 & 0 & 0 \end{vmatrix}$

61. $\begin{vmatrix} 6 & 8 & -12 \\ -1 & 0 & 2 \\ 4 & 0 & -8 \end{vmatrix}$

62. $\begin{vmatrix} 4 & 8 & 0 \\ -1 & -2 & 1 \\ 2 & 4 & 3 \end{vmatrix}$

63. $\begin{vmatrix} -4 & 1 & 4 \\ 2 & 0 & 1 \\ 0 & 2 & 4 \end{vmatrix}$

64. $\begin{vmatrix} 6 & 3 & 2 \\ 1 & 0 & 2 \\ 5 & 7 & 3 \end{vmatrix}$

65. $\begin{vmatrix} 0 & 1 & -3 \\ 7 & 5 & 2 \\ 1 & -2 & 6 \end{vmatrix}$

66. $\begin{vmatrix} 7 & 9 & -3 \\ 7 & -6 & 2 \\ 8 & 1 & 0 \end{vmatrix}$

67. $\begin{vmatrix} 1 & 6 & 7 \\ 0 & 6 & 7 \\ 0 & 0 & 9 \end{vmatrix}$

68. $\begin{vmatrix} 7 & 0 & 0 \\ 1 & 6 & 0 \\ 4 & 2 & 4 \end{vmatrix}$

69. $\begin{vmatrix} 2 & -1 & 3 \\ 6 & 4 & 10 \\ 4 & 5 & 7 \end{vmatrix}$

70. $\begin{vmatrix} 9 & 1 & 7 \\ 12 & 5 & 2 \\ 11 & 4 & 3 \end{vmatrix}$

71. $\begin{vmatrix} -1 & 0 & 2 & 3 \\ 5 & 4 & -3 & 7 \\ 8 & 2 & 9 & -5 \\ 4 & 4 & -1 & 10 \end{vmatrix}$

72. $\begin{vmatrix} 5 & 1 & 4 & 2 \\ 4 & -3 & 7 & -4 \\ 5 & 8 & -3 & 6 \\ 9 & 9 & 0 & 8 \end{vmatrix}$

73. $\begin{vmatrix} 4 & 0 & 0 & 2 \\ -1 & 0 & 3 & 0 \\ 2 & 4 & 0 & 1 \\ 0 & 0 & 1 & 2 \end{vmatrix}$

74. $\begin{vmatrix} -2 & 0 & 4 & 2 \\ 3 & 6 & 0 & 4 \\ 0 & 0 & 0 & 3 \\ 9 & 0 & 2 & -1 \end{vmatrix}$

75. $\begin{vmatrix} 3 & -6 & 5 & -1 \\ 0 & 2 & -1 & 3 \\ -6 & 4 & 2 & 0 \\ -7 & 3 & 1 & 1 \end{vmatrix}$

76. $\begin{vmatrix} 4 & 5 & -1 & -1 \\ 2 & -3 & 1 & 0 \\ -5 & 1 & 3 & 9 \\ 0 & -2 & 1 & 5 \end{vmatrix}$

77. *Concept Check* For the system below, match each determinant in (a)–(d) with its equivalent from choices A–D.

$$4x + 3y - 2z = 1$$
$$7x - 4y + 3z = 2$$
$$-2x + y - 8z = 0$$

(a) D **(b)** D_x **(c)** D_y **(d)** D_z

A. $\begin{vmatrix} 1 & 3 & -2 \\ 2 & -4 & 3 \\ 0 & 1 & -8 \end{vmatrix}$

B. $\begin{vmatrix} 4 & 3 & 1 \\ 7 & -4 & 2 \\ -2 & 1 & 0 \end{vmatrix}$

C. $\begin{vmatrix} 4 & 1 & -2 \\ 7 & 2 & 3 \\ -2 & 0 & -8 \end{vmatrix}$

D. $\begin{vmatrix} 4 & 3 & -2 \\ 7 & -4 & 3 \\ -2 & 1 & -8 \end{vmatrix}$

78. $\{(1, 0, -1)\}$

79. $\{(2, 2)\}$ 80. $\{(-2, 1)\}$

81. $\{(2, -5)\}$ 82. $\{(1, 4)\}$

83. $\{(2, 0)\}$ 84. $\{(0, -2)\}$

85. Cramer's rule does not apply, since $D = 0$; $\emptyset$

86. Cramer's rule does not apply, since $D = 0$; $\emptyset$

87. Cramer's rule does not apply, since $D = 0$; $\left\{\left(\frac{-2y+4}{3}, y\right)\right\}$

88. Cramer's rule does not apply, since $D = 0$; $\left\{\left(\frac{-3y+9}{4}, y\right)\right\}$

89. $\{(-4, 12)\}$ 90. $\{(-16, 6)\}$

91. $\{(-3, 4, 2)\}$ 92. $\{(-4, 3, 5)\}$

93. $\{(0, 0, -1)\}$ 94. $\{(4, 0, 0)\}$

95. Cramer's rule does not apply, since $D = 0$; $\emptyset$

96. Cramer's rule does not apply, since $D = 0$; $\emptyset$

97. Cramer's rule does not apply, since $D = 0$; $\left\{\left(\frac{19z-32}{4}, \frac{-13z+24}{4}, z\right)\right\}$

98. Cramer's rule does not apply, since $D = 0$; $\left\{\left(\frac{6z+5}{11}, \frac{31z+2}{11}, z\right)\right\}$

99. $\{(0, 4, 2)\}$

100. $\{(1, -2, 0)\}$

101. $\left\{\left(\frac{31}{5}, \frac{19}{10}, -\frac{29}{10}\right)\right\}$

102. $\left\{\left(\frac{24}{19}, \frac{63}{38}, \frac{26}{19}\right)\right\}$

103. $W_1 = W_2 = \frac{100\sqrt{3}}{3} \approx 58$ lb

78. *Concept Check* For the following system, $D = -43$, $D_x = -43$, $D_y = 0$, and $D_z = 43$. What is the solution set of the system?

$$\begin{aligned} x + 3y - 6z &= 7 \\ 2x - y + z &= 1 \\ x + 2y + 2z &= -1 \end{aligned}$$

Use Cramer's rule to solve each system of equations. If $D = 0$, use another method to determine the solution set. ***See Examples 5–7.***

79. $\begin{aligned} x + y &= 4 \\ 2x - y &= 2 \end{aligned}$

80. $\begin{aligned} 3x + 2y &= -4 \\ 2x - y &= -5 \end{aligned}$

81. $\begin{aligned} 4x + 3y &= -7 \\ 2x + 3y &= -11 \end{aligned}$

82. $\begin{aligned} 4x - y &= 0 \\ 2x + 3y &= 14 \end{aligned}$

83. $\begin{aligned} 5x + 4y &= 10 \\ 3x - 7y &= 6 \end{aligned}$

84. $\begin{aligned} 3x + 2y &= -4 \\ 5x - y &= 2 \end{aligned}$

85. $\begin{aligned} 1.5x + 3y &= 5 \\ 2x + 4y &= 3 \end{aligned}$

86. $\begin{aligned} 12x + 8y &= 3 \\ 1.5x + y &= 0.9 \end{aligned}$

87. $\begin{aligned} 3x + 2y &= 4 \\ 6x + 4y &= 8 \end{aligned}$

88. $\begin{aligned} 4x + 3y &= 9 \\ 12x + 9y &= 27 \end{aligned}$

89. $\begin{aligned} \frac{1}{2}x + \frac{1}{3}y &= 2 \\ \frac{3}{2}x - \frac{1}{2}y &= -12 \end{aligned}$

90. $\begin{aligned} -\frac{3}{4}x + \frac{2}{3}y &= 16 \\ \frac{5}{2}x + \frac{1}{2}y &= -37 \end{aligned}$

91. $\begin{aligned} 2x - y + 4z &= -2 \\ 3x + 2y - z &= -3 \\ x + 4y + 2z &= 17 \end{aligned}$

92. $\begin{aligned} x + y + z &= 4 \\ 2x - y + 3z &= 4 \\ 4x + 2y - z &= -15 \end{aligned}$

93. $\begin{aligned} 4x - 3y + z + 1 &= 0 \\ 5x + 7y + 2z + 2 &= 0 \\ 3x - 5y - z - 1 &= 0 \end{aligned}$

94. $\begin{aligned} 2x - 3y + z - 8 &= 0 \\ -x - 5y + z + 4 &= 0 \\ 3x - 5y + 2z - 12 &= 0 \end{aligned}$

95. $\begin{aligned} x + 2y + 3z &= 4 \\ 4x + 3y + 2z &= 1 \\ -x - 2y - 3z &= 0 \end{aligned}$

96. $\begin{aligned} 2x - y + 3z &= 1 \\ -2x + y - 3z &= 2 \\ 5x - y + z &= 2 \end{aligned}$

97. $\begin{aligned} -2x - 2y + 3z &= 4 \\ 5x + 7y - z &= 2 \\ 2x + 2y - 3z &= -4 \end{aligned}$

98. $\begin{aligned} 3x - 2y + 4z &= 1 \\ 4x + y - 5z &= 2 \\ -6x + 4y - 8z &= -2 \end{aligned}$

99. $\begin{aligned} 5x - y &= -4 \\ 3x + 2z &= 4 \\ 4y + 3z &= 22 \end{aligned}$

100. $\begin{aligned} 3x + 5y &= -7 \\ 2x + 7z &= 2 \\ 4y + 3z &= -8 \end{aligned}$

101. $\begin{aligned} x + 2y &= 10 \\ 3x + 4z &= 7 \\ -y - z &= 1 \end{aligned}$

102. $\begin{aligned} 5x - 2y &= 3 \\ 4y + z &= 8 \\ x + 2z &= 4 \end{aligned}$

(Modeling) Solve the systems in Exercises 103 and 104.

103. *Roof Trusses* The simplest type of roof truss is a triangle. The truss shown in the figure is used to frame roofs of small buildings. If a 100-pound force is applied at the peak of the truss, then the forces or weights W_1 and W_2 exerted parallel to each rafter of the truss are determined by the following linear system of equations.

$$\frac{\sqrt{3}}{2}(W_1 + W_2) = 100$$

$$W_1 - W_2 = 0$$

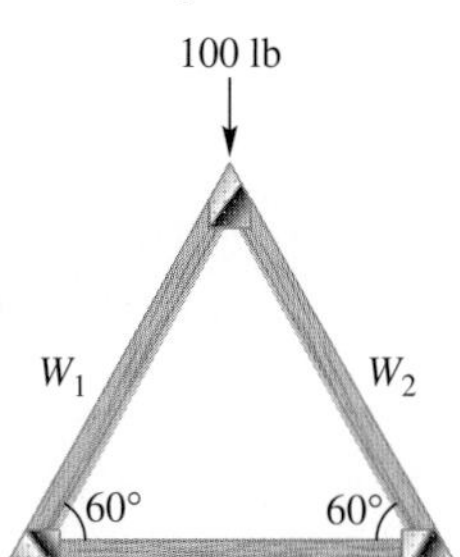

Solve the system to find W_1 and W_2. (*Source:* Hibbeler, R., *Structural Analysis,* Fourth Edition. Reprinted by permission of Pearson Education, Inc., Upper Saddle River, NJ.)

104. $W_1 = \frac{300}{1 + \sqrt{3}} \approx 110$ lb; $W_2 = \frac{300\sqrt{3}}{\sqrt{6} + \sqrt{2}} \approx 134$ lb

105. $\{(-a - b, a^2 + ab + b^2)\}$

106. $\{(-\frac{1}{a}, \frac{a + b}{ab})\}$

107. $\{(1, 0)\}$

108. $\{(\frac{b - ab}{1 - ab}, \frac{a - ab}{1 - ab})\}$

109. $\{(-1, 2)\}$

110. $\{(z - 2, -2z + 3, z)\}$

104. *Roof Trusses* **(Refer to Exercise 103.)** Use the following system of equations to determine the forces or weights W_1 and W_2 exerted on each rafter for the truss shown in the figure.

$$W_1 + \sqrt{2}W_2 = 300$$
$$\sqrt{3}W_1 - \sqrt{2}W_2 = 0$$

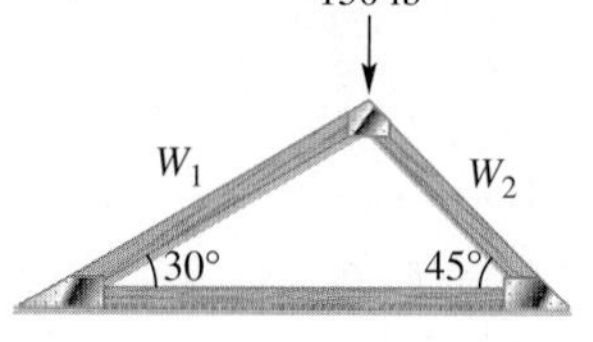

Solve each system for x and y using Cramer's rule. Assume a and b are nonzero constants.

105. $bx + y = a^2$
$ax + y = b^2$

106. $ax + by = \frac{b}{a}$
$x + y = \frac{1}{b}$

107. $b^2x + a^2y = b^2$
$ax + by = a$

108. $x + by = b$
$ax + y = a$

109. Use Cramer's Rule to find the solution set if a, b, c, d, e, and f are consecutive integers.

$$ax + by = c$$
$$dx + ey = f$$

110. In the following system, $a, b, c, \ldots, l$ are consecutive integers. Express the solution set in terms of z.

$$ax + by + cz = d$$
$$ex + fy + gz = h$$
$$ix + jy + kz = l$$

9.4 Partial Fractions

- Decomposition of Rational Expressions
- Distinct Linear Factors
- Repeated Linear Factors
- Distinct Linear and Quadratic Factors
- Repeated Quadratic Factors

Decomposition of Rational Expressions The sums of rational expressions are found by combining two or more rational expressions into one rational expression. Here, the reverse process is considered:

Given one rational expression, express it as the sum of two or more rational expressions.

A special type of sum involving rational expressions is a **partial fraction decomposition**—each term in the sum is a **partial fraction.** The technique of decomposing a rational expression into partial fractions is useful in calculus and other areas of mathematics.

Add rational expressions →

$$\frac{2}{x + 1} + \frac{3}{x} = \frac{5x + 3}{x(x + 1)}$$

← Partial fraction decomposition

LOOKING AHEAD TO CALCULUS
In calculus, partial fraction decomposition provides a powerful technique for determining integrals of rational functions.

Teaching Tip Start by applying the steps for finding partial fraction decomposition to numerical examples such as

$$\frac{5}{6} = \frac{A}{2} + \frac{B}{3}$$

and $\frac{3}{8} = \frac{A}{2} + \frac{B}{2^2} + \frac{C}{2^3}$.

Partial Fraction Decomposition of $\frac{f(x)}{g(x)}$

To form a partial fraction decomposition of a rational expression, follow these steps.

Step 1 If $\frac{f(x)}{g(x)}$ is not a proper fraction (a fraction with the numerator of lesser degree than the denominator), divide $f(x)$ by $g(x)$. For example,

$$\frac{x^4 - 3x^3 + x^2 + 5x}{x^2 + 3} = x^2 - 3x - 2 + \frac{14x + 6}{x^2 + 3}.$$

Then apply the following steps to the remainder, which is a proper fraction.

Step 2 Factor the denominator $g(x)$ completely into factors of the form $(ax + b)^m$ or $(cx^2 + dx + e)^n$, where $cx^2 + dx + e$ is irreducible and m and n are positive integers.

Step 3 **(a)** For each distinct linear factor $(ax + b)$, the decomposition must include the term $\frac{A}{ax + b}$.

(b) For each repeated linear factor $(ax + b)^m$, the decomposition must include the terms

$$\frac{A_1}{ax + b} + \frac{A_2}{(ax + b)^2} + \cdots + \frac{A_m}{(ax + b)^m}.$$

Step 4 **(a)** For each distinct quadratic factor $(cx^2 + dx + e)$, the decomposition must include the term $\frac{Bx + C}{cx^2 + dx + e}$.

(b) For each repeated quadratic factor $(cx^2 + dx + e)^n$, the decomposition must include the terms

$$\frac{B_1x + C_1}{cx^2 + dx + e} + \frac{B_2x + C_2}{(cx^2 + dx + e)^2} + \cdots + \frac{B_nx + C_n}{(cx^2 + dx + e)^n}.$$

Step 5 Use algebraic techniques to solve for the constants in the numerators of the decomposition.

Teaching Tip Have students identify the initial setups for several examples.

To find the constants in Step 5, the goal is to get a system of equations with as many equations as there are unknowns in the numerators.

Distinct Linear Factors

Classroom Example 1
Find the partial fraction decomposition of

$$\frac{3x^3 - 3x^2 + 7x - 4}{x^2 - x}.$$

Answer: $3x + \frac{4}{x} + \frac{3}{x - 1}$

EXAMPLE 1 Finding a Partial Fraction Decomposition

Find the partial fraction decomposition of $\frac{2x^4 - 8x^2 + 5x - 2}{x^3 - 4x}$.

SOLUTION The given fraction is not a proper fraction—the numerator has greater degree than the denominator. Perform the division.

$$\begin{array}{r} 2x \\ x^3 - 4x \overline{)\,2x^4 - 8x^2 + 5x - 2} \\ \underline{2x^4 - 8x^2 } \\ 5x - 2 \end{array}$$ (Section R.3)

The quotient is $$\frac{2x^4 - 8x^2 + 5x - 2}{x^3 - 4x} = 2x + \frac{5x - 2}{x^3 - 4x}.$$

Now, work with the remainder fraction. Factor the denominator as

$$x^3 - 4x = x(x + 2)(x - 2).$$

The factors are distinct linear factors. Use Step 3(a) to write the decomposition as

$$\frac{5x - 2}{x^3 - 4x} = \frac{A}{x} + \frac{B}{x + 2} + \frac{C}{x - 2}, \quad (1)$$

where A, B, and C are constants that need to be found. Multiply each side of equation (1) by $x(x + 2)(x - 2)$ to obtain

$$5x - 2 = A(x + 2)(x - 2) + Bx(x - 2) + Cx(x + 2). \quad (2)$$

Equation (1) is an identity since each side represents the same rational expression. Thus, equation (2) is also an identity. Equation (1) holds for all values of x except 0, -2, and 2. However, equation (2) holds for all values of x. In particular, substituting 0 for x in equation (2) gives $-2 = -4A$, so $A = \frac{1}{2}$. Similarly, choosing $x = -2$ gives $-12 = 8B$, so $B = -\frac{3}{2}$. Finally, choosing $x = 2$ gives $8 = 8C$, so $C = 1$. The remainder rational expression can be written as the following sum of partial fractions.

$$\frac{5x - 2}{x^3 - 4x} = \frac{1}{2x} + \frac{-3}{2(x + 2)} + \frac{1}{x - 2}$$

The given rational expression can be written as follows.

$$\frac{2x^4 - 8x^2 + 5x - 2}{x^3 - 4x} = 2x + \frac{1}{2x} + \frac{-3}{2(x + 2)} + \frac{1}{x - 2}$$

Check the work by combining the terms on the right.

✔ *Now Try Exercise 15.*

Repeated Linear Factors

Classroom Example 2
Find the partial fraction decomposition of

$$\frac{x^2 - 4x + 7}{(x - 3)^3}.$$

Answer:
$$\frac{1}{x - 3} + \frac{2}{(x - 3)^2} + \frac{4}{(x - 3)^3}$$

EXAMPLE 2 Finding a Partial Fraction Decomposition

Find the partial fraction decomposition of $\frac{2x}{(x - 1)^3}$.

SOLUTION This is a proper fraction. The denominator is already factored with repeated linear factors. Write the decomposition as shown, by using Step 3(b).

$$\frac{2x}{(x - 1)^3} = \frac{A}{x - 1} + \frac{B}{(x - 1)^2} + \frac{C}{(x - 1)^3}$$

Clear the denominators by multiplying each side of this equation by $(x - 1)^3$.

$$2x = A(x - 1)^2 + B(x - 1) + C \quad \text{(Section R.5)}$$

Substituting 1 for x leads to $C = 2$.

$$2x = A(x - 1)^2 + B(x - 1) + 2 \quad (1)$$

The only root has been substituted, and values for A and B still need to be found. However, *any* number can be substituted for x. For example, when we choose $x = -1$ (because it is easy to substitute), equation (1) becomes the following.

$$-2 = 4A - 2B + 2 \quad \text{Let } x = -1 \text{ in (1).}$$

$$-4 = 4A - 2B \quad \text{Subtract 2.}$$

$$-2 = 2A - B \quad \text{Divide by 2.} \quad (2)$$

Substituting 0 for x in equation (1) gives another equation in A and B.

$$0 = A - B + 2 \quad \text{Let } x = 0 \text{ in (1).}$$

$$2 = -A + B \quad \text{Subtract 2. Multiply by } -1. \quad (3)$$

$$-2 = 2A - B \quad (2)$$

$$2 = -A + B \quad (3)$$

$$0 = A \quad \text{Add.}$$

If $A = 0$, then using equation (3),

$$2 = 0 + B$$

$$2 = B.$$

Now, solve the system of equations (2) and (3) as shown in the margin to get $A = 0$ and $B = 2$. Substitute these values for A and B and 2 for C to obtain the partial fraction decomposition.

$$\frac{2x}{(x-1)^3} = \frac{2}{(x-1)^2} + \frac{2}{(x-1)^3}$$

We needed three substitutions because there were three constants to evaluate: A, B, and C. To check this result, we could combine the terms on the right.

Now Try Exercise 11.

Distinct Linear and Quadratic Factors

Classroom Example 3
Find the partial fraction decomposition of

$$\frac{8x^2 + 10x - 3}{(x+4)(x^2+1)}.$$

Answer: $\frac{5}{x+4} + \frac{3x-2}{x^2+1}$

EXAMPLE 3 Finding a Partial Fraction Decomposition

Find the partial fraction decomposition of $\frac{x^2 + 3x - 1}{(x+1)(x^2+2)}$.

SOLUTION The denominator $(x + 1)(x^2 + 2)$ has distinct linear and quadratic factors, where neither is repeated. Since $x^2 + 2$ cannot be factored, it is irreducible. The partial fraction decomposition is of the following form.

$$\frac{x^2 + 3x - 1}{(x+1)(x^2+2)} = \frac{A}{x+1} + \frac{Bx + C}{x^2+2}$$

Multiply each side by $(x + 1)(x^2 + 2)$.

$$x^2 + 3x - 1 = A(x^2 + 2) + (Bx + C)(x + 1) \quad (1)$$

First, substitute -1 for x.

Use parentheses around substituted values to avoid errors.

$$(-1)^2 + 3(-1) - 1 = A[(-1)^2 + 2] + 0$$

$$-3 = 3A$$

$$A = -1$$

Replace A with -1 in equation (1) and substitute any value for x. Let $x = 0$.

$$0^2 + 3(0) - 1 = -1(0^2 + 2) + (B \cdot 0 + C)(0 + 1)$$

$$-1 = -2 + C$$

$$C = 1$$

Now, letting $A = -1$ and $C = 1$, substitute again in equation (1), using another value for x. Let $x = 1$.

$$3 = -3 + (B + 1)(2)$$

$$6 = 2B + 2$$

$$B = 2$$

Use $A = -1$, $B = 2$, and $C = 1$ to find the partial fraction decomposition.

$$\frac{x^2 + 3x - 1}{(x+1)(x^2+2)} = \frac{-1}{x+1} + \frac{2x + 1}{x^2+2} \quad \text{Check by combining the terms on the right.}$$

Now Try Exercise 21.

For fractions with denominators that have quadratic factors, another method is often more convenient. The system of equations is formed by equating coefficients of like terms on each side of the partial fraction decomposition. For instance, in **Example 3,** equation (1) was

$$x^2 + 3x - 1 = A(x^2 + 2) + (Bx + C)(x + 1). \quad (1)$$

Multiply on the right and collect like terms.

$$x^2 + 3x - 1 = Ax^2 + 2A + Bx^2 + Bx + Cx + C$$

$$1x^2 + 3x - 1 = (A + B)x^2 + (B + C)x + (C + 2A)$$

Now, equate the coefficients of like powers of x to obtain three equations.

$$1 = A + B$$

$$3 = B + C$$

$$-1 = C + 2A$$

Solving this system for A, B, and C gives the partial fraction decomposition.

Repeated Quadratic Factors

Classroom Example 4
Find the partial fraction decomposition of

$$\frac{x^4 + 3x^3 + 4x^2 + 2x + 4}{x(x^2 + 2)^2}.$$

Answer:
$\frac{1}{x} + \frac{3}{x^2 + 2} + \frac{-4}{(x^2 + 2)^2}$

EXAMPLE 4 Finding a Partial Fraction Decomposition

Find the partial fraction decomposition of $\dfrac{2x}{(x^2 + 1)^2(x - 1)}$.

SOLUTION This expression has both a linear factor and a repeated quadratic factor. Use Steps 3(a) and 4(b) from the beginning of this section.

$$\frac{2x}{(x^2 + 1)^2(x - 1)} = \frac{Ax + B}{x^2 + 1} + \frac{Cx + D}{(x^2 + 1)^2} + \frac{E}{x - 1}$$

Multiply each side by $(x^2 + 1)^2(x - 1)$.

$$2x = (Ax + B)(x^2 + 1)(x - 1) + (Cx + D)(x - 1) + E(x^2 + 1)^2 \quad (1)$$

If $x = 1$, then equation (1) reduces to $2 = 4E$, or $E = \frac{1}{2}$. Substitute $\frac{1}{2}$ for E in equation (1), and expand and combine like terms on the right.

$$2x = \left(A + \frac{1}{2}\right)x^4 + (-A + B)x^3 + (A - B + C + 1)x^2$$
$$+ (-A + B + D - C)x + \left(-B - D + \frac{1}{2}\right) \quad (2)$$

To get additional equations involving the unknowns, equate the coefficients of like powers of x on the two sides of equation (2). Setting corresponding coefficients of x^4 equal, $0 = A + \frac{1}{2}$, or $A = -\frac{1}{2}$. From the corresponding coefficients of x^3, $0 = -A + B$. Since $A = -\frac{1}{2}$, $B = -\frac{1}{2}$.

Using the coefficients of x^2, $0 = A - B + C + 1$. Since $A = -\frac{1}{2}$ and $B = -\frac{1}{2}$, $C = -1$. Finally, from the coefficients of x, $2 = -A + B + D - C$. Substituting for A, B, and C gives $D = 1$. With $A = -\frac{1}{2}$, $B = -\frac{1}{2}$, $C = -1$, $D = 1$, and $E = \frac{1}{2}$, the given fraction has the following partial fraction decomposition.

$$\frac{2x}{(x^2 + 1)^2(x - 1)} = \frac{-\frac{1}{2}x - \frac{1}{2}}{x^2 + 1} + \frac{-x + 1}{(x^2 + 1)^2} + \frac{\frac{1}{2}}{x - 1},$$

or $$\frac{2x}{(x^2 + 1)^2(x - 1)} = \frac{-(x + 1)}{2(x^2 + 1)} + \frac{-x + 1}{(x^2 + 1)^2} + \frac{1}{2(x - 1)}$$

Simplify complex fractions. (Section R.5)

✔ Now Try Exercise 25.

In summary, to solve for the constants in the numerators of a partial fraction decomposition, use either of the following methods or a combination of the two.

Techniques for Decomposition into Partial Fractions

Method 1 **For Linear Factors**

Step 1 Multiply each side of the resulting rational equation by the common denominator.

Step 2 Substitute the zero of each factor in the resulting equation. For repeated linear factors, substitute as many other numbers as necessary to find all the constants in the numerators. The number of substitutions required will equal the number of constants $A, B, \ldots$.

Method 2 **For Quadratic Factors**

Step 1 Multiply each side of the resulting rational equation by the common denominator.

Step 2 Collect like terms on the right side of the equation.

Step 3 Equate the coefficients of like terms to get a system of equations.

Step 4 Solve the system to find the constants in the numerators.

9.4 Exercises

Find the partial fraction decomposition for each rational expression. ***See Examples 1–4.***

1. $\dfrac{5}{3x(2x+1)}$ **2.** $\dfrac{3x-1}{x(x+1)}$ **3.** $\dfrac{4x+2}{(x+2)(2x-1)}$

4. $\dfrac{x+2}{(x+1)(x-1)}$ **5.** $\dfrac{x}{x^2+4x-5}$ **6.** $\dfrac{5x-3}{x^2-2x-3}$

7. $\dfrac{4}{x(1-x)}$ **8.** $\dfrac{9}{x(x-3)}$

9. $\dfrac{4x^2-x-15}{x(x+1)(x-1)}$ **10.** $\dfrac{3}{(x+1)(x+3)}$

11. $\dfrac{2x+1}{(x+2)^3}$ **12.** $\dfrac{2x}{(x+1)(x+2)^2}$

13. $\dfrac{x^2}{x^2+2x+1}$ **14.** $\dfrac{2}{x^2(x+3)}$

15. $\dfrac{2x^5+3x^4-3x^3-2x^2+x}{2x^2+5x+2}$ **16.** $\dfrac{6x^5+7x^4-x^2+2x}{3x^2+2x-1}$

17. $\dfrac{x^3+4}{9x^3-4x}$ **18.** $\dfrac{x^3+2}{x^3-3x^2+2x}$

19. $\dfrac{-3}{x^2(x^2+5)}$ **20.** $\dfrac{1}{x^2(x^2-2)}$

21. $\dfrac{3x-2}{(x+4)(3x^2+1)}$ **22.** $\dfrac{2x+1}{(x+1)(x^2+2)}$

23. $\dfrac{1}{x(2x+1)(3x^2+4)}$ **24.** $\dfrac{3}{x(x+1)(x^2+1)}$

1. $\frac{5}{3x}+\frac{-10}{3(2x+1)}$ **2.** $\frac{-1}{x}+\frac{4}{x+1}$

3. $\frac{6}{5(x+2)}+\frac{8}{5(2x-1)}$

4. $\frac{-1}{2(x+1)}+\frac{3}{2(x-1)}$

5. $\frac{5}{6(x+5)}+\frac{1}{6(x-1)}$

6. $\frac{2}{x+1}+\frac{3}{x-3}$

7. $\frac{4}{x}+\frac{4}{1-x}$ **8.** $\frac{-3}{x}+\frac{3}{x-3}$

9. $\frac{15}{x}+\frac{-5}{x+1}+\frac{-6}{x-1}$

10. $\frac{-3}{2(x+3)}+\frac{3}{2(x+1)}$

11. $\frac{2}{(x+2)^2}+\frac{-3}{(x+2)^3}$

12. $\frac{-2}{x+1}+\frac{2}{x+2}+\frac{4}{(x+2)^2}$

13. $1+\frac{-2}{x+1}+\frac{1}{(x+1)^2}$

14. $\frac{-2}{9x}+\frac{2}{3x^2}+\frac{2}{9(x+3)}$

15. $x^3-x^2+\frac{-1}{3(2x+1)}+\frac{2}{3(x+2)}$

16. $2x^3+x^2+\frac{1}{2(3x-1)}+\frac{1}{2(x+1)}$

17. $\frac{1}{9}+\frac{-1}{x}+\frac{25}{18(3x+2)}+\frac{29}{18(3x-2)}$

18. $1+\frac{1}{x}+\frac{5}{x-2}+\frac{-3}{x-1}$

19. $\frac{-3}{5x^2}+\frac{3}{5(x^2+5)}$

20. $\frac{-1}{2x^2} + \frac{1}{2(x^2 - 2)}$
21. $\frac{-2}{7(x + 4)} + \frac{6x - 3}{7(3x^2 + 1)}$
22. $\frac{-1}{3(x + 1)} + \frac{x + 5}{3(x^2 + 2)}$
23. $\frac{1}{4x} + \frac{-8}{19(2x + 1)} + \frac{-9x - 24}{76(3x^2 + 4)}$
24. $\frac{3}{x} + \frac{-3}{2(x + 1)} + \frac{-3(x + 1)}{2(x^2 + 1)}$
25. $\frac{-1}{x} + \frac{2x}{2x^2 + 1} + \frac{2x + 3}{(2x^2 + 1)^2}$
26. $\frac{1}{x} + \frac{-2x}{(x^2 + 1)^2}$
27. $\frac{-1}{x + 2} + \frac{3}{(x^2 + 4)^2}$
28. $\frac{3}{x - 1} + \frac{1}{x^2 + 1} + \frac{-2}{(x^2 + 1)^2}$

25. $\dfrac{3x - 1}{x(2x^2 + 1)^2}$

26. $\dfrac{x^4 + 1}{x(x^2 + 1)^2}$

27. $\dfrac{-x^4 - 8x^2 + 3x - 10}{(x + 2)(x^2 + 4)^2}$

28. $\dfrac{3x^4 + x^3 + 5x^2 - x + 4}{(x - 1)(x^2 + 1)^2}$

29. $\dfrac{5x^5 + 10x^4 - 15x^3 + 4x^2 + 13x - 9}{x^3 + 2x^2 - 3x}$

30. $\dfrac{3x^6 + 3x^4 + 3x}{x^4 + x^2}$

31. $\dfrac{x^2}{x^4 - 1}$

32. $\dfrac{-2x^2 - 24}{x^4 - 16}$

33. $\dfrac{4x^2 - 3x - 4}{x^3 + x^2 - 2x}$

34. $\dfrac{2x + 4}{x^3 - 2x^2}$

29. $5x^2 + \frac{3}{x} + \frac{-1}{x + 3} + \frac{2}{x - 1}$ 30. $3x^2 + \frac{3}{x} + \frac{-3x}{x^2 + 1}$ 31. $\frac{-1}{4(x + 1)} + \frac{1}{4(x - 1)} + \frac{1}{2(x^2 + 1)}$
32. $\frac{1}{x + 2} + \frac{-1}{x - 2} + \frac{2}{x^2 + 4}$ 33. $\frac{2}{x} + \frac{-1}{x - 1} + \frac{3}{x + 2}$ 34. $\frac{-2}{x} + \frac{-2}{x^2} + \frac{2}{x - 2}$

Chapter 9 Quiz (Sections 9.1–9.4)

Solve each system, using the method indicated, if possible.

1. (Substitution)
$$\begin{aligned} 2x + y &= -4 \\ -x + 2y &= 2 \end{aligned}$$

2. (Substitution)
$$\begin{aligned} 5x + 10y &= 10 \\ x + 2y &= 2 \end{aligned}$$

3. (Elimination)
$$\begin{aligned} x - y &= 6 \\ x - y &= 4 \end{aligned}$$

4. (Elimination)
$$\begin{aligned} 2x - 3y &= 18 \\ 5x + 2y &= 7 \end{aligned}$$

5. (Gauss-Jordan)
$$\begin{aligned} 3x + 5y &= -5 \\ -2x + 3y &= 16 \end{aligned}$$

6. (Cramer's rule)
$$\begin{aligned} 5x + 2y &= -3 \\ 4x - 3y &= -30 \end{aligned}$$

7. (Elimination)
$$\begin{aligned} x + y + z &= 1 \\ -x + y + z &= 5 \\ y + 2z &= 5 \end{aligned}$$

8. (Gauss-Jordan)
$$\begin{aligned} 2x + 4y + 4z &= 4 \\ x + 3y + z &= 4 \\ -x + 3y + 2z &= -1 \end{aligned}$$

9. (Cramer's rule)
$$\begin{aligned} 7x + y - z &= 4 \\ 2x - 3y + z &= 2 \\ -6x + 9y - 3z &= -6 \end{aligned}$$

Solve each problem.

10. *Spending on Food* In 2008, the amount spent by a typical American household on food was about \$6450. For every \$10 spent on food away from home, about \$14 was spent on food at home. Find the amount of household spending on food in each category. (*Source:* U.S. Bureau of Labor Statistics.)

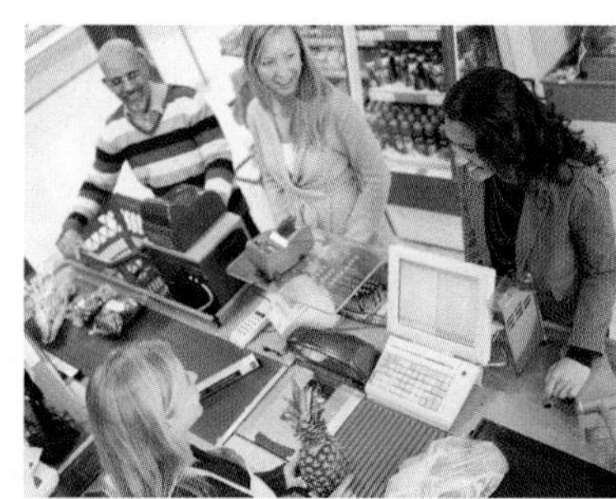

11. *Investments* A sum of \$5000 is invested in three accounts that pay 3%, 4%, and 6% interest rates. The amount of money invested in the account paying 6% equals the total amount of money invested in the other two accounts, and the total annual interest from all three investments is \$240. Find the amount invested at each rate.

12. Let $A = \begin{bmatrix} -5 & 4 \\ 2 & -1 \end{bmatrix}$. Find $|A|$.

13. Evaluate $\begin{vmatrix} -1 & 2 & 4 \\ -3 & -2 & -3 \\ 2 & -1 & 5 \end{vmatrix}$. Use determinant theorems if desired.

Find the partial fraction decomposition for each rational expression.

14. $\dfrac{10x + 13}{x^2 - x - 20}$

15. $\dfrac{2x^2 - 15x - 32}{(x - 1)(x^2 + 6x + 8)}$

[9.1]
1. $\{(-2, 0)\}$
2. $\left\{\left(x, \frac{-x + 2}{2}\right)\right\}$, or $\{(-2y + 2, y)\}$
3. $\emptyset$ 4. $\{(3, -4)\}$

[9.2]
5. $\{(-5, 2)\}$

[9.3]
6. $\{(-3, 6)\}$

[9.1]
7. $\{(-2, 1, 2)\}$

[9.2]
8. $\{(2, 1, -1)\}$

[9.3]
9. Cramer's rule does not apply, since $D = 0$; $\left\{\left(\frac{2y + 6}{9}, y, \frac{23y + 6}{9}\right)\right\}$

[9.1]
10. at home: \$3762.50; away from home: \$2687.50
11. \$1000 at 3%; \$1500 at 4%; \$2500 at 6%

[9.3]
12. -3
13. 59

[9.4]
14. $\frac{7}{x - 5} + \frac{3}{x + 4}$
15. $\frac{-1}{x + 2} + \frac{6}{x + 4} + \frac{-3}{x - 1}$

9.5 Nonlinear Systems of Equations

- Solving Nonlinear Systems with Real Solutions
- Solving Nonlinear Systems with Nonreal Complex Solutions
- Applying Nonlinear Systems

Solving Nonlinear Systems with Real Solutions A system of equations in which at least one equation is *not* linear is a **nonlinear system.**

$$x^2 - y = 4 \qquad x^2 + y^2 = 16$$
$$x + y = -2 \qquad |x| + y = 4 \qquad \text{Nonlinear systems}$$

The substitution method works well for solving many such systems, particularly when one of the equations is linear, as in the next example.

Classroom Example 1
Solve the system.

$$x^2 + y = 6$$
$$x - y = -4$$

Answer: $\{(-2, 2), (1, 5)\}$

EXAMPLE 1 Solving a Nonlinear System by Substitution

Solve the system.

$$x^2 - y = 4 \quad (1)$$
$$x + y = -2 \quad (2)$$

ALGEBRAIC SOLUTION

When one of the equations in a nonlinear system is linear, it is usually best to begin by solving the linear equation for one of the variables.

$$y = -2 - x \quad \text{Solve equation (2) for } y.$$

Substitute this result for y in equation (1).

$$x^2 - y = 4 \quad (1)$$
$$x^2 - (-2 - x) = 4 \quad \text{Let } y = -2 - x. \text{ (Section 9.1)}$$
$$x^2 + 2 + x = 4 \quad \text{Distributive property (Section R.2)}$$
$$x^2 + x - 2 = 0 \quad \text{Standard form (Section 1.4)}$$
$$(x + 2)(x - 1) = 0 \quad \text{Factor. (Section R.4)}$$
$$x + 2 = 0 \quad \text{or} \quad x - 1 = 0 \quad \text{Zero-factor property (Section 1.4)}$$
$$x = -2 \quad \text{or} \quad x = 1$$

Substituting -2 for x in equation (2) gives $y = 0$. If $x = 1$, then $y = -3$. The solution set of the given system is $\{(-2, 0), (1, -3)\}$. A graph of the system is shown in **Figure 9.**

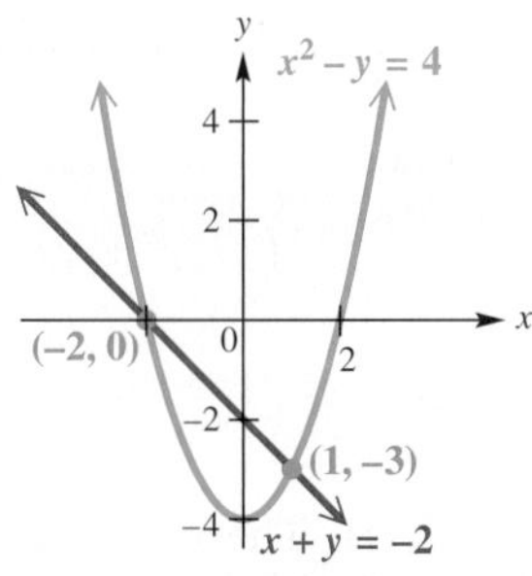

Figure 9

GRAPHING CALCULATOR SOLUTION

Solve each equation for y and graph them in the same viewing window. We obtain

$$Y_1 = X^2 - 4$$

and

$$Y_2 = -X - 2.$$

The screens in **Figure 10,** which indicate that the points of intersection are $(-2, 0)$ and $(1, -3)$, support the solution found algebraically.

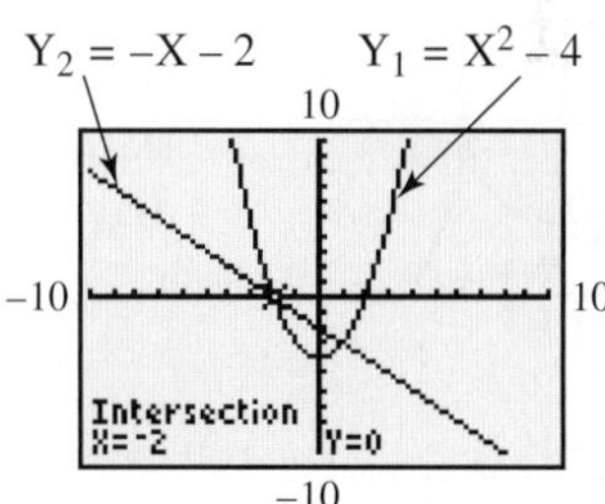

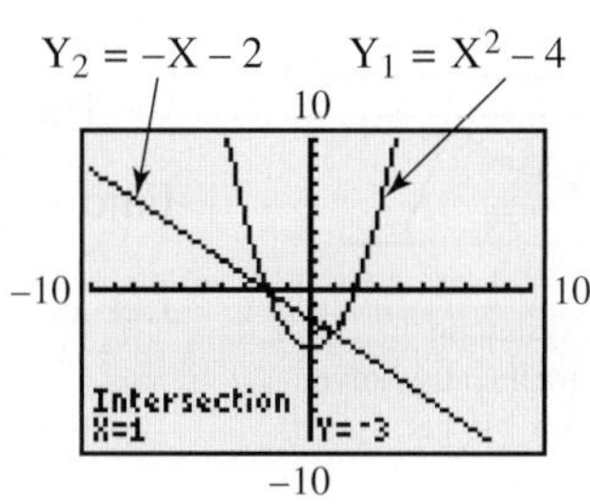

Figure 10

Now Try Exercise 9.

LOOKING AHEAD TO CALCULUS

In calculus, finding the maximum and minimum points for a function of several variables usually requires solving a nonlinear system of equations.

CAUTION If we had solved for x in equation (2) to begin the algebraic solution in **Example 1,** we would have found $y = 0$ or $y = -3$. Substituting $y = 0$ into equation (1) gives $x^2 = 4$, so $x = 2$ or $x = -2$, leading to the ordered pairs $(2, 0)$ and $(-2, 0)$. The ordered pair $(2, 0)$ does not satisfy equation (2), however. ***This illustrates the necessity of checking by substituting all proposed solutions into each equation of the system.***

Visualizing the types of graphs involved in a nonlinear system helps predict the possible numbers of ordered pairs of real numbers that may be in the solution set of the system. For example, a line and a parabola may have 0, 1, or 2 points of intersection, as shown in **Figure 11.**

Teaching Tip Remind students to take the positive *and* negative square roots when solving for variables raised to even powers, as in the solution to **Example 2.**

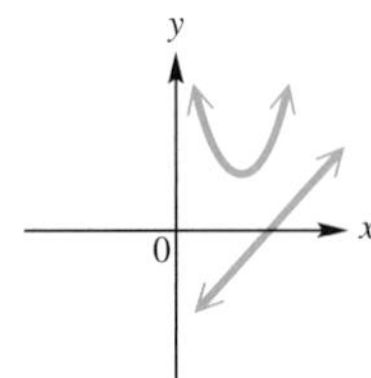

No points of intersection

One point of intersection

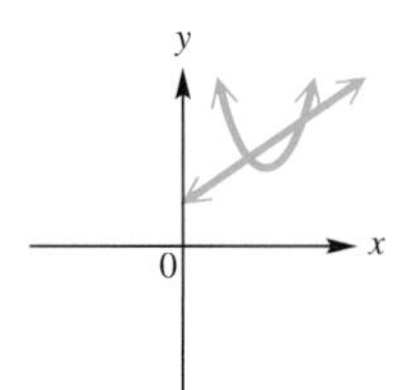

Two points of intersection

Figure 11

Nonlinear systems where both variables are squared in both equations are best solved by elimination, as shown in the next example.

Classroom Example 2
Solve the system.

$$x^2 + y^2 = 9$$
$$9x^2 + 4y^2 = 36$$

Answer: $\{(0, -3), (0, 3)\}$

EXAMPLE 2 Solving a Nonlinear System by Elimination

Solve the system.

$$x^2 + y^2 = 4 \quad (1)$$
$$2x^2 - y^2 = 8 \quad (2)$$

SOLUTION The graph of equation (1) is a circle **(Section 2.2)**, and, as we will learn in the next chapter, the graph of equation (2) is a *hyperbola.* These graphs may intersect in 0, 1, 2, 3, or 4 points. We add to eliminate y^2.

$$x^2 + y^2 = 4 \quad (1)$$
$$2x^2 - y^2 = 8 \quad (2)$$
$$3x^2 = 12 \quad \text{Add. (Section 9.1)}$$
$$x^2 = 4 \quad \text{Divide by 3.}$$
$$x = \pm 2 \quad \text{Square root property (Section 1.4)}$$

Remember to find *both* square roots.

Find y by substituting the values of x in either equation (1) or equation (2).

Let $x = 2$.

$$2^2 + y^2 = 4 \quad (1)$$
$$y^2 = 0$$
$$y = 0$$

Let $x = -2$.

$$(-2)^2 + y^2 = 4 \quad (1)$$
$$y^2 = 0$$
$$y = 0$$

The proposed solutions are $(2, 0)$ and $(-2, 0)$. These satisfy both equations, confirming that the solution set is $\{(2, 0), (-2, 0)\}$.

Now Try Exercise 17.

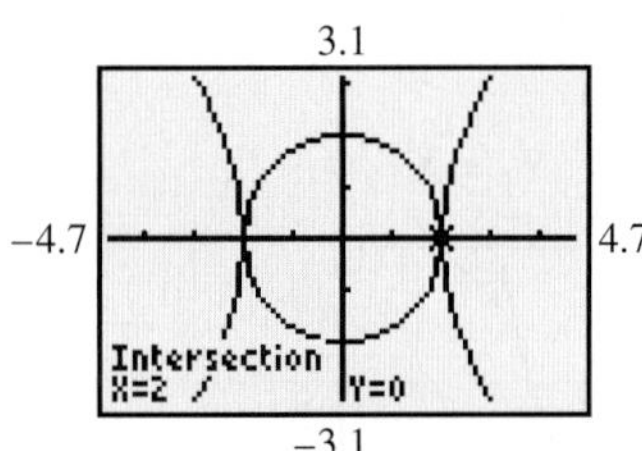

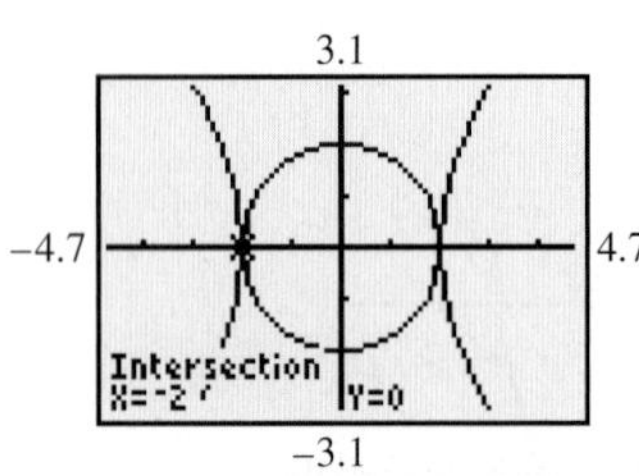

To solve the system in **Example 2** graphically, solve equation (1) for y to get

$$y = \pm\sqrt{4 - x^2}.$$

Similarly, equation (2) yields

$$y = \pm\sqrt{-8 + 2x^2}.$$

Graph these *four* functions to find the two solutions.

NOTE The elimination method works with the system in **Example 2** since the system can be thought of as a system of linear equations where the variables are x^2 and y^2. To see this, substitute u for x^2 and v for y^2. The resulting system is linear in u and v.

Teaching Tip Present several systems and have students determine the best method for solving each one. Then have students predict the number of solutions that can be obtained.

Sometimes a combination of the elimination method and the substitution method is effective in solving a system, as illustrated in **Example 3.**

Classroom Example 3
Solve the system.

$$x^2 + xy + y^2 = 21$$
$$x^2 - xy + y^2 = 9$$

Answer:
$\{(\sqrt{3}, 2\sqrt{3}), (-\sqrt{3}, -2\sqrt{3}), (2\sqrt{3}, \sqrt{3}), (-2\sqrt{3}, -\sqrt{3})\}$

EXAMPLE 3 Solving a Nonlinear System by a Combination of Methods

Solve the system.

$$x^2 + 3xy + y^2 = 22 \quad (1)$$
$$x^2 - xy + y^2 = 6 \quad (2)$$

SOLUTION Begin as with the elimination method.

$$\begin{aligned} x^2 + 3xy + y^2 &= 22 && (1) \\ -x^2 + xy - y^2 &= -6 && \text{Multiply (2) by } -1. \\ \hline 4xy &= 16 && \text{Add.} \quad (3) \\ y &= \frac{4}{x} && \text{Solve for } y \ (x \neq 0). \quad (4) \end{aligned}$$

Now substitute $\frac{4}{x}$ for y in either equation (1) or equation (2). We use equation (2).

$$\begin{aligned} x^2 - x\left(\frac{4}{x}\right) + \left(\frac{4}{x}\right)^2 &= 6 && \text{Let } y = \tfrac{4}{x} \text{ in (2).} \\ x^2 - 4 + \frac{16}{x^2} &= 6 && \text{Multiply and square.} \\ x^4 - 4x^2 + 16 &= 6x^2 && \text{Multiply by } x^2 \text{ to clear fractions. (Section 1.6)} \\ x^4 - 10x^2 + 16 &= 0 && \text{Subtract } 6x^2. \\ (x^2 - 2)(x^2 - 8) &= 0 && \text{Factor.} \end{aligned}$$

This equation is quadratic in form.

$$\begin{aligned} x^2 - 2 = 0 \quad &\text{or} \quad x^2 - 8 = 0 && \text{Zero-factor property} \\ x^2 = 2 \quad &\text{or} \quad x^2 = 8 && \text{Solve each equation.} \\ x = \pm\sqrt{2} \quad &\text{or} \quad x = \pm 2\sqrt{2} && \text{Square root property; } \pm\sqrt{8} = \pm\sqrt{4}\cdot\sqrt{2} = \pm 2\sqrt{2} \text{ (Section R.7)} \end{aligned}$$

For each equation, include *both* square roots.

Substitute these x-values into equation (4) to find corresponding values of y.

Let $x = \sqrt{2}$.	Let $x = -\sqrt{2}$.	Let $x = 2\sqrt{2}$.	Let $x = -2\sqrt{2}$.
$y = \frac{4}{\sqrt{2}} = 2\sqrt{2}$	$y = \frac{4}{-\sqrt{2}} = -2\sqrt{2}$	$y = \frac{4}{2\sqrt{2}} = \sqrt{2}$	$y = \frac{4}{-2\sqrt{2}} = -\sqrt{2}$

The solution set of the system is

$$\{(\sqrt{2}, 2\sqrt{2}), (-\sqrt{2}, -2\sqrt{2}), (2\sqrt{2}, \sqrt{2}), (-2\sqrt{2}, -\sqrt{2})\}.$$

Verify these solutions by substitution in the original system.

Now Try Exercise 35.

Classroom Example 4
Solve the system.

$$x^2 + y^2 = 4$$
$$-|x| + y = 0$$

Answer:
$\{(\sqrt{2}, \sqrt{2}), (-\sqrt{2}, \sqrt{2})\}$

EXAMPLE 4 Solving a Nonlinear System with an Absolute Value Equation

Solve the system.

$$x^2 + y^2 = 16 \quad (1)$$
$$|x| + y = 4 \quad (2)$$

SOLUTION Use the substitution method. Begin by solving equation (2) for $|x|$.

$$|x| = 4 - y \quad (3)$$

In equation (1), the first term is x^2, which is the same as $|x|^2$. Therefore, we substitute $4 - y$ for x in equation (1).

$$x^2 + y^2 = 16 \quad (1)$$
$$(4 - y)^2 + y^2 = 16 \quad \text{Let } x = 4 - y.$$
$$(16 - 8y + y^2) + y^2 = 16 \quad \text{Square the binomial. (Section R.3)}$$
$$2y^2 - 8y = 0 \quad \text{Combine like terms.}$$
$$y^2 - 4y = 0 \quad \text{Divide by 2.}$$
$$y(y - 4) = 0 \quad \text{Factor.}$$
$$y = 0 \quad \text{or} \quad y - 4 = 0 \quad \text{Zero-factor property}$$
$$y = 4 \quad \text{Add 4.}$$

Remember the middle term.

To solve for the corresponding values of x, use either equation (1) or equation (2).

Let $y = 0$.

$$x^2 + 0^2 = 16 \quad (1)$$
$$x^2 = 16$$
$$x = \pm 4$$

Let $y = 4$.

$$x^2 + 4^2 = 16 \quad (1)$$
$$x^2 = 0$$
$$x = 0$$

The solution set, $\{(4, 0), (-4, 0), (0, 4)\}$, includes the points of intersection shown in **Figure 12.** Check the solutions in the original system.

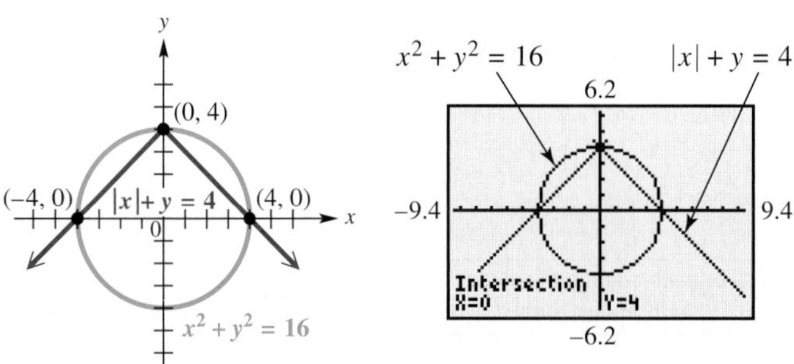

Figure 12

Now Try Exercise 39.

NOTE After solving for y in **Example 4,** the corresponding values of x can found by using equation (2) instead of equation (1). As shown below, the same values for x result.

Let $y = 0$.

$$|x| + 0 = 4 \quad (2)$$
$$|x| = 4$$
$$x = \pm 4$$

Let $y = 4$.

$$|x| + 4 = 4 \quad (2)$$
$$|x| = 0$$
$$x = 0$$

Solving Nonlinear Systems with Nonreal Complex Solutions

Classroom Example 5
Solve the system.

$$x^2 + y^2 = 6$$
$$3x^2 + 2y^2 = 8$$

Answer:
$\{(2i, \sqrt{10}), (2i, -\sqrt{10}), (-2i, \sqrt{10}), (-2i, -\sqrt{10})\}$

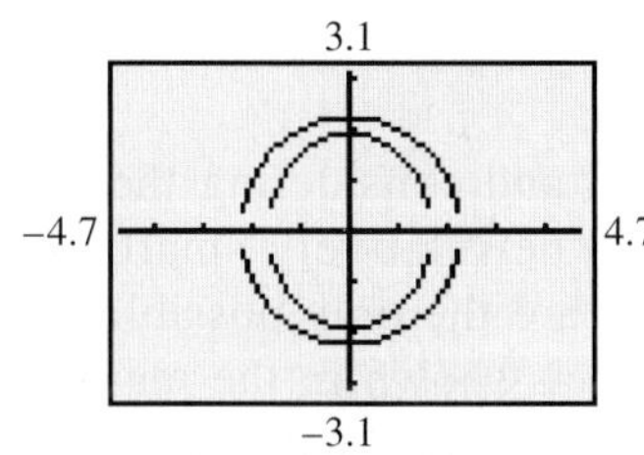

The graphs of the two equations in **Example 5** do not intersect, as seen here. The graphs are obtained by graphing

$$y = \pm\sqrt{5 - x^2}$$

and

$$y = \pm\sqrt{\frac{11 - 4x^2}{3}}.$$

EXAMPLE 5 Solving a Nonlinear System with Nonreal Complex Numbers as Solutions

Solve the system.

$$x^2 + y^2 = 5 \quad (1)$$
$$4x^2 + 3y^2 = 11 \quad (2)$$

SOLUTION Begin by eliminating a variable.

$$-3x^2 - 3y^2 = -15 \quad \text{Multiply (1) by } -3.$$
$$4x^2 + 3y^2 = 11 \quad (2)$$
$$x^2 = -4 \quad \text{Add.}$$
$$x = \pm\sqrt{-4} \quad \text{Square root property}$$
$$x = \pm 2i \quad \sqrt{-4} = i\sqrt{4} = 2i \text{ (Section 1.3)}$$

To find the corresponding values of y, substitute into equation (1).

Let $x = 2i$.

$$(2i)^2 + y^2 = 5 \quad (1)$$
$$-4 + y^2 = 5 \quad i^2 = -1$$
$$y^2 = 9$$
$$y = \pm 3$$

Let $x = -2i$.

$$(-2i)^2 + y^2 = 5 \quad (1)$$
$$-4 + y^2 = 5$$
$$y^2 = 9$$
$$y = \pm 3$$

Checking the proposed solutions confirms the following solution set.

$$\{(2i, 3), (2i, -3), (-2i, 3), (-2i, -3)\}$$

Note that these solutions with nonreal complex number components do not appear as intersection points on the graph of the system.

Now Try Exercise 37.

Applying Nonlinear Systems

Classroom Example 6
A box with an open top has a square base and four sides of equal height. The volume is 384 m^3, and the surface area is 256 m^2. Find the dimensions of the box. Round to the nearest thousandth as necessary.

Answer:
First answer:
length = width = 8 m;
height = 6 m

Second answer:
length = width ≈ 10.422 m;
height ≈ 3.535 m

EXAMPLE 6 Using a Nonlinear System to Find the Dimensions of a Box

A box with an open top has a square base and four sides of equal height. The volume of the box is 75 in.^{3}, and the surface area is 85 in.^{2}. Find the dimensions of the box.

SOLUTION

Step 1 **Read** the problem. We must find the box width, length, and height.

Step 2 **Assign variables.** Let x represent the length and width of the square base, and let y represent the height. See **Figure 13.**

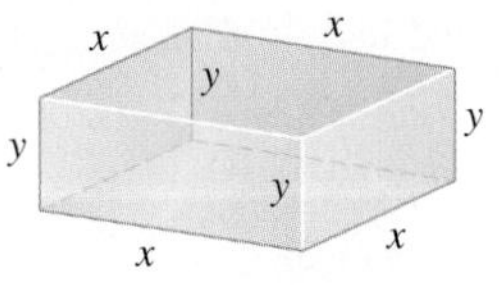

Figure 13

Step 3 **Write a system of equations.** Use the formula for the volume of a box, $V = LWH$, to write one equation using the given volume, 75 in.^{3}.

$$x^2y = 75 \quad \text{Volume formula}$$

The surface consists of the base, whose area is x^2, and four sides, each having area xy. The total surface area of 85 in.^{2} is used to write a second equation.

$$x^2 + 4xy = 85 \quad \text{Sum of areas of base and sides}$$

The two equations form this system.

$$x^2y = 75 \quad (1)$$

$$x^2 + 4xy = 85 \quad (2)$$

Step 4 **Solve** the system. Solve equation (1) for y to get $y = \frac{75}{x^2}$, and substitute into equation (2).

$$x^2 + 4x\left(\frac{75}{x^2}\right) = 85 \qquad \text{Let } y = \tfrac{75}{x^2} \text{ in (2).}$$

$$x^2 + \frac{300}{x} = 85 \qquad \text{Multiply.}$$

$$x^3 + 300 = 85x \qquad \text{Multiply by } x, \text{ where } x \neq 0.$$

$$x^3 - 85x + 300 = 0 \qquad \text{Subtract } 85x.$$

5)	1	0	−85	300
		5	25	−300
	1	5	−60	0

Coefficients of a quadratic polynomial factor (Section 3.2)

We are restricted to positive values for x, and considering the nature of the problem, any solution should be relatively small. By the rational zeros theorem, factors of 300 are the only possible rational solutions. Using synthetic division, as shown in the margin, we see that 5 is a solution. Therefore, one value of x is 5, and $y = \frac{75}{5^2} = 3$. We must now solve

$$x^2 + 5x - 60 = 0$$

for any other possible positive solutions. Use the quadratic formula to find the positive solution.

$$x = \frac{-5 + \sqrt{5^2 - 4(1)(-60)}}{2(1)} \approx 5.639$$

Quadratic formula with $a = 1, b = 5, c = -60$ (Section 1.4)

This value of x leads to $y \approx 2.359$.

Step 5 **State the answer.** There are two possible answers.

First answer: length = width = 5 in.; height = 3 in.

Second answer: length = width ≈ 5.639 in.; height ≈ 2.359 in.

Step 6 **Check.** The check is left for **Exercise 63.**

Now Try Exercises 61 and 63.

9.5 Exercises

7. Consider the graphs; a line and a parabola cannot intersect in more than two points.
8. no
9. $\{(1, 1), (-2, 4)\}$
10. $\{(1, 1), (-2, -2)\}$
11. $\{(2, 1), (\frac{1}{3}, \frac{4}{9})\}$
12. $\{(-4, 1), (-\frac{5}{2}, \frac{1}{4})\}$
13. $\{(2, 12), (-4, 0)\}$

Concept Check In Exercises 1–6 a nonlinear system is given, along with the graphs of both equations in the system. Verify that the points of intersection specified on the graph are solutions of the system by substituting directly into both equations.

1. $x^2 = y - 1$
$y = 3x + 5$

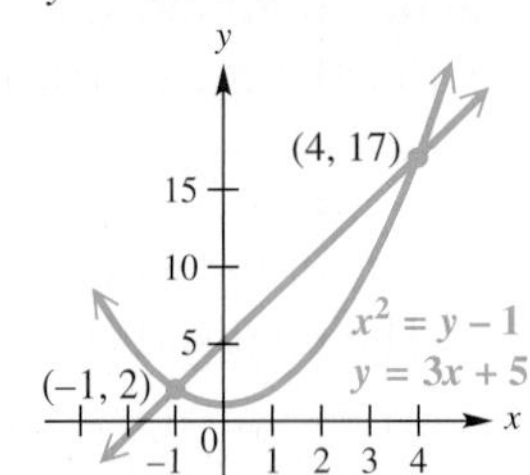

2. $2x^2 = 3y + 23$
$y = 2x - 5$

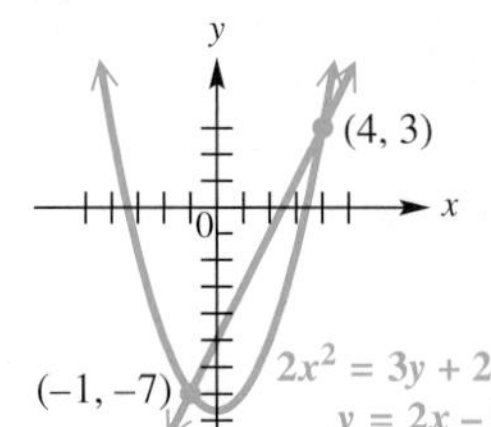

14. $\{(-3,-9),(1,7)\}$
15. $\{(-\frac{3}{5},\frac{7}{5}),(-1,1)\}$
16. $\{(-2,-1),(\frac{38}{25},\frac{41}{25})\}$
17. $\{(2,2),(2,-2),(-2,2),(-2,-2)\}$
18. $\{(3,1),(3,-1),(-3,1),(-3,-1)\}$
19. $\{(0,0)\}$
20. $\{(0,0)\}$
21. $\{(i,\sqrt{6}),(-i,\sqrt{6}),(i,-\sqrt{6}),(-i,-\sqrt{6})\}$
22. $\{(\sqrt{41},4i),(\sqrt{41},-4i),(-\sqrt{41},4i),(-\sqrt{41},-4i)\}$
23. $\{(1,-1),(-1,1),(1,1),(-1,-1)\}$
24. $\{(2,1),(-2,1),(2,-1),(-2,-1)\}$
25. $\emptyset$
26. $\emptyset$
27. $\{(\sqrt{6},0),(-\sqrt{6},0)\}$
28. $\{(-\sqrt{5},0),(\sqrt{5},0)\}$
29. $\{(-3,5),(\frac{15}{4},-4)\}$
30. $\{(-4,-2),(-\frac{4}{3},-6)\}$
31. $\{(4,-\frac{1}{8}),(-2,\frac{1}{4})\}$
32. $\{(6,\frac{1}{15}),(-1,-\frac{2}{5})\}$
33. $\{(3,4),(-3,-4),(\frac{4\sqrt{3}}{3}i,-3i\sqrt{3}),(-\frac{4\sqrt{3}}{3}i,3i\sqrt{3})\}$
34. $\{(\frac{2\sqrt{5}}{5}i,-i\sqrt{5}),(-\frac{2\sqrt{5}}{5}i,i\sqrt{5}),(\sqrt{2},\sqrt{2}),(-\sqrt{2},-\sqrt{2})\}$
35. $\{(\sqrt{5},0),(-\sqrt{5},0),(\sqrt{5},\sqrt{5}),(-\sqrt{5},-\sqrt{5})\}$
36. $\{(\sqrt{3},2\sqrt{3}),(-\sqrt{3},-2\sqrt{3}),(\frac{6\sqrt{7}}{7},-\frac{3\sqrt{7}}{7}),(-\frac{6\sqrt{7}}{7},\frac{3\sqrt{7}}{7})\}$
37. $\{(3,5),(-3,-5),(5i,-3i),(-5i,3i)\}$
38. $\{(2,4),(-2,-4),(4i,-2i),(-4i,2i)\}$

3. $x^2 + y^2 = 5$
$-3x + 4y = 2$

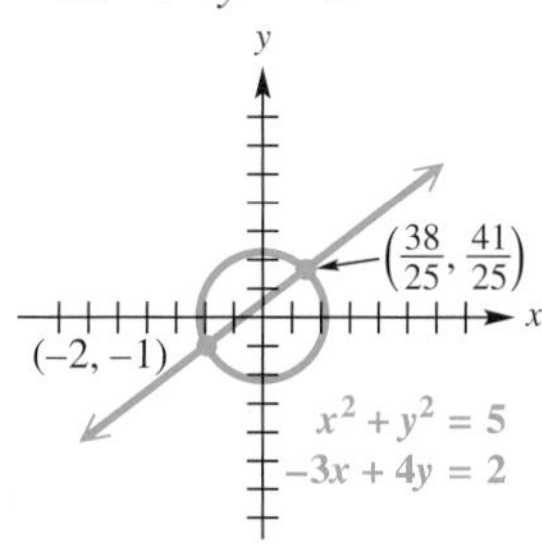

4. $x + y = -3$
$x^2 + y^2 = 45$

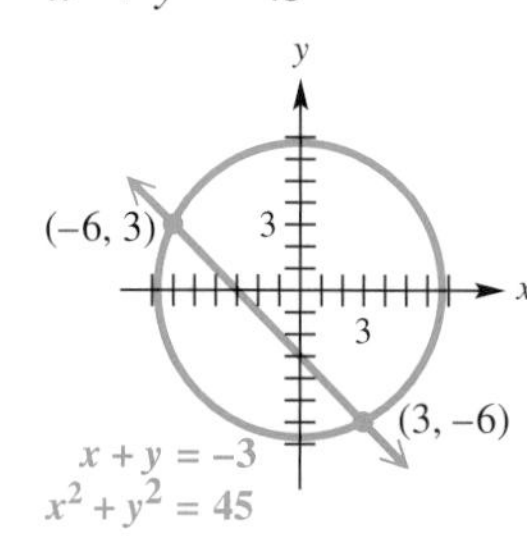

5. $y = \log x$
$x^2 - y^2 = 4$

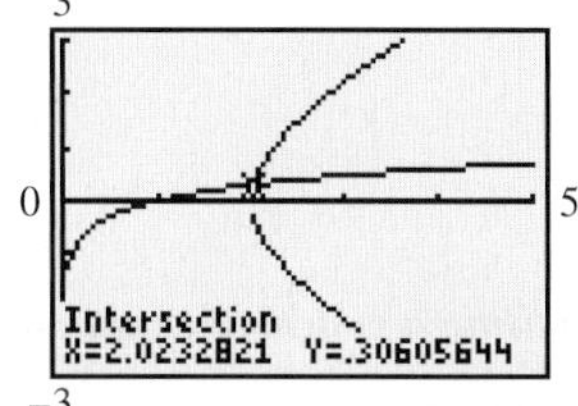

6. $y = \ln(2x + 3)$
$y = \sqrt{2 - 0.5x^2}$

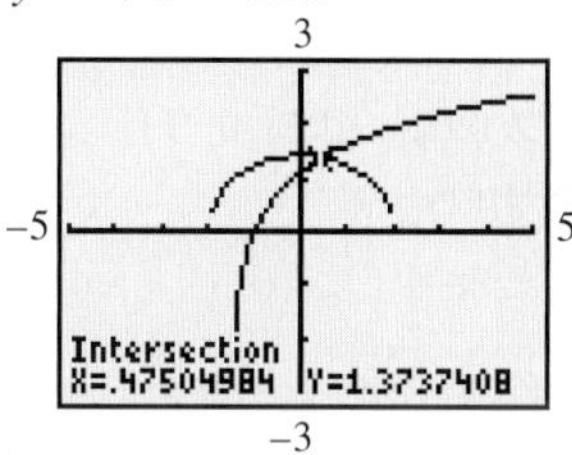

7. *Concept Check* In **Example 1,** we solved the following system. How can we tell, before doing any work, that this system cannot have more than two solutions?

$$x^2 - y = 4$$
$$x + y = -2$$

8. *Concept Check* In **Example 5,** there were four solutions to the system, but there were no points of intersection of the graphs. If a nonlinear system has nonreal complex numbers as components of its solutions, will they appear as intersection points of the graphs?

Give all solutions of each nonlinear system of equations, including those with nonreal complex components. ***See Examples 1–5.***

9. $x^2 - y = 0$
$x + y = 2$

10. $x^2 + y = 2$
$x - y = 0$

11. $y = x^2 - 2x + 1$
$x - 3y = -1$

12. $y = x^2 + 6x + 9$
$x + 2y = -2$

13. $y = x^2 + 4x$
$2x - y = -8$

14. $y = 6x + x^2$
$4x - y = -3$

15. $3x^2 + 2y^2 = 5$
$x - y = -2$

16. $x^2 + y^2 = 5$
$-3x + 4y = 2$

17. $x^2 + y^2 = 8$
$x^2 - y^2 = 0$

18. $x^2 + y^2 = 10$
$2x^2 - y^2 = 17$

19. $5x^2 - y^2 = 0$
$3x^2 + 4y^2 = 0$

20. $x^2 + y^2 = 0$
$2x^2 - 3y^2 = 0$

21. $3x^2 + y^2 = 3$
$4x^2 + 5y^2 = 26$

22. $x^2 + 2y^2 = 9$
$x^2 + y^2 = 25$

23. $2x^2 + 3y^2 = 5$
$3x^2 - 4y^2 = -1$

24. $3x^2 + 5y^2 = 17$
$2x^2 - 3y^2 = 5$

25. $2x^2 + 2y^2 = 20$
$4x^2 + 4y^2 = 30$

26. $x^2 + y^2 = 4$
$5x^2 + 5y^2 = 28$

27. $2x^2 - 3y^2 = 12$
$6x^2 + 5y^2 = 36$

28. $5x^2 - 2y^2 = 25$
$10x^2 + y^2 = 50$

29. $xy = -15$
$4x + 3y = 3$

30. $xy = 8$
$3x + 2y = -16$

31. $2xy + 1 = 0$
$x + 16y = 2$

32. $-5xy + 2 = 0$
$x - 15y = 5$

33. $3x^2 - y^2 = 11$
$xy = 12$

34. $5x^2 - 2y^2 = 6$
$xy = 2$

35. $x^2 - xy + y^2 = 5$
$2x^2 + xy - y^2 = 10$

36. $3x^2 + 2xy - y^2 = 9$
$x^2 - xy + y^2 = 9$

37. $x^2 + 2xy - y^2 = 14$
$x^2 - y^2 = -16$

38. $x^2 + 3xy - y^2 = 12$
$x^2 - y^2 = -12$

39. $\{(5, 0), (-5, 0), (0, -5)\}$
40. $\{(3, 0), (-3, 0), (0, 3)\}$
41. $\{(3, -3), (3, 3)\}$
42. $\{(1, 2), (1, -2)\}$
43. $\{(2, 2), (-2, -2), (2, -2), (-2, 2)\}$
44. $\left\{\left(-\frac{3\sqrt{2}}{2}, -\frac{3\sqrt{2}}{2}\right), \left(-\frac{3\sqrt{2}}{2}, \frac{3\sqrt{2}}{2}\right), \left(\frac{3\sqrt{2}}{2}, -\frac{3\sqrt{2}}{2}\right), \left(\frac{3\sqrt{2}}{2}, \frac{3\sqrt{2}}{2}\right)\right\}$
45. $\{(-0.79, 0.62), (0.88, 0.77)\}$
46. $\{(0.47, 2.13)\}$
47. $\{(0.06, 2.88)\}$
48. $\{(-1.68, -1.78), (2.12, -1.24)\}$
49. 14 and 3
50. 6 and 4
51. 8 and 6, 8 and -6, -8 and 6, -8 and -6
52. 5 and 13, 5 and -13, -5 and 13, -5 and -13
53. 27 and 6, -27 and -6
54. 8 and 6, -8 and -6
55. 5 m and 12 m
56. 20 ft and 21 ft
57. yes
58. $b = \pm 3\sqrt{5}$
59. $y = 9$

39. $x^2 + y^2 = 25$
$|x| - y = 5$

40. $x^2 + y^2 = 9$
$|x| + y = 3$

41. $x = |y|$
$x^2 + y^2 = 18$

42. $2x + |y| = 4$
$x^2 + y^2 = 5$

43. $2x^2 - y^2 = 4$
$|x| = |y|$

44. $x^2 + y^2 = 9$
$|x| = |y|$

Many nonlinear systems cannot be solved algebraically, so graphical analysis is the only way to determine the solutions of such systems. Use a graphing calculator to solve each nonlinear system. Give x- and y-coordinates to the nearest hundredth.

45. $y = \log(x + 5)$
$y = x^2$

46. $y = 5^x$
$xy = 1$

47. $y = e^{x+1}$
$2x + y = 3$

48. $y = \sqrt[3]{x - 4}$
$x^2 + y^2 = 6$

Solve each problem using a system of equations in two variables. ***See Example 6.***

49. *Unknown Numbers* Find two numbers whose sum is 17 and whose product is 42.

50. *Unknown Numbers* Find two numbers whose sum is 10 and whose squares differ by 20.

51. *Unknown Numbers* Find two numbers whose squares have a sum of 100 and a difference of 28.

52. *Unknown Numbers* Find two numbers whose squares have a sum of 194 and a difference of 144.

53. *Unknown Numbers* Find two numbers whose ratio is 9 to 2 and whose product is 162.

54. *Unknown Numbers* Find two numbers whose ratio is 4 to 3 and are such that the sum of their squares is 100.

55. *Triangle Dimensions* The longest side of a right triangle is 13 m in length. One of the other sides is 7 m longer than the shortest side. Find the lengths of the two shorter sides of the triangle.

56. *Triangle Dimensions* The longest side of a right triangle is 29 ft in length. One of the other two sides is 1 ft longer than the shortest side. Find the lengths of the two shorter sides of the triangle.

57. Does the straight line $3x - 2y = 9$ intersect the circle $x^2 + y^2 = 25$? (*Hint:* To find out, solve the system formed by these two equations.)

58. For what value(s) of b will the line $x + 2y = b$ touch the circle $x^2 + y^2 = 9$ in only one point?

59. Find the equation of the line passing through the points of intersection of the graphs of $y = x^2$ and $x^2 + y^2 = 90$.

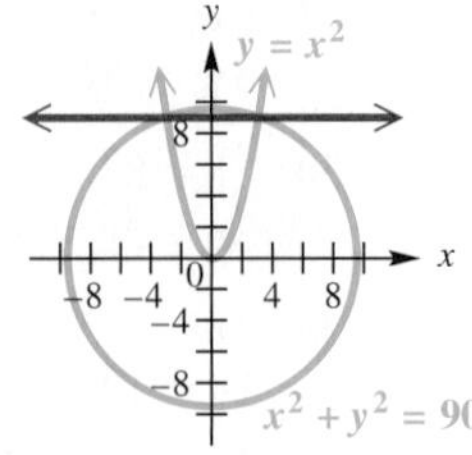

61. *First answer:*
length = width: 6 ft;
height: 10 ft
Second answer:
length = width: 12.780 ft;
height: 2.204 ft
62. radius: 1.538 in.;
height: 6.724 in.
64. (a) 1000 units (b) $2
65. (a) 160 units (b) $3
66. $R \approx 187$; $R_t \approx 645$

60. Suppose you are given the equations of two circles that are known to intersect in exactly two points. Explain how you would find the equation of the only chord common to these circles.

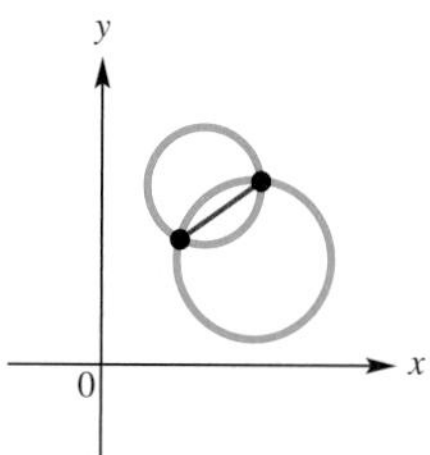

61. *Dimensions of a Box* A box with an open top has a square base and four sides of equal height. The volume of the box is 360 ft^3. If the surface area is 276 ft^2, find the dimensions of the box. (Round answers to the nearest thousandth, if necessary.)

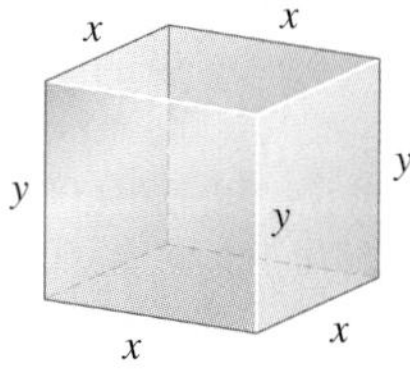

62. *Dimensions of a Cylinder* Find the radius and height (to the nearest thousandth) of an open-ended cylinder with volume 50 in.^{3} and lateral surface area 65 in.^{2}.

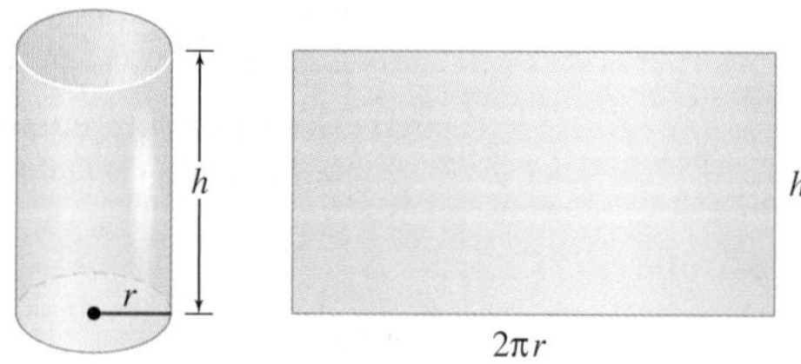

63. *Checking Answers* Check the two answers in **Example 6.**

64. *(Modeling) Equilibrium Demand and Price* The supply and demand equations for a certain commodity are given.

$$\text{supply: } p = \frac{2000}{2000 - q} \quad \text{and} \quad \text{demand: } p = \frac{7000 - 3q}{2q}$$

(a) Find the equilibrium demand.
(b) Find the equilibrium price (in dollars).

65. *(Modeling) Equilibrium Demand and Price* The supply and demand equations for a certain commodity are given.

$$\text{supply: } p = \sqrt{0.1q + 9} - 2 \quad \text{and} \quad \text{demand: } p = \sqrt{25 - 0.1q}$$

(a) Find the equilibrium demand.
(b) Find the equilibrium price (in dollars).

66. *(Modeling) Circuit Gain* In electronics, circuit gain is modeled by

$$G = \frac{Bt}{R + R_t},$$

where R is the value of a resistor, t is temperature, R_t is the value of R at temperature t, and B is a constant. The sensitivity of the circuit to temperature is modeled by

$$S = \frac{BR}{(R + R_t)^2}.$$

If $B = 3.7$ and t is 90 K (kelvins), find the values of R and R_t that will result in the values $G = 0.4$ and $S = 0.001$.

67. (a) The carbon dioxide emissions from consumption of fossil fuels are increasing in both countries, but they are increasing much more rapidly in China than in the United States.
(b) 1996 to 2000
(c) The emissions were equal in 2006, when the levels were approximately 6000 million metric tons in each country.
(d) year: 2007; emissions level: approximately 6107 million metric tons

68. Translate the graph of $y = |x|$ one unit to the right.

69. Translate the graph of $y = x^2$ four units down.

70. $y = \begin{cases} x - 1 & \text{if } x \geq 1 \\ 1 - x & \text{if } x < 1 \end{cases}$

71. $x^2 - 4 = x - 1 \ (x \geq 1)$; $x^2 - 4 = 1 - x \ (x < 1)$

72. $\frac{1 + \sqrt{13}}{2}; \frac{-1 - \sqrt{21}}{2}$

73. $\left\{\left(\frac{1 + \sqrt{13}}{2}, \frac{-1 + \sqrt{13}}{2}\right), \left(\frac{-1 - \sqrt{21}}{2}, \frac{3 + \sqrt{21}}{2}\right)\right\}$

67. *(Modeling) Carbon Dioxide Emissions* The emissions of carbon dioxide from consumption of fossil fuels from 1990 through 2007 are modeled in the graph for both the United States and China.

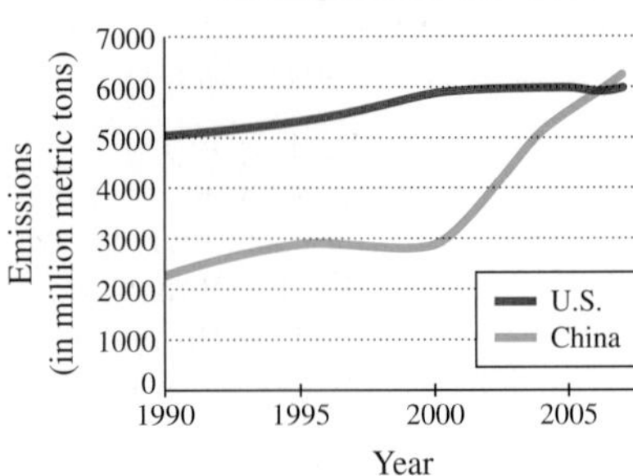

Source: U.S. Energy Information Administration.

(a) Interpret this graph. How are emissions changing with time?

(b) During what years were the carbon dioxide emissions in China decreasing?

(c) Use the graph to estimate the year and the amount of emissions when the carbon emissions for both countries were equal.

(d) The following equations model carbon dioxide emissions in the United States and in China. Use the equations to determine the year and emissions levels when $U = C$.

$$U = 5092(1.0105)^{(x-1990)} \quad \text{United States}$$

$$C = 2082(1.0638)^{(x-1990)} \quad \text{China}$$

Relating Concepts

For individual or collaborative investigation *(Exercises 68–74)*

Consider the following nonlinear system.

$$y = |x - 1|$$
$$y = x^2 - 4$$

Work Exercises 68–74 in order, *to see how concepts from previous chapters are related to the graphs and the solutions of this system.*

68. How is the graph of $y = |x - 1|$ obtained by transforming the graph of $y = |x|$?

69. How is the graph of $y = x^2 - 4$ obtained by transforming the graph of $y = x^2$?

70. Use the definition of absolute value to write $y = |x - 1|$ as a piecewise-defined function.

71. Write two quadratic equations that will be used to solve the system. (*Hint:* Set both parts of the piecewise-defined function in **Exercise 70** equal to $x^2 - 4$.)

72. Use the quadratic formula to solve both equations from **Exercise 71.** Pay close attention to the restriction on x.

73. Use the values of x found in **Exercise 72** to find the solution set of the system.

74. Check the two answers in **Exercise 73.**

Summary Exercises on Systems of Equations

This chapter has introduced methods for solving systems of equations, including substitution and elimination, and matrix methods such as the Gauss-Jordan method and Cramer's rule. Use each method at least once when solving the systems below. Include solutions with nonreal complex number components. For systems with infinitely many solutions, write the solution set using an arbitrary variable.

1. $\begin{aligned} 2x + 5y &= 4 \\ 3x - 2y &= -13 \end{aligned}$

2. $\begin{aligned} x - 3y &= 7 \\ -3x + 4y &= -1 \end{aligned}$

3. $\begin{aligned} 2x^2 + y^2 &= 5 \\ 3x^2 + 2y^2 &= 10 \end{aligned}$

4. $\begin{aligned} 2x - 3y &= -2 \\ x + y &= -16 \\ 3x - 2y + z &= 7 \end{aligned}$

5. $\begin{aligned} 6x - y &= 5 \\ xy &= 4 \end{aligned}$

6. $\begin{aligned} 4x + 2z &= -12 \\ x + y - 3z &= 13 \\ -3x + y - 2z &= 13 \end{aligned}$

7. $\begin{aligned} x + 2y + z &= 5 \\ y + 3z &= 9 \end{aligned}$

8. $\begin{aligned} x - 4 &= 3y \\ x^2 + y^2 &= 8 \end{aligned}$

9. $\begin{aligned} 3x + 6y - 9z &= 1 \\ 2x + 4y - 6z &= 1 \\ 3x + 4y + 5z &= 0 \end{aligned}$

10. $\begin{aligned} x + 2y + z &= 0 \\ x + 2y - z &= 6 \\ 2x - y &= -9 \end{aligned}$

11. $\begin{aligned} x^2 + y^2 &= 4 \\ y &= x + 6 \end{aligned}$

12. $\begin{aligned} x + 5y &= -23 \\ 4y - 3z &= -29 \\ 2x + 5z &= 19 \end{aligned}$

13. $\begin{aligned} y + 1 &= x^2 + 2x \\ y + 2x &= 4 \end{aligned}$

14. $\begin{aligned} 3x + 6z &= -3 \\ y - z &= 3 \\ 2x + 4z &= -1 \end{aligned}$

15. $\begin{aligned} 2x + 3y + 4z &= 3 \\ -4x + 2y - 6z &= 2 \\ 4x + 3z &= 0 \end{aligned}$

16. $\begin{aligned} \frac{3}{x} + \frac{4}{y} &= 4 \\ \frac{1}{x} + \frac{2}{y} &= \frac{2}{3} \end{aligned}$

17. $\begin{aligned} -5x + 2y + z &= 5 \\ -3x - 2y - z &= 3 \\ -x + 6y &= 1 \end{aligned}$

18. $\begin{aligned} x + 5y + 3z &= 9 \\ 2x + 9y + 7z &= 5 \end{aligned}$

19. $\begin{aligned} 2x^2 + y^2 &= 9 \\ 3x - 2y &= -6 \end{aligned}$

20. $\begin{aligned} 2x - 4y - 6 &= 0 \\ -x + 2y + 3 &= 0 \end{aligned}$

21. $\begin{aligned} x + y - z &= 0 \\ 2y - z &= 1 \\ 2x + 3y - 4z &= -4 \end{aligned}$

22. $\begin{aligned} 2y &= 3x - x^2 \\ x + 2y &= 12 \end{aligned}$

23. $\begin{aligned} 4x - z &= -6 \\ \frac{3}{5}y + \frac{1}{2}z &= 0 \\ \frac{1}{3}x + \frac{2}{3}z &= -5 \end{aligned}$

24. $\begin{aligned} x - y + 3z &= 3 \\ -2x + 3y - 11z &= -4 \\ x - 2y + 8z &= 6 \end{aligned}$

25. $\begin{aligned} x^2 + 3y^2 &= 28 \\ y - x &= -2 \end{aligned}$

26. $\begin{aligned} 5x - 2z &= 8 \\ 4y + 3z &= -9 \\ \frac{1}{2}x + \frac{2}{3}y &= -1 \end{aligned}$

27. $\begin{aligned} 2x^2 + 3y^2 &= 20 \\ x^2 + 4y^2 &= 5 \end{aligned}$

28. $\begin{aligned} x + y + z &= -1 \\ 2x + 3y + 2z &= 3 \\ 2x + y + 2z &= -7 \end{aligned}$

29. $\begin{aligned} x + 2z &= 9 \\ y + z &= 1 \\ 3x - 2y &= 9 \end{aligned}$

30. $\begin{aligned} x^2 - y^2 &= 15 \\ x - 2y &= 2 \end{aligned}$

31. $\begin{aligned} -x + y &= -1 \\ x + z &= 4 \\ 6x - 3y + 2z &= 10 \end{aligned}$

32. $\begin{aligned} 2x - y - 2z &= -1 \\ -x + 2y + 13z &= 12 \\ 3x + 9z &= 6 \end{aligned}$

33. $\begin{aligned} xy &= -3 \\ x + y &= -2 \end{aligned}$

34. $\begin{aligned} -3x + 2z &= 1 \\ 4x + y - 2z &= -6 \\ x + y + 4z &= 3 \end{aligned}$

35. $\begin{aligned} y &= x^2 + 6x + 9 \\ x + y &= 3 \end{aligned}$

36. $\begin{aligned} x^2 + y^2 &= 9 \\ 2x - y &= 3 \end{aligned}$

37. $\begin{aligned} 2x - 3y &= -2 \\ x + y - 4z &= -16 \\ 3x - 2y + z &= 7 \end{aligned}$

38. $\begin{aligned} 3x - y &= -2 \\ y + 5z &= -4 \\ -2x + 3y - z &= -8 \end{aligned}$

39. $\begin{aligned} y &= (x - 1)^2 + 2 \\ y &= 2x - 1 \end{aligned}$

1. $\{(-3, 2)\}$
2. $\{(-5, -4)\}$
3. $\{(0, \sqrt{5}), (0, -\sqrt{5})\}$
4. $\{(-10, -6, 25)\}$
5. $\{(\frac{4}{3}, 3), (-\frac{1}{2}, -8)\}$
6. $\{(-1, 2, -4)\}$
7. $\{(5z - 13, -3z + 9, z)\}$
8. $\{(-2, -2), (\frac{14}{5}, -\frac{2}{5})\}$
9. $\emptyset$
10. $\{(-3, 3, -3)\}$
11. $\{(-3 + i\sqrt{7}, 3 + i\sqrt{7}), (-3 - i\sqrt{7}, 3 - i\sqrt{7})\}$
12. $\{(2, -5, 3)\}$
13. $\{(1, 2), (-5, 14)\}$
14. $\emptyset$
15. $\{(0, 1, 0)\}$
16. $\{(\frac{3}{8}, -1)\}$
17. $\{(-1, 0, 0)\}$
18. $\{(-8z - 56, z + 13, z)\}$
19. $\{(0, 3), (-\frac{36}{17}, -\frac{3}{17})\}$
20. $\{(2y + 3, y)\}$
21. $\{(1, 2, 3)\}$
22. $\{(2 + 2i\sqrt{2}, 5 - i\sqrt{2}), (2 - 2i\sqrt{2}, 5 + i\sqrt{2})\}$
23. $\{(-3, 5, -6)\}$
24. $\emptyset$
25. $\{(4, 2), (-1, -3)\}$
26. $\{(2, -3, 1)\}$
27. $\{(\sqrt{13}, i\sqrt{2}), (-\sqrt{13}, i\sqrt{2}), (\sqrt{13}, -i\sqrt{2}), (-\sqrt{13}, -i\sqrt{2})\}$
28. $\{(-z - 6, 5, z)\}$
29. $\{(1, -3, 4)\}$
30. $\{(-\frac{16}{3}, -\frac{11}{3}), (4, 1)\}$
31. $\{(-1, -2, 5)\}$
32. $\emptyset$
33. $\{(1, -3), (-3, 1)\}$
34. $\{(1, -6, 2)\}$
35. $\{(-6, 9), (-1, 4)\}$
36. $\{(0, -3), (\frac{12}{5}, \frac{9}{5})\}$
37. $\{(2, 2, 5)\}$ **38.** $\{(-2, -4, 0)\}$ **39.** $\{(2, 3)\}$

9.6 Systems of Inequalities and Linear Programming

- Solving Linear Inequalities
- Solving Systems of Inequalities
- Linear Programming

Solving Linear Inequalities A line divides a plane into three sets of points: the points of the line itself and the points belonging to the two regions determined by the line. Each of these two regions is a **half-plane.** In **Figure 14,** line r divides the plane into three different sets of points: line r, half-plane P, and half-plane Q. The points on r belong neither to P nor to Q. Line r is the **boundary** of each half-plane.

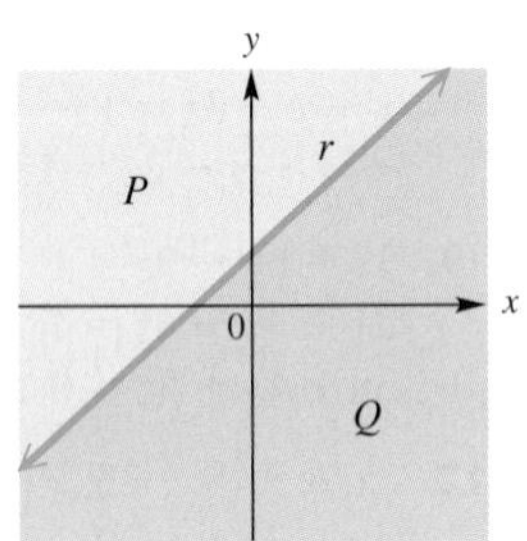

Figure 14

Linear Inequality in Two Variables

A **linear inequality in two variables** is an inequality of the form

$$Ax + By \leq C,$$

where A, B, and C are real numbers, with A and B not both equal to 0. (The symbol $\leq$ could be replaced with $\geq$, $<$, or $>$.)

The graph of a linear inequality is a half-plane, perhaps with its boundary.

Classroom Example 1
Graph $-4x - 3y \geq 12$.

Answer:

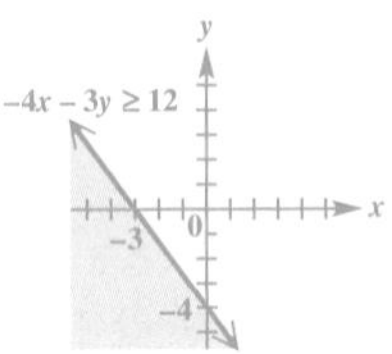

EXAMPLE 1 Graphing a Linear Inequality

Graph $3x - 2y \leq 6$.

SOLUTION First graph the boundary, $3x - 2y = 6$, as shown in **Figure 15.** Since the points of the line $3x - 2y = 6$ satisfy $3x - 2y \leq 6$, this line is part of the solution set. To decide which half-plane (the one above the line $3x - 2y = 6$ or the one below the line) is part of the solution set, solve the original inequality for y.

$$3x - 2y \leq 6$$

$$-2y \leq -3x + 6 \quad \text{Subtract } 3x.$$

$$y \geq \frac{3}{2}x - 3 \quad \text{Divide by } -2. \text{ (Section 1.7)}$$

Reverse the inequality symbol when dividing by a negative number.

For a particular value of x, the inequality will be satisfied by all values of y that are *greater than* or equal to $\frac{3}{2}x - 3$. Thus, the solution set contains the half-plane *above* the line, as shown in **Figure 16.**

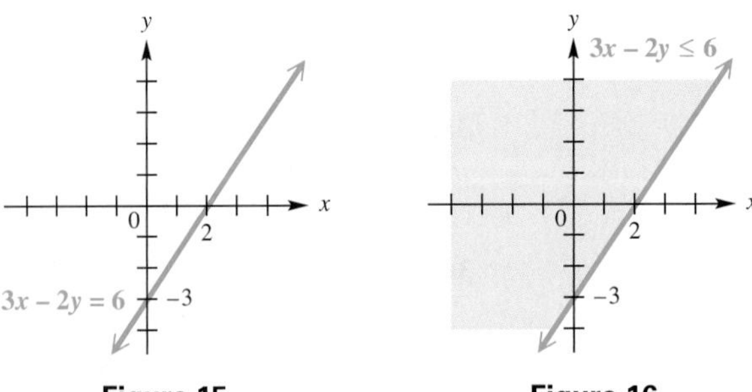

Figure 15 Figure 16

Coordinates for x and y from the solution set (the shaded region) satisfy the original inequality, while coordinates outside the solution set do not.

Now Try Exercise 1.

CAUTION ***A linear inequality must be in slope-intercept form (solved for y) to determine, from the presence of a < symbol or a > symbol, whether to shade the lower or the upper half-plane.*** In **Figure 16,** the upper half-plane is shaded, even though the inequality is $3x - 2y \le 6$ (with a < symbol) in standard form. Only when we write the inequality as

$$y \ge \frac{3}{2}x - 3 \quad \text{Slope-intercept form}$$

does the > symbol indicate to shade the upper half-plane.

Classroom Example 2
Graph $3x - 2y > 6$.

Answer:

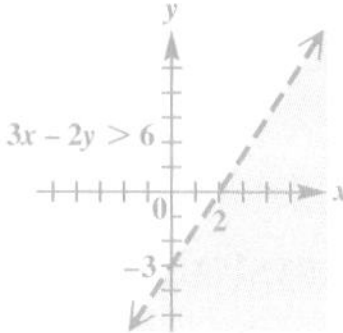

EXAMPLE 2 Graphing a Linear Inequality

Graph $x + 4y > 4$.

SOLUTION The boundary of the graph is the straight line $x + 4y = 4$. Since points on this line do *not* satisfy $x + 4y > 4$, it is customary to make the line dashed, as in **Figure 17.**

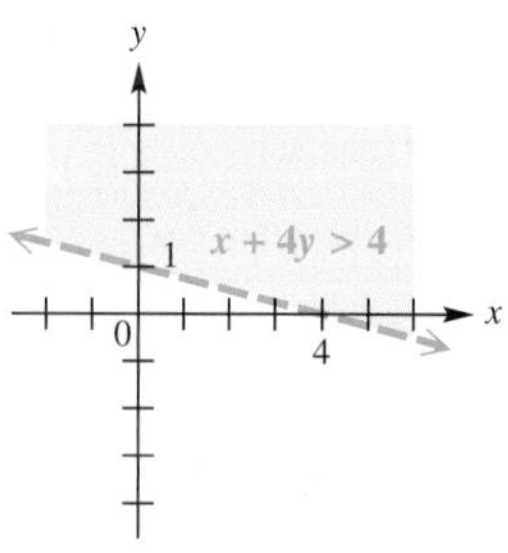

Figure 17

To decide which half-plane represents the solution set, solve for y.

$$x + 4y > 4$$

$$4y > -x + 4 \quad \text{Subtract } x.$$

$$y > -\frac{1}{4}x + 1 \quad \text{Divide by 4.}$$

Since y is *greater than* $-\frac{1}{4}x + 1$, the graph of the solution set is the half-plane *above* the boundary, as shown in **Figure 17.**

Alternatively, or as a check, choose a test point not on the boundary line and substitute into the inequality. The point $(0, 0)$ is a good choice if it does not lie on the boundary, since the substitution is easily done.

$$x + 4y > 4 \quad \text{Original inequality}$$

$$0 + 4(0) \overset{?}{>} 4 \quad \text{Use } (0, 0) \text{ as a test point.}$$

$$0 > 4 \quad \text{False}$$

Since the point $(0, 0)$ is below the boundary, the points that satisfy the inequality must be above the boundary, which agrees with the result above.

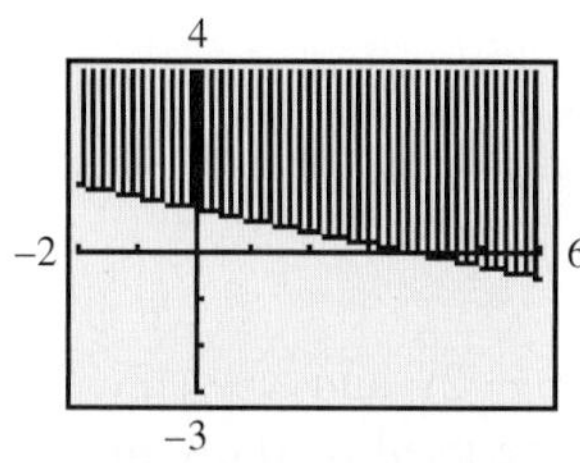

To graph the inequality in **Example 2** using a graphing calculator, solve for y, and then direct the calculator to shade *above* the boundary line, $y = -\frac{1}{4}x + 1$. The calculator graph does not distinguish between solid boundary lines and dashed boundary lines. We must understand the mathematics to interpret a calculator graph correctly.

✔ *Now Try Exercise 5.*

An inequality containing < or > is a **strict inequality** and does not include the boundary in its solution set. This is indicated with a dashed boundary, as shown in **Example 2.** A **nonstrict inequality** contains ≤ or ≥ and does include its boundary in the solution set. This is indicated with a solid boundary, as shown in **Example 1.**

This discussion is summarized in the box on the next page.

Teaching Tip Remind students that the method of shading *above* or *below* a boundary can be applied only to inequalities solved for y. Encourage the use of $(0, 0)$ as a test point to simplify calculations.

Graphing an Inequality in Two Variables

Method 1 If the inequality is or can be solved for y, then the following hold.

- The graph of $y < f(x)$ consists of all the points that are *below* the graph of $y = f(x)$.
- The graph of $y > f(x)$ consists of all the points that are *above* the graph of $y = f(x)$.

Method 2 If the inequality is not or cannot be solved for y, choose a test point not on the boundary.

- If the test point satisfies the inequality, the graph includes all points on the *same* side of the boundary as the test point.
- If the test point does not satisfy the inequality, the graph includes all points on the *other* side of the boundary.

Classroom Example 3
Graph the solution set of each system.

(a) $y < 4 - x^2$
$x < y - 1$

(b) $|y| \le 1$
$x \ge 0$
$y > 2|x| + 1$

Answers:

(a) [graph: $y < 4 - x^2$, $x < y - 1$]

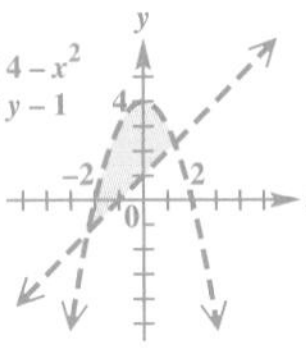

(b) $\emptyset$

Solving Systems of Inequalities

The solution set of a **system of inequalities,** such as

$$x > 6 - 2y$$
$$x^2 < 2y,$$

is the intersection of the solution sets of its members. We find this intersection by graphing the solution sets of all inequalities on the same coordinate axes and identifying, by shading, the region common to all graphs.

EXAMPLE 3 Graphing Systems of Inequalities

Graph the solution set of each system.

(a) $x > 6 - 2y$
$x^2 < 2y$

(b) $|x| \le 3$
$y \le 0$
$y \ge |x| + 1$

SOLUTION

(a) Figures 18(a) and (b) show the graphs of $x > 6 - 2y$ and $x^2 < 2y$. The methods presented earlier in this chapter can be used to show that the boundaries intersect at the points $(2, 2)$ and $\left(-3, \frac{9}{2}\right)$.

The solution set of the system is shown in **Figure 18(c).** Since the points on the boundaries of $x > 6 - 2y$ and $x^2 < 2y$ do not belong to the graph of the solution set, the boundaries are dashed.

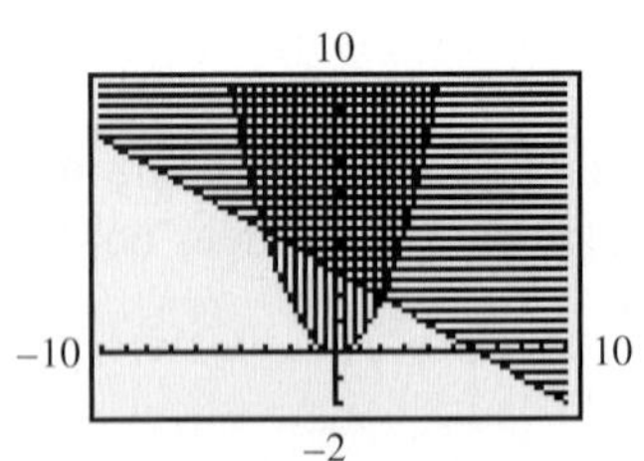

The cross-hatched region is the solution set of the system in **Example 3(a).**

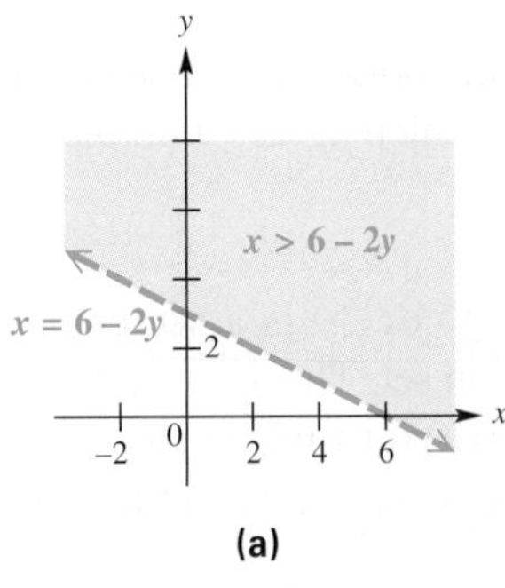

(a)

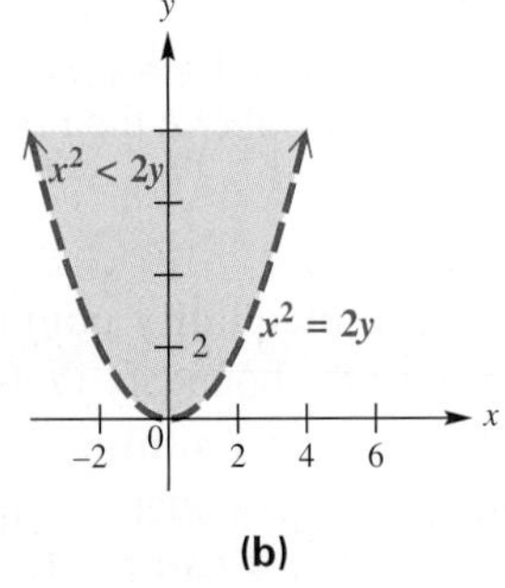

(b)

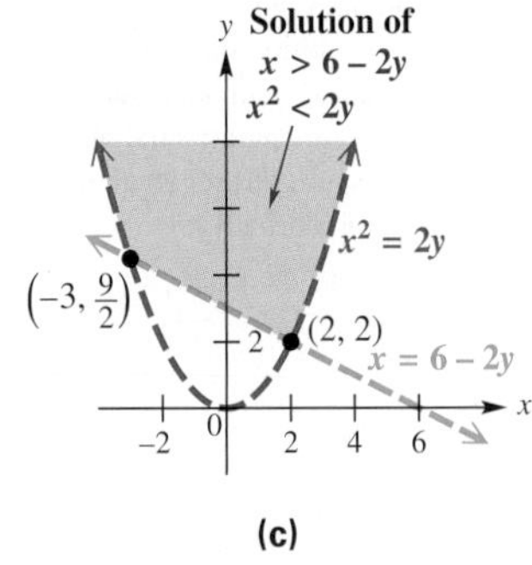

(c)

Figure 18

(b) Writing $|x| \leq 3$ as $-3 \leq x \leq 3$ shows that this inequality is satisfied by points in the region between and including $x = -3$ and $x = 3$. See **Figure 19(a).** The set of points that satisfies $y \leq 0$ includes the points below or on the x-axis. See **Figure 19(b).**

Graph $y = |x| + 1$ and use a test point to verify that the solutions of $y \geq |x| + 1$ are on or above the boundary. See **Figure 19(c).** Since the solution sets of $y \leq 0$ and $y \geq |x| + 1$ shown in **Figures 19(b) and (c)** have no points in common, ***the solution set of the system is ∅.***

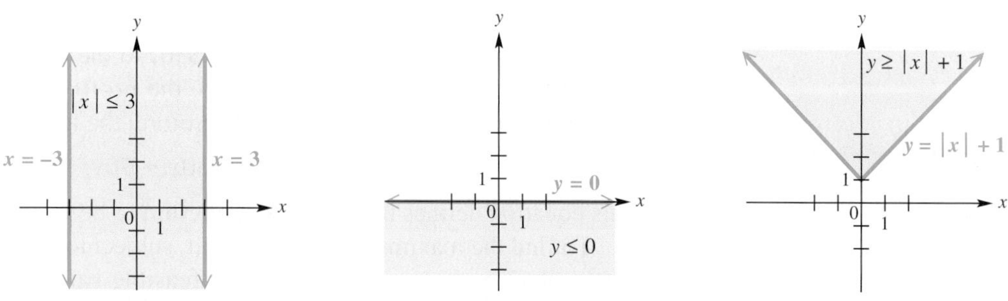

The solution set of the system is ∅, because there are no points common to all three regions.

(a) (b) (c)

Figure 19

✔ *Now Try Exercises 35, 41, and 43.*

Teaching Tip Mention that it is not enough for two out of three inequalities to have a common solution set, as in **Example 3(b).** The solution set must be common to *all* the inequalities of the system.

NOTE While we gave three graphs in the solutions of **Example 3,** in practice we usually give only a final graph showing the solution set of the system.

Linear Programming An important application of mathematics is *linear programming.* We use **linear programming** to find an optimum value such as minimum cost or maximum profit. It was first developed to solve problems in allocating supplies for the U.S. Air Force during World War II.

Classroom Example 4
Refer to **Example 4.** Suppose that the company's profits have changed to \$50 for each MP3 player and \$20 for each DVD player. Also, the company must now manufacture at least 40, but no more than 60, MP3 players per day, and the number of DVD players cannot exceed 75 per day. How many of each should the company manufacture to obtain maximum profit?

Answer:
60 MP3 players;
75 DVD players

EXAMPLE 4 Finding a Maximum Profit

The Charlson Company makes two products: MP3 players and DVD players. Each MP3 player gives a profit of \$30, while each DVD player produces \$70 profit. The company must manufacture at least 10 MP3 players per day to satisfy one of its customers, but it can manufacture no more than 50 per day because of production restrictions. The number of DVD players produced cannot exceed 60 per day, and the number of MP3 players cannot exceed the number of DVD players. How many of each should the company manufacture to obtain maximum profit?

SOLUTION First we translate the statement of the problem into symbols.

Let $x =$ the number of MP3 players to be produced daily,

and $y =$ the number of DVD players to be produced daily.

The company must produce at least 10 MP3 players (10 or more), so

$$x \geq 10.$$

Since no more than 50 MP3 players may be produced,

$$x \leq 50.$$

No more than 60 DVD players may be made in one day, so

$$y \leq 60.$$

The number of MP3 players may not exceed the number of DVD players, so

$$x \leq y.$$

The numbers of MP3 players and of DVD players cannot be negative, so

$$x \geq 0 \quad \text{and} \quad y \geq 0.$$

These restrictions, or **constraints,** form the following system of inequalities.

$$x \geq 10, \quad x \leq 50, \quad y \leq 60, \quad x \leq y, \quad x \geq 0, \quad y \geq 0$$

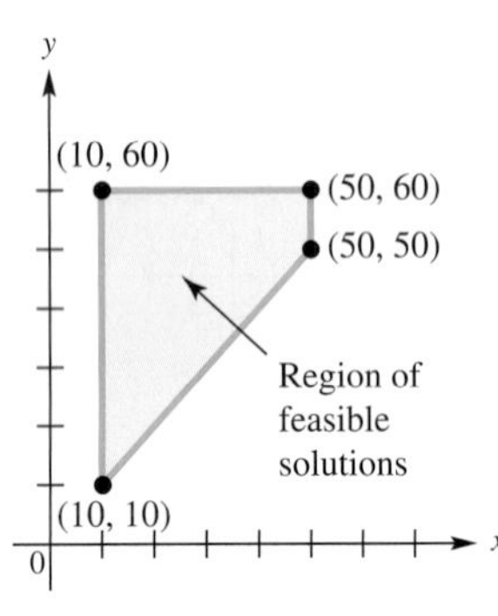

Figure 20

Each MP3 player gives a profit of \$30, so the daily profit from production of x MP3 players is $30x$ dollars. Also, the profit from production of y DVD players will be $70y$ dollars per day. Therefore, the total daily profit is

$$\text{profit} = 30x + 70y.$$

This equation defines the **objective function** to be maximized.

To find the maximum possible profit, subject to these constraints, we sketch the graph of each constraint. The only feasible values of x and y are those that satisfy all constraints—that is, the values that lie in the intersection of the graphs. The intersection is shown in **Figure 20.** Any point lying inside the shaded region or on the boundary in the figure satisfies the restrictions as to the number of MP3 and DVD players that can be produced. (For practical purposes, however, only points with integer coefficients are useful.) This region is called the **region of feasible solutions.**

The **vertices** (singular **vertex**), or **corner points,** of the region of feasible solutions have coordinates

$$(10, 10), \quad (10, 60), \quad (50, 50), \quad \text{and} \quad (50, 60).$$

We must find the value of the objective function $30x + 70y$ for each vertex. We want the vertex that produces the maximum possible value of

$$30x + 70y. \quad \text{Objective function}$$

$$(10, 10): \quad 30(10) + 70(10) = 1000$$

$$(10, 60): \quad 30(10) + 70(60) = 4500$$

$$(50, 50): \quad 30(50) + 70(50) = 5000$$

$$(50, 60): \quad 30(50) + 70(60) = 5700 \leftarrow \text{Maximum}$$

The maximum profit, obtained when 50 MP3 players and 60 DVD players are produced each day, will be

$$30(50) + 70(60) = \$5700 \text{ per day.}$$

✔ *Now Try Exercise 77.*

To justify the procedure used in **Example 4,** consider the following. The Charlson Company needed to find values of x and y in the shaded region of **Figure 20** that produce the maximum profit—that is, the maximum value of $30x + 70y$. To locate the point (x, y) that gives the maximum profit, add to the graph of **Figure 20** lines corresponding to arbitrarily chosen profits of \$0, \$1000, \$3000, and \$7000:

$$30x + 70y = 0, \quad 30x + 70y = 1000,$$

$$30x + 70y = 3000, \quad \text{and} \quad 30x + 70y = 7000.$$

For instance, each point on the line $30x + 70y = 3000$ corresponds to production values that yield a profit of \$3000.

Figure 21 shows the region of feasible solutions, together with these lines. The lines are parallel, and the higher the line, the greater the profit. The line $30x + 70y = 7000$ yields the greatest profit but does not contain any points of the region of feasible solutions. To find the feasible solution of greatest profit, lower the line $30x + 70y = 7000$ until it contains a feasible solution—that is, until it just touches the region of feasible solutions. This occurs at point A, a vertex of the region. See **Figure 22.**

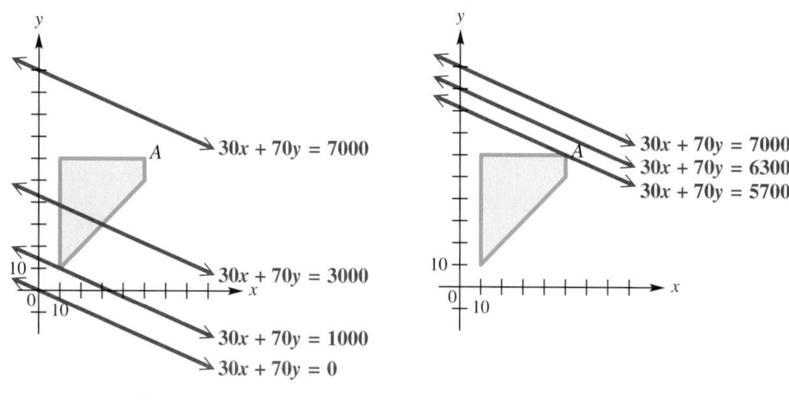

Figure 21 **Figure 22**

This discussion can be generalized.

Fundamental Theorem of Linear Programming

If an optimal value for a linear programming problem exists, it occurs at a vertex of the region of feasible solutions.

Use the following method to solve a linear programming problem.

Solving a Linear Programming Problem

Step 1 Write the objective function and all necessary constraints.

Step 2 Graph the region of feasible solutions.

Step 3 Identify all vertices (corner points).

Step 4 Find the value of the objective function at each vertex.

Step 5 The solution is given by the vertex producing the optimal value of the objective function.

Classroom Example 5
Refer to **Example 5.** If the prices of Robin's vitamins change to \$0.20 per capsule and \$0.10 per chewable tablet, how many of each should she take daily to minimize the cost and yet fulfill her daily requirements?

Answer:
1 capsule;
4 chewable tablets

EXAMPLE 5 Finding a Minimum Cost

Robin takes multivitamins each day. She wants at least 16 units of vitamin A, at least 5 units of vitamin B_1, and at least 20 units of vitamin C. Capsules, costing \$0.10 each, contain 8 units of A, 1 of B_1, and 2 of C. Chewable tablets, costing \$0.20 each, contain 2 units of A, 1 of B_1, and 7 of C. How many of each should she take each day to minimize her cost and yet fulfill her daily requirements?

SOLUTION

Step 1 Let x represent the number of capsules to take each day, and let y represent the number of chewable tablets to take. Then the cost in *pennies* per day is

$$\text{cost} = 10x + 20y. \quad \text{Objective function}$$

Robin takes x of the \$0.10 capsules and y of the \$0.20 chewable tablets, and she gets 8 units of vitamin A from each capsule and 2 units of vitamin A from each tablet. Altogether she gets $8x + 2y$ units of A per day. Since she wants at least 16 units,

$$8x + 2y \geq 16.$$

Each capsule and each tablet supplies 1 unit of vitamin B_1. Robin wants at least 5 units per day, so

$$x + y \geq 5.$$

For vitamin C, the inequality is

$$2x + 7y \geq 20.$$

Since Robin cannot take negative numbers of multivitamins,

$$x \geq 0 \quad \text{and} \quad y \geq 0.$$

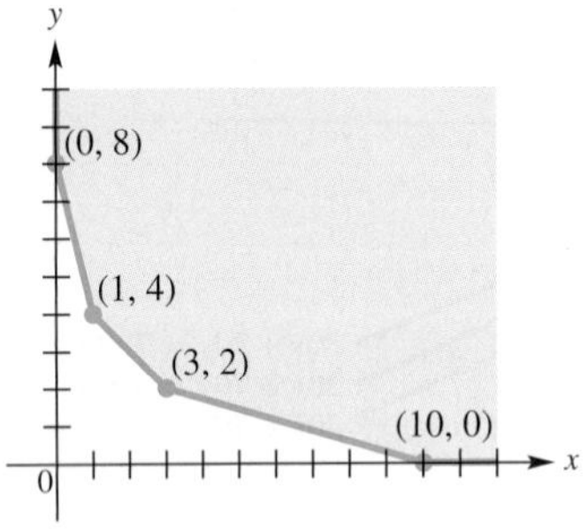

Figure 23

Step 2 **Figure 23** shows the intersection of the graphs of

$$8x + 2y \geq 16, \quad x + y \geq 5, \quad 2x + 7y \geq 20, \quad x \geq 0, \quad \text{and} \quad y \geq 0.$$

Step 3 The vertices are $(0, 8)$, $(1, 4)$, $(3, 2)$, and $(10, 0)$.

Steps 4 and 5 See the table. The minimum cost occurs at $(3, 2)$.

Point	Cost $= 10x + 20y$	
$(0, 8)$	$10(0) + 20(8) = 160$	
$(1, 4)$	$10(1) + 20(4) = 90$	
$(3, 2)$	$10(3) + 20(2) = 70$	← Minimum
$(10, 0)$	$10(10) + 20(0) = 100$	

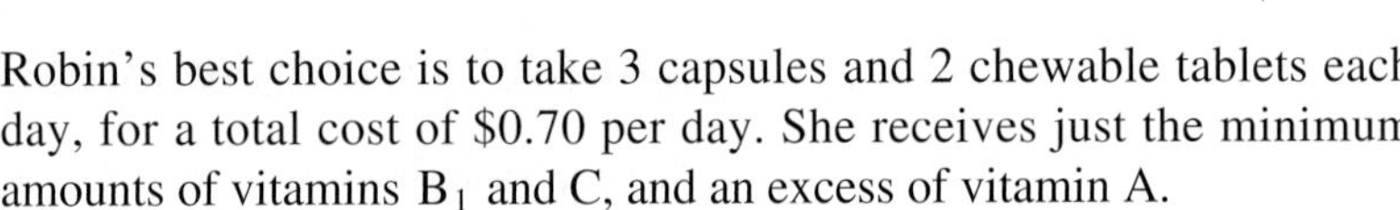

Robin's best choice is to take 3 capsules and 2 chewable tablets each day, for a total cost of \$0.70 per day. She receives just the minimum amounts of vitamins B_1 and C, and an excess of vitamin A.

Now Try Exercises 71 and 81.

9.6 Exercises

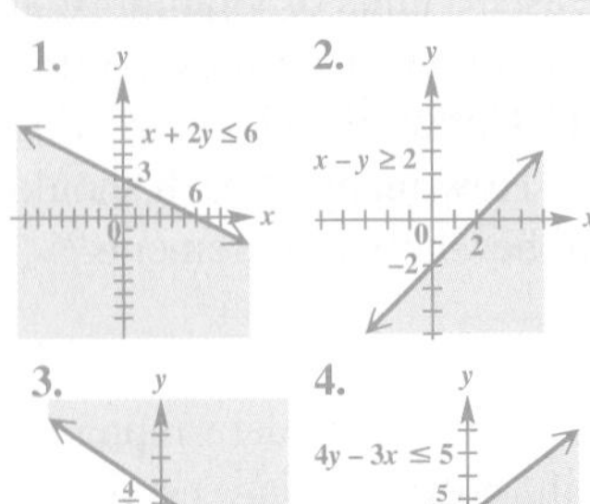

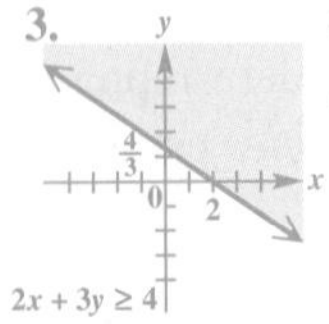

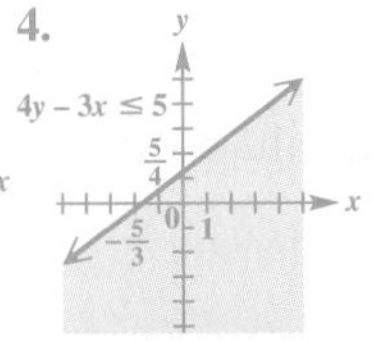

Graph each inequality. ***See Examples 1–3.***

1. $x + 2y \leq 6$
2. $x - y \geq 2$
3. $2x + 3y \geq 4$
4. $4y - 3x \leq 5$
5. $3x - 5y > 6$
6. $x < 3 + 2y$
7. $5x < 4y - 2$
8. $2x > 3 - 4y$
9. $x \leq 3$
10. $y \leq -2$

5. 6.

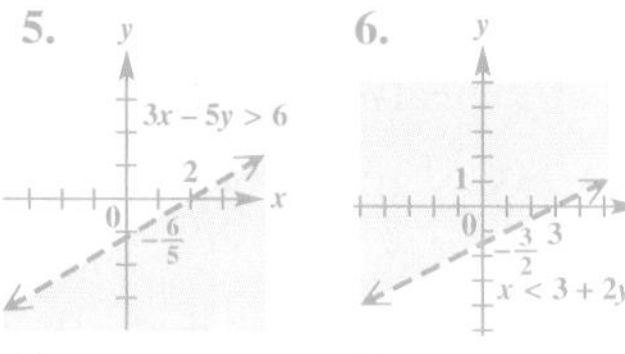

7. 8.

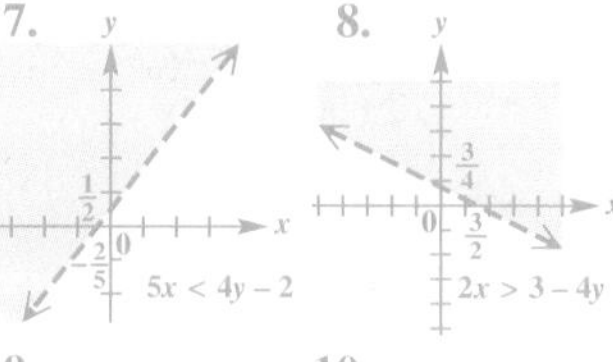

9. 10.

x ≤ 3
y ≤ −2

11. 12.

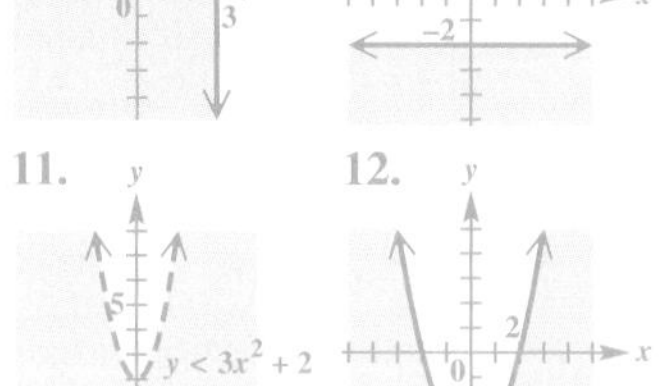

13. 14.

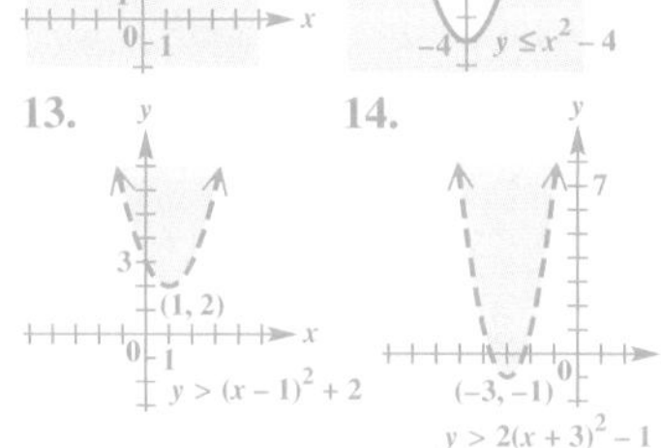

15. 16.

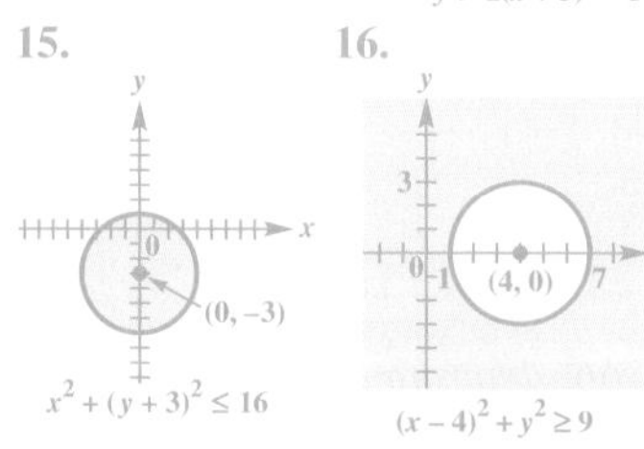

17. 18.

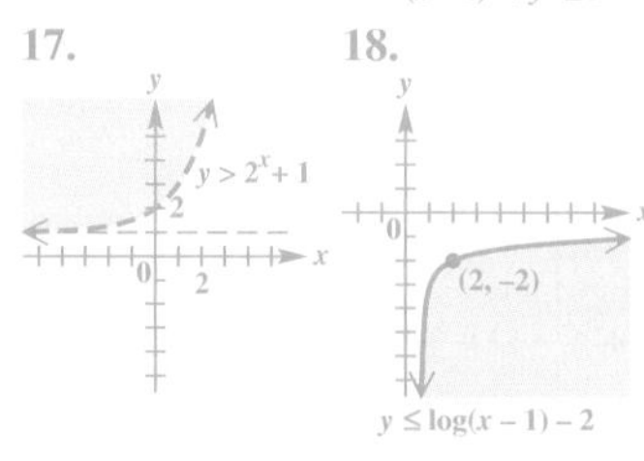

21. above
22. below
23. B
24. $y < 3x + 5$

25. C
26. B
27. A
28. D

29. 30.

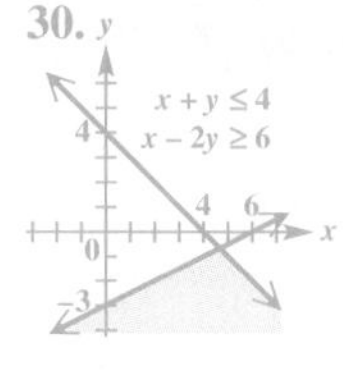

11. $y < 3x^2 + 2$

12. $y \le x^2 - 4$

13. $y > (x - 1)^2 + 2$

14. $y > 2(x + 3)^2 - 1$

15. $x^2 + (y + 3)^2 \le 16$

16. $(x - 4)^2 + y^2 \ge 9$

17. $y > 2^x + 1$

18. $y \le \log(x - 1) - 2$

19. Explain how to determine whether the boundary of the graph of an inequality is solid or dashed.

20. When graphing $y \le 3x - 6$, would you shade above or below the line $y = 3x - 6$? Explain your answer.

Concept Check *Work each problem.*

21. For $Ax + By \ge C$, if $B > 0$, would you shade above or below the line?

22. For $Ax + By \ge C$, if $B < 0$, would you shade above or below the line?

23. Which one of the following is a description of the graph of the inequality

$$(x - 5)^2 + (y - 2)^2 < 4?$$

A. the region inside a circle with center $(-5, -2)$ and radius 2

B. the region inside a circle with center $(5, 2)$ and radius 2

C. the region inside a circle with center $(-5, -2)$ and radius 4

D. the region outside a circle with center $(5, 2)$ and radius 4

24. Find an example of a linear inequality in two variables whose graph does not intersect the graph of $y \ge 3x + 5$.

Concept Check *In Exercises 25–28, match each inequality with the appropriate calculator graph. Do not use your calculator. Instead, use your knowledge of the concepts involved in graphing inequalities.*

25. $y \le 3x - 6$

26. $y \ge 3x - 6$

27. $y \le -3x - 6$

28. $y \ge -3x - 6$

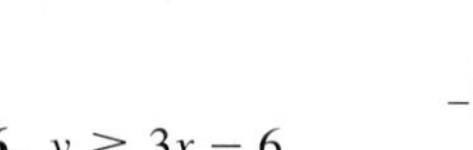

A.

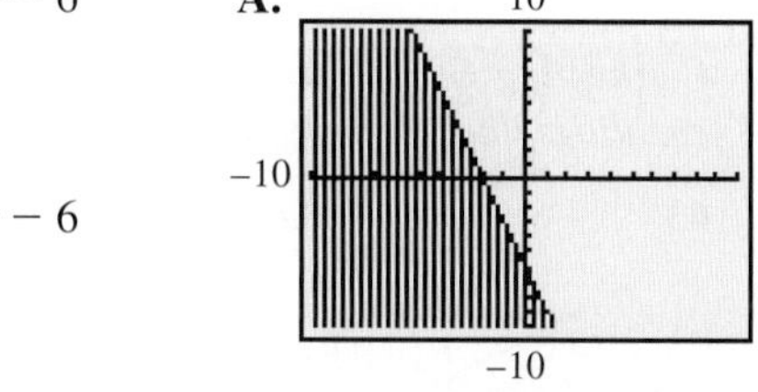

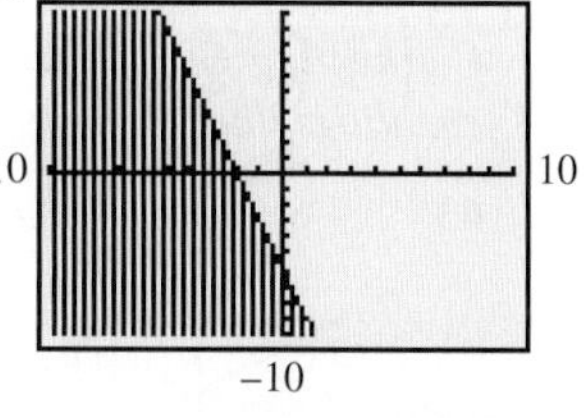

B.

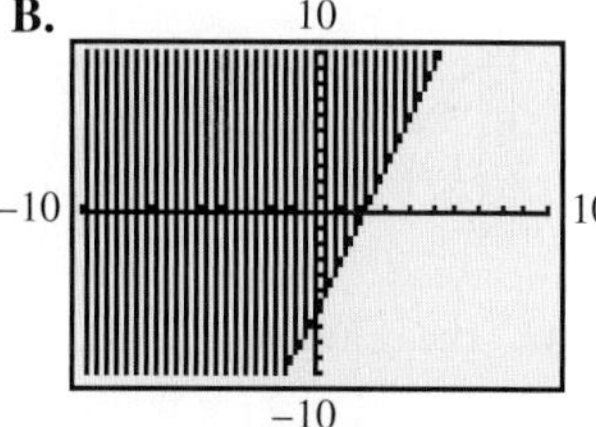

C.

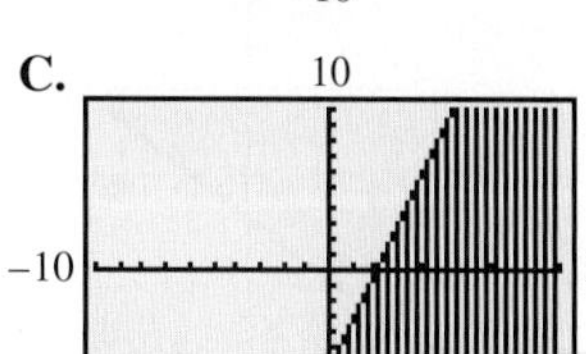

D.

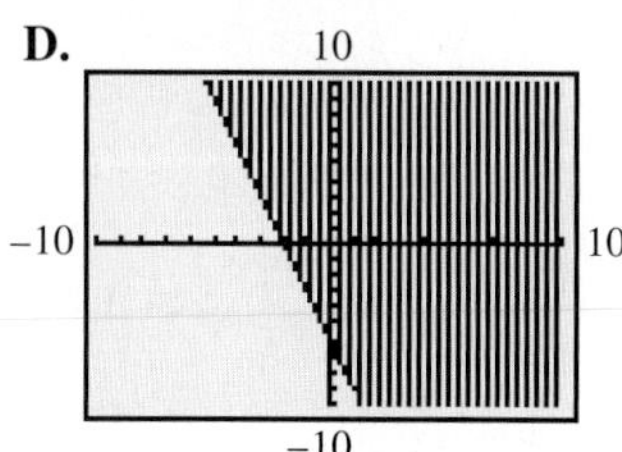

Graph the solution set of each system of inequalities. ***See Example 3.***

29. $x + y \ge 0$
$2x - y \ge 3$

30. $x + y \le 4$
$x - 2y \ge 6$

31. $2x + y > 2$
$x - 3y < 6$

32. $4x + 3y < 12$
$y + 4x > -4$

33. $3x + 5y \le 15$
$x - 3y \ge 9$

34. $y \le x$
$x^2 + y^2 < 1$

35. $4x - 3y \le 12$
$y \le x^2$

36. $y \le -x^2$
$y \ge x^2 - 6$

37. $x + 2y \le 4$
$y \ge x^2 - 1$

31. 32.

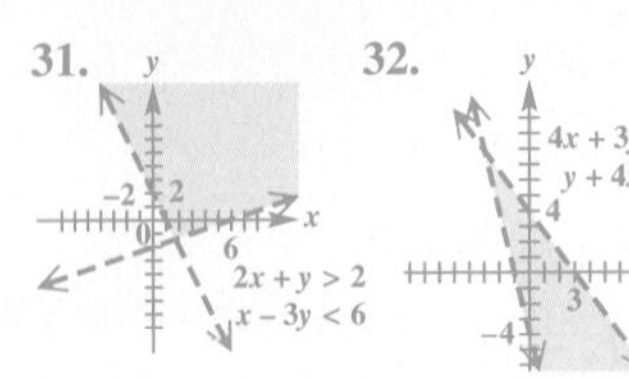

33. 34.

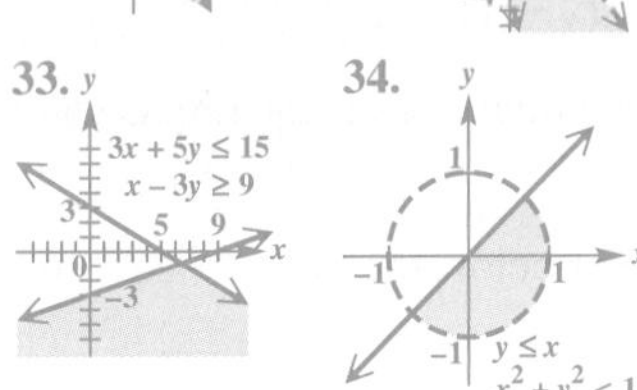

35. 36.

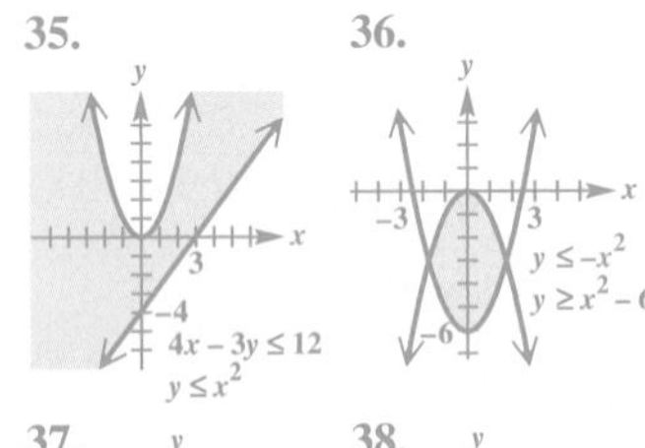

37. 38.

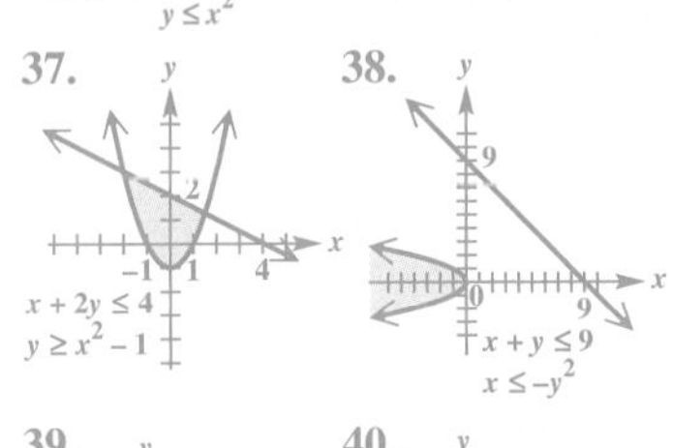

39. 40.

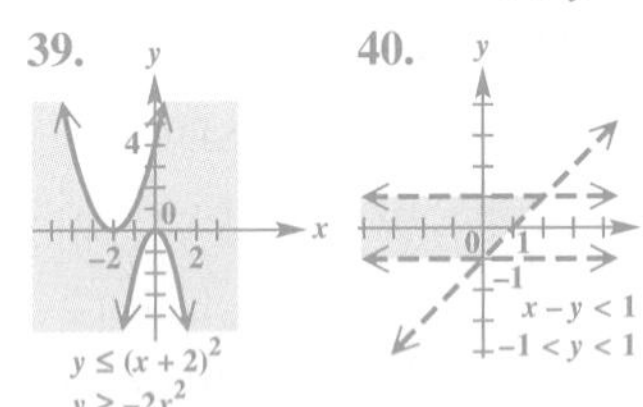
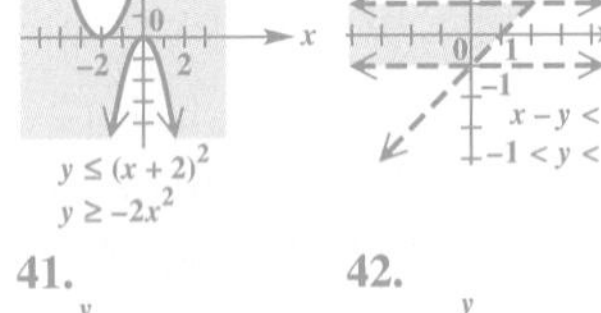

41. 42.

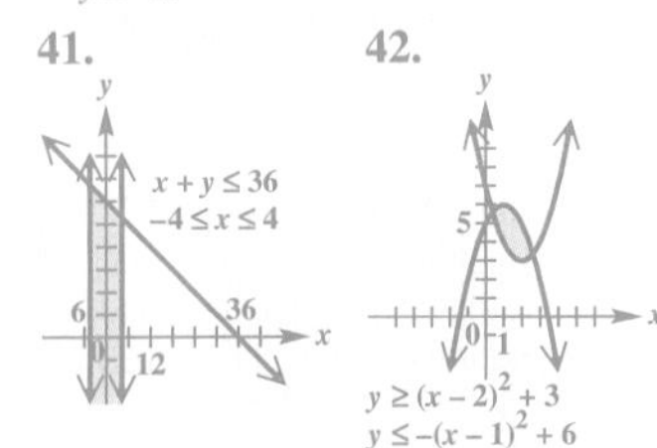

43. The solution set is $\emptyset$.

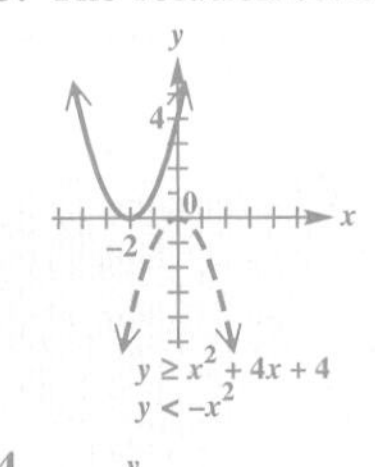

44.

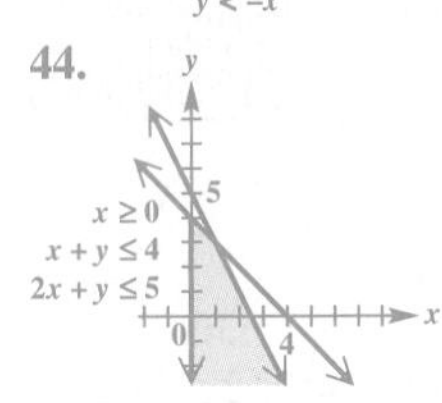

45. 46.

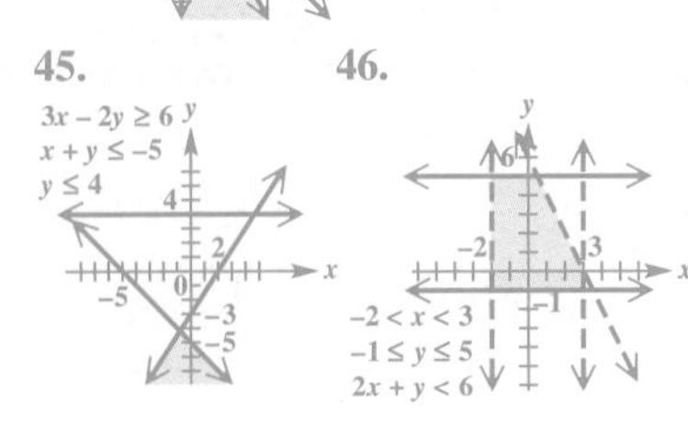

38. $x + y \le 9$, $x \le -y^2$

39. $y \le (x + 2)^2$, $y \ge -2x^2$

40. $x - y < 1$, $-1 < y < 1$

41. $x + y \le 36$, $-4 \le x \le 4$

42. $y \ge (x - 2)^2 + 3$, $y \le -(x - 1)^2 + 6$

43. $y \ge x^2 + 4x + 4$, $y < -x^2$

44. $x \ge 0$, $x + y \le 4$, $2x + y \le 5$

45. $3x - 2y \ge 6$, $x + y \le -5$, $y \le 4$

46. $-2 < x < 3$, $-1 \le y \le 5$, $2x + y < 6$

47. $-2 < x < 2$, $y > 1$, $x - y > 0$

48. $x + y \le 4$, $x - y \le 5$, $4x + y \le -4$

49. $x \le 4$, $x \ge 0$, $y \ge 0$, $x + 2y \ge 2$

50. $2y + x \ge -5$, $y \le 3 + x$, $x \le 0$, $y \le 0$

51. $2x + 3y \le 12$, $2x + 3y > -6$, $3x + y < 4$, $x \ge 0$, $y \ge 0$

52. $y \ge 3^x$, $y \ge 2$

53. $y \le \left(\frac{1}{2}\right)^x$, $y > 4$

54. $\ln x - y \ge 1$, $x^2 - 2x - y \le 1$

55. $y \le \log x$, $y \ge |x - 2|$

56. $e^{-x} - y \le 1$, $x - 2y \ge 4$

57. $y > x^3 + 1$, $y \ge -1$

58. $y \le x^3 - x$, $y > -3$

Use the shading capabilities of your graphing calculator to graph each inequality or system of inequalities.

59. $3x + 2y \ge 6$

60. $y \le x^2 + 5$

61. $x + y \ge 2$, $x + y \le 6$

62. $y \ge |x + 2|$, $y \le 6$

Connecting Graphs with Equations *In Exercises 63–66, determine the system of inequalities illustrated in each graph. Write inequalities in standard form.*

63.

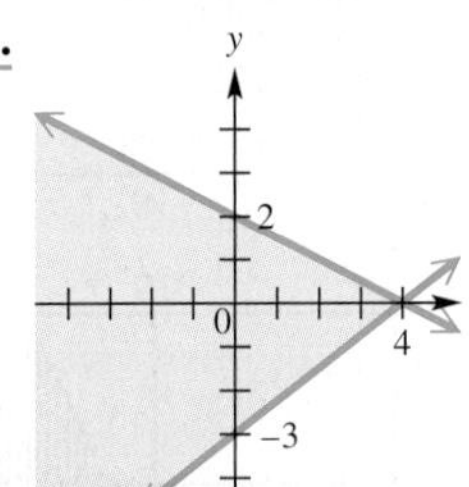
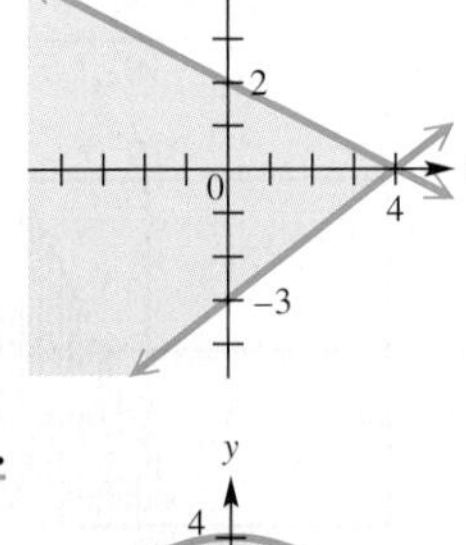

64.

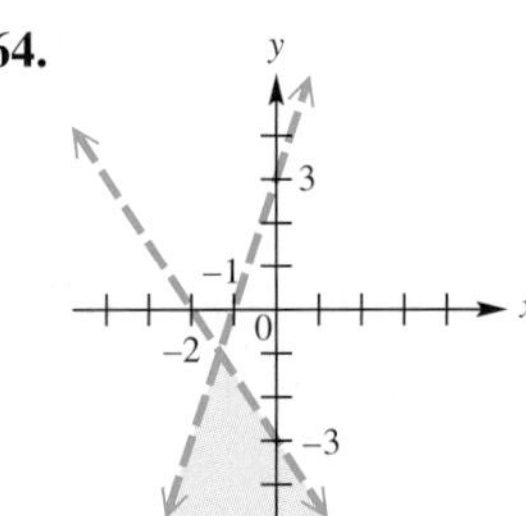

65.

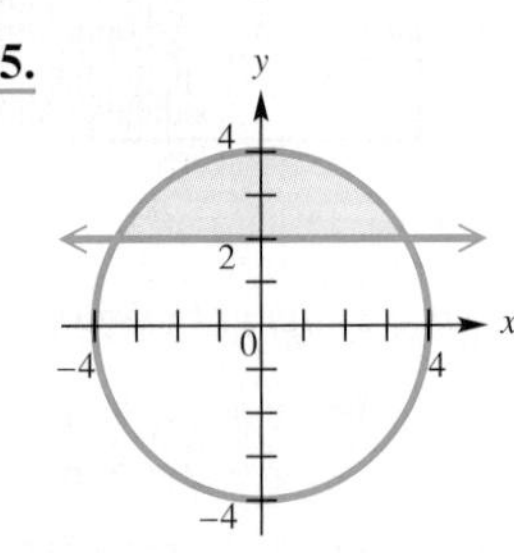

66.

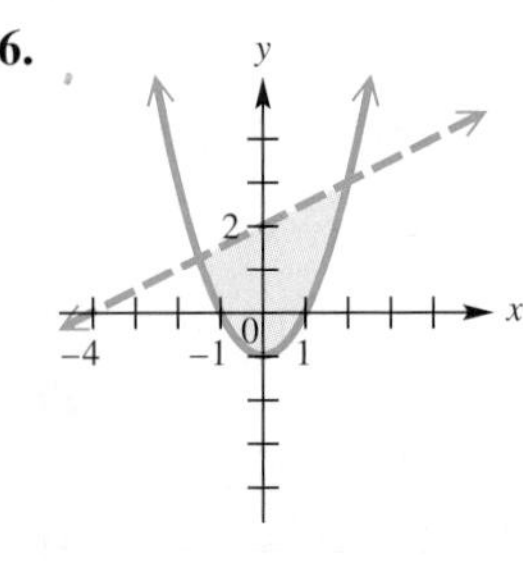

67. *Concept Check* Write a system of inequalities for which the graph is the region in the first quadrant inside and including the circle with radius 2 centered at the origin, and above (not including) the line that passes through the points $(0, -1)$ and $(2, 2)$.

47. 48.

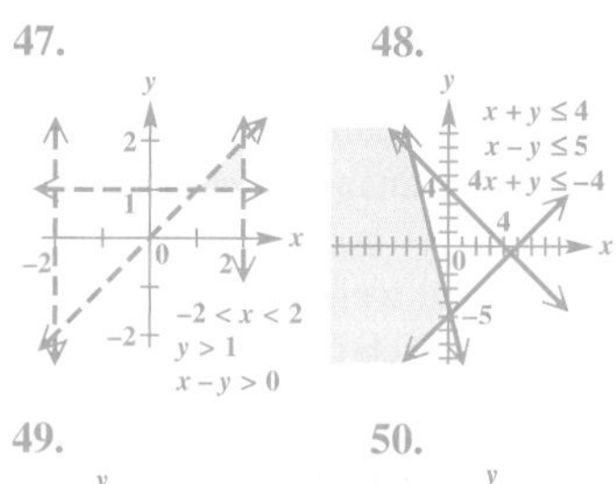

49. 50.

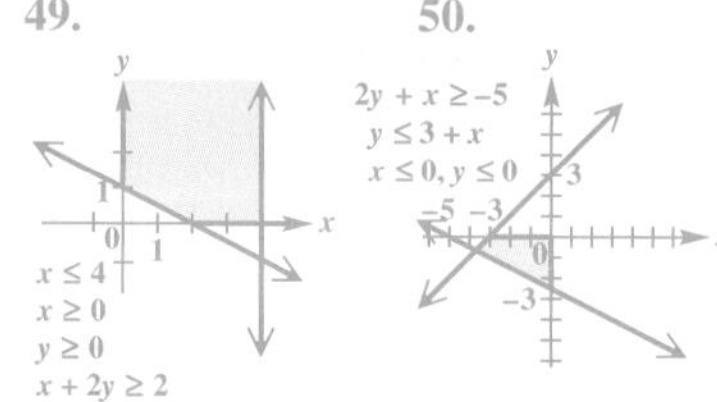

51. 52.

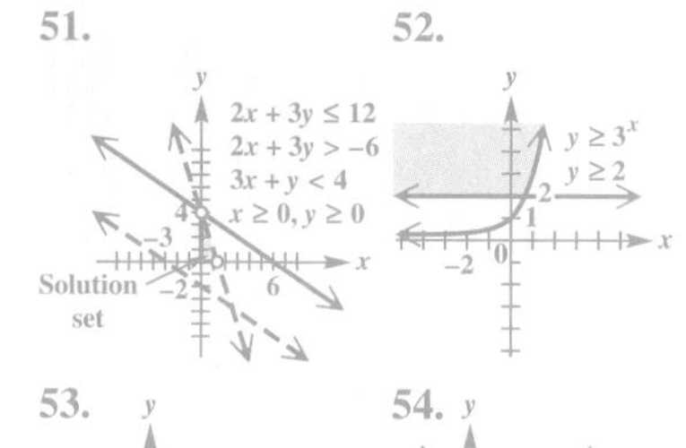

53. 54.

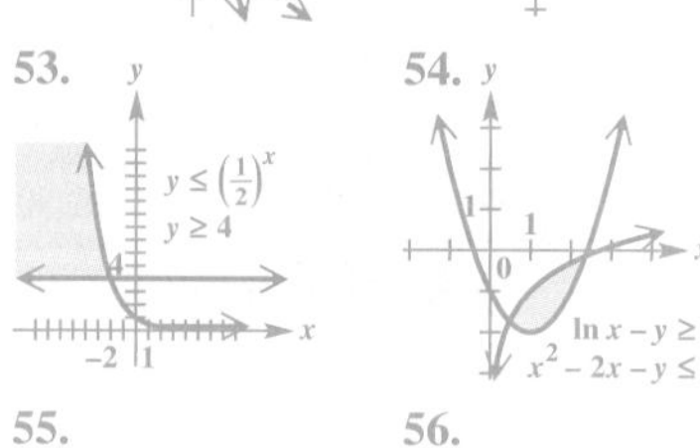

55. 56.

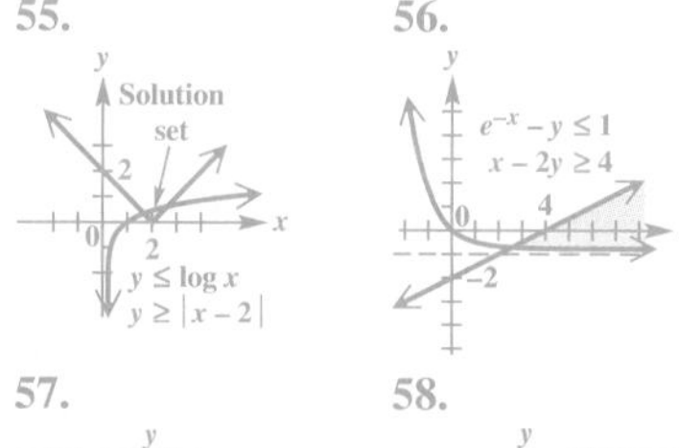

57. 58.

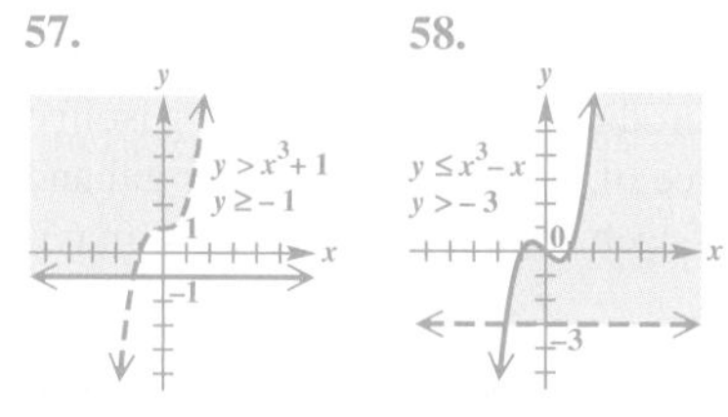

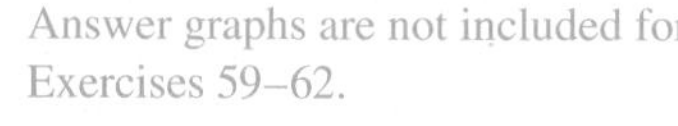
Answer graphs are not included for Exercises 59–62.

63. $x + 2y \le 4$
$3x - 4y \le 12$

64. $3x - y > -3$
$3x + 2y < -6$

65. $x^2 + y^2 \le 16$
$y \ge 2$

66. $y \ge x^2 - 1$
$x - 2y > -4$

67. $x > 0$
$y > 0$
$x^2 + y^2 \le 4$
$y > \frac{3}{2}x - 1$

68. (a) $y = -\frac{1}{2}x + \frac{c}{20}$
(b) up (c) C

68. *Cost of Vitamins* The figure shows the region of feasible solutions for the vitamin problem of **Example 5** and the straight-line graph of all combinations of capsules and chewable tablets for which the cost is \$0.40.

(a) The cost function is $10x + 20y$. Give the linear equation (in slope-intercept form) of the line of constant cost c.

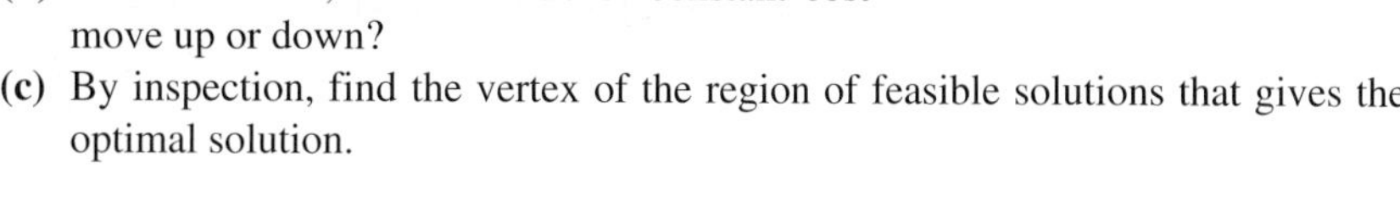

(b) As c increases, does the line of constant cost move up or down?

(c) By inspection, find the vertex of the region of feasible solutions that gives the optimal solution.

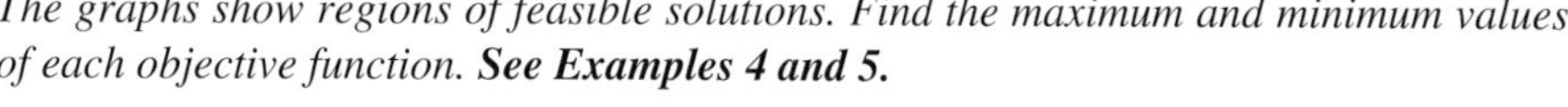

*The graphs show regions of feasible solutions. Find the maximum and minimum values of each objective function. **See Examples 4 and 5.***

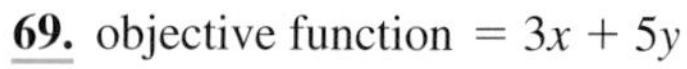

69. objective function $= 3x + 5y$

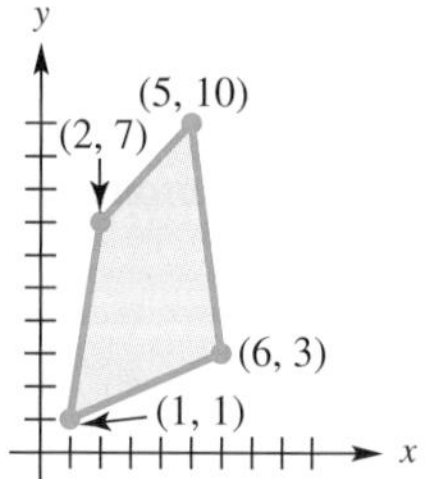

70. objective function $= 6x + y$

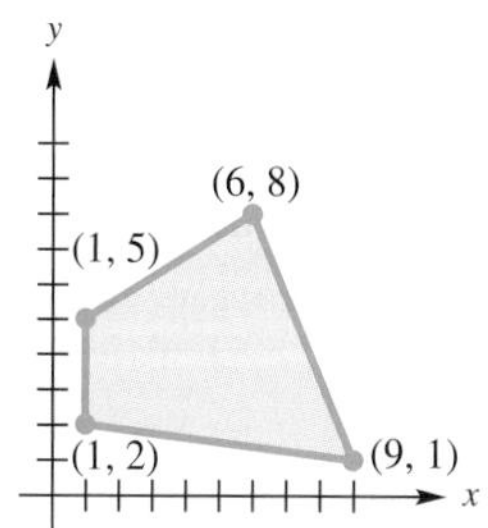

*Find the maximum and minimum values of each objective function over the region of feasible solutions shown at the right. **See Examples 4 and 5.***

71. objective function $= 3x + 5y$

72. objective function $= 5x + 5y$

73. objective function $= 10y$

74. objective function $= 3x - y$

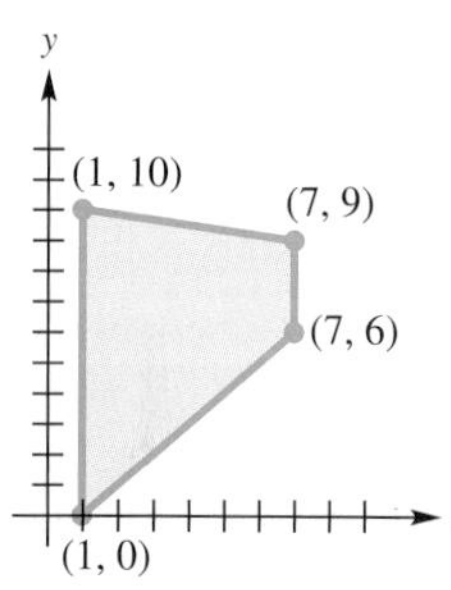

*Write a system of inequalities for each problem, and then graph the region of feasible solutions of the system. **See Examples 4 and 5.***

75. *Vitamin Requirements* Jane Lukas was given the following advice. She should supplement her daily diet with at least 6000 USP units of vitamin A, at least 195 mg of vitamin C, and at least 600 USP units of vitamin D. She finds that Mason's Pharmacy carries Brand X and Brand Y vitamins. Each Brand X pill contains 3000 USP units of A, 45 mg of C, and 75 USP units of D, while the Brand Y pills contain 1000 USP units of A, 50 mg of C, and 200 USP units of D.

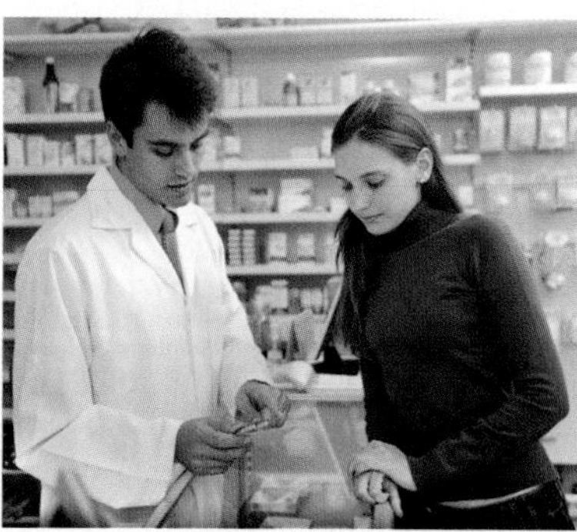

76. *Shipping Requirements* The California Almond Growers have 2400 boxes of almonds to be shipped from their plant in Sacramento to Des Moines and San Antonio. The Des Moines market needs at least 1000 boxes, while the San Antonio market must have at least 800 boxes.

69. maximum of 65 at $(5, 10)$; minimum of 8 at $(1, 1)$

70. maximum of 55 at $(9, 1)$; minimum of 8 at $(1, 2)$

71. maximum of 66 at $(7, 9)$; minimum of 3 at $(1, 0)$

72. maximum of 80 at $(7, 9)$; minimum of 5 at $(1, 0)$

73. maximum of 100 at $(1, 10)$; minimum of 0 at $(1, 0)$

74. maximum of 15 at $(7, 6)$; minimum of -7 at $(1, 10)$

75. Let $x =$ the number of Brand X pills and $y =$ the number of Brand Y pills. Then $3000x + 1000y \geq 6000$, $45x + 50y \geq 195$, $75x + 200y \geq 600$, $x \geq 0$, $y \geq 0$.

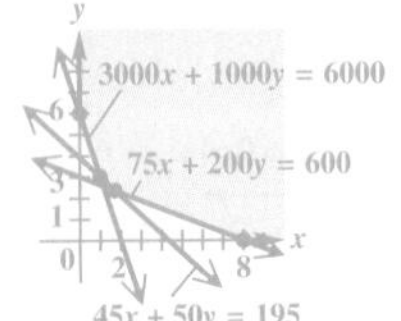

76. Let $x =$ the number of boxes to Des Moines and $y =$ the number of boxes to San Antonio. Then $x \geq 1000$, $y \geq 800$, $x + y \leq 2400$.

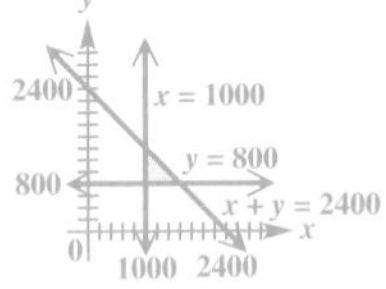

77. (a) 300 cartons of food and 400 cartons of clothes (b) 6200 people

78. (a) no medical kits and 4000 containers of water (b) 4000 medical kits and 2000 containers of water

79. 8 of A and 3 of B, for a maximum storage capacity of 100 ft^3

80. 6.4 million gal of gasoline and 3.2 million gal of fuel oil, for a maximum revenue of \$26,560,000

81. $3\frac{3}{4}$ servings of A and $1\frac{7}{8}$ servings of B, for a minimum cost of \$1.69

82. 800 bargain models and 300 deluxe models, for a maximum profit of \$125,000

Solve each linear programming problem. ***See Examples 4 and 5.***

77. *Aid to Disaster Victims* An agency wants to ship food and clothing to tsunami victims in Japan. Commercial carriers have volunteered to transport the packages, provided they fit in the available cargo space. Each 20-ft^3 box of food weighs 40 lb and each 30-ft^3 box of clothing weighs 10 lb. The total weight cannot exceed 16,000 lb, and the total volume must be at most 18,000 ft^3. Each carton of food will feed 10 people, and each carton of clothing will help 8 people.

(a) How many cartons of food and clothing should be sent to maximize the number of people assisted?

(b) What is the maximum number assisted?

78. *Aid to Disaster Victims* Earthquake victims in Haiti need medical supplies and bottled water. Each medical kit measures 1 ft^3 and weighs 10 lb. Each container of water is also 1 ft^3 but weighs 20 lb. The plane can carry only 80,000 lb with a total volume of 6000 ft^3. Each medical kit will aid 4 people, and each container of water will serve 10 people.

(a) How many of each should be sent to maximize the number of people helped?

(b) If each medical kit could aid 6 people instead of 4, how would the results from part (a) change?

79. *Storage Capacity* An office manager wants to buy some filing cabinets. He knows that cabinet A costs \$10 each, requires 6 ft^2 of floor space, and holds 8 ft^3 of files. Cabinet B costs \$20 each, requires 8 ft^2 of floor space, and holds 12 ft^3 of files. He can spend no more than \$140 due to budget limitations, and his office has room for no more than 72 ft^2 of cabinets. He wants to maximize storage capacity within the limits imposed by funds and space. How many of each type of cabinet should he buy?

80. *Gasoline Revenues* The manufacturing process requires that oil refineries manufacture at least 2 gal of gasoline for each gallon of fuel oil. To meet the winter demand for fuel oil, at least 3 million gal per day must be produced. The demand for gasoline is no more than 6.4 million gal per day. If the price of gasoline is \$2.90 per gal and the price of fuel oil is \$2.50 per gal, how much of each should be produced to maximize revenue?

81. *Diet Requirements* Theo, who is dieting, requires two food supplements, I and II. He can get these supplements from two different products, A and B, as shown in the table. Theo's physician has recommended that he include at least 15 g of each supplement in his daily diet. If product A costs \$0.25 per serving and product B costs \$0.40 per serving, how can he satisfy his requirements most economically?

Supplement (g/serving)	I	II
Product *A*	3	2
Product *B*	2	4

82. *Profit from Televisions* Seall Manufacturing Company makes television monitors. It produces a bargain monitor that sells for \$100 profit and a deluxe monitor that sells for \$150 profit. On the assembly line the bargain monitor requires 3 hr, and the deluxe monitor takes 5 hr. The cabinet shop spends 1 hr on the cabinet for the bargain monitor and 3 hr on the cabinet for the deluxe monitor. Both models require 2 hr of time for testing and packing. On a particular production run, the Seall Company has available 3900 work hours on the assembly line, 2100 work hours in the cabinet shop, and 2200 work hours in the testing and packing department. How many of each model should it produce to make the maximum profit? What is the maximum profit?

9.7 Properties of Matrices

- Basic Definitions
- Adding Matrices
- Special Matrices
- Subtracting Matrices
- Multiplying Matrices
- Applying Matrix Algebra

We used matrix notation to solve a system of linear equations earlier in this chapter. In this section and the next, we discuss algebraic properties of matrices.

Basic Definitions It is customary to use capital letters to name matrices and to use subscript notation to name elements of a matrix, as in the following matrix A.

$$A = \begin{bmatrix} a_{11} & a_{12} & a_{13} & \cdots & a_{1n} \\ a_{21} & a_{22} & a_{23} & \cdots & a_{2n} \\ a_{31} & a_{32} & a_{33} & \cdots & a_{3n} \\ \vdots & \vdots & \vdots & & \vdots \\ a_{m1} & a_{m2} & a_{m3} & \cdots & a_{mn} \end{bmatrix}$$

The first row, first column element is a_{11} (read "a-sub-one-one"); the second row, third column element is a_{23}; and in general, the ith row, jth column element is a_{ij}.

Certain matrices have special names. An $n \times n$ matrix is a **square matrix** because the number of rows is equal to the number of columns. A matrix with just one row is a **row matrix,** and a matrix with just one column is a **column matrix.**

Two matrices are equal if they have the same dimension and if corresponding elements, position by position, are equal. Using this definition, the matrices

$$\begin{bmatrix} 2 & 1 \\ 3 & -5 \end{bmatrix} \quad \text{and} \quad \begin{bmatrix} 1 & 2 \\ -5 & 3 \end{bmatrix} \quad \text{are } \textit{not} \text{ equal}$$

(even though they contain the same elements and have the same dimension), since the corresponding elements differ.

Classroom Example 1
Find the values of the variables for which each statement is true, if possible.

(a) $\begin{bmatrix} a & b \\ -5 & 0 \end{bmatrix} = \begin{bmatrix} -3 & 9 \\ c & d \end{bmatrix}$

(b) $\begin{bmatrix} x & y \end{bmatrix} = \begin{bmatrix} 3 \\ -5 \end{bmatrix}$

Answers:
(a) $a = -3,\ b = 9,\ c = -5,\ d = 0$
(b) This statement can never be true.

EXAMPLE 1 Finding Values to Make Two Matrices Equal

Find the values of the variables for which each statement is true, if possible.

(a) $\begin{bmatrix} 2 & 1 \\ p & q \end{bmatrix} = \begin{bmatrix} x & y \\ -1 & 0 \end{bmatrix}$ **(b)** $\begin{bmatrix} x \\ y \end{bmatrix} = \begin{bmatrix} 1 \\ 4 \\ 0 \end{bmatrix}$

SOLUTION

(a) From the definition of equality given above, the only way that the statement can be true is if $2 = x,\ 1 = y,\ p = -1,$ and $q = 0$.

(b) This statement can never be true since the two matrices have different dimensions. (One is 2×1 and the other is 3×1.)

✔ *Now Try Exercises 1 and 7.*

Adding Matrices Addition of matrices is defined as follows.

Addition of Matrices

To add two matrices of the same dimension, add corresponding elements. ***Only matrices of the same dimension can be added.***

It can be shown that matrix addition satisfies the commutative, associative, closure, identity, and inverse properties. **(See Exercises 87 and 88.)**

EXAMPLE 2 Adding Matrices

Find each sum, if possible.

(a) $\begin{bmatrix} 5 & -6 \\ 8 & 9 \end{bmatrix} + \begin{bmatrix} -4 & 6 \\ 8 & -3 \end{bmatrix}$ **(b)** $\begin{bmatrix} 2 \\ 5 \\ 8 \end{bmatrix} + \begin{bmatrix} -6 \\ 3 \\ 12 \end{bmatrix}$

(c) $A + B$, if $A = \begin{bmatrix} 5 & 8 \\ 6 & 2 \end{bmatrix}$ and $B = \begin{bmatrix} 3 & 9 & 1 \\ 4 & 2 & 5 \end{bmatrix}$

ALGEBRAIC SOLUTION

(a) $\begin{bmatrix} 5 & -6 \\ 8 & 9 \end{bmatrix} + \begin{bmatrix} -4 & 6 \\ 8 & -3 \end{bmatrix}$

$$= \begin{bmatrix} 5 + (-4) & -6 + 6 \\ 8 + 8 & 9 + (-3) \end{bmatrix}$$

$$= \begin{bmatrix} 1 & 0 \\ 16 & 6 \end{bmatrix}$$

(b) $\begin{bmatrix} 2 \\ 5 \\ 8 \end{bmatrix} + \begin{bmatrix} -6 \\ 3 \\ 12 \end{bmatrix} = \begin{bmatrix} -4 \\ 8 \\ 20 \end{bmatrix}$

(c) The matrices

$$A = \begin{bmatrix} 5 & 8 \\ 6 & 2 \end{bmatrix}$$

and $B = \begin{bmatrix} 3 & 9 & 1 \\ 4 & 2 & 5 \end{bmatrix}$

have different dimensions, so A and B cannot be added. The sum $A + B$ does not exist.

GRAPHING CALCULATOR SOLUTION

(a) Figure 24(b) shows the sum of matrices A and B as defined in **Figure 24(a).**

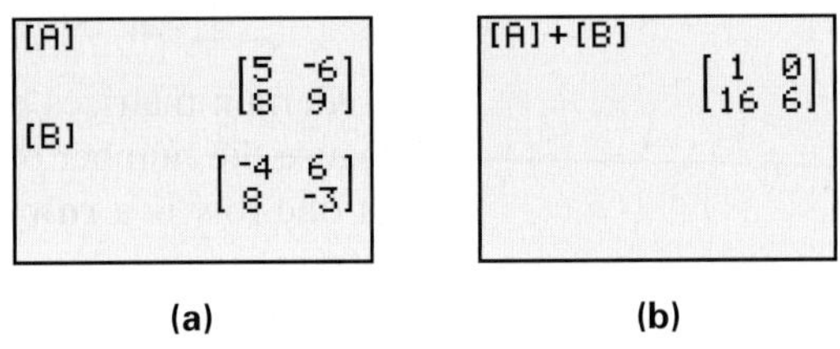

(a) (b)

Figure 24

(b) The screen in **Figure 25** shows how the sum of two column matrices entered directly on the home screen is displayed.

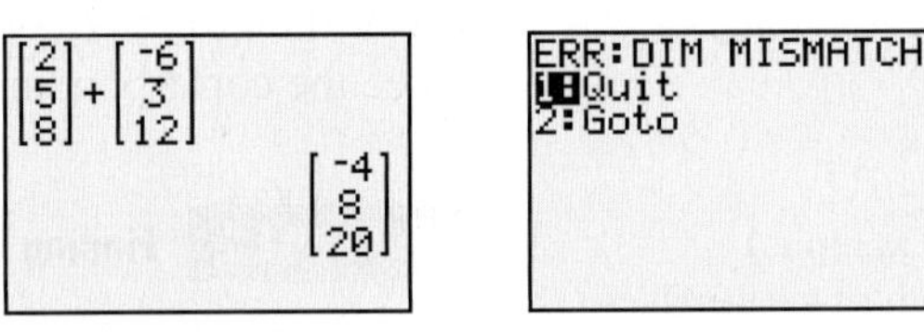

Figure 25 **Figure 26**

(c) A graphing calculator will return an ERROR message if it is directed to perform an operation on matrices that is not possible because of dimension mismatch. See **Figure 26.**

Now Try Exercises 21, 23, and 25.

Classroom Example 2

Find each sum, if possible.

(a) $\begin{bmatrix} 3 & -8 \\ -4 & 6 \end{bmatrix} + \begin{bmatrix} 7 & -5 \\ 10 & -6 \end{bmatrix}$

(b) $\begin{bmatrix} -9 \\ 7 \\ 6 \\ -3 \end{bmatrix} + \begin{bmatrix} -9 \\ 5 \\ -6 \\ 10 \end{bmatrix}$

(c) $A + B$, if

$A = \begin{bmatrix} 3 & -1 & 6 \\ 0 & 7 & -2 \end{bmatrix}$

and $B = \begin{bmatrix} 4 & 8 \\ 3 & -1 \\ -5 & 2 \end{bmatrix}$

Answers:

(a) $\begin{bmatrix} 10 & -13 \\ 6 & 0 \end{bmatrix}$ (b) $\begin{bmatrix} -18 \\ 12 \\ 0 \\ 7 \end{bmatrix}$

(c) A and B cannot be added.

Special Matrices A matrix containing only zero elements is a **zero matrix.** A zero matrix can be written with any dimension.

1×3 zero matrix $\quad$ 2×3 zero matrix

$$O = \begin{bmatrix} 0 & 0 & 0 \end{bmatrix} \qquad O = \begin{bmatrix} 0 & 0 & 0 \\ 0 & 0 & 0 \end{bmatrix}$$

By the additive inverse property, each real number has an additive inverse: If a is a real number, then there is a real number $-a$ such that

$$a + (-a) = 0 \quad \text{and} \quad -a + a = 0. \quad \text{(Section R.2)}$$

Given matrix A, there is a matrix $-A$ such that $A + (-A) = O$. The matrix $-A$ has as elements the additive inverses of the elements of A. (Remember, each element of A is a real number and therefore has an additive inverse.) For example, if

$$A = \begin{bmatrix} -5 & 2 & -1 \\ 3 & 4 & -6 \end{bmatrix}, \quad \text{then} \quad -A = \begin{bmatrix} 5 & -2 & 1 \\ -3 & -4 & 6 \end{bmatrix}.$$

CHECK Test that $A + (-A)$ equals the zero matrix, O.

$$A + (-A) = \begin{bmatrix} -5 & 2 & -1 \\ 3 & 4 & -6 \end{bmatrix} + \begin{bmatrix} 5 & -2 & 1 \\ -3 & -4 & 6 \end{bmatrix} = \begin{bmatrix} 0 & 0 & 0 \\ 0 & 0 & 0 \end{bmatrix} = O \checkmark$$

Matrix $-A$ is the **additive inverse,** or **negative,** of matrix A. Every matrix has an additive inverse.

Subtracting Matrices The real number b is subtracted from the real number a, written $a - b$, by adding a and the additive inverse of b.

$$a - b = a + (-b) \quad \text{Real number subtraction}$$

The same definition applies to subtraction of matrices.

Subtraction of Matrices

If A and B are two matrices of the same dimension, then

$$A - B = A + (-B).$$

Classroom Example 3

Find each difference, if possible.

(a) $\begin{bmatrix} 8 & -9 \\ -6 & 2 \end{bmatrix} - \begin{bmatrix} -8 & 4 \\ -7 & 11 \end{bmatrix}$

(b) $\begin{bmatrix} 9 \\ -3 \\ 6 \end{bmatrix} - \begin{bmatrix} 18 \\ 12 \\ -6 \end{bmatrix}$

(c) $A - B$, if $A = \begin{bmatrix} 4 & 5 & 0 \\ -2 & 3 & 1 \end{bmatrix}$ and $B = \begin{bmatrix} 4 & 5 \\ 3 & -2 \end{bmatrix}$

Answers:

(a) $\begin{bmatrix} 16 & -13 \\ 1 & -9 \end{bmatrix}$ (b) $\begin{bmatrix} -9 \\ -15 \\ 12 \end{bmatrix}$

(c) A and B cannot be subtracted.

In practice, the difference of two matrices of the same dimension is found by subtracting corresponding elements.

EXAMPLE 3 Subtracting Matrices

Find each difference, if possible.

(a) $\begin{bmatrix} -5 & 6 \\ 2 & 4 \end{bmatrix} - \begin{bmatrix} -3 & 2 \\ 5 & -8 \end{bmatrix}$ **(b)** $\begin{bmatrix} 8 & 6 & -4 \end{bmatrix} - \begin{bmatrix} 3 & 5 & -8 \end{bmatrix}$

(c) $A - B$, if $A = \begin{bmatrix} -2 & 5 \\ 0 & 1 \end{bmatrix}$ and $B = \begin{bmatrix} 3 \\ 5 \end{bmatrix}$

SOLUTION

(a) $\begin{bmatrix} -5 & 6 \\ 2 & 4 \end{bmatrix} - \begin{bmatrix} -3 & 2 \\ 5 & -8 \end{bmatrix} = \begin{bmatrix} -5-(-3) & 6-2 \\ 2-5 & 4-(-8) \end{bmatrix}$

$$= \begin{bmatrix} -2 & 4 \\ -3 & 12 \end{bmatrix}$$

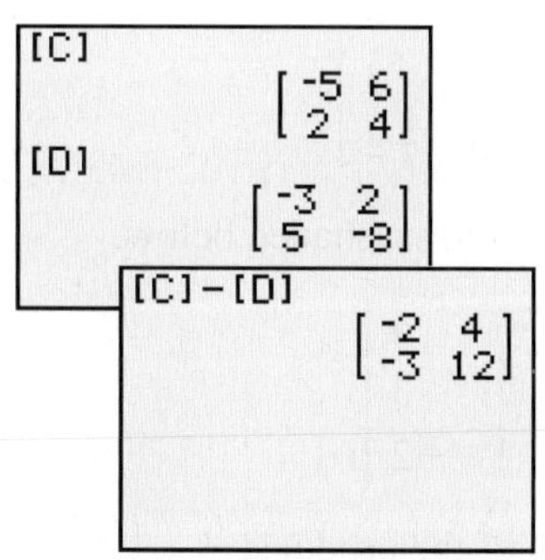

These screens support the result in **Example 3(a).**

(b) $\begin{bmatrix} 8 & 6 & -4 \end{bmatrix} - \begin{bmatrix} 3 & 5 & -8 \end{bmatrix} = \begin{bmatrix} 5 & 1 & 4 \end{bmatrix}$

(c) The matrices

$$A = \begin{bmatrix} -2 & 5 \\ 0 & 1 \end{bmatrix} \quad \text{and} \quad B = \begin{bmatrix} 3 \\ 5 \end{bmatrix}$$

have different dimensions and cannot be subtracted, so the difference $A - B$ does not exist.

✔ Now Try Exercises 27, 29, and 31.

Multiplying Matrices In work with matrices, a real number is called a **scalar** to distinguish it from a matrix. The product of a scalar k and a matrix X is the matrix kX, each of whose elements is k times the corresponding element of X.

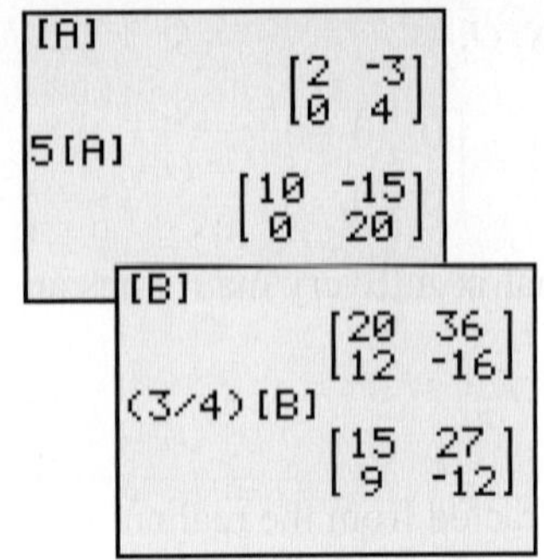

These screens support the results in **Example 4.**

EXAMPLE 4 Multiplying Matrices by Scalars

Find each product.

(a) $5\begin{bmatrix} 2 & -3 \\ 0 & 4 \end{bmatrix}$ **(b)** $\frac{3}{4}\begin{bmatrix} 20 & 36 \\ 12 & -16 \end{bmatrix}$

SOLUTION

(a) $5\begin{bmatrix} 2 & -3 \\ 0 & 4 \end{bmatrix} = \begin{bmatrix} 5(2) & 5(-3) \\ 5(0) & 5(4) \end{bmatrix}$ Multiply each element of the matrix by the scalar 5.

$$= \begin{bmatrix} 10 & -15 \\ 0 & 20 \end{bmatrix}$$

(b) $\frac{3}{4}\begin{bmatrix} 20 & 36 \\ 12 & -16 \end{bmatrix} = \begin{bmatrix} \frac{3}{4}(20) & \frac{3}{4}(36) \\ \frac{3}{4}(12) & \frac{3}{4}(-16) \end{bmatrix}$ Multiply each element of the matrix by the scalar $\frac{3}{4}$.

$$= \begin{bmatrix} 15 & 27 \\ 9 & -12 \end{bmatrix}$$

✔ *Now Try Exercises 37 and 39.*

Classroom Example 4
Find each product.

(a) $-3\begin{bmatrix} 2 & -5 \\ -4 & 0 \end{bmatrix}$

(b) $\frac{4}{5}\begin{bmatrix} -25 & 10 \\ 15 & -45 \end{bmatrix}$

Answers:

(a) $\begin{bmatrix} -6 & 15 \\ 12 & 0 \end{bmatrix}$

(b) $\begin{bmatrix} -20 & 8 \\ 12 & -36 \end{bmatrix}$

The proofs of the following properties of scalar multiplication are left for **Exercises 91–94.**

Properties of Scalar Multiplication

Let A and B be matrices of the same dimension, and let c and d be scalars. Then these properties hold.

$$(c + d)A = cA + dA \qquad (cA)d = (cd)A$$
$$c(A + B) = cA + cB \qquad (cd)A = c(dA)$$

We have seen how to multiply a real number (scalar) and a matrix. To find the product of two matrices, such as

$$A = \begin{bmatrix} -3 & 4 & 2 \\ 5 & 0 & 4 \end{bmatrix} \quad \text{and} \quad B = \begin{bmatrix} -6 & 4 \\ 2 & 3 \\ 3 & -2 \end{bmatrix},$$

first locate *row* 1 of A and *column* 1 of B, which are shown shaded below.

$$A = \begin{bmatrix} -3 & 4 & 2 \\ 5 & 0 & 4 \end{bmatrix} \qquad B = \begin{bmatrix} -6 & 4 \\ 2 & 3 \\ 3 & -2 \end{bmatrix}$$

Multiply corresponding elements, and find the sum of the products.

$$-3(-6) + 4(2) + 2(3) = 32$$

This result is the element for row 1, column 1 of the product matrix.

Now use *row* 1 of A and *column* 2 of B to determine the element in row 1, column 2 of the product matrix.

$$\begin{bmatrix} -3 & 4 & 2 \\ 5 & 0 & 4 \end{bmatrix}\begin{bmatrix} -6 & 4 \\ 2 & 3 \\ 3 & -2 \end{bmatrix} \qquad -3(4) + 4(3) + 2(-2) = -4$$

Next, use *row* 2 of A and *column* 1 of B. This will give the row 2, column 1 entry of the product matrix.

$$\begin{bmatrix} -3 & 4 & 2 \\ 5 & 0 & 4 \end{bmatrix} \begin{bmatrix} -6 & 4 \\ 2 & 3 \\ 3 & -2 \end{bmatrix} \qquad 5(-6) + 0(2) + 4(3) = -18$$

Finally, use *row* 2 of A and *column* 2 of B to find the entry for row 2, column 2 of the product matrix.

$$\begin{bmatrix} -3 & 4 & 2 \\ 5 & 0 & 4 \end{bmatrix} \begin{bmatrix} -6 & 4 \\ 2 & 3 \\ 3 & -2 \end{bmatrix} \qquad 5(4) + 0(3) + 4(-2) = 12$$

The product matrix can now be written.

$$\begin{bmatrix} -3 & 4 & 2 \\ 5 & 0 & 4 \end{bmatrix} \begin{bmatrix} -6 & 4 \\ 2 & 3 \\ 3 & -2 \end{bmatrix} = \begin{bmatrix} 32 & -4 \\ -18 & 12 \end{bmatrix} \qquad \text{Product of } A \text{ and } B$$

NOTE As seen here, the product of a 2×3 matrix and a 3×2 matrix is a 2×2 matrix. The dimension of a product matrix AB is given by the number of rows of A and the number of columns of B, respectively.

By definition, the product AB of an $m \times n$ matrix A and an $n \times p$ matrix B is found as follows. Multiply each element of the first row of A by the corresponding element of the first column of B. The sum of these n products is the first row, first column element of AB. Also, the sum of the products found by multiplying the elements of the first row of A times the corresponding elements of the second column of B gives the first row, second column element of AB, and so on.

To find the ith row, jth column element of AB, multiply each element in the ith row of A by the corresponding element in the jth column of B. (Note the shaded areas in the matrices below.) The sum of these products will give the row i, column j element of AB.

$$A = \begin{bmatrix} a_{11} & a_{12} & a_{13} & \cdots & a_{1n} \\ a_{21} & a_{22} & a_{23} & \cdots & a_{2n} \\ \vdots & & & & \\ a_{i1} & a_{i2} & a_{i3} & \cdots & a_{in} \\ \vdots & & & & \\ a_{m1} & a_{m2} & a_{m3} & \cdots & a_{mn} \end{bmatrix} \qquad B = \begin{bmatrix} b_{11} & b_{12} & \cdots & b_{1j} & \cdots & b_{1p} \\ b_{21} & b_{22} & \cdots & b_{2j} & \cdots & b_{2p} \\ \vdots & & & & & \\ b_{n1} & b_{n2} & \cdots & b_{nj} & \cdots & b_{np} \end{bmatrix}$$

Teaching Tip Encourage students to first determine the dimension of the product matrix. Then have them identify the symbolic form, c_{ij}, of each element of the product matrix. This will help them realize that they need to use the ith row of the first matrix and the jth column of the second matrix to find c_{ij}.

Matrix Multiplication

The number of columns of an $m \times n$ matrix A is the same as the number of rows of an $n \times p$ matrix B (i.e., both n). The element c_{ij} of the product matrix $C = AB$ is found as follows.

$$c_{ij} = a_{i1}b_{1j} + a_{i2}b_{2j} + \cdots + a_{in}b_{nj}$$

Matrix AB will be an $m \times p$ matrix.

Classroom Example 5
Suppose C is a 2×5 matrix and D is a 4×2 matrix.
(a) Can the product CD be calculated?
(b) If CD can be calculated, what is its dimension?
(c) Can DC be calculated?
(d) If DC can be calculated, what is its dimension?

Answers:
(a) no
(b) CD cannot be calculated.
(c) yes
(d) 4×5

EXAMPLE 5 Deciding Whether Two Matrices Can Be Multiplied

Suppose A is a 3×2 matrix, while B is a 2×4 matrix.

(a) Can the product AB be calculated?

(b) If AB can be calculated, what is its dimension?

(c) Can BA be calculated?

(d) If BA can be calculated, what is its dimension?

SOLUTION

(a) The following diagram shows that AB can be calculated, because the number of columns of A is equal to the number of rows of B. (Both are 2.)

Matrix A Matrix B

3×2 2×4

must match

dimension of AB

3×4

(b) As indicated in the diagram above, the product AB is a 3×4 matrix.

(c) The diagram below shows that BA cannot be calculated.

Matrix B Matrix A

2×4 3×2

different

(d) The product BA cannot be calculated, because B has 4 columns and A has only 3 rows.

✔ Now Try Exercises 45 and 47.

Classroom Example 6
Let $A = \begin{bmatrix} 3 & -4 & 5 \\ 1 & 0 & -6 \end{bmatrix}$ and $B = \begin{bmatrix} 4 & 1 \\ -3 & 8 \end{bmatrix}$. Find each product, if possible.
(a) AB (b) BA

Answers:
(a) AB cannot be calculated.
(b) $\begin{bmatrix} 13 & -16 & 14 \\ -1 & 12 & -63 \end{bmatrix}$

EXAMPLE 6 Multiplying Matrices

Let $A = \begin{bmatrix} 1 & -3 \\ 7 & 2 \end{bmatrix}$ and $B = \begin{bmatrix} 1 & 0 & -1 & 2 \\ 3 & 1 & 4 & -1 \end{bmatrix}$. Find each product, if possible.

(a) AB **(b)** BA

SOLUTION

(a) First decide whether AB can be found. Since A is 2×2 and B is 2×4, the product can be found and will be a 2×4 matrix.

$$AB = \begin{bmatrix} 1 & -3 \\ 7 & 2 \end{bmatrix} \begin{bmatrix} 1 & 0 & -1 & 2 \\ 3 & 1 & 4 & -1 \end{bmatrix}$$

$$= \begin{bmatrix} 1(1) + (-3)3 & 1(0) + (-3)1 & 1(-1) + (-3)4 & 1(2) + (-3)(-1) \\ 7(1) + 2(3) & 7(0) + 2(1) & 7(-1) + 2(4) & 7(2) + 2(-1) \end{bmatrix}$$

Use the definition of matrix multiplication.

$$= \begin{bmatrix} -8 & -3 & -13 & 5 \\ 13 & 2 & 1 & 12 \end{bmatrix}$$

Perform the operations.

(b) Since B is a 2×4 matrix, and A is a 2×2 matrix, the number of columns of B (here 4) does not equal the number of rows of A (here 2). Therefore, the product BA cannot be found.

✔ Now Try Exercises 65 and 69.

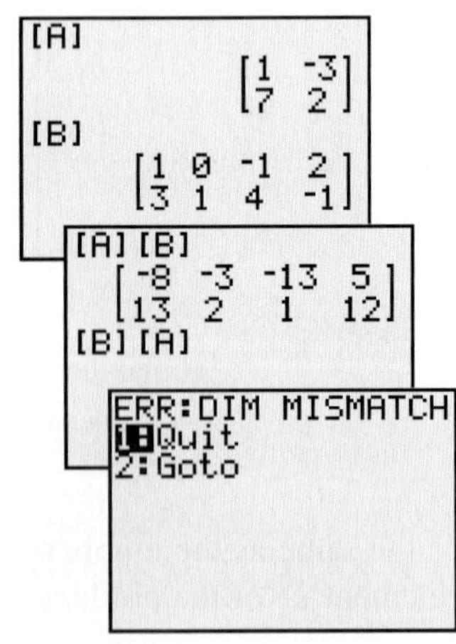

The three screens here support the results of the matrix multiplication in **Example 6.** The final screen indicates that the product BA cannot be found.

Classroom Example 7

Let $C = \begin{bmatrix} 2 & -5 \\ -6 & 1 \end{bmatrix}$ and $D = \begin{bmatrix} 3 & -4 \\ 2 & 3 \end{bmatrix}$. Find each product.

(a) CD (b) DC

Answers:

(a) $\begin{bmatrix} -4 & -23 \\ -16 & 27 \end{bmatrix}$

(b) $\begin{bmatrix} 30 & -19 \\ -14 & -7 \end{bmatrix}$

EXAMPLE 7 Multiplying Square Matrices in Different Orders

Let $A = \begin{bmatrix} 1 & 3 \\ -2 & 5 \end{bmatrix}$ and $B = \begin{bmatrix} -2 & 7 \\ 0 & 2 \end{bmatrix}$. Find each product.

(a) AB **(b)** BA

SOLUTION

(a) $$AB = \begin{bmatrix} 1 & 3 \\ -2 & 5 \end{bmatrix}\begin{bmatrix} -2 & 7 \\ 0 & 2 \end{bmatrix}$$

$$= \begin{bmatrix} 1(-2) + 3(0) & 1(7) + 3(2) \\ -2(-2) + 5(0) & -2(7) + 5(2) \end{bmatrix}$$ Multiply elements of each row of A by elements of each column of B.

$$= \begin{bmatrix} -2 & 13 \\ 4 & -4 \end{bmatrix}$$

(b) $$BA = \begin{bmatrix} -2 & 7 \\ 0 & 2 \end{bmatrix}\begin{bmatrix} 1 & 3 \\ -2 & 5 \end{bmatrix}$$

$$= \begin{bmatrix} -2(1) + 7(-2) & -2(3) + 7(5) \\ 0(1) + 2(-2) & 0(3) + 2(5) \end{bmatrix}$$ Multiply elements of each row of B by elements of each column of A.

$$= \begin{bmatrix} -16 & 29 \\ -4 & 10 \end{bmatrix}$$

Note that $AB \neq BA$.

✔ *Now Try Exercise 75.*

When multiplying matrices, it is important to pay special attention to the dimensions of the matrices as well as the order in which they are to be multiplied. **Examples 5 and 6** showed that the order in which two matrices are to be multiplied may determine whether their product can be found. **Example 7** showed that even when both products AB and BA can be found, they may not be equal.

In general, if A and B are matrices, then $AB \neq BA$. ***Matrix multiplication is not commutative.***

Matrix multiplication does, however, satisfy the associative and distributive properties.

Properties of Matrix Multiplication

If A, B, and C are matrices such that all the following products and sums exist, then these properties hold.

$$(AB)C = A(BC), \quad A(B + C) = AB + AC, \quad (B + C)A = BA + CA$$

For proofs of the first two results for the special cases when A, B, and C are square matrices, see **Exercises 89 and 90.** The identity and inverse properties for matrix multiplication are discussed in the next section.

Applying Matrix Algebra

Classroom Example 8
Suppose that plans for the subdivision in **Example 8** change so that the contractor builds a total of 150 houses. Matrix M shows the new distribution of houses by model and style.

$$\begin{array}{c} \\ \text{Model A} \\ \text{Model B} \\ \text{Model C} \end{array} \begin{array}{c} \begin{array}{cc} \text{Colonial} & \text{Ranch} \end{array} \\ \begin{bmatrix} 10 & 30 \\ 15 & 25 \\ 25 & 45 \end{bmatrix} \end{array} = M$$

Using this new information, answer each of the questions asked in **Example 8.**

Answers:

(a) Model A: \$79,000; Model B: \$69,900; Model C: \$124,600

(b) concrete: 5500 units; lumber: 200 units; brick: 2000 units; shingles: 300 units

(c) \$273,500

Teaching Tip Before multiplying matrices, have students assign labels to the rows and columns of each factor. For example, to multiply matrix P and matrix Q, write

$$\underbrace{\left(\frac{\text{models}}{\text{styles}}\right)}_{P}\underbrace{\left(\frac{\text{styles}}{\text{material used}}\right)}_{Q} = \underbrace{(\text{models}/\text{material used})}_{PQ}.$$

EXAMPLE 8 Using Matrix Multiplication to Model Plans for a Subdivision

A contractor builds three kinds of houses, models A, B, and C, with a choice of two styles, colonial or ranch. Matrix P below shows the number of each kind of house the contractor is planning to build for a new 100-home subdivision. The amounts for each of the main materials used depend on the style of the house. These amounts are shown in matrix Q, while matrix R gives the cost in dollars for each kind of material. Concrete is measured here in cubic yards, lumber in 1000 board feet, brick in 1000s, and shingles in 100 square feet.

$$\begin{array}{c} \\ \text{Model A} \\ \text{Model B} \\ \text{Model C} \end{array} \begin{array}{c} \begin{array}{cc} \text{Colonial} & \text{Ranch} \end{array} \\ \begin{bmatrix} 0 & 30 \\ 10 & 20 \\ 20 & 20 \end{bmatrix} \end{array} = P$$

$$\begin{array}{c} \\ \text{Colonial} \\ \text{Ranch} \end{array} \begin{array}{c} \begin{array}{cccc} \text{Concrete} & \text{Lumber} & \text{Brick} & \text{Shingles} \end{array} \\ \begin{bmatrix} 10 & 2 & 0 & 2 \\ 50 & 1 & 20 & 2 \end{bmatrix} \end{array} = Q \qquad \begin{array}{c} \\ \text{Concrete} \\ \text{Lumber} \\ \text{Brick} \\ \text{Shingles} \end{array} \begin{array}{c} \text{Cost per Unit} \\ \begin{bmatrix} 20 \\ 180 \\ 60 \\ 25 \end{bmatrix} \end{array} = R$$

(a) What is the total cost of materials for all houses of each model?

(b) How much of each of the four kinds of material must be ordered?

(c) What is the total cost of the materials?

SOLUTION

(a) To find the materials cost for each model, first find matrix PQ, which will show the total amount of each material needed for all houses of each model.

$$PQ = \begin{bmatrix} 0 & 30 \\ 10 & 20 \\ 20 & 20 \end{bmatrix}\begin{bmatrix} 10 & 2 & 0 & 2 \\ 50 & 1 & 20 & 2 \end{bmatrix} = \begin{array}{c} \begin{array}{cccc} \text{Concrete} & \text{Lumber} & \text{Brick} & \text{Shingles} \end{array} \\ \begin{bmatrix} 1500 & 30 & 600 & 60 \\ 1100 & 40 & 400 & 60 \\ 1200 & 60 & 400 & 80 \end{bmatrix} \end{array} \begin{array}{c} \\ \text{Model A} \\ \text{Model B} \\ \text{Model C} \end{array}$$

Multiplying PQ and the cost matrix R gives the total cost of materials for each model.

$$(PQ)R = \begin{bmatrix} 1500 & 30 & 600 & 60 \\ 1100 & 40 & 400 & 60 \\ 1200 & 60 & 400 & 80 \end{bmatrix}\begin{bmatrix} 20 \\ 180 \\ 60 \\ 25 \end{bmatrix} = \begin{array}{c} \text{Cost} \\ \begin{bmatrix} 72{,}900 \\ 54{,}700 \\ 60{,}800 \end{bmatrix} \end{array} \begin{array}{c} \\ \text{Model A} \\ \text{Model B} \\ \text{Model C} \end{array}$$

(b) To find how much of each kind of material to order, refer to the columns of matrix PQ. The sums of the elements of the columns will give a matrix whose elements represent the total amounts of all materials needed for the subdivision. Call this matrix T, and write it as a row matrix.

$$T = \begin{bmatrix} 3800 & 130 & 1400 & 200 \end{bmatrix}$$

(c) The total cost of all the materials is given by the product of matrix T, the total amounts matrix, and matrix R, the cost matrix. To multiply these matrices and get a 1×1 matrix, representing the total cost, requires multiplying a 1×4 matrix and a 4×1 matrix. This is why in part (b) a row matrix was written rather than a column matrix.

The total materials cost is given by TR, so

$$TR = \begin{bmatrix} 3800 & 130 & 1400 & 200 \end{bmatrix} \begin{bmatrix} 20 \\ 180 \\ 60 \\ 25 \end{bmatrix} = \begin{bmatrix} 188{,}400 \end{bmatrix}.$$

The total cost of materials is \$188,400. This total may also be found by summing the elements of the column matrix $(PQ)R$.

✔ Now Try Exercise 81.

9.7 Exercises

1. $a = 0, b = 4, c = -3, d = 5$
2. $w = 9, x = 17, y = 8, z = -12$
3. $x = -4, y = 14, z = 3, w = -2$
4. $a = 1, b = -4, c = 9, d = 13$
5. This cannot be true.
6. This cannot be true.
7. $w = 5, x = 6, y = -2, z = 8$
8. $x = 6, y = 2, z = -2, w = 5$
9. $z = 18, r = 3, s = 3, p = 3, a = \frac{3}{4}$
10. $a = 2, z = -3, m = 8, k = 1$
11. dimension; equal
12. dimension
13. 2×2; square
14. 2×3
15. 3×4
16. 1×5; row
17. 2×1; column
18. 1×1; square; row; column

Find the values of the variables for which each statement is true, if possible. ***See Examples 1 and 2.***

1. $\begin{bmatrix} -3 & a \\ b & 5 \end{bmatrix} = \begin{bmatrix} c & 0 \\ 4 & d \end{bmatrix}$

2. $\begin{bmatrix} w & x \\ 8 & -12 \end{bmatrix} = \begin{bmatrix} 9 & 17 \\ y & z \end{bmatrix}$

3. $\begin{bmatrix} x+2 & y-6 \\ z-3 & w+5 \end{bmatrix} = \begin{bmatrix} -2 & 8 \\ 0 & 3 \end{bmatrix}$

4. $\begin{bmatrix} 6 & a+3 \\ b+2 & 9 \end{bmatrix} = \begin{bmatrix} c-3 & 4 \\ -2 & d-4 \end{bmatrix}$

5. $\begin{bmatrix} x & y & z \end{bmatrix} = \begin{bmatrix} 21 & 6 \end{bmatrix}$

6. $\begin{bmatrix} p \\ q \\ r \end{bmatrix} = \begin{bmatrix} 3 \\ -9 \end{bmatrix}$

7. $\begin{bmatrix} 0 & 5 & x \\ -1 & 3 & y+2 \\ 4 & 1 & z \end{bmatrix} = \begin{bmatrix} 0 & w & 6 \\ -1 & 3 & 0 \\ 4 & 1 & 8 \end{bmatrix}$

8. $\begin{bmatrix} 5 & x-4 & 9 \\ 2 & -3 & 8 \\ 6 & 0 & 5 \end{bmatrix} = \begin{bmatrix} y+3 & 2 & 9 \\ z+4 & -3 & 8 \\ 6 & 0 & w \end{bmatrix}$

9. $\begin{bmatrix} -7+z & 4r & 8s \\ 6p & 2 & 5 \end{bmatrix} + \begin{bmatrix} -9 & 8r & 3 \\ 2 & 5 & 4 \end{bmatrix} = \begin{bmatrix} 2 & 36 & 27 \\ 20 & 7 & 12a \end{bmatrix}$

10. $\begin{bmatrix} a+2 & 3z+1 & 5m \\ 8k & 0 & 3 \end{bmatrix} + \begin{bmatrix} 3a & 2z & 5m \\ 2k & 5 & 6 \end{bmatrix} = \begin{bmatrix} 10 & -14 & 80 \\ 10 & 5 & 9 \end{bmatrix}$

11. *Concept Check* Two matrices are equal if they have the same ______ and if corresponding elements are ______.

12. *Concept Check* In order to add two matrices, they must have the same ______.

Concept Check Find the dimension of each matrix. Identify any square, column, or row matrices.

13. $\begin{bmatrix} -4 & 8 \\ 2 & 3 \end{bmatrix}$

14. $\begin{bmatrix} -9 & 6 & 2 \\ 4 & 1 & 8 \end{bmatrix}$

15. $\begin{bmatrix} -6 & 8 & 0 & 0 \\ 4 & 1 & 9 & 2 \\ 3 & -5 & 7 & 1 \end{bmatrix}$

16. $\begin{bmatrix} 8 & -2 & 4 & 6 & 3 \end{bmatrix}$

17. $\begin{bmatrix} 2 \\ 4 \end{bmatrix}$

18. $\begin{bmatrix} -9 \end{bmatrix}$

19. Your friend missed the lecture on adding matrices. Explain to him how to add two matrices.

20. Explain how to subtract two matrices.

21. $\begin{bmatrix} -2 & -5 \\ 17 & 4 \end{bmatrix}$

22. $\begin{bmatrix} 6 & 6 \\ -12 & 9 \end{bmatrix}$

23. $\begin{bmatrix} -2 & -7 & 7 \\ 10 & -2 & 7 \end{bmatrix}$

24. $\begin{bmatrix} 13 & -13 \\ 7 & 7 \\ -7 & 14 \end{bmatrix}$

25. They cannot be added.

26. They cannot be added.

27. $\begin{bmatrix} -6 & 8 \\ 4 & 2 \end{bmatrix}$ 28. $\begin{bmatrix} 11 & -12 \\ -4 & 14 \end{bmatrix}$

29. $\begin{bmatrix} 4 \\ -5 \\ 4 \end{bmatrix}$ 30. $\begin{bmatrix} 12 & -9 & 3 \end{bmatrix}$

31. They cannot be subtracted.

32. They cannot be subtracted.

33. $\begin{bmatrix} -\sqrt{3} & -13 \\ 4 & -2\sqrt{5} \\ -1 & -\sqrt{2} \end{bmatrix}$

34. $\begin{bmatrix} 3 & -4\sqrt{7} \\ 4\sqrt{7} & -8 \end{bmatrix}$

35. $\begin{bmatrix} 5x + y & y \\ 6x + 2y & x + 3y \end{bmatrix}$

36. $\begin{bmatrix} -k - 14y \\ 4z - 8x \\ -2k - a \\ -8m + 4n \end{bmatrix}$

37. $\begin{bmatrix} -4 & 8 \\ 0 & 6 \end{bmatrix}$ 38. $\begin{bmatrix} 18 & -6 \\ -12 & 0 \end{bmatrix}$

39. $\begin{bmatrix} -9 & 3 \\ 6 & 0 \end{bmatrix}$ 40. $\begin{bmatrix} 3 & -6 \\ 0 & -4.5 \end{bmatrix}$

41. $\begin{bmatrix} 14 & 2 \\ -12 & 6 \end{bmatrix}$ 42. $\begin{bmatrix} -20 & 0 \\ 16 & -6 \end{bmatrix}$

43. $\begin{bmatrix} -1 & -3 \\ 2 & -3 \end{bmatrix}$ 44. $\begin{bmatrix} \frac{9}{2} & 1 \\ -4 & \frac{9}{4} \end{bmatrix}$

45. yes; 2×5 46. yes; 5×3

47. no 48. no

49. yes; 3×2 50. no

51. $\begin{bmatrix} 13 \\ 25 \end{bmatrix}$ 52. $\begin{bmatrix} 4 \\ 42 \end{bmatrix}$

53. $\begin{bmatrix} -17 \\ -1 \end{bmatrix}$ 54. $\begin{bmatrix} 27 \\ -1 \end{bmatrix}$

55. $\begin{bmatrix} 17\sqrt{2} & -4\sqrt{2} \\ 35\sqrt{3} & 26\sqrt{3} \end{bmatrix}$

56. $\begin{bmatrix} -7\sqrt{5} \\ 3\sqrt{5} \end{bmatrix}$

57. $\begin{bmatrix} 3 + 4\sqrt{3} & -3\sqrt{2} \\ 2\sqrt{15} + 12\sqrt{6} & -2\sqrt{30} \end{bmatrix}$

Perform each operation when possible. ***See Examples 2 and 3.***

21. $\begin{bmatrix} -4 & 3 \\ 12 & -6 \end{bmatrix} + \begin{bmatrix} 2 & -8 \\ 5 & 10 \end{bmatrix}$

22. $\begin{bmatrix} 9 & 4 \\ -8 & 2 \end{bmatrix} + \begin{bmatrix} -3 & 2 \\ -4 & 7 \end{bmatrix}$

23. $\begin{bmatrix} 6 & -9 & 2 \\ 4 & 1 & 3 \end{bmatrix} + \begin{bmatrix} -8 & 2 & 5 \\ 6 & -3 & 4 \end{bmatrix}$

24. $\begin{bmatrix} 4 & -3 \\ 7 & 2 \\ -6 & 8 \end{bmatrix} + \begin{bmatrix} 9 & -10 \\ 0 & 5 \\ -1 & 6 \end{bmatrix}$

25. $\begin{bmatrix} 2 & 4 & 6 \end{bmatrix} + \begin{bmatrix} -2 \\ -4 \\ -6 \end{bmatrix}$

26. $\begin{bmatrix} 3 \\ 1 \\ 0 \end{bmatrix} + \begin{bmatrix} 2 \\ -6 \end{bmatrix}$

27. $\begin{bmatrix} -6 & 8 \\ 0 & 0 \end{bmatrix} - \begin{bmatrix} 0 & 0 \\ -4 & -2 \end{bmatrix}$

28. $\begin{bmatrix} 11 & 0 \\ -4 & 0 \end{bmatrix} - \begin{bmatrix} 0 & 12 \\ 0 & -14 \end{bmatrix}$

29. $\begin{bmatrix} 12 \\ -1 \\ 3 \end{bmatrix} - \begin{bmatrix} 8 \\ 4 \\ -1 \end{bmatrix}$

30. $\begin{bmatrix} 10 & -4 & 6 \end{bmatrix} - \begin{bmatrix} -2 & 5 & 3 \end{bmatrix}$

31. $\begin{bmatrix} -4 & 3 \end{bmatrix} - \begin{bmatrix} 5 & 8 & 2 \end{bmatrix}$

32. $\begin{bmatrix} 4 & 6 \end{bmatrix} - \begin{bmatrix} 2 \\ 3 \end{bmatrix}$

33. $\begin{bmatrix} \sqrt{3} & -4 \\ 2 & -\sqrt{5} \\ -8 & \sqrt{8} \end{bmatrix} - \begin{bmatrix} 2\sqrt{3} & 9 \\ -2 & \sqrt{5} \\ -7 & 3\sqrt{2} \end{bmatrix}$

34. $\begin{bmatrix} 2 & \sqrt{7} \\ 3\sqrt{28} & -6 \end{bmatrix} - \begin{bmatrix} -1 & 5\sqrt{7} \\ 2\sqrt{7} & 2 \end{bmatrix}$

35. $\begin{bmatrix} 3x + y & -2y \\ x + 2y & 3y \end{bmatrix} + \begin{bmatrix} 2x & 3y \\ 5x & x \end{bmatrix}$

36. $\begin{bmatrix} 4k - 8y \\ 6z - 3x \\ 2k + 5a \\ -4m + 2n \end{bmatrix} - \begin{bmatrix} 5k + 6y \\ 2z + 5x \\ 4k + 6a \\ 4m - 2n \end{bmatrix}$

Let $A = \begin{bmatrix} -2 & 4 \\ 0 & 3 \end{bmatrix}$ *and* $B = \begin{bmatrix} -6 & 2 \\ 4 & 0 \end{bmatrix}$. *Find each of the following.* ***See Example 4.***

37. $2A$ 38. $-3B$ 39. $\frac{3}{2}B$ 40. $-\frac{3}{2}A$

41. $2A - 3B$ 42. $-2A + 4B$ 43. $-A + \frac{1}{2}B$ 44. $\frac{3}{4}A - B$

Suppose that matrix A has dimension 2×3*, B has dimension* 3×5*, and C has dimension* 5×2*. Decide whether the given product can be calculated. If it can, determine its dimension.* ***See Example 5.***

45. AB 46. CA 47. BA 48. AC 49. BC 50. CB

Find each matrix product when possible. ***See Examples 5–7.***

51. $\begin{bmatrix} 1 & 2 \\ 3 & 4 \end{bmatrix}\begin{bmatrix} -1 \\ 7 \end{bmatrix}$

52. $\begin{bmatrix} -1 & 5 \\ 7 & 0 \end{bmatrix}\begin{bmatrix} 6 \\ 2 \end{bmatrix}$

53. $\begin{bmatrix} 3 & -4 & 1 \\ 5 & 0 & 2 \end{bmatrix}\begin{bmatrix} -1 \\ 4 \\ 2 \end{bmatrix}$

54. $\begin{bmatrix} -6 & 3 & 5 \\ 2 & 9 & 1 \end{bmatrix}\begin{bmatrix} -2 \\ 0 \\ 3 \end{bmatrix}$

55. $\begin{bmatrix} \sqrt{2} & \sqrt{2} & -\sqrt{18} \\ \sqrt{3} & \sqrt{27} & 0 \end{bmatrix}\begin{bmatrix} 8 & -10 \\ 9 & 12 \\ 0 & 2 \end{bmatrix}$

56. $\begin{bmatrix} -9 & 2 & 1 \\ 3 & 0 & 0 \end{bmatrix}\begin{bmatrix} \sqrt{5} \\ \sqrt{20} \\ -2\sqrt{5} \end{bmatrix}$

57. $\begin{bmatrix} \sqrt{3} & 1 \\ 2\sqrt{5} & 3\sqrt{2} \end{bmatrix}\begin{bmatrix} \sqrt{3} & -\sqrt{6} \\ 4\sqrt{3} & 0 \end{bmatrix}$

58. $\begin{bmatrix} \sqrt{7} & 0 \\ 2 & \sqrt{28} \end{bmatrix}\begin{bmatrix} 2\sqrt{3} & -\sqrt{7} \\ 0 & -6 \end{bmatrix}$

59. $\begin{bmatrix} -3 & 0 & 2 & 1 \\ 4 & 0 & 2 & 6 \end{bmatrix}\begin{bmatrix} -4 & 2 \\ 0 & 1 \end{bmatrix}$

60. $\begin{bmatrix} -1 & 2 & 4 & 1 \\ 0 & 2 & -3 & 5 \end{bmatrix}\begin{bmatrix} 1 & 2 & 4 \\ -2 & 5 & 1 \end{bmatrix}$

58. $\begin{bmatrix} 2\sqrt{21} & -7 \\ 4\sqrt{3} & -14\sqrt{7} \end{bmatrix}$

59. They cannot be multiplied.

60. They cannot be multiplied.

61. $\begin{bmatrix} 2 & 7 & -4 \end{bmatrix}$

62. $\begin{bmatrix} 4 & 8 & -10 \end{bmatrix}$

63. $\begin{bmatrix} -15 & -16 & 3 \\ -1 & 0 & 9 \\ 7 & 6 & 12 \end{bmatrix}$

64. $\begin{bmatrix} -2 & 5 & 0 \\ 6 & 6 & 1 \\ 12 & 2 & -3 \end{bmatrix}$

65. $\begin{bmatrix} 23 & -9 \\ -6 & -2 \\ 33 & 1 \end{bmatrix}$

66. $\begin{bmatrix} -20 & 10 & -8 \\ -15 & 15 & 9 \end{bmatrix}$

67. $\begin{bmatrix} -25 & 23 & 11 \\ 0 & -6 & -12 \\ -15 & 33 & 45 \end{bmatrix}$

68. $\begin{bmatrix} -22 & -6 \\ 18 & 36 \end{bmatrix}$

69. They cannot be multiplied.

70. They cannot be multiplied.

71. $\begin{bmatrix} 10 & -10 \\ 15 & -5 \end{bmatrix}$

72. $\begin{bmatrix} 10 & -30 \\ 45 & -35 \end{bmatrix}$

73. $BA \neq AB$, $BC \neq CB$, $AC \neq CA$

75. (a) $\begin{bmatrix} 38 & -8 \\ -7 & -2 \end{bmatrix}$
(b) $\begin{bmatrix} 18 & 24 \\ 19 & 18 \end{bmatrix}$

76. (a) $\begin{bmatrix} 25 & -20 \\ -22 & 12 \end{bmatrix}$
(b) $\begin{bmatrix} 4 & -17 \\ -16 & 33 \end{bmatrix}$

77. (a) $\begin{bmatrix} 0 & 1 & -1 \\ 0 & 1 & 0 \\ 0 & 0 & 1 \end{bmatrix}$
(b) $\begin{bmatrix} 0 & 1 & -1 \\ 0 & 1 & 0 \\ 0 & 0 & 1 \end{bmatrix}$

78. (a) $\begin{bmatrix} 1 & 0 & -1 \\ 1 & 1 & 0 \\ 0 & -1 & -1 \end{bmatrix}$
(b) $\begin{bmatrix} -1 & -1 & 0 \\ 0 & 1 & 1 \\ -1 & 0 & 1 \end{bmatrix}$

61. $\begin{bmatrix} -2 & 4 & 1 \end{bmatrix} \begin{bmatrix} 3 & -2 & 4 \\ 2 & 1 & 0 \\ 0 & -1 & 4 \end{bmatrix}$

62. $\begin{bmatrix} 0 & 3 & -4 \end{bmatrix} \begin{bmatrix} -2 & 6 & 3 \\ 0 & 4 & 2 \\ -1 & 1 & 4 \end{bmatrix}$

63. $\begin{bmatrix} -2 & -3 & -4 \\ 2 & -1 & 0 \\ 4 & -2 & 3 \end{bmatrix} \begin{bmatrix} 0 & 1 & 4 \\ 1 & 2 & -1 \\ 3 & 2 & -2 \end{bmatrix}$

64. $\begin{bmatrix} -1 & 2 & 0 \\ 0 & 3 & 2 \\ 0 & 1 & 4 \end{bmatrix} \begin{bmatrix} 2 & -1 & 2 \\ 0 & 2 & 1 \\ 3 & 0 & -1 \end{bmatrix}$

Given $A = \begin{bmatrix} 4 & -2 \\ 3 & 1 \end{bmatrix}$, $B = \begin{bmatrix} 5 & 1 \\ 0 & -2 \\ 3 & 7 \end{bmatrix}$, *and* $C = \begin{bmatrix} -5 & 4 & 1 \\ 0 & 3 & 6 \end{bmatrix}$, *find each product when possible.* ***See Examples 5–7.***

65. BA **66.** AC **67.** BC **68.** CB

69. AB **70.** CA **71.** A^2 **72.** A^3 (*Hint:* $A^3 = A^2 \cdot A$)

73. *Concept Check* Compare the answers to **Exercises 65 and 69, 67 and 68,** and **66 and 70.** How do they show that matrix multiplication is not commutative?

74. Your friend missed the lecture on multiplying matrices. Explain to her the process of matrix multiplication.

For each pair of matrices A and B, find ***(a)*** *AB and* ***(b)*** *BA.* ***See Example 7.***

75. $A = \begin{bmatrix} 3 & 4 \\ -2 & 1 \end{bmatrix}$, $B = \begin{bmatrix} 6 & 0 \\ 5 & -2 \end{bmatrix}$

76. $A = \begin{bmatrix} 0 & -5 \\ -4 & 2 \end{bmatrix}$, $B = \begin{bmatrix} 3 & -1 \\ -5 & 4 \end{bmatrix}$

77. $A = \begin{bmatrix} 0 & 1 & -1 \\ 0 & 1 & 0 \\ 0 & 0 & 1 \end{bmatrix}$, $B = \begin{bmatrix} 1 & 0 & 0 \\ 0 & 1 & 0 \\ 0 & 0 & 1 \end{bmatrix}$

78. $A = \begin{bmatrix} -1 & 0 & 1 \\ 0 & 1 & 1 \\ -1 & -1 & 0 \end{bmatrix}$, $B = \begin{bmatrix} 0 & 0 & 1 \\ 0 & 1 & 0 \\ 1 & 0 & 0 \end{bmatrix}$

79. *Concept Check* In **Exercise 77,** $AB = A$ and $BA = A$. For this pair of matrices, B acts the same way for matrix multiplication as the number ________ acts for multiplication of real numbers.

80. *Concept Check* Find AB and BA for

$$A = \begin{bmatrix} a & b \\ c & d \end{bmatrix} \quad \text{and} \quad B = \begin{bmatrix} 1 & 0 \\ 0 & 1 \end{bmatrix}.$$

What do you notice? Matrix B acts as the multiplicative ________ element for 2×2 square matrices.

Solve each problem. ***See Example 8.***

81. *Income from Yogurt* Yagel's Yogurt sells three types of yogurt: nonfat, regular, and super creamy, at three locations. Location I sells 50 gal of nonfat, 100 gal of regular, and 30 gal of super creamy each day. Location II sells 10 gal of nonfat, and Location III sells 60 gal of nonfat each day. Daily sales of regular yogurt are 90 gal at Location II and 120 gal at Location III. At Location II, 50 gal of super creamy are sold each day, and 40 gal of super creamy are sold each day at Location III.

(a) Write a 3×3 matrix that shows the sales figures for the three locations, with the rows representing the three locations.

(b) The incomes per gallon for nonfat, regular, and super creamy are \$12, \$10, and \$15, respectively. Write a 1×3 or 3×1 matrix displaying the incomes.

(c) Find a matrix product that gives the daily income at each of the three locations.

(d) What is Yagel's Yogurt's total daily income from the three locations?

79. 1 80. identity

81. (a) $\begin{bmatrix} 50 & 100 & 30 \\ 10 & 90 & 50 \\ 60 & 120 & 40 \end{bmatrix}$

(b) $\begin{bmatrix} 12 \\ 10 \\ 15 \end{bmatrix}$ (c) $\begin{bmatrix} 2050 \\ 1770 \\ 2520 \end{bmatrix}$

(d) \$6340

82. (a) $\begin{bmatrix} 47.5 & 57.75 \\ 27 & 33.75 \\ 81 & 95 \\ 12 & 15 \end{bmatrix}$

(b) $\begin{bmatrix} 20 & 200 & 50 & 60 \end{bmatrix}$; $\begin{bmatrix} 220 & 890 & 105 & 125 & 70 \end{bmatrix}$

(c) $\begin{bmatrix} 11{,}120 & 13{,}555 \end{bmatrix}$

82. *Purchasing Costs* The Bread Box, a small neighborhood bakery, sells four main items: sweet rolls, bread, cakes, and pies. The amount of each ingredient (in cups, except for eggs) required for these items is given by matrix A.

	Eggs	Flour	Sugar	Shortening	Milk
Rolls (doz)	1	4	$\frac{1}{4}$	$\frac{1}{4}$	1
Bread (loaf)	0	3	0	$\frac{1}{4}$	0
Cake	4	3	2	1	1
Pie (crust)	0	1	0	$\frac{1}{3}$	0

$= A$

The cost (in cents) for each ingredient when purchased in large lots or small lots is given by matrix B.

	Cost: Large Lot	Small Lot
Eggs	5	5
Flour	8	10
Sugar	10	12
Shortening	12	15
Milk	5	6

$= B$

(a) Use matrix multiplication to find a matrix giving the comparative cost per bakery item for the two purchase options.

(b) Suppose a day's orders consist of 20 dozen sweet rolls, 200 loaves of bread, 50 cakes, and 60 pies. Write the orders as a 1×4 matrix, and, using matrix multiplication, write as a matrix the amount of each ingredient needed to fill the day's orders.

(c) Use matrix multiplication to find a matrix giving the costs under the two purchase options to fill the day's orders.

83. *(Modeling) Northern Spotted Owl Population* Several years ago, mathematical ecologists created a model to analyze population dynamics of the endangered northern spotted owl in the Pacific Northwest. The ecologists divided the female owl population into three categories: juvenile (up to 1 yr old), subadult (1 to 2 yr old), and adult (over 2 yr old). They concluded that the change in the makeup of the northern spotted owl population in successive years could be described by the following matrix equation.

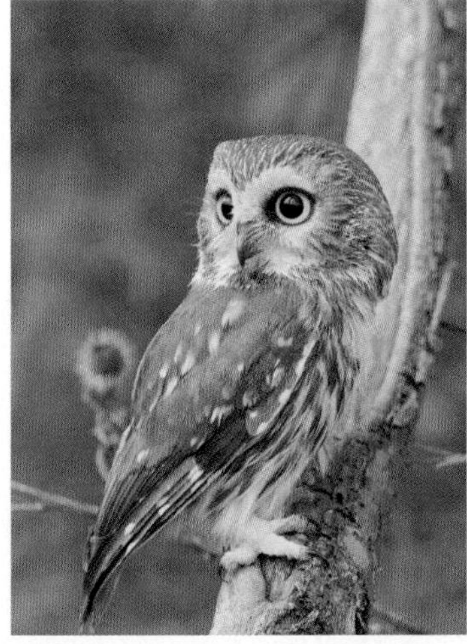

$$\begin{bmatrix} j_{n+1} \\ s_{n+1} \\ a_{n+1} \end{bmatrix} = \begin{bmatrix} 0 & 0 & 0.33 \\ 0.18 & 0 & 0 \\ 0 & 0.71 & 0.94 \end{bmatrix} \begin{bmatrix} j_n \\ s_n \\ a_n \end{bmatrix}$$

The numbers in the column matrices give the numbers of females in the three age groups after n years and $n + 1$ years. Multiplying the matrices yields the following.

$j_{n+1} = 0.33a_n$ Each year 33 juvenile females are born for each 100 adult females.

$s_{n+1} = 0.18j_n$ Each year 18% of the juvenile females survive to become subadults.

$a_{n+1} = 0.71s_n + 0.94a_n$ Each year 71% of the subadults survive to become adults, and 94% of the adults survive.

(*Source:* Lamberson, R. H., R. McKelvey, B. R. Noon, and C. Voss, "A Dynamic Analysis of Northern Spotted Owl Viability in a Fragmented Forest Landscape," *Conservation Biology,* Vol. 6, No. 4.)

83. Answers will vary a little if intermediate steps are rounded.
(a) 2940, 2909, 2861, 2814, 2767
(b) The northern spotted owl will become extinct.
(c) 3023, 3051, 3079, 3107, 3135

84. (a) $d_{n+1} = -0.05m_n + 1.05d_n$; 1.05
(b) after 1 yr: 3020 mountain lions, 515,000 deer; after 2 yr: 3600 mountain lions, 525,650 deer

(a) Suppose there are currently 3000 female northern spotted owls made up of 690 juveniles, 210 subadults, and 2100 adults. Use the matrix equation on the preceding page to determine the total number of female owls for each of the next 5 yr.

(b) Using advanced techniques from linear algebra, we can show that in the long run,

$$\begin{bmatrix} j_{n+1} \\ s_{n+1} \\ a_{n+1} \end{bmatrix} \approx 0.98359 \begin{bmatrix} j_n \\ s_n \\ a_n \end{bmatrix}.$$

What can we conclude about the long-term fate of the northern spotted owl?

(c) In the model, the main impediment to the survival of the northern spotted owl is the number 0.18 in the second row of the 3×3 matrix. This number is low for two reasons.

- The first year of life is precarious for most animals living in the wild.
- In addition, juvenile owls must eventually leave the nest and establish their own territory. If much of the forest near their original home has been cleared, then they are vulnerable to predators while searching for a new home.

Suppose that, thanks to better forest management, the number 0.18 can be increased to 0.3. Rework part (a) under this new assumption.

84. *(Modeling) Predator-Prey Relationship* In certain parts of the Rocky Mountains, deer provide the main food source for mountain lions. When the deer population is large, the mountain lions thrive. However, a large mountain lion population reduces the size of the deer population. Suppose the fluctuations of the two populations from year to year can be modeled with the matrix equation

$$\begin{bmatrix} m_{n+1} \\ d_{n+1} \end{bmatrix} = \begin{bmatrix} 0.51 & 0.4 \\ -0.05 & 1.05 \end{bmatrix} \begin{bmatrix} m_n \\ d_n \end{bmatrix}.$$

The numbers in the column matrices give the numbers of animals in the two populations after n years and $n + 1$ years, where the number of deer is measured in hundreds.

(a) Give the equation for d_{n+1} obtained from the second row of the square matrix. Use this equation to determine the rate at which the deer population will grow from year to year if there are no mountain lions.

(b) Suppose we start with a mountain lion population of 2000 and a deer population of 500,000 (that is, 5000 hundred deer). How large would each population be after 1 yr? 2 yr?

(c) Consider part (b) but change the initial mountain lion population to 4000. Show that the populations would both grow at a steady annual rate of 1.01.

85. *Northern Spotted Owl Population* Refer to **Exercise 83(b).** Show that the number 0.98359 is an approximate zero of the polynomial represented by

$$\begin{vmatrix} -x & 0 & 0.33 \\ 0.18 & -x & 0 \\ 0 & 0.71 & 0.94 - x \end{vmatrix}.$$

86. *Predator-Prey Relationship* Refer to **Exercise 84(c).** Show that the number 1.01 is a zero of the polynomial represented by

$$\begin{vmatrix} 0.51 - x & 0.4 \\ -0.05 & 1.05 - x \end{vmatrix}.$$

For Exercises 87–94, let

$$A = \begin{bmatrix} a_{11} & a_{12} \\ a_{21} & a_{22} \end{bmatrix}, \quad B = \begin{bmatrix} b_{11} & b_{12} \\ b_{21} & b_{22} \end{bmatrix}, \quad \text{and} \quad C = \begin{bmatrix} c_{11} & c_{12} \\ c_{21} & c_{22} \end{bmatrix},$$

where all the elements are real numbers. Use these matrices to show that each statement is true for 2×2 *matrices.*

87. $A + B = B + A$
(commutative property)

88. $A + (B + C) = (A + B) + C$
(associative property)

89. $(AB)C = A(BC)$
(associative property)

90. $A(B + C) = AB + AC$
(distributive property)

91. $c(A + B) = cA + cB$,
for any real number c.

92. $(c + d)A = cA + dA$,
for any real numbers c and d.

93. $(cA)d = (cd)A$, for any real numbers c and d.

94. $(cd)A = c(dA)$, for any real numbers c and d.

9.8 Matrix Inverses

- Identity Matrices
- Multiplicative Inverses
- Solving Systems Using Inverse Matrices

In the previous section, we saw several parallels between the set of real numbers and the set of matrices. Another similarity is that both sets have identity and inverse elements for multiplication.

Identity Matrices

By the identity property for real numbers,

$$a \cdot 1 = a \quad \text{and} \quad 1 \cdot a = a \quad \text{(Section R.2)}$$

for any real number a. If there is to be a multiplicative **identity matrix** I, such that

$$AI = A \quad \text{and} \quad IA = A,$$

for any matrix A, then A and I must be square matrices of the same dimension.

2×2 Identity Matrix

If I_2 represents the 2×2 identity matrix, then

$$I_2 = \begin{bmatrix} 1 & 0 \\ 0 & 1 \end{bmatrix}.$$

To verify that I_2 is the 2×2 identity matrix, we must show that $AI = A$ and $IA = A$ for any 2×2 matrix A. Let

$$A = \begin{bmatrix} x & y \\ z & w \end{bmatrix}.$$

Then

$$AI = \begin{bmatrix} x & y \\ z & w \end{bmatrix}\begin{bmatrix} 1 & 0 \\ 0 & 1 \end{bmatrix} = \begin{bmatrix} x \cdot 1 + y \cdot 0 & x \cdot 0 + y \cdot 1 \\ z \cdot 1 + w \cdot 0 & z \cdot 0 + w \cdot 1 \end{bmatrix} = \begin{bmatrix} x & y \\ z & w \end{bmatrix} = A,$$

and

$$IA = \begin{bmatrix} 1 & 0 \\ 0 & 1 \end{bmatrix}\begin{bmatrix} x & y \\ z & w \end{bmatrix} = \begin{bmatrix} 1 \cdot x + 0 \cdot z & 1 \cdot y + 0 \cdot w \\ 0 \cdot x + 1 \cdot z & 0 \cdot y + 1 \cdot w \end{bmatrix} = \begin{bmatrix} x & y \\ z & w \end{bmatrix} = A.$$

Generalizing, there is an $n \times n$ identity matrix for every $n \times n$ square matrix. The $n \times n$ identity matrix has 1s on the main diagonal and 0s elsewhere.

$n \times n$ Identity Matrix

The $n \times n$ identity matrix is I_n, where

$$I_n = \begin{bmatrix} 1 & 0 & \cdots & 0 \\ 0 & 1 & \cdots & 0 \\ \vdots & \vdots & a_{ij} & \vdots \\ 0 & 0 & \cdots & 1 \end{bmatrix}.$$

The element $a_{ij} = 1$ when $i = j$ (the diagonal elements), and $a_{ij} = 0$ otherwise.

Classroom Example 1

Let $B = \begin{bmatrix} 3 & -5 & 8 \\ 2 & 0 & 6 \\ -4 & 1 & -7 \end{bmatrix}$. Show that $I_3B = B$.

Answer:

$$I_3B = \begin{bmatrix} 1 & 0 & 0 \\ 0 & 1 & 0 \\ 0 & 0 & 1 \end{bmatrix}\begin{bmatrix} 3 & -5 & 8 \\ 2 & 0 & 6 \\ -4 & 1 & -7 \end{bmatrix} = \begin{bmatrix} 3 & -5 & 8 \\ 2 & 0 & 6 \\ -4 & 1 & -7 \end{bmatrix} = B$$

EXAMPLE 1 Verifying the Identity Property of I_3

Let $A = \begin{bmatrix} -2 & 4 & 0 \\ 3 & 5 & 9 \\ 0 & 8 & -6 \end{bmatrix}$. Give the 3×3 identity matrix I_3 and show that $AI_3 = A$.

ALGEBRAIC SOLUTION

The 3×3 identity matrix is

$$I_3 = \begin{bmatrix} 1 & 0 & 0 \\ 0 & 1 & 0 \\ 0 & 0 & 1 \end{bmatrix}.$$

By the definition of matrix multiplication,

$$AI_3 = \begin{bmatrix} -2 & 4 & 0 \\ 3 & 5 & 9 \\ 0 & 8 & -6 \end{bmatrix}\begin{bmatrix} 1 & 0 & 0 \\ 0 & 1 & 0 \\ 0 & 0 & 1 \end{bmatrix}$$

$$= \begin{bmatrix} -2 & 4 & 0 \\ 3 & 5 & 9 \\ 0 & 8 & -6 \end{bmatrix} = A.$$

GRAPHING CALCULATOR SOLUTION

The calculator screen in **Figure 27(a)** shows the identity matrix for $n = 3$. The screens in **Figures 27(b) and (c)** support the algebraic result.

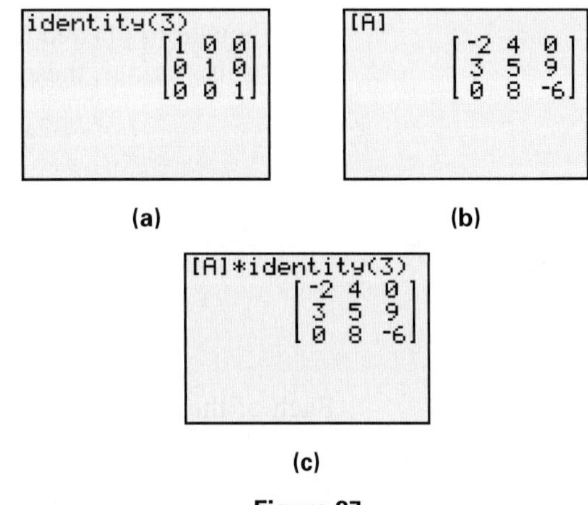

Figure 27

✔ *Now Try Exercise 1.*

Multiplicative Inverses For every *nonzero* real number a, there is a multiplicative inverse $\frac{1}{a}$ that satisfies both of the following.

$$a \cdot \frac{1}{a} = 1 \quad \text{and} \quad \frac{1}{a} \cdot a = 1 \quad \text{(Section R.2)}$$

(Recall: $\frac{1}{a}$ is also written a^{-1}.) In a similar way, if A is an $n \times n$ matrix, then its **multiplicative inverse,** written A^{-1}, must satisfy both of the following.

$$AA^{-1} = I_n \quad \text{and} \quad A^{-1}A = I_n$$

This means that only a square matrix can have a multiplicative inverse.

CAUTION Although $a^{-1} = \frac{1}{a}$ for any nonzero real number a, if A is a matrix, then $A^{-1} \neq \frac{1}{A}$. ***We do not use the symbol $\frac{1}{A}$, since 1 is a number and A is a matrix.***

To find the matrix A^{-1}, we use row transformations, introduced earlier in this chapter. As an example, we find the inverse of

$$A = \begin{bmatrix} 2 & 4 \\ 1 & -1 \end{bmatrix}.$$

Let the unknown inverse matrix be symbolized as follows.

$$A^{-1} = \begin{bmatrix} x & y \\ z & w \end{bmatrix}$$

By the definition of matrix inverse, $AA^{-1} = I_2$.

$$AA^{-1} = \begin{bmatrix} 2 & 4 \\ 1 & -1 \end{bmatrix}\begin{bmatrix} x & y \\ z & w \end{bmatrix} = \begin{bmatrix} 1 & 0 \\ 0 & 1 \end{bmatrix}$$

Use matrix multiplication for the product above.

$$\begin{bmatrix} 2x + 4z & 2y + 4w \\ x - z & y - w \end{bmatrix} = \begin{bmatrix} 1 & 0 \\ 0 & 1 \end{bmatrix} \quad \text{(Section 9.7)}$$

Set the corresponding elements equal to obtain a system of equations.

$$2x + 4z - 1 \quad (1)$$
$$2y + 4w = 0 \quad (2)$$
$$x - z = 0 \quad (3)$$
$$y - w = 1 \quad (4)$$

Since equations (1) and (3) involve only x and z, while equations (2) and (4) involve only y and w, these four equations lead to two systems of equations.

$$\begin{aligned} 2x + 4z &= 1 \\ x - z &= 0 \end{aligned} \quad \text{and} \quad \begin{aligned} 2y + 4w &= 0 \\ y - w &= 1 \end{aligned}$$

Write the two systems as augmented matrices.

$$\left[\begin{array}{cc|c} 2 & 4 & 1 \\ 1 & -1 & 0 \end{array}\right] \quad \text{and} \quad \left[\begin{array}{cc|c} 2 & 4 & 0 \\ 1 & -1 & 1 \end{array}\right] \quad \text{(Section 9.2)}$$

Teaching Tip Review the Gauss-Jordan method of solving a system using matrices. Remind students to proceed column by column, working from left to right.

Each of these systems can be solved by the Gauss-Jordan method. However, since the elements to the left of the vertical bar are identical, the two systems can be combined into one matrix.

$$\left[\begin{array}{cc|c} 2 & 4 & 1 \\ 1 & -1 & 0 \end{array}\right] \quad \text{and} \quad \left[\begin{array}{cc|c} 2 & 4 & 0 \\ 1 & -1 & 1 \end{array}\right] \quad \text{yields} \quad \left[\begin{array}{cc|cc} 2 & 4 & 1 & 0 \\ 1 & -1 & 0 & 1 \end{array}\right]$$

We can solve simultaneously using matrix row transformations. We need to change the numbers on the left of the vertical bar to the 2×2 identity matrix.

$$\left[\begin{array}{cc|cc} 1 & -1 & 0 & 1 \\ 2 & 4 & 1 & 0 \end{array}\right] \quad \text{Interchange R1 and R2 to get 1 in the upper left-hand corner. (Section 9.2)}$$

$$\left[\begin{array}{cc|cc} 1 & -1 & 0 & 1 \\ 0 & 6 & 1 & -2 \end{array}\right] \quad -2\text{R1} + \text{R2}$$

$$\left[\begin{array}{cc|cc} 1 & -1 & 0 & 1 \\ 0 & 1 & \frac{1}{6} & -\frac{1}{3} \end{array}\right] \quad \tfrac{1}{6}\text{R2}$$

$$\left[\begin{array}{cc|cc} 1 & 0 & \frac{1}{6} & \frac{2}{3} \\ 0 & 1 & \frac{1}{6} & -\frac{1}{3} \end{array}\right] \quad \text{R2} + \text{R1}$$

The numbers in the first column to the right of the vertical bar in the final matrix give the values of x and z. The second column gives the values of y and w. That is,

$$\left[\begin{array}{cc|cc} 1 & 0 & x & y \\ 0 & 1 & z & w \end{array}\right] = \left[\begin{array}{cc|cc} 1 & 0 & \frac{1}{6} & \frac{2}{3} \\ 0 & 1 & \frac{1}{6} & -\frac{1}{3} \end{array}\right]$$

so that
$$A^{-1} = \begin{bmatrix} x & y \\ z & w \end{bmatrix} = \begin{bmatrix} \frac{1}{6} & \frac{2}{3} \\ \frac{1}{6} & -\frac{1}{3} \end{bmatrix}.$$

To check, multiply A by A^{-1}. The result should be I_2.

$$\begin{aligned} AA^{-1} &= \begin{bmatrix} 2 & 4 \\ 1 & -1 \end{bmatrix}\begin{bmatrix} \frac{1}{6} & \frac{2}{3} \\ \frac{1}{6} & -\frac{1}{3} \end{bmatrix} \\ &= \begin{bmatrix} \frac{1}{3} + \frac{2}{3} & \frac{4}{3} - \frac{4}{3} \\ \frac{1}{6} - \frac{1}{6} & \frac{2}{3} + \frac{1}{3} \end{bmatrix} \\ &= \begin{bmatrix} 1 & 0 \\ 0 & 1 \end{bmatrix} = I_2 \end{aligned}$$

Thus,
$$A^{-1} = \begin{bmatrix} \frac{1}{6} & \frac{2}{3} \\ \frac{1}{6} & -\frac{1}{3} \end{bmatrix}.$$

This process is summarized below.

Finding an Inverse Matrix

To obtain A^{-1} for any $n \times n$ matrix A for which A^{-1} exists, follow these steps.

Step 1 Form the augmented matrix $[A|I_n]$, where I_n is the $n \times n$ identity matrix.

Step 2 Perform row transformations on $[A|I_n]$ to obtain a matrix of the form $[I_n|B]$.

Step 3 Matrix B is A^{-1}.

NOTE To confirm that two $n \times n$ matrices A and B are inverses of each other, it is sufficient to show that $AB = I_n$. It is not necessary to show also that $BA = I_n$.

Classroom Example 2

Find B^{-1} if $B = \begin{bmatrix} -4 & 2 & 0 \\ 1 & -1 & 2 \\ 0 & 1 & 4 \end{bmatrix}$.

Answer:
$$\begin{bmatrix} -\frac{3}{8} & -\frac{1}{2} & \frac{1}{4} \\ -\frac{1}{4} & -1 & \frac{1}{2} \\ \frac{1}{16} & \frac{1}{4} & \frac{1}{8} \end{bmatrix}$$

EXAMPLE 2 Finding the Inverse of a 3 × 3 Matrix

Find A^{-1} if $A = \begin{bmatrix} 1 & 0 & 1 \\ 2 & -2 & -1 \\ 3 & 0 & 0 \end{bmatrix}$.

SOLUTION Use row transformations as follows.

Step 1 Write the augmented matrix $[A|I_3]$.

$$\left[\begin{array}{ccc|ccc} 1 & 0 & 1 & 1 & 0 & 0 \\ 2 & -2 & -1 & 0 & 1 & 0 \\ 3 & 0 & 0 & 0 & 0 & 1 \end{array}\right]$$

Step 2 Since 1 is already in the upper left-hand corner as desired, begin by using the row transformation that will result in 0 for the first element in the second row. Multiply the elements of the first row by -2, and add the result to the second row.

$$\left[\begin{array}{ccc|ccc} 1 & 0 & 1 & 1 & 0 & 0 \\ 0 & -2 & -3 & -2 & 1 & 0 \\ 3 & 0 & 0 & 0 & 0 & 1 \end{array}\right] \quad -2\text{R1} + \text{R2}$$

To get 0 for the first element in the third row, multiply the elements of the first row by -3 and add to the third row.

$$\left[\begin{array}{ccc|ccc} 1 & 0 & 1 & 1 & 0 & 0 \\ 0 & -2 & -3 & -2 & 1 & 0 \\ 0 & 0 & -3 & -3 & 0 & 1 \end{array}\right] \quad -3\text{R1} + \text{R3}$$

To get 1 for the second element in the second row, multiply the elements of the second row by $-\frac{1}{2}$.

$$\left[\begin{array}{ccc|ccc} 1 & 0 & 1 & 1 & 0 & 0 \\ 0 & 1 & \frac{3}{2} & 1 & -\frac{1}{2} & 0 \\ 0 & 0 & -3 & -3 & 0 & 1 \end{array}\right] \quad -\tfrac{1}{2}\text{R2}$$

To get 1 for the third element in the third row, multiply the elements of the third row by $-\frac{1}{3}$.

$$\left[\begin{array}{ccc|ccc} 1 & 0 & 1 & 1 & 0 & 0 \\ 0 & 1 & \frac{3}{2} & 1 & -\frac{1}{2} & 0 \\ 0 & 0 & 1 & 1 & 0 & -\frac{1}{3} \end{array}\right] \quad -\tfrac{1}{3}\text{R3}$$

To get 0 for the third element in the first row, multiply the elements of the third row by -1 and add to the first row.

$$\left[\begin{array}{ccc|ccc} 1 & 0 & 0 & 0 & 0 & \frac{1}{3} \\ 0 & 1 & \frac{3}{2} & 1 & -\frac{1}{2} & 0 \\ 0 & 0 & 1 & 1 & 0 & -\frac{1}{3} \end{array}\right] \quad -1\text{R3} + \text{R1}$$

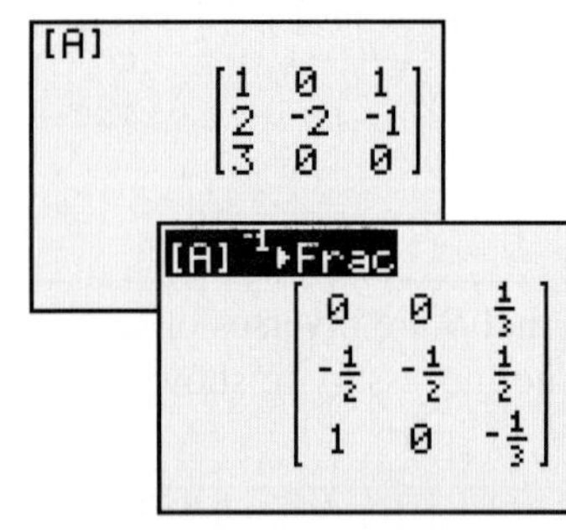

A graphing calculator can be used to find the inverse of a matrix. The screens support the result in **Example 2.** The elements of the inverse are expressed as fractions, so it is easier to compare with the inverse matrix found in the example.

To get 0 for the third element in the second row, multiply the elements of the third row by $-\frac{3}{2}$ and add to the second row.

$$\left[\begin{array}{ccc|ccc} 1 & 0 & 0 & 0 & 0 & \frac{1}{3} \\ 0 & 1 & 0 & -\frac{1}{2} & -\frac{1}{2} & \frac{1}{2} \\ 0 & 0 & 1 & 1 & 0 & -\frac{1}{3} \end{array}\right] \quad -\tfrac{3}{2}\text{R3} + \text{R2}$$

Step 3 The last transformation shows that the inverse is

$$A^{-1} = \begin{bmatrix} 0 & 0 & \frac{1}{3} \\ -\frac{1}{2} & -\frac{1}{2} & \frac{1}{2} \\ 1 & 0 & -\frac{1}{3} \end{bmatrix}.$$

Confirm this by forming the product $A^{-1}A$ or AA^{-1}, each of which should equal the matrix I_3.

Now Try Exercises 11 and 19.

As illustrated by the examples, the most efficient order for the transformations in Step 2 is to make the changes column by column from left to right, so that for each column the required 1 is the result of the first change. Next, perform the steps that obtain the 0s in that column. Then proceed to the next column.

EXAMPLE 3 Identifying a Matrix with No Inverse

Find A^{-1}, if possible, given that $A = \begin{bmatrix} 2 & -4 \\ 1 & -2 \end{bmatrix}$.

ALGEBRAIC SOLUTION

Using row transformations to change the first column of the augmented matrix

$$\left[\begin{array}{cc|cc} 2 & -4 & 1 & 0 \\ 1 & -2 & 0 & 1 \end{array}\right]$$

results in the following matrices.

$$\left[\begin{array}{cc|cc} 1 & -2 & \frac{1}{2} & 0 \\ 1 & -2 & 0 & 1 \end{array}\right] \text{ and } \left[\begin{array}{cc|cc} 1 & -2 & \frac{1}{2} & 0 \\ 0 & 0 & -\frac{1}{2} & 1 \end{array}\right]$$

(We multiplied the elements in row one by $\frac{1}{2}$ in the first step, and in the second step we added the negative of row one to row two.) At this point, the matrix should be changed so that the second row, second element will be 1. Since that element is now 0, there is no way to complete the desired transformation, so A^{-1} does not exist for this matrix A. Just as there is no multiplicative inverse for the real number 0, not every matrix has a multiplicative inverse. Matrix A is an example of such a matrix.

GRAPHING CALCULATOR SOLUTION

If the inverse of a matrix does not exist, the matrix is called **singular,** as shown in **Figure 28** for matrix [A]. This occurs when the determinant of the matrix is 0. **(See Exercises 29–34.)**

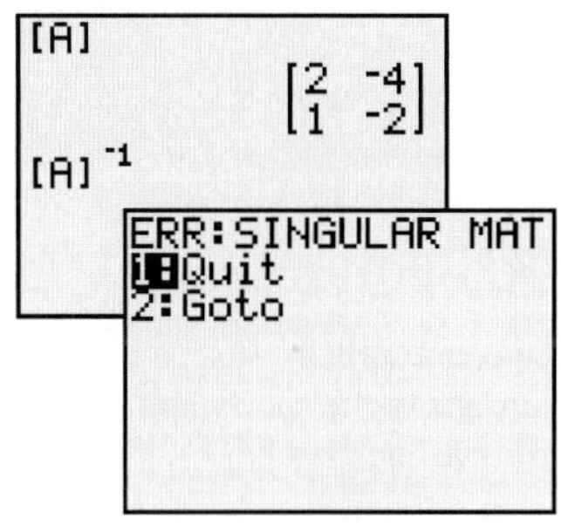

Figure 28

Now Try Exercise 15.

Classroom Example 3
Find A^{-1}, if possible, given that

$$A = \begin{bmatrix} 4 & -2 & 5 \\ 0 & 1 & 0 \\ -8 & 4 & -10 \end{bmatrix}.$$

Answer: The inverse does not exist.

If the inverse of a matrix exists, it is unique. That is, any given square matrix has no more than one inverse. The proof of this is left as **Exercise 70.**

Solving Systems Using Inverse Matrices Matrix inverses can be used to solve square linear systems of equations. (A square system has the same number of equations as variables.) For example, consider the following linear system of three equations with three variables.

$$a_{11}x + a_{12}y + a_{13}z = b_1$$
$$a_{21}x + a_{22}y + a_{23}z = b_2$$
$$a_{31}x + a_{32}y + a_{33}z = b_3$$

The definition of matrix multiplication can be used to rewrite the system using matrices.

$$\begin{bmatrix} a_{11} & a_{12} & a_{13} \\ a_{21} & a_{22} & a_{23} \\ a_{31} & a_{32} & a_{33} \end{bmatrix} \cdot \begin{bmatrix} x \\ y \\ z \end{bmatrix} = \begin{bmatrix} b_1 \\ b_2 \\ b_3 \end{bmatrix} \quad (1)$$

(To see this, multiply the matrices on the left.)

$$\text{If } A = \begin{bmatrix} a_{11} & a_{12} & a_{13} \\ a_{21} & a_{22} & a_{23} \\ a_{31} & a_{32} & a_{33} \end{bmatrix}, \quad X = \begin{bmatrix} x \\ y \\ z \end{bmatrix}, \quad \text{and} \quad B = \begin{bmatrix} b_1 \\ b_2 \\ b_3 \end{bmatrix},$$

then the system given in (1) becomes $AX = B$. If A^{-1} exists, then both sides of $AX = B$ can be multiplied on the left as shown on the next page.

Teaching Tip Caution students to evaluate $A^{-1}B$, not BA^{-1}, when solving $AX = B$, since matrix multiplication is not commutative. Also, point out the analogy between solving a matrix equation and solving an ordinary linear equation, such as $3x = 5$. In each situation, multiply both sides of the equation by the inverse of the coefficient of the unknown quantity.

$$A^{-1}(AX) = A^{-1}B$$

$$(A^{-1}A)X = A^{-1}B \quad \text{Associative property (Section 9.7)}$$

$$I_3X = A^{-1}B \quad \text{Inverse property}$$

$$X = A^{-1}B \quad \text{Identity property}$$

Matrix $A^{-1}B$ gives the solution of the system.

Solution of the Matrix Equation $AX = B$

If A is an $n \times n$ matrix with inverse A^{-1}, X is an $n \times 1$ matrix of variables, and B is an $n \times 1$ matrix, then the matrix equation

$$AX = B$$

has the solution $\quad X = A^{-1}B.$

Classroom Example 4
Use the inverse of the coefficient matrix to solve each system.

(a) $5x + 2y = -1$
$2x + 3y = 15$

(b) $-4x + 2y = 12$
$x - y + 2z = 7$
$y + 4z = 20$

Answers:
(a) $\{(-3, 7)\}$ (b) $\{(-3, 0, 5)\}$

This method of using matrix inverses to solve systems of equations is useful when the inverse is already known or when many systems of the form $AX = B$ must be solved and only B changes.

EXAMPLE 4 Solving Systems of Equations Using Matrix Inverses

Use the inverse of the coefficient matrix to solve each system.

(a) $2x - 3y = 4$
$x + 5y = 2$

(b) $x + z = -1$
$2x - 2y - z = 5$
$3x = 6$

ALGEBRAIC SOLUTION

(a) The system can be written in matrix form as

$$\begin{bmatrix} 2 & -3 \\ 1 & 5 \end{bmatrix}\begin{bmatrix} x \\ y \end{bmatrix} = \begin{bmatrix} 4 \\ 2 \end{bmatrix},$$

where

$$A = \begin{bmatrix} 2 & -3 \\ 1 & 5 \end{bmatrix}, \quad X = \begin{bmatrix} x \\ y \end{bmatrix}, \quad \text{and} \quad B = \begin{bmatrix} 4 \\ 2 \end{bmatrix}.$$

An equivalent matrix equation is $AX = B$ with solution $X = A^{-1}B$. Use the methods described in this section to determine that

$$A^{-1} = \begin{bmatrix} \frac{5}{13} & \frac{3}{13} \\ -\frac{1}{13} & \frac{2}{13} \end{bmatrix}.$$

Now, find $A^{-1}B$.

$$A^{-1}B = \begin{bmatrix} \frac{5}{13} & \frac{3}{13} \\ -\frac{1}{13} & \frac{2}{13} \end{bmatrix}\begin{bmatrix} 4 \\ 2 \end{bmatrix} = \begin{bmatrix} 2 \\ 0 \end{bmatrix}$$

Since $X = A^{-1}B$, $\quad X = \begin{bmatrix} x \\ y \end{bmatrix} = \begin{bmatrix} 2 \\ 0 \end{bmatrix}.$

The final matrix shows that the solution set of the system is $\{(2, 0)\}$.

GRAPHING CALCULATOR SOLUTION

(a) Enter [A] and [B], and then find the product $[A]^{-1}[B]$ as shown in **Figure 29.**

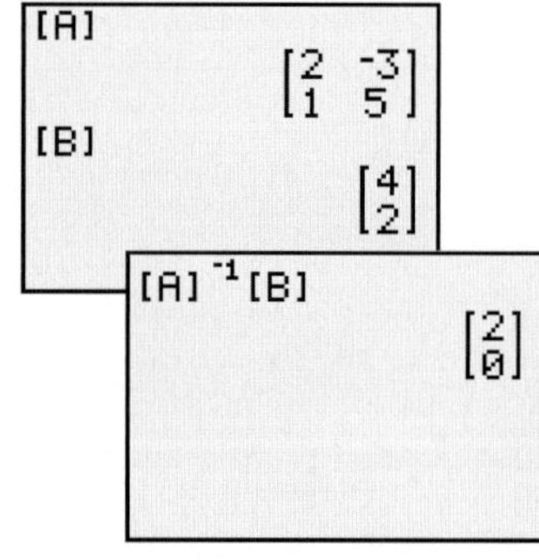

Figure 29

The column matrix indicates the solution set, $\{(2, 0)\}$.

(b) The coefficient matrix A for this system is

$$A = \begin{bmatrix} 1 & 0 & 1 \\ 2 & -2 & -1 \\ 3 & 0 & 0 \end{bmatrix},$$

and its inverse A^{-1} was found in **Example 2.** Let

$$X = \begin{bmatrix} x \\ y \\ z \end{bmatrix} \quad \text{and} \quad B = \begin{bmatrix} -1 \\ 5 \\ 6 \end{bmatrix}.$$

Since $X = A^{-1}B$, we have

$$\begin{bmatrix} x \\ y \\ z \end{bmatrix} = \underbrace{\begin{bmatrix} 0 & 0 & \frac{1}{3} \\ -\frac{1}{2} & -\frac{1}{2} & \frac{1}{2} \\ 1 & 0 & -\frac{1}{3} \end{bmatrix}}_{A^{-1} \text{ from Example 2}} \begin{bmatrix} -1 \\ 5 \\ 6 \end{bmatrix}$$

$$= \begin{bmatrix} 2 \\ 1 \\ -3 \end{bmatrix}.$$

The solution set is $\{(2, 1, -3)\}$.

(b) **Figure 30** shows the coefficient matrix [A] and the column matrix of constants [B]. Note that when $[A]^{-1}[B]$ is determined, it is not necessary to display $[A]^{-1}$. As long as the inverse exists, the product will be computed.

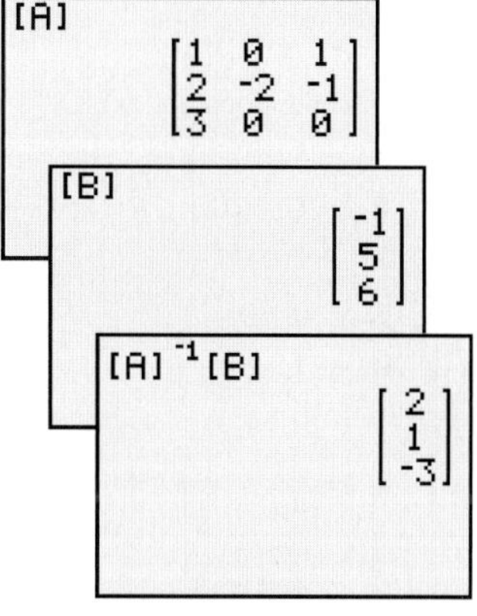

Figure 30

Now Try Exercises 37 and 51.

9.8 Exercises

3. yes 4. yes 5. no 6. no 7. no 8. no 9. yes 10. yes

1. Show that for

$$A = \begin{bmatrix} -2 & 4 & 0 \\ 3 & 5 & 9 \\ 0 & 8 & -6 \end{bmatrix} \quad \text{and} \quad I_3 = \begin{bmatrix} 1 & 0 & 0 \\ 0 & 1 & 0 \\ 0 & 0 & 1 \end{bmatrix},$$

$I_3A = A$. (This result, along with that of **Example 1,** illustrates that the commutative property holds when one of the matrices is an identity matrix.)

2. Let $A = \begin{bmatrix} a & b \\ c & d \end{bmatrix}$ and $I_2 = \begin{bmatrix} 1 & 0 \\ 0 & 1 \end{bmatrix}$. Show that $AI_2 = I_2A = A$, thus proving that I_2 is the identity element for matrix multiplication for 2×2 square matrices.

Decide whether or not the given matrices are inverses of each other. (Hint: Check to see whether their products are the identity matrix I_n.)

3. $\begin{bmatrix} 5 & 7 \\ 2 & 3 \end{bmatrix}$ and $\begin{bmatrix} 3 & -7 \\ -2 & 5 \end{bmatrix}$

4. $\begin{bmatrix} 2 & 3 \\ 1 & 1 \end{bmatrix}$ and $\begin{bmatrix} -1 & 3 \\ 1 & -2 \end{bmatrix}$

5. $\begin{bmatrix} -1 & 2 \\ 3 & -5 \end{bmatrix}$ and $\begin{bmatrix} -5 & -2 \\ -3 & -1 \end{bmatrix}$

6. $\begin{bmatrix} 2 & 1 \\ 3 & 2 \end{bmatrix}$ and $\begin{bmatrix} 2 & 1 \\ -3 & 2 \end{bmatrix}$

7. $\begin{bmatrix} 0 & 1 & 0 \\ 0 & 0 & -2 \\ 1 & -1 & 0 \end{bmatrix}$ and $\begin{bmatrix} 1 & 0 & 1 \\ 1 & 0 & 0 \\ 0 & -1 & 0 \end{bmatrix}$

8. $\begin{bmatrix} 1 & 2 & 0 \\ 0 & 1 & 0 \\ 0 & 1 & 0 \end{bmatrix}$ and $\begin{bmatrix} 1 & -2 & 0 \\ 0 & 1 & 0 \\ 0 & -1 & 1 \end{bmatrix}$

9. $\begin{bmatrix} 1 & 0 & 0 \\ 0 & -1 & 0 \\ 1 & 0 & 1 \end{bmatrix}$ and $\begin{bmatrix} 1 & 0 & 0 \\ 0 & -1 & 0 \\ -1 & 0 & 1 \end{bmatrix}$

10. $\begin{bmatrix} 1 & 3 & 3 \\ 1 & 4 & 3 \\ 1 & 3 & 4 \end{bmatrix}$ and $\begin{bmatrix} 7 & -3 & -3 \\ -1 & 1 & 0 \\ -1 & 0 & 1 \end{bmatrix}$

11. $\begin{bmatrix} -\frac{1}{5} & -\frac{2}{5} \\ \frac{2}{5} & -\frac{1}{5} \end{bmatrix}$

12. $\begin{bmatrix} 0 & \frac{1}{2} \\ -1 & \frac{1}{2} \end{bmatrix}$

13. $\begin{bmatrix} 2 & 1 \\ -\frac{3}{2} & -\frac{1}{2} \end{bmatrix}$

14. $\begin{bmatrix} 2 & 1 \\ 5 & 3 \end{bmatrix}$

15. The inverse does not exist.

16. The inverse does not exist.

17. $\begin{bmatrix} -1 & 1 & 1 \\ 0 & -1 & 0 \\ 2 & -1 & -1 \end{bmatrix}$

18. $\begin{bmatrix} 1 & 0 & 0 \\ 0 & -1 & 0 \\ -1 & 0 & 1 \end{bmatrix}$

19. $\begin{bmatrix} \frac{7}{8} & -\frac{3}{8} & -\frac{3}{8} \\ -\frac{1}{8} & \frac{5}{8} & -\frac{3}{8} \\ -\frac{1}{8} & -\frac{3}{8} & \frac{5}{8} \end{bmatrix}$

20. $\begin{bmatrix} -\frac{4}{5} & \frac{2}{5} & \frac{3}{5} \\ -\frac{11}{10} & \frac{4}{5} & \frac{1}{5} \\ \frac{2}{5} & -\frac{1}{5} & \frac{1}{5} \end{bmatrix}$

21. $\begin{bmatrix} -\frac{15}{4} & -\frac{1}{4} & -3 \\ \frac{5}{4} & \frac{1}{4} & 1 \\ -\frac{3}{2} & 0 & -1 \end{bmatrix}$

22. $\begin{bmatrix} \frac{7}{4} & \frac{5}{2} & 3 \\ -\frac{1}{4} & -\frac{1}{2} & 0 \\ -\frac{1}{4} & -\frac{1}{2} & -1 \end{bmatrix}$

23. $\begin{bmatrix} \frac{1}{2} & 0 & \frac{1}{2} & -1 \\ \frac{1}{10} & -\frac{2}{5} & \frac{3}{10} & -\frac{1}{5} \\ -\frac{7}{10} & \frac{4}{5} & -\frac{11}{10} & \frac{12}{5} \\ \frac{1}{5} & \frac{1}{5} & -\frac{2}{5} & \frac{3}{5} \end{bmatrix}$

24. $\begin{bmatrix} \frac{1}{2} & \frac{1}{2} & -\frac{1}{4} & \frac{1}{2} \\ -1 & 4 & -\frac{1}{2} & -2 \\ -\frac{1}{2} & \frac{5}{2} & -\frac{1}{4} & -\frac{3}{2} \\ \frac{1}{2} & -\frac{1}{2} & \frac{1}{4} & \frac{1}{2} \end{bmatrix}$

25. $\begin{bmatrix} 0 & -\frac{1}{3} & \frac{1}{3} & \frac{2}{3} \\ \frac{1}{3} & \frac{2}{3} & -\frac{1}{3} & -1 \\ \frac{2}{3} & 1 & -\frac{4}{3} & -\frac{4}{3} \\ -\frac{1}{3} & \frac{1}{3} & \frac{1}{3} & 0 \end{bmatrix}$

27. $\begin{bmatrix} 4 & 7 \\ -2 & 2 \end{bmatrix}$ 28. $\begin{bmatrix} 1 & 0 & 1 \\ -1 & 0 & 2 \\ -2 & 1 & 3 \end{bmatrix}$

Find the inverse, if it exists, for each matrix. ***See Examples 2 and 3.***

11. $\begin{bmatrix} -1 & 2 \\ -2 & -1 \end{bmatrix}$ **12.** $\begin{bmatrix} 1 & -1 \\ 2 & 0 \end{bmatrix}$ **13.** $\begin{bmatrix} -1 & -2 \\ 3 & 4 \end{bmatrix}$

14. $\begin{bmatrix} 3 & -1 \\ -5 & 2 \end{bmatrix}$ **15.** $\begin{bmatrix} 5 & 10 \\ -3 & -6 \end{bmatrix}$ **16.** $\begin{bmatrix} -6 & 4 \\ -3 & 2 \end{bmatrix}$

17. $\begin{bmatrix} 1 & 0 & 1 \\ 0 & -1 & 0 \\ 2 & 1 & 1 \end{bmatrix}$ **18.** $\begin{bmatrix} 1 & 0 & 0 \\ 0 & -1 & 0 \\ 1 & 0 & 1 \end{bmatrix}$ **19.** $\begin{bmatrix} 2 & 3 & 3 \\ 1 & 4 & 3 \\ 1 & 3 & 4 \end{bmatrix}$

20. $\begin{bmatrix} -2 & 2 & 4 \\ -3 & 4 & 5 \\ 1 & 0 & 2 \end{bmatrix}$ **21.** $\begin{bmatrix} 2 & 2 & -4 \\ 2 & 6 & 0 \\ -3 & -3 & 5 \end{bmatrix}$ **22.** $\begin{bmatrix} 2 & 4 & 6 \\ -1 & -4 & -3 \\ 0 & 1 & -1 \end{bmatrix}$

23. $\begin{bmatrix} 1 & 1 & 0 & 2 \\ 2 & -1 & 1 & -1 \\ 3 & 3 & 2 & -2 \\ 1 & 2 & 1 & 0 \end{bmatrix}$ **24.** $\begin{bmatrix} 1 & -2 & 3 & 0 \\ 0 & 1 & -1 & 1 \\ -2 & 2 & -2 & 4 \\ 0 & 2 & -3 & 1 \end{bmatrix}$ **25.** $\begin{bmatrix} 3 & 2 & 0 & -1 \\ 2 & 0 & 1 & 2 \\ 1 & 2 & -1 & 0 \\ 2 & -1 & 1 & 1 \end{bmatrix}$

26. Explain the process for finding the inverse of a matrix.

Each graphing calculator screen shows A^{-1} for some matrix A. Find each matrix A. (Hint: $(A^{-1})^{-1} = A$.)

27.

[A]$^{-1}$▸Frac

$\begin{bmatrix} \frac{1}{11} & -\frac{7}{22} \\ \frac{1}{11} & \frac{2}{11} \end{bmatrix}$

28.

[A]$^{-1}$▸Frac

$\begin{bmatrix} \frac{2}{3} & -\frac{1}{3} & 0 \\ \frac{1}{3} & -\frac{5}{3} & 1 \\ \frac{1}{3} & \frac{1}{3} & 0 \end{bmatrix}$

Relating Concepts

For individual or collaborative investigation *(Exercises 29–34)*

It can be shown that the inverse of matrix $A = \begin{bmatrix} a & b \\ c & d \end{bmatrix}$ *is*

$$A^{-1} = \begin{bmatrix} \frac{d}{ad - bc} & \frac{-b}{ad - bc} \\ \frac{-c}{ad - bc} & \frac{a}{ad - bc} \end{bmatrix}.$$

Work Exercises 29–34 in order, *to discover a connection between the inverse and the determinant of a 2×2 matrix.*

29. With respect to the matrix $A = \begin{bmatrix} a & b \\ c & d \end{bmatrix}$, what do we call $ad - bc$?

30. Refer to A^{-1} as given above, and write it using determinant notation.

31. Write A^{-1} using scalar multiplication, where the scalar is $\frac{1}{|A|}$.

32. Explain how the inverse of matrix A can be found using a determinant.

33. Use the method described here to find the inverse of $A = \begin{bmatrix} 4 & 2 \\ 7 & 3 \end{bmatrix}$.

34. Complete the following statement: The inverse of a 2×2 matrix A does not exist if the determinant of A has value ______. (*Hint:* Look at the denominators in A^{-1} as given earlier.)

29. $ad - bc$ is the determinant of A.

30. $\begin{bmatrix} \frac{d}{|A|} & \frac{-b}{|A|} \\ \frac{-c}{|A|} & \frac{a}{|A|} \end{bmatrix}$

31. $A^{-1} = \frac{1}{|A|}\begin{bmatrix} d & -b \\ -c & a \end{bmatrix}$

33. $A^{-1} = \begin{bmatrix} -\frac{3}{2} & 1 \\ \frac{7}{2} & -2 \end{bmatrix}$

34. zero

35. $\{(2, 3)\}$ 36. $\{(2, 3)\}$

37. $\{(-2, 4)\}$ 38. $\{(3, -5)\}$

39. $\{(-2, \frac{3}{4})\}$ 40. $\{(\frac{7}{2}, -1)\}$

41. $\{(-\frac{1}{2}, \frac{2}{3})\}$

42. $\{(-\frac{1}{5}, -\frac{1}{3})\}$

43. $\{(4, -9)\}$ 44. $\{(2, -1)\}$

45. $\{(6, -\frac{5}{6})\}$ 46. $\{(13, -\frac{7}{5})\}$

47. The inverse of $\begin{bmatrix} 7 & -2 \\ 14 & -4 \end{bmatrix}$ does not exist.

48. The inverse of $\begin{bmatrix} 1 & -2 & 3 \\ 2 & -4 & 6 \\ 3 & -6 & 9 \end{bmatrix}$ does not exist.

49. $\{(3, 1, 2)\}$

50. $\{(3, -1, 4)\}$

51. $\{(\frac{5}{4}, \frac{1}{4}, -\frac{3}{4})\}$

52. $\{(-\frac{4}{5}, -\frac{21}{10}, \frac{7}{5})\}$

53. $\{(11, -1, 2)\}$

54. $\{(15, -5, -1)\}$

55. $\{(1, 0, 2, 1)\}$

56. $\{(-2, 0, 1, 0)\}$

57.
(a) $602.7 = a + 5.543b + 37.14c$
$656.7 = a + 6.933b + 41.30c$
$778.5 = a + 7.638b + 45.62c$
(b) $a \approx -490.547$, $b = -89$, $c = 42.71875$
(c) $S = -490.547 - 89A + 42.71875B$

Solve each system by using the inverse of the coefficient matrix. ***See Example 4.***

35. $-x + y = 1$
$2x - y = 1$

36. $x + y = 5$
$x - y = -1$

37. $2x - y = -8$
$3x + y = -2$

38. $x + 3y = -12$
$2x - y = 11$

39. $3x + 4y = -3$
$-5x + 8y = 16$

40. $2x - 3y = 10$
$2x + 2y = 5$

41. $6x + 9y = 3$
$-8x + 3y = 6$

42. $5x - 3y = 0$
$10x + 6y = -4$

43. $0.2x + 0.3y = -1.9$
$0.7x - 0.2y = 4.6$

44. $0.5x + 0.2y = 0.8$
$0.3x - 0.1y = 0.7$

45. $\frac{1}{2}x + \frac{1}{3}y = \frac{49}{18}$
$\frac{1}{2}x + 2y = \frac{4}{3}$

46. $\frac{1}{5}x + \frac{1}{7}y = \frac{12}{5}$
$\frac{1}{10}x + \frac{1}{3}y = \frac{5}{6}$

Concept Check Show that the matrix inverse method cannot be used to solve each system.

47. $7x - 2y = 3$
$14x - 4y = 1$

48. $x - 2y + 3z = 4$
$2x - 4y + 6z = 8$
$3x - 6y + 9z = 14$

Solve each system by using the inverse of the coefficient matrix. For Exercises 51–56, the inverses were found in ***Exercises 19–24. See Example 4.***

49. $x + y + z = 6$
$2x + 3y - z = 7$
$3x - y - z = 6$

50. $2x + 5y + 2z = 9$
$4x - 7y - 3z = 7$
$3x - 8y - 2z = 9$

51. $2x + 3y + 3z = 1$
$x + 4y + 3z = 0$
$x + 3y + 4z = -1$

52. $-2x + 2y + 4z = 3$
$-3x + 4y + 5z = 1$
$x + 2z = 2$

53. $2x + 2y - 4z = 12$
$2x + 6y = 16$
$-3x - 3y + 5z = -20$

54. $2x + 4y + 6z = 4$
$-x - 4y - 3z = 8$
$y - z = -4$

55. $x + y + 2w = 3$
$2x - y + z - w = 3$
$3x + 3y + 2z - 2w = 5$
$x + 2y + z = 3$

56. $x - 2y + 3z = 1$
$y - z + w = -1$
$-2x + 2y - 2z + 4w = 2$
$2y - 3z + w = -3$

(Modeling) Solve each problem.

57. *Plate-Glass Sales* The amount of plate-glass sales S (in millions of dollars) can be affected by the number of new building contracts B issued (in millions) and automobiles A produced (in millions). A plate-glass company in California wants to forecast future sales by using the past three years of sales. The totals for the three years are given in the table. To describe the relationship among these variables, we can use the equation

S	A	B
602.7	5.543	37.14
656.7	6.933	41.30
778.5	7.638	45.62

$$S = a + bA + cB,$$

where the coefficients a, b, and c are constants that must be determined before the equation can be used. (*Source:* Makridakis, S., and S. Wheelwright, *Forecasting Methods for Management,* John Wiley and Sons.)

(a) Substitute the values for S, A, and B for each year from the table into the equation $S = a + bA + cB$, and obtain three linear equations involving a, b, and c.

(b) Use a graphing calculator to solve this linear system for a, b, and c. Use matrix inverse methods.

(c) Write the equation for S using these values for the coefficients.

57. (d) $S \approx 843.5$
(e) $S \approx 1547.5$

58.
(a) $10{,}170 = a + 112.9b + 307.5c$
$15{,}305 = a + 132.9b + 621.63c$
$21{,}289 = a + 155.2b + 1937.13c$
(b) $a \approx -18{,}425$, $b \approx 252.54$, $c \approx 0.26778$
(c) $S = -18{,}425 + 252.54R + 0.26778I$
(d) approximately 11,357
(e) approximately 18,112

59. Answers will vary.

60. $\begin{bmatrix} 0.0543058761 & -0.0543058761 \\ 1.846399787 & 0.153600213 \end{bmatrix}$

61. $\begin{bmatrix} -0.1215875322 & 0.0491390161 \\ 1.544369078 & -0.046799063 \end{bmatrix}$

62. See the next page.

63. $\begin{bmatrix} 2 & -2 & 0 \\ -4 & 0 & 4 \\ 3 & 3 & -3 \end{bmatrix}$

64. $\{(-3.542308934, -4.343268299)\}$

65. $\{(1.68717058, -1.306990242)\}$

66. $\{(-0.9704156959, 1.391914631, 0.1874077432)\}$

67. $\{(13.58736702, 3.929011993, -5.342780076)\}$

(d) For the next year it is estimated that $A = 7.752$ and $B = 47.38$. Predict S. (The actual value for S was 877.6.)

(e) It is predicted that in 6 yr, $A = 8.9$ and $B = 66.25$. Find the value of S in this situation and discuss its validity.

58. *Tire Sales* The number of automobile tire sales is dependent on several variables. In one study the relationship among annual tire sales S (in thousands of dollars), automobile registrations R (in millions), and personal disposable income I (in millions of dollars) was investigated. The results for three years are given in the table. To describe the relationship among these variables, we can use the equation

S	R	I
10,170	112.9	307.5
15,305	132.9	621.63
21,289	155.2	1937.13

$$S = a + bR + cI,$$

where the coefficients a, b, and c are constants that must be determined before the equation can be used. (*Source:* Jarrett, J., *Business Forecasting Methods,* Basil Blackwell, Ltd.)

(a) Substitute the values for S, R, and I for each year from the table into the equation $S = a + bR + cI$, and obtain three linear equations involving a, b, and c.

(b) Use a graphing calculator to solve this linear system for a, b, and c. Use matrix inverse methods.

(c) Write the equation for S using these values for the coefficients.

(d) If $R = 117.6$ and $I = 310.73$, predict S. (The actual value for S was 11,314.)

(e) If $R = 143.8$ and $I = 829.06$, predict S. (The actual value for S was 18,481.)

59. *Social Security Numbers* In **Exercise 89 of Section 3.4,** construction of your own personal Social Security polynomial was discussed. It is also possible to find a polynomial that goes through a given set of points in the plane by using a process called **polynomial interpolation.**

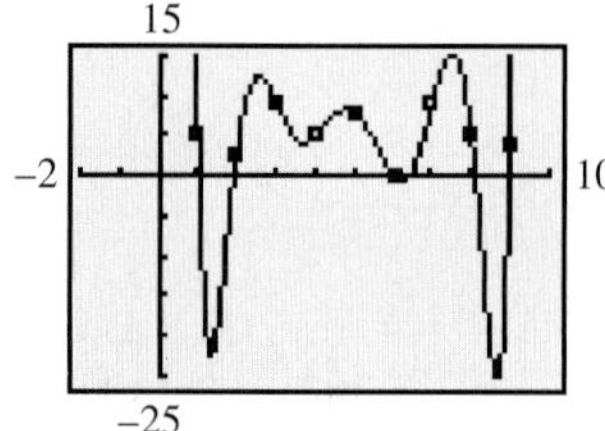

Recall that three points define a second-degree polynomial, four points define a third-degree polynomial, and so on. The only restriction on the points, since polynomials define functions, is that no two distinct points can have the same x-coordinate. Using the same SSN (539-58-0954) as in that exercise, we can find an eighth-degree polynomial that lies on the nine points with x-coordinates 1 through 9 and y-coordinates that are digits of the SSN: $(1, 5)$, $(2, 3)$, $(3, 9)$, $\ldots$, $(9, 4)$. This is done by writing a system of nine equations with nine variables, which is then solved by the inverse matrix method. The graph of this polynomial is shown. Find such a polynomial using your own SSN.

Use a graphing calculator to find the inverse of each matrix. Give as many decimal places as the calculator shows. ***See Example 2.***

60. $\begin{bmatrix} \sqrt{2} & 0.5 \\ -17 & \frac{1}{2} \end{bmatrix}$

61. $\begin{bmatrix} \frac{2}{3} & 0.7 \\ 22 & \sqrt{3} \end{bmatrix}$

62. $\begin{bmatrix} 1.4 & 0.5 & 0.59 \\ 0.84 & 1.36 & 0.62 \\ 0.56 & 0.47 & 1.3 \end{bmatrix}$

63. $\begin{bmatrix} \frac{1}{2} & \frac{1}{4} & \frac{1}{3} \\ 0 & \frac{1}{4} & \frac{1}{3} \\ \frac{1}{2} & \frac{1}{2} & \frac{1}{3} \end{bmatrix}$

Use a graphing calculator and the method of matrix inverses to solve each system. Give as many decimal places as the calculator shows. ***See Example 4.***

64. $x - \sqrt{2}y = 2.6$
$0.75x + y = -7$

65. $2.1x + y = \sqrt{5}$
$\sqrt{2}x - 2y = 5$

66. $\pi x + ey + \sqrt{2}z = 1$
$ex + \pi y + \sqrt{2}z = 2$
$\sqrt{2}x + ey + \pi z = 3$

67. $(\log 2)x + (\ln 3)y + (\ln 4)z = 1$
$(\ln 3)x + (\log 2)y + (\ln 8)z = 5$
$(\log 12)x + (\ln 4)y + (\ln 8)z = 9$

71. $A = \begin{bmatrix} 1 & 0 \\ 1 & 1 \end{bmatrix}, B = \begin{bmatrix} 1 & 1 \\ 0 & 1 \end{bmatrix}$
(Other answers are possible.)

73. $\begin{bmatrix} \frac{1}{a} & 0 & 0 \\ 0 & \frac{1}{b} & 0 \\ 0 & 0 & \frac{1}{c} \end{bmatrix}$

74. $A^{-1} = A^2 = \begin{bmatrix} 1 & 0 & 0 \\ 0 & -1 & 1 \\ 0 & -1 & 0 \end{bmatrix}$

75. $I_n, -A^{-1}, \frac{1}{k}A^{-1}$

Let $A = \begin{bmatrix} a & b \\ c & d \end{bmatrix}$, *and let O be the* 2×2 *zero matrix. Show that the statements in Exercises 68 and 69 are true.*

68. $A \cdot O = O \cdot A = O$

69. For square matrices A and B of the same dimension, if $AB = O$ and if A^{-1} exists, then $B = O$.

70. Prove that any square matrix has no more than one inverse.

71. Give an example of two matrices A and B, where $(AB)^{-1} \neq A^{-1}B^{-1}$.

72. Suppose A and B are matrices, where A^{-1}, B^{-1}, and AB all exist. Show that $(AB)^{-1} = B^{-1}A^{-1}$.

73. Let $A = \begin{bmatrix} a & 0 & 0 \\ 0 & b & 0 \\ 0 & 0 & c \end{bmatrix}$, where a, b, and c are nonzero real numbers. Find A^{-1}.

74. Let $A = \begin{bmatrix} 1 & 0 & 0 \\ 0 & 0 & -1 \\ 0 & 1 & -1 \end{bmatrix}$. Show that $A^3 = I_3$, and use this result to find the inverse of A.

75. What are the inverses of I_n, $-A$ (in terms of A), and kA (k a scalar)?

76. Discuss the similarities and differences between solving the linear equation $ax = b$ and solving the matrix equation $AX = B$.

62. $\begin{bmatrix} 0.9987635516 & -0.252092087 & -0.3330564627 \\ -0.5037783375 & 1.007556675 & -0.2518891688 \\ -0.2481013617 & -0.2556769758 & 1.003768868 \end{bmatrix}$

Chapter 9 Test Prep

Key Terms

9.1 linear equation (first-degree equation) in n unknowns
system of equations
solutions of a system of equations
system of linear equations (linear system)
consistent system
independent equations
inconsistent system
dependent equations
equivalent systems
ordered triple

9.2 matrix (matrices)
element (of a matrix)
augmented matrix
dimension (of a matrix)

9.3 determinant
minor
cofactor
expansion by a row or column
Cramer's rule

9.4 partial fraction decomposition
partial fraction

9.5 nonlinear system

9.6 half-plane
boundary
linear inequality in two variables
system of inequalities
linear programming
constraints
objective function
region of feasible solutions
vertex (corner point)

9.7 square matrix
row matrix
column matrix
zero matrix
additive inverse (negative) of a matrix
scalar

9.8 identity matrix
multiplicative inverse (of a matrix)

New Symbols

(a, b, c)	ordered triple
$[A]$	matrix A (graphing calculator symbolism)
$\lvert A \rvert$	determinant of matrix A
a_{ij}	the element in row i, column j, of a matrix
D, D_x, D_y	determinants used in Cramer's rule
I_2, I_3	identity matrices
A^{-1}	multiplicative inverse of matrix A

(continued)

Quick Review

Concepts	Examples

9.1 Systems of Linear Equations

Transformations of a Linear System

1. Interchange any two equations of the system.
2. Multiply or divide any equation of the system by a nonzero real number.
3. Replace any equation of the system by the sum of that equation and a multiple of another equation in the system.

Systems may be solved by substitution, elimination, or a combination of the two methods.

Substitution Method
Use one equation to find an expression for one variable in terms of the other, and then substitute into the other equation of the system.

Solve by substitution.

$$4x - y = 7 \quad (1)$$
$$3x + 2y = 30 \quad (2)$$

Solve for y in equation (1).

$$y = 4x - 7$$

Substitute $4x - 7$ for y in equation (2), and solve for x.

$$\begin{aligned} 3x + 2(4x - 7) &= 30 & \\ 3x + 8x - 14 &= 30 & \text{Distribute.} \\ 11x - 14 &= 30 & \text{Combine like terms.} \\ 11x &= 44 & \text{Add 14.} \\ x &= 4 & \text{Divide by 11.} \end{aligned}$$

Substitute 4 for x in the equation $y = 4x - 7$ to find that $y = 9$. The solution set is $\{(4, 9)\}$.

Elimination Method
Use multiplication and addition to eliminate a variable from one equation. To eliminate a variable, the coefficients of that variable in the equations must be additive inverses.

Solve the system.

$$x + 2y - z = 6 \quad (1)$$
$$x + y + z = 6 \quad (2)$$
$$2x + y - z = 7 \quad (3)$$

Add equations (1) and (2); z is eliminated and the result is $2x + 3y = 12$.

Eliminate z again by adding equations (2) and (3) to get $3x + 2y = 13$. Now solve the system

$$2x + 3y = 12 \quad (4)$$
$$3x + 2y = 13. \quad (5)$$

$$\begin{aligned} -6x - 9y &= -36 & \text{Multiply (4) by } -3. \\ 6x + 4y &= 26 & \text{Multiply (5) by 2.} \\ \hline -5y &= -10 & \text{Add.} \\ y &= 2 & \text{Divide by } -5. \end{aligned}$$

Let $y = 2$ in equation (4).

$$\begin{aligned} 2x + 3(2) &= 12 & \text{(4) with } y = 2 \\ 2x + 6 &= 12 & \text{Multiply.} \\ 2x &= 6 & \text{Subtract 6.} \\ x &= 3 & \text{Divide by 2.} \end{aligned}$$

Let $y = 2$ and $x = 3$ in any of the original equations to find $z = 1$. The solution set is $\{(3, 2, 1)\}$.

Concepts	Examples

9.2 Matrix Solution of Linear Systems

Matrix Row Transformations
For any augmented matrix of a system of linear equations, the following row transformations will result in the matrix of an equivalent system.

1. Interchange any two rows.
2. Multiply or divide the elements of any row by a nonzero real number.
3. Replace any row of the matrix by the sum of the elements of that row and a multiple of the elements of another row.

Gauss-Jordan Method
The Gauss-Jordan method is a systematic technique for applying matrix row transformations in an attempt to reduce a matrix to diagonal form, with 1s along the diagonal.

Solve the system.

$$\begin{aligned} x + 3y &= 7 \\ 2x + \ \ y &= 4 \end{aligned}$$

$$\left[\begin{array}{cc|c} 1 & 3 & 7 \\ 2 & 1 & 4 \end{array}\right] \quad \text{Augmented matrix}$$

$$\left[\begin{array}{cc|c} 1 & 3 & 7 \\ 0 & -5 & -10 \end{array}\right] \quad -2\text{R1} + \text{R2}$$

$$\left[\begin{array}{cc|c} 1 & 3 & 7 \\ 0 & 1 & 2 \end{array}\right] \quad -\tfrac{1}{5}\text{R2}$$

$$\left[\begin{array}{cc|c} 1 & 0 & 1 \\ 0 & 1 & 2 \end{array}\right] \quad -3\text{R2} + \text{R1}$$

This leads to

$$x = 1$$
$$y = 2,$$

and the solution set is $\{(1, 2)\}$.

9.3 Determinant Solution of Linear Systems

Determinant of a 2×2 Matrix

If $A = \begin{bmatrix} a_{11} & a_{12} \\ a_{21} & a_{22} \end{bmatrix}$, then

$$|A| = \begin{vmatrix} a_{11} & a_{12} \\ a_{21} & a_{22} \end{vmatrix} = a_{11}a_{22} - a_{21}a_{12}.$$

Evaluate.

$$\begin{vmatrix} 3 & 5 \\ -2 & 6 \end{vmatrix} = 3(6) - (-2)5 = 28$$

Determinant of a 3×3 Matrix

If $A = \begin{bmatrix} a_{11} & a_{12} & a_{13} \\ a_{21} & a_{22} & a_{23} \\ a_{31} & a_{32} & a_{33} \end{bmatrix}$, then

$$|A| = \begin{vmatrix} a_{11} & a_{12} & a_{13} \\ a_{21} & a_{22} & a_{23} \\ a_{31} & a_{32} & a_{33} \end{vmatrix} = (a_{11}a_{22}a_{33} + a_{12}a_{23}a_{31} + a_{13}a_{21}a_{32}) - (a_{31}a_{22}a_{13} + a_{32}a_{23}a_{11} + a_{33}a_{21}a_{12}).$$

In practice, we usually evaluate determinants by expansion by minors.

Evaluate by expanding about the second column.

$$\begin{vmatrix} 2 & -3 & -2 \\ -1 & -4 & -3 \\ -1 & 0 & 2 \end{vmatrix} = -(-3)\begin{vmatrix} -1 & -3 \\ -1 & 2 \end{vmatrix} + (-4)\begin{vmatrix} 2 & -2 \\ -1 & 2 \end{vmatrix} - 0\begin{vmatrix} 2 & -2 \\ -1 & -3 \end{vmatrix}$$

$$= 3(-5) - 4(2) - 0(-8)$$
$$= -15 - 8 + 0$$
$$= -23$$

Properties of Determinants
See the discussion in **Section 9.3.**

(continued)

Concepts	Examples
Cramer's Rule for Two Equations in Two Variables Given the system $$a_1x + b_1y = c_1$$ $$a_2x + b_2y = c_2,$$ if $D \neq 0$, then the system has the unique solution $$x = \frac{D_x}{D} \quad \text{and} \quad y = \frac{D_y}{D},$$ where $D = \begin{vmatrix} a_1 & b_1 \\ a_2 & b_2 \end{vmatrix}$, $D_x = \begin{vmatrix} c_1 & b_1 \\ c_2 & b_2 \end{vmatrix}$, and $D_y = \begin{vmatrix} a_1 & c_1 \\ a_2 & c_2 \end{vmatrix}$.	Solve using Cramer's rule. $$x - 2y = -1$$ $$2x + 5y = 16$$ $$x = \frac{\begin{vmatrix} -1 & -2 \\ 16 & 5 \end{vmatrix}}{\begin{vmatrix} 1 & -2 \\ 2 & 5 \end{vmatrix}} = \frac{-5+32}{5+4} = \frac{27}{9} = 3$$ $$y = \frac{\begin{vmatrix} 1 & -1 \\ 2 & 16 \end{vmatrix}}{\begin{vmatrix} 1 & -2 \\ 2 & 5 \end{vmatrix}} = \frac{16+2}{5+4} = \frac{18}{9} = 2$$ The solution set is $\{(3, 2)\}$.
General Form of Cramer's Rule Let an $n \times n$ system have linear equations of the form $$a_1x_1 + a_2x_2 + a_3x_3 + \cdots + a_nx_n = b.$$ Define D as the determinant of the $n \times n$ matrix of coefficients of the variables. Define D_{x_1} as the determinant obtained from D by replacing the entries in column 1 of D with the constants of the system. Define D_{x_i} as the determinant obtained from D by replacing the entries in column i with the constants of the system. If $D \neq 0$, the unique solution of the system is $$x_1 = \frac{D_{x_1}}{D}, \quad x_2 = \frac{D_{x_2}}{D}, \quad x_3 = \frac{D_{x_3}}{D}, \quad \ldots, \quad x_n = \frac{D_{x_n}}{D}.$$	Solve using Cramer's rule. $$3x + 2y + z = -5$$ $$x - y + 3z = -5$$ $$2x + 3y + z = 0$$ Using the method of expansion by minors, it can be shown that $D_x = 45$, $D_y = -30$, $D_z = 0$, and $D = -15$. Thus, $$x = \frac{D_x}{D} = \frac{45}{-15} = -3, \quad y = \frac{D_y}{D} = \frac{-30}{-15} = 2,$$ $$z = \frac{D_z}{D} = \frac{0}{-15} = 0.$$ The solution set is $\{(-3, 2, 0)\}$.
9.4 Partial Fractions To solve for the constants in the numerators of a partial fraction decomposition, use either of the following methods or a combination of the two. (See **Section 9.4** for more details.) ***Method 1* For Linear Factors** ***Step 1*** Multiply each side by the common denominator. ***Step 2*** Substitute the zero of each factor in the resulting equation. For repeated linear factors, substitute as many other numbers as necessary to find all the constants in the numerators. The number of substitutions required will equal the number of constants $A, B, \ldots$. ***Method 2* For Quadratic Factors** ***Step 1*** Multiply each side by the common denominator. ***Step 2*** Collect like terms on the right side of the resulting equation. ***Step 3*** Equate the coefficients of like terms to get a system of equations. ***Step 4*** Solve the system to find the constants in the numerators.	Decompose $\frac{9}{2x^2 + 9x + 9}$ into partial fractions. $$\frac{9}{2x^2+9x+9} = \frac{9}{(2x+3)(x+3)}$$ $$\frac{9}{(2x+3)(x+3)} = \frac{A}{2x+3} + \frac{B}{x+3} \quad *$$ Multiply by $(2x + 3)(x + 3)$. $$9 = A(x+3) + B(2x+3)$$ $$9 = Ax + 3A + 2Bx + 3B$$ $$9 = (A+2B)x + (3A+3B)$$ Now solve the system $$A + 2B = 0$$ $$3A + 3B = 9$$ to obtain $A = 6$ and $B = -3$. $$\frac{9}{2x^2+9x+9} = \frac{6}{2x+3} + \frac{-3}{x+3}$$ Substitute into *.

Concepts	Examples

9.5 Nonlinear Systems of Equations

Solving a Nonlinear System
A nonlinear system can be solved by the substitution method, the elimination method, or a combination of the two.

Solve the system.

$$x^2 + 2xy - y^2 = 14 \quad (1)$$
$$x^2 - y^2 = -16 \quad (2)$$

$$x^2 + 2xy - y^2 = 14$$
$$-x^2 \qquad + y^2 = 16 \quad \text{Multiply (2) by } -1.$$
$$2xy \qquad = 30 \quad \text{Add to eliminate } x^2 \text{ and } y^2.$$
$$xy = 15 \quad \text{Divide by 2.}$$

Solve for y to obtain $y = \frac{15}{x}$. Substitute into equation (2).

$$x^2 - \left(\frac{15}{x}\right)^2 = -16 \quad (2)$$
$$x^2 - \frac{225}{x^2} = -16 \quad \text{Square.}$$
$$x^4 + 16x^2 - 225 = 0 \quad \text{Multiply by } x^2 \text{ and add } 16x^2.$$
$$(x^2 - 9)(x^2 + 25) = 0 \quad \text{Factor.}$$
$$x = \pm 3 \quad \text{or} \quad x = \pm 5i \quad \text{Zero-factor property}$$

Find corresponding y-values to obtain the solution set.

$$\{(3, 5), (-3, -5), (5i, -3i), (-5i, 3i)\}$$

9.6 Systems of Inequalities and Linear Programming

Graphing an Inequality

Method 1

If the inequality is or can be solved for y, then the following hold.

- The graph of $y < f(x)$ consists of all the points that are *below* the graph of $y = f(x)$.
- The graph of $y > f(x)$ consists of all the points that are *above* the graph of $y = f(x)$.

Method 2

If the inequality is not or cannot be solved for y, choose a test point not on the boundary.

- If the test point satisfies the inequality, the graph includes all points on the *same* side of the boundary as the test point.
- If the test point does not satisfy the inequality, the graph includes all points on the *other* side of the boundary.

Solving Systems of Inequalities
To solve a system of inequalities, graph all inequalities on the same axes, and find the intersection of their solution sets.

Graph $y \geq x^2 - 2x + 3$.

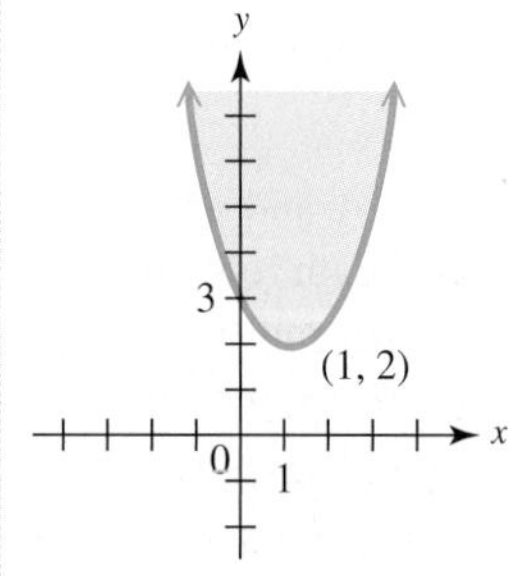

Graph the solution set of the system.

$$3x - 5y > -15$$
$$x^2 + y^2 \leq 25$$

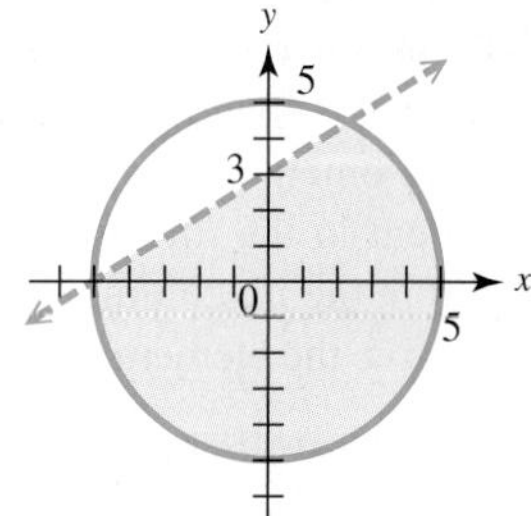

(continued)

Concepts	Examples

Fundamental Theorem of Linear Programming
The optimal value for a linear programming problem occurs at a vertex of the region of feasible solutions.

Solving a Linear Programming Problem

Step 1 Write the objective function and all necessary constraints.

Step 2 Graph the region of feasible solutions.

Step 3 Identify all vertices (corner points).

Step 4 Find the value of the objective function at each vertex.

Step 5 The solution is given by the vertex producing the optimal value of the objective function.

The region of feasible solutions for

$$x + 2y \le 14$$
$$3x + 4y \le 36$$
$$x \ge 0$$
$$y \ge 0$$

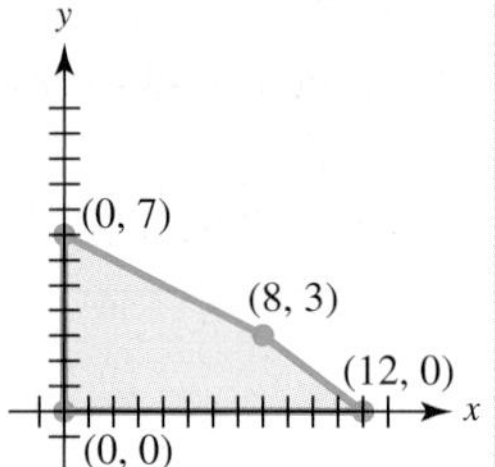

is given in the figure. Maximize the objective function $8x + 12y$.

Vertex Point	Value of $8x + 12y$
(0, 0)	0
(0, 7)	84
(8, 3)	100 ← Maximum
(12, 0)	96

The objective function is maximized for $x = 8$ and $y = 3$.

9.7 Properties of Matrices

Addition and Subtraction of Matrices
To add (subtract) matrices of the same dimension, add (subtract) the corresponding elements.

Find the sum or difference.

$$\begin{bmatrix} 2 & 3 & -1 \\ 0 & 4 & 9 \end{bmatrix} + \begin{bmatrix} -8 & 12 & 1 \\ 5 & 3 & -3 \end{bmatrix} = \begin{bmatrix} -6 & 15 & 0 \\ 5 & 7 & 6 \end{bmatrix}$$

$$\begin{bmatrix} 5 & -1 \\ -8 & 8 \end{bmatrix} - \begin{bmatrix} -2 & 4 \\ 3 & -6 \end{bmatrix} = \begin{bmatrix} 7 & -5 \\ -11 & 14 \end{bmatrix}$$

Scalar Multiplication
To multiply a matrix by a scalar, multiply each element of the matrix by the scalar.

Find the scalar product.

$$3\begin{bmatrix} 6 & 2 \\ 1 & -2 \\ 0 & 8 \end{bmatrix} = \begin{bmatrix} 18 & 6 \\ 3 & -6 \\ 0 & 24 \end{bmatrix}$$

Multiplication of Matrices
The product AB of an $m \times n$ matrix A and an $n \times p$ matrix B is found as follows. To get the ith row, jth column element of the $m \times p$ matrix AB, multiply each element in the ith row of A by the corresponding element in the jth column of B. The sum of these products will give the element of row i, column j of AB.

Find the matrix product.

$$\underset{3 \times 3}{\begin{bmatrix} 1 & -2 & 3 \\ 5 & 0 & 4 \\ -8 & 7 & -7 \end{bmatrix}} \underset{3 \times 1}{\begin{bmatrix} 1 \\ -2 \\ 3 \end{bmatrix}} = \begin{bmatrix} 1(1) + (-2)(-2) + 3(3) \\ 5(1) + 0(-2) + 4(3) \\ -8(1) + 7(-2) + (-7)(3) \end{bmatrix}$$

$$= \underset{3 \times 1}{\begin{bmatrix} 14 \\ 17 \\ -43 \end{bmatrix}}$$

Concepts	Examples
9.8 Matrix Inverses	
Finding an Inverse Matrix To obtain A^{-1} for any $n \times n$ matrix A for which A^{-1} exists, follow these steps. ***Step 1*** Form the augmented matrix $[A\,\vert\,I_n]$, where I_n is the $n \times n$ identity matrix. ***Step 2*** Perform row transformations on $[A\,\vert\,I_n]$ to obtain a matrix of the form $[I_n\,\vert\,B]$. ***Step 3*** Matrix B is A^{-1}.	Find A^{-1} if $A = \begin{bmatrix} 5 & 2 \\ 2 & 1 \end{bmatrix}$. $\left[\begin{array}{cc\|cc} 5 & 2 & 1 & 0 \\ 2 & 1 & 0 & 1 \end{array}\right]$ $\left[\begin{array}{cc\|cc} 1 & 0 & 1 & -2 \\ 2 & 1 & 0 & 1 \end{array}\right]$ $\quad -2R2 + R1$ $\left[\begin{array}{cc\|cc} 1 & 0 & 1 & -2 \\ 0 & 1 & -2 & 5 \end{array}\right]$ $\quad -2R1 + R2$ $I_2 \qquad A^{-1}$ Therefore, $A^{-1} = \begin{bmatrix} 1 & -2 \\ -2 & 5 \end{bmatrix}$.

Chapter 9 Review Exercises

Use the substitution or elimination method to solve each linear system. Identify any inconsistent systems or systems with infinitely many solutions. If the system has infinitely many solutions, write the solution set with y arbitrary.

1. $2x + 6y = 6$
$5x + 9y = 9$

2. $3x - 5y = 7$
$2x + 3y = 30$

3. $x + 5y = 9$
$2x + 10y = 18$

4. $\frac{1}{6}x + \frac{1}{3}y = 8$
$\frac{1}{4}x + \frac{1}{2}y = 12$

5. $y = -x + 3$
$2x + 2y = 1$

6. $0.2x + 0.5y = 6$
$0.4x + y = 9$

7. $3x - 2y = 0$
$9x + 8y = 7$

8. $6x + 10y = -11$
$9x + 6y = -3$

9. $2x - 5y + 3z = -1$
$x + 4y - 2z = 9$
$-x + 2y + 4z = 5$

10. $4x + 3y + z = -8$
$3x + y - z = -6$
$x + y + 2z = -2$

11. $5x - y = 26$
$4y + 3z = -4$
$3x + 3z = 15$

12. $x + z = 2$
$2y - z = 2$
$-x + 2y = -4$

13. *Concept Check* Create your own inconsistent system of two equations.

14. *Connecting Graphs with Equations* Determine the system of equations illustrated in the graph. Write equations in standard form.

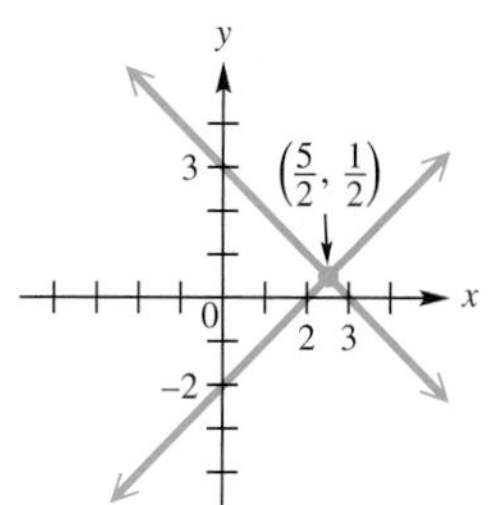

Answers:
1. $\{(0, 1)\}$ 2. $\{(9, 4)\}$
3. $\{(-5y + 9, y)\}$; infinitely many solutions
4. $\{(-2y + 48, y)\}$; infinitely many solutions
5. $\emptyset$; inconsistent system
6. $\emptyset$; inconsistent system
7. $\{(\frac{1}{3}, \frac{1}{2})\}$ 8. $\{(\frac{2}{3}, -\frac{3}{2})\}$
9. $\{(3, 2, 1)\}$ 10. $\{(-2, 0, 0)\}$
11. $\{(5, -1, 0)\}$
12. $\{(4, 0, -2)\}$
13. One possible answer is $x + y = 2$, $x + y = 3$.
14. $x - y = 2$, $x + y = 3$

15. $\frac{1}{3}$ cup of rice, $\frac{1}{5}$ cup of soybeans
16. 80 recordable CDs, 20 play-only CDs
17. 5 blankets, 3 rugs, 8 skirts
18. (a) during 2006
(b) 13.1%
(c) Y = 0.255X + 9.01

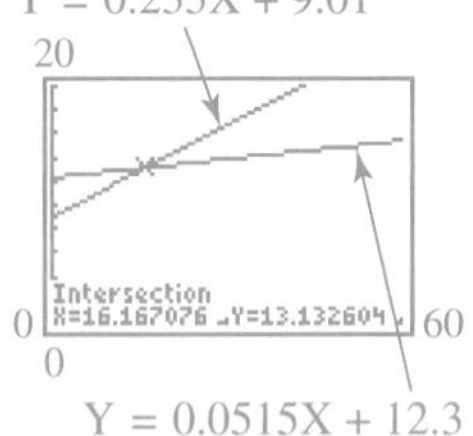

Y = 0.0515X + 12.3

(d) Hispanic

19. $x \approx 177.1$, $y \approx 174.9$; If an athlete's maximum heart rate is 180 beats per minute (bpm), then it will be about 177 bpm 5 sec after stopping and 175 bpm 10 sec after stopping.

20. (a)

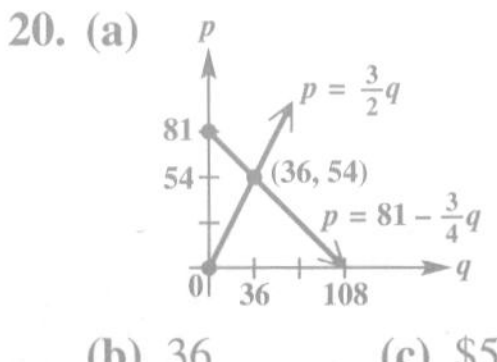

(b) 36 (c) $54

Write a system of linear equations, and then use the system to solve the problem.

15. *Meal Planning* A cup of uncooked rice contains 15 g of protein and 810 calories. A cup of uncooked soybeans contains 22.5 g of protein and 270 calories. How many cups of each should be used for a meal containing 9.5 g of protein and 324 calories?

16. *Order Quantities* A company sells recordable CDs for $0.80 each and play-only CDs for $0.60 each. The company receives $76.00 for an order of 100 CDs. However, the customer neglected to specify how many of each type to send. Determine the number of each type of CD that should be sent.

17. *Indian Weavers* The Waputi Indians make woven blankets, rugs, and skirts. Each blanket requires 24 hr for spinning the yarn, 4 hr for dyeing the yarn, and 15 hr for weaving. Rugs require 30, 5, and 18 hr and skirts 12, 3, and 9 hr, respectively. If there are 306, 59, and 201 hr available for spinning, dyeing, and weaving, respectively, how many of each item can be made? (*Hint:* Simplify the equations you write, if possible, before solving the system.)

18. *(Modeling) Populations of Minorities in the United States* The current and estimated resident populations (in percent) of blacks and Hispanics in the United States for the years 1990–2050 are modeled by the following linear functions.

$$y = 0.0515x + 12.3 \quad \text{Blacks}$$
$$y = 0.255x + 9.01 \quad \text{Hispanics}$$

In each case, x represents the number of years since 1990. (*Source:* U.S. Census Bureau.)

(a) Solve the system to find the year when these population percents will be equal.
(b) What percent of the U.S. resident population will be black or Hispanic in the year found in part (a)?
(c) Use a calculator graph of the system to support your algebraic solution.
(d) Which population is increasing more rapidly? (*Hint:* Consider the slopes of the lines.)

19. *(Modeling) Heart Rate* In a study, a group of athletes was exercised to exhaustion. Let x represent an athlete's heart rate 5 sec after stopping exercise and y this rate 10 sec after stopping. It was found that the maximum heart rate H for these athletes satisfied the two equations

$$H = 0.491x + 0.468y + 11.2$$
$$H = -0.981x + 1.872y + 26.4.$$

If an athlete had maximum heart rate $H = 180$, determine x and y graphically. Interpret your answer. (*Source:* Thomas, V., *Science and Sport,* Faber and Faber.)

20. *(Modeling) Equilibrium Supply and Demand* Let the supply and demand equations for units of backpacks be

$$\text{supply: } p = \frac{3}{2}q \quad \text{and} \quad \text{demand: } p = 81 - \frac{3}{4}q.$$

(a) Graph these equations on the same axes.
(b) Find the equilibrium demand.
(c) Find the equilibrium price.

21. *(Modeling)* Find the equation of the vertical parabola that passes through the points shown in the table.

X	Y1
1	-2.3
2	-1.3
3	4.5
4	15.1
5	30.5
6	50.7
7	75.7

X=1

22. *(Modeling)* Knowing that the equation of a circle may be written in the form

$$x^2 + y^2 + ax + by + c = 0,$$

find the equation of the circle passing through the points $(-3, -7)$, $(4, -8)$, and $(1, 1)$.

Find solutions for each system with the specified arbitrary variable.

23. $\begin{aligned} 3x - 4y + z &= 2 \\ 2x + y &= 1 \end{aligned}$; x

24. $\begin{aligned} 2x - 6y + 4z &= 5 \\ 5x + y - 3z &= 1 \end{aligned}$; z

Use the Gauss-Jordan method to solve each system.

25. $\begin{aligned} 5x + 2y &= -10 \\ 3x - 5y &= -6 \end{aligned}$

26. $\begin{aligned} 2x + 3y &= 10 \\ -3x + y &= 18 \end{aligned}$

27. $\begin{aligned} 3x + y &= -7 \\ x - y &= -5 \end{aligned}$

28. $\begin{aligned} 2x - y + 4z &= -1 \\ -3x + 5y - z &= 5 \\ 2x + 3y + 2z &= 3 \end{aligned}$

29. $\begin{aligned} x - z &= -3 \\ y + z &= 6 \\ 2x - 3z &= -9 \end{aligned}$

30. $\begin{aligned} 2x - y + z &= 4 \\ x + 2y - z &= 0 \\ 3x + y - 2z &= 1 \end{aligned}$

Solve each problem by using the Gauss-Jordan method to solve a system of equations.

31. **Mixing Teas** Three kinds of tea worth \$4.60, \$5.75, and \$6.50 per lb are to be mixed to get 20 lb of tea worth \$5.25 per lb. The amount of \$4.60 tea used is to be equal to the total amount of the other two kinds together. How many pounds of each tea should be used?

32. **Mixing Solutions** A 5% solution of a drug is to be mixed with some 15% solution and some 10% solution to get 20 ml of 8% solution. The amount of 5% solution used must be 2 ml more than the sum of the other two solutions. How many milliliters of each solution should be used?

33. *(Modeling)* **Master's Degrees** During the period 1970–2008, the numbers of master's degrees awarded to both males and females grew, but degrees earned by females grew at a greater rate. If $x = 0$ represents 1970 and $x = 38$ represents 2008, the number of master's degrees earned (in thousands) are closely modeled by the following system.

$$y = 2.81x + 121 \quad \text{Males}$$
$$y = 7.25x + 65.5 \quad \text{Females}$$

Solve the system to find the year in which males and females earned the same number of master's degrees. What was the total number of master's degrees earned in that year? (*Source:* U.S. Census Bureau.)

21. $y = \frac{12}{5}x^2 - \frac{31}{5}x + \frac{3}{2}$, or $y = 2.4x^2 - 6.2x + 1.5$
22. $x^2 + y^2 - 2x + 8y - 8 = 0$
23. $\{(x, -2x + 1, -11x + 6)\}$
24. $\{(\frac{7}{16}z + \frac{11}{32}, \frac{13}{16}z - \frac{23}{32}, z)\}$
25. $\{(-2, 0)\}$
26. $\{(-4, 6)\}$
27. $\{(-3, 2)\}$
28. $\{(0, 1, 0)\}$
29. $\{(0, 3, 3)\}$
30. $\{(\frac{13}{10}, \frac{1}{10}, \frac{3}{2})\}$
31. 10 lb of \$4.60 tea, 8 lb of \$5.75 tea, 2 lb of \$6.50 tea
32. 11 ml of 5%, 3 ml of 15%, 6 ml of 10%
33. 1982; approximately 312 thousand

34. $1\frac{2}{3}$ yr; \$841.67

35. -25 36. -6
37. 0 38. -44
39. -1 40. 0

41. $\left\{\frac{8}{19}\right\}$ 42. $\left\{\frac{1}{31}\right\}$

43. $\{(-4, 2)\}$ 44. $\{(-2, 5)\}$

45. $\emptyset$; inconsistent system

46. $\{(-4, 6, 2)\}$

47. $\{(14, -15, 35)\}$

48. $\left\{\left(\frac{16}{9}, \frac{-9z+8}{18}, z\right)\right\}$; infinitely many solutions

49. $\frac{2}{x-1} - \frac{6}{3x-2}$

50. $\frac{3}{(x-4)^2} - \frac{2}{x-4}$

51. $\frac{1}{x-1} - \frac{x+3}{x^2+2}$

52. $\frac{x+2}{x^2-2} + \frac{2x+1}{(x^2-2)^2}$

53. $\{(-3, 4), (1, 12)\}$

54. $\{(1, 2), (-3, 6)\}$

55. $\{(-4, 1), (-4, -1), (4, -1), (4, 1)\}$

56. $\{(-3, 2), (-3, -2), (3, -2), (3, 2)\}$

57. $\left\{(5, -2), \left(-4, \frac{5}{2}\right)\right\}$

58. $\{(-1, 2), (-2, 1)\}$

59. $\{(-2, 0), (1, 1)\}$

60. $\left\{(3, 2), \left(-7, \frac{24}{7}\right)\right\}$

61. $\left\{\left(\sqrt{3}, 2i\right), \left(\sqrt{3}, -2i\right), \left(-\sqrt{3}, 2i\right), \left(-\sqrt{3}, -2i\right)\right\}$

34. *(Modeling) Comparing Prices* One refrigerator sells for \$700 and uses \$85 worth of electricity per year. A second refrigerator is \$100 more expensive but costs only \$25 per year to run. Assuming that there are no repair costs, the costs to run the refrigerators over a 10-yr period are given by the following system of equations. Here, y represents the total cost in dollars, and x is time in years.

$$y = 700 + 85x$$
$$y = 800 + 25x$$

In how many years will the costs for the two refrigerators be equal? What are the equivalent costs at that time?

Evaluate each determinant.

35. $\begin{vmatrix} -1 & 8 \\ 2 & 9 \end{vmatrix}$ **36.** $\begin{vmatrix} -2 & 4 \\ 0 & 3 \end{vmatrix}$ **37.** $\begin{vmatrix} x & 4x \\ 2x & 8x \end{vmatrix}$

38. $\begin{vmatrix} -2 & 4 & 1 \\ 3 & 0 & 2 \\ -1 & 0 & 3 \end{vmatrix}$ **39.** $\begin{vmatrix} -1 & 2 & 3 \\ 4 & 0 & 3 \\ 5 & -1 & 2 \end{vmatrix}$ **40.** $\begin{vmatrix} -3 & 2 & 7 \\ 6 & -4 & -14 \\ 7 & 1 & 4 \end{vmatrix}$

Solve each determinant equation.

41. $\begin{vmatrix} 3x & 7 \\ -x & 4 \end{vmatrix} = 8$ **42.** $\begin{vmatrix} 6x & 2 & 0 \\ 1 & 5 & 3 \\ x & 2 & -1 \end{vmatrix} = 2x$

Solve each system by Cramer's rule if possible. Identify any inconsistent systems or systems with infinitely many solutions. (Use another method if Cramer's rule cannot be used.)

43. $\begin{aligned} 3x + 7y &= 2 \\ 5x - y &= -22 \end{aligned}$ **44.** $\begin{aligned} 3x + y &= -1 \\ 5x + 4y &= 10 \end{aligned}$ **45.** $\begin{aligned} 6x + y &= -3 \\ 12x + 2y &= 1 \end{aligned}$

46. $\begin{aligned} 3x + 2y + z &= 2 \\ 4x - y + 3z &= -16 \\ x + 3y - z &= 12 \end{aligned}$ **47.** $\begin{aligned} x + y &= -1 \\ 2y + z &= 5 \\ 3x - 2z &= -28 \end{aligned}$ **48.** $\begin{aligned} 5x - 2y - z &= 8 \\ -5x + 2y + z &= -8 \\ x - 4y - 2z &= 0 \end{aligned}$

Find the partial fraction decomposition for each rational expression.

49. $\frac{2}{3x^2 - 5x + 2}$ **50.** $\frac{11 - 2x}{x^2 - 8x + 16}$

51. $\frac{5 - 2x}{(x^2 + 2)(x - 1)}$ **52.** $\frac{x^3 + 2x^2 - 3}{x^4 - 4x^2 + 4}$

Solve each nonlinear system.

53. $\begin{aligned} y &= 2x + 10 \\ x^2 + y &= 13 \end{aligned}$ **54.** $\begin{aligned} x^2 &= 2y - 3 \\ x + y &= 3 \end{aligned}$ **55.** $\begin{aligned} x^2 + y^2 &= 17 \\ 2x^2 - y^2 &= 31 \end{aligned}$

56. $\begin{aligned} 2x^2 + 3y^2 &= 30 \\ x^2 + y^2 &= 13 \end{aligned}$ **57.** $\begin{aligned} xy &= -10 \\ x + 2y &= 1 \end{aligned}$ **58.** $\begin{aligned} xy + 2 &= 0 \\ y - x &= 3 \end{aligned}$

59. $\begin{aligned} x^2 + 2xy + y^2 &= 4 \\ x - 3y &= -2 \end{aligned}$ **60.** $\begin{aligned} x^2 + 2xy &= 15 + 2x \\ xy - 3x + 3 &= 0 \end{aligned}$ **61.** $\begin{aligned} 2x^2 - 3y^2 &= 18 \\ 2x^2 - 2y^2 &= 14 \end{aligned}$

62. $b = \pm 5\sqrt{10}$
63. yes; $\left(\frac{8 - 8\sqrt{41}}{5}, \frac{16 + 4\sqrt{41}}{5}\right)$, $\left(\frac{8 + 8\sqrt{41}}{5}, \frac{16 - 4\sqrt{41}}{5}\right)$
64. $y = 4$
65.

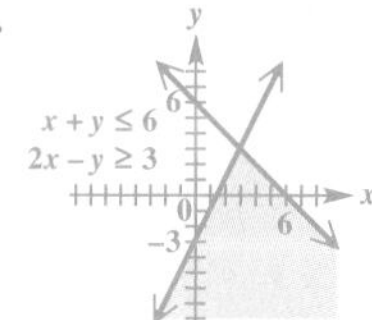

66.

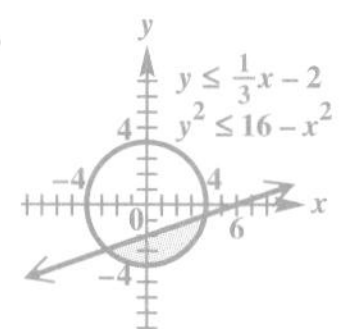

67. maximum of 24 at (0, 6)
68. $x^2 + y \le 1$
$x - 2y < 4$
69. 3 units of food A and 4 units of food B; Minimum cost is \$1.02 per serving.
70. 12 pigs and no geese; Maximum profit is \$240.
71. $a = 5,\ x = \frac{3}{2},\ y = 0,\ z = 9$
72. $k = -8,\ y = 7,\ a = 2,\ m = 3,\ p = 1,\ r = -5$
73. $\begin{bmatrix} -4 \\ 6 \\ 1 \end{bmatrix}$
74. $\begin{bmatrix} 9 & -14 & 13 \\ 21 & -4 & 28 \end{bmatrix}$
75. They cannot be subtracted.
76. $\begin{bmatrix} 11 & 20 \\ 14 & 40 \end{bmatrix}$

62. Find all values of b such that the straight line $3x - y = b$ touches the circle $x^2 + y^2 = 25$ at only one point.

63. Do the circle $x^2 + y^2 = 144$ and the line $x + 2y = 8$ have any points in common? If so, what are they?

64. Find the equation of the line passing through the points of intersection of the graphs of $x^2 + y^2 = 20$ and $x^2 - y = 0$.

Graph the solution set of each system of inequalities.

65. $x + y \le 6$
$2x - y \ge 3$

66. $y \le \frac{1}{3}x - 2$
$y^2 \le 16 - x^2$

67. Find $x \ge 0$ and $y \ge 0$ such that

$$3x + 2y \le 12$$
$$5x + y \ge 5$$

and $2x + 4y$ is maximized.

68. *Connecting Graphs with Equations* Determine the system of inequalities illustrated in the graph. Write them in standard form.

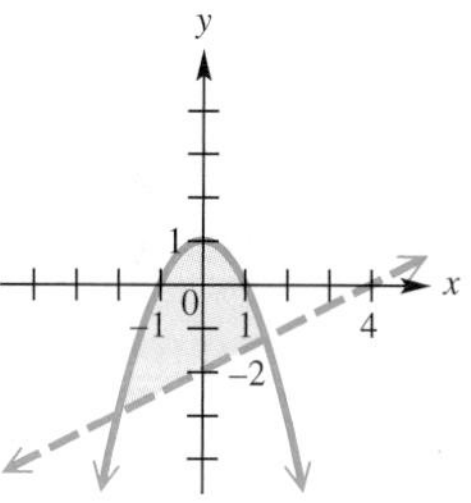

Solve each linear programming problem.

69. *Cost of Nutrients* Certain laboratory animals must have at least 30 g of protein and at least 20 g of fat per feeding period. These nutrients come from food A, which costs \$0.18 per unit and supplies 2 g of protein and 4 g of fat; and from food B, which costs \$0.12 per unit and has 6 g of protein and 2 g of fat. Food B is purchased under a long-term contract requiring that at least 2 units of B be used per serving. How much of each food must be purchased to produce the minimum cost per serving? What is the minimum cost?

70. *Profit from Farm Animals* A 4-H member raises only geese and pigs. She wants to raise no more than 16 animals, including no more than 10 geese. She spends \$5 to raise a goose and \$15 to raise a pig, and she has \$180 available for this project. Each goose produces \$6 in profit, and each pig produces \$20 in profit. How many of each animal should she raise to maximize her profit? What is her maximum profit?

Find the value of each variable.

71. $\begin{bmatrix} 5 & x + 2 \\ -6y & z \end{bmatrix} = \begin{bmatrix} a & 3x - 1 \\ 5y & 9 \end{bmatrix}$

72. $\begin{bmatrix} -6 + k & 2 & a + 3 \\ -2 + m & 3p & 2r \end{bmatrix} + \begin{bmatrix} 3 - 2k & 5 & 7 \\ 5 & 8p & 5r \end{bmatrix} = \begin{bmatrix} 5 & y & 6a \\ 2m & 11 & -35 \end{bmatrix}$

Perform each operation when possible.

73. $\begin{bmatrix} 3 \\ 2 \\ 5 \end{bmatrix} - \begin{bmatrix} 8 \\ -4 \\ 6 \end{bmatrix} + \begin{bmatrix} 1 \\ 0 \\ 2 \end{bmatrix}$

74. $4\begin{bmatrix} 3 & -4 & 2 \\ 5 & -1 & 6 \end{bmatrix} + \begin{bmatrix} -3 & 2 & 5 \\ 1 & 0 & 4 \end{bmatrix}$

75. $\begin{bmatrix} 2 & 5 & 8 \\ 1 & 9 & 2 \end{bmatrix} - \begin{bmatrix} 3 & 4 \\ 7 & 1 \end{bmatrix}$

76. $\begin{bmatrix} -3 & 4 \\ 2 & 8 \end{bmatrix}\begin{bmatrix} -1 & 0 \\ 2 & 5 \end{bmatrix}$

77. $\begin{bmatrix} 3 & -4 \\ 4 & 48 \end{bmatrix}$

78. $\begin{bmatrix} 2 & 12 \\ 12 & 24 \\ -29 & -38 \end{bmatrix}$

79. $\begin{bmatrix} -9 & 3 \\ 10 & 6 \end{bmatrix}$ 80. $\begin{bmatrix} -3 \\ 10 \end{bmatrix}$

81. $\begin{bmatrix} -2 & 5 & -3 \\ 3 & 4 & -4 \\ 6 & -1 & -2 \end{bmatrix}$

82. They cannot be multiplied.

83. $\begin{bmatrix} 3 & -1 \\ -5 & 2 \end{bmatrix}$

84. $\begin{bmatrix} -\frac{1}{4} & \frac{1}{6} \\ 0 & \frac{1}{3} \end{bmatrix}$

85. $\begin{bmatrix} \frac{2}{3} & 0 & -\frac{1}{3} \\ \frac{1}{3} & 0 & -\frac{2}{3} \\ -\frac{2}{3} & 1 & \frac{1}{3} \end{bmatrix}$

86. The inverse does not exist.

87. $\{(\frac{2}{3}, \frac{7}{12})\}$ 88. $\{(2, 1)\}$

89. $\{(-3, 2, 0)\}$ 90. $\{(-1, 0, 2)\}$

77. $\begin{bmatrix} -1 & 0 \\ 2 & 5 \end{bmatrix}\begin{bmatrix} -3 & 4 \\ 2 & 8 \end{bmatrix}$

78. $\begin{bmatrix} 1 & 2 \\ 3 & 0 \\ -6 & 5 \end{bmatrix}\begin{bmatrix} 4 & 8 \\ -1 & 2 \end{bmatrix}$

79. $\begin{bmatrix} 3 & 2 & -1 \\ 4 & 0 & 6 \end{bmatrix}\begin{bmatrix} -2 & 0 \\ 0 & 2 \\ 3 & 1 \end{bmatrix}$

80. $\begin{bmatrix} 1 & -2 & 4 & 2 \\ 0 & 1 & -1 & 8 \end{bmatrix}\begin{bmatrix} -1 \\ 2 \\ 0 \\ 1 \end{bmatrix}$

81. $\begin{bmatrix} -2 & 5 & 5 \\ 0 & 1 & 4 \\ 3 & -4 & -1 \end{bmatrix}\begin{bmatrix} 1 & 0 & -1 \\ -1 & 0 & 0 \\ 1 & 1 & -1 \end{bmatrix}$

82. $\begin{bmatrix} 0 & 1 & 4 \\ 7 & -2 & 9 \\ 10 & 0 & 1 \end{bmatrix}\begin{bmatrix} 3 & 2 & 1 \\ -4 & 7 & 6 \end{bmatrix}$

Find the inverse of each matrix that has an inverse.

83. $\begin{bmatrix} 2 & 1 \\ 5 & 3 \end{bmatrix}$

84. $\begin{bmatrix} -4 & 2 \\ 0 & 3 \end{bmatrix}$

85. $\begin{bmatrix} 2 & -1 & 0 \\ 1 & 0 & 1 \\ 1 & -2 & 0 \end{bmatrix}$

86. $\begin{bmatrix} 2 & 3 & 5 \\ -2 & -3 & -5 \\ 1 & 4 & 2 \end{bmatrix}$

Use the method of matrix inverses to solve each system.

87. $5x - 4y = 1$
$x + 4y = 3$

88. $2x + y = 5$
$3x - 2y = 4$

89. $3x + 2y + z = -5$
$x - y + 3z = -5$
$2x + 3y + z = 0$

90. $x + y + z = 1$
$2x - y = -2$
$3y + z = 2$

Chapter 9 Test

[9.1]

1. $\{(4, 3)\}$

2. $\{(\frac{-3y-7}{2}, y)\}$; infinitely many solutions

3. $\{(1, 2)\}$

4. $\emptyset$; inconsistent system

5. $\{(2, 0, -1)\}$

[9.2]

6. $\{(5, 1)\}$

7. $\{(5, 3, 6)\}$

Use substitution or elimination to solve each system. Identify any system that is inconsistent or has infinitely many solutions. If a system has infinitely many solutions, express the solution set with y arbitrary.

1. $3x - y = 9$
$x + 2y = 10$

2. $6x + 9y = -21$
$4x + 6y = -14$

3. $\frac{1}{4}x - \frac{1}{3}y = -\frac{5}{12}$
$\frac{1}{10}x + \frac{1}{5}y = \frac{1}{2}$

4. $x - 2y = 4$
$-2x + 4y = 6$

5. $2x + y + z = 3$
$x + 2y - z = 3$
$3x - y + z = 5$

Use the Gauss-Jordan method to solve each system.

6. $3x - 2y = 13$
$4x - y = 19$

7. $3x - 4y + 2z = 15$
$2x - y + z = 13$
$x + 2y - z = 5$

[9.1]
8. $y = 2x^2 - 8x + 11$
9. 22 units from Toronto, 56 units from Montreal, 22 units from Ottawa

[9.3]
10. -58
11. -844
12. $\{(-6, 7)\}$
13. $\{(1, -2, 3)\}$

[9.4]
14. $\frac{2}{x} + \frac{-2}{x+1} + \frac{-1}{(x+1)^2}$

[9.5]
15. $y = x^2 - 4x + 4$
$x - 3y = -2$
16. $\{(1, 2), (-1, 2), (1, -2), (-1, -2)\}$
17. $\{(3, 4), (4, 3)\}$
18. 5 and -6

[9.6]
19.

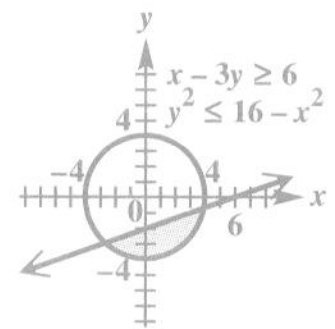

20. maximum of 42 at (12, 6)
21. 0 VIP rings and 24 SST rings; Maximum profit is \$960.

8. *Connecting Graphs with Equations* Find the equation that defines the parabola shown.

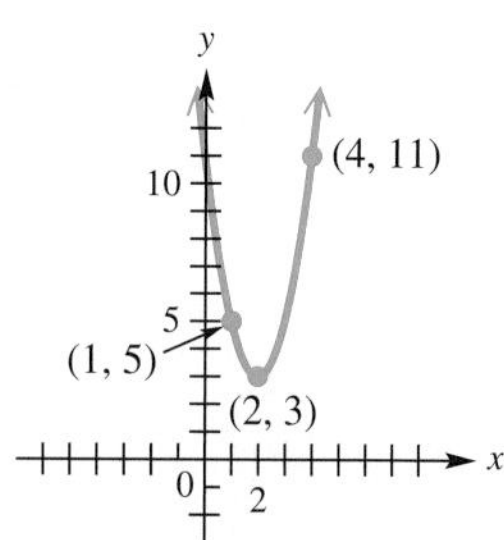

9. *Ordering Supplies* A knitting shop orders yarn from three suppliers in Toronto, Montreal, and Ottawa. One month the shop ordered a total of 100 units of yarn from these suppliers. The delivery costs were \$80, \$50, and \$65 per unit for the orders from Toronto, Montreal, and Ottawa, respectively, with total delivery costs of \$5990. The shop ordered the same amount from Toronto and Ottawa. How many units were ordered from each supplier?

Evaluate each determinant.

10. $\begin{vmatrix} 6 & 8 \\ 2 & -7 \end{vmatrix}$

11. $\begin{vmatrix} 2 & 0 & 8 \\ -1 & 7 & 9 \\ 12 & 5 & -3 \end{vmatrix}$

Solve each system by Cramer's rule.

12. $2x - 3y = -33$
$4x + 5y = 11$

13. $x + y - z = -4$
$2x - 3y - z = 5$
$x + 2y + 2z = 3$

14. Find the partial fraction decomposition of $\dfrac{x+2}{x^3 + 2x^2 + x}$.

15. *Connecting Graphs with Equations* Determine the system of equations illustrated in the graph. Write equations in standard form.

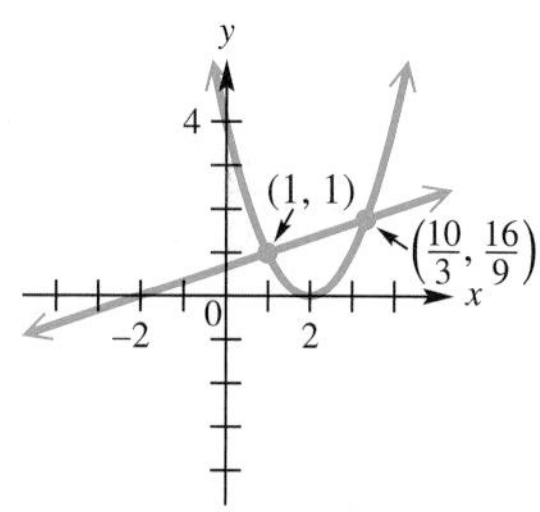

Solve each nonlinear system of equations.

16. $2x^2 + y^2 = 6$
$x^2 - 4y^2 = -15$

17. $x^2 + y^2 = 25$
$x + y = 7$

18. *Unknown Numbers* Find two numbers such that their sum is -1 and the sum of their squares is 61.

19. Graph the solution set.

$$x - 3y \geq 6$$
$$y^2 \leq 16 - x^2$$

20. Find $x \geq 0$ and $y \geq 0$ such that

$$x + 2y \leq 24$$
$$3x + 4y \leq 60$$

and $2x + 3y$ is maximized.

21. *Jewelry Profits* The J. J. Gravois Company designs and sells two types of rings: the VIP and the SST. The company can produce up to 24 rings each day using up to 60 total hours of labor. It takes 3 hr to make one VIP ring and 2 hr to make one SST ring. How many of each type of ring should be made daily in order to maximize the company's profit, if the profit on one VIP ring is \$30 and the profit on one SST ring is \$40? What is the maximum profit?

22. Find the value of each variable in the equation $\begin{bmatrix} 5 & x+6 \\ 0 & 4 \end{bmatrix} = \begin{bmatrix} y-2 & 4-x \\ 0 & w+7 \end{bmatrix}$.

Perform each operation when possible.

23. $3\begin{bmatrix} 2 & 3 \\ 1 & -4 \\ 5 & 9 \end{bmatrix} - \begin{bmatrix} -2 & 6 \\ 3 & -1 \\ 0 & 8 \end{bmatrix}$

24. $\begin{bmatrix} 1 \\ 2 \end{bmatrix} + \begin{bmatrix} 4 \\ -6 \end{bmatrix} + \begin{bmatrix} 2 & 8 \\ -7 & 5 \end{bmatrix}$

25. $\begin{bmatrix} 2 & 1 & -3 \\ 4 & 0 & 5 \end{bmatrix}\begin{bmatrix} 1 & 3 \\ 2 & 4 \\ 3 & -2 \end{bmatrix}$

26. $\begin{bmatrix} 2 & -4 \\ 3 & 5 \end{bmatrix}\begin{bmatrix} 4 \\ 2 \\ 7 \end{bmatrix}$

27. *Concept Check* Which of the following properties does not apply to multiplication of matrices?

A. commutative **B.** associative **C.** distributive **D.** identity

Find the inverse, if it exists, of each matrix.

28. $\begin{bmatrix} -8 & 5 \\ 3 & -2 \end{bmatrix}$

29. $\begin{bmatrix} 4 & 12 \\ 2 & 6 \end{bmatrix}$

30. $\begin{bmatrix} 1 & 3 & 4 \\ 2 & 7 & 8 \\ -2 & -5 & -7 \end{bmatrix}$

Use matrix inverses to solve each system.

31. $2x + y = -6$
$3x - y = -29$

32. $x + y = 5$
$y - 2z = 23$
$x + 3z = -27$

[9.7]

22. $x = -1; y = 7; w = -3$

23. $\begin{bmatrix} 8 & 3 \\ 0 & -11 \\ 15 & 19 \end{bmatrix}$

24. The matrices cannot be added.

25. $\begin{bmatrix} -5 & 16 \\ 19 & 2 \end{bmatrix}$

26. The matrices cannot be multiplied.

27. A

[9.8]

28. $\begin{bmatrix} -2 & -5 \\ -3 & -8 \end{bmatrix}$

29. The inverse does not exist.

30. $\begin{bmatrix} -9 & 1 & -4 \\ -2 & 1 & 0 \\ 4 & -1 & 1 \end{bmatrix}$

31. $\{(-7, 8)\}$

32. $\{(0, 5, -9)\}$

10 Analytic Geometry

When a portion of this *circular* ferris wheel is viewed from an angle, it appears to be *parabolic*, and these shapes are examples of *conic sections*.

10.1 Parabolas

- Conic Sections
- Horizontal Parabolas
- Geometric Definition and Equations of Parabolas
- An Application of Parabolas

Conic Sections *Parabolas, circles, ellipses,* and *hyperbolas* form a group of curves known as **conic sections,** because they are the results of intersecting a cone with a plane. See **Figure 1.** We studied circles and some parabolas (those that open up or down—that is, vertical parabolas) in **Chapters 2 and 3.**

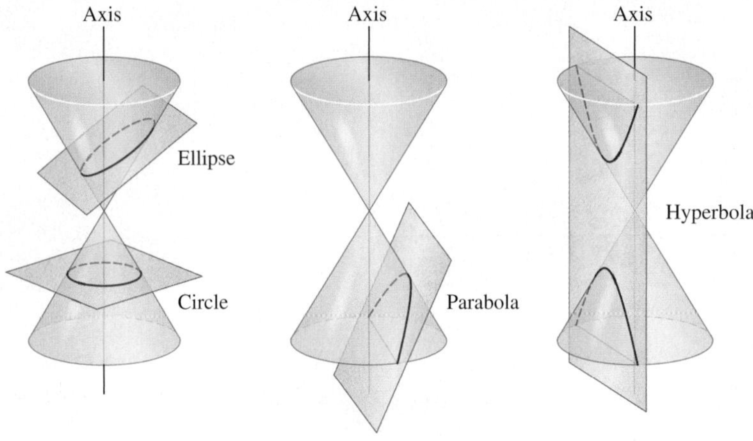

Figure 1

Horizontal Parabolas We know that the graph of the equation

$$y = a(x - h)^2 + k \quad \text{(Section 3.1)}$$

is a parabola with vertex (h, k) and the vertical line $x = h$ as its axis of symmetry. If we subtract k from each side and interchange the roles of $x - h$ and $y - k$, the new equation also has a parabola as its graph.

$$y - k = a(x - h)^2 \quad \text{Subtract } k. \quad (1)$$

$$x - h = a(y - k)^2 \quad \text{Interchange the roles of } x - h \text{ and } y - k. \quad (2)$$

While the graph of $y - k = a(x - h)^2$ has a *vertical* axis of symmetry, the graph of $x - h = a(y - k)^2$ has a *horizontal* axis of symmetry. The graph of the first equation is the graph of a function (specifically a quadratic function), while the graph of the second equation is not. Its graph fails the vertical line test.

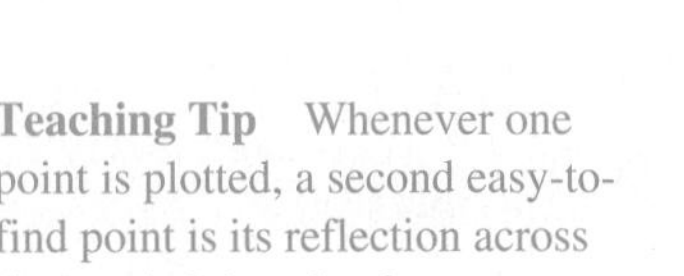

Teaching Tip Whenever one point is plotted, a second easy-to-find point is its reflection across the parabola's axis of symmetry. For example, after the x-intercept in **Example 1** is found and the point $(1, 0)$ is plotted, the point $(1, 4)$ can be obtained using symmetry with respect to the axis $y = 2$.

Parabola with Horizontal Axis of Symmetry

The parabola with vertex (h, k) and the horizontal line $y = k$ as axis of symmetry has an equation of the following form.

$$x - h = a(y - k)^2$$

The parabola opens to the right if $a > 0$ and to the left if $a < 0$.

NOTE When the vertex (h, k) is $(0, 0)$ and $a = 1$ in

$$y - k = a(x - h)^2 \quad (1)$$

and

$$x - h = a(y - k)^2, \quad (2)$$

the equations $y = x^2$ and $x = y^2$, respectively, result. See **Figure 2.** Notice that the graphs are mirror images of each other with respect to the line $y = x$.

$y = x^2$, $y = x$, $x = y^2$

Figure 2

Classroom Example 1
Graph $x - 4 = -(y + 2)^2$. Give the domain and range.

Answer:

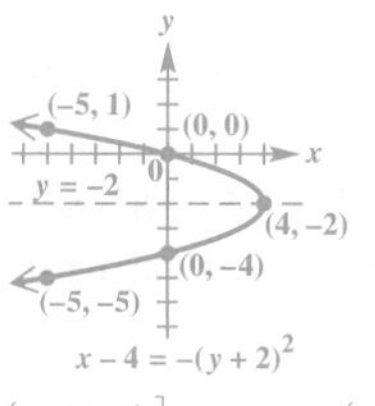

domain: $(-\infty, 4]$; range: $(-\infty, \infty)$

EXAMPLE 1 Graphing a Parabola with Horizontal Axis of Symmetry

Graph $x + 3 = (y - 2)^2$. Give the domain and range.

SOLUTION The graph of $x + 3 = (y - 2)^2$, or $x - (-3) = (y - 2)^2$, has vertex $(-3, 2)$ and opens to the right because $a = 1$, and $1 > 0$. Plotting a few additional points gives the graph shown in **Figure 3.** Note that the graph is symmetric about its axis, $y = 2$. The domain is $[-3, \infty)$, and the range is $(-\infty, \infty)$.

x	y
−3	2
−2	3
−2	1
1	4
1	0

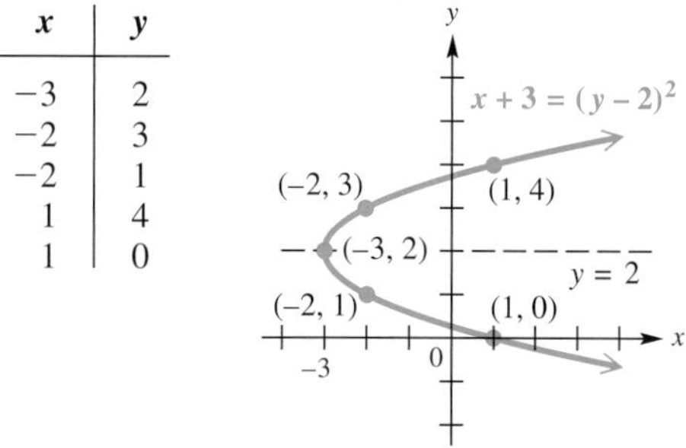

Figure 3

Now Try Exercise 7.

Classroom Example 2 is on the next page.

EXAMPLE 2 Graphing a Parabola with Horizontal Axis of Symmetry

Graph $x = 2y^2 + 6y + 5$. Give the domain and range.

ALGEBRAIC SOLUTION

$$x = 2y^2 + 6y + 5$$

$$x = 2(y^2 + 3y \qquad) + 5 \quad \text{Factor out 2.}$$

$$x = 2\left(y^2 + 3y + \frac{9}{4} - \frac{9}{4}\right) + 5$$

Complete the square; $\left[\frac{1}{2}(3)\right]^2 = \frac{9}{4}$. **(Section 1.4)**

$$x = 2\left(y^2 + 3y + \frac{9}{4}\right) + 2\left(-\frac{9}{4}\right) + 5$$

Distributive property **(Section R.2)**

$$x = 2\left(y + \frac{3}{2}\right)^2 + \frac{1}{2} \quad \text{Factor (Section R.4); simplify.}$$

$$x - \frac{1}{2} = 2\left(y + \frac{3}{2}\right)^2 \quad \text{Subtract } \tfrac{1}{2}. \ (*)$$

The vertex of the parabola is $\left(\frac{1}{2}, -\frac{3}{2}\right)$. The axis is the horizontal line $y = -\frac{3}{2}$. Using the vertex and the axis and plotting a few additional points gives the graph in **Figure 4.** Let $y = 0$ to find that the x-intercept is $(5, 0)$, and because of symmetry, the point $(5, -3)$ also lies on the graph. The domain is $\left[\frac{1}{2}, \infty\right)$, and the range is $(-\infty, \infty)$.

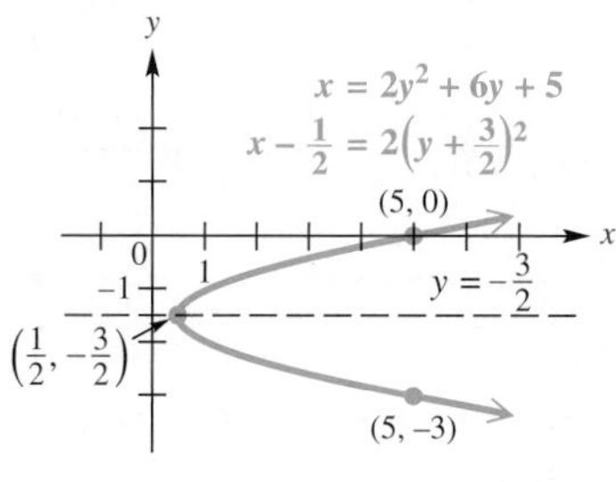

Figure 4

GRAPHING CALCULATOR SOLUTION

Since a horizontal parabola is *not* the graph of a function, to graph it using a graphing calculator in function mode we can write two equations by solving for y.

$$x - \frac{1}{2} = 2\left(y + \frac{3}{2}\right)^2 \quad \text{(*) from algebraic solution}$$

$$x - 0.5 = 2(y + 1.5)^2 \quad \text{Write with decimals.}$$

$$\frac{x - 0.5}{2} = (y + 1.5)^2 \quad \text{Divide by 2.}$$

$$\pm\sqrt{\frac{x - 0.5}{2}} = y + 1.5 \quad \text{Take the square root on each side. (Section 1.4)}$$

$$y = -1.5 \pm \sqrt{\frac{x - 0.5}{2}} \quad \text{Subtract 1.5 and rewrite.}$$

Figure 5 shows the graphs of the two functions defined in the final equation. Their union is the graph of $x = 2y^2 + 6y + 5$.

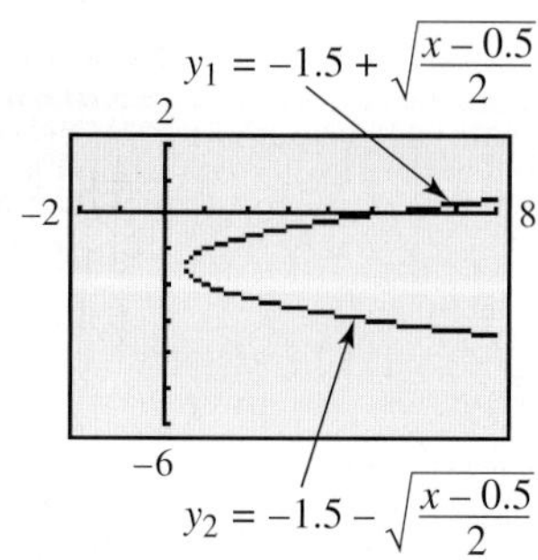

Figure 5

Now Try Exercise 15.

Classroom Example 2
Graph $x = \frac{1}{2}y^2 - 4y + 6$. Give the domain and range.

Answer:

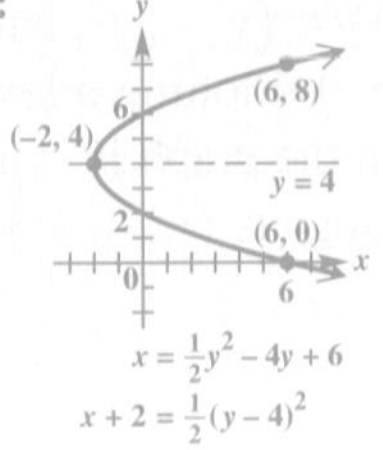

domain: $[-2, \infty)$; range: $(-\infty, \infty)$

Geometric Definition and Equations of Parabolas The equation of a parabola comes from the geometric definition of a parabola as a set of points.

Parabola

A **parabola** is the set of points in a plane equidistant from a fixed point and a fixed line. The fixed point is the **focus,** and the fixed line is the **directrix** of the parabola.

As shown in **Figure 6,** the axis of symmetry of a parabola passes through the focus and is perpendicular to the directrix. The vertex is the midpoint of the line segment joining the focus and directrix on the axis.

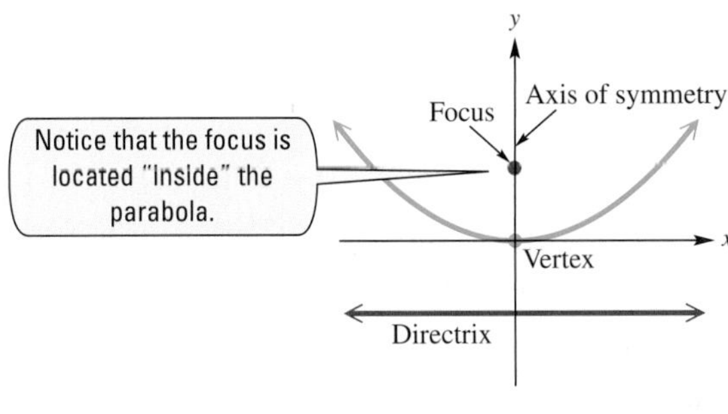

Figure 6

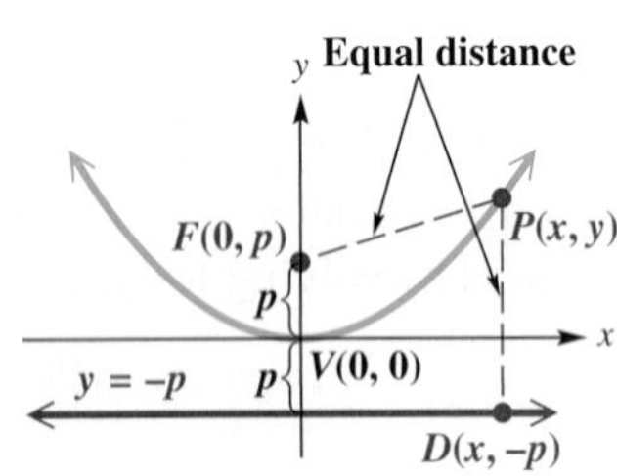

$d(P, F) = d(P, D)$ for all P on the parabola.

Figure 7

We can find an equation of a parabola from the preceding definition. **Let p represent the directed distance from the vertex to the focus.** Then the directrix is the line $y = -p$ and the focus is the point $F(0, p)$, as shown in **Figure 7.** To find the equation of the set of points that are the same distance from the line $y = -p$ and the point $(0, p)$, choose one such point P and give it coordinates (x, y). Since $d(P, F)$ and $d(P, D)$ must be equal, using the distance formula gives

$$d(P, F) = d(P, D)$$

$$\sqrt{(x - 0)^2 + (y - p)^2} = \sqrt{(x - x)^2 + (y - (-p))^2} \quad \text{Distance formula (Section 2.1)}$$

$$\sqrt{x^2 + (y - p)^2} = \sqrt{(y + p)^2} \quad \text{Simplify.}$$

$$x^2 + y^2 - 2yp + p^2 = y^2 + 2yp + p^2 \quad \text{Square each side (Section 1.6) and multiply. (Section R.3)}$$

Remember the middle terms.

$$x^2 = 4py. \quad \text{Simplify.}$$

From this result, if the given form of the equation is $y = ax^2$, then $a = \frac{1}{4p}$.

Parabola with Vertical Axis of Symmetry and Vertex (0, 0)

The parabola with focus $(0, p)$ and directrix $y = -p$ has the following equation.

$$x^2 = 4py$$

This parabola has vertical axis of symmetry $x = 0$ and opens up if $p > 0$ or down if $p < 0$.

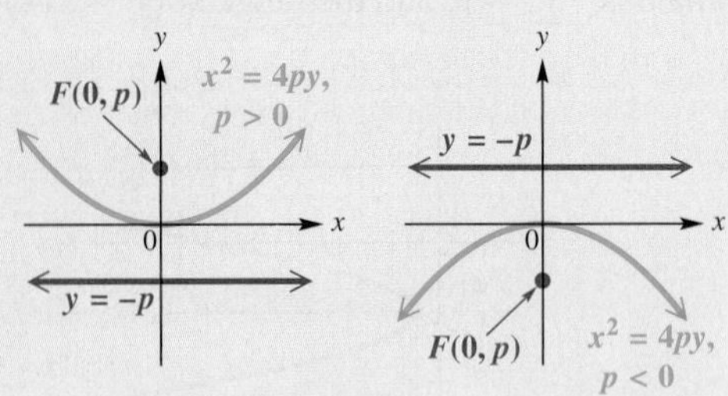

If the directrix is the line $x = -p$ and the focus is $(p, 0)$, a similar procedure leads to the equation of a parabola with a horizontal axis of symmetry.

Parabola with Horizontal Axis of Symmetry and Vertex (0, 0)

The parabola with focus $(p, 0)$ and directrix $x = -p$ has the following equation.

$$y^2 = 4px$$

This parabola has horizontal axis of symmetry $y = 0$ and opens to the right if $p > 0$ or to the left if $p < 0$.

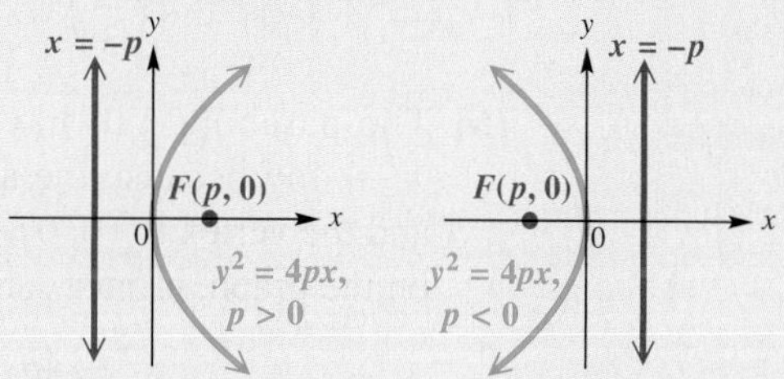

Classroom Example 3
Find the focus, directrix, vertex, and axis of symmetry of each parabola. Then use this information to graph the parabola.
(a) $x^2 = -4y$ (b) $y^2 = 12x$

Answers:
(a) focus: $(0, -1)$; directrix: $y = 1$; vertex: $(0, 0)$; axis: $x = 0$

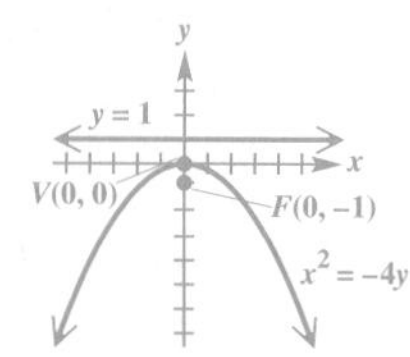

(b) focus: $(3, 0)$; directrix: $x = -3$; vertex: $(0, 0)$; axis: $y = 0$

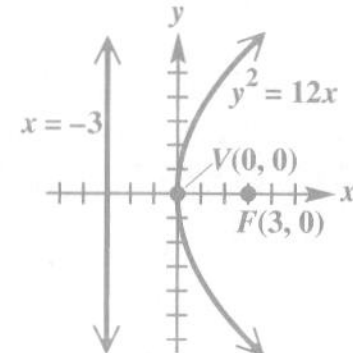

EXAMPLE 3 Graphing Parabolas

Find the focus, directrix, vertex, and axis of symmetry of each parabola. Then use this information to graph the parabola.

(a) $x^2 = 8y$ (b) $y^2 = -28x$

SOLUTION

(a) The equation $x^2 = 8y$ has the form

$$x^2 = 4py, \quad \text{with } 4p = 8, \text{ so} \quad p = 2.$$

Since the x-term is squared, the parabola is vertical, with focus $(0, p) = (0, 2)$ and directrix $y = -2$. The vertex is $(0, 0)$, and the axis of symmetry is the y-axis. See **Figure 8.**

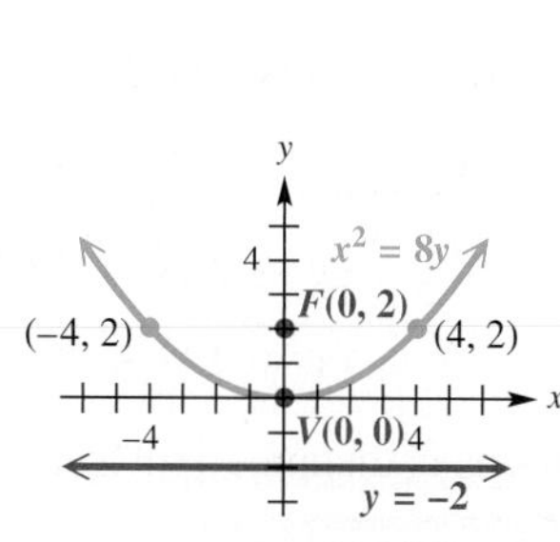

Figure 8

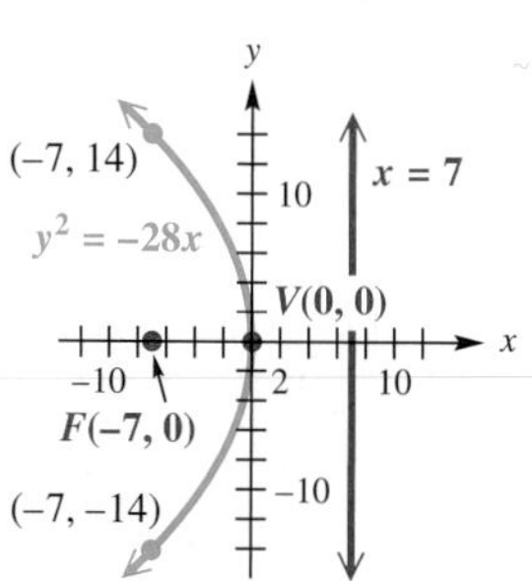

Figure 9

(b) The equation $y^2 = -28x$ has the form

$$y^2 = 4px, \quad \text{with } 4p = -28, \text{ so} \quad p = -7.$$

The parabola is horizontal, with focus $(-7, 0)$, directrix $x = 7$, vertex $(0, 0)$, and the x-axis as axis of symmetry. Since p is negative, the graph opens to the left, as shown in **Figure 9.**

✔ *Now Try Exercises 21 and 25.*

Classroom Example 4
Write an equation for each parabola.
(a) focus $\left(0, \frac{3}{4}\right)$ and vertex at the origin
(b) horizontal axis of symmetry, vertex at the origin, through the point $(-4, -8)$

Answers:
(a) $x^2 = 3y$, or $y = \frac{1}{3}x^2$
(b) $y^2 = -16x$

EXAMPLE 4 Writing Equations of Parabolas

Write an equation for each parabola.

(a) focus $\left(\frac{2}{3}, 0\right)$ and vertex at the origin

(b) vertical axis of symmetry, vertex at the origin, through the point $(-2, 12)$

SOLUTION

(a) Since the focus $\left(\frac{2}{3}, 0\right)$ and the vertex $(0, 0)$ are both on the x-axis, the parabola is horizontal. It opens to the right because $p = \frac{2}{3}$ is positive. See **Figure 10.** The equation, which will have the form $y^2 = 4px$, is

$$y^2 = 4\left(\frac{2}{3}\right)x, \quad \text{or} \quad y^2 = \frac{8}{3}x.$$

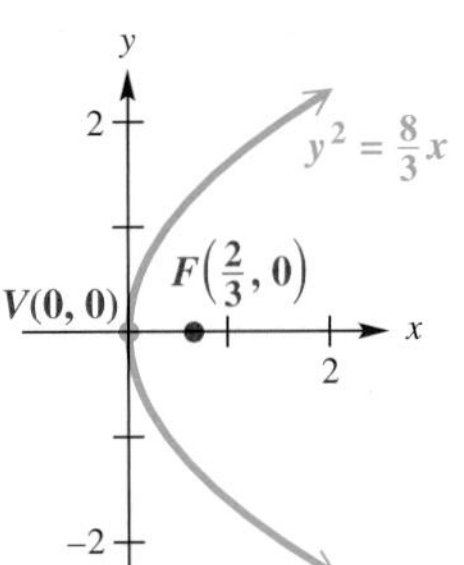

Figure 10

(b) The parabola will have an equation of the form $x^2 = 4py$ because the axis of symmetry is vertical and the vertex is $(0, 0)$. Since the point $(-2, 12)$ is on the graph, it must satisfy the equation.

$$x^2 = 4py$$

$$(-2)^2 = 4p(12) \quad \text{Let } x = -2 \text{ and } y = 12.$$

$$4 = 48p \quad \text{Apply the exponent; multiply.}$$

$$p = \frac{1}{12} \quad \text{Solve for } p. \text{ (Section 1.1)}$$

Then,

$$x^2 = 4\left(\frac{1}{12}\right)y \quad \text{Let } p = \tfrac{1}{12} \text{ in the form } x^2 = 4py.$$

$$x^2 = \frac{1}{3}y, \quad \text{or} \quad y = 3x^2.$$

✔ *Now Try Exercises 33 and 37.*

Teaching Tip When writing an equation of a parabola, encourage students to draw a sketch to represent the given information. This will help them determine the form of the equation that applies.

Classroom Example 5
Write an equation for the parabola with vertex $(-2, 3)$ and focus $(-2, 2)$, and graph it. Give the domain and range.

Answer: $(x + 2)^2 = -4(y - 3)$

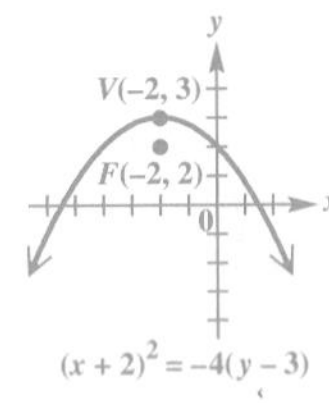

domain: $(-\infty, \infty)$; range: $(-\infty, 3]$

The equations $x^2 = 4py$ and $y^2 = 4px$ can be extended to parabolas having vertex (h, k) by replacing x and y with $x - h$ and $y - k$, respectively.

Equation Forms for Translated Parabolas

A parabola with vertex (h, k) has an equation of the following form.

$$(x - h)^2 = 4p(y - k) \quad \text{Vertical axis of symmetry}$$

or

$$(y - k)^2 = 4p(x - h) \quad \text{Horizontal axis of symmetry}$$

The focus is distance $|p|$ from the vertex.

EXAMPLE 5 Writing an Equation of a Parabola

Write an equation for the parabola with vertex $(1, 3)$ and focus $(-1, 3)$, and graph it. Give the domain and range.

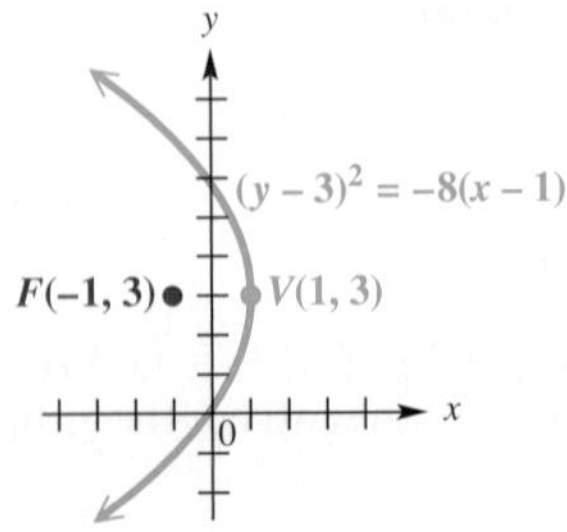

Figure 11

SOLUTION Since the focus is to the left of the vertex, the axis of symmetry is horizontal and the parabola opens to the left. See **Figure 11.** The directed distance between the vertex and the focus is $-1 - 1$, or -2, so $p = -2$ (since the parabola opens to the left).

The equation of the parabola is found as follows.

$$(y-k)^2 = 4p(x-h) \quad \text{Parabola with horizontal axis of symmetry}$$

$$(y-3)^2 = 4(-2)(x-1) \quad \text{Substitute for } p, h, \text{ and } k.$$

$$(y-3)^2 = -8(x-1) \quad \text{Multiply.}$$

The domain is $(-\infty, 1]$, and the range is $(-\infty, \infty)$.

✔ Now Try Exercise 41.

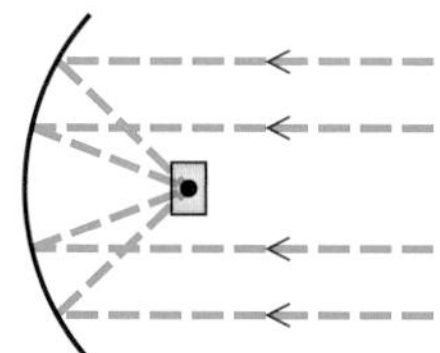

Solar furnace

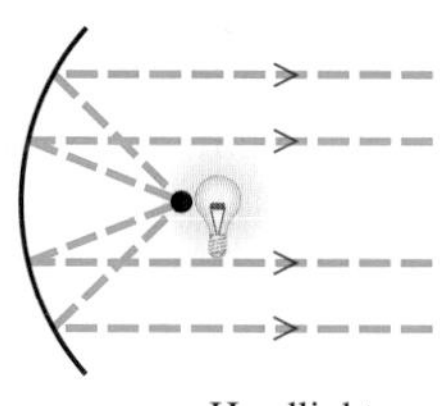

Headlight

An Application of Parabolas Parabolas have a special reflecting property that makes them useful in the design of telescopes, radar equipment, auto headlights, and solar furnaces. When a ray of light or a sound wave traveling parallel to the axis of symmetry of a parabolic shape bounces off the parabola, it passes through the focus.

For example, in a solar furnace, a parabolic mirror collects light at the focus and thereby generates intense heat at that point. If a light source is placed at the focus, then the reflected light rays will be directed straight ahead.

EXAMPLE 6 Modeling the Reflective Property of Parabolas

The Parkes radio telescope has a parabolic dish shape with diameter 210 ft and depth 32 ft. Because of this parabolic shape, distant rays hitting the dish will be reflected directly toward the focus. A cross section of the dish is shown in **Figure 12.** (*Source:* Mar, J., and H. Liebowitz, *Structure Technology for Large Radio and Radar Telescope Systems,* The MIT Press, Massachusetts Institute of Technology.)

Focus
210 ft
32 ft

Figure 12

(a) Determine an equation that models this cross section by placing the vertex at the origin with the parabola opening up.

(b) The receiver must be placed at the focus of the parabola. How far from the vertex of the parabolic dish should the receiver be located?

SOLUTION

(a) Locate the vertex at the origin as shown in **Figure 13.** The form of the parabola is $x^2 = 4py$. The parabola must pass through the point $\left(\frac{210}{2}, 32\right)$, or $(105, 32)$. Use this information to solve for p.

$$x^2 = 4py \quad \text{Parabola with vertical axis of symmetry}$$

$$(105)^2 = 4p(32) \quad \text{Let } x = 105 \text{ and } y = 32.$$

$$11{,}025 = 128p \quad \text{Multiply.}$$

$$p = \frac{11{,}025}{128} \quad \text{Solve for } p.$$

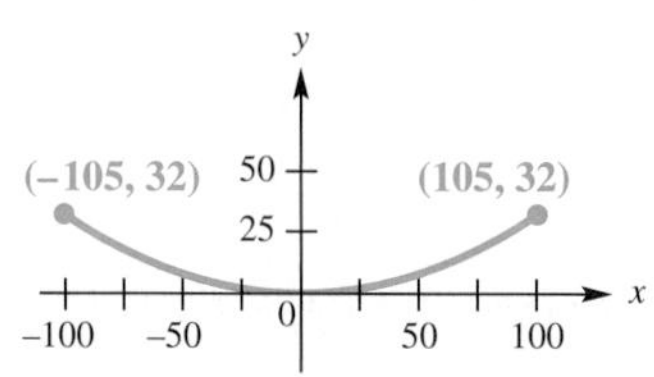

Figure 13

The cross section can be modeled by the following equation.

$$x^2 = 4py$$

$$x^2 = 4\left(\frac{11{,}025}{128}\right)y \quad \text{Substitute for } p.$$

$$x^2 = \frac{11{,}025}{32}y \quad \text{Simplify.}$$

(b) The distance between the vertex and the focus is p. In part (a), we found $p = \frac{11{,}025}{128} \approx 86.1$, so the receiver should be located at $(0, 86.1)$ or 86.1 ft above the vertex.

✔ Now Try Exercise 53.

Classroom Example 6

(a) Using the coordinate system shown in **Figure 13,** find the equation of the directrix of the parabola.

(b) What is the width of the parabolic dish 25 ft above the vertex?

Answers:

(a) $y = -86.1$ **(b)** about 185.6 ft

10.1 Exercises

1. (a) D (b) B (c) C (d) A (e) F (f) H (g) E (h) G
2. (a) B (b) D (c) A (d) C

In Exercises 3–18, we give the domain and then the range.

3. $[-4, \infty)$; $(-\infty, \infty)$
4. $[2, \infty)$; $(-\infty, \infty)$

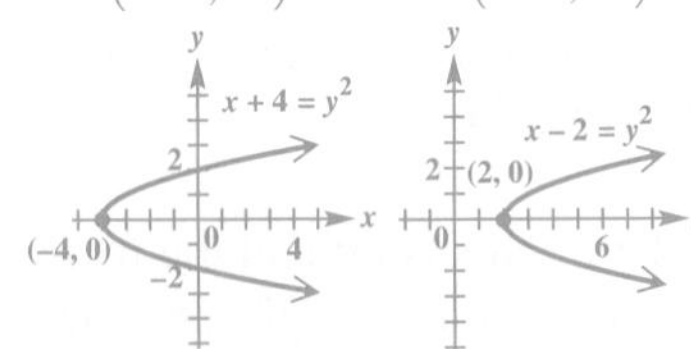

5. $[0, \infty)$; $(-\infty, \infty)$
6. $[0, \infty)$; $(-\infty, \infty)$

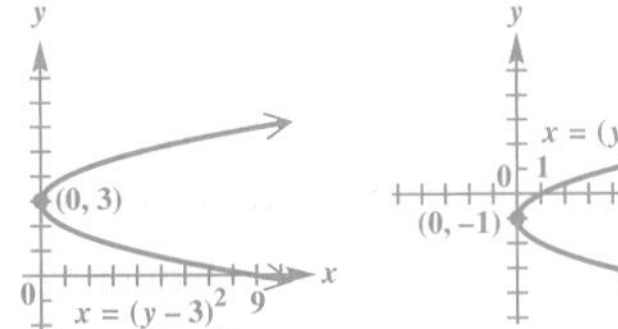

7. $[2, \infty)$; $(-\infty, \infty)$
8. $[-1, \infty)$; $(-\infty, \infty)$

y
(2, 4)
0
$x - 2 = (y - 4)^2$
x
y
$x + 1 = (y + 2)^2$
0
3
(−1, −2)
x

9. $(-\infty, 2]$; $(-\infty, \infty)$
10. $[4, \infty)$; $(-\infty, \infty)$

y
(2, 1)
−1
0
x
$x - 2 = -3(y - 1)^2$
y
(4, 1)
0
x
$x - 4 = \frac{1}{2}(y - 1)^2$

11. $(-\infty, 0]$; $(-\infty, \infty)$

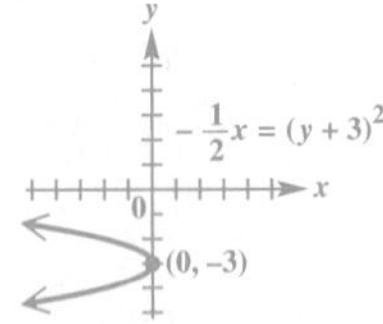

12. $(-\infty, 0]$; $(-\infty, \infty)$

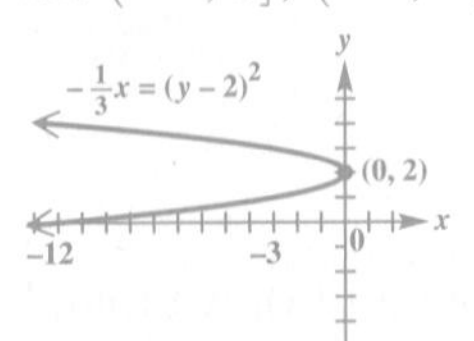

1. *Concept Check* Match each equation of a parabola in Column I with its description in Column II.

I	II
(a) $y - 2 = (x + 4)^2$	A. vertex $(-2, 4)$; opens down
(b) $y - 4 = (x + 2)^2$	B. vertex $(-2, 4)$; opens up
(c) $y - 2 = -(x + 4)^2$	C. vertex $(-4, 2)$; opens down
(d) $y - 4 = -(x + 2)^2$	D. vertex $(-4, 2)$; opens up
(e) $x - 2 = (y + 4)^2$	E. vertex $(2, -4)$; opens left
(f) $x - 4 = (y + 2)^2$	F. vertex $(2, -4)$; opens right
(g) $x - 2 = -(y + 4)^2$	G. vertex $(4, -2)$; opens left
(h) $x - 4 = -(y + 2)^2$	H. vertex $(4, -2)$; opens right

2. *Concept Check* Match each equation of a parabola in Column I with the appropriate description in Column II.

I	II
(a) $y = 2x^2 + 3x + 9$	A. opens right
(b) $y = -3x^2 + 4x - 2$	B. opens up
(c) $x = 2y^2 - 3y + 9$	C. opens left
(d) $x = -3y^2 - 4y + 2$	D. opens down

Graph each horizontal parabola, and give the domain and range. **See Examples 1 and 2.**

3. $x + 4 = y^2$
4. $x - 2 = y^2$
5. $x = (y - 3)^2$
6. $x = (y + 1)^2$
7. $x - 2 = (y - 4)^2$
8. $x + 1 = (y + 2)^2$
9. $x - 2 = -3(y - 1)^2$
10. $x - 4 = \frac{1}{2}(y - 1)^2$
11. $-\frac{1}{2}x = (y + 3)^2$
12. $-\frac{1}{3}x = (y - 2)^2$
13. $x = y^2 + 4y + 2$
14. $x = 2y^2 - 4y + 6$
15. $x = -4y^2 - 4y + 3$
16. $x = -2y^2 + 2y - 3$
17. $2x - y^2 + 4y - 6 = 0$
18. $x + 3y^2 + 18y + 22 = 0$
19. $-x = 3y^2 + 6y + 2$
20. $-x = 2y^2 + 4y - 1$

Give the focus, directrix, and axis of symmetry for each parabola. **See Example 3.**

21. $x^2 = 24y$
22. $x^2 = \frac{1}{8}y$
23. $y = -4x^2$
24. $y = -\frac{1}{9}x^2$
25. $y^2 = -4x$
26. $y^2 = 16x$
27. $x = -32y^2$
28. $x = -16y^2$
29. $(y - 3)^2 = 12(x - 1)$
30. $(y - 2)^2 = 24(x - 3)$
31. $(x - 7)^2 = 16(y + 5)$
32. $(x + 2)^2 = 20(y - 2)$

Write an equation for each parabola with vertex at the origin. **See Example 4.**

33. focus $(5, 0)$
34. focus $\left(-\frac{1}{2}, 0\right)$
35. directrix $y = -\frac{1}{4}$
36. directrix $y = \frac{1}{3}$
37. through $\left(\sqrt{3}, 3\right)$, opening up
38. through $\left(-2, -2\sqrt{2}\right)$, opening left

13. $[-2, \infty)$; $(-\infty, \infty)$

14. $[4, \infty)$; $(-\infty, \infty)$

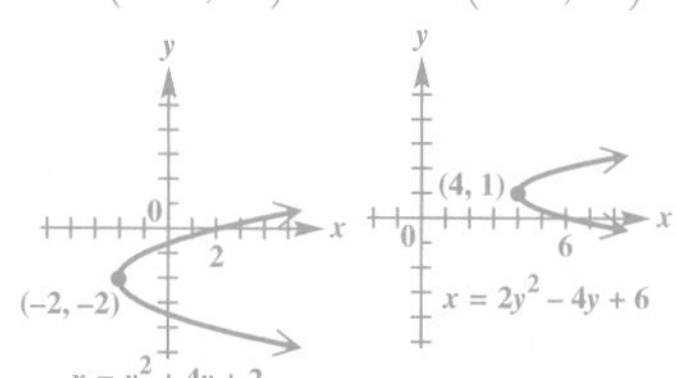

15. $(-\infty, 4]$; $(-\infty, \infty)$

16. $\left(-\infty, -\frac{5}{2}\right]$; $(-\infty, \infty)$

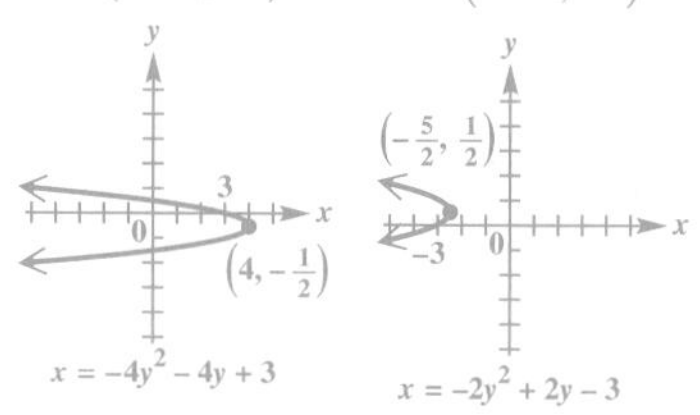

17. $[1, \infty)$; $(-\infty, \infty)$

18. $(-\infty, 5]$; $(-\infty, \infty)$

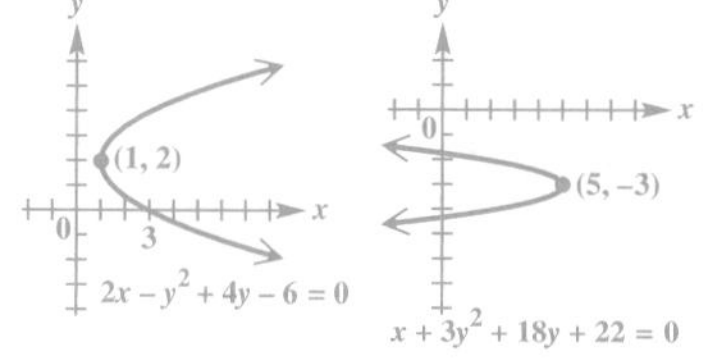

19. $(-\infty, 1]$; $(-\infty, \infty)$

20. $(-\infty, 3]$; $(-\infty, \infty]$

y, 3, 0, x, −11, −2, (1, −1), −3, $-x = 3y^2 + 6y + 2$

y, 1, 0, x, −5, (3, −1), $-x = 2y^2 + 4y - 1$

21. $(0, 6)$, $y = -6$, y-axis

22. $\left(0, \frac{1}{32}\right)$, $y = -\frac{1}{32}$, y-axis

23. $\left(0, -\frac{1}{16}\right)$, $y = \frac{1}{16}$, y-axis

24. $\left(0, -\frac{9}{4}\right)$, $y = \frac{9}{4}$, y-axis

25. $(-1, 0)$, $x = 1$, x-axis

26. $(4, 0)$, $x = -4$, x-axis

27. $\left(-\frac{1}{128}, 0\right)$, $x = \frac{1}{128}$, x-axis

28. $\left(-\frac{1}{64}, 0\right)$, $x = \frac{1}{64}$, x-axis

29. $(4, 3)$, $x = -2$, $y = 3$

30. $(9, 2)$, $x = -3$, $y = 2$

31. $(7, -1)$, $y = -9$, $x = 7$

32. $(-2, 7)$, $y = -3$, $x = -2$

33. $y^2 = 20x$ 34. $y^2 = -2x$

35. $x^2 = y$ 36. $x^2 = -\frac{4}{3}y$

37. $x^2 = y$ 38. $y^2 = -4x$

39. $y^2 = \frac{4}{3}x$ 40. $x^2 = -y$

39. through $(3, 2)$, symmetric with respect to the x-axis

40. through $(2, -4)$, symmetric with respect to the y-axis

Write an equation for each parabola. ***See Example 5.***

41. vertex $(4, 3)$, focus $(4, 5)$

42. vertex $(-2, 1)$, focus $(-2, -3)$

43. vertex $(-5, 6)$, directrix $x = -12$

44. vertex $(1, 2)$, directrix $x = 4$

Determine the two equations necessary to graph each horizontal parabola using a graphing calculator, and graph it in the viewing window specified. ***See Example 2.***

45. $x = 3y^2 + 6y - 4$; $[-10, 2]$ by $[-4, 4]$

46. $x = -2y^2 + 4y + 3$; $[-10, 6]$ by $[-4, 4]$

47. $x + 2 = -(y + 1)^2$; $[-10, 2]$ by $[-4, 4]$

48. $x - 5 = 2(y - 2)^2$; $[-2, 12]$ by $[-2, 6]$

Relating Concepts

For individual or collaborative investigation *(Exercises 49–52)*

(Modeling) Given three noncollinear points, we can find an equation of the form $x = ay^2 + by + c$ of the horizontal parabola joining them by solving a system of equations. ***Work Exercises 49–52 in order,*** *to find the equation of the horizontal parabola containing the points $(-5, 1)$, $(-14, -2)$, and $(-10, 2)$.*

49. Write three equations in a, b, and c, by substituting the given values of x and y into the equation $x = ay^2 + by + c$.

50. Solve the system of three equations determined in **Exercise 49.**

51. Does the horizontal parabola open to the left or to the right? Why?

52. Write the equation of the horizontal parabola.

Solve each problem. ***See Example 6.***

53. ***(Modeling) Radio Telescope Design*** The U.S. Naval Research Laboratory designed a giant radio telescope that had diameter 300 ft and maximum depth 44 ft. (*Source:* Mar, J., and H. Liebowitz, *Structure Technology for Large Radio and Radar Telescope Systems,* The MIT Press.)

(a) Find the equation of a parabola that models the cross section of the dish if the vertex is placed at the origin and the parabola opens up.

(b) The receiver must be placed at the focus of the parabola. How far from the vertex should the receiver be located?

54. ***(Modeling) Radio Telescope Design*** Suppose the telescope in **Exercise 53** had diameter 400 ft and maximum depth 50 ft.

(a) Find the equation of this parabola.

(b) The receiver must be placed at the focus of the parabola. How far from the vertex should the receiver be located?

41. $(x-4)^2 = 8(y-3)$
42. $(x+2)^2 = -16(y-1)$
43. $(y-6)^2 = 28(x+5)$
44. $(y-2)^2 = -12(x-1)$

In the answers for Exercises 45–48, we give only the equations necessary to graph each parabola.

45. $y = -1 \pm \sqrt{\frac{x+7}{3}}$
46. $y = 1 \pm \sqrt{\frac{x-5}{-2}}$
47. $y = -1 \pm \sqrt{-x-2}$
48. $y = 2 \pm \sqrt{\frac{x-5}{2}}$

49. $a + b + c = -5;$
$4a - 2b + c = -14;$
$4a + 2b + c = -10$

50. $\{(-2, 1, -4)\}$
51. It opens to the left, because $a = -2$, and $-2 < 0$.
52. $x = -2y^2 + y - 4$
53. (a) $y = \frac{11}{5625}x^2$ (b) 127.8 ft
54. (a) $y = \frac{1}{800}x^2$ (b) 200 ft
55. 6 ft
56. 60 ft
57. (a) 2000 ft
(b) $y = 1000 - 0.00025x^2$
(c) no
58. (a) Earth: $y = x - \frac{16.1}{961}x^2$;
Mars: $y = x - \frac{6.3}{961}x^2$

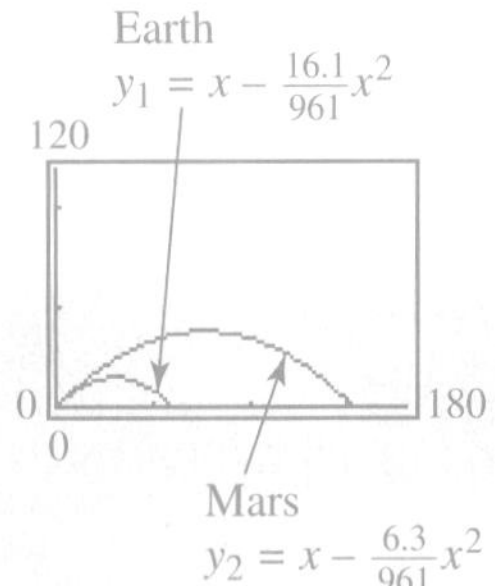

(b) approximately 93 ft

55. *Parabolic Arch* An arch in the shape of a parabola has the dimensions shown in the figure. How wide is the arch 9 ft up?

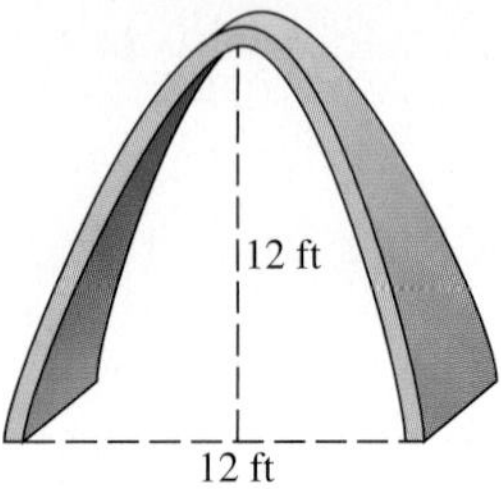

56. *Height of Bridge Cable Supports* The cable in the center portion of a bridge is supported as shown in the figure to form a parabola. The center vertical cable is 10 ft high, the supports are 210 ft high, and the distance between the two supports is 400 ft. Find the height of the remaining vertical cables, if the vertical cables are evenly spaced. (Ignore the width of the supports and cables.)

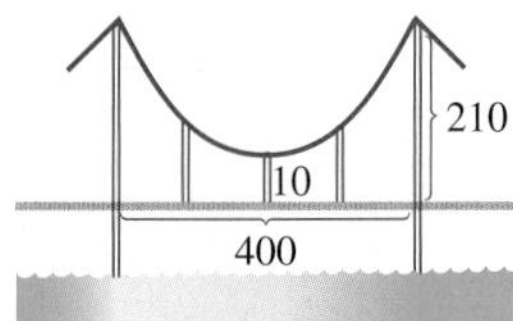

57. *(Modeling) Path of a Cannon Shell* The physicist Galileo observed that certain projectiles follow a parabolic path. For instance, if a cannon fires a shell at a 45° angle with a speed of v feet per second, then the path of the shell (see the top figure) is modeled by the following equation.

$$y = x - \frac{32}{v^2}x^2$$

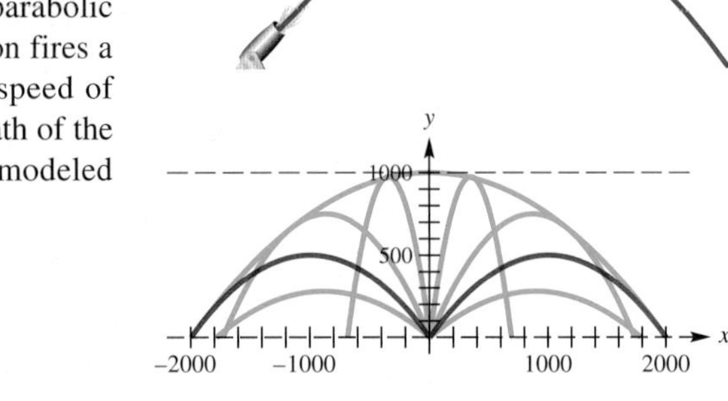

The bottom figure shows the paths of shells all fired at the same speed but at different angles. The greatest distance is achieved with a 45° angle. The outline, or **envelope,** of this family of curves is another parabola with the cannon as focus. The horizontal line through the vertex of the envelope parabola is a directrix for all the other parabolas. Suppose all the shells are fired at a speed of 252.982 ft per sec.

(a) What is the greatest distance that a shell can be fired?
(b) What is the equation of the envelope parabola?
(c) Can a shell reach a helicopter 1500 ft due east of the cannon flying at a height of 450 ft?

58. *(Modeling) Path of an Object* When an object moves under the influence of a constant force (without air resistance), its path is parabolic. This would occur if a ball were thrown near the surface of a planet or other celestial object. Suppose two balls are simultaneously thrown upward at a 45° angle on two different planets. If their initial velocities are both 30 mph, then their xy-coordinates in feet at time x in seconds can be modeled by the following equation.

$$y = x - \frac{g}{1922}x^2$$

Here g is the acceleration due to gravity. The value of g will vary depending on the mass and size of the planet. (*Source:* Zeilik, M., and S. Gregory, *Introductory Astronomy and Astrophysics,* Fourth Edition, Brooks/Cole.)

(a) For Earth $g = 32.2$, while for Mars $g = 12.6$. Find the two equations, and graph on the same screen of a graphing calculator the paths of the two balls thrown on Earth and Mars. Use the window $[0, 180]$ by $[0, 120]$. (*Hint:* If possible, set the mode on your graphing calculator to simultaneous.)
(b) Determine the difference in the horizontal distances traveled by the two balls.

59. (a)

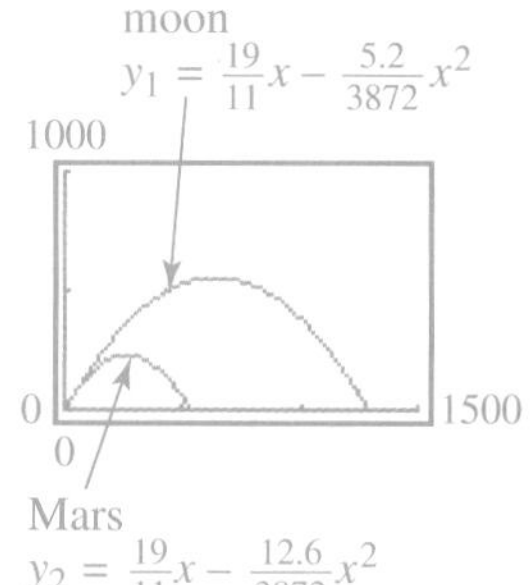

(b) Mars: approximately 229 ft; moon: approximately 555 ft

59. *(Modeling) Path of an Object* (Refer to **Exercise 58.**) Suppose the two balls are now thrown upward at a 60° angle on Mars and the moon. If their initial velocity is 60 mph, then their xy-coordinates in feet at time x in seconds can be modeled by the following equation.

$$y = \frac{19}{11}x - \frac{g}{3872}x^2$$

(*Source:* Zeilik, M., and S. Gregory, *Introductory Astronomy and Astrophysics,* Fourth Edition, Brooks/Cole.)

(a) Graph on the same coordinate axes the paths of the balls if $g = 5.2$ for the moon. Use the window $[0, 1500]$ by $[0, 1000]$.

(b) Determine the maximum height of each ball to the nearest foot.

60. Explain how you can tell, just by looking at the equation of a parabola, whether it has a horizontal or a vertical axis of symmetry.

61. Prove that the parabola with focus $(p, 0)$ and directrix $x = -p$ has the equation $y^2 = 4px$.

62. Write a short paragraph describing the appearances of parabolic shapes in your everyday surroundings.

10.2 Ellipses

- Equations and Graphs of Ellipses
- Translated Ellipses
- Eccentricity
- Applications of Ellipses

Equations and Graphs of Ellipses

Like the circle and the parabola, the ellipse is defined as a set of points in a plane.

Ellipse

An **ellipse** is the set of all points in a plane the sum of whose distances from two fixed points is constant. Each fixed point is a **focus** (plural, **foci**) of the ellipse.

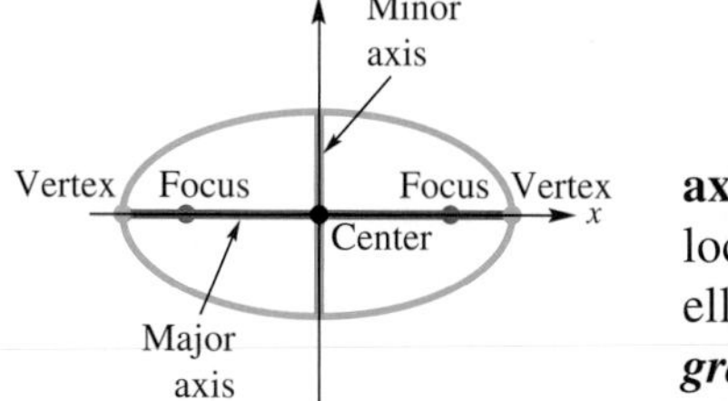

Figure 14

As shown in **Figure 14,** an ellipse has two axes of symmetry, the **major axis** (the longer one) and the **minor axis** (the shorter one). The foci are always located on the major axis. The midpoint of the major axis is the **center** of the ellipse, and the endpoints of the major axis are the **vertices** of the ellipse. ***The graph of an ellipse is not the graph of a function.*** It fails the vertical line test.

The ellipse in **Figure 15** has its center at the origin, foci $F(c, 0)$ and $F'(-c, 0)$, and vertices $V(a, 0)$ and $V'(-a, 0)$. From **Figure 15,** the distance from V to F is $a - c$ and the distance from V to F' is $a + c$. The sum of these distances is $2a$. Since V is on the ellipse, this sum is the constant referred to in the definition of an ellipse. Thus, for any point $P(x, y)$ on the ellipse,

$$d(P, F) + d(P, F') = 2a.$$

y
(0, b) B
P(x, y)
(−a, 0) V′
V (a, 0)
x
F′(−c, 0)
F(c, 0)
(0, −b) B′

Figure 15

By the distance formula,

$$d(P, F) = \sqrt{(x - c)^2 + y^2}$$

and $$d(P, F') = \sqrt{[x - (-c)]^2 + y^2} = \sqrt{(x + c)^2 + y^2}.$$

Thus, we have the following.

$$\sqrt{(x-c)^2+y^2}+\sqrt{(x+c)^2+y^2}=2a \qquad d(P,F)+d(P,F')=2a$$

$$\sqrt{(x-c)^2+y^2}=2a-\sqrt{(x+c)^2+y^2} \qquad \text{Isolate } \sqrt{(x-c)^2+y^2}.$$

$$(x-c)^2+y^2=4a^2-4a\sqrt{(x+c)^2+y^2}+(x+c)^2+y^2 \qquad \text{Square each side. (Sections R.3, 1.6)}$$

Be careful when squaring.

$$x^2-2cx+c^2+y^2=4a^2-4a\sqrt{(x+c)^2+y^2}+x^2+2cx+c^2+y^2 \qquad \text{Square } x-c. \text{ Square } x+c.$$

$$4a\sqrt{(x+c)^2+y^2}=4a^2+4cx \qquad \text{Isolate } 4a\sqrt{(x+c)^2+y^2}.$$

Divide *each* term by 4.

$$a\sqrt{(x+c)^2+y^2}=a^2+cx \qquad \text{Divide by 4.}$$

$$a^2(x^2+2cx+c^2+y^2)=a^4+2ca^2x+c^2x^2 \qquad \text{Square each side.}$$

$$a^2x^2+2ca^2x+a^2c^2+a^2y^2=a^4+2ca^2x+c^2x^2 \qquad \text{Distributive property (Section R.2)}$$

$$a^2x^2+a^2c^2+a^2y^2=a^4+c^2x^2 \qquad \text{Subtract } 2ca^2x.$$

$$a^2x^2-c^2x^2+a^2y^2=a^4-a^2c^2 \qquad \text{Rearrange terms.}$$

$$(a^2-c^2)x^2+a^2y^2=a^2(a^2-c^2) \qquad \text{Factor. (Section R.4)}$$

$$\frac{x^2}{a^2}+\frac{y^2}{a^2-c^2}=1 \quad (1) \qquad \text{Divide by } a^2(a^2-c^2).$$

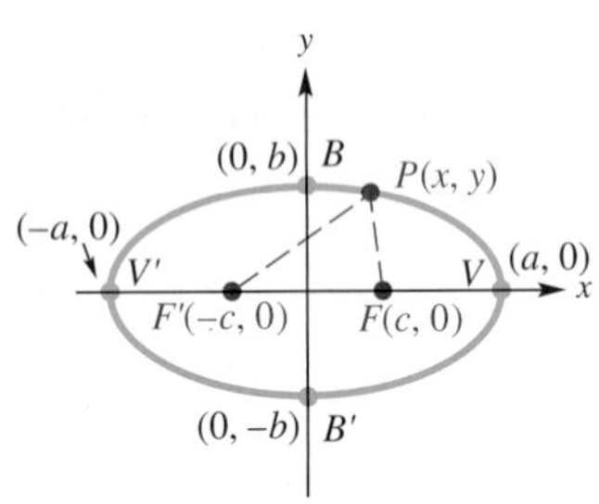

Figure 15 (repeated)

Since $B(0, b)$ is on the ellipse in **Figure 15,** we have the following.

$$d(B,F)+d(B,F')=2a$$

$$\sqrt{(-c)^2+b^2}+\sqrt{c^2+b^2}=2a \qquad \text{Substitute.}$$

$$2\sqrt{c^2+b^2}=2a \qquad \text{Combine like terms.}$$

$$\sqrt{c^2+b^2}=a \qquad \text{Divide by 2.}$$

$$c^2+b^2=a^2 \qquad \text{Square each side.}$$

$$b^2=a^2-c^2 \qquad \text{Subtract } c^2.$$

Replacing a^2-c^2 with b^2 in equation (1) gives the standard form of the equation of an ellipse centered at the origin with foci on the x-axis.

$$\frac{x^2}{a^2}+\frac{y^2}{b^2}=1$$

If the vertices and foci were on the y-axis, an almost identical derivation could be used to get the following standard form.

$$\frac{x^2}{b^2}+\frac{y^2}{a^2}=1$$

Teaching Tip An ellipse can be easily drawn by fixing a loose piece of string to two points, applying tension to the string with a pencil, and sweeping a curve. (See **Exercise 41.**)

LOOKING AHEAD TO CALCULUS

Methods of calculus can be used to solve problems involving ellipses. For example, differentiation is used to find the slope of the tangent line at a point on the ellipse, and integration is used to find the length of any arc of the ellipse.

Standard Forms of Equations for Ellipses

The ellipse with center at the origin and equation

$$\frac{x^2}{a^2} + \frac{y^2}{b^2} = 1 \quad (\text{where } a > b)$$

has vertices $(\pm a, 0)$, endpoints of the minor axis $(0, \pm b)$, and foci $(\pm c, 0)$, where $c^2 = a^2 - b^2$.

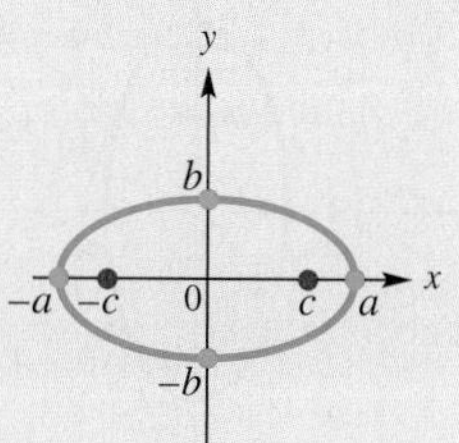

Major axis on x-axis

The ellipse with center at the origin and equation

$$\frac{x^2}{b^2} + \frac{y^2}{a^2} = 1 \quad (\text{where } a > b)$$

has vertices $(0, \pm a)$, endpoints of the minor axis $(\pm b, 0)$, and foci $(0, \pm c)$, where $c^2 = a^2 - b^2$.

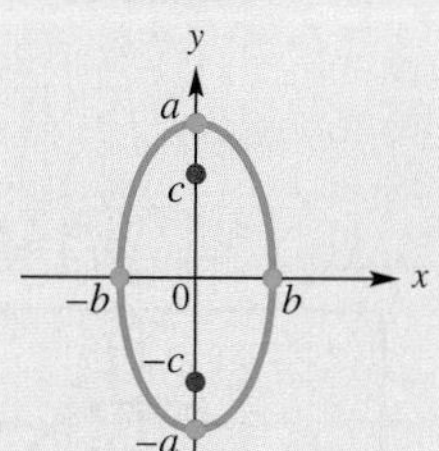

Major axis on y-axis

Classroom Example 1
Graph each ellipse, and find the coordinates of the foci. Give the domain and range.

(a) $4x^2 + 25y^2 = 100$

(b) $49x^2 + 9y^2 = 441$

Answers:

(a)

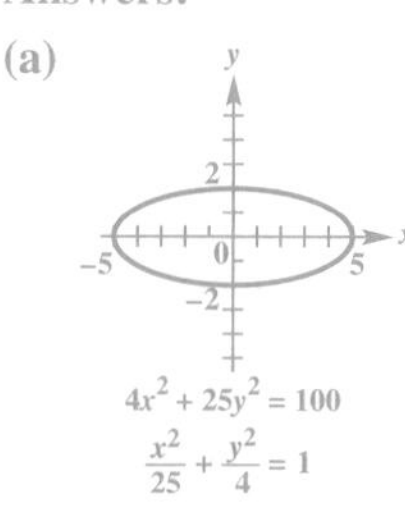

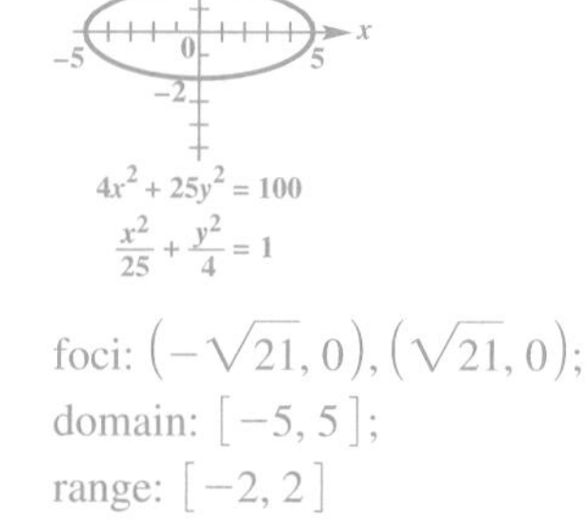

foci: $(-\sqrt{21}, 0), (\sqrt{21}, 0)$;
domain: $[-5, 5]$;
range: $[-2, 2]$

(b)

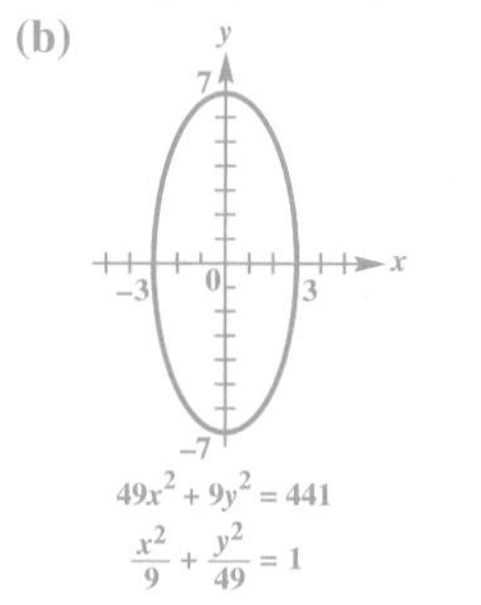

foci: $(0, -2\sqrt{10}), (0, 2\sqrt{10})$;
domain: $[-3, 3]$; range: $[-7, 7]$

Do not be confused by the two standard forms.

- In the first form, a^2 is associated with x^2.
- In the second form, a^2 is associated with y^2.

In practice it is necessary only to find the intercepts of the graph—if the positive x-intercept is greater than the positive y-intercept, then the major axis is horizontal; otherwise, it is vertical.

When using the relationship $b^2 + c^2 = a^2$, or $a^2 - b^2 = c^2$, choose a^2 and b^2 so that $a^2 > b^2$.

EXAMPLE 1 Graphing Ellipses Centered at the Origin

Graph each ellipse, and find the coordinates of the foci. Give the domain and range.

(a) $4x^2 + 9y^2 = 36$ **(b)** $4x^2 = 64 - y^2$

SOLUTION

(a) Divide each side of $4x^2 + 9y^2 = 36$ by 36 to get

Divide *each* term by 36.
$$\frac{x^2}{9} + \frac{y^2}{4} = 1. \quad \text{Standard form of an ellipse}$$

Thus, the x-intercepts are ± 3, and the y-intercepts are ± 2. The graph of the ellipse is shown in **Figure 16.**

Since $9 > 4$, we find the foci of the ellipse by letting $a^2 = 9$ and $b^2 = 4$ in $c^2 = a^2 - b^2$.

$$c^2 = 9 - 4 = 5, \quad \text{so} \quad c = \sqrt{5} \quad \text{By definition, } c > 0.$$

The major axis is along the x-axis. Thus, the foci have coordinates $(-\sqrt{5}, 0)$ and $(\sqrt{5}, 0)$. The domain of this relation is $[-3, 3]$, and the range is $[-2, 2]$.

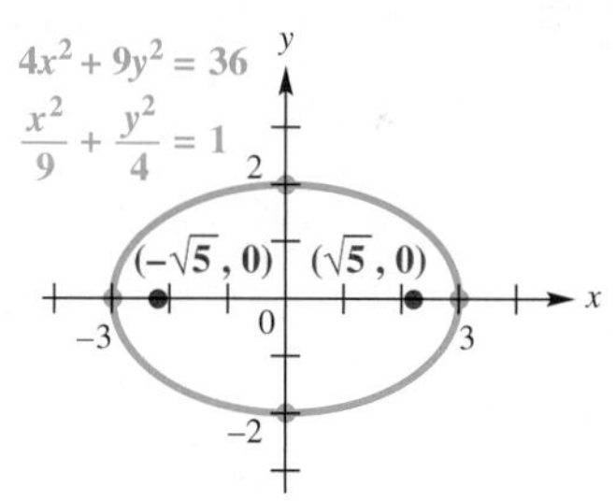

Figure 16

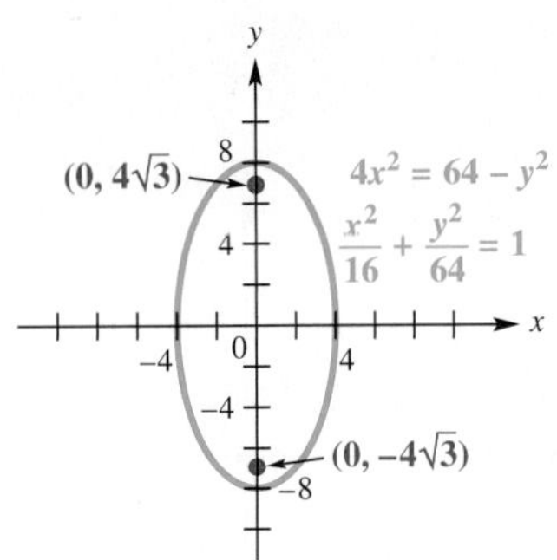

Figure 17

(b) Write the equation $4x^2 = 64 - y^2$ as $4x^2 + y^2 = 64$. Then express it in standard form.

Divide *each* term by 64. $$\frac{x^2}{16} + \frac{y^2}{64} = 1 \quad \text{Standard form of an ellipse}$$

The x-intercepts are ± 4, and the y-intercepts are ± 8. See the graph in **Figure 17.** Here $64 > 16$, so $a^2 = 64$ and $b^2 = 16$. Use $c^2 = a^2 - b^2$.

$$c^2 = 64 - 16 = 48, \quad \text{so} \quad c = \sqrt{48} = 4\sqrt{3} \quad \text{(Section R.7)}$$

The major axis is on the y-axis, which means the coordinates of the foci are $(0, -4\sqrt{3})$ and $(0, 4\sqrt{3})$. The domain of the relation is $[-4, 4]$, and the range is $[-8, 8]$.

✔ *Now Try Exercises 7 and 9.*

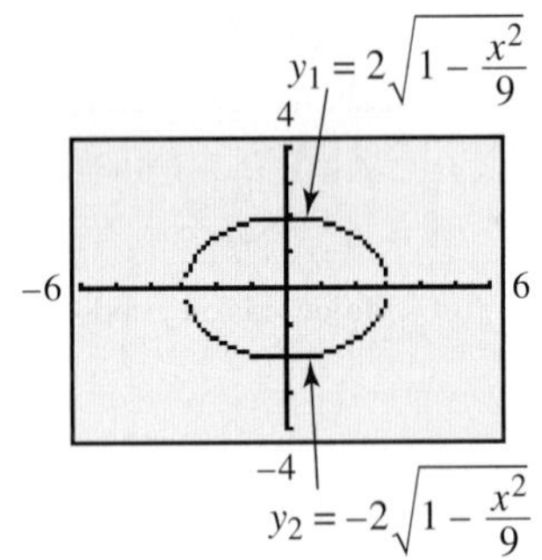

Figure 18

The graph of an ellipse is ***not*** the graph of a function. To graph the ellipse in **Example 1(a)** with a graphing calculator, solve for y in $4x^2 + 9y^2 = 36$ to get equations of thc two functions shown in **Figure 18.**

$$y = 2\sqrt{1 - \frac{x^2}{9}} \quad \text{and} \quad y = -2\sqrt{1 - \frac{x^2}{9}}.$$

Their union is the graph of $4x^2 + 9y^2 = 36$. ■

EXAMPLE 2 Writing the Equation of an Ellipse

Write the equation of the ellipse having center at the origin, foci at $(0, 3)$ and $(0, -3)$, and major axis of length 8 units.

Classroom Example 2
Write the equation of the ellipse having center at the origin, foci at $(-5, 0)$ and $(5, 0)$, and major axis of length 18 units.

Answer: $\frac{x^2}{81} + \frac{y^2}{56} = 1$

SOLUTION Since the major axis is 8 units long, $2a = 8$ and $a = 4$. To find b^2, use the relationship $a^2 - b^2 = c^2$, with $a = 4$ and $c = 3$.

$$a^2 - b^2 = c^2$$
$$4^2 - b^2 = 3^2 \quad \text{Substitute for } a \text{ and } c.$$
$$16 - b^2 = 9 \quad \text{Apply the exponents.}$$
$$b^2 = 7 \quad \text{Solve for } b^2. \text{ (Section 1.1)}$$

Since the foci are on the y-axis, we use the larger intercept, a, to find the denominator for y^2, giving the equation in standard form.

$$\frac{x^2}{7} + \frac{y^2}{16} = 1 \quad \text{Use } \frac{x^2}{b^2} + \frac{y^2}{a^2} = 1.$$

A graph of this ellipse is shown in **Figure 19.** The domain of this relation is $[-\sqrt{7}, \sqrt{7}]$, and the range is $[-4, 4]$.

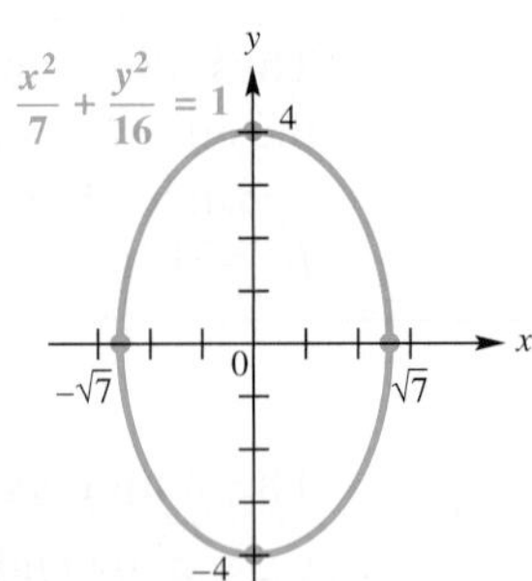

Figure 19

✔ *Now Try Exercise 17.*

Classroom Example 3

Graph $\frac{y}{3} = -\sqrt{1 - \frac{x^2}{4}}$. Give the domain and range.

Answer:

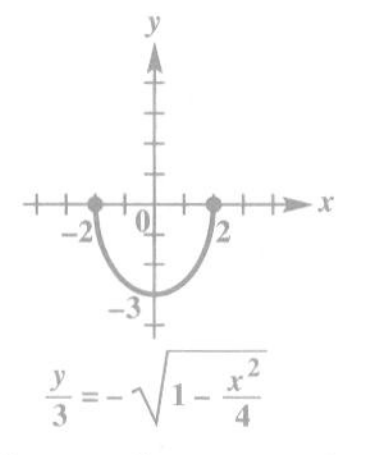

domain: $[-2, 2]$; range: $[-3, 0]$

EXAMPLE 3 Graphing a Half-Ellipse

Graph $\frac{y}{4} = \sqrt{1 - \frac{x^2}{25}}$. Give the domain and range.

SOLUTION Square each side to get

$$\frac{y^2}{16} = 1 - \frac{x^2}{25}, \quad \text{or} \quad \frac{x^2}{25} + \frac{y^2}{16} = 1,$$

the equation of an ellipse with x-intercepts ± 5 and y-intercepts ± 4. In the original equation, the radical expression

$$\sqrt{1 - \frac{x^2}{25}}$$

represents a nonnegative number, so the only possible values of y are those that give the half-ellipse shown in **Figure 20.** This is the graph of a function with

$$\text{domain } [-5, 5] \text{ and range } [0, 4].$$

Now Try Exercise 29.

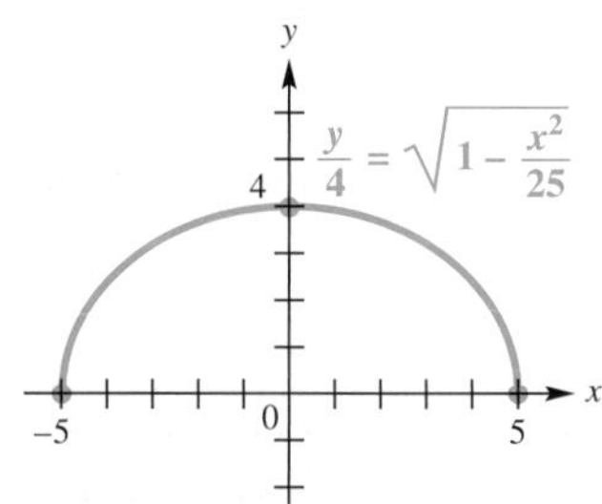

Figure 20

Translated Ellipses Just as a circle need not have its center at the origin, an ellipse may also have its center translated away from the origin.

Standard Forms for Ellipses Centered at (h, k)

An ellipse with center at (h, k) and either a horizontal or a vertical major axis of length $2a$ satisfies one of the following equations, where $a > b > 0$ and $c^2 = a^2 - b^2$ with $c > 0$.

$$\frac{(x-h)^2}{a^2} + \frac{(y-k)^2}{b^2} = 1$$

Major axis: horizontal; foci: $(h \pm c, k)$; vertices: $(h \pm a, k)$

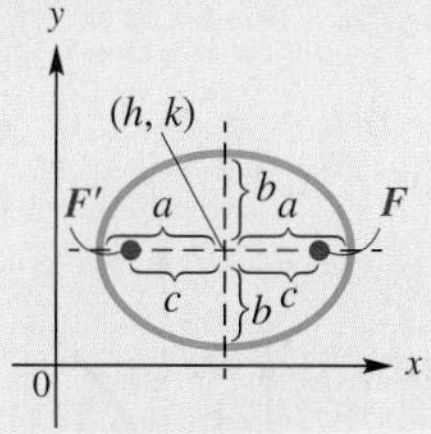

Horizontal major axis

$$\frac{(x-h)^2}{b^2} + \frac{(y-k)^2}{a^2} = 1$$

Major axis: vertical; foci: $(h, k \pm c)$; vertices: $(h, k \pm a)$

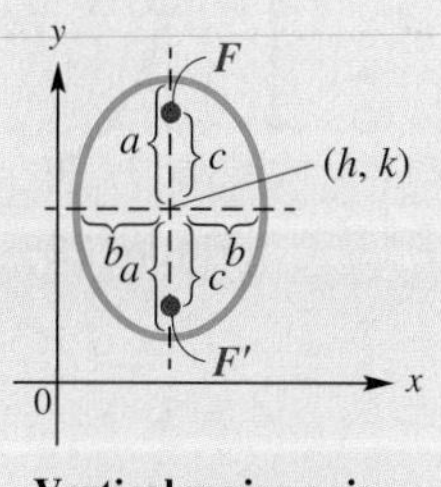

Vertical major axis

When graphing such ellipses, remember that the location of a^2 (the greater denominator) determines whether the ellipse has a horizontal or a vertical major axis.

Classroom Example 4

Graph $\frac{(x-3)^2}{36} + \frac{y^2}{9} = 1$. Give the domain and range.

Answer:

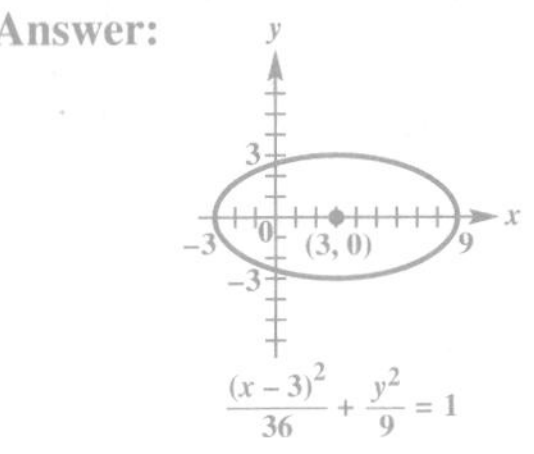

domain: $[-3, 9]$; range: $[-3, 3]$

Teaching Tip In **Example 4**, students may find it easier to graph

$$\frac{x^2}{9} + \frac{y^2}{16} = 1$$

and then apply a horizontal shift 2 units to the right and a vertical shift 1 unit down.

EXAMPLE 4 Graphing an Ellipse Translated Away from the Origin

Graph $\frac{(x-2)^2}{9} + \frac{(y+1)^2}{16} = 1$. Give the domain and range.

SOLUTION The graph of this equation is an ellipse centered at $(2, -1)$. As mentioned earlier, ellipses always have $a > b$. For this ellipse, $a = 4$ and $b = 3$. Since $a = 4$ is associated with y^2, the vertices of the ellipse are on the vertical line through $(2, -1)$.

To find the vertices, locate two points on the vertical line through $(2, -1)$, one 4 units up from $(2, -1)$ and one 4 units down. This gives the vertices $(2, 3)$ and $(2, -5)$. Two other points on the ellipse are on the horizontal line through $(2, -1)$, one 3 units to the right and one 3 units to the left.

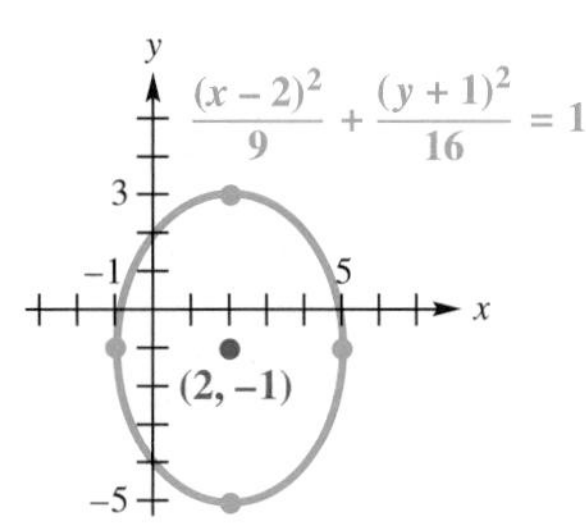

Figure 21

See the graph in **Figure 21.** The domain is $[-1, 5]$, and the range is $[-5, 3]$.

Now Try Exercise 13.

NOTE As suggested by the graphs in this section, an ellipse is symmetric with respect to its major axis, its minor axis, and its center. ***If $a = b$ in the equation of an ellipse, then the graph is a circle.***

Eccentricity

All conics can be characterized by one general definition.

Conic

A **conic** is the set of all points $P(x, y)$ in a plane such that the ratio of the distance from P to a fixed point and the distance from P to a fixed line is constant.

For a parabola, the fixed line is the directrix, and the fixed point is a focus. In **Figure 22,** the focus is $F(c, 0)$, and the directrix is the line $x = -c$. The constant ratio is the **eccentricity** of the conic, written ***e***. **(*This is not the same e as the base of natural logarithms.*)** If the conic is a parabola, then by definition, the distances $d(P, F)$ and $d(P, D)$ in **Figure 22** are equal. ***Thus, every parabola has eccentricity* 1.**

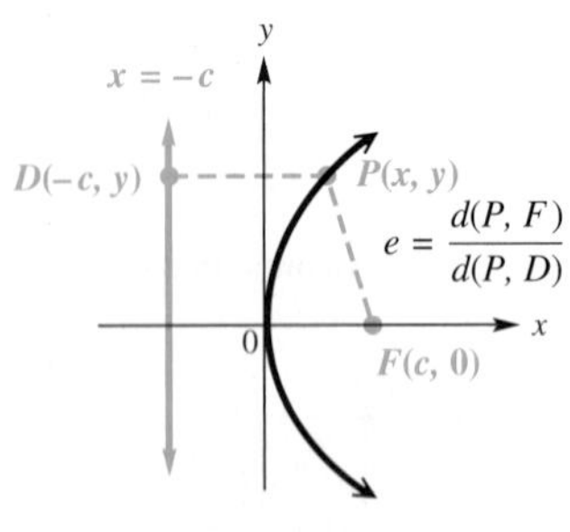

Figure 22

For an ellipse, eccentricity is a measure of its "roundness." The constant ratio in the definition is $e = \frac{c}{a}$, where (as before) c is the distance from the center of the figure to a focus, and a is the distance from the center to a vertex. By the definition of an ellipse, $a^2 > b^2$ and $c = \sqrt{a^2 - b^2}$. Thus, for the ellipse, we have the following.

$$0 < c < a$$

$$0 < \frac{c}{a} < 1 \quad \text{Divide by } a.$$

$$0 < e < 1 \quad e = \frac{c}{a}$$

If a is constant, letting c approach 0 would force the ratio $\frac{c}{a}$ to approach 0, which also forces b to approach a (so that $\sqrt{a^2 - b^2} = c$ would approach 0).

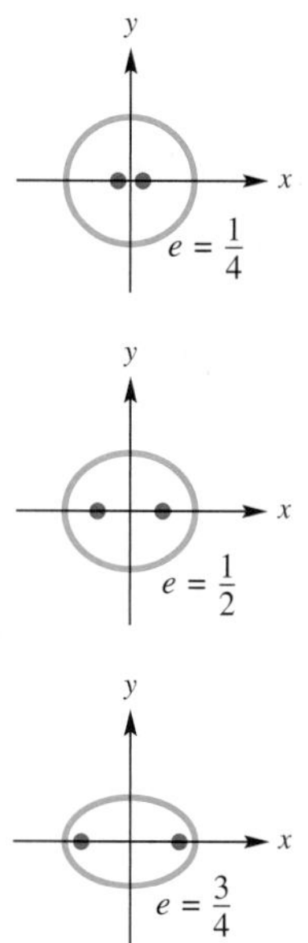

Figure 23

Since b determines the endpoints of the minor axis, this means that the lengths of the major and minor axes are almost the same, producing an ellipse very close in shape to a circle when e is very close to 0. In a similar manner, if e approaches 1, then b will approach 0.

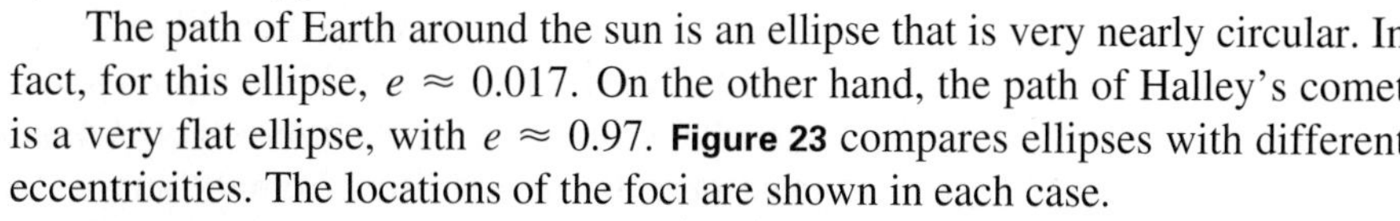

The path of Earth around the sun is an ellipse that is very nearly circular. In fact, for this ellipse, $e \approx 0.017$. On the other hand, the path of Halley's comet is a very flat ellipse, with $e \approx 0.97$. **Figure 23** compares ellipses with different eccentricities. The locations of the foci are shown in each case.

The equation of a circle with center (h, k) and radius r can be written as follows.

$$(x-h)^2 + (y-k)^2 = r^2$$

$$\frac{(x-h)^2}{r^2} + \frac{(y-k)^2}{r^2} = 1 \qquad \text{Divide by } r^2.$$

In a circle, the foci coincide with the center, so $a = b$, $c = \sqrt{a^2 - b^2} = 0$, and $e = 0$. ***Thus, every circle has eccentricity* 0.**

EXAMPLE 5 Finding Eccentricity from Equations of Ellipses

Find the eccentricity of each ellipse.

(a) $\dfrac{x^2}{9} + \dfrac{y^2}{16} = 1$ **(b)** $5x^2 + 10y^2 = 50$

Classroom Example 5
Find the eccentricity of each ellipse.

(a) $\dfrac{x^2}{81} + \dfrac{y^2}{100} = 1$

(b) $\dfrac{x^2}{11} + \dfrac{y^2}{6} = 1$

Answers:
(a) 0.44 (b) 0.67

SOLUTION

(a) Since $16 > 9$, $a^2 = 16$ and thus $a = 4$. To find c, use $c = \sqrt{a^2 - b^2}$.

$$c = \sqrt{a^2 - b^2}$$

$$c = \sqrt{16 - 9} \qquad a^2 = 16,\ b^2 = 9$$

$$c = \sqrt{7} \qquad \text{Subtract.}$$

To find the eccentricity e, use $e = \frac{c}{a}$.

$$e = \frac{\sqrt{7}}{4} \approx 0.66$$

(b) Start by dividing each term of the given equation by 50.

$$5x^2 + 10y^2 = 50$$

$$\frac{5x^2}{50} + \frac{10y^2}{50} = \frac{50}{50} \qquad \text{Divide by 50.}$$

$$\frac{x^2}{10} + \frac{y^2}{5} = 1 \qquad \text{Write in lowest terms.}$$

For this ellipse, $a^2 = 10$ and thus $a = \sqrt{10}$. Find c as in part (a).

$$c = \sqrt{10 - 5} \qquad a^2 = 10,\ b^2 = 5$$

$$c = \sqrt{5} \qquad \text{Subtract.}$$

Finally, find e.

$$e = \frac{\sqrt{5}}{\sqrt{10}} \approx 0.71 \qquad e = \frac{c}{a}$$

✔ *Now Try Exercises 37 and 39.*

Applications of Ellipses

EXAMPLE 6 Applying the Equation of an Ellipse to the Orbit of a Planet

The orbit of the planet Mars is an ellipse with the sun at one focus. The eccentricity of the ellipse is 0.0935, and the closest distance that Mars comes to the sun is 128.5 million mi. (*Source: World Almanac and Book of Facts.*) Find the maximum distance of Mars from the sun.

SOLUTION **Figure 24** shows the orbit of Mars with the origin at the center of the ellipse and the sun at one focus. Mars is closest to the sun when Mars is at the right endpoint of the major axis and farthest from the sun when Mars is at the left endpoint. Therefore, the least distance is $a - c$, and the greatest distance is $a + c$.

$$\frac{a - 128.5}{a} = 0.0935 \qquad \text{Since } a - c = 128.5 \text{, it follows that } c = a - 128.5 \text{. Use } e = \tfrac{c}{a}.$$

$$a - 128.5 = 0.0935a \qquad \text{Multiply by } a.$$

$$a \approx 141.8 \qquad \text{Solve for } a.$$

Then $c = 141.8 - 128.5 = 13.3$ and $a + c = 141.8 + 13.3 = 155.1$.

The maximum distance of Mars from the sun is about 155.1 million mi.

NOT TO SCALE

Figure 24

✔ *Now Try Exercise 45.*

Classroom Example 6
The orbit of Jupiter is an ellipse with the sun at one focus. The eccentricity of the ellipse is 0.0489, and the maximum distance of Jupiter from the sun is 507.4 million mi. (*Source: World Almanac and Book of Facts.*) Find the closest distance that Jupiter comes to the sun.

Answer: 460.1 million mi

When a ray of light or sound emanating from one focus of an ellipse bounces off the ellipse, it passes through the other focus. See **Figure 25.** This reflecting property is responsible for "whispering galleries." In a whispering gallery, a person whispering at a certain point in the room, though barely audible close by, can be heard clearly at another point across the room. The U.S. statesman John Quincy Adams was able to listen in on his opponents' conversations in the old House Chamber (Statuary Hall) because his desk was positioned at one of the foci beneath the ellipsoidal ceiling and his opponents were located across the room at the other focus.

A lithotripter is a machine used to crush kidney stones using shock waves. The patient is placed in an elliptical tub with the kidney stone at one focus of the ellipse. A beam is projected from the other focus to the tub so that it reflects to hit the kidney stone. See **Figure 26** below and **Figure 27** on the next page.

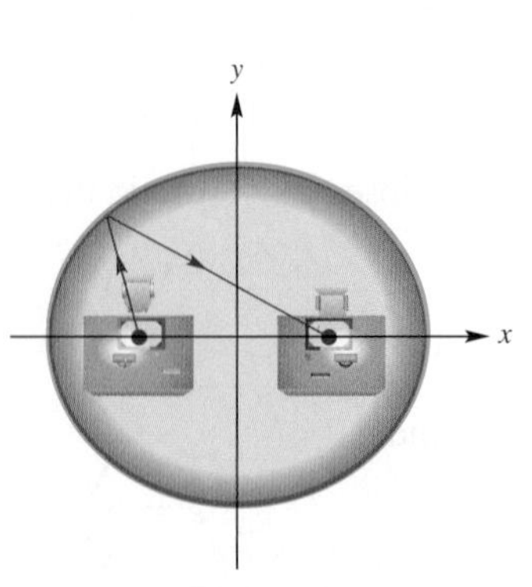

Aerial view of Old House Chamber

Figure 25

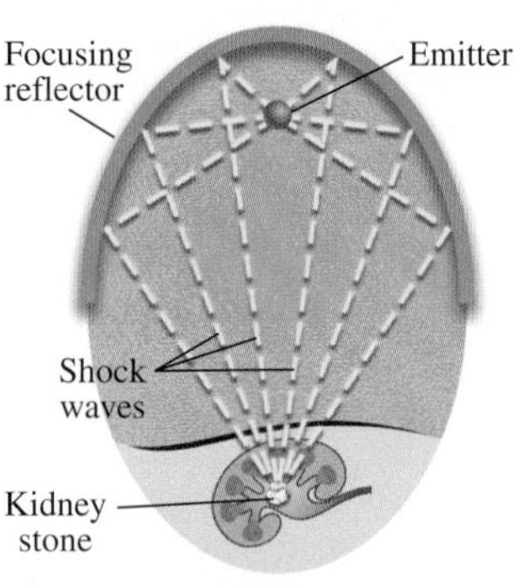

The top of an ellipse is illustrated in this depiction of how a lithotripter crushes a kidney stone.

Figure 26

Classroom Example 7
Repeat **Example 7** for a lithotripter based on the following ellipse.

$$\frac{x^2}{40} + \frac{y^2}{24} = 1$$

Answer: 4 units

EXAMPLE 7 **Modeling the Reflective Property of Ellipses**

If a lithotripter is based on the ellipse

$$\frac{x^2}{36} + \frac{y^2}{27} = 1,$$

determine how many units both the kidney stone and the source of the beam must be placed from the center of the ellipse.

SOLUTION The kidney stone and the source of the beam must be placed at the foci, $(c, 0)$ and $(-c, 0)$. Here $a^2 = 36$ and $b^2 = 27$.

$$c = \sqrt{a^2 - b^2} \quad \text{Equation for } c$$

$$c = \sqrt{36 - 27} \quad \text{Substitute.}$$

$$c = \sqrt{9}, \text{ or } 3 \quad \text{Subtract and simplify.}$$

Thus, the foci are $(3, 0)$ and $(-3, 0)$. The kidney stone and the source both must be placed on a line 3 units from the center. See **Figure 27.**

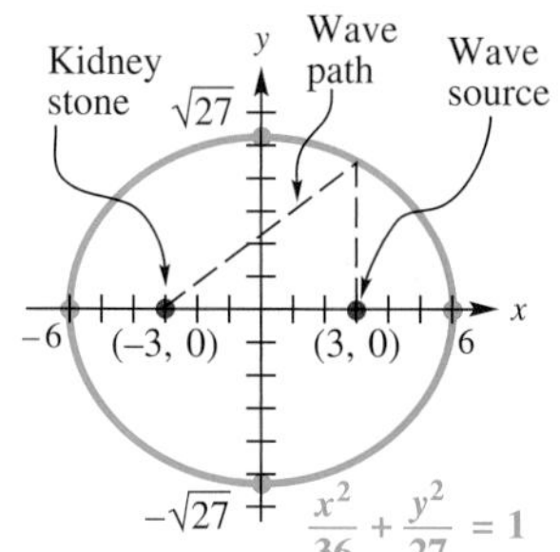

Figure 27

✔ *Now Try Exercise 49.*

1. (a) A **(b)** C **(c)** D **(d)** B **2. (a)** This is an ellipse. **(b)** This is not an ellipse. It is a circle. **(c)** This is not an ellipse. It is a parabola. **(d)** This is not an ellipse. It is a line.

10.2 Exercises

1. *Concept Check* Match each equation of an ellipse in Column I with the appropriate description in Column II.

I	II
(a) $36x^2 + 9y^2 = 324$	**A.** x-intercepts ± 3; y-intercepts ± 6
(b) $9x^2 + 36y^2 = 324$	**B.** x-intercepts ± 4; y-intercepts ± 5
(c) $\frac{x^2}{25} + \frac{y^2}{16} = 1$	**C.** x-intercepts ± 6; y-intercepts ± 3
(d) $\frac{x^2}{16} + \frac{y^2}{25} = 1$	**D.** x-intercepts ± 5; y-intercepts ± 4

2. *Concept Check* Determine whether or not each equation is that of an ellipse. If it is not, state the kind of graph the equation has.

(a) $x^2 + 4y^2 = 4$ **(b)** $x^2 + y^2 = 4$ **(c)** $x^2 + y = 4$ **(d)** $\frac{x}{4} + \frac{y}{25} = 1$

Graph each ellipse. Identify the domain, range, center, vertices, endpoints of the minor axis, and foci in each figure. ***See Examples 1 and 4.***

3. $\frac{x^2}{25} + \frac{y^2}{9} = 1$ **4.** $\frac{x^2}{16} + \frac{y^2}{25} = 1$ **5.** $\frac{x^2}{9} + y^2 = 1$

6. $\frac{x^2}{36} + \frac{y^2}{16} = 1$ **7.** $9x^2 + y^2 = 81$ **8.** $4x^2 + 16y^2 = 64$

9. $4x^2 = 100 - 25y^2$ **10.** $4x^2 = 16 - y^2$

11. $\frac{(x-2)^2}{25} + \frac{(y-1)^2}{4} = 1$ **12.** $\frac{(x+2)^2}{16} + \frac{(y+1)^2}{9} = 1$

13. $\frac{(x+3)^2}{16} + \frac{(y-2)^2}{36} = 1$ **14.** $\frac{(x-1)^2}{9} + \frac{(y+3)^2}{25} = 1$

3. $[-5, 5]$; $[-3, 3]$; $(0, 0)$; $(-5, 0)$, $(5, 0)$; $(0, -3)$, $(0, 3)$; $(-4, 0)$, $(4, 0)$

4. $[-4, 4]$; $[-5, 5]$; $(0, 0)$; $(0, -5)$, $(0, 5)$; $(-4, 0)$, $(4, 0)$; $(0, -3)$, $(0, 3)$

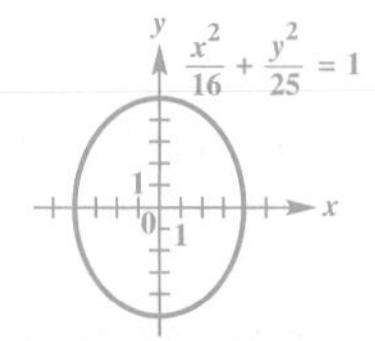

5. $[-3, 3]$; $[-1, 1]$; $(0, 0)$; $(-3, 0)$, $(3, 0)$; $(0, -1)$, $(0, 1)$; $(-2\sqrt{2}, 0)$, $(2\sqrt{2}, 0)$

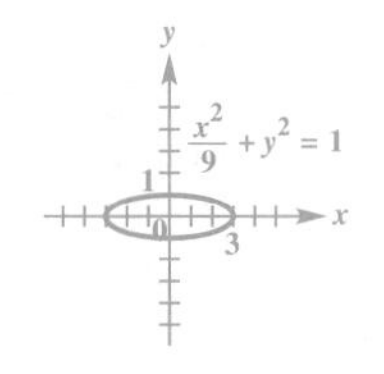

6. $[-6, 6]$; $[-4, 4]$; $(0, 0)$; $(-6, 0)$, $(6, 0)$; $(0, -4)$, $(0, 4)$; $(-2\sqrt{5}, 0)$, $(2\sqrt{5}, 0)$

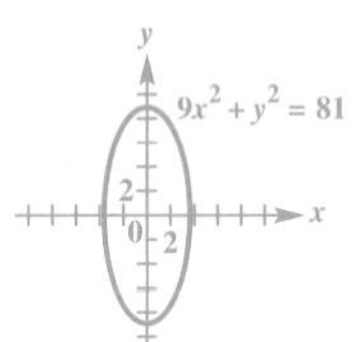

7. $[-3, 3]$; $[-9, 9]$; $(0, 0)$; $(0, -9)$, $(0, 9)$; $(-3, 0)$, $(3, 0)$; $(0, -6\sqrt{2})$, $(0, 6\sqrt{2})$

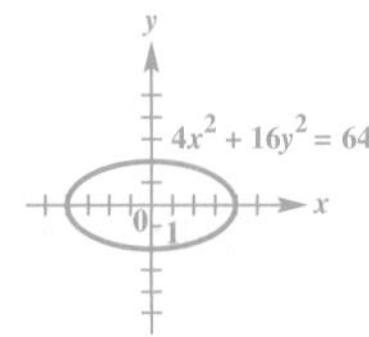

8. $[-4, 4]$; $[-2, 2]$; $(0, 0)$; $(-4, 0)$, $(4, 0)$; $(0, -2)$, $(0, 2)$; $(-2\sqrt{3}, 0)$, $(2\sqrt{3}, 0)$

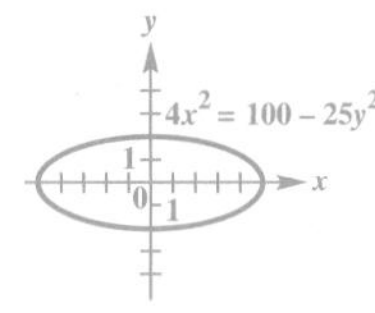

9. $[-5, 5]$; $[-2, 2]$; $(0, 0)$; $(-5, 0)$, $(5, 0)$; $(0, -2)$, $(0, 2)$; $(-\sqrt{21}, 0)$, $(\sqrt{21}, 0)$

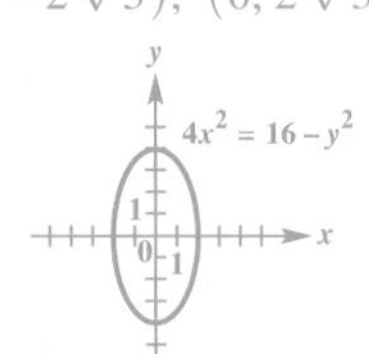

10. $[-2, 2]$; $[-4, 4]$; $(0, 0)$; $(0, -4)$, $(0, 4)$; $(-2, 0)$, $(2, 0)$; $(0, -2\sqrt{3})$, $(0, 2\sqrt{3})$

Write an equation for each ellipse. ***See Example 2.***

15. x-intercepts ± 5; y-intercepts ± 4
16. x-intercepts $\pm \sqrt{15}$; y-intercepts ± 4
17. major axis with length 6; foci at $(0, 2)$, $(0, -2)$
18. minor axis with length 4; foci at $(-5, 0)$, $(5, 0)$
19. center at $(3, 1)$; minor axis vertical, with length 8; $c = 3$
20. center at $(-2, 7)$; major axis vertical, with length 10; $c = 2$
21. foci at $(0, 4)$, $(0, -4)$; sum of distances from foci to point on ellipse is 10 (*Hint:* Consider one of the vertices.)
22. foci at $(-4, 0)$, $(4, 0)$; sum of distances from foci to point on ellipse is 9
23. foci at $(0, -3)$, $(0, 3)$; the point $(8, 3)$ on ellipse
24. foci at $(-3, -3)$, $(7, -3)$; the point $(2, -7)$ on ellipse
25. eccentricity $\frac{4}{5}$; vertices at $(-5, 0)$, $(5, 0)$
26. eccentricity $\frac{1}{2}$; vertices at $(-4, 0)$, $(4, 0)$
27. eccentricity $\frac{3}{4}$; foci at $(0, -2)$, $(0, 2)$
28. eccentricity $\frac{2}{3}$; foci at $(0, -9)$, $(0, 9)$

Graph each equation. Give the domain and range. Identify any that are graphs of functions. ***See Example 3.***

29. $\frac{y}{2} = \sqrt{1 - \frac{x^2}{25}}$
30. $\frac{x}{4} = \sqrt{1 - \frac{y^2}{9}}$
31. $x = -\sqrt{1 - \frac{y^2}{64}}$
32. $y = -\sqrt{1 - \frac{x^2}{100}}$

Determine the two equations necessary to graph each ellipse with a graphing calculator, and graph it in the viewing window indicated. See **Figure 18.**

33. $\frac{x^2}{16} + \frac{y^2}{4} = 1$; $[-4.7, 4.7]$ by $[-3.1, 3.1]$
34. $\frac{x^2}{4} + \frac{y^2}{25} = 1$; $[-9.4, 9.4]$ by $[-6.2, 6.2]$
35. $\frac{(x-3)^2}{25} + \frac{y^2}{9} = 1$; $[-9.4, 9.4]$ by $[-6.2, 6.2]$
36. $\frac{x^2}{36} + \frac{(y+4)^2}{4} = 1$; $[-9.4, 9.4]$ by $[-6.2, 6.2]$

Find the eccentricity of each ellipse. If applicable, round to the nearest hundredth. ***See Example 5.***

37. $\frac{x^2}{3} + \frac{y^2}{4} = 1$
38. $\frac{x^2}{8} + \frac{y^2}{4} = 1$
39. $4x^2 + 7y^2 = 28$
40. $x^2 + 25y^2 = 25$

41. Draftspeople often use the method shown in the sketch to draw an ellipse. Explain why the method works.

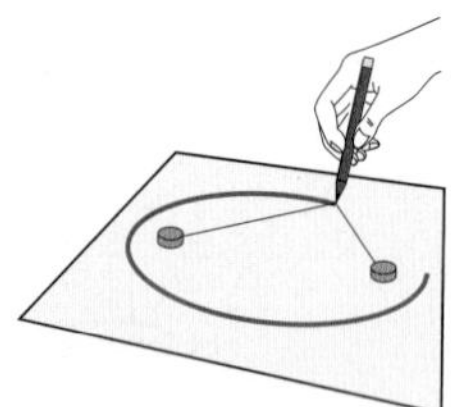

42. Explain how the method of **Exercise 41** can be modified to draw a circle.

11. $[-3, 7]$; $[-1, 3]$; $(2, 1)$; $(-3, 1)$, $(7, 1)$; $(2, -1)$, $(2, 3)$; $(2 - \sqrt{21}, 1)$, $(2 + \sqrt{21}, 1)$

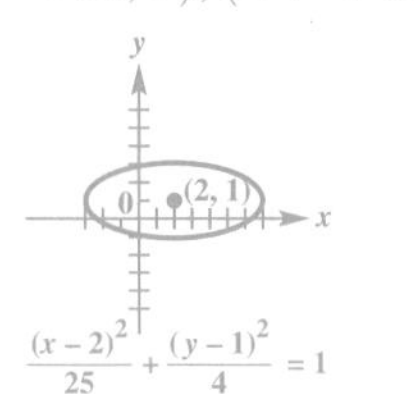

$$\frac{(x-2)^2}{25} + \frac{(y-1)^2}{4} = 1$$

12. $[-6, 2]$; $[-4, 2]$; $(-2, -1)$; $(-6, -1)$, $(2, -1)$; $(-2, -4)$, $(-2, 2)$; $(-2 - \sqrt{7}, -1)$, $(-2 + \sqrt{7}, -1)$

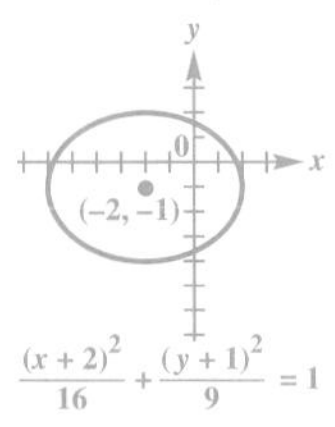

$$\frac{(x+2)^2}{16} + \frac{(y+1)^2}{9} = 1$$

13. $[-7, 1]$; $[-4, 8]$; $(-3, 2)$; $(-3, -4)$, $(-3, 8)$; $(-7, 2)$, $(1, 2)$; $(-3, 2 - 2\sqrt{5})$, $(-3, 2 + 2\sqrt{5})$

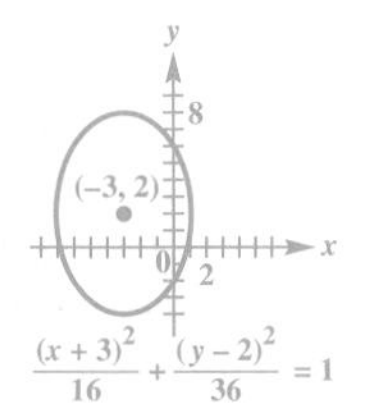

$$\frac{(x+3)^2}{16} + \frac{(y-2)^2}{36} = 1$$

14. $[-2, 4]$; $[-8, 2]$; $(1, -3)$; $(1, -8)$, $(1, 2)$; $(-2, -3)$, $(4, -3)$; $(1, -7)$, $(1, 1)$

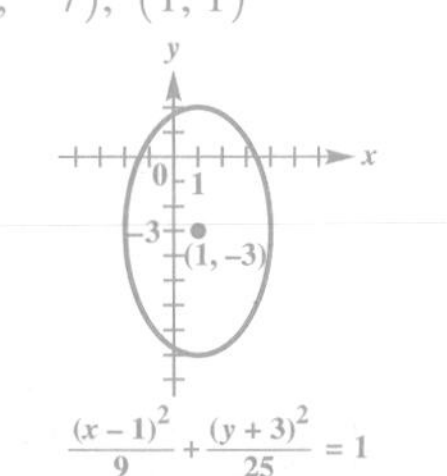

$$\frac{(x-1)^2}{9} + \frac{(y+3)^2}{25} = 1$$

15. $\frac{x^2}{25} + \frac{y^2}{16} = 1$ 16. $\frac{x^2}{15} + \frac{y^2}{16} = 1$

17. $\frac{x^2}{5} + \frac{y^2}{9} = 1$ 18. $\frac{x^2}{29} + \frac{y^2}{4} = 1$

19. $\frac{(x-3)^2}{25} + \frac{(y-1)^2}{16} = 1$

20. $\frac{(x+2)^2}{21} + \frac{(y-7)^2}{25} = 1$

21. $\frac{x^2}{9} + \frac{y^2}{25} = 1$ 22. $\frac{4x^2}{81} + \frac{4y^2}{17} = 1$

23. $\frac{x^2}{72} + \frac{y^2}{81} = 1$

24. $\frac{(x-2)^2}{41} + \frac{(y+3)^2}{16} = 1$

Solve each problem. ***See Examples 6 and 7.***

43. *Height of an Overpass* A one-way road passes under an overpass in the shape of half an ellipse, 15 ft high at the center and 20 ft wide. Assuming a truck is 12 ft wide, what is the tallest truck that can pass under the overpass?

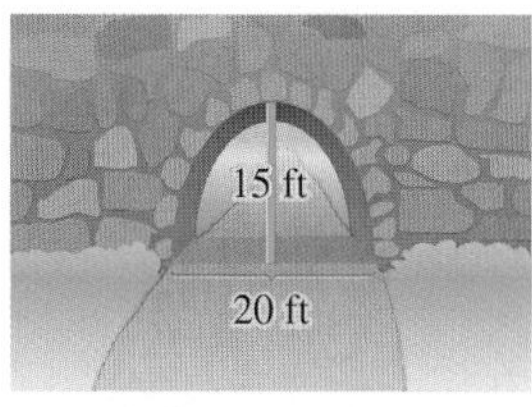

NOT TO SCALE

44. *Height and Width of an Overpass* An arch has the shape of half an ellipse. The equation of the ellipse is $100x^2 + 324y^2 = 32{,}400$, where x and y are in meters.

(a) How high is the center of the arch?

(b) How wide is the arch across the bottom?

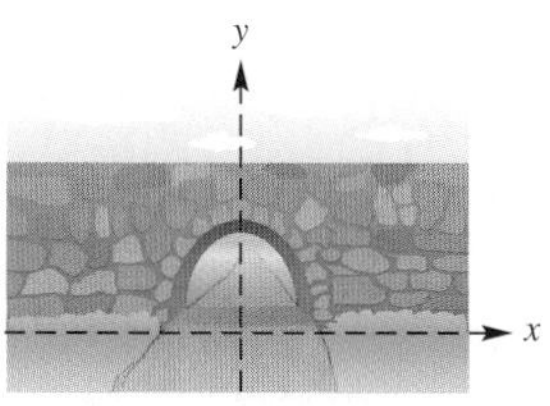

NOT TO SCALE

45. *Orbit of Halley's Comet* The famous Halley's comet last passed by Earth in February 1986 and will next return in 2062. It has an elliptical orbit of eccentricity 0.9673 with the sun at one focus. The greatest distance of the comet from the sun is 3281 million mi. Find the least distance between Halley's comet and the sun. (*Source: World Almanac and Book of Facts.*)

46. *(Modeling) Orbit of a Satellite* The coordinates in miles for the orbit of the artificial satellite Explorer VII can be modeled by the equation

$$\frac{x^2}{a^2} + \frac{y^2}{b^2} = 1,$$

where $a = 4465$ and $b = 4462$. Earth's center is located at one focus of the elliptical orbit. (*Source:* Loh, W., *Dynamics and Thermodynamics of Planetary Entry,* Prentice-Hall; Thomson, W., *Introduction to Space Dynamics,* John Wiley and Sons.)

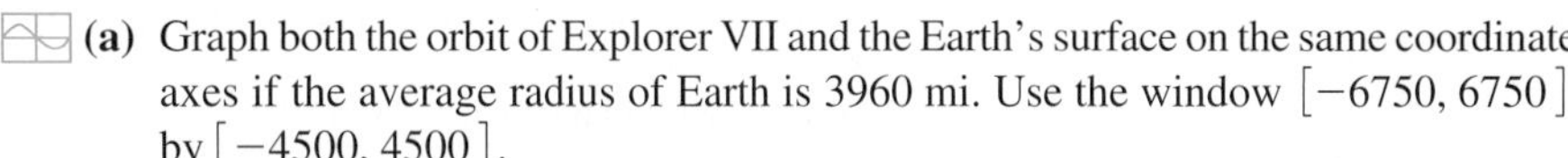

(a) Graph both the orbit of Explorer VII and the Earth's surface on the same coordinate axes if the average radius of Earth is 3960 mi. Use the window $[-6750, 6750]$ by $[-4500, 4500]$.

(b) Find the maximum and minimum heights of the satellite above Earth's surface.

47. *(Modeling) Orbits of Satellites* Neptune and Pluto both have elliptical orbits with the sun at one focus. Neptune's orbit has $a = 30.1$ astronomical units (AU) with an eccentricity of $e = 0.009$, whereas Pluto's orbit has $a = 39.4$ and $e = 0.249$. (*Source:* Zeilik, M., S. Gregory, and E. Smith, *Introductory Astronomy and Astrophysics,* Fourth Edition, Saunders College Publishers.)

(a) Position the sun at the origin and determine equations that model each orbit.

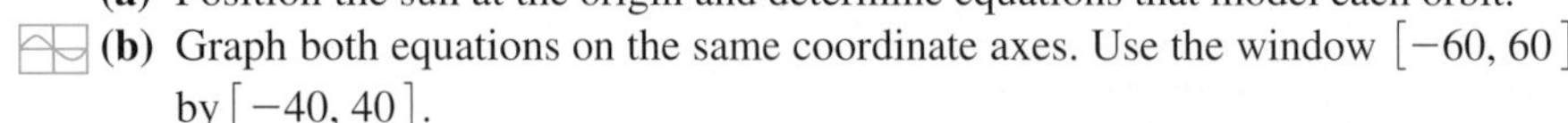

(b) Graph both equations on the same coordinate axes. Use the window $[-60, 60]$ by $[-40, 40]$.

48. *(Modeling) The Roman Colosseum*

(a) The Roman Colosseum is an ellipse with major axis 620 ft and minor axis 513 ft. Find the distance between the foci of this ellipse.

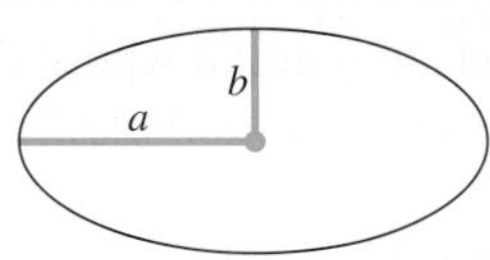

(b) A formula for the approximate perimeter of an ellipse is

$$P \approx 2\pi\sqrt{\frac{a^2 + b^2}{2}},$$

where a and b are the lengths shown in the figure. Use this formula to find the approximate perimeter of the Roman Colosseum.

25. $\frac{x^2}{25} + \frac{y^2}{9} = 1$

26. $\frac{x^2}{16} + \frac{y^2}{12} = 1$

27. $\frac{9x^2}{28} + \frac{9y^2}{64} = 1$

28. $\frac{4x^2}{405} + \frac{4y^2}{729} = 1$

Answers for the remaining exercises are given in the answer section at the back of the text.

49. *Design of a Lithotripter* Suppose a lithotripter is based on the ellipse with equation

$$\frac{x^2}{36} + \frac{y^2}{9} = 1.$$

How far from the center of the ellipse must the kidney stone and the source of the beam be placed? Give the exact answer.

50. *Design of a Lithotripter* Rework **Exercise 49** if the equation of the ellipse is $9x^2 + 4y^2 = 36$.

Chapter 10 Quiz (Sections 10.1–10.2)

[10.1, 10.2]
1. (a) B, D (b) A (c) E (d) C, E (e) D

[10.1]
2. $(y-2)^2 = 12(x+1)$
3. $y = -\frac{1}{2}x^2$

[10.2]
4. $\frac{(x-3)^2}{16} + \frac{(y+2)^2}{25} = 1$
5. $\frac{(x+3)^2}{9} + \frac{(y-7)^2}{25} = 1$

[10.1]
6. parabola; vertex: $(-3, -4)$; focus: $\left(-3, -\frac{15}{4}\right)$; directrix: $y = -\frac{17}{4}$; axis: $x = -3$

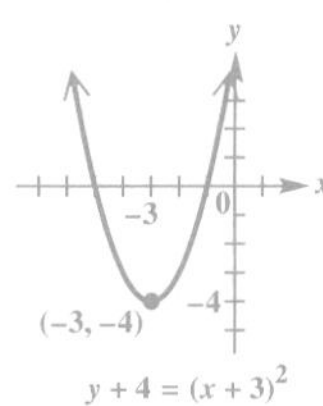

$y + 4 = (x+3)^2$

Answers for the remaining exercises are given in the answer section at the back of the text.

1. *Concept Check* Match each equation of a conic section in Column I with the appropriate description or descriptions in Column II.

I	II
(a) $x + 3 = 4(y-1)^2$	**A.** circle; center $(-3, 1)$
(b) $(x+3)^2 + (y-1)^2 = 81$	**B.** parabola; opens right
(c) $25(x-2)^2 + (y-1)^2 = 100$	**C.** ellipse; major axis horizontal
(d) $\frac{(x-2)^2}{16} + \frac{(y-1)^2}{9} = 1$	**D.** parabola; vertex $(-3, 1)$
(e) $-2(x+3)^2 + 1 = y$	**E.** ellipse; center $(2, 1)$

Write an equation for each conic section.

2. parabola with vertex $(-1, 2)$ and focus $(2, 2)$
3. parabola with vertex at the origin; through $\left(\sqrt{10}, -5\right)$; opening downward
4. ellipse with center $(3, -2)$; $a = 5$; $c = 3$; major axis vertical
5. ellipse with foci at $(-3, 3)$ and $(-3, 11)$; major axis of length 10

Identify and then graph each conic section. If it is a parabola, give the vertex, focus, directrix, and axis of symmetry. If it is an ellipse, give the center, vertices, and foci.

6. $y + 4 = (x+3)^2$
7. $4x^2 + 9y^2 = 36$
8. $8(x+1) = (y+3)^2$
9. $\frac{(x+3)^2}{25} + \frac{(y+2)^2}{36} = 1$
10. $x = -4y^2 - 4y - 3$

10.3 Hyperbolas

- Equations and Graphs of Hyperbolas
- Translated Hyperbolas
- Eccentricity

Equations and Graphs of Hyperbolas

An ellipse was defined as the set of all points in a plane the sum of whose distances from two fixed points is a constant. A *hyperbola* is defined similarly.

Hyperbola

A **hyperbola** is the set of all points in a plane such that the absolute value of the difference of the distances from two fixed points is constant. The two fixed points are the **foci** of the hyperbola.

Suppose a hyperbola has center at the origin and foci at $F'(-c, 0)$ and $F(c, 0)$. See **Figure 28.** The midpoint of the segment $F'F$ is the center of the hyperbola and the points $V'(-a, 0)$ and $V(a, 0)$ are the **vertices** of the hyperbola. The line segment $V'V$ is the **transverse axis** of the hyperbola.

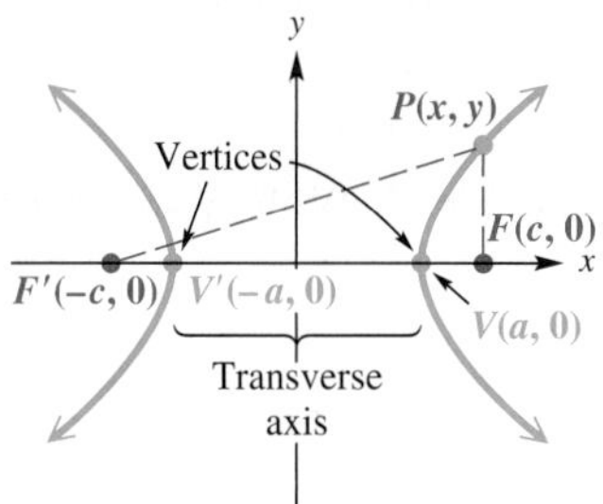

Figure 28

For hyperbolas,

$$d(V, F') - d(V, F) = (c + a) - (c - a) = 2a,$$

so the constant in the definition is $2a$, and

$$|d(P, F') - d(P, F)| = 2a$$

for any point $P(x, y)$ on the hyperbola. The distance formula and algebraic manipulation similar to that used for finding an equation for an ellipse (**Exercise 62**) produce the following result.

$$\frac{x^2}{a^2} - \frac{y^2}{c^2 - a^2} = 1$$

Replacing $c^2 - a^2$ with b^2 gives

$$\mathbf{\frac{x^2}{a^2} - \frac{y^2}{b^2} = 1}$$

as an equation of the hyperbola in **Figure 28.** Letting $y = 0$ shows that the x-intercepts are $\pm a$. If $x = 0$, the equation becomes $y^2 = -b^2$, which has no real number solutions, showing that this hyperbola has no y-intercepts. Again, see **Figure 28.**

Start with the equation for a hyperbola and solve for y.

$$\frac{x^2}{a^2} - \frac{y^2}{b^2} = 1$$

$$\frac{x^2}{a^2} - 1 = \frac{y^2}{b^2} \qquad \text{Subtract 1 and add } \frac{y^2}{b^2}.$$

$$\frac{x^2 - a^2}{a^2} = \frac{y^2}{b^2} \qquad \text{Write the left side as a single fraction.}$$

Remember both the positive and negative square roots.

$$y = \pm\frac{b}{a}\sqrt{x^2 - a^2} \qquad \text{Take the square root on each side (Section 1.6); multiply by } b.$$

If x^2 is very large in comparison to a^2, the difference $x^2 - a^2$ is very close to x^2. If this happens, then the points satisfying the final equation above are very close to one of the lines

$$\mathbf{y = \pm\frac{b}{a}x.}$$

Thus, as $|x|$ increases without bound, the points of the hyperbola $\frac{x^2}{a^2} - \frac{y^2}{b^2} = 1$ approach the lines $y = \pm\frac{b}{a}x$. These lines are **asymptotes** of the hyperbola and are useful when sketching the graph.

EXAMPLE 1 Using Asymptotes to Graph a Hyperbola

Graph $\frac{x^2}{25} - \frac{y^2}{49} = 1$. Sketch the asymptotes, and find the coordinates of the vertices and foci. Give the domain and range.

ALGEBRAIC SOLUTION

For this hyperbola, $a = 5$ and $b = 7$. With these values,

$$y = \pm \frac{b}{a}x \quad \text{becomes} \quad y = \pm \frac{7}{5}x. \quad \text{Asymptotes}$$

If we choose $x = 5$, then $y = \pm 7$. Choosing $x = -5$ also gives $y = \pm 7$. These four ordered pairs—$(5, 7)$, $(5, -7)$, $(-5, 7)$, and $(-5, -7)$—are the coordinates of the corners of the rectangle shown in **Figure 29.**

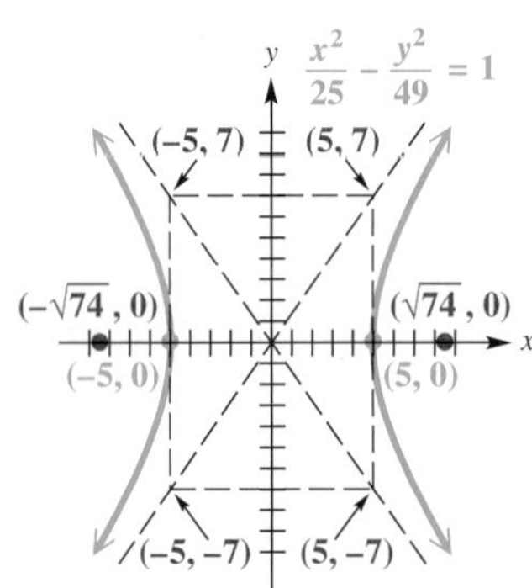

Figure 29

The extended diagonals of this rectangle, called the **fundamental rectangle,** are the asymptotes of the hyperbola. Since $a = 5$, the vertices of the hyperbola are $(5, 0)$ and $(-5, 0)$, as shown in **Figure 29.** We find the foci by letting

$$c^2 = a^2 + b^2 = 25 + 49 = 74, \quad \text{so} \quad c = \sqrt{74}.$$

Therefore, the foci are $\left(\sqrt{74}, 0\right)$ and $\left(-\sqrt{74}, 0\right)$. The domain is $(-\infty, -5] \cup [5, \infty)$, and the range is $(-\infty, \infty)$.

GRAPHING CALCULATOR SOLUTION

The graph of a hyperbola is not the graph of a function. We solve for y in $\frac{x^2}{25} - \frac{y^2}{49} = 1$ to get equations of the ***two*** functions

$$y = \pm \frac{7}{5}\sqrt{x^2 - 25}.$$

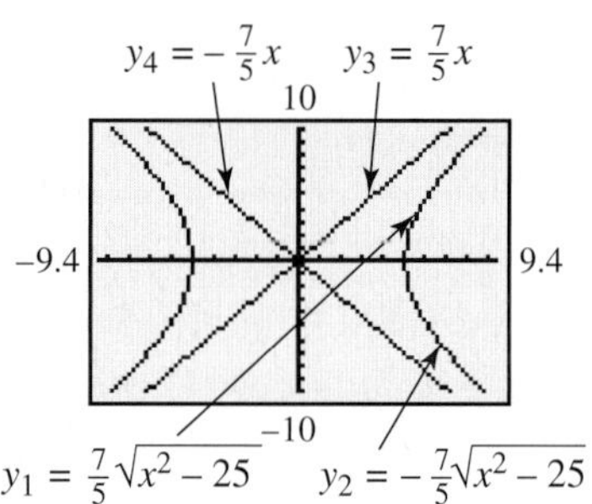

Figure 30

The graph of y_1 is the upper portion of each branch of the hyperbola shown in **Figure 30,** and the graph of y_2 is the lower portion of each branch. We could enter $y_2 = -y_1$ to get the part of the graph below the x-axis.

The asymptotes are also shown. We can use tracing to observe how the branches of the hyperbola approach the asymptotes.

Now Try Exercise 5.

Classroom Example 1

Graph $\frac{x^2}{4} - \frac{y^2}{25} = 1$. Sketch the asymptotes, and find the coordinates of the vertices and foci. Give the domain and range.

Answer:

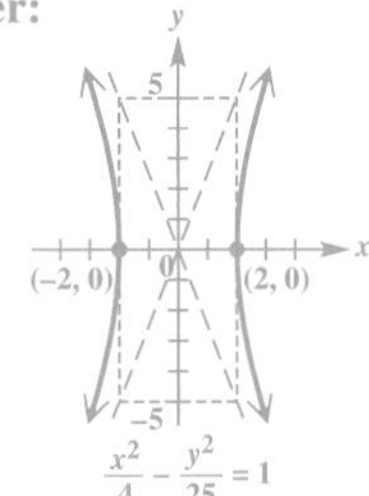

vertices: $(-2, 0)$, $(2, 0)$;

foci: $\left(-\sqrt{29}, 0\right)$, $\left(\sqrt{29}, 0\right)$;

domain: $(-\infty, -2] \cup [2, \infty)$;

range: $(-\infty, \infty)$

NOTE When graphing hyperbolas, remember that the fundamental rectangle and the asymptotes are not actually parts of the graph. They are simply aids in sketching the graph.

While $a > b$ for an ellipse, examples show that for hyperbolas, it is possible that $a > b$, $a < b$, or $a = b$. If the foci of a hyperbola are on the y-axis, the equation of the hyperbola has the form

$$\frac{y^2}{a^2} - \frac{x^2}{b^2} = 1, \quad \text{with asymptotes} \quad y = \pm \frac{a}{b}x.$$

As justified previously, when foci are on the x-axis an asymptote has equation $y = \frac{b}{a}x$, which is equivalent to $ay = bx$, or $ay - bx = 0$. If we divide through by ab, we obtain

$$\frac{y}{b} - \frac{x}{a} = 0. \quad \text{Equation of asymptote}$$

Teaching Tip Show students that if (x_1, y_1) is a point on the graph of a hyperbola centered at the origin, then the points $(-x_1, y_1)$, $(x_1, -y_1)$, and $(-x_1, -y_1)$ are also on the graph.

Now compare this to the original hyperbola equation

$$\frac{y^2}{b^2} - \frac{x^2}{a^2} = 1, \quad \text{Equation of hyperbola}$$

and notice the similarities: The equation of the asymptote has the two squares replaced by the square roots, and 1 replaced by 0.

A similar observation is made for the equation of the other asymptote $y = -\frac{b}{a}x$, and for the case where the foci are on the y-axis.

Standard Forms of Equations for Hyperbolas

The hyperbola with center at the origin and equation

$$\frac{x^2}{a^2} - \frac{y^2}{b^2} = 1$$

has vertices $(\pm a, 0)$, asymptotes $y = \pm \frac{b}{a}x$, and foci $(\pm c, 0)$, where $c^2 = a^2 + b^2$.

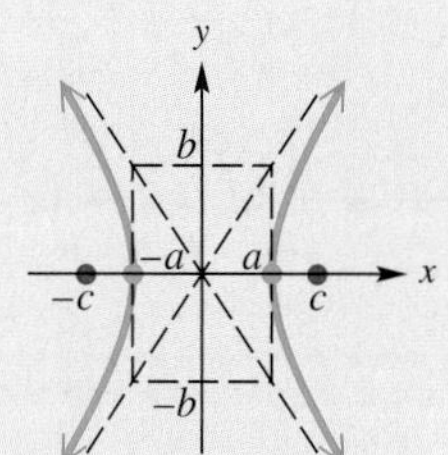

Transverse axis on x-axis

The hyperbola with center at the origin and equation

$$\frac{y^2}{a^2} - \frac{x^2}{b^2} = 1$$

has vertices $(0, \pm a)$, asymptotes $y = \pm \frac{a}{b}x$, and foci $(0, \pm c)$, where $c^2 = a^2 + b^2$.

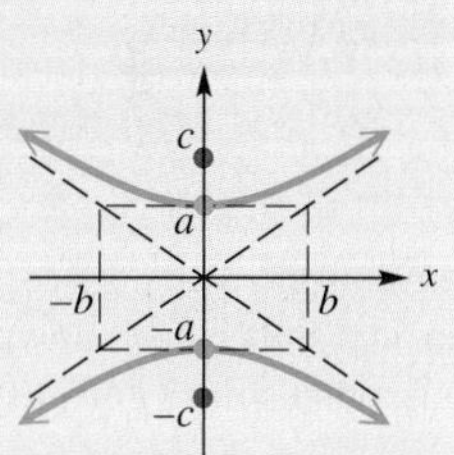

Transverse axis on y-axis

Classroom Example 2
Graph $16y^2 - 25x^2 = 400$. Give the domain and range.

Answer:

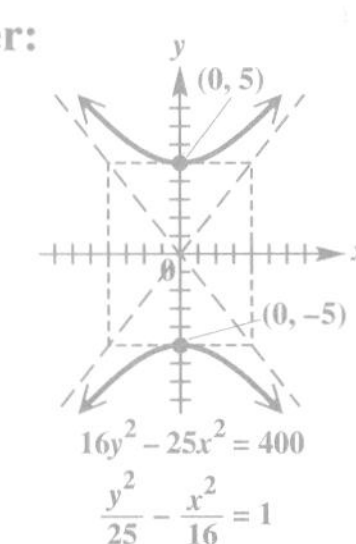

domain: $(-\infty, \infty)$;
range: $(-\infty, -5] \cup [5, \infty)$

EXAMPLE 2 Graphing a Hyperbola

Graph $25y^2 - 4x^2 = 100$. Give the domain and range.

SOLUTION $\frac{y^2}{4} - \frac{x^2}{25} = 1$ Divide by 100, and write in standard form.

This hyperbola is centered at the origin, has foci on the y-axis, and has vertices $(0, 2)$ and $(0, -2)$. One way to find the equations of the asymptotes is to use the forms discussed above.

$$\frac{y^2}{4} - \frac{x^2}{25} = 0 \quad \text{Equation of hyperbola with 1 replaced by 0}$$

$$\frac{y^2}{4} = \frac{x^2}{25} \quad \text{Add } \tfrac{x^2}{25}.$$

$$y^2 = \frac{4x^2}{25} \quad \text{Multiply by 4.}$$

$$y = \pm\frac{2}{5}x \quad \text{Square root property **(Section 1.4)**}$$

Remember both the positive and negative square roots.

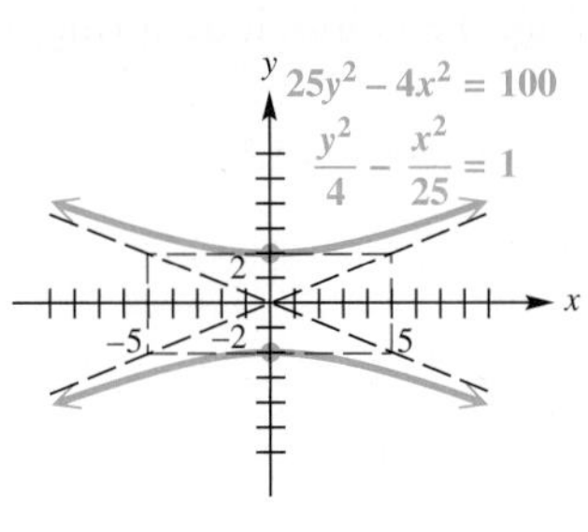

Figure 31

To graph the asymptotes, use the points $(5, 2)$, $(5, -2)$, $(-5, 2)$, and $(-5, -2)$ to determine the fundamental rectangle. The diagonals of this rectangle are the asymptotes for the graph, as shown in **Figure 31.** The domain of the relation is $(-\infty, \infty)$, and the range is $(-\infty, -2] \cup [2, \infty)$.

Now Try Exercise 13.

Translated Hyperbolas Like an ellipse, a hyperbola can have its center translated away from the origin.

Standard Forms for Hyperbolas Centered at (h, k)

A hyperbola with center (h, k) and either a horizontal or vertical transverse axis satisfies one of the following equations, where $c^2 = a^2 + b^2$.

$$\frac{(x-h)^2}{a^2} - \frac{(y-k)^2}{b^2} = 1$$

Transverse axis: horizontal;
vertices: $(h \pm a, k)$;
foci: $(h \pm c, k)$;
asymptotes: $y = \pm \frac{b}{a}(x-h) + k$

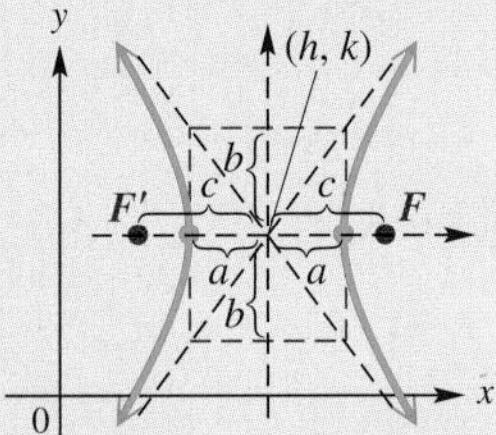

Horizontal transverse axis

$$\frac{(y-k)^2}{a^2} - \frac{(x-h)^2}{b^2} = 1$$

Transverse axis: vertical;
vertices: $(h, k \pm a)$;
foci: $(h, k \pm c)$;
asymptotes: $y = \pm \frac{a}{b}(x-h) + k$

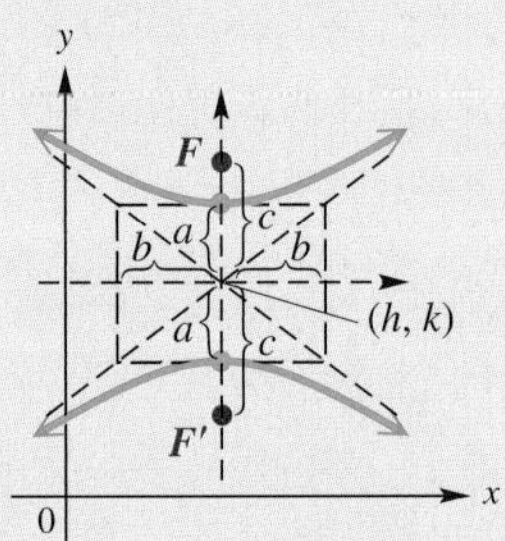

Vertical transverse axis

Teaching Tip Students may find it easier to graph

$$\frac{y^2}{9} - \frac{x^2}{4} = 1$$

in **Example 3** and then apply a vertical shift 2 units down and a horizontal shift 3 units to the left.

Classroom Example 3
Graph $\frac{(x-2)^2}{16} - (y+4)^2 = 1$. Give the domain and range.

Answer:

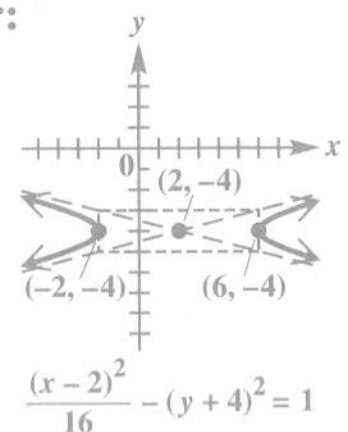

domain: $(-\infty, -2] \cup [6, \infty)$;
range: $(-\infty, \infty)$

NOTE The asymptotes for a hyperbola *always* pass through the center (h, k). By the point-slope form of a line, the equation of any asymptote is $y = m(x-h) + k$. If the transverse axis is horizontal, then $m = \pm\frac{b}{a}$. If it is vertical, then $m = \pm\frac{a}{b}$.

EXAMPLE 3 Graphing a Hyperbola Translated Away from the Origin

Graph $\frac{(y+2)^2}{9} - \frac{(x+3)^2}{4} = 1$. Give the domain and range.

SOLUTION This equation represents a hyperbola centered at $(-3, -2)$. For this vertical hyperbola, $a = 3$ and $b = 2$. The x-values of the vertices are -3. Locate the y-values of the vertices by taking the y-value of the center, -2, and adding and subtracting 3. Thus, the vertices are $(-3, 1)$ and $(-3, -5)$.

The asymptotes have slopes $\pm\frac{3}{2}$ and pass through the center $(-3, -2)$. The equations of the asymptotes can be found by using the point-slope form.

$$[y-(-2)] = \pm\frac{3}{2}[x-(-3)] \quad \text{Point-slope form (Section 2.5)}$$

$$y = \pm\frac{3}{2}(x+3) - 2 \quad \text{Solve for } y.$$

The graph is shown in **Figure 32.** The domain of the relation is $(-\infty, \infty)$, and the range is $(-\infty, -5] \cup [1, \infty)$.

Now Try Exercise 17.

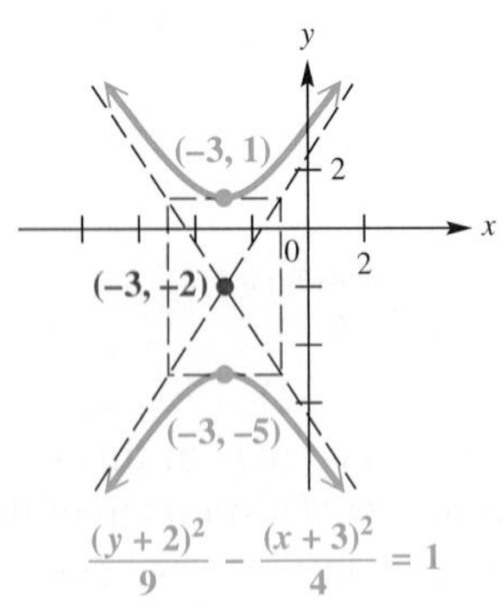

Figure 32

Eccentricity If we apply the definition of eccentricity from the previous section to the hyperbola, we get the following.

$$e = \frac{\sqrt{a^2 + b^2}}{a} = \frac{c}{a}$$

Since $c > a$, we have $e > 1$. ***Narrow hyperbolas have e near 1, and wide hyperbolas have large e.*** See **Figure 33.**

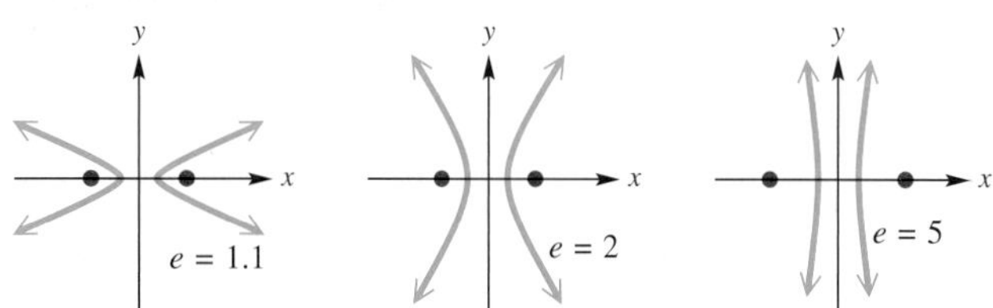

Figure 33

Classroom Example 4
Find the eccentricity of the following hyperbola.

$$\frac{y^2}{100} - \frac{x^2}{81} = 1$$

Answer: 1.3

EXAMPLE 4 Finding Eccentricity from the Equation of a Hyperbola

Find the eccentricity of the hyperbola $\frac{x^2}{9} - \frac{y^2}{4} = 1$.

SOLUTION Here, $a^2 = 9$ and thus $a = 3$. Also, $b^2 = 4$.

$c^2 = a^2 + b^2$ Relationship for hyperbolas

$c = \sqrt{a^2 + b^2}$ Take the positive square root, since $c > 0$.

$c = \sqrt{9 + 4}$ Substitute.

$c = \sqrt{13}$ Add.

The eccentricity e is given by $e = \frac{c}{a}$.

$$e = \frac{\sqrt{13}}{3} \approx 1.2$$

Now Try Exercise 27.

Classroom Example 5
Find the equation of the hyperbola with eccentricity 3 and foci at $(-2, 5)$ and $(-2, -3)$.

Answer:

$$\frac{(y-1)^2}{\frac{16}{9}} - \frac{(x+2)^2}{\frac{128}{9}} = 1, \text{ or}$$

$$\frac{9(y-1)^2}{16} - \frac{9(x+2)^2}{128} = 1$$

EXAMPLE 5 Finding the Equation of a Hyperbola

Find the equation of the hyperbola with eccentricity 2 and foci at $(-9, 5)$ and $(-3, 5)$.

SOLUTION Since the foci have the same y-coordinate, the line through them, and therefore the hyperbola, is horizontal. The center of the hyperbola is halfway between the two foci at $(-6, 5)$. The distance from each focus to the center is $c = 3$ and therefore $c^2 = 9$. Since $e = \frac{c}{a}$, we have $a = \frac{c}{e} = \frac{3}{2}$ and $a^2 = \frac{9}{4}$.

$b^2 = c^2 - a^2$ Relationship for hyperbolas

$b^2 = 9 - \frac{9}{4}$ Substitute.

$b^2 = \frac{27}{4}$ Subtract.

The equation of the hyperbola is

$$\frac{(x+6)^2}{\frac{9}{4}} - \frac{(y-5)^2}{\frac{27}{4}} = 1, \quad \text{or} \quad \frac{4(x+6)^2}{9} - \frac{4(y-5)^2}{27} = 1.$$

Simplify complex fractions. **(Section R.5)**

Now Try Exercise 45.

The following chart summarizes our discussion of eccentricity in this chapter.

Conic	Eccentricity
Parabola	$e = 1$
Circle	$e = 0$
Ellipse	$e = \frac{c}{a}$ and $0 < e < 1$
Hyperbola	$e = \frac{c}{a}$ and $e > 1$

10.3 Exercises

1. C 2. B 3. D 4. A

5. $(-\infty, -4] \cup [4, \infty)$; $(-\infty, \infty)$; $(0, 0)$; $(-4, 0), (4, 0)$; $(-5, 0), (5, 0)$; $y = \pm\frac{3}{4}x$

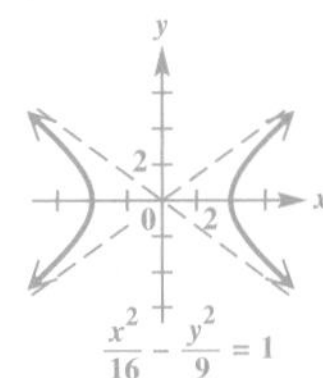

6. $(-\infty, -5] \cup [5, \infty)$; $(-\infty, \infty)$; $(0, 0)$; $(-5, 0), (5, 0)$; $(-13, 0), (13, 0)$; $y = \pm\frac{12}{5}x$

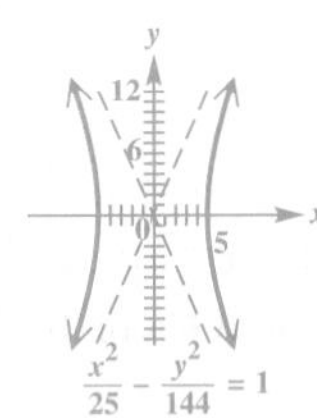

7. $(-\infty, \infty)$; $(-\infty, -5] \cup [5, \infty)$; $(0, 0)$; $(0, -5), (0, 5)$; $(0, -\sqrt{74}), (0, \sqrt{74})$; $y = \pm\frac{5}{7}x$

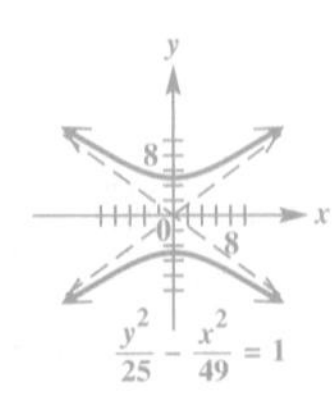

Concept Check Match each equation with the correct graph.

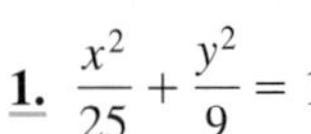

1. $\frac{x^2}{25} + \frac{y^2}{9} = 1$

2. $\frac{x^2}{9} + \frac{y^2}{25} = 1$

3. $\frac{x^2}{9} - \frac{y^2}{25} = 1$

4. $\frac{x^2}{25} - \frac{y^2}{9} = 1$

A.

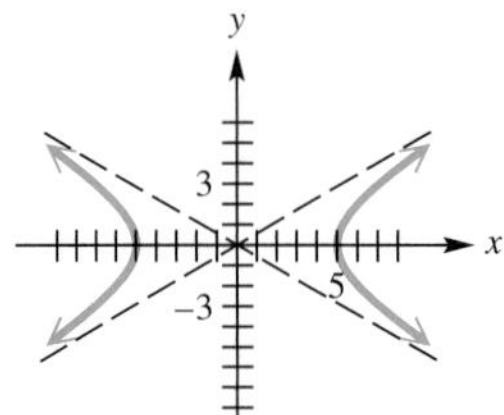

B.

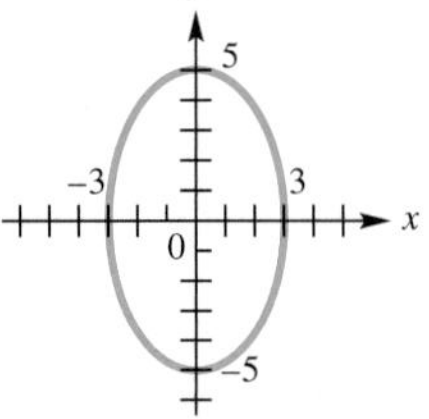

C.

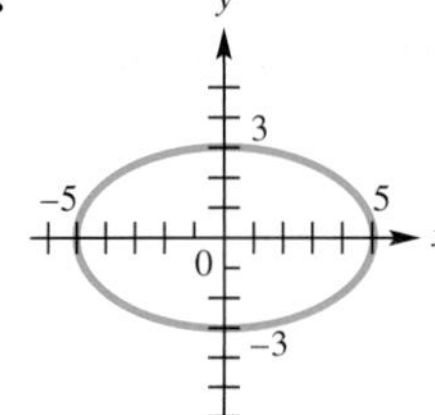

D.

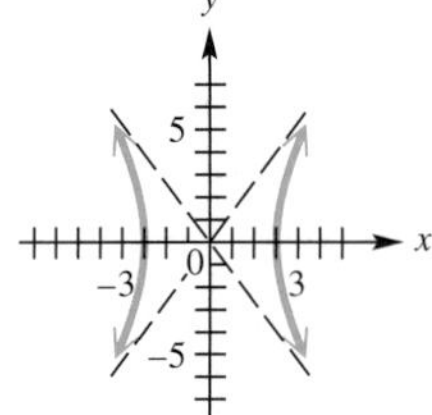

Graph each hyperbola. Give the domain, range, center, vertices, foci, and equations of the asymptotes for each figure. ***See Examples 1–3.***

5. $\frac{x^2}{16} - \frac{y^2}{9} = 1$

6. $\frac{x^2}{25} - \frac{y^2}{144} = 1$

7. $\frac{y^2}{25} - \frac{x^2}{49} = 1$

8. $\frac{y^2}{64} - \frac{x^2}{4} = 1$

9. $x^2 - y^2 = 9$

10. $x^2 - 4y^2 = 64$

11. $9x^2 - 25y^2 = 225$

12. $4y^2 - 16x^2 = 64$

13. $4y^2 - 25x^2 = 100$

14. $x^2 - 4y^2 = 16$

15. $9x^2 - 4y^2 = 1$

16. $25y^2 - 9x^2 = 1$

17. $\frac{(y-7)^2}{36} - \frac{(x-4)^2}{64} = 1$

18. $\frac{(x+6)^2}{144} - \frac{(y+4)^2}{81} = 1$

19. $\frac{(x+3)^2}{16} - \frac{(y-2)^2}{9} = 1$

20. $\frac{(y+5)^2}{4} - \frac{(x-1)^2}{16} = 1$

21. $16(x+5)^2 - (y-3)^2 = 1$

22. $4(x+9)^2 - 25(y+6)^2 = 100$

8. $(-\infty, \infty)$;
$(-\infty, -8] \cup [8, \infty)$;
$(0, 0)$; $(0, -8)$, $(0, 8)$;
$(0, -2\sqrt{17})$, $(0, 2\sqrt{17})$;
$y = \pm 4x$

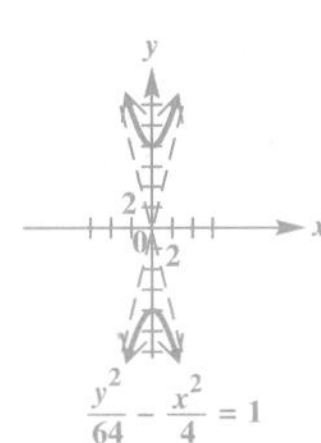

9. $(-\infty, -3] \cup [3, \infty)$;
$(-\infty, \infty)$; $(0, 0)$;
$(-3, 0)$, $(3, 0)$;
$(-3\sqrt{2}, 0)$, $(3\sqrt{2}, 0)$;
$y = \pm x$

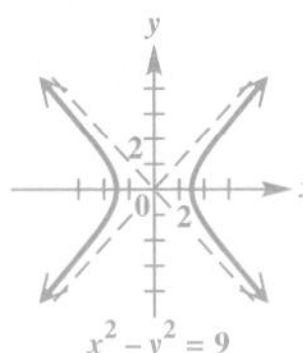

10. $(-\infty, -8] \cup [8, \infty)$;
$(-\infty, \infty)$; $(0, 0)$;
$(-8, 0)$, $(8, 0)$;
$(-4\sqrt{5}, 0)$, $(4\sqrt{5}, 0)$;
$y = \pm \frac{1}{2}x$

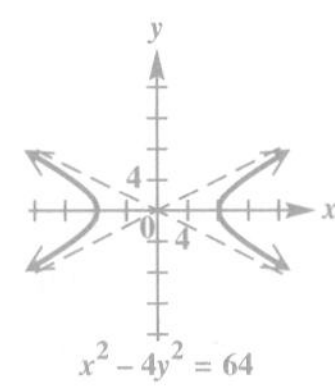

11. $(-\infty, -5] \cup [5, \infty)$;
$(-\infty, \infty)$; $(0, 0)$;
$(-5, 0)$, $(5, 0)$;
$(-\sqrt{34}, 0)$, $(\sqrt{34}, 0)$;
$y = \pm \frac{3}{5}x$

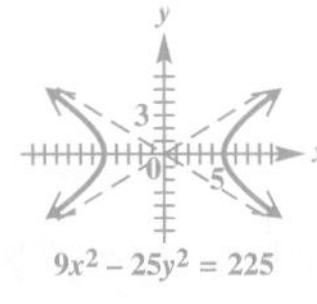

Graph each equation. Give the domain and range. Identify any that are graphs of functions. ***See Example 3 in the previous section.***

23. $\frac{y}{3} = \sqrt{1 + \frac{x^2}{16}}$

24. $\frac{x}{3} = -\sqrt{1 + \frac{y^2}{25}}$

25. $5x = -\sqrt{1 + 4y^2}$

26. $3y = \sqrt{4x^2 - 16}$

Find the eccentricity of each hyperbola to the nearest tenth. ***See Example 4.***

27. $\frac{x^2}{8} - \frac{y^2}{8} = 1$

28. $\frac{x^2}{2} - \frac{y^2}{18} = 1$

29. $16y^2 - 8x^2 = 16$

30. $8y^2 - 2x^2 = 16$

Write an equation for each hyperbola. ***See Examples 4 and 5.***

31. x-intercepts ± 3; foci at $(-5, 0)$, $(5, 0)$

32. y-intercepts ± 12; foci at $(0, -15)$, $(0, 15)$

33. vertices at $(0, 6)$, $(0, -6)$; asymptotes $y = \pm \frac{1}{2}x$

34. vertices at $(-10, 0)$, $(10, 0)$; asymptotes $y = \pm 5x$

35. vertices at $(-3, 0)$, $(3, 0)$; passing through $(-6, -1)$

36. vertices at $(0, 5)$, $(0, -5)$; passing through $(-3, 10)$

37. foci at $(0, \sqrt{13})$, $(0, -\sqrt{13})$; asymptotes $y = \pm 5x$

38. foci at $(-3\sqrt{5}, 0)$, $(3\sqrt{5}, 0)$; asymptotes $y = \pm 2x$

39. vertices at $(4, 5)$, $(4, 1)$; asymptotes $y = \pm 7(x - 4) + 3$

40. vertices at $(5, -2)$, $(1, -2)$; asymptotes $y = \pm \frac{3}{2}(x - 3) - 2$

41. center at $(1, -2)$; focus at $(-2, -2)$; vertex at $(-1, -2)$

42. center at $(9, -7)$; focus at $(9, -17)$; vertex at $(9, -13)$

43. eccentricity 3; center at $(0, 0)$; vertex at $(0, 7)$

44. eccentricity $\frac{5}{3}$; center at $(8, 7)$; focus at $(3, 7)$

45. eccentricity $\frac{25}{9}$; foci at $(9, -1)$, $(-11, -1)$

46. eccentricity $\frac{5}{4}$; vertices at $(2, 10)$, $(2, 2)$

Determine the two equations necessary to graph each hyperbola with a graphing calculator, and graph it in the viewing window indicated. ***See Example 1.***

47. $\frac{x^2}{4} - \frac{y^2}{16} = 1$; $[-9.4, 9.4]$ by $[-10, 10]$

48. $\frac{x^2}{25} - \frac{y^2}{49} = 1$; $[-9.4, 9.4]$ by $[-10, 10]$

49. $4y^2 - 36x^2 = 144$; $[-10, 10]$ by $[-15, 15]$

50. $y^2 - 9x^2 = 9$; $[-10, 10]$ by $[-10, 10]$

12. $(-\infty, \infty)$; $(-\infty, -4] \cup [4, \infty)$; $(0, 0)$; $(0, -4)$, $(0, 4)$; $(0, -2\sqrt{5})$, $(0, 2\sqrt{5})$; $y = \pm 2x$

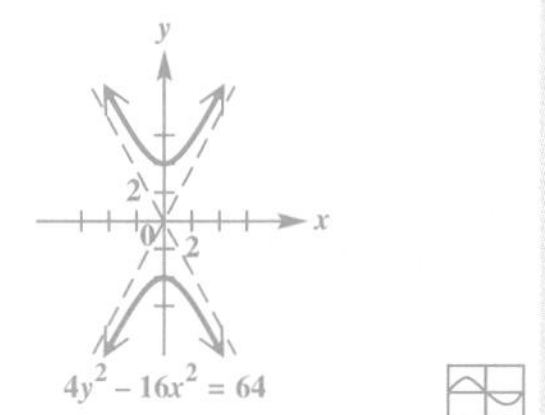

13. $(-\infty, \infty)$; $(-\infty, -5] \cup [5, \infty)$; $(0, 0)$; $(0, -5)$, $(0, 5)$; $(0, -\sqrt{29})$, $(0, \sqrt{29})$; $y = \pm \frac{5}{2}x$

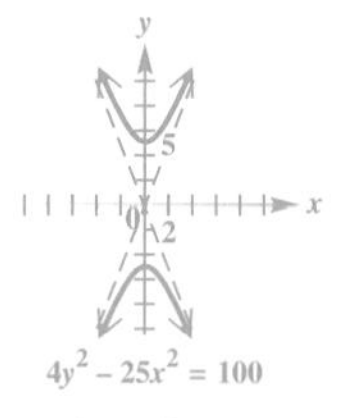

14. $(-\infty, -4] \cup [4, \infty)$; $(-\infty, \infty)$; $(0, 0)$; $(-4, 0)$, $(4, 0)$; $(-2\sqrt{5}, 0)$, $(2\sqrt{5}, 0)$; $y = \pm \frac{1}{2}x$

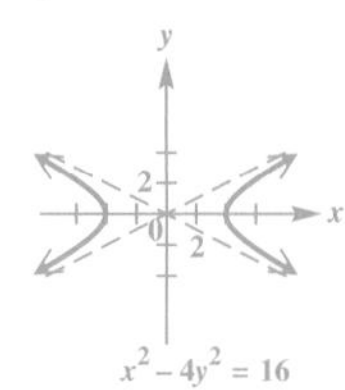

15. $(-\infty, -\frac{1}{3}] \cup [\frac{1}{3}, \infty)$; $(-\infty, \infty)$; $(0, 0)$; $(-\frac{1}{3}, 0)$, $(\frac{1}{3}, 0)$; $\left(-\frac{\sqrt{13}}{6}, 0\right)$, $\left(\frac{\sqrt{13}}{6}, 0\right)$; $y = \pm \frac{3}{2}x$

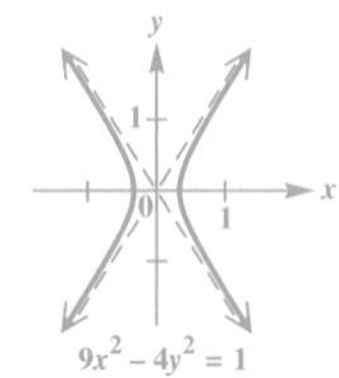

16. $(-\infty, \infty)$; $(-\infty, -\frac{1}{5}] \cup [\frac{1}{5}, \infty)$; $(0, 0)$; $(0, -\frac{1}{5})$, $(0, \frac{1}{5})$; $\left(0, -\frac{\sqrt{34}}{15}\right)$, $\left(0, \frac{\sqrt{34}}{15}\right)$; $y = \pm \frac{3}{5}x$

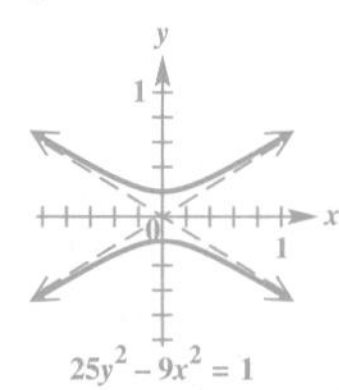

Relating Concepts

For individual or collaborative investigation *(Exercises 51–56)*

The graph of $\frac{x^2}{4} - y^2 = 1$ is a hyperbola. We know that the graph of this hyperbola approaches its asymptotes as $|x|$ increases without bound. ***Work Exercises 51–56 in order,*** *to see the relationship between the hyperbola and one of its asymptotes.*

51. Solve $\frac{x^2}{4} - y^2 = 1$ for y, and choose the positive square root.

52. Find the equation of the asymptote with positive slope.

53. Use a calculator to evaluate the y-coordinate of the point where $x = 50$ on the graph of the portion of the hyperbola represented by the equation obtained in **Exercise 51.** Round your answer to the nearest hundredth.

54. Find the y-coordinate of the point where $x = 50$ on the graph of the asymptote found in **Exercise 52.**

55. Compare your results in **Exercises 53 and 54.** How do they support the following statement?

    When $x = 50$, the graph of the function defined by the equation found in **Exercise 51** lies *below* the graph of the asymptote found in **Exercise 52.**

56. What do you think will happen if we choose x-values greater than 50?

Solve each problem.

57. *(Modeling) Atomic Structure* In 1911, Ernest Rutherford discovered the basic structure of the atom by "shooting" positively charged alpha particles with a speed of 10^7 m per sec at a piece of gold foil 6×10^{-7} m thick. Only a small percentage of the alpha particles struck a gold nucleus head-on and were deflected directly back toward their source. The rest of the particles often followed a hyperbolic trajectory because they were repelled by positively charged gold nuclei. As a result of this famous experiment, Rutherford proposed that the atom was composed mostly of empty space with a small and dense nucleus.

    The figure shows an alpha particle A initially approaching a gold nucleus N and being deflected at an angle $\theta = 90°$. N is located at a focus of the hyperbola, and the trajectory of A passes through a vertex of the hyperbola. (*Source:* Semat, H., and J. Albright, *Introduction to Atomic and Nuclear Physics,* Fifth Edition, International Thomson Publishing.)

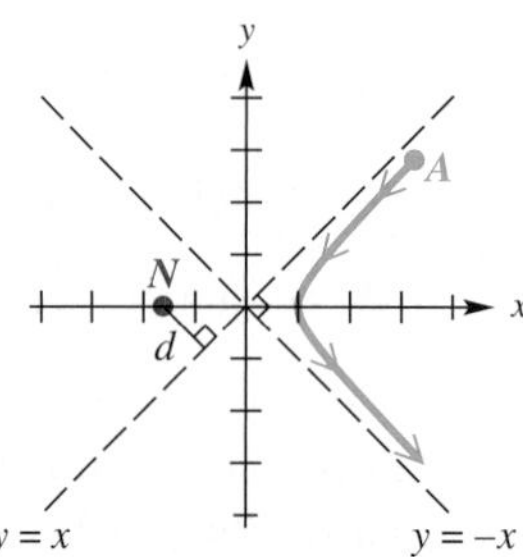

    (a) Determine the equation of the trajectory of the alpha particle if $d = 5 \times 10^{-14}$ m.
    (b) What was the minimum distance between the centers of the alpha particle and the gold nucleus?

58. *LORAN System* Ships and planes often use a location-finding system called LORAN. With this system, a radio transmitter at M in the figure on the next page sends out a series of pulses. When each pulse is received at transmitter S, it then sends out a pulse. A ship at P receives pulses from both M and S. A receiver on the ship measures the difference in the arrival times of the pulses. The navigator then consults a special map showing hyperbolas that correspond to the differences in arrival times (which give the distances d_1 and d_2 in the figure). In this way the ship can be located as lying on a branch of a particular hyperbola.

17. $(-\infty, \infty)$; $(-\infty, 1] \cup [13, \infty)$; $(4, 7)$; $(4, 1)$, $(4, 13)$; $(4, -3)$, $(4, 17)$; $y = \pm\frac{3}{4}(x-4)+7$

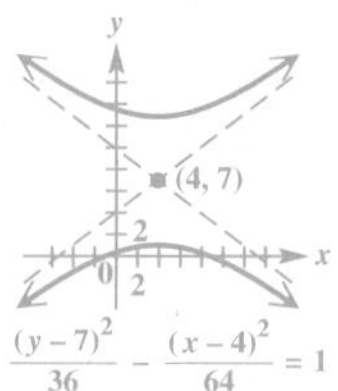

18. $(-\infty, -18] \cup [6, \infty)$; $(-\infty, \infty)$; $(-6, -4)$; $(-18, -4)$, $(6, -4)$; $(-21, -4)$, $(9, -4)$; $y = \pm\frac{3}{4}(x+6)-4$

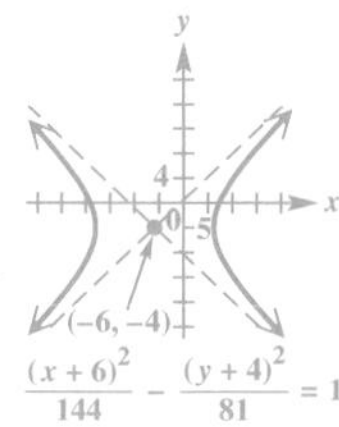

19. $(-\infty, -7] \cup [1, \infty)$; $(-\infty, \infty)$; $(-3, 2)$; $(-7, 2)$, $(1, 2)$; $(-8, 2)$, $(2, 2)$; $y = \pm\frac{3}{4}(x+3)+2$

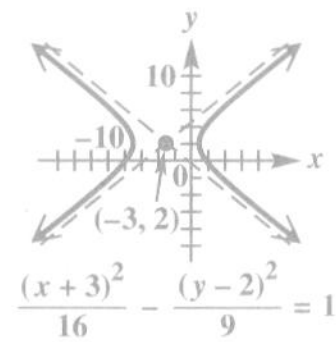

20. $(-\infty, \infty)$; $(-\infty, -7] \cup [-3, \infty)$; $(1, -5)$; $(1, -7)$, $(1, -3)$; $(1, -5-2\sqrt{5})$, $(1, -5+2\sqrt{5})$; $y = \pm\frac{1}{2}(x-1)-5$

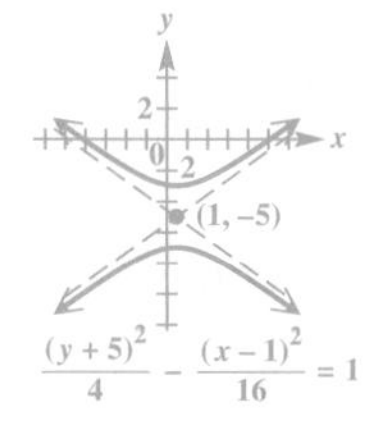

21. $(-\infty, -\frac{21}{4}] \cup [-\frac{19}{4}, \infty)$; $(-\infty, \infty)$; $(-5, 3)$; $(-\frac{21}{4}, 3)$, $(-\frac{19}{4}, 3)$; $\left(-5 - \frac{\sqrt{17}}{4}, 3\right)$, $\left(-5 + \frac{\sqrt{17}}{4}, 3\right)$; $y = \pm 4(x+5)+3$

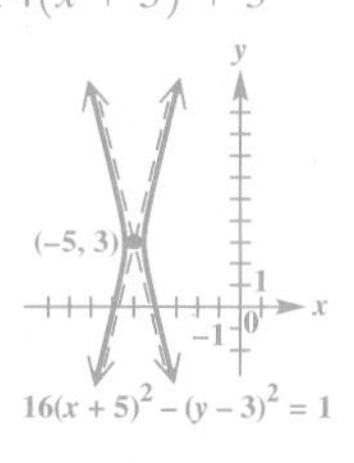

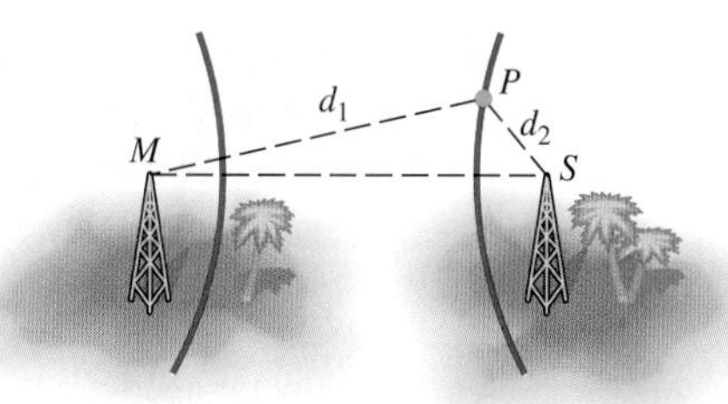

Suppose that in the figure, $d_1 = 80$ mi, $d_2 = 30$ mi, and the distance MS between the transmitters is 100 mi. Use the definition of a hyperbola to find an equation of the hyperbola on which the ship is located.

59. *Sound Detection* Microphones are placed at points $(-c, 0)$ and $(c, 0)$. An explosion occurs at point $P(x, y)$ having positive x-coordinate. See the figure. The sound is detected at the closer microphone t seconds before being detected at the farther microphone. Assume that sound travels at a speed of 330 m per sec, and show that P must be on the hyperbola

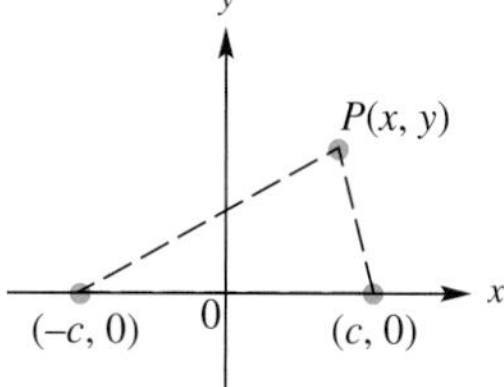

$$\frac{x^2}{330^2t^2} - \frac{y^2}{4c^2 - 330^2t^2} = \frac{1}{4}.$$

60. *Rugby Algebra* A rugby field is similar to a modern football field except that the goalpost, which is 18.5 ft wide, is located on the goal line instead of at the back of the endzone. The rugby equivalent of a touchdown, called a **try,** is scored by touching the ball down beyond the goal line. After a try is scored, the scoring team can earn extra points by kicking the ball through the goalposts. The ball must be placed somewhere on the line perpendicular to the goal line and passing through the point where the try was scored. See the figure below on the left.

If that line passes through the goalposts, then the kicker should place the ball at whatever distance is most comfortable. If the line passes outside the goalposts, then the player might choose the point on the line where angle θ in the figure on the left is as large as possible. The problem of determining this optimal point is similar to a problem posed in 1471 by the astronomer Regiomontanus. (*Source:* Maor, E., *Trigonometric Delights,* Princeton, NJ: Princeton University Press.)

The figure on the right shows a vertical line segment AB, where A and B are a and b units above the horizontal axis, respectively. If point P is located on the axis at a distance of x units from point Q, then angle θ is greatest when $x = \sqrt{ab}$.

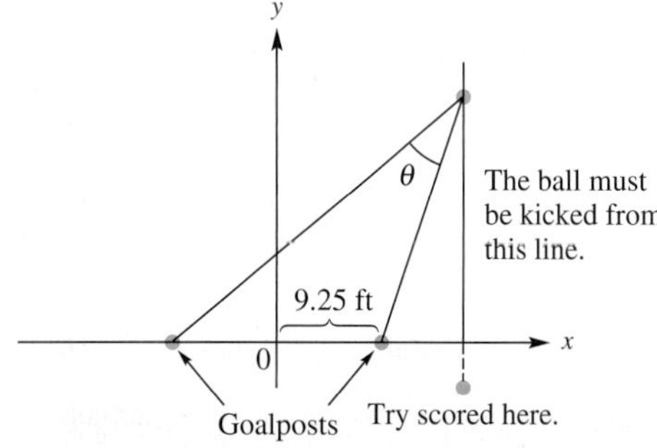

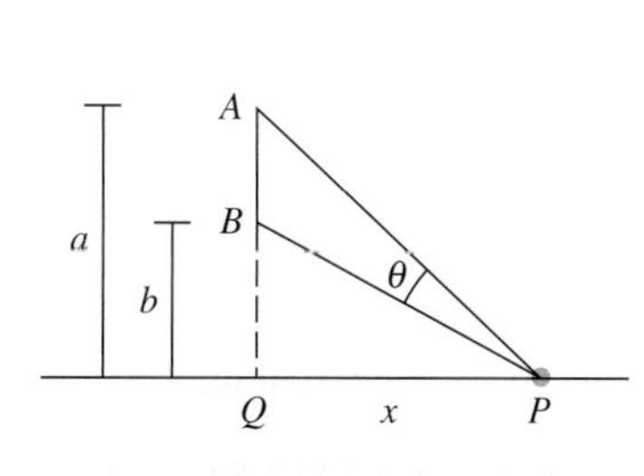

(a) Use the result from Regiomontanus' problem to show that when the line is outside the goalposts, the optimal location to kick the rugby ball lies on the hyperbola

$$x^2 - y^2 = 9.25^2.$$

(b) If the line on which the ball must be kicked is 10 ft to the right of the goalpost, how far from the goal line should the ball be placed to maximize angle θ?

(c) Rugby players find it easier to kick the ball from the hyperbola's asymptote. When the line on which the ball must be kicked is 10 ft to the right of the goalpost, how far will this point differ from the exact optimal location?

22. $(-\infty, -14] \cup [-4, \infty)$; $(-\infty, \infty)$; $(-9, -6)$; $(-14, -6)$, $(-4, -6)$; $(-9 - \sqrt{29}, -6)$, $(-9 + \sqrt{29}, -6)$; $y = \pm\frac{2}{5}(x + 9) - 6$

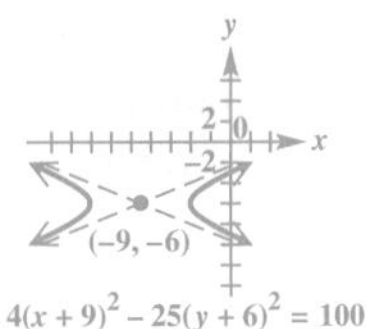

$4(x + 9)^2 - 25(y + 6)^2 = 100$

Answers for the remaining exercises are given in the answer section at the back of the text.

61. *(Modeling) Design of a Sports Complex* Two buildings in a sports complex are shaped and positioned like a portion of the branches of the hyperbola

$$400x^2 - 625y^2 = 250{,}000,$$

where x and y are in meters.

(a) How far apart are the buildings at their closest point?

(b) Find the distance d in the figure.

50 m

d

NOT TO SCALE

62. Suppose a hyperbola has center at the origin, foci at $F'(-c, 0)$ and $F(c, 0)$, and

$$|d(P, F') - d(P, F)| = 2a.$$

Let $b^2 = c^2 - a^2$, and show that an equation of the hyperbola is

$$\frac{x^2}{a^2} - \frac{y^2}{b^2} = 1.$$

10.4 Summary of the Conic Sections

- Characteristics
- Identifying Conic Sections
- Geometric Definition of Conic Sections

Characteristics The graphs of parabolas, circles, ellipses, and hyperbolas are called conic sections since each graph can be obtained by cutting a cone with a plane, as suggested by **Figure 1** at the beginning of the chapter. All conic sections of the types presented in this chapter have equations of the general form

$$Ax^2 + Cy^2 + Dx + Ey + F = 0,$$

where either A or C must be nonzero. The special characteristics of the general equations of the conic sections presented earlier are summarized below.

Teaching Tip When identifying conic sections, caution students to closely regard the sign on each term. A sign can mean the difference between an equation of an ellipse or a hyperbola, as well as between an equation that has no graph, that represents a single point, or that is indeed the graph of a conic section. See **Examples 1 and 2.**

Conic Section	Characteristic	Example
Parabola	Either $A = 0$ or $C = 0$, but not both.	$x^2 - y - 4 = 0$ $y^2 - x - 4y = 0$
Circle	$A = C \neq 0$	$x^2 + y^2 - 16 = 0$
Ellipse	$A \neq C, AC > 0$	$25x^2 + 16y^2 - 400 = 0$
Hyperbola	$AC < 0$	$x^2 - y^2 - 1 = 0$

The following chart summarizes our work with conic sections.

Equation	Graph	Description	Identification
$(x - h)^2 = 4p(y - k)$ or $y - k = a(x - h)^2$	y, x, 0, (h, k) Parabola	Graph opens up if $p > 0$ (or $a > 0$), down if $p < 0$ (or $a < 0$). Vertex is (h, k). Axis of symmetry is $x = h$.	There is an x^2-term. y is not squared.

Equation	Graph	Description	Identification
$(y-k)^2 = 4p(x-h)$ **or** $x - h = a(y-k)^2$	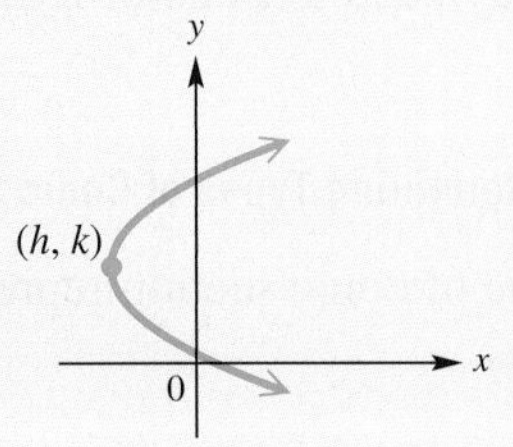 Parabola	Graph opens to the right if $p > 0$ (or $a > 0$), to the left if $p < 0$ (or $a < 0$). Vertex is (h, k). Axis of symmetry is $y = k$.	There is a y^2-term. x is not squared.
$(x-h)^2 + (y-k)^2 = r^2$	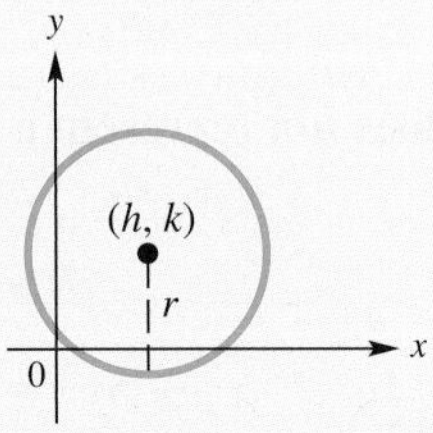 Circle	Center is (h, k), radius is r.	x^2- and y^2-terms have the same positive coefficient.
$\dfrac{(x-h)^2}{a^2} + \dfrac{(y-k)^2}{b^2} = 1 \quad (a > b)$	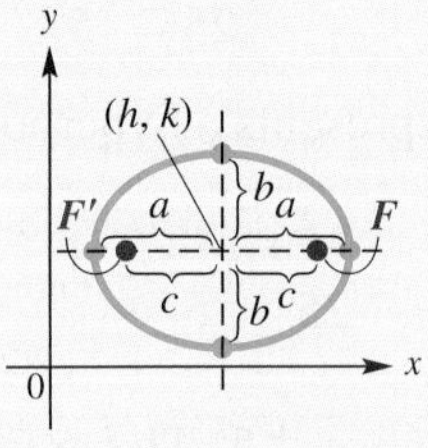 Ellipse	Horizontal major axis, length $= 2a$. $c^2 = a^2 - b^2$ Center is (h, k).	x^2- and y^2-terms have different positive coefficients.
$\dfrac{(x-h)^2}{b^2} + \dfrac{(y-k)^2}{a^2} = 1 \quad (a > b)$	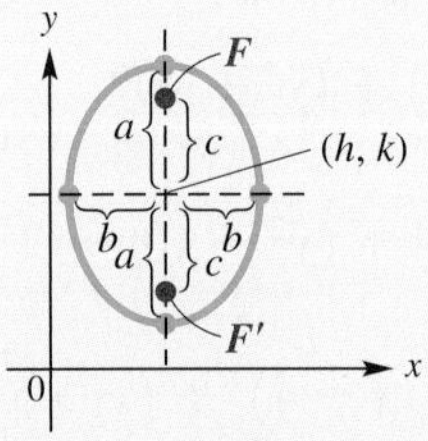 Ellipse	Vertical major axis, length $= 2a$. $c^2 = a^2 - b^2$ Center is (h, k).	x^2- and y^2-terms have different positive coefficients.
$\dfrac{(x-h)^2}{a^2} - \dfrac{(y-k)^2}{b^2} = 1$	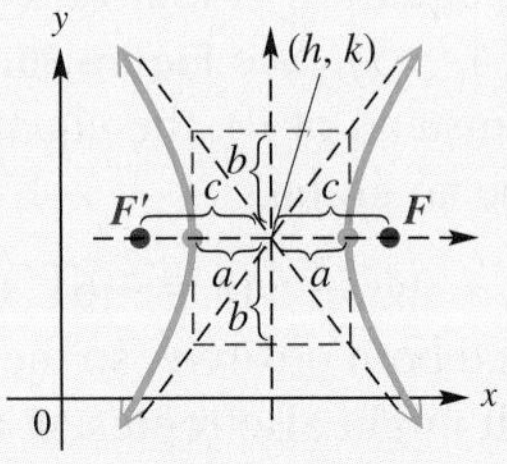 Hyperbola	Graph has horizontal transverse axis. $c^2 = a^2 + b^2$ Asymptotes are $y = \pm\frac{b}{a}(x-h) + k$. Center is (h, k).	x^2-term has a positive coefficient. y^2-term has a negative coefficient.
$\dfrac{(y-k)^2}{a^2} - \dfrac{(x-h)^2}{b^2} = 1$	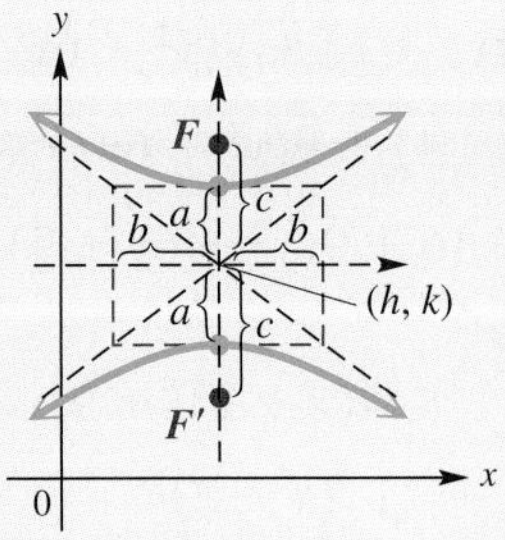 Hyperbola	Graph has vertical transverse axis. $c^2 = a^2 + b^2$ Asymptotes are $y = \pm\frac{a}{b}(x-h) + k$. Center is (h, k).	y^2-term has a positive coefficient. x^2-term has a negative coefficient.

Teaching Tip Encourage students to eliminate some conics by initial inspection. For instance, in **Examples 1(a)–(c),** the parabola can be ruled out because both variables are squared. This initial screening will help students determine how to proceed.

Identifying Conic Sections To recognize the type of graph that a given conic section has, we may need to transform the equation into a more familiar form.

EXAMPLE 1 Determining Types of Conic Sections

Determine the type of conic section represented by each equation, and graph it.

(a) $x^2 = 25 + 5y^2$ **(b)** $x^2 - 8x + y^2 + 10y = -41$

(c) $4x^2 - 16x + 9y^2 + 54y = -61$ **(d)** $x^2 - 6x + 8y - 7 = 0$

SOLUTION

(a) This equation does *not* represent a parabola because both of its variables are squared.

$$x^2 = 25 + 5y^2$$

$$x^2 - 5y^2 = 25 \quad \text{Subtract } 5y^2.$$

$$\frac{x^2}{25} - \frac{y^2}{5} = 1 \quad \text{Divide by 25.}$$

Divide *each* term by 25.

The equation represents a hyperbola centered at the origin, with asymptotes

$$\frac{x^2}{25} - \frac{y^2}{5} = 0, \quad \text{or} \quad y = \pm\frac{\sqrt{5}}{5}x.$$

Remember both the positive and negative square roots.

The x-intercepts are ± 5. The graph is shown in **Figure 34.**

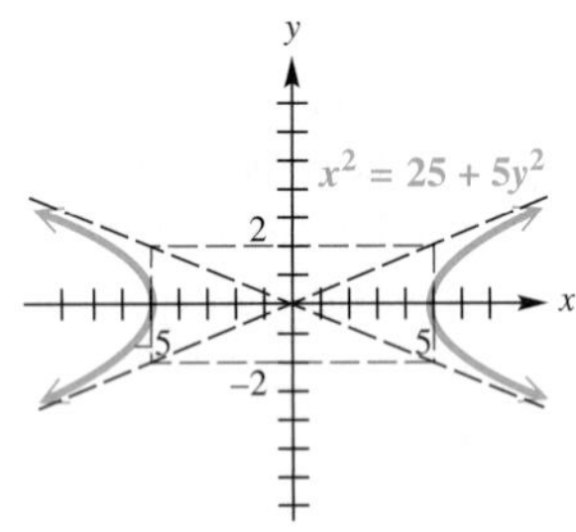

Figure 34

(b)

$$x^2 - 8x + y^2 + 10y = -41$$

$$(x^2 - 8x \quad\quad) + (y^2 + 10y \quad\quad) = -41$$

Rewrite in anticipation of completing the square.

$$(x^2 - 8x + 16) + (y^2 + 10y + 25) = -41 + 16 + 25$$

Complete the square on both x and y. (Section 1.4)

$$(x - 4)^2 + (y + 5)^2 = 0$$

Factor (Section R.4) and add.

The resulting equation is that of a "circle" with radius 0. Its graph is the single point $(4, -5)$. See **Figure 35.** If we had obtained a negative number on the right (instead of 0), the equation would have no solution at all, and there would be no graph.

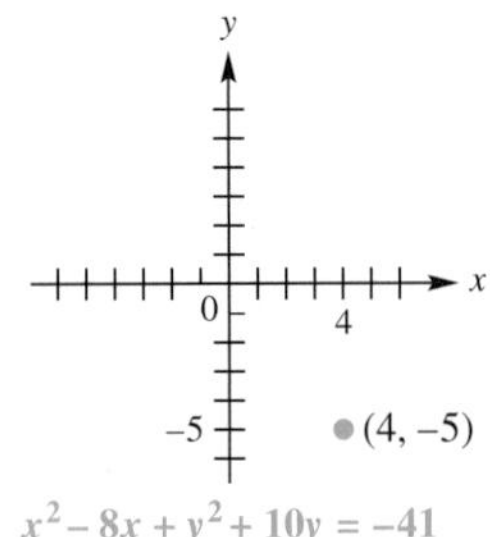

Figure 35

(c) In $4x^2 - 16x + 9y^2 + 54y = -61$, the coefficients of the x^2- and y^2-terms are unequal and both positive, so the equation might represent an ellipse but not a circle. (It might also represent a single point or no points at all.)

$$4x^2 - 16x + 9y^2 + 54y = -61$$

$$4(x^2 - 4x \quad\quad) + 9(y^2 + 6y \quad\quad) = -61 \quad \text{Factor out 4 and factor out 9.}$$

$$4(x^2 - 4x + 4 - 4) + 9(y^2 + 6y + 9 - 9) = -61 \quad \text{Complete the square.}$$

$$4(x^2 - 4x + 4) - 16 + 9(y^2 + 6y + 9) - 81 = -61 \quad \text{Distributive property (Section R.2)}$$

Multiply $4(-4) = -16$.

$$4(x - 2)^2 + 9(y + 3)^2 = 36 \quad \text{Factor; add 97 on each side.}$$

$$\frac{(x - 2)^2}{9} + \frac{(y + 3)^2}{4} = 1 \quad \text{Divide by 36.}$$

This equation represents an ellipse having center $(2, -3)$ and graph as shown in **Figure 36** on the next page.

Classroom Example 1
Determine the type of conic section represented by each equation, and graph it.

(a) $x^2 - 4x + y^2 + 2y = -5$

(b) $9x^2 = 144 - 16y^2$

(c) $2x + y^2 - 10y + 17 = 0$

(d) $-4x^2 - 24x + 25y^2 + 100y = 36$

Answers:

(a) point $(2, -1)$ (b) ellipse

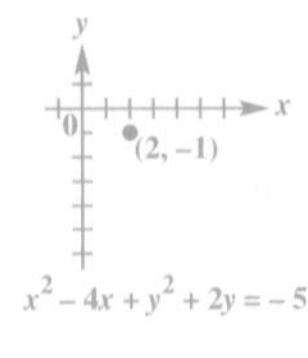

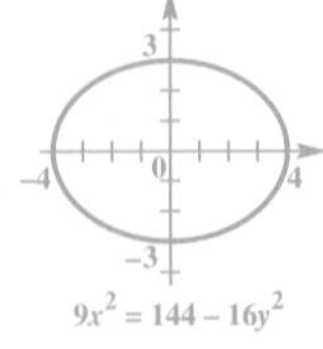

(c) parabola (d) hyperbola

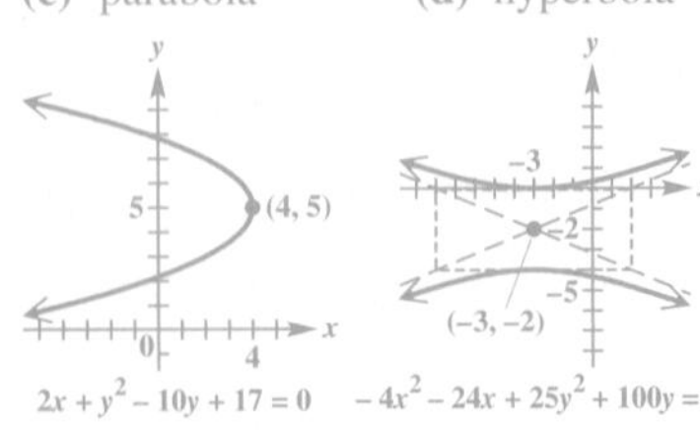

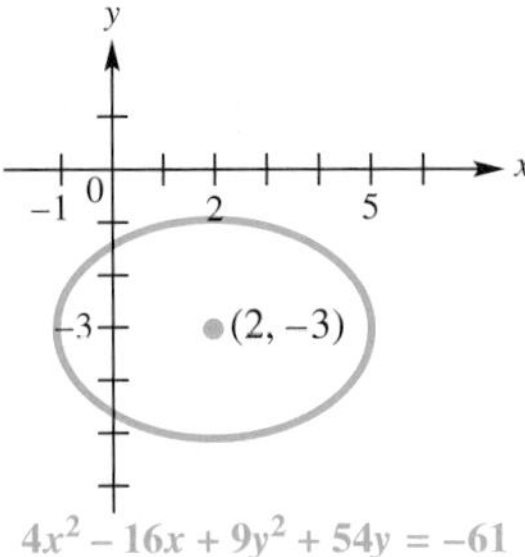

Figure 36

(d) Since only one variable in $x^2 - 6x + 8y - 7 = 0$ is squared (x, and not y), the equation represents a parabola with a vertical axis of symmetry. Isolate the term with y (the variable that is not squared) on one side.

$$x^2 - 6x + 8y - 7 = 0$$

$$8y = -x^2 + 6x + 7 \quad \text{Isolate the } y\text{-term.}$$

$$8y = -(x^2 - 6x \quad\quad) + 7 \quad \text{Regroup terms and factor out } -1.$$

$$8y = -(x^2 - 6x + 9 - 9) + 7 \quad \text{Complete the square.}$$

$$8y = -(x^2 - 6x + 9) + 9 + 7 \quad \text{Distributive property; } -(-9) = +9$$

$$8y = -(x - 3)^2 + 16 \quad \text{Factor and add.}$$

Multiply *each* term by $\frac{1}{8}$. Here, $\frac{1}{8}(16) = 2$.

$$y = -\frac{1}{8}(x - 3)^2 + 2 \quad \text{Multiply by } \tfrac{1}{8}.$$

$$y - 2 = -\frac{1}{8}(x - 3)^2 \quad \text{Subtract 2.}$$

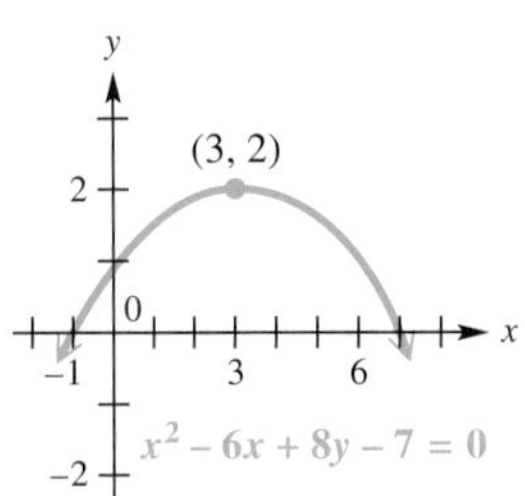

Figure 37

The parabola has vertex $(3, 2)$ and opens down, as shown in **Figure 37.** An equivalent form for this parabola is

$$(x - 3)^2 = -8(y - 2),$$

as seen in **Section 10.1.**

✔ *Now Try Exercises 25, 27, 29, and 33.*

The next example is designed to serve as a warning about a common error.

Classroom Example 2
Identify and sketch the graph of

$9x^2 - 72x + 25y^2 - 100y = -19.$

Answer: ellipse

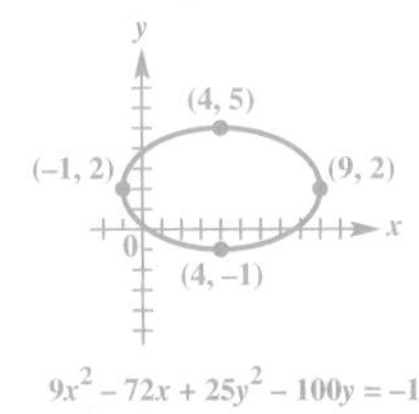

EXAMPLE 2 Determining the Type of Conic Section

Identify and sketch the graph of $4y^2 - 16y - 9x^2 + 18x = -43$.

SOLUTION Once again we complete the square in both variables x and y.

$$4y^2 - 16y - 9x^2 + 18x = -43$$

$$4(y^2 - 4y \quad\quad) - 9(x^2 - 2x \quad\quad) = -43 \quad \text{Factor out 4 and factor out } -9.$$

$$4(y^2 - 4y + 4 - 4) - 9(x^2 - 2x + 1 - 1) = -43 \quad \text{Complete the square.}$$

$$4(y^2 - 4y + 4) - 16 - 9(x^2 - 2x + 1) + 9 = -43 \quad \text{Distributive property}$$

$$4(y - 2)^2 - 9(x - 1)^2 = -36 \quad \text{Factor. Add 16 and subtract 9.}$$

Because of the -36, we might think that this equation does not have a graph. However, the minus sign in the middle on the left shows that the graph is that of a hyperbola.

Be careful here!

$$\frac{(x - 1)^2}{4} - \frac{(y - 2)^2}{9} = 1 \quad \text{Divide by } -36 \text{ and rearrange terms.}$$

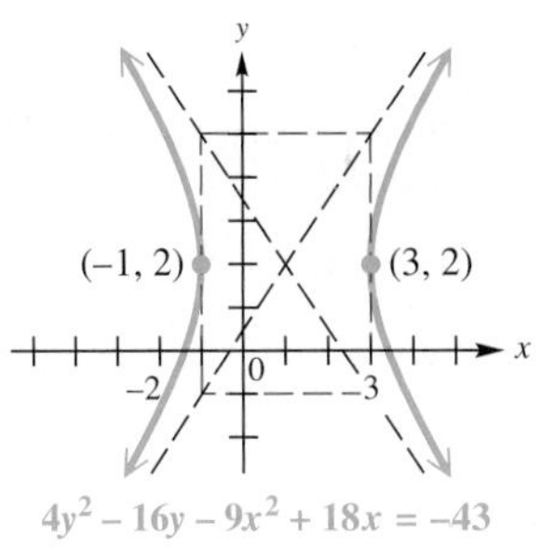

Figure 38

This hyperbola has center $(1, 2)$. The graph is shown in **Figure 38.**

✔ *Now Try Exercise 35.*

Geometric Definition of Conic Sections In **Section 10.1,** a parabola was defined as the set of points in a plane equidistant from a fixed point (focus) and a fixed line (directrix). A parabola has eccentricity 1. This definition can be generalized to apply to the ellipse and the hyperbola. **Figure 39** shows an ellipse with $a = 4$, $c = 2$, and $e = \frac{1}{2}$. The line $x = 8$ is shown also. For any point P on the ellipse,

$$\text{distance of } P \text{ from the focus} = \frac{1}{2}\left[\text{distance of } P \text{ from the line}\right].$$

Figure 40 shows a hyperbola with $a = 2$, $c = 4$, and $e = 2$, along with the line $x = 1$. For any point P on the hyperbola,

$$\text{distance of } P \text{ from the focus} = 2\left[\text{distance of } P \text{ from the line}\right].$$

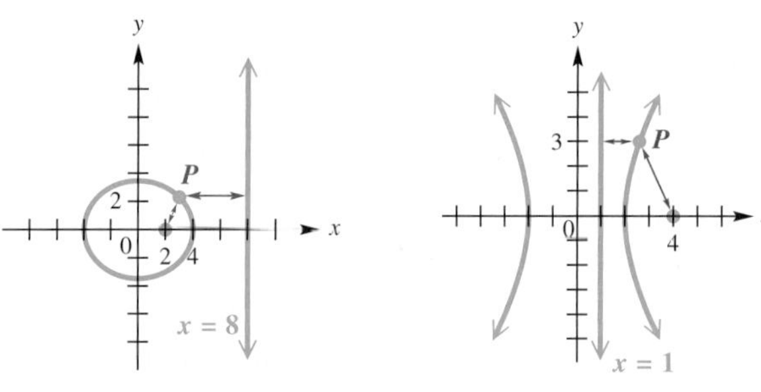

Figure 39 **Figure 40**

The following geometric definition applies to all conic sections except circles, which have $e = 0$.

Geometric Definition of a Conic Section

Given a fixed point F (focus), a fixed line L (directrix), and a positive number e, the set of all points P in the plane such that

$$\textbf{distance of } P \textbf{ from } F = e \cdot [\textbf{distance of } P \textbf{ from } L]$$

is a conic section of eccentricity e. ***The conic section is a parabola when $e = 1$, an ellipse when $0 < e < 1$, and a hyperbola when $e > 1$.***

10.4 Exercises

1. circle
2. circle
3. parabola
4. parabola
5. parabola
6. ellipse
7. ellipse
8. hyperbola
9. hyperbola
10. ellipse
11. hyperbola
12. parabola
13. ellipse
14. hyperbola
15. circle
16. circle
17. parabola
18. parabola
19. point
20. circle
21. parabola
22. parabola
23. point
24. no graph

The equation of a conic section is given in a familiar form. Identify the type of graph that each equation has, without actually graphing. ***See Examples 1 and 2.***

1. $x^2 + y^2 = 144$

2. $(x - 2)^2 + (y + 3)^2 = 25$

3. $y = 2x^2 + 3x - 4$

4. $x = 3y^2 + 5y - 6$

5. $x - 1 = -3(y - 4)^2$

6. $\dfrac{x^2}{25} + \dfrac{y^2}{36} = 1$

7. $\dfrac{x^2}{49} + \dfrac{y^2}{100} = 1$

8. $x^2 - y^2 = 1$

25. no graph

26. circle

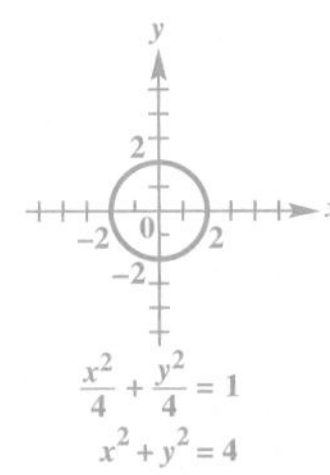

27. hyperbola

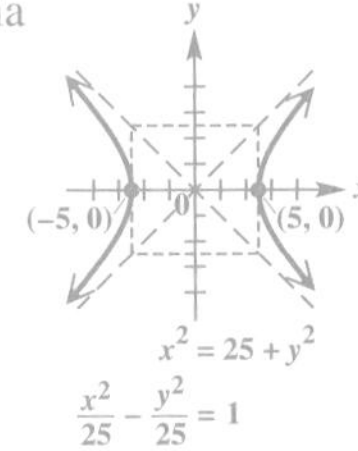

28. ellipse

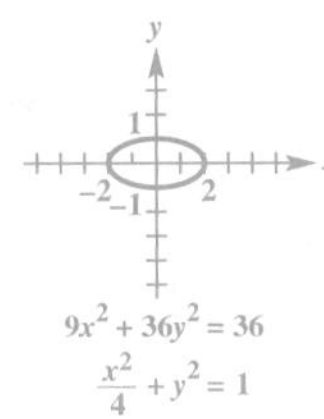

29. parabola

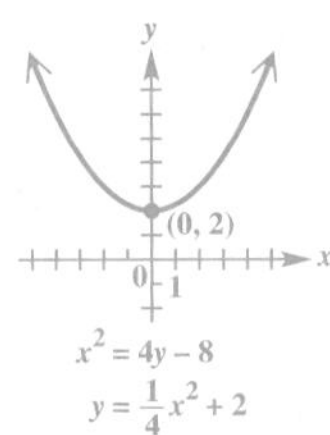

30. point $(4, -1)$

31. parabola

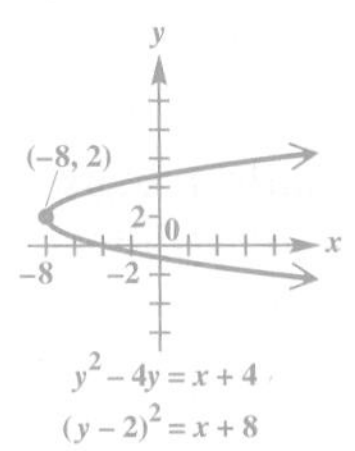

32. no graph

33. circle

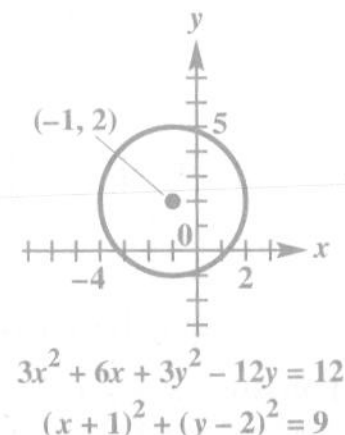

34. hyperbola

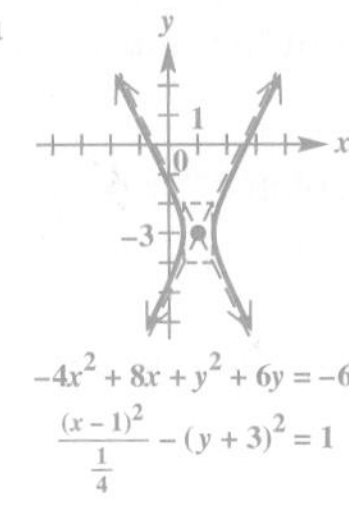

$-4x^2 + 8x + y^2 + 6y = -6$
$\frac{(x-1)^2}{\frac{1}{4}} - (y + 3)^2 = 1$

9. $\frac{x^2}{4} - \frac{y^2}{16} = 1$

10. $\frac{(x + 2)^2}{9} + \frac{(y - 4)^2}{16} = 1$

11. $\frac{x^2}{25} - \frac{y^2}{25} = 1$

12. $y + 7 = 4(x + 3)^2$

13. $\frac{x^2}{4} = 1 - \frac{y^2}{9}$

14. $\frac{x^2}{4} = 1 + \frac{y^2}{9}$

15. $\frac{(x + 3)^2}{16} + \frac{(y - 2)^2}{16} = 1$

16. $x^2 = 25 - y^2$

17. $x^2 - 6x + y = 0$

18. $11 - 3x = 2y^2 - 8y$

19. $4(x - 3)^2 + 3(y + 4)^2 = 0$

20. $2x^2 - 8x + 2y^2 + 20y = 12$

21. $x - 4y^2 - 8y = 0$

22. $x^2 + 2x = -4y$

23. $4x^2 - 24x + 5y^2 + 10y + 41 = 0$

24. $6x^2 - 12x + 6y^2 - 18y + 25 = 0$

Determine the type of conic section represented by each equation, and graph it. ***See Examples 1 and 2.***

25. $\frac{x^2}{4} + \frac{y^2}{4} = -1$

26. $\frac{x^2}{4} + \frac{y^2}{4} = 1$

27. $x^2 = 25 + y^2$

28. $9x^2 + 36y^2 = 36$

29. $x^2 = 4y - 8$

30. $\frac{(x - 4)^2}{8} + \frac{(y + 1)^2}{2} = 0$

31. $y^2 - 4y = x + 4$

32. $(x + 7)^2 + (y - 5)^2 + 4 = 0$

33. $3x^2 + 6x + 3y^2 - 12y = 12$

34. $-4x^2 + 8x + y^2 + 6y = -6$

35. $4x^2 - 8x + 9y^2 - 36y = -4$

36. $3x^2 + 12x + 3y^2 = 0$

In Exercises 37–40, identify which type of conic section is described.

37. The conic section that consists of the set of all points in the plane for which the sum of the distances from the points $(5, 0)$ and $(-5, 0)$ is 14

38. The conic section that consists of the set of all points in the plane for which the absolute value of the difference of the distances from the points $(3, 0)$ and $(-3, 0)$ is 2

39. The conic section that consists of the set of all points in the plane for which the distance from the point $(3, 0)$ is one and one-half times the distance from the line $x = \frac{4}{3}$

40. The conic section that consists of the set of all points in the plane for which the distance from the point $(2, 0)$ is one-third of the distance from the line $x = 10$

Find the eccentricity of each conic section. The point shown on the x-axis is a focus, and the line shown is a directrix.

41.

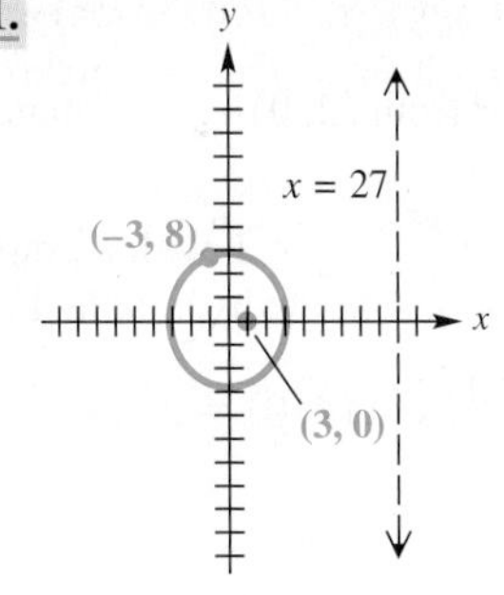

42.

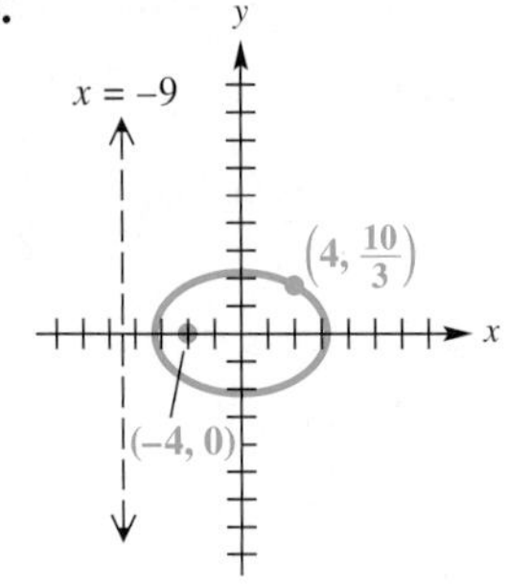

35. ellipse

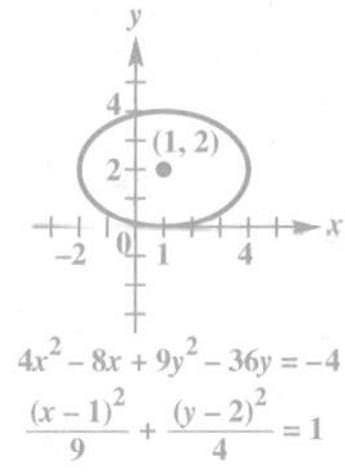

36. circle

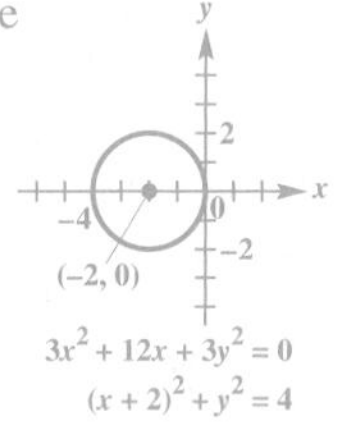

37. ellipse
38. hyperbola
39. hyperbola
40. ellipse
41. $\frac{1}{3}$
42. $\frac{2}{3}$
43. 1
44. 2.3
45. 1.5
46. 1
47. elliptical
48. The minimum increase is $4326 - 2090 = 2236$ m per sec.
50. $\left(-\frac{D}{2A}, -\frac{E}{2C}\right)$

43.

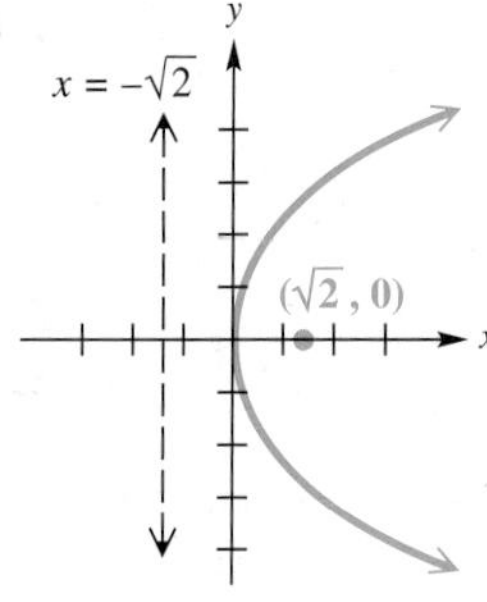

44.

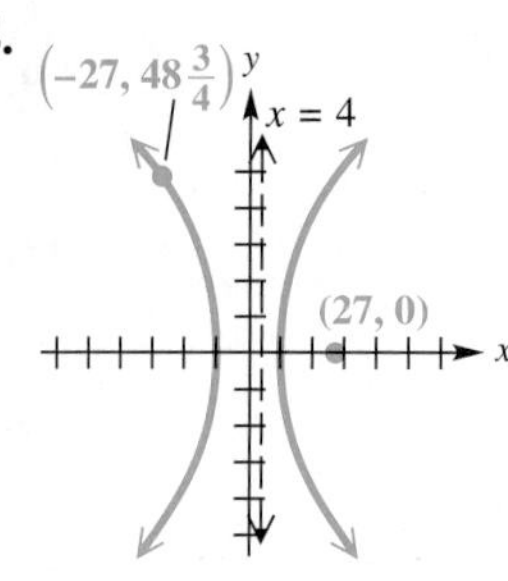

45.

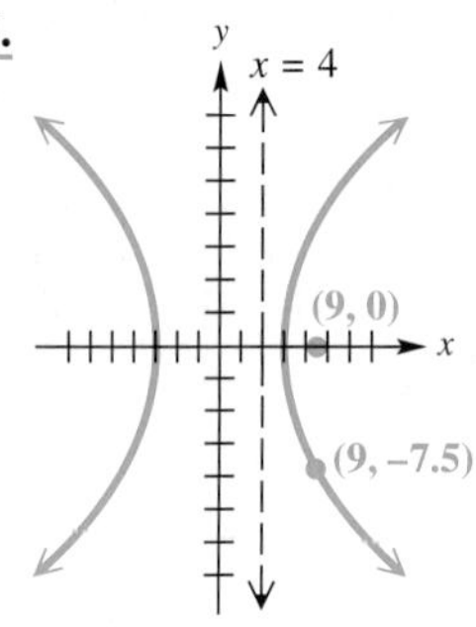

46.

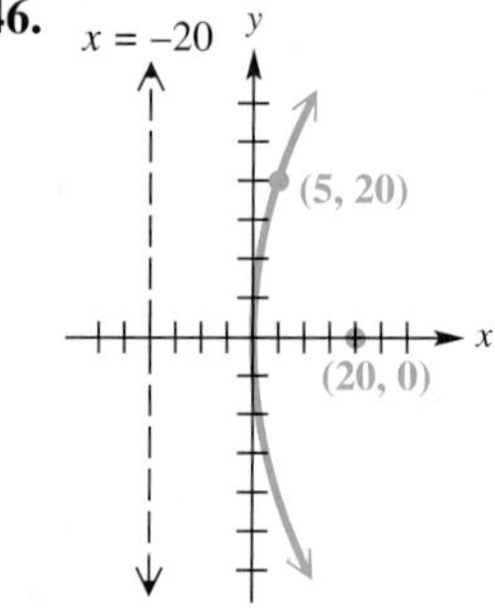

Satellite Trajectory *When a satellite is near Earth, its orbital trajectory may trace out a hyperbola, a parabola, or an ellipse. The type of trajectory depends on the satellite's velocity V in meters per second. It will be hyperbolic if* $V > \frac{k}{\sqrt{D}}$, *parabolic if* $V = \frac{k}{\sqrt{D}}$, *and elliptical if* $V < \frac{k}{\sqrt{D}}$, *where* $k = 2.82 \times 10^7$ *is a constant and D is the distance in meters from the satellite to the center of Earth. Use this information in Exercises 47–49. (Source:* Loh, W., *Dynamics and Thermodynamics of Planetary Entry,* Prentice-Hall, *and* Thomson, W., *Introduction to Space Dynamics,* John Wiley and Sons.)

47. When the artificial satellite Explorer IV was at a maximum distance D of 42.5×10^6 m from Earth's center, it had a velocity V of 2090 m per sec. Determine the shape of its trajectory.

48. If a satellite is scheduled to leave Earth's gravitational influence, its velocity must be increased so that its trajectory changes from elliptical to hyperbolic. Determine the minimum increase in velocity necessary for Explorer IV to escape Earth's gravitational influence when $D = 42.5 \times 10^6$ m.

49. Explain why it is easier to change a satellite's trajectory from an ellipse to a hyperbola when D is maximum rather than minimum.

Solve each problem.

50. If $Ax^2 + Cy^2 + Dx + Ey + F = 0$ is the general equation of an ellipse, find its center point by completing the square.

51. Graph the ellipse $\frac{x^2}{16} + \frac{y^2}{12} = 1$ with a graphing calculator. Trace to find the coordinates of several points on the ellipse. For each of these points P, verify that

$$\text{distance of } P \text{ from } (2, 0) = \frac{1}{2}\left[\text{distance of } P \text{ from the line } x = 8\right].$$

52. Graph the hyperbola $\frac{x^2}{4} - \frac{y^2}{12} = 1$ with a graphing calculator. Trace to find the coordinates of several points on the hyperbola. For each of these points P, verify that

$$\text{distance of } P \text{ from } (4, 0) = 2\left[\text{distance of } P \text{ from the line } x = 1\right].$$

Chapter 10 Test Prep

Key Terms

10.1 conic sections parabola focus directrix	10.2 ellipse foci major axis minor axis	center vertices conic eccentricity	10.3 hyperbola transverse axis asymptotes fundamental rectangle

New Symbols

e eccentricity

Quick Review

Concepts | **Examples**

10.1 Parabolas

Parabola with Vertical Axis and Vertex (0, 0)
The parabola with focus $(0, p)$ and directrix $y = -p$ has equation $x^2 = 4py$. The parabola has vertical axis of symmetry $x = 0$ and opens up if $p > 0$ or down if $p < 0$.

Parabola with Horizontal Axis and Vertex (0, 0)
The parabola with focus $(p, 0)$ and directrix $x = -p$ has equation $y^2 = 4px$. The parabola has horizontal axis of symmetry $y = 0$ and opens to the right if $p > 0$ or to the left if $p < 0$.

Equation Forms for Translated Parabolas
A parabola with vertex (h, k) has an equation of the form

$$(x - h)^2 = 4p(y - k) \quad \text{Vertical axis of symmetry}$$

or $$(y - k)^2 = 4p(x - h), \quad \text{Horizontal axis of symmetry}$$

where p is the directed distance from the vertex to the focus.

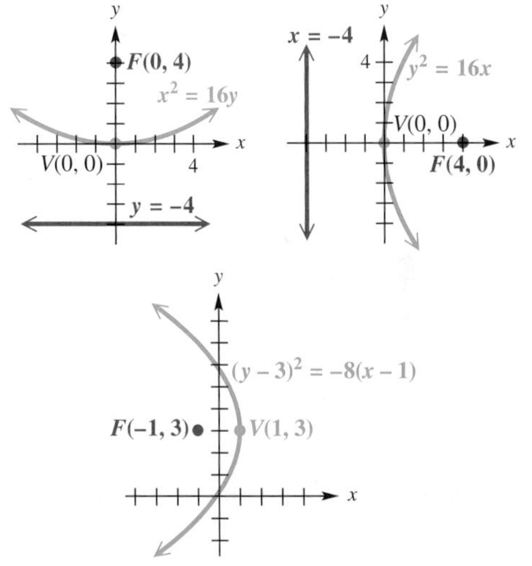

10.2 Ellipses

Standard Forms of Equations for Ellipses
The ellipse with center at the origin and equation

$$\frac{x^2}{a^2} + \frac{y^2}{b^2} = 1 \quad (\text{where } a > b)$$

has vertices $(\pm a, 0)$, endpoints of the minor axis $(0, \pm b)$, and foci $(\pm c, 0)$, where $c^2 = a^2 - b^2$.

The ellipse with center at the origin and equation

$$\frac{x^2}{b^2} + \frac{y^2}{a^2} = 1 \quad (\text{where } a > b)$$

has vertices $(0, \pm a)$, endpoints of the minor axis $(\pm b, 0)$, and foci $(0, \pm c)$, where $c^2 = a^2 - b^2$.

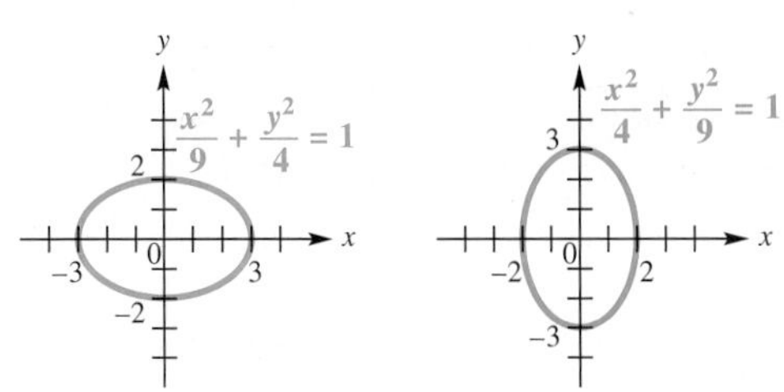

(continued)

Concepts	Examples

Translated Ellipses
The preceding equations can be extended to ellipses having center (h, k) by replacing x and y with $x - h$ and $y - k$, respectively.

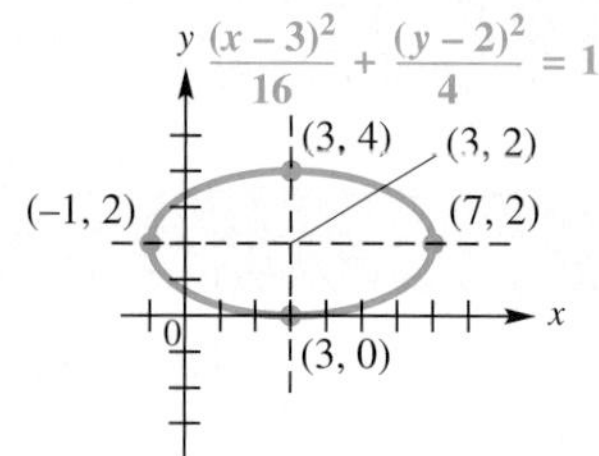

10.3 Hyperbolas

Standard Forms of Equations for Hyperbolas
The hyperbola with center at the origin and equation

$$\frac{x^2}{a^2} - \frac{y^2}{b^2} = 1$$

has vertices $(\pm a, 0)$, asymptotes $y = \pm \frac{b}{a}x$, and foci $(\pm c, 0)$, where $c^2 = a^2 + b^2$.

The hyperbola with center at the origin and equation

$$\frac{y^2}{a^2} - \frac{x^2}{b^2} = 1$$

has vertices $(0, \pm a)$, asymptotes $y = \pm \frac{a}{b}x$, and foci $(0, \pm c)$, where $c^2 = a^2 + b^2$.

Translated Hyperbolas
The preceding equations can be extended to hyperbolas having center (h, k) by replacing x and y with $x - h$ and $y - k$, respectively.

$\frac{x^2}{4} - \frac{y^2}{4} = 1$

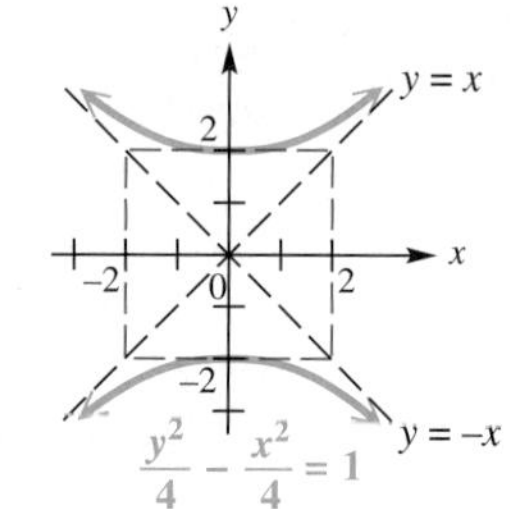

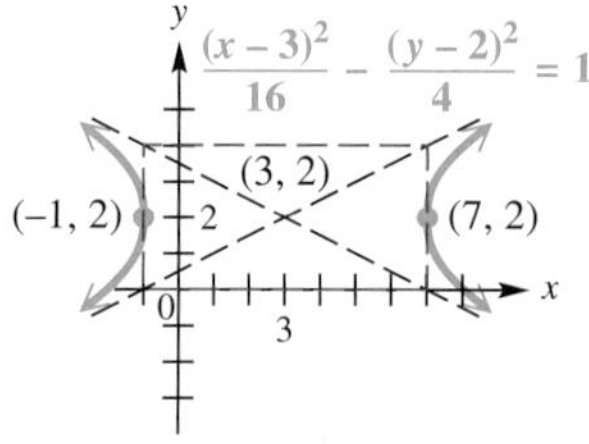

10.4 Summary of the Conic Sections

Conic sections in this chapter have equations that can be written in the following form.

$$Ax^2 + Cy^2 + Dx + Ey + F = 0$$

Conic Section	Characteristic	Example
Parabola	Either $A = 0$ or $C = 0$, but not both.	$x^2 - y - 4 = 0$ $y^2 - x - 4y = 0$
Circle	$A = C \neq 0$	$x^2 + y^2 - 16 = 0$
Ellipse	$A \neq C, AC > 0$	$25x^2 + 16y^2 - 400 = 0$
Hyperbola	$AC < 0$	$x^2 - y^2 - 1 = 0$

See the summary chart in **Section 10.4.**

$y^2 - 4x - 10y + 21 = 0$, or $(y - 5)^2 = 4(x + 1)$, is a parabola with vertex $(-1, 5)$, opening to the right.

$$x^2 - 4x + y^2 + 2y - 4 = 0,$$

or

$$(x - 2)^2 + (y + 1)^2 = 9,$$

is a circle centered at $(2, -1)$ with radius 3.

$$4x^2 + y^2 - 16 = 0, \quad \text{or} \quad \frac{x^2}{4} + \frac{y^2}{16} = 1,$$

is an ellipse centered at the origin with vertical major axis.

$$4x^2 - y^2 - 8x - 4y - 16 = 0,$$

or

$$\frac{(x-1)^2}{4} - \frac{(y+2)^2}{16} = 1,$$

is a hyperbola centered at $(1, -2)$ with horizontal transverse axis.

Chapter 10 Review Exercises

1. $[2, \infty)$; $(-\infty, \infty)$; $(2, 5)$; $y = 5$

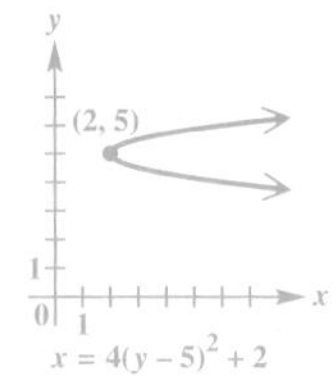

2. $(-\infty, -7]$; $(-\infty, \infty)$; $(-7, -1)$; $y = -1$

(−7, −1)
$x = -(y + 1)^2 - 7$

3. $\left[\frac{7}{4}, \infty\right)$; $(-\infty, \infty)$; $\left(\frac{7}{4}, \frac{1}{2}\right)$; $y = \frac{1}{2}$

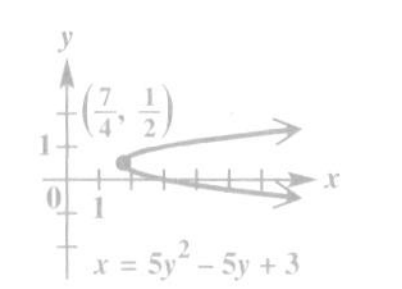

4. $[-1, \infty)$; $(-\infty, \infty)$; $(-1, 1)$; $y = 1$

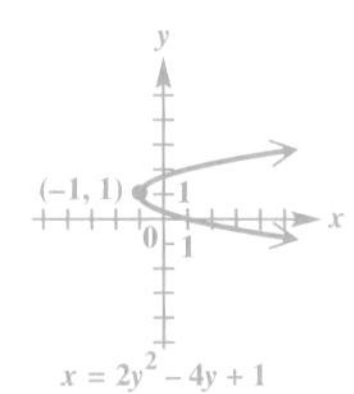

5. $(-\infty, 0]$; $(-\infty, \infty)$; $\left(-\frac{1}{6}, 0\right)$; $x = \frac{1}{6}$; x-axis

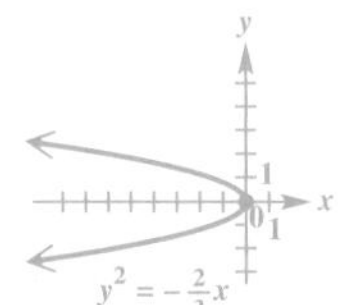

6. $[0, \infty)$; $(-\infty, \infty)$; $\left(\frac{1}{2}, 0\right)$; $x = -\frac{1}{2}$; x-axis

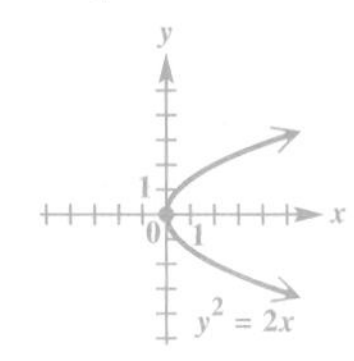

Graph each parabola. In Exercises 1–4, give the domain, range, vertex, and axis of symmetry. In Exercises 5–8, give the domain, range, focus, directrix, and axis of symmetry.

1. $x = 4(y - 5)^2 + 2$ **2.** $x = -(y + 1)^2 - 7$ **3.** $x = 5y^2 - 5y + 3$

4. $x = 2y^2 - 4y + 1$ **5.** $y^2 = -\frac{2}{3}x$ **6.** $y^2 = 2x$

7. $3x^2 = y$ **8.** $x^2 + 2y = 0$

Write an equation for each parabola with vertex at the origin.

9. focus $(4, 0)$ **10.** focus $(0, -3)$

11. through $(-3, 4)$, opening up **12.** through $(2, 5)$, opening right

An equation of a conic section is given. Identify the type of conic section. It may be necessary to transform the equation into a more familiar form.

13. $y^2 + 9x^2 = 9$ **14.** $9x^2 - 16y^2 = 144$ **15.** $3y^2 - 5x^2 = 30$

16. $y^2 + x = 4$ **17.** $4x^2 - y = 0$ **18.** $x^2 + y^2 = 25$

19. $4x^2 - 8x + 9y^2 + 36y = -4$ **20.** $9x^2 - 18x - 4y^2 - 16y - 43 = 0$

Concept Check *Match each equation in Exercises 21–26 with its calculator graph in choices A–F. In all cases except choice B,* Xscl = Yscl = 1.

21. $4x^2 + y^2 = 36$ **22.** $x = 2y^2 + 3$

23. $(x - 2)^2 + (y + 3)^2 = 36$ **24.** $\frac{x^2}{36} + \frac{y^2}{9} = 1$

25. $(y - 1)^2 - (x - 2)^2 = 36$ **26.** $y^2 = 36 + 4x^2$

A.

3.2
−9.4 9.4
−9.2

B.

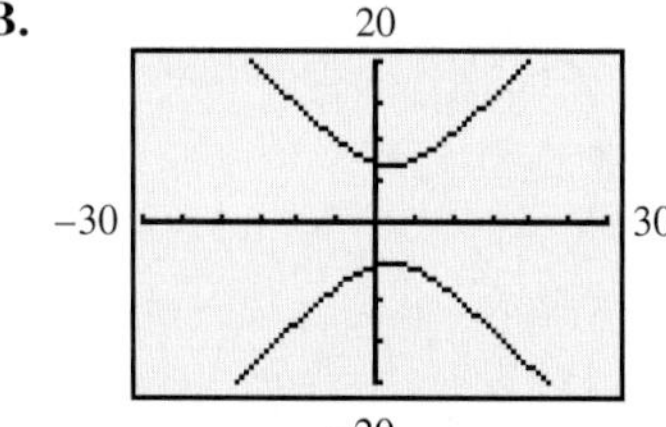

In this screen, Xscl = Yscl = 5.

C.

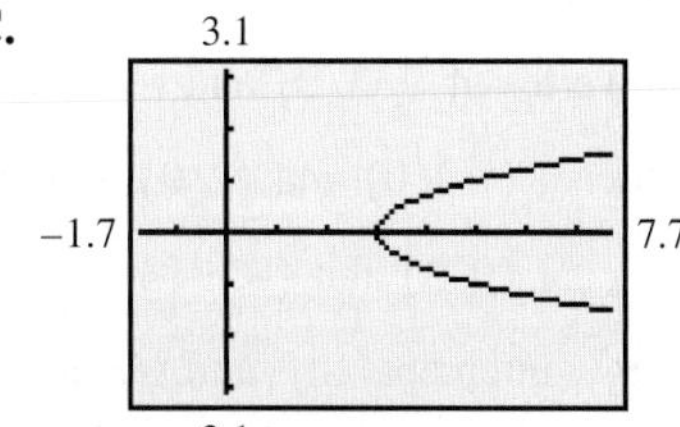

D.

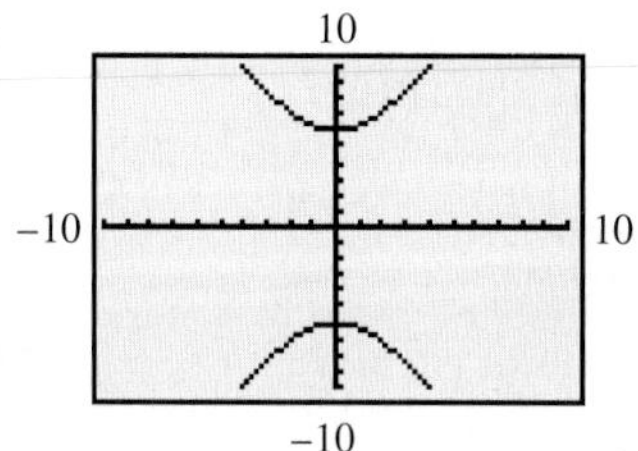

E.

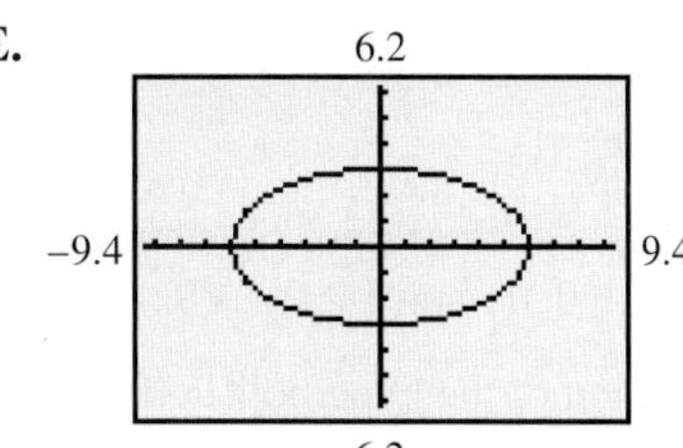

F.

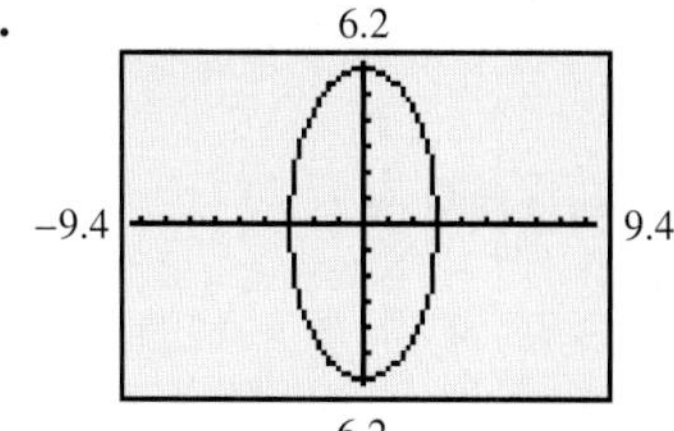

7. $(-\infty, \infty)$; $[0, \infty)$; $\left(0, \frac{1}{12}\right)$; $y = -\frac{1}{12}$; y-axis

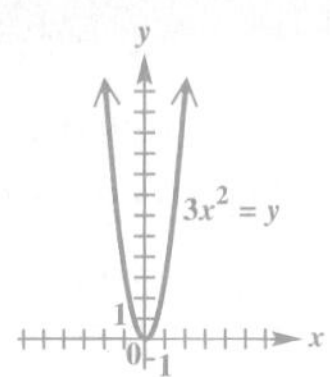

8. $(-\infty, \infty)$; $(-\infty, 0]$; $\left(0, -\frac{1}{2}\right)$; $y = \frac{1}{2}$; y-axis

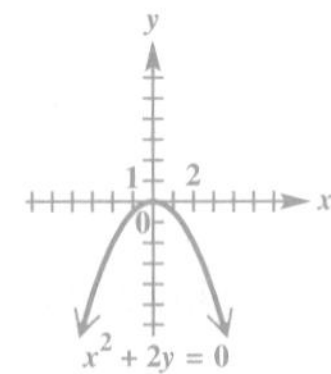

9. $y^2 = 16x$
10. $x^2 = -12y$
11. $x^2 = \frac{9}{4}y$
12. $y^2 = \frac{25}{2}x$
13. ellipse
14. hyperbola
15. hyperbola
16. parabola
17. parabola
18. circle
19. ellipse
20. hyperbola
21. F
22. C
23. A
24. E
25. B
26. D

27. ellipse; $[-2, 2]$; $[-3, 3]$; $(0, -3)$, $(0, 3)$

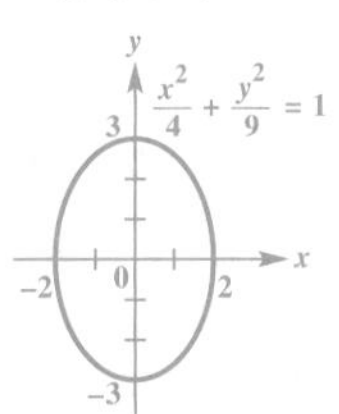

28. ellipse; $[-4, 4]$; $[-2, 2]$; $(-4, 0)$, $(4, 0)$

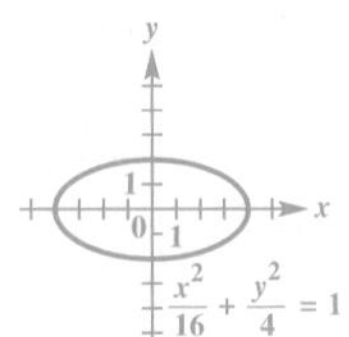

29. hyperbola; $(-\infty, -8] \cup [8, \infty)$; $(-\infty, \infty)$; $(-8, 0)$, $(8, 0)$; $y = \pm\frac{3}{4}x$

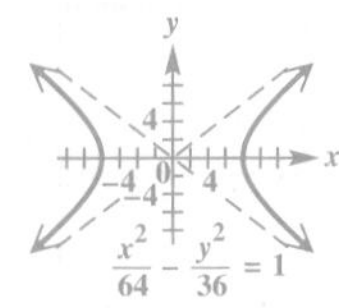

Graph each relation and identify the graph. Give the domain, range, coordinates of the vertices for each ellipse or hyperbola, and equations of the asymptotes for each hyperbola. Give the domain and range for each circle.

27. $\dfrac{x^2}{4} + \dfrac{y^2}{9} = 1$

28. $\dfrac{x^2}{16} + \dfrac{y^2}{4} = 1$

29. $\dfrac{x^2}{64} - \dfrac{y^2}{36} = 1$

30. $\dfrac{y^2}{25} - \dfrac{x^2}{9} = 1$

31. $\dfrac{(x+1)^2}{16} + \dfrac{(y-1)^2}{16} = 1$

32. $(x-3)^2 + (y+2)^2 = 9$

33. $4x^2 + 9y^2 = 36$

34. $x^2 = 16 + y^2$

35. $\dfrac{(x-3)^2}{4} + (y+1)^2 = 1$

36. $\dfrac{(x-2)^2}{9} + \dfrac{(y+3)^2}{4} = 1$

37. $\dfrac{(y+2)^2}{4} - \dfrac{(x+3)^2}{9} = 1$

38. $\dfrac{(x+1)^2}{16} - \dfrac{(y-2)^2}{4} = 1$

39. $x^2 - 4x + y^2 + 6y = -12$

40. $4x^2 + 8x + 25y^2 - 250y = -529$

41. $5x^2 + 20x + 2y^2 - 8y = -18$

42. $-4x^2 + 8x + 4y^2 + 8y = 16$

Graph each relation. Give the domain and range, and identify any that are graphs of functions.

43. $\dfrac{x}{3} = -\sqrt{1 - \dfrac{y^2}{16}}$

44. $x = -\sqrt{1 - \dfrac{y^2}{36}}$

45. $y = -\sqrt{1 + x^2}$

46. $y = -\sqrt{1 - \dfrac{x^2}{25}}$

Write an equation for each conic section with center at the origin.

47. ellipse; vertex at $(0, -4)$, focus at $(0, -2)$

48. ellipse; x-intercept 6, focus at $(2, 0)$

49. hyperbola; focus at $(0, 5)$, transverse axis of length 8

50. hyperbola; y-intercept -2, passing through $(2, 3)$

Write an equation for each conic section satisfying the given conditions.

51. parabola with focus at $(3, 2)$ and directrix $x = -3$

52. parabola with vertex at $(-3, 2)$ and y-intercepts 5 and -1

53. ellipse with foci at $(-2, 0)$ and $(2, 0)$ and major axis of length 10

54. ellipse with foci at $(0, 3)$ and $(0, -3)$ and vertex at $(0, -7)$

55. hyperbola with x-intercepts ± 3 and foci at $(-5, 0)$ and $(5, 0)$

56. hyperbola with foci at $(0, 12)$ and $(0, -12)$ and asymptotes $y = \pm x$

Solve each problem.

57. Find the equation of the ellipse consisting of all points in the plane the sum of whose distances from $(0, 0)$ and $(4, 0)$ is 8.

58. Find the equation of the hyperbola consisting of all points in the plane for which the absolute value of the difference of the distances from $(0, 0)$ and $(0, 4)$ is 2.

30. hyperbola; $(-\infty, \infty)$; $(-\infty, -5] \cup [5, \infty)$; $(0, -5)$, $(0, 5)$; $y = \pm\frac{5}{3}x$

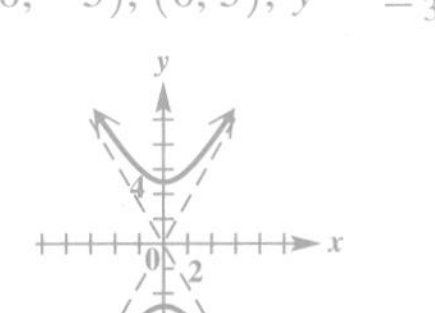

31. circle; $[-5, 3]$; $[-3, 5]$

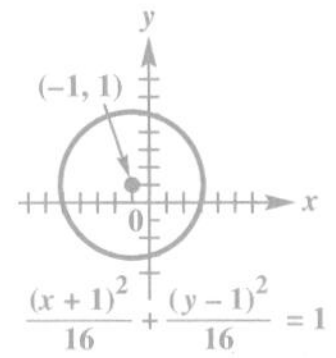

32. circle; $[0, 6]$; $[-5, 1]$

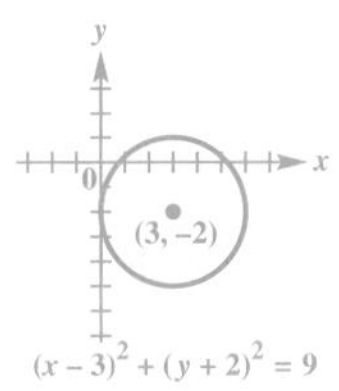

Answers for the remaining exercises are given in the answer section at the back of the text.

59. Calculator graphs are shown in choices A–D. Arrange the graphs so that their eccentricities are in increasing order.

A.

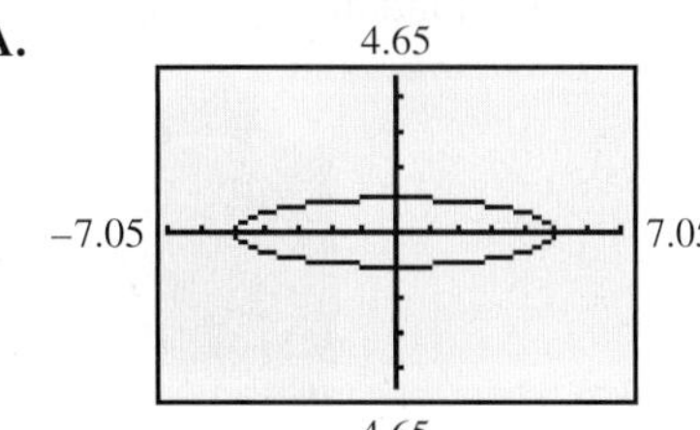

B.

3.1, –4.7, 4.7, –3.1

C.

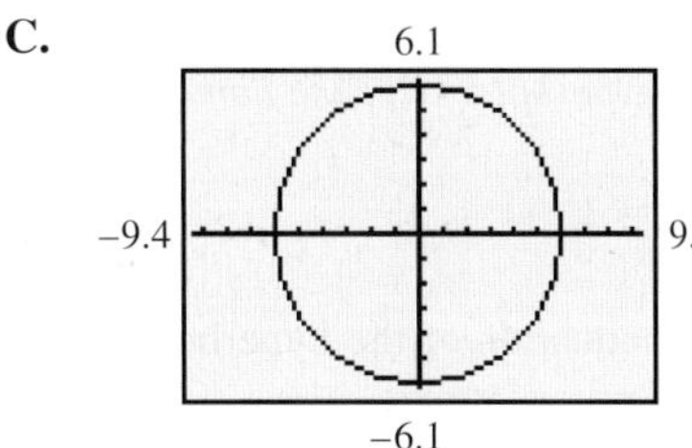

D.

6.2, –9.4, 9.4, –6.2

60. *Orbit of Venus* The orbit of Venus is an ellipse with the sun at one of the foci. The eccentricity of the orbit is $e = 0.006775$, and the major axis has length 134.5 million mi. (*Source: World Almanac and Book of Facts.*) Find the least and greatest distances of Venus from the sun.

61. *Orbit of a Comet* The comet Swift-Tuttle has an elliptical orbit of eccentricity $e = 0.964$, with the sun at one of the foci. Find the equation of the comet given that the closest it comes to the sun is 89 million mi.

62. Find the equation of the hyperbola consisting of all points P in the plane for which the absolute value of the difference of the distances of P from $(-5, 0)$ and $(5, 0)$ is 8. Then graph the hyperbola with a graphing calculator and trace to find the coordinates of several points on the graph of the hyperbola. For each of these points, verify that the absolute value of the differences of the distances is indeed 8.

Chapter 10 Test

[10.1]

1. $(-\infty, \infty)$; $(-\infty, 9]$; $(3, 9)$; $x = 3$

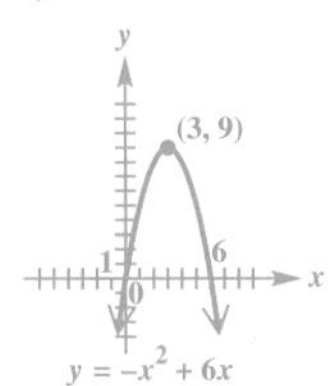

2. $[-4, \infty)$; $(-\infty, \infty)$; $(-4, -1)$; $y = -1$

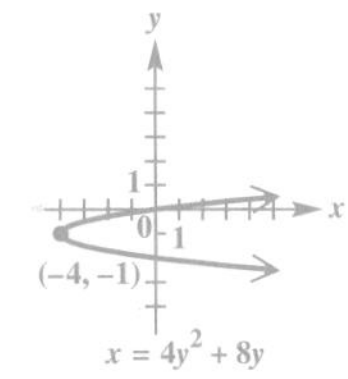

Graph each parabola. Give the domain, range, vertex, and axis of symmetry.

1. $y = -x^2 + 6x$

2. $x = 4y^2 + 8y$

3. Give the coordinates of the focus and the equation of the directrix for the parabola with equation $x = 8y^2$.

4. Write an equation for the parabola with vertex $(2, 3)$, passing through the point $(-18, 1)$, and opening to the left.

5. Explain how to determine just by looking at the equation whether a parabola has a vertical or a horizontal axis of symmetry, and whether it opens up, down, to the left, or to the right.

Graph each ellipse. Give the domain and range.

6. $\frac{(x-8)^2}{100} + \frac{(y-5)^2}{49} = 1$

7. $16x^2 + 4y^2 = 64$

8. Graph $y = -\sqrt{1 - \frac{x^2}{36}}$. Tell whether the graph is that of a function.

9. Write an equation for the ellipse centered at the origin having horizontal major axis with length 6 and minor axis with length 4.

3. $\left(\frac{1}{32}, 0\right)$; $x = -\frac{1}{32}$
4. $x - 2 = -5(y - 3)^2$

[10.2]
6. $[-2, 18]$; $[-2, 12]$

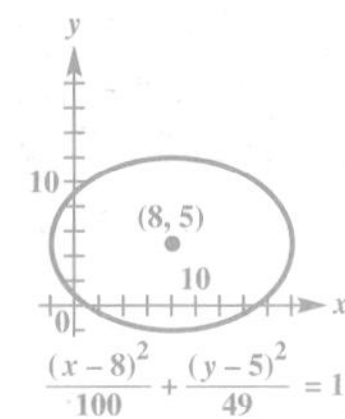

7. $[-2, 2]$; $[-4, 4]$

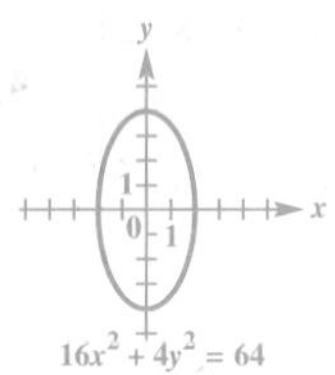

8. It is the graph of a function.

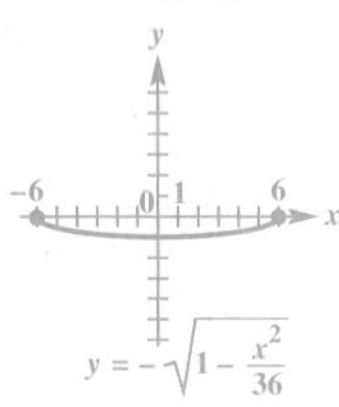

9. $\frac{x^2}{9} + \frac{y^2}{4} = 1$
10. $\frac{x^2}{400} + \frac{y^2}{144} = 1$; approximately 10.39 ft

10. *Height of the Arch of a Bridge* An arch of a bridge has the shape of the top half of an ellipse. The arch is 40 ft wide and 12 ft high at the center. Find the equation of the complete ellipse. Find the height of the arch 10 ft from the center of the bottom.

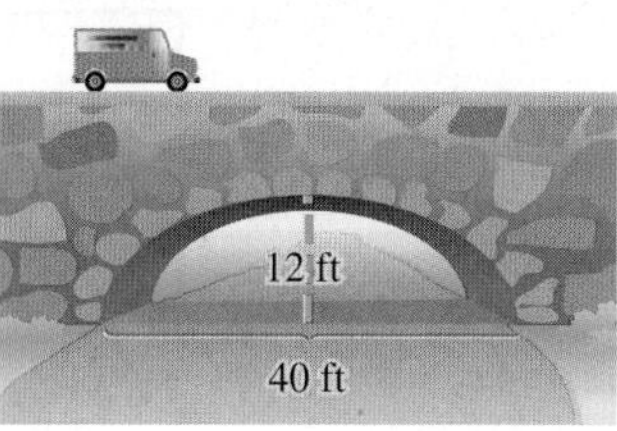

Graph each hyperbola. Give the domain, range, and equations of the asymptotes.

11. $\dfrac{x^2}{4} - \dfrac{y^2}{4} = 1$

12. $9x^2 - 4y^2 = 36$

13. Find the equation of the hyperbola with y-intercepts ± 5 and foci at $(0, -6)$ and $(0, 6)$.

Identify the type of graph, if any, defined by each equation.

14. $x^2 + 8x + y^2 - 4y + 2 = 0$

15. $5x^2 + 10x - 2y^2 - 12y - 23 = 0$

16. $3x^2 + 10y^2 - 30 = 0$

17. $x^2 - 4y = 0$

18. $(x + 9)^2 + (y - 3)^2 = 0$

19. $x^2 + 4x + y^2 - 6y + 30 = 0$

20. The screen shown here gives the graph of

$$\frac{x^2}{25} - \frac{y^2}{49} = 1$$

as generated by a graphing calculator. What two functions y_1 and y_2 were used to obtain the graph?

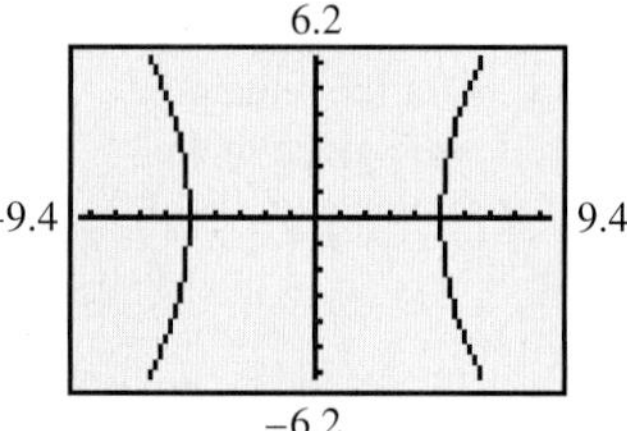

[10.3]
11. $(-\infty, -2] \cup [2, \infty)$; $(-\infty, \infty)$; $y = \pm x$

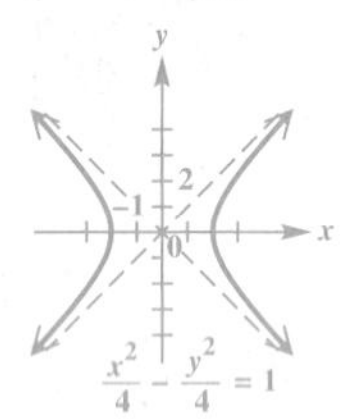

12. $(-\infty, -2] \cup [2, \infty)$; $(-\infty, \infty)$; $y = \pm\frac{3}{2}x$

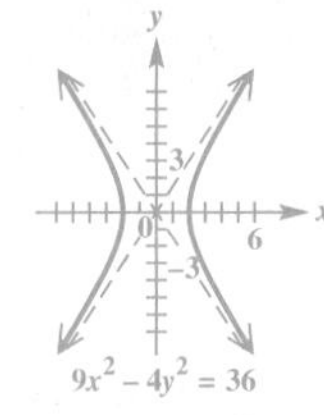

13. $\frac{y^2}{25} - \frac{x^2}{11} = 1$ [10.4] 14. circle 15. hyperbola 16. ellipse 17. parabola 18. point 19. no graph
[10.3] 20. $y_1 = 7\sqrt{\frac{x^2}{25} - 1}$, $y_2 = -7\sqrt{\frac{x^2}{25} - 1}$

11 Further Topics in Algebra

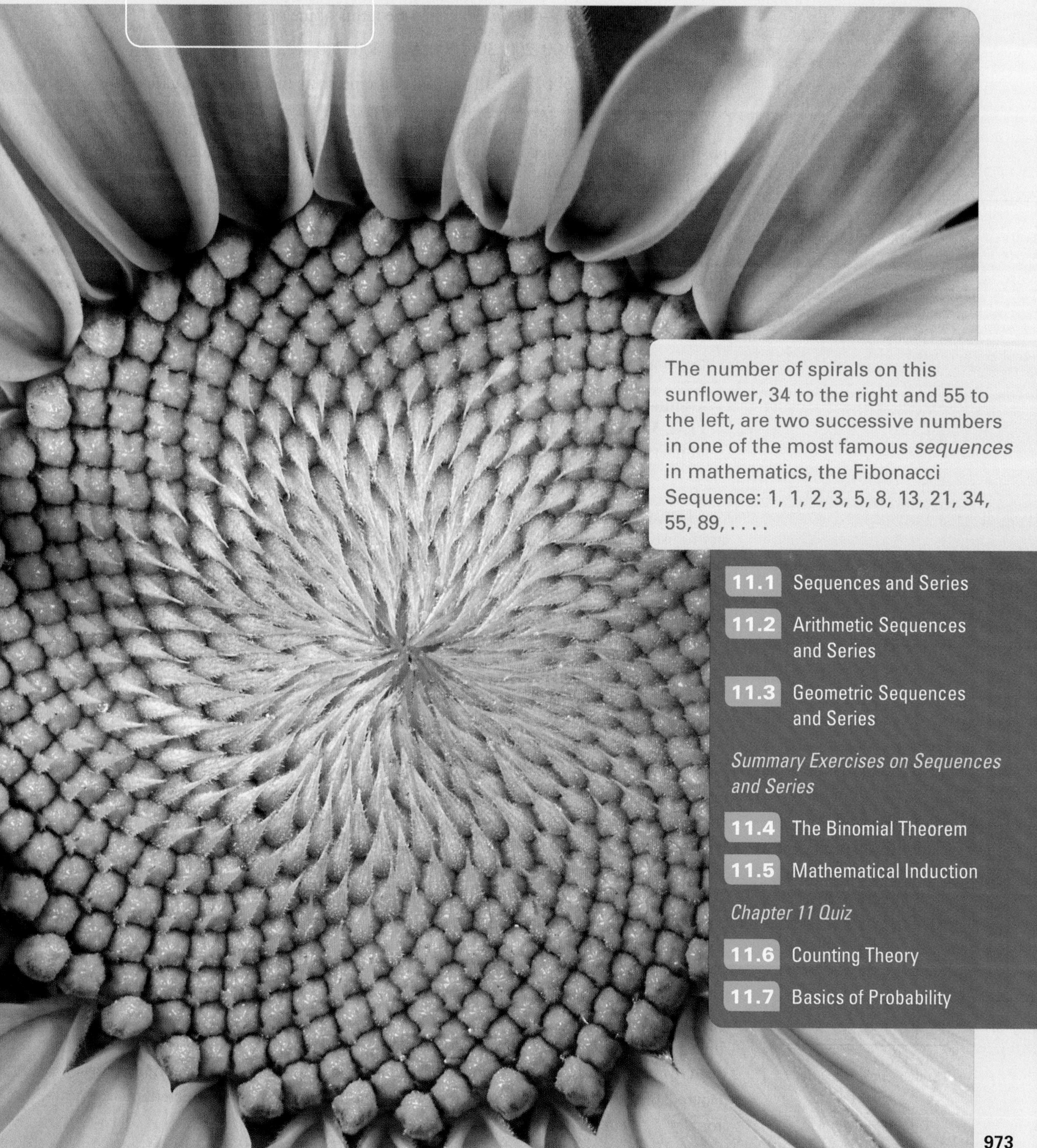

The number of spirals on this sunflower, 34 to the right and 55 to the left, are two successive numbers in one of the most famous *sequences* in mathematics, the Fibonacci Sequence: 1, 1, 2, 3, 5, 8, 13, 21, 34, 55, 89, ...

11.1 Sequences and Series

- Sequences
- Series and Summation Notation
- Summation Properties

Sequences A *sequence* is a function that computes an ordered list. For example, the average person in the United States uses 100 gallons of water each day. The function $f(n) = 100n$ generates the terms of the sequence

$$100, 200, 300, 400, 500, 600, 700, \ldots,$$

when $n = 1, 2, 3, 4, 5, 6, 7, \ldots$. This function represents the number of gallons of water used by the average person after n days.

A second example of a sequence involves investing money. If \$100 is deposited into a savings account paying 3% interest compounded annually, then the function $g(n) = 100(1.03)^n$ calculates the account balance after n years. The terms of the sequence are

$$g(1), g(2), g(3), g(4), g(5), g(6), g(7), \ldots,$$

and can be approximated as

$$103, 106.09, 109.27, 112.55, 115.93, 119.41, 122.99, \ldots.$$

Sequence

A **finite sequence** is a function that has a set of natural numbers of the form $\{1, 2, 3, \ldots, n\}$ as its domain. An **infinite sequence** has the set of natural numbers as its domain.

Teaching Tip Point out the relationship between the graph of a function $y = f(x)$ and the corresponding sequence $a_n = f(n)$.

Instead of using function notation $f(x)$ to indicate a sequence, it is customary to use a_n, where $a_n = f(n)$. ***The letter n is used instead of x as a reminder that n represents a natural number.*** The elements in the range of a sequence, called the **terms** of the sequence, are $a_1, a_2, a_3, \ldots$. The elements of both the domain and the range of a sequence are *ordered.* The first term is found by letting $n = 1$, the second term is found by letting $n = 2$, and so on. The **general term,** or ***n*th term,** of the sequence is a_n.

Figure 1 shows graphs of $f(x) = 2x$ and $a_n = 2n$. Notice that $f(x)$ is a continuous function, and a_n consists of discrete points. To graph a_n, we plot points of the form $(n, 2n)$ for $n = 1, 2, 3, \ldots$. We show only the results for $n = 1, 2, 3, 4,$ and 5.

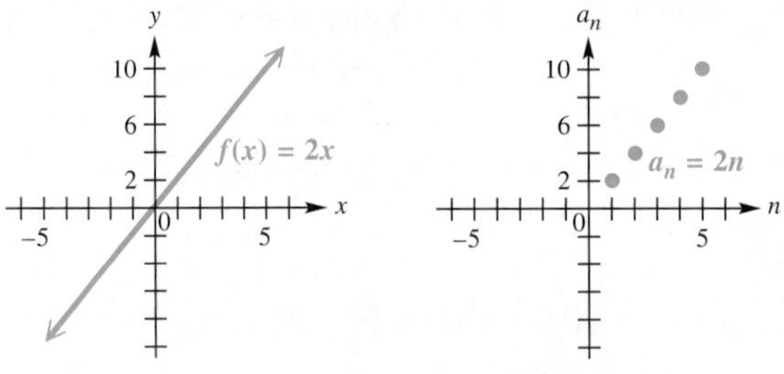

Figure 1

A graphing calculator can list the terms in a sequence. Using sequence mode to list the first 10 terms of the sequence with general term $a_n = n + \frac{1}{n}$ produces the result shown in **Figure 2(a).** Additional terms of the sequence can be seen by scrolling to the right. Sequences can also be graphed in sequence mode. **Figure 2(b)** shows a calculator screen with the graph of $a_n = n + \frac{1}{n}$. Notice that for $n = 5$, the term is $5 + \frac{1}{5} = 5.2$. ■

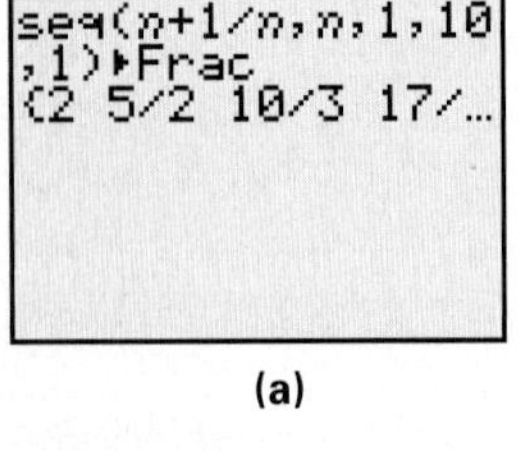

(a)

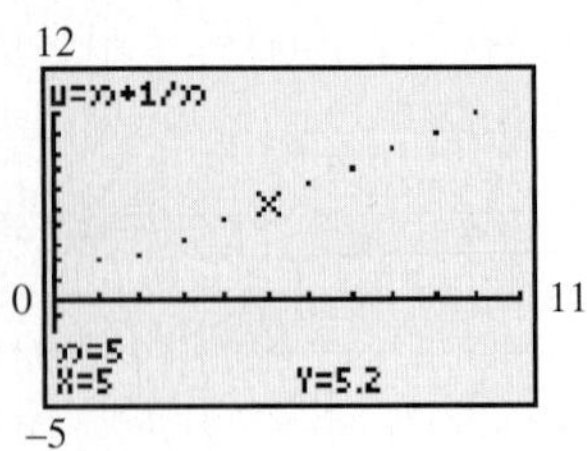

The fifth term is 5.2.

(b)

Figure 2

Classroom Example 1
Write the first five terms for each sequence.

(a) $a_n = 5n - 2$
(b) $a_n = (-1)^{n+1} \cdot n^2$
(c) $a_n = \frac{3n+2}{n^2+3}$

Answers:
(a) 3, 8, 13, 18, 23
(b) 1, −4, 9, −16, 25
(c) $\frac{5}{4}, \frac{8}{7}, \frac{11}{12}, \frac{14}{19}, \frac{17}{28}$

EXAMPLE 1 Finding Terms of Sequences

Write the first five terms for each sequence.

(a) $a_n = \frac{n+1}{n+2}$ **(b)** $a_n = (-1)^n \cdot n$ **(c)** $a_n = \frac{2n+1}{n^2+1}$

SOLUTION

(a) Replace n in $a_n = \frac{n+1}{n+2}$, with 1, 2, 3, 4, and 5.

$$n = 1: \quad a_1 = \frac{1+1}{1+2} = \frac{2}{3}$$

$$n = 2: \quad a_2 = \frac{2+1}{2+2} = \frac{3}{4}$$

$$n = 3: \quad a_3 = \frac{3+1}{3+2} = \frac{4}{5}$$

$$n = 4: \quad a_4 = \frac{4+1}{4+2} = \frac{5}{6}$$

$$n = 5: \quad a_5 = \frac{5+1}{5+2} = \frac{6}{7}$$

(b) Replace n in $a_n = (-1)^n \cdot n$ with 1, 2, 3, 4, and 5.

$$n = 1: \quad a_1 = (-1)^1 \cdot 1 = -1$$

$$n = 2: \quad a_2 = (-1)^2 \cdot 2 = 2$$

$$n = 3: \quad a_3 = (-1)^3 \cdot 3 = -3$$

$$n = 4: \quad a_4 = (-1)^4 \cdot 4 = 4$$

$$n = 5: \quad a_5 = (-1)^5 \cdot 5 = -5$$

(c) For $a_n = \frac{2n+1}{n^2+1}$, we have the following.

$$a_1 = \frac{3}{2}, \quad a_2 = 1, \quad a_3 = \frac{7}{10}, \quad a_4 = \frac{9}{17}, \quad \text{and} \quad a_5 = \frac{11}{26}$$

✔ *Now Try Exercises 3, 7, and 9.*

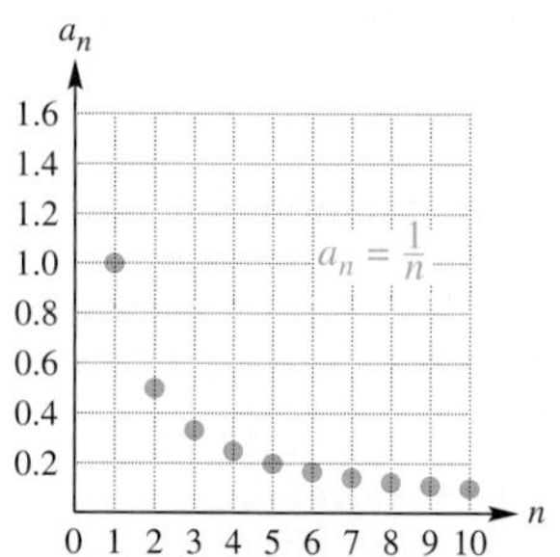

Figure 3

If the terms of an infinite sequence get closer and closer to some real number, the sequence is said to be **convergent** and to **converge** to that real number. For example, the sequence defined by $a_n = \frac{1}{n}$ approaches 0 as n becomes large. Thus, a_n is a convergent sequence that converges to 0. A graph of this sequence for $n = 1, 2, 3, \ldots, 10$ is shown in **Figure 3.** The terms of a_n approach the horizontal axis.

A sequence that does not converge to any number is **divergent.** The terms of the sequence $a_n = n^2$ are

$$1, 4, 9, 16, 25, 36, 49, 64, 81, \ldots.$$

This sequence is divergent because as n becomes large, the values of a_n do not approach a fixed number; rather, they increase without bound.

Some sequences are defined by a **recursive definition,** one in which each term after the first term or first few terms is defined as an expression involving the previous term or terms. On the other hand, the sequences in **Example 1** were defined *explicitly,* with a formula for a_n that does not depend on a previous term.

Leonardo of Pisa (Fibonacci) (1170–1250)

One of the most famous sequences in mathematics is the **Fibonacci sequence,** 1, 1, 2, 3, 5, 8, 13, 21, 34, 55, . . . , named for the Italian mathematician Leonardo of Pisa, who was also known as Fibonacci. The Fibonacci sequence may be defined using a recursion formula.

EXAMPLE 2 Using a Recursion Formula

Find the first four terms of each sequence.

(a) $a_1 = 4$
$a_n = 2 \cdot a_{n-1} + 1$, if $n > 1$

(b) $a_1 = 2$
$a_n = a_{n-1} + n - 1$, if $n > 1$

SOLUTION

(a) This is a recursive definition. We know $a_1 = 4$. Use $a_n = 2 \cdot a_{n-1} + 1$.

$$a_1 = 4$$
$$a_2 = 2 \cdot a_1 + 1 = 2 \cdot 4 + 1 = 9$$
$$a_3 = 2 \cdot a_2 + 1 = 2 \cdot 9 + 1 = 19$$
$$a_4 = 2 \cdot a_3 + 1 = 2 \cdot 19 + 1 = 39$$

(b) In this recursive definition, $a_1 = 2$ and $a_n = a_{n-1} + n - 1$.

$$a_1 = 2$$
$$a_2 = a_1 + 2 - 1 = 2 + 1 = 3$$
$$a_3 = a_2 + 3 - 1 = 3 + 2 = 5$$
$$a_4 = a_3 + 4 - 1 = 5 + 3 = 8$$

✔ Now Try Exercises 23 and 27.

Classroom Example 2
Find the first four terms of each sequence.
(a) $a_1 = 2$
$a_n = 3 \cdot a_{n-1} - 1$, if $n > 1$
(b) $a_1 = 5$
$a_n = a_{n-1} + 2n$, if $n > 1$

Answers:
(a) 2, 5, 14, 41 (b) 5, 9, 15, 23

EXAMPLE 3 Modeling Insect Population Growth

Frequently the population of a particular insect does not continue to grow indefinitely. Instead, its population grows rapidly at first and then levels off because of competition for limited resources. In one study, the behavior of the winter moth was modeled with a sequence similar to the following, where a_n represents the population density, in thousands per acre, during year n. (*Source:* Varley, G. and G. Gradwell, "Population models for the winter moth," Symposium of the Royal Entomological Society of London.)

$$a_1 = 1$$
$$a_n = 2.85a_{n-1} - 0.19a_{n-1}^2, \quad \text{for } n \geq 2$$

(a) Give a table of values for $n = 1, 2, 3, \ldots, 10$.

(b) Graph the sequence. Describe what happens to the population density.

SOLUTION

(a) Evaluate $a_1, a_2, a_3, \ldots, a_{10}$ recursively. We are given $a_1 = 1$.

$$a_2 = 2.85a_1 - 0.19a_1^2 = 2.85(1) - 0.19(1)^2 = 2.66$$
$$a_3 = 2.85a_2 - 0.19a_2^2 = 2.85(2.66) - 0.19(2.66)^2 \approx 6.24$$

Approximate values for $n = 1, 2, 3, \ldots, 10$ are shown in the table. **Figure 4** shows the computation of the sequence, denoted by $u(n)$ rather than a_n, using a calculator.

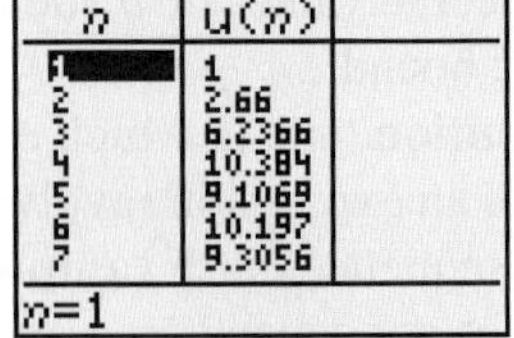

Figure 4

n	1	2	3	4	5	6	7	8	9	10
a_n	1	2.66	6.24	10.4	9.11	10.2	9.31	10.1	9.43	9.98

Classroom Example 3
Suppose an insect population density, in thousands per acre, during year n can be modeled by the recursively defined sequence

$a_1 = 5$
$a_n = 2.6a_{n-1} - 0.25a_{n-1}^2$, for $n > 1$.

(a) Find the population for $n = 1, 2, 3, 4, 5, 6$.
(b) Describe what happens to the population density as n increases.

Answers:
(a) $a_1 = 5$, $a_2 = 6.75$, $a_3 \approx 6.16$, $a_4 \approx 6.53$, $a_5 \approx 6.32$, $a_6 \approx 6.45$
(b) The population density oscillates above and below 6.4 thousand insects per acre and may stabilize near 6.4 thousand per acre.

(b) The graph of a sequence is a set of discrete points. Plot the points $(1, 1)$, $(2, 2.66)$, $(3, 6.24)$, ..., $(10, 9.98)$, as shown in **Figure 5(a).** At first, the insect population increases rapidly, and then it oscillates about the line $y = 9.7$. (See the Note following this example.) The oscillations become smaller as n increases, indicating that the population density may stabilize near 9.7 thousand per acre. In **Figure 5(b),** the first 20 terms have been plotted with a calculator.

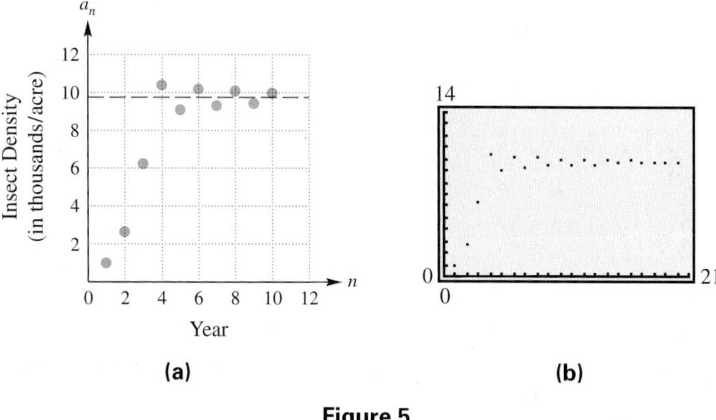

Figure 5

Now Try Exercise 87.

NOTE In **Example 3,** the insect population stabilizes near the value $k = 9.7$ thousand per acre. This value of k can be found by solving the quadratic equation $k = 2.85k - 0.19k^2$, which equates the values of a_n for consecutive years.

Series and Summation Notation Suppose a person has a starting salary of \$30,000 and receives a \$2000 raise each year. Then,

$$30{,}000, \quad 32{,}000, \quad 34{,}000, \quad 36{,}000, \quad 38{,}000$$

are terms of the sequence that describe this person's salaries over a 5-year period. The total earned is given by the *finite series*

$$30{,}000 + 32{,}000 + 34{,}000 + 36{,}000 + 38{,}000,$$

whose sum is \$170,000. A sequence can be used to define a *series*. For example, the infinite sequence

$$1, \frac{1}{3}, \frac{1}{9}, \frac{1}{27}, \frac{1}{81}, \frac{1}{243}, \ldots$$

defines the terms of the *infinite series*

$$1 + \frac{1}{3} + \frac{1}{9} + \frac{1}{27} + \frac{1}{81} + \frac{1}{243} + \cdots.$$

If a sequence has terms $a_1, a_2, a_3, \ldots$, then S_n is defined as the sum of the first n terms. That is,

$$S_n = a_1 + a_2 + a_3 + \cdots + a_n.$$

The sum of the terms of a sequence, called a **series,** is written using **summation notation.** The Greek capital letter **sigma** Σ is used to indicate a sum.

LOOKING AHEAD TO CALCULUS

An infinite series converges if the sequence of partial sums $S_1, S_2, S_3, \ldots$ converges. For example, it can be shown that

$$1 + \frac{1}{2} + \frac{1}{3} + \frac{1}{4} + \cdots \text{ diverges,}$$

while

$$1 - \frac{1}{2} + \frac{1}{3} - \frac{1}{4} + \cdots \text{ converges.}$$

Teaching Tip Tell students that they will use **sigma notation** again in statistics, finite math, and calculus courses.

Series

A **finite series** is an expression of the form

$$S_n = a_1 + a_2 + a_3 + \cdots + a_n = \sum_{i=1}^{n} a_i,$$

and an **infinite series** is an expression of the form

$$S_\infty = a_1 + a_2 + a_3 + \cdots + a_n + \cdots = \sum_{i=1}^{\infty} a_i.$$

The letter i is the **index of summation.**

CAUTION ***Do not confuse this use of i with the use of i as the imaginary unit.*** Other letters, such as k and j, may be used for the index of summation.

Classroom Example 4
Evaluate the series

$$\sum_{j=1}^{5} (j^2 + j + 1).$$

Answer: 75

EXAMPLE 4 Using Summation Notation

Evaluate the series $\sum_{k=1}^{6} (2^k + 1)$.

ALGEBRAIC SOLUTION

Write each of the six terms, and then evaluate the sum.

$$\begin{aligned}\sum_{k=1}^{6} (2^k + 1) &= (2^1 + 1) + (2^2 + 1) + (2^3 + 1) \\ &\quad + (2^4 + 1) + (2^5 + 1) + (2^6 + 1) \\ &= (2 + 1) + (4 + 1) + (8 + 1) \\ &\quad + (16 + 1) + (32 + 1) + (64 + 1) \\ &= 3 + 5 + 9 + 17 + 33 + 65 \\ &= 132\end{aligned}$$

GRAPHING CALCULATOR SOLUTION

A graphing calculator can list the terms of a sequence and then compute the sum of the terms. See **Figure 6.**

```
seq(2^K+1,K,1,6)
{3 5 9 17 33 65}
Σ(K=1 to 6) (2^K+1)
                  132
```

Figure 6

✔ *Now Try Exercise 43.*

Classroom Example 5
Write the terms for each series. Evaluate each sum, if possible.

(a) $\sum_{i=1}^{5} b_i$

(b) $\sum_{j=1}^{4} (2x_j - 5)$ if $x_1 = 3, x_2 = 5, x_3 = 7, x_4 = 9$

(c) $\sum_{k=1}^{3} f(x_k)\Delta x$, if $f(x) = \frac{1}{x}, x_1 = 1, x_2 = 3, x_3 = 5$, and $\Delta x = 4$

Answers:
(a) $b_1 + b_2 + b_3 + b_4 + b_5$
(b) $1 + 5 + 9 + 13 = 28$
(c) $4 + \frac{4}{3} + \frac{4}{5} = \frac{92}{15}$

EXAMPLE 5 Using Summation Notation with Subscripts

Write the terms for each series. Evaluate each sum, if possible.

(a) $\sum_{j=3}^{6} a_j$

(b) $\sum_{i=1}^{3} (6x_i - 2)$, if $x_1 = 2$, $x_2 = 4$, and $x_3 = 6$

(c) $\sum_{i=1}^{4} f(x_i)\Delta x$, if $f(x) = x^2$, $x_1 = 0$, $x_2 = 2$, $x_3 = 4$, $x_4 = 6$, and $\Delta x = 2$

SOLUTION

(a) $\sum_{j=3}^{6} a_j = a_3 + a_4 + a_5 + a_6$ Replace j in a_j with 3, 4, 5, and 6.

(b) $\sum_{i=1}^{3} (6x_i - 2) = (6x_1 - 2) + (6x_2 - 2) + (6x_3 - 2)$ Let $i = 1$, 2, and 3, respectively.

$= (6 \cdot 2 - 2) + (6 \cdot 4 - 2) + (6 \cdot 6 - 2)$ Substitute the given values for x_1, x_2, and x_3.

Use the order of operations.

$= 10 + 22 + 34$ Simplify.

$= 66$ Add.

(c) $\sum_{i=1}^{4} f(x_i)\Delta x = f(x_1)\Delta x + f(x_2)\Delta x + f(x_3)\Delta x + f(x_4)\Delta x$ Let $i = 1, 2, 3,$ and 4.

$= {x_1}^2\Delta x + {x_2}^2\Delta x + {x_3}^2\Delta x + {x_4}^2\Delta x$ $f(x) = x^2$

$= 0^2(2) + 2^2(2) + 4^2(2) + 6^2(2)$ Substitute the given values for x_1, x_2, x_3, and x_4, with $\Delta x = 2$.

$= 0 + 8 + 32 + 72$ Simplify.

$= 112$ Add.

✔ ***Now Try Exercises 51, 53, and 63.***

LOOKING AHEAD TO CALCULUS
Summation notation is used in calculus to describe the area under a curve, to describe the volume of a figure rotated about an axis, and in many other applications, as well as in the definition of integral. In the definition of the definite integral, Σ is replaced with an elongated S:

$$\int_a^b f(x)\,dx = \lim_{n\to\infty} \sum_{i=1}^{n} f(x_i)\,\Delta x_i.$$

In some cases, the definite integral can be interpreted as the sum of the areas of rectangles.

Summation Properties Properties of summation provide useful shortcuts for evaluating series.

Summation Properties

If $a_1, a_2, a_3, \ldots, a_n$ and $b_1, b_2, b_3, \ldots, b_n$ are two sequences, and c is a constant, then for every positive integer n, the following hold.

(a) $\sum_{i=1}^{n} c = nc$ **(b)** $\sum_{i=1}^{n} ca_i = c\sum_{i=1}^{n} a_i$

(c) $\sum_{i=1}^{n} (a_i + b_i) = \sum_{i=1}^{n} a_i + \sum_{i=1}^{n} b_i$ **(d)** $\sum_{i=1}^{n} (a_i - b_i) = \sum_{i=1}^{n} a_i - \sum_{i=1}^{n} b_i$

To prove Property (a), expand the series to obtain

$$c + c + c + c + \cdots + c,$$

where there are n terms of c, so the sum is nc.

Property (c) also can be proved by first expanding the series.

$$\sum_{i=1}^{n} (a_i + b_i) = (a_1 + b_1) + (a_2 + b_2) + \cdots + (a_n + b_n)$$

$$= (a_1 + a_2 + \cdots + a_n) + (b_1 + b_2 + \cdots + b_n)$$

Commutative and associative properties (Section R.2)

$$= \sum_{i=1}^{n} a_i + \sum_{i=1}^{n} b_i$$

Proofs of the other two properties are similar. The following results can be proved by mathematical induction. (See **Section 11.5.**)

Summation Rules

$$\sum_{i=1}^{n} i = 1 + 2 + \cdots + n = \frac{n(n+1)}{2}$$

$$\sum_{i=1}^{n} i^2 = 1^2 + 2^2 + \cdots + n^2 = \frac{n(n+1)(2n+1)}{6}$$

$$\sum_{i=1}^{n} i^3 = 1^3 + 2^3 + \cdots + n^3 = \frac{n^2(n+1)^2}{4}$$

Classroom Example 6
Use the summation properties to find each sum.

(a) $\sum_{i=1}^{50} 12$ (b) $\sum_{i=1}^{35} 3i$

(c) $\sum_{i=1}^{10} (3i^2 + 5)$

Answers:
(a) 600 (b) 1890
(c) 1205

EXAMPLE 6 Using the Summation Properties

Use the summation properties to find each sum.

(a) $\sum_{i=1}^{40} 5$ **(b)** $\sum_{i=1}^{22} 2i$ **(c)** $\sum_{i=1}^{14} (2i^2 - 3)$

SOLUTION

(a) $\sum_{i=1}^{40} 5 = 40(5) = 200$ Property (a) with $n = 40$ and $c = 5$

(b) $\sum_{i=1}^{22} 2i = 2\sum_{i=1}^{22} i$ Property (b) with $c = 2$ and $a_i = i$

$= 2 \cdot \frac{22(22+1)}{2}$ Summation rules

$= 506$ Evaluate.

(c) $\sum_{i=1}^{14} (2i^2 - 3) = \sum_{i=1}^{14} 2i^2 - \sum_{i=1}^{14} 3$ Property (d) with $a_i = 2i^2$ and $b_i = 3$

$= 2\sum_{i=1}^{14} i^2 - \sum_{i=1}^{14} 3$ Property (b) with $c = 2$ and $a_i = i^2$

$= 2 \cdot \frac{14(14+1)(2 \cdot 14 + 1)}{6} - 14(3)$ Summation rules and Property (a)

$= 2030 - 42$ Simplify.

$= 1988$ Subtract.

✔ *Now Try Exercises 67, 69, and 71.*

Classroom Example 7
Evaluate $\sum_{i=1}^{5} (i^3 + 2i - 8)$.

Answer: 215

EXAMPLE 7 Using the Summation Properties

Evaluate $\sum_{i=1}^{6} (i^2 + 3i + 5)$.

SOLUTION

$\sum_{i=1}^{6} (i^2 + 3i + 5) = \sum_{i=1}^{6} i^2 + \sum_{i=1}^{6} 3i + \sum_{i=1}^{6} 5$ Property (c)

$= \sum_{i=1}^{6} i^2 + 3\sum_{i=1}^{6} i + \sum_{i=1}^{6} 5$ Property (b)

$= \sum_{i=1}^{6} i^2 + 3\sum_{i=1}^{6} i + 6(5)$ Property (a)

$= \frac{6(6+1)(2 \cdot 6 + 1)}{6} + 3\left[\frac{6(6+1)}{2}\right] + 6(5)$ Summation rules

$= 91 + 63 + 30$ Simplify.

$= 184$ Add.

✔ *Now Try Exercise 73.*

NOTE It is possible to evaluate the sums in **Examples 6 and 7** without using the summation properties and rules; however, this can be tedious.

11.1 Exercises

1. 14, 18, 22, 26, 30
2. 3, 9, 15, 21, 27
3. $\frac{6}{5}, \frac{7}{6}, \frac{8}{7}, \frac{9}{8}, \frac{10}{9}$
4. $\frac{6}{5}, \frac{5}{4}, \frac{4}{3}, \frac{3}{2}, 2$
5. $0, \frac{1}{9}, \frac{2}{27}, \frac{1}{27}, \frac{4}{243}$
6. $\frac{1}{2}, \frac{1}{2}, \frac{3}{8}, \frac{1}{4}, \frac{5}{32}$
7. $-2, 4, -6, 8, -10$
8. $2, -3, 4, -5, 6$
9. $1, \frac{7}{6}, 1, \frac{5}{6}, \frac{19}{27}$
10. $0, \frac{3}{5}, \frac{4}{5}, \frac{15}{17}, \frac{12}{13}$
11. 3, 4, 7, 12, 19
12. 7, 7, 9, 13, 19

14. A sequence is a function with a domain consisting of natural numbers.

15. finite 16. finite
17. finite 18. finite
19. infinite 20. infinite
21. infinite 22. infinite

23. $-2, 1, 4, 7$
24. $-1, -5, -9, -13$
25. 1, 1, 2, 3
26. 1, 3, 4, 7
27. 2, 4, 12, 48
28. $-3, -12, -72, -576$

29. 35 30. 51
31. $\frac{25}{12}$ 32. $\frac{29}{20}$
33. 288 34. 700
35. 3 36. 28
37. -18 38. 135
39. $\frac{728}{9}$ 40. $\frac{75}{2}$
41. 28 42. 176
43. 343 44. -2
45. 30 46. 254

Write the first five terms of each sequence. ***See Example 1.***

1. $a_n = 4n + 10$ **2.** $a_n = 6n - 3$ **3.** $a_n = \frac{n+5}{n+4}$

4. $a_n = \frac{n-7}{n-6}$ **5.** $a_n = \left(\frac{1}{3}\right)^n (n-1)$ **6.** $a_n = \left(\frac{1}{2}\right)^n (n)$

7. $a_n = (-1)^n(2n)$ **8.** $a_n = (-1)^{n-1}(n+1)$ **9.** $a_n = \frac{4n-1}{n^2+2}$

10. $a_n = \frac{n^2-1}{n^2+1}$ **11.** $a_n = \frac{n^3+8}{n+2}$ **12.** $a_n = \frac{n^3+27}{n+3}$

13. Your friend does not understand what is meant by the nth term, or general term, of a sequence. How would you explain this idea?

14. *Concept Check* How are sequences related to functions?

Concept Check Decide whether each sequence is finite *or* infinite.

15. The sequence of days of the week **16.** The sequence of pages in a book

17. 1, 2, 3, 4, 5 **18.** $-1, -2, -3, -4, -5$

19. 1, 2, 3, 4, 5, . . . **20.** $-1, -2, -3, -4, -5, \ldots$

21. $a_1 = 4$
$a_n = 4 \cdot a_{n-1}$, if $n \geq 2$

22. $a_1 = 2$
$a_2 = 5$
$a_n = a_{n-1} + a_{n-2}$, if $n \geq 3$

Find the first four terms of each sequence. ***See Example 2.***

23. $a_1 = -2$
$a_n = a_{n-1} + 3$, if $n > 1$

24. $a_1 = -1$
$a_n = a_{n-1} - 4$, if $n > 1$

25. $a_1 = 1$
$a_2 = 1$
$a_n = a_{n-1} + a_{n-2}$, if $n \geq 3$
(This is the Fibonacci sequence.)

26. $a_1 = 1$
$a_2 = 3$
$a_n = a_{n-1} + a_{n-2}$, if $n \geq 3$
(This is the Lucas sequence.)

27. $a_1 = 2$
$a_n = n \cdot a_{n-1}$, if $n > 1$

28. $a_1 = -3$
$a_n = 2n \cdot a_{n-1}$, if $n > 1$

Evaluate each series. ***See Example 4.***

29. $\sum_{i=1}^{5} (2i + 1)$ **30.** $\sum_{i=1}^{6} (3i - 2)$ **31.** $\sum_{j=1}^{4} j^{-1}$

32. $\sum_{i=1}^{5} (i + 1)^{-1}$ **33.** $\sum_{i=1}^{4} i^i$ **34.** $\sum_{k=1}^{4} (k + 1)^k$

35. $\sum_{k=1}^{6} (-1)^k \cdot k$ **36.** $\sum_{i=1}^{7} (-1)^{i+1} \cdot i^2$ **37.** $\sum_{i=2}^{5} (6 - 3i)$

38. $\sum_{i=3}^{7} (5i + 2)$ **39.** $\sum_{i=-2}^{3} 2(3)^i$ **40.** $\sum_{i=-1}^{2} 5(2)^i$

41. $\sum_{i=-1}^{5} (i^2 - 2i)$ **42.** $\sum_{i=3}^{6} (2i^2 + 1)$ **43.** $\sum_{i=1}^{5} (3^i - 4)$

44. $\sum_{i=1}^{4} [(-2)^i - 3]$ **45.** $\sum_{i=1}^{3} (i^3 - i)$ **46.** $\sum_{i=1}^{4} (i^4 - i^3)$

Use a graphing calculator to evaluate each series. **See Example 4.**

47. $\sum_{i=1}^{10} (4i^2 - 5)$ **48.** $\sum_{i=1}^{10} (i^3 - 6)$

49. $\sum_{j=3}^{9} (3j - j^2)$ **50.** $\sum_{k=5}^{10} (k^2 - 4k + 7)$

Write the terms for each series. Evaluate the sum, given that $x_1 = -2$, $x_2 = -1$, $x_3 = 0$, $x_4 = 1$, *and* $x_5 = 2$. **See Examples 5(a) and 5(b).**

51. $\sum_{i=1}^{5} x_i$ **52.** $\sum_{i=1}^{5} -x_i$ **53.** $\sum_{i=1}^{5} (2x_i + 3)$

54. $\sum_{i=1}^{4} (-3x_i - 2)$ **55.** $\sum_{i=1}^{3} (3x_i - x_i^2)$ **56.** $\sum_{i=1}^{3} (x_i^2 + x_i)$

57. $\sum_{i=2}^{5} \frac{x_i + 1}{x_i + 2}$ **58.** $\sum_{i=1}^{5} \frac{x_i}{x_i + 3}$ **59.** $\sum_{i=1}^{4} \frac{x_i^3 + 1000}{x_i + 10}$

60. Explain how factoring can make the work in **Exercises 11, 12, and 59** easier.

Write the terms of $\sum_{i=1}^{4} f(x_i)\Delta x$, *with* $x_1 = 0$, $x_2 = 2$, $x_3 = 4$, $x_4 = 6$, *and* $\Delta x = 0.5$, *for each function. Evaluate the sum.* **See Example 5(c).**

61. $f(x) = 4x - 7$ **62.** $f(x) = 6 + 2x$ **63.** $f(x) = 2x^2$

64. $f(x) = x^2 - 1$ **65.** $f(x) = \frac{-2}{x + 1}$ **66.** $f(x) = \frac{5}{2x - 1}$

Use the summation properties and rules to evaluate each series. **See Examples 6 and 7.**

67. $\sum_{i=1}^{100} 6$ **68.** $\sum_{i=1}^{20} 5$ **69.** $\sum_{i=1}^{15} i^2$ **70.** $\sum_{i=1}^{50} 2i^3$

71. $\sum_{i=1}^{5} (5i + 3)$ **72.** $\sum_{i=1}^{5} (8i - 1)$ **73.** $\sum_{i=1}^{5} (4i^2 - 2i + 6)$

74. $\sum_{i=1}^{6} (2 + i - i^2)$ **75.** $\sum_{i=1}^{4} (3i^3 + 2i - 4)$ **76.** $\sum_{i=1}^{6} (i^2 + 2i^3)$

*Concept Check Use summation notation to write each series.**

77. $\frac{1}{3(1)} + \frac{1}{3(2)} + \frac{1}{3(3)} + \cdots + \frac{1}{3(9)}$ **78.** $\frac{5}{1+1} + \frac{5}{1+2} + \frac{5}{1+3} + \cdots + \frac{5}{1+15}$

79. $1 - \frac{1}{2} + \frac{1}{4} - \frac{1}{8} + \cdots - \frac{1}{128}$ **80.** $1 - \frac{1}{4} + \frac{1}{9} - \frac{1}{16} + \cdots - \frac{1}{400}$

Use the sequence feature of a graphing calculator to graph the first ten terms of each sequence as defined. Use the graph to make a conjecture as to whether the sequence converges or diverges. If you think it converges, determine the number to which it converges.

81. $a_n = \frac{n + 4}{2n}$ **82.** $a_n = \frac{1 + 4n}{2n}$ **83.** $a_n = 2e^n$

84. $a_n = n(n + 2)$ **85.** $a_n = \left(1 + \frac{1}{n}\right)^n$ **86.** $a_n = (1 + n)^{1/n}$

*These exercises were suggested by Joe Lloyd Harris, Gulf Coast Community College.

Answers (margin)

47. 1490 **48.** 2965
49. −154 **50.** 217
51. $-2 + (-1) + 0 + 1 + 2;\ 0$
52. $2 + 1 + 0 + (-1) + (-2);\ 0$
53. $-1 + 1 + 3 + 5 + 7;\ 15$
54. $4 + 1 - 2 - 5;\ -2$
55. $-10 - 4 + 0;\ -14$
56. $2 + 0 + 0;\ 2$
57. $0 + \frac{1}{2} + \frac{2}{3} + \frac{3}{4};\ \frac{23}{12}$
58. $-2 - \frac{1}{2} + 0 + \frac{1}{4} + \frac{2}{5};\ -\frac{37}{20}$
59. $124 + 111 + 100 + 91;\ 426$
61. $-3.5 + 0.5 + 4.5 + 8.5;\ 10$
62. $3 + 5 + 7 + 9;\ 24$
63. $0 + 4 + 16 + 36;\ 56$
64. $-0.5 + 1.5 + 7.5 + 17.5;\ 26$
65. $-1 - \frac{1}{3} - \frac{1}{5} - \frac{1}{7};\ -\frac{176}{105}$
66. $-\frac{5}{2} + \frac{5}{6} + \frac{5}{14} + \frac{5}{22};\ -\frac{250}{231}$
67. 600 **68.** 100
69. 1240 **70.** 3,251,250
71. 90 **72.** 115
73. 220 **74.** −58
75. 304 **76.** 973

There are other acceptable forms of the answers in Exercises 77–80.

77. $\sum_{i=1}^{9} \frac{1}{3i}$
78. $\sum_{i=1}^{15} \frac{5}{1+i}$
79. $\sum_{k=1}^{8} \left(-\frac{1}{2}\right)^{k-1}$
80. $\sum_{j=1}^{20} (-1)^{j-1}\frac{1}{j^2}$
81. converges to $\frac{1}{2}$
82. converges to 2
83. diverges
84. diverges
85. converges to $e \approx 2.71828$
86. converges to 1

87. (a) $a_1 = 8$ thousand per acre, $a_2 = 10.4$ thousand per acre, $a_3 = 8.528$ thousand per acre
(b) The population density oscillates above and below 9.5 thousand per acre (approximately).

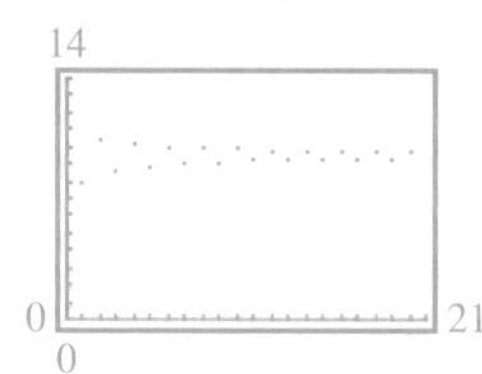

88.

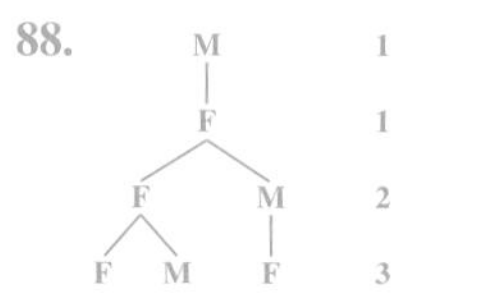

89. (a) $N_{j+1} = 2N_j$ for $j \geq 1$
(b) 1840
(c) 15,000

0 10
0

90. (a)

j	N_j	j	N_j
1	230	11	4901
2	440	12	4950
3	808	13	4975
4	1392	14	4987
5	2178	15	4994
6	3034	16	4997
7	3776	17	4998
8	4303	18	4999
9	4625	19	5000
10	4805	20	5000

(b) 6000

0 20
0

Solve each problem involving sequences and series. ***See Example 3.***

87. *(Modeling) Insect Population* Suppose an insect population density, in thousands per acre, during year n can be modeled by the recursively defined sequence

$$a_1 = 8$$

$$a_n = 2.9a_{n-1} - 0.2a_{n-1}^2, \quad \text{for } n > 1.$$

(a) Find the population for $n = 1, 2, 3$.

(b) Graph the sequence for $n = 1, 2, 3, \ldots, 20$. Use the window $[0, 21]$ by $[0, 14]$. Interpret the graph.

88. *Male Bee Ancestors* One of the most famous sequences in mathematics is the Fibonacci sequence,

$$1, 1, 2, 3, 5, 8, 13, 21, 34, 55, \ldots.$$

(Also see **Exercise 25.**) Male honeybees hatch from eggs that have not been fertilized, so a male bee has only one parent, a female. On the other hand, female honeybees hatch from fertilized eggs, so a female has two parents, one male and one female. The number of ancestors in consecutive generations of bees follows the Fibonacci sequence. Draw a tree showing the number of ancestors of a male bee in each generation following the description given above.

89. *(Modeling) Bacteria Growth* If certain bacteria are cultured in a medium with sufficient nutrients, they will double in size and then divide every 40 minutes. Let N_1 be the initial number of bacteria cells, N_2 the number after 40 minutes, N_3 the number after 80 minutes, and N_j the number after $40(j-1)$ minutes. (*Source:* Hoppensteadt, F. and C. Peskin, *Mathematics in Medicine and the Life Sciences,* Springer-Verlag.)

(a) Write N_{j+1} in terms of N_j for $j \geq 1$.

(b) Determine the number of bacteria after 2 hr if $N_1 = 230$.

(c) Graph the sequence N_j for $j = 1, 2, 3, \ldots, 7$, where $N_1 = 230$. Use the window $[0, 10]$ by $[0, 15{,}000]$.

(d) Describe the growth of these bacteria when there are unlimited nutrients.

90. *(Modeling) Verhulst's Model for Bacteria Growth* Refer to **Exercise 89.** If the bacteria are not cultured in a medium with sufficient nutrients, competition will ensue and growth will slow. According to Verhulst's model, the number of bacteria N_j at time $40(j-1)$ in minutes can be determined by the sequence

$$N_{j+1} = \left[\frac{2}{1 + \frac{N_j}{K}}\right] N_j,$$

where K is a constant and $j \geq 1$. (*Source:* Hoppensteadt, F. and C. Peskin, *Mathematics in Medicine and the Life Sciences,* Springer-Verlag.)

(a) If $N_1 = 230$ and $K = 5000$, make a table of N_j for $j = 1, 2, 3, \ldots, 20$. Round values in the table to the nearest integer.

(b) Graph the sequence N_j for $j = 1, 2, 3, \ldots, 20$. Use the window $[0, 20]$ by $[0, 6000]$.

(c) Describe the growth of these bacteria when there are limited nutrients.

(d) Make a conjecture as to why K is called the **saturation constant.** Test your conjecture by changing the value of K in the given formula.

91. (a) 0.0198026273; $\ln 1.02 \approx 0.0198026273$
(b) -0.0304592075; $\ln 0.97 \approx -0.0304592075$
92. $\pi \approx 3.1407$; This is accurate to three decimal places when rounded.
93. (a) 2.718254; $e \approx 2.718282$
(b) 0.367857; $e^{-1} \approx 0.367879$
94. (a) $a_6 \approx 1.414213562$; $\sqrt{2} \approx 1.414213562$
(b) $a_6 \approx 3.316624805$; $\sqrt{11} \approx 3.31662479$

91. *Approximating* $\ln(1+x)$ The series

$$x - \frac{x^2}{2} + \frac{x^3}{3} - \frac{x^4}{4} + \cdots$$

can be used to approximate the value of $\ln(1+x)$ for values of x in $(-1, 1]$. Use the first six terms of this series to approximate each expression. Compare this approximation with the value obtained on a calculator.

(a) $\ln 1.02$ $(x = 0.02)$ **(b)** $\ln 0.97$ $(x = -0.03)$

92. *Approximating* π Find the sum of the first six terms of the series

$$\frac{\pi^4}{90} = \frac{1}{1^4} + \frac{1}{2^4} + \frac{1}{3^4} + \frac{1}{4^4} + \frac{1}{5^4} + \cdots + \frac{1}{n^4} + \cdots.$$

Multiply this result by 90, and take the fourth root to obtain an approximation of π. Compare your answer to the actual decimal approximation of π.

93. *Approximating Powers of e* The series

$$e^a \approx 1 + a + \frac{a^2}{2!} + \frac{a^3}{3!} + \cdots + \frac{a^n}{n!},$$

where $n! = 1 \cdot 2 \cdot 3 \cdot 4 \cdot \cdots \cdot n$, can be used to approximate the value of e^a for any real number a. Use the first eight terms of this series to approximate each expression. Compare this approximation with the value obtained on a calculator.

(a) e **(b)** e^{-1}

94. *Approximating Square Roots* The recursively defined sequence

$$a_1 = k$$
$$a_n = \frac{1}{2}\left(a_{n-1} + \frac{k}{a_{n-1}}\right), \quad \text{if } n > 1$$

can be used to compute $\sqrt{k}$ for any positive number k. This sequence was known to Sumerian mathematicians 4000 years ago, and it is still used today. Use this sequence to approximate the given square root by finding a_6. Compare your result with the actual value. (*Source:* Heinz-Otto, P., *Chaos and Fractals,* Springer-Verlag.)

(a) $\sqrt{2}$ **(b)** $\sqrt{11}$

11.2 Arithmetic Sequences and Series

- Arithmetic Sequences
- Arithmetic Series

Teaching Tip Point out that the equation

$$d = a_{n+1} - a_n$$

can be written as

$$a_{n+1} = a_n + d.$$

The second form of the equation is used in **Example 2** to generate successive terms of an arithmetic sequence recursively.

Arithmetic Sequences An **arithmetic sequence** (or **arithmetic progression**) is a sequence in which each term after the first differs from the preceding term by a fixed constant, called the **common difference.** The sequence

$$5, 9, 13, 17, 21, \ldots$$

is an arithmetic sequence since each term after the first is obtained by adding 4 to the previous term. That is,

$$9 = 5 + 4$$
$$13 = 9 + 4$$
$$17 = 13 + 4$$
$$21 = 17 + 4, \quad \text{and so on.}$$

The common difference is 4.

If the common difference of an arithmetic sequence is d, then by the definition of an arithmetic sequence,

$$d = a_{n+1} - a_n, \quad \text{Common difference } d$$

for every positive integer n in the domain of the sequence.

Classroom Example 1
Find the common difference, d, for the arithmetic sequence

$$20, 13, 6, -1, -8, \ldots .$$

Answer: -7

EXAMPLE 1 Finding the Common Difference

Find the common difference, d, for the arithmetic sequence

$$-9, -7, -5, -3, -1, \ldots .$$

SOLUTION We find d by choosing any two adjacent terms and subtracting the first from the second. Choosing -7 and -5 gives

$$d = -5 - (-7) = 2.$$

Be careful when subtracting a negative number.

Choosing -9 and -7 would give $d = -7 - (-9) = 2$, the same result.

Now Try Exercise 1.

Classroom Example 2
Find the first five terms for each arithmetic sequence.
(a) The first term is -14, and the common difference is 6.
(b) $a_1 = 1.5, d = -0.5$

Answers:
(a) $-14, -8, -2, 4, 10$
(b) $1.5, 1.0, 0.5, 0, -0.5$

EXAMPLE 2 Finding Terms Given a_1 and d

Find the first five terms for each arithmetic sequence.

(a) The first term is 7, and the common difference is -3.

(b) $a_1 = -12, d = 5$

SOLUTION

(a)

$a_1 = 7$ Start with $a_1 = 7$.

$a_2 = 7 + (-3) = 4$ Add $d = -3$.

$a_3 = 4 + (-3) = 1$ Add -3.

$a_4 = 1 + (-3) = -2$ Add -3.

$a_5 = -2 + (-3) = -5$ Add -3.

(b)

$a_1 = -12$ Start with $a_1 = -12$.

$a_2 = -12 + 5 = -7$ Add $d = 5$.

$a_3 = -7 + 5 = -2$ Add 5.

$a_4 = -2 + 5 = 3$ Add 5.

$a_5 = 3 + 5 = 8$ Add 5.

Now Try Exercises 7 and 9.

If a_1 is the first term of an arithmetic sequence and d is the common difference, then the terms of the sequence are given as follows.

$$a_1 = a_1$$
$$a_2 = a_1 + d$$
$$a_3 = a_2 + d = a_1 + d + d = a_1 + 2d$$
$$a_4 = a_3 + d = a_1 + 2d + d = a_1 + 3d$$
$$a_5 = a_1 + 4d$$
$$a_6 = a_1 + 5d$$

By this pattern, $a_n = a_1 + (n - 1)d$.

This result can be proved by mathematical induction. (See **Section 11.5.**)

nth Term of an Arithmetic Sequence

In an arithmetic sequence with first term a_1 and common difference d, the nth term, a_n, is given by the following.

$$a_n = a_1 + (n - 1)d$$

Classroom Example 3
Find a_{16} and a_n for the arithmetic sequence 23, 20, 17, 14, ...

Answer:
$a_{16} = -22$; $a_n = -3n + 26$

EXAMPLE 3 Finding Terms of an Arithmetic Sequence

Find a_{13} and a_n for the arithmetic sequence $-3, 1, 5, 9, \ldots$.

SOLUTION Here $a_1 = -3$ and $d = 1 - (-3) = 4$. To find a_{13}, substitute 13 for n in the formula for the nth term.

$$a_n = a_1 + (n - 1)d$$

Work inside the parentheses first.

$$a_{13} = a_1 + (13 - 1)d \quad n = 13$$
$$a_{13} = -3 + (12)4 \quad \text{Let } a_1 = -3 \text{ and } d = 4.$$
$$a_{13} = -3 + 48 \quad \text{Multiply.}$$
$$a_{13} = 45 \quad \text{Add.}$$

Find a_n by substituting values for a_1 and d in the formula for a_n.

$$a_n = a_1 + (n - 1)d \quad \text{Formula for } a_n$$
$$a_n = -3 + (n - 1) \cdot 4 \quad \text{Let } a_1 = -3 \text{ and } d = 4.$$
$$a_n = -3 + 4n - 4 \quad \text{Distributive property (Section R.2)}$$
$$a_n = 4n - 7 \quad \text{Simplify.}$$

✔ Now Try Exercise 13.

Classroom Example 4
Find a_n and a_{10} for the arithmetic sequence having $a_4 = 18$ and $a_5 = 22$.

Answer:
$a_n = 4n + 2$; $a_{10} = 42$

EXAMPLE 4 Finding Terms of an Arithmetic Sequence

Find a_n and a_{18} for the arithmetic sequence having $a_2 = 9$ and $a_3 = 15$.

SOLUTION Find d first. Then find a_1.

$$d = a_3 - a_2 = 15 - 9 = 6$$

So,
$$a_2 = a_1 + d \quad \text{Definition of arithmetic sequence}$$
$$9 = a_1 + 6 \quad \text{Let } a_2 = 9 \text{ and } d = 6.$$
$$a_1 = 3. \quad \text{Subtract 6 and interchange sides.}$$

Now,
$$a_n = a_1 + (n - 1)d \quad \text{Formula for } a_n$$
$$a_n = 3 + (n - 1)6 \quad \text{Let } a_1 = 3 \text{ and } d = 6.$$
$$a_n = 3 + 6n - 6 \quad \text{Distributive property}$$
$$a_n = 6n - 3, \quad \text{Simplify.}$$

and
$$a_{18} = 6(18) - 3 \quad \text{Let } n = 18.$$
$$a_{18} = 105. \quad \text{Multiply and then subtract.}$$

✔ Now Try Exercise 17.

Classroom Example 5
Suppose that an arithmetic sequence has $a_6 = 9$ and $a_{15} = -36$. Find a_1.

Answer: 34

EXAMPLE 5 Finding the First Term of an Arithmetic Sequence

Suppose that an arithmetic sequence has $a_8 = -16$ and $a_{16} = -40$. Find a_1.

SOLUTION We obtain a_{16} by adding the common difference to a_8 eight times.

$$a_{16} = a_8 + 8d$$

$$-40 = -16 + 8d \quad \text{Let } a_{16} = -40 \text{ and } a_8 = -16.$$

$$-24 = 8d \quad \text{Add 16. (Section 1.1)}$$

$$d = -3 \quad \text{Divide by 8 and interchange sides.}$$

To find a_1, use the formula for a_n.

$$a_n = a_1 + (n-1)d \quad \text{Formula for } a_n$$

$$-16 = a_1 + (8-1)(-3) \quad \text{Let } a_n = -16,\ n = 8, \text{ and } d = -3.$$

$$-16 = a_1 - 21 \quad \text{Simplify.}$$

$$a_1 = 5 \quad \text{Add 21 and interchange sides.}$$

Now Try Exercise 23.

To determine the characteristics of the graph of an arithmetic sequence, start by rewriting the formula for the nth term.

$$a_n = a_1 + (n-1)d \quad \text{Formula for the } n\text{th term}$$

$$a_n = a_1 + nd - d \quad \text{Distributive property}$$

$$a_n = dn + (a_1 - d) \quad \text{Commutative and associative properties (Section R.2)}$$

$$a_n = dn + c \quad \text{Let } c = a_1 - d.$$

The points in the graph of an arithmetic sequence are determined by

$$a_n = dn + c,$$

where n is a natural number. Thus, the *discrete* points on the graph of the sequence must lie on the *continuous* linear graph

$$y = dx + c. \quad \text{(Section 2.5)}$$

(Slope: d; y-intercept: c)

For example, the sequence a_n shown in **Figure 7(a)** is an arithmetic sequence because the points that make up its graph are collinear (lie on a line). The slope determined by these points is 2, so the common difference d equals 2. On the other hand, the sequence b_n shown in **Figure 7(b)** is not an arithmetic sequence because the points are not collinear.

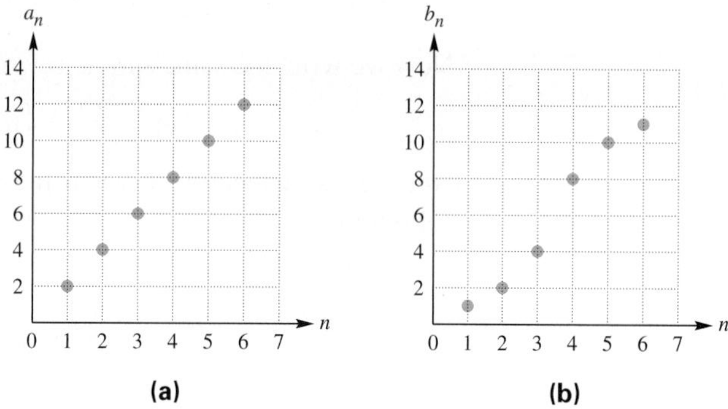

Figure 7

EXAMPLE 6 Finding the *n*th Term from a Graph

Find a formula for the nth term of the sequence a_n shown in **Figure 8.** What are the domain and range of this sequence?

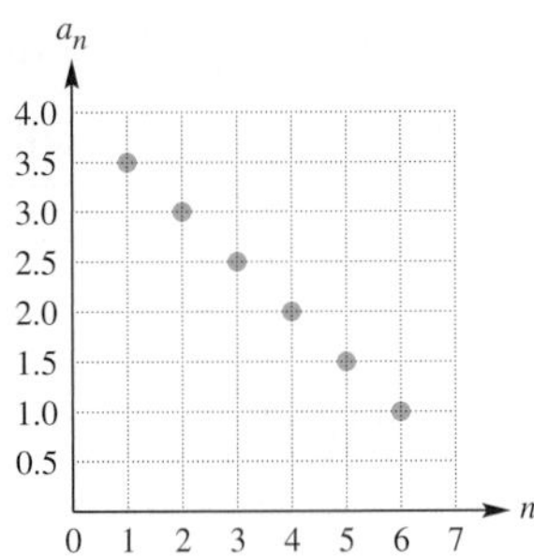

Figure 8

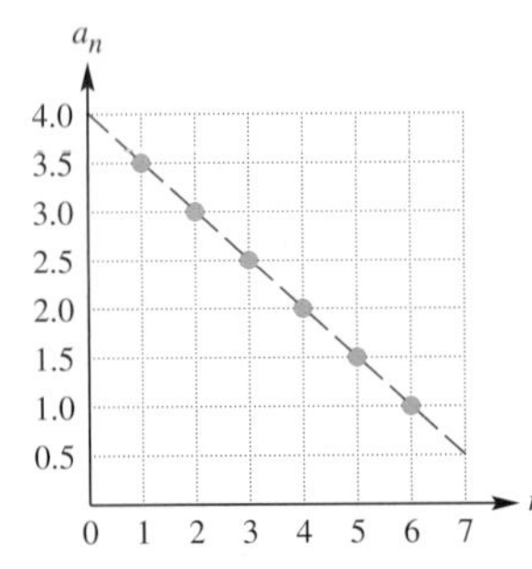

Figure 9

SOLUTION The points in **Figure 8** lie on a line, so the sequence is arithmetic. The equation of the dashed line shown in **Figure 9** is $y = -0.5x + 4$, so the nth term of this sequence is determined by

$$a_n = -0.5n + 4.$$

The sequence consists of the points

$$(1, 3.5), (2, 3), (3, 2.5), (4, 2), (5, 1.5), (6, 1).$$

Thus, the domain of the sequence is given by $\{1, 2, 3, 4, 5, 6\}$, and the range is given by $\{3.5, 3, 2.5, 2, 1.5, 1\}$.

Now Try Exercise 27.

Classroom Example 6

Find a formula for the nth term of the sequence a_n shown in the graph. What are the domain and range of this sequence?

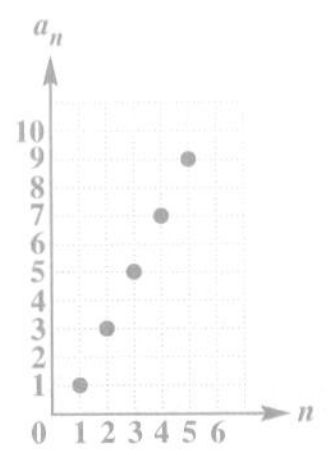

Answer: $a_n = 2n - 1$; domain: $\{1, 2, 3, 4, 5\}$; range: $\{1, 3, 5, 7, 9\}$

Arithmetic Series The sum of the terms of an arithmetic sequence is an **arithmetic series.** To illustrate, suppose that a person borrows \$3000 and agrees to pay \$100 per month plus interest of 1% per month on the unpaid balance until the loan is paid off. The first month, \$100 is paid to reduce the loan, plus interest of $(0.01)3000 = 30$ dollars. The second month, another \$100 is paid toward the loan, and $(0.01)2900 = 29$ dollars is paid for interest. Since the loan is reduced by \$100 each month, interest payments decrease by $(0.01)100 = 1$ dollar each month, forming the arithmetic sequence

$$30, 29, 28, \ldots, 3, 2, 1.$$

The total amount of interest paid is given by the sum of the terms of this sequence. Now we develop a formula to find this sum without adding all 30 numbers directly. Since the sequence is arithmetic, we can write the sum of the first n terms as

$$S_n = a_1 + [a_1 + d] + [a_1 + 2d] + \cdots + [a_1 + (n-1)d]. \quad \text{Formula for the general term}$$

Now we write the same sum in reverse order, beginning with a_n and *subtracting d.*

$$S_n = a_n + [a_n - d] + [a_n - 2d] + \cdots + [a_n - (n-1)d]$$

Adding the respective sides of these two equations term by term, we obtain the following.

$$S_n + S_n = (a_1 + a_n) + (a_1 + a_n) + \cdots + (a_1 + a_n)$$

$$2S_n = n(a_1 + a_n) \quad \text{There are } n \text{ terms of } a_1 + a_n \text{ on the right.}$$

$$S_n = \frac{n}{2}(a_1 + a_n) \quad \text{Solve for } S_n.$$

Teaching Tip You may want to justify the formula for S_n numerically by evaluating the sum of the first six natural numbers as

$$\begin{array}{r} 1+2+3+4+5+6 \\ +6+5+4+3+2+1 \\ \hline 7+7+7+7+7+7=42, \end{array}$$

and $\frac{42}{2}=21$.

Teaching Tip Point out that the formula

$$S_n=\frac{n}{2}(a_1+a_n)$$

can also be written

$$S_n=\left(\frac{a_1+a_n}{2}\right)\cdot n$$

and interpreted as the average of the first and last terms multiplied by the number of terms.

Using the formula $a_n=a_1+(n-1)d$, we can also write this result for S_n as

$$S_n=\frac{n}{2}[a_1+a_1+(n-1)d],$$

or $$S_n=\frac{n}{2}[2a_1+(n-1)d].$$ Alternative formula for the sum of the first n terms

Sum of the First n Terms of an Arithmetic Sequence

If an arithmetic sequence has first term a_1 and common difference d, then the sum S_n of the first n terms is given by these formulas.

$$S_n=\frac{n}{2}(a_1+a_n), \quad \text{or} \quad S_n=\frac{n}{2}[2a_1+(n-1)d]$$

The first formula is used when the first and last terms are known; otherwise, the second formula is used.

For example, in the sequence of interest payments discussed earlier, $n=30$, $a_1=30$, and $a_n=1$.

$$S_n=\frac{n}{2}(a_1+a_n) \quad \text{First formula for } S_n$$

$$S_{30}=\frac{30}{2}(30+1) \quad \text{Substitute known quantities.}$$

$$S_{30}=15(31) \quad \text{Simplify.}$$

$$S_{30}=465 \quad \text{Multiply.}$$

A total of \$465 interest will be paid over the 30 months.

Classroom Example 7
Consider the arithmetic sequence

48, 44, 40, 36, ...

(a) Evaluate S_{21}.

(b) Use a formula for S_n to evaluate the sum of the first 200 positive integers.

Answers:
(a) 168 **(b)** 20,100

EXAMPLE 7 Using the Sum Formulas

Consider the arithmetic sequence $-9, -5, -1, 3, 7, \ldots$.

(a) Evaluate S_{12}.

(b) Use a formula for S_n to evaluate the sum of the first 60 positive integers.

SOLUTION

(a) We want the sum of the first 12 terms.

$$S_n=\frac{n}{2}[2a_1+(n-1)d] \quad \text{Second formula for } S_n$$

$$S_{12}=\frac{12}{2}[2(-9)+11(4)], \quad \text{or} \quad 156$$ Let $a_1=-9$, $n=12$, and $d=4$ and evaluate.

(b) The first 60 positive integers form the arithmetic sequence $1, 2, 3, 4, \ldots, 60$. To find the sum, use $n=60$, $a_1=1$, and $a_{60}=60$.

$$S_n=\frac{n}{2}(a_1+a_n) \quad \text{First formula for } S_n$$

$$S_{60}=\frac{60}{2}(1+60), \quad \text{or} \quad 1830 \quad \text{Substitute and evaluate.}$$

✔ *Now Try Exercises 33 and 43.*

Classroom Example 8
The sum of the first 15 terms of an arithmetic sequence is 345. If $a_{15} = 65$, find a_1 and d.

Answer: $a_1 = -19$; $d = 6$

EXAMPLE 8 Using the Sum Formulas

The sum of the first 17 terms of an arithmetic sequence is 187. If $a_{17} = -13$, find a_1 and d.

SOLUTION

$$S_{17} = \frac{17}{2}(a_1 + a_{17})$$ Use the first formula for S_n, with $n = 17$.

$$187 = \frac{17}{2}(a_1 - 13)$$ Let $S_{17} = 187$ and $a_{17} = -13$.

$$22 = a_1 - 13$$ Multiply by $\frac{2}{17}$.

$$a_1 = 35$$ Add 13 and interchange sides.

Now find d.

$$a_{17} = a_1 + (17 - 1)d$$ Formula for the nth term

$$-13 = 35 + 16d$$ Let $a_{17} = -13$ and $a_1 = 35$.

$$-48 = 16d$$ Subtract 35.

$$d = -3$$ Divide by 16 and interchange sides.

✔ *Now Try Exercise 49.*

Any sum of the form

$$\sum_{i=1}^{n}(di + c), \quad \text{where } d \text{ and } c \text{ are real numbers,}$$

represents the sum of the terms of an arithmetic sequence having first term $a_1 = d(1) + c = d + c$ and common difference d. These sums can be evaluated using the formulas in this section.

Classroom Example 9
Evaluate each sum.

(a) $\sum_{i=5}^{12}(5i - 10)$

(b) $\sum_{k=5}^{10}(8 - 2k)$

Answers:
(a) 260 **(b)** -42

EXAMPLE 9 Using Summation Notation

Evaluate each sum.

(a) $\sum_{i=1}^{10}(4i + 8)$ **(b)** $\sum_{k=3}^{9}(4 - 3k)$

SOLUTION

(a) This sum contains the first 10 terms of the arithmetic sequence having

$$a_1 = 4 \cdot 1 + 8 = 12, \quad \text{First term}$$

and

$$a_{10} = 4 \cdot 10 + 8 = 48. \quad \text{Last term}$$

Thus,
$$\sum_{i=1}^{10}(4i + 8) = S_{10} = \frac{10}{2}(12 + 48) = 5(60) = 300.$$

(b) The first few terms are

$$[4 - 3(3)] + [4 - 3(4)] + [4 - 3(5)] + \cdots = -5 + (-8) + (-11) + \cdots.$$

Thus, $a_1 = -5$ and $d = -3$. If the sequence started with $k = 1$, there would be nine terms. Since it starts at 3, two of those terms are missing, so there are seven terms and $n = 7$. Use the second formula for S_n.

$$\sum_{k=3}^{9}(4 - 3k) = \frac{7}{2}[2(-5) + 6(-3)] = -98$$

✔ *Now Try Exercises 55 and 57.*

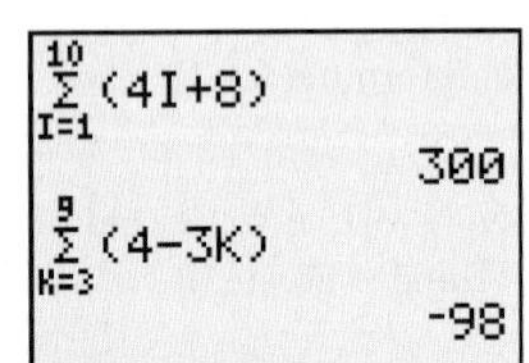

As shown in the previous section, the TI-83/84 Plus will give the sum of a sequence without having to first store the sequence. The screen here illustrates this method for the sequences in **Example 9.**

11.2 Exercises

1. 3 2. 6
3. -5 4. -4
5. $x + 2y$ 6. $-5t^2 + q$

7. 8, 14, 20, 26, 32
8. -2, 10, 22, 34, 46
9. 5, 3, 1, -1, -3
10. 4, 7, 10, 13, 16
11. $10 + \sqrt{7}$, 10, $10 - \sqrt{7}$, $10 - 2\sqrt{7}$, $10 - 3\sqrt{7}$
12. $3 - \sqrt{2}$, 3, $3 + \sqrt{2}$, $3 + 2\sqrt{2}$, $3 + 3\sqrt{2}$

13. $a_8 = 19$; $a_n = 2n + 3$
14. $a_8 = -31$; $a_n = -4n + 1$
15. $a_8 = \frac{85}{3}$; $a_n = \frac{10}{3}n + \frac{5}{3}$
16. $a_8 = 31$; $a_n = 5n - 9$
17. $a_8 = -3$; $a_n = \frac{9}{2}n - 39$
18. $a_8 = 29$; $a_n = -3n + 53$
19. $a_8 = x + 21$; $a_n = x + 3n - 3$
20. $a_8 = y - 17$; $a_n = y - 3n + 7$
21. $a_8 = s + 14p$; $a_n = s + 2pn - 2p$

22. $a_2 = \frac{a_1 + a_3}{2}$

23. 3 24. 27
25. 5 26. 10

In Exercises 27–32, D is the domain and R is the range.
27. $a_n = n - 3$; D: $\{1, 2, 3, 4, 5, 6\}$; R: $\{-2, -1, 0, 1, 2, 3\}$
28. $a_n = -n + 2$; D: $\{1, 2, 3, 4, 5\}$; R: $\{-3, -2, -1, 0, 1\}$
29. $a_n = -\frac{1}{2}n + 3$; D: $\{1, 2, 3, 4, 5, 6\}$; R: $\{0, 0.5, 1, 1.5, 2, 2.5\}$
30. $a_n = 5n - 10$; D: $\{1, 2, 3, 4, 5\}$; R: $\{-5, 0, 5, 10, 15\}$
31. $a_n = -20n + 30$; D: $\{1, 2, 3, 4, 5\}$; R: $\{-70, -50, -30, -10, 10\}$
32. $a_n = 2n - 4$; D: $\{1, 2, 3, 4, 5\}$; R: $\{-2, 0, 2, 4, 6\}$

33. 215 34. 90
35. 230 36. -10

Find the common difference d for each arithmetic sequence. ***See Example 1.***

1. 2, 5, 8, 11, ...
2. 4, 10, 16, 22, ...
3. 3, -2, -7, -12, ...
4. -8, -12, -16, -20, ...
5. $x + 3y$, $2x + 5y$, $3x + 7y$, ...
6. $t^2 + q$, $-4t^2 + 2q$, $-9t^2 + 3q$, ...

Write the first five terms of each arithmetic sequence. ***See Example 2.***

7. The first term is 8, and the common difference is 6.
8. The first term is -2, and the common difference is 12.
9. $a_1 = 5$, $d = -2$
10. $a_1 = 4$, $d = 3$
11. $a_1 = 10 + \sqrt{7}$, $a_2 = 10$
12. $a_1 = 3 - \sqrt{2}$, $a_2 = 3$

Find a_8 and a_n for each arithmetic sequence. ***See Examples 3 and 4.***

13. 5, 7, 9, ...
14. -3, -7, -11, ...
15. $a_1 = 5$, $a_4 = 15$
16. $a_1 = -4$, $a_5 = 16$
17. $a_{10} = 6$, $a_{11} = 10.5$
18. $a_{15} = 8$, $a_{16} = 5$
19. $a_1 = x$, $a_2 = x + 3$
20. $a_2 = y + 1$, $d = -3$
21. $a_4 = s + 6p$, $d = 2p$
22. *Concept Check* If a_1, a_2, a_3 represents an arithmetic sequence, express a_2 in terms of a_1 and a_3.

Find a_1 for each arithmetic sequence. ***See Examples 5 and 8.***

23. $a_5 = 27$, $a_{15} = 87$
24. $a_{12} = 60$, $a_{20} = 84$
25. $S_{16} = -160$, $a_{16} = -25$
26. $S_{28} = 2926$, $a_{28} = 199$

Find a formula for the nth term of the finite arithmetic sequence shown in each graph. Then state the domain and range of the sequence. ***See Example 6.***

27.
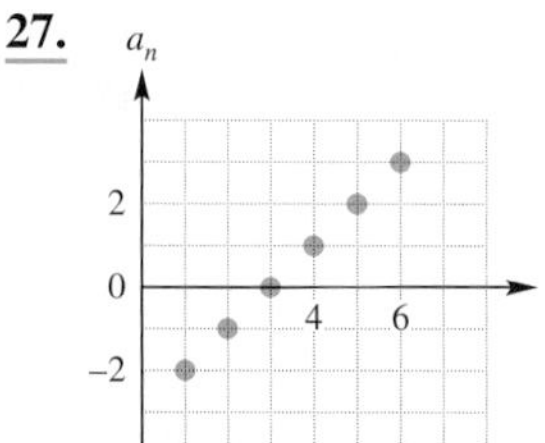

28.
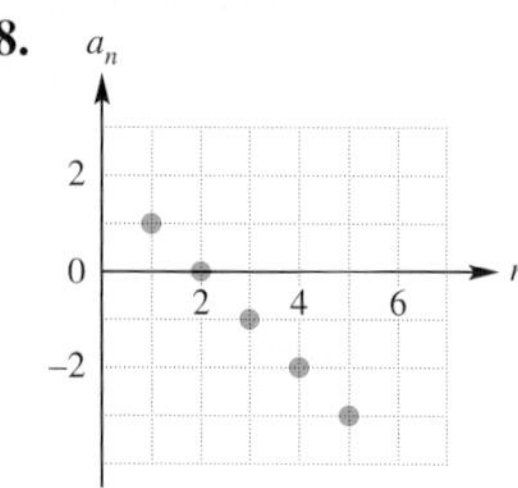

29.
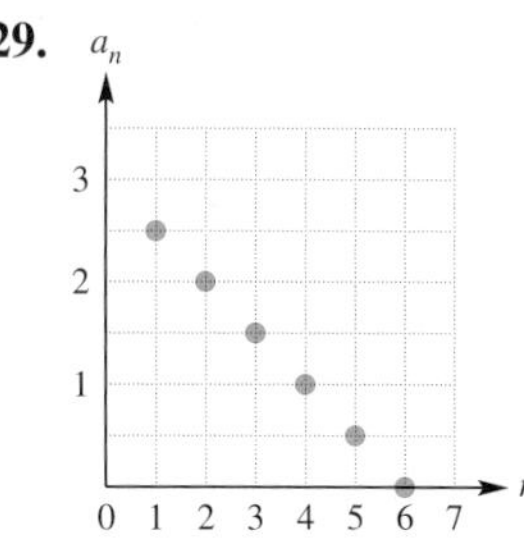

30.
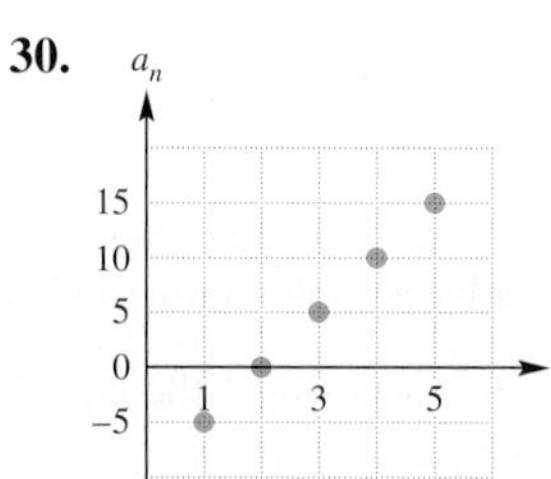

31.
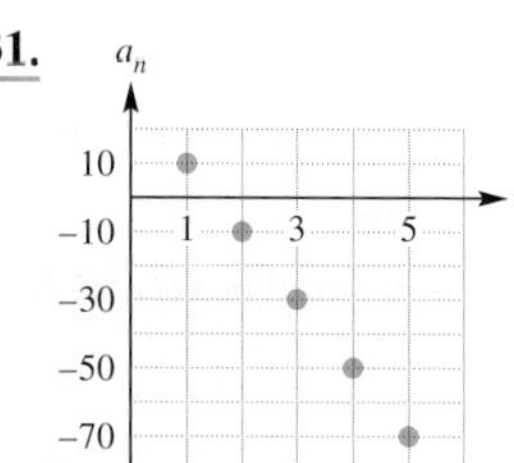

32.
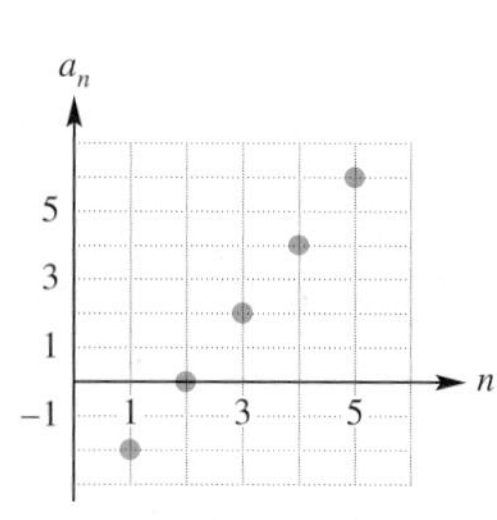

Evaluate S_{10}, the sum of the first ten terms, of each arithmetic sequence. ***See Example 7(a).***

33. 8, 11, 14, ...
34. -9, -5, -1, ...
35. 5, 9, 13, ...
36. 8, 6, 4, ...
37. $a_2 = 9$, $a_4 = 13$
38. $a_3 = 5$, $a_4 = 8$
39. $a_1 = 10$, $a_{10} = 5.5$
40. $a_1 = -8$, $a_{10} = -1.25$
41. $a_1 = \pi$, $a_{10} = 10\pi$

37. 160 38. 125
39. 77.5 40. −46.25
41. 55π

42. yes

43. 3240 44. 7260
45. 2500 46. 8100
47. 3660 48. 4970

49. $a_1 = 7,\ d = 5$
50. $a_1 = 0,\ d = 12$
51. $a_1 = 1,\ d = -\frac{20}{11}$
52. $a_1 = -10,\ d = 3$

53. 18 54. −25
55. 140 56. 465
57. −621 58. −765
59. 500,500 60. 2,001,000
61. −5050

63. 328.3 64. 824.9
65. 172.884 66. −60.850

67. 1281 68. 429
69. 4680 70. 465
71. 54,800 72. 170 m
73. 713 in.

42. *Concept Check* Is this statement accurate? *To find the sum of the first n positive integers, find half the product of n and n + 1.*

Find each sum as described. ***See Example 7(b).***

43. the sum of the first 80 positive integers

44. the sum of the first 120 positive integers

45. the sum of the first 50 positive odd integers

46. the sum of the first 90 positive odd integers

47. the sum of the first 60 positive even integers

48. the sum of the first 70 positive even integers

Find a_1 and d for each arithmetic series. ***See Example 8.***

49. $S_{20} = 1090,\ a_{20} = 102$

50. $S_{31} = 5580,\ a_{31} = 360$

51. $S_{12} = -108,\ a_{12} = -19$

52. $S_{25} = 650,\ a_{25} = 62$

Evaluate each sum. ***See Example 9.***

53. $\sum_{i=1}^{3} (i + 4)$

54. $\sum_{i=1}^{5} (i - 8)$

55. $\sum_{j=1}^{10} (2j + 3)$

56. $\sum_{j=1}^{15} (5j - 9)$

57. $\sum_{i=4}^{12} (-5 - 8i)$

58. $\sum_{k=5}^{19} (-3 - 4k)$

59. $\sum_{i=1}^{1000} i$

60. $\sum_{k=1}^{2000} k$

61. $\sum_{k=1}^{100} -k$

62. Explain the basic difference between a sequence and a series.

Use the summation feature of a graphing calculator to evaluate the sum of the first ten terms of each arithmetic series with a_n defined as shown. In Exercises 65 and 66, round to the nearest thousandth.

63. $a_n = 4.2n + 9.73$

64. $a_n = 8.42n + 36.18$

65. $a_n = \sqrt{8}n + \sqrt{3}$

66. $a_n = -\sqrt[3]{4}n + \sqrt{7}$

Solve each problem.

67. *Integer Sum* Find the sum of all the integers from 51 to 71.

68. *Integer Sum* Find the sum of all the integers from −8 to 30.

69. *Clock Chimes* If a clock strikes the proper number of chimes each hour on the hour, how many times will it chime in a month of 30 days?

70. *Telephone Pole Stack* A stack of telephone poles has 30 in the bottom row, 29 in the next, and so on, with one pole in the top row. How many poles are in the stack?

71. *Population Growth* Five years ago, the population of a city was 49,000. Each year the zoning commission permits an increase of 580 in the population. What will the maximum population be 5 yr from now?

72. *Slide Supports* A super slide of uniform slope is to be built on a level piece of land. There are to be 20 equally spaced vertical supports, with the longest support 15 m long and the shortest 2 m long. Find the total length of all the supports.

73. *Rungs of a Ladder* How much material will be needed for the rungs of a ladder of 31 rungs, if the rungs taper uniformly from 18 in. to 28 in.?

74. (a) 5.8 cm (b) 127.2 cm
75. yes

74. *(Modeling) Children's Growth Pattern* The normal growth pattern for children aged 3–11 follows that of an arithmetic sequence. An increase in height of about 6 cm per year is expected. Thus, 6 would be the common difference of the sequence. For example, a child who measures 96 cm at age 3 would have his expected height in subsequent years represented by the sequence 102, 108, 114, 120, 126, 132, 138, 144. Each term differs from the adjacent terms by the common difference, 6.

(a) If a child measures 98.2 cm at age 3 and 109.8 cm at age 5, what would be the common difference of the arithmetic sequence describing his yearly height?

(b) What would we expect his height to be at age 8?

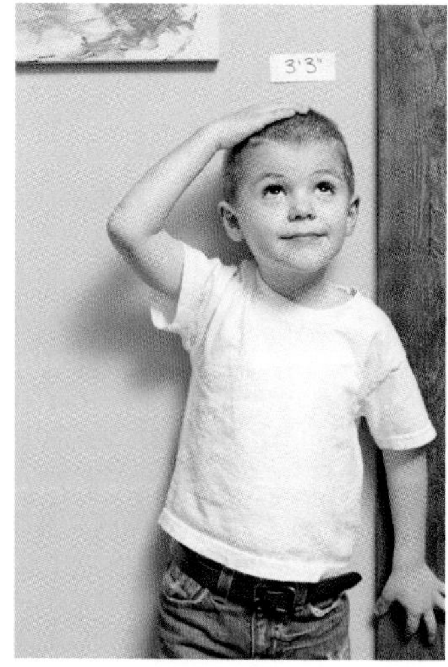

75. *Concept Check* Suppose that $a_1, a_2, a_3, a_4, a_5, \ldots$ is an arithmetic sequence. Is $a_1, a_3, a_5, \ldots$ an arithmetic sequence?

76. Explain why the sequence log 2, log 4, log 8, log 16, ... is arithmetic.

11.3 Geometric Sequences and Series

- Geometric Sequences
- Geometric Series
- Infinite Geometric Series
- Annuities

Geometric Sequences Suppose you agreed to work for \$0.01 the first day, \$0.02 the second day, \$0.04 the third day, \$0.08 the fourth day, and so on, with your wages doubling each day. How much will you earn on day 20, after working 5 days a week for a month? How much will you have earned altogether in 20 days? These questions will be answered in this section.

A **geometric sequence** (or **geometric progression**) is a sequence in which each term after the first is obtained by multiplying the preceding term by a fixed nonzero real number, called the **common ratio.** The sequence discussed above,

$$1, 2, 4, 8, 16, \ldots, \quad \text{In cents}$$

is a geometric sequence in which the first term is 1 and the common ratio is 2.

Notice that if we divide any term after the first term by the preceding term, we obtain the common ratio $r = 2$.

$$\frac{a_2}{a_1} = \frac{2}{1} = 2; \quad \frac{a_3}{a_2} = \frac{4}{2} = 2; \quad \frac{a_4}{a_3} = \frac{8}{4} = 2; \quad \frac{a_5}{a_4} = \frac{16}{8} = 2$$

Teaching Tip Explain that for $r > 1$, the sequence values increase, while for $0 < r < 1$, the sequence values decrease. Point out that the equation

$$r = \frac{a_{n+1}}{a_n}$$

can be written as

$$a_{n+1} = r \cdot a_n.$$

The second form of the equation can be used to generate successive terms of a geometric sequence recursively.

If the common ratio of a geometric sequence is r, then by the definition of a geometric sequence,

$$\boldsymbol{r = \frac{a_{n+1}}{a_n}}, \quad \text{Common ratio } r$$

for every positive integer n. ***Therefore, we find the common ratio by choosing any term after the first and dividing it by the preceding term.***

In the geometric sequence 2, 8, 32, 128, ..., $r = 4$. Notice that

$$8 = 2 \cdot 4$$
$$32 = 8 \cdot 4 = (2 \cdot 4) \cdot 4 = 2 \cdot 4^2$$
$$128 = 32 \cdot 4 = (2 \cdot 4^2) \cdot 4 = 2 \cdot 4^3.$$

To generalize this, assume that a geometric sequence has first term a_1 and common ratio r. The second term is $a_2 = a_1 r$, the third is $a_3 = a_2 r = (a_1 r)r = a_1 r^2$, and so on. Following this pattern, the nth term is $a_n = a_1 r^{n-1}$. Again, this result can be proved by mathematical induction. (See **Section 11.5.**)

nth Term of a Geometric Sequence

In a geometric sequence with first term a_1 and common ratio r, the nth term, a_n, is given by the following.

$$a_n = a_1 r^{n-1}$$

Classroom Example 1
Suppose that you receive a gift on the first day of each month for a year, starting with \$50 on January 1, with the amount doubling each month. How much will you receive on December 1?

Answer: \$102,400

EXAMPLE 1 Finding the nth Term of a Geometric Sequence

Use the formula for the nth term of a geometric sequence to answer the first question posed at the beginning of this section. How much will be earned on day 20 if daily wages follow the sequence 1, 2, 4, 8, 16, ... cents?

SOLUTION

$a_n = a_1 r^{n-1}$ Formula for a_n

$a_{20} = 1(2)^{20-1}$ Let $n = 20$, $a_1 = 1$, and $r = 2$.

$a_{20} = 524{,}288$ cents, or \$5242.88 Evaluate.

Now Try Exercise 1(a).

Classroom Example 2
Find a_5 and a_n for the geometric sequence

6400, 1600, 400, 100, ...

Answer:

$a_5 = 25$; $a_n = 6400\left(\frac{1}{4}\right)^{n-1}$

EXAMPLE 2 Finding Terms of a Geometric Sequence

Find a_5 and a_n for the geometric sequence 4, 12, 36, 108, ...

SOLUTION The first term, a_1, is 4. Find r by choosing any term after the first and dividing it by the preceding term. For example, $r = \frac{36}{12} = 3$.

$a_5 = a_4 \cdot r$ Definition of geometric sequence

$a_5 = 108 \cdot 3$ Let $a_4 = 108$ and $r = 3$.

$a_5 = 324$ Multiply.

We could also find the fifth term by using the formula $a_n = a_1 r^{n-1}$ and replacing n with 5, r with 3, and a_1 with 4.

$$a_5 = 4 \cdot (3)^{5-1} = 4 \cdot 3^4 = 324$$

Thus, $a_n = 4 \cdot 3^{n-1}$. Formula for a_n

Now Try Exercise 11.

Classroom Example 3
Find r and a_1 for the geometric sequence with second term -18 and fifth term 486.

Answer: $r = -3$; $a_1 = 6$

EXAMPLE 3 Finding Terms of a Geometric Sequence

Find r and a_1 for the geometric sequence with third term 20 and sixth term 160.

SOLUTION We obtain a_6 by multiplying a_3 by the common ratio three times.

$a_6 = a_3 r^3$ Definition of geometric sequence

$160 = 20r^3$ Let $a_6 = 160$ and $a_3 = 20$.

$8 = r^3$ Divide by 20.

$r = 2$ Take cube roots and interchange sides. (Section 1.6)

Now use this value of r and the fact that $a_3 = 20$ to find the first term, a_1.

$$a_n = a_1 r^{n-1} \quad \text{Formula for } a_n$$

$$20 = a_1(2)^{3-1} \quad \text{Let } a_n = 20,\ r = 2, \text{ and } n = 3.$$

$$20 = a_1(4) \quad \text{Apply the exponent.}$$

$$a_1 = 5 \quad \text{Divide by 4 and interchange sides.}$$

✔ *Now Try Exercise 17.*

Classroom Example 4
A population of fruit flies is growing in such a way that each generation is 1.75 times as large as the last generation. Suppose there are 250 insects in the first generation. How many would there be in the sixth generation?

Answer: about 4103 insects

EXAMPLE 4 Modeling a Population of Fruit Flies

A population of fruit flies is growing in such a way that each generation is 1.5 times as large as the last generation. Suppose there are 100 insects in the first generation. How many would there be in the fourth generation?

SOLUTION Write the list of populations as a geometric sequence with a_1 as the first-generation population, a_2 the second-generation population, and so on. Then the fourth-generation population is a_4. Use the formula for a_n, with $n = 4$, $r = 1.5$, and $a_1 = 100$.

$$a_4 = a_1 r^3 \quad \text{Definition of geometric sequence}$$

$$a_4 = 100(1.5)^3 \quad \text{Substitute.}$$

$$a_4 = 100(3.375) \quad \text{Apply the exponent.}$$

$$a_4 = 337.5 \quad \text{Multiply.}$$

In the fourth generation, the population will number about 338 insects.

✔ *Now Try Exercise 61.*

Geometric Series A **geometric series** is the sum of the terms of a geometric sequence. In applications, it may be necessary to find the sum of the terms of such a sequence. For example, a scientist might want to know the total number of insects in four generations of the population discussed in **Example 4.** This population would equal $a_1 + a_2 + a_3 + a_4$, or

$$100 + 100(1.5) + 100(1.5)^2 + 100(1.5)^3 = 812.5 \approx 813 \text{ insects.}$$

To find a formula for the sum of the first n terms of a geometric sequence, S_n, first write the sum as

$$S_n = a_1 + a_2 + a_3 + \cdots + a_n$$

or

$$S_n = a_1 + a_1 r + a_1 r^2 + \cdots + a_1 r^{n-1}. \quad (1)$$

If $r = 1$, then $S_n = na_1$, which is a correct formula for this case. If $r \neq 1$, then multiply both sides of equation (1) by r to obtain

$$rS_n = a_1 r + a_1 r^2 + a_1 r^3 + \cdots + a_1 r^n. \quad (2)$$

Now subtract equation (2) from equation (1), and solve for S_n.

$$S_n = a_1 + a_1 r + a_1 r^2 + \cdots + a_1 r^{n-1} \quad (1)$$

$$rS_n = \quad a_1 r + a_1 r^2 + \cdots + a_1 r^{n-1} + a_1 r^n \quad (2)$$

$$S_n - rS_n = a_1 \qquad - a_1 r^n \quad \text{Subtract.}$$

$$S_n(1 - r) = a_1(1 - r^n) \quad \text{Factor. (Section R.4)}$$

$$S_n = \frac{a_1(1 - r^n)}{1 - r}, \quad \text{where } r \neq 1 \quad \text{Divide by } 1 - r. \text{ (Section 1.1)}$$

Sum of the First n Terms of a Geometric Sequence

If a geometric sequence has first term a_1 and common ratio r, then the sum S_n of the first n terms is given by the following.

$$S_n = \frac{a_1(1 - r^n)}{1 - r}, \quad \text{where } r \neq 1$$

We can use a geometric series to find the total fruit fly population in **Example 4** over the four-generation period.

$$S_n = \frac{a_1(1 - r^n)}{1 - r} \qquad \text{Formula for } S_n$$

$$S_4 = \frac{100(1 - 1.5^4)}{1 - 1.5} \qquad \text{Let } n = 4,\ a_1 = 100, \text{ and } r = 1.5.$$

$$S_4 = 812.5 \qquad \text{Evaluate.}$$

$$S_4 \approx 813 \text{ insects} \qquad \text{This agrees with our previous result.}$$

Classroom Example 5
For the situation described in **Classroom Example 1,** what is the total amount of the gifts you will receive throughout the year?

Answer: \$204,750

EXAMPLE 5 Finding the Sum of the First n Terms

At the beginning of this section we asked you to imagine working at a job where you were paid in an unorthodox manner. Then we posed the following question: How much will you have earned altogether after 20 days? Answer this question.

SOLUTION We must find the total amount earned in 20 days with daily wages of 1, 2, 4, 8, ... cents. Use the formula for S_n with $n = 20$, $a_1 = 1$, and $r = 2$.

$$S_{20} = \frac{1(1 - 2^{20})}{1 - 2} = \frac{1 - 1{,}048{,}576}{-1} = 1{,}048{,}575 \text{ cents}, \quad \text{or} \quad \$10{,}485.75$$

✔ *Now Try Exercise 1(b).*

Classroom Example 6
Find $\sum_{i=1}^{8} 4 \cdot 5^i$.

Answer: 1,953,120

EXAMPLE 6 Finding the Sum of the First n Terms

Find $\sum_{i=1}^{6} 2 \cdot 3^i$.

SOLUTION This series is the sum of the first six terms of a geometric sequence having $a_1 = 2 \cdot 3^1 = 6$ and $r = 3$.

$$S_n = \frac{a_1(1 - r^n)}{1 - r} \qquad \text{Formula for } S_n$$

$$S_6 = \frac{6(1 - 3^6)}{1 - 3} \qquad \text{Substitute.}$$

$$S_6 = 2184 \qquad \text{Evaluate.}$$

✔ *Now Try Exercise 29.*

Infinite Geometric Series We extend our discussion of sums of sequences to include infinite geometric sequences such as

$$2, 1, \frac{1}{2}, \frac{1}{4}, \frac{1}{8}, \frac{1}{16}, \ldots, \quad \text{with first term 2 and common ratio } \frac{1}{2}.$$

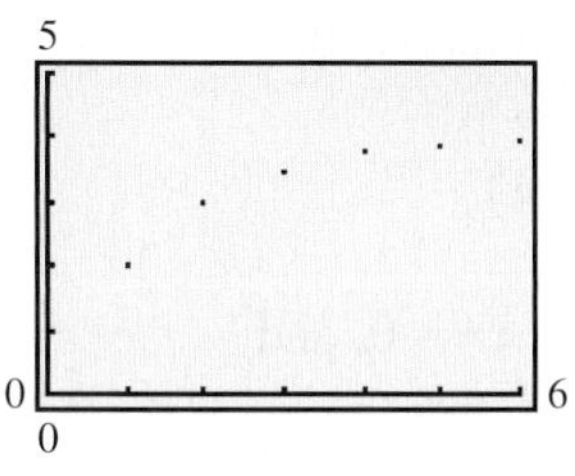

As n gets larger, S_n approaches 4.

Figure 10

Using the formula for S_n gives the following sequence of sums.

$$S_1 = 2, \quad S_2 = 3, \quad S_3 = \frac{7}{2}, \quad S_4 = \frac{15}{4}, \quad S_5 = \frac{31}{8}, \quad S_6 = \frac{63}{16}, \ldots$$

The formula for S_n can also be written as

$$S_n = 4 - \left(\frac{1}{2}\right)^{n-2}.$$

As this formula and **Figure 10** suggest, these sums seem to be getting closer and closer to the number 4. For no value of n is $S_n = 4$. However, if n is large enough, then S_n is as close to 4 as desired. As mentioned earlier, we say the sequence converges to 4. This is expressed as

$$\lim_{n \to \infty} S_n = 4.$$

LOOKING AHEAD TO CALCULUS

In the discussion of

$$\lim_{n \to \infty} S_n = 4,$$

we used the phrases "large enough" and "as close as desired." This description is made more precise in a standard calculus course.

(Read this as: **"The limit of S_n as n increases without bound is 4."**) Since $\lim_{n \to \infty} S_n = 4$, the number 4 is the *sum of the terms* of the infinite geometric sequence

$$2, 1, \frac{1}{2}, \frac{1}{4}, \ldots,$$

and

$$2 + 1 + \frac{1}{2} + \frac{1}{4} + \cdots = 4.$$

Classroom Example 7

Evaluate $4 + \frac{4}{5} + \frac{4}{25} + \cdots$.

Answer: 5

EXAMPLE 7 Evaluating an Infinite Geometric Series

Evaluate $1 + \frac{1}{3} + \frac{1}{9} + \frac{1}{27} + \cdots$.

SOLUTION Use the formula for the sum of the first n terms of a geometric sequence to obtain the following.

$$S_1 = 1, \quad S_2 = \frac{4}{3}, \quad S_3 = \frac{13}{9}, \quad S_4 = \frac{40}{27}$$

In general, $$S_n = \frac{1\left[1 - \left(\frac{1}{3}\right)^n\right]}{1 - \frac{1}{3}}.$$ Let $a_1 = 1$ and $r = \frac{1}{3}$.

The table shows the value of $\left(\frac{1}{3}\right)^n$ for larger and larger values of n.

n	1	10	100	200
$\left(\frac{1}{3}\right)^n$	$\frac{1}{3}$	1.69×10^{-5}	1.94×10^{-48}	3.76×10^{-96}

This graph of the first six values of S_n in **Example 7** suggests that its value is approaching $\frac{3}{2}$. (The y-scale here is $\frac{1}{2}$.)

As n increases without bound, $\left(\frac{1}{3}\right)^n$ approaches 0. That is,

$$\lim_{n \to \infty} \left(\frac{1}{3}\right)^n = 0,$$

making it reasonable that

$$\lim_{n \to \infty} S_n = \lim_{n \to \infty} \frac{1\left[1 - \left(\frac{1}{3}\right)^n\right]}{1 - \frac{1}{3}} = \frac{1(1 - 0)}{1 - \frac{1}{3}} = \frac{1}{\frac{2}{3}} = \frac{3}{2}.$$ Simplify the complex fraction. (Section R.5)

Hence, $$1 + \frac{1}{3} + \frac{1}{9} + \frac{1}{27} + \cdots = \frac{3}{2}.$$

Now Try Exercise 39.

LOOKING AHEAD TO CALCULUS

In calculus, functions are sometimes defined in terms of infinite series. Here are three functions we studied earlier in the text defined that way.

$$e^x = \frac{x^0}{0!} + \frac{x^1}{1!} + \frac{x^2}{2!} + \frac{x^3}{3!} + \cdots$$

$$\ln(1+x) = x - \frac{x^2}{2} + \frac{x^3}{3} - \cdots$$

for x in $(-1, 1)$

$$\frac{1}{1+x} = 1 - x + x^2 - x^3 + \cdots$$

for x in $(-1, 1)$

Teaching Tip Ask students to explain why the sum of an arithmetic series does not converge. Challenge students to determine what happens to the sum of an infinite geometric series when $r = -1$.

Show students how this formula corroborates the fact that $0.333\ldots = \frac{1}{3}$, with $a_1 = 0.3$ and $r = 0.1$.

If a geometric series has first term a_1 and common ratio r, then

$$S_n = \frac{a_1(1 - r^n)}{1 - r} \quad (\text{where } r \neq 1),$$

for every positive integer n. If $-1 < r < 1$, then $\lim_{n \to \infty} r^n = 0$, and

$$\lim_{n \to \infty} S_n = \frac{a_1(1 - 0)}{1 - r} = \frac{a_1}{1 - r}.$$

The resulting quotient, $\frac{a_1}{1 - r}$, is the **sum of the terms of an infinite geometric sequence.**

The limit $\boldsymbol{\lim_{n \to \infty} S_n}$ can be expressed as $\boldsymbol{S_\infty}$ or $\boldsymbol{\sum_{i=1}^{\infty} a_i}$.

Sum of the Terms of an Infinite Geometric Sequence

The sum S_∞ of the terms of an infinite geometric sequence with first term a_1 and common ratio r, where $-1 < r < 1$, is given by the following.

$$\boldsymbol{S_\infty = \frac{a_1}{1 - r}}$$

If $|r| > 1$, then the terms get larger and larger in absolute value, so there is no limit as $n \to \infty$.

Hence, if $|r| > 1$, the terms of the sequence will not have a sum.

Classroom Example 8

Find each sum.

(a) $\sum_{i=1}^{\infty} \left(\frac{2}{5}\right)\left(-\frac{1}{3}\right)^{i-1}$

(b) $\sum_{j=1}^{\infty} (0.9)^j$

Answers:

(a) $\frac{3}{10}$ (b) 9

EXAMPLE 8 Evaluating Infinite Geometric Series

Find each sum.

(a) $\sum_{i=1}^{\infty} \left(-\frac{3}{4}\right)\left(-\frac{1}{2}\right)^{i-1}$ **(b)** $\sum_{i=1}^{\infty} \left(\frac{3}{5}\right)^i$

SOLUTION

(a) Here, $a_1 = -\frac{3}{4}$ and $r = -\frac{1}{2}$. Since $-1 < r < 1$, the preceding formula applies.

$$S_\infty = \frac{a_1}{1 - r} = \frac{-\frac{3}{4}}{1 - \left(-\frac{1}{2}\right)} = \frac{-\frac{3}{4}}{\frac{3}{2}} = -\frac{3}{4} \div \frac{3}{2} = -\frac{3}{4} \cdot \frac{2}{3} = -\frac{1}{2}$$

(b) $\sum_{i=1}^{\infty} \left(\frac{3}{5}\right)^i = \frac{\frac{3}{5}}{1 - \frac{3}{5}} = \frac{\frac{3}{5}}{\frac{2}{5}} = \frac{3}{5} \div \frac{2}{5} = \frac{3}{5} \cdot \frac{5}{2} = \frac{3}{2}$ $\quad a_1 = \frac{3}{5}, r = \frac{3}{5}$

Now Try Exercises 49 and 51.

Annuities A sequence of equal payments made after equal periods of time, such as car payments or house payments, is an **annuity.** If the payments are accumulated in an account (with no withdrawals), the sum of the payments and interest on the payments is the **future value** of the annuity.

Classroom Example 9
To save money for a new car, Clayton James deposited \$3500 at the *end* of each year for 6 yr in an account paying 5% interest, compounded annually. Find the future value of this annuity.

Answer: \$23,806.69

EXAMPLE 9 Finding the Future Value of an Annuity

To save money for a trip, Jacqui Porter deposited \$1000 at the *end* of each year for 4 yr in an account paying 3% interest, compounded annually. Find the future value of this annuity.

SOLUTION We use the formula for interest compounded annually.

$$A = P(1 + r)^t \quad \text{(Section 4.2)}$$

The first payment earns interest for 3 yr, the second payment for 2 yr, and the third payment for 1 yr. The last payment earns no interest.

$$1000(1.03)^3 + 1000(1.03)^2 + 1000(1.03) + 1000 \quad \text{Total amount}$$

This is the sum of the terms of a geometric sequence with first term (starting at the end of the sum as written above) $a_1 = 1000$ and common ratio $r = 1.03$.

$$S_n = \frac{a_1(1 - r^n)}{1 - r} \quad \text{Formula for } S_n$$

$$S_4 = \frac{1000[1 - (1.03)^4]}{1 - 1.03} \quad \text{Substitute.}$$

$$S_4 \approx 4183.63. \quad \text{Evaluate.}$$

The future value of the annuity is \$4183.63.

Now Try Exercise 71.

The general formula for the future value of an annuity can be stated as follows. (See **Exercise 79.**)

Future Value of an Annuity

The formula for the future value of an annuity is given by the following.

$$S = R\left[\frac{(1 + i)^n - 1}{i}\right]$$

Here S is future value, R is payment at the end of each period, i is interest rate per period, and n is number of periods.

11.3 Exercises

1. (a) \$5.12 (b) \$10.23
2. (a) \$20.48 (b) \$40.95
3. (a) \$163.84 (b) \$327.67
4. (a) \$1310.72 (b) \$2621.43

In Exercises 5–16, there may be other ways to express a_n.

5. $a_5 = 80;\ a_n = 5(-2)^{n-1}$
6. $a_5 = 5000;\ a_n = 8(-5)^{n-1}$
7. $a_5 = -108;\ a_n = -\frac{4}{3}(3)^{n-1}$
8. $a_5 = -32;\ a_n = -\frac{1}{8}(4)^{n-1}$
9. $a_5 = -729;\ a_n = -9(-3)^{n-1}$

Refer to the first sentence of this section, which describes a method of payment for a job. Determine **(a)** *how much you will earn on the day indicated and* **(b)** *how much you will have earned altogether after your wages are paid on the day indicated.* ***See Examples 1 and 5.***

1. day 10 **2.** day 12 **3.** day 15 **4.** day 18

Find a_5 and a_n for each geometric sequence. ***See Example 2.***

5. $a_1 = 5, r = -2$ **6.** $a_1 = 8, r = -5$ **7.** $a_2 = -4, r = 3$

8. $a_3 = -2, r = 4$ **9.** $a_4 = 243, r = -3$ **10.** $a_4 = 18, r = 2$

11. $-4, -12, -36, -108, \ldots$ **12.** $-2, 6, -18, 54, \ldots$

10. $a_5 = 36$; $a_n = \frac{9}{4}(2)^{n-1}$
11. $a_5 = -324$; $a_n = -4(3)^{n-1}$
12. $a_5 = -162$; $a_n = -2(-3)^{n-1}$
13. $a_5 = \frac{125}{4}$; $a_n = \frac{4}{5}\left(\frac{5}{2}\right)^{n-1}$
14. $a_5 = \frac{128}{81}$; $a_n = \frac{1}{2}\left(\frac{4}{3}\right)^{n-1}$
15. $a_5 = \frac{5}{8}$; $a_n = 10\left(-\frac{1}{2}\right)^{n-1}$
16. $a_5 = \frac{243}{256}$; $a_n = 3\left(-\frac{3}{4}\right)^{n-1}$

17. 2; -3 18. $\frac{1}{5}$; 125
19. ± 0.1; 5000 20. $\frac{1}{2}$; -2

21. 682 22. 1364
23. $\frac{99}{8}$ 24. $\frac{244}{27}$
25. 860.95 26. -14.82

27. -183 28. 10
29. $\frac{189}{4}$ 30. 422
31. 2032 32. 29,511

33. The sum exists if $|r| < 1$.
34. The sum is 1, which is contrary to the intuition of some people.

35. 2; does not converge
36. $\frac{1}{5}$
37. $\frac{1}{2}$
38. -5; does not converge
40. $\frac{3}{2}$

41. 27 42. $\frac{1000}{9}$
43. $\frac{3}{20}$ 44. $\frac{8}{3}$
45. 4 46. 4
47. $\frac{1}{2}$ 48. $\frac{1}{9}$
49. $-\frac{8}{15}$ 50. $-\frac{1}{7}$
51. $\frac{3}{4}$ 52. $\frac{5}{4}$

13. $\frac{4}{5}, 2, 5, \frac{25}{2}, \ldots$ **14.** $\frac{1}{2}, \frac{2}{3}, \frac{8}{9}, \frac{32}{27}, \ldots$

15. $10, -5, \frac{5}{2}, -\frac{5}{4}, \ldots$ **16.** $3, -\frac{9}{4}, \frac{27}{16}, -\frac{81}{64}, \ldots$

Find r and a_1 for each geometric sequence. ***See Example 3.***

17. $a_2 = -6$, $a_7 = -192$ **18.** $a_3 = 5$, $a_8 = \frac{1}{625}$

19. $a_3 = 50$, $a_7 = 0.005$ **20.** $a_4 = -\frac{1}{4}$, $a_9 = -\frac{1}{128}$

Use the formula for S_n to find the sum of the first five terms of each geometric sequence. In Exercises 25 and 26, round to the nearest hundredth. ***See Example 5.***

21. 2, 8, 32, 128, ... **22.** 4, 16, 64, 256, ...

23. $18, -9, \frac{9}{2}, -\frac{9}{4}, \ldots$ **24.** $12, -4, \frac{4}{3}, -\frac{4}{9}, \ldots$

25. $a_1 = 8.423$, $r = 2.859$ **26.** $a_1 = -3.772$, $r = -1.553$

Find each sum. ***See Example 6.***

27. $\sum_{i=1}^{5} (-3)^i$ **28.** $\sum_{i=1}^{4} (-2)^i$ **29.** $\sum_{j=1}^{6} 48\left(\frac{1}{2}\right)^j$

30. $\sum_{j=1}^{5} 243\left(\frac{2}{3}\right)^j$ **31.** $\sum_{k=4}^{10} 2^k$ **32.** $\sum_{k=3}^{9} 3^k$

33. *Concept Check* Under what conditions does the sum of an infinite geometric series exist?

34. The number 0.999... can be written as the sum of the terms of an infinite geometric sequence: $0.9 + 0.09 + 0.009 + \cdots$. Here we have $a_1 = 0.9$ and $r = 0.1$. Use the formula for S_∞ to find this sum. Does your intuition indicate that your answer is correct?

Find r for each infinite geometric sequence. Identify any whose sum does not converge.

35. 12, 24, 48, 96, ... **36.** 625, 125, 25, 5, ...

37. $-48, -24, -12, -6, \ldots$ **38.** $2, -10, 50, -250, \ldots$

39. Use $\lim_{n\to\infty} S_n$ to show that $2 + 1 + \frac{1}{2} + \frac{1}{4} + \cdots$ converges to 4. **See Example 7.**

40. In **Example 7,** we determined that $1 + \frac{1}{3} + \frac{1}{9} + \frac{1}{27} + \cdots$ converges to $\frac{3}{2}$ using an argument involving limits. Use the formula for the sum of the terms of an infinite geometric sequence to obtain the same result.

Find each sum that converges. ***See Example 8.***

41. $18 + 6 + 2 + \frac{2}{3} + \cdots$ **42.** $100 + 10 + 1 + \cdots$ **43.** $\frac{1}{4} - \frac{1}{6} + \frac{1}{9} - \frac{2}{27} + \cdots$

44. $\frac{4}{3} + \frac{2}{3} + \frac{1}{3} + \cdots$ **45.** $\sum_{i=1}^{\infty} 3\left(\frac{1}{4}\right)^{i-1}$ **46.** $\sum_{i=1}^{\infty} 5\left(-\frac{1}{4}\right)^{i-1}$

47. $\sum_{k=1}^{\infty} 3^{-k}$ **48.** $\sum_{k=1}^{\infty} 10^{-k}$ **49.** $\sum_{i=1}^{\infty} \left(-\frac{2}{3}\right)\left(-\frac{1}{4}\right)^{i-1}$

50. $\sum_{i=1}^{\infty} \left(-\frac{1}{5}\right)\left(-\frac{2}{5}\right)^{i-1}$ **51.** $\sum_{i=1}^{\infty} \left(\frac{3}{7}\right)^i$ **52.** $\sum_{i=1}^{\infty} \left(\frac{5}{9}\right)^i$

53. 97.739
54. −3012.622
55. 0.212
56. 0.016

57. (a) $a_1 = 1169$; $r = 0.916$
(b) $a_{10} = 531$; $a_{20} = 221$; A person who is 10 yr from retirement should have savings of 531% of his or her annual salary. A person 20 yr from retirement should have savings of 221% of his or her annual salary.
58. (a) $a_1 = 1195$; $r = 0.935$
(b) $a_{10} = 653$; $a_{20} = 333$; A person 10 yr from retirement should have savings of 653% of his or her annual salary. A person 20 yr from retirement needs savings of 333% of his or her annual salary.
(c) A conservative investment strategy will accept less risk and, therefore, earn a smaller interest rate than a moderate investment strategy. Thus, more needs to be invested to accumulate the same amount.
59. (a) $a_n = a_1 \cdot 2^{n-1}$
(b) 15 (rounded from 14.28)
(c) 560 min, or 9 hr, 20 min
60. 0.000016%
61. about 488
62. $26,214.40
63. $\frac{10{,}000}{9}$ units

Use the summation feature of a graphing calculator to evaluate each sum. Round to the nearest thousandth.

53. $\sum_{i=1}^{10} (1.4)^i$ **54.** $\sum_{j=1}^{6} -(3.6)^j$ **55.** $\sum_{j=3}^{8} 2(0.4)^j$ **56.** $\sum_{i=4}^{9} 3(0.25)^i$

Solve each problem. ***See Examples 1–8.***

57. *(Modeling) Investment for Retirement* According to T. Rowe Price Associates, a person with a moderate investment strategy and n years to retirement should have accumulated savings of a_n percent of his or her annual salary. The geometric sequence

$$a_n = 1276(0.916)^n$$

gives the appropriate percent for each year n.

(a) Find a_1 and r. Round a_1 to the nearest whole number.
(b) Find and interpret the terms a_{10} and a_{20}. Round to the nearest whole number.

58. *(Modeling) Investment for Retirement* Refer to **Exercise 57.** For someone who has a conservative investment strategy with n years to retirement, the geometric sequence is $a_n = 1278(0.935)^n$. (*Source:* T. Rowe Price Associates.)

(a) Repeat part (a) of **Exercise 57.** **(b)** Repeat part (b) of **Exercise 57.**
(c) Why are the answers in parts (a) and (b) larger than those in **Exercise 57?**

59. *(Modeling) Bacterial Growth* The strain of bacteria described in **Exercise 89** in **Section 11.1** will double in size and then divide every 40 minutes. Let a_1 be the initial number of bacteria cells, a_2 the number after 40 minutes, and a_n the number after $40(n-1)$ minutes. (*Source:* Hoppensteadt, F. and C. Peskin, *Mathematics in Medicine and the Life Sciences,* Springer-Verlag.)

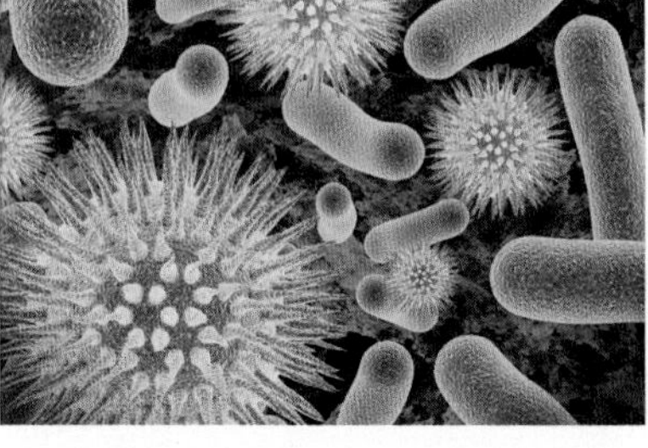

(a) Write a formula for the nth term a_n of the geometric sequence

$$a_1, a_2, a_3, \ldots, a_n, \ldots.$$

(b) Determine the first value for n where $a_n > 1{,}000{,}000$ if $a_1 = 100$.
(c) How long does it take for the number of bacteria to exceed one million?

60. *Photo Processing* For students taking a course in black-and-white photography, the final step in processing a print is to immerse it in a chemical fixer. The print is then washed in running water. Under certain conditions, 98% of the fixer in a print will be removed with 15 min of washing. How much of the original fixer would be left after 1 hr of washing?

61. *(Modeling) Fruit Flies Population* A population of fruit flies is growing in such a way that each generation is 1.25 times as large as the last generation. Suppose there were 200 insects in the first generation. How many would there be in the fifth generation?

62. *Depreciation* Each year a machine loses 20% of the value it had at the beginning of the year. Find the value of the machine at the end of 6 yr if it cost $100,000 new.

63. *Sugar Processing* A sugar factory receives an order for 1000 units of sugar. The production manager thus orders production of 1000 units of sugar. He forgets, however, that the production of sugar requires some sugar (to prime the machines, for example), and so he ends up with only 900 units of sugar. He then orders an additional 100 units and receives only 90 units. A further order for 10 units produces 9 units. Finally, the manager decides to try mathematics. He views the production process as an infinite geometric progression with $a_1 = 1000$ and $r = 0.1$. Using this, find the number of units of sugar that he should have ordered originally.

64. 70 m
65. 62; 2046
66. (b) 3 mg
67. $\frac{1}{64}$ m
68. 12 m; $\frac{4\sqrt{3}}{3}$ m^2
69. Option 2 pays better.

64. *Height of a Dropped Ball* Alicia drops a ball from a height of 10 m and notices that on each bounce the ball returns to about $\frac{3}{4}$ of its previous height. About how far will the ball travel before it comes to rest? (*Hint:* Consider the sum of two sequences.)

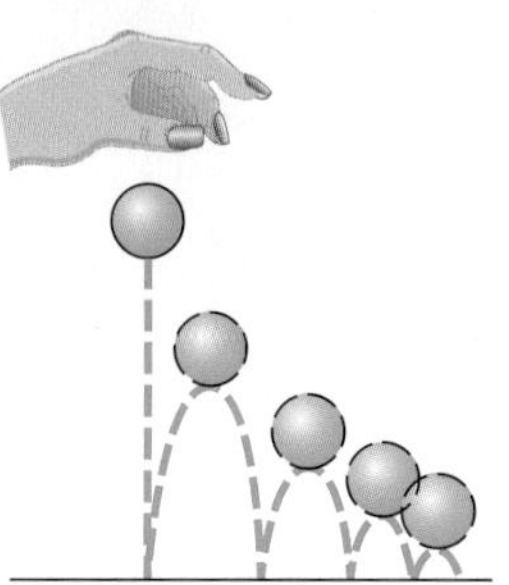

65. *Number of Ancestors* Each person has two parents, four grandparents, eight great-grandparents, and so on. What is the total number of ancestors a person has, going back five generations? ten generations?

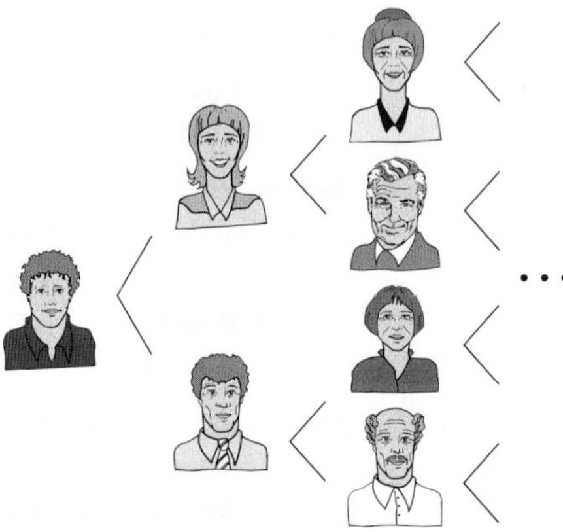

66. *Drug Dosage* Certain medical conditions are treated with a fixed dose of a drug administered at regular intervals. Suppose a person is given 2 mg of a drug each day and that during each 24-hr period, the body utilizes 40% of the amount of drug that was present at the beginning of the period.

(a) Show that the amount of the drug present in the body at the end of n days is

$$\sum_{i=1}^{n} 2(0.6)^i.$$

(b) What will be the approximate quantity of the drug in the body at the end of each day after the treatment has been administered for a long period of time?

67. *Side Length of a Triangle* A sequence of equilateral triangles is constructed. The first triangle has sides 2 m in length. To get the second triangle, midpoints of the sides of the original triangle are connected. What is the length of each side of the eighth such triangle? See the figure.

 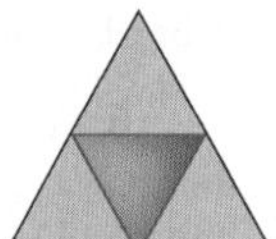 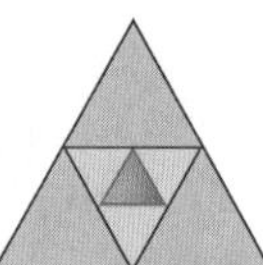 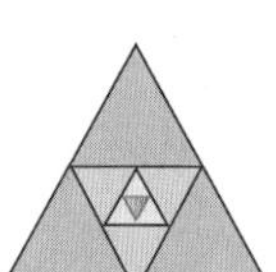

68. *Perimeter and Area of Triangles* In **Exercise 67,** if the process could be continued indefinitely, what would be the total perimeter of all the triangles? What would be the total area of all the triangles, disregarding the overlapping?

69. *Salaries* You are offered a 6-week summer job and are asked to select one of the following salary options.

Option 1: \$5000 for the first day with a \$10,000 raise each day for the remaining 29 days (that is, \$15,000 for day 2, \$25,000 for day 3, and so on)

Option 2: \$0.01 for the first day with the pay doubled each day (that is, \$0.02 for day 2, \$0.04 for day 3, and so on)

Which option would you choose?

70. The correct value is at most $8190 = \sum_{i=1}^{12} 2^i$.

71. $10,159.11 72. $10,729.67
73. $25,423.18 74. $9113.25
75. $28,107.41 76. $10,658.82
77. $72,918.53 78. $76,633.97

81. yes

70. *Number of Ancestors* Suppose a genealogical Web site allows you to identify all your ancestors that lived during the last 300 yr. Assuming that each generation spans about 25 yr, guess the number of ancestors that would be found during the 12 generations. Then use the formula for a geometric series to find the correct value.

Future Value of an Annuity Find the future value of each annuity. ***See Example 9.***

71. There are payments of $1000 at the end of each year for 9 yr at 3% interest compounded annually.

72. There are payments of $800 at the end of each year for 12 yr at 2% interest compounded annually.

73. There are payments of $2430 at the end of each year for 10 yr at 1% interest compounded annually.

74. There are payments of $1500 at the end of each year for 6 yr at 0.5% interest compounded annually.

75. Refer to **Exercise 73.** Use the answer and recursion to find the balance after 11 yr.

76. Refer to **Exercise 74.** Use the answer and recursion to find the balance after 7 yr.

77. *Individual Retirement Account* Starting on his 40th birthday, Michael Branson deposits $2000 per year in an Individual Retirement Account until age 65 (last payment at age 64). Find the total amount in the account if he had a guaranteed interest rate of 3% compounded annually.

78. *Retirement Savings* To save for retirement, Mort put $3000 at the end of each year into an ordinary annuity for 20 yr at 2.5% annual interest. At the end of the 20 yr, what was the amount of the annuity?

79. Show that the formula for future value of an annuity gives the correct answer when compared to the solution in **Example 9.**

80. The screen here shows how the TI-83/84 Plus calculator computes the future value of the annuity described in **Example 9.** Use a calculator with this capability to support your answers in **Exercises 71–78.**

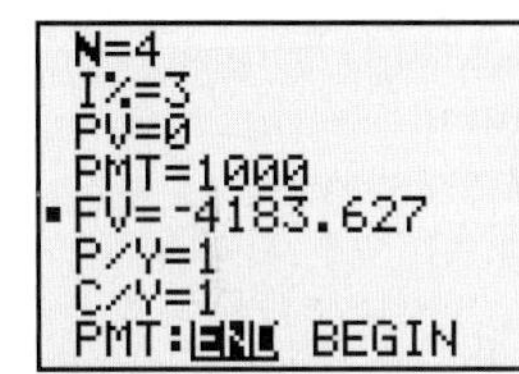

81. *Concept Check* Suppose that $a_1, a_2, a_3, a_4, a_5, \dots$ is a geometric sequence. Is $a_1, a_3, a_5, \dots$ a geometric sequence?

82. Explain why the sequence log 6, log 36, log 1296, log 1,679, 616, ... is geometric.

Summary Exercises on Sequences and Series

Use the following guidelines in Exercises 1–16 on the next page.

Given a sequence $a_1, a_2, a_3, a_4, a_5, \dots,$

- If the differences $a_2 - a_1,\ a_3 - a_2,\ a_4 - a_3,\ a_5 - a_4, \dots$ are all equal to the same number d, then the sequence is *arithmetic,* and d is the *common difference.*
- If the ratios $\frac{a_2}{a_1}, \frac{a_3}{a_2}, \frac{a_4}{a_3}, \frac{a_5}{a_4}, \dots$ are all equal to the same number r, then the sequence is *geometric,* and r is the *common ratio.*

1. geometric; $r = 2$
2. arithmetic; $d = 3$
3. arithmetic; $d = -\frac{5}{2}$
4. neither
5. geometric; $r = \frac{4}{3}$
6. geometric; $r = -3$
7. neither
8. arithmetic; $d = -3$
9. neither

In Exercises 1–9, determine whether the sequence is arithmetic, geometric, *or* neither. *If the sequence is arithmetic, give its common difference d. If the sequence is geometric, give its common ratio r.*

1. $2, 4, 8, 16, 32, \ldots$ **2.** $1, 4, 7, 10, 13, \ldots$ **3.** $3, \frac{1}{2}, -2, -\frac{9}{2}, -7, \ldots$

4. $1, 1, 2, 3, 5, 8, \ldots$ **5.** $\frac{3}{4}, 1, \frac{4}{3}, \frac{16}{9}, \frac{64}{27}, \ldots$ **6.** $4, -12, 36, -108, 324, \ldots$

7. $\frac{1}{2}, \frac{1}{3}, \frac{1}{4}, \frac{1}{5}, \frac{1}{6}, \ldots$ **8.** $5, 2, -1, -4, -7, \ldots$ **9.** $1, 9, 10, 19, 29, \ldots$

10. $d = 0$; $r = 1$

10. *Concept Check* For the sequence $5, 5, 5, \ldots$, find d and r.

11. geometric; $3(2)^{n-1}$; 3069
12. arithmetic; $4n - 2$; 200
13. arithmetic; $-\frac{3}{2}n + \frac{11}{2}$; $-\frac{55}{2}$
14. geometric; $\frac{3}{2}\left(\frac{2}{3}\right)^{n-1}$, or $\left(\frac{2}{3}\right)^{n-2}$; $\frac{58{,}025}{13{,}122}$
15. geometric; $3(-2)^{n-1}$; -1023
16. arithmetic; $-3n - 2$; -185

In Exercises 11–16, determine whether the given sequence is either arithmetic *or* geometric. *Then find a_n and* $\sum_{i=1}^{10} a_i$.

11. $3, 6, 12, 24, 48, \ldots$ **12.** $2, 6, 10, 14, 18, \ldots$ **13.** $4, \frac{5}{2}, 1, -\frac{1}{2}, -2, \ldots$

14. $\frac{3}{2}, 1, \frac{2}{3}, \frac{4}{9}, \frac{8}{27}, \ldots$ **15.** $3, -6, 12, -24, 48, \ldots$ **16.** $5, -8, -11, -14, -17, \ldots$

17. diverges 18. $\frac{1111}{500}$
19. -1850 20. 1092
21. $-\frac{4}{3}$ 22. diverges
23. 144 24. $\frac{1}{4}$
25. diverges

26. $0.3 + 0.03 + 0.003 + \cdots$; $\frac{1}{3}$

Evaluate each sum, where possible. Identify any that diverge.

17. $\sum_{i=1}^{\infty} \frac{1}{3}(-2)^{i-1}$ **18.** $\sum_{j=1}^{4} 2\left(\frac{1}{10}\right)^{j-1}$ **19.** $\sum_{i=1}^{25} (4 - 6i)$

20. $\sum_{i=1}^{6} 3^i$ **21.** $\sum_{i=1}^{\infty} 4\left(-\frac{1}{2}\right)^i$ **22.** $\sum_{i=1}^{\infty} (3i - 2)$

23. $\sum_{j=1}^{12} (2j - 1)$ **24.** $\sum_{k=1}^{\infty} 5^{-k}$ **25.** $\sum_{i=1}^{\infty} 1.0001^i$

26. Write 0.333... as an infinite geometric series. Find the sum.

11.4 The Binomial Theorem

- A Binomial Expansion Pattern
- Pascal's Triangle
- *n*-Factorial
- Binomial Coefficients
- The Binomial Theorem
- *k*th Term of a Binomial Expansion

A Binomial Expansion Pattern In this section, we introduce a method for writing the expansion of expressions of the form $(x + y)^n$, where n is a natural number. Some expansions for various nonnegative integer values of n follow.

$$(x + y)^0 = 1$$
$$(x + y)^1 = x + y$$
$$(x + y)^2 = x^2 + 2xy + y^2$$
$$(x + y)^3 = x^3 + 3x^2y + 3xy^2 + y^3$$
$$(x + y)^4 = x^4 + 4x^3y + 6x^2y^2 + 4xy^3 + y^4$$
$$(x + y)^5 = x^5 + 5x^4y + 10x^3y^2 + 10x^2y^3 + 5xy^4 + y^5$$

Notice that after the special case $(x + y)^0 = 1$, each expansion begins with x raised to the same power as the binomial itself. That is, the expansion of $(x + y)^1$ has a first term of x^1, $(x + y)^2$ has a first term of x^2, $(x + y)^3$ has a first term of x^3, and so on. Also, the last term in each expansion is y to the same power as the binomial. Thus, the expansion of $(x + y)^n$ should begin with the term x^n and end with the term y^n.

LOOKING AHEAD TO CALCULUS

Students taking calculus study the binomial series, which follows from Isaac Newton's extension to the case where the exponent is no longer a positive integer. His result led to a series for $(1 + x)^k$, where k is a real number and $|x| < 1$.

Notice that the exponent on x decreases by 1 in each term after the first, while the exponent on y, beginning with y in the second term, increases by 1 in each succeeding term. That is, the *variables* in the terms of the expansion of $(x + y)^n$ have the following pattern.

$$x^n, x^{n-1}y, x^{n-2}y^2, x^{n-3}y^3, \dots, xy^{n-1}, y^n$$

This pattern suggests that the sum of the exponents on x and y in each term is n. For example, the third term in the list above is $x^{n-2}y^2$, and the sum of the exponents is $n - 2 + 2 = n$.

Pascal's Triangle Now, examine the *coefficients* in the terms of the expansion of $(x + y)^n$. Writing the coefficients alone gives the following pattern.

Pascal's Triangle

	Row
1	0
1 1	1
1 2 1	2
1 3 3 1	3
1 4 6 4 1	4
1 5 10 10 5 1	5

With the coefficients arranged in this way, each number in the triangle is the sum of the two numbers directly above it (one to the right and one to the left). For example, in row four, 1 is the sum of 1 (the only number above it), 4 is the sum of 1 and 3, 6 is the sum of 3 and 3, and so on. This triangular array of numbers is called **Pascal's triangle,** in honor of the seventeenth-century mathematician Blaise Pascal. It was, however, known long before his time.

To find the coefficients for $(x + y)^6$, we need to include row six in Pascal's triangle. Adding adjacent numbers, we find that row six is

$$1 \quad 6 \quad 15 \quad 20 \quad 15 \quad 6 \quad 1.$$

Using these coefficients, we obtain the expansion of $(x + y)^6$.

$$(x + y)^6 = x^6 + 6x^5y + 15x^4y^2 + 20x^3y^3 + 15x^2y^4 + 6xy^5 + y^6$$

Blaise Pascal (1623–1662)

Pascal, a French mathematician, made mathematical contributions in the areas of calculus, geometry, and probability theory. At age 19, he invented the first adding machine, a precursor to our modern-day calculator.

***n*-Factorial** Although it is possible to use Pascal's triangle to find the coefficients of $(x + y)^n$ for any positive integer n, this calculation becomes impractical for large values of n because of the need to write all the preceding rows. A more efficient way of finding these coefficients uses **factorial notation.** The number $\boldsymbol{n!}$ (read **"*n*-factorial"**) is defined as follows.

n-Factorial

For any positive integer n,

$$\boldsymbol{n! = n(n-1)(n-2)\cdots(3)(2)(1).}$$

By definition, $\boldsymbol{0! = 1.}$

```
5!
                120
7!
               5040
2!
                  2
```

For example,

$$5! = 5 \cdot 4 \cdot 3 \cdot 2 \cdot 1 = 120,$$

$$7! = 7 \cdot 6 \cdot 5 \cdot 4 \cdot 3 \cdot 2 \cdot 1 = 5040,$$

and $$2! = 2 \cdot 1 = 2.$$

Teaching Tip Caution students that $3! \cdot 2! \neq 6!$ and that, in general,

$$n! \cdot (n - r)! \neq [n(n - r)]!.$$

Binomial Coefficients Now look at the coefficients of the expansion

$$(x + y)^5 = x^5 + 5x^4y + 10x^3y^2 + 10x^2y^3 + 5xy^4 + y^5.$$

The coefficient of the second term, $5x^4y$, is 5, and the exponents on the variables are 4 and 1. Note that

$$5 = \frac{5!}{4!\,1!}.$$

The coefficient of the third term, $10x^3y^2$, is 10, with exponents of 3 and 2 on the variables, and

$$10 = \frac{5!}{3!\,2!}.$$

The last term (the sixth term) can be written as $y^5 = 1x^0y^5$, with coefficient 1 and exponents of 0 and 5. Since $0! = 1$,

$$1 = \frac{5!}{0!\,5!}.$$

Generalizing from these examples, the coefficient for the term of the expansion of $(x + y)^n$ in which the variable part is $x^r y^{n-r}$ (where $r \leq n$) is

$$\frac{n!}{r!(n - r)!}.$$

This number, called a **binomial coefficient,** is often symbolized $\binom{n}{r}$ or ${}_nC_r$ (read **"*n* choose *r*"**).

Binomial Coefficient

For nonnegative integers n and r, with $r \leq n$, the binomial coefficient is defined as follows.

$${}_nC_r = \binom{n}{r} = \frac{n!}{r!(n - r)!}$$

Classroom Example 1 Evaluate each binomial coefficient.

(a) $\binom{9}{5}$ (b) $\binom{25}{25}$

(c) $\binom{16}{0}$ (d) ${}_{15}C_8$

Answers:

(a) 126 (b) 1

(c) 1 (d) 6435

The binomial coefficients are numbers from Pascal's triangle. For example, $\binom{3}{0}$ is the first number in row three, and $\binom{7}{4}$ is the fifth number in row seven.

EXAMPLE 1 Evaluating Binomial Coefficients

Evaluate each binomial coefficient.

(a) $\binom{6}{2}$ **(b)** $\binom{8}{0}$ **(c)** $\binom{10}{10}$ **(d)** ${}_{12}C_{10}$

ALGEBRAIC SOLUTION

(a) $\binom{6}{2} = \frac{6!}{2!(6-2)!} = \frac{6!}{2!4!}$

$= \frac{6 \cdot 5 \cdot 4 \cdot 3 \cdot 2 \cdot 1}{2 \cdot 1 \cdot 4 \cdot 3 \cdot 2 \cdot 1} = 15$

(b) $\binom{8}{0} = \frac{8!}{0!(8-0)!} = \frac{8!}{0!8!} = \frac{8!}{1 \cdot 8!} = 1 \quad 0! = 1$

(c) $\binom{10}{10} = \frac{10!}{10!(10-10)!} = \frac{10!}{10!0!} = 1 \quad 0! = 1$

(d) ${}_{12}C_{10} = \frac{12!}{10!(12-10)!} = \frac{12!}{10!2!} = 66$

GRAPHING CALCULATOR SOLUTION

Graphing calculators calculate binomial coefficients using the notation ${}_nC_r$. For the TI-83/84 Plus, this function is found in the MATH menu. **Figure 11** shows the values of the binomial coefficients found in parts (a), (b), and (c).

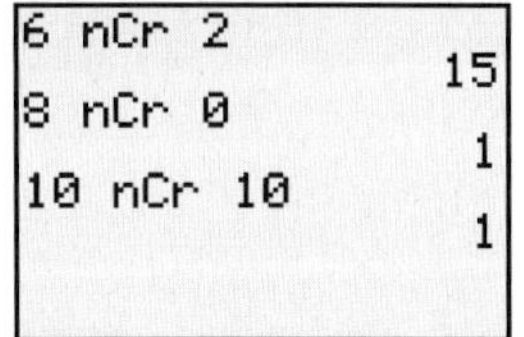

Figure 11

✔ *Now Try Exercises 5, 9, and 17.*

Refer again to Pascal's triangle. Notice the symmetry in each row. This suggests that binomial coefficients should have the same property. That is,

$$\binom{n}{r} = \binom{n}{n-r}.$$

This is true, since

$$\binom{n}{r} = \frac{n!}{r!(n-r)!} \quad \text{and} \quad \binom{n}{n-r} = \frac{n!}{(n-r)!r!}.$$

Teaching Tip Show students that the binomial coefficient $\binom{n}{r}$ can also be calculated by first writing the expansion of $r!$ in the denominator, and then writing, in the numerator, an integer from the descending expansion of $n!$ above each integer in the denominator. For instance,

$$\binom{6}{2} = \frac{6 \cdot 5}{2 \cdot 1}.$$

The formula $\binom{n}{r} = \binom{n}{n-r}$ combined with this tip simplifies the computation of $\binom{n}{r}$ for large values of r.

The Binomial Theorem Our observations about the expansion of $(x + y)^n$ are summarized as follows.

1. There are $n + 1$ terms in the expansion.
2. The first term is x^n, and the last term is y^n.
3. In each succeeding term, the exponent on x decreases by 1 and the exponent on y increases by 1.
4. The sum of the exponents on x and y in any term is n.
5. The coefficient of the term with $x^r y^{n-r}$ or $x^{n-r} y^r$ is $\binom{n}{r}$.

Teaching Tip Encourage students to identify patterns related to the terms of a binomial expansion.

These observations about the expansion of $(x + y)^n$ for any positive integer value of n suggest the **binomial theorem.** A proof of the binomial theorem using mathematical induction is given in **Section 11.5.**

LOOKING AHEAD TO CALCULUS

The binomial theorem is used to show that the derivative of $f(x) = x^n$ is given by the function $f'(x) = nx^{n-1}$. This fact is used extensively in calculus.

Binomial Theorem

For any positive integer n and any complex numbers x and y, $(x + y)^n$ is expanded as follows.

$$(x + y)^n = x^n + \binom{n}{1}x^{n-1}y + \binom{n}{2}x^{n-2}y^2 + \binom{n}{3}x^{n-3}y^3 + \cdots + \binom{n}{r}x^{n-r}y^r + \cdots + \binom{n}{n-1}xy^{n-1} + y^n$$

NOTE The binomial theorem may also be written as a series with summation notation.

$$(x+y)^n = \sum_{r=0}^{n} \binom{n}{r} x^{n-r} y^r$$

In agreement with Pascal's triangle, the coefficients of the first and last terms are both 1. That is,

$$\binom{n}{0} = \binom{n}{n} = 1.$$

Classroom Example 2
Write the binomial expansion of $(x+y)^8$.

Answer: $x^8 + 8x^7y + 28x^6y^2 + 56x^5y^3 + 70x^4y^4 + 56x^3y^5 + 28x^2y^6 + 8xy^7 + y^8$

EXAMPLE 2 Applying the Binomial Theorem

Write the binomial expansion of $(x+y)^9$.

SOLUTION Apply the binomial theorem.

$$(x+y)^9 = x^9 + \binom{9}{1}x^8y + \binom{9}{2}x^7y^2 + \binom{9}{3}x^6y^3 + \binom{9}{4}x^5y^4 + \binom{9}{5}x^4y^5 + \binom{9}{6}x^3y^6 + \binom{9}{7}x^2y^7 + \binom{9}{8}xy^8 + y^9$$

Now evaluate each of the binomial coefficients.

$$\begin{aligned}(x+y)^9 &= x^9 + \frac{9!}{1!8!}x^8y + \frac{9!}{2!7!}x^7y^2 + \frac{9!}{3!6!}x^6y^3 + \frac{9!}{4!5!}x^5y^4 + \frac{9!}{5!4!}x^4y^5 \\ &\quad + \frac{9!}{6!3!}x^3y^6 + \frac{9!}{7!2!}x^2y^7 + \frac{9!}{8!1!}xy^8 + y^9 \\ &= x^9 + 9x^8y + 36x^7y^2 + 84x^6y^3 + 126x^5y^4 + 126x^4y^5 \\ &\quad + 84x^3y^6 + 36x^2y^7 + 9xy^8 + y^9\end{aligned}$$

✔ *Now Try Exercise 23.*

Classroom Example 3
Expand $\left(\frac{m}{3} - n\right)^4$.

Answer: $\frac{m^4}{81} - \frac{4}{27}m^3n + \frac{2}{3}m^2n^2 - \frac{4}{3}mn^3 + n^4$

EXAMPLE 3 Applying the Binomial Theorem

Expand $\left(a - \frac{b}{2}\right)^5$.

SOLUTION Write the binomial as follows.

$$\left(a - \frac{b}{2}\right)^5 = \left(a + \left(-\frac{b}{2}\right)\right)^5$$

Now apply the binomial theorem with $x = a$, $y = -\frac{b}{2}$, and $n = 5$.

$$\begin{aligned}\left(a - \frac{b}{2}\right)^5 &= a^5 + \binom{5}{1}a^4\left(-\frac{b}{2}\right) + \binom{5}{2}a^3\left(-\frac{b}{2}\right)^2 + \binom{5}{3}a^2\left(-\frac{b}{2}\right)^3 + \binom{5}{4}a\left(-\frac{b}{2}\right)^4 + \left(-\frac{b}{2}\right)^5 \\ &= a^5 + 5a^4\left(-\frac{b}{2}\right) + 10a^3\left(-\frac{b}{2}\right)^2 + 10a^2\left(-\frac{b}{2}\right)^3 + 5a\left(-\frac{b}{2}\right)^4 + \left(-\frac{b}{2}\right)^5 \\ &= a^5 - \frac{5}{2}a^4b + \frac{5}{2}a^3b^2 - \frac{5}{4}a^2b^3 + \frac{5}{16}ab^4 - \frac{1}{32}b^5\end{aligned}$$

In this expansion, the signs of the terms alternate because $y = -\frac{b}{2}$ is a negative term.

✔ *Now Try Exercise 35.*

NOTE As **Example 3** illustrates, ***an expansion of the difference of two terms (that is, $x - y$) has alternating signs.***

Classroom Example 4

Expand $\left(\frac{2}{r^3} + \frac{4}{\sqrt{r}}\right)^3$.

Answer:

$\frac{8}{r^9} + \frac{48}{r^{13/2}} + \frac{96}{r^4} + \frac{64}{r^{3/2}}$

EXAMPLE 4 Applying the Binomial Theorem

Expand $\left(\frac{3}{m^2} - 2\sqrt{m}\right)^4$. (Assume $m > 0$.)

SOLUTION Apply the binomial theorem.

$$\left(\frac{3}{m^2} - 2\sqrt{m}\right)^4 = \left(\frac{3}{m^2}\right)^4 + \binom{4}{1}\left(\frac{3}{m^2}\right)^3(-2\sqrt{m}) + \binom{4}{2}\left(\frac{3}{m^2}\right)^2(-2\sqrt{m})^2 + \binom{4}{3}\left(\frac{3}{m^2}\right)(-2\sqrt{m})^3 + (-2\sqrt{m})^4$$

$$= \frac{81}{m^8} + 4\left(\frac{27}{m^6}\right)(-2m^{1/2}) + 6\left(\frac{9}{m^4}\right)(4m) + 4\left(\frac{3}{m^2}\right)(-8m^{3/2}) + 16m^2 \qquad \sqrt{m} = m^{1/2} \text{ (Section R.7)}$$

$$= \frac{81}{m^8} - \frac{216}{m^{11/2}} + \frac{216}{m^3} - \frac{96}{m^{1/2}} + 16m^2$$

Now Try Exercise 37.

kth Term of a Binomial Expansion Earlier in this section, we wrote the binomial theorem in summation notation as $\sum_{r=0}^{n}\binom{n}{r}x^{n-r}y^r$, which gives the form of each term. We can use this form to write any particular term of a binomial expansion without writing out the entire expansion.

For example, to find the tenth term of $(x + y)^n$, where $n \geq 9$, first notice that in the tenth term y is raised to the ninth power (since y has the power 1 in the second term, the power 2 in the third term, and so on). Because the exponents on x and y in any term must have a sum of n, the exponent on x in the tenth term is $n - 9$. Thus, the tenth term of the expansion is

$$\binom{n}{9}x^{n-9}y^9 = \frac{n!}{9!(n-9)!}x^{n-9}y^9.$$

kth Term of the Binomial Expansion

The kth term of the binomial expansion of $(x + y)^n$, where $n \geq k - 1$, is given as follows.

$$\binom{n}{k-1}x^{n-(k-1)}y^{k-1}$$

To find the kth term of the binomial expansion, use the following steps.

Step 1 Find $k - 1$. This is the exponent on the second part of the binomial.

Step 2 Subtract the exponent found in Step 1 from n to get the exponent on the first part of the binomial.

Step 3 Determine the coefficient by using the exponents found in the first two steps and n.

Classroom Example 5
Find the fourth term of $(2c - d)^{12}$.

Answer: $-112{,}640c^9d^3$

EXAMPLE 5 Finding a Particular Term of a Binomial Expansion

Find the seventh term of $(a + 2b)^{10}$.

SOLUTION In the seventh term, $2b$ has an exponent of $7 - 1$, or 6, while a has an exponent of $10 - 6$, or 4.

$$\binom{10}{6}a^4(2b)^6 = 210a^4(64b^6) \quad \text{Seventh term of a binomial expansion}$$

$$= 13{,}440a^4b^6 \quad \text{Multiply.}$$

Now Try Exercise 41.

11.4 Exercises

1. 20 2. 10
3. 35 4. 56
5. 56 6. 35
7. 45 8. 84
9. 1 10. 1
11. n 12. $\frac{n(n-1)}{2}$
13. 56 14. 36
15. 4950 16. 15,504
17. 1 18. 5
19. 12 20. 1
21. $16x^4$; $81y^4$ 22. $\binom{9}{4} = 126$
23. $x^6 + 6x^5y + 15x^4y^2 + 20x^3y^3 + 15x^2y^4 + 6xy^5 + y^6$
24. $m^4 + 4m^3n + 6m^2n^2 + 4mn^3 + n^4$
25. $p^5 - 5p^4q + 10p^3q^2 - 10p^2q^3 + 5pq^4 - q^5$
26. $a^7 - 7a^6b + 21a^5b^2 - 35a^4b^3 + 35a^3b^4 - 21a^2b^5 + 7ab^6 - b^7$
27. $r^{10} + 5r^8s + 10r^6s^2 + 10r^4s^3 + 5r^2s^4 + s^5$
28. $m^4 + 4m^3n^2 + 6m^2n^4 + 4mn^6 + n^8$
29. $p^4 + 8p^3q + 24p^2q^2 + 32pq^3 + 16q^4$
30. $729r^6 + 1458r^5s + 1215r^4s^2 + 540r^3s^3 + 135r^2s^4 + 18rs^5 + s^6$
31. $2401p^4 - 2744p^3q + 1176p^2q^2 - 224pq^3 + 16q^4$
32. $1024a^5 - 6400a^4b + 16{,}000a^3b^2 - 20{,}000a^2b^3 + 12{,}500ab^4 - 3125b^5$

Evaluate each expression. In Exercises 11 and 12, leave answers in terms of n. ***See Example 1.***

1. $\frac{6!}{3!3!}$ 2. $\frac{5!}{2!3!}$ 3. $\frac{7!}{3!4!}$ 4. $\frac{8!}{5!3!}$

5. $\binom{8}{5}$ 6. $\binom{7}{3}$ 7. $\binom{10}{2}$ 8. $\binom{9}{3}$

9. $\binom{14}{14}$ 10. $\binom{15}{15}$ 11. $\binom{n}{n-1}$ 12. $\binom{n}{n-2}$

13. ${}_8C_3$ 14. ${}_9C_7$ 15. ${}_{100}C_{98}$ 16. ${}_{20}C_5$

17. ${}_9C_0$ 18. ${}_5C_1$ 19. ${}_{12}C_1$ 20. ${}_4C_0$

21. *Concept Check* What are the first and last terms in the expansion of $(2x + 3y)^4$?

22. *Concept Check* Determine the binomial coefficient for the fifth term in the expansion of $(x + y)^9$.

Write the binomial expansion for each expression. ***See Examples 2–4.***

23. $(x + y)^6$ 24. $(m + n)^4$ 25. $(p - q)^5$

26. $(a - b)^7$ 27. $(r^2 + s)^5$ 28. $(m + n^2)^4$

29. $(p + 2q)^4$ 30. $(3r + s)^6$ 31. $(7p - 2q)^4$

32. $(4a - 5b)^5$ 33. $(3x - 2y)^6$ 34. $(7k - 9j)^4$

35. $\left(\frac{m}{2} - 1\right)^6$ 36. $\left(3 - \frac{y}{3}\right)^5$ 37. $\left(\sqrt{2}r + \frac{1}{m}\right)^4$

38. $\left(\frac{1}{k} + \sqrt{3}p\right)^3$ 39. $\left(\frac{1}{x^4} + x^4\right)^4$ 40. $\left(\frac{1}{y^5} + y^5\right)^5$

Write the indicated term of each binomial expansion. ***See Example 5.***

41. Sixth term of $(4h - j)^8$ 42. Eighth term of $(2c - 3d)^{14}$

43. Seventeenth term of $(a^2 + b)^{22}$ 44. Twelfth term of $(2x + y^2)^{16}$

45. Fifteenth term of $(x - y^3)^{20}$ 46. Tenth term of $(a^3 + 3b)^{11}$

Concept Check Work Exercises 47–50.

47. Find the middle term of $(3x^7 + 2y^3)^8$.

48. Find the two middle terms of $(-2m^{-1} + 3n^{-2})^{11}$.

33. $729x^6 - 2916x^5y + 4860x^4y^2 - 4320x^3y^3 + 2160x^2y^4 - 576xy^5 + 64y^6$
34. $2401k^4 - 12{,}348k^3j + 23{,}814k^2j^2 - 20{,}412kj^3 + 6561j^4$
35. $\frac{m^6}{64} - \frac{3m^5}{16} + \frac{15m^4}{16} - \frac{5m^3}{2} + \frac{15m^2}{4} - 3m + 1$
36. $243 - 135y + 30y^2 - \frac{10y^3}{3} + \frac{5y^4}{27} - \frac{y^5}{243}$
37. $4r^4 + \frac{8\sqrt{2}r^3}{m} + \frac{12r^2}{m^2} + \frac{4\sqrt{2}r}{m^3} + \frac{1}{m^4}$
38. $\frac{1}{k^3} + \frac{3\sqrt{3p}}{k^2} + \frac{9p^2}{k} + 3\sqrt{3}p^3$
39. $\frac{1}{x^{16}} + \frac{4}{x^8} + 6 + 4x^8 + x^{16}$
40. $\frac{1}{y^{25}} + \frac{5}{y^{15}} + \frac{10}{y^5} + 10y^5 + 5y^{15} + y^{25}$
41. $-3584h^3j^5$
42. $-3432(6^7)c^7d^7$
43. $74{,}613a^{12}b^{16}$
44. $139{,}776x^5y^{22}$
45. $38{,}760x^6y^{42}$
46. $55(3^9)a^6b^9$
47. $90{,}720x^{28}y^{12}$
48. $462(2^6)(3^5)m^{-6}n^{-10}$; $-462(2^5)(3^6)m^{-5}n^{-12}$
49. 11
50. $4455x^4$
51. exact: 3,628,800; approximate: 3,598,695.619
52. about 0.830%
53. exact: 479,001,600; approximate: 475,687,486.5; about 0.692%
54. exact: 6,227,020,800; approximate: 6,187,239,475; about 0.639%; As n gets larger, the percent error decreases.

49. Find the value of n for which the coefficients of the fifth and eighth terms in the expansion of $(x + y)^n$ are the same.

50. Find the term(s) in the expansion of $\left(3 + \sqrt{x}\right)^{11}$ that contain(s) x^4.

Relating Concepts

For individual or collaborative investigation *(Exercises 51–54)*

In this section, we saw how the factorial of a positive integer n can be computed as a product:

$$n! = 1 \cdot 2 \cdot 3 \cdot \cdots \cdot n.$$

Calculators and computers can evaluate factorials very quickly. Before the days of modern technology, mathematicians developed ***Stirling's formula*** *for approximating large factorials. Interestingly enough, the formula involves the irrational numbers* π *and e.*

$$n! \approx \sqrt{2\pi n} \cdot n^n \cdot e^{-n}$$

As an example, the exact value of 5! is 120, and Stirling's formula gives the approximation as 118.019168 with a graphing calculator. This is "off" by less than 2, an error of only 1.65%. ***Work Exercises 51–54 in order.***

51. Use a calculator to find the exact value of 10! and its approximation, using Stirling's formula.

52. Subtract the smaller value from the larger value in **Exercise 51.** Divide it by 10! and convert to a percent. What is the percent error?

53. Repeat **Exercises 51 and 52** for $n = 12$.

54. Repeat **Exercises 51 and 52** for $n = 13$. What seems to happen as n gets larger?

It can be shown that

$$(1 + x)^n = 1 + nx + \frac{n(n-1)}{2!}x^2 + \frac{n(n-1)(n-2)}{3!}x^3 + \cdots$$

is true for any real number n (not just positive integer values) and any real number x, where $|x| < 1$. *Use this series to approximate the given number to the nearest thousandth.*

55. $(1.02)^{-3}$ **56.** $(1.04)^{-5}$ **57.** $(1.01)^{1.5}$ **58.** $(1.03)^2$

55. 0.942 56. 0.822 57. 1.015 58. 1.061

11.5 Mathematical Induction

- Proof by Mathematical Induction
- Proving Statements
- Generalized Principle of Mathematical Induction
- Proof of the Binomial Theorem

Proof by Mathematical Induction Many results in mathematics are claimed true for every positive integer. Any of these results could be checked for $n = 1$, $n = 2$, $n = 3$, and so on, but since the set of positive integers is infinite, it would be impossible to check every possible case. For example, let S_n represent the statement that the sum of the first n positive integers is $\frac{n(n+1)}{2}$.

$$S_n\colon\ 1 + 2 + 3 + \cdots + n = \frac{n(n+1)}{2}$$

Teaching Tip Students may initially think that testing a large number of numerical values is sufficient to prove a statement. Explain that numerical analysis enables us to create conjectures, whereas mathematical proofs are conclusive.

The truth of this statement is easily verified for the first few values of n:

If $n = 1$, then S_1 is $\quad 1 = \frac{1(1+1)}{2}$, which is true.

If $n = 2$, then S_2 is $\quad 1 + 2 = \frac{2(2+1)}{2}$, which is true.

If $n = 3$, then S_3 is $\quad 1 + 2 + 3 = \frac{3(3+1)}{2}$, which is true.

If $n = 4$, then S_4 is $\quad 1 + 2 + 3 + 4 = \frac{4(4+1)}{2}$, which is true.

Continuing in this way for any amount of time would still not prove that S_n is true for *every* positive integer value of n. To prove that such statements are true for every positive integer value of n, the following principle is often used.

Principle of Mathematical Induction

Let S_n be a statement concerning the positive integer n. Suppose that

1. S_1 is true;
2. for any positive integer k, $k \le n$, if S_k is true, then S_{k+1} is also true.

Then S_n is true for every positive integer value of n.

A proof by mathematical induction can be explained as follows. By assumption (1) above, the statement is true when $n = 1$. By assumption (2) above, the fact that the statement is true for $n = 1$ implies that it is true for $n = 1 + 1 = 2$. Using (2) again, the statement is thus true for $2 + 1 = 3$, for $3 + 1 = 4$, for $4 + 1 = 5$, and so on. Continuing in this way shows that the statement must be true for *every* positive integer.

The situation is similar to that of a number of dominoes lined up as shown in **Figure 12.** If the first domino is pushed over, it pushes the next, which pushes the next, and so on until all are down.

Figure 12

Another example of the principle of mathematical induction is the concept of an infinite ladder. Suppose the rungs are spaced so that whenever you are on a rung, you know you can move to the next rung. Then *if* you can get to the first rung, you can go as high up the ladder as you wish.

Two separate steps are required for a proof by mathematical induction.

Proof by Mathematical Induction

Step 1 Prove that the statement is true for $n = 1$.

Step 2 Show that, for any positive integer k, $k \le n$, if S_k is true, then S_{k+1} is also true.

Teaching Tip Students may have difficulty determining how to proceed with mathematical induction. Give some general examples of how to transform terms containing k into terms containing $k + 1$, such as:

(a) a^k (Multiply by a.)

(b) $a \cdot k$ (Add a.)

Proving Statements Mathematical induction is used in the next example to prove the statement S_n mentioned at the beginning of this section.

Classroom Example 1
Let S_n represent the statement

$$2 + 5 + 8 + \cdots + (3n - 1) = \frac{n(3n + 1)}{2}.$$

Prove that S_n is true for every positive integer n.

Answer: The proof appears in the *Solutions to Classroom Examples.*

EXAMPLE 1 Proving an Equality Statement

Let S_n represent the following statement.

$$1 + 2 + 3 + \cdots + n = \frac{n(n + 1)}{2}$$

Prove that S_n is true for every positive integer n.

SOLUTION The proof by mathematical induction is as follows.

Step 1 Show that the statement is true when $n = 1$. If $n = 1$, S_1 becomes

$$1 = \frac{1(1 + 1)}{2}. \quad \text{True}$$

Step 2 Show that S_k implies S_{k+1}, where S_k is the statement

$$1 + 2 + 3 + \cdots + k = \frac{k(k + 1)}{2},$$

and S_{k+1} is the statement

$$1 + 2 + 3 + \cdots + k + (k + 1) = \frac{(k + 1)[(k + 1) + 1]}{2}.$$

Start with S_k and assume it is a true statement.

$$1 + 2 + 3 + \cdots + k = \frac{k(k + 1)}{2}$$

Add $k + 1$ to each side of this equation to obtain S_{k+1}.

Teaching Tip Explain to students that after adding the $(k + 1)$st term to each side, the goal is to perform algebra on the right side to obtain S_{k+1}.

$$1 + 2 + 3 + \cdots + k + (k + 1) = \frac{k(k + 1)}{2} + (k + 1)$$

$$= (k + 1)\left(\frac{k}{2} + 1\right) \quad \text{Factor out } k + 1. \text{ (Section R.4)}$$

$$= (k + 1)\left(\frac{k + 2}{2}\right) \quad \text{Add inside the parentheses. (Section R.5)}$$

$$1 + 2 + 3 + \cdots + k + (k + 1) = \frac{(k + 1)[(k + 1) + 1]}{2} \quad \text{Multiply; } k + 2 = (k + 1) + 1$$

This final result is the statement for $n = k + 1$. It has been shown that if S_k is true, then S_{k+1} is also true.

The two steps required for a proof by mathematical induction have been completed, so the statement S_n is true for every positive integer value of n.

Now Try Exercise 1.

CAUTION Notice that the left side of the statement S_n always includes *all* the terms up to the nth term, as well as the nth term.

Classroom Example 2

Prove: $\left(\frac{a}{b}\right)^n = \frac{a^n}{b^n}$. (Assume a and b are constants, with $b \neq 0$.)

Answer: The proof appears in the *Solutions to Classroom Examples.*

EXAMPLE 2 Proving an Inequality Statement

Prove: If x is a real number between 0 and 1, then for every positive integer n, $0 < x^n < 1$.

SOLUTION

Step 1 Here S_1 is the statement

$$\text{if } 0 < x < 1, \text{ then } 0 < x^1 < 1. \quad \text{True}$$

Step 2 S_k is the statement

$$\text{if } 0 < x < 1, \text{ then } 0 < x^k < 1.$$

To show that this implies that S_{k+1} is true, multiply each of the three parts of $0 < x^k < 1$ by x to get

$$x \cdot 0 < x \cdot x^k < x \cdot 1. \quad \text{(Section 1.7)}$$

(Here the fact that $0 < x$ is used.) Simplify to obtain

$$0 < x^{k+1} < x.$$

Since $x < 1$,

$$0 < x^{k+1} < x < 1$$

and thus $$0 < x^{k+1} < 1. \quad \text{True}$$

This work shows that if S_k is true, then S_{k+1} is true.

Since both steps for a proof by mathematical induction have been completed, the given statement is true for every positive integer n.

Now Try Exercise 23.

Generalized Principle of Mathematical Induction Some statements S_n are not true for the first few values of n but *are* true for all values of n that are greater than or equal to some fixed integer j. The following slightly generalized form of the principle of mathematical induction takes care of these cases.

Generalized Principle of Mathematical Induction

Let S_n be a statement concerning the positive integer n. Let j be a fixed positive integer. Suppose that

Step 1 S_j is true;

Step 2 for any positive integer k, $k \geq j$, S_k implies S_{k+1}.

Then S_n is true for all positive integers n, where $n \geq j$.

Classroom Example 3
Let S_n represent the statement $4^n > 4n$. Show that S_n is true for all values of n such that $n \geq 2$.

Answer: The proof appears in the *Solutions to Classroom Examples.*

EXAMPLE 3 Using the Generalized Principle

Let S_n represent the statement $2^n > 2n + 1$. Show that S_n is true for all values of n such that $n \geq 3$.

SOLUTION (Check that S_n is false for $n = 1$ and $n = 2$.)

Step 1 Show that S_n is true for $n = 3$. If $n = 3$, then S_n is

$$2^3 > 2 \cdot 3 + 1,$$

or $$8 > 7. \quad \text{True}$$

Thus, S_3 is true.

Step 2 Now show that S_k implies S_{k+1}, where $k \geq 3$, and where

$$S_k \quad \text{is} \quad 2^k > 2k + 1,$$

and $$S_{k+1} \text{ is } 2^{k+1} > 2(k + 1) + 1.$$

Start with S_k and assume it is a true statement.

$$2^k > 2k + 1$$

$$2 \cdot 2^k > 2(2k+1)$$ Multiply each side by 2. (Section 1.7)

$2 = 2^1$, so add the exponents.

$$2^{k+1} > 4k+2$$ Product rule (Section R.3); distributive property (Section R.2)

$$2^{k+1} > 2k+2+2k$$ Rewrite $4k$.

$$2^{k+1} > 2(k+1)+2k$$ Factor $2k+2$.

Since $2k > 1$ for positive integers $k \geq 3$, replacing $2k$ with 1 will maintain the truth value of this inequality.

$$2^{k+1} > 2(k+1)+1 \quad \text{True}$$

Thus, S_k implies S_{k+1}, and this, together with the fact that S_3 is true, shows that S_n is true for every positive integer value of n greater than or equal to 3.

✔ ***Now Try Exercise 21.***

Proof of the Binomial Theorem The binomial theorem can be proved by mathematical induction.

$$(x+y)^n = x^n + \binom{n}{1}x^{n-1}y + \binom{n}{2}x^{n-2}y^2 + \binom{n}{3}x^{n-3}y^3 + \cdots + \binom{n}{r}x^{n-r}y^r + \cdots + \binom{n}{n-1}xy^{n-1} + y^n \quad (1)$$

For any positive integer n and any complex numbers x and y (Section 11.4)

Proof Let S_n be statement (1). Begin by verifying S_n for $n = 1$.

$$S_1: \quad (x+y)^1 = x^1 + y^1 \quad \text{True}$$

Now assume that S_n is true for the positive integer k. Statement S_k becomes

$$S_k: \quad (x+y)^k = x^k + \frac{k!}{1!(k-1)!}x^{k-1}y + \frac{k!}{2!(k-2)!}x^{k-2}y^2 + \cdots + \frac{k!}{(k-1)!\,1!}xy^{k-1} + y^k. \quad (2)$$

Definition of the binomial coefficient (Section 11.4)

Multiply each side of equation (2) by $x + y$.

$$(x+y)^k \cdot (x+y)$$

$$= x(x+y)^k + y(x+y)^k \quad \text{Distributive property}$$

$$= \left[x \cdot x^k + \frac{k!}{1!(k-1)!}x^k y + \frac{k!}{2!(k-2)!}x^{k-1}y^2 + \cdots + \frac{k!}{(k-1)!\,1!}x^2y^{k-1} + xy^k\right]$$

$$+ \left[x^k \cdot y + \frac{k!}{1!(k-1)!}x^{k-1}y^2 + \cdots + \frac{k!}{(k-1)!\,1!}xy^k + y \cdot y^k\right]$$

Rearrange terms to get

$$(x+y)^{k+1} = x^{k+1} + \left[\frac{k!}{1!(k-1)!} + 1\right]x^k y + \left[\frac{k!}{2!(k-2)!} + \frac{k!}{1!(k-1)!}\right]x^{k-1}y^2 + \cdots + \left[1 + \frac{k!}{(k-1)!\,1!}\right]xy^k + y^{k+1}. \quad (3)$$

The first expression in brackets in equation (3) simplifies to $\binom{k+1}{1}$. To see this, note the following.

$$\binom{k+1}{1} = \frac{(k+1)(k)(k-1)(k-2)\cdots 1}{1\cdot(k)(k-1)(k-2)\cdots 1} = k+1$$

Also, $$\frac{k!}{1!(k-1)!} + 1 = \frac{k(k-1)!}{1(k-1)!} + 1 = k+1.$$

The second expression becomes $\binom{k+1}{2}$, the last $\binom{k+1}{k}$, and so on. The result of equation (3) is just equation (2) with every k replaced by $k+1$. Thus, the truth of S_k implies the truth of S_{k+1}, which completes the proof of the theorem by mathematical induction.

11.5 Exercises

1. S_1: $1 = 1^2$;
   S_2: $1 + 3 = 2^2$;
   S_3: $1 + 3 + 5 = 3^2$;
   S_4: $1 + 3 + 5 + 7 = 4^2$;
   S_5: $1 + 3 + 5 + 7 + 9 = 5^2$
2. S_1: $2 = 1(1 + 1)$;
   S_2: $2 + 4 = 2(2 + 1)$;
   S_3: $2 + 4 + 6 = 3(3 + 1)$;
   S_4: $2 + 4 + 6 + 8 = 4(4 + 1)$;
   S_5: $2 + 4 + 6 + 8 + 10 = 5(5 + 1)$

Although we do not usually give proofs, the answers for Exercises 3 and 11 are given in the answer section at the back of the text.

Write out in full and verify the statements S_1, S_2, S_3, S_4, and S_5 for the following. Then use mathematical induction to prove that each statement is true for every positive integer n. ***See Example 1.***

1. $1 + 3 + 5 + \cdots + (2n - 1) = n^2$
2. $2 + 4 + 6 + \cdots + 2n = n(n + 1)$

Assume that n is a positive integer. Use mathematical induction to prove each statement S by following these steps. ***See Example 1.***

(a) *Verify the statement for $n = 1$.*
(b) *Write the statement for $n = k$.*
(c) *Write the statement for $n = k + 1$.*
(d) *Assume the statement is true for $n = k$. Use algebra to change the statement in part (b) to the statement in part (c).*
(e) *Write a conclusion based on Steps (a)–(d).*

3. $3 + 6 + 9 + \cdots + 3n = \dfrac{3n(n+1)}{2}$
4. $5 + 10 + 15 + \cdots + 5n = \dfrac{5n(n+1)}{2}$
5. $2 + 4 + 8 + \cdots + 2^n = 2^{n+1} - 2$
6. $3 + 9 + 27 + \cdots + 3^n = \dfrac{1}{2}(3^{n+1} - 3)$
7. $1^2 + 2^2 + 3^2 + \cdots + n^2 = \dfrac{n(n+1)(2n+1)}{6}$
8. $1^3 + 2^3 + 3^3 + \cdots + n^3 = \dfrac{n^2(n+1)^2}{4}$
9. $5 \cdot 6 + 5 \cdot 6^2 + 5 \cdot 6^3 + \cdots + 5 \cdot 6^n = 6(6^n - 1)$
10. $7 \cdot 8 + 7 \cdot 8^2 + 7 \cdot 8^3 + \cdots + 7 \cdot 8^n = 8(8^n - 1)$
11. $\dfrac{1}{1\cdot 2} + \dfrac{1}{2\cdot 3} + \dfrac{1}{3\cdot 4} + \cdots + \dfrac{1}{n(n+1)} = \dfrac{n}{n+1}$
12. $\dfrac{1}{1\cdot 4} + \dfrac{1}{4\cdot 7} + \dfrac{1}{7\cdot 10} + \cdots + \dfrac{1}{(3n-2)(3n+1)} = \dfrac{n}{3n+1}$
13. $\dfrac{1}{2} + \dfrac{1}{2^2} + \dfrac{1}{2^3} + \cdots + \dfrac{1}{2^n} = 1 - \dfrac{1}{2^n}$
14. $\dfrac{4}{5} + \dfrac{4}{5^2} + \dfrac{4}{5^3} + \cdots + \dfrac{4}{5^n} = 1 - \dfrac{1}{5^n}$

15. $n = 1$ or 2
16. $n = 1$
17. $n = 2, 3,$ or 4
18. $n = 1, 2,$ or 3

Find all natural number values for n for which the given statement is false.

15. $2^n > 2n$
16. $3^n > 2n + 1$
17. $2^n > n^2$
18. $n! > 2n$

For Exercises 19–28, we show only the proof for Exercise 19.

19. (a) $(a^m)^1 = a^m$ and $a^{m(1)} = a^m$, so S is true for $n = 1$.
(b) $(a^m)^k = a^{mk}$
(c) $(a^m)^{(k+1)} = a^{m(k+1)}$
(d) $(a^m)^k \cdot (a^m)^1 = a^{mk} \cdot (a^m)^1$
$(a^m)^{(k+1)} = a^{(mk+m)}$ Product rule
$(a^m)^{(k+1)} = a^{m(k+1)}$ Factor.
(e) Since S is true for $n = 1$ and S is true for $n = k + 1$ when it is true for $n = k$, S is true for every positive integer n.

Prove each statement by mathematical induction. ***See Examples 1–3.***

19. $(a^m)^n = a^{mn}$ (Assume a and m are constant.)

20. $(ab)^n = a^n b^n$ (Assume a and b are constant.)

21. $2^n > 2n$, if $n \geq 3$

22. $3^n > 2n + 1$, if $n \geq 2$

23. If $a > 1$, then $a^n > 1$.

24. If $a > 1$, then $a^n > a^{n-1}$.

25. If $0 < a < 1$, then $a^n < a^{n-1}$.

26. $2^n > n^2$, for $n \geq 5$

27. If $n \geq 4$, then $n! > 2^n$, where $n! = n(n-1)(n-2)\cdots(3)(2)(1)$.

28. $4^n > n^4$, for $n \geq 5$

30. $3 \cdot 4^{n-1}$
31. $\frac{4^{n-1}}{3^{n-2}}$, or $3\left(\frac{4}{3}\right)^{n-1}$
33. $2^n - 1$

Solve each problem.

29. *Number of Handshakes* Suppose that each of the n (for $n \geq 2$) people in a room shakes hands with everyone else, but not with himself or herself. Show that the number of handshakes is $\frac{n^2 - n}{2}$.

30. *Sides of a Polygon* The series of sketches below starts with an equilateral triangle having sides of length 1. In the following steps, equilateral triangles are constructed on each side of the preceding figure. The length of the sides of each new triangle is $\frac{1}{3}$ the length of the sides of the preceding triangles. Develop a formula for the number of sides of the nth figure. Use mathematical induction to prove your answer.

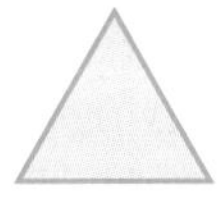   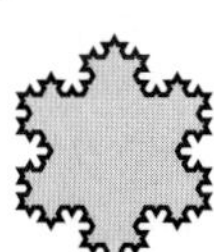

31. *Perimeter* Find the perimeter of the nth figure in **Exercise 30.**

32. *Area* Show that the area of the nth figure in **Exercise 30** is

$$\sqrt{3}\left[\frac{2}{5} - \frac{3}{20}\left(\frac{4}{9}\right)^{n-1}\right].$$

33. *Tower of Hanoi* A pile of n rings, each ring smaller than the one below it, is on a peg. Two other pegs are attached to a board with this peg. In the game called the *Tower of Hanoi* puzzle, all the rings must be moved to a different peg, with only one ring moved at a time, and with no ring ever placed on top of a smaller ring. Find the least number of moves that would be required. Prove your result with mathematical induction.

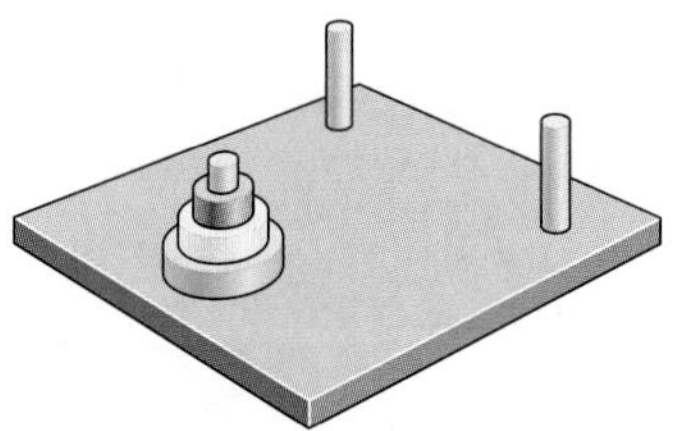

Chapter 11 Quiz (Sections 11.1–11.5)

[11.1–11.3]
1. $-2, -6, -10, -14, -18$; arithmetic
2. $1, -\frac{1}{2}, \frac{1}{4}, -\frac{1}{8}, \frac{1}{16}$; geometric
3. 5, 3, 18, 27, 81; neither

Write the first five terms of each sequence. State whether the sequence is arithmetic, geometric, *or* neither.

1. $a_n = -4n + 2$

2. $a_n = -2\left(-\frac{1}{2}\right)^n$

3. $a_1 = 5, a_2 = 3, a_n = a_{n-1} + 3a_{n-2}$, for $n \geq 3$

4. A certain arithmetic sequence has $a_1 = -6$ and $a_9 = 18$. Find a_7.

5. Find the sum of the first ten terms of each series.

(a) arithmetic, $a_1 = -20$, $d = 14$ (b) geometric, $a_1 = -20$, $r = -\frac{1}{2}$

6. Evaluate each sum that exists.

(a) $\sum_{i=1}^{30}(-3i + 6)$ (b) $\sum_{i=1}^{\infty} 2^i$ (c) $\sum_{i=1}^{\infty}\left(\frac{3}{4}\right)^i$

7. Use the binomial theorem to expand $(x - 3y)^5$.

8. Find the fifth term of the expansion of $\left(4x - \frac{1}{2}y\right)^5$.

9. Evaluate each expression.

(a) $9!$ (b) $\binom{10}{4}$

10. Use mathematical induction to prove that for all positive integers n,

$$6 + 12 + 18 + \cdots + 6n = 3n(n + 1).$$

[11.2]
4. 12

[11.2, 11.3]
5. (a) 430 (b) $-\frac{1705}{128}$
6. (a) -1215
(b) The sum does not exist.
(c) 3

[11.4]
7. $x^5 - 15x^4y + 90x^3y^2 - 270x^2y^3 + 405xy^4 - 243y^5$
8. $\frac{5}{4}xy^4$
9. (a) 362,880 (b) 210

11.6 Counting Theory

- Fundamental Principle of Counting
- Permutations
- Combinations
- Distinguishing between Permutations and Combinations

Fundamental Principle of Counting Consider the following problem.

If there are 3 roads from Albany to Baker and 2 roads from Baker to Creswich, in how many ways can one travel from Albany to Creswich by way of Baker?

For each of the 3 roads from Albany to Baker, there are 2 different roads from Baker to Creswich. Hence, there are $3 \cdot 2 = 6$ different ways to make the trip, as shown in the **tree diagram** in **Figure 13.**

Here, each choice of road is an example of an *event.* Two events are **independent events** if neither influences the outcome of the other. The opening example illustrates the fundamental principle of counting with independent events.

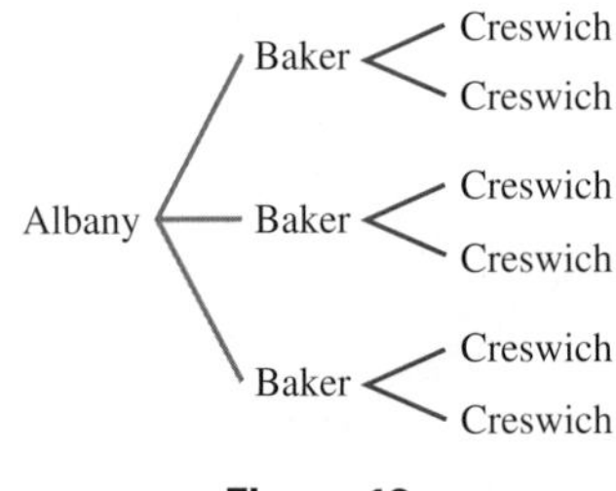

Figure 13

Fundamental Principle of Counting

If n independent events occur, with m_1 ways for event 1 to occur, m_2 ways for event 2 to occur, ... and m_n ways for event n to occur, then there are

$m_1 \cdot m_2 \cdot \cdots \cdot m_n$ different ways for all n events to occur.

Classroom Example 1
Cathy Raftery is shopping for a desktop computer system. She has found 6 computers, 4 monitors, and 3 printers that meet her needs. Find the number of possible computer systems that can be assembled from these choices.

Answer: 72 possible systems

EXAMPLE 1 Using the Fundamental Principle of Counting

A restaurant offers a choice of 3 salads, 5 main dishes, and 2 desserts. Use the fundamental principle of counting to find the number of different 3-course meals that can be selected.

SOLUTION Three events are involved: selecting a salad, selecting a main dish, and selecting a dessert. The first event can occur in 3 ways, the second event can occur in 5 ways, and the third event can occur in 2 ways. Thus, there are

$$3 \cdot 5 \cdot 2 = 30 \text{ possible meals.}$$

Now Try Exercise 1.

Classroom Example 2
Ethel Dorfman has 9 different pictures that she wants to hang in a row on a wall in her new retirement apartment. How many different arrangements of the pictures are possible?

Answer: 362,880 arrangements

EXAMPLE 2 Using the Fundamental Principle of Counting

A teacher has 5 different books that he wishes to arrange in a row. How many different arrangements are possible?

SOLUTION Five events are involved: selecting a book for the first spot, selecting a book for the second spot, and so on. For the first spot the teacher has 5 choices. After a choice has been made, the teacher has 4 choices for the second spot. Continuing in this manner, there are 3 choices for the third spot, 2 for the fourth spot, and 1 for the fifth spot. By the fundamental principle of counting, there are

$$5 \cdot 4 \cdot 3 \cdot 2 \cdot 1 = 120 \text{ different arrangements.}$$

Now Try Exercise 5.

In using the fundamental principle of counting, products such as

$$5 \cdot 4 \cdot 3 \cdot 2 \cdot 1$$

occur often. We use the symbol $n!$ (read **"n-factorial"**), for any counting number n, as follows.

$$\mathbf{n! = n(n-1)(n-2)\cdots(3)(2)(1)} \quad \text{(Section 11.4)}$$

Thus,

$$5 \cdot 4 \cdot 3 \cdot 2 \cdot 1 = 5! \quad \text{and} \quad 3 \cdot 2 \cdot 1 = 3!.$$

By the definition of $n!$, $n[(n-1)!] = n!$ for all natural numbers $n \geq 2$. It is convenient to have this relation hold also for $n = 1$, so, by definition,

$$\mathbf{0! = 1.} \quad \text{(Section 11.4)}$$

Classroom Example 3
In **Classroom Example 2,** suppose that Ethel decides to hang only 6 of the 9 pictures in a row. How many arrangements are possible?

Answer: 60,480 arrangements

EXAMPLE 3 Arranging r of n Items ($r < n$)

Suppose the teacher in **Example 2** wishes to place only 3 of the 5 books in a row. How many arrangements of 3 books are possible?

SOLUTION The teacher still has 5 ways to fill the first spot, 4 ways to fill the second spot, and 3 ways to fill the third. Since only 3 books will be used, there are only 3 spots to be filled (3 events) instead of 5, with

$$5 \cdot 4 \cdot 3 = 60 \text{ arrangements.}$$

Now Try Exercise 7.

Permutations Since each ordering of three books is considered a different *arrangement,* the number 60 in the preceding example is called the number of *permutations* of 5 things taken 3 at a time, written $P(5, 3) = 60$. The number of ways of arranging 5 elements from a set of 5 elements, written $P(5, 5) = 120$, was found in **Example 2.**

A **permutation** of n elements taken r at a time is one of the *arrangements* of r elements from a set of n elements. Generalizing from the examples above, the number of permutations of n elements taken r at a time, denoted by $P(n, r)$, is

$$\begin{aligned} P(n, r) &= n(n-1)(n-2)\cdots(n-r+1) \\ &= \frac{n(n-1)(n-2)\cdots(n-r+1)(n-r)(n-r-1)\cdots(2)(1)}{(n-r)(n-r-1)\cdots(2)(1)} \\ &= \frac{n!}{(n-r)!}. \end{aligned}$$

Teaching Tip Point out that $P(n, r)$ is the product of the first r factors in the descending expansion of n-factorial.

Permutations of n Elements Taken r at a Time

If $P(n, r)$ denotes the number of permutations of n elements taken r at a time, with $r \leq n$, then the following holds.

$$P(n, r) = \frac{n!}{(n - r)!}$$

Alternative notations for $P(n, r)$ are P_r^n and ${}_nP_r$.

Classroom Example 4
Evaluate.

(a) The number of permutations of the digits 0, 2, 4, 6, and 8

(b) The number of permutations of 4 of the digits 0, 2, 4, 6, and 8

Answers:
(a) 120 **(b)** 120

EXAMPLE 4 Using the Permutations Formula

Evaluate.

(a) The number of permutations of the letters L, M, and N

(b) The number of permutations of 2 of the letters L, M, and N

SOLUTION

(a) Use the formula for $P(n, r)$, with $n = 3$ and $r = 3$.

$$P(3, 3) = \frac{3!}{(3 - 3)!} = \frac{3!}{0!} = \frac{3 \cdot 2 \cdot 1}{1} = 6$$

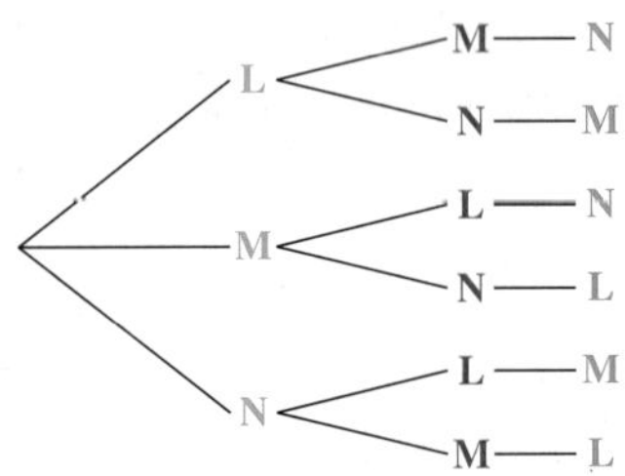

Figure 14

As shown in the tree diagram in **Figure 14,** the 6 permutations are

LMN, LNM, MLN, MNL, NLM, NML.

(b) Evaluate $P(3, 2)$.

$$P(3, 2) = \frac{3!}{(3 - 2)!} = \frac{3!}{1!} = \frac{3 \cdot 2 \cdot 1}{1} = 6$$

This result is the same as the answer in part (a). After the first two choices are made, the third is already determined since only one letter is left.

✔ Now Try Exercise 45.

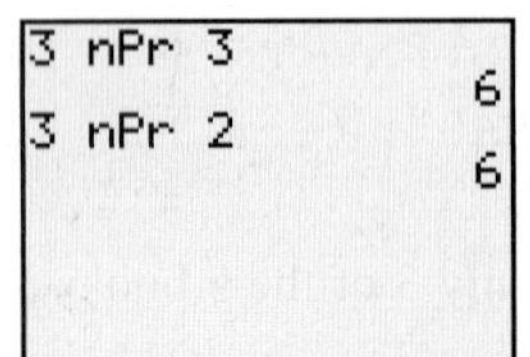

This screen shows how the TI-83/84 Plus calculates $P(3, 3)$ and $P(3, 2)$. **See Example 4.**

Classroom Example 5
A club with 25 members is electing 4 officers: a president, a vice-president, a secretary, and a treasurer. In how many different ways can the 4 officers be chosen?

Answer: 303,600 ways

EXAMPLE 5 Using the Permutations Formula

Suppose 8 people enter an event in a swim meet. In how many ways could the gold, silver, and bronze medals be awarded?

SOLUTION Using the fundamental principle of counting, there are 3 events, giving

$$8 \cdot 7 \cdot 6 = 336 \text{ ways.}$$

We can also use the formula for $P(n, r)$ to obtain the same result.

$$P(8, 3) = \frac{8!}{(8 - 3)!} = \frac{8!}{5!} \quad \text{Formula for permutations}$$

$$= \frac{8 \cdot 7 \cdot 6 \cdot 5 \cdot 4 \cdot 3 \cdot 2 \cdot 1}{5 \cdot 4 \cdot 3 \cdot 2 \cdot 1} \quad \text{Expand the factorials.}$$

$$= 8 \cdot 7 \cdot 6 \quad \text{Divide out common factors.}$$

$$= 336 \text{ ways} \quad \text{Multiply.}$$

✔ Now Try Exercise 43.

Classroom Example 6
In how many ways can a company install 10 different billboards in the 10 locations that the company has rented along a road?

Answer: 3,628,800 ways

EXAMPLE 6 Using the Permutations Formula

In how many ways can 6 students be seated in a row of 6 desks?

SOLUTION Use $P(n, r)$ with $n = 6$ and $r = 6$.

$$P(6, 6) = 6! = 6 \cdot 5 \cdot 4 \cdot 3 \cdot 2 \cdot 1 = 720 \text{ ways}$$

✔ Now Try Exercise 39.

Combinations In **Example 3** we saw that there are 60 ways in which a teacher can arrange 3 of 5 different books in a row. That is, there are 60 permutations of 5 things taken 3 at a time. Suppose now that the teacher does not wish to arrange the books in a row but rather wishes to choose, without regard to order, any 3 of the 5 books to donate to a book sale. In how many ways can the teacher do this?

The number 60 counts all possible *arrangements* of 3 books chosen from 5. The following 6 arrangements, however, would all lead to the same set of 3 books being given to the book sale.

mystery-biography-textbook	biography-textbook-mystery
mystery-textbook-biography	textbook-biography-mystery
biography-mystery-textbook	textbook-mystery-biography

The list shows 6 different *arrangements* of 3 books but only one *set* of 3 books. A subset of items selected *without regard to order* is a **combination.** The number of combinations of 5 things taken 3 at a time is written $\binom{5}{3}$, ${}_5C_3$, or $C(5, 3)$.

To evaluate $\binom{5}{3}$, or $C(5, 3)$, start with the $5 \cdot 4 \cdot 3$ *permutations* of 5 things taken 3 at a time. Since order does not matter, and each subset of 3 items from the set of 5 items can have its elements rearranged in $3 \cdot 2 \cdot 1 = 3!$ ways, we find $\binom{5}{3}$ by dividing the number of permutations by 3!.

$$\binom{5}{3} = \frac{5 \cdot 4 \cdot 3}{3!} = \frac{5 \cdot 4 \cdot 3}{3 \cdot 2 \cdot 1} = 10$$

The teacher can choose 3 books for the book sale in 10 ways.

Generalizing this discussion gives the following formula for the number of combinations of n elements taken r at a time:

$$C(n, r) = \binom{n}{r} = \frac{P(n, r)}{r!}.$$

An alternative version of this formula is found as follows.

$$C(n, r) = \binom{n}{r} = \frac{P(n, r)}{r!} = \frac{n!}{(n-r)!} \cdot \frac{1}{r!} = \frac{n!}{(n-r)!\,r!}$$

This version is most useful for calculation and is the one we used earlier to calculate binomial coefficients.

Combinations of *n* Elements Taken *r* at a Time

If $C(n, r)$, or $\binom{n}{r}$, represents the number of combinations of n elements taken r at a time, with $r \leq n$, then the following holds.

$$C(n, r) = \binom{n}{r} = \frac{n!}{(n-r)!\,r!}, \quad \text{or} \quad \frac{n!}{r!\,(n-r)!}$$

Classroom Example 7
A college club with 22 members needs to choose 4 members to attend a regional conference. In how many ways can this be done?

Answer: 7315 ways

NOTE The formula for $C(n, r)$ given on the preceding page is equivalent to the binomial coefficient formula given in **Section 11.4.**

EXAMPLE 7 Using the Combinations Formula

How many different committees of 3 people can be chosen from a group of 8 people?

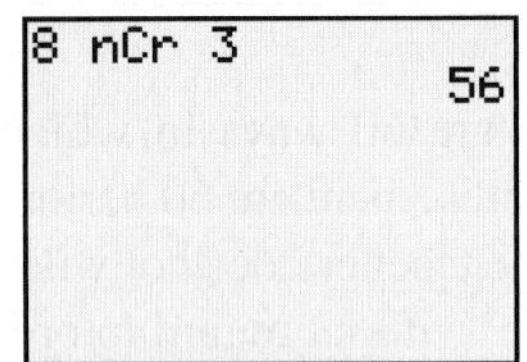

This screen shows how the TI-83/84 Plus calculates $C(8, 3)$. **See Example 7.**

SOLUTION Since a committee is an unordered set, use combinations with $n = 8$ and $r = 3$.

$$C(8, 3) = \binom{8}{3} = \frac{8!}{3!5!} = \frac{8 \cdot 7 \cdot 6 \cdot 5 \cdot 4 \cdot 3 \cdot 2 \cdot 1}{3 \cdot 2 \cdot 1 \cdot 5 \cdot 4 \cdot 3 \cdot 2 \cdot 1} = 56 \text{ committees}$$

✔ *Now Try Exercise 47.*

Classroom Example 8
A local business employs 10 surveyors and 18 engineers. Five employees are to be selected to serve on a committee.

(a) In how many different ways can the committee members be selected?

(b) In how many ways can the committee members be selected if 3 must be surveyors and 2 must be engineers?

Answers:
(a) 98,280 ways
(b) 18,360 ways

EXAMPLE 8 Using the Combinations Formula

A group of stockbrokers consists of 11 women and 19 men. Four will be selected to work on a special project.

(a) In how many different ways can the stockbrokers be selected?

(b) In how many ways can the group of 4 be selected if 2 must be women and 2 must be men?

SOLUTION

(a) Here we wish to know the number of 4-element combinations that can be formed from a set of $11 + 19 = 30$ elements. (We want combinations, not permutations, since order within the group does not matter.)

$$C(30, 4) = \binom{30}{4} = \frac{30!}{4!26!} = 27{,}405$$

There are 27,405 ways to select the project group.

(b) Since order is not important, we use combinations to select 2 of the 11 women and 2 of the 19 men. As a last step, we apply the fundamental principle of counting to multiply these two outcomes.

$$C(11, 2) \cdot C(19, 2) = \frac{11!}{2!9!} \cdot \frac{19!}{2!17!} \quad \text{Use combinations and the fundamental principle of counting.}$$

$$= 55 \cdot 171 \quad \text{Evaluate.}$$

$$= 9405 \quad \text{Multiply.}$$

In this case, the project group can be selected in 9405 ways.

✔ *Now Try Exercise 55.*

Distinguishing between Permutations and Combinations Students often have difficulty determining whether to use permutations or combinations in solving problems. The chart on the next page lists some of the similarities and differences between these two concepts.

These are combinations. The order of the cards in the hands is *not* important.

Permutations	Combinations
These are selections of r items from n items.	
Repetitions are not allowed.	
Order is important.	Order is not important.
These are arrangements of r items from a set of n items.	These are subsets of r items from a set of n items.
$P(n, r) = \frac{n!}{(n-r)!}$	$C(n, r) = \binom{n}{r} = \frac{n!}{r!(n-r)!}$
Clue words: arrangement, schedule, order	Clue words: group, committee, sample, selection

Classroom Example 9
Should permutations or combinations be used to solve each problem?

(a) In how many ways can 5 oranges be chosen from a bin containing 20 oranges?

(b) A Web site requires each user to select a five-letter password in which no letter may be repeated. How many such passwords are available?

(c) In how many ways can 15 employees be assigned to work in 15 cubicles?

(d) In a contest, 3 winners are to be selected at random from among 100 entries. In how many ways can this be done?

Answers:
(a) combinations
(b) permutations
(c) permutations
(d) combinations

EXAMPLE 9 Distinguishing between Permutations and Combinations

Should permutations or combinations be used to solve each problem?

(a) How many 4-digit codes are possible if no digits are repeated?

(b) A sample of 4 light bulbs is randomly selected from a batch of 15 bulbs to be packaged and sold. How many different samples are possible?

(c) In a basketball tournament with 8 teams, how many games must be played so that each team plays every other team exactly once?

(d) In how many ways can 4 stockbrokers be assigned to 6 offices so that each broker has a private office?

SOLUTION

(a) Since changing the order of the 4 digits results in a different code, permutations should be used.

(b) The order in which the 4 light bulbs are selected is not important. The sample is unchanged if the items are rearranged, so combinations should be used.

(c) Selection of 2 teams for a game creates an *unordered* subset of 2 from the set of 8 teams. Use combinations.

(d) The office assignments are an *ordered* selection of 4 offices from the 6 offices. Exchanging the offices of any 2 brokers within a selection of 4 offices gives a different assignment, so permutations should be used.

Now Try Exercise 29.

To illustrate the distinctions between permutations and combinations using tree diagrams, suppose we want to select 2 cans of soup from 4 cans on a shelf:

noodle (N), bean (B), mushroom (M), and tomato (T).

As shown in **Figure 15(a)** on the next page, there are 12 ways to select 2 cans from the 4 cans if order matters (if noodle first and bean second is considered different from bean, then noodle, for example). On the other hand, if order is unimportant, then there are 6 ways to choose 2 cans of soup from the 4 cans, as illustrated in **Figure 15(b).**

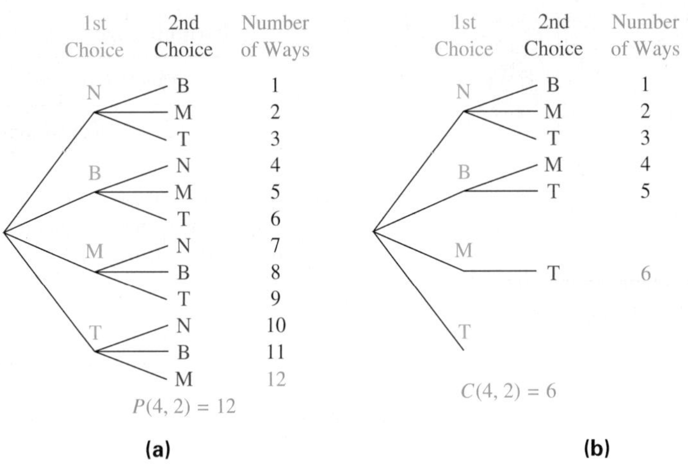

Figure 15

CAUTION Not all counting problems lend themselves to either permutations or combinations. Whenever the fundamental principle of counting or a tree diagram can be used directly, as in the soup example, use it.

11.6 Exercises

1. 24
2. 12
3. 160
4. 128
5. 5040
6. 10,000
7. 3360
8. 303,600
9. 19,958,400
10. 120
11. 72
12. 3,628,800
13. 5
14. 1
15. 6
16. 84
17. 1
18. 8
19. 495
20. 560

Fundamental Counting Principle Use the fundamental principle of counting to solve each problem. ***See Examples 1–3.***

1. On a business trip, Terry took 3 pairs of pants, 4 shirts, 1 jacket, and two pairs of shoes. Determine the number of outfits that Terry can choose.

2. When saddling her horse, Judy can choose from 2 saddles, 3 blankets, and 2 cinches. Find the number of possible choices for saddling Judy's horse.

3. A conference schedule offers 2 main sessions, 20 break-out sessions, and 4 mini-courses. In how many ways can an attendee choose 1 of each to attend?

4. A convenience store offers 16 types of soda with 4 options for flavoring and either crushed or cubed ice. Determine the total number of drink options available for selecting 1 soda with 1 flavor and 1 type of ice.

5. A college has 7 portraits of past college presidents to arrange in a row on a wall. How many different arrangements are possible?

6. A telephone messaging system requires a 4-digit security code. How many security codes are possible if numbers may be repeated.

7. In how many ways can judges select a 1st-place winner, a 2nd-place winner, and a 3rd-place winner from 16 desserts entered in a cooking contest?

8. In how many different ways can 4 different boys be selected from a group of 25 boys on a track team to receive 4 different awards?

Evaluate each expression. ***See Examples 4–8.***

9. $P(12, 8)$ **10.** $P(5, 5)$ **11.** $P(9, 2)$ **12.** $P(10, 9)$

13. $P(5, 1)$ **14.** $P(6, 0)$ **15.** $C(4, 2)$ **16.** $C(9, 3)$

17. $C(6, 0)$ **18.** $C(8, 1)$ **19.** $\binom{12}{4}$ **20.** $\binom{16}{3}$

21. 1,860,480
22. 9,034,502,400
23. 259,459,200
24. 863,040
25. 15,504
26. 75,287,520
27. 6435
28. 35,960

29. (a) permutation (b) permutation (c) combination (d) combination (e) permutation (f) permutation (g) combination

31. 40
32. 840
33. (a) 27,600 (b) 35,152 (c) 1104
34. 120
35. 5040
36. 15
37. (a) 17,576,000 (b) 17,576,000 (c) 456,976,000

Use a calculator to evaluate each expression. ***See Examples 4 and 7.***

21. $_{20}P_5$ **22.** $_{100}P_5$ **23.** $_{15}P_8$ **24.** $_{32}P_4$

25. $_{20}C_5$ **26.** $_{100}C_5$ **27.** $\binom{15}{8}$ **28.** $\binom{32}{4}$

29. Decide whether the situation described involves a permutation or a combination of objects. **See Example 9.**

(a) a telephone number **(b)** a Social Security number
(c) a hand of cards in poker **(d)** a committee of politicians
(e) the "combination" on a padlock **(f)** an automobile license plate
(g) a lottery choice of six numbers where order does not matter

30. Explain the difference between a permutation and a combination. What should you look for in a problem to decide which is an appropriate method of solution?

Use the fundamental principle of counting or permutations to solve each problem. ***See Examples 1–6.***

31. *Home Plan Choices* How many different types of homes are available if a builder offers a choice of 5 basic plans, 4 roof styles, and 2 exterior finishes?

32. *Auto Varieties* An auto manufacturer produces 7 models, each available in 6 different colors, with 4 different upholstery fabrics, and 5 interior colors. How many varieties of the auto are available?

33. *Radio-Station Call Letters* How many different 4-letter radio-station call letters can be made

(a) if the first letter must be K or W and no letter may be repeated?
(b) if repetitions are allowed (but the first letter is K or W)?
(c) if the first letter must be K or W, the last letter must be R, and repetitions are not allowed?

34. *Meal Choices* A menu offers a choice of 3 salads, 8 main dishes, and 5 desserts. How many different 3-course meals (salad, main dish, dessert) are possible?

35. *Arranging Blocks* Baby Finley is arranging 7 blocks in a row. How many different arrangements can he make?

36. *Names for a Baby* A couple has narrowed down the choice of a name for their new baby to 5 first names and 3 middle names. How many different first- and middle-name combinations are possible?

37. *License Plates* For many years, the state of California used 3 letters followed by 3 digits on its automobile license plates.

(a) How many different license plates are possible with this arrangement?
(b) When the state ran out of new plates, the order was reversed to 3 digits followed by 3 letters. How many additional plates were then possible?
(c) When the plates described in part (b) were also used up, the state then issued plates with 1 letter followed by 3 digits and then 3 letters. How many plates does this scheme provide?

38. (a) 78,125 (b) 900,000 (c) 90,000 (d) 10,000 (e) 544,320
39. 362,880
40. 604,800
41. 120
42. 20,523,714,120
43. 2730
44. 60,949,324,800
45. (a) 120 (b) 6
46. 120; 30,240
47. 3,838,380
48. 495; 495
49. 12,650
50. (a) 10 (b) 950
51. (a) 15 (b) 10
52. 10
53. 105; 1365

38. *Telephone Numbers* How many 7-digit telephone numbers are possible if the first digit cannot be 0 and

(a) only odd digits may be used?
(b) the telephone number must be a multiple of 10 (that is, it must end in 0)?
(c) the telephone number must be a multiple of 100?
(d) the first 3 digits are 481?
(e) no repetitions are allowed?

39. *Seating People in a Row* In an experiment on social interaction, 9 people will sit in 9 seats in a row. In how many ways can this be done?

40. *Genetics Experiment* In how many ways can 7 of 10 rats be arranged in a row for a genetics experiment?

41. *Course Schedule Arrangement* A business school offers courses in keyboarding, spreadsheets, transcription, business English, technical writing, and accounting. In how many ways can a student arrange a schedule if 3 courses are taken?

42. *Course Schedule Arrangement* If your college offers 400 courses, 20 of which are in mathematics, and your counselor arranges your schedule of 4 courses by random selection, how many schedules are possible that do not include a math course?

43. *Club Officer Choices* In a club with 15 members, in how many ways can a slate of 3 officers consisting of president, vice-president, and secretary/treasurer be chosen?

44. *Batting Orders* A baseball team has 20 players. How many 9-player batting orders are possible?

45. *Letter Arrangement* Consider the word BRUCE.

(a) In how many ways can all the letters of the word BRUCE be arranged?
(b) In how many ways can the first 3 letters of the word BRUCE be arranged?

46. *Basketball Positions* In how many ways can 5 players be assigned to the 5 positions on a basketball team, assuming that any player can play any position? In how many ways can 10 players be assigned to the 5 positions?

Solve each problem involving combinations. ***See Examples 7 and 8.***

47. *Seminar Presenters* A banker's association has 40 members. If 6 members are selected at random to present a seminar, how many different groups of 6 are possible?

48. *Financial Planners* Four financial planners are to be selected from a group of 12 to participate in a special program. In how many ways can this be done? In how many ways can the group that will not participate be selected?

49. *Apple Samples* How many different samples of 4 apples can be drawn from a crate of 25 apples?

50. *Apple Samples* Suppose in **Exercise 49** that there are 5 rotten apples in the crate.

(a) How many samples of 3 could be drawn in which all 3 are rotten?
(b) How many samples of 3 could be drawn in which there are 2 good apples and 1 rotten apple?

51. *Hamburger Choices* Howard's Hamburger Heaven sells hamburgers with cheese, relish, lettuce, tomato, mustard, or ketchup.

(a) How many different hamburgers can be made that use any 4 of the extras?
(b) How many different hamburgers can be made if one of the 4 extras must be cheese?

52. *Card Combinations* Five cards are marked with the numbers 1, 2, 3, 4, and 5, shuffled, and 2 cards are then drawn. How many different 2-card hands are possible?

53. *Marble Samples* If a bag contains 15 marbles, how many samples of 2 marbles can be drawn from it? How many samples of 4 marbles can be drawn?

54. 28
55. (a) 84 (b) 10
(c) 40 (d) 28
56. (a) 21 (b) 6
(c) 11
57. 1680
58. 220
59. 15
60. 210
61. 6,227,020,800
62. (a) 330 (b) 150
63. (a) 56 (b) 462
(c) 3080 (d) 8526
64. 5040
65. 1,000,000

54. *Marble Samples* In **Exercise 53,** if the bag contains 3 yellow, 4 white, and 8 blue marbles, how many samples of 2 can be drawn in which both marbles are blue?

55. *Convention Delegation Choices* A city council is composed of 5 liberals and 4 conservatives. Three members are to be selected randomly as delegates to an urban convention.

(a) How many delegations are possible?
(b) How many delegations could have all liberals?
(c) How many delegations could have 2 liberals and 1 conservative?
(d) If 1 member of the council serves as mayor, how many delegations are possible that include the mayor?

56. *Delegation Choices* Seven workers decide to send a delegation of 2 to their supervisor to discuss their grievances.

(a) How many different delegations are possible?
(b) If it is decided that a certain employee must be in the delegation, how many different delegations are possible?
(c) If there are 2 women and 5 men in the group, how many delegations would include at least 1 woman?

Use any or all of the methods described in this section to solve each problem. ***See Examples 1–9.***

57. *Course Schedule Arrangement* If Dwight Johnston has 8 courses to choose from, how many ways can he arrange his schedule if he must pick 4 of them?

58. *Pineapple Samples* How many samples of 9 pineapples can be drawn from a crate of 12?

59. *Soup Ingredients* Velma specializes in making different vegetable soups with carrots, celery, beans, peas, mushrooms, and potatoes. How many different soups can she make with any 4 ingredients?

60. *Secretary/Manager Assignments* From a pool of 7 secretaries, 3 are selected to be assigned to 3 managers, 1 secretary to each manager. In how many ways can this be done?

61. *Musical Chairs Seatings* In a game of musical chairs, 13 children will sit in 12 chairs. (1 will be left out.) How many seating arrangements are possible?

62. *Plant Samples* In an experiment on plant hardiness, a researcher gathers 6 wheat plants, 3 barley plants, and 2 rye plants. She wishes to select 4 plants at random.

(a) In how many ways can this be done?
(b) In how many ways can this be done if exactly 2 wheat plants must be included?

63. *Committee Choices* In a club with 8 women and 11 men members, how many 5-member committees can be chosen that have the following?

(a) all women **(b)** all men
(c) 3 women and 2 men **(d)** no more than 3 men

64. *Committee Choices* From 10 names on a ballot, 4 will be elected to a political party committee. In how many ways can the committee of 4 be formed if each person will have a different responsibility?

65. *Combination Lock* A briefcase has 2 locks. The combination to each lock consists of a 3-digit number, where digits may be repeated. How many combinations are possible? (*Hint:* The word *combination* is a misnomer. Lock combinations are permutations where the arrangement of the numbers is important.)

66. *Combination Lock* A typical "combination" for a padlock consists of 3 numbers from 0 to 39. Find the number of "combinations" that are possible with this type of lock, if a number may be repeated.

67. *Garage Door Openers* The code for some garage door openers consists of 12 electrical switches that can be set to either 0 or 1 by the owner. With this type of opener, how many codes are possible? (*Source:* Promax.)

68. *Lottery* To win the jackpot in a lottery game, a person must pick 4 numbers from 0 to 9 in the correct order. If a number can be repeated, how many ways are there to play the game?

69. *Keys* In how many distinguishable ways can 4 keys be put on a circular key ring?

70. *Sitting at a Round Table* In how many different ways can 8 people sit at a round table? Assume that "a different way" means that at least 1 person is sitting next to someone different.

66. 64,000
67. 4096
68. 10,000
69. 6
70. 5040

Prove each statement for positive integers n and r, with $r \leq n$. (Hint: Use the definitions of permutations and combinations.)

71. $P(n, n-1) = P(n, n)$ **72.** $P(n, 1) = n$ **73.** $P(n, 0) = 1$

74. $P(n, n) = n!$ **75.** $\binom{n}{n} = 1$ **76.** $\binom{n}{0} = 1$

77. $\binom{0}{0} = 1$ **78.** $\binom{n}{n-1} = n$ **79.** $\binom{n}{n-r} = \binom{n}{r}$

80. Explain why the restriction $r \leq n$ is needed in the formula for $P(n, r)$.

11.7 Basics of Probability

- Basic Concepts
- Complements and Venn Diagrams
- Odds
- Union of Two Events
- Binomial Probability

Basic Concepts Consider an experiment that has one or more possible **outcomes,** all of which are equally likely to occur. For example, the experiment of tossing a fair coin has 2 equally likely possible outcomes: landing heads up (H) or landing tails up (T). Also, the experiment of rolling a fair die has 6 equally likely outcomes: landing so the face that is up shows 1, 2, 3, 4, 5, or 6 dots.

The set S of all possible outcomes of a given experiment is the **sample space** for the experiment. (In this text, all sample spaces are finite.) The table gives several examples.

Experiment	Sample Space
Toss a coin.	$S = \{H, T\}$
Roll a die.	$S = \{1, 2, 3, 4, 5, 6\}$
Toss two coins.	$S = \{(H, H), (H, T), (T, H), (T, T)\}$
Answer a true/false question.	$S = \{\text{true}, \text{false}\}$

Use set notation. (Section R.1)

Any subset of the sample space is an **event.** In the experiment with the die, for example, "the number showing is a 3" is an event, say E_1, such that $E_1 = \{3\}$. "The number showing is greater than 3" is also an event, say E_2, such that $E_2 = \{4, 5, 6\}$. To represent the number of outcomes that belong to event E, the notation $n(E)$ is used. Then $n(E_1) = 1$ and $n(E_2) = 3$.

Teaching Tip For a review of sets and set operations, refer students to **Section R.1.**

The notation $P(E)$ is used for the *probability* of an event E. If the outcomes in the sample space for an experiment are equally likely, then the probability of event E occurring is found as follows.

Probability of Event *E*

In a sample space with equally likely outcomes, the **probability** of an event E, written $\mathbf{P(E)}$, is the ratio of the number of outcomes in sample space S that belong to event E, $n(E)$, to the total number of outcomes in sample space S, $n(S)$.

$$P(E) = \frac{n(E)}{n(S)}$$

To use this definition to find the probability of the event E_1 in the die experiment, start with the sample space, $S = \{1, 2, 3, 4, 5, 6\}$, and the desired event, $E_1 = \{3\}$. Use $n(E_1) = 1$ and $n(S) = 6$.

$$P(E_1) = \frac{n(E_1)}{n(S)} = \frac{1}{6}$$

Classroom Example 1
Repeat **Example 1** for the following events.

(a) E_1: the number showing is a multiple of 3
(b) E_2: the number showing is less than 6
(c) E_3: the number showing is 0
(d) E_4: the number showing is greater than 0

Answers:
(a) $E_1 = \{3, 6\}$; $P(E_1) = \frac{1}{3}$
(b) $E_2 = \{1, 2, 3, 4, 5\}$; $P(E_2) = \frac{5}{6}$
(c) $E_3 = \emptyset$; $P(E_3) = 0$
(d) $E_4 = \{1, 2, 3, 4, 5, 6\}$; $P(E_4) = 1$

EXAMPLE 1 Finding Probabilities of Events

A single die is rolled. Write each event in set notation and give the probability of the event.

(a) E_3: the number showing is even

(b) E_4: the number showing is greater than 4

(c) E_5: the number showing is less than 7

(d) E_6: the number showing is 7

SOLUTION

(a) Since $E_3 = \{2, 4, 6\}$, $n(E_3) = 3$. As given earlier, $n(S) = 6$.

$$P(E_3) = \frac{3}{6} = \frac{1}{2}$$

(b) Again $n(S) = 6$. Event $E_4 = \{5, 6\}$, with $n(E_4) = 2$.

$$P(E_4) = \frac{2}{6} = \frac{1}{3}$$

(c) $E_5 = \{1, 2, 3, 4, 5, 6\}$ and $P(E_5) = \frac{6}{6} = 1.$

(d) $E_6 = \emptyset$ and $P(E_6) = \frac{0}{6} = 0.$

✔ *Now Try Exercises 7 and 9.*

In **Example 1(c),** $E_5 = S$. Therefore, the event E_5 is certain to occur every time the experiment is performed. An event that is certain to occur always has probability 1. In **Example 1(d),** $E_6 = \emptyset$ and $P(E_6) = 0$. The probability of an impossible event, such as E_6, is always 0, since none of the outcomes in the sample space satisfies the event. ***For any event E, P(E) is between 0 and 1, inclusive.***

Complements and Venn Diagrams The set of all outcomes in the sample space that do *not* belong to event E is the **complement** of E, written E'. For example, in the experiment of drawing a single card from a standard deck of 52 cards, let E be the event "the card is an ace." Then E' is the event "the card is not an ace." From the definition of E', for an event E,

$$E \cup E' = S \quad \text{and} \quad E \cap E' = \emptyset.^*$$

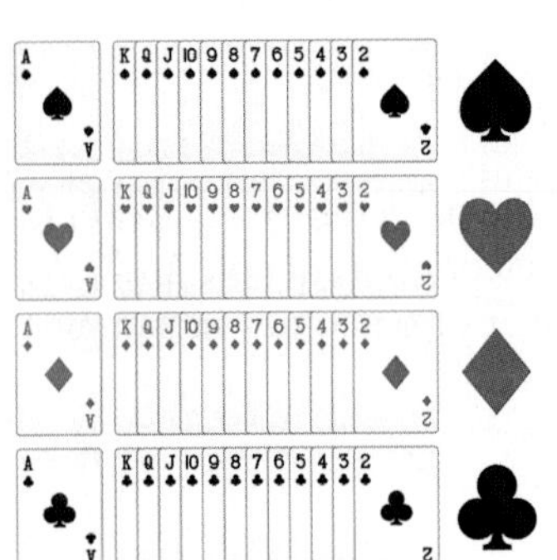

NOTE A standard deck of 52 cards has four suits: hearts ♥, diamonds ♦, spades ♠, and clubs ♣, with 13 cards of each suit. Each suit has a jack, a queen, and a king (sometimes called the "face cards"), an ace, and cards numbered from 2 to 10. The hearts and diamonds are red, and the spades and clubs are black. We will refer to this standard deck of cards in this section.

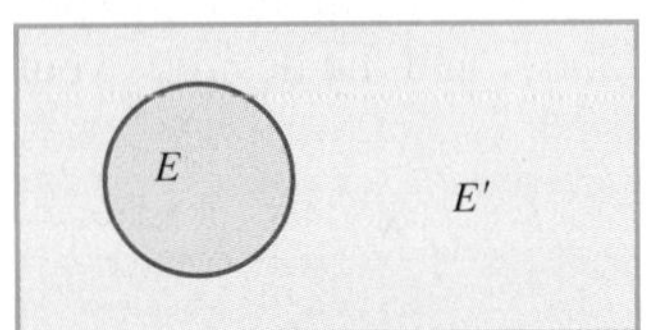

Figure 16

Probability concepts can be illustrated using **Venn diagrams,** as shown in **Figure 16.** The rectangle there represents the sample space in an experiment. The area inside the circle represents event E, and the area inside the rectangle, but outside the circle, represents event E'.

EXAMPLE 2 Using the Complement

In the experiment of drawing a card from a well-shuffled deck, find the probabilities of event E, "the card is an ace," and of event E'.

Classroom Example 2
In the experiment of drawing a card from a well-shuffled deck, find the probabilities of event E, "the card is a heart," and of event E'.

Answer: $P(E) = \frac{1}{4}$; $P(E') = \frac{3}{4}$

SOLUTION Since there are 4 aces in the deck of 52 cards, $n(E) = 4$ and $n(S) = 52$.

$$P(E) = \frac{n(E)}{n(S)} \qquad \text{Definition of probability}$$

$$= \frac{4}{52}, \text{ or } \frac{1}{13} \qquad \text{Write in lowest terms.}$$

Of the 52 cards, 48 are not aces.

$$P(E') = \frac{n(E')}{n(S)} \qquad \text{Definition of probability}$$

$$= \frac{48}{52}, \text{ or } \frac{12}{13} \qquad \text{Write in lowest terms.}$$

✔ *Now Try Exercises 17(a) and (b).*

In **Example 2,** $P(E) + P(E') = \frac{1}{13} + \frac{12}{13} = 1$. This is always true for any event E and its complement E'.

Teaching Tip Mention that the formula $P(E) = 1 - P(E')$ is frequently used when the number of outcomes that belong to event E' is easier to calculate than the number of outcomes that belong to event E.

Rule for Complementary Events

If events E and E' are complements, then all of the following are true.

$$P(E) + P(E') = 1 \qquad P(E) = 1 - P(E') \qquad P(E') = 1 - P(E)$$

---

*The **union** of two sets A and B is the set $A \cup B$ of all elements from either A or B, or both. The **intersection** of sets A and B, written $A \cap B$, includes all elements that belong to both sets. (Section R.1)

These equations suggest an alternative way to compute the probability of an event. For example, if it is known that $P(E) = \frac{1}{13}$, then

$$P(E') = 1 - \frac{1}{13} = \frac{12}{13}.$$

Odds Sometimes probability statements are expressed in terms of odds, a comparison of $P(E)$ with $P(E')$. The **odds** in favor of an event E are expressed as the ratio of $P(E)$ to $P(E')$, or as the quotient $\frac{P(E)}{P(E')}$. For example, if the probability of rain can be established as $\frac{1}{3}$, the odds that it will rain are

$$P(\text{rain}) \text{ to } P(\text{no rain}) = \frac{1}{3} \text{ to } \frac{2}{3} = \frac{\frac{1}{3}}{\frac{2}{3}} = \frac{1}{2}, \quad \text{or} \quad 1 \text{ to } 2.$$

On the other hand, the odds that it will not rain are 2 to 1 $\left(\text{or } \frac{2}{3} \text{ to } \frac{1}{3}\right)$. If the odds in favor of an event are, say, 3 to 5, then the probability of the event is $\frac{3}{8}$, and the probability of the complement of the event is $\frac{5}{8}$.

Rules for Odds

If m represents the number of outcomes in event E and n represents the number of outcomes in event E', then the following are true.

$$P(E) = \frac{m}{m+n} \quad \text{and} \quad P(E') = \frac{n}{m+n}$$

The odds in favor of event E are $\frac{P(E)}{P(E')} = \frac{m}{n}$, or m **to** n.

The odds against event E are $\frac{P(E')}{P(E)} = \frac{n}{m}$, or n **to** m.

Classroom Example 3
A marble is drawn at random from a bowl containing 6 green, 4 blue, and 8 yellow marbles. Find the odds in favor of a yellow marble being drawn.

Answer: $\frac{4}{5}$, or 4 to 5

EXAMPLE 3 Finding Odds in Favor of an Event

A shirt is selected at random from a dark closet containing 6 blue shirts and 4 shirts that are not blue. Find the odds in favor of a blue shirt being selected.

SOLUTION Let E represent "a blue shirt is selected." Then

$$P(E) = \frac{6}{6+4} = \frac{6}{10}, \quad \text{or} \quad \frac{3}{5} \qquad \text{and} \qquad P(E') = \frac{4}{6+4} = \frac{4}{10}, \quad \text{or} \quad \frac{2}{5}.$$

Therefore, the odds in favor of a blue shirt being selected are

$$\frac{P(E)}{P(E')} = \frac{\frac{3}{5}}{\frac{2}{5}} \qquad \text{Formula for odds}$$

$$= \frac{3}{5} \cdot \frac{5}{2} \qquad \text{Division of fractions (Section R.5)}$$

$$= \frac{3}{2}, \quad \text{or} \quad 3 \text{ to } 2.$$

Since the odds in favor of selecting a blue shirt are 3 to 2, if desired we can quickly determine that the odds *against* selecting a blue shirt are 2 to 3.

Now Try Exercise 17(e).

Union of Two Events Suppose a fair die is rolled. Let H be the event "the result is a 3," and K the event "the result is an even number." From earlier in this section,

$$H = \{3\} \qquad K = \{2, 4, 6\} \qquad H \cup K = \{2, 3, 4, 6\}$$

$$P(H) = \frac{1}{6} \qquad P(K) = \frac{3}{6} = \frac{1}{2} \qquad P(H \cup K) = \frac{4}{6} = \frac{2}{3}.$$

Notice that $P(H) + P(K) = P(H \cup K)$.

Consider another event G for this experiment, "the result is a 2."

$$G = \{2\} \qquad K = \{2, 4, 6\} \qquad G \cup K = \{2, 4, 6\}$$

$$P(G) = \frac{1}{6} \qquad P(K) = \frac{3}{6} = \frac{1}{2} \qquad P(G \cup K) = \frac{3}{6} = \frac{1}{2}$$

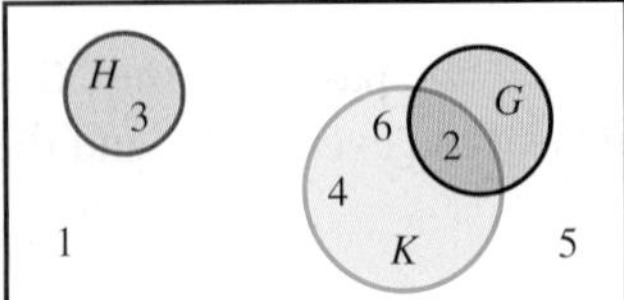

Figure 17

In this case, $P(G) + P(K) \neq P(G \cup K)$. As **Figure 17** suggests, the difference in the two examples above comes from the fact that events H and K cannot occur simultaneously. Such events are **mutually exclusive events.** In fact, $H \cap K = \emptyset$, which is always true for mutually exclusive events. Events G and K, however, can occur simultaneously. Both are satisfied if the result of the roll is a 2, the element in their intersection $(G \cap K = \{2\})$. This example suggests the following property.

Probability of the Union of Two Events

For any events E and F, the following holds.

$$P(E \text{ or } F) = P(E \cup F) = P(E) + P(F) - P(E \cap F)$$

Classroom Example 4
Repeat **Example 4** for the following outcomes.
(a) The card is red or a king.
(b) The card is a club or a diamond.

Answers:
(a) $\frac{7}{13}$ (b) $\frac{1}{2}$

EXAMPLE 4 Finding Probabilities of Unions

One card is drawn from a well-shuffled deck of 52 cards. What is the probability of the following outcomes?

(a) The card is an ace or a spade. **(b)** The card is a 3 or a king.

SOLUTION

(a) The events "drawing an ace" and "drawing a spade" are not mutually exclusive since it is possible to draw the ace of spades, an outcome satisfying both events.

$$P(\text{ace or spade}) = P(\text{ace}) + P(\text{spade}) - P(\text{ace and spade})$$ Probability of the union of two events

$$= \frac{4}{52} + \frac{13}{52} - \frac{1}{52}$$ Substitute known probabilities.

$$= \frac{16}{52}, \text{ or } \frac{4}{13}$$ Add and subtract. Write in lowest terms.

(b) "Drawing a 3" and "drawing a king" are mutually exclusive events because it is impossible to draw one card that is both a 3 and a king.

$$P(3 \text{ or } K) = P(3) + P(K) - P(3 \text{ and } K)$$ Probability of the union of two events

$$= \frac{4}{52} + \frac{4}{52} - 0$$ Substitute known probabilities.

$$= \frac{8}{52}, \text{ or } \frac{2}{13}$$ Add and subtract. Write in lowest terms.

Now Try Exercise 17(d).

Classroom Example 5
Suppose two fair dice are rolled. Find each probability.

(a) The first die shows a 5, or the sum of the two dice is 11 or 12.

(b) The sum of the dots showing is at least 9.

Answers:
(a) $\frac{2}{9}$ (b) $\frac{5}{18}$

EXAMPLE 5 Finding Probabilities of Unions

Suppose two fair dice are rolled. Find each probability.

(a) The first die shows a 2, or the sum of the two dice is 6 or 7.

(b) The sum of the dots showing is at most 4.

SOLUTION

(a) Think of the two dice as being distinguishable—one red and one green, for example. (Actually, the sample space is the same even if they are not apparently distinguishable.) A sample space with equally likely outcomes is shown in **Figure 18,** where $(1, 1)$ represents the event "the first die (red) shows a 1 and the second die (green) shows a 1," $(1, 2)$ represents "the first die shows a 1 and the second die shows a 2," and so on.

(1, 1)	(1, 2)	(1, 3)	(1, 4)	(1, 5)	(1, 6)
(2, 1)	(2, 2)	(2, 3)	(2, 4)	(2, 5)	(2, 6)
(3, 1)	(3, 2)	(3, 3)	(3, 4)	(3, 5)	(3, 6)
(4, 1)	(4, 2)	(4, 3)	(4, 4)	(4, 5)	(4, 6)
(5, 1)	(5, 2)	(5, 3)	(5, 4)	(5, 5)	(5, 6)
(6, 1)	(6, 2)	(6, 3)	(6, 4)	(6, 5)	(6, 6)

Event *A* (row 2); Event *B* (diagonal band)

Figure 18

Let *A* represent the event "the first die shows a 2," and *B* represent the event "the sum of the two dice is 6 or 7." See **Figure 18.** Event *A* has 6 elements, event *B* has 11 elements, and the sample space has 36 elements.

$$P(A) = \frac{6}{36}, \quad P(B) = \frac{11}{36}, \quad \text{and} \quad P(A \cap B) = \frac{2}{36},$$

$$P(A \cup B) = P(A) + P(B) - P(A \cap B) \quad \text{Probability of the union of two events}$$

$$= \frac{6}{36} + \frac{11}{36} - \frac{2}{36} \quad \text{Substitute known probabilities.}$$

$$= \frac{15}{36}, \text{ or } \frac{5}{12} \quad \text{Add and subtract. Write in lowest terms.}$$

(b) "At most 4" can be written as

"2 or 3 or 4."

(A sum of 1 is meaningless here.) Since the events represented by "2," "3," and "4" are mutually exclusive,

$$P(\text{at most } 4) = P(2 \text{ or } 3 \text{ or } 4) = P(2) + P(3) + P(4). \quad (1)$$

The sample space for this experiment includes the 36 possible pairs of numbers shown in **Figure 18.** The pair $(1, 1)$ is the only one with a sum of 2, so $P(2) = \frac{1}{36}$. Also $P(3) = \frac{2}{36}$ since both $(1, 2)$ and $(2, 1)$ give a sum of 3. The pairs $(1, 3)$, $(2, 2)$, and $(3, 1)$ have a sum of 4, so $P(4) = \frac{3}{36}$.

$$P(\text{at most } 4) = \frac{1}{36} + \frac{2}{36} + \frac{3}{36} \quad \text{Substitute into equation (1) above.}$$

$$= \frac{6}{36}, \text{ or } \frac{1}{6} \quad \text{Add and write in lowest terms.}$$

Teaching Tip Challenge students to find the probability "at least a sum of 11 occurs when two dice are rolled" using complementary probability.

✔ *Now Try Exercise 17(c).*

The properties of probability are summarized as follows.

Properties of Probability

Let E and F represent events.

1. $0 \le P(E) \le 1$ **2.** $P(\text{a certain event}) = 1$

3. $P(\text{an impossible event}) = 0$ **4.** $P(E') = 1 - P(E)$

5. $P(E \text{ or } F) = P(E \cup F) = P(E) + P(F) - P(E \cap F)$

CAUTION ***When finding the probability of a union, remember to subtract the probability of the intersection from the sum of the probabilities of the individual events.***

Binomial Probability Many probability experiments consist of a repeated number of independent trials (n) with only two possible outcomes. Consider the example of tossing a coin 5 times and observing the number of tails. In this experiment there are $n = 5$ independent trials, or coin tosses, and there are two possible outcomes, head or tail, for each trial. It is common to consider "obtaining a tail" as a success because it is the outcome of interest, so "obtaining a head" would be considered a failure.

When trials such as coin tosses are independent, the probability of a success (tail) and the probability of a failure (head) remain constant. That is, $P(\text{tail}) = \frac{1}{2}$ and $P(\text{head}) = \frac{1}{2}$ for each coin toss. When a probability experiment satisfies these conditions, it is considered a *binomial experiment.*

Binomial Probability

If a probability experiment consists of n independent trials with two possible outcomes for each trial, and the probabilities remain constant for each trial, then it is a **binomial experiment.** Let p represent the probability of a success, and let $q = 1 - p$ represent the probability of a failure. Then the probability of obtaining exactly r successes in n trials is found as follows.

$$P(r \text{ successes in } n \text{ trials}) = \binom{n}{r} p^r q^{n-r}$$

Classroom Example 6

An experiment consists of rolling a die 12 times. Find each probability.

(a) The probability that in exactly 3 of the rolls, the result is a 5

(b) The probability that in exactly 8 of the rolls, the result is not a 5

Answers:

(a) 0.197 (b) 0.089

Suppose that we want to determine the probability of getting exactly 3 tails in 5 coin tosses. Here $n = 5$, $r = 3$, $p = P(\text{tail}) = \frac{1}{2}$, and $q = P(\text{head}) = 1 - \frac{1}{2} = \frac{1}{2}$.

$$P(3 \text{ tails in 5 coin tosses}) = \binom{5}{3}\left(\frac{1}{2}\right)^3\left(\frac{1}{2}\right)^{5-3}$$ Use the binomial probability formula.

$$= \frac{5!}{3!\,2!}\left(\frac{1}{2}\right)^3\left(\frac{1}{2}\right)^2$$ Apply the combinations formula and subtract.

$$= 10\left(\frac{1}{2}\right)^3\left(\frac{1}{2}\right)^2$$ Evaluate factorials and divide.

$$= 0.3125$$ Apply the exponents and multiply.

EXAMPLE 6 Finding Probabilities in a Binomial Experiment

An experiment consists of rolling a die 10 times and observing the number of 3s.

(a) Find the probability of getting exactly 4 threes.

(b) Find the probability that the result is not a 3 in exactly 9 of the rolls.

ALGEBRAIC SOLUTION

(a) There are $n = 10$ independent trials with $p = P(3) = \frac{1}{6}$ and $q = 1 - \frac{1}{6} = \frac{5}{6}$.

$$P\text{ (4 threes in 10 rolls)} = \binom{10}{4}\left(\frac{1}{6}\right)^4\left(\frac{5}{6}\right)^{10-4}$$

$$= 210\left(\frac{1}{6}\right)^4\left(\frac{5}{6}\right)^6$$

$$\approx 0.054$$

(b) Here $n = 10$, $p = P(\text{not a } 3) = \frac{5}{6}$, and $q = \frac{1}{6}$.

$$P(\text{9 non-threes in 10 rolls}) = \binom{10}{9}\left(\frac{5}{6}\right)^9\left(\frac{1}{6}\right)^1$$

$$\approx 0.323$$

GRAPHING CALCULATOR SOLUTION

Graphing calculators, such as the TI-83/84 Plus, that have statistical distribution functions give binomial probabilities. **Figure 19** shows the results for parts (a) and (b). The numbers in parentheses separated by commas represent n, p, and r, respectively.

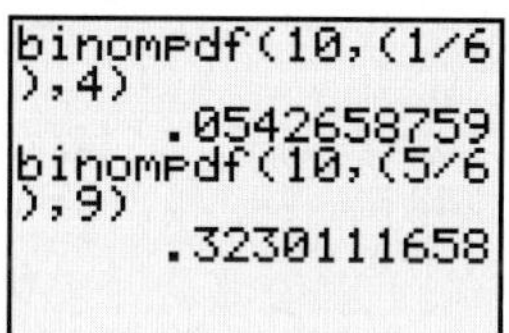

Figure 19

✔ Now Try Exercise 35.

11.7 Exercises

1. $\{H\}$
2. $\{(H, H), (H, T), (T, H), (T, T)\}$
3. $\{(H, H, H), (H, H, T), (H, T, H), (T, H, H), (H, T, T), (T, H, T), (T, T, H), (T, T, T)\}$
4. $\{(1, 2), (1, 3), (1, 4), (2, 1), (2, 3), (2, 4), (3, 1), (3, 2), (3, 4), (4, 1), (4, 2), (4, 3)\}$
5. $\{(1, 1), (1, 2), (1, 3), (2, 1), (2, 2), (2, 3), (3, 1), (3, 2), (3, 3)\}$
6. $\{(1, H), (2, H), (3, H), (4, H), (5, H), (6, H), (1, T), (2, T), (3, T), (4, T), (5, T), (6, T)\}$
7. (a) $\{H\}$; 1 (b) $\emptyset$; 0
8. (a) $\{(H, H), (T, T)\}$; $\frac{1}{2}$ (b) $\{(H, H), (H, T), (T, H)\}$; $\frac{3}{4}$

Write a sample space with equally likely outcomes for each experiment.

1. A two-headed coin is tossed once.

2. Two ordinary coins are tossed.

3. Three ordinary coins are tossed.

4. Slips of paper marked with the numbers 1, 2, 3, and 4 are placed in a box. A slip is drawn and set aside, its number is recorded, and then a second slip is drawn.

5. The spinner shown here is spun twice.

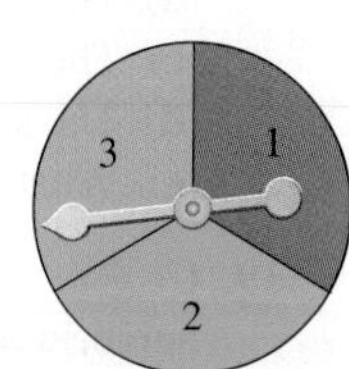

6. A die is rolled and then a coin is tossed.

Write each event in set notation and give the probability of the event. ***See Example 1.***

7. Refer to **Exercise 1.**

(a) The result of the toss is heads. **(b)** The result of the toss is tails.

8. Refer to **Exercise 2.**

(a) Both coins show the same face. **(b)** At least one coin is a head.

9. (a) $\{(H, H, H), (T, T, T)\}; \frac{1}{4}$
(b) $\{(H, T, T), (T, H, T), (T, T, H), (T, T, T)\}; \frac{1}{2}$

10. (a) $\{(2, 4), (4, 2)\}; \frac{1}{6}$
(b) $\{(1, 3), (3, 1)\}; \frac{1}{6}$
(c) $\emptyset$; 0
(d) $\{(1, 2), (1, 4), (2, 1), (2, 3), (3, 2), (3, 4), (4, 1), (4, 3)\}; \frac{2}{3}$

11. (a) $\{(1, 1), (2, 2), (3, 3)\}; \frac{1}{3}$
(b) $\{(1, 1), (1, 3), (2, 1), (2, 3), (3, 1), (3, 3)\}; \frac{2}{3}$
(c) $\{(2, 1), (2, 3)\}; \frac{2}{9}$

13. 0.143

14. (a) F (b) D (c) A (d) F (e) C (f) B (g) E

15. 3 to 7

16. 499 to 1

17. (a) $\frac{1}{4}$ (b) $\frac{3}{4}$ (c) $\frac{1}{2}$ (d) $\frac{11}{26}$ (e) 1 to 3

18. (a) $\frac{1}{6}$ (b) $\frac{5}{6}$ (c) $\frac{1}{3}$ (d) $\frac{1}{6}$ (e) 5 to 1

19. (a) 0.273 (b) 0.869 (c) 0.169 (d) 20,150 to 17,811

9. Refer to **Exercise 3.**

(a) All three coins show the same face. **(b)** At least two coins are tails.

10. Refer to **Exercise 4.**

(a) Both slips are marked with even numbers.
(b) Both slips are marked with odd numbers.
(c) Both slips are marked with the same number.
(d) One slip is marked with an odd number, the other with an even number.

11. Refer to **Exercise 5.**

(a) The result is a repeated number. **(b)** The second number is 1 or 3.
(c) The first number is even and the second number is odd.

12. A student gives the probability of an event in a problem as $\frac{6}{5}$. Explain why this answer must be incorrect.

13. *Concept Check* If the probability of an event is 0.857, what is the probability that the event will not occur?

14. *Concept Check* Associate each probability in parts (a)–(g) with one of the statements in choices A–F. Choices may be used more than once.

(a) $P(E) = -0.1$ **(b)** $P(E) = 0.01$ **(c)** $P(E) = 1$ **(d)** $P(E) - 2$
(e) $P(E) = 0.99$ **(f)** $P(E) = 0$ **(g)** $P(E) = 0.5$

A. The event is certain to occur. **B.** The event cannot occur.
C. The event is very likely to occur. **D.** The event is very unlikely to occur.
E. The event is just as likely to occur as not to occur.
F. The event is impossible.

Work each problem. ***See Examples 1–6.***

15. *Batting Average* A baseball player with a batting average of .300 comes to bat. What are the odds in favor of the ball player getting a hit?

16. *Small Business Loan* The probability that a bank with assets greater than or equal to \$30 billion will make a loan to a small business is 0.002. What are the odds against such a bank making a small business loan? (*Source: The Wall Street Journal* analysis of *CA1 Reports* filed with federal banking authorities.)

17. *Drawing a Card* A card is drawn at random from a standard deck of 52 cards. Find the probabilities in parts (a)–(d).

(a) The card is a spade. **(b)** The card is not a spade.
(c) The card is a spade or a heart. **(d)** The card is a spade or a face card.
(e) What are the odds in favor of drawing a spade?

18. *Dice Rolls* Two dice are rolled. Find the probabilities in parts (a)–(d).

(a) The sum of the dots is at least 10.
(b) The sum of the dots is less than 10.
(c) The sum of the dots is either 7 or at least 10.
(d) The sum of the dots is 2, or the dice both show the same number.
(e) What are the odds against rolling a 7?

19. *Origins of Foreign-Born Population* The numbers (in thousands) of foreign-born people who were living in the United States in 2008, according to region of birth, are given in the table. Find the probability that a foreign-born U.S. resident in 2008 satisfied the following.

(a) born in Asia **(b)** not born in Europe
(c) born in Africa or Europe
(d) What are the odds that a randomly selected foreign-born U.S. resident was born in Latin America?

Region	Number (in thousands)
Africa	1436
Asia	10,356
Europe	4969
Latin America	20,150
Other	1050

Source: U.S. Census Bureau.

20. (a) 0.225 (b) 0.235
(c) 0.419 (d) 0.602
(e) 1937 to 1133

21. $\frac{1}{28,561} \approx 0.0000350$

22. $\frac{48}{28,561} \approx 0.00168$

23. (a) 0.750 (b) 0.250
(c) 0.264 (d) 0.938

24. (a) 0.72 (b) 0.70
(c) 0.79

20. *U.S. Population by Region* The U.S. resident population by region (in millions) for selected years is given in the table. Find the probability that a U.S. resident selected at random satisfies the following.

Region	1995	2000	2009
Northeast	51.4	53.6	55.3
Midwest	61.8	64.4	66.8
South	91.8	100.2	113.3
West	57.7	63.2	71.6

Source: U.S. Census Bureau.

(a) lived in the West in 2000
(b) lived in the Midwest in 1995
(c) lived in the Northeast or Midwest in 2000
(d) lived in the South or West in 2009
(e) What are the odds that a randomly selected U.S. resident in 2009 was not from the South?

21. *State Lottery* One game in a state lottery requires you to pick 1 heart, 1 club, 1 diamond, and 1 spade, in that order, from the 13 cards in each suit. What is the probability of getting all four picks correct and winning $5000?

22. *State Lottery* If three of the four selections in **Exercise 21** are correct, the player wins $200. Find the probability of this outcome.

23. *Male Life Table* The table is an abbreviated version of the 2006 **period life table** used by the Office of the Chief Actuary of the Social Security Administration. (The actual table includes every age, not just every tenth age.) Theoretically, this table follows a group of 100,000 males at birth and gives the number still alive at each age.

Exact Age	Number of Lives	Exact Age	Number of Lives
0	100,000	60	85,026
10	99,067	70	71,586
20	98,519	80	47,073
30	97,079	90	15,051
40	95,431	100	657
50	92,041	110	1

Source: Office of the Actuary, Social Security Administration.

(a) What is the probability that a 40-year-old man will live 30 more years?
(b) What is the probability that a 40-year-old man will not live 30 more years?
(c) Consider a group of five 40-year-old men. What is the probability that exactly three of them survive to age 70? (*Hint:* The longevities of the individual men can be considered as independent trials.)
(d) Consider two 40-year-old men. What is the probability that at least one of them survives to age 70? (*Hint:* The complement of *at least one* is *none*.)

24. *Opinion Survey* The management of a firm wishes to survey the opinions of its workers, classified as follows for the purpose of an interview:

30% have worked for the company 5 or more years,
28% are female,
65% contribute to a voluntary retirement plan, and 50% of the female workers contribute to the retirement plan.

Find each probability if a worker is selected at random.

(a) A male worker is selected.
(b) A worker is selected who has worked for the company less than 5 yr.
(c) A worker is selected who contributes to the retirement plan or is female.

25. 0.3 26. 0.1
27. 0.238 28. 0.044
29. 0.121 30. 0.026
31. 0.38 32. 0.86
33. 0.62 34. 0
35. (a) 0.048 (b) 0.977
(c) 0.103 (d) 0.897

25. *Growth in Stock Value* A financial analyst has determined the possibilities (and their probabilities) for the growth in value of a certain stock during the next year. (Assume these are the only possibilities.) See the table. For instance, the probability of a 5% growth is 0.15. If you invest $10,000 in the stock, what is the probability that the stock will be worth at least $11,400 by the end of the year?

Percent Growth	Probability
5	0.15
8	0.20
10	0.35
14	0.20
18	0.10

26. *Growth in Stock Value* Refer to **Exercise 25.** Suppose the percents and probabilities in the table are estimates of annual growth during the next 3 yr. What is the probability that an investment of $10,000 will grow in value to *at least* $15,000 during the next 3 yr? (*Hint:* Use the formula for (annual) compound interest discussed in **Section 4.2.**)

Insufficient Sleep The table gives the results of a 2008 survey of Americans aged 18–24 in which the respondents were asked, "During the past 30 days, for about how many days have you felt that you did not get enough sleep?"

Number of Days	0	1–13	14–29	30
Percent (as a decimal)	0.23	0.45	0.20	0.12

Source: U.S. Centers for Disease Control and Prevention.

Using the percents as probabilities, find the probability that, out of 10 *respondents in the 18–24 age group selected at random, the following were true.*

27. Exactly 4 did not get enough sleep on 1–13 days.

28. Exactly 5 got enough sleep every night.

29. Fewer than 2 did not get enough sleep on 14 or more days.

30. No more than 3 did not get enough sleep on 1–29 days.

College Applications The table gives the results of a survey of 219,864 *first-year students from the class of 2013 at* 297 *of the nation's four-year colleges and universities.*

Number of Colleges Applied to	1	2 or 3	4–6	7 or more
Percent (as a decimal)	0.14	0.24	0.39	0.23

Source: Higher Education Research Institute, UCLA, 2009.

Using the percents as probabilities, find the probability of each event for a randomly selected student.

31. The student applied to fewer than 4 colleges.

32. The student applied to at least 2 colleges.

33. The student applied to more than 3 colleges.

34. The student applied to no colleges.

Solve each problem.

35. *Color-Blind Males* The probability that a male will be color-blind is 0.042. Find the probabilities that in a group of 53 men, the following are true.

(a) Exactly 5 are color-blind.
(b) No more than 5 are color-blind.
(c) None are color-blind.
(d) At least 1 is color-blind.

36. (a) The probabilities, in order, are 0.125, 0.375, 0.375, and 0.125.
(b) The probabilities, in order, are 0.015625, 0.09375, 0.234375, 0.3125, 0.234375, 0.09375, and 0.015625.
37. (a) about 0.404
(b) about 0.047
(c) about 0.002
38. $p \approx 0.155$ and $p \approx 0.465$

36. The screens illustrate how the TABLE feature of a graphing calculator can be used to find the probabilities of having 0, 1, 2, 3, or 4 girls in a family of 4 children. (Note that 0 appears for X = 5 and X = 6 because these events are impossible.)

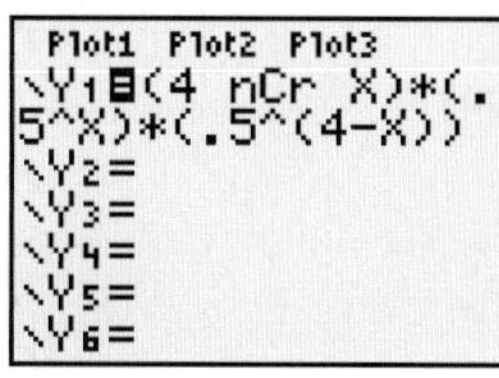

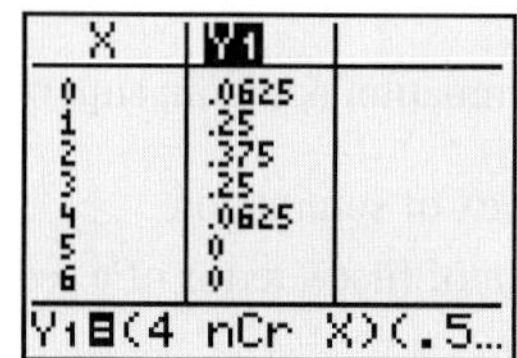

Use this approach for the following.

(a) Find the probabilities of having 0, 1, 2, or 3 boys in a family of 3 children.
(b) Find the probabilities of having 0, 1, 2, 3, 4, 5, or 6 girls in a family of 6 children.

37. *(Modeling) Spread of Disease* What will happen when an infectious disease is introduced into a family? Suppose a family has I infected members and S members who are not infected but are susceptible to contracting the disease. The probability P of exactly k people not contracting the disease during a 1-week period can be calculated by the formula

$$P = \binom{S}{k} q^k (1 - q)^{S-k},$$

where $q = (1 - p)^I$, and p is the probability that a susceptible person contracts the disease from an infected person. For example, if $p = 0.5$, then there is a 50% chance that a susceptible person exposed to 1 infected person for 1 week will contract the disease. (*Source:* Hoppensteadt, F. and C. Peskin, *Mathematics in Medicine and the Life Sciences,* Springer-Verlag.)

(a) Compute the probability P of 3 family members not becoming infected within 1 week if there are currently 2 infected and 4 susceptible members. Assume that $p = 0.1$. (*Hint:* To use the formula, first determine the values of k, I, S, and q.)
(b) A highly infectious disease can have $p = 0.5$. Repeat part (a) with this value of p.
(c) Determine the probability that everyone will become sick in a large family if, initially, $I = 1$, $S = 9$, and $p = 0.5$. Discuss the results.

38. *(Modeling) Spread of Disease* (Refer to **Exercise 37.**) Suppose that in a family $I = 2$ and $S = 4$. If the probability P is 0.25 of there being $k = 2$ uninfected members after 1 week, estimate graphically the possible values of p. (*Hint:* Write P as a function of p.)

Chapter 11 Test Prep

Key Terms

11.1 finite sequence
infinite sequence
terms of a sequence
general term (nth term)
convergent sequence
divergent sequence
recursive definition
Fibonacci sequence
series
summation notation
finite series
infinite series
index of summation

11.2 arithmetic sequence (arithmetic progression)
common difference
arithmetic series

11.3 geometric sequence (geometric progression)
common ratio
geometric series
annuity
future value of an annuity

11.4 Pascal's triangle
factorial notation
binomial coefficient
binomial theorem (general binomial expansion)

11.6 tree diagram
independent events
permutation
combination

11.7 outcome
sample space
event
probability
complement
Venn diagram
odds
mutually exclusive events
binomial experiment

New Symbols

Symbol	Meaning
a_n	nth term of a sequence
$\sum_{i=1}^{n} a_i$	summation notation; sum of n terms
i	index of summation
S_n	sum of first n terms of a sequence
$\sum_{i=1}^{\infty} a_i$	sum of an infinite number of terms
$\lim_{n\to\infty} S_n$	limit of S_n as n increases without bound
$n!$	n-factorial
${}_nC_r$ or $\binom{n}{r}$	binomial coefficient (combinations of n elements taken r at a time)
$P(n, r)$ or ${}_nP_r$	permutations of n elements taken r at a time
$C(n, r)$ or $\binom{n}{r}$	combinations of n elements taken r at a time
$n(E)$	number of outcomes that belong to event E
$P(E)$	probability of event E
E'	complement of event E

Quick Review

Concepts	Examples
11.1 Sequences and Series	
A finite sequence is a function that has a set of natural numbers of the form $\{1, 2, 3, \ldots, n\}$ as its domain. An infinite sequence has the set of natural numbers as its domain. The nth term of a sequence is symbolized a_n. A series is an indicated sum of the terms of a sequence.	The sequence $1, \frac{1}{2}, \frac{1}{3}, \frac{1}{4}, \ldots, \frac{1}{n}$ has general term $a_n = \frac{1}{n}$. The corresponding series is the *sum* $1 + \frac{1}{2} + \frac{1}{3} + \frac{1}{4} + \cdots + \frac{1}{n}$.
Summation Properties If $a_1, a_2, a_3, \ldots, a_n$ and $b_1, b_2, b_3, \ldots, b_n$ are sequences and c is a constant, then for every positive integer n, the following hold.	
(a) $\sum_{i=1}^{n} c = nc$	$\sum_{i=1}^{6} 5 = 6 \cdot 5 = 30$
(b) $\sum_{i=1}^{n} ca_i = c\sum_{i=1}^{n} a_i$	$\sum_{i=1}^{4} 3(2i+1) = 3\sum_{i=1}^{4}(2i+1) = 3(3+5+7+9) = 72$
(c) $\sum_{i=1}^{n}(a_i \pm b_i) = \sum_{i=1}^{n} a_i \pm \sum_{i=1}^{n} b_i$	$\sum_{i=1}^{3}(5i+6i^2) = \sum_{i=1}^{3} 5i + \sum_{i=1}^{3} 6i^2 = (5+10+15)+(6+24+54) = 30+84 = 114$
11.2 Arithmetic Sequences and Series	
Assume a_1 is the first term, a_n is the nth term, and d is the common difference.	The arithmetic sequence $2, 5, 8, 11, \ldots$ has $a_1 = 2$.
Common Difference $d = a_{n+1} - a_n$	$d = 5 - 2 = 3$ (Any two successive terms could have been used.)
***n*th Term** $a_n = a_1 + (n-1)d$	Suppose that $n = 10$. Then the 10th term is $a_{10} = 2 + (10-1)3 = 2 + 9 \cdot 3 = 29.$

Concepts	Examples
Sum of the First n Terms $$S_n = \frac{n}{2}(a_1 + a_n)$$ or $$S_n = \frac{n}{2}[2a_1 + (n-1)d]$$	The sum of the first 10 terms is found as follows. $$S_{10} = \frac{10}{2}(a_1 + a_{10}) = 5(2+29) = 155$$ or $$S_{10} = \frac{10}{2}[2(2) + (10-1)3] = 5(4 + 9 \cdot 3) = 5(4+27) = 155$$
11.3 Geometric Sequences and Series	
Assume a_1 is the first term, a_n is the nth term, and r is the common ratio.	The geometric sequence 1, 2, 4, 8, … has $a_1 = 1$.
Common Ratio $r = \frac{a_{n+1}}{a_n}$	$r = \frac{8}{4} = 2$
***n*th Term** $a_n = a_1 r^{n-1}$	(Any two successive terms could have been used.) Suppose that $n = 6$. Then the sixth term is $$a_6 = (1)(2)^{6-1} = 1(2)^5 = 32.$$
Sum of the First n Terms $$S_n = \frac{a_1(1-r^n)}{1-r} \quad (\text{where } r \neq 1)$$	The sum of the first six terms is found as follows. $$S_6 = \frac{1(1-2^6)}{1-2} = \frac{1-64}{-1} = 63$$
Sum of the Terms of an Infinite Geometric Sequence with $\lvert r \rvert < 1$ $$S_\infty = \frac{a_1}{1-r}$$	The sum of the terms of the infinite geometric sequence $$\sum_{k=0}^{\infty}\left(\frac{1}{2}\right)^k = 1 + \frac{1}{2} + \frac{1}{4} + \cdots$$ is found as follows. $$S_\infty = \frac{1}{1-\frac{1}{2}} = \frac{1}{\frac{1}{2}} = 2$$
11.4 The Binomial Theorem	
For any positive integer n, $$n! = n(n-1)(n-2)\cdots(3)(2)(1).$$	$4! = 4 \cdot 3 \cdot 2 \cdot 1 = 24$
By definition, $0! = 1$.	
Binomial Coefficient For nonnegative integers n and r, with $r \leq n$, $${}_nC_r = \binom{n}{r} = \frac{n!}{r!(n-r)!}.$$	$${}_5C_3 = \frac{5!}{3!(5-3)!} = \frac{5!}{3!2!} = \frac{5 \cdot 4 \cdot 3 \cdot 2 \cdot 1}{3 \cdot 2 \cdot 1 \cdot 2 \cdot 1} = 10$$

(continued)

Concepts	Examples
Binomial Theorem For any positive integer n and any complex numbers x and y, $(x+y)^n$ is expanded as follows. $$(x+y)^n = x^n + \binom{n}{1}x^{n-1}y + \binom{n}{2}x^{n-2}y^2 + \binom{n}{3}x^{n-3}y^3 + \cdots + \binom{n}{r}x^{n-r}y^r + \cdots + \binom{n}{n-1}xy^{n-1} + y^n$$	$$(2m+3)^4 = (2m)^4 + \frac{4!}{1!\,3!}(2m)^3(3) + \frac{4!}{2!\,2!}(2m)^2(3)^2 + \frac{4!}{3!\,1!}(2m)(3)^3 + 3^4$$ $$= 2^4m^4 + 4(2)^3m^3(3) + 6(2)^2m^2(9) + 4(2m)(27) + 81$$ $$= 16m^4 + 12(8)m^3 + 54(4)m^2 + 216m + 81$$ $$= 16m^4 + 96m^3 + 216m^2 + 216m + 81$$
***k*th Term of the Binomial Expansion of $(x+y)^n$** $$\binom{n}{k-1}x^{n-(k-1)}y^{k-1} \quad (\text{where } n \geq k-1)$$	The eighth term of $(a-2b)^{10}$ is $$\binom{10}{7}a^3(-2b)^7 = \frac{10!}{7!\,3!}a^3(-2)^7b^7$$ $$= 120(-128)a^3b^7$$ $$= -15{,}360a^3b^7.$$
11.5 Mathematical Induction **Principle of Mathematical Induction** Let S_n be a statement concerning the positive integer n. Suppose that **1.** S_1 is true; **2.** for any positive integer k, $k \leq n$, if S_k is true, then S_{k+1} is also true. Then S_n is true for every positive integer value of n.	See **Examples 1 and 2** in **Section 11.5.** **Example 3** in **Section 11.5** illustrates the generalized principle of mathematical induction.
11.6 Counting Theory **Fundamental Principle of Counting** If n independent events occur, with m_1 ways for event 1 to occur, m_2 ways for event 2 to occur, ... and m_n ways for event n to occur, then there are $$m_1 \cdot m_2 \cdot \cdots \cdot m_n$$ different ways for all n events to occur.	If there are 2 ways to choose a pair of socks and 5 ways to choose a pair of shoes, then there are $$2 \cdot 5 = 10$$ ways to choose socks and shoes.
Permutations Formula If $P(n, r)$ denotes the number of permutations of n elements taken r at a time, with $r \leq n$, then the following holds. $$P(n, r) = \frac{n!}{(n-r)!}$$	How many ways are there to arrange the letters of the word *triangle* using 5 letters at a time? Because this is an arrangement, use the permutations formula with $n = 8$ and $r = 5$. $$P(8, 5) = \frac{8!}{(8-5)!} = \frac{8!}{3!} = 6720$$

Concepts	Examples
Combinations Formula The number of combinations of n elements taken r at a time, with $r \leq n$, is determined as follows. $$C(n,r) = \binom{n}{r} = \frac{n!}{(n-r)!\,r!}, \quad \text{or} \quad \frac{n!}{r!(n-r)!}$$	How many committees of 4 senators can be formed from a group of 9 senators? Since the arrangement of senators does not matter, this is a combinations problem with $n = 9$ and $r = 4$. $$C(9, 4) = \binom{9}{4} = \frac{9!}{4!\,5!} = 126$$

11.7 Basics of Probability

Concepts	Examples
Probability of an Event E In a sample space S with equally likely outcomes, the probability of an event E is determined as follows. $$P(E) = \frac{n(E)}{n(S)}$$	A number is chosen at random from $S = \{1, 2, 3, 4, 5, 6\}$. What is the probability that the number is less than 3? The event is $E = \{1, 2\}$, $n(S) = 6$, and $n(E) = 2$. $$P(E) = \frac{2}{6} = \frac{1}{3}$$
Properties of Probability Let E and F represent events. **1.** $0 \leq P(E) \leq 1$ **2.** $P(\text{a certain event}) = 1$ **3.** $P(\text{an impossible event}) = 0$ **4.** $P(E') = 1 - P(E)$ **5.** $P(E \text{ or } F) = P(E \cup F)$ $= P(E) + P(F) - P(E \cap F)$	What is the probability that the number is 3 or more? This event is E'. $$P(E') = 1 - \frac{1}{3} = \frac{2}{3}$$
Binomial Probability In a binomial experiment, let p represent the probability of a success, and let $q = 1 - p$ represent the probability of a failure. Then the probability of obtaining exactly r successes in n trials is found as follows. $$P(r \text{ successes in } n \text{ trials}) = \binom{n}{r} p^r q^{n-r}$$	An experiment consists of rolling a die 8 times. Find the probability that exactly 5 rolls result in a 2. Here, we have $n = 8$, $r = 5$, $p = \frac{1}{6}$, and $q = \frac{5}{6}$. $$P(5 \text{ twos in 8 rolls}) = \binom{8}{5}\left(\frac{1}{6}\right)^5\left(\frac{5}{6}\right)^{8-5}$$ $$= 56\left(\frac{1}{6}\right)^5\left(\frac{5}{6}\right)^3 \approx 0.004$$

Chapter 11 Review Exercises

Write the first five terms of each sequence. State whether the sequence is arithmetic, geometric, *or* neither.

1. $a_n = \dfrac{n}{n+1}$ **2.** $a_n = (-2)^n$ **3.** $a_n = 2(n+3)$

4. $a_n = n(n+1)$ **5.** $a_1 = 5$, $a_n = a_{n-1} - 3$, if $n \geq 2$ **6.** $a_1 = 1$, $a_2 = 3$, $a_n = a_{n-2} + a_{n-1}$, if $n \geq 3$

7. *Concept Check* Write an arithmetic sequence that consists of five terms, with first term 4, having the sum of the five terms equal to 25.

In Exercises 8–11, write the first five terms of the sequence described.

8. arithmetic, $a_2 = 10$, $d = -2$ **9.** arithmetic, $a_3 = \pi$, $a_4 = 1$

10. geometric, $a_1 = 6$, $r = 2$ **11.** geometric, $a_1 = -5$, $a_2 = -1$

Answers:

1. $\frac{1}{2}, \frac{2}{3}, \frac{3}{4}, \frac{4}{5}, \frac{5}{6}$; neither
2. $-2, 4, -8, 16, -32$; geometric
3. 8, 10, 12, 14, 16; arithmetic
4. 2, 6, 12, 20, 30; neither
5. $5, 2, -1, -4, -7$; arithmetic
6. 1, 3, 4, 7, 11; neither
7. 4, 4.5, 5, 5.5, 6
8. 12, 10, 8, 6, 4
9. $3\pi - 2, 2\pi - 1, \pi, 1, -\pi + 2$
10. 6, 12, 24, 48, 96
11. $-5, -1, -\frac{1}{5}, -\frac{1}{25}, -\frac{1}{125}$

12. -11; $2n - 13$
13. ± 1; $-8\left(\frac{1}{2}\right)^{n-1} = -\left(\frac{1}{2}\right)^{n-4}$ or $-8\left(-\frac{1}{2}\right)^{n-1} = \left(-\frac{1}{2}\right)^{n-4}$
14. 20
15. $-x + 61$
16. 222
17. 612
18. -162
19. $\frac{4}{25}$
20. 45
21. -40
22. $\frac{13}{36}$
23. 1
24. 70
25. $\frac{73}{12}$
26. 125
27. 3,126,250
28. 248
29. $\frac{4}{3}$
30. The sum does not exist.
31. $-\frac{4}{5}$
32. $\frac{3}{2} + \frac{9}{8} + \frac{27}{32} + \cdots$
33. 36
34. $-\frac{9}{20}$
35. diverges
36. 1
37. -10
38. 2.7

In Exercises 39–42, other answers are possible.

39. $\sum_{i=1}^{15} (-5i + 9)$
40. $\sum_{i=1}^{20} (4i + 6)$
41. $\sum_{i=1}^{6} 4(3)^{i-1}$
42. $\sum_{i=5}^{12} \frac{i}{i+1}$
43. $x^4 + 8x^3y + 24x^2y^2 + 32xy^3 + 16y^4$
44. $27z^3 - 135z^2w + 225zw^2 - 125w^3$
45. $243x^{5/2} - 405x^{3/2} + 270x^{1/2} - 90x^{-1/2} + 15x^{-3/2} - x^{-5/2}$
46. $m^{12} - 4m^7 + 6m^2 - 4m^{-3} + m^{-8}$
47. $-3584x^3y^5$
48. $3003(-3)^6m^8n^6$
49. $x^{12} + 24x^{11} + 264x^{10} + 1760x^9$
50. $480 \cdot 5^{14}a^2b^{14} + 32 \cdot 5^{15}ab^{15} + 5^{16}b^{16}$

12. An arithmetic sequence has $a_5 = -3$ and $a_{15} = 17$. Find a_1 and a_n.

13. A geometric sequence has $a_1 = -8$ and $a_7 = -\frac{1}{8}$. Find a_4 and a_n.

Find a_8 for each arithmetic sequence.

14. $a_1 = 6, d = 2$

15. $a_1 = 6x - 9, a_2 = 5x + 1$

Find S_{12} for each arithmetic sequence.

16. $a_1 = 2, d = 3$

17. $a_2 = 6, d = 10$

Find a_5 for each geometric sequence.

18. $a_1 = -2, r = 3$

19. $a_3 = 4, r = \dfrac{1}{5}$

Find S_4 for each geometric sequence.

20. $a_1 = 3, r = 2$

21. $a_1 = -1, r = 3$

22. $\dfrac{3}{4}, -\dfrac{1}{2}, \dfrac{1}{3}, \ldots$

Evaluate each sum that exists.

23. $\sum_{i=1}^{7} (-1)^{i-1}$

24. $\sum_{i=1}^{5} (i^2 + i)$

25. $\sum_{i=1}^{4} \dfrac{i+1}{i}$

26. $\sum_{j=1}^{10} (3j - 4)$

27. $\sum_{j=1}^{2500} j$

28. $\sum_{i=1}^{5} 4 \cdot 2^i$

29. $\sum_{i=1}^{\infty} \left(\dfrac{4}{7}\right)^i$

30. $\sum_{i=1}^{\infty} -2\left(\dfrac{6}{5}\right)^i$

31. $\sum_{i=1}^{\infty} 2\left(-\dfrac{2}{3}\right)^i$

32. *Concept Check* Find an infinite geometric series having common ratio $\frac{3}{4}$ and sum 6.

Evaluate each series that converges. If the series diverges, say so.

33. $24 + 8 + \dfrac{8}{3} + \dfrac{8}{9} + \cdots$

34. $-\dfrac{3}{4} + \dfrac{1}{2} - \dfrac{1}{3} + \dfrac{2}{9} - \cdots$

35. $\dfrac{1}{12} + \dfrac{1}{6} + \dfrac{1}{3} + \dfrac{2}{3} + \cdots$

36. $0.9 + 0.09 + 0.009 + 0.0009 + \cdots$

Evaluate each sum where $x_1 = 0$, $x_2 = 1$, $x_3 = 2$, $x_4 = 3$, $x_5 = 4$, and $x_6 = 5$.

37. $\sum_{i=1}^{4} (x_i^2 - 6)$

38. $\sum_{i=1}^{6} f(x_i)\,\Delta x$; $f(x) = (x-2)^3$, $\Delta x = 0.1$

Write each sum using summation notation.

39. $4 - 1 - 6 - \cdots - 66$

40. $10 + 14 + 18 + \cdots + 86$

41. $4 + 12 + 36 + \cdots + 972$

42. $\dfrac{5}{6} + \dfrac{6}{7} + \dfrac{7}{8} + \cdots + \dfrac{12}{13}$

Use the binomial theorem to expand each expression.

43. $(x + 2y)^4$

44. $(3z - 5w)^3$

45. $\left(3\sqrt{x} - \dfrac{1}{\sqrt{x}}\right)^5$

46. $(m^3 - m^{-2})^4$

Find the indicated term or terms for each expansion.

47. sixth term of $(4x - y)^8$

48. seventh term of $(m - 3n)^{14}$

49. first four terms of $(x + 2)^{12}$

50. last three terms of $(2a + 5b)^{16}$

55. 72 56. 1
57. 56 58. 362,880
59. 252 60. 3,628,800

61. \$1,800,100
62. \$2,996,971
63. \$1,196,871; Answers will vary.
64. 48
65. 90
66. 24
67. (a) 84 (b) 45
68. 504
69. 456,976,000; 258,336,000
70. (a) 72 (b) 36

Use mathematical induction to prove that each statement is true for every positive integer n.

51. $1 + 3 + 5 + 7 + \cdots + (2n - 1) = n^2$

52. $2 + 6 + 10 + 14 + \cdots + (4n - 2) = 2n^2$

53. $2 + 2^2 + 2^3 + \cdots + 2^n = 2(2^n - 1)$

54. $1^3 + 3^3 + 5^3 + \cdots + (2n - 1)^3 = n^2(2n^2 - 1)$

Evaluate each expression.

55. $P(9, 2)$ **56.** $P(6, 0)$ **57.** $\binom{8}{3}$

58. $9!$ **59.** $C(10, 5)$ **60.** $10 \cdot 9!$

Solve each problem.

61. *Median Annual Earnings* In 2008 the median annual earnings of a high school graduate with no college attendance was \$32,136. This amount is expected to increase by about \$268 per year. How much will a person earning the median amount earn until retirement if he or she joins the work force at age 18 and works until age 65? (*Source:* U.S. Bureau of Labor Statistics.)

62. *Median Annual Earnings* In 2008 the median annual earnings of a person with 4 yr of college was \$52,624. This amount is expected to increase by about \$813 per year. How much will a person earning the median amount earn until retirement if he or she joins the work force at age 22 and works until age 65? (*Source:* U.S. Bureau of Labor Statistics.)

63. *Median Annual Earnings* Refer to **Exercises 61 and 62.** How much more will a person with 4 yr of college who earns the median amount make during his or her career than a person with no college attendance who earns the median amount during his or her career? If the expenses of a 4-yr college degree are estimated at \$130,000, is earning a 4-yr college degree worth it? (*Source:* U.S. Bureau of Labor Statistics.)

64. *Wedding Plans* Two people are planning their wedding. They can select from 2 different chapels, 4 soloists, 3 organists, and 2 ministers. How many different wedding arrangements are possible?

65. *Couch Styles* Bob Schiffer, who is furnishing his apartment, wants to buy a new couch. He can select from 5 different styles, each available in 3 different fabrics, with 6 color choices. How many different couches are available?

66. *Summer Job Assignments* Four students are to be assigned to 4 different summer jobs. Each student is qualified for all 4 jobs. In how many ways can the jobs be assigned?

67. *Conference Delegations* A student council consists of 6 seniors and 3 juniors. Three members are to be selected to attend a conference.

(a) How many different such delegations are possible?

(b) How many are possible if 2 seniors and 1 junior must attend?

68. *Tournament Outcomes* Nine football teams are competing for first-, second-, and third-place titles in a statewide tournament. In how many ways can the winners be determined?

69. *License Plates* How many different license plates can be formed with a letter followed by 3 digits and then 3 letters? How many such license plates have no repeats?

70. *Racetrack Bets* Most racetracks have "compound" bets on 2 or more horses. An *exacta* is a bet in which the first and second finishers in a race are specified in order. A *quinella* is a bet on the first 2 finishers in a race, with order not specified.

(a) In a field of 9 horses, how many different exacta bets can be placed?

(b) How many different quinella bets can be placed in a field of 9 horses?

71. (a) $\frac{4}{15}$ (b) $\frac{2}{3}$
(c) 0 (d) 3 to 2
72. (a) $\frac{1}{26}$ (b) $\frac{4}{13}$
(c) $\frac{4}{13}$ (d) $\frac{3}{4}$
(e) 1 to 12
73. (a) 0.249 (b) 0.116
(c) 0.719
74. (a) 0.86 (b) 0.44
(c) 0
75. 0.296
76. 0.205

71. *Drawing a Marble* A marble is drawn at random from a box containing 4 green, 5 black, and 6 white marbles. Find the following probabilities.

(a) A green marble is drawn.
(b) A marble that is not black is drawn.
(c) A blue marble is drawn.
(d) What are the odds in favor of drawing a marble that is not white?

72. *Drawing a Card* A card is drawn from a standard deck of 52 cards. Find the following probabilities.

(a) A black king is drawn.
(b) A face card or an ace is drawn.
(c) An ace or a diamond is drawn.
(d) A card that is not a diamond is drawn.
(e) What are the odds in favor of drawing an ace?

73. *Master's Degrees* There were 625,023 master's degrees awarded in the United States in 2008. The table shows the numbers of degrees awarded in several fields of study.

Field of Study	Number of Master's Degrees
Business	155,637
Education	175,880
Health professions and related clinical studies	58,120
Visual and performing arts	14,164
Other	221,222

Source: U.S. National Center for Education Statistics.

(a) What is the probability that a randomly selected student who earned a master's degree in 2008 earned a degree in business?
(b) What is the probability that a randomly selected student who earned a master's degree in 2008 earned a degree in either health professions and related clinical studies or the visual and performing arts?
(c) What is the probability that a randomly selected student who earned a master's degree in 2008 earned a degree that was not in education?

74. *Defective Toaster Ovens* A sample shipment of 5 toaster ovens is chosen. The probability of exactly 0, 1, 2, 3, 4, or 5 toaster ovens being defective is given in the table.

Number Defective	0	1	2	3	4	5
Probability	0.31	0.25	0.18	0.12	0.08	0.06

Find the probability that the given number of toaster ovens are defective.

(a) no more than 3 **(b)** at least 2 **(c)** more than 5

75. *Rolling a Die* A die is rolled 12 times. Find the probability (to three decimal places) that exactly 2 of the rolls result in a 5.

76. *Tossing a Coin* A coin is tossed 10 times. Find the probability (to three decimal places) that exactly 4 of the tosses result in a tail.

Chapter 11 Test

Write the first five terms of each sequence. State whether the sequence is arithmetic, geometric, *or* neither.

1. $a_n = (-1)^n(n^2 + 2)$

2. $a_n = -3\left(\frac{1}{2}\right)^n$

3. $a_1 = 2,\ a_2 = 3,\ a_n = a_{n-1} + 2a_{n-2},$ for $n \geq 3$

4. A certain arithmetic sequence has $a_1 = 1$ and $a_3 = 25$. Find a_5.

5. A certain geometric sequence has $a_1 = 81$ and $r = -\frac{2}{3}$. Find a_6.

Find the sum of the first ten terms of each series.

6. arithmetic, $a_1 = -43,\ d = 12$

7. geometric, $a_1 = 5,\ r = -2$

Evaluate each sum that exists.

8. $\sum_{i=1}^{30}(5i + 2)$

9. $\sum_{i=1}^{5}(-3 \cdot 2^i)$

10. $\sum_{i=1}^{\infty}(2^i) \cdot 4$

11. $\sum_{i=1}^{\infty}54\left(\frac{2}{9}\right)^i$

Use the binomial theorem to expand each expression.

12. $(x + y)^6$

13. $(2x - 3y)^4$

14. Find the third term in the expansion of $(w - 2y)^6$.

Evaluate each expression.

15. $8!$

16. $C(10, 2)$

17. $\binom{7}{3}$

18. $P(11, 3)$

19. Use mathematical induction to prove that for all positive integers n,

$$1 + 7 + 13 + \cdots + (6n - 5) = n(3n - 2).$$

Solve each problem.

20. *Athletic Shoe Styles* A shoe manufacturer makes athletic shoes in 4 different styles. Each style comes in 3 different colors, and each color comes in 2 different shades. How many different types of shoes can be made?

21. *Seminar Attendees* A mortgage company has 10 loan officers: 4 women and 6 men. In how many ways can 4 of these officers be selected to attend a seminar? In how many ways can 2 women and 2 men be selected to attend the seminar?

22. *Course Schedule Arrangement* A student must select 4 courses from 15 that are offered in a semester. How many different arrangements of the 4 courses are possible?

23. *Drawing Cards* A card is drawn from a standard deck of 52 cards. Find the following probabilities.

(a) A red three is drawn.
(b) A card that is not a face card is drawn.
(c) A king or a spade is drawn.
(d) What are the odds in favor of drawing a face card?

[11.1–11.3]
1. $-3, 6, -11, 18, -27$; neither
2. $-\frac{3}{2}, -\frac{3}{4}, -\frac{3}{8}, -\frac{3}{16}, -\frac{3}{32}$; geometric
3. 2, 3, 7, 13, 27; neither

[11.2]
4. 49

[11.3]
5. $-\frac{32}{3}$

[11.2]
6. 110

[11.3]
7. -1705

[11.2]
8. 2385

[11.3]
9. -186
10. The sum does not exist.
11. $\frac{108}{7}$

[11.4]
12. $x^6 + 6x^5y + 15x^4y^2 + 20x^3y^3 + 15x^2y^4 + 6xy^5 + y^6$
13. $16x^4 - 96x^3y + 216x^2y^2 - 216xy^3 + 81y^4$
14. $60w^4y^2$

[11.4, 11.6]
15. 40,320 **16.** 45
17. 35 **18.** 990

[11.6]
20. 24
21. 210; 90
22. 32,760

[11.7]
23. (a) $\frac{1}{26}$ **(b)** $\frac{10}{13}$
(c) $\frac{4}{13}$ **(d)** 3 to 10

24. 0.92
25. 0.201

24. *Defective Light Bulbs* A sample of 4 light bulbs is chosen. The probability of exactly 0, 1, 2, 3, or 4 light bulbs being defective is given in the table. Find the probability that at most 2 are defective.

Number Defective	0	1	2	3	4
Probability	0.19	0.43	0.30	0.07	0.01

25. *Rolling a Die* Find the probability (to three decimal places) of obtaining 5 on exactly two of six rolls of a single die.

Appendices

A Polar Form of Conic Sections

- Equations and Graphs
- Converting from Polar to Rectangular Form

Equations and Graphs Until now we have worked with equations of conic sections in rectangular form. If the focus of a conic section is at the pole, the polar form of its equation is

$$r = \frac{ep}{1 \pm e \cdot f(\theta)},$$

where f is either the sine or cosine function.

Polar Forms of Conic Sections

A polar equation of the form

$$r = \frac{ep}{1 \pm e \cos \theta} \quad \text{or} \quad r = \frac{ep}{1 \pm e \sin \theta}$$

has a conic section as its graph. The eccentricity is e (where $e > 0$), and $|p|$ is the distance between the pole (focus) and the directrix.

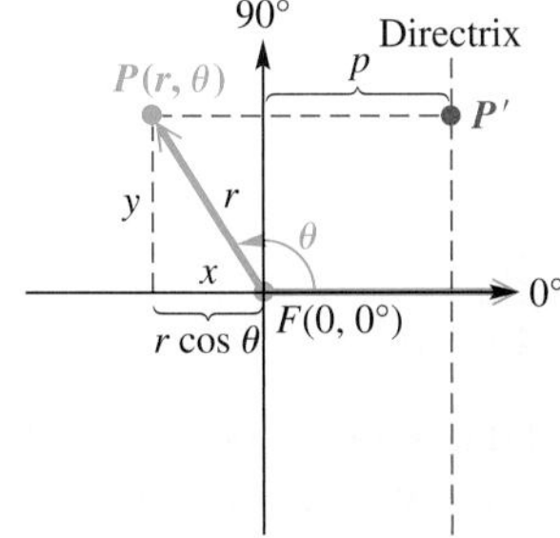

Figure 1

We can verify that $r = \frac{ep}{1 + e \cos \theta}$ does indeed satisfy the definition of a conic section. Consider **Figure 1,** where the directrix is vertical and is p units (where $p > 0$) to the right of the focus $F(0, 0°)$. Let $P(r, \theta)$ be a point on the graph. Then the distance between P and the directrix is found as follows.

$$PP' = |p - x|$$

$$= |p - r \cos \theta| \qquad x = r \cos \theta \text{ (Section 8.5)}$$

$$= \left| p - \left(\frac{ep}{1 + e \cos \theta} \right) \cos \theta \right| \qquad \text{Use the equation for } r.$$

$$= \left| \frac{p(1 + e \cos \theta) - ep \cos \theta}{1 + e \cos \theta} \right| \qquad \text{Write with a common denominator. (Section R.5)}$$

$$= \left| \frac{p + ep \cos \theta - ep \cos \theta}{1 + e \cos \theta} \right| \qquad \text{Distributive property (Section R.2)}$$

$$PP' = \left| \frac{p}{1 + e \cos \theta} \right| \qquad \text{Simplify.}$$

Since $$r = \frac{ep}{1 + e \cos \theta},$$

we multiply each side by $\frac{1}{e}$.

$$\frac{r}{e} = \frac{p}{1 + e \cos \theta}$$

Substitute $\frac{r}{e}$ for the expression in the absolute value bars for PP'.

$$PP' = \left| \frac{p}{1 + e \cos \theta} \right| = \left| \frac{r}{e} \right| = \frac{|r|}{|e|} = \frac{|r|}{e}$$

The distance between the pole and P is $PF = |r|$, so the ratio of PF to PP' is

$$\frac{PF}{PP'} = \frac{|r|}{\frac{|r|}{e}} = e.$$ Simplify the complex fraction. (Section R.5)

Thus, by the definition, the graph has eccentricity e and must be a conic.

In the preceding discussion, we assumed a vertical directrix to the right of the pole. There are three other possible situations, and all four are summarized in the table.

If the equation is:	then the directrix is:
$r = \frac{ep}{1 + e \cos \theta}$	*vertical, p* units to the *right* of the pole.
$r = \frac{ep}{1 - e \cos \theta}$	*vertical, p* units to the *left* of the pole.
$r = \frac{ep}{1 + e \sin \theta}$	*horizontal, p* units *above* the pole.
$r = \frac{ep}{1 - e \sin \theta}$	*horizontal, p* units *below* the pole.

Classroom Example 1

Graph $r = \frac{6}{2 - 4 \cos \theta}$.

Answer:

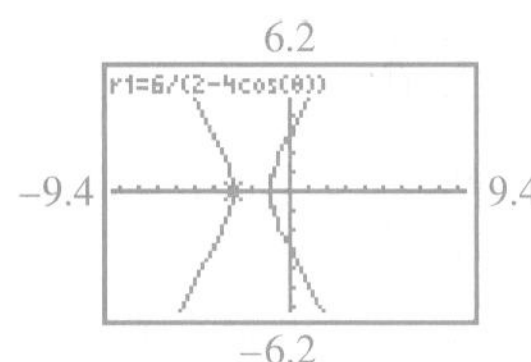

EXAMPLE 1 Graphing a Conic in Polar Form

Graph $r = \frac{8}{4 + 4 \sin \theta}$.

ALGEBRAIC SOLUTION

Divide both numerator and denominator by 4 to get

$$r = \frac{2}{1 + \sin \theta}.$$

Based on the preceding table, this is the equation of a conic with $ep = 2$ and $e = 1$. Thus, $p = 2$. Since $e = 1$, the graph is a parabola. The focus is at the pole, and the directrix is horizontal, 2 units *above* the pole.

The vertex must have polar coordinates $(1, 90°)$. Letting $\theta = 0°$ and $\theta = 180°$ gives the additional points $(2, 0°)$ and $(2, 180°)$. See **Figure 2.**

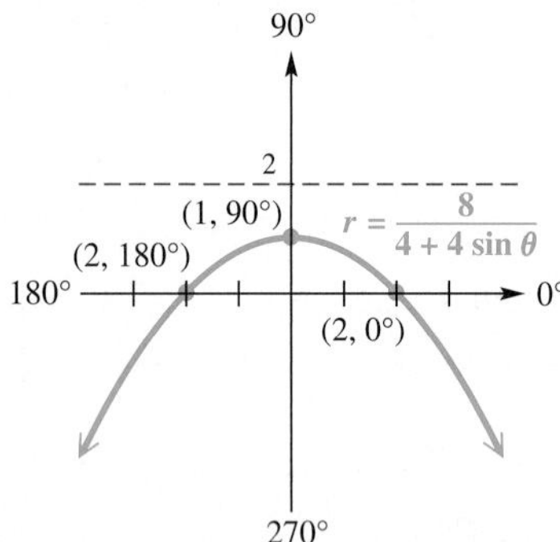

Figure 2

As expected, the graph is a parabola, and it opens downward because the directrix is above the pole.

GRAPHING CALCULATOR SOLUTION

Enter

$$r_1 = \frac{8}{4 + 4 \sin \theta},$$

with the calculator in polar and degree modes. The first two screens in **Figure 3** show the window settings, and the third screen shows the graph. Notice that the point $(1, 90°)$ is indicated at the bottom of the third screen.

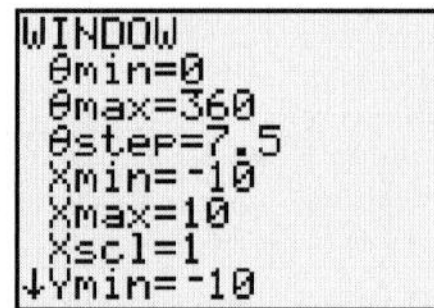

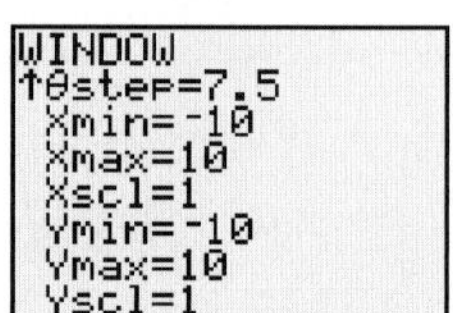

This is a continuation of the screen to the left.

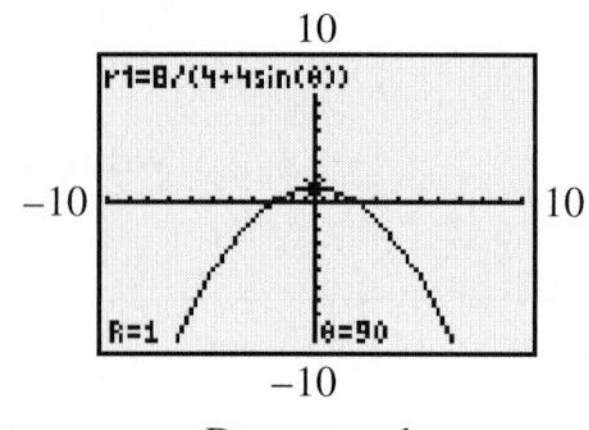

Degree mode

Figure 3

Now Try Exercise 1.

Classroom Example 2
Find the polar equation of a parabola with focus at the pole and horizontal directrix 7 units above the pole.

Answer: $r = \frac{7}{1 + \sin\theta}$

EXAMPLE 2 Finding a Polar Equation

Find the polar equation of a parabola with focus at the pole and vertical directrix 3 units to the left of the pole.

SOLUTION The eccentricity e must be 1, p must equal 3, and the equation must be of the following form.

$$r = \frac{ep}{1 - e\cos\theta}$$

$$r = \frac{1 \cdot 3}{1 - 1\cos\theta} \quad \text{Substitute.}$$

$$r = \frac{3}{1 - \cos\theta} \quad \text{Multiply.}$$

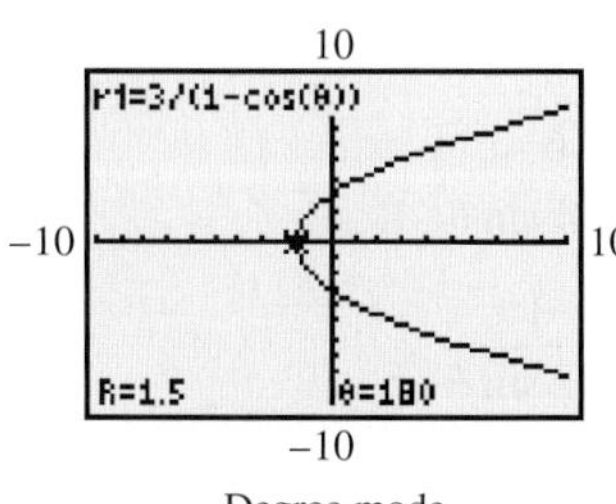

Degree mode

Figure 4

The calculator graph in **Figure 4** supports our result. When $\theta = 180°$, $r = 1.5$. The distance from $F(0, 0°)$ to the directrix is $2r = 2(1.5) = 3$ units, as required.

Now Try Exercise 11.

Converting from Polar to Rectangular Form

Classroom Example 3
Identify the type of conic represented by

$$r = \frac{-4}{2 + 3\sin\theta}.$$

Then convert the equation to rectangular form.

Answer: hyperbola; $4x^2 - 5y^2 - 24y - 16 = 0$

EXAMPLE 3 Identifying and Converting from Polar Form to Rectangular Form

Identify the type of conic represented by $r = \frac{8}{2 - \cos\theta}$. Then convert the equation to rectangular form.

SOLUTION To identify the type of conic, we divide both the numerator and the denominator on the right side by 2.

$$r = \frac{4}{1 - \frac{1}{2}\cos\theta}$$

From the table, we see that this is a conic that has a vertical directrix, with $e = \frac{1}{2}$, making it an ellipse. To convert to rectangular form, we start with the given equation.

$$r = \frac{8}{2 - \cos\theta} \quad \text{Given equation}$$

$$r(2 - \cos\theta) = 8 \quad \text{Multiply by } 2 - \cos\theta.$$

$$2r - r\cos\theta = 8 \quad \text{Distributive property}$$

$$2r = r\cos\theta + 8 \quad \text{Add } r\cos\theta \text{ to each side.}$$

$$(2r)^2 = (r\cos\theta + 8)^2 \quad \text{Square each side. (Section 1.6)}$$

$$(2r)^2 = (x + 8)^2 \quad r\cos\theta = x$$

$$4r^2 = x^2 + 16x + 64 \quad \text{Multiply. (Section R.3)}$$

$$4(x^2 + y^2) = x^2 + 16x + 64 \quad r^2 = x^2 + y^2 \text{ (Section 8.5)}$$

$$4x^2 + 4y^2 = x^2 + 16x + 64 \quad \text{Distributive property}$$

$$3x^2 + 4y^2 - 16x - 64 = 0 \quad \text{Standard form (Section 10.2)}$$

The coefficients of x^2 and y^2 are both positive and are not equal, further supporting our assertion that the graph is an ellipse.

Now Try Exercise 19.

Appendix A Exercises

Graph each conic whose equation is given in polar form. ***See Example 1.***

1. $r = \dfrac{6}{3 + 3 \sin \theta}$ 2. $r = \dfrac{10}{5 + 5 \sin \theta}$

3. $r = \dfrac{-4}{6 + 2 \cos \theta}$ 4. $r = \dfrac{-8}{4 + 2 \cos \theta}$

5. $r = \dfrac{2}{2 - 4 \sin \theta}$ 6. $r = \dfrac{6}{2 - 4 \sin \theta}$

7. $r = \dfrac{-1}{1 + 2 \sin \theta}$ 8. $r = \dfrac{-1}{1 - 2 \sin \theta}$

9. $r = \dfrac{-1}{2 + \cos \theta}$ 10. $r = \dfrac{-1}{2 - \cos \theta}$

Find a polar equation of the parabola with focus at the pole, satisfying the given conditions. ***See Example 2.***

11. The vertical directrix is 3 units to the right of the pole.

12. The vertical directrix is 4 units to the left of the pole.

13. The horizontal directrix is 5 units below the pole.

14. The horizontal directrix is 6 units above the pole.

Find a polar equation for the conic with focus at the pole, satisfying the given conditions. Also identify the type of conic represented. ***See Example 2.***

15. $e = \frac{4}{5}$, and the vertical directrix is 5 units to the right of the pole.

16. $e = \frac{2}{3}$, and the vertical directrix is 6 units to the left of the pole.

17. $e = \frac{5}{4}$, and the horizontal directrix is 8 units below the pole.

18. $e = \frac{3}{2}$, and the horizontal directrix is 4 units above the pole.

Identify the type of conic represented, and convert the equation to rectangular form. ***See Example 3.***

19. $r = \dfrac{6}{3 - \cos \theta}$ 20. $r = \dfrac{8}{4 - \cos \theta}$

21. $r = \dfrac{-2}{1 + 2 \cos \theta}$ 22. $r = \dfrac{-3}{1 + 3 \cos \theta}$

23. $r = \dfrac{-6}{4 + 2 \sin \theta}$ 24. $r = \dfrac{-12}{6 + 3 \sin \theta}$

25. $r = \dfrac{10}{2 - 2 \sin \theta}$ 26. $r = \dfrac{12}{4 - 4 \sin \theta}$

1.
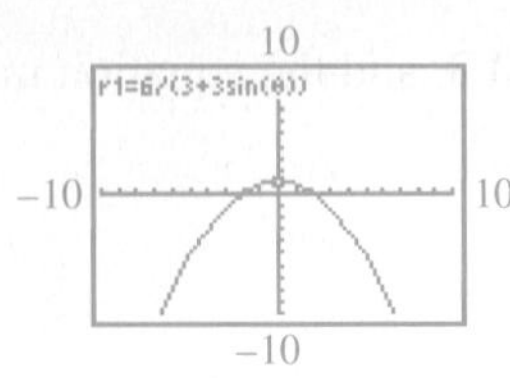

2.
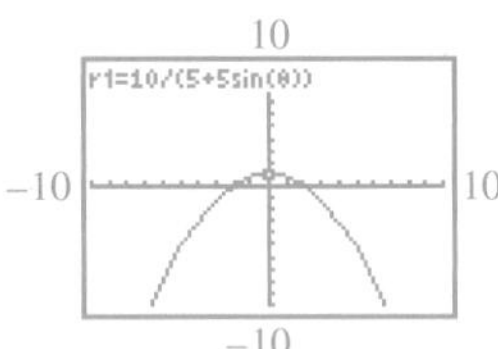

3.
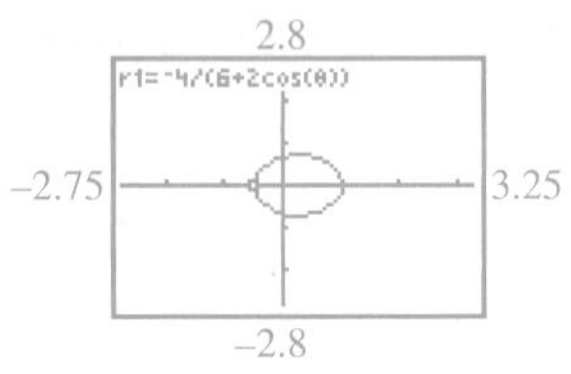

4.
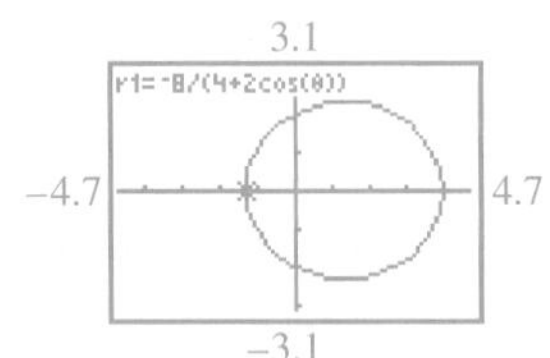

5.
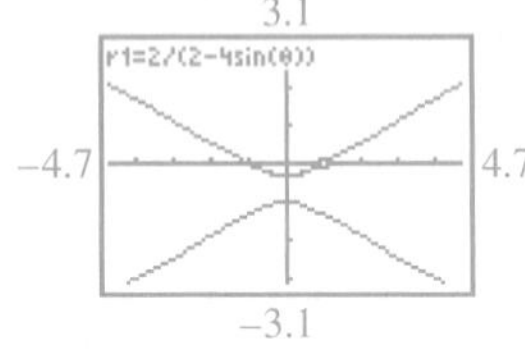

6.
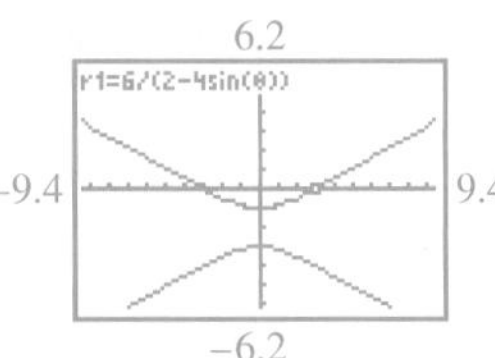

7.
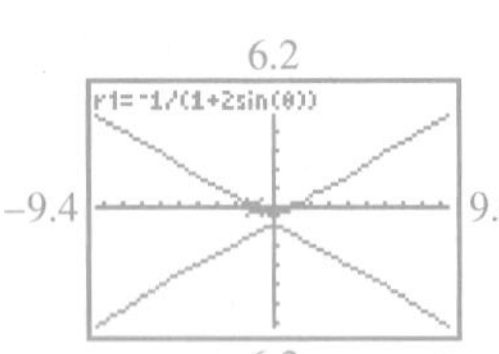

8.
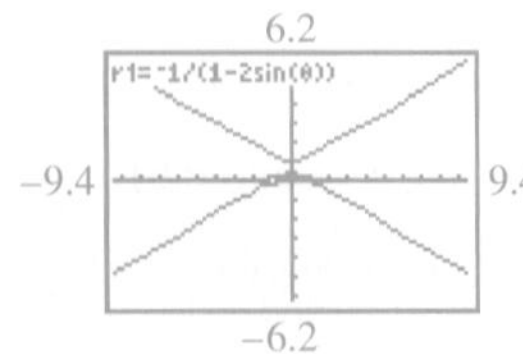

9.
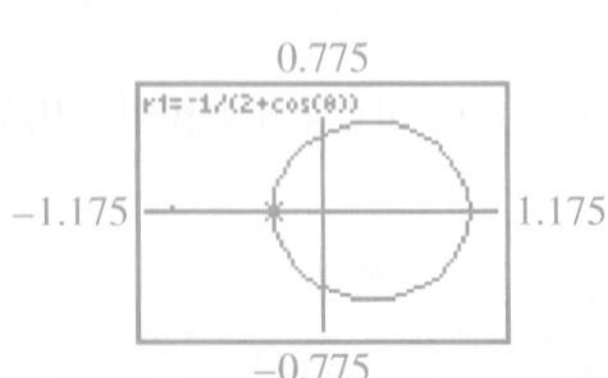

10.
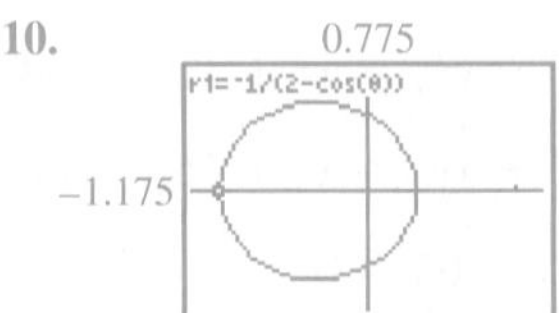

11. $r = \frac{3}{1 + \cos \theta}$ 12. $r = \frac{4}{1 - \cos \theta}$ 13. $r = \frac{5}{1 - \sin \theta}$
14. $r = \frac{6}{1 + \sin \theta}$ 15. $r = \frac{20}{5 + 4 \cos \theta}$; ellipse
16. $r = \frac{12}{3 - 2 \cos \theta}$; ellipse 17. $r = \frac{40}{4 - 5 \sin \theta}$; hyperbola
18. $r = \frac{12}{2 + 3 \sin \theta}$; hyperbola

19. ellipse; $8x^2 + 9y^2 - 12x - 36 = 0$ 20. ellipse; $15x^2 + 16y^2 - 16x - 64 = 0$
21. hyperbola; $3x^2 - y^2 + 8x + 4 = 0$ 22. hyperbola; $8x^2 - y^2 + 18x + 9 = 0$
23. ellipse; $4x^2 + 3y^2 - 6y - 9 = 0$ 24. ellipse; $4x^2 + 3y^2 - 8y - 16 = 0$
25. parabola; $x^2 - 10y - 25 = 0$ 26. parabola; $x^2 - 6y - 9 = 0$

B Rotation of Axes

- Derivation of Rotation Equations
- Applying a Rotation Equation

LOOKING AHEAD TO CALCULUS

Rotation of axes is a topic traditionally covered in calculus texts, in conjunction with parametric equations and polar coordinates. The coverage in calculus is typically the same as that seen in this section.

Derivation of Rotation Equations If we begin with an xy-coordinate system having origin O and rotate the axes about O through an angle θ, the new coordinate system is a **rotation** of the xy-system. Trigonometric identities can be used to obtain equations for converting the coordinates of a point from the xy-system to the rotated $x'y'$-system.

Let P be any point other than the origin, with coordinates (x, y) in the xy-system and (x', y') in the $x'y'$-system. See **Figure 1.**

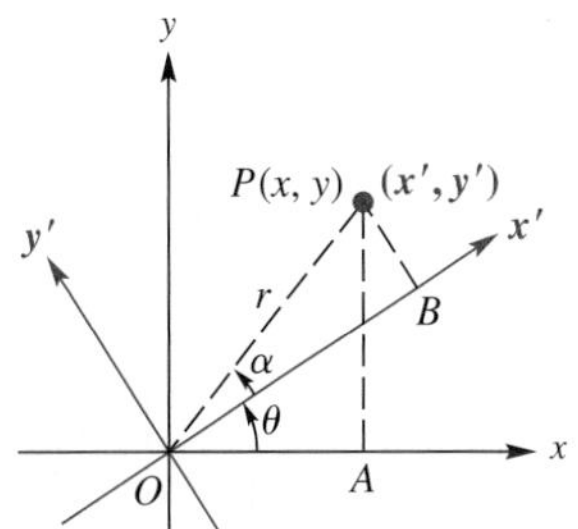

Figure 1

Let $OP = r$, and let α represent the angle made by OP and the x'-axis. **Figure 1** suggests that the following hold.

$$\cos(\theta + \alpha) = \frac{OA}{r} = \frac{x}{r}, \qquad \sin(\theta + \alpha) = \frac{AP}{r} = \frac{y}{r},$$

(Section 5.2)

$$\cos \alpha = \frac{OB}{r} = \frac{x'}{r}, \qquad \sin \alpha = \frac{PB}{r} = \frac{y'}{r}$$

These four statements can be written as follows.

$$x = r\cos(\theta + \alpha), \qquad y = r\sin(\theta + \alpha),$$

$$x' = r\cos \alpha, \qquad y' = r\sin \alpha$$

The trigonometric identity for the cosine of the sum of two angles gives the following equation.

$$x = r\cos(\theta + \alpha)$$

$$= r(\cos\theta\cos\alpha - \sin\theta\sin\alpha) \quad \text{(Section 7.3)}$$

$$= (r\cos\alpha)\cos\theta - (r\sin\alpha)\sin\theta \quad \text{Distributive property (Section R.2)}$$

$$= x'\cos\theta - y'\sin\theta \quad \text{Substitute.}$$

In the same way, the identity for the sine of the sum of two angles gives

$$y = x'\sin\theta + y'\cos\theta.$$

This proves the following result.

Rotation Equations

If the rectangular coordinate axes are rotated about the origin through an angle θ, and if the coordinates of a point P are (x, y) and (x', y') with respect to the xy-system and the $x'y'$-system, respectively, then the **rotation equations** are as follows.

$$\mathbf{x = x'\cos\theta - y'\sin\theta} \quad \text{and} \quad \mathbf{y = x'\sin\theta + y'\cos\theta}$$

Applying a Rotation Equation

Classroom Example 1
The equation of a curve is

$$2x^2 - xy + 2y^2 = 25.$$

Find the resulting equation if the axes are rotated 45°. Graph the equation.

Answer: $\frac{3x'^2}{50} + \frac{y'^2}{10} = 1$

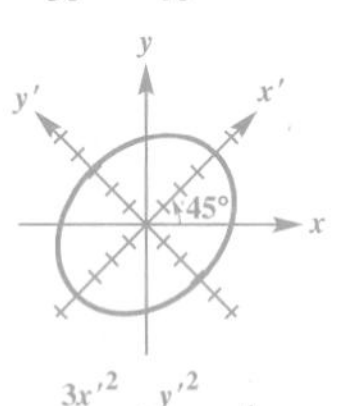

$\frac{3x'^2}{50} + \frac{y'^2}{10} = 1$

EXAMPLE 1 Finding an Equation after a Rotation

The equation of a curve is

$$x^2 + y^2 + 2xy + 2\sqrt{2}x - 2\sqrt{2}y = 0.$$

Find the resulting equation if the axes are rotated 45°. Graph the equation.

SOLUTION If $\theta = 45°$, then $\sin\theta = \frac{\sqrt{2}}{2}$ and $\cos\theta = \frac{\sqrt{2}}{2}$, and the rotation equations become

$$x = \frac{\sqrt{2}}{2}x' - \frac{\sqrt{2}}{2}y' \quad \text{and} \quad y = \frac{\sqrt{2}}{2}x' + \frac{\sqrt{2}}{2}y'.$$

Substitute these values into the given equation.

$$x^2 + y^2 + 2xy + 2\sqrt{2}x - 2\sqrt{2}y = 0$$

$$\left[\frac{\sqrt{2}}{2}x' - \frac{\sqrt{2}}{2}y'\right]^2 + \left[\frac{\sqrt{2}}{2}x' + \frac{\sqrt{2}}{2}y'\right]^2 + 2\left[\frac{\sqrt{2}}{2}x' - \frac{\sqrt{2}}{2}y'\right]\left[\frac{\sqrt{2}}{2}x' + \frac{\sqrt{2}}{2}y'\right]$$

$$+ 2\sqrt{2}\left[\frac{\sqrt{2}}{2}x' - \frac{\sqrt{2}}{2}y'\right] - 2\sqrt{2}\left[\frac{\sqrt{2}}{2}x' + \frac{\sqrt{2}}{2}y'\right] = 0$$

$$\frac{1}{2}x'^2 - x'y' + \frac{1}{2}y'^2 + \frac{1}{2}x'^2 + x'y' + \frac{1}{2}y'^2 + x'^2 - y'^2 + 2x' - 2y' - 2x' - 2y' = 0 \quad \text{Expand terms.}$$

$$2x'^2 - 4y' = 0 \quad \text{Combine like terms.}$$

$$x'^2 - 2y' = 0 \quad \text{Divide by 2.}$$

$$x'^2 = 2y' \quad \text{(Section 10.1)}$$

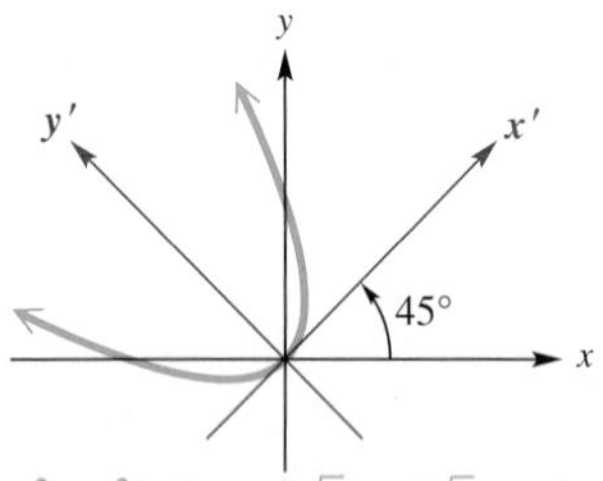

$x^2 + y^2 + 2xy + 2\sqrt{2}x - 2\sqrt{2}y = 0$
$x'^2 = 2y'$

Figure 2

This is the equation of a parabola. The graph is shown in **Figure 2.**

Now Try Exercise 13.

We have graphed equations written in the general form

$$Ax^2 + Cy^2 + Dx + Ey + F = 0. \quad \text{(Section 10.4)}$$

To graph an equation that has an xy-term by hand, it is necessary to find an appropriate **angle of rotation** to eliminate the xy-term. The necessary angle of rotation can be determined by using the following result. The proof is quite lengthy and is not presented here.

Angle of Rotation

The xy-term is removed from the general equation

$$\mathbf{Ax^2 + Bxy + Cy^2 + Dx + Ey + F = 0}$$

by a rotation of the axes through an angle θ, $0° < \theta < 90°$, where

$$\cot 2\theta = \frac{A - C}{B}.$$

To find the rotation equations, first find $\sin\theta$ and $\cos\theta$. **Example 2** illustrates a way to obtain $\sin\theta$ and $\cos\theta$ from $\cot 2\theta$ without first identifying angle θ.

Classroom Example 2
Remove the xy-term from

$$x^2 + xy + y^2 = 3$$

by performing a suitable rotation, and graph the equation.

Answer: $\frac{x'^2}{2} + \frac{y'^2}{6} = 1$

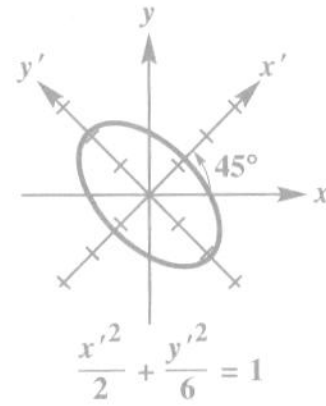

EXAMPLE 2 Rotating and Graphing

Remove the xy-term from $52x^2 - 72xy + 73y^2 = 200$ by performing a suitable rotation, and graph the equation.

SOLUTION Here $A = 52$, $B = -72$, and $C = 73$.

$$\cot 2\theta = \frac{A - C}{B} = \frac{52 - 73}{-72} = \frac{-21}{-72} = \frac{7}{24}$$ Substitute into the angle of rotation equation, and simplify.

To find $\sin\theta$ and $\cos\theta$, use these trigonometric identities.

$$\sin\theta = \sqrt{\frac{1 - \cos 2\theta}{2}} \quad \text{and} \quad \cos\theta = \sqrt{\frac{1 + \cos 2\theta}{2}}$$ (Section 7.4)

Sketch a right triangle as in **Figure 3,** to see that $\cos 2\theta = \frac{7}{25}$. (In the two quadrants for which we are concerned, cosine and cotangent have the same sign.)

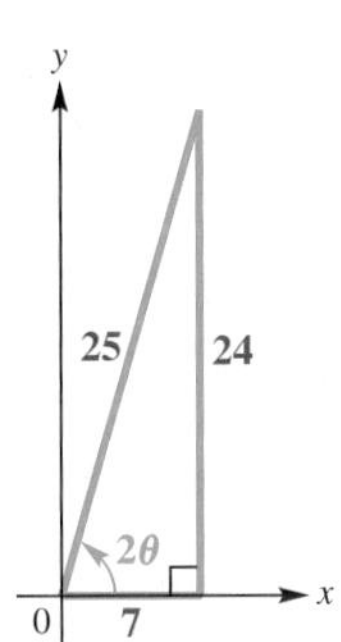

Figure 3

$$\sin\theta = \sqrt{\frac{1 - \frac{7}{25}}{2}} = \sqrt{\frac{9}{25}} = \frac{3}{5} \quad \text{and} \quad \cos\theta = \sqrt{\frac{1 + \frac{7}{25}}{2}} = \sqrt{\frac{16}{25}} = \frac{4}{5}$$

Use these values for $\sin\theta$ and $\cos\theta$ to obtain the following.

$$x = \frac{4}{5}x' - \frac{3}{5}y' \quad \text{and} \quad y = \frac{3}{5}x' + \frac{4}{5}y'$$

Substitute these expressions for x and y into the original equation.

$$52\left[\frac{4}{5}x' - \frac{3}{5}y'\right]^2 - 72\left[\frac{4}{5}x' - \frac{3}{5}y'\right]\left[\frac{3}{5}x' + \frac{4}{5}y'\right] + 73\left[\frac{3}{5}x' + \frac{4}{5}y'\right]^2 = 200$$

$$52\left[\frac{16}{25}x'^2 - \frac{24}{25}x'y' + \frac{9}{25}y'^2\right] - 72\left[\frac{12}{25}x'^2 + \frac{7}{25}x'y' - \frac{12}{25}y'^2\right] + 73\left[\frac{9}{25}x'^2 + \frac{24}{25}x'y' + \frac{16}{25}y'^2\right] = 200$$

$$25x'^2 + 100y'^2 = 200$$ Combine like terms.

$$\frac{x'^2}{8} + \frac{y'^2}{2} = 1$$ Divide by 200.

This is an equation of an ellipse having x'-intercepts $\pm 2\sqrt{2}$ and y'-intercepts $\pm\sqrt{2}$. The graph is shown in **Figure 4.** To find θ, use the following.

$$\frac{\sin\theta}{\cos\theta} = \frac{\frac{3}{5}}{\frac{4}{5}} = \frac{3}{4} = \tan\theta$$ (Section 7.1)

Use a calculator to find $\tan^{-1}\left(\frac{3}{4}\right) \approx 36.87°$.

✔ *Now Try Exercise 17.*

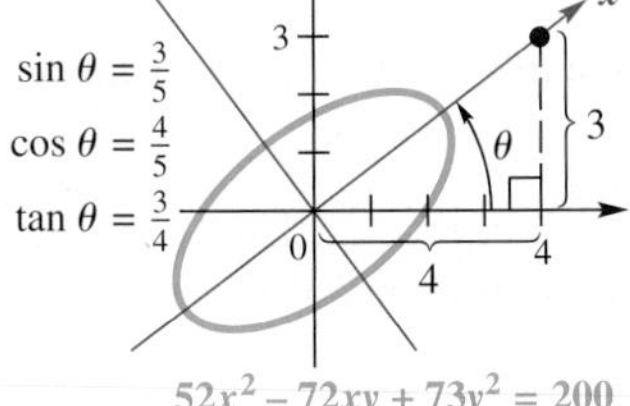

Figure 4

Equation of a Conic with an xy-Term

If the general second-degree equation

$$Ax^2 + Bxy + Cy^2 + Dx + Ey + F = 0$$

has a graph, it will be one of the following:

(a) a circle or an ellipse or a point if $B^2 - 4AC < 0$;

(b) a parabola or one line or two parallel lines if $B^2 - 4AC = 0$;

(c) a hyperbola or two intersecting lines if $B^2 - 4AC > 0$;

(d) a straight line if $A = B = C = 0$, and $D \neq 0$ or $E \neq 0$.

Appendix B Exercises

*Use the summary following **Example 2** to predict the type of graph of each second-degree equation.*

1. $4x^2 + 3y^2 + 2xy - 5x = 8$
2. $x^2 + 2xy - 3y^2 + 2y = 12$
3. $2x^2 + 3xy - 4y^2 = 0$
4. $x^2 - 2xy + y^2 + 4x - 8y = 0$
5. $4x^2 + 4xy + y^2 + 15 = 0$
6. $x^2 - 2xy + y^2 - 16 = 0$

Find the angle of rotation θ that will remove the xy-term in each equation.

7. $2x^2 + \sqrt{3}xy + y^2 + x = 5$
8. $4\sqrt{3}x^2 + xy + 3\sqrt{3}y^2 = 10$
9. $3x^2 + \sqrt{3}xy + 4y^2 + 2x - 3y = 12$
10. $4x^2 + 2xy + 2y^2 + x = 7$
11. $x^2 - 4xy + 5y^2 = 18$
12. $3\sqrt{3}x^2 - 2xy + \sqrt{3}y^2 = 25$

*Find the resulting equation if the axes are rotated through angle θ. Graph the equation. **See Example 1.***

13. $x^2 - xy + y^2 - 6;\ \theta = 45°$
14. $5y^2 + 12xy = 10;\ \theta = \sin^{-1}\left(\frac{3\sqrt{13}}{13}\right)$

*Remove the xy-term from each equation by performing a suitable rotation. Graph each equation. **See Example 2.***

15. $x^2 - 4xy + y^2 = -5$
16. $3x^2 - 2xy + 3y^2 = 8$
17. $7x^2 + 6\sqrt{3}xy + 13y^2 = 64$
18. $x^2 + 2xy + y^2 + 4\sqrt{2}x = 4\sqrt{2}y$
19. $3x^2 - 2\sqrt{3}xy + y^2 - 2x = 2\sqrt{3}y$
20. $7x^2 + 2\sqrt{3}xy + 5y^2 = 24$

Remove the xy-term by rotation. Then translate the axes and sketch the graph.

21. $x^2 + 3xy + y^2 - 5\sqrt{2}y - 15 = 0$
22. $x^2 - \sqrt{3}xy + 2\sqrt{3}x - 3y - 3 = 0$
23. $4x^2 + 4xy + y^2 - 24x + 38y - 19 = 0$
24. $12x^2 + 24xy + 19y^2 - 12x - 40y + 31 = 0$
25. $16x^2 + 24xy + 9y^2 - 130x + 90y = 0$
26. $9x^2 - 6xy + y^2 - 12\sqrt{10}x - 36\sqrt{10}y = 0$
27. *Concept Check* Look at the box titled "Angle of Rotation." Why is no rotation applicable if the value of B is 0?
28. *Concept Check* Look at the equation involving $\cot 2\theta$ in the box titled "Angle of Rotation." Why must the angle of rotation be 45° if the coefficients of x^2 and y^2 are equal, and $B \neq 0$?

---

1. circle or ellipse or a point
2. hyperbola or two intersecting lines
3. hyperbola or two intersecting lines
4. parabola or one line or two parallel lines
5. parabola or one line or two parallel lines
6. parabola or one line or two parallel lines

7. 30° 8. 15°
9. 60° 10. 22.5°
11. 22.5° 12. 75°

13. 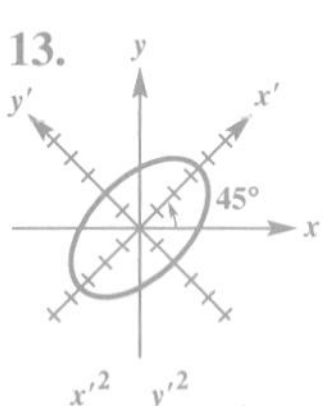

14. 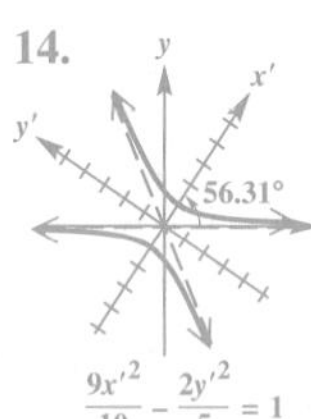

15. 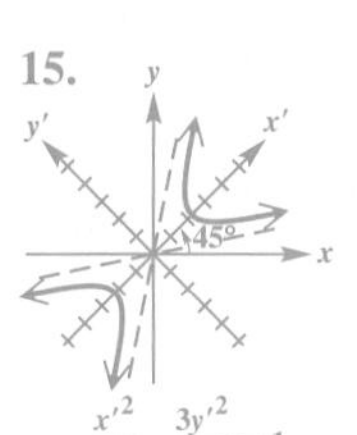

16. 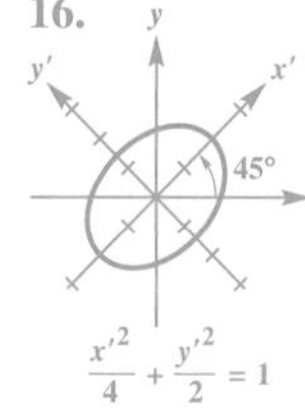

17. 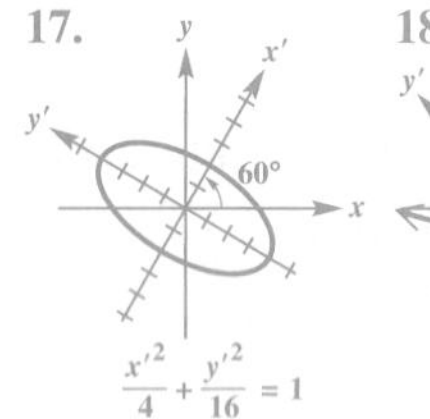

18. 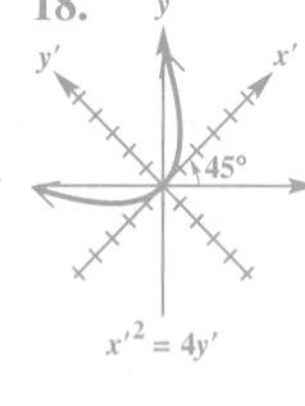

19. 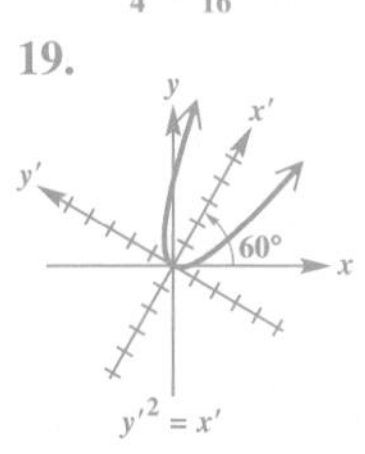

20.
30°
$\frac{x'^2}{3} + \frac{y'^2}{6} = 1$

21. 22. 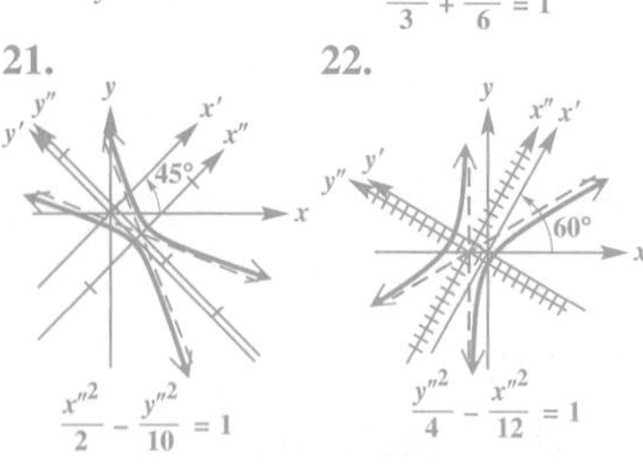

23. 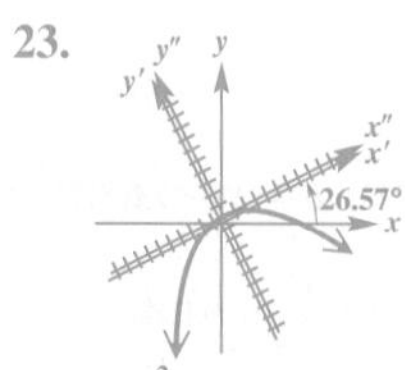

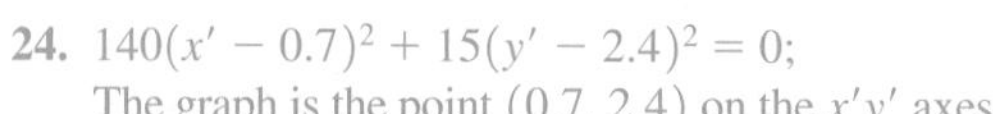
24. $140(x' - 0.7)^2 + 15(y' - 2.4)^2 = 0$; The graph is the point $(0.7, 2.4)$ on the $x'y'$ axes.

25. 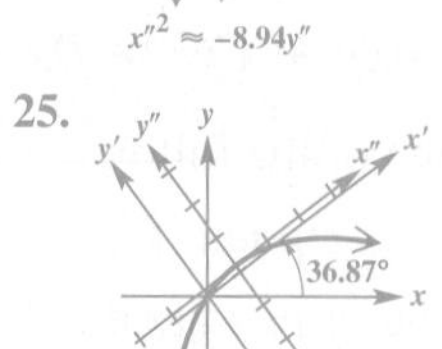

26. 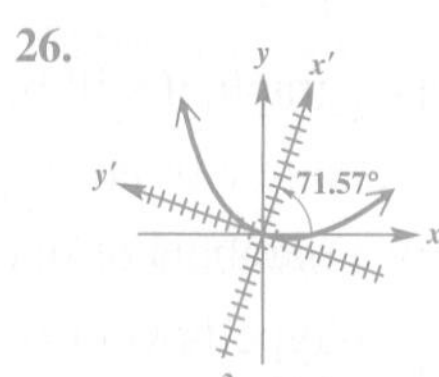

27. If $B = 0$, $\cot 2\theta$ is undefined. The graph may be translated but is not rotated.
28. If $A = C$, then $A - C = 0$, $\cot 2\theta = 0$, $2\theta = 90°$, and $\theta = 45°$.

C Geometry Formulas

Square

Perimeter: $P = 4s$

Area: $\mathcal{A} = s^2$

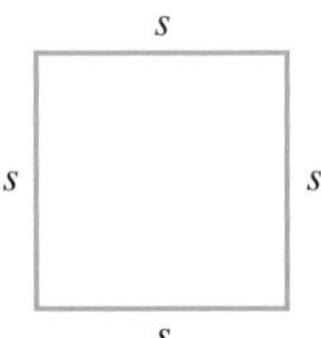

Rectangle

Perimeter: $P = 2L + 2W$

Area: $\mathcal{A} = LW$

Triangle

Perimeter: $P = a + b + c$

Area: $\mathcal{A} = \frac{1}{2}bh$

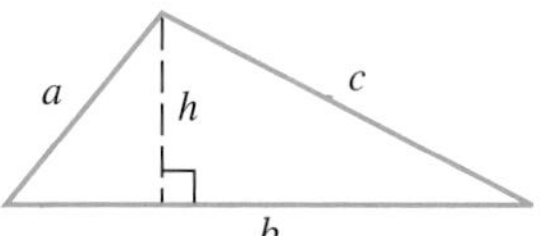

Parallelogram

Perimeter: $P = 2a + 2b$

Area: $\mathcal{A} = bh$

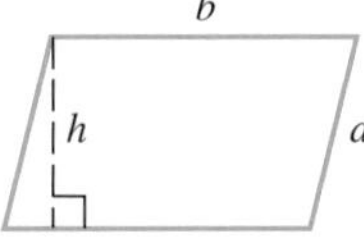

Trapezoid

Perimeter: $P = a + b + c + B$

Area: $\mathcal{A} = \frac{1}{2}h(B + b)$

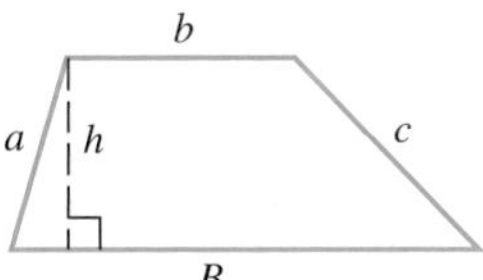

Circle

Diameter: $d = 2r$

Circumference: $C = 2\pi r = \pi d$

Area: $\mathcal{A} = \pi r^2$

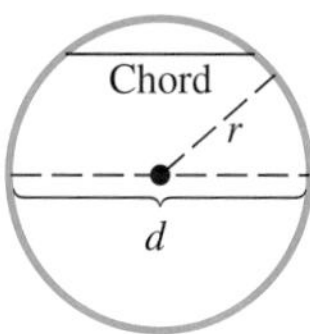

Cube

Volume: $V = e^3$

Surface area: $S = 6e^2$

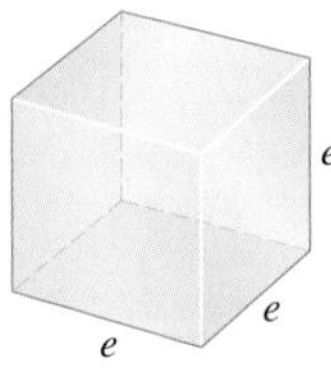

Rectangular Solid

Volume: $V = LWH$

Surface area: $S = 2HW + 2LW + 2LH$

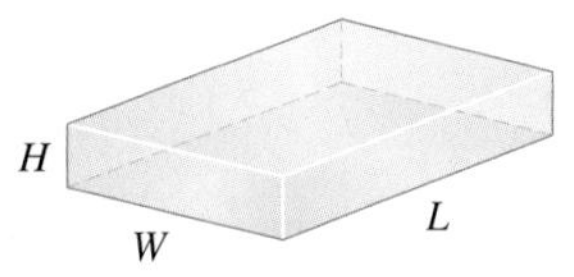

Sphere

Volume: $V = \frac{4}{3}\pi r^3$

Surface area: $S = 4\pi r^2$

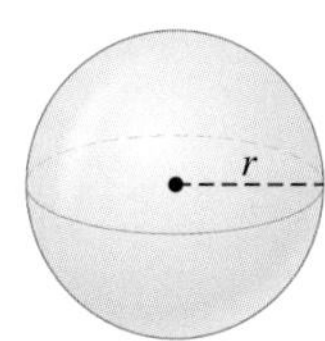

Cone

Volume: $V = \frac{1}{3}\pi r^2 h$

Surface area: $S = \pi r\sqrt{r^2 + h^2}$
(excludes the base)

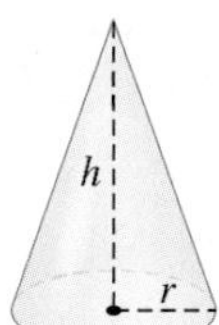

Right Circular Cylinder

Volume: $V = \pi r^2 h$

Surface area: $S = 2\pi rh + 2\pi r^2$
(includes top and bottom)

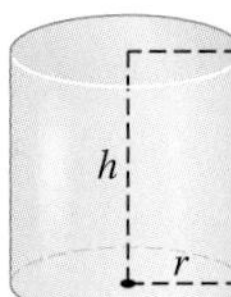

Right Pyramid

Volume: $V = \frac{1}{3}Bh$

B = area of the base

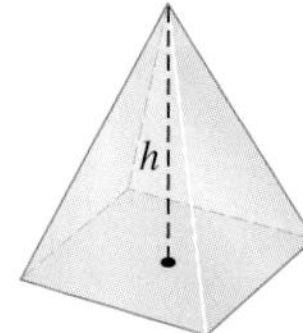

Glossary

For a more complete discussion, see the section(s) in parentheses.

A

abscissa The x-value of a point may be called its abscissa. (Section 6.6)

absolute value The absolute value of a real number is the distance between 0 and the number on a number line. (Sections R.2, 1.8)

absolute value (modulus) of a complex number When a complex number is written in trigonometric (or polar) form as $r(\cos\theta + i\sin\theta)$, the number r is the absolute value (or modulus) of the complex number. (Section 8.5)

acute angle An acute angle is an angle measuring between 0° and 90°. (Section 5.1)

addition of ordinates Addition of ordinates is a method for graphing a function that is the sum of two other functions, using addition of the y-values of the two functions at selected x-values. (Section 6.6)

additive inverse (negative) of a matrix If the sum of two matrices is the zero matrix, then those two matrices are additive inverses (negatives) of each other. (Section 9.7)

airspeed In air navigation, the airspeed of a plane is its speed relative to the air. (Section 8.4)

algebraic expression Any collection of numbers or variables joined by the basic operations of addition, subtraction, multiplication, or division (except by 0), or by the operations of raising to powers or taking roots, formed according to the rules of algebra, is an algebraic expression. (Section R.3)

ambiguous case The situation in which the lengths of two sides of a triangle and the measure of the angle opposite one of them are given (SSA) is the ambiguous case of the law of sines. Depending on the given measurements, this combination of given parts may result in 0, 1, or 2 possible triangles. (Section 8.1)

amplitude The amplitude of a periodic function is half the difference between the maximum and minimum values of the function. (Section 6.3)

angle An angle is formed by rotating a ray around its endpoint. (Section 5.1)

angle of depression The angle of depression from point X to point Y (below X) is the acute angle formed by ray XY and a horizontal ray with endpoint at X. (Section 5.4)

angle of elevation The angle of elevation from point X to point Y (above X) is the acute angle formed by ray XY and a horizontal ray with endpoint at X. (Section 5.4)

angle of rotation The angle through which the axes of an xy-coordinate system are rotated to obtain a new coordinate system is the angle of rotation. (Appendix B)

Angle-Side-Angle (ASA) The Angle-Side-Angle (ASA) congruence axiom states that if two angles and the included side of one triangle are equal, respectively, to two angles and the included side of a second triangle, then the triangles are congruent. (Section 8.1)

angle in standard position An angle is in standard position if its vertex is at the origin and its initial side is along the positive x-axis. (Section 5.1)

angle between two vectors The angle between two vectors is defined to be the angle θ, for $0° \le \theta \le 180°$, having the two vectors as its sides. (Section 8.3)

angular speed ω Angular speed ω (omega) measures the speed of rotation and is defined by $\omega = \frac{\theta}{t}$, where θ is the angle of rotation in radians and t is time. (Section 6.2)

annuity An annuity is a sequence of equal payments made at equal periods of time. (Section 11.3)

argument of a complex number When a complex number is written in trigonometric (or polar) form as $r(\cos\theta + i\sin\theta)$, the angle θ is the argument of the complex number. (Section 8.5)

argument of a function The argument of a function is the expression containing the independent variable of the function. For example, in the function $y = f(x - d)$, the expression $x - d$ is the argument. (Section 6.4)

argument of a logarithm In the expression $\log_a x$, the argument is x. (Section 4.3)

arithmetic sequence (arithmetic progression) An arithmetic sequence is a sequence in which each term after the first differs from the preceding term by a fixed constant. (Section 11.2)

arithmetic series An arithmetic series is the sum of the terms of an arithmetic sequence. (Section 11.2)

asymptotes of a hyperbola The two intersecting straight lines that the branches of a hyperbola approach are the asymptotes of the hyperbola. (Section 10.3)

augmented matrix An augmented matrix is a matrix whose elements are the coefficients of the variables and the constants of a system of equations. An augmented matrix is written with a vertical bar that separates the coefficients of the variables from the constants. (Section 9.2)

average rate of change The slope of a line gives the average rate of change in y, where the value of y depends on the value of x. (Section 2.4)

axis (of symmetry) of a parabola The line of symmetry for a parabola is the axis of the parabola. (Sections 3.1, 10.1)

base of an exponential The base is the number that is a repeated factor in exponential notation. In the expression a^n, the base is a. (Section R.2)

base of a logarithm In the expression $\log_a x$, the base is a. (Section 4.3)

bearing Bearing is used to identify angles in navigation. One method for expressing bearing uses a single angle, with bearing measured in a clockwise direction from due north. A second method for expressing bearing starts with a north-south line and uses an acute angle to show the direction, either east or west, from this line. (Sections 5.4, 8.4)

binomial A binomial is a polynomial containing exactly two terms. (Section R.3)

binomial coefficient For nonnegative integers n and r, with $r \le n$, the binomial coefficient is the value of $\frac{n!}{r!(n-r)!}$. These values are used in calculating the coefficients of the terms of a binomial expansion. (Section 11.4)

binomial experiment In probability, an experiment consisting of n independent trials, with two possible outcomes for each trial, and constant probabilities for each trial, is a binomial experiment. (Section 11.7)

binomial theorem (general binomial expansion) The binomial theorem is used to expand a binomial raised to a power. (Section 11.4)

boundary A line that separates a plane into two half-planes is the boundary of each half-plane. (Section 9.6)

break-even point The break-even point is the point where the revenue from selling a product is equal to the cost of producing it. (Sections 1.7, 2.4)

C

cardioid A cardioid is a heart-shaped curve that is the graph of a polar equation of the form $r = a \pm b \sin \theta$ or $r = a \pm b \cos \theta$, where $\left|\frac{a}{b}\right| = 1$. (Section 8.7)

center of a circle The center of a circle is the given point that is a given distance from all points on the circle. (Section 2.2)

center of an ellipse The center of an ellipse is the midpoint of the major axis. (Section 10.2)

center of a hyperbola The center of a hyperbola is the midpoint of the transverse axis. (Section 10.3)

change in x The change in x is the horizontal difference (the difference in x-coordinates) between two points on a line. (Section 2.4)

change in y The change in y is the vertical difference (the difference in y-coordinates) between two points on a line. (Section 2.4)

circle A circle is the set of all points in a plane that lie a given distance from a given point. (Section 2.2)

circular functions The trigonometric functions of arc lengths, or real numbers, are the circular functions. (Section 6.2)

closed interval A closed interval is an interval that includes both of its endpoints. (Section 1.7)

coefficient (numerical coefficient) The real number factor in a term of an algebraic expression is the coefficient of the other factors. (Section R.3)

cofactor The product of a minor of an element of a square matrix and the number $+1$ (if the sum of the row number and column number is even) or -1 (if the sum of the row number and column number is odd) is a cofactor. (Section 9.3)

cofunctions The function pairs sine and cosine, tangent and cotangent, and secant and cosecant are cofunctions. (Section 5.3)

collinear Two points are collinear if they lie on the same line. (Section 2.1)

column matrix A matrix with exactly one column is a column matrix. (Section 9.7)

combination A subset of items selected *without regard to order* is a combination. (Section 11.6)

combined variation Variation in which one variable depends on more than one other variable is combined variation. (Section 3.6)

common difference In an arithmetic sequence, the fixed number that is added to each term to obtain the next term is the common difference. (Section 11.2)

common logarithm A base 10 logarithm is a common logarithm. (Section 4.4)

common ratio In a geometric sequence, the fixed number by which each term is multiplied to obtain the next term is the common ratio. (Section 11.3)

complement of an event In probability, the set of all outcomes in a sample space that do *not* belong to an event E is the complement of E, written E'. (Section 11.7)

complement of a set The set of all elements in the universal set U that do not belong to set A is the complement of A, written A'. (Sections R.1, 11.7)

complementary angles (complements) Two positive angles are complementary angles (or complements) if the sum of their measures is 90°. (Section 5.1)

completing the square The process of adding to a binomial the expression that makes it a perfect square trinomial is called completing the square. (Sections 1.4, 3.1)

complex conjugates The complex numbers $a + bi$ and $a - bi$ are complex conjugates. (Section 1.3)

complex fraction A complex fraction is a quotient of two rational expressions. (Section R.5)

complex number A complex number is a number of the form $a + bi$, where a and b are real numbers and $i = \sqrt{-1}$. (Section 1.3)

complex plane The complex plane is a two-dimensional representation of the complex numbers in which the horizontal axis is the real axis and the vertical axis is the imaginary axis. (Section 8.5)

composite function (composition) If f and g are functions, then the composite function, or composition, of g and f is symbolized by $(g \circ f)(x)$, and defined to be $g(f(x))$. (Section 2.8)

compound amount In an investment paying compound interest, the compound amount is the balance *after* interest has been earned. (The compound amount is sometimes called the *future value.*) (Section 4.2)

compound interest For compound interest, interest is paid both on the principal and previously earned interest. (Section 4.2)

conditional equation An equation that is satisfied by some numbers but not by others is a conditional equation. (Section 1.1)

congruent triangles Triangles that are both the same size and the same shape are congruent triangles. (Section 8.1)

conic (conic section) (geometric definition) A conic is the set of all points $P(x, y)$ in a plane such that the ratio of the distance from P to a fixed point and the distance from P to a fixed line is constant. (Sections 10.2, 10.4)

conic sections (conics) When a plane intersects a portion of a cone, the figure formed by the intersection is a conic section. (Section 10.1)

conjugates The expressions $a - b$ and $a + b$ are conjugates. (Section R.7)

consistent system A consistent system is a system of equations with at least one solution. (Section 9.1)

constant function A function f is constant on an interval I if, for every x_1 and x_2 in I, $f(x_1) = f(x_2)$. (Section 2.3)

constant of variation In a variation equation, such as $y = kx$, $y = \frac{k}{x}$, or $y = kx^n z^m$, the real number k is the constant of variation. (Section 3.6)

constraints In linear programming, the inequalities that represent restrictions on a particular situation are the constraints. (Section 9.6)

continuous compounding Continuous compounding of money involves the computation of interest as the frequency of compounding approaches infinity, leading to the formula $A = Pe^{rt}$. (Section 4.2)

continuous function (informal definition) A function is continuous over an interval of its domain if its hand-drawn graph over that interval can be sketched without lifting the pencil from the paper. (Section 2.6)

contradiction An equation that has no solution is a contradiction. (Section 1.1)

convergent sequence An infinite sequence is convergent if its terms get closer and closer to some real number. (Section 11.1)

coordinate (on a number line) A number that corresponds to a particular point on a number line is the coordinate of the point. (Section R.2)

coordinate plane (*xy*-plane) The plane into which the rectangular coordinate system is introduced is the coordinate plane (or *xy*-plane). (Section 2.1)

coordinates (in the *xy*-plane) The coordinates of a point in the *xy*-plane are the numbers in the ordered pair that correspond to that point. (Section 2.1)

coordinate system (on a number line) The correspondence between points on a line and the real numbers is a coordinate system. (Section R.2)

cosecant Let $P(x, y)$ be a point other than the origin on the terminal side of an angle θ in standard position. Let $r = \sqrt{x^2 + y^2}$ represent the distance from the origin to P. Then the cosecant function is defined by $\csc \theta = \frac{r}{y}$ $(y \neq 0)$. (Section 5.2)

cosine Let $P(x, y)$ be a point other than the origin on the terminal side of an angle θ in standard position. Let $r = \sqrt{x^2 + y^2}$ represent the distance from the origin to P. Then the cosine function is defined by $\cos \theta = \frac{x}{r}$. (Section 5.2)

cotangent Let $P(x, y)$ be a point other than the origin on the terminal side of an angle θ in standard position. Let $r = \sqrt{x^2 + y^2}$ represent the distance from the origin to P. Then the cotangent function is defined by $\cot \theta = \frac{x}{y}$ $(y \neq 0)$. (Section 5.2)

coterminal angles Two angles that have the same initial side and the same terminal side, but different amounts of rotation, are coterminal angles. The measures of coterminal angles differ by a multiple of 360°. (Section 5.1)

Cramer's rule Cramer's rule is a method of using determinants to solve systems of linear equations. (Section 9.3)

cycloid A cycloid is a curve that represents the path traced by a fixed point on the circumference of a circle rolling along a line. (Section 8.8)

D

damped oscillatory motion Damped oscillatory motion is oscillatory motion that has been slowed down (damped) by the force of friction. Friction causes the amplitude of the motion to diminish gradually until the weight comes to rest. (Section 6.7)

decreasing function A function f is decreasing on an interval I if, whenever $x_1 < x_2$ in I, $f(x_1) > f(x_2)$. (Section 2.3)

degree The degree is a unit of measure for angles. One degree, written 1°, represents $\frac{1}{360}$ of a rotation. (Section 5.1)

degree of a polynomial The greatest degree of any term in a polynomial is the degree of the polynomial. (Section R.3)

degree of a term The degree of a term is the sum of the exponents on the variable factors in the term. (Section R.3)

dependent equations Two equations are dependent if any solution of one equation is also a solution of the other. (Section 9.1)

dependent variable If the value of the variable y depends on the value of the variable x, then y is the dependent variable. (Section 2.3)

determinant Every $n \times n$ matrix A is associated with a single real number called the determinant of A, written $|A|$. (Section 9.3)

difference quotient If f is a function and h is a positive number, then the expression $\frac{f(x + h) - f(x)}{h}$ is the difference quotient. (Section 2.8)

dimension of a matrix The dimension of a matrix indicates the number of rows and columns that the matrix has, with the number of rows given first. (Section 9.2)

direction angle The positive angle between the *x*-axis and a position vector is the direction angle for the vector. (Section 8.3)

directrix A directrix is a fixed line that, together with a focus, is used to determine the points that form a conic section. (Sections 10.1, 10.4)

discriminant The quantity under the radical in the quadratic formula, $b^2 - 4ac$, is the discriminant of the expression $ax^2 + bx + c$. (Section 1.4)

disjoint sets Two sets that have no elements in common are disjoint sets. (Section R.1)

divergent sequence If an infinite sequence does not converge to some number, then it is a divergent sequence. (Section 11.1)

domain of a rational expression The domain of a rational expression is the set of real numbers for which the expression is defined. (Section R.5)

domain of a relation In a relation, the set of all values of the independent variable (x) is the domain. (Section 2.3)

dominating term In a polynomial function, the dominating term is the term of greatest degree. (Section 3.4)

dot product The dot product of two vectors is the sum of the product of their first components and the product of their second components. The dot product of the two vectors $\mathbf{u} = \langle a, b \rangle$ and $\mathbf{v} = \langle c, d \rangle$ is denoted $\mathbf{u} \cdot \mathbf{v}$ and given by $\mathbf{u} \cdot \mathbf{v} = ac + bd$. (Section 8.3)

doubling time The amount of time that it takes for a quantity that grows exponentially to become twice its initial amount is its doubling time. (Section 4.6)

E

eccentricity The eccentricity of a parabola, ellipse, or hyperbola is the fixed ratio of the distance from a point P to a focus compared to the distance from the same point to the directrix. (The eccentricity of a circle is 0). (Section 10.2)

element of a matrix Each entry in a matrix is an element of the matrix. (Section 9.2)

elements (members) The objects that belong to a set are the elements (members) of the set. (Section R.1)

ellipse An ellipse is the set of all points in a plane such that the sum of the distances from two fixed points is constant. (Section 10.2)

empty set (null set) The empty set (or null set), written $\emptyset$ or $\{\ \}$, is the set containing no elements. (Sections R.1, 1.1)

end behavior The end behavior of a graph of a polynomial function describes how the values of y increase or decrease as $|x|$ increases without bound. (Section 3.4)

endpoint of a ray In a given ray AB, point A is the endpoint of the ray. (Section 5.1)

equation An equation is a statement that two expressions are equal. (Section 1.1)

equation with a rational exponent An equation that contains a variable or variable expression, raised to an exponent that is a rational number is an equation with a rational exponent. (Section 1.6)

equilibrant The opposite vector of the resultant of two vectors is the equilibrant. (Section 8.4)

equivalent equations Equations with the same solution set are equivalent equations. (Section 1.1)

equivalent systems Equivalent systems are systems of equations that have the same solution set. (Section 9.1)

even function A function f is an even function if, for all x in the domain of f, $f(-x) = f(x)$. The graph of an even function is symmetric with respect to the *y*-axis. (Section 2.7)

event In probability, any subset of the sample space is an event. (Section 11.7)

exact number A number that represents the result of counting, or a number that results from theoretical work and is not the result of a measurement, is an exact number. (Section 5.4)

expansion by a row or column Expansion by a row or column is a method for evaluating a determinant of a 3×3 or larger matrix. This process involves multiplying each element of any row or column of the matrix by its cofactor and then adding these products. (Section 9.3)

exponent In the expression a^n, the exponent n indicates the number of times that the base a is used as a factor. (Section R.2)

exponential equation An exponential equation is an equation with a variable in an exponent. (Section 4.2)

exponential function If $a > 0$ and $a \neq 1$, then $f(x) = a^x$ defines the exponential function with base a. (Section 4.2)

exponential growth or decay function An exponential growth or decay function models a situation in which a quantity changes at a rate proportional to the amount present. Such a function is defined by an equation of the form $y = y_0e^{kt}$, where y_0 is the amount or number present at time $t = 0$ and k is a constant. (Section 4.6)

F

factored completely A polynomial is factored completely when it is written as a product of prime polynomials. (Section R.4)

factored form A polynomial is written in factored form when it is written as a product of polynomials. (Section R.4)

factorial notation Factorial notation is a compact way of writing a product of consecutive natural numbers. The symbol $n!$, read "n-factorial," is defined as follows: For any positive integer n, $n! = n(n-1)(n-2) \cdots (3)(2)(1)$, and $0! = 1$. (Section 11.4)

factoring The process of finding polynomials whose product equals a given polynomial is called factoring. (Section R.4)

factoring by grouping Factoring by grouping is a method of grouping the terms of a polynomial in such a way that the polynomial can be factored even though the greatest common factor of its terms is 1. (Section R.4)

Fibonacci sequence The Fibonacci sequence is the sequence 1, 1, 2, 3, 5, 8, 13, ... In this sequence, each term starting with the third term is the sum of the previous two terms. (Section 11.1)

finite sequence A sequence is a finite sequence if its domain is the set $\{1, 2, 3, \ldots, n\}$, where n is a natural number. (Section 11.1)

finite series A finite series is an expression of the form $S_n = a_1 + a_2 + a_3 + \cdots + a_n = \sum_{i=1}^{n} a_i$. (Section 11.1)

finite set A finite set is a set that has a limited number of elements. (Section R.1)

foci (singular, **focus**) Foci are fixed points used to determine the points that form a parabola, an ellipse, or a hyperbola. A parabola has one focus, and an ellipse or a hyperbola has two foci. (Sections 10.1, 10.2, 10.3)

four-leaved rose A four-leaved rose is a curve that is the graph of a polar equation of the form $r = a \sin 2\theta$ or $r = a \cos 2\theta$. (Section 8.7)

frequency In simple harmonic motion, the frequency is the number of cycles per unit of time, or the reciprocal of the period. (Section 6.7)

function A function is a relation (set of ordered pairs) in which, for each value of the first component of the ordered pairs, there is *exactly one* value of the second component. (Section 2.3)

function notation Function notation $f(x)$ (read "f of x") represents the y-value of the function f for the indicated x-value. (Section 2.3)

fundamental rectangle The fundamental rectangle is used as a guide in sketching the graph of a hyperbola. The extended diagonals of this rectangle are the asymptotes of the hyperbola. (Section 10.3)

future value In an investment paying compound interest, the future value is the balance *after* interest has been earned. (The future value is sometimes called the *compound amount*.) (Section 4.2)

future value of an annuity If the payments on an annuity are accumulated in an account (with no withdrawals), then the sum of the payments and interest on the payments is the future value of the annuity. (Section 11.3)

G

general term (*n*th term) In the sequence $a_1, a_2, a_3, \ldots$, the general term (or nth term) is symbolized a_n. (Section 11.1)

geometric sequence (geometric progression) A geometric sequence is a sequence in which each term after the first is obtained by multiplying the preceding term by a constant nonzero real number. (Section 11.3)

geometric series A geometric series is the sum of the terms of a geometric sequence. (Section 11.3)

graph of an equation The graph of an equation is the set of all points that correspond to all of the ordered pairs that satisfy the equation. (Section 2.1)

ground speed In air navigation, the ground speed of a plane is its speed relative to the ground. (Section 8.4)

half-life The amount of time that it takes for a quantity that decays exponentially to become half its initial amount is its half-life. (Section 4.6)

half-plane A line separates a plane into two regions, each of which is a half-plane. (Section 9.6)

horizontal asymptote A horizontal line that a graph approaches as $|x|$ increases without bound is a horizontal asymptote. The line $y = b$ is a horizontal asymptote if $y \to b$ as $|x| \to \infty$. (Section 3.5)

horizontal component When a vector **u** is expressed as an ordered pair in the form $\mathbf{u} = \langle a, b \rangle$, the number a is the horizontal component of the vector. (Section 8.3)

hyperbola A hyperbola is the set of all points in a plane such that the absolute value of the difference of the distances from two fixed points is constant. (Section 10.3)

identity An equation satisfied by every number that is a meaningful replacement for the variable is an identity. (Sections 1.1, 7.1)

identity matrix (multiplicative identity matrix) An identity matrix is an $n \times n$ matrix with 1s on the main diagonal and 0s everywhere else. (Section 9.8)

imaginary axis In the complex plane, the vertical axis is the imaginary axis. (Section 8.5)

imaginary part In the complex number $a + bi$, the imaginary part is b. (Section 1.3)

imaginary unit The number i, defined by $i = \sqrt{-1}$ (and thus $i^2 = -1$), is the imaginary unit. (Section 1.3)

inconsistent system An inconsistent system is a system of equations with no solution. (Section 9.1)

increasing function A function f is increasing on an interval I if, whenever $x_1 < x_2$ in I, $f(x_1) < f(x_2)$. (Section 2.3)

independent events Two events are independent events if neither influences the outcomes of the other. (Section 11.6)

independent variable If the value of the variable y depends on the value of the variable x, then x is the independent variable. (Section 2.3)

index of a radical In a radical of the form $\sqrt[n]{a}$, the index is n. (Section R.7)

index of summation In the summation $\sum_{i=1}^{n} a_i$, the letter i is the index of summation. Other letters may also be used, such as j and k. (Section 11.1)

inequality An inequality says that one expression is greater than, greater than or equal to, less than, or less than or equal to, another. (Sections R.2, 1.7)

infinite sequence A sequence is an infinite sequence if its domain is the set of *all* natural numbers. (Section 11.1)

infinite series An infinite series is an expression of the form $S_\infty = a_1 + a_2 + a_3 + \cdots + a_n + \cdots = \sum_{i=1}^{\infty} a_i$. (Section 11.1)

infinite set (This is an informal definition.) An infinite set is a set that has an unending list of distinct elements. (Section R.1)

initial point When two letters are used to name a vector, the first letter indicates the initial (starting) point of the vector. (Section 8.3)

initial side When a ray is rotated around its endpoint to form an angle, the ray in its starting position is the initial side of the angle. (Section 5.1)

integers The set of integers is $\{\ldots, -3, -2, -1, 0, 1, 2, 3, \ldots\}$. (Section R.2)

intersection The intersection of sets A and B, written $A \cap B$, is the set of elements that belong to both A and B. (Sections R.1, 11.7)

interval An interval is a portion of the real number line, which may or may not include its endpoint(s). (Section 1.7)

interval notation Interval notation is a simplified notation for writing intervals. It uses parentheses and brackets to show whether the endpoints are included. (Section 1.7)

inverse function Let f be a one-to-one function. Then g is the inverse function of f if $(f \circ g)(x) = x$ for every x in the domain of g, and $(g \circ f)(x) = x$ for every x in the domain of f. (Section 4.1)

irrational numbers Real numbers that cannot be represented as quotients of integers are irrational numbers. (Section R.2)

L

latitude Latitude gives the measure of a central angle with vertex at Earth's center whose initial side goes through the equator and whose terminal side goes through the given location. (Section 6.1)

leading coefficient In a polynomial function of degree n, the leading coefficient is a_n. It is the coefficient of the term of greatest degree. (Section 3.1)

lemniscate A lemniscate is a figure-eight-shaped curve that is the graph of a polar equation of the form $r^2 = a^2 \sin 2\theta$ or $r^2 = a^2 \cos 2\theta$. (Section 8.7)

like radicals Radicals with the same radicand and the same index are like radicals. (Section R.7)

like terms Terms with the same variables each raised to the same powers are like terms. (Section R.3)

limaçon A limaçon is the graph of a polar equation of the form $r = a \pm b \sin \theta$ or $r = a \pm b \cos \theta$. If $\left|\frac{a}{b}\right| = 1$, then the limaçon is a cardioid. (Section 8.7)

line Two distinct points A and B determine the line AB. (Section 5.1)

line segment (segment) Line segment AB is the portion of line AB between A and B, including A and B. (Section 5.1)

linear cost function A linear cost function is a linear function that has the form $C(x) = mx + b$, where x represents the number of items produced, m represents the cost per item, and b represents the fixed cost. (Section 2.4)

linear equation (first-degree equation) in *n* unknowns Any equation of the form $a_1x_1 + a_2x_2 + \cdots + a_nx_n = b$, for real numbers $a_1, a_2, \ldots, a_n$ (not all of which are 0) and b, is a linear equation, or a first-degree equation, in n unknowns. (Section 9.1)

linear equation (first-degree equation) in one variable A linear equation in one variable is an equation that can be written in the form $ax + b = 0$, where a and b are real numbers with $a \neq 0$. (Section 1.1)

linear function A function f of the form $f(x) = ax + b$, for real numbers a and b, is a linear function. (Section 2.4)

linear inequality in one variable A linear inequality in one variable is an inequality that can be written in the form $ax + b > 0$, where a and b are real numbers with $a \neq 0$. (Any of the symbols $<$, $\geq$, and $\leq$ may also be used.) (Section 1.7)

linear inequality in two variables A linear inequality in two variables is an inequality of the form $Ax + By \leq C$, where A, B, and C are real numbers with A and B not both equal to 0. (Any of the symbols $\geq$, $<$, and $>$ may also be used.) (Section 9.6)

linear programming Linear programming, an application of mathematics to business and social science, is a method for finding an optimum value, such as minimum cost or maximum profit. (Section 9.6)

linear speed *v* The linear speed v measures the distance traveled per unit of time. (Section 6.2)

literal equation A literal equation is an equation that relates two or more variables (letters). (Section 1.1)

logarithm A logarithm is an exponent. The expression $\log_a x$ is the power to which the base a must be raised to obtain x. (Section 4.3)

logarithmic equation A logarithmic equation is an equation with a logarithm of a variable expression in at least one term. (Section 4.3)

logarithmic function If $a > 0$, $a \neq 1$, and $x > 0$, then $f(x) = \log_a x$ defines the logarithmic function with base a. (Section 4.3)

lowest terms A rational expression is in lowest terms when the greatest common factor of its numerator and its denominator is 1. (Section R.5)

M

magnitude The length of a vector represents the magnitude of the vector quantity. (Section 8.3)

major axis The major axis of an ellipse is its longer axis of symmetry. (Section 10.2)

mathematical induction Mathematical induction is a method for proving that a statement S_n is true for every positive integer n. (Section 11.5)

mathematical model A mathematical model is an equation (or inequality) that describes the relationship between two or more quantities. (Section 1.2)

matrix (plural, **matrices**) A matrix is a rectangular array of numbers enclosed in brackets. (Section 9.2)

minor In an $n \times n$ matrix with $n \geq 3$, the minor of a particular element is the determinant of the $(n - 1) \times (n - 1)$ matrix that results when the row and column that contain the chosen element are eliminated. (Section 9.3)

minor axis The minor axis of an ellipse is its shorter axis of symmetry. (Section 10.2)

minute One minute, written 1′, is $\frac{1}{60}$ of a degree. (Section 5.1)

monomial A monomial is a polynomial containing exactly one term. (Section R.3).

multiplicative inverse of a matrix If A is an $n \times n$ matrix, then its multiplicative inverse, written A^{-1}, is the matrix that satisfies both $AA^{-1} = I_n$ and $A^{-1}A = I_n$. It is possible that A^{-1} does not exist. (Section 9.8)

mutually exclusive events In probability, two events that cannot occur simultaneously are mutually exclusive events. (Section 11.7)

natural logarithm A base e logarithm is a natural logarithm. (Section 4.4)

natural numbers (counting numbers) The natural numbers, or counting numbers, form the set of numbers $\{1, 2, 3, 4, \dots\}$. (Sections R.1, R.2)

negative angle A negative angle is an angle that is formed by clockwise rotation around its endpoint. (Section 5.1)

nonlinear system A system of equations in which at least one equation is *not* linear is called a nonlinear system. (Section 9.5)

nonreal complex number A complex number $a + bi$ in which $b \neq 0$ is a nonreal complex number. (Section 1.3)

nonstrict inequality An inequality in which the symbol is either $\leq$ or $\geq$ is a nonstrict inequality. (Sections 1.7, 9.6)

***n*th root of a complex number** For a positive integer n, the complex number $a + bi$ is an nth root of the complex number $x + yi$ if $(a + bi)^n = x + yi$. (Section 8.6)

objective function In linear programming, the function to be maximized or minimized is the objective function. (Section 9.6)

oblique asymptote A nonvertical, nonhorizontal line that a graph approaches as $|x|$ increases without bound is an oblique asymptote. (Section 3.5)

oblique triangle A triangle that is not a right triangle is an oblique triangle. (Section 8.1)

obtuse angle An obtuse angle is an angle measuring more than 90° but less than 180°. (Section 5.1)

odd function A function f is an odd function if, for all x in the domain of f, $f(-x) = -f(x)$. The graph of an odd function is symmetric with respect to the origin. (Section 2.7)

odds The odds in favor of an event are expressed as the ratio of the probability of the event to the probability of the complement of the event. (Section 11.7)

one-to-one function If a function is defined so that each range element is used only once, then it is a one-to-one function. (Section 4.1)

open interval An open interval is an interval that does not include its endpoint(s). (Section 1.7)

opposite of a vector The opposite of a vector **v** is a vector −**v** that has the same magnitude as **v** but opposite direction. (Section 8.3)

ordered pair An ordered pair consists of two components, written inside parentheses. Ordered pairs are used to identify points in the coordinate plane. (Section 2.1)

ordered triple An ordered triple consists of three components, written inside parentheses. Ordered triples are used to identify points in space. (Section 9.1)

ordinate The y-value of a point may be called its ordinate. (Section 6.6)

origin (of a rectangular coordinate system) The point of intersection of the x-axis and the y-axis of a rectangular coordinate system is the origin. (Section 2.1)

orthogonal vectors Orthogonal vectors are vectors that are perpendicular, meaning that the angle between the two vectors is 90°. (Section 8.3)

outcome In probability, a possible result of each trial in an experiment is an outcome of the experiment. (Section 11.7)

parabola (function definition) A parabola is a curve that is the graph of a quadratic function over the set of all real numbers. (Sections 2.6, 3.1)

parabola (geometric definition) A parabola is the set of all points in a plane equidistant from a fixed point (the focus) and a fixed line (the directrix). (Section 10.1)

parallelogram rule The parallelogram rule is a geometric interpretation of the sum of two vectors. If the two vectors are placed so that their initial points coincide and a parallelogram is completed that has these two vectors as two of its sides, then the diagonal vector of the parallelogram that has the same initial point as the two vectors is their sum. (Section 8.3)

parameter A parameter is a variable in terms of which two or more other variables are expressed. In a pair of parametric equations $x = f(t)$ and $y = g(t)$, the variable t is the parameter. (Section 8.8)

parametric equations of a plane curve A pair of equations $x = f(t)$ and $y = g(t)$ are parametric equations of a plane curve. (Section 8.8)

partial fraction Each term in the decomposition of a rational expression is a partial fraction. (Section 9.4)

partial fraction decomposition When one rational expression is expressed as the sum of two or more rational expressions, the sum is the partial fraction decomposition. (Section 9.4)

Pascal's triangle Pascal's triangle is a triangular array of numbers that is helpful in expanding binomials. The numbers in the triangle are the binomial coefficients. (Section 11.4)

period For a periodic function such that $f(x) = f(x + np)$, the least possible positive value of p is the period of the function. (Section 6.3)

periodic function A periodic function is a function f such that $f(x) = f(x + np)$, for every real number x in the domain of f, every integer n, and some positive real number p. (Section 6.3)

permutation A permutation of n elements taken r at a time is one of the *arrangements* of r elements from a set of n elements. (Section 11.6)

phase shift For trigonometric functions, a horizontal translation is a phase shift. (Section 6.4)

piecewise-defined function A piecewise-defined function is a function that is defined by different rules over different intervals of its domain. (Section 2.6)

plane curve A plane curve is a set of points (x, y) such that $x = f(t)$ and $y = g(t)$, and f and g are both defined on an interval I. (Section 8.8)

point-slope form The point-slope form of the equation of the line with slope m through the point (x_1, y_1) is $y - y_1 = m(x - x_1)$. (Section 2.5)

polar axis The polar axis is a specific ray in the polar coordinate system that has the pole as its endpoint. The polar axis is usually drawn in the direction of the positive x-axis. (Section 8.7)

polar coordinates In the polar coordinate system, the ordered pair (r, θ) gives polar coordinates of point P, where r is the directed distance from the pole to P and θ is the directed angle from the positive x-axis to ray OP. (Section 8.7)

polar coordinate system The polar coordinate system is a coordinate system based on a point (the pole) and a ray (the polar axis). (Section 8.7)

polar equation A polar equation is an equation that uses polar coordinates. The variables are r and θ. (Section 8.7)

pole The pole is the single fixed point in the polar coordinate system that is the endpoint of the polar axis. The pole is usually placed at the origin of a rectangular coordinate system. (Section 8.7)

polynomial A polynomial is a term or a finite sum of terms, with only positive or zero integer exponents permitted on the variables. (Section R.3)

polynomial function of degree *n* A polynomial function of degree n, where n is a nonnegative integer, is a function defined by an expression of the form $f(x) = a_n x^n + a_{n-1}x^{n-1} + \cdots + a_1 x + a_0$, where $a_n, a_{n-1}, \ldots, a_1$, and a_0 are real numbers, with $a_n \neq 0$. (Section 3.1)

position vector A vector with its initial point at the origin is a position vector. (Section 8.3)

positive angle A positive angle is an angle that is formed by counterclockwise rotation around its endpoint. (Section 5.1)

power (exponential expression, exponential) An expression of the form a^n is a power, an exponential expression, or an exponential. (Section R.2)

present value In an investment paying compound interest, the principal is also called the present value. (Section 4.2)

prime polynomial A polynomial with variable terms that cannot be written as a product of two polynomials of lesser degree is a prime polynomial. (Section R.4)

principal *n*th root For even values of n (square roots, fourth roots, and so on), when a is positive, there are two real nth roots, one positive and one negative. In such cases, the notation $\sqrt[n]{a}$ represents the positive root, or principal nth root. (Section R.7)

probability of an event In a sample space with equally likely outcomes, the probability of an event is the ratio of the number of outcomes in the sample space that belong to the event to the number of outcomes in the sample space. (Section 11.7)

pure imaginary number A complex number $a + bi$ in which $a = 0$ and $b \neq 0$ is a pure imaginary number. (Section 1.3)

Pythagorean theorem The Pythagorean theorem states that in a right triangle, the sum of the squares of the lengths of the legs is equal to the square of the length of the hypotenuse. (Section 1.5)

quadrantal angle A quadrantal angle is an angle that, when placed in standard position, has its terminal side along the x-axis or the y-axis. (Section 5.1)

quadrants The quadrants of a rectangular coordinate system are the four regions into which the x-axis and y-axis divide the plane. (Section 2.1)

quadratic equation (second-degree equation) An equation that can be written in the form $ax^2 + bx + c = 0$, where a, b, and c are real numbers with $a \neq 0$, is a quadratic equation. (Section 1.4)

quadratic in form An equation is quadratic in form if it can be written as $au^2 + bu + c = 0$, where $a \neq 0$ and u is some algebraic expression. (Section 1.6)

quadratic formula The quadratic formula $x = \frac{-b \pm \sqrt{b^2 - 4ac}}{2a}$ is a general formula that can be used to solve a quadratic equation of the form $ax^2 + bx + c = 0$, with $a \neq 0$. (Section 1.4)

quadratic function A function f of the form $f(x) = ax^2 + bx + c$, where a, b, and c are real numbers, with $a \neq 0$, is a quadratic function. (Section 3.1)

quadratic inequality A quadratic inequality is an inequality that can be written in the form $ax^2 + bx + c < 0$, for real numbers a, b, and c, with $a \neq 0$. (The symbol $<$ can be replaced with $>$, $\leq$, and $\geq$.) (Section 1.7)

radian A radian is a unit of measure for angles. An angle with its vertex at the center of a circle that intercepts an arc on the circle equal in length to the radius of the circle has a measure of 1 radian. (Section 6.1)

radicand The number or expression under a radical symbol is the radicand. (Section R.7)

radius The radius of a circle is the given distance between the center and any point on the circle. (Section 2.2)

range In a relation, the set of all values of the dependent variable (y) is the range. (Section 2.3)

rational equation A rational equation is an equation that has a rational expression for one or more terms. (Section 1.6)

rational expression The quotient of two polynomials P and Q, with $Q \neq 0$, is a rational expression. (Section R.5)

rational function A function f of the form $f(x) = \frac{p(x)}{q(x)}$, where $p(x)$ and $q(x)$ are polynomials, with $q(x) \neq 0$, is a rational function. (Section 3.5)

rational inequality A rational inequality is an inequality in which one or both sides contain rational expressions. (Section 1.7)

rationalizing the denominator Rationalizing a denominator is the process of writing a radical expression so that there are no radicals in the denominator. (Section R.7)

rational numbers The rational numbers are the set of numbers $\frac{p}{q}$, where p and q are integers and $q \neq 0$. (Section R.2)

ray The portion of line AB that starts at A and continues through B, and on past B, is ray AB. (Section 5.1)

real axis In the complex plane, the horizontal axis is the real axis. (Section 8.5)

real numbers The set of all numbers that correspond to points on a number line is the real numbers. (Section R.2)

real part In the complex number $a + bi$, the real part is a. (Section 1.3)

reciprocal function The function $f(x) = \frac{1}{x}$ is the reciprocal function. (Section 3.5)

rectangular (Cartesian) coordinate system Two perpendicular number lines intersecting at their origins form a rectangular coordinate system. (Section 2.1)

rectangular (Cartesian) equation A rectangular (or Cartesian) equation is an equation that uses rectangular coordinates. If it is an equation in two variables, the variables are x and y. (Section 8.7)

rectangular form (standard form) of a complex number The rectangular form (or standard form) of a complex number is $a + bi$, where a and b are real numbers. (Section 8.5)

recursive definition A sequence is defined by a recursive definition if each term after the first term or first few terms is defined as an expression involving the previous term or terms. (Section 11.1)

reference angle The reference angle for an angle θ, written θ', is the positive acute angle made by the terminal side of angle θ and the x-axis. (Section 5.3)

reference arc The reference arc for a point on the unit circle is the shortest arc from the point itself to the nearest point on the x-axis. (Section 6.2)

region of feasible solutions In linear programming, the region of feasible solutions is the region of the graph that satisfies all of the constraints. (Section 9.6)

relation A relation is a set of ordered pairs. (Section 2.3)

resultant If **A** and **B** are vectors, the vector sum **A** + **B** is the resultant of vectors **A** and **B**. (Section 8.3)

right angle A right angle is an angle measuring exactly 90°. (Section 5.1)

root (solution) of an equation A zero k of $f(x)$ is a number such that $f(k) = 0$ is true. (Section 3.2)

rose curve A rose curve is a member of a family of curves that resemble flowers. It is the graph of a polar equation of the form $r = a \sin n\theta$ or $r = a \cos n\theta$. (Section 8.7)

rotation of axes If the axes in an xy-coordinate system having origin O are rotated about O through an angle θ, the new coordinate system is called a rotation of the xy-system. (Appendix B)

row matrix A matrix with exactly one row is a row matrix. (Section 9.7)

S

sample space In probability, the set of all possible outcomes of a given experiment is the sample space. (Section 11.7)

scalar A scalar is a quantity that involves magnitude and can be represented by a real number. (Sections 8.3, 9.7)

scalar product The scalar product of a real number (or scalar) k and a vector $\mathbf{u}$ is the vector $k \cdot \mathbf{u}$, which has magnitude $|k|$ times the magnitude of $\mathbf{u}$. (Section 8.3)

scatter diagram (of ordered pairs) A scatter diagram is a graph of specific ordered pairs of data. (Section 2.5)

secant Let $P(x, y)$ be a point other than the origin on the terminal side of an angle θ in standard position. Let $r = \sqrt{x^2 + y^2}$ represent the distance from the origin to P. Then the secant function is defined by $\sec \theta = \frac{r}{x}\ (x \neq 0)$. (Section 5.2)

second One second, written $1''$, is $\frac{1}{60}$ of a minute. (Section 5.1)

sector of a circle A sector of a circle is the portion of the interior of a circle intercepted by a central angle. (Section 6.1)

semiperimeter The semiperimeter is half the sum of the lengths of the three sides of a triangle. (Section 8.2)

sequence A sequence is a function that has a set of natural numbers of the form $\{1, 2, 3, \ldots, n\}$ or $\{1, 2, 3, \ldots, n, \ldots\}$ as its domain. (Section 11.1)

series A series is the indicated sum of the terms of a sequence. (Section 11.1)

set A set is a collection of objects. (Section R.1)

set-builder notation Set-builder notation uses the form $\{x \mid x$ has a certain property$\}$ to describe a set without having to list all of it elements. (Section R.1)

set operations The processes of finding the complement of a set, the intersection of two sets, and the union of two sets are set operations. (Section R.1)

side of an angle One of the two rays (or line segments) with a common endpoint that form an angle is a side of the angle. (Section 5.1)

Side-Angle-Side (SAS) The Side-Angle-Side (SAS) congruence axiom states that if two sides and the included angle of one triangle are equal, respectively, to two sides and the included angle of a second triangle, then the triangles are congruent. (Section 8.1)

Side-Side-Side (SSS) The Side-Side-Side (SSS) congruence axiom states that if three sides of one triangle are equal, respectively, to three sides of a second triangle, then the triangles are congruent. (Section 8.1)

significant digit A significant digit is a digit obtained by actual measurement. (Section 5.4)

simple harmonic motion Simple-harmonic motion is oscillatory motion about an equilibrium position. If friction is neglected, this motion can be described by a sinusoid. (Section 6.7)

simple interest In computing simple interest, interest is paid only on the principal, and not on previously earned interest. (Section 1.1)

sine Let $P(x, y)$ be a point other than the origin on the terminal side of an angle θ in standard position. Let $r = \sqrt{x^2 + y^2}$ represent the distance from the origin to P. Then the sine function is defined by $\sin \theta = \frac{y}{r}$. (Section 5.2)

sine wave (sinusoid) The graph of a sine function is a sine wave (or sinusoid). (Section 6.3)

slope The slope of a nonvertical line is the ratio of the change in y to the change in x. (Section 2.4)

slope-intercept form The slope-intercept form of the equation of the line with slope m and y-intercept b is $y = mx + b$. (Section 2.5)

solution (root) A solution (or root) of an equation is a number that makes the equation a true statement. (Section 1.1)

solution set The solution set of an equation is the set of all numbers that satisfy the equation. (Section 1.1)

solution of a system of equations A solution of a system of equations must satisfy every equation in the system. (Section 9.1).

spiral of Archimedes A spiral of Archimedes is an infinite curve that is the graph of a polar equation of the form $r = n\theta$. (Section 8.7)

square matrix A matrix with the same number of columns as rows is a square matrix. (Section 9.7)

standard form of a complex number A complex number written in the form $a + bi$ (or $a + ib$) is in standard form. (Section 1.3)

standard form of a linear equation The form $Ax + By = C$ is the standard form of a linear equation. (Section 2.4)

step function A step function is a function that is defined using the greatest integer function, and whose graph resembles a series of steps. (Section 2.6)

straight angle A straight angle is an angle measuring exactly 180°. (Section 5.1)

strict inequality An inequality in which the symbol is either $<$ or $>$ is a strict inequality. (Sections 1.7, 9.6)

subset If every element of set A is also an element of set B, then A is a subset of B, written $A \subseteq B$. (Section R.1)

summation notation Summation notation is a compact way of writing a series using the general term of the corresponding sequence and the Greek capital letter sigma Σ. (Section 11.1)

supplementary angles (supplements) Two positive angles are supplementary angles (or supplements) if the sum of their measures is 180°. (Section 5.1)

synthetic division Synthetic division is a shortcut method of dividing a polynomial by a binomial of the form $x - k$. (Section 3.2)

system of equations A group of equations that are considered at the same time is a system of equations. (Section 9.1)

system of inequalities A group of inequalities that are considered at the same time is a system of inequalities. (Section 9.6)

system of linear equations (linear system) If all the equations in a system are linear, then the system is a system of linear equations, or a linear system. (Section 9.1)

T

tangent Let $P(x, y)$ be a point other than the origin on the terminal side of an angle θ in standard position. Let $r = \sqrt{x^2 + y^2}$ represent the distance from the origin to P. Then the tangent function is defined by $\tan \theta = \frac{y}{x}\ (x \neq 0)$. (Section 5.2)

term The product of a real number and one or more variables raised to powers is a term. (Section R.3)

terminal point When two letters are used to name a vector, the second letter indicates the terminal (ending) point of the vector. (Section 8.3)

terminal side When a ray is rotated around its endpoint to form an angle, the ray in its location after rotation is the terminal side of the angle. (Section 5.1)

terms of a sequence The elements in the range of a sequence are the terms of the sequence. (Section 11.1)

translation A translation is a horizontal or vertical shift of a graph. (Sections 2.7, 6.4)

transverse axis The line segment that has the vertices of a hyperbola as endpoints is the transverse axis of the hyperbola. (Section 10.3)

tree diagram A tree diagram is a diagram with branches that is used to systematically list all the outcomes of a counting situation or probability experiment. (Section 11.6)

trigonometric (polar) form of a complex number The expression $r(\cos\theta + i\sin\theta)$ is the trigonometric form (or polar form) of the complex number $x + yi$. The expression $\cos\theta + i\sin\theta$ is sometimes abbreviated as cis θ. (Section 8.5)

trinomial A trinomial is a polynomial containing exactly three terms. (Section R.3)

turning points The points on the graph of a function where the function changes from increasing to decreasing or from decreasing to increasing are turning points of the graph. (Section 3.4)

U

union The union of sets A and B, written $A \cup B$, is the set of all elements that belong to set A or set B (or both). (Sections R.1, 1.7, 11.7)

unit circle The unit circle is the circle with center at the origin and radius 1. (Section 6.2)

unit vector A unit vector is a vector that has magnitude 1. Two useful unit vectors are $\mathbf{i} = \langle 1, 0 \rangle$ and $\mathbf{j} = \langle 0, 1 \rangle$. (Section 8.3)

universal set The universal set, written U, contains all the elements under discussion in a particular situation. (Section R.1)

V

varies directly (is directly proportional to) y varies directly as x, or y is directly proportional to x, if there exists a nonzero real number k such that $y = kx$. (Section 3.6)

varies inversely (is inversely proportional to) y varies inversely as x, or y is inversely proportional to x, if there exists a nonzero real number k such that $y = \frac{k}{x}$. (Section 3.6)

varies jointly In joint variation, a variable depends on the product of two or more other variables. If m and n are real numbers, then y varies jointly as the nth power of x and the mth power of z if there exists a nonzero real number k such that $y = kx^n z^m$. (Section 3.6)

vector A vector is a directed line segment that represents a vector quantity with direction and magnitude. (Section 8.3)

vector quantities Quantities that involve both magnitude and direction are vector quantities. (Section 8.3)

Venn diagram A Venn diagram is a diagram used to illustrate relationships among sets or probability concepts. (Sections R.1, 11.7)

vertex (corner point) In linear programming, a vertex, or corner point, is one of the vertices of the region of feasible solutions. (Section 9.6)

vertex of an angle The vertex of an angle is the endpoint of the ray that is rotated to form the angle. (Section 5.1)

vertex of a parabola The vertex of a parabola is the point where the axis of symmetry intersects the parabola. For the graph of a quadratic function, this is the turning point of the parabola. (Sections 2.6, 3.1, 10.1)

vertical asymptote A vertical line that a graph approaches, but never touches or intersects, is a vertical asymptote. The line $x = a$ is a vertical asymptote for a function f if $|f(x)| \to \infty$ as $x \to a$. (Sections 3.5, 6.5)

vertical component When a vector $\mathbf{u}$ is expressed as an ordered pair in the form $\mathbf{u} = \langle a, b \rangle$, the number b is the vertical component of the vector. (Section 8.3)

vertices of an ellipse The vertices of an ellipse are the endpoints of the major axis. (Section 10.2)

vertices of a hyperbola The vertices of a hyperbola are the two points on the hyperbola that are closest to the center. (Section 10.3)

W

whole numbers The set of whole numbers $\{0, 1, 2, 3, 4, \ldots\}$ is the union of the set of natural numbers and $\{0\}$. (Section R.2)

X

x-axis The horizontal number line in a rectangular coordinate system is the x-axis. (Section 2.1)

x-intercept An x-intercept is the x-value of a point where the graph of an equation intersects the x-axis. (Section 2.1)

Y

y-axis The vertical number line in a rectangular coordinate system is the y-axis. (Section 2.1)

y-intercept A y-intercept is the y-value of a point where the graph of an equation intersects the y-axis. (Section 2.1)

Z

zero-factor property The zero-factor property states that if the product of two (or more) complex numbers is 0, then at least one of the numbers must be 0. (Section 1.4)

zero matrix A matrix containing only zero elements is a zero matrix. (Section 9.7)

zero of multiplicity n A polynomial function has a zero k of multiplicity n if the zero occurs n times. The polynomial has exactly n factors of $x - k$. (Section 3.3)

zero polynomial The function f defined by $f(x) = 0$ is the zero polynomial. (Section 3.1)

zero of a polynomial function A zero of a polynomial function f is a number k such that $f(k) = 0$. (Section 3.2)

zero vector The zero vector is the vector with magnitude 0. (Section 8.3)

Solutions to Selected Exercises

Chapter R Review of Basic Concepts

R.2 Exercises *(pages 17–20)*

61. No; in general $a - b \neq b - a$.

Examples:

$a = 15, b = 0$: $a - b = 15 - 0 = 15$, but
$b - a = 0 - 15 = -15$.

$a = 12, b = 7$: $a - b = 12 - 7 = 5$, but
$b - a = 7 - 12 = -5$.

$a = -6, b = 4$: $a - b = -6 - 4 = -10$, but
$b - a = 4 - (-6) = 10$.

$a = -18, b = -3$: $a - b = -18 - (-3) = -15$, but
$b - a = -3 - (-18) = 15$.

73. The process in your head should be like the following.

$$\begin{aligned} 72 \cdot 17 + 28 \cdot 17 &= 17(72 + 28) \\ &= 17(100) \\ &= 1700 \end{aligned}$$

113. $\frac{x^3}{y} > 0$

The quotient of two numbers is positive if they have the same sign (both positive or both negative). The sign of x^3 is the same as the sign of x. Therefore, $\frac{x^3}{y} > 0$ if x and y have the same sign.

R.3 Exercises *(pages 28–31)*

15. $$\begin{aligned} -(4m^3n^0)^2 &= -[4^2(m^3)^2(n^0)^2] && \text{Power rule 2} \\ &= -(4^2)m^6n^0 && \text{Power rule 1} \\ &= -(4^2)m^6 \cdot 1 && \text{Zero exponent} \\ &= -4^2m^6, \quad \text{or} \quad -16m^6 \end{aligned}$$

43. $$\begin{aligned} &(6m^4 - 3m^2 + m) - (2m^3 + 5m^2 + 4m) + (m^2 - m) \\ &= (6m^4 - 3m^2 + m) + (-2m^3 - 5m^2 - 4m) \\ &\quad + (m^2 - m) \\ &= 6m^4 - 2m^3 + (-3 - 5 + 1)m^2 + (1 - 4 - 1)m \\ &= 6m^4 - 2m^3 - 7m^2 - 4m \end{aligned}$$

67. $[(2p - 3) + q]^2$

$= (2p - 3)^2 + 2(2p - 3)(q) + q^2$
Square of a binomial, treating $(2p - 3)$ as one term

$= (2p)^2 - 2(2p)(3) + 3^2 + 2(2p - 3)(q) + q^2$
Square the binomial $(2p - 3)$.

$= 4p^2 - 12p + 9 + 4pq - 6q + q^2$

87. $$\begin{aligned} &\frac{-4x^7 - 14x^6 + 10x^4 - 14x^2}{-2x^2} \\ &= \frac{-4x^7}{-2x^2} + \frac{-14x^6}{-2x^2} + \frac{10x^4}{-2x^2} + \frac{-14x^2}{-2x^2} \\ &= 2x^5 + 7x^4 - 5x^2 + 7 \end{aligned}$$

R.4 Exercises *(pages 38–40)*

39. $$\begin{aligned} &24a^4 + 10a^3b - 4a^2b^2 \\ &= 2a^2(12a^2 + 5ab - 2b^2) && \text{Factor out the GCF, } 2a^2. \\ &= 2a^2(4a - b)(3a + 2b) && \text{Factor the trinomial.} \end{aligned}$$

47. $(a - 3b)^2 - 6(a - 3b) + 9$

$= [(a - 3b) - 3]^2$ Factor the perfect square trinomial.

$= (a - 3b - 3)^2$

69. $27 - (m + 2n)^3$

$= 3^3 - (m + 2n)^3$ Write as a difference of cubes.

$= [3 - (m + 2n)][3^2 + 3(m + 2n) + (m + 2n)^2]$
Factor the difference of cubes.

$= (3 - m - 2n)(9 + 3m + 6n + m^2 + 4mn + 4n^2)$
Distributive property; square the binomial $(m + 2n)$.

81. $$\begin{aligned} &9(a - 4)^2 + 30(a - 4) + 25 \\ &= 9u^2 + 30u + 25 && \text{Replace } a - 4 \text{ with } u. \\ &= (3u)^2 + 2(3u)(5) + 5^2 \\ &= (3u + 5)^2 && \text{Factor the perfect square trinomial.} \\ &= [3(a - 4) + 5]^2 && \text{Replace } u \text{ with } a - 4. \\ &= (3a - 12 + 5)^2 && \text{Distributive property} \\ &= (3a - 7)^2 && \text{Add.} \end{aligned}$$

R.5 Exercises *(pages 47–49)*

19. $$\begin{aligned} \frac{8m^2 + 6m - 9}{16m^2 - 9} &= \frac{(2m + 3)(4m - 3)}{(4m + 3)(4m - 3)} && \text{Factor.} \\ &= \frac{2m + 3}{4m + 3} && \text{Lowest terms} \end{aligned}$$

35. $$\begin{aligned} &\frac{x^3 + y^3}{x^3 - y^3} \cdot \frac{x^2 - y^2}{x^2 + 2xy + y^2} \\ &= \frac{(x + y)(x^2 - xy + y^2)}{(x - y)(x^2 + xy + y^2)} \cdot \frac{(x + y)(x - y)}{(x + y)(x + y)} && \text{Factor.} \\ &= \frac{x^2 - xy + y^2}{x^2 + xy + y^2} && \text{Lowest terms} \end{aligned}$$

57. $\dfrac{4}{x+1}+\dfrac{1}{x^2-x+1}-\dfrac{12}{x^3+1}$

$=\dfrac{4}{x+1}+\dfrac{1}{x^2-x+1}-\dfrac{12}{(x+1)(x^2-x+1)}$ Factor the sum of cubes.

$=\dfrac{4(x^2-x+1)}{(x+1)(x^2-x+1)}+\dfrac{1(x+1)}{(x+1)(x^2-x+1)}-\dfrac{12}{(x+1)(x^2-x+1)}$ Write each fraction with the common denominator.

$=\dfrac{4(x^2-x+1)+1(x+1)-12}{(x+1)(x^2-x+1)}$ Add and subtract numerators.

$=\dfrac{4x^2-4x+4+x+1-12}{(x+1)(x^2-x+1)}$ Distributive property

$=\dfrac{4x^2-3x-7}{(x+1)(x^2-x+1)}$ Combine like terms.

$=\dfrac{(4x-7)(x+1)}{(x+1)(x^2-x+1)}$ Factor the numerator.

$=\dfrac{4x-7}{x^2-x+1}$ Lowest terms

75. $\dfrac{\frac{1}{x+h}-\frac{1}{x}}{h}$

Multiply both numerator and denominator by the LCD of all the fractions, $x(x+h)$.

$\dfrac{\frac{1}{x+h}-\frac{1}{x}}{h}=\dfrac{x(x+h)\left(\frac{1}{x+h}-\frac{1}{x}\right)}{x(x+h)(h)}$

$=\dfrac{x(x+h)\left(\frac{1}{x+h}\right)-x(x+h)\left(\frac{1}{x}\right)}{x(x+h)(h)}$ Distributive property

$=\dfrac{x-(x+h)}{x(x+h)(h)}$ Divide common factors in the numerator.

$=\dfrac{-h}{x(x+h)(h)}$ Subtract.

$=\dfrac{-1}{x(x+h)}$ Lowest terms

R.6 Exercises *(pages 55–58)*

43. $\left(-\dfrac{64}{27}\right)^{1/3}=-\dfrac{4}{3}$ because $\left(-\dfrac{4}{3}\right)^3=-\dfrac{64}{27}$.

71. $\dfrac{p^{1/5}p^{7/10}p^{1/2}}{(p^3)^{-1/5}}=\dfrac{p^{1/5+7/10+1/2}}{p^{-3/5}}$ Product rule; power rule 1

$=\dfrac{p^{2/10+7/10+5/10}}{p^{-6/10}}$ Write all fractions with the LCD, 10.

$=\dfrac{p^{14/10}}{p^{-6/10}}$ Add exponents.

$=p^{(14/10)-(-6/10)}$ Quotient rule

$=p^{20/10}$ Subtract and write exponent in lowest terms.

$=p^2$

81. $(r^{1/2}-r^{-1/2})^2=(r^{1/2})^2-2(r^{1/2})(r^{-1/2})+(r^{-1/2})^2$ Square of a binomial

$=r-2r^0+r^{-1}$ Power rule 1; product rule

$=r-2\cdot 1+r^{-1}$ Zero exponent

$=r-2+r^{-1}$, or $r-2+\dfrac{1}{r}$ Negative exponent

103. $\dfrac{x-9y^{-1}}{(x-3y^{-1})(x+3y^{-1})}$

$=\dfrac{x-\frac{9}{y}}{\left(x-\frac{3}{y}\right)\left(x+\frac{3}{y}\right)}$ Definition of negative exponent

$=\dfrac{x-\frac{9}{y}}{x^2-\frac{9}{y^2}}$ Multiply in the denominator.

$=\dfrac{y^2\left(x-\frac{9}{y}\right)}{y^2\left(x^2-\frac{9}{y^2}\right)}$ Multiply numerator and denominator by the LCD, y^2.

$=\dfrac{y^2x-9y}{y^2x^2-9}$ Distributive property

$=\dfrac{y(xy-9)}{x^2y^2-9}$ Factor the numerator.

R.7 Exercises *(pages 66–69)*

61. $\sqrt[4]{\dfrac{g^3h^5}{9r^6}}=\dfrac{\sqrt[4]{g^3h^5}}{\sqrt[4]{9r^6}}$ Quotient rule

$=\dfrac{\sqrt[4]{h^4(g^3h)}}{\sqrt[4]{r^4(9r^2)}}$ Factor out perfect fourth powers.

$=\dfrac{h\sqrt[4]{g^3h}}{r\sqrt[4]{9r^2}}$ Remove all perfect fourth powers from the radicals.

$=\dfrac{h\sqrt[4]{g^3h}}{r\sqrt[4]{9r^2}}\cdot\dfrac{\sqrt[4]{9r^2}}{\sqrt[4]{9r^2}}$ Rationalize the denominator.

$=\dfrac{h\sqrt[4]{9g^3hr^2}}{r\sqrt[4]{81r^4}}$ Product rule

$=\dfrac{h\sqrt[4]{9g^3hr^2}}{3r^2}$ Simplify.

79. $\left(\sqrt[3]{11}-1\right)\left(\sqrt[3]{11^2}+\sqrt[3]{11}+1\right)$

This product is a difference of cubes.

$$(x-y)(x^2+xy+y^2)=x^3-y^3$$

Thus,

$$\begin{aligned}\left(\sqrt[3]{11}-1\right)\left(\sqrt[3]{11^2}+\sqrt[3]{11}+1\right)&=\left(\sqrt[3]{11}\right)^3-1^3\\&=11-1\\&=10.\end{aligned}$$

81. $\left(\sqrt{3}+\sqrt{8}\right)^2=\left(\sqrt{3}\right)^2+2\left(\sqrt{3}\right)\left(\sqrt{8}\right)+\left(\sqrt{8}\right)^2$ Square of a binomial

$$\begin{aligned}&=3+2\sqrt{24}+8\\&=3+2\sqrt{4\cdot 6}+8\\&=3+2\left(2\sqrt{6}\right)+8 \quad \text{Product rule}\\&=11+4\sqrt{6}\end{aligned}$$

91. $\dfrac{-4}{\sqrt[3]{3}}+\dfrac{1}{\sqrt[3]{24}}-\dfrac{2}{\sqrt[3]{81}}=\dfrac{-4}{\sqrt[3]{3}}+\dfrac{1}{\sqrt[3]{8\cdot 3}}-\dfrac{2}{\sqrt[3]{27\cdot 3}}$

$=\dfrac{-4}{\sqrt[3]{3}}+\dfrac{1}{2\sqrt[3]{3}}-\dfrac{2}{3\sqrt[3]{3}}$

Simplify radicals.

$=\dfrac{-4\cdot 6}{\sqrt[3]{3}\cdot 6}+\dfrac{1\cdot 3}{2\sqrt[3]{3}\cdot 3}-\dfrac{2\cdot 2}{3\sqrt[3]{3}\cdot 2}$

Write fractions with common denominator, $6\sqrt[3]{3}$.

$=\dfrac{-24}{6\sqrt[3]{3}}+\dfrac{3}{6\sqrt[3]{3}}-\dfrac{4}{6\sqrt[3]{3}}$

$=\dfrac{-25}{6\sqrt[3]{3}}$

$=\dfrac{-25}{6\sqrt[3]{3}}\cdot\dfrac{\sqrt[3]{3^2}}{\sqrt[3]{3^2}}$

Rationalize the denominator.

$=\dfrac{-25\sqrt[3]{9}}{6\sqrt[3]{27}}$

$=\dfrac{-25\sqrt[3]{9}}{6\cdot 3}$

$=\dfrac{-25\sqrt[3]{9}}{18}$

95. $\dfrac{\sqrt{7}-1}{2\sqrt{7}+4\sqrt{2}}$

$=\dfrac{\sqrt{7}-1}{2\sqrt{7}+4\sqrt{2}}\cdot\dfrac{2\sqrt{7}-4\sqrt{2}}{2\sqrt{7}-4\sqrt{2}}$

Multiply numerator and denominator by the conjugate of the denominator.

$=\dfrac{\left(\sqrt{7}-1\right)\left(2\sqrt{7}-4\sqrt{2}\right)}{\left(2\sqrt{7}+4\sqrt{2}\right)\left(2\sqrt{7}-4\sqrt{2}\right)}$

Multiply numerators. Multiply denominators.

$=\dfrac{\sqrt{7}\cdot 2\sqrt{7}-\sqrt{7}\cdot 4\sqrt{2}-1\cdot 2\sqrt{7}+1\cdot 4\sqrt{2}}{\left(2\sqrt{7}\right)^2-\left(4\sqrt{2}\right)^2}$

Use FOIL in the numerator. Find the product of the sum and difference of two terms in the denominator.

$=\dfrac{2\cdot 7-4\sqrt{14}-2\sqrt{7}+4\sqrt{2}}{4\cdot 7-16\cdot 2}$

$=\dfrac{14-4\sqrt{14}-2\sqrt{7}+4\sqrt{2}}{-4}$

$=\dfrac{-2\left(-7+2\sqrt{14}+\sqrt{7}-2\sqrt{2}\right)}{-2\cdot 2}$

Factor the numerator and denominator.

$=\dfrac{-7+2\sqrt{14}+\sqrt{7}-2\sqrt{2}}{2}$

Chapter 1 Equations and Inequalities

1.1 Exercises *(pages 84–86)*

25.

$$\begin{aligned}0.5x+\frac{4}{3}x&=x+10\\\frac{1}{2}x+\frac{4}{3}x&=x+10 && \text{Change decimal to fraction.}\\6\left(\frac{1}{2}x+\frac{4}{3}x\right)&=6(x+10) && \text{Multiply by the LCD, 6.}\\3x+8x&=6x+60 && \text{Distributive property}\\11x&=6x+60 && \text{Combine like terms.}\\5x&=60 && \text{Subtract } 6x.\\x&=12 && \text{Divide by 5.}\end{aligned}$$

Solution set: $\{12\}$

33.

$$\begin{aligned}0.3(x+2)-0.5(x+2)&=-0.2x-0.4\\10[0.3(x+2)-0.5(x+2)]&=10(-0.2x-0.4) && \text{Multiply by 10 to clear decimals.}\\3(x+2)-5(x+2)&=-2x-4 && \text{Distributive property}\\3x+6-5x-10&=-2x-4 && \text{Distributive property}\\-2x-4&=-2x-4 && \text{Combine like terms.}\\0&=0 && \text{Add } 2x. \text{ Add } 4.\end{aligned}$$

$0=0$ is a true statement, so the equation is an identity. Solution set: {all real numbers}

57.

$$\begin{aligned}3x&=(2x-1)(m+4), && \text{for } x\\3x&=2xm+8x-m-4 && \text{FOIL}\\m+4&=2xm+5x && \text{Add } m+4. \text{ Subtract } 3x.\\m+4&=x(2m+5) && \text{Factor out } x.\end{aligned}$$

$\dfrac{m+4}{2m+5}=x,\quad$ or $\quad x=\dfrac{m+4}{2m+5}\quad$ Divide by $2m+5$.

1.2 Exercises *(pages 92–97)*

17. Let $h =$ the height of the box. Use the formula for the surface area of a rectangular box.

$S = 2lw + 2wh + 2hl$

$496 = 2 \cdot 18 \cdot 8 + 2 \cdot 8 \cdot h + 2 \cdot h \cdot 18$ Let $S = 496,\ l = 18,\ w = 8.$

$496 = 288 + 16h + 36h$ Multiply.

$496 = 288 + 52h$ Combine like terms.

$208 = 52h$ Subtract 288.

$4 = h$ Divide by 52.

The height of the box is 4 ft.

21. Let $x =$ David's biking speed.

Then $x + 4.5 =$ David's driving speed.

Summarize the given information in a table, using the equation $d = rt$. Because the speeds are given in miles per hour, the times must be changed from minutes to hours.

	r	t	d
Driving	$x + 4.5$	$\frac{1}{3}$	$\frac{1}{3}(x + 4.5)$
Biking	x	$\frac{3}{4}$	$\frac{3}{4}x$

Distance driving = Distance biking

$\frac{1}{3}(x + 4.5) = \frac{3}{4}x$

$12\left(\frac{1}{3}(x + 4.5)\right) = 12\left(\frac{3}{4}x\right)$

Multiply by the LCD, 12.

$4(x + 4.5) = 9x$ Multiply.

$4x + 18 = 9x$ Distributive property

$18 = 5x$ Subtract $4x$.

$\frac{18}{5} = x$ Divide by 5.

Now find the distance.

$$d = rt = \frac{3}{4}x = \frac{3}{4}\left(\frac{18}{5}\right) = \frac{27}{10} = 2.7$$

David travels 2.7 mi to work.

33. Let $x =$ the number of milliliters of water to be added.

Strength	Milliliters of Solution	Milliliters of Salt
6%	8	$0.06(8)$
0%	x	$0(x)$
4%	$8 + x$	$0.04(8 + x)$

The number of milliliters of salt in the 6% solution plus the number of milliliters of salt in the water (0% solution) must equal the number of milliliters of salt in the 4% solution.

$$0.06(8) + 0(x) = 0.04(8 + x)$$
$$0.48 = 0.32 + 0.04x$$
$$0.16 = 0.04x$$
$$4 = x$$

To reduce the saline concentration to 4%, 4 mL of water should be added.

43. (a) Since 1 ft^2 of plywood flooring can emit 140 μg/hr of formaldehyde, 100 ft^2 of flooring can emit

$$100 \text{ ft}^2 \times \frac{140\ \mu\text{g/hr}}{1 \text{ ft}^2} = 14{,}000\ \mu\text{g/hr}.$$

Thus, the linear equation that computes the amount of formaldehyde, in micrograms, emitted into the room in x hours is

$$F = 14{,}000x.$$

(b) The amount of formaldehyde necessary to produce a concentration of 33 $\mu\text{g/ft}^3$ in a room with 800 ft^3 of air is

$$800 \text{ ft}^3 \times 33\ \mu\text{g/ft}^3 = 26{,}400\ \mu\text{g}.$$

In the linear equation from part (a), substitute this value for F and solve for x.

$$F = 14{,}000x$$
$$26{,}400 = 14{,}000x$$
$$x \approx 1.9 \text{ hr}$$

1.3 Exercises *(pages 102–104)*

49. $-i\sqrt{2} - 2 - (6 - 4i\sqrt{2}) - (5 - i\sqrt{2})$

$= (-2 - 6 - 5) + [-\sqrt{2} - (-4\sqrt{2}) - (-\sqrt{2})]i$

$= -13 + 4i\sqrt{2}$

67. $(2 + i)(2 - i)(4 + 3i)$

$= [(2 + i)(2 - i)](4 + 3i)$ Associative property

$= (2^2 - i^2)(4 + 3i)$ Product of the sum and difference of two terms

$= [4 - (-1)](4 + 3i)$ $i^2 = -1$

$= 5(4 + 3i)$

$= 20 + 15i$ Distributive property

95. $\frac{1}{i^{-11}} = i^{11}$ Negative exponent rule

$= i^8 \cdot i^3$ $a^{m+n} = a^m \cdot a^n$

$= (i^4)^2 \cdot i^3$ $a^{mn} = (a^m)^n$

$= 1(-i)$ $(i^4)^2 = 1^2 = 1$

$= -i$ Multiply.

99.

$$\left(\frac{\sqrt{2}}{2}+\frac{\sqrt{2}}{2}i\right)^2=\left(\frac{\sqrt{2}}{2}\right)^2+2\cdot\frac{\sqrt{2}}{2}\cdot\frac{\sqrt{2}}{2}i+\left(\frac{\sqrt{2}}{2}i\right)^2$$

$$=\frac{2}{4}+2\cdot\frac{2}{4}i+\frac{2}{4}i^2$$ Square of a binomial; Apply exponents and multiply.

$$=\frac{1}{2}+i+\frac{1}{2}i^2$$ Simplify.

$$=\frac{1}{2}+i+\frac{1}{2}(-1)$$ $i^2=-1$

$$=\frac{1}{2}+i-\frac{1}{2}$$ Multiply and combine real parts.

$$=i$$

Thus, $\frac{\sqrt{2}}{2}+\frac{\sqrt{2}}{2}i$ is a square root of i.

1.4 Exercises *(pages 111–113)*

17. $-4x^2+x=-3$

$$0=4x^2-x-3$$ Standard form

$$0=(4x+3)(x-1)$$ Factor.

$4x+3=0$ or $x-1=0$ Zero-factor property

$x=-\frac{3}{4}$ or $x=1$

Solution set: $\left\{-\frac{3}{4},1\right\}$

59. $\frac{1}{2}x^2+\frac{1}{4}x-3=0$

$$4\left(\frac{1}{2}x^2+\frac{1}{4}x-3\right)=4\cdot 0$$ Multiply by the LCD, 4.

$$2x^2+x-12=0$$ Distributive property; standard form

$$x=\frac{-b\pm\sqrt{b^2-4ac}}{2a}$$ Quadratic formula

$$x=\frac{-1\pm\sqrt{1^2-4(2)(-12)}}{2(2)}$$ $a=2,\ b=1,\ c=-12$

$$x=\frac{-1\pm\sqrt{97}}{4}$$

Solution set: $\left\{\frac{-1\pm\sqrt{97}}{4}\right\}$

79. $4x^2-2xy+3y^2=2$

(a) Solve for x in terms of y.

$$4x^2-2yx+3y^2-2=0$$ Standard form

$$4x^2-(2y)x+(3y^2-2)=0$$ $a=4,\ b=-2y,\ c=3y^2-2$

$$x=\frac{-b\pm\sqrt{b^2-4ac}}{2a}$$

$$=\frac{-(-2y)\pm\sqrt{(-2y)^2-4(4)(3y^2-2)}}{2(4)}$$

$$=\frac{2y\pm\sqrt{4y^2-16(3y^2-2)}}{8}$$

$$=\frac{2y\pm\sqrt{4y^2-48y^2+32}}{8}$$

$$=\frac{2y\pm\sqrt{32-44y^2}}{8}$$

$$=\frac{2y\pm\sqrt{4(8-11y^2)}}{8}$$

$$=\frac{2y\pm 2\sqrt{8-11y^2}}{8}$$

$$=\frac{2\left(y\pm\sqrt{8-11y^2}\right)}{2(4)}$$

$$x=\frac{y\pm\sqrt{8-11y^2}}{4}$$

(b) Solve for y in terms of x.

$$3y^2-2xy+4x^2-2=0$$ Standard form

$$3y^2-(2x)y+(4x^2-2)=0$$ $a=3,\ b=-2x,\ c=4x^2-2$

$$y=\frac{-b\pm\sqrt{b^2-4ac}}{2a}$$

$$=\frac{-(-2x)\pm\sqrt{(-2x)^2-4(3)(4x^2-2)}}{2(3)}$$

$$=\frac{2x\pm\sqrt{4x^2-12(4x^2-2)}}{6}$$

$$=\frac{2x\pm\sqrt{4x^2-48x^2+24}}{6}$$

$$=\frac{2x\pm\sqrt{24-44x^2}}{6}$$

$$=\frac{2x\pm\sqrt{4(6-11x^2)}}{6}$$

$$=\frac{2x\pm 2\sqrt{6-11x^2}}{6}$$

$$=\frac{2\left(x\pm\sqrt{6-11x^2}\right)}{2(3)}$$

$$y=\frac{x\pm\sqrt{6-11x^2}}{3}$$

95. $x=4$ or $x=5$

$x-4=0$ or $x-5=0$ Zero-factor property in reverse

$$(x-4)(x-5)=0$$

$$x^2-9x+20=0$$ Multiply.

$a=1,\ b=-9,\ c=20$

(Any nonzero constant multiple of these numbers will also work.)

1.5 Exercises *(pages 118–124)*

25.
$S = 2\pi rh + 2\pi r^2$ Surface area of a cylinder
$8\pi = 2\pi r \cdot 3 + 2\pi r^2$ Let $S = 8\pi$, $h = 3$.
$8\pi = 6\pi r + 2\pi r^2$ Multiply.
$0 = 2\pi r^2 + 6\pi r - 8\pi$ Standard form
$0 = r^2 + 3r - 4$ Divide by 2π.
$0 = (r + 4)(r - 1)$ Factor.
$r + 4 = 0$ or $r - 1 = 0$ Zero-factor property
$r = -4$ or $r = 1$ Solve each equation.

Because r represents the radius of a cylinder, -4 is not reasonable. The radius is 1 ft.

33. Let $r =$ the radius of the circle.

$\mathcal{A} = s^2$ Area of square
$800 = s^2$ Let $\mathcal{A} = 800$.
$800 = (2r)^2$ Let $s = 2r$.
$800 = 4r^2$ $(ab)^2 = a^2b^2$
$200 = r^2$ Divide by 4.
$r = \pm\sqrt{200} = \pm 10\sqrt{2}$

Because r represents the radius of a circle, $-10\sqrt{2}$ is not reasonable. The radius is $10\sqrt{2}$ ft.

55. Let $x =$ number of passengers in excess of 75.
Then $225 - 5x =$ the cost per passenger (in dollars)
and $75 + x =$ the number of passengers.

(Number of passengers)(Cost per passenger) = Revenue

$(75 + x)(225 - 5x) = 16{,}000$
$16{,}875 - 150x - 5x^2 = 16{,}000$
$0 = 5x^2 + 150x - 875$
$0 = x^2 + 30x - 175$
$0 = (x + 35)(x - 5)$
$x + 35 = 0$ or $x - 5 = 0$
$x = -35$ or $x = 5$

The negative solution is not meaningful. Since there are 5 passengers in excess of 75, the total number of passengers is 80.

1.6 Exercises *(pages 133–136)*

13. $\dfrac{4}{x^2 + x - 6} - \dfrac{1}{x^2 - 4} = \dfrac{2}{x^2 + 5x + 6}$

$\dfrac{4}{(x + 3)(x - 2)} - \dfrac{1}{(x + 2)(x - 2)} = \dfrac{2}{(x + 3)(x + 2)}$ Factor denominators.

$(x+3)(x-2)(x+2)\left(\dfrac{4}{(x+3)(x-2)} - \dfrac{1}{(x+2)(x-2)}\right)$
$= (x + 3)(x - 2)(x + 2)\left(\dfrac{2}{(x + 3)(x + 2)}\right)$

Multiply by the LCD, $(x + 3)(x - 2)(x + 2)$, where $x \neq -3$, $x \neq 2$, $x \neq -2$.

$4(x + 2) - 1(x + 3) = 2(x - 2)$ Simplify each side.
$4x + 8 - x - 3 = 2x - 4$ Distributive property
$3x + 5 = 2x - 4$ Combine like terms.
$x = -9$ Subtract $2x$ and subtract 5.

The restrictions $x \neq -3$, $x \neq 2$, $x \neq -2$ do not affect the proposed solution.

Solution set: $\{-9\}$

31. Let $x =$ the number of hours to fill the pool with both pipes open.

	Rate	Time	Part of the Job Accomplished
Inlet Pipe	$\frac{1}{5}$	x	$\frac{1}{5}x$
Outlet Pipe	$\frac{1}{8}$	x	$\frac{1}{8}x$

Filling the pool is 1 whole job, but because the outlet pipe empties the pool, its contribution should be subtracted from the contribution of the inlet pipe.

$\dfrac{1}{5}x - \dfrac{1}{8}x = 1$
$40\left(\dfrac{1}{5}x - \dfrac{1}{8}x\right) = 40 \cdot 1$ Multiply by the LCD, 40.
$8x - 5x = 40$ Distributive property
$3x = 40$ Combine like terms.
$x = \dfrac{40}{3} = 13\dfrac{1}{3}$ Divide by 3.

It took $13\frac{1}{3}$ hr to fill the pool.

53.
$\sqrt{2\sqrt{7x + 2}} = \sqrt{3x + 2}$
$\left(\sqrt{2\sqrt{7x + 2}}\right)^2 = \left(\sqrt{3x + 2}\right)^2$ Square each side.
$2\sqrt{7x + 2} = 3x + 2$
$\left(2\sqrt{7x + 2}\right)^2 = (3x + 2)^2$ Square each side again.
$4(7x + 2) = 9x^2 + 12x + 4$ Square the binomial on the right.
$28x + 8 = 9x^2 + 12x + 4$
$0 = 9x^2 - 16x - 4$
$0 = (9x + 2)(x - 2)$
$9x + 2 = 0$ or $x - 2 = 0$
$x = -\dfrac{2}{9}$ or $x = 2$

Check both proposed solutions by substituting first $-\frac{2}{9}$ and then 2 in the *original* equation. These checks will verify that both of these numbers are solutions.

Solution set: $\{-\frac{2}{9}, 2\}$

89. $x^{-2/3} + x^{-1/3} - 6 = 0$

Since $(x^{-1/3})^2 = x^{-2/3}$, let $u = x^{-1/3}$.

$u^2 + u - 6 = 0$ Substitute.

$(u + 3)(u - 2) = 0$ Factor.

$u + 3 = 0$ or $u - 2 = 0$

$u = -3$ or $u = 2$

Now replace u with $x^{-1/3}$.

$x^{-1/3} = -3$ or $x^{-1/3} = 2$

$(x^{-1/3})^{-3} = (-3)^{-3}$ or $(x^{-1/3})^{-3} = 2^{-3}$

Raise each side of each equation to the -3 power.

$x = \frac{1}{(-3)^3}$ or $x = \frac{1}{2^3}$

$x = -\frac{1}{27}$ or $x = \frac{1}{8}$

A check will show that both $-\frac{1}{27}$ and $\frac{1}{8}$ satisfy the original equation.

Solution set: $\left\{-\frac{1}{27}, \frac{1}{8}\right\}$

99. $m^{3/4} + n^{3/4} = 1$, for m

$m^{3/4} = 1 - n^{3/4}$

$(m^{3/4})^{4/3} = (1 - n^{3/4})^{4/3}$ Raise each side to the $\frac{4}{3}$ power.

$m = (1 - n^{3/4})^{4/3}$

1.7 Exercises *(pages 145–149)*

51. $x^2 - 2x \leq 1$

Step 1 Solve the corresponding quadratic equation.

$$x^2 - 2x = 1$$

$$x^2 - 2x - 1 = 0$$

The trinomial $x^2 - 2x - 1$ cannot be factored, so solve this equation with the quadratic formula.

$$x = \frac{-b \pm \sqrt{b^2 - 4ac}}{2a}$$

$$x = \frac{-(-2) \pm \sqrt{(-2)^2 - 4(1)(-1)}}{2(1)}$$

$a = 1,\ b = -2,\ c = -1$

Simplify to obtain $x = 1 \pm \sqrt{2}$—that is, $x = 1 - \sqrt{2}$ or $x = 1 + \sqrt{2}$.

Step 2 Identify the intervals determined by the solutions of the equation. The values $1 - \sqrt{2}$ and $1 + \sqrt{2}$ divide a number line into three intervals: $\left(-\infty, 1 - \sqrt{2}\right)$, $\left(1 - \sqrt{2}, 1 + \sqrt{2}\right)$, and $\left(1 + \sqrt{2}, \infty\right)$. Use solid circles on $1 - \sqrt{2}$ and $1 + \sqrt{2}$ because the inequality symbol includes equality. (Note that $1 - \sqrt{2} \approx -0.4$ and $1 + \sqrt{2} \approx 2.4$.)

Interval A $(-\infty, 1-\sqrt{2})$ | Interval B $(1-\sqrt{2}, 1+\sqrt{2})$ | Interval C $(1+\sqrt{2}, \infty)$

$1-\sqrt{2}$ 0 $1+\sqrt{2}$

Step 3 Use a test value from each interval to determine which intervals form the solution set.

Interval	Test Value	Is $x^2 - 2x \leq 1$ True or False?
A: $\left(-\infty, 1 - \sqrt{2}\right)$	-1	$(-1)^2 - 2(-1) \stackrel{?}{\leq} 1$ $3 \leq 1$ False
B: $\left(1 - \sqrt{2}, 1 + \sqrt{2}\right)$	0	$0^2 - 2(0) \stackrel{?}{\leq} 1$ $0 \leq 1$ True
C: $\left(1 + \sqrt{2}, \infty\right)$	3	$3^2 - 2(3) \stackrel{?}{\leq} 1$ $3 \leq 1$ False

Only Interval B makes the inequality true. Both endpoints are included because the given inequality is a nonstrict inequality.

Solution set: $\left[1 - \sqrt{2}, 1 + \sqrt{2}\right]$

61. $4x - x^3 \geq 0$

Step 1 $4x - x^3 = 0$ Corresponding equation

$x(4 - x^2) = 0$ Factor out the GCF, x.

$x(2 - x)(2 + x) = 0$

Factor the difference of squares.

$x = 0$ or $2 - x = 0$ or $2 + x = 0$

Zero-factor property

$x = 0$ or $x = 2$ or $x = -2$

Step 2 The values -2, 0, and 2 divide the number line into four intervals.

Interval A: $(-\infty, -2)$; Interval B: $(-2, 0)$; Interval C: $(0, 2)$; Interval D: $(2, \infty)$

Step 3 Use a test value from each interval to determine which intervals form the solution set.

Interval	Test Value	Is $4x - x^3 \geq 0$ True or False?
A: $(-\infty, -2)$	-3	$4(-3) - (-3)^3 \stackrel{?}{\geq} 0$ $15 \geq 0$ True
B: $(-2, 0)$	-1	$4(-1) - (-1)^3 \stackrel{?}{\geq} 0$ $-3 \geq 0$ False
C: $(0, 2)$	1	$4(1) - 1^3 \stackrel{?}{\geq} 0$ $3 \geq 0$ True
D: $(2, \infty)$	3	$4(3) - 3^3 \stackrel{?}{\geq} 0$ $-15 \geq 0$ False

Intervals A and C make the inequality true. The endpoints -2, 0, and 2 are all included because the given inequality is a nonstrict inequality.

Solution set: $(-\infty, -2] \cup [0, 2]$

79. $\dfrac{7}{x+2} \geq \dfrac{1}{x+2}$

Step 1 Rewrite the inequality so that 0 is on one side and there is a single fraction on the other side.

$$\frac{7}{x+2} - \frac{1}{x+2} \geq 0$$

$$\frac{6}{x+2} \geq 0$$

Step 2 Determine the values that will cause either the numerator or the denominator to equal 0.

Since $6 \neq 0$, the numerator is never equal to 0. The denominator is equal to 0 when $x + 2 = 0$, or $x = -2$.

-2 divides a number line into two intervals, $(-\infty, -2)$ and $(-2, \infty)$. Use an open circle on -2 because it makes the denominator 0.

Interval A $(-\infty, -2)$ | Interval B $(-2, \infty)$; -2, 0

Step 3 Use a test value from each interval to determine which intervals form the solution set.

Interval	Test Value	Is $\frac{7}{x+2} \geq \frac{1}{x+2}$ True or False?
A: $(-\infty, -2)$	-3	$\frac{7}{-3+2} \overset{?}{\geq} \frac{1}{-3+2}$ $-7 \geq -1$ False
B: $(-2, \infty)$	0	$\frac{7}{0+2} \overset{?}{\geq} \frac{1}{0+2}$ $\frac{7}{2} \geq \frac{1}{2}$ True

Interval B satisfies the original inequality. The endpoint -2 is not included because it makes the denominator 0.

Solution set: $(-2, \infty)$

87. $\dfrac{2x-3}{x^2+1} \geq 0$

The inequality already has 0 on one side, so set the numerator and denominator equal to 0. If $2x - 3 = 0$, then $x = \frac{3}{2}$. $x^2 + 1 = 0$ has no real solutions. $\frac{3}{2}$ divides a number line into two intervals, $\left(-\infty, \frac{3}{2}\right)$ and $\left(\frac{3}{2}, \infty\right)$.

Interval	Test Value	Is $\frac{2x-3}{x^2+1} \geq 0$ True or False?
A: $\left(-\infty, \frac{3}{2}\right)$	0	$\frac{2(0)-3}{0^2+1} \overset{?}{\geq} 0$ $-3 \geq 0$ False
B: $\left(\frac{3}{2}, \infty\right)$	2	$\frac{2(2)-3}{2^2+1} \overset{?}{\geq} 0$ $\frac{1}{5} \geq 0$ True

Interval B satisfies the original inequality, along with the endpoint $\frac{3}{2}$.

Solution set: $\left[\frac{3}{2}, \infty\right)$

1.8 Exercises *(pages 153–155)*

33.

$$4|x-3| > 12$$

$|x - 3| > 3$ Divide by 4.

$x - 3 < -3$ or $x - 3 > 3$ Case 3

$x < 0$ or $x > 6$ Add 3.

Solution set: $(-\infty, 0) \cup (6, \infty)$

49. $\left|5x + \dfrac{1}{2}\right| - 2 < 5$

$\left|5x + \frac{1}{2}\right| < 7$ Add 2.

$-7 < 5x + \frac{1}{2} < 7$ Case 2

$2(-7) < 2\left(5x + \frac{1}{2}\right) < 2(7)$ Multiply each part by 2.

$-14 < 10x + 1 < 14$ Distributive property

$-15 < 10x < 13$ Subtract 1 from each part.

$\frac{-15}{10} < x < \frac{13}{10}$ Divide each part by 10.

$-\frac{3}{2} < x < \frac{13}{10}$ Lowest terms

Solution set: $\left(-\frac{3}{2}, \frac{13}{10}\right)$

63. $|2x + 1| \leq 0$

Since the absolute value of a number is always non-negative, $|2x + 1| < 0$ is never true, so $|2x + 1| \leq 0$ is true only when $|2x + 1| = 0$. Solve this equation.

$|2x + 1| = 0$

$2x + 1 = 0$ If $|a| = 0$, then $a = 0$.

$x = -\frac{1}{2}$ Solve.

Solution set: $\left\{-\frac{1}{2}\right\}$

75. $|x^2 + 1| - |2x| = 0$

$|x^2 + 1| = |2x|$

If $|a| = |b|$, then $a = b$ or $a = -b$.

$x^2 + 1 = 2x$ or $x^2 + 1 = -2x$

$x^2 - 2x + 1 = 0$ or $x^2 + 2x + 1 = 0$

$(x - 1)^2 = 0$ or $(x + 1)^2 = 0$

$x - 1 = 0$ or $x + 1 = 0$

$x = 1$ or $x = -1$

Solution set: $\{-1, 1\}$

Chapter 2 Graphs and Functions

2.1 Exercises *(pages 178–180)*

17. $P(3\sqrt{2}, 4\sqrt{5}), Q(\sqrt{2}, -\sqrt{5})$

(a) $d(P, Q) = \sqrt{(\sqrt{2} - 3\sqrt{2})^2 + (-\sqrt{5} - 4\sqrt{5})^2}$

Let $x_1 = 3\sqrt{2}, y_1 = 4\sqrt{5}, x_2 = \sqrt{2}, y_2 = -\sqrt{5}$.

$= \sqrt{(-2\sqrt{2})^2 + (-5\sqrt{5})^2}$

$= \sqrt{8 + 125}$ $\quad (ab)^2 = a^2b^2; (\sqrt{a})^2 = a$

$= \sqrt{133}$

(b) The midpoint of segment PQ has these coordinates.

$$\left(\frac{3\sqrt{2} + \sqrt{2}}{2}, \frac{4\sqrt{5} + (-\sqrt{5})}{2}\right) = \left(2\sqrt{2}, \frac{3\sqrt{5}}{2}\right)$$

2.2 Exercises *(pages 185–187)*

49. Let $P(x, y)$ be a point whose distance from $A(1, 0)$ is $\sqrt{10}$ and whose distance from $B(5, 4)$ is also $\sqrt{10}$.

$d(P, A) = \sqrt{10}$, so

$\sqrt{(1 - x)^2 + (0 - y)^2} = \sqrt{10}$ $\quad$ Distance formula

$(1 - x)^2 + y^2 = 10.$ $\quad$ Square each side.

$d(P, B) = \sqrt{10}$, so

$\sqrt{(5 - x)^2 + (4 - y)^2} = \sqrt{10}$

$(5 - x)^2 + (4 - y)^2 = 10.$

Set these distances equal to each other. Remember the middle term when squaring each binomial.

$(1 - x)^2 + y^2 = (5 - x)^2 + (4 - y)^2$

$1 - 2x + x^2 + y^2 = 25 - 10x + x^2 + 16 - 8y + y^2$

$1 - 2x = 41 - 10x - 8y$ $\quad$ Subtract x^2 and y^2. Add.

$8y = 40 - 8x$ $\quad$ Solve for y.

$y = 5 - x$ $\quad$ Divide by 8.

Substitute $5 - x$ for y in the equation $(1 - x)^2 + y^2 = 10$ and solve for x.

$(1 - x)^2 + (5 - x)^2 = 10$

$1 - 2x + x^2 + 25 - 10x + x^2 = 10$

$2x^2 - 12x + 26 = 10$

$2x^2 - 12x + 16 = 0$

$x^2 - 6x + 8 = 0$

$(x - 2)(x - 4) = 0$

$x - 2 = 0$ or $x - 4 = 0$

$x = 2$ or $x = 4$

To find the corresponding values of y, substitute in the equation $y = 5 - x$.

If $x = 2$, then $y = 5 - 2 = 3$.

If $x = 4$, then $y = 5 - 4 = 1$.

The points satisfying the given conditions are $(2, 3)$ and $(4, 1)$.

2.4 Exercises *(pages 211–216)*

49. $5x - 2y = 10$

(a) Write the equation in slope-intercept form.

$5x - 2y = 10$

$-2y = -5x + 10$ $\quad$ Subtract $5x$.

$y = \frac{5}{2}x - 5$ $\quad$ Divide by -2.

The slope is $\frac{5}{2}$, the coefficient of x.

(b) The y-intercept b is -5, so plot $(0, -5)$. The slope m is $\frac{5}{2}$, so move 5 units up and 2 units to the right to locate a second point, $(2, 0)$. Draw a line through those two points.

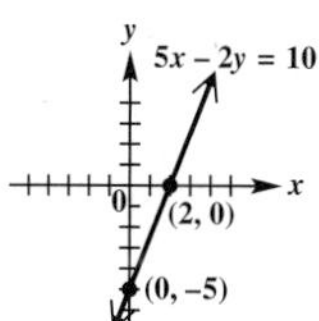

89. fixed cost = \$1650; cost = \$400; price of item = \$305

(a) $C(x) = 400x + 1650$ $\quad m = 400, b = 1650$

(b) $R(x) = 305x$

(c) $P(x) = R(x) - C(x)$

$P(x) = 305x - (400x + 1650)$

$P(x) = -95x - 1650$

(d) $C(x) = R(x)$

$400x + 1650 = 305x$

$95x = -1650$

$x \approx -17.4$

This result indicates a negative "break-even point," but the number of units produced must be a positive number. A calculator graph of the lines $y = 400x + 1650$ and $y = 305x$ on the same screen or solving the inequality $305x < 400x + 1650$ will show that $R(x) < C(x)$ for all positive values of x (in fact, whenever x is greater than about -17.4). Do not produce the product since it is impossible to make a profit.

2.5 Exercises *(pages 226–230)*

15. Since the x-intercept is 3 and the y-intercept is -2, the line passes through the points $(3, 0)$ and $(0, -2)$. Use these points to find the slope.

$$m = \frac{-2 - 0}{0 - 3} = \frac{-2}{-3} = \frac{2}{3}$$

The slope is $\frac{2}{3}$ and the y-intercept is -2, so the equation of the line in slope-intercept form is as follows.

$$y = \frac{2}{3}x - 2$$

53. **(a)** Find the slope of the line $3y + 2x = 6$.

$$3y + 2x = 6$$
$$3y = -2x + 6$$
$$y = -\frac{2}{3}x + 2$$

Thus, $m = -\frac{2}{3}$. A line parallel to $3y + 2x = 6$ will also have slope $-\frac{2}{3}$. Use the points $(4, -1)$ and $(k, 2)$ and the definition of slope.

$$\frac{2-(-1)}{k-4} = -\frac{2}{3}$$

Solve this equation for k.

$$\frac{3}{k-4} = -\frac{2}{3}$$
$$3(k-4)\left(\frac{3}{k-4}\right) = 3(k-4)\left(-\frac{2}{3}\right)$$
$$9 = -2(k-4)$$
$$9 = -2k + 8$$
$$2k = -1$$
$$k = -\frac{1}{2}$$

(b) Find the slope of the line $2y - 5x = 1$.

$$2y - 5x = 1$$
$$2y = 5x + 1$$
$$y = \frac{5}{2}x + \frac{1}{2}$$

Thus, $m = \frac{5}{2}$. A line perpendicular to $2y - 5x = 1$ will have slope $-\frac{2}{5}$ since $\frac{5}{2}\left(-\frac{2}{5}\right) = -1$. Use the points $(4, -1)$ and $(k, 2)$ and the definition of slope.

$$\frac{2-(-1)}{k-4} = -\frac{2}{5}$$

Solve this equation for k.

$$\frac{3}{k-4} = -\frac{2}{5}$$
$$5(k-4)\left(\frac{3}{k-4}\right) = 5(k-4)\left(-\frac{2}{5}\right)$$
$$15 = -2(k-4)$$
$$15 = -2k + 8$$
$$2k = -7$$
$$k = -\frac{7}{2}$$

77. $A(-1, 4), B(-2, -1), C(1, 14)$

For A and B, $m = \frac{-1-4}{-2-(-1)} = \frac{-5}{-1} = 5.$

For B and C, $m = \frac{14-(-1)}{1-(-2)} = \frac{15}{3} = 5.$

For A and C, $m = \frac{14-4}{1-(-1)} = \frac{10}{2} = 5.$

All three slopes are the same, so the three points are collinear.

2.6 Exercises *(pages 238–242)*

35. First, notice that for all x-values less than or equal to 0, the function value is -1. Next, notice that for all x-values greater than 0, the function value is 1. Thus, we have a function f that is defined in two pieces.

$$f(x) = \begin{cases} -1 & \text{if } x \le 0 \\ 1 & \text{if } x > 0 \end{cases}$$

The domain is $(-\infty, \infty)$ and the range is $\{-1, 1\}$.

2.7 Exercises *(pages 253–257)*

91. $f(x) = 2x + 5$

Translate the graph of $f(x)$ up 2 units to obtain the graph of

$$t(x) = (2x + 5) + 2$$
$$= 2x + 7.$$

Now translate the graph of $t(x)$ left 3 units to obtain the graph of

$$g(x) = 2(x + 3) + 7$$
$$= 2x + 6 + 7$$
$$= 2x + 13.$$

(Note that if the original graph is first translated left 3 units and then up 2 units, the final result is the same.)

2.8 Exercises *(pages 266–271)*

25. **(a)** From the graph, $f(-1) = 0$ and $g(-1) = 3$.

$$(f + g)(-1) = f(-1) + g(-1)$$
$$= 0 + 3$$
$$= 3$$

(b) From the graph, $f(-2) = -1$ and $g(-2) = 4$.

$$(f - g)(-2) = f(-2) - g(-2)$$
$$= -1 - 4$$
$$= -5$$

(c) From the graph, $f(0) = 1$ and $g(0) = 2$.

$$(fg)(0) = f(0) \cdot g(0)$$
$$= 1 \cdot 2$$
$$= 2$$

(d) From the graph, $f(2) = 3$ and $g(2) = 0$.

$$\left(\frac{f}{g}\right)(2) = \frac{f(2)}{g(2)} = \frac{3}{0}, \quad \text{which is undefined.}$$

39. $f(x) = \frac{1}{x}$

(a) $f(x + h) = \frac{1}{x + h}$

(b) $f(x + h) - f(x) = \frac{1}{x+h} - \frac{1}{x}$ Substitute.

$$= \frac{x - (x + h)}{x(x + h)}$$

Use a common denominator.

$$= \frac{-h}{x(x + h)}$$ Simplify.

(c) $\dfrac{f(x+h)-f(x)}{h} = \dfrac{-h}{x(x+h)} \cdot \dfrac{1}{h}$ Substitute in the numerator. Multiply by the reciprocal of the denominator.

$= \dfrac{-1}{x(x+h)}$ Multiply. Write in lowest terms.

79. $g(f(2)) = g(1) = 2$; $g(f(3)) = g(2) = 5$
Since $g(f(1)) = 7$ and $f(1) = 3$, $g(3) = 7$.
Completed table:

x	$f(x)$	$g(x)$	$g(f(x))$
1	3	2	7
2	1	5	2
3	2	7	5

Chapter 3 Polynomial and Rational Functions

3.1 Exercises *(pages 292–300)*

3. $f(x) = -2(x+3)^2 + 2$

(a) domain: $(-\infty, \infty)$
range: $(-\infty, 2]$

(b) vertex: $(h, k) = (-3, 2)$

(c) axis: $x = -3$

(d) $y = -2(0+3)^2 + 2$ Let $x = 0$.
$y = -16$
y-intercept: -16

(e) $0 = -2(x+3)^2 + 2$ Set $f(x) = 0$.
$2(x+3)^2 = 2$ Add $2(x+3)^2$.
$(x+3)^2 = 1$ Divide by 2.
$x + 3 = \pm\sqrt{1}$ Take square roots.
$x + 3 = 1$ or $x + 3 = -1$
$x = -2$ or $x = -4$
x-intercepts: $-4, -2$

59. $y = \dfrac{-16x^2}{0.434v^2} + 1.15x + 8$

(a) $10 = \dfrac{-16(15)^2}{0.434v^2} + 1.15(15) + 8$ Let $y = 10$, $x = 15$.

$10 = \dfrac{-3600}{0.434v^2} + 17.25 + 8$

$\dfrac{3600}{0.434v^2} = 15.25$ Isolate the variable term.

$3600 = 6.6185v^2$ Multiply by $0.434v^2$.

$v^2 = \dfrac{3600}{6.6185}$ Divide by 6.6185.

$v = \pm\sqrt{\dfrac{3600}{6.6185}} \approx \pm 23.32$ Take square roots.

Because v represents velocity, only the positive square root is meaningful. The basketball should have an initial velocity of approximately 23.32 ft per sec.

(b) $y = \dfrac{-16x^2}{0.434(23.32)^2} + 1.15x + 8$

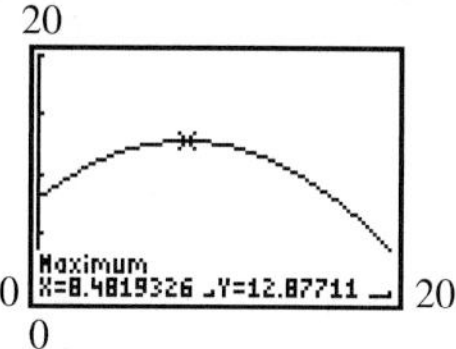

Since the coefficient of x^2 is negative, the graph of this function is a parabola that opens downward. Find the coordinates of the vertex to answer the question.

$$x = -\frac{b}{2a} = -\frac{1.15}{2\left(\dfrac{-16}{0.434(23.32)^2}\right)} \approx 8.48$$

The y-value of the vertex is found by evaluating the function at $x = 8.48$.

$$f(8.48) = \frac{-16(8.48)^2}{0.434(23.32)^2} + 1.15(8.48) + 8$$
$$\approx 12.88$$

The maximum height of the basketball is approximately 12.88 ft. This is verified by the calculator screen.

69. $y = x^2 - 10x + c$
An x-intercept occurs where $y = 0$, or

$$0 = x^2 - 10x + c.$$

There will be exactly one x-intercept if this equation has exactly one solution, meaning that the discriminant is 0.

$$b^2 - 4ac = 0$$
$$(-10)^2 - 4(1)c = 0$$
$$100 = 4c$$
$$c = 25$$

3.2 Exercises *(pages 305–306)*

7. $\dfrac{x^5 + 3x^4 + 2x^3 + 2x^2 + 3x + 1}{x+2}$

$x + 2 = x - (-2)$

$$\begin{array}{r|rrrrrr} -2 & 1 & 3 & 2 & 2 & 3 & 1 \\ & & -2 & -2 & 0 & -4 & 2 \\ \hline & 1 & 1 & 0 & 2 & -1 & 3 \end{array}$$

In the last row of the synthetic division, all numbers except the last one give the coefficients of the quotient, and the last number gives the remainder. The quotient is

$$1x^4 + 1x^3 + 0x^2 + 2x - 1, \quad \text{or} \quad x^4 + x^3 + 2x - 1,$$

and the remainder is 3. Thus,

$$\frac{x^5 + 3x^4 + 2x^3 + 2x^2 + 3x + 1}{x+2} = x^4 + x^3 + 2x - 1 + \frac{3}{x+2}.$$

19. $f(x) = 2x^3 + x^2 + x - 8; \quad k = -1$

$$\begin{array}{r|rrrr} -1 & 2 & 1 & 1 & -8 \\ & & -2 & 1 & -2 \\ \hline & 2 & -1 & 2 & -10 \end{array}$$

$$f(x) = (x + 1)(2x^2 - x + 2) + (-10)$$
$$f(x) = (x + 1)(2x^2 - x + 2) - 10$$

27. $f(x) = x^2 + 5x + 6; \quad k = -2$

$$\begin{array}{r|rrr} -2 & 1 & 5 & 6 \\ & & -2 & -6 \\ \hline & 1 & 3 & 0 \end{array}$$

The last number in the bottom row of the synthetic division gives the remainder, 0. Therefore, by the remainder theorem, $f(-2) = 0$.

3.3 Exercises *(pages 315–317)*

23. $f(x) = x^3 + (7 - 3i)x^2 + (12 - 21i)x - 36i; \quad k = 3i$

$$\begin{array}{r|rrrr} 3i & 1 & 7-3i & 12-21i & -36i \\ & & 3i & 21i & 36i \\ \hline & 1 & 7 & 12 & 0 \end{array}$$

The quotient is $x^2 + 7x + 12$, so

$$f(x) = (x - 3i)(x^2 + 7x + 12)$$
$$f(x) = (x - 3i)(x + 4)(x + 3).$$

45. $f(x) = 3x(x - 2)(x + 3)(x^2 - 1)$

$0 = 3x(x - 2)(x + 3)(x^2 - 1)$ Set $f(x) = 0$.

$0 = 3x(x - 2)(x + 3)(x - 1)(x + 1)$ Factor the difference of squares.

$x = 0$ or $x - 2 = 0$ or $x + 3 = 0$ or $x - 1 = 0$ or $x + 1 = 0$ Zero-factor property

$x = 0$ or $x = 2$ or $x = -3$ or $x = 1$ or $x = -1$

The zeros are 0, 2, -3, 1, and -1, all of multiplicity 1.

51. Zeros of -2, 1, and 0; $f(-1) = -1$

The factors of $f(x)$ are $x - (-2) = x + 2$, $x - 1$, and $x - 0 = x$.

$$f(x) = a(x + 2)(x - 1)(x)$$
$$f(-1) = a(-1 + 2)(-1 - 1)(-1) = -1$$
$$a(1)(-2)(-1) = -1$$
$$2a = -1$$
$$a = -\frac{1}{2}$$

Therefore,

$$f(x) = -\frac{1}{2}(x + 2)(x - 1)(x)$$
$$f(x) = -\frac{1}{2}(x^2 + x - 2)(x)$$
$$f(x) = -\frac{1}{2}(x^3 + x^2 - 2x)$$
$$f(x) = -\frac{1}{2}x^3 - \frac{1}{2}x^2 + x.$$

3.4 Exercises *(pages 330–337)*

37. $f(x) = x^3 + 5x^2 - x - 5$

$f(x) = x^2(x + 5) - 1(x + 5)$

$f(x) = (x + 5)(x^2 - 1)$ Factor by grouping.

$f(x) = (x + 5)(x + 1)(x - 1)$ Factor the difference of squares.

Find the real zeros of f. (All three zeros of this function are real.)

$x + 5 = 0$ or $x + 1 = 0$ or $x - 1 = 0$ Zero-factor property

$x = -5$ or $x = -1$ or $x = 1$

The zeros of f are -5, -1, and 1, so the x-intercepts of the graph are also -5, -1, and 1. Plot the points

$$(-5, 0), (-1, 0), \text{ and } (1, 0).$$

Now find the y-intercept.

$$f(0) = 0^3 + 5 \cdot 0^2 - 0 - 5 = -5$$

The y-intercept is -5. Plot the point $(0, -5)$.

The dominating term of $f(x)$ is x^3, so the end behavior of the graph is ↙ ↗. Each zero is of multiplicity 1, indicating that the graph will cross the x-axis at each zero. Begin at either end of the graph with the appropriate end behavior and draw a smooth curve that crosses the x-axis at each zero, has a turning point between successive zeros, and passes through the y-intercept.

Additional points may be used to verify whether the graph is above or below the x-axis between the zeros and to add detail to the sketch of the graph. A typical selection of test points and the results are shown in the table.

Interval	Test Point	Value of $f(x)$	Sign of $f(x)$	Graph Above or Below x-axis
$(-\infty, -5)$	-6	-35	Negative	Below
$(-5, -1)$	-2	9	Positive	Above
$(-1, 1)$	0	-5	Negative	Below
$(1, \infty)$	2	21	Positive	Above

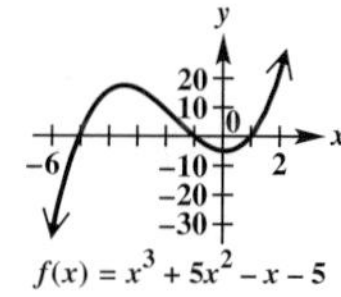

83. $f(x) = x^3 + 4x^2 - 8x - 8;\quad [-3.8, -3]$
Graph this function in a window that will produce a comprehensive graph, such as $[-10, 10]$ by $[-50, 50]$, with X-scale = 1, Y-scale = 10.

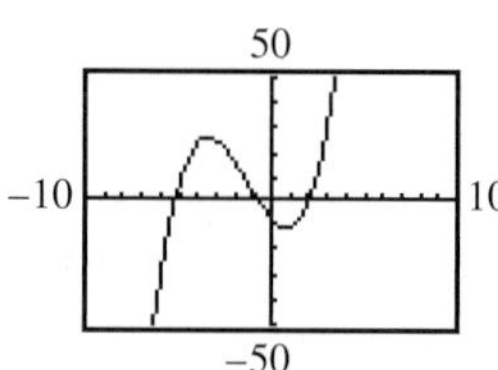

From this graph, we can see that the turning point in the interval $[-3.8, -3]$ is a maximum. Use "Maximum" from the CALC menu to approximate the coordinates of this turning point.

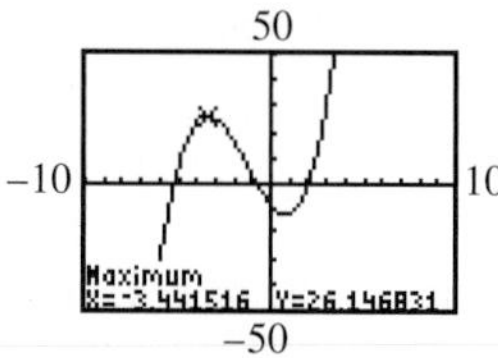

To the nearest hundredth, the turning point in the interval $[-3.8, -3]$ is $(-3.44, 26.15)$.

101. Use the following volume formulas:

$$V_{\text{cylinder}} = \pi r^2 h$$

$$V_{\text{hemisphere}} = \frac{1}{2}V_{\text{sphere}} = \frac{1}{2}\left(\frac{4}{3}\pi r^3\right) = \frac{2}{3}\pi r^3$$

$$\pi r^2 h + 2\left(\frac{2}{3}\pi r^3\right) = \text{Total volume of tank}$$

$$\pi x^2(12) + \frac{4}{3}\pi x^3 = 144\pi \qquad \text{Let } V = 144\pi,\ h = 12, \text{ and } r = x.$$

$$\frac{4}{3}\pi x^3 + 12\pi x^2 - 144\pi = 0 \qquad \text{Subtract } 144\pi \text{ and rewrite.}$$

$$\frac{4}{3}x^3 + 12x^2 - 144 = 0 \qquad \text{Divide by } \pi.$$

$$4x^3 + 36x^2 - 432 = 0 \qquad \text{Multiply by 3.}$$

$$x^3 + 9x^2 - 108 = 0 \qquad \text{Divide by 4.}$$

Synthetic division or graphing $f(x) = x^3 + 9x^2 - 108$ with a graphing calculator will show that this function has zeros of -6 (multiplicity 2) and 3. Because x represents the radius of the hemispheres, a negative solution for the equation $x^3 + 9x^2 - 108 = 0$ is not meaningful. In order to get a volume of 144π ft^3, a radius of 3 ft should be used.

3.5 Exercises *(pages 351–359)*

81. $f(x) = \dfrac{1}{x^2 + 1}$

Because $x^2 + 1 = 0$ has no real solutions, the denominator is never 0, so there are no vertical asymptotes. As $|x| \to \infty$, $y \to 0$, and thus the line $y = 0$ (the x-axis) is the horizontal asymptote.

$f(0) = \dfrac{1}{0^2 + 1} = 1$, so the y-intercept is 1.

The equation $f(x) = 0$ has no solutions (notice that the numerator has no zeros), so there are no x-intercepts.

$f(-x) = \dfrac{1}{(-x)^2 + 1} = \dfrac{1}{x^2 + 1} = f(x)$, so the graph is symmetric with respect to the y-axis.

Use a table of values to obtain several additional points on the graph. Use the asymptote, the y-intercept, and these additional points (which reflect the symmetry of the graph) to sketch the graph of the function.

x	y
± 0.5	0.8
± 1	0.5
± 2	0.2

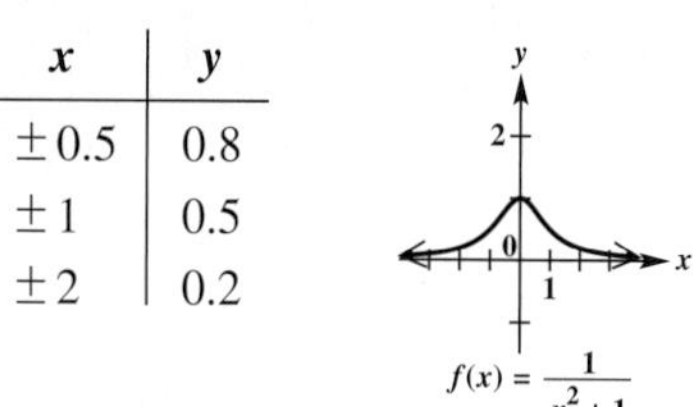

89. $f(x) = \dfrac{x^2 + 2x}{2x - 1}$

Step 1 Find any vertical asymptotes.

$$2x - 1 = 0$$
$$x = \frac{1}{2}$$

Step 2 Find any horizontal or oblique asymptotes. Because the numerator has degree exactly one more than the denominator, there is an oblique asymptote. Because $2x - 1$ is not of the form $x - a$, use polynomial long division rather than synthetic division to find the equation of this asymptote.

$$\begin{array}{r} \frac{1}{2}x + \frac{5}{4} \\ 2x - 1\overline{)x^2 + 2x + 0} \\ \underline{x^2 - \frac{1}{2}x\quad} \\ \frac{5}{2}x + 0 \\ \underline{\frac{5}{2}x - \frac{5}{4}} \\ \frac{5}{4} \end{array}$$

Disregard the remainder. The equation of the oblique asymptote is $y = \frac{1}{2}x + \frac{5}{4}$.

Step 3 Find the y-intercept.

$$f(0) = \frac{0^2 + 2(0)}{2(0) - 1} = \frac{0}{-1} = 0,$$

so the y-intercept is 0.

Step 4 Find the x-intercepts, if any, by finding the zeros of the numerator.

$$x^2 + 2x = 0$$
$$x(x + 2) = 0$$
$$x = 0 \quad \text{or} \quad x = -2$$

There are two x-intercepts, -2 and 0.

Step 5 The graph does not intersect its oblique asymptote because the following equation has no solution.

$$\frac{x^2 + 2x}{2x - 1} = \frac{1}{2}x + \frac{5}{4}$$

$$x^2 + 2x = x^2 + 2x - \frac{5}{4} \quad \text{Multiply by } (2x - 1).$$

$$0 = -\frac{5}{4} \quad \text{False}$$

Step 6

x	y
-3	$-\frac{3}{7} \approx -0.43$
-1	$\frac{1}{3} \approx 0.33$
$\frac{1}{4} = 0.25$	$-\frac{9}{8} = -1.125$
1	3
3	3

Use the asymptotes, the intercepts, and these additional points to sketch the graph.

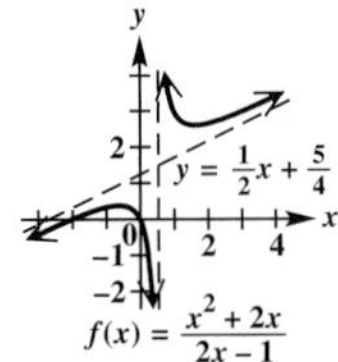

101. The graph has one vertical asymptote, $x = 2$, so $x - 2$ is a factor of the denominator of the rational expression. There is a point of discontinuity ("hole") in the graph at $x = -2$, so there is a factor of $x + 2$ in both numerator and denominator. There is one x-intercept, 3, so 3 is a zero of the numerator, which means that $x - 3$ is a factor of the numerator. Putting all of this information together, we have a possible function for the graph:

$$f(x) = \frac{(x - 3)(x + 2)}{(x - 2)(x + 2)}, \quad \text{or} \quad f(x) = \frac{x^2 - x - 6}{x^2 - 4}.$$

Note: From the second form of the function, we can see that the graph of this function has a horizontal asymptote of $y = 1$, which is consistent with the given graph.

3.6 Exercises *(pages 365–369)*

19. ***Step 1*** Write the general relationship among the variables as an equation. Use the constant k.

$$a = \frac{kmn^2}{y^3}$$

Step 2 Substitute $a = 9$, $m = 4$, $n = 9$, and $y = 3$ to find k.

$$9 = \frac{k \cdot 4 \cdot 9^2}{3^3}$$
$$9 = 12k$$
$$k = \frac{3}{4}$$

Step 3 Substitute this value of k into the equation from Step 1, obtaining a specific formula.

$$a = \frac{3}{4} \cdot \frac{mn^2}{y^3}$$

$$a = \frac{3mn^2}{4y^3}$$

Step 4 Substitute $m = 6$, $n = 2$, and $y = 5$ and solve for a.

$$a = \frac{3mn^2}{4y^3} = \frac{3 \cdot 6 \cdot 2^2}{4 \cdot 5^3} = \frac{18}{125}$$

35. ***Step 1*** Let F represent the force of the wind, $\mathscr{A}$ represent the area of the surface, and v represent the velocity of the wind.

$$F = k\mathscr{A}v^2$$

Step 2 $50 = k\left(\frac{1}{2}\right)(40)^2$ Let $F = 50$, $\mathscr{A} = \frac{1}{2}$, $v = 40$.

$50 = 800k$

$k = \frac{50}{800}$, or $\frac{1}{16}$

Step 3 $F = \frac{1}{16}\mathscr{A}v^2$

Step 4 $F = \frac{1}{16} \cdot 2 \cdot 80^2 = 800$

The force of the wind would be 800 lb.

43. Let t represent the Kelvin temperature.

$R = kt^4$

$213.73 = k \cdot 293^4$ Let $R = 213.73$, $t = 293$.

$k = \frac{213.73}{293^4} \approx 2.9 \times 10^{-8}$

Thus, $R = (2.9 \times 10^{-8})t^4$.

If $t = 335$, then

$R = (2.9 \times 10^{-8})(335^4) \approx 365.24$.

Chapter 4 Inverse, Exponential, and Logarithmic Functions

4.1 Exercises *(pages 393–397)*

47. $f(x) = \frac{2}{x+6}$, $\quad g(x) = \frac{6x+2}{x}$

$(f \circ g)(x) = f(g(x))$ Definition

$= f\left(\frac{6x+2}{x}\right)$ Replace $g(x)$ with $\frac{6x+2}{x}$.

$= \frac{2}{\frac{6x+2}{x} + 6}$ Use $f(x) = \frac{2}{x+6}$.

$= \frac{2}{\frac{6x+2+6x}{x}}$ Add terms in the denominator.

$= \frac{2}{1} \cdot \frac{x}{12x+2}$ Definition of division

$= \frac{2x}{12x+2}$ Multiply.

$= \frac{2x}{2(6x+1)}$ Factor.

$= \frac{x}{6x+1}$ Lowest terms

$(f \circ g)(x) \neq x$

Since $(f \circ g)(x) \neq x$, the functions are not inverses. It is not necessary to check $(g \circ f)(x)$.

69. $f(x) = \frac{1}{x-3}$, $x \neq 3$

(a) $y = \frac{1}{x-3}$, $x \neq 3$ $\quad y = f(x)$

$x = \frac{1}{y-3}$, $y \neq 3$ $\quad$ Interchange x and y.

$x(y-3) = 1$ $\quad$ Solve for y.

$xy - 3x = 1$

$xy = 1 + 3x$

$y = \frac{1+3x}{x}$

$f^{-1}(x) = \frac{1+3x}{x}$, $x \neq 0$ $\quad$ Replace y with $f^{-1}(x)$.

(b)

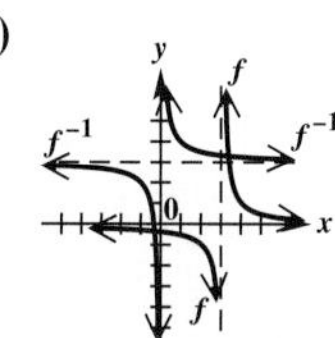

(c) For both f and f^{-1}, the domain contains all real numbers except those for which the denominator is equal to 0.

Domain of f = range of $f^{-1} = (-\infty, 3) \cup (3, \infty)$

Domain of f^{-1} = range of $f = (-\infty, 0) \cup (0, \infty)$

4.2 Exercises *(pages 409–413)*

7. $g(x) = \left(\frac{1}{4}\right)^x$

$g(-2) = \left(\frac{1}{4}\right)^{-2} = 4^2 = 16$

73. $\left(\frac{1}{e}\right)^{-x} = \left(\frac{1}{e^2}\right)^{x+1}$

$(e^{-1})^{-x} = (e^{-2})^{x+1}$ $\quad$ Definition of negative exponent

$e^x = e^{-2(x+1)}$ $\quad$ $(a^m)^n = a^{mn}$

$e^x = e^{-2x-2}$ $\quad$ Distributive property

$x = -2x - 2$ $\quad$ Property (b)

$3x = -2$ $\quad$ Add $2x$.

$x = -\frac{2}{3}$ $\quad$ Divide by 3.

Solution set: $\left\{-\frac{2}{3}\right\}$

75. $(\sqrt{2})^{x+4} = 4^x$

$(2^{1/2})^{x+4} = (2^2)^x$ Definition of $a^{1/n}$; Write each side as a power of a common base.

$2^{(1/2)(x+4)} = 2^{2x}$ $(a^m)^n = a^{mn}$

$2^{(1/2)x+2} = 2^{2x}$ Distributive property

$\frac{1}{2}x + 2 = 2x$ Property (b)

$2 = \frac{3}{2}x$ Subtract $\frac{1}{2}x$.

$\frac{4}{3} = x$ Multiply by $\frac{2}{3}$.

Solution set: $\{\frac{4}{3}\}$

77. $\frac{1}{27} = x^{-3}$

$3^{-3} = x^{-3}$ Write $\frac{1}{27}$ as a power of 3.

$x = 3$ Property of logarithms

Solution set: $\{3\}$

Alternative solution:

$\frac{1}{27} = x^{-3}$

$\frac{1}{27} = \frac{1}{x^3}$ Definition of negative exponent

$27 = x^3$ Take reciprocals.

$x = \sqrt[3]{27}$ Find cube roots.

$x = 3$ Evaluate.

Solution set: $\{3\}$

4.3 Exercises *(pages 422–425)*

23. $\log_x 25 = -2$

We must consider only positive values for x, $x \neq 1$, because it is the base of the logarithm.

$x^{-2} = 25$ Write in exponential form.

$(x^{-2})^{-1/2} = 25^{-1/2}$ Raise each side to the same power.

$x = \frac{1}{25^{1/2}}$ Definition of negative exponent

$x = \frac{1}{5}$ $25^{1/2} = \sqrt{25} = 5$

Solution set: $\{\frac{1}{5}\}$

83. $-\frac{2}{3}\log_5 5m^2 + \frac{1}{2}\log_5 25m^2$

$= \log_5(5m^2)^{-2/3} + \log_5(25m^2)^{1/2}$ Power property

$= \log_5[(5m^2)^{-2/3} \cdot (25m^2)^{1/2}]$ Product property

$= \log_5(5^{-2/3}m^{-4/3} \cdot 5m)$ Properties of exponents

$= \log_5(5^{-2/3} \cdot 5^1 \cdot m^{-4/3} \cdot m^1)$

$= \log_5(5^{1/3} \cdot m^{-1/3})$ $a^m \cdot a^n = a^{m+n}$

$= \log_5 \frac{5^{1/3}}{m^{1/3}}$, or $\log_5 \sqrt[3]{\frac{5}{m}}$

91. $\log_{10} \sqrt{30} = \log_{10} 30^{1/2}$ Definition of $a^{1/n}$

$= \frac{1}{2}\log_{10} 30$ Power property

$= \frac{1}{2}\log_{10}(10 \cdot 3)$

$= \frac{1}{2}(\log_{10} 10 + \log_{10} 3)$ Product property

$= \frac{1}{2}(1 + 0.4771)$ $\log_a a = 1$; Substitute.

$= \frac{1}{2}(1.4771)$

≈ 0.7386

4.4 Exercises *(pages 434–439)*

87. $\log_{\sqrt{13}} 12 = \frac{\ln 12}{\ln \sqrt{13}}$ Change-of-base theorem

$= \frac{\ln 12}{\ln 13^{1/2}}$ Definition of $a^{1/n}$

$= \frac{\ln 12}{0.5 \ln 13}$ Power property

≈ 1.9376 Use a calculator and round answer to four decimal places.

The required logarithm can also be found by entering $\ln \sqrt{13}$ into the calculator directly.

$\log_{\sqrt{13}} 12 = \frac{\ln 12}{\ln \sqrt{13}}$ Change-of-base theorem

≈ 1.9376 Use a calculator and round answer to four decimal places.

91. $\ln(b^4\sqrt{a}) = \ln(b^4a^{1/2})$ Definition of $a^{1/n}$

$= \ln b^4 + \ln a^{1/2}$ Product property

$= 4 \ln b + \frac{1}{2}\ln a$ Power property

$= 4v + \frac{1}{2}u$ Substitute v for $\ln b$ and u for $\ln a$.

4.5 Exercises *(pages 446–450)*

23. $3(2)^{x-2} + 1 = 100$

$3(2)^{x-2} = 99$ Subtract 1.

$2^{x-2} = 33$ Divide by 3.

$\log 2^{x-2} = \log 33$ Take the logarithm on each side.

$(x-2)\log 2 = \log 33$ Power property

$x \log 2 - 2\log 2 = \log 33$ Distributive property

$x \log 2 = \log 33 + 2\log 2$ Add 2 log 2.

$x = \frac{\log 33 + 2\log 2}{\log 2}$ Divide by log 2.

$x \approx 7.044$

Solution set: $\{7.044\}$

75. $\log_2(\log_2 x) = 1$

$\log_2 x = 2^1$ Write in exponential form.

$x = 2^2$ Write in exponential form.

$x = 4$ Evaluate.

Solution set: $\{4\}$

77. $\log x^2 = (\log x)^2$

$2 \log x = (\log x)^2$ Power property

$(\log x)^2 - 2 \log x = 0$ Subtract $2 \log x$ and rewrite.

$(\log x)(\log x - 2) = 0$ Factor.

$\log x = 0$ or $\log x - 2 = 0$ Zero-factor property

$\log x = 2$

$x = 10^0$ or $x = 10^2$ Write in exponential form.

$x = 1$ or $x = 100$

Solution set: $\{1, 100\}$

81. $p = a + \dfrac{k}{\ln x}$, for x

$p - a = \dfrac{k}{\ln x}$ Subtract a.

$(\ln x)(p - a) = k$ Multiply by $\ln x$.

$\ln x = \dfrac{k}{p - a}$ Divide by $p - a$.

$x = e^{k/(p-a)}$ Change to exponential form.

4.6 Exercises *(pages 457–462)*

29. $A = Pe^{rt}$ Continuous compounding formula

$3P = Pe^{0.05t}$ Let $A = 3P$ and $r = 0.05$.

$3 = e^{0.05t}$ Divide by P.

$\ln 3 = \ln e^{0.05t}$ Take the logarithm on each side.

$\ln 3 = 0.05t$ $\ln e^x = x$, for all x.

$\dfrac{\ln 3}{0.05} = t$ Divide by 0.05.

$t \approx 21.97$ Use a calculator.

It will take about 21.97 yr for the investment to triple.

43. $f(t) = 200(0.90)^t$

Find t when $f(t) = 50$.

$50 = 200(0.90)^t$ Let $f(t) = 50$.

$0.25 = (0.90)^t$ Divide by 200.

$\ln 0.25 = \ln(0.90)^t$ Take the logarithm on each side.

$\ln 0.25 = t \ln 0.90$ Power property

$\dfrac{\ln 0.25}{\ln 0.90} = t$ Divide by ln 0.90.

$t \approx 13.2$ Use a calculator.

The initial dose will reach a level of 50 mg in about 13.2 hr.

Chapter 5 Trigonometric Functions

5.1 Exercises *(pages 481–484)*

47. $90° - 72° 58' 11''$

$89° 59' 60''$ Write 90° as $89° 59' 60''$.

$\underline{-72° 58' 11''}$

$17° 01' 49''$

Thus, $90° - 72° 58' 11'' = 17° 01' 49''$.

133. 600 rotations per min

$= \dfrac{600}{60}$ rotations per sec

$= 10$ rotations per sec

$= 5$ rotations per $\frac{1}{2}$ sec

$= 5(360°)$ per $\frac{1}{2}$ sec

$= 1800°$ per $\frac{1}{2}$ sec

A point on the edge of the tire will move 1800° in $\frac{1}{2}$ sec.

5.2 Exercises *(pages 496–499)*

65. Evaluate $\tan 360° + 4 \sin 180° + 5 \cos^2 180°$.

$$\tan 360° = \tan 0° = \frac{y}{x} = \frac{0}{1} = 0$$

$$\sin 180° = \frac{y}{r} = \frac{0}{1} = 0$$

$$\cos 180° = \frac{x}{r} = \frac{-1}{1} = -1$$

$\tan 360° + 4 \sin 180° + 5 \cos^2 180° = 0 + 4(0) + 5(-1)^2$ Substitute; $\cos^2 x = (\cos x)^2$.

$= 5$

127. We are given $\tan \theta = -\frac{15}{8}$, with θ in quadrant II. Draw θ in standard position in quadrant II. Because $\tan \theta = \frac{y}{x}$ and θ is in quadrant II, we can use the values $y = 15$ and $x = -8$ for a point on its terminal side.

$r = \sqrt{x^2 + y^2} = \sqrt{(-8)^2 + 15^2} = \sqrt{64 + 225}$

$= \sqrt{289} = 17$

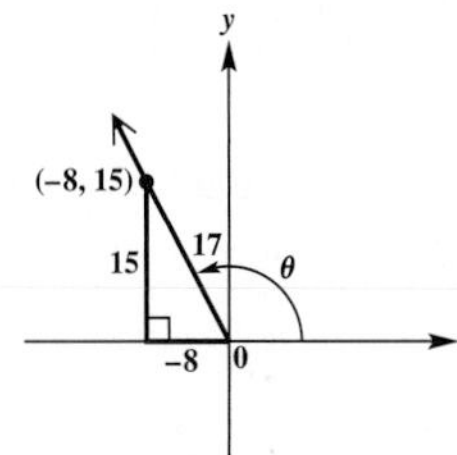

Use the values of x, y, and r and the definitions of the trigonometric functions to find the six trigonometric function values for θ.

$\sin \theta = \dfrac{y}{r} = \dfrac{15}{17}$ $\csc \theta = \dfrac{r}{y} = \dfrac{17}{15}$

$\cos \theta = \dfrac{x}{r} = \dfrac{-8}{17} = -\dfrac{8}{17}$ $\sec \theta = \dfrac{r}{x} = \dfrac{17}{-8} = -\dfrac{17}{8}$

$\tan \theta = \dfrac{y}{x} = \dfrac{15}{-8} = -\dfrac{15}{8}$ $\cot \theta = \dfrac{x}{y} = \dfrac{-8}{15} = -\dfrac{8}{15}$

143. Multiply the compound inequality $90° < \theta < 180°$ by 2 to find that $180° < 2\theta < 360°$. Thus, 2θ must lie in quadrant III or quadrant IV. In both of these quadrants, the sine function is negative, so $\sin 2\theta$ must be negative.

5.3 Exercises *(pages 508–513)*

51. One point on the line $y = \sqrt{3}x$ is the origin, $(0, 0)$. Let (x, y) be any other point on this line. Then, by the definition of slope, $m = \frac{y-0}{x-0} = \frac{y}{x} = \sqrt{3}$, but also, by the definition of tangent, $\tan\theta = \frac{y}{x}$. Thus, $\tan\theta = \sqrt{3}$. Because $\tan 60° = \sqrt{3}$, the line $y = \sqrt{3}x$ makes a $60°$ angle with the positive x-axis. (See **Exercise 50.**)

53. Apply the relationships among the lengths of the sides of a $30°$–$60°$ right triangle first to the triangle on the left to find the values of x and y, and then to the triangle on the right to find the values of z and w. In a $30°$–$60°$ right triangle, the side opposite the $30°$ angle is $\frac{1}{2}$ the length of the hypotenuse. The longer leg is $\sqrt{3}$ times the shorter leg.

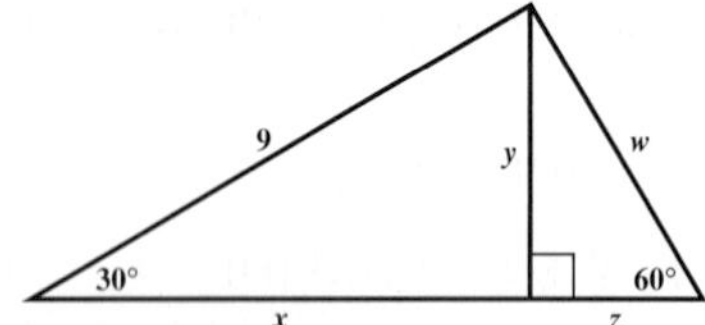

Thus,

$$y = \frac{1}{2}(9) = \frac{9}{2} \quad \text{and} \quad x = y\sqrt{3} = \frac{9\sqrt{3}}{2}.$$

Since $y = z\sqrt{3}$,

$$z = \frac{y}{\sqrt{3}} = \frac{\frac{9}{2}}{\sqrt{3}} = \frac{9}{2\sqrt{3}} \cdot \frac{\sqrt{3}}{\sqrt{3}} = \frac{9\sqrt{3}}{6} = \frac{3\sqrt{3}}{2},$$

and

$$w = 2z = 2\left(\frac{3\sqrt{3}}{2}\right) = 3\sqrt{3}.$$

83. To find the reference angle for $-300°$, sketch this angle in standard position.

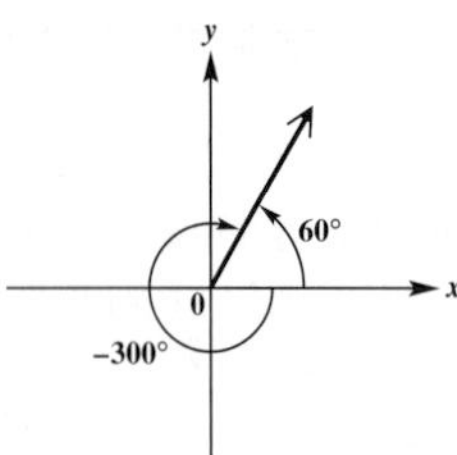

The reference angle for $-300°$ is

$$-300° + 360° = 60°.$$

Because $-300°$ is in quadrant I, the values of all its trigonometric functions are positive, so these values will be identical to the trigonometric function values for $60°$. (See the Function Values of Special Angles table that follows **Example 3** in **Section 5.3.**)

$$\sin(-300°) = \frac{\sqrt{3}}{2} \qquad \csc(-300°) = \frac{2\sqrt{3}}{3}$$

$$\cos(-300°) = \frac{1}{2} \qquad \sec(-300°) = 2$$

$$\tan(-300°) = \sqrt{3} \qquad \cot(-300°) = \frac{\sqrt{3}}{3}$$

147. For parts (a) and (b), $\theta = 3°$, $g = 32.2$, and $f = 0.14$.

(a) Use the fact that 45 mph = 66 ft per sec.

$$R = \frac{V^2}{g(f + \tan\theta)} = \frac{66^2}{32.2(0.14 + \tan 3°)} \approx 703 \text{ ft}$$

(b) Use the fact that 70 mph $= \frac{70(5280)}{3600}$ ft per sec = 102.67 ft per sec.

$$R = \frac{V^2}{g(f + \tan\theta)} = \frac{102.67^2}{32.2(0.14 + \tan 3°)} \approx 1701 \text{ ft}$$

(c) Intuitively, increasing θ would make it easier to negotiate the curve at a higher speed much as is done at a race track. Mathematically, a larger value of θ (acute) will lead to a larger value for $\tan\theta$. If $\tan\theta$ increases, then the ratio determining R will *decrease*. Thus, the radius can be smaller and the curve sharper if θ is increased.

$$R = \frac{V^2}{g(f + \tan\theta)} = \frac{66^2}{32.2(0.14 + \tan 4°)} \approx 644 \text{ ft}$$

$$R = \frac{V^2}{g(f + \tan\theta)} = \frac{102.67^2}{32.2(0.14 + \tan 4°)} \approx 1559 \text{ ft}$$

As predicted, both values are less.

5.4 Exercises *(pages 523–530)*

15. Solve the right triangle with $B = 73.0°$, $b = 128$ in., and $C = 90°$.

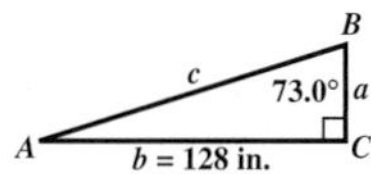

$$A = 90° - 73.0° = 17.0°$$

$$\tan 73.0° = \frac{128}{a} \qquad \tan B = \frac{b}{a}$$

$$a = \frac{128}{\tan 73.0°} \approx 39.1 \text{ in.} \qquad \text{Three significant digits}$$

$$\sin 73.0° = \frac{128}{c} \qquad \sin B = \frac{b}{c}$$

$$c = \frac{128}{\sin 73.0°} \approx 134 \text{ in.} \qquad \text{Three significant digits}$$

33. Let x represent the horizontal distance between the two buildings and y represent the height of the portion of the building across the street that is higher than the window.

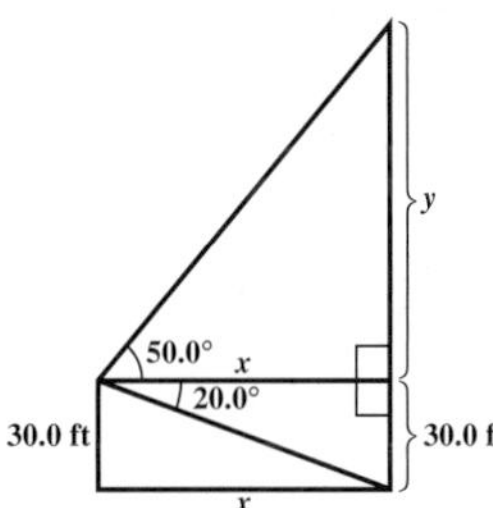

$$\tan 20.0° = \frac{30.0}{x} \qquad \text{Tangent ratio}$$

$$x = \frac{30.0}{\tan 20.0°} \approx 82.4 \qquad \text{Solve for } x.$$

$$\tan 50.0° = \frac{y}{x} \qquad \text{Tangent ratio}$$

$$y = x \tan 50.0° = \left(\frac{30.0}{\tan 20.0°}\right) \tan 50.0° \approx 98.2$$

Solve for y.

$$\text{height} = y + 30.0 = \left(\frac{30.0}{\tan 20.0°}\right) \tan 50.0° + 30.0 \approx 128$$

Three significant digits

The height of the building across the street is about128 ft.

39. Let h represent the height of the tower.

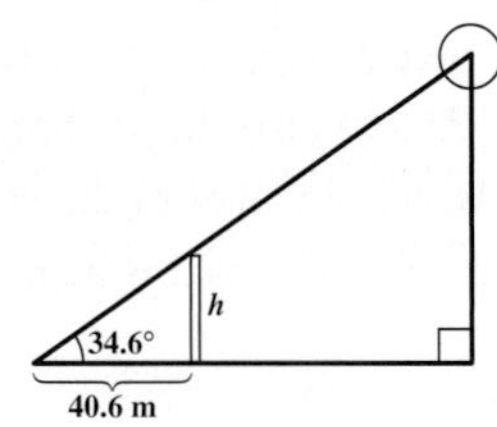

$$\tan 34.6° = \frac{h}{40.6} \qquad \text{Tangent ratio}$$

$$h = 40.6 \tan 34.6° \approx 28.0$$

Three significant digits

The height of the tower is about 28.0 m.

63. Let x = the distance between the two ships. The angle between the bearings of the ships is

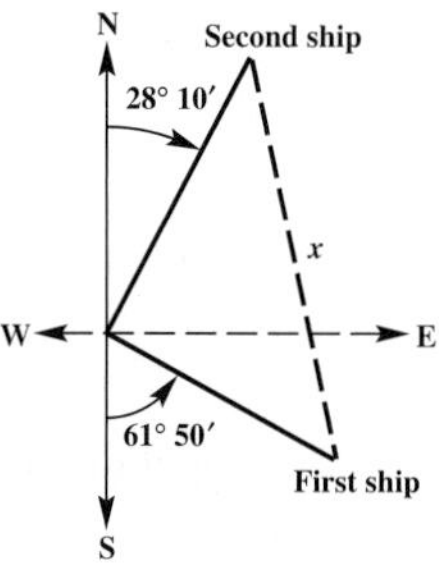

$$180° - (28° \, 10' + 61° \, 50') = 90°.$$

The triangle formed is a right triangle.

Distance traveled at 24.0 mph:

$$(4 \text{ hr})(24.0 \text{ mph}) = 96 \text{ mi}$$

Distance traveled at 28.0 mph:

$$(4 \text{ hr})(28.0 \text{ mph}) = 112 \text{ mi}$$

Applying the Pythagorean theorem gives the following.

$$x^2 = 96^2 + 112^2$$

$$x^2 = 21{,}760$$

$$x \approx 148$$

The ships are 148 mi apart.

71. Let x = the distance from the closer point on the ground to the base of height h of the pyramid.

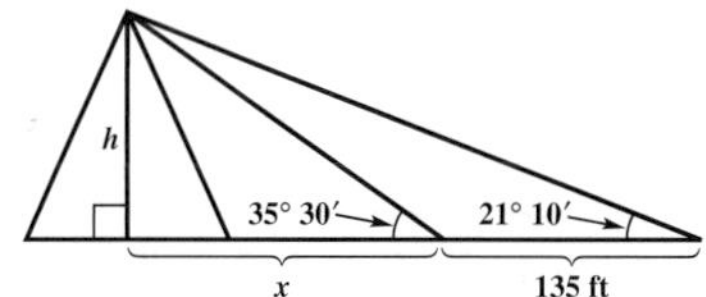

In the larger right triangle,

$$\tan 21° \, 10' = \frac{h}{135 + x}$$

$$h = (135 + x) \tan 21° \, 10'.$$

In the smaller right triangle,

$$\tan 35° \, 30' = \frac{h}{x}$$

$$h = x \tan 35° \, 30'.$$

Substitute for h in this equation, and solve for x.

$$(135 + x) \tan 21° \, 10' = x \tan 35° \, 30'$$

Substitute $(135 + x) \tan 21° 10'$ for h.

$$135 \tan 21° \, 10' + x \tan 21° \, 10' = x \tan 35° \, 30'$$

Distributive property

$$135 \tan 21° \, 10' = x \tan 35° \, 30' - x \tan 21° \, 10'$$

Write the x-terms on one side.

$$135 \tan 21° \, 10' = x(\tan 35° \, 30' - \tan 21° \, 10')$$

Factor out x.

$$\frac{135 \tan 21° \, 10'}{\tan 35° \, 30' - \tan 21° \, 10'} = x$$

Divide by the coefficient of x.

Then substitute for x in the equation for the smaller triangle.

$$h = \left(\frac{135 \tan 21° 10'}{\tan 35° 30' - \tan 21° 10'}\right) \tan 35° 30'$$

$$h \approx 114$$

The height of the pyramid is about 114 ft.

Chapter 6 The Circular Functions and Their Graphs

6.1 Exercises *(pages 548–555)*

101. For the large gear and pedal,

$$s = r\theta = 4.72\pi. \quad 180° = \pi \text{ radians}$$

Thus, the chain moves 4.72π in. Find the angle through which the small gear rotates.

$$\theta = \frac{s}{r} = \frac{4.72\pi}{1.38} \approx 3.42\pi$$

The angle θ for the wheel and for the small gear are the same, so for the wheel,

$$s = r\theta = 13.6(3.42\pi) \approx 146 \text{ in.}$$

The bicycle will move about 146 in.

133. (a)

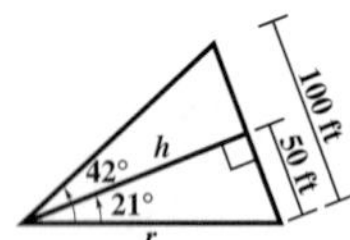

The triangle formed by the sides of the central angle and the chord is isosceles. Therefore, the bisector of the central angle is also the perpendicular bisector of the chord and divides the larger triangle into two congruent right triangles.

$$\sin 21° = \frac{50}{r}$$

$$r = \frac{50}{\sin 21°} \approx 140 \text{ ft}$$

The radius of the curve is about 140 ft.

(b) $r = \dfrac{50}{\sin 21°}$; $\theta = 42°$

$$42° = 42\left(\frac{\pi}{180} \text{ radian}\right) = \frac{7\pi}{30} \text{ radian}$$

$$s = r\theta = \frac{50}{\sin 21°} \cdot \frac{7\pi}{30} = \frac{35\pi}{3 \sin 21°} \approx 102 \text{ ft}$$

The length of the arc determined by the 100-ft chord is about 102 ft.

(c) The portion of the circle bounded by the arc and the 100-ft chord is the shaded region in the figure below.

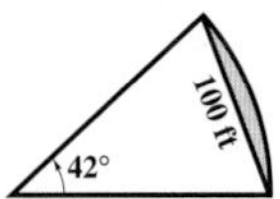

The area of the portion of the circle can be found by subtracting the area of the triangle from the area of the sector. From the figure in part (a),

$$\tan 21° = \frac{50}{h}, \quad \text{so} \quad h = \frac{50}{\tan 21°}.$$

$$\mathscr{A}_{\text{sector}} = \frac{1}{2} r^2\theta$$

$$= \frac{1}{2}\left(\frac{50}{\sin 21°}\right)^2\left(\frac{7\pi}{30}\right) \quad \text{From part (b), } 42° = \tfrac{7\pi}{30}.$$

$$\approx 7135 \text{ ft}^2$$

$$\mathscr{A}_{\text{triangle}} = \frac{1}{2}bh = \frac{1}{2}(100)\left(\frac{50}{\tan 21°}\right)$$

$$\approx 6513 \text{ ft}^2$$

$$\mathscr{A}_{\text{portion}} = \mathscr{A}_{\text{sector}} - \mathscr{A}_{\text{triangle}}$$

$$\approx 7135 \text{ ft}^2 - 6513 \text{ ft}^2$$

$$= 622 \text{ ft}^2$$

The area of the portion is about 622 ft^2.

135. Use the Pythagorean theorem to find the hypotenuse of the right triangle, which is also the radius of the sector of the circle.

$$r^2 = 30^2 + 40^2 = 900 + 1600 = 2500$$

$$r = \sqrt{2500} = 50$$

$$\mathscr{A}_{\text{triangle}} = \frac{1}{2}bh = \frac{1}{2}(30)(40)$$

$$= 600 \text{ yd}^2$$

$$\mathscr{A}_{\text{sector}} = \frac{1}{2}r^2\theta$$

$$= \frac{1}{2}(50)^2 \cdot \frac{\pi}{3} \quad 60° = \tfrac{\pi}{3}$$

$$= \frac{1250\pi}{3} \text{ yd}^2$$

$$\text{Total area} = \mathscr{A}_{\text{triangle}} + \mathscr{A}_{\text{sector}}$$

$$= 600 \text{ yd}^2 + \frac{1250\pi}{3} \text{ yd}^2$$

$$\approx 1900 \text{ yd}^2$$

The area of the lot is about 1900 yd^2.

6.2 Exercises *(pages 564–568)*

45. $\cos 2$

$\frac{\pi}{2} \approx 1.57$ and $\pi \approx 3.14$, so $\frac{\pi}{2} < 2 < \pi$. Thus, an angle of 2 radians is in quadrant II. (The figure for **Exercises 35–44** also shows that 2 radians is in quadrant II.) Because values of the cosine function are negative in quadrant II, $\cos 2$ is negative.

63. $\left[\pi, \dfrac{3\pi}{2}\right]$; $\tan s = \sqrt{3}$

Recall that $\tan \frac{\pi}{3} = \sqrt{3}$ and that in quadrant III, $\tan s$ is positive.

$$\tan\left(\pi + \frac{\pi}{3}\right) = \tan \frac{4\pi}{3} = \sqrt{3}$$

Thus, $s = \frac{4\pi}{3}$.

101. The hour hand of a clock moves through an angle of 2π radians (one complete revolution) in 12 hr. Find ω as follows.

$$\omega = \frac{\theta}{t} = \frac{2\pi}{12} = \frac{\pi}{6} \text{ radian per hr}$$

111. At 215 revolutions per min, the bicycle tire is moving $215(2\pi) = 430\pi$ radians per min. This is the angular velocity ω. Find v as follows.

$$v = r\omega = 13(430\pi) = 5590\pi \text{ in. per min}$$

Convert this velocity to miles per hour.

$$v = \frac{5590\pi \text{ in.}}{1 \text{ min}} \cdot \frac{60 \text{ min}}{1 \text{ hr}} \cdot \frac{1 \text{ ft}}{12 \text{ in.}} \cdot \frac{1 \text{ mi}}{5280 \text{ ft}} \approx 16.6 \text{ mph}$$

6.3 Exercises *(pages 578–583)*

55. $E = 5 \cos 120\pi t$

(a) The amplitude is $|5| = 5$, and the period is $\frac{2\pi}{120\pi} = \frac{1}{60}$.

(b) Since the period is $\frac{1}{60}$, one cycle is completed in $\frac{1}{60}$ sec. Therefore, in 1 sec, 60 cycles are completed.

(c) For $t = 0$, $E = 5 \cos 120\pi(0) = 5 \cos 0 = 5$.
For $t = 0.03$, $E = 5 \cos 120\pi(0.03) \approx 1.545$.
For $t = 0.06$, $E = 5 \cos 120\pi(0.06) \approx -4.045$.
For $t = 0.09$, $E \approx -4.045$.
For $t = 0.12$, $E \approx 1.545$.

(d)

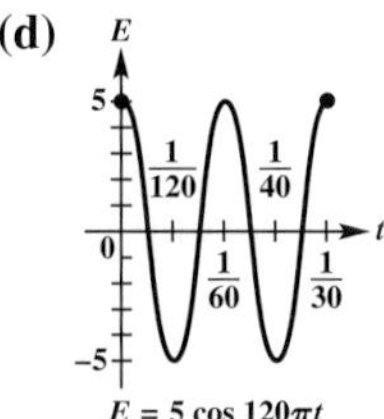

6.4 Exercises *(pages 590–593)*

55. $y = \frac{1}{2} + \sin\left[2\left(x + \frac{\pi}{4}\right)\right]$

This equation has the form $y = c + a \sin[b(x - d)]$ with $c = \frac{1}{2}$, $a = 1$, $b = 2$, and $d = -\frac{\pi}{4}$. Start with the graph of $y = \sin x$ and modify it to take into account the amplitude, period, and translations required to obtain the desired graph.

Amplitude: $|a| = 1$

Period: $\frac{2\pi}{b} = \frac{2\pi}{2} = \pi$

Vertical translation: $\frac{1}{2}$ unit up

Phase shift (horizontal translation): $\frac{\pi}{4}$ unit to the left

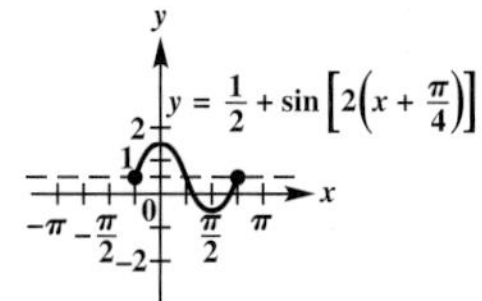

6.5 Exercises *(pages 601–603)*

29. $y = -1 + \frac{1}{2}\cot(2x - 3\pi)$

$$y = -1 + \frac{1}{2}\cot\left[2\left(x - \frac{3\pi}{2}\right)\right] \quad \text{Rewrite } 2x - 3\pi \text{ as } 2\left(x - \tfrac{3\pi}{2}\right).$$

Period: $\frac{\pi}{b} = \frac{\pi}{2}$

Vertical translation: 1 unit down

Phase shift (horizontal translation): $\frac{3\pi}{2}$ units to the right

Because the function is to be graphed over a two-period interval, locate three adjacent vertical asymptotes. Because asymptotes of the graph of $y = \cot x$ occur at multiples of π, the following equations can be solved to locate asymptotes.

$$2\left(x - \frac{3\pi}{2}\right) = -2\pi, \quad 2\left(x - \frac{3\pi}{2}\right) = -\pi, \quad \text{and}$$
$$2\left(x - \frac{3\pi}{2}\right) = 0$$

Solve each of these equations.

$$2\left(x - \frac{3\pi}{2}\right) = -2\pi$$
$$x - \frac{3\pi}{2} = -\pi \quad \text{Divide by 2.}$$
$$x = -\pi + \frac{3\pi}{2} \quad \text{Add } \tfrac{3\pi}{2}.$$
$$x = \frac{\pi}{2}$$

$$2\left(x - \frac{3\pi}{2}\right) = -\pi$$
$$x - \frac{3\pi}{2} = -\frac{\pi}{2}$$
$$x = -\frac{\pi}{2} + \frac{3\pi}{2}$$
$$x = \frac{2\pi}{2}, \quad \text{or} \quad \pi$$

$$2\left(x - \frac{3\pi}{2}\right) = 0$$
$$x - \frac{3\pi}{2} = 0$$
$$x = \frac{3\pi}{2}$$

Divide the interval $\left(\frac{\pi}{2}, \pi\right)$ into four equal parts to obtain the following key x-values.

first-quarter value: $\frac{5\pi}{8}$; middle value: $\frac{3\pi}{4}$;

third-quarter value: $\frac{7\pi}{8}$

Evaluating the given function at these three key x-values gives the following points.

$$\left(\frac{5\pi}{8}, -\frac{1}{2}\right), \left(\frac{3\pi}{4}, -1\right), \left(\frac{7\pi}{8}, -\frac{3}{2}\right)$$

Connect these points with a smooth curve and continue the graph to approach the asymptotes $x = \frac{\pi}{2}$ and $x = \pi$ to complete one period of the graph. Sketch an identical curve between the asymptotes $x = \pi$ and $x = \frac{3\pi}{2}$ to complete a second period of the graph.

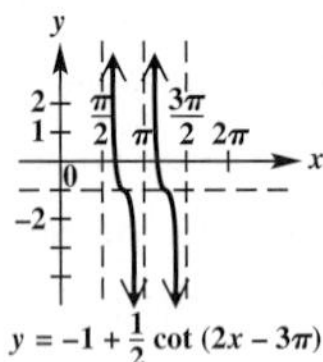

45. $\tan(-x) = \dfrac{\sin(-x)}{\cos(-x)}$ Quotient identity

$= \dfrac{-\sin x}{\cos x}$ Negative-angle identities

$= -\dfrac{\sin x}{\cos x}$ $\frac{-a}{b} = -\frac{a}{b}$

$= -\tan x$ Quotient identity

6.6 Exercises *(pages 609–611)*

31. $\sec(-x) = \dfrac{1}{\cos(-x)}$ Reciprocal identity

$= \dfrac{1}{\cos x}$ Negative-angle identity

$= \sec x$ Reciprocal identity

6.7 Exercises *(pages 614–615)*

19. (a) We will use a model of the form $s(t) = a \cos \omega t$ with $a = -3$. Since

$$s(0) = -3 \cos(\omega \cdot 0) = -3 \cos 0 = -3 \cdot 1 = -3,$$

using a cosine function rather than a sine function will avoid the need for a phase shift.

The frequency of $\frac{6}{\pi}$ cycles per sec is the reciprocal of the period.

$\dfrac{6}{\pi} = \dfrac{\omega}{2\pi}$ Frequency $= \frac{1}{\text{period}}$

$6 \cdot 2 = \omega$ Multiply by 2π.

$\omega = 12$ Multiply and rewrite.

Therefore, a model for the position of the weight at time t seconds is

$$s(t) = -3 \cos 12t.$$

(b) Period $= \dfrac{1}{\frac{6}{\pi}} = 1 \div \dfrac{6}{\pi} = 1 \cdot \dfrac{\pi}{6} = \dfrac{\pi}{6}$ sec

Chapter 7 Trigonometric Identities and Equations

7.1 Exercises *(pages 633–636)*

35. $\cot \theta = \frac{4}{3}$, $\sin \theta > 0$

Because $\cot \theta > 0$ and $\sin \theta > 0$, θ is in quadrant I, so all the function values are positive.

$\tan \theta = \dfrac{1}{\cot \theta} = \dfrac{1}{\frac{4}{3}} = \dfrac{3}{4}$ Reciprocal identity

$\sec^2 \theta = \tan^2 \theta + 1$ Pythagorean identity

$= \left(\dfrac{3}{4}\right)^2 + 1 = \dfrac{9}{16} + \dfrac{16}{16} = \dfrac{25}{16}$

$\sec \theta = \sqrt{\dfrac{25}{16}} = \dfrac{5}{4}$ $\sec \theta > 0$

$\cos \theta = \dfrac{1}{\sec \theta} = \dfrac{1}{\frac{5}{4}} = \dfrac{4}{5}$ Reciprocal identity

$\sin^2 \theta = 1 - \cos^2 \theta$ Alternative form of Pythagorean identity

$= 1 - \left(\dfrac{4}{5}\right)^2 = \dfrac{9}{25}$

$\sin \theta = \sqrt{\dfrac{9}{25}} = \dfrac{3}{5}$ $\sin \theta > 0$

$\csc \theta = \dfrac{1}{\sin \theta} = \dfrac{1}{\frac{3}{5}} = \dfrac{5}{3}$ Reciprocal identity

Thus, $\sin \theta = \frac{3}{5}$, $\cos \theta = \frac{4}{5}$, $\tan \theta = \frac{3}{4}$, $\sec \theta = \frac{5}{4}$, and $\csc \theta = \frac{5}{3}$.

57. $\csc x = \dfrac{1}{\sin x}$ Reciprocal identity

$= \dfrac{1}{\pm\sqrt{1 - \cos^2 x}}$ Alternative form of Pythagorean identity

$= \dfrac{\pm 1}{\sqrt{1 - \cos^2 x}}$ Redistribute signs.

$= \dfrac{\pm 1}{\sqrt{1 - \cos^2 x}} \cdot \dfrac{\sqrt{1 - \cos^2 x}}{\sqrt{1 - \cos^2 x}}$ Rationalize the denominator.

$\csc x = \dfrac{\pm\sqrt{1 - \cos^2 x}}{1 - \cos^2 x}$ Multiply.

73. $\sec\theta - \cos\theta = \dfrac{1}{\cos\theta} - \cos\theta$

$= \dfrac{1}{\cos\theta} - \dfrac{\cos^2\theta}{\cos\theta}$ Use a common denominator.

$= \dfrac{1-\cos^2\theta}{\cos\theta}$ Subtract fractions.

$= \dfrac{\sin^2\theta}{\cos\theta}$ $1-\cos^2\theta = \sin^2\theta$

$= \dfrac{\sin\theta}{\cos\theta}\cdot\sin\theta$ $\sin^2\theta = \sin\theta\cdot\sin\theta$

$= \tan\theta\sin\theta$ $\frac{\sin\theta}{\cos\theta} = \tan\theta$

85. Since $\cos x = \frac{1}{5} > 0$, x is in quadrant I or quadrant IV.

$$\sin x = \pm\sqrt{1-\cos^2 x} = \pm\sqrt{1-\left(\frac{1}{5}\right)^2}$$

$$= \pm\sqrt{\frac{24}{25}} = \pm\frac{2\sqrt{6}}{5}$$

$$\tan x = \frac{\sin x}{\cos x} = \frac{\pm\frac{2\sqrt{6}}{5}}{\frac{1}{5}} = \pm 2\sqrt{6}$$

$$\sec x = \frac{1}{\cos x} = \frac{1}{\frac{1}{5}} = 5$$

Quadrant I:

$$\frac{\sec x - \tan x}{\sin x} = \frac{5-2\sqrt{6}}{\frac{2\sqrt{6}}{5}} = \frac{5(5-2\sqrt{6})}{2\sqrt{6}}$$

$$= \frac{25-10\sqrt{6}}{2\sqrt{6}}\cdot\frac{\sqrt{6}}{\sqrt{6}}$$

$$= \frac{25\sqrt{6}-60}{12}$$

Quadrant IV:

$$\frac{\sec x - \tan x}{\sin x} = \frac{5-(-2\sqrt{6})}{-\frac{2\sqrt{6}}{5}} = \frac{5(5+2\sqrt{6})}{-2\sqrt{6}}$$

$$= \frac{25+10\sqrt{6}}{-2\sqrt{6}}\cdot\frac{-\sqrt{6}}{-\sqrt{6}}$$

$$= \frac{-25\sqrt{6}-60}{12}$$

7.2 Exercises *(pages 642–644)*

11. $$\frac{1}{1+\cos x} - \frac{1}{1-\cos x} = \frac{1(1-\cos x) - 1(1+\cos x)}{(1+\cos x)(1-\cos x)}$$

$$= \frac{1-\cos x - 1 - \cos x}{1-\cos^2 x}$$

$$= \frac{-2\cos x}{\sin^2 x}$$

$$= -\frac{2\cos x}{\sin^2 x}$$

$$= -2\left(\frac{\cos x}{\sin x}\right)\left(\frac{1}{\sin x}\right)$$

$$= -2\cot x\csc x$$

15. $(\sin x + 1)^2 - (\sin x - 1)^2$

$= [(\sin x + 1) + (\sin x - 1)][(\sin x + 1) - (\sin x - 1)]$ Factor the difference of squares.

$= [2\sin x][\sin x + 1 - \sin x + 1]$ Simplify.

$= [2\sin x][2]$ Simplify again.

$= 4\sin x$ Multiply.

59. Verify that $\dfrac{\tan^2 t - 1}{\sec^2 t} = \dfrac{\tan t - \cot t}{\tan t + \cot t}$ is an identity.

Work with the right hand side.

$$\frac{\tan t - \cot t}{\tan t + \cot t} = \frac{\tan t - \frac{1}{\tan t}}{\tan t + \frac{1}{\tan t}}$$ $\cot t = \frac{1}{\tan t}$

$$= \frac{\tan t}{\tan t}\left(\frac{\tan t - \frac{1}{\tan t}}{\tan t + \frac{1}{\tan t}}\right)$$

Multiply numerator and denominator of the complex fraction by the LCD, $\tan t$.

$= \dfrac{\tan^2 t - 1}{\tan^2 t + 1}$ Distributive property

$= \dfrac{\tan^2 t - 1}{\sec^2 t}$ $\tan^2 t + 1 = \sec^2 t$

87. Show that $\sin(\csc t) = 1$ is not an identity.

We need find only one value for which the statement is false. Let $t = 2$. Use a calculator to find that $\sin(\csc 2) \approx 0.891094$, which is not equal to 1. Thus, $\sin(\csc t) = 1$ is not true for *all* real numbers t, so it is not an identity.

7.3 Exercises *(pages 654–658)*

35. $\sec \theta = \csc\left(\frac{\theta}{2} + 20°\right)$

By a cofunction identity, $\sec \theta = \csc(90° - \theta)$.

$$\csc\left(\frac{\theta}{2} + 20°\right) = \csc(90° - \theta) \quad \text{Substitute.}$$

$$\frac{\theta}{2} + 20° = 90° - \theta \quad \text{Set the angle measures equal.}$$

$$\frac{3\theta}{2} = 70° \quad \text{Add } \theta \text{ and subtract } 20°.$$

$$\theta = \frac{2}{3}(70°) = \frac{140°}{3} \quad \text{Multiply by } \tfrac{2}{3}.$$

65. $\tan \frac{11\pi}{12} = \tan\left(\frac{3\pi}{4} + \frac{\pi}{6}\right)$ $\quad \frac{3\pi}{4} = \frac{9\pi}{12}; \frac{\pi}{6} = \frac{2\pi}{12}$

$$= \frac{\tan \frac{3\pi}{4} + \tan \frac{\pi}{6}}{1 - \tan \frac{3\pi}{4} \tan \frac{\pi}{6}} \quad \text{Tangent sum identity}$$

$$= \frac{-1 + \frac{\sqrt{3}}{3}}{1 - (-1)\left(\frac{\sqrt{3}}{3}\right)} \quad \tan \tfrac{3\pi}{4} = -1 \text{ and } \tan \tfrac{\pi}{6} = \tfrac{\sqrt{3}}{3}$$

$$= \frac{-1 + \frac{\sqrt{3}}{3}}{1 + \frac{\sqrt{3}}{3}} \quad \text{Simplify.}$$

$$= \frac{-1 + \frac{\sqrt{3}}{3}}{1 + \frac{\sqrt{3}}{3}} \cdot \frac{3}{3} \quad \text{Multiply numerator and denominator by 3.}$$

$$= \frac{-3 + \sqrt{3}}{3 + \sqrt{3}} \quad \text{Distributive property}$$

$$= \frac{-3 + \sqrt{3}}{3 + \sqrt{3}} \cdot \frac{3 - \sqrt{3}}{3 - \sqrt{3}} \quad \text{Rationalize the denominator.}$$

$$= \frac{-9 + 6\sqrt{3} - 3}{9 - 3} \quad \text{FOIL}$$

$$= \frac{-12 + 6\sqrt{3}}{6} \quad \text{Subtract.}$$

$$= \frac{6(-2 + \sqrt{3})}{6} \quad \text{Factor the numerator.}$$

$$= -2 + \sqrt{3} \quad \text{Lowest terms}$$

93. $\cos s = -\frac{8}{17}$ and $\cos t = -\frac{3}{5}$, s and t in quadrant III

In order to substitute into sum and difference identities, we need to find the values of $\sin s$ and $\sin t$, and also the values of $\tan s$ and $\tan t$. Because s and t are both in quadrant III, the values of $\sin s$ and $\sin t$ will be negative, and $\tan s$ and $\tan t$ will be positive.

$$\sin s = -\sqrt{1 - \cos^2 s} = -\sqrt{1 - \left(-\frac{8}{17}\right)^2} = -\sqrt{\frac{225}{289}} = -\frac{15}{17}$$

$$\sin t = -\sqrt{1 - \cos^2 t} = -\sqrt{1 - \left(-\frac{3}{5}\right)^2} = -\sqrt{\frac{16}{25}} = -\frac{4}{5}$$

$$\tan s = \frac{\sin s}{\cos s} = \frac{-\frac{15}{17}}{-\frac{8}{17}} = \frac{15}{8}$$

$$\tan t = \frac{\sin t}{\cos t} = \frac{-\frac{4}{5}}{-\frac{3}{5}} = \frac{4}{3}$$

(a) $\sin(s + t) = \sin s \cos t + \cos s \sin t$

$$= \left(-\frac{15}{17}\right)\left(-\frac{3}{5}\right) + \left(-\frac{8}{17}\right)\left(-\frac{4}{5}\right)$$

$$= \frac{45}{85} + \frac{32}{85} = \frac{77}{85}$$

(b) $\tan(s + t) = \dfrac{\tan s + \tan t}{1 - \tan s \tan t} = \dfrac{\frac{15}{8} + \frac{4}{3}}{1 - \left(\frac{15}{8}\right)\left(\frac{4}{3}\right)}$

$$= \frac{\frac{45}{24} + \frac{32}{24}}{1 - \frac{60}{24}} = \frac{\frac{77}{24}}{-\frac{36}{24}} = -\frac{77}{36}$$

(c) From parts (a) and (b), $\sin(s + t) > 0$ and $\tan(s + t) < 0$. The only quadrant in which values of sine are positive and values of tangent are negative is quadrant II. Thus, $s + t$ is in quadrant II.

107. Verify that $\dfrac{\sin(x - y)}{\sin(x + y)} = \dfrac{\tan x - \tan y}{\tan x + \tan y}$ is an identity.

Work with the left hand side.

$$\frac{\sin(x - y)}{\sin(x + y)} = \frac{\sin x \cos y - \cos x \sin y}{\sin x \cos y + \cos x \sin y} \quad \text{Sine sum and difference identities}$$

$$= \frac{\frac{\sin x \cos y}{\cos x \cos y} - \frac{\cos x \sin y}{\cos x \cos y}}{\frac{\sin x \cos y}{\cos x \cos y} + \frac{\cos x \sin y}{\cos x \cos y}} \quad \text{Divide numerator and denominator by } \cos x \cos y.$$

$$= \frac{\frac{\sin x}{\cos x} \cdot 1 - 1 \cdot \frac{\sin y}{\cos y}}{\frac{\sin x}{\cos x} \cdot 1 + 1 \cdot \frac{\sin y}{\cos y}} \quad \text{Divide.}$$

$$= \frac{\tan x - \tan y}{\tan x + \tan y} \quad \text{Tangent quotient identity}$$

7.4 Exercises *(pages 668–671)*

25. $\frac{1}{4} - \frac{1}{2}\sin^2 47.1° = \frac{1}{4}(1 - 2\sin^2 47.1°)$ Factor out $\frac{1}{4}$.

$= \frac{1}{4}\cos 2(47.1°)$

$\cos 2A = 1 - 2\sin^2 A$

$= \frac{1}{4}\cos 94.2°$

31. $\tan 3x = \tan(2x + x)$

$= \frac{\tan 2x + \tan x}{1 - \tan 2x \tan x}$ Tangent sum identity

$= \frac{\frac{2\tan x}{1 - \tan^2 x} + \tan x}{1 - \frac{2\tan x}{1 - \tan^2 x} \cdot \tan x}$ Tangent double-angle identity

$= \frac{\frac{2\tan x + (1 - \tan^2 x)\tan x}{1 - \tan^2 x}}{\frac{1 - \tan^2 x - 2\tan^2 x}{1 - \tan^2 x}}$ Add and subtract using the common denominator.

$= \frac{2\tan x + \tan x - \tan^3 x}{1 - \tan^2 x - 2\tan^2 x}$

Multiply numerator and denominator by $1 - \tan^2 x$.

$\tan 3x = \frac{3\tan x - \tan^3 x}{1 - 3\tan^2 x}$ Combine like terms.

59. Find $\tan \frac{\theta}{2}$, given $\sin \theta = \frac{3}{5}$, with $90° < \theta < 180°$.
To find $\tan \frac{\theta}{2}$, we need the values of $\sin \theta$ and $\cos \theta$.
We know $\sin \theta = \frac{3}{5}$.

$\cos \theta = \pm\sqrt{1 - \sin^2 \theta}$ Fundamental identity

$= \pm\sqrt{1 - \left(\frac{3}{5}\right)^2}$ Substitute.

$= \pm\sqrt{\frac{16}{25}}$ Simplify.

$\cos \theta = -\frac{4}{5}$ θ is in quadrant II.

Thus,

$\tan \frac{\theta}{2} = \frac{\sin \theta}{1 + \cos \theta}$ Half-angle identity

$= \frac{\frac{3}{5}}{1 - \frac{4}{5}}$ Substitute.

$= 3.$ Simplify.

89. Verify that $\sin 4x = 4\sin x \cos x \cos 2x$ is an identity.
Work with the left hand side.

$\sin 4x = \sin 2(2x)$ Factor: $4 = 2 \cdot 2$.

$= 2\sin 2x \cos 2x$ Sine double-angle identity

$= 2(2\sin x \cos x)\cos 2x$ Sine double-angle identity

$= 4\sin x \cos x \cos 2x$ Multiply.

93. Verify that $\sec^2 \frac{x}{2} = \frac{2}{1 + \cos x}$ is an identity.

Work with the left hand side.

$\sec^2 \frac{x}{2} = \frac{1}{\cos^2 \frac{x}{2}}$ Reciprocal identity

$= \frac{1}{\left(\pm\sqrt{\frac{1 + \cos x}{2}}\right)^2}$ Cosine half-angle identity

$= \frac{1}{\frac{1 + \cos x}{2}}$ Apply the exponent.

$= \frac{2}{1 + \cos x}$ Divide.

7.5 Exercises *(pages 683–687)*

87. $\sin\left(2\cos^{-1}\frac{1}{5}\right)$

Let $\theta = \cos^{-1}\frac{1}{5}$, so $\cos \theta = \frac{1}{5}$. The inverse cosine function yields values only in quadrants I and II, and since $\frac{1}{5}$ is positive, θ is in quadrant I. Sketch θ in quadrant I, and label the sides of a right triangle. By the Pythagorean theorem, the length of the side opposite θ will be

$$\sqrt{5^2 - 1^2} = \sqrt{24} = 2\sqrt{6}.$$

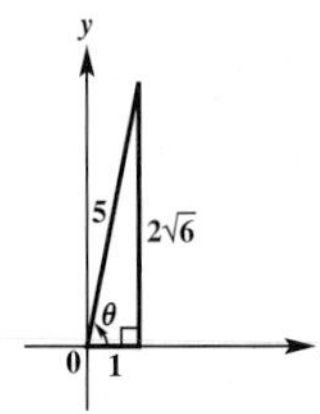

From the figure, $\sin \theta = \frac{2\sqrt{6}}{5}$.

$\sin\left(2\cos^{-1}\frac{1}{5}\right) = \sin 2\theta$

$= 2\sin \theta \cos \theta$

Sine double-angle identity

$= 2\left(\frac{2\sqrt{6}}{5}\right)\left(\frac{1}{5}\right)$

$= \frac{4\sqrt{6}}{25}$

93. $\sin\left(\sin^{-1}\frac{1}{2}+\tan^{-1}(-3)\right)$

Let $\sin^{-1}\frac{1}{2}=A$ and $\tan^{-1}(-3)=B$. Then $\sin A=\frac{1}{2}$ and $\tan B=-3$. Sketch angle A in quadrant I and angle B in quadrant IV, and use the Pythagorean theorem to find the unknown side in each triangle.

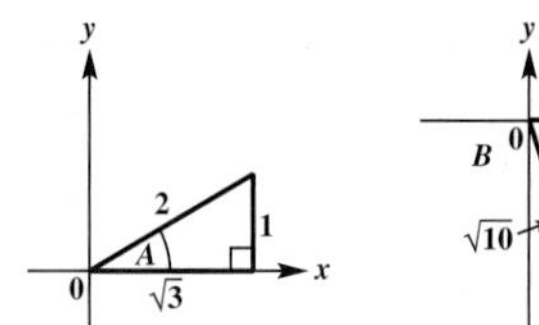

$$\sin\left(\sin^{-1}\frac{1}{2}+\tan^{-1}(-3)\right)=\sin(A+B)$$

$$=\sin A\cos B+\cos A\sin B$$

Sine sum identity

$$=\frac{1}{2}\cdot\frac{1}{\sqrt{10}}+\frac{\sqrt{3}}{2}\cdot\frac{-3}{\sqrt{10}}$$

$$=\frac{1-3\sqrt{3}}{2\sqrt{10}},\quad\text{or}\quad\frac{\sqrt{10}-3\sqrt{30}}{20}$$

7.6 Exercises *(pages 695–700)*

15. $\tan^2 x+3=0$, so $\tan^2 x=-3$.

The square of a real number cannot be negative, so this equation has no solution. Solution set: $\emptyset$

25.

$2\sin\theta-1=\csc\theta$ — Original equation

$2\sin\theta-1=\dfrac{1}{\sin\theta}$ — Reciprocal identity

$2\sin^2\theta-\sin\theta=1$ — Multiply by $\sin\theta$.

$2\sin^2\theta-\sin\theta-1=0$ — Subtract 1.

$(2\sin\theta+1)(\sin\theta-1)=0$ — Factor.

$2\sin\theta+1=0$ or $\sin\theta-1=0$ — Zero-factor property

$\sin\theta=-\dfrac{1}{2}$ or $\sin\theta=1$

Over the interval $[0°, 360°)$, the equation $\sin\theta=-\frac{1}{2}$ has two solutions, the angles in quadrants III and IV that have reference angle 30°. These are 210° and 330°. In the same interval, the only angle θ for which $\sin\theta=1$ is 90°. All three of these check.

Solution set: $\{90°, 210°, 330°\}$

57.

$\dfrac{2\tan\theta}{3-\tan^2\theta}=1$ — Original equation

$2\tan\theta=3-\tan^2\theta$ — Multiply by $3-\tan^2\theta$.

$\tan^2\theta+2\tan\theta-3=0$ — Write in standard quadratic form.

$(\tan\theta-1)(\tan\theta+3)=0$ — Factor.

$\tan\theta-1=0$ or $\tan\theta+3=0$ — Zero-factor property

$\tan\theta=1$ or $\tan\theta=-3$

Over the interval $[0°, 360°)$, the equation $\tan\theta=1$ has two solutions, 45° and 225°. Over the same interval, the equation $\tan\theta=-3$ has two solutions that are approximately $-71.6°+180°=108.4°$ and $-71.6°+360°=288.4°$. All of these check.

The period of the tangent function is 180°, so the solution set is $\{45°+180°n,\ 108.4°+180°n$, where n is any integer$\}$.

81. $\cos 2x+\cos x=0$

We choose the identity for $\cos 2x$ that involves only the cosine function.

$\cos 2x+\cos x=0$ — Original equation

$2\cos^2 x-1+\cos x=0$ — Cosine double-angle identity

$2\cos^2 x+\cos x-1=0$ — Standard quadratic form

$(2\cos x-1)(\cos x+1)=0$ — Factor.

$2\cos x-1=0$ or $\cos x+1=0$ — Zero-factor property

$2\cos x=1$

$\cos x=\dfrac{1}{2}$ or $\cos x=-1$ — Solve for $\cos x$.

Over the interval $[0, 2\pi)$, the equation $\cos x=\frac{1}{2}$ has two solutions, $\frac{\pi}{3}$ and $\frac{5\pi}{3}$. Over the same interval, the equation $\cos x=-1$ has only one solution, π.

Solution set: $\left\{\frac{\pi}{3}, \pi, \frac{5\pi}{3}\right\}$

89.

$2\sin\theta=2\cos 2\theta$ — Original equation

$\sin\theta=\cos 2\theta$ — Divide by 2.

$\sin\theta=1-2\sin^2\theta$ — Cosine double-angle identity

$2\sin^2\theta+\sin\theta-1=0$ — Standard quadratic form

$(2\sin\theta-1)(\sin\theta+1)=0$ — Factor.

$2\sin\theta-1=0$ or $\sin\theta+1=0$ — Zero-factor property

$\sin\theta=\dfrac{1}{2}$ or $\sin\theta=-1$ — Solve for $\sin\theta$.

Over the interval $[0°, 360°)$, the equation $\sin\theta = \frac{1}{2}$ has two solutions, 30° and 150°. Over the same interval, the equation $\sin\theta = -1$ has one solution, 270°.

The period of the sine function is 360°, so the solution set is $\{30° + 360°n, 150° + 360°n, 270° + 360°n$, where n is any integer$\}$.

7.7 Exercises *(pages 704–706)*

15. $y = \cos(x + 3)$, for x in $[-3, \pi - 3]$ — Original equation

$x + 3 = \arccos y$ — Definition of arccosine

$x = -3 + \arccos y$ — Subtract 3.

37. $\arccos x + 2 \arcsin \frac{\sqrt{3}}{2} = \pi$ — Original equation

$\arccos x = \pi - 2 \arcsin \frac{\sqrt{3}}{2}$ — Isolate arccos x.

$\arccos x = \pi - 2\left(\frac{\pi}{3}\right)$ — $\arcsin \frac{\sqrt{3}}{2} = \frac{\pi}{3}$

$\arccos x = \pi - \frac{2\pi}{3}$ — Multiply.

$\arccos x = \frac{\pi}{3}$ — Subtract.

$x = \cos \frac{\pi}{3}$ — Rewrite.

$x = \frac{1}{2}$ — Evaluate.

Solution set: $\{\frac{1}{2}\}$

41. $\cos^{-1} x + \tan^{-1} x = \frac{\pi}{2}$ — Original equation

$\cos^{-1} x = \frac{\pi}{2} - \tan^{-1} x$ — Subtract $\tan^{-1} x$.

$x = \cos\left(\frac{\pi}{2} - \tan^{-1} x\right)$ — Definition of $\cos^{-1} x$

$x = \cos \frac{\pi}{2} \cdot \cos(\tan^{-1} x) + \sin \frac{\pi}{2} \cdot \sin(\tan^{-1} x)$ — Cosine difference identity

$x = 0 \cdot \cos(\tan^{-1} x) + 1 \cdot \sin(\tan^{-1} x)$ — $\cos \frac{\pi}{2} = 0$ and $\sin \frac{\pi}{2} = 1$

$x = \sin(\tan^{-1} x)$

Let $u = \tan^{-1} x$, so $\tan u = x$.

y
$\sqrt{1+x^2}$
x
u
0
1
x

From the triangle, we find that $\sin u = \frac{x}{\sqrt{1 + x^2}}$, so the equation $x = \sin(\tan^{-1} x)$ becomes

$$x = \frac{x}{\sqrt{1 + x^2}}.$$

Solve this equation.

$x = \frac{x}{\sqrt{1 + x^2}}$

$x\sqrt{1 + x^2} = x$ — Multiply by $\sqrt{1 + x^2}$.

$x\sqrt{1 + x^2} - x = 0$ — Subtract x.

$x\left(\sqrt{1 + x^2} - 1\right) = 0$ — Factor.

$x = 0$ or $\sqrt{1 + x^2} - 1 = 0$ — Zero-factor property

$\sqrt{1 + x^2} = 1$ — Isolate the radical.

$1 + x^2 = 1$ — Square each side.

$x^2 = 0$ — Subtract 1.

$x = 0$ — Take square roots.

Solution set: $\{0\}$

Chapter 8 Applications of Trigonometry

8.1 Exercises *(pages 727–733)*

31. $\frac{\sin B}{b} = \frac{\sin A}{a}$ — Alternative form of the law of sines

$\frac{\sin B}{2} = \frac{\sin 60°}{\sqrt{6}}$ — Substitute values from the figure.

$\sin B = \frac{2 \sin 60°}{\sqrt{6}}$ — Multiply by 2.

$\sin B = \frac{2 \cdot \frac{\sqrt{3}}{2}}{\sqrt{6}}$ — $\sin 60° = \frac{\sqrt{3}}{2}$

$\sin B = \frac{\sqrt{3}}{\sqrt{6}} = \sqrt{\frac{1}{2}} = \frac{\sqrt{2}}{2}$ — Simplify and rationalize.

$B = 45°$ — Use the inverse sine function.

There is another angle between 0° and 180° whose sine is $\frac{\sqrt{2}}{2}$: $180° - 45° = 135°$. However, this is too large because $A = 60°$ and $60° + 135° = 195°$. Since $195° > 180°$, there is only one solution, $B = 45°$.

39. $A = 142.13°$, $b = 5.432$ ft, $a = 7.297$ ft

$$\frac{\sin B}{b} = \frac{\sin A}{a} \quad \text{Alternative form of the law of sines}$$

$$\sin B = \frac{b \sin A}{a} \quad \text{Multiply by } b.$$

$$\sin B = \frac{5.432 \sin 142.13°}{7.297} \quad \text{Substitute given values.}$$

$$\sin B \approx 0.45697580 \quad \text{Simplify.}$$

$$B \approx 27.19° \quad \text{Use the inverse sine function.}$$

Because angle A is obtuse, angle B must be acute, so this is the only possible value for B and there is one triangle with the given measurements.

$$C = 180° - A - B \quad \text{Angle sum formula, solved for } C$$

$$C \approx 180° - 142.13° - 27.19°$$

$$C \approx 10.68°$$

Thus, $B \approx 27.19°$ and $C \approx 10.68°$.

63. We cannot find θ directly because the length of the side opposite angle θ is not given. Redraw the triangle shown in the figure, and label the third angle as α.

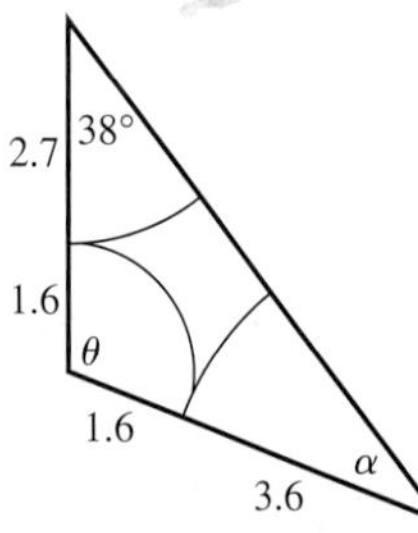

$$\frac{\sin \alpha}{1.6 + 2.7} = \frac{\sin 38°}{1.6 + 3.6} \quad \text{Alternative form of the law of sines}$$

$$\frac{\sin \alpha}{4.3} = \frac{\sin 38°}{5.2} \quad \text{Add in the denominators.}$$

$$\sin \alpha = \frac{4.3 \sin 38°}{5.2} \approx 0.50910468$$

$$\alpha \approx 31° \quad \text{Use the inverse sine function.}$$

Then $\theta \approx 180° - 38° - 31°$

$\theta \approx 111°$.

73. To find the area of the triangle, use $\mathscr{A} = \frac{1}{2}bh$, with $b = 1$ and $h = \sqrt{2}$.

$$\mathscr{A} = \frac{1}{2}(1)(\sqrt{2}) = \frac{\sqrt{2}}{2}$$

Now use $\mathscr{A} = \frac{1}{2}ab \sin C$, with $a = 2$, $b = 1$, and $C = 45°$.

$$\mathscr{A} = \frac{1}{2}(2)(1) \sin 45° = \sin 45° = \frac{\sqrt{2}}{2}$$

Both formulas show that the area is $\frac{\sqrt{2}}{2}$ sq unit.

8.2 Exercises *(pages 740–746)*

21. $C = 45.6°$, $b = 8.94$ m, $a = 7.23$ m

First find c.

$$c^2 = a^2 + b^2 - 2ab \cos C \quad \text{Law of cosines}$$

$$c^2 = 7.23^2 + 8.94^2 - 2(7.23)(8.94) \cos 45.6° \quad \text{Substitute given values.}$$

$$c^2 \approx 41.7493 \quad \text{Use a calculator.}$$

$$c \approx 6.46 \quad \text{Square root property}$$

Find A next since angle A is smaller than angle B (because $a < b$), and thus angle A must be acute.

$$\frac{\sin A}{a} = \frac{\sin C}{c} \quad \text{Alternative form of the law of sines}$$

$$\sin A = \frac{a \sin C}{c} \quad \text{Multiply by } a.$$

$$\sin A = \frac{7.23 \sin 45.6°}{6.46} \quad \text{Substitute.}$$

$$\sin A \approx 0.79963428 \quad \text{Simplify.}$$

$$A \approx 53.1° \quad \text{Use the inverse sine function.}$$

Finally, find B.

$$B = 180° - C - A$$

$$B \approx 180° - 45.6° - 53.1°$$

$$B \approx 81.3°$$

Thus, $c \approx 6.46$ m, $A \approx 53.1°$, and $B \approx 81.3°$.

43. Find AC, or b, in this figure.

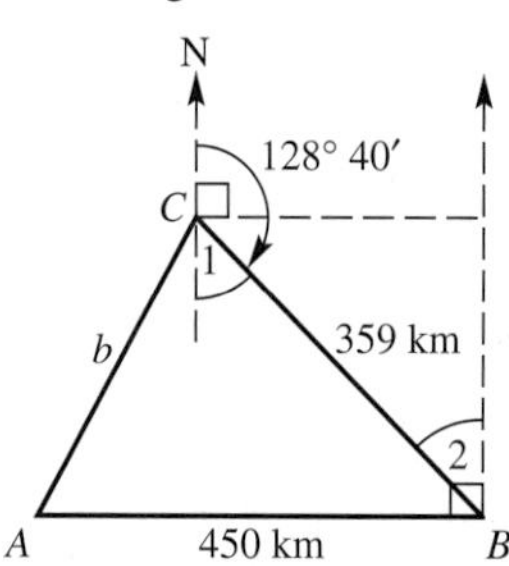

Angle 1 = $180° - 128° 40' = 51° 20'$

Angles 1 and 2 are alternate interior angles formed when two parallel lines (the north lines) are cut by a transversal, line BC, so angle 2 = angle 1 = $51° 20'$.

angle $ABC = 90° -$ angle 2 $= 90° - 51° 20' = 38° 40'$ Complementary angles

$$b^2 = a^2 + c^2 - 2ac \cos B \quad \text{Law of cosines}$$

$$b^2 = 359^2 + 450^2 - 2(359)(450) \cos 38° 40' \quad \text{Substitute values from the figure.}$$

$$b^2 \approx 79{,}106 \quad \text{Use a calculator.}$$

$$b \approx 281 \quad \text{Square root property}$$

C is about 281 km from A.

8.3 Exercises (pages 755–758)

19. Use the figure to find the components of **u** and **v**: $\mathbf{u} = \langle -8, 8\rangle$ and $\mathbf{v} = \langle 4, 8\rangle$.

(a) $\mathbf{u} + \mathbf{v} = \langle -8, 8\rangle + \langle 4, 8\rangle = \langle -8+4, 8+8\rangle$
$= \langle -4, 16\rangle$

(b) $\mathbf{u} - \mathbf{v} = \langle -8, 8\rangle - \langle 4, 8\rangle = \langle -8-4, 8-8\rangle$
$= \langle -12, 0\rangle$

(c) $-\mathbf{u} = -\langle -8, 8\rangle = \langle 8, -8\rangle$

47. $\mathbf{v} = \langle a, b\rangle = \langle 5\cos(-35°), 5\sin(-35°)\rangle$
$= \langle 4.0958, -2.8679\rangle$

81. First write the given vectors in component form.

$$3\mathbf{i} + 4\mathbf{j} = \langle 3, 4\rangle; \quad \mathbf{j} = \langle 0, 1\rangle$$

$$\cos\theta = \frac{\langle 3, 4\rangle \cdot \langle 0, 1\rangle}{|\langle 3, 4\rangle|\,|\langle 0, 1\rangle|}$$

$$\cos\theta = \frac{3(0) + 4(1)}{\sqrt{9+16}\cdot\sqrt{0+1}} = \frac{4}{5} = 0.8$$

$$\theta = \cos^{-1} 0.8 \approx 36.87°$$

8.4 Exercises (pages 762–765)

5. Use the parallelogram rule. In the figure, **x** represents the second force and **v** is the resultant.

$$\alpha = 180° - 78°\,50' = 101°\,10'$$

$$\beta = 78°\,50' - 41°\,10' = 37°\,40'$$

$$\frac{|\mathbf{x}|}{\sin 41°\,10'} = \frac{176}{\sin 37°\,40'} \quad \text{Law of sines}$$

$$|\mathbf{x}| = \frac{176 \sin 41°\,10'}{\sin 37°\,40'} \approx 190$$

$$\frac{|\mathbf{v}|}{\sin 101°\,10'} = \frac{176}{\sin 37°\,40'} \quad \text{Law of sines}$$

$$|\mathbf{v}| = \frac{176 \sin 101°\,10'}{\sin 37°\,40'} \approx 283$$

Thus, the magnitude of the second force is about 190 lb, and the magnitude of the resultant is about 283 lb.

27. Let **v** represent the airspeed vector.

The ground speed is $\dfrac{400 \text{ mi}}{2.5 \text{ hr}} = 160$ mph.

angle $BAC = 328° - 180° = 148°$

$$|\mathbf{v}|^2 = 11^2 + 160^2 - 2(11)(160)\cos 148° \quad \text{Law of cosines}$$

$$|\mathbf{v}|^2 \approx 28{,}706$$

$$|\mathbf{v}| \approx 169.4$$

$$\frac{\sin B}{11} = \frac{\sin 148°}{169.4} \quad \text{Law of sines}$$

$$\sin B = \frac{11 \sin 148°}{169.4} \approx 0.03441034$$

$$B \approx 2°$$

The bearing must be approximately $360° - 2° = 358°$.

The airspeed must be approximately 170 mph.

8.5 Exercises (pages 774–777)

31. $3 \text{ cis } 150°$

$= 3(\cos 150° + i \sin 150°)$

$= 3\left(-\dfrac{\sqrt{3}}{2} + i \cdot \dfrac{1}{2}\right)$ $\cos 150° = -\frac{\sqrt{3}}{2}$; $\sin 150° = \frac{1}{2}$

$= -\dfrac{3\sqrt{3}}{2} + \dfrac{3}{2}i$ Rectangular form

43. $-5 - 5i$

Sketch the graph of $-5 - 5i$ in the complex plane.

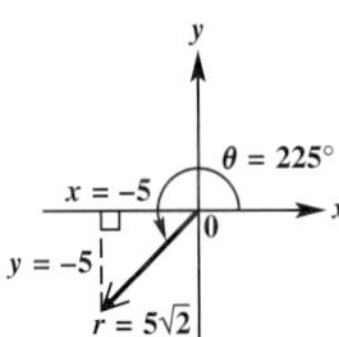

Since $x = -5$ and $y = -5$,

$$r = \sqrt{x^2 + y^2} = \sqrt{(-5)^2 + (-5)^2} = \sqrt{50} = 5\sqrt{2}$$

and $$\tan\theta = \frac{y}{x} = \frac{-5}{-5} = 1.$$

Since $\tan\theta = 1$, the reference angle for θ is 45°. The graph shows that θ is in quadrant III, so

$$\theta = 180° + 45° = 225°.$$

Use these results.

$$-5 - 5i = 5\sqrt{2}(\cos 225° + i \sin 225°)$$

69. $[4(\cos 60° + i \sin 60°)][6(\cos 330° + i \sin 330°)]$

$= 4 \cdot 6[\cos(60° + 330°) + i \sin(60° + 330°)]$ Product theorem

$= 24(\cos 390° + i \sin 390°)$ Multiply and add.

$= 24(\cos 30° + i \sin 30°)$ 390° and 30° are coterminal angles.

$= 24\left(\dfrac{\sqrt{3}}{2} + i \cdot \dfrac{1}{2}\right)$ $\cos 30° = \frac{\sqrt{3}}{2}$; $\sin 30° = \frac{1}{2}$

$= 12\sqrt{3} + 12i$ Rectangular form

85. $\dfrac{-i}{1+i}$

Numerator: $-i = 0 - 1i$

$r = \sqrt{0^2 + (-1)^2} = 1$

$\theta = 270°$ since $\cos 270° = 0$ and $\sin 270° = -1$. Thus $-i = 1 \text{ cis } 270°$.

Denominator: $1 + i = 1 + 1i$

$$r = \sqrt{1^2 + 1^2} = \sqrt{2}$$

$$\tan \theta = \frac{y}{x} = \frac{1}{1} = 1$$

Since x and y are both positive, θ is in quadrant I, and $\theta = \tan^{-1} 1 = 45°$. Thus, $1 + i = \sqrt{2} \text{ cis } 45°$.

$\dfrac{-i}{1+i}$

$= \dfrac{1 \text{ cis } 270°}{\sqrt{2} \text{ cis } 45°}$ Substitute.

$= \dfrac{1}{\sqrt{2}} \text{cis}(270° - 45°)$ Quotient theorem

$= \dfrac{\sqrt{2}}{2} \text{ cis } 225°$ Rationalize and subtract.

$= \dfrac{\sqrt{2}}{2}(\cos 225° + i \sin 225°)$ Equivalent form

$= \dfrac{\sqrt{2}}{2}\left(-\dfrac{\sqrt{2}}{2} - i \cdot \dfrac{\sqrt{2}}{2}\right)$ $\cos 225° = -\frac{\sqrt{2}}{2}$; $\sin 225° = -\frac{\sqrt{2}}{2}$

$= -\dfrac{1}{2} - \dfrac{1}{2}i$ Rectangular form

8.6 Exercises *(pages 782–784)*

11. $(-2 - 2i)^5$

First write $-2 - 2i$ in trigonometric form.

$$r = \sqrt{(-2)^2 + (-2)^2} = \sqrt{8} = 2\sqrt{2}$$

$$\tan \theta = \frac{y}{x} = \frac{-2}{-2} = 1$$

Because x and y are both negative, θ is in quadrant III. Thus $\theta = 225°$.

$-2 - 2i = 2\sqrt{2}(\cos 225° + i \sin 225°)$

$(-2 - 2i)^5 = \left[2\sqrt{2}(\cos 225° + i \sin 225°)\right]^5$

$= (2\sqrt{2})^5[\cos(5 \cdot 225°) + i \sin(5 \cdot 225°)]$ De Moivre's theorem

$= 32 \cdot 4\sqrt{2}(\cos 1125° + i \sin 1125°)$

$= 128\sqrt{2}(\cos 1125° + i \sin 1125°)$

$= 128\sqrt{2}(\cos 45° + i \sin 45°)$ 1125° and 45° are coterminal.

$= 128\sqrt{2}\left(\dfrac{\sqrt{2}}{2} + i \cdot \dfrac{\sqrt{2}}{2}\right)$ $\cos 45° = \frac{\sqrt{2}}{2}$; $\sin 45° = \frac{\sqrt{2}}{2}$

$= 128 + 128i$ Rectangular form

41. $x^3 - (4 + 4i\sqrt{3}) = 0$

$$x^3 = 4 + 4i\sqrt{3}$$

$$r = \sqrt{4^2 + (4\sqrt{3})^2} = \sqrt{16 + 48} = \sqrt{64} = 8$$

$$\tan \theta = \frac{4\sqrt{3}}{4} = \sqrt{3}$$

θ is in quadrant I, so $\theta = 60°$.

$$x^3 = 4 + 4i\sqrt{3}$$

$$x^3 = 8\left(\frac{1}{2} + i\frac{\sqrt{3}}{2}\right)$$

$$r^3(\cos 3\alpha + i \sin 3\alpha) = 8(\cos 60° + i \sin 60°)$$

$r^3 = 8$, so $r = 2$.

$\alpha = \dfrac{60°}{3} + \dfrac{360° \cdot k}{3}$, k any integer *n*th root theorem

$\alpha = 20° + 120° \cdot k$, k any integer

If $k = 0$, then $\alpha = 20° + 0° = 20°$.

If $k = 1$, then $\alpha = 20° + 120° = 140°$.

If $k = 2$, then $\alpha = 20° + 240° = 260°$.

Solution set: $\{2(\cos 20° + i \sin 20°),$ $2(\cos 140° + i \sin 140°), 2(\cos 260° + i \sin 260°)\}$

8.7 Exercises *(pages 793–797)*

57. $r = 2 \sin \theta$

$r^2 = 2r \sin \theta$ Multiply by r.

$x^2 + y^2 = 2y$ $r^2 = x^2 + y^2$, $r \sin \theta = y$

$x^2 + y^2 - 2y = 0$ Subtract $2y$.

$x^2 + y^2 - 2y + 1 = 1$ Add 1 to complete the square on y.

$x^2 + (y - 1)^2 = 1$ Factor the perfect square trinomial.

The graph is a circle with center $(0, 1)$ and radius 1.

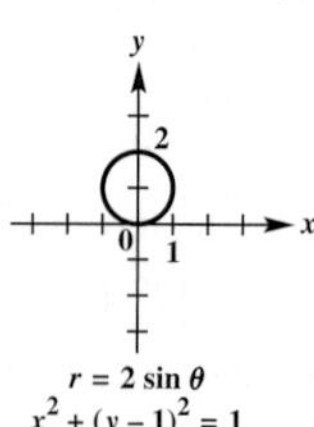

63. $r = 2 \sec \theta$

$r = \dfrac{2}{\cos \theta}$ Reciprocal identity

$r \cos \theta = 2$ Multiply by $\cos \theta$.

$x = 2$ $r \cos \theta = x$

The graph is the vertical line through $(2, 0)$.

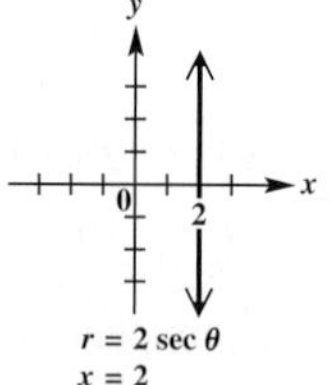

8.8 Exercises *(pages 803–806)*

9. $x = t^3 + 1,\ y = t^3 - 1,\quad$ for t in $(-\infty, \infty)$

(a)

t	x	y
-2	-7	-9
-1	0	-2
0	1	-1
1	2	0
2	9	7
3	28	26

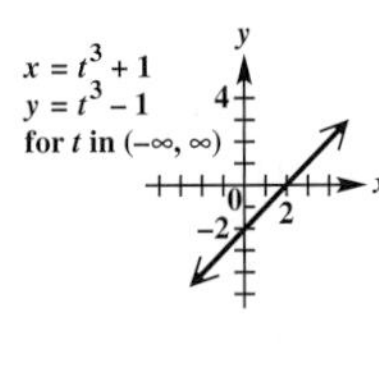

(b)

$$\begin{aligned} x &= t^3 + 1 \\ y &= t^3 - 1 \\ \hline x - y &= 2 \quad \text{Subtract equations to eliminate } t. \\ y &= x - 2 \quad \text{Solve for } y. \end{aligned}$$

The rectangular equation is $y = x - 2$, for x in $(-\infty, \infty)$. The graph is a line with slope 1 and y-intercept -2.

13. $x = 3 \tan t,\ y = 2 \sec t,\quad$ for t in $\left(-\frac{\pi}{2}, \frac{\pi}{2}\right)$

(a)

t	x	y
$-\frac{\pi}{3}$	$-3\sqrt{3} \approx -5.2$	4
$-\frac{\pi}{6}$	$-\sqrt{3} \approx -1.7$	$\frac{4\sqrt{3}}{3} \approx 2.3$
0	0	2
$\frac{\pi}{6}$	$\sqrt{3} \approx 1.7$	$\frac{4\sqrt{3}}{3} \approx 2.3$
$\frac{\pi}{3}$	$3\sqrt{3} \approx 5.2$	4

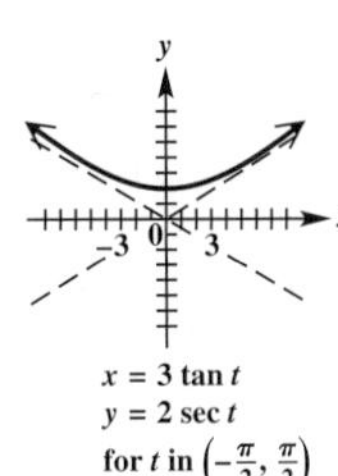

(b) $x = 3 \tan t$, so $\frac{x}{3} = \tan t$.

$y = 2 \sec t$, so $\frac{y}{2} = \sec t$.

$1 + \tan^2 t = \sec^2 t$ — Pythagorean identity

$1 + \left(\frac{x}{3}\right)^2 = \left(\frac{y}{2}\right)^2$ — Substitute expressions for $\tan t$ and $\sec t$.

$1 + \frac{x^2}{9} = \frac{y^2}{4}$ — Apply the exponents.

$y^2 = 4\left(1 + \frac{x^2}{9}\right)$ — Multiply by 4. Rewrite.

$y = 2\sqrt{1 + \frac{x^2}{9}}$ — Use the positive square root because $y > 0$ in the given interval for t.

The rectangular equation is $y = 2\sqrt{1 + \frac{x^2}{9}}$, for x in $(-\infty, \infty)$. The graph is the upper half of a hyperbola.

Chapter 9 Systems and Matrices

9.1 Exercises *(pages 832–838)*

27. $\frac{x}{2} + \frac{y}{3} = 4 \quad (1)$

$\frac{3x}{2} + \frac{3y}{2} = 15 \quad (2)$

It is easier to work with equations that do not contain fractions. To clear fractions, multiply each equation by the LCD of all the fractions in the equation.

$6\left(\frac{x}{2} + \frac{y}{3}\right) = 6(4)$ — Multiply by the LCD, 6.

$2\left(\frac{3x}{2} + \frac{3y}{2}\right) = 2(15)$ — Multiply by the LCD, 2.

This gives the system

$$3x + 2y = 24 \quad (3)$$
$$3x + 3y = 30. \quad (4)$$

To solve this system by elimination, multiply equation (3) by -1 and add the result to equation (4), eliminating x.

$$\begin{aligned} -3x - 2y &= -24 \\ 3x + 3y &= 30 \\ \hline y &= 6 \end{aligned}$$

Substitute 6 for y in equation (1).

$\frac{x}{2} + \frac{6}{3} = 4$ — Let $y = 6$.

$\frac{x}{2} + 2 = 4$ — $\frac{6}{3} = 2$

$\frac{x}{2} = 2$ — Subtract 2.

$x = 4$ — Multiply by 2.

Solution set: $\{(4, 6)\}$

69. $\frac{2}{x} + \frac{1}{y} = \frac{3}{2} \quad (1)$

$\frac{3}{x} - \frac{1}{y} = 1 \quad (2)$

Let $\frac{1}{x} = t$ and $\frac{1}{y} = u$. With these substitutions, the system becomes

$$2t + u = \frac{3}{2} \quad (3)$$
$$3t - u = 1. \quad (4)$$

Add these equations, eliminating u, and solve for t.

$5t = \frac{5}{2}$

$t = \frac{1}{2}$ — Multiply by $\frac{1}{5}$.

Substitute $\frac{1}{2}$ for t in equation (3) and solve for u.

$$2\left(\frac{1}{2}\right) + u = \frac{3}{2} \quad \text{Let } t = \tfrac{1}{2}.$$

$$1 + u = \frac{3}{2} \quad \text{Multiply.}$$

$$u = \frac{1}{2} \quad \text{Subtract 1.}$$

Now find the values of x and y, the variables in the original system. Since $t = \frac{1}{x}$, $tx = 1$, and $x = \frac{1}{t}$. Likewise, $y = \frac{1}{u}$.

$$x = \frac{1}{t} = \frac{1}{\frac{1}{2}} = 2 \quad \text{Let } t = \tfrac{1}{2}.$$

$$y = \frac{1}{u} = \frac{1}{\frac{1}{2}} = 2 \quad \text{Let } u = \tfrac{1}{2}.$$

Solution set: $\{(2, 2)\}$

9.2 Exercises *(pages 845–850)*

47. $\dfrac{1}{(x-1)(x+1)} = \dfrac{A}{x-1} + \dfrac{B}{x+1}$

$$\frac{1}{(x-1)(x+1)}$$

$$= \frac{A(x+1)}{(x-1)(x+1)} + \frac{B(x-1)}{(x-1)(x+1)} \quad \text{Write terms on the right with the LCD, } (x-1)(x+1).$$

$$= \frac{A(x+1) + B(x-1)}{(x-1)(x+1)} \quad \text{Combine terms on the right.}$$

$1 = A(x+1) + B(x-1)$ Denominators are equal, so numerators must be equal.

$1 = Ax + A + Bx - B$ Distributive property

$1 = (Ax + Bx) + (A - B)$ Group like terms.

$1 = (A + B)x + (A - B)$ Factor $Ax + Bx$.

Since $1 = 0x + 1$, we can equate the coefficients of like powers of x to obtain the following system.

$$A + B = 0$$
$$A - B = 1$$

Solve this system by the Gauss-Jordan method.

$$\left[\begin{array}{cc|c} 1 & 1 & 0 \\ 1 & -1 & 1 \end{array}\right]$$

$$\left[\begin{array}{cc|c} 1 & 1 & 0 \\ 0 & -2 & 1 \end{array}\right] \quad -\text{R1} + \text{R2}$$

$$\left[\begin{array}{cc|c} 1 & 1 & 0 \\ 0 & 1 & -\frac{1}{2} \end{array}\right] \quad -\tfrac{1}{2}\text{R2}$$

$$\left[\begin{array}{cc|c} 1 & 0 & \frac{1}{2} \\ 0 & 1 & -\frac{1}{2} \end{array}\right] \quad -\text{R2} + \text{R1}$$

From the final matrix, $A = \frac{1}{2}$ and $B = -\frac{1}{2}$.

55. Let $x =$ number of cubic centimeters of 2% solution and $y =$ number of cubic centimeters of 7% solution. From the given information, we can write the system

$$\begin{aligned} x + \quad\;\; y &= 40 \\ 0.02x + 0.07y &= 0.032(40), \end{aligned}$$

or

$$\begin{aligned} x + \quad\;\; y &= 40 \\ 0.02x + 0.07y &= 1.28. \end{aligned}$$

Solve this system by the Gauss-Jordan method.

$$\left[\begin{array}{cc|c} 1 & 1 & 40 \\ 0.02 & 0.07 & 1.28 \end{array}\right]$$

$$\left[\begin{array}{cc|c} 1 & 1 & 40 \\ 0 & 0.05 & 0.48 \end{array}\right] \quad -0.02\text{R1} + \text{R2}$$

$$\left[\begin{array}{cc|c} 1 & 1 & 40 \\ 0 & 1 & 9.6 \end{array}\right] \quad \tfrac{1}{0.05}\text{R2 (or 20R2)}$$

$$\left[\begin{array}{cc|c} 1 & 0 & 30.4 \\ 0 & 1 & 9.6 \end{array}\right] \quad -\text{R2} + \text{R1}$$

From the final matrix, $x = 30.4$ and $y = 9.6$, so the chemist should mix 30.4 cm^3 of the 2% solution with 9.6 cm^3 of the 7% solution.

9.3 Exercises *(pages 858–862)*

63. $\begin{vmatrix} -4 & 1 & 4 \\ 2 & 0 & 1 \\ 0 & 2 & 4 \end{vmatrix} = \begin{vmatrix} 0 & 1 & 6 \\ 2 & 0 & 1 \\ 0 & 2 & 4 \end{vmatrix}$ Use determinant theorem 6; 2R2 + R1.

$$= 0\begin{vmatrix} 0 & 1 \\ 2 & 4 \end{vmatrix} - 2\begin{vmatrix} 1 & 6 \\ 2 & 4 \end{vmatrix} + 0\begin{vmatrix} 1 & 6 \\ 0 & 1 \end{vmatrix}$$

Expand about column 1.

$$= 0 - 2(4 - 12) + 0$$
$$= -2(-8)$$
$$= 16$$

85. $1.5x + 3y = 5$ (1)

$2x + 4y = 3$ (2)

$$D = \begin{vmatrix} 1.5 & 3 \\ 2 & 4 \end{vmatrix} = 1.5(4) - 2(3) = 6 - 6 = 0$$

Because $D = 0$, Cramer's rule does not apply. To determine whether the system is inconsistent or has infinitely many solutions, use the elimination method.

$6x + 12y = 20$ Multiply equation (1) by 4.

$-6x - 12y = -9$ Multiply equation (2) by -3.

$0 = 11$ False

The system is inconsistent.

Solution set: $\emptyset$

9.4 Exercises *(pages 867–868)*

13. $\dfrac{x^2}{x^2 + 2x + 1}$

This is not a proper fraction; the numerator has degree greater than or equal to that of the denominator. Divide the numerator by the denominator.

$$\begin{array}{r} 1 \\ x^2 + 2x + 1 \overline{)x^2 } \\ \underline{x^2 + 2x + 1} \\ -2x - 1 \end{array}$$

Find the partial fraction decomposition for

$$\frac{-2x - 1}{x^2 + 2x + 1} = \frac{-2x - 1}{(x + 1)^2}.$$

$\dfrac{-2x - 1}{(x + 1)^2} = \dfrac{A}{x + 1} + \dfrac{B}{(x + 1)^2}$ Factor the denominator of the given fraction.

$-2x - 1 = A(x + 1) + B$ Multiply by the LCD, $(x + 1)^2$.

Use this equation to find the value of B.

$-2(-1) - 1 = A(-1 + 1) + B$ Let $x = -1$.

$1 = B$

Now use the same equation and the value of B to find the value of A.

$-2(2) - 1 = A(2 + 1) + 1$ Let $x = 2$ and $B = 1$.

$-5 = 3A + 1$

$-6 = 3A$

$A = -2$

Thus,

$$\frac{x^2}{x^2 + 2x + 1} = 1 + \frac{-2}{x + 1} + \frac{1}{(x + 1)^2}.$$

23. $\dfrac{1}{x(2x + 1)(3x^2 + 4)}$

The denominator contains two linear factors and one quadratic factor. All factors are distinct, and $3x^2 + 4$ cannot be factored, so it is irreducible. The partial fraction decomposition is of the form

$$\frac{1}{x(2x + 1)(3x^2 + 4)} = \frac{A}{x} + \frac{B}{2x + 1} + \frac{Cx + D}{3x^2 + 4}.$$

We need to find the values of A, B, C, and D.

$1 = A(2x + 1)(3x^2 + 4) + Bx(3x^2 + 4) + (Cx + D)(x)(2x + 1)$ Multiply by the LCD, $x(2x + 1)(3x^2 + 4)$.

$1 = A(6x^3 + 3x^2 + 8x + 4) + B(3x^3 + 4x) + C(2x^3 + x^2) + D(2x^2 + x)$ Multiply on the right.

$1 = 6Ax^3 + 3Ax^2 + 8Ax + 4A + 3Bx^3 + 4Bx + 2Cx^3 + Cx^2 + 2Dx^2 + Dx$ Distributive property

$1 = (6A + 3B + 2C)x^3 + (3A + C + 2D)x^2 + (8A + 4B + D)x + 4A$ Combine like terms.

Since $1 = 0x^3 + 0x^2 + 0x + 1$, equating the coefficients of like powers of x produces the following system of equations:

$$\begin{aligned} 6A + 3B + 2C &= 0 \\ 3A + C + 2D &= 0 \\ 8A + 4B + D &= 0 \\ 4A &= 1. \end{aligned}$$

Any of the methods from this chapter can be used to solve this system. If the Gauss-Jordan method is used, begin by representing the system with the augmented matrix

$$\left[\begin{array}{cccc|c} 6 & 3 & 2 & 0 & 0 \\ 3 & 0 & 1 & 2 & 0 \\ 8 & 4 & 0 & 1 & 0 \\ 4 & 0 & 0 & 0 & 1 \end{array}\right].$$

After performing a series of row operations, we obtain the following final matrix:

$$\left[\begin{array}{cccc|c} 1 & 0 & 0 & 0 & \frac{1}{4} \\ 0 & 1 & 0 & 0 & -\frac{8}{19} \\ 0 & 0 & 1 & 0 & -\frac{9}{76} \\ 0 & 0 & 0 & 1 & -\frac{6}{19} \end{array}\right],$$

from which we read the values of the four variables: $A = \frac{1}{4}$, $B = -\frac{8}{19}$, $C = -\frac{9}{76}$, $D = -\frac{6}{19}$.

Substitute these values into the form given at the beginning of this solution for the partial fraction decomposition.

$$\frac{1}{x(2x + 1)(3x^2 + 4)} = \frac{A}{x} + \frac{B}{2x + 1} + \frac{Cx + D}{3x^2 + 4}$$

$$\frac{1}{x(2x + 1)(3x^2 + 4)} = \frac{\frac{1}{4}}{x} + \frac{-\frac{8}{19}}{2x + 1} + \frac{-\frac{9}{76}x - \frac{6}{19}}{3x^2 + 4}$$

$$= \frac{\frac{1}{4}}{x} + \frac{-\frac{8}{19}}{2x + 1} + \frac{-\frac{9}{76}x - \frac{24}{76}}{3x^2 + 4}$$

Get a common denominator for the numerator of the last term.

$$= \frac{1}{4x} + \frac{-8}{19(2x + 1)} + \frac{-9x - 24}{76(3x^2 + 4)}$$

Simplify the complex fractions.

9.5 Exercises *(pages 874–878)*

53. Let x and y represent the two numbers.

$\dfrac{x}{y} = \dfrac{9}{2}$ (1)

$xy = 162$ (2)

Solve equation (1) for x.

$y\left(\dfrac{x}{y}\right) = y\left(\dfrac{9}{2}\right)$ Multiply by y.

$x = \dfrac{9}{2}y$

Substitute $\frac{9}{2}y$ for x in equation (2).

$$\left(\frac{9}{2}y\right)y = 162$$

$$\frac{9}{2}y^2 = 162 \quad \text{Multiply.}$$

$$\frac{2}{9}\left(\frac{9}{2}y^2\right) = \frac{2}{9}(162) \quad \text{Multiply by } \tfrac{2}{9}.$$

$$y^2 = 36$$

$$y = \pm 6 \quad \text{Take square roots.}$$

If $y = 6$, then $x = \frac{9}{2}(6) = 27$.

If $y = -6$, then $x = \frac{9}{2}(-6) = -27$. The two numbers are either 27 and 6, or -27 and -6.

65. supply: $p = \sqrt{0.1q + 9} - 2$

demand: $p = \sqrt{25 - 0.1q}$

(a) Equilibrium occurs when supply = demand, so solve the system formed by the supply and demand equations. This system can be solved by substitution. Substitute $\sqrt{0.1q + 9} - 2$ for p in the demand equation, and solve the resulting equation for q.

$$\sqrt{0.1q + 9} - 2 = \sqrt{25 - 0.1q}$$

$$\left(\sqrt{0.1q + 9} - 2\right)^2 = \left(\sqrt{25 - 0.1q}\right)^2$$

Square each side.

$$0.1q + 9 - 4\sqrt{0.1q + 9} + 4 = 25 - 0.1q$$

$(x - y)^2 = x^2 - 2xy + y^2$

$$0.2q - 12 = 4\sqrt{0.1q + 9}$$

Combine like terms, and isolate the radical.

$$(0.2q - 12)^2 = \left(4\sqrt{0.1q + 9}\right)^2$$

Square each side again.

$$0.04q^2 - 4.8q + 144 = 16(0.1q + 9)$$

$$0.04q^2 - 4.8q + 144 = 1.6q + 144$$

Distributive property

$$0.04q^2 - 6.4q = 0 \quad \text{Combine like terms.}$$

$$0.04q(q - 160) = 0 \quad \text{Factor.}$$

$$q = 0 \quad \text{or} \quad q = 160 \quad \text{Zero-factor property}$$

Disregard an equilibrium demand of 0. The equilibrium demand is 160 units.

(b) Substitute 160 for q in either equation and solve for p. Using the supply equation, we obtain the following.

$$p = \sqrt{0.1q + 9} - 2$$

$$= \sqrt{0.1(160) + 9} - 2 \quad \text{Let } q = 160.$$

$$= \sqrt{16 + 9} - 2$$

$$= \sqrt{25} - 2$$

$$= 5 - 2$$

$$= 3$$

The equilibrium price is \$3.

9.6 Exercises *(pages 886–890)*

55. $y \le \log x$

$y \ge |x - 2|$

Graph $y = \log x$ as a solid curve because $y \le \log x$ is a nonstrict inequality.

(Recall that "$\log x$" means $\log_{10} x$.) This graph contains the points $(0.1, -1)$, $(1, 0)$, and $(10, 1)$ because $10^{-1} = 0.1$, $10^0 = 1$, and $10^1 = 10$. Because the symbol is $\le$, shade the region *below* the curve.

Now graph $y = |x - 2|$. Make this boundary solid because $y \ge |x - 2|$ is also a nonstrict inequality. This graph can be obtained by translating the graph of $y = |x|$ to the right 2 units. It contains the points $(0, 2)$, $(2, 0)$, and $(4, 2)$. Because the symbol is $\ge$, shade the region *above* the absolute value graph.

The solution set of the system is the intersection (or overlap) of the two shaded regions, which is shown in the final graph.

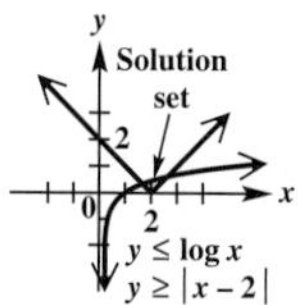

79. Let x = the number of cabinet A and y = the number of cabinet B.

The cost constraint is

$$10x + 20y \le 140.$$

The space constraint is

$$6x + 8y \le 72.$$

Since the numbers of cabinets cannot be negative, we also have the constraints $x \ge 0$ and $y \ge 0$. We want to maximize the objective function,

$$\text{storage capacity} = 8x + 12y.$$

Find the region of feasible solutions by graphing the system of inequalities that is made up of the constraints.

To graph $10x + 20y \le 140$, draw the line with x-intercept 14 and y-intercept 7 as a solid line. Because the symbol is $\le$, shade the region *below* the line.

To graph $6x + 8y \le 72$, draw the line with x-intercept 12 and y-intercept 9 as a solid line. Because the symbol is $\le$, shade the region *below* the line.

The constraints $x \ge 0$ and $y \ge 0$ restrict the graph to the first quadrant. The graph of the feasible region is the intersection of the regions that are the graphs of the individual constraints.

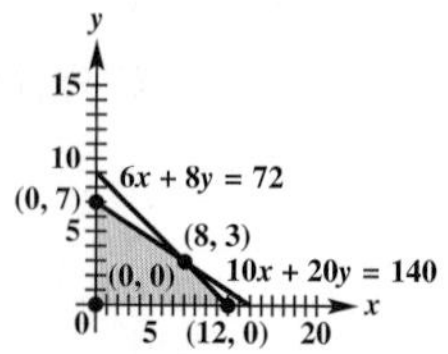

From the graph, observe that three of the vertices are $(0, 0)$, $(0, 7)$, and $(12, 0)$. The fourth vertex is the intersection point of the lines $10x + 20y = 140$ and $6x + 8y = 72$. To find this point, solve the system

$$\begin{aligned} 10x + 20y &= 140 \\ 6x + 8y &= 72 \end{aligned}$$

to obtain the ordered pair $(8, 3)$. Evaluate the objective function at each vertex.

Point	Storage Capacity $= 8x + 12y$
$(0, 0)$	$8(0) + 12(0) = 0$
$(0, 7)$	$8(0) + 12(7) = 84$
$(8, 3)$	$8(8) + 12(3) = 100$
$(12, 0)$	$8(12) + 12(0) = 96$

The maximum value of $8x + 12y$ occurs at the vertex $(8, 3)$, so the office manager should buy 8 of cabinet A and 3 of cabinet B for a maximum storage capacity of 100 ft^3.

9.8 Exercises *(pages 911–915)*

23. $A = \begin{bmatrix} 1 & 1 & 0 & 2 \\ 2 & -1 & 1 & -1 \\ 3 & 3 & 2 & -2 \\ 1 & 2 & 1 & 0 \end{bmatrix}$

$$\left[\begin{array}{cccc|cccc} 1 & 1 & 0 & 2 & 1 & 0 & 0 & 0 \\ 2 & -1 & 1 & -1 & 0 & 1 & 0 & 0 \\ 3 & 3 & 2 & -2 & 0 & 0 & 1 & 0 \\ 1 & 2 & 1 & 0 & 0 & 0 & 0 & 1 \end{array}\right] \quad \text{Write the augmented matrix } [A|I_4].$$

$$\left[\begin{array}{cccc|cccc} 1 & 1 & 0 & 2 & 1 & 0 & 0 & 0 \\ 0 & -3 & 1 & -5 & -2 & 1 & 0 & 0 \\ 0 & 0 & 2 & -8 & -3 & 0 & 1 & 0 \\ 0 & 1 & 1 & -2 & -1 & 0 & 0 & 1 \end{array}\right] \quad \begin{matrix} \\ -2\text{R1} + \text{R2} \\ -3\text{R1} + \text{R3} \\ -1\text{R1} + \text{R4} \end{matrix}$$

$$\left[\begin{array}{cccc|cccc} 1 & 1 & 0 & 2 & 1 & 0 & 0 & 0 \\ 0 & 1 & -\frac{1}{3} & \frac{5}{3} & \frac{2}{3} & -\frac{1}{3} & 0 & 0 \\ 0 & 0 & 2 & -8 & -3 & 0 & 1 & 0 \\ 0 & 1 & 1 & -2 & -1 & 0 & 0 & 1 \end{array}\right] \quad \begin{matrix} \\ -\frac{1}{3}\text{R2} \\ \\ \\ \end{matrix}$$

$$\left[\begin{array}{cccc|cccc} 1 & 0 & \frac{1}{3} & \frac{1}{3} & \frac{1}{3} & \frac{1}{3} & 0 & 0 \\ 0 & 1 & -\frac{1}{3} & \frac{5}{3} & \frac{2}{3} & -\frac{1}{3} & 0 & 0 \\ 0 & 0 & 2 & -8 & -3 & 0 & 1 & 0 \\ 0 & 0 & \frac{4}{3} & -\frac{11}{3} & -\frac{5}{3} & \frac{1}{3} & 0 & 1 \end{array}\right] \quad \begin{matrix} -1\text{R2} + \text{R1} \\ \\ \\ -1\text{R2} + \text{R4} \end{matrix}$$

$$\left[\begin{array}{cccc|cccc} 1 & 0 & \frac{1}{3} & \frac{1}{3} & \frac{1}{3} & \frac{1}{3} & 0 & 0 \\ 0 & 1 & -\frac{1}{3} & \frac{5}{3} & \frac{2}{3} & -\frac{1}{3} & 0 & 0 \\ 0 & 0 & 1 & -4 & -\frac{3}{2} & 0 & \frac{1}{2} & 0 \\ 0 & 0 & \frac{4}{3} & -\frac{11}{3} & -\frac{5}{3} & \frac{1}{3} & 0 & 1 \end{array}\right] \quad \begin{matrix} \\ \\ \frac{1}{2}\text{R3} \\ \\ \end{matrix}$$

$$\left[\begin{array}{cccc|cccc} 1 & 0 & 0 & \frac{5}{3} & \frac{5}{6} & \frac{1}{3} & -\frac{1}{6} & 0 \\ 0 & 1 & 0 & \frac{1}{3} & \frac{1}{6} & -\frac{1}{3} & \frac{1}{6} & 0 \\ 0 & 0 & 1 & -4 & -\frac{3}{2} & 0 & \frac{1}{2} & 0 \\ 0 & 0 & 0 & \frac{5}{3} & \frac{1}{3} & \frac{1}{3} & -\frac{2}{3} & 1 \end{array}\right] \quad \begin{matrix} -\frac{1}{3}\text{R3} + \text{R1} \\ \frac{1}{3}\text{R3} + \text{R2} \\ \\ -\frac{4}{3}\text{R3} + \text{R4} \end{matrix}$$

$$\left[\begin{array}{cccc|cccc} 1 & 0 & 0 & \frac{5}{3} & \frac{5}{6} & \frac{1}{3} & -\frac{1}{6} & 0 \\ 0 & 1 & 0 & \frac{1}{3} & \frac{1}{6} & -\frac{1}{3} & \frac{1}{6} & 0 \\ 0 & 0 & 1 & -4 & -\frac{3}{2} & 0 & \frac{1}{2} & 0 \\ 0 & 0 & 0 & 1 & \frac{1}{5} & \frac{1}{5} & -\frac{2}{5} & \frac{3}{5} \end{array}\right] \quad \begin{matrix} \\ \\ \\ \frac{3}{5}\text{R4} \end{matrix}$$

$$\left[\begin{array}{cccc|cccc} 1 & 0 & 0 & 0 & \frac{1}{2} & 0 & \frac{1}{2} & -1 \\ 0 & 1 & 0 & 0 & \frac{1}{10} & -\frac{2}{5} & \frac{3}{10} & -\frac{1}{5} \\ 0 & 0 & 1 & 0 & -\frac{7}{10} & \frac{4}{5} & -\frac{11}{10} & \frac{12}{5} \\ 0 & 0 & 0 & 1 & \frac{1}{5} & \frac{1}{5} & -\frac{2}{5} & \frac{3}{5} \end{array}\right] \quad \begin{matrix} -\frac{5}{3}\text{R4} + \text{R1} \\ -\frac{1}{3}\text{R4} + \text{R2} \\ 4\text{R4} + \text{R3} \\ \\ \end{matrix}$$

$$A^{-1} = \begin{bmatrix} \frac{1}{2} & 0 & \frac{1}{2} & -1 \\ \frac{1}{10} & -\frac{2}{5} & \frac{3}{10} & -\frac{1}{5} \\ -\frac{7}{10} & \frac{4}{5} & -\frac{11}{10} & \frac{12}{5} \\ \frac{1}{5} & \frac{1}{5} & -\frac{2}{5} & \frac{3}{5} \end{bmatrix}$$

43. $0.2x + 0.3y = -1.9$
$0.7x - 0.2y = 4.6$

$$A = \begin{bmatrix} 0.2 & 0.3 \\ 0.7 & -0.2 \end{bmatrix}, \quad X = \begin{bmatrix} x \\ y \end{bmatrix}, \quad B = \begin{bmatrix} -1.9 \\ 4.6 \end{bmatrix}$$

Find A^{-1}.

$$\left[\begin{array}{cc|cc} 0.2 & 0.3 & 1 & 0 \\ 0.7 & -0.2 & 0 & 1 \end{array}\right] \quad \text{Write the augmented matrix } [A|I_2].$$

$$\left[\begin{array}{cc|cc} 1 & 1.5 & 5 & 0 \\ 0.7 & -0.2 & 0 & 1 \end{array}\right] \quad 5\text{R1}$$

$$\left[\begin{array}{cc|cc} 1 & 1.5 & 5 & 0 \\ 0 & -1.25 & -3.5 & 1 \end{array}\right] \quad -0.7\text{R1} + \text{R2}$$

$$\left[\begin{array}{cc|cc} 1 & 1.5 & 5 & 0 \\ 0 & 1 & 2.8 & -0.8 \end{array}\right] \quad \frac{1}{-1.25}\text{R2, or } -0.8\text{R2}$$

$$\left[\begin{array}{cc|cc} 1 & 0 & 0.8 & 1.2 \\ 0 & 1 & 2.8 & -0.8 \end{array}\right] \quad -1.5\text{R2} + \text{R1}$$

$$A^{-1} = \begin{bmatrix} 0.8 & 1.2 \\ 2.8 & -0.8 \end{bmatrix}$$

$$X = A^{-1}B = \begin{bmatrix} 0.8 & 1.2 \\ 2.8 & -0.8 \end{bmatrix} \begin{bmatrix} -1.9 \\ 4.6 \end{bmatrix} = \begin{bmatrix} 4 \\ -9 \end{bmatrix}$$

Solution set: $\{(4, -9)\}$

Chapter 10 Analytic Geometry

10.1 Exercises (pages 936–939)

31. $(x-7)^2 = 16(y+5)$

This equation can be rewritten as

$$(x-7)^2 = 16[y-(-5)].$$

Thus, it has the form

$$(x-h)^2 = 4p(y-k),$$

with $h = 7$, $k = -5$, and $4p = 16$ and thus $p = 4$. The graph of the given equation is a parabola with vertical axis of symmetry. The vertex (h, k) is $(7, -5)$. Because this parabola has a vertical axis and $p > 0$, the parabola opens up, so the focus is distance $p = 4$ units above the vertex. Thus, the focus is the point $(7, -1)$.

The directrix is a horizontal line $p = 4$ units below the vertex, so the directrix is the line $y = -9$. The axis of symmetry is the vertical line through the vertex, so the equation of the axis is $x = 7$.

39. Write an equation for the parabola through $(3, 2)$, symmetric with respect to the x-axis.

This parabola has a horizontal axis of symmetry (the x-axis) because of the symmetry and vertex $(0, 0)$, so its equation can be written in the form $y^2 = 4px$. Use this equation with the coordinates of the point $(3, 2)$ to find the value of p.

$y^2 = 4px$

$2^2 = 4p \cdot 3$ Let $x = 3$ and $y = 2$.

$4 = 12p$ Apply the exponent. Multiply.

$p = \frac{1}{3}$ Divide by 4. Rewrite.

Thus, the equation of the parabola is

$$y^2 = 4px = 4\left(\frac{1}{3}\right)x = \frac{4}{3}x.$$

55. Place the parabola that represents the arch on a coordinate system with the center of the bottom of the arch at the origin. Then the vertex will be at $(0, 12)$, and the points $(-6, 0)$ and $(6, 0)$ will also be on the parabola.

Because the axis of symmetry of the parabola is the y-axis and the vertex is $(0, 12)$, the equation will have the form $x^2 = 4p(y-12)$. Use the coordinates of the point $(6, 0)$ to find the value of p.

$x^2 = 4p(y-12)$

$6^2 = 4p(0-12)$ Let $x = 6$ and $y = 0$.

$36 = -48p$ Multiply.

$p = -\frac{3}{4}$ Divide by 36. Rewrite in lowest terms.

Thus, the equation of the parabola is

$$x^2 = 4\left(-\frac{3}{4}\right)(y-12), \quad \text{or} \quad x^2 = -3(y-12).$$

Now find the x-coordinate of a point whose y-coordinate is 9 and whose x-coordinate is positive.

$x^2 = -3(y-12)$

$x^2 = -3(9-12)$ Let $y = 9$.

$x^2 = 9$ Subtract and multiply.

$x = \sqrt{9} = 3$ $x > 0$

Using symmetry, the width of the arch 9 ft up is $2(3 \text{ ft}) = 6$ ft.

10.2 Exercises (pages 947–950)

21. The center is located halfway between the foci at $(0, 0)$. The distance from the center to either focus is $c = 4$. The sum of the distances from the foci to a point on the ellipse is $2a = 10$, so $a = 5$. The form of the equation of the ellipse with center at the origin and vertical major axis is

$$\frac{x^2}{b^2} + \frac{y^2}{a^2} = 1.$$

$c^2 = a^2 - b^2$, so $b^2 = a^2 - c^2$.

$$b^2 = 5^2 - 4^2 = 25 - 16 = 9$$

Thus, the equation of the ellipse is

$$\frac{x^2}{9} + \frac{y^2}{25} = 1.$$

43. Place the half-ellipse that represents the overpass on a coordinate system with the center of the bottom of the overpass at the origin. If the complete ellipse were drawn, the center of the ellipse would also be at $(0, 0)$. Then the half-ellipse will include the points $(0, 15)$, $(-10, 0)$, and $(10, 0)$. Thus, for the complete ellipse, $a = 15$ and $b = 10$.

$\frac{x^2}{b^2} + \frac{y^2}{a^2} = 1$

$\frac{x^2}{10^2} + \frac{y^2}{15^2} = 1$ Let $a = 15$ and $b = 10$.

$\frac{x^2}{100} + \frac{y^2}{225} = 1$ Apply the exponents.

To find the equation of the half-ellipse, solve this equation for y and use the positive square root since the overpass is represented by the upper half of the ellipse.

$\frac{y^2}{225} = 1 - \frac{x^2}{100}$

$y^2 = 225\left(1 - \frac{x^2}{100}\right)$ Multiply by 225.

$y = \sqrt{225\left(1 - \frac{x^2}{100}\right)}$ Take square roots; $y \geq 0$.

$y = 15\sqrt{1 - \frac{x^2}{100}}$ Product rule: $\sqrt{ab} = \sqrt{a} \cdot \sqrt{b}$

Find the y-coordinate of the point whose x-coordinate is $\frac{1}{2}(12) = 6$.

$$y = 15\sqrt{1 - \frac{x^2}{100}} \quad \text{Equation of half-ellipse}$$

$$y = 15\sqrt{1 - \frac{6^2}{100}} \quad \text{Let } x = 6.$$

$$y = 15\sqrt{1 - \frac{36}{100}} = 15\sqrt{\frac{64}{100}} = 15(0.8) = 12$$

The tallest truck that can pass under the overpass is 12 ft tall.

10.3 Exercises *(pages 956–960)*

23. $\frac{y}{3} = \sqrt{1 + \frac{x^2}{16}}$

Square each side to get

$$\frac{y^2}{9} = 1 + \frac{x^2}{16}, \quad \text{or} \quad \frac{y^2}{9} - \frac{x^2}{16} = 1.$$

This is the equation of a hyperbola with center $(0, 0)$, vertices $(0, 3)$ and $(0, -3)$, and asymptotes $y = \pm\frac{3}{4}x$. The original equation represents the top half of the hyperbola. The domain is $(-\infty, \infty)$, and the range is $[3, \infty)$. The vertical line test shows that this is the graph of a function.

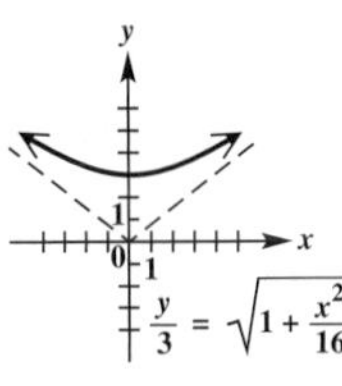

33. vertices at $(0, 6)$ and $(0, -6)$; asymptotes $y = \pm\frac{1}{2}x$
From the given vertices, the equation is of the following form.

$$\frac{y^2}{a^2} - \frac{x^2}{b^2} = 1$$

$$\frac{y^2}{6^2} - \frac{x^2}{b^2} = 1 \quad \text{Let } a = 6.$$

$$\frac{y^2}{36} - \frac{x^2}{b^2} = 1 \quad \text{Apply the exponent.}$$

The slopes of the asymptotes are $\pm\frac{1}{2}$. Use the positive slope to find the value of b.

$$\frac{a}{b} = \frac{1}{2}$$

$$\frac{6}{b} = \frac{1}{2} \quad \text{Let } a = 6.$$

$$b = 12 \quad \text{Multiply by } 2b. \text{ Rewrite.}$$

Use the values of a and b to write the equation of the hyperbola.

$$\frac{y^2}{6^2} - \frac{x^2}{12^2} = 1 \quad \text{Let } a = 6 \text{ and } b = 12.$$

$$\frac{y^2}{36} - \frac{x^2}{144} = 1 \quad \text{Apply the exponents.}$$

43. eccentricity 3; center at $(0, 0)$; vertex at $(0, 7)$

Since the center and the given vertex lie on the y-axis, the equation is of the form

$$\frac{y^2}{a^2} - \frac{x^2}{b^2} = 1.$$

The distance between the center and a vertex is 7 units, so $a = 7$. Use the given eccentricity to find the value of c.

$$e = \frac{c}{a}$$

$$3 = \frac{c}{7} \quad \text{Let } e = 3 \text{ and } a = 7.$$

$$c = 21 \quad \text{Multiply by 7. Rewrite.}$$

Now find the value of b^2. Since $c^2 = a^2 + b^2$,

$$b^2 = c^2 - a^2 = 21^2 - 7^2 = 441 - 49 = 392.$$

The equation of the hyperbola is

$$\frac{y^2}{49} - \frac{x^2}{392} = 1.$$

10.4 Exercises *(pages 964–966)*

31. $y^2 - 4y = x + 4$

$$y^2 - 4y + 4 = x + 4 + 4 \quad \text{Add 4 to complete the square.}$$

$$(y - 2)^2 = x + 8 \quad \text{Factor on the left and add on the right.}$$

The equation is a parabola of the form $(y - k)^2 = 4p(x - h)$ with $p = \frac{1}{4}$, $h = -8$, and $k = 2$. Thus, the vertex is $(-8, 2)$ and the parabola opens to the right.

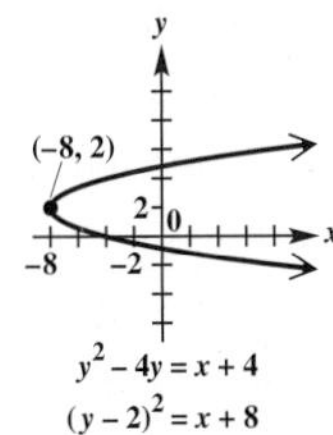

41. From the graph, the coordinates of P (a point on the graph) are $(-3, 8)$, the coordinates of F (a focus) are $(3, 0)$, and the equation of L (the directrix) is $x = 27$. By the distance formula, the distance from P to F is

$$\sqrt{(x_2 - x_1)^2 + (y_2 - y_1)^2} = \sqrt{[3 - (-3)]^2 + (0 - 8)^2}$$
$$= \sqrt{6^2 + (-8)^2}$$
$$= \sqrt{36 + 64} = \sqrt{100} = 10.$$

The distance between a point and a line is defined as the perpendicular distance, so the distance from P to L is $|27 - (-3)| = 30$.

$$e = \frac{\text{distance from } P \text{ to } F}{\text{distance from } P \text{ to } L} = \frac{10}{30} = \frac{1}{3}$$

Chapter 11 Further Topics in Algebra

11.2 Exercises *(pages 991–993)*

25. $S_{16} = -160,\ a_{16} = -25$

$$S_n = \frac{n}{2}(a_1 + a_n) \quad \text{Use the first formula for } S_n.$$

$$S_{16} = \frac{16}{2}(a_1 + a_{16}) \quad \text{Let } n = 16.$$

$$-160 = 8(a_1 - 25) \quad \text{Let } S_{16} = -160 \text{ and } a_{16} = -25.$$

$$-20 = a_1 - 25 \quad \text{Divide by 8.}$$

$$a_1 = 5 \quad \text{Add 25. Interchange sides.}$$

47. The positive even integers form the arithmetic sequence 2, 4, 6, 8, . . . , with $a_1 = 2$ and $d = 2$. Find the sum of the first 60 terms of this sequence.

$$S_n = \frac{n}{2}[2a_1 + (n-1)d] \quad \text{Use the second formula for } S_n.$$

$$S_{60} = \frac{60}{2}(2a_1 + 59d) \quad \text{Let } n = 60.$$

$$= \frac{60}{2}(2 \cdot 2 + 59 \cdot 2) \quad \text{Let } a_1 = 2 \text{ and } d = 2.$$

$$= 30(4 + 118) \quad \text{Multiply inside parentheses.}$$

$$= 30(122) \quad \text{Add inside parentheses.}$$

$$S_{60} = 3660 \quad \text{Multiply.}$$

11.3 Exercises *(pages 999–1003)*

31. $\sum_{k=4}^{10} 2^k$

This series is the sum of the fourth through tenth terms of a geometric sequence with $a_1 = 2^1 = 2$ and $r = 2$. One way to find this sum is to find the difference between the sum of the first ten terms and the sum of the first three terms.

$$\sum_{k=4}^{10} 2^k = \sum_{k=1}^{10} 2^k - \sum_{k=1}^{3} 2^k$$

$$= \frac{2(1-2^{10})}{1-2} - \frac{2(1-2^3)}{1-2}$$

$$= \frac{2(1-1024)}{-1} - \frac{2(1-8)}{-1}$$

$$= \frac{2(-1023)}{-1} - \frac{2(-7)}{-1}$$

$$= 2046 - 14$$

$$\sum_{k=4}^{10} 2^k = 2032$$

43. $\frac{1}{4} - \frac{1}{6} + \frac{1}{9} - \frac{2}{27} + \cdots$

For this infinite geometric series, $a_1 = \frac{1}{4}$.

$$r = \frac{-\frac{1}{6}}{\frac{1}{4}} = -\frac{1}{6} \cdot \frac{4}{1} = -\frac{2}{3}$$

Because $-1 < r < 1$, this series converges.

$$S_\infty = \frac{a_1}{1-r} = \frac{\frac{1}{4}}{1-\left(-\frac{2}{3}\right)} = \frac{\frac{1}{4}}{\frac{5}{3}} = \frac{1}{4} \cdot \frac{3}{5} = \frac{3}{20} \quad \text{Let } a_1 = \tfrac{1}{4},\ r = -\tfrac{2}{3}.$$

69. Option 1 is modeled by the arithmetic sequence $a_n = 5000 + 10{,}000(n-1)$ with sum S_{30}.

$$S_n = \frac{n}{2}[2a_1 + (n-1)d]$$

$$S_{30} = \frac{30}{2}[2a_1 + 29d] \quad \text{Let } n = 30.$$

$$= \frac{30}{2}(2 \cdot 5000 + 29 \cdot 10{,}000) \quad \text{Let } a_1 = 5000 \text{ and } d = 10{,}000.$$

$$= 15(10{,}000 + 290{,}000)$$

$$S_{30} = 15(300{,}000) = 4{,}500{,}000$$

Thus, Option 1 pays you a total of \$4,500,000.00.

Option 2 is modeled by the geometric sequence $a_n = 0.01(2)^{n-1}$ with sum S_{30}.

$$S_n = \frac{a_1(1-r^n)}{1-r}$$

$$S_{30} = \frac{0.01(1-2^{30})}{1-2} = 10{,}737{,}418.23 \quad \text{Let } n = 30,\ a_1 = 0.01,\ r = 2.$$

Thus, Option 2 pays you a total of \$10,737,418.23. You should choose Option 2.

11.4 Exercises *(pages 1010–1011)*

11. $$\binom{n}{n-1} = \frac{n!}{(n-1)![n-(n-1)]!}$$

$$= \frac{n!}{(n-1)!\,1!} = \frac{n!}{(n-1)!}$$

$$= \frac{n(n-1)!}{(n-1)!} = n$$

45. The fifteenth term of $(x - y^3)^{20}$ is found using $n = 20$ and $k = 15$. Thus $k - 1 = 14$ and $n - (k-1) = 6$. The fifteenth term of the expansion is

$$\binom{20}{14}x^6(-y^3)^{14} = 38{,}760x^6y^{42}.$$

47. To find the middle term of $(3x^7 + 2y^3)^8$, note that this expansion has nine terms, so the middle term is the fifth term. Here, $n = 8$ and $k = 5$. Thus $k - 1 = 4$ and $n - (k-1) = 4$. The fifth term of the expansion is

$$\binom{8}{4}(3x^7)^4(2y^3)^4 = 70(81x^{28})(16y^{12}) = 90{,}720x^{28}y^{12}.$$

11.5 Exercises *(pages 1016–1017)*

29. Let S_n represent the statement that the number of handshakes for n people is $\frac{n^2 - n}{2}$.

We need to prove that this statement is true for every positive integer $n \geq 2$, so we will use the generalized principle of mathematical induction.

Step 1 Show that the statement is true when $n = 2$.

S_2 is the statement that for 2 people, the number of handshakes is

$$\frac{2^2 - 2}{2} = \frac{4 - 2}{2} = \frac{2}{2} = 1. \quad \text{True}$$

Step 2 Show that S_k implies S_{k+1}, where S_k is the statement that the number of handshakes for k people is $\frac{k^2 - k}{2}$, and S_{k+1} is the statement that the number of handshakes for $(k + 1)$ people is $\frac{(k + 1)^2 - (k + 1)}{2}$.

Start with S_k and assume it is a true statement:

For k people, there are $\frac{k^2 - k}{2}$ handshakes.

If one more person enters the room that already contains k people, this person will shake hands once with each of the k people who were already in the room, so there will be k additional handshakes. Thus, the number of handshakes for $(k + 1)$ people is as follows.

$$\frac{k^2 - k}{2} + k = \frac{k^2 - k}{2} + \frac{2k}{2}$$
Get a common denominator.
$$= \frac{k^2 - k + 2k}{2}$$
Add rational expressions.
$$= \frac{k^2 + k}{2}$$
Combine like terms in the numerator.
$$= \frac{(k^2 + 2k + 1) - k - 1}{2}$$
Write k as $2k - k$, add 1 to complete the square, and subtract 1.
$$= \frac{(k + 1)^2 - (k + 1)}{2}$$
Factor out -1 in the numerator.

This work shows that if S_k is true, then S_{k+1} is true.

Since both steps for a proof by the generalized principle of mathematical induction have been completed, the given statement is true for every positive integer $n \geq 2$.

11.6 Exercises *(pages 1024–1028)*

33. (a) There are two choices for the first letter, K and W.

The second letter can be any of the 26 letters of the alphabet except for the one chosen for the first letter, so there are 25 choices. The third letter can be any of the remaining 24 letters of the alphabet, so there are 24 choices. The fourth letter can be any of the remaining 23 letters of the alphabet, so there are 23 choices.

Therefore, by the fundamental principle of counting, the number of possible 4-letter radio-station call letters is

$$2 \cdot 25 \cdot 24 \cdot 23 = 27{,}600.$$

(b) There are two choices for the first letter.

Because repetitions are allowed, there are 26 choices for each of the remaining 3 letters.

Therefore, by the fundamental principle of counting, the number of possible 4-letter radio-station call letters is

$$2 \cdot 26 \cdot 26 \cdot 26 = 35{,}152.$$

(c) There are two choices for the first letter.

There are 24 choices for the second letter since it cannot repeat the first letter and cannot be R. There are 23 choices for the third letter since it cannot repeat either of the first two letters and cannot be R. There is only one choice for the last letter since it must be R.

Therefore, by the fundamental principle of counting, the number of possible 4-letter radio-station call letters is

$$2 \cdot 24 \cdot 23 \cdot 1 = 1104.$$

63. (c) Choosing 3 women and 2 men involves two independent events, each of which involves combinations.

First, select the women. There are 8 women in the club, so the number of ways to do this is

$$C(8, 3) = \binom{8}{3} = \frac{8!}{3!5!} = 56.$$

Now select the men. There are 11 men in the club, so the number of ways to do this is

$$C(11, 2) = \binom{11}{2} = \frac{11!}{2!9!} = 55.$$

To find the number of committees, use the fundamental principle of counting. The number of committees with 3 women and 2 men is

$$56 \cdot 55 = 3080.$$

69. Because the keys are arranged in a circle, there is no "first" key. The number of distinguishable arrangements is the number of ways to arrange the other three keys in relation to any one of the keys.

$$P(3, 3) = 3! = 6$$

11.7 Exercises *(pages 1035–1039)*

23. (a) A 40-yr-old man who lives 30 more yr would be 70 yr old.

Let E be the event "selected man will live to be 70"; then $n(E) = 71{,}586$. For this situation, the sample space S is the set of all 40-yr-old men, so $n(S) = 95{,}431$. Thus, the probability that a 40-yr-old man will live 30 more yr is

$$P(E) = \frac{n(E)}{n(S)} = \frac{71{,}586}{95{,}431} \approx 0.750.$$

(b) Using the notation and result from part (a), the probability that a 40-yr-old man will not live 30 more yr is

$$P(E') = 1 - P(E) = 1 - 0.750 = 0.250.$$

(c) Use the notation and results from parts (a) and (b). In this binomial experiment, we call "a 40-yr-old man survives to age 70" a success. Then,

$$p = P(E) = 0.750$$

and $\quad q = 1 - p = P(E') = 0.250.$

There are 5 independent trials and we need the probability of 3 successes, so $n = 5$ and $r = 3$. The probability that exactly 3 of the 40-yr-old men survive to age 70 is

$$\binom{n}{r}p^r q^{n-r} = \binom{5}{3}(0.750)^3(0.250)^2$$
$$= 10(0.750)^3(0.250)^2$$
$$\approx 0.264.$$

(d) Let F be the event "at least one man survives to age 70." The easiest way to find $P(F)$ is to first find the probability of the complementary event F': "neither man survives to age 70."

$$P(F') = P(E') \cdot P(E') = (0.250)^2 = 0.0625$$

Then,

$$P(F) = 1 - P(F') = 1 - 0.0625 \approx 0.938.$$

29. P(not enough sleep on 14 or more days)
$= P$(not enough sleep on 14–29 days or on 30 days)
$= P$(not enough sleep on 14–29 days) $+$ P(not enough sleep on 30 days)
$= 0.20 + 0.12$
$= 0.32$

In this binomial experiment, call "not enough sleep on 14 or more days" a success. Then $p = 0.32$ and $q = 1 - p = 0.68$. "Fewer than 2" means 0 or 1.

P(0 did not get enough sleep on 14 or more days) $+$
P(1 did not get enough sleep on 14 or more days)

$$= \binom{10}{0}(0.32)^0(0.68)^{10} + \binom{10}{1}(0.32)^1(0.68)^9$$
$$\approx 0.021139 + 0.099479$$
$$\approx 0.121$$

Answers to Selected Exercises

3.3 Exercises *(pages 315–317)*

79.

Positive	Negative	Nonreal Complex
1	1	2

80.

Positive	Negative	Nonreal Complex
1	3	0
1	1	2

81.

Positive	Negative	Nonreal Complex
4	0	0
2	0	2
0	0	4

82.

Positive	Negative	Nonreal Complex
0	4	0
0	2	2
0	0	4

83.

Positive	Negative	Nonreal Complex
2	3	0
2	1	2
0	3	2
0	1	4

84.

Positive	Negative	Nonreal Complex
4	1	0
2	1	2
0	1	4

85.

Positive	Negative	Nonreal Complex
0	5	0
0	3	2
0	1	4

86.

Positive	Negative	Nonreal Complex
5	0	0
3	0	2
1	0	4

87.

Positive	Negative	Nonreal Complex
2	3	0
2	1	2
0	3	2
0	1	4

88.

Positive	Negative	Nonreal Complex
3	2	0
3	0	2
1	2	2
1	0	4

89.

Positive	Negative	Nonreal Complex
4	2	0
4	0	2
2	2	2
2	0	4
0	2	4
0	0	6

90.

Positive	Negative	Nonreal Complex
2	4	0
2	2	2
2	0	4
0	4	2
0	2	4
0	0	6

91. $-5, 3, \pm 2i\sqrt{3}$ **92.** $2, 2, -3$ **93.** $1, 1, 1, -4$
94. $-4, 2 \pm \sqrt{2}$ **95.** $-3, -3, 0, \frac{1 \pm i\sqrt{31}}{4}$
96. $5, \frac{-3 \pm 2\sqrt{3}}{3}$ **97.** $2, 2, 2, \pm i\sqrt{2}$
98. $-\frac{5}{3}, -\frac{5}{3}, \pm 2i\sqrt{6}$ **99.** $-\frac{1}{2}, 1, \pm 2i$
100. $\frac{1}{4}, -\frac{3}{8}, 3, 3$ **101.** $-\frac{1}{5}, 1 \pm i\sqrt{5}$
102. $-\frac{3}{4}, \pm i\sqrt{2}$ **103.** $\pm 2i, \pm 5i$
104. $-1, -1, -1, -1$ **105.** $\pm i, \pm i$
106. $2, 2, 2, 2$ **107.** $0, 0, 3 \pm \sqrt{2}$ **108.** $\pm\frac{1}{2}, \pm 4$
109. $3, 3, 1 \pm i\sqrt{7}$ **110.** $\frac{1}{4}, -\frac{2}{3}, 2 \pm i\sqrt{2}$
111. $\pm 2, \pm 3, \pm 2i$ **112.** $1, -2, 3, -5, 2 \pm \sqrt{3}$
117. 2 **118.** 4

Summary Exercises on Polynomial Functions, Zeros, and Graphs *(pages 338–339)*

11. (a)

Positive	Negative	Nonreal Complex
1	3	0
1	1	2

(b) $\pm 1, \pm 2, \pm 3, \pm 6$ **(c)** $-3, -1$ (multiplicity 2), 2
(d) no other real zeros **(e)** no other complex zeros
(f) $-3, -1, 2$ **(g)** -6 **(h)** $f(4) = 350$; $(4, 350)$

11. (i) **(j)**

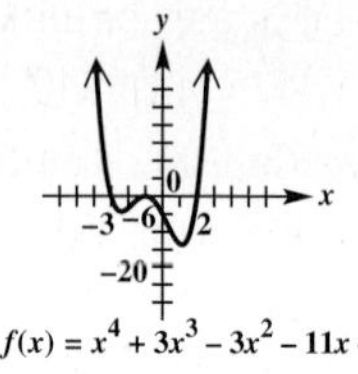

$f(x) = x^4 + 3x^3 - 3x^2 - 11x - 6$

12. (a)

Positive	Negative	Nonreal Complex
3	2	0
3	0	2
1	2	2
1	0	4

(b) $\pm 1, \pm 3, \pm 5, \pm 9, \pm 15, \pm 45, \pm\frac{1}{2}, \pm\frac{3}{2}, \pm\frac{5}{2}, \pm\frac{9}{2}, \pm\frac{15}{2}, \pm\frac{45}{2}$ **(c)** $-3, \frac{1}{2}, 5$ **(d)** $-\sqrt{3}, \sqrt{3}$ **(e)** no other complex zeros **(f)** $-3, \frac{1}{2}, 5, -\sqrt{3}, \sqrt{3}$ **(g)** 45 **(h)** $f(4) = 637$; $(4, 637)$ **(i)**

(j)

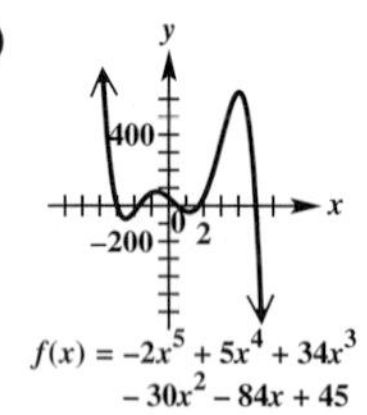

$f(x) = -2x^5 + 5x^4 + 34x^3 - 30x^2 - 84x + 45$

13. (a)

Positive	Negative	Nonreal Complex
4	1	0
2	1	2
0	1	4

(b) $\pm 1, \pm 5, \pm\frac{1}{2}, \pm\frac{5}{2}$ **(c)** 5 **(d)** $-\frac{\sqrt{2}}{2}, \frac{\sqrt{2}}{2}$ **(e)** $-i, i$ **(f)** $-\frac{\sqrt{2}}{2}, \frac{\sqrt{2}}{2}, 5$ **(g)** 5 **(h)** $f(4) = -527$; $(4, -527)$ **(i)** **(j)**

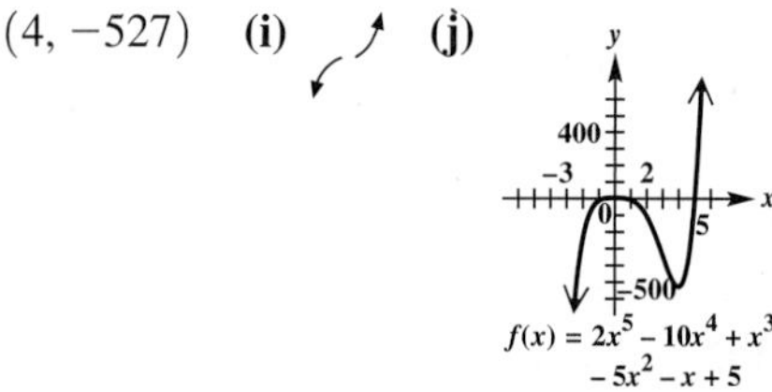

$f(x) = 2x^5 - 10x^4 + x^3 - 5x^2 - x + 5$

14. (a)

Positive	Negative	Nonreal Complex
2	2	0
2	0	2
0	2	2
0	0	4

(b) $\pm 1, \pm 2, \pm 3, \pm 6, \pm 9, \pm 18, \pm\frac{1}{3}, \pm\frac{2}{3}$ **(c)** $-\frac{2}{3}, 3$ **(d)** $\frac{-1+\sqrt{13}}{2}, \frac{-1-\sqrt{13}}{2}$ **(e)** no other complex zeros **(f)** $-\frac{2}{3}, 3, \frac{-1 \pm \sqrt{13}}{2}$ **(g)** 18 **(h)** $f(4) = 238$; $(4, 238)$ **(i)**

(j)

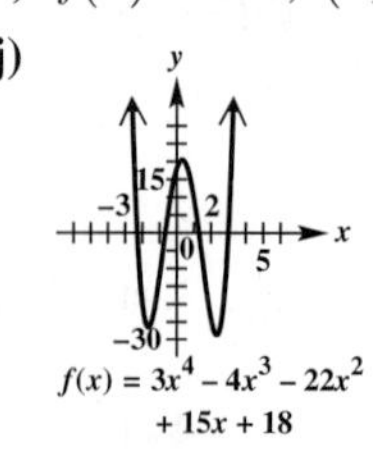

$f(x) = 3x^4 - 4x^3 - 22x^2 + 15x + 18$

15. (a)

Positive	Negative	Nonreal Complex
1	3	0
1	1	2

(b) $\pm 1, \pm 2, \pm\frac{1}{2}$ **(c)** $-1, 1$ **(d)** no other real zeros **(e)** $-\frac{1}{4} + \frac{\sqrt{15}}{4}i, -\frac{1}{4} - \frac{\sqrt{15}}{4}i$ **(f)** $-1, 1$ **(g)** 2 **(h)** $f(4) = -570$; $(4, -570)$ **(i)**

(j)

$f(x) = -2x^4 - x^3 + x + 2$

16. (a)

Positive	Negative	Nonreal Complex
0	4	0
0	2	2
0	0	4

(b) $0, \pm 1, \pm 3, \pm 9, \pm 27, \pm\frac{1}{2}, \pm\frac{3}{2}, \pm\frac{9}{2}, \pm\frac{27}{2}, \pm\frac{1}{4}, \pm\frac{3}{4}, \pm\frac{9}{4}, \pm\frac{27}{4}$ **(c)** $0, -\frac{3}{2}$ (multiplicity 2) **(d)** no other real zeros **(e)** $\frac{1}{2} + \frac{\sqrt{11}}{2}i, \frac{1}{2} - \frac{\sqrt{11}}{2}i$ **(f)** $0, -\frac{3}{2}$ **(g)** 0 **(h)** $f(4) = 7260$; $(4, 7260)$ **(i)** **(j)**

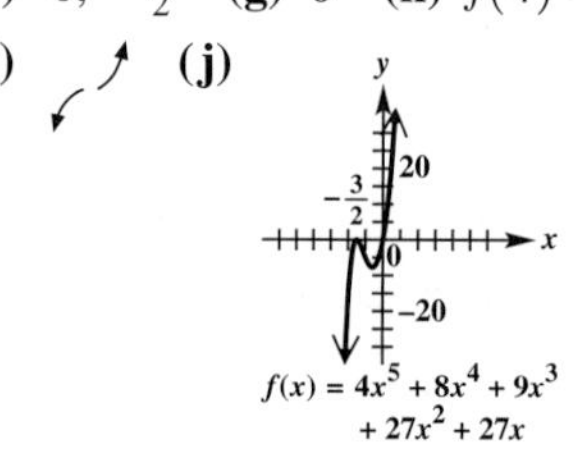

$f(x) = 4x^5 + 8x^4 + 9x^3 + 27x^2 + 27x$

17. (a)

Positive	Negative	Nonreal Complex
1	1	2

(b) $\pm 1, \pm 5, \pm\frac{1}{3}, \pm\frac{5}{3}$ **(c)** no rational zeros **(d)** $-\sqrt{5}, \sqrt{5}$ **(e)** $-\frac{\sqrt{3}}{3}i, \frac{\sqrt{3}}{3}i$ **(f)** $-\sqrt{5}, \sqrt{5}$ **(g)** -5 **(h)** $f(4) = 539$; $(4, 539)$ **(i)**

(j)

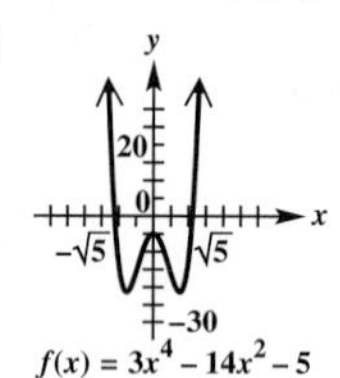

$f(x) = 3x^4 - 14x^2 - 5$

18. (a)

Positive	Negative	Nonreal Complex
2	3	0
2	1	2
0	3	2
0	1	4

(b) $\pm 1, \pm 3, \pm 9$ **(c)** $-3, -1$ (multiplicity 2), $1, 3$ **(d)** no other real zeros **(e)** no other complex zeros

18. (f) $-3, -1, 1, 3$ **(g)** -9 **(h)** $f(4) = -525$; $(4, -525)$ **(i)** ↖ ↘ **(j)**

$f(x) = -x^5 - x^4 + 10x^3 + 10x^2 - 9x - 9$

19. (a) Positive	**Negative**	**Nonreal Complex**
4	0	0
2	0	2
0	0	4

(b) $\pm 1, \pm 2, \pm 3, \pm 4, \pm 6, \pm 12, \pm\frac{1}{3}, \pm\frac{2}{3}, \pm\frac{4}{3}$
(c) $\frac{1}{3}$, 2 (multiplicity 2), 3 **(d)** no other real zeros
(e) no other complex zeros **(f)** $\frac{1}{3}, 2, 3$ **(g)** -12
(h) $f(4) = -44$; $(4, -44)$ **(i)** ↙ ↘

19. (j)

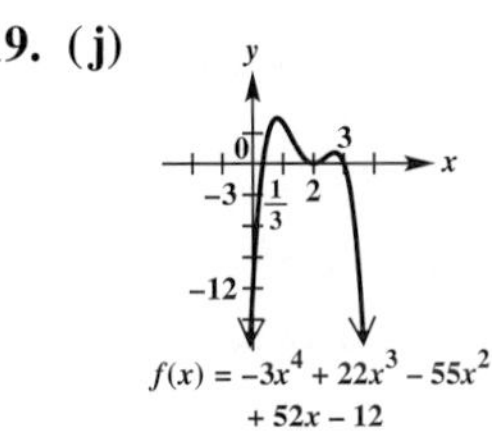

20. For the function in **Exercise 12:** ± 1.732; for the function in **Exercise 13:** ± 0.707; for the function in **Exercise 14:** $-2.303, 1.303$; for the function in **Exercise 17:** ± 2.236

3.5 Exercises *(pages 351–359)*

Calculator graphs are not included with the answers for **Exercises 109–113, 115, 117, and 118.**

109. $f(1.25) = -0.8\overline{1}$ **110.** $f(1.25) = -2.\overline{2}$
111. $f(1.25) = 2.708\overline{3}$ **112.** $f(1.25) = -1.75$
113. (a) 26 per min **(b)** 5 park attendants
114. (a) The waiting time is negative. **(b)** 12 **(c)** 15 min **(d)** 4.25 min per customer; An assistant could be hired to handle menial tasks, or a more efficient register system could be installed.
115. (a) approximately 52.1 mph
(b)

x	$d(x)$	x	$d(x)$
20	**34**	50	**273**
25	**56**	55	**340**
30	**85**	60	**415**
35	**121**	65	**499**
40	**164**	70	**591**
45	**215**		

116. (a) $D(0.05) \approx 238$; The braking distance for a car traveling at 50 mph on a wet 5% uphill grade is about 238 ft. **(c)** 7.9%

117. All answers are given in tens of millions.
(a) \$65.5 **(b)** \$64 **(c)** \$60 **(d)** \$40 **(e)** \$0
118. All answers are given in tens of millions.
(a) \$42.9 **(b)** \$40 **(c)** \$30 **(d)** \$0
119. $y = 1$ **120.** $(x + 4)(x + 1)(x - 3)(x - 5)$
121. (a) $(x - 1)(x - 2)(x + 2)(x - 5)$
(b) $f(x) = \frac{(x + 4)(x + 1)(x - 3)(x - 5)}{(x - 1)(x - 2)(x + 2)(x - 5)}$
122. (a) $x - 5$ **(b)** 5 **123.** $-4, -1, 3$ **124.** -3
125. $x = 1,\ x = 2,\ x = -2$
126. $\left(\frac{7 + \sqrt{241}}{6}, 1\right), \left(\frac{7 - \sqrt{241}}{6}, 1\right)$
127.

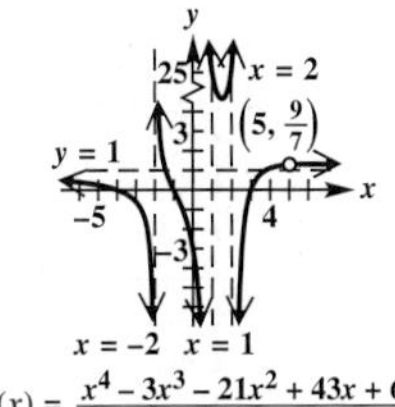

128. (a) $(-4, -2) \cup (-1, 1) \cup (2, 3)$
(b) $(-\infty, -4) \cup (-2, -1) \cup (1, 2) \cup (3, 5) \cup (5, \infty)$

Chapter 3 Quiz *(page 359)*

[3.4] **6.**

$f(x) = x(x - 2)^3(x + 2)^2$

7.

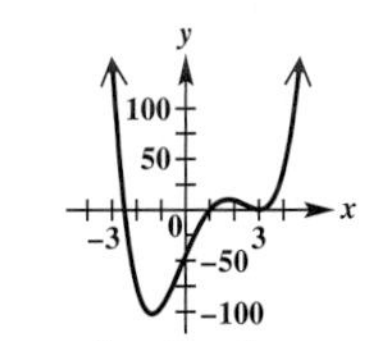

8.

$f(x) = -4x^5 + 16x^4 + 13x^3 - 76x^2 - 3x + 18$
$= -(x + 2)(2x + 1)(2x - 1)(x - 3)^2$

[3.5] **9.**

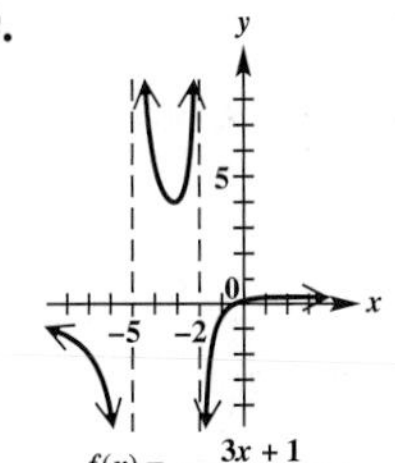

10.

$y = x + 3$

$f(x) = \frac{x^2 + 2x + 1}{x - 1}$

5.2 Exercises *(pages 496–499)*

130. $-\frac{\sqrt{3}}{2}; -\frac{1}{2}; \sqrt{3}; \frac{\sqrt{3}}{3}; -2; -\frac{2\sqrt{3}}{3}$ **131.** $\frac{8\sqrt{67}}{67}; \frac{\sqrt{201}}{67}; \frac{8\sqrt{3}}{3}; \frac{\sqrt{3}}{8}; \frac{\sqrt{201}}{3}; \frac{\sqrt{67}}{8}$ **132.** $\frac{1}{2}; -\frac{\sqrt{3}}{2}; -\frac{\sqrt{3}}{3}; -\sqrt{3}; -\frac{2\sqrt{3}}{3}; 2$ **133.** $\frac{\sqrt{2}}{6}; -\frac{\sqrt{34}}{6}; -\frac{\sqrt{17}}{17}; -\sqrt{17}; -\frac{3\sqrt{34}}{17}; 3\sqrt{2}$ **134.** $-\frac{\sqrt{59}}{8}; \frac{\sqrt{5}}{8}; -\frac{\sqrt{295}}{5}; -\frac{\sqrt{295}}{59}; \frac{8\sqrt{5}}{5}; -\frac{8\sqrt{59}}{59}$ **135.** $\frac{\sqrt{15}}{4}; -\frac{1}{4}; -\sqrt{15}; -\frac{\sqrt{15}}{15}; -4; \frac{4\sqrt{15}}{15}$

136. $-\frac{1}{3}$; $\frac{2\sqrt{2}}{3}$; $-\frac{\sqrt{2}}{4}$; $-2\sqrt{2}$; $\frac{3\sqrt{2}}{4}$; -3 **137.** 0.164215; -0.986425; -0.166475; -6.00691; -1.01376; 6.08958 **138.** -0.555762; 0.831342; -0.668512; -1.49586; 1.20287; -1.79933 **141.** This statement is false. For example, $\sin 180° + \cos 180° = 0 + (-1) = -1 \neq 1$. **142.** This statement is false. There is no value for θ for which $\sin \theta = 2$. **143.** negative **144.** negative **145.** positive **146.** positive **147.** positive **148.** positive **149.** positive **150.** positive

Summary Exercises on Graphing Circular Functions *(page 611)*

1.

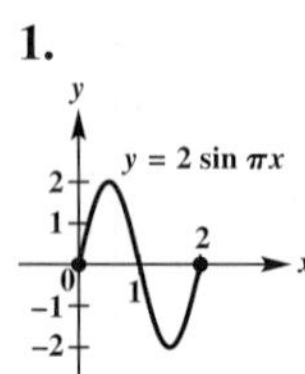

2.

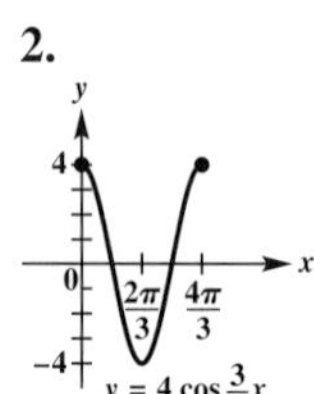

3.

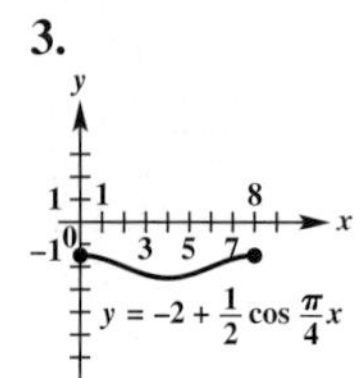

4.

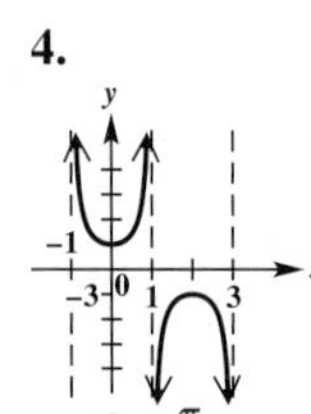

5.

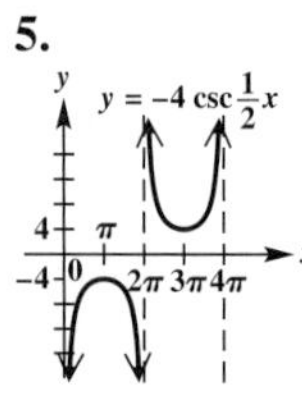

6.

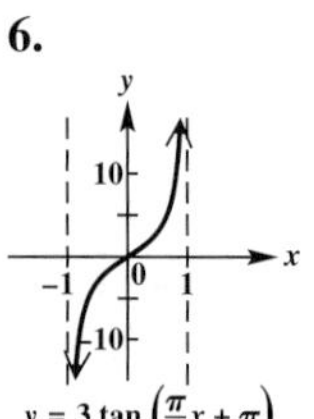

7.

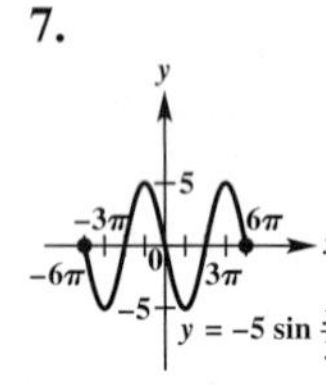

8.

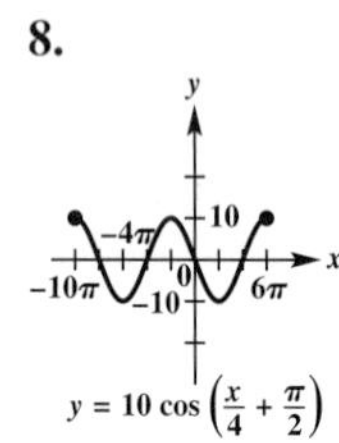

9.

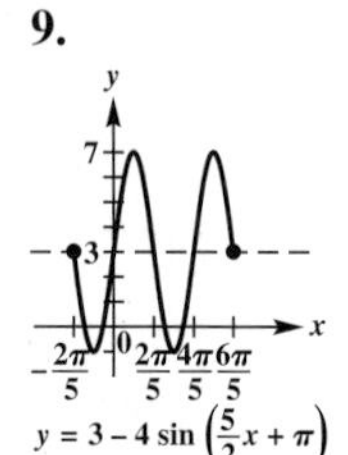

10.

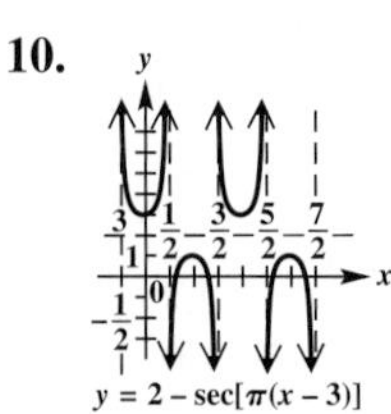

8.6 Exercises *(pages 782–784)*

31. $\{\cos 0° + i \sin 0°, \cos 120° + i \sin 120°, \cos 240° + i \sin 240°\}$ **32.** $\{\cos 60° + i \sin 60°, \cos 180° + i \sin 180°, \cos 300° + i \sin 300°\}$ **33.** $\{\cos 90° + i \sin 90°, \cos 210° + i \sin 210°, \cos 330° + i \sin 330°\}$ **34.** $\{\cos 67.5° + i \sin 67.5°, \cos 157.5° + i \sin 157.5°, \cos 247.5° + i \sin 247.5°, \cos 337.5° + i \sin 337.5°\}$ **35.** $\{2(\cos 0° + i \sin 0°), 2(\cos 120° + i \sin 120°), 2(\cos 240° + i \sin 240°)\}$ **36.** $\{3(\cos 60° + i \sin 60°), 3(\cos 180° + i \sin 180°), 3(\cos 300° + i \sin 300°)\}$ **37.** $\{\cos 45° + i \sin 45°, \cos 135° + i \sin 135°, \cos 225° + i \sin 225°, \cos 315° + i \sin 315°\}$ **38.** $\{2(\cos 45° + i \sin 45°), 2(\cos 135° + i \sin 135°), 2(\cos 225° + i \sin 225°), 2(\cos 315° + i \sin 315°)\}$ **39.** $\{\cos 22.5° + i \sin 22.5°, \cos 112.5° + i \sin 112.5°, \cos 202.5° + i \sin 202.5°, \cos 292.5° + i \sin 292.5°\}$ **40.** $\{\cos 18° + i \sin 18°, \cos 90° + i \sin 90°, \cos 162° + i \sin 162°, \cos 234° + i \sin 234°, \cos 306° + i \sin 306°\}$ **41.** $\{2(\cos 20° + i \sin 20°), 2(\cos 140° + i \sin 140°), 2(\cos 260° + i \sin 260°)\}$ **42.** $\{2(\cos 15° + i \sin 15°), 2(\cos 105° + i \sin 105°), 2(\cos 195° + i \sin 195°), 2(\cos 285° + i \sin 285°)\}$ **43.** $1, -\frac{1}{2} + \frac{\sqrt{3}}{2}i, -\frac{1}{2} - \frac{\sqrt{3}}{2}i$ **44.** $-3, \frac{3}{2} + \frac{3\sqrt{3}}{2}i, \frac{3}{2} - \frac{3\sqrt{3}}{2}i$ **45.** $\cos 2\theta + i \sin 2\theta$ **46.** $(\cos^2 \theta - \sin^2 \theta) + i(2 \cos \theta \sin \theta) = \cos 2\theta + i \sin 2\theta$ **47.** $\cos 2\theta = \cos^2 \theta - \sin^2 \theta$ **48.** $\sin 2\theta = 2 \sin \theta \cos \theta$ **49.** (a) yes (b) no (c) yes **50.** (a) blue (b) red (c) yellow **51.** $1, 0.30901699 + 0.95105652i, -0.809017 + 0.58778525i, -0.809017 - 0.5877853i, 0.30901699 - 0.9510565i$ **52.** $1, 0.80901699 + 0.58778525i, 0.30901699 + 0.95105652i$ **53.** $-4, 2 - 2i\sqrt{3}$ **54.** $\{0.3436 + 1.4553i, -0.3436 - 1.4553i\}$ **55.** $\{-1.8174 + 0.5503i, 1.8174 - 0.5503i\}$ **56.** $\{1.3606 + 1.2637i, -1.7747 + 0.5464i, 0.4141 - 1.8102i\}$ **57.** $\{0.8771 + 0.9492i, -0.6317 + 1.1275i, -1.2675 - 0.2524i, -0.1516 - 1.2835i, 1.1738 - 0.5408i\}$ **58.** 64; 2; 62 **59.** false **60.** false

8.7 Exercises *(pages 793–797)*

55.

$r = 2 \sin \theta \tan \theta$

56.

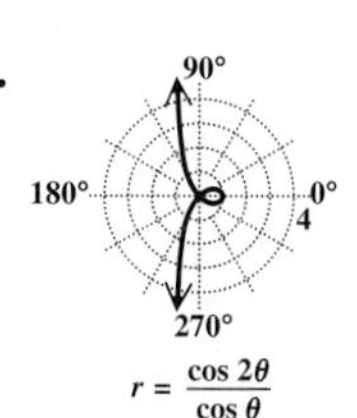

57. $x^2 + (y - 1)^2 = 1$

$r = 2 \sin \theta$
$x^2 + (y - 1)^2 = 1$

58. $(x - 1)^2 + y^2 = 1$

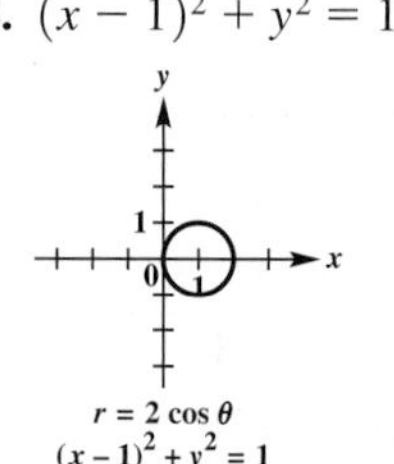

59. $y^2 = 4(x + 1)$

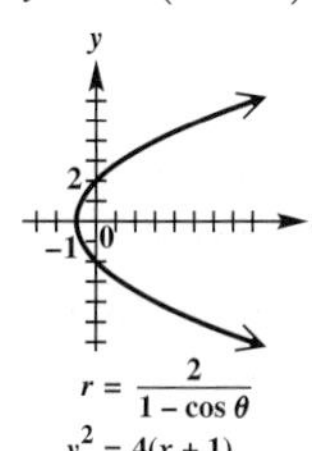

60. $x^2 = 6\left(y + \frac{3}{2}\right)$

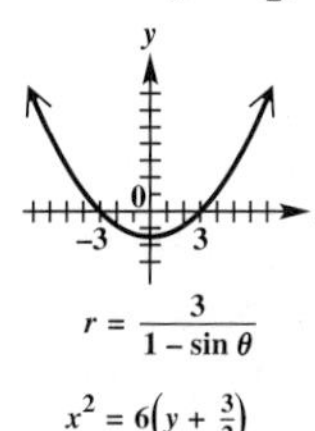

61. $(x + 1)^2 + (y + 1)^2 = 2$ **62.** $4x - y = 3$

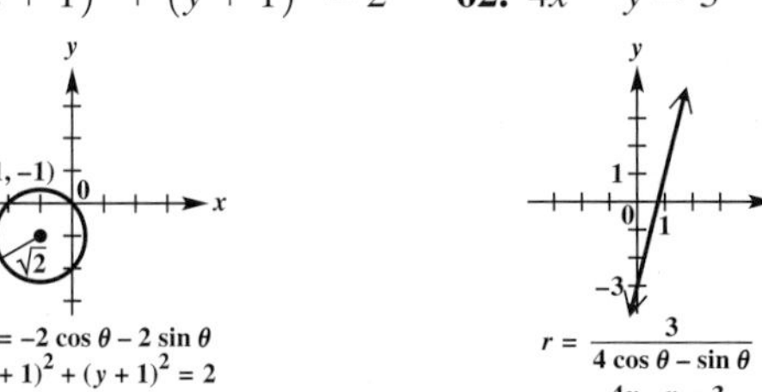

63. $x = 2$

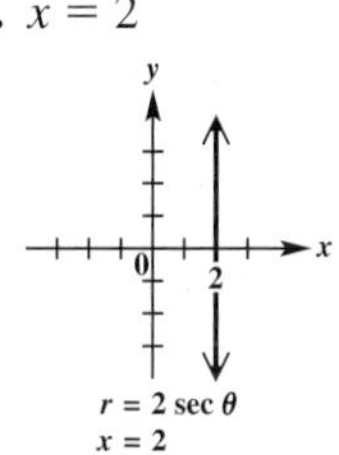

64. $y = -5$

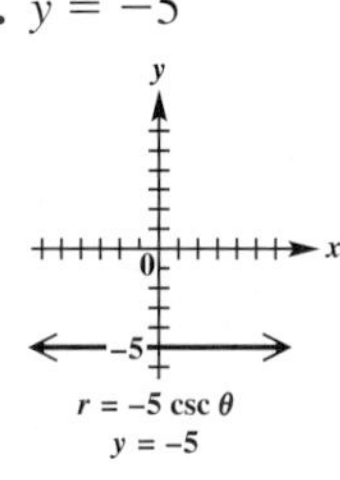

65. $x + y = 2$

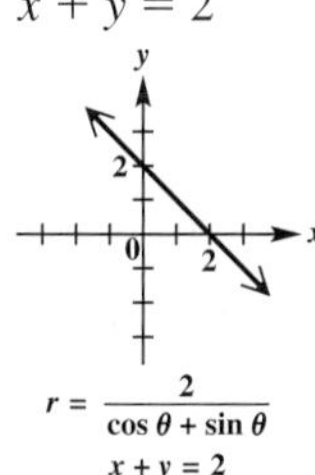

66. $2x + y = 2$

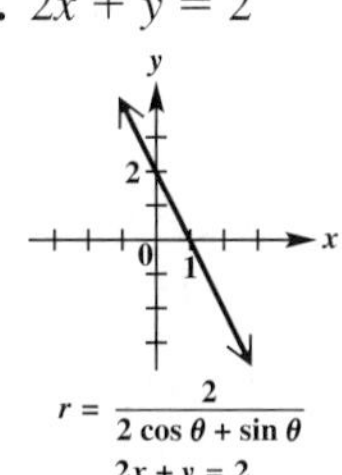

67.

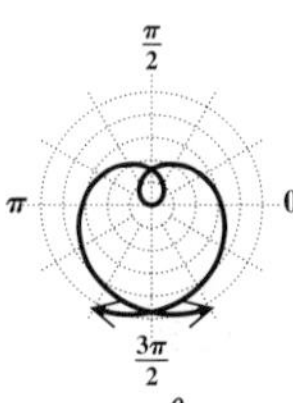

68.

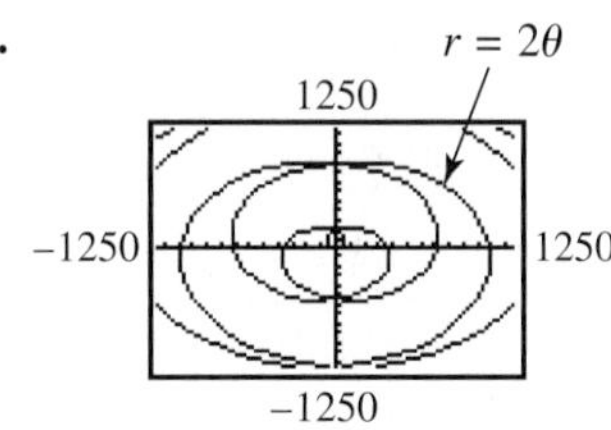

Degree mode

69. $r = \frac{2}{2\cos\theta + \sin\theta}$

71. (a) $(r, -\theta)$ **(b)** $(r, \pi - \theta)$ or $(-r, -\theta)$ **(c)** $(r, \pi + \theta)$ or $(-r, \theta)$ **72. (a)** $-\theta$ **(b)** $\pi - \theta$ **(c)** $-r$; $-\theta$ **(d)** $-r$ **(e)** $\pi + \theta$ **(f)** the polar axis **(g)** the line $\theta = \frac{\pi}{2}$

73.

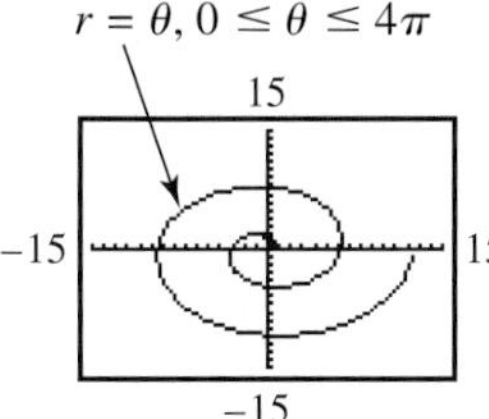

74.

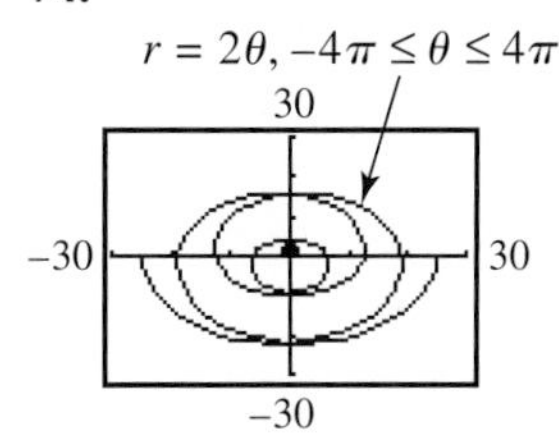

75.

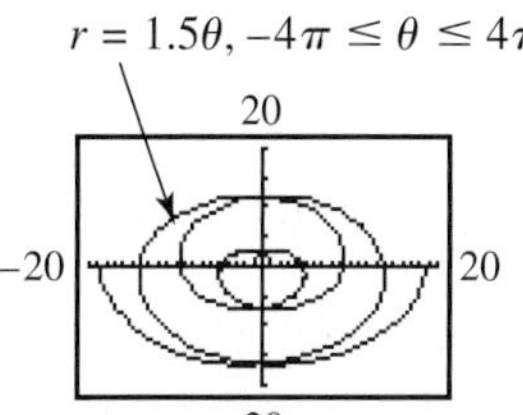

76.

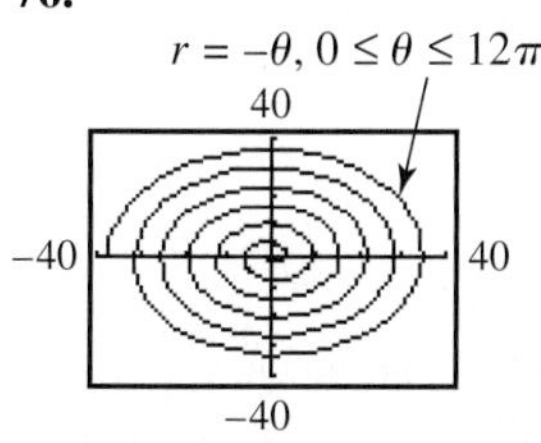

77. $\left(2, \frac{\pi}{6}\right), \left(2, \frac{5\pi}{6}\right), (0, 0)$ **78.** $(3, 60°), (3, 300°)$

79. $\left(\frac{4 + \sqrt{2}}{2}, \frac{\pi}{4}\right), \left(\frac{4 - \sqrt{2}}{2}, \frac{5\pi}{4}\right)$ **80.** $\left(1, \frac{\pi}{4}\right), \left(0, \frac{\pi}{2}\right), \left(-1, \frac{3\pi}{4}\right)$

81. (a)

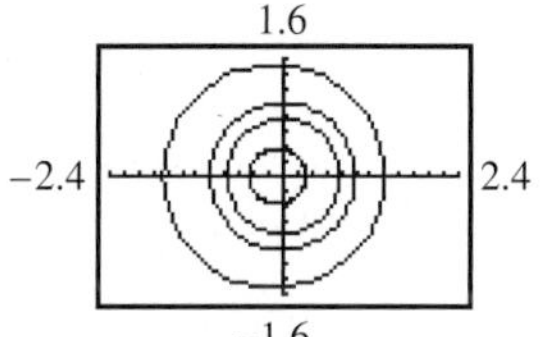

(b)

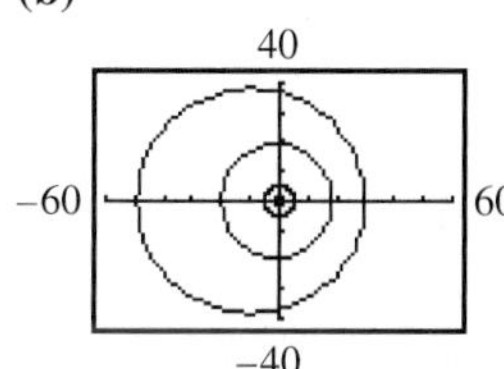

Earth is closest to the sun.

(c) no

82. (a) Inside the "figure eight" the radio signal can be received. This region is generally in an east-west direction from the two radio towers with maximum distance 200 mi.

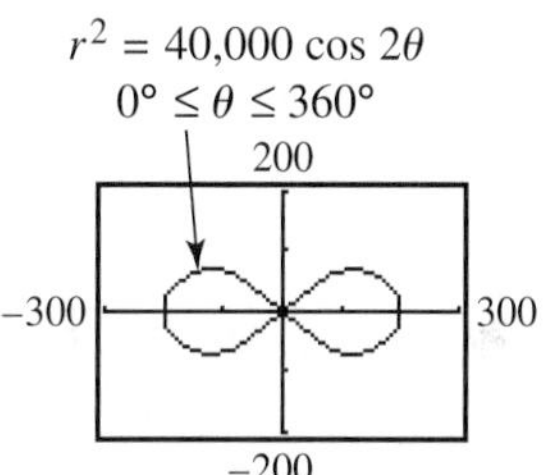

(b) The radio signal is received inside the "figure eight." This region is generally in a southwest-northeast direction from the two towers with maximum distance 150 mi.

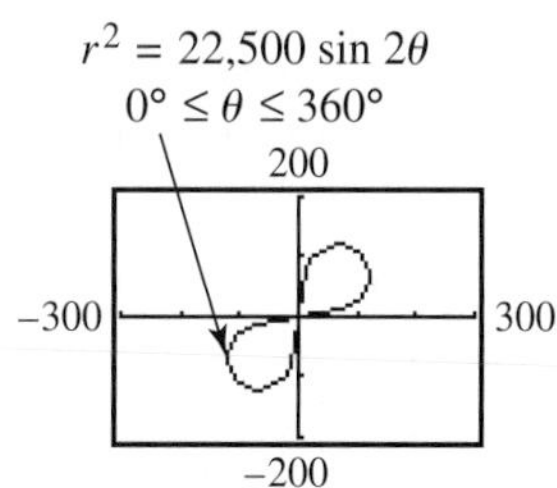

8.8 Exercises *(pages 803–806)*

38.

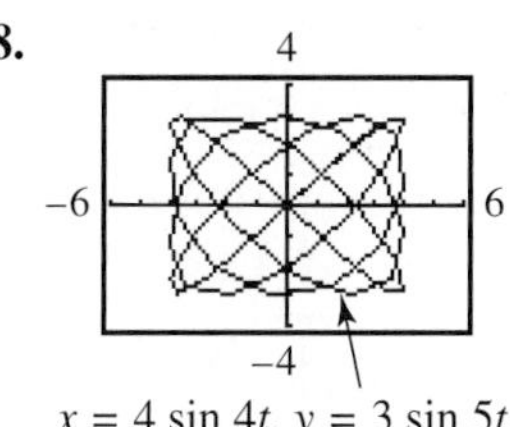

$x = 4\sin 4t$, $y = 3\sin 5t$, for t in $[0, 6.5]$

39. **(a)** $x = 24t,\ y = -16t^2 + 24\sqrt{3}t$
(b) $y = -\frac{1}{36}x^2 + \sqrt{3}x$ **(c)** 2.6 sec; 62 ft
40. **(a)** $x = 75t,\ y = -16t^2 + 75\sqrt{3}t$
(b) $y = -\frac{16}{5625}x^2 + \sqrt{3}x$ **(c)** 8.1 sec; 609 ft
41. **(a)** $x = (88 \cos 20°)t,\ y = 2 - 16t^2 + (88 \sin 20°)t$
(b) $y = 2 - \frac{x^2}{484 \cos^2 20°} + (\tan 20°)x$ **(c)** 1.9 sec; 161 ft
42. **(a)** $x = (136 \cos 29°)t,\ y = 2.5 - 16t^2 + (136 \sin 29°)t$
(b) $y = 2.5 - \frac{x^2}{1156 \cos^2 29°} + (\tan 29°)x$ **(c)** 4.2 sec; 495 ft
43. **(a)** $y = -\frac{1}{256}x^2 + \sqrt{3}x + 8$; parabolic path
(b) approximately 7 sec; approximately 448 ft
44. about 1456 ft
45. **(a)** $x = 32t,\ y = 32\sqrt{3}t - 16t^2 + 3$ **(b)** about 112.6 ft
(c) 51 ft maximum height; The ball had traveled horizontally about 55.4 ft. **(d)** yes
46. **(a)**
$x = 82.69295063t$
$y = -16t^2 + 30.09777261t$

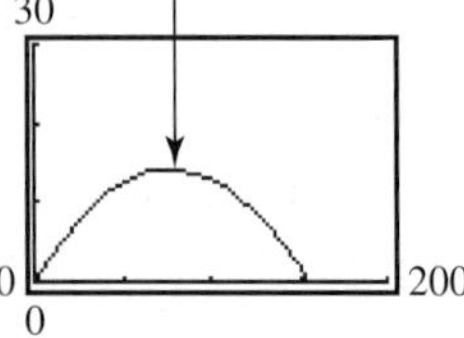

(b) 20.0°
(c) $x = (88 \cos 20.0°)t$,
$y = -16t^2 + (88 \sin 20.0°)t$

47. **(a)**
$x = 56.56530965t$
$y = -16t^2 + 67.41191099t$

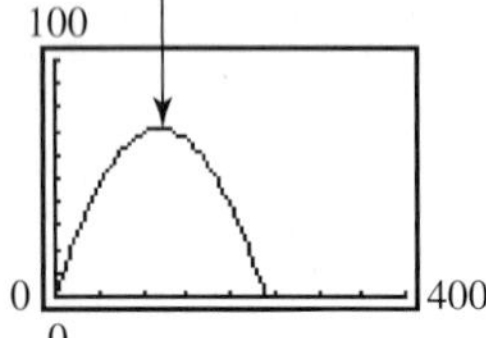

(b) 50.0°
(c) $x = (88 \cos 50.0°)t$,
$y = -16t^2 + (88 \sin 50.0°)t$

48. Many answers are possible, two of which are $x = t$, $y = m(t - x_1) + y_1$ and $x = t + x_1,\ y = mt + y_1$.
49. Many answers are possible; for example, $y = a(t - h)^2 + k,\ x = t$ and $y = at^2 + k,\ x = t + h$.
50. Many answers are possible; for example, $x = a \sec \theta$, $y = b \tan \theta$ and $x = t,\ y^2 = \frac{b^2}{a^2}(t^2 - a^2)$.
51. Many answers are possible; for example, $x = a \sin t$, $y = b \cos t$ and $x = t,\ y^2 = b^2\left(1 - \frac{t^2}{a^2}\right)$.
54. the pair $x = \cos t,\ y = -\sin t$ **55.** The graph is translated c units to the right. **56.** The graph is translated d units up.

10.2 Exercises *(pages 947–950)*

29. $[-5, 5]; [0, 2]$; function

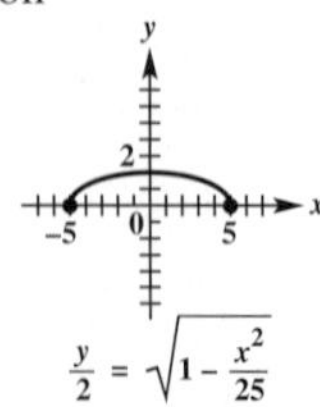

30. $[0, 4]; [-3, 3]$

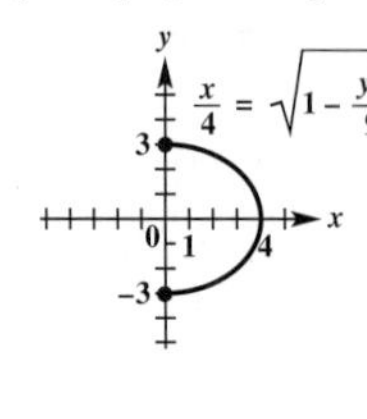

31. $[-1, 0]; [-8, 8]$

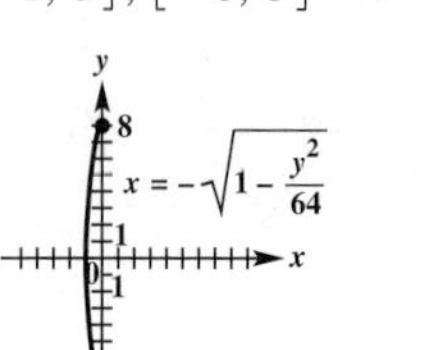

32. $[-10, 10]; [-1, 0]$; function

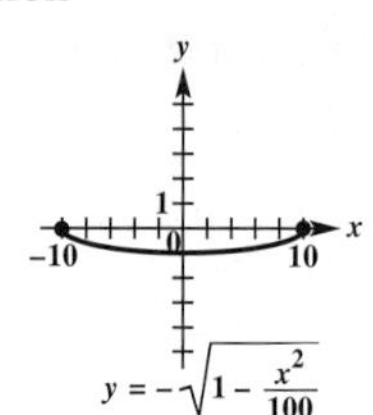

In the answers for Exercises 33–36, we give only the equations necessary to graph each ellipse.

33. $y = \pm 2\sqrt{1 - \frac{x^2}{16}}$ **34.** $y = \pm 5\sqrt{1 - \frac{x^2}{4}}$
35. $y = \pm 3\sqrt{1 - \frac{(x - 3)^2}{25}}$ **36.** $y = -4 \pm 2\sqrt{1 - \frac{x^2}{36}}$
37. $\frac{1}{2}$ **38.** 0.71 **39.** 0.65 **40.** 0.98 **43.** 12 ft tall
44. **(a)** 10 m **(b)** 36 m **45.** approximately 55 million mi
46. **(a)**

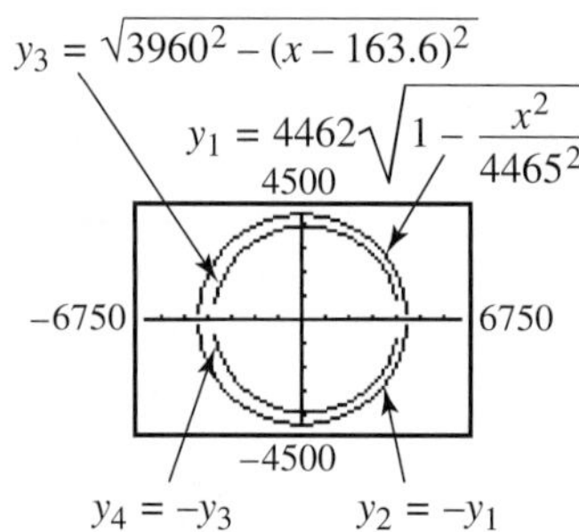

(b) minimum: approximately 341 mi; maximum: approximately 669 mi
47. **(a)** Neptune: $\frac{(x - 0.2709)^2}{30.1^2} + \frac{y^2}{30.1^2} = 1$;
Pluto: $\frac{(x - 9.8106)^2}{39.4^2} + \frac{y^2}{38.16^2} = 1$
(b)

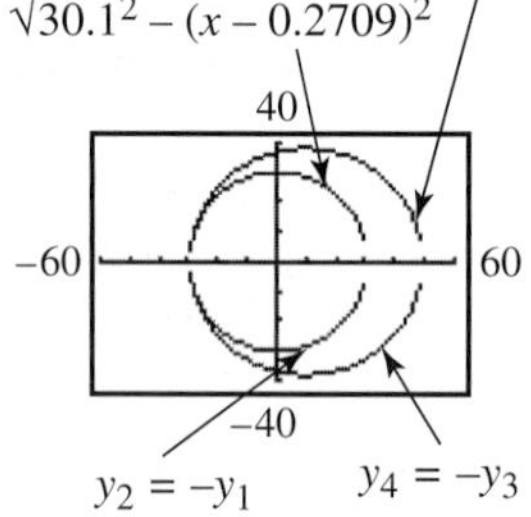

48. **(a)** 348.2 ft **(b)** 1787.6 ft **49.** $3\sqrt{3}$ units
50. $\sqrt{5}$ units

Chapter 10 Quiz *(page 950)*

[10.1, 10.2]

7. ellipse; center: $(0, 0)$; vertices: $(-3, 0), (3, 0)$; foci: $\left(-\sqrt{5}, 0\right), \left(\sqrt{5}, 0\right)$

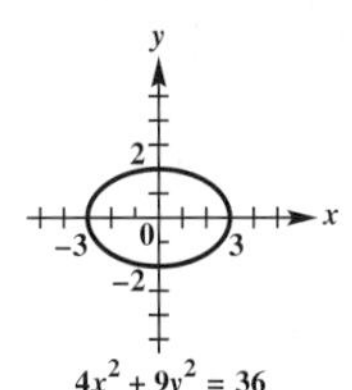

8. parabola; vertex: $(-1, -3)$; focus: $(1, -3)$; directrix: $x = -3$; axis: $y = -3$

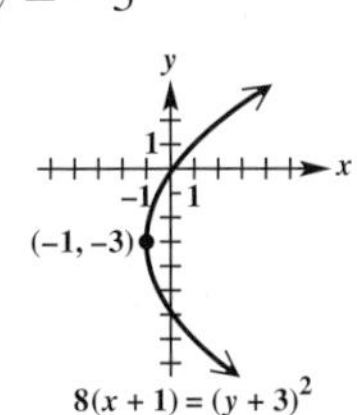

9. ellipse; center: $(-3, -2)$; vertices: $(-3, 4), (-3, -8)$; foci: $\left(-3, -2 + \sqrt{11}\right)$, $\left(-3, -2 - \sqrt{11}\right)$

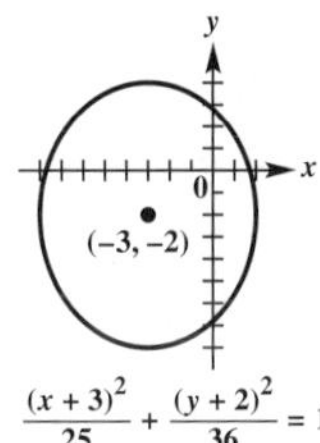

10. parabola; vertex: $\left(-2, -\frac{1}{2}\right)$; focus: $\left(-\frac{33}{16}, -\frac{1}{2}\right)$; directrix: $x = -\frac{31}{16}$; axis: $y = -\frac{1}{2}$

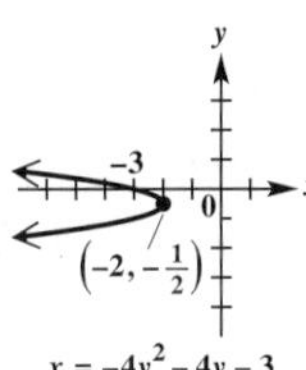

10.3 Exercises *(pages 956–960)*

23. $(-\infty, \infty)$; $[3, \infty)$; function

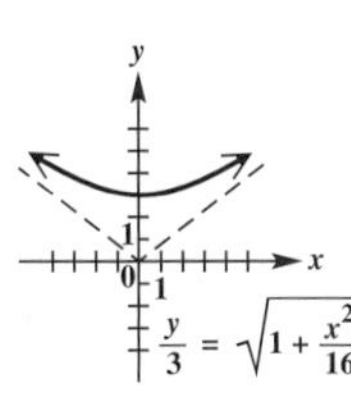

24. $(-\infty, -3]$; $(-\infty, \infty)$

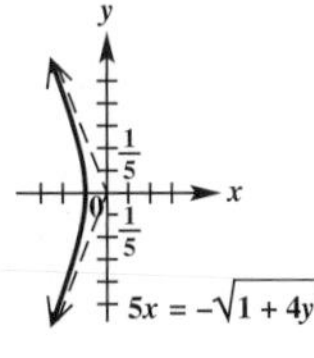

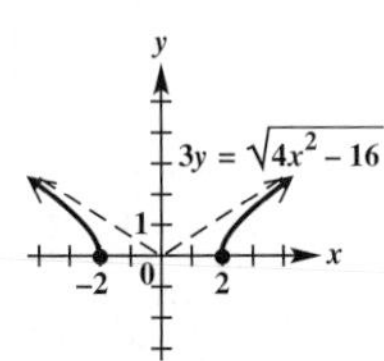

25. $\left(-\infty, -\frac{1}{5}\right]$; $(-\infty, \infty)$

26. $(-\infty, -2] \cup [2, \infty)$; $[0, \infty)$; function

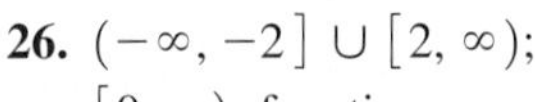

27. 1.4 **28.** 3.2 **29.** 1.7 **30.** 2.2 **31.** $\frac{x^2}{9} - \frac{y^2}{16} = 1$

32. $\frac{y^2}{144} - \frac{x^2}{81} = 1$ **33.** $\frac{y^2}{36} - \frac{x^2}{144} = 1$ **34.** $\frac{x^2}{100} - \frac{y^2}{2500} = 1$

35. $\frac{x^2}{9} - 3y^2 = 1$ **36.** $\frac{y^2}{25} - \frac{x^2}{3} = 1$ **37.** $\frac{2y^2}{25} - 2x^2 = 1$

38. $\frac{x^2}{9} - \frac{y^2}{36} = 1$ **39.** $\frac{(y-3)^2}{4} - \frac{49(x-4)^2}{4} = 1$

40. $\frac{(x-3)^2}{4} - \frac{(y+2)^2}{9} = 1$ **41.** $\frac{(x-1)^2}{4} - \frac{(y+2)^2}{5} = 1$

42. $\frac{(y+7)^2}{36} - \frac{(x-9)^2}{64} = 1$ **43.** $\frac{y^2}{49} - \frac{x^2}{392} = 1$

44. $\frac{(x-8)^2}{9} - \frac{(y-7)^2}{16} = 1$ **45.** $\frac{25(x+1)^2}{324} - \frac{25(y+1)^2}{2176} = 1$

46. $\frac{(y-6)^2}{16} - \frac{(x-2)^2}{9} = 1$

In the answers for Exercises 47–50, we give only the equations necessary to graph each hyperbola.

47. $y = \pm 2\sqrt{x^2 - 4}$ **48.** $y = \pm\frac{7}{5}\sqrt{x^2 - 25}$

49. $y = \pm 3\sqrt{x^2 + 4}$ **50.** $y = \pm 3\sqrt{x^2 + 1}$

51. $y = \frac{1}{2}\sqrt{x^2 - 4}$ **52.** $y = \frac{1}{2}x$ **53.** $y \approx 24.98$

54. $y = 25$ **55.** Because $24.98 < 25$, the graph of $y = \frac{1}{2}\sqrt{x^2 - 4}$ lies below the graph of $y = \frac{1}{2}x$ when $x = 50$.

56. The y-values on the hyperbola will approach the y-values on the asymptote. **57. (a)** $x = \sqrt{y^2 + 2.5 \times 10^{-27}}$ **(b)** approximately 1.2×10^{-13} m **58.** $\frac{x^2}{625} - \frac{y^2}{1875} = 1$

60. (b) approximately 16.88 ft **(c)** approximately 2.4 ft

61. (a) 50 m **(b)** 69.3 m

Chapter 10 Review Exercises *(pages 969–971)*

33. ellipse; $[-3, 3]$; $[-2, 2]$; $(-3, 0), (3, 0)$

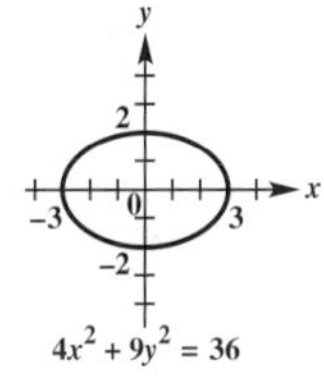

34. hyperbola; $(-\infty, -4] \cup [4, \infty)$; $(-\infty, \infty)$; $(-4, 0), (4, 0)$; $y = \pm x$

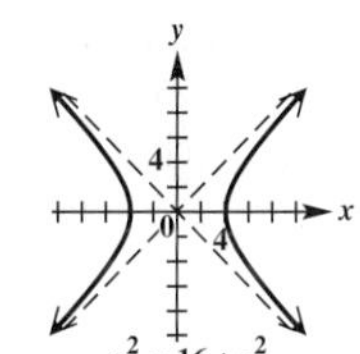

35. ellipse; $[1, 5]$; $[-2, 0]$; $(1, -1), (5, -1)$

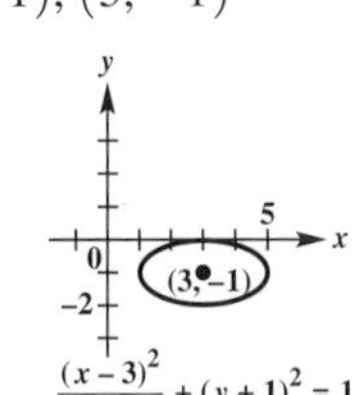

36. ellipse; $[-1, 5]$; $[-5, -1]$; $(-1, -3), (5, -3)$

37. hyperbola; $(-\infty, \infty)$; $(-\infty, -4] \cup [0, \infty)$; $(-3, -4), (-3, 0)$; $y = \pm\frac{2}{3}(x + 3) - 2$

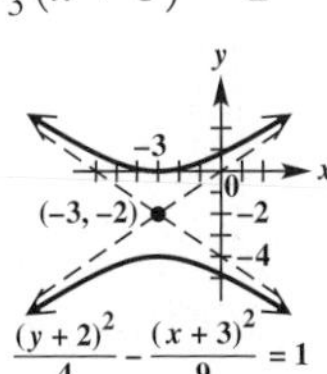

38. hyperbola; $(-\infty, -5] \cup [3, \infty)$; $(-\infty, \infty)$; $(-5, 2), (3, 2)$; $y = \pm\frac{1}{2}(x + 1) + 2$

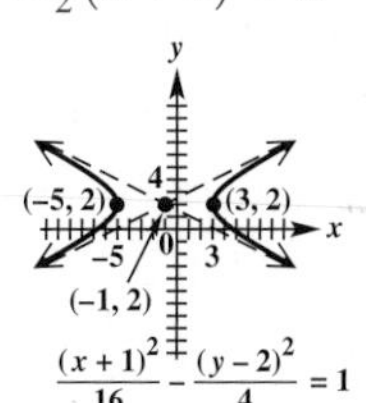

39. circle; $[1, 3]$; $[-4, -2]$

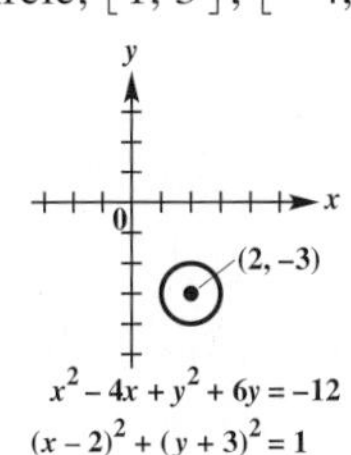

40. ellipse; $[-6, 4]$; $[3, 7]$; $(-6, 5), (4, 5)$

(−1, 5)
$4x^2 + 8x + 25y^2 - 250y = -529$
$\frac{(x+1)^2}{25} + \frac{(y-5)^2}{4} = 1$

41. ellipse; $\left[-2-\sqrt{2}, -2+\sqrt{2}\right]$; $\left[2-\sqrt{5}, 2+\sqrt{5}\right]$; $\left(-2, 2-\sqrt{5}\right)$, $\left(-2, 2+\sqrt{5}\right)$

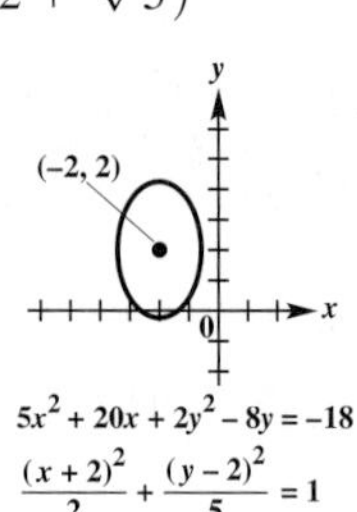

42. hyperbola; $(-\infty, \infty)$; $(-\infty, -3] \cup [1, \infty)$; $(1, -3)$, $(1, 1)$; $y = \pm(x-1)-1$

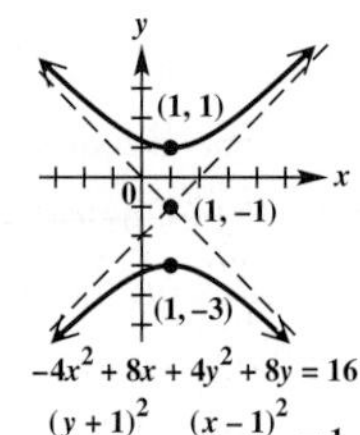

43. $[-3, 0]$; $[-4, 4]$

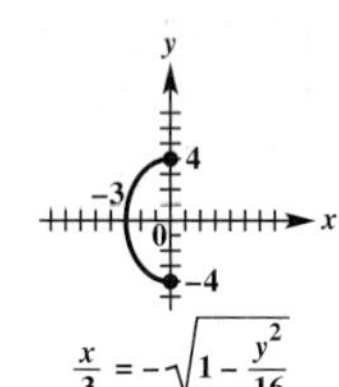

44. $[-1, 0]$; $[-6, 6]$

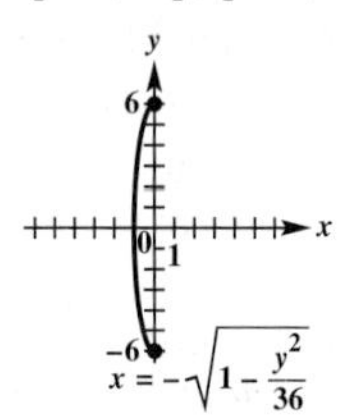

45. $(-\infty, \infty)$; $(-\infty, -1]$; function

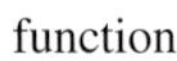

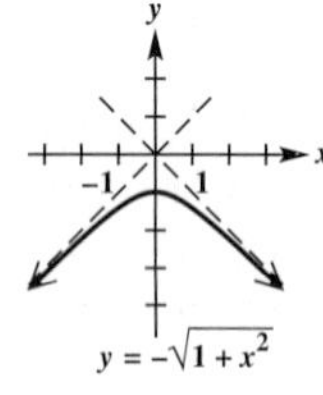

46. $[-5, 5]$; $[-1, 0]$; function

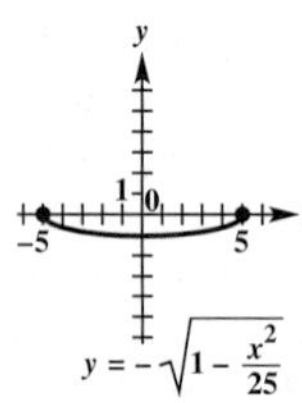

47. $\frac{x^2}{12}+\frac{y^2}{16}=1$ **48.** $\frac{x^2}{36}+\frac{y^2}{32}=1$ **49.** $\frac{y^2}{16}-\frac{x^2}{9}=1$

50. $\frac{y^2}{4}-\frac{5x^2}{16}=1$ **51.** $(y-2)^2=12x$ **52.** $(y-2)^2=3(x+3)$ **53.** $\frac{x^2}{25}+\frac{y^2}{21}=1$ **54.** $\frac{x^2}{40}+\frac{y^2}{49}=1$ **55.** $\frac{x^2}{9}-\frac{y^2}{16}=1$ **56.** $\frac{y^2}{72}-\frac{x^2}{72}=1$ **57.** $\frac{(x-2)^2}{16}+\frac{y^2}{12}=1$ **58.** $(y-2)^2-\frac{x^2}{3}=1$ **59.** C, A, B, D **60.** 66.8 and 67.7 million mi **61.** $\frac{x^2}{6{,}111{,}883}+\frac{y^2}{432{,}135}=1$ **62.** $\frac{x^2}{16}-\frac{y^2}{9}=1$

11.5 Exercises *(pages 1016–1017)*

Although we do not usually give proofs, the answers for Exercises 3 and 11 are given here.

3. (a) $3(1)=3$ and $\frac{3(1)(1+1)}{2}=\frac{6}{2}=3$, so S is true for $n=1$. **(b)** $3+6+9+\cdots+3k=\frac{3(k)(k+1)}{2}$ **(c)** $3+6+9+\cdots+3(k+1)=\frac{3(k+1)[(k+1)+1]}{2}$ **(d)** Add $3(k+1)$ to each side of the equation in part (b). Simplify the expression on the right side to match the right side of the equation in part (c). **(e)** Since S is true for $n=1$ and S is true for $n=k+1$ when it is true for $n=k$, S is true for every positive integer n.

11. (a) $\frac{1}{1\cdot 2}=\frac{1}{2}$ and $\frac{1}{1+1}=\frac{1}{2}$, so S is true for $n=1$. **(b)** $\frac{1}{1\cdot 2}+\frac{1}{2\cdot 3}+\frac{1}{3\cdot 4}+\cdots+\frac{1}{k(k+1)}=\frac{k}{k+1}$ **(c)** $\frac{1}{1\cdot 2}+\frac{1}{2\cdot 3}+\cdots+\frac{1}{(k+1)[(k+1)+1]}=\frac{k+1}{(k+1)+1}$ **(d)** Add the last term on the left of the equation in part (c) to each side of the equation in part (b). Simplify the right side until it matches the right side in part (c). **(e)** Since S is true for $n=1$ and S is true for $n=k+1$ when it is true for $n=k$, S is true for every positive integer n.

Photo Credits

page 1: Diseñador/Fotolia **page 15:** Brian Maudsley/Shutterstock **page 19:** Don Emmert/AFP/Getty Images/Newscom **page 20:** Photo from the collection of John Hornsby **page 31:** Gregory Johnston/Shutterstock **page 49:** Bestimagesevercom/Shutterstock **page 57:** Jami Garrison/Shutterstock **page 69:** Danshutter/Shutterstock **page 74:** Corepics/Shutterstock **page 75:** David Anderson/Shutterstock **page 79:** Gunter Marx/Dorling Kindersley, Ltd. **page 86:** Barry Blackburn/Shutterstock **page 88:** AP Images **page 91:** Michaeljung/Shutterstock **page 93:** (cake) Tarzan Wong/EyePress/Newscom; (bin) Scanrail/Fotolia **page 95:** Digital Vision/Thinkstock **page 96:** Armadillo Stock/Shutterstock **page 115:** Photodisc/Digital Vision/Thinkstock **page 116:** Photos/Getty Images/Thinkstock **page 117:** Medioimages/Photodisc/Thinkstock **page 118:** (parking lot) David Phillips/Shutterstock; (picture) Thomas Bethge/ Shutterstock **page 119:** (carpet) Marco Mayer/Shutterstock; (floor) Kutlayev Dmitry/Shutterstock **page 123:** Thomas Barrat/Shutterstock **page 134:** Columbia Pictures/Everett Collection **page 148:** (Avatar) Trademark™ and copyright © 20th Century Fox. All Rights Reserved/Everett Collection; (refuse) Picsfive/Shutterstock **page 155:** Kapu/Shutterstock **page 161:** Discpicture/Shutterstock **page 162:** (lawn mower) Suzanne Tucker/Shutterstock; (U-Haul truck) Beth Anderson/Pearson Education, Inc. **page 167:** Sergei Butorin/Shutterstock **page 169:** Peter Radacsi/Shutterstock **page 171:** Photos/Getty Images/Thinkstock **page 174:** Mipstudio/Shutterstock **page 184:** Evlakhov Valeriy/Shutterstock **page 187:** Photo from the collection of John Hornsby **page 190:** Courtesy of John Hornsby **page 224:** Karen Roach/Shutterstock **page 228:** Wolfgang Kloehr/Shutterstock **page 237:** Samuel Acosta/Shutterstock **page 263:** Imagemore/Getty Images **page 271:** Dudarev Mikhail/Shutterstock **page 283:** Cloudrain/Fotolia **page 291:** Avava/Shutterstock **page 299:** Yuri Arcurs/Shutterstock **page 309:** Bilwissedition Ltd. & Co. KG/Alamy **page 328:** EML/Shutterstock **page 351:** Michael Fritzen/Fotolia **page 366:** Ocean Image Photography/Shutterstock **page 367:** Poznukhov Yuriy/Shutterstock **page 369:** Cristi Lucaci/Shutterstock **page 379:** Stocker1970/Shutterstock **page 383:** Lee Prince/ Shutterstock **page 412:** Guy J. Sagi/Shutterstock **page 421:** Dave King/The Science Museum, London, Dorling Kindersley, Ltd. **page 429:** Etienjones/Shutterstock **page 433:** Kletr/Shutterstock **page 436:** Jiri Flogel/Shutterstock **page 448:** Lukich/Shutterstock **page 453:** Dariush M/Shutterstock **page 455:** Gosphotodesign/Shutterstock **page 457:** Konstantin Mironov/Shutterstock **page 459:** Steven Frame/Shutterstock **page 460:** Crok Photography/Shutterstock **page 462:** Courtesy of B. K. Tomlin **page 470:** Dgareri/Dreamstime **page 475:** Carroteater/Shutterstock **page 512:** VladKol/Shutterstock **page 514:** Manfredxy/Shutterstock **page 525:** Holbox/Shutterstock **page 541:** Christopher Meder Photography/Shutterstock **page 548:** Comstock/Thinkstock **page 562:** Pixland/Thinkstock **page 567:** J. Bell, Cornell University/M. Wolff, Space Science Center/NASA **page 581:** TextArt/Fotolia **page 588:** Natalia Bratslavsky/Shutterstock **page 593:** Lisa F. Young/Shutterstock **page 614:** Alexander Sakhatovsky/Shutterstock **page 621:** Trinacria Photo/Shutterstock **page 625:** Photos/Thinkstock **page 629:** Photos/Thinkstock **page 641:** Paul Matthew Photography/Shutterstock **page 658:** (whispering) Lifesize/Thinkstock; (bending) Iconogenic/iStockphoto **page 670:** Digital Vision/Thinkstock **page 671:** Comstock/Thinkstock **page 683:** Engine Images/Fotolia **page 694:** Ingrid Balabanova/Shutterstock **page 698:** Alenavlad/Shutterstock **page 716:** Pete Saloutos/Shutterstock **page 717:** Kadmy/Fotolia **page 733:** Shutterstock **page 763:** (pyramids) Sculpies/Shutterstock; (motorboat) Mypokcik/Shutterstock **page 770:** Ross Anderson **page 778:** North Wind Picture Archives/Alamy **page 783:** (both fractals) Ross Anderson **page 797:** NatUlrich/Shutterstock **page 817:** Xymm/Fotolia **page 821:** Haak78/Shutterstock **page 826:** Ampyang/Shutterstock **page 836:** Courtesy of Callie Daniels **page 838:** James Thew/Shutterstock **page 846:** Christina Richards/Shutterstock **page 847:** Stockbyte/Thinkstock **page 848:** Yuri Arcurs/Shutterstock **page 849:** Jeff Banke/Shutterstock **page 850:** U.S. Dept. of Energy's Ames Laboratory **page 868:** Tyler Olson/Shutterstock **page 889:** Mangostock/Shutterstock **page 890:** Snyderdf/Dreamstime **page 898:** Chad McDermott/Shutterstock **page 902:** Pix2go/Shutterstock **page 903:** Tom Reichner/Shutterstock **page 922:** Andrey Yurlov/Shutterstock **page 924:** U.S. Environmental Protection Agency **page 929:** Biczó Zsolt/Shutterstock **page 937:** Ilya Genkin/Shutterstock **page 946:** Kcphotos/Dreamstime **page 949:** SF Photo/Shutterstock **page 973:** John Foxx/Stockbyte/Thinkstock **page 976:** (Fibonacci) Science Source/Photo Researchers, Inc.; (moth) Mikhail Melnikov/Shutterstock **page 983:** Michael Avory/Shutterstock **page 993:** Michael Drager/Shutterstock **page 995:** Liew Weng Keong/Shutterstock **page 1001:** Tischenko Irina/Shutterstock **page 1005:** Georgios Kollidas/Fotolia **page 1021:** Courtesy of John Hornsby **page 1023:** Courtesy of Callie Daniels **page 1025:** Courtesy of Margaret Westmoreland **page 1030:** Xiver/Shutterstock **page 1045:** Creatas/Thinkstock

Index

C

F

O

P

T

Identities and Formulas

6.1 Conversion of Angular Measure

Degree/Radian Relationship: $180° = \pi$ radians

Conversion Formulas:

From	To	Multiply by
Degrees	Radians	$\frac{\pi}{180}$
Radians	Degrees	$\frac{180°}{\pi}$

6.1 Applications of Radian Measure

Arc Length: $s = r\theta$, θ in radians

Area of Sector: $\mathcal{A} = \frac{1}{2}r^2\theta$, θ in radians

6.2

Angular Speed	Linear Speed
$\omega = \frac{\theta}{t}$	$v = \frac{s}{t}$
(ω in radians per unit time, θ in radians)	$v = \frac{r\theta}{t}$
	$v = r\omega$

7.1 Fundamental Identities

$\cot\theta = \frac{1}{\tan\theta}$ $\quad\sec\theta = \frac{1}{\cos\theta}$ $\quad\csc\theta = \frac{1}{\sin\theta}$

$\tan\theta = \frac{\sin\theta}{\cos\theta}$ $\quad\cot\theta = \frac{\cos\theta}{\sin\theta}$

$\sin^2\theta + \cos^2\theta = 1$ $\quad\tan^2\theta + 1 = \sec^2\theta$ $\quad 1 + \cot^2\theta = \csc^2\theta$

$\sin(-\theta) = -\sin\theta$ $\quad\cos(-\theta) = \cos\theta$ $\quad\tan(-\theta) = -\tan\theta$

$\csc(-\theta) = -\csc\theta$ $\quad\sec(-\theta) = \sec\theta$ $\quad\cot(-\theta) = -\cot\theta$

7.3 Sum and Difference Identities

$\cos(A + B) = \cos A\cos B - \sin A\sin B$

$\cos(A - B) = \cos A\cos B + \sin A\sin B$

$\sin(A + B) = \sin A\cos B + \cos A\sin B$

$\sin(A - B) = \sin A\cos B - \cos A\sin B$

$$\tan(A + B) = \frac{\tan A + \tan B}{1 - \tan A\tan B}$$

$$\tan(A - B) = \frac{\tan A - \tan B}{1 + \tan A\tan B}$$

7.4 Product-to-Sum and Sum-to-Product Identities

$$\cos A\cos B = \frac{1}{2}[\cos(A + B) + \cos(A - B)]$$

$$\sin A\sin B = \frac{1}{2}[\cos(A - B) - \cos(A + B)]$$

$$\sin A\cos B = \frac{1}{2}[\sin(A + B) + \sin(A - B)]$$

$$\cos A\sin B = \frac{1}{2}[\sin(A + B) - \sin(A - B)]$$

$$\sin A + \sin B = 2\sin\left(\frac{A + B}{2}\right)\cos\left(\frac{A - B}{2}\right)$$

$$\sin A - \sin B = 2\cos\left(\frac{A + B}{2}\right)\sin\left(\frac{A - B}{2}\right)$$

$$\cos A + \cos B = 2\cos\left(\frac{A + B}{2}\right)\cos\left(\frac{A - B}{2}\right)$$

$$\cos A - \cos B = -2\sin\left(\frac{A + B}{2}\right)\sin\left(\frac{A - B}{2}\right)$$

7.3 Cofunction Identities

$\cos(90° - \theta) = \sin\theta$

$\sin(90° - \theta) = \cos\theta$

$\tan(90° - \theta) = \cot\theta$

$\cot(90° - \theta) = \tan\theta$

$\sec(90° - \theta) = \csc\theta$

$\csc(90° - \theta) = \sec\theta$

7.4 Double-Angle and Half-Angle Identities

$\cos 2A = \cos^2 A - \sin^2 A$ $\quad\cos 2A = 1 - 2\sin^2 A$

$\cos 2A = 2\cos^2 A - 1$ $\quad\sin 2A = 2\sin A\cos A$

$$\tan 2A = \frac{2\tan A}{1 - \tan^2 A} \qquad \cos\frac{A}{2} = \pm\sqrt{\frac{1 + \cos A}{2}}$$

$$\sin\frac{A}{2} = \pm\sqrt{\frac{1 - \cos A}{2}} \qquad \tan\frac{A}{2} = \pm\sqrt{\frac{1 - \cos A}{1 + \cos A}}$$

$$\tan\frac{A}{2} = \frac{\sin A}{1 + \cos A} \qquad \tan\frac{A}{2} = \frac{1 - \cos A}{\sin A}$$

8.1 Law of Sines

In any triangle ABC, with sides a, b, and c,

$$\frac{a}{\sin A} = \frac{b}{\sin B}, \quad \frac{a}{\sin A} = \frac{c}{\sin C}, \quad \text{and} \quad \frac{b}{\sin B} = \frac{c}{\sin C}.$$

Area of a Triangle

The area $\mathcal{A}$ of a triangle is given by half the product of the lengths of two sides and the sine of the angle between the two sides.

$$\mathcal{A} = \frac{1}{2}bc\sin A, \quad \mathcal{A} = \frac{1}{2}ab\sin C, \quad \mathcal{A} = \frac{1}{2}ac\sin B$$

8.2 Law of Cosines

In any triangle ABC, with sides a, b, and c,

$$a^2 = b^2 + c^2 - 2bc\cos A, \quad b^2 = a^2 + c^2 - 2ac\cos B,$$

and

$$c^2 = a^2 + b^2 - 2ab\cos C.$$

Heron's Area Formula

If a triangle has sides of lengths a, b, and c, with semiperimeter $s = \frac{1}{2}(a + b + c)$, then the area $\mathcal{A}$ of the triangle is

$$\mathcal{A} = \sqrt{s(s - a)(s - b)(s - c)}.$$

Graphs of Functions

2.6 Identity Function

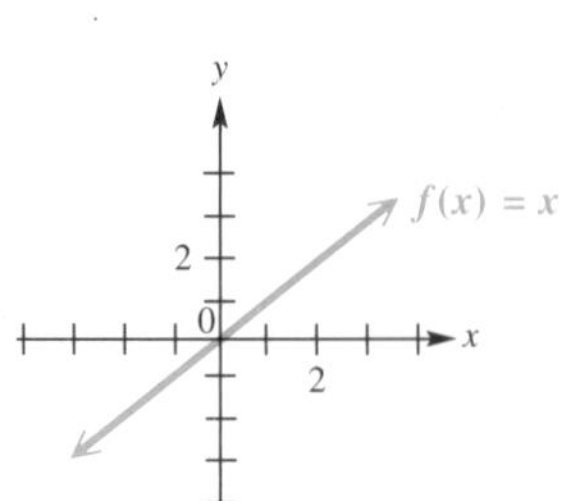

2.6 Squaring Function

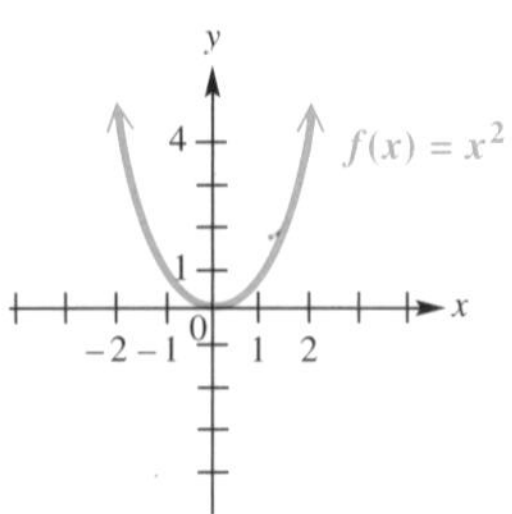

2.6 Cubing Function

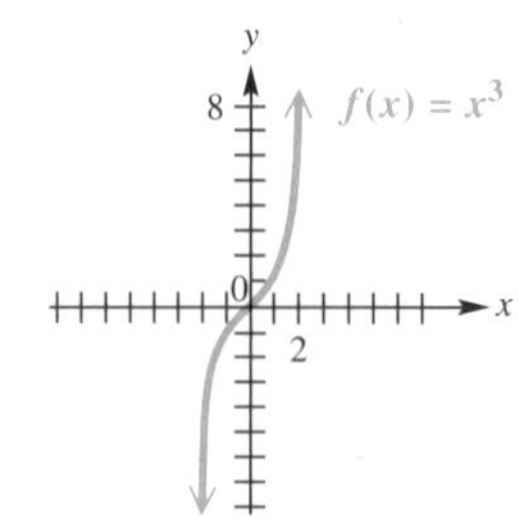

2.6 Square Root Function

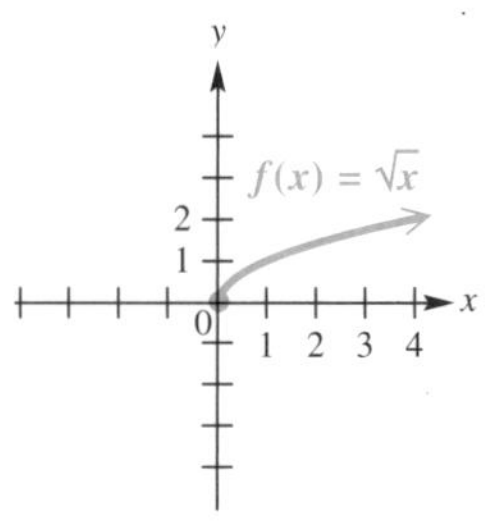

2.6 Cube Root Function

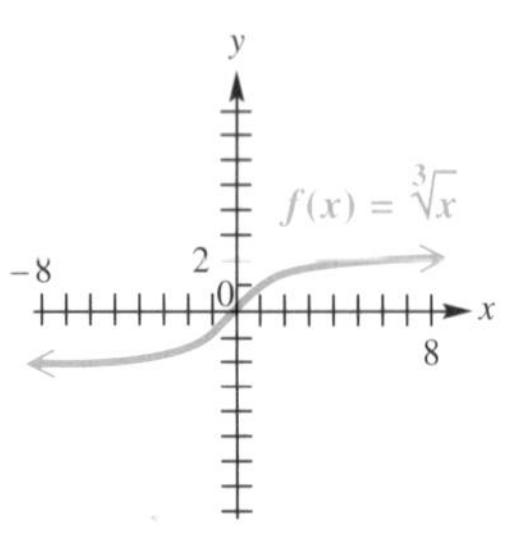

2.6 Absolute Value Function

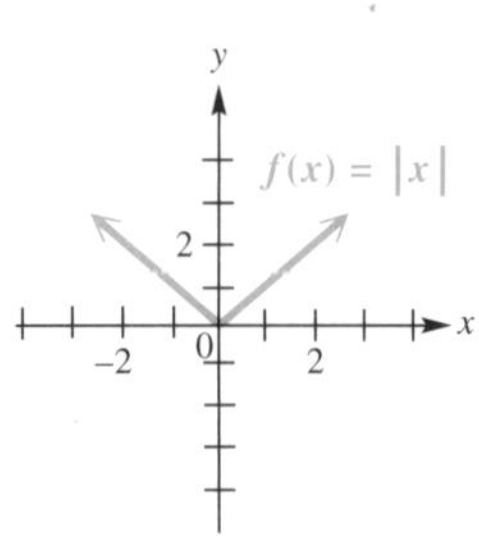

2.6 Greatest Integer Function

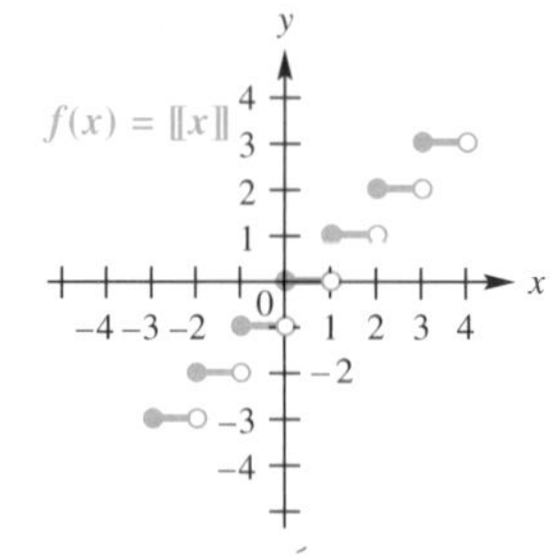

3.5 Reciprocal Function

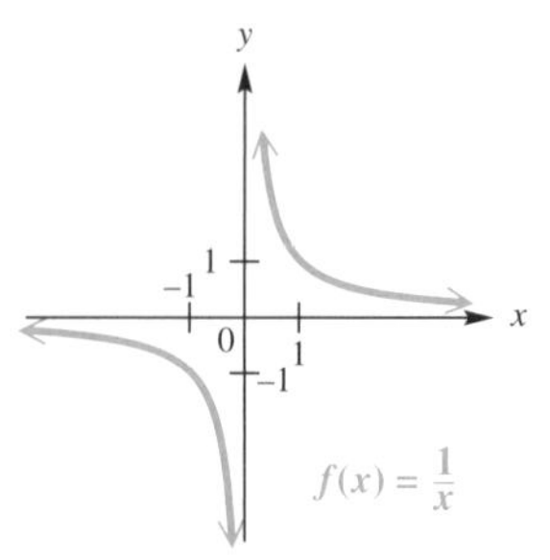

3.4 Polynomial Functions

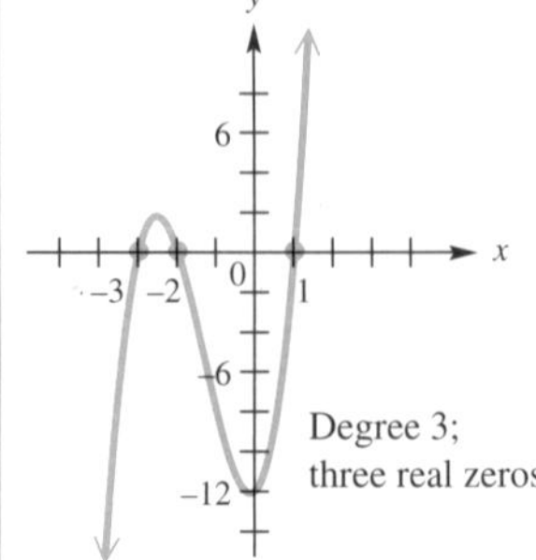

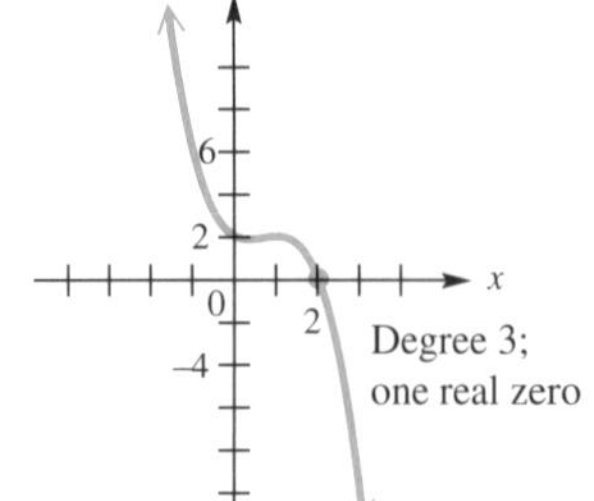

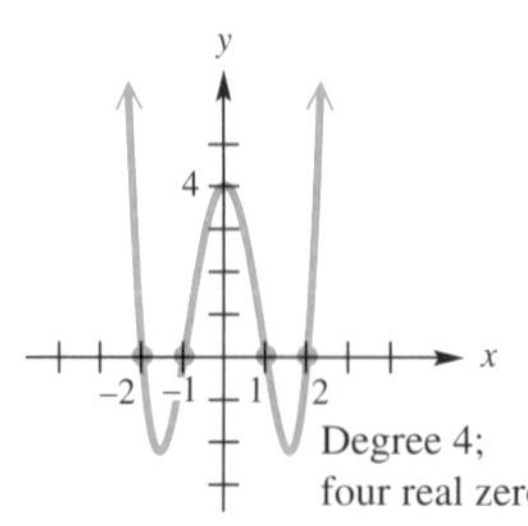

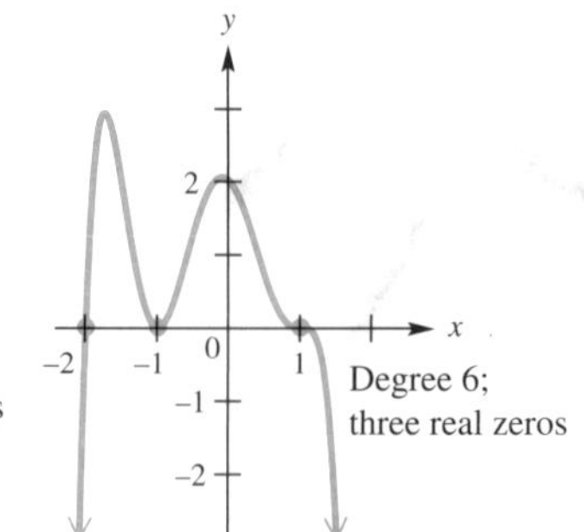

4.2 Exponential Functions

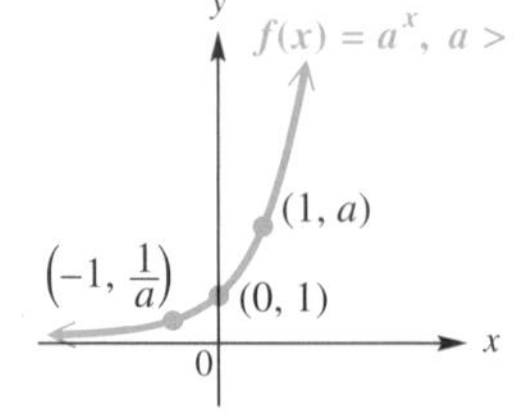

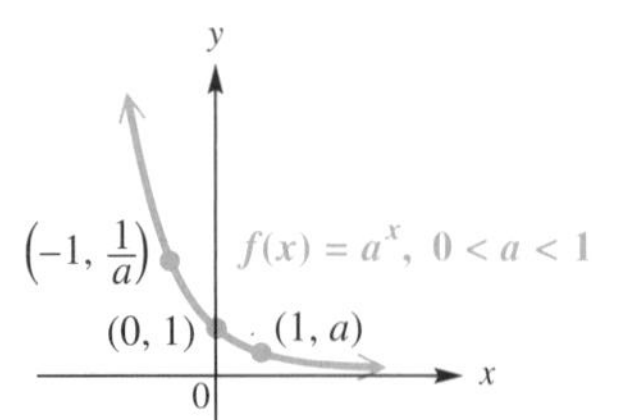

4.3 Logarithmic Functions

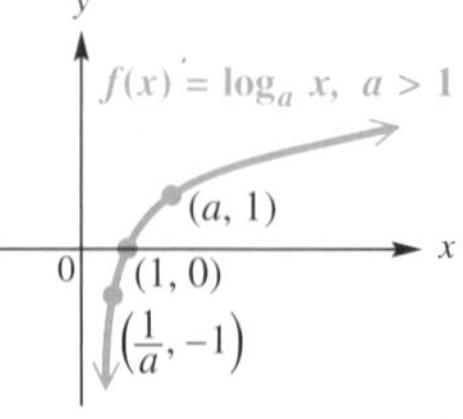

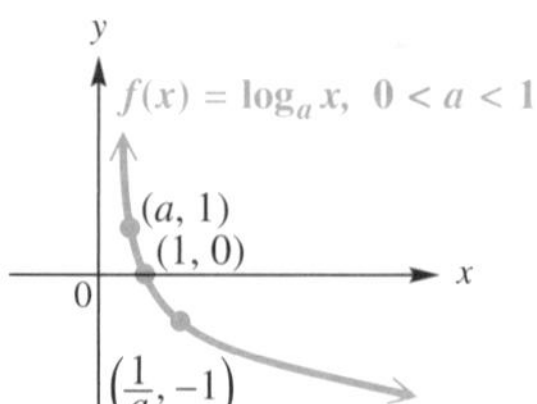